UNDERSTAND! Biochemistry Interactive CD-ROM
Lehninger Principles of Biochemistry 3/e Version
The Mona Group LLC

The *UNDERSTAND! Biochemistry* CD is a self-paced study tool that allows students to review, visualize, and test their mastery of biochemistry. At the core of the CD are 65 "Minicourses" organized as self-contained tutorials on key subject areas in biochemistry. Offered below is a listing of Minicourses and Minicourse Topics. In parentheses after each Topic are the most closely related chapter(s) in the Lehninger textbook.

Background
- Chemical Fundamentals I (4, 5)
- Chemical Fundamentals II (3, 9)
- Fundamentals of Cell Architecture (2)
- Fundamentals of Gene Action (24, 26, 27)

Molecules of Life
- Important Biomolecules (5, 9–11, 14)
- Macromolecules: Proteins and Nucleic Acids (3, 6, 10)
- An Interactive Gallery of Molecular Models (5, 9–11, 14–18, 20–23)
- An Interactive Gallery of Protein Structures (5–9, 12, 13, 15–20, 22–26, 28)

Proteins in Action
- Catalysis and Regulation (5–9, 18, 26)
- Nonallosteric Enzyme Kinetics (8)
- Allosteric Enzymes (7, 8)
- Molecules of the Immune Systems (7, 25)
- The Immune Responses (7)

Bioenergetics
- Energetics (6, 12, 14)
- Energetics of Metabolism (1, 14, 19)
- Glycolysis and Fermentation (15)
- Glycogen (9, 15, 20)
- Gluconeogenesis (20)
- Pentose Phosphate Pathway (15, 20)
- Citric Acid Cycle (16)
- Electron Transport – Oxidative Phosphorylation (14, 19)
- Fatty Acid Breakdown (17)
- Photophosphorylation (19)
- Calvin Cycle (20)
- C4 Photosynthesis (2, 20, 21)

Biosynthesis and Catabolism
- Fatty Acid Biosynthesis (21)
- Steroid Biosynthesis (21)
- Nitrogen Metabolism (8, 18, 22)
- Amino Acid Degradation (18)
- Nucleotide Biosynthesis and Breakdown (22)

Nucleic Acids and Their Expression
- DNA Replication (25)
- Mutation (10, 25, 27)
- Mutagens (10, 25)
- DNA Repair (25, 28)
- DNA Binding Proteins (26, 28)
- Transcription (26, 28)
- Translation (26, 27)
- Gene Regulation - Prokaryotes (28)
- Gene Regulation - Phage Lambda (25, 28)
- RNA Processing (26, 27)
- DNA Topology (24)
- Recombination (25)

Cellular Architecture and Traffic
- The Cytoskeleton (2, 12)
- Motor Proteins and Movement (2)
- The Extracellular Matrix (6, 9, 12)
- Protein Targeting (27)
- Protein Sorting in the ER and Golgi Apparatus (2, 27)
- Nuclear Transport (2, 27)
- Vesicle Transport (27)
- Sorting of Lysosomal Enzymes (2, 27)
- Protein Modification (2, 12, 27)
- Cell Signaling (13)
- Phototransduction (13)
- Neuronal Signaling (13)
- Receptor Kinase Signaling (13)
- G Protein Signaling (13)
- Second Messenger Molecules (13)

The Dividing Cell
- Cell Division Cycle (13)
- Cancer (2, 19, 25)
- Proto-Oncogenes and Tumor Suppressors (13)
- Viruses (2, 12, 26)

Some Important Techniques
- DNA Cloning and Sequencing (10, 29)
- Nucleic Acid Analysis (5, 10, 29)
- Protein Separation and Analysis (5, 7, 29)
- Protein Structure Determination (5, 6)

For more information on the CD and other supplements supporting the textbook, please see page viii of the textbook Preface.

Visit our Web site at: www.worthpublishers.com/lehninger

The site offers a variety of online resources designed to expand and enhance your understanding of the textbook's topics and features the *Lehninger 3-D Structure Tutorials* built using the Chemscape Chime plug-in. Additional site features include up-to-date annotated Web links to guide students to useful sites related to the topics in the textbook, stereoscopic views of many of the textbook's molecular illustrations, and support for students working the new edition's Biochemistry on the Internet problems. For more information on the Web site see page ix of the textbook's Preface.

THIRD EDITION

Lehninger Principles of Biochemistry

David L. Nelson
Professor of Biochemistry
University of Wisconsin–Madison

Michael M. Cox
Professor of Biochemistry
University of Wisconsin–Madison

WORTH PUBLISHERS

Lehninger Principles of Biochemistry Third Edition

David L. Nelson and Michael M. Cox

Printed in the United States of America

Library of Congress Cataloging-in-Publication Data

Nelson, David L.
Lehninger principles of biochemistry / David L. Nelson, Michael M. Cox.— 3rd ed.
p. cm.
Includes index.
ISBN 1-57259-153-6
1. Biochemistry. I. Nelson, David L. (David Lee), 1942- II. Cox, Michael M. III. Title.

QD415 .L44 2000
572—dc21 99-049137

Printing: 5 4 3 2 1 Year: 04 03 02 01 00

Development Editor: Morgan Ryan, with Linda Strange and Valerie Neal
Project Editor: Elizabeth Geller
Art Director: Barbara Rusin
Design: Paul Lacy
Production Supervisor: Bernadine Richey
Layout: York Graphic Services and Paul Lacy
Photo Editor: Deborah Goodsite
Illustrations: Susan Tilberry (with Alan Landau and Joan Waites), J.B. Woolsey & Associates, Laura Pardi Duprey, and York Graphic Services
Molecular Graphics: Jean-Yves Sgro
Composition: York Graphic Services
Printing and Binding: R.R. Donnelley and Sons

Cover (from top to bottom): Cut-away view of GroEL, a protein complex involved in protein folding; cut-away view of tobacco mosaic virus, an RNA virus; ribbon model of a β-barrel structural domain from UDP *N*-acetylglucosamine acyltransferase; cut-away view of the F_1 subunit of ATP synthase, with bound ATP shown as a stick structure; mesh surface image of the electron-transfer protein cytochrome *c*, with its heme group shown as a stick structure.

Cover images created by Jean-Yves Sgro.

Illustration credits begin on p. IC-1 and constitute a continuation of the copyright page.

Worth Publishers
41 Madison Avenue
New York, NY 10010

Preface

Lehninger Principles of Biochemistry, Third Edition, is an introduction to our favorite subject—and our attempt to make it yours. It has been designed for one- and two-semester courses for undergraduates majoring in biochemistry and related disciplines, as well as for graduate students who require a broad introduction to biochemistry, and for students in medical, dental, veterinary, and pharmacy programs. In revising, we have been encouraged and aided by the helpful comments of many students and teachers around the world who used the second edition of our book in one of the many languages in which it was published.

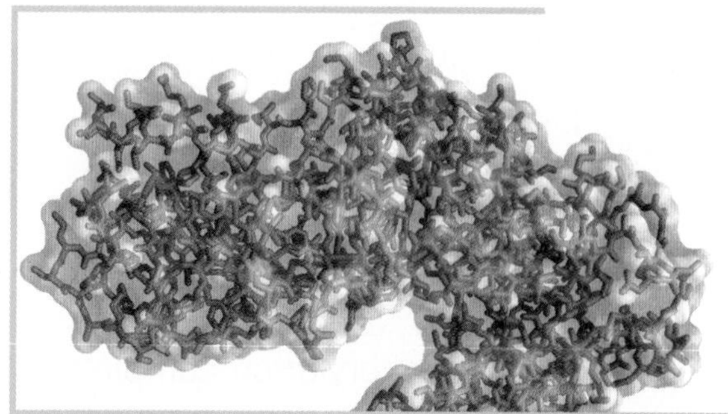

page 276

Biochemistry has undergone profound changes in the years since the last (1993) edition of this book. The third edition reflects those changes. Every chapter has been revised. There are new and completely reorganized chapters, new or greatly expanded sections covering areas of recent progress, and much new art, including an entirely new molecular graphics program. Altogether there are more than a thousand full-color illustrations (diagrams, photos, molecular graphics) in this edition. However, our ambitions in this new edition extend well beyond a thorough updating. We have expended at least as much effort on enhancing the presentation of fundamentals.

New and reorganized chapters, new or expanded sections covering recent progress, entirely new molecular graphics program

Changes in the New Edition: Revised, Refined, Up-to-Date Content

The third edition evolved gradually during those rare quiet moments in our offices, on walks along Lake Mendota, on long plane trips around the world, at halftime during Badger football games, in conversations with colleagues, during "free" afternoons at scientific meetings. Evolution is not revolution. While we have changed the content extensively, our goal has been to maintain the excellent qualities that led so many users to cherish the textbooks of Albert Lehninger since the first edition of *Biochemistry* appeared thirty years ago: clear prose and good illustrations, logical organization and development of subjects, challenging study questions. The organization of the book remains faithful to an order of presentation that has consistently proven successful through multiple editions of this book, and which is now familiar to instructors and students around the world. Part I provides a series of background and introductory chapters. Part II introduces the major classes of molecules in living cells. Intermediary metabolism is presented in Part III. The book finishes with a discussion of information pathways in Part IV.

Maintains the tradition of Albert Lehninger's cherished textbooks

Nevertheless, there are some notable changes. One is in the coverage of proteins and enzymes in Chapters 5–8. Chapters 5 and 6 from the second edition have been merged into the new and streamlined Chapter 5, which now introduces proteins and some of the important methods used to

Old Chapters 5 and 6 merged in a new and streamlined Chapter 5

Structure, folding, and denaturation in Chapter 6

New Chapter 7 on protein function

New chapter on biosignaling surveys one of today's most active research fields

Oxidative phosphorylation completely revised, including the remarkable mechanism of ATP synthase

Comprehensive revision of information pathways

New material on NMR, SELEX, DNA microarrays and other important methods

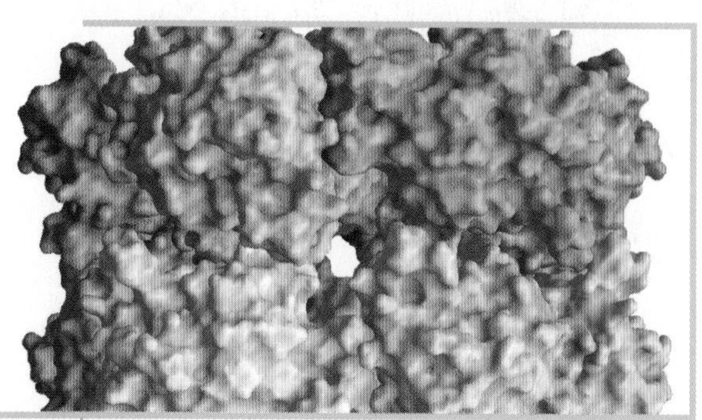

page 824

analyze them. Chapter 6 is a thoroughly revised presentation of protein structure, folding, and denaturation. Chapter 7 focuses on protein function. Developed from a supplement published for the second edition, this new chapter features an expanded treatment of the principles underlying the reversible binding of ligands to proteins, within an updated discussion of the function of myoglobin, hemoglobin, immunoglobulins, and muscle. Chapter 8, which focuses on enzyme function, has also been extensively revised and updated.

Following a much-expanded chapter on membranes and transport, a new chapter (Chapter 13; Biosignaling) covers one of the most active areas of research in biochemistry today: the signal transductions by which cells detect and respond to external cues such as hormones, neurotransmitters, growth factors, and environmental stimuli. This chapter emphasizes the universal features of receptor and transducer mechanisms, including those for vertebrate vision, olfaction, and gustation. The chapter also describes the mechanisms that regulate the cell cycle, and it details the effects of altered regulatory proteins encoded in oncogenes and tumor supressor genes. Finally, the biosignaling chapter provides the bridge from biomolecular structure to the next part of the book, on intermediary metabolism.

The chapters in Part III (Metabolism) have been updated to incorporate new developments in metabolic regulation, in the structure of enzymes and enzyme complexes, and new examples of human diseases that result from defective metabolism. The chapter on oxidative phosphorylation and photophosphorylation (Chapter 19) has been thoroughly reworked, and now includes recent information on the structures of membrane-bound enzyme complexes and the remarkable mechanism of ATP synthase. We have also used, in this chapter and throughout the book, the widely accepted nonintegral stoichiometry for ATP production in oxidative phosphorylation—2.5 ATP per electron pair from NADH, 1.5 ATP from $FADH_2$.

The chapters on information pathways in Part IV were systematically revised to reflect the fast pace of advance in many of the research areas covered, and to improve the order and flow of the presentation. Few areas went untouched. Highlights include a new presentation of DNA metabolism that includes a range of recent experimental insights, expanded and updated coverage of eukaryotic transcription and regulation of gene expression, descriptions of new information concerning protein synthesis and protein sorting, and an updating of the technology coverage in the final chapter.

A Closer Look at Experimental Methods

An understanding of modern biochemistry is impossible without an introduction to the experimental methods that made possible each major advance. We consider the methodology to be so intrinsic to the results that we have woven descriptions of many of these important techniques into the presentations of the concepts and principles they have revealed. There are new sections on methods such as NMR, mass spectrometry, SELEX, and DNA microarrays. The presentations of x-ray crystallography, DNA and protein sequencing, hybridization methods, nucleic acid and peptide synthesis, protein purification, PCR, and many other methods have been improved and updated. A complete listing of the experimental methods described in the book can be found in the index, under "Techniques."

New Pedagogic Artwork

The most immediately visible difference between this and the previous edition is the many new illustrations of macromolecular structures generated by Jean-Yves Sgro, biophysicist and master of molecular graphics. Armrest-to-armrest for many, many hours in front of his computers, we described what we wanted to see and Jean-Yves conjured more than we had hoped for. Sgro has been our valued partner in this revision, producing clear and beautiful images that are closely coordinated with the text and other art. All of the more than 200 molecular graphics images are new, original, and unique to this book. Our ability to illustrate biochemical principles with known structures has been greatly enhanced by the dramatic improvements in the methods for solving macromolecular structures by x-ray crystallography and nuclear magnetic resonance spectroscopy. The past few years of research in structural biology have yielded a profusion of protein and nucleic acid structures, rich in molecular detail. The progress is strikingly apparent when the present edition is compared to the first (1970) edition of Lehninger's *Biochemistry,* which featured a drawing of only one protein (myoglobin) in molecular detail!

Molecular graphics and diagrams all follow a consistent color scheme throughout the text

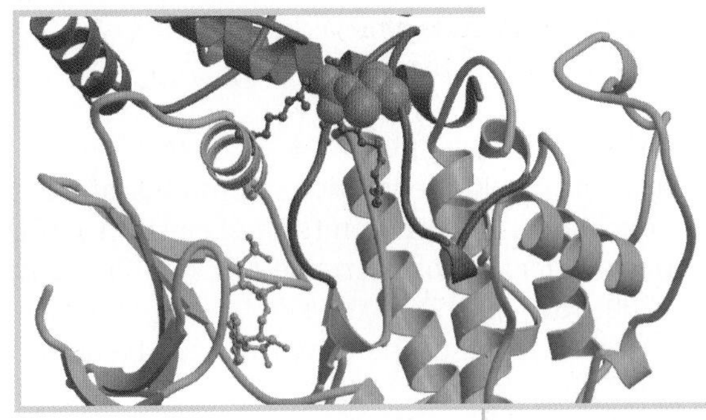

page 470

Emphasis on Fundamentals

To help students navigate the ocean of information in biochemistry, we wrote with these goals:

- to introduce the language of biochemistry, with careful explanations of the meaning, origin, and significance of terms;
- to provide a balanced understanding of the physical, chemical, and biological context in which each biomolecule, reaction, or pathway operates;
- to project a clear and repeated emphasis on major themes, especially those relating to evolution, thermodynamics, regulation, and the relationship between structure and function;
- to explain and to place in context the most important techniques that have brought us to our current understanding of biochemistry; and
- to sustain the student's interest by developing topics in a logical and stepwise manner; taking every opportunity to point out connections between processes; identifying gaps in our knowledge that promise to challenge future generations of scientists; supplying the historical context of selected major discoveries, when such context is useful; and providing with each chapter challenging problems that require the student to work with real data or real situations.

The four parts of *Lehninger Principles of Biochemistry,* Third Edition, were planned with these goals in mind. Themes that unify groups of chapters are introduced in the opening text for each part and reinforced in individual chapters. The organization within and across chapters will help students to maintain the focus on major themes and essential information and so will enhance student understanding. We have written introductions to each chapter that allow that chapter to stand alone, so that an instructor using this book is not obliged to teach his or her course in exactly the sequence of the book. This is especially true of the middle chapters on metabolism, where there is considerable variation in the sequence used by different teachers.

The appendices include an extensive list of abbreviations found in the biochemical literature and brief solutions to the end-of-chapter problems. The Glossary has been revised and expanded to provide definitions of more than 800 important terms.

In presenting this new edition of *Principles of Biochemistry,* we welcome your criticisms, suggestions, and comments of all kinds.

New Boxes and Problems on Practical Applications

To encourage students to go deeper into the applications and implications of biochemical research, this edition features:

- More boxed features on the relevance of biochemistry to medicine, biotechnology, and other aspects of daily life.
- New end-of-chapter problems that challenge students to apply biochemical principles to questions from the real world.
- *Biochemistry on the Internet* problems, which pose questions and present strategies for finding answers using biochemistry resources on the World Wide Web.

Tools and Techniques

The text describes many techniques used in biochemical research, most of which are interwoven with the results of their application. Among the techniques described in this book are:

Working with cells and animals

Working with polypeptides and proteins

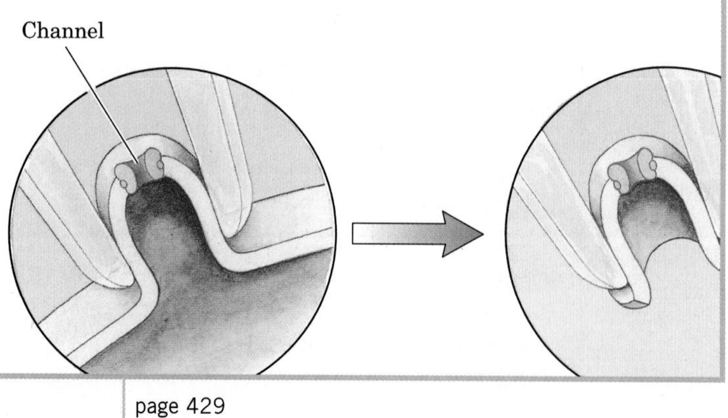

page 429

Working with DNA and RNA

Working with carbohydrates, lipids, and membranes

Physical techniques for structure determination

Applications and Explanations in Boxes

Boxed features throughout the text give supplementary information on applications of biochemistry, biochemical explanations of intriguing biological phenomena, and other types of student enrichment. Additional applications of biochemistry relating to health and disease, the environment, industry, agriculture, and other fields appear throughout the pages of this book.

Media and Supplements Supporting *Lehninger Principles of Biochemistry,* Third Edition

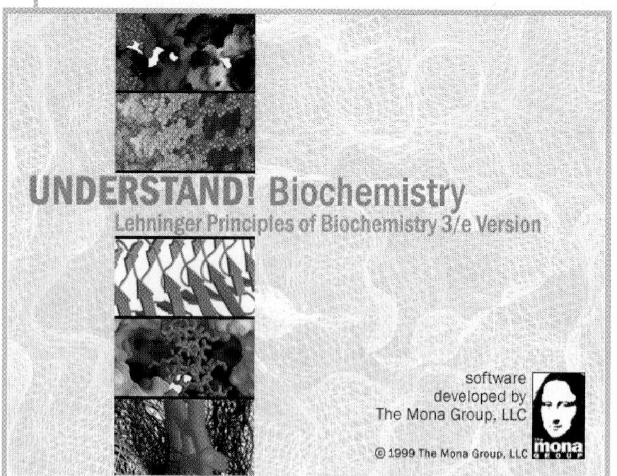

UNDERSTAND! Biochemistry
Lehninger Principles of Biochemistry 3/e Version
software developed by The Mona Group, LLC
© 1999 The Mona Group, LLC
mona

Worth Publishers offers supplements of the highest quality for students and instructors learning and teaching from this textbook. Lists of the Minicourses and animations on the *UNDERSTAND! Biochemistry* CD-ROM can be found on the front endpapers. In addition to those features, the CD contains more than 120 original animations that enable students to see important biochemical processes in action. For a complete list of the animations, see our Web site. The CD also includes self-quizzes that give students immediate feedback; interactive three-dimensional models of macromolecular structures; spoken pronunciations of over 400 biochemical terms; 900 original illustrations; and an index function that allows students to jump instantly to brief descriptions and explanations of important terms.

The Absolute, Ultimate Guide to Lehninger Principles of Biochemistry, Third Edition Study Guide and Solutions Manual

Marcy Osgood and Karen Ocorr, University of Michigan, Ann Arbor

The *Guide* combines the features of an innovative study guide and a detailed solutions manual in one convenient volume. The study guide features have been thoroughly class-tested and have proven to help students grasp the fundamentals of the course. For each textbook chapter, the *Guide* includes:

- **Expanded solutions** for all end-of-chapter problems
- **What to Review,** recapping key points from previous chapters
- Open-ended **Discussion Questions** for each chapter section designed for study groups, classroom discussions, or individual review
- A **Self-Test** focusing on terms, facts, and the application of facts, plus **Answers**
- For some chapters, the *Guide* also offers **Biochemistry on the Internet problems,** which teach students how to use online databases and biochemical resources.
- A poster-sized **Cellular Metabolic Map** is packaged with each copy of the *Guide*. On the map, students can draw the reactions and pathways of metabolism in their proper compartments within the cell. This exercise helps students to see important connections between cellular pathways and puts reaction pathways in their proper cellular context.

Supplements for Instructors

Traditional ideas and the latest technology have been combined to create innovative supplements to support instructors as they teach from the textbook.

Instructor's Resource CD-ROM

The instructor's CD offers all the figures from the textbook in JPEG format, plus all the animations from the *UNDERSTAND! Biochemistry* CD. Also included are stereoscopic views of molecular structures featured in the textbook, plus a database of PDB coordinates used to create the textbook's molecular illustrations. The CD can be used with any presentation software or can be preloaded into bundled Presentation Manager Pro software. Presentation Manager Pro makes it easy for instructors to create lecture slide shows that combine textbook illustrations with additional videos, animations, and illustrations from local or Internet sources.

Overhead Transparencies

A set of 152 acetates of the most important figures from the textbook.

Test Bank

By Gerald Frenkel, Rutgers University; David L. Nelson and Brook Chase Soltvedt, University of Wisconsin–Madison. Printed as well as computerized (Windows and Macintosh versions on a hybrid disk) versions are available. The *Test Bank* contains about 50 multiple-choice and short-answer questions per chapter, keyed to the corresponding section of the textbook and rated by level of difficulty. The questions in the computerized questions can be downloaded, edited, amended, and resequenced to suit each instructor's needs.

Online Testing

Instructors can create online exams and quizzes using questions from the Test Bank, with the option of inserting multimedia, graphics, or three-dimensional molecular representations into the questions. High-level security allows instructors to restrict individual testing capability to individual computers, and to specify a time block during which a test can be taken. The *Online Testing* package also has a suite of grade book and question analysis features.

Lehninger on the Web

www.worthpublishers.com/lehninger

Students and instructors will both benefit from the online offerings of the Lehninger Principles of Biochemistry Web site. The Web site features the *Lehninger 3-D Structure Tutorials* created by Timothy Driscoll, Frieda Reichsman, and Eric Martz, University of Massachusetts, Amherst. These self-paced, interactive tutorials use three-dimensional graphics to show how molecular structure relates to function, and allow students to more fully explore basic and advanced topics covered in the textbook. New topics will be added as the tutorials develop, so check back often. Tutorial topics will include:

ATP synthase
Bacteriorhodopsin
Chymotrypsin
Cyclooxygenases (COX-1/COX-2)
DNA-binding motifs
G proteins (heterotrimeric and monomeric)
Glycogen phosphorylase
Hammerhead ribozymes
HIV protease inhibitors
Lac repressor (the three operators)
MHC proteins
Oxygen-binding proteins
Protein architecture
Protein kinases
Restriction enzymes

Additional Web site features include:

- **Annotated Web links** for each chapter, to guide students to useful sites related to topics covered throughout the textbook
- Online support for the **Biochemistry on the Internet problems** in the textbook, including links to databases and research tools on the Internet, as well as instructions for downloading the three-dimensional molecular viewing utilities Rasmol and Chemscape Chime
- **Database of PDB** coordinates used to create molecular textbook illustrations
- **Stereoscopic views** of many of the textbook illustrations

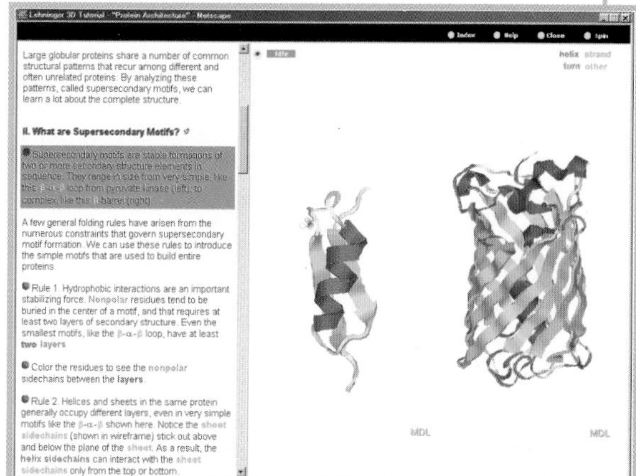

The 3D Structure Tutorials allow students to more fully explore basic and advanced topics covered in *Lehninger Principles of Biochemistry*—interactively and in 3 dimensions.

Acknowledgments

Our writing has been supported by many people whose advice and encouragement were critical throughout the years of revising, and we are indebted to all of them. At Worth Publishers, the overall organization of the project and management of its development were overseen in the early phase by Judith Wilson, and in the main phase by Morgan Ryan, with timely help on some chapters from Linda Strange and Valerie Neal. Most critical to the overall effort was the enormous infusion of constructive criticism, queries, encouragement, and inspiration contributed by Morgan Ryan. His deft editorial touch improved the book in innumerable ways. The final manuscript benefited from the superb copyediting of Linda Strange (who has been involved in the writing of all five editions of Lehninger's biochemistry texts!). Our project editor, Elizabeth Geller, kept all the elements of the book moving in the right directions throughout the production process, showing us sometimes velvet, sometimes steel. Her energy and focus kept the project on track. In Madison, we relied heavily on the invaluable editorial and organizational input, involving every phase of the project, provided by Brook Soltvedt. Notable clerical assistance was provided by Shannon Baruth. It was a great privilege to work with so talented and energetic a team, and they have all had much to do with the quality of the book before you.

The art for this edition includes work by Susan Tilberry, Laura Duprey, and J.B. Woolsey and Associates. Paul Lacy, Photoshop wizard and page layout artist, repeatedly amazed us with pages in which illustrations and text are in perfect synch. We have been impressed and gratified to witness the process by which our often indecipherable scribbles have been transformed into the elegant figures that grace this book.

We have already noted the invaluable contribution of Jean-Yves Sgro, who did all of the molecular graphics. The photographs used in this edition were tracked down by the indefatigable Deborah Goodsite, with assistance from Joan Peterson, Connie Gardner, and Jennifer MacMillan. Yuna Lee arranged for the many reviewers of this book, and coordinated the development of the printed supplements. The development of the book support Web site was under the editorial supervision of Sonia DiVittorio. Marcy Osgood and Karen Ocorr of University of Michigan, Ann Arbor, wrote many new end-of-chapter problems, including all the Biochemistry on the Internet problems, and made helpful comments on the manuscript as they revised their study guide for this book. Lou Pech of Carroll College researched and wrote five of the new boxes for this edition. Paul Marrione helped us select problems from Paul Van Eikeren's guide to the first edition.

One of the special advantages we have had is our access to the extraordinary community of researchers at the University of Wisconsin–Madison. More often than not, the expertise we needed to tie down a loose end or develop a new topic was right at hand. Our colleagues provided us with timely data, critiqued sections of the text, helped us develop figures, answered our questions, and helped in innumerable other ways. Space precludes a detailing of every contribution, but we are especially grateful to Laurens Anderson, Wayne Becker, Brian Fox, Hazel Holden, Ivan Rayment, and Brian Volkman, each of whom spent many hours helping us.

These other Madison colleagues also made important contributions to some aspect of the book: Rick Amasino, Alan Attie, Sebastian Bednarek, Susan Eichhorn, Jerry Ensign, Ray Evert, Bob Fillingame, Perry Frey, Rick Gourse, Colleen Hayes, Judith Kimble, Bob Landick, Paul Ludden, John Markley, Anant Menon, James Ntambi, Ann Palmenberg, Ron Raines, Tom Record, George Reed, Bill Reznikoff, Ruth Saecker, Heinrich Schnoes, Tom Sharkey, Lloyd Smith, Sue Smith, Gary Splitter, Bill Sugden, Mike Sussman, and Marv Wickens.

We were also able to draw on contacts from around the world. For their patience in answering our questions, often including an informal reading and critique of passages, we thank: Andre Adoutté (Université d'Orsay), Marlene Belfort (Wadsworth Center, New York State Department of Health), Terry Beveridge (University of Guelph), Pat Brown (Stanford University), Peter Burgers (Washington University School of Medicine), Joseph Clark (Oxford University), Ron Conaway (University of Oklahoma), Herbert Friedmann (University of Chicago), Barry Ganong (Mansfield University), Myron Goodman (University of Southern California), Jack Griffith (University of North Carolina), Carol Gross (University of California, San Francisco), Peter Hinkle (Cornell University), Denis Lynn (University of Guelph), Lynn Margulis (University of Massachusetts, Amherst), Ken Marians (Memorial Sloan-Kettering Cancer Center), Stanley Miller (University of California, San Diego), Paul Modrich (Duke University), Chris Raetz (Duke University), Aziz Sancar (University of North Carolina at Chapel Hill), William Schopf (University of California–Los Angeles), Sue Travis (University of Iowa), Claire Walczak (Indiana University), and Richard Wolfenden (University of North Carolina).

The book has benefited enormously from the combined wisdom and experience of many reviewers. Every chapter was critiqued at multiple stages by biochemists with interests and expertise in both the chapter topic and biochemical education. These formal reviews helped to shape this new edition in many ways. For their indispensible advice and criticism, we thank:

Reviewers of the third edition:

Sankar Adhya
National Cancer Institute, National Institutes of Health

Tom Alber
University of California, Berkeley

Laurens Anderson
University of Wisconsin–Madison

Norman Arnheim
University of Southern California

Alan Attie
University of Wisconsin–Madison

Tania Baker
Massachusetts Institute of Technology

Vahe Bandarian
University of Wisconsin–Madison

Wayne Becker
University of Wisconsin–Madison

Robert W. Bernlohr
Penn State University

Michael R. Borenstein
Temple University

Ross D. Brown, Jr.
University of Florida, Gainesville

John Browse
Washington State University

Peter Burgers
Washington University School of Medicine

Thomas R. Cech
University of Colorado at Boulder

Michael J. Chamberlin
University of California, Berkeley

Ron Conaway
University of Oklahoma

Joseph Clark
Oxford University

Jaleh Daie
University of Wisconsin–Madison

H. Garry Dallmann
University of Colorado Health Sciences Center

Louis T. J. Delbaere
University of Saskatchewan

William Dowhan
University of Texas–Houston Medical School

Diana Downs
University of Wisconsin–Madison

Jeffrey D. Esko
University of Alabama at Birmingham

Gerald W. Feigenson
Cornell University

Bob Fillingame
University of Wisconsin–Madison

Hartmut Follmann
Universität Kassel, Kassel, Germany

Brian Fox
University of Wisconsin–Madison

Maxim D. Frank-Kamenetskii
Boston University

Herbert Friedmann
University of Chicago

David Goodman
Princeton University

Myron Goodman
University of Southern California

Jack Gorski
University of Wisconsin–Madison

Rick Gourse
University of Wisconsin–Madison

Lawrence M. Gracz
Massachusetts College of Pharmacy and Allied Health Sciences

Rachel Green
Johns Hopkins University School of Medicine

Carol Gross
University of California, San Francisco

Lawrence Grossman
Johns Hopkins University

Richard I. Gumport
University of Illinois at Urbana-Champaign

Mitchell F. Halperin
St. Michael's Hospital, University of Toronto

F. Ulrich Hartl
Max-Planck-Institut für Biochemie

John W. B. Hershey
University of California, Davis School of Medicine

Lowell E. Hokin
University of Wisconsin Medical School

Jon M. Kaguni
Michigan State University

Pierre Kamoun
Hôpital Necker-Enfants Malades

Harold Kasinsky
University of British Columbia

Judith Kimble
University of Wisconsin–Madison

Roy L. Kisliuk
Tufts University

Randy D. Krauss
Boston University School of Medicine

Bob Landick
University of Wisconsin–Madison

Bob LaRossa
DuPont Company

Michael Lieberman
University of Cincinnati, College of Medicine

Janet E. Lindsley
University of Utah School of Medicine

Stuart Linn
University of California, Berkeley

Elsebet Lund
University of Wisconsin–Madison

Michael J. MacDonald
University of Wisconsin Medical School

T. F. J. Martin
University of Wisconsin–Madison

Anant Menon
University of Wisconsin–Madison

Julie T. Millard
Colby College

Cynthia J. Moore
Washington University in St. Louis

Pierre Morell
University of North Carolina at Chapel Hill

Ronald L. Niece
University of California at Irvine

James Ntambi
University of Wisconsin–Madison

Michael O'Donnell
The Rockefeller University

James Ofengand
University of Miami School of Medicine

Archie R. Portis, Jr.
USDA Agricultural Research Service

Jack Preiss
Michigan State University

Frank Pugh
The Pennsylvania State University

George Reed
University of Wisconsin–Madison

David Reibstein
University of Pennsylvania

Bill Reznikoff
University of Wisconsin–Madison

Daniel H. Rich
University of Wisconsin–Madison

Peter J. Roach
Indiana University School of Medicine

Gary Roberts
University of Wisconsin–Madison

Jeff Roberts
Cornell University

Francis Rolleston
Medical Research Council of Canada

Douglas W. Russell
Dalhousie University, Nova Scotia, Canada

Lisa M. Salati
West Virginia University School of Medicine

Aziz Sancar
University of North Carolina at Chapel Hill

Paul Schimmel
The Scripps Research Institute

Bob Schleif
Johns Hopkins University

Herbert P. Schweizer
Colorado State University

Tom Sharkey
University of Wisconsin–Madison

Richard R. Sinden
Institute of Biosciences and Technology, Texas A & M University

Cassandra L. Smith
Boston University

Lloyd M. Smith
University of Wisconsin–Madison

Gary Splitter
University of Wisconsin–Madison

Howard Sprecher
Ohio State University College of Medicine and Public Health

William Tapprich
University of Nebraska at Omaha

Jeremy Thorner
University of California, Berkeley

Bruce Tiberiis
University of British Columbia

Harald Tschesche
Universität Bielefeld, Bielefeld, Germany

Thomas L. Vandergon
Pepperdine University

Désirée Vanderwel
University of Winnipeg

Alejandro J. Vera
University of British Columbia

Jon A. Wolff
University of Wisconsin Medical School

William Wolodko
University of Alberta

We lack the space here to acknowledge all of the other individuals whose special efforts went into this book. We offer instead our sincere thanks, and the finished book that they helped guide to completion. We, of course, assume full responsibility for errors of fact or emphasis.

We are grateful to our students at the University of Wisconsin–Madison, who provided inspiration (especially Jason Celitti, who provided valuable criticism, and Erik Mikkelson, an excellent proofreader of chemistry); to the students and staff of our research groups, who helped us balance the competing demands of research, teaching, administration, and textbook writing; and to our colleagues in the Department of Biochemistry at Madison, who patiently answered our questions, corrected our misconceptions, and in many cases reviewed chapters. We have received many letters over the past eight years from students and professors who used our book and who pointed out places it could be made better, for which we are grateful. (We hope future users will continue to tell us how we can improve the book). We also wish to thank the Lehninger Team at Worth Publishers, who allowed us the freedom we needed and exhibited extraordinary patience as we added late-breaking advances to "finalized" chapters, and whose dedication to producing a fine-quality book inspired our best efforts and made this a rewarding experience.

Finally, we express our deepest appreciation to our wives, Brook and Beth, and our children, who endured the long evenings and weekends we devoted to book writing with extraordinary grace and provided constant encouragement.

Madison, Wisconsin
October 1999

David L. Nelson
Michael M. Cox

About the Authors

Michael M. Cox and David L. Nelson

David L. Nelson, born in Fairmont, Minnesota, received his BS in Chemistry and Biology from St. Olaf College in 1964. He earned his PhD in Biochemistry at Stanford Medical School under Arthur Kornberg, and was a postdoctoral fellow at the Harvard Medical School with Eugene P. Kennedy, who was one of Lehninger's first graduate students. Nelson went to the University of Wisconsin–Madison in 1971 and became a full professor of biochemistry in 1982.

Nelson's thesis research at Stanford was on the intermediary metabolism of sporulating and germinating bacteria. At Harvard he studied the energetics, genetics, and biochemistry of ion transport in *E. coli.* At Wisconsin his research has focused on the signal transductions that regulate ciliary motion and exocytosis in the protozoan *Paramecium.* The enzymes of signal transductions, including a variety of protein kinases, are primary targets of study. His research group uses enzyme purification, immunological techniques, electron microscopy, genetics, molecular biology, and electrophysiology to study these processes.

Dr. Nelson has a distinguished record as a lecturer and research supervisor. For 30 years he has taught an intensive survey of biochemistry for advanced biochemistry undergraduates and graduate students in the life sciences (using Lehninger's *Biochemistry* and *Principles of Biochemistry* for much of that time). He has also taught a survey of biochemistry for nursing students, a graduate course on membrane structure and function, and a graduate seminar on membranes and sensory transductions. He has sponsored numerous PhD, MS, and undergraduate honors theses, and has received awards for his outstanding teaching, including the Dreyfus Teacher–Scholar Award and the Atwood Distinguished Professorship. In 1991–1992 he was a visiting professor of chemistry and biology at Spelman College.

Michael M. Cox was born in Wilmington, Delaware. In his first biochemistry course, Lehninger's *Biochemistry* was a major influence in refocusing his fascination with biology and inspiring him to pursue a career in biochemistry. After graduating from the University of Delaware in 1974, Cox went to Brandeis University to do his doctoral work with William Jencks, and then to Stanford in 1979 for postdoctoral study with I. Robert Lehman, moving to the University of Wisconsin–Madison in 1983. He became a full professor of biochemistry in 1992.

His doctoral research was on general acid and base catalysis as a model for enzyme-catalyzed reactions. At Stanford, Cox began work on the enzymes involved in genetic recombination, designing still-used purification and assay methods, illuminating the process of DNA branch migration, and cloning the gene for a site-specific recombinase from yeast. Exploration of the enzymes of genetic recombination has remained the central theme of his research.

Dr. Cox has coordinated a large and active research team at Wisconsin, investigating the enzymology, topology, and energetics of genetic recombination. A primary focus has been the mechanism of DNA strand exchange and the role of ATP in the RecA system. The research team has also concentrated on the FLP recombinase of yeast and the process it controls, and is developing chromosomal targeting systems based on the FLP recombinase. For the past 15 years he has taught (with Dave Nelson) the survey of biochemistry and has lectured in graduate courses on DNA structure and topology, protein–DNA interactions, and the biochemistry of recombination. He has received awards for both his teaching and his research, including the Dreyfus Teacher–Scholar Award and the 1989 Eli Lilly Award in Biological Chemistry. His hobbies include gardening, wine collecting, and assisting in the design of laboratory buildings.

To our teachers:

Paul R. Burton

Albert Finholt

William P. Jencks

Eugene P. Kennedy

Homer Knoss

Arthur Kornberg

I. Robert Lehman

Earl K. Nelson

David E. Sheppard

Harold B. White

Contents in Brief

Contents

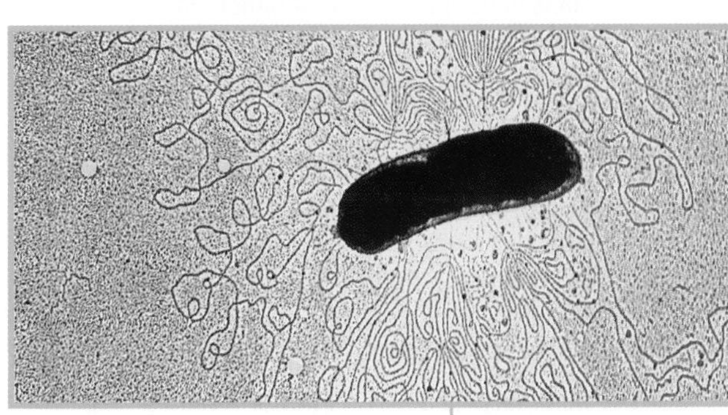

page 13

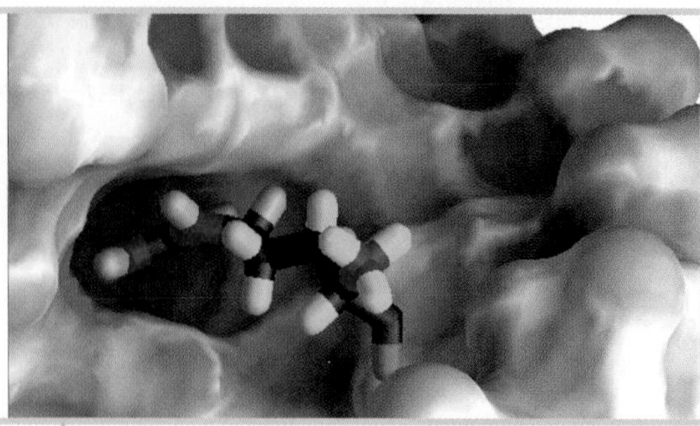

page 62

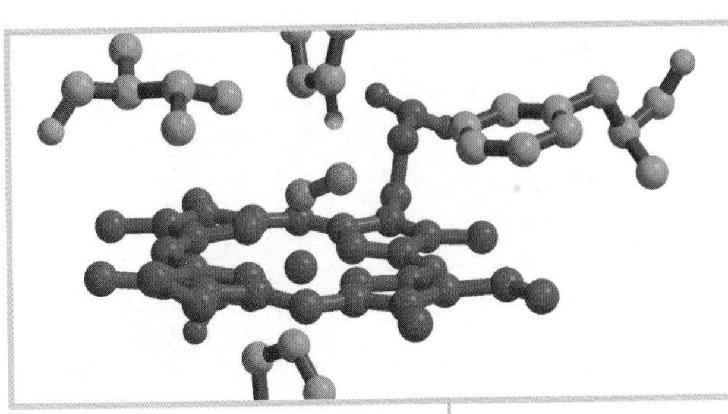
page 209

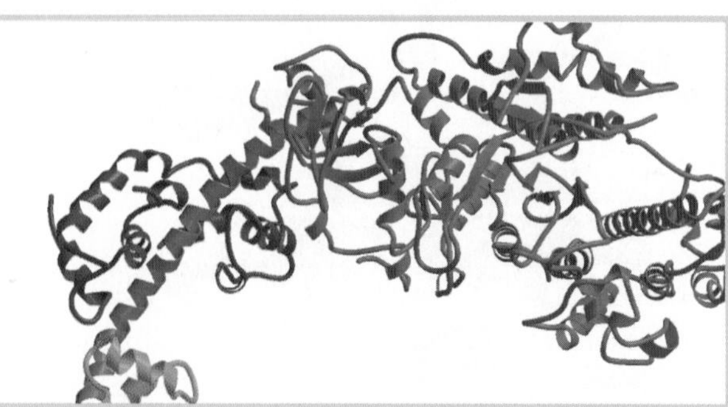

page 233

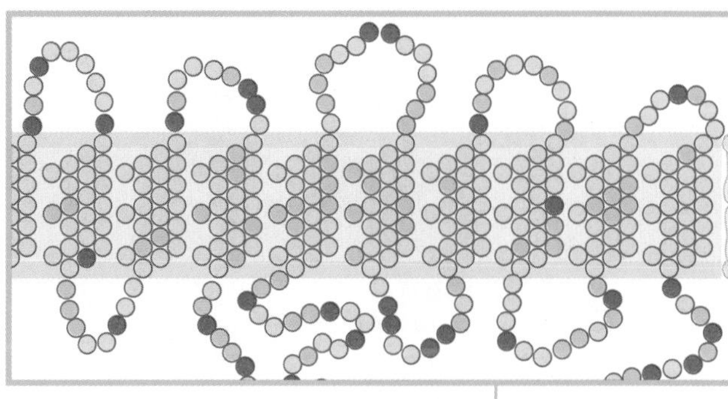

page 411

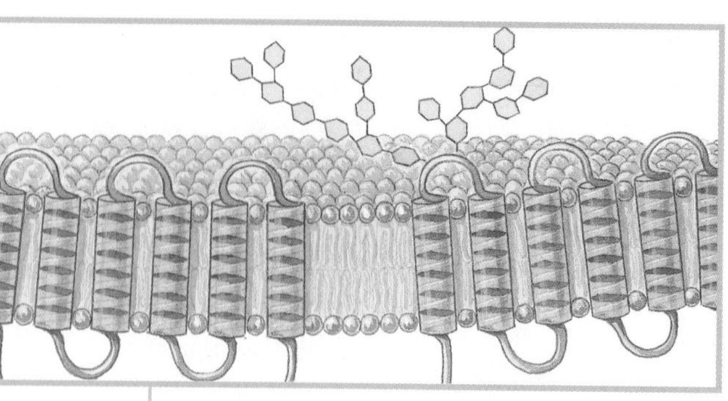

page 418

page 509

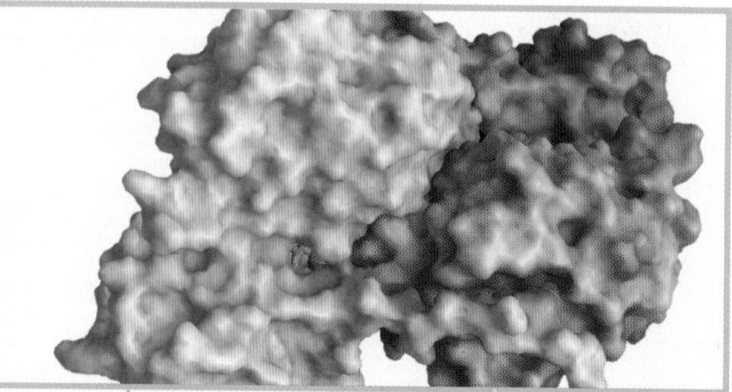

page 629

page 681

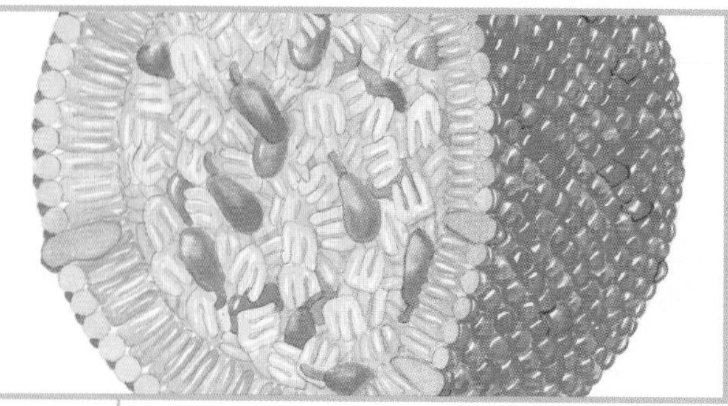
page 805

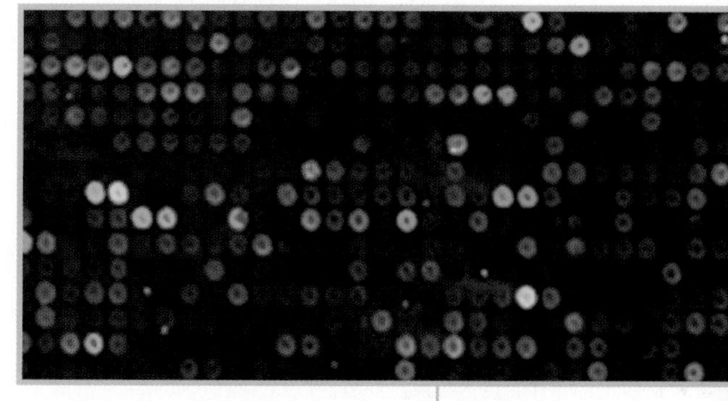

page 904

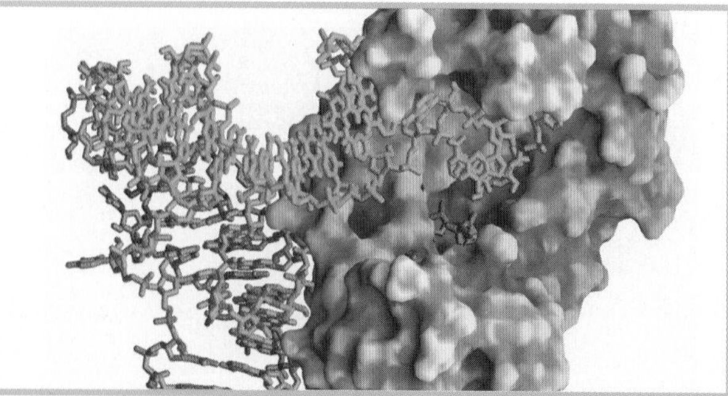

page 1043

page 1145

part

I

Foundations of Biochemistry

1 The Molecular Logic of Life
2 Cells
3 Biomolecules
4 Water

Fifteen to twenty billion years ago, the universe arose as a cataclysmic eruption of hot, energy-rich subatomic particles. Within seconds, the simplest elements (hydrogen and helium) were formed. As the universe expanded and cooled, material condensed under the influence of gravity to form stars. Some stars became enormous and then exploded as supernovae, releasing the energy needed to fuse simpler atomic nuclei into the more complex elements. Thus were produced, over billions of years, the earth itself and the chemical elements found on the earth today. About four billion years ago, life arose—simple microorganisms with the ability to extract energy from organic compounds or from sunlight, which they used to make a vast array of more complex **biomolecules** from the simple elements and compounds on the earth's surface. Biochemistry asks how the thousands of different biomolecules interact with each other to confer the remarkable properties of living organisms.

In Part I, we will summarize the biological and chemical background to biochemistry. Living organisms obey the same physical laws that apply to all natural processes, and we begin by discussing those laws and several axioms that flow from them (Chapter 1). These axioms make up the molecular logic of life. They define the means by which cells transform energy to accomplish work, catalyze chemical transformations, assemble complex molecules from simpler subunits, form supramolecular structures that are the machinery of life, and store and pass on the instructions for the assembly of all future generations of organisms from simple, nonliving precursors.

Cells, the units of all living organisms, share certain features; but the cells of different organisms, and the various cell types within a single organism, are remarkably diverse in structure and function. Chapter 2 is a brief description of the common features and the diverse specializations of cells, and of the evolutionary processes that have led to such diversity.

Nearly all of the organic compounds from which living organisms are constructed are products of biological activity. These molecules were selected during the course of biological evolution for their fitness in performing specific biochemical and cellular functions. Biomolecules can be characterized and understood in the same terms that apply to molecules of inanimate matter: the types of bonds between atoms, the factors that contribute to bond formation and bond strength, the three-dimensional

facing page
The Orion Nebula, a tremendous cloud of gas in which many hot, young stars are evolving rapidly toward cataclysmic cosmic explosions called supernovae. Energy released by nuclear explosions in such supernovae brought about the fusion of simple atomic nuclei, forming the more complex elements of which the earth, its atmosphere, and all living things are composed.

structures of molecules, and chemical reactivities. Three-dimensional structure is especially important in biochemistry. Biological interactions, such as those between enzyme and substrate, antibody and antigen, hormone and receptor, are highly specific, and this specificity is achieved by steric and electrostatic complementarity between molecules. Prominent among the forces that stabilize three-dimensional structure are noncovalent interactions, individually weak but with significant cumulative effects. Chapter 3 provides the chemical basis for later discussions of the structure, catalysis, and metabolic interconversions of individual classes of biomolecules.

Water is the medium in which the first cells arose, and it is the solvent in which most biochemical transformations occur. The properties of water have shaped the course of evolution, and the structure and interactions of biomolecules are profoundly influenced by the aqueous solution in which biomolecules reside. The weak interactions within and between biomolecules are strongly affected by the solvent properties of water. Even water-insoluble components of cells, such as membrane lipids, interact with each other in ways dictated by the polar properties of water. In Chapter 4 we consider the properties of water, the weak noncovalent interactions that occur in aqueous solutions of biomolecules, and the ionization of water and of solutes in aqueous solution.

These initial chapters are intended to provide a chemical backdrop for the later discussions of biochemical structures and reactions, so whatever your background in chemistry or biology, you can immediately begin to follow, and to enjoy, the action.

chapter

1

The Molecular Logic of Life

Living organisms are composed of lifeless molecules. When these molecules are isolated and examined individually, they conform to all the physical and chemical laws that describe the behavior of inanimate matter. Yet living organisms possess extraordinary attributes not exhibited by any random collection of molecules. In this chapter, we first consider the properties of living organisms that distinguish them from other collections of matter, and then we describe a set of principles that characterize all living organisms. These principles underlie the organization of organisms and their cells, and they provide the framework for this book. They will help you to keep the larger picture in mind while exploring the illustrative examples presented in the text.

(a)

(b)

(c)

figure 1–1

Some characteristics of living matter. (a) Microscopic complexity and organization are apparent in this colorized thin section of vertebrate muscle tissue, viewed with the electron microscope. **(b)** A prairie falcon acquires nutrients by consuming a smaller bird. **(c)** Biological reproduction occurs with near-perfect fidelity.

The Chemical Unity of Diverse Living Organisms

What distinguishes living organisms from inanimate objects? First is their degree of chemical complexity and organization. Thousands of different molecules make up a cell's intricate internal structures (Fig. 1–1a). By contrast, inanimate matter—clay, sand, rocks, seawater—usually consists of mixtures of relatively simple chemical compounds.

Second, living organisms extract, transform, and use energy from their environment (Fig. 1–1b), usually in the form of chemical nutrients or sunlight. This energy enables organisms to build and maintain their intricate structures and to do mechanical, chemical, osmotic, and other types of work. Inanimate matter does not use energy in a systematic, dynamic way to maintain structure or to do work; rather, it tends to decay toward a more disordered state, to come to equilibrium with its surroundings.

The third attribute of living organisms is the capacity for precise self-replication and self-assembly, a property that is the quintessence of the living state (Fig. 1–1c). A single bacterial cell placed in a sterile nutrient medium can give rise to a billion identical "daughter" cells in 24 hours. Each of the cells contains thousands of different molecules, some extremely complex; yet each bacterium is a faithful copy of the original, its construction directed entirely from information contained within the genetic material of the original cell.

Although the ability to self-replicate has no true analog in the nonliving world, there is an instructive analogy in the growth of crystals in saturated solutions. Crystallization produces more material identical in lattice structure to the original "seed" crystal. Crystals are much less complex than the simplest living organisms, and their structure is static, not dynamic as are living cells. Nevertheless, the ability of crystals to "reproduce" themselves

Erwin Schrödinger
1887–1961

led the physicist Erwin Schrödinger to propose in his famous essay "What Is Life?" that the genetic material of cells must have some of the properties of a crystal. Schrödinger's 1944 notion (years before our modern understanding of gene structure) describes rather accurately some of the properties of deoxyribonucleic acid, the material of genes.

Each component of a living organism has a specific function. This is true not only of macroscopic structures, such as leaves and stems or hearts and lungs, but also of microscopic intracellular structures such as the nucleus or chloroplast and of individual chemical compounds. The interplay among the chemical components of a living organism is dynamic; changes in one component cause coordinating or compensating changes in another, with the whole ensemble displaying a character beyond that of its individual constituents. The collection of molecules carries out a program, the end result of which is reproduction of the program and self-perpetuation of that collection of molecules; in short, life.

Biochemistry Explains Diverse Forms of Life in Unifying Chemical Terms

If living organisms are composed of molecules that are intrinsically inanimate, how do these molecules confer the remarkable combination of characteristics we call life? How can a living organism be more than the sum of its inanimate parts? Philosophers once answered that living organisms are endowed with a mysterious and divine life force, but this doctrine, called vitalism, has been firmly rejected by modern science. The study of biochemistry shows how the collections of inanimate molecules that constitute living organisms interact to maintain and perpetuate life animated solely by the chemical laws that govern the nonliving universe.

Living organisms are enormously diverse (Fig. 1–2). In appearance and function, birds and beasts, trees, grasses, and microscopic organisms differ

figure 1–2
Diverse living organisms share common chemical features. Birds, beasts, plants, and soil microorganisms share with humans the same basic structural units (cells) and the same kinds of macromolecules (DNA, RNA, proteins) made up of the same kinds of monomeric subunits (nucleotides, amino acids). They utilize the same pathways for synthesis of cellular components, share the same genetic code, and derive from the same evolutionary ancestors. ("The Garden of Eden" (detail), by Jan van Kessel, the Younger (1626–1679).)

greatly. Yet, biochemical research has revealed that all organisms are remarkably alike at the cellular and chemical levels. Biochemistry describes in molecular terms the structures, mechanisms, and chemical processes shared by all organisms, and provides organizing principles that underlie life in all of its diverse forms, principles we shall refer to collectively as *the molecular logic of life.* Although biochemistry provides important insights and practical applications in medicine, agriculture, nutrition, and industry, its ultimate concern is with the wonder of life itself.

Despite the fundamental unity of life, very few generalizations about living organisms are absolutely correct for every organism under every condition. The range of habitats in which organisms live, from hot springs to Arctic tundra, from animal intestines to college dormitories, is matched by a correspondingly wide range of specific biochemical adaptations, achieved within a common chemical framework. For the sake of clarity, we will sometimes risk certain generalizations, which, though not perfect, remain useful; we will also frequently point out the exceptions that illuminate scientific generalizations.

All Macromolecules Are Constructed from a Few Simple Compounds

Most of the molecular constituents of living systems are composed of carbon atoms covalently joined with other carbon atoms and with hydrogen, oxygen, or nitrogen. The special bonding properties of carbon permit the formation of a great variety of molecules. Organic compounds of molecular weight (also called relative molecular mass, M_r)[1] less than about 500, such as amino acids, nucleotides, and monosaccharides, serve as **monomeric subunits** of **macromolecules:** proteins, nucleic acids, and polysaccharides. A single protein molecule may have 1,000 or more amino acids, and deoxyribonucleic acid has millions of nucleotides.

Each cell of the bacterium *Escherichia coli* (*E. coli*) contains several thousand kinds of organic compounds, including a thousand different proteins, a similar number of different nucleic acid molecules, and hundreds of types of carbohydrates and lipids. In humans there may be tens of thousands of different proteins, as well as many types of polysaccharides (chains of simple sugars), a variety of lipids, and many other compounds of lower molecular weight.

To purify and to characterize thoroughly all of these molecules would be an insuperable task were it not for the fact that each class of macromolecules (proteins, nucleic acids, polysaccharides) is composed of a small, common set of monomeric subunits. These monomeric subunits can be covalently linked in a virtually limitless variety of sequences (Fig. 1–3), just as the 26 letters of the English alphabet can be arranged into a limitless number of words, sentences, and books.

Deoxyribonucleic acids (DNA) are constructed from only four different kinds of simple monomeric subunits, the deoxyribonucleotides. **Ribonucleic acids (RNA)** are composed of just four types of ribonucleotides. **Proteins** are composed of 20 different kinds of amino acids. The eight nucleotides from which all nucleic acids are built and the 20 different amino acids from which all proteins are built are identical in all living

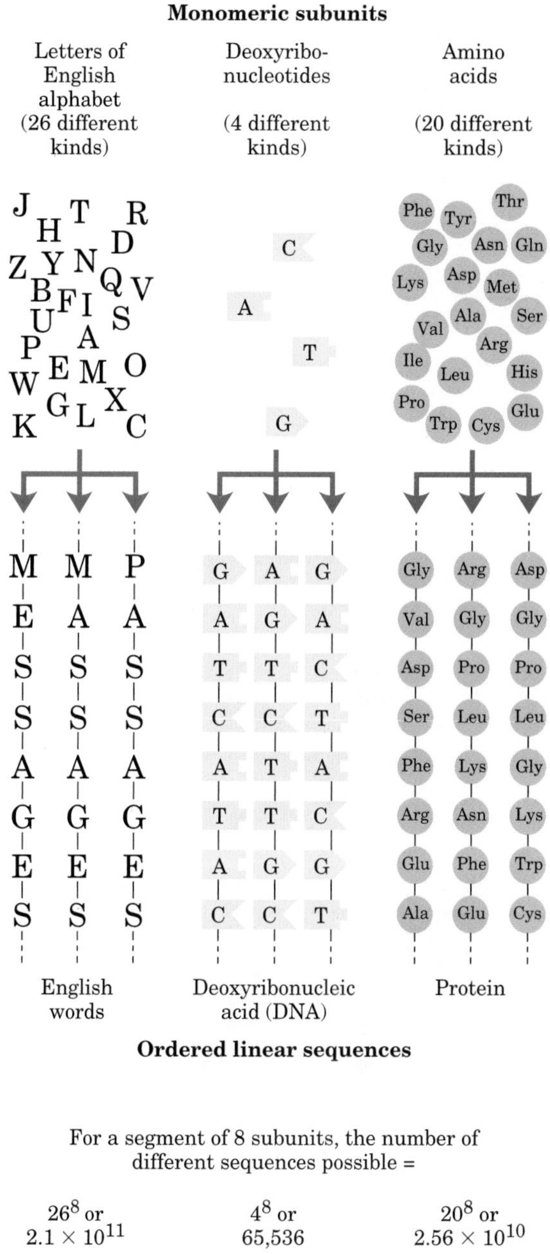

figure 1–3

Monomeric subunits in linear sequences can spell infinitely complex messages. The number of different sequences possible (S) depends on the number of different kinds of subunits (N) and the length of the linear sequence (L): $S = N^L$. For an average-sized protein ($L \approx 400$), S is 20^{400}—an astronomical number.

[1] The terms used to indicate the size of a molecule are often confused. We use molecular weight or M_r, relative molecular mass, a dimensionless ratio of the mass of a molecule to one-twelfth the mass of ^{12}C. The size of a molecule can also be correctly given in terms of molecular mass (m), which has units of daltons (Da) or atomic mass units (amu). A molecule should never be described as having a molecular weight or M_r (a dimensionless property) expressed in daltons or atomic mass units.

organisms. The specific sequence of monomeric subunits together with their arrangement in space shapes macromolecules for their particular biological functions as genes, catalysts, hormones, and so on.

Most of the monomeric subunits from which all macromolecules are constructed serve more than one function in living cells. Nucleotides serve not only as subunits of nucleic acids but also as energy-carrying molecules. Amino acids are subunits of protein molecules and are also precursors of hormones, neurotransmitters, pigments, and many other kinds of biomolecules.

We can now set out some of the principles in the molecular logic of life:

> **All living organisms build molecules from the same kinds of monomeric subunits.**
>
> **The structure of a macromolecule determines its specific biological function.**
>
> **Each genus and species is defined by its distinctive set of macromolecules.**

Energy Production and Consumption in Metabolism

Energy is a central theme in biochemistry: cells and organisms depend on a constant supply of energy to oppose the inexorable tendency in nature for a system to decay to its lowest energy state. The storage and expression of information costs energy, without which structures rich in information inevitably become disordered and meaningless. The synthetic reactions that occur within cells, like the synthetic processes in any factory, require the input of energy. Energy is consumed in the motion of a bacterium or an Olympic sprinter, in the flashing of a firefly or the electrical discharge of an eel. Cells have evolved highly efficient mechanisms for coupling the energy obtained from sunlight or fuels to the many energy-consuming processes they carry out.

Organisms Are Never at Equilibrium with Their Surroundings

One of the first developments in biological evolution must have been an oily membrane that enclosed the water-soluble molecules of the primitive cell, segregating them and allowing them to accumulate to relatively high concentrations. The molecules and ions contained within a living organism differ in kind and in concentration from those in the organism's surroundings. For example, the cells of a freshwater fish contain certain inorganic ions at concentrations far different from those in the surrounding water (Fig. 1–4). Proteins, nucleic acids, sugars, and fats are present in the fish but are essentially absent from the surrounding medium, which contains only simpler molecules such as carbon dioxide, molecular oxygen, and water. Only by continuously expending energy can the fish establish and maintain its constituents at concentrations distinct from those of the surroundings. When the fish dies, its components eventually come to equilibrium with its surroundings.

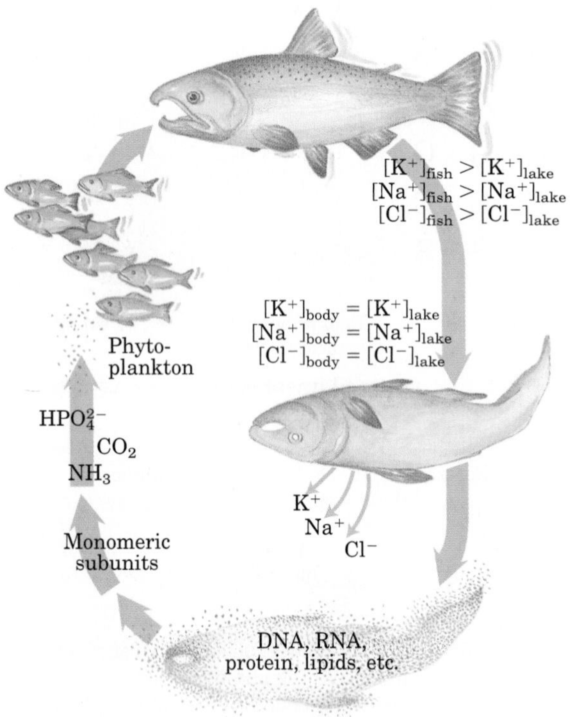

figure 1–4
Living organisms are not at equilibrium with their surroundings. Death and decay restore the equilibrium. During life, the fish uses energy from food to build complex molecules and to concentrate ions from the surroundings. When it dies, it no longer derives energy from food and thus cannot maintain concentration gradients; ions leak out. Inexorably, macromolecular components decay to simpler compounds. These simple compounds serve as nutritional sources for microscopic plants and algae (the phytoplankton), which are then eaten by larger organisms. (By convention, square brackets denote concentration—in this case, of ionic species.)

Molecular Composition Reflects a Dynamic Steady State

Although the chemical composition of an organism may be almost constant through time, the population of molecules within a cell or organism is far from static. Molecules are synthesized and then broken down by continuous chemical reactions, involving a constant flux of mass and energy through the system. The hemoglobin molecules carrying oxygen from your lungs to

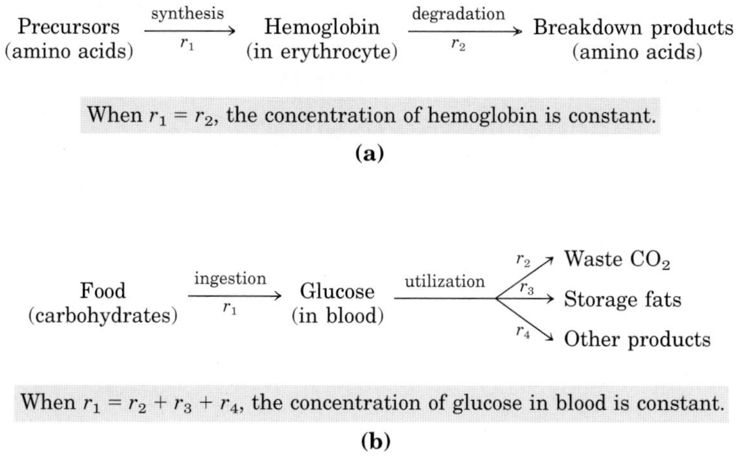

$$\underset{\text{(amino acids)}}{\text{Precursors}} \xrightarrow[r_1]{\text{synthesis}} \underset{\text{(in erythrocyte)}}{\text{Hemoglobin}} \xrightarrow[r_2]{\text{degradation}} \underset{\text{(amino acids)}}{\text{Breakdown products}}$$

When $r_1 = r_2$, the concentration of hemoglobin is constant.

(a)

$$\underset{\text{(carbohydrates)}}{\text{Food}} \xrightarrow[r_1]{\text{ingestion}} \underset{\text{(in blood)}}{\text{Glucose}} \xrightarrow{\text{utilization}} \begin{cases} \xrightarrow{r_2} \text{Waste } CO_2 \\ \xrightarrow{r_3} \text{Storage fats} \\ \xrightarrow{r_4} \text{Other products} \end{cases}$$

When $r_1 = r_2 + r_3 + r_4$, the concentration of glucose in blood is constant.

(b)

figure 1–5
The dynamic steady state. A dynamic steady state results when the rate of appearance of a cellular component is exactly matched by the rate of its disappearance. In this scheme, r_1, r_2, and so forth, represent the rates of the various processes. In **(a),** a protein (hemoglobin) is synthesized, then degraded. In **(b),** glucose derived from food (or from carbohydrate stores) enters the bloodstream in some tissues (intestine, liver), then leaves the blood to be consumed by metabolic processes in other tissues (heart, brain, skeletal muscle). The dynamic steady-state concentrations of hemoglobin and glucose are maintained by complex mechanisms regulating the relative rates of the processes shown here.

your brain at this moment were synthesized within the past month; by next month they will have been degraded and replaced by new molecules. The glucose you ingested with your most recent meal is now circulating in your bloodstream; before the day is over these particular glucose molecules will have been converted into something else, such as carbon dioxide or fat, and will have been replaced with a fresh supply of glucose. The amounts of hemoglobin and glucose in the blood remain nearly constant because the rate of synthesis or intake of each just balances the rate of its breakdown, consumption, or conversion into some other product (Fig. 1–5). The constancy of concentration is the result of a **dynamic steady state.**

Organisms Transform Energy and Matter from Their Surroundings

Living cells and organisms must perform work to stay alive and to reproduce themselves. The continual synthesis of cellular components requires chemical work; the accumulation and retention of salts and various organic compounds against a concentration gradient involves osmotic work; and the contraction of a muscle or the motion of a bacterial flagellum represents mechanical work. Biochemistry examines the processes by which energy is extracted, channeled, and consumed, so it is essential to develop an understanding of the fundamental principles of **bioenergetics**—the energy transformations and exchanges on which all living organisms depend.

For chemical reactions occurring in solution, we can define a **system** as all of the reactants and products present, the solvent, and the immediate atmosphere—in short, everything within a defined region of space. The system and its surroundings together constitute the **universe.** If the system exchanges neither matter nor energy with its surroundings, it is said to be **closed.** If the system exchanges energy but not matter with its surroundings, it is an **isolated** system; if it exchanges both energy and material with its surroundings, it is an **open** system.

A living organism is an open system; it exchanges both matter and energy with its surroundings. Living organisms use either of two strategies to derive energy from their surroundings: (1) they take up chemical fuels from the environment and extract energy by oxidizing them; or (2) they absorb energy from sunlight.

> Living organisms create and maintain their complex, orderly structures using energy extracted from fuels or sunlight.

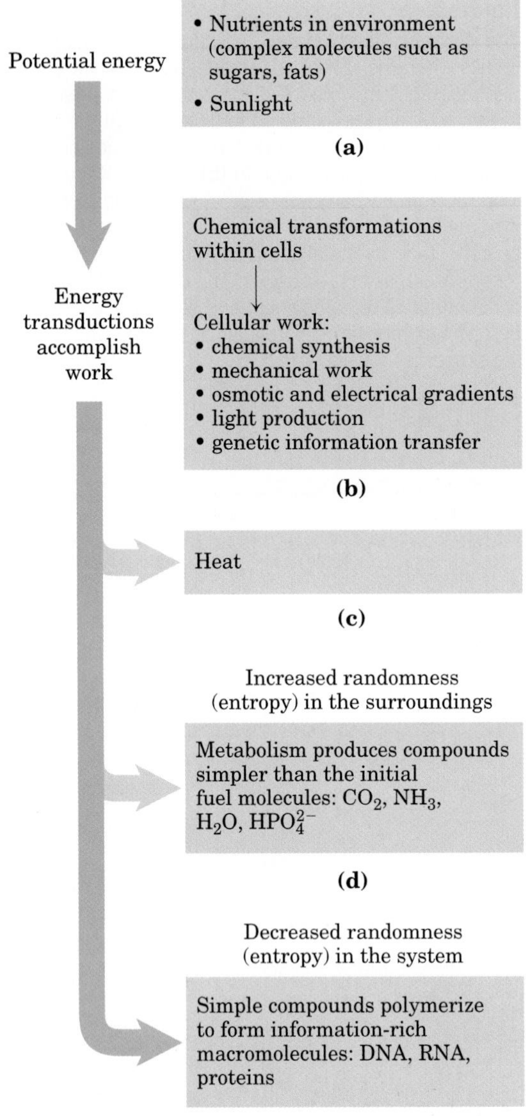

figure 1–6
During metabolic transductions, the randomness of the system plus surroundings (expressed quantitatively as entropy) increases as the potential energy of complex nutrient molecules decreases. Living organisms **(a)** extract energy from their environment; **(b)** convert some of it into useful forms of energy to produce work; **(c)** return some energy to the environment as heat; and **(d)** release end-product molecules that are less well organized than the starting fuel, increasing the entropy of the universe. One effect of all these transformations is **(e)** increased order (decreased randomness) in the form of complex macromolecules. We shall return to a quantitative treatment of entropy in Chapter 14.

The first law of thermodynamics, developed from physics and chemistry but fully valid for biological systems as well, describes the principle of the conservation of energy:

> **In any physical or chemical change, the total amount of energy in the universe remains constant, although the form of the energy may change.**

Cells are consummate transducers of energy, capable of interconverting chemical, electromagnetic, mechanical, and osmotic energy with great efficiency (Fig. 1–6). Biological energy transducers differ from many familiar machines that depend on temperature or pressure differences. The steam engine, for example, converts the chemical energy of fuel into heat, raising the temperature of water to its boiling point to produce steam pressure that drives a mechanical device. The internal combustion engine, similarly, depends upon changes in temperature and pressure. By contrast, all parts of a living organism must operate at about the same temperature and pressure, and heat flow is therefore not a useful source of energy.

> **Living cells are chemical engines that function at constant temperature.**

The Flow of Electrons Provides Energy for Organisms

Nearly all living organisms derive their energy, directly or indirectly, from the radiant energy of sunlight, which arises from thermonuclear fusion reactions occurring in the sun (Fig. 1–7). Photosynthetic cells absorb light energy and use it to drive electrons from water to carbon dioxide, forming energy-rich products such as starch and sucrose and releasing molecular oxygen into the atmosphere (Fig. 1–8). Nonphotosynthetic cells and organisms obtain the energy they need by oxidizing the energy-rich products of photosynthesis and then passing electrons to atmospheric oxygen to form water, carbon dioxide, and other end products, which are recycled in the environment. Virtually all energy transductions in cells can be traced to

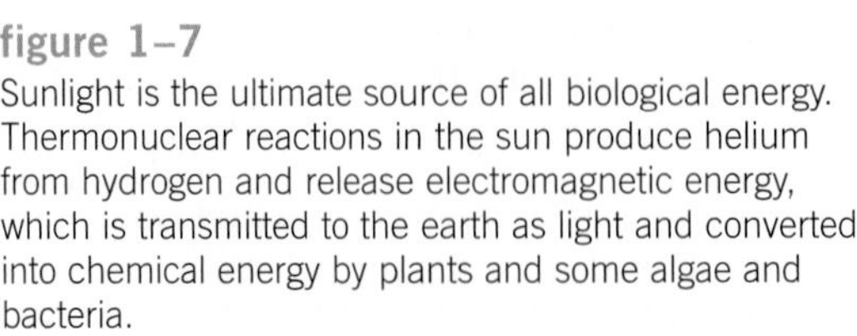

figure 1–7
Sunlight is the ultimate source of all biological energy. Thermonuclear reactions in the sun produce helium from hydrogen and release electromagnetic energy, which is transmitted to the earth as light and converted into chemical energy by plants and some algae and bacteria.

this flow of electrons from one molecule to another, in a "downhill" flow from higher to lower electrochemical potential; as such, it is formally analogous to the flow of electrons in a battery-driven electric circuit. All these reactions involving electron flow are **oxidation-reduction reactions;** some reactant is oxidized (loses electrons) as another is reduced (gains electrons).

> The energy needs of virtually all organisms are provided, directly or indirectly, by solar energy.
>
> The flow of electrons in oxidation-reduction reactions underlies energy transductions in living cells.
>
> Living organisms are interdependent, exchanging energy and matter via the environment.

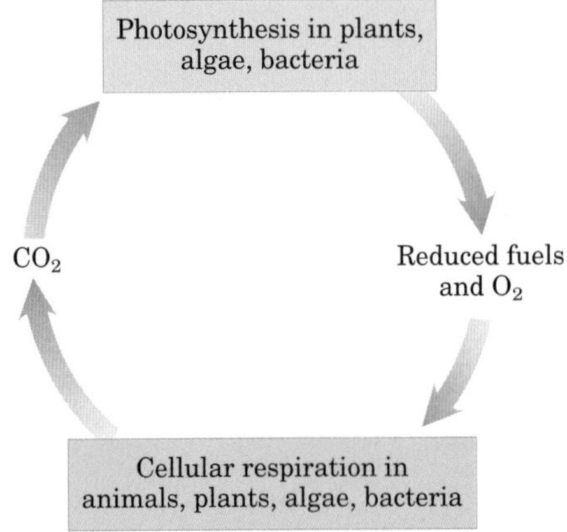

figure 1–8
Photosynthetic organisms (plants, some algae, and some bacteria) are the ultimate providers of fuels—reduced, energy-rich compounds—in the biosphere. The energy of sunlight drives the synthesis of fuels such as sucrose and starch, with O_2 as a by-product. These fuels, or the photosynthetic organisms themselves, are then a source of food for animals, which oxidize the sucrose and starch (using O_2 and producing CO_2) to supply energy. This process of fuel oxidation—cellular respiration—is the energy source for metabolism in both photosynthetic and nonphotosynthetic organisms.

Energy Coupling Links Reactions in Biology

The central issue in bioenergetics is the means by which energy from fuel metabolism or light capture is coupled to energy-requiring reactions. It is instructive to consider the simple mechanical example of energy coupling shown in Figure 1–9a. An object at the top of an inclined plane has a certain amount of potential energy as a result of its elevation. It tends spontaneously to slide down the plane, losing its potential energy of position as it approaches the ground. When an appropriate string-and-pulley device couples the falling object to another, smaller object, the spontaneous downward motion of the larger can lift the smaller, accomplishing a certain amount of work. The amount of energy actually available to do work, called the **free energy, *G*,** will always be somewhat less than the theoretical amount of energy released, because some energy is dissipated as the heat of friction. The greater the elevation of the larger object relative to its final position, the greater is the release of energy as it slides downward, and the greater the amount of work that can be accomplished.

Chemical reactions can also be coupled so that an energy-releasing reaction drives an energy-requiring one. Chemical reactions in closed systems proceed spontaneously until **equilibrium** is reached. When a system is at equilibrium, the rate of product formation exactly equals the rate at which

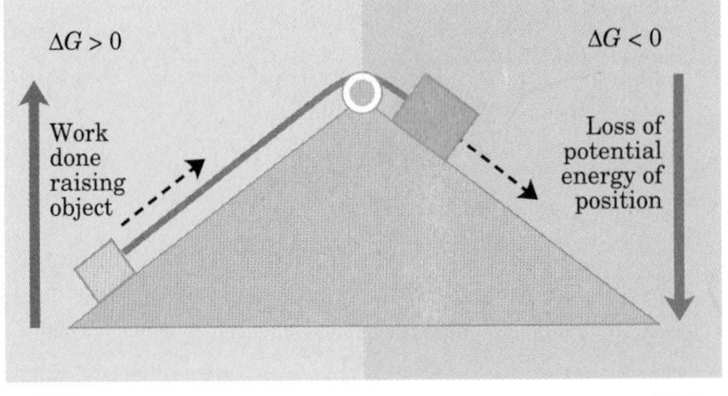

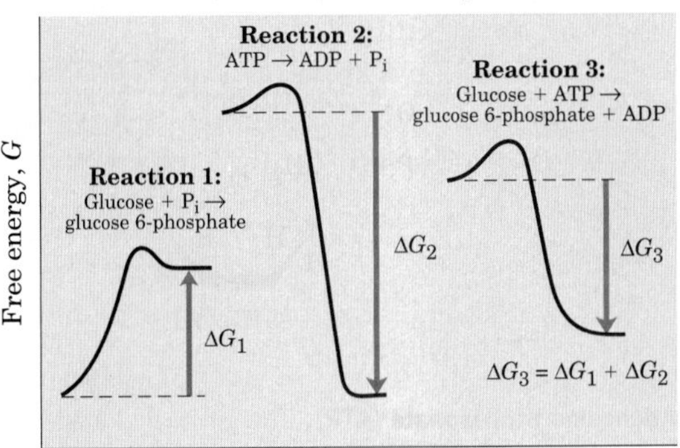

figure 1–9
Energy coupling in mechanical and chemical processes. (a) The downward motion of an object releases potential energy that can do mechanical work. The potential energy made available by spontaneous downward motion, an exergonic process (pink), can be coupled to the endergonic upward movement of another object (blue). **(b)** In reaction 1, the formation of glucose 6-phosphate from glucose and inorganic phosphate, P_i, yields a product of higher energy than the two reactants. For this endergonic reaction, ΔG is positive. In reaction 2, the exergonic breakdown of adenosine triphosphate (ATP; see Fig. 1–10) can drive an endergonic reaction when the two reactions are coupled. The exergonic reaction has a large, negative free-energy change (ΔG_2), and the endergonic reaction has a smaller, positive free-energy change (ΔG_1). The third reaction accomplishes the sum of reactions 1 and 2, and the free-energy change, ΔG_3, is the arithmetic sum of ΔG_1 and ΔG_2. Because the value of ΔG_3 is negative, the overall reaction is exergonic and proceeds spontaneously.

product is converted to reactant. Thus there is no net change in the concentration of reactants and products; a "steady state" is achieved. The energy change as the system moves from its initial state to equilibrium, with no changes in temperature or pressure, is given by the **free-energy change, ΔG.** The magnitude of ΔG depends on the particular chemical reaction and on how far from equilibrium the system is initially. Each compound involved in a chemical reaction contains a certain amount of potential energy, related to the kind and number of its bonds. In reactions that occur spontaneously, the products have less free energy than the reactants, thus the reaction releases free energy, which is then available to do work. Such reactions are **exergonic;** the decline in free energy from reactants to products is expressed as a negative value. **Endergonic** reactions require an input of energy, and their ΔG values are therefore positive. As in mechanical processes, only part of the energy released in exergonic biochemical reactions can be used to accomplish work. In living systems some energy is dissipated as heat or lost to increasing entropy, a measure of randomness, which we will define more rigorously in Chapter 14.

In living organisms, as in the mechanical example in Figure 1–9a, an exergonic reaction can be coupled to an endergonic reaction or process to drive otherwise unfavorable reactions. Figure 1–9b illustrates this principle for the case of glucose 6-phosphate synthesis, a reaction occurring in muscle cells. The simplest way to produce glucose 6-phosphate would be reaction 1, which is endergonic. (P_i is an abbreviation for inorganic phosphate, HPO_4^{2-}. Don't be concerned about the structure of these compounds now; we will describe them in detail later.)

$$\text{Reaction 1: Glucose} + P_i \longrightarrow \text{glucose 6-phosphate} \quad (\text{endergonic}, \Delta G \text{ is positive})$$

In this reaction, the product contains more energy than the reactants.

A second, very exergonic reaction can occur in living cells.

$$\text{Reaction 2: ATP} \longrightarrow \text{ADP} + P_i \quad (\text{exergonic}, \Delta G \text{ is negative})$$

In this reaction, the products contain *less* energy than the reactant—the reaction releases energy. The two chemical reactions share a common intermediate, P_i, which is consumed in reaction 1 and produced in reaction 2. The two reactions can be coupled in the form of a third reaction, which we can write as the sum of reactions 1 and 2, with the common intermediate P_i omitted from both sides of the equation:

$$\text{Reaction 3: Glucose} + \text{ATP} \longrightarrow \text{glucose 6-phosphate} + \text{ADP}$$

Because more energy is released in reaction 2 than is consumed in reaction 1, reaction 3 is exergonic: some energy is released (ΔG_3 in Fig. 1–9b). Living cells thus make glucose 6-phosphate by catalyzing a direct reaction between glucose and ATP, in effect coupling reaction 1 to reaction 2.

The coupling of exergonic reactions with endergonic ones is absolutely central to the energy exchanges in living systems. The mechanism by which energy coupling occurs in biological reactions is via a shared intermediate. We will see that reaction 2 in Figure 1–9b, the breakdown of **adenosine triphosphate (ATP),** is the exergonic reaction that drives many endergonic processes in cells. In fact, ATP (Fig. 1–10) is the major carrier of chemical energy in all cells, coupling endergonic processes to exergonic ones. The terminal phosphoryl group of ATP, shaded pink in Figure 1–10, is transferred to a variety of acceptor molecules, which are thereby activated for further chemical transformation. The adenosine diphosphate (ADP) that remains is recycled (phosphorylated) to ATP, at the expense of either chemical energy (during oxidation of fuels) or solar energy (in photosynthetic cells).

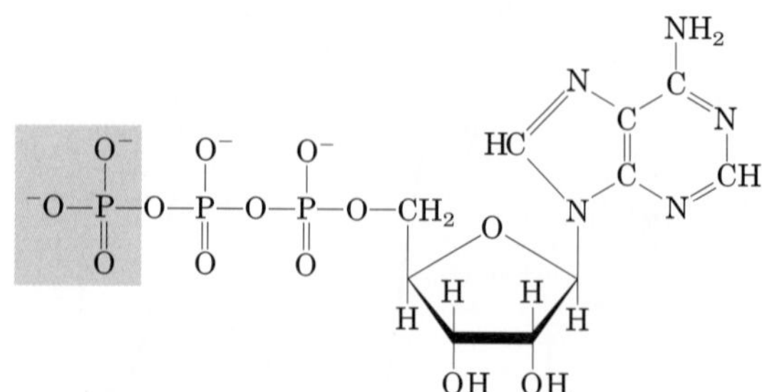

figure 1–10
Adenosine triphosphate (ATP). The removal of the terminal phosphoryl of ATP (shaded pink) is highly exergonic, and this reaction is coupled to many endergonic reactions in the cell as in the example described in Figure 1–9b.

> Endergonic cellular reactions are driven by coupling them to exergonic chemical or photochemical processes through shared chemical intermediates.

Enzymes Promote Sequences of Chemical Reactions

An exergonic reaction does not necessarily proceed rapidly. The path from reactant(s) to product(s) almost invariably involves an energy barrier, called the activation barrier (Fig. 1–11), that must be surmounted for any reaction to occur. The breaking of existing bonds and formation of new ones generally requires the distortion of the existing bonds, creating a **transition state** of higher free energy than either reactant or product. The highest point in the reaction coordinate diagram represents the transition state.

Virtually every cellular chemical reaction occurs at a measurable rate only because of the presence of **enzymes**—biocatalysts that, like all other catalysts, greatly enhance the rate of specific chemical reactions without being consumed in the process. Enzymes lower the energy barrier between reactant and product. The **activation energy** ($\Delta G^{\ddagger}$; Fig. 1–11) required to overcome this energy barrier could in principle be supplied by heating the reaction mixture, thereby increasing the kinetic energy of the molecules, the frequency with which they collide, and the likelihood that they will react. However, this option is not available in living cells, which generally maintain a constant temperature. In fact, many cell components (proteins, membranes) are inactivated by temperatures only a few degrees above an organism's normal internal temperature. Instead, enzymes speed reactions by taking advantage of binding effects. Two or more reactants bind to the enzyme's surface close to each other and with stereospecific orientations that favor the reaction between them. Through this combination of proximity and orientation, the probability of productive collisions between reactants is increased by orders of magnitude relative to the uncatalyzed process, when reactants are randomly oriented and distributed throughout an aqueous solution. Furthermore, the reactants themselves, in the process of binding to the enzyme, undergo changes in shape that distort them toward the transition state, thereby lowering the activation energy and enormously accelerating the rate of the reaction (Fig. 1–12). The relationship between the activation energy and reaction rate is exponential; a small decrease in $\Delta G^{\ddagger}$ results in a very large increase in reaction rate. Enzyme-catalyzed reactions commonly proceed at rates up to 10^{10} to 10^{14} times faster than uncatalyzed reactions.

Metabolic catalysts are, with a few exceptions, proteins. (In a few cases, RNA molecules have catalytic roles, as discussed in Chapter 26.) Again with a few exceptions, each enzyme protein catalyzes a specific reaction, and each reaction in a cell is catalyzed by a different enzyme. Thousands of different enzymes are therefore required by each cell. The multiplicity of enzymes, their specificity (the ability to discriminate between reactants), and their susceptibility to regulation give cells the capacity to lower activation barriers selectively. This selectivity is crucial for the effective regulation of cellular processes.

The thousands of enzyme-catalyzed chemical reactions in cells are functionally organized into many different sequences of consecutive reactions called **pathways,** in which the product of one reaction becomes the reactant in the next (Fig. 1–13). Some pathways degrade organic nutrients

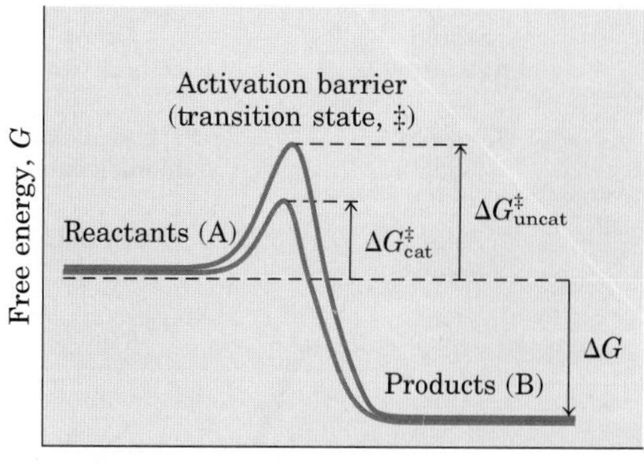

figure 1–11
Energy changes during a chemical reaction. An activation barrier, representing the transition state, must be overcome in the conversion of reactants (A) into products (B), even though the products are more stable than the reactants, as indicated by a large, negative free-energy change (ΔG). The energy required to overcome the activation barrier is the activation energy ($\Delta G^{\ddagger}$). Enzymes catalyze reactions by lowering the activation barrier. They bind the transition-state intermediates tightly, and the binding energy of this interaction effectively reduces the activation energy from $\Delta G^{\ddagger}_{uncat}$ to $\Delta G^{\ddagger}_{cat}$. (Note that the activation energy is unrelated to the free-energy change of the reaction, ΔG.)

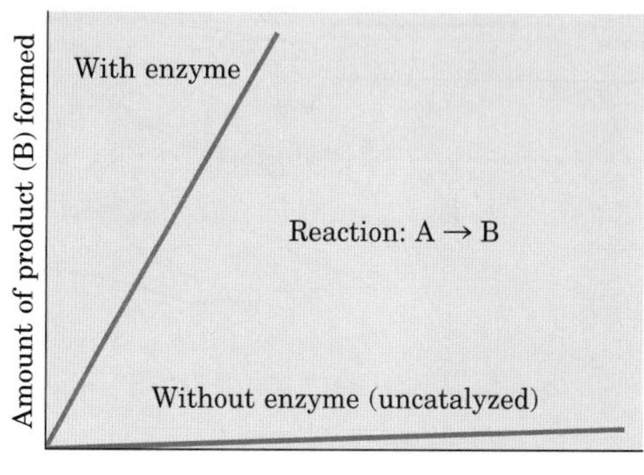

figure 1–12
An enzyme increases the rate of a specific chemical reaction. In the presence of an enzyme specific for the conversion of reactant A into product B, the rate of the reaction may increase by many orders of magnitude (powers of ten) over that of the uncatalyzed reaction. Like all catalysts, the enzyme is not consumed in the process; one enzyme molecule can act repeatedly to convert many molecules of A to B.

$$A \xrightarrow{\text{enzyme 1}} B \xrightarrow{\text{enzyme 2}} C \xrightarrow{\text{enzyme 3}} D \xrightarrow{\text{enzyme 4}} E \xrightarrow{\text{enzyme 5}} F$$

figure 1–13
A linear metabolic pathway. In this pathway, the reactant A is converted in five steps into the product F, with each step catalyzed by an enzyme specific for that reaction.

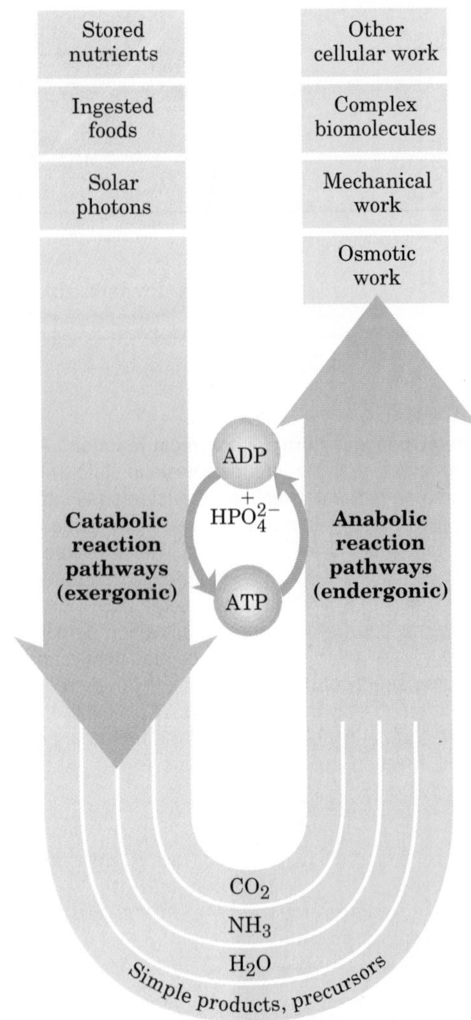

figure 1–14
ATP is the shared chemical intermediate linking energy-releasing to energy-requiring cell processes. Its role in the cell is analogous to that of money in an economy: it is "earned/produced" in exergonic reactions and "spent/consumed" in endergonic ones.

into simple end products in order to extract chemical energy and convert it into a form useful to the cell. Together these degradative, free-energy-yielding reactions are designated **catabolism.** Other pathways start with small precursor molecules and convert them to progressively larger and more complex molecules, including proteins and nucleic acids. Such synthetic pathways invariably require the input of energy and, taken together, represent **anabolism.** The overall network of enzyme-catalyzed pathways constitutes cellular **metabolism.** ATP is the major connecting link (the shared intermediate) between the catabolic and anabolic components of this network (Fig. 1–14). The linked systems of enzyme-catalyzed reactions that act on the main constituents of cells—proteins, fats, sugars, and nucleic acids—are virtually identical in all living organisms.

[ATP is the universal carrier of metabolic energy, linking catabolic and anabolic pathways.]

Metabolism Is Regulated to Achieve Balance and Economy

Not only do living cells simultaneously synthesize thousands of different kinds of carbohydrate, fat, protein, and nucleic acid molecules and their simpler subunits, they do so in the precise proportions required by the cell. For example, during rapid cell growth, the precursors of proteins and nucleic acids must be made in large quantities, whereas in nongrowing cells the requirement for these precursors is much reduced. Key enzymes in each metabolic pathway are regulated so that each type of precursor molecule is produced in a quantity appropriate to the current requirements of the cell. Consider the pathway that leads to the synthesis of isoleucine, one of the amino acids, the monomeric subunits of proteins (Fig. 1–15). If a cell begins to produce more isoleucine than is needed for protein synthesis, the unused isoleucine accumulates. High concentrations of isoleucine inhibit the catalytic activity of the first enzyme in the pathway, immediately slowing the production of the amino acid. Such **feedback inhibition** keeps the production and utilization of each metabolic intermediate in balance.

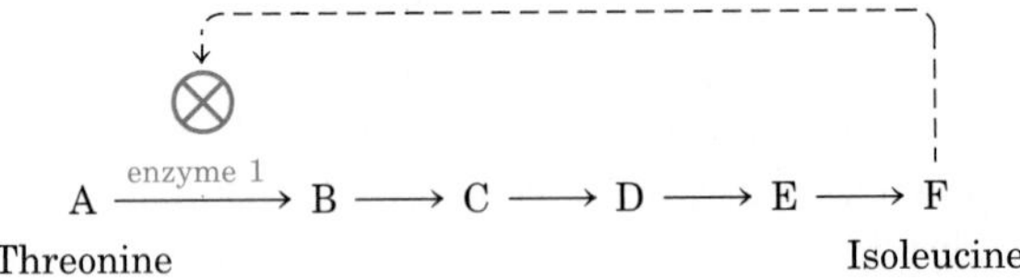

figure 1–15
Feedback inhibition. Regulation by feedback inhibition in a typical synthetic (anabolic) pathway. In the bacterium *E. coli,* the amino acid threonine is converted to another amino acid, isoleucine, in five steps, each catalyzed by a separate enzyme. (The letters A through F represent the compounds, or intermediates, in this pathway.) The accumulation of the product isoleucine (F) causes inhibition of the first reaction in the pathway by binding to the enzyme catalyzing this reaction and reducing its activity.

Living cells also regulate the synthesis of their own catalysts, the enzymes. Thus a cell can switch off the synthesis of an enzyme required to make a given product whenever that product is adequately supplied. These self-adjusting and self-regulating properties allow cells to maintain themselves in a dynamic steady state, despite fluctuations in the external environment.

[Living cells are self-regulating chemical engines, continually adjusting for maximum economy.]

Biological Information Transfer

The continued existence of a biological species requires that its genetic information be maintained in a stable form and, at the same time, be expressed with very few errors. Effective storage and accurate expression of the genetic message defines individual species, distinguishes them from one another, and assures their continuity over successive generations.

Among the seminal discoveries of twentieth-century biology are the chemical nature and the three-dimensional structure of the genetic material, deoxyribonucleic acid, or DNA. The sequence of deoxyribonucleotides in this linear polymer encodes the instructions for forming all other cellular components and provides a template for the production of identical DNA molecules to be distributed to progeny when a cell divides.

Genetic Continuity Is Vested in DNA Molecules

Perhaps the most remarkable of all the properties of living cells and organisms is their ability to reproduce themselves with nearly perfect fidelity for countless generations. This continuity of inherited traits implies constancy, over thousands or millions of years, in the structure of the molecules that contain the genetic information. Very few historical records of civilization, even those etched in copper or carved in stone, have survived for a thousand years. But there is good evidence that the genetic instructions in living organisms have remained nearly unchanged over very much longer periods; many bacteria have nearly the same size, shape, and internal structure and contain the same kinds of precursor molecules and enzymes as those that lived a billion years ago (Fig. 1–16).

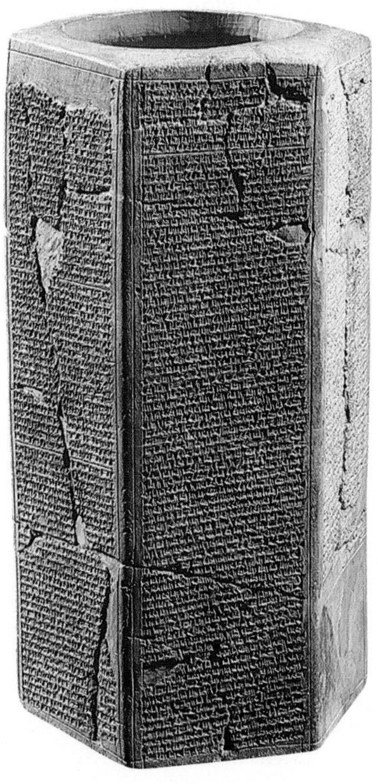

(a)

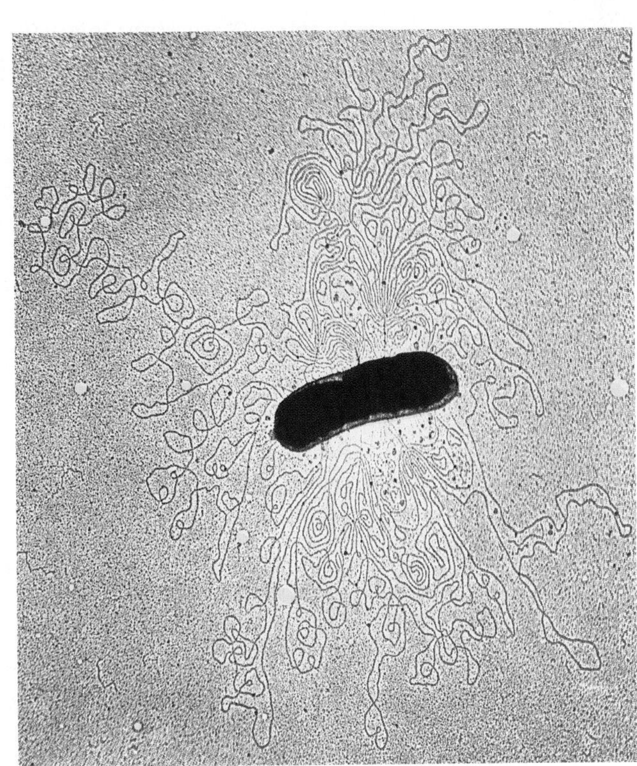

(b)

figure 1–16
Two ancient scripts. **(a)** The Prism of Sennacherib, inscribed in about 700 B.C., describes in characters of the Assyrian language some historical events during the reign of King Sennacherib. The Prism contains about 20,000 characters, weighs about 50 kg, and has survived almost intact for about 2,700 years. **(b)** The single DNA molecule of the bacterium *E. coli*, seen leaking out of a disrupted cell, is hundreds of times longer than the cell itself and contains all of the encoded information necessary to specify the cell's structure and functions. The bacterial DNA contains about 10 million characters (nucleotides), weighs less than 10^{-10} g, and has undergone only relatively minor changes during the past several million years. The yellow spots and dark specks in this colorized electron micrograph are artifacts of the preparation.

Hereditary information is preserved in DNA, a long, thin organic polymer so fragile that it will fragment from the shear forces arising in a solution that is stirred or pipetted. A human sperm or egg, carrying the accumulated hereditary information of millions of years of evolution, transmits these instructions in the form of DNA molecules, in which the linear sequence of covalently linked nucleotide subunits encodes the genetic message.

The Structure of DNA Allows for Its Repair and Replication with Near-Perfect Fidelity

The capacity of living cells to preserve their genetic material and to duplicate it for the next generation results from the structural complementarity between the two halves of the DNA molecule (Fig. 1–17). The basic unit of DNA is a linear polymer of four different monomeric subunits, **deoxyribonucleotides** (Fig. 1–3), arranged in a precise linear sequence. It is this linear sequence that encodes the genetic information. Two of these polymeric strands are twisted about each other to form the DNA double helix, in which each monomeric subunit in one strand pairs specifically with a complementary subunit in the opposite strand. Before a cell divides, the two DNA strands separate and each serves as a template for the synthesis of a new complementary strand, generating two identical double-helical molecules, one for each daughter cell. If one strand is damaged, continuity of information is assured by the information present in the other strand, which acts as a template for repair of the damage.

> Genetic information is encoded in the linear sequence of four kinds of subunits of DNA.
>
> The double-helical DNA molecule contains an internal template for its own replication and repair.

Changes in the Hereditary Instructions Allow Evolution

Despite the near-perfect fidelity of genetic replication, infrequent, unrepaired mistakes in the replication process produce changes in the nucleotide sequence of DNA, representing a genetic **mutation** (Fig. 1–18). Incorrectly repaired damage to one of the DNA strands has the same effect. Mutations can change the instructions for producing cellular components. Many mutations are harmful or even lethal to the organism; they may, for example, cause the synthesis of a defective enzyme that is not able to catalyze an essential metabolic reaction. Occasionally, a mutation better equips an organism or cell to survive in its environment. The mutant enzyme

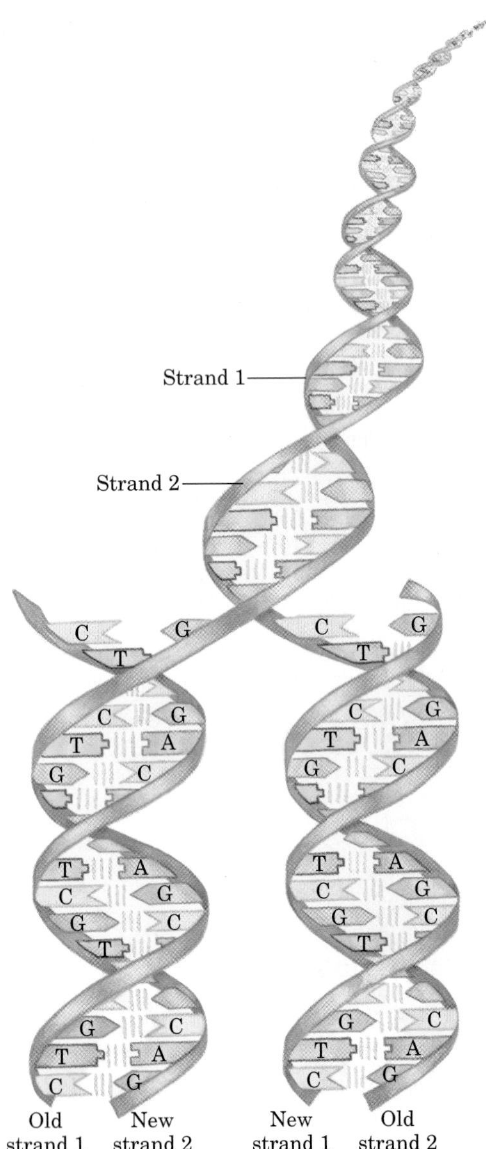

figure 1–17

The complementary structure of DNA. Complementarity between the two strands accounts for the accurate replication essential for genetic continuity. DNA is a linear polymer of covalently joined subunits, the four deoxyribonucleotides: deoxyadenylate (A), deoxyguanylate (G), deoxycytidylate (C), and deoxythymidylate (T). Each nucleotide has the intrinsic ability, due to its precise three-dimensional structure, to associate very specifically but noncovalently with one other nucleotide in the complementary chain: A always associates with its complement T, and G with its complement C. Thus, in the double-stranded DNA molecule, the entire sequence of nucleotides in one strand is **complementary** to the sequence in the other; wherever G occurs in strand 1, C occurs in strand 2; wherever A occurs in strand 1, T occurs in strand 2. The two strands of the DNA, held together by a large number of hydrogen bonds (represented here by vertical blue lines) between the pairs of complementary nucleotides, twist about each other to form the DNA double helix. In DNA replication, prior to cell division, the two strands of the original DNA separate and two new strands are synthesized, each with a sequence complementary to one of the original strands. The result is two double-helical DNA molecules, each identical to the original DNA.

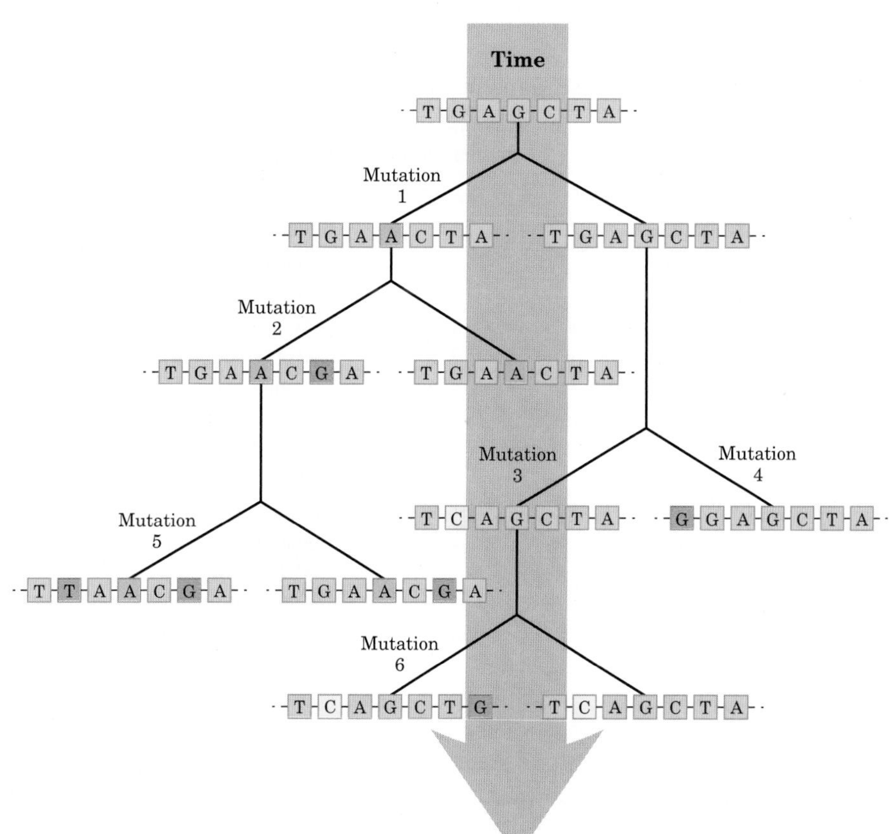

figure 1–18
Role of mutation in evolution. The gradual accumulation of mutations over long periods of time results in new biological species, each with a unique DNA sequence. At the top is shown a short segment of a gene in a hypothetical progenitor organism. With the passage of time, changes in nucleotide sequence (mutations, indicated here by colored boxes) occur, *one nucleotide at a time,* resulting in progeny with different DNA sequences. These mutant progeny themselves undergo occasional mutations, yielding their own progeny differing by two or more nucleotides from the original sequence. When two lineages have changed so much by this mechanism that they can no longer interbreed, a new species has been created.

might, for example, have acquired a slightly different specificity, so that it is now able to use as a reactant some compound that the cell was previously unable to metabolize. If a population of cells were to find itself in an environment where that compound was the only available source of fuel, the mutant cell would have an advantage over the other, unmutated **(wild-type)** cells in the population. The mutant cell and its progeny would survive in the new environment, whereas wild-type cells would starve and be eliminated.

Chance genetic variations in individuals in a population, combined with natural selection (survival and reproduction of the fittest individuals in a challenging or changing environment), have resulted in the evolution of an enormous variety of organisms, each adapted to life in a particular ecological niche.

Carolus Linnaeus
1707–1778

Molecular Anatomy Reveals Evolutionary Relationships

The eighteenth-century naturalist Carolus Linnaeus recognized the anatomic similarities and differences among living organisms and provided a framework for assessing the relatedness of species. Charles Darwin, in the nineteenth century, gave us a unifying hypothesis to explain the phylogeny of modern organisms—the origin of different species from a common ancestor. Now biochemical research in the twentieth century has revealed the molecular anatomy of cells of different species—the subunit sequences and the three-dimensional structures of individual nucleic acids and proteins. Biochemists have an enormously rich and increasing treasury of evidence that can be used to analyze evolutionary relationships and to refine evolutionary theory. The nucleotide sequences of entire genomes (the genome is

Charles Darwin
1809–1882

table 1–1

Some Organisms Whose Genomes Have Been Completely Sequenced

Organism	Genome size (million bases)	Biological interest
Mycoplasma pneumoniae	0.8	Causes pneumonia
Treponema pallidum	1.1	Causes syphilis
Borrelia burgdorferi	1.3	Causes Lyme disease
Helicobacter pylori	1.7	Causes gastric ulcers
Methanococcus jannaschii	1.7	Grows at 85 °C!
Haemophilus influenzae	1.8	Causes bacterial influenza
Methanobacterium thermoautotrophicum	1.8	Member of the Archaea
Archaeoglobus fulgidus	2.2	High-temperature methanogen
Synechocystis sp.	3.6	Cyanobacterium
Bacillus subtilis	4.2	Common soil bacterium
Escherichia coli	4.6	Some strains cause toxic shock syndrome
Saccharomyces cerevisiae	12.1	Unicellular eukaryote
Caenorhabditis elegans	97	Multicellular roundworm

the complete genetic endowment of an organism) have been determined (Table 1–1). The genomic sequences of a number of eubacteria and an archaebacterium, a eukaryotic microorganism (*Saccharomyces cerevisiae*), and a multicellular animal (*Caenorhabditis elegans*) have been determined; those of a plant (*Arabidopsis thaliana*) and even of *Homo sapiens* will soon be known. With such sequences in hand, highly detailed and quantitative comparisons among species will provide deep insight into the evolutionary process. Thus far, the molecular phylogeny derived from gene sequences is consistent with, but in many cases more precise than, the classical phylogeny based on macroscopic structures. Molecular structures and mechanisms have been conserved in evolution even though organisms have continuously diverged at the level of gross anatomy. At the molecular level, the basic unity of life is readily apparent; crucial molecular structures and mechanisms are remarkably similar from the simplest to the most complex organisms. Biochemistry makes possible the discovery of the unifying features common to all life.

The Linear Sequence in DNA Encodes Proteins with Three-Dimensional Structures

The information in DNA is encoded as a linear (one-dimensional) sequence of the nucleotide subunits, but the expression of this information results in a three-dimensional cell. This change from one to three dimensions occurs in two phases. A linear sequence of deoxyribonucleotides in DNA codes (through an intermediary, RNA) for the production of a protein with a corresponding linear sequence of amino acids (Fig. 1–19). The protein folds into a particular three-dimensional shape, determined by its amino acid sequence and stabilized primarily by noncovalent interactions. Although the final shape of the folded protein is dictated by its amino acid sequence, the folding process is aided by proteins that act as "molecular chaperones," discouraging incorrect folding. The precise three-dimensional structure, or **native conformation,** is crucial to the protein's function.

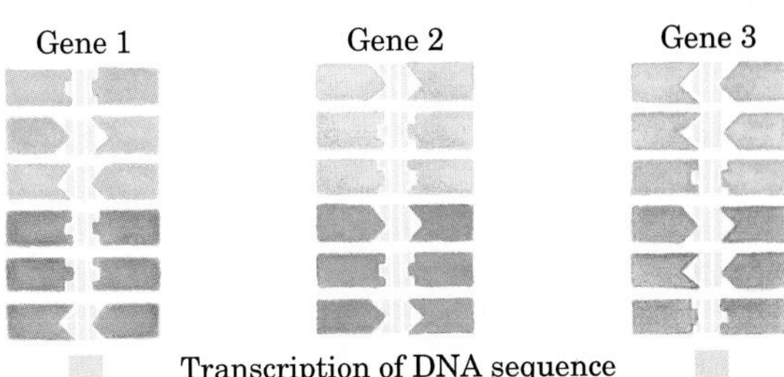

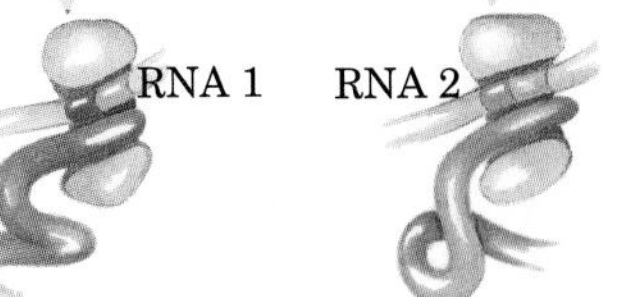

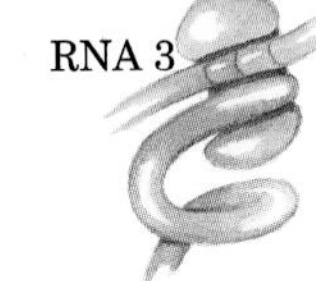

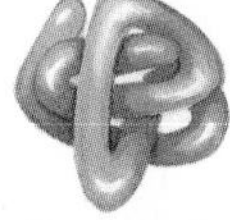

figure 1–19
Linear sequences of deoxyribonucleotides in DNA, arranged into units known as genes, are transcribed into ribonucleic acid (RNA) molecules with complementary ribonucleotide sequences. The RNA sequences are then translated into linear protein chains, which fold into their native three-dimensional shapes, often aided by other proteins called molecular chaperones. Individual proteins commonly associate with other proteins to form supramolecular complexes, stabilized by numerous weak interactions.

[**The linear sequence of amino acids in a protein leads to the acquisition of a unique three-dimensional structure.**]

Once a protein has folded into its native conformation, it may associate noncovalently with other proteins, or with nucleic acids or lipids, to form supramolecular complexes such as chromosomes, ribosomes, and membranes. These complexes are in many cases self-assembling. The individual molecules of these complexes have specific, high-affinity binding sites for each other, and within the cell they spontaneously form functional complexes.

[**Individual macromolecules with specific affinity for other macromolecules self-assemble into supramolecular complexes.**]

Noncovalent Interactions Stabilize Three-Dimensional Structures

The forces that provide stability and specificity to the three-dimensional structures of macromolecules and supramolecular complexes are mostly noncovalent interactions. These interactions, individually weak but collectively strong, include hydrogen bonds, ionic interactions among charged groups, van der Waals interactions, and hydrophobic interactions among nonpolar groups. These weak interactions are transient; individually they form and break in small fractions of a second. The transient nature of noncovalent interactions gives macromolecules a flexibility that is critical to their function. Furthermore, the large number of noncovalent interactions in a single macromolecule makes it unlikely that at any given moment all the interactions will be broken; thus macromolecular structures are stable over time.

[**Three-dimensional biological structures combine the properties of flexibility and stability.**]

For example, the double-helical DNA molecule, with its complementary strands held together by many weak interactions, has enough flexibility to allow strand separation during DNA replication (Fig. 1–17), yet enough stability to ensure genetic continuity.

Noncovalent interactions are also central to the specificity and catalytic efficiency of enzymes. Enzymes bind transition-state intermediates through numerous weak but precisely oriented interactions. Because the weak interactions are flexible, the enzyme-substrate complex survives the structural distortions that occur as the reactant is converted into product.

Noncovalent interactions provide the energy for self-assembly of macromolecules by stabilizing their native conformations relative to their unfolded, random forms. A protein will assume this more stable shape, its native conformation, when the energetic advantages of forming weak interactions outweigh the tendency of the protein chain to assume random forms.

The Physical Roots of the Biochemical World

We can now summarize the various principles of the molecular logic of life:

A living cell is a self-contained, self-assembling, self-adjusting, self-perpetuating constant-temperature system of molecules that extracts free energy and raw materials from its environment.

The cell uses this energy to maintain itself in a dynamic steady state, far from equilibrium with its surroundings.

The many chemical transformations within cells are organized into a network of reaction pathways, promoted at each step by specific catalysts, called enzymes, which the cell itself produces. A great economy of parts and processes is achieved by regulation of the activity of key enzymes.

Self-replication through many generations is ensured by the self-repairing, linear information-coding system. Genetic information encoded as sequences of nucleotide subunits in DNA and RNA specifies the sequence of amino acids in each distinct protein, which ultimately determines the three-dimensional structure and function of each protein.

Many weak (noncovalent) interactions, acting cooperatively, stabilize the three-dimensional structures of biological macromolecules and supramolecular complexes, while allowing sufficient flexibility for biological actions.

The chemical reactions and regulatory processes of cells have been highly refined over the course of billions of years of evolution. Nevertheless, no matter how complex it may seem, the organic machinery of living cells functions within the same set of physical laws that governs the operation of inanimate machines.

This set of principles has been most thoroughly validated in studies of unicellular organisms (such as the bacterium *E. coli*), which are exceptionally amenable to biochemical and genetic investigation. Multicellular organisms must solve certain problems not encountered by unicellular organisms, such as the differentiation of the fertilized egg into specialized cell types. Yet here, too, the same principles have been found to apply. Can such simple and mechanical statements apply to humans as well, with their extraordinary capacity for thought, language, and creativity? The pace of recent biochemical progress toward understanding such processes as gene regulation, cellular differentiation, communication among cells, and neural function has been extraordinarily fast and is accelerating. The success of biochemical methods in solving and redefining these problems justifies the hope that the most complex functions of the most highly developed organisms will eventually be explicable in molecular terms.

The relevant facts of biochemistry are many; the student approaching this subject for the first time may occasionally feel overwhelmed. Perhaps the most encouraging development in twentieth-century biology is the realization that, for all of the enormous diversity in the biological world, there is a fundamental unity and simplicity to life. The organizing principles, the biochemical unity, and the evolutionary perspective of diversity provided at the molecular level will serve as helpful frames of reference for the study of biochemistry.

further reading

Asimov, I. (1962) *Life and Energy: An Exploration of the Physical and Chemical Basis of Modern Biology,* Doubleday & Co., Inc., New York.

An engaging account of the role of energy transformations in biology, written for the intelligent layperson by a biochemist and superb writer.

Blum, H.F. (1968) *Time's Arrow and Evolution,* 3rd edn, Princeton University Press, Princeton, NJ.

An excellent discussion of the way the second law of thermodynamics has influenced biological evolution.

Darwin, C. (1964) *On the Origin of Species. A Facsimile of the First Edition (published in 1859),* Harvard University Press, Cambridge, MA.

One of the most influential scientific works ever published.

Dulbecco, R. (1987) *The Design of Life,* Yale University Press, New Haven, CT.

An unusual and excellent introduction to biology.

Fruton, J.S. (1972) *Molecules and Life: Historical Essays on the Interplay of Chemistry and Biology,* Wiley-Interscience, New York.

This series of essays describes the development of biochemistry from Pasteur's studies of fermentation to the present studies of metabolism and information transfer. You may want to refer to these essays as you progress through this textbook.

Fruton, J.S. (1992) *A Skeptical Biochemist,* Harvard University Press, Cambridge, MA.

Jacob, F. (1973) *The Logic of Life: A History of Heredity,* Pantheon Books, Inc., New York. Originally published (1970) as *La logique du vivant: une histoire de l'hérédité,* Editions Gallimard, Paris.

A fascinating historical and philosophical account of the route by which we came to the present molecular understanding of life.

Judson, H.F. (1979) *The Eighth Day of Creation: The Makers of the Revolution in Biology,* Jonathan Cape, London.

A highly readable and authoritative account of the rise of biochemistry and molecular biology in the twentieth century.

Kornberg, A. (1987) The two cultures: chemistry and biology. *Biochemistry* **26,** 6888–6891.

The importance of applying chemical tools to biological problems, described by an eminent practitioner.

Mayr, E. (1997) *This Is Biology: The Science of the Living World,* Belknap Press, Cambridge, MA.

A history of the development of science, with special emphasis on Darwinian evolution, by an eminent Darwin scholar.

Monod, J. (1971) *Chance and Necessity,* Alfred A. Knopf, Inc., New York. [Paperback version (1972) Vintage Books, New York.] Originally published (1970) as *Le hasard et la nécessité,* Editions du Seuil, Paris.

An exploration of the philosophical implications of biological knowledge.

Schrödinger, E. (1944) *What Is Life?* Cambridge University Press, New York. [Reprinted (1956) in *What Is Life? and Other Scientific Essays,* Doubleday Anchor Books, Garden City, NY.]

A thought-provoking look at life, written by a prominent physical chemist.

chapter

2 Cells

Cells are the structural and functional units of all living organisms. The smallest organisms consist of single cells and are microscopic, whereas larger organisms are multicellular. The human body, for example, contains at least 10^{14} cells. Unicellular organisms are found in great variety throughout virtually every environment from Antarctica to hot springs to the inner recesses of larger organisms. Multicellular organisms contain many different types of cells, which vary in size, shape, and specialized function. Yet no matter how large and complex the organism, each of its cells retains some individuality and independence.

Despite their many differences, cells of all kinds share certain structural features (Fig. 2–1). The **plasma membrane** defines the periphery of the cell, separating its contents from the surroundings. It is composed of enormous numbers of lipid and protein molecules, held together primarily by noncovalent hydrophobic interactions (p. 17), forming a thin, tough, pliable, hydrophobic layer around the cell. The membrane is a barrier to the free passage of inorganic ions and most other charged or polar compounds. Transport proteins in the plasma membrane allow the passage of certain ions and molecules. Other membrane proteins include receptors that transmit signals from the outside to the inside of the cell and enzymes that participate in membrane-associated reaction pathways.

Because the individual lipids and proteins of the plasma membrane are not covalently linked, the entire structure is remarkably flexible, allowing changes in the shape and size of the cell. As a cell grows, newly made lipid and protein molecules are inserted into its plasma membrane; cell division produces two cells, each with its own membrane. Growth and fission occur without loss of membrane integrity. In a reversal of the fission process, two separate membrane surfaces can fuse, also without loss of integrity. Membrane fusion and fission are central to mechanisms of transport into and out of cells known as endocytosis and exocytosis, respectively.

The internal volume bounded by the plasma membrane, the **cytoplasm,** is composed of an aqueous solution, the **cytosol,** and a variety of insoluble, suspended particles (Fig. 2–1). The cytosol is a highly concentrated aqueous solution with a complex composition and gel-like consis-

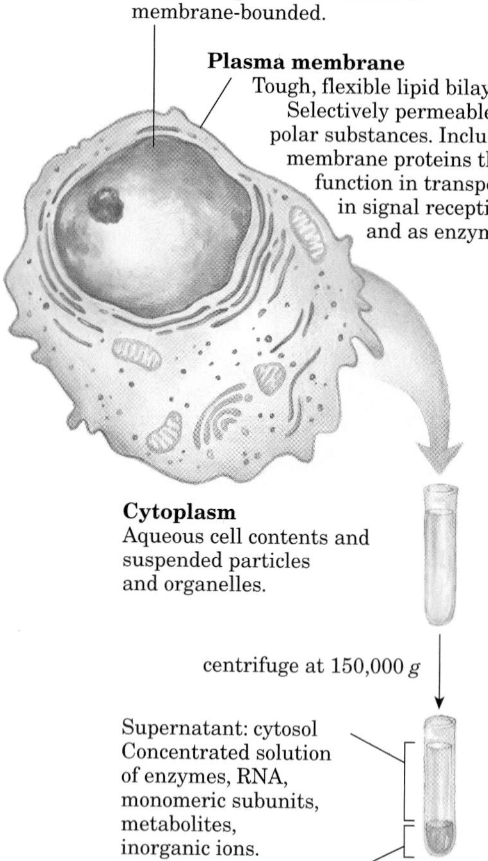
Nucleus (eukaryotes) or nucleoid (bacteria)
Contains genetic material–DNA and associated proteins. Nucleus is membrane-bounded.
Plasma membrane
Tough, flexible lipid bilayer. Selectively permeable to polar substances. Includes membrane proteins that function in transport, in signal reception, and as enzymes.
Cytoplasm
Aqueous cell contents and suspended particles and organelles.
centrifuge at 150,000 g
Supernatant: cytosol
Concentrated solution of enzymes, RNA, monomeric subunits, metabolites, inorganic ions.
Pellet: particles and organelles
Ribosomes, storage granules, mitochondria, chloroplasts, lysosomes, endoplasmic reticulum.

figure 2–1
The universal features of living cells. All cells have a nucleus or nucleoid, a plasma membrane, and cytoplasm. The cytosol is defined operationally as that portion of the cytoplasm that remains in the supernatant after centrifugation of a cell extract at 150,000 *g* for 1 hour.

tency. Dissolved in the cytosol are many enzymes and the RNA molecules that encode them; the monomeric subunits (amino acids and nucleotides) from which these macromolecules are assembled; hundreds of small organic molecules called **metabolites,** intermediates in biosynthetic and degradative pathways; **coenzymes,** compounds of M_r 200 to 1,000 that are essential participants in many enzyme-catalyzed reactions; and inorganic ions.

Among the particles suspended in the cytosol are supramolecular complexes and, in almost all nonbacterial cells, a variety of membrane-bounded organelles containing specialized metabolic machinery. **Ribosomes,** small particles 18 to 22 nm in diameter (1 nm is 10^{-9} m) that are composed of over 50 different protein and RNA molecules, are the sites at which protein synthesis occurs. Ribosomes engaged in protein synthesis often occur in clusters called **polysomes** (polyribosomes) held together by a strand of messenger RNA. Also present in the cytoplasm of many cells are granules or droplets containing stored nutrients such as starch and fat.

All living cells have, for at least some part of their life, either a **nucleus** or a **nucleoid,** in which the **genome** (the complete set of genes, composed of DNA) is stored and replicated. The DNA molecules are always far longer than the cells themselves and are tightly folded and packed within the nucleus or nucleoid as supramolecular complexes of DNA with specific proteins. The bacterial nucleoid is not separated from the cytoplasm by a membrane, but in higher organisms the nuclear material is enclosed within a double membrane, the nuclear envelope. Cells with nuclear envelopes are called **eukaryotes** (Greek *eu,* "true," and *karyon,* "nucleus"); those without nuclear envelopes—bacterial cells—are **prokaryotes** (Greek *pro,* "before").

Unlike bacteria, eukaryotes have a variety of other membrane-bounded organelles in their cytoplasm, including mitochondria, endoplasmic reticulum, Golgi complexes, lysosomes and vacuoles (related organelles found in animal and plant cells, respectively), and, in photosynthetic cells, chloroplasts.

In this chapter we briefly review the evolutionary relationships among some commonly studied cells and organisms and the structural features that distinguish cells of various types. Our main focus is on eukaryotic cells. Also discussed in brief are the cellular parasites known as viruses.

Cellular Dimensions

Most cells are microscopic, invisible to the unaided eye. Animal and plant cells are typically 5 to 100 μm in diameter, and many bacteria are only 1 to 2 μm long (1 μm is 10^{-6} m).

What limits the dimensions of a cell? The lower limit is probably set by the minimum number of each type of biomolecule required by the cell. The smallest cells, certain bacteria known as mycoplasmas, are 300 nm in diameter and have a volume of about 10^{-14} mL. A single bacterial ribosome is about 20 nm in its longest dimension, so a few ribosomes take up a substantial fraction of the volume in a mycoplasmal cell. In a cell of this size, a 1 μM solution of a compound (a typical concentration for some small metabolites) represents only 6,000 molecules.

The upper limit of cell size is probably set by the rate of diffusion of solute molecules in aqueous systems. A bacterial cell that depends upon oxygen-consuming reactions for energy production (an **aerobic** cell) must obtain molecular oxygen (O_2) from the surrounding medium by diffusion through its plasma membrane. The cell is so small, and the ratio of its surface area to its volume is so large, that every part of its cytoplasm is easily

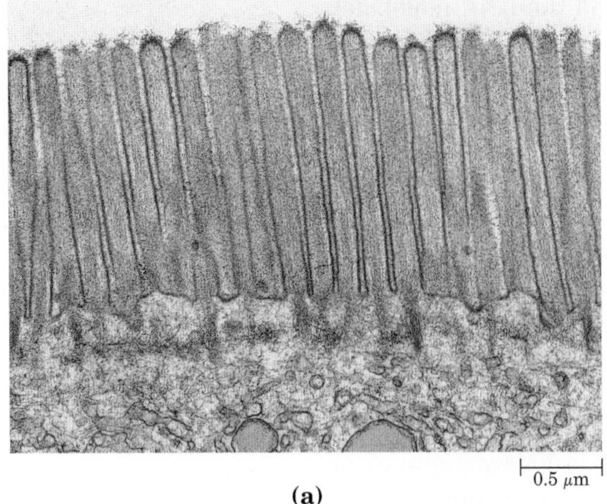

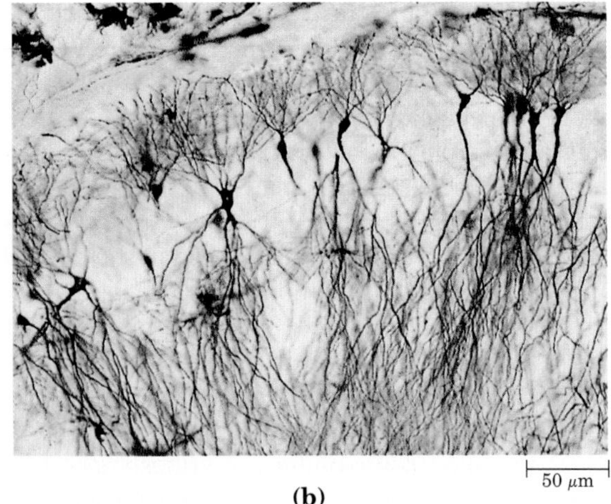

figure 2–2
Convolutions of the plasma membrane, or long, thin extensions of the cytoplasm, increase the surface-to-volume ratio of cells. **(a)** In cells of the intestinal mucosa (the inner lining of the small intestine), the plasma membrane facing the intestinal lumen is folded into microvilli, increasing the area for absorption of nutrients from the intestine. **(b)** Neurons of the hippocampus of the rat brain are several millimeters long, but the long extensions (axons) are only about 10 nm wide.

reached by O_2 diffusing into the cell. As cell size increases, however, surface-to-volume ratio decreases, until metabolism consumes O_2 faster than diffusion can supply it. Aerobic metabolism thus becomes impossible as cell size increases beyond a certain point, placing a theoretical upper limit on the size of the aerobic cell.

There are interesting exceptions to the generalization that cells must be small. The green alga *Nitella* has giant cells several centimeters long. To assure the delivery of nutrients, metabolites, and genetic information (RNA) to all of its parts, each cell is vigorously "stirred" by active cytoplasmic streaming (see Fig. 2–18). The shape of a cell can also help to compensate for its large size. A smooth sphere has the smallest surface-to-volume ratio possible for a given volume. Many large cells, although roughly spherical, have highly convoluted surfaces (Fig. 2–2a), creating larger surface areas for the same volume and thus facilitating the uptake of fuels and nutrients and the release of waste products to the surrounding medium. Other large cells (neurons, for example) have large surface-to-volume ratios because they are long and thin, star-shaped, or highly branched (Fig. 2–2b), rather than spherical.

Cells and Tissues Used in Biochemical Studies

Because all living cells have evolved from the same progenitors, they share certain fundamental similarities. Careful biochemical study of just a few types of cells, however different in biochemical details and varied in superficial appearance, should therefore yield general principles applicable to all cells and organisms. The burgeoning of biological knowledge over the past 150 years has repeatedly supported these propositions. Certain cells, tissues, and organisms have proved more amenable to experimental studies than others. Knowledge in biochemistry is derived primarily from a few representative tissues and organisms, such as the bacterium *Escherichia coli,* the yeast *Saccharomyces cerevisiae*, photosynthetic algae such as *Chlamydomonas*, spinach leaves, rat liver, and the skeletal muscle of several vertebrates. Some biochemical studies focus on the isolation, purification, and characterization of cellular components; other research investigates the metabolic and genetic pathways of living cells.

An experimenter ideally begins the isolation of enzymes and other cellular components with a plentiful and homogeneous source of the material. The component of interest (such as an enzyme or nucleic acid) often represents only a miniscule fraction of the total material, and grams or even kilograms of starting material are needed to obtain a few micrograms of the purified component. A homogeneous source of an enzyme or nucleic acid, in which all the cells are genetically and biochemically identical, leaves no doubt about which cell type yielded the purified component and makes it safer to extrapolate the results of in vitro studies to the situation in vivo. A large culture of bacterial or protistan cells (*E. coli, S. cerevisiae,* or *Chlamydomonas,* for example), all derived by division from the same parent and therefore genetically identical, meets the requirement for a plentiful and homogeneous source. Individual tissues from laboratory animals (rat liver, pig brain, rabbit muscle) are plentiful sources of similar, though not identical, cells. Some animal and plant cells proliferate in cell culture, producing populations of identical (cloned) cells in quantities suitable for biochemical analysis.

Genetic mutants in which a defect in a single gene produces a defective protein, which causes a specific functional defect in the cell or organism, are extremely useful in establishing that a certain protein is essential to a particular cellular function. Because it is technically much simpler to produce and detect mutants in bacteria and yeast, these organisms (*E. coli* and *S. cerevisiae,* for example) have been favorite experimental targets for biochemical geneticists. Once the gene for a protein has been isolated, it can often be inserted into a bacterial or yeast cell, which then acts as a biological factory, overproducing the protein. With genetic engineering techniques, experimenters can introduce specific mutations into such genes and determine their effects on protein structure and function.

An organism that is easy to culture in the laboratory, and has a short generation time, offers significant advantages to the research biochemist. An organism that requires only a few simple precursor molecules in its growth medium can be cultured in the presence of a radioisotopically labeled precursor, and the metabolic fate of that precursor can then be conveniently traced by following the incorporation of the radioactive atoms into its metabolic products. The short generation time of microorganisms (minutes or hours) allows the investigator to follow a labeled precursor or a genetic defect through many generations in a few days. In organisms with generation times of months or years, this is virtually impossible.

Some highly specialized tissues of multicellular organisms are remarkably enriched in some particular component related to their specialized function. For studies on such specific components or functions, biochemists commonly choose the specialized tissue for their experimental systems. For example, vertebrate skeletal muscle is a rich source of actin and myosin; pancreatic secretory cells contain high concentrations of rough endoplasmic reticulum; sperm cells are rich in DNA; liver contains high concentrations of many enzymes of biosynthetic pathways; and spinach leaves contain large numbers of chloroplasts.

Sometimes it is simplicity of structure or function that makes a particular cell or organism attractive as an experimental system. For studies of plasma membrane structure and function, the mature erythrocyte (red blood cell) has been a favorite, because it has no internal membranes to

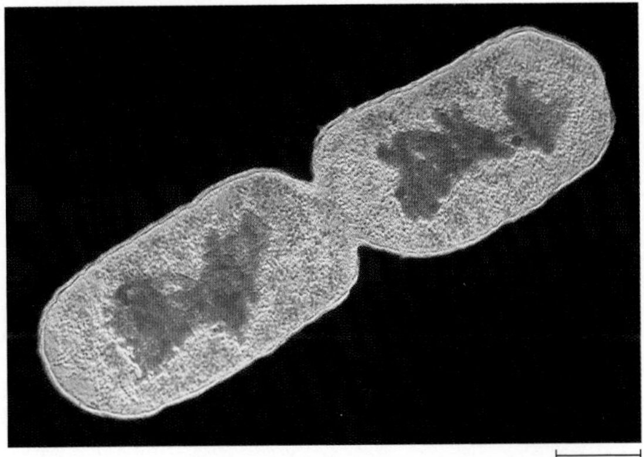

A dividing *Escherichia coli* cell.

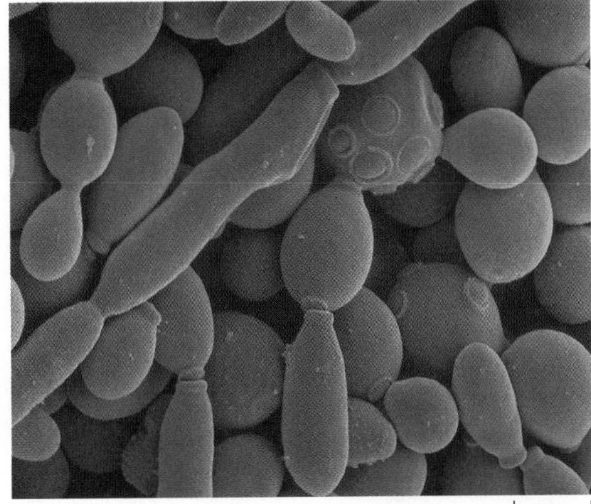

Dividing *Saccharomyces cerevisiae* (baker's yeast) cells.

complicate purification of the plasma membrane. Some bacterial viruses (bacteriophages) have few genes. Their DNA molecules are therefore smaller and much simpler than those of humans or maize plants. It has proved easier to study DNA replication with these viruses than with eukaryotic chromosomes because when a virus infects a bacterial cell, there is a synchronous burst of DNA synthesis, often accompanied by increased levels of the enzymes of DNA replication.

The biochemical description of living cells in this book is a composite, based on studies of many types of cells. Biochemists must always exercise caution in generalizing from results obtained in studies of selected cells, tissues, and organisms and in relating what is observed in vitro to what happens within the living cell.

Evolution and Structure of Prokaryotic Cells

Two large groups of extant prokaryotes can be distinguished on biochemical grounds: **archaebacteria** (Greek *archē,* "origin") and **eubacteria** (Greek *eu,* "true"). Eubacteria inhabit soils, surface waters, and the tissues of other living or decaying organisms. Most common and well-studied bacteria, including *Escherichia coli*, are eubacteria. The archaebacteria are more recently discovered and less well characterized biochemically. Most inhabit more extreme environments—salt lakes, hot springs, bogs, and the ocean depths. The available evidence suggests that the archaebacteria and eubacteria diverged early in evolution and constitute two separate ur-kingdoms or domains, sometimes called Bacteria and Archaea. All eukaryotic organisms, which constitute the third domain, Eukarya, evolved from the same branch that gave rise to the Archaea; archaebacteria are therefore more closely related to eukaryotes than to eubacteria. As complete genomic sequences have become available for archaebacteria (such as *Methanococcus jannaschii*) and eubacteria (*E. coli*), the extent of the divergence between these domains of life has become starkly apparent: less than half the genes of *M. jannaschii* have recognizable homologs in *E. coli*! Furthermore, the genes that encode the proteins required for DNA replication, RNA transcription, and protein synthesis in the archaebacterium *M. jannaschii* are of the same general type as those found in eukaryotes and distinctly different from those involved in the same processes in eubacteria.

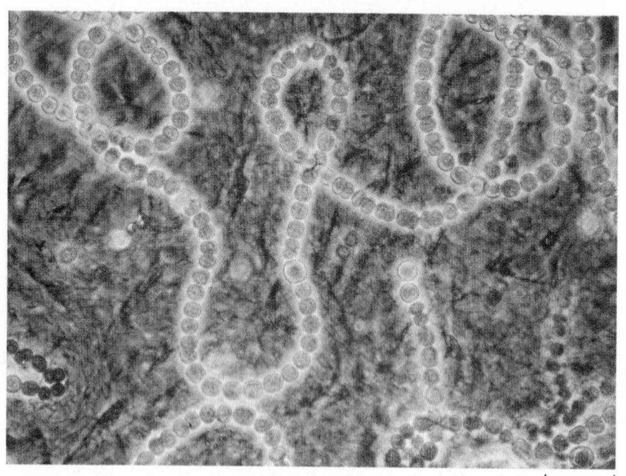

Nostoc sp., a photosynthetic cyanobacterium. This light micrograph shows long strings of the individual round cells.

Within the domains of Bacteria and Archaea are subgroups distinguished by the habitats in which they live. In aerobic habitats with a plentiful supply of oxygen, some resident organisms live by aerobic metabolism; their catabolic processes ultimately result in the transfer of electrons from fuel molecules to oxygen. Other environments are **anaerobic,** virtually devoid of oxygen, and microorganisms adapted to these environments carry out catabolism without it. These bacteria transfer electrons to nitrate (forming N_2), sulfate (forming H_2S), or CO_2 (forming methane, CH_4). Many organisms that have evolved in anaerobic environments are *obligate anaerobes*; they die when exposed to oxygen.

Organisms can be divided into two broad categories according to their energy sources: **phototrophs** (Greek *trophē*, "nourishment") trap sunlight, whereas **chemotrophs** derive their energy from oxidation of a fuel. The phototrophs can be further divided into those that can obtain all needed carbon from CO_2 **(autotrophs)** and those that require organic nutrients **(heterotrophs).** No chemotroph can get its carbon atoms exclusively from CO_2 (that is, there are no autotrophs in this group), but the chemotrophs may be further classified according to a different criterion: whether the fuels they oxidize are inorganic **(lithotrophs)** or organic **(organotrophs).** Most known organisms fall within one of these four broad categories—autotrophs or heterotrophs among the photosynthesizers; lithotrophs or

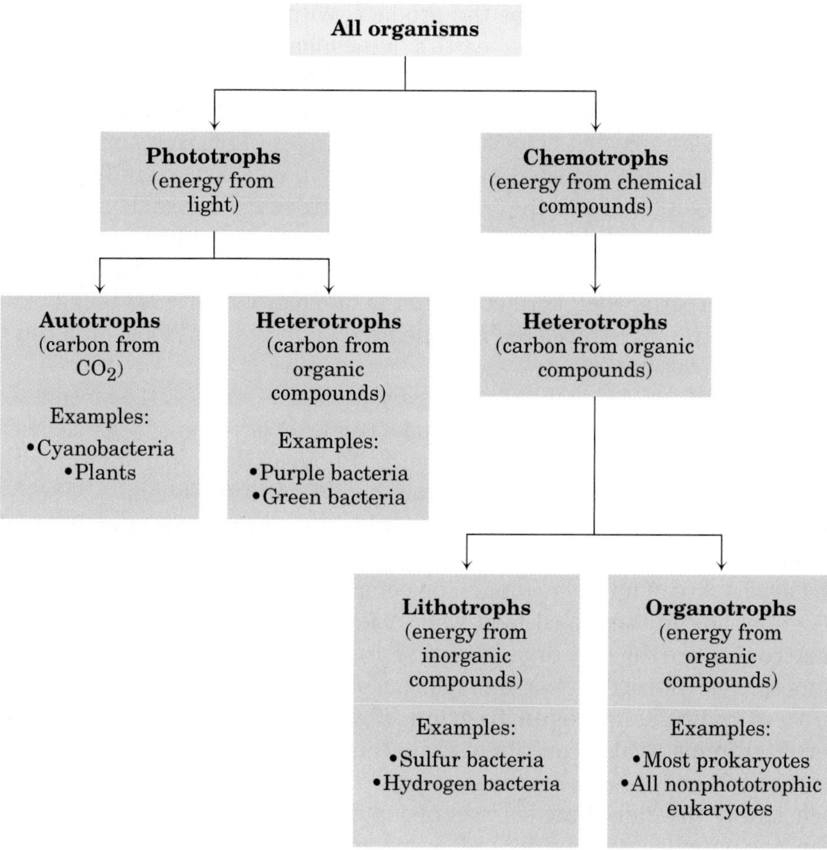

figure 2–3
Organisms can be classified according to their source of energy (sunlight or oxidizable chemical compounds) and their source of carbon for the synthesis of cellular material.

organotrophs among the chemical oxidizers (Fig. 2–3). Thus the prokaryotes have several general modes of obtaining carbon and energy. *E. coli,* for example, is a chemoorganoheterotroph; it requires organic compounds from its environment as fuel and as a source of carbon. Cyanobacteria are photolithoautotrophs; they use sunlight as an energy source and convert CO_2 into biomolecules.

As shown in Figure 2–4, the earliest cells arose about 3.5 billion (3.5×10^9) years ago in the rich mixture of organic compounds, the "primordial soup," of prebiotic times; they were almost certainly chemoheterotrophs. The organic compounds they required were originally synthesized from such components of the early earth's atmosphere as CO, CO_2, N_2, and CH_4 by the nonbiological actions of volcanic heat and lightning (Chapter 3). Early heterotrophs gradually acquired the ability to derive energy from certain compounds in their environment and to use that energy to synthesize more and more of their own precursor molecules, thereby becoming less dependent on outside sources of these compounds. A very significant evolutionary event was the development of pigments capable of capturing visible light from the sun, allowing the cell to use light energy to reduce or "fix" CO_2 into more complex, organic compounds. The original electron donor for these **photosynthetic** organisms was probably H_2S, yielding elemental sulfur or sulfate (SO_4^{2-}) as the byproduct, but later cells developed the enzymatic capacity to use H_2O as the electron donor in photosynthetic reactions, eliminating O_2 as waste. Cyanobacteria are the modern descendants of these early photosynthetic oxygen producers.

The atmosphere of the earth in the earliest stages of biological evolution was nearly devoid of oxygen, and the earliest cells were therefore anaerobic. Under these conditions, chemoheterotrophs could oxidize organic compounds to CO_2 by passing electrons not to O_2, but to acceptors

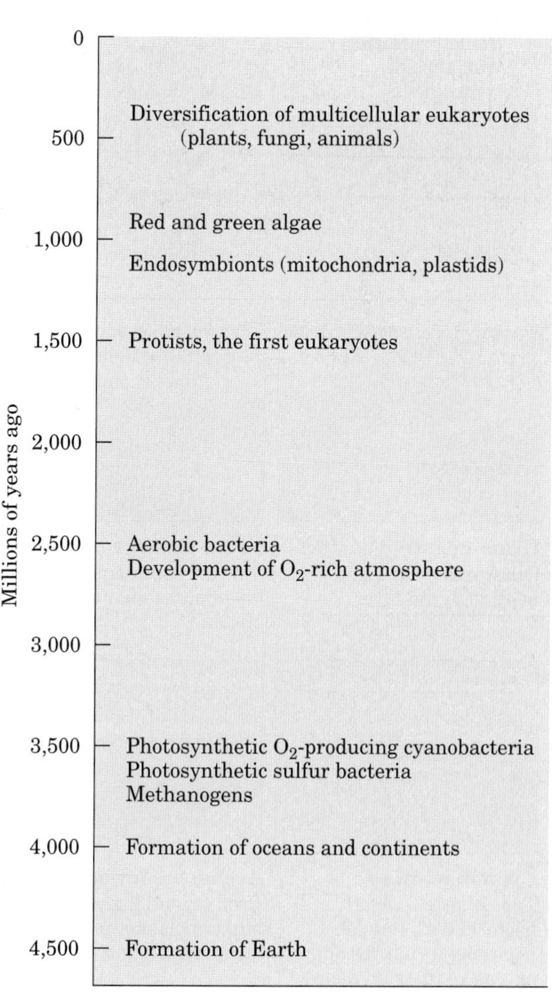

figure 2–4
Landmarks in the evolution of life on Earth.

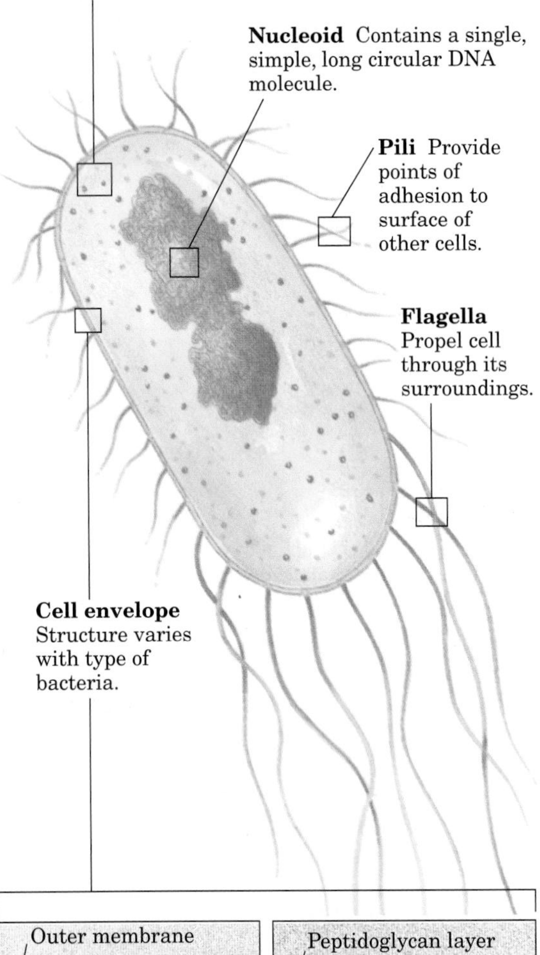
Ribosomes Bacterial ribosomes are smaller than eukaryotic ribosomes, but serve the same function—protein synthesis from an RNA message.
Nucleoid Contains a single, simple, long circular DNA molecule.
Pili Provide points of adhesion to surface of other cells.
Flagella Propel cell through its surroundings.
Cell envelope Structure varies with type of bacteria.

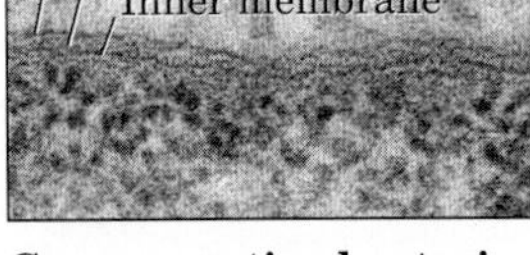
Outer membrane
Peptidoglycan layer
Inner membrane

Gram-negative bacteria Outer membrane and peptidoglycan layer

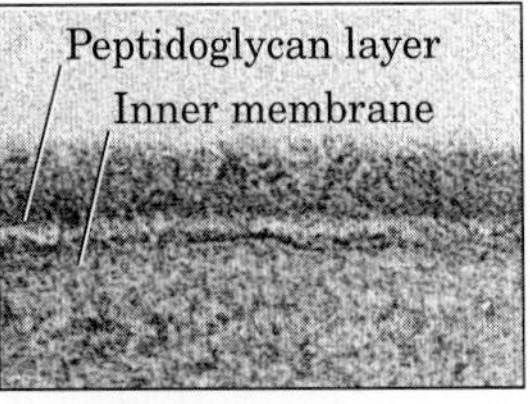
Peptidoglycan layer
Inner membrane

Gram-positive bacteria Thicker peptidoglycan layer; outer membrane absent

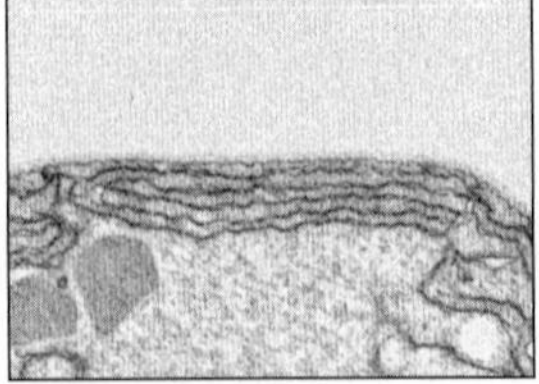

Cyanobacteria Type of gram-negative bacteria with tougher peptidoglycan layer and extensive internal membrane system containing photosynthetic pigments

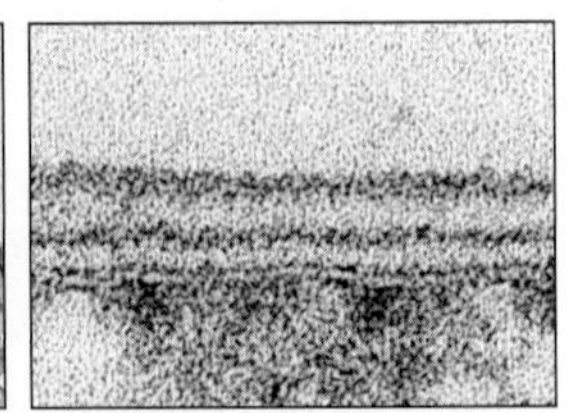

Archaebacteria Pseudopeptidoglycan layer outside plasma membrane; outer membrane absent

such as SO_4^{2-}, yielding H_2S as the product. With the rise of O_2-producing photosynthetic bacteria, the earth's atmosphere became progressively richer in oxygen—a powerful oxidant and deadly poison to anaerobes adapted to a milder environment. Responding to the evolutionary pressure of the so-called "oxygen holocaust," some lineages of microorganisms gave rise to aerobes that obtained energy by passing electrons from fuel molecules to oxygen. Because the transfer of electrons from organic molecules to O_2 releases a great deal of energy (the reaction is strongly exergonic; see Chapter 1), aerobic organisms had an energetic advantage over their anaerobic counterparts when both competed in an environment containing oxygen. This advantage translated into the predominance of aerobic organisms in O_2-rich environments.

Modern bacteria inhabit almost every ecological niche in the biosphere, and there are bacteria capable of using virtually every type of organic compound as a source of carbon and energy. Photosynthetic bacteria in both fresh and marine waters trap solar energy and use it to generate carbohydrates and all other cell constituents, which are in turn used as food by other forms of life. A potential limit to growth in the rest of the biosphere is the availability of nitrogen-containing compounds, and here bacteria are an essential link in the global food web. A few strains of bacteria, called **diazatrophs,** are the only organisms on Earth that can metabolically convert atmospheric nitrogen (N_2) into biologically necessary compounds, in a process known as **nitrogen fixation.** Lightning-driven reactions and the fertilizer industry also contribute significantly to the global budget of fixed nitrogen. However, these ultimate sources of bioavailable nitrogen are not necessarily the immediate sources of supply for the biosphere. Most nitrogen compounds taken up by organisms are recycled from organic waste, and here again, bacteria play an essential role in the global food web by acting as the ultimate consumers, degrading the organic material of dead plants and animals and recycling the end products to the environment.

Escherichia coli Is the Best-Studied Prokaryotic Cell

Bacterial cells share certain common structural features, but also show group-specific specializations (Fig. 2–5). *E. coli* is a usually harmless inhabitant of the human intestinal tract. The *E. coli* cell is about 2 μm long and a little less than 1 μm in diameter. It has a protective outer membrane and an inner plasma membrane that encloses the cytoplasm and the nucleoid. Between the inner and outer membranes is a thin but strong layer of peptidoglycans (sugar polymers cross-linked by amino acids), which gives the cell its shape and rigidity. The plasma membrane and the layers outside it constitute the **cell envelope.** Differences in the cell envelope among bacterial species account for the different affinities for the dye gentian violet,

figure 2–5
Common structural features of bacterial cells. Because of differences in cell envelope structure, some eubacteria (gram-positive bacteria) retain Gram's stain, and others (gram-negative bacteria) do not. *E. coli* is gram-negative. Cyanobacteria are also eubacteria but are distinguished by their extensive internal membrane system, in which photosynthetic pigments are localized. Although the cell envelopes of archaebacteria and gram-positive eubacteria look similar under the electron microscope, the structures of the membrane lipids and the polysaccharides of the cell envelope are distinctly different in these organisms.

which is the basis for Gram's stain; gram-positive bacteria retain the dye, gram-negative bacteria do not. The outer membrane of *E. coli,* like that of other gram-negative eubacteria, is similar to the plasma membrane in structure but is different in composition. Gram-positive bacteria (*Bacillus subtilis* and *Staphylococcus aureus,* for example) lack an outer membrane, and the peptidoglycan layer surrounding the plasma membrane is much thicker than that in gram-negative bacteria. In the Archaea, rigidity is conferred by a different type of cross-linked sugar polymer ("pseudopeptidoglycan"). The plasma membranes of eubacteria consist of a thin bilayer of lipid molecules penetrated by proteins. Archaebacterial membranes have a similar architecture, although their lipids differ strikingly from those of the eubacteria.

The plasma membrane contains proteins capable of transporting ions and compounds into and out of the cell. Also in the plasma membrane of most eubacteria are electron-carrying proteins (cytochromes) essential in the formation of ATP from ADP (Chapter 1). In photosynthetic bacteria, internal membranes derived from the plasma membrane contain chlorophyll and other light-trapping pigments.

From the outer membrane of *E. coli* cells and some other eubacteria protrude short, hairlike structures called **pili,** by which cells adhere to the surfaces of other cells. Strains of *E. coli* and other motile bacteria have one or more long **flagella** (singular, **flagellum**), which can propel the bacterium through its aqueous surroundings. Bacterial flagella are thin, rigid, helical rods, 10 to 20 nm thick and up to several hundred micrometers long. Each is attached to a rotary motor, a protein structure that spins in the cell envelope, rotating the flagellum.

The cytoplasm of *E. coli* contains about 15,000 ribosomes, thousands of copies of each of about 1,000 different enzymes, numerous metabolites and cofactors, and a variety of inorganic ions. Under some conditions, granules of polysaccharides or droplets of lipid accumulate. The nucleoid contains a single, circular molecule of DNA. Although the DNA molecule of an *E. coli* cell is 1,000 times longer than the cell itself, it is packaged with proteins and tightly folded into the nucleoid, which is less than 1 μm in its longest dimension. As in all bacteria, no membrane surrounds the genetic material. In addition to the DNA in the nucleoid, the cytoplasm of most bacteria contains one or more smaller, circular segments of DNA called **plasmids.** In nature, some plasmids confer resistance to toxins and antibiotics in the environment. In the laboratory, these DNA segments, because they are nonessential, are especially amenable to experimental manipulation and are extremely useful to molecular geneticists.

There is a division of labor within the bacterial cell. The cell envelope, which includes the plasma membrane, regulates the flow of materials into and out of the cell and protects the cell from noxious environmental agents. The plasma membrane and the cytoplasm contain a variety of enzymes essential to energy metabolism and the synthesis of precursor molecules; the ribosomes manufacture proteins; and the nucleoid stores and transmits genetic information. Most bacteria lead existences that are nearly independent of other cells, but in some bacterial species, cells tend to associate in clusters or filaments, and a few (the myxobacteria, for example) demonstrate simple social behavior.

Evolution of Eukaryotic Cells

All fossils older than 1.5 billion years are the remains of small and relatively simple organisms, similar in size and shape to modern prokaryotes. Starting about 1.5 billion years ago, the fossil record begins to show evidence of larger and more complex organisms, probably the earliest eukaryotic cells (Fig. 2–4). Details of the evolutionary path from prokaryotes to eukaryotes cannot be deduced from the fossil record alone, but morphological and biochemical comparisons of modern organisms have suggested a reasonable sequence of events consistent with the fossil evidence.

Eukaryotic Cells Evolved from Prokaryotes in Several Stages

Three major changes must have occurred as prokaryotes gave rise to eukaryotes (Fig. 2–6). First, as cells acquired more DNA, the mechanisms required to fold it compactly into discrete complexes with specific proteins and to divide it equally between daughter cells at cell division became more elaborate. These DNA-protein complexes, **chromosomes** (Greek *chroma*, "color," and *soma*, "body"), become especially compact at the time of cell division, when they can be visualized with the light microscope as threads of **chromatin.**

Second, as cells became larger, a system of intracellular membranes developed, including a double membrane surrounding the DNA. This membrane segregated the nuclear process of RNA synthesis on a DNA template from the cytoplasmic process of protein synthesis on ribosomes. Finally,

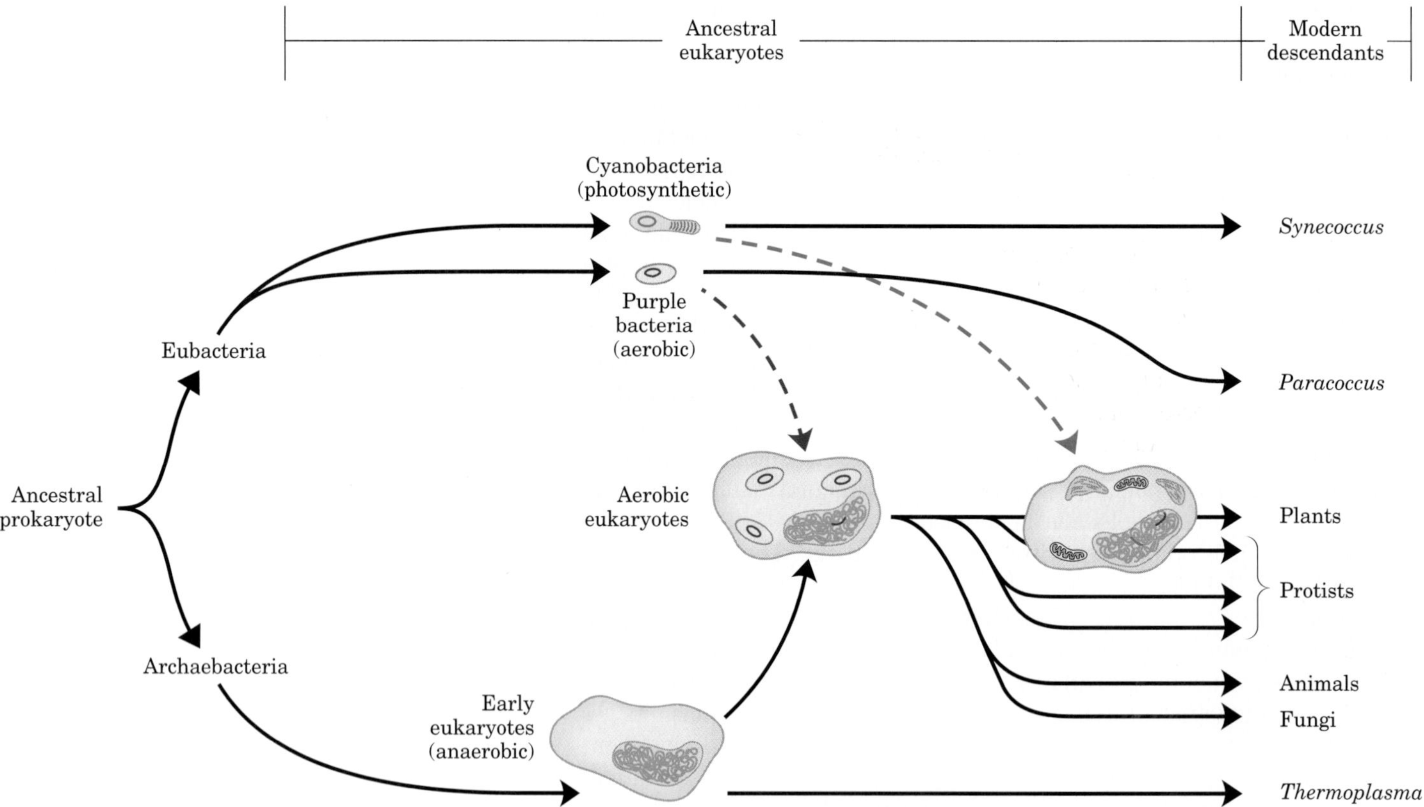

figure 2–6
Evolution of eukaryotes. Modern organisms may have derived from a common ancestral prokaryote by a series of endosymbiotic associations. The early anaerobic eukaryote derived its nuclear structures (red) from an archaebacterium and its motile apparatus (not shown) from an anaerobic eubacterium with which it fused. This early eukaryote later acquired endosymbiotic purple bacteria (yellow), which brought their capacity for aerobic catabolism and became, over time, mitochondria. When photosynthetic cyanobacteria (green) subsequently became endosymbionts of some aerobic eukaryotes, these cells became the photosynthetic precursors of modern green algae and plants.

table 2–1

Comparison of Prokaryotic and Eukaryotic Cells

Characteristic	Prokaryotic cell	Eukaryotic cell
Size	Generally small (1–10 μm)	Generally large (5–100 μm)
Genome	DNA with nonhistone protein; genome in nucleoid, not surrounded by membrane	DNA complexed with histone and nonhistone proteins in chromosomes; chromosomes in nucleus with membranous envelope
Cell division	Fission or budding; no mitosis	Mitosis including mitotic spindle; centrioles in many species
Membrane-bounded organelles	Absent	Mitochondria, chloroplasts (in plants, some algae), endoplasmic reticulum, Golgi complexes, lysosomes (in animals), etc.
Nutrition	Absorption; some photosynthesis	Absorption, ingestion; photosynthesis in some species
Energy metabolism	No mitochondria; oxidative enzymes bound to plasma membrane; great variation in metabolic pattern	Oxidative enzymes packaged in mitochondria; more unified pattern of oxidative metabolism
Cytoskeleton	None	Complex, with microtubules, intermediate filaments, actin filaments
Intracellular movement	None	Cytoplasmic streaming, endocytosis, phagocytosis, mitosis, vesicle transport

Source: Modified from Hickman, C.P., Roberts, L.S., & Hickman, F.M. (1990) *Biology of Animals,* 5th edn, p. 30, Mosby–Yearbook, Inc., St. Louis, MO.

early eukaryotic cells, which were incapable of photosynthesis or aerobic metabolism, enveloped aerobic bacteria or photosynthetic bacteria to form **endosymbiotic** associations that became permanent. Some aerobic bacteria evolved into the mitochondria of modern eukaryotes, and some photosynthetic cyanobacteria became plastids, such as the chloroplasts of green algae, the likely ancestors of modern plant cells. Prokaryotic and eukaryotic cells are compared in Table 2–1.

Early Eukaryotic Cells Gave Rise to Diverse Protists

With the rise of early eukaryotic cells, further evolution led to a tremendous diversity of unicellular eukaryotic organisms **(protists).** Some of these (those with chloroplasts) resembled modern photosynthetic protists such as *Euglena* and *Chlamydomonas;* other, nonphotosynthetic protists were more like *Paramecium* or *Dictyostelium.* Unicellular eukaryotes are abundant, and the cells of all multicellular organisms—animals, plants, and fungi—are eukaryotic.

Major Structural Features of Eukaryotic Cells

Typical eukaryotic cells (Fig. 2–7) are much larger than prokaryotic cells—commonly 5 to 100 μm in diameter, with cell volumes a thousand to a million times larger than those of bacteria. The distinguishing characteristic of eukaryotes is the nucleus, which has a complex internal structure surrounded by a double membrane. Another striking difference between eukaryotes and prokaryotes is that eukaryotes contain a number of other membrane-bounded organelles. The following sections describe the structures and roles of the components of eukaryotic cells in more detail.

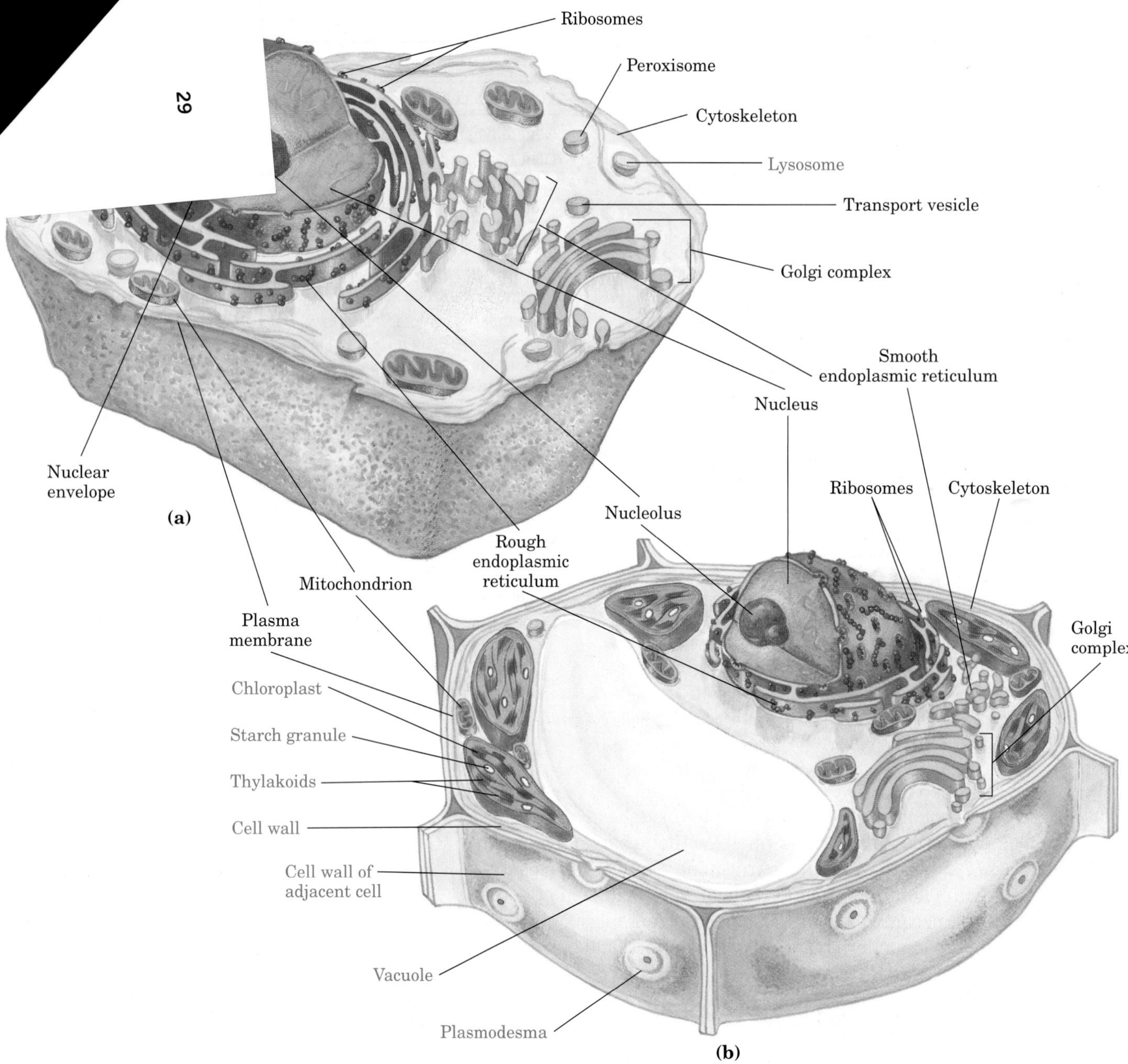

figure 2–7
Schematic illustrations of the two major types of eukaryotic cell: a representative animal cell **(a)** and a representative plant cell **(b).** Plant cells are usually 10 to 100 μm in diameter—larger than animal cells, which typically range from 5 to 30 μm. Structures labeled in red are unique to either animal or plant cells.

The Plasma Membrane Contains Transporters and Receptors

The external surface of a cell is in contact with other cells, the extracellular fluid, and the solutes, nutrient molecules, hormones, neurotransmitters, and antigens in that fluid. The plasma membranes of all cells contain many **transporters,** proteins that span the membrane and carry nutrients into the cell and various products out. Cells also have surface membrane proteins **(signal receptors)** with highly specific binding sites for extracellular signaling molecules (receptor ligands). When an external ligand binds to its specific receptor, the receptor protein transduces the signal carried by that ligand into an intracellular message (Fig. 2–8). For example, some surface receptors are associated with **ion channels** that open when the receptor is occupied, permitting entry of specific ions; others activate or inhibit cellular enzymes on the inner membrane surface. Whatever the mode of **signal**

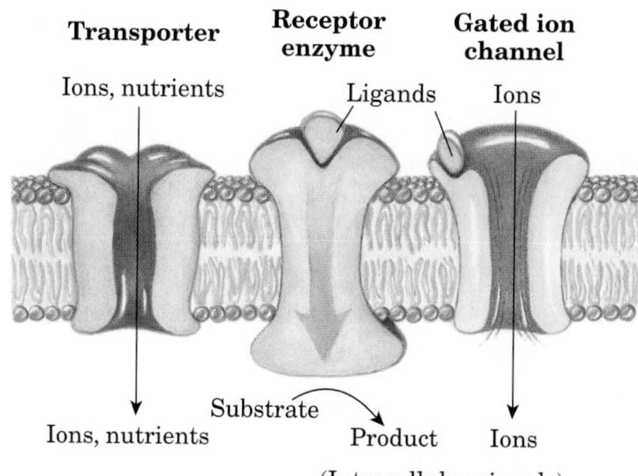

figure 2–8
Proteins in the plasma membrane serve as transporters, signal receptors, and ion channels. Transporters carry substances into and out of the cell; some transporters use energy to pump ions and compounds against a concentration gradient. Extracellular signals are amplified by receptors: binding of a single ligand molecule to the surface receptor causes the formation of many molecules of an intracellular messenger or the flow of many ions through an opened channel.

transduction, surface receptors characteristically act as signal amplifiers—a single ligand molecule bound to a single receptor may cause the flux of thousands of ions through an opened channel or the synthesis of thousands of molecules of an intracellular messenger molecule by an activated enzyme.

Some surface receptors recognize ligands of low molecular weight, and others recognize macromolecules. For example, binding of acetylcholine (M_r 146) to its receptor begins a cascade of cellular events that underlie the transmission of signals for muscle contraction. Blood proteins (M_r >20,000) that carry lipids (lipoproteins) are recognized by specific cell surface receptors, which mediate lipid entry into the cells. Antigens (proteins, viruses, or bacteria, recognized by the immune system as foreign) bind to specific receptors and trigger the production of antibodies. During the development of multicellular organisms, neighboring cells influence each other's developmental paths, as signal molecules from one cell type react with receptors of other cells. Thus the surface membrane of a cell is a complex mosaic of different kinds of highly specific "molecular antennae" through which cells receive, amplify, and react to external signals.

Most cells of higher plants have a **cell wall** outside the plasma membrane (Fig. 2–7b), which serves as a rigid, protective shell. The cell wall, composed of cellulose and other carbohydrate polymers, is thick but porous. It allows water and small molecules to pass readily, but swelling of the cell due to the accumulation of water is resisted by the rigidity of the wall.

Endocytosis and Exocytosis Carry Traffic across the Plasma Membrane

Endocytosis is a mechanism for transporting components of the surrounding medium deep into the cytoplasm. In this process (Fig. 2–9), a region of the plasma membrane invaginates, enclosing a small volume of extracellular fluid within a bud that pinches off inside the cell by membrane fission. The resulting small vesicle **(endosome)** can move into the interior of the cell, delivering its contents to another organelle bounded by a single membrane (a lysosome, for example; see p. 33) by fusion of the two membranes. The endosome thus serves as an intracellular extension of the plasma membrane, effectively allowing intimate contact between components of the extracellular medium and regions deep within the cytoplasm, which could not be reached by diffusion alone. **Phagocytosis** is a special case of endocytosis in which the material carried into the cell (within a phagosome) is particulate, such as a cell fragment or even another, smaller

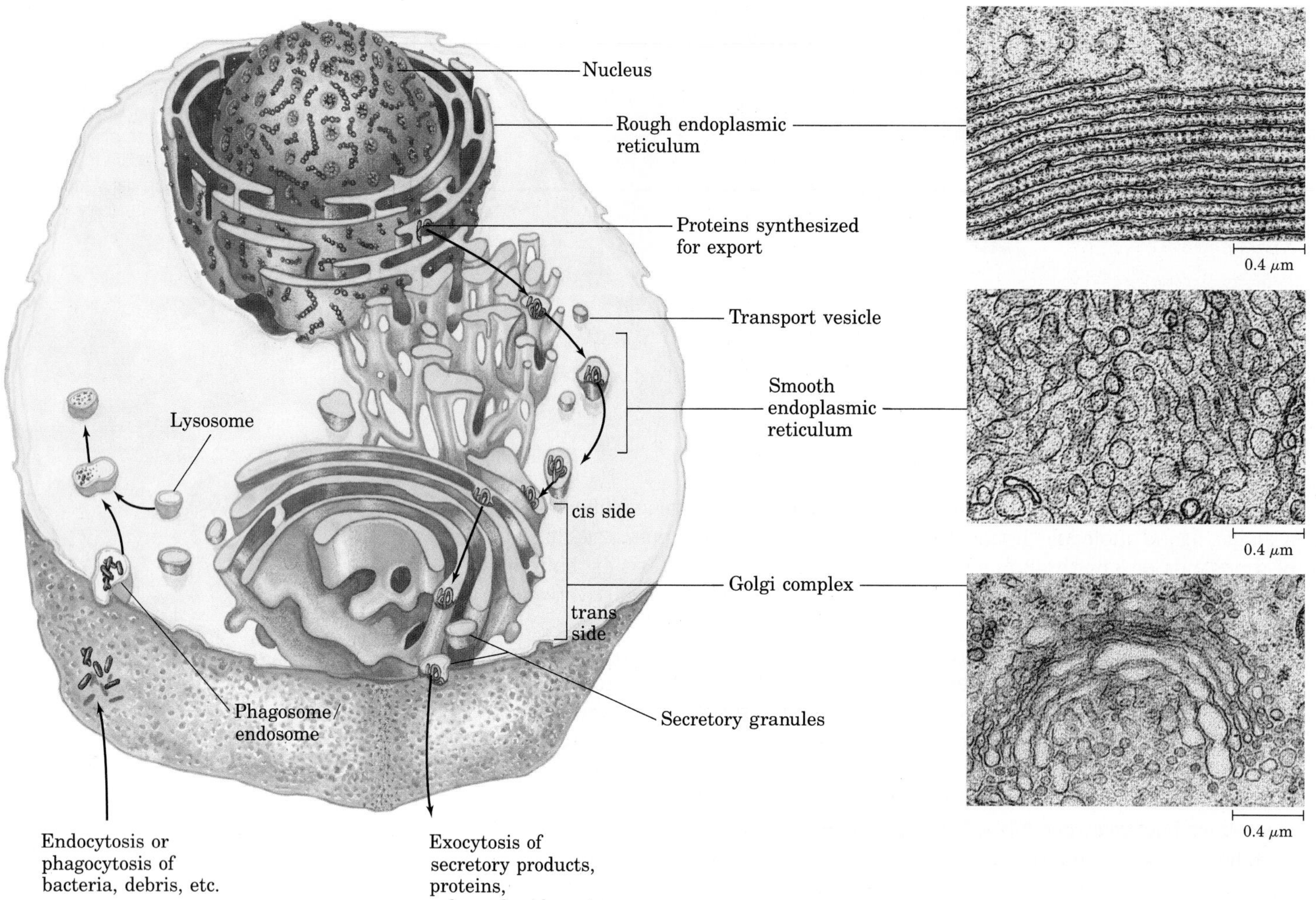

figure 2–9

The endomembrane system. This system includes the nuclear envelope, endoplasmic reticulum, Golgi complex, and several types of small vesicles. It encloses a compartment (lumen) distinct from the cytosol. Contents of the lumen move from one region of the endomembrane system to another as small transport vesicles bud from one component and fuse with another. High-magnification electron micrographs of a sectioned cell show rough endoplasmic reticulum studded with ribosomes, smooth endoplasmic reticulum, and the Golgi complex. (The size of the Golgi complex is exaggerated in the diagram for clarity.)

The endomembrane system is dynamic; newly synthesized proteins move into the lumen of the rough endoplasmic reticulum and thus to the smooth endoplasmic reticulum, then to the Golgi complex via transport vesicles. The cis portion of the Golgi complex faces the nucleus; the trans portion is that nearer the plasma membrane. In the Golgi complex, molecular "addresses" are added to specific proteins to direct them to the cell surface, lysosomes, or secretory granules. The contents of secretory granules are released from the cell by exocytosis. Endocytosis and phagocytosis bring extracellular materials into the cell. Fusion of endosomes (or phagosomes) with lysosomes, which contain digestive enzymes, results in degradation of the extracellular materials.

cell. The inverse of endocytosis is **exocytosis** (Fig. 2–9), in which a vesicle in the cytoplasm moves to the inside surface of the plasma membrane, fuses with it, then releases the vesicular contents outside the membrane. Many proteins destined for secretion into the extracellular space are packaged into vesicles called secretory granules then released by exocytosis.

The Endoplasmic Reticulum Organizes the Synthesis of Proteins and Lipids

The small transport vesicles moving to and from the plasma membrane in exocytosis and endocytosis are parts of a dynamic system of intracellular membranes that includes the endoplasmic reticulum, the Golgi complex, the nuclear envelope, and a variety of small vesicles such as lysosomes and peroxisomes (Fig. 2–9). Although generally represented as discrete and static elements, these structures are in fact in constant flux, with membrane vesicles continually budding off, moving through the cell, and merging with membranous structures elsewhere.

The **endoplasmic reticulum** (ER) is a highly convoluted, three-dimensional network of membrane-enclosed spaces extending throughout the cytoplasm and enclosing a subcellular compartment (the lumen of the ER) separate from the cytoplasm. The many flattened branches (cisternae) of this compartment are continuous with each other and with the nuclear envelope. In cells specialized for the secretion of proteins, such as the pancreatic cells that secrete the hormone insulin, the ER is particularly prominent. The ribosomes that synthesize proteins destined for export attach to the outer (cytoplasmic) surface of the ER, and the secretory proteins are passed through the membrane into the lumen as they are synthesized. Digestive enzymes that will be sequestered within lysosomes or proteins destined for insertion into the nuclear or plasma membranes are also synthesized on ribosomes attached to the ER. By contrast, proteins that will remain and function within the cytosol are synthesized on cytoplasmic ribosomes unassociated with the ER.

The attachment of thousands of ribosomes (usually in regions of large cisternae) gives the **rough endoplasmic reticulum** its granular appearance (Fig. 2–9) and thus its name. In other regions of the cell, the ER is free of ribosomes. This **smooth endoplasmic reticulum,** which is physically continuous with the rough ER, is the site of lipid biosynthesis and a variety of other important processes, including the metabolism of certain drugs and toxic compounds. Smooth ER is generally tubular, in contrast to the long, flattened cisternae typical of rough ER. In some tissues (skeletal muscle, for example), the ER is specialized for the storage and rapid release of calcium ions. Release of Ca^{2+} is the trigger for many cellular events, including muscle contraction.

The Golgi Complex Processes and Sorts Proteins

Nearly all eukaryotic cells have **Golgi complexes,** systems of membranous sacs, or cisternae, arranged as flattened stacks (Fig. 2–9). Named after its discoverer, Camillo Golgi, the Golgi complex is asymmetric, structurally and functionally. The cis side faces the rough endoplasmic reticulum (and the nucleus), and the trans side faces the plasma membrane; between these are the medial elements. Proteins, during their synthesis on ribosomes bound to the rough ER, are inserted into the interior (lumen) of the ER cisternae. Small membrane vesicles containing the newly synthesized proteins bud from the ER and move to the Golgi complex, fusing with the cis side. As the proteins pass through the Golgi complex to the trans side, enzymes in the complex modify the protein molecules by adding sulfate, carbohydrate, or lipid moieties to side chains of certain amino acids. One of the functions of this modification of a newly synthesized protein is to “address” it to its proper destination as it leaves the Golgi complex in a transport vesicle budding from the trans side. Certain proteins are enclosed in secretory granules, eventually to be released from the cell by exocytosis. Others are targeted for intracellular organelles such as lysosomes or for incorporation into the plasma membrane during cell growth.

Lysosomes Are the Sites of Degradative Reactions

Lysosomes, found only in animal cells, are spherical vesicles bounded by a single membrane bilayer (Fig. 2–9). They are usually about 1 μm in diameter. Lysosomes contain enzymes capable of digesting proteins, polysaccharides, nucleic acids, and lipids. They function as cellular recycling centers, breaking down complex molecules brought into the cell by endocytosis, fragments of foreign cells brought in by phagocytosis, or worn-out organelles from the cell's own cytoplasm. These materials selectively enter the lysosome by fusion of the lysosomal membrane with endosomes, phagosomes, or defective organelles, and are then degraded to their simple components (amino acids, monosaccharides, fatty acids, etc.), which are released into the cytosol to be recycled into new cellular components or further catabolized.

The degradative enzymes within a lysosome would be free to act on all cellular components were they not confined by the lysosomal membrane. A second line of defense against unwanted destruction of cytosolic macromolecules by lysosomal enzymes is the difference in pH between the lysosome and the cytosol, maintained by the action of an ATP-fueled proton pump in the lysosomal membrane. The lysosomal compartment is more acidic ($pH \leq 5$) than the cytosol ($pH \approx 7$), and lysosomal enzymes are much less active at the higher pH of the cytosol.

figure 2–10
The vacuole of a plant cell contains high concentrations of Ca^{2+} and a variety of stored compounds and waste products. Water enters the vacuole, increasing the vacuolar volume and pressing the cytoplasm against the plasma membrane, creating turgor pressure. The rigidity of the cell wall prevents expansion and rupture of the plasma membrane.

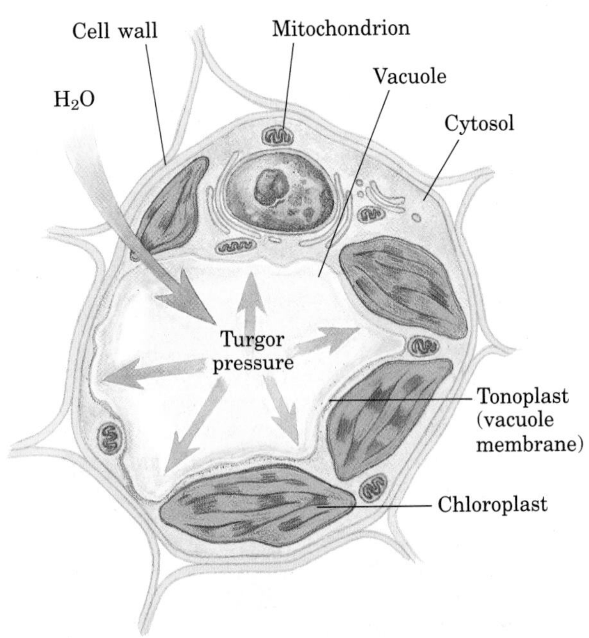

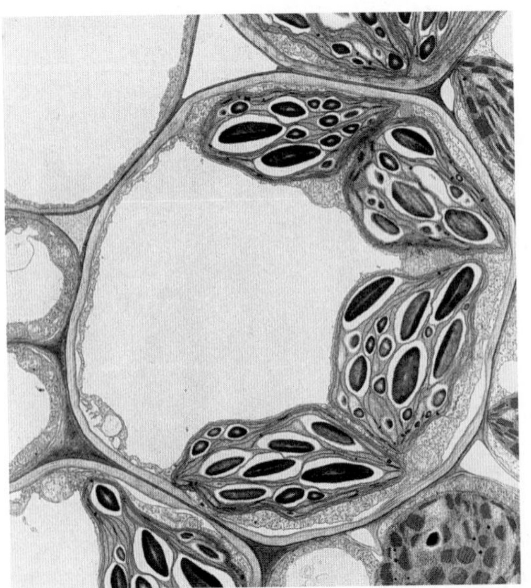

Vacuoles of Plant Cells Play Several Important Roles

Plant cells do not have lysosomes, but their **vacuoles** carry out similar degradative reactions as well as other functions not found in animal cells. Growing plant cells contain several small vacuoles, vesicles bounded by a single membrane bilayer. As the cell matures, the vacuoles fuse and become one large central vacuole (Fig. 2–10; see also Fig. 2–7b). The vacuole may represent as much as 90% of the total cell volume in a mature cell, pressing the cytoplasm into a thin layer between the vacuole and the plasma membrane. The membrane surrounding the vacuole, called the **tonoplast,** regulates the entry of ions, metabolites, and cellular structures destined for degradation, and the liquid within the vacuole contains digestive enzymes that degrade and recycle macromolecular components. As in the lysosome, the pH within the vacuole is generally lower than the pH of the surrounding cytosol. In some plant cells, the vacuole contains high concentrations of pigments (anthocyanins) that give flowers and fruits their deep purple and red colors. In addition to its role in storage and degradation of cellular components, the vacuole also provides physical support to the plant cell. Because the concentration of solutes (salts, ions, degradation products) is greater in the vacuole than in the cytosol, water passes osmotically into the vacuole, establishing, at equilibrium, an outward **turgor pressure** on the cytoplasm and the cell wall that stiffens the plant tissue (Fig. 2–10).

Peroxisomes Destroy Hydrogen Peroxide, and Glyoxysomes Convert Fats to Carbohydrates

Some of the oxidative reactions in the breakdown of amino acids and fats produce free radicals and hydrogen peroxide (H_2O_2), very reactive chemical species that could damage cellular machinery. To protect the cell from these destructive byproducts, such reactions are segregated within small membrane-bounded vesicles called **peroxisomes.** The hydrogen peroxide is degraded by catalase, an enzyme present at high concentration in peroxisomes; it catalyzes the reaction

$$2H_2O_2 \longrightarrow 2H_2O + O_2.$$

Glyoxysomes are specialized peroxisomes found in certain plant cells. They contain high concentrations of the enzymes of the **glyoxylate cycle,** a metabolic pathway unique to plants that converts stored fats to carbohydrates during seed germination. Lysosomes, peroxisomes, and glyoxysomes are sometimes referred to collectively as **microbodies.**

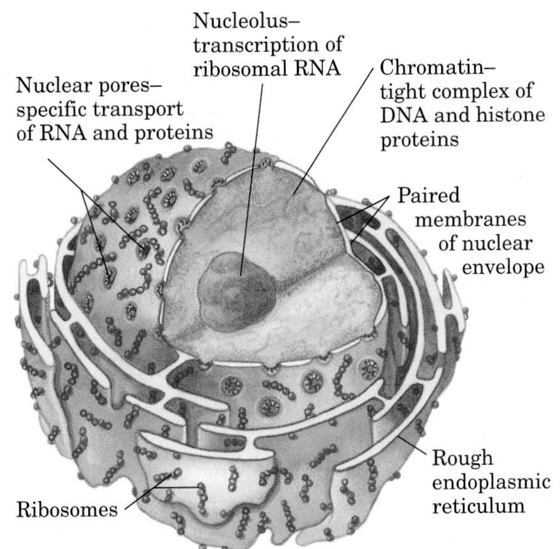

figure 2–11
The nucleus and nuclear envelope.

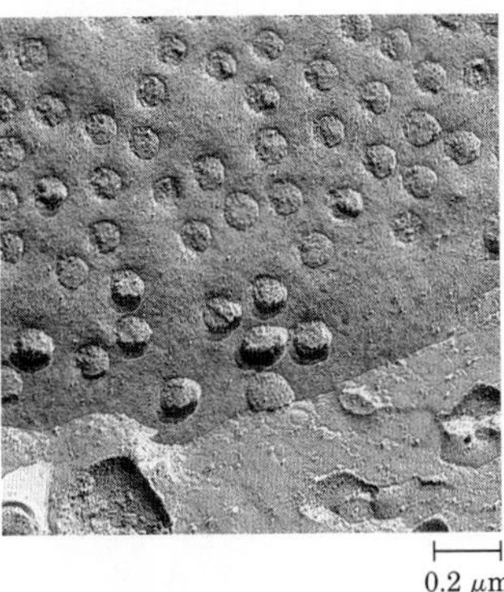

(a) Scanning electron micrograph of the surface of the nuclear envelope, showing numerous nuclear pores.

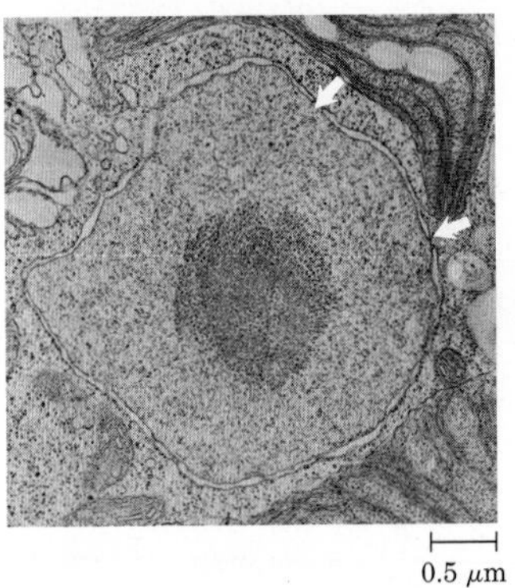

(b) Electron micrograph of the nucleus of the alga *Chlamydomonas.* The dark body in the center is the nucleolus, and the granular material that fills the rest of the nucleus is chromatin. Two nuclear pores piercing the paired membranes of the nuclear envelope are shown by arrows.

The Nucleus Contains the Genome

The eukaryotic nucleus is quite complex in structure and biological activity compared with the relatively simple nucleoid of prokaryotes. The nucleus contains nearly all of the cell's DNA, which can be thousands of times more than is present in a bacterial cell; a small amount of DNA is also present in mitochondria and chloroplasts. The nucleus is surrounded by a **nuclear envelope,** composed of two membrane bilayers separated by a narrow space and continuous with the rough endoplasmic reticulum (Fig. 2–11; see also Fig. 2–9). At intervals the inner and outer nuclear membranes are pinched together around openings **(nuclear pores),** which have a diameter of about 90 nm. Associated with the pores are protein structures called nuclear pore complexes, specific transporters that allow certain macromolecules to pass between the cytoplasm and the aqueous phase of the nucleus (the **nucleoplasm**). Traffic into the nucleus through the nuclear pore complexes includes enzymes and other proteins synthesized in the cytoplasm and required in the nucleoplasm for DNA replication and repair, transcription, and RNA processing. Passing out through the nuclear pores are messenger RNA precursors, with associated proteins, which will be translated on ribosomes in the cytoplasm.

The nucleus of an interphase (nondividing) cell is filled with a diffuse material called **chromatin,** so called because early microscopists found that it stained brightly with certain dyes. Chromatin consists of DNA and proteins bound tightly together and is the substance of the chromosomes, which do not condense and become individually visible until the cell is ready to divide. The **nucleolus** is a specific region of the nucleus in which the DNA contains many copies of the genes encoding ribosomal RNA. To produce the large number of ribosomes needed by the cell, these genes are continually transcribed into RNA. The nucleolus appears dense in electron micrographs (Fig. 2–11b) because of its high RNA content. Ribosomal RNA produced in the nucleolus enters the cytoplasm through the nuclear pores.

Nuclear division **(mitosis)** occurs before cell division **(cytokinesis).** The double-helical DNA of the chromatin is replicated, then in the first

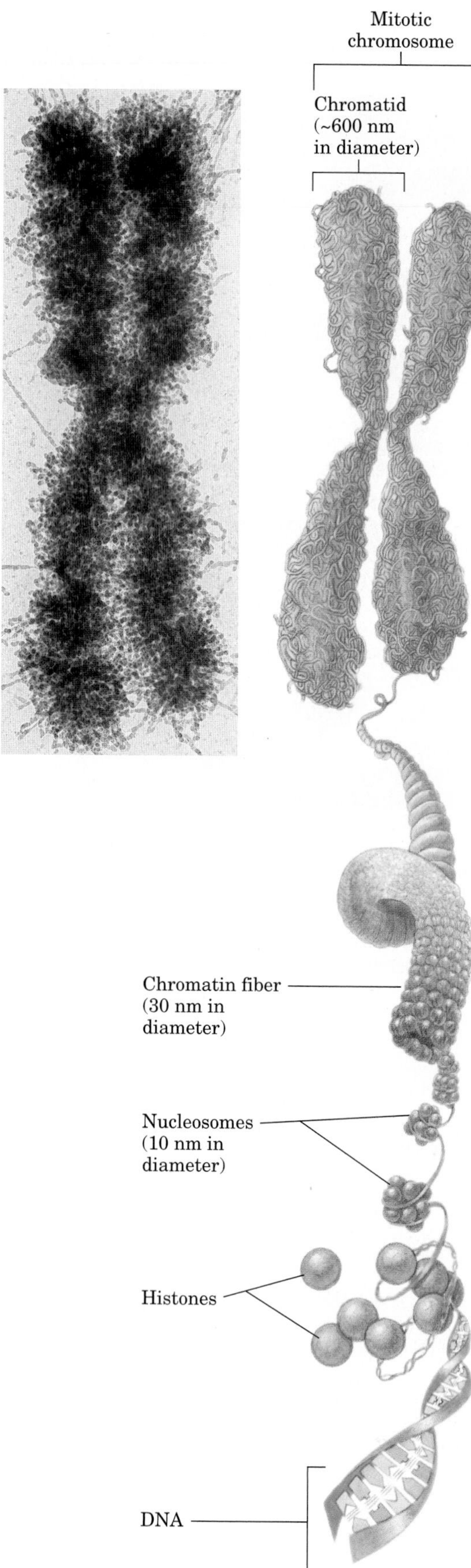

figure 2–12
Chromosomes are visible microscopically during mitosis. Shown here is an electron micrograph of one of the 46 human chromosomes in a diploid (somatic) cell. Every mitotic chromosome is composed of two chromatids, each consisting of tightly folded chromatin fibers. Each chromatin fiber is in turn formed by the packaging of a DNA molecule wrapped about histone proteins to form a series of nucleosomes. (Adapted from Becker, W.M. & Deamer, D.W. (1991) *The World of the Cell,* 2nd edn, Fig. 13–20, The Benjamin/Cummings Publishing Company, Menlo Park, CA.)

phase of mitosis the chromatin fibers condense into discrete bodies, the **chromosomes,** each consisting of two identical **chromatids** (Fig. 2–12). The two sister chromatids separate, one moving to each pole of the cell, where they become part of the newly formed nucleus of each daughter cell.

Cells of each species have a characteristic number of chromosomes of specific sizes and shapes. For example, the protist *Tetrahymena pyriformis* has 5 pairs; cabbage has 9, humans have 23, and the fern *Ophioglossum reticulatum* has about 630! The cells that make up most of the body of a multicellular organism, the somatic cells, have two copies of each chromosome and are said to be **diploid (2*n*).** Gametes (egg and sperm, for example), produced by meiosis (Chapter 25) and having only one copy of each chromosome, are **haploid (*n*).** During sexual reproduction, two haploid gametes combine to regenerate a diploid cell in which each chromosome pair consists of a maternal and a paternal chromosome.

The DNA of chromatin and chromosomes is bound tightly to a family of positively charged proteins, the **histones,** which associate strongly with the many negatively charged phosphate groups in DNA. About half the mass of chromatin is DNA and half is histones. When DNA replicates prior to cell division, large quantities of histones are also synthesized to maintain this 1:1 mass ratio. The histones and DNA associate in complexes called **nucleosomes,** in which the DNA strand winds around a core of histone molecules (Fig. 2–12). The DNA of a single human chromosome forms about a million nucleosomes; nucleosomes associate to form very regular and compact supramolecular complexes. The resulting chromatin fibers, about 30 nm in diameter, condense further by forming a series of looped regions, which cluster with adjacent looped regions to form the chromosomes visible during cell division. This tight packing of DNA into nucleosomes achieves a remarkable condensation of the DNA molecules. The DNA in the chromosomes of a single diploid human cell would have a combined length of about 2 m if fully extended as a DNA double helix, but the combined length of all 46 chromosomes is only about 200 μm.

Mitochondria Are the Power Plants of Aerobic Eukaryotic Cells

Mitochondria (singular, **mitochondrion**) are very conspicuous in the cytoplasm of most eukaryotic cells when viewed by electron microscopy (Fig. 2–13). These membrane-bounded organelles vary in size, but typically have a diameter of about 1 μm, similar to that of bacterial cells. Mitochondria also vary widely in shape, number, and location, depending on the cell type or tissue function. Most plant and animal cells contain several hundred to a thousand mitochondria. Generally, cells in more metabolically active tissues devote a larger proportion of their volume to mitochondria.

Each mitochondrion has two membranes. The outer membrane is unwrinkled and completely surrounds the organelle. The inner membrane has infoldings called **cristae,** which give it a large surface area. Enclosed by the inner membrane is the **matrix,** a very concentrated aqueous solution of enzymes and chemical intermediates involved in energy-yielding metabolism. Mitochondrial enzymes catalyze the oxidation of organic nutrients by molecular oxygen (O_2); some of these enzymes are in the matrix and some are embedded in the inner membrane. The chemical energy released in mitochondrial oxidations is used to generate ATP, the major energy-carrying molecule of cells. In aerobic cells, mitochondria are the principal producers of ATP, which diffuses to all parts of the cell and provides the energy for cellular work.

Unlike other membranous structures such as lysosomes, Golgi complexes, and the nuclear envelope, mitochondria are produced only by division of previously existing mitochondria; each mitochondrion contains its own DNA, RNA, and ribosomes. Mitochondrial DNA codes for certain proteins specific to the mitochondrial inner membrane. This and other evidence supports the theory (outlined below) that mitochondria are the descendants of aerobic bacteria that lived endosymbiotically with early eukaryotic cells.

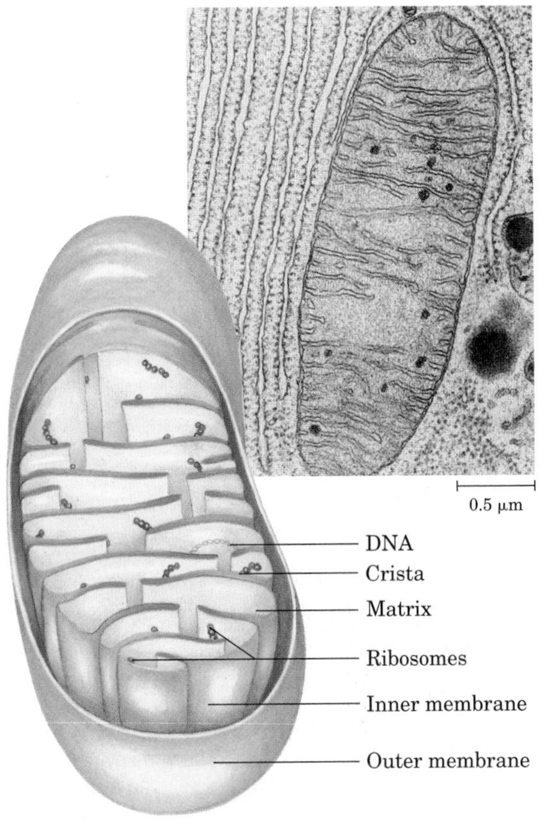

figure 2–13
Structure of a mitochondrion. The electron micrograph shows the extensive infolding of the inner membrane; the folds are called cristae. (Note the rough endoplasmic reticulum surrounding the mitochondrion.)

Chloroplasts Convert Solar Energy into Chemical Energy

The cytoplasm of plants contains **plastids,** specialized organelles surrounded by envelopes consisting of two membranes. Most conspicuous of the plastids and characteristically present in the photosynthetic cells of plants and algae are the **chloroplasts** (Fig. 2–14). Like mitochondria, the chloroplasts may be considered power plants, with the important difference that chloroplasts use solar energy, whereas mitochondria use the chemical energy of oxidizable compounds. Pigment molecules in chloroplasts absorb the energy of light and use it to make ATP and, ultimately, to reduce carbon dioxide to form carbohydrates such as starch and sucrose. Photosynthesis in eukaryotes and in cyanobacteria produces O_2 as a byproduct of the light-capturing reactions. Photosynthetic plant cells contain both chloroplasts and mitochondria. Chloroplasts produce ATP only in the light; mitochondria function independently of light, oxidizing carbohydrates generated by photosynthesis during daylight hours.

Chloroplasts are generally larger (diameter 5 μm) than mitochondria and have various shapes. Because chloroplasts contain a high concentration of the pigment **chlorophyll,** photosynthetic cells are usually green, but their color depends on the relative amounts of other pigments present. Chlorophyll and other pigment molecules, which together can absorb light

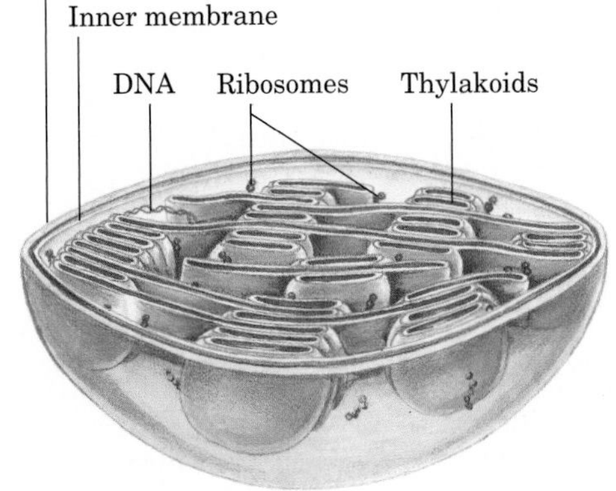

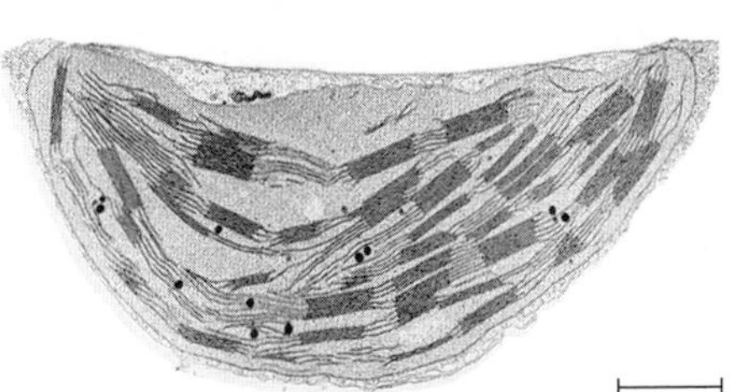

figure 2–14
Structure of a chloroplast. The thylakoids are flattened membranous sacs that contain chlorophyll, the light-harvesting pigment.

energy over much of the visible spectrum, are localized in the internal membranes of the chloroplast; these membranes form stacks of closed cisternae known as **thylakoids** (Fig. 2–14). Like mitochondria, chloroplasts contain DNA, RNA, and ribosomes.

Mitochondria and Chloroplasts Probably Evolved from Endosymbiotic Bacteria

Several independent lines of evidence suggest that the mitochondria and chloroplasts of modern eukaryotes were derived during evolution from aerobic bacteria and cyanobacteria that took up endosymbiotic residence in early eukaryotic cells (Fig. 2–15; see also Fig. 2–6). Mitochondria are always derived from preexisting mitochondria, and chloroplasts from chloroplasts, by simple fission, just as bacteria multiply by fission. Mitochondria and chloroplasts are in fact semiautonomous; they contain DNA, ribosomes, and the enzymatic machinery to synthesize proteins encoded in their DNA. Sequences in mitochondrial DNA are strikingly similar to sequences in certain aerobic bacteria, and chloroplast DNA shows strong sequence similarity to the DNA of certain cyanobacteria. The ribosomes of mitochondria and chloroplasts are more similar in size, overall structure, and RNA sequences to those of bacteria than to those in the cytoplasm of the eukaryotic cell. The enzymes that catalyze protein synthesis in these organelles also more closely resemble those of bacteria.

Despite their complement of DNA and protein-synthesizing machinery, mitochondria and chloroplasts are only semiautonomous. If these or-

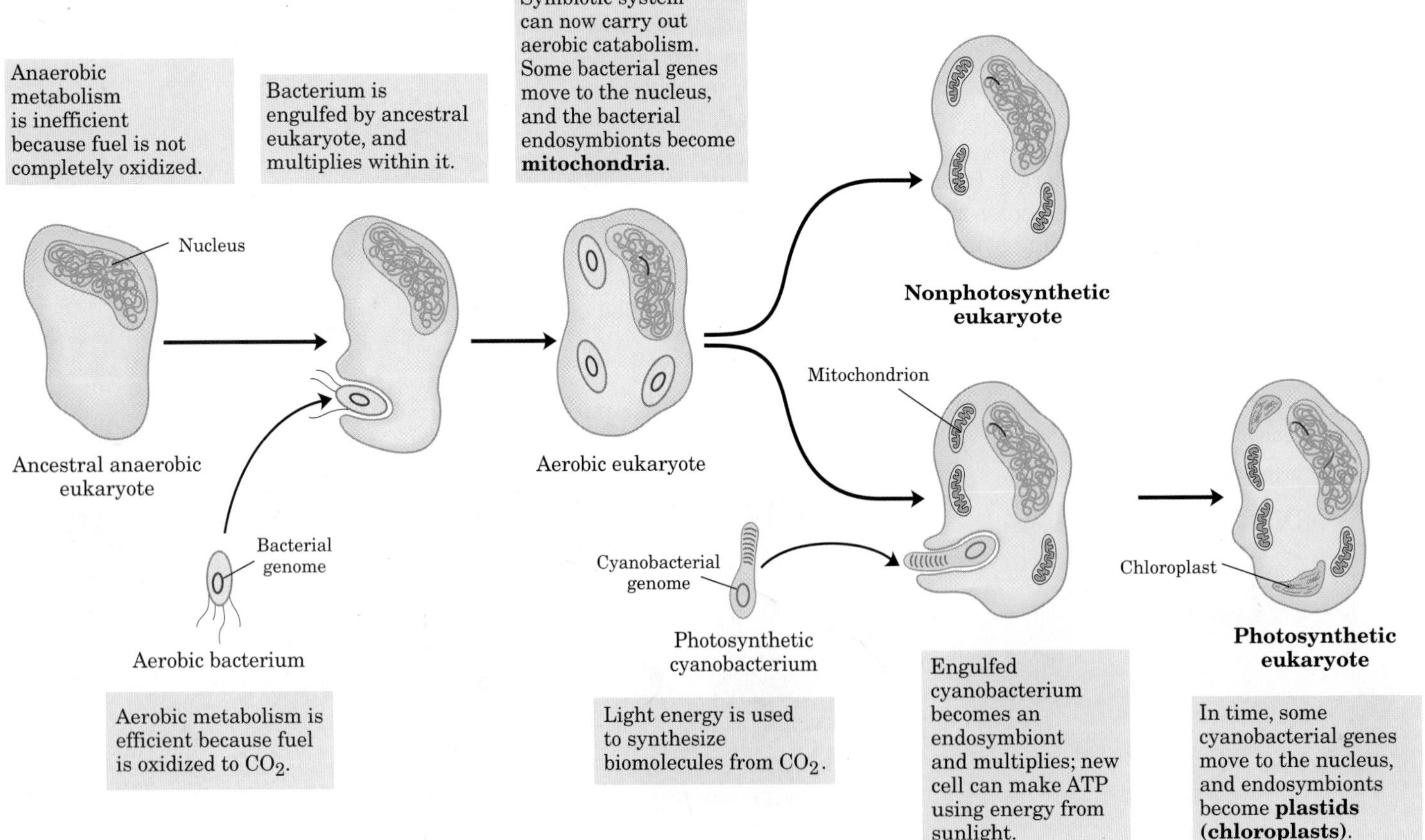

figure 2–15
A plausible theory for the evolutionary origin of mitochondria and chloroplasts. It is based on a number of striking biochemical and genetic similarities between certain aerobic bacteria and mitochondria, and between certain cyanobacteria and chloroplasts. During the evolution of eukaryotic cells, the bacteria became symbiotic within the ancestral anaerobe. Ultimately the cytoplasmic bacteria became the mitochondria and chloroplasts of modern eukaryotes.

ganelles are indeed the descendants of early bacterial endosymbionts, some of the genes present in the original free-living bacteria must have been transferred into the nuclear DNA of the host eukaryote over the course of evolution. Neither mitochondria nor chloroplasts contain all the genes necessary to specify all of their proteins. Most mitochondrial and chloroplast proteins are encoded in nuclear genes, translated on cytoplasmic ribosomes, and subsequently imported into the organelles.

The Cytoskeleton Stabilizes Cell Shape, Organizes the Cytoplasm, and Produces Motion

Several types of protein filaments visible with the electron microscope crisscross the eukaryotic cell, forming an interlocking three-dimensional meshwork, the **cytoskeleton,** that extends throughout the cytoplasm. There are three general types of cytoplasmic filaments: actin filaments, microtubules, and intermediate filaments (Fig. 2–16). They differ in width (from about 6 to 22 nm), composition, and specific function, but all apparently provide structure and organization to the cytoplasm and shape to the cell. Actin filaments and microtubules also help to produce the motion of organelles or of the whole cell.

Each of the cytoskeletal components is composed of simple protein subunits that polymerize to form filaments of uniform thickness. These filaments are not permanent structures; they undergo constant disassembly into their monomeric subunits and reassembly into filaments. Their locations in cells are not rigidly fixed, but may change dramatically with mitosis, cytokinesis, amoeboid motion, or changes in cell shape. All types of filaments associate with other proteins that cross-link filaments to themselves or to other filaments, influence assembly or disassembly, or move cytoplasmic organelles along the filaments.

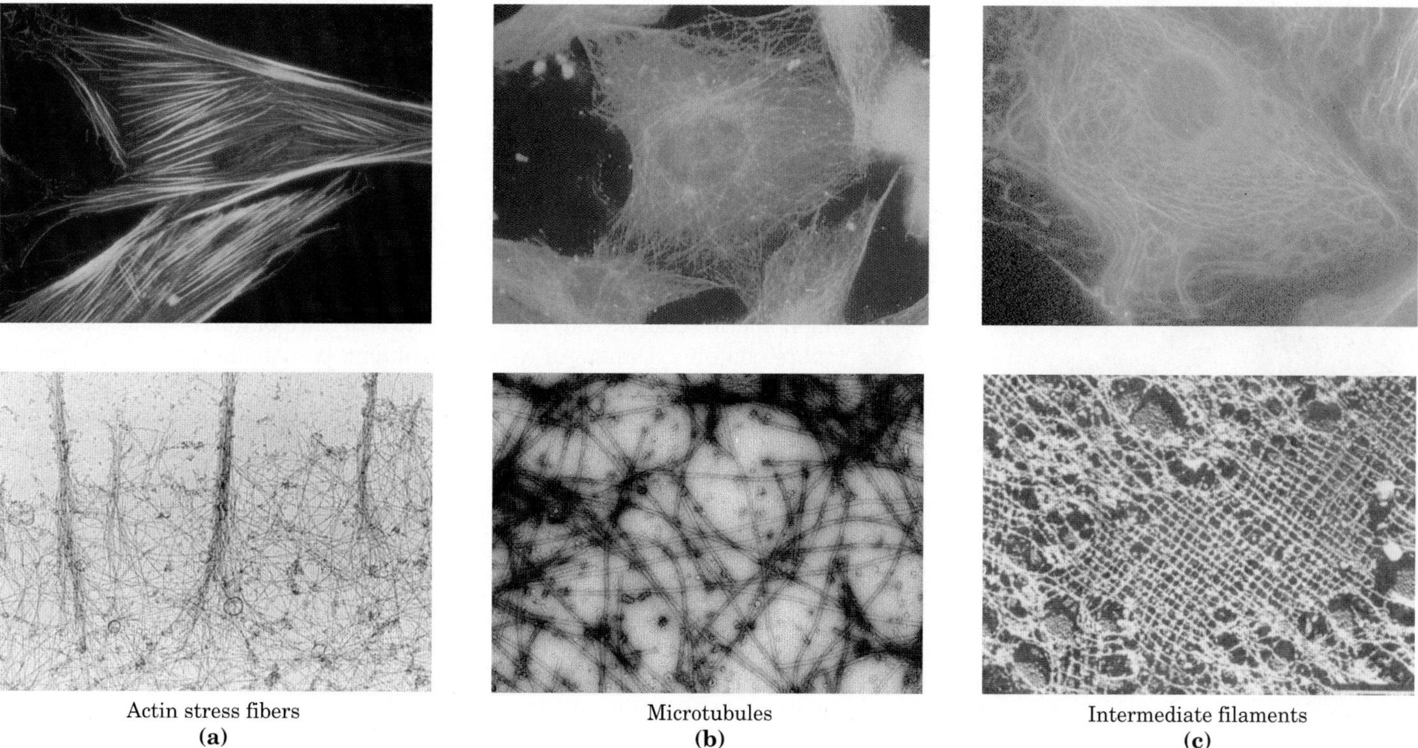

figure 2–16

The three types of cytoskeletal filaments. The upper panels show epithelial cells photographed after treatment with antibodies that bind to and specifically stain **(a)** actin filaments bundled together to form "stress fibers," **(b)** microtubules radiating from the cell center, and **(c)** intermediate filaments extending throughout the cytoplasm. For these experiments, antibodies that specifically recognize actin, tubulin, or intermediate filament proteins are covalently attached to a fluorescent compound. When the cell is viewed with a fluorescence microscope, only the stained structures are visible. The lower panels show each type of filament as visualized by transmission **(a, b)** or scanning **(c)** electron microscopy.

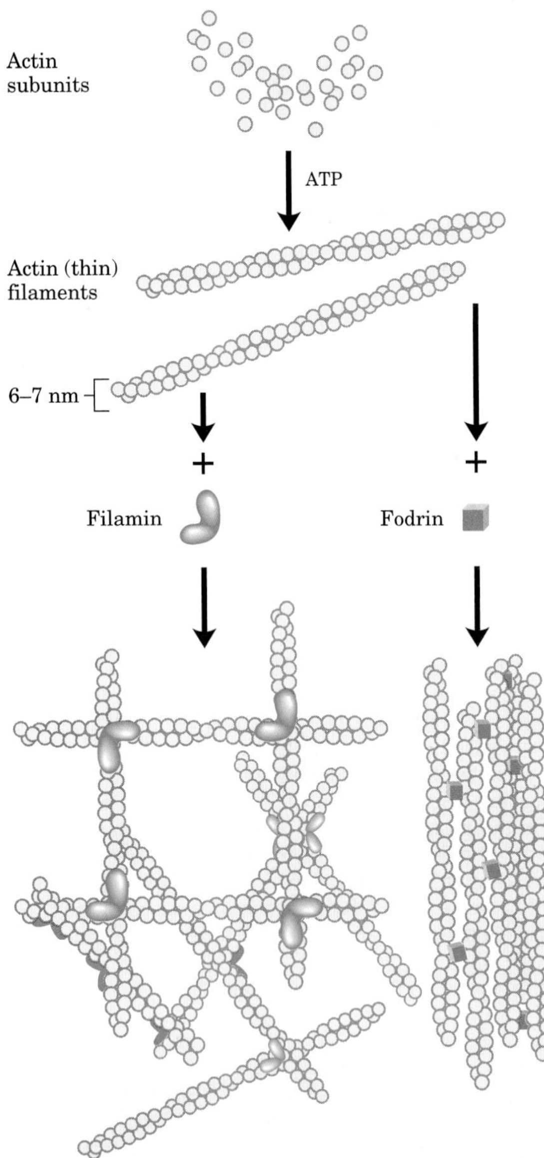

figure 2–17
Individual subunits of actin polymerize to form actin filaments. The protein filamin holds two filaments together where they cross at right angles. The protein fodrin forms cross-links between filaments to create side-by-side aggregates or bundles.

The protein actin, found in virtually all eukaryotic cells, assembles in the presence of ATP into long, helical, noncovalent polymers, 6 to 7 nm in diameter, called **actin filaments** or **microfilaments** (Fig. 2–17). Cells contain a variety of proteins that bind to actin monomers or filaments and influence their localization or state of aggregation. Filamin and fodrin cross-link actin filaments to each other, stabilizing the meshwork and greatly increasing the viscosity of the surrounding medium. Large numbers of actin filaments bound to specific plasma membrane proteins lie just beneath and more or less parallel to the plasma membrane, conferring shape and rigidity on the cell surface (Fig. 2–16a).

Actin filaments also bind to a family of proteins called myosins, molecular motors that convert the chemical energy of ATP into mechanical work, moving themselves along the actin filament. The simplest members of this family, such as myosin I, have a globular head and a short tail. The head binds to and moves along an actin filament, driven by the breakdown of ATP (Fig. 2–18). The tail region binds to the membrane of a cytoplasmic organelle, dragging the organelle behind as the myosin head moves along the actin filament. This motion is readily seen in living cells such as the giant cells of the green alga *Nitella*, in which organelles and vesicles move uniformly around the cell in a process called cytoplasmic streaming (Fig. 2–18). This motion has the effect of mixing the cytoplasmic contents of the enormous algal cell much more efficiently than would occur by diffusion alone.

A larger form of myosin occurs in the contractile systems of a wide variety of organisms, from slime molds to humans. This myosin also has a globular head that binds to and moves along actin filaments in an ATP-driven reaction, but it has a longer tail, which permits the myosin molecules to associate side by side to form thick filaments (see Fig. 7–30). Actin-myosin complexes form the contractile ring that squeezes the cytoplasm in two during cytokinesis in all eukaryotes. The muscle cells of multicellular animals are filled with highly organized arrays of actin (thin) filaments and myosin (thick) filaments, which produce a coordinated contractile force by ATP-driven sliding of actin filaments past stationary myosin filaments.

Like actin filaments, **microtubules** form spontaneously from their monomeric subunits, but the polymeric structure of microtubules is slightly more complex. Dimers of α- and β-tubulin, two similar proteins, form the hollow microtubule, which is about 22 nm in diameter. In cells, most microtubules undergo continual polymerization and depolymerization by addition of tubulin subunits primarily at one end and dissociation at the other. Microtubules are present throughout the cytoplasm, but are concentrated in specific regions at certain times. For example, after sister chromatids separate and move to opposite poles of a cell during mitosis, a highly organized array of microtubules (the mitotic spindle) provides the framework and probably the motive force for the separation of these daughter chromosomes.

Microtubules, like actin filaments, associate with a variety of proteins that move along them, form cross-bridges, or influence their state of polymerization. Kinesin and cytoplasmic dynein, proteins found in the cytoplasm of many cells, bind to and move along microtubules using the energy

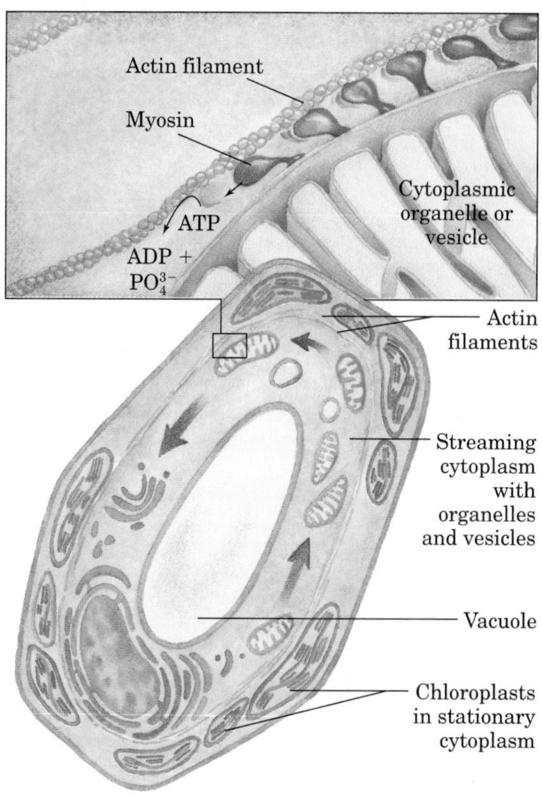

figure 2–18

Organelle transport. Myosin molecules move along actin filaments using energy from ATP. Cytoplasmic streaming is produced in the giant cells of the green alga *Nitella* as myosin pulls organelles around a track of actin filaments. Endoplasmic reticulum, mitochondria, nucleus, and other membrane-bounded organelles and vesicles move uniformly around the cell at 50 to 75 μm/s. The chloroplasts are located in the layer of stationary cytoplasm that lies between the actin filaments and the plasma membrane.

of ATP to drive their motion (Fig. 2–19). Each protein is capable of associating with specific organelles and pulling them along the microtubule over long distances at rates of about 1 μm/s. The beating motion of cilia and eukaryotic flagella also involves dynein and microtubules.

The contraction of skeletal muscle, the propelling action of cilia and flagella, and the intracellular transport of organelles all rely on the same fundamental mechanism: the splitting of ATP by proteins such as kinesin, myosin, and dynein drives sliding motion along microfilaments or microtubules.

Intermediate filaments are a family of structures with dimensions (diameter 8 to 10 nm) intermediate between actin filaments and microtubules. Several different types of monomeric protein subunits reversibly form intermediate filaments. The cytoplasmic distribution of these structures is subject to regulated changes.

One function of intermediate filaments is to provide internal mechanical support for the cell and to position its organelles. For example, vimentin is the monomeric subunit of the intermediate filaments found in the endothelial cells that line blood vessels and in adipocytes (fat cells). Vimentin fibers appear to anchor the nucleus and fat droplets in specific cellular locations. The intermediate filaments composed of keratins, a family of structural proteins, are particularly prominent in certain epidermal cells of vertebrates, forming covalently cross-linked meshworks that persist even after the cell dies. Hair, fingernails, and feathers are among the structures composed primarily of keratins.

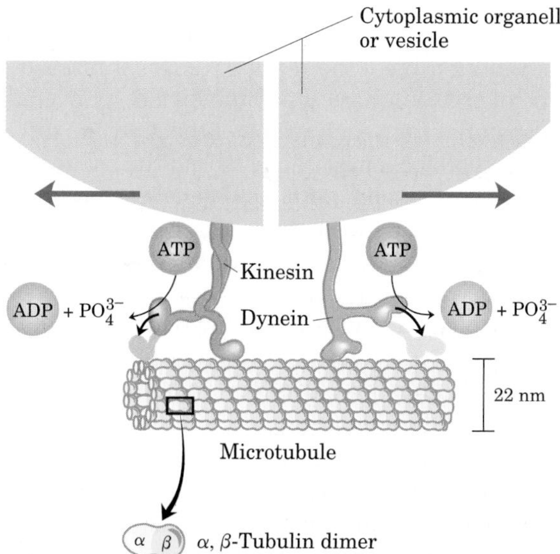

figure 2–19

Kinesin and cytoplasmic dynein are ATP-driven molecular motors that can attach to cytoplasmic organelles or vesicles and drag them along microtubular "rails" at a rate of about 1 μm/s.

The Cytoplasm Is Crowded, Highly Ordered, and Dynamic

The picture that emerges from this brief survey is of a eukaryotic cell with a cytoplasm crisscrossed by a meshwork of structural fibers, throughout which extends a complex system of membrane-bounded compartments (Fig. 2–7). Both the filaments and the organelles are dynamic: the filaments disassemble and then reassemble elsewhere; membranous vesicles bud from one organelle and fuse with another. Organelles move through the cytoplasm along protein filaments, drawn by kinesin, cytoplasmic dynein, myosin, and perhaps other similar proteins. The endomembrane system segregates specific metabolic processes and provides surfaces on which certain enzyme-catalyzed reactions occur. Exocytosis and endocytosis provide paths between the cell interior and the surrounding medium, allowing for the secretion of proteins and other components produced within the cell and the uptake of extracellular substances.

Although complex, this organization of the cytoplasm is far from random. The motion and positioning of organelles and cytoskeletal elements are under tight regulation, and at certain stages in a eukaryotic cell's life, dramatic, finely orchestrated reorganizations occur, such as the events of mitosis. The interactions between the cytoskeleton and organelles are noncovalent, reversible, and subject to regulation in response to various intracellular and extracellular signals.

Study of Cellular Components

Organelles Can Be Isolated by Centrifugation

A major advance in the biochemical study of cells was the development of methods for separating organelles from the cytosol and from each other. In a typical cellular fractionation, cells or tissues are disrupted by gentle homogenization in a medium containing sucrose (about 0.2 M). This treatment ruptures the plasma membrane but leaves most of the organelles intact. (The sucrose creates a medium with an osmotic pressure similar to that within organelles; this prevents diffusion of water into the organelles, which would cause them to swell, burst, and spill their contents.)

Organelles such as nuclei, mitochondria, and lysosomes differ in size and therefore sediment at different rates during centrifugation. They also differ in specific gravity, and they "float" at different levels in a density gradient (Fig. 2–20). Differential centrifugation results in a rough fractionation of the cytoplasmic contents, which may be further purified by isopycnic ("same density") centrifugation. In this procedure, organelles of different buoyant densities (the result of different ratios of lipid and protein in each type of organelle) are separated on a density gradient. By carefully removing material from each region of the gradient and observing it with a microscope, the biochemist can establish the sedimentation position of each organelle and obtain purified organelles for further study. In this way it was established, for example, that lysosomes contain degradative enzymes, mitochondria contain oxidative enzymes, and chloroplasts contain photosynthetic pigments. The isolation of an organelle enriched in a certain enzyme is often the first step in the purification of that enzyme.

In Vitro Studies May Overlook Important Interactions among Molecules

One of the most effective approaches to understanding a biological process is to study purified individual molecules such as enzymes, nucleic acids, or structural proteins. The purified components are amenable to detailed characterization in vitro; their physical properties and catalytic activities can be studied without "interference" from other molecules present in the

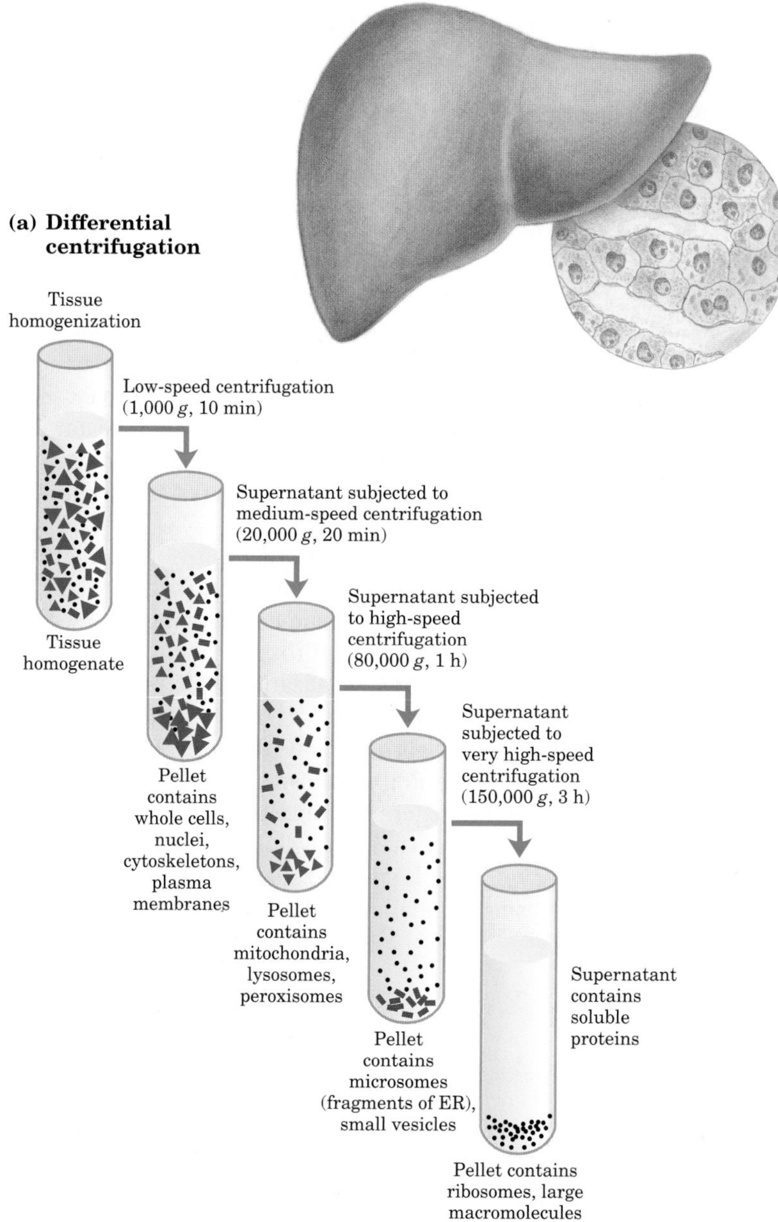
(a) Differential centrifugation
Tissue homogenization
Low-speed centrifugation (1,000 g, 10 min)
Tissue homogenate
Supernatant subjected to medium-speed centrifugation (20,000 g, 20 min)
Pellet contains whole cells, nuclei, cytoskeletons, plasma membranes
Supernatant subjected to high-speed centrifugation (80,000 g, 1 h)
Pellet contains mitochondria, lysosomes, peroxisomes
Supernatant subjected to very high-speed centrifugation (150,000 g, 3 h)
Pellet contains microsomes (fragments of ER), small vesicles
Supernatant contains soluble proteins
Pellet contains ribosomes, large macromolecules

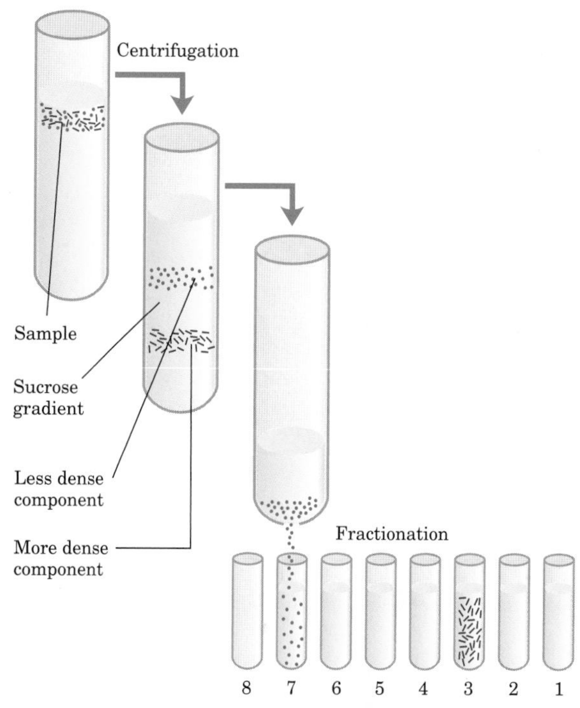
(b) Isopycnic (sucrose-density) centrifugation
Centrifugation
Sample
Sucrose gradient
Less dense component
More dense component
Fractionation
8 7 6 5 4 3 2 1

figure 2–20
Subcellular fractionation of tissue. A tissue such as liver is mechanically homogenized to break cells and disperse their contents in an aqueous buffer. The large and small particles in this suspension can be separated by centrifugation at different speeds **(a),** or particles of different density can be separated by isopycnic centrifugation **(b).** In isopycnic centrifugation, a centrifuge tube is filled with a solution, the density of which increases from top to bottom; a solute such as sucrose is dissolved at different concentrations to produce the density gradient. When a mixture of organelles is layered on top of the density gradient and the tube is centrifuged at high speed, individual organelles sediment until their buoyant density exactly matches that in the gradient. Each layer can be collected separately.

intact cell. Although this approach has been remarkably revealing, we must keep in mind that the inside of a cell is quite different from the inside of a test tube. The "interfering" components eliminated by purification may be critical to the biological function or regulation of the molecule purified. In vitro studies of pure enzymes are commonly done at very low enzyme concentrations in thoroughly stirred aqueous solutions. In the cell, an enzyme is dissolved or suspended in a gel-like cytosol with thousands of other proteins, some of which bind to that enzyme and influence its activity. Some enzymes in cells are parts of multienzyme complexes in which reactants are channeled from one enzyme to another without ever entering the bulk solvent. Diffusion is hindered in the gel-like cytosol, and the cytosolic composition varies in different regions of the cell. In short, a given molecule may function somewhat differently within the cell than it does in vitro. A central challenge of biochemistry is to understand the influences of cellular organization and macromolecular associations on the function of individual enzymes—to understand function in vivo as well as in vitro.

Evolution of Multicellular Organisms and Cellular Differentiation

All modern unicellular eukaryotes—the protists—contain the organelles and mechanisms that we have described, indicating that these organelles and mechanisms must have evolved relatively early. The protists are extraordinarily versatile. The ciliated protist *Paramecium,* for example, moves rapidly through its aqueous surroundings by beating its cilia; senses mechanical, chemical, and thermal stimuli from its environment, and responds by changing its path; finds, engulfs, and digests a variety of food organisms, and excretes the indigestible fragments; eliminates excess water that leaks in through its membrane; and finds and mates with sexual partners. Nonetheless, being unicellular has its limitations. Paramecia probably live out their lives in a very small region of the pond in which they began life, because their motility is limited by the small thrust of their microscopic cilia, and their ability to detect a better environment at a distance is limited by the short range of their sensory apparatus.

At some later stage of evolution, unicellular organisms found it advantageous to cluster together, thereby acquiring greater motility, efficiency, or reproductive success than their free-living single-celled competitors. Further evolution of such clustered organisms led to permanent associations among individual cells and eventually to specialization within the colony—to cellular differentiation.

The advantages of cellular specialization led to the evolution of ever more complex and highly differentiated organisms, in which some cells carried out the sensory functions, others the digestive, photosynthetic, or reproductive functions. Many modern multicellular organisms contain hundreds of different cell types, each specialized for some function that supports the entire organism. Fundamental mechanisms that evolved

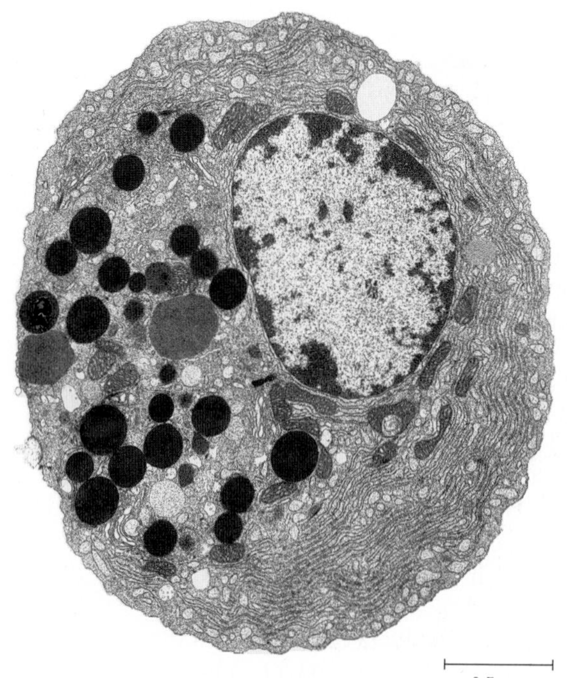

(a)

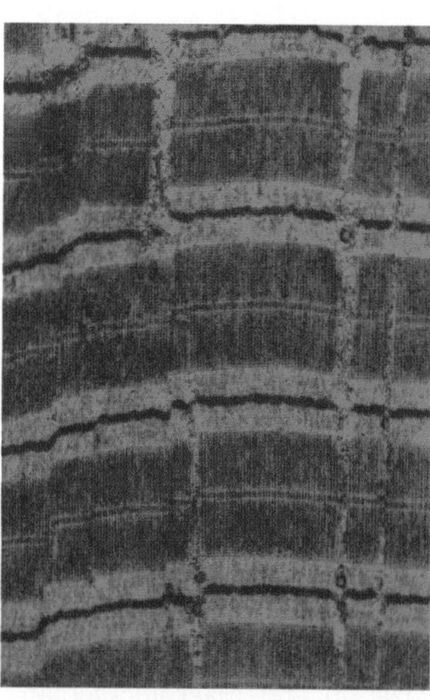
(b)

early have been further refined and embellished through evolution. The simple mechanism responsible for the motion of myosin along actin filaments in slime molds has been conserved and elaborated in vertebrate muscle cells. The same basic structure and mechanism that underlie the beating motion of cilia in *Paramecium* and flagella in *Chlamydomonas* are employed by the highly differentiated vertebrate sperm cell. Figure 2–21 illustrates some of the cellular specializations encountered in multicellular organisms.

figure 2–21

A gallery of structurally and functionally differentiated cells. (a) Secretory cell of the pancreas. Its extensive endoplasmic reticulum is the site of synthesis of the secreted protein(s). **(b)** Portion of a skeletal muscle cell (artificial color). The highly organized actin and myosin filaments slide relative to each other in the ATP-dependent process that produces macroscopic muscle contraction. **(c)** Collenchyma cells of a plant stem. These cells, lacking a rigid cell wall, provide flexible support for the growing stem. **(d)** Human sperm cells (artificial color). The long flagella propel the sperm through the female reproductive tract toward the egg. **(e)** Mature human erythrocytes (artificial color). These cells have no nucleus or endomembrane system; each cell is filled with the soluble oxygen-binding protein hemoglobin and is flexible enough to fit through capillaries of small diameter. **(f)** Human embryo at the two-celled stage. The egg cell from which it was derived was packed with stored fuel and messenger RNA to support the rapid protein synthesis that follows fertilization.

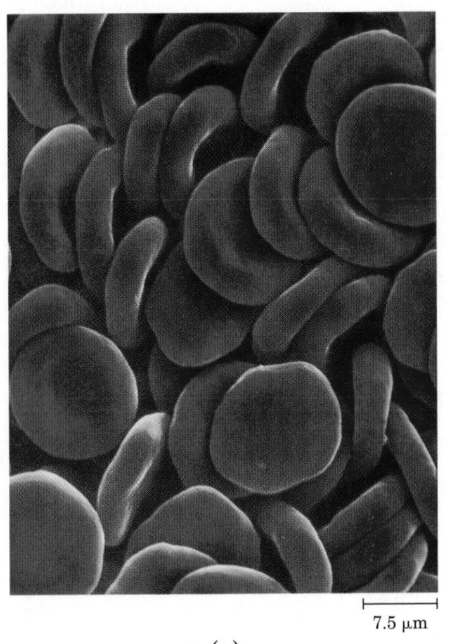

(e)

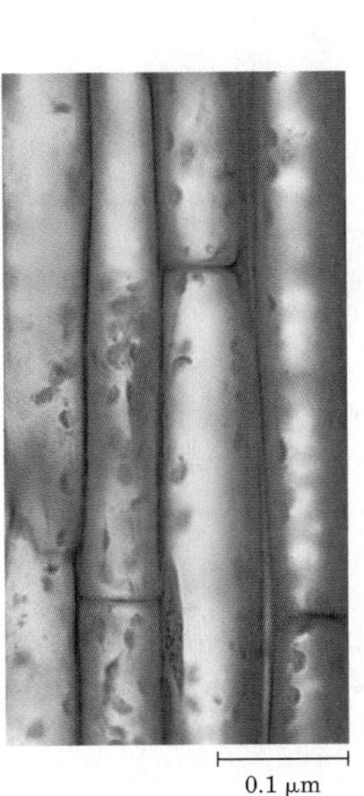

(c)

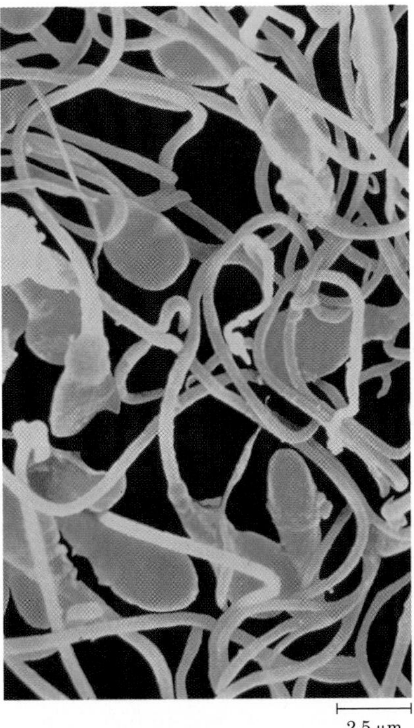

(d)

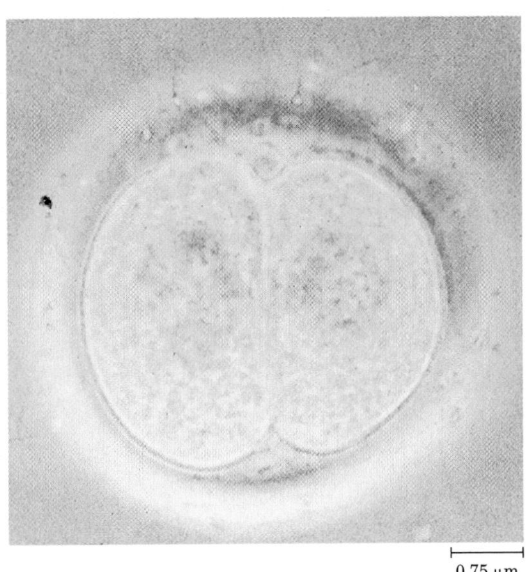

(f)

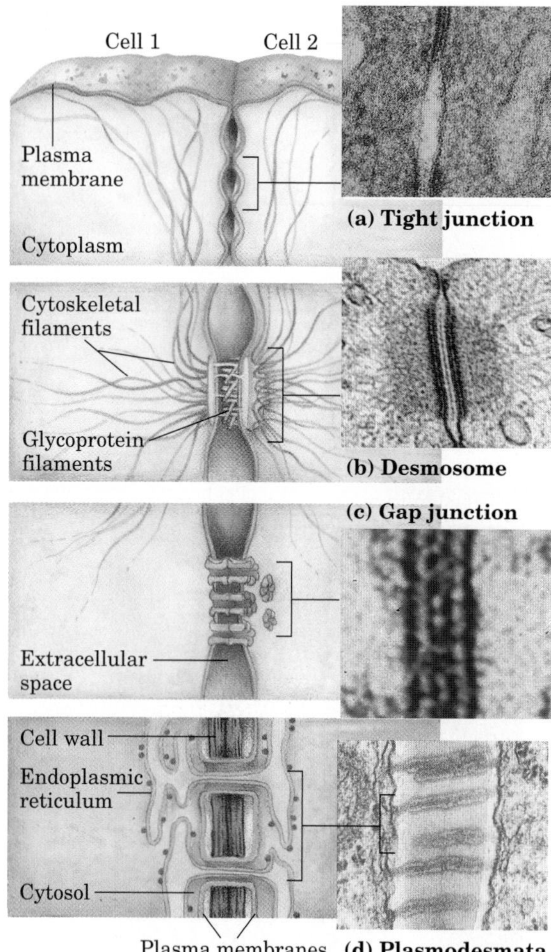

figure 2–22
Cellular connections. Animal cells can be joined by three types of junctions. **(a)** Tight junctions produce a watertight seal between adjacent epithelial cells. **(b)** Desmosomes weld adjacent epithelial cells together and are reinforced by various cytoskeletal elements. **(c)** Gap junctions allow ions and electric currents to flow between adjacent cells. **(d)** In plants, plasmodesmata connect adjacent cells, providing a path through the cell wall for the passage of small metabolites and proteins.

The individual cells of a multicellular organism remain delimited by their plasma membranes, but they have developed specialized surface structures for attachment to and communication with each other (Fig. 2–22). Each type of intercellular junction is reinforced by membrane proteins or cytoskeletal filaments. Animals have three types of junctions that serve different purposes. At **tight junctions,** the plasma membranes of adjacent epithelial cells are closely apposed, with no extracellular fluid separating them. Tight junctions form a belt around the cell, providing a barrier between the tissue and the outside environment. **Desmosomes** are fibrous plaques that weld epithelial cells together; the small extracellular space between the cells is filled with fibrous and adhesive proteins. Desmosomes mechanically strengthen the physical connections between cells, but do not prevent the passage of materials through the extracellular space between the cells they connect. **Gap junctions** provide small, reinforced openings between adjacent cells, through which electric currents, ions, and small molecules can pass. They serve as channels of communication between adjacent cells. Higher plants have **plasmodesmata** (singular, **plasmodesma**), channels functionally similar to gap junctions but structurally quite different, in part because of the presence of the cell wall in plants. Plasmodesmata provide a path through the cell wall and plasma membrane for the movement of metabolites—even some small proteins—between adjacent cells.

Viruses: Parasites of Cells

Viruses are supramolecular complexes that can replicate themselves in appropriate host cells. They consist of a nucleic acid (DNA or RNA) molecule surrounded by a protective shell, or capsid, made up of protein molecules and, in some cases, a membranous envelope. Viruses exist in two states. Outside the host cells that formed them, viruses are simply nonliving particles called **virions,** which can be crystallized. Once a virus or its nucleic acid component gains entry into a specific host cell, it becomes an intracellular parasite. The viral nucleic acid carries the genetic message specifying the structure of the intact virion. It diverts the host cell's enzymes and ribosomes from their normal cellular roles to the manufacture of many new daughter viral particles. As a result, hundreds of progeny viruses may arise from the single virion that infected the host cell. In some host-virus systems, the progeny virions escape through the host cell's plasma membrane. Other viruses cause cell lysis (membrane breakdown and host cell death) as they are released. Much of the pathology associated with viral diseases results from this lysis of the host cell.

A different type of response results from some viral infections, in which viral DNA becomes integrated into a host chromosome and is replicated with the host's own genes. Integrated viral genes may have little or no effect on the host's survival, but, in rare cases, they cause profound changes in the host cell's appearance and activity.

Many hundreds of different viruses are known (Fig. 2–23), each more or less specific for a host cell, which may be an animal, plant, or bacterial cell. Viruses specific for bacteria are known as **bacteriophages,** or simply **phages** (Greek *phagein,* "to eat"). Some viruses contain only one kind of protein in their capsid—the tobacco mosaic virus, for example, a simple plant virus and the first to be crystallized. Other viruses contain as many as a hundred different kinds of proteins. Even some of these large and complex viruses have been crystallized, and their detailed molecular structures are known. Viruses differ greatly in size. Bacteriophage øX174, one of the smallest, has a diameter of 18 nm. Vaccinia virus is one of the largest; its virions are almost as large as the smallest bacteria. Viruses also differ in shape and complexity of structure. The human immunodeficiency virus (HIV) is relatively simple in structure, but devastating in effect; it causes AIDS by destroying cells central to the human immune response. The outbreaks of Hantavirus in the southwestern United States in 1993 and of the Ebola virus in central Africa in 1995 illustrate the extreme pathogenicity of some viruses. Both viruses produce diseases with rapid courses and high mortality. Other viruses that are highly pathogenic in humans cause poliomyelitis, influenza, herpes, hepatitis, the common cold, infectious mononucleosis, shingles, and certain types of cancer.

Biochemistry has profited enormously from the study of viruses, which has provided new information about the structure of the genome, the enzymatic mechanisms of nucleic acid and protein synthesis, and the regulation of the flow of genetic information.

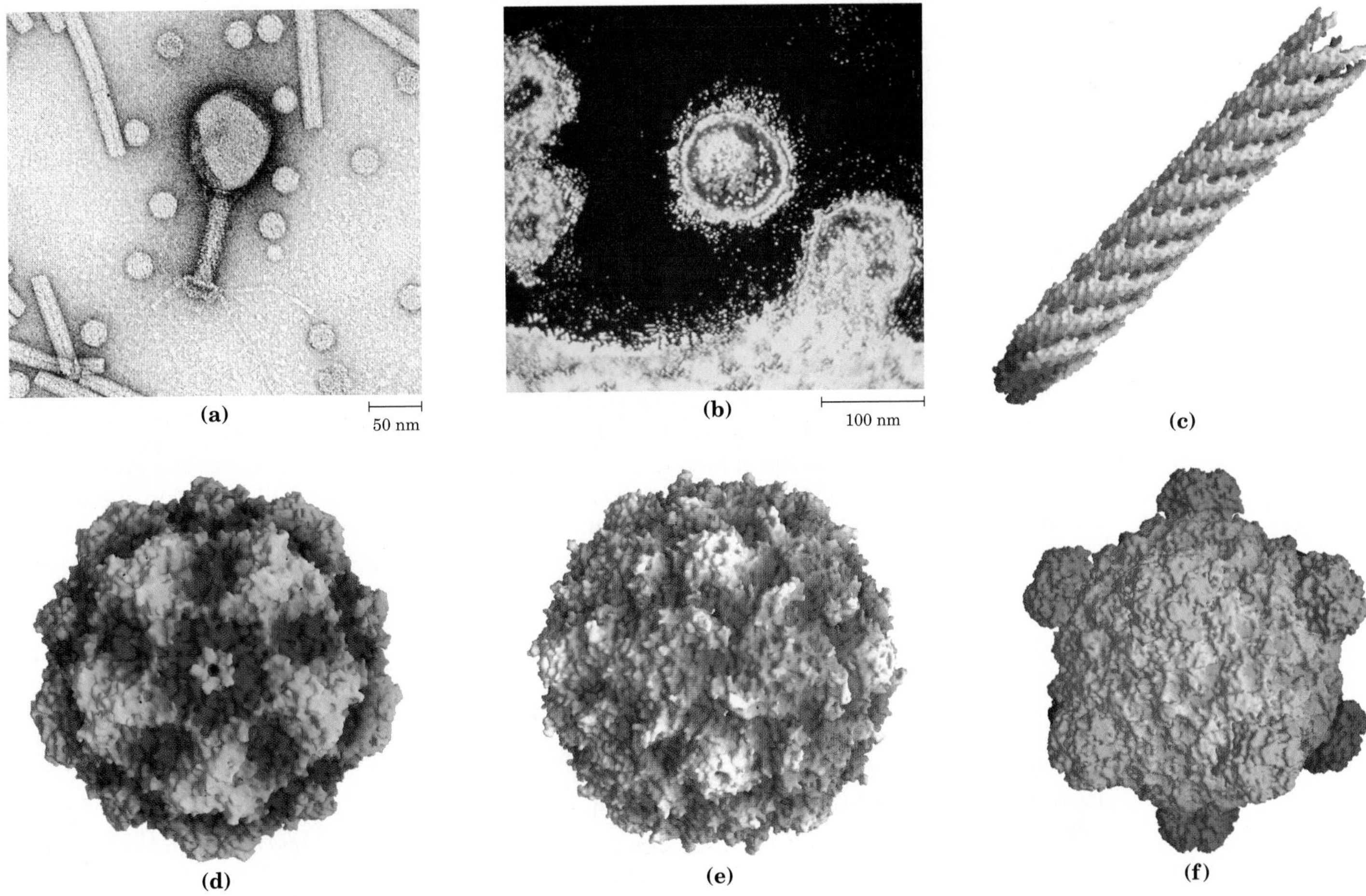

figure 2–23

A gallery of viruses. **(a)** Electron micrograph showing turnip yellow mosaic virus (small spherical particles), tobacco mosaic virus (long cylinders), and bacteriophage T4 (shaped like a hand mirror with spidery legs). **(b)** Electron micrograph (artificial color) showing human immunodeficiency viruses (HIV), the causative agent of AIDS, leaving an infected T lymphocyte of the immune system. **(c)** Molecular surface model of filamentous phage fd. **(d)** Molecular surface model of the canine parvovirus, a serious health hazard to unvaccinated dogs. **(e)** Molecular surface model of human poliovirus (type 2), a picornavirus. Widespread vaccination has nearly eliminated poliovirus as a health hazard in humans. **(f)** Molecular surface model of the bacteriophage ϕX174.

summary

Cells, the structural and functional units of living organisms, are of microscopic dimensions. Their small size, combined with convolutions of their surfaces, results in high surface-to-volume ratios, facilitating the diffusion of fuels, nutrients, and waste products between the cell and its surroundings. All cells share certain features: DNA containing the genetic information, ribosomes, and a plasma membrane that surrounds the cytoplasm. In eukaryotes the genetic material is surrounded by a nuclear envelope; prokaryotes have no such membrane.

The plasma membrane is a tough, flexible permeability barrier, which contains numerous transporters as well as receptors for a variety of extracellular signals. The cytoplasm of eukaryotic cells consists of the cytosol and organelles. The cytosol is a concentrated solution of proteins, RNA, metabolic intermediates and cofactors, and inorganic ions. Ribosomes are supramolecular complexes on which protein synthesis occurs; bacterial ribosomes are slightly smaller than those of eukaryotic cells, but are similar in structure and function.

Certain organisms, tissues, and cells offer advantages for biochemical studies. *E. coli* and yeast can be cultured in large quantities, have short generation times, and are especially amenable to genetic manipulation. The specialized functions of liver, muscle, and fat tissue, and of erythrocytes, make them attractive for the study of specific processes.

The first living cells were prokaryotic and anaerobic; they arose about 3.5 billion years ago, when the atmosphere was devoid of oxygen. With the passage of time, biological evolution led to cells capable of photosynthesis, with O_2 as a byproduct. As O_2 accumulated, prokaryotic cells capable of the aerobic oxidation of fuels evolved. The two major groups of prokaryotes, eubacteria and archaebacteria, diverged early in evolution. The cell envelope of some types of bacteria includes layers outside the plasma membrane that provide rigidity or protection. Some bacteria have flagella for propulsion. The cytoplasm of bacteria has no membrane-bounded organelles but does contain ribosomes and granules of stored fuels, as well as a nucleoid that contains the cell's DNA. Some photosynthetic bacteria have extensive intracellular membranes that contain light-capturing pigments.

About 1.5 billion years ago, eukaryotic cells emerged. They were larger than prokaryotes and their genetic material was more complex. These early cells established symbiotic relationships with prokaryotes that lived in their cytoplasm; modern mitochondria and chloroplasts are derived from these early endosymbionts. Mitochondria and chloroplasts are intracellular organelles surrounded by a double membrane. They are the principal sites of ATP synthesis in eukaryotic, aerobic cells. Chloroplasts are found only in photosynthetic organisms, but mitochondria are ubiquitous among eukaryotes.

Modern eukaryotic cells have a complex system of intracellular membranes. This endomembrane system consists of the nuclear envelope, rough and smooth endoplasmic reticulum, the Golgi complex, transport vesicles, lysosomes, and endosomes. Proteins synthesized on ribosomes bound to the rough endoplasmic reticulum pass into the endomembrane system, traveling through the Golgi complex on their way to organelles or to the cell surface, where they are secreted by exocytosis. Endocytosis brings extracellular materials into the cell, where they can be digested by degradative enzymes in the lysosomes. In plants, the central vacuole is the site of degradative processes; it also serves as a storage depot for pigments and other metabolic products and maintains cell turgor.

The genetic material in eukaryotic cells is organized into chromosomes, highly ordered complexes of DNA and histone proteins. Before cell division (cytokinesis), each chromosome is replicated and the duplicate chromosomes are separated by the process of mitosis.

The cytoskeleton is an intracellular meshwork of actin filaments, microtubules, and intermediate filaments of several types. The cytoskeleton confers shape on the cell, and reorganization of cytoskeletal filaments results in the shape changes accompanying amoeboid movement and cell division. Intracellular organelles move along filaments of the cytoskeleton, propelled by proteins such as kinesin, cytoplasmic dynein, and myosin, using the energy of ATP. Biochemists use differential centrifugation and isopycnic centrifugation to isolate subcellular components for study.

In multicellular organisms, there is a division of labor among different types of cells. The individual epithelial cells in animals can be joined to each other mechanically by tight junctions and desmosomes; communication channels are provided by gap junctions (in animals) and plasmodesmata (in plants). Viruses are parasites of living cells, capable of subverting the cellular machinery for their own replication. They infect animal, plant, and bacterial cells and are responsible for a variety of serious human diseases.

further reading

General

Alberts, B., Bray, D., Lewis, J., Raff, M., Roberts, K., & Watson, J.D. (1994) *Molecular Biology of the Cell,* 3rd edn, Garland Publishing, Inc., New York.

A superb textbook on cell structure and function, covering the topics considered in this chapter, and a useful reference for many of the following chapters.

Becker, W.M., Reece, J.M., & Peonie, M.F. (1995) *The World of the Cell,* 3rd edn, The Benjamin/Cummings Publishing Company, Redwood City, CA.

An excellent introductory textbook of cell biology.

Lodish, H., Baltimore, D., Berk, A., Zipursky, S.L., Matsudaira, P., & Darnell, J. (1995) *Molecular Cell Biology,* 3rd edn, Scientific American Books, Inc., New York.

Like the book by Alberts and coauthors, a superb text useful for this and later chapters.

Margulis, L. (1996) Archaeal-eubacterial mergers in the origin of Eukarya: phylogenetic classification of life. *Proc. Natl. Acad. Sci. USA* **93,** 1071–1076.

The arguments for dividing all living creatures into five kingdoms: Monera, Protoctista, Fungi, Animalia, Plantae.

Margulis, L., Gould, S.J., Schwartz, K.V., & Margulis, A.R. (1998) *Five Kingdoms: An Illustrated Guide to the Phyla of Life on Earth,* 3rd edn, W.H. Freeman and Company, New York.

Description of all major groups of organisms, beautifully illustrated with electron micrographs and drawings.

Purves, W.K., Orians, G.H., Heller, H.C., & Sadava, D. (1998) *Life: The Science of Biology,* 5th edn, Sinauer Associates, Inc., and W.H. Freeman and Company, New York.

A well-written, well-illustrated, up-to-date general biology textbook.

Structure of Cells, Organelles, and Cytoskeleton

Block, S.M. (1998) Leading the procession: new insights into kinesin motors. *J. Cell Biol.* **140,** 1281–1284.

Fawcett, D.W. & Jensh, R.O. (eds) (1997) *Bloom and Fawcett: Concise Histology,* Chapman & Hall, London.

A well-illustrated textbook of cell structure at the microscopic level.

Frontiers in Cell Biology: The Cytoskeleton. (1998) *Science* **279,** 509–533.

This special issue includes the following papers:

Hall, A., Rho GTPases and the actin cytoskeleton (pp. 509–514); **Fuchs, E. & Cleveland, D.W.,** A structural scaffolding of intermediate filaments in health and disease (pp. 514–519); **Hirokawa, N.,** Kinesin and dynein superfamily proteins and the mechanism of organelle transport (pp. 519–526); **Mermall, V., Post, P.L., & Mooseker, M.S.,** Unconventional myosins in cell movement, membrane traffic, and signal transduction (pp. 527–533).

Gelfand, V. & Bershadsky, A.D. (1991) Microtubule dynamics: mechanism, regulation, and function. *Annu. Rev. Cell Biol.* **7,** 93–116.

Organization of the Cytoplasm. (1981) *Cold Spring Harb. Symp. Quant. Biol.* **46.**

More than 90 excellent papers on microtubules, microfilaments, and intermediate filaments and their biological roles.

Rothman, J.E. & Orci, L. (1996) Budding vesicles in living cells. *Sci. Am.* **274** (March), 70–75.

A clear description of the dynamics of the endomembrane system.

Schroer, T.A. & Sheetz, M.P. (1991) Functions of microtubule-based motors. *Annu. Rev. Physiol.* **53,** 629–652.

Spudich, J.A. (1996) Structure-function analysis of the motor domain of myosin. *Annu. Rev. Cell Dev. Biol.* **12,** 543–573.

Takai, Y., Sasaki, T., Tanaka, K., & Nakanishi, H. (1995) Rho as a regulator of the cytoskeleton. *Trends Biochem. Sci.* **20,** 227–231.

Short review of the evidence that the small GTP-binding protein Rho controls the assembly and structure of actin filaments.

Vale, R.D. & Fletterick, R.J. (1997) The design plan of kinesin motors. *Annu. Rev. Cell Dev. Biol.* **13,** 745–777.

Detailed review of the structure and mechanism of the molecular motors in the kinesin superfamily.

Evolution of Cells

de Duve, C. (1995) The beginnings of life on earth. *Am. Sci.* **83,** 428–437.

One scenario for the succession of chemical steps that led to the first living organism.

de Duve, C. (1996) The birth of complex cells. *Sci. Am.* **274** (April), 50–57.

Dyer, B.D. & Obar, R.A. (1994) *Tracing the History of Eukaryotic Cells: The Enigmatic Smile,* Columbia University Press, New York.

Fenchel, T. & Finlay, B.J. (1994) The evolution of life without oxygen. *Am. Sci.* **82**, 22–29.

Discussion of the endosymbiotic hypothesis in the light of modern endosymbiotic anaerobic organisms.

Knoll, A.H. (1991) End of the proterozoic eon. *Sci. Am.* **265** (October), 64–73.

Discussion of the evidence that an increase in atmospheric oxygen led to the development of multicellular organisms, including large animals.

Lazcano, A. & Miller, S.L. (1994) How long did it take for life to begin and evolve to cyanobacteria? *J. Mol. Evol.* **39,** 546–554.

Lazcano, A. & Miller, S.L. (1996) The origin and early evolution of life: prebiotic chemistry, the pre-RNA world, and time. *Cell* **85,** 793–798.

Brief review of recent developments in studies of the origin of life: primitive atmospheres, submarine vents, autotrophic versus heterotrophic origin, the RNA and pre-RNA worlds, and the time required for life to arise.

Margulis, L. (1992) *Symbiosis in Cell Evolution: Microbial Evolution in the Archean and Proterozoic Eons,* 2nd edn, W.H. Freeman and Company, New York.

Clear discussion of the hypothesis that mitochondria and chloroplasts are descendants of bacteria; all eukaryotic cells evolved from microbial symbioses.

Martin, W. & Mueller, M. (1998) The hydrogen hypothesis for the first eukaryote. *Nature* **392,** 37–41.

An interesting new hypothesis for the origin of eukaryotic cells, based on the comparative biochemistry of energy metabolism. It postulates that eukaryotic cells arose from fusion of an H_2-producing eubacterium with a strictly H_2-dependent archaebacterium.

Schopf, J.W. (1992) *Major Events in the History of Life,* Jones and Bartlett Publishers, Boston.

Vidal, G. (1984) The oldest eukaryotic cell. *Sci. Am.* **250** (February), 48–57.

Relationship of Archaea and Eubacteria

Brow, J.R. & Doolittle, W.F. (1997) Archaea and the prokaryote-to-eukaryote transition. *Microbiol. Mol. Biol. Rev.* **61,** 456–502.

A very thorough discussion of the arguments for placing the Archaea on the phylogenetic branch that led to multicellular organisms.

Keeling, P.J. & Doolittle, W.F. (1995) Archaea: narrowing the gap between prokaryotes and eukaryotes. *Proc. Natl. Acad. Sci. USA* **92,** 5761–5764.

Madigan, T. & Marris, B.L. (1997) Extremophiles. *Sci. Am.* **276** (April), 82–87.

A biochemical assessment of archaebacteria that live where it is hottest, coldest, saltiest, most acid, and most alkaline.

Reviews of Archaea. (1997) *Cell* **89.**

This issue contains five reviews of the biochemistry and genomics of the Archaea and their relationship to the eukaryotes:

Olsen, G.J. & Woese, C.R., Archaeal genomics: an overview (pp. 991–994); **Edgell, D.R. & Doolittle, W.F.,** Archaea and the origin(s) of DNA replication proteins (pp. 995–998); **Reeve, J.N., Sandman, K., & Daniels, C.J.,** Archaeal histones, nucleosomes, and transcription initiation (pp. 999–1002); **Belfort, M. & Weiner, A.,** Another bridge between kingdoms: tRNA splicing in Archaea and Eukaryotes (pp. 1003–1006); **Dennis, P.P.,** Ancient ciphers: translation in Archaea (pp. 1007–1010).

problems

Some problems related to the contents of the chapter follow. They involve simple geometrical and numerical relationships concerning cell structure and activities. (In solving end-of-chapter problems, you may wish to refer to the tables printed on the inside of the back cover.) Each problem has a title for easy reference and discussion.

1. The Size of Cells and Their Components

(a) If you were to magnify a cell 10,000 fold (which is typical of the magnification that is achieved using a microscope), how big would it appear? Assume you are viewing a "typical" eukaryotic cell with a cellular diameter of 50 microns.

(b) If this cell were a muscle cell, how many molecules of actin could it hold assuming there were no other cellular components present? (Actin molecules are spherical with a diameter of 3.6 nm; assume the muscle cell is spherical. The volume of a sphere is $4/3\ \pi r^3$.)

(c) If this were a liver cell of the same dimensions, how many mitochondria could it hold assuming there were no other cellular components present? (Assume mitochondria are spherical with a diameter of 1.5 μm; assume the liver cell is spherical. The volume of a sphere is $4/3\ \pi r^3$.)

(d) Glucose is the major energy-yielding nutrient for most cells. Assuming it is present at a concentration of 1 mM, calculate how many molecules of glucose would be present in our hypothetical (and spherical) eukaryotic cell? (Avogadro's number, the number of molecules in 1 mol of a nonionized substance is 6.02×10^{23}.)

(e) Hexokinase is an important enzyme in the metabolism of glucose by cells (see Chapter 15). If the molar concentration of hexokinase in our eukaryotic cell is 20 μM, how many glucose molecules are available for each hexokinase enzyme molecule to metabolize?

2. Components of *E. coli* *E. coli* cells are rod-shaped, about 2 μm long and 0.8 μm in diameter. The volume of a cylinder is $\pi r^2 h$, where h is the height of the cylinder.

(a) If the average density of *E. coli* (mostly water) is 1.1×10^3 g/L, what is the mass of a single cell?

(b) The protective cell wall of *E. coli* is 10 nm thick. What percentage of the total volume of the bacterium does the wall occupy?

(c) *E. coli* is capable of growing and multiplying rapidly because of the inclusion in each cell of some 15,000 spherical ribosomes (diameter 18 nm), which carry out protein synthesis. What percentage of the total cell volume do the ribosomes occupy?

3. Genetic Information in *E. coli* DNA The genetic information contained in DNA consists of a linear sequence of successive coding units, known as codons. Each codon is a specific sequence of three nucleotides (three nucleotide pairs in double-stranded DNA), and each codon codes for a single amino acid unit in a protein. The molecular weight of an *E. coli* DNA molecule is about 3.1×10^9. The average molecular weight of a nucleotide pair is 660, and each nucleotide pair contributes 0.34 nm to the length of DNA.

(a) Calculate the length of an *E. coli* DNA molecule. Compare the length of the DNA molecule with the cell dimensions (see Problem 2). How does the DNA molecule fit into the cell?

(b) Assume that the average protein in *E. coli* consists of a chain of 400 amino acids. What is the maximum number of proteins that can be coded by an *E. coli* DNA molecule?

4. The High Rate of Bacterial Metabolism Bacterial cells have a much higher rate of metabolism than animal cells. Under ideal conditions some bacteria will double in size and divide in 20 min, whereas most animal cells under rapid growth conditions require 24 h. The high rate of bacterial metabolism requires a high ratio of surface area to cell volume.

(a) Why would the surface-to-volume ratio have an effect on the maximum rate of metabolism?

(b) Calculate the surface-to-volume ratio for the spherical bacterium *Neisseria gonorrhoeae* (diameter 0.5 μm), responsible for the disease gonorrhea. Compare it with the surface-to-volume ratio for a globular amoeba, a large eukaryotic cell (diameter 150 μm). The surface area of a sphere is $4\pi r^2$.

5. A Strategy to Increase the Surface Area of Cells Certain cells whose function is to absorb nutrients, such as the cells lining the small intestine or the root hair cells of a plant, are optimally adapted to their role because their exposed surface area is increased by microvilli. Consider a spherical epithelial cell (diameter 20 μm) in the lining of the small intestine. Given that only a part of the cell surface faces the interior of the intestine, assume that a "patch" corresponding to 25% of the cell area is covered with microvilli. Furthermore, assume that the microvilli are cylinders 0.1 μm in diameter, 1.0 μm long, and spaced in a regular grid 0.2 μm on center.

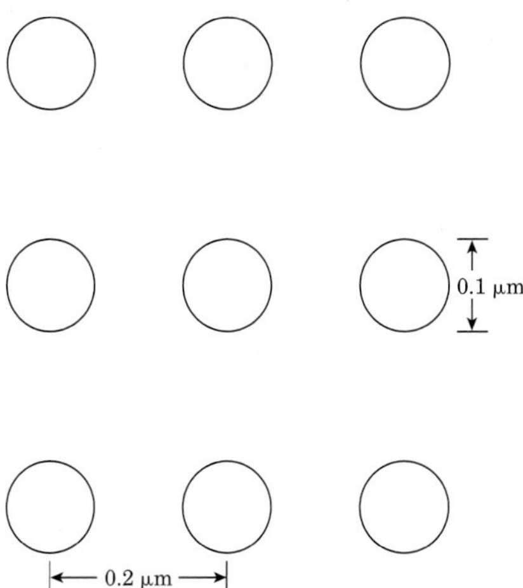

Arrangement of microvilli on the "patch"

(a) Calculate the number of microvilli on the patch.

(b) Calculate the surface area of the patch, assuming it has no microvilli.

(c) Calculate the surface area of the patch, assuming it does have microvilli.

(d) What percentage improvement in absorptive capacity (reflected by the surface-to-volume ratio) does the presence of microvilli provide?

(e) What other organelles and organ systems utilize this strategy to improve absorptive capacity?

6. Fast Axonal Transport Neurons possess long, thin processes called axons, which are structures specialized for conducting signals throughout an organism. Some axonal processes can be as long as 2 m—for example, the axons that originate in the spinal cord and terminate in the muscles of your toes. Small membrane-enclosed vesicles carrying materials essential to axonal function move along microtubules from the cell body to the tips of axons by kinesin-dependent "fast axonal transport."

(a) If the average velocity of a vesicle is 1 μm/sec, how long does it take a vesicle to move from a cell body in the spinal cord to the axonal tip in the toes?

(b) Movement of large molecules in cells by diffusion occurs relatively slowly in cells. (For example, hemoglobin diffuses at a rate of approximately 5 μm/s.) However, the diffusion of sucrose in an aqueous solution occurs at a rate approaching that of fast transport mechanisms (about 4 μm/s). What are some advantages to a cell or an organism of fast, directed transport mechanisms, compared to what a cell could do relying on diffusion alone?

(c) Some of the studies that originally determined the velocity of vesicular movement were performed on microtubules in vitro (in a dish). In order to isolate the microtubules for these studies, intact neurons were initially homogenized (broken) in the presence of 0.2 M sucrose to prevent osmotic swelling and bursting of intracellular organelles. Why is this an important consideration in studies involving cell fractionation?

chapter

3

Biomolecules

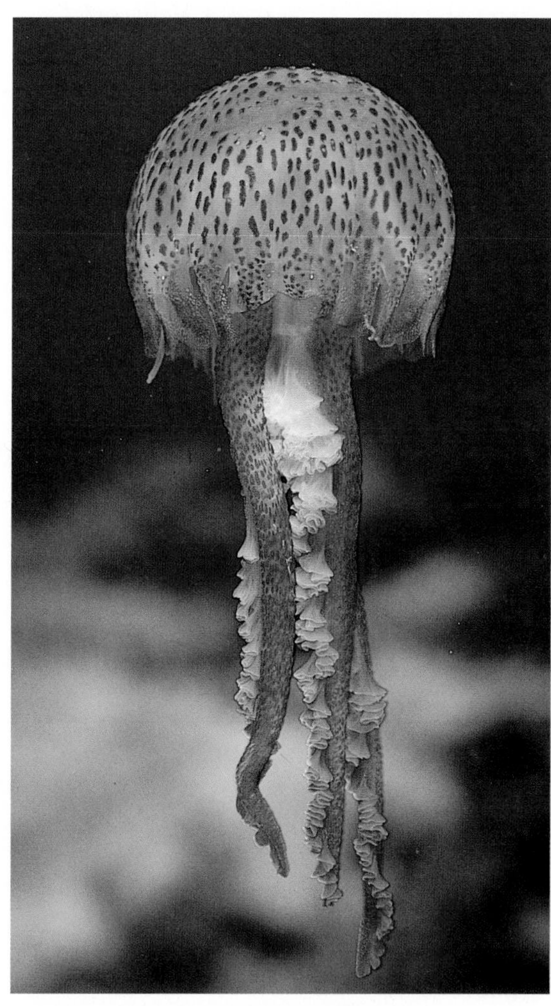

The chemical composition of living material, such as this jellyfish, differs from that of its physical environment, which for this organism is salt water.

Biochemistry aims to explain biological form and function in chemical terms. One of the most fruitful approaches to understanding biological phenomena has been to purify an individual chemical component, such as a protein, from a living organism and to characterize its chemical structure or catalytic activity. As we begin our study of biomolecules and their interactions in living cells, some basic questions come naturally to mind. What kinds of molecules are present in living organisms, and in what proportions? What are the structures of these molecules, and what forces stabilize their structures? What are their chemical properties and reactivities in isolation? How do they interact with each other? How and where did the biomolecules of the first living cells originate?

In this chapter, we review some of the chemical principles that govern the properties of biological molecules: the covalent bonding of carbon with itself and with other elements, the functional groups that occur in common biological molecules, the three-dimensional structure and stereochemistry of carbon compounds, the effects of chemical structure on reactivity, and the common classes of chemical reactions that occur in living organisms. Next, we discuss the monomeric subunits from which macromolecules are constructed and the energetics of their polymerization. Finally, we consider the origin of these monomeric subunits from simple compounds in the earth's atmosphere during prebiological times—that is, prebiotic evolution.

Chemical Composition and Bonding

By the late eighteenth century, it had become clear to chemists that the composition of living matter is strikingly different from that of the inanimate world. Antoine Lavoisier (1743–1794) noted the relative chemical simplicity of the "mineral world" and contrasted it with the complexity of the "plant and animal worlds"; the latter, he knew, were composed of compounds rich in the elements carbon, oxygen, nitrogen, and phosphorus.

Only about 30 of the more than 90 naturally occurring chemical elements are essential to living organisms. Most of the elements in living matter have relatively low atomic numbers; only five have atomic numbers above that of selenium, 34 (Fig. 3–1). The four most abundant elements in living organisms, in terms of the percentage of the total number of atoms, are hydrogen, oxygen, nitrogen, and carbon, which together make up over 99% of the mass of most cells. They are the lightest elements capable of forming one, two, three, and four bonds, respectively (Fig. 3–2). In general, the lightest elements form the strongest bonds. The trace elements (Fig. 3–1) represent a miniscule fraction of the weight of the human body, but all

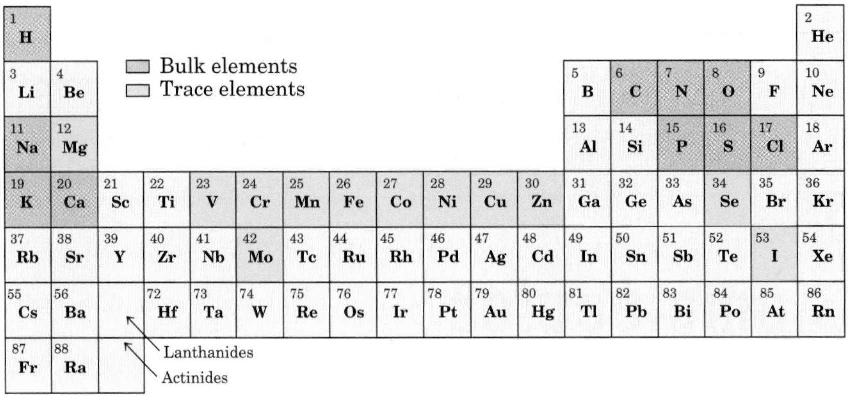

figure 3–1

Elements essential to animal life and health. Bulk elements (shaded orange) are structural components of cells and tissues and are required in the diet in gram quantities daily. For trace elements (shaded bright yellow), the requirements are much smaller: for humans, a few milligrams per day of Fe, Cu, and Zn, even less of the others. The elemental requirements for plants and microorganisms are similar to those shown here.

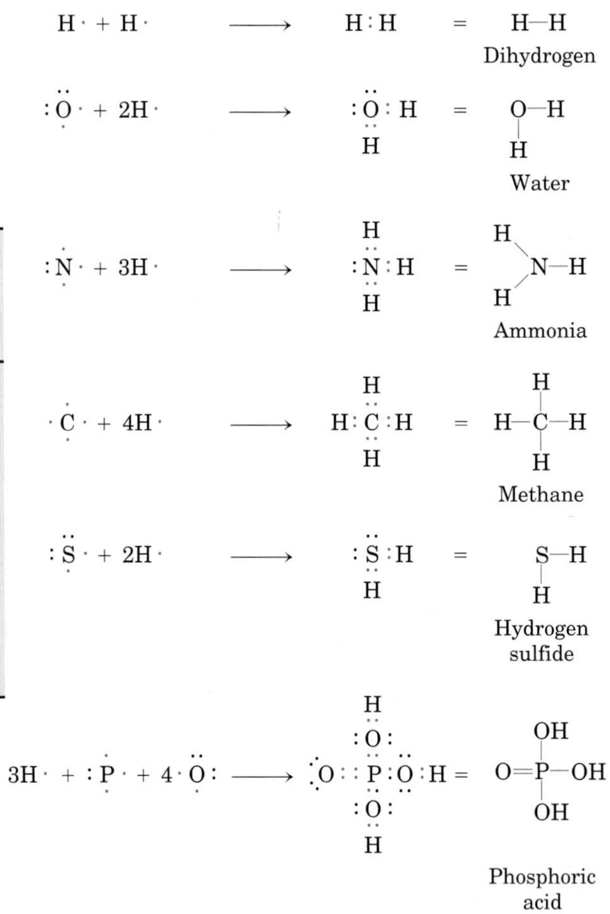

figure 3–2

Covalent bonding. Two atoms with unpaired electrons in their outer shells can form covalent bonds with each other by sharing electron pairs. Atoms participating in covalent bonding tend to fill their outer electron shells.

Atom	Number of unpaired electrons (in red)	Number of electrons in complete outer shell
H·	1	2
O	2	8
N	3	8
C	4	8
S	2	8
P	3	8

H· + H· ⟶ H:H = H—H Dihydrogen

O + 2H· ⟶ H:O:H = H—O—H Water

N + 3H· ⟶ H:N:H with H = NH_3 Ammonia

C + 4H· ⟶ H:C:H with 2 H = CH_4 Methane

S + 2H· ⟶ H:S:H = H—S—H Hydrogen sulfide

3H· + P + 4O ⟶ O::P(OH)₃ = O=P(OH)—OH Phosphoric acid

are essential to life, usually because they are essential to the function of specific proteins, including enzymes. The oxygen-transporting capacity of the hemoglobin molecule, for example, is absolutely dependent on four iron ions that make up only 0.3% of its mass.

Biomolecules Are Compounds of Carbon

The chemistry of living organisms is organized around carbon, which accounts for more than half the dry weight of cells. Carbon can form single bonds with hydrogen atoms, and both single and double bonds with oxygen and nitrogen atoms (Fig. 3–3). Of greatest significance in biology is the ability of carbon atoms to share electron pairs with each other to form very stable carbon–carbon single bonds. Each carbon atom can form single bonds with one, two, three, or four other carbon atoms. Two carbon atoms also can share two (or three) electron pairs, thus forming double (or triple) bonds (Fig. 3–3).

The four single covalent bonds that can be formed by a carbon atom are arranged tetrahedrally, with an angle of about 109.5° between any two bonds (Fig. 3–4) and an average length of 0.154 nm. There is free rotation around each single bond unless very large or highly charged groups are attached to both carbon atoms, in which case rotation may be restricted. A double bond is shorter (about 0.134 nm) and rigid and allows little rotation about its axis.

Covalently linked carbon atoms in biomolecules can form linear chains, branched chains, and cyclic structures. To these carbon skeletons are added groups of other atoms, called **functional groups,** which confer specific chemical properties on the molecule. Molecules with covalently bonded carbon backbones are called **organic compounds;** they occur in limitless variety. Most biomolecules are organic compounds; we can therefore infer that the bonding versatility of carbon was a major factor in the selection of carbon compounds for the molecular machinery of cells during the origin and evolution of living organisms. No other chemical element can form molecules of such widely different sizes and shapes or with such a variety of functional groups.

figure 3–3
Versatility of carbon in forming covalent single, double, and triple bonds (in red), particularly between carbon atoms. Triple bonds occur only rarely in biomolecules.

Electron-dot formation	Structural formula
·Ċ· + ·H ⟶ ·Ċ:H	—C—H
·Ċ· + ·Ö: ⟶ ·Ċ:Ö·	—C—O—
·Ċ· + ·Ö: ⟶ C::O	>C=O
·Ċ· + ·Ṅ: ⟶ ·Ċ:N:	—C—N<
·Ċ· + ·Ṅ: ⟶ C::N·	>C=N—
·Ċ· + ·Ċ· ⟶ ·Ċ:Ċ·	—C—C—
·Ċ· + ·Ċ· ⟶ C::C	>C=C<
·Ċ· + ·Ċ· ⟶ ·C:::C·	—C≡C—

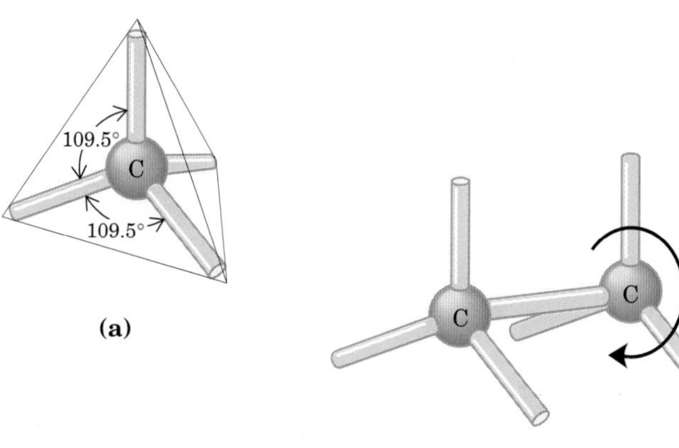

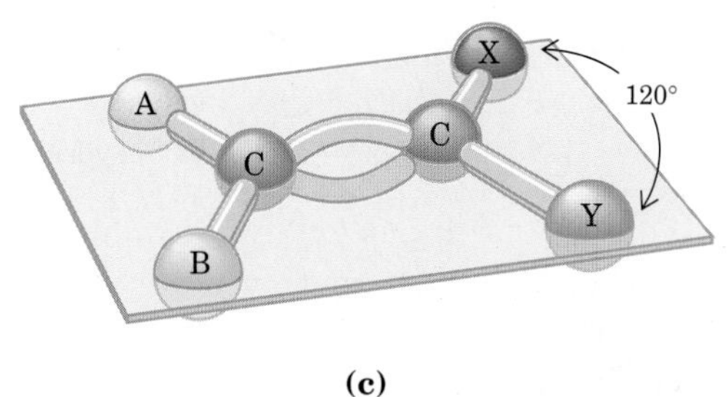

figure 3–4
Geometry of carbon bonding. (a) Carbon atoms have a characteristic tetrahedral arrangement of their four single bonds, which are about 0.154 nm long and at an angle of 109.5° to each other. **(b)** Carbon–carbon single bonds have freedom of rotation, as shown for the compound ethane (CH_3—CH_3). **(c)** Double bonds are shorter and do not allow free rotation. The single bonds on each doubly bonded carbon make an angle of 120° with each other. The two doubly bonded carbons and the atoms designated A, B, X, and Y all lie in the same rigid plane.

Functional Groups Determine Chemical Properties

Most biomolecules can be regarded as derivatives of hydrocarbons, compounds with a covalently linked carbon backbone to which only hydrogen atoms are bonded. The backbones of hydrocarbons are very stable. The hydrogen atoms may be replaced by a variety of functional groups to yield different families of organic compounds. Typical of these are alcohols, which have one or more hydroxyl groups; amines, which have amino groups; aldehydes and ketones, which have carbonyl groups; and carboxylic acids, which have carboxyl groups (Fig. 3–5).

Many biomolecules are polyfunctional, containing two or more different kinds of functional groups (Fig. 3–6), each with its own chemical characteristics and reactions. The chemical "personality" of a compound such as epinephrine or acetyl-coenzyme A is determined by the chemistry of its functional groups and their disposition in three-dimensional space.

figure 3–5

Some common functional groups of biomolecules. All groups are shown in their uncharged (nonionized) form. In this figure and throughout the book, we use R to represent "any substituent." It may be as simple as a hydrogen atom, but typically it is a carbon-containing moiety. When two or more substituents are shown in a molecule, we designate them R^1, R^2, and so forth.

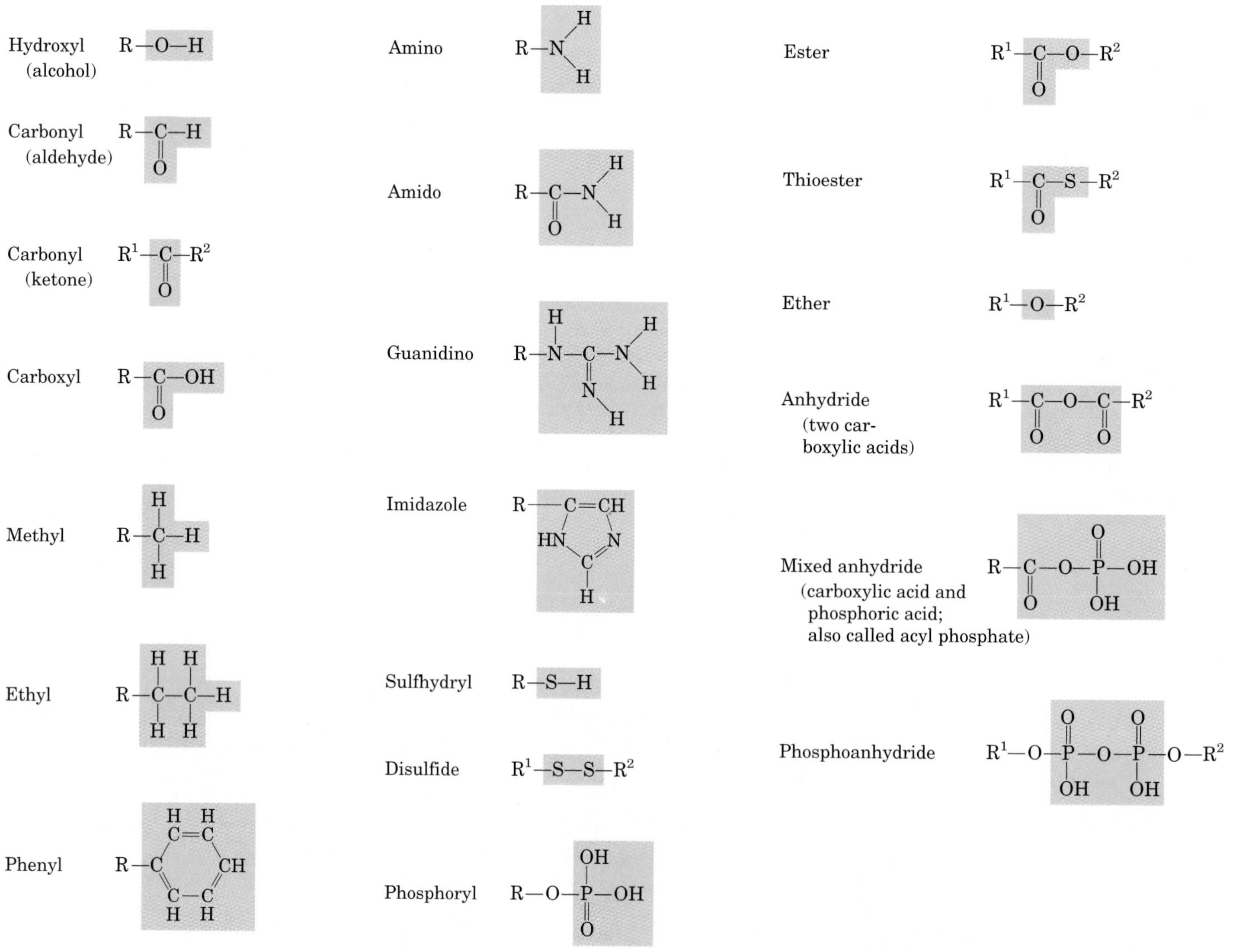

Hydroxyl (alcohol) R—O—H

Carbonyl (aldehyde) R—C(=O)—H

Carbonyl (ketone) R^1—C(=O)—R^2

Carboxyl R—C(=O)—OH

Methyl R—CH₃ (R—C(H)(H)—H)

Ethyl R—C(H)(H)—C(H)(H)—H

Phenyl R—C₆H₅ (R—C, ring of HC=CH, CH, HC—CH)

Amino R—N(H)—H

Amido R—C(=O)—N(H)—H

Guanidino R—N(H)—C(=NH)—N(H)—H

Imidazole R—C=CH, HN, N, C—H (ring)

Sulfhydryl R—S—H

Disulfide R^1—S—S—R^2

Phosphoryl R—O—P(OH)(=O)—OH

Ester R^1—C(=O)—O—R^2

Thioester R^1—C(=O)—S—R^2

Ether R^1—O—R^2

Anhydride (two carboxylic acids) R^1—C(=O)—O—C(=O)—R^2

Mixed anhydride (carboxylic acid and phosphoric acid; also called acyl phosphate) R—C(=O)—O—P(=O)(OH)—OH

Phosphoanhydride R^1—O—P(=O)(OH)—O—P(=O)(OH)—O—R^2

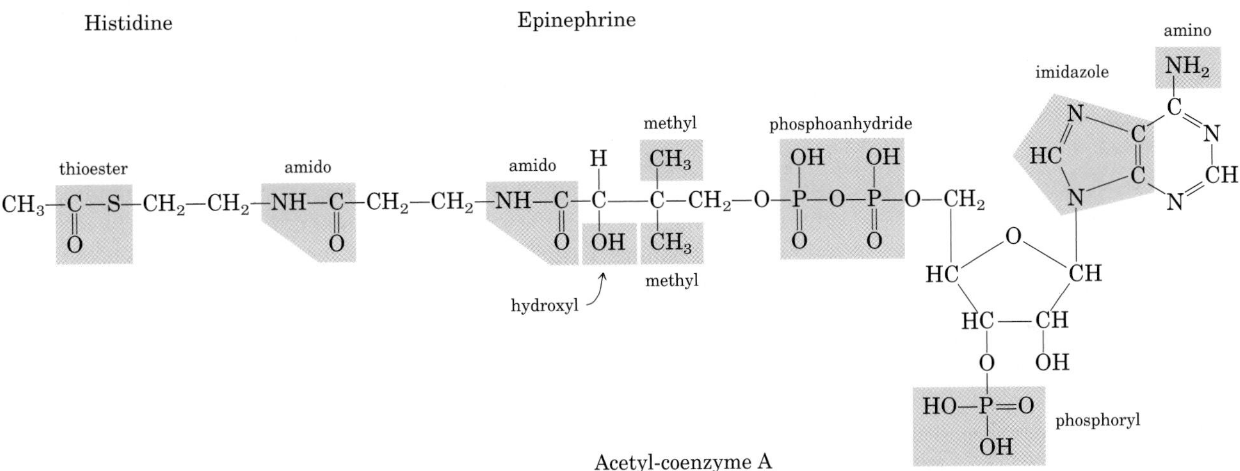

figure 3–6

Common functional groups in biomolecules. Histidine is one of the amino acids found in proteins; epinephrine is a hormone; acetyl-coenzyme A is a carrier of acetyl groups in some enzymatic reactions. Note that the amino group in epinephrine is a secondary (*s*) amino group, with one of its amino hydrogens replaced by another group.

Three-Dimensional Structure: Configuration and Conformation

Although the covalent bonds and functional groups of a biomolecule are central to its function, the arrangement of the molecule's constituent atoms in three-dimensional space—its stereochemistry—is also crucially important. Compounds of carbon commonly exist as **stereoisomers,** different molecules in which the order of bonding is the same, but the spatial relationship among the atoms is different. Molecular interactions between biomolecules are invariably stereospecific; that is, they require specific stereochemistry in the interacting molecules.

Figure 3–7 shows three ways to illustrate the stereochemical configuration of simple molecules. The perspective diagram specifies configuration unambiguously, but bond angles and center-to-center bond lengths are better represented with ball-and-stick models. In space-filling models, the radius of each atom is proportional to its van der Waals radius (Table 3–1), and the contours of the molecule represent the outer limits of the region from which atoms of other molecules are excluded.

table 3–1

Van der Waals Radii and Covalent (Single-Bond) Radii of Some Elements*

Element	Van der Waals radius (nm)	Covalent radius for single bond (nm)
H	0.1	0.030
O	0.14	0.074
N	0.15	0.073
C	0.17	0.077
S	0.18	0.103
P	0.19	0.110
I	0.22	0.133

*Van der Waals radii describe the space-filling dimensions of atoms. When two atoms are joined covalently, the atomic radii at the point of bonding are less than the van der Waals radii, because the joined atoms are pulled together by the shared electron pair. The distance between nuclei in a van der Waals interaction or in a covalent bond is about equal to the sum of the van der Waals radii or the covalent radii, respectively, for the two atoms. Thus the length of a carbon–carbon single bond is about 0.077 nm + 0.077 nm = 0.154 nm.

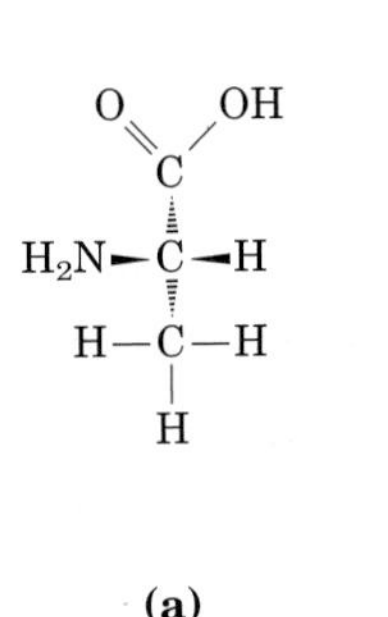

(a)

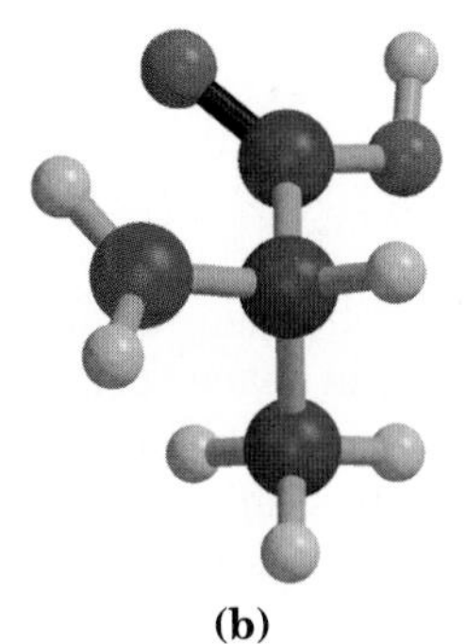

(b)

(c)

figure 3–7

Three ways to represent the structure of the amino acid alanine. **(a)** Structural formula in perspective form: a solid wedge (◀) represents a bond in which the atom at the wide end projects out of the plane of the paper, toward the reader; a dashed wedge represents a bond extending behind the plane of the paper. **(b)** Ball-and-stick model, showing relative bond lengths and the bond angles. **(c)** Space-filling model, in which each atom is shown with its correct relative van der Waals radius (see Table 3–1).

The Configuration of a Molecule Is Changed Only by Breaking a Bond

Configuration denotes the fixed spatial arrangement of atoms in an organic molecule that is conferred by the presence of either (1) double bonds, around which there is no freedom of rotation, or (2) chiral centers, around which substituent groups are arranged in a specific sequence. The identifying characteristic of configurational isomers is that they cannot be interconverted without temporarily breaking one or more covalent bonds.

Figure 3–8a shows the configurations of maleic acid and its isomer, fumaric acid. These compounds are **geometric** or **cis-trans isomers;** they differ in the arrangement of their substituent groups with respect to the nonrotating double bond. Maleic acid is the cis isomer and fumaric acid the trans isomer; each is a well-defined compound that can be separated from the other, and each has its own unique chemical properties. A binding site (on an enzyme, for example) that is complementary to one of these molecules would not be a suitable binding site for the other, which explains why these compounds have distinct biological roles despite their similar chemistry.

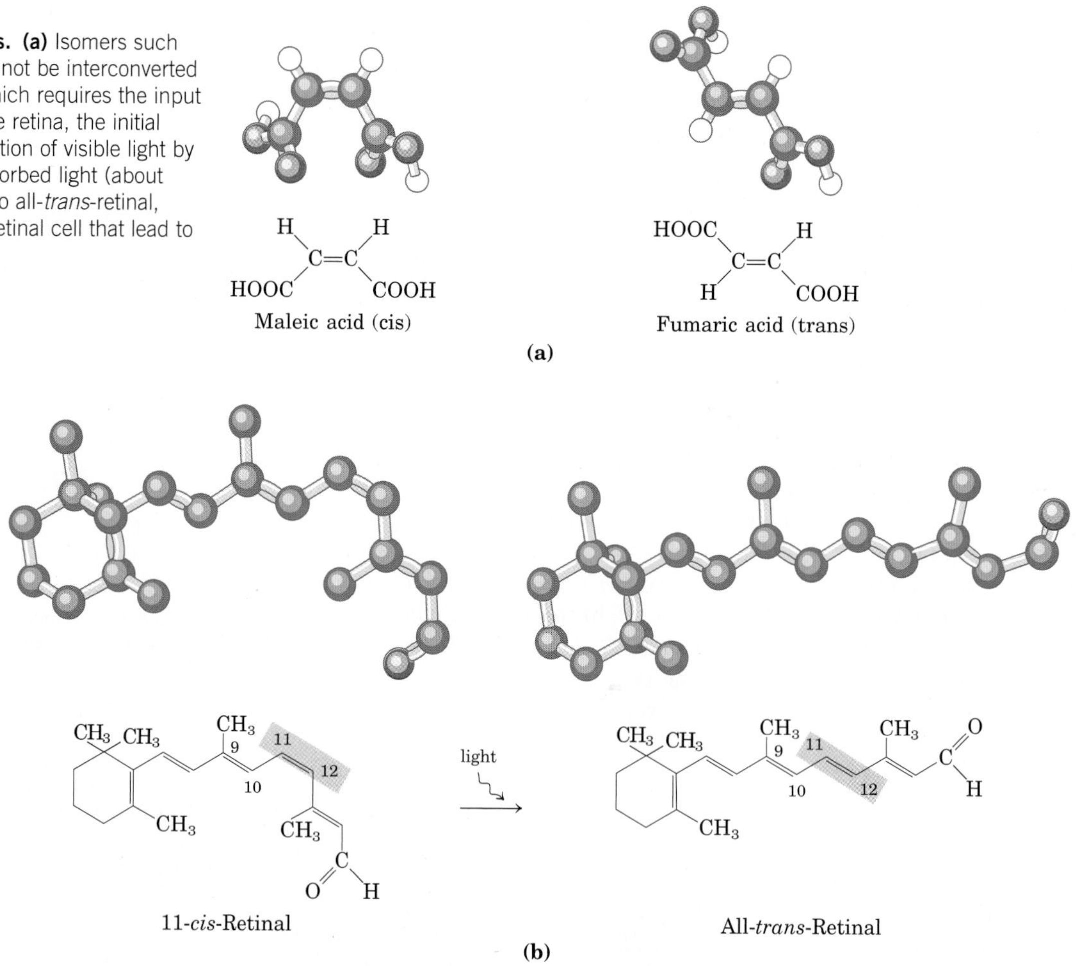

figure 3–8
Configurations of geometric isomers. (a) Isomers such as maleic acid and fumaric acid cannot be interconverted without breaking covalent bonds, which requires the input of much energy. **(b)** In the vertebrate retina, the initial event in light detection is the absorption of visible light by 11-*cis*-retinal. The energy of the absorbed light (about 250 kJ/mol) converts 11-*cis*-retinal to all-*trans*-retinal, triggering electrical changes in the retinal cell that lead to a nerve impulse.

Four different substituents bonded to a tetrahedral carbon atom may be arranged two different ways in space (i.e., have two configurations; Fig. 3–9), yielding two stereoisomers with similar or identical chemical properties, but differing in certain physical and biological properties. A carbon atom with four different substituents is said to be asymmetric, and asym-

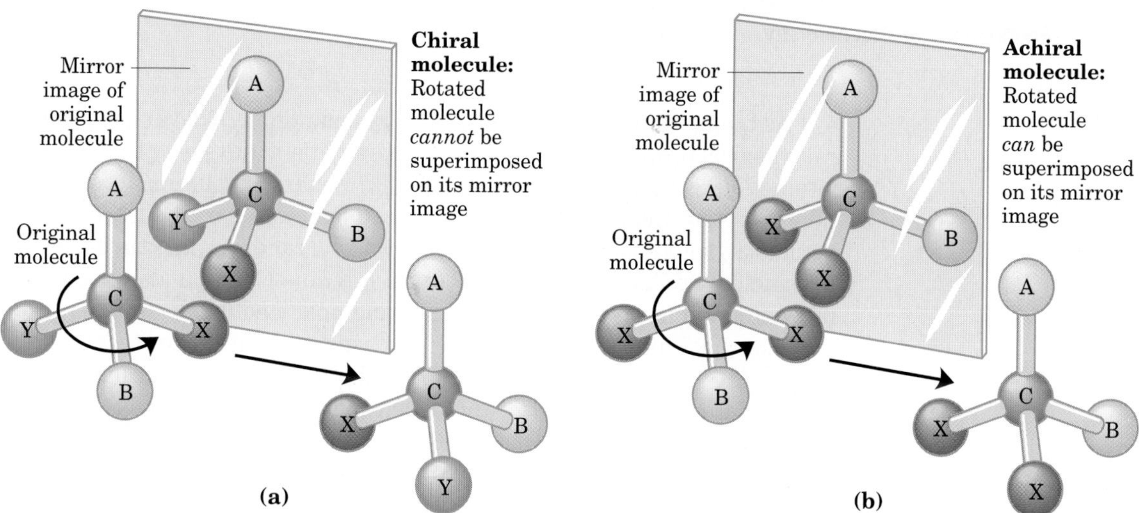

figure 3–9

Molecular asymmetry: chiral and achiral molecules. **(a)** When a carbon atom has four different substituent groups (A, B, X, Y), they can be arranged in two ways that represent nonsuperimposable mirror images of each other (enantiomers). Such a carbon atom is asymmetric and is called a chiral atom or chiral center. **(b)** When a tetrahedral carbon has only three dissimilar groups (i.e., the same group occurs twice), only one configuration is possible and the molecule is symmetric, or achiral. In this case the molecule is superimposable on its mirror image: the molecule on the left can be rotated counterclockwise (when looking down the vertical bond from A to C) to create the molecule in the mirror.

metric carbons are called **chiral centers** (Greek *chiros,* "hand"; some stereoisomers are related structurally as the right hand is to the left). A molecule with only one chiral carbon can have only two stereoisomers, but when two or more (n) chiral carbons are present, there can be 2^n stereoisomers. Some stereoisomers are mirror images of each other; they are called **enantiomers** (Fig. 3–9). Pairs of stereoisomers that are not mirror images of each other are called **diastereomers** (Fig. 3–10).

figure 3–10

Two types of stereoisomers. There are four different 2,3-disubstituted butanes (n = 2 asymmetric carbons, hence 2^n = 4 stereoisomers). Each is shown in a box as a perspective formula and a ball-and-stick model, which has been rotated to allow the reader to view all the groups. Some pairs of stereoisomers are mirror images of each other, and thus enantiomers. Other pairs are not mirror images; these are diastereomers. (Adapted from Carroll, F. (1998) *Perspectives on Structure and Mechanism in Organic Chemistry,* p. 63, Brooks/Cole Publishing Co., Pacific Grove, CA.)

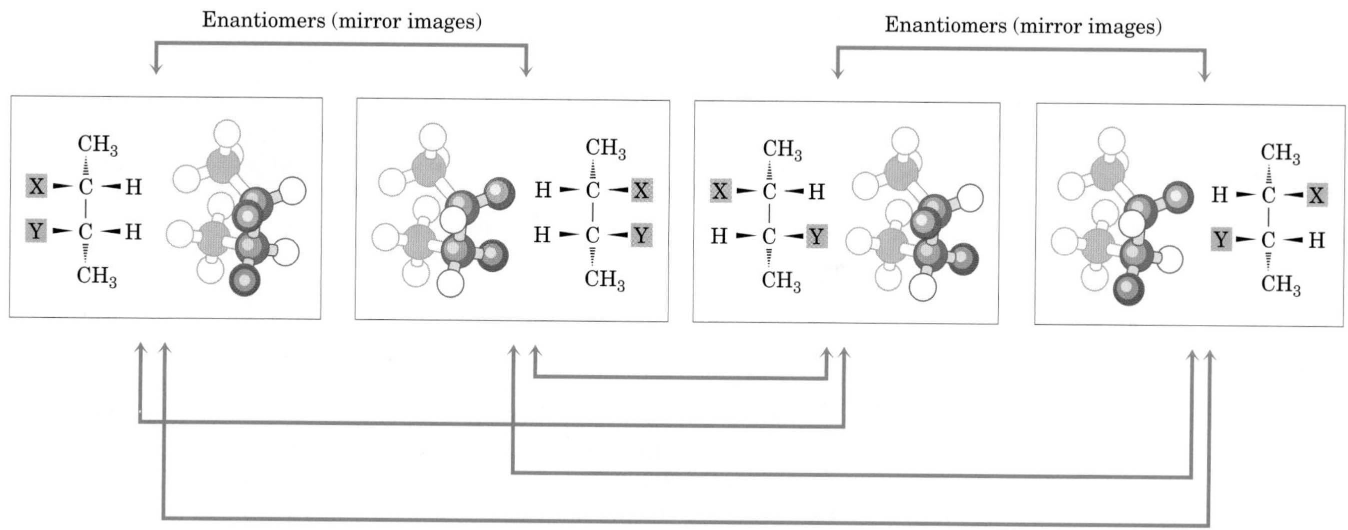

As Louis Pasteur observed (Box 3–1), enantiomers have nearly identical chemical properties but differ in a characteristic physical property, their interaction with plane-polarized light. In separate solutions, two enantiomers rotate the plane of plane-polarized light in opposite directions, but equimolar solutions of the two enantiomers (**racemic mixtures,** in the terminology of Pasteur) show no optical rotation. Compounds without chiral centers do not rotate the plane of plane-polarized light.

Biological interactions (between enzyme and substrate, receptor and hormone, or antibody and antigen, for example) are stereospecific: the "fit" in such interactions must be stereochemically correct. We must therefore name and represent the structure of a biomolecule so as to make its stereochemistry unambiguous. For compounds with more than one chiral center, the RS system of nomenclature is often more useful than the D and L system described in Chapter 5. In the RS system, each group attached to a chiral carbon is assigned a *priority.* The priorities of some common substituents are

$$—OCH_2 > —OH > —NH_2 > —COOH > —CHO > —CH_2OH > —CH_3 > —H$$

The chiral atom is viewed with the group of lowest priority (4) pointing away from the viewer. If the priority of the other three groups (1 to 3) decreases in clockwise order, the configuration is (*R*) (Latin *rectus*, "right"); if in counterclockwise order, the configuration is (*S*) (Latin *sinister*, "left").

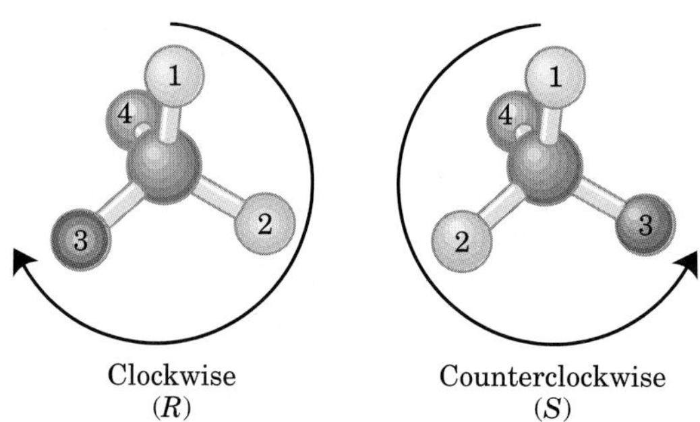

In this way each chiral carbon is designated as either (*R*) or (*S*), and the inclusion of these designations in the name of the compound provides an unambiguous description of the stereochemistry at each chiral center.

Molecular Conformation Is Changed by Rotation about Single Bonds

Molecular **conformation** refers to the spatial arrangement of substituent groups that, without breaking any bonds, are free to assume different positions in space because of the freedom of bond rotation. In the simple hydrocarbon ethane, for example, there is nearly complete freedom of rotation around the C—C bond. Many different, interconvertible conformations of the ethane molecule are therefore possible, depending on the degree of rotation (Fig. 3–11). Two conformations are of special interest: the staggered, which is more stable than all others and thus predominates, and the eclipsed, which is least stable. It is not possible to isolate either of these conformational forms, because they are freely interconvertible. However, when one or more of the hydrogen atoms on each carbon is replaced by a functional group that is either very large or electrically charged, freedom of rotation around the C—C bond is hindered. This limits the number of stable conformations of the ethane derivative.

box 3–1 Louis Pasteur and Optical Activity: *In Vino, Veritas*

Louis Pasteur
1822–1895

Louis Pasteur encountered the phenomenon of **optical activity** in 1843, during his investigation of the crystalline sediment that accumulated in wine casks ("paratartaric acid," also called racemic acid, from Latin *racemus,* "bunch of grapes"). He used fine forceps to separate two types of crystals identical in shape, but mirror images of each other. Both types proved to have all the chemical properties of tartaric acid, but in solution one type rotated polarized light to the left (levorotatory), the other to the right (dextrorotatory). Pasteur later described the experiment and its interpretation:

> In isomeric bodies, the elements and the proportions in which they are combined are the same, only the arrangement of the atoms is different. . . . We know, on the one hand, that the molecular arrangements of the two tartaric acids are asymmetric, and, on the other hand, that these arrangements are absolutely identical, excepting that they exhibit asymmetry in opposite directions. Are the atoms of the dextro acid grouped in the form of a right-handed spiral, or are they placed at the apex of an irregular tetrahedron, or are they disposed according to this or that asymmetric arrangement? We do not know.*

Now we do know. X-ray crystallographic studies in 1951 confirmed that the levorotatory and dextrorotatory forms of tartaric acid are mirror images of each other at the molecular level, and established the absolute configuration of each (Fig. 1). The same approach has been used to demonstrate that although the amino acid alanine has two stereoisomeric forms (designated D and L), alanine in proteins exists exclusively in one form (the L isomer; see Chapter 5).

HOOC[1] [4]COOH
C[2]—C[3]
H OH / H OH

(2*R*,3*R*)-Tartaric acid
(dextrorotatory)

HOOC[1] [4]COOH
C[2]—C[3]
HO H / H OH

(2*S*,3*S*)-Tartaric acid
(levorotatory)

figure 1
Pasteur separated crystals of two stereoisomers of tartaric acid and showed that solutions of the separated forms rotated polarized light to the same extent but in opposite directions. These dextrorotatory and levorotatory forms were later shown to be the (*R,R*) and (*S,S*) isomers represented here. The RS system of nomenclature is described in the text.

* From Pasteur's lecture to the Société Chimique de Paris in 1883, quoted in DuBos, R. (1976) *Louis Pasteur: Free Lance of Science,* p. 95, Charles Scribner's Sons, New York.

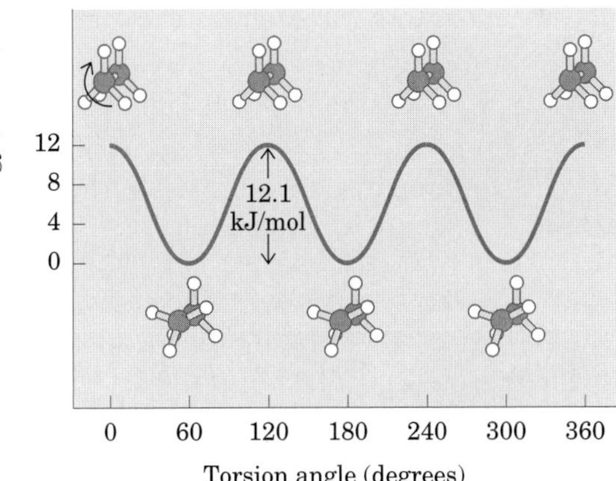

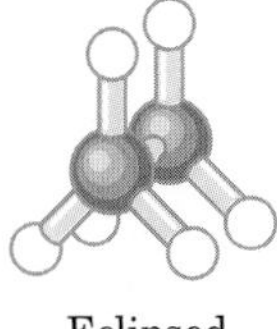

Eclipsed

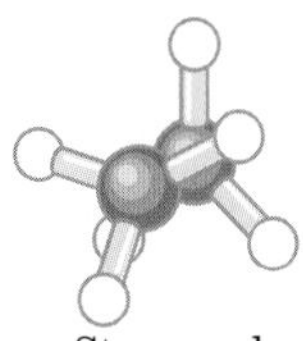

Staggered

figure 3–11
Many conformations of ethane are possible because of freedom of rotation around the C—C bond. When the front carbon atom (as viewed by the reader) with its three attached hydrogens is rotated relative to the rear carbon atom, the potential energy of the molecule rises in the fully eclipsed conformation (torsion angle 0°, 120°, etc.), then falls in the fully staggered conformation (torsion angle 60°, 180°, etc.). Because the energy differences are small enough to allow rapid interconversion of the two forms (millions of times per second), the eclipsed and staggered forms cannot be separately isolated.

Configuration and Conformation Define Biomolecular Structures

The three-dimensional structure of biomolecules large and small—the combination of configuration and conformation—is of the utmost importance in their biological interactions. For example, in the binding of a substrate (reactant) to the catalytic site of an enzyme, the two molecules must complement each other closely for effective catalysis. Such complementarity is also required in the binding of a hormone molecule to its receptor on a cell surface, or in the recognition and binding of an antigen by a specific antibody (Fig. 3–12).

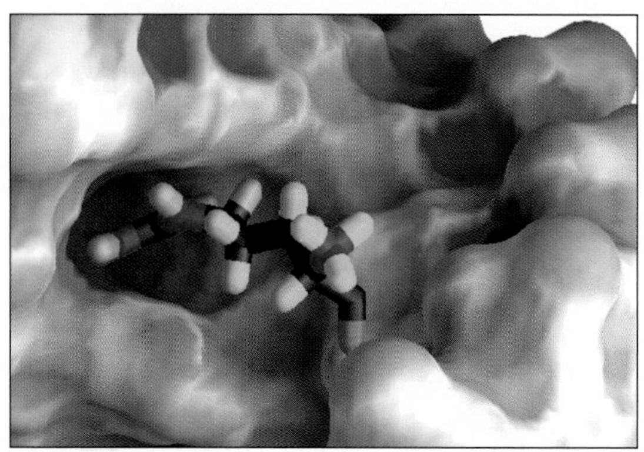

figure 3–12
Complementary fit of a macromolecule and a small molecule. Shown here are a segment of RNA from the regulatory region TAR of the HIV genome (gray) and argininamide (colored), representing one residue of a protein that binds to this region. The argininamide fits into a pocket on the RNA surface and is held in this orientation by several noncovalent interactions with the RNA. This representation of the RNA molecule is produced with the computer program GRASP©, which can calculate the shape of the outer surface of a macromolecule, defined either by the van der Waals radii of all of the atoms in the molecule or by the "solvent exclusion volume," beyond which a water molecule cannot penetrate.

The study of biomolecular stereochemistry with precise physical methods is an important part of modern research on cell structure and biochemical function. To date, the most productive method for structural investigations has been x-ray crystallography. For a compound that can be crystallized, the diffraction of x rays by crystals can be used to determine with great precision the position of every atom in the molecule relative to every other atom. The structures of most small biomolecules (those with fewer than about 50 atoms) and of hundreds of proteins and small nucleic acids have been determined by this means.

X-ray crystallography yields a static picture of the molecule within the confines of the crystal. However, biomolecules almost never exist as crystals within cells; rather, they are dissolved in the cytosol or associated with some other component(s) of the cell. Molecules have more freedom of intramolecular motion in solution than in a crystal. In a large molecule such as a protein, the small variations allowed in the conformations of its monomeric subunits add up to extensive flexibility; a protein may have several alternative stable conformations. Techniques such as nuclear magnetic resonance (NMR) spectroscopy complement x-ray crystallography by providing information about the three-dimensional structures of biomolecules in solution, including the flexibility of protein segments that might be held rigid in the crystal. With advances in the technology for producing very strong magnetic fields, NMR spectroscopy is increasingly used to determine the structure of larger molecules, including proteins (Fig. 3–13). The techniques of x-ray crystallography and NMR spectroscopy are described further in Box 6–3.

figure 3–13
Structure of the protein brazzein as determined by NMR spectroscopy. Brazzein, isolated from the fruit of *Pentadiplandra,* a plant of western Africa, is gram-for-gram 2,000 times sweeter than sucrose. In this representation, only the backbone of the protein is shown, without the individual amino acid side chains. NMR spectroscopy shows that in solution the protein can assume up to 43 very similar conformations, each shown here in a different color. This kind of conformational variability is not seen by x-ray crystallography, which requires that a protein be essentially frozen into one conformation in its crystalline form.

Interactions between Biomolecules Are Stereospecific

In living organisms, chiral molecules are usually present in only one of their chiral forms. For example, the amino acids occur in proteins only as the L isomers. Glucose, the monomeric subunit of starch, occurs biologically in only one of its chiral forms, the D isomer. (The conventions for naming stereoisomers of the amino acids are described in Chapter 5; those for sugars, in Chapter 9.) In contrast, when a compound with an asymmetric carbon atom is chemically synthesized in the laboratory, the nonbiological reactions usually produce all possible chiral forms in an equimolar mixture that does not rotate polarized light (a racemic mixture). The chiral forms in such a mixture can be separated only by painstaking physical methods (recall that Pasteur separated crystals with forceps). Chiral compounds in living cells are produced in only one chiral form because the enzymes that synthesize them are also chiral molecules.

Stereospecificity, the ability to distinguish between stereoisomers, is a property of enzymes and other proteins and a characteristic feature of the molecular logic of living cells. If the binding site on a protein is complementary to one isomer of a chiral compound, it will not be complementary to the other isomer, for the same reason that a left glove does not fit a right hand. Two striking examples of the ability of biological systems to distinguish stereoisomers are shown in Figure 3–14. Specific sensory receptors (for smell and taste) easily distinguish the members of each pair of diastereomers.

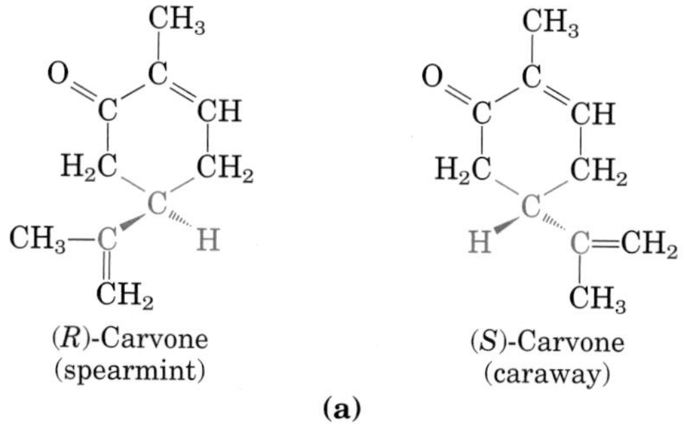

(*R*)-Carvone
(spearmint)

(*S*)-Carvone
(caraway)

(a)

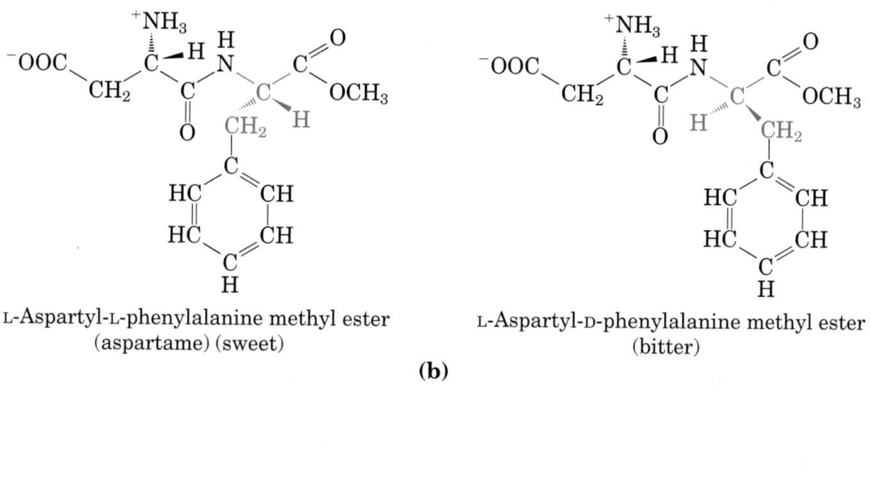

L-Aspartyl-L-phenylalanine methyl ester
(aspartame) (sweet)

L-Aspartyl-D-phenylalanine methyl ester
(bitter)

(b)

figure 3–14
Stereoisomers distinguishable by smell and taste in humans. **(a)** Two stereoisomers of carvone: (*R*)-carvone (isolated from spearmint oil) has the characteristic fragrance of spearmint; (*S*)-carvone (from caraway seed oil) smells like caraway. **(b)** Aspartame, the artificial sweetener sold under the trade name NutraSweet, is easily distinguishable by taste receptors from its bitter-tasting stereoisomer, although the two differ only in the configuration at one of the two chiral carbon atoms.

Chemical Reactivity

The mechanisms of biochemical reactions are not fundamentally different from those of other chemical reactions. They may be understood and predicted from the nature of the functional groups of the reactants. Functional groups alter the electron distribution and the geometry of neighboring atoms and thus affect the chemical reactivity of the entire molecule. Although a large number of different chemical reactions occur in a typical cell, these reactions are of only a few general types. We will briefly review the basic facts about chemical bonding and reactivity, and then summarize five reaction types commonly encountered in biochemistry. Later we will consider specific reaction types in more detail.

Bond Strength Is Related to the Properties of the Bonded Atoms

In chemical reactions, bonds are broken and new ones are formed. The strength of a chemical bond depends on the relative electronegativities—the relative affinities for electrons—of the bonding elements (Table 3–2), the distance of the bonding electrons from each nucleus, and the nuclear charge of each atom. The number of electrons shared also influences bond strength; double bonds are stronger than single bonds, and triple bonds are stronger yet (Table 3–3). The strength of a bond is expressed as bond energy, in joules. (Biochemists have often used the calorie as a unit of energy—bond energy and free energy, for example, are often given in kcal/mol. The joule is the unit of energy in the International System of Units and has replaced the calorie in most biochemical usage; we use it throughout this book. For conversions, 1 cal equals 4.184 J.) Bond energy can be thought of as either the amount of energy required to break a bond or the amount of energy gained by the surroundings when the bond forms. One way to put energy into a system is to heat it, which gives the molecules more kinetic energy, raising the fraction of molecules with kinetic energies

table 3–2

The Electronegativities of Some Elements

Element	Electronegativity*
F	4.0
O	3.5
Cl	3.0
N	3.0
Br	2.8
S	2.5
C	2.5
I	2.5
Se	2.4
P	2.1
H	2.1
Cu	1.9
Fe	1.8
Co	1.8
Ni	1.8
Mo	1.8
Zn	1.6
Mn	1.5
Mg	1.2
Ca	1.0
Li	1.0
Na	0.9
K	0.8

*The higher the number, the more electronegative (the greater the electron affinity of) the element.

table 3–3

Strengths of Bonds Common in Biomolecules

Type of bond	Bond dissociation energy* (kJ/mol)	Type of bond	Bond dissociation energy (kJ/mol)
Single bonds		**Double bonds**	
O—H	461	C═O	712
H—H	435	C═N	615
P—O	419	C═C	611
C—H	414	P═O	502
N—H	389		
C—O	352	**Triple bonds**	
C—C	348	C≡C	816
S—H	339	N≡N	930
C—N	293		
C—S	260		
N—O	222		
S—S	214		

*The greater the energy required for bond dissociation (breakage), the stronger the bond.

high enough to react. When molecular motion is sufficiently violent, intramolecular vibrations and intermolecular collisions sometimes break chemical bonds and allow new ones to form.

When bonds are broken and formed in a chemical reaction, the difference between the energy extracted from the surroundings to break bonds and the energy released to the surroundings during the formation of new bonds can be approximated as the **enthalpy change, ΔH,** for the reaction. If heat energy is absorbed by the system as the change occurs (i.e., if the reaction is endothermic), then H has, by definition, a positive value; if heat is given off, the reaction is exothermic and H is negative. In short, the change in enthalpy for a chemical reaction reflects the kinds and numbers of bonds that are made and broken. As we shall see later in this chapter, the enthalpy change is one of three factors that determine the free-energy change for a reaction; the other two are the temperature and the change in entropy.

Five General Types of Chemical Transformations Occur in Cells

Most cells have the capacity to carry out thousands of specific, enzyme-catalyzed reactions: for example, transformation of a simple nutrient such as glucose into amino acids, nucleotides, or lipids; extraction of energy from fuels by oxidation; or polymerization of monomeric subunits into macromolecules. Fortunately for the student of biochemistry, there are patterns within this multitude of reactions; you do not need to learn all of these reactions to comprehend the molecular logic of life. Most of the reactions in living cells fall into one of five general categories: (1) oxidation-reductions, (2) cleavage and formation of carbon–carbon bonds, (3) internal rearrangements, (4) group transfers, and (5) condensation reactions in which monomeric subunits are joined, with the elimination of a molecule of water. Reactions within each general category usually occur by similar mechanisms.

All Oxidation-Reduction Reactions Involve Electron Transfer

When the two atoms sharing electrons in a covalent bond have the same electronegativity, as in the case of two carbon atoms, the bond is nonpolar. When two elements that differ in electronegativity form a covalent bond (e.g., C and O), that bond is polarized; the shared electrons are more likely to be in the region of the more electronegative atom (O in this case) than of the less electronegative atom (C in this case; see Table 3–2). In the extreme case of two elements of very different electronegativity, such as Na and Cl, one atom gives up its electron(s) to the other atom, resulting in the formation of ions and ionic interactions such as those in solid NaCl.

In carbon–hydrogen bonds, the more electronegative C "owns" the two electrons shared with H, but in carbon–oxygen bonds, electron sharing is unequal in favor of oxygen. Thus, in going from $—CH_3$ (an alkane) to $—CH_2OH$ (an alcohol), the carbon atom has effectively lost electrons—which is, by definition, oxidation. As Figure 3–15 shows, carbon atoms encountered in biochemistry can exist in five oxidation states, depending on the elements with which carbon shares electrons.

In many biological oxidations, a compound loses two electrons and two hydrogen ions (i.e., two hydrogen atoms); these reactions are commonly called **dehydrogenations** and the enzymes that catalyze them are called **dehydrogenases** (Fig. 3–16). In some, but not all, biological oxidations, a carbon atom becomes covalently bonded to an oxygen atom. The enzymes that catalyze these oxidations are generally called **oxidases** or, if the oxygen atom is derived directly from molecular oxygen (O_2), **oxygenases.**

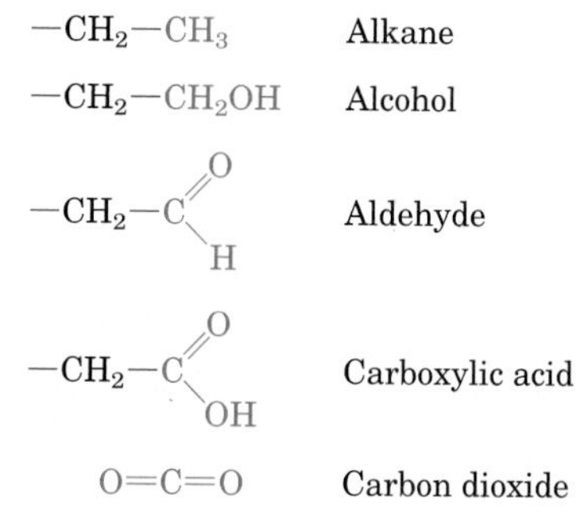

$—CH_2—CH_3$	Alkane
$—CH_2—CH_2OH$	Alcohol
$—CH_2—CHO$	Aldehyde
$—CH_2—COOH$	Carboxylic acid
$O{=}C{=}O$	Carbon dioxide

figure 3–15

The oxidation states of carbon in biomolecules. Each compound is formed by oxidation of the red carbon in the compound listed above it. Carbon dioxide is the most highly oxidized form of carbon found in living systems.

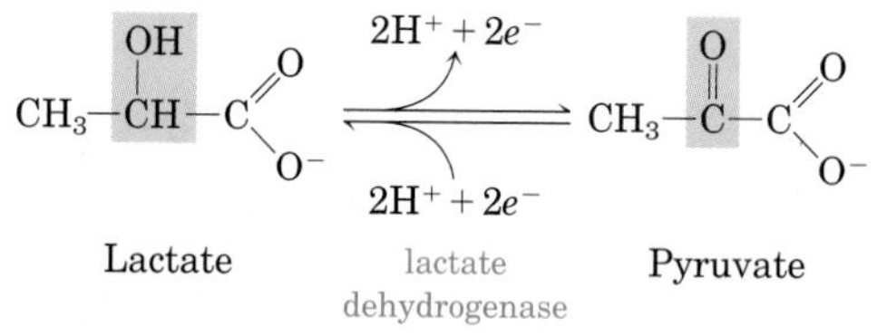

$$CH_3—CH(OH)—COO^- \underset{2H^+ + 2e^-}{\overset{2H^+ + 2e^-}{\rightleftharpoons}} CH_3—CO—COO^-$$

Lactate — lactate dehydrogenase — Pyruvate

figure 3–16

An oxidation-reduction reaction. Shown here is the oxidation of lactate to pyruvate. In this dehydrogenation, two electrons and two hydrogen ions (the equivalent of two hydrogen atoms) are removed from the C-2 of lactate, a ketone. In cells the reaction is catalyzed by lactate dehydrogenase and the electrons are transferred to a cofactor called nicotinamide adenine dinucleotide. This reaction is fully reversible; pyruvate can be reduced by electrons from the cofactor. We will discuss the factors that determine the direction of a reaction in Chapter 14.

Every oxidation must be accompanied by a reduction, in which an electron acceptor acquires the electrons removed by oxidation. Oxidation reactions generally release energy (think of camp fires, in which various compounds in wood are oxidized by oxygen molecules in the air). Most living cells obtain the energy needed for cellular work by oxidizing metabolic fuels such as carbohydrate or fat; photosynthetic organisms can also trap and use the energy of sunlight. The catabolic (energy-yielding) pathways described in Chapters 15 to 19 are oxidative reaction sequences that result in the transfer of electrons from fuel molecules through a series of electron carriers and finally to oxygen. The high affinity of O_2 for electrons makes the overall electron-transfer process highly exergonic, providing the energy that drives ATP synthesis—the central goal of catabolism.

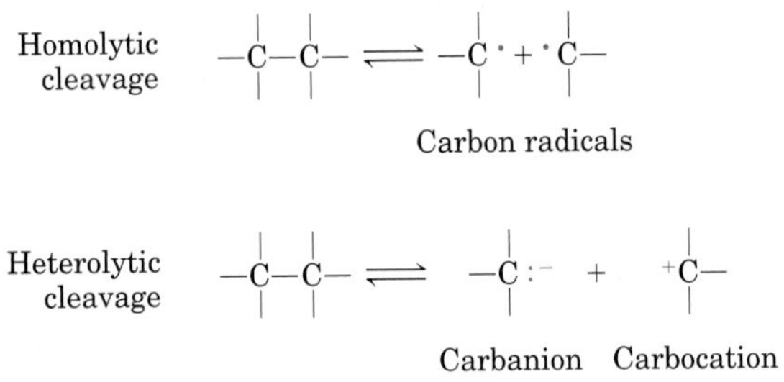

Homolytic cleavage $-\overset{|}{\underset{|}{C}}-\overset{|}{\underset{|}{C}}- \rightleftharpoons -\overset{|}{\underset{|}{C}}\cdot + \cdot\overset{|}{\underset{|}{C}}-$

Carbon radicals

Heterolytic cleavage $-\overset{|}{\underset{|}{C}}-\overset{|}{\underset{|}{C}}- \rightleftharpoons -\overset{|}{\underset{|}{C}}:^- + {}^+\overset{|}{\underset{|}{C}}-$

Carbanion Carbocation

figure 3–17
Two mechanisms for cleavage of a C—C bond. In homolytic cleavages, each carbon atom keeps one of the bonding electrons, resulting in two carbon radicals (i.e., carbons having unpaired electrons). In heterolytic cleavages, one of the two carbon atoms keeps both bonding electrons, producing a carbanion; the other becomes a carbocation.

Carbon–Carbon Bonds Are Cleaved and Formed by Nucleophilic Substitution Reactions

A covalent bond can be broken in two general ways (Fig. 3–17). In **homolytic** cleavage, each atom leaves the bond as a radical, carrying one of the two electrons (now unpaired) that held the bonded atoms together. Homolytic reactions occur only rarely in living organisms (but see Fig. 22–39 for an example). More common are **heterolytic** cleavages in which one atom keeps both bonding electrons (forming an anion), leaving the other atom one electron short (a cation). When a second electron-rich group replaces the departing anion, a **nucleophilic substitution** occurs (Fig. 3–18). Many biochemical reactions involve interactions between **nucleophiles,** functional groups rich in electrons and capable of donating them, and **electrophiles,** electron-deficient functional groups that seek electrons. Nucleophiles combine with, and give up electrons to, electrophiles. Functional groups containing oxygen, nitrogen, and sulfur are important biological nucleophiles (Table 3–4). Positively charged hydrogen atoms (hydrogen ions, or protons) and positively charged metals (cations) frequently act as electrophiles in cells. A carbon atom can act as either a nucleophile or an electrophile, depending on which bonds and functional groups surround it.

There are two general mechanisms by which one nucleophile can replace another in the formation of carbon–carbon bonds. In the first (Fig. 3–19a), the leaving group (the nucleophile W; see Fig. 3–18) departs with its electrons, leaving the former partner as a relatively unstable carbocation (positively charged carbon, an electrophile), before the substituting group (Z, a nucleophile) comes on the scene. This mechanism is called an **SN1 reaction,** (SN1 indicating substitution nucleophilic, unimolecular). In the second type of nucleophilic substitution, an attacking nucleophile (Z) arrives before the leaving group (W) departs, and a pentacovalent intermediate forms transiently (Fig. 3–19b). This is an **SN2 reaction** (substitution nucleophilic, bimolecular). As Figure 3–19 suggests, SN2 reactions typically result in an inversion of configuration around the attacked carbon upon departure of the leaving group, whereas SN1 reactions usually result in either retention of the original configuration or racemization. In general, weaker nucleophiles make better leaving groups and stronger nucleophiles are better attacking species.

The aldol condensation catalyzed by aldolase (see Fig. 15–4) is an example of a nucleophilic substitution employed to form carbon–carbon bonds in cells. These reactions are reversible; aldolase can join two three-carbon moieties to form a six-carbon sugar, or it can split the six-carbon sugar to yield two three-carbon moieties.

table 3–4

Some Functional Groups Active as Nucleophiles within Cells*

Water	$H\ddot{O}H$
Hydroxide ion	$H\ddot{O}:^-$
Hydroxyl (alcohol)	$R\ddot{O}H$
Alkoxyl	$R\ddot{O}:^-$
Sulfhydryl	RSH
Sulfide	$R\ddot{S}^-$
Amino	$R\ddot{N}H_2$
Carboxylate	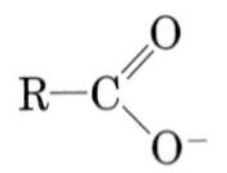 $R-C(=O)-O^-$
Imidazole	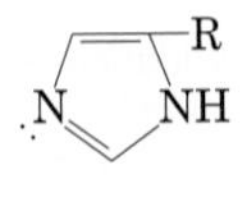 (imidazole ring with R; :N, NH)
Inorganic orthophosphate	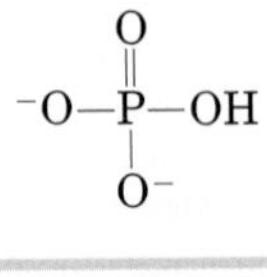 $^-O-P(=O)(O^-)-OH$

*Listed in order of decreasing strength. Weaker nucleophiles make better leaving groups.

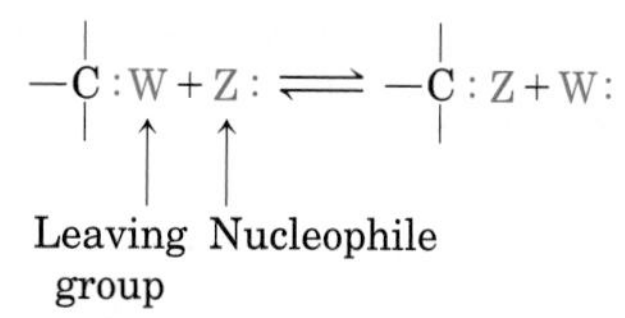

figure 3–18

A nucleophilic substitution reaction. An electron-rich nucleophile (Z) attacks an electron-poor center (a carbon atom, for example) and displaces a nucleophilic group (W), which is called the leaving group.

—C:W + Z: ⇌ —C:Z + W:
Leaving group
Nucleophile

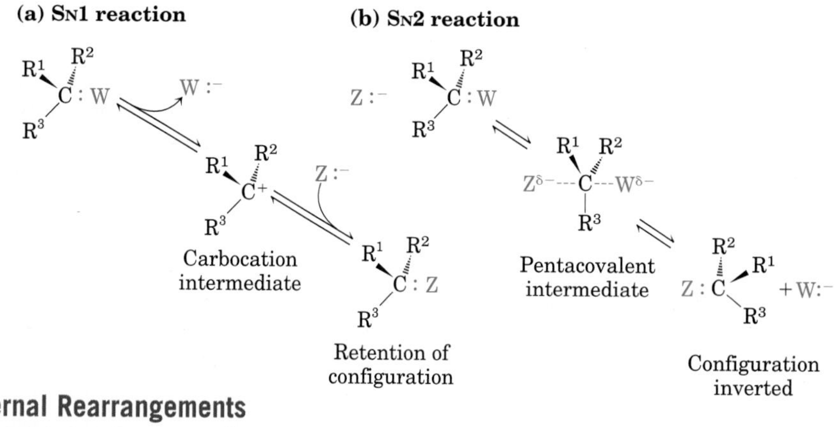

figure 3–19

Two classes of nucleophilic substitution reactions. **(a)** SN1: The leaving group (W) departs with a bonding electron, leaving a carbocation, before the attacking nucleophile (Z) arrives. **(b)** SN2: The attacking nucleophile (Z) approaches one side of the electrophilic carbon while the leaving group (W) remains bonded to the other side, resulting in a transient pentacovalent intermediate. The departure of W leaves the substituted compound with a completely inverted configuration at the reacting carbon atom.

(a) SN1 reaction
Carbocation intermediate
Retention of configuration
(b) SN2 reaction
Pentacovalent intermediate
Configuration inverted

Electron Transfers within a Molecule Produce Internal Rearrangements

Another common cellular reaction type is intramolecular rearrangement, in which redistribution of electrons results in isomerization, transposition of double bonds, and cis-trans rearrangements of double bonds. An example of isomerization is the formation of fructose 6-phosphate from glucose 6-phosphate during sugar metabolism (Chapter 15). In this reaction (Fig. 3–20a), C-1 is reduced (from aldehyde to alcohol) and C-2 is oxidized (from alcohol to ketone). In Figure 3–20b, which shows the details of the electron movements that result in isomerization, we have employed the convention of "electron-pushing" diagrams, which we will use to indicate reaction mechanisms throughout the book. Curved blue arrows show the movement of electrons as the reaction proceeds.

A simple transposition of one C=C bond occurs during metabolism of the common fatty acid oleic acid (see Fig. 17–9), and we will see spectacular examples of double-bond repositioning in the synthesis of cholesterol (see Fig. 21–35).

figure 3–20

An isomerization reaction. (a) The conversion of glucose 6-phosphate to fructose 6-phosphate, a reaction of sugar metabolism catalyzed by phosphohexose isomerase. **(b)** This reaction proceeds through an enediol intermediate. The curved blue arrows represent the movement of bonding electrons from nucleophile (red) to electrophile (blue). B_1 and B_2 are basic groups on the enzyme; they are capable of donating and accepting hydrogen ions (protons) as the reaction progresses.

(a)
Glucose 6-phosphate
phosphohexose isomerase
Fructose 6-phosphate
(b)
① B_1 abstracts a proton.
② This allows the formation of a C=C double bond.
③ Electrons from carbonyl form an O—H bond with the hydrogen ion.
④ B_2 abstracts a proton, allowing the formation of a C=O bond.
⑤ An electron leaves the C=C bond to form a C—H bond with the proton donated by B_1.
Enediol intermediate

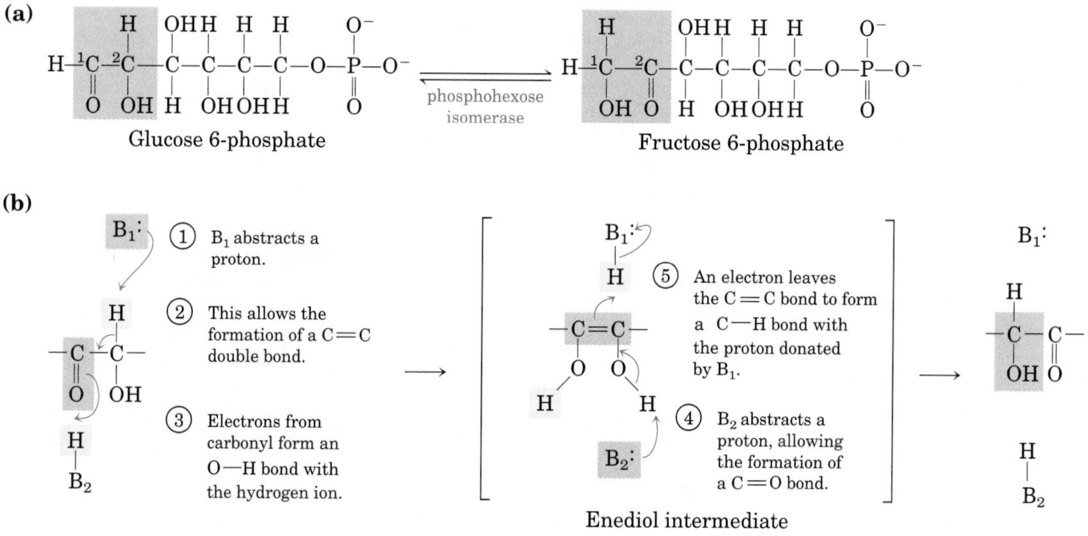

Group Transfer Reactions Activate Metabolic Intermediates

A general theme in metabolism is the attachment of a good leaving group to a metabolic intermediate to "activate" the intermediate for subsequent reaction. Among the better leaving groups in nucleophilic substitution reactions (Table 3–4) are inorganic orthophosphate (the ionized form of H_3PO_4 at neutral pH, a mixture of $H_2PO_4^-$ and HPO_4^{2-}, commonly abbreviated P_i) and inorganic pyrophosphate ($P_2O_7^{6-}$, abbreviated PP_i). Esters and anhydrides of phosphoric acid play central roles in cellular chemistry. Nucleophilic substitutions in which the phosphoryl group ($—PO_3^{2-}$) serves as a leaving group occur in hundreds of metabolic reactions; nucleophilic substitution is made more favorable by the attachment of a phosphoryl group to an otherwise poor leaving group such as —OH.

Phosphorus can form five covalent bonds. The conventional representation of P_i (Fig. 3–21a) with three P—O bonds and one P═O bond is not an accurate picture. In P_i, four equivalent P—O bonds share some double-bond character, and the anion has a tetrahedral structure (Fig. 3–21b). As oxygen is more electronegative than phosphorus, the sharing of electrons is unequal. The central phosphorus bears a positive charge and can therefore act as an electrophile. In a very large number of metabolic reactions, a phosphoryl group ($—PO_3^{2-}$) is transferred from ATP to an alcohol (forming a phosphate ester) (Fig. 3–21c) or to a carboxylic acid (forming a mixed anhydride; see Fig. 3–5). When a nucleophile attacks the electrophilic phosphorus atom in ATP, a relatively stable pentacovalent structure is formed as a reaction intermediate (Fig. 3–21d). With departure of the leaving group (ADP), the transfer of a phosphoryl group is complete. The large family of enzymes that catalyze phosphoryl group transfers with ATP as donor are called kinases (Greek *kinein*, "to move"). Hexokinase, for example, "moves" a phosphoryl group from ATP to glucose.

Phosphoryl groups are not the only activators of this type. Thioalcohols (thiols), in which the oxygen atom of an alcohol is replaced with a sulfur atom, are also good leaving groups. Thiols activate carboxylic acids by forming thioesters (thiol esters) with them (Fig. 3–5). We will encounter a number of cases, including the reactions catalyzed by the fatty acyl transferases in lipid synthesis (see Fig. 21–2), in which nucleophilic substitution at the carbonyl carbon of a thioester results in transfer of the acyl group to another moiety.

(a)

$^-O—P(O^-)(O^-)═O \longleftrightarrow {}^-O—P(═O)(O^-)—O^-$

$\updownarrow \qquad\qquad \updownarrow$

$^-O—P(O^-)(═O)—O^- \longleftrightarrow O═P(O^-)(O^-)—O^-$

(b)

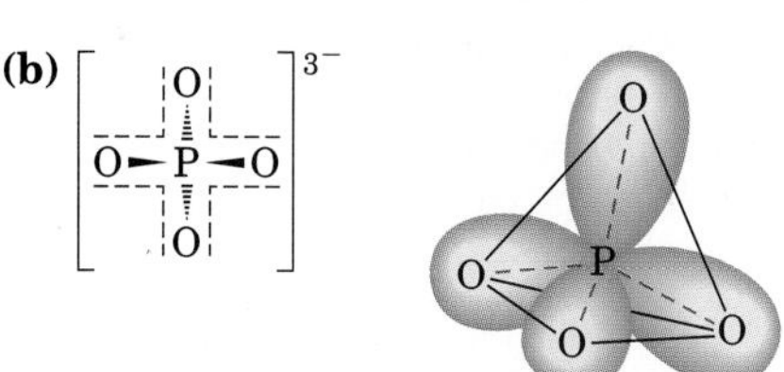

(c) Adenine—Ribose—O—P(═O)(O⁻)—O—P(═O)(O⁻)—O—P(═O)(O⁻)—O⁻ + HÖ—R (Glucose)

ATP

↓

Adenine—Ribose—O—P(═O)(O⁻)—O—P(═O)(O⁻)—O⁻ + ⁻O—P(═O)(O⁻)—O—R

ADP — Glucose 6-phosphate, a phosphate ester

(d) Z---P(O)(O)(O)---W

figure 3–21

Alternative ways of showing the structure of inorganic orthophosphate. (a) In one (inadequate) representation, three oxygens are single-bonded to phosphorus, and the fourth is double-bonded, allowing the four different resonance structures shown. **(b)** The four resonance structures can be represented more accurately by showing all four P—O bonds with some double-bond character; the hybrid orbitals so represented are arranged in a tetrahedron with P at its center. **(c)** When a nucleophile Z (in this case, the —OH on C-6 of glucose) attacks ATP, it displaces ADP (W). In this S_N2 reaction, a pentacovalent intermediate **(d)** forms transiently.

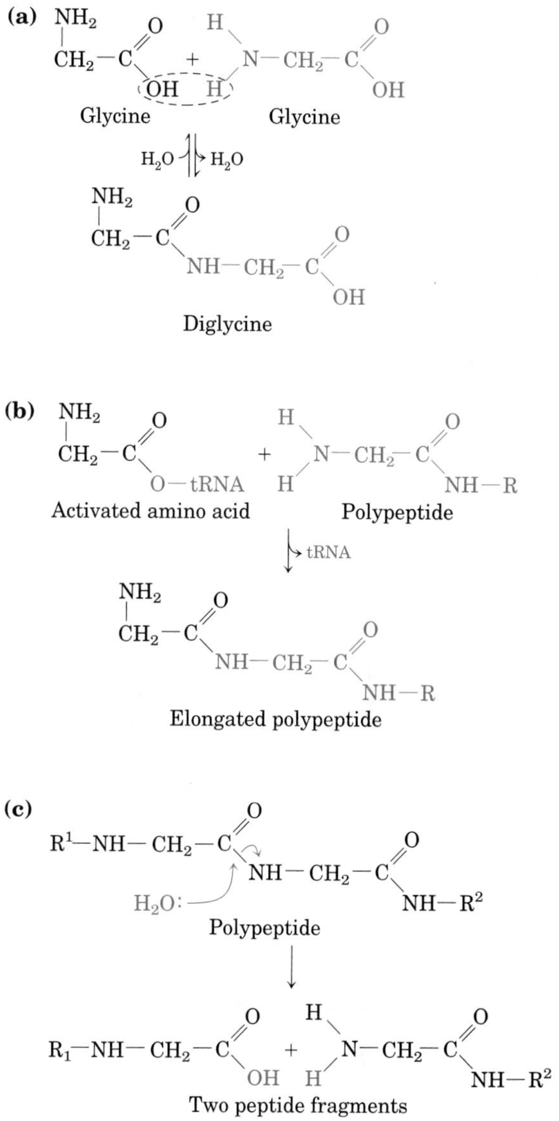

figure 3–22

Condensation and hydrolysis. Shown here are formation and hydrolysis of a peptide bond. **(a)** Removal of the elements of water from two molecules of the amino acid glycine produces a peptide bond, but because —OH is not a good leaving group, this reaction is unfavorable. **(b)** In cells, amino acids are activated prior to polymerization by attachment of a better leaving group than —OH, a short RNA (transfer RNA or tRNA) that forms an oxygen ester with the α-carboxyl group. **(c)** The hydrolysis of a peptide bond (shown here in a polypeptide) is essentially the reverse of the reaction in **(a)**: H_2O makes a nucleophilic attack on the carbonyl carbon, displacing the nitrogen of the α-amino group.

Biopolymers Are Formed by Condensations

The monomeric subunits that make up proteins, nucleic acids, and polysaccharides are joined by nucleophilic displacement reactions that replace a good leaving group. For example, the joining of two amino acid molecules to form a dipeptide could occur by the simple mechanism in Figure 3–22a. However, —OH is a poor leaving group, and the reaction by this mechanism is not efficient. Cells solve this problem by first attaching a better leaving group, a small RNA molecule (transfer RNA, about 75 nucleotides long), in ester linkage to the α-carboxyl group of the amino acid. This activates the carboxyl group for condensation with the α-amino group of another amino acid (Fig. 3–22b). Similar strategies are employed in the biosynthesis of nucleic acids and polysaccharides.

Macromolecules can be broken down by hydrolysis reactions, in which H_2O is the attacking nucleophile, displacing a monomeric subunit or a smaller polymer fragment (Fig. 3–22c). Enzymes that catalyze hydrolysis of biopolymers (hydrolases) are essential in the digestive process and serve also to regulate the level of such critical macromolecules as messenger RNA.

Macromolecules and Their Monomeric Subunits

Many biological molecules are macromolecules, polymers of high molecular weight assembled from relatively simple precursors. Polysaccharides, proteins, and nucleic acids are produced by the polymerization of relatively small compounds with molecular weights of 500 or less. The total number of polymerized units can range from tens to millions. Synthesis of macromolecules is a major energy-consuming activity of cells. Macromolecules themselves may be further assembled into supramolecular complexes, forming functional units such as ribosomes, which are constructed of about 70 different proteins and several different RNA molecules.

Macromolecules Are the Major Constituents of Cells

Table 3–5 shows the major classes of biomolecules in the bacterium *Escherichia coli.* Water is the most abundant compound in *E. coli* and in virtually all other cells and organisms. Nearly all of the solid matter in cells is organic and is present in four forms: proteins, nucleic acids, polysaccharides, and lipids. Inorganic salts and mineral elements constitute a very small fraction of the total dry weight.

Proteins, long polymers of amino acids, constitute the largest fraction (besides water) of cells. Some proteins have catalytic activity and function as enzymes; others serve as structural elements, signal receptors, or transporters that carry specific substances into or out of cells. Proteins are per-

table 3–5

Molecular Components of an *E. coli* Cell

	Percentage of total weight of cell	Approximate number of different molecular species
Water	70	1
Proteins	15	3,000
Nucleic acids		
DNA	1	1
RNA	6	>3,000
Polysaccharides	3	5
Lipids	2	20
Monomeric subunits and intermediates	2	500
Inorganic ions	1	20

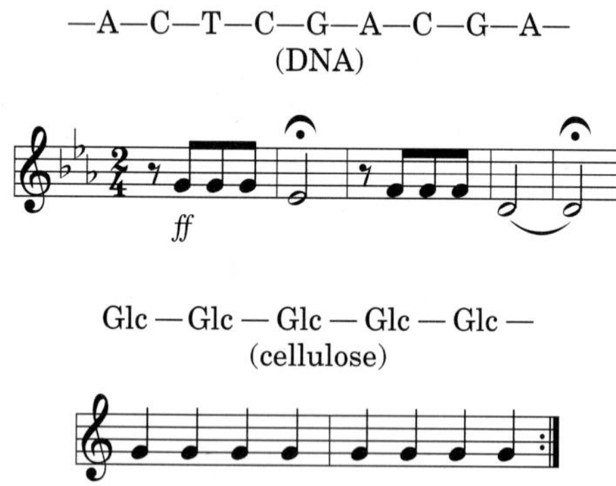

figure 3–23

Informational and structural macromolecules. A, T, C, and G represent the four subunits of DNA, and glucose (Glc) is the repeating subunit of cellulose. The variety of sequences that can be made from the four subunits of DNA is limitless, as is the number of melodies possible with a few musical notes. Cellulose, a polymer of one subunit type, is information-poor and monotonous.

haps the most versatile of all biomolecules. The **nucleic acids,** DNA and RNA, are polymers of nucleotides. They store and transmit genetic information, and some RNA molecules have structural roles in macromolecular complexes. The **polysaccharides,** polymers of simple sugars such as glucose, have two major functions: as energy-yielding fuel stores and as extracellular structural elements. Shorter polymers of sugars (oligosaccharides) attached to proteins or lipids at the cell surface serve as specific cellular signals. The **lipids,** greasy or oily hydrocarbon derivatives, serve as structural components of membranes, energy-rich fuel stores, pigments, and intracellular signals. Proteins, nucleotides, polysaccharides, and lipids are synthesized in condensation reactions. In the first three categories, the number of monomeric subunits in the polymer is very large. Proteins have molecular weights in the range of 5,000 to over 1 million; nucleic acids have molecular weights ranging up to several billion; and polysaccharides, such as starch, have molecular weights into the millions. Individual lipid molecules are much smaller (M_r 750 to 1,500) and are not classified as macromolecules. However, when large numbers of lipid molecules associate noncovalently, very large structures result. Cellular membranes are built of enormous aggregates containing millions of lipid molecules.

Macromolecules Are Composed of Monomeric Subunits

Although living organisms contain a very large number of different proteins and different nucleic acids, a fundamental simplicity underlies their structure (Chapter 1). The simple monomeric units from which all proteins and all nucleic acids are constructed are few in number and identical in all living species. Proteins and nucleic acids are **informational macromolecules:** each protein and each nucleic acid has a characteristic information-rich subunit sequence (Fig. 3–23).

Polysaccharides with only a single kind of subunit, or with two different alternating units, are not informational molecules in the same sense as are proteins and nucleic acids. However, short sugar polymers made up of six or more different kinds of sugars connected in branched chains have the structural and stereochemical variety to carry information recognizable by other macromolecules.

Monomeric Subunits Have Simple Structures

Figure 3–24 shows the structures of some of the monomeric units of the large biomolecules, arranged in families. Twenty different amino acids are found in proteins; all have an amino group (an imino group in the case of proline) and a carboxyl group attached to the same carbon atom, designated the α carbon. These α-amino acids differ from each other only in their side chains (Fig. 3–24a).

The recurring structural units of all nucleic acids are eight different nucleotides; four kinds of nucleotides are the structural units of DNA, and four others are the units of RNA (Fig. 3–24b). Each nucleotide is made up of three components: a nitrogenous organic base and a five-carbon sugar (which combined are called a nucleoside), to which is attached a phosphate group. The eight different nucleotides of DNA and RNA are built from five different organic bases and two different sugars.

Lipids also are constructed from relatively few kinds of compounds. Most lipid molecules contain one or more long-chain fatty acids, of which palmitic acid and oleic acid are parent compounds (Fig. 3–24c). Many lipids also contain an alcohol, such as glycerol, and some contain phosphate.

The most abundant polysaccharides in nature, starch and cellulose, consist of repeating units of D-glucose (Fig. 3–24d). Other polysaccharides are composed of a variety of sugar molecules derived from glucose.

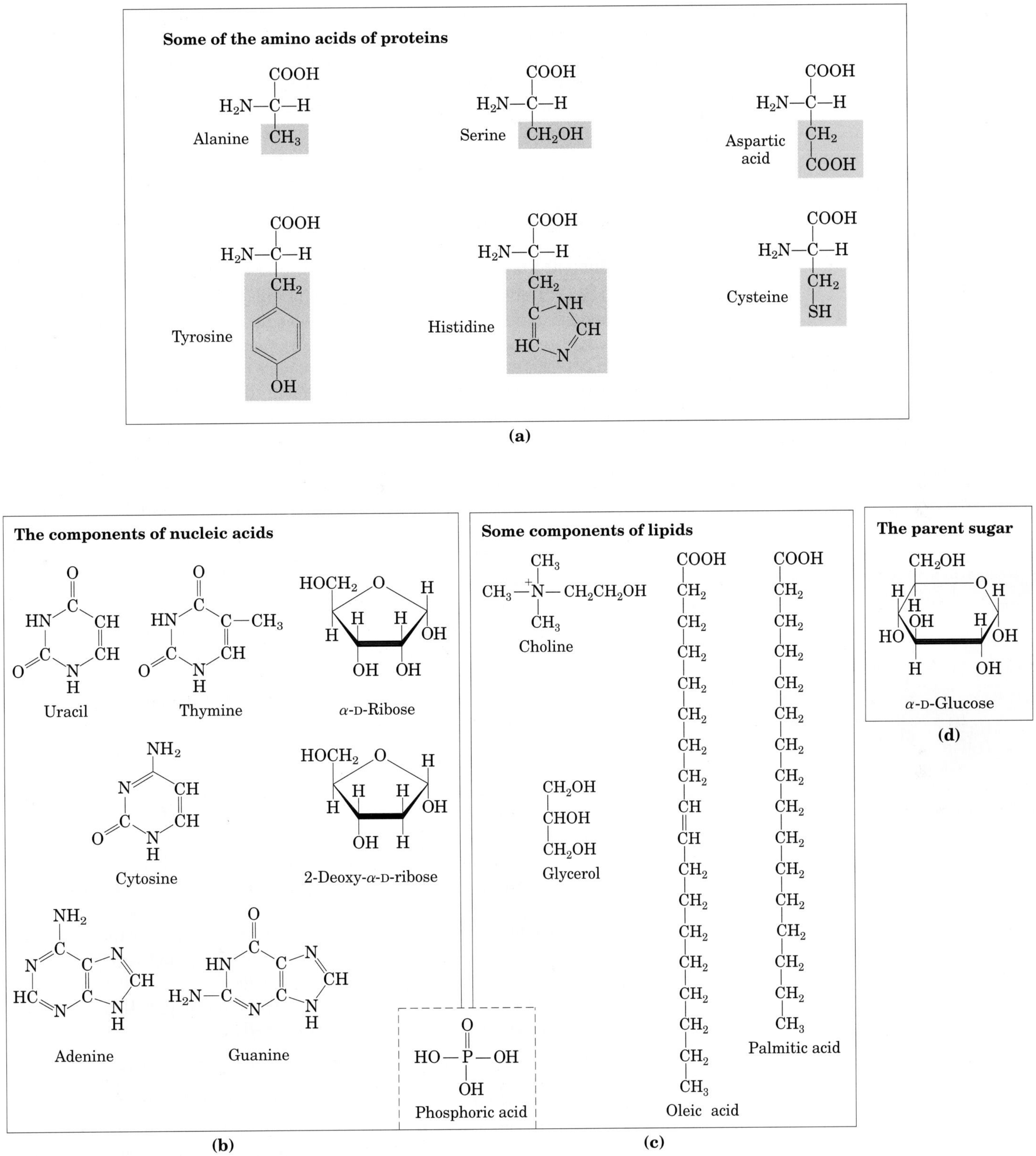

figure 3–24

The organic compounds from which most cellular materials are constructed: the ABCs of biochemistry. Shown here are **(a)** six of the 20 amino acids from which all proteins are built (the side chains are shaded red); **(b)** the five nitrogenous bases, two five-carbon sugars, and phosphoric acid from which all nucleic acids are built; **(c)** five components of many membrane lipids, and **(d)** D-glucose, the parent sugar from which most carbohydrates are derived. Note that phosphoric acid is a component of both nucleic acids and membrane lipids. All compounds are shown in their nonionized form.

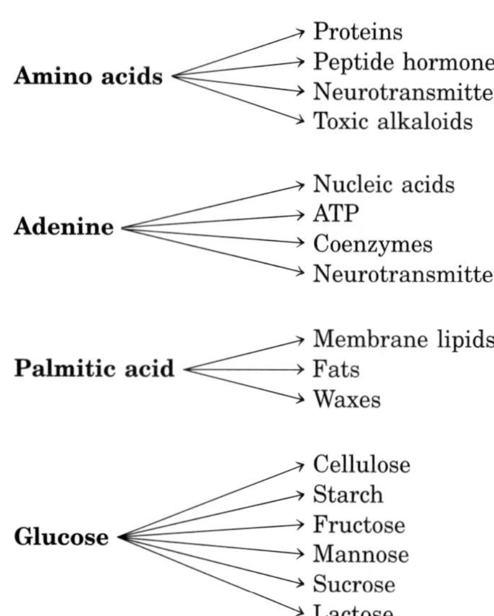

figure 3–25
Each compound in Figure 3–24 is a precursor of many other kinds of biomolecules.

J. Willard Gibbs
1839–1903

Thus, only three dozen organic compounds are the parents of most biomolecules. Each of the compounds in Figure 3–24 has multiple functions in living organisms (Fig. 3–25). Amino acids are not only the monomeric subunits of proteins; some also act as neurotransmitters and as precursors of hormones and toxins. The nitrogenous base adenine serves both as a subunit of nucleic acids and ATP and as a neurotransmitter. Fatty acids serve as components of membrane lipids and fuel-storage fats, and also as precursors of a group of potent signaling molecules, the eicosanoids. D-Glucose is the monomeric subunit of starch and cellulose and is also the precursor of other sugars such as D-mannose and sucrose.

Subunit Condensation Creates Order and Requires Energy

It is extremely improbable that amino acids in a mixture would spontaneously condense into a protein molecule with a unique sequence. This would represent increased order in a population of molecules; but according to the second law of thermodynamics (Chapter 14), the tendency in nature is toward ever-greater disorder in the universe. To bring about the synthesis of macromolecules from their monomeric units, free energy must be supplied to the system, in this case the cell (Chapter 1).

The randomness of the components of a chemical system is expressed as **entropy, *S*.** Any change in randomness of the system is expressed as entropy change, ΔS, which has a positive value when randomness increases. J. Willard Gibbs, who developed the theory of energy changes during chemical reactions, showed that the free-energy content (G) of any closed system can be defined in terms of three quantities: enthalpy (H), reflecting the number and kinds of bonds (p. 65); entropy (S); and the absolute temperature (T in degrees Kelvin). The definition of free energy is: $G = H - TS$. When a chemical reaction occurs at constant temperature, the free-energy change, ΔG, is determined by ΔH, reflecting the kinds and numbers of chemical bonds and noncovalent interactions broken and formed, and ΔS, the change in the system's randomness:

$$\Delta G = \Delta H - T\,\Delta S$$

Recall from Chapter 1 that a process tends to occur spontaneously only if ΔG is negative. Yet cells depend on many molecules, such as proteins and nucleic acids, for which the free energy of formation is positive: they are less stable and more highly ordered than a mixture of their monomeric components. To overcome the free-energy deficit of thermodynamically unfavorable (endergonic) reactions, cells couple these reactions to other reactions that liberate free energy (exergonic reactions), so that the overall process is exergonic: the *sum* of the free-energy changes is negative. The usual source of free energy in coupled biological reactions is the energy released by hydrolysis of phosphoanhydride bonds such as those connecting phosphate groups (represented as Ⓟ) in ATP:

$$\text{Amino acids} \longrightarrow \text{polymer} \qquad \Delta G_1 \text{ is positive (endergonic)}$$

$$-Ⓟ-Ⓟ \longrightarrow Ⓟ + Ⓟ \qquad \Delta G_2 \text{ is negative (exergonic)}$$

When these reactions are coupled, the sum of ΔG_1 and ΔG_2 is negative (the overall process is exergonic). By this strategy, cells are able to synthesize and maintain the information-rich polymers (DNA, RNA, and protein) essential to life.

Cells Have a Structural Hierarchy

The monomeric units in Figure 3–24 are much smaller than biological macromolecules. A molecule of alanine is less than 0.5 nm long. Hemoglobin, the oxygen-carrying protein of erythrocytes, consists of nearly 600 amino acid units covalently linked into four long chains, which are folded into globular shapes and associated in a tetrameric structure with a diameter of 5.5 nm. Protein molecules in turn are small compared with ribosomes (about 20 nm in diameter), which are in turn much smaller than organelles such as mitochondria, typically 1,000 nm in diameter. It is a long jump from simple biomolecules to cellular structures that can be seen with the light microscope. Figure 3–26 illustrates the structural hierarchy in cellular organization.

In proteins, nucleic acids, and polysaccharides, the individual monomeric subunits are joined by covalent bonds. In supramolecular complexes, however, macromolecules are held together by noncovalent interactions—much weaker, individually, than covalent bonds. Among these are hydrogen bonds (between polar groups), ionic interactions (between charged groups), hydrophobic interactions (among nonpolar groups in aqueous solution), and van der Waals interactions, all of which have energies substantially smaller than those of covalent bonds (Table 3–3). The nature of these noncovalent interactions is described in the next chapter. The large numbers of weak interactions between macromolecules in supramolecular complexes stabilize these noncovalent assemblies, producing their unique "native" structures.

Although the monomeric subunits of macromolecules are much smaller than cells and organelles, they influence the shape and function of these much larger structures. In sickle-cell anemia, a hereditary human disorder, the hemoglobin molecule is defective: valine occurs in two of the four polypeptide chains (the two β chains) of hemoglobin at a position normally occupied by glutamic acid. This single difference in the sequence of the 146 amino acids of the β chain affects only a tiny portion of the hemoglobin molecule, yet, as explained in Chapter 7, it causes the molecules to form large aggregates within the erythrocytes, which become deformed (sickled) and functionally abnormal.

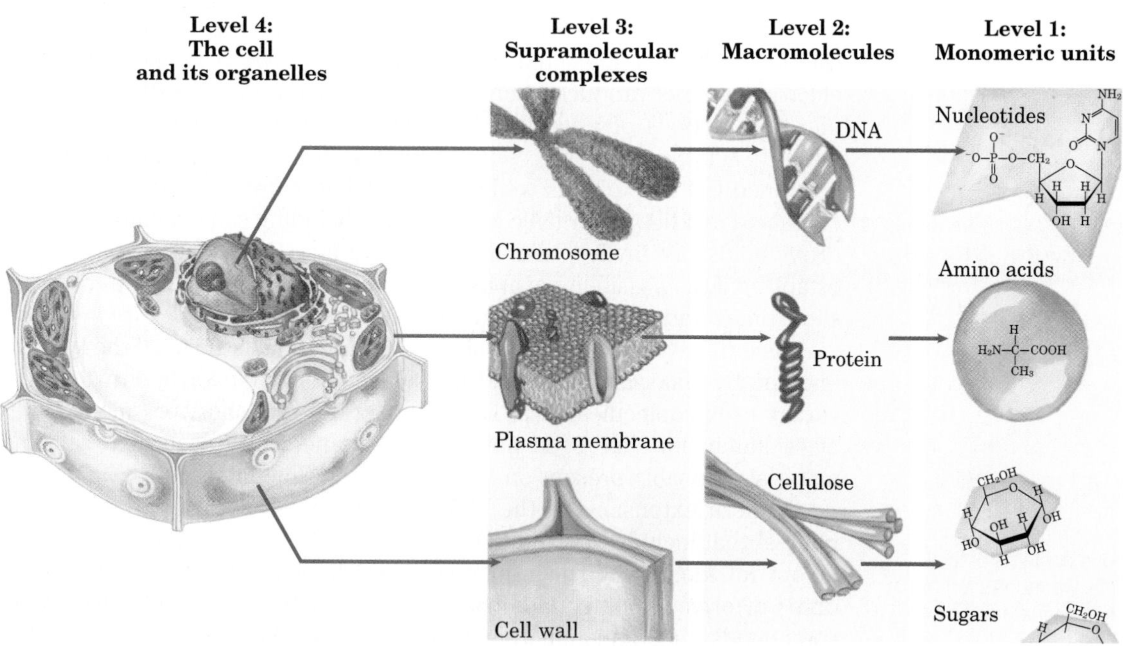

figure 3–26

Structural hierarchy in the molecular organization of cells. In this plant cell, the nucleus is an organelle containing several types of supramolecular complexes, including chromosomes. Chromosomes consist of macromolecules—DNA and many different proteins. Each type of macromolecule is constructed from simple subunits—DNA from deoxyribonucleotides, for example.

Prebiotic Evolution

The finding that all biological macromolecules in all organisms are made from the same three dozen subunits has provided strong evidence that modern organisms are descended from a single primordial cell line whose fundamental chemistry would be recognizable even today. Furthermore, several billion years of adaptive selection have refined cellular systems to take maximum advantage of the chemical and physical properties of these molecular raw materials for carrying out the basic energy-transforming and self-replicating features of a living cell.

Biomolecules First Arose by Chemical Evolution

We come now to a puzzle. Apart from their occurrence in living organisms, organic compounds, including the basic biomolecules such as amino acids and carbohydrates, occur in only trace amounts in the earth's crust, the sea, and the atmosphere. How did the first living organisms acquire their characteristic organic building blocks? In 1922, the biochemist Aleksandr I. Oparin proposed a theory for the origin of life early in the history of the earth, postulating that the atmosphere then was very different from that of today. Rich in methane, ammonia, and water, and essentially devoid of oxygen, it was a reducing atmosphere, in contrast to the oxidizing environment of our era. In Oparin's theory, electrical energy from lightning discharges or heat energy from volcanoes caused ammonia, methane, water vapor, and other components of the primitive atmosphere to react, forming simple organic compounds. These compounds then dissolved in the ancient seas, which over many millenia became enriched with a large variety of simple organic substances. In this warm solution (the "primordial soup"), some organic molecules had a greater tendency than others to associate into larger complexes. Over millions of years, these in turn assembled spontaneously to form membranes and catalysts (enzymes), which came together to become precursors of the earliest cells. For many years, Oparin's views remained speculative and appeared untestable, until a surprising experiment was conducted using simple equipment on a desktop.

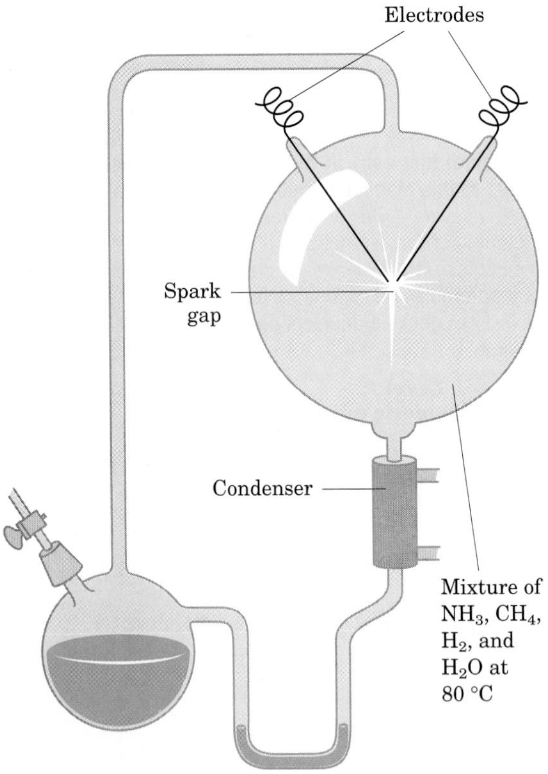

Electrodes

Spark gap

Condenser

Mixture of NH_3, CH_4, H_2, and H_2O at 80 °C

figure 3–27
Spark-discharge apparatus of the type used by Miller and Urey in experiments demonstrating abiotic formation of organic compounds under primitive atmospheric conditions. After subjection of the gaseous contents of the system to electrical sparks, products were collected by condensation. Biomolecules such as amino acids were among the products (see Table 3–6).

Chemical Evolution Can Be Simulated in the Laboratory

A classic experiment on the abiotic (nonbiological) origin of organic biomolecules was carried out in 1953 by Stanley Miller in the laboratory of Harold Urey. Miller subjected gaseous mixtures of NH_3, CH_4, H_2O, and H_2 to electrical sparks produced across a pair of electrodes (to simulate lightning) for periods of a week or more, then analyzed the contents of the closed reaction vessel (Fig. 3–27). The gas phase of the resulting mixture contained CO and CO_2, as well as the starting materials. The water phase contained a variety of organic compounds, including some amino acids, hydroxy acids, aldehydes, and hydrogen cyanide (HCN). This experiment established the possibility of abiotic production of biomolecules in relatively short times under relatively mild conditions.

Several developments have allowed more refined studies of the type pioneered by Miller and Urey and have yielded strong evidence that a wide variety of biomolecules, including polypeptides and short polymers of nucleosides, could have been produced spontaneously from simple starting materials probably present on the earth at the time life arose.

Modern extensions of the Miller experiments have employed "atmospheres" that include CO_2 and CO, with N_2 and H_2, and much-improved technology for identifying small quantities of products. The formation of hundreds of organic compounds has been demonstrated (Table 3–6). These compounds include common amino acids, a variety of mono-, di-, and tri-

table 3–6

Some Products Formed under Prebiotic Conditions

Carboxylic acids	Nucleic acid bases	Amino acids	Sugars
Formic acid	Adenine	Glycine	Straight and branched pentoses and hexoses
Acetic acid	Guanine	Alanine	
Propionic acid	Xanthine	α-Aminobutyric acid	
Straight and branched fatty acids (C_4–C_{10})	Hypoxanthine	Valine	
	Cytosine	Leucine	
Glycolic acid	Uracil	Isoleucine	
Lactic acid		Proline	
Succinic acid		Aspartic acid	
		Glutamic acid	
		Serine	
		Threonine	

Source: From Miller, S.L. (1987) Which organic compounds could have occurred on the prebiotic earth? *Cold Spring Harb. Symp. Quant. Biol.* **52,** 17–27.

carboxylic acids, fatty acids, adenine, and formaldehyde. Under certain conditions, formaldehyde polymerizes to form sugars containing three, four, five, and six carbons. In addition to the many small molecules that form in these experiments, polymers of nucleotides (nucleic acids) and of amino acids (proteins) also form. Some products of the self-condensation of HCN are effective promoters of such polymerization reactions, and inorganic ions present in the earth's crust (Cu^{2+}, Ni^{2+}, and Zn^{2+}) also enhance the rate of polymerization. The sources of energy that are effective in bringing about the formation of these compounds include heat, visible and ultraviolet (UV) light, x rays, gamma radiation, ultrasound and shock waves, and bombardment with alpha and beta particles.

In short, laboratory experiments on the spontaneous formation of biomolecules under prebiotic conditions have provided good evidence that many of the chemical components of living cells, including polypeptides and RNA-like molecules, can form under these conditions. Short polymers of RNA can act as catalysts in biologically significant reactions (Chapter 26), and RNA probably played a crucial role in prebiotic evolution, both as catalyst and as information repository.

If life evolved on Earth by this chemical evolution process, it is likely that life arose also on suitable planets of other solar systems. Many prebiotic compounds such as HCN, formic acid, and cyanoacetylene have been found in comets, in the atmospheres of Jupiter, Saturn, and Titan (a moon of Saturn), and in the dust clouds of interstellar space. Analysis of the Murchison meteorite, which fell to Earth in 1969, revealed the presence of amino acids, hydroxy acids, purines, and pyrimidines. It is therefore conceivable that the organic precursors for the evolution of life on Earth originated elsewhere in the solar system.

RNA or Related Precursors May Have Been the First Genes and Catalysts

In modern organisms, nucleic acids encode the genetic information that specifies the structure of enzymes, and enzymes have the ability to catalyze the replication and repair of nucleic acids. The mutual dependence of these two classes of biomolecules brings up the perplexing question: which came first, DNA or protein?

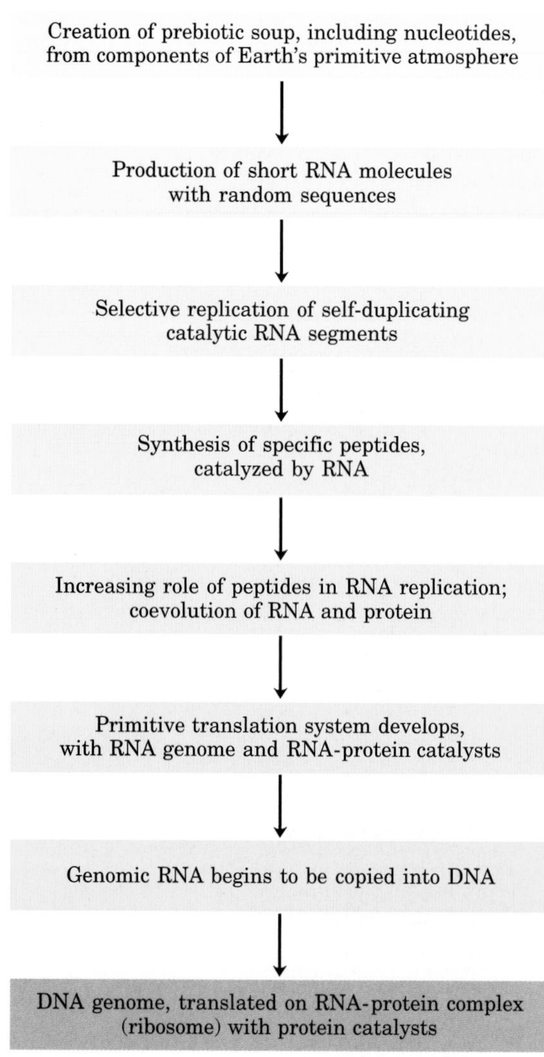
Creation of prebiotic soup, including nucleotides, from components of Earth's primitive atmosphere
Production of short RNA molecules with random sequences
Selective replication of self-duplicating catalytic RNA segments
Synthesis of specific peptides, catalyzed by RNA
Increasing role of peptides in RNA replication; coevolution of RNA and protein
Primitive translation system develops, with RNA genome and RNA-protein catalysts
Genomic RNA begins to be copied into DNA
DNA genome, translated on RNA-protein complex (ribosome) with protein catalysts

figure 3–28
One possible "RNA world" scenario, showing the transition from the prebiotic RNA world (shades of yellow) to the biotic DNA world (orange).

The answer may be: neither. The discovery that RNA molecules can act as catalysts in their own formation suggests that RNA or a similar molecule may have been the first gene *and* the first catalyst. According to this scenario (Fig. 3–28), one of the earliest stages of biological evolution was the chance formation, in the primordial soup, of an RNA molecule that had the ability to catalyze the formation of other RNA molecules of the same sequence—a self-replicating, self-perpetuating RNA. The concentration of a self-replicating RNA molecule would increase exponentially, as one molecule formed two, two formed four, and so on. The fidelity of self-replication was presumably less than perfect, so the process would generate variants of the RNA, some of which might be even better able to self-replicate. In the competition for nucleotides, the most efficient of the self-replicating sequences would win, and less efficient replicators would fade from the population.

The division of function between DNA (genetic information storage) and protein (catalysis) was, according to the "RNA world" hypothesis, a later development. New variants of self-replicating RNA molecules developed, with the additional ability to catalyze the condensation of amino acids into peptides. Occasionally, the peptide(s) thus formed would reinforce the self-replicating ability of the RNA, and the pair—RNA molecule and helping peptide—could undergo further modifications in sequence, generating even more efficient self-replicating systems. Some time after the evolution of this primitive protein-synthesizing system, there was a further development: DNA molecules with sequences complementary to the self-replicating RNA molecules took over the function of conserving the "genetic" information, and RNA molecules evolved to play roles in protein synthesis. Proteins proved to be versatile catalysts and, over time, took over that function. Lipidlike compounds in the primordial soup formed relatively impermeable layers around self-replicating collections of molecules. The concentration of proteins and nucleic acids within these lipid enclosures favored the molecular interactions required in self-replication.

This "RNA world" hypothesis is plausible but by no means universally accepted. The hypothesis does make testable predictions, and to the extent that experimental tests are possible within finite times, the hypothesis will be tested and refined.

Biological Evolution Began More Than Three and a Half Billion Years Ago

Earth was formed about 4.5 billion years ago, and the first evidence of life dates to more than 3.5 billion years ago. An international group of scientists showed in 1980 that certain ancient rock formations (stromatolites) in Western Australia contained fossils of primitive microorganisms (Fig. 3–29). In 1996, scientists working in Greenland found not fossil remains but chemical evidence of life from as far back as 3.85 billion years ago, forms of carbon embedded in rock that appear to have a distinctly biological origin. Somewhere on Earth during its first billion years there arose the first simple organism, capable of replicating its own structure from a template (RNA?) that was the first genetic material. Because the terrestrial atmosphere at the dawn of life was nearly devoid of oxygen, and because there were few microorganisms to scavenge organic compounds formed by natural processes, these compounds were relatively stable. Given this stability and eons of time, the improbable became inevitable: the organic compounds were incorporated into evolving cells to produce more and more effective self-reproducing catalysts. The process of biological evolution had begun.

Organisms developed mechanisms to harness the energy of sunlight through photosynthesis, to make sugars and other organic molecules from carbon dioxide, and to incorporate molecular nitrogen from the atmosphere into nitrogenous biomolecules such as amino acids. By developing their own capacities to synthesize biomolecules, cells became independent of the random processes by which such compounds had first appeared on Earth. As evolution proceeded, organisms began to interact and to derive mutual benefits from each other's products, forming increasingly complex ecological systems.

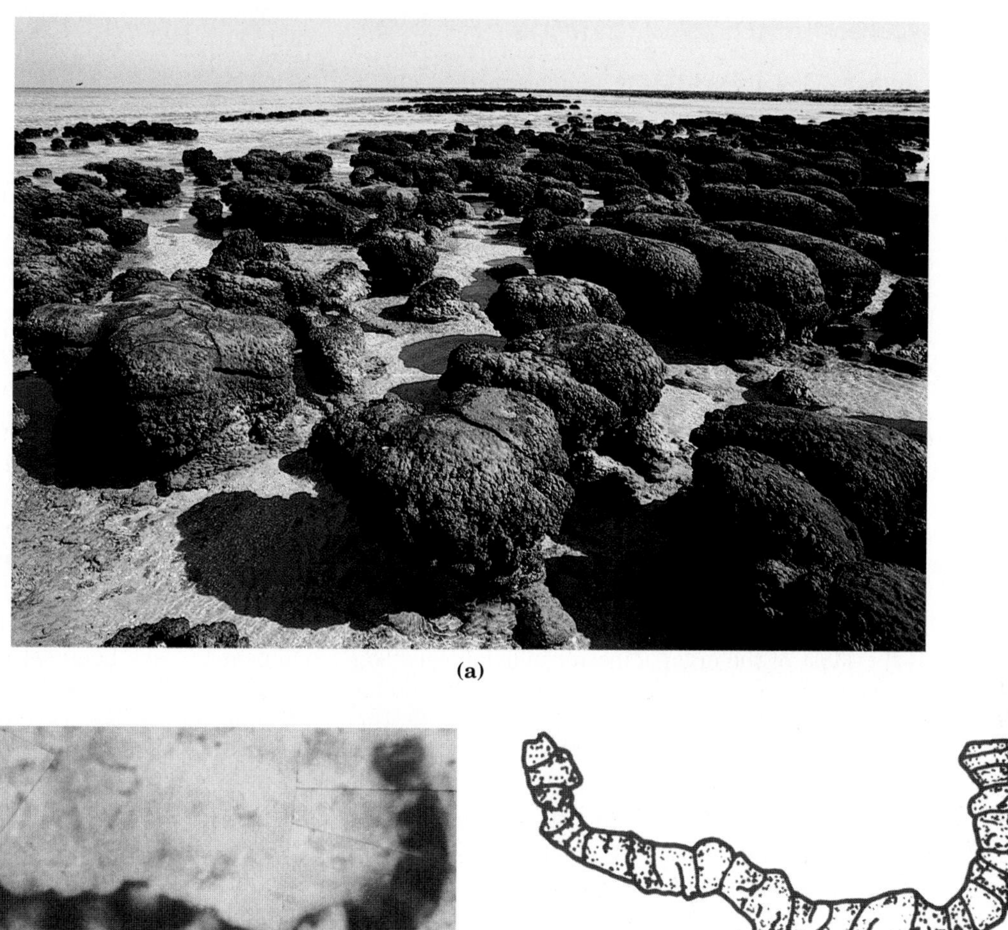

figure 3–29
Ancient reefs in Australia contain fossil evidence of microbial life in the sea of 3.5 billion years ago. Bits of sand and limestone became trapped in the sticky extracellular coats of cyanobacteria, gradually building up these stromatolites found in Hamelin Bay, Western Australia **(a).** Microscopic examination of thin sections of such ancient rock reveals microfossils of filamentous bacteria **(b),** interpreted as shown in the drawing **(c).**

summary

Most of the dry weight of living organisms consists of organic compounds, molecules containing covalently bonded carbon backbones to which other carbon, hydrogen, oxygen, or nitrogen atoms may be attached. The different kinds of functional groups attached to the backbone determine the chemical properties of the molecule. Organic biomolecules also have characteristic shapes (configurations and conformations) in three dimensions. Many biomolecules occur in asymmetric or chiral forms called enantiomers, stereoisomers that are nonsuperimposable mirror images of each other. Usually, only one of a pair of enantiomers has biological activity.

The strength of covalent chemical bonds, measured in joules, depends on the electronegativities and sizes of the atoms that share electrons. The enthalpy change (ΔH) for a chemical reaction reflects the numbers and kinds of bonds made and broken. For endothermic reactions, ΔH is positive; for exothermic reactions, negative. The many different chemical reactions that occur within a cell fall into five general categories: oxidation-reduction reactions, breakage or formation of carbon–carbon bonds, rearrangements of the bonds around carbon atoms, group transfers, and condensations.

Most of the organic matter in living cells consists of macromolecules: nucleic acids, proteins, and polysaccharides. Lipid molecules, another important component of cells, are small molecules that form large aggregates. Biological macromolecules are composed of small, covalently linked monomeric subunits of relatively few kinds. Proteins are polymers of 20 different kinds of amino acids, nucleic acids are polymers of different nucleotide units (four kinds in DNA, four in RNA), and polysaccharides are polymers of recurring sugar units. Nucleic acids and proteins are informational macromolecules; the characteristic sequences of their subunits constitute the genetic individuality of a species. Simple polysaccharides act as structural components; some complex polysaccharides are informational macromolecules, their specific sequence and stereochemistry allowing them to be recognized with high specificity by other molecules.

There is a structural hierarchy in the molecular organization of cells. Cells contain organelles, such as nuclei, mitochondria, and chloroplasts, which in turn contain supramolecular complexes, such as membranes and ribosomes, and these consist in turn of clusters of macromolecules bound together by many relatively weak, noncovalent forces. Macromolecules consist of covalently linked subunits. The formation of macromolecules from simple subunits creates order (decreases entropy); this synthesis requires energy and therefore must be coupled to exergonic reactions.

The small biomolecules such as amino acids and sugars probably first arose spontaneously from atmospheric gases and water under the influence of electrical energy (lightning) during the early history of the earth. The same process can be simulated in the laboratory. The monomeric components of cellular macromolecules appear to have been selected early in biological evolution. These subunit molecules are relatively few in number, yet very versatile; evolution has combined small biomolecules to yield macromolecules of immense diversity. The first macromolecules may have been RNA-like molecules capable of catalyzing their own replication. Later in evolution, DNA took over the function of storing genetic information, proteins became the cellular catalysts, and RNA mediated between DNA and protein, allowing the expression of genetic information as proteins.

further reading

General

Frausto da Silva, J.J.R. & Williams, R.J.P. (1994) *The Biological Chemistry of the Elements: The Inorganic Chemistry of Life,* Clarendon Press, Oxford.

An excellent, highly readable text on the role of inorganic elements in biochemistry. Clear diagrams, good references.

Frieden, E. (1972) The chemical elements of life. *Sci. Am.* **227** (July), 52–61.

The Molecules of Life. (1985) *Sci. Am.* **253** (October).

An entire issue devoted to the structure and function of biomolecules. It includes articles on DNA, RNA, and proteins, and their subunits.

Chemistry and Stereochemistry

Barta, N.S. & Stille, J.R. (1994) Grasping the concepts of stereochemistry. *J. Chem. Educ.* **71,** 20–23.

A clear description of the RS system for naming stereoisomers, with practical suggestions for determining and remembering the "handedness" of isomers.

Brewster, J.H. (1986) Stereochemistry and the origins of life. *J. Chem. Educ.* **63,** 667–670.

An interesting and lucid discussion of the ways in which evolution could have selected only one of two stereoisomers for the construction of proteins and other molecules.

Hegstrom, R.A. & Kondepudi, D.K. (1990) The handedness of the universe. *Sci. Am.* **262** (January), 108–115.

Stereochemistry and the asymmetry of biomolecules, viewed in the context of the universe.

Kotz, J.C. & Treichel, P., Jr. (1998) *Chemistry and Chemical Reactivity,* Saunders College Publishing, Fort Worth, TX.

An excellent, comprehensive introduction to chemistry.

Loudon, M. (1995) *Organic Chemistry,* 3rd edn, The Benjamin/Cummings Publishing Company, Menlo Park, CA.

This and the following two books provide details on stereochemistry and the chemical reactivity of functional groups. All are excellent textbooks.

Morrison, R.T. & Boyd, R.N. (1999) *Organic Chemistry,* 7th edn, Allyn & Bacon, Inc., Boston, MA.

Streitwieser, A., Jr., Heathcock, C.H., & Kosower, E.M. (1998) *Introduction to Organic Chemistry,* 4th edn, revised printing, Prentice-Hall, Upper Saddle River, NJ.

Wagner, G. (1997) An account of NMR in structural biology. *Nature Struct. Biol.,* NMR Suppl. (October), 841–844.

A short, clear account of the development of NMR as a biochemical tool, current applications, and future prospects.

Prebiotic Evolution

Cavalier-Smith, T. (1987) The origin of cells: a symbiosis between genes, catalysts, and membranes. *Cold Spring Harb. Symp. Quant. Biol.* **52,** 805–824.

Darnell, J.E. & Doolittle, W.F. (1986) Speculations on the early course of evolution. *Proc. Natl. Acad. Sci. USA* **83,** 1271–1275.

A clear statement of the RNA world scenario.

de Duve, C. (1995) The beginnings of life on earth. *Am. Sci.* **83,** 428–437.

One scenario for the succession of chemical steps that led to the first living organism.

Evolution of Catalytic Function. (1987) *Cold Spring Harb. Symp. Quant. Biol.* **52.**

A collection of almost 100 articles on all aspects of prebiotic and early biological evolution; probably the single best source on molecular evolution.

Ferris, J.P. (1984) The chemistry of life's origin. *Chem. Eng. News* **62,** 21–35.

A short, clear description of the experimental evidence for the synthesis of biomolecules under prebiotic conditions.

Gesteland, R.F. & Atkins, J.F. (eds) (1993) *The RNA World,* Cold Spring Harbor Laboratory Press, Cold Spring Harbor, NY.

A collection of stimulating reviews on a wide range of topics related to the RNA world scenario.

Hager, A.J., Pollard, J.D., & Szostak, J.W. (1996) Ribozymes: aiming at RNA replication and protein synthesis. *Chem. Biol.* **3,** 717–725.

A short review of studies on RNA catalysis of RNA synthesis and amide bond synthesis, and the relevance of these findings to the RNA world scenario.

Hirao, I. & Ellington, A.D. (1995) Re-creating the RNA world. *Curr. Biol.* **5,** 1017–1022.

This and the article by Robertson and Ellington (1997), listed below, describe research aimed at reproducing in the laboratory the evolutionary rise of catalytic RNA.

Horgan, J. (1991) In the beginning . . . *Sci. Am.* **264** (February), 116–125.

A brief, clear statement of current theories on prebiotic evolution.

Lazcano, A. & Miller, S.L. (1996) The origin and early evolution of life: prebiotic chemistry, the pre-RNA world, and time. *Cell* **85,** 793–798.

Brief review of recent developments in studies of the origin of life: primitive atmospheres, submarine vents, autotrophic versus heterotrophic origin, the RNA and pre-RNA worlds, and the time required for life to arise.

Miller, S.L. (1987) Which organic compounds could have occurred on the prebiotic earth? *Cold Spring Harb. Symp. Quant. Biol.* **52,** 17–27.

Summary of laboratory experiments on chemical evolution, by the person who did the original Miller–Urey experiment.

Miller, S.L. & Orgel, L.E. (1974) *The Origins of Life on the Earth,* Prentice-Hall, Inc., Englewood Cliffs, NJ.

The New Age of RNA. (1993) *FASEB J.* **7** (1).

A collection of about 15 short articles related to the RNA world scenario.

Robertson, M.P. & Ellington, A.D. (1997) Ribozymes: red in tooth and claw. *Curr. Biol.* **7,** R376–R379.

Schopf, J.W. (ed.) (1983) *Earth's Earliest Biosphere,* Princeton University Press, Princeton, NJ.

A comprehensive discussion of geologic history and its relation to the development of life.

problems

1. Vitamin C: Is the Synthetic Vitamin as Good as the Natural One? A claim sometimes put forth by purveyors of health foods is that vitamins obtained from natural sources are more healthful than those obtained by chemical synthesis. For example, pure L-ascorbic acid (vitamin C) extracted from rose hips is better than pure L-ascorbic acid manufactured in a chemical plant. Are the vitamins from the two sources different? Can the body distinguish a vitamin's source?

2. Identification of Functional Groups Figures 3–5 and 3–6 show some common functional groups of biomolecules. Because the properties and biological activities of biomolecules are largely determined by their functional groups, it is important to be able to identify them. In each of the compounds below, circle and identify by name each constituent functional group.

```
   H  H
   |  |
H2N—C—C—OH
   |  |
   H  H
```

Ethanolamine

(a)

```
   H
   |
H—C—OH
   |
H—C—OH
   |
H—C—OH
   |
   H
```

Glycerol

(b)

```
      O
      ‖
  HO—P—O⁻
      |
 H    O
  \   |
   C=C—COOH
  /
 H
```

Phosphoenolpyruvic acid, an intermediate in glucose metabolism

(c)

```
    COOH
     |
H2N—C—H
     |
  H—C—OH
     |
    CH3
```

Threonine, an amino acid

(d)

```
   HO   O
     \ //
      C
      |
     CH2
      |
     CH2
      |
     NH
      |
     C=O
      |
   H—C—OH
      |
CH3—C—CH3
      |
     CH2OH
```

Pantothenic acid, a vitamin

(e)

```
    H   O
     \ //
      C
      |
   H—C—NH2
      |
  HO—C—H
      |
   H—C—OH
      |
   H—C—OH
      |
     CH2OH
```

D-Glucosamine

(f)

Problem 2

3. Drug Activity and Stereochemistry The quantitative differences in biological activity between the two enantiomers of a compound are sometimes quite large. For example, the D isomer of the drug isoproterenol, used to treat mild asthma, is 50 to 80 times more effective as a bronchodilator than the L isomer. Identify the chiral center in isoproterenol. Why do the two enantiomers have such radically different bioactivity?

```
             OH       H  H
             |        |  |
HO—⟨benzene⟩—C—CH2—N—C—CH3
   |         |           |
   HO        H          CH3
```

Isoproterenol

4. Drug Action and Shape of Molecules Some years ago two drug companies marketed a drug under the trade names Dexedrine and Benzedrine. The structure of the drug is shown below.

```
                  H
                  |
⟨benzene⟩—CH2—C—CH3
                  |
                 NH2
```

The physical properties (C, H, and N analysis, melting point, solubility, etc.) of Dexedrine and Benzedrine were identical. The recommended oral dosage of Dexedrine (which is still available) was 5 mg/day, but the recommended dosage of Benzedrine (no longer available) was twice that. Apparently it required considerably more Benzedrine than Dexedrine to yield the same physiological response. Explain this apparent contradiction.

5. Components of Complex Biomolecules Figure 3–24 shows the major components of complex biomolecules. For each of the three important biomolecules below and at right (shown in their ionized forms at physiological pH), identify the constituents.

(a) Guanosine triphosphate (GTP), an energy-rich nucleotide that serves as a precursor to RNA:

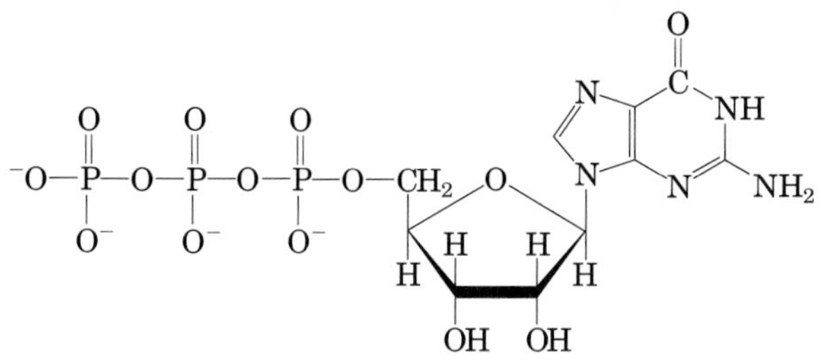

(b) Phosphatidylcholine, a component of many membranes:

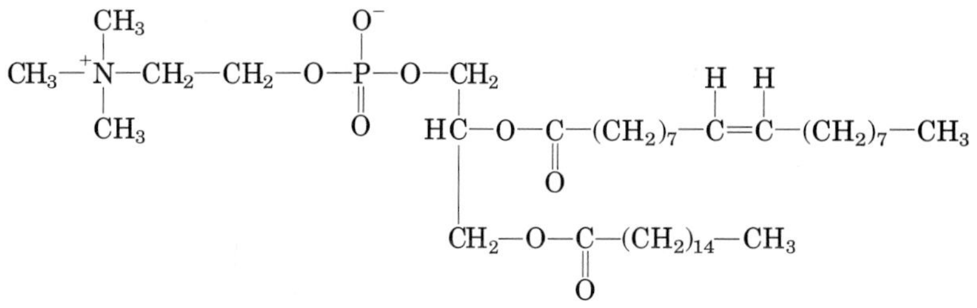

(c) Methionine enkephalin, the brain's own opiate:

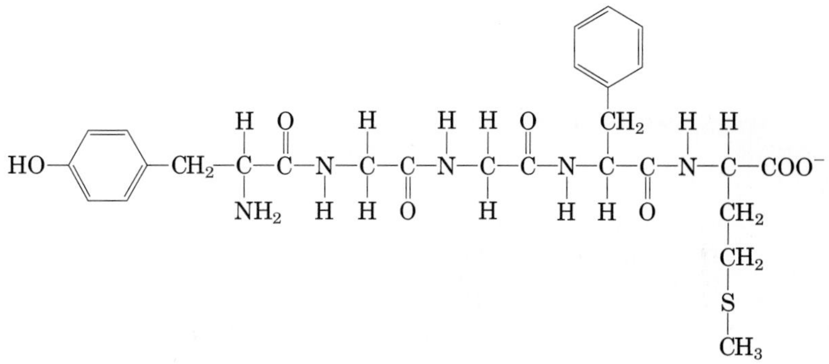

Problem 5

6. Determination of the Structure of a Biomolecule An unknown substance, X, was isolated from rabbit muscle. Its structure was determined from the following observations and experiments. Qualitative analysis showed that X was composed entirely of C, H, and O. A weighed sample of X was completely oxidized, and the H_2O and CO_2 produced were measured; this quantitative analysis revealed that X contained 40.00% C, 6.71% H, and 53.29% O by weight. The molecular mass of X, determined by a mass spectrometer, was 90.00 amu. An infrared spectrum showed that X contained one double bond. X dissolved readily in water to give an acidic solution; the solution demonstrated optical activity when tested in a polarimeter.

(a) Determine the empirical and molecular formula of X.

(b) Draw the possible structures of X that fit the molecular formula and contain one double bond. Consider *only* linear or branched structures and disregard cyclic structures. Note that oxygen makes very poor bonds to itself.

(c) What is the structural significance of the observed optical activity? Which structures in (b) does this observation eliminate? Which structures are consistent with the observation?

(d) What is the structural significance of the observation that a solution of X was acidic? Which structures in (b) are now eliminated? Which structures are consistent with the observation?

(e) What is the structure of X? Is more than one structure consistent with all the data?

7. Separating Biomolecules In laboratory biochemistry, it is first necessary to separate the molecule of interest from the other biomolecules in the sample—that is, to *purify* the protein, nucleic acid, carbohydrate, or lipid. Specific purification techniques will be addressed later in the text. However, just by looking at the monomeric subunits from which the larger biomolecules are made, you should have some ideas as to what characteristics of those biomolecules would allow you to separate them one from another.

(a) What characteristics of amino acids and fatty acids would allow them to be easily separated from each other?

(b) How might nucleotides be separated from glucose molecules?

8. Silicon-Based Life? Silicon is in the same group of the periodic table as carbon and, like carbon, can form up to four single bonds. Many science fiction stories have been written based on the premise of silicon-based life. Is this realistic? What characteristics of silicon make it *less* well adapted to performing as the central organizing element for life? To answer this question, use the information in this chapter about carbon's bonding versatility, and refer to a beginning inorganic chemistry textbook for silicon's bonding properties.

chapter

4 Water

This view of Earth from space shows that most of the planet's surface is covered with water. The seas, where life probably first arose, are today the habitat of countless organisms.

Water is the most abundant substance in living systems, making up 70% or more of the weight of most organisms. The first living organisms doubtless arose in an aqueous environment, and the course of evolution has been shaped by the properties of the aqueous medium in which life began.

This chapter begins with descriptions of the physical and chemical properties of water, to which all aspects of cell structure and function are adapted. The attractive forces between water molecules and the slight tendency of water to ionize are of crucial importance to the structure and function of biomolecules. We will review the topic of ionization in terms of equilibrium constants, pH, and titration curves, and consider how aqueous solutions of weak acids or bases and their salts act as buffers against pH changes in biological systems. The water molecule and its ionization products, H^+ and OH^-, profoundly influence the structure, self-assembly, and properties of all cellular components, including proteins, nucleic acids, and lipids. The noncovalent interactions responsible for the strength and specificity of "recognition" among biomolecules are decisively influenced by the solvent properties of water.

Weak Interactions in Aqueous Systems

Hydrogen bonds between water molecules provide the cohesive forces that make water a liquid at room temperature and that favor the extreme ordering of molecules that is typical of crystalline water (ice). Polar biomolecules dissolve readily in water because they can replace water-water interactions with more energetically favorable water-solute interactions. In contrast, nonpolar biomolecules interfere with water-water interactions but are unable to form water-solute interactions—consequently, nonpolar molecules are poorly soluble in water. In aqueous solutions, nonpolar molecules tend to cluster together.

Hydrogen bonds and ionic, hydrophobic (Greek, "water-fearing"), and van der Waals interactions are individually weak, but collectively they have a very significant influence on the three-dimensional structures of proteins, nucleic acids, polysaccharides, and membrane lipids.

Hydrogen Bonding Gives Water Its Unusual Properties

Water has a higher melting point, boiling point, and heat of vaporization than most other common solvents (Table 4–1). These unusual properties

table 4–1

Melting Point, Boiling Point, and Heat of Vaporization of Some Common Solvents

	Melting point (°C)	Boiling point (°C)	Heat of vaporization (J/g)*
Water	0	100	2,260
Methanol (CH_3OH)	−98	65	1,100
Ethanol (CH_3CH_2OH)	−117	78	854
Propanol ($CH_3CH_2CH_2OH$)	−127	97	687
Butanol ($CH_3(CH_2)_2CH_2OH$)	−90	117	590
Acetone (CH_3COCH_3)	−95	56	523
Hexane ($CH_3(CH_2)_4CH_3$)	−98	69	423
Benzene (C_6H_6)	6	80	394
Butane ($CH_3(CH_2)_2CH_3$)	−135	−0.5	381
Chloroform ($CHCl_3$)	−63	61	247

*The heat energy required to convert 1.0 g of a liquid at its boiling point, at atmospheric pressure, into its gaseous state at the same temperature. It is a direct measure of the energy required to overcome attractive forces between molecules in the liquid phase.

are a consequence of attractions between adjacent water molecules that give liquid water great internal cohesion. A look at the electron structure of the H_2O molecule reveals the cause of these intermolecular attractions.

Each hydrogen atom of a water molecule shares an electron pair with the oxygen atom. The geometry of the molecule is dictated by the shapes of the outer electron orbitals of the oxygen atom, which are similar to the bonding orbitals of carbon (see Fig. 3–4a). These orbitals describe a rough tetrahedron, with a hydrogen atom at each of two corners and unshared electron pairs at the other two corners (Fig. 4–1a). The H—O—H bond angle is 104.5°, slightly less than the 109.5° of a perfect tetrahedron because of crowding by the nonbonding orbitals of the oxygen atom.

The oxygen nucleus attracts electrons more strongly than does the hydrogen nucleus (a proton); oxygen is more electronegative (see Table 3–2). The sharing of electrons between H and O is therefore unequal; the electrons are more often in the vicinity of the oxygen atom than of the hydrogen. The result of this unequal electron sharing is two electric dipoles in the water molecule, one along each of the H—O bonds; the oxygen atom bears a partial negative charge ($2\delta^-$), and each hydrogen a partial positive charge (δ^+). As a result, there is an electrostatic attraction between the oxygen atom of one water molecule and the hydrogen of another (Fig. 4–1c), called a **hydrogen bond.** Throughout this book, we will represent hydrogen bonds with three parallel blue lines, as in Figure 4–1c.

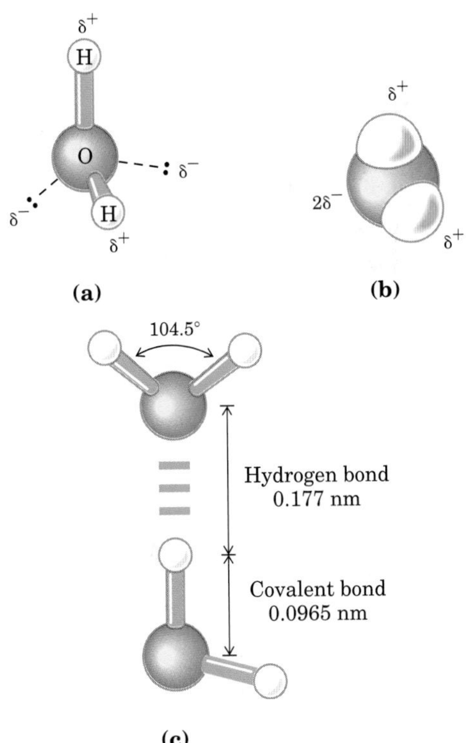

figure 4–1

Structure of the water molecule. The dipolar nature of the H_2O molecule is shown by **(a)** ball-and-stick and **(b)** space-filling models. The dashed lines in **(a)** represent the nonbonding orbitals. There is a nearly tetrahedral arrangement of the outer-shell electron pairs around the oxygen atom; the two hydrogen atoms have localized partial positive charges (δ^+) and the oxygen atom has a partial negative charge ($2\delta^-$). **(c)** Two H_2O molecules joined by a hydrogen bond (designated here, and throughout this book, by three blue lines) between the oxygen atom of the upper molecule and a hydrogen atom of the lower one. Hydrogen bonds are longer and weaker than covalent O—H bonds.

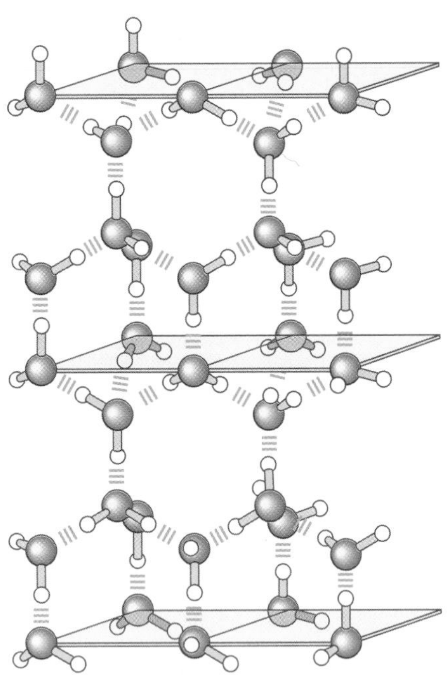

figure 4–2
Hydrogen bonding in ice. Each water molecule forms the maximum of four hydrogen bonds, creating a regular crystal lattice. In liquid water at room temperature and atmospheric pressure, by contrast, each water molecule hydrogen bonds with an average of 3.4 other water molecules. The crystal lattice of ice occupies more space than that occupied by the same number of H_2O molecules in liquid water; ice is less dense than—and thus floats on—liquid water.

Hydrogen bonds are weaker than covalent bonds. The hydrogen bonds in liquid water have a **bond dissociation energy** (the energy required to break a bond) of about 20 kJ/mol, compared with 348 kJ/mol for the covalent C—C bond. At room temperature, the thermal energy of an aqueous solution (the kinetic energy of motion of the individual atoms and molecules) is of the same order of magnitude as that required to break hydrogen bonds. When water is heated, the increase in temperature reflects the faster motion of individual water molecules. Although at any given time most of the molecules in liquid water are engaged in hydrogen bonding, the lifetime of each hydrogen bond is less than 1×10^{-9} s. The apt phrase "flickering clusters" has been applied to the short-lived groups of hydrogen-bonded molecules in liquid water. The sum of all the hydrogen bonds between molecules nevertheless confers great internal cohesion on liquid water.

The nearly tetrahedral arrangement of the orbitals about the oxygen atom (Fig. 4–1a) allows each water molecule to form hydrogen bonds with as many as four neighboring water molecules. In liquid water at room temperature and atmospheric pressure, however, water molecules are disorganized and in continuous motion, so that each molecule forms hydrogen bonds with an average of only 3.4 other molecules. In ice, on the other hand, each water molecule is fixed in space and forms hydrogen bonds with four other water molecules to yield a regular lattice structure (Fig. 4–2). Breakage of a sufficient number of hydrogen bonds to destabilize the crystal lattice of ice requires much thermal energy, which accounts for the relatively high melting point of water (Table 4–1). When ice melts or water evaporates, heat is taken up by the system:

$$H_2O(s) \longrightarrow H_2O(l) \qquad \Delta H = +5.9 \text{ kJ/mol}$$

$$H_2O(l) \longrightarrow H_2O(g) \qquad \Delta H = +44.0 \text{ kJ/mol}$$

During melting or evaporation, the entropy of the aqueous system increases as more highly ordered arrays of water molecules relax into the less orderly hydrogen-bonded arrays in liquid water or the wholly disordered gaseous state. At room temperature, both the melting of ice and the evaporation of water occur spontaneously; the tendency of the water molecules to associate through hydrogen bonds is outweighed by the energetic push toward randomness. Recall that the free-energy change (ΔG) must have a negative value for a process to occur spontaneously: $\Delta G = \Delta H - T\,\Delta S$, where ΔG represents the driving force, ΔH the enthalpy change from making and breaking bonds, and ΔS the change in randomness. Because ΔH is positive for melting and evaporation, it is clearly the increase in entropy (ΔS) that makes ΔG negative and drives these transformations.

Water Forms Hydrogen Bonds with Polar Solutes

Hydrogen bonds are not unique to water. They readily form between an electronegative atom (the hydrogen acceptor, usually oxygen or nitrogen with a lone pair of electrons) and a hydrogen atom covalently bonded to another electronegative atom (the hydrogen donor) in the same or another molecule (Fig. 4–3). Hydrogen atoms covalently bonded to carbon atoms (which are not electronegative) do not participate in hydrogen bonding. The distinction explains why butanol ($CH_3(CH_2)_2CH_2OH$) has a relatively high boiling point of 117 °C, whereas butane ($CH_3(CH_2)_2CH_3$) has a boiling point of only −0.5 °C. Butanol has a polar hydroxyl group and thus can form intermolecular hydrogen bonds.

Uncharged but polar biomolecules such as sugars dissolve readily in water because of the stabilizing effect of hydrogen bonds between the hydroxyl groups or carbonyl oxygen of the sugar and the polar water molecules. Alcohols, aldehydes, ketones, and compounds containing N—H bonds all form hydrogen bonds with water molecules (Fig. 4–4) and tend to be soluble in water.

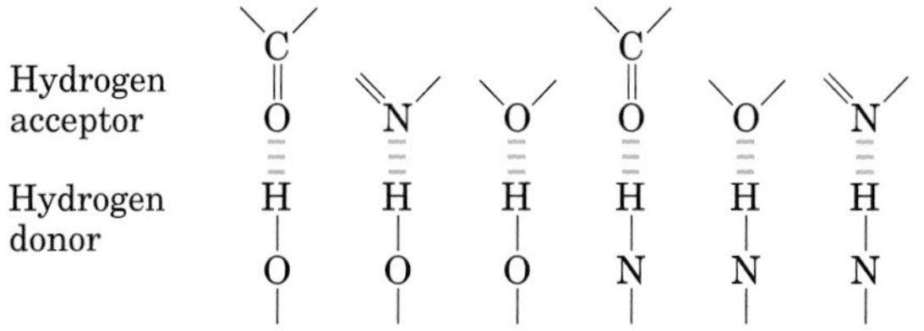

Hydrogen acceptor

Hydrogen donor

figure 4–3
Common hydrogen bonds in biological systems. The hydrogen acceptor is usually oxygen or nitrogen.

Between the hydroxyl group of an alcohol and water

Between the carbonyl group of a ketone and water

Between peptide groups in polypeptides

Between complementary bases of DNA

Thymine

Adenine

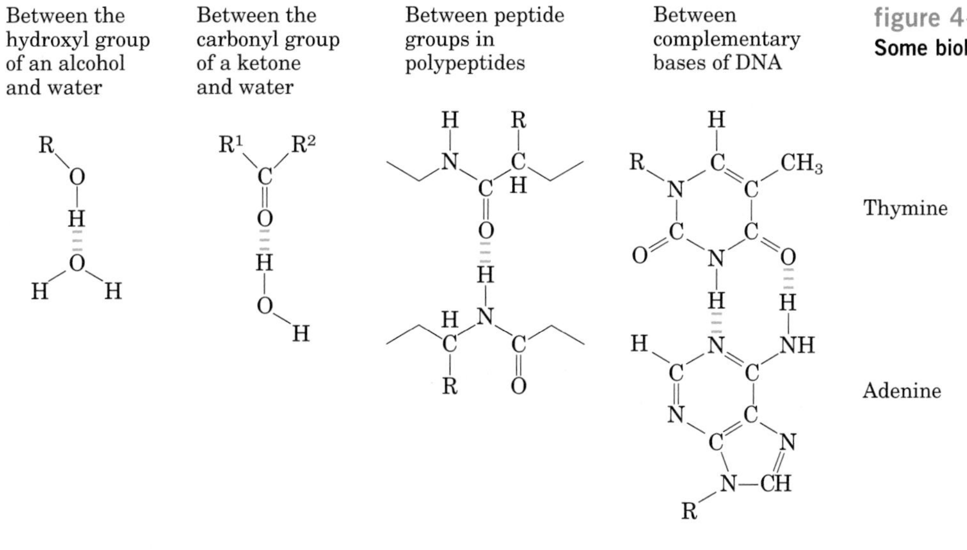

figure 4–4
Some biologically important hydrogen bonds.

Hydrogen bonds are strongest when the bonded molecules are oriented to maximize electrostatic interaction, which occurs when the hydrogen atom and the two atoms that share it are in a straight line—that is, when the acceptor atom is in line with the covalent bond between the donor atom and H (Fig. 4–5). Hydrogen bonds are thus highly directional and capable of holding two hydrogen-bonded molecules or groups in a specific geometric arrangement. As we shall see later, this property of hydrogen bonds confers very precise three-dimensional structures on protein and nucleic acid molecules, which have many intramolecular hydrogen bonds.

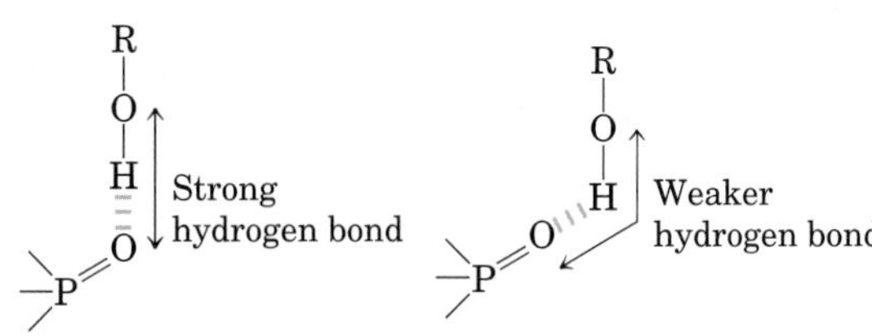

figure 4–5
Directionality of the hydrogen bond. The attraction between the partial electric charges (see Fig. 4–1) is greatest when the three atoms involved (in this case O, H, and O) lie in a straight line. When the hydrogen-bonded moieties are structurally constrained (as when they are parts of a single protein molecule, for example), this ideal geometry may not be possible and the resulting hydrogen bond is weaker.

Water Interacts Electrostatically with Charged Solutes

Water is a polar solvent. It readily dissolves most biomolecules, which are generally charged or polar compounds (Table 4–2); compounds that dissolve easily in water are **hydrophilic** (Greek, "water-loving"). In contrast, nonpolar solvents such as chloroform and benzene are poor solvents for polar biomolecules but easily dissolve those that are **hydrophobic**—nonpolar molecules such as lipids and waxes.

Water dissolves salts such as NaCl by hydrating and stabilizing the Na^+ and Cl^- ions, weakening the electrostatic interactions between them and thus counteracting their tendency to associate in a crystalline lattice (Fig.

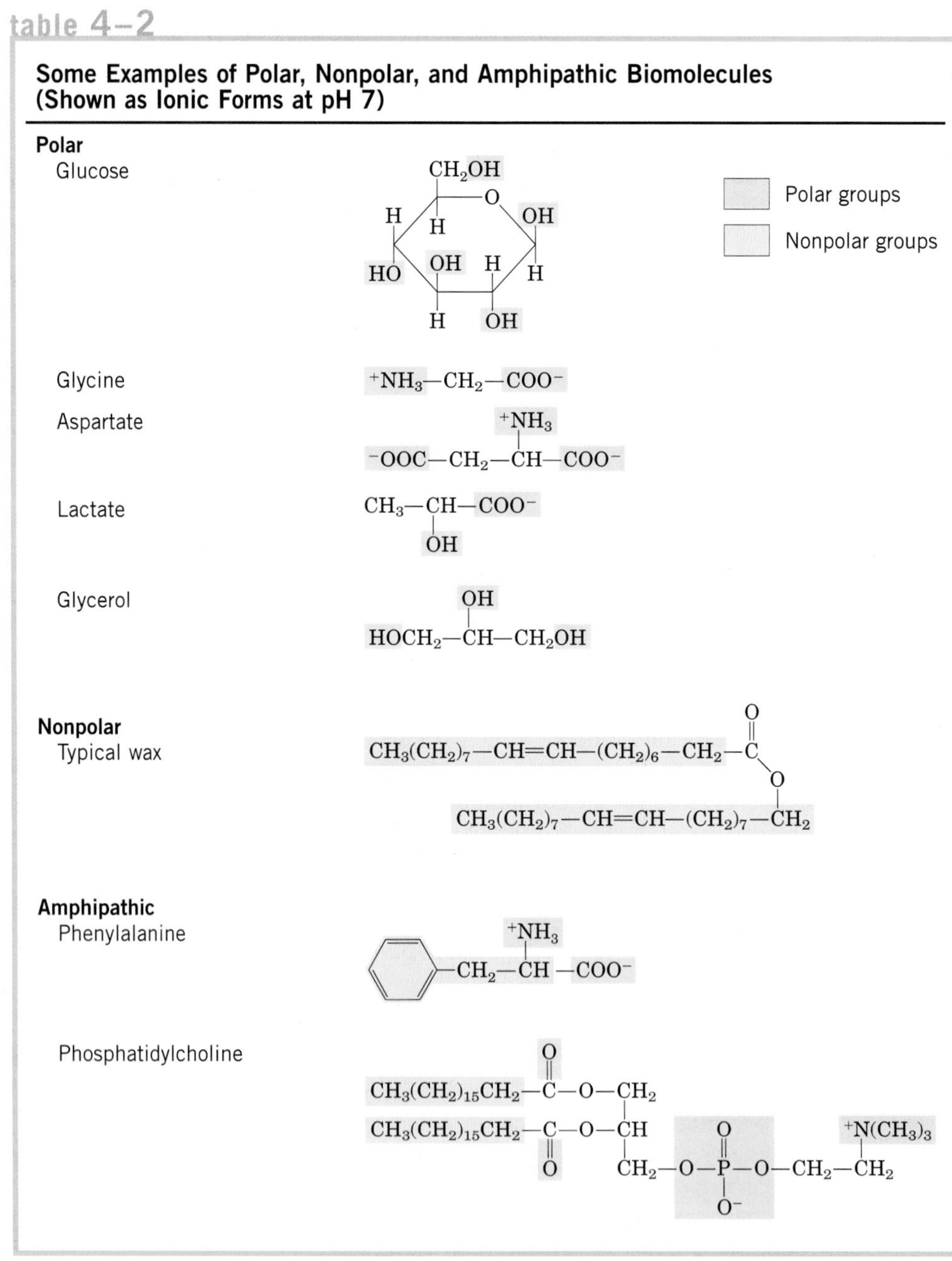

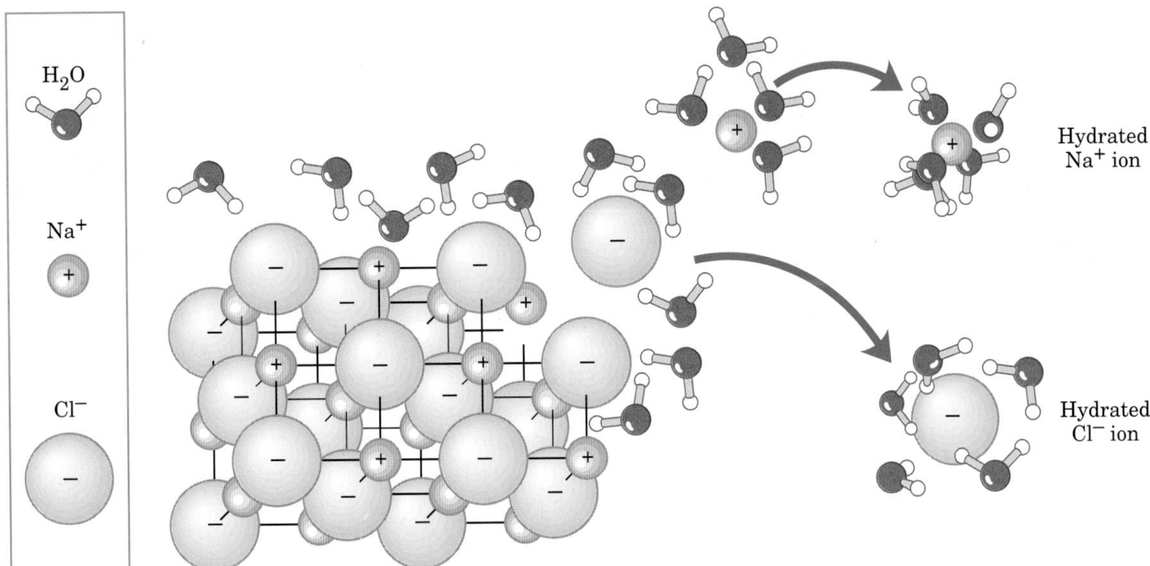

figure 4–6
Water dissolves many crystalline salts by hydrating their component ions. The NaCl crystal lattice is disrupted as water molecules cluster about the Cl^- and Na^+ ions. The ionic charges are partially neutralized, and the electrostatic attractions necessary for lattice formation are weakened.

4–6). The same factors apply to charged biomolecules, compounds with functional groups such as ionized carboxylic acids ($—COO^-$), protonated amines ($—NH_3^+$), and phosphate esters or anhydrides. Water readily dissolves such compounds by replacing solute-solute hydrogen bonds with solute-water hydrogen bonds, thus screening the electrostatic interactions between solute molecules.

Water is especially effective in screening the electrostatic interactions between dissolved ions because of its high dielectric constant, a physical property reflecting the number of dipoles in a solvent. The strength, or force (F), of ionic interactions in a solution depends upon the magnitude of the charges (Q), the distance between the charged groups (r), and the dielectric constant (ϵ) of the solvent in which the interactions occur:

$$F = \frac{Q_1 Q_2}{\epsilon r^2}$$

For water at 25 °C, ϵ (which is dimensionless) is 78.5, and for the very nonpolar solvent benzene, ϵ is 4.6. Thus, ionic interactions are much stronger in less polar environments. The dependence on r^2 is such that ionic attractions or repulsions operate only over short distances—in the range of 10 to 40 nm (depending on the electrolyte concentration) when the solvent is water.

Entropy Increases as Crystalline Substances Dissolve

As a salt such as NaCl dissolves, the Na^+ and Cl^- ions leaving the crystal lattice acquire far greater freedom of motion (Fig. 4–6). The resulting increase in the entropy (randomness) of the system is largely responsible for the ease of dissolving salts such as NaCl in water. In thermodynamic terms, formation of the solution occurs with a favorable change in free energy: $\Delta G = \Delta H - T\,\Delta S$, where ΔH has a small positive value and $T\,\Delta S$ a large positive value; thus ΔG is negative.

Nonpolar Gases Are Poorly Soluble in Water

The molecules of the biologically important gases CO_2, O_2, and N_2 are nonpolar. In O_2 and N_2, electrons are shared equally by both atoms. In CO_2, each C=O bond is polar, but the two dipoles are oppositely directed and cancel each other (Table 4–3). The movement of molecules from the disordered gas phase into aqueous solution constrains their motion and the motion of water molecules and therefore represents a decrease in entropy. The nonpolar nature of these gases and the decrease in entropy when they enter solution combine to make them very poorly soluble in water (Table 4–3). Some organisms have water-soluble carrier proteins (hemoglobin and myoglobin, for example) that facilitate the transport of O_2. Carbon dioxide forms carbonic acid (H_2CO_3) in aqueous solution and is transported as the HCO_3^- (bicarbonate) ion, either free—bicarbonate is very soluble in water (~100 g/L at 25 °C)—or bound to hemoglobin.

Two other gases, NH_3 and H_2S, also have biological roles in some organisms; these gases are polar and dissolve readily in water.

table 4–3

Solubilities of Some Gases in Water

Gas	Structure*	Polarity	Solubility in water (g/L)†
Nitrogen	N≡N	Nonpolar	0.018 (40 °C)
Oxygen	O=O	Nonpolar	0.035 (50 °C)
Carbon dioxide	$\overset{\delta^-}{\longleftarrow}\ \overset{\delta^-}{\longrightarrow}$ O=C=O	Nonpolar	0.97 (45 °C)
Ammonia	H H H / N, dipole arrow δ^-	Polar	900 (10 °C)
Hydrogen sulfide	H H / S, dipole arrow δ^-	Polar	1,860 (40 °C)

*The arrows represent electric dipoles; there is a partial negative charge (δ^-) at the head of the arrow, a partial positive charge (δ^+; not shown here) at the tail.

†Note that polar molecules dissolve far better even at low temperatures than do nonpolar molecules at relatively high temperatures.

Nonpolar Compounds Force Energetically Unfavorable Changes in the Structure of Water

When water is mixed with benzene or hexane, two phases form; neither liquid is soluble in the other. Nonpolar compounds such as benzene and hexane are hydrophobic—they are unable to undergo energetically favorable interactions with water molecules, and they actually interfere with the hydrogen bonding among water molecules. All molecules or ions in aqueous solution interfere with the hydrogen bonding of some water molecules in their immediate vicinity, but polar or charged solutes (such as NaCl) compensate for lost water-water hydrogen bonds by forming new solute-water interactions. The net change in enthalpy (ΔH) for dissolving these solutes is generally small. Hydrophobic solutes, however, offer no such compensation, and their addition to water may therefore result in a small gain of enthalpy; the breaking of hydrogen bonds between water molecules takes up

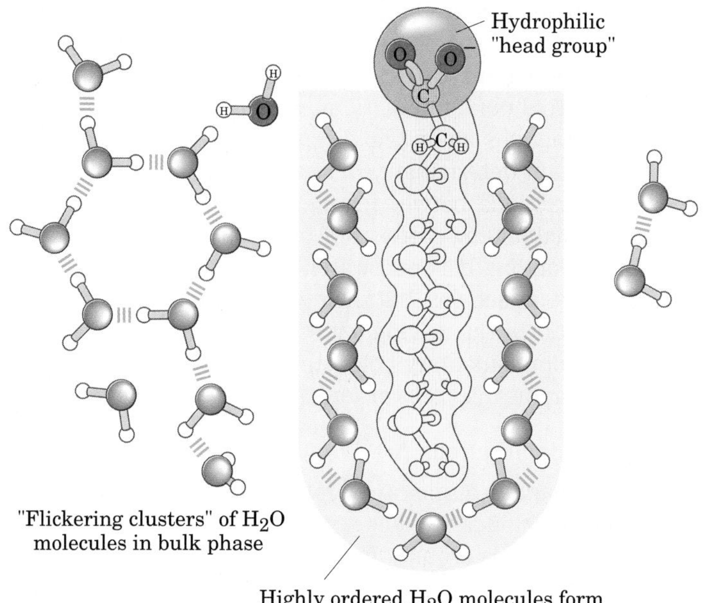

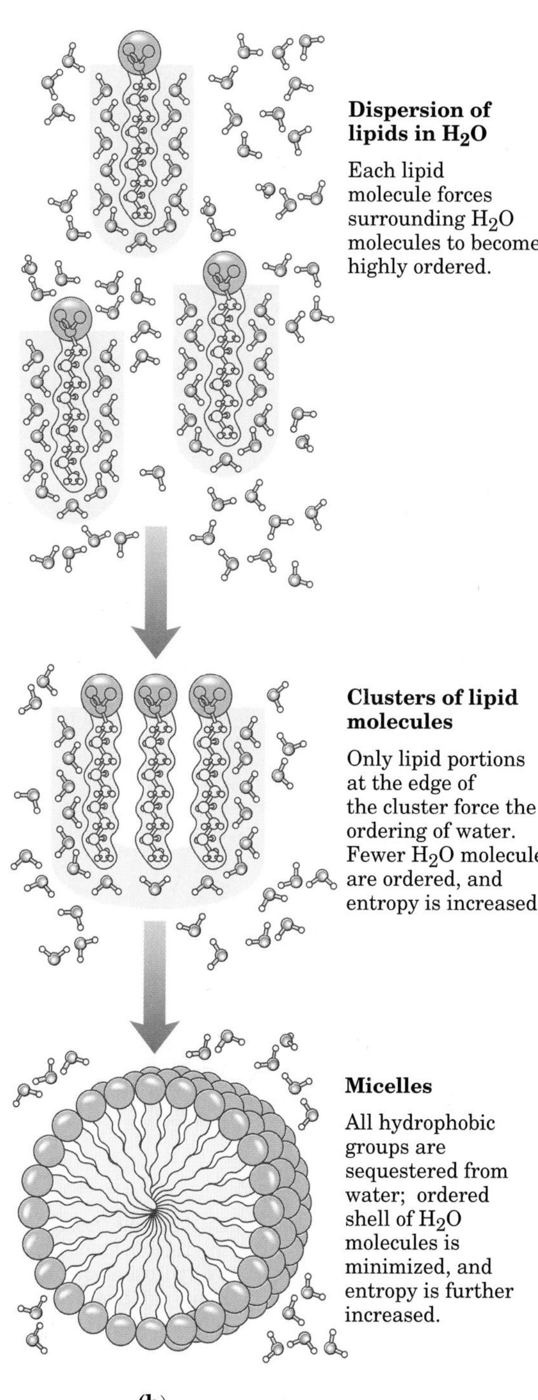

Dispersion of lipids in H_2O
Each lipid molecule forces surrounding H_2O molecules to become highly ordered.
Clusters of lipid molecules
Only lipid portions at the edge of the cluster force the ordering of water. Fewer H_2O molecules are ordered, and entropy is increased.
Micelles
All hydrophobic groups are sequestered from water; ordered shell of H_2O molecules is minimized, and entropy is further increased.
(b)

figure 4–7
Amphipathic compounds in aqueous solution. (a) Long-chain fatty acids have very hydrophobic alkyl chains, each of which is surrounded by a layer of highly ordered water molecules. **(b)** By clustering together in micelles, the fatty acid molecules expose the smallest possible hydrophobic surface area to the water, and fewer water molecules are required in the shell of ordered water. The energy gained by freeing immobilized water molecules stabilizes the micelle.

energy from the system. Furthermore, dissolving hydrophobic compounds in water produces a measurable decrease in entropy. Water molecules in the immediate vicinity of a nonpolar solute are constrained in their possible orientations as they form a highly ordered cagelike shell around each solute molecule. These water molecules are not as highly ordered as those in the crystalline compound of a nonpolar solute and water (a **clathrate**), but the effect is the same in both cases: the ordering of water molecules reduces entropy. The number of ordered water molecules, and therefore the magnitude of the entropy decrease, is proportional to the surface area of the hydrophobic solute enclosed within the cage of water molecules. The free-energy change for dissolving a nonpolar solute in water is thus unfavorable: $\Delta G = \Delta H - T\,\Delta S$, where ΔH has a positive value, ΔS has a negative value, and ΔG is positive.

Amphipathic compounds contain regions that are polar (or charged) and regions that are nonpolar (Table 4–2). When an amphipathic compound is mixed with water, the polar, hydrophilic region interacts favorably with the solvent and tends to dissolve, but the nonpolar, hydrophobic region tends to avoid contact with the water (Fig. 4–7a). The nonpolar regions of the molecules cluster together to present the smallest hydrophobic area to the aqueous solvent, and the polar regions are arranged to maximize their interaction with the solvent (Fig. 4–7b). These stable structures of amphipathic compounds in water, called **micelles,** may contain hundreds or thousands of molecules. The forces that hold the nonpolar regions of the molecules together are called **hydrophobic interactions.** The strength of hydrophobic interactions is not due to any intrinsic attraction between nonpolar moieties. Rather, it results from the system's achieving greatest thermodynamic stability by minimizing the number of ordered water molecules required to surround hydrophobic portions of the solute molecules.

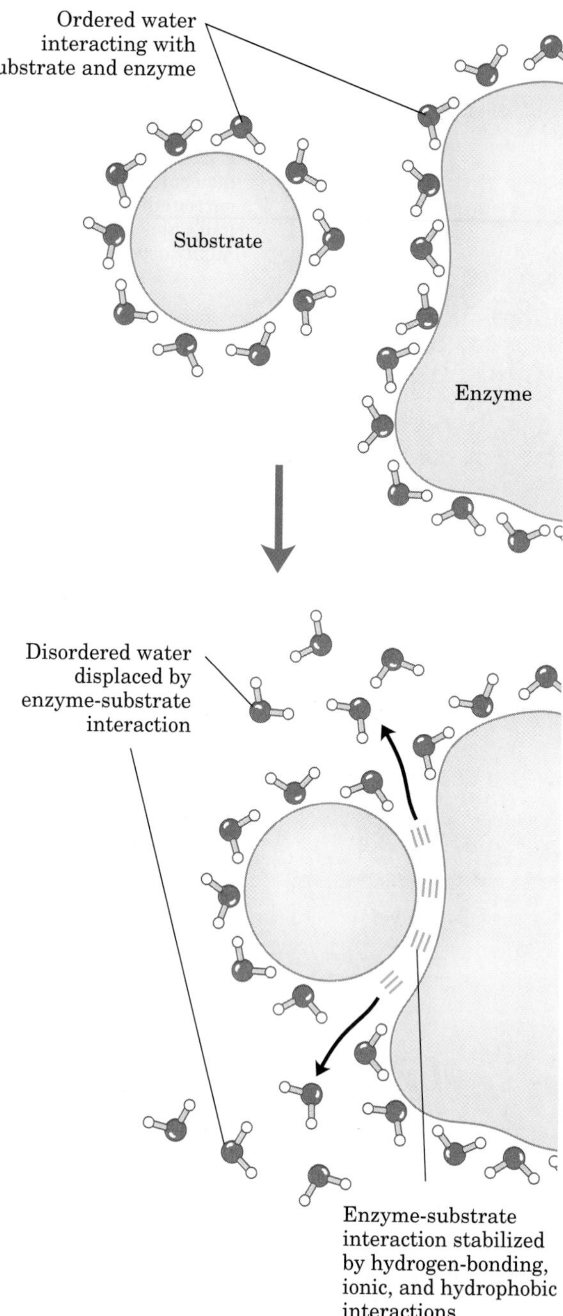

figure 4–8

Release of ordered water favors formation of an enzyme-substrate complex. While separate, both enzyme and substrate force neighboring water molecules into an ordered shell. Binding of substrate to enzyme releases some of the ordered water, and the resulting increase in entropy provides a thermodynamic push toward formation of the enzyme-substrate complex.

Many biomolecules are amphipathic; proteins, pigments, certain vitamins, and the sterols and phospholipids of membranes all have polar and nonpolar surface regions. Structures composed of these molecules are stabilized by hydrophobic interactions among the nonpolar regions. Hydrophobic interactions among lipids, and between lipids and proteins, are the most important determinants of structure in biological membranes. Hydrophobic interactions between nonpolar amino acids also stabilize the three-dimensional folding patterns of proteins.

Hydrogen bonding between water and polar solutes also causes some ordering of water molecules, but the effect is less significant than with nonpolar solutes. Part of the driving force for binding of a polar substrate (reactant) to the complementary polar surface of an enzyme is the entropy increase as the enzyme displaces ordered water from the substrate (Fig. 4–8).

Van der Waals Interactions Are Weak Interatomic Attractions

When two uncharged atoms are brought very close together, their surrounding electron clouds influence each other. Random variations in the positions of the electrons around one nucleus may create a transient electric dipole, which induces a transient, opposite electric dipole in the nearby atom. The two dipoles weakly attract each other, bringing the two nuclei closer. These weak attractions are called **van der Waals interactions.** As the two nuclei draw closer together, their electron clouds begin to repel each other. At the point when the van der Waals attraction exactly balances this repulsive force, the nuclei are said to be in van der Waals contact. Each atom has a characteristic **van der Waals radius,** a measure of how close that atom will allow another to approach (see Table 3–1). In the "space-filling" molecular models shown throughout this book (e.g., Fig. 3–7c) the atoms are depicted in sizes proportional to their van der Waals radii.

Weak Interactions Are Crucial to Macromolecular Structure and Function

The noncovalent interactions we have described (hydrogen bonds and ionic, hydrophobic, and van der Waals interactions) (Table 4–4) are much weaker than covalent bonds. An input of about 350 kJ of energy is required to break a mole of (6×10^{23}) C—C single bonds, and about 410 kJ to break a mole of C—H bonds, but as little as 4 kJ is sufficient to disrupt a mole of typical van der Waals interactions. Hydrophobic interactions are also much weaker than covalent bonds, although they are substantially strengthened by a highly polar solvent (a concentrated salt solution, for example). Ionic interactions and hydrogen bonds are variable in strength, depending on the polarity of the solvent, but they are always significantly weaker than covalent bonds. In aqueous solvent at 25 °C, the available thermal energy can be of the same order of magnitude as the strength of these weak interactions, and the interaction between solute and solvent (water) molecules is nearly as favorable as solute-solute interactions. Consequently, hydrogen bonds and ionic, hydrophobic, and van der Waals interactions are continually formed and broken.

Although these four types of interactions are individually weak relative to covalent bonds, the cumulative effect of many such interactions with a protein or nucleic acid can be very significant. For example, the noncovalent binding of an enzyme to its substrate may involve several hydrogen bonds and one or more ionic interactions, as well as hydrophobic and van der Waals interactions. The formation of each of these weak bonds contributes to a net decrease in the free energy of the system. The stability of a noncovalent interaction such as that of a small molecule hydrogen-bonded to its macromolecular partner is calculable from the binding energy. Stabil-

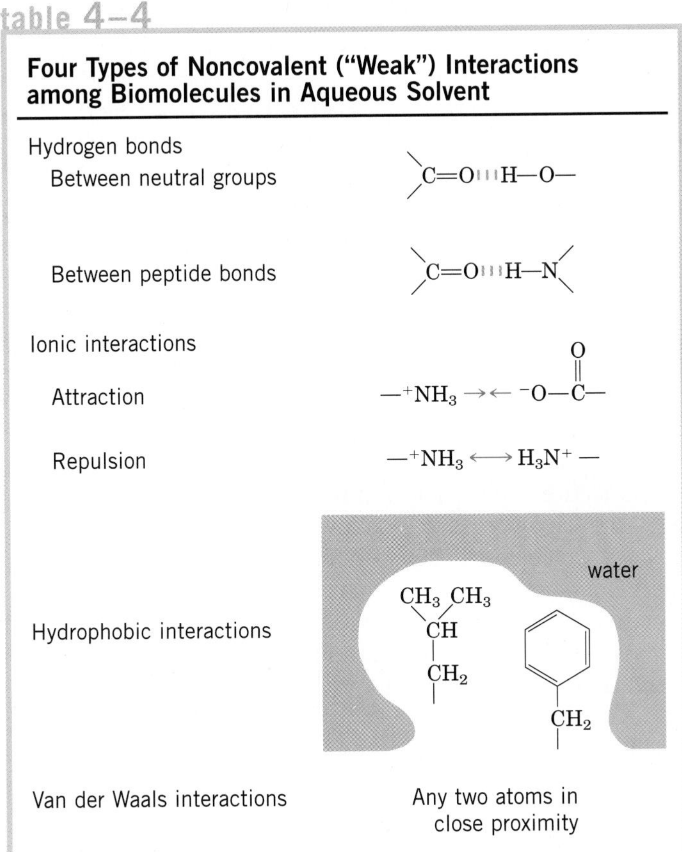

table 4–4

Four Types of Noncovalent ("Weak") Interactions among Biomolecules in Aqueous Solvent

Hydrogen bonds	
Between neutral groups	$>C{=}O \cdots H{-}O{-}$
Between peptide bonds	$>C{=}O \cdots H{-}N<$
Ionic interactions	
Attraction	$-{}^{+}NH_3 \rightarrow\leftarrow {}^{-}O{-}C(=O){-}$
Repulsion	$-{}^{+}NH_3 \longleftrightarrow H_3N^{+}-$
Hydrophobic interactions	water; CH_3 CH_3 / CH / CH_2; CH_2
Van der Waals interactions	Any two atoms in close proximity

ity, as measured by the equilibrium constant (see below) of the binding reaction, varies *exponentially* with binding energy. The dissociation of two biomolecules associated noncovalently by multiple weak interactions (such as an enzyme and its bound substrate) requires all these interactions to be disrupted at the same time. Because the interactions fluctuate randomly, such simultaneous disruptions are very unlikely. The molecular stability bestowed by two or five or 20 weak interactions is therefore much greater than would be expected intuitively from a simple summation of small binding energies.

Macromolecules such as proteins, DNA, and RNA contain so many sites of potential hydrogen bonding or ionic, van der Waals, or hydrophobic interactions that the cumulative effect of the many small binding forces is enormous. For macromolecules, the most stable (native) structure is usually that in which weak-bonding possibilities are maximized. The folding of a single polypeptide or polynucleotide chain into its three-dimensional shape is determined by this principle. The binding of an antigen to a specific antibody depends on the cumulative effects of many weak interactions. As noted earlier, the energy released when an enzyme binds noncovalently to its substrate is the main source of the enzyme's catalytic power. The binding of a hormone or a neurotransmitter to its cellular receptor protein is the result of weak interactions. One consequence of the large size of enzymes and receptors is that their extensive surfaces provide many opportunities for weak interactions. At the molecular level, the complementarity between interacting biomolecules reflects the complementarity and weak interactions between polar, charged, and hydrophobic groups on the surfaces of the molecules.

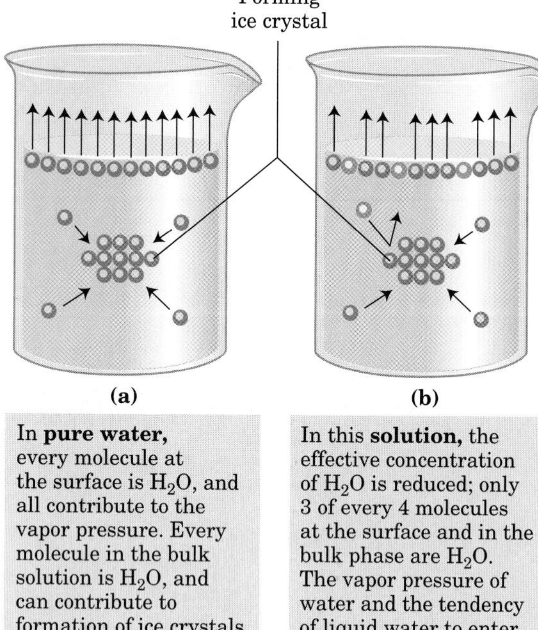

= H_2O
= Solute
Forming ice crystal
(a)
(b)
In **pure water,** every molecule at the surface is H_2O, and all contribute to the vapor pressure. Every molecule in the bulk solution is H_2O, and can contribute to formation of ice crystals.
In this **solution,** the effective concentration of H_2O is reduced; only 3 of every 4 molecules at the surface and in the bulk phase are H_2O. The vapor pressure of water and the tendency of liquid water to enter a crystal are reduced proportionately.

figure 4–9

Solutes alter the colligative properties of aqueous solutions. (a) At 101 kPa (1 atm) pressure, pure water boils at 100 °C and freezes at 0 °C. **(b)** The presence of solute molecules reduces the probability of a water molecule leaving the solution and entering the gas phase, thereby reducing the vapor pressure of the solution and increasing the boiling point. Similarly, the probability of a water molecule colliding with and joining a forming ice crystal is reduced when some of the molecules colliding with the crystal are solute, not water, molecules. The effect is depression of the freezing point.

Solutes Affect the Colligative Properties of Aqueous Solutions

Dissolved solutes of all kinds alter certain physical properties of the solvent, water: its vapor pressure, boiling point, melting point (freezing point), and osmotic pressure. These are called **colligative** ("tied together") **properties** because the effect of solutes on all four properties has the same basis: the concentration of water is lower in solutions than in pure water. The effect of solute concentration on the colligative properties of water is independent of the chemical properties of the solute; it depends only on the *number* of solute particles (molecules, ions) in a given amount of water. A compound such as NaCl, which dissociates in solution, has twice the effect on osmotic pressure, for example, as an equal number of moles of a nondissociating solute such as glucose.

Dissolved solutes alter the colligative properties of aqueous solutions by lowering the effective concentration of water. For example, when a significant fraction of the molecules at the surface of an aqueous solution are not water but solute, the tendency of water molecules to escape into the vapor phase—the vapor pressure—is lowered (Fig. 4–9). Similarly, the tendency of water molecules to move from the aqueous phase to the surface of a forming ice crystal is reduced when some of the molecules that collide with the crystal are solute, not water. In that case, the solution will freeze more slowly than pure water and at a lower temperature. For a 1.00 molal aqueous solution (1.00 mol of solute per 1,000 g water) of an ideal, nonvolatile, and nondissociating solute at 101 kPa (1 atm) of pressure, the freezing point is 1.86 °C lower and the boiling point is 0.543 °C higher than for pure water. For 0.100 molal solutions of the same solute, the changes are one-tenth as large.

Water molecules tend to move from a region of higher water concentration to one of lower water concentration. When two different aqueous solutions are separated by a semipermeable membrane (one that allows the passage of water but not solute molecules), water molecules diffusing from the region of higher water concentration to that of lower water concentration produce osmotic pressure (Fig. 4–10). This pressure, Π, measured as the force necessary to resist water movement (Fig. 4–10c), is approximated by the van't Hoff equation:

$$\Pi = icRT$$

in which R is the gas constant and T is the absolute temperature. The term ic is the **osmolarity** of the solution, the product of the solute's molar concentration c and the van't Hoff factor i, which is a measure of the extent to

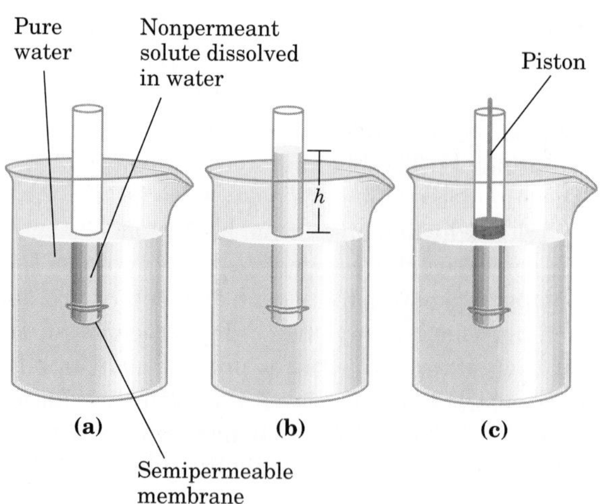

figure 4–10

Osmosis and the measurement of osmotic pressure. (a) The initial state. The tube contains an aqueous solution, the beaker contains pure water, and the semipermeable membrane allows the passage of water but not solute. Water flows from the beaker into the tube to equalize its concentration across the membrane. **(b)** The final state. Water has moved into the solution of the nonpermeant compound, diluting it and raising the column of water within the tube. At equilibrium, the force of gravity operating on the solution in the tube exactly balances the tendency of water to move into the tube, where its concentration is lower. **(c)** Osmotic pressure (Π) is measured as the force that must be applied to return the solution in the tube to the level of that in the beaker. This force is proportional to the height, h, of the column in **(b).**

which the solute dissociates into two or more ionic species. In dilute NaCl solutions, the solute completely dissociates into Na^+ and Cl^-, doubling the number of solute particles, and $i = 2$. For nonionizing solutes, i is always 1. For solutions of several (n) solutes, Π is the sum of the contributions of each species:

$$\Pi = RT(i_1c_1 + i_2c_2 + \cdots + i_nc_n)$$

Osmosis, water movement across a semipermeable membrane driven by differences in osmotic pressure, is an important factor in the life of most cells. Plasma membranes are more permeable to water than to most other small molecules, ions, and macromolecules. This permeability is due partly to simple diffusion of water through the lipid bilayer and partly to protein channels (aquaporins) in the membrane that selectively permit the passage of water. Solutions of equal osmolarity are said to be **isotonic.** Surrounded by an isotonic solution, a cell neither gains nor loses water (Fig. 4–11). In a **hypertonic** solution, one with higher osmolarity than the cytosol, the cell shrinks as water flows out. In **hypotonic** solution (of lower osmolarity), the cell swells and, if unsupported by a cell wall, eventually bursts. Cells generally contain higher concentrations of biomolecules and ions than their surroundings, so osmotic pressure tends to drive water into cells. If not somehow counterbalanced, this inward movement of water would distend the plasma membrane and eventually cause explosion of the cell (osmotic lysis).

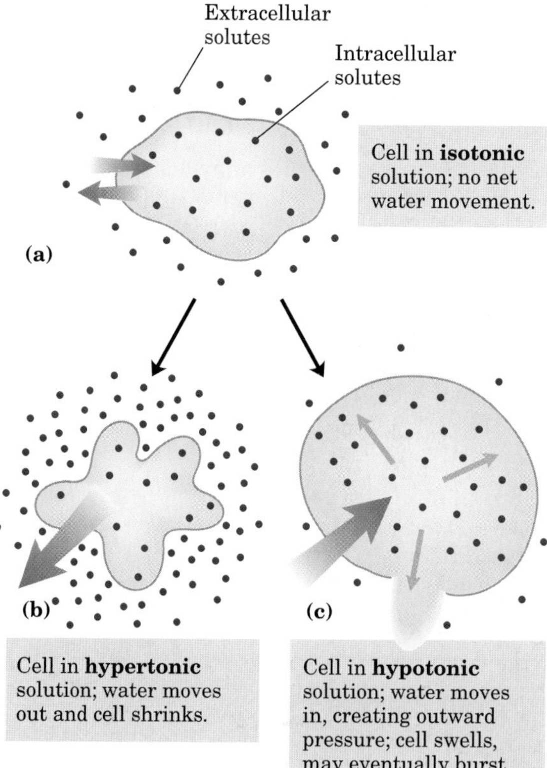
Extracellular solutes
Intracellular solutes
Cell in **isotonic** solution; no net water movement.
(a)
(b)
(c)
Cell in **hypertonic** solution; water moves out and cell shrinks.
Cell in **hypotonic** solution; water moves in, creating outward pressure; cell swells, may eventually burst.

figure 4–11
The effect of extracellular osmolarity on water movement across a plasma membrane. When a cell in osmotic balance with its surrounding medium (that is, in an isotonic medium) **(a)** is transferred into a hypertonic solution **(b)** or hypotonic solution **(c)**, water moves across the plasma membrane in the direction that tends to equalize osmolarity outside and inside the cell.

Three mechanisms have evolved to prevent this catastrophe. In bacteria and plants, the plasma membrane is surrounded by a nonexpandable cell wall of sufficient rigidity and strength to resist osmotic pressure and prevent osmotic lysis. Certain freshwater protozoans, which live in a highly hypotonic medium, have an organelle (contractile vacuole) that pumps water out of the cell. In multicellular animals, blood plasma and interstitial fluid (the extracellular fluid of tissues) are maintained at an osmolarity close to that of the cytosol. The high concentration of albumin and other proteins in blood plasma contributes to its osmolarity. Cells also actively pump out ions such as Na^+ into the interstitial fluid to stay in osmotic balance with their surroundings.

Because the effect of solutes on osmolarity depends on the *number* of dissolved particles, not their *masses*, macromolecules (proteins, nucleic acids, polysaccharides) have far less effect on the osmolarity of a solution than would an equal mass of their monomeric components. For example, a *gram* of a polysaccharide composed of 1,000 glucose units has the same effect on osmolarity as a *milligram* of glucose. One effect of storing fuel as polysaccharides (starch or glycogen) rather than as glucose or other simple sugars is prevention of an enormous increase in osmotic pressure within the storage cell.

Plants use osmotic pressure to achieve mechanical rigidity. The very high solute concentration in the vacuole draws water into the cell (see Fig. 2–10). The resulting osmotic pressure against the cell wall (turgor pressure) stiffens the cell, the tissue, and the plant body. When the lettuce in your salad wilts, it is because loss of water has reduced turgor pressure. Dramatic alterations in turgor pressure produce the movement of plant parts seen in touch-sensitive plants such as the Venus flytrap and mimosa (Box 4–1).

Osmosis also has consequences for laboratory protocols. Mitochondria, chloroplasts, and lysosomes, for example, are bounded by semipermeable membranes. In isolating these organelles from broken cells (see Fig. 2–20), biochemists must perform the fractionations in isotonic solutions. Buffers used in cellular fractionations commonly contain sufficient concentrations (about 0.2 M) of sucrose or some other inert solute to protect the organelles from osmotic lysis.

box 4–1 Touch Response in Plants: An Osmotic Event

The highly specialized leaves of the Venus flytrap (*Dionaea muscipula*) rapidly fold together in response to a light touch by an unsuspecting insect, entrapping the insect for later digestion. Attracted by nectar on the leaf surface, the insect touches three mechanically sensitive hairs, triggering the traplike closing of the leaf (Fig. 1). This leaf movement is produced by sudden (within 0.5 s) changes of turgor pressure in mesophyll cells (the inner cells of the leaf), probably achieved by the release of K^+ ions from the cells and the resulting efflux, by osmosis, of water. Digestive glands in the leaf's surface release enzymes that extract nutrients from the insect.

The sensitive plant (*Mimosa pudica*) also undergoes a remarkable change in leaf shape triggered by mechanical touch (Fig. 2). A light touch or vibration produces a sudden drooping of the leaves, a result of a dramatic reduction in turgor pressure in cells at the base of each leaflet and leaf. As in the Venus flytrap, the drop in turgor pressure results from K^+ release followed by the efflux of water.

(a)

(b)

figure 1
Touch response in the Venus flytrap. A fly approaching an open leaf **(a)** is trapped for digestion by the plant **(b)**.

(a)

(b)

figure 2
The feathery leaflets of the sensitive plant **(a)** close and drop **(b)** to protect the plant from structural damage by wind.

Ionization of Water, Weak Acids, and Weak Bases

Although many of the solvent properties of water can be explained in terms of the uncharged H_2O molecule, the small degree of ionization of water to hydrogen ions (H^+) and hydroxide ions (OH^-) must also be taken into account. Like all reversible reactions, the ionization of water can be described by an equilibrium constant. When weak acids are dissolved in water, they contribute H^+ by ionizing; bases consume H^+ by being protonated. These processes are also governed by equilibrium constants. The total hydrogen ion concentration from all sources is experimentally measurable, and is expressed as the pH of the solution. To predict the state of ionization of solutes in water, we must take into account the relevant equilibrium constants for each ionization reaction. We therefore turn now to a brief discussion of the ionization of water, and of weak acids and bases dissolved in water.

Pure Water Is Slightly Ionized

Water molecules have a slight tendency to undergo reversible ionization to yield a hydrogen ion (proton) and a hydroxide ion, giving the equilibrium

$$H_2O \rightleftharpoons H^+ + OH^- \qquad (4\text{–}1)$$

Although we commonly show the dissociation product of water as H^+, free protons do not exist in solution; hydrogen ions formed in water are immediately hydrated to **hydronium ions** (H_3O^+). Hydrogen bonding between water molecules makes the hydration of dissociating protons virtually instantaneous:

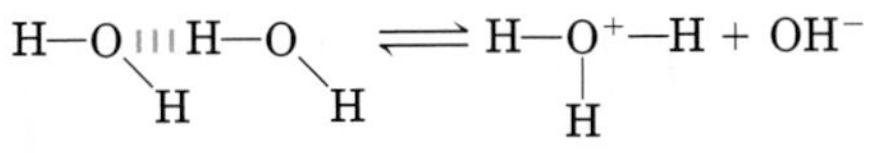

$$\mathrm{H{-}O{\cdots}H{-}O{-}H} \rightleftharpoons \mathrm{H{-}O^+(H){-}H} + OH^-$$

The ionization of water can be measured by its electrical conductivity; pure water carries electrical current as H^+ migrates toward the cathode and OH^- toward the anode. The movement of hydronium and hydroxide ions in the electric field is anomalously fast compared with that of other ions such as Na^+, K^+, and Cl^-. This high ionic mobility results from the kind of "proton hopping" shown in Figure 4–12. No individual proton moves very far through the bulk solution, but a series of proton hops between hydrogen-bonded water molecules causes the net movement of a proton over a long distance in a remarkably short time. As a result of the high ionic mobility of H^+ (and of OH^-, which also moves rapidly by proton hopping, but in the opposite direction), acid-base reactions in aqueous solutions are generally exceptionally fast. Proton hopping very likely also plays a role in biological proton transfer reactions.

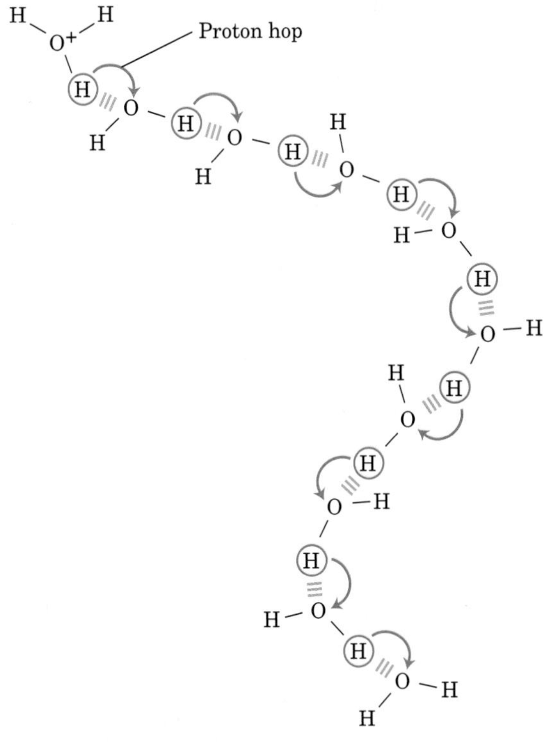

Proton hop

figure 4–12
Proton hopping. Short "hops" of protons between a series of hydrogen-bonded water molecules effects an extremely rapid net movement of a proton over a long distance. As a hydronium ion (upper left) gives up a proton, a water molecule some distance away (lower right) acquires one, becoming a hydronium ion. Proton hopping is much faster than true diffusion and explains the remarkably high ionic mobility of hydrogen ions compared with other monovalent cations such as Na^+ or K^+.

Because reversible ionization is crucial to the role of water in cellular function, we must have a means of expressing the extent of ionization of water in quantitative terms. A brief review of some properties of reversible chemical reactions will show how this can be done.

The position of equilibrium of any chemical reaction is given by its **equilibrium constant, K_{eq}** (sometimes expressed simply as K). For the generalized reaction

$$A + B \rightleftharpoons C + D \tag{4–2}$$

an equilibrium constant can be defined in terms of the concentrations of reactants (A and B) and products (C and D) at equilibrium:

$$K_{eq} = \frac{[C][D]}{[A][B]}$$

Strictly speaking, the concentration terms should be the *activities*, or effective concentrations in nonideal solutions, of each species. Except in very accurate work, the equilibrium constant may be approximated by measuring the *concentrations* at equilibrium. For reasons beyond the scope of this discussion, equilibrium constants are dimensionless. However, we have generally retained the concentration units (M) in the equilibrium expressions used in this book to remind you that molarity is the unit of concentration used in calculating K_{eq}.

The equilibrium constant is fixed and characteristic for any given chemical reaction at a specified temperature. It defines the composition of the final equilibrium mixture, regardless of the starting amounts of reactants and products. Conversely, one can calculate the equilibrium constant for a given reaction at a given temperature if the equilibrium concentrations of all its reactants and products are known. As we will show in Chapter 14, the standard free-energy change ($\Delta G°$) is directly related to K_{eq}.

The Ionization of Water Is Expressed by an Equilibrium Constant

The degree of ionization of water at equilibrium (Eqn 4–1) is small; at 25 °C only about one of every 10^7 molecules in pure water is ionized at any instant. The equilibrium constant for the reversible ionization of water (Eqn 4–1) is

$$K_{eq} = \frac{[H^+][OH^-]}{[H_2O]} \tag{4–3}$$

In pure water at 25 °C, the concentration of water is 55.5 M (grams of H_2O in 1 L divided by its gram molecular weight: (1,000 g/L)/(18.015 g/mol)) and is essentially constant in relation to the very low concentrations of H^+ and OH^-, namely, 1×10^{-7} M. Accordingly, we can substitute 55.5 M in the equilibrium constant expression (Eqn 4–3) to yield

$$K_{eq} = \frac{[H^+][OH^-]}{55.5 \text{ M}},$$

which, on rearranging, becomes

$$(55.5 \text{ M})(K_{eq}) = [H^+][OH^-] = K_w \tag{4–4}$$

where K_w designates the product $(55.5 \text{ M})(K_{eq})$, the **ion product of water** at 25 °C.

The value for K_{eq}, determined by electrical-conductivity measurements of pure water, is 1.8×10^{-16} M at 25 °C. Substituting this value for K_{eq} in Equation 4–4 gives the ion product of water:

$$K_w = [H^+][OH^-] = (55.5 \text{ M})(1.8 \times 10^{-16} \text{ M}) = 1.0 \times 10^{-14} \text{ M}^2$$

box 4–2 The Ion Product of Water: Two Illustrative Problems

The ion product of water makes it possible to calculate the concentration of H^+, given the concentration of OH^-, and vice versa; the following problems demonstrate this.

1. What is the concentration of H^+ in a solution of 0.1 M NaOH?

$$K_w = [H^+][OH^-]$$

Solving for $[H^+]$ gives

$$[H^+] = \frac{K_w}{[OH^-]} = \frac{1 \times 10^{-14}\ \text{M}^2}{0.1\ \text{M}} = \frac{10^{-14}\ \text{M}^2}{10^{-1}\ \text{M}}$$
$$= 10^{-13}\ \text{M} \quad (\textit{answer})$$

2. What is the concentration of OH^- in a solution in which the H^+ concentration is 0.00013 M?

$$K_w = [H^+][OH^-]$$

Solving for $[OH^-]$ gives

$$[OH^-] = \frac{K_w}{[H^+]} = \frac{1 \times 10^{-14}\ \text{M}^2}{0.00013\ \text{M}} = \frac{10^{-14}\ \text{M}^2}{1.3 \times 10^{-4}\ \text{M}}$$
$$= 7.7 \times 10^{-11}\ \text{M} \quad (\textit{answer})$$

Thus the product $[H^+][OH^-]$ in aqueous solutions at 25 °C always equals 1×10^{-14} M^2. When there are exactly equal concentrations of both H^+ and OH^-, as in pure water, the solution is said to be at **neutral pH.** At this pH, the concentration of H^+ and OH^- can be calculated from the ion product of water as follows:

$$K_w = [H^+][OH^-] = [H^+]^2$$

Solving for $[H^+]$ gives

$$[H^+] = \sqrt{K_w} = \sqrt{1 \times 10^{-14}\ \text{M}^2}$$
$$[H^+] = [OH^-] = 10^{-7}\ \text{M}$$

As the ion product of water is constant, whenever $[H^+]$ is greater than 1×10^{-7} M, $[OH^-]$ must become less than 1×10^{-7} M, and vice versa. When $[H^+]$ is very high, as in a solution of hydrochloric acid, $[OH^-]$ must be very low. From the ion product of water we can calculate $[H^+]$ if we know $[OH^-]$, and vice versa (Box 4–2).

The pH Scale Designates the H^+ and OH^- Concentrations

The ion product of water, K_w, is the basis for the **pH scale** (Table 4–5). It is a convenient means of designating the concentration of H^+ (and thus of OH^-) in any aqueous solution in the range between 1.0 M H^+ and 1.0 M OH^-. The term **pH** is defined by the expression

$$\text{pH} = \log \frac{1}{[H^+]} = -\log\,[H^+]$$

The symbol p denotes "negative logarithm of." For a precisely neutral solution at 25 °C, in which the concentration of hydrogen ions is 1.0×10^{-7} M, the pH can be calculated as follows:

$$\text{pH} = \log \frac{1}{1.0 \times 10^{-7}} = \log\,(1.0 \times 10^{7}) = \log 1.0 + \log 10^{7} = 0 + 7 = 7$$

The value of 7 for the pH of a precisely neutral solution is not an arbitrarily chosen figure; it is derived from the absolute value of the ion product of water at 25 °C, which by convenient coincidence is a round number. Solutions

table 4–5

The pH Scale

$[H^+]$ (M)	pH	$[OH^-]$ (M)	pOH*
10^0 (1)	0	10^{-14}	14
10^{-1}	1	10^{-13}	13
10^{-2}	2	10^{-12}	12
10^{-3}	3	10^{-11}	11
10^{-4}	4	10^{-10}	10
10^{-5}	5	10^{-9}	9
10^{-6}	6	10^{-8}	8
10^{-7}	7	10^{-7}	7
10^{-8}	8	10^{-6}	6
10^{-9}	9	10^{-5}	5
10^{-10}	10	10^{-4}	4
10^{-11}	11	10^{-3}	3
10^{-12}	12	10^{-2}	2
10^{-13}	13	10^{-1}	1
10^{-14}	14	10^0 (1)	0

*The expression pOH is sometimes used to describe the basicity, or OH^- concentration, of a solution; pOH is defined by the expression pOH $= -\log\,[OH^-]$, which is analogous to the expression for pH. Note that in all cases, pH + pOH = 14.

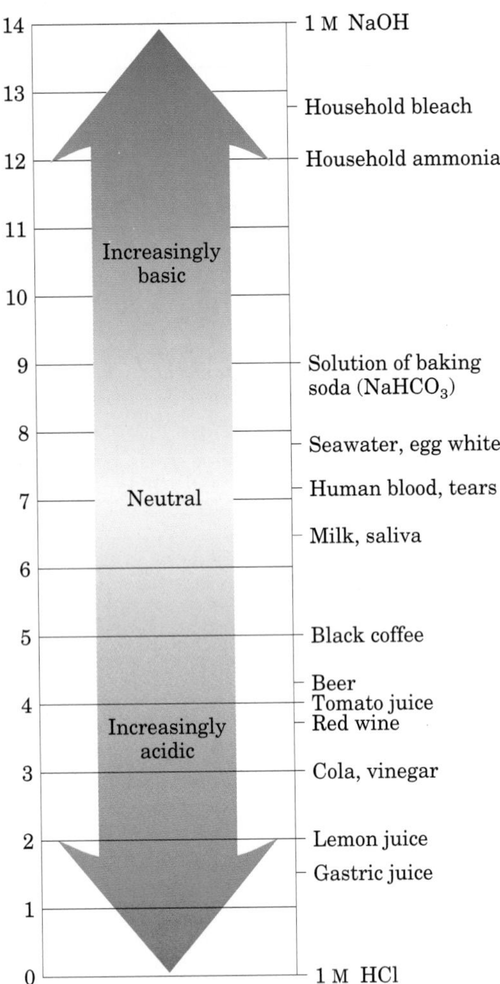

figure 4–13
The pH of some aqueous fluids.

having a pH greater than 7 are alkaline or basic; the concentration of OH^- is greater than that of H^+. Conversely, solutions having a pH less than 7 are acidic.

Note that the pH scale is logarithmic, not arithmetic. To say that two solutions differ in pH by 1 pH unit means that one solution has ten times the H^+ concentration of the other, but it does not tell us the absolute magnitude of the difference. Figure 4–13 gives the pH of some common aqueous fluids. A cola drink (pH 3.0) or red wine (pH 3.7) has an H^+ concentration approximately 10,000 times that of blood (pH 7.4).

The pH of an aqueous solution can be approximately measured using various indicator dyes, including litmus, phenolphthalein, and phenol red, which undergo color changes as a proton dissociates from the dye molecule. Accurate determinations of pH in the chemical or clinical laboratory are made with a glass electrode that is selectively sensitive to H^+ concentration but insensitive to Na^+, K^+, and other cations. In a pH meter the signal from such an electrode is amplified and compared with the signal generated by a solution of accurately known pH.

Measurement of pH is one of the most important and frequently used procedures in biochemistry. The pH affects the structure and activity of biological macromolecules; for example, the catalytic activity of enzymes is strongly dependent on pH (see Fig. 4–19). Measurements of the pH of blood and urine are commonly used in medical diagnoses. The pH of the blood plasma of severely diabetic people, for example, is often below the normal value of 7.4; this condition is called acidosis. In certain other disease states the pH of the blood is higher than normal, the condition of alkalosis.

Weak Acids and Bases Have Characteristic Dissociation Constants

Hydrochloric, sulfuric, and nitric acids, commonly called strong acids, are completely ionized in dilute aqueous solutions; the strong bases NaOH and KOH are also completely ionized. Of more interest to biochemists is the behavior of weak acids and bases—those not completely ionized when dissolved in water. These are common in biological systems and play important roles in metabolism and its regulation. The behavior of aqueous solutions of weak acids and bases is best understood if we first define some terms.

Acids may be defined as proton donors and bases as proton acceptors. A proton donor and its corresponding proton acceptor make up a **conjugate acid-base pair** (Fig. 4–14). Acetic acid (CH_3COOH), a proton donor, and the acetate anion (CH_3COO^-), the corresponding proton acceptor, constitute a conjugate acid-base pair, related by the reversible reaction

$$CH_3COOH \rightleftharpoons H^+ + CH_3COO^-$$

Each acid has a characteristic tendency to lose its proton in an aqueous solution. The stronger the acid, the greater its tendency to lose its proton. The tendency of any acid (HA) to lose a proton and form its conjugate base (A^-) is defined by the equilibrium constant (K_{eq}) for the reversible reaction

$$HA \rightleftharpoons H^+ + A^-$$

which is

$$K_{eq} = \frac{[H^+][A^-]}{[HA]} = K_a$$

Equilibrium constants for ionization reactions are usually called ionization or **dissociation constants,** often designated K_a. The dissociation constants of some acids are given in Figure 4–14. Stronger acids, such as phosphoric and carbonic acids, have larger dissociation constants; weaker acids, such as

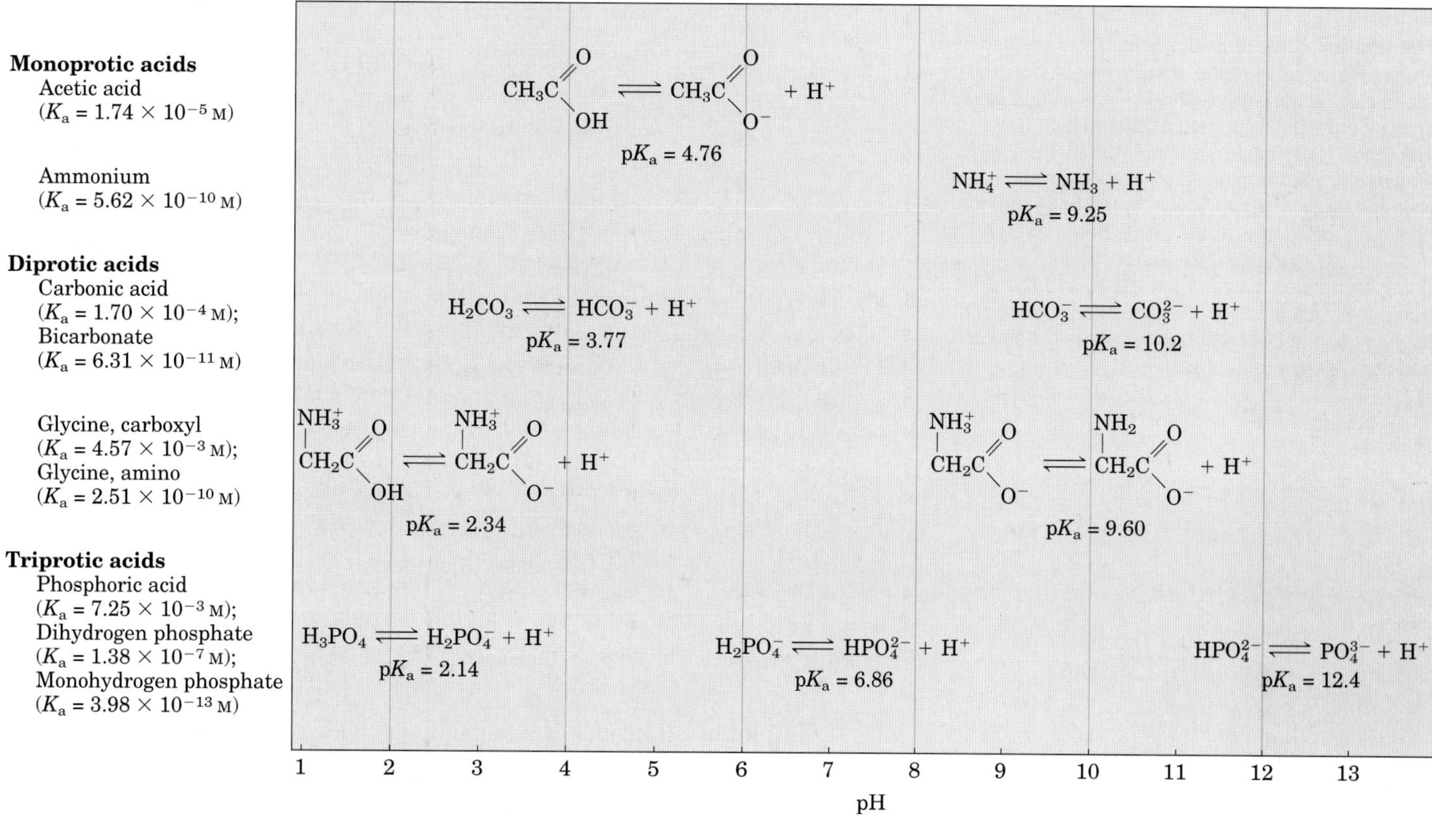

figure 4–14

Conjugate acid-base pairs consist of a proton donor and a proton acceptor. Some compounds, such as acetic acid and ammonia, are monoprotic; they can give up only one proton. Others are diprotic (H_2CO_3 and glycine) or triprotic (H_3PO_4). The dissociation reactions for each pair are shown where they occur along a pH gradient. The equilibrium or dissociation constant (K_a) and its negative logarithm, the pK_a, are shown for each reaction.

monohydrogen phosphate (HPO_4^{2-}), have smaller dissociation constants.

Also included in Figure 4–14 are values of **pK_a,** which is analogous to pH and is defined by the equation

$$pK_a = \log \frac{1}{K_a} = -\log K_a$$

The stronger the tendency to dissociate a proton, the stronger is the acid and the lower its pK_a. As we shall now see, the pK_a of any weak acid can be determined quite easily.

Titration Curves Reveal the pK_a of Weak Acids

Titration is used to determine the amount of an acid in a given solution. A measured volume of the acid is titrated with a solution of a strong base, usually sodium hydroxide (NaOH), of known concentration. The NaOH is added in small increments until the acid is consumed (neutralized), as determined with an indicator dye or a pH meter. The concentration of the acid in the original solution can be calculated from the volume and concentration of NaOH added.

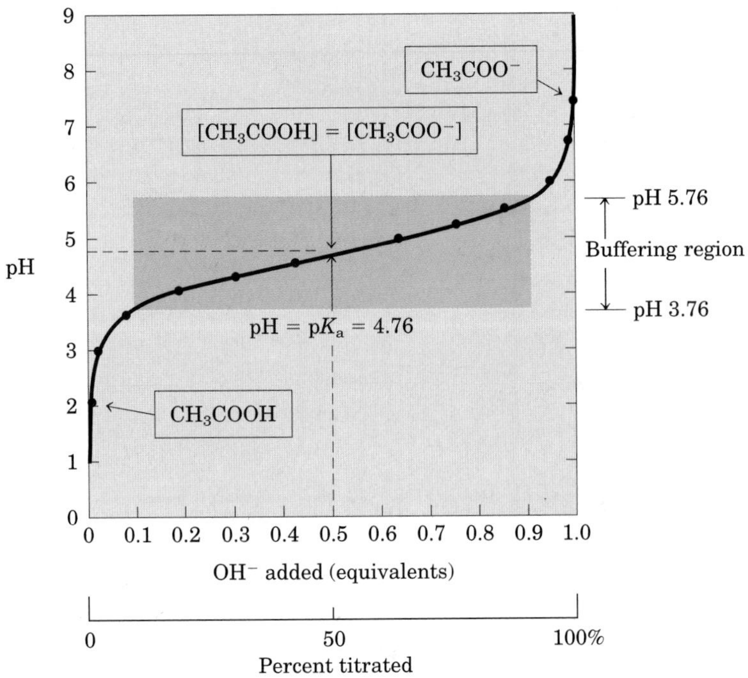

figure 4–15

The titration curve of acetic acid. After addition of each increment of NaOH to the acetic acid solution, the pH of the mixture is measured. This value is plotted against the amount of NaOH expressed as the fraction of the total amount of NaOH required to convert all the acetic acid to its deprotonated form, acetate. The points so obtained yield the titration curve. Shown in the boxes are the predominant ionic forms at the points designated. At the midpoint of the titration, the concentrations of the proton donor and proton acceptor are equal, and the pH is numerically equal to the pK_a. The shaded zone is the useful region of buffering power, generally between 10% and 90% titration of the weak acid.

A plot of pH against the amount of NaOH added (a **titration curve**) reveals the pK_a of the weak acid. Consider the titration of a 0.1 M solution of acetic acid (for simplicity denoted as HAc) with 0.1 M NaOH at 25 °C (Fig. 4–15). Two reversible equilibria are involved in the process:

$$H_2O \rightleftharpoons H^+ + OH^- \qquad (4\text{–}5)$$

$$HAc \rightleftharpoons H^+ + Ac^- \qquad (4\text{–}6)$$

The equilibria must simultaneously conform to their characteristic equilibrium constants, which are, respectively,

$$K_w = [H^+][OH^-] = 1 \times 10^{-14}\ \text{M}^2 \qquad (4\text{–}7)$$

$$K_a = \frac{[H^+][Ac^-]}{[HAc]} = 1.74 \times 10^{-5}\ \text{M} \qquad (4\text{–}8)$$

At the beginning of the titration, before any NaOH is added, the acetic acid is already slightly ionized, to an extent that can be calculated from its dissociation constant (Eqn 4–8).

As NaOH is gradually introduced, the added OH^- combines with the free H^+ in the solution to form H_2O, to an extent that satisfies the equilibrium relationship in Equation 4–7. As free H^+ is removed, HAc dissociates further to satisfy its own equilibrium constant (Eqn 4–8). The net result as the titration proceeds is that more and more HAc ionizes, forming Ac^-, as the NaOH is added. At the midpoint of the titration, at which exactly 0.5 equivalent of NaOH has been added, one-half of the original acetic acid has undergone dissociation, so that the concentration of the proton donor, [HAc], now equals that of the proton acceptor, $[Ac^-]$. At this midpoint a very important relationship holds: the pH of the equimolar solution of acetic acid and acetate is exactly equal to the pK_a of acetic acid ($pK_a = 4.76$; see Figs 4–14, 4–15). The basis for this relationship, which holds for all weak acids, will soon become clear.

As the titration is continued by adding further increments of NaOH, the remaining undissociated acetic acid is gradually converted into acetate. The end point of the titration occurs at about pH 7.0: all the acetic acid has lost its protons to OH^-, to form H_2O and acetate. Throughout the titration the

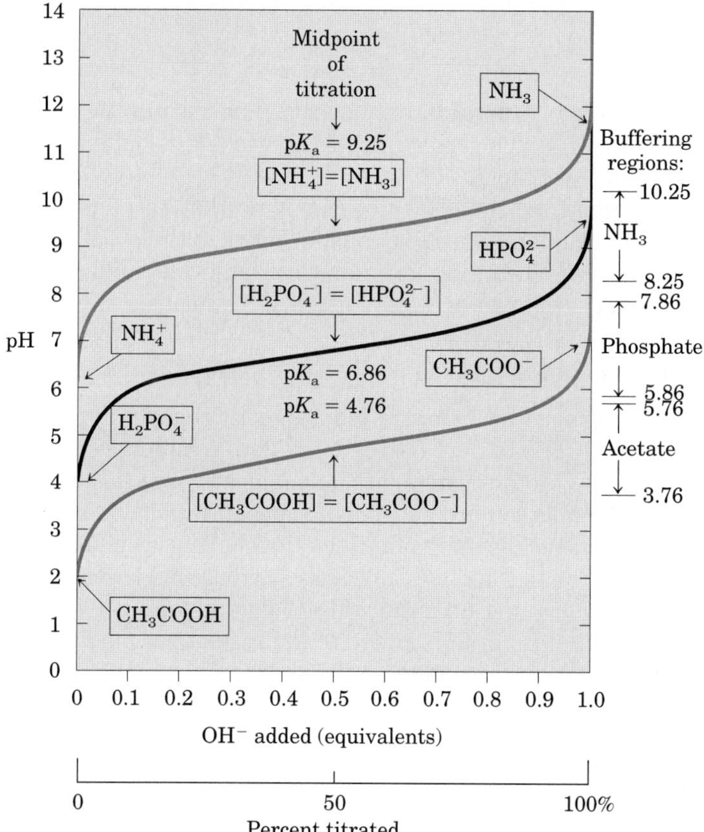

figure 4–16
Comparison of the titration curves of three weak acids, CH_3COOH, $H_2PO_4^-$, and NH_4^+. The predominant ionic forms at designated points in the titration are given in boxes. The regions of buffering capacity are indicated at the right. Conjugate acid-base pairs are effective buffers between approximately 10% and 90% neutralization of the proton-donor species.

two equilibria (Eqns 4–5, 4–6) coexist, each always conforming to its equilibrium constant.

Figure 4–16 compares the titration curves of three weak acids with very different dissociation constants: acetic acid (pK_a = 4.76); dihydrogen phosphate, $H_2PO_4^-$ (pK_a = 6.86); and ammonium ion, NH_4^+ (pK_a = 9.25). Although the titration curves of these acids have the same shape, they are displaced along the pH axis because the three acids have different strengths. Acetic acid is the strongest (loses its proton most readily) because its K_a is highest (pK_a lowest) of the three. Acetic acid is already half dissociated at pH 4.76. Dihydrogen phosphate loses a proton less readily, being half dissociated at pH 6.86. Ammonium ion is the weakest acid of the three and does not become half dissociated until pH 9.25.

The most important point about the titration curve of a weak acid is that it shows graphically that a weak acid and its anion—a conjugate acid-base pair—can act as a buffer.

Buffering against pH Changes in Biological Systems

Almost every biological process is pH dependent; a small change in pH produces a large change in the rate of the process. This is true not only for the many reactions in which the H^+ ion is a direct participant, but also for those in which there is no apparent role for H^+ ions. The enzymes that catalyze cellular reactions, and many of the molecules on which they act, contain ionizable groups with characteristic pK_a values. The protonated amino and carboxyl groups of amino acids and the phosphate groups of nucleotides, for example, function as weak acids; their ionic state depends on the pH of the surrounding medium. As we noted above, ionic interactions are among the forces that stabilize a protein molecule and allow an enzyme to recognize and bind its substrate.

Cells and organisms maintain a specific and constant cytosolic pH, keeping biomolecules in their optimal ionic state, usually near pH 7. In multicellular organisms, the pH of extracellular fluids is also tightly regulated. Constancy of pH is achieved primarily by biological buffers: mixtures of weak acids and their conjugate bases.

We describe here the ionization equilibria that account for buffering, and we show the quantitative relationship between the pH of a buffered solution and the pK_a of the buffer. Biological buffering is illustrated by the phosphate and carbonate buffering systems of humans.

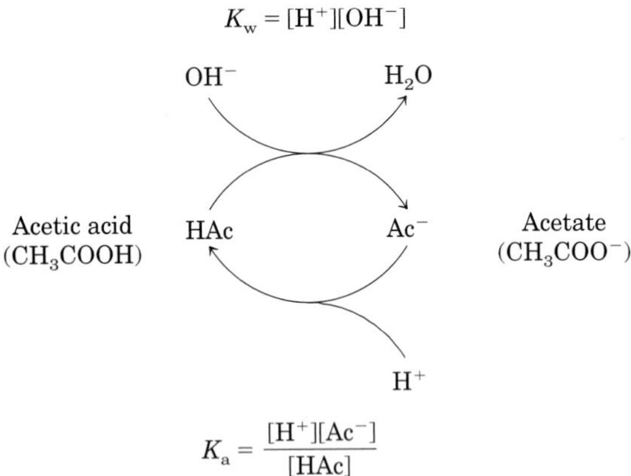

figure 4–17
The acetic acid–acetate pair as a buffer system. The system is capable of absorbing either H^+ or OH^- through the reversibility of the dissociation of acetic acid. The proton donor, acetic acid (HAc), contains a reserve of bound H^+, which can be released to neutralize an addition of OH^- to the system, forming H_2O. This happens because the product $[H^+][OH^-]$ transiently exceeds K_w (1×10^{-14} M^2). The equilibrium quickly adjusts so that this product equals 1×10^{-14} M^2 (at 25 °C), thus transiently reducing the concentration of H^+. But now the quotient $[H^+][Ac^-]/[HAc]$ is less then K_a, so HAc dissociates further to restore equilibrium. Similarly, the conjugate base, Ac^-, can react with H^+ ions added to the system; again, the two ionization reactions simultaneously come to equilibrium. Thus a conjugate acid-base pair, such as acetic acid and acetate ion, tends to resist a change in pH when small amounts of acid or base are added. Buffering action is simply the consequence of two reversible reactions taking place simultaneously and reaching their points of equilibrium as governed by their equilibrium constants, K_w and K_a.

Buffers Are Mixtures of Weak Acids and Their Conjugate Bases

Buffers are aqueous systems that tend to resist changes in pH when small amounts of acid (H^+) or base (OH^-) are added. A buffer system consists of a weak acid (the proton donor) and its conjugate base (the proton acceptor). As an example, a mixture of equal concentrations of acetic acid and acetate ion, found at the midpoint of the titration curve in Figure 4–15, is a buffer system. The titration curve of acetic acid has a relatively flat zone extending about 1 pH unit on either side of its midpoint pH of 4.76. In this zone, an amount of H^+ or OH^- added to the system has much less effect on pH than the same amount added outside the buffer range. This relatively flat zone is the buffering region of the acetic acid–acetate buffer pair. At the midpoint of the buffering region, where the concentration of the proton donor (acetic acid) exactly equals that of the proton acceptor (acetate), the buffering power of the system is maximal; that is, its pH changes least on addition of H^+ or OH^-. The pH at this point in the titration curve of acetic acid is equal to its pK_a. The pH of the acetate buffer system does change slightly when a small amount of H^+ or OH^- is added, but this change is very small compared with the pH change that would result if the same amount of H^+ (or OH^-) were added to pure water or to a solution of the salt of a strong acid and strong base, such as NaCl, which has no buffering power.

Buffering results from two reversible reaction equilibria occurring in a solution of nearly equal concentrations of a proton donor and its conjugate proton acceptor. Figure 4–17 explains how a buffer system works. Whenever H^+ or OH^- is added to a buffer, the result is a small change in the ratio of the relative concentrations of the weak acid and its anion and thus a small change in pH. The decrease in concentration of one component of the system is balanced exactly by an increase in the other. The sum of the buffer components does not change, only their ratio.

Each conjugate acid-base pair has a characteristic pH zone in which it is an effective buffer (Fig. 4–16). The $H_2PO_4^-/HPO_4^{2-}$ pair has a pK_a of 6.86 and thus can serve as an effective buffer system between approximately pH 5.9 and pH 7.9; the NH_4^+/NH_3 pair, with a pK_a of 9.25, can act as a buffer between approximately pH 8.3 and pH 10.3.

A Simple Expression Relates pH, p*K*, and Buffer Concentration

The titration curves of acetic acid, $H_2PO_4^-$, and NH_4^+ (Fig. 4–16) have nearly identical shapes, suggesting that these curves reflect a fundamental law or relationship. This is indeed the case. The shape of the titration curve of any weak acid is described by the **Henderson-Hasselbalch equation,** which is important for understanding buffer action and acid-base balance in the blood and tissues of vertebrates. This equation is simply a useful way of restating the expression for the dissociation constant of an acid. For the dissociation of a weak acid HA into H^+ and A^-, the Henderson-Hasselbalch equation can be derived as follows:

$$K_a = \frac{[H^+][A^-]}{[HA]}$$

First solve for $[H^+]$:

$$[H^+] = K_a \frac{[HA]}{[A^-]}$$

Then take the negative logarithm of both sides:

$$-\log [H^+] = -\log K_a - \log \frac{[HA]}{[A^-]}$$

Substitute pH for $-\log [H^+]$ and pK_a for $-\log K_a$:

$$pH = pK_a - \log \frac{[HA]}{[A^-]}$$

Now invert $-\log [HA]/[A^-]$, which involves changing its sign, to obtain the Henderson-Hasselbalch equation:

$$pH = pK_a + \log \frac{[A^-]}{[HA]}$$

Stated more generally,

$$pH = pK_a + \log \frac{[\text{proton acceptor}]}{[\text{proton donor}]}$$

This equation fits the titration curve of all weak acids and enables us to deduce a number of important quantitative relationships. For example, it shows why the pK_a of a weak acid is equal to the pH of the solution at the midpoint of its titration. At that point, [HA] equals $[A^-]$, and

$$pH = pK_a + \log 1.0 = pK_a + 0 = pK_a$$

As shown in Box 4–3, the Henderson-Hasselbalch equation also makes it possible to (1) calculate pK_a, given pH and the molar ratio of proton donor and acceptor; (2) calculate pH, given pK_a and the molar ratio of proton donor and acceptor; and (3) calculate the molar ratio of proton donor and acceptor, given pH and pK_a.

box 4–3 Solving Problems Using the Henderson-Hasselbalch Equation

1. Calculate the pK_a of lactic acid, given that when the concentration of lactic acid is 0.010 M and the concentration of lactate is 0.087 M, the pH is 4.80.

$$pH = pK_a + \log \frac{[\text{lactate}]}{[\text{lactic acid}]}$$

$$pK_a = pH - \log \frac{[\text{lactate}]}{[\text{lactic acid}]}$$

$$= 4.80 - \log \frac{0.087}{0.010} = 4.80 - \log 8.7$$

$$= 4.80 - 0.94 = 3.86 \quad \textit{(answer)}$$

2. Calculate the pH of a mixture of 0.1 M acetic acid and 0.2 M sodium acetate. The pK_a of acetic acid is 4.76.

$$pH = pK_a + \log \frac{[\text{acetate}]}{[\text{acetic acid}]}$$

$$= 4.76 + \log \frac{0.2}{0.1} = 4.76 + 0.301$$

$$= 5.06 \quad \textit{(answer)}$$

3. Calculate the ratio of the concentrations of acetate and acetic acid required in a buffer system of pH 5.30.

$$pH = pK_a + \log \frac{[\text{acetate}]}{[\text{acetic acid}]}$$

$$\log \frac{[\text{acetate}]}{[\text{acetic acid}]} = pH - pK_a$$

$$= 5.30 - 4.76 = 0.54$$

$$\frac{[\text{acetate}]}{[\text{acetic acid}]} = \text{antilog } 0.54 = 3.47 \quad \textit{(answer)}$$

Weak Acids or Bases Buffer Cells and Tissues against pH Changes

The intracellular and extracellular fluids of multicellular organisms have a characteristic and nearly constant pH. The organism's first line of defense against changes in internal pH is provided by buffer systems. The cytoplasm of most cells contains high concentrations of proteins, which contain many amino acids with functional groups that are weak acids or weak bases. For example, the side chain of histidine (Fig. 4–18) has a pK_a of 6.0; proteins containing histidine residues therefore buffer effectively near neutral pH. Nucleotides such as ATP, as well as many low molecular weight metabolites, contain ionizable groups that can contribute buffering power to the cytoplasm. Some highly specialized organelles and extracellular compartments have high concentrations of compounds that contribute buffering capacity: organic acids buffer the vacuoles of plant cells; ammonia buffers urine.

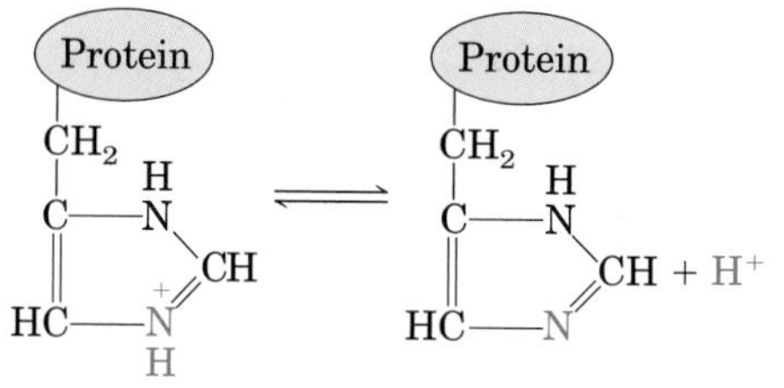

Protein–CH_2–(imidazolium ring, C–NH, CH, +N–H, HC) $\rightleftharpoons$ Protein–CH_2–(imidazole ring, C–NH, CH, N, HC) + H^+

figure 4–18
The amino acid histidine, a component of proteins, is a weak acid. The pK_a of the protonated nitrogen of the side chain is 6.0.

Two especially important biological buffers are the phosphate and bicarbonate systems. The phosphate buffer system, which acts in the cytoplasm of all cells, consists of $H_2PO_4^-$ as proton donor and HPO_4^{2-} as proton acceptor:

$$H_2PO_4^- \rightleftharpoons H^+ + HPO_4^{2-}$$

The phosphate buffer system is maximally effective at a pH close to its pK_a of 6.86 (Figs 4–14, 4–16) and thus tends to resist pH changes in the range between about 5.9 and 7.9. It is therefore an effective buffer in biological fluids; in mammals, for example, extracellular fluids and most cytoplasmic compartments have a pH in the range of 6.9 to 7.4.

Blood plasma is buffered in part by the bicarbonate system, consisting of carbonic acid (H_2CO_3) as proton donor and bicarbonate (HCO_3^-) as proton acceptor:

$$H_2CO_3 \rightleftharpoons H^+ + HCO_3^-$$

$$K_1 = \frac{[H^+][HCO_3^-]}{[H_2CO_3]}$$

This buffer system is more complex than other conjugate acid-base pairs because one of its components, carbonic acid (H_2CO_3), is formed from dissolved (d) carbon dioxide and water, in a reversible reaction:

$$CO_2(d) + H_2O \rightleftharpoons H_2CO_3$$

$$K_2 = \frac{[H_2CO_3]}{[CO_2(d)][H_2O]}$$

Carbon dioxide is a gas under normal conditions, and the concentration of dissolved CO_2 is the result of equilibration with CO_2 of the gas phase:

$$CO_2(g) \rightleftharpoons CO_2(d)$$

$$K_3 = \frac{[CO_2(d)]}{[CO_2(g)]}$$

box 4–4 Blood, Lungs, and Buffer: The Bicarbonate Buffer System

In animals with lungs, the bicarbonate buffer system is an effective physiological buffer near pH 7.4 because the H_2CO_3 of blood plasma is in equilibrium with a large reserve capacity of $CO_2(g)$ in the air space of the lungs. This buffer system involves three reversible equilibria between gaseous CO_2 in the lungs and bicarbonate (HCO_3^-) in the blood plasma (Fig. 1). When H^+ (from lactic acid produced in muscle tissue during vigorous exercise, for example) is added to blood as it passes through the tissues, reaction 1 proceeds toward a new equilibrium, in which the concentration of H_2CO_3 is increased. This increases the concentration of $CO_2(d)$ in the blood plasma (reaction 2) and thus increases the pressure of $CO_2(g)$ in the air space of the lungs (reaction 3); the extra CO_2 is exhaled.

Conversely, when the pH of blood plasma is raised (by NH_3 production during protein catabolism, for example), the opposite events occur: the H^+ concentration of blood plasma is lowered, causing more H_2CO_3 to dissociate into H^+ and HCO_3^-. This in turn causes more $CO_2(g)$ from the lungs to dissolve in the blood plasma. The rate of breathing—that is, the rate of inhaling and exhaling CO_2—can quickly adjust these equilibria to keep the blood pH nearly constant.

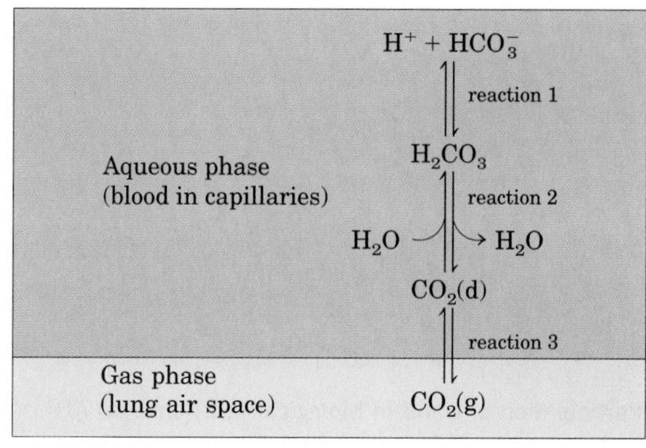

figure 1
The CO_2 in the air space of the lungs is in equilibrium with the bicarbonate buffer in the blood plasma passing through the lung capillaries. Because the concentration of dissolved CO_2 can be adjusted rapidly through changes in the rate of breathing, the bicarbonate buffer system of the blood is in near-equilibrium with a large potential reservoir of CO_2.

The pH of a bicarbonate buffer system depends on the concentration of H_2CO_3 and HCO_3^-, the proton donor and acceptor components. The concentration of H_2CO_3 in turn depends on the concentration of dissolved CO_2, which in turn depends on the concentration of CO_2 in the gas phase, called the **partial pressure** of CO_2. Thus the pH of a bicarbonate buffer exposed to a gas phase is ultimately determined by the concentration of HCO_3^- in the aqueous phase and the partial pressure of CO_2 in the gas phase (Box 4–4).

Human blood plasma normally has a pH close to 7.4. Should the pH-regulating mechanisms fail or be overwhelmed, as may happen in severe uncontrolled diabetes when an overproduction of metabolic acids causes acidosis, the pH of the blood can fall to 6.8 or below, leading to irreparable cell damage and death. In other diseases the pH may rise to lethal levels. Although many aspects of cell structure and function are influenced by pH, it is the catalytic activity of enzymes that is especially sensitive. Enzymes typically show maximal catalytic activity at a characteristic pH, called the **pH optimum** (Fig. 4–19). On either side of the optimum pH their catalytic activity often declines sharply. Thus, a small change in pH can make a large difference in the rate of some crucial enzyme-catalyzed reactions. Biological control of the pH of cells and body fluids is therefore of central importance in all aspects of metabolism and cellular activities.

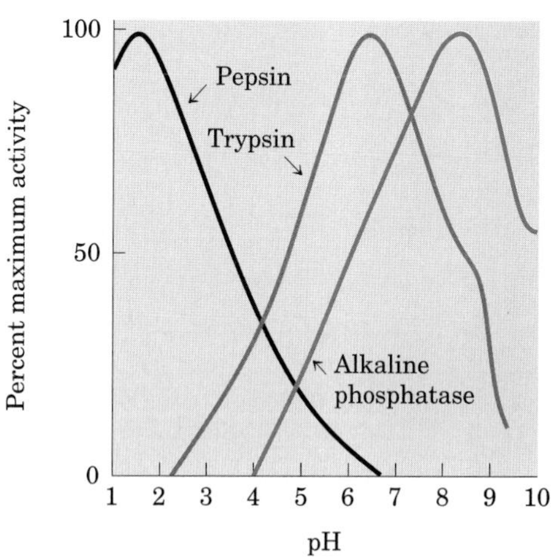

figure 4–19
The pH optima of some enzymes. Pepsin is a digestive enzyme secreted into gastric juice; trypsin, a digestive enzyme that acts in the small intestine; alkaline phosphatase of bone tissue, a hydrolytic enzyme thought to aid in bone mineralization.

Water as a Reactant

Water is not only the solvent in which the chemical reactions of living cells occur; it is very often a direct participant in those reactions. The formation of ATP from ADP and inorganic phosphate is an example of a **condensation reaction** (p. 69) in which the elements of water are eliminated (Fig. 4–20a). The reverse of this reaction—cleavage accompanied by the addition of the elements of water—is a **hydrolysis reaction.** Hydrolysis reactions are also responsible for the enzymatic depolymerization of proteins, carbohydrates, and nucleic acids. Hydrolysis reactions, catalyzed by enzymes called **hydrolases,** are almost invariably exergonic. The formation of cellular polymers from their subunits by simple reversal of hydrolysis would be endergonic and therefore does not occur. As we shall see, cells circumvent this thermodynamic obstacle by coupling endergonic condensation reactions to exergonic processes, such as breakage of the anhydride bond in ATP.

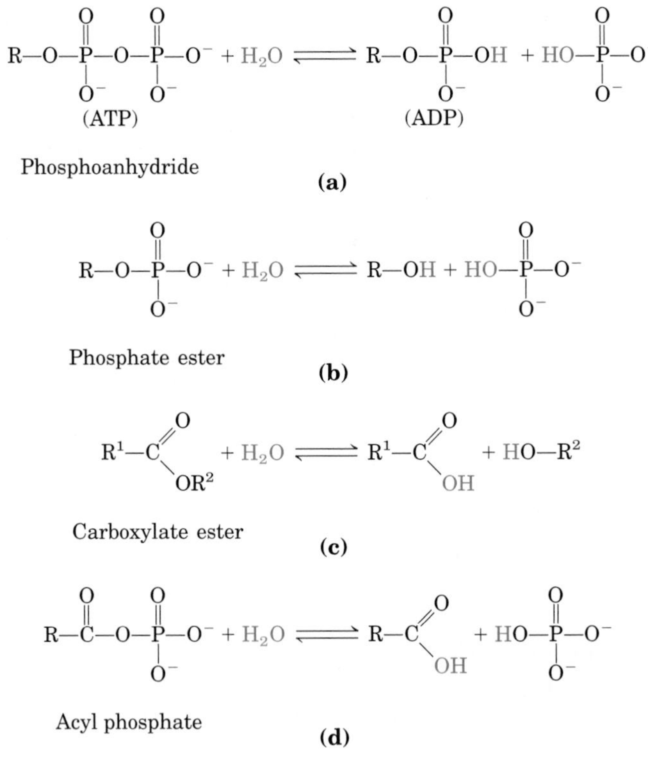

figure 4–20
Participation of water in biological reactions. (a) ATP is a phosphoanhydride formed by a condensation reaction (loss of the elements of water) between ADP and phosphate. R represents adenosine monophosphate (AMP). This condensation reaction requires energy. The hydrolysis of (addition of the elements of water to) ATP releases an equivalent amount of energy. Also shown are some other condensation and hydrolysis reactions common in biological systems **(b), (c), (d).**

$$R\text{—O—}P(=O)(O^-)\text{—O—}P(=O)(O^-)\text{—}O^- + H_2O \rightleftharpoons R\text{—O—}P(=O)(O^-)\text{—OH} + HO\text{—}P(=O)(O^-)\text{—}O^-$$

(ATP) (ADP)

Phosphoanhydride

(a)

$$R\text{—O—}P(=O)(O^-)\text{—}O^- + H_2O \rightleftharpoons R\text{—OH} + HO\text{—}P(=O)(O^-)\text{—}O^-$$

Phosphate ester

(b)

$$R^1\text{—}C(=O)\text{—}OR^2 + H_2O \rightleftharpoons R^1\text{—}C(=O)\text{—OH} + HO\text{—}R^2$$

Carboxylate ester

(c)

$$R\text{—}C(=O)\text{—O—}P(=O)(O^-)\text{—}O^- + H_2O \rightleftharpoons R\text{—}C(=O)\text{—OH} + HO\text{—}P(=O)(O^-)\text{—}O^-$$

Acyl phosphate

(d)

You are (we hope!) consuming oxygen as you read. Water and carbon dioxide are the end products of the oxidation of fuels such as glucose. The overall reaction can be summarized as

$$\underset{\text{Glucose}}{C_6H_{12}O_6} + 6O_2 \rightleftharpoons 6CO_2 + 6H_2O$$

The "metabolic water" thus formed from solid food and stored fuels is actually enough to allow some animals in very dry habitats (gerbils, kangaroo rats, camels) to survive without drinking water for extended periods.

Green plants and algae use the energy of sunlight to split water in the process of photosynthesis:

$$2H_2O + 2A \xrightarrow{\text{light}} O_2 + 2AH_2$$

In this reaction, A is an electron-accepting species, which varies with the type of photosynthetic organism.

The Fitness of the Aqueous Environment for Living Organisms

Organisms have effectively adapted to their aqueous environment and have even evolved means of exploiting the unusual properties of water. The high specific heat of water (the heat energy required to raise the temperature of 1 g of water by 1 °C) is useful to cells and organisms because it allows water to act as a "heat buffer," permitting the temperature of an organism to remain relatively constant as the temperature of the air fluctuates and as heat is generated as a byproduct of metabolism. Furthermore, some vertebrates exploit the high heat of vaporization of water (Table 4–1) by using (thus losing) excess body heat to evaporate sweat. The high degree of internal cohesion of liquid water, due to hydrogen bonding, is exploited by plants as a means of transporting dissolved nutrients from the roots to the leaves during the process of transpiration. Even the density of ice, lower than that of liquid water, has important biological consequences in the life cycles of aquatic organisms. Ponds freeze from the top down, and the layer of ice at the top insulates the water below from frigid air, preventing the pond (and the organisms in it) from freezing solid. Most fundamental to all living organisms is the fact that many physical and biological properties of cell macromolecules, particularly the proteins and nucleic acids, derive from their interactions with water molecules of the surrounding medium. The influence of water on the course of biological evolution has been profound and determinative. If life forms have evolved elsewhere in the universe, it is unlikely that they resemble those of Earth unless their extraterrestrial origin is also a place in which plentiful liquid water is available.

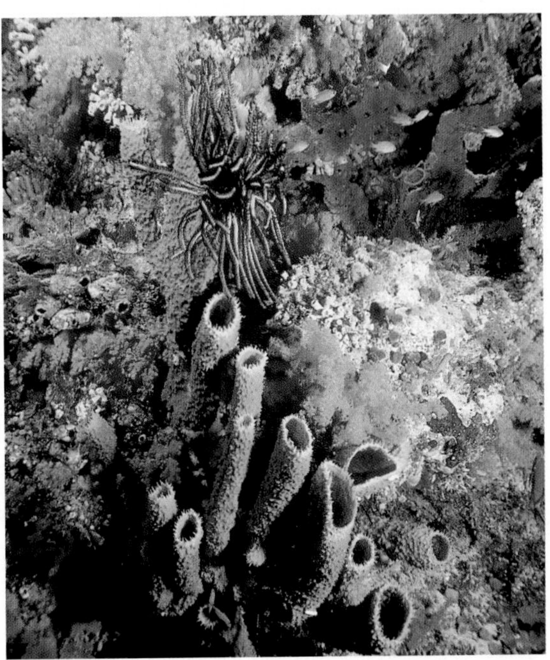

Aqueous environments support a myriad of species. Soft corals, sponges, bryozoans, and algae compete for space on this reef substrate off the Philippine Islands.

summary

Water is the most abundant compound in living organisms. Its relatively high freezing point, boiling point, and heat of vaporization are the result of strong intermolecular attractions in the form of hydrogen bonding between adjacent water molecules. Liquid water has considerable short-range order and consists of short-lived hydrogen-bonded clusters. The polarity and hydrogen-bonding properties of water make it a potent solvent for many ionic compounds and other polar molecules. Nonpolar compounds, including the gases CO_2, O_2, and N_2, are poorly soluble in water.

Four types of weak interactions occur within and between biomolecules in an aqueous solvent: ionic, hydrophobic, and van der Waals interactions, and hydrogen bonds. Although weak individually, these interactions collectively create a very strong stabilizing force for proteins, nucleic acids, and membranes. Weak (noncovalent) interactions are also at the heart of enzyme catalysis, antibody function, and receptor-ligand interactions.

When aqueous solutions of different concentration are separated by a semipermeable membrane, water crosses the membrane in the direction of lower water concentration. This tendency toward movement of water across a semipermeable membrane (osmosis) creates osmotic pressure. For cells in hypotonic solutions, inward osmotic movement of water across the plasma membrane causes swelling, producing turgor pressure or, if the cell is not protected by a rigid wall, osmotic lysis. The colligative properties of aqueous solutions (melting and boiling points, vapor pressure, and osmotic pressure) depend on the number of dissolved particles (ions, molecules), not on their molecular mass or chemical properties.

Water ionizes very slightly to form H^+ and OH^- ions. The rapid hopping of protons along strings of hydrogen-bonded water molecules gives the appearance of exceptionally fast diffusion of protons in water. In dilute aqueous solutions, the concentrations of H^+ and OH^- ions are inversely related by the expression $K_w = [H^+][OH^-] = 1 \times 10^{-14}$ M^2 (at 25 °C). The hydrogen-ion concentration of biological systems is usually expressed in terms of pH, defined as $\text{pH} = -\log [H^+]$.

Acids are defined as proton donors and bases as proton acceptors. A conjugate acid-base pair consists of a proton donor (HA) and its corresponding proton acceptor (A^-). The tendency of an acid HA to donate protons is expressed by its dissociation constant ($K_a = [H^+][A^-]/[HA]$) or by the function pK_a, defined as $-\log K_a$, which can be determined from an experimental titration curve. The pH of a solution of a weak acid is quantitatively related to its pK_a and to the ratio of the concentrations of its proton-donor and proton-acceptor species by the Henderson-Hasselbalch equation.

A conjugate acid-base pair can act as a buffer and resist changes in pH; its capacity to do so is greatest at a pH equal to its pK_a. Many types of biomolecules have functional groups that contribute buffering capacity. Proteins, H_2CO_3/HCO_3^-, and $H_2PO_4^-/HPO_4^{2-}$ are important biological buffers. The catalytic activity of enzymes is strongly influenced by pH, and the environments in which enzymes function must be buffered against large pH changes.

Water is not only the solvent in which metabolic reactions occur; it participates directly in many biochemical processes, including hydrolysis and condensation reactions. The physical and chemical properties of water are central to biological structure and function. The evolution of life on Earth has been influenced greatly by both the solvent and reactant properties of water.

further reading

General

Denny, M.W. (1993) *Air and Water: The Biology and Physics of Life's Media,* Princeton University Press, Princeton, NJ.

A wonderful investigation of the biological relevance of the properties of water.

Eisenberg, D. & Kauzmann, W. (1969) *The Structure and Properties of Water,* Oxford University Press, New York.

An advanced treatment of the physical chemistry of water.

Franks, F. & Mathias, S.F. (eds) (1982) *Biophysics of Water,* John Wiley & Sons, Inc., New York.

A large collection of papers on the structure of pure water and of the cytoplasm.

Gerstein, M. & Levitt, M. (1998) Simulating water and the molecules of life. *Sci. Am.* **279** (November), 100–105.

A well-illustrated description of the use of computer simulation to study the biologically important association of water with proteins and nucleic acids.

Gronenborn, A. & Clore, M. (1997) Water in and around proteins. *The Biochemist* **19** (3), 18–21.

A brief discussion of protein-bound water as detected by crystallography and NMR.

Kornblatt, J. & Kornblatt, J. (1997) The role of water in recognition and catalysis by enzymes. *The Biochemist* **19** (3), 14–17.

A short, useful summary of the ways in which bound water influences the structure and activity of proteins.

Kuntz, I.D. & Zipp, A. (1977) Water in biological systems. *N. Engl. J. Med.* **297,** 262–266.

A brief review of the physical state of cytosolic water and its interactions with dissolved biomolecules.

Ladbury, J. (1996) Just add water! The effect of water on the specificity of protein-ligand binding sites and its potential application to drug design. *Chem. Biol.* **3,** 973–980.

Stillinger, F.H. (1980) Water revisited. *Science* **209,** 451–457.

A short review of the physical structure of water, including the importance of hydrogen bonding and the nature of hydrophobic interactions.

Westhof, E. (ed.) (1993) *Water and Biological Macromolecules,* CRC Press, Inc., Boca Raton, FL.

Fourteen chapters, each by a different author, cover (at an advanced level) the structure of water and its interactions with proteins, nucleic acids, polysaccharides, and lipids.

Wiggins, P.M. (1990) Role of water in some biological processes. *Microbiol. Rev.* **54,** 432–449.

A review of water in biology, including discussion of the physical structure of liquid water, its interaction with biomolecules, and the state of water in living cells.

Weak Interactions in Aqueous Systems

Fersht, A.R. (1987) The hydrogen bond in molecular recognition. *Trends Biochem. Sci.* **12,** 301–304.

A clear, brief, quantitative discussion of the contribution of hydrogen bonding to molecular recognition and enzyme catalysis.

Frieden, E. (1975) Non-covalent interactions: key to biological flexibility and specificity. *J. Chem. Educ.* **52,** 754–761.

Review of the four kinds of weak interactions that stabilize macromolecules and confer biological specificity, with clear examples.

Jeffrey, G.A. (1997) *An Introduction to Hydrogen Bonding,* Oxford University Press, New York.

A detailed and advanced discussion of the structure and properties of hydrogen bonds, including those in water and biomolecules.

Schwabe, J.W.R. (1997) The role of water in protein-DNA interactions. *Curr. Opin. Struct. Biol.* **7,** 126–134.

Examines the important role of water in both the specificity and affinity of protein-DNA interactions.

Tanford, C. (1978) The hydrophobic effect and the organization of living matter. *Science* **200,** 1012–1018.

A review of the chemical and energetic basis for hydrophobic interactions between biomolecules in aqueous solutions.

Weak Acids, Weak Bases, and Buffers: Problems for Practice

Segel, I.H. (1976) *Biochemical Calculations,* 2nd edn, John Wiley & Sons, Inc., New York.

problems

1. Simulated Vinegar One way to make vinegar (*not* the preferred way) is to prepare a solution of acetic acid, the sole acid component of vinegar, at the proper pH (see Fig. 4–13) and add appropriate flavoring agents. Acetic acid (M_r 60) is a liquid at 25 °C with a density of 1.049 g/mL. Calculate the volume that must be added to distilled water to make 1 L of simulated vinegar (see Fig. 4–14).

2. Acidity of Gastric HCl In a hospital laboratory, a 10.0 mL sample of gastric juice, obtained several hours after a meal, was titrated with 0.1 M NaOH to neutrality; 7.2 mL of NaOH was required. The patient's stomach contained no ingested food or drink, thus assume that no buffers were present. What was the pH of the gastric juice?

3. Measurement of Acetylcholine Levels by pH Changes The concentration of acetylcholine (a neurotransmitter) in a sample can be determined from the pH changes that accompany its hydrolysis. When the sample is incubated with the enzyme acetylcholinesterase, acetylcholine is quantitatively converted into choline and acetic acid, which dissociates to yield acetate and a hydrogen ion:

$$\underset{\text{Acetylcholine}}{CH_3{-}\overset{\overset{O}{\|}}{C}{-}O{-}CH_2{-}CH_2{-}{}^{+}\underset{\underset{CH_3}{|}}{\overset{\overset{CH_3}{|}}{N}}{-}CH_3} \xrightarrow{H_2O}$$

$$\underset{\text{Choline}}{HO{-}CH_2{-}CH_2{-}{}^{+}\underset{\underset{CH_3}{|}}{\overset{\overset{CH_3}{|}}{N}}{-}CH_3} + \underset{\text{Acetate}}{CH_3{-}\underset{\underset{O}{\|}}{C}{-}O^-} + H^+$$

In a typical analysis, 15 mL of an aqueous solution containing an unknown amount of acetylcholine had a pH of 7.65. When incubated with acetylcholinesterase, the pH of the solution decreased to 6.87. Assuming that there was no buffer in the assay mixture, determine the number of moles of acetylcholine in the 15 mL sample.

4. Osmotic Balance in a Marine Frog The crab-eating frog of Southeast Asia, *Rana cancrivora*, is born and matures in fresh water but searches for its food in coastal mangrove swamps (80% to full-strength seawater). Consequently, when the frog moves from its freshwater home to seawater it experiences a large change in the osmolarity of its environment (from hypotonic to hypertonic).

(a) Eighty percent seawater contains 460 mM NaCl, 10 mM KCl, 10 mM $CaCl_2$ and 50 mM $MgCl_2$. What are the concentrations of the various ionic species in this seawater? Assuming that these salts account for nearly all the solutes in seawater, what is the osmolarity of the seawater?

(b) The chart below lists the cytoplasmic concentrations of ions in *Rana cancrivora*. Ignoring dissolved proteins, amino acids, nucleic acids, and other small metabolites, what is the osmolarity of the frog's cells based solely on the ionic concentrations given below?

	Na^+ (mM)	K^+ (mM)	Cl^- (mM)	Ca^{2+} (mM)	Mg^{2+} (mM)
Rana cancrivora	122	10	100	2	1

(c) Like all frogs, the crab-eating frog can exchange gases through its permeable skin, allowing it to stay underwater for long periods of time without breathing. How does the high permeability of frog skin affect the frog's cells when it moves from fresh water to seawater?

(d) The crab-eating frog uses two mechanisms to maintain its cells in osmotic balance with its environment. First, it allows the Na^+ and Cl^- concentrations in its cells to slowly increase as the ions diffuse down their concentration gradients. Second, like many elasmobranchs (sharks), it retains the waste product urea in its cells. The addition of both NaCl and urea increases the osmolarity of the cytosol to a value that is nearly equal to that of the surrounding environment.

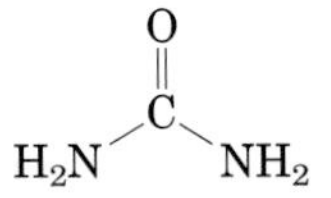

$H_2N-C(=O)-NH_2$

Urea (CH_4N_2O)

Assuming the volume of water in a typical frog is 100 mL, how many grams of NaCl (formula weight (FW) 58.44) does the frog need to take up in order to make its tissues isotonic with seawater?

(e) How many grams of urea (FW 60) must it retain to accomplish the same thing?

5. Properties of a Buffer The amino acid glycine is often used as the main ingredient of a buffer in biochemical experiments. The amino group of glycine, which has a pK_a of 9.6, can exist either in the protonated form ($-NH_3^+$) or as the free base ($-NH_2$) because of the reversible equilibrium

$$R-NH_3^+ \rightleftharpoons R-NH_2 + H^+$$

(a) In what pH range can glycine be used as an effective buffer due to its amino group?

(b) In a 0.1 M solution of glycine at pH 9.0, what fraction of glycine has its amino group in the $-NH_3^+$ form?

(c) How much 5 M KOH must be added to 1.0 L of 0.1 M glycine at pH 9.0 to bring its pH to exactly 10.0?

(d) When 99% of the glycine is in its $-NH_3^+$ form, what is the numerical relation between the pH of the solution and the pK_a of the amino group?

6. The Effect of pH on Solubility The strongly polar, hydrogen-bonding properties of water make it an excellent solvent for ionic (charged) species. By contrast, nonionized, nonpolar organic molecules, such as benzene, are relatively insoluble in water. In principle, the aqueous solubility of any organic acid or base can be increased by conversion of the molecules to charged species. For example, the solubility of benzoic acid in water is low. The addition of sodium bicarbonate to a mixture of water and benzoic acid raises the pH and deprotonates the benzoic acid to form benzoate ion, which is quite soluble in water.

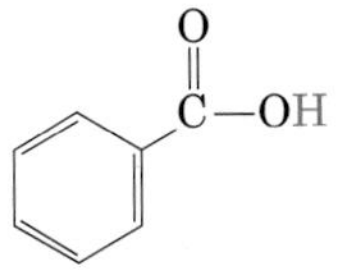

Benzoic acid ($C_6H_5-C(=O)-OH$)
$pK_a \approx 5$

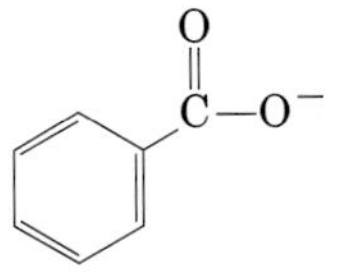

Benzoate ion ($C_6H_5-C(=O)-O^-$)

Are the following compounds more soluble in an aqueous solution of 0.1 M NaOH or 0.1 M HCl? (The dissociable protons are shown in red.)

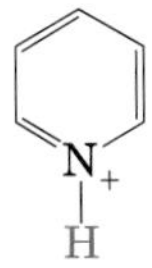

(a) Pyridine ion, $pK_a \approx 5$

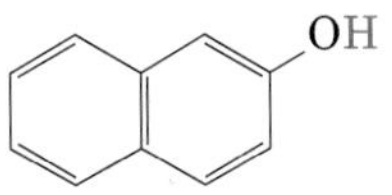

(b) β-Naphthol, $pK_a \approx 10$

(c) *N*-Acetyltyrosine methyl ester, $pK_a \approx 10$

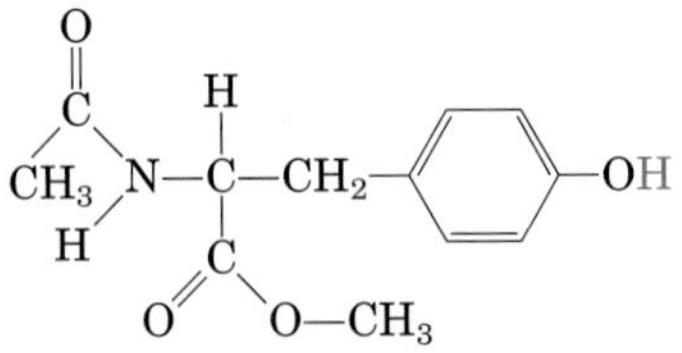

7. Treatment of Poison Ivy Rash The components of poison ivy and poison oak that produce the characteristic itchy rash are catechols substituted with long-chain alkyl groups.

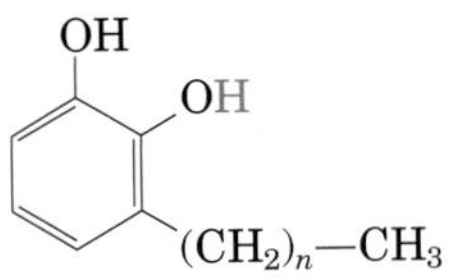

OH
OH
$(CH_2)_n—CH_3$
$pK_a \approx 8$

If you were exposed to poison ivy, which of the treatments below would you apply to the affected area? Justify your choice.

(a) Wash the area with cold water.

(b) Wash the area with dilute vinegar or lemon juice.

(c) Wash the area with soap and water.

(d) Wash the area with soap, water, and baking soda (sodium bicarbonate).

8. pH and Drug Absorption Aspirin is a weak acid with a pK_a of 3.5.

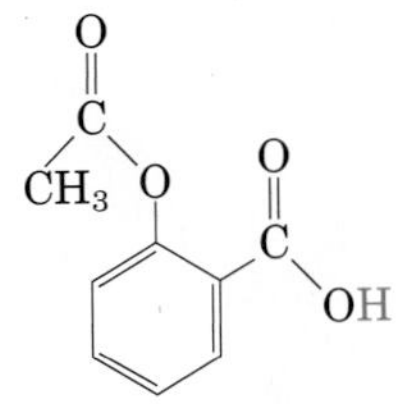

O
C
CH_3 O
O
C
OH

It is absorbed into the blood through the cells lining the stomach and the small intestine. Absorption requires passage through the plasma membrane, the rate of which is determined by the polarity of the molecule: charged and highly polar molecules pass slowly, whereas neutral hydrophobic ones pass rapidly. The pH of the stomach contents is about 1.5, and the pH of the contents of the small intestine is about 6. Is more aspirin absorbed into the bloodstream from the stomach or from the small intestine? Clearly justify your choice.

9. Preparation of Standard Buffer for Calibration of a pH Meter The glass electrode used in commercial pH meters gives an electrical response proportional to the concentration of hydrogen ion. To convert these responses into pH, glass electrodes must be calibrated against standard solutions of known H^+ concentration. Determine the weight in grams of sodium dihydrogen phosphate ($NaH_2PO_4 \cdot H_2O$; formula weight (FW) 138.01) and disodium hydrogen phosphate (Na_2HPO_4; FW 141.98) needed to prepare 1 L of a standard buffer at pH 7.00 with a total phosphate concentration of 0.100 M (see Fig. 4–14).

10. Control of Blood pH by Respiration Rate

(a) The partial pressure of CO_2 in the lungs can be varied rapidly by the rate and depth of breathing. For example, a common remedy to alleviate hiccups is to increase the concentration of CO_2 in the lungs. This can be achieved by holding one's breath, by very slow and shallow breathing (hypoventilation), or by breathing in and out of a paper bag. Under such conditions, the partial pressure of CO_2 in the air space of the lungs rises above normal. Qualitatively explain the effect of these procedures on the blood pH.

(b) A common practice of competitive short-distance runners is to breathe rapidly and deeply (hyperventilation) for about half a minute to remove CO_2 from their lungs just before running in, say, a 100 m dash. Their blood pH may rise to 7.60. Explain why the blood pH increases.

(c) During a short-distance run the muscles produce a large amount of lactic acid ($CH_3CH(OH)COOH$, $K_a = 1.38 \times 10^{-4}$) from their glucose stores. In view of this fact, why might hyperventilation before a dash be useful?

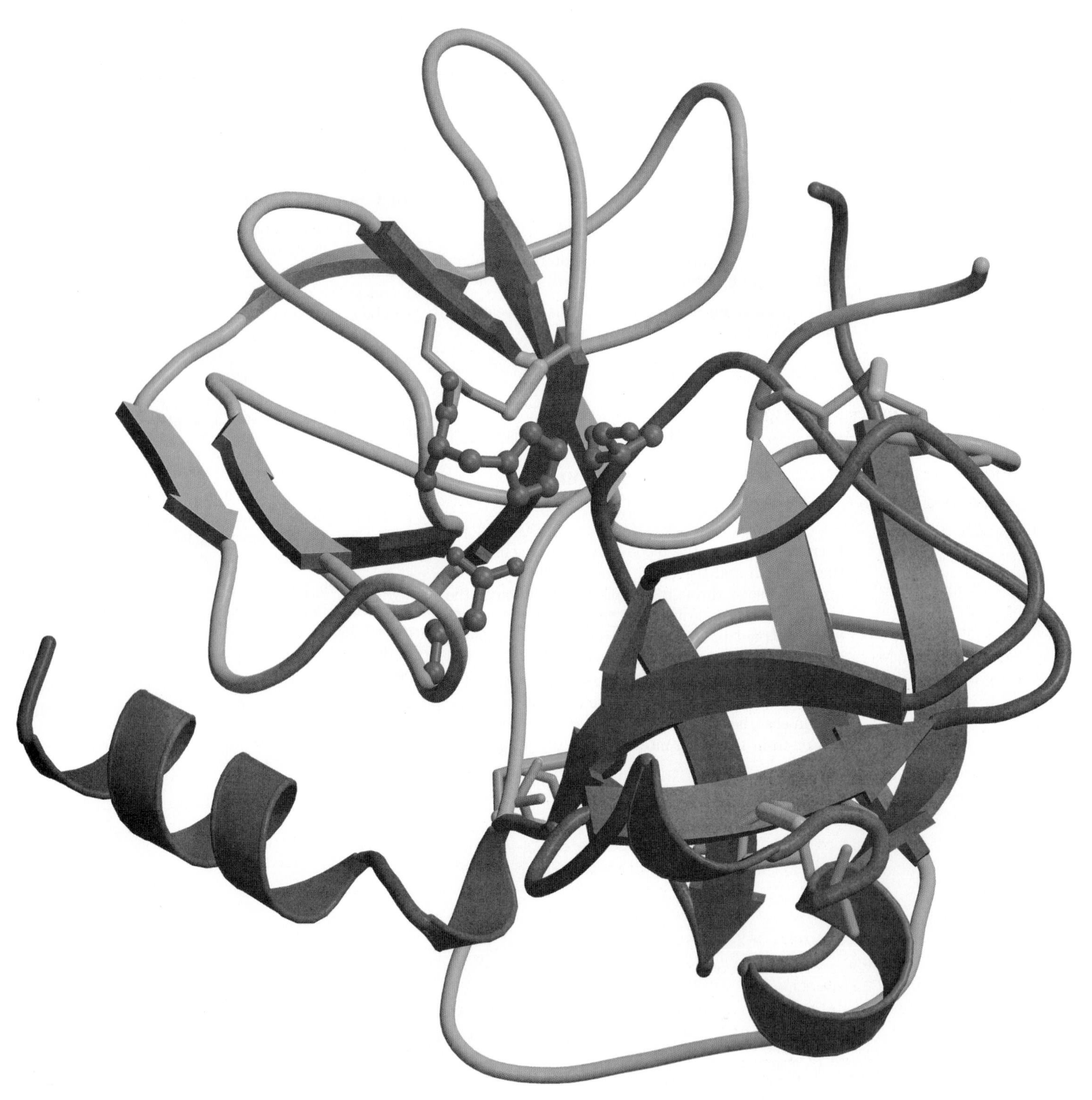

part

II

Structure and Catalysis

In Part I we contrasted the complex structure and function of living cells with the relative simplicity of the monomeric units from which the macromolecules, supramolecular complexes, and organelles of the cells are constructed. Part II is devoted to the structure and function of the major classes of cellular constituents: amino acids and proteins (Chapters 5 through 8), sugars and polysaccharides (Chapter 9), nucleotides and nucleic acids (Chapter 10), fatty acids and lipids (Chapter 11), and, finally, membranes and membrane signaling proteins (Chapters 12 and 13). We begin in each case by considering the covalent structure of the simple subunits (amino acids, monosaccharides, nucleotides, and fatty acids). These subunits are a major part of the language of biochemistry; familiarity with them is a prerequisite for understanding more advanced topics covered in this book, as well as the rapidly growing and exciting literature of biochemistry.

After describing the covalent chemistry of the monomeric units, we consider the structure of the macromolecules and supramolecular complexes derived from them. An overriding theme is that the polymeric macromolecules in living systems, though large, are highly ordered chemical entities, with specific sequences of monomeric subunits giving rise to discrete structures and functions. This fundamental theme can be broken down into three interrelated principles: (1) the unique structure of each macromolecule determines its function; (2) noncovalent interactions play a critical role in the structure and thus the function of macromolecules; and (3) the monomeric subunits in polymeric macromolecules occur in specific sequences, representing a form of information upon which the ordered living state depends.

The relationship between structure and function is especially evident in proteins, which exhibit an extraordinary diversity of functions. One particular polymeric sequence of amino acids produces a strong, fibrous structure

facing page
A view of the enzyme chymotrypsin, one of the best-understood biological catalysts.

found in hair and wool; another produces a protein that transports oxygen in the blood; a third binds other proteins and catalyzes the cleavage of the bonds between their amino acids. Similarly, the special functions of polysaccharides, nucleic acids, and lipids can be understood as a direct manifestation of their chemical structure, with their characteristic monomeric subunits linked in precise functional polymers. Sugars linked together become energy stores and structural fibers; nucleotides strung together in DNA or RNA provide the blueprint for an entire organism; and aggregated lipids form membrane bilayers. Chapter 13 unifies the discussion of biomolecule function, describing how specific signaling systems regulate the activities of biomolecules within a cell and between organs to keep an organism in homeostasis.

As we move from monomeric units to larger and larger polymers, the chemical focus shifts from covalent bonds to noncovalent interactions. The properties of covalent bonds, both in the monomeric units and in the bonds that connect them in polymers, place constraints on the shapes assumed by large molecules. It is the numerous noncovalent interactions, however, that dictate the stable native conformations of large molecules while permitting the flexibility necessary for their biological function. We will see that noncovalent interactions are essential to the catalytic power of enzymes, the critical interaction of complementary base pairs in nucleic acids, the arrangement and properties of lipids in membranes, and the interaction of a hormone or growth factor with its membrane receptor.

The principle that sequences of monomeric subunits are information-rich emerges fully in the discussion of nucleic acids in Chapter 10. However, proteins and some short polymers of sugars (oligosaccharides) are also information-rich molecules. The amino acid sequence is a form of information that directs the folding of the protein into its unique three-dimensional structure, and ultimately determines the function of the protein. Some oligosaccharides also have unique sequences and three-dimensional structures that are recognized by other macromolecules.

For each class of molecules we find a similar structural hierarchy, in which subunits of fixed structure are connected by bonds of limited flexibility to form macromolecules with three-dimensional structures determined by noncovalent interactions. These macromolecules then interact to form the supramolecular structures and organelles that allow a cell to carry out its many metabolic functions. Together, the molecules described in Part II are the "stuff" of life. We begin with the amino acids and proteins.

chapter

5

Amino Acids, Peptides, and Proteins

Proteins are the most abundant biological macromolecules, occurring in all cells and all parts of cells. Proteins also occur in great variety; thousands of different kinds, ranging in size from relatively small peptides to huge polymers with molecular weights in the millions, may be found in a single cell. Moreover, proteins exhibit enormous diversity of biological function and are the most important final products of the information pathways discussed in Part IV of this book. Proteins are the molecular instruments through which genetic information is expressed. It is appropriate to begin our study of biological macromolecules with the proteins, whose name derives from the Greek *prōtos,* meaning "first" or "foremost."

Relatively simple monomeric subunits provide the key to the structure of the thousands of different proteins. All proteins, whether from the most ancient lines of bacteria or from the most complex forms of life, are constructed from the same ubiquitous set of 20 amino acids, covalently linked in characteristic linear sequences. Because each of these amino acids has a side chain with distinctive chemical properties, this group of 20 precursor molecules may be regarded as the alphabet in which the language of protein structure is written.

What is most remarkable is that cells can produce proteins with strikingly different properties and activities by joining the same 20 amino acids in many different combinations and sequences. From these building blocks different organisms can make such widely diverse products as enzymes, hormones, antibodies, transporters, muscle, the lens protein of the eye, feathers, spider webs, rhinoceros horn, milk proteins, antibiotics, mushroom poisons, and a myriad of other substances having distinct biological activities (Fig. 5–1). Among these protein products, the enzymes are the most varied and specialized. Virtually all cellular reactions are catalyzed by enzymes.

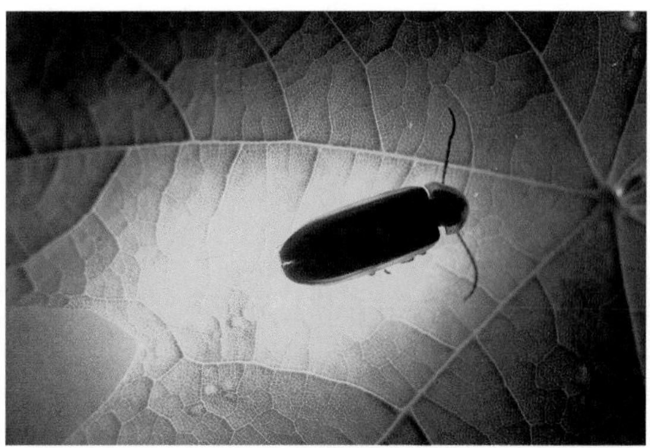

(a)

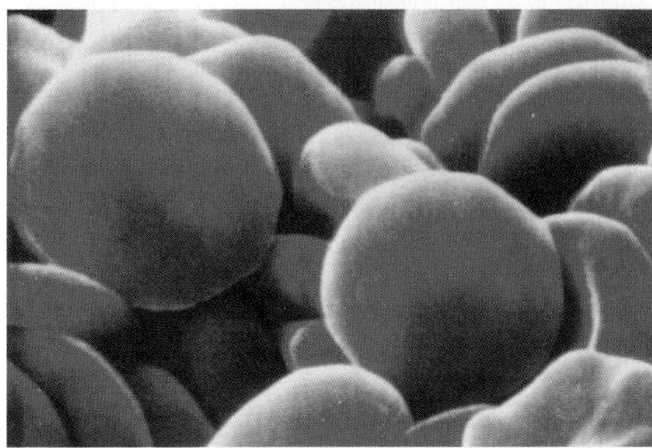

(b)

(c)

figure 5–1

Some functions of proteins. (a) The light produced by fireflies is the result of a reaction involving the protein luciferin and ATP, catalyzed by the enzyme luciferase (see Box 14–3). **(b)** Erythrocytes contain large amounts of the oxygen-transporting protein hemoglobin. **(c)** The protein keratin, formed by all vertebrates, is the chief structural component of hair, scales, horn, wool, nails, and feathers. The black rhinoceros is nearing extinction in the wild because of the myth prevalent in some parts of the world that a powder derived from its horn has aphrodisiac properties. In reality, the chemical properties of powdered rhinoceros horn are no different from those of powdered bovine hooves or human fingernails.

Protein structure and function are the topics of this and the next three chapters. We begin with a description of the fundamental chemical properties of amino acids, peptides, and proteins.

Amino Acids

Proteins are dehydration polymers of amino acids, with each **amino acid residue** joined to its neighbor by a specific type of covalent bond. (The term "residue" reflects the loss of the elements of water when one amino acid is joined to another.) Proteins can be broken down (hydrolyzed) to their constituent amino acids by a variety of methods, and the earliest studies of proteins naturally focused on the free amino acids derived from them. The first to be discovered was asparagine, in 1806. The last of the 20 to be found, threonine, was not identified until 1938. All the amino acids have trivial or common names, in some cases derived from the source from which they were first isolated. Asparagine was first found in asparagus, and glutamate in wheat gluten; tyrosine was first isolated from cheese (its name is derived from the Greek *tyros,* "cheese"); and glycine (Greek *glykos,* "sweet") was so named because of its sweet taste.

Amino Acids Share Common Structural Features

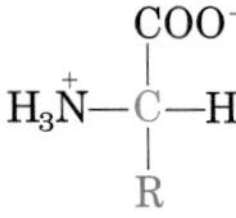

figure 5–2
General structure of an amino acid. This structure is common to all but one of the α-amino acids. (Proline, a cyclic amino acid, is the exception.) The R group or side chain (red) attached to the α carbon (blue) is different in each amino acid.

All 20 standard amino acids found in proteins are α-amino acids. They have a carboxyl group and an amino group bonded to the same carbon atom (the α carbon) (Fig. 5–2). They differ from each other in their side chains, or **R groups,** which vary in structure, size, and electric charge, and which influence the solubility of the amino acids in water. The 20 amino acids of proteins are often referred to as the standard amino acids, to distinguish them from less common amino acids that are residues modified after a protein has been synthesized, and from the many other kinds of amino acids present in living organisms but not in proteins. The standard amino acids have been assigned three-letter abbreviations and one-letter symbols (Table 5–1, p. 118), which are used as shorthand to indicate the composition and sequence of amino acids polymerized in proteins.

In a practice that can be confusing, two conventions are used to identify the carbons within an amino acid. The additional carbons in an R group are commonly designated β, γ, δ, ε, and so forth, proceeding out from the α carbon. For most other organic molecules, carbon atoms are simply numbered from one end, giving highest priority to carbons with substitutions containing atoms with the highest atomic numbers. Within this latter convention, the carboxyl group of an amino acid would be C-1 and the α carbon would be C-2. In some cases, such as amino acids with heterocyclic R groups, the Greek lettering system is ambiguous and the numbering convention is therefore used.

ε 6, δ 5, γ 4, β 3, α 2, 1

$CH_2—CH_2—CH_2—CH_2—CH—COO^-$ (with $^+NH_3$ on C-6 and on C-2)

Lysine

For all the standard amino acids except glycine, the α carbon is bonded to four different groups: a carboxyl group, an amino group, an R group, and a hydrogen atom (Fig. 5–2; in glycine, the R group is another hydrogen atom). The α-carbon atom is thus a **chiral center** (see Fig. 3–9). Because of the tetrahedral arrangement of the bonding orbitals around the α-carbon atom, the four different groups can occupy two different spatial arrangements that are nonsuperimposable mirror images of each other (Fig. 5–3). These two forms represent a class of stereoisomers called **enantiomers** (see Fig. 3–10). All molecules with a chiral center are also **optically active**—that is, they rotate plane-polarized light (see Box 3–1).

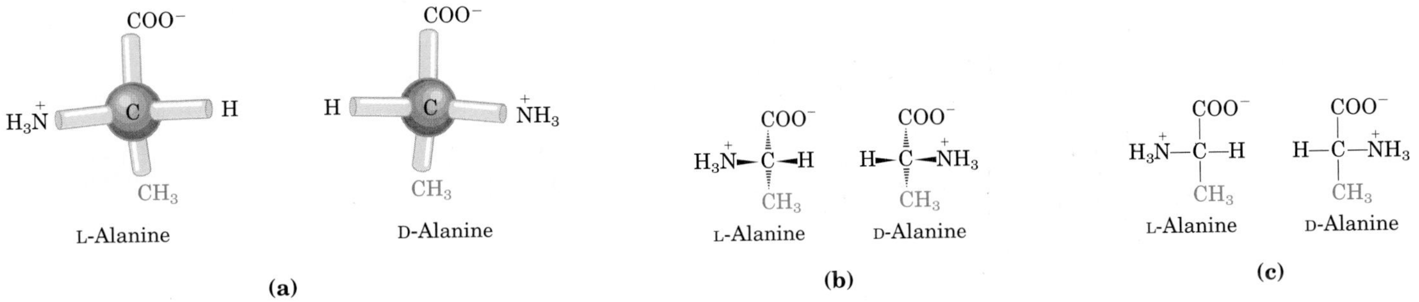

COO^- / $H_3\overset{+}{N}$—C—H / CH_3 L-Alanine; COO^- / H—C—$\overset{+}{N}H_3$ / CH_3 D-Alanine

(a) **(b)** **(c)**

figure 5–3

Stereoisomerism in α-amino acids. (a) The two stereoisomers of alanine, L- and D-alanine, are nonsuperimposable mirror images of each other (enantiomers). **(b, c)** Two different conventions for showing the configurations in space of stereoisomers. In perspective formulas **(b)** the solid wedge-shaped bonds project out of the plane of the paper, the dashed bonds behind it. In projection formulas **(c)** the horizontal bonds are assumed to project out of the plane of the paper, the vertical bonds behind. However, projection formulas are often used casually and are not always intended to portray a specific stereochemical configuration.

Special nomenclature has been developed to specify the **absolute configuration** of the four substituents of asymmetric carbon atoms. The absolute configurations of simple sugars and amino acids are specified by the **D, L system** (Fig. 5–4), based on the absolute configuration of the three-carbon sugar glyceraldehyde, a convention proposed by Emil Fischer in 1891. (Fischer knew what groups surrounded the asymmetric carbon of glyceraldehyde but had to guess at their absolute configuration; his guess was later confirmed by x-ray diffraction analysis.) For all chiral compounds, stereoisomers having a configuration related to that of L-glyceraldehyde are designated L, and stereoisomers related to D-glyceraldehyde are designated D. The functional groups of L-alanine are related to those of L-glyceraldehyde by simple chemical conversions. Thus the carboxyl group of L-alanine occupies the same position about the chiral carbon as does the aldehyde group of L-glyceraldehyde, because an aldehyde is readily converted (oxidized) to a carboxyl group. Historically, the similar *l* and *d* designations were used for levorotatory (rotating light to the left) and dextrorotatory (rotating light to the right) (see Box 3–1). However, not all L-amino acids are levorotatory, and the convention shown in Figure 5–4 was needed to avoid potential ambiguities about absolute configuration. By Fischer's convention, L and D refer *only* to the absolute configuration of the four substituents around the chiral carbon.

Another system of specifying configuration around a chiral center is the **RS system** (explained in Chapter 3), which is used in the systematic nomenclature of organic chemistry and describes more precisely the configuration of molecules with more than one chiral center.

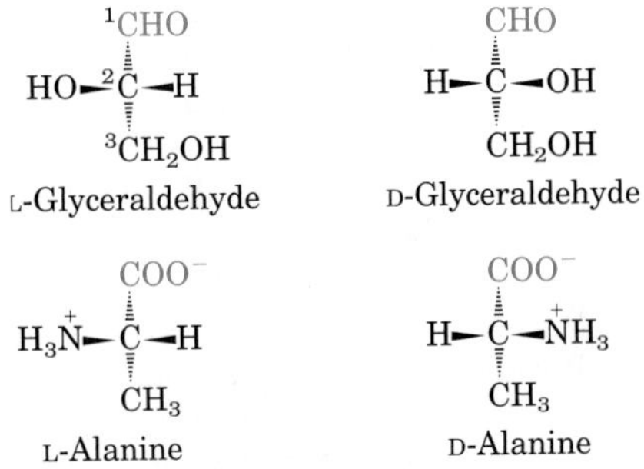

^1CHO / HO—^{2}C—H / $^3CH_2OH$ L-Glyceraldehyde; CHO / H—C—OH / CH_2OH D-Glyceraldehyde

COO^- / $H_3\overset{+}{N}$—C—H / CH_3 L-Alanine; COO^- / H—C—$\overset{+}{N}H_3$ / CH_3 D-Alanine

figure 5–4

Steric relationship of the stereoisomers of alanine to the absolute configuration of L- and D-glyceraldehyde. In these perspective formulas, the carbons are lined up vertically, with the chiral atom in the center. The carbons in these molecules are numbered beginning with the aldehyde or carboxyl carbons on the end (red), 1 to 3 from top to bottom as shown. When presented in this way, the R group of the amino acid (in this case the methyl group of alanine) is always below the α carbon. L-Amino acids are those with the α-amino group on the left, and D-amino acids have the α-amino group on the right.

The Amino Acid Residues in Proteins Are L Stereoisomers

Nearly all biological compounds with a chiral center occur naturally in only one stereoisomeric form, either D or L. The amino acid residues in protein molecules are exclusively L stereoisomers. D-Amino acid residues have been found only in a few, generally small peptides, including some peptides of bacterial cell walls and certain peptide antibiotics.

It is remarkable that all amino acid residues in proteins are L stereoisomers. As we noted in Chapter 3, when chiral compounds are formed by ordinary chemical reactions, the result is a racemic mixture of D and L isomers, which are difficult for a chemist to distinguish and separately isolate. But to a living system, D and L isomers are as different as the right hand and the left. The formation of stable, repeating substructures in proteins (Chapter 6) generally requires that their constituent amino acids be of one stereochemical series. Cells are able to specifically synthesize the L isomers of amino acids because the active sites of enzymes are asymmetric, causing the reactions they catalyze to be stereospecific.

table 5–1

Properties and Conventions Associated with the Standard Amino Acids

Amino acid	Abbreviated names		M_r	pK_a values: pK_1 (—COOH)	pK_2 ($—NH_3^+$)	pK_R (R group)	pI	Hydropathy index*	Occurrence in proteins (%)†
Nonpolar, aliphatic R groups									
Glycine	Gly	G	75	2.34	9.60		5.97	−0.4	7.2
Alanine	Ala	A	89	2.34	9.69		6.01	1.8	7.8
Valine	Val	V	117	2.32	9.62		5.97	4.2	6.6
Leucine	Leu	L	131	2.36	9.60		5.98	3.8	9.1
Isoleucine	Ile	I	131	2.36	9.68		6.02	4.5	5.3
Methionine	Met	M	149	2.28	9.21		5.74	1.9	2.3
Aromatic R groups									
Phenylalanine	Phe	F	165	1.83	9.13		5.48	2.8	3.9
Tyrosine	Tyr	Y	181	2.20	9.11	10.07	5.66	−1.3	3.2
Tryptophan	Trp	W	204	2.38	9.39		5.89	−0.9	1.4
Polar, uncharged R groups									
Serine	Ser	S	105	2.21	9.15		5.68	−0.8	6.8
Proline	Pro	P	115	1.99	10.96		6.48	1.6	5.2
Threonine	Thr	T	119	2.11	9.62		5.87	−0.7	5.9
Cysteine	Cys	C	121	1.96	10.28	8.18	5.07	2.5	1.9
Asparagine	Asn	N	132	2.02	8.80		5.41	−3.5	4.3
Glutamine	Gln	Q	146	2.17	9.13		5.65	−3.5	4.2
Positively charged R groups									
Lysine	Lys	K	146	2.18	8.95	10.53	9.74	−3.9	5.9
Histidine	His	H	155	1.82	9.17	6.00	7.59	−3.2	2.3
Arginine	Arg	R	174	2.17	9.04	12.48	10.76	−4.5	5.1
Negatively charged R groups									
Aspartate	Asp	D	133	1.88	9.60	3.65	2.77	−3.5	5.3
Glutamate	Glu	E	147	2.19	9.67	4.25	3.22	−3.5	6.3

*A scale combining hydrophobicity and hydrophilicity of R groups; it can be used to measure the tendency of an amino acid to seek an aqueous environment (− values) or a hydrophobic environment (+ values). See Chapter 12. From Kyte, J. & Doolittle, R.F. (1982) *J. Mol. Biol.* **157,** 105–132.

†Average occurrence in over 1150 proteins. From Doolittle, R.F. (1989) Redundancies in protein sequences. In *Prediction of Protein Structure and the Principles of Protein Conformation* (Fasman, G.D., ed) Plenum Press, NY, pp. 599–623.

Amino Acids Can Be Classified by R Group

Knowledge of the chemical properties of the standard amino acids is central to an understanding of biochemistry. The topic can be simplified by grouping the amino acids into five main classes based on the properties of their R groups (Table 5–1), in particular, their **polarity** or tendency to interact with water at biological pH (near pH 7.0). The polarity of the R groups varies widely, from totally nonpolar or hydrophobic (water-insoluble) to highly polar or hydrophilic (water-soluble).

The structures of the 20 standard amino acids are shown in Figure 5–5, and some of their properties are listed in Table 5–1. Within each class there are gradations of polarity, size, and shape of the R groups.

Nonpolar, Aliphatic R Groups The R groups in this class of amino acids are nonpolar and hydrophobic. The side chains of **alanine, valine, leucine,** and **isoleucine** tend to cluster together within proteins, stabilizing protein structure by means of hydrophobic interactions. **Glycine** has the simplest

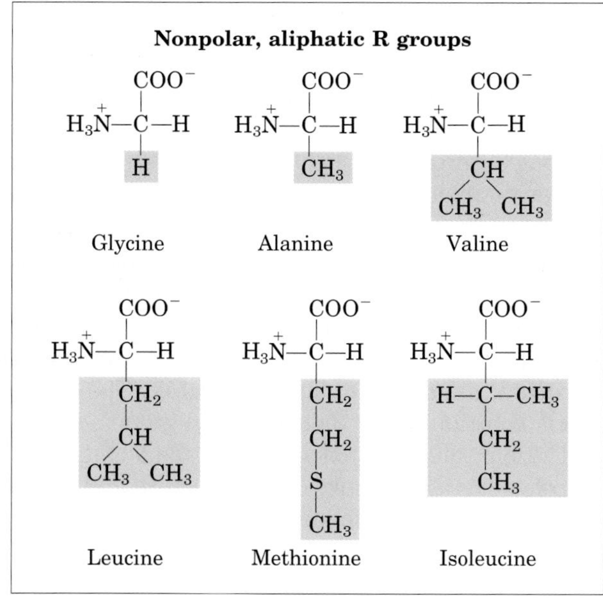

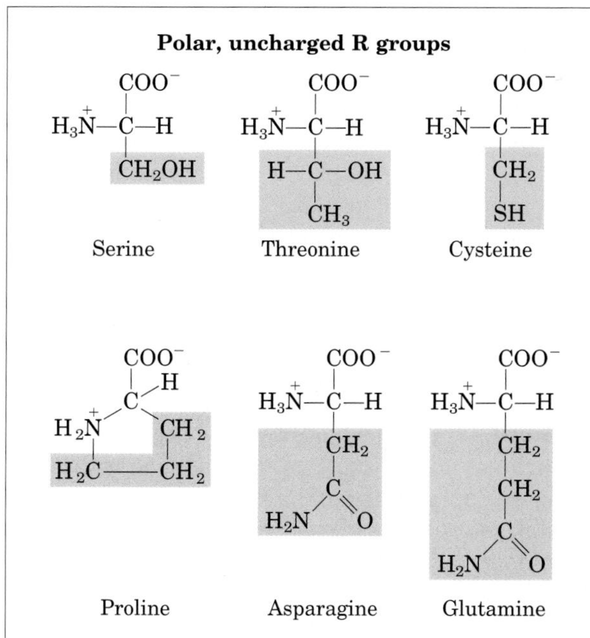

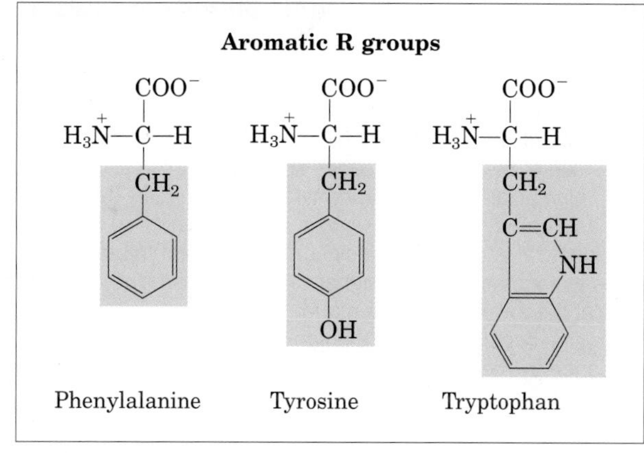

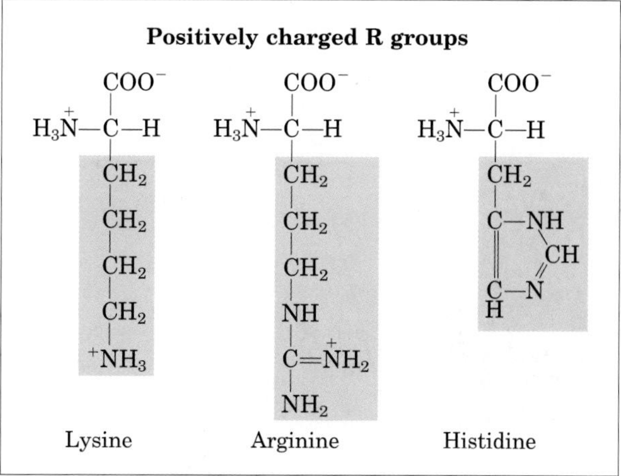

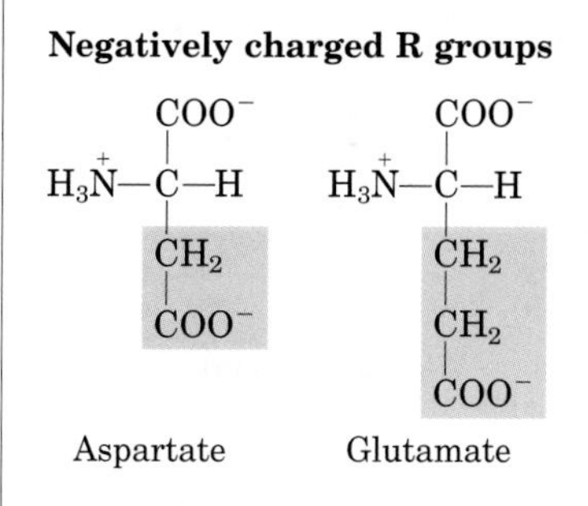

figure 5–5

The 20 standard amino acids of proteins. The structural formulas show the state of ionization that would predominate at pH 7.0. The unshaded portions are those common to all the amino acids; the portions shaded in red are the R groups. Although the R group of histidine is shown uncharged, its pK_a (see Table 5–1) is such that a small but significant fraction of these groups are positively charged at pH 7.0.

structure. Although it is formally nonpolar, its very small side chain makes no real contribution to hydrophobic interactions. **Methionine,** one of the two sulfur-containing amino acids, has a nonpolar thioether group in its side chain.

Aromatic R Groups **Phenylalanine, tyrosine,** and **tryptophan,** with their aromatic side chains, are relatively nonpolar (hydrophobic). All can participate in hydrophobic interactions. The hydroxyl group of tyrosine can form hydrogen bonds, and it is an important functional group in some enzymes. Tyrosine and tryptophan are significantly more polar than phenylalanine because of the tyrosine hydroxyl group and the nitrogen of the tryptophan indole ring.

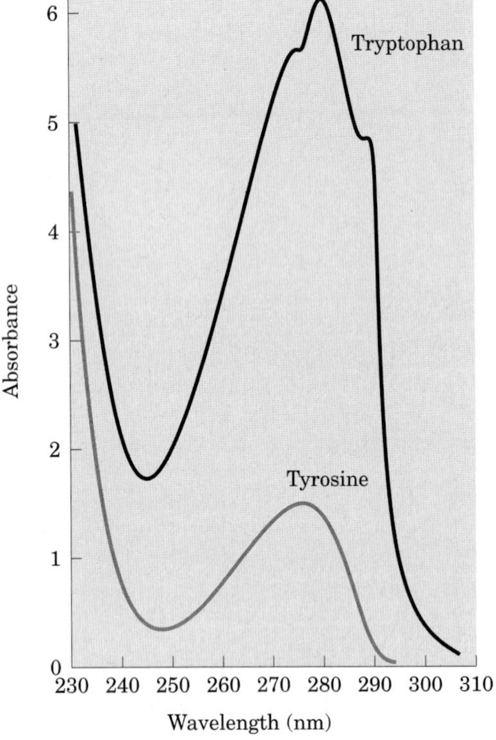

figure 5–6
Absorbance of ultraviolet light by aromatic amino acids. Comparison of the light absorbance spectra of the aromatic amino acids tryptophan and tyrosine at pH 6.0. The amino acids are present in equimolar amounts (10^{-3} M) under identical conditions. The light absorbance of tryptophan is as much as fourfold higher than that of tyrosine. Note that the absorbance maxima for both tryptophan and tyrosine occur near a wavelength of 280 nm. Light absorbance by the third aromatic amino acid, phenylalanine (not shown), generally contributes little to the absorbance properties of proteins.

Tryptophan and tyrosine, and to a much lesser extent phenylalanine, absorb ultraviolet light (Fig. 5–6; Box 5–1). This accounts for the characteristic strong absorbance of light by most proteins at a wavelength of 280 nm, a property exploited by researchers in the characterization of proteins.

Polar, Uncharged R Groups The R groups of these amino acids are more soluble in water, or more hydrophilic, than those of the nonpolar amino acids because they contain functional groups that form hydrogen bonds with water. This class of amino acids includes **serine, threonine, cysteine, proline, asparagine,** and **glutamine.** The polarity of serine and threonine is contributed by their hydroxyl groups; that of cysteine by its sulfhydryl group; and that of asparagine and glutamine by their amide groups. Proline has a distinctive cyclic structure and is only moderately polar. The secondary amino (imino) group of Pro residues is held in a rigid conformation that reduces the structural flexibility of polypeptide regions containing proline.

Asparagine and glutamine are the amides of two other amino acids also found in proteins, aspartate and glutamate, respectively, to which asparagine and glutamine are easily hydrolyzed by acid or base. Cysteine is readily oxidized to form a covalently linked dimeric amino acid called **cystine,** in which two cysteine molecules or residues are joined by a disulfide bond (Fig. 5–7). The disulfide-linked residues are strongly hydrophobic (nonpolar). Disulfide bonds play a special role in the structures of many proteins by forming covalent links between parts of a protein molecule or between two different protein chains.

Positively Charged (Basic) R Groups The most hydrophilic R groups are those that are either positively or negatively charged. The amino acids in which the R groups have significant positive charge at pH 7.0 are **lysine,** which has a second primary amino group at the ϵ position on its aliphatic chain; **arginine,** which has a positively charged guanidino group; and **histidine,** which has an imidazole group. Histidine is the only standard amino acid having an ionizable side chain with a pK_a near neutrality. In many enzyme-catalyzed reactions, a His residue facilitates the reaction by serving as a proton donor/acceptor.

Negatively Charged (Acidic) R Groups The two amino acids having R groups with a net negative charge at pH 7.0 are **aspartate** and **glutamate,** each of which has a second carboxyl group.

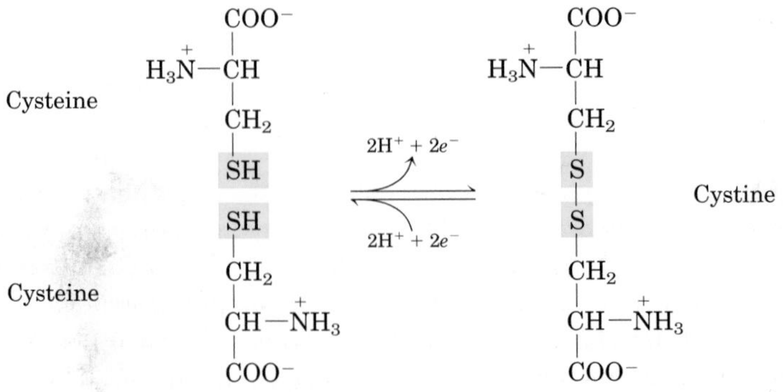

Cysteine: COO^- | $H_3\overset{+}{N}-CH$ | CH_2 | SH
Cysteine: SH | CH_2 | $CH-\overset{+}{N}H_3$ | COO^-

$\rightleftharpoons$ ($2H^+ + 2e^-$)

Cystine: COO^- | $H_3\overset{+}{N}-CH$ | CH_2 | S | S | CH_2 | $CH-\overset{+}{N}H_3$ | COO^-

figure 5–7
Reversible formation of a disulfide bond by the oxidation of two molecules of cysteine. Disulfide bonds between Cys residues stabilize the structures of many proteins.

box 5–1

Absorption of Light by Molecules: The Lambert-Beer Law

A wide range of biomolecules absorb light at characteristic wavelengths, just as tryptophan absorbs light at 280 nm (Fig. 5–6). Measurement of light absorption by a spectrophotometer is used to detect and identify molecules and to measure their concentration in solution. The fraction of the incident light absorbed by a solution at a given wavelength is related to the thickness of the absorbing layer (path length) and the concentration of the absorbing species (Fig. 1). These two relationships are combined into the Lambert-Beer law,

$$\log \frac{I_0}{I} = \epsilon cl$$

where I_0 is the intensity of the incident light, I is the intensity of the transmitted light, ϵ is the molar extinction coefficient (in units of liters per mole-centimeter), c is the concentration of the absorbing species (in moles per liter), and l is the path length of the light-absorbing sample (in centimeters). The Lambert-Beer law assumes that the incident light is parallel and monochromatic (of a single wavelength) and that the solvent and solute molecules are randomly oriented. The expression log (I_0/I) is called the **absorbance,** designated A.

It is important to note that each successive millimeter of path length of absorbing solution in a 1.0 cm cell absorbs not a constant amount but a constant fraction of the light that is incident upon it. However, with an absorbing layer of fixed path length, *the absorbance* A *is directly proportional to the concentration of the absorbing solute.*

The molar extinction coefficient varies with the nature of the absorbing compound, the solvent, and the wavelength, and also with pH if the light-absorbing species is in equilibrium with an ionization state that has different absorbance properties.

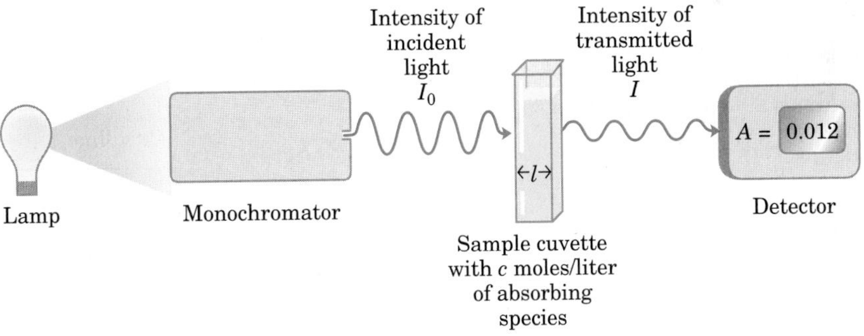

figure 1
The principal components of a spectrophotometer. A light source emits light along a broad spectrum, then the monochromator selects and transmits light of a particular wavelength. The monochromatic light passes through the sample in a cuvette of path length l and is absorbed by the sample in proportion to the concentration of the absorbing species. The transmitted light is measured by a detector.

Nonstandard Amino Acids Also Have Important Functions

In addition to the 20 standard amino acids, proteins may contain nonstandard residues created by modification of standard residues already incorporated into a polypeptide (Fig. 5–8a, p. 122). Among the nonstandard amino acids are **4-hydroxyproline,** a derivative of proline, and **5-hydroxylysine,** derived from lysine. The former is found in plant cell wall proteins, and both are found in collagen, a fibrous protein of connective tissues. **6-*N*-Methyllysine** is a constituent of myosin, a contractile protein of

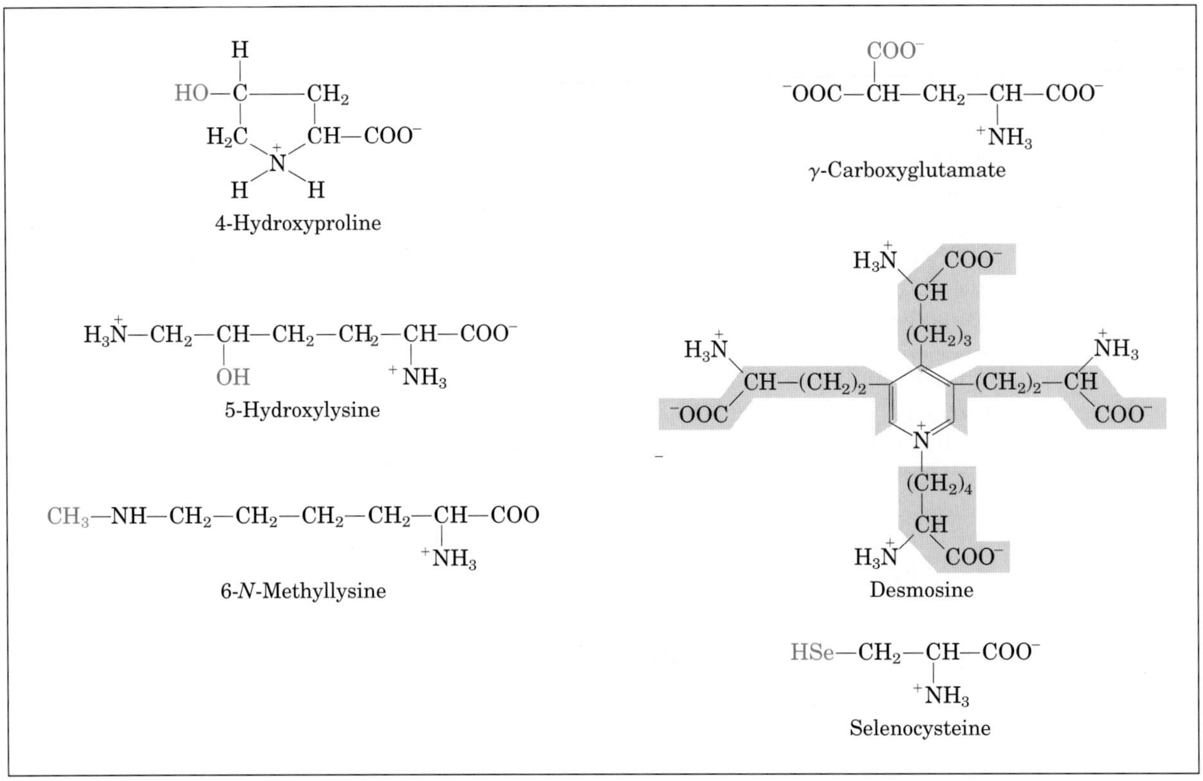

4-Hydroxyproline

γ-Carboxyglutamate

5-Hydroxylysine

Desmosine

6-*N*-Methyllysine

Selenocysteine

(a)

Ornithine

Citrulline

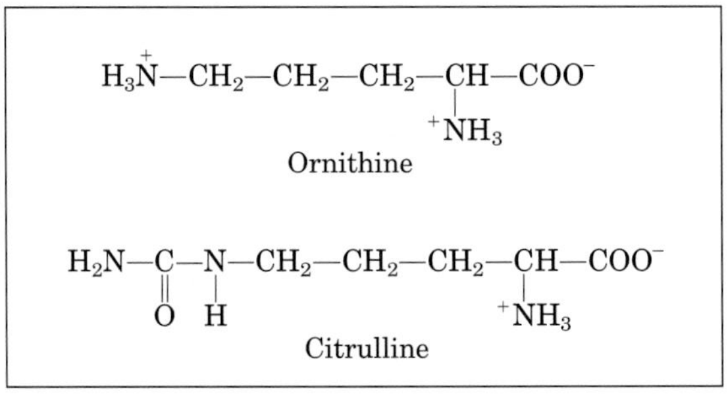

(b)

figure 5–8

Nonstandard amino acids. (a) Some nonstandard amino acids found in proteins. All are derived from standard amino acids. Extra functional groups added by modification reactions are shown in red. Desmosine is formed from four Lys residues (the four carbon backbones are shaded gray). Note the use of both numbers and Greek letters to identify the carbon atoms in these structures. **(b)** Ornithine and citrulline, which are not found in proteins, are intermediates in the biosynthesis of arginine and in the urea cycle.

muscle. Another important nonstandard amino acid is **γ-carboxyglutamate,** found in the blood-clotting protein prothrombin and in certain other proteins that bind Ca^{2+} as part of their biological function. More complicated is **desmosine,** a derivative of four Lys residues, which is found in the fibrous protein elastin.

Selenocysteine is a special case. This rare amino acid residue is introduced during protein synthesis rather than created through a postsynthetic modification. It contains selenium rather than the sulfur of cysteine. Actually derived from serine, selenocysteine is a constituent of just a few known proteins.

Some 300 additional amino acids have been found in cells. They have a variety of functions but are not constituents of proteins. **Ornithine** and **citrulline** (Fig. 5–8b) deserve special note because they are key intermediates in the biosynthesis of arginine (Chapter 22) and in the urea cycle (Chapter 18).

Amino Acids Can Act as Acids and Bases

When an amino acid is dissolved in water, it exists in solution as the dipolar ion, or **zwitterion** (German for "hybrid ion"), shown in Figure 5–9. A zwitterion can act as either an acid (proton donor):

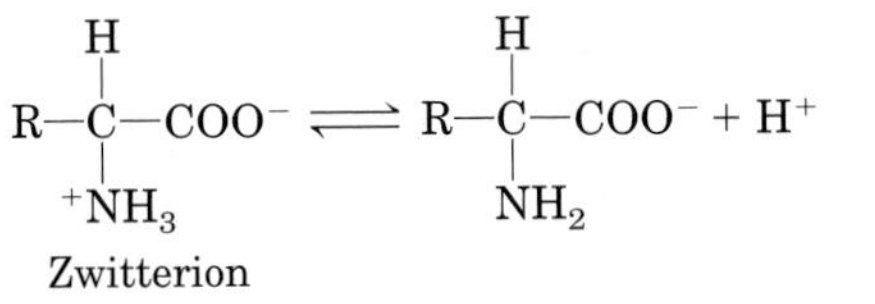

$$\underset{\text{Zwitterion}}{\mathrm{R{-}CH({}^{+}NH_3){-}COO^-}} \rightleftharpoons \mathrm{R{-}CH(NH_2){-}COO^-} + \mathrm{H^+}$$

or a base (proton acceptor):

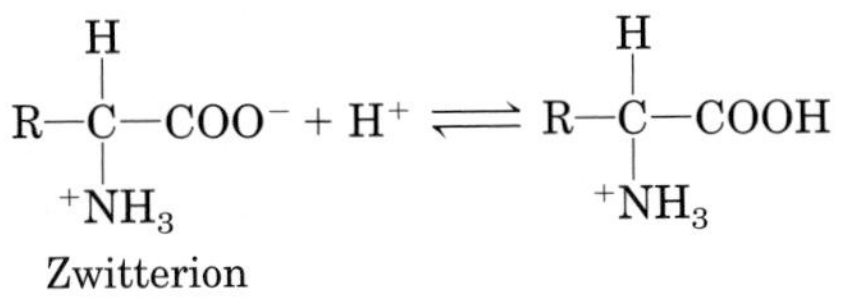

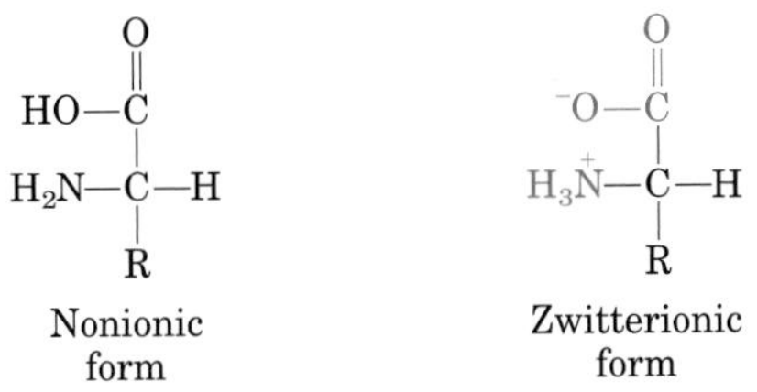

$$\underset{\text{Zwitterion}}{\mathrm{R{-}CH({}^{+}NH_3){-}COO^-}} + \mathrm{H^+} \rightleftharpoons \mathrm{R{-}CH({}^{+}NH_3){-}COOH}$$

Substances having this dual nature are **amphoteric** and are often called **ampholytes** (from "amphoteric electrolytes"). A simple monoamino monocarboxylic α-amino acid, such as alanine, is a diprotic acid when fully protonated—it has two groups, the —COOH group and the —NH_3^+ group, that can yield protons:

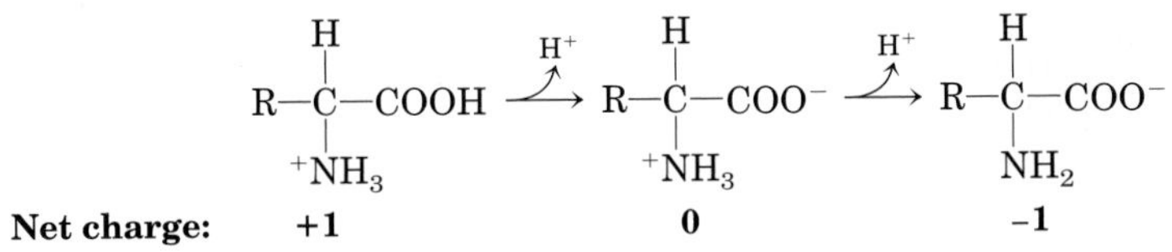

$$\mathrm{R{-}CH({}^{+}NH_3){-}COOH} \xrightarrow{\mathrm{H^+}} \mathrm{R{-}CH({}^{+}NH_3){-}COO^-} \xrightarrow{\mathrm{H^+}} \mathrm{R{-}CH(NH_2){-}COO^-}$$

Net charge: +1, 0, −1

$$\underset{\text{Nonionic form}}{\mathrm{HO{-}C({=}O){-}CH(R){-}H_2N}} \qquad \underset{\text{Zwitterionic form}}{\mathrm{{}^{-}O{-}C({=}O){-}CH(R){-}{}^{+}NH_3}}$$

figure 5–9
Nonionic and zwitterionic forms of amino acids. The nonionic form does not occur in significant amounts in aqueous solutions. The zwitterion predominates at neutral pH.

Amino Acids Have Characteristic Titration Curves

Acid-base titration involves the gradual addition or removal of protons (Chapter 4). Figure 5–10 shows the titration curve of the diprotic form of glycine. The plot has two distinct stages, corresponding to deprotonation of two different groups on glycine. Each of the two stages resembles in shape the titration curve of a monoprotic acid, such as acetic acid (see Fig. 4–15), and can be analyzed in the same way. At very low pH, the predominant ionic species of glycine is $^{+}H_3N{-}CH_2{-}COOH$, the fully protonated form. At the midpoint in the first stage of the titration, in which the —COOH group of glycine loses its proton, equimolar concentrations of the proton-donor ($^{+}H_3N{-}CH_2{-}COOH$) and proton-acceptor ($^{+}H_3N{-}CH_2{-}COO^-$) species are present. At the midpoint of any titration, a point of inflection is reached where the pH is equal to the pK_a of the protonated group being titrated (see Fig. 4–16). For glycine, the pH at the midpoint is 2.34, thus its —COOH group has a pK_a (labeled pK_1 in Fig. 5–10) of 2.34. (Recall from Chapter 4 that pH and pK_a are simply convenient notations for proton concentration and the equilibrium constant for ionization, respectively. The pK_a is a measure of the tendency of a group to give up a proton, with that tendency decreasing tenfold as the pK_a increases by one unit.) As the titration proceeds, another important point is reached at pH 5.97. Here there is another point of inflection, at which removal of the first proton is essentially complete and

figure 5–10
Titration of an amino acid. Shown here is the titration curve of 0.1 M glycine at 25 °C. The ionic species predominating at key points in the titration are shown above the graph. The shaded boxes, centered at about $pK_1 = 2.34$ and $pK_2 = 9.60$, indicate the regions of greatest buffering power.

$$\mathrm{{}^{+}NH_3{-}CH_2{-}COOH} \underset{}{\overset{pK_1}{\rightleftharpoons}} \mathrm{{}^{+}NH_3{-}CH_2{-}COO^-} \underset{}{\overset{pK_2}{\rightleftharpoons}} \mathrm{NH_2{-}CH_2{-}COO^-}$$

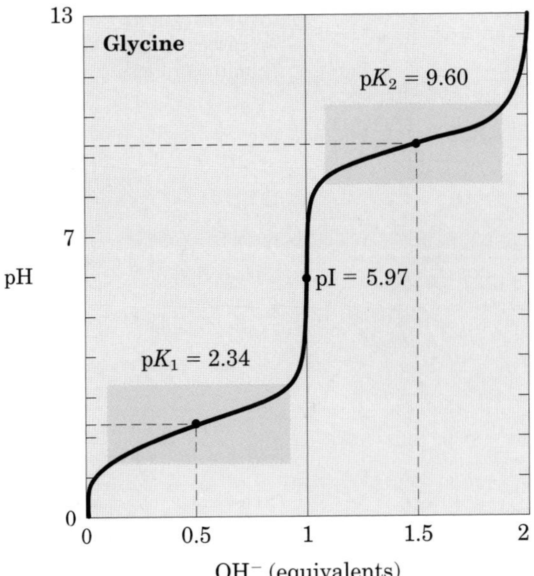

removal of the second has just begun. At this pH glycine is present largely as the dipolar ion $^+H_3N—CH_2—COO^-$. We shall return to the significance of this inflection point in the titration curve (pI in Fig. 5–10) shortly.

The second stage of the titration corresponds to the removal of a proton from the $—NH_3^+$ group of glycine. The pH at the midpoint of this stage is 9.60, equal to the pK_a (labeled pK_2 in Fig. 5–10) for the $—NH_3^+$ group. The titration is essentially complete at a pH of about 12, at which point the predominant form of glycine is $H_2N—CH_2—COO^-$.

From the titration curve of glycine we can derive several important pieces of information. First, it gives a quantitative measure of the pK_a of each of the two ionizing groups: 2.34 for the —COOH group and 9.60 for the $—NH_3^+$ group. Note that the carboxyl group of glycine is over 100 times more acidic (more easily ionized) than the carboxyl group of acetic acid, which, as we saw in Chapter 4, has a pK_a of 4.76, about average for a carboxyl group attached to an otherwise unsubstituted aliphatic hydrocarbon. The perturbed pK_a of glycine is caused by repulsion between the departing proton and the nearby positively charged amino group on the α-carbon atom, as described in Figure 5–11. The opposite charges on the resulting zwitterion are stabilizing, nudging the equilibrium farther to the right. Similarly, the pK_a of the amino group in glycine is perturbed downward relative to the average pK_a of an amino group. This effect is due partly to the electronegative oxygen atoms in the carboxyl groups, which tend to pull electrons toward them regardless of the carboxyl group charge, increasing the tendency of the amino group to give up a proton. Hence, the α-amino group has a pK_a that is lower than that of an aliphatic amine such as methylamine (Fig. 5–11). In short, the pK_a of any functional group is greatly affected by its chemical environment, a phenomenon sometimes exploited in the active sites of enzymes to promote exquisitely adapted reaction mechanisms that depend on the perturbed pK_a values of proton donor/acceptor groups of specific residues.

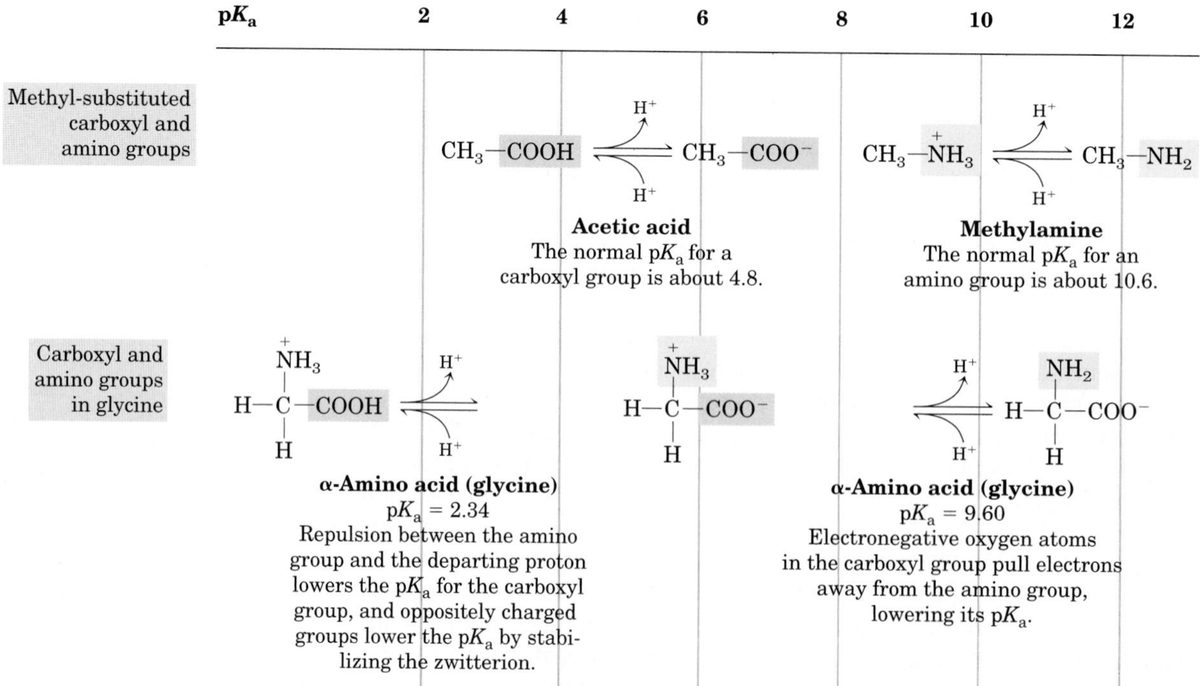

figure 5–11
Effect of the chemical environment on pK_a. The pK_a values for the ionizable groups in glycine are lower than those for simple, methyl-substituted amino and carboxyl groups. These downward perturbations of pK_a are due to intramolecular interactions. Similar effects can be caused by chemical groups that happen to be positioned nearby—for example, in the active site of an enzyme.

The second piece of information provided by the titration curve of glycine (Fig. 5–10) is that this amino acid has *two* regions of buffering power (see Fig. 4–17). One of these is the relatively flat portion of the curve, extending for approximately one pH unit on either side of the first pK_a of 2.34, indicating that glycine is a good buffer near this pH. The other buffering zone is centered around pH 9.60. Note that glycine is not a good buffer at the pH of intracellular fluid or blood, about 7.4. Within the buffering ranges of glycine, the Henderson-Hasselbalch equation (Chapter 4) can be used to calculate the proportions of proton-donor and proton-acceptor species of glycine required to make a buffer at a given pH.

Titration Curves Predict the Electric Charge of Amino Acids

Another important piece of information derived from the titration curve of an amino acid is the relationship between its net electric charge and the pH of the solution. At pH 5.97, the point of inflection between the two stages in its titration curve, glycine is present predominantly as its dipolar form, fully ionized but with no *net* electric charge (Fig. 5–10). The characteristic pH at which the net electric charge is zero is called the **isoelectric point** or **isoelectric pH,** designated **pI.** For glycine, which has no ionizable group in its side chain, the isoelectric point is simply the arithmetic mean of the two pK_a values:

$$\text{pI} = \frac{1}{2}(\text{p}K_1 + \text{p}K_2) = \frac{1}{2}(2.34 + 9.60) = 5.97$$

As is evident in Figure 5–10, glycine has a net negative charge at any pH above its pI and will thus move toward the positive electrode (the anode) when placed in an electric field. At any pH below its pI, glycine has a net positive charge and will move toward the negative electrode (the cathode). The farther the pH of a glycine solution is from its isoelectric point, the greater the net electric charge of the population of glycine molecules. At pH 1.0, for example, glycine exists almost entirely as the form $^+H_3N{-}CH_2{-}COOH$, with a net positive charge of 1.0. At pH 2.34, where there is an equal mixture of $^+H_3N{-}CH_2{-}COOH$ and $^+H_3N{-}CH_2{-}COO^-$, the average or net positive charge is 0.5. The sign and the magnitude of the net charge of any amino acid at any pH can be predicted in the same way.

Amino Acids Differ in Their Acid-Base Properties

The shared properties of many amino acids permit some simplifying generalizations about their acid-base behaviors.

All amino acids with a single α-amino group, a single α-carboxyl group, and an R group that does not ionize have titration curves resembling that of glycine (Fig. 5–10). These amino acids have very similar, although not identical, pK_a values: pK_a of the —COOH group in the range of 1.8 to 2.4, and pK_a of the $-NH_3^+$ group in the range of 8.8 to 11.0 (Table 5–1).

Amino acids with an ionizable R group have more complex titration curves, with *three* stages corresponding to the three possible ionization steps; thus they have three pK_a values. The additional stage for the titration of the ionizable R group merges to some extent with the other two. The titration curves for two amino acids of this type, glutamate and histidine, are shown in Figure 5–12. The isoelectric points reflect the nature of the ionizing R groups present. For example, glutamate has a pI of 3.22, considerably lower than that of glycine. This is due to the presence of two carboxyl groups which, at the average of their pK_a values (3.22), contribute a net negative charge of −1 that balances the +1 contributed by the amino group. Similarly, the pI of histidine, with two groups that are positively charged when protonated, is 7.59 (the average of the pK_a values of the amino and imidazole groups), much higher than that of glycine.

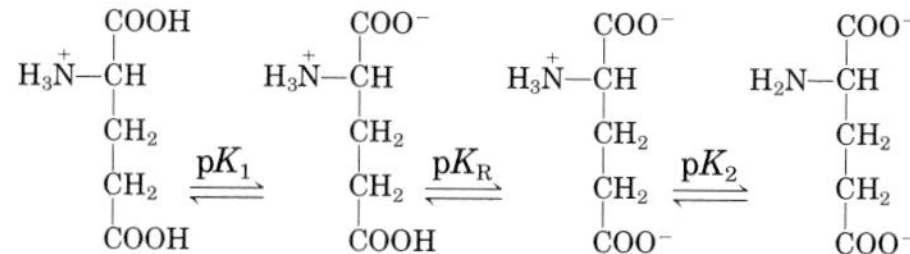

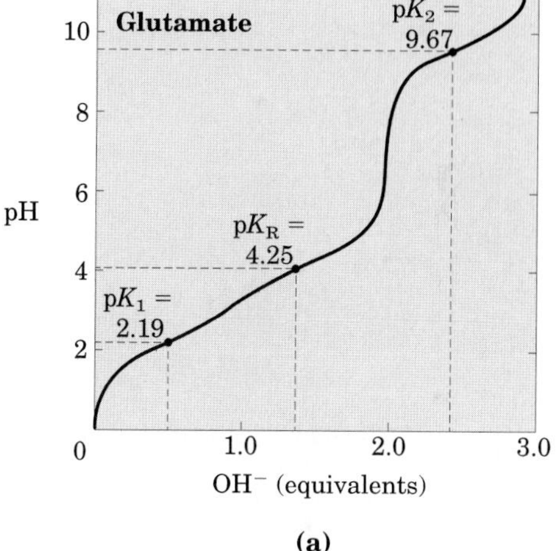

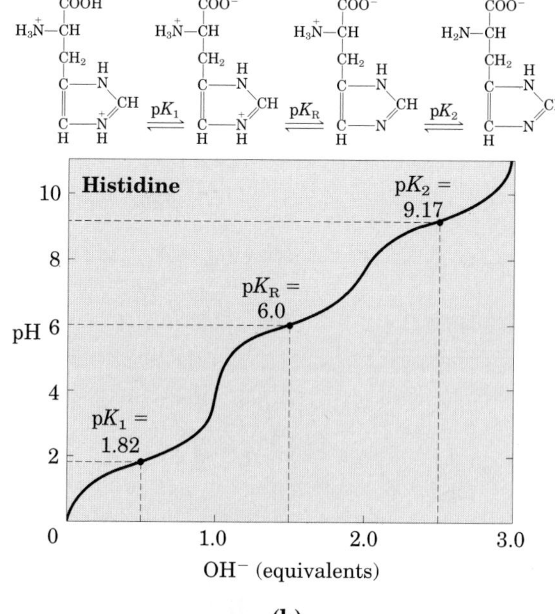

figure 5–12

Titration curves for **(a)** glutamate and **(b)** histidine. The pK_a of the R group is designated here as pK_R.

Another important generalization can be made about the acid-base behavior of the 20 standard amino acids. As pointed out earlier, under the general condition of free and open exposure to the aqueous environment, only histidine has an R group ($pK_a = 6.0$) providing significant buffering power near the neutral pH usually found in the intracellular and intercellular fluids of most animals and bacteria. No other amino acid has an ionizable side chain with a pK_a value near enough to pH 7.0 to be an effective physiological buffer (Table 5–1).

Peptides and Proteins

We now turn to polymers of amino acids, the **peptides** and **proteins.** Biologically occurring peptides range in size from small to very large, consisting of two or three to thousands of linked amino acid residues. The focus here is on the fundamental chemical properties of these polymers.

Peptides Are Chains of Amino Acids

Two amino acid molecules can be covalently joined through a substituted amide linkage, termed a **peptide bond,** to yield a dipeptide. Such a linkage is formed by removal of the elements of water (dehydration) from the α-carboxyl group of one amino acid and the α-amino group of another (Fig. 5–13). Peptide bond formation is an example of a condensation reaction, a common class of reaction in living cells. Under standard biochemical conditions the reaction shown in Figure 5–13 has an equilibrium that favors reactants rather than products. To make the reaction thermodynamically more favorable, the carboxyl group must be chemically modified or activated so that the hydroxyl group can be more readily eliminated. A chemical approach to this problem is outlined later in this chapter. The biological approach to peptide bond formation is a major topic of Chapter 27.

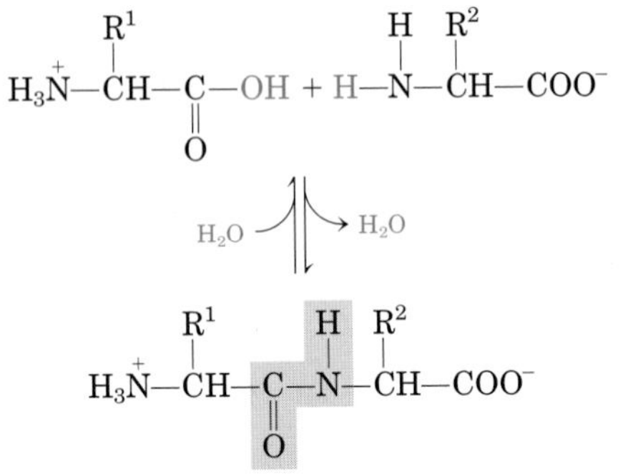

$$H_3\overset{+}{N}-CH(R^1)-C(=O)-OH + H-N(H)-CH(R^2)-COO^-$$

$$\rightleftharpoons \quad (H_2O)$$

$$H_3\overset{+}{N}-CH(R^1)-C(=O)-N(H)-CH(R^2)-COO^-$$

figure 5–13

Formation of a peptide bond by condensation. The α-amino group of one amino acid (with R^2 group) acts as a nucleophile (see Table 3–4) to displace the hydroxyl group of another amino acid (with R^1 group), forming a peptide bond (shaded in gray). Amino groups are good nucleophiles, but the hydroxyl group is a poor leaving group and is not readily displaced. At physiological pH, the reaction shown does not occur to any appreciable extent.

Three amino acids can be joined by two peptide bonds to form a tripeptide; similarly, amino acids can be linked to form tetrapeptides and pentapeptides. When a few amino acids are joined in this fashion, the structure is called an **oligopeptide.** When many amino acids are joined, the product is called a **polypeptide.** Proteins may have thousands of amino acid residues. Although the terms "protein" and "polypeptide" are sometimes used interchangeably, molecules referred to as polypeptides generally have molecular weights below 10,000.

Figure 5–14 shows the structure of a pentapeptide. As already noted, an amino acid unit in a peptide is often called a residue (the part left over after losing a hydrogen atom from its amino group and a hydroxyl moiety from its carboxyl group). In a peptide, the amino acid residue at the end with a free α-amino group is the **amino-terminal** (or *N*-terminal) residue; the residue at the other end, which has a free carboxyl group, is the **carboxyl-terminal** (*C*-terminal) residue.

Although hydrolysis of a peptide bond is an exergonic reaction, it occurs slowly because of its high activation energy. As a result, the peptide bonds in proteins are quite stable, with a half-life ($t_{1/2}$) of about 7 years under most intracellular conditions.

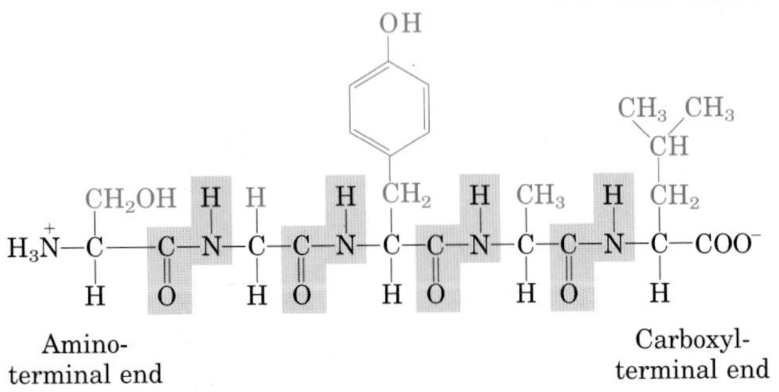

$$H_3\overset{+}{N}-CH(CH_2OH)-C(=O)-NH-CH_2-C(=O)-NH-CH(CH_2C_6H_4OH)-C(=O)-NH-CH(CH_3)-C(=O)-NH-CH(CH_2CH(CH_3)_2)-COO^-$$

Amino-terminal end — Carboxyl-terminal end

figure 5–14

The pentapeptide serylglycyltyrosylalanylleucine, or Ser–Gly–Tyr–Ala–Leu. Peptides are named beginning with the amino-terminal residue, which by convention is placed at the left. The peptide bonds are shaded in gray, the R groups are in red.

Peptides Can Be Distinguished by Their Ionization Behavior

Peptides contain only one free α-amino group and one free α-carboxyl group, one at each end of the chain (Fig. 5–15). These groups ionize as they do in free amino acids, although the ionization constants are different because the oppositely charged group is absent from the α carbon. The α-amino and α-carboxyl groups of all nonterminal amino acids are covalently joined in the form of peptide bonds, which do not ionize and thus do not contribute to the total acid-base behavior of peptides. However, the R groups of some amino acids can ionize (Table 5–1), and in a peptide these contribute to the overall acid-base properties of the molecule (Fig. 5–15). Thus the acid-base behavior of a peptide can be predicted from its free α-amino and α-carboxyl groups as well as the nature and number of its ionizable R groups. Like free amino acids, peptides have characteristic titration curves and a characteristic isoelectric pH (pI) at which they do not move in an electric field. These properties are exploited in some of the techniques used to separate peptides and proteins, as we shall see later in the chapter. It should be emphasized that the pK_a value for an ionizable R group can change somewhat when an amino acid becomes a residue in a peptide. The loss of charge in the α-carboxyl and α-amino groups, interactions with other peptide R groups, and other environmental factors can affect the pK_a. The pK_a values for R groups listed in Table 5–1 can be a useful guide to the pH range in which a given group will ionize, but they cannot be strictly applied to peptides.

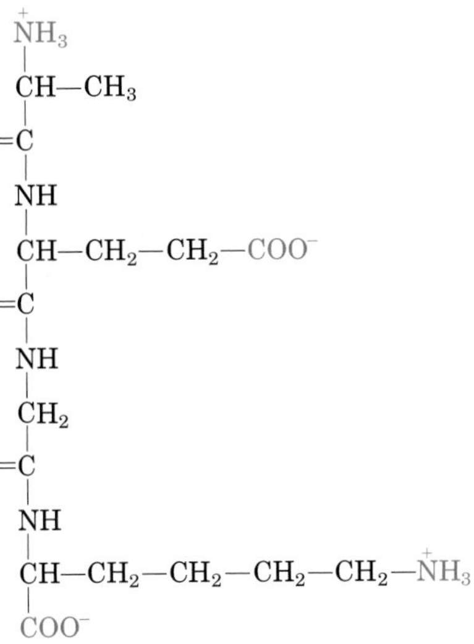

Ala $\overset{+}{N}H_3$–CH–CH_3; O=C–NH
Glu CH–CH_2–CH_2–COO^-; O=C–NH
Gly CH_2; O=C–NH
Lys CH–CH_2–CH_2–CH_2–CH_2–$\overset{+}{N}H_3$; COO^-

figure 5–15

Alanylglutamylglycyllysine. This tetrapeptide has one free α-amino group, one free α-carboxyl group, and two ionizable R groups. The groups ionized at pH 7.0 are in red.

Biologically Active Peptides and Polypeptides Occur in a Vast Range of Sizes

No generalizations can be made about the molecular weights of biologically active peptides and proteins in relation to their function. Naturally occurring peptides range in length from two amino acids to many thousands of residues. Even the smallest peptides can have biologically important effects. Consider the commercially synthesized dipeptide L-aspartyl-L-phenylalanine methyl ester, the artificial sweetener better known as aspartame or NutraSweet.

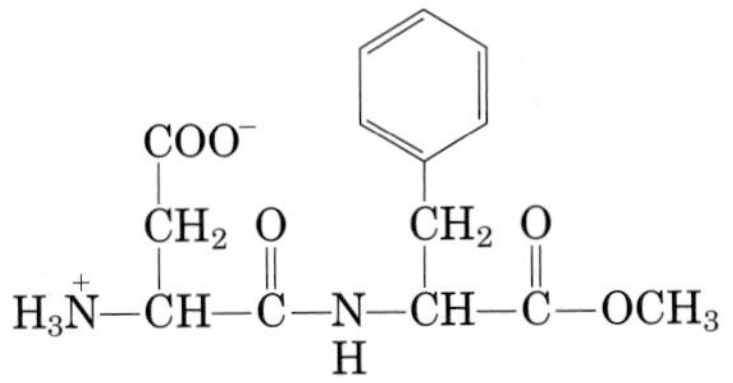

$$H_3\overset{+}{N}-\underset{\underset{COO^-}{|}}{\overset{}{CH}}(CH_2)-\overset{O}{\overset{\|}{C}}-\underset{H}{N}-CH(CH_2C_6H_5)-\overset{O}{\overset{\|}{C}}-OCH_3$$

L-Aspartyl-L-phenylalanine methyl ester (aspartame)

Many small peptides exert their effects at very low concentrations. For example, a number of vertebrate hormones (Chapter 23) are small peptides. These include oxytocin (nine amino acid residues), which is secreted by the posterior pituitary and stimulates uterine contractions; bradykinin (nine residues), which inhibits inflammation of tissues; and thyrotropin-releasing factor (three residues), which is formed in the hypothalamus and stimulates the release of another hormone, thyrotropin, from the anterior pituitary gland. Some extremely toxic mushroom poisons, such as amanitin, are also small peptides, as are many antibiotics.

Slightly larger are small polypeptides and oligopeptides such as the pancreatic hormone insulin, which contains two polypeptide chains, one having 30 amino acid residues and the other 21. Glucagon, another pancreatic hormone, has 29 residues; it opposes the action of insulin. Corticotropin is a 39-residue hormone of the anterior pituitary gland that stimulates the adrenal cortex.

How long are the polypeptide chains in proteins? As Table 5–2 shows, lengths vary considerably. Human cytochrome *c* has 104 amino acid residues linked in a single chain; bovine chymotrypsinogen has 245 residues. At the extreme is titin, a constituent of vertebrate muscle, which has nearly 27,000 amino acid residues and a molecular weight of about 3,000,000. The vast majority of naturally occurring polypeptides are much smaller than this, containing less than 2,000 amino acid residues.

Some proteins consist of a single polypeptide chain, but others, called **multisubunit** proteins, have two or more polypeptides associated

table 5–2

Molecular Data on Some Proteins

	Molecular weight	Number of residues	Number of polypeptide chains
Cytochrome *c* (human)	13,000	104	1
Ribonuclease A (bovine pancreas)	13,700	124	1
Lysozyme (egg white)	13,930	129	1
Myoglobin (equine heart)	16,890	153	1
Chymotrypsin (bovine pancreas)	21,600	241	3
Chymotrypsinogen (bovine)	22,000	245	1
Hemoglobin (human)	64,500	574	4
Serum albumin (human)	68,500	609	1
Hexokinase (yeast)	102,000	972	2
RNA polymerase (*E. coli*)	450,000	4,158	5
Apolipoprotein B (human)	513,000	4,536	1
Glutamine synthetase (*E. coli*)	619,000	5,628	12
Titin (human)	2,993,000	26,926	1

noncovalently (Table 5–2). The individual polypeptide chains in a multisubunit protein may be identical or different. If at least two are identical the protein is said to be **oligomeric,** and identical units (consisting of one or more polypeptide chains) are referred to as **protomers.** Hemoglobin, for example, has four polypeptide subunits: two identical α chains and two identical β chains, all four held together by noncovalent interactions. Each α subunit is paired in an identical way with a β subunit within the structure of this multisubunit protein, so that hemoglobin can be considered either a tetramer of four polypeptide subunits or a dimer of $\alpha\beta$ protomers.

A few proteins contain two or more polypeptide chains linked covalently. For example, the two polypeptide chains of insulin are linked by disulfide bonds. In such cases, the individual polypeptides are not considered subunits, but are commonly referred to simply as chains.

We can calculate the approximate number of amino acid residues in a simple protein containing no other chemical group by dividing its molecular weight by 110. Although the average molecular weight of the 20 standard amino acids is about 138, the smaller amino acids predominate in most proteins; if we take into account the proportions in which the various amino acids occur in proteins (Table 5–1), the average molecular weight is nearer to 128. Because a molecule of water (M_r 18) is removed to create each peptide bond, the average molecular weight of an amino acid residue in a protein is about 128 − 18 = 110.

table 5–3

Amino Acid Composition of Two Proteins*

Amino acid	Number of residues per molecule of protein: Bovine cytochrome *c*	Bovine chymotrypsinogen
Ala	6	22
Arg	2	4
Asn	5	15
Asp	3	8
Cys	2	10
Gln	3	10
Glu	9	5
Gly	14	23
His	3	2
Ile	6	10
Leu	6	19
Lys	18	14
Met	2	2
Phe	4	6
Pro	4	9
Ser	1	28
Thr	8	23
Trp	1	8
Tyr	4	4
Val	3	23
Total	104	245

*Note that standard procedures for the acid hydrolysis of proteins convert Asn and Gln to Asp and Glu, respectively. In addition, Trp is destroyed. Special procedures must be employed to determine the amounts of these amino acids.

Polypeptides Have Characteristic Amino Acid Compositions

Hydrolysis of peptides or proteins with acid yields a mixture of free α-amino acids. When completely hydrolyzed, each type of protein yields a characteristic proportion or mixture of the different amino acids. The 20 standard amino acids almost never occur in equal amounts in a protein. Some amino acids may occur only once per molecule or not at all in a given type of protein; others may occur in large numbers. Table 5–3 shows the composition of the amino acid mixtures obtained on complete hydrolysis of bovine cytochrome *c* and chymotrypsinogen, the inactive precursor of the digestive enzyme chymotrypsin. These two proteins, with very different functions, also differ significantly in the relative numbers of each kind of amino acid they contain.

Some Proteins Contain Chemical Groups Other Than Amino Acids

Many proteins, for example the enzymes ribonuclease and chymotrypsinogen, contain only amino acid residues and no other chemical groups; these are considered simple proteins. However, some proteins contain permanently associated chemical components in addition to amino acids; these are called **conjugated proteins.** The non–amino acid part of a conjugated protein is usually called its **prosthetic group.** Conjugated proteins are classified on the basis of the chemical nature of their prosthetic groups (Table 5–4); for example, **lipoproteins** contain lipids, **glycoproteins** contain sugar groups, and **metalloproteins** contain a specific metal. A number of proteins contain more than one prosthetic group. Usually the prosthetic group plays an important role in the protein's biological function.

table 5–4

Conjugated Proteins

Class	Prosthetic group(s)	Example
Lipoproteins	Lipids	β_1-Lipoprotein of blood
Glycoproteins	Carbohydrates	Immunoglobulin G
Phosphoproteins	Phosphate groups	Casein of milk
Hemoproteins	Heme (iron porphyrin)	Hemoglobin
Flavoproteins	Flavin nucleotides	Succinate dehydrogenase
Metalloproteins	Iron	Ferritin
	Zinc	Alcohol dehydrogenase
	Calcium	Calmodulin
	Molybdenum	Dinitrogenase
	Copper	Plastocyanin

There Are Several Levels of Protein Structure

For large macromolecules such as proteins, the tasks of describing and understanding structure are approached at several levels of complexity, arranged in a kind of conceptual hierarchy. Four levels of protein structure are commonly defined (Fig. 5–16). A description of all covalent bonds (mainly peptide bonds and disulfide bonds) linking amino acid residues in a polypeptide chain is its **primary structure.** The most important element of primary structure is the *sequence* of amino acid residues. **Secondary structure** refers to particularly stable arrangements of amino acid residues giving rise to recurring structural patterns. **Tertiary structure** describes all aspects of the three-dimensional folding of a polypeptide. When a protein has two or more polypeptide subunits, their arrangement in space is referred to as **quaternary structure.**

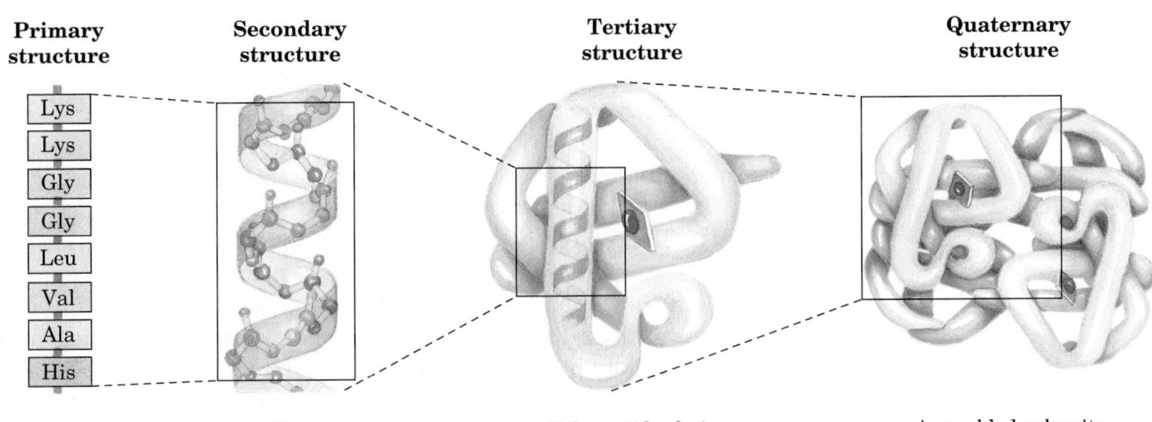

figure 5–16

Levels of structure in proteins. The *primary structure* consists of a sequence of amino acids linked together by peptide bonds and includes any disulfide bonds. The resulting polypeptide can be coiled into units of *secondary structure,* such as an α helix. The helix is a part of the *tertiary structure* of the folded polypeptide, which is itself one of the subunits that make up the *quaternary structure* of the multisubunit protein, in this case hemoglobin.

Working with Proteins

Our understanding of protein structure and function has been derived from the study of many individual proteins. To study a protein in any detail it must be separated from all other proteins, and techniques must be available to determine its properties. The necessary methods come from protein chemistry, a discipline as old as biochemistry itself and one that retains a central position in biochemical research.

Proteins Can Be Separated and Purified

A pure preparation of a protein is essential before its properties, amino acid composition, and sequence can be determined. Given that cells contain thousands of different kinds of proteins, how can one protein be purified? Methods for separating proteins take advantage of properties that vary from one protein to the next. For example, many proteins bind to other biomolecules with great specificity, and such proteins can be separated on the basis of their binding properties.

The source of a protein is generally tissue or microbial cells. The first step in any protein purification procedure is to break open these cells, releasing their proteins into a solution called a **crude extract.** If necessary, differential centrifugation can be used to prepare subcellular fractions or to isolate specific organelles (see Fig. 2–20).

Once the extract or organelle preparation is ready, various methods are available for purifying one or more of the proteins it contains. Commonly, the extract is subjected to treatments that separate the proteins into different fractions based on some property such as size or charge, a process referred to as **fractionation.** Early fractionation steps in a purification utilize differences in protein solubility, which is a complex function of pH, temperature, salt concentration, and other factors. The solubility of proteins is generally lowered at high salt concentrations, an effect called "salting out." The addition of a salt in the right amounts can selectively precipitate some proteins, while others remain in solution. Ammonium sulfate ($(NH_4)_2SO_4$) is often used for this purpose because of its high solubility in water.

A solution containing the protein of interest often must be further altered before subsequent purification steps are possible. For example, **dialysis** is a procedure that separates proteins from solvents by taking advantage of the proteins' larger size. The partially purified extract is placed in a bag or tube made of a semipermeable membrane. When this is suspended in a larger volume of buffered solution of appropriate ionic strength, the membrane allows the exchange of salt and buffer but not proteins. Thus dialysis retains large proteins within the membranous bag or tube while allowing the concentration of other solutes in the protein preparation to change until they come into equilibrium with the solution outside the membrane. Dialysis might be used, for example, to remove ammonium sulfate from the protein preparation.

The most powerful methods for fractionating proteins make use of **column chromatography,** which takes advantage of differences in protein charge, size, binding affinity, and other properties (Fig. 5–17). A porous solid material with appropriate chemical properties (the stationary phase) is held in a column, and a buffered solution (the mobile phase) percolates through it. The protein-containing solution is layered on the top of the column, then also percolates through the solid matrix as an ever-expanding band within the larger mobile phase (Fig. 5–17b). Individual proteins migrate faster or more slowly through the column depending on their properties. For example, in cation-exchange chromatography (Fig. 5–18a), the solid matrix has negatively charged groups. In the mobile phase, proteins

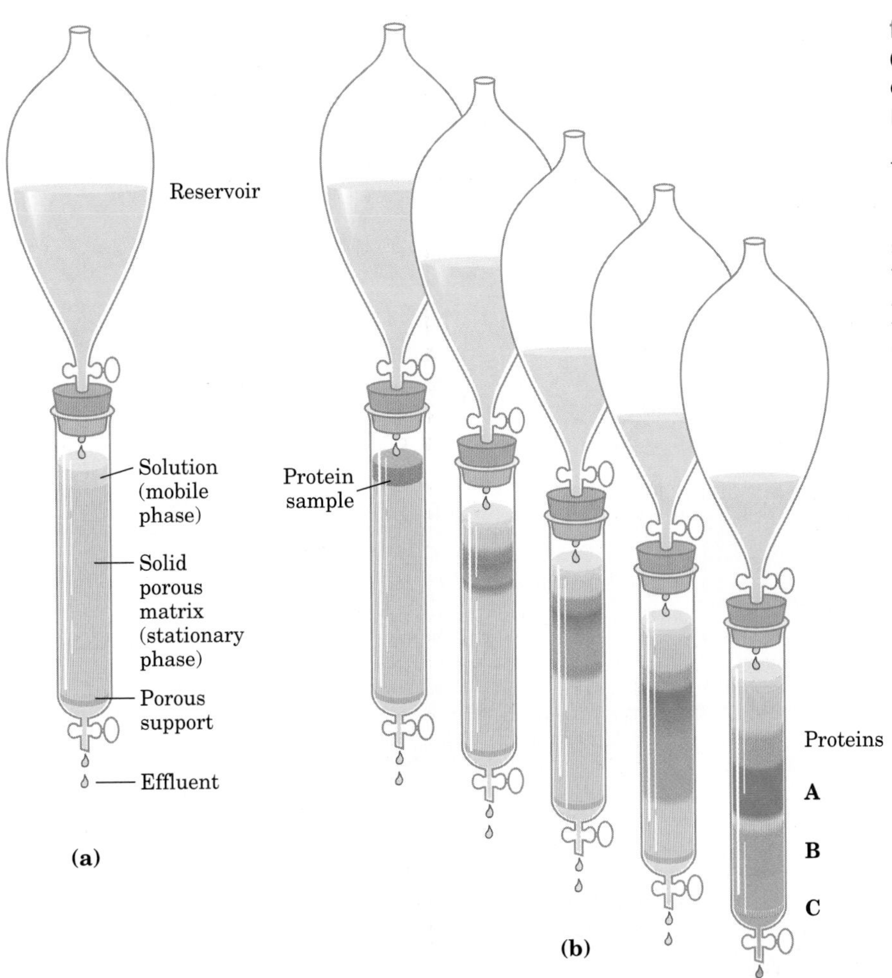

figure 5–17
Column chromatography. (a) The standard elements of a chromatographic column. A solid, porous material is supported inside a column generally made of some form of plastic. The solid material (matrix) makes up the stationary phase through which flows a solution, the mobile phase. The solution that passes out of the column at the bottom (the effluent) is constantly replaced by solution supplied from a reservoir at the top. **(b)** The protein solution to be separated is layered on top of the column and allowed to percolate into the solid matrix. Additional solution is added on top. The protein solution forms a band within the mobile phase that is initially the depth of the protein solution applied to the column. As proteins migrate through the column, they are retarded to different degrees by their different interactions with the matrix material. The overall protein band thus widens as it moves through the column. Individual types of proteins (such as A, B, and C, shown in blue, red, and green) gradually separate from each other, forming bands within the broader protein band. Separation improves (resolution increases) as the length of the column increases. However, each individual protein band also broadens with time due to diffusional spreading, a process that decreases resolution. In this example, protein A is well separated from B and C, but diffusional spreading prevents complete separation of B and C under these conditions.

with a net positive charge migrate through the matrix more slowly than those with a net negative charge, because the migration of the former is retarded more by interaction with the stationary phase. The two types of protein can separate into two distinct bands. The expansion of the protein band in the mobile phase (the protein solution) is caused both by separation of proteins with different properties and by diffusional spreading. As the length of the column increases, the resolution of two types of protein with different net charges generally improves. However, the rate at which the protein solution can flow through the column usually decreases with column length. As the length of time spent on the column increases, the resolution can decline as a result of diffusional spreading within each protein band.

Figure 5–18 shows two other variations of column chromatography in addition to ion exchange.

A modern refinement in chromatographic methods is **HPLC,** or **high-performance liquid chromatography.** HPLC makes use of high-pressure pumps that speed the movement of the protein molecules down the column, as well as higher-quality chromatographic materials that can withstand the crushing force of the pressurized flow. By reducing the transit time on the column, HPLC can limit diffusional spreading of protein bands and thus greatly improve resolution.

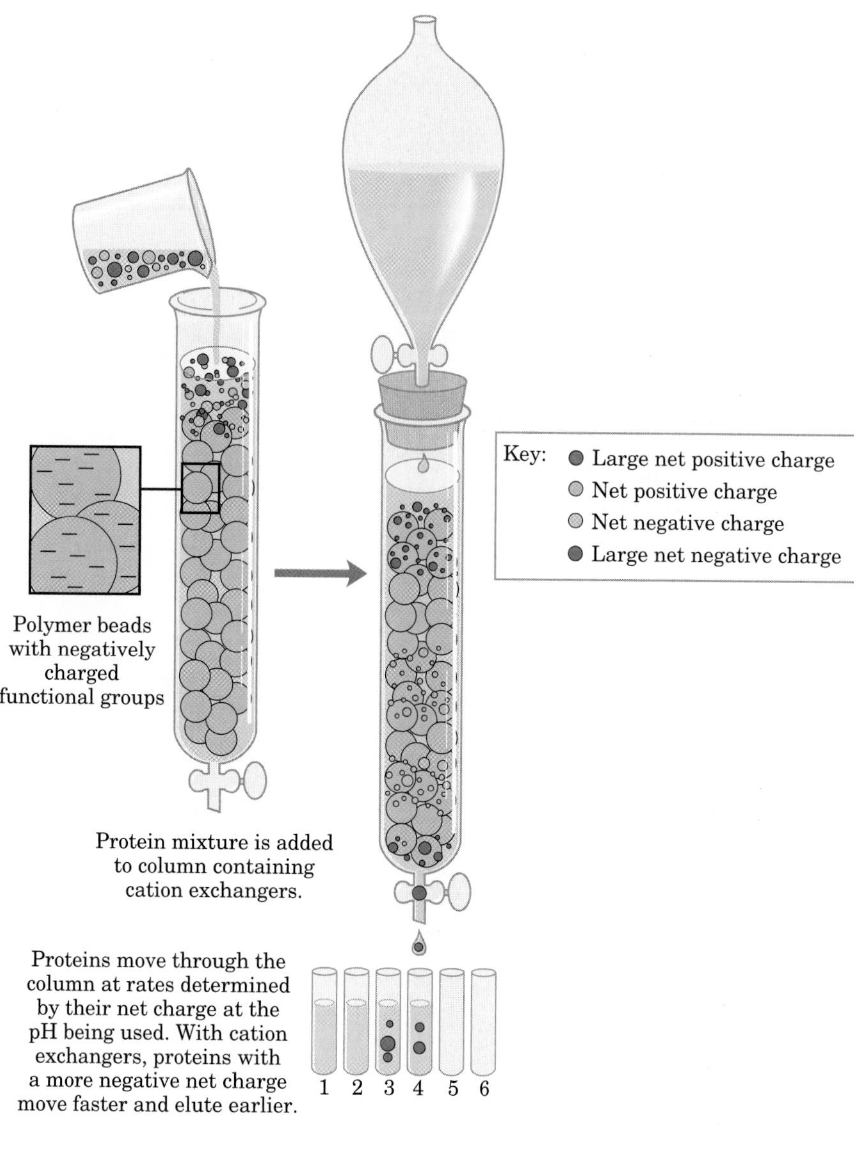
Polymer beads
with negatively
charged
functional groups
Key:
Large net positive charge
Net positive charge
Net negative charge
Large net negative charge
Protein mixture is added
to column containing
cation exchangers.
Proteins move through the
column at rates determined
by their net charge at the
pH being used. With cation
exchangers, proteins with
a more negative net charge
move faster and elute earlier.
1 2 3 4 5 6
(a)

figure 5–18

Three chromatographic methods used in protein purification. (a) Ion-exchange chromatography exploits differences in the sign and magnitude of the net electric charges of proteins at a given pH. The column matrix is a synthetic polymer containing bound charged groups; those with bound anionic groups are called **cation exchangers,** and those with bound cationic groups are called **anion exchangers.** Ion-exchange chromatography on a cation exchanger is shown here. The affinity of each protein for the charged groups on the column is affected by the pH (which determines the ionization state of the molecule) and the concentration of competing free salt ions in the surrounding solution. Separation can be optimized by gradually changing the pH and/or salt concentration of the mobile phase so as to create a pH or salt gradient. **(b) Size-exclusion chromatography,** also called gel filtration, separates proteins according to size. The column matrix is a cross-linked polymer with pores of selected size. Larger proteins migrate faster than smaller ones because they are too large to enter the pores in the beads and hence take a more direct route through the column. The smaller proteins enter the pores and are slowed by the more labyrinthine path they take through the column. **(c) Affinity chromatography** separates proteins by their binding specificities. The proteins retained on the column are those that bind specifically to a ligand cross-linked to the beads. (In biochemistry, the term "ligand" is used to refer to a group or molecule that binds to a macromolecule such as a protein.) After proteins that do not bind to the ligand are washed through the column, the bound protein of particular interest is eluted (washed out of the column) by a solution containing free ligand.

The approach to the purification of a protein that has not been previously isolated is guided both by established precedents and by common sense. In most cases, several different methods must be used sequentially to purify a protein completely. The choice of method is somewhat empirical, and many protocols may be tried before the most effective one is found. Trial and error can often be minimized by basing the procedure on purification techniques developed for similar proteins. Published purification protocols are available for many thousands of proteins. Common sense dictates that inexpensive procedures such as "salting out" be used first, when the total volume and number of contaminants is greatest. Chromatographic methods are often impractical at early stages because the amount of chromatographic medium needed increases with sample size. As each purification step is completed, the sample size generally becomes smaller (Table 5–5), making it feasible to use more sophisticated (and expensive) chromatographic procedures at later stages.

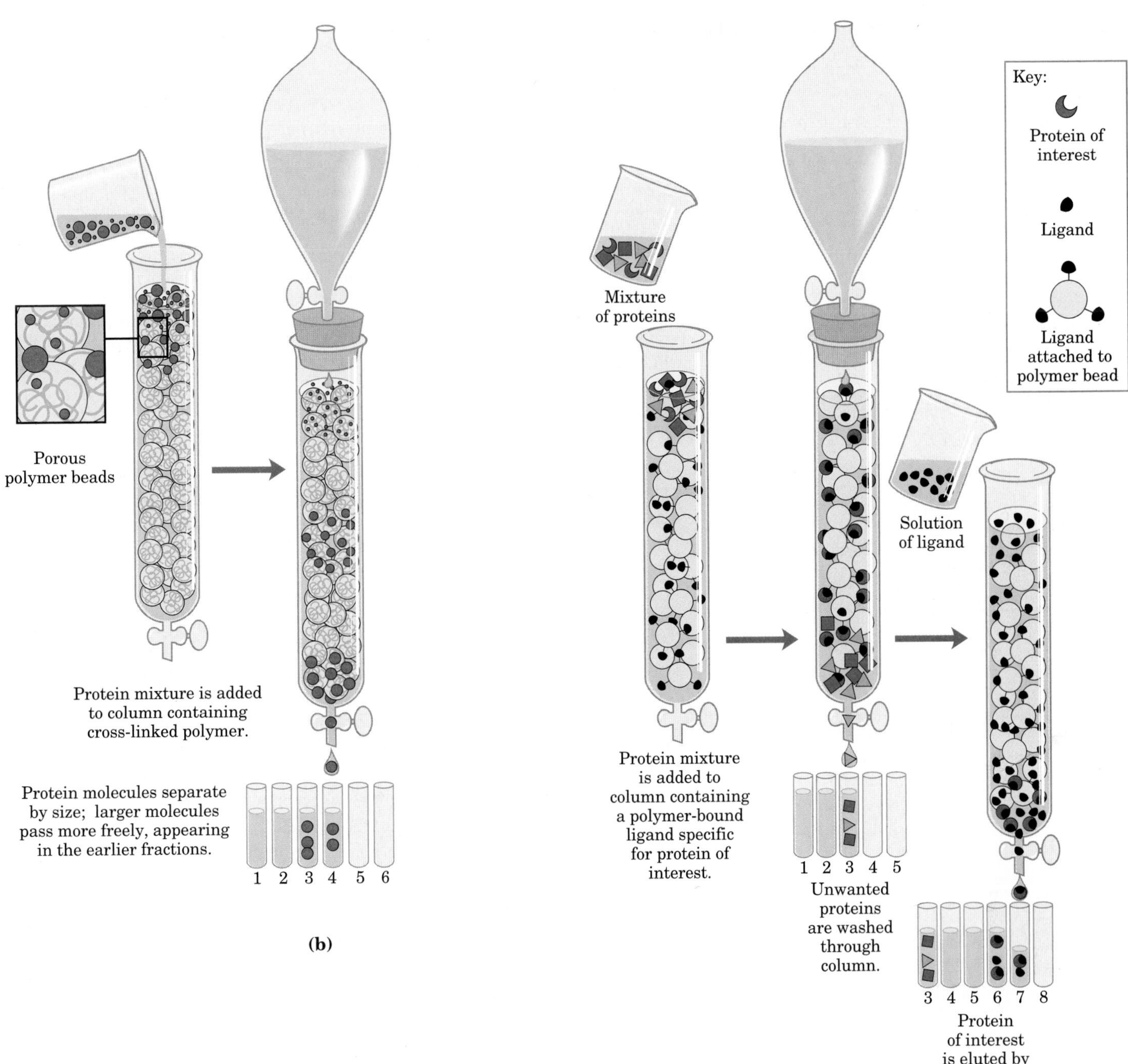
Key:
Protein of interest
Ligand
Ligand attached to polymer bead
Porous polymer beads
Protein mixture is added to column containing cross-linked polymer.
Protein molecules separate by size; larger molecules pass more freely, appearing in the earlier fractions.
1 2 3 4 5 6
Mixture of proteins
Protein mixture is added to column containing a polymer-bound ligand specific for protein of interest.
1 2 3 4 5
Unwanted proteins are washed through column.
Solution of ligand
3 4 5 6 7 8
Protein of interest is eluted by ligand solution.

(b)

(c)

table 5–5

A Purification Table for a Hypothetical Enzyme*

Procedure or step	Fraction volume (ml)	Total protein (mg)	Activity (units)	Specific activity (units/mg)
1. Crude cellular extract	1,400	10,000	100,000	10
2. Precipitation with ammonium sulfate	280	3,000	96,000	32
3. Ion-exchange chromatography	90	400	80,000	200
4. Size-exclusion chromatography	80	100	60,000	600
5. Affinity chromatog-raphy	6	3	45,000	15,000

*All data represent the status of the sample *after* the designated procedure has been carried out. Activity and specific activity are defined on page 137.

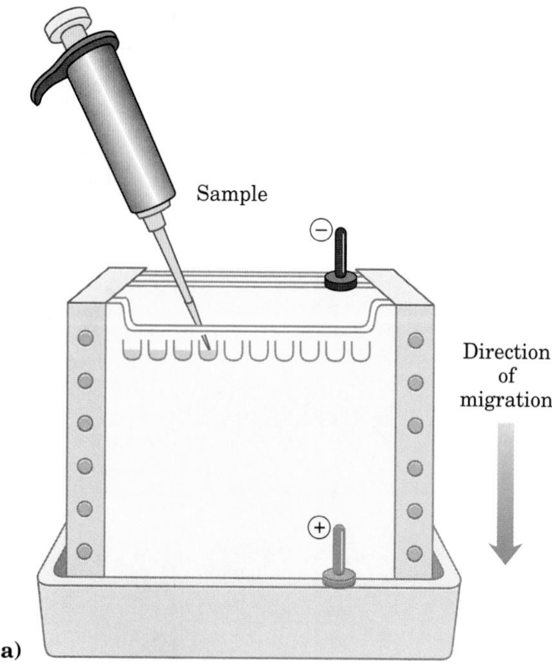

(a)

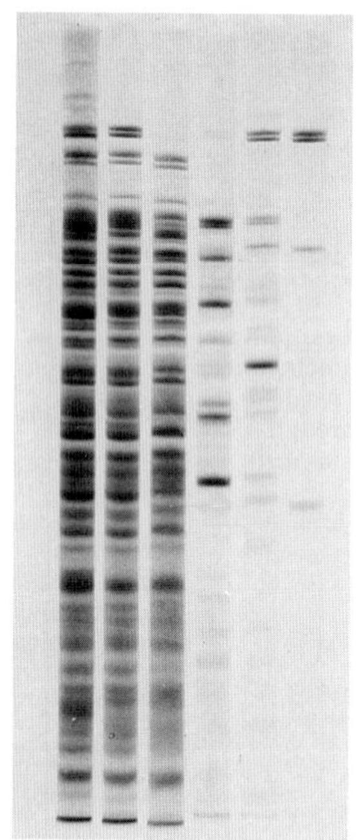

(b)

figure 5–19

Electrophoresis. (a) Different samples are loaded in wells or depressions at the top of the polyacrylamide gel. The proteins move into the gel when an electric field is applied. The gel minimizes convection currents caused by small temperature gradients, and it minimizes protein movements other than those induced by the electric field. **(b)** Proteins can be visualized after electrophoresis by treating the gel with a stain such as Coomassie blue, which binds to the proteins but not to the gel itself. Each band on the gel represents a different protein (or protein subunit); smaller proteins move through the gel more rapidly than larger proteins and therefore are found nearer the bottom of the gel. This gel illustrates the purification of the enzyme RNA polymerase from the bacterium *E. coli*. The first lane shows the proteins present in the crude cellular extract. Successive lanes (left to right) show the proteins present after each purification step. The purified protein contains four subunits, as seen in the last lane on the right.

Proteins Can Be Separated and Characterized by Electrophoresis

Another important technique for the separation of proteins is based on the migration of charged proteins in an electric field, a process called **electrophoresis.** These procedures are not generally used to purify proteins in large amounts because simpler alternatives are usually available and electrophoretic methods often adversely affect the structure and thus the function of proteins. Electrophoresis is, however, especially useful as an analytical method. Its advantage is that proteins can be visualized as well as separated, permitting a researcher to estimate quickly the number of different proteins in a mixture or the degree of purity of a particular protein preparation. Also, electrophoresis allows determination of crucial properties of a protein such as its isoelectric point and approximate molecular weight.

Electrophoresis of proteins is generally carried out in gels made up of the cross-linked polymer polyacrylamide (Fig. 5–19). The polyacrylamide gel acts as a molecular sieve, slowing the migration of proteins approximately in proportion to their charge-to-mass ratio. Migration may also be affected by protein shape. In electrophoresis, the force moving the macromolecule is the electrical potential, E. The electrophoretic mobility of the molecule, μ, is the ratio of the velocity of the particle, V, to the electrical potential. Electrophoretic mobility is also equal to the net charge of the molecule, Z, divided by the frictional coefficient, f, which reflects in part a protein's shape. Thus:

$$\mu = \frac{V}{E} = \frac{Z}{f}$$

The migration of a protein in a gel during electrophoresis is therefore a function of its size and its shape.

An electrophoretic method commonly employed for estimation of purity and molecular weight makes use of the detergent **sodium dodecyl sulfate (SDS).** SDS binds to most proteins (probably by hydrophobic interactions; see Chapter 4) in amounts roughly proportional to the molecular weight of the protein, about one molecule of SDS for every two amino acid residues. The bound SDS contributes a large net negative charge, rendering the intrinsic charge of the protein insignificant and conferring on each protein a similar charge-to-mass ratio. In addition, the native conformation of a protein is altered when SDS is bound, and most proteins assume a similar shape. Electrophoresis in the presence of SDS therefore separates proteins almost exclusively on the basis of mass (molecular weight), with smaller polypeptides migrating more rapidly. After electrophoresis, the proteins are visualized by adding a dye such as Coomassie blue, which binds to proteins but not to the gel itself (Fig. 5–19b). Thus one can monitor the progress of a protein purification procedure, because the number of protein bands visible on the gel should decrease after each new fractionation step. When compared with the positions to which proteins of known molecular weight migrate in the gel, the position of an unidentified protein can provide an excellent measure of its molecular weight (Fig. 5–20). If the protein has two

$$Na^+\ {}^-O-\overset{\overset{O}{\|}}{\underset{\underset{O}{\|}}{S}}-O-(CH_2)_{11}CH_3$$

Sodium dodecyl sulfate (SDS)

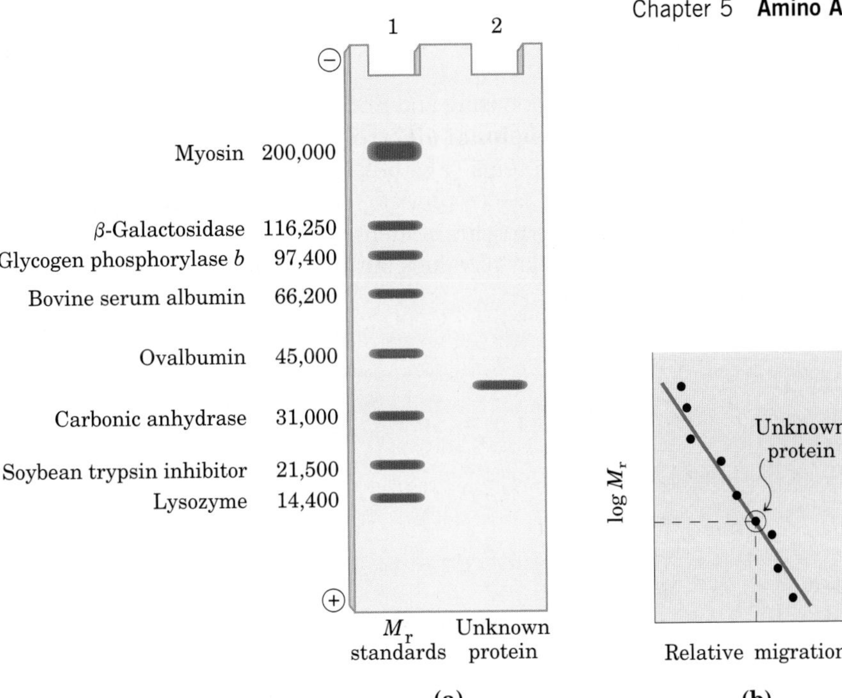

figure 5–20

Estimating the molecular weight of a protein. The electrophoretic mobility of a protein on an SDS polyacrylamide gel is related to its molecular weight, M_r. **(a)** Standard proteins of known molecular weight are subjected to electrophoresis (lane 1). These marker proteins can be used to estimate the molecular weight of an unknown protein (lane 2). **(b)** A plot of log M_r of the marker proteins versus relative migration during electrophoresis is linear, which allows the molecular weight of the unknown protein to be read from the graph.

or more different subunits, the subunits will generally be separated by the SDS treatment and a separate band will appear for each.

Isoelectric focusing is a procedure used to determine the isoelectric point (pI) of a protein (Fig. 5–21). A pH gradient is established by allowing a mixture of low molecular weight organic acids and bases (ampholytes; see p. 123) to distribute themselves in an electric field generated across the gel. When a protein mixture is applied, each protein migrates until it reaches the pH that matches its pI (Table 5–6). Proteins with different isoelectric points are thus distributed differently throughout the gel.

table 5–6

The Isoelectric Points of Some Proteins

Protein	pI
Pepsin	~1.0
Egg albumin	4.6
Serum albumin	4.9
Urease	5.0
β-Lactoglobulin	5.2
Hemoglobin	6.8
Myoglobin	7.0
Chymotrypsinogen	9.5
Cytochrome *c*	10.7
Lysozyme	11.0

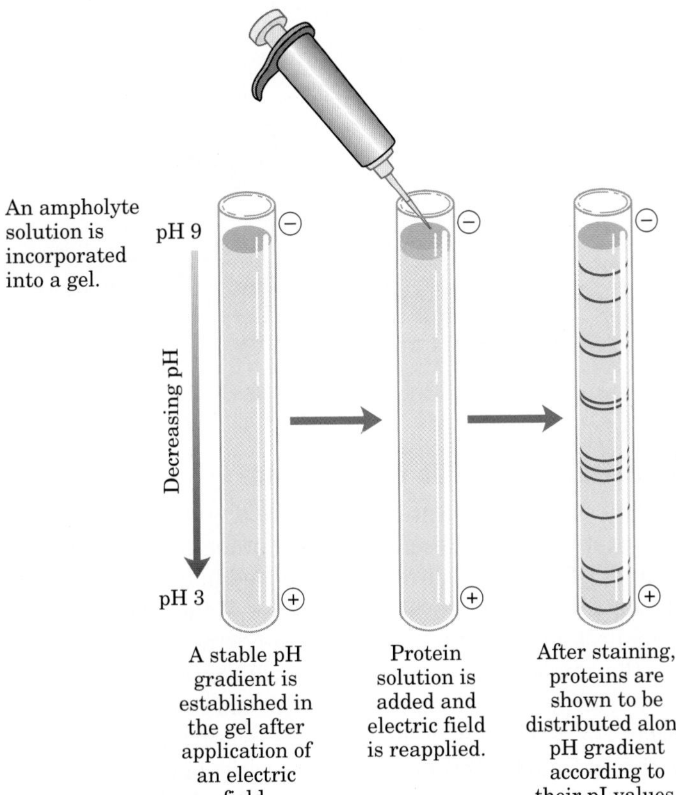

figure 5–21

Isoelectric focusing. This technique separates proteins according to their isoelectric points. A stable pH gradient is established in the gel by the addition of appropriate ampholytes. A protein mixture is placed in a well on the gel. With an applied electric field, proteins enter the gel and migrate until each reaches a pH equivalent to its pI. Remember that when pH = pI, the net charge of a protein is zero.

Combining isoelectric focusing and SDS electrophoresis sequentially in a process called **two-dimensional electrophoresis** permits the resolution of complex mixtures of proteins (Fig. 5–22). This is a more sensitive analytical method than either electrophoretic method alone. Two-dimensional electrophoresis separates proteins of identical molecular weight that differ in pI, or proteins with similar pI values but different molecular weights.

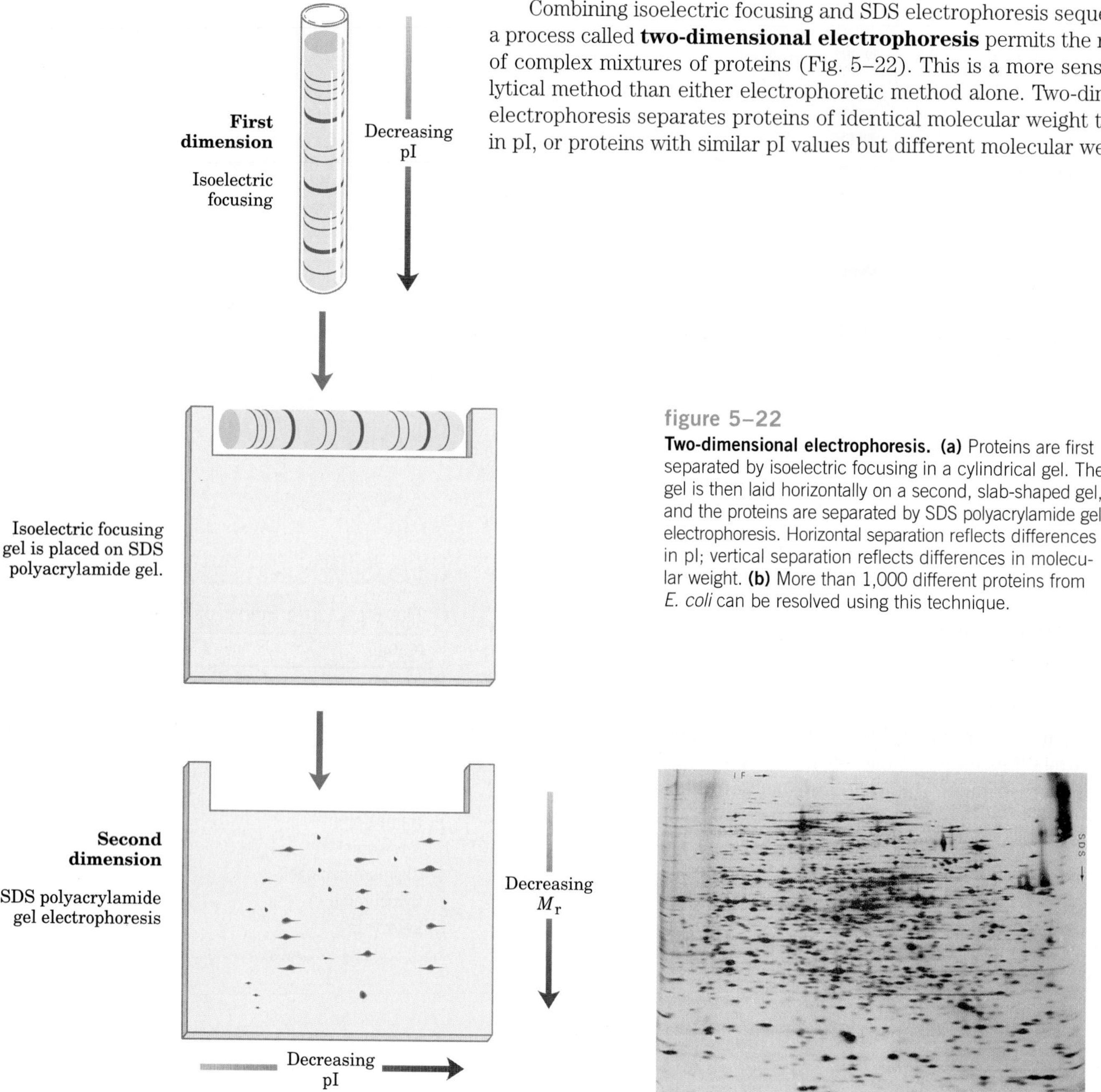

figure 5–22

Two-dimensional electrophoresis. (a) Proteins are first separated by isoelectric focusing in a cylindrical gel. The gel is then laid horizontally on a second, slab-shaped gel, and the proteins are separated by SDS polyacrylamide gel electrophoresis. Horizontal separation reflects differences in pI; vertical separation reflects differences in molecular weight. **(b)** More than 1,000 different proteins from *E. coli* can be resolved using this technique.

Unseparated Proteins Can Be Quantified

To purify a protein, it is essential to have a way of detecting and quantifying that protein in the presence of many other proteins at each stage of the procedure. Often, purification must proceed in the absence of any information about the size and physical properties of the protein, or the fraction of the total protein mass it represents in the extract. For proteins that are enzymes, the amount in a given solution or tissue extract can be measured or assayed in terms of the catalytic effect the enzyme produces, that is, the *increase* in the rate at which its substrate is converted to reaction products when the enzyme is present. For this purpose one must know (1) the overall equation of the reaction catalyzed, (2) an analytical procedure for determining the disappearance of the substrate or the appearance of a reaction product, (3) whether the enzyme requires cofactors such as metal ions or coenzymes, (4) the dependence of the enzyme activity on substrate concentration, (5) the optimum pH, and (6) a temperature zone in which the enzyme is stable and has high activity. Enzymes are usually assayed at their optimum pH and at some convenient temperature within the range 25 to 38 °C. Also, very high substrate concentrations are generally required so that the initial reaction rate, measured experimentally, is proportional to enzyme concentration (Chapter 8).

By international agreement, 1.0 unit of enzyme activity is defined as the amount of enzyme causing transformation of 1.0 μmol of substrate per minute at 25 °C under optimal conditions of measurement. The term **activity** refers to the total units of enzyme in a solution. The **specific activity** is the number of enzyme units per milligram of total protein (Fig. 5–23). The specific activity is a measure of enzyme purity: it increases during purification of an enzyme and becomes maximal and constant when the enzyme is pure (Table 5–5).

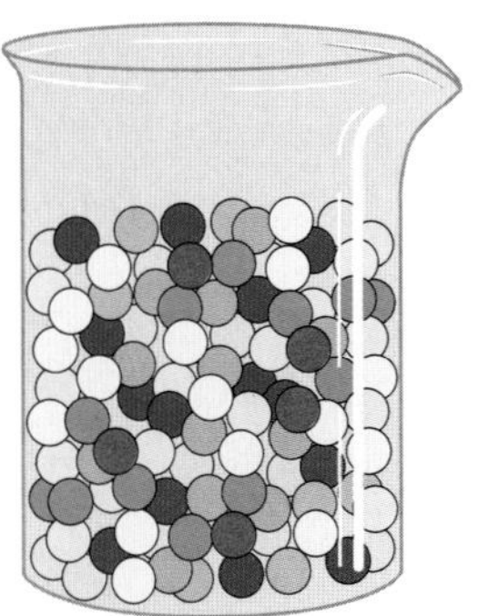

figure 5–23
Activity versus specific activity. The difference between these two terms can be illustrated by considering two beakers of marbles. The beakers contain the same number of red marbles, but different numbers of marbles of other colors. If the marbles represent proteins, both beakers contain the same *activity* of the protein represented by the red marbles. The second beaker, however, has the higher *specific activity* because here the red marbles represent a much higher fraction of the total.

After each purification step, the activity of the preparation (in units) is assayed, the total amount of protein is determined independently, and their ratio gives the specific activity. Activity and total protein generally decrease with each step. Activity decreases because some loss always occurs due to inactivation or nonideal interactions with chromatographic materials or other molecules in the solution. Total protein decreases because the objective is to remove as much unwanted or nonspecific protein as possible. In a successful step, the loss of nonspecific protein is much greater than the loss of activity; therefore, specific activity increases even as total activity falls. The data are then assembled in a purification table similar to Table 5–5. A protein is generally considered pure when further purification steps fail to increase specific activity and when only a single protein species can be detected (for example, by electrophoresis).

For proteins that are not enzymes, other quantification methods are required. Transport proteins can be assayed by their binding to the molecule they transport, and hormones and toxins by the biological effect they produce; for example, growth hormones will stimulate the growth of certain cultured cells. Some structural proteins represent such a large fraction of a tissue mass that they can be readily extracted and purified without a functional assay. The approaches are as varied as the proteins themselves.

The Covalent Structure of Proteins

Purification of a protein is usually only a prelude to a detailed biochemical dissection of its structure and function. What is it that makes one protein an enzyme, another a hormone, another a structural protein, and still another an antibody? How do they differ chemically? The most obvious distinctions are structural, and these distinctions can be approached at every level of structure defined in Figure 5–16.

The differences in primary structure can be especially informative. Each protein has a distinctive number and sequence of amino acid residues. As we shall see in Chapter 6, the primary structure of a protein determines how it folds up into a unique three-dimensional structure, and this in turn determines the function of the protein. Primary structure now becomes the focus of the remainder of the chapter. We first consider empirical clues that amino acid sequence and protein function are closely linked, then describe how amino acid sequence is determined, and finally outline the many uses to which this information can be put.

The Function of a Protein Depends on Its Amino Acid Sequence

The bacterium *E. coli* produces more than 3,000 different proteins; a human being produces 50,000 to 100,000. In both cases, each type of protein has a unique three-dimensional structure and this structure confers a unique function. Each type of protein also has a unique amino acid sequence. Intuition suggests that the amino acid sequence must play a fundamental role in determining the three-dimensional structure of the protein, and ultimately its function, but is this expectation correct? A quick survey of proteins and how they vary in amino acid sequence provides a number of empirical clues that help substantiate the important relationship between amino acid sequence and biological function. First, as we have already noted, proteins with different functions always have different amino acid sequences. Second, thousands of human genetic diseases have been traced to the production of defective proteins. Perhaps one-third of these proteins are defective because of a single change in their amino acid sequence; hence, if the primary structure is altered, the function of the protein may also be changed. Finally, on comparing functionally similar proteins from different species, we find that these proteins often have similar amino acid sequences (Box 5–2). An extreme case is ubiquitin, a 76-residue protein involved in regulating the degradation of other proteins. The amino acid sequence of ubiquitin is identical in species as disparate as fruit flies and humans.

Is the amino acid sequence absolutely fixed, or invariant, for a particular protein? No; some flexibility is possible. An estimated 20% to 30% of the proteins in humans are **polymorphic,** having amino acid sequence variants in the human population. Many of these variations in sequence have little or no effect on the function of the protein. Furthermore, proteins that carry out a broadly similar function in distantly related species can differ greatly in overall size and amino acid sequence.

Proteins often contain crucial regions within their amino acid sequence that are essential to their biological functions. The amino acid sequence in other regions might vary considerably without affecting these functions. The fraction of the sequence that is critical varies from protein to protein, complicating the task of relating sequence to three-dimensional structure, and structure to function. Before we can consider this problem further, however, we must examine how sequence information is obtained.

The Amino Acid Sequences of Numerous Proteins Have Been Determined

Two major discoveries in 1953 were of crucial importance in the history of biochemistry. In that year James D. Watson and Francis Crick deduced the double-helical structure of DNA and proposed a structural basis for its precise replication (Chapter 10). Their proposal illuminated the molecular reality behind the idea of a gene. In that same year, Frederick Sanger worked out the sequence of amino acid residues in the polypeptide chains of the hormone insulin (Fig. 5–24), surprising many researchers who had long thought that elucidation of the amino acid sequence of a polypeptide would be a hopelessly difficult task. It quickly became evident that the nucleotide

A chain	B chain
$\overset{+}{N}H_3$	$\overset{+}{N}H_3$
Gly	Phe
Ile	Val
Val	Asn
Gln	Gln
5 Gln	5 His
Cys (S—S to A-chain Cys 11)	Leu
Cys—S—S—Cys	Cys
Ala	Gly
Ser	Ser
10 Val	10 His
Cys	Leu
Ser	Val
Leu	Glu
Tyr	Ala
15 Gln	15 Leu
Leu	Tyr
Glu	Leu
Asn	Val
Tyr	S—Cys
20 Cys—S	20 Gly
Asn	Glu
COO^-	Arg
	Gly
	Phe
	25 Phe
	Tyr
	Thr
	Pro
	Lys
	30 Ala
	COO^-

figure 5–24
Amino acid sequence of bovine insulin. The two polypeptide chains are joined by disulfide cross-linkages. The A chain is identical in human, pig, dog, rabbit, and sperm whale insulins. The B chains of the cow, pig, dog, goat, and horse are identical. Such identities between similar proteins of different species are discussed in Box 5–2.

box 5–2 Protein Homology among Species

Homologous proteins are proteins that are evolutionarily related. They usually perform the same function in different species; an example is **cytochrome *c*,** an iron-containing mitochondrial protein that transfers electrons during biological oxidations in eukaryotic cells. Homologous proteins from different species may have polypeptide chains that are identical or nearly identical in length. Many positions in the amino acid sequence are occupied by the same residue in all species and are thus called **invariant residues.** Other positions show considerable variation in the amino acid residue from one species to another; these are called **variable residues.**

The functional significance of sequence homology is well illustrated by cytochrome *c* (M_r ~13,000), which has about 100 amino acid residues in most species. The amino acid sequences of cytochrome *c* molecules from many different species have been determined, and 27 positions in the chain are invariant in all species tested (Fig. 1), suggesting that they are the most important residues specifying the biological activity of this protein. The residues in other positions exhibit some interspecies variation. There are clear gradations in the number of differences observed in the variable residues. In some positions, most substitutions involve similar amino acid residues (for example, positively charged Arg might replace positively charged Lys); these are called **conservative substitutions.** At other positions the substitutions are less restricted (nonconservative). As we will show in the next chapter, the polypeptide chains of proteins are folded into characteristic and specific conformations, which depend on amino acid sequence. Clearly, the invariant residues are more critical to the structure and function of a protein than the variable ones. Recognizing which residues fall into each category is an important step in deciphering the complicated question of how amino acid sequence is translated into a specific three-dimensional structure.

The variable amino acids provide information of another sort. Phylogenetic (evolutionary) relationships based on taxonomic methods have been tested and experimentally confirmed through biochemistry. The examination of sequences of cytochrome *c* and other homologous proteins has led to an important conclusion: the number of residues that differ in homologous proteins from any two species is in proportion to the phylogenetic difference between those species. For example, 48 amino acid residues differ in the cytochrome *c* molecules of the horse and of yeast, which are very widely separated species, whereas only two residues differ in the cytochrome *c* molecules of the much more closely related duck and chicken. In fact, cytochrome *c* has identical amino acid sequences in the chicken and the turkey, and in the pig, cow,

figure 1

Amino acid sequence of human cytochrome *c*. Amino acid substitutions in other species are listed below the individual residues. Invariant amino acids are shaded in yellow, conservative substitutions are shaded in blue, and nonconservative substitutions are unshaded. X is an unusual amino acid, trimethyllysine. The one-letter abbreviations for amino acids are given in Table 5–1.

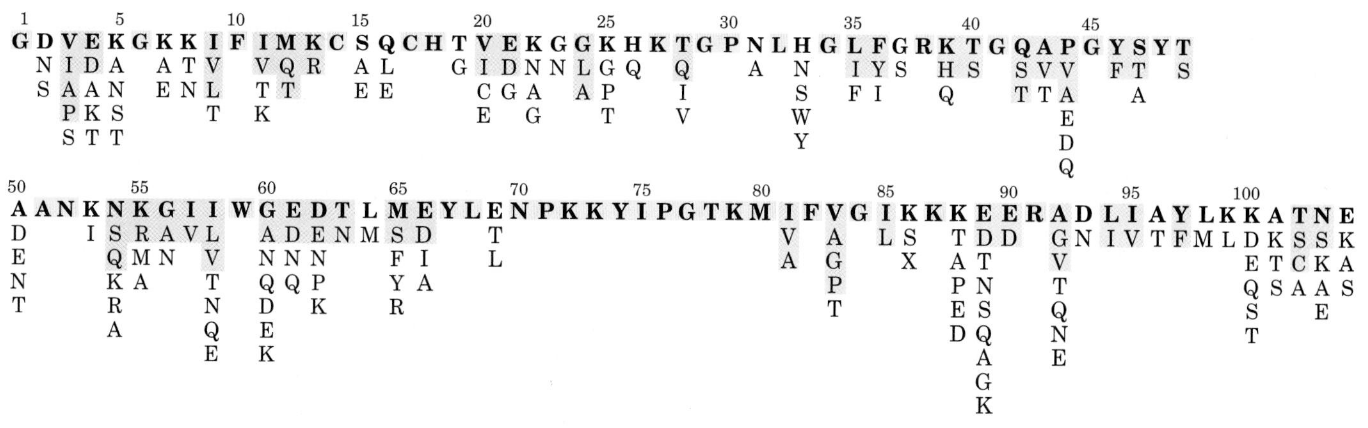

Continued on next page

and sheep. Information on the number of residue differences between homologous proteins of different species allows the construction of evolutionary trees that show the origin and sequence of appearance of different species during the course of evolution (Fig. 2). The relationships established by anatomic and biochemical taxonomy are in close agreement.

figure 2

Main branches of the eukaryotic evolutionary tree constructed from the number of amino acid differences between cytochrome *c* molecules of various species. The numbers represent the number of residues by which the cytochrome *c* of a given line of organism differs from its ancestor. Branch points reflect a common ancestor.

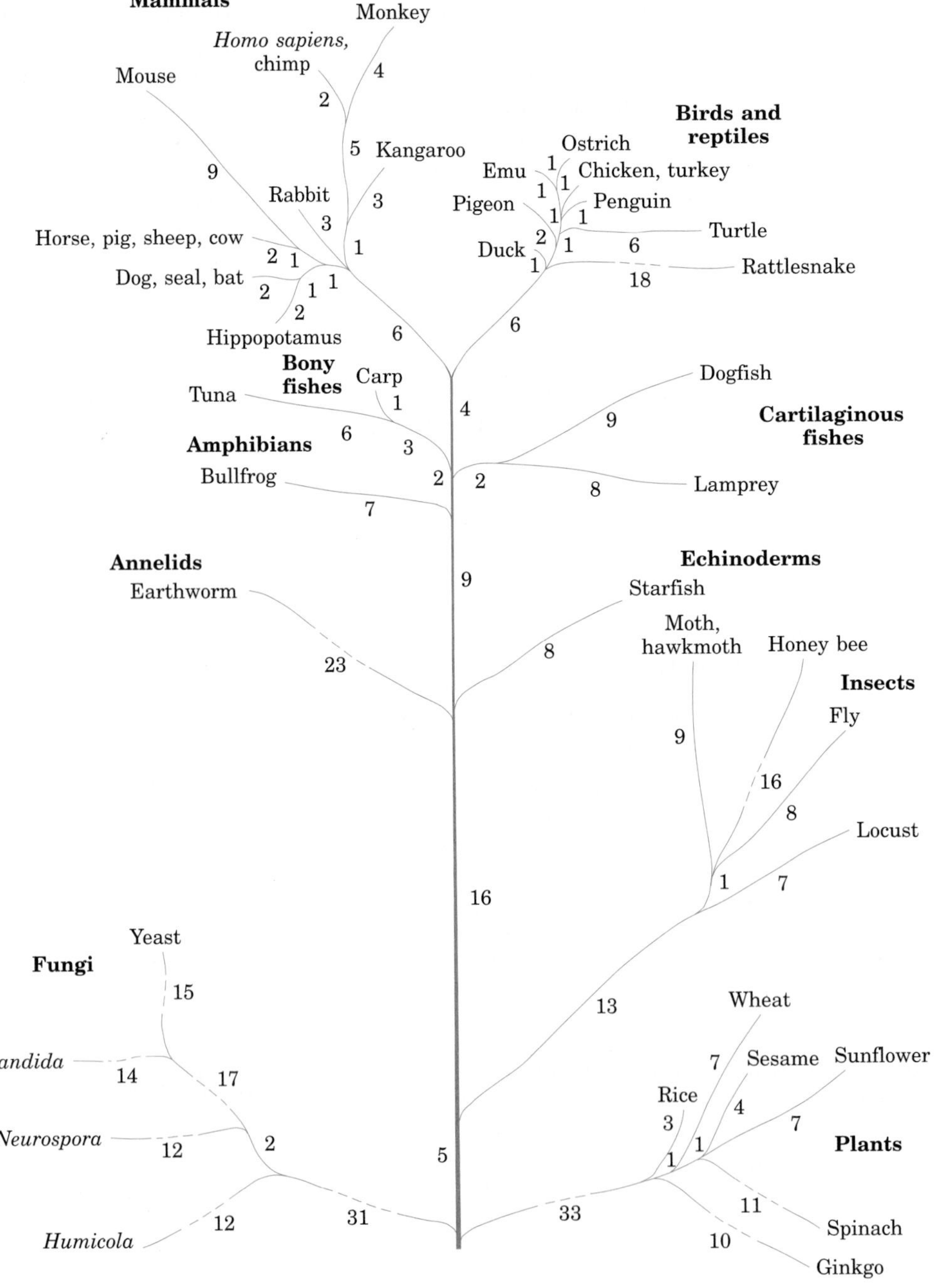

sequence in DNA and the amino acid sequence in proteins were somehow related. Barely a decade after these discoveries, the role of DNA nucleotide sequence in determining the amino acid sequence of protein molecules had been revealed (Chapter 27).

The amino acid sequences of thousands of different proteins from many species have been determined using principles first developed by Sanger. These methods are still in use, although with many variations and improvements in detail. Chemical protein sequencing now complements a growing list of newer methods, providing multiple avenues to obtain amino acid sequence data. Such data are now critical to every area of biochemical investigation.

Short Polypeptides Are Sequenced Using Automated Procedures

Various procedures are used to analyze protein primary structure. One is to hydrolyze the protein and determine its amino acid composition (Fig. 5–25a). This information is often valuable in interpreting the results of other procedures. Because amino acid composition differs from one protein to the next, it can serve as a kind of fingerprint. It can be used, for example, to help determine whether proteins isolated by different laboratories are the same or different. Hydrolysis alone, however, cannot be used to determine the *sequence* of amino acids in a protein.

A procedure that is often used in conjunction with hydrolysis is to label and identify the amino-terminal amino acid residue (Fig. 5–25b). For this purpose Sanger developed the reagent 1-fluoro-2,4-dinitrobenzene (FDNB).

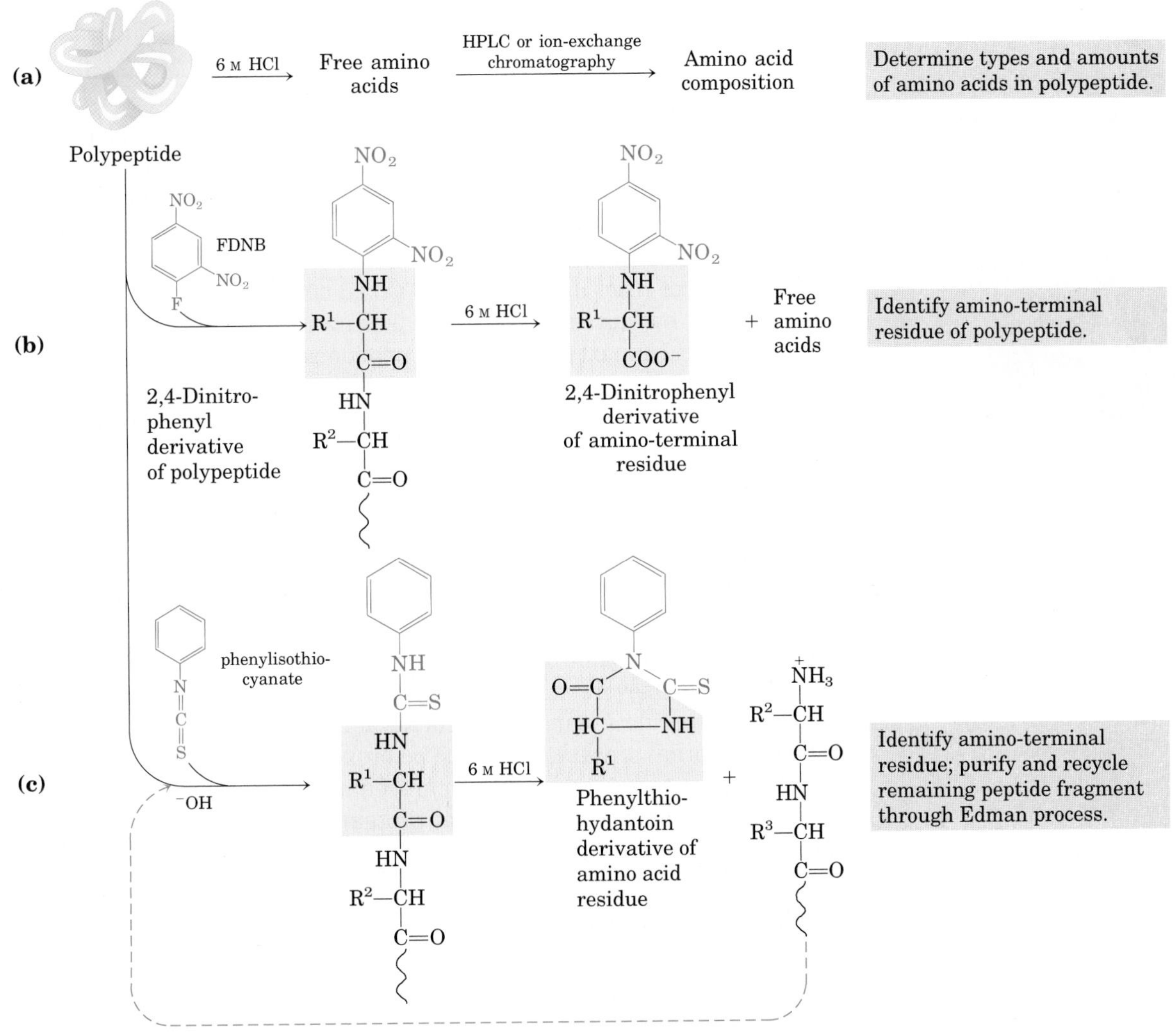

figure 5–25
Steps in sequencing a polypeptide. (a) Determination of amino acid composition and **(b)** identification of the amino-terminal residue are often the first steps in sequencing a polypeptide. Sanger's method for identifying the amino-terminal residue is shown here. The Edman degradation procedure **(c)** reveals the entire sequence of a peptide. For shorter peptides, this method alone readily yields the entire sequence, and steps **(a)** and **(b)** are often omitted. The latter procedures are useful in the case of larger polypeptides, which are often fragmented into smaller peptides for sequencing (see Fig. 5–27).

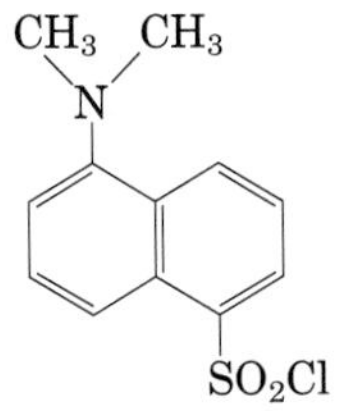

Dansyl chloride

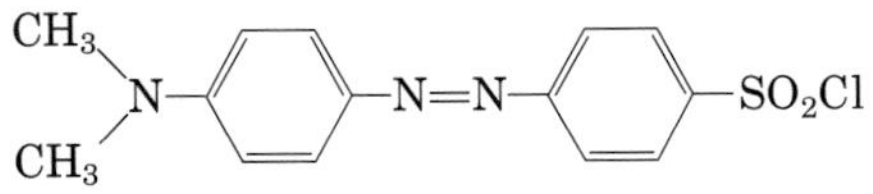

Dabsyl chloride

Other reagents used to label the amino-terminal residue, dansyl chloride and dabsyl chloride, yield derivatives that are more easily detectable than the dinitrophenyl derivatives. After the amino-terminal residue is labeled with one of these reagents, the polypeptide is hydrolyzed to its constituent amino acids and the labeled amino acid is identified. Because the hydrolysis stage destroys the polypeptide, this procedure cannot be used to sequence a polypeptide beyond its amino-terminal residue. However, it can help determine the number of chemically distinct polypeptides in a protein, provided each has a different amino-terminal residue. For example, two residues—Phe and Gly—would be labeled if insulin (Fig. 5–24) were subjected to this procedure.

To sequence an entire polypeptide, a chemical method devised by Pehr Edman is usually employed. The **Edman degradation** procedure labels and removes only the amino-terminal residue from a peptide, leaving all other peptide bonds intact (Fig. 5–25c). The peptide is reacted with phenylisothiocyanate, and the amino-terminal residue is ultimately removed as a phenylthiohydantoin derivative. After removal and identification of the amino-terminal residue, the *new* amino-terminal residue so exposed can be labeled, removed, and identified through the same series of reactions. This procedure is repeated until the entire sequence is determined. The Edman degradation is carried out on a machine, called a **sequenator,** which mixes reagents in the proper proportions, separates the products, identifies them, and records the results. These methods are extremely sensitive. Often, the complete amino acid sequence can be determined starting with only a few micrograms of protein.

The length of polypeptide that can be accurately sequenced by the Edman degradation depends on the efficiency of the individual chemical steps. Consider a peptide beginning with the sequence Gly–Pro–Lys– at its amino terminus. If glycine were removed with 97% efficiency, 3% of the polypeptide molecules in the solution would retain a Gly residue at their amino terminus. In the second Edman cycle, 97% of the liberated amino acids would be proline, and 3% glycine, while 3% of the polypeptide molecules would retain Gly (0.1%) or Pro (2.9%) residues at their amino terminus. At each cycle, peptides that did not react in earlier cycles would contribute amino acids to an ever-increasing background, eventually making it impossible to determine which amino acid is next in the original peptide sequence. Modern sequenators achieve efficiencies of better than 99% per cycle, permitting the sequencing of more than 50 contiguous amino acid residues in a polypeptide. The primary structure of insulin, worked out by Sanger and colleagues over a period of 10 years, could now be completely determined in a day or two.

Frederick Sanger

Large Proteins Must Be Sequenced in Smaller Segments

The overall accuracy of amino acid sequencing generally declines as the length of the polypeptide increases. The very large polypeptides found in proteins must be broken down into smaller pieces to be sequenced efficiently. There are several steps in this process. First, the protein is cleaved into a set of specific fragments by chemical or enzymatic methods. If any disulfide bonds are present, they must be broken. Each fragment is then purified, then sequenced by the Edman procedure. Finally, the order in which the fragments appear in the original protein is determined and disulfide bonds (if any) are located.

Breaking Disulfide Bonds Disulfide bonds interfere with the sequencing procedure. A cystine residue (Fig. 5–7) that has one of its peptide bonds cleaved by the Edman procedure may remain attached to another polypep-

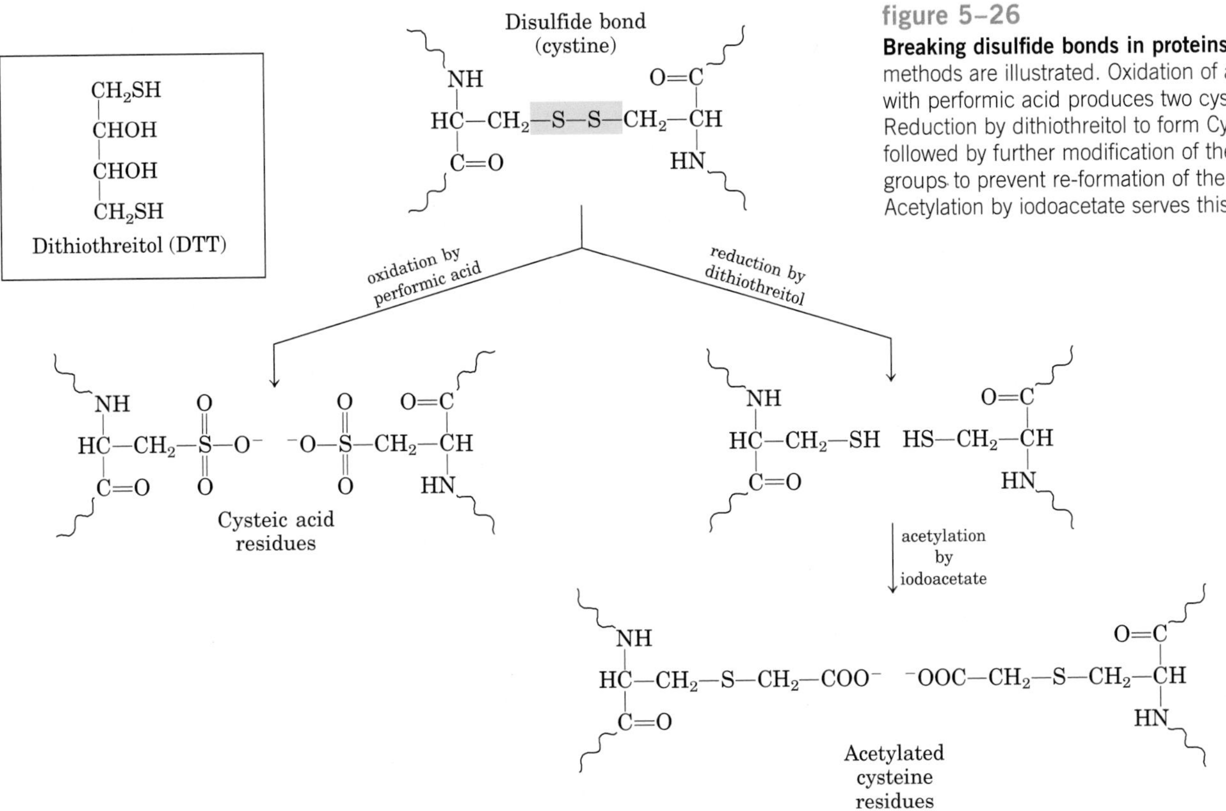

figure 5–26

Breaking disulfide bonds in proteins. Two common methods are illustrated. Oxidation of a cystine residue with performic acid produces two cysteic acid residues. Reduction by dithiothreitol to form Cys residues must be followed by further modification of the reactive —SH groups to prevent re-formation of the disulfide bond. Acetylation by iodoacetate serves this purpose.

tide strand via its disulfide bond. Disulfide bonds also interfere with the enzymatic or chemical cleavage of the polypeptide. Two approaches to irreversible breakage of disulfide bonds are outlined in Figure 5–26.

Cleaving the Polypeptide Chain Several methods can be used for fragmenting the polypeptide chain. Enzymes called **proteases** catalyze the hydrolytic cleavage of peptide bonds. Some proteases cleave only the peptide bond adjacent to particular amino acid residues (Table 5–7) and thus fragment a polypeptide chain in predictable and reproducible ways. A number of chemical reagents also cleave the peptide bond adjacent to specific residues.

Among proteases, the digestive enzyme trypsin catalyzes the hydrolysis of only those peptide bonds in which the carbonyl group is contributed by either a Lys or an Arg residue, regardless of the length or amino acid sequence of the chain. The number of smaller peptides produced by trypsin cleavage can thus be predicted from the total number of Lys or Arg residues in the original polypeptide, as determined by hydrolysis of an intact sample (Fig. 5–27). A polypeptide with five Lys and/or Arg residues will usually yield six smaller peptides on cleavage with trypsin. Moreover, all except one of these will have a carboxyl-terminal Lys or Arg. The fragments produced by trypsin (or other enzyme or chemical) action are then separated by chromatographic or electrophoretic methods.

Sequencing of Peptides Each peptide fragment resulting from the action of trypsin is sequenced separately by the Edman procedure.

Ordering Peptide Fragments The order of the "trypsin fragments" in the original polypeptide chain must now be determined. Another sample of the intact polypeptide is cleaved into fragments using a different enzyme or reagent, one that cleaves peptide bonds at points other than those

table 5–7

The Specificity of Some Common Methods for Fragmenting Polypeptide Chains

Treatment*	Cleavage points†
Trypsin	Lys, Arg (C)
Submaxillarus protease	Arg (C)
Chymotrypsin	Phe, Trp, Tyr (C)
Staphylococcus aureus V8 protease	Asp, Glu (C)
Asp-*N*-protease	Asp, Glu (N)
Pepsin	Phe, Trp, Tyr (N)
Endoproteinase Lys C	Lys (C)
Cyanogen bromide	Met (C)

*All except cyanogen bromide are proteases. All are available from commercial sources.

†Residues furnishing the primary recognition point for the protease or reagent; peptide bond cleavage occurs on either the carbonyl (C) or the amino (N) side of the indicated amino acid residues.

cleaved by trypsin. For example, cyanogen bromide cleaves only those peptide bonds in which the carbonyl group is contributed by Met. The fragments resulting from this second procedure are then separated and sequenced as before.

The amino acid sequences of each fragment obtained by the two cleavage procedures are examined, with the objective of finding peptides from the second procedure whose sequences establish continuity, because of overlaps, between the fragments obtained by the first cleavage procedure (Fig. 5–27). Overlapping peptides obtained from the second fragmentation yield the correct order of the peptide fragments produced in the first. If the amino-terminal amino acid has been identified before the original cleavage of the protein, this information can be used to establish which fragment is derived from the amino terminus. The two sets of fragments can be compared for possible errors in determining the amino acid sequence of each fragment.

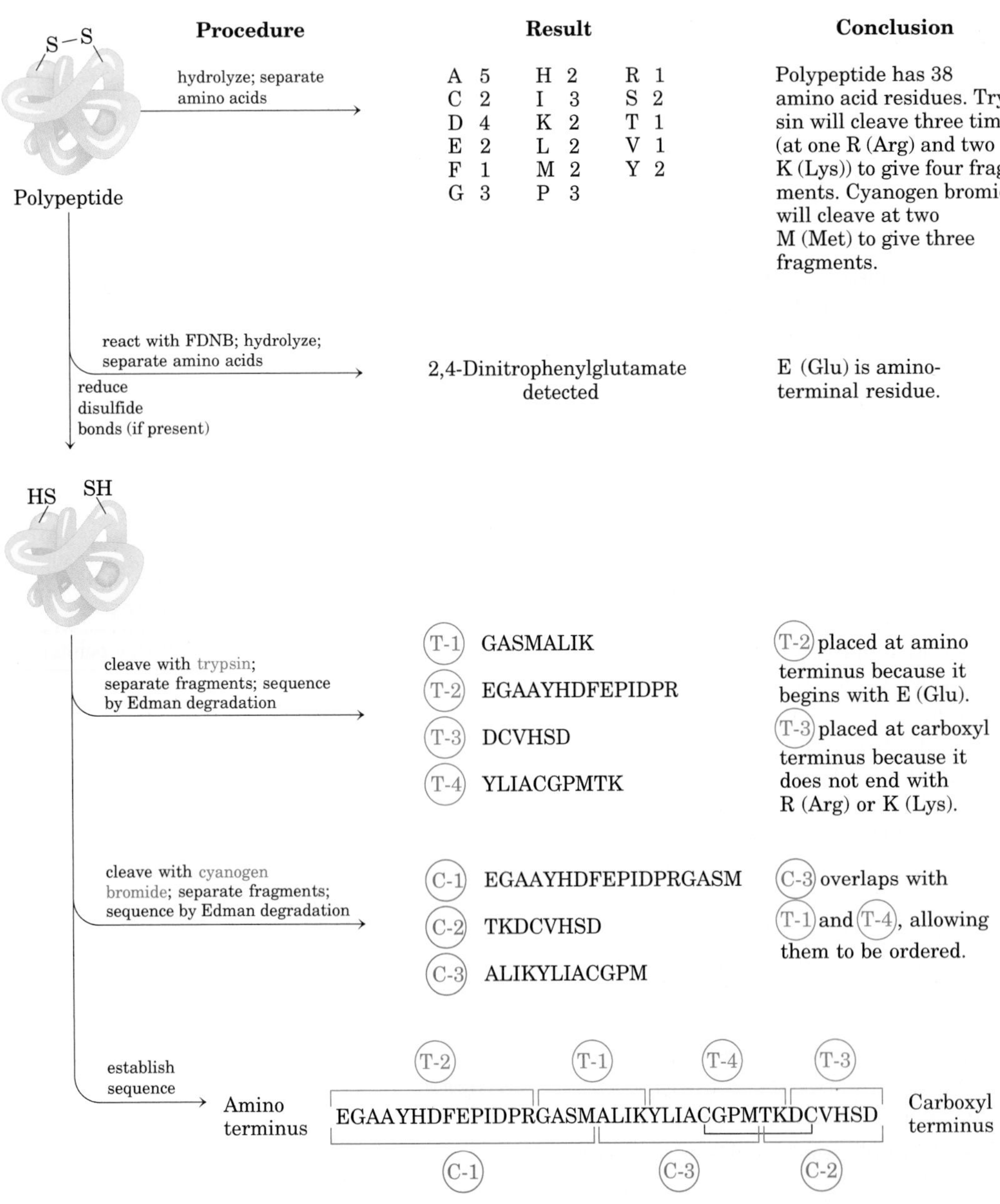

S–S
Procedure
Result
Conclusion
hydrolyze; separate amino acids
A 5 H 2 R 1
C 2 I 3 S 2
D 4 K 2 T 1
E 2 L 2 V 1
F 1 M 2 Y 2
G 3 P 3
Polypeptide has 38 amino acid residues. Trypsin will cleave three times (at one R (Arg) and two K (Lys)) to give four fragments. Cyanogen bromide will cleave at two M (Met) to give three fragments.
Polypeptide
react with FDNB; hydrolyze; separate amino acids
2,4-Dinitrophenylglutamate detected
E (Glu) is amino-terminal residue.
reduce disulfide bonds (if present)
HS SH
cleave with trypsin; separate fragments; sequence by Edman degradation
T-1 GASMALIK
T-2 EGAAYHDFEPIDPR
T-3 DCVHSD
T-4 YLIACGPMTK
T-2 placed at amino terminus because it begins with E (Glu).
T-3 placed at carboxyl terminus because it does not end with R (Arg) or K (Lys).
cleave with cyanogen bromide; separate fragments; sequence by Edman degradation
C-1 EGAAYHDFEPIDPRGASM
C-2 TKDCVHSD
C-3 ALIKYLIACGPM
C-3 overlaps with T-1 and T-4, allowing them to be ordered.
establish sequence
T-2 T-1 T-4 T-3
Amino terminus
EGAAYHDFEPIDPRGASMALIKYLIACGPMTKDCVHSD
Carboxyl terminus
C-1 C-3 C-2

figure 5–27
Cleaving proteins and sequencing and ordering the peptide fragments. First, the amino acid composition and amino-terminal residue of an intact sample are determined. Then any disulfide bonds are broken prior to fragmenting so that sequencing can proceed efficiently. In this example, there are only two Cys (C) residues, and thus only one possibility for location of the disulfide bond. In polypeptides with three or more Cys residues, the position of disulfide bonds can be determined as described in the text. (The one-letter abbreviations for amino acids are given in Table 5–1.)

If the second cleavage procedure fails to establish continuity between all peptides from the first cleavage, a third or even a fourth cleavage method must be used to obtain a set of peptides that can provide the necessary overlap(s).

Locating Disulfide Bonds If the primary structure includes disulfide bonds, their locations are determined in an additional step after sequencing is completed. A sample of the protein is again cleaved with a reagent such as trypsin, this time without first breaking the disulfide bonds. When the resulting peptides are separated by electrophoresis and compared with the original set of peptides generated by trypsin, for each disulfide bond, two of the original peptides will be missing and a new, larger peptide will appear. The two missing peptides represent the regions of the intact polypeptide that are linked by the disulfide bond.

Amino Acid Sequences Can Also Be Deduced by Other Methods

The approach outlined above is not the only way to determine amino acid sequences. New methods based on mass spectrometry permit the sequencing of short polypeptides (20 to 30 amino acid residues) in just a few minutes (Box 5–3, pp. 146–149). In addition, the development of rapid DNA sequencing methods (Chapter 10), the elucidation of the genetic code (Chapter 27), and the development of techniques for isolating genes (Chapter 29) make it possible to deduce the sequence of a polypeptide by determining the sequence of nucleotides in the gene that codes for it (Fig. 5–28). The techniques used to determine protein and DNA sequences are complementary. When the gene is available, sequencing the DNA can be faster and more accurate than sequencing the protein. Most proteins are now sequenced in this indirect way. If the gene has not been isolated, direct sequencing of peptides is necessary, and this can provide information (e.g., the location of disulfide bonds) not available in a DNA sequence. In addition, a knowledge of the amino acid sequence of even a part of a polypeptide can greatly facilitate the isolation of the corresponding gene (Chapter 29).

The array of methods now available to analyze both proteins and nucleic acids is ushering in a new discipline of whole cell biochemistry. The complete sequence of an organism's DNA, its genome, is now available for organisms ranging from viruses to bacteria to multicellular eukaryotes (see Table 1–1). Genes are being discovered by the thousands, including many that encode proteins with no known function. To describe the entire protein complement encoded by an organism's DNA, researchers have coined the term **proteome.** Analysis of a cell's proteome is an increasingly important and informative adjunct to the completion of its genomic sequence. Proteins from a cell are separated and displayed by two-dimensional gel electrophoresis (Fig. 5–22). Individual protein spots can be extracted from such a gel. Small peptides derived from the proteins are sequenced by mass spectrometry (Box 5–3), and these sequences are compared with the genomic sequence to identify the protein. Often, knowledge of the sequence of a segment of six to eight amino acid residues is enough to pinpoint the gene encoding the entire protein. Inevitably, some of these proteins are already known and have well-studied functions; others are more mysterious. Once most of the proteins are matched to a gene, changes in a cell's protein complement brought on by the environment, nutritional changes, stress, or disease can be examined. This work can provide clues to the role of proteins whose functions are as yet unknown. Eventually, such studies will complement work carried out on cellular intermediary metabolism and nucleic acid metabolism to provide a new and increasingly complete picture of biochemistry at the level of cells and even organisms.

Amino acid sequence (protein)	Gln–Tyr–Pro–Thr–Ile–Trp
DNA sequence (gene)	CAGTATCCTACGATTTGG

figure 5–28
Correspondence of DNA and amino acid sequences. Each amino acid is encoded by a specific sequence of three nucleotides in DNA. The genetic code is described in detail in Chapter 27.

box 5–3 Investigating Proteins with Mass Spectrometry

The mass spectrometer has long been an indispensable tool in chemistry. Molecules to be analyzed, referred to as **analytes,** are first ionized in a vacuum. When the newly charged molecules are introduced into an electric and/or magnetic field, their paths through the field are a function of their mass-to-charge ratio, m/z. This measured property of the ionized species can be used to deduce the mass (M) of the analyte with very high precision.

Although mass spectrometry has been in use for many years, it could not be applied to macromolecules such as proteins and nucleic acids. The m/z measurements are made on molecules in the gas phase, and the heating or other treatment needed to bring a macromolecule into the gas phase usually caused its rapid decomposition. In 1988, two different techniques were developed to overcome this problem. In one, proteins are placed in a light-absorbing matrix. With a short pulse of laser light, the proteins are ionized and then desorbed from the matrix into the vacuum system. This process, known as **matrix-assisted laser desorption/ionization mass spectrometry,** or **MALDI MS,** has been successfully used to measure the mass of a wide range of macromolecules. In a second and equally successful method, macromolecules in solution are forced directly from the liquid to gas phase. A solution of analytes is passed through a charged needle that is kept at a high electrical potential, dispersing the solution into a fine mist of charged microdroplets. The solvent surrounding the macromolecules rapidly evaporates and the resulting multiply charged macromolecular ions are thus introduced nondestructively into the gas phase. This technique is called **electrospray ionization mass spectrometry,** or **ESI MS.** Protons added during passage through the needle give additional charge to the macromolecule. The m/z of the molecule can be analyzed in the vacuum chamber.

Mass spectrometry provides a wealth of information for proteome research, enzymology, and protein chemistry in general. The techniques require only miniscule amounts of sample, so they can be readily applied to the small amounts of protein that can be extracted from a two-dimensional electrophoretic gel. The accurately measured molecular mass of a protein is one of the critical parameters in its identification. Once the mass of a protein is accurately known, mass spectrometry is a convenient and accurate method for detecting changes in mass due to the presence of bound cofactors, bound metal ions, covalent modifications, and so on.

The process for determining the molecular mass of a protein with ESI MS is illustrated in Figure 1. As it is injected into the gas phase, a protein acquires a variable number of protons,

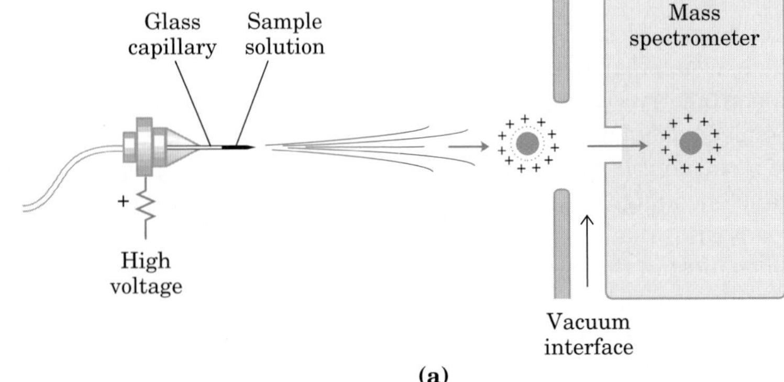

figure 1
Electrospray mass spectrometry of a protein. (a) A protein solution is dispersed into highly charged droplets by passage through a needle under the influence of a high-voltage electric field. The droplets evaporate, and the ions (with added protons in this case) enter the mass spectrometer for m/z measurement. The spectrum generated **(b)** is a family of peaks, with each successive peak (from right to left) corresponding to a charged species increased by 1 in both mass and charge. A computer-generated transformation of this spectrum is shown in the inset.

and thus positive charges, from the solvent. This creates a spectrum of species with different mass-to-charge ratios. Each successive peak corresponds to a species that differs from that of its neighboring peak by a charge difference of 1 and a mass difference of 1 (1 proton). The mass of the protein can be determined from any two neighboring peaks. The measured *m/z* of one peak is

$$(m/z)_2 = \frac{M + n_2 X}{n_2}$$

where M is the mass of the protein, n_2 is the number of charges, and X is the mass of the added groups (protons in this case). Similarly for the neighboring peak,

$$(m/z)_1 = \frac{M + (n_2 + 1)X}{n_2 + 1}$$

We now have two unknowns (M and n_2) and two equations. We can solve first for n_2 and then for M:

$$n_2 = \frac{(m/z)_2 - X}{(m/z)_2 - (m/z)_1}$$

$$M = n_2[(m/z)_2 - X]$$

This calculation using the *m/z* values for any two peaks in a spectrum such as that shown in Figure 1b will usually provide the mass of the protein (in this case, aerolysin k; 47,342 Da) with an error of only ±0.01%. Generating several sets of peaks, repeating the calculation, and averaging the results generally provides an even more accurate value for M. Computer algorithms can transform the *m/z* spectrum into a single peak that also provides a very accurate mass measurement (Fig. 1b, inset).

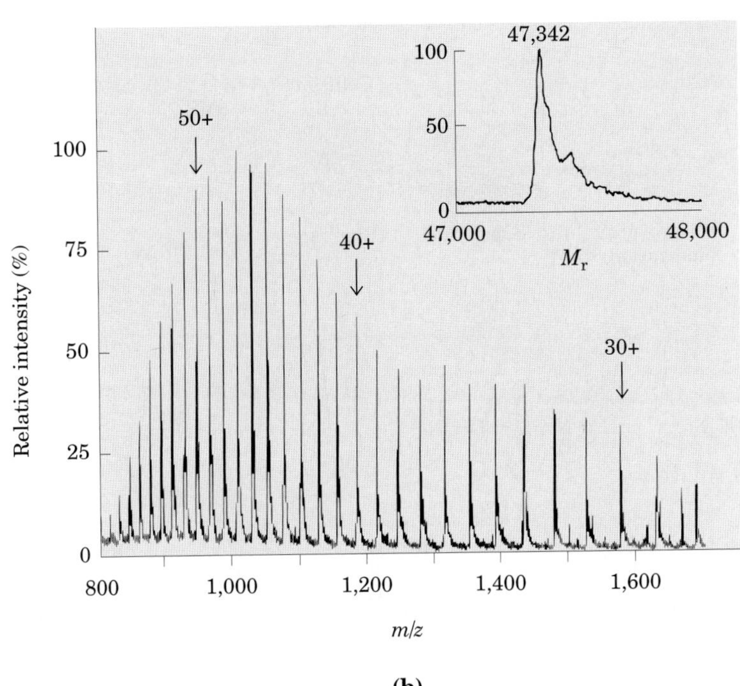

(b)

Continued on next page

Mass spectrometry can also be used to sequence short stretches of polypeptide, an application that has emerged as an invaluable tool for quickly identifying unknown proteins. Sequence information is extracted using a technique called **tandem MS,** or **MS/MS.** A solution containing the protein under investigation is first treated with a protease or chemical reagent to reduce it by hydrolytic cleavage to a mixture of shorter peptides. The mixture is then injected into a device that is essentially two mass spectrometers in tandem (Fig. 2a, top). In the first, the peptide mixture is sorted and the ionized fragments are manipulated so that only one of the several types of peptides produced by cleavage emerges at the other end. The sample of the selected peptide, each molecule of which has a charge somewhere along its length, then travels through a vacuum chamber between the two mass spectrometers. In this collision cell, the peptide is further fragmented by high-energy impact with a "collision gas," a small amount of a noble gas such as helium or argon that is bled into the vacuum chamber. This procedure is designed to fragment many of the peptide molecules in the sample, with each individual peptide broken in only one place on average. Most breaks occur at peptide bonds. This fragmentation does not involve the addition of water (it is done in a near-vacuum), so the products may include molecular ion radicals such as carbonyl radicals (Fig. 2a, bottom). The charge on the original peptide is retained on one of the fragments generated from it.

The second mass spectrometer then measures the m/z ratios of all the charged fragments (uncharged fragments are not detected). This generates one or more sets of peaks. A given set of peaks (Fig. 2b) consists of all the charged fragments that were generated by breaking the same type of bond (but at different points in the peptide), and derived from the same side of the bond breakage, either the carboxyl-terminal or amino-terminal side. Each successive peak in a given set has one less amino acid than the peak before. The difference in mass from peak to peak identifies the amino acid that was lost in each

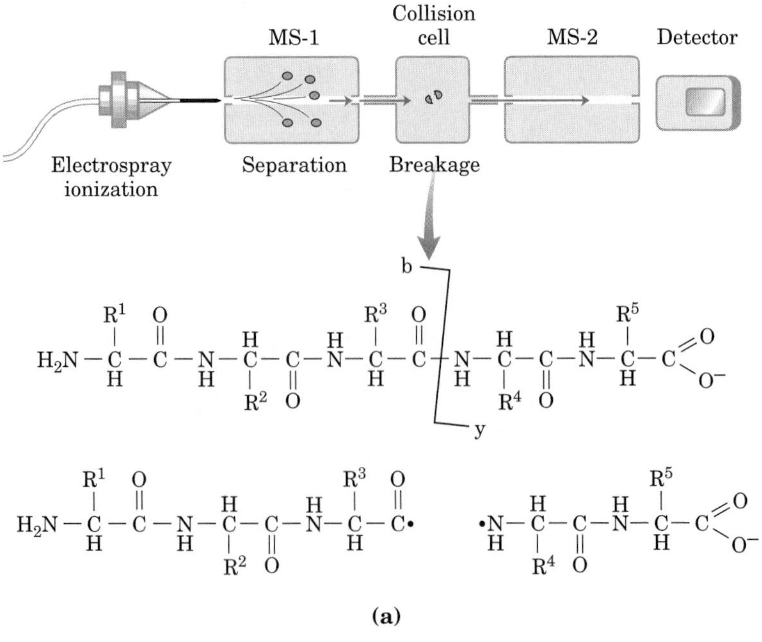

figure 2

Obtaining protein sequence information with tandem mass spectrometry. (a) After proteolytic hydrolysis, a protein solution is injected into a mass spectrometer (MS-1). The different peptides are sorted so that only one type is selected for further analysis. The selected peptide is further fragmented in a chamber between the two mass spectrometers, and m/z for each fragment is measured in the second mass spectrometer (MS-2). Many of the ions generated during this second fragmentation result from breakage of the peptide bond, as shown. These are called b-type or y-type ions, depending on whether the charge is retained on the amino- or carboxyl-terminal side, respectively. **(b)** A typical spectrum with peaks representing the peptide fragments generated from a sample of one small peptide (10 residues). The labeled peaks are y-type ions. The large peak next to y_5'' is a doubly charged ion and is not part of the y set. The successive peaks differ by the mass of a particular amino acid in the original peptide. In this case, the deduced sequence was Phe–Pro–Gly–Gln–(Ile/Leu)–Asn–Ala–Asp–(Ile/Leu)–Arg. Note the ambiguity about Ile and Leu residues because they have the same molecular mass. In this example, the set of peaks derived from y-type ions predominates, and the spectrum is greatly simplified as a result. This is because an Arg residue occurs at the carboxyl terminus of the peptide, and most of the positive charges are retained on this residue.

case, thus revealing the sequence of the peptide. The only ambiguities involve leucine and isoleucine, which have the same mass.

The charge on the peptide can be retained on either the carboxyl- or amino-terminal fragment, and bonds other than the peptide bond can be broken in the fragmentation process, with the result that multiple sets of peaks are usually generated. The two most prominent sets generally consist of charged fragments derived from breakage of the peptide bonds. The set consisting of the carboxyl-terminal fragments can be unambiguously distinguished from that consisting of amino-terminal fragments. Because the bond breaks generated between the spectrometers (in the collision cell) do not yield full carboxyl and amino groups at the sites of the breaks, the only intact α-amino and α-carboxyl groups on the peptide fragments are those at the very ends (Fig. 2a). The two sets of fragments can thereby be assigned by the resulting slight differences in mass. The amino acid sequence derived from one set can be confirmed by the other, improving the confidence in the sequence information obtained.

Even a short sequence is often enough to permit unambiguous association of a protein with its gene, if the gene sequence is known. Sequencing by mass spectrometry cannot replace the Edman degradation procedure for the sequencing of long polypeptides, but it is ideal for proteome research aimed at cataloging the hundreds of cellular proteins that might be separated on a two-dimensional gel. In the coming decades, detailed genomic sequence data will be available from hundreds, eventually thousands, of organisms. The ability to rapidly associate proteins with genes using mass spectrometry will greatly facilitate the exploitation of this extraordinary information resource.

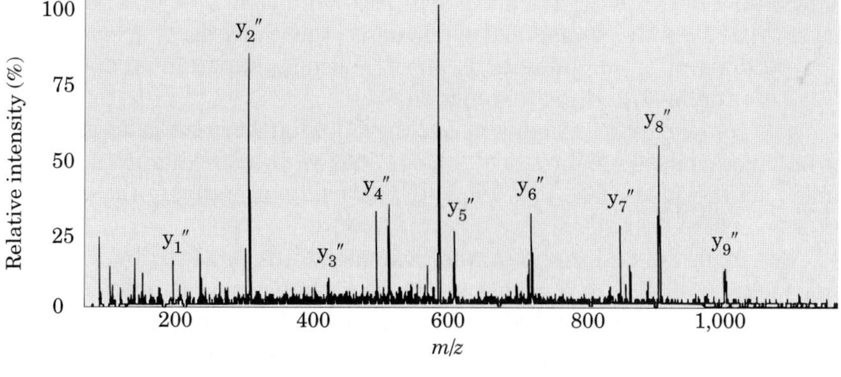

(b)

Amino Acid Sequences Provide Important Biochemical Information

Knowledge of the sequence of amino acids in a protein can offer insights into its three-dimensional structure and its function, cellular location, and evolution. Most of these insights are derived by searching for similarities with other known sequences. Thousands of sequences are known and available in databases accessible through the Internet. The comparison of a newly obtained sequence with this large bank of stored sequences often reveals relationships both surprising and enlightening.

Exactly how the amino acid sequence determines three-dimensional structure is not understood in detail, nor can we always predict function from sequence. However, protein families that have some shared structural or functional features can be readily identified on the basis of amino acid sequence similarities. Individual proteins are assigned to families based on the degree of similarity in amino acid sequence. Members of a family are usually identical across 25% or more of their sequences, and proteins in these families generally share at least some structural and functional characteristics. Some families are defined, however, by identities involving only a few amino acid residues that are critical to a certain function. A number of similar substructures (to be defined in Chapter 6 as "domains") occur in many functionally unrelated proteins. These domains often fold up into structural configurations that have an unusual degree of stability or that are specialized for a certain environment. Evolutionary relationships can also be inferred from the structural and functional similarities within protein families.

Certain amino acid sequences often serve as signals that determine the cellular location, chemical modification, and half-life of a protein. Special signal sequences, usually at the amino terminus, are used to target certain proteins for export from the cell, while other proteins are targeted for distribution to the nucleus, the cell surface, the cytosol, and other cellular locations. Other sequences act as attachment sites for prosthetic groups, such as sugar groups in glycoproteins and lipids in lipoproteins. Some of these signals are well characterized and are easily recognized if they occur in the sequence of a newly characterized protein.

Small Peptides and Proteins Can Be Chemically Synthesized

Many peptides are potentially useful as pharmacologic agents, and their production is of considerable commercial importance. There are three ways to obtain a peptide: (1) purification from tissue, a task often made difficult by the vanishingly low concentrations of some peptides; (2) genetic engineering (Chapter 29); or (3) direct chemical synthesis. Powerful techniques now make direct chemical synthesis an attractive option in many cases. In addition to commercial applications, the synthesis of specific peptide portions of larger proteins is an increasingly important tool for the study of protein structure and function.

R. Bruce Merrifield

The complexity of proteins makes the traditional synthetic approaches of organic chemistry impractical for peptides with more than four or five amino acid residues. One problem is the difficulty of purifying the product after each step.

The major breakthrough in this technology was provided by R. Bruce Merrifield in 1962. His innovation involved synthesizing a peptide while keeping it attached at one end to a solid support. The support is an insoluble polymer (resin) contained within a column, similar to that used for chromatographic procedures. The peptide is built up on this support one amino acid at a time using a standard set of reactions in a repeating cycle (Fig. 5–29). At each successive step in the cycle, protective chemical groups block unwanted reactions.

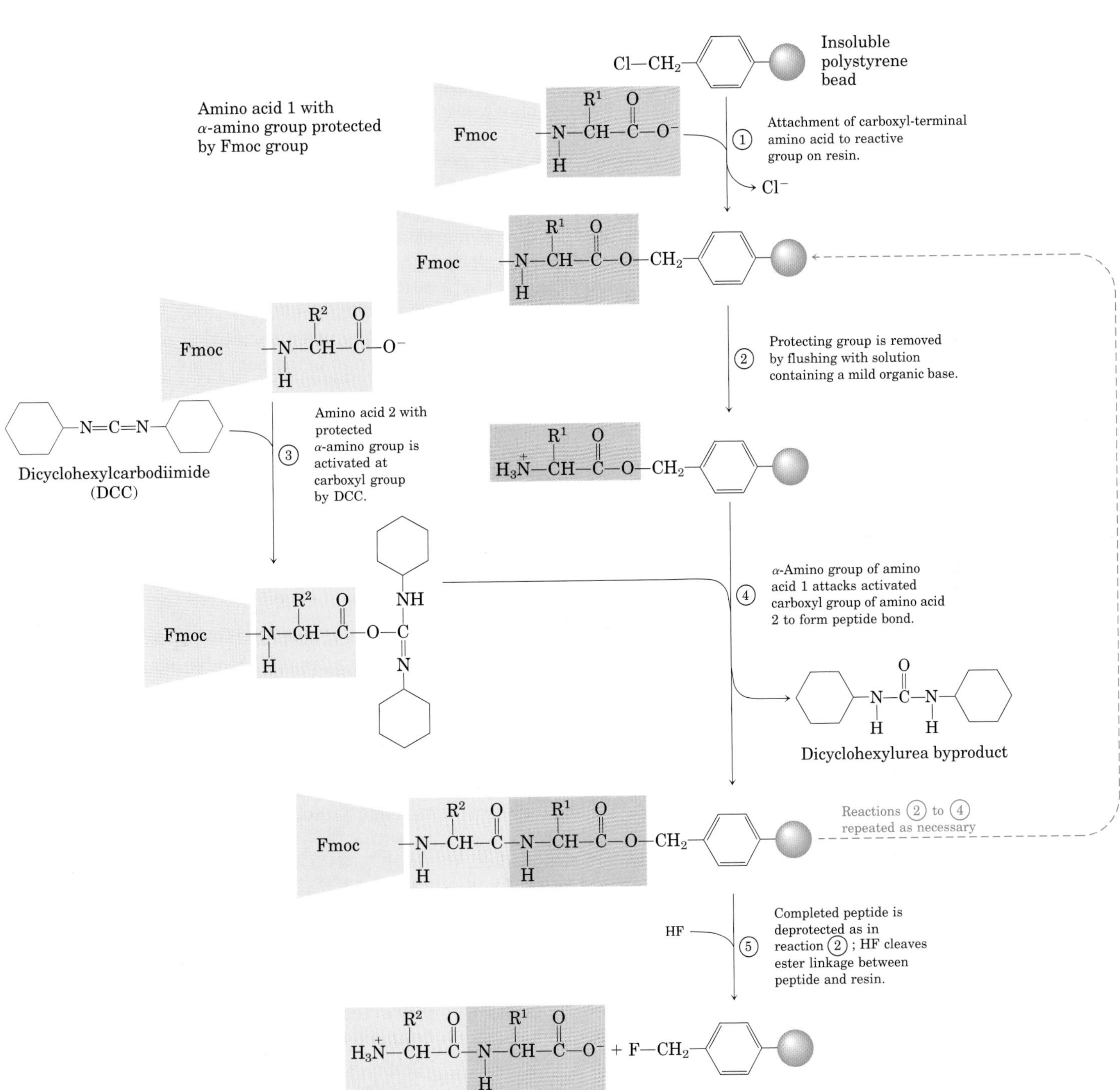

figure 5–29
Chemical synthesis of a peptide on an insoluble polymer support. Reactions ① through ④ are necessary for the formation of each peptide bond. The 9-fluorenylmethoxycarbonyl (Fmoc) group (shaded blue) prevents unwanted reactions at the α-amino group of the residue (shaded red). Chemical synthesis proceeds from the carboxyl terminus to the amino terminus, the reverse of the direction of protein synthesis in vivo (Chapter 27).

table 5–8

Effect of Stepwise Yield on Overall Yield in Peptide Synthesis

Number of residues in the final polypeptide	Overall yield of final peptide (%) when the yield of each step is: 96.0%	99.8%
11	66	98
21	44	96
31	29	94
51	13	90
100	1.7	82

The technology for chemical peptide synthesis is automated, and several kinds of commercial instruments are now available. The most important limitation of the process is the efficiency of each chemical cycle, as can be seen by calculating the overall yields of peptides of various lengths when the yield for addition of each new amino acid is 96.0% versus 99.8% (Table 5–8). Incomplete reaction at one stage can lead to formation of an impurity (in the form of a shorter peptide) in the next. The chemistry has been optimized to permit the synthesis of proteins of 100 amino acid residues in a few days in reasonable yield. A very similar approach is used to synthesize nucleic acids (see Fig. 10–37). It is worth noting that this technology, impressive as it is, still pales when compared with biological processes. The same 100 amino acid protein would be synthesized with exquisite fidelity in about 5 seconds in a bacterial cell.

A variety of new methods for the efficient ligation (joining together) of peptides has made possible the assembly of synthetic peptides into larger proteins. With these methods, novel forms of proteins can be created with precisely positioned chemical groups, including those that might not normally be found in a cellular protein. These novel forms provide new ways to test theories of enzyme catalysis, to create proteins with new chemical properties, and to design protein sequences that will fold into particular structures. This last application provides the ultimate test of our increasing ability to relate the primary structure of a peptide to the three-dimensional structure that it takes up in solution. It is the higher-order structures of proteins that we turn to in the next chapter.

summary

The 20 standard amino acids commonly found as residues in proteins contain an α-carboxyl group, an α-amino group, and a distinctive R group substituted on the α-carbon atom. The α-carbon atom of all amino acids except glycine is asymmetric, and thus they can exist in at least two stereoisomeric forms. Only the L stereoisomers, which are related to the absolute configuration of the reference molecule L-glyceraldehyde, are found in proteins. Amino acids are classified on the basis of the polarity and charge (at pH 7) of their R groups. The nonpolar, aliphatic class includes alanine, glycine, isoleucine, leucine, methionine, and valine. Phenylalanine, tryptophan, and tyrosine have aromatic side chains and are relatively hydrophobic. The polar, uncharged class includes asparagine, cysteine, glutamine, proline, serine, and threonine. The negatively charged (acidic) amino acids are aspartate and glutamate; the positively charged (basic) ones are arginine, histidine, and lysine. Nonstandard amino acids also exist, either as constituents of proteins (through modification of standard amino acid residues after protein synthesis) or as free metabolites.

Monoamino monocarboxylic amino acids (with nonionizable R groups) are diprotic acids ($^{+}H_3NCH(R)COOH$) at low pH. As the pH is raised, a proton is lost from the carboxyl group to form the dipolar or zwitterionic species $^{+}H_3NCH(R)COO^{-}$, which is electrically neutral. Further increase in pH causes loss of the second proton to yield the ionic species $H_2NCH(R)COO^{-}$. Amino acids with ionizable R groups may exist as additional ionic species, depending on the pH and the pK_a of the R group. Thus amino acids vary in their acid-base properties.

Amino acids can be joined covalently through peptide bonds to form peptides and proteins. Cells generally contain thousands of different proteins, each with a different function or biological activity. Proteins can be very long polypeptide chains of 100 to several thousand amino acid residues. However, some naturally occurring peptides have only a few amino acid residues. Some proteins are composed of several noncovalently associated polypeptide chains, which are referred to as subunits. Simple proteins yield only amino acids on hydrolysis; conjugated proteins contain in addition some other component, such as a metal ion or organic prosthetic group.

There are four generally recognized levels of protein structure. Primary structure refers to the amino acid sequence and the location of disulfide bonds. Secondary structure is the spatial relationship of adjacent amino acids in localized stretches. Tertiary structure is the three-dimensional conformation of an entire polypeptide chain. Quaternary structure involves the spatial relationship of multiple polypeptide chains (subunits) that are stably associated.

Proteins are purified by taking advantage of various properties in which they differ. Proteins can be selectively precipitated by the addition of certain salts. A wide range of chromatographic procedures make use of differences in size, binding affinities, charge, and other properties. Electrophoresis can separate proteins on the basis of mass or charge. All purification procedures require a method for quantifying or assaying the protein of interest in the presence of other proteins.

Differences in protein function result from differences in amino acid composition and sequence. Amino acid sequences are deduced by fragmenting polypeptides into smaller peptides using reagents known to cleave specific peptide bonds, determining the amino acid sequence of each fragment by the automated Edman degradation procedure, then ordering the peptide fragments by finding sequence overlaps between fragments generated by different reagents. A protein sequence can also be deduced from the nucleotide sequence of its corresponding gene in DNA. Comparison of a protein's amino acid sequence with the thousands of known sequences often provides insights into the structure, function, cellular location, and evolution of the protein.

Short proteins and peptides (up to about 100 residues long) can be chemically synthesized. The peptide is built up, one amino acid residue at a time, while remaining tethered to a solid support.

further reading

General

Creighton, T.E. (1993) *Proteins: Structures and Molecular Properties,* 2nd edn, W.H. Freeman and Company, New York.

Very useful general source.

Sanger, F. (1988) Sequences, sequences, sequences. *Annu. Rev. Biochem.* **57,** 1–28.

A nice historical account of the development of sequencing methods.

Amino Acids

Greenstein, J.P. & Winitz, M. (1961) *Chemistry of the Amino Acids,* 3 Vols, John Wiley & Sons, New York.

Kreil, G. (1997) D-Amino acids in animal peptides. *Annu. Rev. Biochem.* **66,** 337–345.

An update on the occurrence of these unusual stereoisomers of amino acids.

Meister, A. (1965) *Biochemistry of the Amino Acids,* 2nd edn, Vols 1 and 2, Academic Press, Inc., New York.

Encyclopedic treatment of the properties, occurrence, and metabolism of amino acids.

Peptides and Proteins

Doolittle, R.F. (1985) Proteins. *Sci. Am.* **253** (October), 88–99.

An overview that highlights evolutionary relationships.

Working with Proteins

Dunn, M.J. (1997) Quantitative two-dimensional gel electrophoresis: from proteins to proteomes. *Biochem. Soc. Trans.* **25,** 248–254.

Dunn, M.J. & Corbett, J.M. (1996) Two-dimensional polyacrylamide gel electrophoresis. *Methods Enzymol.* **271,** 177–203.

A detailed description of the technology.

Kornberg, A. (1990) Why purify enzymes? *Methods Enzymol.* **182,** 1–5.

The critical role of classical biochemical methods in a new age.

Scopes, R.K. (1994) *Protein Purification: Principles and Practice,* 3rd edn, Springer-Verlag, New York.

A good source for more complete descriptions of the principles underlying chromatography and other methods.

Covalent Structure of Proteins

Andersen, J.S., Svensson, B., & Roepstorff, P. (1996) Electrospray ionization and matrix assisted laser desorption/ionization mass spectrometry: powerful analytical tools in recombinant protein chemistry. *Nat. Biotechnol.* **14,** 449–457.

A summary emphasizing applications.

Bork, P. & Koonin, E.V. (1998) Predicting functions from protein sequences—where are the bottlenecks? *Nat. Genet.* **18,** 313–318.

A good description of the technology and the roadblocks still limiting its use.

Dongre, A.R., Eng, J.K., & Yates, J.R. III. (1997) Emerging tandem-mass-spectrometry techniques for the rapid identification of proteins. *Trends Biotechnol.* **15,** 418–425.

A detailed description of methods.

Gibney, B.R., Rabanal, F., & Dutton, P.L. (1997) Synthesis of novel proteins. *Curr. Opin. Chem. Biol.* **1,** 537–542.

Koonin, E.V., Tatusov, R.L., & Galperin, M.Y. (1998) Beyond complete genomes: from sequence to structure and function. *Curr. Opin. Struct. Biol.* **8,** 355–363.

A good discussion of what we will do with the tremendous amount of protein sequence information becoming available.

Mann, M. & Wilm, M. (1995) Electrospray mass spectrometry for protein characterization. *Trends Biochem. Sci.* **20,** 219–224.

An approachable summary for beginners.

Wallace, C.J. (1995) Peptide ligation and semisynthesis. *Curr. Opin. Biotechnol.* **6,** 403–410.

Good summary of methods available for peptide ligation. Includes some case studies

Wilken, J. & Kent, S.B. (1998) Chemical protein synthesis. *Curr. Opin. Biotechnol.* **9,** 412–426.

A good overview of chemical synthesis, focusing on peptide ligation methods and applications.

problems

1. Absolute Configuration of Citrulline The citrulline isolated from watermelons has the structure shown below. Is it a D- or L-amino acid? Explain.

$$\begin{array}{c} CH_2(CH_2)_2NH-\underset{\underset{O}{\|}}{C}-NH_2 \\ H-\overset{|}{\underset{|}{C}}-\overset{+}{N}H_3 \\ COO^- \end{array}$$

2. Relationship between the Titration Curve and the Acid-Base Properties of Glycine A 100 mL solution of 0.1 M glycine at pH 1.72 was titrated with 2 M NaOH solution. The pH was monitored and the results were plotted on a graph, as shown at right. The key points in the titration are designated I to V. For each of the statements (a) to (o), *identify* the appropriate key point in the titration and *justify* your choice.

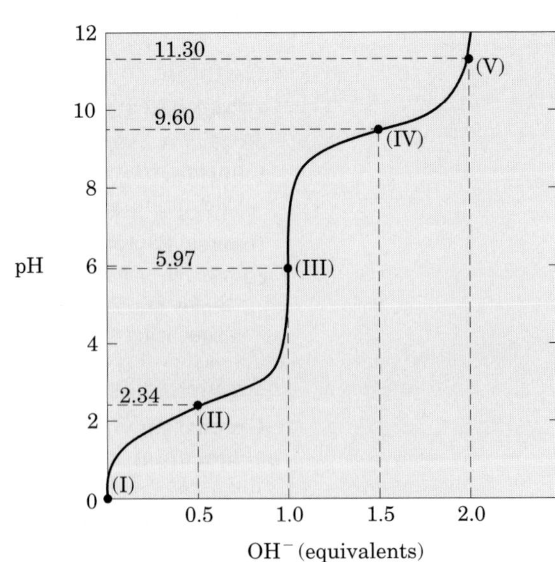

(a) Glycine is present predominantly as the species $^{+}H_3N—CH_2—COOH$.

(b) The *average* net charge of glycine is $+\frac{1}{2}$.

(c) Half of the amino groups are ionized.

(d) The pH is equal to the pK_a of the carboxyl group.

(e) The pH is equal to the pK_a of the protonated amino group.

(f) Glycine has its maximum buffering capacity.

(g) The *average* net charge of glycine is zero.

(h) The carboxyl group has been completely titrated (first equivalence point).

(i) Glycine is completely titrated (second equivalence point).

(j) The predominant species is $^{+}H_3N—CH_2—COO^-$.

(k) The *average* net charge of glycine is -1.

(l) Glycine is present predominantly as a 50:50 mixture of $^{+}H_3N—CH_2—COOH$ and $^{+}H_3N—CH_2—COO^-$.

(m) This is the isoelectric point.

(n) This is the end of the titration.

(o) These are the *worst* pH regions for buffering power.

3. How Much Alanine Is Present as the Completely Uncharged Species? At a pH equal to the isoelectric point of alanine, the *net* charge on alanine is zero. Two structures can be drawn that have a net charge of zero, but the predominant form of alanine at its pI is zwitterionic.

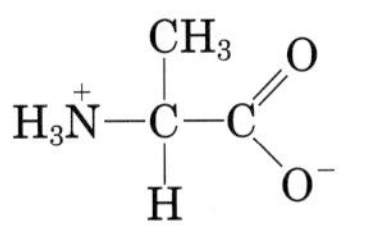

CH_3 O
$H_3\overset{+}{N}—C—C$
H O^-
Zwitterionic

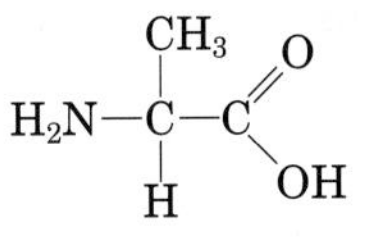

CH_3 O
$H_2N—C—C$
H OH
Uncharged

(a) Why is alanine predominantly zwitterionic rather than completely uncharged at its pI?

(b) What fraction of alanine is in the completely uncharged form at its pI? Justify your assumptions.

4. Ionization State of Amino Acids Each ionizable group of an amino acid can exist in one of two states, charged or neutral. The electric charge on the functional group is determined by the relationship between its pK_a and the pH of the solution. This relationship is described by the Henderson-Hasselbalch equation.

(a) Histidine has three ionizable functional groups. Write the equilibrium equations for its three ionizations and assign the proper pK_a for each ionization. Draw the structure of histidine in each ionization state. What is the net charge on the histidine molecule in each ionization state?

(b) Draw the structures of the predominant ionization state of histidine at pH 1, 4, 8, and 12. Note that the ionization state can be approximated by treating each ionizable group independently.

(c) What is the net charge of histidine at pH 1, 4, 8, and 12? For each pH, will histidine migrate toward the anode (+) or cathode (−) when placed in an electric field?

5. Separation of Amino Acids by Ion-Exchange Chromatography Mixtures of amino acids are analyzed by first separating the mixture into its components through ion-exchange chromatography. Amino acids placed on a cation-exchange resin containing sulfonate groups (see Fig. 5–18) flow down the column at different rates because of two factors that influence their movement: (1) ionic attraction between the $—SO_3^-$ residues on the column and positively charged functional groups on the amino acids, and (2) hydrophobic interactions between amino acid side chains and the strongly hydrophobic backbone of the polystyrene resin. For each pair of amino acids listed, determine which will be eluted first from an ion-exchange column using a pH 7.0 buffer.

(a) Asp and Lys

(b) Arg and Met

(c) Glu and Val

(d) Gly and Leu

(e) Ser and Ala

6. Naming the Stereoisomers of Isoleucine The structure of the amino acid isoleucine is

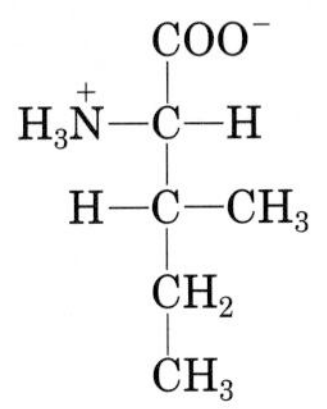

COO^-
$H_3\overset{+}{N}—C—H$
$H—C—CH_3$
CH_2
CH_3

(a) How many chiral centers does it have?

(b) How many optical isomers?

(c) Draw perspective formulas for all the optical isomers of isoleucine.

7. Comparing the pK_a Values of Alanine and Polyalanine The titration curve of alanine shows the ionization of two functional groups with pK_a values of 2.34 and 9.69, corresponding to the ionization of the carboxyl and the protonated amino groups, respectively. The titration of di-, tri-, and larger oligopeptides of alanine also shows the ionization of only two functional groups, although the experimental pK_a values are different. The trend in pK_a values is summarized in the table.

Amino acid or peptide	pK_1	pK_2
Ala	2.34	9.69
Ala–Ala	3.12	8.30
Ala–Ala–Ala	3.39	8.03
Ala–(Ala)$_n$–Ala, $n \geq 4$	3.42	7.94

(a) Draw the structure of Ala–Ala–Ala. Identify the functional groups associated with pK_1 and pK_2.

(b) Why does the value of pK_1 *increase* with each addition of an Ala residue to the Ala oligopeptide?

(c) Why does the value of pK_2 *decrease* with each addition of an Ala residue to the Ala oligopeptide?

8. The Size of Proteins What is the approximate molecular weight of a protein with 682 amino acid residues in a single polypeptide chain?

9. The Number of Tryptophan Residues in Bovine Serum Albumin A quantitative amino acid analysis reveals that bovine serum albumin (BSA) contains 0.58% tryptophan (M_r 204) by weight.

(a) Calculate the *minimum* molecular weight of BSA (i.e., assuming there is only one tryptophan residue per protein molecule).

(b) Gel filtration of BSA gives a molecular weight estimate of 70,000. How many tryptophan residues are present in a molecule of serum albumin?

10. Net Electric Charge of Peptides A peptide has the sequence

Glu – His – Trp – Ser – Gly – Leu – Arg – Pro – Gly

(a) What is the net charge of the molecule at pH 3, 8, and 11? (Use pK_a values for side chains and terminal amino and carboxyl groups as given in Table 5–1.)

(b) Estimate the pI for this peptide.

11. Isoelectric Point of Pepsin Pepsin is the name given to several digestive enzymes that are secreted (as larger precursor proteins) by glands that line the stomach. These glands also secrete hydrochloric acid, which dissolves the particulate matter in food, allowing pepsin to enzymatically cleave individual protein molecules. The resulting mixture of food, HCl, and digestive enzymes is known as chyme and has a pH near 1.5. What pI would you predict for the pepsin proteins? What functional groups must be present to confer this pI on pepsin? Which amino acids in the proteins would contribute such groups?

12. The Isoelectric Point of Histones Histones are proteins found in eukaryotic cell nuclei, tightly bound to DNA, which has many phosphate groups. The pI of histones is very high, about 10.8. What amino acid residues must be present in relatively large numbers in histones? In what way do these residues contribute to the strong binding of histones to DNA?

13. Solubility of Polypeptides One method for separating polypeptides makes use of their differential solubilities. The solubility of large polypeptides in water depends upon the relative polarity of their R groups, particularly on the number of ionized groups: the more ionized groups there are, the more soluble the polypeptide. Which of each pair of polypeptides below is more soluble at the indicated pH?

(a) $(Gly)_{20}$ or $(Glu)_{20}$ at pH 7.0

(b) $(Lys\text{-}Ala)_3$ or $(Phe\text{-}Met)_3$ at pH 7.0

(c) $(Ala\text{-}Ser\text{-}Gly)_5$ or $(Asn\text{-}Ser\text{-}His)_5$ at pH 6.0

(d) $(Ala\text{-}Asp\text{-}Gly)_5$ or $(Asn\text{-}Ser\text{-}His)_5$ at pH 3.0

14. Purification of an Enzyme A biochemist discovers and purifies a new enzyme, generating the purification table below.

Procedure	Total protein (mg)	Activity (units)
1. Crude extract	20,000	4,000,000
2. Precipitation (salt)	5,000	3,000,000
3. Precipitation (pH)	4,000	1,000,000
4. Ion-exchange chromatography	200	800,000
5. Affinity chromatography	50	750,000
6. Size-exclusion chromatography	45	675,000

(a) From the information given in the table, calculate the specific activity of the enzyme solution after each purification procedure.

(b) Which of the purification procedures used for this enzyme is most effective (i.e., gives the greatest relative increase in purity)?

(c) Which of the purification procedures is least effective?

(d) Is there any indication based on the results shown in the table that the enzyme after step 6 is now pure? What else could be done to estimate the purity of the enzyme preparation?

15. Sequence Determination of the Brain Peptide Leucine Enkephalin A group of peptides that influence nerve transmission in certain parts of the brain has been isolated from normal brain tissue. These peptides are known as opioids, because they bind to specific receptors that also bind opiate drugs, such as morphine and naloxone. Opioids thus mimic some of the properties of opiates. Some researchers consider these peptides to be the brain's own pain killers. Using the information below, determine the amino acid sequence of the opioid leucine enkephalin. Explain how your structure is consistent with each piece of information.

(a) Complete hydrolysis by 6 M HCl at 110 °C followed by amino acid analysis indicated the presence of Gly, Leu, Phe, and Tyr, in a 2:1:1:1 molar ratio.

(b) Treatment of the peptide with 1-fluoro-2,4-dinitrobenzene followed by complete hydrolysis and chromatography indicated the presence of the 2,4-dinitrophenyl derivative of tyrosine. No free tyrosine could be found.

(c) Complete digestion of the peptide with pepsin followed by chromatography yielded a dipeptide containing Phe and Leu, plus a tripeptide containing Tyr and Gly in a 1:2 ratio.

16. Structure of a Peptide Antibiotic from *Bacillus brevis* Extracts from the bacterium *Bacillus brevis* contain a peptide with antibiotic properties. This peptide forms complexes with metal ions and apparently disrupts ion transport across the cell membranes of other bacterial species, killing them. The structure of the peptide has been determined from the following observations.

(a) Complete acid hydrolysis of the peptide followed by amino acid analysis yielded equimolar amounts of Leu, Orn, Phe, Pro, and Val. Orn is ornithine, an amino acid not present in proteins but present in some peptides. It has the structure

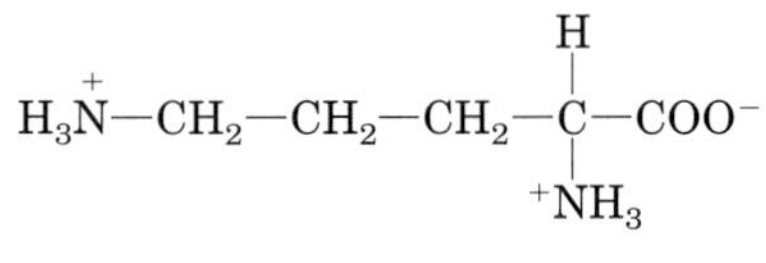

$$H_3\overset{+}{N}{-}CH_2{-}CH_2{-}CH_2{-}\underset{\overset{|}{{}^{+}NH_3}}{\overset{\overset{H}{|}}{C}}{-}COO^-$$

(b) The molecular weight of the peptide was estimated as about 1,200.

(c) The peptide failed to undergo hydrolysis when treated with the enzyme carboxypeptidase. This enzyme catalyzes the hydrolysis of the carboxyl-terminal residue of a polypeptide unless the residue is Pro or does not contain a free carboxyl group for some reason.

(d) Treatment of the intact peptide with 1-fluoro-2,4-dinitrobenzene, followed by complete hydrolysis and chromatography, yielded only free amino acids and the following derivative:

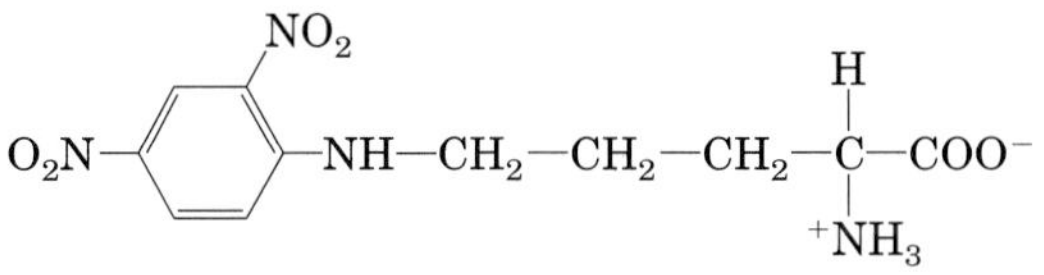

$$O_2N{-}C_6H_3(NO_2){-}NH{-}CH_2{-}CH_2{-}CH_2{-}\underset{\overset{|}{{}^{+}NH_3}}{\overset{\overset{H}{|}}{C}}{-}COO^-$$

(Hint: Note that the 2,4-dinitrophenyl derivative involves the amino group of a side chain rather than the α-amino group.)

(e) Partial hydrolysis of the peptide followed by chromatographic separation and sequence analysis yielded the following di- and tripeptides (the amino-terminal amino acid is always at the left):

Leu–Phe Phe–Pro Orn–Leu Val–Orn
Val–Orn–Leu Phe–Pro–Val Pro–Val–Orn

Given the above information, deduce the amino acid sequence of the peptide antibiotic. Show your reasoning. When you have arrived at a structure, demonstrate that it is consistent with *each* experimental observation.

17. Efficiency in Peptide Sequencing A peptide with the primary structure Lys–Arg–Pro–Leu–Ile–Asp–Gly–Ala is sequenced by the Edman procedure. If each Edman cycle were 96% efficient, what percentage of the amino acids liberated in the fourth cycle would be leucine? Do the calculation a second time, but assume a 99% efficiency for each cycle.

18. Biochemistry Protocols: Your First Protein Purification As the newest and least experienced student in a biochemistry research lab, your first few weeks are spent washing glassware and labeling test tubes. You then graduate to making buffers and stock solutions for use in various laboratory procedures. Finally, you are given responsibility for purifying a protein. It is a citric acid cycle enzyme, citrate synthase, located in the mitochondrial matrix. Following a protocol for the purification, you proceed through the steps below. As you work, a more experienced student questions you about the rationale for each procedure. Supply the answers. (Hint: See Chapter 2 for information on separation of organelles from cells, and Chapter 4 for information about osmolarity).

(a) You pick up 20 kg of beef hearts from a nearby slaughterhouse. You transport the hearts on ice, and perform each step of the purification in a walk-in cold room or on ice. You homogenize the beef heart tissue in a high-speed blender in a medium containing ~0.2 M sucrose, buffered to a pH of 7.2. *Why do you use beef heart tissue, and in such large quantity? What is the purpose of keeping the tissue cold and suspending it in 0.2 M sucrose, at pH 7.2? What happens to the tissue when it is homogenized?*

(b) You subject the resulting heart homogenate, which is dense and opaque, to a series of differential centrifugation steps. *What does this accomplish?*

(c) You proceed with the purification using the supernatant fraction that contains mostly intact mitochondria. Next you osmotically lyse the mitochondria. The lysate, which is less dense than the homogenate, but still opaque, consists primarily of mitochondrial membranes and internal mitochondrial contents. To this lysate you add ammonium sulfate, a highly soluble salt, to a specific concentration. You centrifuge the solution, decant the supernatant, and discard the pellet. To the supernatant, which is clearer than the lysate, you add *more* ammonium sulfate. Once again, you centrifuge the sample, but this time you save the pellet because it contains the protein of interest. *What is the rationale for the two-step addition of the salt?*

(d) You solubilize the ammonium sulfate pellet containing the mitochondrial proteins and dialyze it overnight against large volumes of buffered (pH 7.2) solution. *Why isn't ammonium sulfate included in the dialysis buffer? Why do you use the buffer solution instead of water?*

(e) You run the dialyzed solution over a size-exclusion chromatographic column. Following the protocol, you collect the *first* protein fraction that exits the column, and discard the rest of the fractions that elute from the column later. You detect the protein by measuring UV absorption (at 280 nm) in the fractions. *What does the instruction to collect the first fraction tell you about the protein? Why is UV absorption at 280 nm a good way to monitor for the presence of protein in the eluted fractions?*

(f) You place the fraction collected in (e) on a cation-exchange chromatographic column. After discarding the initial solution that exits the column (the flowthrough), you add a washing solution of higher pH to the column and collect the protein fraction that immediately elutes. *Explain what you are doing.*

(g) You run a small sample of your fraction, now very reduced in volume and quite clear (though tinged pink) on an isoelectric focusing gel. When stained, the gel shows three sharp bands. According to the protocol, the protein of interest is the one with the pI of 5.6, but you decide to do one more assay of the protein's purity. You cut out the pI 5.6 band and subject it to SDS-polyacrylamide gel electrophoresis. The protein resolves as a single band. *Why were you unconvinced of the purity of the "single" protein band on your IEF gel? What did the results of the SDS gel tell you? Why is it important to do the SDS gel electrophoresis* after *the isoelectric focusing?*

chapter

6

The Three-Dimensional Structure of Proteins

The covalent backbone of a typical protein contains hundreds of individual bonds. Because free rotation is possible around many of these bonds, the protein can assume an unlimited number of conformations. However, each protein has a specific chemical or structural function, strongly suggesting that each has a unique three-dimensional structure (Fig. 6–1). By the late 1920s, several proteins had been crystallized, including hemoglobin (M_r 64,500) and the enzyme urease (M_r 483,000). Given that the ordered array of molecules in a crystal can generally form only if the molecular units are identical, the simple fact that many proteins can be crystallized provides strong evidence that even very large proteins are discrete chemical entities with unique structures. This conclusion revolutionized thinking about proteins and their functions.

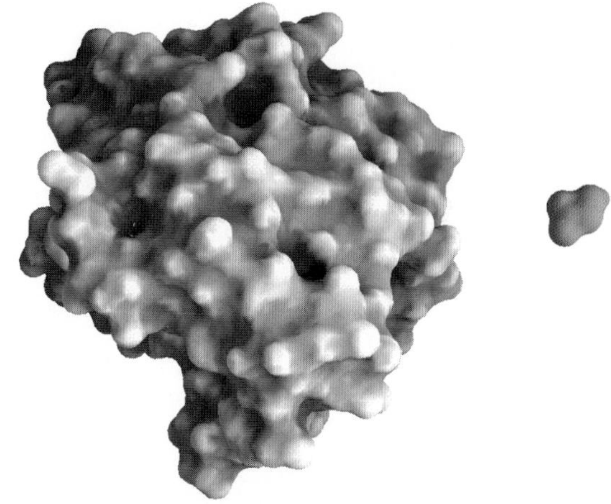

figure 6–1
Structure of the enzyme chymotrypsin, a globular protein. Proteins are large molecules, and we will see that each has a unique structure. A molecule of glycine (blue) is shown for size comparison.

In this chapter, we will explore the three-dimensional structure of proteins, emphasizing five themes. First, the three-dimensional structure of a protein is determined by its amino acid sequence. Second, the function of a protein depends on its structure. Third, an isolated protein has a unique, or nearly unique, structure. Fourth, the most important forces stabilizing the specific structure maintained by a given protein are noncovalent interactions. Finally, amid the huge number of unique protein structures, we can recognize some common structural patterns that help us organize our understanding of protein architecture.

These themes should not be taken to imply that proteins have static, unchanging three-dimensional structures. Protein function often entails an interconversion between two or more structural forms. The dynamic aspects of protein structure will be explored in Chapters 7 and 8.

The relationship between the amino acid sequence of a protein and its three-dimensional structure is an intricate puzzle that is gradually yielding to techniques used in modern biochemistry. An understanding of structure, in turn, is essential to the discussion of function in succeeding chapters. We can find and understand the patterns within the biochemical labyrinth of protein structure by applying fundamental principles of chemistry and physics.

Overview of Protein Structure

The spatial arrangement of atoms in a protein is called its **conformation.** The possible conformations of a protein include any structural state that can be achieved without breaking covalent bonds. A change in conformation could occur, for example, by rotation about single bonds. Of the numerous conformations that are theoretically possible in a protein containing hundreds of single bonds, one or a few generally predominate under biological conditions. The conformation existing under a given set of conditions is usually the one that is thermodynamically the most stable, having the lowest Gibbs free energy (G). Proteins in any of their functional, folded conformations are called **native** proteins.

What principles determine the most stable conformation of a protein? An understanding of protein conformation can be built stepwise from the discussion of primary structure in Chapter 5, through a consideration of secondary, tertiary, and quaternary structure. To this traditional approach must be added a new emphasis on supersecondary structures, a growing set of known and classifiable protein folding patterns that provides an important organizational context to this complex endeavor. We begin by introducing some guiding principles.

A Protein's Conformation Is Stabilized Largely by Weak Interactions

In the context of protein structure, the term **stability** can be defined as the tendency to maintain a native conformation. Native proteins are only marginally stable; the ΔG separating the folded and unfolded states in typical proteins under physiological conditions is in the range of only 20 to 65 kJ/mol. A given polypeptide chain can theoretically assume countless different conformations, and as a result the unfolded state of a protein is characterized by a high degree of conformational entropy. This entropy, and the hydrogen-bonding interactions of many groups in the polypeptide chain with solvent (water), tend to maintain the unfolded state. The chemical interactions that counteract these effects and stabilize the native conformation include disulfide bonds and weak (noncovalent) interactions described in Chapter 4: hydrogen bonds, and hydrophobic and ionic interactions. An appreciation of the role of these weak interactions is especially important to our understanding of how polypeptide chains fold into specific secondary and tertiary structures, and combine with other proteins to form quaternary structures.

About 200 to 460 kJ/mol are required to break a single covalent bond, whereas weak interactions can be disrupted by a mere 4 to 30 kJ/mol. Individual covalent bonds that contribute to the native conformations of proteins, such as disulfide bonds linking separate parts of a single polypeptide chain, are clearly much stronger than individual weak interactions, yet it is weak interactions that predominate as a stabilizing force in protein structure because they are so numerous. In general, the protein conformation with the lowest free energy (i.e., the most stable conformation) is the one with the maximum number of weak interactions.

The stability of a protein is not simply the sum of the free energies of formation of the many weak interactions within it. Every hydrogen-bonding group in a folded polypeptide chain was hydrogen bonded to water prior to folding, and for every hydrogen bond formed in a protein, a hydrogen bond (of similar strength) between the same group and water was broken. The net stability contributed by a given weak interaction, or the *difference* in free energies of the folded and unfolded states, may be close to zero. We must therefore look elsewhere to explain why the native conformation of a protein is favored.

We find that the contribution of weak interactions to protein stability can be understood in terms of the properties of water (Chapter 4). Pure water contains a network of hydrogen-bonded H_2O molecules. No other molecule has the hydrogen-bonding potential of water, and other molecules present in an aqueous solution disrupt the hydrogen bonding of water. When water surrounds a hydrophobic molecule, the optimal arrangement of hydrogen bonds results in a highly structured shell or **solvation layer** of water in the immediate vicinity. The increased order of the water molecules in the solvation layer correlates with an unfavorable decrease in the entropy of the water. However, when nonpolar groups are clustered together, there is a decrease in the extent of the solvation layer because each group no longer presents its entire surface to the solution. The result is a favorable increase in entropy. As described in Chapter 4, this entropy term is the major thermodynamic driving force for the association of hydrophobic groups in aqueous solution. Hydrophobic amino acid side chains therefore tend to be clustered in a protein's interior, away from water.

Under physiological conditions, the formation of hydrogen bonds and ionic interactions in a protein is driven largely by this same entropic effect. Polar groups can generally form hydrogen bonds with water and hence are soluble in water. However, the number of hydrogen bonds per unit mass is generally greater for pure water than for any other liquid or solution, and there are limits to the solubility of even the most polar molecules because their presence causes a net decrease in hydrogen bonding per unit mass. Therefore, a solvation shell of structured water will also form to some extent around polar molecules. Even though the energy of formation of an intramolecular hydrogen bond or ionic interaction between two polar groups in a macromolecule is largely canceled out by the elimination of such interactions between the same groups and water, the release of structured water when the intramolecular interaction is formed provides an entropic driving force for folding. Most of the net change in free energy that occurs when weak interactions are formed within a protein is therefore derived from the increased entropy in the surrounding aqueous solution resulting from the burial of hydrophobic surfaces. This more than counterbalances the large loss of conformational entropy as a polypeptide is constrained into a single folded conformation.

Hydrophobic interactions are clearly important in stabilizing a protein conformation; the interior of a protein is generally a densely packed core of hydrophobic amino acid side chains. It is also important that any polar or charged groups in the protein interior have suitable partners for hydrogen bonding or ionic interactions. One hydrogen bond seems to contribute little to the stability of a native structure, but the presence of hydrogen-bonding or charged groups without partners in the hydrophobic core of a protein can be so *destabilizing* that conformations containing such a group are often thermodynamically untenable. The favorable free-energy change realized by combining such a group with a partner in the surrounding solution can be greater than the difference in free energy between the folded and unfolded states. In addition, hydrogen bonds between groups in proteins form cooperatively. Formation of one hydrogen bond facilitates the formation of additional hydrogen bonds. The overall contribution of hydrogen bonds and other noncovalent interactions to the stabilization of protein conformation is still being evaluated. The interaction of oppositely charged groups that form an ion pair (salt bridge) may also have a stabilizing effect on one or more native conformations of some proteins.

Most of the structural patterns outlined in this chapter reflect two simple rules: (1) hydrophobic residues are largely buried in the protein interior, away from water, and (2) the number of hydrogen bonds within the protein is maximized. Insoluble proteins and proteins within membranes (which we will examine in Chapter 12) follow somewhat different rules because of their function or their environment, but weak interactions are still critical structural elements.

Linus Pauling
1901–1994

The Peptide Bond Is Rigid and Planar

Covalent bonds also place important constraints on the conformation of a polypeptide. In the late 1930s, Linus Pauling and Robert Corey embarked on a series of studies that laid the foundation for our present understanding of protein structure. They began with a careful analysis of the peptide

Robert Corey
1897–1971

bond. The α carbons of adjacent amino acid residues are separated by three covalent bonds, arranged as C_α—C—N—C_α. X-ray diffraction studies of crystals of amino acids and of simple dipeptides and tripeptides demonstrated that the peptide C—N bond is somewhat shorter than the C—N bond in a simple amine and that the atoms associated with the peptide bond are coplanar. This indicated a resonance or partial sharing of two pairs of electrons between the carbonyl oxygen and the amide nitrogen (Fig. 6–2a). The oxygen has a partial negative charge and the nitrogen a partial positive charge, setting up a small electric dipole. The six atoms of the **peptide group** lie in a single plane, with the oxygen atom of the carbonyl group and the hydrogen atom of the amide nitrogen trans to each other. From these findings Pauling and Corey concluded that the peptide C—N bonds are unable to rotate freely because of their partial double-bond character. Rotation is permitted about the N—C_α and the C_α—C bonds. The backbone of a polypeptide chain can thus be pictured as a series of rigid planes with consecutive planes sharing a common point of rotation at C_α (Fig. 6–2b). The

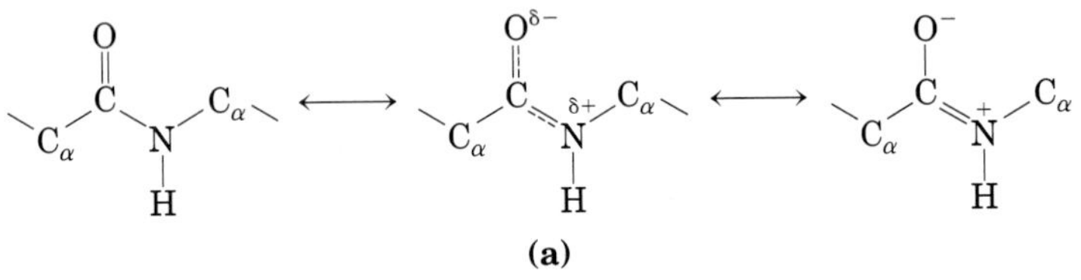

(a)

The carbonyl oxygen has a partial negative charge and the amide nitrogen a partial positive charge, setting up a small electric dipole. Virtually all peptide bonds in proteins occur in this trans configuration; an exception is noted in Figure 6–8b.

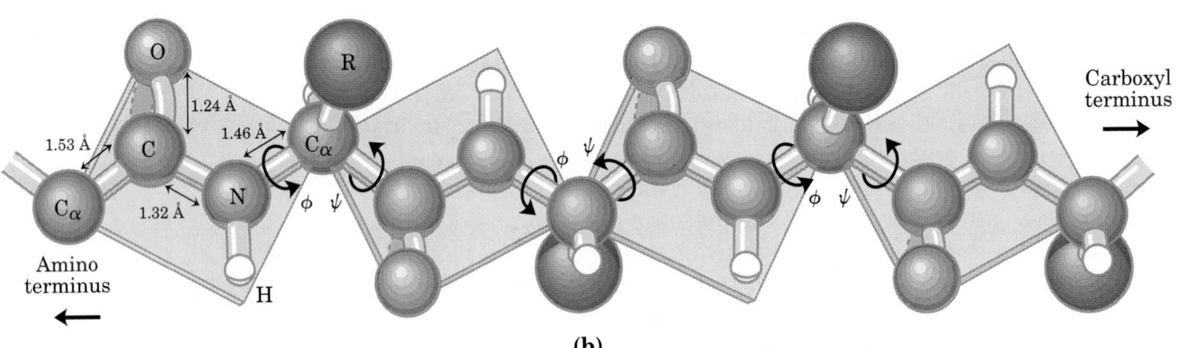

(b)

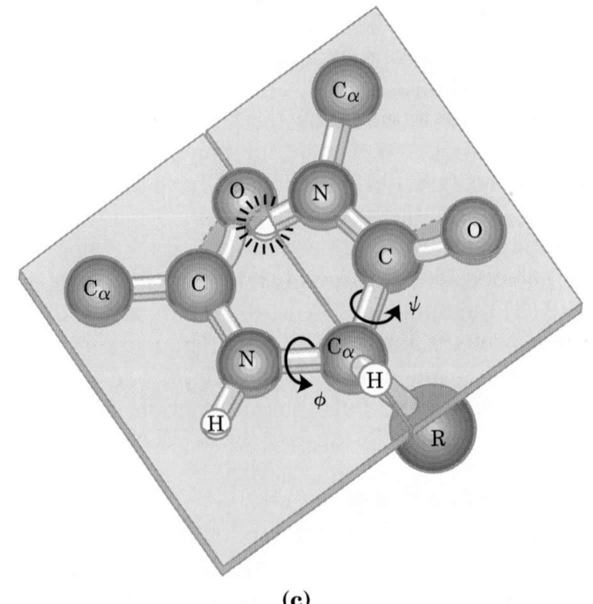

(c)

figure 6–2

The planar peptide group. (a) Each peptide bond has some double-bond character due to resonance and cannot rotate. **(b)** Three bonds separate sequential α carbons in a polypeptide chain. The N—C_α and C_α—C bonds can rotate, with bond angles designated ϕ and ψ, respectively. The peptide C—N bond is not free to rotate. Other single bonds in the backbone may also be rotationally hindered, depending on the size and charge of the R groups. **(c)** By convention, ϕ and ψ are both defined as 0° when the two peptide bonds flanking that α carbon are in the same plane and positioned as shown. In a protein, this conformation is prohibited by steric overlap between an α-carbonyl oxygen and an α-amino hydrogen atom. To illustrate the bonds between atoms, the balls representing each atom are smaller than the van der Waals radii for this scale. 1 Å = 0.1 nm.

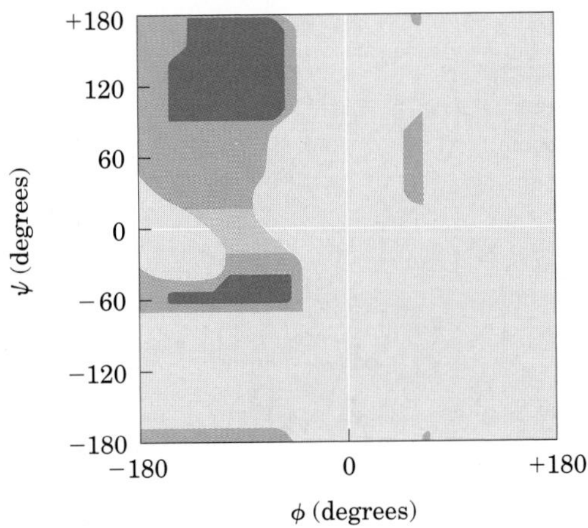

figure 6–3

Ramachandran plot for L-Ala residues. The conformations of peptides are defined by the values of ϕ and ψ. Conformations deemed possible are those that involve little or no steric interference, based on calculations using known van der Waals radii and bond angles. The areas shaded dark blue reflect conformations that involve no steric overlap and thus are fully allowed; medium blue indicates conformations allowed at the extreme limits for unfavorable atomic contacts; the lightest blue area reflects conformations that are permissible if a little flexibility is allowed in the bond angles. The asymmetry of the plot results from the L stereochemistry of the amino acid residues. The plots for other L-amino acid residues with unbranched side chains are nearly identical. The allowed ranges for branched amino acid residues such as Val, Ile, and Thr are somewhat smaller than for Ala. The Gly residue, which is less sterically hindered, exhibits a much broader range of allowed conformations. The range for Pro residues is greatly restricted because ϕ is limited by the cyclic side chain to the range of −35° to −85°.

rigid peptide bonds limit the range of conformations that can be assumed by a polypeptide chain.

By convention the bond angles resulting from rotations at C_α are labeled ϕ (phi) for the N—C_α bond and ψ (psi) for the C_α—C bond. Again by convention, both ϕ and ψ are defined as 180° when the polypeptide is in its fully extended conformation and all peptide groups are in the same plane (Fig. 6–2b). In principle, ϕ and ψ can have any value between −180° and +180°, but many values are prohibited by steric interference between atoms in the polypeptide backbone and amino acid side chains. The conformation in which both ϕ and ψ are 0° (Fig. 6–2c) is prohibited for this reason; this is used merely as a reference point for describing the angles of rotation. Allowed values for ϕ and ψ are graphically revealed when ψ is plotted versus ϕ in a **Ramachandran plot** (Fig. 6–3), introduced by G.N. Ramachandran.

Protein Secondary Structure

The term **secondary structure** refers to the local conformation of some part of the polypeptide. The discussion of secondary structure most usefully focuses on common regular folding patterns of the polypeptide backbone. A few types of secondary structure are particularly stable and occur widely in proteins. The most prominent are the α helix and β conformations described below. Using fundamental chemical principles and a few experimental observations, Pauling and Corey predicted the existence of these secondary structures in 1951, several years before the first complete protein structure was elucidated.

The α Helix Is a Common Protein Secondary Structure

Pauling and Corey were aware of the importance of hydrogen bonds in orienting polar chemical groups such as the C=O and N—H groups of the peptide bond. They also had the experimental results of William Astbury, who in the 1930s had conducted pioneering x-ray studies of proteins. Astbury demonstrated that the protein that makes up hair and porcupine quills (the fibrous protein α-keratin) has a regular structure that repeats every 5.15 to 5.2 Å. (The angstrom, Å (named after the physicist Anders J. Ångström), is equal to 0.1 nm. Although not an SI unit, it is used universally by structural biologists to describe atomic distances.) With this information and their data on the peptide bond, and with the help of precisely constructed models, Pauling and Corey set out to determine the likely conformations of protein molecules.

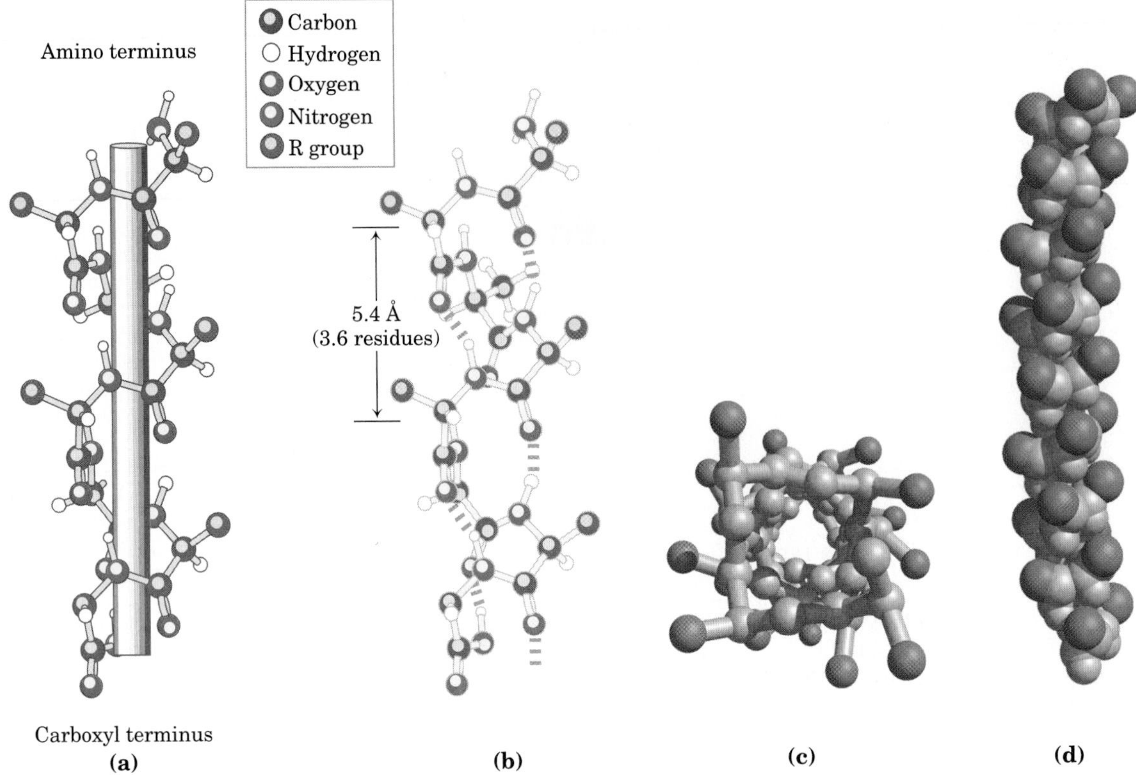

figure 6–4

Four models of the α helix, showing different aspects of its structure. (a) Formation of a right-handed α helix. The planes of the rigid peptide bonds are parallel to the long axis of the helix, depicted here as a vertical rod. **(b)** Ball-and-stick model of a right-handed α helix, showing the intrachain hydrogen bonds. The repeat unit is a single turn of the helix, 3.6 residues. **(c)** The α helix as viewed from one end, looking down the longitudinal axis. Note the positions of the R groups, represented by purple spheres. This ball-and-stick model, used to emphasize the helical arrangement, gives the false impression that the helix is hollow because the balls do not represent the van der Waals radii of the individual atoms. As the space-filling model **(d)** shows, the atoms in the center of the α helix are in very close contact.

The simplest arrangement the polypeptide chain could assume with its rigid peptide bonds (but other single bonds free to rotate) is a helical structure, which Pauling and Corey called the ***α* helix** (Fig. 6–4). In this structure the polypeptide backbone is tightly wound around an imaginary axis drawn longitudinally through the middle of the helix, and the R groups of the amino acid residues protrude outward from the helical backbone. The repeating unit is a single turn of the helix, which extends about 5.4 Å along the long axis, slightly greater than the periodicity Astbury observed on x-ray analysis of hair keratin. The amino acid residues in an α helix have conformations with $\psi = -45°$ to $-50°$ and $\phi = -60°$, and each helical turn includes 3.6 amino acid residues. The helical twist of the α helix found in all proteins is right-handed (Box 6–1). The α helix proved to be the predominant structure in α-keratins. More generally, about one-fourth of all amino acid residues in polypeptides are found in α helices, the exact fraction varying greatly from one protein to the next.

Why does the α helix form more readily than many other possible conformations? The answer is, in part, that an α helix makes optimal use of internal hydrogen bonds. The structure is stabilized by a hydrogen bond between the hydrogen atom attached to the electronegative nitrogen atom of a peptide linkage and the electronegative carbonyl oxygen atom of the fourth amino acid on the amino-terminal side of that peptide bond (Fig. 6–4b). Within the α helix, every peptide bond (except those close to each end of the helix) participates in such hydrogen bonding. Each successive turn of the α helix is held to adjacent turns by three to four hydrogen bonds. All the hydrogen bonds combined give the entire helical structure considerable stability.

Further model-building experiments have shown that an α helix can form in polypeptides consisting of either L- or D-amino acids. However, all residues must be of one stereoisomeric series; a D-amino acid will disrupt a

box 6–1 **Knowing the Right Hand from the Left**

There is a simple method for determining whether a helical structure is right-handed or left-handed. Make fists of your two hands with thumbs outstretched and pointing straight up. Looking at your right hand, think of a helix spiraling up your right thumb in the direction in which the other four fingers are curled as shown (counterclockwise). The resulting helix is right-handed. Your left hand will demonstrate a left-handed helix, which rotates in the clockwise direction as it spirals up your thumb.

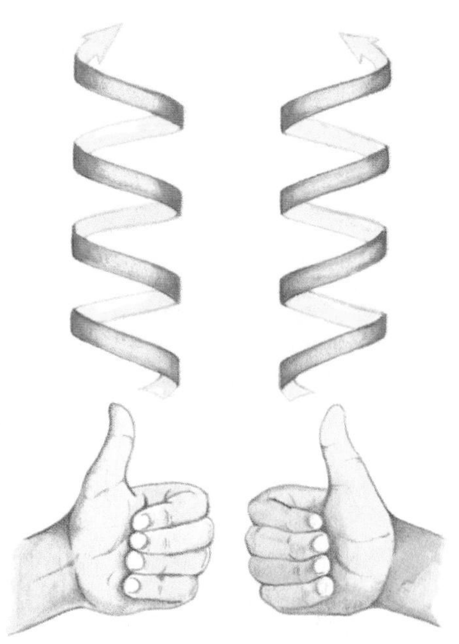

regular structure consisting of L-amino acids, and vice versa. Naturally occurring L-amino acids can form either right- or left-handed α helices, but extended left-handed helices have not been observed in proteins.

Amino Acid Sequence Affects α Helix Stability

Not all polypeptides can form a stable α helix. Interactions between amino acid side chains can stabilize or destabilize this structure. For example, if a polypeptide chain has a long block of Glu residues, this segment of the chain will not form an α helix at pH 7.0. The negatively charged carboxyl groups of adjacent Glu residues repel each other so strongly that they overcome the stabilizing influence of hydrogen bonds on the α helix. For the same reason, if there are many adjacent Lys and/or Arg residues, which have positively charged R groups at pH 7.0, they will also repel each other and prevent formation of the α helix. The bulk and shape of Asn, Ser, Thr, and Leu residues can also destabilize an α helix if they are close together in the chain.

The twist of an α helix ensures that critical interactions occur between an amino acid side chain and the side chain three (and sometimes four) residues away on either side of it (Fig. 6–5). Positively charged amino acids are often found three residues away from negatively charged amino acids, permitting the formation of an ion pair. Two aromatic amino acid residues are often similarly spaced, resulting in a hydrophobic interaction.

A constraint on the formation of the α helix is the presence of Pro or Gly residues. In proline, the nitrogen atom is part of a rigid ring (see Fig. 6–8b), and rotation about the N—C_α bond is not possible. Thus, a Pro residue introduces a destabilizing kink in an α helix. In addition, the nitrogen atom of a Pro residue in peptide linkage has no substituent hydrogen to participate in hydrogen bonds with other residues. For these reasons, proline is only

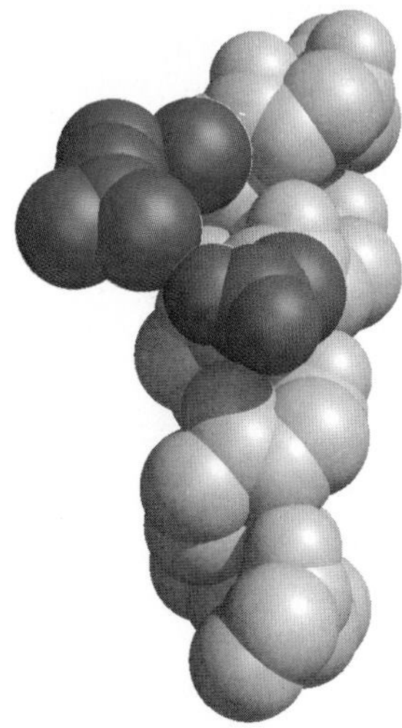

figure 6–5
Interactions between R groups of amino acids three residues apart in an α helix. An ionic interaction between Asp^{100} and Arg^{103} in an α-helical region of the protein troponin C, a calcium-binding protein associated with muscle, is shown in this space-filling model. The polypeptide backbone (carbons, α-amino nitrogens, and α-carbonyl oxygens) is shown in gray for a helix segment 13 residues long. The only side chains represented here are the interacting Asp (red) and Arg (blue) side chains.

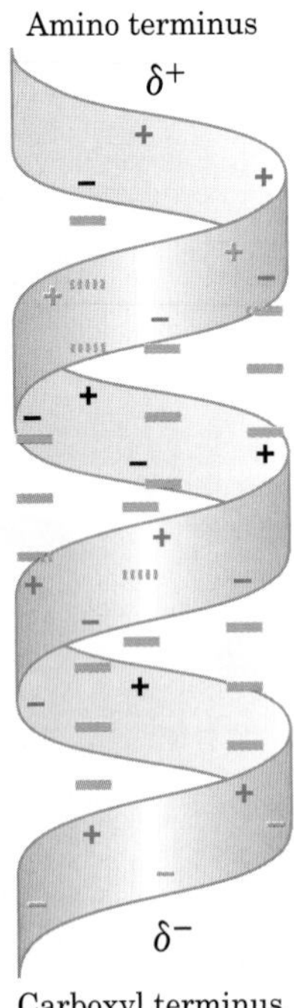

figure 6–6
The electric dipole of a peptide bond (see Fig. 6–2a) is transmitted along an α-helical segment through the intrachain hydrogen bonds, resulting in an overall helix dipole. In this illustration, the amino and carbonyl constituents of each peptide bond are indicated by + and − symbols, respectively. Non-hydrogen-bonded amino and carbonyl constituents in the peptide bonds near each end of the α-helical region are shown in red.

rarely found within an α helix. Glycine occurs infrequently in α helices for a different reason: it has more conformational flexibility than the other amino acid residues. Polymers of glycine tend to take up coiled structures quite different from an α helix.

A final factor affecting the stability of an α helix in a polypeptide is the identity of the amino acid residues near the ends of the α-helical segment. A small electric dipole exists in each peptide bond (Fig. 6–2a). These dipoles are connected through the hydrogen bonds of the helix, resulting in a net dipole extending down the helix that increases with helix length (Fig. 6–6). The four amino acid residues at each end of the helix do not participate fully in the helix hydrogen bonds. The partial positive and negative charges of the helix dipole actually reside on the peptide amino and carbonyl groups near the amino-terminal and carboxyl-terminal ends of the helix, respectively. For this reason, negatively charged amino acids are often found near the amino terminus of the helical segment, where they have a stabilizing interaction with the positive charge of the helix dipole; a positively charged amino acid at the amino-terminal end is destabilizing. The opposite is true at the carboxyl-terminal end of the helical segment.

Thus five different kinds of constraints affect the stability of an α helix: (1) the electrostatic repulsion (or attraction) between successive amino acid residues with charged R groups, (2) the bulkiness of adjacent R groups, (3) the interactions between amino acid side chains spaced three (or four) residues apart, (4) the occurrence of Pro and Gly residues, and (5) the interaction between amino acid residues at the ends of the helical segment and the electric dipole inherent to the α helix. Hence, the tendency of a given segment of a polypeptide chain to fold up as an α helix depends on the identity and sequence of amino acid residues within the segment.

The β Conformation Organizes Polypeptide Chains into Sheets

Pauling and Corey predicted a second type of repetitive structure, the **β conformation.** This is a more extended conformation of polypeptide chains, and its structure has been confirmed by x-ray analysis. In the β conformation, the backbone of the polypeptide chain is extended into a zigzag rather than helical structure (Fig. 6–7). The zigzag polypeptide chains can be arranged side by side to form a structure resembling a series of pleats. In this arrangement, called a **β sheet,** hydrogen bonds are formed between adjacent segments of polypeptide chain. The individual segments that form a β sheet are usually nearby on the polypeptide chain, but can also be quite distant from each other in the linear sequence of the polypeptide; they may even be segments in different polypeptide chains. The R groups of adjacent amino acids protrude from the zigzag structure in opposite directions, creating an alternating pattern as seen in the side views in Figure 6–7.

The adjacent polypeptide chains in a β sheet can be either parallel or antiparallel (having the same or opposite amino-to-carboxyl orientations, respectively). The structures are somewhat similar, although the repeat period is shorter for the parallel conformation (6.5 Å, versus 7 Å for antiparallel) and the hydrogen-bonding patterns are different.

Some protein structures limit the kinds of amino acids that can occur in the β sheet. When two or more β sheets are layered closely together within a protein, the R groups of the amino acid residues on the touching surfaces must be relatively small. β-Keratins such as silk fibroin and the fibroin of spider webs have a very high content of Gly and Ala residues, the two amino acids with the smallest R groups. Indeed, in silk fibroin Gly and Ala alternate over large parts of the sequence.

(a) Antiparallel

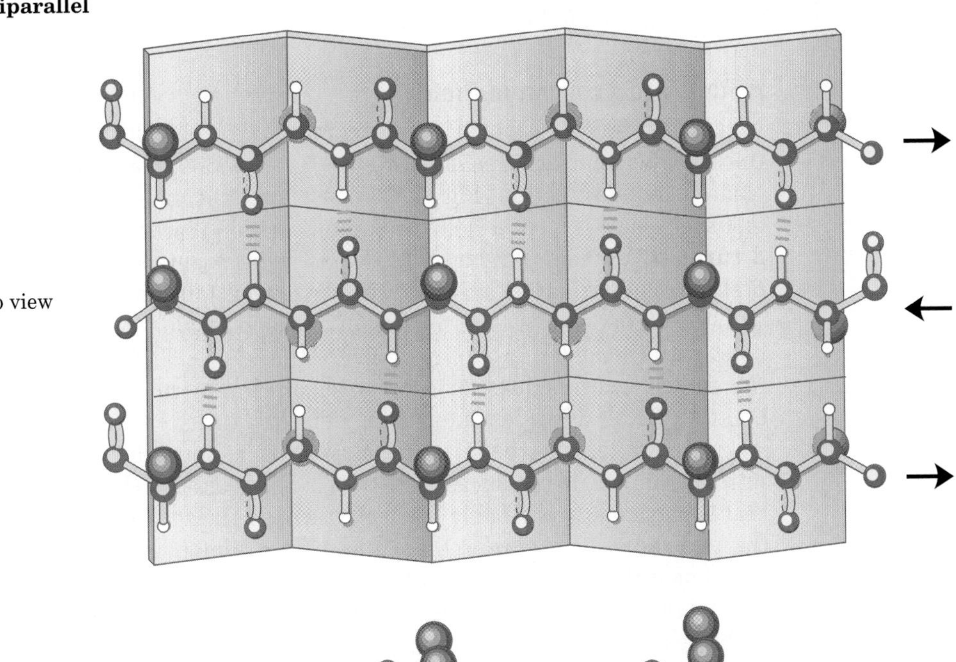

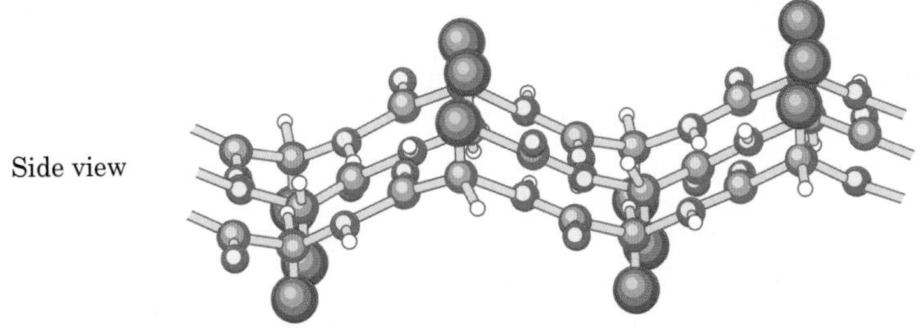

(b) Parallel

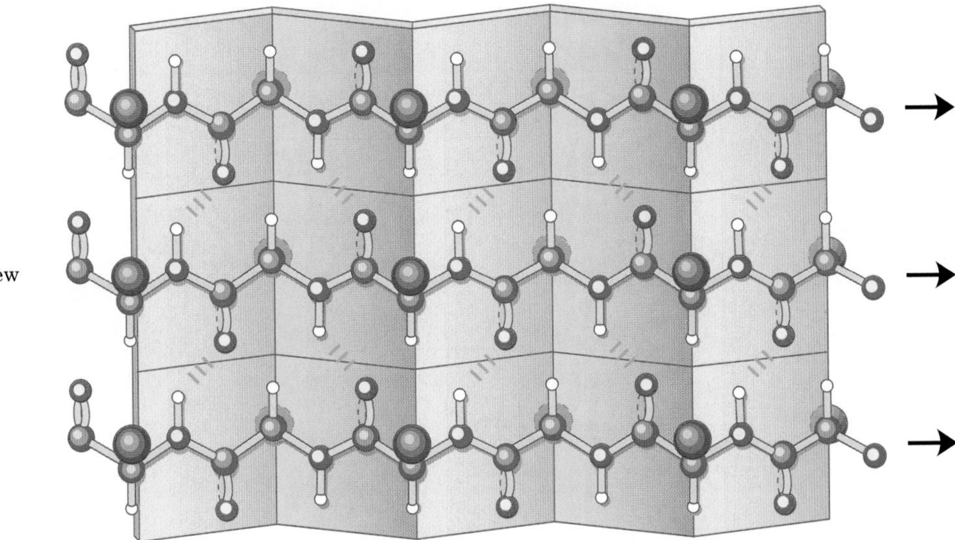

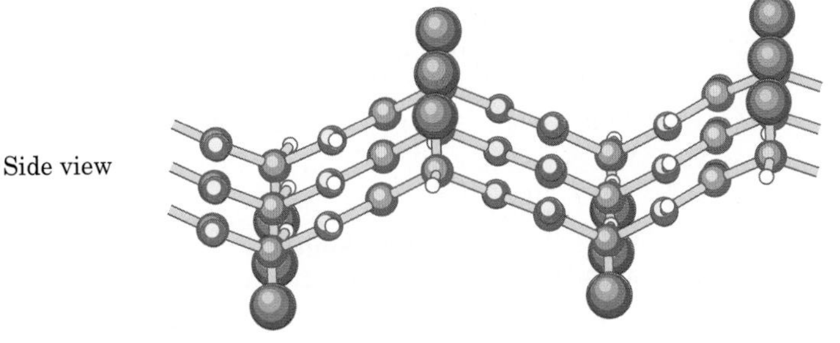

figure 6–7

The β conformation of polypeptide chains. These top and side views reveal the R groups extending out from the β sheet and emphasize the pleated shape described by the planes of the peptide bonds. (An alternate name for this structure is β-pleated sheet.) Hydrogen-bond cross-links between adjacent chains are also shown. **(a)** Antiparallel β sheet, in which the amino-terminal to carboxyl-terminal orientation of adjacent chains (arrows) is inverse. **(b)** Parallel β sheet.

β Turns Are Common in Proteins

In globular proteins, which have a compact folded structure, nearly one-third of the amino acid residues are in turns or loops where the polypeptide chain reverses direction (Fig. 6–8). These are the connecting elements that link successive runs of α helix or β conformation. Particularly common are **β turns** that connect the ends of two adjacent segments of an antiparallel β sheet. The structure is a 180° turn involving four amino acid residues, with the carbonyl oxygen of the first amino acid residue forming a hydrogen bond with the amino-group hydrogen of the fourth. The peptide groups of the central two residues do not participate in any interresidue hydrogen bonding. Gly and Pro residues often occur in β turns, the former because it is small and flexible, the latter because peptide bonds involving the imino nitrogen of proline readily assume the cis configuration (Fig. 6–8b), a form that is particularly amenable to a tight turn. Of the several types of β turns, the two shown in Figure 6–8 are the most common. Beta turns are often found near the surface of a protein, where the peptide groups of the central two amino acid residues in the turn can hydrogen bond with water. Considerably less common is the γ turn, a three-residue turn with a hydrogen bond between the first and third residues.

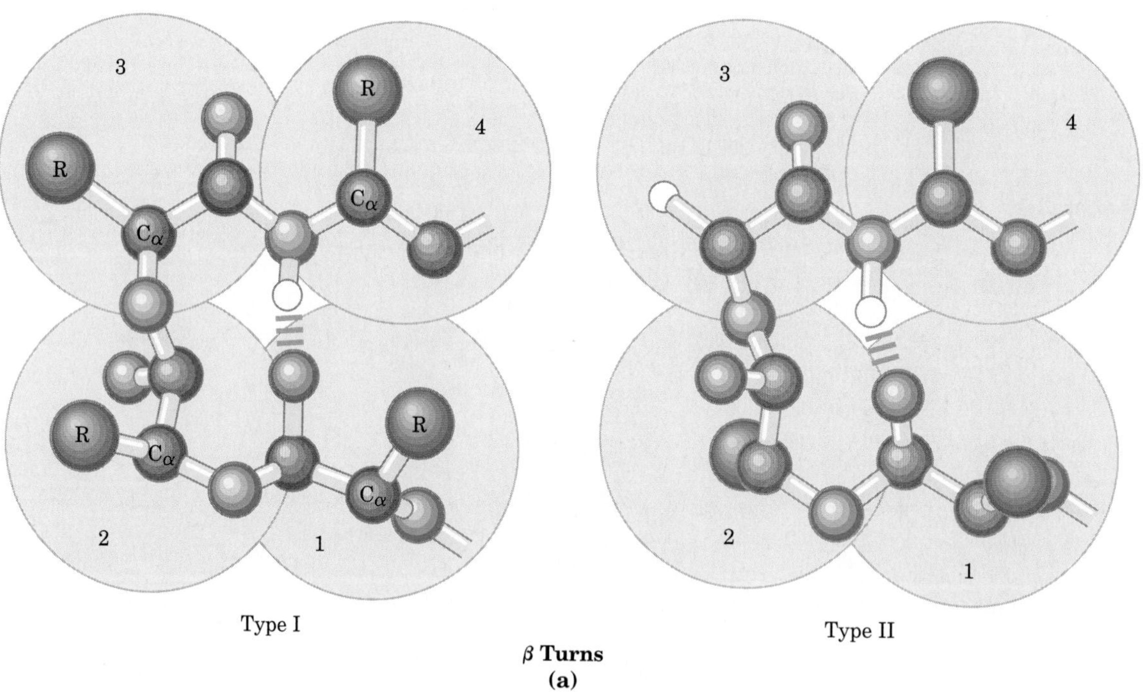

figure 6–8

Structures of β turns. (a) Type I and type II β turns are most common; type I turns occur more than twice as frequently as type II. Type II β turns always have Gly as the third residue. Note the hydrogen bond between the peptide groups of the first and fourth residues of the bends. (Individual amino acid residues are framed by large blue circles.) **(b)** The trans and cis isomers of a peptide bond involving the imino nitrogen of proline. Of the peptide bonds between amino acid residues other than Pro, over 99.95% are in the trans configuration. For peptide bonds involving the imino nitrogen of proline, however, about 6% are in the cis configuration; many of these occur at β turns.

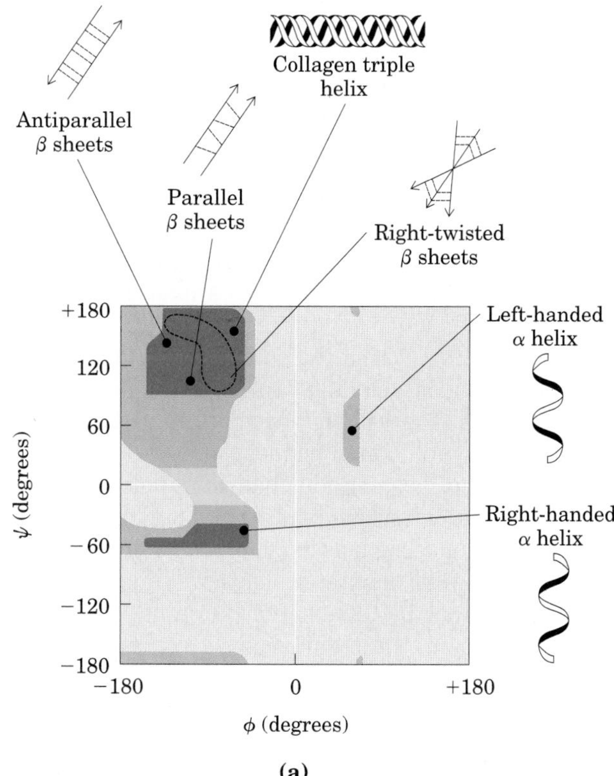

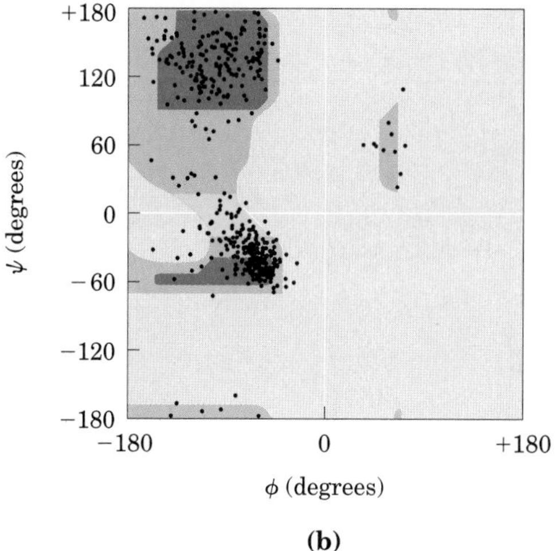

figure 6–9

Ramachandran plots for a variety of structures. (a) The values of ϕ and ψ for various allowed secondary structures are overlaid on the plot from Figure 6–3. Although left-handed α helices extending over several amino acid residues are theoretically possible, they have not been observed in proteins. **(b)** The values of ϕ and ψ for all the amino acid residues except Gly in the enzyme pyruvate kinase (isolated from rabbit) are overlaid on the plot of theoretically allowed conformations (Fig. 6–3). The small, flexible Gly residues were excluded because they frequently fall outside the expected ranges (blue).

Common Secondary Structures Have Characteristic Bond Angles and Amino Acid Content

The α helix and the β conformation are the major repetitive secondary structures in a wide variety of proteins, although other repetitive structures do exist in some specialized proteins (an example is collagen; see Fig. 6–13).

Every type of secondary structure can be completely described by the bond angles ϕ and ψ at each residue. As shown by a Ramachandran plot, the α helix and β conformation fall within a relatively restricted range of sterically allowed structures (Fig. 6–9a). Most values of ϕ and ψ taken from known protein structures fall into the expected regions, with high concentrations near the α helix and β conformation values as predicted (Fig. 6–9b). The only amino acid residue often found in a conformation outside these regions is glycine. Because its side chain, a single hydrogen atom, is small, a Gly residue can take part in many conformations that are sterically forbidden for other amino acids.

Some amino acids are accommodated better than others in the different types of secondary structures. An overall summary is presented in Figure 6–10. Some biases, such as the common presence of Pro and Gly residues in β turns and their relative absence in α helices, is readily explained by the known constraints on the different secondary structures. Other evident biases may be explained by taking into account the sizes or charges of side chains, but not all the trends in Figure 6–10 are understood.

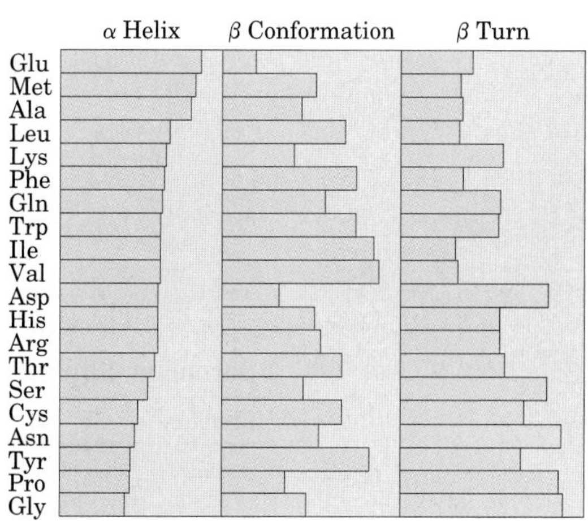

figure 6–10

Relative probabilities that a given amino acid will occur in the three common types of secondary structure.

Protein Tertiary and Quaternary Structures

The overall three-dimensional arrangement of all atoms in a protein is referred to as the protein's **tertiary structure.** Whereas the term secondary structure refers to the spatial arrangement of amino acid residues that are adjacent in the primary structure, tertiary structure includes *longer-range* aspects of amino acid sequence. Amino acids that are far apart in the polypeptide sequence and that reside in different types of secondary structure may interact within the completely folded structure of a protein. The location of bends (including β turns) in the polypeptide chain and the direction and angle of these bends are determined by the number and location of specific bend-producing residues, such as Pro, Thr, Ser, and Gly. Interacting segments of polypeptide chains are held in their characteristic tertiary positions by different kinds of weak-bonding interactions (and sometimes by covalent bonds such as disulfide cross-links) between the segments.

Some proteins contain two or more separate polypeptide chains or subunits, which may be identical or different. The arrangement of these protein subunits in three-dimensional complexes constitutes **quaternary structure.**

In considering these higher levels of structure, it is useful to classify proteins into two major groups: **fibrous proteins,** having polypeptide chains arranged in long strands or sheets, and **globular proteins,** having polypeptide chains folded into a spherical or globular shape. The two groups are structurally distinct: fibrous proteins usually consist largely of a single type of secondary structure; globular proteins often contain several types of secondary structure. The groups differ functionally in that the structures that provide support, shape, and external protection to vertebrates are made of fibrous proteins, whereas most enzymes and regulatory proteins are globular proteins. Certain fibrous proteins played a key role in the development of our modern understanding of protein structure and provide particularly clear examples of the relationship between structure and function. We begin our discussion with fibrous proteins before turning to the more complex folding patterns observed in globular proteins.

Fibrous Proteins Are Adapted for a Structural Function

α-Keratin, collagen, and silk fibroin nicely illustrate the relationship between protein structure and biological function (Table 6–1). Fibrous proteins share properties that give strength and/or flexibility to the structures in which they occur. In each case, the fundamental structural unit is a simple repeating element of secondary structure. All fibrous proteins are insoluble in water, a property conferred by a high concentration of hydrophobic amino acid residues both in the interior of the protein and on its surface. These hydrophobic surfaces are largely buried by packing many similar polypeptide chains together to form elaborate supramolecular complexes.

table 6–1

Secondary Structures and Properties of Fibrous Proteins

Structure	Characteristics	Examples of occurrence
α Helix, cross-linked by disulfide bonds	Tough, insoluble protective structures of varying hardness and flexibility	α-Keratin of hair, feathers, and nails
β Conformation	Soft, flexible filaments	Silk fibroin
Collagen triple helix	High tensile strength, without stretch	Collagen of tendons, bone matrix

The underlying structural simplicity of fibrous proteins makes them particularly useful for illustrating some of the fundamental principles of protein structure discussed above.

α-Keratin The α-keratins have evolved for strength. Found in mammals, these proteins constitute almost the entire dry weight of hair, wool, nails, claws, quills, horns, hooves, and much of the outer layer of skin. The α-keratins are part of a broader family of proteins called intermediate filament (IF) proteins. Other IF proteins are found in the cystoskeletons of animal cells. All IF proteins have a structural function and share structural features exemplified by the α-keratins.

The α-keratin helix is a right-handed α helix, the same helix found in many other proteins. Francis Crick and Linus Pauling in the early 1950s independently suggested that the α helices of keratin were arranged as a coiled coil. Two strands of α-keratin, oriented in parallel (with their amino termini at the same end) are wrapped about each other to form a supertwisted coiled coil. The supertwisting amplifies the strength of the overall structure, just as strands are twisted to make a strong rope (Fig. 6–11). The twisting of the axis of an α helix to form a coiled coil explains the discrepancy between the 5.4 Å per turn predicted for an α helix by Pauling and Corey and the 5.15 to 5.2 Å repeating structure observed in the x-ray diffraction of hair (p. 163). The helical path of the supertwists is left-handed, opposite in sense to the α helix. The surfaces where the two α helices touch are made up of hydrophobic amino acid residues, their R groups meshed together in a regular interlocking pattern. This permits a close packing of the polypeptide chains within the left-handed supertwist. Not surprisingly, α-keratin is rich in the hydrophobic residues Ala, Val, Leu, Ile, Met, and Phe.

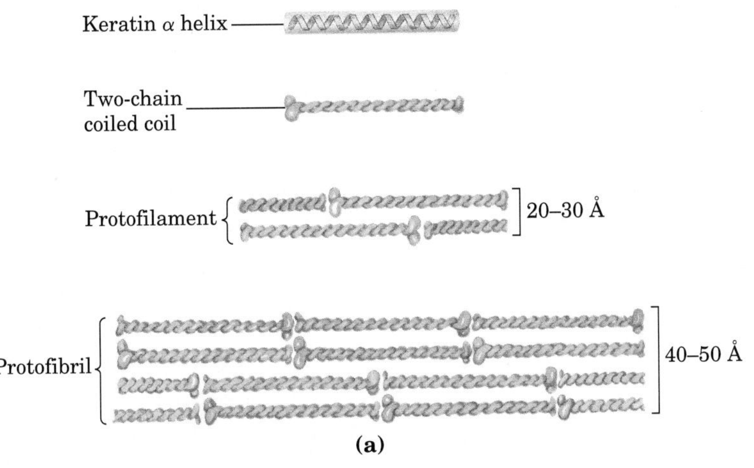

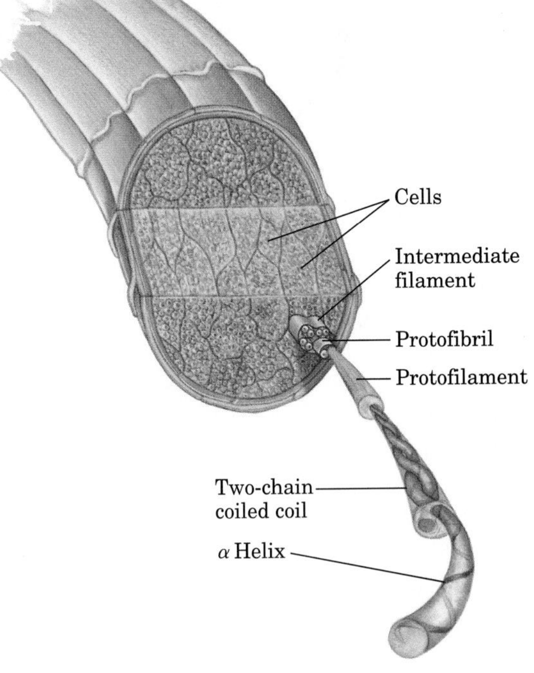

figure 6–11
Structure of hair. (a) Hair α-keratin is an elongated α helix with somewhat thicker elements near the amino and carboxyl termini. Pairs of these helices are interwound in a left-handed sense to form two-chain coiled coils. These then combine in higher-order structures called protofilaments and protofibrils. About four protofibrils—32 strands of α-keratin altogether—combine to form an intermediate filament. The individual two-chain coiled coils in the various substructures also appear to be interwound, but the handedness of the interwinding and other structural details are unknown. **(b)** A hair is an array of many α-keratin filaments, made up of the substructures shown in **(a)**.

An individual polypeptide in the α-keratin coiled coil has a relatively simple tertiary structure, dominated by an α-helical secondary structure with its helical axis twisted in a left-handed superhelix. The intertwining of the two α-helical polypeptides is an example of quaternary structure. Coiled coils of this type are common structural elements in filamentous proteins and in the muscle protein myosin (see Fig. 7–29). The quaternary structure of α-keratin can be quite complex. Many coiled coils can be assembled into large supramolecular complexes, such as the arrangement of α-keratin to form the intermediate filament of hair (Fig. 6–11b).

The strength of fibrous proteins is enhanced by covalent cross-links between polypeptide chains within the multihelical "ropes" and between adjacent chains in a supramolecular assembly. In α-keratins, the cross-links stabilizing quaternary structure are disulfide bonds (Box 6–2). In the hardest and toughest α-keratins, such as those of rhinoceros horn, up to 18% of the residues are cysteines involved in disulfide bonds.

box 6–2 Permanent Waving Is Biochemical Engineering

When hair is exposed to moist heat, it can be stretched. At the molecular level, the α helices in the α-keratin of hair are stretched out until they arrive at the fully extended β conformation. On cooling they spontaneously revert to the α-helical conformation. The characteristic "stretchability" of α-keratins, as well as their numerous disulfide cross-linkages, are the basis of permanent waving. The hair to be waved or curled is first bent around a form of appropriate shape. A solution of a reducing agent, usually a compound containing a thiol or sulfhydryl group (—SH), is then applied with heat. The reducing agent cleaves the cross-linkages by reducing each disulfide bond to form two Cys residues. The moist heat breaks hydrogen bonds and causes the α-helical structure of the polypeptide chains to uncoil. After a time the reducing solution is removed, and an oxidizing agent is added to establish *new* disulfide bonds between pairs of Cys residues of adjacent polypeptide chains, but not the same pairs as before the treatment. After the hair is washed and cooled, the polypeptide chains revert to their α-helical conformation. The hair fibers now curl in the desired fashion because the new disulfide cross-linkages exert some torsion or twist on the bundles of α-helical coils in the hair fibers. A permanent wave is not truly permanent because the hair grows; in the new hair replacing the old, the α-keratin has the natural, nonwavy pattern of disulfide bonds.

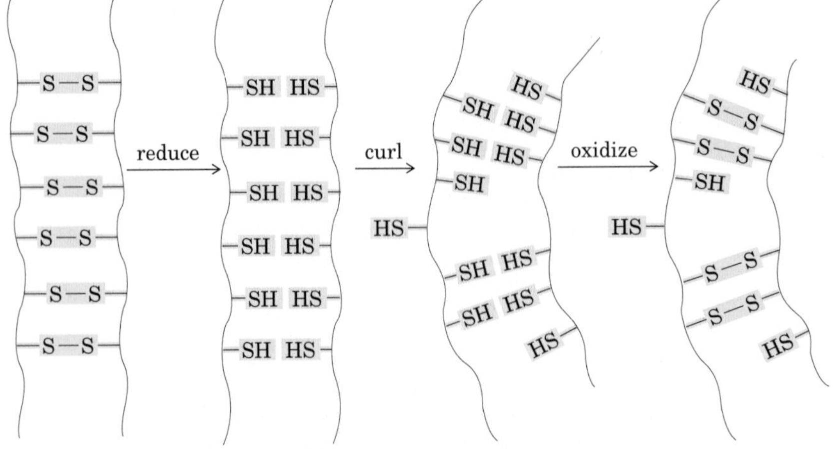

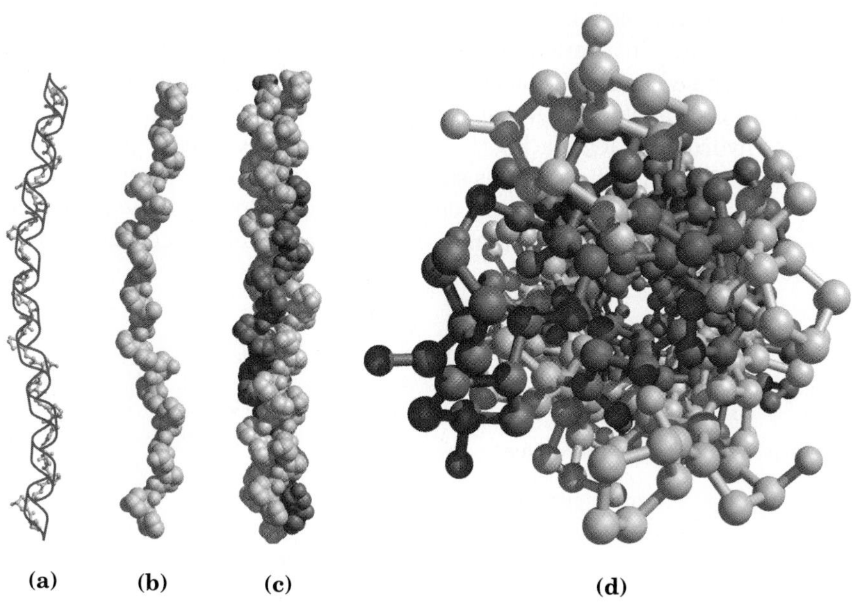

figure 6–12

Structure of collagen. (a) The α chain of collagen has a repeating secondary structure unique to this protein. The repeating tripeptide sequence Gly–X–Pro or Gly–X–HyPro adopts a left-handed helical structure with three residues per turn. The repeating sequence used to generate this model is Gly–Pro–HyPro. **(b)** Space-filling model of the same α chain. **(c)** Three of these helices (shown here in gray, light blue, and dark blue) wrap around one another with a right-handed twist. **(d)** The three-stranded collagen superhelix shown from one end, in a ball-and-stick representation. Gly residues are shown in red. Glycine, because of its small size, is required at the tight junction where the three chains are in contact. The balls in this illustration do not represent the van der Waals radii of the individual atoms. The center of the three-stranded superhelix is not hollow as it appears here, but is very tightly packed.

Collagen Like the α-keratins, collagen has evolved to provide strength. It is found in connective tissue such as tendons, cartilage, the organic matrix of bone, and the cornea of the eye. The collagen helix is a unique secondary structure quite distinct from the α helix. It is left-handed and has three amino acid residues per turn (Fig. 6–12). Collagen is also a coiled coil, but one with distinct tertiary and quaternary structure: three separate polypeptides, called α chains (not to be confused with α helices), are supertwisted about each other (Fig. 6–12c). The superhelical twisting is right-handed in collagen, opposite in sense to the left-handed helix of the α chains.

Collagen is 35% Gly, 11% Ala, and 21% Pro and HyPro (hydroxyproline, a nonstandard amino acid; see Fig. 5–8). The food product gelatin, derived from collagen, has little nutritional value as a protein because collagen is extremely low in many amino acids that are essential in the human diet. The unusual amino acid content of collagen is related to structural constraints unique to the collagen helix. The amino acid sequence in collagen is generally a repeating tripeptide unit, Gly–X–Pro or Gly–X–HyPro, where X can be any amino acid residue. Only Gly residues can be accommodated at the very tight junctions between the individual α chains (Fig. 6–12d); Pro residues permit the sharp twisting of the collagen helix. The amino acid sequence and the supertwisted quaternary structure of collagen allow a very close packing of its three polypeptides.

The tight wrapping of the α chains in the collagen triple helix provides tensile strength greater than that of a steel wire of equal cross section. Collagen fibrils (Fig. 6–13) are supramolecular assemblies consisting of triple-

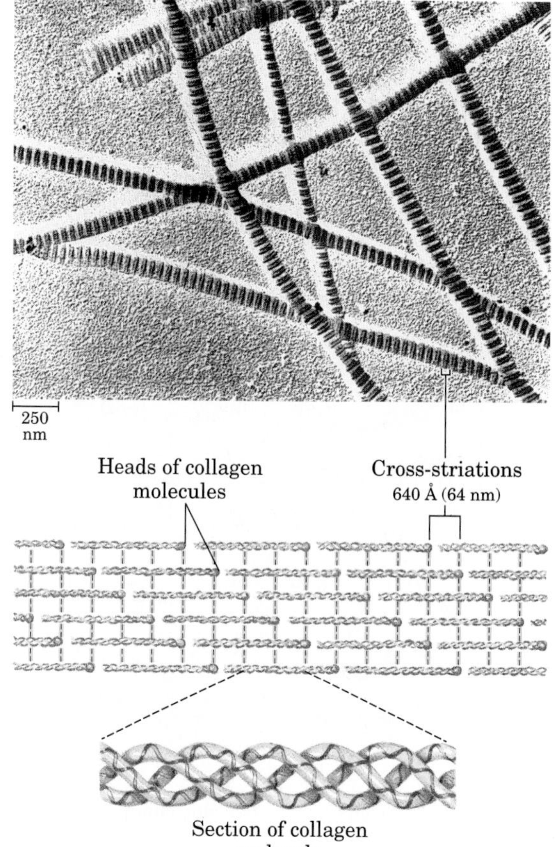

figure 6–13

Structure of collagen fibrils. Collagen (M_r 300,000) is a rod-shaped molecule, about 3,000 Å long and only 15 Å thick. Its three helically intertwined α chains may have different sequences, but each has about 1,000 amino acid residues. Collagen fibrils are made up of collagen molecules aligned in a staggered fashion and cross-linked for strength. The specific alignment and degree of cross-linking vary with the tissue and produce characteristic cross-striations in an electron micrograph. In the example shown here, alignment of the head groups of every fourth molecule produces striations 640 Å apart.

helical collagen molecules (sometimes referred to as tropocollagen molecules) associated in a variety of ways to provide different degrees of tensile strength. The α chains of collagen molecules and the collagen molecules of fibrils are cross-linked by unusual types of covalent bonds involving Lys, HyLys (hydroxylysine; see Fig. 5–8), or His residues. These links create nonstandard amino acid residues such as dehydrohydroxylysinonorleucine. The increasingly rigid and brittle character of connective tissue as people age results from an accumulation of covalent cross-links in collagen fibrils.

H—N ... N—H

$\mathrm{CH{-}CH_2{-}CH_2{-}CH_2{-}CH{=}N{-}CH_2{-}CH{-}CH_2{-}CH_2{-}CH}$

O=C ... OH ... C=O

Polypeptide chain | Lys residue minus ϵ-amino group (norleucine) | HyLys residue | Polypeptide chain

Dehydrohydroxylysinonorleucine

A typical mammal has more than 30 different structural variants of collagen that occur in particular tissues. Each is somewhat different in sequence and function. Some human genetic defects in collagen structure illustrate the close relationship between amino acid sequence and three-dimensional structure in this protein. Osteogenesis imperfecta is characterized by abnormal bone formation in babies; Ehlers-Danlos syndrome by loose joints. Both conditions can be lethal, and both result from the substitution of an amino acid residue with a larger R group (such as Cys or Ser) for a single Gly residue in each α chain (a different Gly residue in each disorder). These single-residue substitutions have a catastrophic effect on collagen function because they disrupt the Gly–X–Pro repeat that gives collagen its unique helical structure. Given its role in the collagen triple helix (Fig. 6–12d), Gly cannot be replaced by another amino acid residue without substantial deleterious effects on the structure.

Silk Fibroin Fibroin, the protein of silk, is produced by insects and spiders. Its polypeptide chains are predominantly in the β conformation. Fibroin is rich in Ala and Gly residues, permitting a close packing of β sheets and an interlocking arrangement of R groups (Fig. 6–14). The overall structure is

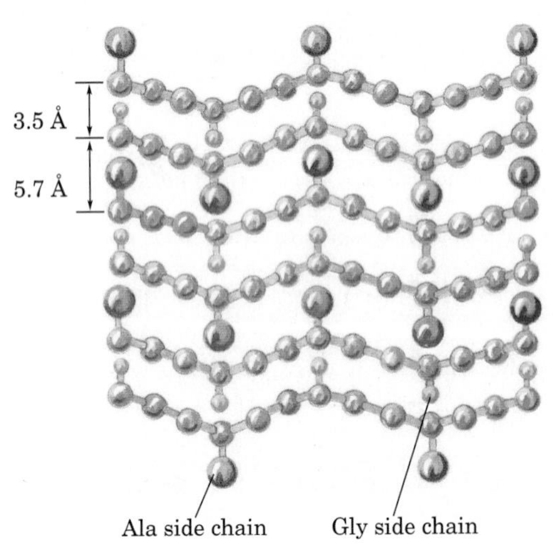

(a)

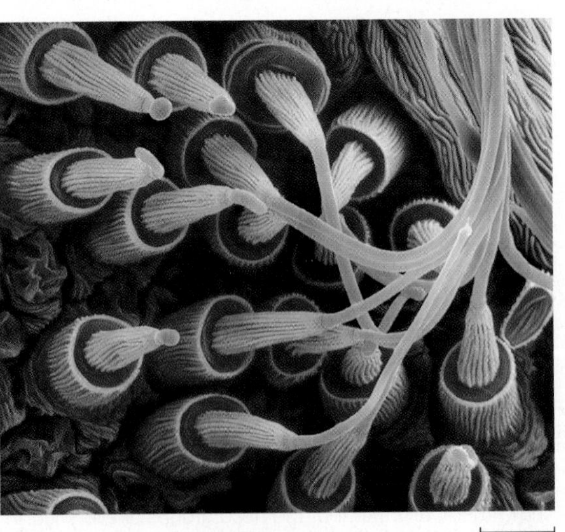

(b)

figure 6–14
Structure of silk. The fibers used to make silk cloth or a spider web are made up of the protein fibroin. **(a)** Fibroin consists of layers of antiparallel β sheets rich in Ala (purple) and Gly (yellow) residues. The small side chains interdigitate and allow close packing of each layered sheet, as shown in this side view. **(b)** Strands of fibroin (blue) emerge from the spinnerets of a spider in this colorized electron micrograph.

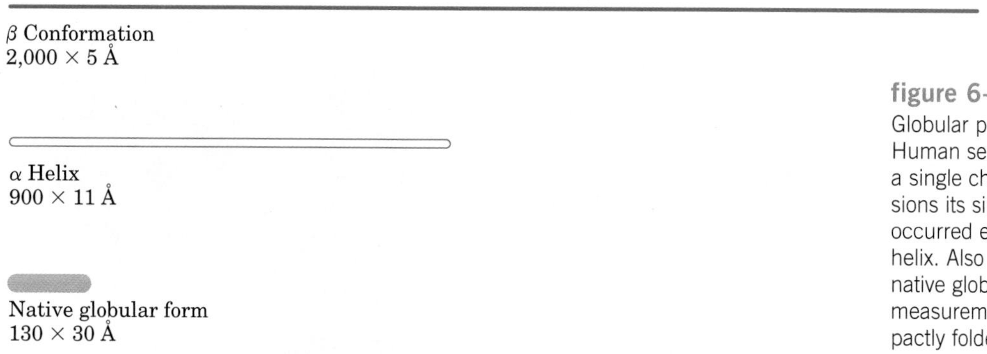

β Conformation
2,000 × 5 Å

α Helix
900 × 11 Å

Native globular form
130 × 30 Å

figure 6–15
Globular protein structures are compact and varied. Human serum albumin (M_r 64,500) has 585 residues in a single chain. Given here are the approximate dimensions its single polypeptide chain would have if it occurred entirely in extended β conformation or as an α helix. Also shown is the actual size of the protein in its native globular form, as determined by physicochemical measurements; the polypeptide chain must be very compactly folded to fit into these dimensions.

stabilized by extensive hydrogen bonding between all peptide linkages in the polypeptides of each β sheet and by the optimization of van der Waals interactions between sheets. Silk does not stretch because the β conformation is already highly extended (Fig. 6–7; see also Fig. 6–15). However, the structure is flexible because the sheets are held together by numerous weak interactions rather than by covalent bonds such as the disulfide bonds in α-keratins.

Structural Diversity Reflects Functional Diversity in Globular Proteins

In a globular protein, different segments of a polypeptide chain (or multiple polypeptide chains) fold back on each other. As illustrated in Figure 6–15, this folding generates a compact form relative to polypeptides in a fully extended conformation. The folding also provides the structural diversity necessary for proteins to carry out a wide array of biological functions. Globular proteins include enzymes, transport proteins, motor proteins, regulatory proteins, immunoglobulins, and proteins with many other functions.

As a new millenium begins, the number of known three-dimensional protein structures is in the thousands and more than doubles every two years. This wealth of structural information is revolutionizing our understanding of protein structure, the relation of structure to function, and even the evolutionary paths by which proteins arrived at their present state, which can be glimpsed in the family resemblances among proteins that are revealed as protein databases are sifted and sorted. The sheer variety of structures can seem daunting, yet as new protein structures become available it is becoming increasingly clear that they are manifestations of a finite set of recognizable, stable folding patterns.

Our discussion of globular protein structure begins with the principles gleaned from the earliest protein structures to be elucidated. This is followed by a detailed description of protein substructure and comparative categorization. Such discussions are only possible because of the vast amount of information available over the Internet from resources such as the Protein Data Bank (PDB), an archive of experimentally determined three-dimensional structures of biological macromolecules.

Myoglobin Provided Early Clues about the Complexity of Globular Protein Structure

The first breakthrough in understanding the three-dimensional structure of a globular protein came from x-ray diffraction studies of myoglobin carried out by John Kendrew and his colleagues in the 1950s. Myoglobin is a relatively small (M_r 16,700), oxygen-binding protein of muscle cells. It functions both to store oxygen and to facilitate oxygen diffusion in rapidly contracting muscle tissue. Myoglobin contains a single polypeptide chain of 153 amino acid residues of known sequence and a single iron protoporphyrin, or

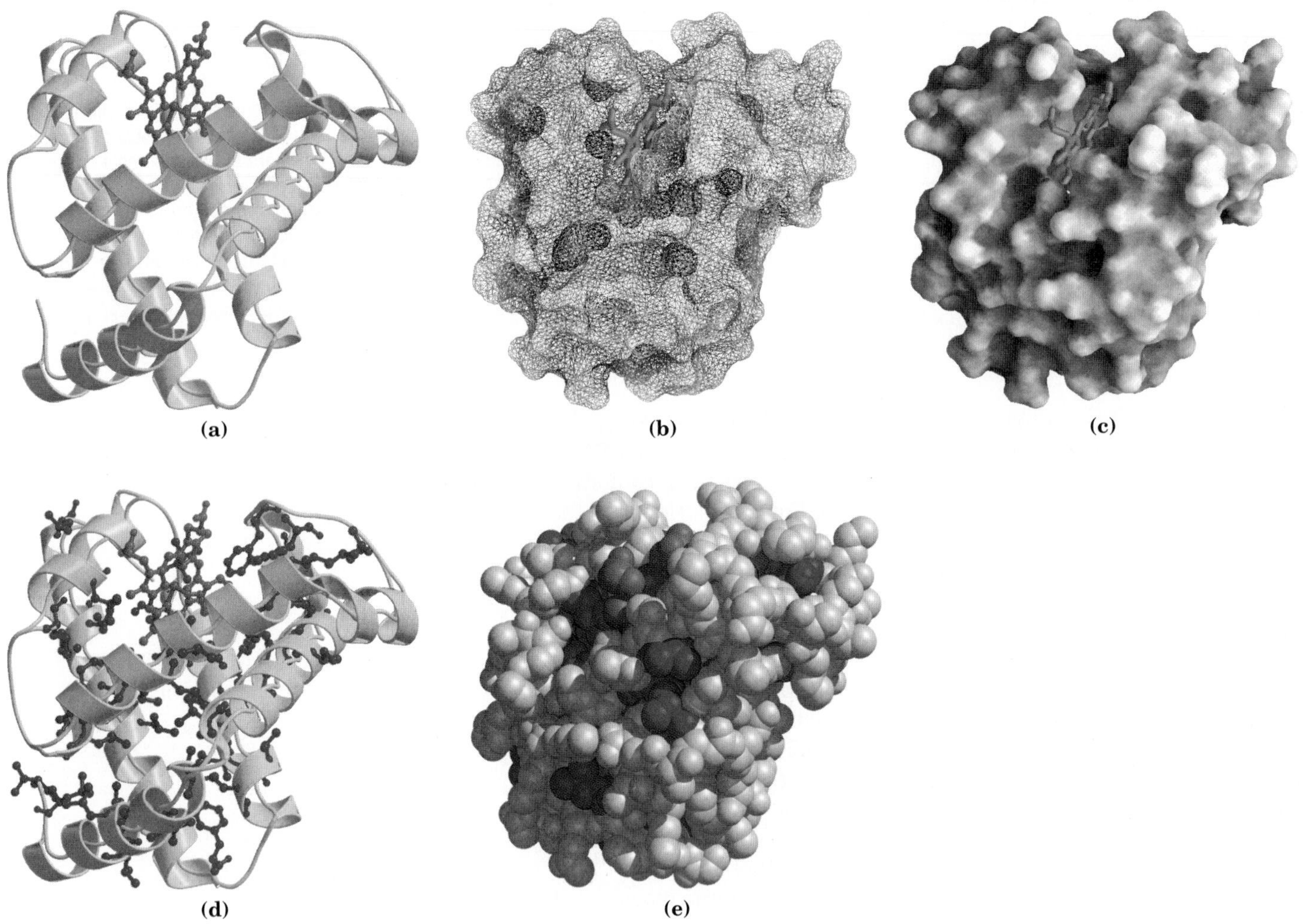

figure 6–16

Tertiary structure of sperm whale myoglobin. The orientation of the protein is similar in all panels; the heme group is shown in red. In addition to illustrating the myoglobin structure, this figure provides examples of several different ways to display protein structure. **(a)** The polypeptide backbone, shown in a ribbon representation of a type introduced by Jane Richardson, which highlights regions of secondary structure. The α-helical regions are evident. **(b)** A "mesh" image emphasizes the protein surface. **(c)** A surface contour image is useful for visualizing pockets in the protein where other molecules might bind. **(d)** A ribbon representation, including side chains (blue) for the hydrophobic residues Leu, Ile, Val, and Phe. **(e)** A space-filling model with all amino acid side chains. Each atom is represented by a sphere encompassing its van der Waals radius. The hydrophobic residues are again shown in blue; most are not visible because they are buried in the interior of the protein.

heme, group. The same heme group is found in hemoglobin, the oxygen-binding protein of erythrocytes, and is responsible for the deep red-brown color of both myoglobin and hemoglobin. Myoglobin is particularly abundant in the muscles of diving mammals such as the whale, seal, and porpoise, whose muscles are so rich in this protein that they are brown. Storage and distribution of oxygen by muscle myoglobin permits these animals to remain submerged for long periods of time.

Figure 6–16 shows several structural representations of myoglobin, illustrating how the polypeptide chain is folded in three dimensions—its tertiary structure. The red group surrounded by protein is heme. The backbone of the myoglobin molecule is made up of eight relatively straight segments of α helix interrupted by bends, some of which are β turns. The longest α helix has 23 amino acid residues and the shortest only seven; all are right-handed. More than 70% of the amino acid residues in myoglobin are in these α-helical regions. X-ray analysis has revealed the precise position of each of the R groups, which occupy nearly all the space within the folded chain.

Many important conclusions were drawn from the structure of myoglobin. The positioning of amino acid side chains reflects a structure that derives much of its stability from hydrophobic interactions. Most of the hydrophobic R groups are in the interior of the myoglobin molecule, hidden from exposure to water. All but two of the polar R groups are located on the outer surface of the molecule, and all are hydrated. The myoglobin molecule is so compact that its interior has room for only four molecules of

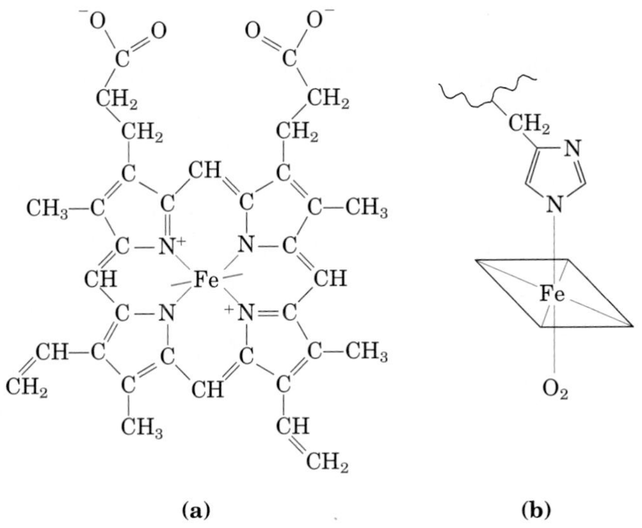

(a) (b)

figure 6–17

The heme group. This group is present in myoglobin, hemoglobin, cytochromes, and many other heme proteins. **(a)** Heme consists of a complex organic ring structure, protoporphyrin, to which is bound an iron atom in its ferrous (Fe^{2+}) state. The iron atom has six coordination bonds, four in the plane of, and bonded to, the flat porphyrin molecule and two perpendicular to it. **(b)** In myoglobin and hemoglobin, one of the perpendicular coordination bonds is bound to a nitrogen atom of a His residue. The other is "open" and serves as the binding site for an O_2 molecule.

water. This dense hydrophobic core is typical of globular proteins. The fraction of space occupied by atoms in an organic liquid is 0.4 to 0.6; in a typical crystal the fraction is 0.70 to 0.78, near the theoretical maximum. In a globular protein the fraction is about 0.75, comparable to that in a crystal. In this closely packed environment, weak interactions strengthen and reinforce each other. For example, the nonpolar side chains in the core are so close together that short-range van der Waals interactions make a significant contribution to stabilizing hydrophobic interactions.

Deduction of the structure of myoglobin confirmed some expectations and introduced some new elements of secondary structure. As predicted by Pauling and Corey, all the peptide bonds are in the planar trans configuration. The α helices in myoglobin provided the first direct experimental evidence for the existence of this type of secondary structure. Three of the four Pro residues of myoglobin are found at bends (recall that proline, with its fixed ϕ bond angle and lack of a peptide-bond N—H group for participation in hydrogen bonds, is largely incompatible with α-helical structure). The fourth Pro residue occurs within an α helix, where it creates a kink necessary for tight helix packing. Other bends contain Ser, Thr, and Asn residues, which are among the amino acids whose bulk and shape tend to make them incompatible with α-helical structure if they are in close proximity in the amino acid sequence (p. 165).

The flat heme group rests in a crevice, or pocket, in the myoglobin molecule. The iron atom in the center of the heme group has two bonding (coordination) positions perpendicular to the plane of the heme (Fig. 6–17). One of these is bound to the R group of the His residue at position 93; the other is the site at which an O_2 molecule binds. Within this pocket, the accessibility of the heme group to solvent is highly restricted. This is important for function because free heme groups in an oxygenated solution are rapidly oxidized from the ferrous (Fe^{2+}) form, which is active in the reversible binding of O_2, to the ferric (Fe^{3+}) form, which does not bind O_2.

Knowledge of the structure of myoglobin allowed researchers for the first time to understand in detail the correlation between the structure and function of a protein. Hundreds of proteins have been subjected to similar analysis since then. Today, techniques such as NMR spectroscopy supplement x-ray diffraction data, providing more information on a protein's structure (Box 6–3). The ongoing sequencing of genomic DNA from many organisms (Chapter 29) has identified thousands of genes that encode proteins of known sequence but unknown function. Our first insight into what these proteins do often comes from our still-limited understanding of how primary structure determines tertiary structure, and how tertiary structure determines function.

box 6–3 Methods for Determining the Three-Dimensional Structure of a Protein

X-Ray Diffraction

The spacing of atoms in a crystal lattice can be determined by measuring the locations and intensities of spots produced on photographic film by a beam of x rays of given wavelength, after the beam has been diffracted by the electrons of the atoms. For example, x-ray analysis of sodium chloride crystals shows that Na^+ and Cl^- ions are arranged in a simple cubic lattice. The spacing of the different kinds of atoms in complex organic molecules, even very large ones such as proteins, can also be analyzed by x-ray diffraction methods. However, the technique for analyzing crystals of complex molecules is far more laborious than for simple salt crystals. When the repeating pattern of the crystal is a molecule as large as, say, a protein, the numerous atoms in the molecule yield thousands of diffraction spots that must be analyzed by computer.

The process may be understood at an elementary level by considering how images are generated in a light microscope. Light from a point source is focused on an object. The light waves are scattered by the object, and these scattered waves are recombined by a series of lenses to generate an enlarged image of the object. The smallest object whose structure can be determined by such a system (i.e., the resolving power of the microscope) is determined by the wavelength of the light—in this case, visible light, with wavelengths in the range of 400 to 700 nm. Objects smaller than half the wavelength of the incident light cannot be resolved. To resolve objects as small as proteins we must use x rays, with wavelengths in the range of 0.7 to 1.5 Å (0.07 to 0.15 nm). However, there are no lenses that can recombine x rays to form an image; instead the pattern of diffracted x rays is collected directly and an image is reconstructed by mathematical techniques.

The amount of information obtained from x-ray crystallography depends on the degree of structural order in the sample. Some important structural parameters were obtained from early studies of the diffraction patterns of the fibrous proteins arranged in fairly regular arrays in hair and wool. However, the orderly bundles formed by fibrous proteins are not crystals—the molecules are aligned side by side, but not all are oriented in the same direction. More detailed three-dimensional structural information about proteins requires a highly ordered protein crystal. Protein crystallization is something of an empirical science, and the structures of many important proteins are not yet known simply because they have proved difficult to crystallize. Practitioners have compared making protein crystals to holding together a stack of bowling balls with cellophane tape.

Operationally, there are several steps in x-ray

(a)
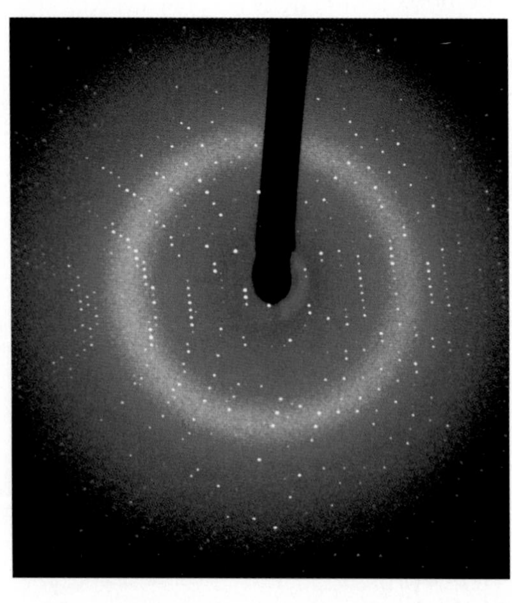

(b)
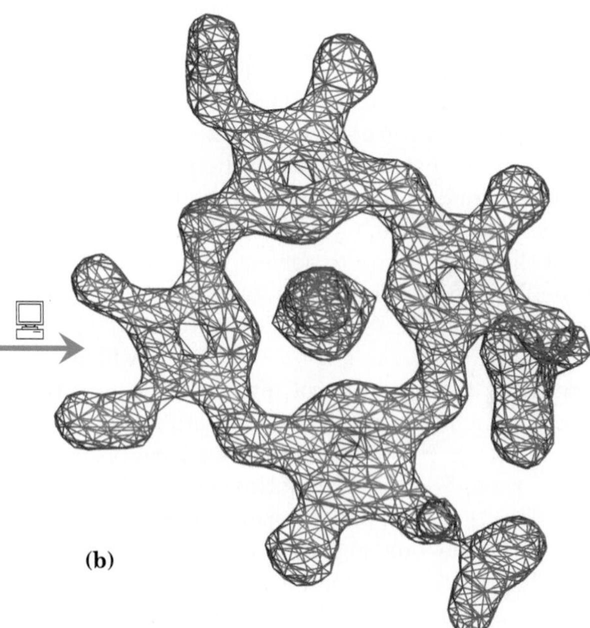

structural analysis (Fig. 1). Once a crystal is obtained, it is placed in an x-ray beam between the x-ray source and a detector, and a regular array of spots called reflections is generated. The spots are created by the diffracted x-ray beam, and each atom in a molecule makes a contribution to each spot. An electron-density map of the protein is reconstructed from the overall diffraction pattern of spots by using a mathematical technique called a Fourier transform. In effect, the computer acts as a "computational lens." A model for the structure is then built that is consistent with the electron-density map.

John Kendrew found that the x-ray diffraction pattern of crystalline myoglobin (isolated from muscles of the sperm whale) is very complex, with nearly 25,000 reflections. Computer analysis of these reflections took place in stages. The resolution improved at each stage, until in 1959 the positions of virtually all the non-hydrogen atoms in the protein had been determined. The amino acid sequence of the protein, obtained by chemical analysis, was consistent with the molecular model. The structures of thousands of proteins, many of them much more complex than myoglobin, have since been determined to a similar level of resolution.

The physical environment within a crystal is not identical to that in solution or in a living cell. A crystal imposes a space and time average on the structure deduced from its analysis, and x-ray diffraction studies provide little information about molecular motion within the protein. The conformation of proteins in a crystal could in principle also be affected by nonphysiological factors such as incidental protein-protein contacts within the crystal. However, when structures derived from the analysis of crystals are compared with structural information obtained by other means (such as NMR, as described below), the crystal-derived structure almost always represents a functional conformation of the protein. X-ray crystallography can be applied successfully to proteins too large to be structurally analyzed by NMR.

figure 1
Steps in the determination of the structure of sperm whale myoglobin by x-ray crystallography. **(a)** X-ray diffraction patterns are generated from a crystal of the protein. **(b)** Data extracted from the diffraction patterns are used to calculate a three-dimensional electron-density map of the protein. The electron density of only part of the structure, the heme, is shown. **(c)** Regions of greatest electron density reveal the location of atomic nuclei, and this information is used to piece together the final structure. Here, the heme structure is modeled into its electron density map. **(d)** The completed structure of sperm whale myoglobin, including the heme.

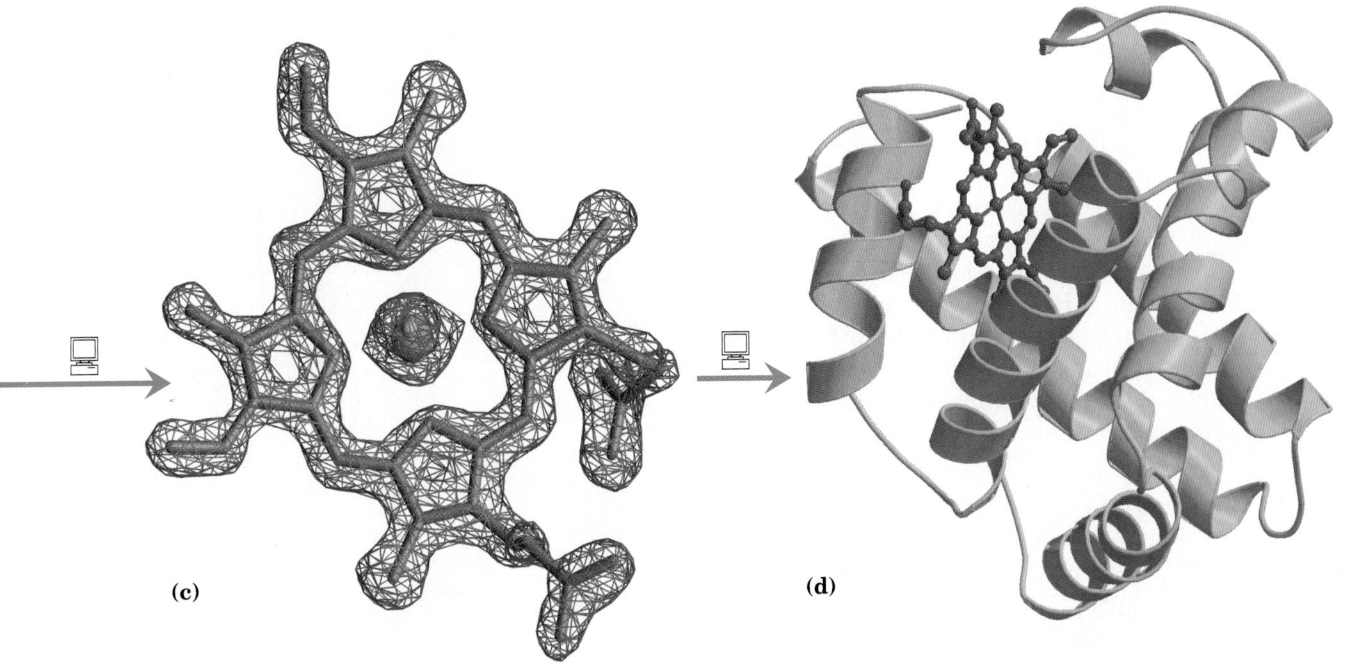

Continued on next page

Nuclear Magnetic Resonance

An important complementary method for determining the three-dimensional structures of macromolecules is nuclear magnetic resonance (NMR). Modern NMR techniques are being used to determine the structures of ever-larger macromolecules, including carbohydrates, nucleic acids, and small to average-sized proteins. An advantage of NMR studies is that they are carried out on macromolecules in solution, whereas x-ray crystallography is limited to molecules that can be crystallized. NMR can also illuminate the dynamic side of protein structure, including conformational changes, protein folding, and interactions with other molecules.

NMR is a manifestation of nuclear spin angular momentum, a quantum mechanical property of atomic nuclei. Only certain atoms, including ^{1}H, ^{13}C, ^{15}N, ^{19}F, and ^{31}P, possess the kind of nuclear spin that gives rise to an NMR signal. Nuclear spin generates a magnetic dipole. When a strong, static magnetic field is applied to a solution containing a single type of macromolecule, the magnetic dipoles are aligned in the field in one of two orientations, parallel (low energy) or antiparallel (high energy). A short ($\sim$10 μs) pulse of electromagnetic energy of suitable frequency (the resonant frequency, which is in the radio frequency range) is applied at right angles to the nuclei aligned in the magnetic field. Some energy is absorbed as nuclei switch to the high-energy state, and the absorption spectrum that results contains information about the identity of the nuclei and their immediate chemical environment. The data from many such experiments performed on a sample are averaged, increasing the signal-to-noise ratio, and an NMR spectrum such as that in Figure 2 is generated.

^{1}H is particularly important in NMR experiments because of its high sensitivity and natural abundance. For macromolecules, ^{1}H NMR spectra can become quite complicated. Even a small protein has hundreds of ^{1}H atoms, typically resulting in a one-dimensional NMR spectrum too complex for analysis. Structural analysis of proteins became possible with the advent of two-dimensional NMR techniques (Fig. 3). These methods allow measurement of distance-dependent coupling of nuclear spins in nearby atoms through space (the nuclear Overhauser effect (NOE), in a method dubbed NOESY) or the coupling of nuclear spins in atoms connected by covalent bonds (total correlation spectroscopy, or TOCSY).

Translating a two-dimensional NMR spectrum into a complete three-dimensional structure can be a laborious process. The NOE signals provide some information about the distances between individual atoms, but for these distance constraints to be useful, the atoms giving rise to each signal must be identified. Complementary TOCSY experiments can help identify which NOE signals reflect atoms that are linked by covalent bonds. Certain patterns of NOE signals have been associated with secondary structures such as α helices. Modern genetic engineering (Chapter 29) can be used to prepare proteins that contain the rare isotopes ^{13}C or ^{15}N. The new NMR signals produced by these atoms, and the coupling with ^{1}H signals resulting from these substitutions, help in the assignment of individual ^{1}H NOE signals. The process is also aided by a knowledge of the amino acid sequence of the polypeptide.

To generate a three-dimensional structure, the distance constraints are fed into a computer along with known geometric constraints such as chirality, van der Waals radii, and bond lengths and angles. The computer generates a family of closely related structures that represent the

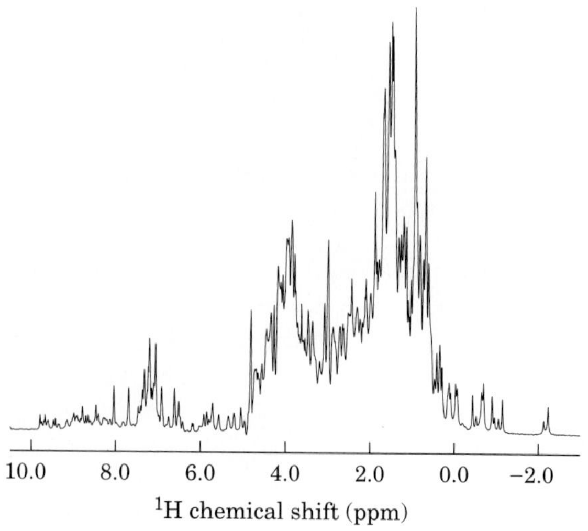

figure 2

A one-dimensional NMR spectrum of a globin from a marine blood worm. This protein and sperm whale myoglobin are very close structural analogs, belonging to the same protein structural family and sharing an oxygen-transport function.

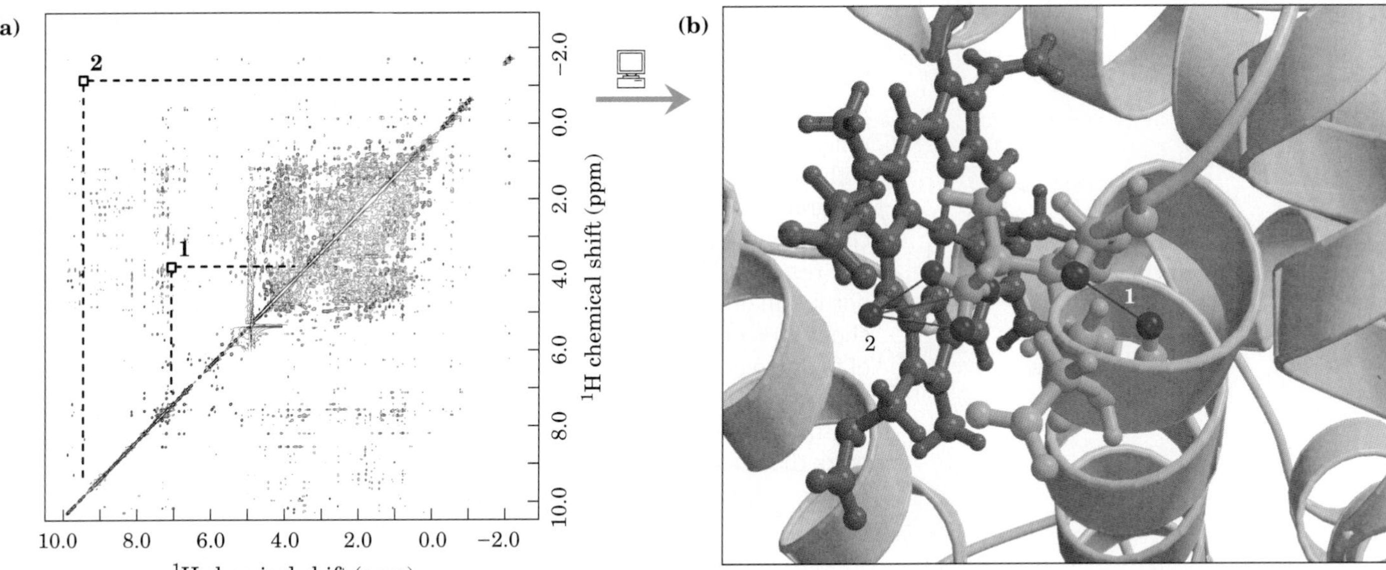

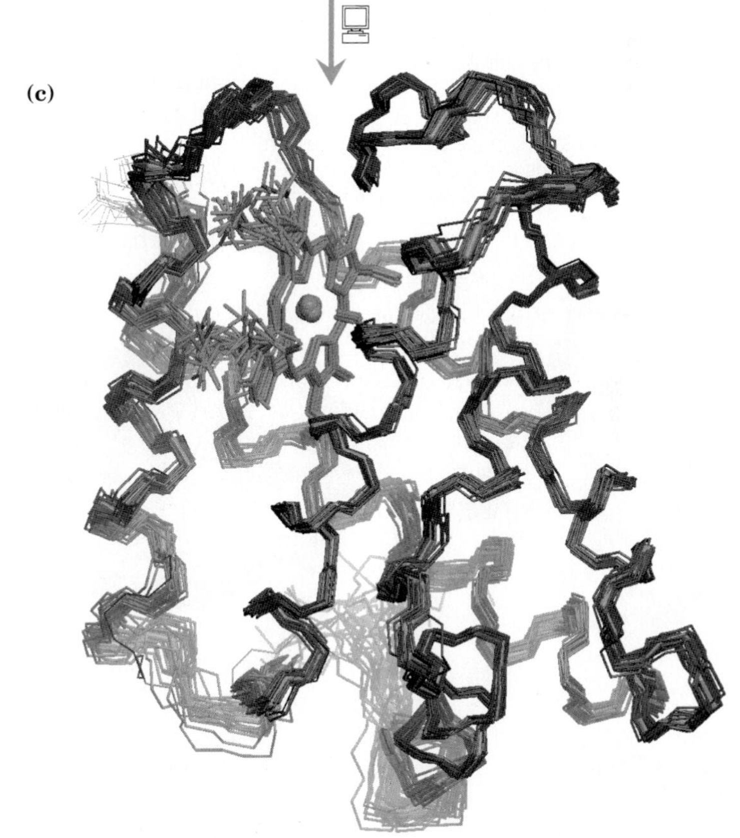

figure 3
The use of two-dimensional NMR to generate a three-dimensional structure of a globin, the same protein used to generate the data in Figure 2. The diagonal in a two-dimensional NMR spectrum is equivalent to a one-dimensional spectrum. The off-diagonal peaks are NOE signals generated by close-range interactions of ^{1}H atoms that may generate signals quite distant in the one-dimensional spectrum. Two such interactions are identified in **(a),** and their identities are shown with blue lines in **(b).** Three lines are drawn for interaction 2 between a methyl group in the protein and a hydrogen on the heme. The methyl group rotates rapidly such that each of its three hydrogens contributes equally to the interaction and the NMR signal. Such information is used to determine the complete three-dimensional structure, as in **(c).** The multiple lines shown for the protein backbone represent the family of structures consistent with the distance constraints in the NMR data. The structural similarity with myoglobin (see Fig. 1) is evident. The proteins are oriented in the same way in both figures.

range of conformations consistent with the NOE distance constraints (Fig. 3c). The uncertainty in structures generated by NMR is in part a reflection of the molecular vibrations (breathing) within a protein structure in solution, discussed in more detail in Chapter 7. Normal experimental uncertainty can also play a role.

When a protein structure has been determined by both x-ray crystallography and NMR, the structures generally agree well. In some cases, the precise locations of particular amino acid side chains on the protein exterior are different, often because of effects related to the packing of adjacent protein molecules in a crystal. The two techniques together are at the heart of the rapid increase in the availability of structural information about the macromolecules of living cells.

Globular Proteins Have a Variety of Tertiary Structures

With elucidation of the tertiary structures of hundreds of other globular proteins by x-ray analysis, it became clear that myoglobin represents only one of many ways in which a polypeptide chain can be folded. In Figure 6–18 the structures of cytochrome *c*, lysozyme, and ribonuclease are compared. All have different amino acid sequences and different tertiary structures, reflecting differences in function. All are relatively small and easy to work with, facilitating structural analysis. Cytochrome *c* is a component of the respiratory chain of mitochondria (Chapter 19). Like myoglobin, cytochrome *c* is a heme protein. It contains a single polypeptide chain of about 100 residues (M_r 12,400) and a single heme group. In this case, the protoporphyrin of the heme group is covalently attached to the polypeptide. Only about 40% of the polypeptide is in α-helical segments, compared with 70% of the myoglobin chain. The rest of the cytochrome *c* chain contains β turns and irregularly coiled and extended segments.

Lysozyme (M_r 14,600) is an enzyme abundant in egg white and human tears that catalyzes the hydrolytic cleavage of polysaccharides in the protective cell walls of some families of bacteria. Lysozyme, because it can lyse, or degrade, bacterial cell walls, serves as a bactericidal agent. As in cytochrome *c*, about 40% of its 129 amino acid residues are in α-helical segments, but the arrangement is different and some β-sheet structure is also present (Fig. 6–18). Four disulfide bonds contribute stability to this structure. The α helices line a long crevice in the side of the molecule, called the active site, which is the site of substrate binding and catalysis. The bacterial polysaccharide that is the substrate for lysozyme fits into this crevice.

Ribonuclease, another small globular protein (M_r 13,700), is an enzyme secreted by the pancreas into the small intestine, where it catalyzes the hydrolysis of certain bonds in the ribonucleic acids present in ingested food. Its tertiary structure, determined by x-ray analysis, shows that little of its

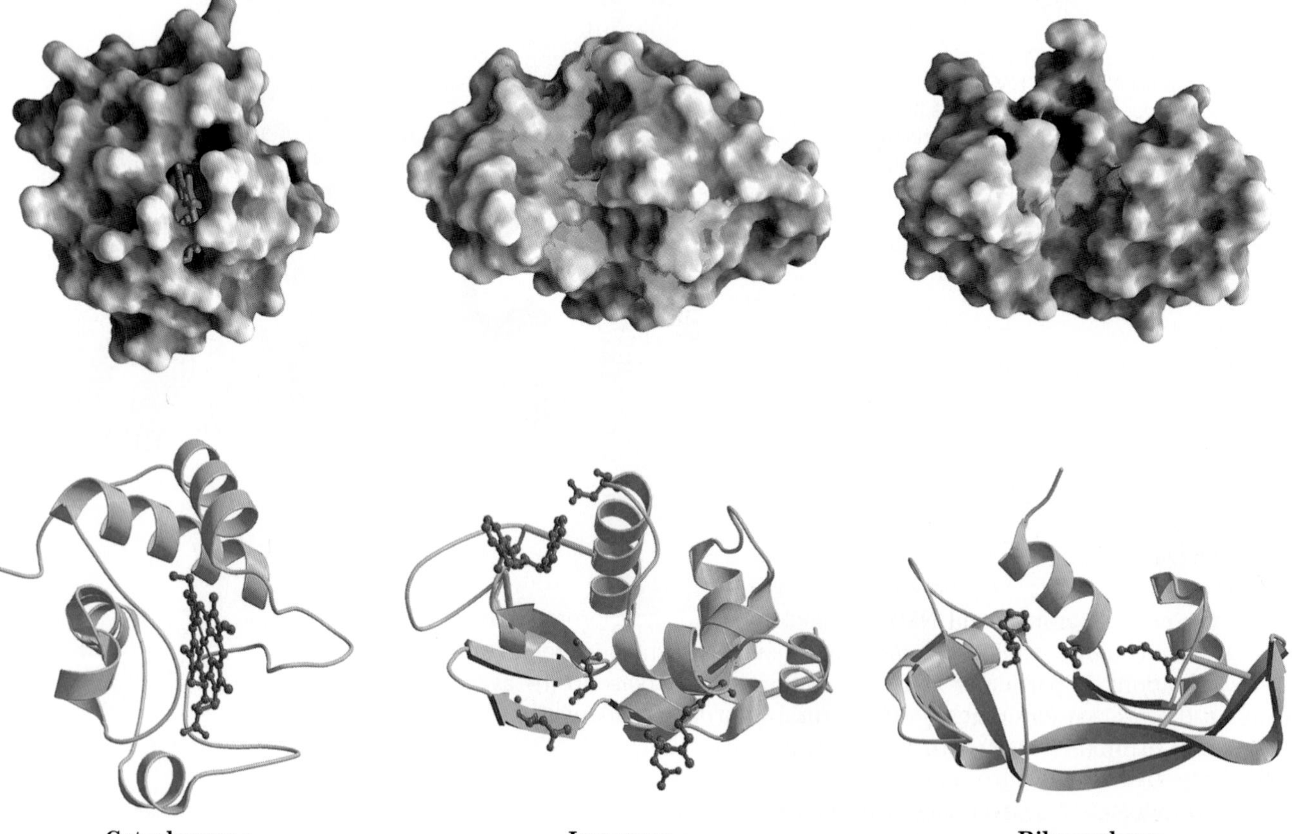

figure 6–18
Three-dimensional structures of some small proteins. Shown here are cytochrome *c*, lysozyme, and ribonuclease. Key functional groups (the heme in cytochrome *c*; amino acid side chains in the active site of lysozyme and ribonuclease) are shown in red. Disulfide bonds are shown in yellow. Each protein is shown in surface contour and in a ribbon representation, in the same orientation. In the ribbon depictions, regions in the β conformation are represented by flat arrows and the α helices are represented by spiral ribbons.

table 6–2

Approximate Amounts of α Helix and β Conformation in Some Single-Chain Proteins*

Protein (total residues)	Residues (%)	
	α Helix	β Conformation
Chymotrypsin (247)	14	45
Ribonuclease (124)	26	35
Carboxypeptidase (307)	38	17
Cytochrome *c* (104)	39	0
Lysozyme (129)	40	12
Myoglobin (153)	78	0

Source: Data from Cantor, C.R. & Schimmel, P.R. (1980) *Biophysical Chemistry,* Part I: *The Conformation of Biological Macromolecules,* p. 100, W.H. Freeman and Company, New York.

*Portions of the polypeptide chains that are not accounted for by α helix or β conformation consist of bends and irregularly coiled or extended stretches. Segments of α helix and β conformation sometimes deviate slightly from their normal dimensions and geometry.

124 amino acid polypeptide chain is in an α-helical conformation, but it contains many segments in the β conformation (Fig. 6–18). Like lysozyme, ribonuclease has four disulfide bonds between loops of the polypeptide chain.

In small proteins, hydrophobic residues are less likely to be sheltered in a hydrophobic interior—simple geometry dictates that the smaller the protein, the lower the ratio of volume to surface area. Small proteins also have fewer potential weak interactions available to stabilize them. This explains why many smaller proteins such as those in Figure 6–18 are stabilized by a number of covalent bonds. Lysozyme and ribonuclease, for example, have disulfide linkages, and the heme group in cytochrome *c* is covalently linked to the protein on two sides, providing significant stabilization of the entire protein structure.

Table 6–2 shows the proportions of α helix and β conformation (expressed as percentage of residues in each secondary structure) in several small, single-chain, globular proteins. Each of these proteins has a distinct structure, adapted for its particular biological function, but together they share several important properties. Each is folded compactly, and in each case the hydrophobic amino acid side chains are oriented toward the interior (away from water) and the hydrophilic side chains are on the surface. The structures are also stabilized by a multitude of hydrogen bonds and some ionic interactions.

Analysis of Many Globular Proteins Reveals Common Structural Patterns

For the beginning student, the very complex tertiary structures of globular proteins much larger than those shown in Figure 6–18 are best approached by focusing on structural patterns that recur in different and often unrelated proteins. The three-dimensional structure of a typical globular protein can be considered an assemblage of polypeptide segments in the α-helix and β-sheet conformations, linked by connecting segments. The structure can then be described to a first approximation by defining how these segments stack on one another, and how the segments that connect them are arranged. This formalism has led to the development of databases that allow informative comparisons of protein structures, complementing other databases that permit comparisons of protein sequences.

An understanding of a complete three-dimensional structure is built upon an analysis of its parts. We begin by defining terms used to describe protein substructures, then turn to the folding rules elucidated from analysis of the structures of many proteins.

Supersecondary structures, also called **motifs** or simply **folds,** are particularly stable arrangements of several elements of secondary structure and the connections between them. There is no universal agreement among

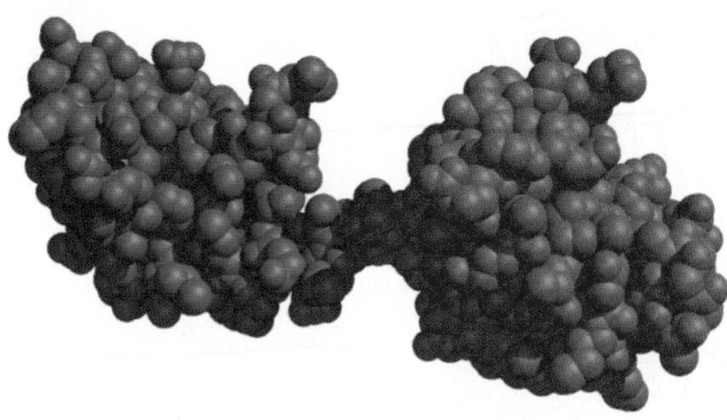

figure 6–19
Structural domains in the polypeptide troponin C. This calcium-binding protein associated with muscle has separate calcium-binding domains, indicated in blue and purple.

biochemists on the application of the three terms, and they are often used interchangeably. The terms are also applied to a wide range of structures. Recognized motifs range from simple to complex, sometimes appearing in repeating units or combinations. A single large motif may comprise the entire protein. We have already encountered one well-studied motif, the coiled coil of α-keratin, also found in a number of other proteins.

Polypeptides with more than a few hundred amino acid residues often fold into two or more stable, globular units called **domains.** In many cases, a domain from a large protein will retain its correct three-dimensional structure even when it is separated (for example, by proteolytic cleavage) from the remainder of the polypeptide chain. A protein with multiple domains may appear to have a distinct globular lobe for each domain (Fig. 6–19), but, more commonly, extensive contacts between domains make individual domains hard to discern. Different domains often have distinct functions, such as the binding of small molecules or interaction with other proteins. Small proteins usually have only one domain (the domain *is* the protein).

Folding of polypeptides is subject to an array of physical and chemical constraints. A sampling of the prominent folding rules that have emerged provides an opportunity to introduce some simple motifs.

1. Hydrophobic interactions make a large contribution to the stability of protein structures. Burial of hydrophobic amino acid R groups so as to exclude water requires at least two layers of secondary structure. Two simple motifs, the **β-α-β loop** and the **α-α corner** (Fig. 6–20a), create two layers.
2. Where they occur together in proteins, α helices and β sheets generally are found in different structural layers. This is because the backbone of a polypeptide segment in the β conformation (Fig. 6–7) cannot readily hydrogen bond to an α helix aligned with it.
3. Polypeptide segments adjacent to each other in the primary sequence are usually stacked adjacent to each other in the folded structure. Although distant segments of a polypeptide may come together in the tertiary structure, this is not the norm.
4. Connections between elements of secondary structure cannot cross or form knots (Fig. 6–20b).
5. The β conformation is most stable when the individual segments are twisted slightly in a right-handed sense. This influences both the arrangement of β sheets relative to one another and the path of the polypeptide connection between them. Two parallel β strands, for example, must be connected by a crossover strand (Fig. 6–20c). In principle, this crossover could have a right- or left-handed conformation, but in proteins it is almost always right-handed. Right-handed connections tend to be shorter than left-handed connections and tend to bend through smaller angles, making them easier to form. The twisting of β sheets also leads to a characteristic twisting of the structure formed when many segments are put together. Two examples of resulting structures are the β barrel and twisted β sheet (Fig. 6–20d), which form the core of many larger structures.

Following these rules, complex motifs can be built up from simple ones. For example, a series of β-α-β loops, arranged so that the β strands form a barrel, creates a particularly stable and common motif called the ***α/β* barrel** (Fig. 6–21). In this structure, each parallel β segment is attached to its neighbor by an α-helical segment. All connections are right-handed. The α/β barrel is found in many enzymes, often with a binding site for a cofactor or substrate in the form of a pocket near one end of the barrel. Note that domains exhibiting similar folding patterns are said to have the same motif even though their constituent α helices and β sheets may differ in length.

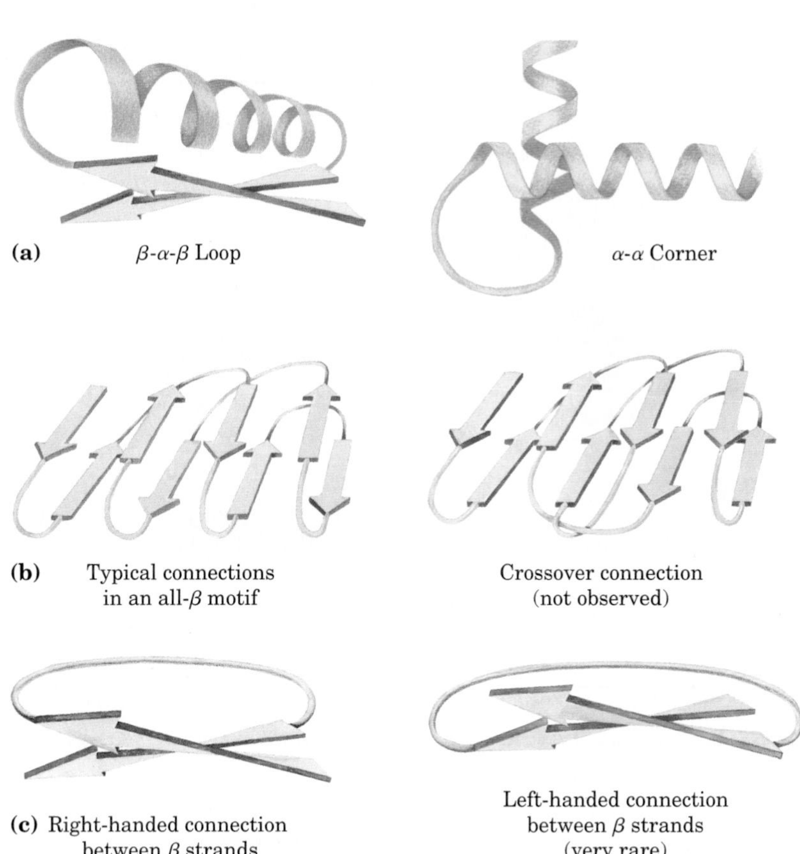

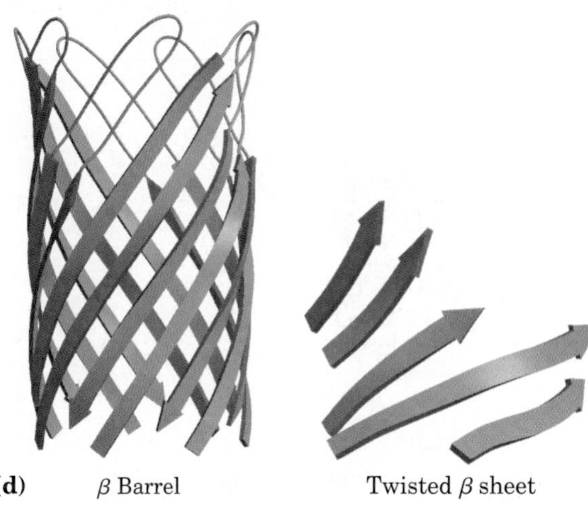

figure 6–20

Stable folding patterns in proteins. (a) Two simple and common motifs that provide two layers of secondary structure. Amino acid side chains at the interface between elements of secondary structure are shielded from water. Note that the β strands in the β-α-β loop tend to twist in a right-handed fashion. **(b)** Connections between β strands in layered β sheets. The strands are shown from one end, with no twisting included in the schematic. Thick connections are those at the ends nearest the viewer; thin connections are at the far ends of the β strands. The connections on a given end (e.g., near the viewer) do not cross each other. **(c)** Because of the twist in β strands, connections between strands are generally right-handed. Left-handed connections must traverse sharper angles and are harder to form. **(d)** Two arrangements of β strands stabilized by the tendency of the strands to twist. This β barrel is a single domain of α-hemolysin from the bacterium *Staphylococcus aureus.* The twisted β sheet is from a domain of photolyase from *E. coli.*

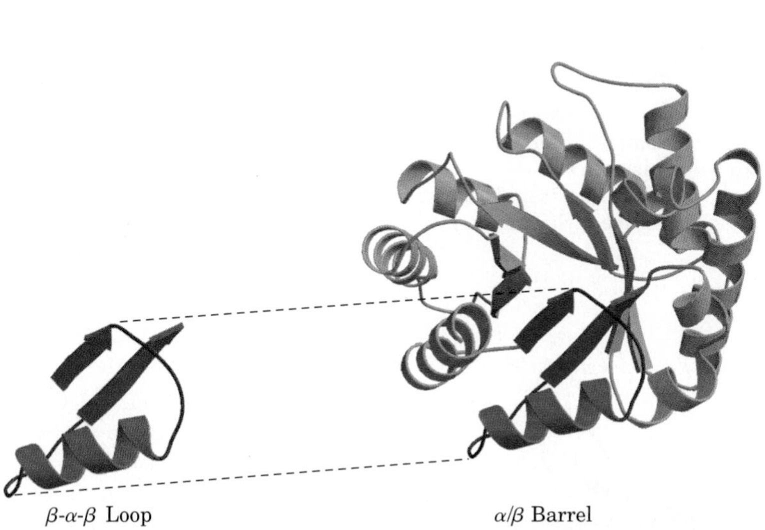

figure 6–21

Constructing large motifs from smaller ones. The α/β barrel is a common motif constructed from repetitions of the simpler β-α-β loop motif. This α/β barrel is a domain of the enzyme pyruvate kinase from rabbit.

Protein Motifs Are the Basis for Protein Structural Classification

As we have seen, the complexities of tertiary structure are decreased by considering substructures. Taking this idea further, researchers have organized the complete contents of databases according to hierarchical levels of structure. The Structural Classification of Proteins (SCOP) database offers a good example of this very important trend in biochemistry. At the highest level of classification, the SCOP database borrows a scheme already in common use, in which protein structures are divided into four classes: all α, all β, α/β (in which the α and β segments are interspersed or alternate), and $\alpha + \beta$ (in which the α and β regions are somewhat segregated) (Fig. 6–22).

All α

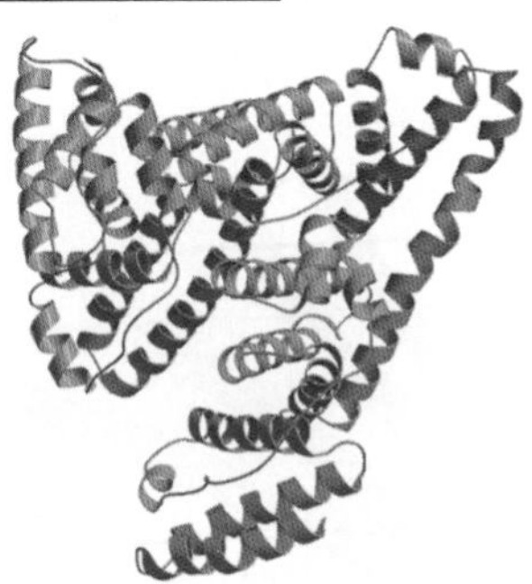

1ao6
Serum albumin
Serum albumin
Serum albumin
Serum albumin
Human (*Homo sapiens*)

1bcf
Ferritin-like
Ferritin-like
Ferritin
Bacterioferritin (cytochrome b_1)
Escherichia coli

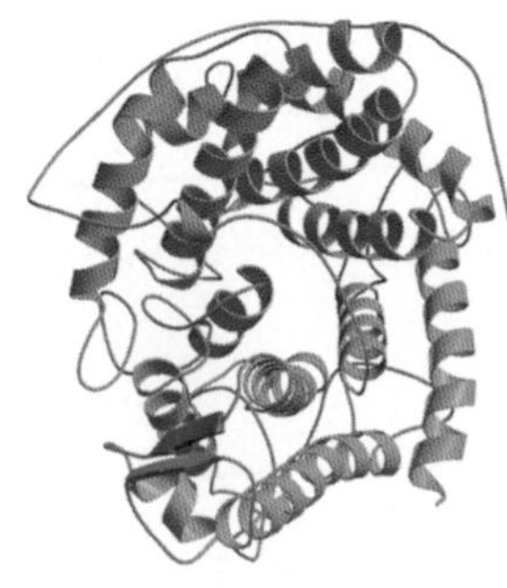

1gai
α/α toroid
Glycosyltransferases of the superhelical fold
Glucoamylase
Glucoamylase
Aspergillus awamori, variant x100

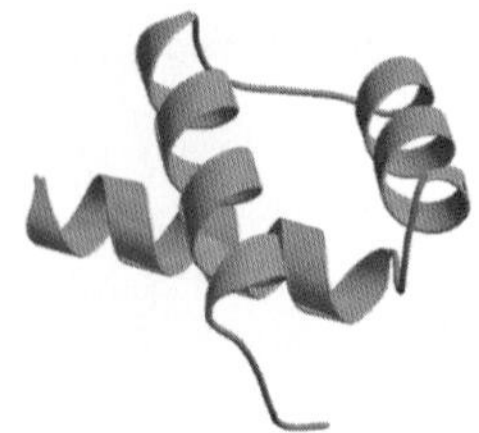

1enh
DNA-binding 3-helical bundle
Homeodomain-like
Homeodomain
engrailed Homeodomain
Drosophila melanogaster

All β

1hoe
α-Amylase inhibitor
α-Amylase inhibitor
α-Amylase inhibitor
HOE-467A
Streptomyces tendae 4158

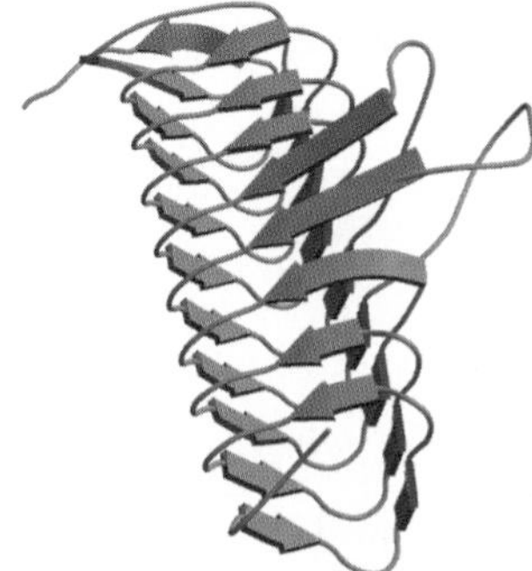

1lxa
Single-stranded left-handed β helix
Trimeric LpxA-like enzymes
UDP *N*-acetylglucosamine acyltransferase
UDP *N*-acetylglucosamine acyltransferase
Escherichia coli

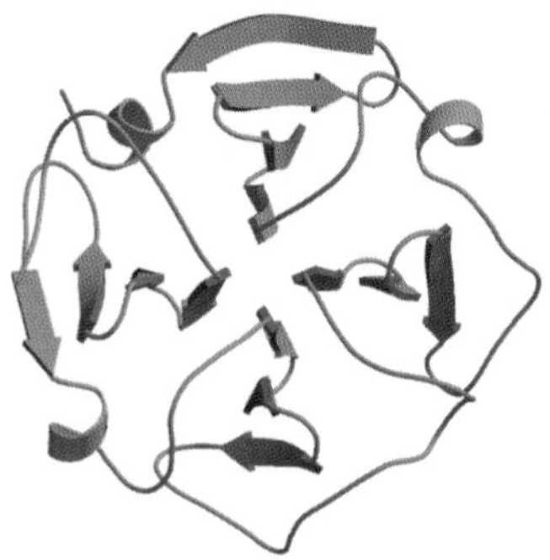

1pex
Four-bladed β propeller
Hemopexin-like domain
Hemopexin-like domain
Collagenase-3 (MMP-13), carboxyl-terminal domain
Human (*Homo sapiens*)

1jpc
β-Prism II
α-D-Mannose-specific plant lectins
α-D-Mannose-specific plant lectins
Lectin (agglutinin)
Snowdrop (*Galanthus nivalis*)

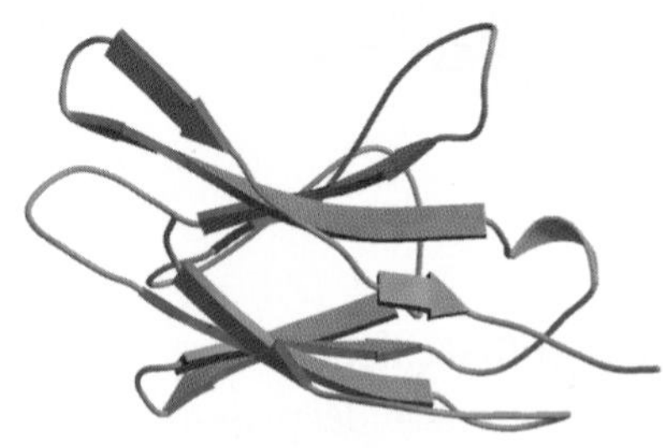

1cd8
Immunoglobulin-like β sandwich
Immunoglobulin
Antibody variable domain-like
CD8
Human (*Homo sapiens*)

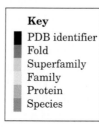

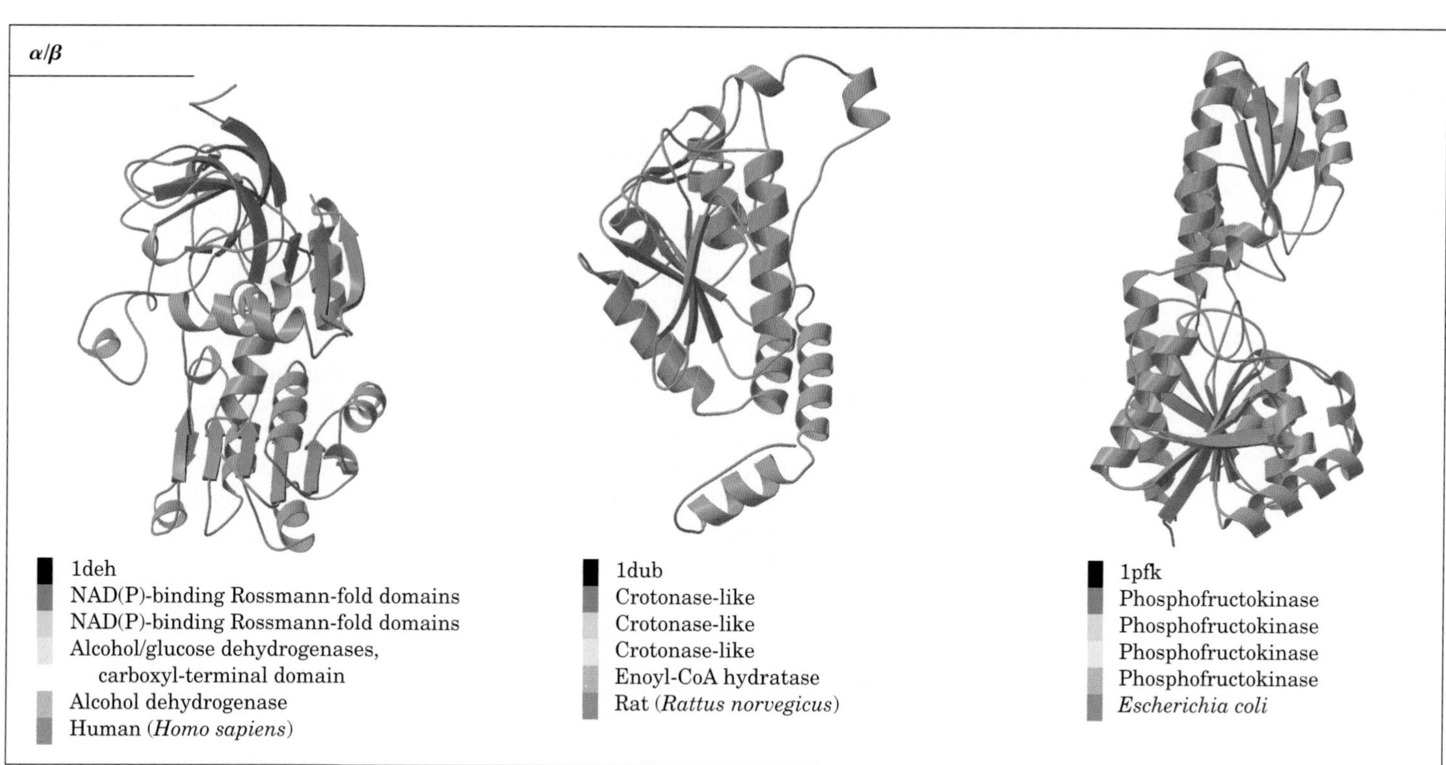

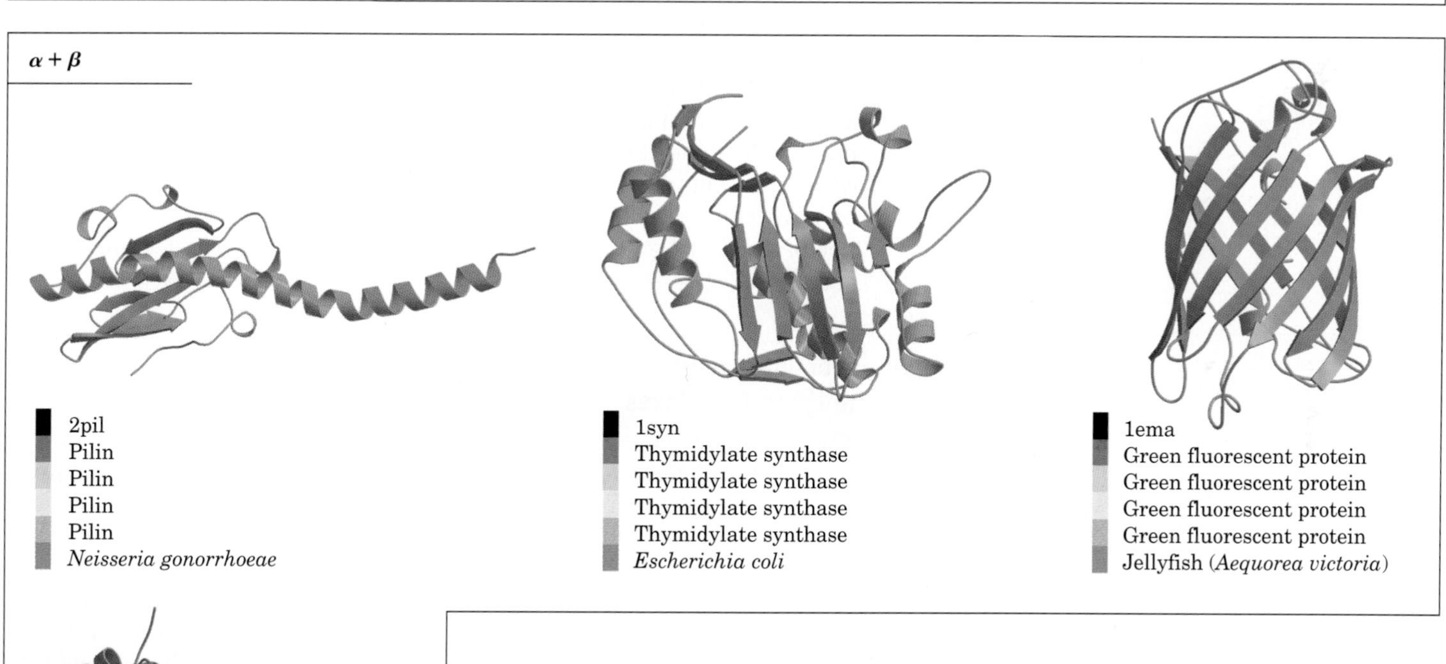

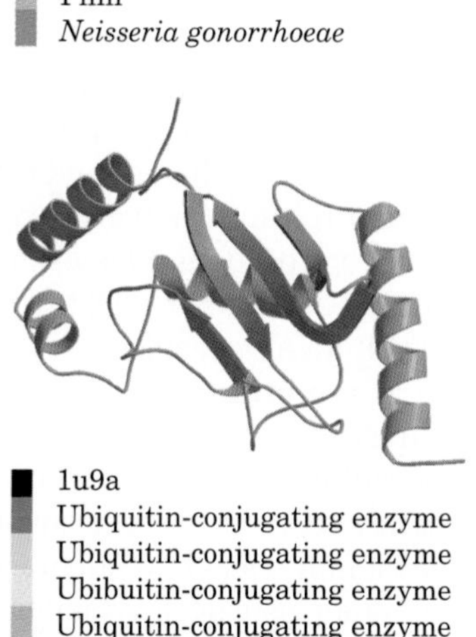

Key
- PDB identifier
- Fold
- Superfamily
- Family
- Protein
- Species

figure 6–22

Organization of proteins based on motifs. Shown here are just a small number of the hundreds of known stable motifs. They are divided into four classes: all α, all β, α/β, and $\alpha + \beta$. Structural classification data from the SCOP database, readily accessible on the Web, are also provided. The PDB identifier is the unique number given each structure archived in the Protein Data Bank. The α/β barrel, shown in Figure 6–21, is another particularly common α/β motif.

Within each class are tens to hundreds of different folding arrangements, built up from increasingly identifiable substructures. Some of the substructure arrangements are very common, others have been found in just one protein. Figure 6–22 displays a variety of motifs arrayed among the four classes of protein structure. Those illustrated are just a minute sample of the hundreds of known motifs. The number of folding patterns is not infinite, however. As the rate at which new protein structures are elucidated has increased, the fraction of those structures containing a new motif has steadily declined. Fewer than 1,000 different folds or motifs may exist in all proteins. Figure 6–22 also shows how real proteins can be organized based on the presence of the motifs just discussed. The top two levels, **class** and **fold,** are purely structural. Below that level, categorization is based on evolutionary relationships.

Proteins with significant primary sequence similarity, and/or with demonstrably similar structure and function, are said to be in the same protein **family.** A strong evolutionary relationship is usually evident within a protein family. For example, the globin family has many different proteins with both structural and sequence similarity to myoglobin (as seen in the proteins used as examples in Box 6–3 and again in the next chapter).

Two or more families with little primary sequence similarity sometimes make use of the same major structural motif and have functional similarities; these families are grouped as **superfamilies.** An evolutionary relationship between the families in a superfamily is considered probable, even though time and functional distinctions, hence different adaptive pressures, may have erased many of the telltale sequence relationships.

Structural motifs become especially important in defining protein families and superfamilies. Improved classification and comparison systems for proteins lead inevitably to the elucidation of new functional relationships. Given the central role of proteins in living systems, these structural comparisons can help illuminate every aspect of biochemistry, from the evolution of individual proteins to the evolutionary history of complete metabolic pathways.

Protein Quaternary Structures Range from Simple Dimers to Large Complexes

Many proteins have multiple polypeptide subunits. The association of polypeptide chains can serve a variety of functions. Many multisubunit proteins have regulatory roles; the binding of small molecules may affect the interaction between subunits, causing large changes in the protein's activity in response to small changes in the concentration of substrate or regulatory molecules (Chapter 8). In other cases, separate subunits can take on separate but related functions, such as catalysis and regulation. Some associations, such as the fibrous proteins considered earlier in this chapter and the coat proteins of viruses, serve primarily structural roles. Some very large protein assemblies are the site of complex, multistep reactions. One example is the ribosome, site of protein synthesis, which incorporates dozens of protein subunits along with a number of structural RNA molecules.

A multisubunit protein is also referred to as a **multimer.** Multimeric proteins can have from two to hundreds of subunits. A multimer with just a few subunits is often called an **oligomer.** If a multimer is composed of a number of nonidentical subunits, the overall structure of the protein can be asymmetric and quite complicated. However, most multimers have identical subunits or repeating groups of nonidentical subunits, usually in symmetric arrangements. The repeating structural unit in such a multimeric protein, whether it is a single subunit or a group of subunits, is called a **protomer.**

The first oligomeric protein for which the three-dimensional structure was determined was hemoglobin (M_r 64,500), which contains four polypeptide chains and four heme prosthetic groups, in which the iron atoms are in the ferrous (Fe^{2+}) state (Fig. 6–17). The protein portion, called globin, consists of two α chains (141 residues each) and two β chains (146 residues each). Note that in this case α and β do not refer to secondary structures. Because hemoglobin is four times as large as myoglobin, much more time and effort were required to solve its three-dimensional structure by x-ray analysis, finally achieved by Max Perutz, John Kendrew, and their colleagues in 1959. The subunits of hemoglobin are arranged in symmetric pairs (Fig. 6–23), each pair having one α and one β subunit. Hemoglobin can therefore be described either as a tetramer or as a dimer of $\alpha\beta$ protomers.

Identical subunits of multimeric proteins are generally arranged in one or a limited set of symmetric patterns. A description of the structure of these proteins requires an understanding of conventions used to define symmetries. Oligomers can have either **rotational symmetry** or **helical symmetry;** that is, individual subunits can be superimposed on others (brought to coincidence) by rotation about one or more rotational axes, or by a helical rotation. In proteins with rotational symmetry, the subunits pack about the rotational axes to form closed structures. Proteins with helical symmetry tend to form structures that are more open-ended, with subunits added in a spiraling array.

Max Perutz (left)
John Kendrew, 1917–1997 (right)

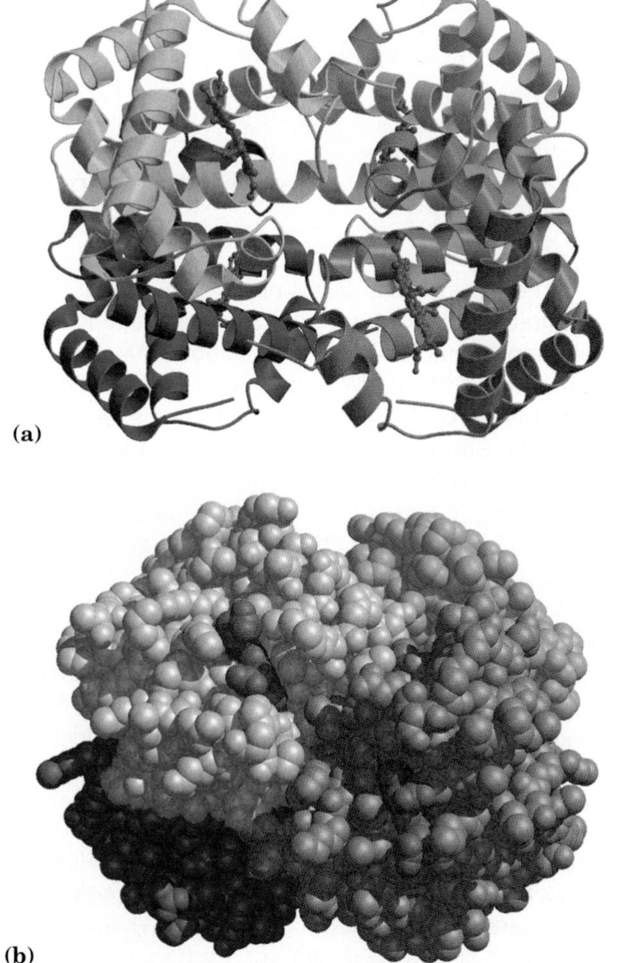

figure 6–23
The quaternary structure of deoxyhemoglobin. X-ray diffraction analysis of deoxyhemoglobin (hemoglobin without oxygen molecules bound to the heme groups) shows how the four polypeptide subunits are packed together. **(a)** A ribbon representation. **(b)** A space-filling model. The α subunits are shown in gray and light blue; the β subunits in pink and dark blue. Note that the heme groups (red) are relatively far apart.

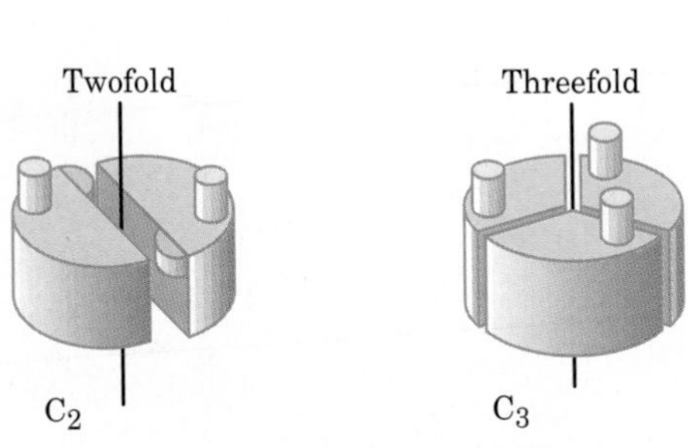

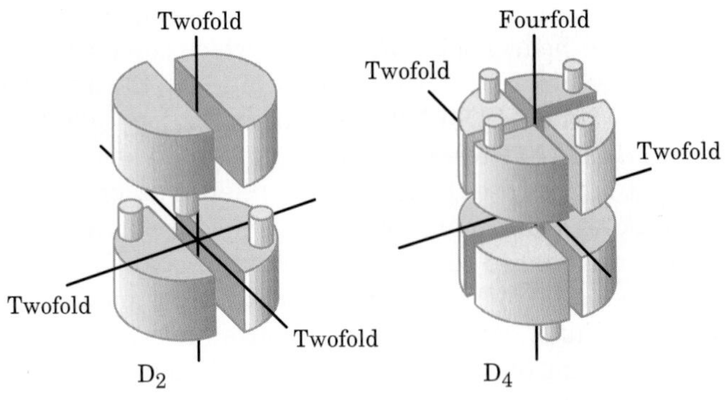

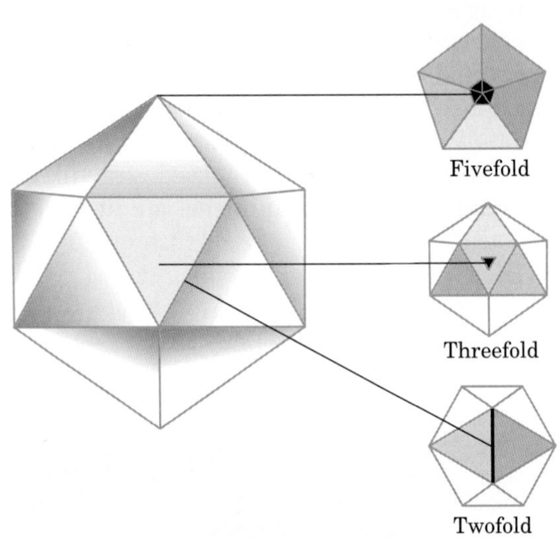

figure 6–24

Rotational symmetry in proteins. (a) In cyclic symmetry, subunits are related by rotation about a single *n*-fold axis, where *n* is the number of subunits so related. The axes are shown as black lines; the numbers are values of *n*. Only two of many possible C_n arrangements are shown. **(b)** In dihedral symmetry, all subunits can be related by rotation about one or both of two axes, one of which is twofold. D_2 symmetry is most common. **(c)** Icosahedral symmetry. Relating all 20 triangular faces of an icosahedron requires rotation about one or more of three separate rotational axes: twofold, threefold, and fivefold. An end-on view of each of these axes is shown at the right.

There are several forms of rotational symmetry. The simplest is **cyclic symmetry,** involving rotation about a single axis (Fig. 6–24a). If subunits can be superimposed by rotation about a single axis, the protein has a symmetry defined by convention as C_n (C for cyclic, n for the number of subunits related by the axis). The axis itself is described as an n-fold rotational axis. The $\alpha\beta$ protomers of hemoglobin (Fig. 6–23) are related by C_2 symmetry. A somewhat more complicated rotational symmetry is **dihedral symmetry,** in which a twofold rotational axis intersects an n-fold axis at right angles. The symmetry is defined as D_n (Fig. 6–24b). A protein with dihedral symmetry has $2n$ protomers.

Proteins with cyclic or dihedral symmetry are particularly common. More complex rotational symmetries are possible, but only a few are regularly encountered. One example is icosahedral symmetry. An icosahedron is a regular 12-cornered polyhedron having 20 equilateral triangular faces (Fig. 6–24c). Each face can be brought to coincidence with another by rotation about one or more of three rotational axes. This is a common structure in virus coats, or capsids. The human poliovirus has an icosahedral capsid (Fig. 6–25a). Each triangular face is made up of three protomers, each protomer containing single copies of four different polypeptide chains, three of which are accessible at the outer surface. Sixty protomers form the 20 faces of the icosahedral shell enclosing the genetic material (RNA).

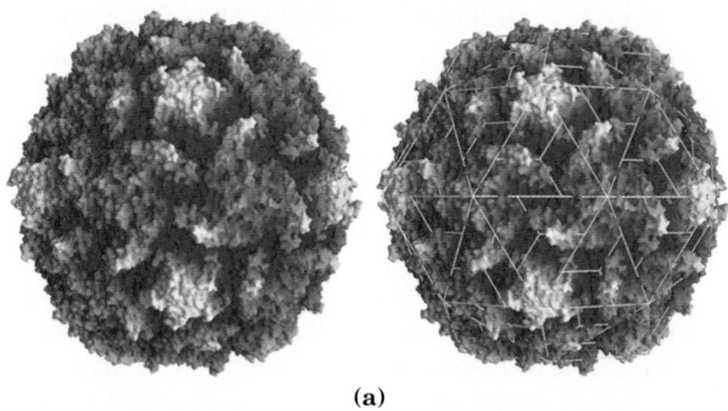

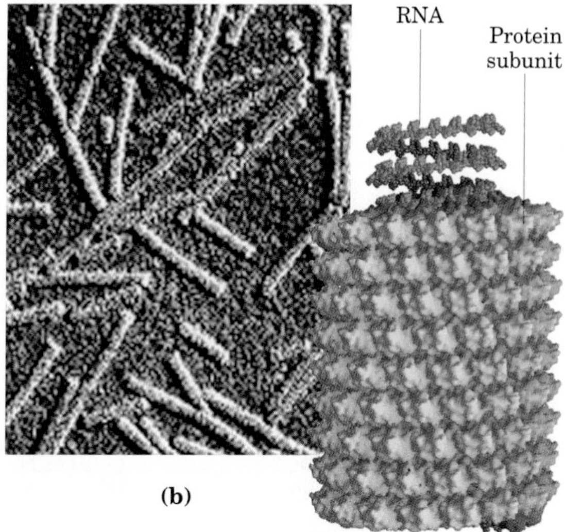

figure 6–25
Viral capsids. (a) Poliovirus. The coat proteins of poliovirus assemble into an icosahedron 300 Å in diameter. Icosahedral symmetry is a type of rotational symmetry (see Fig. 6–24c). On the left is a surface contour image of the poliovirus capsid. In the image on the right, lines have been superimposed to show the axes of symmetry. **(b)** Tobacco mosaic virus. This rod-shaped virus (as shown in the electron micrograph) is 3,000 Å long and 180 Å in diameter; it has helical symmetry.

The other major type of symmetry found in oligomers, helical symmetry, also occurs in capsids. Tobacco mosaic virus is a right-handed helical filament made up of 2,130 identical subunits (Fig. 6–25b). This cylindrical structure encloses the viral RNA. Proteins with subunits arranged in helical filaments can also form long, fibrous structures such as the actin filaments of muscle (Fig. 7–30).

There Are Limits to the Size of Proteins

The relatively large size of proteins reflects their functions. The function of an enzyme, for example, requires a stable structure containing a pocket large enough to bind its substrate and catalyze a reaction. Protein size has limits, however, imposed by two factors: the genetic coding capacity of nucleic acids and the accuracy of the protein biosynthetic process. The use of many copies of one or a few proteins to make a large enclosing structure (capsid) is important for viruses because this strategy conserves genetic material. Remember that there is a linear correspondence between the sequence of a gene in the nucleic acid and the amino acid sequence of the protein for which it codes (see Box 5–2). The nucleic acids of viruses are much too small to encode the information required for a protein shell made of a single polypeptide. By using many copies of much smaller polypeptides, a much shorter nucleic acid is needed for coding the capsid subunits, and this nucleic acid can be efficiently used over and over again. Cells also use large complexes of polypeptides in muscle, cilia, the cytoskeleton, and other structures. It is simply more efficient to make many copies of a small polypeptide than one copy of a very large protein. In fact, most proteins with a molecular weight greater than 100,000 have multiple subunits, identical or different. The second factor limiting the size of proteins is the error frequency during protein biosynthesis. The error frequency is low (about 1 mistake per 10,000 amino acid residues added), but even this low rate results in a high probability of a damaged protein if the protein is very large. Simply put, the potential for incorporating a "wrong" amino acid in a protein is greater for a large protein than for a small one.

Protein Denaturation and Folding

All proteins begin their existence on a ribosome as a linear sequence of amino acid residues (Chapter 27). This polypeptide must fold during and following synthesis to take up its native conformation. We have seen that a native protein conformation is only marginally stable. Modest changes in the protein's environment can bring about structural changes that can affect function. We now explore the transition that occurs between the folded and unfolded states.

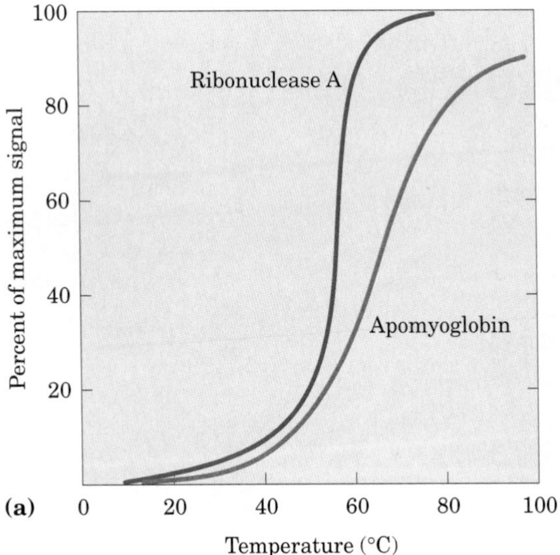

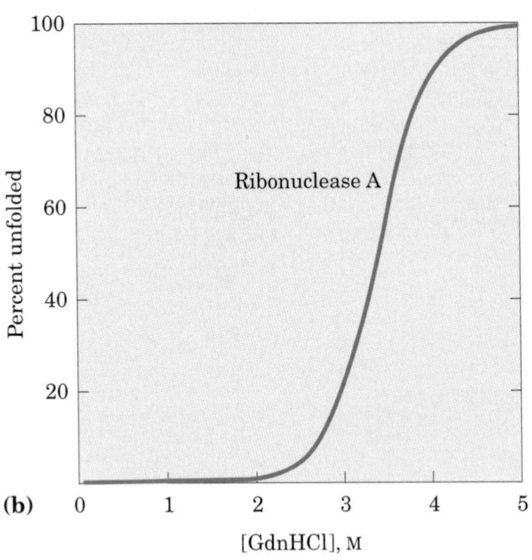

figure 6–26

Protein denaturation. Results are shown for proteins denatured by two different environmental changes. In each case, the transition from the folded to unfolded state is fairly abrupt, suggesting cooperativity in the unfolding process. **(a)** Thermal denaturation of horse apomyoglobin (myoglobin without the heme prosthetic group) and ribonuclease A (with its disulfide bonds intact; see Fig. 6–27). The midpoint of the temperature range over which denaturation occurs is called the melting temperature, or T_m. The denaturation of apomyoglobin was monitored by circular dichroism, a technique that measures the amount of helical structure in a macromolecule. Denaturation of ribonuclease A was tracked by monitoring changes in the intrinsic fluorescence of the protein, which is affected by changes in the environment of Trp residues. **(b)** Denaturation of disulfide-intact ribonuclease A by guanidine hydrochloride (GdnHCl), monitored by circular dichroism.

Loss of Protein Structure Results in Loss of Function

Protein structures have evolved to function in particular cellular environments. Conditions different from those in the cell can result in protein structural changes, large and small. A loss of three-dimensional structure sufficient to cause loss of function is called **denaturation.** The denatured state does not necessarily equate with complete unfolding of the protein and randomization of conformation. Under most conditions, denatured proteins exist in a set of partially folded states that are poorly understood.

Most proteins can be denatured by heat, which affects the weak interactions in a protein (primarily hydrogen bonds) in a complex manner. If the temperature is increased slowly, a protein's conformation generally remains intact until an abrupt loss of structure (and function) occurs over a narrow temperature range (Fig. 6–26). The abruptness of the change suggests that unfolding is a cooperative process: loss of structure in one part of the protein destabilizes other parts. The effects of heat on proteins are not yet readily predictable. The very heat-stable proteins of thermophilic bacteria have evolved to function at the temperature of hot springs (~100 °C). Yet, the structures of these proteins often differ only slightly from those of homologous proteins derived from bacteria such as *Escherichia coli.* How these small differences promote structural stability at high temperatures is not yet understood.

Proteins can be denatured not only by heat but by extremes of pH, by certain miscible organic solvents such as alcohol or acetone, by certain solutes such as urea and guanidine hydrochloride, or by detergents. Each of these denaturing agents represents a relatively mild treatment in the sense that no covalent bonds in the polypeptide chain are broken. Organic solvents, urea, and detergents act primarily by disrupting the hydrophobic interactions that make up the stable core of globular proteins; extremes of pH alter the net charge on the protein, causing electrostatic repulsion and the disruption of some hydrogen bonding. The denatured states obtained with these various treatments need not be equivalent.

Amino Acid Sequence Determines Tertiary Structure

The tertiary structure of a globular protein is determined by its amino acid sequence. The most important proof of this came from experiments showing that denaturation of some proteins is reversible. Certain globular proteins denatured by heat, extremes of pH, or denaturing reagents will regain their native structure and their biological activity if returned to conditions in which the native conformation is stable. This process is called **renaturation.**

A classic example is the denaturation and renaturation of ribonuclease. Purified ribonuclease can be completely denatured by exposure to a concentrated urea solution in the presence of a reducing agent. The reducing agent cleaves the four disulfide bonds to yield eight Cys residues, and the urea disrupts the stabilizing hydrophobic interactions, thus freeing the entire polypeptide from its folded conformation. Denaturation of ribonuclease is accompanied by a complete loss of catalytic activity. When the urea and the reducing agent are removed, the randomly coiled, denatured ribonuclease spontaneously refolds into its correct tertiary structure, with full restoration of its catalytic activity (Fig. 6–27). The refolding of ribonuclease is so accurate that the four intrachain disulfide bonds are re-formed in the same positions in the renatured molecule as in the native ribonuclease. As calculated mathematically, the eight Cys residues could recombine at random to form up to four disulfide bonds in 105 different ways. In fact, an essentially random distribution of disulfide bonds was obtained when the

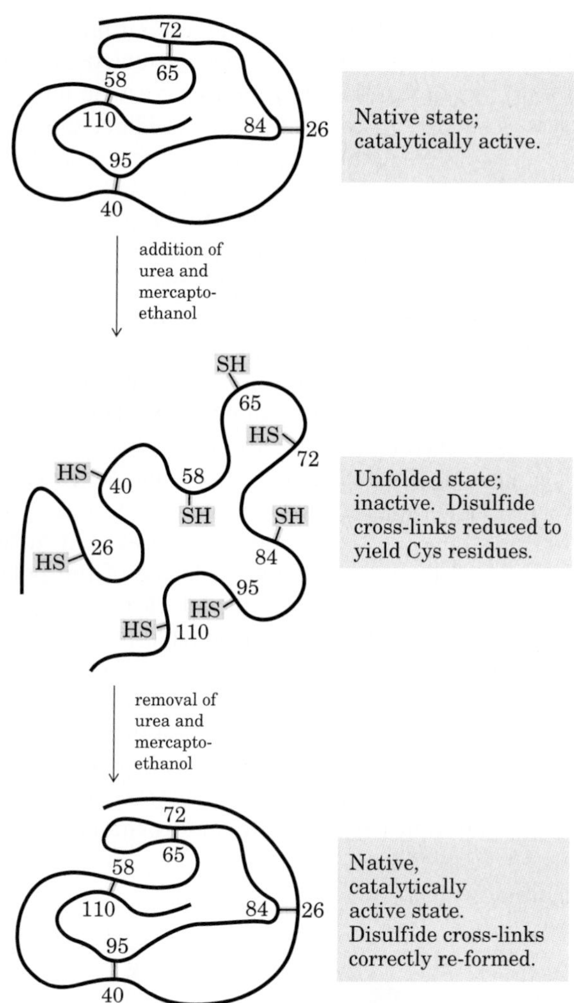

figure 6–27

Renaturation of unfolded, denatured ribonuclease. Urea is used to denature ribonuclease, and mercaptoethanol ($HOCH_2CH_2SH$) to reduce and thus cleave the disulfide bonds to yield eight Cys residues. Renaturation involves reestablishment of the correct disulfide cross-links.

disulfides were allowed to re-form in the presence of denaturant, indicating that weak bonding interactions are required for correct positioning of disulfide bonds and assumption of the native conformation.

This classic experiment, carried out by Christian Anfinsen in the 1950s, provided the first evidence that the amino acid sequence of a polypeptide chain contains all the information required to fold the chain into its native, three-dimensional structure. Later, similar results were obtained using chemically synthesized, catalytically active ribonuclease. This eliminated the possibility that some minor contaminant in Anfinsen's purified ribonuclease preparation might have contributed to the renaturation of the enzyme, thus dispelling any remaining doubt that this enzyme folds spontaneously.

Polypeptides Fold Rapidly by a Stepwise Process

In living cells, proteins are assembled from amino acids at a very high rate. For example, *E. coli* cells can make a complete, biologically active protein molecule containing 100 amino acid residues in about 5 s at 37 °C. How does such a polypeptide chain arrive at its native conformation? Let's assume conservatively that each of the amino acid residues could take up 10 different conformations on average, giving 10^{100} different conformations for the polypeptide. Let's also assume that the protein folds itself spontaneously by a random process in which it tries out all possible conformations around every single bond in its backbone until it finds its native, biologically active form. If each conformation were sampled in the shortest possible time ($\sim 10^{-13}$ s, or the time required for a single molecular vibration), it would take about 10^{77} yr to sample all possible conformations. Thus protein folding cannot be a completely random, trial-and-error process. There must be shortcuts. This problem was first pointed out by Cyrus Levinthal in 1968, and is sometimes called Levinthal's paradox.

The folding pathway of a large polypeptide chain is unquestionably complicated, and not all the principles that guide the process have been worked out. However, extensive study has led to the development of several plausible models. In one, the folding process is envisioned as hierarchical. Local secondary structures form first. Certain amino acid sequences fold readily into α helices or β sheets, guided by constraints we have reviewed in our discussion of secondary structure. This is followed by longer-range interactions between, say, two α helices that come together to form stable supersecondary structures. The process continues until complete domains form and the entire polypeptide is folded (Fig. 6–28). In an alternative model, folding is initiated by a spontaneous collapse of the polypeptide into a compact state, mediated by hydrophobic interactions among nonpolar residues. The state resulting from this "hydrophobic collapse" may have a high content of secondary structure, but many amino acid side chains are not entirely fixed. The collapsed state is often referred to as a **molten globule.**

Most proteins probably fold by a process that incorporates features of both models. Instead of following a single pathway, a population of peptide molecules may take a variety of routes to the same end point, with the number of different partly folded conformational species decreasing as folding nears completion.

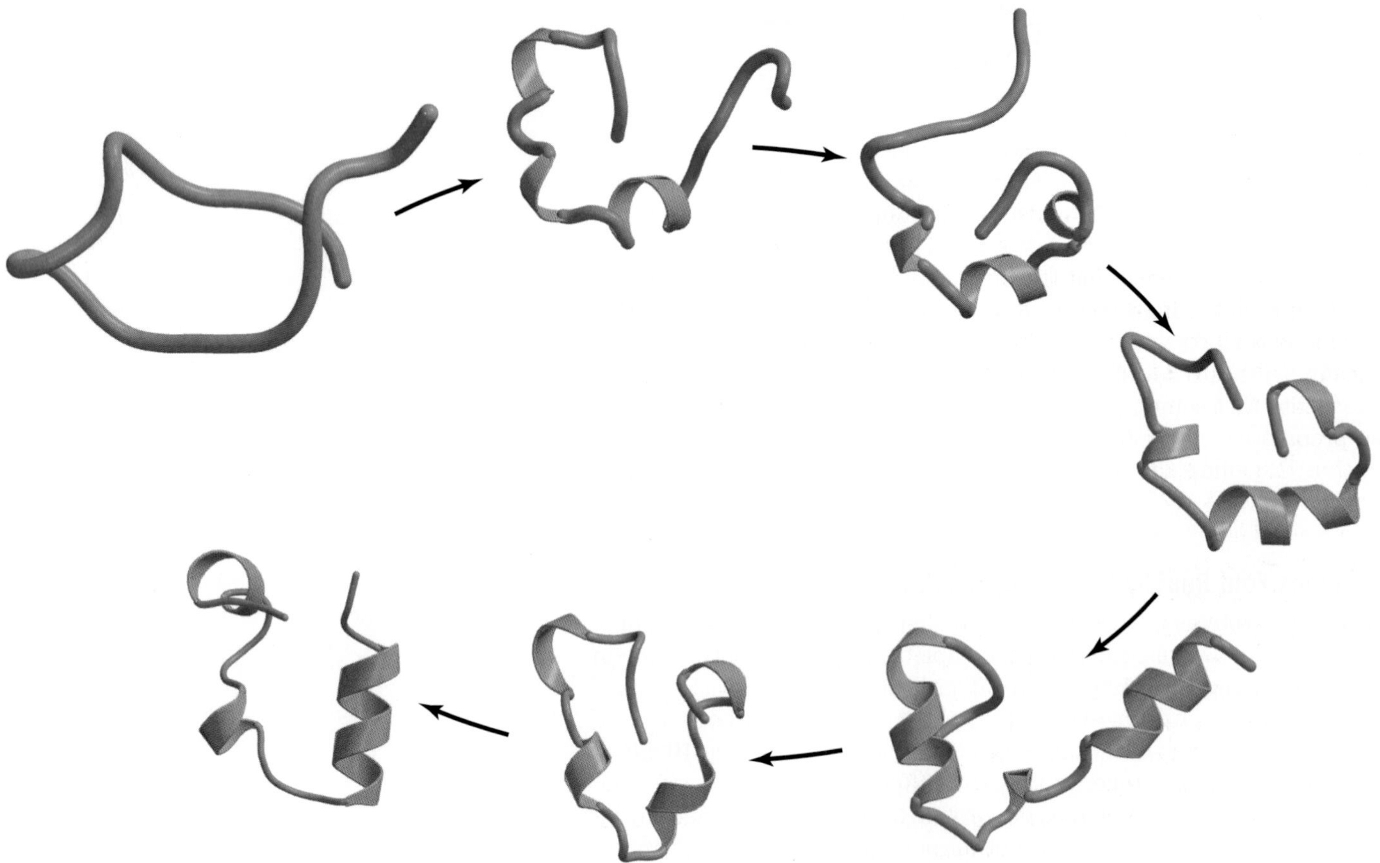

figure 6–28

A simulated folding pathway. The folding pathway of a 36-residue subdomain of the protein villin was simulated by computer. The process started with the randomly coiled peptide and 3,000 surrounding water molecules in a virtual "water box." The molecular motions of the peptide and the effects of the water molecules were taken into account in mapping the most likely paths to the final structure among the countless alternatives. The simulated folding took place in a theoretical timespan of 1 μs; however, the calculation required half a billion integration steps on two Cray supercomputers, each running for two months.

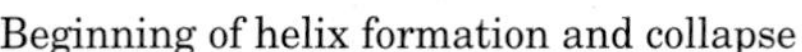

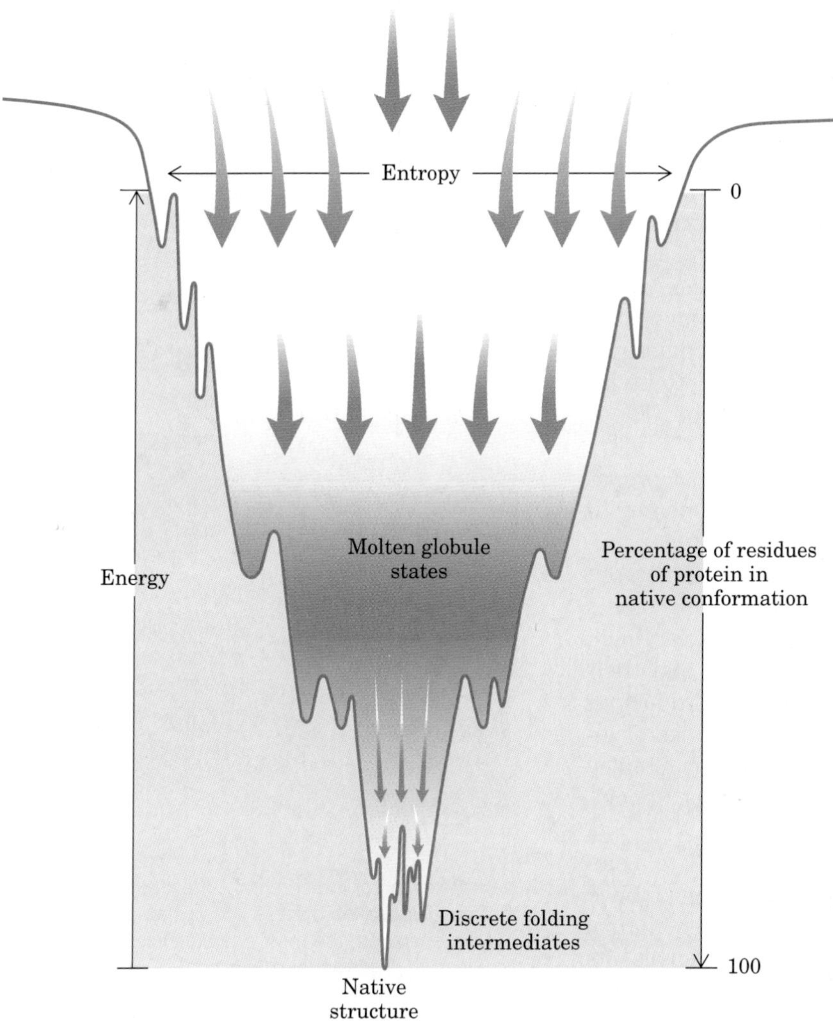

figure 6–29

The thermodynamics of protein folding depicted as a free-energy funnel. At the top, the number of conformations, and hence the conformational entropy, is large. Only a small fraction of the intramolecular interactions that will exist in the native conformation are present. As folding progresses, the thermodynamic path down the funnel reduces the number of states present (decreases entropy), increases the amount of protein in the native conformation, and decreases the free energy. Depressions on the sides of the funnel represent semistable folding intermediates, which may, in some cases, slow the folding process.

Thermodynamically, the folding process can be viewed as a kind of free-energy funnel (Fig. 6–29). The unfolded states are characterized by a high degree of conformational entropy and relatively high free energy. As folding proceeds, the narrowing of the funnel represents a decrease in the number of conformational species present. Small depressions along the sides of the free-energy funnel represent semistable intermediates that can briefly slow the folding process. At the bottom of the funnel, an ensemble of folding intermediates has been reduced to a single native conformation (or one of a small set of native conformations).

Defects in protein folding may be the molecular basis for a wide range of human genetic disorders. For example, cystic fibrosis is caused by defects in a membrane-bound protein called cystic fibrosis transmembrane conductance regulator (CFTR), which acts as a channel for chloride ions. The most common cystic fibrosis–causing mutation is the deletion of a Phe residue at position 508 in CFTR, which causes improper protein folding (see Box 12–3). Many of the disease-related mutations in collagen (p. 174) also cause defective folding. Improved understanding of protein folding may lead to new therapies for these and many other diseases (Box 6–4).

box 6–4 Death by Misfolding: The Prion Diseases

A misfolded protein appears to be the causative agent of a number of rare degenerative brain diseases in mammals. Perhaps the best known of these is mad cow disease, an outbreak of which made international headlines in the spring of 1996. Related diseases include kuru and Creutzfeldt-Jakob disease in humans and scrapie in sheep. The diseases are sometimes referred to as spongiform encephalopathies, so named because the diseased brain frequently becomes riddled with holes (Fig. 1). Typical symptoms include dementia and loss of coordination. These diseases are fatal.

In the 1960s, investigators found that preparations of the disease-causing agents appeared to lack nucleic acids. At this time, Tikvah Alper suggested that the agent was a protein. Initially, the idea seemed heretical. All disease-causing agents known up to that time—viruses, bacteria, fungi, and so on—contained nucleic acids, and their virulence was related to genetic reproduction and propagation. However, three decades of investigations, pursued most notably by Stanley Prusiner, have provided evidence that spongiform encephalopathies are different.

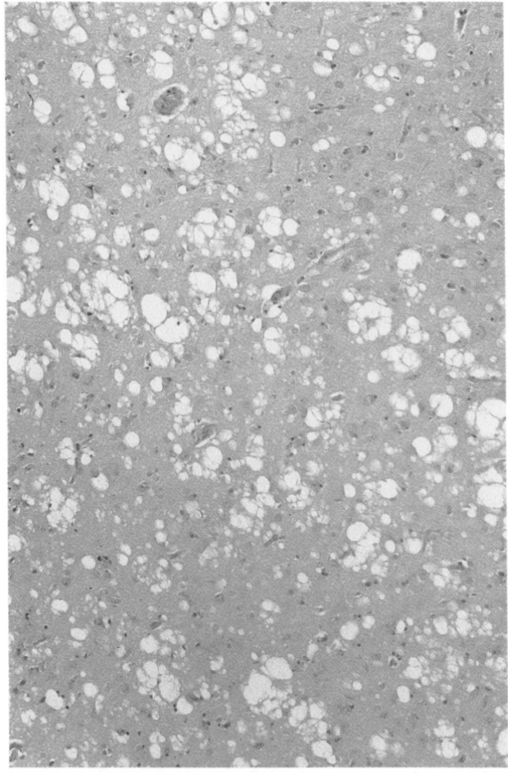

figure 1
A stained section of the cerebral cortex from a patient with Creutzfeldt-Jakob disease shows spongiform (vacuolar) degeneration, the most characteristic neurohistological feature. The vacuoles (white spots) are intracellular and occur mostly in pre- and postsynaptic processes of neurons. The vacuoles in this section vary in diameter from 20 to 100 μm.

The infectious agent has been traced to a single protein (M_r 28,000), which Prusiner dubbed prion protein (PrP). Prion protein is a normal constituent of brain tissue in all mammals. Its function is not known. Strains of mice lacking the gene for PrP (and thus the protein itself) appear to suffer no ill effects. Illness occurs only when the normal cellular PrP, or PrP^C, occurs in an altered conformation called PrP^{Sc} (Sc denotes scrapie). The interaction of PrP^{Sc} with PrP^C converts the latter to PrP^{Sc}, initiating a domino effect in which more and more of the cellular protein converts to the disease-causing form. The mechanism by which the presence of PrP^{Sc} leads to spongiform encephalopathy is not understood.

In inherited forms of prion diseases, a mutation in the gene encoding PrP produces a change in one amino acid residue that is believed to make the conversion of PrP^C to PrP^{Sc} more likely. A complete understanding of prion diseases awaits new information about how prion protein affects brain function, as well as more detailed structural information about both forms of PrP.

Some Proteins Undergo Assisted Folding

Not all proteins fold spontaneously as they are synthesized in the cell. Folding for many proteins is facilitated by the action of specialized proteins. **Molecular chaperones** are proteins that interact with partially folded or improperly folded polypeptides, facilitating correct folding pathways or providing microenvironments in which folding can occur. Two classes of molecular chaperones have been well-studied. Both are found in organisms ranging

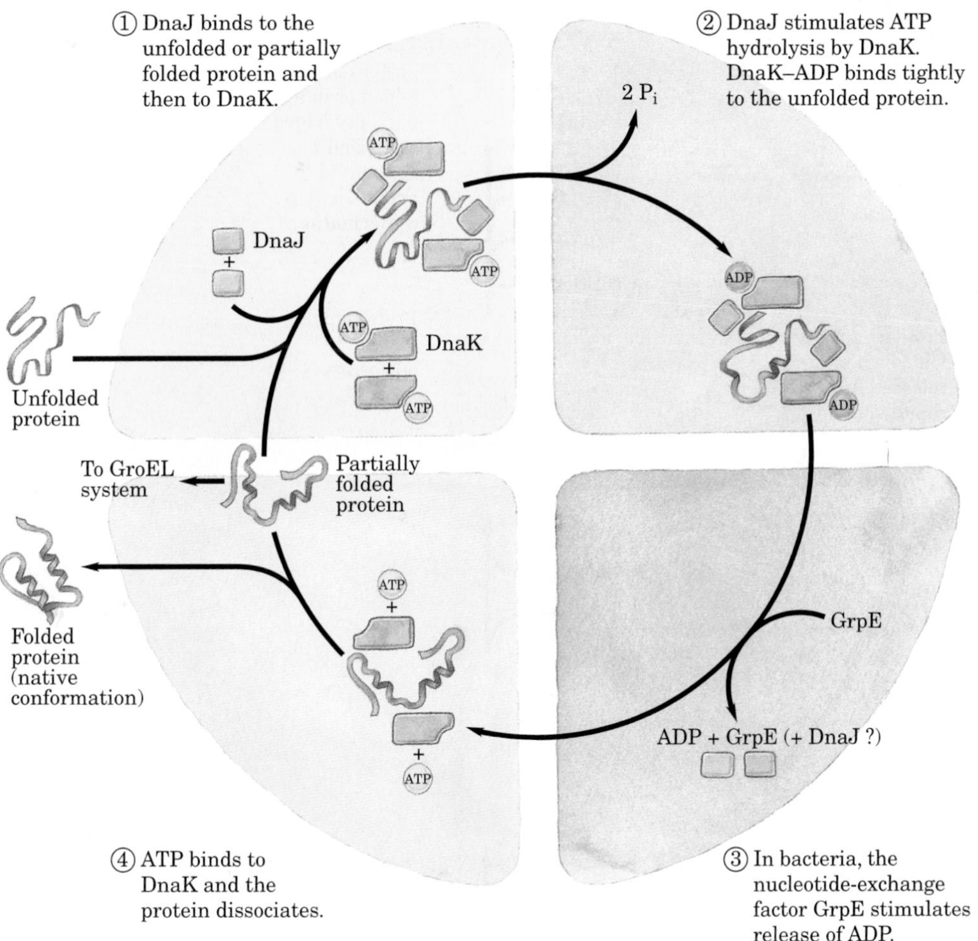
① DnaJ binds to the unfolded or partially folded protein and then to DnaK.
② DnaJ stimulates ATP hydrolysis by DnaK. DnaK–ADP binds tightly to the unfolded protein.
2 P_i
DnaJ
DnaK
ATP
ADP
Unfolded protein
To GroEL system
Partially folded protein
Folded protein (native conformation)
GrpE
ADP + GrpE (+ DnaJ ?)
④ ATP binds to DnaK and the protein dissociates.
③ In bacteria, the nucleotide-exchange factor GrpE stimulates release of ADP.

figure 6–30

Chaperones in protein folding. The cyclic pathway by which chaperones bind and release polypeptides is illustrated for the *E. coli* chaperone proteins DnaK and DnaJ, homologs of the eukaryotic chaperones Hsp70 and Hsp40. The chaperones do not actively promote the folding of the substrate protein, but instead prevent aggregation of unfolded peptides. For a population of polypeptides, some fraction of the polypeptides released at the end of the cycle are in the native conformation. The remainder are rebound by DnaK or are diverted to the chaperonin system (GroEL; see Fig. 6–31). In bacteria, a protein called GrpE interacts transiently with DnaK late in the cycle, promoting dissociation of ADP and possibly DnaJ. No eukaryotic analog of GrpE is known.

from bacteria to humans. The first class, a family of proteins called **Hsp70,** generally have a molecular weight near 70,000 and are more abundant in cells stressed by elevated temperatures (hence, *h*eat *s*hock *p*roteins of M_r 70,000, or Hsp70). Hsp70 binds to regions of unfolded polypeptides that are rich in hydrophobic residues, preventing inappropriate aggregation. These chaperones thus "protect" proteins that have been denatured by heat and peptides that are being synthesized (and are still unfolded). Hsp70 proteins also block the folding of certain proteins that must remain unfolded until they have been translocated across membranes (described in Chapter 27). Some chaperones also facilitate the quaternary assembly of oligomeric proteins. The Hsp70 proteins bind to and release polypeptides in a cycle that also involves several other proteins (including a class called Hsp40) and ATP hydrolysis. Figure 6–30 illustrates chaperone-assisted folding as elucidated for the chaperones DnaK and DnaJ in *E. coli,* homologs of the eukaryotic Hsp70 and Hsp40. The DnaK and DnaJ proteins were first identified as proteins required for in vitro replication of certain viral DNA molecules (hence the "Dna" designation).

The second class of chaperones are called **chaperonins.** These are elaborate protein complexes required for the folding of a number of cellular proteins that do not fold spontaneously. In *E. coli* an estimated 10% to 15% of cellular proteins require the resident chaperonin system, called GroEL/GroES, for folding under normal conditions (up to 30% require this assistance when the cells are heat stressed). These proteins first became known when they were found to be necessary for the growth of certain bacterial viruses (hence the designation "Gro"). Unfolded proteins are bound

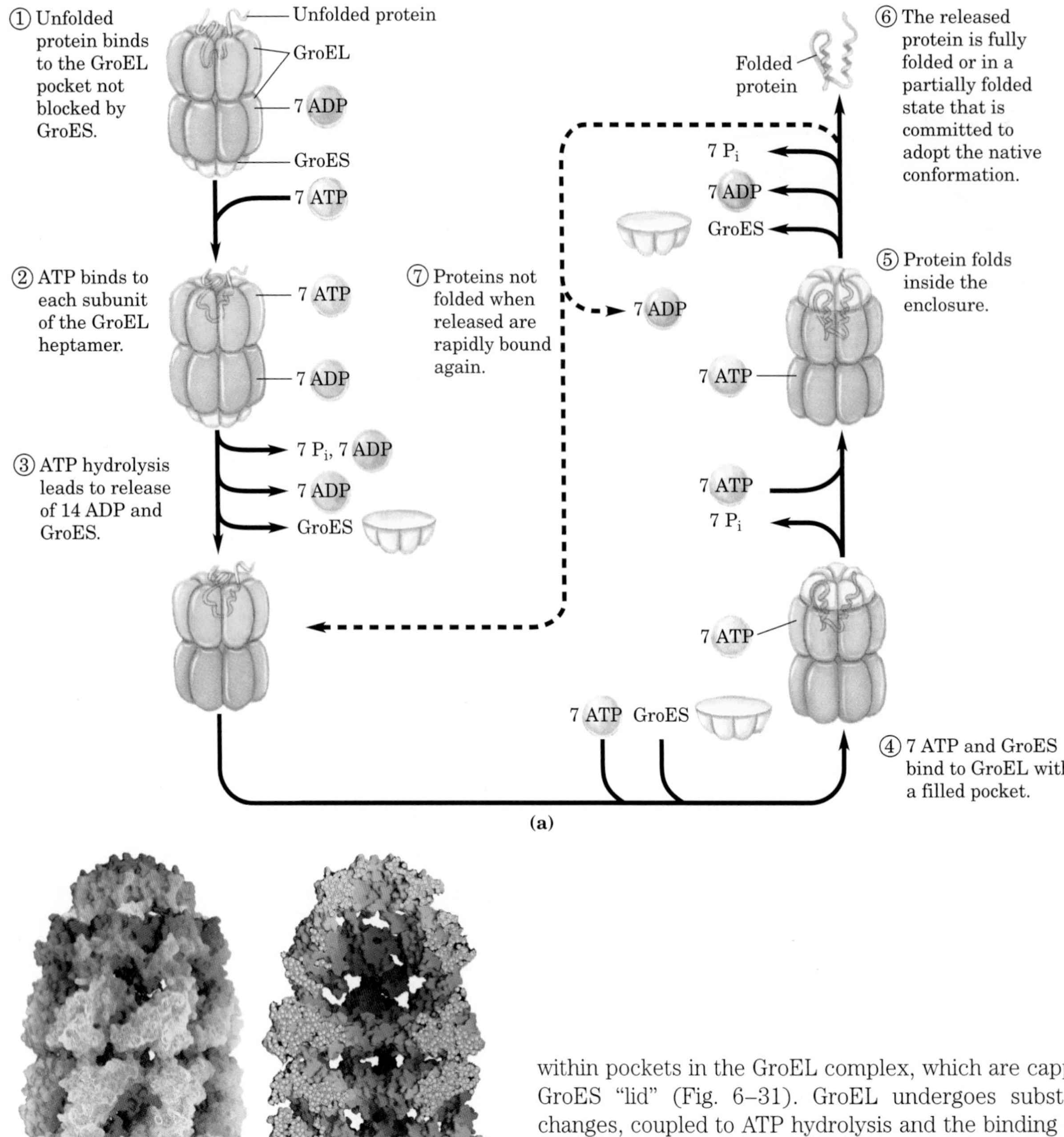

① Unfolded protein binds to the GroEL pocket not blocked by GroES.
Unfolded protein
GroEL
7 ADP
GroES
7 ATP
② ATP binds to each subunit of the GroEL heptamer.
7 ATP
7 ADP
③ ATP hydrolysis leads to release of 14 ADP and GroES.
7 P_i, 7 ADP
7 ADP
GroES
⑦ Proteins not folded when released are rapidly bound again.
7 ADP
Folded protein
⑥ The released protein is fully folded or in a partially folded state that is committed to adopt the native conformation.
7 P_i
7 ADP
GroES
⑤ Protein folds inside the enclosure.
7 ATP
7 ATP
7 P_i
7 ATP
7 ATP
GroES
④ 7 ATP and GroES bind to GroEL with a filled pocket.
(a)
(b)

figure 6–31

Chaperonins in protein folding. **(a)** A proposed pathway for the action of the *E. coli* chaperonins GroEL (a member of the Hsp60 protein family) and GroES. Each GroEL complex consists of two large pockets formed by two heptameric rings (each subunit M_r 57,000). GroES, also a heptamer (subunits M_r 10,000), blocks one of the GroEL pockets. **(b)** Surface and cut-away images of the GroEL/GroES complex. The cut-away illustrates the large interior space within which other proteins are bound.

within pockets in the GroEL complex, which are capped transiently by the GroES "lid" (Fig. 6–31). GroEL undergoes substantial conformational changes, coupled to ATP hydrolysis and the binding and release of GroES, which promote folding of the bound polypeptide. Although the structure of the GroEL/GroES chaperonin is known, many details of its mechanism of action remain unresolved.

Finally, the folding pathways of a number of proteins require two enzymes that catalyze isomerization reactions. **Protein disulfide isomerase (PDI)** is a widely distributed enzyme that catalyzes the interchange or shuffling of disulfide bonds until the bonds of the native conformation are formed. Among its functions, PDI catalyzes the elimination of folding intermediates with inappropriate disulfide cross-links. **Peptide prolyl cis-trans isomerase (PPI)** catalyzes the interconversion of the cis and trans isomers of proline peptide bonds (Fig. 6–8b), which can be a slow step in the folding of proteins that contain some bonds in the cis conformation.

Protein folding is likely to be a more complex process in the densely packed cellular environment than in the test tube. More classes of proteins that facilitate protein folding may be discovered as the biochemical dissection of the folding process continues.

summary

Every protein has a unique three-dimensional structure that reflects its function. Protein structure is stabilized by multiple weak interactions. Hydrophobic interactions provide the major contribution to stabilizing the globular form of most soluble proteins; hydrogen bonds and ionic interactions are optimized in the specific structure that is thermodynamically most stable.

The nature of the covalent bonds in the polypeptide chain places constraints on structure. The peptide bond exhibits partial double-bond character that keeps the entire peptide group in a rigid planar configuration. The N—C_α and C_α—C bonds can rotate with bond angles ϕ and ψ, respectively. Secondary structure can be defined completely if the ϕ and ψ angles are known for all amino acid residues in that polypeptide segment.

Tertiary structure, the complete three-dimensional structure of a polypeptide chain, can be understood by examining common, stable substructures variably called supersecondary structures, motifs, or folds. Motifs range from simple to very complex. The thousands of known protein structures are generally assembled from a repertoire of only a few hundred motifs, some of which are very common. Regions of a polypeptide chain that can fold stably and independently are called domains. Small proteins generally have only a single domain, whereas large proteins may have several.

There are two general classes of proteins: fibrous and globular. Fibrous proteins, which serve mainly structural roles, have simple repeating elements of secondary structure and were models for early studies of protein structure. Two major types of secondary structure were predicted by model building based on information obtained from fibrous proteins: the α helix and the β conformation. Both are characterized by optimal hydrogen bonding between peptide bonds in the polypeptide backbone. The stability of these structures within a protein is influenced by their amino acid content and by the relative placement of amino acid residues in the sequence. Another type of secondary structure common in proteins is the β turn.

In fibrous proteins such as keratins and collagen, a single type of secondary structure predominates. The polypeptide chains are supercoiled into ropes and then combined in larger bundles to provide strength. The β sheets of silk fibroin are stacked to build a strong but flexible structure.

Globular proteins have more complicated tertiary structures, often containing several types of secondary structure in the same polypeptide chain. The first globular protein structure to be determined, using x-ray diffraction methods, was that of myoglobin. This structure confirmed that a predicted secondary structure (α helix) occurs in proteins; that hydrophobic amino acid residues are located in the protein interior; and that globular proteins are compact. Subsequent research on the structure of many globular proteins has reinforced these conclusions while demonstrating that great variety can be found in tertiary structure.

The complex structures of globular proteins can be analyzed by examining substructures, including motifs and domains. In protein structural databases, structures are commonly organized into four major classes: all α, all β, α/β, and $\alpha + \beta$. Specific proteins in each class are grouped into families and superfamilies based on correlations in sequence, structure, and function.

Quaternary structure refers to the interaction between the subunits of multisubunit (multimeric) proteins or large protein assemblies. Some multimeric proteins have a repeated unit consisting of a single subunit or a group of subunits referred to as a protomer. The protomers are usually related by rotational or helical symmetry. The best-studied multimeric protein is hemoglobin.

The three-dimensional structure of proteins can be destroyed by treatments that disrupt weak interactions, a process called denaturation. Denaturation destroys protein function, demonstrating a relationship between structure and function. Some denatured proteins (e.g., ribonuclease) can renature spontaneously to form biologically active protein, showing that the tertiary structure of a protein is determined by its amino acid sequence.

Protein folding in cells probably involves multiple pathways. Initially, regions of secondary structure may form, followed by folding into supersecondary structures. Large ensembles of folding intermediates are rapidly brought to a single native conformation. For many proteins, folding is facilitated by Hsp70 chaperones and by chaperonins. Disulfide bond formation and the cis-trans isomerization of proline peptide bonds are catalyzed by specific enzymes.

further reading

General

Anfinsen, C.B. (1973) Principles that govern the folding of protein chains. *Science* **181,** 223–230.

The author reviews his classic work on ribonuclease.

Branden, C. & Tooze, J. (1991) *Introduction to Protein Structure*, Garland Publishing, Inc., New York.

Creighton, T.E. (1993) *Proteins: Structures and Molecular Properties,* 2nd edn, W.H. Freeman and Company, New York.

A comprehensive and authoritative source.

Evolution of Catalytic Function. (1987) *Cold Spring Harb. Symp. Quant. Biol.* **52.**

A source of excellent articles on many topics, including protein structure, folding, and function.

Kendrew, J.C. (1961) The three-dimensional structure of a protein molecule. *Sci. Am.* **205** (December), 96–111.

Describes how the structure of myoglobin was determined and what was learned from it.

Richardson, J.S. (1981) The anatomy and taxonomy of protein structure. *Adv. Prot. Chem.* **34,** 167–339.

An outstanding summary of protein structural patterns and principles; the author originated the very useful "ribbon" representations of protein structure.

Secondary, Tertiary, and Quaternary Structure

Brenner, S.E., Chothia, C., & Hubbard, T.J.P. (1997) Population statistics of protein structures: lessons from structural classifications. *Curr. Opin. Struct. Biol.* **7,** 369–376.

Chothia, C., Hubbard, T., Brenner, S., Barns, H., & Murzin, A. (1997) Protein folds in the all-β and all-α classes. *Annu. Rev. Physiol. Biomol. Struct.* **26,** 597–627.

Fuchs, E. & Cleveland, D.W. (1998) A structural scaffolding of intermediate filaments in health and disease. *Science* **279,** 514–519.

McPherson, A. (1989) Macromolecular crystals. *Sci. Am.* **260** (March), 62–69.

Describes how macromolecules such as proteins are crystallized.

Prockop, D.J. & Kivirikko, K.I. (1995) Collagens, molecular biology, diseases, and potentials for therapy. *Annu. Rev. Biochem.* **64,** 403–434.

Shoeman, R.L. & Traub, P. (1993) Assembly of intermediate filaments. *Bioessays* **15,** 605–611.

Protein Denaturation and Folding

Aurora, R., Creamer, T.P., Srinivasan, R., & Rose, G.D. (1997) Local interactions in protein folding: lessons from the α-helix. *J. Biol. Chem.* **272,** 1413–1416.

Baldwin, R.L. (1994) Matching speed and stability. *Nature* **369,** 183–184.

Creighton, T.E., Darby, N.J., & Kemmink, J. (1996) The roles of partly folded intermediates in protein folding. *FASEB J.* **10,** 110–118.

Dill, K.A. & Chan, H.S. (1997) From Levinthal to pathways to funnels. *Nat. Struct. Biol.* **4,** 10–19.

Johnson, J.L. & Craig, E.A. (1997) Protein folding *in vivo*: unraveling complex pathways. *Cell* **90,** 201–204.

Netzer, W.J. & Hartl, F.U. (1998) Protein folding in the cytosol: chaperonin-dependent and independent mechanisms. *Trends Biochem. Sci.* **23,** 68–73.

Prusiner, S.B. (1995) The prion diseases. *Sci. Am.* **272** (January), 48–57.

Prusiner, S.B., Scott, M.R., DeArmond, S.J., & Cohen, F.E. (1998) Prion protein biology. *Cell* **93,** 337–348.

Richardson, A., Landry, S.J., & Georgopolous, C. (1998) The ins and outs of a molecular chaperone machine. *Trends Biochem. Sci.* **23,** 138–143.

Ruddon, R.R. & Bedows, E. (1997) Assisted protein folding. *J. Biol. Chem.* **272,** 3125–3128.

Thomas, P.J., Qu, B-H., & Pederson, P.L. (1995) Defective protein folding as a basis of human disease. *Trends Biochem. Sci.* **20,** 456–459.

problems

1. Properties of the Peptide Bond In x-ray studies of crystalline peptides, Linus Pauling and Robert Corey found that the C—N bond in the peptide link is intermediate in length (1.32 Å) between a typical C—N single bond (1.49 Å) and a C═N double bond (1.27 Å). They also found that the peptide bond is planar (all four atoms attached to the C—N group are located in the same plane) and that the two α-carbon atoms attached to the C—N are always trans to each other (on opposite sides of the peptide bond):

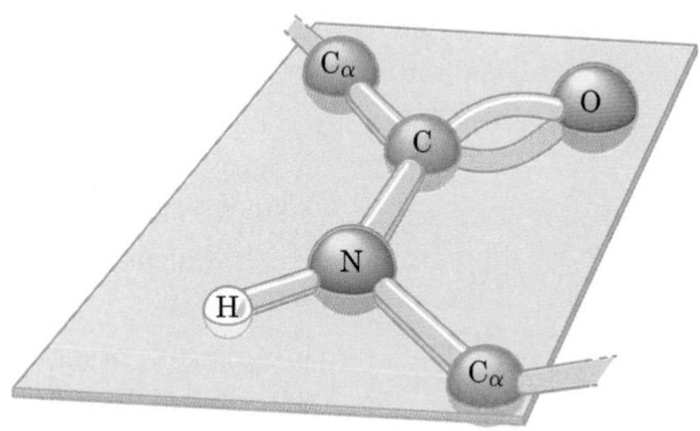

(a) What does the length of the C—N bond in the peptide linkage indicate about its strength and its bond order (i.e., whether it is single, double, or triple)?

(b) What do the observations of Pauling and Corey tell us about the ease of rotation about the C—N peptide bond?

2. Structural and Functional Relationships in Fibrous Proteins William Astbury discovered that the x-ray pattern of wool shows a repeating structural unit spaced about 5.2 Å along the direction of the wool fiber. When he steamed and stretched the wool, the x-ray pattern showed a new repeating structural unit at a spacing of 7.0 Å. Steaming and stretching the wool and then letting it shrink gave an x-ray pattern consistent with the original spacing of about 5.2 Å. Although these observations provided important clues to the molecular structure of wool, Astbury was unable to interpret them at the time.

(a) Given our current understanding of the structure of wool, interpret Astbury's observations.

(b) When wool sweaters or socks are washed in hot water or heated in a dryer, they shrink. Silk, on the other hand, does not shrink under the same conditions. Explain.

3. Rate of Synthesis of Hair α-Keratin Hair grows at a rate of 15 to 20 cm/yr. All this growth is concentrated at the base of the hair fiber, where α-keratin filaments are synthesized inside living epidermal cells and assembled into ropelike structures (see Fig. 6–11). The fundamental structural element of α-keratin is the α helix, which has 3.6 amino acid residues per turn and a rise of 5.4 Å per turn (see Fig. 6–4b). Assuming that the biosynthesis of α-helical keratin chains is the rate-limiting factor in the growth of hair, calculate the rate at which peptide bonds of α-keratin chains must be synthesized (peptide bonds per second) to account for the observed yearly growth of hair.

4. The Effect of pH on the Conformation of α-Helical Secondary Structures The unfolding of the α helix of a polypeptide to a randomly coiled conformation is accompanied by a large decrease in a property called its specific rotation, a measure of a solution's capacity to rotate plane-polarized light. Polyglutamate, a polypeptide made up of only L-Glu residues, has the α-helical conformation at pH 3. However, when the pH is raised to 7, there is a large decrease in the specific rotation of the solution. Similarly, polylysine (L-Lys residues) is an α helix at pH 10, but when the pH is lowered to 7 the specific rotation also decreases, as shown by the following graph.

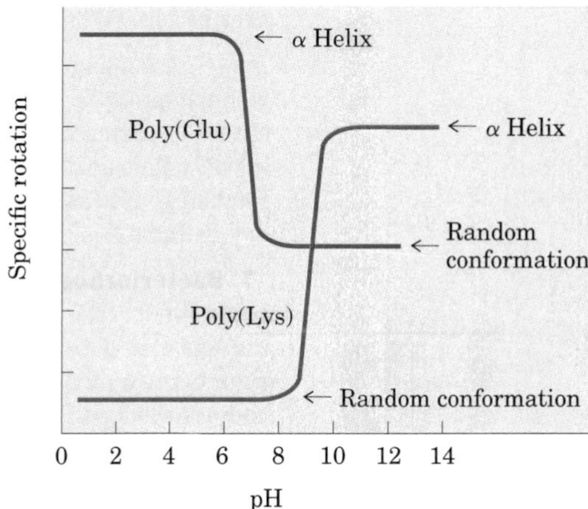

What is the explanation for the effect of the pH changes on the conformations of poly(Glu) and poly(Lys)? Why does the transition occur over such a narrow range of pH?

5. Disulfide Bonds Determine the Properties of Many Proteins A number of natural proteins are very rich in disulfide bonds, and their mechanical properties (tensile strength, viscosity, hardness, etc.) are correlated with the degree of disulfide bonding. For example, glutenin, a wheat protein rich in disulfide bonds, is responsible for the cohesive and elastic character of dough made from wheat flour. Similarly, the hard, tough nature of tortoise shell is due to the extensive disulfide bonding in its α-keratin.

(a) What is the molecular basis for the correlation between disulfide-bond content and mechanical properties of the protein?

(b) Most globular proteins are denatured and lose their activity when briefly heated to 65 °C. However, globular proteins that contain multiple disulfide bonds often must be heated longer at higher temperatures to denature them. One such protein is bovine pancreatic trypsin inhibitor (BPTI), which has 58 amino acid residues in a single chain and contains three disulfide bonds. On cooling a solution of denatured BPTI, the activity of the protein is restored. What is the molecular basis for this property?

6. Amino Acid Sequence and Protein Structure Our growing understanding of how proteins fold allows researchers to make predictions about protein structure based on primary amino acid sequence data.

1 2 3 4 5 6 7 8 9 10
Ile – Ala – His – Thr – Tyr – Gly – Pro – Phe – Glu – Ala –

11 12 13 14 15 16 17 18 19 20
Ala – Met – Cys – Lys – Trp – Glu – Ala – Gln – Pro – Asp –

21 22 23 24 25 26 27 28
Gly – Met – Glu – Cys – Ala – Phe – His – Arg

(a) Based on the amino acid sequence above, where would you predict that bends or β turns would occur?

(b) Where might intrachain disulfide cross-linkages be formed?

(c) Assuming that this sequence is part of a larger globular protein, indicate the probable location (the external surface or interior of the protein) of the following amino acid residues: Asp, Ile, Thr, Ala, Gln, Lys. Explain your reasoning. (Hint: See the hydropathy index in Table 5–1).

7. Bacteriorhodopsin in Purple Membrane Proteins Under the proper environmental conditions, the salt-loving bacterium *Halobacterium halobium* synthesizes a membrane protein (M_r 26,000) known as bacteriorhodopsin, which is purple because it contains retinal. Molecules of this protein aggregate into "purple patches" in the cell membrane. Bacteriorhodopsin acts as a light-activated proton pump that provides energy for cell functions. X-ray analysis of this protein reveals that it consists of seven parallel α-helical segments, each of which traverses the bacterial cell membrane (thickness 45 Å). Calculate the minimum number of amino acid residues necessary for one segment of α helix to traverse the membrane completely. Estimate the fraction of the bacteriorhodopsin protein that is involved in membrane-spanning helices. (Use an average amino acid residue weight of 110.)

8. Pathogenic Action of Bacteria That Cause Gas Gangrene The highly pathogenic anaerobic bacterium *Clostridium perfringens* is responsible for gas gangrene, a condition in which animal tissue structure is destroyed. This bacterium secretes an enzyme that efficiently catalyzes the hydrolysis of the peptide bond indicated in red in the sequence:

$$—X—Gly—Pro—Y— \xrightarrow{H_2O} —X—COO^- + H_3\overset{+}{N}—Gly—Pro—Y—$$

where X and Y are any of the 20 standard amino acids. How does the secretion of this enzyme contribute to the invasiveness of this bacterium in human tissues? Why does this enzyme not affect the bacterium itself?

9. Number of Polypeptide Chains in a Multisubunit Protein A sample (660 mg) of an oligomeric protein of M_r 132,000 was treated with an excess of 1-fluoro-2,4-dinitrobenzene (Sanger's reagent) under slightly alkaline conditions until the chemical reaction was complete. The peptide bonds of the protein were then completely hydrolyzed by heating it with concentrated HCl. The hydrolysate was found to contain 5.5 mg of the following compound:

$CH_3\ CH_3$ / CH / NO_2 / O_2N—(benzene ring)—NH—C—COOH / H

However, 2,4-dinitrophenyl derivatives of the α-amino groups of other amino acids could not be found.

(a) Explain how this information can be used to determine the number of polypeptide chains in an oligomeric protein.

(b) Calculate the number of polypeptide chains in this protein.

(c) What other protein analysis technique could you employ to determine whether the polypeptide chains in this protein are similar or different?

Biochemistry on the Internet

10. Protein Modeling on the Internet A group of patients suffering from Crohn's disease (an inflammatory bowel disease) underwent biopsies of their intestinal mucosa in an attempt to identify the causative agent. A protein was identified that was expressed at higher levels in patients with Crohn's disease than in patients with an unrelated inflammatory bowel disease or in unaffected controls. The protein was isolated and the following *partial* amino acid sequence was obtained (reads left to right):

EAELCPDRCI HSFQNLGIQC VKKRDLEQAI
SQRIQTNNNP FQVPIEEQRG DYDLNAVRLC
FQVTVRDPSG RPLRLPPVLP HPIFDNRAPN
TAELKICRVN RNSGSCLGGD EIFLLCDKVQ
KEDIEVYFTG PGWEARGSFS QADVHRQVAI
VFRTPPYADP SLQAPVRVSM QLRRPSDREL
SEPMEFQYLP DTDDRHRIEE KRKRTYETFK
SIMKKSPFSG PTDPRPPPRR IAVPSRSSAS
VPKPAPQPYP

(a) You can identify this protein using a protein database on the Internet. Some good places to start include PIR-International Protein Sequence Database, Structural Classification of Proteins (SCOP), and Prosite. For the current URLs of these and other protein database sites, use an Internet search engine or go to the *Principles of Biochemistry,* 3/e site at http://www.worthpublishers.com/lehninger.

At these sites, follow links to locate the sequence comparison "engine." Enter about 30 residues from the sequence of the protein in the appropriate search field and submit it for analysis. What does this analysis tell you about the identity of the protein?

(b) Try using different portions of the protein amino acid sequence. Do you always get the same result?

(c) A variety of Web sites provide information about the three-dimensional structure of proteins. Find information about the protein's secondary, tertiary, and quaternary structure using database sites such as the Protein Data Bank (PDB) or SCOP.

(d) In the course of your Web searches try to find information about the cellular function of the protein.

chapter

7

Protein Function

Knowing the three-dimensional structure of a protein is an important part of understanding how the protein functions. However, the structure shown in two dimensions on a page is deceptively static. Proteins are dynamic molecules whose functions almost invariably depend on interactions with other molecules, and these interactions are affected in physiologically important ways by sometimes subtle, sometimes striking changes in protein conformation.

In this chapter, we explore how proteins interact with other molecules and how their interactions are related to dynamic protein structure. The importance of molecular interactions to a protein's function can hardly be overemphasized. In Chapter 6, we saw that the function of fibrous proteins as structural elements of cells and tissues depends on stable, long-term quaternary interactions between identical polypeptide chains. As we will see in this chapter, the functions of many other proteins involve interactions with a variety of different molecules. Most of these interactions are fleeting, though they may be the basis of complex physiological processes such as oxygen transport, immune function, and muscle contraction, the topics we examine in detail in this chapter. The proteins that carry out these processes illustrate the following key principles of protein function, some of which will be familiar from the previous chapter:

The functions of many proteins involve the reversible binding of other molecules. A molecule bound reversibly by a protein is called a **ligand.** A ligand may be any kind of molecule, including another protein. The transient nature of protein-ligand interactions is critical to life, allowing an organism to respond rapidly and reversibly to changing environmental and metabolic circumstances.

A ligand binds at a site on the protein called the **binding site,** which is complementary to the ligand in size, shape, charge, and hydrophobic or hydrophilic character. Furthermore, the interaction is specific: the protein can discriminate among the thousands of different molecules in its environment and selectively bind only one or a few. A given protein may have separate binding sites for several different ligands. These specific molecular interactions are crucial in maintaining the high degree of order in a living system. (This discussion excludes the binding of water, which may interact weakly and nonspecifically with many parts of a protein. In Chapter 8, we consider water as a specific ligand for many enzymes.)

Proteins are flexible. Changes in conformation may be subtle, reflecting molecular vibrations and small movements of amino acid residues

throughout the protein. A protein flexing in this way is sometimes said to "breathe." Changes in conformation may also be quite dramatic, with major segments of the protein structure moving as much as several nanometers. Specific conformational changes are frequently essential to a protein's function.

The binding of a protein and ligand is often coupled to a conformational change in the protein that makes the binding site more complementary to the ligand, permitting tighter binding. The structural adaptation that occurs between protein and ligand is called **induced fit.**

In a multisubunit protein, a conformational change in one subunit often affects the conformation of other subunits.

Interactions between ligands and proteins may be regulated, usually through specific interactions with one or more additional ligands. These other ligands may cause conformational changes in the protein that affect the binding of the first ligand.

Enzymes represent a special case of protein function. Enzymes bind and chemically transform other molecules—they catalyze reactions. The molecules acted upon by enzymes are called reaction **substrates** rather than ligands, and the ligand-binding site is called the **catalytic site** or **active site.** In this chapter we emphasize the noncatalytic functions of proteins. In Chapter 8 we consider catalysis by enzymes, a central topic in biochemistry. You will see that the themes of this chapter—binding, specificity, and conformational change—are continued in the next chapter, with the added element of proteins acting as reactants in chemical transformations.

Reversible Binding of a Protein to a Ligand: Oxygen-Binding Proteins

Myoglobin and hemoglobin may be the most-studied and best-understood proteins. They were the first proteins for which three-dimensional structures were determined, and our current understanding of myoglobin and hemoglobin is garnered from the work of thousands of biochemists over several decades. Most important, they illustrate almost every aspect of that most central of biochemical processes: the reversible binding of a ligand to a protein. This classic model of protein function will tell us a great deal about how proteins work.

Oxygen Can Be Bound to a Heme Prosthetic Group

Oxygen is poorly soluble in aqueous solutions (see Table 4–3) and cannot be carried to tissues in sufficient quantity if it is simply dissolved in blood serum. Diffusion of oxygen through tissues is also ineffective over distances greater than a few millimeters. The evolution of larger, multicellular animals depended on the evolution of proteins that could transport and store oxygen. However, none of the amino acid side chains in proteins is suited for the reversible binding of oxygen molecules. This role is filled by certain transition metals, among them iron and copper, that have a strong tendency to bind oxygen. Multicellular organisms exploit the properties of metals, most commonly iron, for oxygen transport. However, free iron promotes the formation of highly reactive oxygen species such as hydroxyl radicals that can damage DNA and other macromolecules. Iron used in cells is therefore bound in forms that sequester it and/or make it less reactive. In multicellular organisms—especially those in which iron, in its oxygen-carrying capacity, must be transported over large distances—iron is often incorporated into a protein-bound prosthetic group called **heme.** (A prosthetic group is a compound permanently associated with a protein that contributes to the protein's function.)

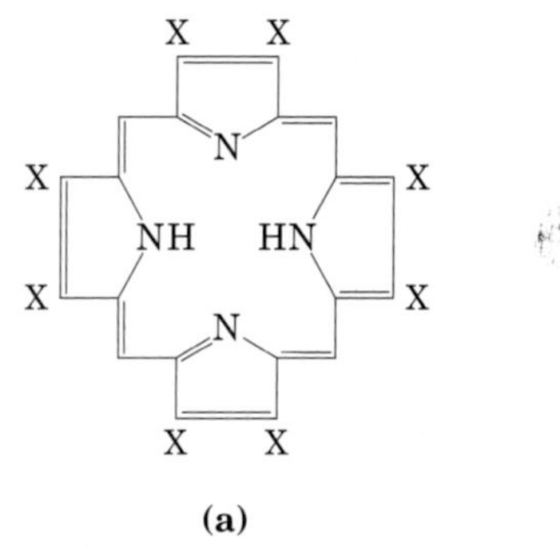

figure 7–1

Heme. The heme group is present in myoglobin, hemoglobin, and many other proteins, designated heme proteins. Heme consists of a complex organic ring structure, protoporphyrin IX, to which is bound an iron atom in its ferrous (Fe^{2+}) state. Porphyrins, of which protoporphyrin IX is only one example, consist of four pyrrole rings linked by methene bridges **(a),** with substitutions at one or more of the positions denoted X. Two representations of heme are shown in **(b)** and **(c).** The iron atom of heme has six coordination bonds: four in the plane of, and bonded to, the flat porphyrin ring system, and two perpendicular to it **(d).**

(b)

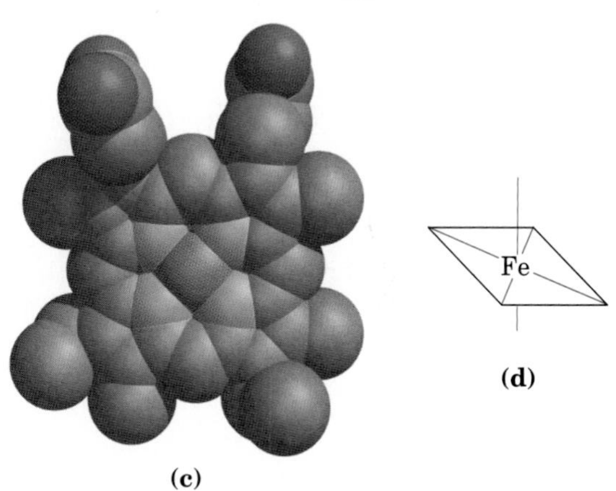

Heme (or haem) consists of a complex organic ring structure, **protoporphyrin,** to which is bound a single iron atom in its ferrous (Fe^{2+}) state (Fig. 7–1). The iron atom has six coordination bonds, four to nitrogen atoms that are part of the flat **porphyrin ring** system and two perpendicular to the porphyrin. The coordinated nitrogen atoms (which have an electron-donating character) help prevent conversion of the heme iron to the ferric (Fe^{3+}) state. Iron in the Fe^{2+} state binds oxygen reversibly; in the Fe^{3+} state it does not bind oxygen. Heme is found in a number of oxygen-transporting proteins, as well as in some proteins, such as the cytochromes, that participate in oxidation-reduction (electron transfer) reactions (Chapter 19).

In free heme molecules, reaction of oxygen at one of the two "open" coordination bonds of iron (perpendicular to the plane of the porphyrin molecule, above and below) can result in irreversible conversion of Fe^{2+} to Fe^{3+}. In heme-containing proteins, this reaction is prevented by sequestering the heme deep within a protein structure where access to the two open coordination bonds is restricted. One of these two coordination bonds is occupied by a side-chain nitrogen of a His residue. The other is the binding site for molecular oxygen (O_2) (Fig. 7–2). When oxygen binds, the electronic properties of heme iron change; this accounts for the change in color from the dark purple of oxygen-depleted venous blood to the bright red of oxygen-rich arterial blood. Some small molecules, such as carbon monoxide (CO) and nitric oxide (NO), coordinate to heme iron with greater affinity than does O_2. When a molecule of CO is bound to heme, O_2 is excluded, which is why CO is highly toxic to aerobic organisms. By surrounding and sequestering heme, oxygen-binding proteins regulate the access of CO and other small molecules to heme iron.

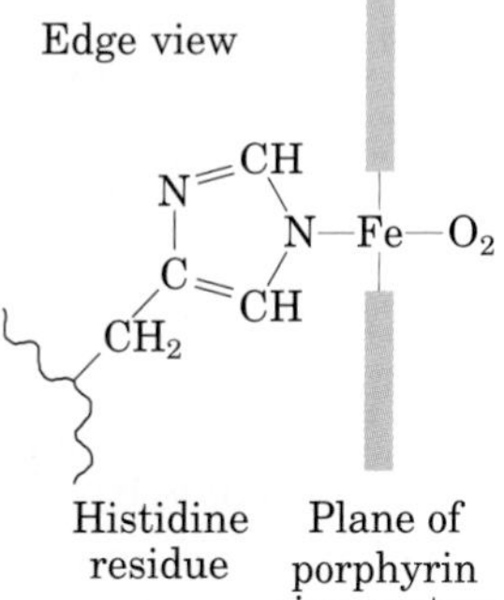

figure 7–2

The heme group viewed from the side. This view shows the two coordination bonds to Fe^{2+} perpendicular to the porphyrin ring system. One of these two bonds is occupied by a His residue, sometimes called the proximal His. The other is the binding site for oxygen. The remaining four coordination bonds are in the plane of, and bonded to, the flat porphyrin ring system.

Myoglobin Has a Single Binding Site for Oxygen

Myoglobin (M_r 16,700; abbreviated Mb) is a relatively simple oxygen-binding protein found in almost all mammals, primarily in muscle tissue. It is particularly abundant in the muscles of diving mammals such as seals and whales that must store enough oxygen for prolonged excursions undersea. Proteins very similar to myoglobin are widely distributed, occurring even in some single-celled organisms. Myoglobin stores oxygen for periods when energy demands are high and facilitates its distribution to oxygen-starved tissues.

Myoglobin is a single polypeptide of 153 amino acid residues with one molecule of heme. It is typical of the family of proteins called **globins,** which have similar primary and tertiary structures. The polypeptide is made up of eight α-helical segments connected by bends (Fig. 7–3). About 78% of the amino acid residues in the protein are found in these α helices.

Any detailed discussion of protein function inevitably involves protein structure. Our treatment of myoglobin will be facilitated by introducing some structural conventions peculiar to globins. As seen in Figure 7–3, the helical segments are labeled A through H. An individual amino acid residue may be designated either by its position in the amino acid sequence or by its location within the sequence of a particular α-helical segment. For example, the His residue coordinated to the heme in myoglobin, His^{93} (the 93rd amino acid residue from the amino-terminal end of the myoglobin polypeptide sequence), is also called His F8 (the 8th residue in α helix F). The bends in the structure are labeled AB, CD, EF, and so forth, reflecting the α-helical segments they connect.

Protein-Ligand Interactions Can Be Described Quantitatively

The function of myoglobin depends on the protein's ability not only to bind oxygen, but also to release it when and where it is needed. Function in biochemistry often revolves around a reversible protein-ligand interaction of this type. A quantitative description of this interaction is therefore a central part of many biochemical investigations.

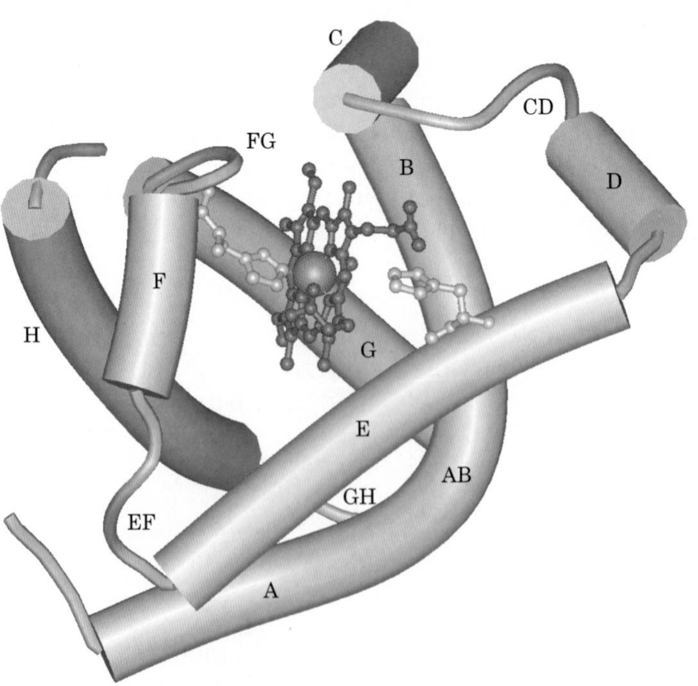

figure 7–3

The structure of myoglobin. The eight α-helical segments (shown here as cylinders) are labeled A through H. Nonhelical residues in the bends that connect them are labeled AB, CD, EF, and so forth, indicating the segments they interconnect. A few bends, including BC and DE, are abrupt and do not contain any residues; these are not normally labeled. (The short segment visible between D and E is an artifact of the computer representation.) The heme is bound in a pocket made up largely of the E and F helices, although amino acid residues from other segments of the protein also participate.

In general, the reversible binding of a protein (P) to a ligand (L) can be described by a simple **equilibrium expression:**

$$\mathrm{P} + \mathrm{L} \rightleftharpoons \mathrm{PL} \tag{7–1}$$

The reaction is characterized by an equilibrium constant, K_a, such that

$$K_a = \frac{[\mathrm{PL}]}{[\mathrm{P}][\mathrm{L}]} \tag{7–2}$$

The term $\boldsymbol{K_a}$ is an **association constant** (not to be confused with the K_a that denotes an acid dissociation constant; see p. 98). The association constant provides a measure of the affinity of the ligand L for the protein. K_a has units of M^{-1}; a higher value of K_a corresponds to a higher affinity of the ligand for the protein. A rearrangement of Equation 7–2 shows that the ratio of bound to free protein is directly proportional to the concentration of free ligand:

$$K_a[\mathrm{L}] = \frac{[\mathrm{PL}]}{[\mathrm{P}]} \tag{7–3}$$

When the concentration of the ligand is much greater than the concentration of ligand-binding sites, the binding of the ligand by the protein does not appreciably change the concentration of free (unbound) ligand—that is, [L] remains constant. This condition is broadly applicable to most ligands that bind to proteins in cells and simplifies our description of the binding equilibrium.

Thus we can consider the binding equilibrium from the standpoint of the fraction, θ (theta), of ligand-binding sites on the protein that are occupied by ligand:

$$\theta = \frac{\text{binding sites occupied}}{\text{total binding sites}} = \frac{[\mathrm{PL}]}{[\mathrm{PL}] + [\mathrm{P}]} \tag{7–4}$$

Substituting $K_a[\mathrm{L}][\mathrm{P}]$ for [PL] (see Eqn 7–3) and rearranging terms gives

$$\theta = \frac{K_a[\mathrm{L}][\mathrm{P}]}{K_a[\mathrm{L}][\mathrm{P}] + [\mathrm{P}]} = \frac{K_a[\mathrm{L}]}{K_a[\mathrm{L}] + 1} = \frac{[\mathrm{L}]}{[\mathrm{L}] + \dfrac{1}{K_a}} \tag{7–5}$$

The term K_a can be determined from a plot of θ versus the concentration of free ligand, [L] (Fig. 7–4a). Any equation of the form $x = y/(y + z)$ describes a hyperbola, and θ is thus found to be a hyperbolic function of [L]. The fraction of ligand-binding sites occupied approaches saturation asymptotically as [L] increases. The [L] at which half of the available ligand-binding sites are occupied (at $\theta = 0.5$) corresponds to $1/K_a$.

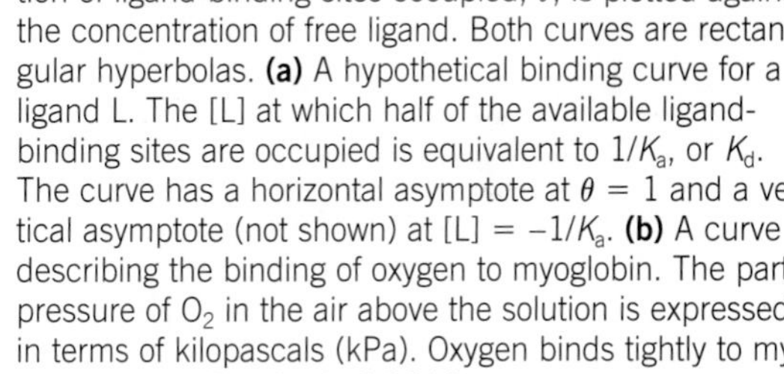

figure 7–4

Graphical representations of ligand binding. The fraction of ligand-binding sites occupied, θ, is plotted against the concentration of free ligand. Both curves are rectangular hyperbolas. **(a)** A hypothetical binding curve for a ligand L. The [L] at which half of the available ligand-binding sites are occupied is equivalent to $1/K_a$, or K_d. The curve has a horizontal asymptote at $\theta = 1$ and a vertical asymptote (not shown) at $[\mathrm{L}] = -1/K_a$. **(b)** A curve describing the binding of oxygen to myoglobin. The partial pressure of O_2 in the air above the solution is expressed in terms of kilopascals (kPa). Oxygen binds tightly to myoglobin with a P_{50} of only 0.26 kPa.

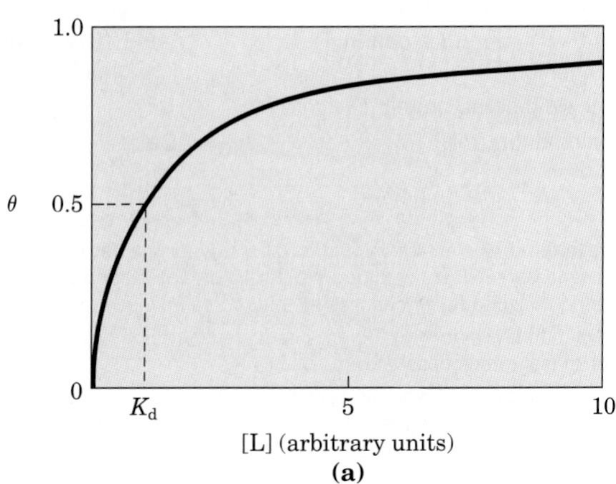

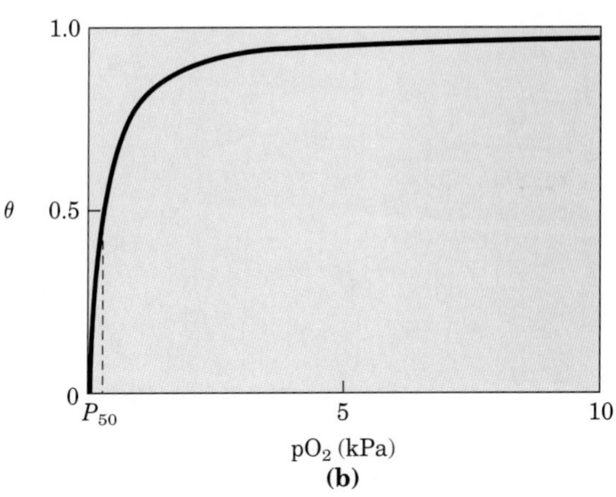

It is sometimes intuitively simpler to consider the **dissociation constant, K_d,** which is the reciprocal of K_a ($K_d = 1/K_a$) and is given in units of molar concentration (M). K_d is the equilibrium constant for the release of ligand. The relevant expressions change to

$$K_d = \frac{[P][L]}{[PL]} \tag{7–6}$$

$$[PL] = \frac{[P][L]}{K_d} \tag{7–7}$$

$$\theta = \frac{[L]}{[L] + K_d} \tag{7–8}$$

When [L] is equal to K_d, half of the ligand-binding sites are occupied. When [L] is lower than K_d, little ligand binds to the protein. In order for 90% of the available ligand-binding sites to be occupied, [L] must be nine times greater than K_d. In practice, K_d is used much more often than K_a to express the affinity of a protein for a ligand. Note that a lower value of K_d corresponds to a higher affinity of ligand for the protein. The mathematics can be reduced to simple statements: K_d is the molar concentration of ligand at which half of the available ligand-binding sites are occupied. At this point, the protein is said to have reached half saturation with respect to ligand binding. The more tightly a protein binds a ligand, the lower the concentration of ligand required for half the binding sites to be occupied, and thus the lower the value of K_d. Some representative dissociation constants are given in Table 7–1.

The binding of oxygen to myoglobin follows the patterns discussed above, but because oxygen is a gas, we must make some minor adjustments to the equations. We can simply substitute the concentration of dissolved oxygen for [L] in Equation 7–8 to give

$$\theta = \frac{[O_2]}{[O_2] + K_d} \tag{7–9}$$

As for any ligand, K_d is equal to the $[O_2]$ at which half of the available ligand-binding sites are occupied, or $[O_2]_{0.5}$. Equation 7–9 becomes

$$\theta = \frac{[O_2]}{[O_2] + [O_2]_{0.5}} \tag{7–10}$$

table 7–1

Some Protein Dissociation Constants

Protein	Ligand	K_d (M)*
Avidin (egg white)[†]	Biotin	1×10^{-15}
Insulin receptor (human)	Insulin	1×10^{-10}
Anti-HIV immunoglobulin (human)[‡]	gp41 (HIV-1 surface protein)	4×10^{-10}
Nickel-binding protein (*E. coli*)	Ni^{2+}	1×10^{-7}
Calmodulin (rat)[§]	Ca^{2+}	3×10^{-6} 2×10^{-5}

*A reported dissociation constant is valid only for the particular solution conditions under which it was measured. K_d values for a protein-ligand interaction can be altered, sometimes by several orders of magnitude, by changes in solution salt concentration, pH, or other variables.

[†]Interaction of avidin with the enzymatic cofactor biotin is among the strongest noncovalent biochemical interactions known.

[‡]This immunoglobulin was isolated as part of an effort to develop a vaccine against HIV. Immunoglobulins (described later in the chapter) are highly variable, and the K_d reported here should not be considered characteristic of all immunoglobulins.

[§]Calmodulin has four binding sites for calcium. The values shown reflect the highest- and lowest-affinity binding sites observed in one set of measurements.

The concentration of a volatile substance in solution, however, is always proportional to its partial pressure in the gas phase above the solution. In experiments using oxygen as a ligand, it is the partial pressure of oxygen, pO_2, that is varied because this is easier to measure than the concentration of dissolved oxygen. If we define the partial pressure of oxygen at $[O_2]_{0.5}$ as P_{50}, substitution in Equation 7–10 gives

$$\theta = \frac{pO_2}{pO_2 + P_{50}} \quad (7\text{–}11)$$

A binding curve for myoglobin that relates θ to pO_2 is shown in Figure 7–4b.

Protein Structure Affects How Ligands Bind

The binding of a ligand to a protein is rarely as simple as the above equations would suggest. The interaction is greatly affected by protein structure and is often accompanied by conformational changes. For example, the specificity with which heme binds its various ligands is altered when the heme is a component of myoglobin. CO binds to free heme molecules over 20,000 times better than does O_2 (the K_d or P_{50} for CO binding is more than 20,000 times lower than that for O_2) but binds only about 200 times better when the heme is bound in myoglobin. The difference is partly explained by steric hindrance. When O_2 binds to free heme, the axis of the oxygen molecule is positioned at an angle to the Fe—O bond (Fig. 7–5a). In contrast, when CO binds to free heme, the Fe, C, and O atoms lie in a straight line (Fig. 7–5b). In both cases, the binding reflects the geometry of hybrid orbitals in each ligand. In myoglobin, His^{64} (His E7), on the O_2-binding side of the heme, is too far away to coordinate with the heme iron, but it does interact with a ligand bound to heme. This residue, called the *distal His,* does not affect the binding of O_2 (Fig. 7–5c) but may preclude the linear binding of CO, providing one explanation for the diminished binding of CO to heme in myoglobin (and hemoglobin). This effect on CO binding is physiologically important, because CO is a low-level byproduct of cellular metabolism. Other factors, not yet well-defined, also seem to modulate the interaction of heme with CO in these proteins.

The binding of O_2 to the heme in myoglobin also depends on molecular motions, or "breathing," in the protein structure. The heme molecule is deeply buried in the folded polypeptide, with no direct path for oxygen to go from the surrounding solution to the ligand-binding site. If the protein were rigid, O_2 could not enter or leave the heme pocket at a measurable rate. However, rapid molecular flexing of the amino acid side chains produces transient cavities in the protein structure, and O_2 evidently makes its way in and out by moving through these cavities. Computer simulations of rapid structural fluctuations in myoglobin suggest that there are many such pathways. One major route is provided by rotation of the side chain of the distal His (His^{64}), which occurs on a nanosecond (10^{-9} s) time scale. Even subtle conformational changes can be critical for protein activity.

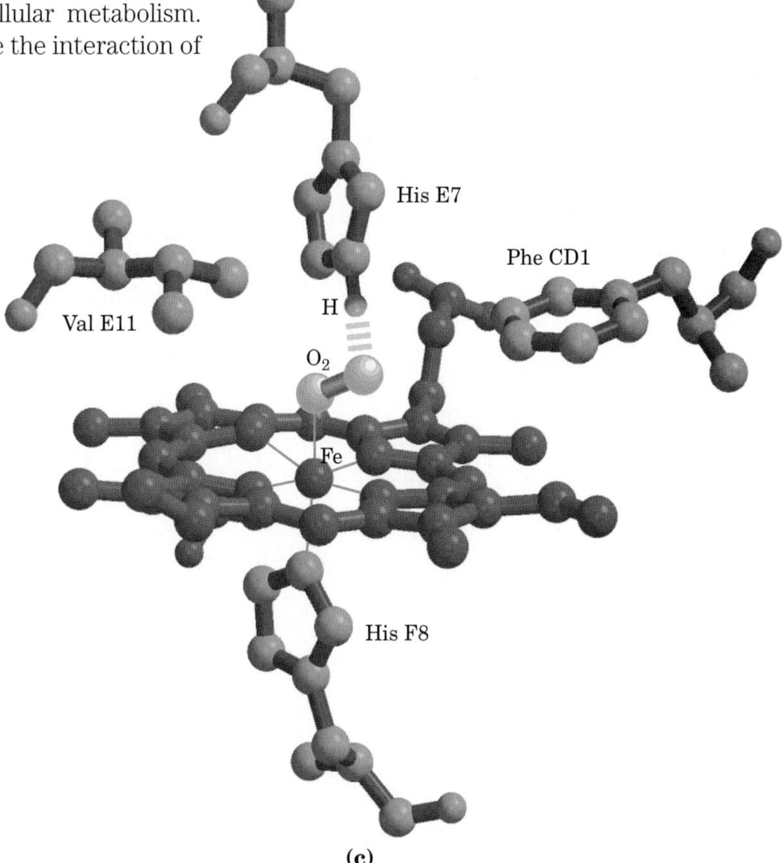

figure 7–5

Steric effects on the binding of ligands to the heme of myoglobin. (a) Oxygen binds to heme with the O_2 axis at an angle, a binding conformation readily accommodated by myoglobin. **(b)** Carbon monoxide binds to free heme with the CO axis perpendicular to the plane of the porphyrin ring. CO binding to the heme in myoglobin is forced to adopt a slight angle because the perpendicular arrangement is sterically blocked by His E7, the distal His. This effect weakens the binding of CO to myoglobin. **(c)** Another view showing the arrangement of key amino acid residues around the heme of myoglobin. The bound O_2 is hydrogen-bonded to the distal His, His E7 (His^{64}), further facilitating the binding of O_2.

Oxygen Is Transported in Blood by Hemoglobin

Nearly all the oxygen carried by whole blood in animals is bound and transported by hemoglobin in erythrocytes (red blood cells). Normal human erythrocytes are small (6 to 9 μm in diameter), biconcave disks. They are formed from precursor stem cells called **hemocytoblasts.** In the maturation process, the stem cell produces daughter cells that form large amounts of hemoglobin and then lose their intracellular organelles—nucleus, mitochondria, and endoplasmic reticulum. Erythrocytes are thus incomplete, vestigial cells, unable to reproduce and, in humans, destined to survive for only about 120 days. Their main function is to carry hemoglobin, which is dissolved in the cytosol at a very high concentration (~34% by weight).

In arterial blood passing from the lungs through the heart to the peripheral tissues, hemoglobin is about 96% saturated with oxygen. In the venous blood returning to the heart, hemoglobin is only about 64% saturated. Thus, each 100 mL of blood passing through a tissue releases about one-third of the oxygen it carries, or 6.5 mL of O_2 gas at atmospheric pressure and body temperature.

Myoglobin, with its hyperbolic binding curve for oxygen (Fig. 7–4b), is relatively insensitive to small changes in the concentration of dissolved oxygen and so functions well as an oxygen-storage protein. Hemoglobin, with its multiple subunits and O_2-binding sites, is better suited to oxygen transport. As we will see, interactions between the subunits of a multimeric protein can permit a highly sensitive response to small changes in ligand concentration. Interactions among the subunits in hemoglobin cause conformational changes that alter the affinity of the protein for oxygen. The modulation of oxygen binding allows the O_2-transport protein to respond to changes in oxygen demand by tissues.

Hemoglobin Subunits Are Structurally Similar to Myoglobin

Hemoglobin (M_r 64,500; abbreviated Hb) is roughly spherical, with a diameter of nearly 5.5 nm. It is a tetrameric protein containing four heme prosthetic groups, one associated with each polypeptide chain. Adult hemoglobin contains two types of globin, two α chains (141 residues each) and two β chains (146 residues each). Although fewer than half of the amino acid residues in the polypeptide sequences of the α and β subunits are identical, the three-dimensional structures of the two types of subunits are very similar. Furthermore, their structures are very similar to that of myoglobin (Fig. 7–6), even though the amino acid sequences of the three polypeptides are identical at only 27 positions (Fig. 7–7). All three polypeptides are

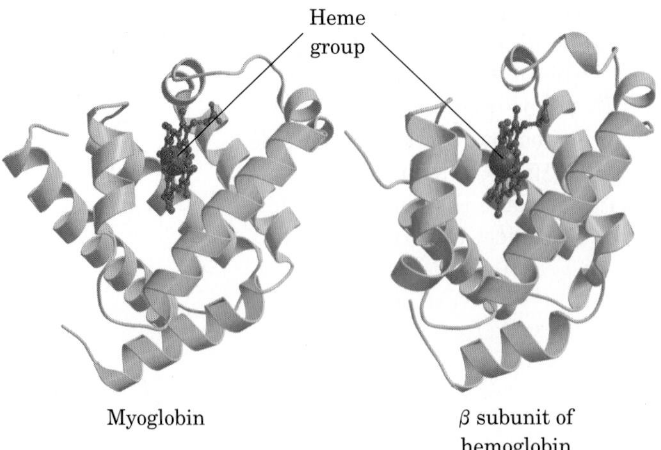

figure 7–6
A comparison of the structures of myoglobin and the β subunit of hemoglobin.

	Mb	Hbα	Hbβ
NA1---	1 V	1 V	1 V
	—	—	H
	L	L	L
A1 ---	S	S	T ---
	E	P	P
	G	A	E
	E	D	E
	W	K	K
	Q	T	S
	L	N	A
	V	V	V
	L	K	T
	H	A	A
	V	A	L
	W	W	W
	A	G	G
	K	K	K
	V	V	V
A16 ---	E	G	— ---
	A	A	—
B1 --	20 D	20 H	N ---
	V	A	20 V
	A	G	D
	G	E	E
	H	Y	V
	G	G	G
	Q	A	G
	D	E	E
	I	A	A
	L	L	L
	I	E	G
	R	R	R
	L	M	L
	F	F	L
	K	L	V
B16 ---	S	S	V ---
C1 ---	H	F	Y ---
	P	P	P
	E	T	W
	T	T	T
	40 L	40 K	Q
	E	T	40 R
C7 ---	K	Y	F ---
	F	F	F
	D	P	E
	R	H	S
	F	F	F
	K	—	G
	H	D	D
	L	L	L
	K	S	S
D1 ---	T	H	T ---

	Mb	Hbα	Hbβ
	E	—	P
	A	—	D
	E	—	A
	M	—	V
	K	—	M
D7 ---	A	G	G ---
E1 ---	S	S	N ---
	E	A	P
	60 D	Q	K
	L	V	60 V
	K	K	K
	K	G	A
Distal His **E7**	H	H	H
	G	G	G
	V	60 K	K
	T	K	K
	V	V	V
	L	A	L
	T	D	G
	A	A	A
	L	L	F
	G	T	S
	A	N	D
	I	A	G
E19 ---	L	V	L ---
	K	A	A
	K	H	H
	K	V	L
	80 G	D	D
	H	D	80 N
	H	M	L
	E	P	K
	A	N	G
	E	A	T
F1 ---	L	80 L	F ---
	K	S	A
	P	A	T
	L	L	L
	A	S	S
	Q	D	E
	S	L	L
Proximal His **F8**	H	H	H
F9 ---	A	A	C ---
	T	H	D
	K	K	K
	H	L	L
	K	R	H
	I	V	V
G1--	100 P	D	D ---
	I	P	100 P
	K	V	E
	Y	N	N

	Mb	Hbα	Hbβ	
	L	F	F	
	E	K	R	
	F	100 L	L	
	I	L	L	
	S	S	G	
	E	H	N	
	A	C	V	
	I	L	L	
	I	L	V	
	H	V	C	
	V	T	V	
	L	L	L	
	H	A	A	
	S	A	H	
G19 ---	R	H	H ---	
	H	L	F	
	120 P	P	G	
	G	A	120 K	
	D	E	E	
	F	F	F	
H1 ---	G	T	T ---	
	A	P	P	
	D	120 A	P	
	A	V	V	
	Q	H	Q	
	G	A	A	
	A	S	A	
	M	L	Y	
	N	D	Q	
	K	K	K	
	A	F	V	
	L	L	V	
	E	A	A	
	L	S	G	
	F	V	V	
	R	S	A	
	140 K	T	N	
	D	V	140 A	
	I	L	L	
	A	T	A	
H21 ---	A	S	H ---	
	K	K	K --- HC1	Hbα and Hbβ only
	Y	140 Y	Y --- HC2	
	K	141 R	146 H --- HC3	
	E			
H26 ---	L ------------			
	G			
	Y			
	Q			
	153 G			

figure 7–7

The amino acid sequences of whale myoglobin and the α and β chains of human hemoglobin. Dashed lines mark helix boundaries. To align the sequences optimally, short breaks must be incorporated into both Hb sequences where a few amino acids are present in the other sequences. With the exception of the missing D helix in Hbα, this alignment permits the use of the helix lettering convention that emphasizes the common positioning of amino acid residues that are identical in all three structures (shaded). Residues shaded in red are conserved in all known globins. Note that a common letter-and-number designation for amino acids in two or three different structures does not necessarily correspond to a common position in the linear sequence of amino acids in the polypeptides. For example, the distal His residue is His E7 in all three structures, but corresponds to His^{64}, His^{58}, and His^{63} in the linear sequences of Mb, Hbα, and Hbβ, respectively. Nonhelical residues at the amino and carboxyl termini, beyond the first (A) and last (H) α-helical segments, are labeled NA and HC, respectively.

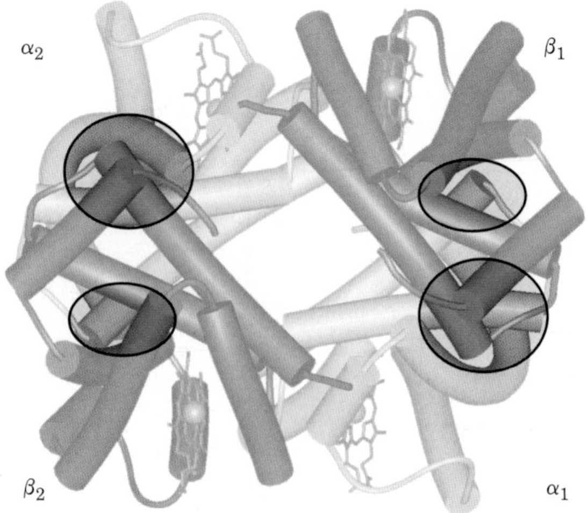

figure 7–8
Dominant interactions between hemoglobin subunits. In this representation, α subunits are light and β subunits are dark. The strongest subunit interactions, highlighted, occur between unlike subunits. When oxygen binds, the $\alpha_1\beta_1$ contact changes little, but there is a large change at the $\alpha_1\beta_2$ contact, with several ion pairs broken.

members of the globin family of proteins. The helix-naming convention described for myoglobin is also applied to the hemoglobin polypeptides, except that the α subunit lacks the short D helix. The heme-binding pocket is made up largely of the E and F helices.

The quaternary structure of hemoglobin features strong interactions between unlike subunits. The $\alpha_1\beta_1$ interface (and its $\alpha_2\beta_2$ counterpart) involves over 30 residues and is sufficiently strong that although mild treatment of hemoglobin with urea tends to cause the tetramer to disassemble into $\alpha\beta$ dimers, the dimers remain intact. The $\alpha_1\beta_2$ (and $\alpha_2\beta_1$) interface involves 19 residues (Fig. 7–8). Hydrophobic interactions predominate at the interfaces, but there are also many hydrogen bonds and a few ion pairs (sometimes referred to as salt bridges), whose importance is discussed below.

Hemoglobin Undergoes a Structural Change on Binding Oxygen

X-ray analysis has revealed two major conformations of hemoglobin: the **R state** and the **T state.** Although oxygen binds to hemoglobin in either state, it has a significantly higher affinity for hemoglobin in the R state. Oxygen binding stabilizes the R state. When oxygen is absent experimentally, the T state is more stable and is thus the predominant conformation of **deoxyhemoglobin.** T and R originally denoted "tense" and "relaxed," respectively, because the T state is stabilized by a greater number of ion pairs, many of which lie at the $\alpha_1\beta_2$ (and $\alpha_2\beta_1$) interface (Fig. 7–9). The binding

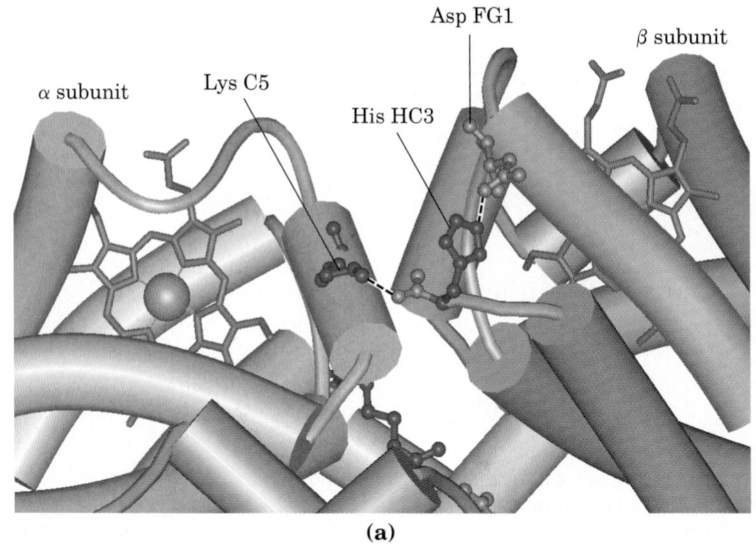

(a)

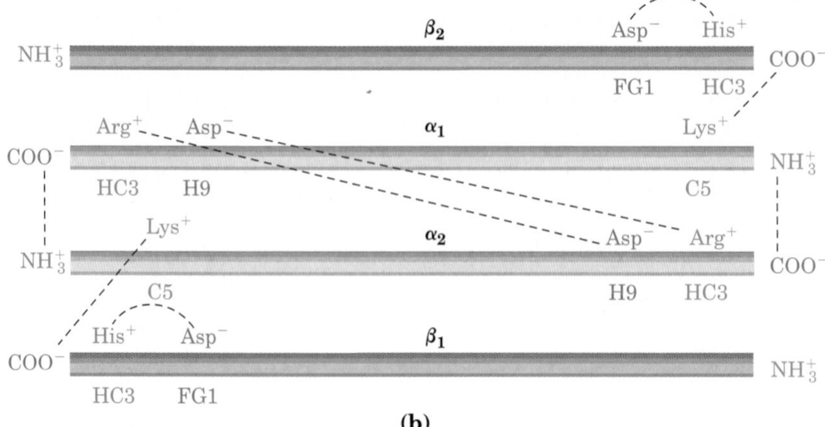

(b)

figure 7–9
Some ion pairs that stabilize the T state of deoxyhemoglobin. (a) A close-up view of a portion of a deoxyhemoglobin molecule in the T state. Interactions between the ion pairs His HC3 and Asp FG1 of the β subunit (blue) and between Lys C5 of the α subunit (gray) and the α-carboxyl group of His HC3 of the β subunit are shown with dashed lines. (Recall that HC3 is the carboxyl-terminal residue of the β subunit.) **(b)** The interactions between these ion pairs and others not shown in **(a)** are schematized in this representation of the extended polypeptide chains of hemoglobin.

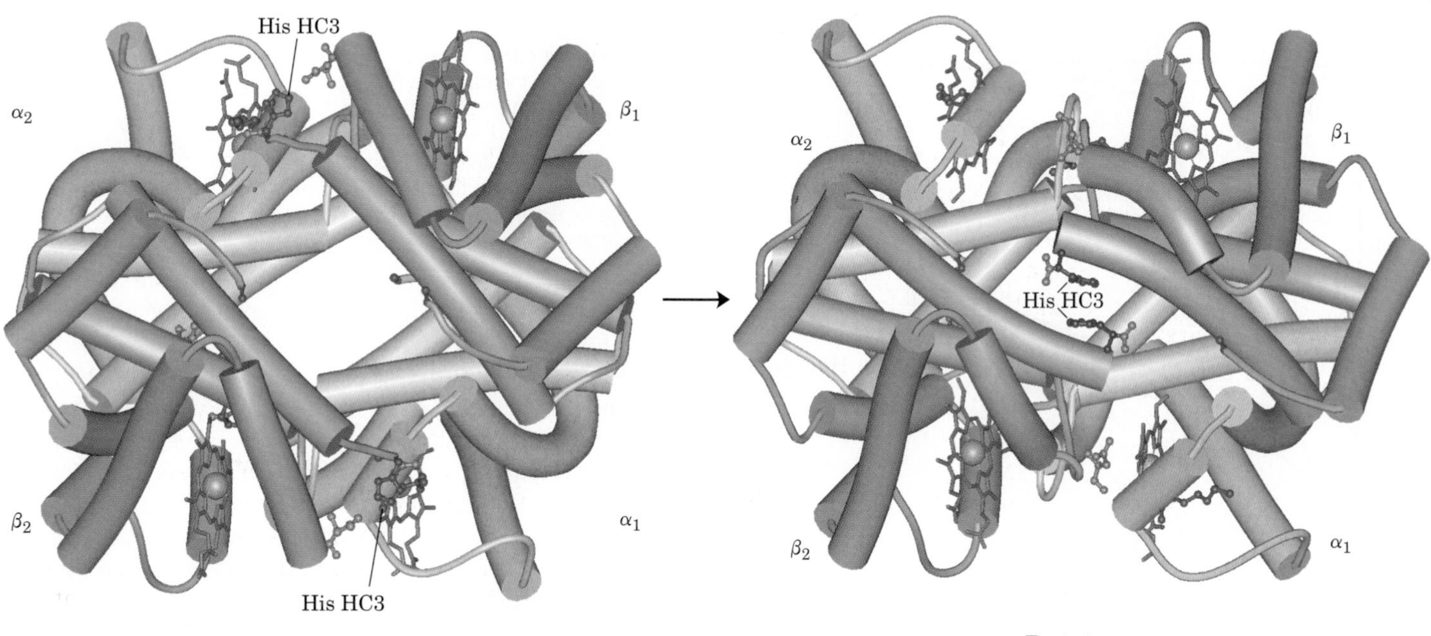

figure 7–10
The T ⟶ R transition. In these depictions of deoxyhemoglobin, as in Figure 7–9, the β subunits are light blue and the α subunits are gray. Positively charged side chains and chain termini involved in ion pairs are shown in blue, their negatively charged partners in pink. The Lys C5 of each α subunit and Asp FG1 of each β subunit are visible but not labeled (compare Fig. 7–9a). Note that the molecule is oriented slightly differently than in Figure 7–9. The transition from the T state to the R state shifts the subunit pairs substantially, affecting certain ion pairs. Most noticeably, the His HC3 residues at the carboxyl termini of the β subunits, which are involved in ion pairs in the T state, rotate in the R state toward the center of the molecule where they are no longer in ion pairs. Another dramatic result of the T ⟶ R transition is a narrowing of the pocket between the β subunits.

of O_2 to a hemoglobin subunit in the T state triggers a change in conformation to the R state. When the entire protein undergoes this transition, the structures of the individual subunits change little, but the $\alpha\beta$ subunit pairs slide past each other and rotate, narrowing the pocket between the β subunits (Fig. 7–10). In this process, some of the ion pairs that stabilize the T state are broken and some new ones are formed.

Max Perutz proposed that the T ⟶ R transition is triggered by changes in the positions of key amino acid side chains surrounding the heme. In the T state, the porphyrin is slightly puckered, causing the heme iron to protrude somewhat on the proximal His (His F8) side. The binding of O_2 causes the heme to assume a more planar conformation, shifting the position of the proximal His and the attached F helix (Fig. 7–11). Also, a Val residue in the E helix (Val E11) partially blocks the heme in the T state and must swing out of the way for oxygen to bind (Fig. 7–10). These changes lead to adjustments in the ion pairs at the $\alpha_1\beta_2$ interface.

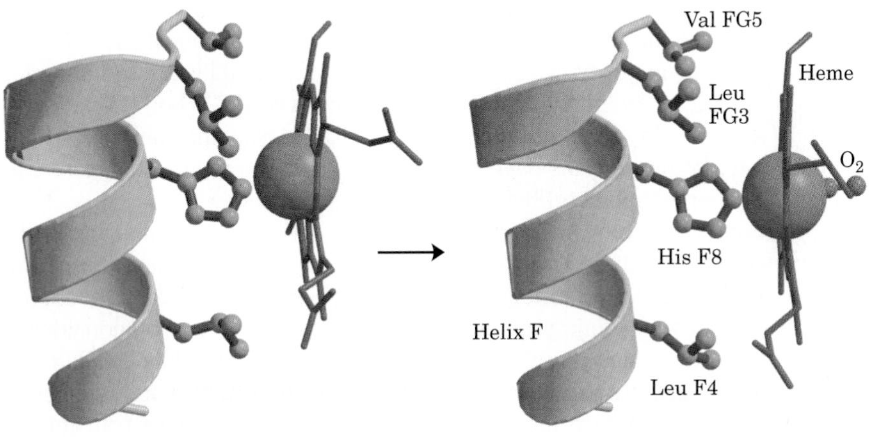

figure 7–11
Changes in conformation near heme on O_2 binding. The shift in the position of the F helix when heme binds O_2 is one of the adjustments that is believed to trigger the T ⟶ R transition.

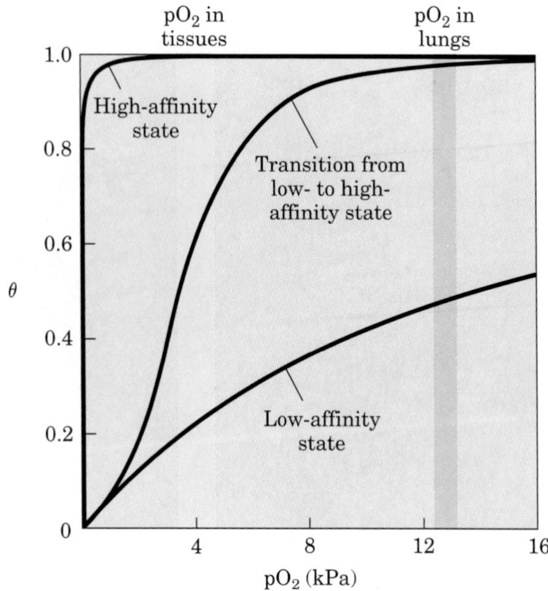

figure 7–12

A sigmoid (cooperative) binding curve. A sigmoid binding curve can be viewed as a hybrid curve reflecting a transition from a low-affinity to a high-affinity state. Cooperative binding, as manifested by a sigmoid binding curve, renders hemoglobin more sensitive to the small differences in O_2 concentration between the tissues and the lungs, allowing hemoglobin to bind oxygen in the lungs where pO_2 is high and release it in the tissues where pO_2 is low.

Hemoglobin Binds Oxygen Cooperatively

Hemoglobin must bind oxygen efficiently in the lungs, where the pO_2 is about 13.3 kPa, and release oxygen in the tissues, where the pO_2 is about 4 kPa. Myoglobin, or any protein that binds oxygen with a hyperbolic binding curve, would be ill-suited to this function, for the reason illustrated in Figure 7–12. A protein that bound O_2 with high affinity would bind it efficiently in the lungs but would not release much of it in the tissues. If the protein bound oxygen with a sufficiently low affinity to release it in the tissues, it would not pick up much oxygen in the lungs.

Hemoglobin solves the problem by undergoing a transition from a low-affinity state (the T state) to a high-affinity state (the R state) as more O_2 molecules are bound. As a result, hemoglobin has a hybrid S-shaped, or sigmoid, binding curve for oxygen (Fig. 7–12). A single-subunit protein with a single ligand-binding site cannot produce a sigmoid binding curve—even if binding elicits a conformational change—because each molecule of ligand binds independently and cannot affect the binding of another molecule. In contrast, O_2 binding to individual subunits of hemoglobin can alter the affinity for O_2 in adjacent subunits. The first molecule of O_2 that interacts with deoxyhemoglobin binds weakly, because it binds to a subunit in the T state. Its binding, however, leads to conformational changes that are communicated to adjacent subunits, making it easier for additional molecules of O_2 to bind. In effect, the T $\longrightarrow$ R transition occurs more readily in the second subunit once O_2 is bound to the first subunit. The last (fourth) O_2 molecule binds to a heme in a subunit that is already in the R state, and hence it binds with much higher affinity than the first molecule.

An **allosteric protein** is one in which the binding of a ligand to one site affects the binding properties of another site on the same protein. The term allosteric derives from the Greek *allos,* "other," and *stereos,* "solid" or "shape." Allosteric proteins are those having "other shapes" or conformations induced by the binding of ligands referred to as modulators. The conformational changes induced by the modulator(s) interconvert more-active and less-active forms of the protein. The modulators for allosteric proteins may be either inhibitors or activators. When the normal ligand and modulator are identical, the interaction is termed **homotropic.** When the modulator is a molecule other than the normal ligand the interaction is **heterotropic.** Some proteins have two or more modulators and therefore can have both homotropic and heterotropic interactions.

Cooperative binding of a ligand to a multimeric protein, such as we observe with the binding of O_2 to hemoglobin, is a form of allosteric binding often observed in multimeric proteins. The binding of one ligand affects the affinities of any remaining unfilled binding sites, and O_2 can be considered as both a normal ligand and an activating homotropic modulator. There is only one binding site for O_2 on each subunit, so the allosteric effects giving rise to cooperativity are mediated by conformational changes transmitted from one subunit to another by subunit-subunit interactions. A sigmoid bonding curve is diagnostic of cooperative binding. It permits a much more sensitive response to ligand concentration and is important to the function of many multisubunit proteins. The principle of allostery extends readily to regulatory enzymes, as we will see in Chapter 8.

Cooperative Ligand Binding Can Be Described Quantitatively

Cooperative binding of oxygen by hemoglobin was first analyzed by Archibald Hill in 1910. For a protein with n binding sites, the equilibrium of Equation 7–1 becomes

$$P + nL \rightleftharpoons PL_n \tag{7–12}$$

and the expression for the association constant becomes

$$K_a = \frac{[PL_n]}{[P][L]^n} \tag{7–13}$$

The expression for θ (see Eqn 7–8) is

$$\theta = \frac{[L]^n}{[L]^n + K_d} \tag{7–14}$$

Rearranging, then taking the log of both sides, yields

$$\frac{\theta}{1-\theta} = \frac{[L]^n}{K_d} \tag{7–15}$$

$$\log\left(\frac{\theta}{1-\theta}\right) = n \log [L] - \log K_d \tag{7–16}$$

Equation 7–16 is the **Hill equation,** and a plot of log $[\theta/(1-\theta)]$ versus log [L] is called a **Hill plot.** Based on the equation, the Hill plot should have a slope of n. However, the experimentally determined slope actually reflects not the number of binding sites, but the degree of interaction between them. The slope of a Hill plot is therefore denoted n_H, the **Hill coefficient,** which is a measure of the degree of cooperativity. If n_H equals 1, ligand binding is not cooperative, a situation that can arise even in a multisubunit protein if the subunits do not communicate. An n_H of greater than 1 indicates positive cooperativity in ligand binding. This is the situation observed in hemoglobin, in which the binding of one molecule of ligand facilitates the binding of others. The theoretical upper limit for n_H is reached when $n_H = n$. In this case the binding would be completely cooperative: all binding sites on the protein would bind ligand simultaneously, and no protein molecules partially saturated with ligand would be present under any conditions. This limit is never reached in practice, and the measured value of n_H is always less than the actual number of ligand-binding sites in the protein.

An n_H of less than 1 indicates negative cooperativity, in which the binding of one molecule of ligand *impedes* the binding of others. Well-documented cases of negative cooperativity are rare.

To adapt the Hill equation to the binding of oxygen to hemoglobin we must again substitute pO_2 for [L] and P_{50} for K_d:

$$\log\left(\frac{\theta}{1-\theta}\right) = n \log pO_2 - \log P_{50} \tag{7–17}$$

Hill plots for myoglobin and hemoglobin are given in Figure 7–13.

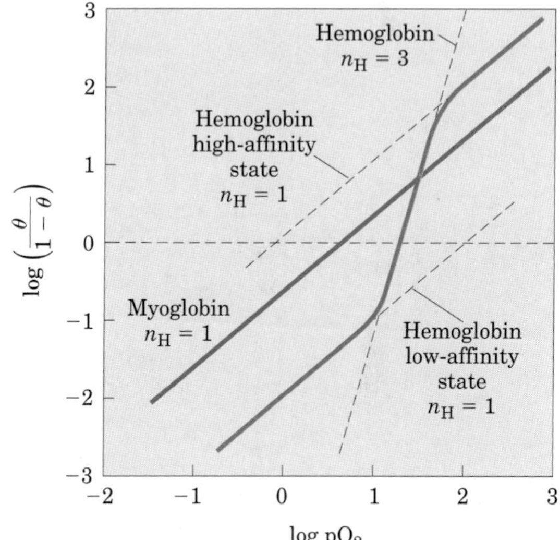

figure 7–13

Hill plots for the binding of oxygen to myoglobin and hemoglobin. When $n_H = 1$, there is no evident cooperativity. The maximum degree of cooperativity observed for hemoglobin corresponds approximately to $n_H = 3$. Note that while this indicates a high level of cooperativity, n_H is less than n, the number of O_2-binding sites in hemoglobin. This is normal for a protein that exhibits allosteric binding behavior.

Two Models Suggest Mechanisms for Cooperative Binding

Biochemists now know a great deal about the T and R states of hemoglobin, but much remains to be learned about how the T $\longrightarrow$ R transition occurs. Two models for the cooperative binding of ligands to proteins with multiple binding sites have greatly influenced thinking about this problem.

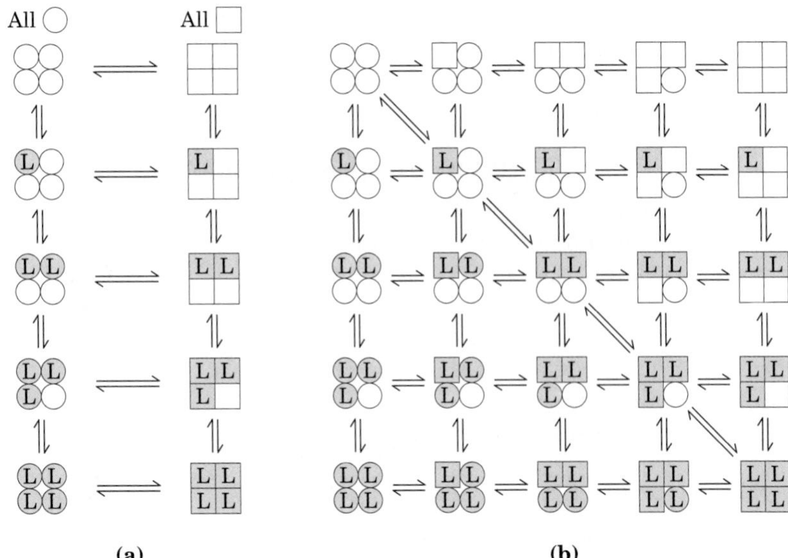

figure 7–14

Two general models for the interconversion of inactive and active forms of cooperative ligand-binding proteins. Although the models may be applied to any protein—including any enzyme (Chapter 8)—that exhibits cooperative binding, four subunits are shown because the model was originally proposed for hemoglobin. In the concerted, or all-or-none, model **(a)** all the subunits are postulated to be in the same conformation, either all ○ (low affinity or inactive) or all □ (high affinity or active). Depending on the equilibrium, K_1, between ○ and □ forms, the binding of one or more ligand molecules (L) will pull the equilibrium toward the □ form. Subunits with bound L are shaded. In the sequential model **(b)** each individual subunit can be in either the ○ or □ form. A very large number of conformations is thus possible.

The first model was proposed by Jacques Monod, Jeffries Wyman, and Jean-Pierre Changeux in 1965, and is called the **MWC model** or the **concerted model** (Fig. 7–14a). The concerted model assumes that the subunits of a cooperatively binding protein are functionally identical, that each subunit can exist in (at least) two conformations, and that all subunits undergo the transition from one conformation to the other simultaneously. In this model, no protein has individual subunits in different conformations. The two conformations are in equilibrium. The ligand can bind to either conformation, but binds each with different affinity. Successive binding of ligand molecules to the low-affinity conformation (which is more stable in the absence of ligand) makes a transition to the high-affinity conformation more likely.

In the second model, the **sequential model** (Fig. 7–14b), proposed in 1966 by Daniel Koshland and colleagues, ligand binding can induce a change of conformation in an individual subunit. A conformational change in one subunit makes a similar change in an adjacent subunit, as well as the binding of a second ligand molecule, more likely. There are more potential intermediate states in this model than in the concerted model. The two models are not mutually exclusive; the concerted model may be viewed as the "all-or-none" limiting case of the sequential model. In Chapter 8 we will use these models when we investigate allosteric enzymes.

Hemoglobin Also Transports H^+ and CO_2

In addition to carrying nearly all the oxygen required by cells from the lungs to the tissues, hemoglobin carries two end products of cellular respiration—H^+ and CO_2—from the tissues to the lungs and the kidneys, where they are excreted. The CO_2, produced by oxidation of organic fuels in mitochondria, is hydrated to form bicarbonate:

$$CO_2 + H_2O \rightleftharpoons H^+ + HCO_3^-$$

This reaction is catalyzed by **carbonic anhydrase,** an enzyme particularly abundant in erythrocytes. Carbon dioxide is not very soluble in aqueous solution, and bubbles of CO_2 would form in the tissues and blood if it were not converted to bicarbonate. As you can see from the equation, the hydration of CO_2 results in an increase in the H^+ concentration (a decrease in pH) in the tissues. The binding of oxygen by hemoglobin is profoundly influenced by pH and CO_2 concentration, so the interconversion of CO_2 and bicarbonate is of great importance to the regulation of oxygen binding and release in the blood.

Hemoglobin transports about 20% of the total H^+ and CO_2 formed in the tissues to the lungs and the kidneys. The binding of H^+ and CO_2 is inversely related to the binding of oxygen. At the relatively low pH and high CO_2 concentration of peripheral tissues, the affinity of hemoglobin for oxygen decreases as H^+ and CO_2 are bound, and O_2 is released to the tissues. Conversely, in the capillaries of the lung, as CO_2 is excreted and the blood pH consequently rises, the affinity of hemoglobin for oxygen increases and the protein binds more O_2 for transport to the peripheral tissues. This effect of pH and CO_2 concentration on the binding and release of oxygen by hemoglobin is called the **Bohr effect,** after Christian Bohr, the Danish physiologist (and father of physicist Niels Bohr) who discovered it in 1904.

The binding equilibrium for hemoglobin and one molecule of oxygen can be designated by the reaction

$$Hb + O_2 \rightleftharpoons HbO_2$$

but this is not a complete statement. To account for the effect of H^+ concentration on this binding equilibrium, we rewrite the reaction as

$$HHb^+ + O_2 \rightleftharpoons HbO_2 + H^+$$

where HHb^+ denotes a protonated form of hemoglobin. This equation tells us that the O_2-saturation curve of hemoglobin is influenced by the H^+ concentration (Fig. 7–15). Both O_2 and H^+ are bound by hemoglobin, but with inverse affinity. When the oxygen concentration is high, as in the lungs, hemoglobin binds O_2 and releases protons. When the oxygen concentration is low, as in the peripheral tissues, H^+ is bound and O_2 is released.

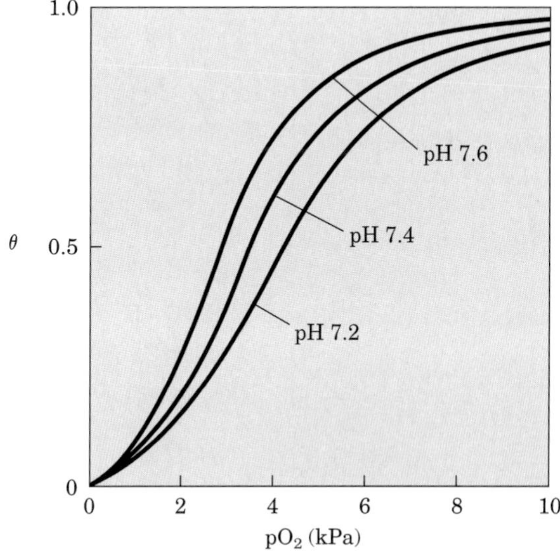

figure 7–15

Effect of pH on the binding of oxygen to hemoglobin. The pH of blood is 7.6 in the lungs and 7.2 in the tissues. Experimental measurements on hemoglobin binding are often performed at pH 7.4.

Oxygen and H^+ are not bound at the same sites in hemoglobin. Oxygen binds to the iron atoms of the hemes, whereas H^+ binds to any of several amino acid residues in the protein. A major contribution to the Bohr effect is made by His^{146} (His HC3) of the β subunits. When protonated, this residue forms one of the ion pairs—to Asp^{94} (Asp FG1)—that helps stabilize deoxyhemoglobin in the T state (Fig. 7–9). The ion pair stabilizes the protonated form of His HC3, giving this residue an abnormally high pK_a in the T state. The pK_a falls to its normal value of 6.0 in the R state because the ion pair cannot form, and this residue is largely unprotonated in oxyhemoglobin at pH 7.6, the blood pH in the lungs. As the concentration of H^+ rises, protonation of His HC3 promotes release of oxygen by favoring a transition to the T state. Protonation of the amino-terminal residues of the α subunits, certain other His residues, and perhaps other groups has a similar effect.

Thus we see that the four polypeptide chains of hemoglobin communicate with each other not only about O_2 binding to their heme groups, but also about H^+ binding to specific amino acid residues. And there is still more to the story. Hemoglobin also binds CO_2, again in a manner inversely related to the binding of oxygen. Carbon dioxide binds as a carbamate group to the α-amino group at the amino-terminal end of each globin chain, forming carbaminohemoglobin:

$$O{=}C{=}O + H_2N{-}\underset{R}{\overset{H}{C}}{-}\overset{O}{\overset{\|}{C}}{-} \xrightarrow{H^+} {}^{-}O{-}\overset{O}{\overset{\|}{C}}{-}\overset{H}{N}{-}\underset{R}{\overset{H}{C}}{-}\overset{O}{\overset{\|}{C}}{-}$$

Amino-terminal residue — Carbamino-terminal residue

This reaction produces H^+, contributing to the Bohr effect. The bound carbamates also form additional salt bridges (not shown in Fig. 7–9) that help to stabilize the T state and promote the release of oxygen.

When the concentration of carbon dioxide is high, as in peripheral tissues, some CO_2 binds to hemoglobin and the affinity for O_2 decreases, causing its release. Conversely, when hemoglobin reaches the lungs, the high oxygen concentration promotes binding of O_2 and release of CO_2. It is the capacity to communicate ligand-binding information from one polypeptide subunit to the others that makes the hemoglobin molecule so beautifully adapted to integrating the transport of O_2, CO_2, and H^+ by erythrocytes.

Oxygen Binding to Hemoglobin Is Regulated by 2,3-Bisphosphoglycerate

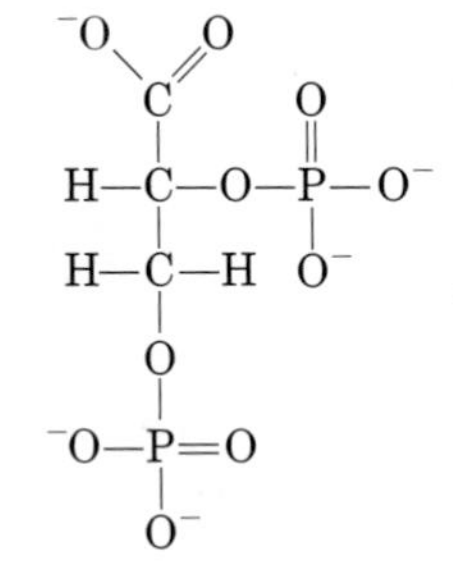

2,3-Bisphosphoglycerate

The interaction of **2,3-bisphosphoglycerate** (BPG) with hemoglobin provides an example of heterotropic allosteric modulation. BPG is present in relatively high concentrations in erythrocytes. When hemoglobin is isolated, it contains substantial amounts of bound BPG, which can be difficult to remove completely. In fact, the O_2-binding curves for hemoglobin that we have examined to this point were obtained in the presence of bound BPG. 2,3-Bisphosphoglycerate is known to greatly reduce the affinity of hemoglobin for oxygen—there is an inverse relationship between the binding of O_2 and the binding of BPG. We can therefore describe another binding process for hemoglobin:

$$\text{HbBPG} + O_2 \rightleftharpoons \text{HbO}_2 + \text{BPG}$$

BPG binds at a site distant from the oxygen-binding site and regulates the O_2-binding affinity of hemoglobin in relation to the pO_2 in the lungs. BPG plays an important role in the physiological adaptation to the lower pO_2 available at high altitudes. For a healthy human strolling by the ocean, the binding of O_2 to hemoglobin is regulated such that the amount of O_2 delivered to the tissues is equivalent to nearly 40% of the maximum that could be carried by the blood (Fig. 7–16). If the same person is quickly transported to a mountainside at an altitude of 4,500 meters, where the pO_2 is considerably lower, the delivery of O_2 to the tissues is reduced. However,

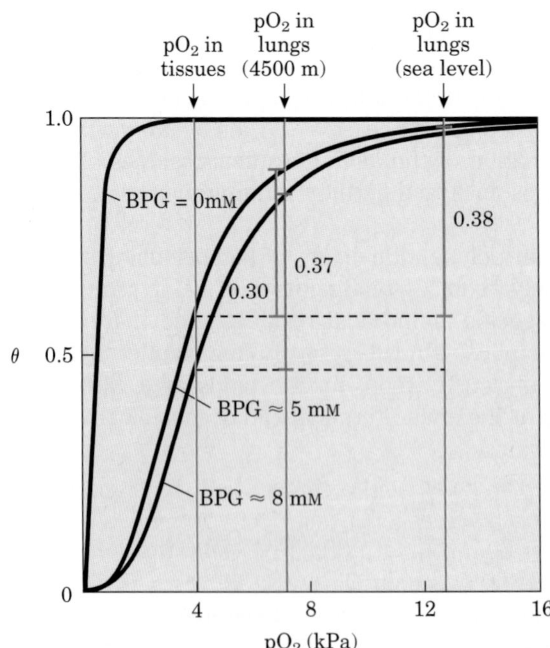

figure 7–16

Effect of BPG on the binding of oxygen to hemoglobin. The BPG concentration in normal human blood is about 5 mM at sea level and about 8 mM at high altitudes. Note that hemoglobin binds to oxygen quite tightly when BPG is entirely absent, and the binding curve appears to be hyperbolic. In reality, the measured Hill coefficient for O_2-binding cooperativity decreases only slightly (from 3 to about 2.5) when BPG is removed from hemoglobin, but the rising part of the sigmoid curve is confined to a very small region close to the origin. At sea level, hemoglobin is nearly saturated with O_2 in the lungs, but only 60% saturated in the tissues, so that the amount of oxygen released in the tissues is close to 40% of the maximum that can be carried in the blood. At high altitudes, O_2 delivery declines by about one-fourth, to 30% of maximum. An increase in BPG concentration, however, decreases the affinity of hemoglobin for O_2 so that nearly 40% of what can be carried is again delivered to the tissues.

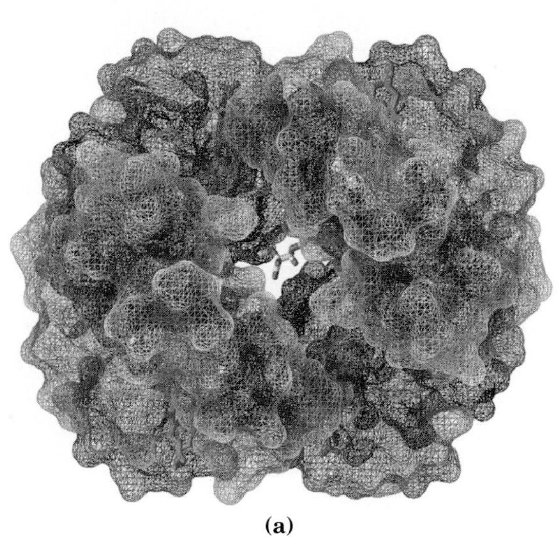
(a)

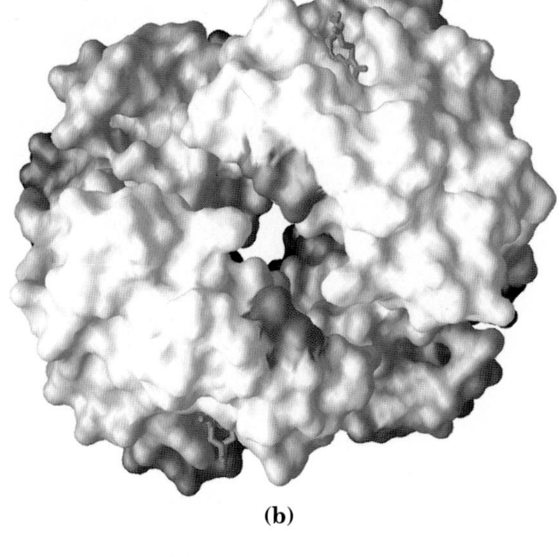
(b)

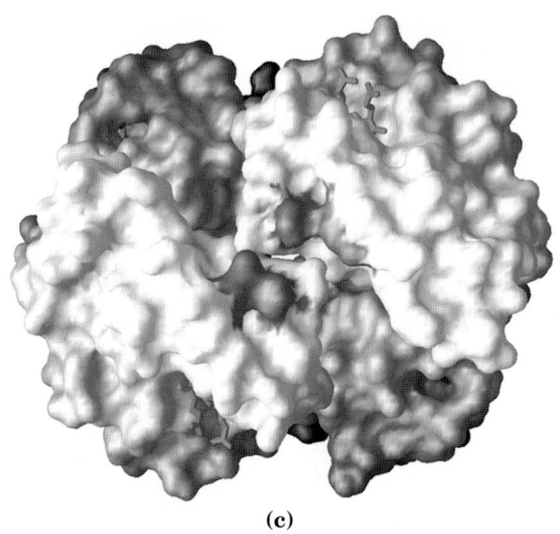
(c)

figure 7–17
Binding of BPG to deoxyhemoglobin. (a) BPG binding stabilizes the T state of deoxyhemoglobin, shown here as a mesh surface image. **(b)** The negative charges of BPG interact with several positively charged groups (shown in blue in this GRASP surface image) that surround the pocket between the β subunits in the T state. **(c)** The binding pocket for BPG disappears on oxygenation, following transition to the R state. (Compare **(b)** and **(c)** with Fig. 7–10.)

after just a few hours at the higher altitude, the BPG concentration in the blood has begun to rise, leading to a decrease in the affinity of hemoglobin for oxygen. This adjustment in the BPG level has only a small effect on the binding of O_2 in the lungs but a considerable effect on the release of O_2 in the tissues. As a result, the delivery of oxygen to the tissues is restored to nearly 40% of that which can be transported by the blood. The situation is reversed when the person returns to sea level. The BPG concentration in erythrocytes also increases in people suffering from **hypoxia,** lowered oxygenation of peripheral tissues due to inadequate function of the lungs or circulatory system.

BPG binds to hemoglobin in the cavity between the β subunits in the T state (Fig. 7–17). This cavity is lined with positively charged amino acid residues that interact with the negatively charged groups of BPG. Unlike O_2, only one molecule of BPG is bound to each hemoglobin tetramer. BPG lowers hemoglobin's affinity for oxygen by stabilizing the T state. The transition to the R state narrows the binding pocket for BPG, precluding BPG binding. In the absence of BPG, hemoglobin is converted to the R state more easily.

Regulation of oxygen binding to hemoglobin by BPG has an important role in fetal development. Because a fetus must extract oxygen from its mother's blood, fetal hemoglobin must have greater affinity than the maternal hemoglobin for O_2. In fetuses, γ subunits are synthesized rather than β subunits, and $\alpha_2\gamma_2$ hemoglobin is formed. This tetramer has a much lower affinity for BPG than normal adult hemoglobin, and a correspondingly higher affinity for O_2.

Sickle-Cell Anemia Is a Molecular Disease of Hemoglobin

The great importance of the amino acid sequence in determining the secondary, tertiary, and quaternary structures of globular proteins, and thus their biological functions, is strikingly demonstrated by the hereditary human disease sickle-cell anemia. More than 300 genetic variants of hemoglobin are known to occur in the human population. Most of these variations consist of differences in a single amino acid residue. The effects on hemoglobin structure and function are often minor but can sometimes be extraordinary. Each hemoglobin variation is the product of an altered gene. The variant genes are called alleles. Because humans generally have two copies of each gene, an individual may have two copies of one allele (thus being homozygous for that gene) or one copy of each of two different alleles

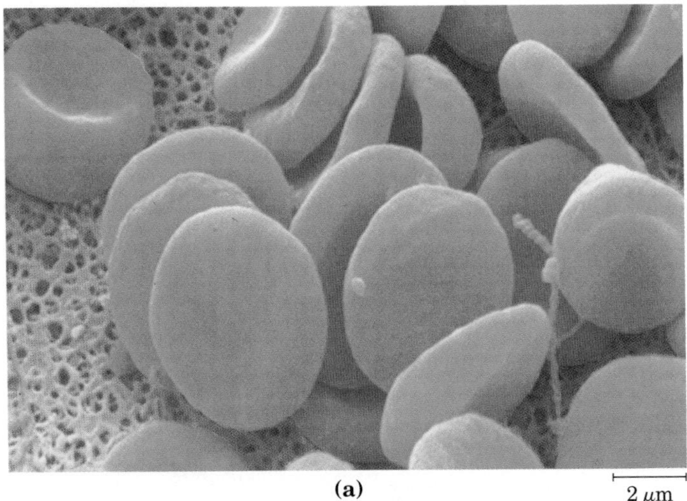

(a)

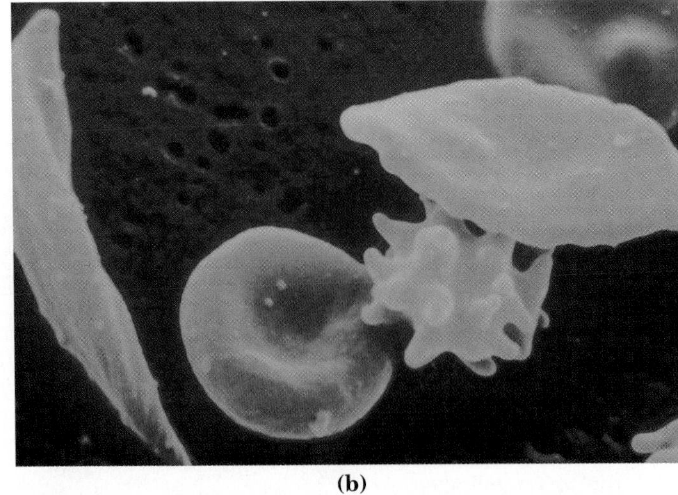

(b)

figure 7–18
A comparison of uniform, cup-shaped, normal erythrocytes **(a)** with the variably shaped erythrocytes seen in sickle-cell anemia **(b).** These cells range from normal to spiny or sickle-shaped.

(heterozygous). Sickle-cell anemia is a genetic disease in which an individual has inherited the allele for sickle-cell hemoglobin from both parents. The erythrocytes of these individuals are fewer and also abnormal. In addition to an unusually large number of immature cells, the blood contains many long, thin, crescent-shaped erythrocytes that look like the blade of a sickle (Fig. 7–18). When hemoglobin from sickle cells (called hemoglobin S) is deoxygenated, it becomes insoluble and forms polymers that aggregate into tubular fibers (Fig. 7–19). Normal hemoglobin (hemoglobin A) remains sol-

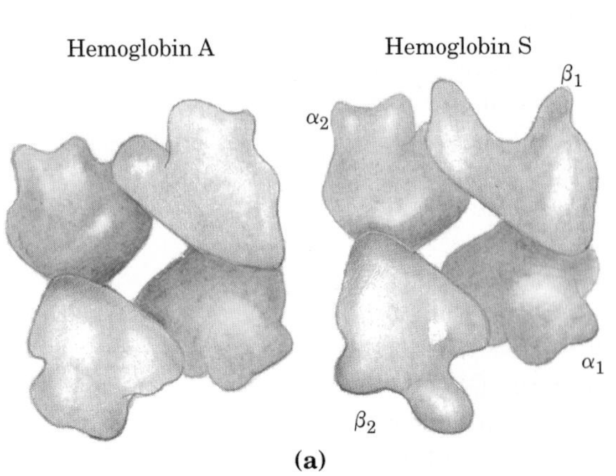

(a)

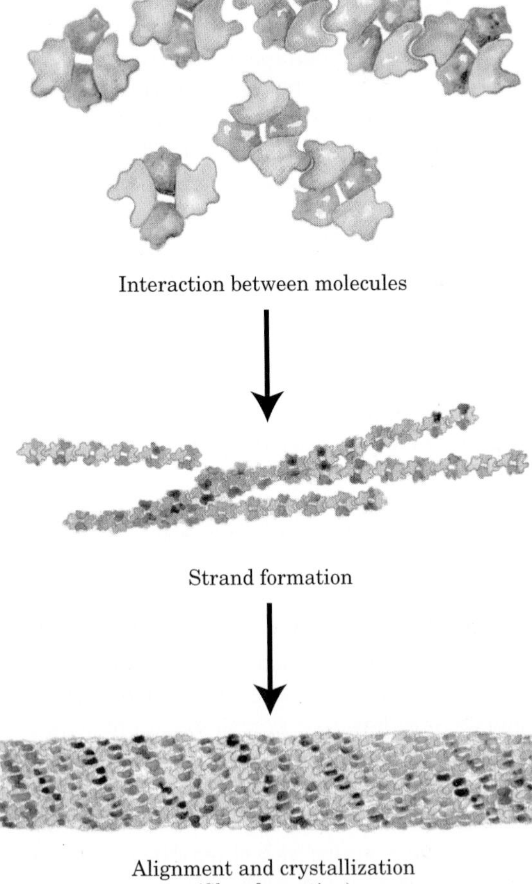

(b)

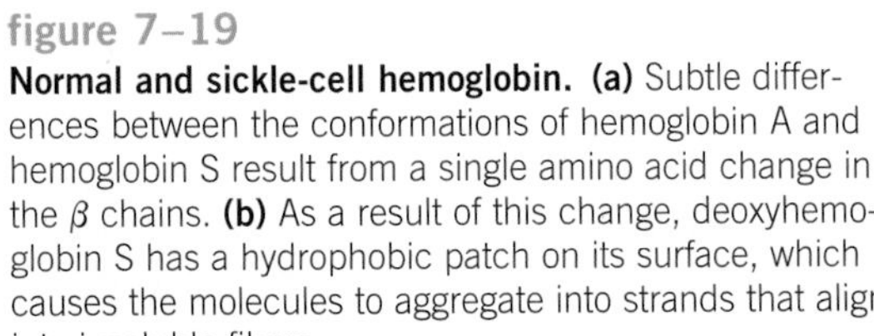

figure 7–19
Normal and sickle-cell hemoglobin. (a) Subtle differences between the conformations of hemoglobin A and hemoglobin S result from a single amino acid change in the β chains. **(b)** As a result of this change, deoxyhemoglobin S has a hydrophobic patch on its surface, which causes the molecules to aggregate into strands that align into insoluble fibers.

uble on deoxygenation. The insoluble fibers of deoxygenated hemoglobin S are responsible for the deformed sickle shape of the erythrocytes, and the proportion of sickled cells increases greatly as blood is deoxygenated.

The altered properties of hemoglobin S result from a single amino acid substitution, a Val instead of a Glu residue at position 6 in the two β chains. The R group of valine has no electric charge, whereas glutamate has a negative charge at pH 7.4. Hemoglobin S therefore has two fewer negative charges than hemoglobin A, one for each of the two β chains. Replacement of the Glu residue by Val creates a "sticky" hydrophobic contact point at position 6 of the β chain, which is on the outer surface of the molecule. These sticky spots cause deoxyhemoglobin S molecules to associate abnormally with each other, forming the long, fibrous aggregates characteristic of this disorder.

Sickle-cell anemia occurs in individuals homozygous for the sickle-cell allele of the gene encoding the β subunit of hemoglobin. Individuals who receive the sickle-cell allele from only one parent and are thus heterozygous experience a milder condition called sickle-cell trait; only about 1% of their erythrocytes become sickled on deoxygenation. These individuals may live completely normal lives if they avoid vigorous exercise or other stresses on the circulatory system.

People with sickle-cell anemia suffer from repeated crises brought on by physical exertion. They become weak, dizzy, and short of breath, and they also experience heart murmurs and an increased pulse rate. The hemoglobin content of their blood is only about half the normal value of 15 to 16 g/100 mL because sickled cells are very fragile and rupture easily; this results in anemia ("lack of blood"). An even more serious consequence is that capillaries become blocked by the long, abnormally shaped cells, causing severe pain and interfering with normal organ function—a major factor in the early death of many people with the disease.

Without medical treatment, people with sickle-cell anemia usually die in childhood. Nevertheless, the sickle-cell allele is surprisingly common in certain parts of Africa. Investigation into the persistence of an allele that is so obviously deleterious in homozygous individuals led to the finding that the allele confers a small but significant resistance to lethal forms of malaria in heterozygous individuals. Natural selection has resulted in an allele population that balances the deleterious effects of the homozygous condition against the resistance to malaria afforded by the heterozygous condition.

Complementary Interactions between Proteins and Ligands: The Immune System and Immunoglobulins

Our discussion of oxygen-binding proteins showed how the conformations of these proteins affect and are affected by the binding of small ligands (O_2 or CO) to the heme group. However, most protein-ligand interactions do not involve a prosthetic group. Instead, the binding site for a ligand is more often like the hemoglobin binding site for BPG—a cleft in the protein lined with amino acid residues, arranged to render the binding interaction highly specific. Effective discrimination between ligands is the norm at binding sites, even when the ligands have only minor structural differences.

All vertebrates have an immune system capable of distinguishing molecular "self" from "nonself" and then destroying those entities identified as nonself. In this way, the immune system eliminates viruses, bacteria, and other pathogens and molecules that may pose a threat to the organism. On a physiological level, the response of the immune system to an invader is an intricate and coordinated set of interactions among many classes of proteins, molecules, and cell types. However, at the level of individual proteins, the immune response demonstrates how an acutely sensitive and specific biochemical system is built upon the reversible binding of ligands to proteins.

The Immune Response Features a Specialized Array of Cells and Proteins

Immunity is brought about by a variety of **leukocytes** (white blood cells), including **macrophages** and **lymphocytes,** all arising from undifferentiated stem cells in the bone marrow. Leukocytes can leave the bloodstream and patrol the tissues, each cell producing one or more proteins capable of recognizing and binding to molecules that might signal an infection.

The immune response consists of two complementary systems, the humoral and cellular immune systems. The **humoral immune system** (Latin *humor,* "fluid") is directed at bacterial infections and extracellular viruses (those found in the body fluids), but can also respond to individual proteins introduced into the organism. The **cellular immune system** destroys host cells infected by viruses and also destroys some parasites and foreign tissues.

The proteins at the heart of the humoral immune response are soluble proteins called **antibodies** or **immunoglobulins,** often abbreviated Ig. Immunoglobulins bind bacteria, viruses, or large molecules identified as foreign and target them for destruction. Making up 20% of blood protein, the immunoglobulins are produced by **B lymphocytes** or **B cells,** so named because they complete their development in the *b*one marrow.

The agents at the heart of the cellular immune response are a class of **T lymphocytes** or **T cells** (so called because the latter stages of their development occur in the *t*hymus) known as **cytotoxic T cells (T_C cells,** also called killer T cells). Recognition of infected cells or parasites involves proteins called **T-cell receptors** on the surface of T_C cells. Recall from Chapter 2 (p. 30) that receptors are proteins, usually found on the outer surface of cells and extending through the plasma membrane; they recognize and bind extracellular ligands, triggering changes inside the cell.

In addition to cytotoxic T cells, there are **helper T cells (T_H cells),** whose function it is to produce soluble signaling proteins called cytokines, which include the interleukins. T_H cells interact with macrophages. Table 7–2 summarizes the functions of the various leukocytes of the immune system.

table 7–2

Some Types of Leukocytes Associated with the Immune System

Cell type	Function
Macrophages	Ingest large particles and cells by phagocytosis
B lymphocytes (B cells)	Produce and secrete antibodies
T lymphocytes (T cells)	
Cytotoxic (killer) T cells (T_C)	Interact with infected host cells through receptors on T-cell surface
Helper T cells (T_H)	Interact with macrophages and secrete cytokines (interleukins) that stimulate T_C, T_H, and B cells to proliferate.

Each recognition protein of the immune system, either an antibody produced by a B cell or a receptor on the surface of a T cell, specifically binds some particular chemical structure, distinguishing it from virtually all others. Humans are capable of producing over 10^8 different antibodies with distinct binding specificities. This extraordinary diversity makes it likely that any chemical structure on the surface of a virus or invading cell will be recognized and bound by one or more antibodies. Antibody diversity is derived from random reassembly of a set of immunoglobulin gene segments via genetic recombination mechanisms that are discussed in Chapter 25.

Some properties of the interactions between antibodies or T-cell receptors and the molecules they bind are unique to the immune system, and a specialized lexicon is used to describe them. Any molecule or pathogen capable of eliciting an immune response is called an **antigen.** An antigen may be a virus, a bacterial cell wall, or an individual protein or other macromolecule. A complex antigen may be bound by a number of different antibodies. An individual antibody or T-cell receptor binds only a particular molecular structure within the antigen, called its **antigenic determinant** or **epitope.**

It would be unproductive for the immune system to respond to small molecules that are common intermediates and products of cellular metabolism. Molecules of M_r <5,000 are generally not antigenic. However, small molecules can be covalently attached to large proteins in the laboratory, and in this form they may elicit an immune response. These small molecules are called **haptens.** The antibodies produced in response to protein-linked haptens will then bind to the same small molecules when they are free. Such antibodies are sometimes used in the development of analytical tests described later in this chapter or as catalytic antibodies (described in Box 8–3).

The interactions of antibody and antigen are much better understood than are the binding properties of T-cell receptors. However, before focusing on antibodies, we need to look at the humoral and cellular immune systems in more detail to put the fundamental biochemical interactions into their proper context.

Self Is Distinguished from Nonself by the Display of Peptides on Cell Surfaces

The immune system must identify and destroy pathogens, but it must also recognize and *not* destroy the normal proteins and cells of the host organism—the "self." Detection of protein antigens in the host is mediated by **MHC (major histocompatibility complex) proteins.** MHC proteins bind peptide fragments of proteins digested in the cell and present them on the outside surface of the cell. These peptides normally come from the digestion of typical cellular proteins, but during a viral infection viral proteins are also digested and presented by MHC proteins. Peptide fragments from foreign proteins that are displayed by MHC proteins are the antigens the immune system recognizes as nonself. T-cell receptors bind these fragments and launch the subsequent steps of the immune response. There are two classes of MHC proteins (Fig. 7–20), which differ in their distribution among cell types and in the source of digested proteins whose peptides they display.

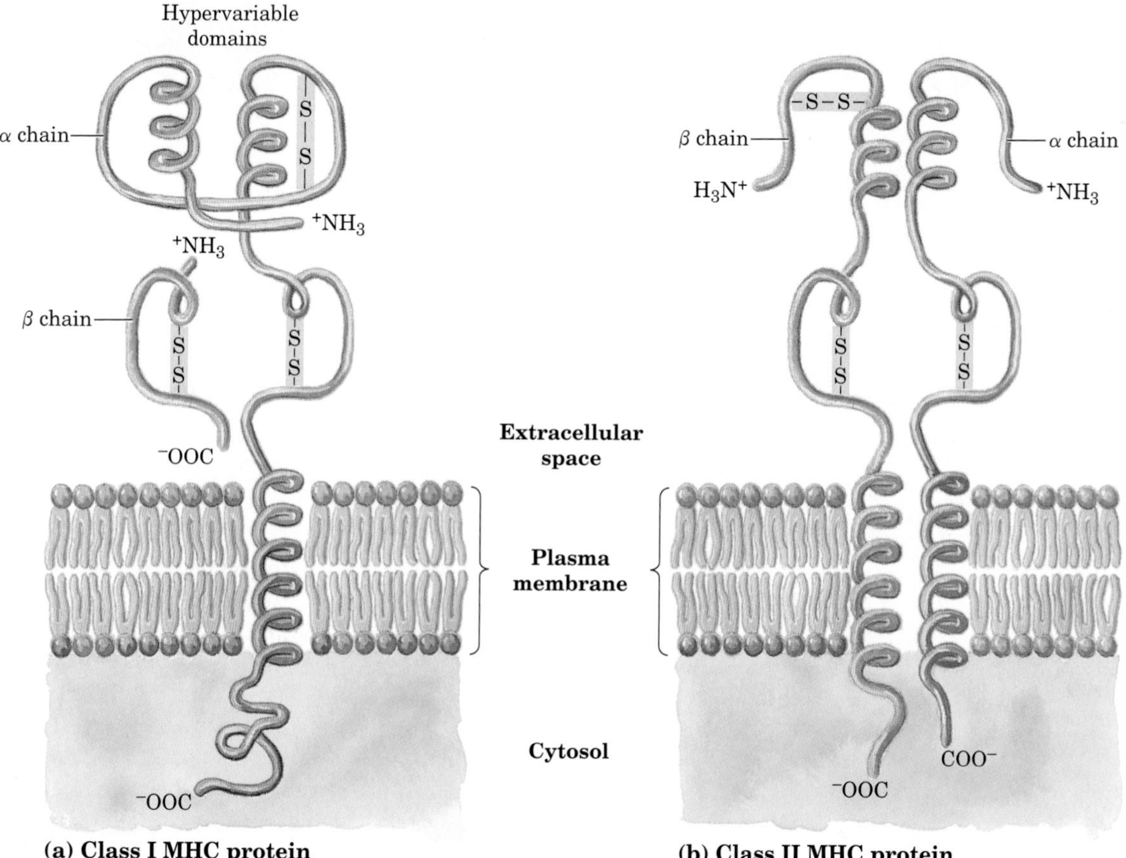

figure 7–20
MHC proteins These proteins consist of α and β chains. In class I MHC proteins **(a),** the small β chain is invariant but the amino acid sequence of the α chain exhibits a high degree of variability, localized in specific domains of the protein that appear on the outside of the cell. Each human produces up to six different α chains for class 1 MHC proteins. In class II MHC proteins **(b),** both the α and β chains have regions of relatively high variability near their amino-terminal ends.

Class I MHC proteins (Fig. 7–21) are found on the surface of virtually all vertebrate cells. There are countless variants in the human population, placing them among the most polymorphic of proteins. Because individuals produce up to six class I MHC protein variants, any two individuals are unlikely to have the same set. Class I MHC proteins bind and display peptides derived from the proteolytic degradation and turnover of proteins that occurs randomly within the cell. These complexes of peptides and class I MHC proteins are the recognition targets of the T-cell receptors of the T_C cells in the cellular immune system. The general pattern of immune system recognition was first described by Rolf Zinkernagel and Peter Doherty in 1974.

Each T_C cell has many copies of only one T-cell receptor that is specific for a particular class I MHC protein–peptide complex. To avoid creating a

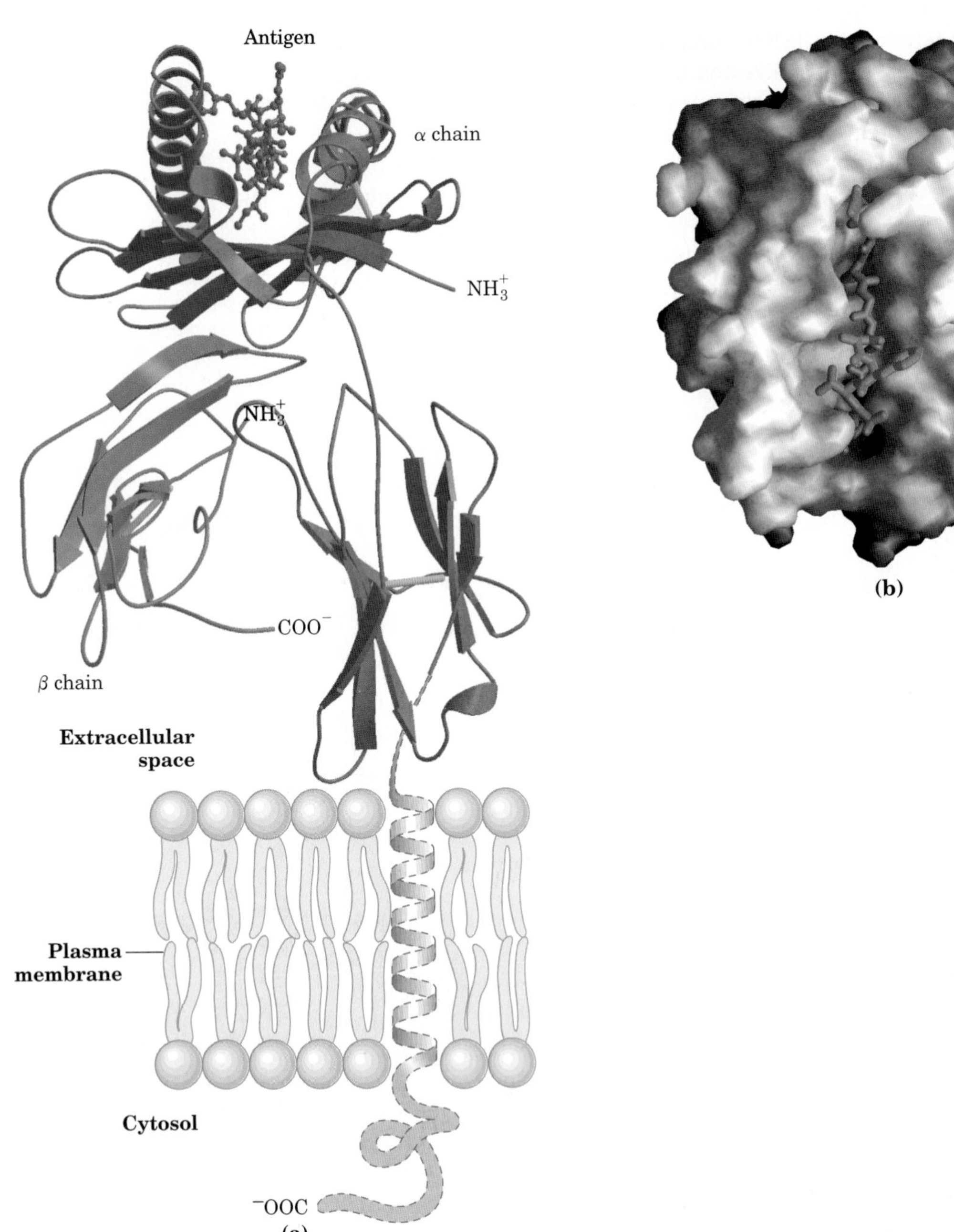

figure 7–21
Structure of a human class I MHC protein. (a) This image is derived in part from the determined structure of the extracellular portion of the protein. The α chain of MHC is shown in gray; the small β chain is blue; the disulfide bonds are yellow. A bound ligand, a peptide derived from HIV, is shown in red. **(b)** Top view showing a surface contour image of the site where peptides are bound and displayed. The HIV peptide (red) occupies the site. This part of the class I MHC protein interacts with T-cell receptors.

legion of T_C cells that would set upon and destroy normal cells, the maturation of T_C cells in the thymus includes a stringent selection process that eliminates more than 95% of the developing T_C cells, including those that might recognize and bind class I MHC proteins displaying peptides from cellular proteins of the organism itself. The T_C cells that survive and mature are those with T-cell receptors that do not bind to the organism's own proteins. The result is a population of cells that bind foreign peptides bound to class I MHC proteins of the host cell. These binding interactions lead to the destruction of parasites and virus-infected cells. When an organ is transplanted, its foreign class I MHC proteins are also bound by T_C cells, leading to tissue rejection.

Class II MHC proteins occur on the surfaces of a few types of specialized cells that take up foreign antigens, including macrophages and B lymphocytes. Like class I MHC proteins, the class II proteins are highly polymorphic, with many variants in the human population. Each human is capable of producing up to 12 variants, and thus it is unlikely that any two individuals have an identical set of variants. The class II MHC proteins bind and display peptides derived not from cellular proteins but from external proteins ingested by the cells. The resulting class II MHC protein–peptide complexes are the binding targets of the T-cell receptors of the various helper T cells. T_H cells, like T_C cells, undergo a stringent selection process in the thymus, eliminating those that recognize the individual's own cellular proteins.

Despite the elimination of most T_C and T_H cells during the selection process in the thymus, a very large number survive, and these provide the immune response. Each survivor has a single type of T-cell receptor that can bind to one particular chemical structure. The T cells patrolling the bloodstream and the tissues carry millions of different binding specificities in the T-cell receptors. Within the highly varied T-cell population there is almost always a contingent of cells that can specifically bind any antigen that might appear. The vast majority of these cells never encounter a foreign antigen to which they can bind and typically die within a few days, replaced by new generations of T cells endlessly patrolling in search of the interaction that will launch the full immune response.

Molecular Interactions at Cell Surfaces Trigger the Immune Response

A new antigen is often the harbinger of an infection—a signal to the immune system that a virus or other parasite may be rapidly growing in the organism. Those few T cells and B cells possessing receptors or antibodies that can bind the antigen must be rapidly and selectively propagated to eliminate the infection. A hypothetical viral infection illustrates how this occurs.

When a virus invades a cell, it makes use of cellular functions and resources to replicate its nucleic acid and make viral proteins. Once inside the cell, viral macromolecules are relatively inaccessible to the antibodies of the humoral immune system. However, some of the class I MHC proteins that find their way to the surface of an infected cell will generally display peptide fragments from viral proteins, which can then be recognized by T_C lymphocytes. Mature viruses become vulnerable to the humoral immune system when they are released from the infected cell and are present for a time in the extracellular environment. Some are then ingested by macrophages (which ingest only those antigens that are recognized by the antibodies produced by a particular B cell). Viral peptide fragments will be displayed on the surfaces of the macrophages and B cells, complexed to class II MHC proteins, and the peptide antigens will trigger a multi-pronged response involving B cells, T_C cells, and T_H cells (Fig. 7–22).

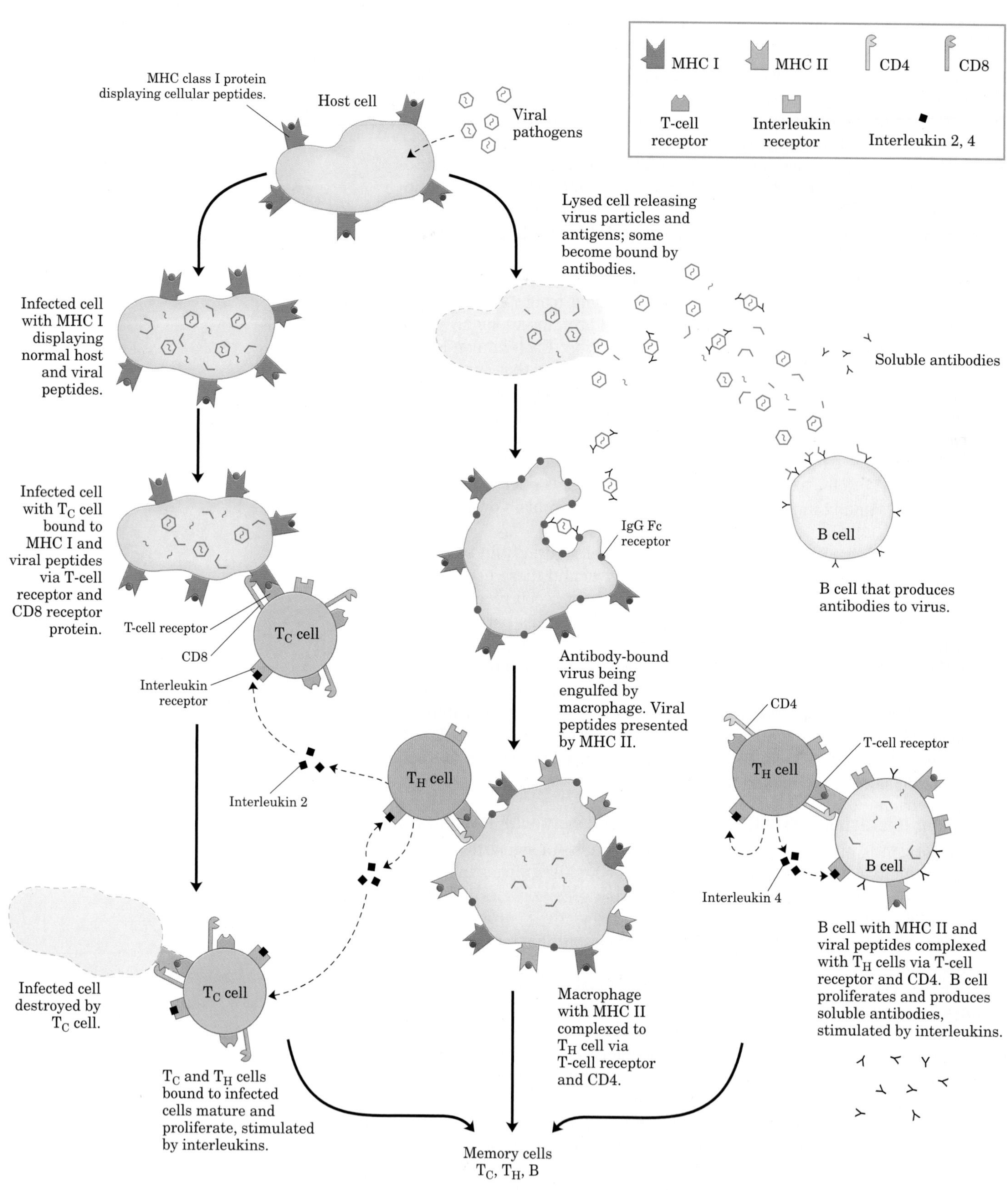
MHC I
MHC II
CD4
CD8
T-cell receptor
Interleukin receptor
Interleukin 2, 4
MHC class I protein displaying cellular peptides.
Host cell
Viral pathogens
Lysed cell releasing virus particles and antigens; some become bound by antibodies.
Soluble antibodies
Infected cell with MHC I displaying normal host and viral peptides.
Infected cell with T_C cell bound to MHC I and viral peptides via T-cell receptor and CD8 receptor protein.
T-cell receptor
CD8
Interleukin receptor
T_C cell
IgG Fc receptor
B cell
B cell that produces antibodies to virus.
Antibody-bound virus being engulfed by macrophage. Viral peptides presented by MHC II.
CD4
T-cell receptor
T_H cell
Interleukin 2
T_H cell
B cell
Interleukin 4
B cell with MHC II and viral peptides complexed with T_H cells via T-cell receptor and CD4. B cell proliferates and produces soluble antibodies, stimulated by interleukins.
Infected cell destroyed by T_C cell.
T_C cell
Macrophage with MHC II complexed to T_H cell via T-cell receptor and CD4.
T_C and T_H cells bound to infected cells mature and proliferate, stimulated by interleukins.
Memory cells
T_C, T_H, B

figure 7–22

Overview of the immune response to a viral infection. The individual steps are described in the text.

The class I MHC protein–peptide complexes on infected cells are recognized as foreign and bound by those T_C cells with T-cell receptors having the appropriate binding specificity. The T-cell receptors respond only to peptide antigens that are complexed to class I MHC proteins. The T_C cells have an additional receptor, **CD8,** also called a coreceptor, that enhances the binding interactions of T-cell receptors and MHC proteins (Fig. 7–22, middle left). The T_C cells live up to the name killer T cells by destroying the virally infected cell to which they are complexed through their T-cell receptors. Cell death is brought about by a number of mechanisms, not all well understood. One mechanism involves the release of a protein called **perforin,** which binds to and aggregates in the plasma membrane of the target cell, forming molecular pores that destroy the capacity of that cell to regulate its interior environment. T_C cells also induce a process called **programmed cell death,** or **apoptosis** (most commonly pronounced app′-a-toe′-sis), in which the cells complexed to T_C cells undergo metabolic changes that rapidly lead to the demise of the cell.

T_C cells with the proper specificity must proliferate selectively if large numbers of virus-infected cells are to be destroyed. To this end, T_C cells complexed to an infected cell generate cell-surface receptors for signaling proteins called **interleukins.** Interleukins, secreted by a variety of cells, stimulate the proliferation of only those T and B cells bearing the required interleukin receptors. Because T and B cells produce interleukin receptors only when they are complexed with an antigen, the only immune system cells that proliferate are those few that can respond to the antigen. The process of producing a population of cells by stimulated reproduction of a particular ancestor cell is called **clonal selection.**

The peptides complexed to class II MHC proteins and displayed on the surface of macrophages and B lymphocytes are similarly bound by the appropriate T-cell receptors of T_H cells. The T_H cells also have a coreceptor, called **CD4,** that enhances the binding interactions of the T-cell receptors. This overall binding interaction, in concert with secondary molecular signals that are currently being identified, activates the T_H cells. A subpopulation of activated T_H cells secrete a small signal protein called interleukin-2 (IL-2; M_r 15,000), which stimulates proliferation of nearby T_C cells and T_H cells having the appropriate interleukin receptors. This greatly increases the number of available immune system cells capable of recognizing and responding to the antigen. Another subpopulation of activated T_H cells complexed to macrophages or B lymphocytes secrete interleukin-4 (IL-4; M_r 20,000), which stimulates the proliferation of B cells that recognize the antigen (Fig. 7–22, bottom right). Proliferation of the responding B, T_C, and T_H cells continues as long as the appropriate antigen is present.

The proliferating B cells promote the destruction of any extracellular viruses or bacterial cells. They first secrete large amounts of soluble antibody that binds to the antigen. This bound antibody recruits a cellular system of about 20 proteins collectively called **complement** because they complement and enhance the action of the antibodies. The complement proteins disrupt the coats of many viruses or, in bacterial infections, produce holes in the cell walls of bacteria, causing them to swell and burst by osmotic shock.

Unlike T cells, B cells do not undergo selection in the thymus to eliminate those producing antibodies that recognize host (self) proteins. However, B cells do not contribute significantly to an immune response unless they are stimulated to proliferate by T_H cells. The T_H cells *do* undergo selection in the thymus, leaving no T_H cells capable of stimulating B cells that produce antibodies potentially dangerous to the host.

The T_H cells themselves participate only indirectly in the destruction of infected cells and pathogens, but their role is critical to the entire immune response. This is dramatically illustrated by the epidemic produced by HIV (human immunodeficiency virus), the virus that causes AIDS (acquired immune deficiency syndrome). The primary targets of HIV infection are T_H cells. Elimination of these cells progressively incapacitates the entire immune system.

Once antigen is depleted, activated immune cells generally die in a matter of days by programmed cell death. However, a few of the stimulated B and T cells mature into **memory cells.** These are long-lived cells that do not participate directly in the primary immune response when the antigen is first encountered. Instead they become permanent residents of the blood, ready to respond to a reappearance of the same antigen. Memory cells, when subsequently challenged by the antigen, can mount a secondary immune response that is generally much more rapid and vigorous than the primary response because of prior clonal expansion. By this mechanism, vertebrates once exposed to a virus or other pathogen can respond quickly to the pathogen when exposed again. This is the basis of the long-term immunity conferred by vaccines and the natural immunity to repeated infections by the same strain of a virus.

Antibodies Have Two Identical Antigen-Binding Sites

Immunoglobulin G (IgG) is the major class of antibody molecule and one of the most abundant proteins in the blood serum. IgG has four polypeptide chains: two large ones, called heavy chains, and two light chains, linked by noncovalent and disulfide bonds into a complex of M_r 150,000. The heavy chains of an IgG molecule interact at one end, then branch to interact separately with the light chains, forming a Y-shaped molecule (Fig. 7–23). At the "hinges" separating the base of an IgG molecule from its branches, the immunoglobulin can be cleaved with proteases. Cleavage with the protease papain liberates the basal fragment, called **Fc** because it usually *c*rystallizes readily, and the two branches, which are called **Fab,** the *a*ntigen-*b*inding fragments. Each branch has a single antigen-binding site.

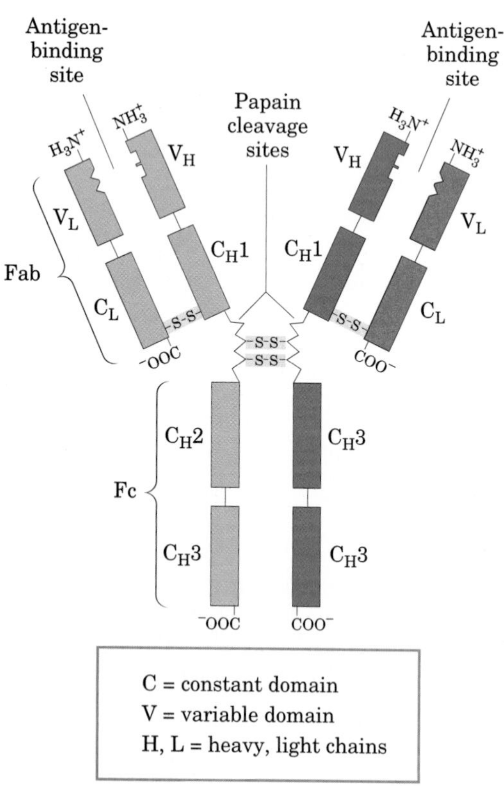

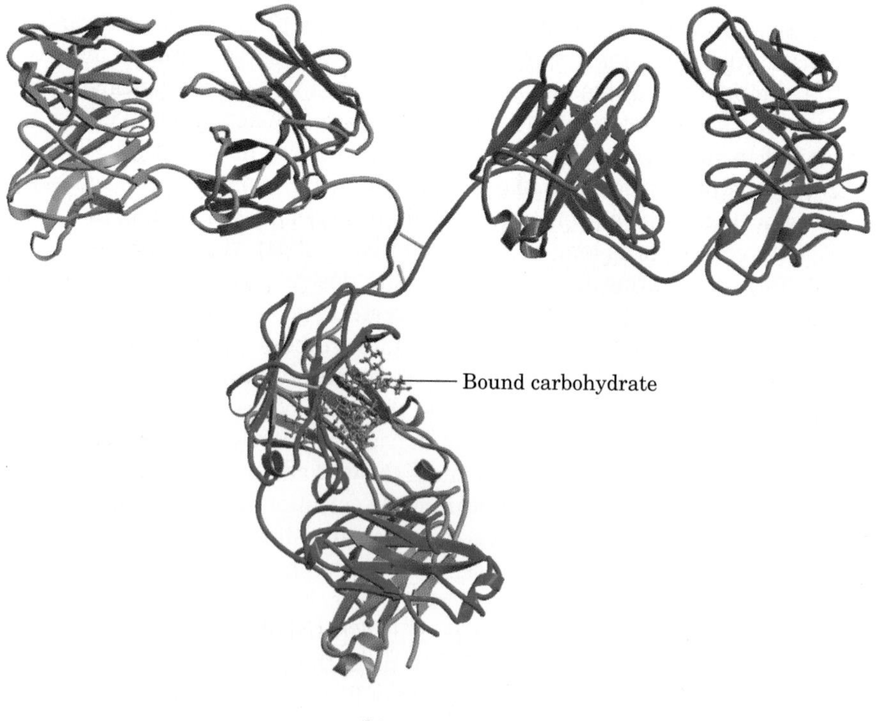

figure 7–23
The structure of immunoglobulin G. (a) Pairs of heavy and light chains combine to form a Y-shaped molecule. Two antigen-binding sites are formed by the combination of variable domains from one light (V_L) and one heavy (V_H) chain. Cleavage with papain separates the Fab and Fc portions of the protein in the hinge region. The Fc portion of the molecule also contains bound carbohydrate. **(b)** A ribbon model of the first complete IgG molecule to be crystallized and structurally analyzed. Although the molecule contains two identical heavy chains (two shades of blue) and two identical light chains (two shades of red), it crystallized in the asymmetric conformation shown. Conformational flexibility may be important to the function of immunoglobulins.

The fundamental structure of immunoglobulins was first established by Gerald Edelman and Rodney Porter. Each chain is made up of identifiable domains; some are constant in sequence and structure from one IgG to the next, others are variable. The constant domains have a characteristic structure known as the **immunoglobulin fold,** a well-conserved structural motif in the all-β class. There are three of these constant domains in each heavy chain and one in each light chain. The heavy and light chains also have one variable domain each, in which most of the variability in amino acid residue sequence is found. The variable domains associate to create the antigen-binding site (Fig. 7–24).

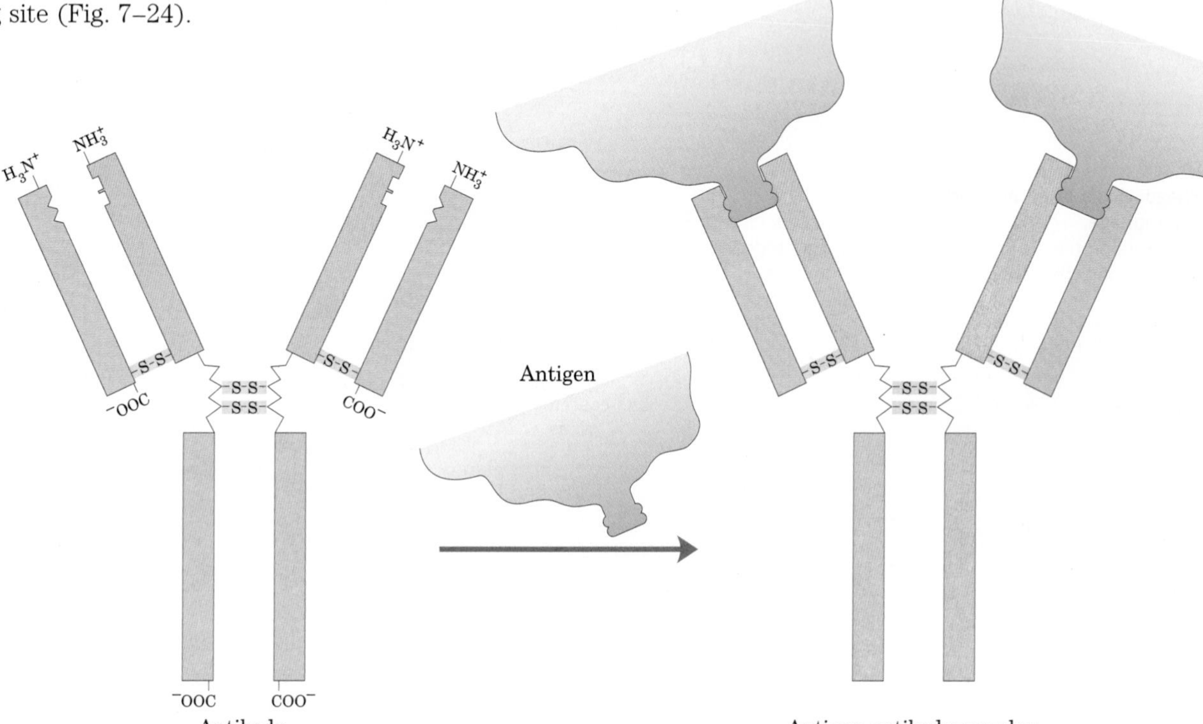

figure 7–24
Binding of IgG to an antigen. To generate an optimal fit for the antigen, the binding sites of IgG often undergo slight conformational changes. Such induced fit is common to many protein-ligand interactions.

In many vertebrates, IgG is only one of five classes of immunoglobulins. Each class has a characteristic type of heavy chain, denoted α, δ, ϵ, γ, and μ for IgA, IgD, IgE, IgG, and IgM, respectively. Two types of light chain, κ and λ, occur in all classes of immunoglobulins. The overall structures of **IgD** and **IgE** are similar to that of IgG. **IgM** occurs in either a monomeric, membrane-bound form or a secreted form that is a cross-linked pentamer of this basic structure (Fig. 7–25). **IgA,** found principally in secretions such as saliva, tears, and milk, can be a monomer, dimer, or trimer. IgM is the first antibody to be made by B lymphocytes and is the major antibody in the early stages of a primary immune response. Some B cells soon begin to produce IgD (with the same antigen-binding site as the IgM produced by the same cell), but the unique function of IgD is less clear.

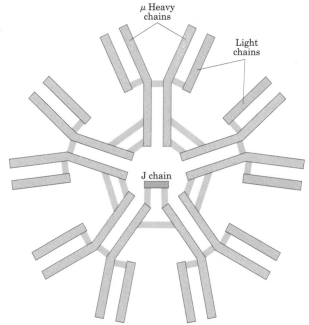

figure 7–25
IgM pentamer of immunoglobulin units. The pentamer is cross-linked with disulfide bonds. The J chain is a polypeptide of M_r 20,000 found in both IgA and IgM.

The IgG described above is the major antibody in secondary immune responses, which are initiated by memory B cells. As part of the organism's ongoing immunity to antigens already encountered and dealt with, IgG is the most abundant immunoglobulin in the blood. When IgG binds to an invading bacterium or virus, it not only activates the complement system, but also activates certain leukocytes such as macrophages to engulf and destroy the invader. Yet another class of receptors on the cell surface of macrophages recognizes and binds the Fc region of IgG. When these Fc receptors bind an antibody-pathogen complex, the macrophage engulfs the complex by phagocytosis (Fig. 7–26).

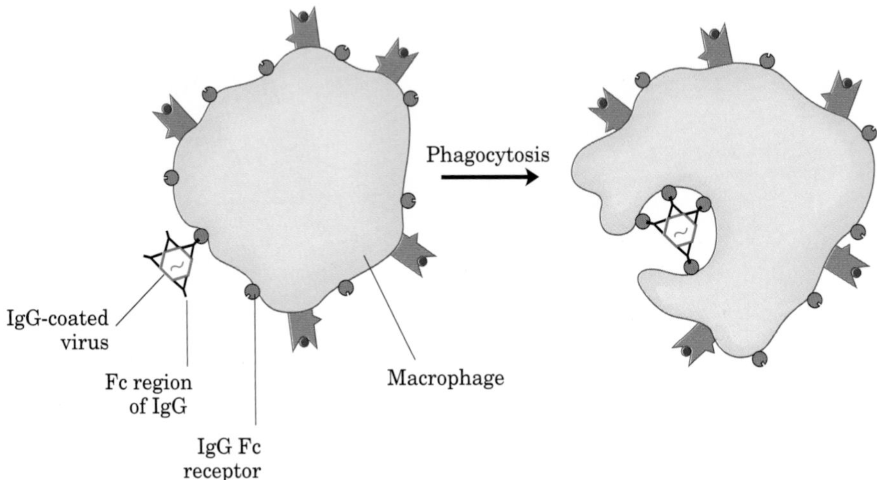

figure 7–26
Phagocytosis of an antibody-bound virus by a macrophage. The Fc regions of the antibodies bind to Fc receptors on the surface of the macrophage, triggering the macrophage to engulf and destroy the virus.

IgE plays an important role in the allergic response, interacting with basophils (phagocytic leukocytes) in the blood and histamine-secreting cells called mast cells that are widely distributed in tissues. This immunoglobulin binds, through its Fc region, to special Fc receptors on the basophils or mast cells. In this form, IgE serves as a kind of receptor for antigen. If antigen is bound, the cells are induced to secrete histamine and other biologically active amines that cause dilation and increased permeability of blood vessels. These effects on the blood vessels are thought to facilitate the movement of immune system cells and proteins to sites of inflammation. They also produce the symptoms normally associated with allergies. Pollen or other allergens are recognized as foreign, triggering an immune response normally reserved for pathogens.

Antibodies Bind Tightly and Specifically to Antigen

The binding specificity of an antibody is determined by the amino acid residues in the variable domains of its heavy and light chains. Many residues in these domains are variable, but not equally so. Some, particularly those lining the antigen-binding site, are hypervariable—especially likely to differ. Specificity is conferred by chemical complementarity between the antigen and its specific binding site, in terms of shape and the location of charged, nonpolar, and hydrogen-bonding groups. For example, a binding site with a negatively charged group may bind an antigen with a positive charge in the complementary position. In many instances, complementarity is achieved interactively as the structures of antigen and binding site are influenced by each other during the approach of the ligand. Conformational changes in the antibody and/or the antigen then occur that allow the complementary groups to interact fully. This is an example of induced fit (Fig. 7–27).

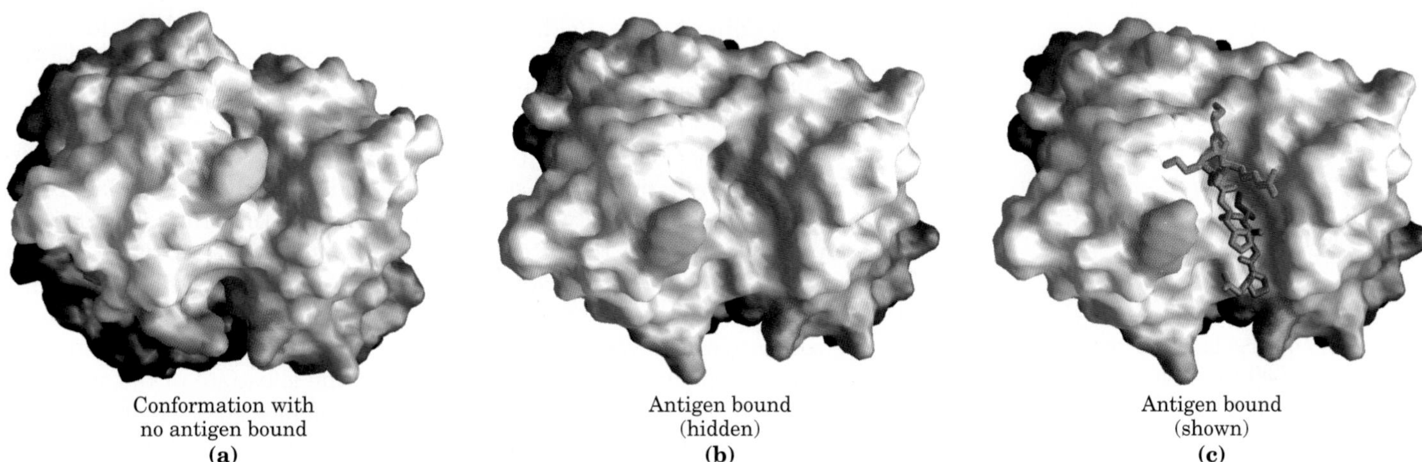

figure 7–27
Induced fit in the binding of an antigen to IgG. The molecule, shown in surface contour, is the Fab fragment of an IgG. The antigen this IgG binds is a small peptide derived from HIV. Two residues from the heavy chain (blue) and one from the light chain (pink) are colored to provide visual points of reference. **(a)** View of the Fab fragment looking down on the antigen-binding site. **(b)** The same view, but here the Fab fragment is in the "bound" conformation; the antigen has been omitted from the image to provide an unobstructed view of the altered binding site. Note how the binding cavity has enlarged and several groups have shifted position. **(c)** The same view as in **(b),** but with the antigen pictured in the binding site as a red stick structure.

A typical antibody-antigen interaction is quite strong, characterized by K_d values as low as 10^{-10} M (recall that a lower K_d corresponds to a stronger binding interaction). The K_d reflects the energy derived from the various ionic, hydrogen-bonding, hydrophobic, and van der Waals interactions that stabilize the binding. The binding energy required to produce a K_d of 10^{-10} M is about 65 kJ/mol.

A complex of a peptide derived from HIV (a model antigen) and an Fab molecule illustrates some of these properties (Fig. 7–27). The changes in structure observed on antigen binding are particularly striking in this example.

The Antibody-Antigen Interaction Is the Basis for a Variety of Important Analytical Procedures

The extraordinary binding affinity and specificity of antibodies makes them valuable analytical reagents. Two types of antibody preparations are in use: polyclonal and monoclonal. **Polyclonal antibodies** are those produced by many different B lymphocytes responding to one antigen, such as a protein injected into an animal. Cells in the population of B lymphocytes produce antibodies that bind specific, different epitopes within the antigen. Thus, polyclonal preparations contain a mixture of antibodies that recognize different parts of the protein. **Monoclonal antibodies,** in contrast, are synthesized by a population of identical B cells (a **clone**) grown in cell culture. These antibodies are homogeneous, all recognizing the same epitope. The techniques for producing monoclonal antibodies were developed by Georges Köhler and Cesar Milstein.

Georges Köhler

Cesar Milstein

The specificity of antibodies has practical uses. A selected antibody can be covalently attached to a resin and used in a chromatography column of the type shown in Figure 5–18c. When a mixture of proteins is added to the column, the antibody will specifically bind its target protein and retain it on the column while other proteins are washed through. The target protein can then be eluted from the resin by a salt solution or some other agent. This is a powerful tool for protein purification.

In another versatile analytical technique, an antibody is attached to a radioactive label or some other reagent that makes it easy to detect. When the antibody binds the target protein, the label reveals the presence of the protein in a solution or its location in a gel or even a living cell. Several variations of this procedure are illustrated in Figure 7–28.

An **ELISA** (enzyme-linked immunosorbent assay) allows for rapid screening and quantification of the presence of an antigen in a sample (Fig. 7–28b). Proteins in a sample are adsorbed to an inert surface, usually a 96-well polystyrene plate. The surface is washed with a solution of an inexpensive nonspecific protein (often casein from nonfat dry milk powder) to block proteins in subsequent steps from also adsorbing to these surfaces. The surface is then treated with a solution containing the primary antibody—an antibody against the protein of interest. Unbound antibody is washed away and the surface is treated with a solution containing antibodies against the primary antibody. These secondary antibodies have been linked to an enzyme that catalyzes a reaction that forms a colored product. After unbound secondary antibody is washed away, the substrate of the antibody-linked enzyme is added. Product formation (monitored as color intensity) is proportional to the concentration of the protein of interest in the sample.

① Coat surface with sample (antigens).
② Block unoccupied sites with nonspecific protein.
③ Incubate with primary antibody against specific antigen.
④ Incubate with antibody-enzyme complex that binds primary antibody.
⑤ Add substrate.
⑥ Formation of colored product indicates presence of specific antigen.

In an **immunoblot assay** (Fig. 7–28c), proteins that have been separated by gel electrophoresis are transferred electrophoretically to a nitrocellulose membrane. The membrane is blocked (as described above for ELISA), then treated successively with primary antibody, secondary antibody linked to enzyme, and substrate. A colored precipitate forms only along the band containing the protein of interest. The immunoblot allows

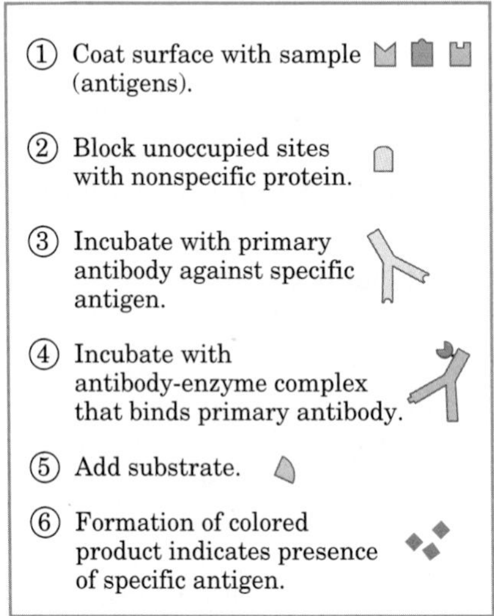

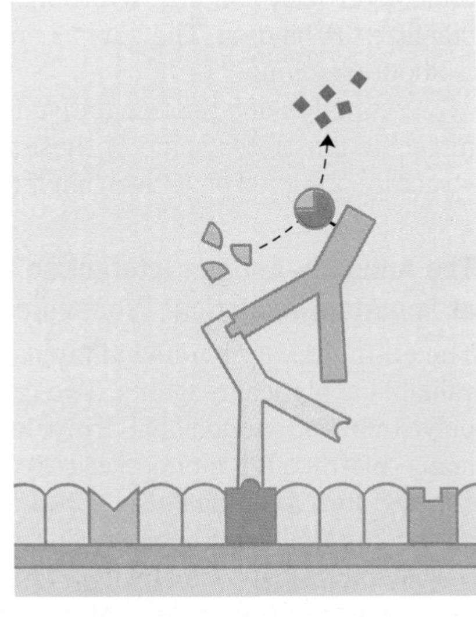

(a)

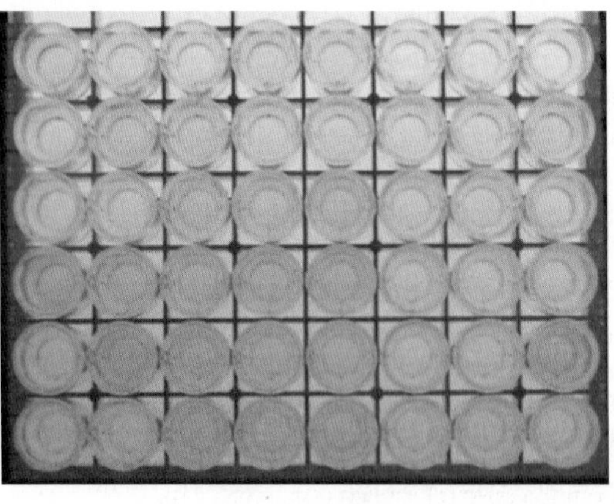

ELISA assay
(b)

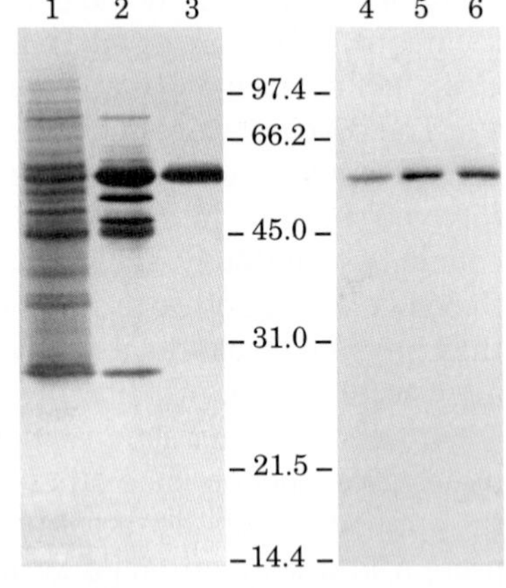

(c)

figure 7–28
Antibody techniques. The specific reaction of an antibody with its antigen is the basis of several techniques that identify and quantify a specific protein in a complex sample. **(a)** A schematic representation of the general method. **(b)** An ELISA testing for the presence of herpes simplex virus (HSV) antibodies in blood samples. Wells were coated with an HSV antigen, to which antibodies against HSV in a patient's blood will bind. The second antibody is anti–human IgG linked to horseradish peroxidase. Blood samples with greater amounts of HSV antibody turn brighter yellow. **(c)** An immunoblot. Lanes 1 to 3 are from an SDS gel; samples from successive stages in the purification of a protein kinase have been separated and stained with Coomassie blue. Lanes 4 to 6 show the same samples, but these were electrophoretically transferred to a nitrocellulose membrane after separation on an SDS gel. The membrane was then "probed" with antibody against the protein kinase. The numbers between the gel and the immunoblot indicate M_r ($\times 10^{-3}$).

the detection of a minor component in a sample and provides an approximation of its molecular weight.

We will encounter other aspects of antibodies in later chapters. They are extremely important in medicine and can tell us much about the structure of proteins and the action of genes.

Protein Interactions Modulated by Chemical Energy: Actin, Myosin, and Molecular Motors

Organisms move. Cells move. Organelles and macromolecules within cells move. Most of these movements arise from the activity of the fascinating class of protein-based molecular motors. Fueled by chemical energy, usually derived from ATP, large aggregates of motor proteins undergo cyclic conformational changes that accumulate into a unified, directional force—the tiny force that pulls apart chromosomes in a dividing cell, and the immense force that levers a pouncing, quarter-ton jungle cat into the air.

The interactions among motor proteins, as you might predict, feature complementary arrangements of ionic, hydrogen-bonding, hydrophobic, and van der Waals interactions at protein binding sites. In motor proteins, however, these interactions achieve exceptionally high levels of spatial and temporal organization.

Motor proteins underlie the contraction of muscles, the migration of organelles along microtubules, the rotation of bacterial flagella, and the movement of some proteins along DNA. As we noted in Chapter 2, proteins called kinesins and dyneins move along microtubules in cells, pulling along organelles or reorganizing chromosomes during cell division (see Fig. 2–19). An interaction of dynein with microtubules brings about the motion of eukaryotic flagella and cilia. Flagellar motion in bacteria involves a complex rotational motor at the base of the flagellum (see Fig. 19–32). Helicases, polymerases, and other proteins move along DNA as they carry out their functions in DNA metabolism (Chapter 25). Here, we focus on the well-studied example of the contractile proteins of vertebrate skeletal muscle as a paradigm for how proteins translate chemical energy into motion.

The Major Proteins of Muscle Are Myosin and Actin

The contractile force of muscle is generated by the interaction of two proteins, myosin and actin. These proteins are arranged in filaments that undergo transient interactions and slide past each other to bring about contraction. Together, actin and myosin make up over 80% of the protein mass of muscle.

Myosin (M_r 540,000) has six subunits: two heavy chains (M_r 220,000) and four light chains (M_r 20,000). The heavy chains account for much of the overall structure. At their carboxyl termini, they are arranged as extended α helices, wrapped around each other in a fibrous, left-handed coiled coil similar to that of α-keratin (Fig. 7–29a). At its amino termini, each heavy chain has a large globular domain containing a site where ATP is hydrolyzed. The light chains are associated with the globular domains.

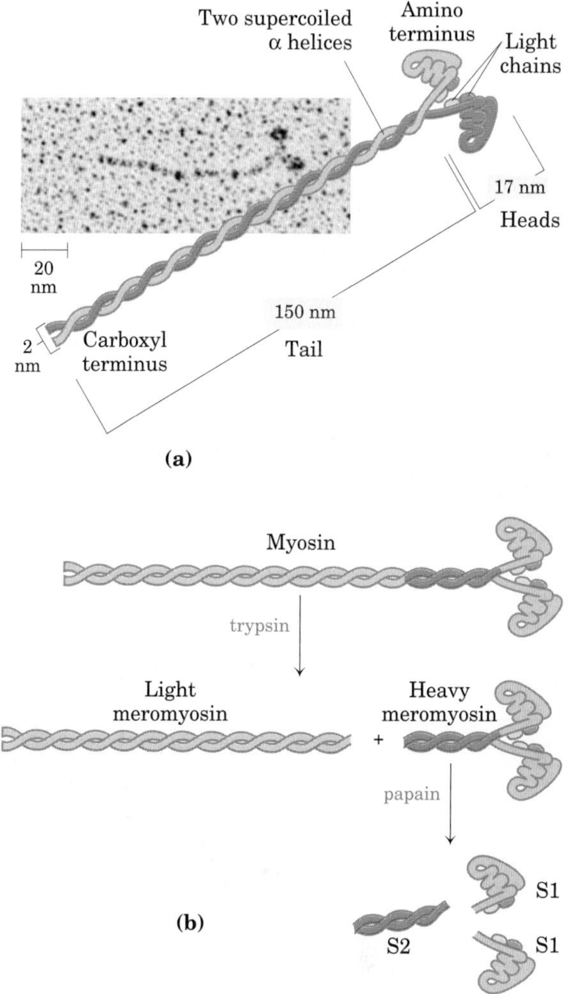

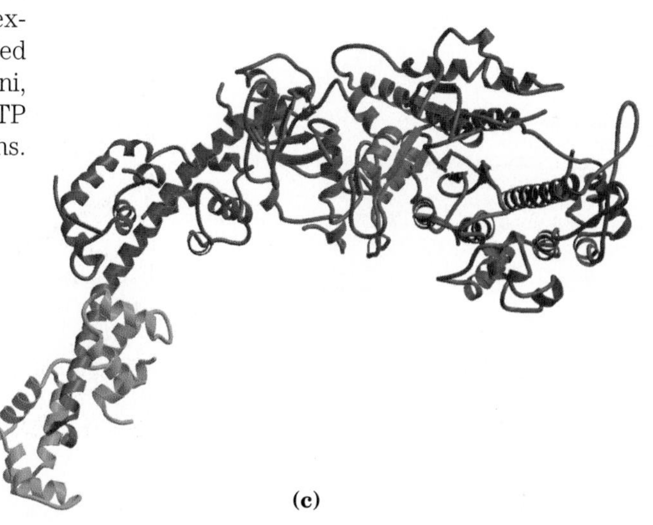

figure 7–29

Myosin. (a) Myosin has two heavy chains (in two shades of pink), the carboxyl termini forming an extended coiled coil (tail) and the amino termini having globular domains (heads). Two light chains (blue) are associated with each myosin head. **(b)** Cleavage with trypsin and papain separates the myosin heads (S1 fragments) from the tails. **(c)** Ribbon representation of the myosin S1 fragment. The heavy chain is in gray, the two light chains in two shades of blue.

When myosin is treated briefly with the protease trypsin, much of the fibrous tail is cleaved off, dividing the protein into components called light and heavy meromyosin (Fig. 7–29b). The globular domain, called myosin subfragment 1, or S1, or simply the myosin head group, is liberated from heavy meromyosin by cleavage with papain. The S1 fragment produced by this procedure is the motor domain that makes muscle contraction possible. S1 fragments can be crystallized, and their structure has been determined. The overall structure of the S1 fragment as determined by Ivan Rayment and Hazel Holden is shown in Figure 7–29c.

In muscle cells, molecules of myosin aggregate to form structures called **thick filaments** (Fig. 7–30a). These rodlike structures serve as the core of the contractile unit. Within a thick filament, several hundred myosin molecules are arranged with their fibrous "tails" associated to form a long bipolar structure. The globular domains project from either end of this structure, in regular stacked arrays.

The second major muscle protein, **actin,** is abundant in almost all eukaryotic cells. In muscle, molecules of monomeric actin, called G-actin (*g*lobular actin; M_r 42,000), associate to form a long polymer called F-actin (*f*ilamentous actin). The **thin filament** (Fig. 7–30b) consists of F-actin, along with the proteins troponin and tropomyosin. The filamentous parts of thin filaments assemble as successive monomeric actin molecules add to one end. On addition, each monomer binds ATP, then hydrolyzes it to ADP, so all actin molecules in the filament are complexed to ADP. However, this ATP hydrolysis by actin functions only in the assembly of the filaments; it does not contribute directly to the energy expended in muscle contraction. Each actin monomer in the thin filament can bind tightly and specifically to one myosin head group (Fig. 7–30c).

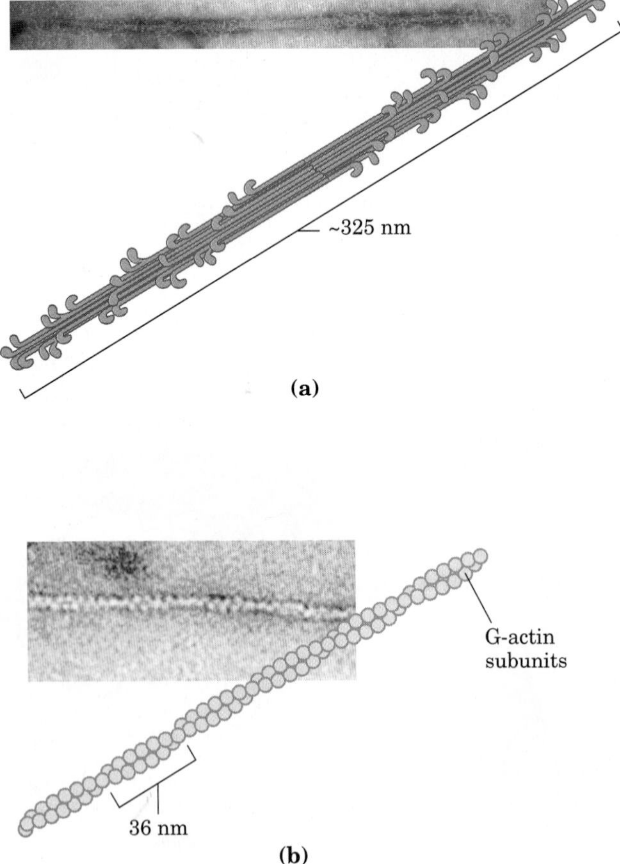

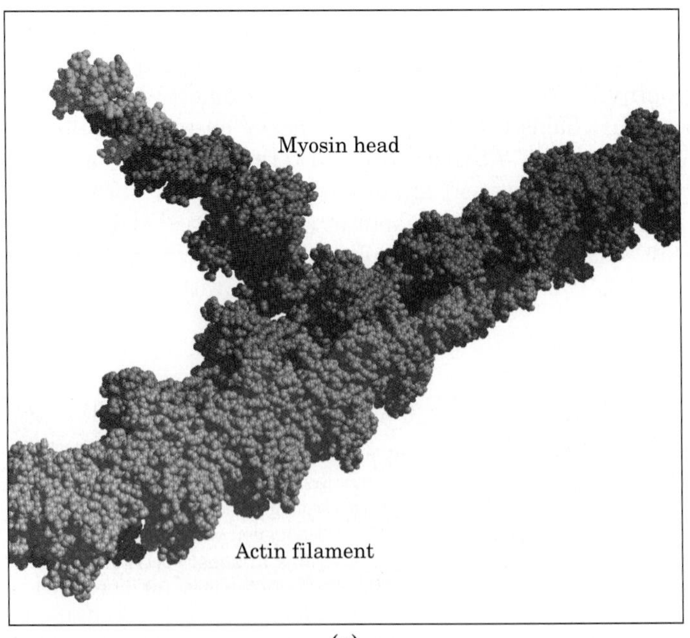

figure 7–30

The major components of muscle. (a) Myosin aggregates to form a bipolar structure called a thick filament. **(b)** F-actin is a filamentous assemblage of G-actin monomers that polymerize two by two, giving the appearance of two filaments spiraling about one another in a right-handed fashion. An electron micrograph and a model of F-actin are shown. **(c)** Space-filling model of an actin filament (red) with one myosin head (gray and two shades of blue) bound to an actin monomer within the filament.

Additional Proteins Organize the Thin and Thick Filaments into Ordered Structures

Skeletal muscle consists of parallel bundles of **muscle fibers,** each fiber a single, very large, multinucleated cell, 20 to 100 μm in diameter, formed from many cells fused together and often spanning the length of the muscle. Each fiber, in turn, contains about 1,000 **myofibrils,** 2 μm in diameter, each consisting of a vast number of regularly arrayed thick and thin filaments complexed to other proteins (Fig. 7–31). A system of flat membranous vesicles called the **sarcoplasmic reticulum** surrounds each myofibril. Examined under the electron microscope, muscle fibers reveal alternating regions of high and low electron density, called the **A** and **I bands** (Fig. 7–31b,c). The A and I bands arise from the arrangement of

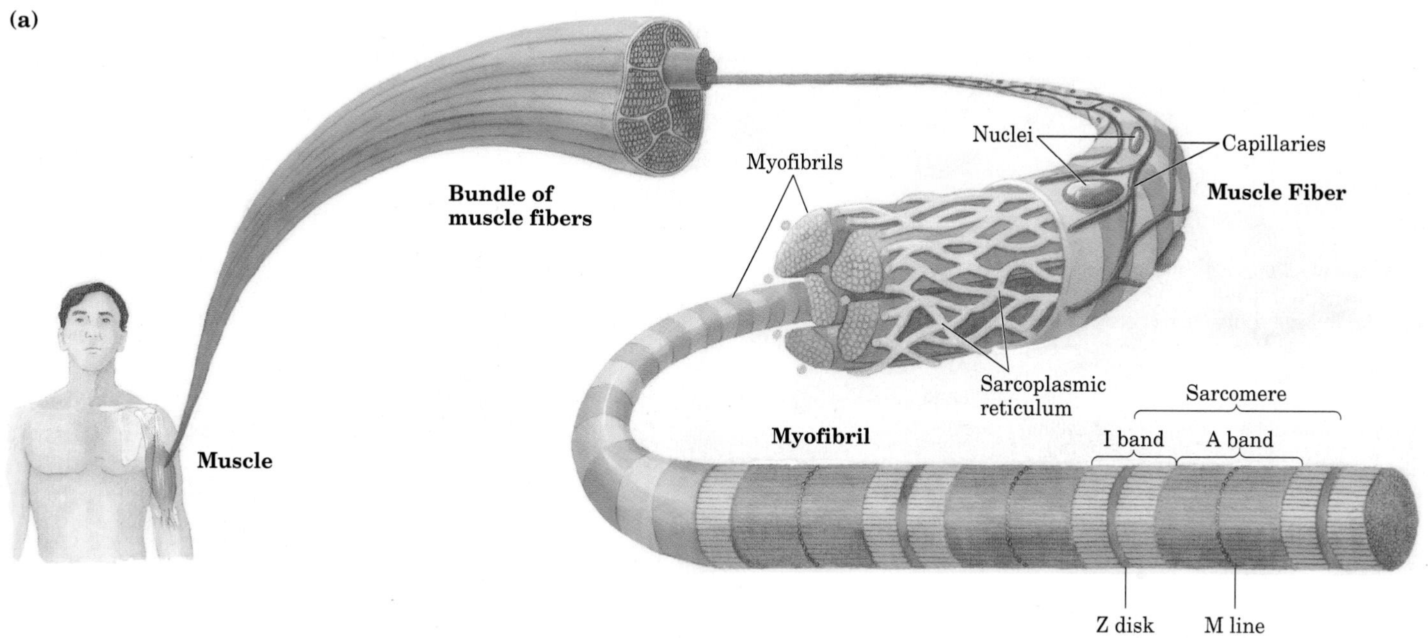

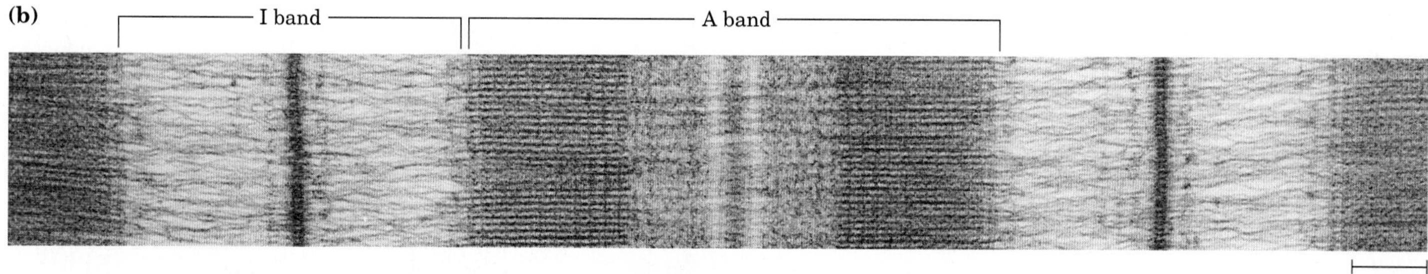

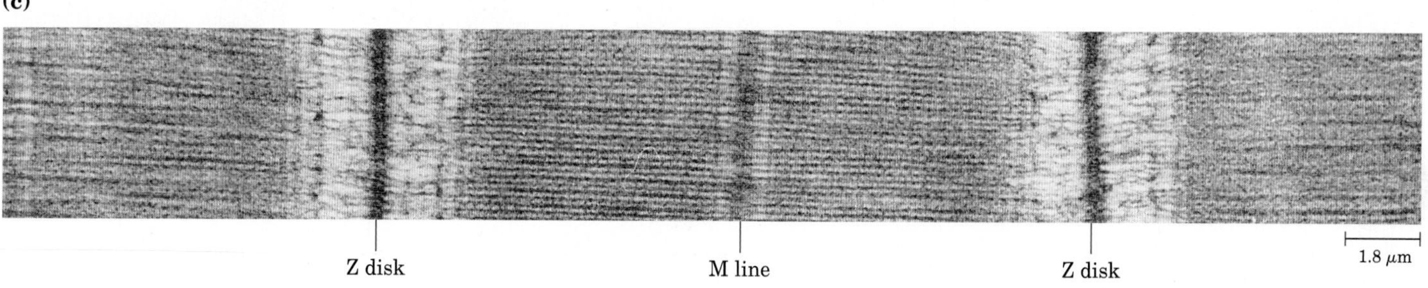

figure 7–31

Structure of skeletal muscle. (a) Muscle fibers consist of single, elongated, multinucleated cells that arise from the fusion of many precursor cells. Within the fibers are many myofibrils (only six are shown here for simplicity) surrounded by the membranous sarcoplasmic reticulum. The organization of thick and thin filaments in the myofibril gives it a striated appearance. When muscle contracts, the I bands narrow and the Z disks come closer together, as seen in electron micrographs of relaxed **(b)** and contracted **(c)** muscle.

thick and thin filaments, which are aligned and partially overlapping. The I band is the region of the bundle that in cross section would contain only thin filaments. The darker A band stretches the length of the thick filament and includes the region where parallel thick and thin filaments overlap. Bisecting the I band is a thin structure called the **Z disk,** perpendicular to the thin filaments and serving as an anchor to which the thin filaments are attached. The A band too is bisected by a thin line, the **M line** or M disk, a region of high electron density in the middle of the thick filaments. The entire contractile unit, consisting of bundles of thick filaments interleaved at either end with bundles of thin filaments, is called the **sarcomere.** The arrangement of interleaved bundles allows the thick and thin filaments to slide past each other (by a mechanism discussed below), causing a progressive shortening of each sarcomere (Fig. 7–32).

The thin actin filaments are attached at one end to the Z disk in a regular pattern. The assembly includes the minor muscle proteins **α-actinin, desmin,** and **vimentin.** Thin filaments also contain a large protein called **nebulin** (~7,000 amino acid residues), thought to be structured as an α helix long enough to span the length of the filament. The M line similarly

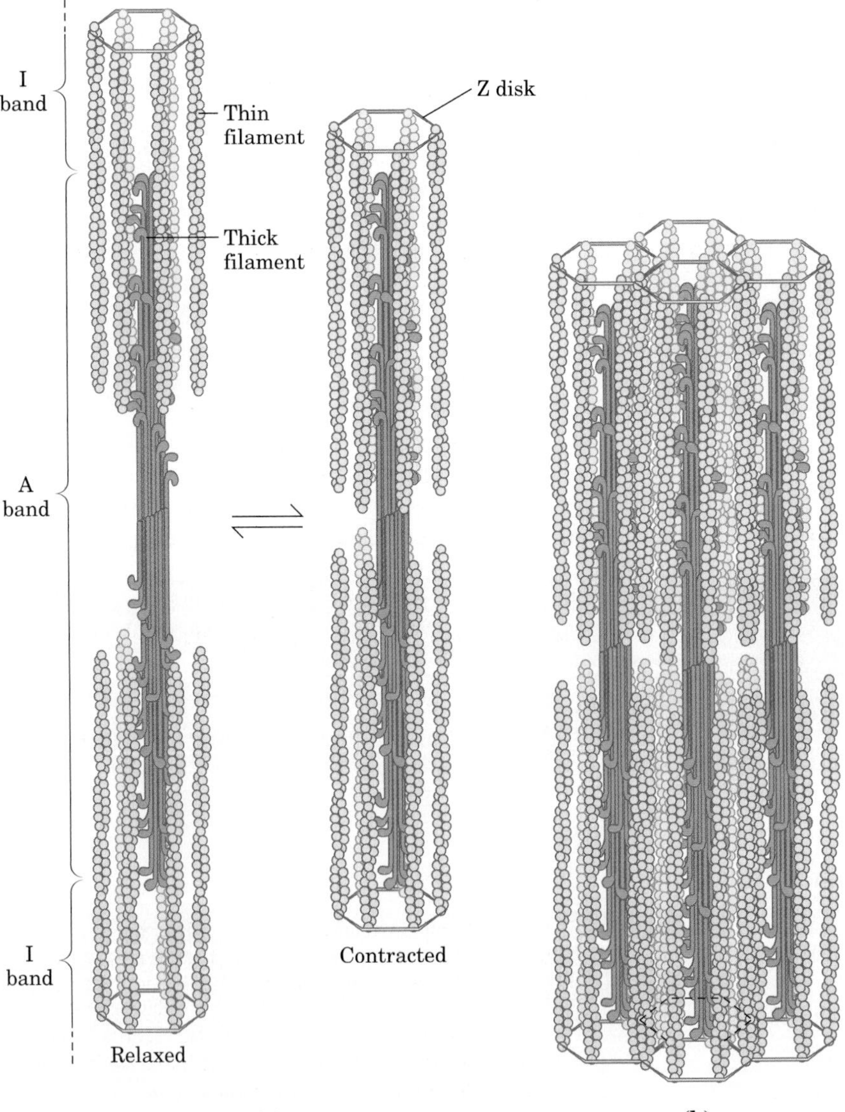

figure 7–32
Muscle contraction. Thick filaments are bipolar structures created by the association of many myosin molecules. **(a)** Muscle contraction occurs by the sliding of the thick and thin filaments past each other so that the Z disks in neighboring I bands approach each other. **(b)** The thick and thin filaments are interleaved such that each thick filament is surrounded by six thin filaments.

organizes the thick filaments. It contains the proteins **paramyosin, C-protein,** and **M-protein.** Another class of proteins called **titins,** the largest known single polypeptide chains (the titin of human cardiac muscle has 26,926 amino acid residues), link the thick filaments to the Z disk, providing additional organization to the overall structure. Among their structural functions, the proteins nebulin and titin are believed to act as "molecular rulers," regulating the length of the thin and thick filaments, respectively. Titin extends from the Z disk to the M line, regulating the length of the sarcomere itself and preventing overextension of the muscle. The characteristic sarcomere length varies from one muscle tissue to the next in a vertebrate organism, attributed in large part to the expression of different titin variants.

Myosin Thick Filaments Slide along Actin Thin Filaments

The interaction between actin and myosin, like that between all proteins and ligands, involves weak bonds. When ATP is not bound to myosin, a face on the myosin head group binds tightly to actin (Fig. 7–33). When ATP binds to myosin and is hydrolyzed to ADP and phosphate, a coordinated and cyclic series of conformational changes occur in which myosin releases the F-actin subunit and binds another subunit farther along the thin filament.

The cycle has four major steps (Fig. 7–33). ① ATP binds to myosin, and a cleft in the myosin molecule opens, disrupting the actin-myosin interaction so that the bound actin is released. ATP is then hydrolyzed (step ②), causing a conformational change in the protein to a "high-energy" state that moves the myosin head and changes its orientation in relation to the actin thin filament. Myosin then binds weakly to an F-actin subunit closer to the Z disk than the one just released. As the phosphate product of ATP hydrolysis is released from myosin in step ③, another conformational change occurs in which the myosin cleft closes, strengthening the myosin-actin binding. This is followed quickly by the final step, ④, a "power stroke" during which the conformation of the myosin head returns to the original resting state, its orientation relative to the bound actin changing so as to pull the tail of the myosin toward the Z disk. ADP is then released to complete the cycle. Each cycle generates about 3 to 4 pN (piconewtons) of force and moves the thick filament 5 to 10 nm relative to the thin filament.

Because there are many myosin heads in a thick filament, at any given moment some (probably 1% to 3%) are bound to the thin filaments. This prevents the thick filaments from slipping backward when an individual myosin head releases the actin subunit to which it was bound. The thick filament thus actively slides forward past the adjacent thin filaments. This process, coordinated among the many sarcomeres in a muscle fiber, brings about muscle contraction.

The interaction between actin and myosin must be regulated so that contraction occurs only in response to appropriate signals from the nervous

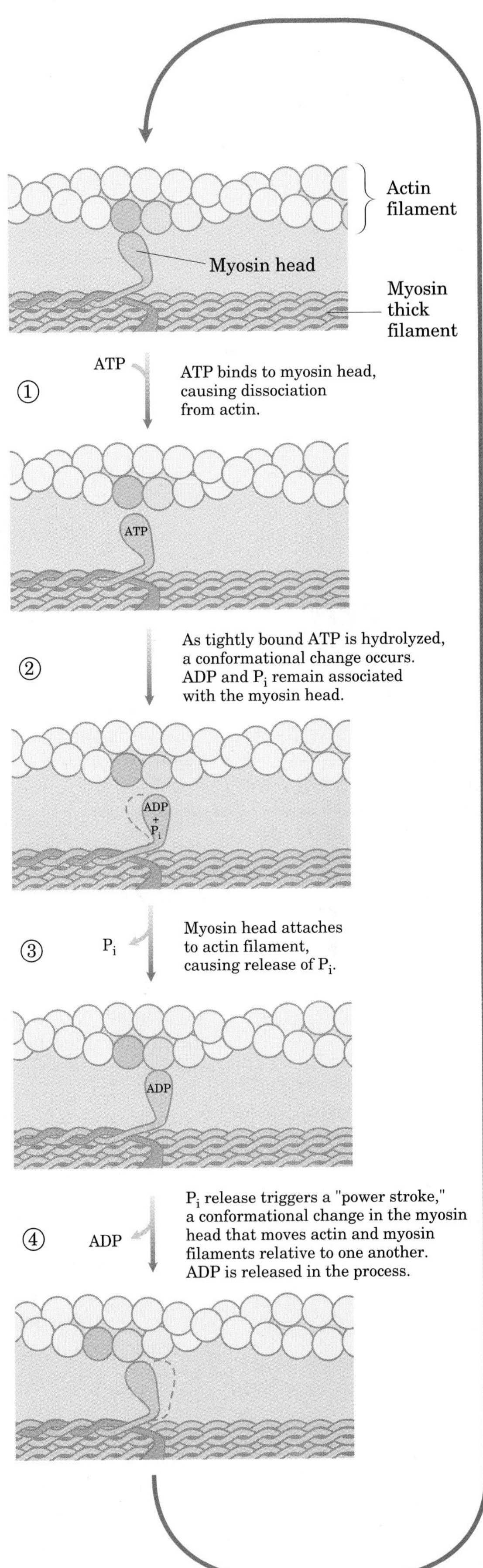
Actin filament
Myosin head
Myosin thick filament
ATP
①
ATP binds to myosin head, causing dissociation from actin.
ATP
②
As tightly bound ATP is hydrolyzed, a conformational change occurs. ADP and P_i remain associated with the myosin head.
ADP + P_i
③
P_i
Myosin head attaches to actin filament, causing release of P_i.
ADP
④
ADP
P_i release triggers a "power stroke," a conformational change in the myosin head that moves actin and myosin filaments relative to one another. ADP is released in the process.

figure 7–33
Molecular mechanism of muscle contraction. Conformational changes in the myosin head that are coupled to stages in the ATP hydrolytic cycle cause myosin to successively dissociate from one actin subunit, then associate with another farther along the actin filament. In this way the myosin heads slide along the thin filaments, drawing the thick filament array into the thin filament array (see Fig. 7–32).

system. The regulation is mediated by a complex of two proteins, **tropomyosin** and **troponin.** Tropomyosin binds to the thin filament, blocking the attachment sites for the myosin head groups. Troponin is a Ca^{2+}-binding protein. A nerve impulse causes release of Ca^{2+} from the sarcoplasmic reticulum. The released Ca^{2+} binds to troponin (another protein-ligand interaction) and causes a conformational change in the tropomyosin-troponin complexes, exposing the myosin-binding sites on the thin filaments. Contraction follows.

Working skeletal muscle requires two types of molecular functions that are common in proteins—binding and catalysis. The actin-myosin interaction, a protein-ligand interaction like that of immunoglobulins with antigens, is reversible and leaves the participants unchanged. When ATP binds myosin, however, it is hydrolyzed to ADP and P_i. Myosin is not only an actin-binding protein, it is also an ATPase—an enzyme. The function of enzymes in catalyzing chemical transformations is the topic of the next chapter.

summary

Protein function often entails interactions with other molecules. A molecule bound by a protein is called a ligand, and the site on the protein to which it binds is called the binding site. Proteins are not rigid and may undergo conformational changes when a ligand binds, a process called induced fit. In a multisubunit protein, the binding of a ligand to one subunit may affect ligand binding to other subunits. Ligand binding can be regulated. Oxygen-binding proteins, immune system proteins, and motor proteins are useful models with which to illustrate these principles.

Myoglobin and hemoglobin contain a prosthetic group called heme to which oxygen binds. Heme consists of a single atom of Fe^{2+} iron coordinated within a porphyrin. Some other small molecules, such as CO and NO, can also bind heme. Myoglobin is a single polypeptide with eight α-helical regions connected by bends. It has a single molecule of heme, located in a pocket deep within the polypeptide. Oxygen binds to myoglobin reversibly. Simple reversible binding can be described by an association constant K_a or a dissociation constant K_d. For a monomeric protein, the fraction of binding sites occupied by a ligand is a hyperbolic function of ligand concentration. Because O_2 is a gas, the term P_{50}, which describes the partial pressure of oxygen at which an O_2-binding protein is half saturated with bound oxygen, is used in place of K_d. The entry and exit of O_2 depend upon small molecular motions, called "breathing," of the myoglobin molecule.

Normal adult hemoglobin has four heme-containing subunits, two α and two β. They are similar in structure to each other and to myoglobin. Strong interactions occur between unlike (α and β) subunits. Hemoglobin exists in two interchangeable states, called T and R. The T state is stabilized by several salt bridges and is most stable when oxygen is not bound. Oxygen binding promotes a transition to the R state.

Oxygen binding to hemoglobin is both allosteric and cooperative. Binding of O_2 to one binding site of hemoglobin affects binding of O_2 to other such sites, an example of allosteric binding behavior. Conformational changes between the T and R states, mediated by subunit-subunit interactions, give rise to a form of allostery called cooperative binding. Cooperative binding results in a sigmoid binding curve and can be analyzed by a Hill plot. Two major models have been proposed to explain the cooperative binding of ligands to multisubunit proteins. In the concerted model, all subunits are in the same conformation at any given time, and the entire protein undergoes a reversible transition between two possible conformations. Successive binding of ligand molecules to the low-affinity conformation facilitates transition to the high-affinity conformation. In the sequential model, individual subunits can undergo conformational changes. Binding of a ligand to one subunit alters that subunit's conformation, facilitating similar changes in, and binding of additional ligands to, the other subunits.

Hemoglobin also binds H^+ and CO_2. In both cases, binding results in the formation of ion pairs that stabilize the T state and O_2 binding is weakened, a phenomenon called the Bohr effect. The binding of H^+ and CO_2 to hemoglobin in the tissues promotes the release of O_2, and the binding of O_2 to hemoglobin in the lungs promotes the release of H^+ and CO_2. Oxygen binding to hemoglobin is also modulated by 2,3-bisphosphoglycerate, which binds to and stabilizes the T state.

Sickle-cell anemia is a genetic disease caused by a single amino acid substitution (Glu to Val) at position 6 in the β chains of hemoglobin. The change produces a hydrophobic patch on the surface of the protein that causes the hemoglobin molecules to aggregate into bundles of fibers. These bundles give the erythrocytes a sickle shape. This homozygous condition results in serious medical complications.

The immune response is mediated by interactions among an array of specialized leukocytes and their associated proteins. T lymphocytes produce T-cell receptors. B lymphocytes produce immunoglobulins. All cells produce MHC proteins, which display host (self) or antigenic (nonself) peptides on the cell surface. Helper T cells induce the proliferation of those B cells and cytotoxic T cells producing immunoglobulins or T-cell receptors that bind to a specific antigen, a process called clonal selection.

Humans have five classes of immunoglobulins, each with different biological functions. The most abundant is IgG, a Y-shaped protein with two heavy and two light chains. The domains near the upper ends of the Y are hypervariable within the broad population of IgGs and form two antigen-binding sites. A given immunoglobulin generally binds to only a part, called the epitope, of a large antigen. Binding often involves a conformational change in the IgG, an induced fit to the antigen.

Protein-ligand interactions achieve a special degree of spatial and temporal organization in motor proteins. Muscle contraction results from choreographed interactions between myosin and actin, coupled to the hydrolysis of ATP by myosin. Myosin consists of two heavy and four light chains, forming a fibrous coiled coil (tail) domain and a globular (head) domain. Myosin molecules are organized into thick filaments, which slide past thin filaments composed largely of actin. ATP hydrolysis in myosin is coupled to a series of conformational changes in the myosin head, leading to dissociation of myosin from one F-actin subunit and its eventual reassociation with another F-actin subunit farther along the thin filament. The myosin thus slides along the actin filaments. Muscle contraction is stimulated by the release of Ca^{2+} from the sarcoplasmic reticulum. The Ca^{2+} binds to the protein troponin, leading to a conformational change in a troponin-tropomyosin complex that triggers the cycle of actin-myosin interactions.

further reading

Oxygen-Binding Proteins

Ackers, G.K. & Hazzard, J.H. (1993) Transduction of binding energy into hemoglobin cooperativity. *Trends Biochem. Sci.* **18,** 385–390.

Changeux, J.-P. (1993) Allosteric proteins: from regulatory enzymes to receptors—personal recollections. *Bioessays* **15,** 625–634.

An interesting perspective from a leader in the field.

Dickerson, R.E. & Geis, I. (1982) *Hemoglobin: Structure, Function, Evolution, and Pathology,* The Benjamin/Cummings Publishing Company, Redwood City, CA.

di Prisco, G., Condò, S.G., Tamburrini, M., & Giardina, B. (1991) Oxygen transport in extreme environments. *Trends Biochem. Sci.* **16,** 471–474.

A revealing comparison of the oxygen-binding properties of hemoglobins from polar species.

Koshland, D.E., Jr., Nemethy, G., & Filmer, D. (1966) Comparison of experimental binding data and theoretical models in proteins containing subunits. *Biochemistry* **6,** 365–385.

The paper in which the sequential model is introduced.

Monod, J., Wyman, J., & Changeux, J.-P. (1965) On the nature of allosteric transitions: a plausible model. *J. Mol. Biol.* **12,** 88–118.

The concerted model was first proposed in this landmark paper.

Olson, J.S. & Phillips, G.N., Jr. (1996) Kinetic pathways and barriers for ligand binding to myoglobin. *J. Biol. Chem.* **271,** 17,593–17,596.

Perutz, M.F. (1989) Myoglobin and haemoglobin: role of distal residues in reactions with haem ligands. *Trends Biochem. Sci.* **14,** 42–44.

Perutz, M.F., Wilkinson, A.J., Paoli, M., & Dodson, G.G. (1998) The stereochemical mechanism of the cooperative effects in hemoglobin revisited. *Annu. Rev. Biophys. Biomol. Struct.* **27,** 1–34.

Immune System Proteins

Blom, B., Res, P.C., & Spits, H. (1998) T cell precursors in man and mice. *Crit. Rev. Immunol.* **18,** 371–388.

Cohen, I.R. (1988) The self, the world and autoimmunity. *Sci. Am.* **258** (April), 52–60.

Davies, D.R. & Chacko, S. (1993) Antibody structure. *Acc. Chem. Res.* **26,** 421–427.

Davies, D.R., Padlan, E.A., & Sheriff, S. (1990) Antibody-antigen complexes. *Annu. Rev. Biochem.* **59,** 439–473.

Davis, M.M. (1990) T cell receptor gene diversity and selection. *Annu. Rev. Biochem.* **59,** 475–496.

Dutton, R.W., Bradley, L.M., & Swain, S.L. (1998) T cell memory. *Annu. Rev. Immunol.* **16,** 201–223.

Life, Death and the Immune System. (1993) *Sci. Am.* **269** (September).

A special issue on the immune system.

Marrack, P. & Kappler, J. (1987) The T cell receptor. *Science* **238,** 1073–1079.

Müller-Eberhard, H.J. (1988) Molecular organization and function of the complement system. *Annu. Rev. Biochem.* **57,** 321–337.

Parham, P. & Ohta, T. (1996) Population biology of antigen presentation by MHC class I molecules. *Science* **272,** 67–74.

Ploegh, H.L. (1998) Viral strategies of immune evasion. *Science* **280,** 248–253.

Thomsen, A.R., Nansen, A., & Christensen, J.P. (1998) Virus-induced T cell activation and the inflammatory response. *Curr. Top. Microbiol. Immunol.* **231,** 99–123.

Van Parjis, L. & Abbas, A.K. (1998) Homeostasis and self-tolerance in the immune system: turning lymphocytes off. *Science* **280,** 243–248.

York, I.A. & Rock, K.L. (1996) Antigen processing and presentation by the class-I major histocompatibility complex. *Annu. Rev. Immunol.* **14,** 369–396.

Molecular Motors

Finer, J.T., Simmons, R.M., & Spudich, J.A. (1994) Single myosin molecule mechanics: piconewton forces and nanometre steps. *Nature* **368,** 113–119.

Modern techniques reveal the forces affecting individual motor proteins.

Geeves, M.A. & Holmes, K.C. (1999) Structural mechanism of muscle contraction. *Annu. Rev. Biochem.* **68,** 687–728.

Goldman, Y.E. (1998) Wag the tail: structural dynamics of actomyosin. *Cell* **93,** 1–4.

Huxley, H.E. (1998) Getting to grips with contraction: the interplay of structure and biochemistry. *Trends Biochem. Sci.* **23,** 84–87.

An interesting historical perspective on deciphering the mechanism of muscle contraction.

Labeit, S. & Kolmerer, B. (1995) Titins: giant proteins in charge of muscle ultrastructure and elasticity. *Science* **270,** 293–296.

A structural and functional description of some of the largest known proteins.

Rayment, I. (1996) The structural basis of the myosin ATPase activity. *J. Biol. Chem.* **271,** 15,850–15,853.

Examining mechanism from a structural perspective.

Rayment, I. & Holden, H.M. (1994) The three-dimensional structure of a molecular motor. *Trends Biochem. Sci.* **19,** 129–134.

Spudich, J.A. (1994) How molecular motors work. *Nature* **372,** 515–518.

problems

1. Relationship between Affinity and Dissociation Constant Protein A has a binding site for ligand X with a K_d of 10^{-6} M. Protein B has a binding site for ligand X with a K_d of 10^{-9} M. Which protein has a higher affinity for ligand X? Explain your reasoning. Convert the K_d to K_a for both proteins.

2. Negative Cooperativity Which of the following situations would produce a Hill plot with $n_H < 1.0$? Explain your reasoning in each case.

(a) The protein has multiple subunits, each with a single ligand-binding site. Binding of ligand to one site decreases the binding affinity of other sites for the ligand.

(b) The protein is a single polypeptide with two ligand-binding sites, each having a different affinity for the ligand.

(c) The protein is a single polypeptide with a single ligand-binding site. As purified, the protein preparation is heterogeneous, containing some protein molecules that are partially denatured and thus have a lower binding affinity for the ligand.

3. Affinity for Oxygen in Myoglobin and Hemoglobin What is the effect of the following changes on the O_2 affinity of myoglobin and hemoglobin? (a) A drop in the pH of blood plasma from 7.4 to 7.2. (b) A decrease in the partial pressure of CO_2 in the lungs from 6 kPa (holding one's breath) to 2 kPa (normal). (c) An increase in the BPG level from 5 mM (normal altitudes) to 8 mM (high altitudes).

4. Cooperativity in Hemoglobin Under appropriate conditions, hemoglobin dissociates into its four subunits. The isolated α subunit binds oxygen, but the O_2-saturation curve is hyperbolic rather than sigmoid. In addition, the binding of oxygen to the isolated α subunit is not affected by the presence of H^+, CO_2, or BPG. What do these observations indicate about the source of the cooperativity in hemoglobin?

5. Comparison of Fetal and Maternal Hemoglobins Studies of oxygen transport in pregnant mammals have shown that the O_2-saturation curves of fetal and maternal blood are markedly different when measured under the same conditions. Fetal erythrocytes contain a structural variant of hemoglobin, HbF, consisting of two α and two γ subunits ($\alpha_2\gamma_2$), whereas maternal erythrocytes contain HbA ($\alpha_2\beta_2$).

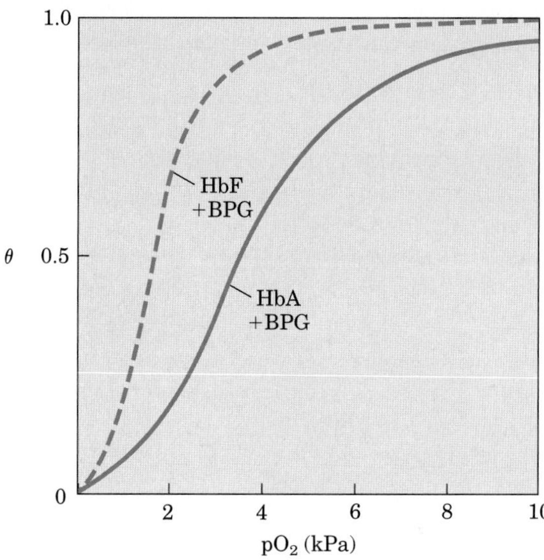

(a) Which hemoglobin has a higher affinity for oxygen under physiological conditions, HbA or HbF? Explain.

(b) What is the physiological significance of the different O_2 affinities?

(c) When all the BPG is carefully removed from samples of HbA and HbF, the measured O_2-saturation curves (and consequently the O_2 affinities) are displaced to the left. However, HbA now has a greater affinity for oxygen than does HbF. When BPG is reintroduced, the O_2-saturation curves return to normal, as shown in the graph. What is the effect of BPG on the O_2 affinity of hemoglobin? How can the above information be used to explain the different O_2 affinities of fetal and maternal hemoglobin?

6. Hemoglobin Variants There are almost 500 naturally occurring variants of hemoglobin. Most are the result of a single amino acid substitution in a globin polypeptide chain. Some variants produce clinical illness, though not all variants have deleterious effects. A brief sample is presented below:

HbS (sickle-cell Hb): substitutes a Val for a Glu on the surface

Hb Cowtown: eliminates an ion pair involved in T-state stabilization

Hb Memphis: substitutes one uncharged polar residue for another of similar size on the surface

Hb Bibba: substitutes a Pro for a Leu involved in an α helix

Hb Milwaukee: substitutes a Glu for a Val

Hb Providence: substitutes an Asn for a Lys that normally projects into the central cavity of the tetramer

Hb Philly: substitutes a Phe for a Tyr, disrupting hydrogen bonding at the $\alpha_1\beta_1$ interface

Explain your choices for each of the following:

(a) The Hb variant *least* likely to cause pathological symptoms.

(b) The variant(s) most likely to show pI values different from that of HbA when run on an isoelectric focusing gel.

(c) The variant(s) most likely to show a decrease in BPG binding and an increase in the overall affinity of the hemoglobin for oxygen.

7. Reversible (but Tight) Binding to an Antibody An antibody binds to an antigen with a K_d of 5×10^{-8} M. At what concentration of antigen will θ be (a) 0.2, (b) 0.5, (c) 0.6, (d) 0.8?

8. Using Antibodies to Probe Structure-Function Relationships in Proteins A monoclonal antibody binds to G-actin but not to F-actin. What does this tell you about the epitope recognized by the antibody?

9. The Immune System and Vaccines A host organism needs time, often days, to mount an immune response against a new antigen, but memory cells permit a rapid response to pathogens previously encountered. A vaccine to protect against a particular viral infection often consists of weakened or killed virus or isolated proteins from a viral protein coat. When injected into a human patient, the vaccine generally does not cause an infection and illness, but it effectively "teaches" the immune system what the viral particles look like, stimulating the production of memory cells. On subsequent infection, these cells can bind to the virus and trigger a rapid immune response. Some pathogens, including HIV, have developed mechanisms to evade the immune system, making it difficult or impossible to develop effective vaccines against them. What strategy could a pathogen use to evade the immune system? Assume that antibodies and/or T-cell receptors are available to bind to any structure that might appear on the surface of a pathogen and that, once bound, the pathogen is destroyed.

10. How We Become a "Stiff" When a higher vertebrate dies, its muscles stiffen as they are deprived of ATP, a state called rigor mortis. Explain the molecular basis of the rigor state.

11. Sarcomeres from Another Point of View The symmetry of thick and thin filaments in a sarcomere is such that six thin filaments ordinarily surround each thick filament in a hexagonal array. Draw a cross section (transverse cut) of a myofibril at the following points: (a) at the M line; (b) through the I band; (c) through the dense region of the A band; (d) through the less dense region of the A band, adjacent to the M line (see Fig. 7–31b).

Biochemistry on the Internet

12. Lysozyme and Antibodies To fully appreciate how proteins function in a cell, it is helpful to have a three-dimensional view of how proteins interact with other cellular components. Fortunately, this is possible using the Internet and on-line protein databases. Go to the biochemistry site at

http://www.worthpublishers.com/lehninger

to learn how to use the Chemscape Chime three-dimensional molecular viewing utility. You can then use the Protein Data Bank and Chemscape Chime to investigate the interactions between antibodies and antigens in more detail.

To examine the interactions between the enzyme lysozyme (Chapter 6) and the Fab portion of the anti-lysozyme antibody, go to the Protein Data Bank Website. Use the PDB identifier 1FDL to retrieve the data page for the IgG1 Fab Fragment-Lysozyme Complex (antibody-antigen complex). Open the structure using Chemscape Chime, and use the different viewing options to answer the following questions:

(a) Which chains in the three-dimensional model correspond to the antibody fragment and which correspond to the antigen, lysozyme?

(b) What secondary structure predominates in this Fab fragment?

(c) How many amino acid residues are in the heavy and light chains of the Fab fragment and in lysozyme? Estimate the percentage of the lysozyme that interacts with the antigen-binding site of the antibody fragment.

(d) Identify the specific amino acid residues in lysozyme and in the variable regions of the heavy and light chains that appear to be situated at the antigen-antibody interface. Are the residues contiguous in the primary sequence of the polypeptide chain?

chapter

8

Enzymes

There are two fundamental conditions for life. One, the living entity must be able to self-replicate (a topic considered in Part IV of this book); two, the organism must be able to catalyze chemical reactions efficiently and selectively. The central importance of catalysis may surprise some beginning students of biochemistry, but it is easy to illustrate. As described in Chapter 1, living systems make use of energy from the environment. Many humans, for example, consume substantial amounts of sucrose—common table sugar—as a kind of fuel—whether in the form of sweetened foods and drinks or as sugar itself. The conversion of sucrose to CO_2 and H_2O in the presence of oxygen is a highly exergonic process, releasing free energy that can be used to think, move, taste, and see. However, a bag of sugar can be stored for years without any obvious conversion to CO_2 and H_2O. Although this chemical process is thermodynamically favorable, it is very slow! Yet when sucrose is consumed by a human (or almost any other organism), it releases its chemical energy in seconds. The difference is catalysis. Without catalysis, the chemical reactions needed to sustain life could not occur on a useful time scale.

We now turn our attention to the reaction catalysts of biological systems: the enzymes, the most remarkable and highly specialized proteins. Enzymes have extraordinary catalytic power, often far greater than that of synthetic or inorganic catalysts. They have a high degree of specificity for their substrates, they accelerate chemical reactions tremendously, and they function in aqueous solutions under very mild conditions of temperature and pH. Few nonbiological catalysts have all these properties.

Enzymes are central to every biochemical process. Acting in organized sequences, they catalyze the hundreds of stepwise reactions by which nutrient molecules are degraded, chemical energy is conserved and transformed, and biological macromolecules are made from simple precursors. Through the action of regulatory enzymes, metabolic pathways are highly coordinated to yield a harmonious interplay among the many different activities necessary to sustain life.

The study of enzymes has immense practical importance. In some diseases, especially inheritable genetic disorders, there may be a deficiency or even a total absence of one or more enzymes. For other disease conditions, an excessive activity of an enzyme may be the cause. Measurements of the activities of enzymes in blood plasma, erythrocytes, or tissue samples are important in diagnosing certain illnesses. Many drugs exert their biological effects through interactions with enzymes. And enzymes are important practical tools, not only in medicine but in the chemical industry, food processing, and agriculture.

We begin with descriptions of the properties of enzymes and the principles underlying their catalytic power, then introduce enzyme kinetics, a discipline that provides much of the framework for any discussion of enzymes. Specific examples of enzyme mechanisms are then provided, illustrating principles introduced earlier in the chapter. We end with a discussion of regulatory enzymes.

An Introduction to Enzymes

Eduard Buchner
1860–1917

Much of the history of biochemistry is the history of enzyme research. Biological catalysis was first recognized and described in the late 1700s, in studies on the digestion of meat by secretions of the stomach, and research continued in the 1800s with examinations of the conversion of starch into sugar by saliva and various plant extracts. In the 1850s, Louis Pasteur concluded that fermentation of sugar into alcohol by yeast is catalyzed by "ferments." He postulated that these ferments were inseparable from the structure of living yeast cells, a view called vitalism that prevailed for many years. Then in 1897 Eduard Buchner discovered that yeast extracts can ferment sugar to alcohol, proving that fermentation was promoted by molecules that continued to function when removed from cells. Frederick W. Kühne called these molecules **enzymes.** As vitalistic notions of life were disproved, the isolation of new enzymes and the investigation of their properties advanced the science of biochemistry.

James Sumner
1887–1955

James Sumner's isolation and crystallization of urease in 1926 provided a breakthrough in early enzyme studies. Sumner found that urease crystals consisted entirely of protein and postulated that all enzymes are proteins. In the absence of other examples, this idea remained controversial for some time. Only in the 1930s was Sumner's conclusion widely accepted, after John Northrop and Moses Kunitz crystallized pepsin, trypsin, and other digestive enzymes and found them also to be proteins. During this period, J.B.S. Haldane wrote a treatise entitled "Enzymes." Although the molecular nature of enzymes was not yet fully appreciated, Haldane made the remarkable suggestion that weak-bonding interactions between an enzyme and its substrate might be used to distort the substrate and catalyze a reaction. This insight lies at the heart of our current understanding of enzymatic catalysis.

In the latter part of the twentieth century, research on the enzymes catalyzing the reactions of cellular metabolism has been intensive. It has led to the purification of thousands of enzymes, elucidation of the structure and chemical mechanism of many of these, and a general understanding of how enzymes work.

J.B.S. Haldane
1892–1964

Most Enzymes Are Proteins

With the exception of a small group of catalytic RNA molecules (Chapter 26), all enzymes are proteins. Their catalytic activity depends on the integrity of their native protein conformation. If an enzyme is denatured or dissociated into subunits, catalytic activity is usually lost. If an enzyme is broken down into its component amino acids, its catalytic activity is always destroyed. Thus the primary, secondary, tertiary, and quaternary structures of protein enzymes are essential to their catalytic activity.

Enzymes, like other proteins, have molecular weights ranging from about 12,000 to over 1 million. Some enzymes require no chemical groups for activity other than their amino acid residues. Others require an additional chemical component called a **cofactor**—either one or more inor-

table 8–1

Some Inorganic Elements That Serve as Cofactors for Enzymes

Cu^{2+}	Cytochrome oxidase
Fe^{2+} or Fe^{3+}	Cytochrome oxidase, catalase, peroxidase
K^{+}	Pyruvate kinase
Mg^{2+}	Hexokinase, glucose 6-phosphatase, pyruvate kinase
Mn^{2+}	Arginase, ribonucleotide reductase
Mo	Dinitrogenase
Ni^{2+}	Urease
Se	Glutathione peroxidase
Zn^{2+}	Carbonic anhydrase, alcohol dehydrogenase, carboxypeptidases A and B

ganic ions, such as Fe^{2+}, Mg^{2+}, Mn^{2+}, or Zn^{2+} (Table 8–1), or a complex organic or metalloorganic molecule called a **coenzyme** (Table 8–2). Some enzymes require *both* a coenzyme and one or more metal ions for activity. A coenzyme or metal ion that is very tightly or even covalently bound to the enzyme protein is called a **prosthetic group.** A complete, catalytically active enzyme together with its bound coenzyme and/or metal ions is called a **holoenzyme.** The protein part of such an enzyme is called the **apoenzyme** or **apoprotein.** Coenzymes function as transient carriers of specific functional groups. They are often derived from vitamins, organic nutrients required in small amounts in the diet. We consider coenzymes in more detail as we encounter them in the metabolic pathways discussed in Part III of this book.

Finally, some enzyme proteins are modified covalently by phosphorylation, glycosylation, and other processes. Many of these alterations are involved in the regulation of enzyme activity.

table 8–2

Some Coenzymes That Serve as Transient Carriers of Specific Atoms or Functional Groups*

Coenzyme	Examples of chemical groups transferred	Dietary precursor in mammals
Biocytin	CO_2	Biotin
Coenzyme A	Acyl groups	Pantothenic acid and other compounds
5′-Deoxyadenosylcobalamin (coenzyme B_{12})	H atoms and alkyl groups	Vitamin B_{12}
Flavin adenine dinucleotide	Electrons	Riboflavin (vitamin B_2)
Lipoate	Electrons and acyl groups	Not required in diet
Nicotinamide adenine dinucleotide	Hydride ion ($:H^-$)	Nicotinic acid (niacin)
Pyridoxal phosphate	Amino groups	Pyridoxine (vitamin B_6)
Tetrahydrofolate	One-carbon groups	Folate
Thiamine pyrophosphate	Aldehydes	Thiamine (vitamin B_1)

*The structure and mode of action of these coenzymes are described in Part III of this book.

table 8–3

International Classification of Enzymes*

No.	Class	Type of reaction catalyzed
1	Oxidoreductases	Transfer of electrons (hydride ions or H atoms)
2	Transferases	Group-transfer reactions
3	Hydrolases	Hydrolysis reactions (transfer of functional groups to water)
4	Lyases	Addition of groups to double bonds, or formation of double bonds by removal of groups
5	Isomerases	Transfer of groups within molecules to yield isomeric forms
6	Ligases	Formation of C—C, C—S, C—O, and C—N bonds by condensation reactions coupled to ATP cleavage

*Most enzymes catalyze the transfer of electrons, atoms, or functional groups. They are therefore classified, given code numbers, and assigned names according to the type of transfer reaction, the group donor, and the group acceptor.

Enzymes Are Classified by the Reactions They Catalyze

Many enzymes have been named by adding the suffix "-ase" to the name of their substrate or to a word or phrase describing their activity. Thus urease catalyzes hydrolysis of urea, and DNA polymerase catalyzes the polymerization of nucleotides to form DNA. Other enzymes, such as pepsin and trypsin, have names that do not denote their substrates or reactions. Sometimes the same enzyme has two or more names, or two different enzymes have the same name. Because of such ambiguities, and the ever-increasing number of newly discovered enzymes, a system for naming and classifying enzymes has been adopted by international agreement. This system divides enzymes into six major classes, each with subclasses, based on the type of reaction catalyzed (Table 8–3). Each enzyme is assigned a four-digit classification number and a systematic name, which identifies the reaction it catalyzes. As an example, the formal systematic name of the enzyme catalyzing the reaction

$$\text{ATP} + \text{D-glucose} \longrightarrow \text{ADP} + \text{D-glucose 6-phosphate}$$

is ATP:glucose phosphotransferase, which indicates that it catalyzes the transfer of a phosphoryl group from ATP to glucose. Its Enzyme Commission number (E.C. number) is 2.7.1.1. The first digit (2) denotes the class name (transferase); the second digit (7), the subclass (phosphotransferase); the third digit (1), a phosphotransferase with a hydroxyl group as acceptor; and the fourth digit (1), D-glucose as the phosphoryl-group acceptor. For many enzymes, a trivial name is more commonly used—in this case hexokinase.

A complete list and description of the thousands of known enzymes is well beyond the scope of this book. This chapter is devoted primarily to principles and properties common to all enzymes.

How Enzymes Work

The enzymatic catalysis of reactions is essential to living systems. Under biologically relevant conditions, uncatalyzed reactions tend to be slow—most biological molecules are quite stable in the neutral-pH, mild-temperature, aqueous environment inside cells. Furthermore, many common reactions in biochemistry entail chemical events that are unfavorable or unlikely in the cellular environment, such as the transient formation of unstable charged intermediates or the collision of two or more molecules in the precise orientation required for reaction. Reactions required to digest food, send nerve signals, or contract a muscle simply do not occur at a useful rate without catalysis.

An enzyme circumvents these problems by providing a specific environment within which a given reaction is energetically more favorable. The distinguishing feature of an enzyme-catalyzed reaction is that it occurs within the confines of a pocket on the enzyme called the **active site** (Fig. 8–1). The molecule that is bound in the active site and acted upon by the enzyme is called the **substrate.** The surface of the active site is lined with amino acid residues whose substituent groups bind the substrate and catalyze its chemical transformation. The enzyme-substrate complex, whose existence was first proposed by Adolphe Wurtz in 1880, is central to the action of enzymes. It is also the starting point for mathematical treatments defining the kinetic behavior of enzyme-catalyzed reactions and for theoretical descriptions of enzyme mechanisms.

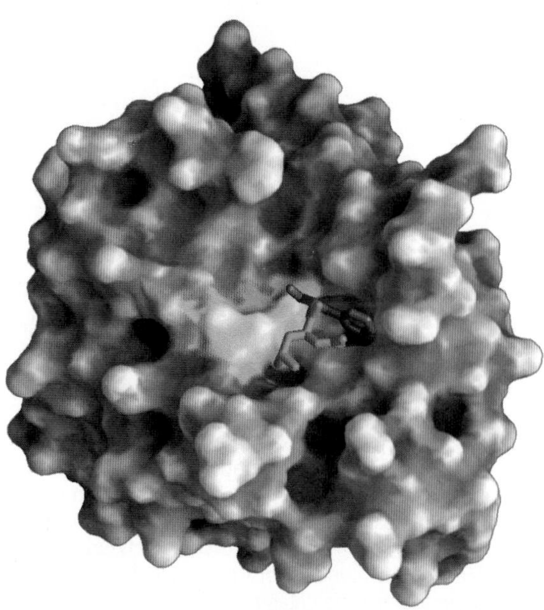

figure 8–1
Binding of a substrate to an enzyme at the active site. The enzyme chymotrypsin is shown with the bound substrate in red. Some key active-site amino acids are shown as red splotches on the enzyme surface.

Enzymes Affect Reaction Rates, Not Equilibria

A simple enzymatic reaction might be written

$$\mathrm{E + S \rightleftharpoons ES \rightleftharpoons EP \rightleftharpoons E + P} \qquad (8\text{–}1)$$

where E, S, and P represent the enzyme, substrate, and product. ES and EP are transient complexes of the enzyme with the substrate and with the product.

To understand catalysis, we must first appreciate the important distinction between reaction equilibria (discussed in Chapter 4) and reaction rates. The function of a catalyst is to increase the *rate* of a reaction. Catalysts do not affect reaction *equilibria.* Any reaction, such as $\mathrm{S \rightleftharpoons P}$, can be described by a reaction coordinate diagram (Fig. 8–2), a picture of the energy changes during the reaction. As we noted in Chapters 1 and 3, energy in biological systems is described in terms of free energy, G. In the coordinate diagram, the free energy of the system is plotted against the progress of the reaction (reaction coordinate). The starting point for either the forward or the reverse reaction is called the **ground state,** the contribution to the free energy of the system by an average molecule (S or P) under a given set of conditions. To describe the free-energy changes for reactions, chemists define a standard set of conditions (temperature 298 K; partial pressure of each gas 1 atm or 101.3 kPa; concentration of each solute 1 M) and express the free-energy change for this reacting system as $\Delta G°$, the **standard free-energy change.** Because biochemical systems commonly involve H^+ concentrations far from 1 M, biochemists define a **biochemical standard free-energy change** $\Delta G'°$, the standard free-energy change *at pH 7.0,* which we will employ throughout the book. A more complete definition of $\Delta G'°$ is given in Chapter 14.

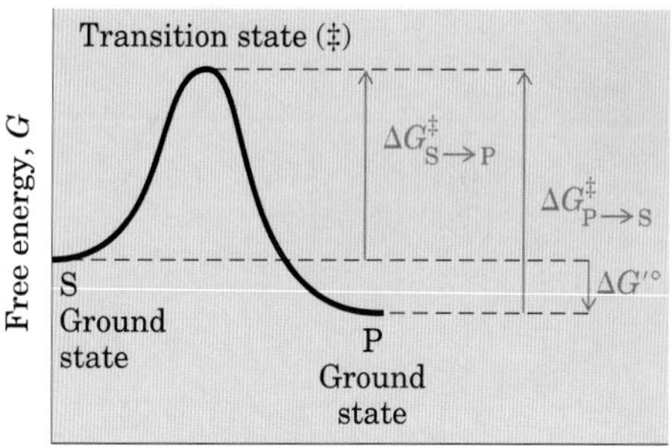

figure 8–2
Reaction coordinate diagram for a chemical reaction. The free energy of the system is plotted against the progress of the reaction $\mathrm{S \rightleftharpoons P}$. A diagram of this kind is a description of the energy changes during the reaction, and the horizontal axis (reaction coordinate) reflects the progressive chemical changes (e.g., bond breakage or formation) as S is converted to P. The activation energies, $\Delta G^‡$, for the $\mathrm{S \rightarrow P}$ and $\mathrm{P \rightarrow S}$ reactions are indicated. $\Delta G'°$ is the overall standard free-energy change in the direction $\mathrm{S \rightarrow P}$.

The equilibrium between S and P reflects the difference in the free energies of their ground states. In the example shown in Figure 8–2, the free energy of the ground state of P is lower than that of S, so $\Delta G'°$ for the reaction is negative and the equilibrium favors P. The position and direction of equilibrium are *not* affected by any catalyst.

A favorable equilibrium does not mean that the $\mathrm{S \rightarrow P}$ conversion will occur at a detectable rate. The rate of a reaction is dependent on an entirely different parameter. There is an energy barrier between S and P, the energy required for alignment of reacting groups, formation of transient unstable charges, bond rearrangements, and other transformations required for the reaction to proceed in either direction. This is illustrated by the energy

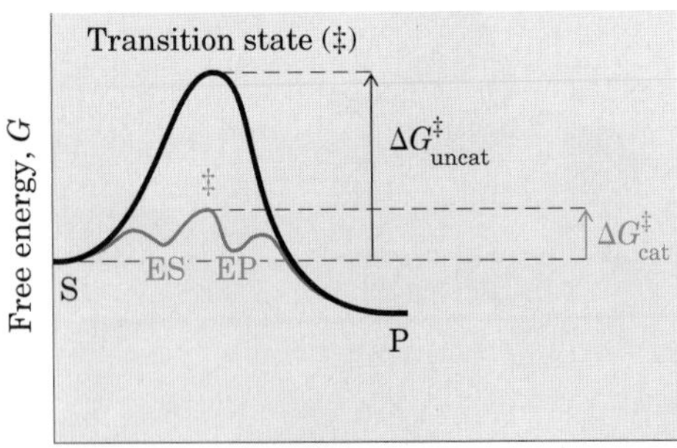

figure 8–3

Reaction coordinate diagram comparing enzyme-catalyzed and uncatalyzed reactions. In the reaction S → P, the ES and EP intermediates occupy minima in the energy progress curve of the enzyme-catalyzed reaction. The terms $\Delta G^{\ddagger}_{uncat}$ and $\Delta G^{\ddagger}_{cat}$ correspond to the activation energies for the uncatalyzed and catalyzed reactions. The activation energy for the overall process is lower when the enzyme catalyzes the reaction.

"hill" in Figures 8–2 and 8–3. To undergo reaction, the molecules must overcome this barrier and therefore must be raised to a higher energy level. At the top of the energy hill is a point at which decay to the S or P state is equally probable (it is downhill either way). This is called the **transition state.** The transition state is not a chemical species with any significant stability and should not be confused with a reaction intermediate (such as ES or EP). It is simply a fleeting molecular moment in which events such as bond breakage, bond formation, and charge development have proceeded to the precise point at which decomposition to either substrate or product is equally likely. The difference between the energy levels of the ground state and the transition state is called the **activation energy** ($\Delta G^{\ddagger}$). The rate of a reaction reflects this activation energy; a higher activation energy corresponds to a slower reaction. Reaction rates can be increased by raising the temperature, thereby increasing the number of molecules with sufficient energy to overcome the energy barrier. Alternatively, the activation energy can be lowered by adding a catalyst (Fig. 8–3). *Catalysts enhance reaction rates by lowering activation energies.*

Enzymes are no exception to the rule that catalysts do not affect reaction equilibria. The bidirectional arrows in Equation 8–1 make this point: any enzyme that catalyzes the reaction S → P also catalyzes the reaction P → S. The role of enzymes is to *accelerate* the interconversion of S and P. The enzyme is not used up in the process, and the equilibrium point is unaffected. However, the reaction reaches equilibrium much faster when the appropriate enzyme is present because the rate of the reaction is increased.

This general principle can be illustrated by considering the conversion of sucrose and oxygen to carbon dioxide and water:

$$C_{12}H_{22}O_{11} + 12O_2 \rightleftharpoons 12CO_2 + 11H_2O$$

This conversion, which takes place through a series of separate reactions, has a very large and negative $\Delta G'^{\circ}$, and at equilibrium the amount of sucrose present is negligible. Yet sucrose is a stable compound because the activation energy barrier that must be overcome before sucrose reacts with oxygen is quite high. It can be stored in a container with oxygen almost indefinitely without reacting. In cells, however, sucrose is readily broken down to CO_2 and H_2O in a series of reactions catalyzed by enzymes. These enzymes not only accelerate the reactions, they organize and control them so that much of the energy released is recovered in other chemical forms and made available to the cell for other tasks. The reaction pathway by which sucrose (and other sugars) is broken down is the primary energy-yielding pathway for cells (see Fig. 16–1), and the enzymes of this pathway allow the reaction sequence to proceed on a biologically useful time scale.

In practice, any reaction may have several steps involving the formation and decay of transient chemical species called **reaction intermediates.*** When the S ⇌ P reaction is catalyzed by an enzyme, the ES and EP complexes are intermediates (Eqn 8–1); they occupy valleys in the reaction coordinate diagram (Fig. 8–3). When several steps occur in a reaction, the overall rate is determined by the step (or steps) with the highest activation

*Note that the terms "step" and "intermediate" in this chapter refer to chemical species occurring on the reaction pathway of a single enzymatically catalyzed reaction. In the context of metabolic pathways involving multiple enzymes (Part III of this book), these terms are used somewhat differently. An entire enzyme reaction is often referred to as a "step" in a pathway, and the product of one enzyme reaction (which is the substrate for the next enzyme in the pathway) is referred to as an "intermediate."

energy; this is called the **rate-limiting step.** In a simple case the rate-limiting step is the highest-energy point in the diagram for interconversion of S and P. In practice, the rate-limiting step can vary with reaction conditions, and for many enzymes several steps may have similar activation energies, which means they are all partially rate-limiting.

As described in Chapter 1, activation energies are energy barriers to chemical reactions; these barriers are crucial to life itself. The stability of a molecule increases with the height of its activation barrier. Without such energy barriers, complex macromolecules would revert spontaneously to much simpler molecular forms, and the complex and highly ordered structures and metabolic processes of cells could not exist. Enzymes have evolved to lower activation energies *selectively* for reactions that are needed for cell survival.

Reaction Rates and Equilibria Have Precise Thermodynamic Definitions

Reaction *equilibria* are inextricably linked to $\Delta G'^{\circ}$ and reaction *rates* are linked to $\Delta G^{\ddagger}$. A basic introduction to these thermodynamic relationships is the next step in understanding how enzymes work.

An equilibrium such as $S \rightleftharpoons P$ is described by an **equilibrium constant,** K_{eq} or simply K (Chapter 4). Under the standard conditions used to compare biochemical processes, an equilibrium constant is denoted K'_{eq} (or K'):

$$K'_{eq} = \frac{[P]}{[S]} \tag{8–2}$$

From thermodynamics, the relationship between K'_{eq} and $\Delta G'^{\circ}$ can be described by the expression

$$\Delta G'^{\circ} = -RT \ln K'_{eq} \tag{8–3}$$

where R is the gas constant, 8.315 J/mol · K, and T is the absolute temperature, 298 K (25 °C). Equation 8–3 is developed and discussed in more detail in Chapter 14. The important point here is that the equilibrium constant is directly related to the overall standard free-energy change for the reaction (Table 8–4). A large negative value for $\Delta G'^{\circ}$ reflects a favorable reaction equilibrium—but as already noted, this does not mean the reaction will proceed at a rapid rate.

table 8–4

Relationship between K'_{eq} and $\Delta G'^{\circ}$ (see Eqn 8–3)

K'_{eq}	$\Delta G'^{\circ}$ (kJ/mol)
10^{-6}	34.2
10^{-5}	28.5
10^{-4}	22.8
10^{-3}	17.1
10^{-2}	11.4
10^{-1}	5.7
1	0.0
10^{1}	−5.7
10^{2}	−11.4
10^{3}	−17.1

The rate of any reaction is determined by the concentration of the reactant (or reactants) and by a **rate constant,** usually denoted by k. For the unimolecular reaction $S \rightarrow P$, the rate or velocity of the reaction, V, representing the amount of S that reacts per unit time, is expressed by a **rate equation:**

$$V = k[S] \tag{8–4}$$

In this reaction, the rate depends only on the concentration of S. This is called a first-order reaction. The factor k is a proportionality constant that reflects the probability of reaction under a given set of conditions (pH, temperature, etc.). Here, k is a first-order rate constant and has units of reciprocal time, such as s^{-1}. If a first-order reaction has a rate constant k of 0.03 s^{-1}, this may be interpreted (qualitatively) to mean that 3% of the available S will be converted to P in 1 s. A reaction with a rate constant of 2,000 s^{-1} will be over in a small fraction of a second. If the reaction rate depends on the concentration of two different compounds, or if the reaction is between two molecules of the same compound, the reaction is second order and k is a second-order rate constant, with units of $\text{M}^{-1}s^{-1}$. The rate equation then becomes

$$V = k[S_1][S_2] \tag{8–5}$$

From transition-state theory, an expression can be derived that relates the magnitude of a rate constant to the activation energy:

$$k = \frac{\mathbf{k}T}{h}e^{-\Delta G^{\ddagger}/RT} \tag{8–6}$$

where **k** is the Boltzmann constant and h is Planck's constant. The important point here is that the relationship between the rate constant k and the activation energy $\Delta G^{\ddagger}$ is inverse and exponential. In simplified terms, this is the basis for the statement that a lower activation energy means a higher reaction rate, and vice versa.

Now we turn from *what* enzymes do to *how* they do it.

A Few Principles Explain the Catalytic Power and Specificity of Enzymes

Enzymes are extraordinary catalysts. The rate enhancements brought about by enzymes are in the range of 5 to 17 orders of magnitude (Table 8–5). Enzymes are also very specific, readily discriminating between substrates with quite similar structures. How can these enormous and highly selective rate enhancements be explained? Where does the energy come from for the dramatic lowering of the activation energies for specific reactions?

The answer to these questions has two distinct but interwoven parts. The first lies in the rearrangements of covalent bonds during an enzyme-catalyzed reaction. Chemical reactions of many types take place between substrates and enzyme functional groups (specific amino acid side chains, metal ions, and coenzymes). Catalytic functional groups on an enzyme may form a transient covalent bond with a substrate and activate it for reaction, or some group may be transiently transferred from the substrate to a group on the enzyme. In many cases, these reactions occur only in the enzyme active site. They lower the activation energy (and thereby accelerate the reaction) by providing an alternative, lower-energy reaction path.

The second part of the explanation lies in the noncovalent interactions between enzyme and substrate. Much of the energy required to lower activation energies is derived from weak, noncovalent interactions between substrate and enzyme. The factor that really sets enzymes apart from most nonenzymatic catalysts is the formation of a specific ES complex. The in-

table 8–5

Some Rate Enhancements Produced by Enzymes

Enzyme	Rate enhancement
Cyclophilin	10^5
Carbonic anhydrase	10^7
Triose phosphate isomerase	10^9
Carboxypeptidase A	10^{11}
Phosphoglucomutase	10^{12}
Succinyl-CoA transferase	10^{13}
Urease	10^{14}
Orotidine monophosphate decarboxylase	10^{17}

teraction between substrate and enzyme in this complex is mediated by the same forces that stabilize protein structure, including hydrogen bonds and hydrophobic and ionic interactions (Chapter 6). Formation of each weak interaction in the ES complex is accompanied by a small release of free energy that provides a degree of stability to the interaction. The energy derived from enzyme-substrate interaction is called **binding energy,** ΔG_B. Its significance extends beyond a simple stabilization of the enzyme-substrate interaction. *Binding energy is a major source of free energy used by enzymes to lower the activation energies of reactions.*

Two fundamental and interrelated principles provide a general explanation for how enzymes use noncovalent binding energy.

1. Much of the catalytic power of enzymes is ultimately derived from the free energy released in forming multiple weak bonds and interactions between an enzyme and its substrate. This binding energy contributes to specificity as well as catalysis.
2. Weak interactions are optimized in the reaction transition state; enzyme active sites are complementary not to the substrates per se, but to the transition states through which substrates pass as they are converted into products during the course of an enzymatic reaction.

These themes are critical to an understanding of enzymes, and they now become our primary focus.

Weak Interactions between Enzyme and Substrate Are Optimized in the Transition State

How does an enzyme use binding energy to lower the activation energy for a reaction? Formation of the ES complex is not the explanation in itself, although some of the earliest considerations of enzyme mechanisms began with this idea. Studies on enzyme specificity carried out by Emil Fischer led him to propose, in 1894, that enzymes were structurally complementary to their substrates, so that they fit together like a "lock and key" (Fig. 8–4).

This elegant idea, that a specific (exclusive) interaction between two biological molecules is mediated by molecular surfaces with complementary shapes, has greatly influenced the development of biochemistry, and such interactions lie at the heart of many biochemical processes. However, the "lock and key" hypothesis can be misleading when applied to enzymatic catalysis. An enzyme completely complementary to its substrate would be a very poor enzyme.

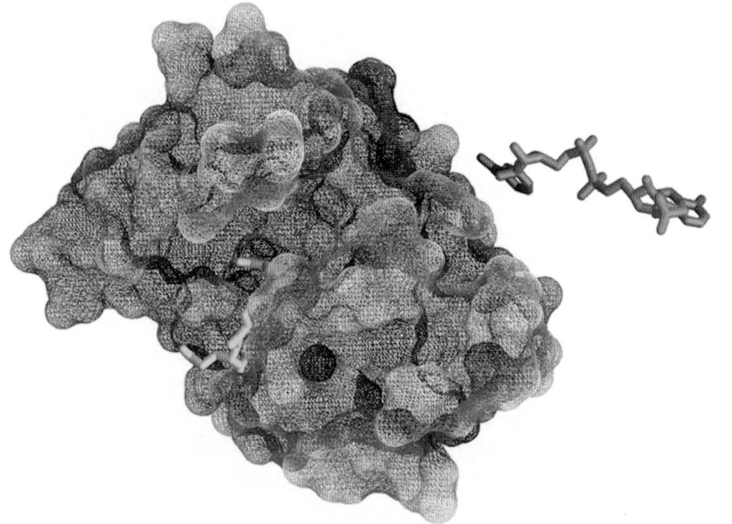

figure 8–4
Complementary shapes of a substrate and its binding site on an enzyme. The enzyme dihydrofolate reductase is shown with its substrate $NADP^+$ (red) unbound (left) and bound (right). Another bound substrate, tetrahydrofolate (yellow), is also visible. The $NADP^+$ binds to a pocket that is complementary to it in shape and ionic properties. In reality, the complementarity between protein and ligand (in this case substrate) is rarely perfect, as we saw in Chapter 7. The interaction of a protein with a ligand often involves changes in the conformation of one or both molecules, a process called induced fit. This *lack* of perfect complementarity between enzyme and substrate (not evident in this figure) is important to enzymatic catalysis.

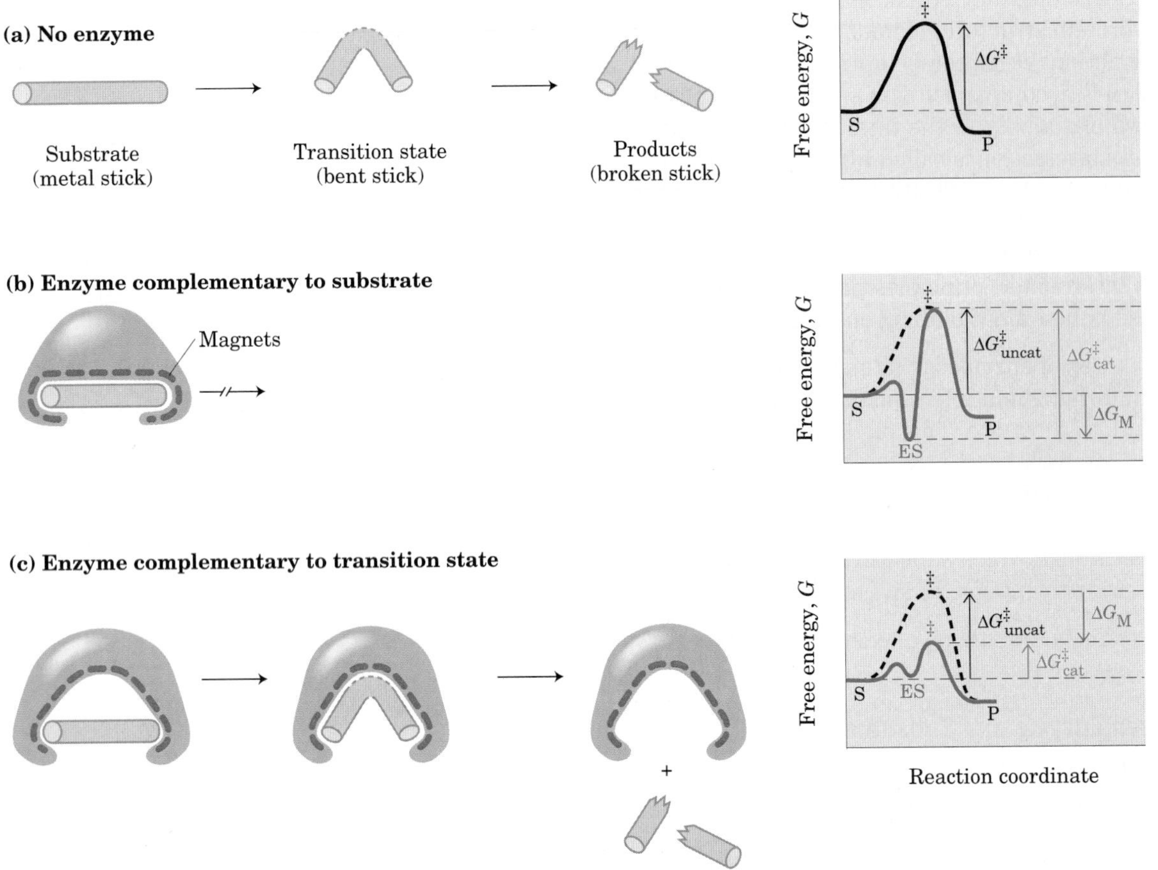

figure 8–5

An imaginary enzyme (stickase) designed to catalyze the breaking of a metal stick. (a) Before the stick is broken, it must first be bent (the transition state). In both stickase examples, magnetic interactions take the place of weak-bonding interactions between enzyme and substrate. **(b)** A stickase with a magnet-lined pocket complementary in structure to the stick (the substrate) stabilizes the substrate. Bending is impeded by the magnetic attraction between stick and stickase. **(c)** An enzyme complementary to the reaction transition state helps to destabilize the stick, contributing to catalysis of the reaction. The binding energy of the magnetic interactions compensates for the increase in free energy required to bend the stick. Reaction coordinate diagrams (right) show the energy consequences of complementarity to substrate versus complementarity to transition state (EP complexes are omitted). ΔG_M represents the difference between the transition-state energies of the uncatalyzed and catalyzed reactions, and is contributed by the magnetic interactions between the stick and stickase. When the enzyme is complementary to the substrate **(b),** the ES complex is more stable and has less free energy in the ground state than substrate alone. The result is an *increase* in the activation energy.

Consider an imaginary reaction, the breaking of a magnetized metal stick. The uncatalyzed reaction is shown in Figure 8–5a. Let's examine two imaginary enzymes—two "stickases"—that catalyze this reaction, both of which employ magnetic forces as a paradigm for the binding energy used by real enzymes. We first design an enzyme perfectly complementary to the substrate (Fig. 8–5b). The active site of this stickase is a pocket lined with magnets. To react (break), the stick must reach the transition state of the reaction, but the stick fits so tightly in the active site that it cannot bend, because bending would eliminate some of the magnetic interactions between stick and enzyme. Such an enzyme *impedes* the reaction, stabilizing the substrate instead. In a reaction coordinate diagram (Fig. 8–5b), this kind of ES complex would correspond to an energy trough from which the substrate would have difficulty escaping. Such an enzyme would be useless.

The modern notion of enzymatic catalysis, first proposed by Haldane in 1930, was elaborated by Linus Pauling in 1946: in order to catalyze reactions, an enzyme must be complementary to the *reaction transition state.* This means that optimal interactions (through weak bonding) between substrate and enzyme occur only in the transition state. Figure 8–5c demonstrates how such an enzyme can work. The metal stick binds, but only a subset of the possible magnetic interactions are used in forming the ES complex. The bound substrate must still undergo the increase in free energy needed to reach the transition state. Now, however, the increase in free energy required to draw the stick into a bent and partially broken conformation is offset or "paid for" by the magnetic interactions (binding energy) that form between the enzyme and substrate in the transition state. Many of these interactions involve parts of the stick that are distant from the point of breakage; thus interactions between the stickase and nonreacting parts of the stick provide some of the energy needed to catalyze stick breakage. This "energy payment" translates into a lower net activation energy and a faster reaction rate.

Real enzymes work on an analogous principle. Some weak interactions are formed in the ES complex, but the full complement of such interactions between substrate and enzyme are formed only when the substrate reaches the transition state. The free energy (binding energy) released by the formation of these interactions partially offsets the energy required to reach the top of the energy hill. The summation of the unfavorable (positive) activation energy $\Delta G^{\ddagger}$ and the favorable (negative) binding energy ΔG_B results in a lower *net* activation energy (Fig. 8–6). Even on the enzyme, the transition state is not a stable species but a brief point in time that the substrate spends atop an energy hill. The enzyme-catalyzed reaction is much faster than the uncatalyzed process, however, because the hill is much smaller. The important principle is that *weak-bonding interactions between the enzyme and the substrate provide a major driving force for enzymatic catalysis.* The groups on the substrate that are involved in these weak interactions can be at some distance from the bonds that are broken or changed. The weak interactions formed only in the transition state are those that make the primary contribution to catalysis.

The requirement for multiple weak interactions to drive catalysis is one reason why enzymes (and some coenzymes) are so large. An enzyme must provide functional groups for ionic, hydrogen-bond, and other interactions, and also must precisely position these groups so that binding energy is optimized in the transition state.

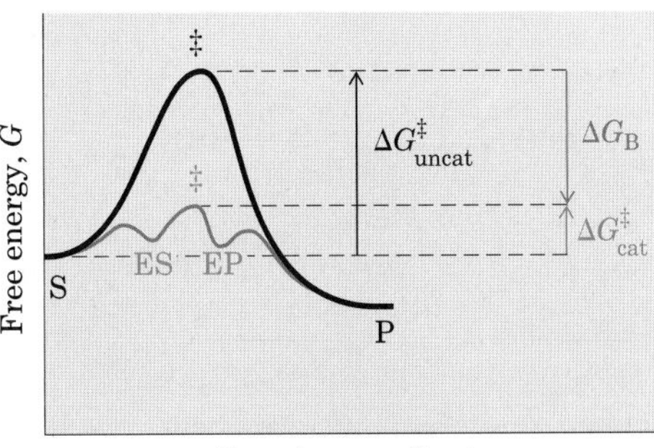

figure 8–6
Role of binding energy in catalysis. To lower the activation energy for a reaction, the system must acquire an amount of energy equivalent to the amount by which $\Delta G^{\ddagger}$ is lowered. Much of this energy comes from binding energy (ΔG_B) contributed by formation of weak noncovalent interactions between substrate and enzyme in the transition state. The role of ΔG_B is analogous to that of ΔG_M in Figure 8–5.

Binding Energy Contributes to Reaction Specificity and Catalysis

Can we demonstrate quantitatively that binding energy accounts for the huge rate accelerations brought about by enzymes? Yes. As a point of reference, Equation 8–6 allows us to calculate that $\Delta G^{\ddagger}$ must be lowered by about 5.7 kJ/mol to accelerate a first-order reaction by a factor of ten under conditions commonly found in cells. The energy available from formation of a single weak interaction is generally estimated to be 4 to 30 kJ/mol. The overall energy available from formation of a number of such interactions is therefore sufficient to lower activation energies by the 60 to 100 kJ/mol required to explain the large rate enhancements observed for many enzymes.

The same binding energy that provides energy for catalysis also gives an enzyme its **specificity,** the ability to discriminate between a substrate and a competing molecule. Conceptually, specificity is easy to distinguish from catalysis, but this distinction is much more difficult to make experimentally because catalysis and specificity arise from the same phenomenon. If an enzyme active site has functional groups arranged optimally to form a variety of weak interactions with a given substrate in the transition state, the enzyme will not be able to interact to the same degree with any other molecule. For example, if the substrate has a hydroxyl group that forms a hydrogen bond with a specific Glu residue on the enzyme, any molecule lacking that particular hydroxyl group would be a poorer substrate for the enzyme. In addition, any molecule with an extra functional group for which the enzyme has no pocket or binding site is likely to be excluded from the enzyme. In general, *specificity* is derived from the formation of multiple weak interactions between the enzyme and its specific substrate molecule.

The general principles outlined above can be illustrated by a variety of recognized catalytic mechanisms. These mechanisms are not mutually exclusive, and a given enzyme might incorporate several in its overall mechanism of action. It is often difficult to quantify the contribution of any one catalytic mechanism to the rate and/or specificity of a particular enzyme-catalyzed reaction.

Binding energy is the dominant driving force in several mechanisms and can be the major, and sometimes the only, contributor to catalysis. Consider what needs to occur for a reaction to take place. Prominent physical and thermodynamic factors contributing to $\Delta G^{\ddagger}$, the barrier to reaction, include (1) the change in entropy, in the form of the freedom of motion of two molecules in solution; (2) the solvation shell of hydrogen-bonded water that surrounds and helps to stabilize most biomolecules in aqueous solution; (3) the distortion of substrates that must occur in many reactions; and (4) the need for proper alignment of catalytic functional groups on the enzyme. Binding energy can be used to overcome all of these barriers.

A large reduction in the relative motions of two substrates that are to react, or **entropy reduction,** is one obvious benefit of binding them to an enzyme. Binding energy holds the substrates in the proper orientation to react—a major contribution to catalysis because productive collisions between molecules in solution can be exceedingly rare. Substrates can be precisely aligned on the enzyme, with a multitude of weak interactions between each substrate and strategically located groups on the enzyme clamping the substrate molecules into the proper positions. Studies have shown that constraining the motion of two reactants can produce rate enhancements of as much as 10^8 M (Fig. 8–7).

Formation of weak bonds between substrate and enzyme also results in **desolvation** of the substrate. Enzyme-substrate interactions replace most or all of the hydrogen bonds between the substrate and water.

Binding energy involving weak interactions formed only in the reaction transition state helps to compensate thermodynamically for any distortion, primarily electron redistribution, that the substrate must undergo to react.

Finally, the enzyme itself usually undergoes a change in conformation when the substrate binds, induced again by multiple weak interactions with the substrate. This is referred to as **induced fit,** a mechanism postulated by Daniel Koshland in 1958. Induced fit serves to bring specific functional groups on the enzyme into the proper position to catalyze the reaction. The conformational change also permits formation of additional weak-bonding

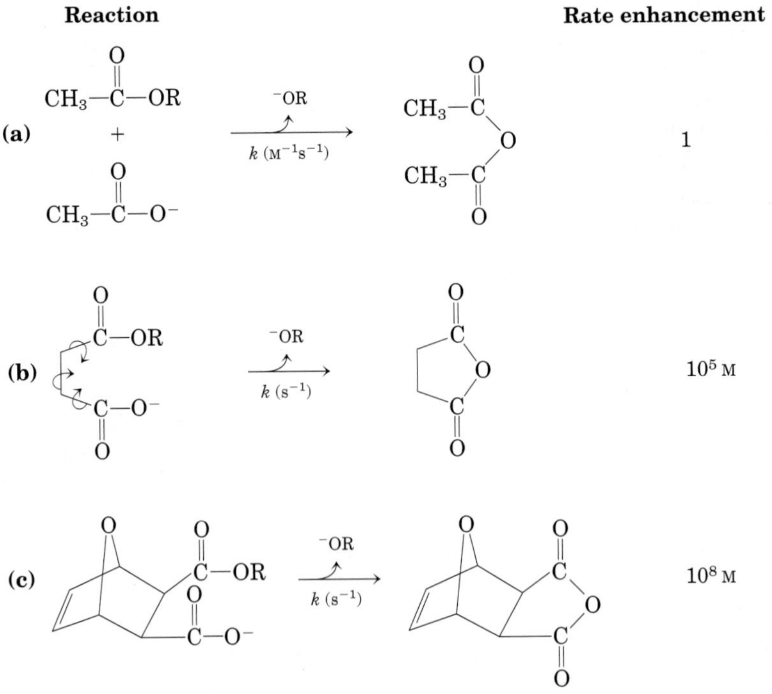

	Reaction	Rate enhancement
(a)	$CH_3-C(=O)-OR$ + $CH_3-C(=O)-O^-$ $\xrightarrow[k\ (M^{-1}s^{-1})]{^-OR}$ $CH_3-C(=O)-O-C(=O)-CH_3$	1
(b)	intramolecular $-C(=O)-OR$ / $-C(=O)-O^-$ $\xrightarrow[k\ (s^{-1})]{^-OR}$ cyclic anhydride	10^5 M
(c)	bicyclic $-C(=O)-OR$ / $-C(=O)-O^-$ $\xrightarrow[k\ (s^{-1})]{^-OR}$ bicyclic cyclic anhydride	10^8 M

figure 8–7

Rate enhancement by entropy reduction. Reactions of an ester with a carboxylate group to form an anhydride are depicted. The R group is the same in each case. **(a)** For this bimolecular reaction, the rate constant *k* is second order with units of $M^{-1}s^{-1}$. **(b)** When the two reacting groups are in a single molecule, the reaction is much faster. For this unimolecular reaction, *k* has units of s^{-1}. Dividing the rate constant for **(b)** by the rate constant for **(a)** gives a rate enhancement of about 10^5 M. (The enhancement has units of molarity because we are comparing a unimolecular and a bimolecular reaction.) Put another way, if the reactant in **(b)** were present at a concentration of 1 M, the reacting groups would *behave* as though they were present at a concentration of 10^5 M. Note that the reactant in **(b)** has freedom of rotation about three bonds (shown with curved arrows), but this still represents a substantial reduction of entropy over **(a)**. If the bonds that rotate in **(b)** are constrained as in **(c)**, the entropy is reduced further and the reaction exhibits a rate enhancement of 10^8 M relative to **(a)**.

interactions in the transition state. In either case the new enzyme conformation has enhanced catalytic properties. As we have seen, induced fit is a common feature of the reversible binding of ligands to proteins (Chapter 7). Induced fit is also important in the interaction of almost every enzyme with its substrate.

Specific Catalytic Groups Contribute to Catalysis

Once a substrate is bound to an enzyme, properly positioned catalytic functional groups aid in bond cleavage and formation by a variety of mechanisms, including general acid-base catalysis, covalent catalysis, and metal ion catalysis. These are distinct from mechanisms based on binding energy because they generally involve transient *covalent* interaction with a substrate or group transfer to or from a substrate.

General Acid-Base Catalysis Many biochemical reactions involve the formation of unstable charged intermediates that tend to break down rapidly to their constituent reactant species, thus impeding the reaction (Fig. 8–8).

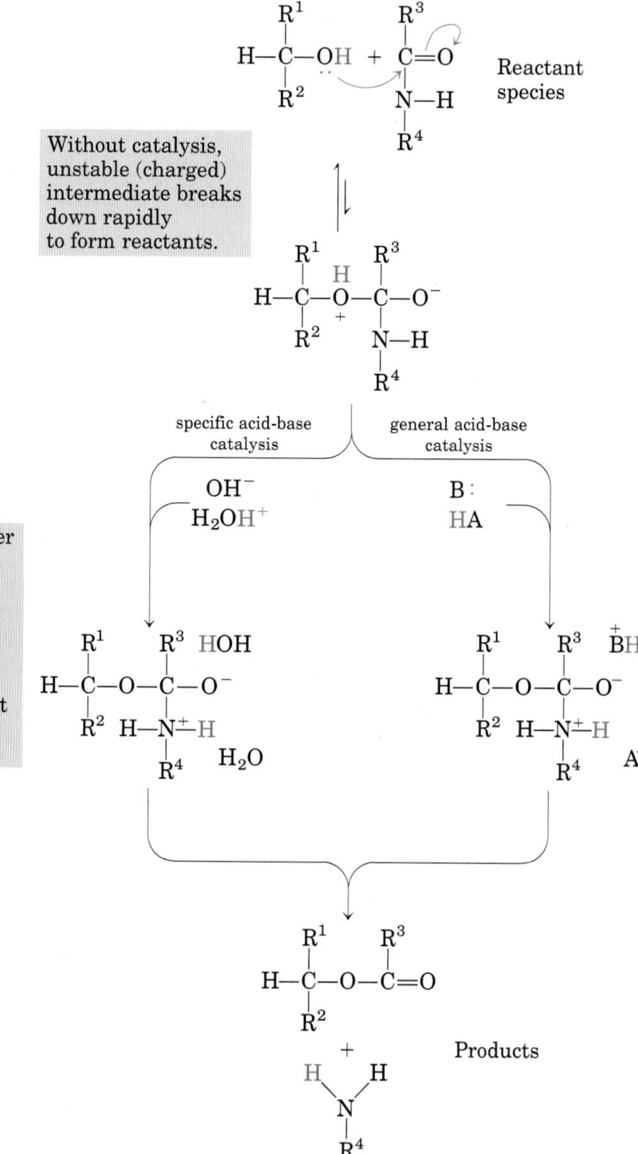

figure 8–8

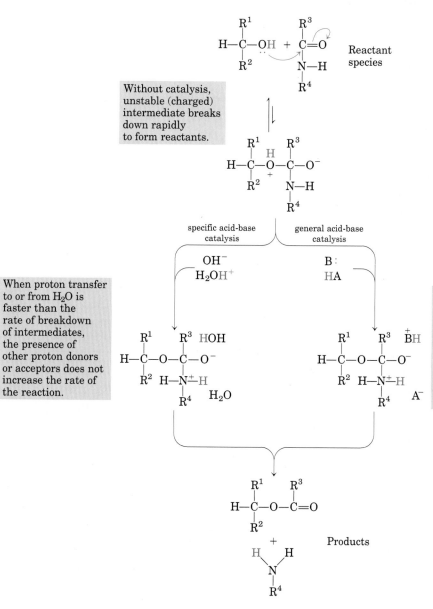

Unfavorable charge development during cleavage of an amide and its circumvention by catalysis. Shown here is hydrolysis of an amide bond, the same reaction as that catalyzed by chymotrypsin and other proteases. Charge development is unfavorable and can be circumvented by donation of a proton by H_3O^+ (specific acid catalysis) or HA (general acid catalysis), where HA represents any acid. Similarly, charge can be neutralized by proton abstraction by OH^- (specific base catalysis) or B**:** (general base catalysis), where B**:** represents any base.

Charged intermediates can often be stabilized by the transfer of protons to or from the substrate or intermediate to form a species that breaks down more readily to products than to reactants. For nonenzymatic reactions, the proton transfers can involve either the constituents of water alone or other weak proton donors or acceptors. Catalysis of this type that uses only the H^+ (H_3O^+) or OH^- ions present in water is referred to as **specific acid-base catalysis.** If protons are transferred between the intermediate and water faster than the intermediate breaks down to reactants, the intermediate is effectively stabilized every time it forms. No additional catalysis mediated by other proton acceptors or donors will occur. In many cases, however, water is not enough. The term **general acid-base catalysis** refers to proton transfers mediated by other classes of molecules. For nonenzymatic reactions in aqueous solutions, this occurs only when the unstable reaction intermediate breaks down to reactants faster than protons can be transferred to or from water. A variety of weak organic acids can supplement water as proton donors in this situation, or weak organic bases can serve as proton acceptors.

In the active site of an enzyme, a number of amino acid side chains can similarly act as proton donors and acceptors (Fig. 8–9). These groups can be precisely positioned in an enzyme active site to allow proton transfers, providing rate enhancements of the order of 10^2 to 10^5. This type of catalysis occurs on the vast majority of enzymes. In fact, proton transfers are the most common biochemical reactions.

Covalent Catalysis In this type of catalysis, a transient covalent bond is formed between the enzyme and the substrate. Consider the hydrolysis of a bond between groups A and B:

$$\text{A—B} \xrightarrow{H_2O} \text{A} + \text{B}$$

In the presence of a covalent catalyst (an enzyme with a nucleophilic group X:) the reaction becomes

$$\text{A—B} + \text{X:} \longrightarrow \text{A—X} + \text{B} \xrightarrow{H_2O} \text{A} + \text{X:} + \text{B}$$

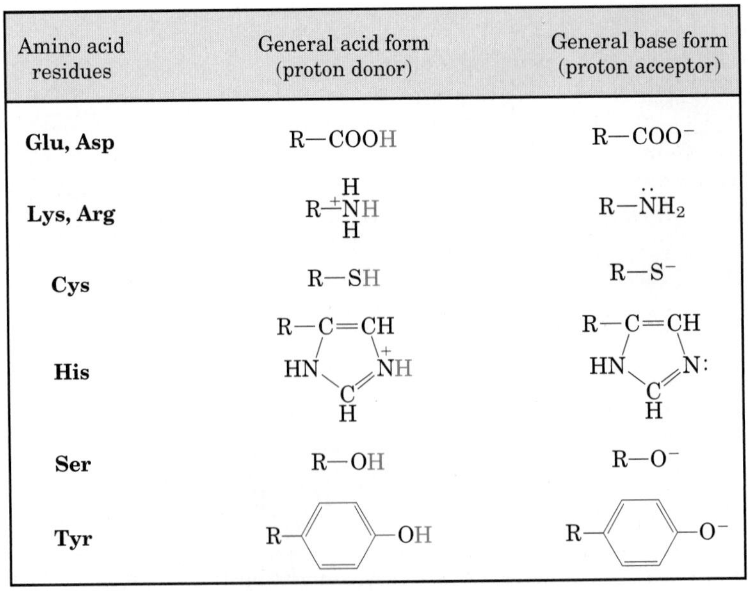

Amino acid residues	General acid form (proton donor)	General base form (proton acceptor)
Glu, Asp	R—COOH	$R—COO^-$
Lys, Arg	$R—\overset{+}{N}H_3$	$R—\ddot{N}H_2$
Cys	R—SH	$R—S^-$
His	R—C=CH, HN–CH=$\overset{+}{N}$H (imidazolium ring)	R—C=CH, HN–CH=N: (imidazole ring)
Ser	R—OH	$R—O^-$
Tyr	R—C₆H₄—OH	$R—C_6H_4—O^-$

figure 8–9

Amino acids in general acid-base catalysis. Many organic reactions are promoted by proton donors (general acids) or proton acceptors (general bases). The active sites of some enzymes contain amino acid functional groups, such as those shown here, that can participate in the catalytic process as proton donors or proton acceptors.

This alters the pathway of the reaction and results in catalysis only when the new pathway has a lower activation energy than the uncatalyzed pathway. Both of the new steps must be faster than the uncatalyzed reaction. A number of amino acid side chains, including all those in Figure 8–9, and the functional groups of some enzyme cofactors can serve as nucleophiles in the formation of covalent bonds with substrates. These covalent complexes always undergo further reaction to regenerate the free enzyme. The covalent bond formed between the enzyme and the substrate can activate a substrate for further reaction in a manner that is usually specific to the particular group or coenzyme.

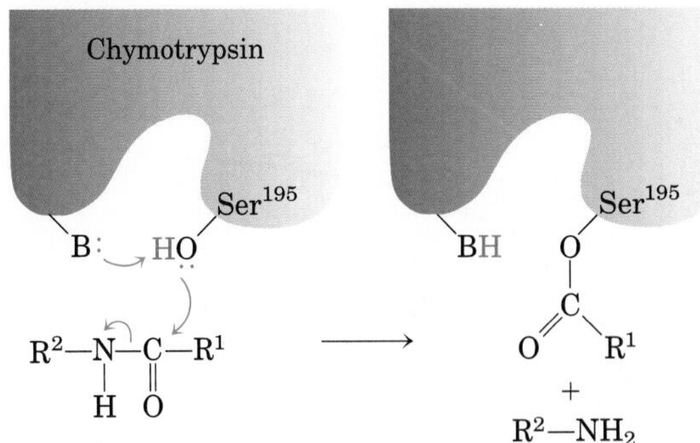

figure 8–10

Covalent and general acid-base catalysis. The first step in the reaction catalyzed by chymotrypsin is the acylation step. The hydroxyl group of Ser^{195} is the nucleophile in a reaction aided by general base catalysis (the base is the side chain of His^{57}). This provides a new pathway for the hydrolytic cleavage of a peptide bond. Catalysis occurs only if each step in the new pathway is faster than the uncatalyzed reaction. The chymotrypsin reaction is described in more detail in Figure 8–19.

Metal Ion Catalysis Metals, whether tightly bound to the enzyme or taken up from solution along with the substrate, can participate in catalysis in several ways. Ionic interactions between an enzyme-bound metal and a substrate can help orient the substrate for reaction or stabilize charged reaction transition states. This use of weak-bonding interactions between metal and substrate is similar to some of the uses of enzyme-substrate binding energy described earlier. Metals can also mediate oxidation-reduction reactions by reversible changes in the metal ion's oxidation state. Nearly a third of all known enzymes require one or more metal ions for catalytic activity.

Most enzymes employ a combination of several catalytic strategies to bring about a rate enhancement. A good example of the use of both covalent catalysis and general acid-base catalysis is the reaction catalyzed by chymotrypsin. The first step is cleavage of a peptide bond, which is accompanied by formation of a covalent linkage between a Ser residue on the enzyme and part of the substrate; the reaction is enhanced by general base catalysis by other groups on the enzyme (Fig. 8–10). The chymotrypsin reaction is described in more detail later in this chapter.

Enzyme Kinetics As an Approach to Understanding Mechanism

Multiple approaches are commonly used to study the mechanism of action of purified enzymes. A knowledge of the three-dimensional structure of the protein provides important information, and the value of structural information is greatly enhanced by classical protein chemistry and modern methods of site-directed mutagenesis (changing the amino acid sequence of a protein by genetic engineering; see Chapter 29). These technologies permit enzymologists to examine the role of individual amino acids in enzyme structure and action. However, the central approach to studying the mechanism of an enzyme-catalyzed reaction is to determine the *rate* of the reaction and how it changes in response to changes in experimental parameters, a discipline known as **enzyme kinetics.** This is the oldest approach to understanding enzyme mechanism and remains the most important today. We provide here a basic introduction to the kinetics of enzyme-catalyzed reactions. More advanced treatments are available in the sources cited at the end of the chapter.

Substrate Concentration Affects the Rate of Enzyme-Catalyzed Reactions

A key factor affecting the rate of a reaction catalyzed by a purified enzyme in vitro is the concentration of substrate, [S]. However, studying the effects of substrate concentration is complicated by the fact that [S] changes during the course of a reaction as substrate is converted to product. One

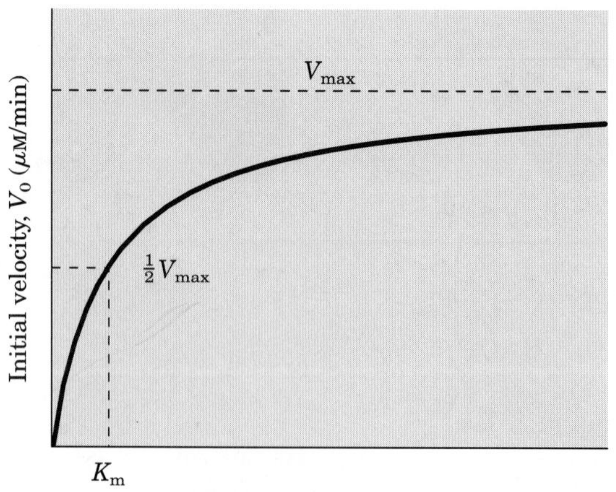

figure 8–11

Effect of substrate concentration on the initial velocity of an enzyme-catalyzed reaction. V_{max} can only be approximated from such a plot, because V_0 approaches but never quite reaches V_{max}. The substrate concentration at which V_0 is half maximal is K_m, the Michaelis constant. The concentration of enzyme in an experiment such as this is generally so low that [S] >> [E] even when [S] is described as low or relatively low. The units shown are typical for enzyme-catalyzed reactions and are given only to help illustrate the meaning of V_0 and [S]. (Note that the curve describes *part* of a rectangular hyperbola, with one asymptote at V_{max}. If the curve were continued below [S] = 0, it would approach a vertical asymptote at [S] = $-K_m$.)

Leonor Michaelis
1875–1949

Maud Menten
1879–1960

simplifying approach in kinetics experiments is to measure the **initial rate** (or **initial velocity**), designated V_0, when [S] is generally much greater than the concentration of enzyme, [E]. Then, if the time is sufficiently short following the start of a reaction, changes in [S] are negligible and [S] can be regarded as a constant.

The effect on V_0 of varying [S] when the enzyme concentration is held constant is shown in Figure 8–11. At relatively low concentrations of substrate, V_0 increases almost linearly with an increase in [S]. At higher substrate concentrations, V_0 increases by smaller and smaller amounts in response to increases in [S]. Finally, a point is reached beyond which increases in V_0 are vanishingly small as [S] increases. This plateau-like V_0 region is close to the **maximum velocity, V_{max}.**

The ES complex is the key to understanding this kinetic behavior, just as it was a starting point for our discussion of catalysis. The kinetic pattern in Figure 8–11 led Victor Henri, following the lead of Wurtz, to propose in 1903 that combination of an enzyme with its substrate molecule to form an ES complex is a necessary step in enzyme catalysis. This idea was expanded into a general theory of enzyme action, particularly by Leonor Michaelis and Maud Menten in 1913. They postulated that the enzyme first combines reversibly with its substrate to form an enzyme-substrate complex in a relatively fast reversible step:

$$E + S \underset{k_{-1}}{\overset{k_1}{\rightleftharpoons}} ES \quad (8\text{–}7)$$

The ES complex then breaks down in a slower second step to yield the free enzyme and the reaction product P:

$$ES \underset{k_{-2}}{\overset{k_2}{\rightleftharpoons}} E + P \quad (8\text{–}8)$$

Because the slower second reaction (Eqn 8–8) must limit the rate of the overall reaction, the overall rate must be proportional to the concentration of the species that reacts in the second step, that is, ES.

At any given instant in an enzyme-catalyzed reaction, the enzyme exists in two forms, the free or uncombined form E and the combined form ES. At low [S], most of the enzyme is in the uncombined form E. Here, the rate is proportional to [S] because the equilibrium of Equation 8–7 is pushed toward formation of more ES as [S] is increased. The maximum initial rate of the catalyzed reaction (V_{max}) is observed when virtually all of the enzyme is present as the ES complex and the concentration of E is vanishingly small. Under these conditions, the enzyme is "saturated" with its substrate, so that further increases in [S] have no effect on rate. This condition exists when [S] is sufficiently high that essentially all the free enzyme has been converted into the ES form. After the ES complex breaks down to yield the product P, the enzyme is free to catalyze reaction of another molecule of substrate. The saturation effect is a distinguishing characteristic of enzyme catalysts and is responsible for the plateau observed in Figure 8–11.

When the enzyme is first mixed with a large excess of substrate, there is an initial period, the **pre-steady state,** during which the concentration of ES builds up. This period is usually too short to be easily observed. The reaction quickly achieves a **steady state** in which [ES] (and the concentrations of any other intermediates) remains approximately constant over time. The concept of a steady state was introduced by G.E. Briggs and Haldane in 1925. The measured V_0 generally reflects the steady state even though V_0 is limited to early in the course of the reaction, and analysis of these initial rates is referred to as **steady-state kinetics.**

The Relationship between Substrate Concentration and Reaction Rate Can Be Expressed Quantitatively

The curve expressing the relationship between [S] and V_0 (Fig. 8–11) has the same general shape for most enzymes (it approaches a rectangular hyperbola), which can be expressed algebraically by the Michaelis-Menten equation. Michaelis and Menten derived this equation starting from their basic hypothesis that the rate-limiting step in enzymatic reactions is the breakdown of the ES complex to product and free enzyme. The equation is

$$V_0 = \frac{V_{max}[S]}{K_m + [S]} \qquad (8\text{–}9)$$

The important terms are [S], V_0, V_{max}, and a constant called the Michaelis constant, K_m. All of these terms are readily measured experimentally.

Here we develop the basic logic and the algebraic steps in a modern derivation of the Michaelis-Menten equation, which includes the steady-state assumption introduced by Briggs and Haldane. The derivation starts with the two basic steps involved in the formation and breakdown of ES (Eqns 8–7 and 8–8). Early in the reaction, the concentration of the product, [P], is negligible, and we make the simplifying assumption that k_{-2} (which describes the reverse reaction from P to S) can be ignored. This assumption is not critical but it simplifies our task. The overall reaction then reduces to

$$E + S \underset{k_{-1}}{\overset{k_1}{\rightleftharpoons}} ES \overset{k_2}{\rightleftharpoons} E + P \qquad (8\text{–}10)$$

V_0 is determined by the breakdown of ES to form product, which is determined by [ES]:

$$V_0 = k_2[ES] \qquad (8\text{–}11)$$

As [ES] in Equation 8–11 is not easily measured experimentally, we must begin by finding an alternative expression for [ES]. First, we introduce the term $[E_t]$, representing the total enzyme concentration (the sum of free and substrate-bound enzyme). Free or unbound enzyme can then be represented by $[E_t] - [ES]$. Also, because [S] is ordinarily far greater than $[E_t]$, the amount of substrate bound by the enzyme at any given time is negligible compared with the total [S]. With these conditions in mind, the following steps lead us to an expression for V_0 in terms of parameters that are easily measured.

Step 1. The rates of formation and breakdown of ES are determined by the steps governed by the rate constants k_1 (formation) and $k_{-1} + k_2$ (breakdown), according to the expressions

$$\text{Rate of ES formation} = k_1([E_t] - [ES])[S] \qquad (8\text{–}12)$$

$$\text{Rate of ES breakdown} = k_{-1}[ES] + k_2[ES] \qquad (8\text{–}13)$$

Step 2. An important assumption is now made that the initial rate of reaction reflects a steady state in which [ES] is constant—that is, the rate of formation of ES is equal to its rate of breakdown. This is called the steady-state assumption. The expressions in Equations 8–12 and 8–13 can be equated for the steady state, giving

$$k_1([E_t] - [ES])[S] = k_{-1}[ES] + k_2[ES] \qquad (8\text{–}14)$$

Step 3. A series of algebraic steps is now taken to solve Equation 8–14 for [ES]. First, the left side is multiplied out and the right side is simplified:

$$k_1[E_t][S] - k_1[ES][S] = (k_{-1} + k_2)[ES] \qquad (8\text{–}15)$$

Adding the term $k_1[\mathrm{ES}][\mathrm{S}]$ to both sides of the equation and simplifying gives

$$k_1[\mathrm{E_t}][\mathrm{S}] = (k_1[\mathrm{S}] + k_{-1} + k_2)[\mathrm{ES}] \tag{8–16}$$

We then solve this equation for [ES]:

$$[\mathrm{ES}] = \frac{k_1[\mathrm{E_t}][\mathrm{S}]}{k_1[\mathrm{S}] + k_{-1} + k_2} \tag{8–17}$$

This can now be simplified further, combining the rate constants into one expression:

$$[\mathrm{ES}] = \frac{[\mathrm{E_t}][\mathrm{S}]}{[\mathrm{S}] + (k_2 + k_{-1})/k_1} \tag{8–18}$$

The term $(k_2 + k_{-1})/k_1$ is defined as the **Michaelis constant, K_m.** Substituting this into Equation 8–18 simplifies the expression to

$$[\mathrm{ES}] = \frac{[\mathrm{E_t}][\mathrm{S}]}{K_m + [\mathrm{S}]} \tag{8–19}$$

Step 4. V_0 can now be expressed in terms of [ES]. Substituting the right side of Equation 8–19 for [ES] in Equation 8–11 gives

$$V_0 = \frac{k_2[\mathrm{E_t}][\mathrm{S}]}{K_m + [\mathrm{S}]} \tag{8–20}$$

This equation can be further simplified. Because the maximum velocity occurs when the enzyme is saturated, with $[\mathrm{ES}] = [\mathrm{E_t}]$, V_{max} can be defined as $k_2[\mathrm{E_t}]$. Substituting this in Equation 8–20 gives Equation 8–9:

$$V_0 = \frac{V_{max}[\mathrm{S}]}{K_m + [\mathrm{S}]}$$

This is the **Michaelis-Menten equation,** the **rate equation** for a one-substrate enzyme-catalyzed reaction. It is a statement of the quantitative relationship between the initial velocity V_0, the maximum initial velocity V_{max}, and the initial substrate concentration [S], all related through the Michaelis constant K_m. Note that K_m has units of concentration. Does the equation fit experimental observations? Yes; we can confirm this by considering the limiting situations where [S] is very high or very low, as shown in Figure 8–12.

An important numerical relationship emerges from the Michaelis-Menten equation in the special case when V_0 is exactly one-half V_{max} (Fig. 8–12). Then

$$\frac{V_{max}}{2} = \frac{V_{max}[\mathrm{S}]}{K_m + [\mathrm{S}]} \tag{8–21}$$

On dividing by V_{max}, we obtain

$$\frac{1}{2} = \frac{[\mathrm{S}]}{K_m + [\mathrm{S}]} \tag{8–22}$$

Solving for K_m, we get $K_m + [\mathrm{S}] = 2[\mathrm{S}]$, or

$$K_m = [\mathrm{S}], \quad \text{when } V_0 = \tfrac{1}{2}V_{max} \tag{8–23}$$

This represents a very useful, practical definition of K_m: K_m is equivalent to the substrate concentration at which V_0 is one-half V_{max}.

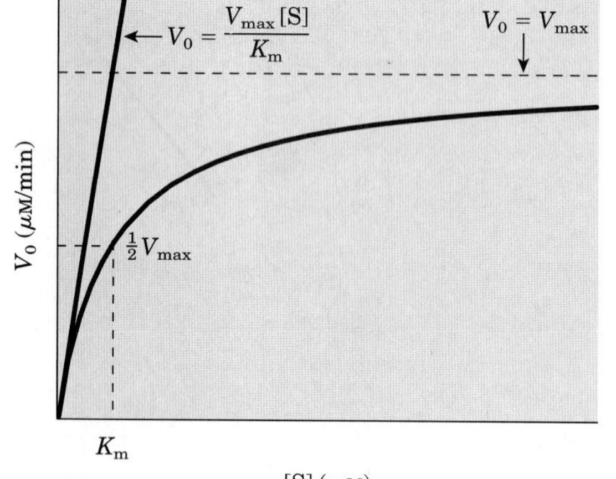

figure 8–12

Dependence of initial velocity on substrate concentration. This graph shows the kinetic parameters that define the limits of the curve at high and low [S]. At low [S], $K_m >> [\mathrm{S}]$ and the [S] term in the denominator of the Michaelis-Menten equation (Eqn 8–9) becomes insignificant. The equation simplifies to $V_0 = V_{max}[\mathrm{S}]/K_m$, and V_0 exhibits a linear dependence on [S], as observed. At high [S], where $[\mathrm{S}] >> K_m$, the K_m term in the denominator of the Michaelis-Menten equation becomes insignificant and the equation simplifies to $V_0 = V_{max}$; this is consistent with the plateau observed at high [S]. The Michaelis-Menten equation is therefore consistent with the observed dependence of V_0 on [S], and the shape of the curve is defined by the terms V_{max}/K_m at low [S] and V_{max} at high [S].

box 8–1 Transformations of the Michaelis-Menten Equation: The Double-Reciprocal Plot

The Michaelis-Menten equation

$$V_0 = \frac{V_{max}[S]}{K_m + [S]}$$

can be algebraically transformed into equations that are more useful in plotting experimental data. One common transformation is derived simply by taking the reciprocal of both sides of the Michaelis-Menten equation:

$$\frac{1}{V_0} = \frac{K_m + [S]}{V_{max}[S]}$$

Separating the components of the numerator on the right side of the equation gives

$$\frac{1}{V_0} = \frac{K_m}{V_{max}[S]} + \frac{[S]}{V_{max}[S]}$$

which simplifies to

$$\frac{1}{V_0} = \frac{K_m}{V_{max}[S]} + \frac{1}{V_{max}}$$

This form of the Michaelis-Menten equation is called the **Lineweaver-Burk equation.** For enzymes obeying the Michaelis-Menten relationship, a plot of $1/V_0$ versus $1/[S]$ (the "double-reciprocal" of the V_0 versus [S] plot we have been using to this point) yields a straight line (Fig. 1). This line has a slope of K_m/V_{max}, an intercept of $1/V_{max}$ on the $1/V_0$ axis, and an intercept of $-1/K_m$ on the $1/[S]$ axis. The double-reciprocal presentation, also called a Lineweaver-Burk plot, has the great advantage of allowing a more accurate determination of V_{max}, which can only be *approximated* from a simple plot of V_0 versus [S] (see Fig. 8–12).

Other transformations of the Michaelis-Menten equation have been derived, each with some particular advantage in analyzing enzyme kinetic data. (See problem 11 on page 291.)

The double-reciprocal plot of enzyme reaction rates is very useful in distinguishing between certain types of enzymatic reaction mechanisms (see Fig. 8–14) and in analyzing enzyme inhibition (see Box 8–2).

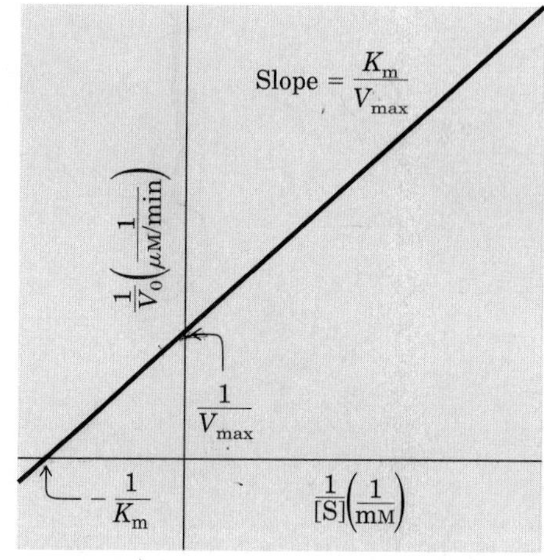

figure 1
A double-reciprocal or Lineweaver-Burk plot.

The Michaelis-Menten equation (Eqn 8–9) can be algebraically transformed into forms that are useful in the practical determination of K_m and V_{max} (Box 8–1) and, as we describe later, in the analysis of inhibitor action (see Box 8–2).

Kinetic Parameters Are Used to Compare Enzyme Activities

It is important to distinguish between the Michaelis-Menten equation and the specific kinetic mechanism on which it was originally based. The equation describes the kinetic behavior of a great many enzymes, and all enzymes that exhibit a hyperbolic dependence of V_0 on [S] are said to follow **Michaelis-Menten kinetics.** The practical rule that $K_m = [S]$ when $V_0 = \frac{1}{2}V_{max}$ (Eqn 8–23) holds for all enzymes that follow Michaelis-Menten kinetics. (The major exceptions to Michaelis-Menten kinetics are the regulatory enzymes, discussed at the end of this chapter.) However, the Michaelis-Menten equation does not depend on the relatively simple two-step reaction mechanism proposed by Michaelis and Menten (Eqn 8–10).

Many enzymes that follow Michaelis-Menten kinetics have quite different reaction mechanisms, and enzymes that catalyze reactions with six or eight identifiable steps often exhibit the same steady-state kinetic behavior. Even though Equation 8–23 holds true for many enzymes, both the magnitude and the real meaning of V_{max} and K_m can differ from one enzyme to the next. This is an important limitation of the steady-state approach to enzyme kinetics. V_{max} and K_m are parameters that can be obtained experimentally for any given enzyme, but by themselves they provide little information about the number, rates, or chemical nature of discrete steps in the reaction. Steady-state kinetics nevertheless represents the standard language by which the catalytic efficiencies of enzymes are characterized and compared. We now turn to the application and interpretation of the terms V_{max} and K_m.

A simple graphical method for obtaining an approximate value for K_m is shown in Figure 8–12. A more convenient procedure, using a **double-reciprocal plot,** is presented in Box 8–1. The K_m can vary greatly from enzyme to enzyme, and even for different substrates of the same enzyme (Table 8–6). The term is sometimes used (often inappropriately) as an indication of the affinity of an enzyme for its substrate. The actual meaning of K_m depends on specific aspects of the reaction mechanism such as the number and relative rates of the individual steps of the reaction. For reactions with two steps,

$$K_m = \frac{k_2 + k_{-1}}{k_1} \tag{8–24}$$

When k_2 is rate-limiting, $k_2 << k_{-1}$ and K_m reduces to k_{-1}/k_1, which is defined as the **dissociation constant,** K_d, of the ES complex. Where these conditions hold, K_m does represent a measure of the affinity of the enzyme for its substrate in the ES complex. However, this scenario does not apply for most enzymes. Sometimes $k_2 >> k_{-1}$, and then $K_m = k_2/k_1$. In other cases, k_2 and k_{-1} are comparable and K_m remains a more complex function of all three rate constants (Eqn 8–24). The Michaelis-Menten equation and the characteristic saturation behavior of the enzyme still apply, but K_m cannot be considered a simple measure of substrate affinity. Even more common are cases in which the reaction goes through multiple steps after formation of ES; K_m can then become a very complex function of many rate constants.

The quantity V_{max} also varies greatly from one enzyme to the next. If an enzyme reacts by the two-step Michaelis-Menten mechanism, V_{max} is equivalent to $k_2[E_t]$, where k_2 is rate-limiting. However, the number of reaction

table 8–6

K_m for Some Enzymes and Substrates

Enzyme	Substrate	K_m (mM)
Catalase	H_2O_2	25
Hexokinase (brain)	ATP	0.4
	D-Glucose	0.05
	D-Fructose	1.5
Carbonic anhydrase	HCO_3^-	26
Chymotrypsin	Glycyltyrosinylglycine	108
	N-Benzoyltyrosinamide	2.5
β-Galactosidase	D-Lactose	4.0
Threonine dehydratase	L-Threonine	5.0

steps and the identity of the rate-limiting step(s) can vary from enzyme to enzyme. For example, consider the quite common situation where product release, EP → E + P, is rate-limiting. Early in the reaction (when [P] is low), the overall reaction can be described by the scheme

$$E + S \underset{k_{-1}}{\overset{k_1}{\rightleftharpoons}} ES \underset{k_{-2}}{\overset{k_2}{\rightleftharpoons}} EP \overset{k_3}{\rightleftharpoons} E + P \quad (8\text{–}25)$$

In this case, most of the enzyme is in the EP form at saturation, and $V_{max} = k_3[E_t]$. It is useful to define a more general rate constant, $\boldsymbol{k_{cat}}$, to describe the limiting rate of any enzyme-catalyzed reaction at saturation. If there are several steps in the reaction and one is clearly rate-limiting, k_{cat} is equivalent to the rate constant for that limiting step. For the simple reaction of Equation 8–10, $k_{cat} = k_2$. For the reaction of Equation 8–25, $k_{cat} = k_3$. When several steps are partially rate-limiting, k_{cat} can become a complex function of several of the rate constants that define each individual reaction step. In the Michaelis-Menten equation, $k_{cat} = V_{max}/[E_t]$, and Equation 8–9 becomes

$$V_0 = \frac{k_{cat}[E_t][S]}{K_m + [S]} \quad (8\text{–}26)$$

The constant k_{cat} is a first-order rate constant and hence has units of reciprocal time. It is also called the **turnover number.** It is equivalent to the number of substrate molecules converted to product in a given unit of time on a single enzyme molecule when the enzyme is saturated with substrate. The turnover numbers of several enzymes are given in Table 8–7.

The kinetic parameters k_{cat} and K_m are generally useful for the study and comparison of different enzymes, whether their reaction mechanisms are simple or complex. Each enzyme has optimum values of k_{cat} and K_m that reflect the cellular environment, the concentration of substrate normally encountered in vivo by the enzyme, and the chemistry of the reaction being catalyzed.

The parameters k_{cat} and K_m also allow us to evaluate the kinetic efficiency of enzymes, but either parameter alone is insufficient for this task. Two enzymes catalyzing different reactions may have the same k_{cat} (turnover number), yet the rates of the uncatalyzed reactions may be different and thus the rate enhancements brought about by the enzymes may differ greatly. Experimentally, the K_m for an enzyme tends to be similar to the cellular concentration of its substrate. An enzyme that acts on a substrate present at a very low concentration in the cell usually has a lower K_m than an enzyme that acts on a substrate that is more abundant.

The best way to compare the catalytic efficiencies of different enzymes or the turnover of different substrates by the same enzyme is to compare the ratio k_{cat}/K_m for the two reactions. This parameter, sometimes called the

table 8–7

Turnover Numbers (k_{cat}) of Some Enzymes

Enzyme	Substrate	k_{cat} (s^{-1})
Catalase	H_2O_2	40,000,000
Carbonic anhydrase	HCO_3^-	400,000
Acetylcholinesterase	Acetylcholine	140,000
β-Lactamase	Benzylpenicillin	2,000
Fumarase	Fumarate	800
RecA protein (an ATPase)	ATP	0.4

specificity constant, is the rate constant for the conversion of E+S to E+P. When [S] << K_m, Equation 8–26 reduces to the form

$$V_0 = \frac{k_{cat}}{K_m}[E_t][S] \qquad (8\text{–}27)$$

V_0 in this case depends on the concentration of two reactants, [E_t] and [S]; therefore this is a second-order rate equation and the constant k_{cat}/K_m is a second-order rate constant with units of $M^{-1}s^{-1}$. There is an upper limit to k_{cat}/K_m, imposed by the rate at which E and S can diffuse together in an aqueous solution. This diffusion-controlled limit is 10^8 to 10^9 $M^{-1}s^{-1}$, and many enzymes have a k_{cat}/K_m near this range (Table 8–8). Such enzymes are said to have achieved catalytic perfection. Note that different values of k_{cat} and K_m can produce the maximum ratio.

table 8–8

Enzymes for Which k_{cat}/K_m Is Close to the Diffusion-Controlled Limit (10^8 to 10^9 $M^{-1}s^{-1}$)

Enzyme	Substrate	k_{cat} (s^{-1})	K_m (M)	k_{cat}/K_m ($M^{-1}s^{-1}$)
Acetylcholinesterase	Acetylcholine	1.4×10^4	9×10^{-5}	1.6×10^8
Carbonic anhydrase	CO_2	1×10^6	1.2×10^{-2}	8.3×10^7
	HCO_3^-	4×10^5	2.6×10^{-2}	1.5×10^7
Catalase	H_2O_2	4×10^7	1.1	4×10^7
Crotonase	Crotonyl-CoA	5.7×10^3	2×10^{-5}	2.8×10^8
Fumarase	Fumarate	8×10^2	5×10^{-6}	1.6×10^8
	Malate	9×10^2	2.5×10^{-5}	3.6×10^7
β-Lactamase	Benzylpenicillin	2.0×10^3	2×10^{-5}	1×10^8
Triose phosphate isomerase	Glyceraldehyde 3-phosphate	4.3×10^3	4.7×10^{-4}	2.4×10^8

Source: Fersht, A. (1999) *Structure and Mechanism in Protein Science,* p. 166, W.H. Freeman and Company, New York.

Many Enzymes Catalyze Reactions with Two or More Substrates

We have seen how [S] affects the rate of a simple enzyme reaction (S → P) with only one substrate molecule. In many enzymatic reactions, however, two (and sometimes more than two) different substrate molecules bind to the enzyme and participate in the reaction. For example, in the reaction catalyzed by hexokinase, ATP and glucose are the substrate molecules, and ADP and glucose 6-phosphate are the products:

$$\text{ATP + glucose} \longrightarrow \text{ADP + glucose 6-phosphate}$$

The rates of such bisubstrate reactions can also be analyzed by the Michaelis-Menten approach. Hexokinase has a characteristic K_m for each of its two substrates (Table 8–6).

Enzymatic reactions with two substrates usually involve transfer of an atom or a functional group from one substrate to the other. These reactions proceed by one of several different pathways. In some cases, both substrates are bound to the enzyme concurrently at some point in the course of the reaction, forming a noncovalent ternary complex (Fig. 8–13a). Such a complex can be formed by substrates binding in a random sequence or in a specific order. No ternary complex is formed if the first substrate is converted to product and dissociates before the second substrate binds. An example of this is the Ping-Pong or double-displacement mechanism

(a) Enzyme reaction involving a ternary complex

Random order

$$E \rightleftharpoons ES_1 \rightleftharpoons ES_1S_2 \longrightarrow E + P_1 + P_2$$
$$E \rightleftharpoons ES_2 \rightleftharpoons ES_1S_2$$

Ordered

$$E + S_1 \rightleftharpoons ES_1 \overset{S_2}{\rightleftharpoons} ES_1S_2 \longrightarrow E + P_1 + P_2$$

(b) Enzyme reaction in which no ternary complex is formed

$$E + S_1 \rightleftharpoons ES_1 \rightleftharpoons E'P_1 \overset{P_1}{\rightleftharpoons} E' \overset{S_2}{\rightleftharpoons} E'S_2 \longrightarrow E + P_2$$

figure 8–13

Common mechanisms for enzyme-catalyzed bisubstrate reactions. In **(a)** the enzyme and both substrates come together to form a ternary complex. In ordered binding, substrate 1 must bind before substrate 2 can bind productively. In random binding, the substrates can bind in either order. In **(b)** an enzyme-substrate complex forms, a product leaves the complex, the altered enzyme forms a second complex with another substrate molecule, and the second product leaves, regenerating the enzyme. Substrate 1 may transfer a functional group to the enzyme (to form the covalently modified E′), which is subsequently transferred to substrate 2. This is a Ping-Pong or double-displacement mechanism.

(Fig. 8–13b). Steady-state kinetics can often help distinguish among these possibilities (Fig. 8–14).

Pre-Steady State Kinetics Can Provide Evidence for Specific Reaction Steps

We have introduced kinetics as the primary method for studying the steps in an enzymatic reaction, and we have also outlined the limitations of the most common kinetic parameters in providing such information. The two most important experimental parameters provided by steady-state kinetics are k_{cat} and k_{cat}/K_m. Variation in these parameters with changes in pH or temperature can provide additional information about steps in a reaction pathway. In the case of bisubstrate reactions, steady-state kinetics can help determine whether a ternary complex is formed during the reaction (Fig. 8–14). A more complete picture generally requires more sophisticated kinetic methods that go beyond the scope of an introductory text. Here, we briefly introduce one of the most important kinetic approaches for studying reaction mechanisms, pre-steady state kinetics.

A complete description of an enzyme-catalyzed reaction requires direct measurement of the rates of individual reaction steps, for example the measurement of the association of enzyme and substrate to form the ES complex. It is during the pre-steady state that the rates of many reaction steps can be measured independently. Reaction conditions are adjusted to facilitate the measurement of events that occur during reaction of a single substrate molecule. Because the pre-steady state phase is generally very short, this often requires specialized techniques for very rapid mixing and sampling. One objective is to gain a complete and quantitative picture of the energy changes during the reaction. As we have already noted, reaction rates and equilibria are related to the free-energy changes that occur during the reaction. Measuring the rate of individual reaction steps reveals how energy is used by a specific enzyme, which is an important component of the overall reaction mechanism. In a number of cases it has proved possible to measure the rates of every individual step in a multistep enzymatic reaction. Some examples of the application of pre-steady state kinetics are included in the descriptions of specific enzymes later in this chapter.

Enzymes Are Subject to Inhibition

Enzyme inhibitors are molecular agents that interfere with catalysis, slowing or halting enzymatic reactions. Enzymes catalyze virtually all cellular processes, so it should not be surprising that enzyme inhibitors are among

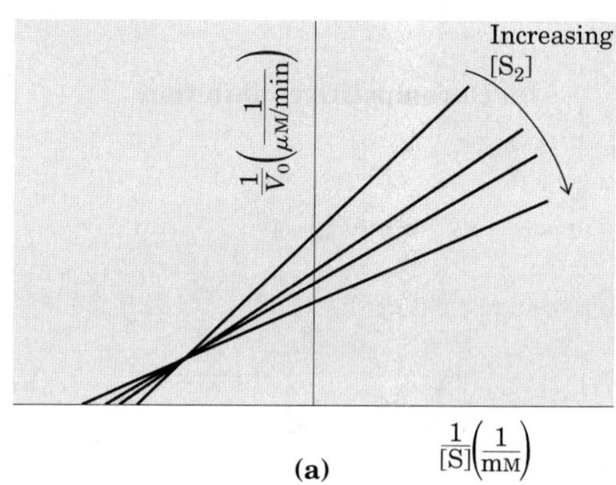

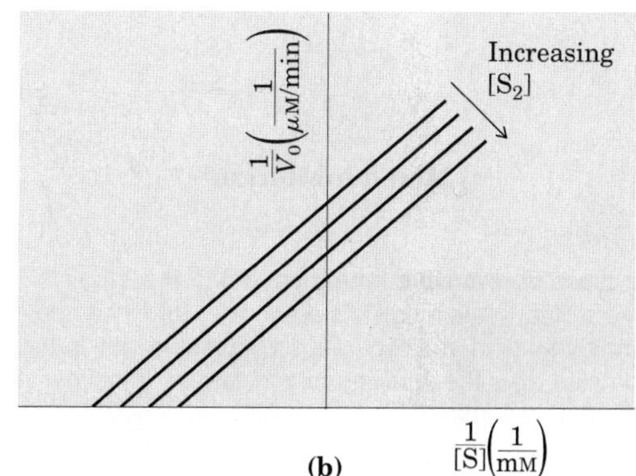

figure 8–14

Steady-state kinetic analysis of bisubstrate reactions. In these double-reciprocal plots (see Box 8–1), the concentration of substrate 1 is varied while the concentration of substrate 2 is held constant. This is repeated for several values of [S_2], generating several separate lines. Intersecting lines indicate that a ternary complex is formed in the reaction **(a);** parallel lines indicate a Ping-Pong or double-displacement pathway **(b).**

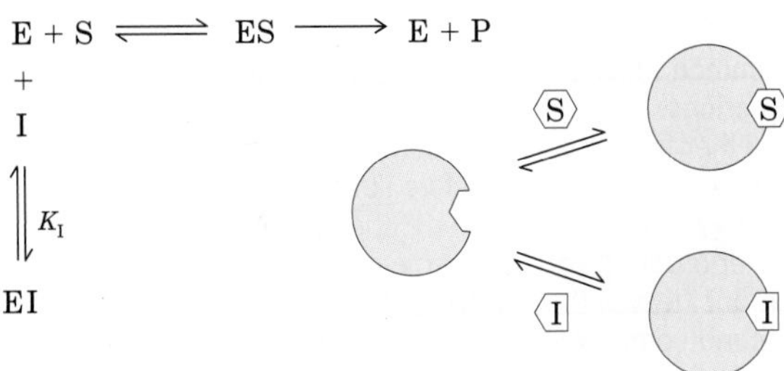

(a) Competitive inhibition

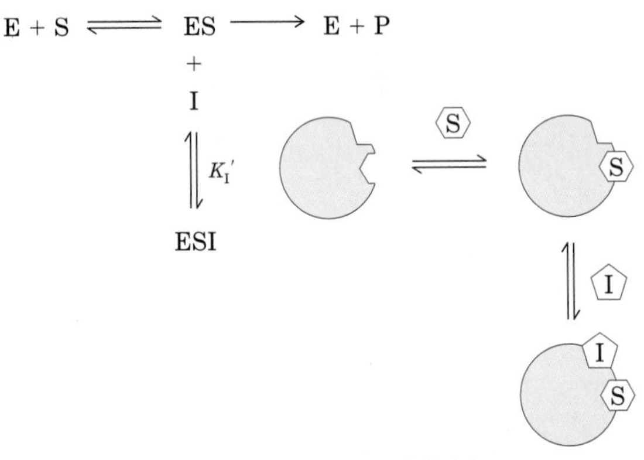

(b) Uncompetitive inhibition

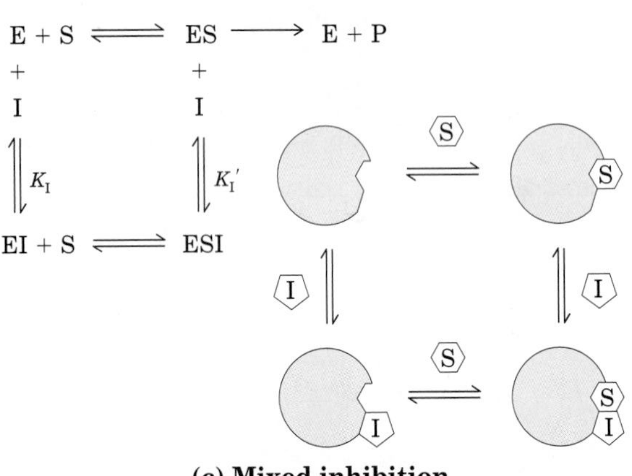

(c) Mixed inhibition

figure 8–15

Three types of reversible inhibition. (a) Competitive inhibitors bind to the enzyme's active site. **(b)** Uncompetitive inhibitors bind at a separate site, but bind only to the ES complex. K_I is the equilibrium constant for inhibitor binding to E, whereas K_I' is the equilibrium constant for inhibitor binding to ES. **(c)** Mixed inhibitors bind at a separate site, but may bind to either E or ES.

the most important pharmaceutical agents known. For example, aspirin (acetylsalicylate) inhibits the enzyme that catalyzes the first step in the synthesis of prostaglandins, compounds involved in many processes including some that produce pain. The study of enzyme inhibitors also has provided valuable information about enzyme mechanisms and has helped define some metabolic pathways. There are two broad classes of enzyme inhibitors: reversible and irreversible.

Reversible Inhibition Can Be Competitive, Uncompetitive, or Mixed

One common type of **reversible inhibition** is called competitive (Fig. 8–15a). A **competitive inhibitor** competes with the substrate for the active site of an enzyme. While the inhibitor (I) occupies the active site it prevents binding of the substrate to the enzyme. Competitive inhibitors are often compounds that resemble the substrate and combine with the enzyme to form an EI complex, but without leading to catalysis. Even fleeting combinations of this type will negatively affect the efficiency of the enzyme. By taking into account the molecular geometry of inhibitors that resemble the substrate, we can often reach conclusions about which parts of the normal substrate bind to the enzyme. Competitive inhibition can be analyzed quantitatively by steady-state kinetics. In the presence of a competitive inhibitor, the Michaelis-Menten equation (Eqn 8–9) becomes

$$V_0 = \frac{V_{max}[S]}{\alpha K_m + [S]} \tag{8–28}$$

where

$$\alpha = 1 + \frac{[I]}{K_I}$$

and

$$K_I = \frac{[E][I]}{[EI]}$$

Equation 8–28 describes the important features of competitive inhibition. The experimentally determined term αK_m, the K_m observed in the presence of the inhibitor, is often called the "apparent" K_m.

Because the inhibitor binds reversibly to the enzyme, the competition can be biased to favor the substrate simply by adding more substrate. When [S] far exceeds [I], the probability that an inhibitor molecule will bind to the enzyme is minimized, and the reaction exhibits a normal V_{max}. However, the [S] at which $V_0 = \frac{1}{2} V_{max}$, the apparent K_m, will increase in the presence of inhibitor by the factor α. This effect on the apparent K_m combined with the absence of an effect on V_{max} is diagnostic of competitive inhibition and is readily revealed in a double-reciprocal plot (Box 8–2). The equilibrium constant for inhibitor binding, K_I, can be obtained from the same plot.

A medical therapy based on competition at the active site is used to treat patients who have ingested methanol, a solvent found in gas-line antifreeze. The liver enzyme alcohol dehydrogenase converts methanol to formaldehyde, which is damaging to many tissues. Blindness is a common result of methanol ingestion because the eyes are particularly sensitive to formaldehyde. Ethanol competes effectively with methanol as an alternative substrate for alcohol dehydrogenase. The effect of ethanol is much like that of a competitive inhibitor, with the distinction that ethanol is also a substrate and its concentration will decrease over time as the enzyme converts it to acetaldehyde. The therapy for methanol poisoning is a gradual intravenous infusion of ethanol at a rate that maintains a controlled concen-

box 8–2 Kinetic Tests for Determining Inhibition Mechanisms

The double-reciprocal plot (see Box 8–1) offers an easy way of determining whether an enzyme inhibitor is competitive, uncompetitive, or mixed. Two sets of rate experiments are carried out, with the enzyme concentration held constant in each set. In the first set, [S] is also held constant, permitting measurement of the effect of increasing inhibitor concentration [I] on the initial rate V_0 (not shown). In the second set, [I] is held constant but [S] is varied. The results are plotted as $1/V_0$ versus 1/[S].

Figure 1 shows a set of double-reciprocal plots, one obtained in the absence of the inhibitor and two at different concentrations of a competitive inhibitor. Increasing [I] results in a family of lines with a common intercept on the $1/V_0$ axis but with different slopes. Because the intercept on the $1/V_0$ axis is equal to $1/V_{max}$, we know that V_{max} is unchanged by the presence of a competitive inhibitor. That is, regardless of the concentration of a competitive inhibitor, a sufficiently high substrate concentration will always displace the inhibitor from the enzyme's active site. Above the graph is the rearrangement of Equation 8–28 on which the plot is based. The value of α can be calculated from the change in slope at any given [I]. Knowing [I] and α, we can calculate K_I from the expression

$$\alpha = 1 + \frac{[I]}{K_I}$$

For uncompetitive and mixed inhibition, similar plots of rate data give the families of lines shown in Figures 2 and 3. Changes in axis intercepts signal changes in V_{max} and K_m.

$$\frac{1}{V_0} = \left(\frac{\alpha K_m}{V_{max}}\right)\frac{1}{[S]} + \frac{1}{V_{max}}$$

$\frac{1}{V_0}\left(\frac{1}{\mu M/min}\right)$
[I]
$\alpha = 3$
$\alpha = 2$
$\alpha = 1$
No inhibitor
$\frac{1}{V_{max}}$
Slope = $\frac{\alpha K_m}{V_{max}}$
$\frac{1}{[S]}\left(\frac{1}{mM}\right)$

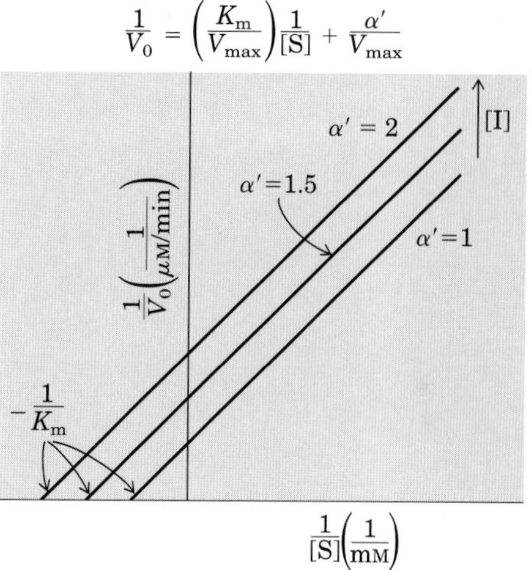

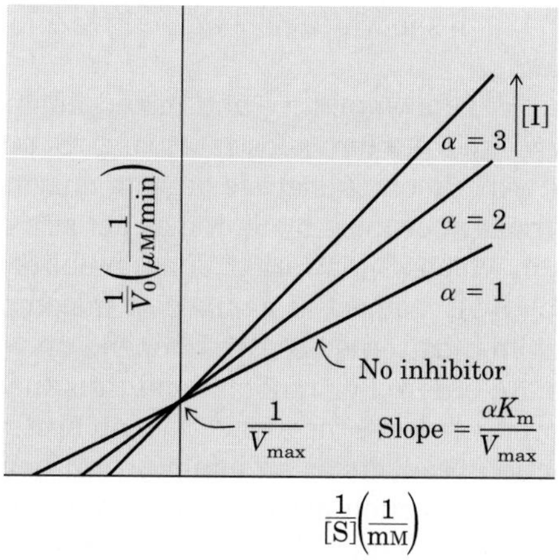

figure 1
Competitive inhibition.

$$\frac{1}{V_0} = \left(\frac{K_m}{V_{max}}\right)\frac{1}{[S]} + \frac{\alpha'}{V_{max}}$$

$\alpha' = 2$
[I]
$\alpha'=1.5$
$\alpha'=1$
$\frac{1}{V_0}\left(\frac{1}{\mu M/min}\right)$
$-\frac{1}{K_m}$
$\frac{1}{[S]}\left(\frac{1}{mM}\right)$

figure 2
Uncompetitive inhibition.

$$\frac{1}{V_0} = \left(\frac{\alpha K_m}{V_{max}}\right)\frac{1}{[S]} + \frac{\alpha'}{V_{max}}$$

[I]
$\frac{1}{V_0}\left(\frac{1}{\mu M/min}\right)$
No inhibitor
$\frac{1}{[S]}\left(\frac{1}{mM}\right)$

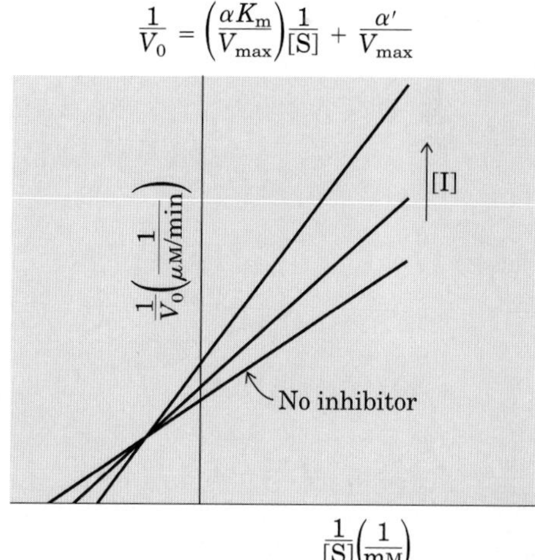

figure 3
Mixed inhibition.

tration in the bloodstream for several hours. This slows the formation of formaldehyde, lessening the danger while the kidneys filter out the methanol to be excreted harmlessly in the urine.

Two other types of reversible inhibition, uncompetitive and mixed, are often defined in terms of one-substrate enzymes but in practice are observed only with enzymes having two or more substrates. An **uncompetitive inhibitor** (Fig. 8–15b) binds at a site distinct from the substrate active site and, unlike a competitive inhibitor, binds only to the ES complex. In the presence of an uncompetitive inhibitor, the Michaelis-Menten equation is altered to

$$V_0 = \frac{V_{max}[S]}{K_m + \alpha'[S]} \tag{8–29}$$

where

$$\alpha' = 1 + \frac{[I]}{K'_I}$$

and

$$K'_I = \frac{[ES][I]}{[ESI]}$$

As described by Equation 8–29, at high concentrations of substrate, V_0 approaches V_{max}/α'. Thus, an uncompetitive inhibitor lowers the measured V_{max}. The apparent K_m also decreases, because the [S] required to reach one-half V_{max} decreases by the factor α'.

A **mixed inhibitor** (Fig. 8–15c) also binds at a site distinct from the substrate active site, but it will bind to either E or ES. The rate equation describing mixed inhibition is

$$V_0 = \frac{V_{max}[S]}{\alpha K_m + \alpha'[S]} \tag{8–30}$$

where α and α' are defined as before. A mixed inhibitor usually affects both K_m and V_{max}. The special case of $\alpha = \alpha'$, rarely encountered in practice, classically has been defined as **noncompetitive inhibition.** Examine Equation 8–30 to see why a noncompetitive inhibitor would affect the V_{max} but not the K_m.

In practice, uncompetitive and mixed inhibition are observed only for enzymes with two or more substrates (e.g., S_1 and S_2), and are very important in the experimental analysis of such enzymes. If an inhibitor binds to the site normally occupied by S_1, it may act as a competitive inhibitor in experiments in which $[S_1]$ is varied. If an inhibitor binds to the site normally occupied by S_2, it may act as a mixed or uncompetitive inhibitor of S_1. The actual inhibition patterns observed depend on whether the S_1 and S_2 binding events are ordered or random, and thus the order in which substrates bind and products leave the active site can be determined. Often, using one of the reaction products as an inhibitor can be particularly informative. If only one of two reaction products is present, no reverse reaction can take place. However, a product will generally bind to some part of the active site and thus serve as an inhibitor. Inhibition studies are often elaborate and can provide a detailed picture of the mechanism of a bisubstrate reaction.

Irreversible Inhibition Is an Important Tool in Enzyme Research and Pharmacology

Irreversible inhibitors are those that combine with or destroy a functional group on an enzyme that is essential for the enzyme's activity, or that form a particularly stable noncovalent association. Formation of a covalent link between an irreversible inhibitor and an enzyme is common. Irreversible inhibitors are another useful tool for studying reaction mechanisms. Amino acids with key catalytic functions in the active site can some-

$$\underset{(\mathrm{Ser}^{195})}{\mathrm{Enz{-}CH_2{-}OH}} + \underset{\mathrm{DIFP}}{\mathrm{F{-}P({=}O)(O{-}CH(CH_3)_2){-}O{-}CH(CH_3)_2}} \xrightarrow{\mathrm{H^+ + F^-}} \mathrm{Enz{-}CH_2{-}O{-}P({=}O)(O{-}CH(CH_3)_2){-}O{-}CH(CH_3)_2}$$

figure 8–16

Irreversible inhibition. Reaction of chymotrypsin with diisopropylfluorophosphate (DIFP) irreversibly inhibits the enzyme. This has led to the conclusion that Ser^{195} is the key active-site serine residue in chymotrypsin.

times be identified by determining which amino acid is covalently linked to an inhibitor after the enzyme is inactivated. An example is shown in Figure 8–16.

A special class of irreversible inhibitors is the **suicide inactivators.** These compounds are relatively unreactive until they bind to the active site of a specific enzyme. A suicide inactivator is designed to carry out the first few chemical steps of the normal enzyme reaction, but instead of being transformed into the normal product, the inactivator is converted to a very reactive compound that combines irreversibly with the enzyme. These compounds are also called **mechanism-based inactivators,** because they utilize the normal enzyme reaction mechanism to inactivate the enzyme. Suicide inactivators play a central role in rational drug design, a modern approach to obtaining new pharmaceutical agents in which novel substrates are synthesized based on our knowledge of known substrates and reaction mechanisms. A well-designed suicide inactivator is specific for a single enzyme and is unreactive until within that enzyme's active site, so drugs based on this approach can offer the important advantage of few side effects (see Box 22–2).

Enzyme Activity Is Affected by pH

Enzymes have an optimum pH (or pH range) at which their activity is maximal (Fig. 8–17); at higher or lower pH, activity decreases. This is not surprising. Amino acid side chains in the active site may act as weak acids and bases with critical functions that depend on their maintaining a certain state of ionization, and elsewhere in the protein, ionized side chains may play an essential role in the interactions that maintain protein structure. Removing a proton from a His residue, for example, might eliminate an ionic interaction essential for stabilization of the active conformation of the enzyme. A less common cause of pH sensitivity is titration of a group on the substrate.

The pH range over which an enzyme undergoes changes in activity can provide a clue to what amino acid is involved (see Table 5–1). A change in activity near pH 7.0, for example, often reflects titration of a His residue. The effects of pH must be interpreted with some caution, however. In the closely packed environment of a protein, the $\mathrm{p}K_\mathrm{a}$ of amino acid side chains can be significantly altered. For example, a nearby positive charge can lower the $\mathrm{p}K_\mathrm{a}$ of a Lys residue, and a nearby negative charge can increase it. Such effects sometimes result in a $\mathrm{p}K_\mathrm{a}$ that is shifted by 2 or more pH units from its normal (free amino acid) value. In the enzyme acetoacetate decarboxylase, one Lys residue (normal $\mathrm{p}K_\mathrm{a}$ 10.5) has a $\mathrm{p}K_\mathrm{a}$ of 6.6 due to electrostatic effects of nearby positive charges.

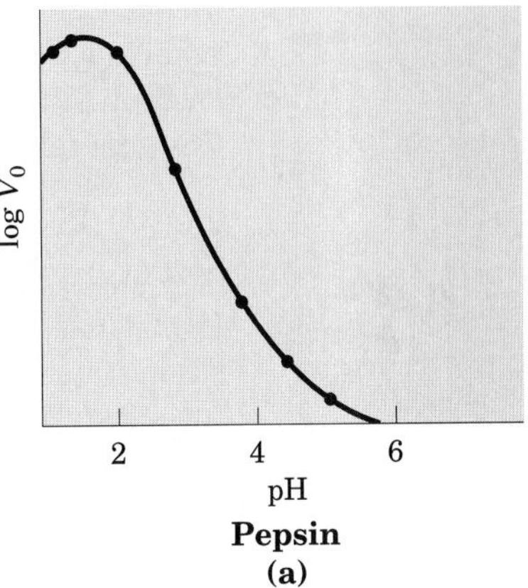

Pepsin
(a)

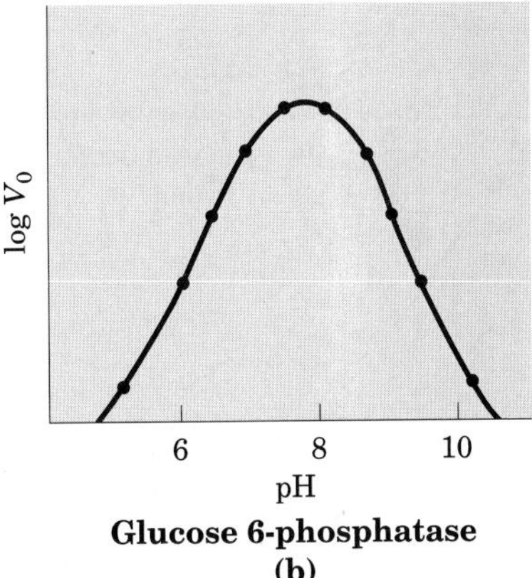

Glucose 6-phosphatase
(b)

figure 8–17

The pH-activity profiles of two enzymes. These curves are constructed from measurements of initial velocities when the reaction is carried out in buffers of different pH. Because pH is a logarithmic scale reflecting tenfold changes in $[\mathrm{H}^+]$, the changes in V_0 are also plotted on a logarithmic scale. The pH optimum for the activity of an enzyme is generally close to the pH of the environment in which the enzyme is normally found. **(a)** Pepsin, which hydrolyzes certain peptide bonds of proteins during digestion in the stomach, has a pH optimum of about 1.6. The pH of gastric juice is between 1 and 2. **(b)** Glucose 6-phosphatase of hepatocytes (liver cells), with a pH optimum of about 7.8, is responsible for releasing glucose into the blood. The normal pH of the cytosol of hepatocytes is about 7.2.

Examples of Enzymatic Reactions

This chapter has focused thus far on the general principles of catalysis and on introducing some of the kinetic parameters used to describe enzyme action. Principles and kinetics are combined in Box 8–3, which describes some of the evidence indicating that binding energy and transition-state complementarity are central to enzymatic catalysis. We now turn to several examples of specific enzyme reaction mechanisms.

box 8–3 Evidence for Enzyme–Transition State Complementarity

The transition state of a reaction is difficult to study because it is so short-lived. To understand enzymatic catalysis, however, we must dissect the interaction between the enzyme and this ephemeral moment in the course of a reaction. Complementarity between an enzyme and the transition state is virtually a requirement for catalysis, because the energy hill upon which the transition state sits is what the enzyme must lower if catalysis is to occur. How can we obtain evidence for enzyme–transition state complementarity? Fortunately, we have a variety of approaches, old and new, to address this problem, each providing compelling evidence in support of this general principle of enzyme action.

Structure-Activity Correlations

If enzymes are complementary to reaction transition states, then some functional groups in both the substrate and the enzyme must interact preferentially in the transition state rather than in the ES complex. Altering these groups should have little effect on formation of the ES complex and hence should not affect kinetic parameters (the dissociation constant, K_d; or sometimes K_m if $K_d = K_m$) that reflect the $E + S \rightleftharpoons ES$ equilibrium. Changing these same groups should have a large effect on the overall rate (k_{cat} or k_{cat}/K_m) of the reaction, however, because the bound substrate lacks potential binding interactions needed to lower the activation energy.

An excellent example of this effect is seen in the kinetics associated with a series of related substrates for the enzyme chymotrypsin (Fig. 1). Chymotrypsin normally catalyzes the hydrolysis of peptide bonds next to aromatic amino acids. The substrates shown in Figure 1 are convenient smaller models for the natural substrates (long polypeptides and proteins; see Chapter 5). The additional chemical groups added in each substrate (A to B to C) are shaded in red. As the

figure 1
Effects of small structural changes in the substrate on kinetic parameters for chymotrypsin-catalyzed amide hydrolysis.

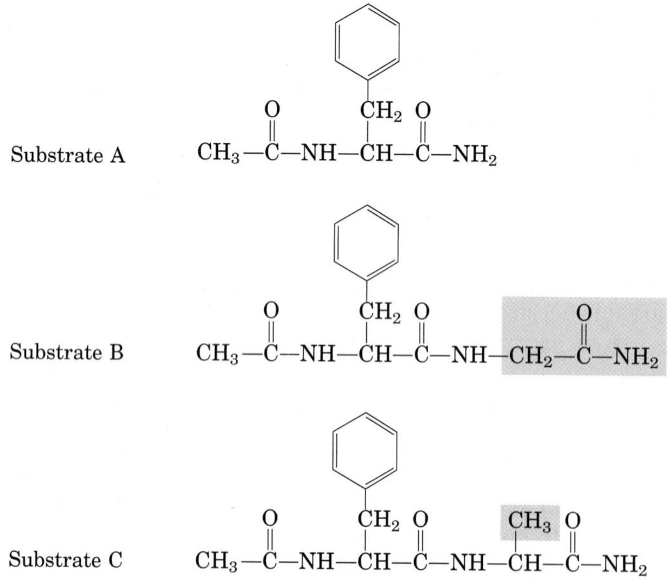

	k_{cat} (s^{-1})	K_m (mM)	k_{cat}/K_m ($M^{-1}s^{-1}$)
Substrate A	0.06	31	2
Substrate B	0.14	15	10
Substrate C	2.8	25	114

table shows, the interaction between the enzyme and these added functional groups has a minimal effect on K_m (taken here as a reflection of K_d) but a large, positive effect on k_{cat} and k_{cat}/K_m. This is what we would expect if the interaction contributed largely to stabilization of the transition state. The results also demonstrate that the rate of a reaction can be affected greatly by enzyme-substrate interactions that are physically remote from the covalent bonds that are altered in the enzyme-catalyzed reaction. Chymotrypsin is described in more detail beginning on page 273.

A complementary experimental approach is to modify the enzyme, eliminating certain enzyme-substrate interactions by replacing specific amino acids through site-directed mutagenesis (see Fig. 29–15). Results from such experiments again demonstrate the importance of binding energy in stabilizing the transition state.

Transition-State Analogs

Even though transition states cannot be observed directly, chemists can often predict the approximate structure of a transition state based on accumulated knowledge about reaction mechanisms. The transition state is by definition transient and so unstable that direct measurement of the binding interaction between this species and the enzyme is impossible. In some cases, however, stable molecules can be designed that resemble transition states. These are called **transition-state analogs.** In principle, they should bind to an enzyme more tightly than does the substrate in the ES complex, because they should fit the active site better (i.e., form a greater number of weak interactions) than the substrate itself. The idea of transition-state analogs was suggested by Pauling in the 1940s, and it has been explored using a number of enzymes. These experiments have the limitation that a transition-state analog cannot perfectly mimic a transition state. Some analogs, however, bind an enzyme 10^2 to 10^6 times more tightly than does the normal substrate, providing good evidence that enzyme active sites are indeed complementary to transition states. The same principle is now used routinely in the pharmaceutical industry to design new drugs. The powerful anti-HIV drugs called protease inhibitors were designed as tight-binding transition-state analogs directed at the active site of HIV protease.

Catalytic Antibodies

If a transition-state analog can be designed for the reaction S $\rightarrow$ P, then an antibody that binds tightly to this analog might be expected to catalyze S $\rightarrow$ P. Antibodies (immunoglobulins; see Fig. 7–23) are key components of the immune response. When a transition-state analog is used as a protein-bound epitope to stimulate the production of antibodies, the antibodies that bind it are potential catalysts of the corresponding reaction. This use of "catalytic antibodies," first suggested by William P. Jencks in 1969, has become practical with the development of laboratory techniques to produce quantities of identical antibodies that bind one specific antigen (monoclonal antibodies; see Chapter 7).

Pioneering work in the laboratories of Richard Lerner and Peter Schultz has resulted in the isolation of a number of monoclonal antibodies that catalyze the hydrolysis of esters or carbonates (Fig. 2). In these reactions, the attack by water (OH^-) on the carbonyl carbon produces a tetrahedral transition state in which a partial negative charge has developed on the carbonyl oxygen. Phosphonate compounds mimic the structure and charge distribution of this transition state in ester hydrolysis, making them good transition-state analogs; phosphate compounds are used for carbonate hydrolysis reactions. Antibodies that bind the phosphonate or phosphate compound tightly have been found to accelerate the corresponding ester or carbonate hydrolysis reaction by factors of 10^3 to 10^4. Structural analyses of a few of these catalytic antibodies have shown that some catalytic amino acid side chains are arranged such that they could interact with the substrate largely in the transition state.

Catalytic antibodies generally do not approach the catalytic efficiency of enzymes, but medical and industrial uses for them are nevertheless emerging. For example, catalytic antibodies designed to degrade cocaine are being investigated as a potential aid in the treatment of cocaine addiction.

Continued on next page

figure 2
The expected transition states for ester or carbonate hydrolysis reactions. Phosphonate and phosphate compounds, respectively, make good transition-state analogs for these reactions.

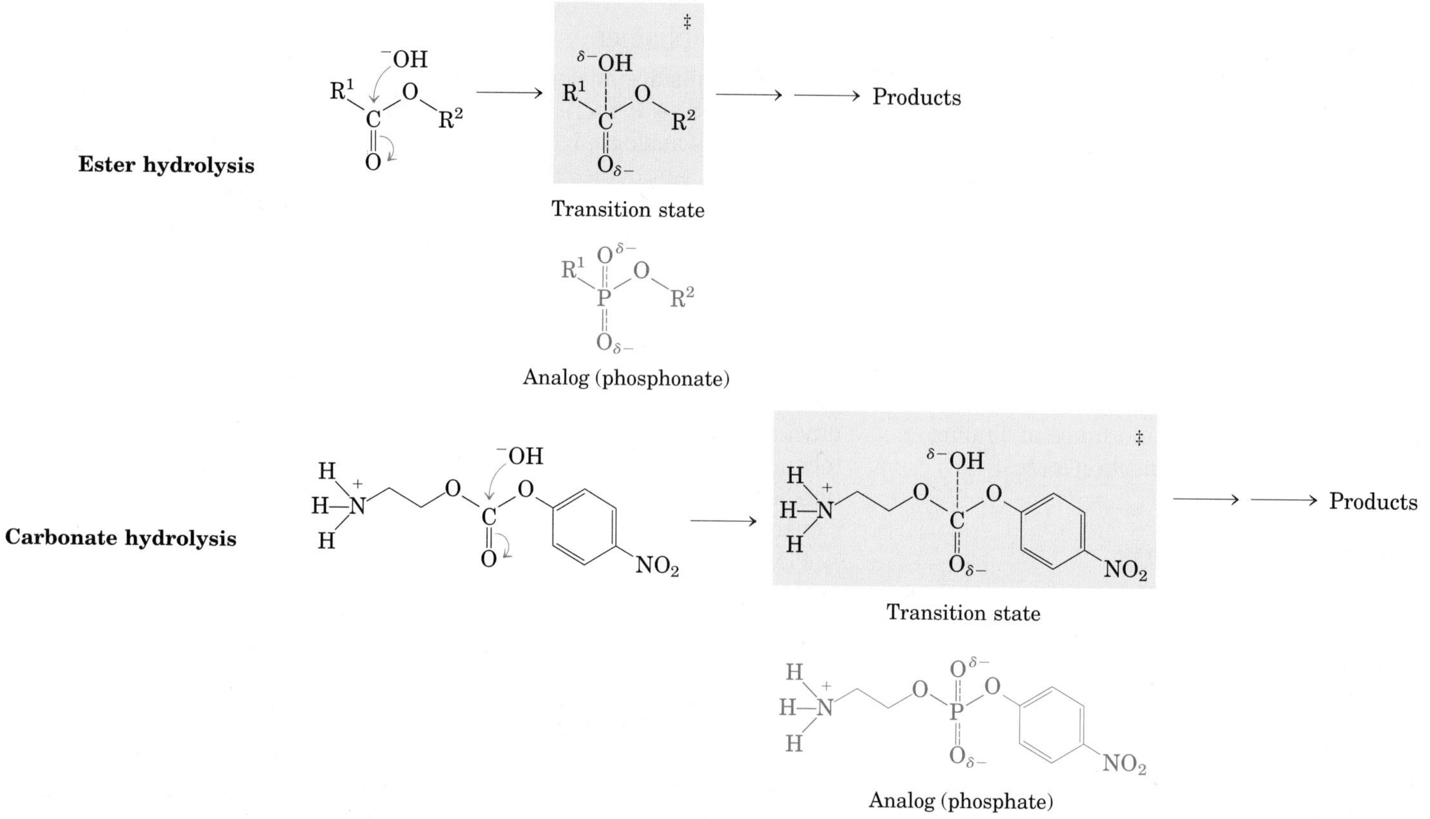

An understanding of the complete mechanism of action of a purified enzyme requires identification of all substrates, cofactors, products, and regulators. Moreover, it requires a knowledge of (1) the temporal sequence in which enzyme-bound reaction intermediates form, (2) the structure of each intermediate and each transition state, (3) the rates of interconversion between intermediates, (4) the structural relationship of the enzyme with each intermediate, and (5) the energy that all reacting and interacting groups contribute to intermediate complexes and transition states. There is probably no enzyme for which our current understanding meets all of these requirements completely. Many decades of research, however, have produced mechanistic information about hundreds of enzymes, and in some cases this information is highly detailed.

Reaction Mechanisms Illustrate Principles

We present here the mechanisms for three enzymes: chymotrypsin, hexokinase, and enolase. These enzymes are chosen not because they are the best understood or because they cover all possible classes of enzyme chemistry,

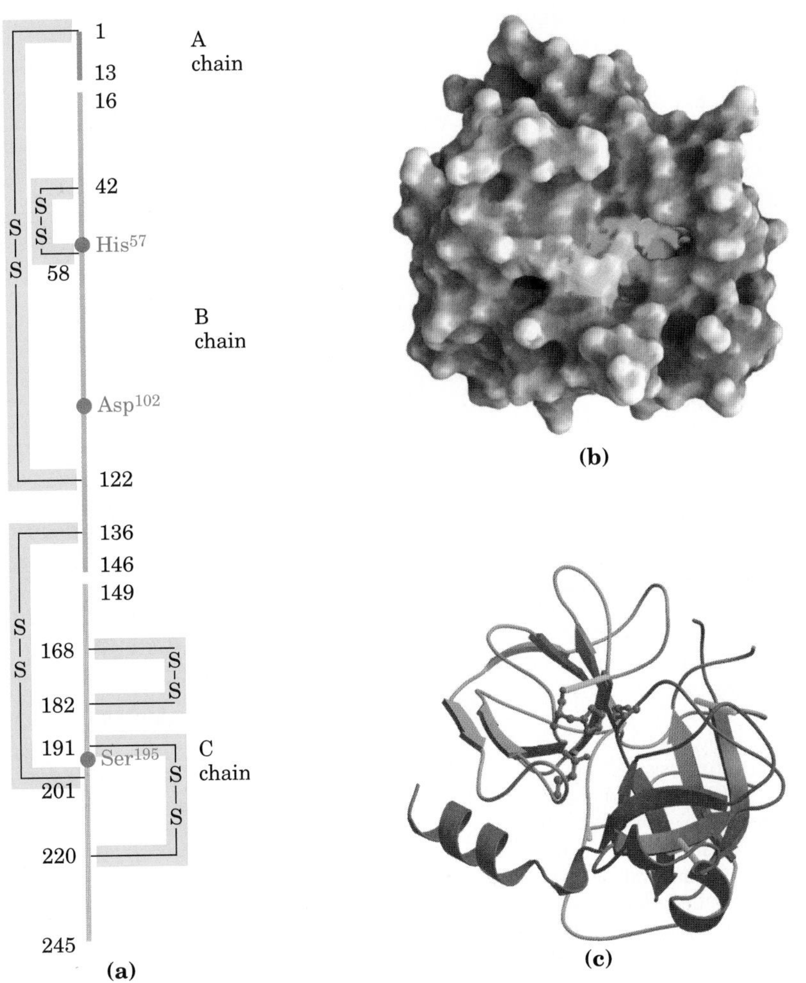

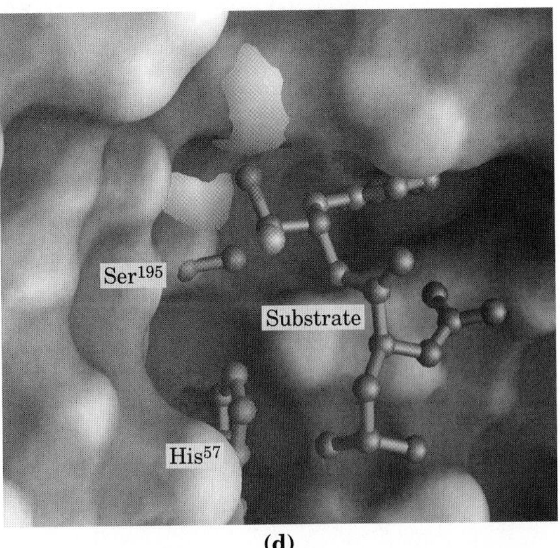

figure 8–18

The structure of chymotrypsin. (a) A representation of primary structure, showing disulfide bonds and the amino acid residues crucial to catalysis. The protein consists of three polypeptide chains linked by disulfide bonds. (The numbering of residues in chymotrypsin (14, 15, 147, and 148 "missing") is explained in Fig. 8–31.) The active-site amino acid residues are grouped together in the three-dimensional structure. **(b)** A depiction of the enzyme emphasizing its surface. The pocket in which the aromatic amino acid side chain of the substrate is bound is shown in green. Key active-site residues, including Ser^{195}, His^{57}, and Asp^{102}, are shown in red. The role of these residues in catalysis is illustrated in Figure 8–19. **(c)** The polypeptide backbone as a ribbon structure. Disulfide bonds are yellow; the A, B, and C chains are colored as in part **(a)**. **(d)** A close-up of the active site with a substrate (mostly green) bound. Two of the active-site residues, Ser^{195} and His^{57} (red), are partly visible. Ser^{195} attacks the carbonyl group of the substrate (the oxygen is in purple); the developing negative charge on the oxygen is stabilized by the oxyanion hole (amide nitrogens in orange), as explained in Figure 8–19. In the substrate, the aromatic amino acid side chain and the amide nitrogen of the peptide bond to be cleaved (protruding toward the viewer and projecting the path of the rest of the substrate polypeptide chain) are in blue.

but because they clearly illustrate some general principles outlined in this chapter. The discussion concentrates on selected principles, along with some key experiments that have helped to bring principles into focus. Much mechanistic detail and experimental evidence is omitted, and in no instance do the mechanisms described provide a complete explanation for the catalytic rate enhancements brought about by these enzymes.

Chymotrypsin Chymotrypsin (M_r 25,000) is a protease, an enzyme that catalyzes the hydrolytic cleavage of peptide bonds. This protease is specific for cleavage of peptide bonds adjacent to aromatic amino acid residues (see Table 5–7). The three-dimensional structure of chymotrypsin is shown in Figure 8–18, with functional groups in the active site emphasized. The reaction catalyzed by this enzyme illustrates the principle of transition-state stabilization and also provides a classic example of general acid-base catalysis and covalent catalysis.

Chymotrypsin enhances the rate of peptide bond hydrolysis by a factor of at least 10^9. The enzyme does not catalyze a direct attack of water on the peptide bond; instead, a transient covalent acyl-enzyme intermediate is formed. The reaction thus has two major phases (Fig. 8–19). In the acylation phase, the peptide bond is cleaved and an ester linkage is formed between the peptide carbonyl carbon and the enzyme. In the deacylation phase, the ester linkage is hydrolyzed and the nonacylated enzyme is

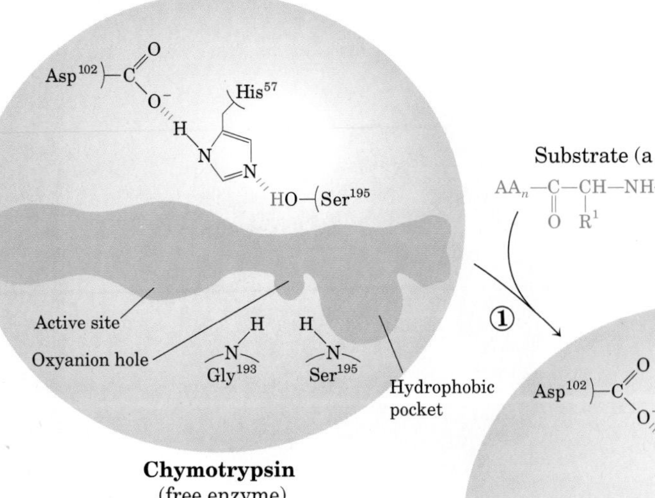

Chymotrypsin
(free enzyme)

Substrate (a polypeptide)

①

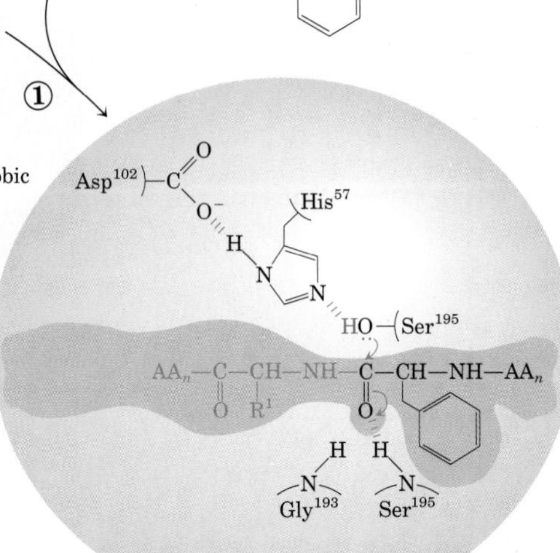

Enzyme-substrate complex

②

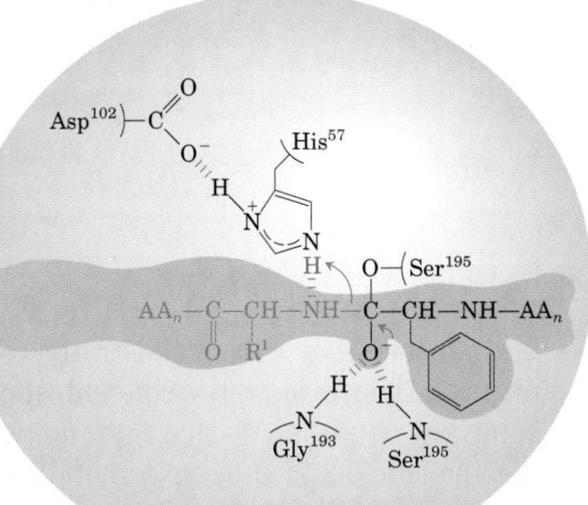

Short-lived intermediate
(acylation)

③

Product 1

Acyl-enzyme intermediate

④

Acyl-enzyme intermediate

⑤

Short-lived intermediate
(deacylation)

⑥

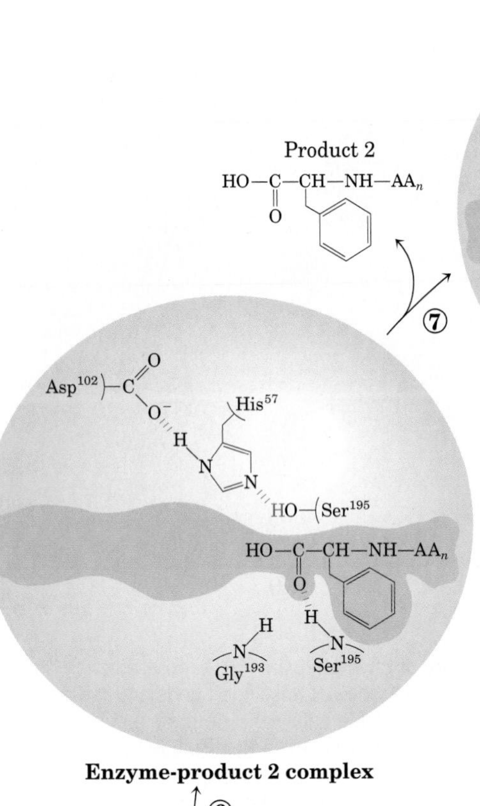

Enzyme-product 2 complex

⑦

figure 8–19

Steps in the hydrolytic cleavage of a peptide bond by chymotrypsin. The substrate (a polypeptide or protein) is bound at the active site. The peptide bond to be cleaved is positioned by the binding of the adjacent hydrophobic amino acid side chain (a Phe residue in this example) in a special hydrophobic pocket on the enzyme. The reaction consists of two phases. In steps ① to ③ formation of a covalent acyl-enzyme intermediate is coupled to cleavage of the peptide bond (the acylation phase). In steps ④ to ⑦ deacylation regenerates the free enzyme (the deacylation phase). In both phases, the carbonyl oxygen of the substrate acquires a negative charge in the tetrahedral intermediate. The charge is stabilized by hydrogen bonding to the amide nitrogens of Gly^{193} and Ser^{195}; the hydrogen bond to Gly^{193} forms only in this short-lived intermediate and in the transition states leading to its formation and breakdown. Deacylation is essentially the reverse of acylation, with water serving in place of the amine component of the substrate. The His and Asp residues cooperate in a catalytic triad, providing general base catalysis of steps ② and ⑤ and general acid catalysis of steps ③ and ⑦.

regenerated. The nucleophile in the acylation phase is the oxygen of Ser^{195}. A serine hydroxyl is normally protonated at neutral pH, but the Ser^{195} of chymotrypsin is hydrogen-bonded to His^{57}, which is further hydrogen-bonded to Asp^{102}. These three amino acid residues are often referred to as the catalytic triad. As the Ser^{195} oxygen attacks the carbonyl carbon of a peptide bond, the hydrogen-bonded His^{57} functions as a general base to remove the serine proton, and the negatively charged Asp^{102} stabilizes the positive charge that forms on the His residue. This prevents development of a very unstable positive charge on the Ser^{195} hydroxyl and makes it more nucleophilic. His^{57} can also act as a proton donor to protonate the amino group in the displaced portion of the substrate (the leaving group). A similar set of proton transfers occurs in the deacylation step.

As the Ser^{195} oxygen attacks the carbonyl group of the substrate, a very short-lived intermediate is formed in which the carbonyl oxygen acquires a negative charge. This charge forms within a pocket on the enzyme called the oxyanion hole, and it is stabilized by hydrogen bonds contributed by the amide nitrogens of two peptide bonds in the chymotrypsin backbone. One of these hydrogen bonds is present only in this intermediate and in the transition states for its formation and breakdown; its presence reduces the energy required to reach these states. This is an example of the use of binding energy in catalysis.

The first evidence for a covalent acyl-enzyme intermediate came from a classic application of pre-steady state kinetics. In addition to its action on polypeptides, chymotrypsin also catalyzes the hydrolysis of small esters and amides. These reactions are much slower than hydrolysis of peptides because less binding energy is available with the smaller substrates, but they are easier to study. Investigations by B.S. Hartley and B.A. Kilby found that hydrolysis of the ester *p*-nitrophenylacetate by chymotrypsin, as measured by release of *p*-nitrophenol, proceeded with a rapid burst before leveling off to a slower rate (Fig. 8–20). By extrapolating back to zero time, they concluded that the burst phase corresponded to just under one molecule of *p*-nitrophenol released for every enzyme molecule present. Hartley and Kilby suggested that this reflected a rapid acylation of all the enzyme molecules (with release of *p*-nitrophenol), and the rate for subsequent turnover of the enzyme was limited by a slow deacylation step. Similar results have since been obtained with many other enzymes. The observation of a burst phase provides yet another example of the use of kinetics to break down a reaction into its constituent steps.

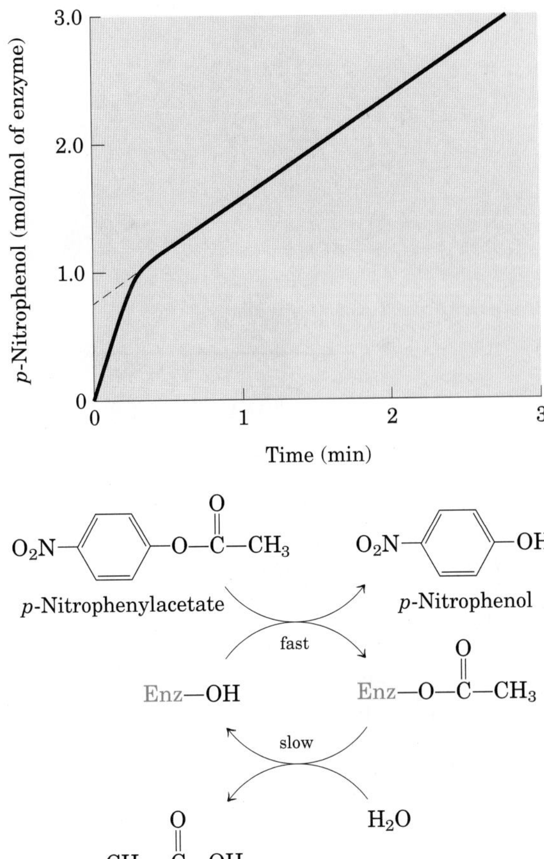

figure 8–20
Pre-steady state kinetic evidence for an acyl-enzyme intermediate. The hydrolysis of *p*-nitrophenylacetate by chymotrypsin is measured by release of *p*-nitrophenol (a colored product). Initially, an amount of *p*-nitrophenol nearly stoichiometric with the amount of enzyme present is released in a rapid burst. This reflects the fast acylation phase of the reaction. The subsequent rate is slower because enzyme turnover is limited by the rate of the slower deacylation phase.

Hexokinase This is a bisubstrate enzyme (M_r 100,000) that catalyzes the interconversion of glucose and ATP with glucose 6-phosphate and ADP. ATP and ADP each bind to enzymes as a complex with the metal ion Mg^{2+}. The hydroxyl at C-6 of glucose (to which the γ-phosphoryl of ATP is transferred) is similar in chemical reactivity to water, and water freely enters the enzyme active site. Yet hexokinase discriminates between glucose and water, with glucose favored by a factor of 10^6.

Mg • ATP + Glucose (CH_2OH) $\xrightleftharpoons{\text{hexokinase}}$ Mg • ADP + Glucose 6-phosphate ($CH_2OPO_3^{2-}$)

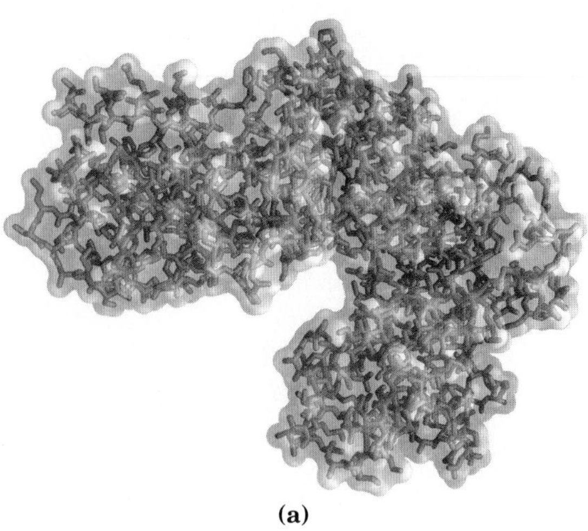

(a)

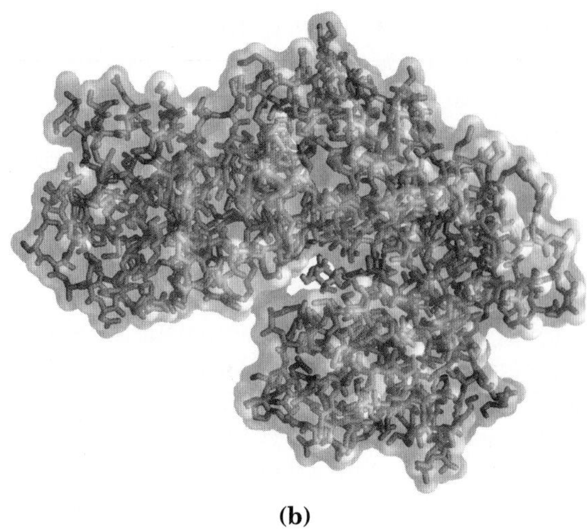

(b)

figure 8–21

Induced fit in hexokinase. The ends of the U-shaped enzyme hexokinase **(a)** pinch toward each other in a conformational change induced by binding D-glucose (red) **(b).**

Hexokinase can discriminate between glucose and water because of a conformational change in the enzyme when the correct substrates are bound (Fig. 8–21). Hexokinase thus provides a good example of induced fit. When glucose is not present, the enzyme is in an inactive conformation with the active-site amino acid side chains out of position for reaction. When glucose (but not water) and Mg · ATP bind, the binding energy derived from this interaction induces a conformational change in hexokinase to the catalytically active form.

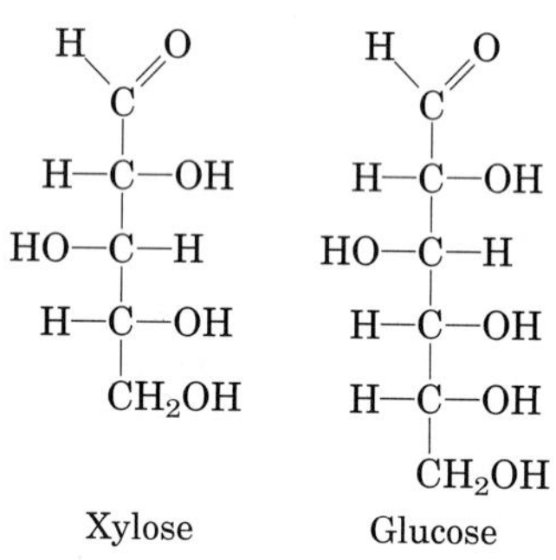

Xylose Glucose

This conclusion has been reinforced by kinetic studies. The five-carbon sugar xylose, stereochemically similar to glucose but one carbon shorter, binds to hexokinase but is in a position where it cannot be phosphorylated. However, addition of xylose to the reaction mixture increases the rate of ATP hydrolysis. Evidently, the binding of xylose is sufficient to induce a change in hexokinase to its active conformation, and the enzyme is thereby "tricked" into phosphorylating water.

The hexokinase reaction also illustrates that enzyme specificity is not always a simple matter of binding one compound but not another. In the case of hexokinase, specificity is observed not in the formation of the ES complex, but in the relative rates of subsequent catalytic steps. Water is not excluded from the active site, but reaction rates increase greatly in the presence of the functional phosphoryl group acceptor (glucose). Induced fit is only one aspect of the catalytic mechanism of hexokinase—like chymotrypsin, hexokinase uses several catalytic strategies. For example, the active-site amino acid residues (those brought into position by the conformational change that follows substrate binding) participate in general acid-base catalysis and transition-state stabilization.

Enolase This enzyme catalyzes a step of glycolysis (Chapter 15), the reversible dehydration of 2-phosphoglycerate to phosphoenolpyruvate:

$$\underset{\text{2-Phosphoglycerate}}{\mathrm{H{-}C(COO^-)(CH_2{-}OH){-}O{-}PO_3^{2-}}} \underset{}{\overset{\text{enolase}}{\rightleftharpoons}} \underset{\text{Phosphoenolpyruvate}}{\mathrm{C(COO^-)(=CH_2){-}O{-}PO_3^{2-}}} + H_2O$$

2-Phosphoglycerate bound to enolase

Enolic intermediate

Phosphoenolpyruvate

(a)

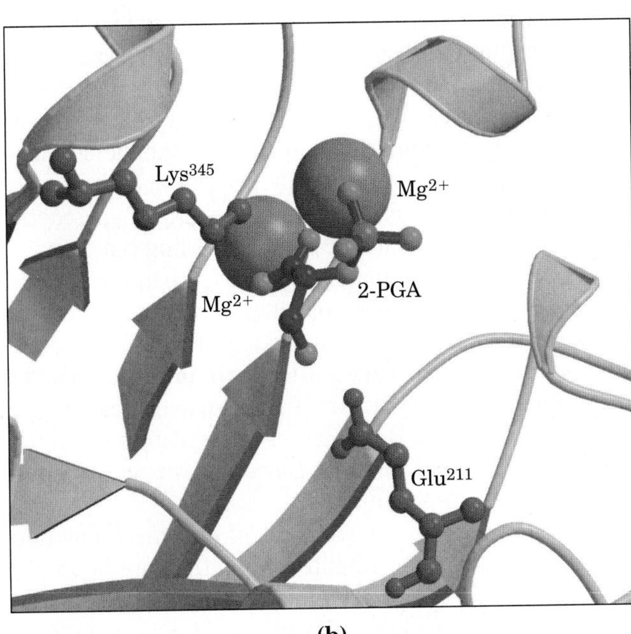

(b)

figure 8–22

The two-step reaction catalyzed by enolase. (a) The mechanism by which 2-phosphoglycerate is converted to phosphoenolpyruvate by enolase. The carboxyl group of 2-phosphoglycerate (2-PGA) is coordinated by two magnesium ions at the active site. A proton is abstracted by general base catalysis (Lys^{345}). The resulting enolic intermediate is stabilized by the two Mg^{2+} ions. Subsequent elimination of the —OH is facilitated by general acid catalysis (Glu^{211}). **(b)** The substrate, 2-PGA, in relation to the magnesium ions, Lys^{345}, and Glu^{211} in the enolase active site. Hydrogen atoms are not shown. All oxygen atoms of 2-PGA are light blue; phosphorus is orange.

Yeast enolase (M_r 96,000) is a dimer with 436 amino acid residues per subunit. The enolase reaction illustrates one type of metal ion catalysis and provides an additional example of general acid-base catalysis and transition-state stabilization. The reaction occurs in two steps (Fig. 8–22a). The residue Lys^{345} acts as a general base catalyst, abstracting a proton from C-2 of 2-phosphoglycerate in the first step; Glu^{211} acts as a general acid catalyst, donating a proton to the —OH leaving group in the second step.

The proton at C-2 of 2-phosphoglycerate is not very acidic and thus is not readily removed. However, in the enzyme active site, 2-phosphoglycerate undergoes strong ionic interactions with two bound Mg^{2+} ions (Fig. 8–22b), making the C-2 proton more acidic (lowering the pK_a) and easier to abstract. Hydrogen bonding to other active-site amino acid residues also contributes to the overall mechanism. The various interactions effectively stabilize both the enolate intermediate and the transition state preceding its formation.

Not included in these discussions of enzyme mechanisms is the special contribution of coenzymes to the catalytic activity of many enzymes. The function of coenzymes is chemically varied, and we will describe each as it is encountered in Part III of this book.

Regulatory Enzymes

In cellular metabolism, groups of enzymes work together in sequential pathways to carry out a given metabolic process, such as the multireaction conversion of glucose to lactate in skeletal muscle or the multireaction synthesis of an amino acid from simpler precursors in a bacterial cell. In such enzyme systems, the reaction product of the first enzyme becomes the substrate of the next, and so on.

Most of the enzymes in each system follow the kinetic patterns we have already described. In each metabolic pathway, however, there is at least one enzyme that sets the rate of the overall sequence because it catalyzes the slowest or rate-limiting reaction. Furthermore, these **regulatory enzymes** exhibit increased or decreased catalytic activity in response to certain signals. Adjustments in the rate of reactions catalyzed by regulatory enzymes, and therefore in the rate of entire metabolic sequences, allow the cell to meet changing needs for energy and for biomolecules required in growth and repair.

In most multienzyme systems, the first enzyme of the sequence is a regulatory enzyme. Catalyzing even the first few reactions of a pathway that leads to an unneeded product diverts energy and metabolites from more important processes. Therefore, an excellent place to regulate the pathway is at the point of commitment to that metabolic sequence. The other enzymes in the sequence are usually present at levels providing an excess of catalytic activity; they can generally promote their reactions as fast as their substrates are made available from preceding reactions.

The activities of regulatory enzymes are modulated in a variety of ways. There are two major classes of regulatory enzymes in metabolic pathways. **Allosteric enzymes** function through reversible, noncovalent binding of regulatory compounds called **allosteric modulators,** which are generally small metabolites or cofactors. Other enzymes are regulated by reversible covalent modification. Both classes of regulatory enzymes tend to be multisubunit proteins, and in some cases the regulatory site(s) and the active site are on separate subunits.

At least two other mechanisms of enzyme regulation occur. Some enzymes are stimulated or inhibited when they are bound by separate regulatory proteins. Others are activated when peptide segments are removed by proteolytic cleavage; unlike effector-mediated regulation, regulation by proteolytic cleavage is irreversible. Important examples of both these mechanisms are found in physiological processes such as digestion, blood clotting, hormone action, and vision.

Cell growth and survival depend on the efficient use of resources made possible by regulatory enzymes. No single rule governs the occurrence of different types of regulation in different systems. To a degree, allosteric (noncovalent) regulation may permit fine-tuning of metabolic pathways that are required continuously but at different levels of activity as cellular conditions change. Regulation by covalent modification may be all or none—usually the case with proteolytic cleavage—or it may allow for subtle changes in activity. Multiple types of regulation are observed in a number of regulatory enzymes. The remainder of this chapter is devoted to a discussion of enzyme regulation.

Allosteric Enzymes Undergo Conformational Changes in Response to Modulator Binding

As we saw in Chapter 7, allosteric proteins are those having "other shapes" or conformations induced by the binding of modulators. The same concept applies to certain regulatory enzymes, as conformational changes induced by one or more modulators interconvert more-active and less-active forms of the enzyme. The modulators for allosteric enzymes may be either inhibitory or stimulatory. An activator is often the substrate itself; regulatory enzymes for which substrate and modulator are identical are called homotropic. The effect is similar to that of O_2 binding to the nonenzymatic protein hemoglobin (Chapter 7); binding of the substrate causes conformational changes that affect the subsequent activity of other sites on the protein. When the modulator is a molecule other than the substrate, the enzyme is said to be heterotropic.

Note that allosteric modulators should not be confused with uncompetitive and mixed inhibitors. Although the latter bind at a second site on the enzyme, they do not necessarily mediate conformational changes between active and inactive forms, and the kinetic effects are distinct.

The properties of allosteric enzymes are significantly different from those of simple nonregulatory enzymes. Some of the differences are structural. In addition to active sites, allosteric enzymes generally have one or more regulatory or allosteric sites for binding the modulator (Fig. 8–23). Just as an enzyme's active site is specific for its substrate, each regulatory site is specific for its modulator. Enzymes with several modulators generally have different specific binding sites for each. In homotropic enzymes, the active site and regulatory site are the same.

Allosteric enzymes are generally larger and more complex than nonallosteric enzymes. Most have two or more polypeptide chains, or subunits. Aspartate transcarbamoylase, which catalyzes the first reaction in the biosynthesis of pyrimidine nucleotides (see Fig. 22–34), has 12 polypeptide chains organized into catalytic and regulatory subunits. Figure 8–24 shows the quaternary structure of this enzyme, deduced from x-ray analysis.

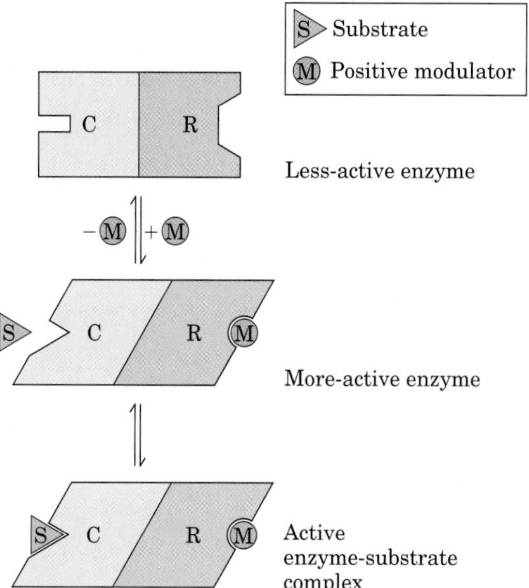

figure 8–23

Subunit interactions in an allosteric enzyme, and interactions with inhibitors and activators. In many allosteric enzymes the substrate binding site and the modulator binding site(s) are on different subunits, the catalytic (C) and regulatory (R) subunits, respectively. Binding of the positive (stimulatory) modulator (M) to its specific site on the regulatory subunit is communicated to the catalytic subunit through a conformational change. This change renders the catalytic subunit active and capable of binding the substrate (S) with higher affinity. On dissociation of the modulator from the regulatory subunit, the enzyme reverts to its inactive or less active form.

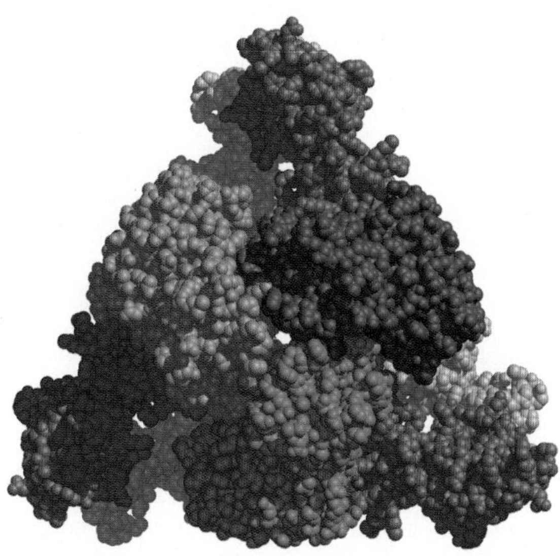

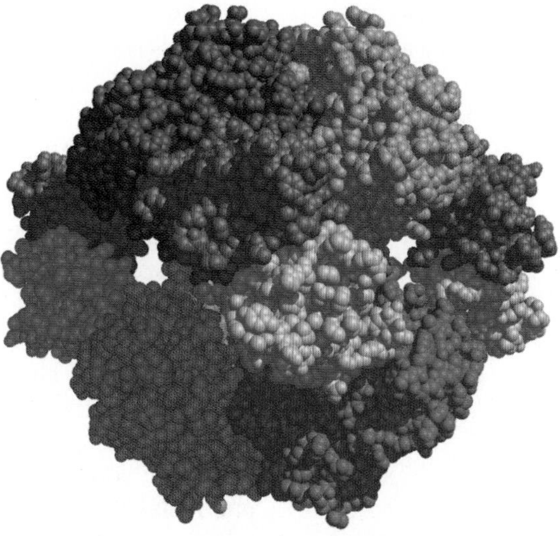

figure 8–24

Two views of the regulatory enzyme aspartate transcarbamoylase. This allosteric regulatory enzyme has two stacked catalytic clusters, each with three catalytic polypeptide chains (in shades of blue and purple), and three regulatory clusters, each with two regulatory polypeptide chains (in red and yellow). The regulatory clusters form the points of a triangle surrounding the catalytic subunits. Binding sites for allosteric modulators are on the regulatory subunits. Modulator binding produces large changes in enzyme conformation and activity. The role of this enzyme in nucleotide synthesis, and details of its regulation, are discussed in Chapter 22.

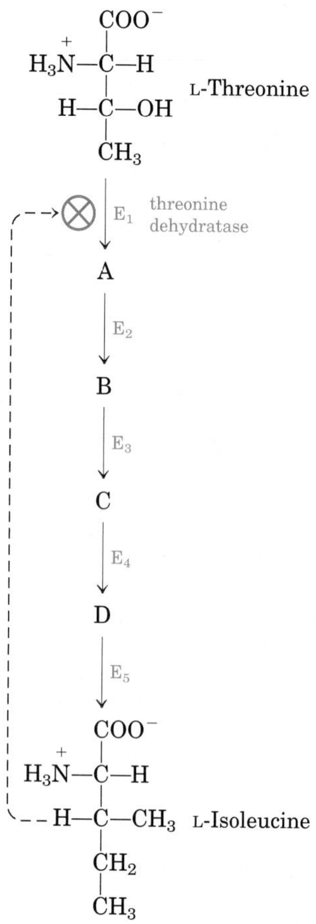

$$\mathrm{COO^-}$$
$$\mathrm{H_3\overset{+}{N}—C—H}$$
$$\mathrm{H—C—OH}$$
$$\mathrm{CH_3}$$
L-Threonine

E_1 threonine dehydratase

A

E_2

B

E_3

C

E_4

D

E_5

$$\mathrm{COO^-}$$
$$\mathrm{H_3\overset{+}{N}—C—H}$$
$$\mathrm{H—C—CH_3}$$
$$\mathrm{CH_2}$$
$$\mathrm{CH_3}$$
L-Isoleucine

figure 8–25

Feedback inhibition. The conversion of L-threonine to L-isoleucine is catalyzed by a sequence of five enzymes (E_1 to E_5). Threonine dehydratase (E_1) is specifically inhibited allosterically by L-isoleucine, the end product of the sequence, but not by any of the four intermediates (A to D). Feedback inhibition is indicated by the dashed feedback line and the ⊗ symbol at the threonine dehydratase reaction arrow, a device that is used throughout this book.

The Regulatory Step in Many Pathways Is Catalyzed by an Allosteric Enzyme

In some multienzyme systems, the regulatory enzyme is specifically inhibited by the end product of the pathway whenever the concentration of the end product exceeds the cell's requirements. When the regulatory enzyme reaction is slowed, all subsequent enzymes operate at reduced rates as their substrates are depleted. The rate of production of the pathway's end product is thereby brought into balance with the cell's needs. This type of regulation is called **feedback inhibition.** Buildup of the pathway's end product ultimately slows the entire pathway.

One of the first examples of allosteric feedback inhibition to be discovered was the bacterial enzyme system that catalyzes the conversion of L-threonine to L-isoleucine in five steps (Fig. 8–25). In this system, the first enzyme, threonine dehydratase, is inhibited by isoleucine, the product of the last reaction of the series. This is an example of heterotropic allosteric inhibition. Isoleucine is quite specific as an inhibitor. No other intermediate in this sequence inhibits threonine dehydratase, nor is any other enzyme in the sequence inhibited by isoleucine. Isoleucine binds not to the active site but to another specific site on the enzyme molecule, the regulatory site. This binding is noncovalent and readily reversible; if the isoleucine concentration decreases, the rate of threonine dehydratase activity increases. Thus threonine dehydratase activity responds rapidly and reversibly to fluctuations in the cellular concentration of isoleucine.

The Kinetic Properties of Allosteric Enzymes Diverge from Michaelis-Menten Behavior

Allosteric enzymes show relationships between V_0 and [S] that differ from Michaelis-Menten kinetics. They do exhibit saturation with the substrate when [S] is sufficiently high, but for some allosteric enzymes, when V_0 is plotted against [S] (Fig. 8–26) a sigmoid saturation curve results, rather than the hyperbolic curve typical of nonregulatory enzymes. Although we can find a value of [S] on the sigmoid saturation curve at which V_0 is half-maximal, we cannot refer to it with the designation K_m because the enzyme does not follow the hyperbolic Michaelis-Menten relationship. Instead the symbol $[S]_{0.5}$ or $K_{0.5}$ is often used to represent the substrate concentration giving half-maximal velocity of the reaction catalyzed by an allosteric enzyme (Fig. 8–26).

Sigmoid kinetic behavior generally reflects cooperative interactions between multiple protein subunits. In other words, changes in the structure of one subunit are translated into structural changes in adjacent subunits, an effect that is mediated by noncovalent interactions at the subunit-subunit interface(s). The principles are particularly well illustrated by O_2 binding to hemoglobin (Chapter 7). Sigmoid kinetic behavior is explained by the concerted and sequential models for subunit interactions (see Fig. 7–14).

Homotropic allosteric enzymes generally have multiple subunits. As noted earlier, the same binding site on each subunit may function as both the active site and the regulatory site. The substrate can be a positive modulator (an activator) because the subunits act cooperatively: the binding of one molecule of substrate to one binding site alters the enzyme's conformation and enhances the binding of subsequent substrate molecules. This accounts for the sigmoid rather than hyperbolic increase in V_0 with increasing [S]. One characteristic of sigmoid kinetics is that small changes in the concentration of a modulator can be associated with large changes in activity. As is evident in Figure 8–26a, a relatively small increase in [S] in the steep part of the curve causes a comparatively large increase in V_0.

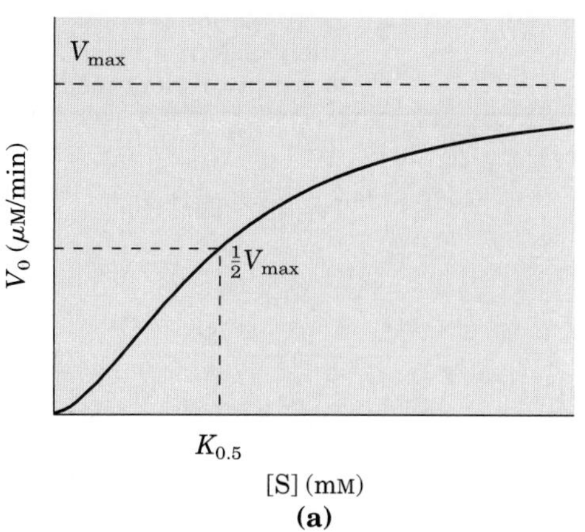

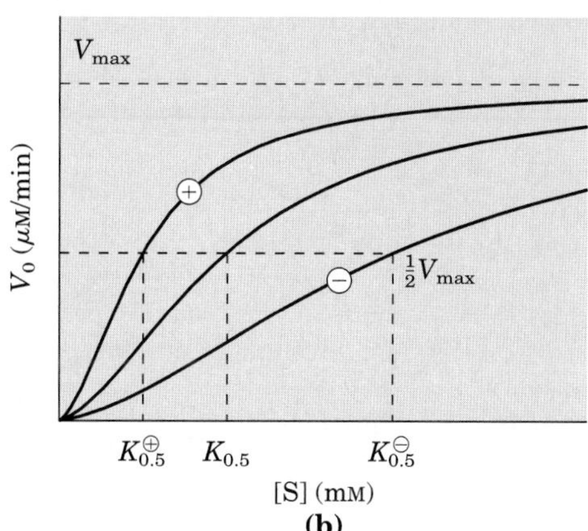

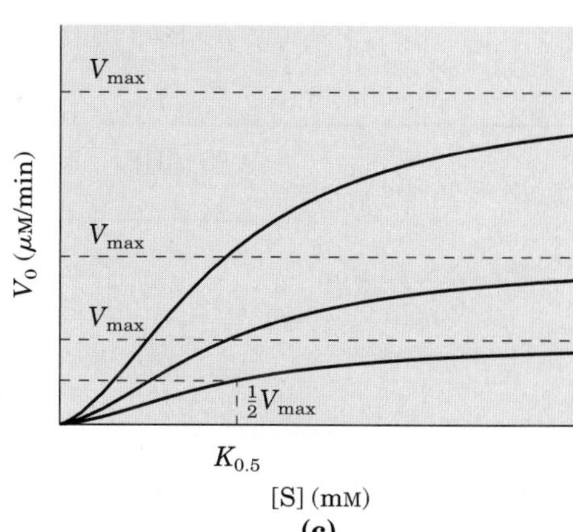

figure 8–26

Substrate-activity curves for representative allosteric enzymes. Three examples of complex responses of allosteric enzymes to their modulators. **(a)** The sigmoid curve of a homotropic enzyme, in which the substrate also serves as a positive (stimulatory) modulator, or activator. Note the resemblance to the oxygen-saturation curve of hemoglobin (see Fig. 7–12). **(b)** The effects of a positive modulator ⊕ and a negative modulator ⊖ on an allosteric enzyme in which $K_{0.5}$ is altered without a change in V_{max}. The central curve shows the substrate-activity relationship without a modulator. **(c)** A less common type of modulation, in which V_{max} is altered and $K_{0.5}$ is nearly constant.

For heterotropic allosteric enzymes, with a modulator that is a metabolite other than the substrate itself, it is difficult to generalize about the shape of the substrate-saturation curve. An activator may cause the curve to become more nearly hyperbolic, with a decrease in $K_{0.5}$ but no change in V_{max}, thus resulting in an increased reaction velocity at a fixed substrate concentration (V_0 is higher for any value of [S]) (Fig. 8–26b, upper curve). Other heterotropic allosteric enzymes respond to an activator by an increase in V_{max} with little change in $K_{0.5}$ (Fig. 8–26c). A negative modulator (an inhibitor) may produce a *more* sigmoid substrate-saturation curve, with an increase in $K_{0.5}$ (Fig. 8–26b, lower curve). Heterotropic allosteric enzymes therefore show different kinds of responses in their substrate-activity curves because some have inhibitory modulators, some have activating modulators, and some have both.

Some Regulatory Enzymes Undergo Reversible Covalent Modification

In another important class of regulatory enzymes, activity is modulated by covalent modification of the enzyme molecule. Modifying groups include phosphoryl, adenylyl, uridylyl, adenosine diphosphate ribosyl, and methyl groups (Fig. 8–27, p. 282). These groups are generally covalently linked to and removed from the regulatory enzyme by separate enzymes.

An example of methylation is the methyl-accepting chemotaxis protein of bacteria. This protein is part of a system that permits a bacterium to swim toward an attractant (such as a sugar) in solution and away from repellent chemicals. The methylating agent is *S*-adenosyl-methionine (adoMet), described in Chapter 18. ADP-ribosylation is an especially interesting reaction observed in only a few proteins. ADP-ribose is derived from nicotinamide adenine dinucleotide (NAD) (see Fig. 10–41). This type of modification occurs for the bacterial enzyme dinitrogenase reductase, resulting in regulation of the important process of biological nitrogen fixation. In addition, both diphtheria toxin and cholera toxin are enzymes that catalyze the ADP-ribosylation (and inactivation) of key cellular enzymes or proteins. Diphtheria toxin acts on and inhibits elongation factor 2, a protein involved in protein biosynthesis. Cholera toxin acts on a specific protein (a G protein that is part of a signaling pathway; see Fig. 13–28), leading ultimately to several physiological responses including a massive loss of body fluids and sometimes death.

Phosphorylations make up the vast majority of known regulatory modifications; one-third to one-half of all proteins in a eukaryotic cell are phosphorylated. Some proteins have only one phosphorylated residue, others

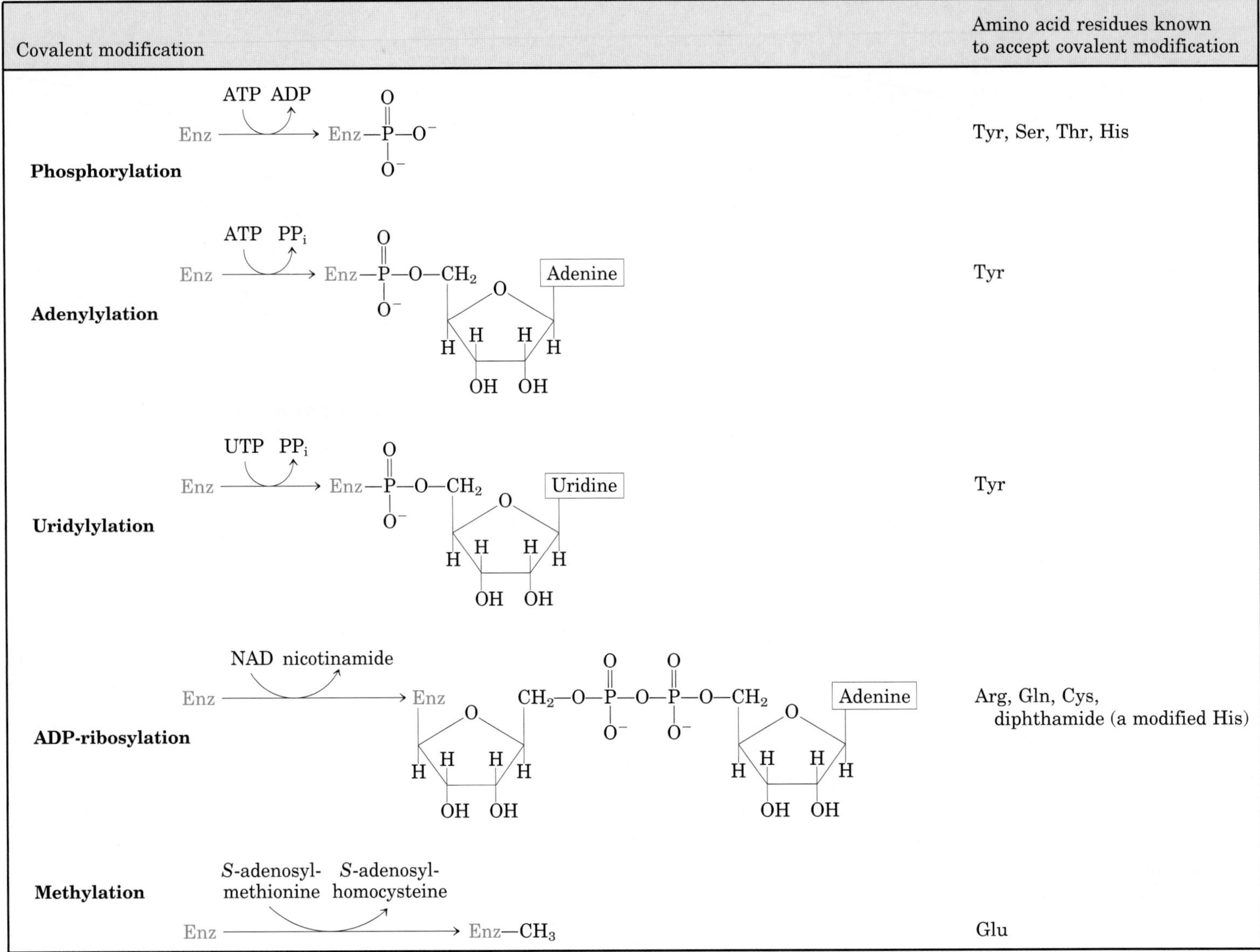

figure 8–27
Examples of enzyme modification reactions.

have several, and a few have dozens of sites for phosphorylation. This mode of covalent modification is central to a large number of regulatory pathways and is therefore treated in considerable detail below.

Phosphoryl Groups Affect the Structure and Catalytic Activity of Proteins

The attachment of phosphoryl groups to specific amino acid residues of a protein is catalyzed by **protein kinases;** removal of phosphoryl groups is catalyzed by protein phosphatases. The addition of a phosphoryl group to a Ser, Thr, or Tyr residue introduces a bulky, charged group into a region that was only moderately polar. The oxygens of a phosphoryl group are capable of hydrogen bonding with one or several groups in a protein, commonly the amide groups of the peptide backbone at the start of an α helix or the charged guanidinium group of the Arg side chain. The double negative charge on a phosphorylated side chain can also repel neighboring negatively charged (Asp or Glu) residues. (No unmodified amino acid side chain has two negative charges.) When the modified side chain is located in a region

table 8–9

Consensus Sequences for Protein Kinases

Protein kinase	Consensus sequence and phosphorylated residue*
Protein kinase A	–X–R–(R/K)–X–(S/T)–B–
Protein kinase G	–X–R–(R/K)–X–(S/T)–X–
Protein kinase C	–(R/K)–(R/K)–X–(S/T)–B–(R/K)–(R/K)–
Protein kinase B	–X–R–X–(S/T)–X–K–
Ca^{2+}/calmodulin kinase I	–B–X–R–X–X–(S/T)–X–X–X–B–
Ca^{2+}/calmodulin kinase II	–B–X–(R/K)–X–X–(S/T)–X–X–
Myosin light chain kinase (smooth muscle)	–K–K–R–X–X–S–X–B–B–
Phosphorylase *b* kinase	*–K–R–K–Q–I–S–V–R–*
Extracellular signal–regulated kinase (ERK)	–P–X–(S/T)–P–P–
Cyclin-dependent protein kinase (cdc2)	–X–(S/T)–P–X–(K/R)–
Casein kinase I	–(Sp/Tp)–X–X–(X)–(S/T)–B
Casein kinase II	–X–(S/T)–X–X–(E/D/Sp/Yp)–X–
β-Adrenergic receptor kinase	–$(D/E)_n$–(S/T)–X–X–X–
Rhodopsin kinase	–X–X–(S/T)–$(E)_n$–
Insulin receptor kinase	–X–E–E–E–Y–M–M–M–M–*K–K–S–R–G–D–Y–M–T–M–Q–I–G–K–K–K–L–P–A–T–G–D–Y–M–N–M–S–P–V–G–D–*
Epidermal growth factor (EGF) receptor kinase	*–E–E–E–E–Y–F–E–L–V–*

Sources: Pinna, L.A. & Ruzzene, M.H. (1996) How do protein kinases recognize their substrates? *Biochim. Biophys. Acta* **1314,** 191–225. Kemp, B.E. & Pearson, R.B. (1990) Protein kinase recognition sequence motifs. *Trends Biochem. Sci.* **15,** 342–346. Kennelly, P.J. & Krebs, E.G. (1991) Consensus sequences as substrate specificity determinants for protein kinases and protein phosphatases. *J. Biol. Chem.* **266,** 15,555–15,558.

*Shown here are deduced consensus sequences (in roman type) and actual sequences from known substrates (italic). The Ser (S), Thr (T), or Tyr (Y) residue to undergo phosphorylation is in red; all amino acid residues are shown as their one-letter abbreviations (see Table 5–1). X represents any amino acid; B, any hydrophobic amino acid; Sp, Tp, and Yp, already phosphorylated Ser, Thr, and Tyr residues.

proline. Primary sequence is not the only important factor in determining whether a given residue will be phosphorylated, however. Protein folding brings together residues that are distant in the primary sequence, and the resulting three-dimensional structure can determine whether a protein kinase has access to a given residue and can recognize it as a substrate. One factor influencing the substrate specificity of certain protein kinases is the proximity of other phosphorylated residues.

Regulation by phosphorylation is often quite complicated. Some proteins have consensus sequences recognized by several different protein kinases, each of which can phosphorylate the protein and alter its enzymatic activity. In some cases, phosphorylation is hierarchical: a certain residue can be phosphorylated only if a neighboring residue has been phosphorylated first. For example, glycogen synthase is inactivated by phosphorylation of specific Ser residues and is also modulated by at least four other

Kinase	Glycogen synthase sites phosphorylated	Degree of synthase inactivation
Protein kinase A	1A, 1B, 2, 4	+
Protein kinase G	1A, 1B, 2	+
Protein kinase C	1A	+
Ca^{2+}/calmodulin kinase	1B, 2	+
Phosphorylase *b* kinase	2	+
Casein kinase I	At least 9 sites	+ + + +
Casein kinase II	5	0
Glycogen synthase kinase 3	3A, 3B, 3C	+ + +
Glycogen synthase kinase 4	2	+

figure 8–30

Multiple regulatory phosphorylations. The enzyme glycogen synthase contains at least nine separate sites in five designated regions susceptible to phosphorylation by one of the cellular protein kinases. The activity of this enzyme is therefore capable of modulation in response to a variety of signals. Thus regulation is a matter not of binary (on/off) switching but of finely tuned modulation of activity over a wide range.

protein kinases that phosphorylate four other sites in the protein (Fig. 8–30). The protein is not a substrate for glycogen synthase kinase 3, for example, until one site has been phosphorylated by casein kinase II. Some phosphorylations inhibit glycogen synthase more than others, and some combinations of phosphorylations are cumulative. These multiple regulatory phosphorylations provide the potential for extremely subtle modulation of enzyme activity.

To serve as an effective regulatory mechanism, phosphorylation must be reversible. In general, phosphoryl groups are added and removed by different enzymes, and the processes can therefore be separately regulated. Cells contain a family of phosphoprotein phosphatases that hydrolyze specific Ⓟ–Ser, Ⓟ–Thr, and Ⓟ–Tyr esters, releasing P_i. The known phosphoprotein phosphatases act only on a subset of phosphoproteins, but they show less substrate specificity than protein kinases. Although phosphoprotein phosphatases are not yet as thoroughly studied as the protein kinases, they are likely to prove just as important in regulating cellular processes and metabolism.

Some Types of Regulation Require Proteolytic Cleavage of an Enzyme Precursor

For some enzymes, an inactive precursor called a **zymogen** is cleaved to form the active enzyme. Many proteolytic enzymes (proteases) of the stomach and pancreas are regulated in this way. Chymotrypsin and trypsin are initially synthesized as chymotrypsinogen and trypsinogen, respectively (Fig. 8–31). Specific cleavage causes conformational changes that expose the enzyme active site. Because this type of activation is irreversible, other mechanisms are needed to inactivate these enzymes. Proteases are inactivated by inhibitor proteins that bind very tightly to the enzyme active site. For example, pancreatic trypsin inhibitor (M_r 6,000) binds to and inhibits trypsin; α_1-antiproteinase (M_r 53,000) primarily inhibits neutrophil elastase (neutrophils are a type of white blood cell). An insufficiency of α_1-antiproteinase, which can be caused by exposure to cigarette smoke, has been associated with lung damage, including emphysema.

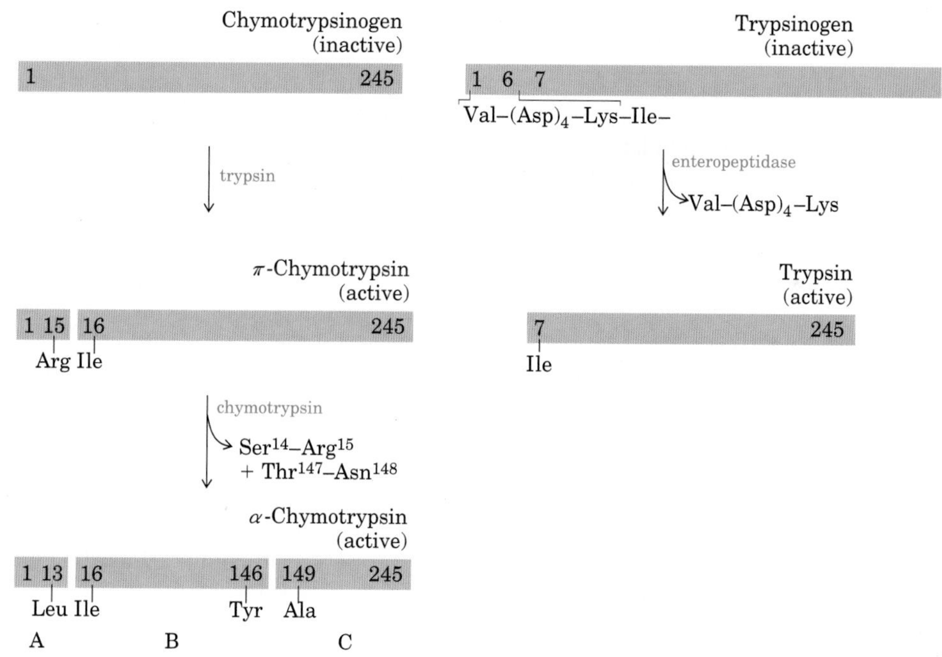

figure 8–31

Activation of zymogens by proteolytic cleavage. Shown here is the formation of chymotrypsin and trypsin from their zymogens. The bars represent the primary sequences of the polypeptide chains. Amino acids at the termini of the polypeptide fragments generated by cleavage are indicated below the bars. The numbering of amino acid residues represents their positions in the primary sequence of the zymogens, chymotrypsinogen or trypsinogen (the amino-terminal residue is number 1). Recall that the three polypeptide chains (A, B, and C) of chymotrypsin are linked by disulfide bonds (see Fig. 8–18).

Proteases are not the only proteins activated by proteolysis. In other cases, however, the precursors are called not zymogens but, more generally, **proproteins** or **proenzymes,** as appropriate. For example, the connective tissue protein collagen is initially synthesized as the soluble precursor procollagen. The blood clotting system provides many examples of the proteolytic activation of proteins. Fibrin, the protein of blood clots, is produced by proteolysis of its inactive proprotein, fibrinogen. The protease responsible for this activation is thrombin (similar in many respects to chymotrypsin), which itself is produced by proteolysis of a proprotein (in this case a zymogen), prothrombin. Blood clotting is mediated by a complicated cascade of zymogen activations.

Some Regulatory Enzymes Use Multiple Regulatory Mechanisms

Glycogen phosphorylase catalyzes the first reaction in a pathway that feeds stored glucose into energy-yielding carbohydrate metabolism (Chapter 15). This is an important metabolic step, and its regulation is correspondingly complex. Although its primary regulation is through covalent modification, as outlined in Figure 8–28, glycogen phosphorylase is also modulated in a noncovalent, allosteric manner by AMP, which is an activator of phosphorylase *b*, and by several other molecules that are inhibitors.

Other complex regulatory enzymes are found at key metabolic crossroads. Bacterial glutamine synthetase, which catalyzes one of the steps that introduce reduced nitrogen into cellular metabolism (Chapter 22), is one of the most complex regulatory enzymes known. It is regulated by allostery (with at least eight different allosteric modulators) and reversible covalent modification. It is also regulated by the association of other regulatory proteins, a mechanism only briefly alluded to in this chapter but examined in detail when we consider the regulation of specific metabolic pathways.

What is the advantage of such complexity in the regulation of enzymatic activity? We began this chapter by stressing the central importance of catalysis to the very existence of life. The control of catalysis is also critical to life. If all possible reactions in a cell were catalyzed simultaneously, macromolecules and metabolites would quickly be broken down to much simpler chemical forms. Instead, only the reactions needed by the cell at a given moment are catalyzed. When chemical resources are plentiful, glucose and other metabolites are synthesized and stored. When chemical resources are scarce, these stores are used to fuel cellular metabolism. Chemical energy is used economically, parceled out to various metabolic pathways as cellular needs dictate. The availability of powerful catalysts, each specific for a given reaction, makes the regulation of these reactions possible. This in turn gives rise to the complex, highly regulated symphony we call life.

summary

Life depends on the existence of powerful and specific catalysts: the enzymes. Virtually every biochemical reaction is catalyzed by an enzyme. With the exception of a few catalytic RNAs, all known enzymes are proteins. Enzymes are extraordinarily effective catalysts, commonly enhancing reaction rates by a factor of 10^5 to 10^{17}. To be active, some enzymes require a chemical cofactor, which can be loosely or tightly bound. Each enzyme is classified according to the specific reaction it catalyzes.

Enzyme-catalyzed reactions are characterized by the formation of a complex between substrate and enzyme (an ES complex). The binding occurs in a pocket on the enzyme called the active site. The function of enzymes and other catalysts is to lower the activation energy for the reaction and thereby enhance the reaction rate. The equilibrium of a reaction is unaffected by the enzyme.

The energy used for enzymatic rate enhancements is derived from weak interactions (hydrogen bonds and hydrophobic and ionic interactions) between substrate and enzyme. The enzyme active site is structured so that some of these weak interactions occur preferentially in the reaction transition state, thus stabilizing the transition state. The energy available from the numerous weak interactions between enzyme and substrate (the binding energy) is substantial and can generally account for observed rate enhancements. The need for multiple interactions is one reason for the large size of enzymes. Binding energy can be used to lower substrate entropy, to strain the substrate, or to cause a conformational change in the enzyme (induced fit). This same binding energy accounts for the exquisite specificity of enzymes for their substrates. Other catalytic mechanisms include general acid-base catalysis and covalent catalysis. Detailed reaction mechanisms have been worked out for many enzymes.

Kinetics is an important method for the study of enzyme mechanisms. Most enzymes have some kinetic properties in common. As the concentration of substrate is increased, the catalytic activity of a fixed concentration of an enzyme increases in a hyperbolic fashion to approach a characteristic maximum rate V_{max} at which essentially all the enzyme is in the form of the ES complex. The substrate concentration giving one-half V_{max} is the Michaelis constant K_m, which is characteristic for each enzyme acting on a given substrate. The Michaelis-Menten equation

$$V_0 = \frac{V_{max}[S]}{K_m + [S]}$$

relates the initial velocity of an enzymatic reaction to the substrate concentration and V_{max} through the constant K_m. Both K_m and V_{max} can be measured; they have different meanings for different enzymes. The limiting rate of an enzyme-catalyzed reaction at saturation is described by the constant k_{cat}, also called the turnover number. The ratio k_{cat}/K_m provides a good measure of catalytic efficiency. The Michaelis-Menten equation is also applicable to bisubstrate reactions, which occur by either ternary-complex or double-displacement (Ping-Pong) pathways. Each enzyme has an optimum pH (or pH range) at which it has maximal activity.

Enzymes can be inactivated by irreversible modification of a functional group essential for catalytic activity. They can also be inhibited by molecules that bind reversibly. Competitive inhibitors compete reversibly with the substrate for binding to the active site but are not transformed by the enzyme. Uncompetitive inhibitors bind only to the ES complex, at a site distinct from the active site. In mixed inhibition, an inhibitor binds to either E or ES, again at a site distinct from that where substrate binds.

Some enzymes regulate the activity of metabolic pathways in cells. In feedback inhibition, the end product of a pathway inhibits the first enzyme of that pathway. The activity of some regulatory enzymes, called allosteric enzymes, is adjusted by reversible binding of a specific modulator to a regulatory site. Such modulators may be inhibitory or stimulatory and may be either the substrate itself or some other metabolite. The kinetic behavior of allosteric enzymes reflects cooperative interactions among the enzyme subunits. Other regulatory enzymes are modulated by covalent modification of a specific functional group necessary for activity. The phosphorylation of specific amino acid residues is a particularly common way to regulate enzyme activity. Many proteolytic enzymes have inactive precursors called zymogens; small inactive peptides are cleaved from zymogens to form the active proteases.

further reading

General

Evolution of Catalytic Function. (1987) *Cold Spring Harb. Symp. Quant. Biol.* **52.**

A collection of excellent papers on fundamentals; continues to be very useful.

Fersht, A. (1999) *Structure and Mechanism in Protein Science: A Guide to Enzyme Catalysis and Protein Folding,* W.H. Freeman and Company, New York.

A clearly written, concise introduction. More advanced.

Friedmann, H. (ed.) (1981) *Benchmark Papers in Biochemistry,* Vol. 1: *Enzymes,* Hutchinson Ross Publishing Company, Stroudsburg, PA.

A collection of classic papers on enzyme chemistry, with historical commentaries by the editor. Extremely interesting.

Jencks, W.P. (1987) *Catalysis in Chemistry and Enzymology,* Dover Publications, Inc., New York.

An outstanding book on the subject. More advanced.

Kornberg, A. (1989) *For the Love of Enzymes: The Odyssey of a Biochemist,* Harvard University Press, Cambridge, MA.

Principles of Catalysis

Hansen, D.E. & Raines, R.T. (1990) Binding energy and enzymatic catalysis. *J. Chem. Educ.* **67,** 483–489.

A good place for the beginning student to acquire a better understanding of principles.

Kraut, J. (1988) How do enzymes work? *Science* **242,** 533–540.

Landry, D.W., Zhao, K., Yang, G.X.-Q., Glickman, M., & Georgiadis, T.M. (1993) Antibody degradation of cocaine. *Science* **259,** 1899–1901.

An interesting application of catalytic antibodies.

Lerner, R.A., Benkovic, S.J., & Schulz, P.G. (1991) At the crossroads of chemistry and immunology: catalytic antibodies. *Science* **252,** 659–667.

Schramm, V.L. (1998) Enzymatic transition states and transition state analog design. *Annu. Rev. Biochem.* **67,** 693–720.

Many good illustrations of the principles introduced in this chapter.

Kinetics

Cleland, W.W. (1977) Determining the chemical mechanisms of enzyme-catalyzed reactions by kinetic studies. *Adv. Enzymol.* **45,** 273–387.

Radzicka, A. & Wolfenden, R. (1995) A proficient enzyme. *Science* **267,** 90–93.

Definitive examination of rate enhancement by an enzyme that accelerates its reaction by a factor of 10^{17}.

Raines, R.T. & Hansen, D.E. (1988) An intuitive approach to steady-state kinetics. *J. Chem. Educ.* **65,** 757–759.

Segel, I.H. (1975) *Enzyme Kinetics: Behavior and Analysis of Rapid Equilibrium and Steady State Enzyme Systems,* John Wiley & Sons, Inc., New York.

A more advanced treatment.

Enzyme Examples

Babbit, P.C. & Gerlt, J.A. (1997) Understanding enzyme superfamilies: chemistry as the fundamental determinant in the evolution of new catalytic activities. *J. Biol. Chem.* **27,** 30,591–30,594.

An interesting description of the evolution of enzymes with different catalytic specificities, making use of a limited repertoire of protein structural motifs.

Babbitt, P.C., Hasson, M.S., Wedekind, J.E., Palmer, D.R.J., Barrett, W.C., Reed, G.H., Rayment, I., Ringe, D., Kenyon, G.L., & Gerlt, J.A. (1996) The enolase superfamily: a general strategy for enzyme-catalyzed abstraction of the α-protons of carboxylic acids. *Biochemistry* **35,** 16,489–16,501.

Warshel, A., Naray-Szabo, G., Sussman, F., & Hwang, J.-K. (1989) How do serine proteases really work? *Biochemistry* **28,** 3629–3637.

Regulatory Enzymes

Barford, D., Das, A.K., & Egloff, M.-P. (1998). The structure and mechanism of protein phosphatases: insights into catalysis and regulation. *Annu. Rev. Biophys. Biomol. Struct.* **27,** 133–164.

Dische, Z. (1976) The discovery of feedback inhibition. *Trends Biochem. Sci.* **1,** 269–270.

Hunter, T. & Plowman, G.D. (1997) The protein kinases of budding yeast: six score and more. *Trends Biochem. Sci.* **22,** 18–22.

Details of the variety of these important enzymes in a model eukaryote.

Johnson, L.N. & Barford, D. (1993) The effects of phosphorylation on the structure and function of proteins. *Annu. Rev. Biophys. Biomol. Struct.* **22,** 199–232.

Koshland, D.E., Jr. & Neet, K.E. (1968) The catalytic and regulatory properties of enzymes. *Annu. Rev. Biochem.* **37,** 359–410.

Monod, J., Changeux, J.-P., & Jacob, F. (1963) Allosteric proteins and cellular control systems. *J. Mol. Biol.* **6,** 306–329.

A classic paper introducing the concept of allosteric regulation.

problems

1. Keeping the Sweet Taste of Corn The sweet taste of freshly picked corn (maize) is due to the high level of sugar in the kernels. Store-bought corn (several days after picking) is not as sweet because about 50% of the free sugar is converted into starch within one day of picking. To preserve the sweetness of fresh corn, the husked ears can be immersed in boiling water for a few minutes ("blanched") then cooled in cold water. Corn processed in this way and stored in a freezer maintains its sweetness. What is the biochemical basis for this procedure?

2. Intracellular Concentration of Enzymes To approximate the actual concentration of enzymes in a bacterial cell, assume that the cell contains equal concentrations of 1,000 different enzymes in solution in the cytosol and that each protein has a molecular weight of 100,000. Assume also that the bacterial cell is a cylinder (diameter 1 μm, height 2.0 μm), that the cytosol (specific gravity 1.20) is 20% soluble protein by weight, and that the soluble protein consists entirely of enzymes. Calculate the *average* molar concentration of each enzyme in this hypothetical cell.

3. Rate Enhancement by Urease The enzyme urease enhances the rate of urea hydrolysis at pH 8.0 and 20 °C by a factor of 10^{14}. If a given quantity of urease can completely hydrolyze a given quantity of urea in 5 min at 20 °C and pH 8.0, how long would it take for this amount of urea to be hydrolyzed under the same conditions in the absence of urease? Assume that both reactions take place in sterile systems so that bacteria cannot attack the urea.

4. Protection of an Enzyme against Denaturation by Heat When enzyme solutions are heated, there is a progressive loss of catalytic activity over time due to denaturation of the enzyme. A solution of the enzyme hexokinase incubated at 45 °C lost 50% of its activity in 12 min, but when incubated at 45 °C in the presence of a very large concentration of one of its substrates, it lost only 3% of its activity in 12 min. Suggest why thermal denaturation of hexokinase was retarded in the presence of one of its substrates.

5. Requirements of Active Sites in Enzymes Carboxypeptidase, which sequentially removes carboxyl-terminal amino acid residues from its peptide substrates, is a single polypeptide of 307 amino acids. The two essential catalytic groups in the active site are furnished by Arg^{145} and Glu^{270}.

(a) If the carboxypeptidase chain were a perfect α helix, how far apart (in Ångstroms) would Arg^{145} and Glu^{270} be? (Hint: See Fig. 6–4b.)

(b) Explain how two amino acids separated by this distance can catalyze a reaction occurring in the space of a few Ångstroms.

6. Quantitative Assay for Lactate Dehydrogenase The muscle enzyme lactate dehydrogenase catalyzes the reaction

$$\underset{\text{Pyruvate}}{CH_3{-}\overset{\overset{\large O}{\|}}{C}{-}COO^-} + NADH + H^+ \longrightarrow \underset{\text{Lactate}}{CH_3{-}\underset{\underset{\large H}{|}}{\overset{\overset{\large OH}{|}}{C}}{-}COO^-} + NAD^+$$

NADH and NAD^+ are the reduced and oxidized forms, respectively, of the coenzyme NAD. Solutions of NADH, but *not* NAD^+, absorb light at 340 nm. This property is used to determine the concentration of NADH in solution by measuring spectrophotometrically the amount of light absorbed at 340 nm by the solution. Explain how these properties of NADH can be used to design a quantitative assay for lactate dehydrogenase.

7. Relation between Reaction Velocity and Substrate Concentration: Michaelis-Menten Equation (a) At what substrate concentration will an enzyme with a k_{cat} of 30 s^{-1} and a K_m of 0.005 M show one-quarter of its maximum rate? (b) Determine the fraction of V_{max} that would occur at the following substrate concentrations: [S] = $\frac{1}{2}K_m$, $2K_m$, and $10K_m$.

8. Estimation of V_{max} and K_m by Inspection Although graphical methods are available for accurate determination of the V_{max} and K_m of an enzyme-catalyzed reaction (see Box 8–1), sometimes these quantities can be quickly estimated by inspecting values of V_0 at increasing [S]. Estimate the V_{max} and K_m of the enzyme-catalyzed reaction for which the following data were obtained:

[S] (M)	V_0 (μM/min)
2.5×10^{-6}	28
4.0×10^{-6}	40
1×10^{-5}	70
2×10^{-5}	95
4×10^{-5}	112
1×10^{-4}	128
2×10^{-3}	139
1×10^{-2}	140

9. Properties of an Enzyme of Prostaglandin Synthesis Prostaglandins are a class of eicosanoids, fatty acid derivatives with a variety of extremely potent actions on vertebrate tissues, whose structure and action will be discussed further in Chapters 11 and 21. Prostaglandins are responsible for producing fever and inflammation and its associated pain. They are derived from the 20-carbon fatty acid arachidonic acid in a reaction catalyzed by the enzyme prostaglandin endoperoxide synthase. This enzyme, a cyclooxygenase, uses oxygen to convert arachidonic acid to PGG_2, the immediate precursor of many different prostaglandins (Chapter 21).

(a) The kinetic data given below are for the reaction catalyzed by prostaglandin endoperoxide synthase. Focusing here on the first two columns, determine the V_{max} and K_m of the enzyme.

[Arachidonic acid] (mM)	Rate of formation of PGG_2 (mM/min)	Rate of formation of PGG_2 with 10 mg/mL ibuprofen (mM/min)
0.5	23.5	16.67
1.0	32.2	25.25
1.5	36.9	30.49
2.5	41.8	37.04
3.5	44.0	38.91

(b) Ibuprofen is an inhibitor of prostaglandin endoperoxide synthase. By inhibiting the synthesis of prostaglandins, ibuprofen reduces inflammation and pain. Using the data in the first and third columns of the table, determine the type of inhibition that ibuprofen exerts on prostaglandin endoperoxide synthase.

10. Graphical Analysis of V_{max} and K_m The following experimental data were collected during a study of the catalytic activity of an intestinal peptidase with the substrate glycylglycine:

$$\text{Glycylglycine} + H_2O \longrightarrow 2\ \text{glycine}$$

[S] (mM)	Product formed (μmol/min)
1.5	0.21
2.0	0.24
3.0	0.28
4.0	0.33
8.0	0.40
16.0	0.45

Use graphical analysis (see Box 8–1) to determine the K_m and V_{max} for this enzyme preparation and substrate.

11. The Eadie-Hofstee Equation One transformation of the Michaelis-Menten equation is the Lineweaver-Burk, or double-reciprocal, equation. Multiplying both sides of the Lineweaver-Burk equation by V_{max} and rearranging gives the Eadie-Hofstee equation:

$$V_0 = (-K_m)\frac{V_0}{[S]} + V_{max}$$

A plot of V_0 vs. $V_0/[S]$ for an enzyme-catalyzed reaction is shown below.

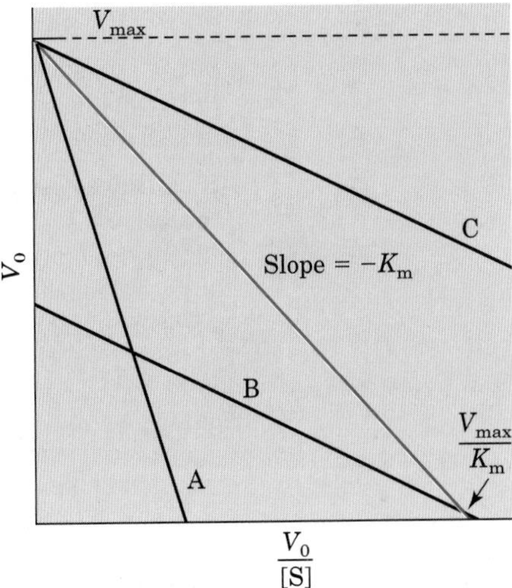

The blue curve was obtained in the absence of inhibitor. Which of the other curves (A, B, or C) shows the enzyme activity when a competitive inhibitor is added to the reaction mixture?

12. The Turnover Number of Carbonic Anhydrase Carbonic anhydrase of erythrocytes (M_r 30,000) has one of the highest turnover numbers among known enzymes. It catalyzes the reversible hydration of CO_2:

$$H_2O + CO_2 \rightleftharpoons H_2CO_3$$

This is an important process in the transport of CO_2 from the tissues to the lungs. If 10 μg of pure carbonic anhydrase catalyzes the hydration of 0.30 g of CO_2 in 1 min at 37 °C at V_{max}, what is the turnover number (k_{cat}) of carbonic anhydrase (in units of min^{-1})?

13. Deriving a Rate Equation for Competitive Inhibition The rate equation for an enzyme subject to competitive inhibition is

$$V_0 = \frac{V_{max}[S]}{\alpha K_m + [S]}$$

Beginning with a new definition of total enzyme as

$$[E]_t = [E] + [EI] + [ES]$$

and the definitions of α and K_I on p. 266, derive the rate equation above. Use the derivation of the Michaelis-Menten equation in the text as a guide.

14. Irreversible Inhibition of an Enzyme Many enzymes are inhibited irreversibly by heavy-metal ions such as Hg^{2+}, Cu^{2+}, or Ag^+, which can react with essential sulfhydryl groups to form mercaptides:

$$\text{Enz–SH} + Ag^+ \longrightarrow \text{Enz–S–Ag} + H^+$$

The affinity of Ag^+ for sulfhydryl groups is so great that Ag^+ can be used to titrate —SH groups quantitatively. To 10 mL of a solution containing 1.0 mg/mL of a pure enzyme, an investigator added just enough $AgNO_3$ to completely inactivate the enzyme. A total of 0.342 μmol of $AgNO_3$ was required. Calculate the *minimum* molecular weight of the enzyme. Why does the value obtained in this way give only the minimum molecular weight?

15. Clinical Application of Differential Enzyme Inhibition Human blood serum contains a class of enzymes known as acid phosphatases, which hydrolyze biological phosphate esters under slightly acidic conditions (pH 5.0):

$$\text{R—O—}\overset{\displaystyle O^-}{\underset{\displaystyle O}{\overset{|}{\underset{\|}{P}}}}\text{—O}^- + H_2O \longrightarrow \text{R—OH} + \text{HO—}\overset{\displaystyle O^-}{\underset{\displaystyle O}{\overset{|}{\underset{\|}{P}}}}\text{—O}^-$$

Acid phosphatases are produced by erythrocytes, the liver, kidney, spleen, and prostate gland. The enzyme from the prostate gland is clinically important because its increased activity in the blood is frequently an indication of prostate cancer. The phosphatase from the prostate gland is strongly inhibited by tartrate ion, but acid phosphatases from other tissues are not. How can this information be used to develop a specific procedure for measuring the activity of the acid phosphatase of the prostate gland in human blood serum?

16. Inhibition of Carbonic Anhydrase by Acetazolamide Carbonic anhydrase is strongly inhibited by the drug acetazolamide, which is used as a diuretic (to increase the production of urine) and to treat glaucoma (to reduce excessively high pressure in the eye due to accumulation of intraocular fluid). Carbonic anhydrase plays an important role in these and other secretory processes because it participates in regulating the pH and bicarbonate content of a number of body fluids. The experimental curve of initial reaction velocity (as percentage of V_{max}) versus [S] for the carbonic anhydrase reaction is illustrated below (upper curve). When the experiment is repeated in the presence of acetazolamide, the lower curve is obtained. From an inspection of the curves and your knowledge of the kinetic properties of competitive and mixed enzyme inhibitors, determine the nature of the inhibition by acetazolamide. Explain.

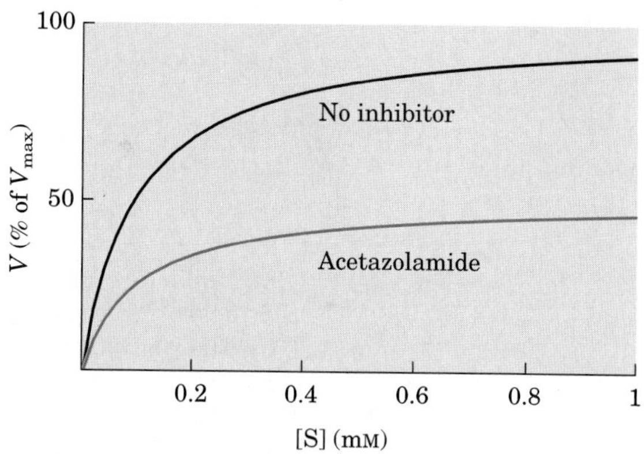

17. pH Optimum of Lysozyme The active site of lysozyme contains two amino acid residues essential for catalysis: Glu^{35} and Asp^{52}. The pK_a values of the carboxyl side chains of these two residues are 5.9 and 4.5, respectively. What is the ionization state (protonated or deprotonated) of each residue at pH 5.2, the pH optimum of lysozyme? How can the ionization states of these residues explain the pH-activity profile of lysozyme shown below?

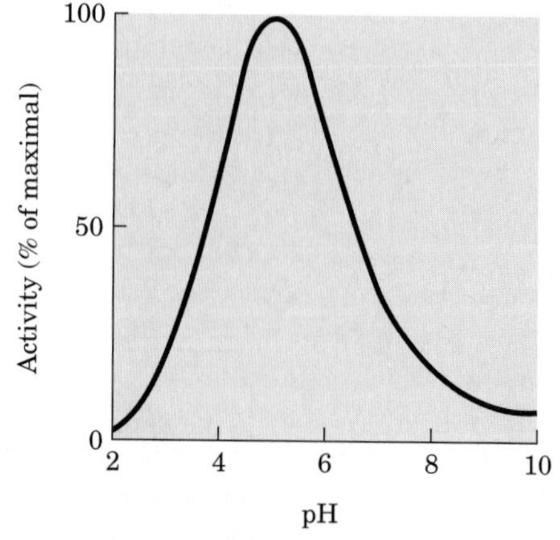

chapter

9

Carbohydrates and Glycobiology

Carbohydrates are the most abundant biomolecules on Earth. Each year, photosynthesis converts more than 100 billion metric tons of CO_2 and H_2O into cellulose and other plant products. Certain carbohydrates (sugar and starch) are a dietary staple in most parts of the world, and the oxidation of carbohydrates is the central energy-yielding pathway in most nonphotosynthetic cells. Insoluble carbohydrate polymers serve as structural and protective elements in the cell walls of bacteria and plants and in the connective tissues of animals. Other carbohydrate polymers lubricate skeletal joints and participate in recognition and adhesion between cells. More complex carbohydrate polymers covalently attached to proteins or lipids act as signals that determine the intracellular location or metabolic fate of these hybrid molecules, called **glycoconjugates**. This chapter introduces the major classes of carbohydrates and glycoconjugates and provides a few examples of their many structural and functional roles.

Carbohydrates are predominantly cyclized polyhydroxy aldehydes or ketones, or substances that yield such compounds on hydrolysis. Many, but not all, carbohydrates have the empirical formula $(CH_2O)_n$; some also contain nitrogen, phosphorus, or sulfur.

There are three major size classes of carbohydrates: monosaccharides, oligosaccharides, and polysaccharides (the word "saccharide" is derived from the Greek *sakcharon,* meaning "sugar"). **Monosaccharides,** or simple sugars, consist of a single polyhydroxy aldehyde or ketone unit. The most abundant monosaccharide in nature is the six-carbon sugar D-glucose, sometimes referred to as dextrose.

Oligosaccharides consist of short chains of monosaccharide units, or residues, joined by characteristic linkages called glycosidic bonds. The most abundant are the **disaccharides,** with two monosaccharide units. Typical is sucrose, or cane sugar, which consists of the six-carbon sugars D-glucose and D-fructose. All common monosaccharides and disaccharides have names ending with the suffix "-ose." In cells, most oligosaccharides having three or more units do not occur as free entities but are joined to nonsugar molecules (lipids or proteins) in glycoconjugates.

Sugar polymers occur in a continuous range of sizes. Those containing more than about 20 monosaccharide units are generally called **polysaccharides.** Polysaccharides may have hundreds or thousands of monosaccharide units. Some polysaccharide molecules, such as cellulose, are linear chains, whereas others, such as glycogen, are branched chains. The plant products starch and cellulose both consist of recurring units of D-glucose, but they differ in the type of glycosidic linkage, and consequently have strikingly different properties and biological roles.

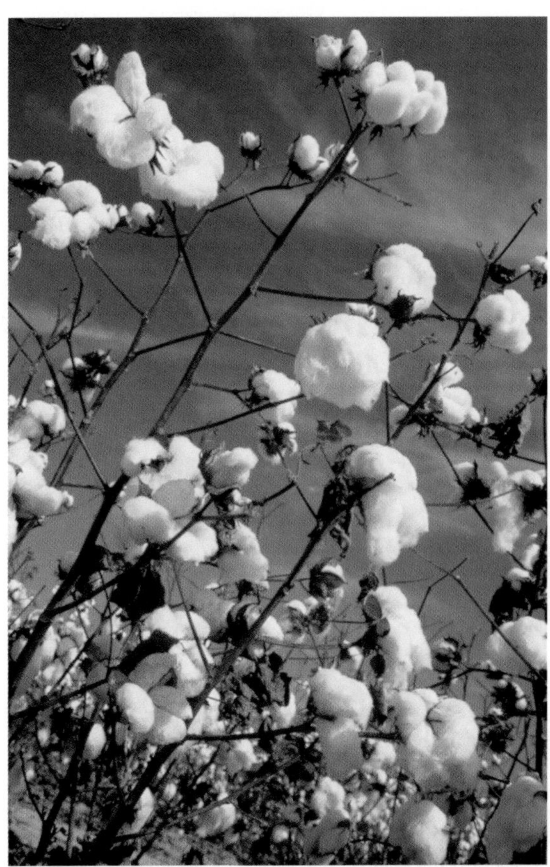

Cotton bolls ready for harvest. The cotton fiber consists of cellulose, a polysaccharide with physical and chemical properties that give cotton textiles their desirable qualities. The cotton plant has been cultivated and used in textile production for thousands of years.

Monosaccharides and Disaccharides

The simplest of the carbohydrates, the monosaccharides, are either aldehydes or ketones with one or more hydroxyl groups; the six-carbon monosaccharides glucose and fructose have five hydroxyl groups. The carbon atoms to which hydroxyl groups are attached are often chiral centers, which give rise to the many sugar stereoisomers found in nature. We begin by describing the families of monosaccharides with backbones of three to seven carbons—their structure and stereoisomeric forms, and the means of representing their three-dimensional structures on paper. We then discuss several chemical reactions of the carbonyl groups of monosaccharides. One such reaction, the addition of a hydroxyl group from within the same molecule, generates the cyclic forms of five- and six-carbon sugars (which predominate in aqueous solution) and creates a new chiral center, adding further stereochemical complexity to this class of compounds. The nomenclature for unambiguously specifying the configuration about each carbon atom in a cyclic form and the means of representing these structures on paper are therefore described in some detail; these will be useful later in our discussion of the metabolism of monosaccharides.

We also introduce here some important monosaccharide derivatives that will be encountered in later chapters.

The Two Families of Monosaccharides Are Aldoses and Ketoses

Monosaccharides are colorless, crystalline solids that are freely soluble in water but insoluble in nonpolar solvents. Most have a sweet taste. The backbones of common monosaccharide molecules are unbranched carbon chains in which all the carbon atoms are linked by single bonds. In the open-chain form, one of the carbon atoms is double-bonded to an oxygen atom to form a carbonyl group; each of the other carbon atoms has a hydroxyl group. If the carbonyl group is at an end of the carbon chain (i.e., in an aldehyde group) the monosaccharide is an **aldose;** if the carbonyl group is at any other position (in a ketone group) the monosaccharide is a **ketose.** The simplest monosaccharides are the two three-carbon trioses: glyceraldehyde, an aldotriose, and dihydroxyacetone, a ketotriose (Fig. 9–1a).

Monosaccharides with four, five, six, and seven carbon atoms in their backbones are called, respectively, tetroses, pentoses, hexoses, and heptoses. There are aldoses and ketoses of each of these chain lengths: aldotetroses and ketotetroses, aldopentoses and ketopentoses, and so on. The hexoses, which include the aldohexose D-glucose and the ketohexose D-fructose (Fig. 9–1b), are the most common monosaccharides in nature. The aldopentoses D-ribose and 2-deoxy-D-ribose (Fig. 9–1c) are components of nucleotides and nucleic acids (Chapter 10).

figure 9–1

Representative monosaccharides. (a) Two trioses, an aldose and a ketose. The carbonyl group in each is shaded. **(b)** Two common hexoses. **(c)** The pentose components of nucleic acids. D-Ribose is a component of ribonucleic acid (RNA), and 2-deoxy-D-ribose is a component of deoxyribonucleic acid (DNA).

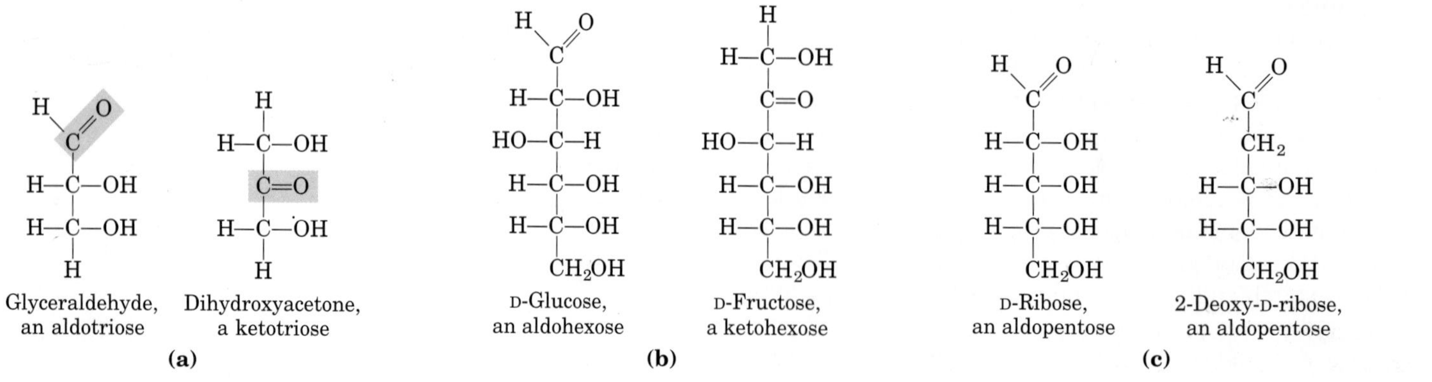

Glyceraldehyde, an aldotriose; Dihydroxyacetone, a ketotriose **(a)**

D-Glucose, an aldohexose; D-Fructose, a ketohexose **(b)**

D-Ribose, an aldopentose; 2-Deoxy-D-ribose, an aldopentose **(c)**

Monosaccharides Have Asymmetric Centers

All the monosaccharides except dihydroxyacetone contain one or more asymmetric (chiral) carbon atoms and thus occur in optically active isomeric forms (pp. 58–61). The simplest aldose, glyceraldehyde, contains one chiral center (the middle carbon atom) and therefore has two different optical isomers, or enantiomers (Fig. 9–2). By convention, one of these two forms is designated the D isomer, the other the L isomer. As for other biomolecules with chiral centers, the absolute configurations of sugars are known from x-ray crystallography. To represent three-dimensional sugar structures on paper, we often use **Fischer projection formulas** (Fig. 9–2).

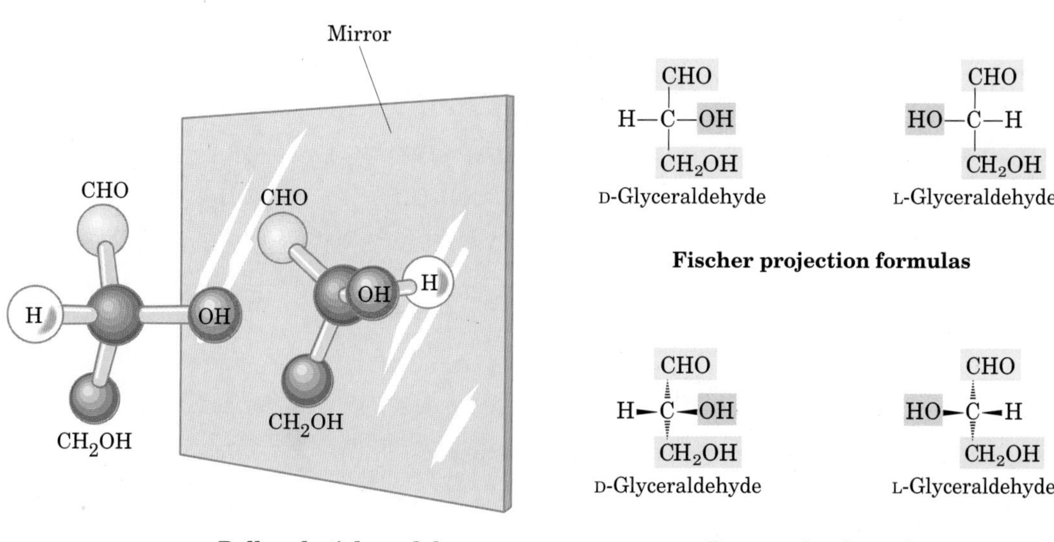

figure 9–2
Three ways to represent the two stereoisomers of glyceraldehyde. The stereoisomers are mirror images of each other. Ball-and-stick models show the actual configuration of molecules. By convention, in Fischer projection formulas, horizontal bonds project out of the plane of the paper, toward the reader; vertical bonds project behind the plane of the paper, away from the reader. Recall (see Fig. 3–7) that in perspective formulas, solid wedge-shaped bonds point toward the reader, dashed wedges point away.

In general, a molecule with n chiral centers can have 2^n stereoisomers. Glyceraldehyde has $2^1 = 2$; the aldohexoses, with four chiral centers, have $2^4 = 16$ stereoisomers. The stereoisomers of monosaccharides of each carbon chain length can be divided into two groups that differ in the configuration about the chiral center *most distant* from the carbonyl carbon. Those in which the configuration at this reference carbon is the same as that of D-glyceraldehyde are designated D isomers, and those with the same configuration as L-glyceraldehyde are L isomers. When the hydroxyl group on the reference carbon is on the right in the projection formula, the sugar is the D isomer; when on the left, it is the L isomer. Of the 16 possible aldohexoses, eight are D forms and eight are L. Most of the hexoses of living organisms are D isomers.

Figure 9–3 shows the structures of the D stereoisomers of all the aldoses and ketoses having three to six carbon atoms. The carbons of a sugar are numbered beginning at the end of the chain nearest the carbonyl group. Each of the eight D-aldohexoses, which differ in the stereochemistry at C-2, C-3, or C-4, has its own name: D-glucose, D-galactose, D-mannose, and so forth (Fig. 9–3a). The four- and five-carbon ketoses are designated by

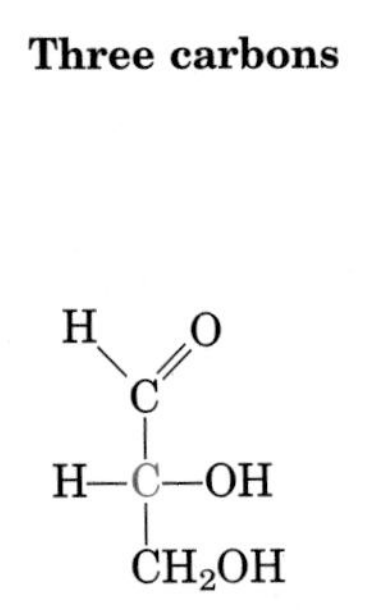

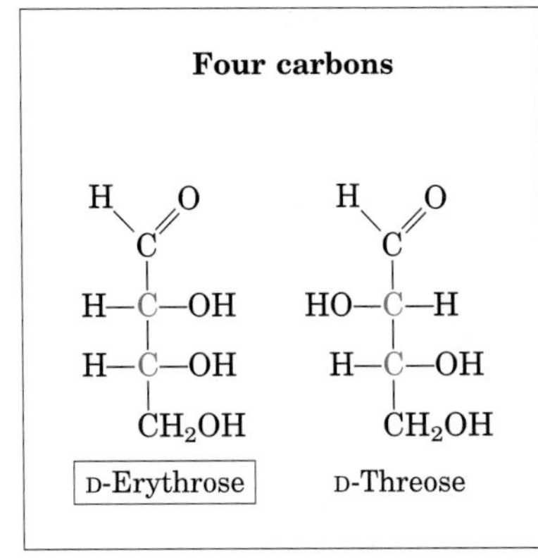

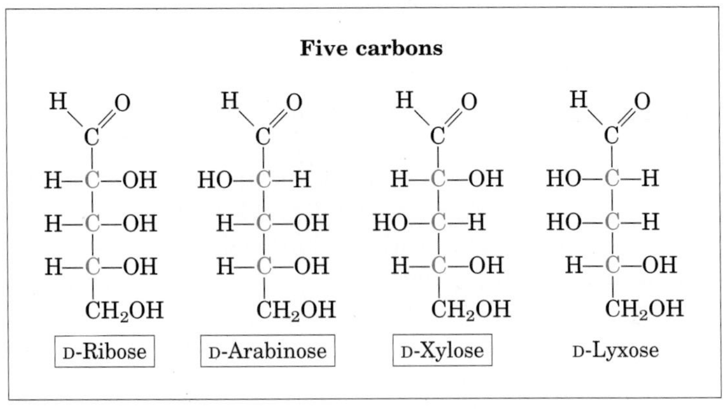

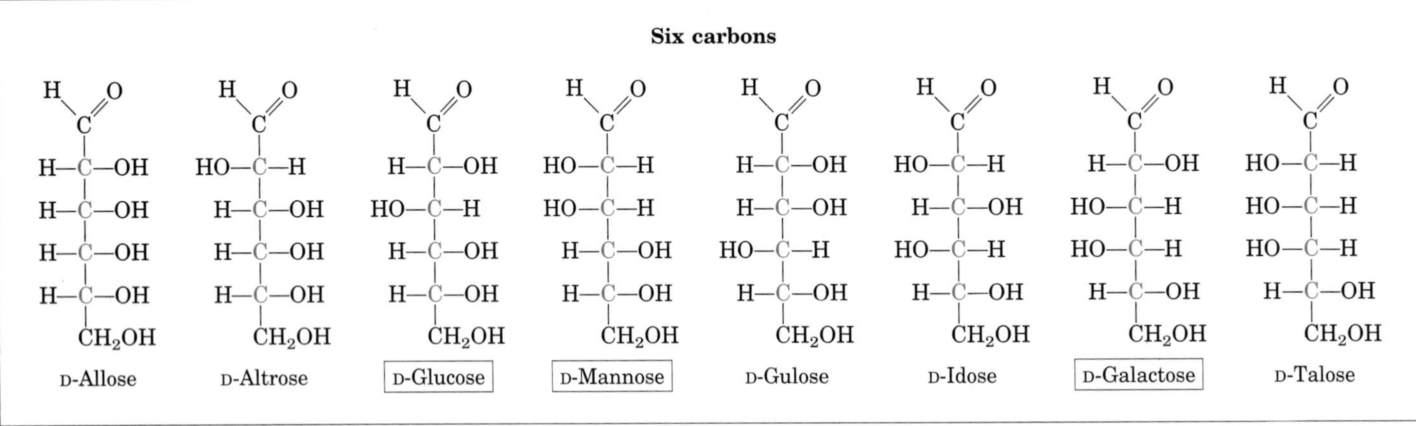

D-Aldoses
(a)

figure 9–3
The series of **(a)** D-aldoses and **(b)** D-ketoses having from three to six carbon atoms, shown as projection formulas. The carbon atoms in red are chiral centers. In all these D isomers, the chiral carbon *most distant from the carbonyl carbon* has the same configuration as the chiral carbon in D-glyceraldehyde. The sugars named in boxes are the most common in nature; we shall encounter these again in this and later chapters.

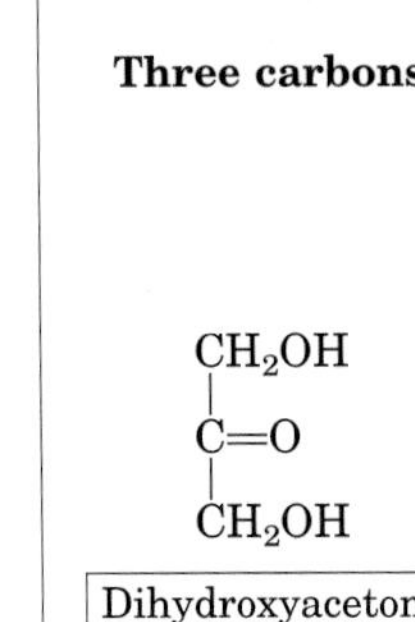

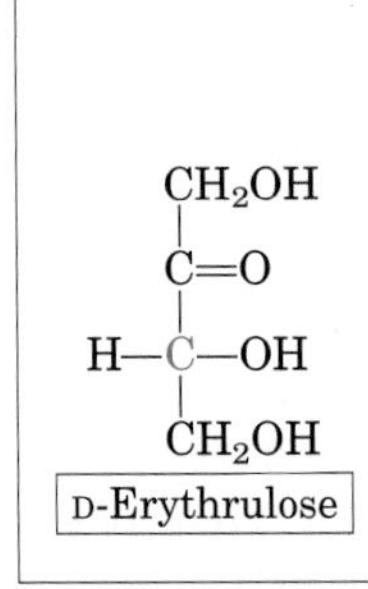

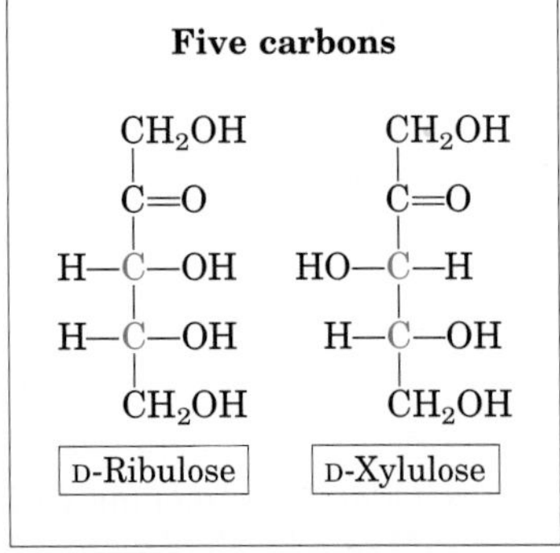

Six carbons

D-Psicose D-Fructose D-Sorbose D-Tagatose

D-Ketoses
(b)

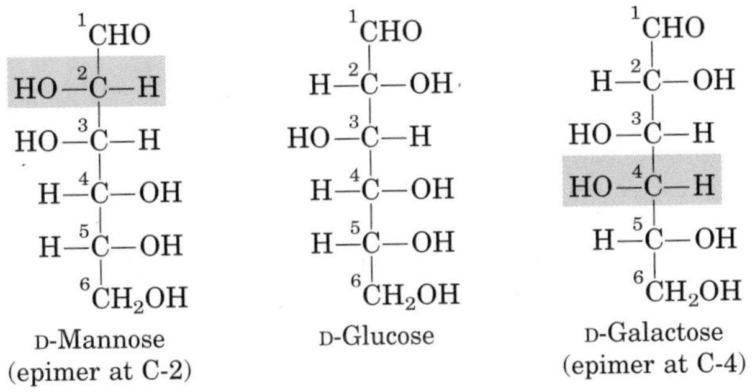

figure 9–4

Epimers. D-Glucose and two of its epimers are shown as projection formulas. Each epimer differs from D-glucose in the configuration at one chiral center (shaded red).

inserting "ul" into the name of a corresponding aldose; for example, D-ribulose is the ketopentose corresponding to the aldopentose D-ribose. The ketohexoses are named otherwise, such as fructose (from Latin *fructus*, meaning "fruit"; fruits are rich in this sugar) and sorbose (from *Sorbus*, the genus of mountain ash, which has berries rich in the related sugar alcohol sorbitol). Two sugars that differ only in the configuration around one carbon atom are called **epimers;** D-glucose and D-mannose, which differ only in the stereochemistry at C-2, are epimers, as are D-glucose and D-galactose (which differ at C-4) (Fig. 9–4).

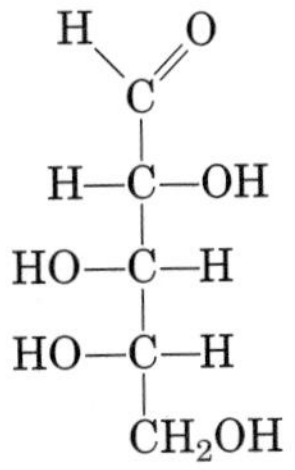

Some sugars occur naturally in their L form; examples are L-arabinose and the L isomers of some sugar derivatives (discussed below) that are common components of glycoconjugates.

The Common Monosaccharides Have Cyclic Structures

For simplicity, we have thus far represented the structures of various aldoses and ketoses as straight-chain forms (Figs. 9–3, 9–4). In fact, in aqueous solution, aldotetroses and all monosaccharides with five or more carbon atoms in the backbone occur predominantly as cyclic (ring) structures in which the carbonyl group has formed a covalent bond with the oxygen of a hydroxyl group along the chain. The formation of these ring structures is the result of a general reaction between aldehydes or ketones and alcohols to form derivatives called **hemiacetals** or **hemiketals** (Fig. 9–5), which contain an additional asymmetric carbon atom and thus can exist in two stereoisomeric forms. For example, D-glucose exists in solution as an intramolecular hemiacetal in which the free hydroxyl group at C-5 has reacted

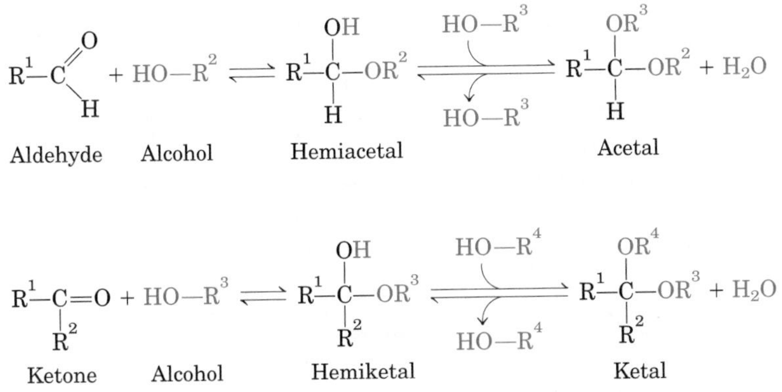

figure 9–5

Formation of hemiacetals and hemiketals. An aldehyde or ketone can react with an alcohol in a 1 : 1 ratio to yield a hemiacetal or hemiketal, respectively, creating a new chiral center at the carbonyl carbon. Substitution of a second alcohol molecule produces an acetal or ketal. When the second alcohol is part of another sugar molecule, the bond produced is a glycosidic bond (p. 301).

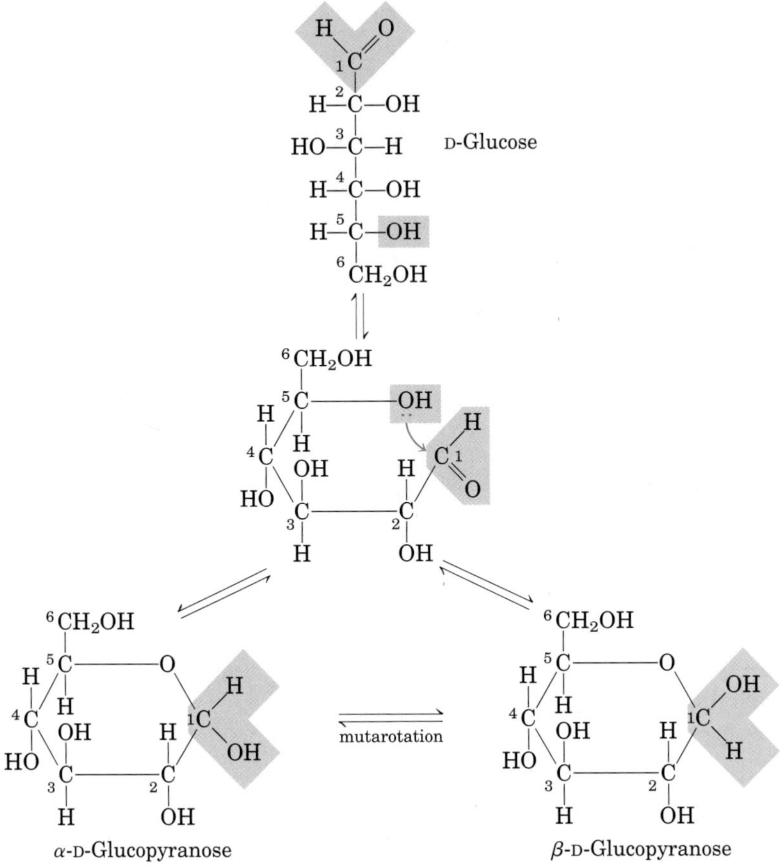

figure 9–6
Formation of the two cyclic forms of D-glucose. Reaction between the aldehyde group at C-1 and the hydroxyl group at C-5 forms a hemiacetal linkage, producing either of two stereoisomers, the α and β anomers, which differ only in the stereochemistry around the hemiacetal carbon. The interconversion of α and β anomers is called mutarotation.

with the aldehydic C-1, rendering the latter asymmetric and producing two stereoisomers, designated α and β (Fig. 9–6). These six-membered ring compounds are called **pyranoses** because they resemble the six-membered ring compound pyran (Fig. 9–7). The systematic names for the two ring forms of D-glucose are α-D-glucopyranose and β-D-glucopyranose.

Aldohexoses also exist in cyclic forms having five-membered rings, which, because they resemble the five-membered ring compound furan, are called **furanoses.** However, the six-membered aldopyranose ring is much more stable than the aldofuranose ring and predominates in aldohexose solutions. Only aldoses having five or more carbon atoms can form pyranose rings.

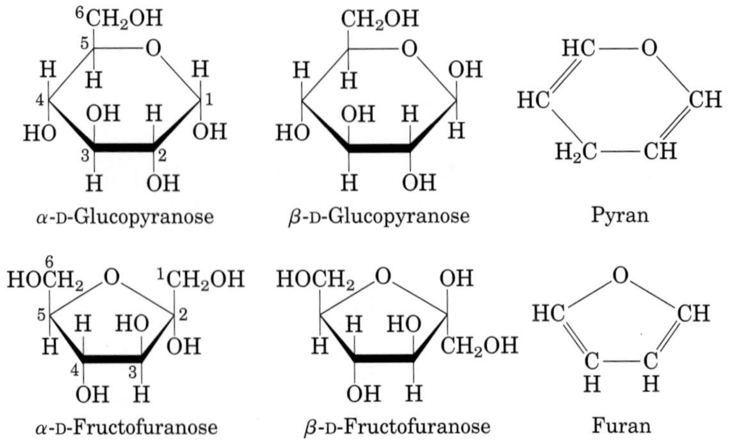

figure 9–7
Pyranoses and furanoses. The pyranose forms of D-glucose and the furanose forms of D-fructose are shown here as Haworth perspective formulas. The edges of the ring nearest the reader are represented by bold lines. Hydroxyl groups below the plane of the ring in these Haworth perspectives would appear at the right side of a Fischer projection (compare with Fig. 9–6). Pyran and furan are shown for comparison.

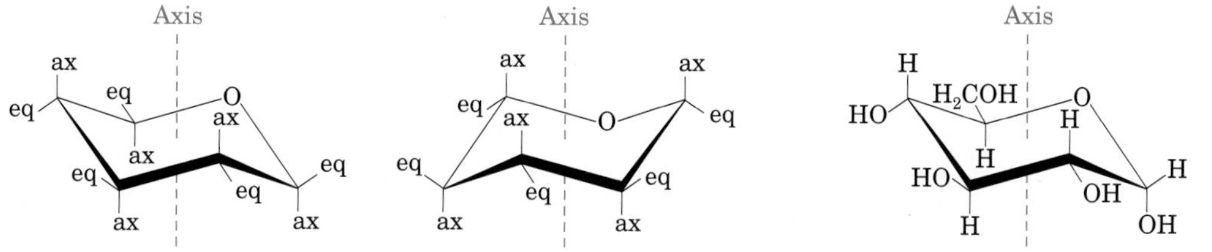

Two possible chair forms
(a)

α-D-Glucopyranose
(b)

figure 9–8

Conformational formulas of pyranoses. (a) Two chair forms of the pyranose ring. Substituents on the ring carbons may be either axial (ax), projecting parallel with the vertical axis through the ring, or equatorial (eq), projecting roughly perpendicular to this axis. Generally, substituents in the equatorial positions are less sterically hindered by neighboring substituents, and conformations with their bulky substituents in equatorial positions are favored. Another conformation, the "boat" (not shown), is only seen in derivatives with very bulky substituents. **(b)** A chair conformation of α-D-glucopyranose.

Isomeric forms of monosaccharides that differ only in their configuration about the hemiacetal or hemiketal carbon atom are called **anomers.** The hemiacetal or carbonyl carbon atom is called the **anomeric carbon.** The α and β anomers of D-glucose interconvert in aqueous solution by a process called **mutarotation.** Thus a solution of α-D-glucose and a solution of β-D-glucose eventually form identical equilibrium mixtures having identical optical properties. This mixture consists of about one-third α-D-glucose, two-thirds β-D-glucose, and very small amounts of the linear and five-membered ring (glucofuranose) forms.

Ketohexoses also occur in α and β anomeric forms. In these compounds the hydroxyl group on C-5 (or C-6) reacts with the keto group at C-2, forming a furanose (or pyranose) ring containing a hemiketal linkage (Fig. 9–5). D-Fructose forms predominantly the furanose ring (Fig. 9–7); the more common anomer in combined forms or derivatives is β-D-fructofuranose.

Haworth perspective formulas like those in Figure 9–7 are commonly used to show the ring forms of monosaccharides. However, the six-membered pyranose ring is not planar, as Haworth perspectives suggest, but tends to assume either of two "chair" conformations (Fig. 9–8). Recall from Chapter 3 that two *conformations* of a molecule are interconvertible without the breakage of covalent bonds, but two *configurations* can be interconverted only by breaking a covalent bond—for example, in the case of α and β configurations, the bond involving the ring oxygen atom. The specific three-dimensional conformations of the monosaccharide units are important in determining the biological properties and functions of some polysaccharides, as we shall see.

Organisms Contain a Variety of Hexose Derivatives

In addition to simple hexoses such as glucose, galactose, and mannose, there are a number of sugar derivatives in which a hydroxyl group in the parent compound is replaced with another substituent, or a carbon atom is oxidized to a carboxyl group (Fig. 9–9). In glucosamine, galactosamine, and mannosamine, the hydroxyl at C-2 of the parent compound is replaced with an amino group. The amino group is nearly always condensed with acetic acid, as in *N*-acetylglucosamine. This glucosamine derivative is part of many structural polymers, including those of the bacterial cell wall. Bacterial cell walls also contain a derivative of glucosamine called *N*-acetylmuramic acid, in which lactic acid (a three-carbon carboxylic acid) is ether-linked to the oxygen at C-3 of *N*-acetylglucosamine. The substitution of a hydrogen for the hydroxyl group at C-6 of L-galactose or L-mannose produces L-fucose or L-rhamnose, respectively; these deoxy sugars are found in plant polysaccharides and in the complex oligosaccharide components of glycoproteins and glycolipids described later.

When the carbonyl (aldehyde) carbon of glucose is oxidized to the carboxyl level, gluconic acid is produced; other aldoses yield other **aldonic acids.** Oxidation of the carbon at the other end of the carbon chain—C-6

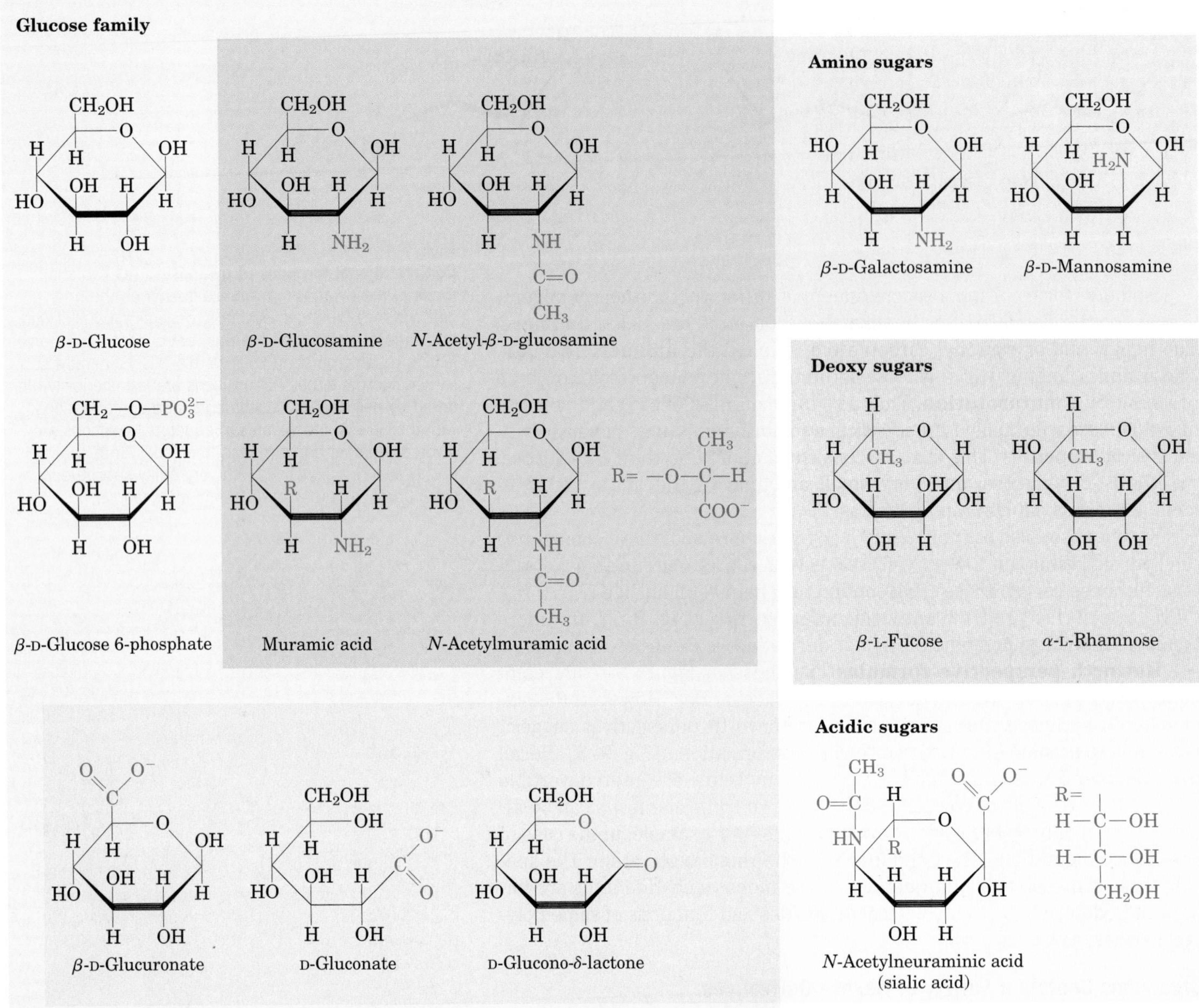

figure 9–9
Some hexose derivatives important in biology. In amino sugars, an —NH_2 group replaces one of the —OH groups in the parent hexose. Substitution of —H for —OH produces a deoxy sugar. Note that the deoxy sugars shown here occur in nature as the L isomers. The acidic sugars contain a carboxylate group, which confers a negative charge at neutral pH. D-Glucono-δ-lactone results from formation of an ester linkage between the C-1 carboxylate group and the C-5 (also known as the δ carbon) hydroxyl group of D-gluconate.

of glucose, galactose, or mannose—forms the corresponding **uronic acid:** glucuronic, galacturonic, or mannuronic acid. Both aldonic and uronic acid form stable intramolecular esters called lactones (Fig. 9–9, lower left). In addition to these acidic hexose derivatives, one nine-carbon acidic sugar deserves mention: *N*-acetylneuraminic acid (sialic acid), a derivative of *N*-acetylmannosamine, is a component of many glycoproteins and glycolipids in animals. The carboxylic acid groups of the acidic sugar derivatives are ionized at pH 7, and the compounds are therefore correctly named as carboxylates—glucuronate, galacturonate, and so forth.

In the synthesis and metabolism of carbohydrates, the intermediates are very often not the sugars themselves but their phosphorylated derivatives. Condensation of phosphoric acid with one of the hydroxyl groups of a sugar forms a phosphate ester, as in glucose 6-phosphate (Fig. 9–9). Sugar phosphates are relatively stable at neutral pH and bear a negative charge. One effect of sugar phosphorylation within cells is to trap the sugar inside the cell; cells do not in general have plasma membrane transporters for phosphorylated sugars. Phosphorylation also activates sugars for subsequent chemical transformation. Several important phosphorylated derivatives of sugars are discussed in the next chapter.

Monosaccharides Are Reducing Agents

Monosaccharides can be oxidized by relatively mild oxidizing agents such as ferric (Fe^{3+}) or cupric (Cu^{2+}) ion (Fig. 9–10a). The carbonyl carbon is oxidized to a carboxyl group. Glucose and other sugars capable of reducing ferric or cupric ion are called **reducing sugars.** This property is the basis of Fehling's reaction, a qualitative test for the presence of reducing sugar. By measuring the amount of oxidizing agent reduced by a solution of a sugar, it is also possible to estimate the concentration of that sugar. For many years, this test was used to detect and measure elevated glucose levels in blood and urine in the diagnosis of diabetes mellitus. Now, more sensitive methods for measuring blood glucose employ an enzyme, glucose oxidase (Fig. 9–10b).

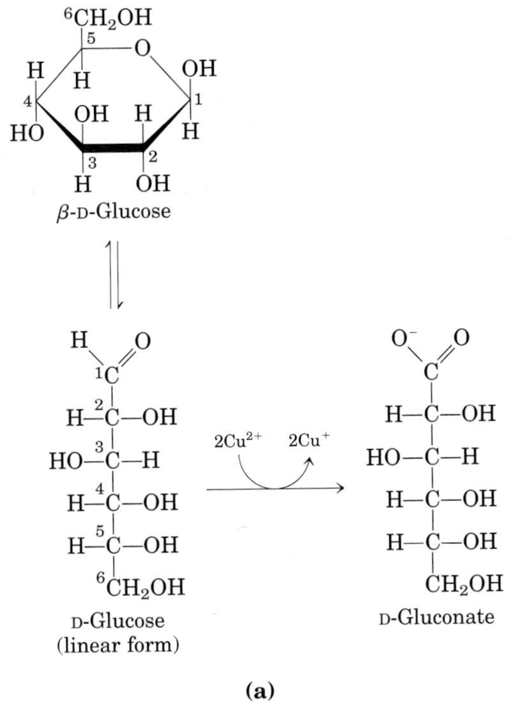

$$\text{D-Glucose} + O_2 \xrightarrow{\text{glucose oxidase}} \text{D-Glucono-}\delta\text{-lactone} + H_2O_2$$

(b)

figure 9–10
Sugars as reducing agents. (a) Oxidation of the anomeric carbon of glucose and other sugars is the basis for Fehling's reaction. The cuprous ion (Cu^+) produced under alkaline conditions forms a red cuprous oxide precipitate. In the hemiacetal (ring) form, C-1 of glucose cannot be oxidized by Cu^{2+}. However, the open-chain form is in equilibrium with the ring form, and eventually the oxidation reaction goes to completion. The reaction with Cu^{2+} is not as simple as the equation here implies; in addition to D-gluconate, a number of shorter-chain acids are produced by the fragmentation of glucose. **(b)** Blood glucose concentration is commonly determined by measuring the amount of H_2O_2 produced in the reaction catalyzed by glucose oxidase. In the reaction mixture, a second enzyme, peroxidase, catalyzes reaction of the H_2O_2 with a colorless compound to produce a colored compound, the amount of which is then measured spectrophotometrically.

Disaccharides Contain a Glycosidic Bond

Disaccharides (such as maltose, lactose, and sucrose) consist of two monosaccharides joined covalently by an ***O*-glycosidic bond,** which is formed when a hydroxyl group of one sugar reacts with the anomeric carbon of the other (Fig. 9–11). This reaction represents the formation of an acetal from a hemiacetal (such as glucopyranose) and an alcohol (a hydroxyl group of a second sugar molecule) (Fig. 9–5). When an anomeric carbon participates

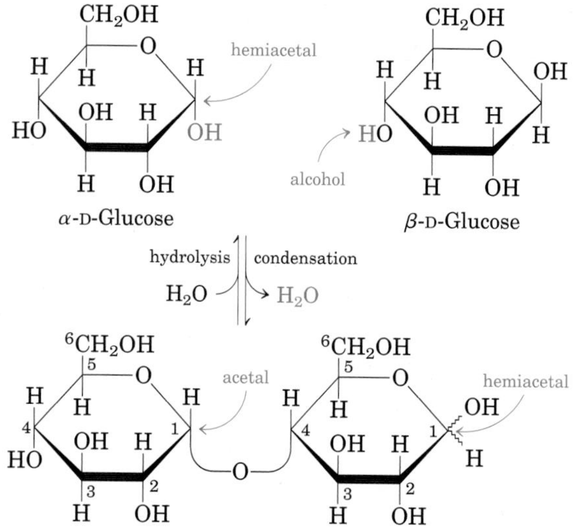

figure 9–11

Formation of maltose. A disaccharide is formed from two monosaccharides (here, two molecules of D-glucose) when an —OH (alcohol) of one glucose molecule (right) condenses with the intramolecular hemiacetal of the other glucose molecule (left), with the elimination of H_2O and formation of an *O*-glycosidic bond. The reversal of this reaction is hydrolysis—attack by H_2O on the glycosidic bond. The maltose molecule retains a reducing hemiacetal at the C-1 not involved in the glycosidic bond. Because mutarotation interconverts the α and β forms of the hemiacetal, the bonds at this position are sometimes depicted with wavy lines to indicate that the structure may be either α or β.

in a glycosidic bond, it cannot be oxidized by cupric or ferric ion. The sugar containing the anomeric carbon atom cannot exist in linear form and no longer acts as a reducing sugar. In describing disaccharides or polysaccharides, the end of a chain with a free anomeric carbon (i.e., not involved in a glycosidic bond) is commonly called the **reducing end.** Glycosidic bonds are readily hydrolyzed by acid, but resist cleavage by base. Thus disaccharides can be hydrolyzed to yield their free monosaccharide components by boiling with dilute acid. Another type of glycosidic bond joins the anomeric carbon of a sugar to a nitrogen atom in glycoproteins (see Fig. 9–25). These *N*-glycosyl bonds are also found in all nucleotides (Chapter 10).

The disaccharide maltose (Fig. 9–11) contains two D-glucose residues joined by a glycosidic linkage between C-1 (the anomeric carbon) of one glucose residue and C-4 of the other. Because the free anomeric carbon (C-1 of the glucose residue on the right in Fig. 9–11) can be oxidized, maltose is a reducing disaccharide. The configuration of the anomeric carbon atom in the glycosidic linkage is α. The glucose residue with the free anomeric carbon is capable of existing in α- and β-pyranose forms.

To name reducing disaccharides such as maltose unambiguously, and especially to name more complex oligosaccharides, several rules are followed. By convention, the name describes the compound with its nonreducing end to the left, and the name is "built up" in the following order. (1) The configuration (α or β) at the anomeric carbon joining the first monosaccharide unit (on the left) to the second is given. (2) The nonreducing residue is named. To distinguish five- and six-membered ring structures, "furano" or "pyrano" is inserted into the name. (3) The two carbon atoms joined by the glycosidic bond are indicated in parentheses, with an arrow connecting the two numbers; for example, (1→4) shows that C-1 of the first-named sugar residue is joined to C-4 of the second. (4) The second residue is named. If there is a third residue, the second glycosidic bond is described next, by the same conventions. (To shorten the description of complex polysaccharides, three-letter abbreviations for each monosaccharide are often used, as given in Table 9–1.) Following this convention for naming oligosaccharides, maltose is α-D-glucopyranosyl-(1→4)-D-glucopyranose. Because most sugars encountered in this book are the D enantiomers and the pyranose form of hexoses predominates, we will generally use a shortened version of the formal name of such compounds, giving the configuration of the anomeric carbon and naming the carbons joined by the glycosidic bond. In this abbreviated nomenclature, maltose is Glc(α1→4)Glc.

table 9–1

Abbreviations for Common Monosaccharides and Some of Their Derivatives

Abequose	Abe	Glucuronic acid	GlcA
Arabinose	Ara	Galactosamine	GalN
Fructose	Fru	Glucosamine	GlcN
Fucose	Fuc	*N*-Acetylgalactosamine	GalNAc
Galactose	Gal	*N*-Acetylglucosamine	GlcNAc
Glucose	Glc	Muramic acid	Mur
Mannose	Man	*N*-Acetylmuramic acid	Mur2Ac
Rhamnose	Rha	*N*-Acetylneuraminic acid (sialic acid)	Neu5Ac
Ribose	Rib		
Xylose	Xyl		

The disaccharide lactose (Fig. 9–12), which yields D-galactose and D-glucose on hydrolysis, occurs naturally only in milk. The anomeric carbon of the glucose residue is available for oxidation, and thus lactose is a reducing disaccharide. Its abbreviated name is Gal(β1→4)Glc. Sucrose (table sugar) is a disaccharide of glucose and fructose. It is formed by plants but not by higher animals. In contrast to maltose and lactose, sucrose contains no free anomeric carbon atom; the anomeric carbons of both monosaccharide units are involved in the glycosidic bond (Fig. 9–12). Sucrose is therefore not a reducing sugar. Nonreducing disaccharides are named as glycosides; the positions joined are the anomeric carbons. In the abbreviated nomenclature, a double-headed arrow connects the symbols specifying the anomeric carbons and their configurations. For example, the abbreviated name of sucrose is either Glc(α1↔2β)Fru or Fru(β2↔1α)Glc. Sucrose is a major intermediate product of photosynthesis; in many plants it is the principal form in which sugar is transported from the leaves to other parts of the plant body. Trehalose, Glc(α1↔1α)Glc (Fig. 9–12), is a disaccharide of D-glucose that, like sucrose, is a nonreducing sugar. It is a major constituent of the circulating fluid (hemolymph) of insects, in which it serves as an energy storage compound.

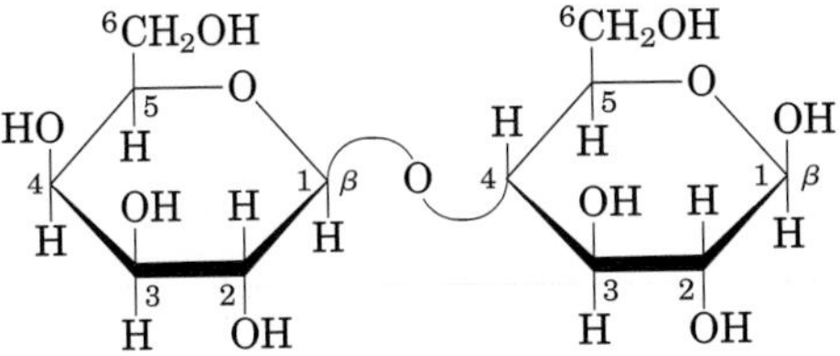

Lactose (β form)
β-D-galactopyranosyl-(1→4)-β-D-glucopyranose
Gal(β1→4)Glc

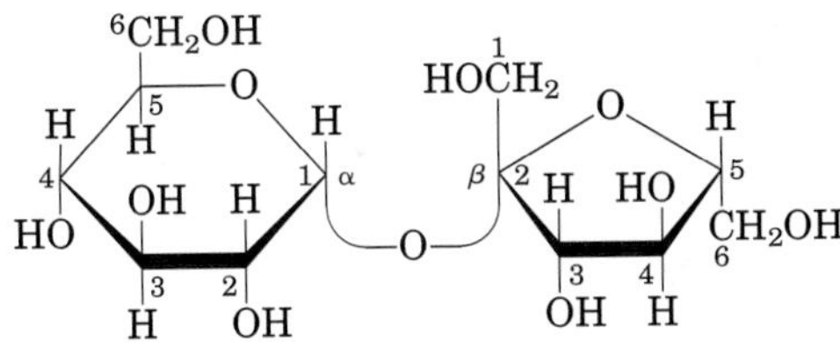

Sucrose
β-D-fructofuranosyl α-D-glucopyranoside
Fru(β2↔1α)Glc

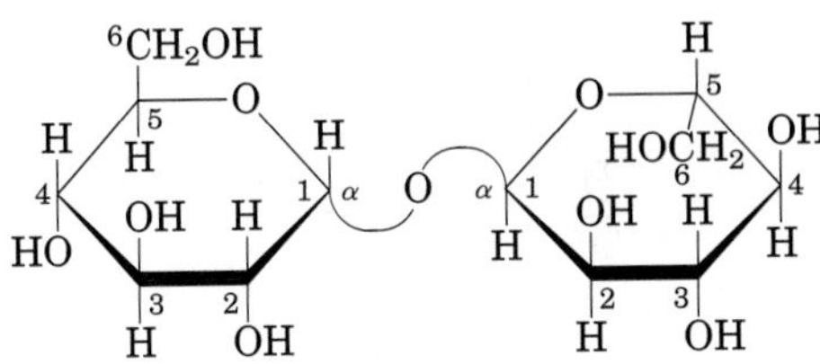

Trehalose
α-D-glucopyranosyl α-D-glucopyranoside
Glc(α1↔1α)Glc

figure 9–12
Some common disaccharides. Like maltose in Figure 9–11, these are shown as Haworth perspectives. The common name, full systematic name, and abbreviation are given for each disaccharide.

Polysaccharides

Most carbohydrates found in nature occur as polysaccharides, polymers of medium to high molecular weight. Polysaccharides, also called **glycans,** differ from each other in the identity of their recurring monosaccharide units, in the length of their chains, in the types of bonds linking the units, and in the degree of branching. **Homopolysaccharides** contain only a single type of monomer; **heteropolysaccharides** contain two or more different kinds (Fig. 9–13). Some homopolysaccharides serve as storage forms of monosaccharides used as fuels; starch and glycogen are homopolysaccharides of this type. Other homopolysaccharides (cellulose and chitin, for example) serve as structural elements in plant cell walls and animal exoskeletons. Heteropolysaccharides provide extracellular support for organisms of all kingdoms. For example, the rigid layer of the bacterial cell envelope (the peptidoglycan) is composed in part of a heteropolysaccharide built from two alternating monosaccharide units. In animal tissues, the extracellular

Homopolysaccharides
Unbranched
Branched
Heteropolysaccharides
Two monomer types, unbranched
Multiple monomer types, branched

figure 9–13
Polysaccharides may be composed of one, two, or several different monosaccharides, in straight or branched chains of varying length.

space is occupied by several types of heteropolysaccharides, which form a matrix that holds individual cells together and provides protection, shape, and support to cells, tissues, and organs.

Unlike proteins, polysaccharides generally do not have definite molecular weights. This difference is a consequence of the mechanisms of assembly of the two types of polymers. As we shall see in Chapter 27, proteins are synthesized on a template (messenger RNA) of defined sequence and length, by enzymes that follow the template exactly. For polysaccharide synthesis, there is no template; rather, the program for polysaccharide synthesis is intrinsic to the enzymes that catalyze the polymerization of the monomeric units.

Starch and Glycogen Are Stored Fuels

The most important storage polysaccharides are starch in plant cells and glycogen in animal cells. Both polysaccharides occur intracellularly as large clusters or granules (Fig. 9–14). Starch and glycogen molecules are heavily hydrated because they have many exposed hydroxyl groups available to hydrogen bond with water. Most plant cells have the ability to form starch, but it is especially abundant in tubers, such as potatoes, and in seeds, such as those of maize (corn).

Starch contains two types of glucose polymer, amylose and amylopectin (Fig. 9–15). The former consists of long, unbranched chains of D-glucose residues connected by (α1→4) linkages. Such chains vary in molecular weight from a few thousand to over a million. Amylopectin also has a high molecular weight (up to 100 million) but unlike amylose is highly branched. The glycosidic linkages joining successive glucose residues in amylopectin chains are (α1→4); the branch points (about one per 24 to 30 residues) are (α1→6) linkages.

Glycogen is the main storage polysaccharide of animal cells. Like amylopectin, glycogen is a polymer of (α1→4)-linked subunits of glucose, with (α1→6)-linked branches, but glycogen is more extensively branched (on average, one branch per 8 to 12 residues) and more compact than starch. Glycogen is especially abundant in the liver, where it may constitute as much as 7% of the wet weight; it is also present in skeletal muscle. In hepatocytes glycogen is found in large granules (Fig. 9–14b), which are themselves clusters of smaller granules composed of single, highly branched glycogen molecules with an average molecular weight of several million. Such glycogen granules also contain, in tightly bound form, the enzymes responsible for the synthesis and degradation of glycogen.

figure 9–14

Electron micrographs of starch and glycogen granules. **(a)** Large starch granules in a single chloroplast. Starch is made in the chloroplast from D-glucose formed photosynthetically. **(b)** Glycogen granules in a hepatocyte. These granules form in the cytosol and are much smaller (~0.1 μm) than starch granules (~1.0 μm).

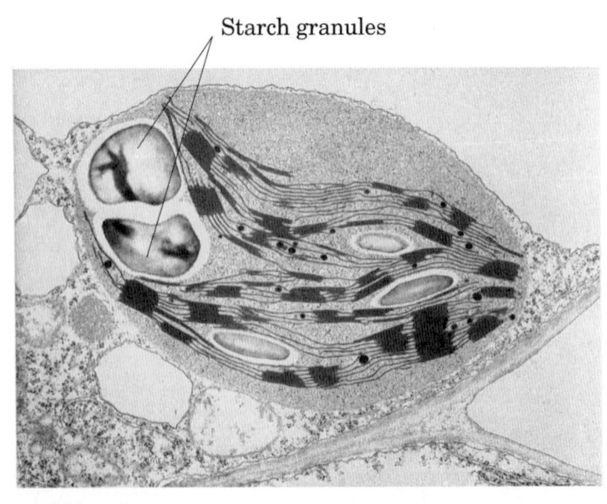

(a)

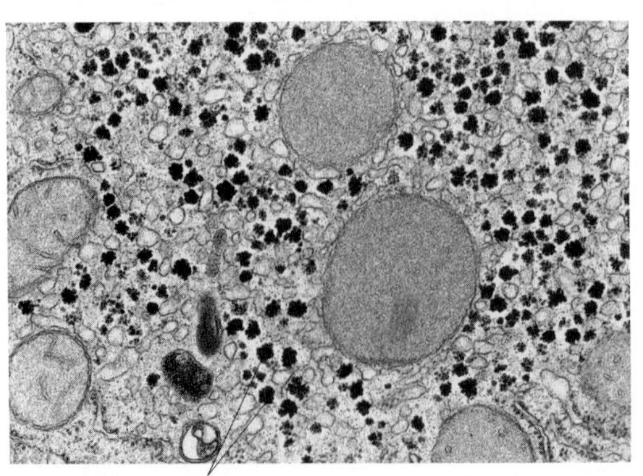

(b)

Nonreducing end

$^6CH_2OH$ CH_2OH CH_2OH CH_2OH

Reducing end

(a)

$^6CH_2OH$

Branch

(α1→6) branch point

$^6CH_2$

Main chain

(b)

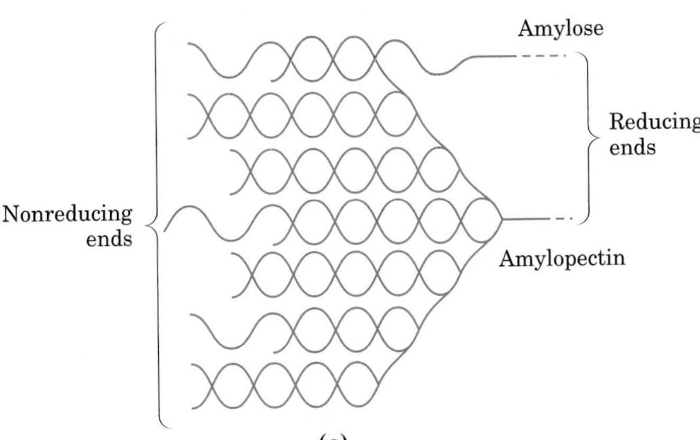

(c)

figure 9–15

Amylose and amylopectin, the polysaccharides of starch. (a) A short segment of amylose, a linear polymer of D-glucose residues in (α1→4) linkage. A single chain can contain several thousand glucose residues. Amylopectin has stretches of similarly linked residues between branch points. **(b)** An (α1→6) branch point of amylopectin. **(c)** A cluster of amylose and amylopectin like that believed to occur in starch granules. Strands of amylopectin (red) form double-helical structures with each other or with amylose strands (blue). Glucose residues at the nonreducing ends of the outer branches are removed enzymatically, one at a time, during the intracellular mobilization of starch for energy production. Glycogen has a similar structure but is more highly branched and more compact.

Because each branch in glycogen ends with a nonreducing sugar (one without a free anomeric carbon), a glycogen molecule has as many nonreducing ends as it has branches, but only one reducing end. When glycogen is used as an energy source, glucose units are removed one at a time from the nonreducing ends. Degradative enzymes that act only at nonreducing ends can work simultaneously on the many branches, speeding the conversion of the polymer to monosaccharides.

Why not store glucose in its monomeric form? It has been calculated that hepatocytes store glycogen equivalent to a glucose concentration of 0.4 M. The actual concentration of glycogen, which is insoluble and contributes little to the osmolarity of the cytosol, is about 0.01 μM. If the cytosol contained 0.4 M glucose, the osmolarity would be threateningly elevated, leading to osmotic entry of water that might rupture the cell (see Fig. 4–11). Furthermore, with an intracellular glucose concentration of 0.4 M and an external concentration of about 5 mM (the concentration in the blood of a mammal), the free-energy change for glucose uptake into cells against this very high concentration gradient would be prohibitively large.

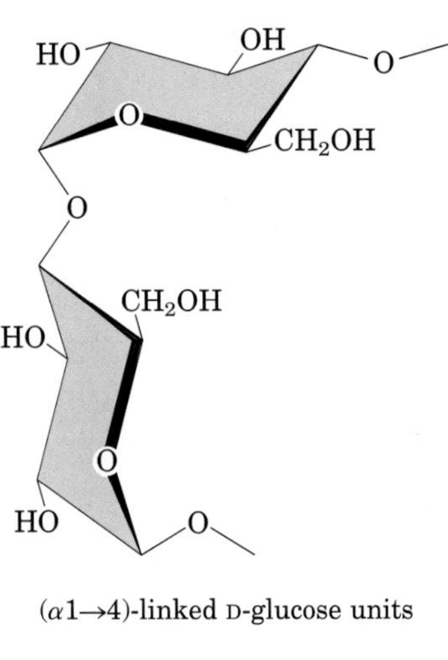

(a)

(b)

figure 9–16
The structure of starch (amylose). (a) In the most stable conformation of adjacent rigid chairs, the polysaccharide chain is curved, rather than linear as in cellulose (see Fig. 9–17). **(b)** Scale drawing of a segment of amylose. The conformation of (α1→4) linkages in amylose, amylopectin, and glycogen causes these polymers to assume tightly coiled helical structures. These compact structures produce the dense granules of stored starch or glycogen seen in many cells (see Fig. 9–14).

The three-dimensional structure of starch is shown in Figure 9–16, and it is compared with the structure of cellulose below.

Cellulose and Chitin Are Structural Homopolysaccharides

Cellulose, a fibrous, tough, water-insoluble substance, is found in the cell walls of plants, particularly in stalks, stems, trunks, and all the woody portions of the plant body. Cellulose constitutes much of the mass of wood, and cotton is almost pure cellulose. Like amylose and the main chains of amylopectin and glycogen, the cellulose molecule is a linear, unbranched homopolysaccharide, consisting of 10,000 to 15,000 D-glucose units. But there is a very important difference: in cellulose the glucose residues have the β configuration (Fig. 9–17), whereas in amylose, amylopectin, and glycogen the glucose is in the α configuration. The glucose residues in cellulose are linked by (β1→4) glycosidic bonds. This difference gives cellulose and amylose very different three-dimensional structures and physical properties.

The three-dimensional structure of carbohydrate macromolecules is based on the same principles as those underlying polypeptide structure: subunits with a more-or-less rigid structure dictated by covalent bonds form three-dimensional macromolecular structures that are stabilized by weak interactions. Because polysaccharides have so many hydroxyl groups, hydrogen bonding has an especially important influence on their structures. Polymers of β-D-glucose, such as cellulose, can be represented as a series of rigid pyranose rings connected by an oxygen atom bridging two carbon atoms (the glycosidic bond). There is free rotation about both C—O bonds linking the two residues (Fig. 9–17a). The most stable conformation for the polymer is that in which each chair is turned 180° relative to its neighbors, yielding a straight, extended chain. With several chains lying side by side, a stabilizing network of interchain and intrachain hydrogen bonds produces straight, stable supramolecular fibers of great tensile strength (Fig. 9–17b). The tensile strength of cellulose has made it a useful substance to civilizations for millenia. Many manufactured products, including paper, cardboard, rayon, insulating tiles, and other packing and building materials, are derived from cellulose. The water content of these materials is low because extensive interchain hydrogen bonding between cellulose molecules satisfies their capacity for hydrogen-bond formation.

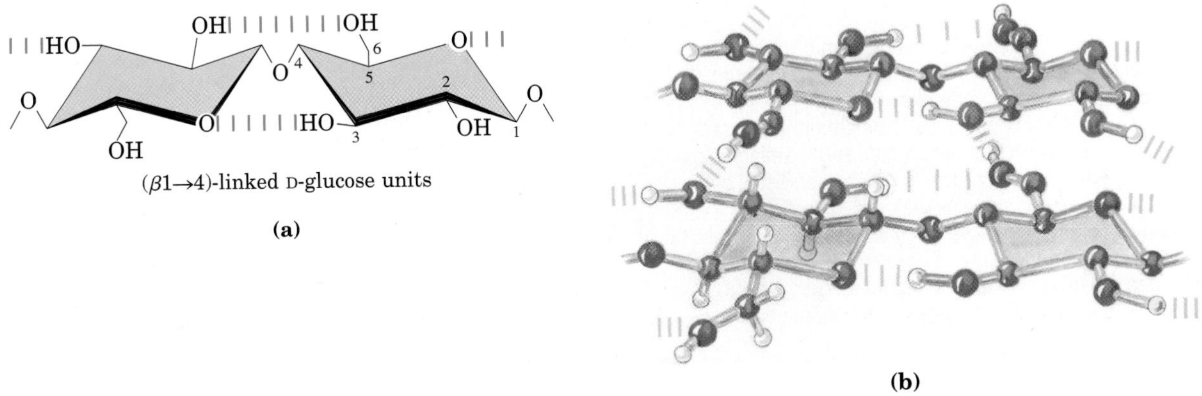

(β1→4)-linked D-glucose units

(a)

(b)

figure 9–17
The structure of cellulose. (a) Two units of a cellulose chain; the D-glucose residues are in (β1→4) linkage. The rigid chair structures can rotate relative to one another. **(b)** Scale drawing of segments of two parallel cellulose chains, showing the conformation of the D-glucose residues and the hydrogen-bond cross-links. In the hexose unit at the lower left, all hydrogen atoms are shown; in the other three hexose units, all hydrogens attached to carbon have been omitted for clarity as they do not participate in hydrogen bonding.

In contrast to the straight fibers produced by (β1→4)-linked polymers such as cellulose, the most favorable conformation for (α1→4)-linked polymers of D-glucose, such as starch and glycogen, is a tightly coiled helical structure stabilized by hydrogen bonds (Fig. 9–16).

Glycogen and starch ingested in the diet are hydrolyzed by α-amylases, enzymes in saliva and intestinal secretions that break (α1→4) glycosidic bonds between glucose units. Most animals cannot use cellulose as a fuel source because they lack an enzyme to hydrolyze the (β1→4) linkages. Termites readily digest cellulose (and therefore wood), but only because their intestinal tract harbors a symbiotic microorganism, *Trichonympha,* that secretes cellulase, an enzyme that hydrolyzes (β1→4) linkages between glucose units. Wood-rot fungi and bacteria also produce cellulase. The only vertebrates able to use cellulose as food are cattle and other ruminants (sheep, goats, camels, giraffes). The extra stomach compartment (rumen) of a ruminant teems with bacteria and protists that secrete cellulase.

Chitin is a linear homopolysaccharide composed of *N*-acetylglucosamine residues in β linkage (Fig. 9–18). The only chemical difference from cellulose is the replacement of the hydroxyl group at C-2 with an acetylated amino group. Chitin forms extended fibers similar to those of cellulose, and like cellulose is indigestible by vertebrate animals. Chitin is the principal component of the hard exoskeletons of nearly a million species of arthropods—insects, lobsters, and crabs, for example—and is probably the second most abundant polysaccharide, next to cellulose, in nature.

figure 9–18
A short segment of chitin, a homopolymer of *N*-acetyl-D-glucosamine units in (β1→4) linkage.

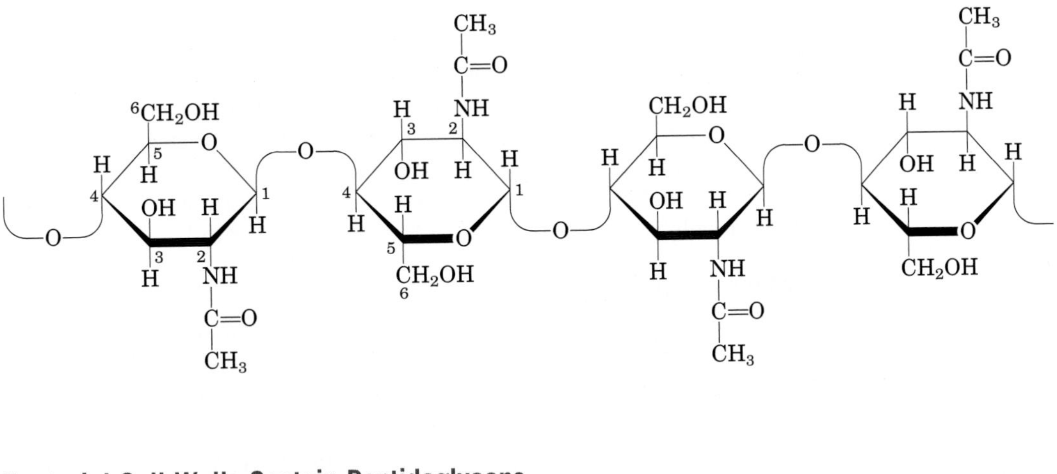

Bacterial Cell Walls Contain Peptidoglycans

The rigid component of bacterial cell walls is a heteropolymer of alternating (β1→4)-linked *N*-acetylglucosamine and *N*-acetylmuramic acid residues (Fig. 9–19). The linear polymers lie side by side in the cell wall, cross-linked

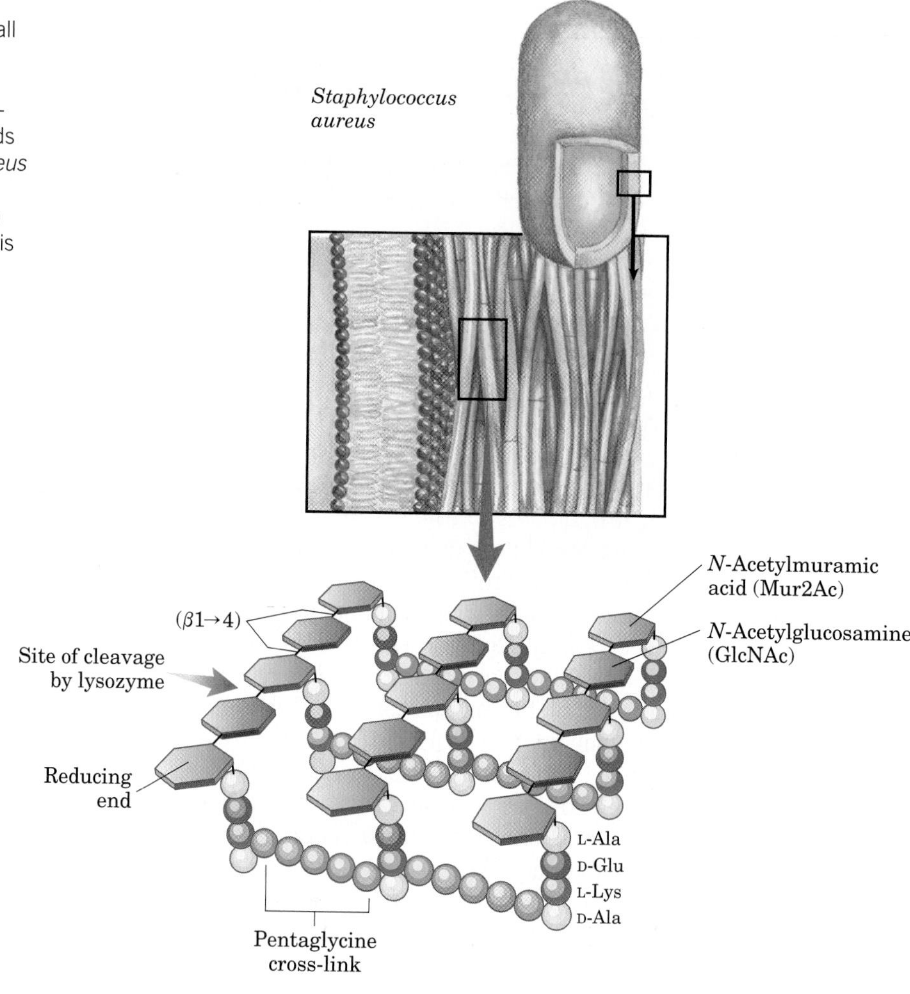

figure 9–19

Peptidoglycan. This is the peptidoglycan of the cell wall of *Staphylococcus aureus,* a gram-positive bacterium. Peptides (strings of colored spheres) covalently link *N*-acetylmuramic acid residues in neighboring polysaccharide chains. Note the mixture of L and D amino acids in the peptides. Gram-positive bacteria such as *S. aureus* have a pentaglycine chain in the cross-link. Gram-negative bacteria such as *E. coli* lack the pentaglycine; instead, the terminal D-Ala residue of one tetrapeptide is attached directly to a neighboring tetrapeptide through either L-Lys or a lysine-like amino acid, diaminopimelic acid.

by short peptides, the exact structure of which depends on the bacterial species. The peptide cross-links weld the polysaccharide chains into a strong sheath that envelops the entire cell and prevents cellular swelling and lysis due to the osmotic entry of water. The enzyme lysozyme kills bacteria by hydrolyzing the (β1→4) glycosidic bond between *N*-acetylglucosamine and *N*-acetylmuramic acid. Lysozyme is notably present in tears, presumably as a defense against bacterial infections of the eye. It is also produced by certain bacterial viruses to ensure their release from the host bacterial cell, an essential step of the viral infection cycle. Penicillin and related antibiotics kill bacteria by preventing the synthesis of cross-links, leaving the cell wall too weak to resist osmotic lysis (see Box 20–1).

Glycosaminoglycans Are Components of the Extracellular Matrix

The extracellular space in the tissues of multicellular animals is filled with a gel-like material, the **extracellular matrix,** also called ground substance, which holds the cells together and provides a porous pathway for the diffusion of nutrients and oxygen to individual cells. The extracellular matrix is composed of an interlocking meshwork of heteropolysaccharides and fibrous proteins such as collagen, elastin, fibronectin, and laminin. The heteropolysaccharides, called **glycosaminoglycans,** are a family of linear polymers composed of repeating disaccharide units (Fig. 9–20). One of the two monosaccharides is always either *N*-acetylglucosamine or *N*-acetylgalactosamine; the other is in most cases a uronic acid, usually D-glucuronic or L-iduronic acid. In some glycosaminoglycans, one or more of the hydroxyls of the amino sugar is esterified with sulfate. The combination of sulfate groups and the carboxylate groups of the uronic acid residues gives glycosaminoglycans a very high density of negative charge. To minimize the repulsive forces among neighboring charged groups, these molecules assume an extended conformation in solution. The specific patterns of sulfated and nonsulfated sugar residues in glycosaminoglycans provide for specific recognition by a variety of protein ligands that bind electrostatically to these molecules. Glycosaminoglycans are attached to extracellular proteins to form **proteoglycans** (discussed below).

The glycosaminoglycan **hyaluronic acid** (hyaluronate at physiological pH) contains alternating residues of D-glucuronic acid and *N*-acetylglucosamine (Fig. 9–20). With up to 50,000 repeats of the basic disaccharide unit, hyaluronates have molecular weights greater than 1 million; they form clear, highly viscous solutions that serve as lubricants in the synovial fluid of joints and give the vitreous humor of the vertebrate eye its jellylike consistency (the Greek *hyalos* means "glass"; hyaluronates can have a glassy

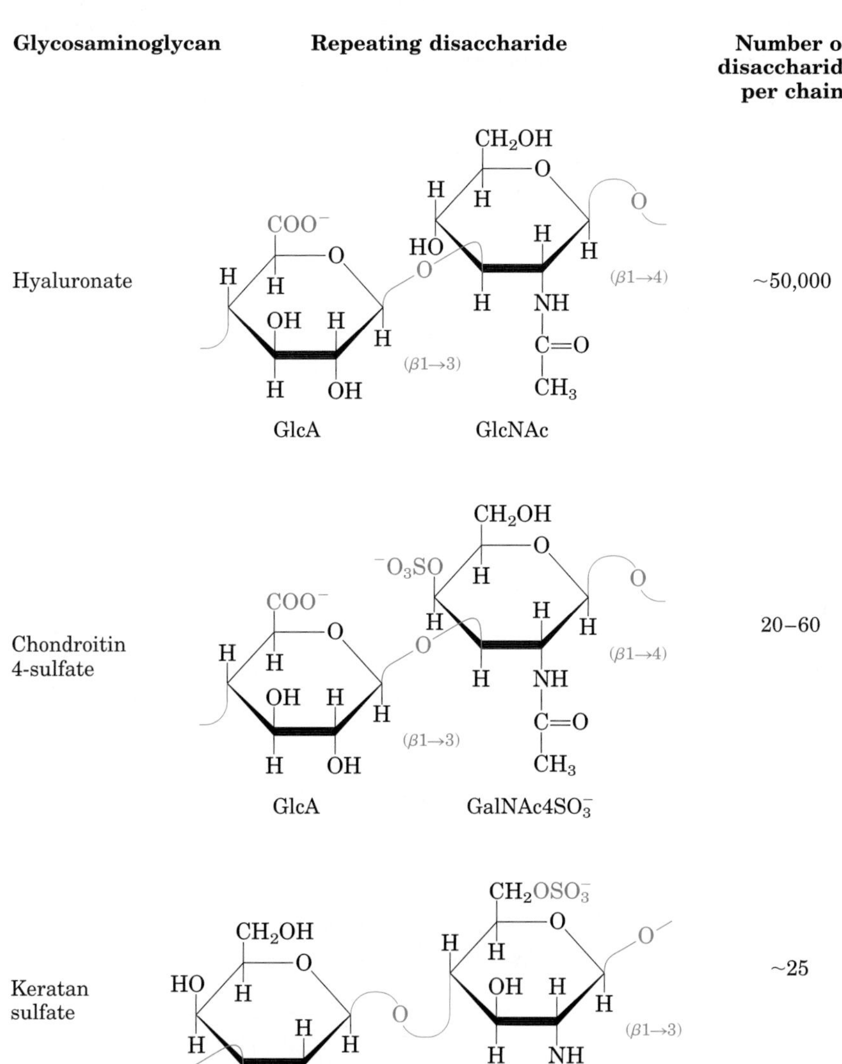

figure 9–20
Repeating units of some common glycosaminoglycans of extracellular matrix. The molecules are copolymers of alternating uronic acid and amino sugar residues, with sulfate esters in any of several positions. The ionized carboxylate and sulfate groups (red) give these polymers their characteristic highly negative charge.

or translucent appearance). Hyaluronate is also an essential component of the extracellular matrix of cartilage and tendons, to which it contributes tensile strength and elasticity as a result of strong interactions with other components of the matrix. Hyaluronidase, an enzyme secreted by some pathogenic (disease-causing) bacteria, can hydrolyze the glycosidic linkages of hyaluronate, rendering tissues more susceptible to bacterial invasion. In many organisms, a similar enzyme in sperm hydrolyzes an outer glycosaminoglycan coat around the ovum, allowing sperm penetration.

Other glycosaminoglycans differ from hyaluronate in two respects: they are generally much shorter polymers and they are covalently linked to specific proteins (proteoglycans, discussed below). Chondroitin sulfate (Greek *chondros,* "cartilage") contributes to the tensile strength of cartilage, tendons, ligaments, and the walls of the aorta. Dermatan sulfate (Greek *derma,* "skin") contributes to the pliability of skin and is also present in blood vessels and heart valves. Keratan sulfates (Greek *keras,* "horn") have no uronic acid and their sulfate content is variable. They are present in cornea, cartilage, bone, and a variety of horny structures formed of dead cells: horn, hair, hoofs, nails, and claws. Heparin (Greek *hēpar,* "liver") is a natural anticoagulant made in mast cells and released into the blood, where it inhibits coagulation by binding to and stimulating the anticoagulant protein antithrombin III. The interaction is strongly electrostatic; heparin has the highest negative charge density of any known biological macromolecule. Purified heparin is routinely added to blood samples obtained for clinical analysis, and to blood donated for transfusion, to prevent clotting.

Table 9–2 summarizes the composition, properties, roles, and occurrence of the polysaccharides described in this section.

table 9–2

Structures and Roles of Some Polysaccharides

Polymer	Type*	Repeating unit[†]	Size (number of monosaccharide units)	Roles
Starch				Energy storage: in plants
Amylose	Homo-	(α1→4)Glc, linear	50–5,000	
Amylopectin	Homo-	(α1→4)Glc, with (α1→6)Glc branches every 24 to 30 residues	Up to 10^6	
Glycogen	Homo-	(α1→4)Glc, with (α1→6)Glc branches every 8 to 12 residues	Up to 50,000	Energy storage: in bacteria and animal cells
Cellulose	Homo-	(β1→4)Glc	Up to 15,000	Structural: in plants, gives rigidity and strength to cell walls
Chitin	Homo-	(β1→4)GlcNAc	Very large	Structural: in insects, spiders, crustaceans, gives rigidity and strength to exoskeletons
Peptidoglycan	Hetero-; peptides attached	4)Mur2Ac(β1→4) GlcNAc(β1	Very large	Structural: in bacteria, gives rigidity and strength to cell envelope
Hyaluronate (a glycosaminoglycan)	Hetero-; acidic	4)GlcA(β1→3) GlcNAc(β1	Up to 100,000	Structural: in vertebrates, extracellular matrix of skin and connective tissue; viscosity and lubrication in joints

* Each polymer is classified as a homopolysaccharide (homo-) or heteropolysaccharide (hetero-).

[†]The abbreviated names for the peptidoglycan and hyaluronate repeating units indicate that the polymer contains repeats of this disaccharide unit, with the GlcNAc of one disaccharide unit linked β(1→4) to the first residue of the next disaccharide unit.

Glycoconjugates: Proteoglycans, Glycoproteins, and Glycolipids

In addition to their important roles as stored fuels (starch, glycogen) and as structural materials (cellulose, chitin, peptidoglycan), polysaccharides (and oligosaccharides) are information carriers: they serve as destination labels for some proteins and as mediators of specific cell-cell interactions and interactions between cells and the extracellular matrix. Specific carbohydrate-containing molecules act in cell-cell recognition and adhesion, cell migration during development, blood clotting, the immune response, and wound healing, to name but a few of their many roles. In most of these cases, the informational carbohydrate is covalently joined to a protein or a lipid to form a **glycoconjugate,** which is the biologically active molecule.

Proteoglycans are macromolecules of the cell surface or extracellular matrix in which one or more glycosaminoglycan chains are joined covalently to a membrane protein or a secreted protein. The glycosaminoglycan moiety commonly forms the greater fraction (by mass) of the proteoglycan molecule, dominates the structure, and is often the main site of biological activity. In many cases the biological activity is the provision of multiple binding sites, rich in opportunities for hydrogen bonding and electrostatic interactions with other proteins of the cell surface or the extracellular matrix. Proteoglycans are major components of connective tissue such as cartilage, in which their many noncovalent interactions with other proteoglycans, proteins, and glycosaminoglycans provide strength and resilience.

Glycoproteins have one or several oligosaccharides of varying complexity joined covalently to a protein. They are found on the outer face of the plasma membrane, in the extracellular matrix, and in the blood. Inside cells they are found in specific organelles such as Golgi complexes, secretory granules, and lysosomes. The oligosaccharide portions of glycoproteins are less monotonous than the glycosaminoglycan chains of proteoglycans; they are rich in information, forming highly specific sites for recognition and high-affinity binding by other proteins.

Glycolipids are membrane lipids in which the hydrophilic head groups are oligosaccharides, which, as in glycoproteins, act as specific sites for recognition by carbohydrate-binding proteins.

Glycobiology, the study of the structure and function of glycoconjugates, is one of the most active and exciting areas of biochemistry and cell biology. Our discussion uses just a few examples to illustrate the diversity of structure and the range of biological activity of the glycoconjugates. In Chapter 20 we discuss the biosynthesis of polysaccharides, including the glycosaminoglycans, and in Chapter 27, the assembly of oligosaccharide chains on glycoproteins.

Proteoglycans Are Glycosaminoglycan-Containing Macromolecules of the Cell Surface and Extracellular Matrix

The basic proteoglycan unit consists of a "core protein" with covalently attached glycosaminoglycan(s). For example, the sheetlike extracellular matrix (basal lamina) that separates organized groups of cells contains a family of core proteins (M_r 20,000 to 40,000), each with several covalently attached heparan sulfate chains. (Heparan sulfate is structurally similar to heparin but has a lower density of sulfate esters.) The point of attachment is commonly a Ser residue, to which the glycosaminoglycan is joined through a trisaccharide bridge (Fig. 9–21). The Ser residue is generally in the sequence –Ser–Gly–X–Gly– (where X is any amino acid residue), although not every protein with this sequence has an attached glycosaminoglycan. Many proteoglycans are secreted into the extracellular matrix, but

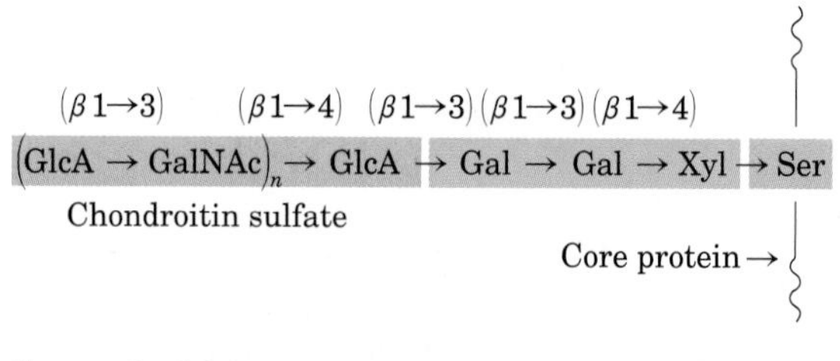

(β1→3) (β1→4) (β1→3) (β1→3) (β1→4)

$(\text{GlcA} \rightarrow \text{GalNAc})_n \rightarrow \text{GlcA} \rightarrow \text{Gal} \rightarrow \text{Gal} \rightarrow \text{Xyl} \rightarrow \text{Ser}$

Chondroitin sulfate

Core protein →

figure 9–21

Proteoglycan structure, showing the trisaccharide bridge. A typical trisaccharide linker (blue) connects a glycosaminoglycan—in this case chondroitin sulfate (orange)—to a Ser residue (red) in the core protein. The xylose residue at the reducing end of the linker is joined by its anomeric carbon to the hydroxyl of the Ser residue.

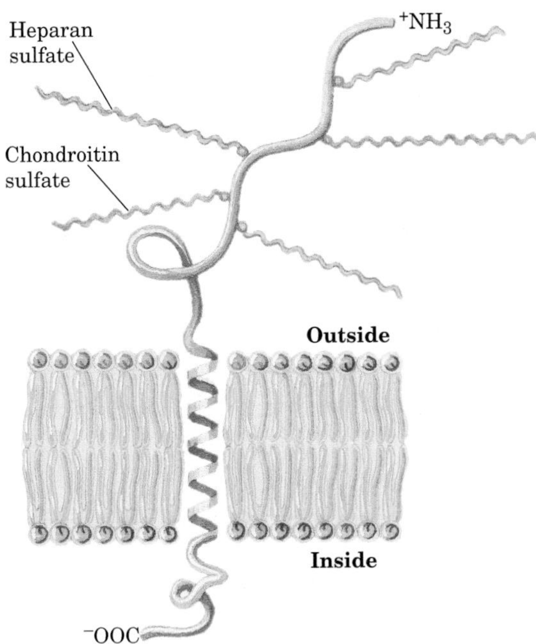

figure 9–22

Proteoglycan structure of an integral membrane protein. This schematic diagram shows syndecan, a core protein of the plasma membrane. The amino-terminal domain on the extracellular side of the membrane is covalently attached (by trisaccharide linkers such as those in Fig. 9–21) to three heparan sulfate chains and two chondroitin sulfate chains.

some are integral membrane proteins. For example, syndecan core protein (M_r 56,000) has a single transmembrane domain and an extracellular domain bearing three chains of heparan sulfate and two of chondroitin sulfate, each attached to a Ser residue (Fig. 9–22). The heparan sulfate moieties bind a variety of extracellular ligands and thereby modulate the ligands' interaction with specific receptors of the cell surface.

Some proteoglycans can form **proteoglycan aggregates,** enormous supramolecular assemblies of many core proteins all bound to a single molecule of hyaluronate. Aggrecan core protein (M_r ~250,000) has multiple chains of chondroitin sulfate and keratan sulfate, joined to Ser residues in the core protein through trisaccharide linkers, to give an aggrecan monomer of M_r ~2×10^6. When a hundred or more of these "decorated" core proteins bind a single, extended molecule of hyaluronate (Fig. 9–23), the resulting proteoglycan aggregate (M_r $>2 \times 10^8$) and its associated water of hydration occupy a volume about equal to that of a bacterial cell! Aggrecan interacts strongly with collagen in the extracellular matrix of cartilage, contributing to its development and tensile strength.

Interwoven with these enormous extracellular proteoglycans are fibrous matrix proteins such as collagen, elastin, and fibronectin, forming a cross-linked meshwork that gives the whole extracellular matrix strength and resilience. Some of these proteins are multiadhesive, a single protein having binding sites for several different matrix molecules. Fibronectin, for example, has separate domains that bind fibrin, heparan sulfate, collagen, and a family of plasma membrane proteins called integrins that mediate signaling between the cell interior and the extracellular matrix (see Fig. 12–19). Integrins, in turn, have binding sites for a number of other extracellular macromolecules. The picture of cell-matrix interactions that emerges (Fig. 9–24) shows an array of interactions between cellular and extracellular molecules. These interactions serve not merely to anchor cells to the extracellular matrix but also to provide paths that direct the migration of cells

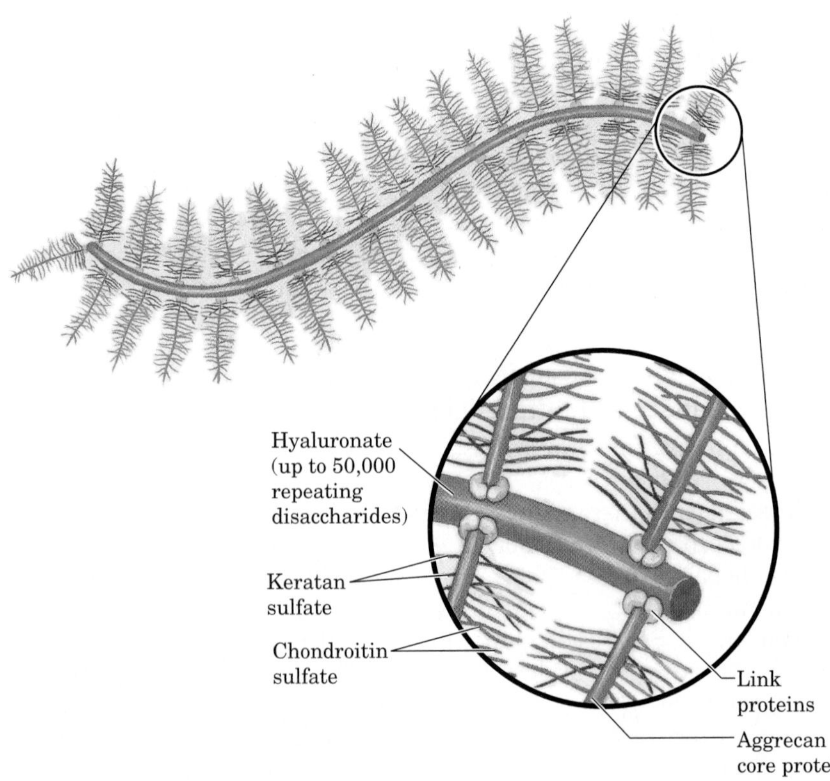

figure 9–23

A proteoglycan aggregate of the extracellular matrix. One very long molecule of hyaluronate is associated noncovalently with about 100 molecules of the core protein aggrecan. Each aggrecan molecule contains many covalently bound chondroitin sulfate and keratan sulfate chains. Link proteins situated at the junction between each core protein and the hyaluronate backbone mediate the core protein–hyaluronate interaction.

in developing tissue and, through integrins, to convey information in both directions across the plasma membrane.

Matrix proteoglycans are essential in the response of cells to certain extracellular growth factors. For example, fibroblast growth factor (FGF), an extracellular protein signal that stimulates cell division, first binds to heparan sulfate moieties of syndecan molecules in the target cell's plasma membrane. Syndecan then "presents" FGF to the specific FGF plasma membrane receptor, and only then can FGF interact productively with its receptor to trigger cell division.

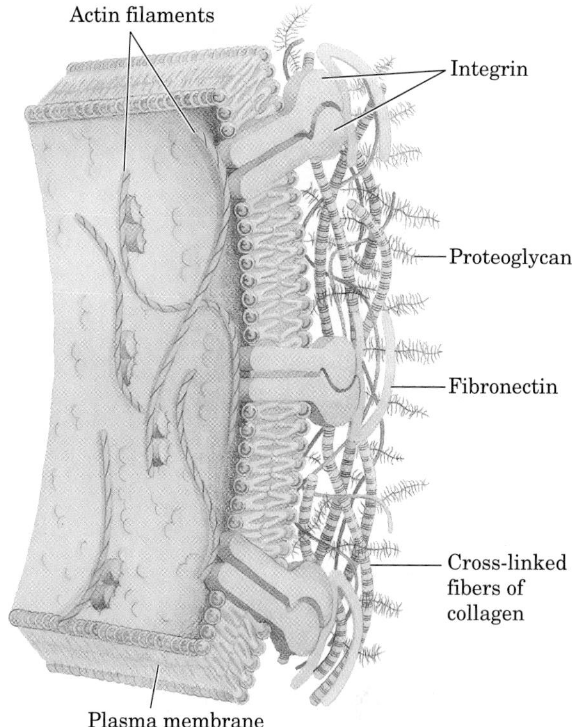

figure 9–24
Interactions between cells and extracellular matrix. The association between cells and the proteoglycan of extracellular matrix is mediated by a membrane protein (integrin) and by an extracellular protein (fibronectin in this example) with binding sites for both integrin and the proteoglycan. Note the close association of collagen fibers with the fibronectin and proteoglycan.

Glycoproteins Are Information-Rich Conjugates Containing Oligosaccharides

Glycoproteins are carbohydrate-protein conjugates in which the carbohydrate moieties are smaller and more structurally diverse than the glycosaminoglycans of proteoglycans. The carbohydrate is attached at its anomeric carbon through a glycosidic link to the —OH of a Ser or Thr residue (*O*-linked), or through an *N*-glycosyl link to the amide nitrogen of an Asn residue (*N*-linked) (Fig. 9–25). Some glycoproteins have a single oligosaccharide chain, but many have more than one; the carbohydrate may constitute from 1% to 70% or more of the glycoprotein by mass. The structures of a large number of *O*- and *N*-linked oligosaccharides from a variety of glycoproteins are known; Figure 9–25 shows a few typical examples.

As we will see in Chapter 12, the external surface of the plasma membrane has many membrane glycoproteins with arrays of covalently attached oligosaccharides of varying complexity. One of the best-characterized membrane glycoproteins is glycophorin A of the erythrocyte membrane (see Fig. 12–10). It contains 60% carbohydrate by mass in the form of 16 oligosaccharide chains (totalling 60 to 70 monosaccharide residues) covalently attached to amino acid residues near the amino terminus of the polypeptide chain. Fifteen of the oligosaccharide chains are *O*-linked to Ser or Thr residues, and one is *N*-linked to an Asn residue.

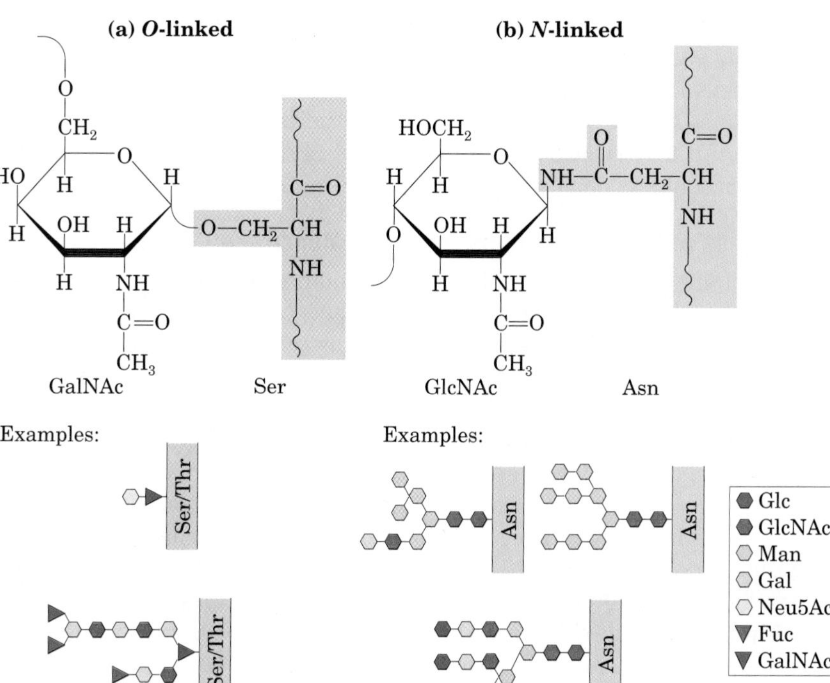

figure 9–25
Oligosaccharide linkages in glycoproteins. (a) *O*-linked oligosaccharides have a glycosidic bond to the hydroxyl group of Ser or Thr residues (shaded pink), illustrated here with GalNAc as the sugar at the reducing end of the oligosaccharide. One simple chain and one complex chain are shown. (Recall that Neu5Ac—*N*-acetylneuraminic acid—is commonly called sialic acid.) **(b)** *N*-linked oligosaccharides have an *N*-glycosyl bond to the amide nitrogen of an Asn residue (shaded green), illustrated here with GlcNAc as the terminal sugar. Three common types of oligosaccharide chains that are *N*-linked in glycoproteins are shown. A complete description of oligosaccharide structure requires specification of the position and stereochemistry (α or β) of each glycosidic linkage.

Many of the proteins secreted by eukaryotic cells are glycoproteins, including most of the proteins of blood. For example, immunoglobulins (antibodies) and certain hormones, such as follicle-stimulating hormone, luteinizing hormone, and thyroid-stimulating hormone, are glycoproteins. Many milk proteins, including lactalbumin, and some of the proteins secreted by the pancreas (e.g., ribonuclease) are glycosylated, as are most of the proteins contained in lysosomes.

The biological advantages of adding oligosaccharides to proteins are not fully understood. The very hydrophilic clusters of carbohydrate alter the polarity and solubility of the proteins with which they are conjugated. Oligosaccharide chains attached to newly synthesized proteins in the Golgi complex may also influence the sequence of polypeptide-folding events that determine the tertiary structure of the protein (see Fig. 27–36); steric interactions between peptide and oligosaccharide may preclude one folding route and favor another. When numerous negatively charged oligosaccharide chains are clustered in a single region of a protein, the charge repulsion among them favors the formation of an extended, rodlike structure in that region. The bulkiness and negative charge of oligosaccharide chains also protect some proteins from attack by proteolytic enzymes.

Beyond these global physical effects on protein structure, there are more specific biological effects of oligosaccharide chains in glycoproteins. We noted earlier the difference between the information-rich linear sequences of nucleic acids and proteins and the monotonous regularity of homopolysaccharides such as cellulose (see Fig. 3–23). The oligosaccharides attached to glycoproteins are generally not monotonous but enormously rich in structural information. Consider the oligosaccharide chains in Figure 9–25, typical of those found in many glycoproteins. The most complex of those shown contains 14 monosaccharide residues of four different kinds, variously linked as (1→2), (1→3), (1→4), (1→6), (2→3), and (2→6), some with the α and some with the β configuration. The number of possible permutations and combinations of monosaccharide types and glycosidic linkages in an oligosaccharide of this size is astronomical. Each of the oligosaccharides in Figure 9–25 therefore presents a unique face, recognizable by the enzymes and receptors that interact with it. A number of cases are known in which the same protein produced in two tissues has different glycosylation patterns. The human protein interferon IFN-β1 produced in ovarian cells, for example, has oligosaccharide chains that differ from those of the same protein produced in breast epithelial cells. The biological significance of these **tissue glycoforms** is not understood, but in some way the oligosaccharide chains represent a tissue-specific mark.

Glycolipids and Lipopolysaccharides Are Membrane Components

Glycoproteins are not the only cellular components that bear complex oligosaccharide chains; some lipids, too, have covalently bound oligosaccharides. **Gangliosides** are membrane lipids of eukaryotic cells in which the polar head group, the part of the lipid that forms the outer surface of the membrane, is a complex oligosaccharide containing sialic acid (Fig. 9–9) and other monosaccharide residues. Some of the oligosaccharide moieties of gangliosides, such as those that determine human blood groups (see Fig. 11–12), are identical with those found in certain glycoproteins, which therefore also contribute to blood group type determination. Like the oligosaccharide moieties of glycoproteins, those of membrane lipids are generally, perhaps always, found on the outer face of the plasma membrane.

Lipopolysaccharides are the dominant surface feature of the outer membrane of gram-negative bacteria such as *E. coli* and *Salmonella typhimurium.* They are prime targets of the antibodies produced by the ver-

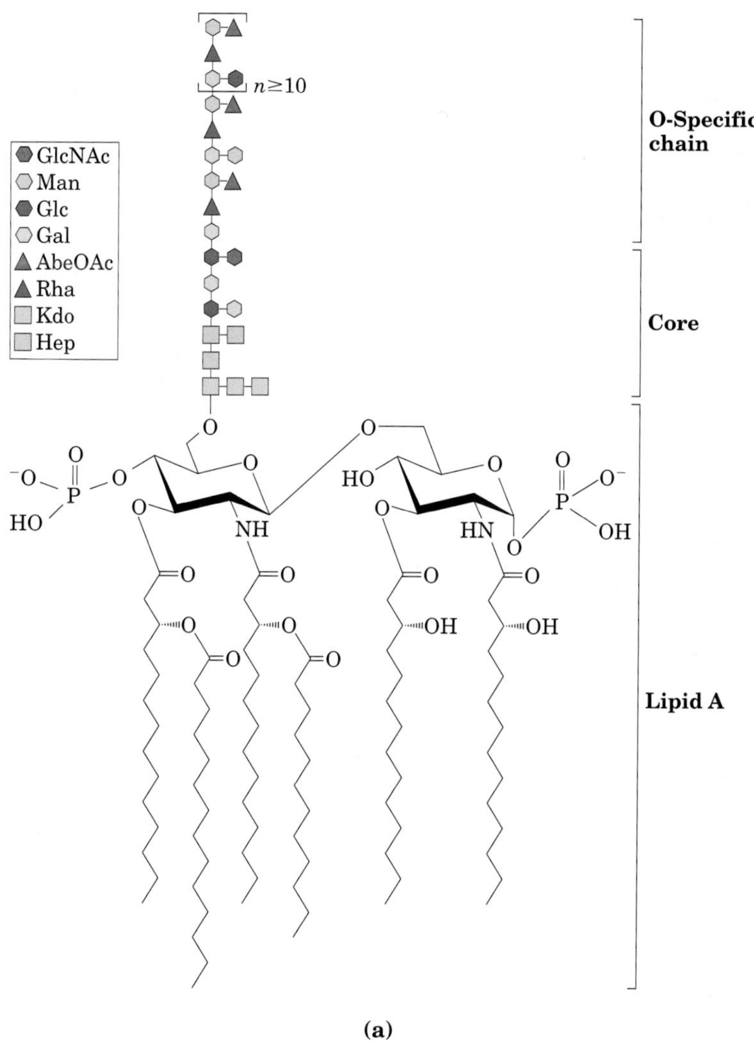

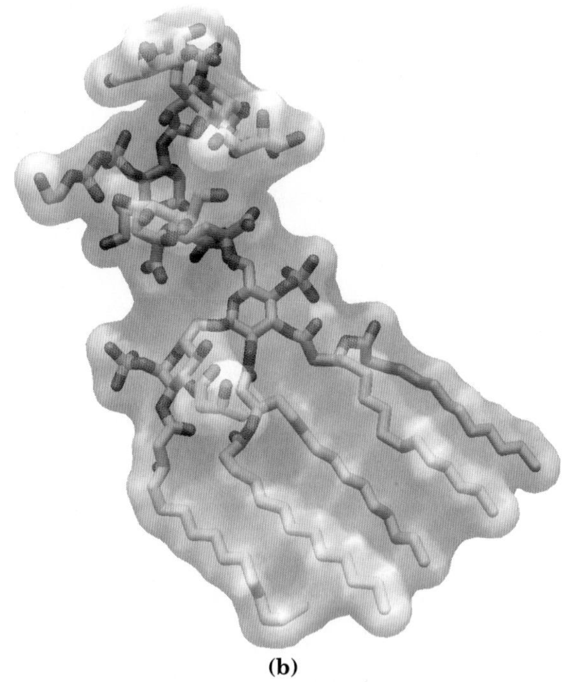

figure 9–26

Bacterial lipopolysaccharides. (a) Schematic diagram of the lipopolysaccharide of the outer membrane of *Salmonella typhimurium.* Kdo is 3-deoxy-D-manno-octonic acid, also called *keto*deoxy*o*ctulosonic acid; Hep is L-glycero-D-mannoheptose; AbeOAc is abequose (a 3,6-dideoxyhexose) acetylated on one of its hydroxyls. There are six fatty acids in the lipid A portion of the molecule. Different bacterial species have subtly different lipopolysaccharide structures, but they have in common a lipid region (lipid A), a core oligosaccharide, and an "O-specific" chain, which is the principal determinant of the serotype (immunological reactivity) of the bacterium. The outer membrane of the gram-negative bacteria *S. typhimurium* and *E. coli* contains so many lipopolysaccharide molecules that the cell surface is virtually covered with O-specific chains. **(b)** Space-filling molecular model of the lipopolysaccharide from *E. coli.*

tebrate immune system in response to bacterial infection and are therefore important determinants of the serotype of bacterial strains (serotypes are strains that are distinguished on the basis of antigenic properties). The lipopolysaccharides of *S. typhimurium* contain six fatty acids bound to two glucosamine residues, one of which is the point of attachment for a complex oligosaccharide (Fig. 9–26). *E. coli* has similar but unique lipopolysaccharides. The lipopolysaccharides of some bacteria are toxic to humans and other animals; for example, they are responsible for the dangerously lowered blood pressure that occurs in toxic shock syndrome resulting from gram-negative bacterial infections.

Oligosaccharide-Lectin Interactions Mediate Many Biological Processes

Lectins, found in all organisms, are proteins that bind carbohydrates with high affinity and specificity. A number of critically situated hydrogen-bonding partners in the carbohydrate recognition domain of each type of protein bind to specific oligosaccharides, and thus lectins easily distinguish between closely similar sugars (Table 9–3). Lectins serve in a wide variety of cell-cell recognition and adhesion processes. Here we discuss just a few examples of the roles of lectins in nature. In the laboratory, purified lectins are useful reagents for detecting and separating glycoproteins with different oligosaccharide moieties.

table 9–3

Lectins and the Oligosaccharide Ligands That They Bind

Lectin family and lectin	Abbreviation	Ligand(s)
Plant		
Concanavalin A	ConA	Manα1—OCH_3
Griffonia simplicifolia lectin 4	GS4	Lewis b (Le^b) tetrasaccharide
Wheat germ agglutinin	WGA	Neu5Ac(α2→3)Gal(β1→4)Glc GlcNAc(β1→4)GlcNAc
Ricin		Gal(β1→4)Glc
Animal		
Galectin-1		Gal(β1→4)Glc
Mannose-binding protein A	MBP-A	High-mannose octasaccharide
Viral		
Influenza virus hemagglutinin	HA	Neu5Ac(α2→6)Gal(β1→4)Glc
Polyoma virus protein 1	VP1	Neu5Ac(α2→3)Gal(β1→4)Glc
Bacterial		
Enterotoxin	LT	Gal
Cholera toxin	CT	GM1 pentasaccharide

Source: Weiss, W.I. & Drickamer, K. (1996) Structural basis of lectin-carbohydrate recognition. *Annu. Rev. Biochem.* **65,** 441–473.

The sialic acid (Neu5Ac) residues situated at the ends of the oligosaccharide chains of many plasma glycoproteins (Fig. 9–25) protect the proteins from uptake and degradation in the liver. For example, ceruloplasmin, a copper-transporting glycoprotein, has several oligosaccharide chains ending in sialic acid. Removal of sialic acid residues by the enzyme sialidase is one way in which the body marks "old" proteins for destruction and replacement. The plasma membrane of hepatocytes has lectin molecules (asialoglycoprotein receptors; "asialo-" indicating "without sialic acid") that specifically bind oligosaccharide chains with galactose residues no longer "protected" by a terminal sialic acid residue. Receptor-ceruloplasmin interaction triggers endocytosis and destruction of the ceruloplasmin.

A similar mechanism is apparently responsible for removing old erythrocytes from the mammalian bloodstream. Newly synthesized erythrocytes have several membrane glycoproteins with oligosaccharide chains that end in sialic acid. When sialic acid residues are removed by withdrawing a sample of blood, treating it with sialidase in vitro, and reintroducing it into the circulation, the treated erythrocytes disappear from the bloodstream within a few hours, whereas those with intact oligosaccharides (erythrocytes withdrawn and reintroduced without sialidase treatment) continue to circulate for days.

Some peptide hormones that circulate in the blood have oligosaccharide moieties that strongly influence their circulatory lifetime. Luteinizing hormone and thyrotropin (polypeptide hormones produced in the adrenal cortex) have *N*-linked oligosaccharides that end with the disaccharide GalNAc4SO_3^-(β1→4)GlcNAc, which is recognized by a lectin (receptor) of hepatocytes. (GalNAc4SO_3^- is *N*-acetylgalactosamine sulfated on the —OH group of C-4.) Receptor-hormone interaction mediates the uptake and destruction of luteinizing hormone and thyrotropin, reducing their concentration in the blood. Thus the blood levels of these hormones undergo a periodic rise (due to secretion by the adrenal cortex) and fall (due to destruction by hepatocytes).

Selectins are a family of lectins, found in plasma membranes, that mediate cell-cell recognition and adhesion in a wide range of cellular processes. One such process is the movement of immune cells (T lymphocytes) through the capillary wall, from blood to tissues, at sites of infection or inflammation (Fig. 9–27). At an infection site, P-selectin on the surface of capillary endothelial cells interacts with a specific oligosaccharide of the glycoproteins of circulating T lymphocytes. This interaction slows the T cells as they adhere to and roll along the endothelial lining of the capillaries. A second interaction, between integrin molecules in the T-cell plasma membrane and an adhesion protein on the endothelial cell surface, now stops the T cell and allows it to move through the capillary wall into the infected tissues to initiate the immune attack. Two other selectins participate in this process: E-selectin on the endothelial cell and L-selectin on the T cell bind their cognate oligosaccharides on the other cell.

Some microbial pathogens have lectins that mediate bacterial adhesion to host cells or toxin entry into cells. The bacterium believed responsible for most gastric ulcers, *Helicobacter pylori*, adheres to the inner surface of the stomach by interactions between bacterial membrane lectins and specific oligosaccharides of membrane glycoproteins of the gastric epithelial cells (Fig. 9–28). Among the binding sites recognized by *H. pylori* is the oligosaccharide Le^b when it is a part of the type O blood group determinant. This observation helps to explain the severalfold greater incidence of gastric ulcers in people of blood type O than those of type A or B. Chemically synthesized analogs of the Le^b oligosaccharide may prove useful in treating this type of ulcer. Administered orally, they could prevent bacterial adhesion (and thus infection) by competing with the gastric glycoproteins for binding to the bacterial lectin.

The cholera toxin molecule (produced by *Vibrio cholerae*) triggers diarrhea after entering intestinal cells responsible for water absorption from the intestine. It attaches to its target cell through the oligosaccharide of ganglioside GM1, a membrane phospholipid (see Box 11–2, Fig. 1), in the surface of intestinal epithelial cells. Similarly, the pertussis toxin produced by *Bordetella pertussis,* the bacterium that causes whooping cough, enters target cells only after interacting with an oligosaccharide (or perhaps several oligosaccharides) with a terminal sialic acid residue. Understanding the details of the oligosaccharide-binding sites of these toxins (lectins) may allow the development of genetically engineered toxin analogs for use in vaccines. Toxin analogs engineered to lack the carbohydrate binding site would be harmless because they could not bind to and enter cells, but they might elicit an immune response that would protect the recipient if exposed to the natural toxin.

Several animal viruses, including the influenza virus, attach to their host cells through interactions with oligosaccharides displayed on the host cell surface. The lectin of the influenza virus is the HA protein, which we will describe in Chapter 12. The HA protein is essential for viral entry and infection (see pp. 405–406 and Fig. 12–21a).

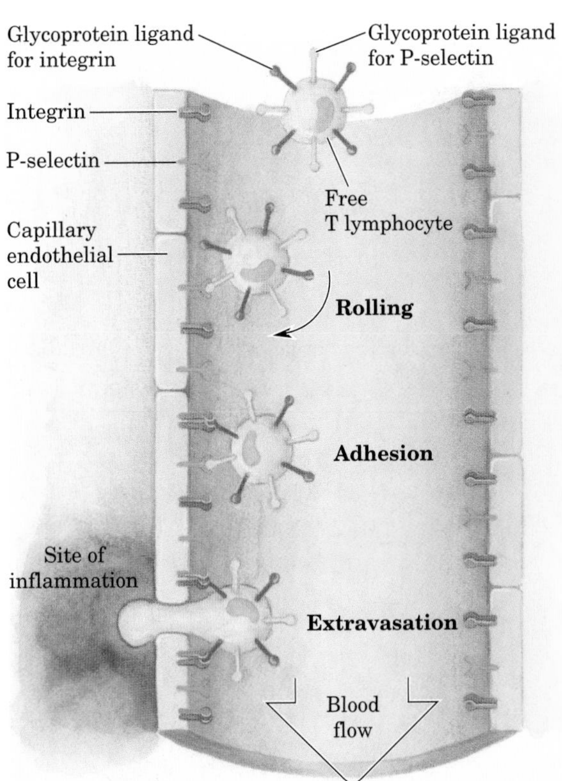

figure 9–27
Role of lectin-ligand interactions in lymphocyte movement to the site of an infection or injury. A T lymphocyte circulating through a capillary is slowed by transient interactions between P-selectin molecules in the plasma membrane of capillary endothelial cells and glycoprotein ligands for P-selectin on the T-cell surface. As it interacts with successive P-selectin molecules, the T cell rolls along the capillary surface. Near a site of inflammation, stronger interactions between integrin in the capillary surface and its ligand in the T-cell surface lead to tight adhesion. The T-cell stops rolling and, under the influence of signals sent out from the site of the inflammation, begins extravasation—escape through the capillary wall—as it moves toward the site of inflammation.

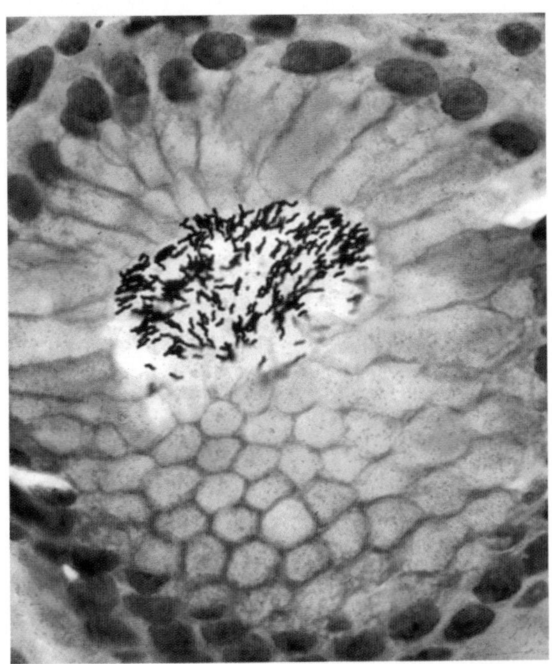

figure 9–28
Helicobacter pylori cells adhering to the gastric surface. This bacterium causes ulcers by interactions between a bacterial surface lectin and the Le^b oligosaccharide (a blood group antigen) of the gastric epithelium.

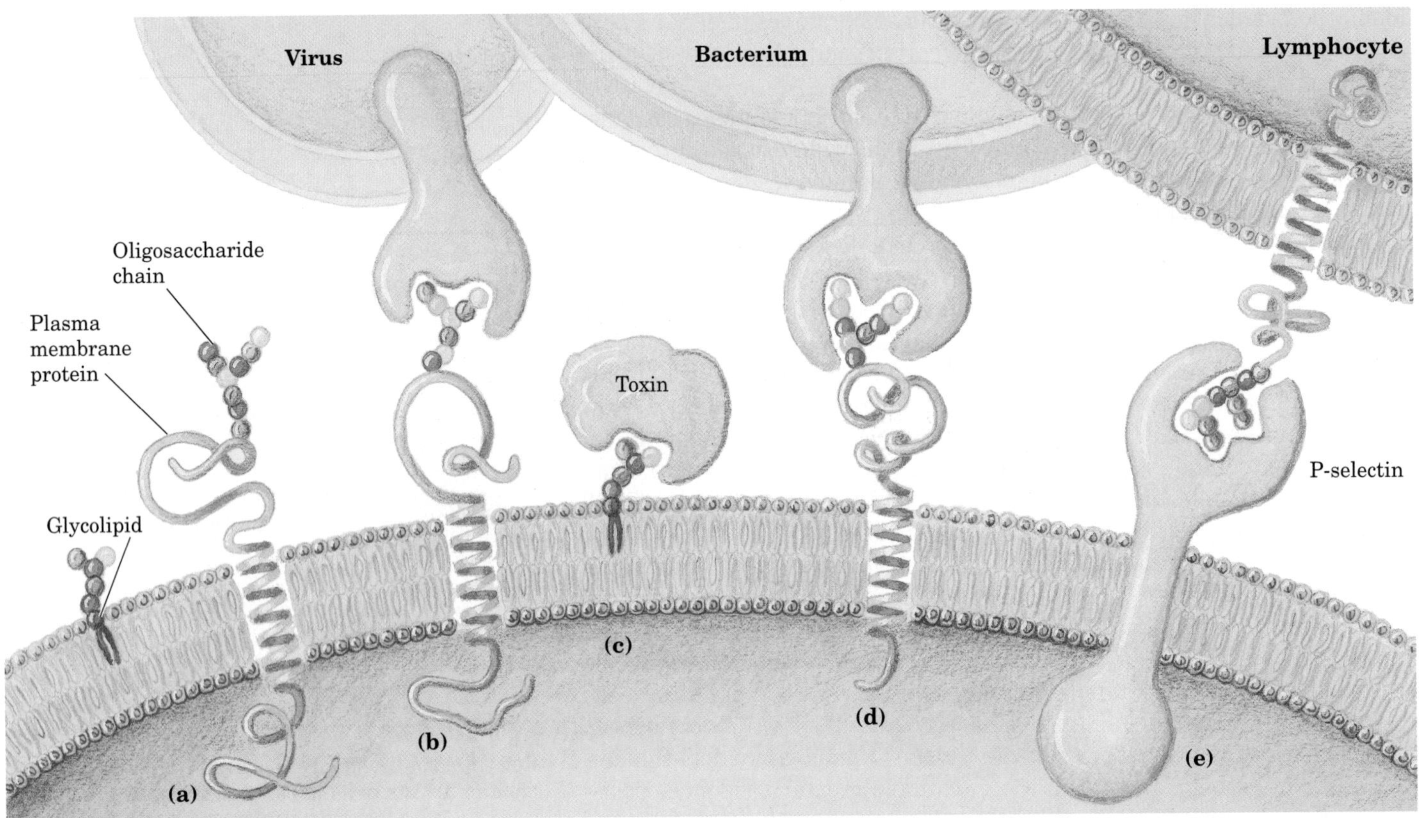

figure 9–29

Roles of oligosaccharides in recognition and adhesion at the cell surface. (a) Oligosaccharides with unique structures (represented as strings of colored balls), components of a variety of glycoproteins or glycolipids on the outer surface of plasma membranes, interact with high specificity and affinity with lectins in the extracellular milieu. **(b)** Viruses that infect animal cells, such as the influenza virus, bind to glycoproteins on the cell as the first step of infection. **(c)** Bacterial toxins, such as the cholera and pertussis toxins, bind to a surface glycolipid before entering a cell. **(d)** Some bacteria, such as *Helicobacter pylori*, adhere to then colonize or infect animal cells. **(e)** Lectins called selectins in the plasma membrane of certain cells mediate cell-cell interactions, such as those of T lymphocytes with the endothelial cells of the capillary wall at an infection site.

Figure 9–29 summarizes the types of biological interactions that are mediated by carbohydrate binding.

Analysis of Carbohydrates

The growing appreciation of the importance of oligosaccharide structure in biological recognition has been the driving force behind the development of methods for analyzing the structure and stereochemistry of complex oligosaccharides. Oligosaccharide analysis is complicated by the fact that, unlike nucleic acids and proteins, oligosaccharides can be branched and are joined by a variety of linkages. For such analysis, oligosaccharides are generally removed from their protein or lipid conjugates, then subjected to stepwise degradation with specific reagents that reveal bond position or stereochemistry. Mass spectrometry and NMR spectroscopy have also become invaluable in deciphering oligosaccharide structure.

The oligosaccharide moieties of glycoproteins or glycolipids can be released by purified enzymes—glycosidases that specifically cleave *O*- or *N*-linked oligosaccharides or lipases that remove lipid head groups. Mixtures of carbohydrates are resolved into their individual components (Fig. 9–30) by some of the same techniques useful in protein and amino acid separation: fractional precipitation by solvents, and ion-exchange and gel filtration (size-exclusion) chromatography (see Fig. 5–18). Highly purified lectins, attached covalently to an insoluble support, are commonly used in affinity chromatography of carbohydrates (see Fig. 5–18c). Hydrolysis of oligosaccharides and polysaccharides in strong acid yields a mixture of monosaccharides, which, after conversion to suitable volatile derivatives, may be separated, identified, and quantified by gas-liquid chromatography (see p. 385) to yield the overall composition of the polymer. For simple, linear polymers such as amylose, the positions of the glycosidic bonds are determined

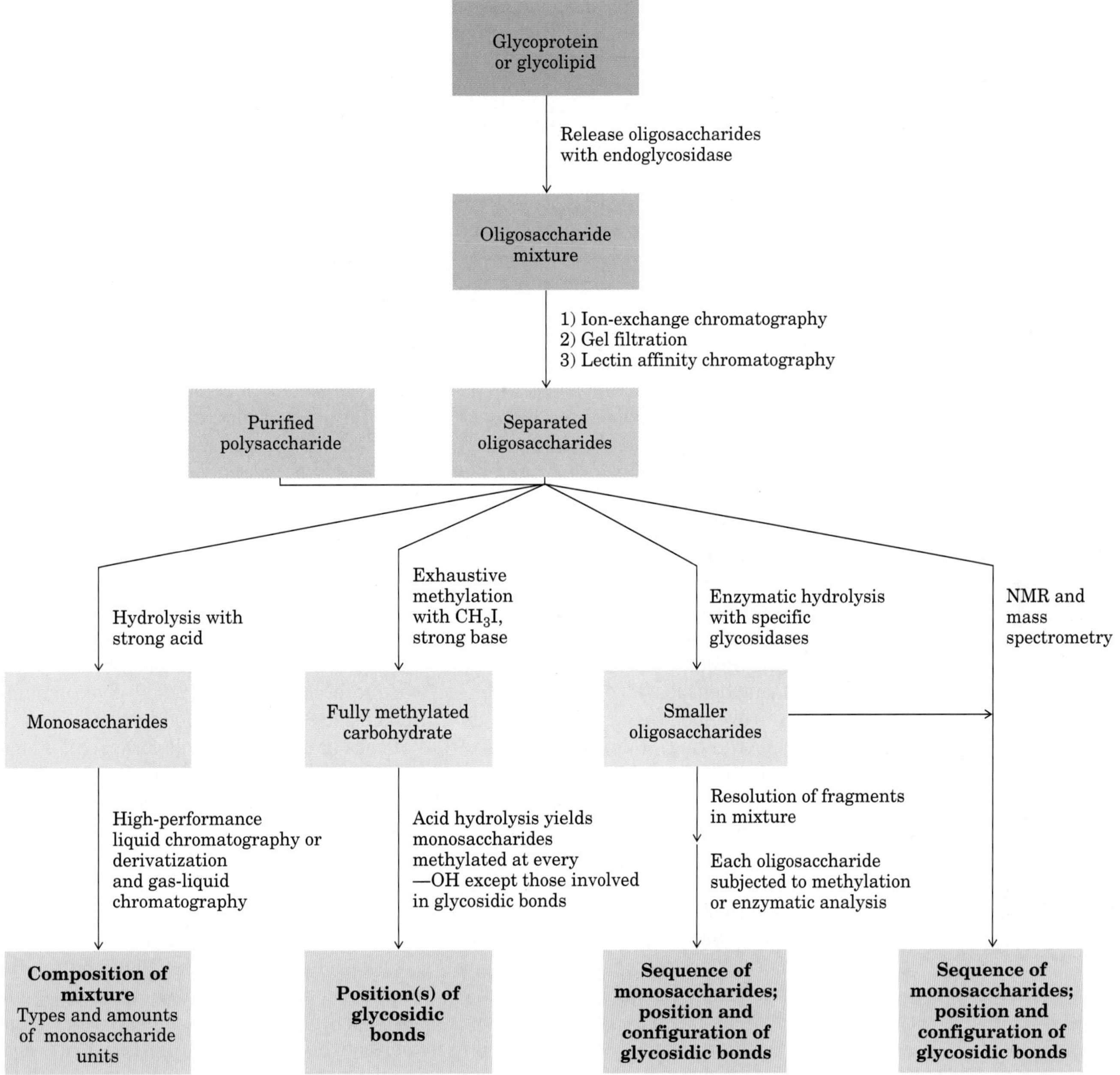

figure 9–30

Methods of carbohydrate analysis. A carbohydrate purified in the first stage of the analysis often requires all four analytical routes for its complete characterization.

by treating the intact polysaccharide with methyl iodide in a strongly basic medium to convert all free hydroxyls to acid-stable methyl ethers, then hydrolyzing the methylated polysaccharide in acid. The only free hydroxyls present in the monosaccharide derivatives produced are those that were involved in glycosidic bonds. To determine the sequence of monosaccharide residues, including branches if they are present, exoglycosidases of known specificity are used to remove residues one at a time from the nonreducing end(s). The specificity of these exoglycosidases often allows deduction of the position and stereochemistry of the linkages.

Oligosaccharide analysis relies increasingly on mass spectrometry and high-resolution NMR spectroscopy. NMR analysis alone, especially for oligosaccharides of moderate size, can yield much information about sequence, linkage position, and anomeric carbon configuration. Polysaccharides and large oligosaccharides can be treated chemically or with endoglycosidases to split specific internal glycosidic bonds, producing several smaller, more easily analyzable oligo-saccharides. Automated procedures and commercial instruments are used for the routine determination of oligosaccharide structure, but the sequencing of branched oligosaccharides joined by more than one type of bond remains a far more formidable problem than determining the linear sequences of proteins and nucleic acids, which have monomers joined by a single bond type.

summary

Carbohydrates are predominantly cyclized polyhydroxy aldehydes or ketones, which occur in nature as monosaccharides (aldoses or ketoses), oligosaccharides (several monosaccharide units), and polysaccharides (large linear or branched molecules containing many monosaccharide units). Monosaccharides have at least one asymmetric carbon atom and thus exist in stereoisomeric forms. Most common, naturally occurring sugars, such as ribose, glucose, fructose, and mannose, are of the D series. Simple sugars having four or more carbon atoms may exist in the form of closed-ring hemiacetals or hemiketals, either furanoses (five-membered ring) or pyranoses (six-membered ring). Furanoses and pyranoses occur in anomeric α and β forms, which are interconverted in the process of mutarotation. Sugars with free, oxidizable anomeric carbons are called reducing sugars. Many derivatives of the simple sugars are found in living cells, including amino sugars and their acetylated derivatives, aldonic acids, and uronic acids. The hydroxyl groups of monosaccharides can also form phosphate and sulfate esters. Disaccharides consist of two monosaccharides joined by a glycosidic bond.

Polysaccharides (glycans) contain many monosaccharide residues in glycosidic linkage. Some (starch and glycogen) function as storage forms of carbohydrate, high molecular weight, branched polymers of glucose having ($\alpha1{\rightarrow}4$) linkages in the main chains and ($\alpha1{\rightarrow}6$) linkages at the branch points. Other polysaccharides play a structural role in cell walls and exoskeletons. Cellulose (in plants) has D-glucose units in ($\beta1{\rightarrow}4$) linkage; chitin (in insect exoskeletons) is a linear polymer of *N*-acetylglucosamine joined by ($\beta1{\rightarrow}4$) linkages. Bacterial cell walls contain peptidoglycans, linear polysaccharides of alternating *N*-acetylmuramic acid and *N*-acetylglucosamine residues, cross-linked by short peptide chains. The extracellular matrix surrounding cells in animal tissues contains a variety of glycosaminoglycans, linear polymers of alternating amino sugar and uronic acid units. These include hyaluronate, a high molecular weight polymer of alternating D-glucuronic acid and *N*-acetyl-D-glucosamine residues, and a variety of shorter, sulfated, very acidic heteropolysaccharides such as heparan sulfate, chondroitin sulfate, keratan sulfate, and dermatan sulfate.

Many of the oligosaccharides in a cell are in glycoconjugates—hybrid molecules in which the carbohydrate moiety is covalently bound to a protein or lipid. Proteoglycans, proteins (such as syndecan and aggrecan core proteins) with one or more covalently attached glycosaminoglycan chains, are generally found on the outer surface of cells or in the extracellular matrix. In most tissues, the extracellular matrix contains a variety of proteoglycans and multiadhesive proteins such as fibronectin, as well as huge proteoglycan aggregates consisting of many proteoglycans noncovalently bound to a single large hyaluronate molecule.

Glycoproteins contain covalently linked oligosaccharides that are smaller but more structurally complex, and therefore more information-rich, than glycosaminoglycans. Many cell surface or extracellular proteins are glycoproteins, as are most secreted proteins. The carbohydrate moieties serve as biological labels, which are "read" by lectins, proteins that bind oligosaccharide chains with high affinity and selectivity. Interactions between lectins and specific oligosaccharides are central to cell-cell recognition and adhesion. In vertebrates, oligosaccharide tags govern the rate of degradation of certain peptide hormones, circulating proteins, and blood cells. Selectins are plasma membrane lectins that bind carbohydrate chains in the extracellular matrix or on the surfaces of other cells, thereby mediating the flow of information between cell and matrix or between cells. The adhesion of bacterial and viral pathogens (including *Helicobacter,* the cholera and pertussis toxins, and the influenza virus) to their animal cell targets occurs through binding of lectins in the pathogens to oligosaccharides of the target cell surface.

The structure of oligosaccharides and polysaccharides is investigated by a combination of methods: specific enzymatic hydrolysis to determine stereochemistry and produce smaller fragments for further analysis; methylation analysis to locate the glycosidic bonds; mass spectrometry, stepwise degradation, and NMR spectroscopy to determine sequences; and high-resolution NMR spectroscopy to establish configurations at anomeric carbons.

further reading

General Background on Carbohydrate Chemistry

Aspinall, G.O. (ed.) (1982, 1983, 1985) *The Polysaccharides,* Vols 1–3, Academic Press, Inc., New York.

Collins, P.M. & Ferrier, R.J. (1995) *Monosaccharides: Their Chemistry and Their Roles in Natural Products,* John Wiley & Sons, Chichester, England.

A comprehensive text at the graduate level.

Fukuda, M. & Hindsgaul, O. (1994) *Molecular Glycobiology,* IRL Press at Oxford University Press, Inc., New York.

Thorough, advanced treatment of the chemistry and biology of cell surface carbohydrates. Good chapters on lectins, carbohydrate recognition in cell-cell interactions, and chemical synthesis of oligosaccharides.

Lehmann, J. (Haines, A.H., trans.) (1998) *Carbohydrates: Structure and Biology,* G. Thieme, Verlag, New York.

The fundamentals of carbohydrate chemistry and biology, presented at a level suitable for advanced undergraduates and graduate students.

Melendez-Hevia, E., Waddell, T.G., & Shelton, E.D. (1993) Optimization of molecular design in the evolution of metabolism: the glycogen molecule. *Biochem. J.* **295,** 477–483.

Morrison, R.T. & Boyd, R.N. (1992) *Organic Chemistry,* 6th edn, Prentice Hall, Inc., Englewood Cliffs, NJ.

Chapters 34 and 35 cover the structure, stereochemistry, nomenclature, and chemical reactions of carbohydrates.

Pigman, W. & Horton, D. (eds) (1970, 1972, 1980) *The Carbohydrates: Chemistry and Biochemistry,* Vols IA, IB, IIA, and IIB, Academic Press, Inc., New York.

Comprehensive treatise on carbohydrate chemistry.

Glycosaminoglycans and Proteoglycans

Carney, S.L. & Muir, H. (1988) The structure and function of cartilage proteoglycans. *Physiol. Rev.* **68,** 858–910.

An advanced review.

Iozzo, R.V. (1998) Matrix proteoglycans: from molecular design to cellular function. *Annu. Rev. Biochem.* **67,** 609–652.

A review focusing on the most recent genetic and molecular biological studies of the matrix proteoglycans. The structure-function relationships of some paradigmatic proteoglycans are discussed in depth and novel aspects of their biology are examined.

Jackson, R.L., Busch, S.J., & Cardin, A.D. (1991) Glycosaminoglycans: molecular properties, protein interactions, and role in physiological processes. *Physiol. Rev.* **71,** 481–539.

An advanced review of the chemistry and biology of glycosaminoglycans.

Glycoproteins

Gahmberg, C.G. & Tolvanen, M. (1996) Why mammalian cell surface proteins are glycoproteins. *Trends Biochem. Sci.* **21,** 308–311.

Kobata, A. (1992) Structures and functions of the sugar chains of glycoproteins. *Eur. J. Biochem.* **209,** 483–501.

Opdenakker, G., Rudd, P., Ponting, C., & Dwek, R. (1993) Concepts and principles of glycobiology. *FASEB J.* **7,** 1330–1337.

A review that considers the genesis of glycoforms, functional roles for glycosylation, and structure-function relationships for several glycoproteins.

Varki, A. (1993) Biological roles of oligosaccharides: all of the theories are correct. *Glycobiology* **3,** 97–130.

Lectins, Recognition, and Adhesion

Aplin, A.E., Howe, A., Alahari, S.K., & Juliano, R.L. (1998) Signal transduction and signal modulation by cell adhesion receptors: the role of integrins, cadherins, immunoglobulin-cell adhesion molecules, and selectins. *Pharmacol. Rev.* **50,** 197–263.

Borén, T., Normark, S., & Falk, P. (1994) *Helicobacter pylori:* molecular basis for host recognition and bacterial adherence. *Trends Microbiol.* **2,** 221–228.

A look at the role of the oligosaccharides that determine blood type in the adhesion of this microorganism to the stomach lining, producing ulcers.

Cornejo, C.J., Winn, R.K., & Harlan, J.M. (1997) Anti-adhesion therapy. *Adv. Pharmacol.* **39,** 99–142.

Analogs of recognition oligosaccharides are used to block adhesion of a pathogen to its host cell target.

Hooper, L.A., Manzella, S.M., & Baenziger, J.U. (1996) From legumes to leukocytes: biological roles for sulfated carbohydrates. *FASEB J.* **10,** 1137–1146.

Evidence for roles of sulfated oligosaccharides in peptide hormone half-life, symbiont interactions in nitrogen-fixing legumes, and lymphocyte homing.

Horwitz, A.F. (1997) Integrins and health. *Sci. Am.* **276** (May), 68–75.

Article on the role of integrins in cell-cell adhesion, and possible roles in arthritis, heart disease, stroke, osteoporosis, and the spread of cancer.

Leahy, D.J. (1997) Implications of atomic-resolution structures for cell adhesion. *Annu. Rev. Cell Dev. Biol.* **13,** 363–393.

This review briefly describes several recently determined structures of cell adhesion molecules, summarizes some of the main findings about each structure, and highlights common features of different cell adhesion systems.

McEver, R.P., Moore, K.L., & Cummings, R.D. (1995) Leukocyte trafficking mediated by selectin-carbohydrate interactions *J. Biol. Chem.* **270,** 11,025–11,028.

A short review that focuses on the interaction of selectins with their carbohydrate ligands.

Sharon, N. & Lis, H. (1993) Carbohydrates in cell recognition. *Sci. Am.* **268** (January), 82–89.

Chemical basis and biological roles of carbohydrate recognition.

Varki, A. (1997) Sialic acids as ligands in recognition phenomena. *FASEB J.* **11,** 248–255.

Weiss, W.I. & Drickamer, K. (1996) Structural basis of lectin–carbohydrate recognition. *Annu. Rev. Biochem.* **65,** 441–473.

Good treatment of the chemical basis of carbohydrate-protein interactions.

Analysis of Carbohydrates

Chaplin, M.F. & Kennedy, J.F. (eds) (1994) *Carbohydrate Analysis: A Practical Approach,* 2nd edn, IRL Press, Oxford.

Very useful manual for analysis of all types of sugar-containing molecules—monosaccharides, polysaccharides and glycosaminoglycans, glycoproteins, proteoglycans, and glycolipids.

Dwek, R.A., Edge, C.J., Harvey, D.J., & Wormald, M.R. (1993) Analysis of glycoprotein-associated oligosaccharides. *Annu. Rev. Biochem.* **62,** 65–100.

Excellent survey of the uses of NMR, mass spectrometry, and enzymatic reagents to determine oligosaccharide structure.

Fukuda, M. & Kobata, A. (1993) *Glycobiology: A Practical Approach,* IRL Press, Oxford.

A how-to manual for the isolation and characterization of the oligosaccharide moieties of glycoproteins, using the whole range of modern techniques.

Jay, A. (1996) The methylation reaction in carbohydrate analysis. *J. Carbohydr. Chem.* **15,** 897–923.

Thorough description of methylation analysis of carbohydrates.

Lennarz, W.J. & Hart, G.W. (eds) (1994) *Guide to Techniques in Glycobiology,* Methods in Enzymology, Vol. 230, Academic Press, Inc., New York.

Practical guide to working with oligosaccharides.

McCleary, B.V. & Matheson, N.K. (1986) Enzymic analysis of polysaccharide structure. *Adv. Carbohydr. Chem. Biochem.* **44,** 147–276.

On the use of purified enzymes in analysis of structure and stereochemistry.

Rudd, P.M., Guile, G.R., Kuester, B., Harvey, D.J., Opdenakker, G., & Dwek, R.A. (1997) Oligosaccharide sequencing technology. *Nature* **388,** 205–207.

problems

1. Determination of an Empirical Formula An unknown substance containing only C, H, and O was isolated from goose liver. A 0.423 g sample produced 0.620 g of CO_2 and 0.254 g of H_2O after complete combustion in excess oxygen. Is the empirical formula of this substance consistent with its being a carbohydrate? Explain.

2. Sugar Alcohols In the monosaccharide derivatives known as sugar alcohols, the carbonyl oxygen is reduced to a hydroxyl group. For example, D-glyceraldehyde can be reduced to glycerol. However, this sugar alcohol is no longer designated D or L. Why?

3. Melting Points of Monosaccharide Osazone Derivatives Many carbohydrates react with phenylhydrazine ($C_6H_5NHNH_2$) to form bright yellow crystalline derivatives known as osazones:

$$
\begin{array}{ccc}
\begin{array}{c}
\mathrm{H}\diagdown\ \diagup\!\!\!\!/\mathrm{O} \\
\mathrm{C} \\
| \\
\mathrm{H-C-OH} \\
| \\
\mathrm{OH-C-H} \\
| \\
\mathrm{H-C-OH} \\
| \\
\mathrm{H-C-OH} \\
| \\
\mathrm{CH_2OH}
\end{array}
&
\xrightarrow{\mathrm{C_6H_5NHNH_2}}
&
\begin{array}{c}
\mathrm{H-C{=}NNHC_6H_5} \\
| \\
\mathrm{C{=}NNHC_6H_5} \\
| \\
\mathrm{OH-C-H} \\
| \\
\mathrm{H-C-OH} \\
| \\
\mathrm{H-C-OH} \\
| \\
\mathrm{CH_2OH}
\end{array}
\\
\text{Glucose} & & \text{Osazone derivative of glucose}
\end{array}
$$

The melting temperatures of these derivatives are easily determined and are characteristic for each osazone. This information was used to help identify monosaccharides before the development of HPLC or gas-liquid chromatography. Listed below are the melting points (MPs) of some aldose-osazone derivatives:

Monosaccharide	MP of anhydrous monosaccharide (°C)	MP of osazone derivative (°C)
Glucose	146	205
Mannose	132	205
Galactose	165–168	201
Talose	128–130	201

As the table shows, certain pairs of derivatives have the same melting points, although the underivatized monosaccharides do not. Why do glucose and mannose, and galactose and talose, form osazone derivatives with the same melting points?

4. Interconversion of D-Galactose Forms A solution of one stereoisomer of a given monosaccharide rotates plane-polarized light to the left (counterclockwise) and is called the levorotatory isomer, designated (−); the other stereoisomer rotates plane-polarized light to the same extent but to the right (clockwise) and is called the dextrorotatory isomer, designated (+). An equimolar mixture of the (+) and (−) forms does not rotate plane-polarized light.

The optical activity of a stereoisomer is expressed quantitatively by its *optical rotation,* the number of degrees by which plane-polarized light is rotated on passage through a given path length of a solution of the compound at a given concentration. The *specific rotation* $[\alpha]_{D}^{25°C}$ of an optically active compound is defined thus:

$$[a]_{D}^{25°C} = \frac{\text{observed optical rotation (°)}}{\text{optical path length (dm)} \times \text{concentration (g/mL)}}$$

The temperature and the wavelength of the light employed (usually the D line of sodium, 589 nm) must be specified in the definition.

A freshly prepared solution of α-D-galactose (1 g/mL in a 10 cm cell) shows an optical rotation of +150.7°. Over time, the observed rotation of the solution gradually decreases and reaches an equilibrium value of +80.2°. In contrast, a freshly prepared solution (1 g/mL) of β-D-galactose shows an optical rotation of only +52.8°. Moreover, the rotation increases over time to an equilibrium value of +80.2°, identical to that reached by α-D-galactose.

(a) Draw the Haworth perspective formulas of the α and β forms of D-galactose. What feature distinguishes the two forms?

(b) Why does the optical rotation of a freshly prepared solution of the α form gradually decrease with time? Why do solutions of the α and β forms (at equal concentrations) reach the same optical rotation at equilibrium?

(c) Calculate the percentage composition of the two forms of D-galactose at equilibrium.

5. A Taste of Honey The fructose in honey is mainly in the β-D-pyranose form. This is one of the sweetest substances known, about twice as sweet as glucose. The β-D-furanose form of fructose is much less sweet. The sweetness of honey gradually decreases at a high temperature. Also, high-fructose corn syrup (a commercial product in which much of the glucose in corn syrup is converted to fructose) is used for sweetening *cold* but not *hot* drinks. What chemical property of fructose could account for both of these observations?

6. Glucose Oxidase in Determination of Blood Glucose The enzyme glucose oxidase isolated from the mold *Penicillium notatum* catalyzes the oxidation of β-D-glucose to D-glucono-δ-lactone. This enzyme is highly specific for the β anomer of glucose and does not affect the α anomer. In spite of this specificity, the reaction catalyzed by glucose oxidase is commonly used in a clinical assay for total blood glucose—that is, for solutions consisting of a mixture of β- and α-D-glucose. How is this possible? Aside from allowing the detection of smaller quantities of glucose, what advantage does glucose oxidase offer over Fehling's reagent for the determination of blood glucose?

7. Invertase "Inverts" Sucrose The hydrolysis of sucrose (specific rotation +66.5°) yields an equimolar mixture of D-glucose (specific rotation +52.5°) and D-fructose (specific rotation –92°).

(a) Suggest a convenient way to determine the rate of hydrolysis of sucrose by an enzyme preparation extracted from the lining of the small intestine.

(b) Explain why an equimolar mixture of D-glucose and D-fructose formed by hydrolysis of sucrose is called invert sugar in the food industry.

(c) The enzyme invertase (its preferred name is now sucrase) is allowed to act on a solution of sucrose until the optical rotation of the solution becomes zero. What fraction of the sucrose has been hydrolyzed?

8. Manufacture of Liquid-Filled Chocolates The manufacture of chocolates containing a liquid center is an interesting application of enzyme engineering. The flavored liquid center consists largely of an aqueous solution of sugars rich in fructose to provide sweetness.

The technical dilemma is the following: the chocolate coating must be prepared by pouring hot melted chocolate over a solid (or almost solid) core, yet the final product must have a liquid, fructose-rich center. Suggest a way to solve this problem. (Hint: Sucrose is much less soluble than a mixture of glucose and fructose.)

9. Anomers of Sucrose? Although lactose exists in two anomeric forms, no anomeric forms of sucrose have been reported. Why?

10. Physical Properties of Cellulose and Glycogen The almost pure cellulose obtained from the seed threads of *Gossypium* (cotton) is tough, fibrous, and completely insoluble in water. In contrast, glycogen obtained from muscle or liver disperses readily in hot water to make a turbid solution. Although they have markedly different physical properties, both substances are composed of (1→4)-linked D-glucose polymers of comparable molecular weight. What structural features of these two polysaccharides underlie their different physical properties? Explain the biological advantages of their respective properties.

11. Growth Rate of Bamboo The stems of bamboo, a tropical grass, can grow at the phenomenal rate of 0.3 m/day under optimal conditions. Given that the stems are composed almost entirely of cellulose fibers oriented in the direction of growth, calculate the number of sugar residues per second that must be added enzymatically to growing cellulose chains to account for the growth rate. Each D-glucose unit in the cellulose molecule is about 0.45 nm long.

12. Glycogen as Energy Storage: How Long Can a Game Bird Fly? Since ancient times it has been observed that certain game birds, such as grouse, quail, and pheasants, are easily fatigued. The Greek historian Xenophon wrote, "The bustards . . . can be caught if one is quick in starting them up, for they will fly only a short distance, like partridges, and soon tire; and their flesh is delicious." The flight muscles of game birds rely almost entirely on the use of glucose 1-phosphate for energy, in the form of ATP (Chapter 15). In game birds, glucose 1-phosphate is formed by the breakdown of stored muscle glycogen, catalyzed by the enzyme glycogen phosphorylase. The rate of ATP production is limited by the rate at which glycogen can be broken down. During a "panic flight," the game bird's rate of glycogen breakdown is quite high, approximately 120 μmol/min of glucose 1-phosphate produced per gram of fresh tissue. Given that the flight muscles usually contain about 0.35% glycogen by weight, calculate how long a game bird can fly.

13. Volume of Chondroitin Sulfate in Solution One critical function of chondroitin sulfate is to act as a lubricant in skeletal joints by creating a gel-like medium that is resilient to friction and shock. This function appears to be related to a distinctive property of chondroitin sulfate: the volume occupied by the molecule is much greater in solution than in the dehydrated solid. Why is the volume occupied by the molecule so much larger in solution?

14. Heparin Interactions Heparin, a highly negatively charged glycosaminoglycan, is used clinically as an anticoagulant. It acts by binding several plasma proteins, including antithrombin III, an inhibitor of blood clotting. The 1:1 binding of heparin to antithrombin III appears to cause a conformational change in the protein that greatly increases its ability to inhibit clotting. What amino acid residues of antithrombin III are likely to interact with heparin?

15. Information Content of Oligosaccharides The carbohydrate portion of some glycoproteins may serve as a cellular recognition site. In order to perform this function, the oligosaccharide moiety of glycoproteins must have the potential to exist in a large variety of forms. Which can produce a greater variety of structures: oligopeptides composed of five different amino acid residues or oligosaccharides composed of five different monosaccharide residues? Explain.

16. Determination of the Extent of Branching in Amylopectin The extent of branching (number of (α1→6) glycosidic bonds) in amylopectin can be determined by the following procedure. A sample of amylopectin is exhaustively treated with a methylating agent (methyl iodide) that replaces all the hydrogens of the sugar hydroxyls with methyl groups, converting —OH to —OCH_3. All the glycosidic bonds in the treated sample are then hydrolyzed in aqueous acid. The amount of 2,3-di-*O*-methylglucose in the hydrolyzed sample is determined.

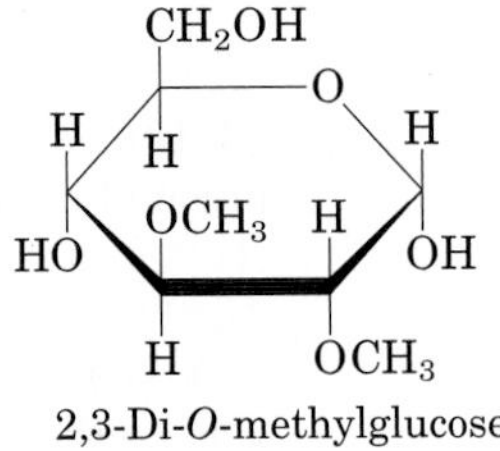

2,3-Di-*O*-methylglucose

(a) Explain the basis of this procedure for determining the number of (α1→6) branch points in amylopectin. What happens to the unbranched glucose residues in amylopectin during the methylation and hydrolysis procedure?

(b) A 258 mg sample of amylopectin treated as described above yielded 12.4 mg of 2,3-di-*O*-methylglucose. Determine what percentage of the glucose residues in amylopectin contain an (α1→6) branch.

17. Structural Analysis of a Polysaccharide A polysaccharide of unknown structure was isolated, subjected to exhaustive methylation, and hydrolyzed. Analysis of the products revealed three methylated sugars in the ratio 20:1:1. The sugars were 2,3,4-tri-*O*-methyl-D-glucose; 2,4-di-*O*-methyl-D-glucose; and 2,3,4,6-tetra-*O*-methyl-D-glucose. What is the structure of the polysaccharide?

chapter

10

Nucleotides and Nucleic Acids

Nucleotides have a variety of roles in cellular metabolism. They are the energy currency in metabolic transactions; the essential chemical links in the response of cells to hormones and other extracellular stimuli; and the structural components of an array of enzyme cofactors and metabolic intermediates. And, last but certainly not least, they are the constituents of nucleic acids: deoxyribonucleic acid (DNA) and ribonucleic acid (RNA), the molecular repositories of genetic information. The structure of every protein, and ultimately of every biomolecule and cellular component, is a product of information programmed into the nucleotide sequence of a cell's nucleic acids. The ability to store and transmit genetic information from one generation to the next is a fundamental condition for life.

This chapter provides an overview of the chemical nature of the nucleotides and nucleic acids found in most cells; a more detailed examination of the function of nucleic acids is the focus of Part IV of this text.

Some Basics

The amino acid sequence of every protein in a cell, and the nucleotide sequence of every RNA, is specified by a nucleotide sequence in the cell's DNA. A segment of a DNA molecule that contains the information required for the synthesis of a functional biological product, whether protein or RNA, is referred to as a **gene.** A cell typically has many thousands of genes, and DNA molecules, not surprisingly, tend to be very large. The storage and transmission of biological information are the only known functions of DNA.

RNAs have a broader range of functions, and several classes are found in cells. **Ribosomal RNAs** (rRNA) are structural components of ribosomes, the complexes that carry out the synthesis of proteins. **Messenger RNAs** (mRNA) are intermediaries, carrying genetic information from one or a few genes to a ribosome, where the corresponding proteins can be synthesized. **Transfer RNAs** (tRNA) are adapter molecules that faithfully translate the information in mRNA into a specific sequence of amino acids. In addition to these major classes there is a wide variety of RNAs with special functions, described in depth in Part IV.

Nucleotides and Nucleic Acids Have Characteristic Bases and Pentoses

Nucleotides have three characteristic components: (1) a nitrogenous (nitrogen-containing) base, (2) a pentose, and (3) a phosphate (Fig. 10–1). The molecule without the phosphate group is called a **nucleoside.** The nitrogenous bases are derivatives of two parent compounds, **pyrimidine**

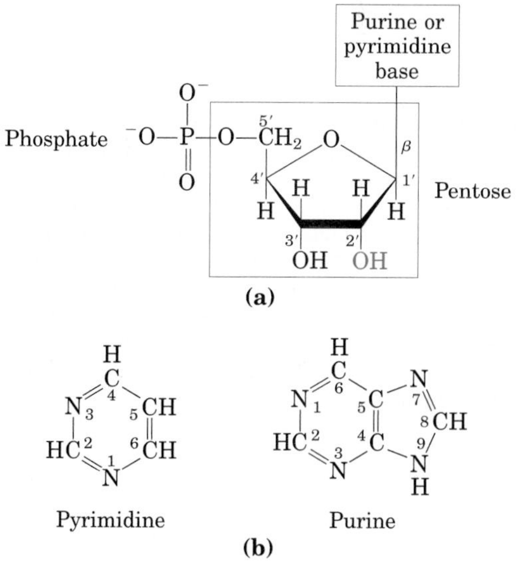

Purine or pyrimidine base

Phosphate

Pentose

(a)

Pyrimidine

Purine

(b)

figure 10–1

Structure of nucleotides. (a) General structure showing the numbering convention for the pentose ring. This is a ribonucleotide. In deoxyribonucleotides the —OH group on the 2′ carbon (in red) is replaced with —H. **(b)** The parent compounds of the pyrimidine and purine bases of nucleotides and nucleic acids, showing the numbering conventions.

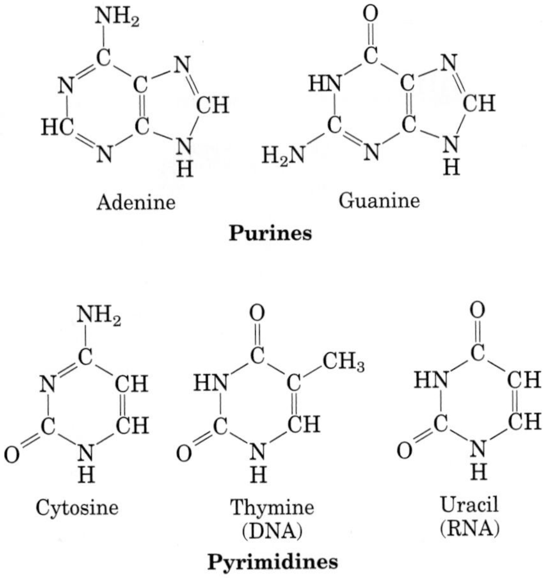

figure 10–2
Major purine and pyrimidine bases of nucleic acids. Some of the common names of these bases reflect the circumstances of their discovery. Guanine, for example, was first isolated from guano (bird manure), and thymine was first isolated from thymus tissue.

and **purine.** The bases and pentoses of the common nucleotides are heterocyclic compounds. The carbon and nitrogen atoms in the parent structures are conventionally numbered to facilitate the naming and identification of the many derivative compounds. The convention for the pentose ring follows rules outlined in Chapter 9, but in the pentoses of nucleotides and nucleosides the carbon numbers are given a prime (′) designation to distinguish them from the numbered atoms of the nitrogenous bases.

The base of a nucleotide is joined covalently (at N-1 of pyrimidines and N-9 of purines) in an *N*-β-glycosyl bond to the 1′ carbon of the pentose, and the phosphate is esterified to the 5′ carbon. The *N*-β-glycosyl bond is formed by removal of the elements of water (a hydroxyl group from the pentose and hydrogen from the base), as in *O*-glycosidic bond formation (see Fig. 9–25).

Both DNA and RNA contain two major purine bases, **adenine** (A) and **guanine** (G), and two major pyrimidines. In both DNA and RNA one of the pyrimidines is **cytosine** (C), but the second major pyrimidine is **thymine** (T) in DNA and **uracil** (U) in RNA. Only rarely does thymine occur in RNA or uracil in DNA. The structures of the five major bases are shown in Figure 10–2, and the nomenclature of their corresponding nucleotides and nucleosides is summarized in Table 10–1 (opposite page).

Nucleic acids have two kinds of pentoses. The recurring deoxyribonucleotide units of DNA contain 2′-deoxy-D-ribose, and the ribonucleotide units of RNA contain D-ribose. In nucleotides, both types of pentoses are in their β-furanose (closed five-membered ring) form. As Figure 10–3 shows, the pentose ring is not planar but occurs in one of a variety of conformations generally described as "puckered."

Figure 10–4 gives the structures and names of the four major **deoxyribonucleotides** (deoxyribonucleoside 5′-monophosphates), the structural units of DNAs, and the four major **ribonucleotides** (ribonucleoside 5′-monophosphates), the structural units of RNAs. Specific long sequences of A, T, G, and C nucleotides in DNA are the repository of genetic information.

figure 10–3
Conformations of ribose. (a) In solution, the straight-chain (aldehyde) and ring (β-furanose) forms of free ribose are in equilibrium. RNA contains only the ring form, β-D-ribofuranose. Deoxyribose undergoes a similar interconversion in solution, but in DNA exists solely as β-2′-deoxy-D-ribofuranose. **(b)** Ribofuranose rings in nucleotides can exist in four different puckered conformations. In all cases, four of the five atoms are in a single plane. The fifth atom (C-2′ or C-3′) is on either the same (endo) or the opposite (exo) side of the plane relative to the C-5′ atom.

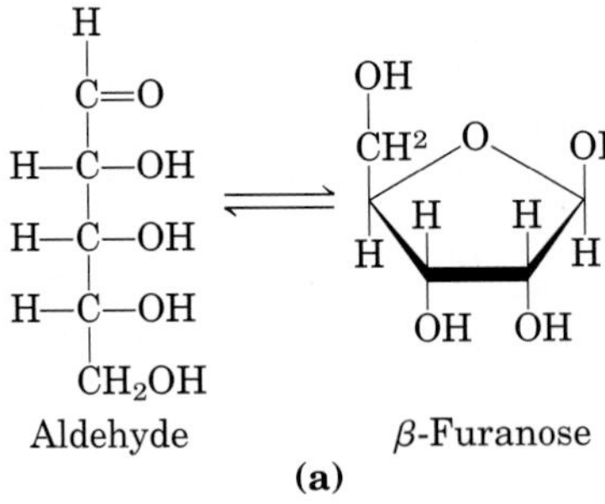

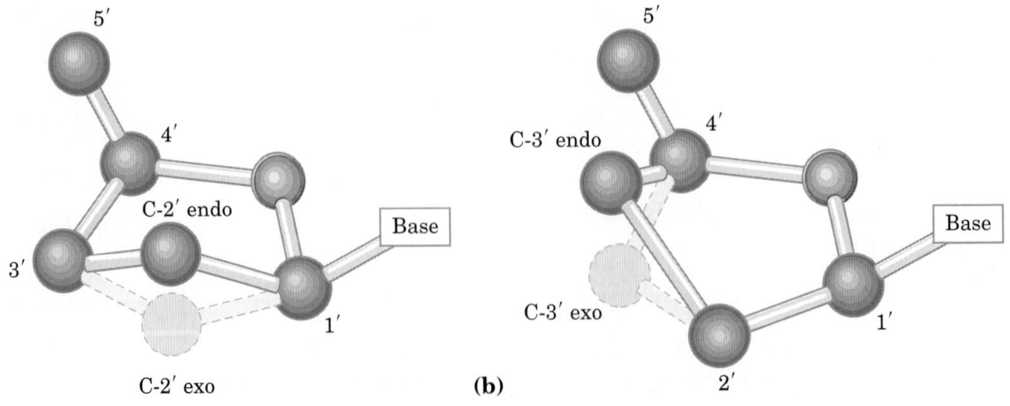

table 10–1

Nucleotide and Nucleic Acid Nomenclature

Base	Nucleoside*	Nucleotide*	Nucleic acid
Purines			
Adenine	Adenosine	Adenylate	RNA
	Deoxyadenosine	Deoxyadenylate	DNA
Guanine	Guanosine	Guanylate	RNA
	Deoxyguanosine	Deoxyguanylate	DNA
Pyrimidines			
Cytosine	Cytidine	Cytidylate	RNA
	Deoxycytidine	Deoxycytidylate	DNA
Thymine	Thymidine or deoxythymidine	Thymidylate or deoxythymidylate	DNA
Uracil	Uridine	Uridylate	RNA

**Nucleoside* and *nucleotide* are generic terms that include both ribo- and deoxyribo- forms. Note that here ribonucleosides and ribonucleotides are designated simply as nucleosides and nucleotides (e.g., riboadenosine as adenosine), and deoxyribonucleosides and deoxyribonucleotides as deoxynucleosides and deoxynucleotides (e.g., deoxyriboadenosine as deoxyadenosine). Both forms of naming are acceptable, but the shortened names are more commonly used. Thymine is an exception; the name ribothymidine is used to describe its unusual occurrence in RNA.

figure 10–4

Deoxyribonucleotides and ribonucleotides of nucleic acids. All nucleotides are shown in their free form at pH 7.0. The nucleotide units of DNA **(a)** are usually symbolized as A, G, T, and C, sometimes as dA, dG, dT, and dC; those of RNA **(b)** as A, G, U, and C. In their free form the deoxyribonucleotides are commonly abbreviated dAMP, dGMP, dTMP, and dCMP; the ribonucleotides, AMP, GMP, UMP, and CMP. For each nucleotide, the more common name is followed by the complete name in parentheses. All abbreviations assume that the phosphate group is at the 5′ position. The nucleoside portion of each molecule is shaded in red. In this and the following illustrations, the ring carbons are not shown.

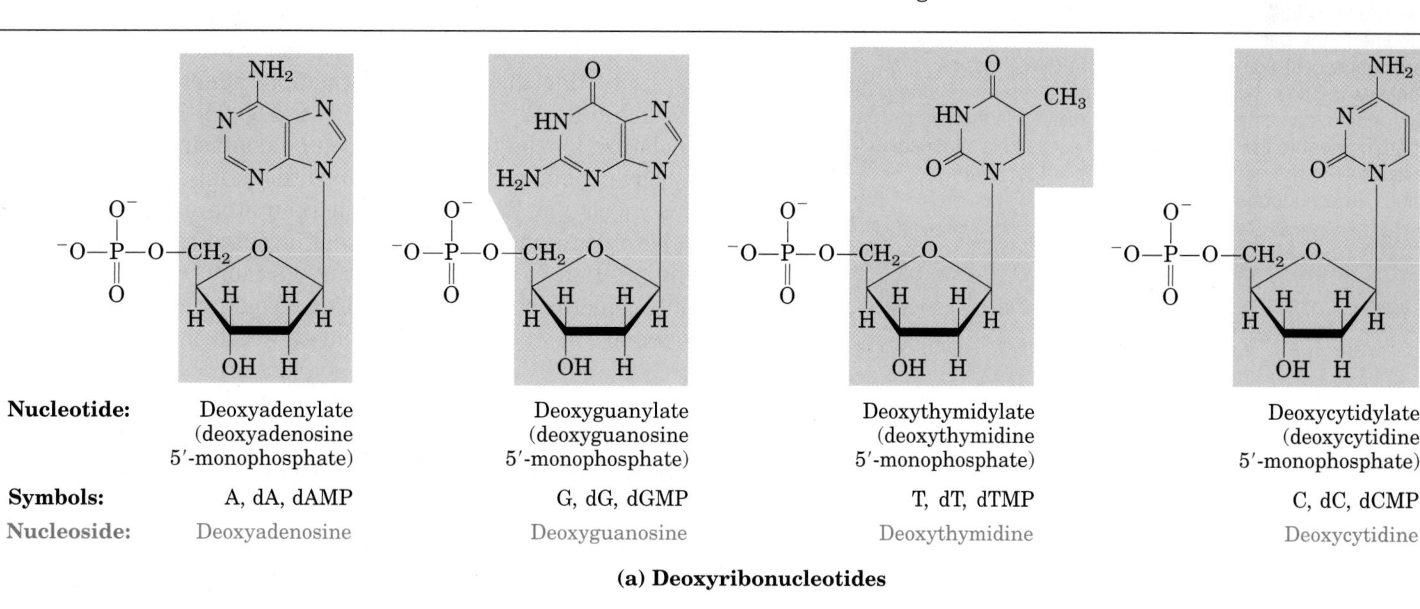

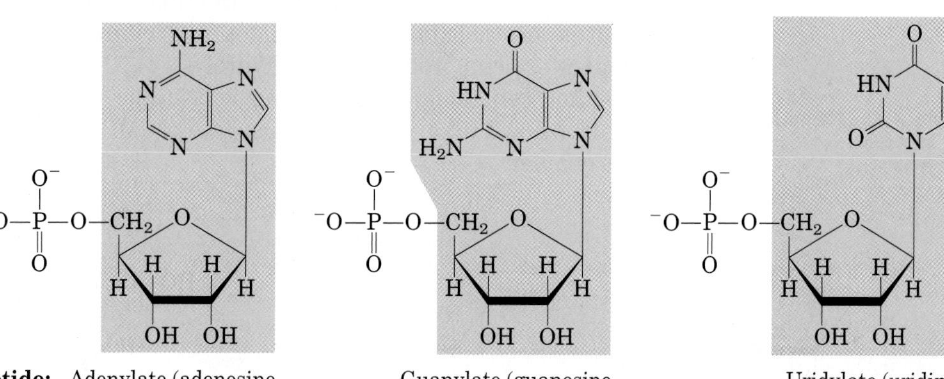

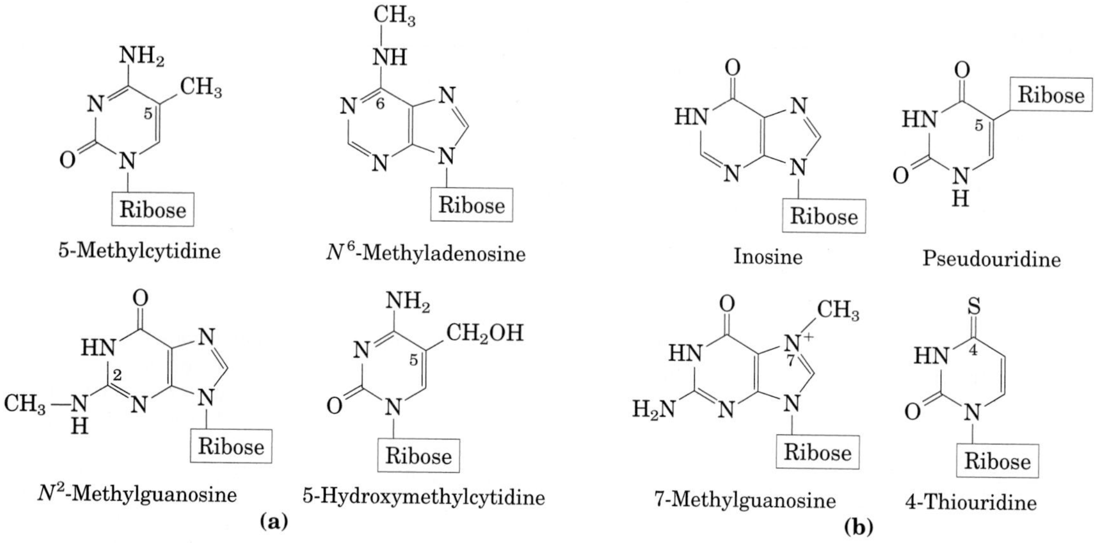

5-Methylcytidine

N^6-Methyladenosine

N^2-Methylguanosine

5-Hydroxymethylcytidine

(a)

Inosine

Pseudouridine

7-Methylguanosine

4-Thiouridine

(b)

figure 10–5

Some minor purine and pyrimidine bases, shown as the nucleosides. (a) Minor bases of DNA. 5-Methylcytidine occurs in the DNA of animals and higher plants, N^6-methyladenosine in bacterial DNA, and 5-hydroxymethylcytidine in the DNA of bacteria infected with certain bacteriophages. **(b)** Some minor bases of tRNAs. Inosine contains the base hypoxanthine. Note that pseudouridine, like uridine, contains uracil; they are distinct in the point of attachment to the ribose—in uridine, uracil is attached through N-1, the normal attachment point for pyrimidines; in pseudouridine, through C-5.

Although nucleotides bearing the major purines and pyrimidines are most common, both DNA and RNA also contain some minor bases (Fig. 10–5). In DNA the most common of these are methylated forms of the major bases; in some viral DNAs, certain bases may be hydroxymethylated or glucosylated. Altered or unusual bases in DNA molecules often have roles in regulating or protecting the genetic information. Minor bases of many types are also found in RNAs, especially in tRNA (Fig. 26–26).

The nomenclature for the minor bases can be confusing. Like the major bases, many have common names—hypoxanthine, for example, shown as its nucleoside inosine in Figure 10–5. When an atom in the purine or pyrimidine ring is substituted, the usual convention (used here) is simply to indicate the ring position of the substituent by its number—for example, 5-methylcytosine, 7-methylguanine, and 5-hydroxymethylcytosine (shown as the nucleosides in Fig. 10–5). The element to which the substituent is attached (N, C, etc.) is not identified. The convention changes when the substituted atom is exocyclic (not within the ring structure), in which case the type of atom is identified and the ring position to which it is attached is denoted with a superscript. The amino nitrogen attached to C-6 of adenine is N^6; similarly, the carbonyl oxygen and amino nitrogen at C-6 and C-2 of guanine are O^6 and N^2, respectively. Examples of this nomenclature are N^6-methyladenosine and N^2-methylguanosine (Fig. 10–5).

Cells also contain nucleotides with phosphate groups in positions other than on the 5′ carbon (Fig. 10–6). **Ribonucleoside 2′,3′-cyclic monophosphates** are isolatable intermediates and **ribonucleoside 3′-monophosphates** are end products of the hydrolysis of RNA by certain ribonucleases. Other variations are adenosine 3′,5′-cyclic monophosphate (cAMP) and guanosine 3′,5′-cyclic monophosphate (cGMP), considered at the end of this chapter.

figure 10–6

Some adenosine monophosphates. Adenosine 2′-monophosphate, 3′-monophosphate, and 2′,3′-cyclic monophosphate are formed by enzymatic and alkaline hydrolysis of RNA.

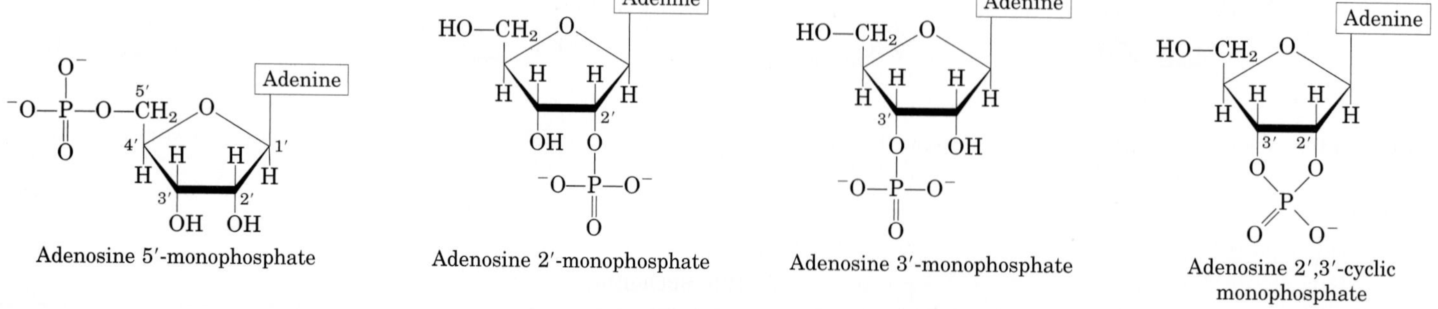

Adenosine 5′-monophosphate

Adenosine 2′-monophosphate

Adenosine 3′-monophosphate

Adenosine 2′,3′-cyclic monophosphate

Phosphodiester Bonds Link Successive Nucleotides in Nucleic Acids

The successive nucleotides of both DNA and RNA are covalently linked through phosphate-group "bridges," in which the 5′-hydroxyl group of one nucleotide unit is joined to the 3′-hydroxyl group of the next nucleotide by a **phosphodiester linkage** (Fig. 10–7). Thus the covalent backbones of nucleic acids consist of alternating phosphate and pentose residues, and the nitrogenous bases may be regarded as side groups joined to the backbone at regular intervals. The backbones of both DNA and RNA are hydrophilic. The hydroxyl groups of the sugar residues form hydrogen bonds with water. The phosphate groups, with a pK_a near 0, are completely ionized and negatively charged at pH 7, and the negative charges are generally neutralized by ionic interactions with positive charges on proteins, metal ions, and polyamines.

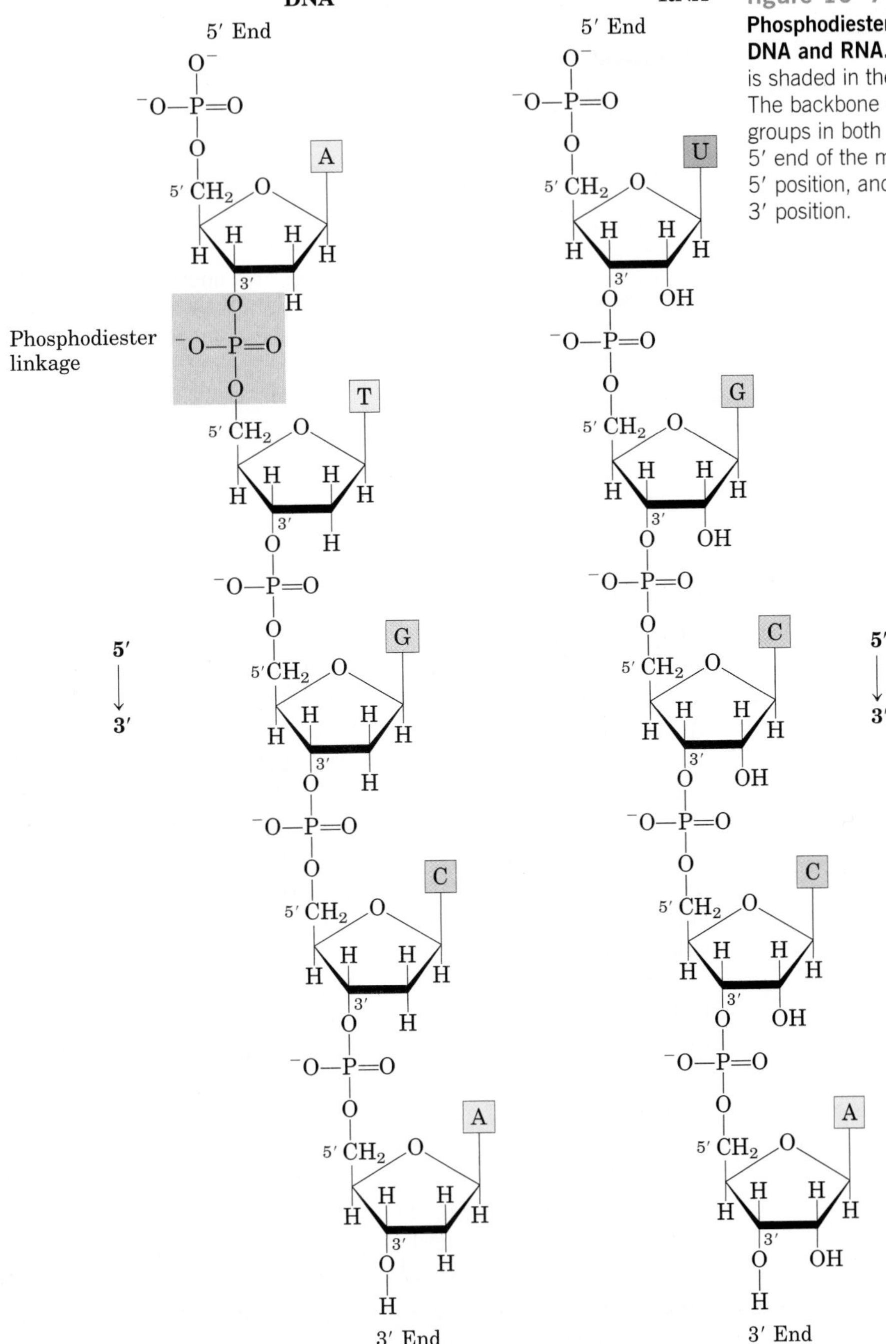

figure 10–7

Phosphodiester linkages in the covalent backbone of DNA and RNA. The phosphodiester bonds (one of which is shaded in the DNA) link successive nucleotide units. The backbone of alternating pentose and phosphate groups in both types of nucleic acid is highly polar. The 5′ end of the macromolecule lacks a nucleotide at the 5′ position, and the 3′ end lacks a nucleotide at the 3′ position.

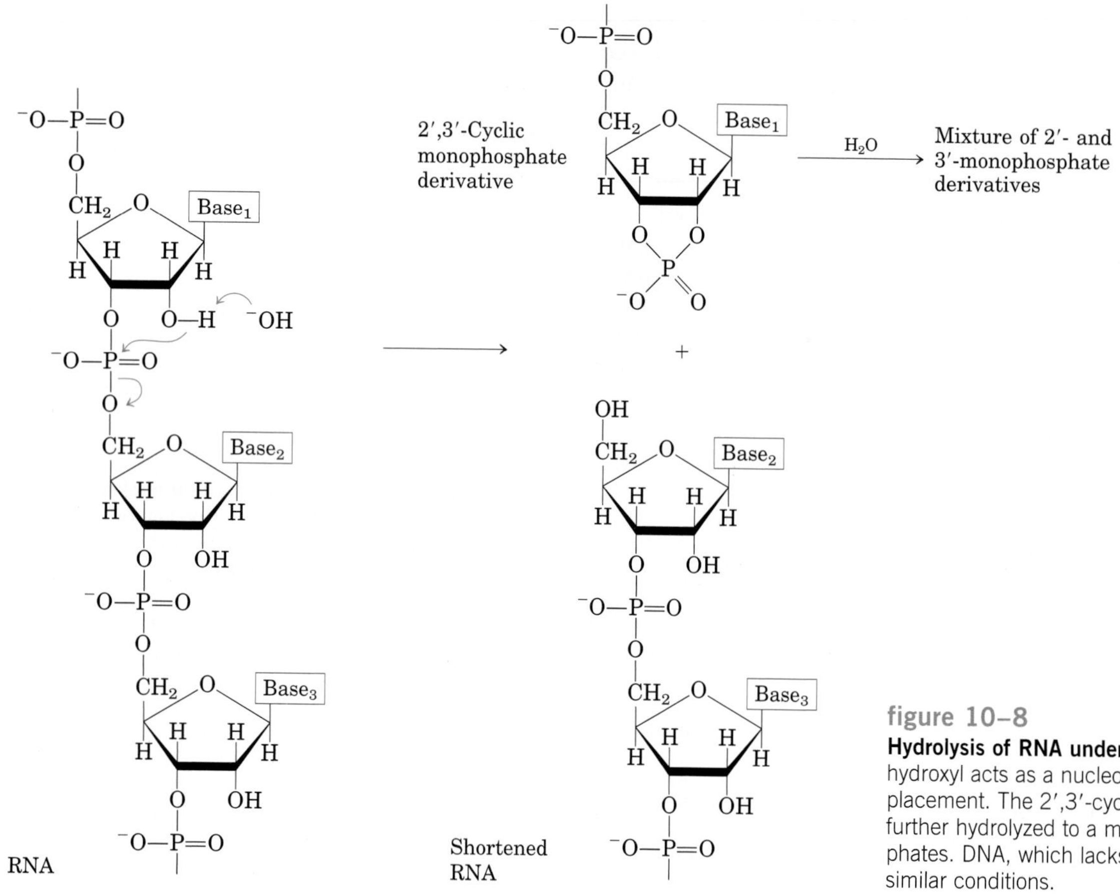

figure 10–8
Hydrolysis of RNA under alkaline conditions. The 2′ hydroxyl acts as a nucleophile in an intramolecular displacement. The 2′,3′-cyclic monophosphate derivative is further hydrolyzed to a mixture of 2′- and 3′-monophosphates. DNA, which lacks 2′ hydroxyls, is stable under similar conditions.

All the phosphodiester linkages have the same orientation along the chain (Fig. 10–7), giving each linear nucleic acid strand a specific polarity and distinct 5′ and 3′ ends. By definition, the **5′ end** lacks a nucleotide at the 5′ position and the **3′ end** lacks a nucleotide at the 3′ position. Other groups (most often one or more phosphates) may be present on one or both ends.

The covalent backbone of DNA and RNA is subject to slow, nonenzymatic hydrolysis of the phosphodiester bonds. In the test tube, RNA is hydrolyzed rapidly under alkaline conditions, but DNA is not; the 2′-hydroxyl groups in RNA (absent in DNA) are directly involved in the process. Cyclic 2′,3′-monophosphates are the first products of the action of alkali on RNA and are rapidly hydrolyzed further to yield a mixture of 2′- and 3′-nucleoside monophosphates (Fig. 10–8).

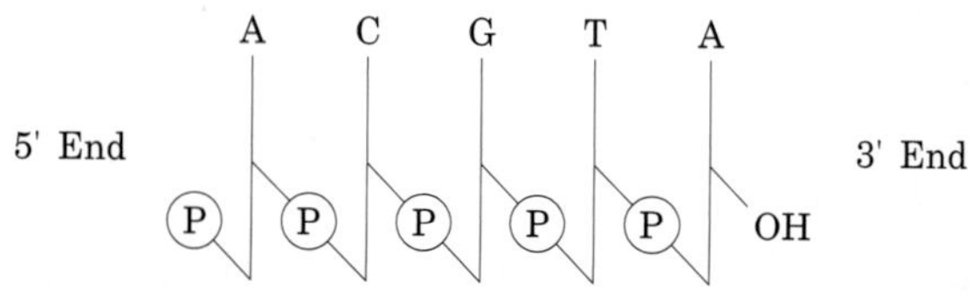

The nucleotide sequences of nucleic acids can be represented schematically, as illustrated on the right by a segment of DNA with five nucleotide units. The phosphate groups are symbolized by Ⓟ and each deoxyribose is symbolized by a vertical line, from C-1′ at the top to C-5′ at the bottom (but keep in mind that the sugar is always in its closed-ring β-furanose form in nucleic acids). The connecting lines between nucleotides (which pass through Ⓟ) are drawn diagonally from the middle (C-3′) of the deoxyribose of one nucleotide to the bottom (C-5′) of the next. By convention, the structure of a single strand of nucleic acid is always written with the 5′ end at the left and the 3′ end at the right—that is, in the 5′→3′ direction. Some simpler representations of this pentadeoxyribonucleotide are pA-C-G-T-A_{OH}, pApCpGpTpA, and pACGTA.

A short nucleic acid is referred to as an **oligonucleotide.** The definition of "short" is somewhat arbitrary, but polymers containing 50 or fewer nucleotides are generally called oligonucleotides. A longer nucleic acid is called a **polynucleotide.**

The Properties of Nucleotide Bases Affect the Three-Dimensional Structure of Nucleic Acids

Free pyrimidines and purines are weakly basic compounds and are thus called bases. They have a variety of chemical properties that affect the structure, and ultimately the function, of nucleic acids. The purines and pyrimidines common in DNA and RNA are highly conjugated molecules (Fig. 10–2), a property with important consequences for the structure, electron distribution, and light absorption of nucleic acids. Resonance among atoms in the ring gives most of the bonds partial double-bond character. One result is that pyrimidines are planar molecules; purines are very nearly planar, with a slight pucker. Free pyrimidine and purine bases may exist in two or more tautomeric forms depending on the pH. Uracil, for example, occurs in lactam, lactim, and double lactim forms (Fig. 10–9). The structures shown in Figure 10–2 are the tautomers that predominate at pH 7.0. As a result of resonance, all nucleotide bases absorb UV light, and nucleic acids are characterized by a strong absorption at wavelengths near 260 nm (Fig. 10–10).

The purine and pyrimidine bases are hydrophobic and relatively insoluble in water at the near-neutral pH of the cell. At acidic or alkaline pH the bases become charged and their solubility in water increases. Hydrophobic stacking interactions in which two or more bases are positioned with the planes of their rings parallel (similar to a stack of coins) are one of two important modes of interaction between bases in nucleic acids. The stacking also involves a combination of van der Waals and dipole-dipole interactions between the bases. Base stacking helps to minimize contact of the bases with water, and base-stacking interactions are very important in stabilizing the three-dimensional structure of nucleic acids, as described later.

Lactam ⇌ Lactim ⇌ Double lactim

figure 10–9
Tautomeric forms of uracil. The lactam form predominates at pH 7.0; the other forms become more prominent as pH decreases. The other free pyrimidines and the free purines also have tautomeric forms, but they are more rarely encountered.

figure 10–10
Absorption spectra of the common nucleotides. The spectra are shown as the variation in molar absorption coefficient with wavelength. The molar absorption coefficients at 260 nm and pH 7.0 (ϵ_{260}) are listed in the table. The spectra of corresponding ribonucleotides and deoxyribonucleotides, as well as the nucleosides, are essentially identical. For mixtures of nucleotides, a wavelength of 260 nm (dashed vertical lines) is used for absorption measurements.

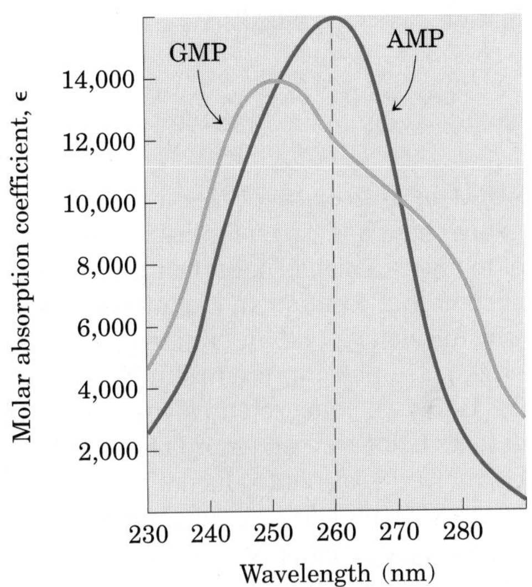

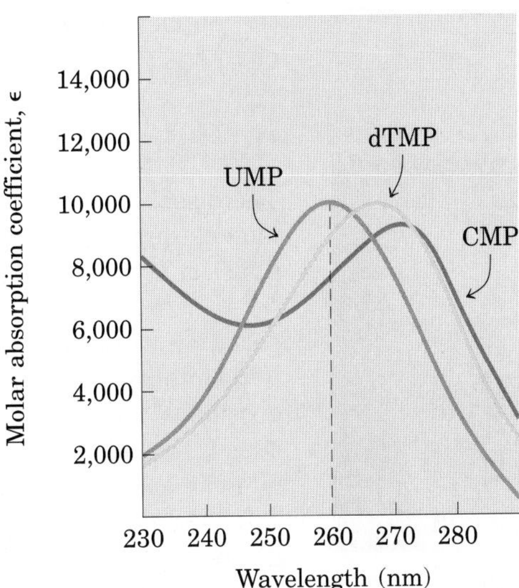

Molar absorption coefficient at 260 nm, ϵ_{260} ($M^{-1}cm^{-1}$)	
AMP	15,400
GMP	11,700
CMP	7,500
UMP	9,900
dTMP	9,200

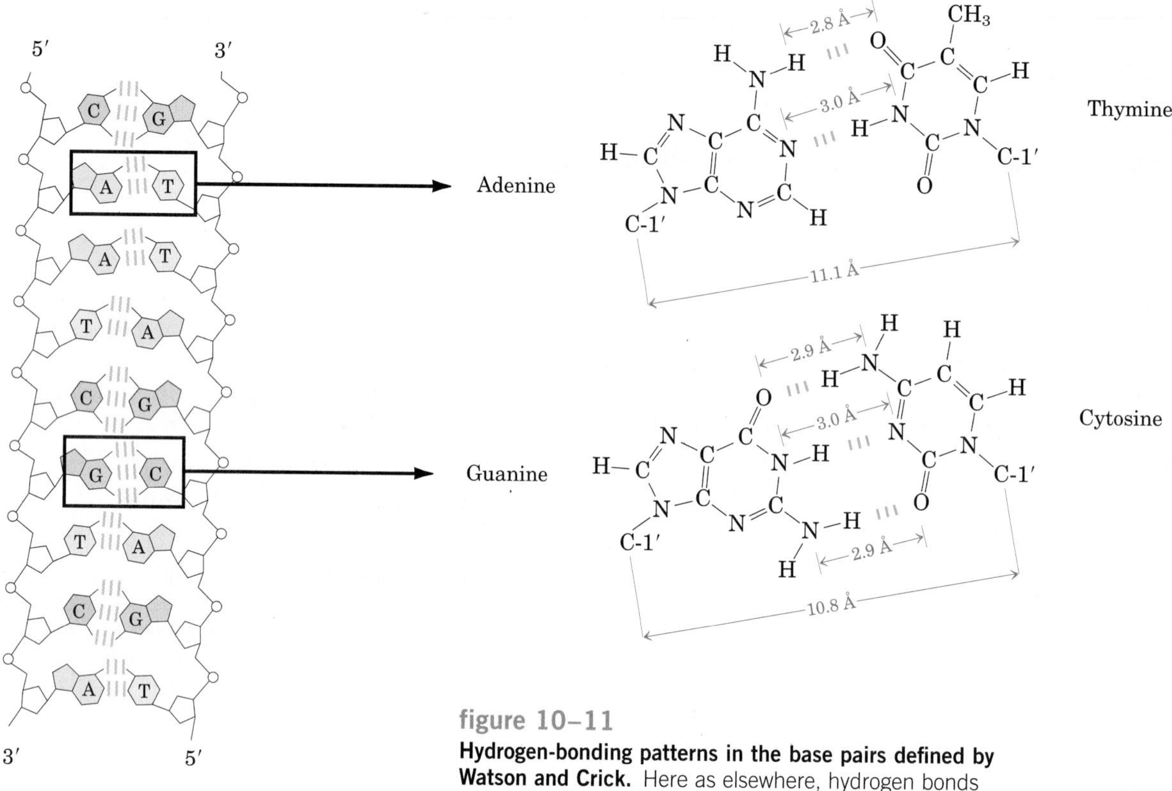

figure 10–11
Hydrogen-bonding patterns in the base pairs defined by Watson and Crick. Here as elsewhere, hydrogen bonds are represented by sets of blue lines.

The most important functional groups of pyrimidines and purines are ring nitrogens, carbonyl groups, and exocyclic amino groups. Hydrogen bonds involving the amino and carbonyl groups are the second important mode of interaction between bases in nucleic acid molecules. Hydrogen bonds between bases permit a complementary association of two (and occasionally three or four) strands of nucleic acid. The most important hydrogen-bonding patterns are those defined by James D. Watson and Francis Crick in 1953, in which A bonds specifically to T (or U) and G bonds to C (Fig. 10–11). These two types of **base pairs** predominate in double-stranded DNA and RNA, and the tautomers shown in Figure 10–2 are responsible for these patterns. It is this specific pairing of bases that permits the duplication of genetic information, as we shall discuss later in this chapter.

James Watson

Francis Crick

Nucleic Acid Structure

The discovery of the structure of DNA by Watson and Crick in 1953 was a momentous event in science, an event that gave rise to entirely new disciplines and influenced the course of many established ones. Our present understanding of the storage and utilization of a cell's genetic information is based on work made possible by this discovery, and an outline of how genetic information is processed by the cell is now a prerequisite for the discussion of any area of biochemistry. Here, we concern ourselves with DNA structure itself, the events that led to its discovery, and more recent refinements in our understanding. RNA structure is also introduced.

As in the case of protein structure (Chapter 6), it is sometimes useful to describe nucleic acid structure in terms of hierarchical levels of complexity (primary, secondary, tertiary). The primary structure of a nucleic acid is its covalent structure and nucleotide sequence. Any regular, stable structure taken up by some or all of the nucleotides in a nucleic acid can be referred to as secondary structure. All structures considered in the remainder of this chapter fall under the heading of secondary structure. The complex folding of large chromosomes within eukaryotic chromatin and bacterial nucleoids is generally considered tertiary structure; this is discussed in Chapter 24.

DNA Stores Genetic Information

The biochemical investigation of DNA began with Friedrich Miescher, who carried out the first systematic chemical studies of cell nuclei. In 1868 Miescher isolated a phosphorus-containing substance, which he called "nuclein," from the nuclei of pus cells (leukocytes) obtained from discarded surgical bandages. He found nuclein to consist of an acidic portion, which we know today as DNA, and a basic portion, protein. Miescher later found a similar acidic substance in the heads of sperm cells from salmon. Although he partially purified nuclein and studied its properties, the covalent (primary) structure of DNA (as shown in Fig. 10–7) was not known with certainty until the late 1940s.

Miescher and many others suspected that nuclein (nucleic acid) was associated in some way with cell inheritance, but the first direct evidence that DNA is the bearer of genetic information came in 1944 through a discovery made by Oswald T. Avery, Colin MacLeod, and Maclyn McCarty. These investigators found that DNA extracted from a virulent (disease-causing) strain of the bacterium *Streptococcus pneumoniae*, also known as pneumococcus, genetically transformed a nonvirulent strain of this organism into a virulent form (Fig. 10–12). Avery and his colleagues concluded that the DNA extracted from the virulent strain carried the inheritable genetic message for virulence. Not everyone accepted these conclusions, because protein impurities present in the DNA could have been the carrier of the genetic information. This possibility was soon eliminated by the finding that treatment of the DNA with proteolytic enzymes did not destroy the transforming activity, but treatment with deoxyribonucleases (DNA-hydrolyzing enzymes) did.

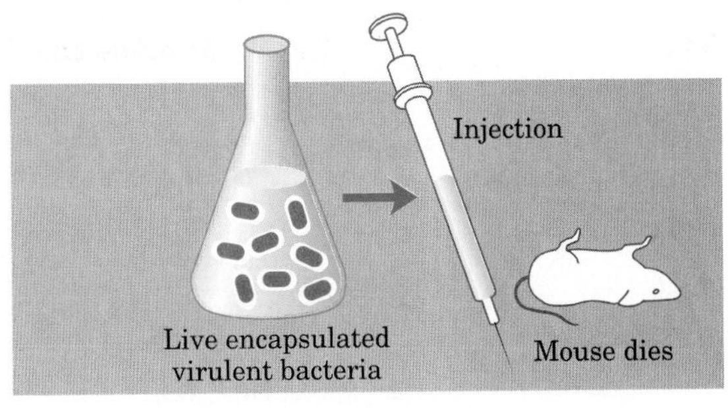

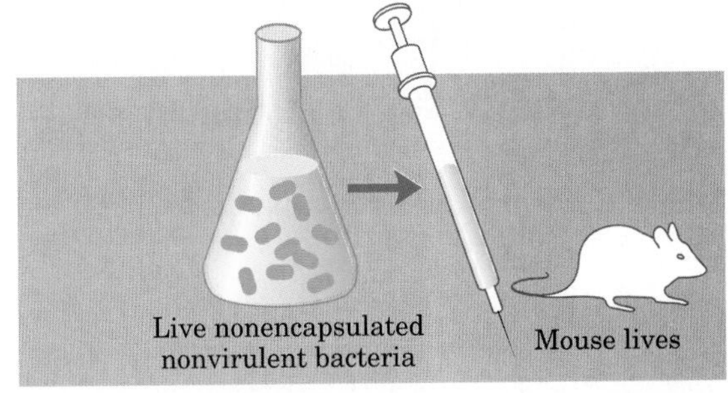

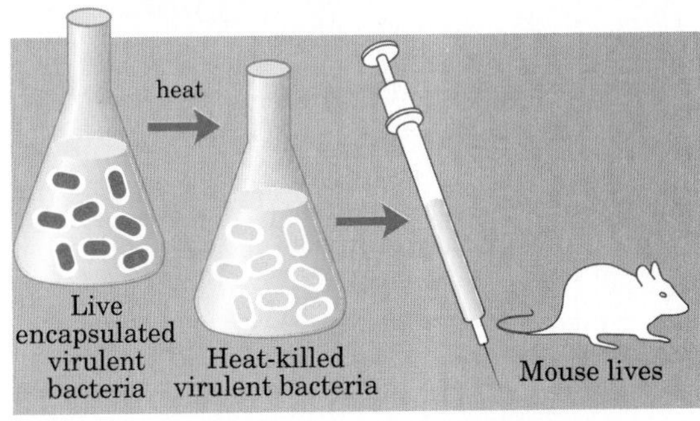

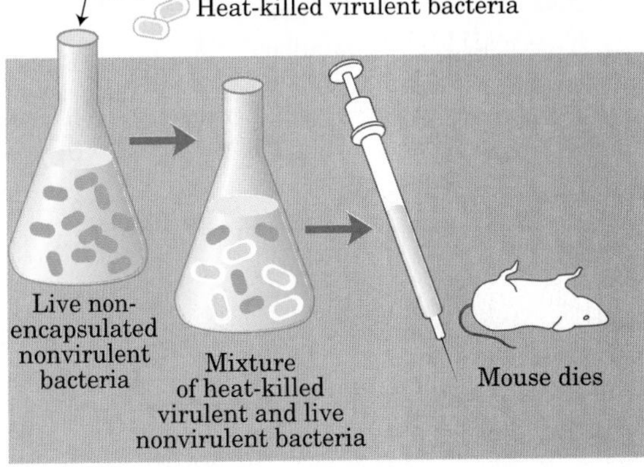

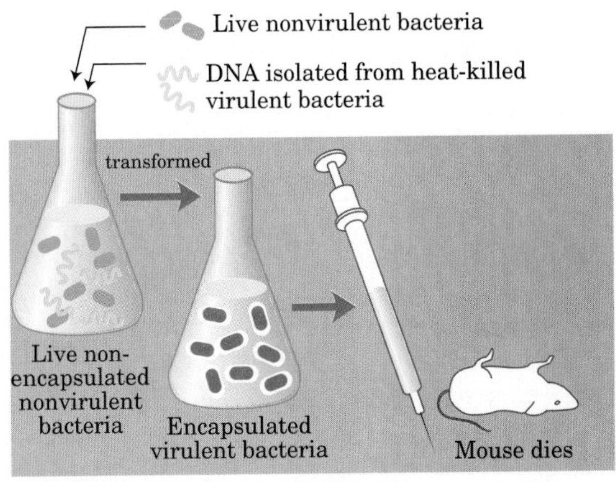

figure 10–12

The Avery-MacLeod-McCarty experiment. When injected into mice, the encapsulated strain of pneumococcus **(a)** is lethal, whereas the nonencapsulated strain **(b),** like the heat-killed encapsulated strain **(c),** is harmless. Earlier research by the bacteriologist Frederick Griffith had shown that adding heat-killed virulent bacteria (harmless to mice) to a live nonvirulent strain permanently transformed the latter into lethal, virulent, encapsulated bacteria **(d).** He concluded that a transforming factor in the heat-killed virulent bacteria had gained entrance into the live nonvirulent bacteria and rendered them virulent and encapsulated.

Avery and his colleagues identified the transforming factor as DNA. They extracted the DNA from heat-killed virulent pneumococci, removing the protein as completely as possible, and added this DNA to nonvirulent bacteria **(e).** The nonvirulent pneumococci were permanently transformed into a virulent strain. The DNA evidently gained entrance into the nonvirulent bacteria, and the genes for virulence and capsule formation became incorporated into the chromosome of the nonvirulent strain. All subsequent generations of these bacteria were virulent and encapsulated.

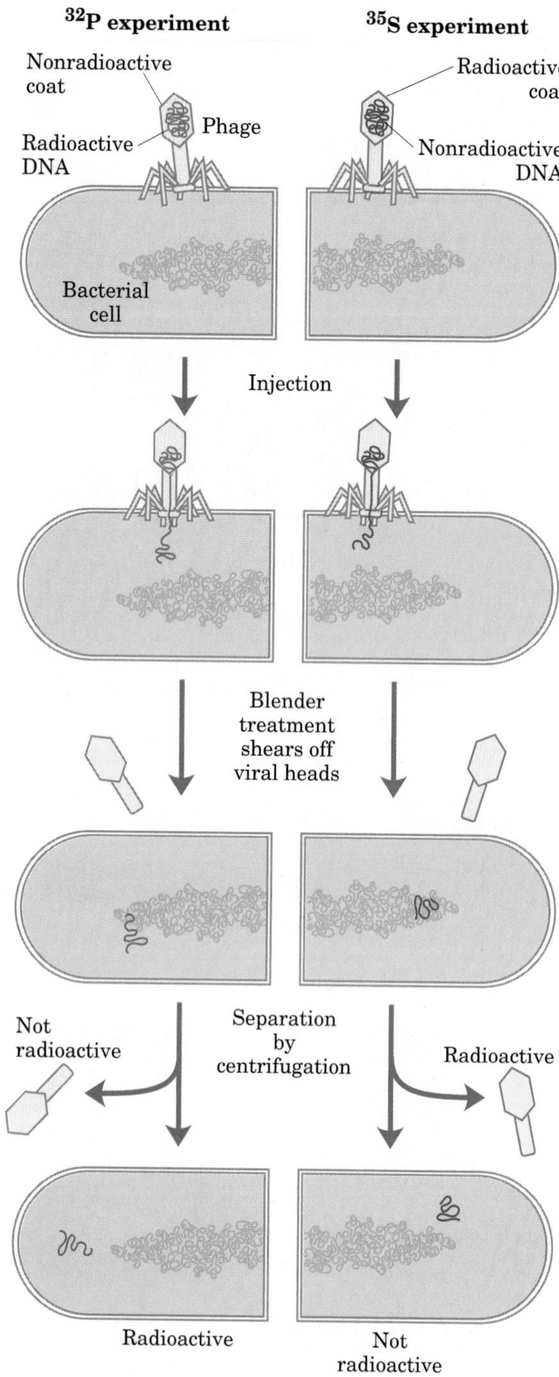

figure 10–13

The Hershey-Chase experiment. Two batches of isotopically labeled bacteriophage T2 particles were prepared. One was labeled with ^{32}P in the phosphate groups of the DNA, the other with ^{35}S in the sulfur-containing amino acids of the protein coats (capsids). (Note that DNA contains no sulfur and viral protein contains no phosphorus.) The two batches of labeled phage were then allowed to infect separate suspensions of unlabeled bacteria. Each suspension of phage-infected cells was agitated in a blender to shear the viral capsids from the bacteria. The bacteria and empty viral coats (called ghosts) were then separated by centrifugation. The cells infected with the ^{32}P-labeled phage were found to contain ^{32}P, indicating that the labeled viral DNA had entered the cells; the viral ghosts contained no radioactivity. The cells infected with ^{35}S-labeled phage were found to have no radioactivity after blender treatment, but the viral ghosts contained ^{35}S. Progeny virus particles (not shown) were produced in both batches of bacteria some time after the viral coats were removed, indicating that the genetic message for their replication had been introduced by viral DNA, not by viral protein.

A second important experiment provided independent evidence that DNA carries genetic information. In 1952 Alfred D. Hershey and Martha Chase used radioactive phosphorus (^{32}P) and radioactive sulfur (^{35}S) tracers to show that when the bacterial virus (bacteriophage) T2 infects its host cell, *Escherichia coli,* it is the phosphorus-containing DNA of the viral particle, not the sulfur-containing protein of the viral coat, that enters the host cell and furnishes the genetic information for viral replication (Fig. 10–13).

These important early experiments and many other lines of evidence have shown that DNA is the exclusive chromosomal component bearing the genetic information of living cells.

DNA Molecules Have Distinctive Base Compositions

A most important clue to the structure of DNA came from the work of Erwin Chargaff and his colleagues in the late 1940s. They found that the four nucleotide bases of DNA occur in different ratios in the DNAs of different organisms and that the amounts of certain bases are closely related. These data, collected from DNAs of a great many different species, led Chargaff to the following conclusions:

1. The base composition of DNA generally varies from one species to another.
2. DNA specimens isolated from different tissues of the same species have the same base composition.
3. The base composition of DNA in a given species does not change with an organism's age, nutritional state, or changing environment.
4. In *all* cellular DNAs, regardless of the species, the number of adenosine residues is equal to the number of thymidine residues (that is, A = T), and the number of guanosine residues is equal to the number of cytidine residues (G = C). From these relationships it follows that the sum of the purine residues equals the sum of the pyrimidine residues; that is, A + G = T + C.

These quantitative relationships, sometimes called "Chargaff's rules," were confirmed by many subsequent researchers. They were a key to establishing the three-dimensional structure of DNA and yielded clues to how genetic information is encoded in DNA and passed from one generation to the next.

DNA Is a Double Helix

To shed more light on the structure of DNA, Rosalind Franklin and Maurice Wilkins used the powerful method of x-ray diffraction (see Box 6–3) to analyze DNA fibers. They showed in the early 1950s that DNA produces a characteristic x-ray diffraction pattern (Fig. 10–14). From this pattern it was deduced that DNA molecules are helical with two periodicities along their long axis, a primary one of 3.4 Å and a secondary one of 34 Å. The problem then was to formulate a three-dimensional model of the DNA molecule that could account not only for the x-ray diffraction data but also for the specific A = T and G = C base equivalences discovered by Chargaff and for the other chemical properties of DNA.

Rosalind Franklin
1920–1958

Maurice Wilkins

In 1953 Watson and Crick postulated a three-dimensional model of DNA structure that accounted for all the available data. It consists of two helical DNA chains wound around the same axis to form a right-handed double helix (see Box 6–1 for an explanation of the right- or left-handed sense of a helical structure). The hydrophilic backbones of alternating deoxyribose and phosphate groups are on the outside of the double helix, facing the surrounding water. The furanose ring of each deoxyribose is in the C-2′ endo conformation. The purine and pyrimidine bases of both strands are stacked inside the double helix, with their hydrophobic and nearly planar ring structures very close together and perpendicular to the long axis. The offset pairing of the two strands creates a **major groove** and **minor groove** on the surface of the duplex (Fig. 10–15, next page). Each nucleotide base of one strand is paired in the same plane with a base of the other strand. Watson and Crick found that the hydrogen-bonded base pairs illustrated in Figure 10–11, G with C and A with T, are those that fit best within the structure, providing a rationale for Chargaff's rule that in any DNA, G = C and A = T. It is important to note that three hydrogen bonds can form between G and C, symbolized G≡C, but only two can form between A and T, symbolized A=T. This is one reason for the finding that separation of paired DNA strands is more difficult the higher the ratio of G≡C to A=T base pairs. Other pairings of bases tend (to varying degrees) to destabilize the double-helical structure.

When Watson and Crick constructed their model, they had to decide at the outset whether the strands of DNA should be **parallel** or **antiparallel**—whether their 5′,3′-phosphodiester bonds should run in the same or opposite directions. Antiparallel produced the most convincing model, and later work with DNA polymerases (Chapter 25) provided experimental evidence that the strands are indeed antiparallel, a finding ultimately confirmed by x-ray analysis.

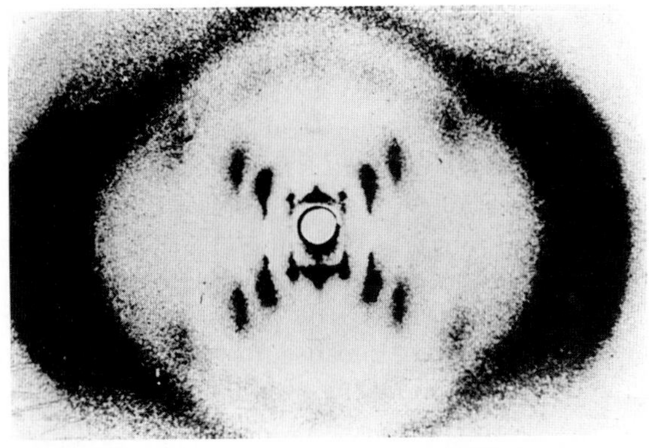

figure 10–14
X-ray diffraction pattern of DNA. The spots forming a cross in the center denote a helical structure. The heavy bands at the left and right arise from the recurring bases.

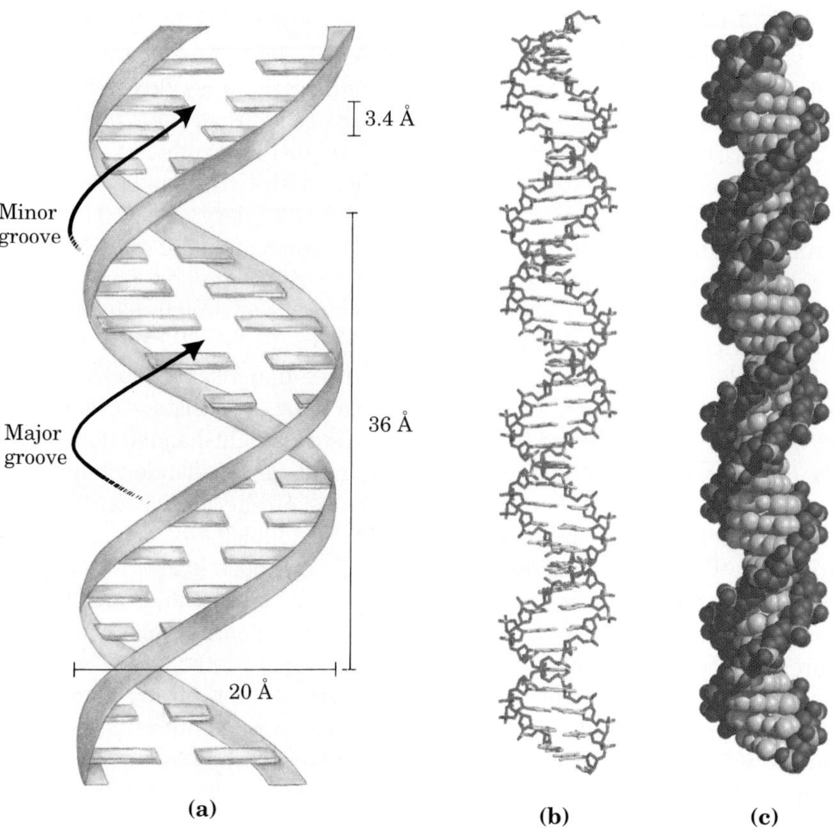

figure 10–15

Watson-Crick model for the structure of DNA. The original model of Watson and Crick proposed 10 base pairs or 34 Å (3.4 nm) per turn of the helix; subsequent measurements revealed 10.5 base pairs or 36 Å (3.6 nm) per turn. **(a)** Schematic representation, showing dimensions of the helix. **(b)** Stick representation showing the backbone and stacking of the bases. **(c)** Space-filling model.

To account for the periodicities observed in the x-ray diffraction patterns of DNA fibers, Watson and Crick manipulated molecular models to arrive at a structure in which the vertically stacked bases inside the double helix would be 3.4 Å apart; the secondary repeat distance of about 34 Å was accounted for by the presence of 10 base pairs in each complete turn of the double helix. In aqueous solution the structure differs slightly from that in fibers, having 10.5 base pairs per helical turn (Fig. 10–15).

As Figure 10–16 shows, the two antiparallel polynucleotide chains of double-helical DNA are not identical in either base sequence or composition. Instead they are **complementary** to each other. Wherever adenine occurs in one chain, thymine is found in the other; similarly, wherever guanine occurs in one chain, cytosine is found in the other.

The DNA double helix or duplex is held together by two forces, as described earlier: hydrogen bonding between complementary base pairs (Fig. 10–11) and base-stacking interactions. The complementarity between the DNA strands is attributable to the hydrogen bonding between base pairs. The base-stacking interactions, which are largely nonspecific with respect to the identity of the stacked bases, make the major contribution to the stability of the double helix.

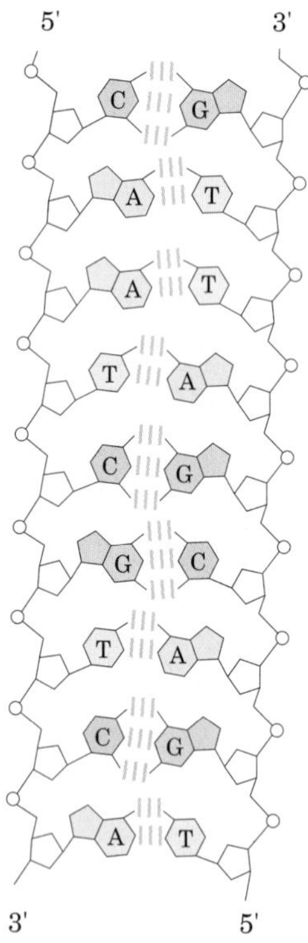

figure 10–16

Complementarity of strands in the DNA double helix. The complementary antiparallel strands of DNA follow the pairing rules proposed by Watson and Crick. The base-paired antiparallel strands differ in base composition: the left strand has the composition A_3 T_2 G_1 C_3; the right, A_2 T_3 G_3 C_1. They also differ in sequence when each chain is read in the 5′→3′ direction. Note the base equivalences: A = T and G = C in the duplex.

The important features of the double-helical model of DNA structure are supported by much chemical and biological evidence. Moreover, the model immediately suggested a mechanism for the transmission of genetic information. The essential feature of the model is the complementarity of the two DNA strands. As Watson and Crick were able to see, well before confirmatory data became available, this structure could logically be replicated by (1) separating the two strands and (2) synthesizing a complementary strand for each. Because nucleotides in each new strand are joined in a sequence specified by the base-pairing rules stated above, each preexisting strand functions as a template to guide the synthesis of one complementary strand (Fig. 10–17). These expectations were experimentally confirmed, inaugurating a revolution in our understanding of biological inheritance.

DNA Can Occur in Different Three-Dimensional Forms

DNA is a remarkably flexible molecule. Considerable rotation is possible around a number of bonds in the sugar–phosphate (phosphodeoxyribose) backbone, and thermal fluctuation can produce bending, stretching, and unpairing (melting) of the strands. Many significant deviations from the Watson-Crick DNA structure are found in cellular DNA, and some or all of these may play important roles in DNA metabolism. These structural variations generally do not affect the key properties of DNA defined by Watson and Crick: strand complementarity, antiparallel strands, and the requirement for A═T and G≡C base pairs.

Structural variation in DNA reflects three things: the different possible conformations of the deoxyribose, rotation about the contiguous bonds that make up the phosphodeoxyribose backbone (Fig. 10–18a), and free rotation about the C-1′–*N*-glycosyl bond (Fig. 10–18b). Because of steric constraints, purines in purine nucleotides are restricted to two stable conformations with respect to deoxyribose, called syn and anti (Fig. 10–18b). Pyrimidines are generally restricted to the anti conformation because of steric interference between the sugar and the carbonyl oxygen at C-2 of the pyrimidine.

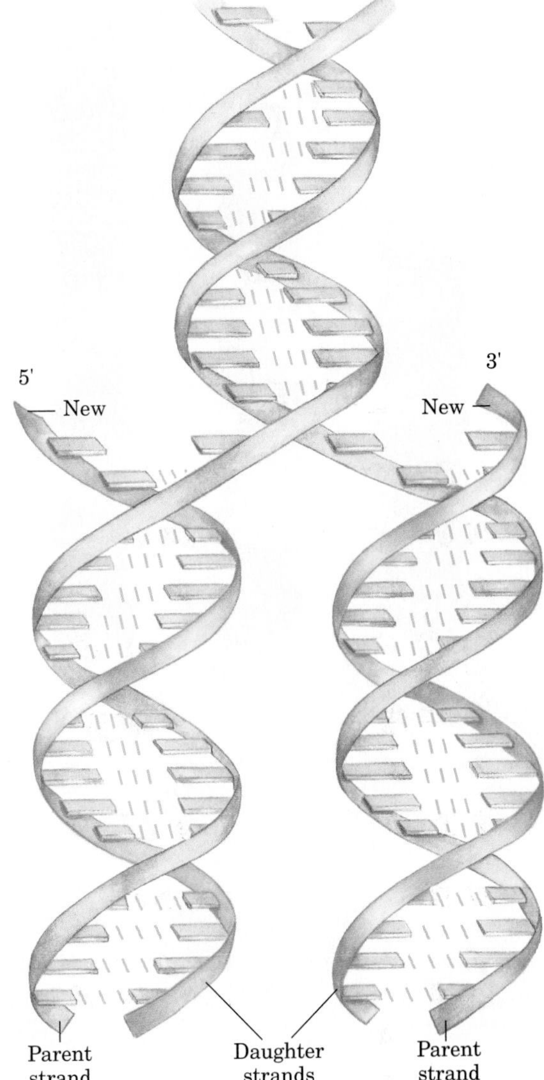

figure 10–17
Replication of DNA as suggested by Watson and Crick. The preexisting or "parent" strands become separated, and each is the template for biosynthesis of a complementary "daughter" strand (in red).

figure 10–18
Structural variation in DNA. (a) The conformation of a nucleotide in DNA is affected by rotation about seven different bonds. Six of the bonds rotate freely. The limited rotation about bond 4 gives rise to ring pucker, in which one of the atoms in the five-membered furanose ring is out of the plane described by the other four. This conformation is described as endo or exo, depending on whether the atom is displaced to the same side of the plane as C-5′ or to the opposite side (see also Fig. 10–3b). **(b)** For purine bases in nucleotides, only two conformations with respect to the attached ribose units are sterically permitted, anti or syn. Pyrimidines generally occur in the anti conformation.

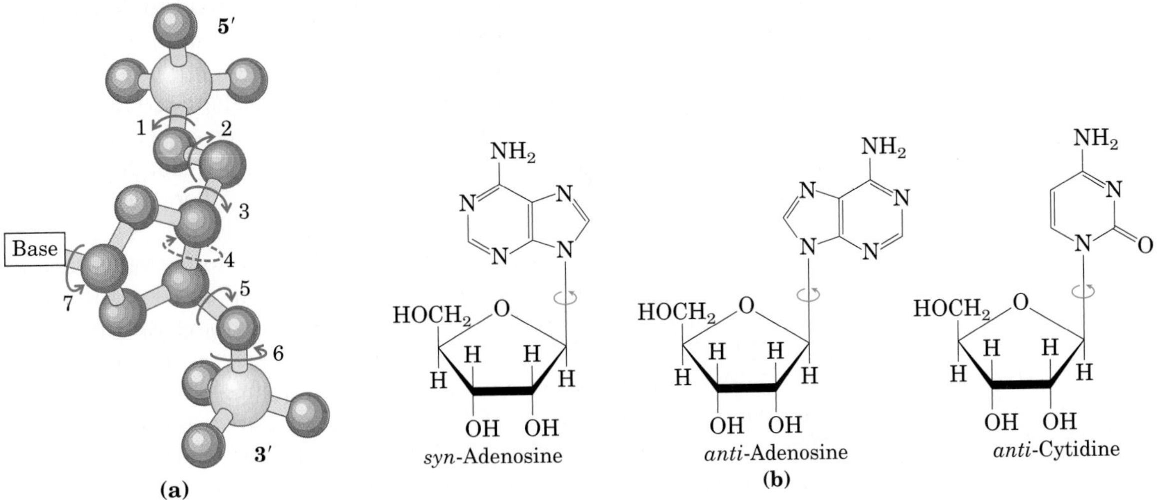

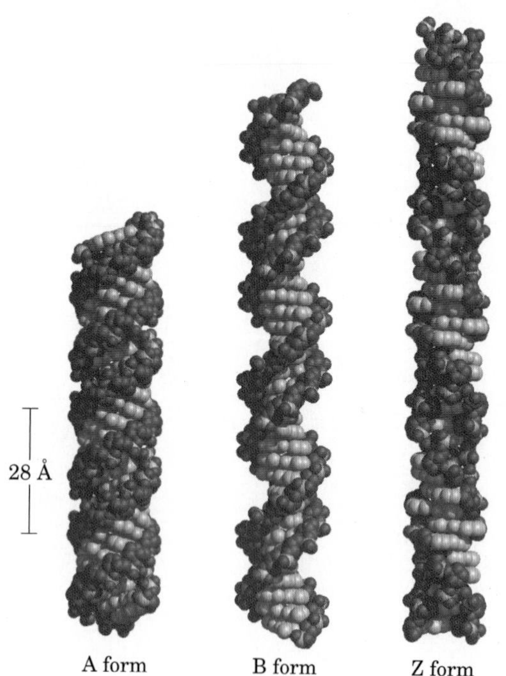

	A form	B form	Z form
Helical sense	Right handed	Right handed	Left handed
Diameter	~26 Å	~20 Å	~18 Å
Base pairs per helical turn	11	10.5	12
Helix rise per base pair	2.6 Å	3.4 Å	3.7 Å
Base tilt normal to the helix axis	20°	6°	7°
Sugar pucker conformation	C-3′ endo	C-2′ endo	C-2′ endo for pyrimidines; C-3′ endo for purines
Glycosyl bond conformation	Anti	Anti	Anti for pyrmidines; syn for purines

figure 10–19
Comparison of A, B, and Z forms of DNA. Each structure shown here has 36 base pairs.The bases are shown in white, the phosphate atoms in yellow, and the riboses and phosphate oxygens in blue. Blue is the standard color used to represent DNA strands in later chapters. The table summarizes some properties of the three forms of DNA.

The Watson-Crick structure is also referred to as **B-form DNA,** or B-DNA. The B form is the most stable structure for a random-sequence DNA molecule under physiological conditions and is therefore the standard point of reference in any study of the properties of DNA. Two structural variants that have been well characterized in crystal structures are the **A** and **Z forms.** These three DNA conformations are shown in Figure 10–19, with a summary of their properties. The A form is favored in many solutions that are relatively devoid of water. The DNA is still arranged in a right-handed double helix, but the helix is wider and the number of base pairs per helical turn is 11, rather than 10.5 as in B-DNA. The plane of the base pairs in A-DNA is tilted about 20° with respect to the helix axis. These structural changes deepen the major groove while making the minor groove shallower. The reagents used to promote crystallization of DNA tend to dehydrate it, and thus most short DNA molecules tend to crystallize in the A form.

Z-form DNA is a more radical departure from the B structure; the most obvious distinction is the left-handed helical rotation. There are 12 base pairs per helical turn, and the structure appears more slender and elongated. The DNA backbone takes on a zigzag appearance. Certain nucleotide sequences fold into left-handed Z helices much more readily than others. Prominent examples are sequences in which pyrimidines alternate with purines, especially alternating C and G or 5-methyl-C and G residues. To form the left-handed helix in Z-DNA, the purine residues flip to the syn conformation, alternating with pyrimidines in the anti conformation. The major groove is barely apparent in Z-DNA, and the minor groove is narrow and deep.

Whether A-DNA occurs in cells is uncertain, but there is evidence for some short stretches (tracts) of Z-DNA in both prokaryotes and eukaryotes. These Z-DNA tracts may play a role (as yet undefined) in the regulation of the expression of some genes or in genetic recombination.

Certain DNA Sequences Adopt Unusual Structures

A number of other sequence-dependent structural variations have been detected within larger chromosomes that may affect the function and metabolism of the DNA segments in their immediate vicinity. For example, bends occur in the DNA helix whenever four or more adenosine residues appear sequentially in one strand. Six adenosines in a row produce a bend of about 18°. The bending observed with this and other sequences may be important in the binding of some proteins to DNA.

A rather common type of DNA sequence is a **palindrome.** A palindrome is a word, phrase, or sentence that is spelled identically reading forward or backward; two examples are ROTATOR and NURSES RUN. The term is applied to regions of DNA with **inverted repeats** of base sequence having twofold symmetry over two strands of DNA (Fig. 10–20). Such sequences are self-complementary within each strand and therefore have the potential to form hairpin or cruciform (cross-shaped) structures (Fig. 10–21). When the inverted repeat occurs within each individual strand of the DNA, the sequence is called a **mirror repeat.** Mirror repeats do not have complementary sequences within the same strand and cannot form hairpin or cruciform structures. Sequences of these types are found in virtually every large DNA molecule and can encompass a few base pairs or thousands. The extent to which palindromes occur as cruciforms in cells is not known, although some cruciform structures have been demonstrated in vivo in *E. coli*. Self-complementary sequences cause isolated single strands of DNA (or RNA) in solution to fold into complex structures containing multiple hairpins.

Several unusual DNA structures involve three or even four DNA strands. These structural variations merit investigation because there is a tendency for many of them to appear at sites where important events in

Palindrome

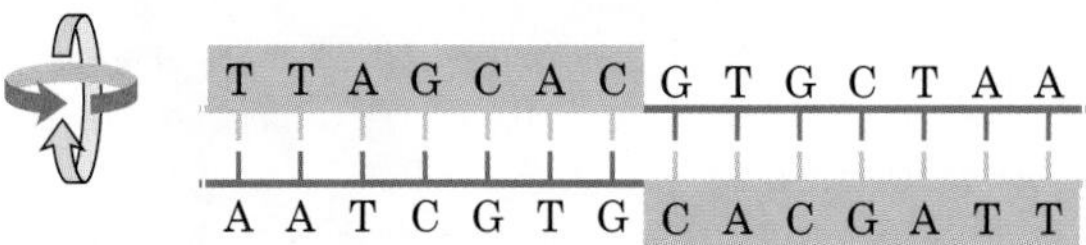

Mirror repeat

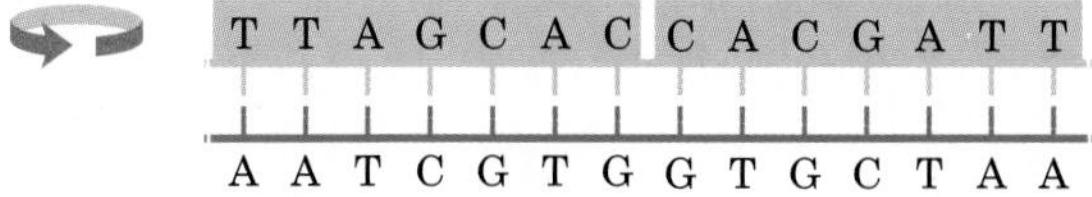

figure 10–20
Palindromes and mirror repeats. Palindromes are sequences of double-stranded nucleic acids with twofold symmetry. In order to superimpose one repeat (shaded sequence) on the other, it must be rotated 180° about the horizontal axis then 180° about the vertical axis, as shown by the colored arrows. A mirror repeat, on the other hand, has a symmetric sequence within each strand. Superimposing one repeat on the other requires only a single 180° rotation about the vertical axis.

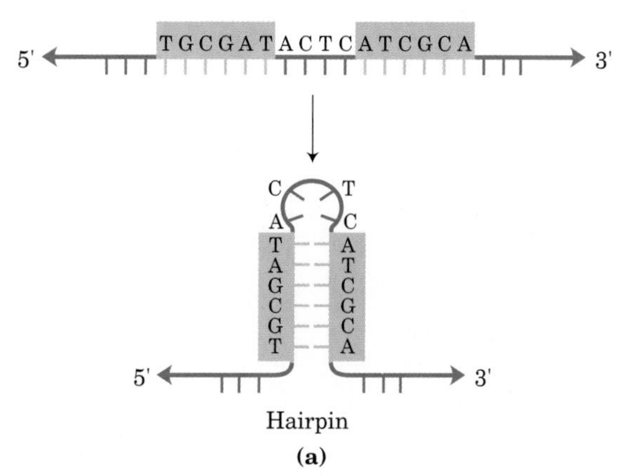

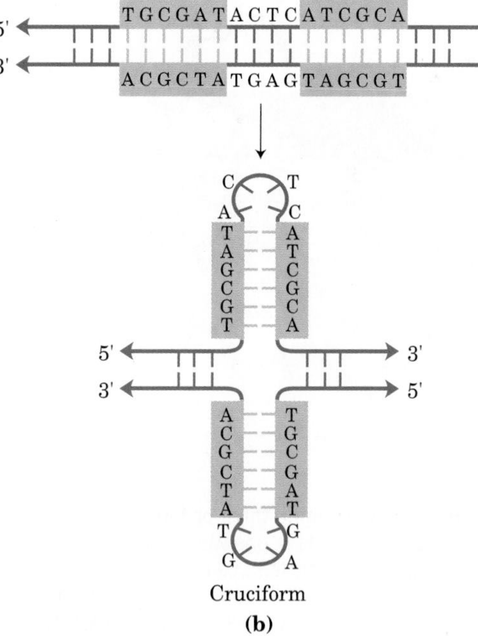

figure 10–21
Hairpins and cruciforms. Palindromic DNA (or RNA) sequences can form alternative structures with intrastrand base pairing. **(a)** When only a single DNA (or RNA) strand is involved, the structure is called a hairpin. **(b)** When both strands of a duplex DNA are involved, it is called a cruciform. Blue shading highlights asymmetric sequences that can pair with the complementary sequence either in the same strand or in the complementary strand.

DNA metabolism (replication, recombination, transcription) are initiated or regulated. Nucleotides participating in a Watson-Crick base pair (Fig. 10–11) can form a number of additional hydrogen bonds, particularly with functional groups arrayed in the major groove. For example, a cytidine residue (if protonated) can pair with the guanosine residue of a G≡C nucleotide pair, and a thymidine can pair with the adenosine of an A═T pair (Fig. 10–22). The N-7, O^6, and N^6 of purines, the atoms that participate in the hydrogen bonding of triplex DNA, are often referred to as **Hoogsteen positions,** and the non-Watson-Crick pairing is called **Hoogsteen pairing,**

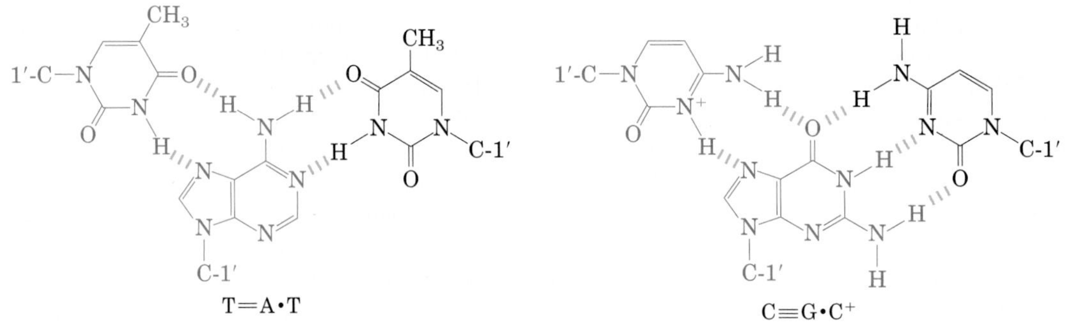

T═A•T

C≡G•C⁺

(a)

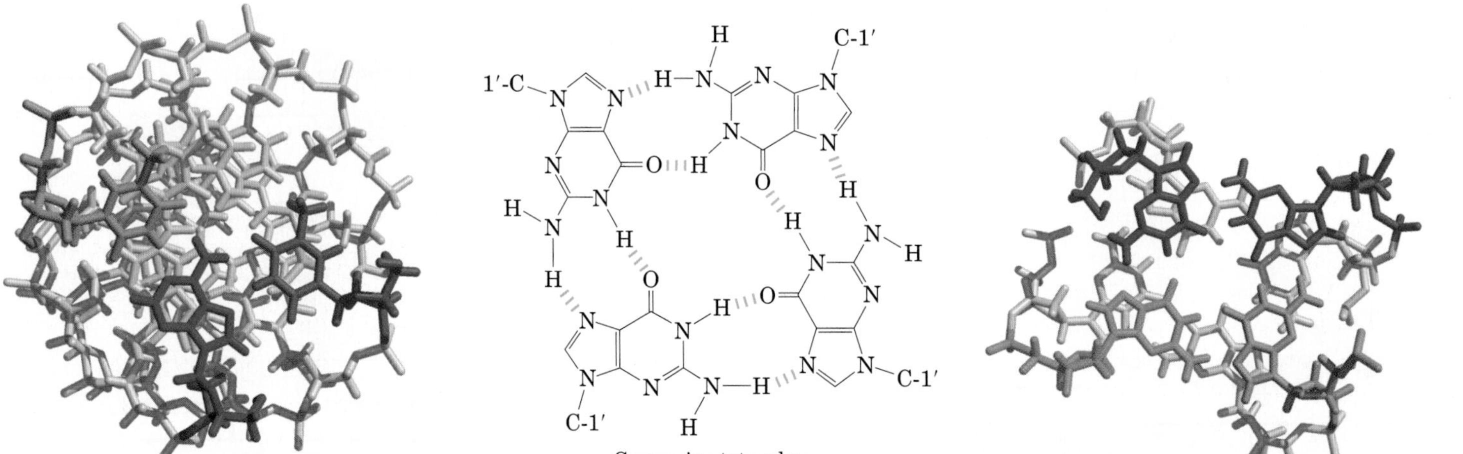

Guanosine tetraplex

(b) (c) (d)

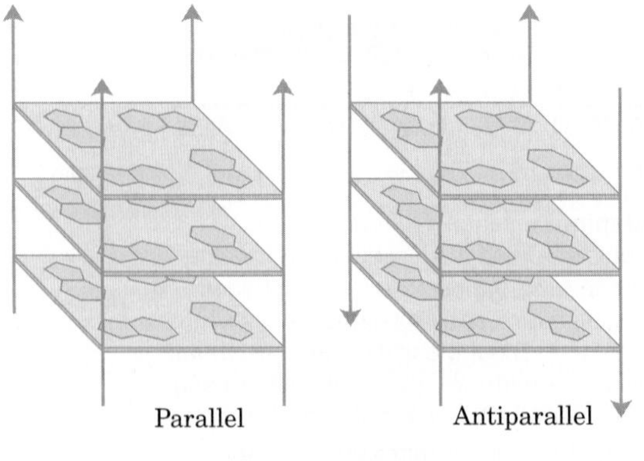

Parallel Antiparallel

(e)

figure 10–22

DNA structures containing three or four DNA strands. **(a)** Base-pairing patterns in one well-characterized form of triplex DNA. The Hoogsteen pair in each case is shown in red. **(b)** Triple-helical DNA containing two pyrimidine strands (poly(T)) and one purine strand (poly(A)). The dark blue and light blue strands are antiparallel and related by normal Watson-Crick base-pairing patterns. The third (all-pyrimidine) strand (purple) is parallel to the purine strand and paired through non-Watson-Crick hydrogen bonds. The triplex is viewed end-on, with five triplets shown. Only the triplet closest to the viewer is colored. **(c)** Base-pairing pattern in the guanosine tetraplex structure. **(d)** Two successive tetraplets from a G tetraplex structure, viewed end-on with the one closest to the viewer in color. **(e)** Possible variants in the orientation of strands in a G tetraplex.

after Karst Hoogsteen who in 1963 first recognized the potential for these unusual pairings. Hoogsteen pairing allows the formation of **triplex DNAs.** The triplexes shown in Figure 10–22a, b, are most stable at low pH, because the C≡G • C^+ triplet requires a protonated cytosine. In the triplex, the pK_a of this cytosine is >7.5, altered from its normal value of 4.2. The triplexes also form most readily within long sequences containing only pyrimidines or only purines in a given strand. Some triplex DNAs contain two pyrimidine strands and one purine strand, others contain two purine strands and one pyrimidine strand.

Four DNA strands can also pair to form a tetraplex (also called a quadruplex), but this occurs readily only for DNA sequences with a very high proportion of guanosine residues (Fig. 10–22c, d). The guanosine or **G tetraplex** is quite stable over a wide range of conditions. The orientation of strands in the tetraplex can vary as shown in Figure 10–22e.

A particularly exotic DNA structure, known as **H-DNA,** is found in polypyrimidine or polypurine tracts that also incorporate a mirror repeat. A simple example is a long stretch of alternating T and C residues (Fig. 10–23). The H-DNA structure features the triple-stranded form illustrated in Figure 10–22a, b. Two of the three strands in the H-DNA triple helix contain pyrimidines and the third contains purines.

In the DNA of living cells, sites recognized by many sequence-specific DNA-binding proteins (Chapter 28) are arranged as palindromes, and polypyrimidine or polypurine sequences that can form triple helices or even H-DNA are found within regions involved in the regulation of expression of some eukaryotic genes. In principle, synthetic DNA strands designed to pair with these sequences to form triplex DNA could disrupt gene expression. This approach to controlling cellular metabolism is of growing commercial interest for its potential application in medicine and agriculture.

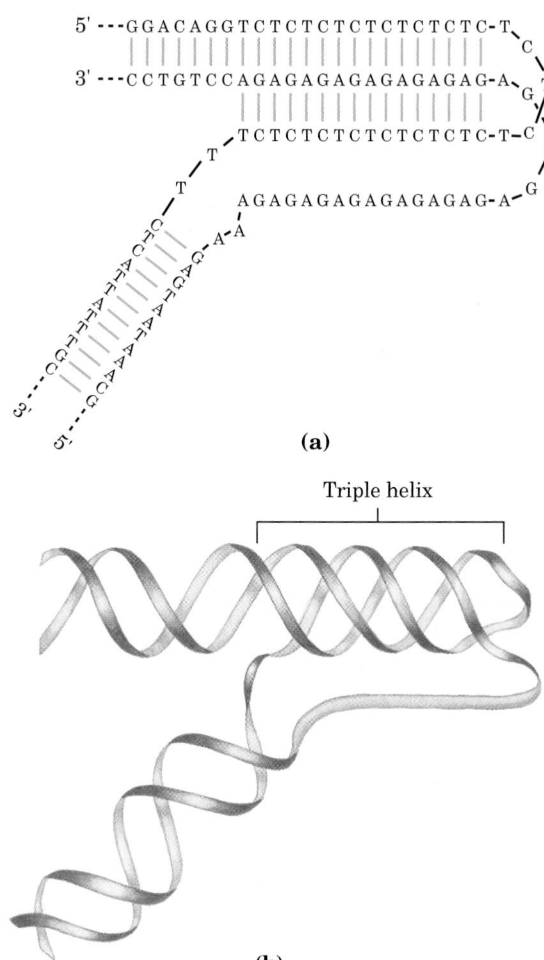

figure 10–23
H-DNA. A sequence of alternating T and C residues can be considered a mirror repeat centered about a central T or C **(a).** These sequences form an unusual structure in which the strands in one half of the mirror repeat are separated and the pyrimidine-containing strand (alternating T and C residues) folds back on the other half of the repeat to form a triple helix **(b).** The purine strand (alternating A and G residues) is left unpaired. This structure produces a sharp bend in the DNA.

Messenger RNAs Code for Polypeptide Chains

We now turn our attention briefly from DNA structure to the expression of the genetic information that it contains. RNA, the second major form of nucleic acid in cells, has many functions. In gene expression, RNA acts as an intermediary by using the information encoded in DNA to specify the amino acid sequence of a functional protein.

Given that the DNA of eukaryotes is largely confined to the nucleus whereas protein synthesis occurs on ribosomes in the cytoplasm, some molecule other than DNA must carry the genetic message from the nucleus to the cytoplasm. As early as the 1950s, RNA was considered the logical candidate: RNA is found in both the nucleus and the cytoplasm, and an increase in protein synthesis is accompanied by an increase in the amount of cytoplasmic RNA and an increase in its rate of turnover. These and other observations led several researchers to suggest that RNA carries genetic information from DNA to the protein biosynthetic machinery of the ribosome. In 1961, François Jacob and Jacques Monod presented a unified (and essentially correct) picture of many aspects of this process. They proposed the name messenger RNA (mRNA) for that portion of the total cellular RNA carrying the genetic information from DNA to the ribosomes, where the messengers provide the templates that specify amino acid sequences in polypeptide chains. Although mRNAs from different genes can vary greatly in length, the mRNAs from a particular gene generally have a defined size. The process of forming mRNA on a DNA template is known as transcription.

In prokaryotes a single mRNA molecule may code for one or several polypeptide chains. If it carries the code for only one polypeptide, the mRNA is **monocistronic;** if it codes for two or more different polypeptides,

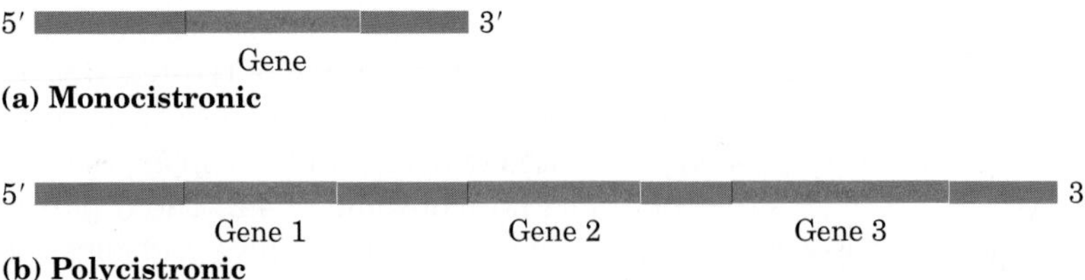

figure 10–24
Prokaryotic mRNA. Schematic diagrams show monocistronic **(a)** and polycistronic **(b)** mRNAs of prokaryotes. Red segments represent RNA coding for a gene product; gray segments represent noncoding RNA. In the polycistronic transcript, noncoding RNA separates the three genes.

the mRNA is **polycistronic.** In eukaryotes, most mRNAs are monocistronic. (For the purposes of this discussion, "cistron" refers to a gene. The term itself has historical roots in the science of genetics, and its formal genetic definition is beyond the scope of this text.) The minimum length of an mRNA is set by the length of the polypeptide chain for which it codes. For example, a polypeptide chain of 100 amino acid residues requires an RNA coding sequence of at least 300 nucleotides, because each amino acid is coded by a nucleotide triplet (this and other details of protein synthesis are discussed in Chapter 27). However, mRNAs transcribed from DNA are always somewhat longer than the length needed simply to code for a polypeptide sequence (or sequences). The additional noncoding RNA includes sequences that regulate protein synthesis. Figure 10–24 summarizes the general structure of prokaryotic mRNAs.

Many RNAs Have More Complex Three-Dimensional Structures

Messenger RNA is only one of several classes of cellular RNA. Transfer RNAs serve as adapter molecules in protein synthesis; covalently linked to an amino acid at one end, they pair with the mRNA in such a way that amino acids are joined to a growing polypeptide in the correct sequence. Ribosomal RNAs are structural components of ribosomes. There is also a wide variety of special-function RNAs, including some (called ribozymes) that have enzymatic activity. All of these are considered in detail in Chapter 26. The diverse and often complex functions of these RNAs reflect a diversity of structure much richer than that observed in DNA molecules.

The product of transcription of DNA is always single-stranded RNA. The single strand tends to assume a right-handed helical conformation dominated by base-stacking interactions (Fig. 10–25), which are stronger between two purines than between a purine and pyrimidine or between two pyrimidines. The purine-purine interaction is so strong that a pyrimidine separating two purines is often displaced from the stacking pattern so that the purines can interact. Any self-complementary sequences in the molecule produce more complex structures. RNA can base-pair with complementary regions of either RNA or DNA. Base pairing matches the pattern for DNA: G pairs with C and A pairs with U (or with the occasional T residue in some RNAs). One difference is that base pairing between G and U residues—unusual in DNA—is fairly common in RNA (see Fig. 10–27). The paired strands in RNA or RNA-DNA duplexes are antiparallel, as in DNA.

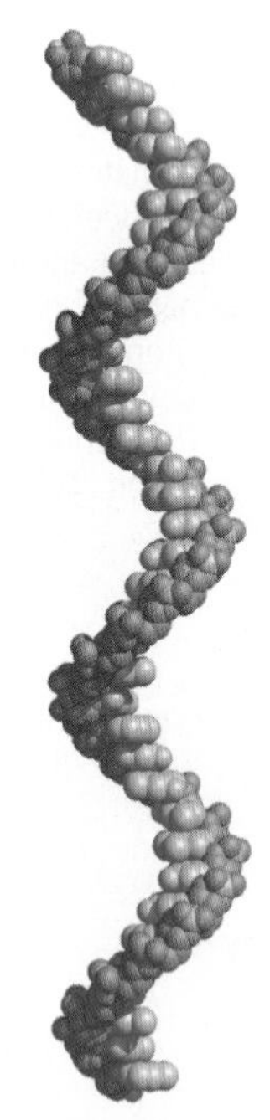

figure 10–25
Typical right-handed stacking pattern of single-stranded RNA. The bases are shown in white, the phosphate atoms in yellow, and the riboses and phosphate oxygens in green. Green is used to represent RNA strands in succeeding chapters, just as blue is used for DNA.

RNA has no simple, regular secondary structure that serves as a reference point, as does the double helix for DNA. The three-dimensional structures of many RNAs, like those of proteins, are complex and unique. Weak interactions, especially base-stacking interactions, play a major role in stabilizing RNA structures, just as they do in DNA. Where complementary sequences are present, the predominant double-stranded structure is an A-form right-handed double helix. Z-form helices have been made in the laboratory (under very high-salt or high-temperature conditions). The B form of RNA has not been observed. Breaks in the regular A-form helix

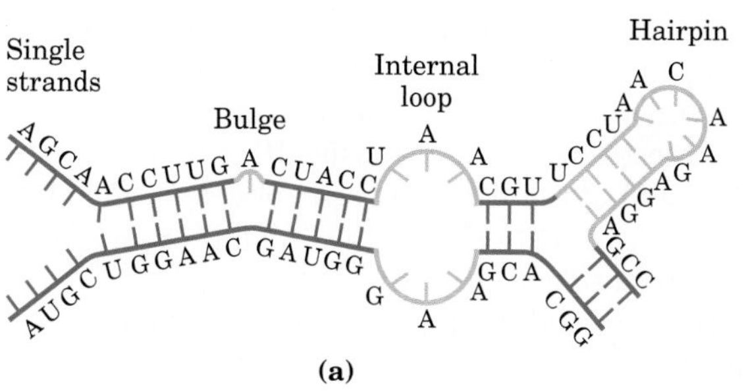

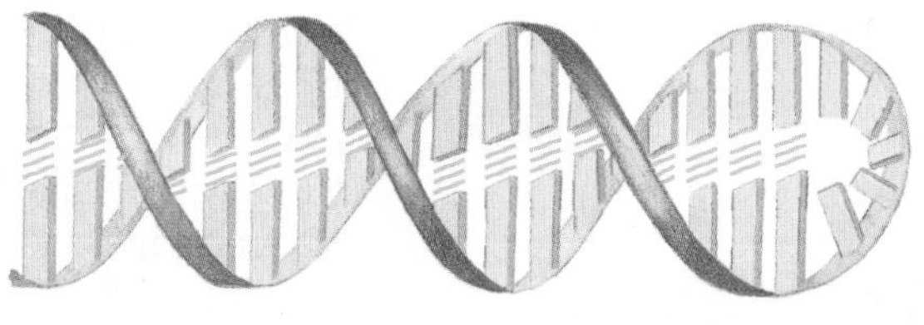

figure 10–26
Secondary structure of RNAs. (a) Bulge, internal loop, and hairpin loop. The paired regions generally have an A-form right-handed helix, as shown for a hairpin **(b).**

caused by mismatched or unmatched bases in one or both strands are common and result in bulges or internal loops (Fig. 10–26). Hairpin loops form between nearby self-complementary sequences. The potential for base-paired helical structures in many RNAs is extensive (Fig. 10–27), and the resulting hairpins are the most common type of secondary structure in RNA. Specific short base sequences (e.g., UUCG) are often found at the ends of RNA hairpins and are known to form particularly tight and stable loops. Such sequences may act as starting points for the folding of an RNA molecule into its precise three-dimensional structure. Important additional

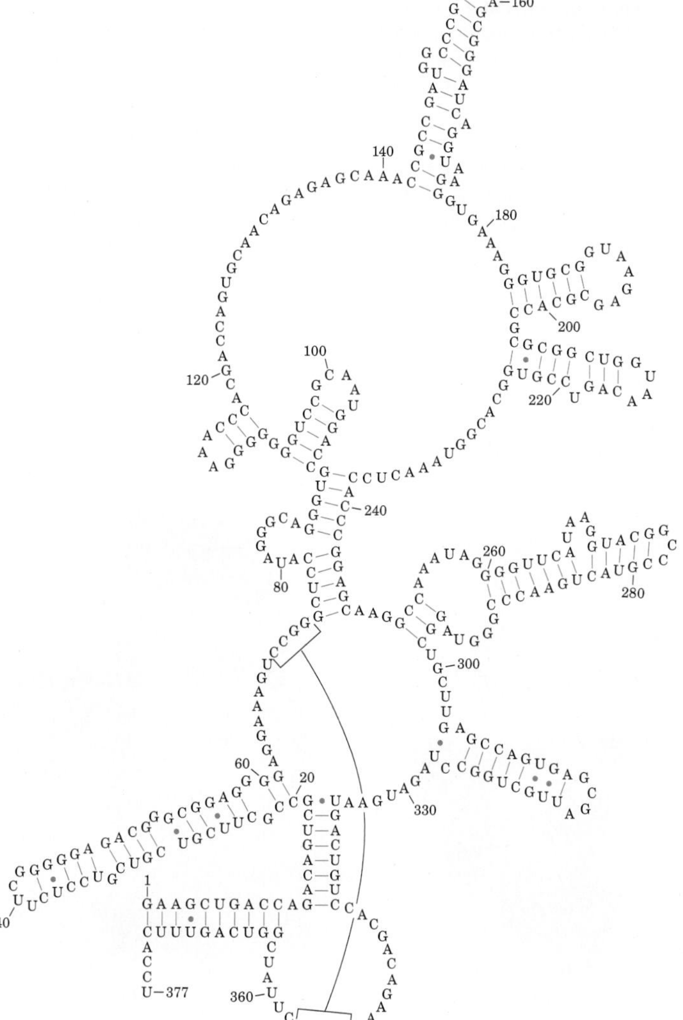

figure 10–27
Base-paired helical structures in an RNA. Shown here is the possible secondary structure of the M1 RNA component of the enzyme RNase P of *E. coli,* with many hairpins. RNase P, which also contains a protein component (not shown), functions in the processing of transfer RNAs (Chapter 26). The two brackets indicate additional complementary sequences that may be paired in the three-dimensional structure. The blue dots indicate non-Watson-Crick G=U base pairs (boxed inset). Note that G=U base pairs are allowed only when presynthesized strands of RNA fold up or anneal with each other. There are no RNA polymerases that insert a U opposite a template G or vice versa during RNA synthesis.

Guanine
Uracil

structural contributions are made by hydrogen bonds that are not part of standard Watson-Crick base pairs. For example, the 2′-hydroxyl group of ribose can hydrogen-bond with other groups. Some of these properties are evident in the structure of the phenylalanine transfer RNA of yeast—the tRNA responsible for inserting Phe residues into polypeptides—and in two RNA enzymes or ribozymes, whose functions, like those of protein enzymes, depend on their three-dimesional structures (Fig. 10–28).

The analysis of RNA structure and the relationship between structure and function is an emerging field of inquiry that has many of the same complexities as the analysis of protein structure. The importance of understanding RNA structure grows as we become increasingly aware of the large number of functional roles for RNA molecules.

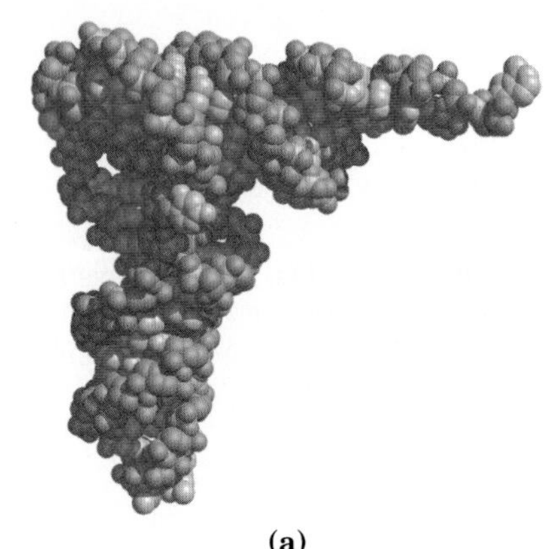

(a)

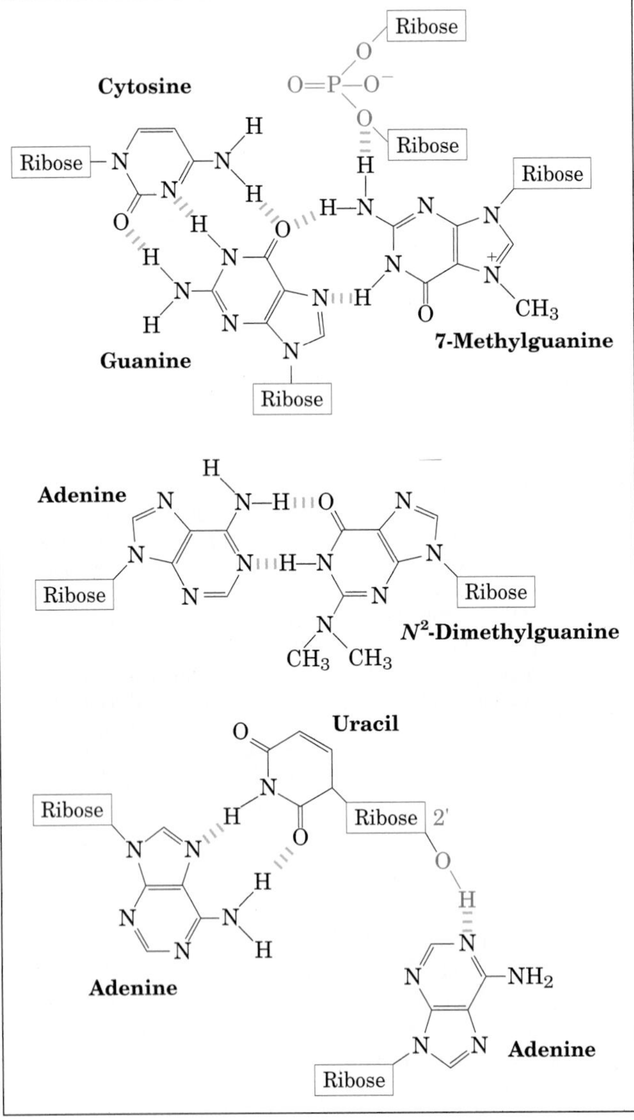

(b)

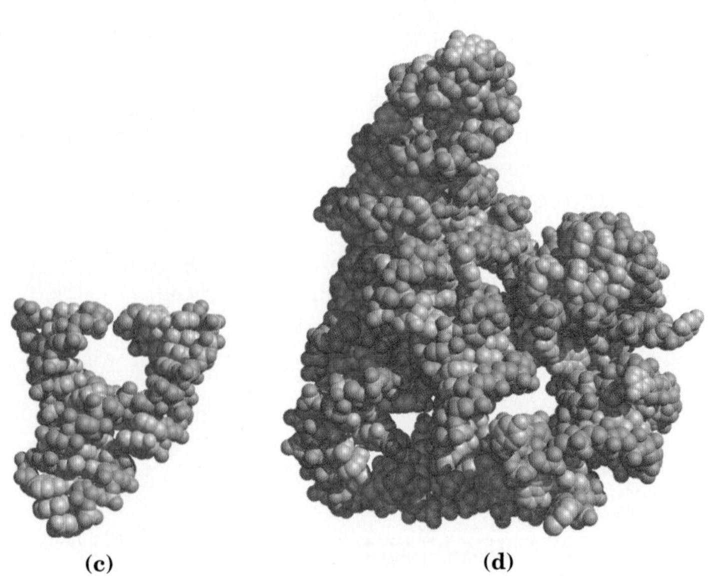

(c) **(d)**

figure 10–28
Three-dimensional structure in RNA. (a) Three-dimensional structure of phenylalanine tRNA of yeast. Some unusual base-pairing patterns are found in this tRNA **(b).** Note also the involvement of the oxygen of a ribose phosphodiester bond in one hydrogen-bonding arrangement, and a ribose 2′-hydroxyl group in another (both in red). **(c)** A hammerhead ribozyme (so named because the secondary structure at the active site looks like the head of a hammer), derived from certain plant viruses. Ribozymes, or RNA enzymes, catalyze a variety of reactions, primarily in RNA metabolism. The complex three-dimensional structures of these RNAs reflect the complexity inherent in catalysis, as described for protein enzymes in Chapter 8. **(d)** A segment of mRNA known as an intron, from the ciliated protozoan *Tetrahymena thermophila.* This intron (a ribozyme) catalyzes its own excision from between exons in an mRNA strand (introns and exons are discussed in Chapter 26).

Nucleic Acid Chemistry

To understand how nucleic acids function, we must understand their chemical properties as well as their structures. The role of DNA as a repository of genetic information depends in part on its inherent stability. The chemical transformations that do occur are generally very slow in the absence of an enzyme catalyst. The long-term storage of information without alteration is so important to a cell, however, that even very slow reactions that alter DNA structure can be physiologically significant. Processes such as carcinogenesis and aging may be intimately linked to slowly accumulating, irreversible alterations of DNA. Other, nondestructive alterations also occur, such as the strand separation that must precede DNA replication or transcription. In addition to providing insights into physiological processes, our understanding of nucleic acid chemistry has given us a powerful array of technologies that have applications in molecular biology, medicine, and forensic science. We now examine the chemical properties of DNA and some of these technologies.

Double-Helical DNA and RNA Can Be Denatured

Solutions of carefully isolated, native DNA are highly viscous at pH 7.0 and room temperature (25 °C). When such a solution is subjected to extremes of pH or to temperatures above 80 °C, its viscosity decreases sharply, indicating that the DNA has undergone a physical change. Just as heat and extremes of pH denature globular proteins, they also cause denaturation or melting of double-helical DNA. Disruption of the hydrogen bonds between paired bases and of base stacking causes unwinding of the double helix to form two single strands, completely separate from each other along the entire length or part of the length (partial denaturation) of the molecule. No covalent bonds in the DNA are broken (Fig. 10–29).

Renaturation of a DNA molecule is a rapid one-step process as long as a double-helical segment of a dozen or more residues still unites the two strands. When the temperature or pH is returned to the range in which most organisms live, the unwound segments of the two strands spontaneously rewind or **anneal** to yield the intact duplex (Fig. 10–29). However, if the two strands are completely separated, renaturation occurs in two steps. In the first, relatively slow step, the two strands "find" each other by random collisions and form a short segment of complementary double helix. The second step is much faster: the remaining unpaired bases successively come into register as base pairs, and the two strands "zipper" themselves together to form the double helix.

The close interaction between stacked bases in a nucleic acid has the effect of decreasing its absorption of UV light relative to that of a solution with the same concentration of free nucleotides, and the absorption is decreased further when two complementary nucleic acids are paired. This is called the hypochromic effect. Denaturation of a double-stranded nucleic acid produces the opposite result, an increase in absorption called the hyperchromic effect. The transition from double-stranded DNA to the single-stranded, denatured form can thus be detected by monitoring the absorption of UV light.

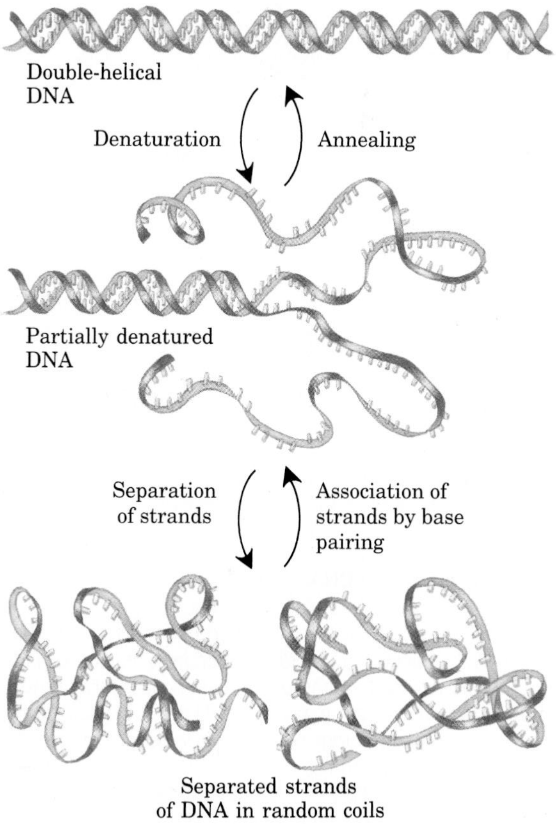

figure 10–29
Reversible denaturation and annealing (renaturation) of DNA.

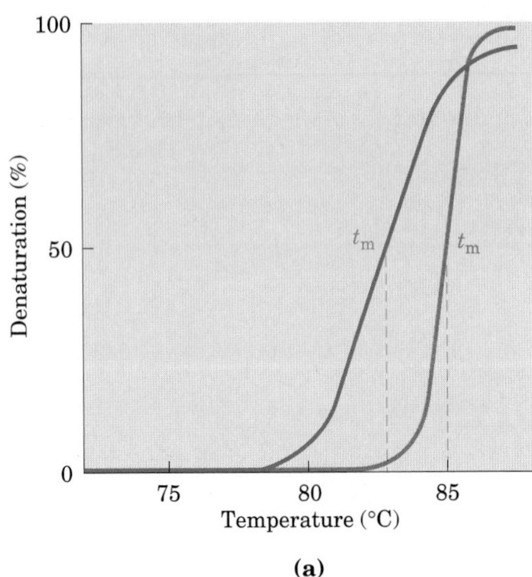

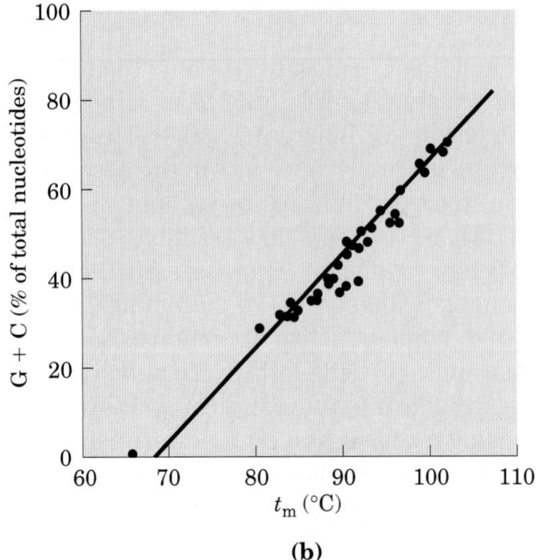

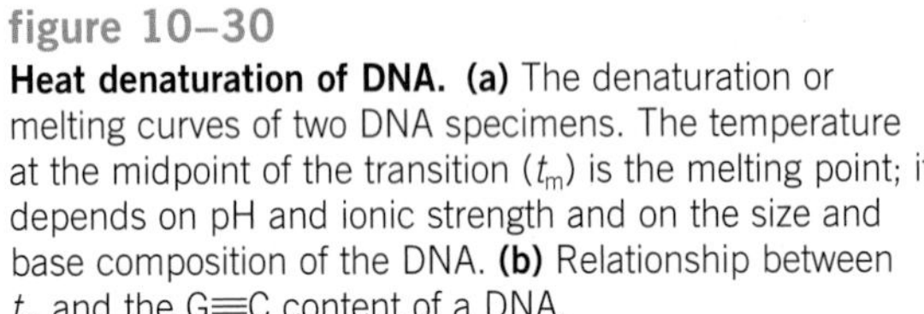
figure 10–30

Heat denaturation of DNA. (a) The denaturation or melting curves of two DNA specimens. The temperature at the midpoint of the transition (t_m) is the melting point; it depends on pH and ionic strength and on the size and base composition of the DNA. **(b)** Relationship between t_m and the G≡C content of a DNA.

Viral or bacterial DNA molecules in solution denature when they are heated slowly (Fig. 10–30). Each species of DNA has a characteristic denaturation temperature or melting point (t_m): the higher its content of G≡C base pairs, the higher the melting point of the DNA. This is because G≡C base pairs, with three hydrogen bonds, require more heat energy to dissociate than A═T base pairs. Careful determination of the melting point of a DNA specimen, under fixed conditions of pH and ionic strength, can yield an estimate of its base composition. If denaturation conditions are carefully controlled, regions that are rich in A═T base pairs will specifically denature while most of the DNA remains double-stranded. Such denatured regions (called bubbles) can be visualized with electron microscopy (Fig. 10–31). Strand separation of DNA *must* occur in vivo during processes such as DNA replication and transcription. As we will see, the DNA sites where these processes are initiated are often rich in A═T base pairs.

Duplexes of two RNA strands or one RNA strand and one DNA strand (RNA-DNA hybrids) can also be denatured. Notably, RNA duplexes are more stable than DNA duplexes. At neutral pH, denaturation of a double-helical RNA often requires temperatures 20 °C or more higher than those required for denaturation of a DNA molecule with a comparable sequence. The stability of an RNA-DNA hybrid is generally intermediate between that of RNA and that of DNA. The physical basis for these differences in stability is not known.

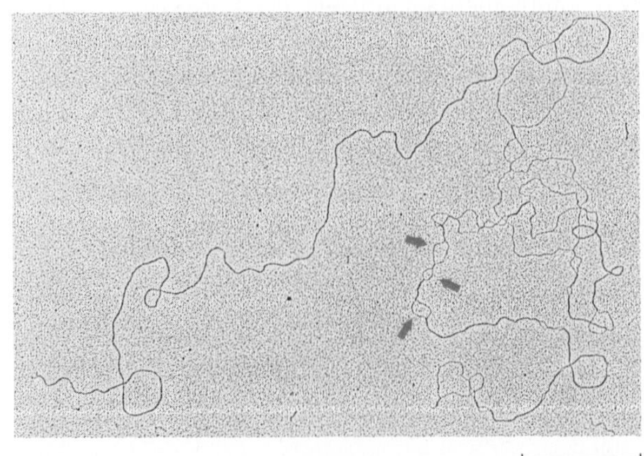

figure 10–31

Partially denatured DNA. This DNA was partially denatured, then fixed to prevent renaturation during sample preparation. The shadowing method used to visualize the DNA in this electron micrograph increases its diameter approximately fivefold and obliterates most details of the helix. However, length measurements can be obtained, and single-stranded regions are readily distinguishable from double-stranded regions. The arrows point to some single-stranded bubbles where denaturation has occurred. The regions that denature are highly reproducible and are rich in A═T base pairs.

Nucleic Acids from Different Species Can Form Hybrids

The ability of two complementary DNA strands to pair with one another can be used to detect similar DNA sequences in two different species or within the genome of a single species. If duplex DNAs isolated from human cells and from mouse cells are completely denatured by heating then mixed and kept at 65 °C for many hours, much of the DNA will anneal. Most of the mouse DNA strands anneal with complementary mouse DNA strands to form mouse duplex DNA; similarly, most human DNA strands anneal with complementary human DNA strands. However, some strands of the mouse DNA will associate with human DNA strands to yield **hybrid duplexes,** in which segments of a mouse DNA strand form base-paired regions with segments of a human DNA strand (Fig. 10–32). This reflects a common evolutionary heritage; different organisms generally have some proteins and RNAs with similar functions and, often, similar structures. In many cases, the DNAs encoding these proteins and RNAs have similar sequences. The closer the evolutionary relationship between two species, the more extensively their DNAs will hybridize. For example, human DNA hybridizes much more extensively with mouse DNA than with DNA from yeast.

The hybridization of DNA strands from different sources forms the basis for a powerful set of techniques essential to the practice of modern molecular genetics. A specific DNA sequence or gene can be detected in the presence of many other sequences if one already has an appropriate complementary DNA strand (usually labeled in some way) to hybridize with it (Chapter 29). The complementary DNA can be from a different species or from the same species, or it can be synthesized chemically in the laboratory using techniques described later in this chapter. Hybridization techniques can be varied to detect a specific RNA rather than DNA. The isolation and identification of specific genes and RNAs rely on these hybridization techniques. Applications of this technology make possible the identification of an individual on the basis of a single hair left at the scene of a crime or the prediction of the onset of a disease decades before symptoms appear (see Box 29–1).

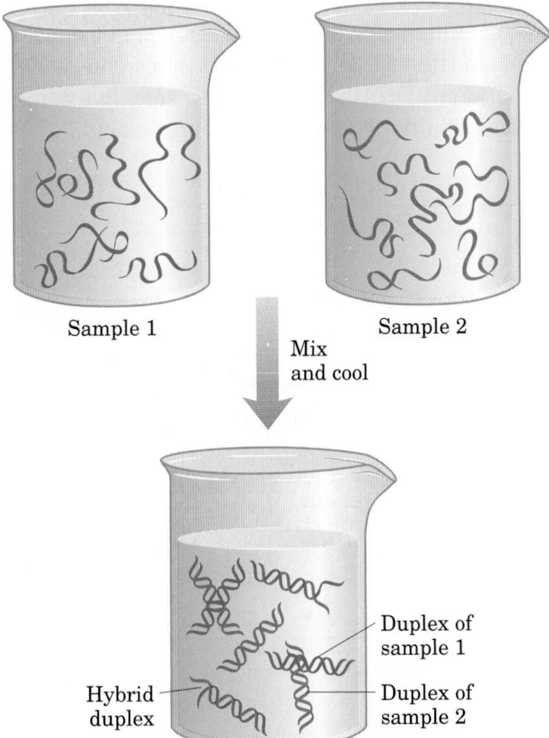

figure 10–32
DNA hybridization. Two DNA samples to be compared are completely denatured by heating. When the two solutions are mixed and slowly cooled, DNA strands of each sample associate with their normal complementary partner and anneal to form duplexes. If the two DNAs have significant sequence similarity, they also tend to form partial duplexes or hybrids with each other: the greater the sequence similarity between the two DNAs, the greater the number of hybrids formed. Hybrid formation can be measured in several ways. One of the DNAs is usually labeled with a radioactive isotope to simplify the measurements.

figure 10–33

Some well-characterized nonenzymatic reactions of nucleotides. (a) Deamination reactions. Only the base is shown. **(b)** Depurination, in which a purine is lost by hydrolysis of the *N*-β-glycosyl bond. The deoxyribose remaining after depurination is readily converted from the β-furanose to the aldehyde form (see Fig. 10–3). Further nonenzymatic reactions are illustrated in Figures 10–34 and 10–35.

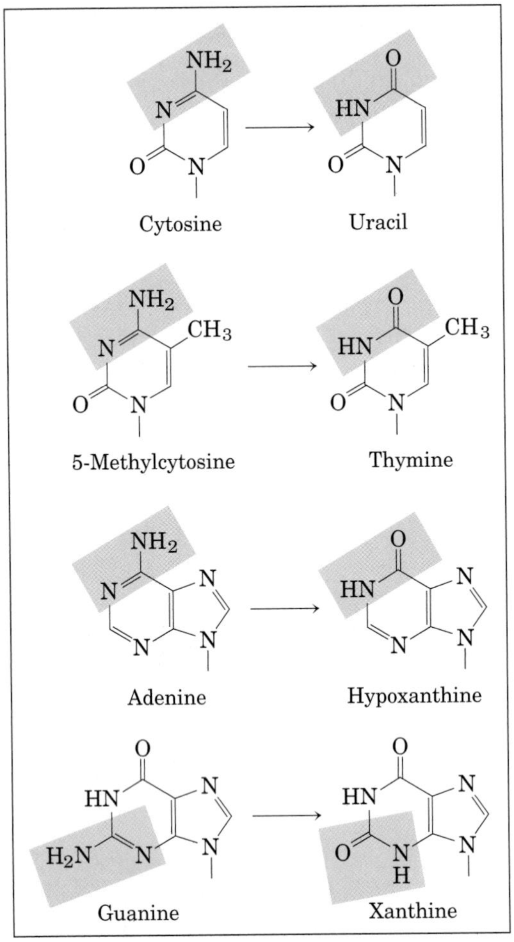

Cytosine → Uracil

5-Methylcytosine → Thymine

Adenine → Hypoxanthine

Guanine → Xanthine

Deamination
(a)

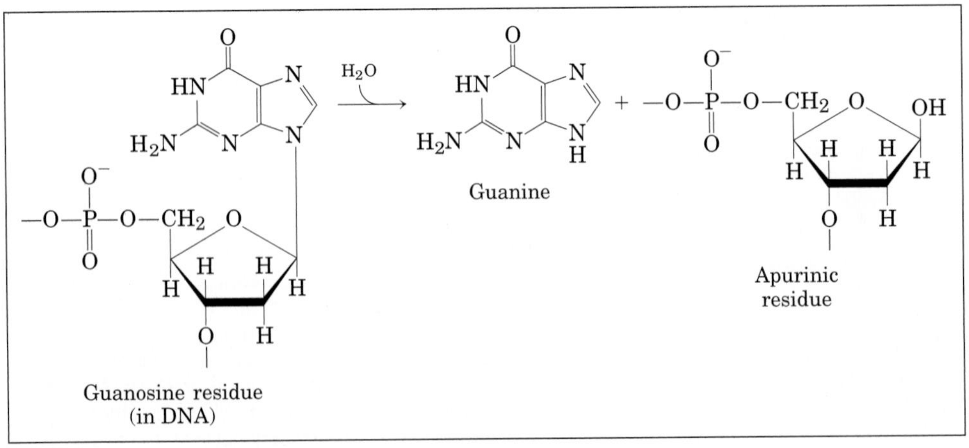

Guanosine residue (in DNA) → (H_2O) Guanine + Apurinic residue

Depurination
(b)

Nucleotides and Nucleic Acids Undergo Nonenzymatic Transformations

Purines and pyrimidines, along with the nucleotides of which they are a part, undergo a number of spontaneous alterations in their covalent structure. The rate of these reactions is generally *very slow,* but they are physiologically significant because of the cell's very low tolerance for alterations in its genetic information. Alterations in DNA structure that produce permanent changes in the genetic information encoded therein are called **mutations,** and much evidence suggests an intimate link between the accumulation of mutations and the processes of aging and carcinogenesis.

Several nucleotide bases undergo spontaneous loss of their exocyclic amino groups (deamination) (Fig. 10–33a). For example, under conditions found in a typical cell, deamination of cytosine (in DNA) to uracil occurs in about one of every 10^7 cytidine residues in 24 hours. This corresponds to about 100 spontaneous events per day, on average, in a mammalian cell. Deamination of adenine and guanine is about 100 times slower.

The slow cytosine deamination reaction seems innocuous enough, but is almost certainly the reason why DNA contains thymine rather than uracil. The product of cytosine deamination (uracil) is readily recognized as foreign in DNA and is removed by a repair system (Chapter 25). If DNA normally contained uracil, recognition of uracils resulting from cytosine deamination would be more difficult, and unrepaired uracils would lead to permanent sequence changes as they were paired with adenines during replication. Cytosine deamination would gradually lead to a decrease in G≡C base pairs and an increase in A=U base pairs in the DNA of all cells. Over the millennia, cytosine deamination could eliminate G≡C base pairs and the genetic code that depends on them. Establishing thymine as one of the four bases in DNA may well have been one of the crucial turning points in evolution, making the long-term storage of genetic information possible.

Another important reaction in deoxyribonucleotides is the hydrolysis of the *N*-β-glycosyl bond between the base and the pentose (Fig. 10–33b). This occurs at a higher rate for purines than for pyrimidines. As many as one in 10^5 purines (10,000 per mammalian cell) are lost from DNA every 24 hours under typical cellular conditions. Depurination of ribonucleotides and RNA is much slower and generally is not considered physiologically significant. In the test tube, loss of purines can be accelerated by dilute acid. Incubation of DNA at pH 3 causes selective removal of the purine bases, resulting in a derivative called apurinic acid.

Other reactions are promoted by radiation. UV light induces the condensation of two ethylene groups to form a cyclobutane ring. In the cell, the same reaction between adjacent pyrimidine bases in nucleic acids forms cyclobutane pyrimidine dimers. This happens most frequently between adjacent thymidine residues on the same DNA strand (Fig. 10–34). A second type of pyrimidine dimer, called a 6-4 photoproduct, is also formed during UV irradiation. Ionizing radiation (x rays and gamma rays) can cause ring opening and fragmentation of bases as well as breaks in the covalent backbone of nucleic acids.

Virtually all forms of life are exposed to energy-rich radiation capable of causing chemical changes in DNA. Near-UV radiation (with wavelengths of 200 to 400 nm), which makes up a significant portion of the solar spectrum, is known to cause pyrimidine dimer formation and other chemical changes in the DNA of bacteria and of human skin cells. We are subject to a constant field of ionizing radiation in the form of cosmic rays, which can penetrate deep into the earth, as well as radiation emitted from radioactive elements, such as radium, plutonium, uranium, radon, ^{14}C, and ^{3}H. X rays used in medical and dental examinations and in radiation therapy of cancer and other diseases are another form of ionizing radiation. It is estimated that UV and ionizing radiations are responsible for about 10% of all DNA damage caused by environmental agents.

DNA also may be damaged by reactive chemicals introduced into the environment as products of industrial activity. Such products may not be injurious per se, but may be metabolized by cells into forms that are. Two

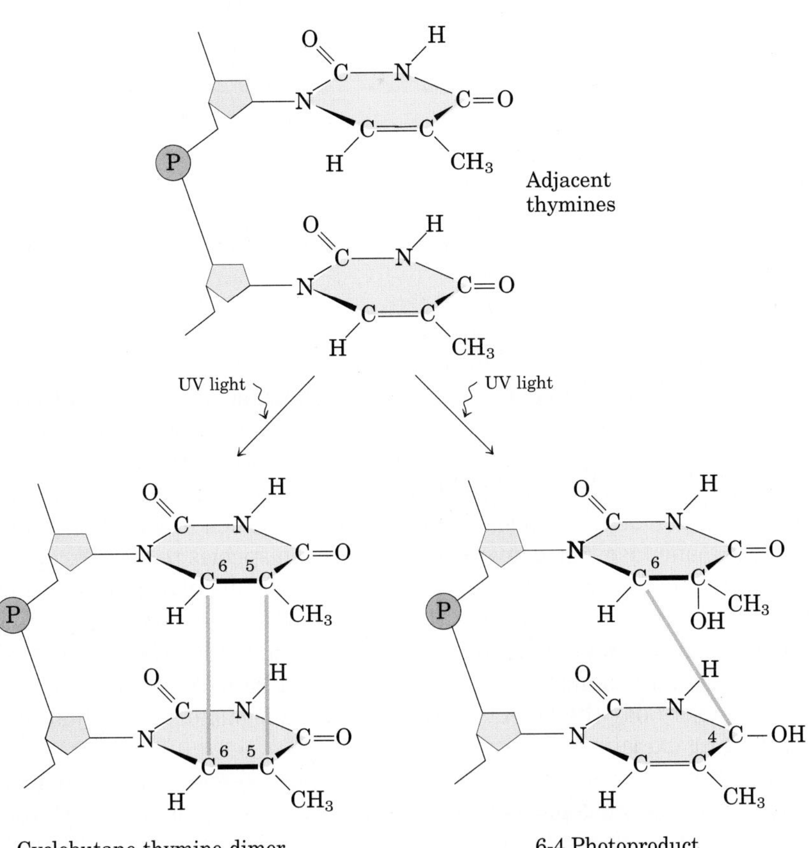

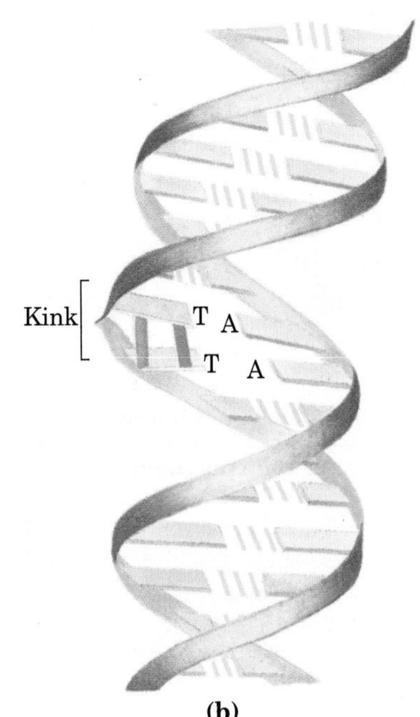

figure 10–34
Formation of pyrimidine dimers induced by UV light. **(a)** One type of reaction (on the left) results in the formation of a cyclobutyl ring involving C-5 and C-6 of adjacent pyrimidine residues. An alternative reaction (on the right) results in a 6-4 photoproduct, with a linkage between C-6 of one pyrimidine and C-4 of its neighbor. **(b)** A bend or kink is introduced into the DNA on formation of a cyclobutane pyrimidine dimer.

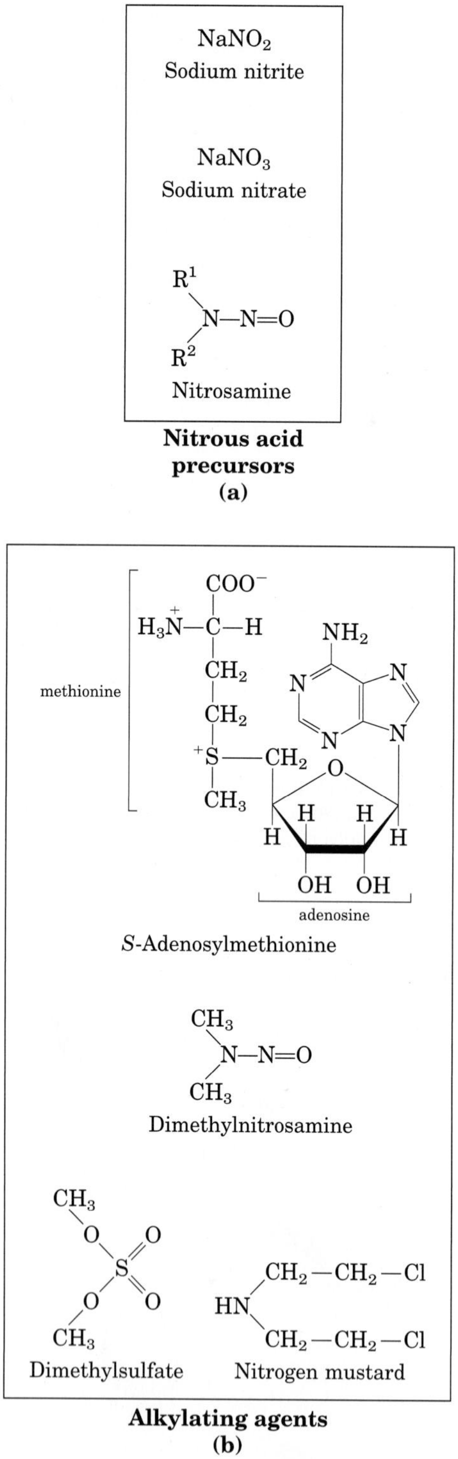

figure 10–35

Chemical agents that cause DNA damage. (a) Precursors of nitrous acid, which promotes deamination reactions. **(b)** Alkylating agents.

$NaNO_2$ Sodium nitrite

$NaNO_3$ Sodium nitrate

R^1 N—N=O R^2 Nitrosamine

Nitrous acid precursors (a)

methionine

adenosine

S-Adenosylmethionine

Dimethylnitrosamine

Dimethylsulfate

Nitrogen mustard

Alkylating agents (b)

prominent classes of such agents (Fig. 10–35) are (1) deaminating agents, particularly nitrous acid (HNO_2) or compounds that can be metabolized to nitrous acid or nitrites, and (2) alkylating agents.

Nitrous acid, formed from organic precursors such as nitrosamines and from nitrite and nitrate salts, is a potent accelerator of the deamination of bases. Bisulfite has similar effects. Both agents are used as preservatives in processed foods to prevent the growth of toxic bacteria. They do not appear to increase cancer risks significantly when used in this way, perhaps because they are used in small amounts and make only a minor contribution to the overall levels of DNA damage. (The potential health risk from food spoilage if these preservatives were not used is much greater.)

Alkylating agents can alter certain bases of DNA. For example, the highly reactive chemical dimethylsulfate (Fig. 10–35b) can methylate a guanine to yield O^6-methylguanine, which cannot base-pair with cytosine.

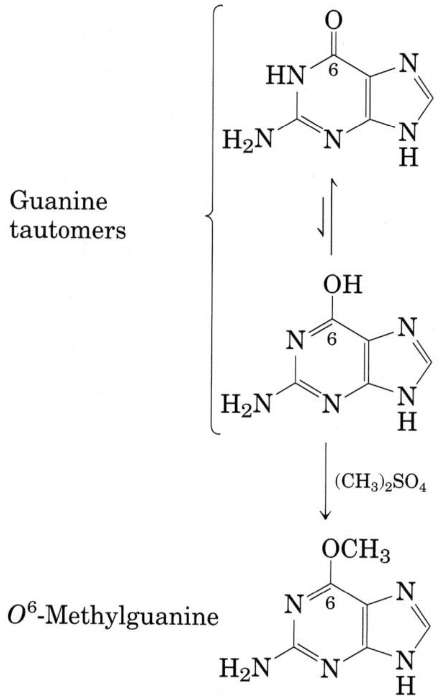

Guanine tautomers

$(CH_3)_2SO_4$

O^6-Methylguanine

Many similar reactions are brought about by alkylating agents normally present in cells, such as *S*-adenosylmethionine.

Possibly the most important source of mutagenic alterations in DNA is oxidative damage. Excited-oxygen species such as hydrogen peroxide, hydroxyl radicals, and superoxide radicals arise during irradiation or as a byproduct of aerobic metabolism. Cells have an elaborate defense system to destroy these reactive species, including enzymes such as catalase and superoxide dismutase that convert reactive oxygen species to harmless products. A fraction of these oxidants inevitably escape cellular defenses, however, and damage to DNA occurs through any of a large, complex group of reactions ranging from oxidation of deoxyribose and base moieties to strand breaks. Accurate estimates for the extent of this damage are not yet available, but every day the DNA of each human cell is subjected to thousands of damaging oxidative reactions.

This is merely a sampling of the best-understood reactions that damage DNA. Many carcinogenic compounds in food, water, or air exert their cancer-causing effects by modifying bases in DNA. Nevertheless, the integrity of DNA as a polymer is better maintained than that of either RNA or protein, because DNA is the only macromolecule that has the benefit of biochemical repair systems. These repair processes (described in Chapter 25) greatly lessen the impact of damage to DNA.

Some Bases of DNA Are Methylated

Certain nucleotide bases in DNA molecules are enzymatically methylated. Adenine and cytosine are methylated more often than guanine and thymine. Methylation is generally confined to certain sequences or regions of a DNA molecule. In some cases the function of methylation is well understood; in others the function remains unclear. All known DNA methylases use *S*-adenosylmethionine as a methyl group donor. *E. coli* has two prominent methylation systems. One serves as part of a defense mechanism that helps the cell to distinguish its DNA from foreign DNA by marking its own DNA with methyl groups and destroying (foreign) DNA without the methyl groups (this is known as a restriction-modification system; see Chapter 29). The other system methylates adenosine residues within the sequence (5′)GATC(3′) to N^6-methyladenosine (Fig. 10–5a). This is mediated by the Dam (*D*NA *a*denine *m*ethylation) methylase, a component of a system that repairs mismatched base pairs formed occasionally during DNA replication (Chapter 25).

In eukaryotic cells, about 5% of cytidine residues in DNA are methylated to 5-methylcytidine (Fig. 10–5a). Methylation is most common at CpG sequences, producing methyl-CpG symmetrically on both strands of the DNA. The extent of methylation of CpG sequences varies by region in large eukaryotic DNA molecules. Methylation suppresses the migration of segments of DNA called transposons, described in Chapter 25. These methylations of cytosine also have structural significance. The presence of 5-methylcytosine in an alternating CpG sequence markedly increases the tendency for that segment of DNA to assume the Z form.

The Sequences of Long DNA Strands Can Be Determined

In its capacity as a repository of information, a DNA molecule's most important property is its nucleotide sequence. Until the late 1970s, determining the sequence of a nucleic acid containing even five or ten nucleotides was difficult and very laborious. The development of two new techniques in 1977, one by Alan Maxam and Walter Gilbert and the other by Frederick Sanger, has made possible the sequencing of ever larger DNA molecules with an ease unimagined just a few decades ago. The techniques depend on an improved understanding of nucleotide chemistry and DNA metabolism, and on electrophoretic methods for separating DNA strands differing in size by only one nucleotide. Electrophoresis of DNA is similar to that of proteins (see Fig. 5–19). Polyacrylamide is often used as the gel matrix when working with short DNA molecules (up to a few hundred nucleotides); agarose is generally used for longer pieces of DNA.

In both Sanger and Maxam-Gilbert sequencing, the general principle is to reduce the DNA to four sets of labeled fragments. The reaction producing each set is base-specific, so that the lengths of the fragments correspond to positions in the DNA sequence where a certain base occurs. For example, for an oligonucleotide with the sequence pAATCGACT, labeled at the 5′ end (the left end), a reaction that breaks the DNA after each C residue will generate two labeled fragments: a four-nucleotide and a seven-nucleotide fragment; a reaction that breaks the DNA after each G will produce only a five-nucleotide fragment. Because the fragments are radioactively labeled at their 5′ ends, only the fragment to the 5′ side of the break

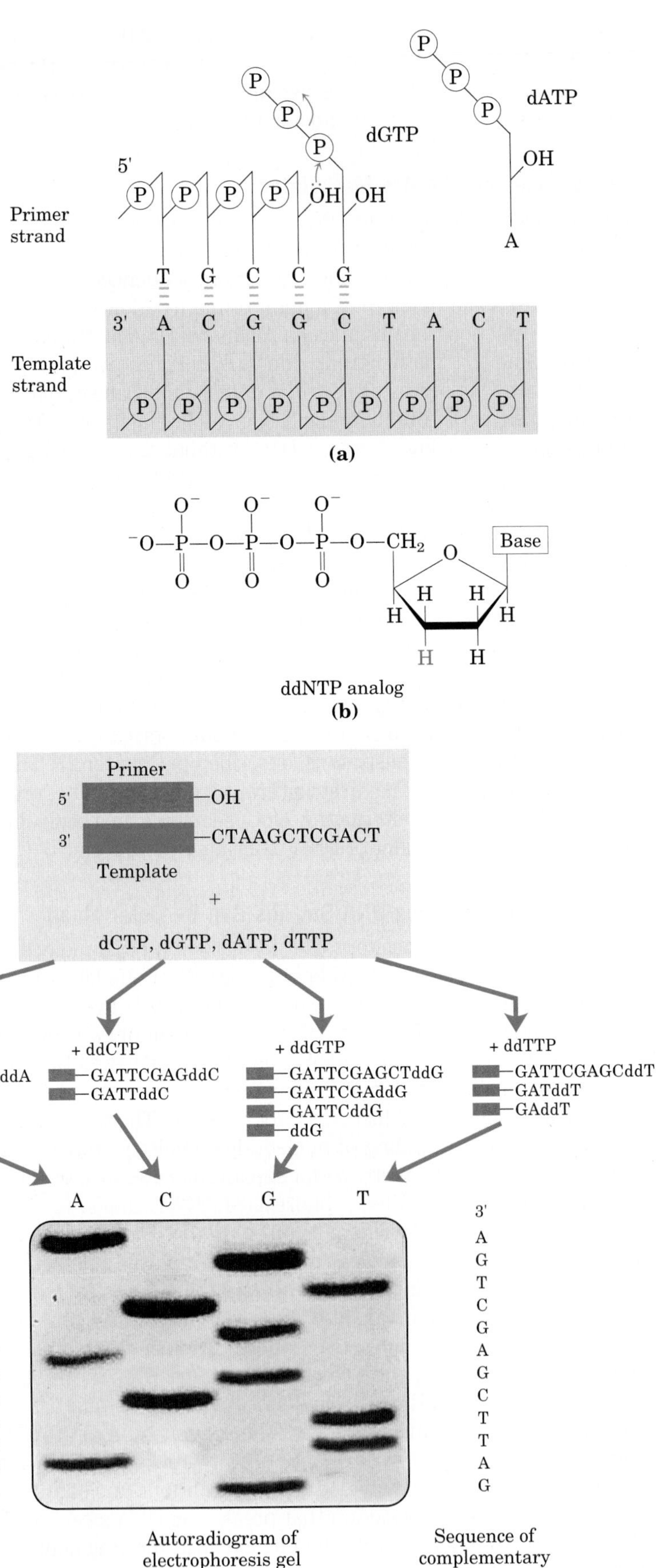

figure 10–36

DNA sequencing by the Sanger method. This method makes use of the mechanism of DNA synthesis by DNA polymerases (Chapter 25). **(a)** DNA polymerases require both a primer (a short oligonucleotide strand), to which nucleotides are added, and a template strand to guide selection of each new nucleotide. In cells, the 3′-hydroxyl group of the primer reacts with an incoming deoxynucleoside triphosphate (dNTP) to form a new phosphodiester bond. The Sanger sequencing procedure uses dideoxynucleoside triphosphate (ddNTP) analogs **(b)** to interrupt DNA synthesis. (The Sanger method is also known as the dideoxy method.) When a ddNTP is inserted in place of a dNTP, strand elongation is halted after the analog is added because it lacks the 3′-hydroxyl group needed for the next step.

(c) The DNA to be sequenced is used as the template strand, and a short primer, radioactively or fluorescently labeled, is annealed to it. By adding small amounts of a single ddNTP, for example ddCTP, to an otherwise normal reaction system, the synthesized strands will be prematurely terminated at some locations where dC normally occurs. Given the excess of dCTP over ddCTP, the chance that the analog will be incorporated whenever a dC is to be added is small. However, ddCTP is present in sufficient amounts to ensure that each new strand has a high probability of acquiring at least one ddC at some point during synthesis. The result is a solution containing a mixture of labeled fragments, each ending with a C residue. Each C residue in the sequence generates a set of fragments of a particular length, such that the different-sized fragments, separated by electrophoresis, reveal the location of C residues. This procedure is repeated separately for each of the four ddNTPs, and the sequence can be read directly from an autoradiogram of the gel. Because shorter DNA fragments migrate faster, the fragments near the bottom of the gel represent the nucleotide positions closest to the primer (the 5′ end), and the sequence is read (in the 5′→3′ direction) from bottom to top. Note that the sequence obtained is that of the strand *complementary* to the strand being analyzed.

is seen on the gel. The fragment sizes correspond to the relative positions of C and G residues in the sequence. When the sets of fragments corresponding to each of the four bases are electrophoretically separated side by side, they produce a ladder of bands from which the sequence can be read directly (Fig. 10–36). We illustrate only the Sanger method because it has proven to be technically easier and is in more widespread use. It requires the enzymatic synthesis of a DNA strand complementary to the strand under analysis, using a radioactively labeled "primer" and dideoxynucleotides.

DNA sequencing is readily automated, using a variation of Sanger's sequencing method in which the dideoxynucleotides used for each reaction are labeled with a differently colored fluorescent tag (Fig. 10–37). This technology allows DNA sequences containing thousands of nucleotides to be determined in a few hours, and very large DNA-sequencing projects are in progress. The most ambitious of these is the Human Genome Initiative, in which all 3 billion base pairs of the DNA in a human cell are being sequenced.

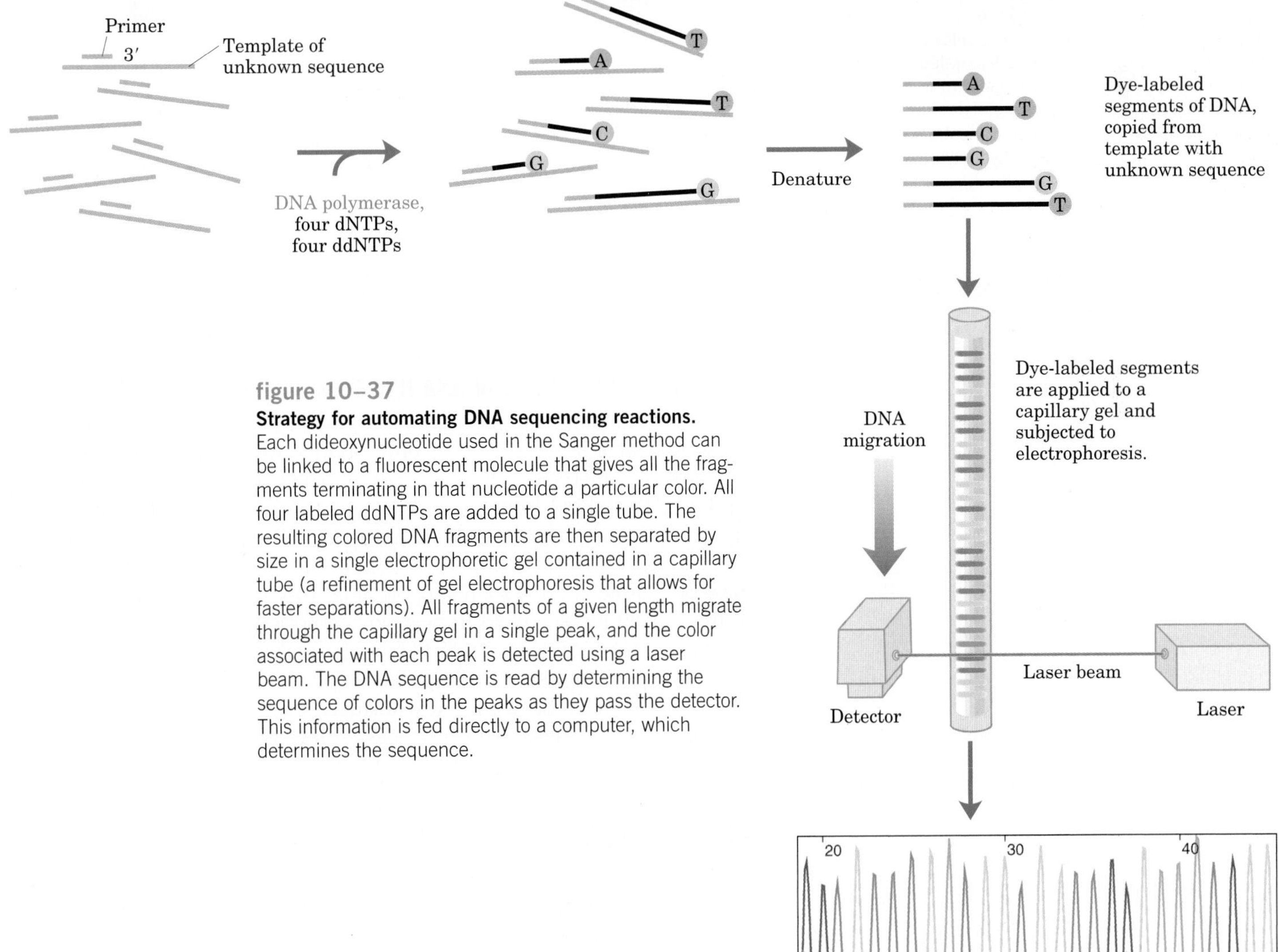

figure 10–37
Strategy for automating DNA sequencing reactions. Each dideoxynucleotide used in the Sanger method can be linked to a fluorescent molecule that gives all the fragments terminating in that nucleotide a particular color. All four labeled ddNTPs are added to a single tube. The resulting colored DNA fragments are then separated by size in a single electrophoretic gel contained in a capillary tube (a refinement of gel electrophoresis that allows for faster separations). All fragments of a given length migrate through the capillary gel in a single peak, and the color associated with each peak is detected using a laser beam. The DNA sequence is read by determining the sequence of colors in the peaks as they pass the detector. This information is fed directly to a computer, which determines the sequence.

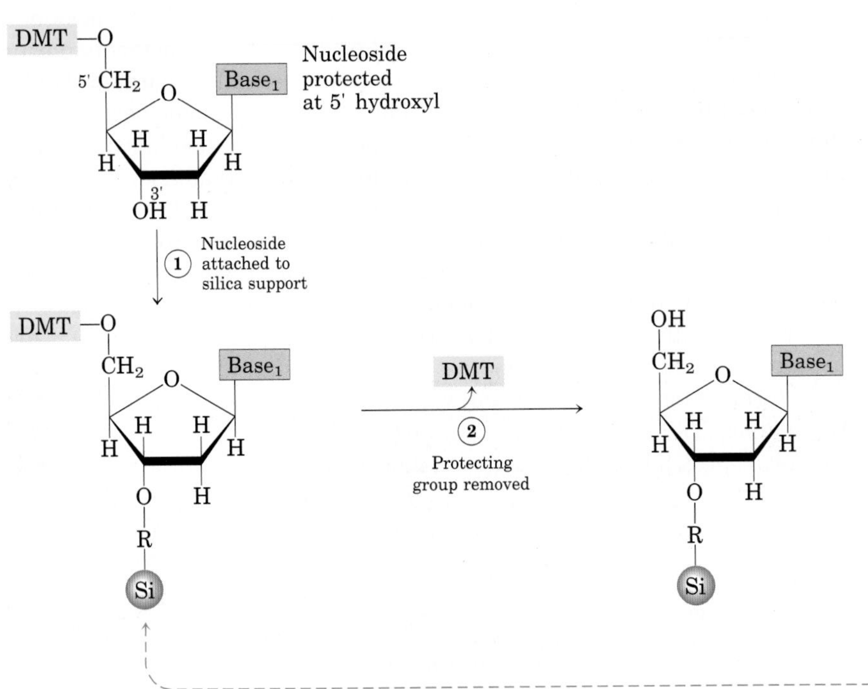

figure 10–38
Chemical synthesis of DNA. Automated DNA synthesis is conceptually similar to the solid-state synthesis of polypeptides. The oligonucleotide is built up on a solid support (silica), one nucleotide at a time, in a repeated series of chemical reactions with suitably protected nucleotide precursors. ① The first nucleoside (which will be the 3′ end) is attached to the silica support at the 3′ hydroxyl (through a linking group, R) and is protected at the 5′ hydroxyl with an acid-labile dimethoxytrityl group (DMT). The reactive groups on all bases are also chemically protected. ② The protecting DMT group is removed by washing the column with acid (the DMT group is colored, so this reaction can be followed spectrophotometrically). ③ The next nucleotide is activated with a diisopropylamino group and reacted with the bound nucleotide to form a 5′,3′ linkage, which in step ④ is oxidized with iodine to produce a phosphotriester linkage. (One of the phosphate oxygens carries a cyanoethyl protecting group.) Reactions ② through ④ are repeated until all nucleotides are added. At each step, excess nucleotide is removed before addition of the next nucleotide. In steps ⑤ and ⑥ the remaining protecting groups on the bases and the phosphates are removed, and in ⑦ the oligonucleotide is separated from the solid support and purified. The chemical synthesis of RNA is somewhat more complicated because of the need to protect the 2′ hydroxyl of ribose without adversely affecting the reactivity of the 3′ hydroxyl.

The Chemical Synthesis of DNA Has Been Automated

Another technology that has paved the way for many biochemical advances is the chemical synthesis of oligonucleotides with any chosen sequence. The chemical methods for synthesizing nucleic acids were developed primarily by H. Gobind Khorana and his colleagues in the 1970s. Refinement and automation of these methods have made possible the rapid and accurate synthesis of DNA strands. The synthesis is carried out with the growing strand attached to a solid support (Fig. 10–38), using principles similar to those used by Merrifield in peptide synthesis (see Fig. 5–29). The efficiency of each addition step is very high, allowing the routine laboratory synthesis of polymers containing 70 or 80 nucleotides and, in some laboratories, much longer strands. The availability of relatively inexpensive DNA polymers with predesigned sequences is having a powerful impact on all areas of biochemistry (Chapter 29).

Other Functions of Nucleotides

In addition to their roles as the subunits of nucleic acids, nucleotides have a variety of other functions in every cell: as energy carriers, components of enzyme cofactors, and chemical messengers.

Nucleotides Carry Chemical Energy in Cells

The phosphate group covalently linked at the 5′ hydroxyl of a ribonucleotide may have one or two additional phosphates attached. The resulting

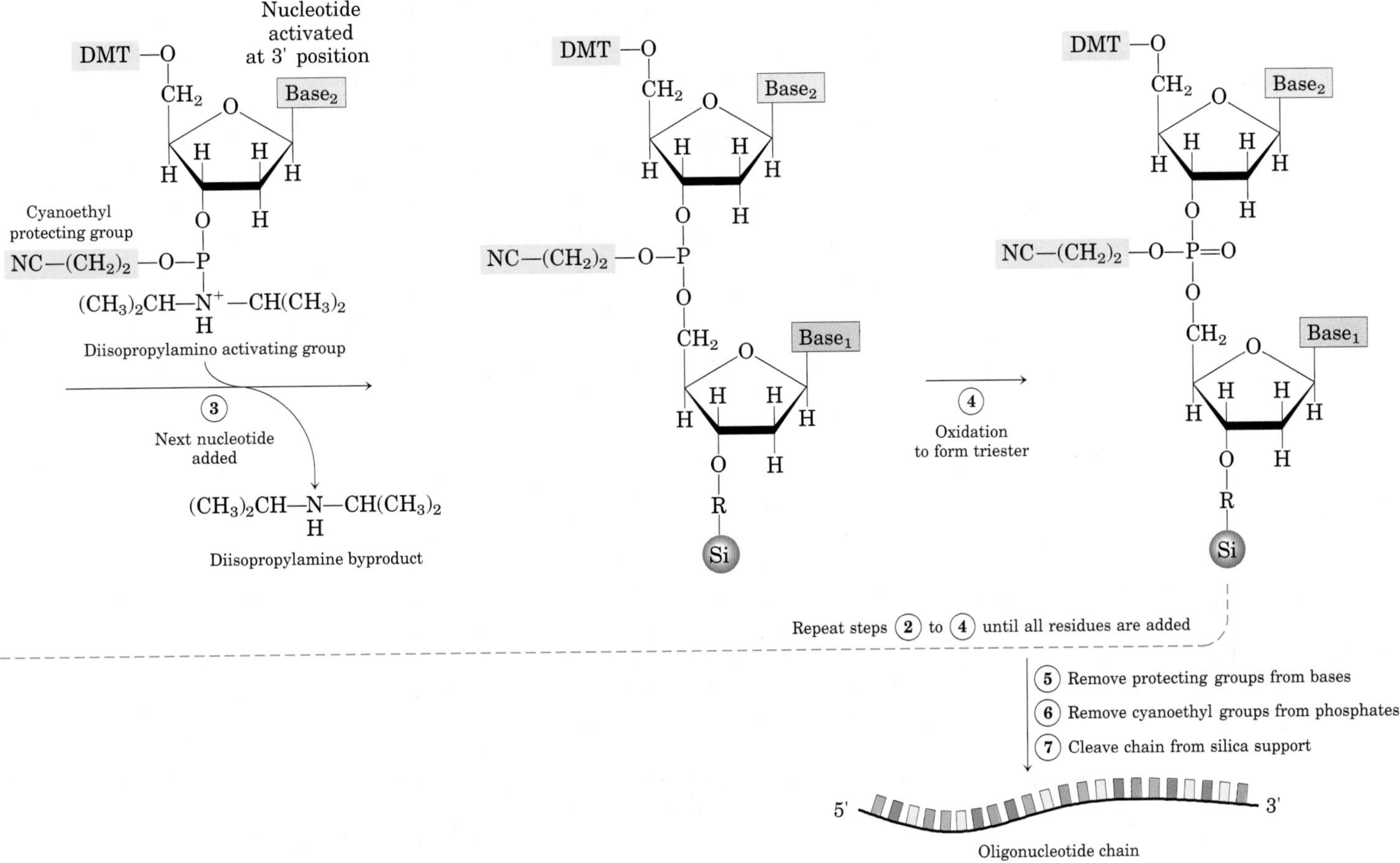

molecules are referred to as nucleoside mono-, di-, and triphosphates (Fig. 10–39). Starting from the ribose, the three phosphates are generally labeled α, β, and γ. Hydrolysis of nucleoside triphosphates provides the chemical energy to drive a wide variety of biochemical reactions. Adenosine 5′-triphosphate, ATP, is by far the most widely used for this purpose, but UTP, GTP, and CTP are also used in some reactions. Nucleoside triphosphates also serve as the activated precursors of DNA and RNA synthesis, as described in Chapters 25 and 26.

figure 10–39
General structure of the nucleoside 5′-mono-, di-, and triphosphates (NMPs, NDPs, and NTPs) and their standard abbreviations. In the deoxyribonucleoside phosphates (dNMPs, dNDPs, and dNTPs), the pentose is 2′-deoxy-D-ribose.

NMP

NDP

NTP

Abbreviations of ribonucleoside 5'-phosphates			
Base	Mono-	Di-	Tri-
Adenine	AMP	ADP	ATP
Guanine	GMP	GDP	GTP
Cytosine	CMP	CDP	CTP
Uracil	UMP	UDP	UTP

Abbreviations of deoxyribonucleoside 5'-phosphates			
Base	Mono-	Di-	Tri-
Adenine	dAMP	dADP	dATP
Guanine	dGMP	dGDP	dGTP
Cytosine	dCMP	dCDP	dCTP
Thymine	dTMP	dTDP	dTTP

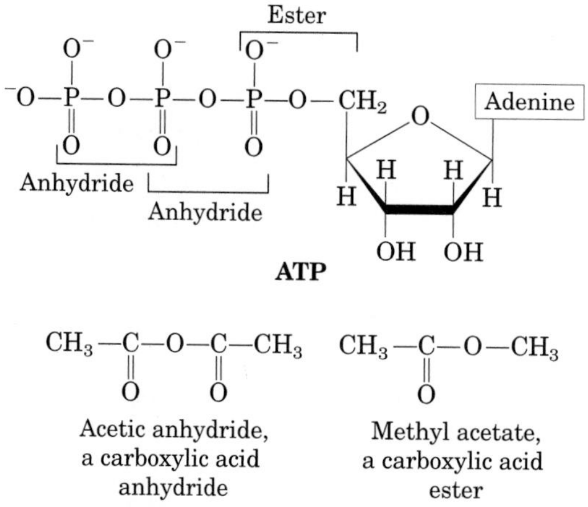

figure 10–40

The phosphate ester and phosphoanhydride bonds of ATP. Hydrolysis of an anhydride bond yields more energy than hydrolysis of the ester. A carboxylic acid anhydride and carboxylic acid ester are shown for comparison.

The energy released by hydrolysis of ATP and the other nucleoside triphosphates is accounted for by the structure of the triphosphate group. The bond between the ribose and the α phosphate is an ester linkage. The α,β and β,γ linkages are phosphoanhydrides (Fig. 10–40). Hydrolysis of the ester linkage yields about 14 kJ/mol under standard conditions, whereas hydrolysis of each anhydride bond yields about 30 kJ/mol. ATP hydrolysis often plays an important thermodynamic role in biosynthesis. When coupled to a reaction with a positive free-energy change, ATP hydrolysis shifts the equilibrium of the overall process to favor product formation (recall the relationship between equilibrium constant and free-energy change described by Equation 8–3 on p. 249).

Adenine Nucleotides Are Components of Many Enzyme Cofactors

A variety of enzyme cofactors serving a wide range of chemical functions include adenosine as part of their structure (Fig. 10–41). They are unrelated structurally except for the presence of adenosine. In none of these cofactors does the adenosine portion participate directly in the primary function, but removal of adenosine generally results in a drastic reduction of cofactor activities. For example, removal of the adenine nucleotide (3′-phosphoadenosine diphosphate) from acetoacetyl-CoA, the coenzyme A derivative of acetoacetate, reduces its reactivity as a substrate for β-ketoacyl-CoA transferase (an enzyme of lipid metabolism) by a factor of 10^6. Although this requirement for adenosine has not been investigated in detail, it must involve the binding energy between enzyme and substrate (or cofactor) that is used both in catalysis and in stabilizing the initial enzyme-substrate complex (Chapter 8). In the case of β-ketoacyl-CoA transferase, the nucleotide moiety of coenzyme A appears to be a binding "handle" that helps to pull the substrate (acetoacetyl-CoA) into the active site. Similar roles may be found for the nucleoside portion of other nucleotide cofactors.

Why is adenosine, rather than some other large molecule, used in these structures? The answer here may involve a form of evolutionary economy. Adenosine is certainly not unique in the amount of potential binding energy it can contribute. The importance of adenosine probably lies not so much in some special chemical characteristic but in the evolutionary advantage of using one compound for multiple roles. Once ATP became the universal source of chemical energy, systems developed to synthesize ATP in greater

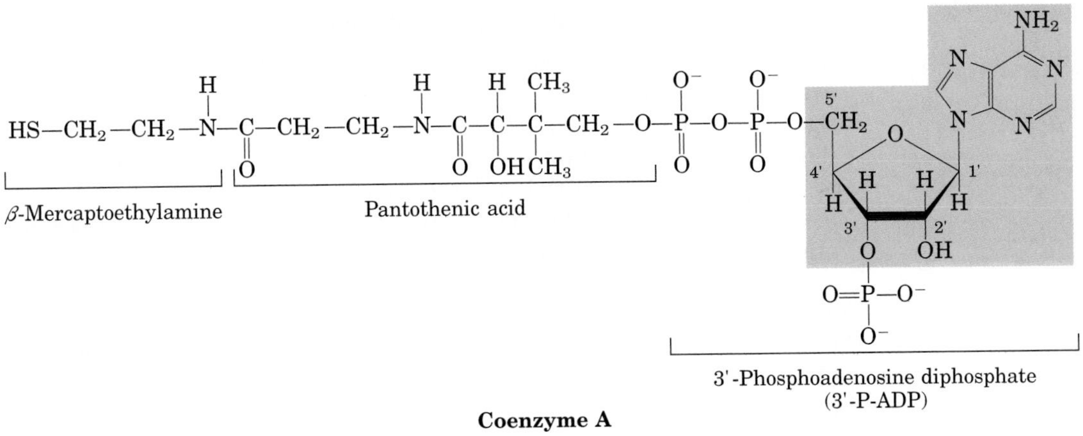

Coenzyme A

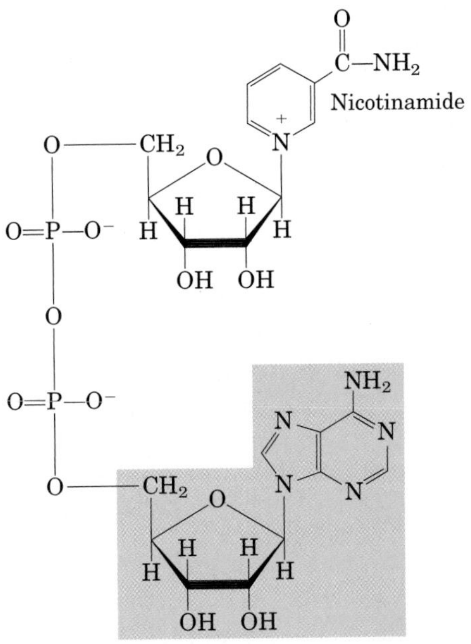

Nicotinamide adenine dinucleotide (NAD^+)

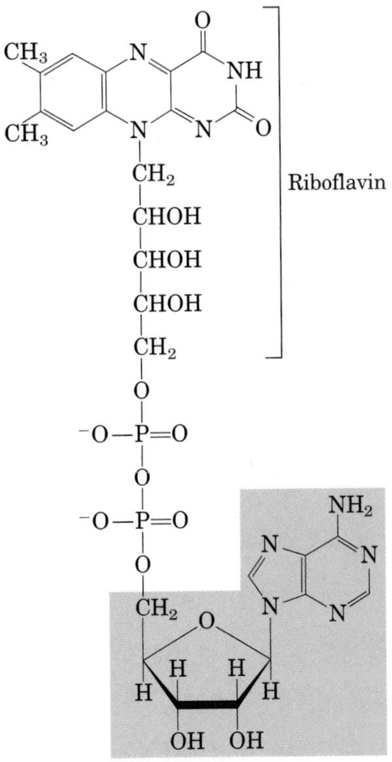

Flavin adenine dinucleotide (FAD)

figure 10–41

Some coenzymes containing adenosine. The adenosine portion is shaded in red. Coenzyme A (CoA) functions in acyl group transfer reactions; the acyl group (such as the acetyl or acetoacetyl group) is attached to the CoA through a thioester linkage to the β-mercaptoethylamine moiety. NAD^+ functions in hydride transfers, and FAD, the active form of vitamin B_2 (riboflavin), in electron transfers. Another coenzyme incorporating adenosine is 5′-deoxyadenosylcobalamin, the active form of vitamin B_{12} (see Box 17–2), which participates in intramolecular group transfers between adjacent carbons.

abundance than the other nucleotides; because it is abundant, it becomes the logical choice for incorporation into a wide variety of structures. The economy extends to protein structure. A protein domain that binds adenosine can be used in a wide variety of enzymes. Such a domain, called a **nucleotide-binding fold,** is found in many enzymes that bind ATP and nucleotide cofactors.

Some Nucleotides Are Regulatory Molecules

Cells respond to their environment by taking cues from hormones or other external chemical signals. The interaction of these extracellular chemical signals (first messengers) with receptors on the cell surface often leads to the production of **second messengers** inside the cell, which in turn lead to adaptive changes in the cell interior (Chapter 13). Often, the second messenger is a nucleotide (Fig. 10–42). One of the most common is **adenosine 3′,5′-cyclic monophosphate (cyclic AMP,** or **cAMP),** formed from ATP in a reaction catalyzed by adenylate cyclase, an enzyme associated with the inner face of the plasma membrane. Cyclic AMP serves regulatory functions in virtually every cell outside the plant kingdom. Guanosine 3′,5′-cyclic monophosphate (cGMP) occurs in many cells and also has regulatory functions.

Another regulatory nucleotide, ppGpp (Fig. 10–42), is produced in bacteria in response to a slowdown in protein synthesis during amino acid starvation. This nucleotide inhibits the synthesis of the rRNA and tRNA molecules (see Fig. 28–26) needed for protein synthesis, preventing the unnecessary production of nucleic acids.

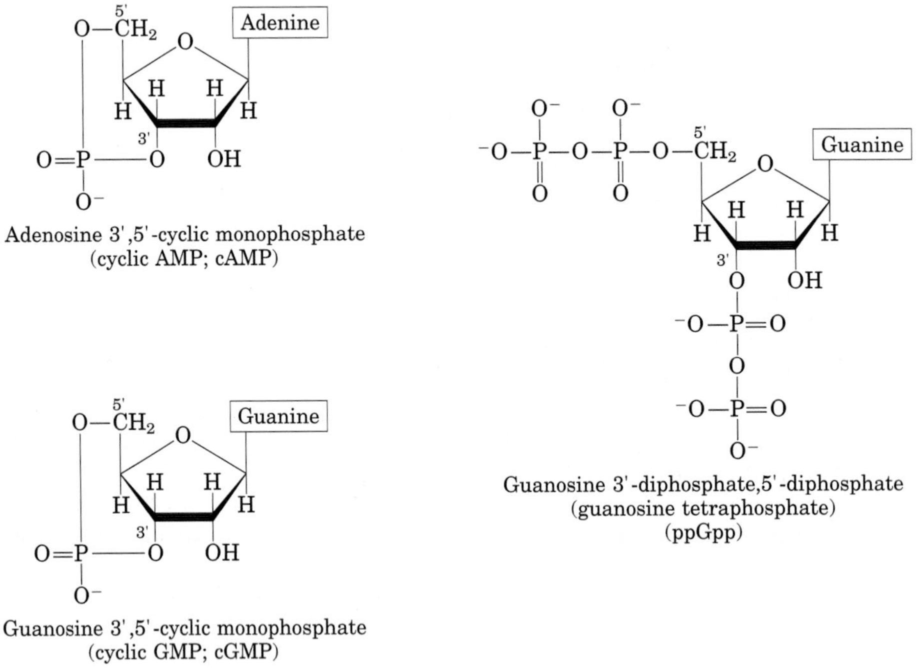

Adenosine 3′,5′-cyclic monophosphate (cyclic AMP; cAMP)

Guanosine 3′,5′-cyclic monophosphate (cyclic GMP; cGMP)

Guanosine 3′-diphosphate,5′-diphosphate (guanosine tetraphosphate) (ppGpp)

figure 10–42
Three regulatory nucleotides.

summary

Nucleotides serve a diverse set of important functions in cells. As subunits of nucleic acids they carry genetic information. They are also the primary carriers of chemical energy in cells, structural components of many enzyme cofactors, and cellular second messengers.

A nucleotide consists of a nitrogenous base (purine or pyrimidine), a pentose sugar, and one or more phosphate groups. Nucleic acids are polymers of nucleotides, joined together by phosphodiester linkages between the 5′ hydroxyl of one pentose and the 3′ hydroxyl of the next. There are two types of nucleic acid: RNA and DNA. The nucleotides in RNA contain ribose, and the common pyrimidine bases are uracil and cytosine. In DNA, the nucleotides contain 2′-deoxyribose, and the pyrimidine bases are generally thymine and cytosine. The primary purines are adenine and guanine in both RNA and DNA.

Many lines of evidence show that DNA bears genetic information. In particular, the Avery-MacLeod-McCarty experiment showed that DNA isolated from one strain of a bacterium can enter and transform the cells of another strain, endowing it with some of the inheritable characteristics of the donor. The Hershey-Chase experiment showed that the DNA of a bacterial virus, but not its protein coat, carries the genetic message for replication of the virus in a host cell.

From Franklin and Wilkins's x-ray diffraction studies of DNA fibers and Chargaff's discovery of the base equivalences in DNA (A = T and G = C), Watson and Crick postulated that native DNA consists of two antiparallel chains in a right-handed double-helical arrangement. Complementary base pairs, A═T and G≡C, are formed by hydrogen bonding within the helix, and the hydrophilic sugar–phosphate backbones are located on the outside. The base pairs are stacked perpendicular to the long axis, 3.4 Å apart; there are 10.5 base pairs in each complete turn of the double helix.

DNA can exist in several structural forms. Two variations of the Watson-Crick B-form DNA, the A and Z forms, have been characterized in DNA crystal structures. The A-form helix is shorter and of greater diameter than a B-form helix with the same sequence. The Z form is a left-handed helix. Some sequence-dependent structural variations cause bends in the DNA. DNA strands with self-complementary inverted repeats can form hairpin or cruciform structures. Some polypyrimidine or polypurine segments can pair with additional strands or duplexes of appropriate sequence to form triplex or tetraplex DNA. Polypyrimidine tracts arranged in mirror repeats can assume a triple-helical structure called H-DNA.

Messenger RNA is the vehicle by which genetic information is transferred from DNA to ribosomes for protein synthesis. Transfer RNA and ribosomal RNA are also involved in protein synthesis. RNA can be structurally complex, with single RNA strands often folded into hairpins, double-stranded regions, and complex loops.

Native DNA undergoes reversible unwinding and separation (melting) of strands on heating or at extremes of pH. Because G≡C base pairs are more stable than A═T pairs, the melting point of DNAs rich in G≡C pairs is higher than that of DNAs rich in A═T pairs. Denatured single-stranded DNAs from two species can form a hybrid duplex, the degree of hybridization depending on the extent of sequence similarity. Hybridization is the basis for important techniques used to study and isolate specific genes and RNAs.

DNA is a relatively stable polymer. Spontaneous reactions such as deamination of certain bases, hydrolysis of base–sugar *N*-glycosyl bonds, radiation-induced formation of pyrimidine dimers, and oxidative damage occur at very low rates, yet are important because of cells' very low tolerance for changes in genetic material. DNA sequences can be determined and DNA polymers synthesized using simple, automated protocols involving chemical and enzymatic methods.

ATP is the central carrier of chemical energy in cells. The presence of an adenosine moiety in a variety of enzyme cofactors may be related to binding-energy requirements. Cyclic AMP, formed from ATP in a reaction catalyzed by adenylate cyclase, is a common second messenger produced in response to hormones and other chemical signals.

further reading

General

Chang, K.Y. & Varani, G. (1997) Nucleic acids structure and recognition. *Nature Struct. Biol.* **4** (Suppl.), 854–858.

Describes the application of NMR to determination of nucleic acid structure.

Friedberg, E.C., Walker, G.C., & Siede, W. (1995) *DNA Repair and Mutagenesis,* W.H. Freeman and Company, New York.

A good source for more information on the chemistry of nucleotides and nucleic acids.

Hecht, S.M. (ed.) (1996) *Bioorganic Chemistry: Nucleic Acids,* Oxford University Press, Oxford.

A very useful set of articles.

Kornberg, A. & Baker, T.A. (1991) *DNA Replication,* 2nd edn, W.H. Freeman and Company, New York.

The best place to start learning more about DNA structure.

Sinden, R.R. (1994) *DNA Structure and Function*, Academic Press, Inc., San Diego.

Good discussion of many topics covered in this chapter.

Variations in DNA Structure

Frank-Kamenetskii, M.D. & Mirkin, S.M. (1995) Triplex DNA structures. *Annu. Rev. Biochem.* **64,** 65–95.

Herbert, A. & Rich, A. (1996) The biology of left-handed Z-DNA. *J. Biol. Chem.* **271,** 11,595–11,598.

Htun, H. & Dahlberg, J.E. (1989) Topology and formation of triple-stranded H-DNA. *Science* **243,** 1571–1576.

Moore, P.B. (1999) Structural motifs in RNA. *Annu. Rev. Biochem.* **68,** 287–300.

Shafer, R.H. (1998) Stability and structure of model DNA triplexes and quadruplexes and their interactions with small ligands. *Prog. Nucleic Acid Res. Mol. Biol.* **59,** 55–94.

Vasquez, K.M. & Wilson, J.H. (1998) Triplex-directed modification of genes and gene activity. *Trends Biochem. Sci.* **23,** 4–9.

Synthetic oligonucleotides with appropriate all-purine or all-pyrimidine sequences can be targeted to complementary sequences in the genome, forming stable triplexes for a variety of purposes in biotechnology.

Wells, R.D. (1988) Unusual DNA structures. *J. Biol. Chem.* **263,** 1095–1098.

Minireview; a concise summary.

ATP As Energy Carrier

Jencks, W.P. (1987) Economics of enzyme catalysis. *Cold Spring Harb. Symp. Quant. Biol.* **52,** 65–73.

A relatively short article, full of insights.

Historical

Judson, H.F. (1996) *The Eighth Day of Creation: Makers of the Revolution in Biology,* expanded edn, Cold Spring Harbor Laboratory Press, Cold Spring Harbor, NY.

Olby, R.C. (1994) *The Path to the Double Helix: The Discovery of DNA*, Dover Publications, Inc., New York.

Sayre, A. (1978) *Rosalind Franklin and DNA,* W.W. Norton & Co., Inc., New York.

Watson, J.D. (1968) *The Double Helix: A Personal Account of the Discovery of the Structure of DNA,* W.W. Norton & Co., Inc., New York. [Paperback edition (1981).]

problems

1. Determination of Protein Concentration in a Solution Containing Proteins and Nucleic Acids The concentration of protein or nucleic acid in a solution containing both can be estimated by using their different light absorption properties: proteins absorb most strongly at 280 nm and nucleic acids at 260 nm. Their respective concentrations in a mixture can be estimated by measuring the absorbance (A) of the solution at 280 nm and 260 nm and using the table below, which gives $R_{280/260}$, the ratio of absorbances at 280 and 260 nm; the percentage of total mass that is nucleic acid; and a factor, F, that corrects the A_{280} reading and gives a more accurate protein estimate. The protein concentration (in mg/ml) $= F \times A_{280}$ (assuming the cuvette is 1 cm wide). Calculate the protein concentration in a solution of $A_{280} = 0.69$ and $A_{260} = 0.94$.

$R_{280/260}$	Proportion of nucleic acid (%)	F
1.75	0.00	1.116
1.63	0.25	1.081
1.52	0.50	1.054
1.40	0.75	1.023
1.36	1.00	0.994
1.30	1.25	0.970
1.25	1.50	0.944
1.16	2.00	0.899
1.09	2.50	0.852
1.03	3.00	0.814
0.979	3.50	0.776
0.939	4.00	0.743
0.874	5.00	0.682
0.846	5.50	0.656
0.822	6.00	0.632
0.804	6.50	0.607
0.784	7.00	0.585
0.767	7.50	0.565
0.753	8.00	0.545
0.730	9.00	0.508
0.705	10.00	0.478
0.671	12.00	0.422
0.644	14.00	0.377
0.615	17.00	0.322
0.595	20.00	0.278

2. Nucleotide Structure Which positions in a purine ring of a purine nucleotide in DNA have the potential to form hydrogen bonds but are not involved in Watson-Crick base pairing?

3. Base Sequence of Complementary DNA Strands One strand of double-helical DNA has the sequence (5′)GCGCAATATTTCTCAAAATATTGCGC (3′). Write the base sequence of the complementary strand. What special type of sequence is contained in this DNA segment? Does the double-stranded DNA have the potential to form any alternative structures?

4. DNA of the Human Body Calculate the weight in grams of a double-helical DNA molecule stretching from the earth to the moon (~320,000 km). The DNA double helix weighs about 1×10^{-18} g per 1,000 nucleotide pairs; each base pair extends 3.4 Å. For an interesting comparison, your body contains about 0.5 g of DNA!

5. DNA Bending Assume that a poly(A) tract five base pairs long produces a bend in a DNA strand of about 20°. Calculate the total (net) bend produced in a DNA if the center base pairs (the third of five) of two successive $(dA)_5$ tracts are located (a) 10 base pairs apart; (b) 15 base pairs apart. Assume 10 base pairs per turn in the DNA double helix.

6. Distinction between DNA Structure and RNA Structure Hairpins may form at palindromic sequences in single strands of either RNA or DNA. How is the helical structure of a long and fully base-paired (except at the end) hairpin in RNA different from that of a similar hairpin in DNA?

7. Nucleotide Chemistry The cells of many eukaryotic organisms have highly specialized systems that specifically repair G–T mismatches in DNA. The mismatch is repaired to form a G≡C (not A=T) base pair. This G–T mismatch repair mechanism occurs in addition to a more general system that repairs virtually all mismatches. Can you suggest why cells might require a specialized system to repair G–T mismatches?

8. Nucleic Acid Structure Explain why the absorption of UV light increases (hyperchromic effect) when double-stranded DNA is denatured.

9. Base Pairing in DNA In samples of DNA isolated from two unidentified species of bacteria, adenine makes up 32% and 17%, respectively, of the total bases. What relative proportions of adenine, guanine, thymine, and cytosine would you expect to find in the two DNA samples? What assumptions have you made? One of these species was isolated from a hot spring (64 °C). Suggest which DNA came from this thermophilic bacterium. What is the basis for your answer?

10. DNA Sequencing The following DNA fragment was sequenced by the Sanger method. The red asterisk indicates a fluorescent label.

```
*5' ———————— 3'-OH
 3' ———————— ATTACGCAAGGACATTAGAC---5'
```

A sample of the DNA was reacted with DNA polymerase and each of the nucleotide mixtures (in an appropriate buffer) listed below. Dideoxynucleotides (ddNTPs) were added in relatively small amounts.

1. dATP, dTTP, dCTP, dGTP, ddTTP
2. dATP, dTTP, dCTP, dGTP, ddGTP
3. dATP, dCTP, dGTP, ddTTP
4. dATP, dTTP, dCTP, dGTP

The resulting DNA was separated by electrophoresis on an agarose gel, and the fluorescent bands on the gel were located. The band pattern resulting from nucleotide mixture 1 is shown. Assuming that all mixtures were run on the same gel, what did the remaining lanes of the gel look like?

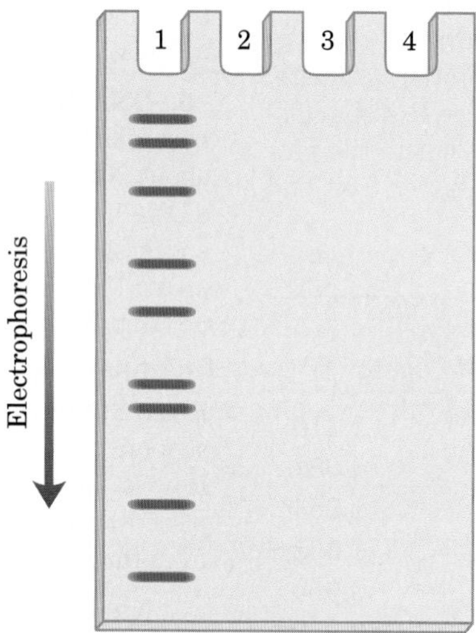

11. Solubility of the Components of DNA Draw the following structures and rate their relative solubilities in water (most soluble to least soluble): deoxyribose, guanine, phosphate. How are these solubilities consistent with the three-dimensional structure of double-stranded DNA?

12. Snake Venom Phosphodiesterase An exonuclease is an enzyme that sequentially cleaves nucleotides from the end of a polynucleotide strand. Snake venom phosphodiesterase, which hydrolyzes nucleotides from the 3′ end of any oligonucleotide with a free 3′-hydroxyl group, cleaves between the 3′ hydroxyl of the ribose or deoxyribose and the phosphoryl group of the next nucleotide. It acts on single-stranded DNA or RNA and has no base specificity. This enzyme was used in sequence determination experiments before the development of modern nucleic acid sequencing techniques. What are the products of partial digestion by snake venom phosphodiesterase of an oligonucleotide with the sequence (5′)GCGCCAUUGC(3′)—OH?

13. Preserving DNA in Bacterial Endospores Bacterial endospores form when the environment is no longer conducive to active cell metabolism. The soil bacterium *Bacillus subtilis,* for example, begins the process of sporulation when one or more nutrients are depleted. The end product is a small, metabolically dormant structure that can survive almost indefinitely with no detectable metabolism. Spores have mechanisms to prevent accumulation of potentially lethal mutations in their DNA over periods of dormancy that can exceed 1,000 years. *Bacillus subtilis* spores are much more resistant than the organism's growing cells to heat, UV radiation, and oxidizing agents, all of which promote mutations.

(a) One factor that prevents potential DNA damage in spores is their greatly decreased water content. How would this affect some types of mutations?

(b) Endospores have a category of proteins called small acid-soluble proteins (SASPs) that bind to their DNA, preventing formation of cyclobutane-type dimers. What causes cyclobutane dimers, and why do bacterial endospores need mechanisms to prevent their formation?

Biochemistry on the Internet

14. The Structure of DNA Elucidation of the three-dimensional structure of DNA helped researchers to understand how this molecule conveys information that can be faithfully replicated from one generation to the next. To see the secondary structure of double-stranded DNA, go to the Protein Data Bank Web site. Use the PDB identifiers listed below to retrieve the data pages for the two forms of DNA. Open the structures using Chime and use the different viewing options to complete the following exercises.

(a) Obtain the file for 141D, a highly conserved DNA sequence from an HIV-1 long terminal repeat region. Display the molecule as a stick or ball-and-stick structure. Identify the sugar-phosphate backbone for each strand of the DNA duplex. Locate and identify individual bases. Which is the 5′ end of this molecule? Locate the major and minor grooves. Is this a right- or left-handed helix?

(b) Obtain the file for 145D, DNA in the Z conformation. Display the molecule as a stick or ball-and-stick structure. Identify the sugar-phosphate backbone for each strand of the DNA duplex. Is this a right- or left-handed helix?

(c) To fully appreciate the secondary structure of DNA, turn on the Stereo option in the viewer. You will see two images of the DNA molecule. Sit with your nose approximately 10 inches from the monitor and focus on the tip of your nose. In the background you should see three images of the DNA helix. Shift your focus from the tip of your nose to the middle image, which should appear three-dimensional. For additional tips, see the Study Guide or http://www.worthpublishers.com/lehninger.

chapter

11

Lipids

Biological lipids are a chemically diverse group of compounds, the common and defining feature of which is their insolubility in water. The biological functions of the lipids are as diverse as their chemistry. Fats and oils are the principal stored forms of energy in many organisms. Phospholipids and sterols are major structural elements of biological membranes. Other lipids, although present in relatively small quantities, play crucial roles as enzyme cofactors, electron carriers, light-absorbing pigments, hydrophobic anchors, emulsifying agents, hormones, and intracellular messengers. This chapter introduces representative lipids of each type, with emphasis on their chemical structure and physical properties.

Storage Lipids

The fats and oils used almost universally as stored forms of energy in living organisms are derivatives of **fatty acids.** The fatty acids are hydrocarbon derivatives, at about the same low oxidation state (that is, as highly reduced) as the hydrocarbons in fossil fuels. The cellular oxidation of fatty acids (to CO_2 and H_2O), like the controlled, rapid burning of fossil fuels in internal combustion engines, is highly exergonic.

We introduce here the structures and nomenclature of the fatty acids most commonly found in living organisms. Two types of fatty acid–containing compounds, triacylglycerols and waxes, are described to illustrate the diversity of structure and physical properties in this family of compounds.

Fatty Acids Are Hydrocarbon Derivatives

Fatty acids are carboxylic acids with hydrocarbon chains ranging from 4 to 36 carbons long (C_4 to C_{36}). In some fatty acids, this chain is fully saturated (contains no double bonds) and unbranched; in others the chain contains one or more double bonds (Table 11–1). A few contain three-carbon rings, hydroxyl groups, or methyl-group branches. A simplified nomenclature for these compounds specifies the chain length and number of double bonds, separated by a colon; the 16-carbon saturated palmitic acid is abbreviated 16:0, and the 18-carbon oleic acid, with one double bond, is 18:1. The positions of any double bonds are specified by superscript numbers following Δ (delta); a 20-carbon fatty acid with one double bond between C-9 and C-10 (C-1 being the carboxyl carbon) and another between C-12 and C-13 is designated 20:2($\Delta^{9,12}$). The most commonly occurring fatty acids have even numbers of carbon atoms in an unbranched chain of 12 to 24 carbons (Table 11–1). As we shall see in Chapter 21, the even number of carbons results

table 11–1

Some Naturally Occurring Fatty Acids

Carbon skeleton	Structure*	Systematic name†	Common name (derivation)	Melting point (°C)	Solubility at 30 °C (mg/g solvent) Water	Benzene
12:0	$CH_3(CH_2)_{10}COOH$	*n*-Dodecanoic acid	Lauric acid (Latin *laurus*, "laurel plant")	44.2	0.063	2,600
14:0	$CH_3(CH_2)_{12}COOH$	*n*-Tetradecanoic acid	Myristic acid (Latin *Myristica*, nutmeg genus)	53.9	0.024	874
16:0	$CH_3(CH_2)_{14}COOH$	*n*-Hexadecanoic acid	Palmitic acid (Latin *palma*, "palm tree")	63.1	0.0083	348
18:0	$CH_3(CH_2)_{16}COOH$	*n*-Octadecanoic acid	Stearic acid (Greek *stear*, "hard fat")	69.6	0.0034	124
20:0	$CH_3(CH_2)_{18}COOH$	*n*-Eicosanoic acid	Arachidic acid (Latin *Arachis*, legume genus)	76.5		
24:0	$CH_3(CH_2)_{22}COOH$	*n*-Tetracosanoic acid	Lignoceric acid (Latin *lignum*, "wood" + *cera*, "wax")	86.0		
16:1(Δ^9)	$CH_3(CH_2)_5CH{=}CH(CH_2)_7COOH$	*cis*-9-Hexadecenoic acid	Palmitoleic acid	−0.5		
18:1(Δ^9)	$CH_3(CH_2)_7CH{=}CH(CH_2)_7COOH$	*cis*-9-Octadecenoic acid	Oleic acid (Latin *oleum*, "oil")	13.4		
18:2($\Delta^{9,12}$)	$CH_3(CH_2)_4CH{=}CHCH_2CH{=}CH(CH_2)_7COOH$	*cis*-,*cis*-9,12-Octadecadienoic acid	Linoleic acid (Greek *linon*, "flax")	−5		
18:3($\Delta^{9,12,15}$)	$CH_3CH_2CH{=}CHCH_2CH{=}CHCH_2CH{=}CH(CH_2)_7COOH$	*cis*-,*cis*-,*cis*-9,12,15-Octadecatrienoic acid	α-Linolenic acid	−11		
20:4($\Delta^{5,8,11,14}$)	$CH_3(CH_2)_4CH{=}CHCH_2CH{=}CHCH_2CH{=}CHCH_2CH{=}CH(CH_2)_3COOH$	*cis*-,*cis*-,*cis*-,*cis*-5,8,11,14-Icosatetraenoic acid	Arachidonic acid	−49.5		

*All acids are shown in their nonionized form. At pH 7, all free fatty acids have an ionized carboxylate. Note that numbering of carbon atoms begins at the carboxyl carbon.

†The prefix *n*- indicates the "normal" unbranched structure. For instance, "dodecanoic" simply indicates 12 carbon atoms, which could be arranged in a variety of branched forms; "*n*-dodecanoic" specifies the linear, unbranched form. For unsaturated fatty acids, the configuration of each double bond is indicated; in biological fatty acids the configuration is almost always cis.

from the mode of synthesis of these compounds, which involves condensation of acetate (two-carbon) units.

There is also a common pattern in the location of double bonds; in most monounsaturated fatty acids the double bond is between C-9 and C-10 (Δ^9), and the other double bonds of polyunsaturated fatty acids are generally Δ^{12} and Δ^{15}. (Arachidonic acid is an exception to this generalization.) The double bonds of polyunsaturated fatty acids are almost never conjugated (alternating single and double bonds, as in —CH=CH—CH=CH—), but are separated by a methylene group (—CH=CH—CH_2—CH=CH—). In nearly all naturally occurring unsaturated fatty acids, the double bonds are in the cis configuration.

The physical properties of the fatty acids, and of compounds that contain them, are largely determined by the length and degree of unsaturation of the hydrocarbon chain. The nonpolar hydrocarbon chain accounts for the poor solubility of fatty acids in water. Lauric acid (12:0, M_r 200), for example, has a solubility in water of 0.063 mg/g—much less than that of glucose

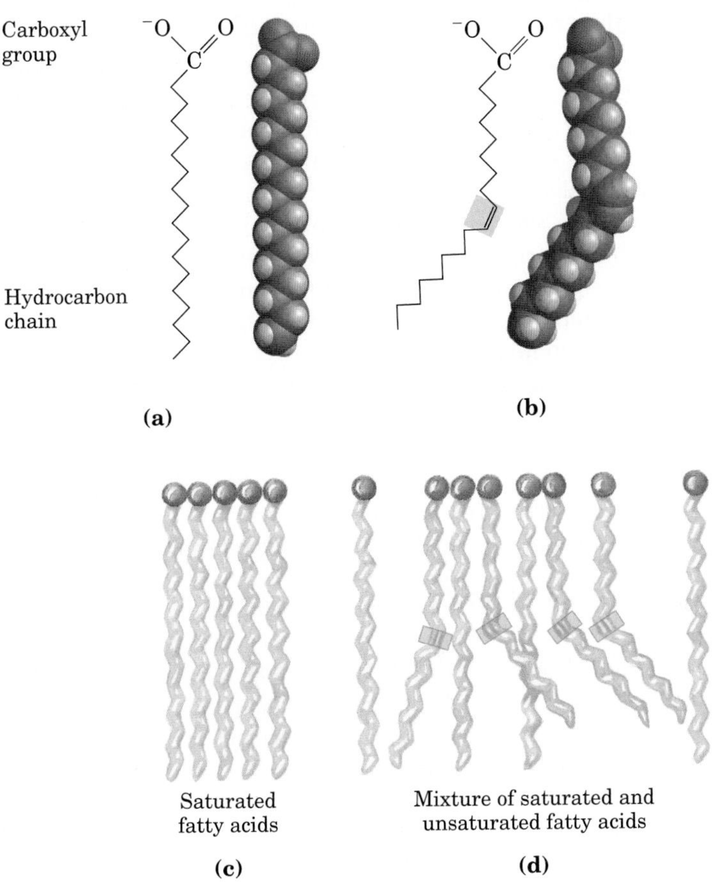

figure 11–1
The packing of fatty acids into stable aggregates. The extent of packing depends on the degree of saturation. **(a)** Two representations of the fully saturated acid stearic acid (stearate at pH 7) in its usual extended conformation. Each line segment of the zig-zag represents a single bond between adjacent carbons. **(b)** The cis double bond (shaded) in oleic acid (oleate) does not permit rotation and introduces a rigid bend in the hydrocarbon tail. All other bonds in the chain are free to rotate. **(c)** Fully saturated fatty acids in the extended form pack into nearly crystalline arrays, stabilized by many hydrophobic interactions. **(d)** The presence of one or more cis double bonds interferes with this tight packing and results in less stable aggregates.

(M_r 180), which is 1,100 mg/g. The longer the fatty acyl chain and the fewer the double bonds, the lower is the solubility in water. The carboxylic acid group is polar (and ionized at neutral pH) and accounts for the slight solubility of short-chain fatty acids in water.

Melting points are also strongly influenced by the length and degree of unsaturation of the hydrocarbon chain. At room temperature (25 °C), the saturated fatty acids from 12:0 to 24:0 have a waxy consistency, whereas unsaturated fatty acids of these lengths are oily liquids. This difference in melting points is due to different degrees of packing of the fatty acid molecules (Fig. 11–1). In the fully saturated compounds, free rotation around each carbon–carbon bond gives the hydrocarbon chain great flexibility; the most stable conformation is the fully extended form, in which the steric hindrance of neighboring atoms is minimized. These molecules can pack together tightly in nearly crystalline arrays, with atoms all along their lengths in van der Waals contact with the atoms of neighboring molecules. In unsaturated fatty acids, a cis double bond forces a kink in the hydrocarbon chain. Fatty acids with one or several such kinks cannot pack together as tightly as fully saturated fatty acids and their interactions with each other are therefore weaker. Because it takes less thermal energy to disorder these poorly ordered arrays of unsaturated fatty acids, they have markedly lower melting points than saturated fatty acids of the same chain length (Table 11–1).

In vertebrates, free fatty acids (unesterified fatty acids having a free carboxylate group) circulate in the blood bound noncovalently to a protein carrier, serum albumin. However, fatty acids are present in blood plasma mostly as carboxylic acid derivatives such as esters or amides. Lacking the charged carboxylate group, these fatty acid derivatives are generally even less soluble in water than are the free fatty acids.

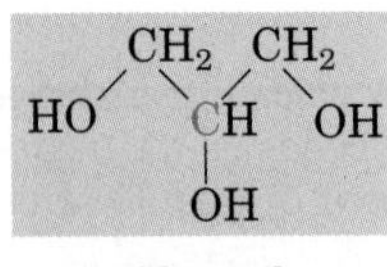

Glycerol

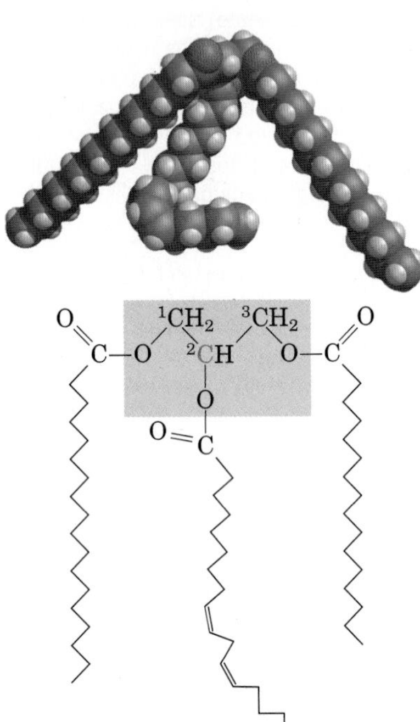

1-Stearoyl, 2-linoleoyl, 3-palmitoyl glycerol, a mixed triacylglycerol

figure 11–2

Glycerol and a triacylglycerol. The mixed triacylglycerol shown here has three different fatty acids attached to the glycerol backbone. When glycerol has two different fatty acids at C-1 and C-3, the C-2 is a chiral center (p. 59).

Triacylglycerols Are Fatty Acid Esters of Glycerol

The simplest lipids constructed from fatty acids are the **triacylglycerols,** also referred to as triglycerides, fats, or neutral fats. Triacylglycerols are composed of three fatty acids each in ester linkage with a single glycerol (Fig. 11–2). Those containing the same kind of fatty acid in all three positions are called simple triacylglycerols and are named after the fatty acid they contain. Simple triacylglycerols of 16:0, 18:0, and 18:1, for example, are tristearin, tripalmitin, and triolein, respectively. Most naturally occurring triacylglycerols are mixed; they contain two or more different fatty acids. To name these compounds unambiguously, the name and position of each fatty acid must be specified.

Because the polar hydroxyls of glycerol and the polar carboxylates of the fatty acids are bound in ester linkages, triacylglycerols are nonpolar, hydrophobic molecules, essentially insoluble in water. Lipids have lower specific gravities than water, which explains why mixtures of oil and water (oil-and-vinegar salad dressing, for example) have two phases: oil, with the lower specific gravity, floats on the aqueous phase.

Triacylglycerols Provide Stored Energy and Insulation

In most eukaryotic cells, triacylglycerols form a separate phase of microscopic, oily droplets in the aqueous cytosol, serving as depots of metabolic fuel (Fig. 11–3). In vertebrates, specialized cells called adipocytes, or fat cells, store large amounts of triacylglycerols as fat droplets that nearly fill the cell. Triacylglycerols are also stored as oils in the seeds of many types of plants, providing energy and biosynthetic precursors during seed germination (Fig. 11–3b). Adipocytes and germinating seeds contain **lipases,** enzymes that catalyze the hydrolysis of stored triacylglycerols, releasing fatty acids for export to sites where they are required as fuel.

There are two significant advantages to using triacylglycerols as stored fuels rather than polysaccharides such as glycogen and starch. First, because the carbon atoms of fatty acids are more reduced than those of sugars, oxidation of triacylglycerols yields more than twice as much energy, gram for gram, as the oxidation of carbohydrates. Second, because triacylglycerols are hydrophobic and therefore unhydrated, the organism that carries fat as fuel does not have to carry the extra weight of water of hydration

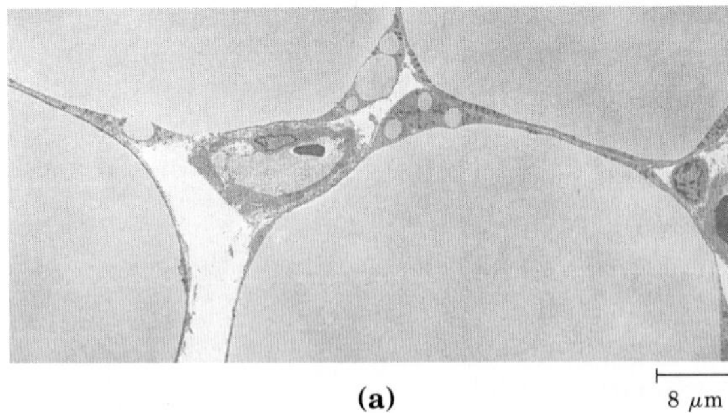

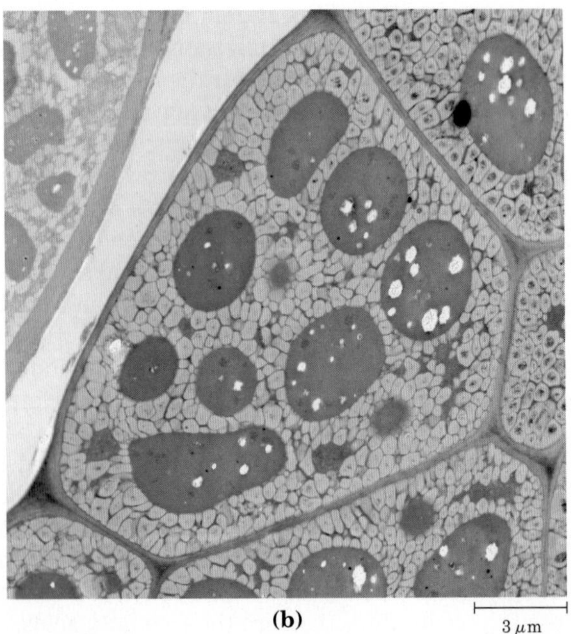

figure 11–3

Fat stores in cells. (a) Cross section of four guinea pig adipocytes, showing huge fat droplets that virtually fill the cells. Also visible are several capillaries in cross section. **(b)** Cross section of a cotyledon cell from a seed of the plant *Arabidopsis*. The large dark structures are protein bodies, which are surrounded by stored oils in the light-colored oil bodies.

box 11–1 Sperm Whales: Fatheads of the Deep

Studies of sperm whales have uncovered another way in which triacylglycerols are biologically useful. The sperm whale's head is very large, accounting for over one-third of its total body weight. About 90% of the weight of the head is made up of the spermaceti organ, a blubbery mass that contains up to 18,000 kg (about 4 tons) of spermaceti oil, a mixture of triacylglycerols and waxes containing an abundance of unsaturated fatty acids. This mixture is liquid at the normal resting body temperature of the whale, about 37 °C, but it begins to crystallize at about 31 °C and becomes solid when the temperature drops several more degrees.

The probable biological function of spermaceti oil has been deduced from research on the anatomy and feeding behavior of the sperm whale. These mammals feed almost exclusively on squid in very deep water. In their feeding dives they descend 1,000 m or more; the deepest recorded dive is 3,000 m (almost 2 miles). At these depths, there are no competitors for the very plentiful squid; the sperm whale rests quietly, waiting for schools of squid to pass.

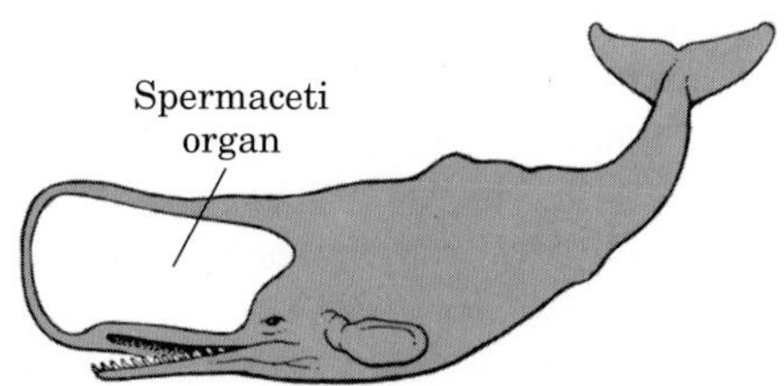

For a marine animal to remain at a given depth without a constant swimming effort, it must have the same density as the surrounding water. The sperm whale undergoes changes in buoyancy to match the density of its surroundings—from the tropical ocean surface to great depths where the water is much colder and thus denser. The key is the freezing point of spermaceti oil. When the temperature of the oil is lowered several degrees during a deep dive, it congeals or crystallizes and becomes denser. Thus the buoyancy of the whale changes to match the density of seawater. Various physiological mechanisms promote rapid cooling of the oil during a dive. During the return to the surface, the congealed spermaceti oil warms and melts, decreasing its density to match that of the surface water. Thus we see in the sperm whale a remarkable anatomical and biochemical adaptation. The triacylglycerols and waxes synthesized by the sperm whale contain fatty acids of the necessary chain length and degree of unsaturation to give the spermaceti oil the proper melting point for the animal's diving habits.

Unfortunately for the sperm whale population, spermaceti oil was at one time considered the finest lamp oil and continues to be commercially valuable as a lubricant. Several centuries of intensive hunting of these mammals have driven sperm whales onto the endangered species list, with a total worldwide population of about 500,000 remaining.

that is associated with stored polysaccharides (2 g per gram of polysaccharide). Humans have fat tissue, which is composed primarily of adipocytes, under the skin, in the abdominal cavity, and in the mammary glands. Moderately obese people with 15 to 20 kg of triacylglycerols deposited in their adipocytes could meet their energy needs for months by drawing on their fat stores. In contrast, the human body can store less than a day's energy supply in the form of glycogen. Carbohydrates such as glucose and glycogen do offer certain advantages as quick sources of metabolic energy, one of which is their ready solubility in water.

In some animals, triacylglycerols stored under the skin serve not only as energy stores but as insulation against low temperatures. Seals, walruses, penguins, and other warm-blooded polar animals are amply padded with triacylglycerols. In hibernating animals (bears, for example), the huge fat reserves accumulated before hibernation serve the dual purposes of insulation and energy storage (see Box 17–1, "Fat Bears Carry on β-Oxidation in Their Sleep"). The low density of triacylglycerols is the basis for another remarkable function of these compounds. In sperm whales, a store of triacylglycerols and waxes allows the animals to match the buoyancy of their bodies to that of their surroundings during deep dives in cold water (Box 11–1).

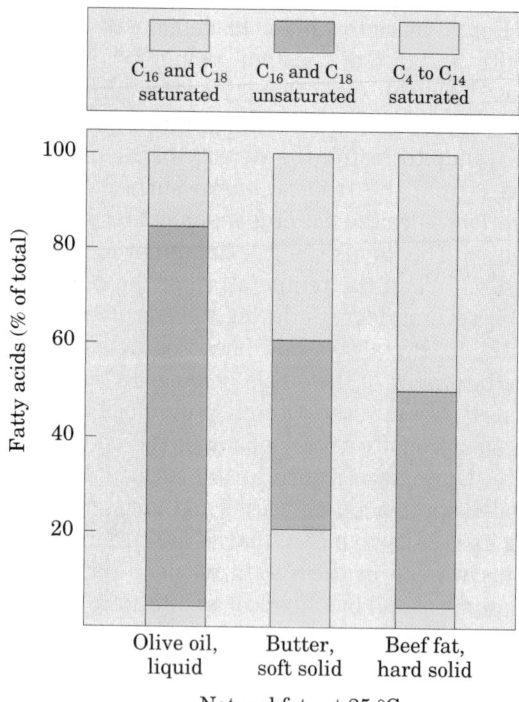

figure 11–4

Fatty acid composition of three food fats. Olive oil, butter, and beef fat consist of mixtures of triacylglycerols, differing in their fatty acid composition. The melting points of these fats—and hence their physical state at room temperature (25 °C)—are a direct function of their fatty acid composition. Olive oil has a high proportion of long-chain (C_{16} and C_{18}) unsaturated fatty acids, which accounts for its liquid state at 25 °C. The higher proportion of long-chain (C_{16} and C_{18}) saturated fatty acids in butter increases its melting point, so butter is a soft solid at room temperature. Beef fat, with an even higher proportion of long-chain saturated fatty acids, is a hard solid.

Many Foods Contain Triacylglycerols

Most natural fats, such as those in vegetable oils, dairy products, and animal fat, are complex mixtures of simple and mixed triacylglycerols. These contain a variety of fatty acids differing in chain length and degree of saturation (Fig. 11–4). Vegetable oils such as corn (maize) and olive oil are composed largely of triacylglycerols with unsaturated fatty acids and thus are liquids at room temperature. They are converted industrially into solid fats by catalytic hydrogenation, which reduces some of their double bonds to single bonds. Triacylglycerols containing only saturated fatty acids, such as tristearin, the major component of beef fat, are white, greasy solids at room temperature.

When lipid-rich foods are exposed too long to the oxygen in air, they may spoil and become rancid. The unpleasant taste and smell associated with rancidity result from the oxidative cleavage of the double bonds in unsaturated fatty acids, which produces aldehydes and carboxylic acids of shorter chain length and therefore higher volatility.

Waxes Serve as Energy Stores and Water Repellents

Biological waxes are esters of long-chain (C_{14} to C_{36}) saturated and unsaturated fatty acids with long-chain (C_{16} to C_{30}) alcohols (Fig. 11–5). Their melting points (60 to 100 °C) are generally higher than those of triacylglycerols. In plankton, the free-floating marine microorganisms at the bottom of the food chain, waxes are the chief storage form of metabolic fuel.

Waxes also serve a diversity of other functions in nature related to their water-repellent properties and their firm consistency. Certain skin glands of vertebrates secrete waxes to protect hair and skin and keep it pliable, lubricated, and waterproof. Birds, particularly waterfowl, secrete waxes from their preen glands to keep their feathers water-repellent. The shiny leaves of holly, rhododendrons, poison ivy, and many tropical plants are coated with a thick layer of waxes, which prevents excessive evaporation of water and protects against parasites.

Biological waxes find a variety of applications in the pharmaceutical, cosmetic, and other industries. Lanolin (from lamb's wool), beeswax (Fig. 11–5), carnauba wax (from a Brazilian palm tree), and wax extracted from spermaceti oil (from whales; see Box 11–1) are widely used in the manufacture of lotions, ointments, and polishes.

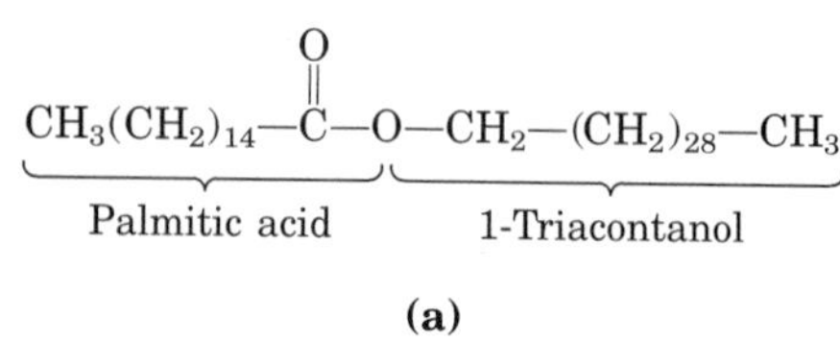

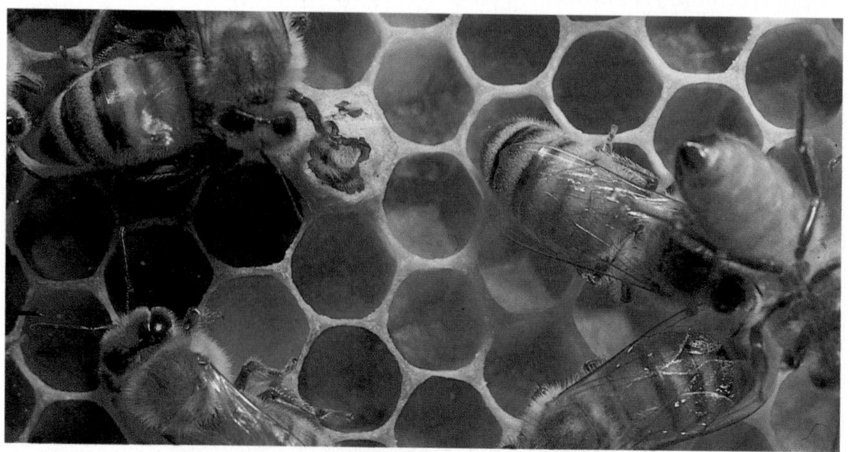

(b)

figure 11–5

Biological wax. (a) Triacontanoylpalmitate, the major component of beeswax, is an ester of palmitic acid with the alcohol triacontanol. **(b)** A honeycomb, constructed of beeswax, is firm at 25 °C and completely impervious to water. The term "wax" originates in the Old English word *weax,* meaning "the material of the honeycomb."

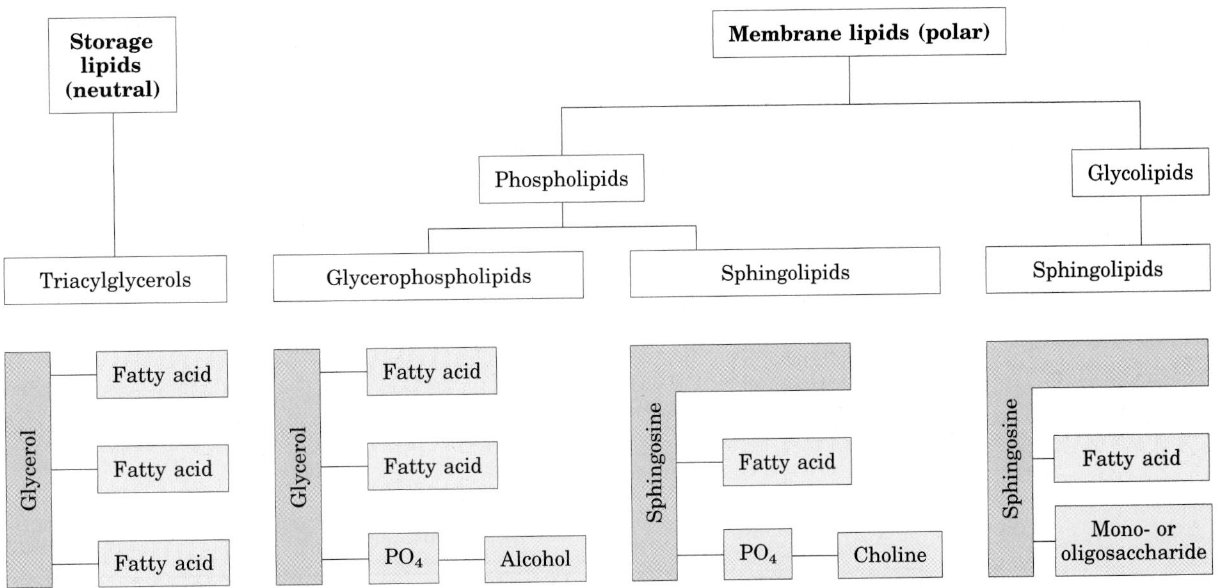

figure 11–6

The principal classes of storage and membrane lipids. All the lipids shown here have either glycerol or sphingosine as the backbone. A third class of membrane lipids, the sterols, is described later (see Fig. 11–14).

Structural Lipids in Membranes

The central architectural feature of biological membranes is a double layer of lipids, which acts as a barrier to the passage of polar molecules and ions. Membrane lipids are amphipathic; one end of the molecule is hydrophobic, the other hydrophilic. Their hydrophobic interactions with each other and their hydrophilic interactions with water direct their packing into sheets called membrane bilayers. In the next sections we describe three general types of membrane lipids: glycerophospholipids, in which the hydrophobic regions are composed of two fatty acids joined to glycerol; sphingolipids, in which a single fatty acid is joined to a fatty amine, sphingosine; and sterols, compounds characterized by a rigid system of four fused hydrocarbon rings. The hydrophilic moieties in these amphipathic compounds may be as simple as a single —OH group at one end of the sterol ring system, or they may be much more complex. In glycerophospholipids and some sphingolipids, a polar head group is joined to the hydrophobic moiety by a phosphodiester linkage; these are the **phospholipids.** Other sphingolipids lack phosphate but have a simple sugar or complex oligosaccharide at their polar ends; these are the **glycolipids** (Fig. 11–6). Within these groups of membrane lipids, enormous diversity results from various combinations of fatty acid "tails" and polar "heads." The arrangement of these lipids in membranes, and their structural and functional roles therein, are considered in the next chapter.

Glycerophospholipids Are Derivatives of Phosphatidic Acid

Glycerophospholipids, also called phosphoglycerides, are membrane lipids in which two fatty acids are attached in ester linkage to the first and second carbons of glycerol, and a highly polar or charged group is attached through a phosphodiester linkage to the third carbon. Glycerol is prochiral; it has no asymmetric carbons, but attachment of phosphate at either end converts it into a chiral compound, which can be correctly named either L-glycerol 3-phosphate or D-glycerol 1-phosphate (Fig. 11–7). Glycerophospholipids are named as derivatives of the parent compound phosphatidic acid (Fig. 11–8), according to the polar alcohol in the head group.

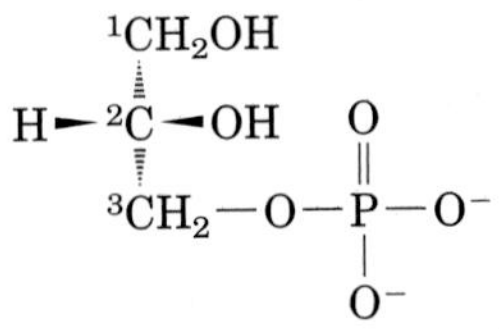

$$
\begin{array}{l}
\quad\ \ {}^{1}CH_2OH \\
H-{}^{2}C-OH \quad\ \ O \\
\qquad\ \ \qquad\qquad\ \ \| \\
\quad\ \ {}^{3}CH_2-O-P-O^- \\
\qquad\qquad\qquad\ \ | \\
\qquad\qquad\qquad O^-
\end{array}
$$

figure 11–7

L-Glycerol 3-phosphate, the backbone of phospholipids. Note that this compound can also be called D-glycerol 1-phosphate.

Phosphatidylcholine and phosphatidylethanolamine have choline and ethanolamine in their polar head groups, for example. In all of these compounds, the head group is joined to glycerol through a phosphodiester bond, in which the phosphate group bears a negative charge at neutral pH. The polar alcohol may be negatively charged (as in phosphatidylinositol

Glycerophospholipid (general structure)

$$\begin{array}{l} {}^{1}CH_2{-}O{-}C(=O){-}\text{(saturated chain)} \\ {}^{2}CH{-}O{-}C(=O){-}\text{(unsaturated chain)} \\ {}^{3}CH_2{-}O{-}P(=O)(O^-){-}O{-}X \end{array}$$

Saturated fatty acid (e.g., palmitic acid)

Unsaturated fatty acid (e.g., oleic acid)

X: Head-group substituent

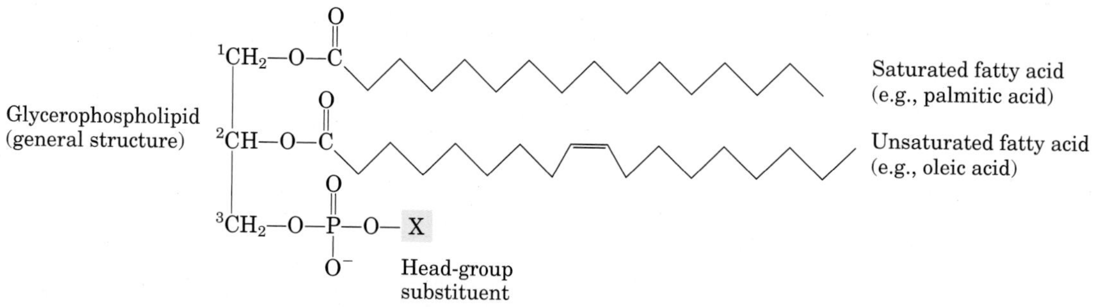

Name of glycerophospholipid	Name of X	Formula of X	Net charge (at pH 7)
Phosphatidic acid	—	$-H$	−1
Phosphatidylethanolamine	Ethanolamine	$-CH_2-CH_2-\overset{+}{N}H_3$	0
Phosphatidylcholine	Choline	$-CH_2-CH_2-\overset{+}{N}(CH_3)_3$	0
Phosphatidylserine	Serine	$-CH_2-CH(COO^-)-\overset{+}{N}H_3$	−1
Phosphatidylglycerol	Glycerol	$-CH_2-CH(OH)-CH_2-OH$	−1
Phosphatidylinositol 4,5-bisphosphate	*myo*-Inositol 4,5-bisphosphate	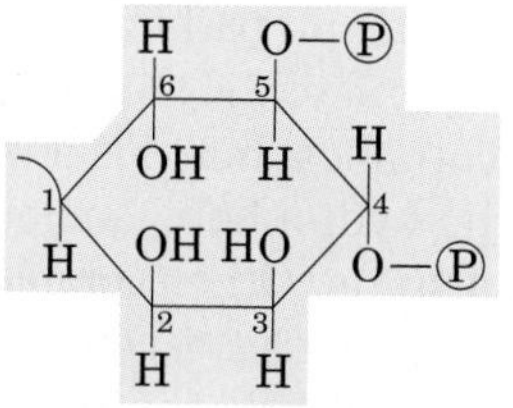 inositol ring (C-1 attached; OH at 2, 3, 6; O–Ⓟ at 4 and 5)	−4
Cardiolipin	Phosphatidyl-glycerol	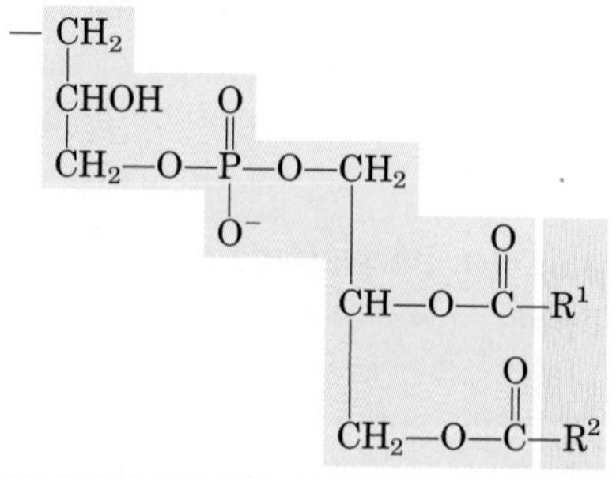 $-CH_2-CHOH-CH_2-O-P(=O)(O^-)-O-CH_2-CH(O-C(=O)-R^1)-CH_2-O-C(=O)-R^2$	−2

figure 11–8

Glycerophospholipids. The common glycerophospholipids are diacylglycerols linked to head-group alcohols through a phosphodiester bond. Phosphatidic acid, a phosphomonoester, is the parent compound. Each derivative is named for the head-group alcohol (X), with the prefix "phosphatidyl-." In cardiolipin, two phosphatidic acids share a single glycerol.

4,5-bisphosphate), neutral (phosphatidylserine), or positively charged (phosphatidylcholine, phosphatidylethanolamine). As we shall see in Chapter 12, these charges contribute significantly to the surface properties of membranes.

The fatty acids in glycerophospholipids can be any of a wide variety, so a given phospholipid (phosphatidylcholine, for example) may consist of a number of molecular species, each with its unique complement of fatty acids. The distribution of molecular species is specific for different organisms, different tissues of the same organism, and different glycerophospholipids in the same cell or tissue. In general, glycerophospholipids contain a C_{16} or C_{18} saturated fatty acid at C-1 and a C_{18} to C_{20} unsaturated fatty acid at C-2. With few exceptions, the biological significance of the variation in fatty acids and head groups is not yet understood.

Some Phospholipids Have Ether-Linked Fatty Acids

Some animal tissues and some unicellular organisms are rich in ether lipids, in which one of the two acyl chains is attached to glycerol in ether, rather than ester, linkage. The ether-linked chain may be saturated, as in the alkyl ether lipids, or may contain a double bond between C-1 and C-2, as in **plasmalogens** (Fig. 11–9). Vertebrate heart tissue is uniquely enriched in ether lipids; about half of the heart phospholipids are plasmalogens. The membranes of halophilic bacteria, ciliated protists, and certain invertebrates also contain high proportions of ether lipids. The functional significance of ether lipids in these membranes is unknown; perhaps their resistance to the phospholipases that cleave ester-linked fatty acids from membrane lipids is important in some roles. At least one ether lipid, **platelet-activating factor,** is a potent molecular signal. It is released from leukocytes called basophils and stimulates platelet aggregation and the release of serotonin (a vasoconstrictor) from platelets. It also exerts a variety of effects on liver, smooth muscle, heart, uterine, and lung tissues, and plays an important role in inflammation and the allergic response.

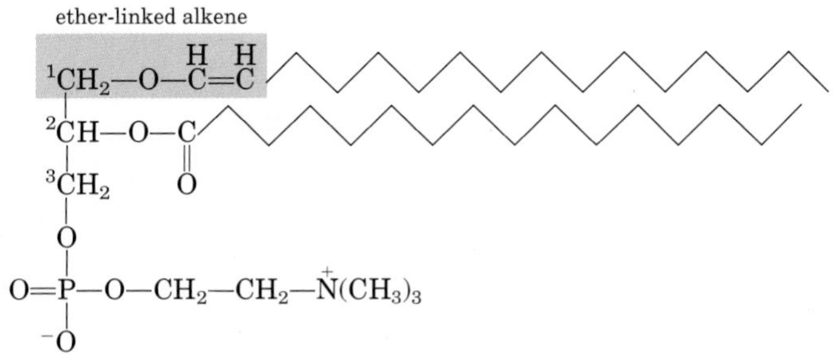

ether-linked alkene

Plasmalogen

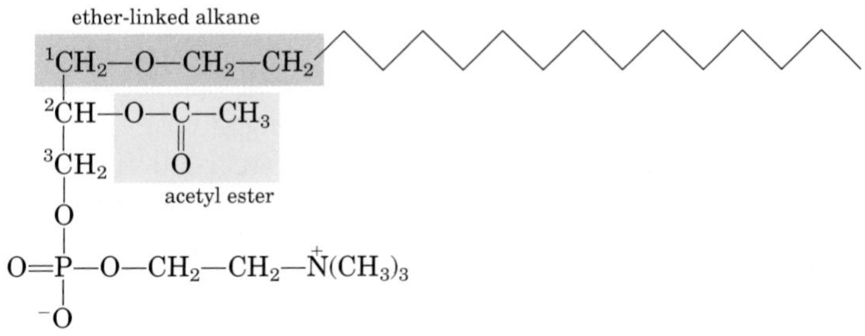

ether-linked alkane

acetyl ester

Platelet-activating factor

figure 11–9

Ether lipids. Plasmalogens have an ether-linked alkenyl chain where most glycerophospholipids have an ester-linked fatty acid (compare Fig. 11–8). Platelet-activating factor has a long ether-linked alkyl chain at C-1 of glycerol, but C-2 is ester-linked to acetic acid, which makes the compound much more water-soluble than most glycerophospholipids and plasmalogens. The head-group alcohol is choline in plasmalogens and in platelet-activating factor.

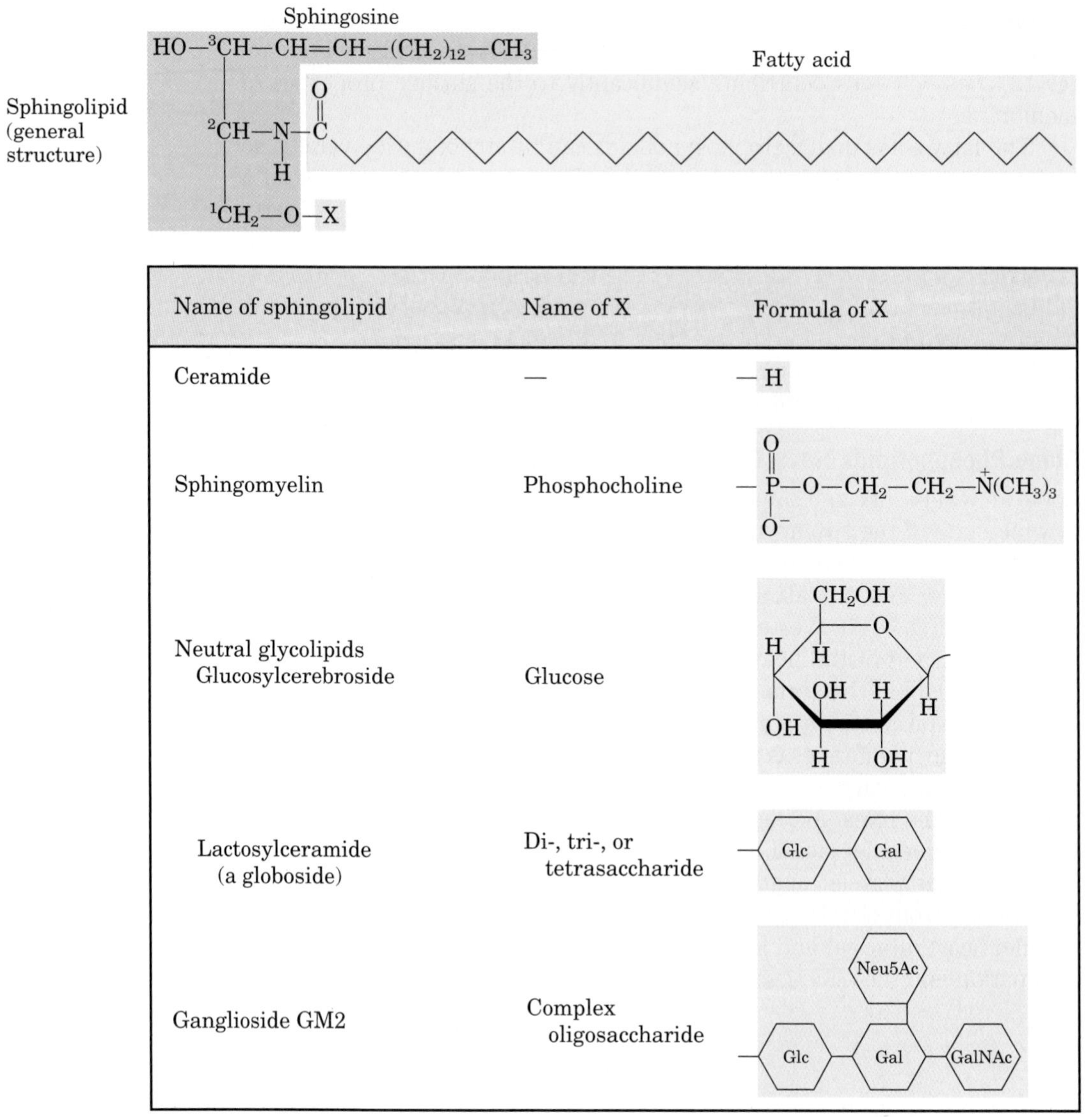

Sphingosine

$HO{-}^{3}CH{-}CH{=}CH{-}(CH_2)_{12}{-}CH_3$

Fatty acid

Sphingolipid (general structure)

$^{2}CH{-}N(H){-}C(=O){-}$

$^{1}CH_2{-}O{-}X$

Name of sphingolipid	Name of X	Formula of X
Ceramide	—	$-H$
Sphingomyelin	Phosphocholine	$-P(=O)(O^-){-}O{-}CH_2{-}CH_2{-}\overset{+}{N}(CH_3)_3$
Neutral glycolipids Glucosylcerebroside	Glucose	(glucose ring: CH_2OH, O, H, OH)
Lactosylceramide (a globoside)	Di-, tri-, or tetrasaccharide	—Glc—Gal
Ganglioside GM2	Complex oligosaccharide	—Glc—Gal(—Neu5Ac)—GalNAc

figure 11–10

Sphingolipids. The first three carbons at the polar end of sphingosine are analogous to the three carbons of glycerol in glycerophospholipids. The amino group at C-2 bears a fatty acid in amide linkage. The fatty acid is usually saturated or monounsaturated, with 16, 18, 22, or 24 carbon atoms. Ceramide is the parent compound for this group. Other sphingolipids differ in the polar head group (X) attached at C-1. Gangliosides have very complex oligosaccharide head groups. Standard abbreviations for sugars are used in this figure: Glc, D-glucose; Gal, D-galactose; GalNAc, *N*-acetyl-D-galactosamine; Neu5Ac, *N*-acetylneuraminic acid (sialic acid).

Sphingolipids Are Derivatives of Sphingosine

Sphingolipids, the second large class of membrane lipids, also have a polar head group and two nonpolar tails, but unlike glycerophospholipids they contain no glycerol. Sphingolipids are composed of one molecule of the long-chain amino alcohol sphingosine (4-sphingenine) or one of its derivatives, one molecule of a long-chain fatty acid, and a polar head group that is joined by a glycosidic linkage in some cases and by a phosphodiester in others (Fig. 11–10).

Carbons C-1, C-2, and C-3 of the sphingosine molecule are structurally analogous to the three carbons of glycerol in glycerophospholipids. When a fatty acid is attached in amide linkage to the $-NH_2$ on C-2, the resulting compound is a **ceramide,** which is structurally similar to a diacylglycerol. Ceramide is the structural parent of all sphingolipids.

There are three subclasses of sphingolipids, all derivatives of ceramide but differing in their head groups: sphingomyelins, neutral (uncharged) glycolipids, and gangliosides. **Sphingomyelins** contain phosphocholine or phosphoethanolamine as their polar head group and are therefore classified along with glycerophospholipids as phospholipids (Fig. 11–6). Indeed, sphingomyelins resemble phosphatidylcholines in their general properties and three-dimensional structure, and in having no net charge on their head

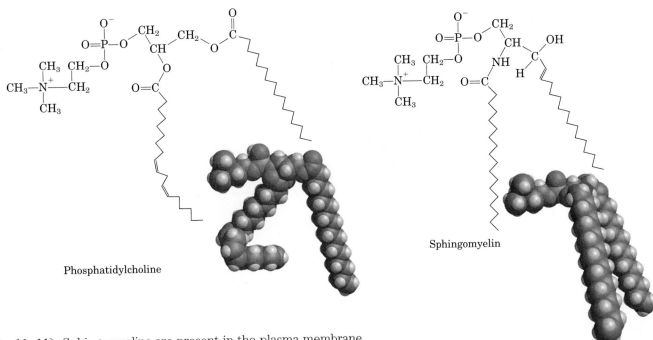

Phosphatidylcholine

Sphingomyelin

figure 11–11
The similarities in shape and in molecular structure of phosphatidylcholine (a glycerophospholipid) and sphingomyelin (a sphingolipid) are clear when their space-filling and structural formulas are drawn as here.

groups (Fig. 11–11). Sphingomyelins are present in the plasma membrane of animal cells and are especially prominent in myelin, a membranous sheath that surrounds and insulates the axons of some neurons—thus the name "sphingomyelins."

Glycosphingolipids, which occur largely in the outer face of plasma membranes, have head groups with one or more sugars connected directly to the —OH at C-1 of the ceramide moiety; they do not contain phosphate. **Cerebrosides** have a single sugar linked to ceramide; those with galactose are characteristically found in the plasma membranes of cells in neural tissue, and those with glucose in the plasma membranes of cells in nonneural tissues. **Globosides** are neutral (uncharged) glycosphingolipids with two or more sugars, usually D-glucose, D-galactose, or *N*-acetyl-D-galactosamine. Cerebrosides and globosides are sometimes called **neutral glycolipids,** as they have no charge at pH 7.

Gangliosides, the most complex sphingolipids, have oligosaccharides as their polar head groups and one or more residues of *N*-acetylneuraminic acid (Neu5Ac), also called sialic acid, at the termini. Sialic acid gives gangliosides the negative charge at pH 7 that distinguishes them from globosides. Gangliosides with one sialic acid residue are in the GM (M for mono-) series, those with two sialic acids are in the GD (D for di-) series, and so on.

N-Acetylneuraminic acid (sialic acid) (Neu5Ac)

Sphingolipids at Cell Surfaces Are Sites of Biological Recognition

When sphingolipids were discovered a century ago by the physician-chemist Johann Thudichum, their biological role seemed as enigmatic as the Sphinx, for which he therefore named them. In humans, at least 60 different sphingolipids have been identified in cellular membranes. Many of these are especially prominent in the plasma membranes of neurons, and some are clearly recognition sites on the cell surface, but for only a few sphingolipids has a specific function been discovered. The carbohydrate moieties of certain sphingolipids define the human blood groups and therefore determine the type of blood that individuals can safely receive in blood transfusions (Fig. 11–12, p. 374). The kinds and amounts of gangliosides in the plasma membrane change dramatically with embryonic development, and tumor formation induces the synthesis of a new complement of gangliosides. Very low concentrations of a specific ganglioside induce differentiation of cultured neuronal tumor cells. Investigation of the biological roles of diverse gangliosides remains fertile ground for future research.

Johann Thudichum
1829–1901

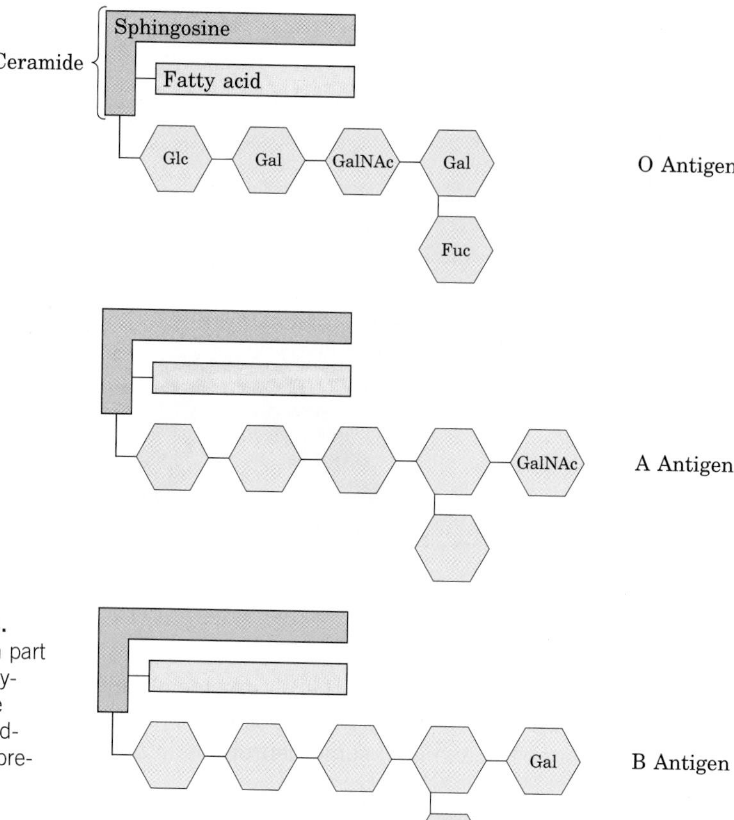

figure 11–12

Glycosphingolipids as determinants of blood groups. The human blood groups (O, A, B) are determined in part by the oligosaccharide head groups (blue) of these glycosphingolipids. The same three oligosaccharides are also found attached to certain blood proteins of individuals of blood types O, A, and B, respectively. (Fuc represents the sugar fucose.)

Phospholipids and Sphingolipids Are Degraded in Lysosomes

Most cells continually degrade and replace their membrane lipids. For each hydrolyzable bond in a glycerophospholipid, there is a specific hydrolytic enzyme in the lysosome (Fig. 11–13). Phospholipases of the A type remove one of the two fatty acids, producing a lysophospholipid. (These esterases do not attack the ether link of plasmalogens.) Lysophospholipases remove the remaining fatty acid.

Gangliosides are degraded by a set of lysosomal enzymes that catalyze the stepwise removal of sugar units, finally yielding a ceramide. A genetic defect in any of these hydrolytic enzymes leads to the accumulation of gangliosides in the cell, with severe medical consequences (Box 11–2).

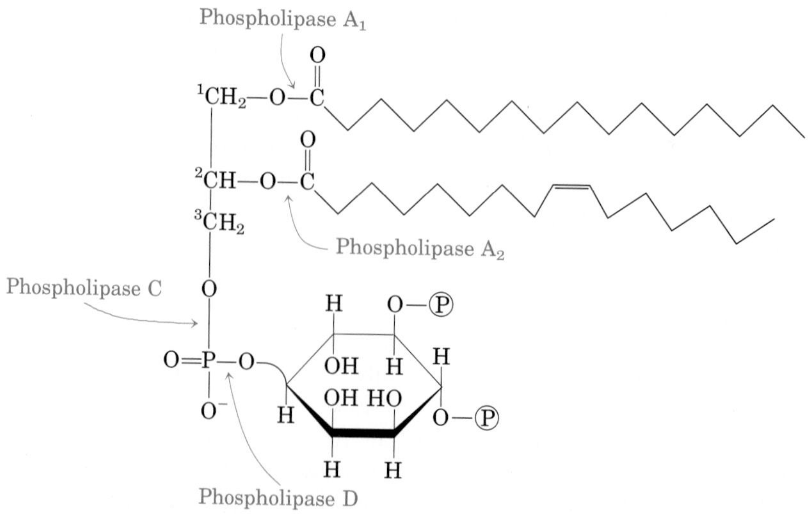

Phospholipase A_1

Phospholipase A_2

Phospholipase C

Phospholipase D

figure 11–13

The specificities of phospholipases. Phospholipases A_1 and A_2 hydrolyze the ester bonds of intact glycerophospholipids at C-1 and C-2 of glycerol, respectively. Phospholipases C and D each split one of the phosphodiester bonds in the head group. Some phospholipases act on only one type of glycerophospholipid, such as phosphatidylinositol 4,5-bisphosphate (shown here) or phosphatidylcholine; others are less specific. When one of the fatty acids has been removed by a type A phospholipase, the second fatty acid is cleaved from the molecule by a lysophospholipase (not shown).

box 11–2 Inherited Human Diseases Resulting from Abnormal Accumulations of Membrane Lipids

The polar lipids of membranes undergo constant metabolic turnover, the rate of their synthesis normally counterbalanced by the rate of breakdown. The breakdown of lipids is promoted by hydrolytic enzymes in lysosomes, each enzyme capable of hydrolyzing a specific bond. When sphingolipid degradation is impaired by a defect in one of these enzymes (Fig. 1), partial breakdown products accumulate in the tissues, causing serious disease.

For example, Niemann-Pick disease is caused by a rare genetic defect in the enzyme sphingomyelinase, which cleaves phosphocholine from sphingomyelin. Sphingomyelin accumulates in the brain, spleen, and liver. The disease becomes evident in infants, and causes mental retardation and early death.

More common is Tay-Sachs disease (Fig. 2), in which ganglioside GM2 accumulates in the brain and spleen owing to lack of the enzyme hexosaminidase A. The symptoms of Tay-Sachs disease are progressive retardation in development, paralysis, blindness, and death by the age of 3 or 4 years.

Genetic counseling can predict and avert many inheritable diseases. Tests on prospective parents can detect abnormal enzymes, then DNA testing can determine the exact nature of the defect and the risk it poses for offspring. Once a pregnancy occurs, fetal cells obtained by sampling a part of the placenta (chorionic villus biopsy) or the fluid surrounding the fetus (amniocentesis) can be tested in the same way.

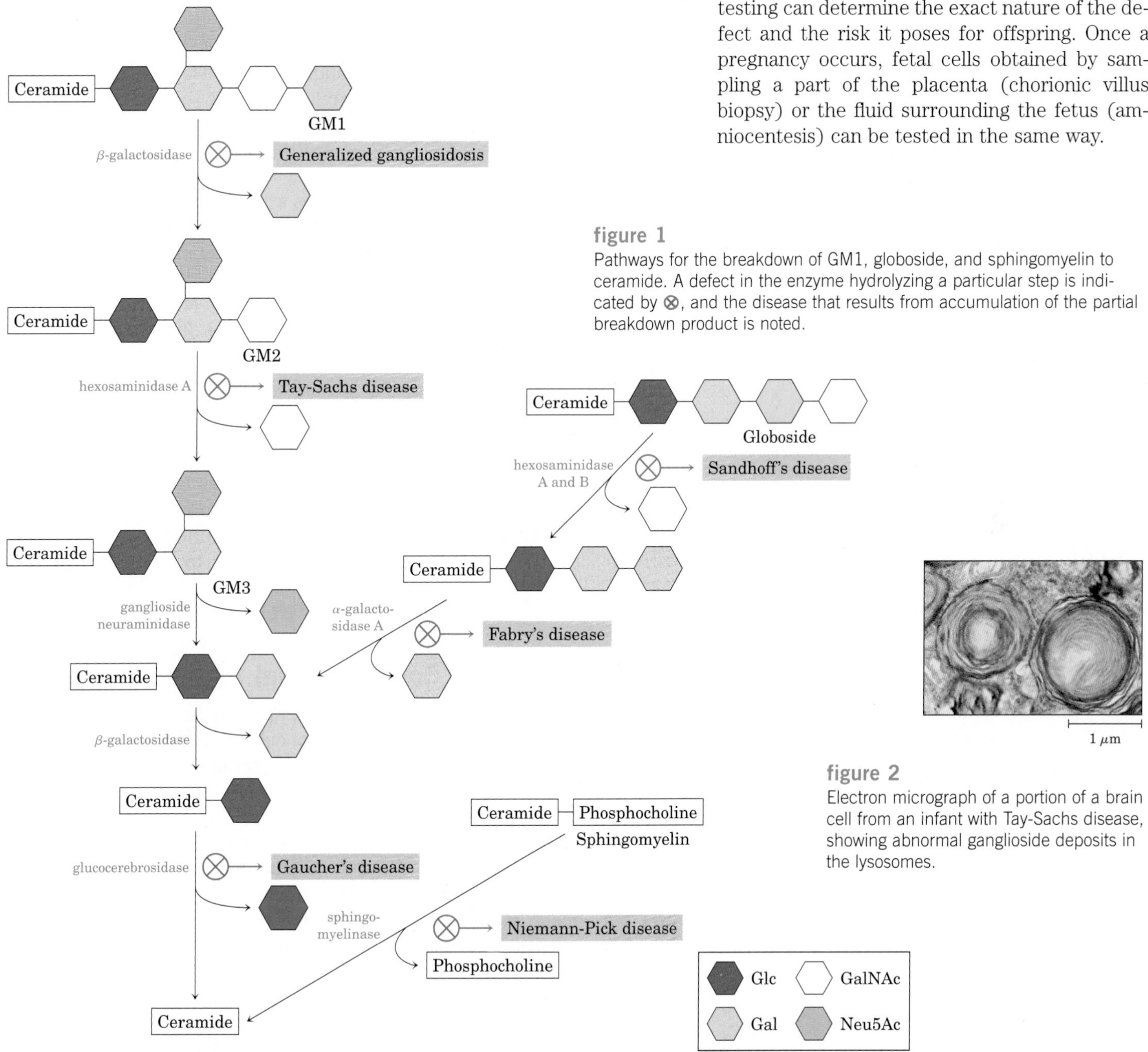

figure 1
Pathways for the breakdown of GM1, globoside, and sphingomyelin to ceramide. A defect in the enzyme hydrolyzing a particular step is indicated by ⊗, and the disease that results from accumulation of the partial breakdown product is noted.

figure 2
Electron micrograph of a portion of a brain cell from an infant with Tay-Sachs disease, showing abnormal ganglioside deposits in the lysosomes.

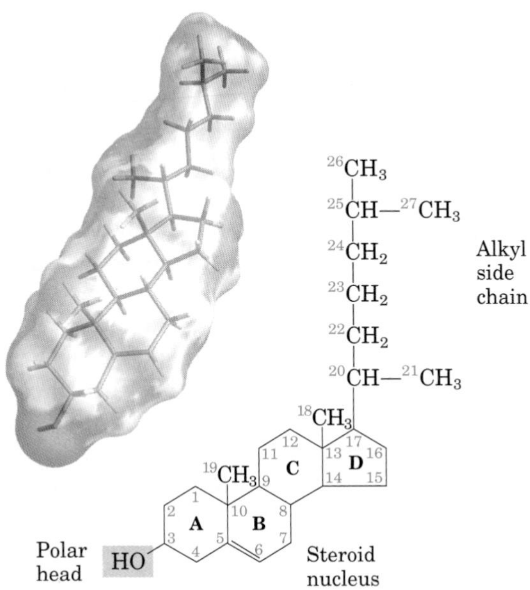

figure 11–14

Cholesterol. The stick structure of cholesterol is visible through a transparent surface contour model of the molecule. In the chemical structure, the rings are labeled A through D to simplify reference to derivatives of the steroid nucleus and the carbon atoms are numbered in blue. The hydroxyl group on C-3 (red in both representations) represents the polar head group. For storage and transport of the sterol, this hydroxyl group condenses with a fatty acid to form a sterol ester.

Sterols Have Four Fused Carbon Rings

Sterols are structural lipids present in the membranes of most eukaryotic cells. Their characteristic structure is the steroid nucleus consisting of four fused rings, three with six carbons and one with five (Fig. 11–14). The steroid nucleus is almost planar and is relatively rigid; the fused rings do not allow rotation about C—C bonds. **Cholesterol,** the major sterol in animal tissues, is amphipathic, with a polar head group (the hydroxyl group at C-3) and a nonpolar hydrocarbon body (the steroid nucleus and the hydrocarbon side chain at C-17) about as long as a 16-carbon fatty acid in its extended form. Similar sterols are found in other eukaryotes: stigmasterol in plants and ergosterol in fungi, for example. Bacteria cannot synthesize sterols; a few bacterial species, however, can incorporate exogenous sterols into their membranes. The sterols of all eukaryotes are synthesized from simple five-carbon isoprene subunits, as are the fat-soluble vitamins, quinones, and dolichols described below.

In addition to their roles as membrane constituents, the sterols serve as precursors for a variety of products with specific biological activities. Steroid hormones, for example, are potent biological signals that regulate gene expression. Bile acids are polar derivatives of cholesterol that act as detergents in the intestine, emulsifying dietary fats to make them more readily accessible to digestive lipases. We shall return to cholesterol and other sterols in later chapters, to consider the structural role of cholesterol, in biological membranes (Chapter 12), signaling by steroid hormones (Chapter 13), the remarkable biosynthetic pathway to cholesterol, and the transport of cholesterol by lipoprotein carriers (Chapter 21).

Taurocholic acid
(a bile acid)

Lipids as Signals, Cofactors, and Pigments

The two functional classes of lipids considered thus far (storage lipids and structural lipids) are major cellular components; membrane lipids make up 5% to 10% of the dry mass of most cells, and storage lipids more than 80% of the mass of an adipocyte. With some important exceptions, these lipids play a *passive* role in the cell; lipid fuels are stored until oxidized by enzymes, and membrane lipids form impermeable barriers that surround cells and cellular compartments. Another group of lipids, present in much smaller amounts, have active roles in the metabolic traffic as metabolites and messengers. Some serve as potent signals—as hormones carried in the blood from one tissue to another, or as intracellular messengers generated in response to an extracellular signal (hormone or growth factor). Others function as enzyme cofactors in electron-transfer reactions in chloroplasts and mitochondria, or in the transfer of sugar moieties in a variety of glycosylation (addition of sugar) reactions. A third group consists of lipids with a system of conjugated double bonds: pigment molecules that absorb visible light. Some of these act as light-capturing pigments in vision and photosynthesis; others produce natural colorations, such as the orange of pumpkins and carrots and the yellow of canary feathers. Specialized lipids such as these are derived from lipids of the plasma membrane or from the fat-soluble vitamins A, D, E, and K. We will briefly describe a few of these biologically active lipids. In later chapters, their synthesis and biological roles are considered in more detail.

Phosphatidylinositols Act as Intracellular Signals

Phosphatidylinositol and its phosphorylated derivatives act at several levels to regulate cell structure and metabolism (Fig. 11–15). Phosphatidylinositol 4,5-bisphosphate (Fig. 11–8) in the cytosolic (inner) face of plasma membranes serves as a specific binding site for certain cytoskeletal proteins and for some soluble proteins involved in membrane fusion during exocytosis. It also serves as a reservoir of messenger molecules that are released inside the cell in response to extracellular signals interacting with specific receptors on the outer surface of the plasma membrane. The signals act through a series of steps (Fig. 11–15) that begins with enzymatic removal of a phospholipid head group and ends with activation of an enzyme (protein kinase C). For example, when the hormone vasopressin binds to plasma membrane receptors on the epithelial cells of the renal collecting duct, a specific phospholipase C is activated. Phospholipase C hydrolyzes the bond between glycerol and phosphate in phosphatidylinositol 4,5-bisphosphate, releasing two products: inositol 1,4,5-trisphosphate (IP_3), which is water-soluble, and diacylglycerol, which remains associated with the plasma membrane. IP_3 triggers release of Ca^{2+} from the endoplasmic reticulum, and the combination of diacylglycerol and elevated cytosolic Ca^{2+} activates the enzyme protein kinase C. This enzyme catalyzes the transfer of a phosphoryl group from ATP to a specific residue in one or more target proteins, thereby altering their activity and consequently the cell's metabolism. Membrane sphingolipids can also serve as sources of intracellular messengers; both ceramide and sphingomyelin (Fig. 11–10) are potent regulators of protein kinases.

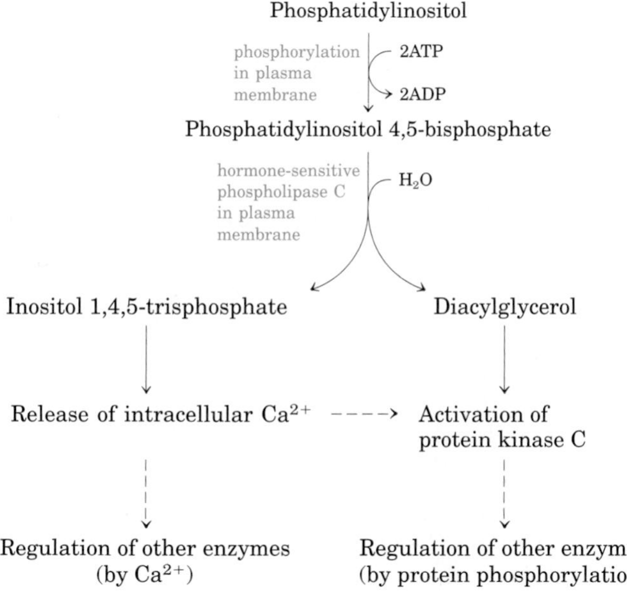

figure 11–15
Phosphatidylinositols in cellular regulation. Phosphatidylinositol 4,5-bisphosphate in the plasma membrane is hydrolyzed by a specific phospholipase C in response to hormonal signals. Both products of hydrolysis act as intracellular messengers.

Eicosanoids Carry Messages to Nearby Cells

Eicosanoids are paracrine hormones, substances that act only on cells near the point of synthesis instead of being transported in the blood to act on cells in other tissues or organs. These fatty acid derivatives have a variety of dramatic effects on vertebrate tissues. They are known to be involved in reproductive function, in the inflammation, fever, and pain associated with injury or disease, in the formation of blood clots and the regulation of blood pressure, in gastric acid secretion, and in a variety of other processes important in human health or disease.

All eicosanoids are derived from the 20-carbon polyunsaturated fatty acid arachidonic acid, 20:4($\Delta^{5,8,11,14}$) (Fig. 11–16), from which they take their general name (Greek *eikosi*, "twenty"). There are three classes of eicosanoids: prostaglandins, thromboxanes, and leukotrienes.

Prostaglandins (PG) contain a five-carbon ring originating from the chain of arachidonic acid. They derive their name from the prostate gland, the tissue from which they were first isolated by Bengt Samuelsson and Sune Bergström. Two groups of prostaglandins were originally defined: PGE, for *e*ther-soluble, and PGF, for phosphate buffer–soluble (*fosfat* in Swedish). Each contains numerous subtypes, named PGE_1, PGE_2, and so forth.

Prostaglandins act in many tissues by regulating the synthesis of the intracellular messenger molecule 3′,5′-cyclic AMP (cAMP). Because cAMP mediates the action of many hormones, the prostaglandins affect a wide range of cellular and tissue functions. Some prostaglandins stimulate contraction of the smooth muscle of the uterus during menstruation and labor. Others affect blood flow to specific organs, the wake-sleep cycle, and the responsiveness of certain tissues to hormones such as epinephrine and glucagon. Prostaglandins in a third group elevate body temperature (producing fever) and cause inflammation and pain.

figure 11–16
Arachidonic acid and some eicosanoid derivatives. **(a)** In response to hormonal signals, phospholipase A_2 cleaves arachidonic acid–containing membrane phospholipids to release arachidonic acid (arachidonate at pH 7), the precursor to various eicosanoids. **(b)** These include prostaglandins such as PGE_1, in which C-8 and C-12 of arachidonate are joined to form the characteristic five-membered ring. In thromboxane A_2, the C-8 and C-12 are joined and an oxygen atom is added to form the six-membered ring. Leukotriene A_4 has a series of three conjugated double bonds. Nonsteroidal antiinflammatory drugs (NSAIDs) such as aspirin, acetaminophen, and ibuprofen block the formation of prostaglandins and thromboxanes from arachidonate by inhibiting the enzyme cyclooxygenase (prostaglandin H_2 synthase).

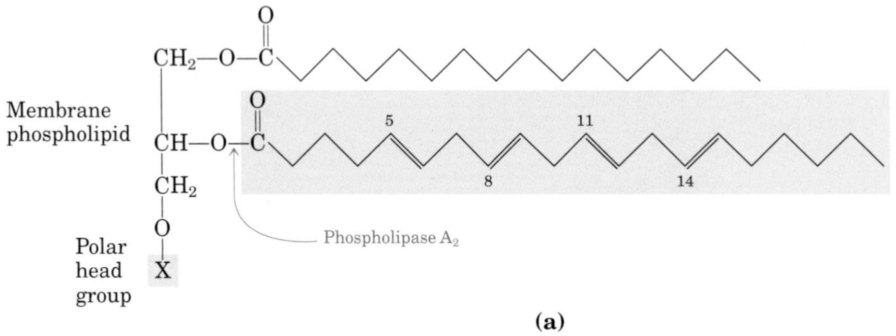

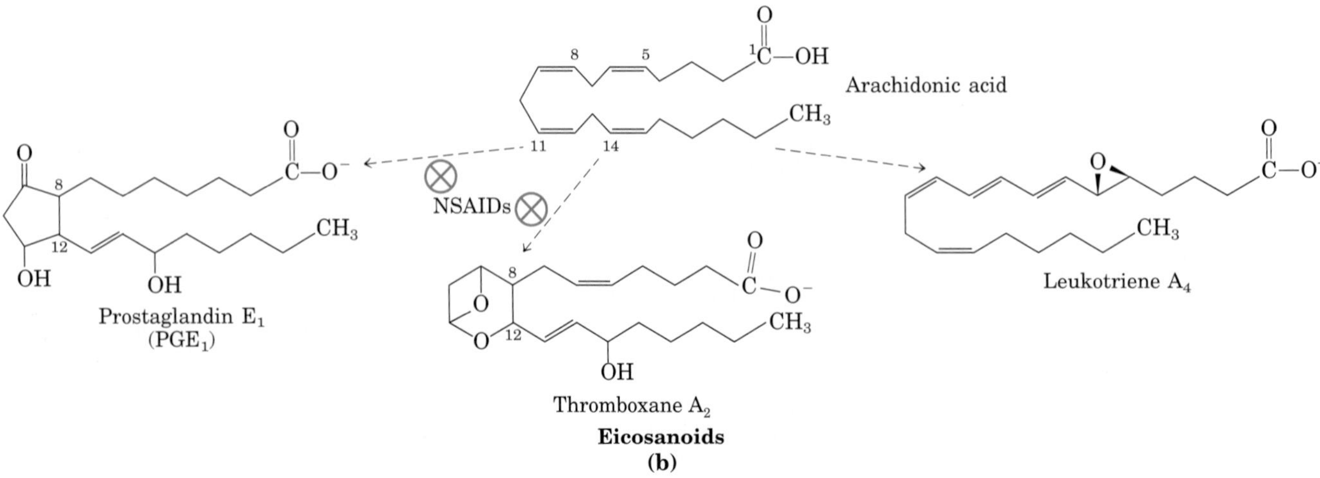

The **thromboxanes** have a six-membered ring containing an ether. They are produced by platelets (thrombocytes) and act in the formation of blood clots and the reduction of blood flow to the site of a clot. The nonsteroidal antiinflammatory drugs (NSAIDs)—aspirin, ibuprofen, acetaminophen, and meclofenamate, for example—were shown by John Vane to inhibit the enzyme prostaglandin H_2 synthase (also called cyclooxygenase or COX), which catalyzes an early step in the pathway from arachidonate to prostaglandins and thromboxanes (Fig. 11–16; see also Box 21–2).

Leukotrienes, found first in leukocytes, contain three conjugated double bonds. They are powerful biological signals. For example, leukotriene D_4, derived from leukotriene A_4, induces contraction of the muscle lining the airways to the lung. Overproduction of leukotrienes causes asthmatic attacks, and leukotriene synthesis is one target of antiasthmatic drugs such as prednisone. The strong contraction of the smooth muscles of the lung that occurs during anaphylactic shock is part of the potentially fatal allergic reaction in individuals hypersensitive to bee stings, penicillin, or various other agents.

John Vane, Sune Bergström, and Bengt Samuelsson

Steroid Hormones Carry Messages between Tissues

Steroids are oxidized derivatives of sterols; they have the sterol nucleus but lack the alkyl chain attached to ring D of cholesterol, and they are more polar than cholesterol. Steroid hormones move through the bloodstream on protein carriers from their site of production to target tissues, where they enter cells, bind to highly specific receptor proteins in the nucleus, and trigger changes in gene expression and metabolism. Because hormones have very high affinity for their receptors, very low concentrations of hormones (nanomolar or less) are sufficient to produce responses in target tissues. The major groups of steroid hormones are the male and female sex hormones and the hormones produced by the adrenal cortex, cortisol and aldosterone (Fig. 11–17). Prednisone and prednisolone are steroid drugs with potent antiinflammatory activities, mediated in part by the inhibition of arachidonate release by phospholipase A_2 (Fig. 11–16) and consequent inhibition of the synthesis of leukotrienes, prostaglandins, and thromboxanes. They have a variety of medical applications, including the treatment of asthma and rheumatoid arthritis.

figure 11–17

Steroids derived from cholesterol. Testosterone, the male sex hormone, is produced in the testes. Estradiol, one of the female sex hormones, is produced in the ovaries and placenta. Cortisol and aldosterone are hormones synthesized in the cortex of the adrenal gland; they regulate glucose metabolism and salt excretion, respectively. Prednisolone and prednisone are steroid drugs used as antiinflammatory agents.

Testosterone

Estradiol

Cortisol

Aldosterone

Prednisolone

Prednisone

Vitamins A and D Are Hormone Precursors

During the first third of the twentieth century, a major focus of research in physiological chemistry was the identification of **vitamins**—compounds that are essential to the health of humans and other vertebrates but cannot be synthesized by these animals and must therefore be obtained in the diet. Early nutritional studies identified two general classes of such compounds: those soluble in nonpolar organic solvents (fat-soluble vitamins) and those that could be extracted from foods with aqueous solvents (water-soluble vitamins). Eventually the fat-soluble group was resolved into the four vitamin groups A, D, E, and K, all of which are isoprenoid compounds synthesized by the condensation of multiple isoprene units. Two of these (D and A) serve as hormone precursors.

$$CH_2{=}C(CH_3){-}CH{=}CH_2$$

Isoprene

Vitamin D_3, also called **cholecalciferol,** is normally formed in the skin from 7-dehydrocholesterol in a photochemical reaction driven by the UV component of sunlight (Fig. 11–18). Vitamin D_3 is not itself biologically active, but is converted by enzymes in the liver and kidney to 1,25-dihydroxycholecalciferol, a hormone that regulates calcium uptake in the intestine and calcium levels in kidney and bone. Deficiency of vitamin D leads to defective bone formation and the disease rickets, for which administration of

7-Dehydrocholesterol

UV light ↓ 2 steps (in skin)

Cholecalciferol (vitamin D_3)

↓ 1 step in the liver

↓ 1 step in the kidney

1,25-Dihydroxycholecalciferol (1,25-dihydroxyvitamin D_3)

(a)

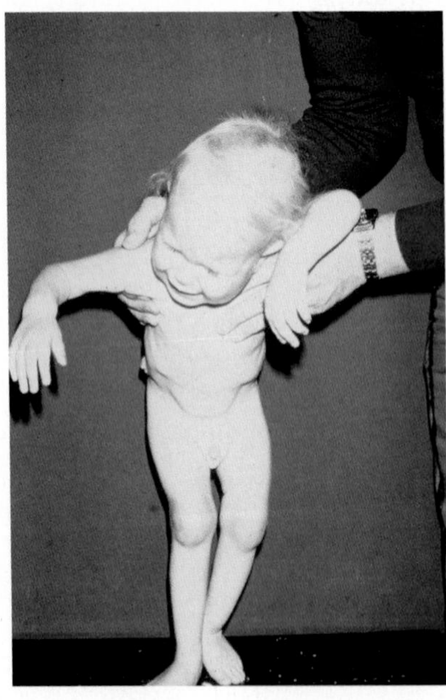

Before vitamin D treatment

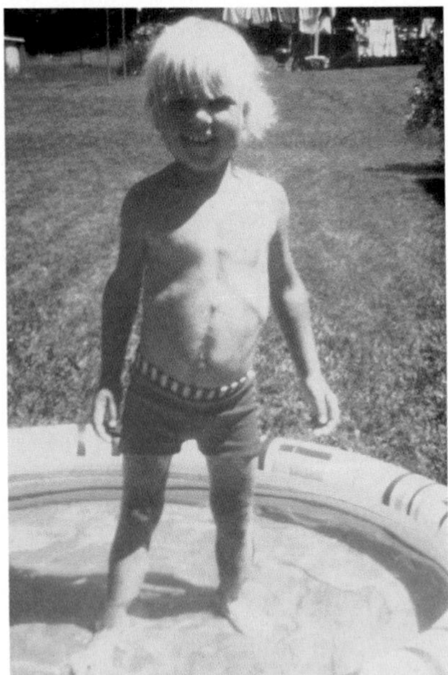

After 14 months of vitamin D treatment

(b)

figure 11–18

Vitamin D_3 production and metabolism. (a) Cholecalciferol (vitamin D_3) is produced in the skin by UV irradiation of 7-dehydrocholesterol. In the liver, a hydroxyl group is added at C-25; in the kidney, a second hydroxylation at C-1 produces the active hormone, 1,25-dihydroxycholecalciferol. This hormone regulates the metabolism of Ca^{2+} in kidney, intestine, and bone. **(b)** Dietary vitamin D prevents rickets, a disease once common in cold climates where heavy clothing blocks the UV component of sunlight necessary for the production of vitamin D_3 in skin. On the left is a 2 1/2-year-old boy with severe rickets; on the right, the same boy at age 5, after 14 months of vitamin D therapy.

vitamin D produces a dramatic cure. Vitamin D_2 (ergocalciferol) is a commercial product formed by UV irradiation of the ergosterol of yeast. Vitamin D_2 is structurally similar to D_3, with slight modification to the side chain attached to the sterol D ring. Both have the same biological effects, and D_2 is commonly added to milk and butter as a dietary supplement. Like steroid hormones, the product of vitamin D metabolism, 1,25-dihydroxycholecalciferol, regulates gene expression—for example, turning on the synthesis of an intestinal Ca^{2+}-binding protein.

Vitamin A (retinol) in its various forms functions as a hormone and as the visual pigment of the vertebrate eye (Fig. 11–19). Acting through receptor proteins in the cell nucleus, the vitamin A derivative retinoic acid regulates gene expression in the development of epithelial tissue, including skin. Retinoic acid is the active ingredient in the drug tretinoin (Retin-A), used in the treatment of severe acne and wrinkled skin. The vitamin A derivative retinal is the pigment that initiates the response of rod and cone cells of the retina to light, producing a neuronal signal to the brain. This role of retinal is described in detail in Chapter 13.

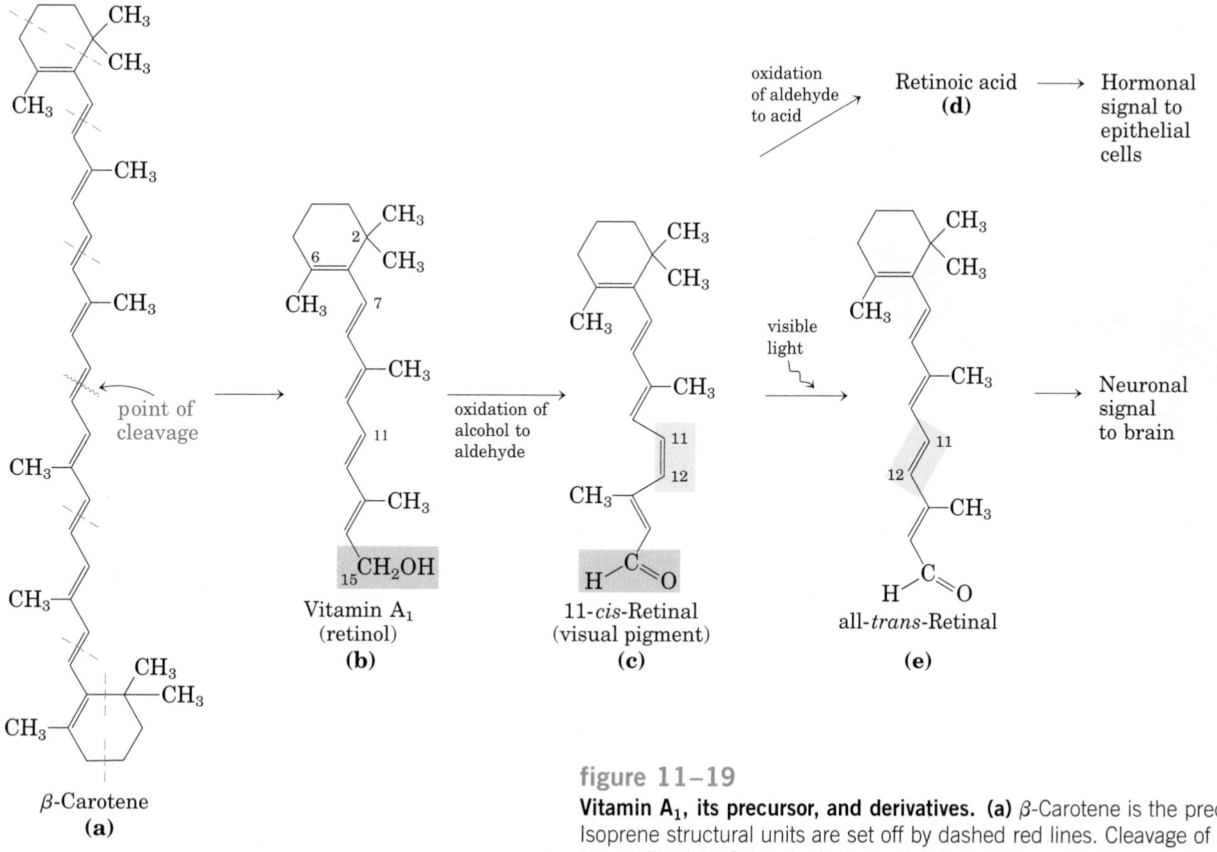

figure 11–19
Vitamin A_1, its precursor, and derivatives. (a) β-Carotene is the precursor of vitamin A_1. Isoprene structural units are set off by dashed red lines. Cleavage of β-carotene yields two molecules of vitamin A_1 (retinol) **(b).** Oxidation at C-15 converts retinol to the aldehyde, retinal **(c),** and further oxidation produces retinoic acid **(d),** a hormone that regulates gene expression in skin. Retinal combines with the protein opsin to form rhodopsin (not shown), a visual pigment widely employed in nature. In the dark, retinal of rhodopsin is in the 11-*cis* form **(c).** When a rhodopsin molecule is excited by visible light, the 11-*cis*-retinal undergoes a series of photochemical reactions that convert it to all-*trans*-retinal **(e),** forcing a change in the shape of the entire rhodopsin molecule. This transformation in the rod cell of the vertebrate retina sends an electrical signal to the brain that is the basis of visual transduction, a topic we address in more detail in Chapter 13.

Vitamin A was first isolated from fish liver oils; liver, eggs, whole milk, and butter are good dietary sources. In vertebrates, β-carotene, the pigment that gives carrots, sweet potatoes, and other yellow vegetables their characteristic color, can be enzymatically converted to vitamin A. Deficiency of vitamin A leads to a variety of symptoms in humans, including dryness of the skin, eyes, and mucous membranes; retarded development and growth; and night blindness, an early symptom commonly used in diagnosing vitamin A deficiency.

Vitamins E and K and the Lipid Quinones Are Oxidation-Reduction Cofactors

Vitamin E is the collective name for a group of closely related lipids called **tocopherols,** all of which contain a substituted aromatic ring and a long isoprenoid side chain (Fig. 11–20a). Because they are hydrophobic, tocopherols associate with cell membranes, lipid deposits, and lipoproteins in the blood. Tocopherols are biological antioxidants. The aromatic ring reacts with and destroys the most reactive forms of oxygen radicals and other free radicals, protecting unsaturated fatty acids from oxidation and preventing oxidative damage to membrane lipids, which can cause cell fragility. Tocopherols are found in eggs and vegetable oils, and are especially abundant in wheat germ. Laboratory animals fed diets depleted of vitamin E develop scaly skin, muscular weakness and wasting, and sterility. Vitamin E deficiency in humans is very rare; the principal symptom is fragile erythrocytes.

Henrik Dam
1895–1976

Edward A. Doisy
1893–1986

The aromatic ring of **vitamin K** (Fig. 11–20b) undergoes a cycle of oxidation and reduction during the formation of active prothrombin, a blood plasma protein essential in blood clot formation. Prothrombin is a proteolytic enzyme that splits peptide bonds in the blood protein fibrinogen to convert it to fibrin, the insoluble fibrous protein that holds blood clots together. Henrik Dam and Edward A. Doisy independently discovered that vitamin K deficiency slows blood clotting, which can be fatal. Vitamin K deficiency is very uncommon in humans, aside from a small percentage of infants who suffer from hemorrhagic disease of the newborn, a potentially fatal disorder. In the United States, newborns are routinely given a 1 mg injection of vitamin K. Vitamin K_1 (phylloquinone) is found in green plant leaves; a related form, vitamin K_2 (menaquinone), is formed by bacteria residing in the vertebrate intestine.

Warfarin (Fig. 11–20c) is a synthetic compound that inhibits the formation of active prothrombin. It is particularly poisonous to rats, causing death by internal bleeding. Ironically, this potent rodenticide is also an invaluable anticoagulant drug for treating humans at risk for excessive blood clotting, such as surgical patients and victims of coronary thrombosis.

Ubiquinone (also called coenzyme Q) and plastoquinone (Fig. 11–20d, e) are isoprenoids that function as lipophilic electron carriers in the oxidation-reduction reactions that drive ATP synthesis in mitochondria and chloroplasts, respectively. Both ubiquinone and plastoquinone can accept either one or two electrons and either one or two protons, as shown in Figure 19–2.

Dolichols Activate Sugar Precursors for Biosynthesis

During assembly of the complex carbohydrates of bacterial cell walls, and during the addition of polysaccharide units to certain proteins (glycoproteins) and lipids (glycolipids) in eukaryotes, the sugar units to be added are chemically activated by attachment to isoprenoid alcohols called **dolichols** (Fig. 11–20f). These compounds have strong hydrophobic interactions with membrane lipids, anchoring the attached sugars to the membrane where they participate in sugar-transfer reactions.

(a) Vitamin E: an antioxidant

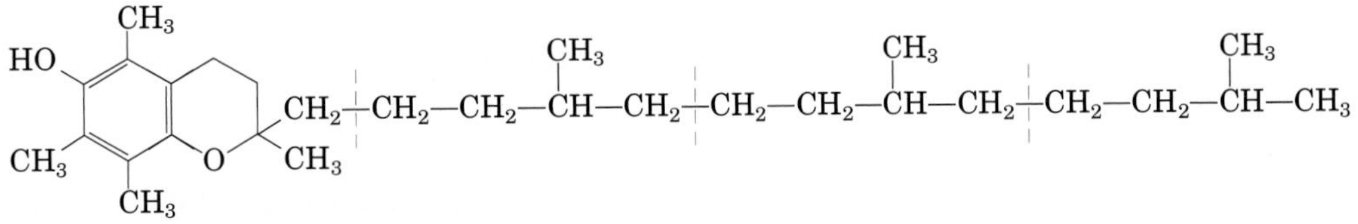

(b) Vitamin K_1: a blood-clotting cofactor (phylloquinone)

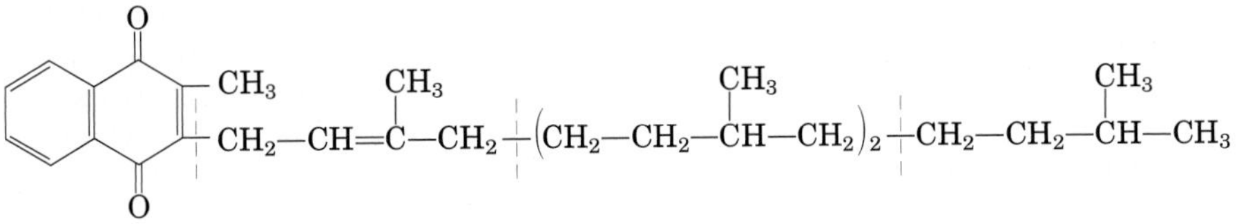

(c) Warfarin: a blood anticoagulant

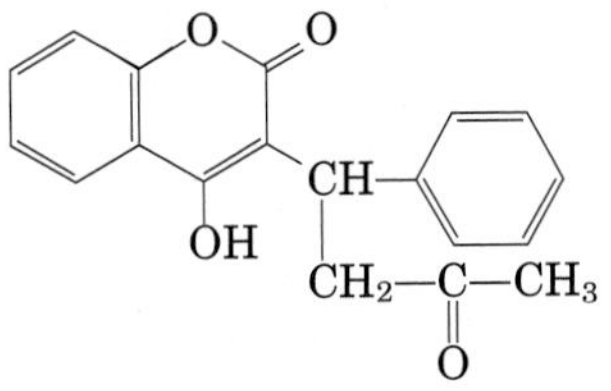

(d) Ubiquinone: a mitochondrial electron carrier (coenzyme Q) (n = 4–8)

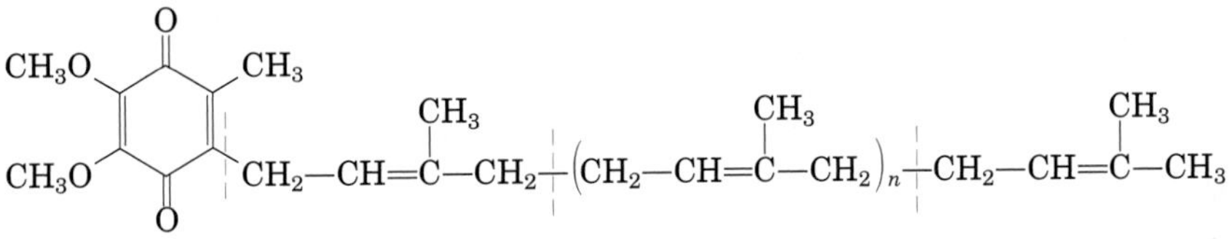

(e) Plastoquinone: a chloroplast electron carrier (n = 4–8)

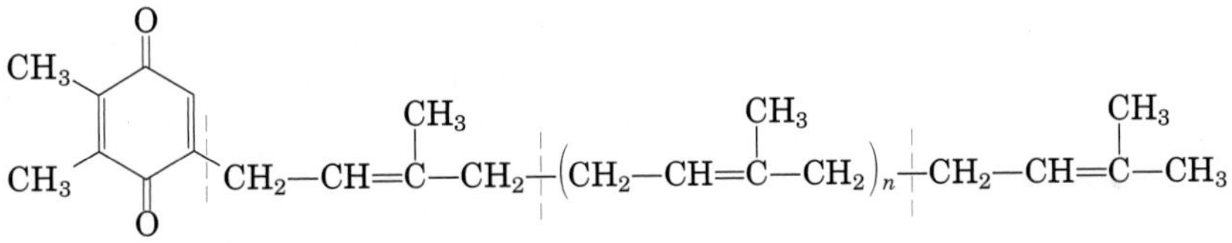

(f) Dolichol: a sugar carrier (n = 9–22)

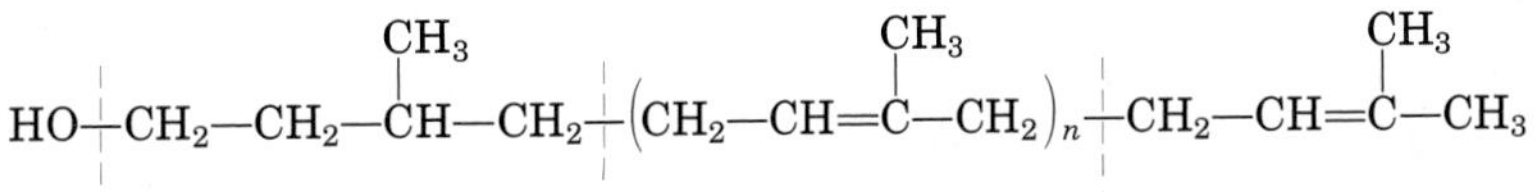

figure 11–20
Some other biologically active isoprenoid compounds or derivatives. Isoprene structural units are set off by dashed red lines. In most mammalian tissues, ubiquinone (also called coenzyme Q) has ten isoprene units. Dolichols of animals have 17 to 21 isoprene units (85 to 105 carbon atoms), bacterial dolichols have 11, and those of plants and fungi have 14 to 24.

Separation and Analysis of Lipids

In exploring the biological role of lipids in cells and tissues, it is useful to know which lipids are present and in what proportions. Because lipids are insoluble in water, their extraction and subsequent fractionation require the use of organic solvents and some techniques not commonly used in the purification of water-soluble molecules such as proteins and carbohydrates. In general, complex mixtures of lipids are separated by differences in the polarity or solubility of the components in nonpolar solvents. Lipids that contain ester- or amide-linked fatty acids can be hydrolyzed by treatment with acid or alkali or with highly specific hydrolytic enzymes (phospholipases,

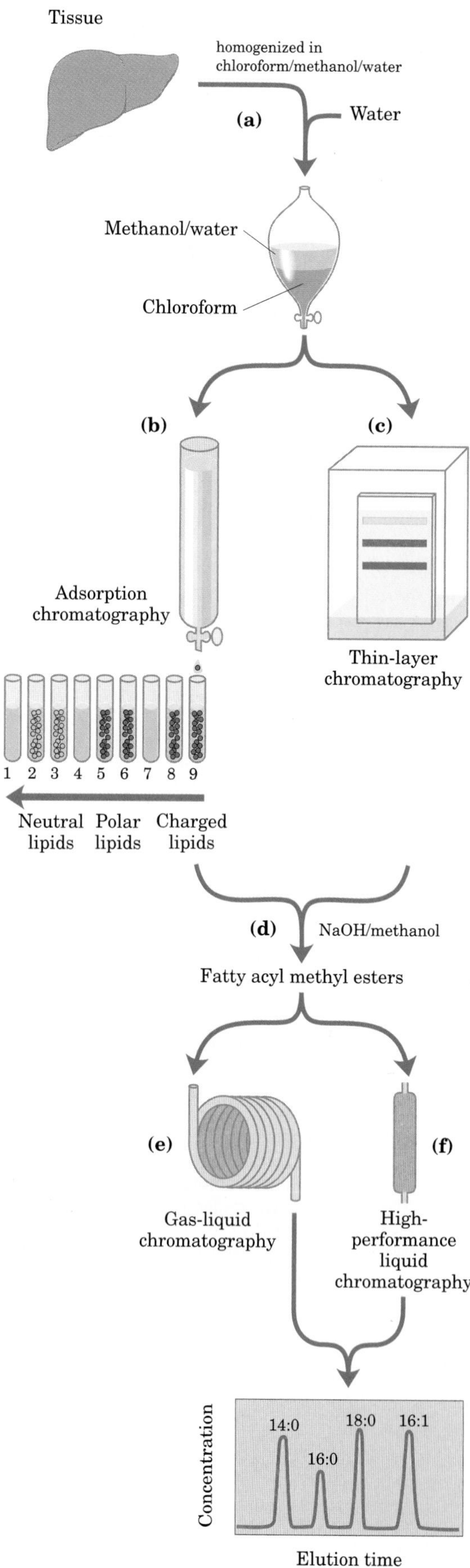

glycosidases) to yield their component parts for analysis. Some methods commonly used in lipid analysis are shown in Figure 11–21 and discussed below.

Lipid Extraction Requires Organic Solvents

Neutral lipids (triacylglycerols, waxes, pigments, etc.) are readily extracted from tissues with ethyl ether, chloroform, or benzene, solvents in which lipid clustering driven by hydrophobic interactions does not occur. Membrane lipids are more effectively extracted by more polar organic solvents, such as ethanol or methanol, which reduce the hydrophobic interactions among lipid molecules while also weakening the hydrogen bonds and electrostatic interactions that bind membrane lipids to membrane proteins. A commonly used extractant is a mixture of chloroform, methanol, and water, initially in volume proportions (1:2:0.8) that are miscible, producing a single phase. After tissue is homogenized in this solvent to extract all lipids, more water is added to the resulting extract and the mixture separates into two phases, methanol/water (top phase) and chloroform (bottom phase). The lipids remain in the chloroform layer, and more polar molecules such as proteins and sugars partition into the methanol/water layer.

Adsorption Chromatography Separates Lipids of Different Polarity

The complex mixture of tissue lipids can be fractionated further by chromatographic procedures based on the different polarities of each class of lipid. In adsorption chromatography, an insoluble, polar material such as silica gel (a form of silicic acid, $Si(OH)_4$) is packed into a glass column, and the lipid mixture (in chloroform solution) is applied to the top of the column. (In high-performance liquid chromatography, the column is of smaller diameter, and solvents are forced through the column under high pressure.) The polar lipids bind tightly to the polar silicic acid, but the neutral lipids pass directly through the column and emerge in the first chloroform wash (Fig. 11–21b). The polar lipids are then eluted, in order of increasing polarity, by washing the column with solvents of progressively higher polarity. Uncharged but polar lipids (cerebrosides, for example) are eluted with acetone, and very polar or charged lipids (such as glycerophospholipids) are eluted with methanol.

figure 11–21

Common procedures in the extraction, separation, and identification of cellular lipids. **(a)** Tissue is homogenized in a chloroform/methanol/water mixture, which on addition of water and removal of unextractable sediment by centrifugation yields two phases. Different types of extracted lipids in the chloroform phase may be separated by **(b)** adsorption chromatography on a column of silica gel, through which solvents of increasing polarity are passed, or **(c)** thin-layer chromatography (TLC), in which lipids are carried up a silica gel–coated plate by a rising solvent front, less polar lipids traveling farther than more polar or charged lipids. TLC with appropriate solvents can also be used to separate closely related lipid species; for example, the charged lipids phosphatidylserine, phosphatidylglycerol, and phosphatidylinositol are easily separated by TLC. For the determination of fatty acid composition, a lipid fraction containing ester-linked fatty acids is transesterified in a warm aqueous solution of NaOH and methanol **(d),** producing a mixture of fatty acyl methyl esters. These methyl esters are then separated on the basis of chain length and degree of saturation by **(e)** gas-liquid chromatography (GLC) or **(f)** high-performance liquid chromatography (HPLC). Precise determination of molecular mass by mass spectrometry (see Box 5–3) allows unambiguous identification of individual lipids.

Thin-layer chromatography on silicic acid employs the same principle. A thin layer of silica gel is spread onto a glass plate, to which it adheres. A small sample of lipids dissolved in chloroform is applied near one edge of the plate, which is dipped in a shallow container of an organic solvent or solvent mixture—all of which is enclosed within a chamber saturated with the solvent vapor. As the solvent rises on the plate by capillary action, it carries lipids with it (Fig. 11–21c). The less polar lipids move farthest, as they have less tendency to bind to the silicic acid. The lipids can be detected after separation by spraying the plate with a dye (rhodamine) that fluoresces when associated with lipids, or by exposing the plate to iodine fumes. Iodine reacts reversibly with the double bonds in fatty acids, such that lipids containing unsaturated fatty acids develop a yellow or brown color. A number of other spray reagents are also useful in detecting specific lipids. For subsequent analysis, regions containing separated lipids can be scraped from the plate and the lipids recovered by extraction with an organic solvent.

Gas-Liquid Chromatography Resolves Mixtures of Volatile Lipid Derivatives

Gas-liquid chromatography separates volatile components of a mixture according to their relative tendencies to dissolve in the inert material packed in the chromatography column and to volatilize and move through the column, carried by a current of an inert gas such as helium. Some lipids are naturally volatile, but most must first be derivatized to increase their volatility (that is, lower their boiling point). For an analysis of the fatty acids in a sample of phospholipids, the lipids are first heated in a methanol/HCl or methanol/NaOH mixture, which converts fatty acids esterified to glycerol into their methyl esters (transesterification). These fatty acyl methyl esters are then loaded onto the gas-liquid chromatography column, and the column is heated to volatilize the compounds. Those fatty acyl esters most soluble in the column material partition into (dissolve in) that material; the less-soluble lipids are carried by the stream of inert gas and emerge first from the column. The order of elution depends on the nature of the solid adsorbant in the column and on the boiling point of the components of the lipid mixture. Using these techniques, mixtures of fatty acids of various chain lengths and various degrees of unsaturation can be completely resolved (Fig. 11–21e).

Specific Hydrolysis Aids in Determination of Lipid Structure

Certain classes of lipids are susceptible to degradation under specific conditions. For example, all ester-linked fatty acids in triacylglycerols, phospholipids, and sterol esters are released by mild acid or alkaline treatment, and somewhat harsher hydrolysis conditions release amide-bound fatty acids from sphingolipids. Enzymes that specifically hydrolyze certain lipids are also useful in the determination of lipid structure. Phospholipases A, C, and D (Fig. 11–13) each split particular bonds in phospholipids and yield products with characteristic solubilities and chromatographic behaviors. Phospholipase C, for example, releases a water-soluble phosphoryl alcohol (such as phosphocholine from phosphatidylcholine) and a chloroform-soluble diacylglycerol, each of which can be characterized separately to determine the structure of the intact phospholipid. The combination of specific hydrolysis with characterization of the products by thin-layer, gas-liquid, or high-performance liquid chromatography often allows determination of a lipid structure.

Mass Spectrometry Reveals Complete Lipid Structure

To establish unambiguously the length of a hydrocarbon chain or the position of double bonds, mass spectral analysis of lipids or their volatile derivatives is invaluable. The chemical properties of similar lipids (for example, two fatty acids unsaturated at different positions, or two isoprenoids with different numbers of isoprene units) are very much alike, and their positions of elution from the various chromatographic procedures often do not distinguish between them. When the output from a chromatography column is constantly sampled by mass spectrometry, however, the components of a lipid mixture can be simultaneously separated and identified.

summary

Lipids are water-insoluble cellular components that can be extracted by nonpolar solvents. Some lipids serve as structural components of membranes and others as storage forms of fuel. Fatty acids, which provide the hydrocarbon components of many lipids, usually have an even number of carbon atoms (usually 12 to 24) and may be saturated or unsaturated; unsaturated fatty acids have double bonds in the cis configuration. In most unsaturated fatty acids, one double bond is at the Δ^9 position (between C-9 and C-10).

Triacylglycerols contain three fatty acid molecules esterified to the three hydroxyl groups of glycerol. Simple triacylglycerols contain only one type of fatty acid, mixed triacylglycerols two or three types. Triacylglycerols are primarily storage fats; they are present in many types of foods.

The polar lipids, with polar heads and nonpolar tails, are major components of membranes. The most abundant are the glycerophospholipids, which contain two fatty acid molecules esterified to two hydroxyl groups of glycerol, and a second alcohol, the head group, esterified to the third hydroxyl of glycerol via a phosphodiester bond. Glycerophospholipids differ in the structure of the head group; common glycerophospholipids are phosphatidylethanolamine and phosphatidylcholine. The polar heads of the glycerophospholipids carry electric charges at pH near 7. The sphingolipids, also membrane components, contain sphingosine, a long-chain aliphatic amino alcohol, but no glycerol. Sphingomyelin has, in addition to phosphoric acid and choline, two long hydrocarbon chains, one contributed by a fatty acid and the other by sphingosine. Three other classes of sphingolipids are cerebrosides, globosides, and gangliosides, which contain sugar components. Cholesterol, a sterol, is a precursor of steroids and an important component of the plasma membranes of animal cells.

Some types of lipids, although present in relatively small quantities, play critical roles as cofactors or signals. Phosphatidylinositol is hydrolyzed to yield two intracellular messengers, diacylglycerol and inositol 1,4,5-trisphosphate. Prostaglandins, thromboxanes, and leukotrienes are extremely potent hormones derived from arachidonate. Steroid hormones are derived from sterols. Vitamins D, A, E, and K are fat-soluble compounds made up of isoprene units. All play essential roles in the metabolism or physiology of animals. Vitamin D is precursor to a hormone that regulates calcium metabolism. Vitamin A furnishes the visual pigment of the vertebrate eye and acts as a regulator of gene expression during epithelial cell growth. Vitamin E functions in the protection of membrane lipids from oxidative damage, and vitamin K is essential in the blood-clotting process. Ubiquinones and plastoquinones, also isoprenoid derivatives, function as electron carriers in mitochondria and chloroplasts, respectively. Dolichols activate and anchor sugars on cellular membranes for use in the synthesis of certain complex carbohydrates, glycolipids, and glycoproteins.

In the determination of lipid composition, lipids are extracted from tissues with organic solvents and separated by thin-layer, gas-liquid, or high-performance liquid chromatography. Individual lipids are identified by their chromatographic behavior, their susceptibility to hydrolysis by specific enzymes, or by mass spectrometry.

further reading

General

Gurr, M.I. & Harwood, J.L. (1990) *Lipid Biochemistry: An Introduction,* 4th edn, Chapman & Hall, London.

A good general resource on lipid structure and metabolism, at the intermediate level.

Harwood, J.L. & Russell, N.J. (1984) *Lipids in Plants and Microbes,* George Allen & Unwin, Ltd., London.

Short, clear descriptions of lipid types, their distribution, metabolism, and function in plants and microbes; intermediate level.

Mead, J.F., Alfin-Slater, R.B., Howton, D.R., & Popják, G. (1986) *Lipids: Chemistry, Biochemistry and Nutrition,* Plenum Press, New York.

An intermediate-level textbook on chemical, metabolic, and nutritional aspects of lipids.

Vance, D.E. & Vance, J.E. (eds) (1996) *Biochemistry of Lipids, Lipoproteins and Membranes,* New Comprehensive Biochemistry, Vol. 31, Elsevier Science Publishing Co., Inc., New York.

An excellent collection of reviews on various aspects of lipid structure, biosynthesis, and function.

Structural Lipids in Membranes

Gravel, R.A., Clarke, J.T.R., Kaback, M.M., Mahuran, D., Sandhoff, K., & Suzuki, K. (1995) The GM_2 gangliosidoses. In *The Metabolic and Molecular Bases of Inherited Disease,* 7th edn (Scriver, C.R., Beaudet, A.L., Sly, W.S., & Valle, D., eds), pp. 2839–2879, McGraw-Hill, Inc., New York.

This article is one of many in a three-volume set that contains definitive descriptions of the clinical, biochemical, and genetic aspects of hundreds of human metabolic diseases—an authoritative source and fascinating reading.

Hakamori, S. (1986) Glycosphingolipids. *Sci. Am.* **254** (May), 44–53.

Hoekstra, D. (ed.) (1994) *Cell Lipids,* Current Topics in Membranes, Vol. 4, Academic Press, Inc., San Diego.

Ostro, M.J. (1987) Liposomes. *Sci. Am.* **256** (January), 102–111.

Sastry, P.S. (1985) Lipids of nervous tissue: composition and metabolism. *Prog. Lipid Res.* **24,** 69–176.

Spector, A.A. & Yorek, M.A. (1985) Membrane lipid composition and cellular function. *J. Lipid Res.* **26,** 1015–1035.

Lipids with Specific Biological Activities

Bell, R.M., Exton, J.H., & Prescott, S.M. (eds) (1996) *Lipid Second Messengers,* Handbook of Lipid Research, Vol. 8, Plenum Press, New York.

Chojnacki, T. & Dallner, G. (1988) The biological role of dolichol. *Biochem. J.* **251,** 1–9.

Machlin, L.J. & Bendich, A. (1987) Free radical tissue damage: protective role of antioxidant nutrients. *FASEB J.* **1,** 441–445.

Brief discussion of tocopherols as antioxidants and their role in preventing damage by oxygen free radicals.

Prescott, S.M., Zimmerman, G.A., & McIntyre, T.M. (1990) Platelet-activating factor. *J. Biol. Chem.* **265,** 17,381–17,384.

Snyder, F., Lee, T.-C., & Blank, M.L. (1989) Platelet-activating factor and related ether lipid mediators: biological activities, metabolism, and regulation. *Ann. N. Y. Acad. Sci.* **568,** 35–43.

Suttie, J.W. (1993) Synthesis of vitamin K-dependent proteins. *FASEB J.* **7,** 445–452.

Vermeer, C. (1990) γ-Carboxyglutamate-containing proteins and the vitamin K-dependent carboxylase. *Biochem. J.* **266,** 625–636.

Describes the biochemical basis for the requirement of vitamin K in blood clotting, and the importance of carboxylation in the synthesis of the blood-clotting protein thrombin.

Viitala, J. & Järnefelt, J. (1985) The red cell surface revisited. *Trends Biochem. Sci.* **10,** 392–395.

Includes discussion of the human A, B, and O blood type determinants.

Separation and Analysis of Lipids

Kates, M. (1986) *Techniques of Lipidology: Isolation, Analysis and Identification of Lipids,* 2nd edn, Laboratory Techniques in Biochemistry and Molecular Biology, Vol. 3, Part 2 (Burdon, R.H. & van Knippenberg, P.H., eds), Elsevier Science Publishing Co., Inc., New York.

Matsubara, T. & Hagashi, A. (1991) FAB/Mass spectrometry of lipids. *Prog. Lipid Res.* **30,** 301–322.

An advanced discussion of the identification of lipids by fast atom bombardment (FAB) mass spectrometry, a powerful technique for structure determination.

problems

1. Operational Definition of Lipids How is the definition of "lipid" different from the types of definitions used for other biomolecules that we have considered, such as amino acids, nucleic acids, and proteins?

2. Melting Points of Lipids The melting points of a series of 18-carbon fatty acids are: stearic acid, 69.6 °C; oleic acid, 13.4 °C; linoleic acid, −5 °C; and linolenic acid, −11 °C. (a) What structural aspect of these 18-carbon fatty acids can be correlated with the melting point? Provide a molecular explanation for the trend in melting points. (b) Draw all of the possible triacylglycerols that can be constructed from glycerol, palmitic acid, and oleic acid. Rank them in order of increasing melting point. (c) Branched-chain fatty acids are found in some bacterial membrane lipids. Would their presence increase or decrease the fluidity of the membranes (that is, make them have a lower or higher melting point)? Why?

3. Preparation of Béarnaise Sauce During the preparation of béarnaise sauce, egg yolks are incorporated into melted butter to stabilize the sauce and avoid separation. The stabilizing agent in the egg yolks is lecithin (phosphatidylcholine). Suggest why this works.

4. Hydrophobic and Hydrophilic Components of Membrane Lipids A common structural feature of membrane lipid molecules is their amphipathic nature. For example, in phosphatidylcholine, the two fatty acid chains are hydrophobic and the phosphocholine head group is hydrophilic. For each of the following membrane lipids, name the components that serve as the hydrophobic and hydrophilic units: (a) phosphatidylethanolamine; (b) sphingomyelin; (c) galactosylcerebroside; (d) ganglioside; (e) cholesterol.

5. Alkali Lability of Triacylglycerols A common procedure for cleaning the grease trap in a sink is to add a product that contains sodium hydroxide. Explain why this works.

6. The Action of Phospholipases The venom of the Eastern diamondback rattler and the Indian cobra contains phospholipase A_2, which catalyzes the hydrolysis of fatty acids at the C-2 position of glycerophospholipids. The phospholipid breakdown product of this reaction is lysolecithin (lecithin is phosphatidylcholine). At high concentrations, this and other lysophospholipids act as detergents, dissolving the membranes of erythrocytes and lysing the cells. Extensive hemolysis may be life-threatening.

(a) Detergents are amphipathic. What are the hydrophilic and hydrophobic portions of lysolecithin?

(b) The pain and inflammation caused by a snake bite can be treated with certain steroids. What is the basis of this treatment?

(c) Though high levels of phospholipase A_2 can be deadly, this enzyme is necessary for a variety of normal metabolic processes. What are these processes?

7. Intracellular Messengers from Phosphatidylinositols When the hormone vasopressin stimulates cleavage of phosphatidylinositol 4,5-bisphosphate by hormone-sensitive phospholipase C, two products are formed. Compare their properties and their solubilities in water, and predict whether either would diffuse readily through the cytosol.

8. Storage of Fat-Soluble Vitamins In contrast to water-soluble vitamins, which must be a part of our daily diet, fat-soluble vitamins can be stored in the body in amounts sufficient for many months. Suggest an explanation for this difference based on solubilities.

9. Hydrolysis of Lipids Name the products of mild hydrolysis with dilute NaOH of (a) 1-stearoyl-2,3-dipalmitoylglycerol; (b) 1-palmitoyl-2-oleoylphosphatidylcholine.

10. Effect of Polarity on Solubility Rank the following in order of increasing solubility in water: a triacylglycerol, a diacylglycerol, and a monoacylglycerol, all containing only palmitic acid.

11. Chromatographic Separation of Lipids A mixture of lipids is applied to a silica gel column, and the column is then washed with increasingly polar solvents. The mixture consists of: phosphatidylserine, cholesteryl palmitate (a sterol ester), phosphatidylethanolamine, phosphatidylcholine, sphingomyelin, palmitate, *n*-tetradecanol, triacylglycerol, and cholesterol. In what order do you expect the lipids to elute from the column?

12. Identification of Unknown Lipids Johann Thudichum, who practiced medicine in London about 100 years ago, also dabbled in lipid chemistry in his spare time. He isolated a variety of lipids from neural tissue, and characterized and named many of them. His carefully sealed and labeled vials of isolated lipids were rediscovered many years later. (a) How would you confirm, using techniques not available to Thudichum, that the vials labeled "sphingomyelin" and "cerebroside" actually contain these compounds? (b) How would you distinguish sphingomyelin from phosphatidylcholine by chemical, physical, or enzymatic tests?

13. Ninhydrin to Detect Lipids on TLC Plates Ninhydrin reacts specifically with primary amines to form a purplish-blue product. A thin-layer chromatogram of rat liver phospholipids is sprayed with ninhydrin, and the color is allowed to develop. Which phospholipids can be detected in this way?

chapter

12

Biological Membranes and Transport

The first cell probably came into being when a membrane formed, enclosing a small volume of aqueous solution and separating it from the rest of the universe. Membranes define the external boundaries of cells and regulate the molecular traffic across that boundary; in eukaryotic cells, they divide the internal space into discrete compartments to segregate processes and components (Fig. 12–1). They organize complex reaction sequences and are central to both biological energy conservation and cell-to-cell communication. The biological activities of membranes flow from their remarkable physical properties. Membranes are flexible, self-sealing, and selectively permeable to polar solutes. Their flexibility permits the shape changes that accompany cell growth and movement (such as amoeboid movement). With their ability to break and reseal, two membranes can fuse, as in exocytosis, or a single membrane-enclosed compartment can undergo fission to yield two sealed compartments, as in endocytosis or cell division, without creating gross leaks through cellular surfaces. Because membranes are selectively permeable, they retain certain compounds and ions within cells and within specific cellular compartments, while excluding others.

Membranes are not merely passive barriers. They include an array of proteins specialized for promoting or catalyzing various cellular processes. At the cell surface, transporters move specific organic solutes and inorganic ions across the membrane; receptors sense extracellular signals and trigger molecular changes in the cell; adhesion molecules hold neighboring cells together. Within the cell, membranes organize cellular processes such as the synthesis of lipids and certain proteins, and the energy transductions in mitochondria and chloroplasts. Membranes consist of just two layers of molecules and are therefore very thin; they are essentially two-dimensional. Because intermolecular collisions are far more probable in this two-dimensional space than in three-dimensional space, the efficiency of enzyme-catalyzed processes organized within membranes is vastly increased.

In this chapter we first describe the composition of cellular membranes and their chemical architecture—the dynamic molecular structures that underlie their biological functions. We describe several classes of membrane proteins defined by the ways in which they associate with the lipid bilayer. Cell adhesion, endocytosis, and the membrane fusion that occurs with viral infection will illustrate the dynamic role of membrane proteins. We then turn to the protein-mediated passage of solutes across membranes via transporters and ion channels. In later chapters we will discuss the role of membranes in signal transduction (Chapters 13 and 23), energy transduction (Chapter 19), lipid synthesis (Chapter 21), and protein synthesis (Chapter 27).

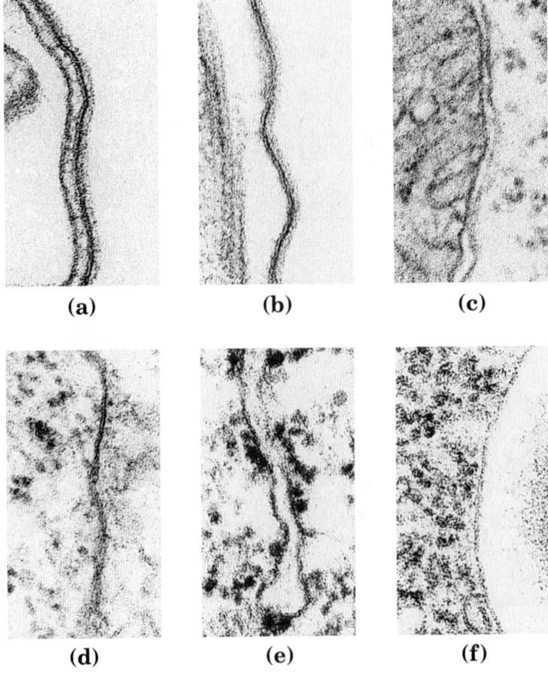

figure 12–1

Biological membranes. Viewed in cross section, all intracellular membranes share a characteristic trilaminar appearance. The protozoan *Paramecium* contains a variety of specialized membrane-bounded organelles. When a thin section of a *Paramecium* is stained with osmium tetroxide to highlight membranes, each of the membranes appears as a three-layer structure, 5 to 8 nm (50 to 80 Å) thick. The trilaminar images consist of two electron-dense layers on the inner and outer surfaces separated by a less dense central region. Above are high-magnification views of the membranes of **(a)** a cell body (plasma and alveolar membranes tightly apposed), **(b)** a cilium, **(c)** a mitochondrion, **(d)** a digestive vacuole, **(e)** the endoplasmic reticulum, and **(f)** a secretory vesicle.

table 12–1

Major Components of Plasma Membranes in Various Organisms

	Components (% by weight)				
	Protein	Phospholipid	Sterol	Sterol type	Other lipids
Human myelin sheath	30	30	19	Cholesterol	Galactolipids, plasmalogens
Mouse liver	45	27	25	Cholesterol	—
Maize leaf	47	26	7	Sitosterol	Galactolipids
Yeast	52	7	4	Ergosterol	Triacylglycerols, steryl esters
Paramecium (ciliated protist)	56	40	4	Stigmasterol	—
E. coli	75	25	0	—	—

The Molecular Constituents of Membranes

One approach to understanding membrane function is to study membrane composition—to determine, for example, which components are common to all membranes and which are unique to membranes with specific functions. Before describing membrane structure and function, we therefore consider the molecular components of membranes: proteins and polar lipids, which account for almost all the mass of biological membranes, and carbohydrate present as part of glycoproteins and glycolipids.

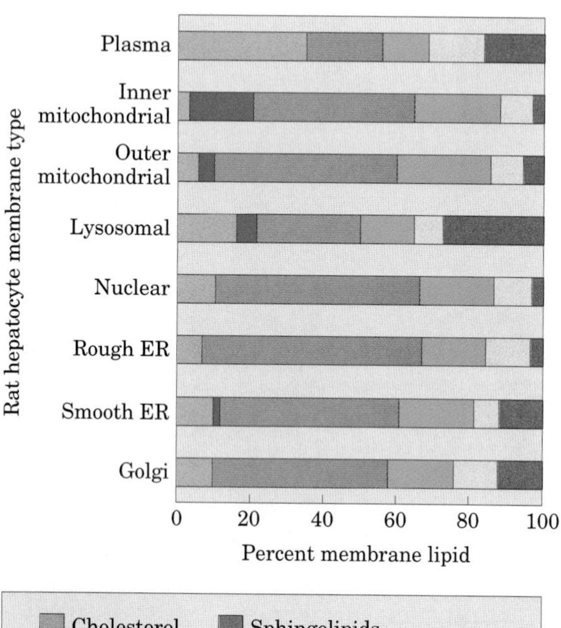

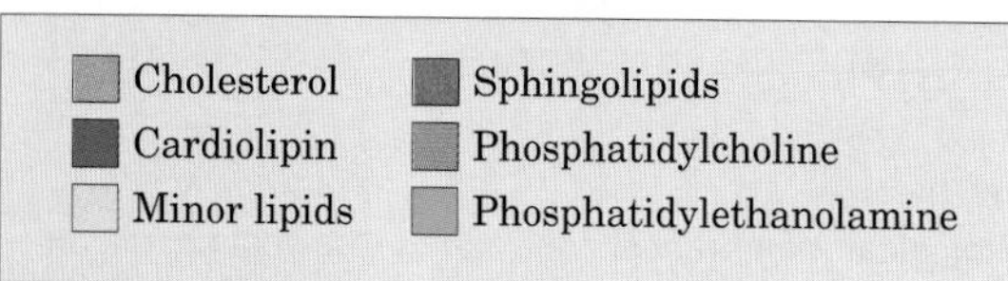

figure 12–2

Lipid composition of the plasma membrane and organelle membranes of a rat hepatocyte. The functional specialization of each membrane type is reflected in its unique lipid composition. Cholesterol (orange) is prominent in plasma membranes but barely detectable in mitochondrial membranes. Cardiolipin (blue) is a major component of the inner mitochondrial membrane but not of the plasma membrane. Phosphatidylserine, phosphatidylinositol, and phosphatidylglycerol (yellow) are relatively minor components of most membranes but may serve critical functions. Phosphatidylinositol and its derivatives, for example, are important in signal transductions triggered by hormones. Sphingolipids (red), phosphatidylcholine (purple), and phosphatidylethanolamine (green) are present in most membranes, but in varying proportions. Glycolipids, which are major components of the chloroplast membranes of plants, are virtually absent from animal cells.

Each Type of Membrane Has Characteristic Lipids and Proteins

The relative proportions of protein and lipid vary with the type of membrane (Table 12–1), reflecting the diversity of biological roles. For example, certain neurons have a myelin sheath, an extended plasma membrane that wraps around the cell many times and acts as a passive electrical insulator. The myelin sheath consists primarily of lipids, whereas the plasma membranes of bacteria and the membranes of mitochondria and chloroplasts, in which many enzyme-catalyzed processes take place, contain more protein than lipid.

For studies of membrane composition, the first task is to isolate a selected membrane. When eukaryotic cells are subjected to mechanical shear, their plasma membranes are torn and fragmented, releasing cytosolic components and membrane-bounded organelles such as mitochondria, chloroplasts, lysosomes, and nuclei. Plasma membrane fragments and intact organelles can be isolated by centrifugal techniques described in Chapter 2 (see Fig. 2–20).

Chemical analysis of membranes isolated from various sources reveals certain common properties. Each kingdom, each species, each tissue or cell type, and the organelles of each cell type have a characteristic set of membrane lipids. The plasma membranes of hepatocytes, for example, are enriched in cholesterol and contain no detectable cardiolipin (Fig. 12–2); in the inner mitochondrial membrane of the hepatocytes, this distribution is reversed. Cells clearly have mechanisms to control the kinds and amounts of membrane lipids they synthesize and to target specific lipids to particular organelles. In many cases, we can surmise the adaptive advantages of distinct combinations of membrane lipids; in other cases, the functional significance of these combinations remains to be discovered.

The protein composition of membranes from different sources varies even more widely than their lipid composition, reflecting functional specialization. In a rod cell of the vertebrate retina, one portion of the cell is highly specialized for the reception of light; more than 90% of the plasma membrane protein in this region is the light-absorbing glycoprotein rhodopsin. The less specialized plasma membrane of the erythrocyte has about 20 prominent types of proteins as well as dozens of minor ones; many of these are transporters, each moving a specific solute across the membrane. The plasma membrane of *Escherichia coli* contains hundreds of different proteins, including transporters and many enzymes involved in energy-conserving metabolism, lipid synthesis, protein export, and cell division. The outer membrane of *E. coli* has a different function (protection) and a different set of proteins.

Some membrane proteins are covalently linked to complex arrays of carbohydrate. For example, in glycophorin, a glycoprotein of the erythrocyte plasma membrane, 60% of the mass consists of complex oligosaccharide units covalently attached to specific amino acid residues. Ser, Thr, and Asn residues are the most common points of attachment (see Fig. 9–25). At the other end of the scale is rhodopsin of the rod cell plasma membrane, which contains a single hexasaccharide. The sugar moieties of surface glycoproteins influence the folding of the protein, as well as its stability and intracellular destination, and they play a significant role in the specific binding of ligands to glycoprotein surface receptors (see Fig. 9–29). Unlike plasma membranes, intracellular membranes such as those of mitochondria and chloroplasts rarely contain covalently bound carbohydrates.

Some membrane proteins are covalently attached to one or more lipids, which serve as hydrophobic anchors that hold the proteins to the membrane, as we shall see.

The Supramolecular Architecture of Membranes

All biological membranes share certain fundamental properties. They are impermeable to most polar or charged solutes, but permeable to nonpolar compounds; they are 5 to 8 nm (50 to 80 Å) thick and appear trilaminar when viewed in cross section with the electron microscope (Fig. 12–1). The combined evidence from electron microscopy and studies of chemical composition, as well as physical studies of permeability and the motion of individual protein and lipid molecules within membranes, led to the development of the **fluid mosaic model** for the structure of biological membranes (Fig. 12–3). Phospholipids and sterols form a lipid bilayer in which the nonpolar regions of the lipid molecules face each other at the core of the bilayer and their polar head groups face outward. In this bilayer sheet, proteins are

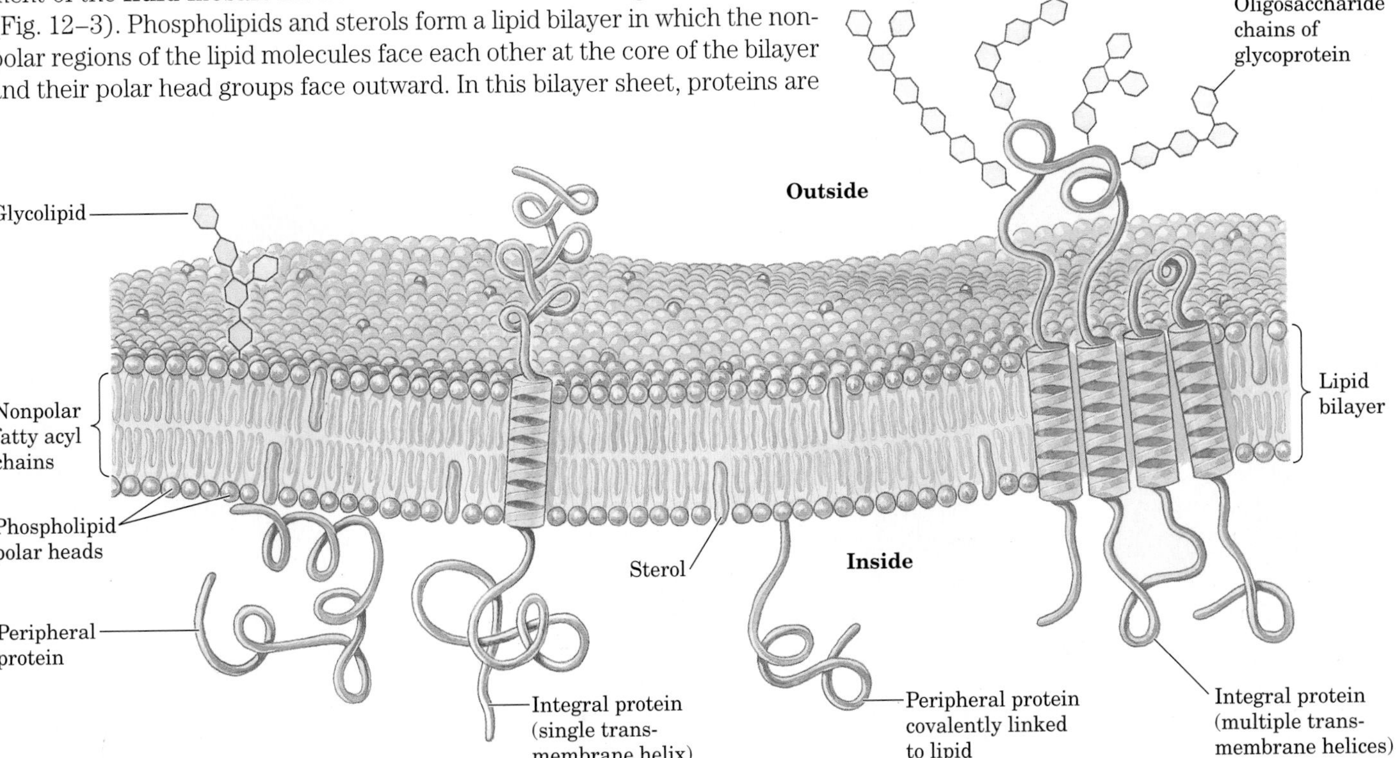

figure 12–3
Fluid mosaic model for membrane structure. The fatty acyl chains in the interior of the membrane form a fluid, hydrophobic region. Integral proteins float in this sea of lipid, held by hydrophobic interactions with their nonpolar amino acid side chains. Both proteins and lipids are free to move laterally in the plane of the bilayer, but movement of either from one face of the bilayer to the other is restricted. The carbohydrate moieties attached to some proteins and lipids of the plasma membrane are exposed on the extracellular face of the membrane.

embedded at irregular intervals, held by hydrophobic interactions between the membrane lipids and hydrophobic domains in the proteins. Some proteins protrude from only one side of the membrane; others have domains exposed on both sides. The orientation of proteins in the bilayer is asymmetric, giving the membrane "sidedness"; the protein domains exposed on one side of the bilayer are different from those exposed on the other side, reflecting functional asymmetry. The individual lipid and protein units in a membrane form a fluid mosaic whose pattern, unlike a mosaic of ceramic tile and mortar, is free to change constantly. The membrane mosaic is fluid because most of the interactions among its components are noncovalent, leaving individual lipid and protein molecules free to move laterally in the plane of the membrane.

We now look at some of these features of the fluid mosaic model in more detail and consider the experimental evidence that confirms the model.

A Lipid Bilayer Is the Basic Structural Element of Membranes

Glycerophospholipids, sphingolipids, and sterols are virtually insoluble in water. When mixed with water, they spontaneously form microscopic lipid aggregates in a phase separate from their aqueous surroundings, clustering together with their hydrophobic moieties in contact with each other and their hydrophilic groups interacting with the surrounding water. Recall that lipid clustering reduces the amount of hydrophobic surface exposed to water and thus minimizes the number of molecules in the shell of ordered water at the lipid-water interface (see Fig. 4–7), resulting in an increase in entropy. Hydrophobic interactions among lipid molecules provide the thermodynamic driving force for the formation and maintenance of these clusters.

Depending on the precise conditions and the nature of the lipids, three types of lipid aggregates can form when amphipathic lipids are mixed with water (Fig. 12–4). **Micelles** are spherical structures containing a few dozen to a few thousand molecules arranged with their hydrophobic regions aggregated in the interior, excluding water, and their hydrophilic head groups at the surface, in contact with water. Micelle formation is favored when the cross-sectional area of the head group is greater than that of the acyl side chain(s), as in free fatty acids, lysophospholipids (phospholipids lacking one fatty acid), and detergents such as sodium dodecyl sulfate (SDS; p. 134).

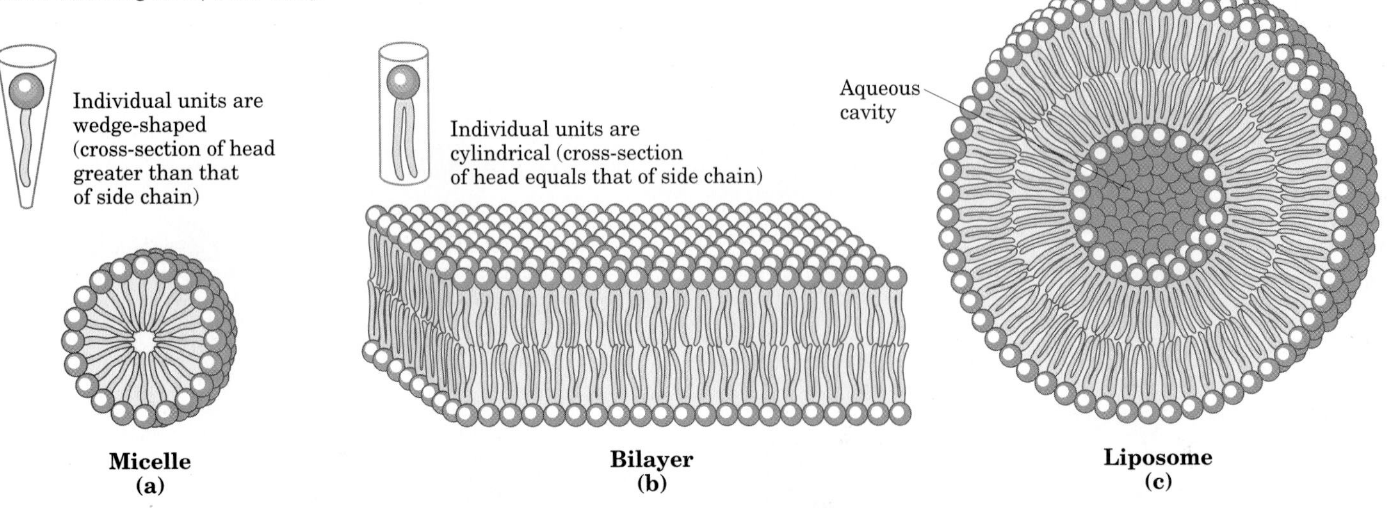

figure 12–4
Amphipathic lipid aggregates that form in water. (a) In micelles, the hydrophobic chains of the fatty acids are sequestered at the core of the sphere. There is virtually no water in the hydrophobic interior. **(b)** In an open bilayer, all acyl side chains except those at the edges of the sheet are protected from interaction with water. **(c)** When a two-dimensional bilayer folds on itself, it forms a closed bilayer, a three-dimensional hollow vesicle (liposome) enclosing an aqueous cavity.

A second type of lipid aggregate in water is the **bilayer,** in which two lipid monolayers form a two-dimensional sheet. Bilayer formation occurs most readily when the cross-sectional areas of the head group and acyl side chain(s) are similar, as in glycerophospholipids and sphingolipids. The hydrophobic portions in each monolayer, excluded from water, interact with each other. The hydrophilic head groups interact with water at each surface of the bilayer. Because the hydrophobic regions at its edges (Fig. 12–4b) are transiently in contact with water, the bilayer sheet is relatively unstable and spontaneously forms a third type of aggregate: it folds back on itself to form a hollow sphere called a vesicle or **liposome.** By forming vesicles, bilayers lose their hydrophobic edge regions, achieving maximal stability in their aqueous environment. These bilayer vesicles enclose water, creating a separate aqueous compartment. It is likely that the first living cells resembled liposomes, their aqueous contents segregated from the rest of the world by a hydrophobic shell.

All evidence indicates that biological membranes are constructed of lipid bilayers. The thickness of membranes, measured by electron microscopy, is about that expected for a lipid bilayer 3 nm (30 Å) thick with proteins protruding on each side. X-ray diffraction studies of membranes show the distribution of electron density expected for a bilayer structure. And liposomes formed in the laboratory show the same relative impermeability to polar solutes as is seen in biological membranes (although the latter, as we shall see, are permeable to solutes for which they have specific transporters).

Membrane lipids are asymmetrically distributed between the two monolayers of the bilayer, although the asymmetry, unlike that of membrane proteins, is not absolute. In the plasma membrane, for example, certain lipids are typically found primarily in the outer monolayer and others in the inner (cytosolic) monolayer, but rarely is a lipid found only on one side, not the other (Fig. 12–5).

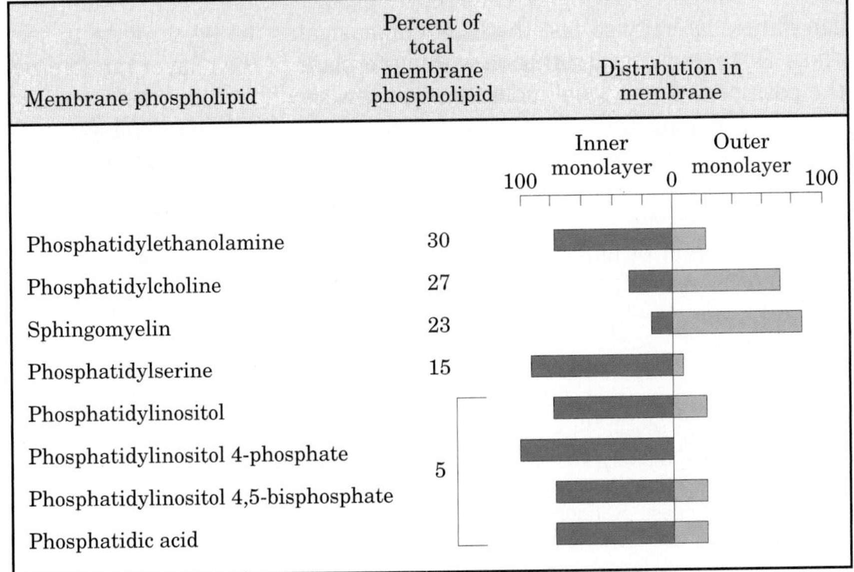

Membrane phospholipid
Percent of total membrane phospholipid
Distribution in membrane
Inner monolayer
Outer monolayer
100
0
100
Phosphatidylethanolamine
30
Phosphatidylcholine
27
Sphingomyelin
23
Phosphatidylserine
15
Phosphatidylinositol
Phosphatidylinositol 4-phosphate
5
Phosphatidylinositol 4,5-bisphosphate
Phosphatidic acid

figure 12–5
Asymmetric distribution of phospholipids between the inner and outer monolayers of the erythrocyte plasma membrane. The distribution of a specific phospholipid is determined by treating the intact cell with phospholipase C, which cannot reach lipids in the inner monolayer but removes the head groups of lipids in the outer monolayer. (Other membrane-impermeant reagents could also be used; see Fig. 12–9.) The proportion of each head group released provides an estimate of the fraction of each lipid in the outer monolayer.

Membrane Lipids Are in Constant Motion

Although the lipid bilayer structure itself is stable, the individual phospholipid and sterol molecules have great freedom of motion within the plane of the membrane (Fig. 12–6). The interior of the bilayer is fluid; individual hydrocarbon chains of fatty acids are in constant motion produced by rotation about the carbon-carbon bonds of the long acyl side chains. The degree of fluidity depends on lipid composition and temperature. At low temperature, relatively little lipid motion occurs and the bilayer exists as a nearly crystalline (paracrystalline) array. Above a certain temperature, lipids can undergo rapid motion. The **transition temperature,** the temperature above which the paracrystalline solid changes to fluid, is characteristic for each membrane and depends on its lipid composition. Saturated fatty acids pack well into a paracrystalline array, but the kinks in unsaturated fatty acids (see Fig. 11–1) interfere with this packing, hindering the formation of a paracrystalline state. The higher the proportion of saturated fatty acids, the higher is the transition temperature of the membrane.

The sterol content of a membrane is another important determinant of transition temperature. The rigid planar structure of the steroid nucleus, inserted between fatty acyl side chains, has two effects on fluidity. Below the membrane's transition temperature, sterol insertion prevents the highly ordered packing of fatty acyl chains and thus increases membrane fluidity. Above the thermal transition point, the rigid ring system of the sterol reduces the freedom of neighboring fatty acyl chains to move by rotation about their carbon-carbon bonds and thus reduces the fluidity in the core of the bilayer. Sterols therefore tend to moderate the extremes of solidity and fluidity of membranes.

Cells regulate their lipid composition to achieve a constant membrane fluidity under various growth conditions. For example, bacteria synthesize more unsaturated fatty acids and fewer saturated ones when cultured at low temperatures than when cultured at higher temperatures (Table 12–2). As a result of this adjustment in lipid composition, membranes of bacteria cultured at high or low temperatures have about the same degree of fluidity.

A second type of lipid motion involves not merely the flexing of fatty acyl chains but the movement of an entire lipid molecule relative to its neighbors (Fig. 12–6b). A molecule in one monolayer, or face, of the lipid bilayer—the outer face of the erythrocyte plasma membrane, for example—can diffuse laterally so fast that it circumnavigates the erythrocyte in seconds. This rapid lateral diffusion within the plane of the bilayer randomizes the positions of individual molecules in a few seconds. The combination of acyl chain flexing and lateral diffusion produces a membrane bilayer with the properties of a liquid crystal: a high degree of regularity in one dimension (perpendicular to the bilayer) and great mobility in the other (the plane of the bilayer).

A third kind of lipid motion, much less probable than conformational motion or lateral diffusion, is transbilayer or "flip-flop" diffusion of a molecule from one face of the bilayer to the other (Fig. 12–6c). This motion

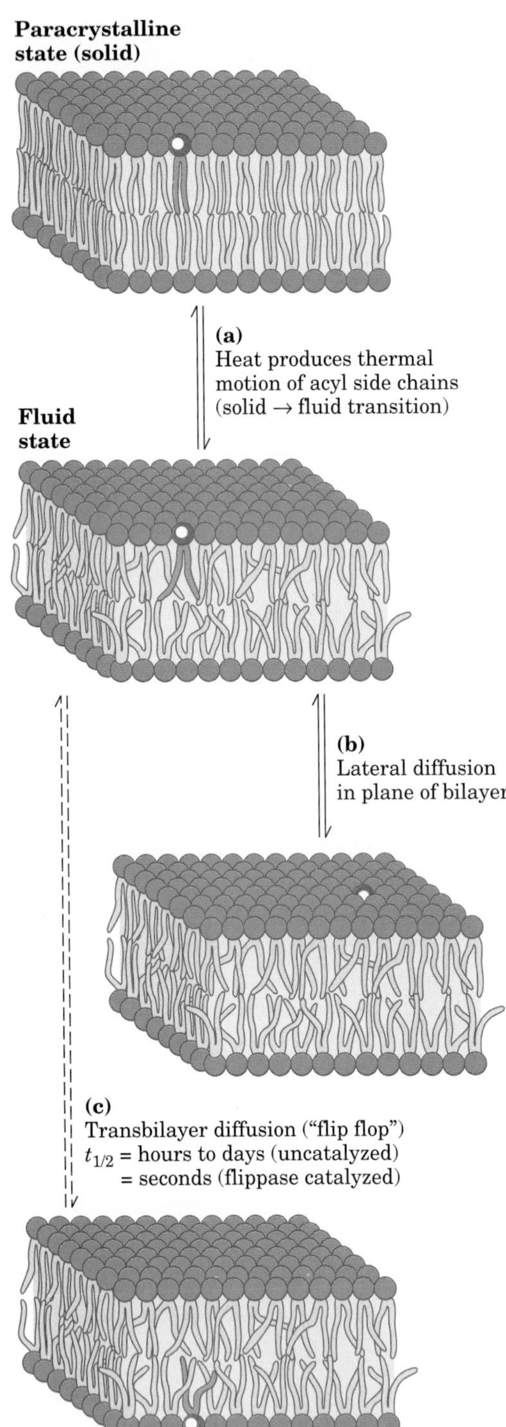

figure 12–6

Motion of membrane lipids. Lipid motion within a bilayer includes **(a)** thermal motion of the fatty acyl groups in the bilayer interior and **(b)** lateral diffusion of individual molecules in one face of the bilayer. A single lipid molecule can diffuse from one end of an *E. coli* cell to the other in about one second at 37 °C. Because of the thermal motion, the bilayer is fluid above a certain transition temperature; as temperature decreases, the bilayer becomes paracrystalline. **(c)** Movement of individual lipid molecules between the two faces of the bilayer (flip-flop diffusion) is very slow unless catalyzed by a specific lipid transporter, or flippase. Random distribution of the lipids in one monolayer between the two monolayers takes hours if uncatalyzed, but only seconds when a flippase is present.

table 12–2

Fatty Acid Composition of *E. coli* Cells Cultured at Different Temperatures

	Percentage of total fatty acids*			
	10 °C	20 °C	30 °C	40 °C
Myristic acid (14:0)	4	4	4	8
Palmitic acid (16:0)	18	25	29	48
Palmitoleic acid (16:1)	26	24	23	9
Oleic acid (18:1)	38	34	30	12
Hydroxymyristic acid	13	10	10	8
Ratio of unsaturated to saturated†	2.9	2.0	1.6	0.38

Source: Data from Marr, A.G. & Ingraham, J.L. (1962) Effect of temperature on the composition of fatty acids in *Escherichia coli*. *J. Bacteriol.* **84,** 1260.

*The exact fatty acid composition depends not only on growth temperature but on growth stage and growth medium composition.

†Calculated as the total percentage of 16:1 plus 18:1 divided by the total percentage of 14:0 plus 16:0. Hydroxymyristic acid was omitted from this calculation.

requires a polar (and possibly charged) head group to leave its aqueous environment and move into the hydrophobic interior of the bilayer, a process with a large, positive free-energy change. Yet this highly endergonic event is sometimes necessary, for example during synthesis of the bacterial plasma membrane, when phospholipids produced on the inside surface of the membrane must undergo flip-flop diffusion to enter the outer face of the bilayer. Similar transbilayer diffusion must also occur in eukaryotic cells as membrane lipids synthesized in one compartment move to other organelles. A family of proteins (flippases) facilitates flip-flop diffusion, providing a transmembrane path that is energetically more favorable than uncatalyzed diffusion.

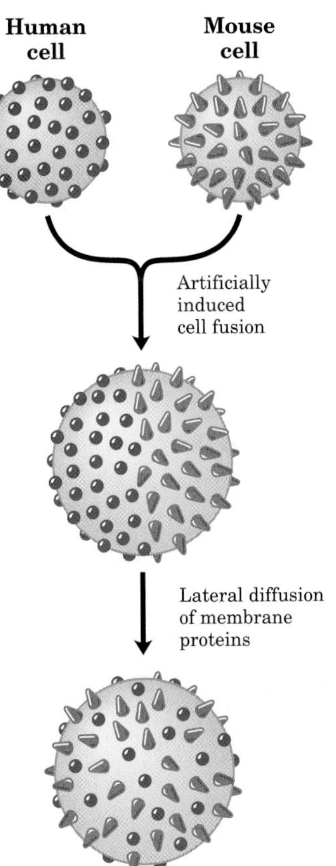

figure 12–7
Demonstration of lateral diffusion of membrane proteins. Fusion of a mouse cell with a human cell results in the randomization of membrane proteins from the two cells. After fusion, the location of each type of membrane protein is determined by staining cells with species-specific antibodies. Anti-mouse and anti-human antibodies are specifically tagged with molecules that fluoresce with different colors. Observed with the fluorescence microscope, the colored antibodies are seen to mix on the surface of the hybrid cell within minutes after fusion, indicating rapid lateral diffusion of the membrane proteins throughout the lipid bilayer.

Membrane Proteins Diffuse Laterally in the Bilayer

Many membrane proteins behave as though they were afloat in a sea of lipids. Like membrane lipids, these proteins are free to diffuse laterally in the plane of the bilayer and are in constant motion, as shown by the experiment diagrammed in Figure 12–7. Some membrane proteins associate to form large aggregates ("patches") on the surface of a cell or organelle, in which individual protein molecules do not move relative to one another; for example, acetylcholine receptors (see Fig. 12–39) form dense patches on neuron plasma membranes at synapses. Other membrane proteins are anchored to internal structures that prevent their free diffusion. In the erythrocyte membrane, both glycophorin and the chloride-bicarbonate exchanger (p. 413) are tethered to a filamentous cytoskeletal protein, spectrin (Fig. 12–8).

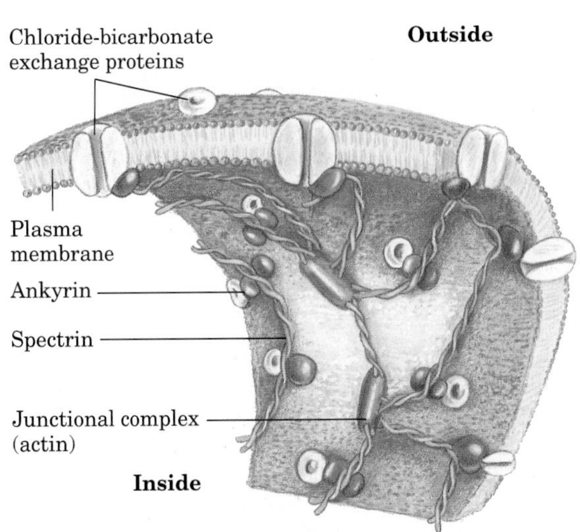

figure 12–8
Restricted motion of the erythrocyte chloride-bicarbonate exchanger. The protein spans the membrane and is tethered to the cytoskeletal protein spectrin by another protein, ankyrin, limiting the exchanger's lateral mobility. Ankyrin is anchored in the membrane by a covalently bound palmitoyl side chain (see Fig. 12–13). Spectrin, a long, filamentous protein, is cross-linked at junctional complexes containing actin. A network of cross-linked spectrin molecules attached to the cytoplasmic face of the plasma membrane stabilizes the membrane against deformation.

box 12–1 Looking at Membranes

Electron microscopy combined with techniques for preparing and staining tissues has been essential to studies of membrane structure. When thin sections of tissue are stained with osmium tetroxide, which accentuates differences in electron density, **transmission electron microscopy** shows the trilaminar structure (see Fig. 12–1). This level of resolution does not reveal individual membrane proteins, however. When a tissue sample is quick-frozen (to avoid the distortion that results when ice crystals form within cells) then cut with a fine knife, fracture lines frequently run along the surface of a membrane or through its center, splitting the bilayer into two monolayers (Fig. 1). The surfaces exposed by this **freeze-fracture** technique can be coated with a thin layer of carbon (evaporated from a carbon electrode under vacuum). **Scanning electron microscopy** of the resulting carbon "replica" shows details in bas relief of the membrane surface and the inside face of one monolayer exposed by splitting of the bilayer (Fig. 2). A variety of particles with the dimensions of single protein molecules or protein complexes protrude from a featureless background (the lipid bilayer). In many membranes, the particles are randomly positioned, but in specialized regions such as the postsynaptic membrane, they are clustered and ordered. In the myelin sheath, the particles are sparsely distributed. In membranes rich in enzyme complexes—the thylakoid membranes of chloroplasts, for instance—the surface is virtually filled with particles.

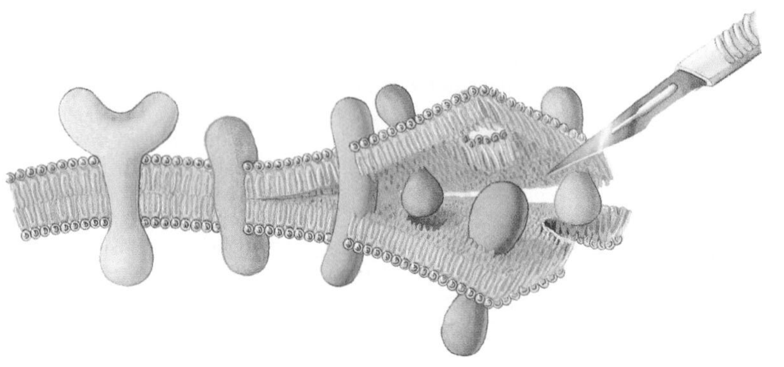

figure 1
Splitting of a membrane bilayer by the freeze-fracture technique.

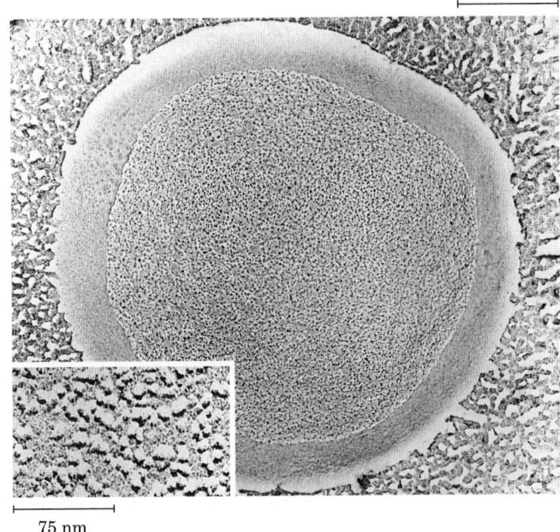

figure 2
An erythrocyte membrane is frozen in water, etched (by sublimation of some water), and a carbon replica is viewed in the electron microscope. The replica shows integral membrane proteins as bumps of various sizes and shapes, which are more readily visible at higher magnification (inset).

Some Membrane Proteins Span the Lipid Bilayer

The individual protein molecules and multiprotein complexes of biological membranes can be visualized by electron microscopy of the freeze-fractured membranes (Box 12–1). Some proteins appear on only one face of the membrane; others, which span the full thickness of the bilayer, protrude from both inner and outer membrane surfaces. Among the latter are proteins that conduct solutes or signals across the membrane.

Membrane protein localization has also been investigated with reagents that react with protein side chains but cannot cross membranes (Fig. 12–9). The human erythrocyte is convenient for such studies because it has no

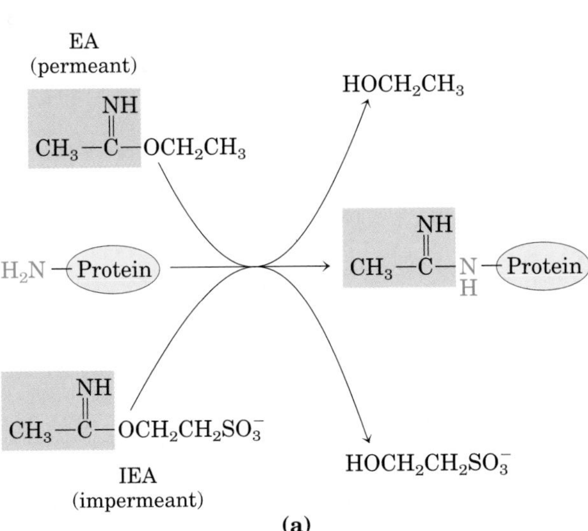
EA
(permeant)
NH
$CH_3-C-OCH_2CH_3$
$HOCH_2CH_3$
H_2N-Protein
NH
CH_3-C-N-Protein
H
NH
$CH_3-C-OCH_2CH_2SO_3^-$
IEA
(impermeant)
$HOCH_2CH_2SO_3^-$
(a)

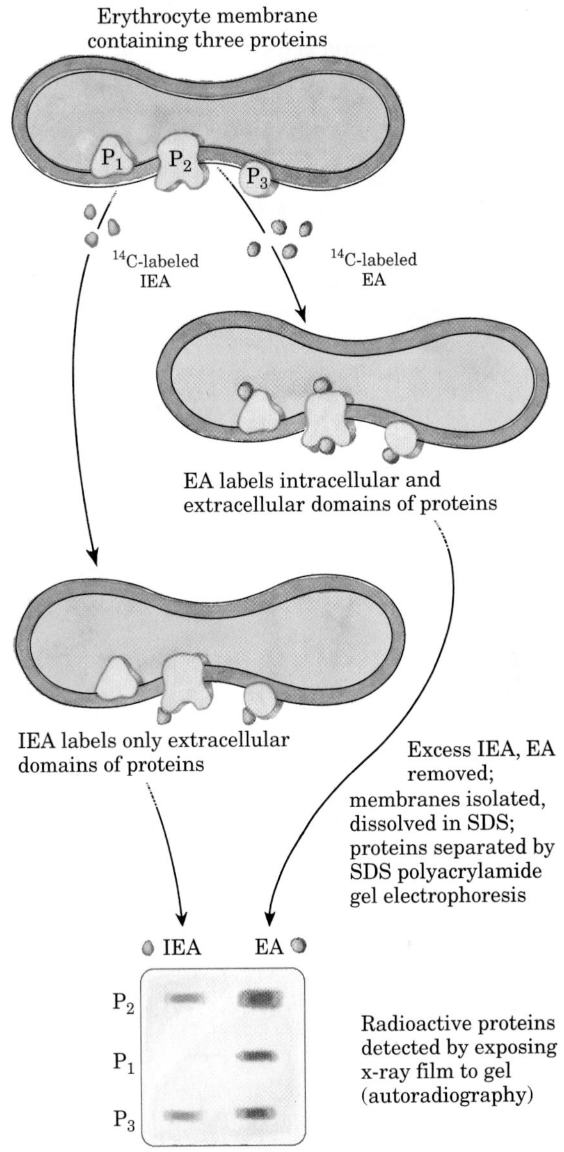
Erythrocyte membrane containing three proteins
P_1
P_2
P_3
^{14}C-labeled IEA
^{14}C-labeled EA
EA labels intracellular and extracellular domains of proteins
IEA labels only extracellular domains of proteins
Excess IEA, EA removed; membranes isolated, dissolved in SDS; proteins separated by SDS polyacrylamide gel electrophoresis
IEA
EA
P_2
P_1
P_3
Radioactive proteins detected by exposing x-ray film to gel (autoradiography)
(b)

figure 12–9
Experiments to determine the transmembrane arrangement of membrane proteins. (a) Both ethylacetimidate (EA) and isoethionylacetimidate (IEA) react with free amino groups in proteins, but only the ethyl derivative (EA) diffuses freely through the membrane. **(b)** Comparison of the labeling patterns after treatment of erythrocytes with the two reagents reveals whether a given protein is exposed only on the outer surface or only on the inner surface. A protein (P2) labeled by both reagents, but more heavily by the permeant reagent (EA), is exposed on both sides and thus spans the membrane. Proteins (P1) not labeled by the impermeant reagent (IEA) are exposed on the inner surface only.

membrane-bounded organelles; the plasma membrane is the only membrane present. If a membrane protein in an intact erythrocyte reacts with a membrane-impermeant reagent (IEA in Fig. 12–9), it must have at least one domain exposed on the outer (extracellular) face of the membrane. Proteases such as trypsin are unable to cross membranes and are commonly used to explore membrane protein topology, the disposition of protein domains relative to the lipid bilayer. Trypsin cleaves extracellular domains but does not affect domains buried within the bilayer or exposed on the inner surface only.

Experiments like those described in Figure 12–9 show that the glycoprotein glycophorin spans the erythrocyte membrane. Its amino-terminal domain (bearing the carbohydrate) is on the outer surface and can be cleaved by trypsin. The carboxyl terminus protrudes on the inside of the

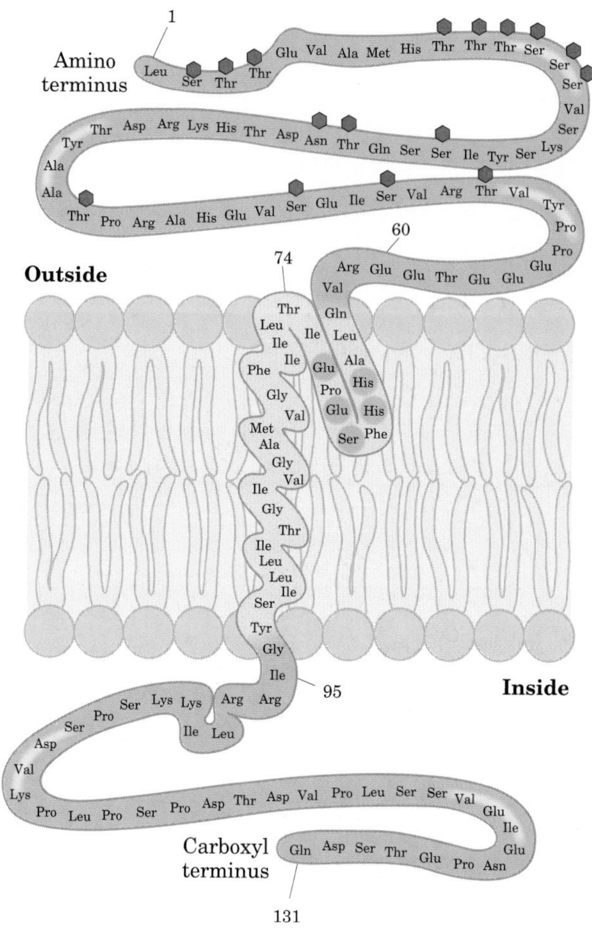

figure 12–10

Transbilayer disposition of glycophorin in the erythrocyte. One hydrophilic domain, containing all the sugar residues, is on the outer surface, and another hydrophilic domain protrudes from the inner face of the membrane. The red hexagons represent a tetrasaccharide (containing two Neu5Ac (sialic acid), Gal, and GalNAc) *O*-linked to a Ser or Thr residue; the blue hexagon represents an oligosaccharide chain *N*-linked to an Asn residue. The relative size of the oligosaccharide units is larger than shown here. A segment of 19 hydrophobic residues (residues 75 to 93) forms an α helix that traverses the membrane bilayer (see Fig. 12–17a). The segment from residue 64 to 74 has some hydrophobic residues and probably penetrates into the outer face of the lipid bilayer.

cell. Both the amino-terminal and carboxyl-terminal domains contain many polar or charged amino acid residues and are therefore quite hydrophilic. However, a long segment in the center of the protein contains mainly hydrophobic amino acid residues. These findings suggest that glycophorin has a transmembrane segment arranged as shown in Figure 12–10.

One further fact may be deduced from the results of the experiments with glycophorin: its disposition in the membrane is asymmetric. Similar studies of other membrane proteins show that each has a specific orientation in the bilayer and that proteins reorient by flip-flop diffusion very slowly, if at all. Furthermore, glycoproteins of the plasma membrane are invariably situated with their sugar residues on the outer surface of the cell. As we shall see, the asymmetric arrangement of membrane proteins results in functional asymmetry. All the molecules of a given ion pump, for example, have the same orientation in the membrane and therefore pump in the same direction.

Peripheral Membrane Proteins Are Easily Solubilized

Membrane proteins may be divided into two operational groups (Fig. 12–11). **Integral** (intrinsic) proteins are very firmly associated with the membrane, removable only by agents that interfere with hydrophobic interactions, such as detergents, organic solvents, or denaturants. **Peripheral** (extrinsic) proteins associate with the membrane through electrostatic interactions and hydrogen bonding with the hydrophilic domains of integral proteins and with the polar head groups of membrane lipids. They can

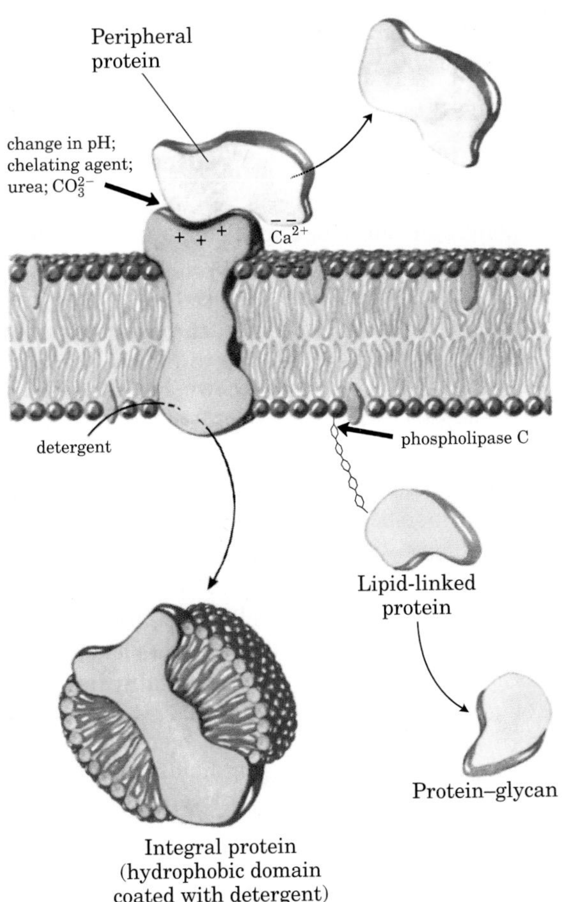

figure 12–11

Peripheral and integral proteins. Membrane proteins can be operationally distinguished by the conditions required to release them from the membrane. Most peripheral proteins can be released by changes in pH or ionic strength, removal of Ca^{2+} by a chelating agent, or addition of urea or carbonate. Peripheral proteins covalently attached to a membrane lipid, such as a glycosyl phosphatidylinositol (see Fig. 12–13), can be released by treatment with phospholipase C. Integral proteins can be extracted with detergents, which disrupt the hydrophobic interactions with the lipid bilayer and form micelle-like clusters around individual protein molecules.

be released by relatively mild treatments that interfere with electrostatic interactions or break hydrogen bonds. Peripheral proteins may serve as regulators of membrane-bound enzymes or may limit the mobility of integral proteins by tethering them to intracellular structures.

Annexins, for example, are a family of peripheral membrane proteins that bind reversibly to acidic membrane phospholipids, such as those that enrich the cytosolic (inner) surface of plasma membranes and the cytosolic (outer) face of organelles. Ca^{2+} ions, which interact simultaneously with negatively charged groups on annexin and the negatively charged head group of a phospholipid, are essential to the annexin-membrane interaction (Fig. 12–12). Roles for various annexins have been suggested in vesicle aggregation, membrane fusion, and interactions between membranes and cytoskeletal elements. A number of other proteins that act at a membrane–aqueous solution interface (blood-clotting proteins, for example) also undergo reversible Ca^{2+}-dependent membrane binding, perhaps by a similar mechanism.

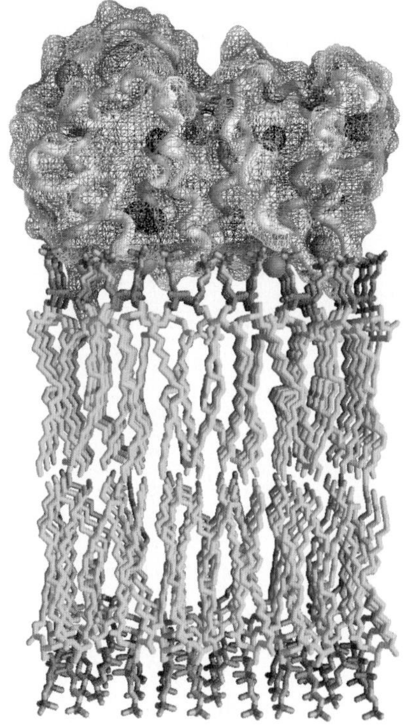

figure 12–12

Proposed structure of annexin V in association with the lipid head groups of a membrane. In this image, the protein backbone of annexin V (derived from crystallographic studies) is seen through a transparent mesh of the surface contour. The membrane modeled here is composed entirely of phosphocholine, with the choline moiety shown in blue, the phosphate groups red, and the fatty acid tails yellow. The bridging Ca^{2+} ions are purple. Electrostatic interactions hold annexin to the membrane.

figure 12–13

Lipid-linked membrane proteins. Covalently attached lipids anchor membrane proteins to the lipid bilayer. A palmitoyl group is shown attached by thioester linkage to a Cys residue; an *N*-myristoyl group is generally attached to an amino-terminal Gly; the farnesyl and geranylgeranyl groups attached to carboxyl-terminal Cys residues are isoprenoids of 15 and 20 carbons, respectively. These three lipid-protein assemblies are found only on the inside surface of the plasma membrane. Glycosyl phosphatidylinositol (GPI) anchors are derivatives of phosphatidylinositol in which the inositol bears a short oligosaccharide covalently joined to the carboxyl-terminal residue of a protein through phosphoethanolamine. GPI-linked proteins are always on the extracellular surface of the plasma membrane.

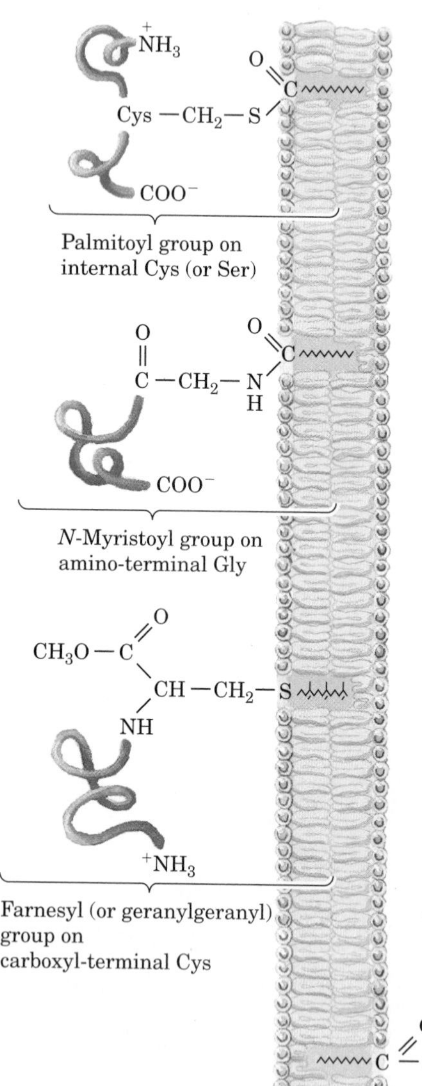

Covalently Attached Lipids Anchor Some Peripheral Membrane Proteins

Some membrane proteins contain one or more covalently linked lipids of several types: long-chain fatty acids, isoprenoids, or *g*lycosylated derivatives of *p*hosphatidyl*i*nositol, GPI (Fig. 12–13). The attached lipid provides a hydrophobic anchor, which inserts into the lipid bilayer and holds the protein at the membrane surface. The strength of the hydrophobic interaction between a bilayer and a single hydrocarbon chain linked to a protein is barely enough to anchor the protein securely. Other interactions, such as ionic attractions between positively charged Lys residues in the protein and negatively charged lipid head groups, probably stabilize the attachment. The association of these lipid-linked proteins with the membrane is certainly weaker than that for integral membrane proteins and is, in at least some cases, reversible.

Beyond merely anchoring a protein to the membrane, the attached lipid may have a more specific role. In the plasma membrane, proteins with GPI anchors are exclusively on the outer (extracellular) face, whereas other types of lipid-linked proteins (Fig. 12–13) are found exclusively on the inner (cytosolic) face. In epithelial cells, GPI-linked membrane proteins are far more common on the apical surface (the surface facing the external environment or interior cavity) than on the basal surface (the adjoining lower-layer epithelium). The attachment of a specific lipid to a membrane protein may therefore have a targeting function, directing the newly synthesized protein to its correct membrane location. The association of lipid-linked proteins with the membrane may be mediated by specific membrane receptors.

Integral Proteins Are Held in the Membrane by Hydrophobic Interactions with Lipids

The firm attachment of integral proteins to membranes is the result of hydrophobic interactions between membrane lipids and hydrophobic domains of the protein. Some proteins have a single hydrophobic sequence in the middle (glycophorin, for example) or at the amino or carboxyl terminus. Others have multiple hydrophobic sequences, each of which, when in the α-helical conformation, is long enough to span the lipid bilayer (Fig. 12–14). The same techniques that have allowed us to determine the three-dimensional

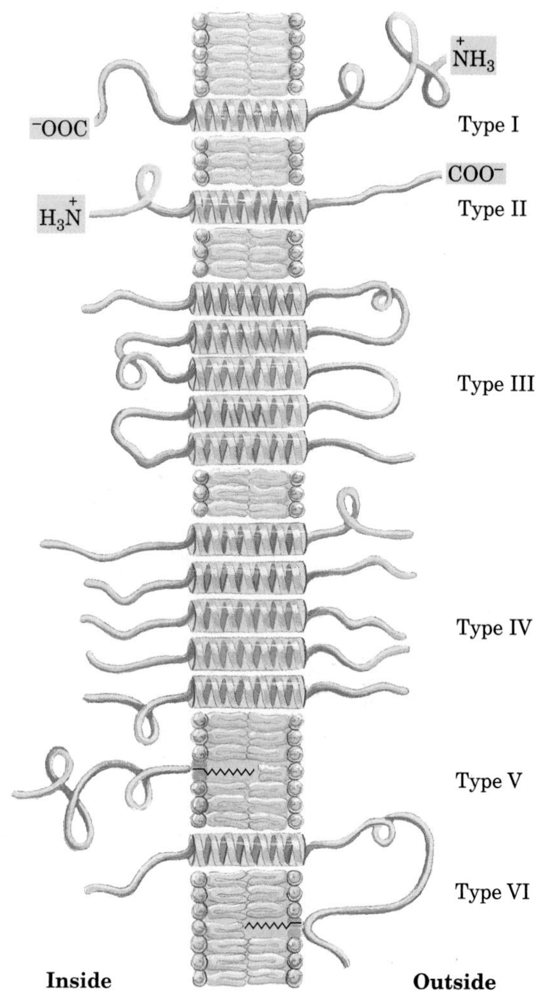

figure 12–14

Integral membrane proteins. For known proteins of the plasma membrane, the spatial relationships of protein domains to the lipid bilayer fall into six categories. Types I and II have only one transmembrane helix; the amino-terminal domain is outside the cell in type I proteins and inside in type II. Type III proteins have multiple transmembrane helices in a single polypeptide. In type IV proteins, transmembrane domains of several different polypeptides assemble to form a channel through the membrane. Type V proteins are held to the bilayer primarily by covalently linked lipids (see Fig. 12–13), and type VI proteins have both transmembrane helices and lipid (GPI) anchors.

In this figure, and in other figures throughout the book, we represent transmembrane protein segments in their most likely conformations: as α helices of six to seven turns. Sometimes these helices will be shown simply as cylinders. As relatively few membrane protein structures have been deduced by x-ray crystallography, our representation of the extra-membrane domains is arbitrary and not necessarily to scale, unless otherwise noted.

structures of many soluble proteins can in principle be applied to membrane proteins. However, generally useful techniques for crystallizing membrane proteins have only recently been developed, and relatively few membrane protein structures have been solved crystallographically.

One of the best-studied membrane-spanning proteins, bacteriorhodopsin, has seven very hydrophobic internal sequences and crosses the lipid bilayer seven times. Bacteriorhodopsin is a light-driven proton pump densely packed in regular arrays in the purple membrane of the bacterium *Halobacterium halobium.* When these arrays are viewed with the electron microscope from several angles, the resulting images allow a three-dimensional reconstruction of the bacteriorhodopsin molecule. X-ray crystallography yields a structure similar in all essential features (Fig. 12–15). Seven α-helical segments, each traversing the lipid bilayer, are connected by nonhelical loops at the inner and outer face of the membrane. In the amino acid sequence of bacteriorhodopsin, seven segments of about 20 hydrophobic residues can be identified, each segment just long enough to make an α helix that spans the bilayer. Hydrophobic interactions between the nonpolar amino acids and the fatty acyl groups of the membrane lipids firmly anchor the protein in the membrane. The seven helices are clustered together and oriented roughly perpendicular to the bilayer plane, providing a transmembrane pathway for proton movement. As we shall see in Chapter 13, this pattern of seven hydrophobic membrane-spanning helices is a common motif in membrane proteins involved in signal reception.

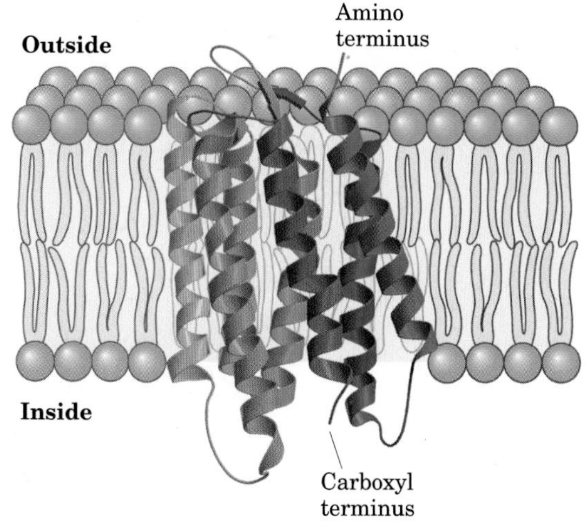

figure 12–15

Bacteriorhodopsin, a membrane-spanning protein. The single polypeptide chain folds into seven hydrophobic α helices, each of which traverses the lipid bilayer roughly perpendicular to the plane of the membrane. The seven transmembrane helices are clustered, and the space around and between them is filled with the acyl chains of membrane lipids. The light-absorbing pigment retinal (see Fig. 11–19) is buried deep in the membrane in contact with several of the helical segments (not shown). The helices are colored to correspond with the hydropathy plot in Figure 12–17b.

The photosynthetic reaction center from a purple bacterium was the first membrane protein structure solved by crystallography. Although a more complex membrane protein than bacteriorhodopsin, it is constructed on the same structural principles. The reaction center has four protein subunits, three of which contain α-helical segments that span the membrane (Fig. 12–16). These segments are rich in nonpolar amino acids, with their hydrophobic side chains oriented toward the outside of the molecule where they interact with membrane lipids. The architecture of the reaction center protein is therefore the inverse of that seen in most water-soluble proteins, in which hydrophobic residues are buried within the protein core and hydrophilic residues are on the surface (recall the structures of myoglobin and hemoglobin, for example). In Chapter 19 we will encounter several complex membrane proteins having multiple transmembrane helical segments in which hydrophobic chains are positioned to interact with the lipid bilayer.

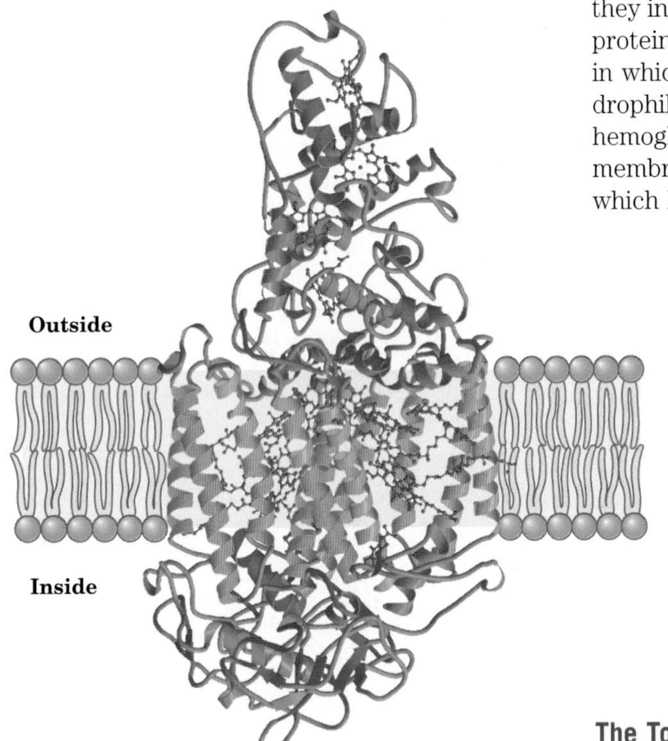

figure 12–16
Three-dimensional structure of the photosynthetic reaction center of *Rhodopseudomonas viridis*, a purple bacterium. This was the first integral membrane protein to have its atomic structure determined by x-ray diffraction methods. Eleven α-helical segments from three of the four subunits span the lipid bilayer, forming a cylinder 45 Å (4.5 nm) long; hydrophobic residues on the exterior of the cylinder interact with lipids of the bilayer. In this ribbon representation, residues that are part of the transmembrane helices are shown in yellow. The prosthetic groups (light-absorbing pigments and electron carriers; see Fig. 19–45) are red.

The Topology of an Integral Membrane Protein Can Sometimes Be Predicted from Its Sequence

Determining the three-dimensional structure of a membrane protein or its topology is generally much more difficult than determining its amino acid sequence, which can be accomplished by sequencing the protein or its gene. More than a thousand sequences are known for membrane proteins, but relatively few three-dimensional structures. The presence of long hydrophobic sequences in a membrane protein is commonly taken as evidence that these sequences traverse the lipid bilayer, acting as hydrophobic anchors or forming transmembrane channels. Virtually all integral proteins have at least one such sequence.

What can we predict about the secondary structure of the membrane-spanning portions of integral proteins? An α-helical sequence of 20 to 25 residues is just long enough to span the thickness (30 Å) of the lipid bilayer (recall that the length of an α helix is 1.5 Å (0.15 nm) per amino acid residue). A polypeptide chain surrounded by lipids, having no water molecules with which to form hydrogen bonds, will tend to form α helices or β sheets, in which intrachain hydrogen bonding is maximized. If the side chains of all amino acids in a helix are nonpolar, hydrophobic interactions with the surrounding lipids further stabilize the helix.

Several simple methods of analyzing amino acid sequences yield reasonably accurate predictions of secondary structure for transmembrane proteins. The relative polarity of each amino acid has been determined experimentally by measuring the free-energy change accompanying the movement of that amino acid's side chain from a hydrophobic solvent into water. This free energy of transfer ranges from very exergonic for charged or polar residues to very endergonic for amino acids with aromatic or

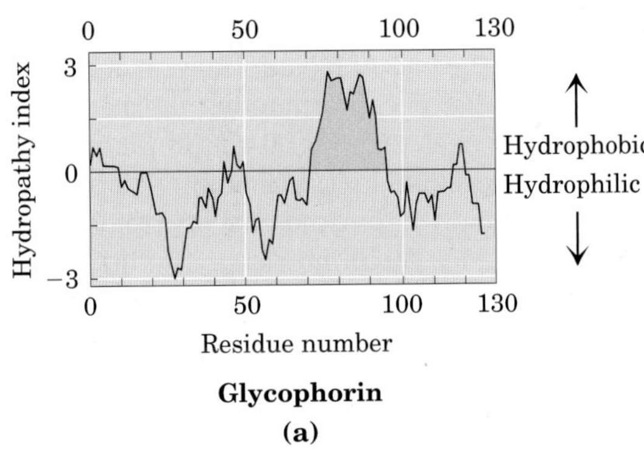

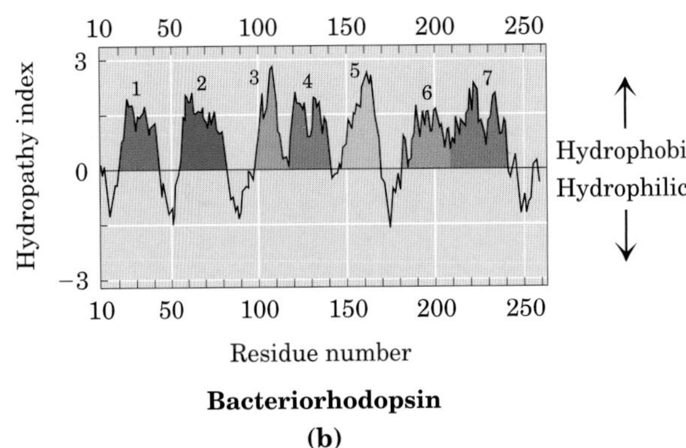

figure 12–17
Hydropathy plots. Hydropathy index (see Table 5–1) is plotted against residue number for two integral membrane proteins. The hydropathy index for each amino acid residue in a sequence of defined length (called the window) is used to calculate the average hydropathy for the residues in that window. The horizontal axis shows the residue number for the middle of the window. **(a)** Glycophorin from human erythrocytes has a single hydrophobic sequence between residues 75 and 93 (yellow); compare this with Figure 12–10. **(b)** Bacteriorhodopsin, known from independent physical studies to have seven transmembrane helices (see Fig. 12–15), has seven hydrophobic regions. Note, however, that the hydropathy plot is ambiguous in the region of segments 6 and 7. Physical studies have confirmed that this region has two transmembrane segments.

aliphatic hydrocarbon side chains. The overall hydrophobicity of a sequence of amino acids is estimated by summing the free energies of transfer for the residues in the sequence, which yields a **hydropathy index** for that region (see Table 5–1). To scan a polypeptide sequence for potential membrane-spanning segments, one calculates the hydropathy index for successive segments (called windows) of a given size, which may be from seven to 20 residues. For a window of seven residues, for example, the indices for residues 1 to 7, 2 to 8, 3 to 9, and so on, are plotted as in Figure 12–17 (plotted for the middle residue in each window—residue 4 for the window 1 to 7, for example). A region with more than 20 residues of high hydropathy index is presumed to be a transmembrane segment. When the sequences of membrane proteins of known three-dimensional structure are scanned in this way, we find a reasonably good correspondence between predicted and known membrane-spanning segments. Hydropathy analysis predicts a single hydrophobic helix for glycophorin (Fig. 12–17a) and seven transmembrane segments for bacteriorhodopsin (Fig. 12–17b)—in agreement with experimental studies.

Many of the transport proteins described in this chapter are believed, on the basis of their amino acid sequences and hydropathy plots, to have multiple membrane-spanning helical regions—that is, they are type III or type IV integral proteins (Fig. 12–14). When predictions are consistent with chemical studies of protein localization (such as those described above for glycophorin and bacteriorhodopsin), the assumption that hydrophobic regions correspond to membrane-spanning domains is better justified.

Not all integral membrane proteins are composed of transmembrane α helices. Another structural motif common in membrane proteins is the β barrel (see Fig. 6–20d), in which 20 or more transmembrane segments form β sheets that line a cylinder (Fig. 12–18). The same factors that favor α-helix formation in the hydrophobic interior of a lipid bilayer also stabilize β barrels. When no water molecules are available to hydrogen bond with the carbonyl oxygen and nitrogen of the peptide bond, maximal intrachain hydrogen bonding gives the most stable conformation. Porins, proteins that allow certain polar solutes to cross the outer membrane of gram-negative bacteria such as *E. coli,* have many-stranded β barrels lining the polar transmembrane passage.

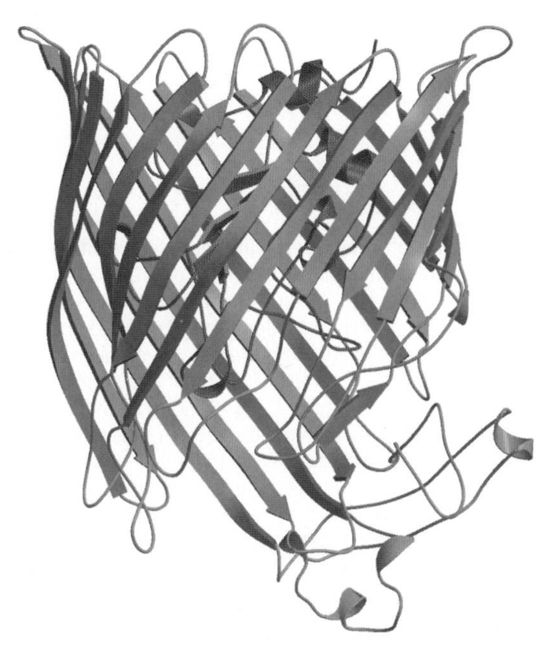

figure 12–18
Porin FhuA, an integral membrane protein with β-barrel structure. This protein of the *E. coli* outer membrane includes a β barrel composed of 22 antiparallel β strands (gray), forming a transmembrane channel through which iron ion bound to the carrier ferrichrome enters from the surrounding medium. The residues on the outside surface of the barrel are hydrophobic and interact with the lipids and lipopolysaccharides in the outer membrane.

A polypeptide is more extended in the β conformation than in an α helix; just seven to nine residues of β conformation are needed to span a membrane. Recall that in the β conformation, alternating side chains project above and below the sheet (see Fig. 6–7). In β strands of membrane proteins, every second residue in the membrane-spanning segment is hydrophobic and interacts with the lipid bilayer; aromatic side chains are commonly found at the lipid-protein interface. The other residues may or may not be hydrophilic. The hydropathy plot is not useful in predicting transmembrane segments for proteins with β barrel motifs, but as the database of known β barrel motifs increases, sequence-based predictions of transmembrane β sheets become more feasible.

Integral Proteins Mediate Cell-Cell Interactions and Adhesion

Several families of integral proteins in the plasma membrane provide specific points of attachment between cells, or between a cell and extracellular matrix proteins. **Integrins** are heterodimeric proteins (with two unlike subunits, α and β) anchored to the plasma membrane by a single hydrophobic transmembrane helix in each subunit (Fig. 12–19). The large extracellular domains of the α and β subunits combine to form a specific binding site for extracellular proteins such as collagen and fibronectin. As there are at least 14 different α subunits and at least eight different β subunits, a wide variety of specificities may be generated from various combinations of α and β. One common determinant of integrin binding present in several extracellular partners of integrins is the sequence Arg–Gly–Asp (RGD).

Integrins are not merely adhesives; they serve as receptors and signal transducers, carrying information across the plasma membrane in both directions. Integrins regulate many processes, including platelet aggregation at the site of a wound, tissue repair, the activity of immune cells, and the invasion of tissue by a tumor (see Fig. 9–27). Mutation in an integrin gene encoding the β subunit known as CD18 is the cause of leukocyte adhesion deficiency in humans, a rare genetic disease in which leukocytes fail to pass out of blood vessels to reach sites of infection. Infants with a severe defect in CD18 commonly die of infections before the age of two.

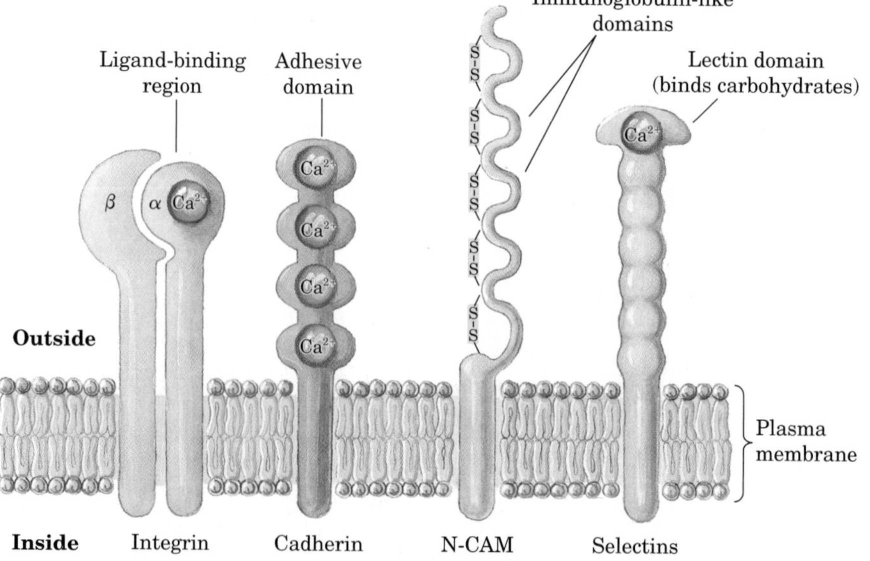

figure 12–19

Four examples of integral protein types that function in cell-cell interactions. Integrins consist of α and β transmembrane polypeptides; their extracellular domains combine to form binding sites for divalent metal ions and proteins of the extracellular matrix (such as collagen and fibronectin) or specific surface proteins of other cells. Cadherin has four extracellular Ca^{2+}-binding domains, the most distal of which contains the site that binds to cadherin on another cell surface. N-CAM (*n*euronal *c*ell *a*dhesion *m*olecule) is one of a family of immunoglobulin-like proteins that mediate Ca^{2+}-independent interactions with surface proteins of nearby cells. Selectins bind tightly to carbohydrate moieties in neighboring cells; the binding is Ca^{2+}-dependent.

At least three other families of plasma membrane proteins are also involved in surface adhesion (Fig. 12–19). **Cadherins** undergo homophilic ("with same kind") interactions with identical cadherins in an adjacent cell. **Immunoglobulin-like proteins** can undergo either homophilic interactions with their identical counterparts on another cell or heterophilic interactions with an integrin on a neighboring cell. **Selectins** have extracellular domains that, in the presence of Ca^{2+}, bind specific polysaccharides on the surface of an adjacent cell. Selectins are present primarily in the various types of blood cells and in the endothelial cells that line blood vessels. They are an essential part of the blood-clotting process.

Integral proteins play a role in many other cellular processes. They serve as transporters and ion channels (discussed later in this chapter) and as receptors for hormones, neurotransmitters, and growth factors (Chapter 13). They are central to oxidative phosphorylation and photosynthesis (Chapter 19) and to cell-cell and antigen-cell recognition in the immune system (Chapter 7). Integral proteins are also major players in the membrane fusion that accompanies exocytosis, endocytosis, and the entry of many types of viruses into host cells.

Membrane Fusion Is Central to Many Biological Processes

A remarkable feature of the biological membrane is its ability to undergo fusion with another membrane without losing its integrity. Although membranes are stable, they are by no means static. Within the endomembrane system (see Fig. 2–9), the membranous compartments constantly reorganize as small vesicles bud from the Golgi complex to carry newly synthesized lipids and proteins to other organelles and to the plasma membrane. Exocytosis, endocytosis, cell division, fusion of egg and sperm cells, and entry of a membrane-enveloped virus into its host cell all involve membrane reorganization in which the fundamental operation is fusion of two membrane segments without loss of continuity (Fig. 12–20).

Specific fusion of two membranes requires that (1) they recognize each other; (2) their surfaces become closely apposed, which requires the removal of water molecules normally associated with the polar head groups of lipids; (3) their bilayer structures become locally disrupted; and (4) the two bilayers fuse to form a single continuous bilayer. Receptor-mediated endocytosis or regulated secretion also requires that (5) the fusion process is triggered at the appropriate time or in response to a specific signal. Integral proteins called **fusion proteins** mediate these events, bringing about specific recognition and a transient, local distortion of the bilayer structure that favors membrane fusion. (Note that these fusion proteins are unrelated to the products of two fused genes, also called fusion proteins, as discussed in Chapter 29).

One of the best-studied cases of membrane fusion occurs during infection of a host cell by a membrane-enveloped virus, such as the influenza virus or HIV. Infection by influenza virus begins when an integral protein of

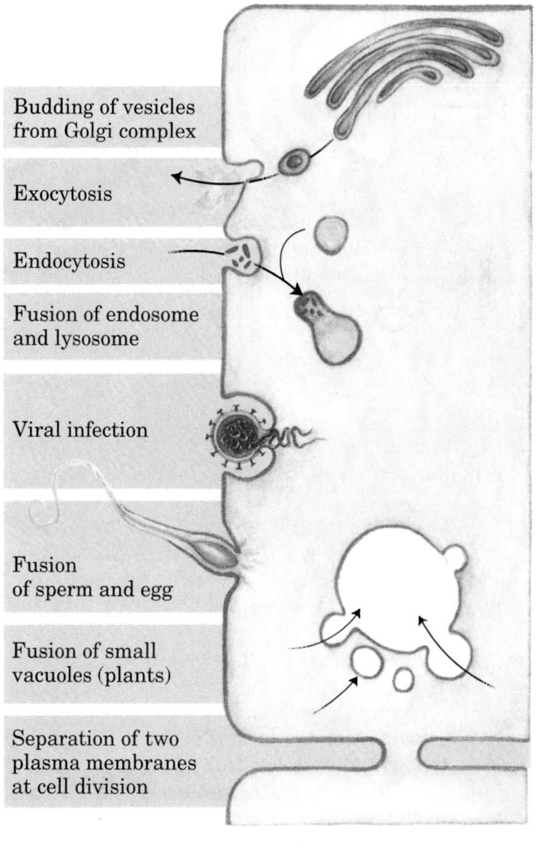

figure 12–20
Membrane fusion. The fusion of two membranes is central to a variety of cellular processes involving both organelles and the plasma membrane.

the virus envelope, the hemagglutination protein, binds to sialic acid (Neu5Ac) residues of glycoproteins or glycolipids in the host cell's plasma membrane (Fig. 12–21a). Endocytosis then conveys the virus into the cell within an endosome. The low pH in the endosome triggers a major alteration in the conformation of the hemagglutination protein, exposing a short hydrophobic peptide called the fusion peptide. Insertion of the fusion peptide into the lipid bilayer of the endosomal membrane produces a bridge between viral and endosomal membranes. This presumably brings about local distortion of both bilayers, leading to membrane fusion and release of the viral contents into the host cell.

In HIV infection, a pair of glycoproteins in the HIV envelope (gp41 and gp120) bind to two "receptors"—CD4 and a coreceptor protein—on the surface of the host cell (Fig. 12–21b). CD4 alone is sufficient to allow binding of HIV to the cell, but for fusion (and thus infection) to occur the host cell also must have a coreceptor such as CCR5. (CCR5 normally functions as a receptor for growth factors called chemokines that direct leukocytes to sites of inflammation.) The requirement for both CD4 and a coreceptor in HIV infection explains the specificity of HIV for T lymphocytes and phagocytes of the immune system; they are the only cell types that have both proteins in their plasma membranes. Presumably, these "receptors" evolved to serve functions useful to the cell, but the virus has evolved to exploit them. Agents that interfere with the interaction of HIV with CD4 or the coreceptor may prove useful in the treatment of HIV infection or in the prevention of progression of HIV infection to AIDS.

According to a current model for HIV entry by fusion with the plasma membrane, as shown in Figure 12–21b, the gp41-gp120 complex is normally anchored to the viral membrane by a hydrophobic sequence near the carboxyl terminus of gp41. Contact with CD4 and the coreceptor in the host cell's surface triggers a large conformational change in gp41, which stretches into an extended structure and exposes the fusion peptide near its amino terminus. This fusion peptide is inserted into the plasma membrane of the host cell; gp41 thus forms a bridge between virus and cell. A second dramatic structural change now occurs: gp41 bends sharply at a hinge region near its center, forming a hairpin-like structure that brings its amino terminus (in the plasma membrane) close to its carboxyl terminus (in the viral membrane). According to this model, the insertion of the fusion peptide into the plasma membrane somehow disorganizes the lipid bilayer locally. Realignment of lipid molecules in this disorganized region results in coalescence of the two membranes, and fusion occurs.

Membrane fusion is central to other cellular processes, too, such as the movement of newly synthesized membrane components through the endomembrane system from the endoplasmic reticulum through the Golgi complex to the plasma membrane via membrane vesicles, and the release of proteins, hormones, or neurotransmitters by exocytosis. The proteins required for these membrane fusions, called SNARES, resemble the viral fusion proteins in several respects. SNAREs (the family name derives from *syn*aptosome-*a*ssociated protein *re*ceptors) form hairpin-shaped rods that bring together protein domains bound to the plasma membrane and the vesicle membrane. SNAREs in the cytosolic face of the intracellular *v*esicles are called v-SNAREs; those in the *t*arget membranes with which the vesicles fuse are t-SNAREs. The extended forms of SNAREs, like those of the viral fusion proteins, have coiled coils, long regions in which several α helices are coiled about each other. These structural similarities very probably reflect a fundamental similarity in fusion mechanisms in viruses and intracellular vesicles.

figure 12–21

Membrane fusion during viral entry into a host cell. (a) Influenza virus adheres to carbohydrate moieties in the host cell's surface (see Fig. 9–29b) and triggers endocytosis, which encloses the virus in an endosome. Proton pumps (p. 416) acidify the endosome contents, causing a pH-dependent change in the structure of the hemagglutination (HA) protein and thus exposing a hydrophobic fusion peptide. The fusion peptide penetrates the endosomal membrane and brings about fusion of the viral and endosomal membranes, releasing the viral genome into the host cell. **(b)** When HIV encounters a T lymphocyte with CD4 and a second coreceptor (cytokine receptor) protein in its plasma membrane, the virus adheres to the cell through two glycoproteins (gp41 and gp120) in its own membrane, releasing the viral contents into the host cell. HIV infects only those cells (lymphocytes) with CD4 in their plasma membrane.

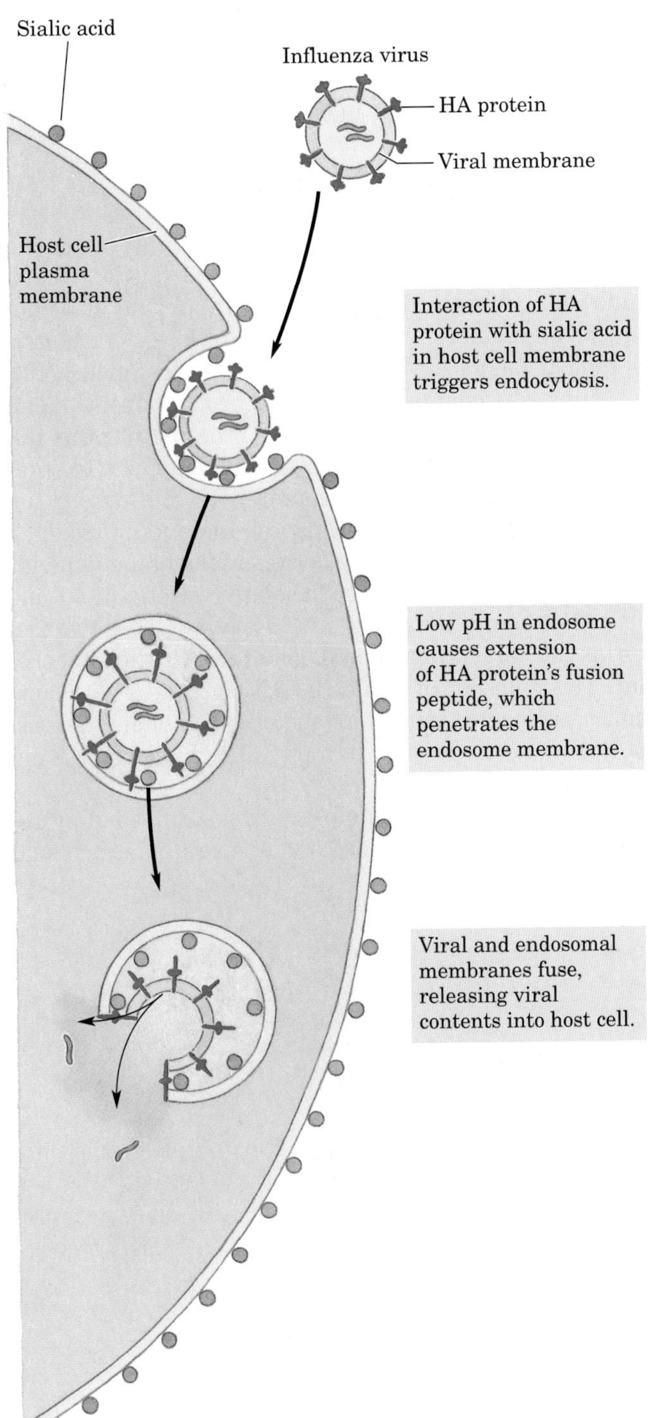

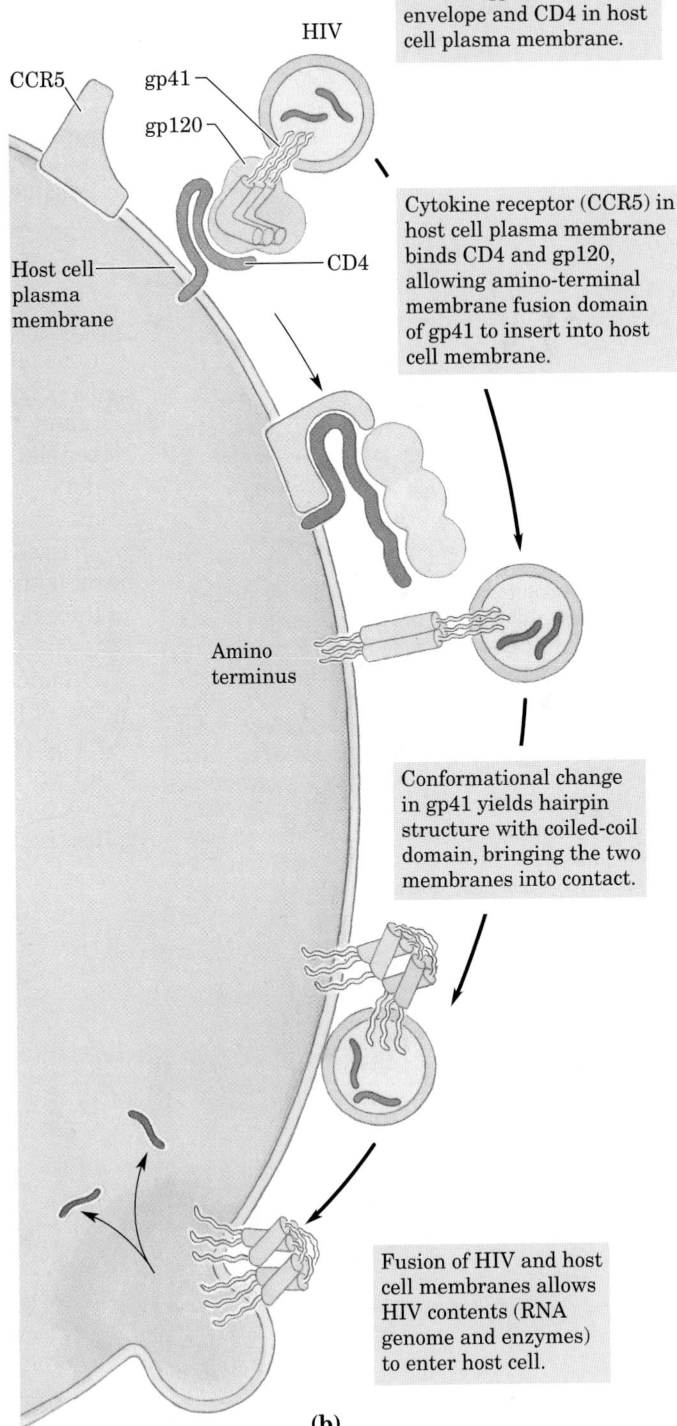

Solute Transport across Membranes

Every living cell must acquire from its surroundings the raw materials for biosynthesis and for energy production, and must release to its environment the byproducts of metabolism. The plasma membrane contains proteins that specifically recognize and carry into the cell such necessities as sugars, amino acids, and inorganic ions. In some cases, these components are imported against a gradient of concentration, electrical charge, or both—they are "pumped" in. Certain other materials are pumped out to keep their concentrations in the cytosol lower than in the surrounding medium. With few exceptions, the traffic of small molecules across the plasma membrane is mediated by proteins such as transmembrane channels, carriers, or pumps. Within the eukaryotic cell, different compartments have different concentrations of metabolic intermediates and products, and these, too, must move across intracellular membranes in tightly regulated, protein-mediated processes.

Passive Transport Is Facilitated by Membrane Proteins

When two aqueous compartments containing unequal concentrations of a soluble compound or ion are separated by a permeable divider (membrane), the solute moves by **simple diffusion** from the region of higher concentration, through the membrane, to the region of lower concentration, until the two compartments have equal solute concentrations (Fig. 12–22a). When ions of opposite charge are separated by a permeable membrane, there is a transmembrane electrical gradient, the **membrane potential, V_m** (expressed in volts or millivolts). This membrane potential produces a force opposing ion movements that increase V_m and driving ion movements that reduce V_m (Fig. 12–22b). Thus, the direction in which a charged solute tends to move spontaneously across a membrane depends on both the chemical gradient (the difference in solute concentration) and the electrical gradient (V_m) across the membrane. Together, these two factors are referred to as the **electrochemical gradient** or the **electrochemical potential.** This behavior of solutes is in accord with the second law of thermodynamics: molecules tend spontaneously to assume the distribution of greatest randomness—that is, entropy will increase and the energy of the system will be minimized.

In living organisms, simple diffusion of most solutes is impeded by **selectively permeable** barriers—membranes that separate intracellular

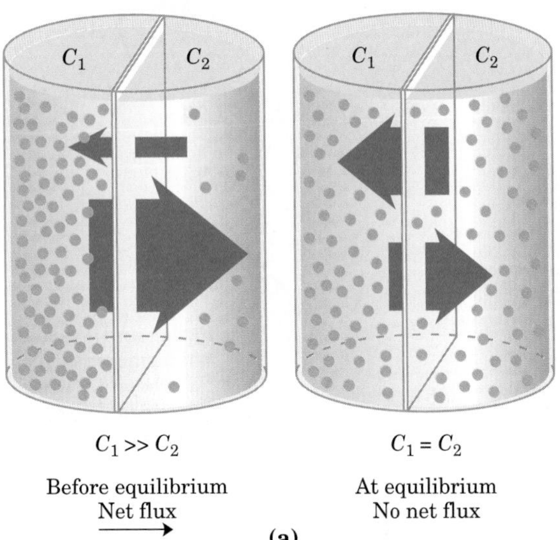

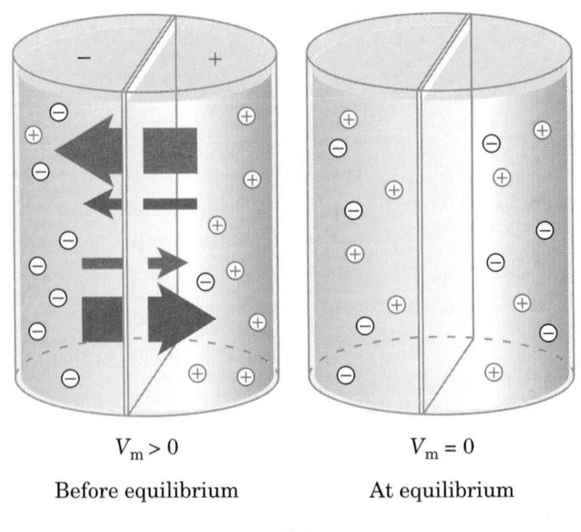

figure 12–22
Movement of solutes across a permeable membrane. (a) Net movement of electrically neutral solutes is toward the side of lower solute concentration until equilibrium is achieved. The solute concentrations on the left and right sides of the membrane are designated C_1 and C_2. The rate of transmembrane movement is proportional to the concentration gradient, C_1/C_2, as indicated by the large arrows. **(b)** Net movement of electrically charged solutes is dictated by a combination of the electrical potential (V_m) and the chemical concentration difference across the membrane; net ion movement continues until this electrochemical potential reaches zero.

compartments and surround cells. To pass through a lipid bilayer, a polar or charged solute must first give up its interactions with the water molecules in its hydration shell, then diffuse about 3 nm (30 Å) through a solvent (lipid) in which it is poorly soluble (Fig. 12–23). The energy used to strip away the hydration shell and to move the polar compound from water into and through lipid is regained as the compound leaves the membrane on the other side and is rehydrated. However, the intermediate stage of transmembrane passage is a high-energy state comparable to the transition state in an enzyme-catalyzed chemical reaction. In both cases, an activation barrier must be overcome to reach the intermediate stage (Fig. 12–23; compare with Fig. 8–3). The energy of activation ($\Delta G^{\ddagger}$) for translocation of a polar solute across the bilayer is so large that pure lipid bilayers are virtually impermeable to polar and charged species over periods of time relevant to cells.

A few biologically important gases can cross membranes by simple diffusion: molecular oxygen, nitrogen, and methane, all of which are relatively nonpolar. Despite its polarity, water crosses some biological membranes slowly by simple diffusion, presumably as a result of its very high concentration (about 55 M). However, for tissues in which rapid transmembrane water movement is essential (in the kidney, for example), water diffuses through channels formed by specific integral proteins, the aquaporins, described below.

Transmembrane passage of polar compounds and ions is made possible by membrane proteins that lower the activation energy for transport by providing an alternative path for specific solutes through the lipid bilayer. Proteins that bring about this **facilitated diffusion** or **passive transport** are not enzymes in the usual sense; their "substrates" are moved from one compartment to another, but are not chemically altered. Membrane proteins that speed the movement of a solute across a membrane by facilitating diffusion are called **transporters** or **permeases.**

The kind of detailed structural information obtained for many soluble proteins by x-ray crystallography is not yet available for most membrane transporters; as a group, these integral membrane proteins are both difficult to purify and difficult to crystallize. However, from studies of the specificity and kinetics of transporters we have learned that their action is closely analogous to that of enzymes. Like enzymes, transporters bind their substrates with stereochemical specificity through many weak, noncovalent interactions. The negative free-energy change associated with these weak interactions, $\Delta G_{\text{binding}}$, counterbalances the positive free-energy change that accompanies loss of the water of hydration from the substrate, $\Delta G_{\text{dehydration}}$, thereby lowering $\Delta G^{\ddagger}$ for transmembrane passage (Fig. 12–23). Transporters span the lipid bilayer at least once, and usually several times, forming a transmembrane channel lined with hydrophilic amino acid side chains. The channel provides an alternative path for a specific substrate to move across the lipid bilayer without its having to dissolve in the bilayer, further lowering $\Delta G^{\ddagger}$ for transmembrane diffusion. The result is an increase of orders of magnitude in the rate of transmembrane passage of the substrate.

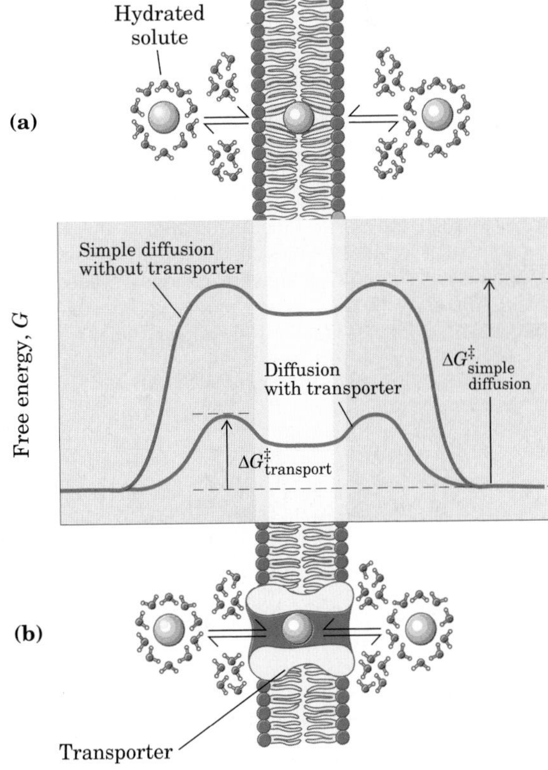

figure 12–23
Energy changes accompanying passage of a hydrophilic solute through the lipid bilayer of a biological membrane. **(a)** In simple diffusion, removal of the hydration shell is highly endergonic, and the energy of activation ($\Delta G^{\ddagger}$) for diffusion through the bilayer is very high. **(b)** A transporter protein reduces the $\Delta G^{\ddagger}$ for transmembrane diffusion of the solute. It does this by forming noncovalent interactions with the dehydrated solute to replace the hydrogen bonding with water, and by providing a hydrophilic transmembrane passageway.

table 12–3

Aquaporins

Aquaporin	Roles and location
AQP-1	Fluid reabsorption in proximal renal tubule; secretion of aqueous humor in eye and cerebrospinal fluid in central nervous system; water homeostasis in lung
AQP-2	Water permeability in renal collecting duct (mutations produce nephrogenic diabetes insipidus)
AQP-3	Water retention in renal collecting duct
AQP-4	Reabsorption of cerebrospinal fluid in central nervous system; regulation of brain edema
AQP-5	Fluid secretion in salivary glands, lachrymal glands, and alveolar epithelium of lung
γ-TIP	Water uptake by plant vacuole, regulating turgor pressure

Source: King, L.S. & Agre, P. (1996) Pathophysiology of the aquaporin water channels. *Annu. Rev. Physiol.* **58,** 619–648.

Aquaporins Form Hydrophilic Transmembrane Channels for the Passage of Water

A family of integral proteins, the **aquaporins** (AQPs), provide channels for rapid movement of water molecules across plasma membranes in a variety of specialized tissues. Erythrocytes, which swell or shrink rapidly in response to abrupt changes in extracellular osmolarity as blood travels through the renal medulla, have a high density of aquaporin in their plasma membranes (2×10^5 copies of AQP-1 per cell). Also rich in aquaporins are the plasma membranes of proximal renal tubule cells, which reabsorb water during urine formation, and the vacuolar membrane of plant cells, which require osmotic movement of water into the vacuole in order to maintain turgor pressure (see Fig. 2–10). Table 12–3 provides some examples of aquaporins.

All aquaporins are type III integral proteins (Fig. 12–14) with six transmembrane helical segments (Fig. 12–24a). In AQP-1, four monomers of M_r 28,000 form a tetrameric transmembrane channel lined with hydrophilic side chains and having a sufficient diameter (~3 Å) to allow passage of water molecules in single file (Fig. 12–24b). Water flows through an AQP-1 channel at the rate of about 5×10^8 molecules per second. For comparison, the highest known turnover number for an enzyme is that for catalase, 4×10^7 s^{-1}, and many enzymes have turnover numbers between 1 s^{-1} and 10^4 s^{-1} (see Table 8–7). The low activation energy for passage of water through aquaporin channels ($\Delta G^{\ddagger} < 15$ kJ/mol) suggests that water moves through the channels in a continuous stream, flowing in the direction dictated

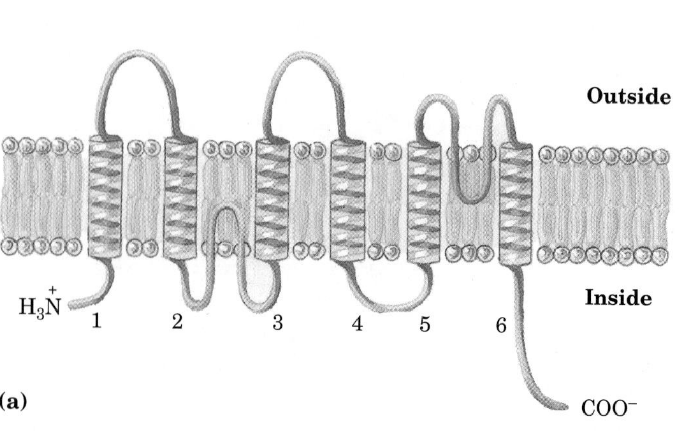

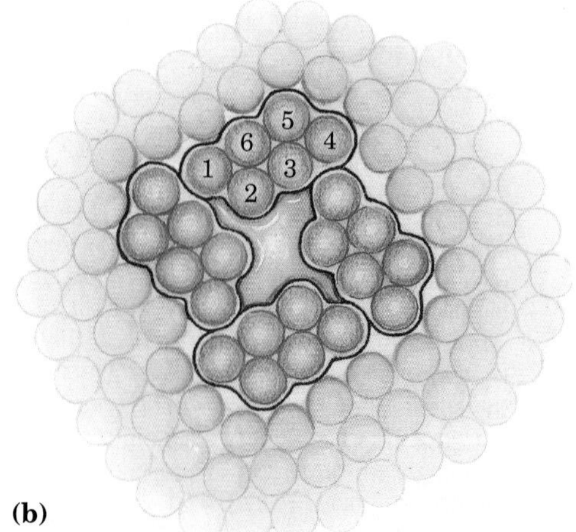

figure 12–24
Likely transmembrane topology of an aquaporin, AQP-1. This protein is also called CHIP-28. **(a)** Each monomer consists of six transmembrane helices. **(b)** In a proposed structure for the aquaporin channel (viewed perpendicular to, or through, the plane of the membrane), four AQP-1 monomers associate side by side, with their 24 transmembrane helices surrounding a central channel (shaded blue). The channel, through which water molecules diffuse one by one, is lined with hydrophilic side chains.

by the osmotic gradient. (For a discussion of osmosis, see p. 93.) Like most aquaporins, AQP-1 does not allow the passage of ions or other small solutes.

The Glucose Transporter of Erythrocytes Mediates Passive Transport

Energy-yielding metabolism in erythrocytes depends on a constant supply of glucose from the blood plasma, where the glucose concentration is maintained at about 5 mM. Glucose enters the erythrocyte by facilitated diffusion via a specific glucose transporter at a rate about 50,000 times greater than the uncatalyzed diffusion rate. This well-studied transporter provides an excellent example of how passive transport is managed in cells.

The glucose transporter of erythrocytes (called GluT1 to distinguish it from related glucose transporters in other tissues) is a type III integral protein (M_r 45,000) with 12 hydrophobic segments, each of which is believed to form a membrane-spanning helix. The detailed structure of GluT1 is not yet known, but one plausible model suggests that the side-by-side assembly of several helices produces a transmembrane channel lined with hydrophilic residues that can hydrogen bond with glucose as it moves through the channel (Fig. 12–25).

Hydrophobic
Polar
Charged
Outside
Inside
$^+NH_3$
COO^-

(a)

–Ser–Leu–Val–Thr–Asn–Phe–Ile–

1 Ser, 2 Leu, 3 Val, 4 Thr, 5 Asn, 6 Phe, 7 Ile

(b)

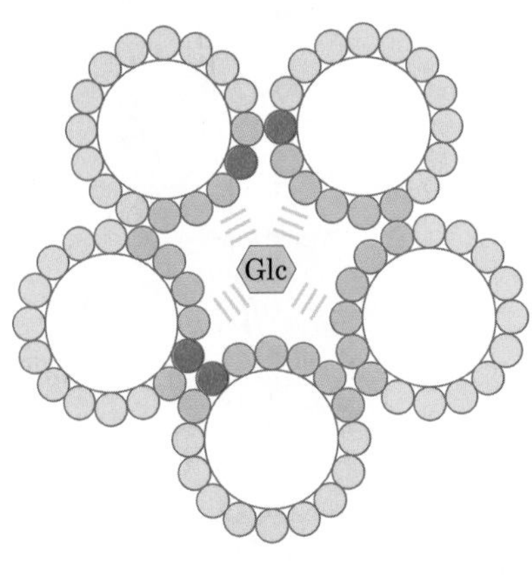

(c)

figure 12–25
Proposed structure of GluT1. (a) Transmembrane helices are represented as oblique (angled) rows of three or four amino acid residues, each row depicting one turn of the α helix. Nine of the 12 helices contain three or more polar or charged amino acid residues, often separated by several hydrophobic residues. **(b)** A helical wheel diagram shows the distribution of polar and nonpolar residues on the surface of a helical segment. The helix is diagrammed as though observed along its axis from the amino terminus. Adjacent residues in the linear sequence are connected with arrows, and each residue is placed around the wheel in the position it occupies in the helix; recall that 3.6 residues are required to make one complete turn of the α helix. In this example, the polar residues (blue) are on one side of the helix and the hydrophobic residues (yellow) on the other. This is, by definition, an amphipathic helix. **(c)** Side-by-side association of five or six amphipathic helices, each with its polar face oriented toward the central cavity, can produce a transmembrane channel lined with polar and charged residues. This channel provides many opportunities for hydrogen bonding with glucose moving through the transporter. The three-dimensional structure of GluT1 has not yet been determined by x-ray crystallography, but researchers expect that the hydrophilic transmembrane channels of this and many other transporters and ion channels will prove to resemble this model.

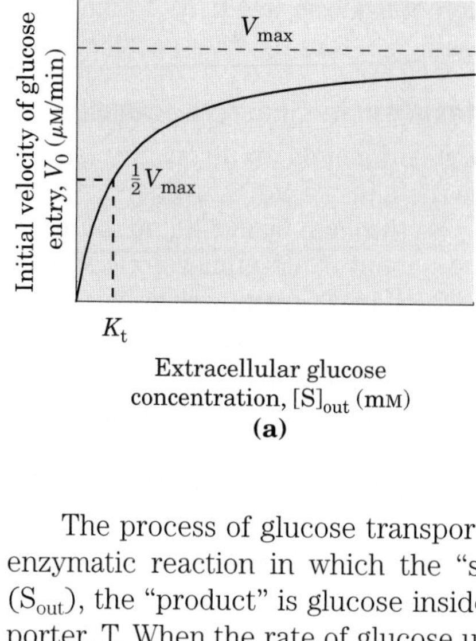

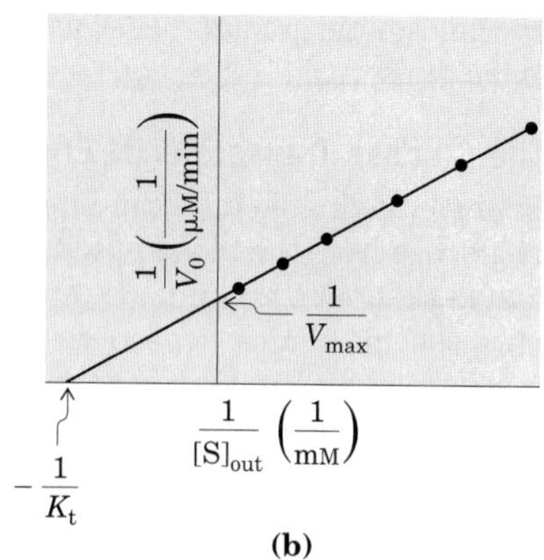

figure 12–26
Kinetics of glucose transport into erythrocytes. (a) The initial rate of glucose entry into an erythrocyte, V_0, depends upon the initial concentration of glucose on the outside, $[S]_{out}$. **(b)** Double-reciprocal plot of the data in **(a).** The kinetics of facilitated diffusion is analogous to the kinetics of an enzyme-catalyzed reaction. Compare these plots with Figure 8–11, and Figure 1 in Box 8–1. Note that K_t is analogous to K_m, the Michaelis constant.

The process of glucose transport can be described by analogy with an enzymatic reaction in which the "substrate" is glucose outside the cell (S_{out}), the "product" is glucose inside (S_{in}), and the "enzyme" is the transporter, T. When the rate of glucose uptake is measured as a function of external glucose concentration (Fig. 12–26), the resulting plot is hyperbolic; at high external glucose concentrations the rate of uptake approaches V_{max}. Formally, such a transport process can be described by the equations

$$S_{out} + T \underset{k_{-1}}{\overset{k_1}{\rightleftharpoons}} S_{out} \cdot T \underset{k_{-2}}{\overset{k_2}{\rightleftharpoons}} S_{in} \cdot T \underset{k_{-3}}{\overset{k_3}{\rightleftharpoons}} S_{in} + T$$

in which k_1, k_{-1}, and so forth, are the forward and reverse rate constants for each step. The steps are summarized in Figure 12–27.

The rate equations for this process can be derived exactly as for enzyme-catalyzed reactions (Chapter 8), yielding an expression analogous to the Michaelis-Menten equation (p. 260):

$$V_0 = \frac{V_{max}[S]_{out}}{K_t + [S]_{out}}$$

in which V_0 is the initial velocity of accumulation of glucose inside the cell when its concentration in the surrounding medium is $[S]_{out}$, and K_t ($K_{transport}$) is a constant analogous to the Michaelis constant, a combination of rate constants characteristic of each transport system. This equation describes the *initial* velocity—the rate observed when $[S]_{in} = 0$. As is the case for enzyme-catalyzed reactions, the slope-intercept form of the equation describes a linear plot of $1/V_0$ against $1/[S]_{out}$, from which we can obtain values of K_t and V_{max} (Fig. 12–26b). K_t, like K_m, is a measure of the affinity of transporter for substrate; the lower the K_t, the higher the affinity.

Because no chemical bonds are made or broken in the conversion of S_{out} to S_{in}, neither "substrate" nor "product" is intrinsically more stable, and the process of entry is therefore fully reversible. As $[S]_{in}$ approaches $[S]_{out}$,

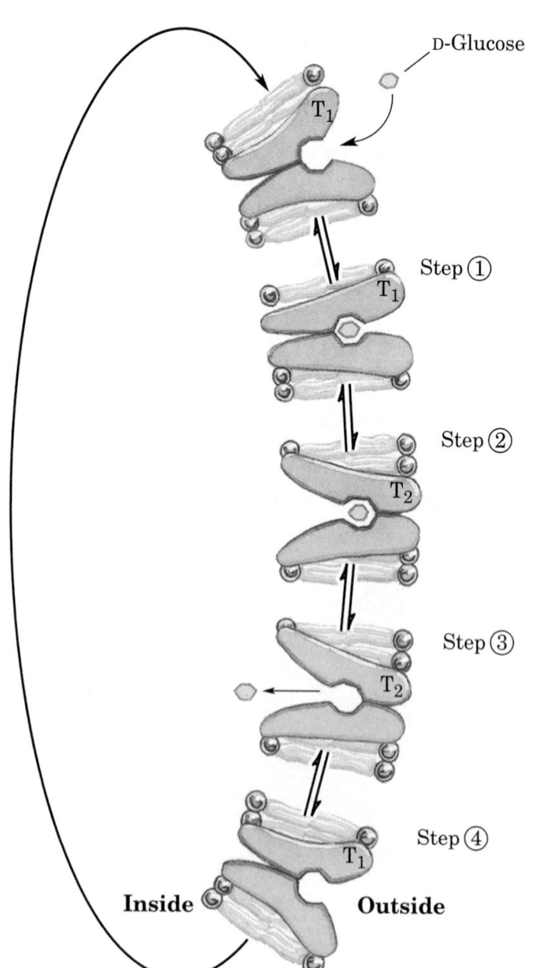

figure 12–27
Model of glucose transport into erythrocytes by GluT1. The transporter exists in two conformations: T_1, with the glucose-binding site exposed on the outer surface of the plasma membrane, and T_2, with the binding site on the inner surface. Glucose transport occurs in four steps. In step ①, glucose in blood plasma binds to a stereospecific site on T_1, thus lowering the activation energy for a conformational change (step ②) from $S_{out} \cdot T_1$ to $S_{in} \cdot T_2$, effecting the transmembrane passage of the glucose. In step ③, glucose is released from T_2 into the cytosol, and in step ④, the transporter returns to the T_1 conformation, ready to transport another glucose molecule.

the rates of entry and exit become equal. Such a system is therefore incapable of accumulating the substrate (glucose) within cells at concentrations above that in the surrounding medium; it simply achieves equilibration of glucose on the two sides of the membrane at a much higher rate than would occur in the absence of a specific transporter. GluT1 is specific for D-glucose, having a measured K_t of 1.5 mM. For the close analogs D-mannose and D-galactose, which differ only in the position of one hydroxyl group, the values of K_t are 20 and 30 mM, respectively; and for L-glucose, K_t exceeds 3,000 mM! Thus, GluT1 shows the three hallmarks of passive transport: high rates of diffusion down a concentration gradient, saturability, and specificity. Because the concentration of blood glucose, 4.5 to 5 mM, is maintained at about three times K_t, GluT1 is nearly saturated with substrate and operates near V_{max}.

In liver, a different glucose transporter, GluT2, transports glucose out of hepatocytes when liver glycogen is broken down to replenish the blood glucose. GluT2 has a K_t of about 66 mM and can therefore respond to increased levels of intracellular glucose by increasing outward transport. Muscle and adipose tissue have yet another glucose transporter, GluT4, which is distinguished by its stimulation by insulin (Box 12–2).

Chloride and Bicarbonate Are Cotransported across the Erythrocyte Membrane

The erythrocyte contains another facilitated diffusion system, an anion exchanger, which is essential in CO_2 transport to the lungs from tissues such as muscle and liver. Waste CO_2 released from respiring tissues into the blood plasma enters the erythrocyte, where it is converted into bicarbonate (HCO_3^-) by the enzyme carbonic anhydrase. The HCO_3^- reenters the blood plasma for transport to the lungs (Fig. 12–28). Because HCO_3^- is much more soluble in blood plasma than is CO_2, this roundabout route increases the blood's capacity to carry carbon dioxide from the tissues to the lungs. In the lungs, HCO_3^- reenters the erythrocyte and is converted to CO_2, which is eventually released into the lung space and exhaled. This shuttle, to be effective, requires very rapid movement of HCO_3^- across the erythrocyte membrane.

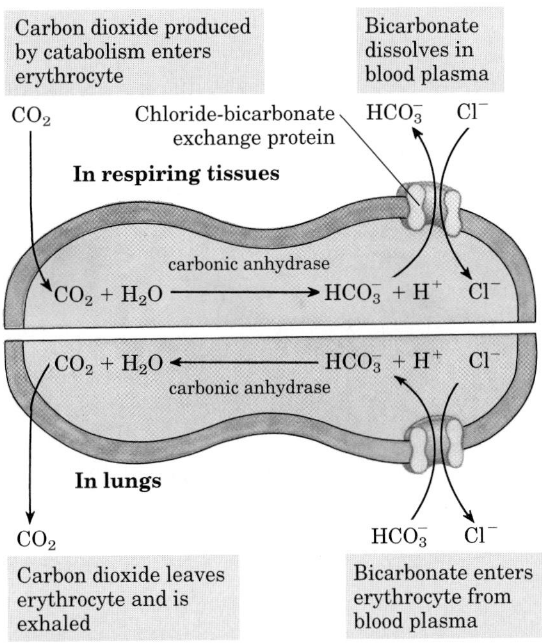

figure 12–28
Chloride-bicarbonate exchanger of the erythrocyte membrane. This cotransport system allows the entry and exit of HCO_3^- without changes in the transmembrane electrical potential. Its role is to increase the CO_2-carrying capacity of the blood.

The **chloride-bicarbonate exchanger,** also called the **anion exchange protein,** increases the permeability of the erythrocyte membrane to HCO_3^- by a factor of more than a million. Like the glucose transporter, it is an integral protein that probably spans the membrane 12 times. This protein mediates the simultaneous movement of two anions; for each HCO_3^- ion that moves in one direction, one Cl^- ion moves in the opposite direction (Fig. 12–28). The coupling of Cl^- and HCO_3^- movement is obligatory; in the absence of chloride, bicarbonate transport stops. In this respect, the anion exchanger is typical of all systems, called **cotransport systems,** that simultaneously carry two solutes across a membrane. When, as in this case, the two substrates move in opposite directions, the process is antiport. In **symport,** two substrates are moved simultaneously in the same direction. Transporters that carry only one substrate, such as the erythrocyte glucose transporter, are **uniport** systems (Fig. 12–29).

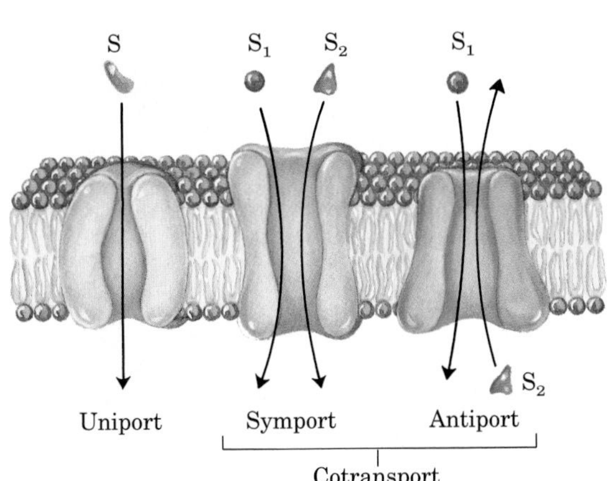

figure 12–29
Three general classes of transport systems. Transporters differ in the number of solutes (substrates) transported and the direction in which each is transported. Examples of all three types of transporters are discussed in the text. Note that this classification tells us nothing about whether these are energy-requiring (active transport) or energy-independent (passive transport) processes.

box 12–2 Defective Glucose and Water Transport in Two Forms of Diabetes

When ingestion of a carbohydrate-rich meal causes blood glucose to exceed the usual concentration between meals (5 mM), excess glucose is taken up by cardiac muscle and skeletal muscle (which store it as glycogen) and by adipocytes (which convert it to triacylglycerols). Glucose uptake into myocytes and adipocytes is mediated by the glucose transporter GluT4. Between meals, the plasma membrane of these cells contains some GluT4, but most is sequestered in the membranes of small intracellular vesicles (Fig. 1). Insulin released from the pancreas in response to high blood glucose triggers the movement of these intracellular vesicles to the plasma membrane, where they fuse, thus exposing GluT4 molecules on the outer surface of the cell. With more GluT4 molecules in action, the rate of glucose uptake increases 15-fold or more. When blood glucose levels return to normal, insulin release slows and most GluT4 molecules are removed from the plasma membranes and stored in vesicles.

In type I (juvenile onset) diabetes mellitus, the inability to release insulin (and thus to mobilize glucose transporters) results in low rates of glucose uptake into muscle and adipose tissue. One consequence is a prolonged period of high blood glucose after a carbohydrate-rich meal. This condition is the basis for the glucose tolerance test used to diagnose diabetes (Chapter 23).

The water permeability of epithelial cells lining the renal collecting duct in the kidney is due to the presence of an aquaporin (AQP-2) in their apical plasma membranes (facing the lumen of the duct). Antidiuretic hormone (ADH) regulates the retention of water by mobilizing AQP-2 molecules stored in vesicle membranes within the epithelial cells, much as insulin mobilizes GluT4 in muscle and adipose tissue. When the vesicles fuse with the epithelial cell plasma membrane, water permeability increases dramatically and more water is reabsorbed from the collecting duct and returned to the blood. When the ADH level drops, AQP-2 is resequestered within vesicles, reducing water retention. In the relatively rare human disease diabetes insipidus, a genetic defect in AQP-2 leads to impaired water reabsorption by the kidney. The result is excretion of copious volumes of very dilute urine.

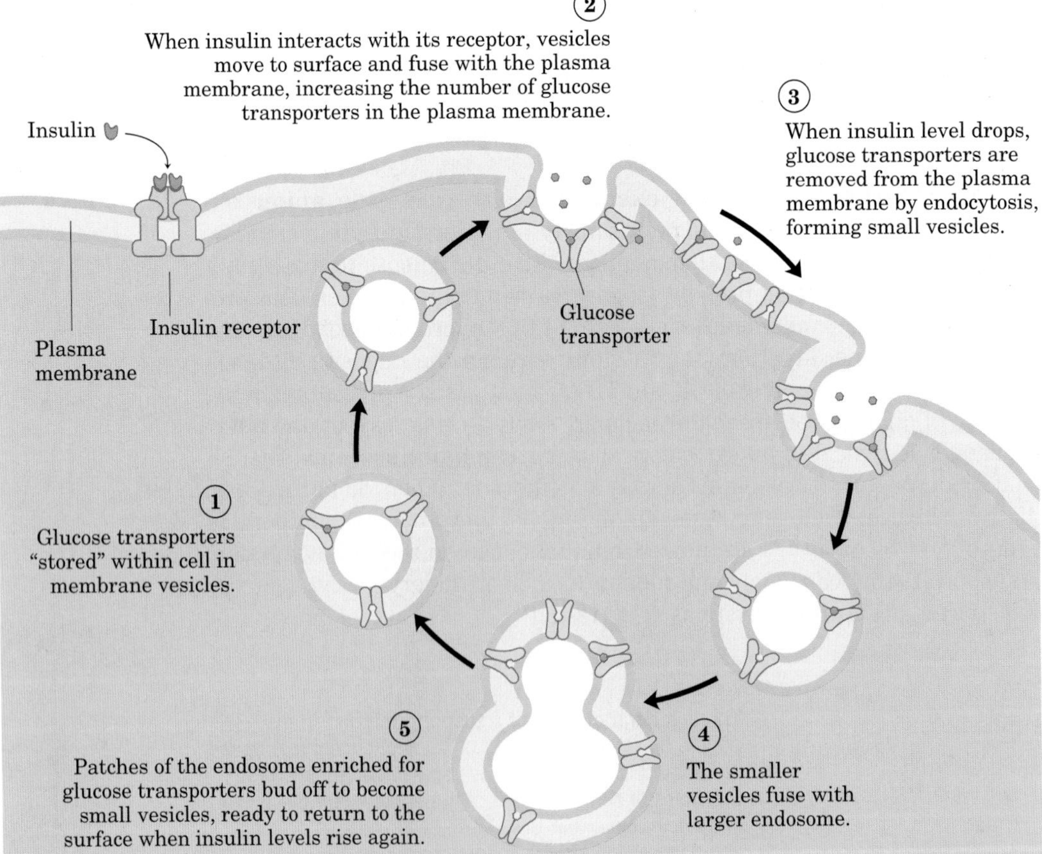

2
When insulin interacts with its receptor, vesicles move to surface and fuse with the plasma membrane, increasing the number of glucose transporters in the plasma membrane.
3
When insulin level drops, glucose transporters are removed from the plasma membrane by endocytosis, forming small vesicles.
Insulin
Insulin receptor
Plasma membrane
Glucose transporter
1
Glucose transporters "stored" within cell in membrane vesicles.
5
Patches of the endosome enriched for glucose transporters bud off to become small vesicles, ready to return to the surface when insulin levels rise again.
4
The smaller vesicles fuse with larger endosome.

figure 1
Regulation by insulin of glucose transport into a myocyte.

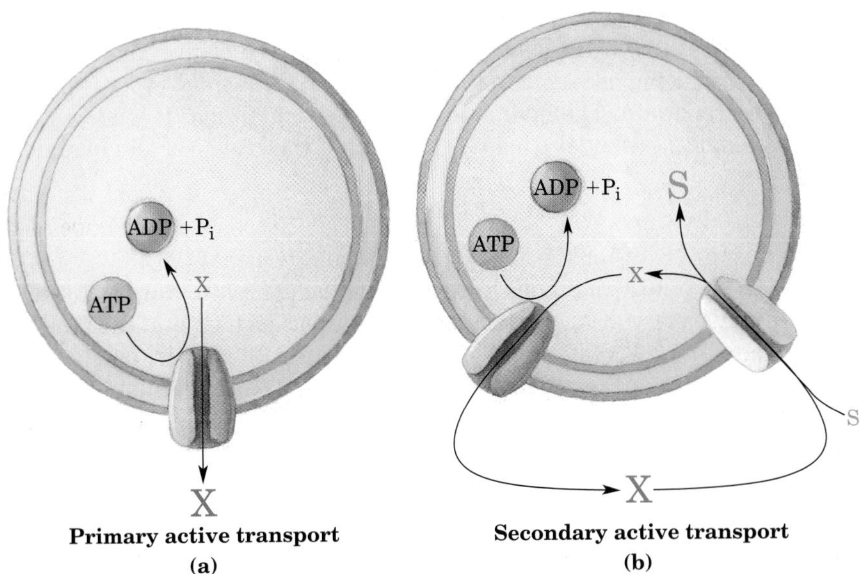

figure 12–30
Two types of active transport. (a) In primary active transport, the energy released by ATP hydrolysis drives solute movement against an electrochemical gradient. **(b)** In secondary active transport, a gradient of an ion X (often Na^+) has been established by primary active transport. Movement of X down its electrochemical gradient then provides the energy to drive cotransport of a second solute (S) against its electrochemical gradient.

Active Transport Results in Solute Movement against a Concentration or Electrochemical Gradient

In passive transport, the transported species always moves down its electrochemical gradient and is not accumulated above the equilibrium point. Active transport, by contrast, results in the accumulation of a solute above the equilibrium point. Active transport is thermodynamically unfavorable (endergonic) and occurs only when coupled (directly or indirectly) to an exergonic process such as the absorption of sunlight, an oxidation reaction, the breakdown of ATP, or the concomitant flow of some other chemical species down its electrochemical gradient. In **primary active transport**, solute accumulation is coupled directly to an exergonic chemical reaction, such as conversion of ATP to ADP + P_i (Fig. 12–30). **Secondary active transport** occurs when endergonic (uphill) transport of one solute is coupled to the exergonic (downhill) flow of a different solute that was originally pumped uphill by primary active transport.

The amount of energy needed for the transport of a solute against a gradient can be calculated from the initial concentration gradient. The general equation for the free-energy change in the chemical process that converts S into P is

$$\Delta G = \Delta G'^{\circ} + RT \ln [\mathrm{P}]/[\mathrm{S}] \qquad (12\text{–}1)$$

where R is the gas constant, 8.315 J/mol·K, and T is the absolute temperature. When the "reaction" is simply transport of a solute from a region where its concentration is C_1 to another region where its concentration is C_2, no bonds are made or broken and the standard free-energy change, $\Delta G'^{\circ}$, is zero. The free-energy change for transport, ΔG_t, is then

$$\Delta G_t = RT \ln (C_2/C_1) \qquad (12\text{–}2)$$

If there is a tenfold difference in concentration between two compartments, the cost of moving 1 mol of an uncharged solute at 25 °C across a membrane separating the compartments is therefore

$$\Delta G_t = (8.315 \text{ J/mol} \cdot \text{K})(298 \text{ K})(\ln 10/1) = 5{,}705 \text{ J/mol} = 5.7 \text{ kJ/mol}$$

Equation 12–2 holds for all uncharged solutes.

When the solute is an ion, its movement without an accompanying counterion results in the endergonic separation of positive and negative charges, producing an electrical potential; such a transport process is said to be **electrogenic.** The energetic cost of moving an ion depends on the electrochemical potential (p. 408), the sum of the chemical and electrical gradients:

$$\Delta G_t = RT \ln (C_2/C_1) + Z\mathcal{J}\Delta\psi \qquad (12\text{–}3)$$

where Z is the charge on the ion, $\mathcal{J}$ is the Faraday constant (96,480 J/V·mol), and $\Delta\psi$ is the transmembrane electrical potential (in volts). Eukaryotic cells typically have electrical potentials across their plasma membranes of the order of 0.05 to 0.2 V (with the interior negative relative to the outside), so the second term of Equation 12–3 can be a significant contribution to the total free-energy change for transporting an ion. Most cells maintain more than tenfold differences in ion concentrations across their plasma or intracellular membranes, and for many cells and tissues, active transport is therefore a major energy-consuming process.

The mechanism of active transport is of fundamental importance in biology. As we shall see in Chapter 19, the formation of ATP in mitochondria and chloroplasts occurs by a mechanism that is essentially ATP-driven ion transport operating in reverse. The energy made available by the spontaneous flow of protons across a membrane is calculable from Equation 12–3; remember that for flow *down* an electrochemical gradient, the sign of ΔG is opposite to that for transport *against* the gradient.

There Are at Least Four General Types of Transport ATPases

In the course of evolution, several distinct types of ATP-dependent active transporters have arisen, differing in structure, mechanism, and localization in specific tissues and intracellular compartments (Table 12–4).

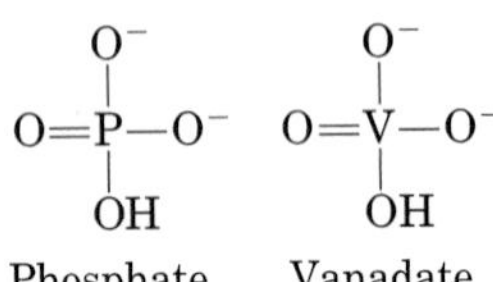

P-type ATPases are ATP-driven cation transporters that are reversibly phosphorylated by ATP as part of the transport cycle. All P-type transport ATPases have similarities in amino acid sequence, especially near the Asp residue that undergoes phosphorylation, and all are sensitive to inhibition by the phosphate analog **vanadate.** Each is an integral protein with multiple membrane-spanning regions in a single polypeptide; some also have a second subunit (Fig. 12–31a). P-type transporters are very widely distributed. In animal tissues, the Na^+K^+ ATPase, an antiporter for Na^+ and K^+, and the Ca^{2+} ATPase, a uniporter for Ca^{2+}, are ubiquitous, well-understood P-type ATPases that maintain disequilibrium in the ionic composition between the cytosol and the extracellular medium. Parietal cells in the lining of the mammalian stomach have a P-type ATPase that pumps H^+ and K^+ across the plasma membrane, thereby acidifying the stomach contents. In higher plants, a P-type ATPase pumps protons out of the cell, establishing a difference of as much as 2 pH units and 250 mV across the plasma membrane. A similar P-type ATPase in the bread mold *Neurospora* pumps protons out of cells to establish an inside-negative membrane potential, which is used to drive the uptake of substrates and ions from the surrounding medium by secondary active transport. Bacteria use P-type ATPases to pump out toxic heavy metal ions such as Cd^{2+} and Cu^{2+}.

A distinctly different class of proton-transporting ATPases is responsible for acidifying intracellular compartments in many organisms. The vacuoles of fungi and higher plants maintain a pH between 3 and 6, well below that of the surrounding cytosol (pH 7.5), by the action of **V-type ATPases**—proton pumps. V-type ATPases (V for vacuolar) are also responsible for the acidification of lysosomes, endosomes, the Golgi complex,

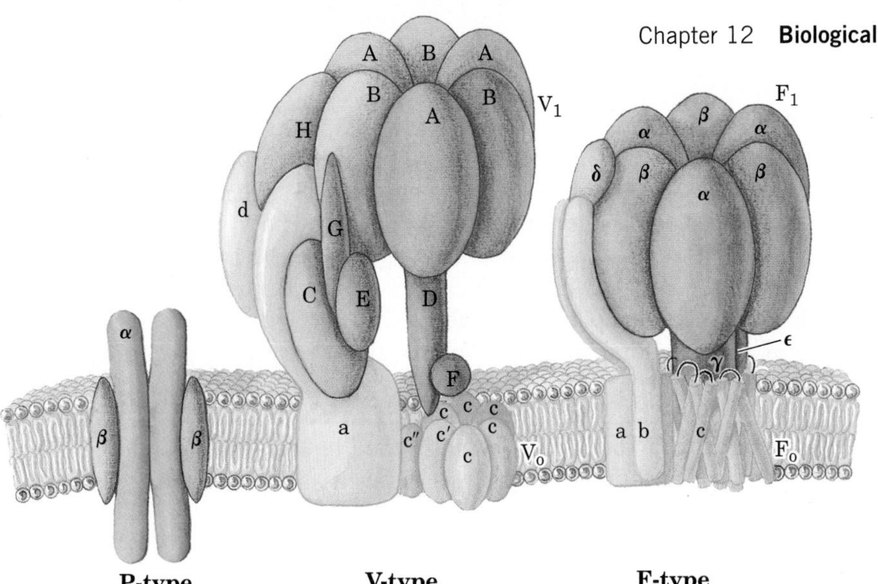

figure 12–31
Subunit structure of three types of ion-transporting ATPases. (a) The P-type ATPases generally have two types of integral protein subunits. The α subunit, which is essential, has the Asp residue that is phosphorylated during transport. **(b)** V-type ATPases have a peripheral domain, V_1 (blue), composed of seven different types of subunits including three A and three B subunits, and an integral domain, V_o (yellow), with three types of subunits including multiple copies of c. **(c)** F-type ATPases have a peripheral domain, F_1 (blue), homologous with V_1 of the V-type ATPases. The integral portion of F-type ATPases, F_o (yellow), also has three types of subunits, with multiple copies of c. P-type pumps such as the Na^+K^+ ATPase move two ions in opposite directions. V-type and F-type proton pumps move protons in one direction—from top to bottom in this diagram—*into* vacuoles for V-type, *out of* mitochondria for F-type.

table 12–4

Four Classes of Transport ATPases

	Organism or tissue	Type of membrane	Role of ATPase
P-type ATPases			
Na^+K^+	Animal tissues	Plasma	Maintains low $[Na^+]$, high $[K^+]$ inside cell; creates transmembrane electrical potential
H^+K^+	Acid-secreting (parietal) cells of mammals	Plasma	Acidifies contents of stomach
H^+	Fungi *(Neurospora)*	Plasma	Create H^+ gradient to drive secondary transport of extracellular solutes into cell
H^+	Higher plants	Plasma	
Ca^{2+}	Animal tissues	Plasma	Maintains low $[Ca^{2+}]$ in cytosol
Ca^{2+}	Myocytes of animals	Sarcoplasmic reticulum (endoplasmic reticulum)	Sequesters intracellular Ca^{2+}, keeping cytosolic $[Ca^{2+}]$ low
Cd^{2+}, Hg^{2+}, Cu^{2+}	Bacteria	Plasma	Pumps heavy metal ions out of cell
V-type ATPases			
H^+	Animals	Lysosomal, endosomal, secretory vesicles	Create low pH in compartment, activating proteases and other hydrolytic enzymes
H^+	Higher plants	Vacuolar	
H^+	Fungi	Vacuolar	
F-type ATPases			
H^+	Eukaryotes	Inner mitochondrial	Catalyze formation of ATP from ADP + P_i
H^+	Higher plants	Thylakoid	
H^+	Prokaryotes	Plasma	
Multidrug transporter			
	Animal tumor cells	Plasma	Removes a wide variety of hydrophobic natural products and synthetic drugs from cytosol, including vinblastine, doxorubicin, actinomycin D, mitomycin, taxol, colchicine, and puromycin

and secretory vesicles in animal cells. Structurally unrelated to P-type ATPases, the V-type ATPases do not undergo cyclic phosphorylation and dephosphorylation and are not inhibited by vanadate. All V-type ATPases have a similar complex structure, with an integral (transmembrane) domain

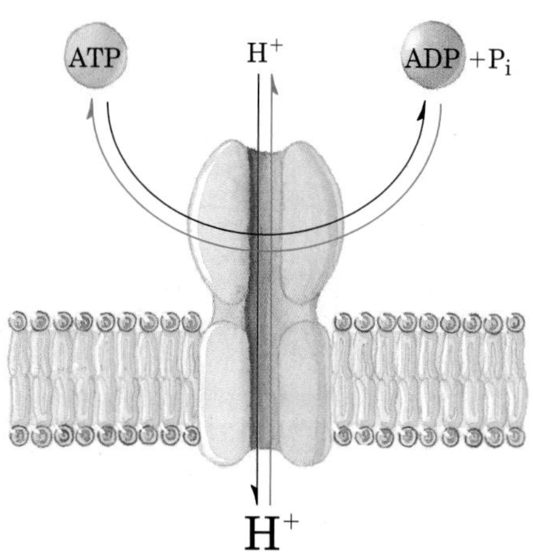

figure 12–32
Reversibility of F-type ATPases. An ATP-driven proton transporter also catalyzes ATP synthesis as protons flow *down* their electrochemical gradient.

(V_o) that serves as a proton channel and a peripheral domain (V_1) that contains the ATP-binding site and the ATPase activity (Fig. 12–31b). The mechanism by which V-type ATPases couple ATP hydrolysis to the uphill transport of protons is not understood in detail. The V-type proton pumps are related in structure, and probably in mechanism, to a third family of proton pumps, the F-type ATPases.

F-type ATPases play a central role in energy-conserving reactions in bacteria, mitochondria, and chloroplasts; that role will be discussed in detail when we describe oxidative phosphorylation and photophosphorylation in Chapter 19. (The *F* in their name originated in their identification as energy-coupling *f*actors.) They catalyze the uphill transmembrane passage of protons driven by ATP hydrolysis, as well as the reverse reaction, in which downhill proton flow drives ATP synthesis (Fig. 12–32). In the second case, the F-type ATPases are more appropriately named **ATP synthases.** The proton gradient in oxidative phosphorylation and photophosphorylation is established by other types of proton pumps powered by substrate oxidation or sunlight. The F-type ATPases/ATP synthases are multisubunit complexes that provide a transmembrane pore (the integral protein F_o) for protons and a molecular machine (the peripheral protein F_1)

box 12–3 A Defective Ion Channel Causes Cystic Fibrosis

Cystic fibrosis (CF) is a serious and relatively common hereditary disease of humans. About 5% of white Americans are carriers, having one defective and one normal copy of the gene. Only individuals with two defective copies show the severe symptoms of the disease: obstruction of the gastrointestinal and respiratory tracts, commonly leading to bacterial infection of the airways and death due to respiratory insufficiency before the age of 30. In CF, the thin layer of mucus that normally coats the internal surfaces of the lungs is abnormally thick, obstructing air flow and providing a haven for pathogenic bacteria, particularly *Staphylococcus aureus* and *Pseudomonas aeruginosa.*

The defective gene in CF patients was discovered in 1989. It encodes a membrane protein called *c*ystic *f*ibrosis *t*ransmembrane conductance *r*egulator, or CFTR. Hydropathy analysis predicted that CFTR has 12 transmembrane helices and is structurally related to the multidrug transporters of drug-resistant tumors (Fig. 1). The normal CFTR protein proved to be an ion

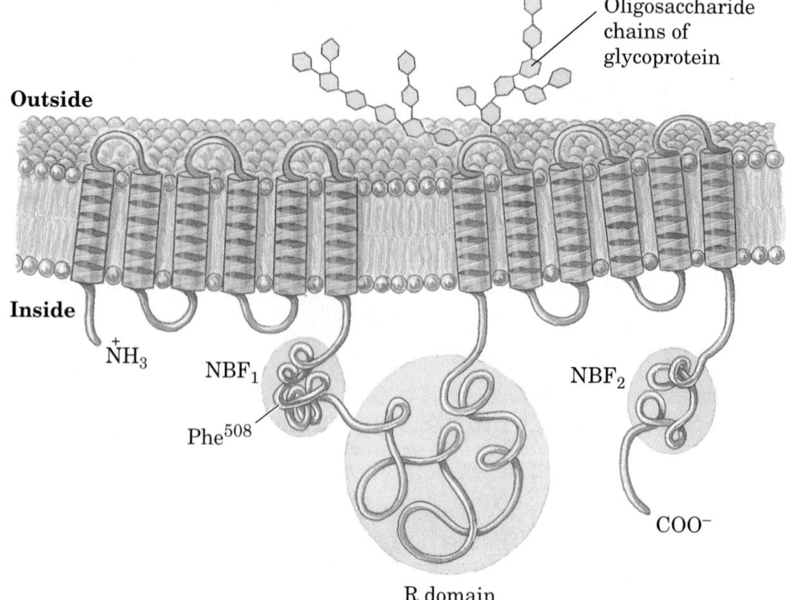

figure 1
Topology of the cystic fibrosis transmembrane conductance regulator, CFTR. There are 12 transmembrane helices, and three functionally significant domains extend from the cytosolic surface: NBF_1 and NBF_2 are nucleotide-binding folds to which ATP binds, and a regulatory domain (R-domain) is the site of phosphorylation by cAMP-dependent protein kinase. Oligosaccharide chains are attached to several residues on the outer surface on the segment between helices 7 and 8. The most commonly occurring mutation leading to CF is the deletion of Phe^{508}, in the NBF_1 domain. The structure of CFTR is very similar to that of the multidrug transporter of tumors, described in the text.

that uses the energy released by downhill proton flow through F_0 to form the phosphoanhydride bonds of ATP. The ATP-synthesizing and ATPase activities reside in the F_1 protein (Fig. 12–31c).

In the mid-1980s it became apparent that some tumors are strikingly resistant to a number of generally effective antitumor compounds. Investigation revealed that the plasma membranes of these tumors contain an ATP-dependent transporter that can export many different drugs, preventing their accumulation within the tumor cells and their growth-inhibitory effects. The transported drugs are chemically dissimilar but are generally hydrophobic. The **multidrug transporter** responsible for removing these drugs from the tumor cell cytosol is an integral protein (M_r 170,000) with 12 transmembrane segments and two ATP-binding sites. Export of the drugs is driven by ATP hydrolysis. The multidrug transporter is also an ion channel that specifically allows transmembrane diffusion of Cl^- down its concentration gradient; this activity is independent of ATP. In the genetic disease cystic fibrosis, the underlying defect is in the gene that encodes a Cl^--transporting protein, the cystic fibrosis conductance transmembrane regulator (CFTR), which may be related to the multidrug transporter of tumor cells (Box 12–3).

channel specific for Cl^-. The Cl^- channel activity increases greatly when phosphoryl groups are transferred from ATP to several side chains of the protein, catalyzed by cAMP-dependent protein kinase (Chapter 13). The mutation responsible for CF in 70% of cases results in deletion of a Phe residue at position 508, with the effect that the mutant protein is not correctly folded and inserted in the plasma membrane. Other mutations yield a protein that is inserted properly but cannot be activated by phosphorylation. In each case, the fundamental problem is a nonfunctional Cl^- channel in the epithelial cells that line the airways (Fig. 2), the digestive tract, and exocrine glands (pancreas, sweat glands, bile ducts, and vas deferens).

Normally, epithelial cells that line the inner surface of the lungs secrete a substance that traps and kills bacteria, and the cilia on the epithelial cells constantly sweep away the resulting debris. According to one hypothesis, the bactericidal activity requires a relatively low NaCl concentration. In CF patients, the defect in the Cl^- channel of CFTR results in surface fluid containing a high concentration of NaCl, and this fluid is much less effective in killing bacteria. Frequent infections by bacteria such as *S. aureus* and *P. aeruginosa* progressively damage the lungs and reduce respiratory efficiency.

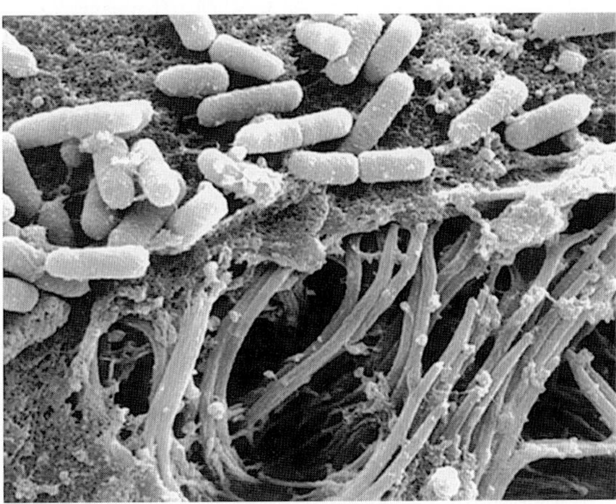

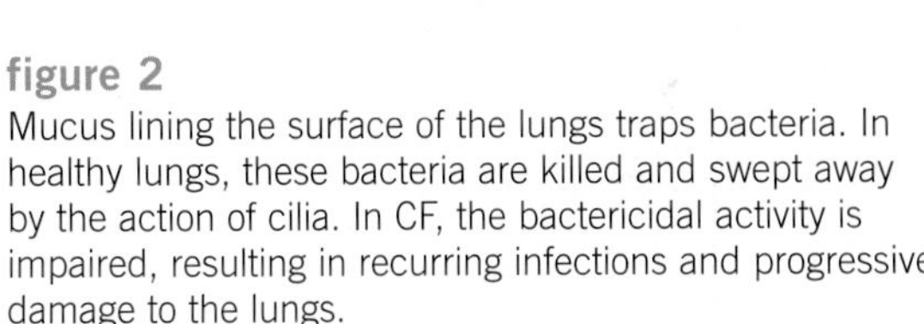

figure 2
Mucus lining the surface of the lungs traps bacteria. In healthy lungs, these bacteria are killed and swept away by the action of cilia. In CF, the bactericidal activity is impaired, resulting in recurring infections and progressive damage to the lungs.

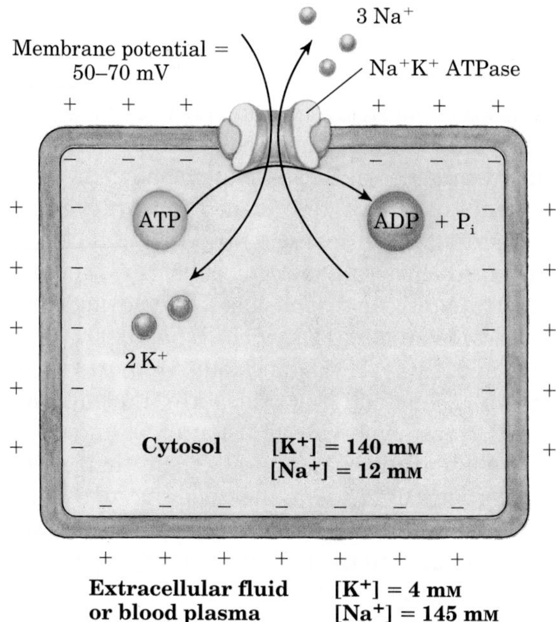

figure 12–33
Na^+K^+ ATPase. In animal cells, this active transport system is primarily responsible for setting and maintaining the intracellular concentrations of Na^+ and K^+ and for generating the transmembrane electrical potential. It does this by moving three Na^+ out of the cell for every two K^+ it moves in. The electrical potential is central to electrical signaling in neurons, and the gradient of Na^+ is used to drive the uphill cotransport of solutes in a variety of cell types.

A P-Type ATPase Catalyzes Active Cotransport of Na^+ and K^+

Jens Skou

In virtually every animal cell, the concentration of Na^+ is lower in the cell than in the surrounding medium, and the concentration of K^+ is higher (Fig. 12–33). This imbalance is established and maintained by a primary active transport system in the plasma membrane. The enzyme **Na^+K^+ ATPase,** discovered by Jens Skou in 1957, couples breakdown of ATP to the simultaneous movement of both Na^+ and K^+ against their electrochemical gradients. For each molecule of ATP converted to ADP and P_i, the transporter moves two K^+ ions inward and three Na^+ ions outward across the plasma membrane. The Na^+K^+ ATPase is an integral protein with two subunits (M_r ~50,000 and ~110,000), both of which span the membrane.

The detailed mechanism by which ATP hydrolysis is coupled to transport awaits determination of the protein's three-dimensional structure, but a current model (Fig. 12–34) supposes that the ATPase cycles between two forms, a phosphorylated form (designated P–Enz_{II}) with high affinity for K^+ and low affinity for Na^+, and a dephosphorylated form (Enz_I) with high affinity for Na^+ and low affinity for K^+. The conversion of ATP to ADP and P_i occurs in two steps catalyzed by the enzyme.

(1) Formation of phosphoenzyme:

$$ATP + Enz_I \longrightarrow ADP + P\text{–}Enz_{II}$$

(2) Hydrolysis of phosphoenzyme:

$$P\text{–}Enz_{II} + H_2O \longrightarrow Enz_I + P_i$$

The net reaction for these two steps is

$$ATP + H_2O \longrightarrow ADP + P_i$$

Because three Na^+ ions move outward for every two K^+ ions that move inward, the process is electrogenic—it creates a net separation of charge across the membrane. The result is a transmembrane potential of −50 to −70 mV (inside negative relative to outside), which is characteristic of most animal cells and essential to the conduction of action potentials in neurons. The central role of the Na^+K^+ ATPase is reflected in the energy invested in this single reaction: about 25% of the total energy consumption of a human at rest! The steroid derivative ouabain (pronounced wah′-bane; from *waa bayyo,* Somali for "arrow poison") is a potent and specific inhibitor of the Na^+K^+ ATPase. Ouabain and another steroid derivative, digitoxigenin, are the active ingredients of digitalis, an extract of the leaves of the foxglove plant. Digitalis has long been used in human medicine to treat congestive heart failure. Inhibition of Na^+ efflux by digitalis leads to higher Na^+ concentration in cells, activating a Na^+-Ca^{2+} antiporter in cardiac muscle. The increased influx of Ca^{2+} through this antiporter produces elevated cytosolic Ca^{2+}, which strengthens the contractions of heart muscle.

Ouabain

ATP-Driven Ca^{2+} Pumps Maintain a Low Concentration of Calcium in the Cytosol

The cytosolic concentration of free Ca^{2+} is generally about 100 nM, far below that in the surrounding medium, whether pond water or blood plasma. The ubiquitous occurrence of inorganic phosphates (P_i and PP_i) at millimolar concentrations in the cytosol necessitates a low cytosolic Ca^{2+} concentration, because inorganic phosphate combines with calcium to form relatively insoluble calcium phosphates. Calcium ions are pumped out of the cytosol by a P-type ATPase, the **plasma membrane Ca^{2+} pump.** Another P-type Ca^{2+} pump in the endoplasmic reticulum moves Ca^{2+} into the ER lumen, a compartment separate from the cytosol. In myocytes, Ca^{2+} is normally sequestered in a specialized form of endoplasmic reticulum called the sarcoplasmic reticulum. The *s*arcoplasmic and *e*ndoplasmic *r*eticulum *ca*lcium (SERCA) pumps are closely related in structure and mechanism, and both are inhibited by the tumor-promoting agent thapsigargin, which does not affect the plasma membrane Ca^{2+} pump.

The plasma membrane Ca^{2+} pump and SERCA pumps are integral proteins that cycle between two conformations in a mechanism similar to that for Na^+K^+ ATPase (Fig. 12–34). The Ca^{2+} pump of the sarcoplasmic reticulum, which comprises 80% of the protein in that membrane, has been thoroughly characterized and is the prototype for Ca^{2+} pumps of the P-type. It consists of a single polypeptide (M_r ~100,000) that spans the membrane ten times. A large cytosolic domain includes a site for ATP binding and an Asp residue that undergoes reversible phosphorylation by ATP. Phosphorylation favors a conformation with a high-affinity Ca^{2+}-binding site exposed on the cytosolic side, and dephosphorylation favors a conformation with a low-affinity Ca^{2+}-binding site on the lumenal side. As a result of the cyclic changes in conformation, the transporter binds Ca^{2+} on the side of the membrane where Ca^{2+} concentration is low and releases it on the side where the concentration is high. In this way, the energy released by hydrolysis of ATP to ADP and P_i during one phosphorylation-dephosphorylation cycle drives Ca^{2+} across the membrane against a large electrochemical gradient.

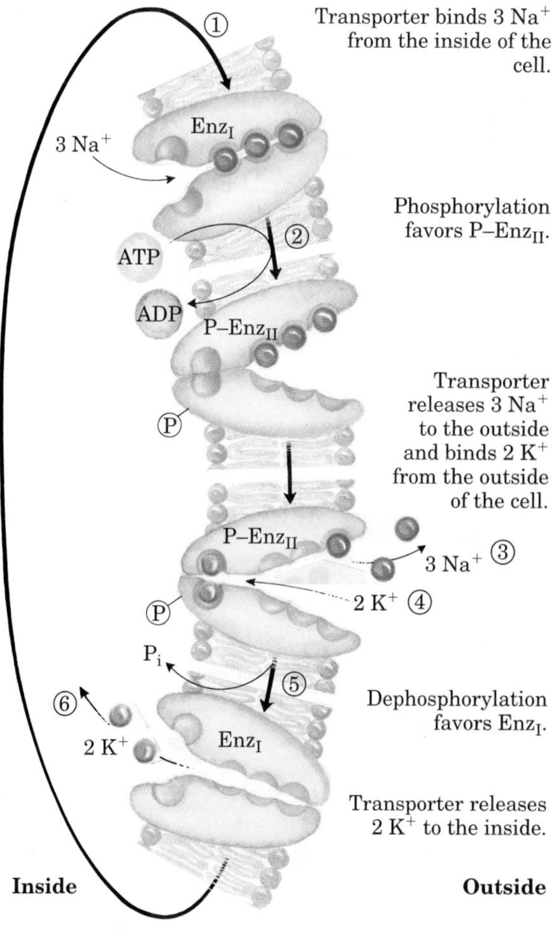

Transporter binds 3 Na^+ from the inside of the cell.
①
3 Na^+
Enz_I
Phosphorylation favors $P–Enz_{II}$.
②
ATP
ADP
$P–Enz_{II}$
Transporter releases 3 Na^+ to the outside and binds 2 K^+ from the outside of the cell.
Ⓟ
$P–Enz_{II}$
3 Na^+ ③
2 K^+ ④
Ⓟ
P_i
⑤
Dephosphorylation favors Enz_I.
⑥
2 K^+
Enz_I
Transporter releases 2 K^+ to the inside.
Inside
Outside

figure 12–34

Postulated mechanism of Na^+ and K^+ transport by the Na^+K^+ ATPase. ① The process begins with the binding of three Na^+ to high-affinity sites on the large subunit of the transport protein on the inner (cytosolic) surface of the membrane. This same part of the large subunit also has the ATP-binding site. Phosphorylation of the transporter changes its conformation ② and decreases its affinity for Na^+, leading to ③ Na^+ release on the outer (extracellular) surface. ④ K^+ on the outside now binds to high-affinity sites on the extracellular portion of the large subunit; ⑤ the enzyme is dephosphorylated, reducing its affinity for K^+; and ⑥ K^+ is discharged on the inside of the cell. The transport protein is now ready for another cycle of Na^+ and K^+ pumping.

table 12–5

Cotransport Systems Driven by Gradients of Na^+ or H^+

Organism or tissue	Transported solute (moving against its gradient)	Cotransported solute (moving down its gradient)	Type of transport
E. coli	Lactose	H^+	Symport
	Proline	H^+	Symport
	Dicarboxylic acids	H^+	Symport
Intestine, kidney of vertebrates	Glucose	Na^+	Symport
	Amino acids	Na^+	Symport
Vertebrate cells (many types)	Ca^{2+}	Na^+	Antiport
Higher plants	K^+	H^+	Antiport
Fungi (*Neurospora*)	K^+	H^+	Antiport

Ion Gradients Provide the Energy for Secondary Active Transport

The ion gradients formed by primary transport of Na^+ or H^+ can themselves provide the driving force for cotransport of other solutes. Many cells contain transport systems that couple the spontaneous, downhill flow of these ions to the simultaneous uphill pumping of another ion, sugar, or amino acid (Table 12–5). The galactoside transporter of *E. coli*, for example, allows the accumulation of lactose to levels 100 times higher than the concentration in the surrounding growth medium (Fig. 12–35). *E. coli* normally has a gradient of protons and charge across its plasma membrane produced by energy-yielding metabolism; protons tend to flow back spontaneously into the cell down this electrochemical gradient. The lipid bilayer is impermeable to protons, but the galactoside transporter provides a route for proton reentry, and lactose is simultaneously carried into the cell by symport. The endergonic accumulation of lactose is thereby coupled to the exergonic flow of protons; the overall free-energy change for the coupled process is negative.

In intestinal epithelial cells, glucose and certain amino acids are accumulated by symport with Na^+, using the Na^+ gradient established by the Na^+K^+ ATPase of the plasma membrane (Fig. 12–36). The apical surface of the intestinal epithelial cell is covered with microvilli, long, thin projections of the plasma membrane that greatly increase the surface area exposed to the intestinal contents. Na^+-glucose symporters in the apical plasma membrane take up glucose from the intestine in a process driven by the downhill flow of Na^+:

$$2Na^+_{out} + glucose_{out} \longrightarrow 2Na^+_{in} + glucose_{in}$$

The energy required for this process comes from two sources: the greater concentration of Na^+ outside than inside (the chemical potential) and the

figure 12–35

Lactose uptake in *E. coli*. (a) The primary transport of H^+ out of the cell, driven by the oxidation of a variety of fuels, establishes both a proton gradient and an electrical potential (inside negative) across the membrane. Secondary active transport of lactose into the cell involves symport of H^+ and lactose by the galactoside transporter. The uptake of lactose against its concentration gradient is entirely dependent on this inflow of H^+, driven by the electrochemical gradient. **(b)** When the energy-yielding oxidation reactions of metabolism are blocked by cyanide (CN^-), the galactoside transporter allows equilibration of lactose inside and outside the cell via passive transport. The dashed line represents the concentration of lactose in the surrounding medium.

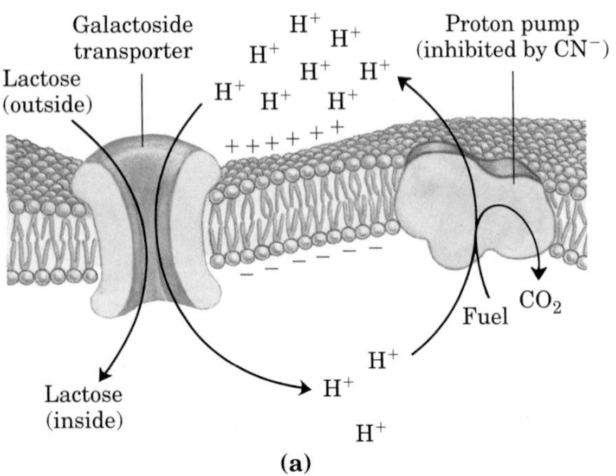

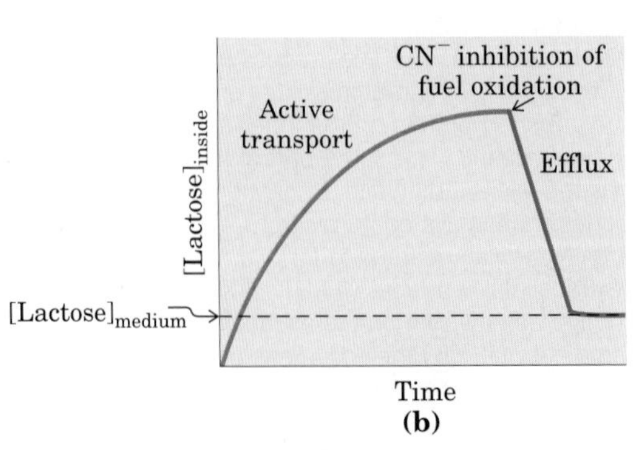

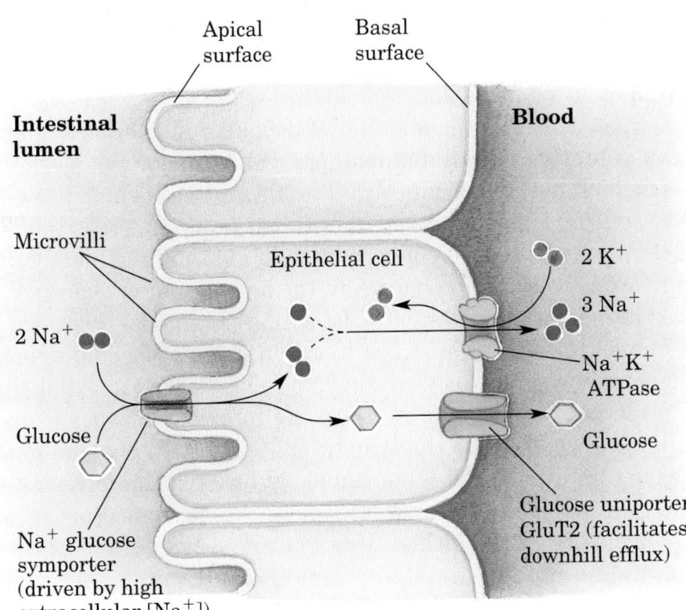

figure 12–36
Glucose transport in intestinal epithelial cells. Glucose is cotransported with Na^+ across the apical plasma membrane into the epithelial cell. It moves through the cell to the basal surface, where it passes into the blood via GluT2, a passive glucose transporter. The Na^+K^+ ATPase continues to pump Na^+ outward to maintain the gradient of Na^+ that drives glucose uptake.

transmembrane potential (the electrical potential), which is inside-negative and therefore draws Na^+ inward. The electrochemical potential of Na^+ is

$$\Delta G = RT \ln \frac{[Na^+]_{in}}{[Na^+]_{out}} + n\mathcal{J}\Delta E$$

where $n = 2$, the number of Na^+ ions cotransported with each glucose molecule. Given the typical membrane potential of −50 mV, an intracellular $[Na^+]$ of 12 mM, and an extracellular $[Na^+]$ of 145 mM, the energy, ΔG, made available as two Na^+ ions reenter the cell is −25 kJ, enough to pump glucose against a large concentration gradient.

$$\Delta G = -25 \text{ kJ} = RT \ln \frac{[\text{glucose}]_{in}}{[\text{glucose}]_{out}}$$

$$\frac{[\text{glucose}]_{in}}{[\text{glucose}]_{out}} = 30{,}000$$

That is, the cotransporter can pump glucose inward until its concentration within the epithelial cell is about 30,000 times that in the intestine. As glucose is pumped from the intestine into the epithelial cell at the apical surface, it is simultaneously moved from the cell into the blood by passive transport through a glucose transporter (GluT2) in the basal surface (Fig. 12–36). The crucial role of Na^+ in symport and antiport systems such as these requires the continued outward pumping of Na^+ to maintain the transmembrane Na^+ gradient.

Because of the essential role of ion gradients in active transport and energy conservation, compounds that collapse the ion gradients across cellular membranes are poisons, and those that are specific for infectious microorganisms can serve as antibiotics. Valinomycin is a small, cyclic peptide that surrounds K^+ and neutralizes its positive charge (Fig. 12–37). The peptide then acts as a shuttle, carrying K^+ across membranes down its concentration gradient and deflating that gradient. Compounds that shuttle ions across membranes in this way are called **ionophores,** literally "ion bearers." Both valinomycin and monensin (a Na^+-carrying ionophore) are antibiotics; they kill microbial cells by disrupting secondary transport processes and energy-conserving reactions.

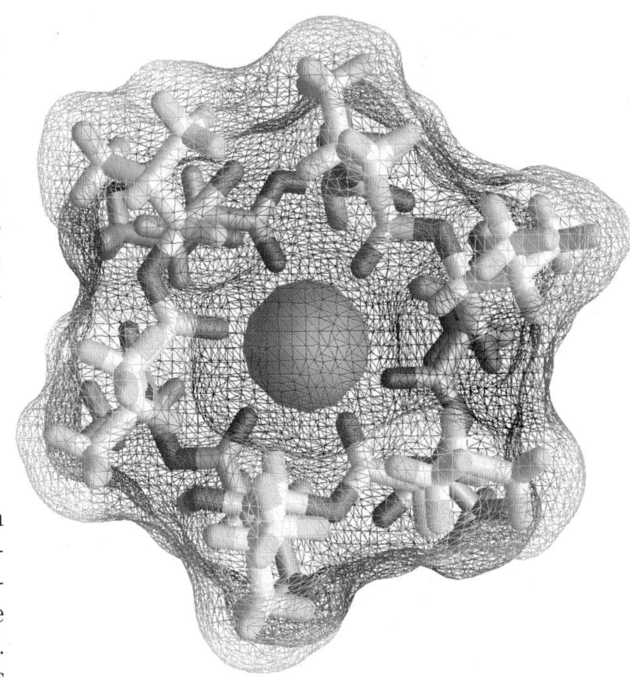

figure 12–37
Valinomycin, a peptide ionophore that binds K^+. In this image, the surface contours are shown as a transparent mesh, through which a stick structure of the peptide and a K^+ atom (green) are visible. The oxygen atoms (red) that bind K^+ are part of a central hydrophilic cavity. Hydrophobic amino acid side chains (yellow) coat the outside of the molecule. Because the exterior of the K^+-valinomycin complex is hydrophobic, the complex readily diffuses through membranes, carrying K^+ down its concentration gradient. The resulting dissipation of the transmembrane ion gradient kills microbial cells, making valinomycin a potent antibiotic.

Ion-Selective Channels Allow Rapid Movement of Ions across Membranes

First recognized in neurons and now known to be present in the plasma membranes of all cells, as well as in the intracellular membranes of eukaryotes, the **ion-selective channel** is another mechanism for moving inorganic ions across membranes. Ion channels determine the plasma membrane's permeability to specific ions and, together with ion pumps such as the Na^+K^+ ATPase, regulate the cytosolic concentration of ions and the membrane potential. In neurons, very rapid changes in the activity of ion channels cause the changes in membrane potential—the action potentials—that carry signals from one end of a neuron to the other. In myocytes, rapid opening of Ca^{2+} channels in the sarcoplasmic reticulum releases the Ca^{2+} that triggers muscle contraction. We discuss the signaling functions of ion channels in Chapter 13. Here we describe the structural basis for ion channel function, using as examples a bacterial K^+ channel, the acetylcholine receptor ion channel, and the voltage-dependent Na^+ channel of neurons.

Ion channels are distinguished from ion transporters in at least three ways. First, the rate of flux through channels can be orders of magnitude greater than the turnover number for a transporter—10^7 to 10^8 ions per channel per second, near the theoretical maximum rate for unrestricted diffusion. Second, ion channels are not saturable; their rates do not approach a maximum at high substrate concentration. Third, they are "gated"—opened or closed in response to some cellular event. In **ligand-gated channels** (which are generally oligomeric), binding of some extracellular or intracellular small molecule forces an allosteric transition in the protein, which opens or closes the channel. In **voltage-gated ion channels,** a charged protein domain moves relative to the membrane in response to a change in transmembrane electrical potential, causing the ion channel to open or close. Gating by either ligands or membrane potential can be very fast. A channel typically opens in a fraction of a millisecond, and may remain open for only milliseconds, making these molecular devices effective for very fast signal transmission in the nervous system.

The Structure of a K^+ Channel Shows the Basis for Its Ion Specificity

The structure of a potassium channel from the bacterium *Streptomyces lividans*, determined by x-ray crystallography in 1998, provides much insight into the way ion channels work. This bacterial ion channel is related in sequence to all other known K^+ channels and serves as the prototype for such channels, including the voltage-gated K^+ channel of neurons described in the next chapter. Among the members of this protein family, the similarities in sequence are greatest in the "pore region," which contains the ion selectivity filter that allows K^+ (radius 1.33 Å) to pass 10,000 times more readily than Na^+ (radius 0.95 Å), at a rate (about 10^8 ions per second) approaching the theoretical limit for unrestricted diffusion.

The K^+ channel consists of four identical subunits that span the membrane and form a cone within a cone surrounding the ion channel, with the cone's wide end facing the extracellular space (Fig. 12–38). Each subunit has two transmembrane α helices as well as a third, shorter helix that contributes to the pore region. The outer cone is formed by one of the transmembrane helices of each subunit. The inner cone, formed by the other four transmembrane helices, surrounds the ion channel and cradles the ion selectivity filter.

Both the ion specificity and the high flux through the channel are understandable from the channel's structure. At the inner and outer plasma membrane surfaces, the entryways to the channel have several negatively charged amino acid residues, which presumably increase the local concentration of cations such as K^+ and Na^+. The ion path through the membrane begins (on the inner surface) as a wide, water-filled channel in which the ion can retain its hydration sphere. Further stabilization is provided by the short α helices in the pore region of each subunit, with their carboxyl termini and the associated partial negative charges pointed at K^+ in the channel. About two-thirds of the way through the membrane, this channel narrows in the region of the selectivity filter, forcing the ion to give up its hydrating water molecules. Carbonyl oxygen atoms in the backbone of the selectivity filter replace the water molecules in the hydration sphere of K^+, forming a series of perfect coordination shells through which the K^+ moves. This favorable interaction with the filter is not possible with Na^+, which is too small to make contact with all the potential oxygen ligands. The preferential stabilization of K^+ is the basis for the ion selectivity of the filter, and mutations that change residues in this part of the protein eliminate the channel's ion selectivity.

K^+ ions pass through the filter in single file. In the crystallographic structure, two K^+ ions are visible, one at each end of the selectivity filter, about 7.5 Å apart. Their mutual electrostatic repulsion most likely just balances the interaction of each with the selectivity filter and keeps them moving through it rapidly.

Other K^+ channels are similar in sequence, and presumably in structure and mechanism, to the *S. lividans* K^+ channel. For example, the product of the Shaker gene, which is a K^+ channel in *Drosophila*, and K^+ channels from the nematode *Caenorhabditis elegans*, the ciliated protozoan *Paramecium*, and the plant *Arabidopsis thaliana* all have sequences very similar to those in the pore region of the *S. lividans* K^+ channel. Furthermore, the amino acid sequences of Na^+ and Ca^{2+} channels suggest that they, too, share some structural and functional similarities with the bacterial K^+ channel. Determination of this channel's structure was therefore a landmark in channel biochemistry.

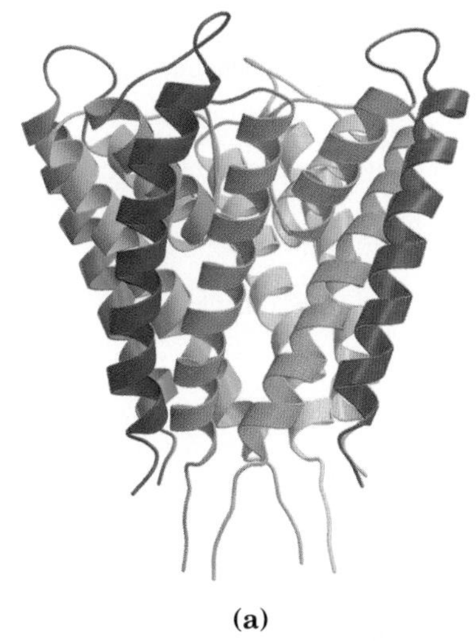

(a)

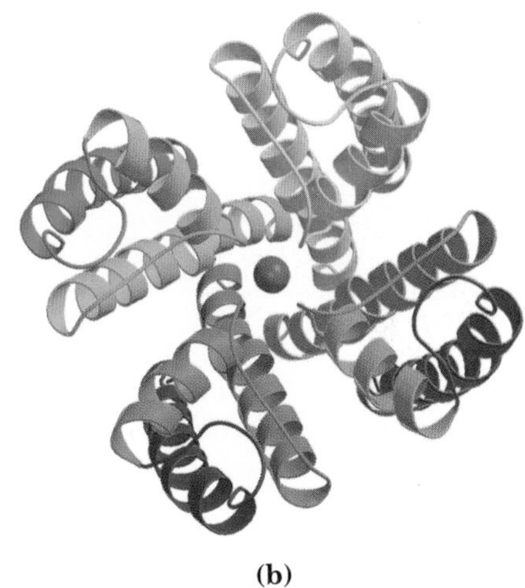

(b)

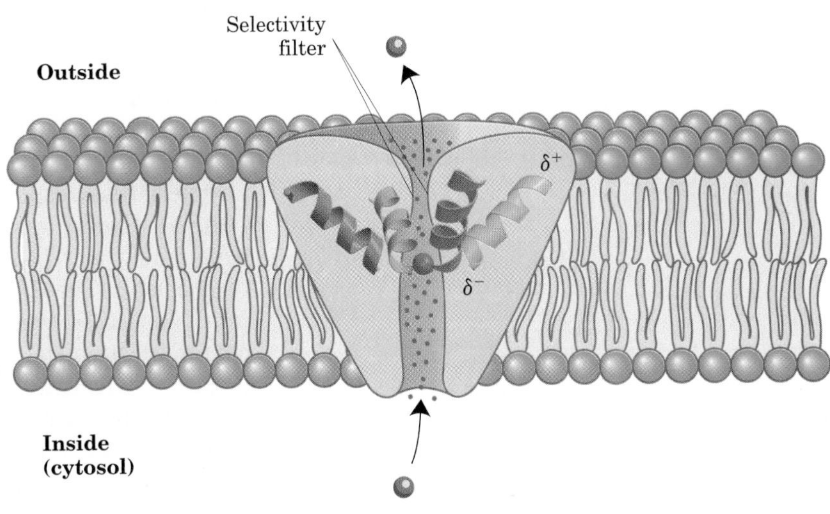

(c)

figure 12–38
Structure of the K^+ channel of *Streptomyces lividans*. **(a)** Viewed in the plane of the membrane, the channel consists of eight transmembrane helices (two from each of the four identical subunits), forming a cone with its wide end toward the extracellular space. The inner helices of the cone (lighter-colored) line the transmembrane channel, and the outer helices interact with the lipid bilayer. Short segments of each subunit converge in the open end of the cone to make a selectivity filter. **(b)** This view perpendicular to the plane of the membrane shows the four subunits arranged around a central channel just wide enough for a single K^+ ion to pass. Carbonyl oxygens of the peptide backbone protrude into the channel, interacting with and stabilizing a K^+ ion passing through. These ligands are perfectly positioned to interact with K^+, but not with the smaller Na^+ ion. This preferential interaction with K^+ is the basis for the channel's ion selectivity. **(c)** Cross section showing the water-filled and relatively broad channel at the cytosolic side (stippled blue). In this region, a K^+ ion (green) is stabilized by hydration and by the partial negative charge at the carboxyl-terminal end of the short helices projecting from each of the four subunits. The channel narrows at the selectivity filter, where the water of hydration is stripped from the K^+ and replaced with carbonyl oxygens of the selectivity filter.

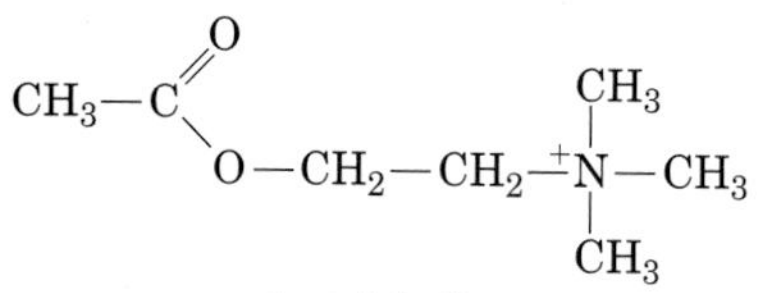

$$CH_3{-}C({=}O){-}O{-}CH_2{-}CH_2{-}\overset{+}{N}(CH_3)_3$$

Acetylcholine

The Acetylcholine Receptor Is a Ligand-Gated Ion Channel

Another very well-studied ion channel is the **nicotinic acetylcholine receptor,** which is essential in the passage of an electrical signal from a motor neuron to a muscle fiber at the neuromuscular junction (signaling the muscle to contract). (Nicotinic receptors were originally distinguished from muscarinic receptors by the sensitivity of the former to nicotine, the latter to the mushroom alkaloid muscarine.) Acetylcholine released by the motor neuron diffuses a few micrometers to the plasma membrane of a myocyte, where it binds to the acetylcholine receptor. This forces a conformational change in the receptor, causing the ion channel intrinsic to the receptor to open. The resulting inward movement of positive charges depolarizes the plasma membrane of the myocyte, triggering contraction. The acetylcholine receptor allows Na^+, Ca^{2+}, and K^+ to pass through with equal ease, but other cations and all anions are unable to pass.

Na^+ movement through an acetylcholine receptor ion channel is unsaturable (its rate is linear with respect to extracellular Na^+ concentration) and very fast—about 2×10^7 Na^+ ions per second under physiological conditions.

This receptor channel is typical of many other ion channels that produce or respond to electrical signals: it has a "gate" that opens in response to stimulation by acetylcholine, and an intrinsic timing mechanism that closes the gate after a split second. Thus the acetylcholine signal is transient—an essential feature of electrical signal conduction. The structural changes underlying gating in the acetylcholine receptor are known, but the mechanism of "desensitization"—of closing the gate even in the continued presence of acetylcholine—is not well understood.

The nicotinic acetylcholine receptor (which is structurally and functionally distinct from the muscarinic acetylcholine receptors) is composed of five subunits: single copies of subunits β, γ, and δ, and two identical α subunits each with an acetylcholine-binding site. All five subunits are related in sequence and tertiary structure, each having four transmembrane helical segments (M1 to M4) (Fig. 12–39a). The five subunits surround a central pore, which is lined with their M2 helices. The pore is about 20 Å wide in the parts of the channel that protrude on the cytosolic and extracellular surfaces, but narrows as it passes through the lipid bilayer. Near the center of the bilayer is a ring of bulky hydrophobic Leu side chains from the M2 helices, positioned so close together that they prevent ions from passing through the channel. Allosteric conformational changes induced by acetylcholine binding to the two α subunits include a slight twisting of the M2 helices (Fig. 12–39b), which draws these hydrophobic side chains away from the center of the channel and opens it to the passage of ions.

Based on similarities between the amino acid sequences of other ligand-gated ion channels and the acetylcholine receptor, the receptor channels that respond to the extracellular signals γ-aminobutyric acid (GABA), glycine, and serotonin are classified in an acetylcholine receptor superfamily and probably share three-dimensional structure and gating mechanisms. The $GABA_A$ and glycine receptors are anion channels specific for Cl^- or HCO_3^-, whereas the serotonin receptor, like the acetylcholine receptor, is cation-specific. The subunits of each of these channels, like those of the acetylcholine receptor, have four transmembrane helical segments and form oligomeric channels.

A second class of ligand-gated ion channels responds to *intracellular* ligands: 3′,5′-cyclic guanosine mononucleotide (cGMP) in the vertebrate eye, cGMP and cAMP in olfactory neurons, and ATP and inositol 1,4,5-

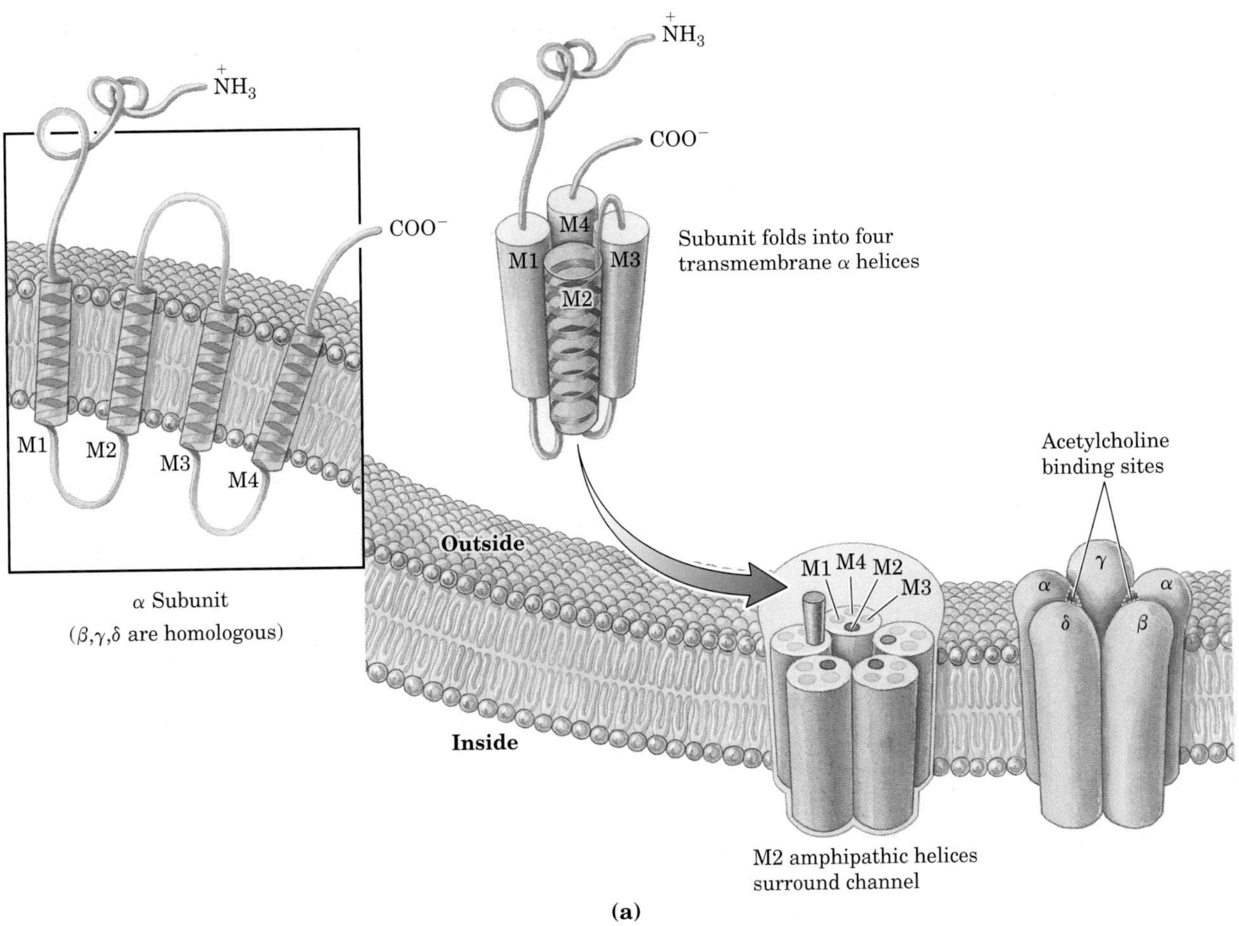

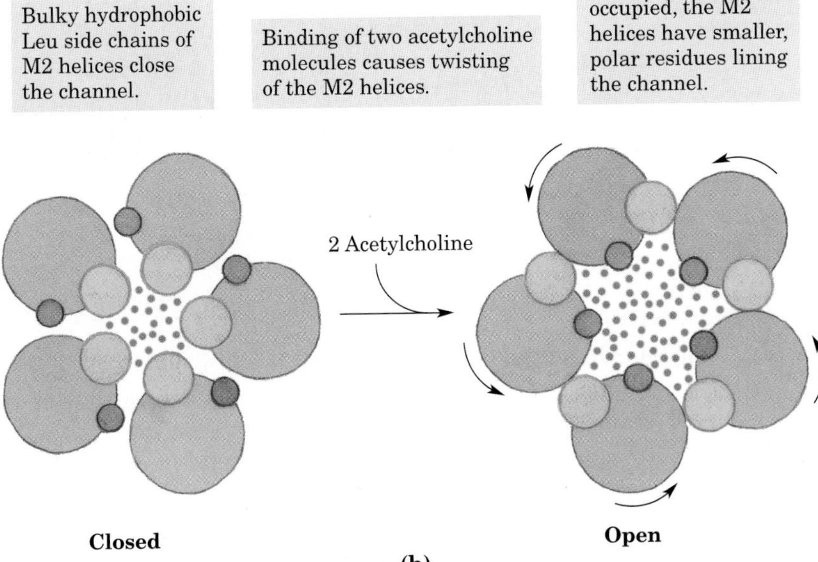
Bulky hydrophobic Leu side chains of M2 helices close the channel.
Binding of two acetylcholine molecules causes twisting of the M2 helices.
With receptor sites occupied, the M2 helices have smaller, polar residues lining the channel.
2 Acetylcholine
Closed
Open
(b)

figure 12–39

Structure of the acetylcholine receptor ion channel. **(a)** Each of the five subunits ($\alpha_2\beta\gamma\delta$) has four transmembrane helices, M1 to M4. The M2 helices are amphipathic; the others have mainly hydrophobic residues. The five subunits are arranged around a central transmembrane channel, which is lined with the polar sides of the M2 helices. At the top and bottom of the channel are rings of negatively charged amino acid residues. Near the middle of the bilayer, five Leu side chains (one from each M2 helix) protrude into the channel, constricting it to a diameter too small to allow passage of ions such as Ca^{2+}, Na^+, and K^+. **(b)** This top view of a cross section through the center of the M2 helices shows the blocking of the channel by the bulky Leu side chains. When both acetylcholine receptor sites (one on each α subunit) are occupied, a conformational change occurs. As the M2 helices twist slightly, the five Leu residues rotate away from the channel and are replaced by smaller, polar residues. This gating mechanism opens the channel, allowing the passage of Ca^{2+}, Na^+, or K^+.

trisphosphate (IP_3) in many cell types. These channels are composed of multiple subunits, each with six transmembrane helical domains. We discuss these channels in the context of their signaling functions in Chapter 13.

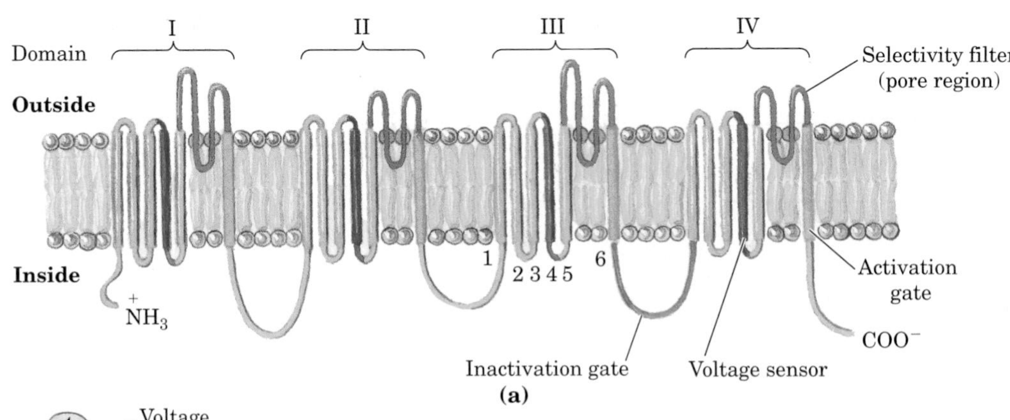

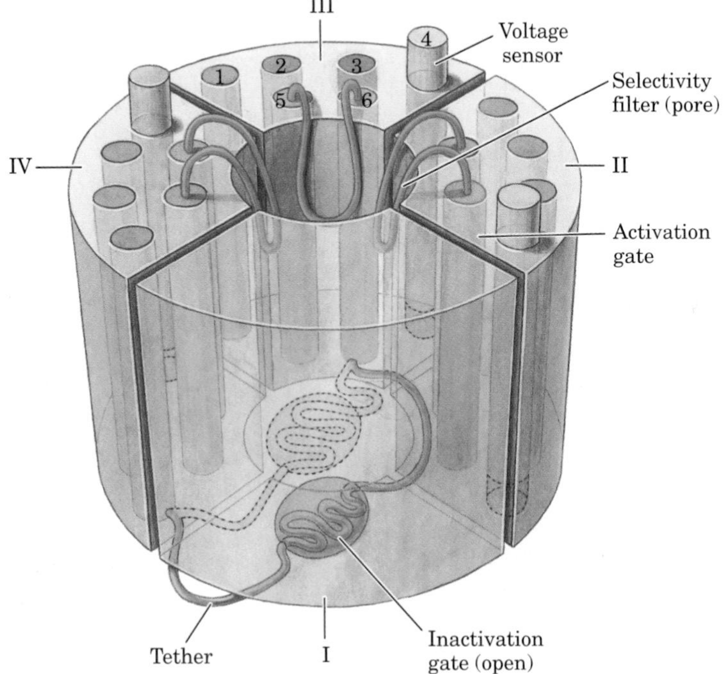

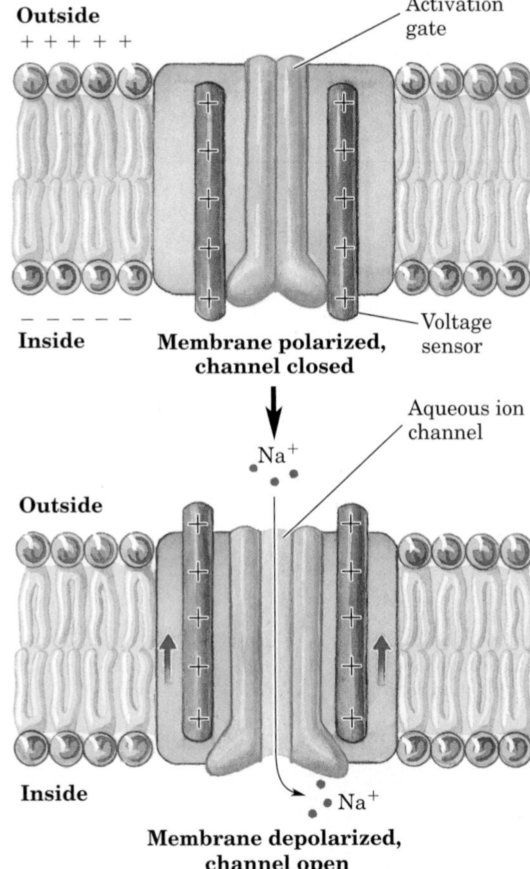

figure 12–40

Voltage-gated Na^+ channels of neurons. Sodium channels of various tissues and organisms have a variety of subunits, but only the principal subunit (α) is essential. **(a)** The α subunit is a large protein with four homologous domains (I to IV), each containing six transmembrane helices. Helix 4 in each domain (blue) is the voltage sensor; helix 6 (orange) is believed to be the activation gate. The segments between helices 5 and 6, the pore region (red), form the selectivity filter, and the segment connecting domains III and IV (green) is the inactivation gate. **(b)** The four domains are wrapped about a central transmembrane channel lined with polar amino acid residues. The segments linking helices 5 and 6 (red) in each domain come together near the extracellular surface to form the selectivity filter, which is conserved in all Na^+ channels. The filter gives the channel its ability to discriminate between Na^+ and other ions of similar size. The inactivation gate (green) closes (dotted lines) soon after the activation gate opens. **(c)** The voltage-sensing mechanism involves movement of helix 4 (blue) perpendicular to the plane of the membrane in response to a change in transmembrane potential. The strong positive charge on helix 4 allows it to be pulled inward in response to the inside-negative membrane potential (V_m). Depolarization lessens this pull, and helix 4 relaxes by moving outward. This movement is communicated to the activation gate (orange), inducing conformational changes that open the channel in response to depolarization.

The Neuronal Na^+ Channel Is a Voltage-Gated Ion Channel

Na^+ channels in the plasma membranes of neurons and of myocytes of heart and skeletal muscle sense electrical gradients across the membrane and respond by opening or closing. These **voltage-gated ion channels** are typically very selective for Na^+ over other monovalent or divalent cations (by factors of 100 or more) and have a very high flux rate ($>10^7$ ions per second). Normally in the closed conformation, Na^+ channels are activated (opened) by reduction in the transmembrane electrical potential. They then undergo very rapid inactivation. Within milliseconds of the opening, the channel closes and remains inactive for many milliseconds. Activation followed by inactivation of Na^+ channels is the basis for signaling by neurons (see Fig. 13–5).

The essential component of Na^+ channels is a single, large polypeptide (1,840 amino acid residues) organized into four domains (Fig. 12–40) clustered around a central channel, providing a path for Na^+ through the membrane. That path is made Na^+-specific by a "pore region" composed of the segments between transmembrane helices 5 and 6 of each domain, which fold into the channel. Helix 4 of each domain has a high density of positively charged residues. This segment is believed to move within the membrane in response to changes in the transmembrane voltage, from the "resting" potential of about −60 mV (inside negative) to about +30 mV. The movement of helix 4 triggers opening of the channel, and this is the basis for voltage gating.

Inactivation of the channel is believed to occur by a ball-and-chain mechanism. A protein domain on the cytosolic surface of the Na^+ channel, called the inactivation gate (the ball), is tethered to the channel by a short segment of the polypeptide (the chain) (Fig. 12–40b). This domain is free to move about when the channel is closed, but when it opens, a site on the inner face of the channel becomes available for the tethered ball to bind, blocking the channel. The length of the tether appears to determine how long an ion channel stays open; the longer the tether, the longer the open period. Inactivation of other ion channels may occur by a similar mechanism.

Ion-Channel Function Is Measured Electrically

A single ion channel typically remains open for only a few milliseconds, beyond the limit of most biochemical measurements. Ion fluxes must therefore be measured electrically, either as changes in V_m (in the millivolt range) or as electrical currents I (in the microampere or picoampere range), using microelectrodes and appropriate amplifiers (Fig. 12–41). Patch-clamping, a technique in which very small currents are measured through a tiny region of the membrane surface containing only one or a few ion channel molecules, reveals that as many as 10^4 ions can move through a single ion channel in one millisecond. This represents a huge amplification of the initial signal; for the acetylcholine receptor, for example, the signal may have been only a few molecules of acetylcholine.

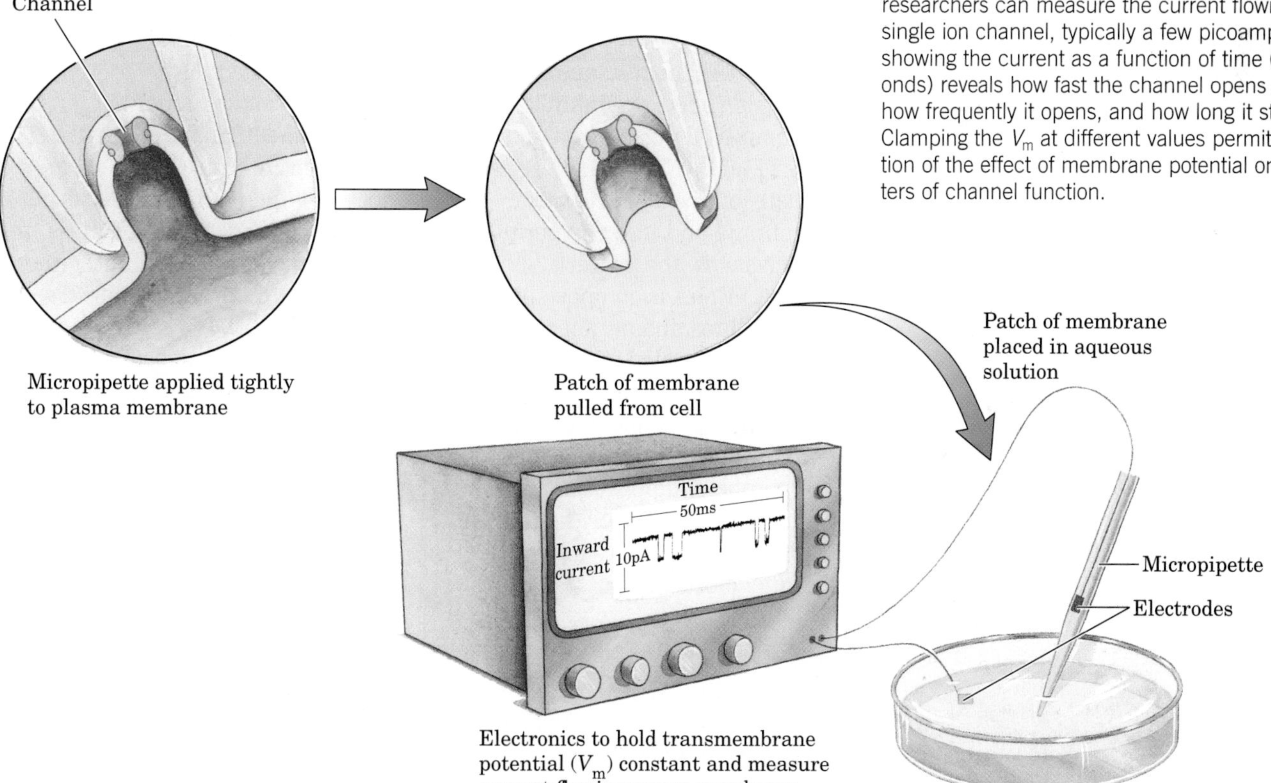

figure 12–41

Electrical measurements of ion channel function. The "activity" of an ion channel is estimated by measuring the flow of ions through it, using the patch-clamping technique. A finely drawn-out pipette (micropipette) is pressed against a cell's surface, and negative pressure in the pipette is used to form a pressure seal between pipette and membrane. When the pipette is pulled away from the cell, it pulls off a tiny patch of membrane (which may contain one or a few ion channels). When the pipette and attached patch are placed in an aqueous solution, channel activity can be measured as the electrical current that flows between the contents of the pipette and the aqueous solution. In practice, a circuit is set up that "clamps" the transmembrane potential at a given value and measures the current that must flow to maintain this voltage. With highly sensitive current detectors, researchers can measure the current flowing through a single ion channel, typically a few picoamperes. The trace showing the current as a function of time (in milliseconds) reveals how fast the channel opens and closes, how frequently it opens, and how long it stays open. Clamping the V_m at different values permits a determination of the effect of membrane potential on these parameters of channel function.

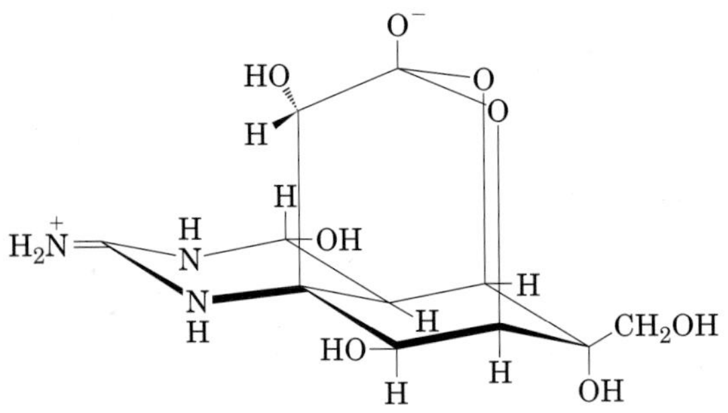

Tetrodotoxin

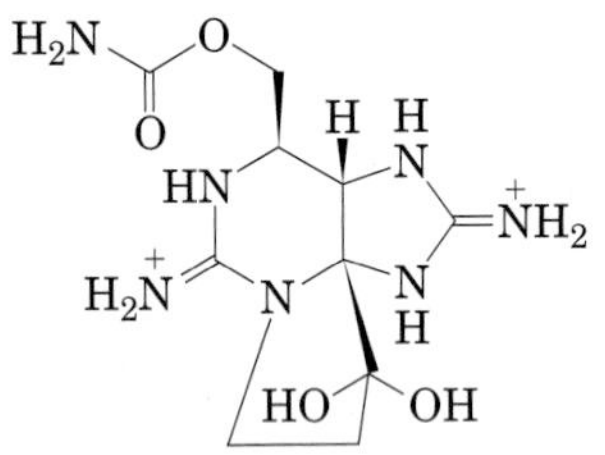

Saxitoxin

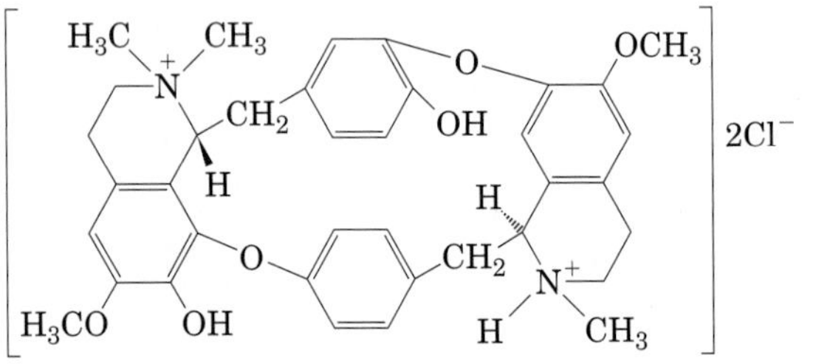

D-Tubocurarine chloride

Defective Ion Channels Can Have Striking Physiological Consequences

The importance of ion channels to physiological processes is clear from the effects of mutations in specific ion channel proteins. Genetic defects in the voltage-gated Na^+ channel of the myocyte plasma membrane result in diseases in which muscles are periodically either paralyzed (as in hyperkalemic periodic paralysis) or stiff (as in paramyotonia congenita). As noted earlier, cystic fibrosis is the result of a mutation that changes one amino acid in the protein CFTR, a chloride ion channel. The defective process here is not neurotransmission, but secretion by various exocrine gland cells whose activities are tied to Cl^- ion fluxes.

Naturally occurring toxins often act on ion channels, and the potency of these toxins further illustrates the importance of normal ion channel function. **Tetrodotoxin** (produced by the puffer fish, *Spheroides rubripes*) and **saxitoxin** (produced by the marine dinoflagellate *Gonyaulax,* which causes the occasional phenomenon of "red tides") are poisons that act by binding to the voltage-gated Na^+ channels of neurons and preventing normal action potentials. Puffer fish is an ingredient of the Japanese delicacy fugu, which may only be prepared by chefs specially trained to separate succulent morsel from deadly poison. Eating shellfish that have fed on *Gonyaulax* can also be fatal; shellfish are not sensitive to saxitoxin, but they concentrate it in their muscles, which become highly poisonous to organisms farther up the food chain. The venom of the black mamba snake contains **dendrotoxin,** which interferes with voltage-gated K^+ channels. **Tubocurarine,** the active component of curare (used as an arrow poison in the Amazon), and two other toxins from snake venoms, **cobrotoxin** and **bungarotoxin**, block the acetylcholine receptor or prevent the opening of its ion channel. By blocking signals from nerves to muscles, all of these toxins cause paralysis and possibly death. On the positive side, the extremely high affinity of bungarotoxin for the acetylcholine receptor ($K_d = 10^{-15}$ M) proved useful when radiolabeled toxin was used to quantify the receptor during its purification.

Porins Are Transmembrane Channels for Small Molecules

In the outer membrane of gram-negative bacteria such as *E. coli,* protein channels called **porins** allow the passage of molecules much larger than ions, but by a mechanism more like a gated channel than a transporter. The porin FhuA (Fig. 12–18) serves in *E. coli* to bring iron (in the form of a chelate of ferrichrome) from the extracellular medium across the outer membrane and into the periplasmic space. (In gram-negative bacteria, the periplasm is the material between the inner (plasma) and outer membranes.) FhuA is composed of a large 22-stranded β-barrel domain of about 560 residues and an amino-terminal "cork" domain of 160 residues, which normally obstructs the barrel and keeps the channel closed (Fig. 12–42). Binding of the ferrichrome complex to a specific site on the outer surface of the cell triggers allosteric changes that move the ferrichrome into the barrel and allow interaction of the FhuA protein with proteins of the inner membrane and periplasm. This interaction moves the cork out of the barrel and allows passage of the ferrichrome through the channel. With the emptying of the ferrichrome binding site, the changes are reversed and the channel closes.

The transport systems discussed in this chapter are summarized in Figure 12–43.

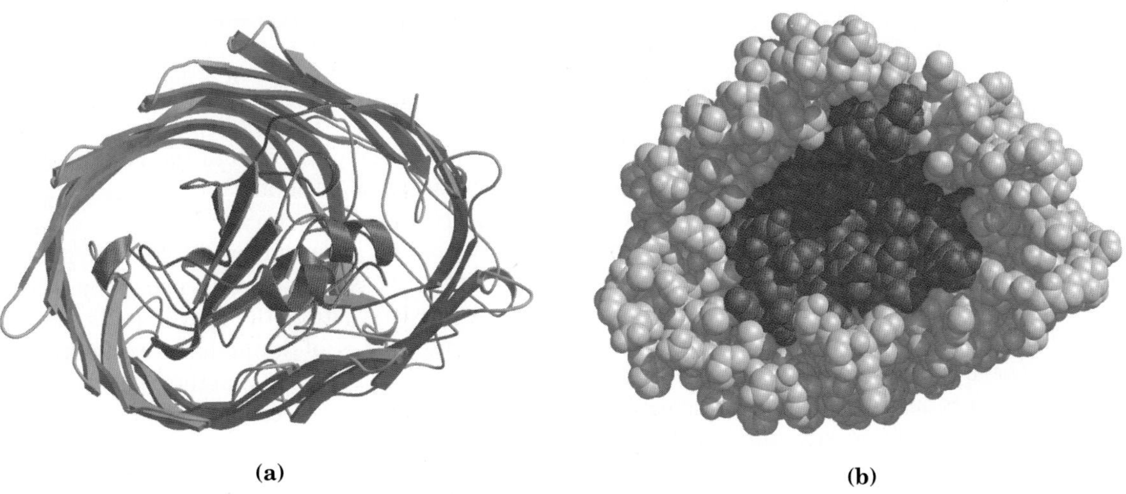

figure 12–42

Structure of FhuA, an iron transporter from *E. coli.* This protein of the *E. coli* outer membrane transports iron ion into the cell in the form of a ferrichrome-iron chelate. Figure 12–18 shows a ribbon representation of the protein viewed parallel to the membrane. Here we see **(a)** ribbon and **(b)** space-filling models, viewed perpendicular to the membrane. The 22-stranded β barrel (gray) is seen as a hollow pipe, plugged at one end by the cork domain (red). The binding of a ferrichrome-iron molecule at a specific site on the outer surface of FhuA drives a conformational change that helps to move the plug out of the channel; the ferrichrome-iron then moves through the channel, and the transporter resumes its closed conformation.

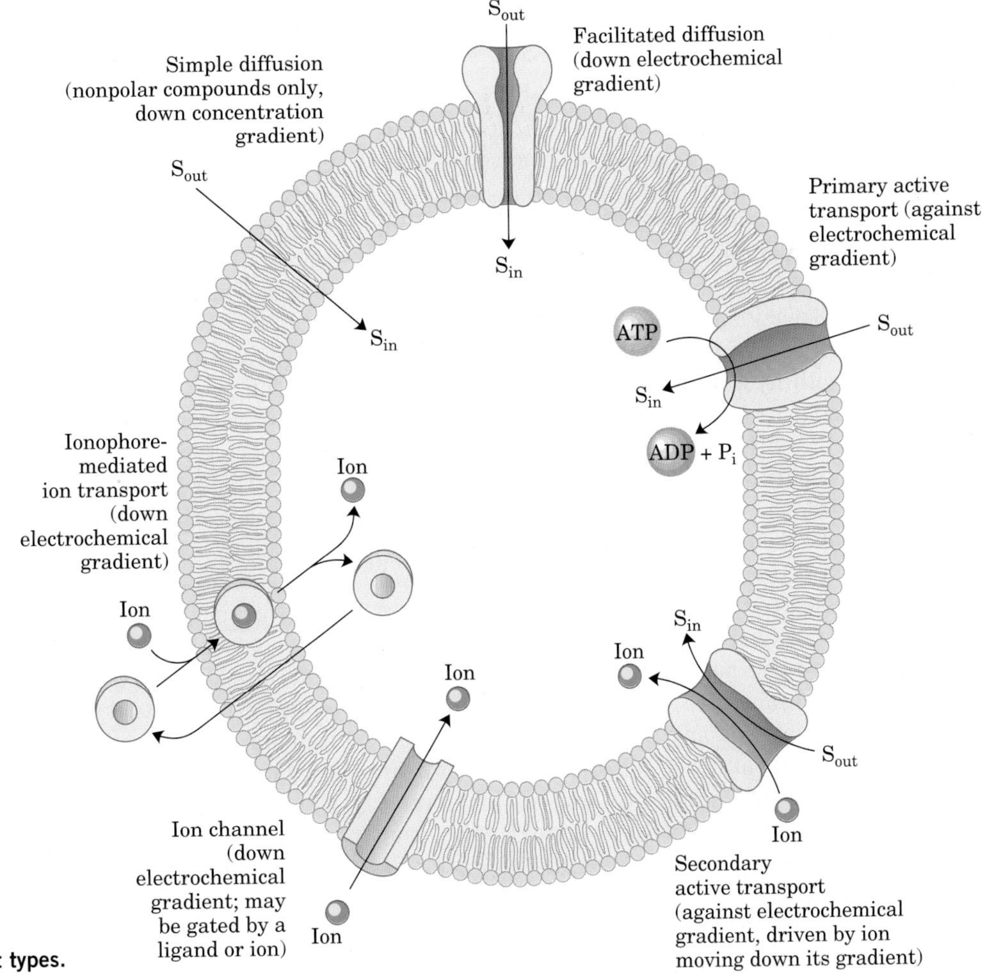

figure 12–43

Summary of transport types.

summary

Biological membranes are central to life. They define cellular boundaries, divide cells into discrete compartments, organize complex reaction sequences, and act in signal reception and energy transformations. Membranes are composed of lipids and proteins in varying combinations particular to each species, cell type, and organelle. The fluid mosaic model describes features common to all biological membranes. The lipid bilayer is the basic structural unit. Fatty acyl chains of phospholipids and the steroid nucleus of sterols are oriented toward the interior of the bilayer; their hydrophobic interactions stabilize the bilayer but allow the structure to be flexible. Lipids and most proteins are free to diffuse laterally within the membrane, and the hydrophobic moieties of the lipids undergo rapid thermal motion, making the interior of the bilayer fluid. Fluidity is affected by temperature, fatty acid composition, and sterol content. Cells regulate their lipid compositions to maintain a constant fluidity when external circumstances change.

Peripheral proteins are loosely associated with the membrane through electrostatic interactions and hydrogen bonds or by covalently attached lipid anchors. Integral proteins associate firmly with membranes by hydrophobic interactions between the lipid bilayer and their nonpolar amino acid side chains, which are oriented toward the outside of the protein molecule. Some membrane proteins span the lipid bilayer several times, with hydrophobic sequences of about 20 amino acid residues forming transmembrane α helices. Such hydrophobic sequences detected in the primary structures of proteins can be used to predict the secondary structure and transmembrane disposition of these molecules. Multistranded β barrels are also common in integral membrane proteins.

The lipids and proteins of membranes are inserted into the bilayer with specific sidedness; thus, membranes are structurally and functionally asymmetric. Many membrane proteins contain covalently attached oligosaccharides of various degrees of complexity. Plasma membrane glycoproteins are always oriented with the carbohydrate-bearing domain on the extracellular surface. A variety of proteins mediate the fusion of two membranes, which accompanies processes such as endocytosis and exocytosis. Fusion requires close apposition of two membranes and local disruption of the lipid bilayers, both processes attributed to fusion proteins called SNAREs.

The lipid bilayer is impermeable to polar substances. Water is an important exception; it can diffuse passively across the bilayer. Cells that require large fluxes of water have water-specific channels called aquaporins in their plasma membranes. Other polar species cross biological membranes only by way of specific membrane proteins—transporters and ion channels.

Transporters (transport proteins), like enzymes, show saturation and substrate specificity. Transport via these systems may be passive (down the electrochemical gradient, hence independent of metabolic energy) or active (against the gradient and thus dependent on metabolic energy). Glucose transport into erythrocytes and water transport through aquaporins is passive. The energy input for active transport may come from light, oxidation reactions, ATP hydrolysis, or cotransport of some other solute. Some transporters carry out symport, the simultaneous passage of two substances in the same direction; others mediate antiport, in which two substances move simultaneously in opposite directions. The lactose transporter of *E. coli* and the glucose transporter of intestinal epithelial cells are symporters, both cotransporting Na^+ along with the sugar. The chloride-bicarbonate exchanger of erythrocytes and the ubiquitous Na^+K^+ ATPase are antiporters. In animal cells, the differences in cytosolic and extracellular concentrations of Na^+ and K^+ are established and maintained by active transport through the Na^+K^+ ATPase. The resulting Na^+ gradient across the plasma membrane is used as an energy source for a variety of secondary active transport processes, both symport and antiport.

In primary active transport, ATP serves directly as the energy source. There are four general types of ATP-driven active transporters. P-type ATPases undergo reversible phosphorylation during their catalytic cycle and are inhibited by the phosphate analog vanadate; examples are the Na^+K^+ ATPase of the plasma membrane and the Ca^{2+} transporters of the sarcoplasmic and endoplasmic reticulum (the SERCA pumps). V-type ATPases produce gradients of protons across a variety of intracellular membranes, including plant vacuolar membranes. F-type proton pumps (ATP synthases) are central to energy-conserving mechanisms in mitochondria and chloroplasts. A multidrug transporter in tumor cells uses ATP to drive the export of a variety of drugs. Ionophores are lipid-soluble molecules that bind specific ions and carry them passively across membranes, dissipating the energy of electrochemical ion gradients.

Ion channels provide hydrophilic pores through which select ions can diffuse, moving down their electrical or chemical concentration gradients. They are characteristically unsaturable and have very high flux rates. Many are highly specific for one ion, and most are gated by either voltage or a ligand. The structure of the K^+ channel is known and serves as the prototype for other channels such as the acetylcholine receptor ion channel and the voltage-gated Na^+ channel. Several α-helical transmembrane domains cluster about a central aqueous channel. Ion passage is restricted by a selectivity filter and by a gate that obstructs the ion channel when it is not activated by voltage or a bound ligand.

Porins are integral membrane proteins that consist of a β barrel with a central transmembrane opening. A globular protein domain hinders passage of ion, but the binding of a specific substrate transiently opens the central pore, allowing uptake of that substrate.

further reading

Membrane Architecture

Devaux, P.F. (1991) Static and dynamic lipid asymmetry in cell membranes. *Biochemistry* **30,** 1163–1173.

Frye, L.D. & Ediden, M. (1970) The rapid intermixing of cell-surface antigens after formation of mouse-human heterokaryons. *J. Cell Sci.* **7,** 319–335.

The classic demonstration of membrane protein mobility.

Haltia, T. & Freire, E. (1995) Forces and factors that contribute to the structural stability of membrane proteins. *Biochim. Biophys. Acta* **1241,** 295–322.

Good discussion of the secondary and tertiary structures of membrane proteins and the factors that stabilize them.

Jacobson, K., Sheets, E.D., & Simson, R. (1995) Revisiting the fluid mosaic model of membranes. *Science* **268,** 1441–1442.

Discussion of the restricted mobility of membrane proteins, updating the model in paper by Singer and Nicolson (below).

Singer, S.J. & Nicolson, G.L. (1972) The fluid mosaic model of the structure of cell membranes. *Science* **175,** 720–731.

The classic statement of the model.

von Heijne, G. (1994) Membrane proteins: from sequence to structure. *Annu. Rev. Biophys. Biomol. Struct.* **23,** 167–192.

A review of the steps required to predict the structure of an integral protein from its sequence.

White, S.H. & Wimley, W.C. (1999) Membrane protein folding and stability: Physical principles. *Annu. Rev. Biophys. Biomol. Struct.* **28,** 319–365.

Membrane Fusion

Nichols, B.J. & Pelham, H.R. (1998) SNAREs and membrane fusion in the Golgi. *Biochim. Biophys. Acta* **1404,** 9–31.

Transporters

Lienhard, F.E., Slot, J.W., James, D.E., & Mueckler, M.M. (1992) How cells absorb glucose. *Sci. Am.* **266** (January), 86–91.

Introduction to the glucose transporter and its regulation by insulin.

Lingrel, J.B. & Kuntzweiler, T. (1994) Na^+, K^+-ATPase. *J. Biol. Chem.* **269,** 19,659–19,662.

Møller, J.V., Juul, B., & LeMaire, M. (1996) Structural organization, ion transport, and energy transduction of P-type ATPases. *Biochim. Biophys. Acta* **1286,** 1–51.

Mueckler, M. (1994) Facilitative glucose transporters. *Eur. J. Biochem.* **219,** 713–725.

Postle, K. (1999) Active transport by customized β-barrels. *Nat. Struct. Biol.* **6,** 3–6.

Introductory description of the structure of the *E. coli* porins and their coupling to metabolic energy.

Sheppard, D.N. & Welsh, M.J. (1999) Structure and function of the CFTR chloride channel. *Physiolog. Rev.* **79,** S23–S46.

This issue of the journal has 11 reviews of the CFTR chloride channel, covering its structure, activity, regulation, biosynthesis, and pathophysiology.

Ion Channels

Aidley, D.J. & Stanfield, P.R. (1996) *Ion Channels: Molecules in Action,* Cambridge University Press.

Excellent chapters on molecular structure, gating, and drug interactions of ion channels, and human diseases that result from defective channels.

Changeux, J.P. (1993) Chemical signaling in the brain. *Sci. Am.* **269** (November), 58–62.

Structure and function of the acetylcholine receptor channel.

Doyle, D.A., Cabral, K.M., Pfuetzner, R.A., Kuo, A., Gulbis, J.M., Cohen, S.L., Chait, B.T., & MacKinnon, R. (1998) The structure of the potassium channel: Molecular basis of K^+ conduction and selectivity. *Science* **280**, 69–77.

The first crystal structure of an ion channel.

Edelstein, S.J. & Changeux, J.P. (1998) Allosteric transitions of the acetylcholine receptor. *Adv. Prot. Chem.* **51,** 121–184.

Advanced discussion of the conformational changes induced by acetylcholine.

Hille, B. (1992) *Ionic Channels of Excitable Membranes.* Sinauer Associates, Sunderland, Massachusetts.

Emphasizes the function of ion channels; good complement to Aidley and Stanfield (above).

Marban, E., Yamagishi, T., & Tomaselli, G.F. (1998) Structure and function of voltage-gated sodium channels. *J. Physiol.* **508,** 647–657.

Review of evolution, structure, gating, drug interactions, and synthesis of voltage-gated Na^+ channels.

Neher, E. & Sakmann, B. (1992) The patch clamp technique. *Sci. Am.* (March) **266,** 44–51.

Clear description of the electrophysiological techniques used to measure the activity of single ion channels, by their Nobel Prize-winning developers.

Perozo, E., Cortes, D.M., & Cuello, L.G. (1999) Structural rearrangements underlying K^+ channel activation gating. *Science* **285,** 73–78.

Physical studies of the mechanism by which the K^+ channel opens and closes.

Unwin, N. (1998) The nicotinic acetylcholine receptor of the *Torpedo* electric ray. *J. Struct. Biol.* **121,** 181–190.

Evidence for the structural change that accompanies channel closing.

problems

1. Determining the Cross-Sectional Area of a Lipid Molecule When phospholipids are layered gently onto the surface of water, they orient at the air-water interface with their head groups in the water and their hydrophobic tails in the air. An experimental apparatus **(a)** has been devised that reduces the surface area available to a layer of lipids. By measuring the force necessary to push the lipids together, it is possible to determine when the molecules are packed tightly in a continuous monolayer; when that area is approached, the force needed to further reduce the surface area increases sharply **(b).** How would you use this apparatus to determine the average area occupied by a single lipid molecule in the monolayer?

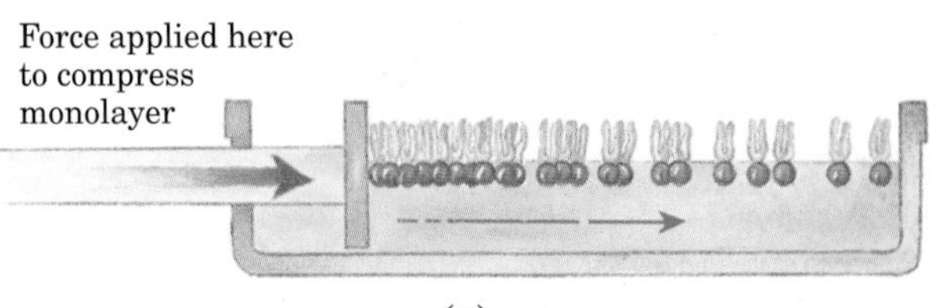

(a)

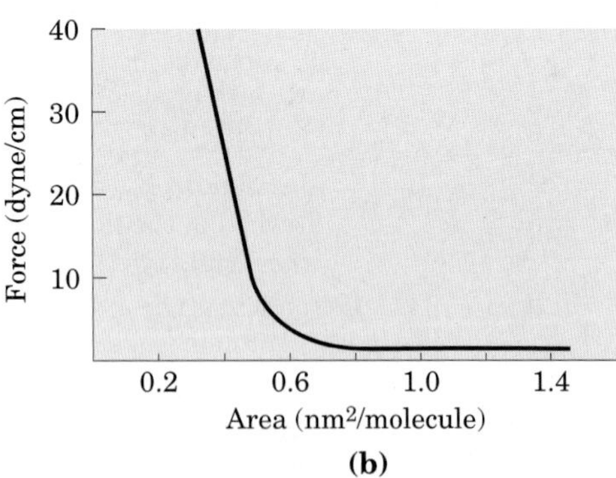

(b)

2. Evidence for Lipid Bilayer In 1925, E. Gorter and F. Grendel used an apparatus like that described in Problem 1 to determine the surface area of a lipid monolayer formed by lipids extracted from erythrocytes of several animal species. They used a microscope to measure the dimensions of individual cells, from which they calculated the average surface area of one erythrocyte. They obtained the data shown on the next page. Were these investigators justified in concluding that "chromocytes [erythrocytes] are covered by a layer of fatty substances that is two molecules thick" (i.e., a lipid bilayer)?

Animal	Volume of packed cells (mL)	Number of cells (per mm^3)	Total surface area of lipid monolayer from cells (m^2)	Total surface area of one cell (μm^2)
Dog	40	8,000,000	62	98
Sheep	10	9,900,000	6.0	29.8
Human	1	4,740,000	0.92	99.4

Source: Data from Gorter, E. & Grendel, F. (1925) On bimolecular layers of lipoids on the chromocytes of the blood. *J. Exp. Med.* **41,** 439–443.

3. Number of Detergent Molecules per Micelle When a small amount of sodium dodecyl sulfate ($Na^+CH_3(CH_2)_{11}OSO_3^-$) is dissolved in water, the detergent ions enter the solution as monomeric species. As more detergent is added, a concentration is reached (the critical micelle concentration) at which the monomers associate to form micelles. The critical micelle concentration of SDS is 8.2 mM. The micelles have an average particle weight (the sum of the molecular weights of the constituent monomers) of 18,000. Calculate the number of detergent molecules in the average micelle.

4. Properties of Lipids and Lipid Bilayers Lipid bilayers formed between two aqueous phases have this important property: they form two-dimensional sheets, the edges of which close upon each other and undergo self-sealing to form liposomes.

(a) What properties of lipids are responsible for this property of bilayers? Explain.

(b) What are the consequences of this property with regard to the structure of biological membranes?

5. Length of a Fatty Acid Molecule The carbon–carbon bond distance for single-bonded carbons such as those in a saturated fatty acyl chain is about 1.5 Å. Estimate the length of a single molecule of palmitate in its fully extended form. If two molecules of palmitate were placed end to end, how would their total length compare with the thickness of the lipid bilayer in a biological membrane?

6. Temperature Dependence of Lateral Diffusion The experiment described in Figure 12–7 was performed at 37 °C. If the experiment were carried out at 10 °C, what effect would you expect on the rate of cell-cell fusion and the rate of membrane protein mixing? Why?

7. Synthesis of Gastric Juice: Energetics Gastric juice (pH 1.5) is produced by pumping HCl from blood plasma (pH 7.4) into the stomach. Calculate the amount of free energy required to concentrate the H^+ in 1 L of gastric juice at 37 °C. Under cellular conditions, how many moles of ATP must be hydrolyzed to provide this amount of free energy? The free-energy change for ATP hydrolysis under cellular conditions is about −58 kJ/mol (as we will explain in Chapter 14). Ignore the effects of the transmembrane electric potential.

8. Energetics of the Na^+K^+ ATPase The concentration of Na^+ inside a vertebrate cell is about 12 mM, and that in blood plasma is about 145 mM. For a typical cell with a transmembrane potential of −0.07 V (inside negative relative to outside), what is the free-energy change for transporting 1 mol of Na^+ out of the cell and into the blood at 37 °C?

9. Action of Ouabain on Kidney Tissue Ouabain specifically inhibits the Na^+K^+ ATPase activity of animal tissues but is not known to inhibit any other enzyme. When ouabain is added to thin slices of living kidney tissue, it inhibits oxygen consumption by 66%. Why? What does this observation tell us about the use of respiratory energy by kidney tissue?

10. Energetics of Symport Suppose that you determined experimentally that a cellular transport system for glucose, driven by symport of Na^+, could accumulate glucose to concentrations 25 times greater than in the external medium, while the external [Na^+] was only ten times greater than the intracellular [Na^+]. Would this violate the laws of thermodynamics? If not, how could you explain this observation?

11. Location of a Membrane Protein The following observations are made on an unknown membrane protein, X. It can be extracted from disrupted erythrocyte membranes into a concentrated salt solution, and the isolated X can be cleaved into fragments by proteolytic enzymes. Treatment of erythrocytes with proteolytic enzymes followed by disruption and extraction of membrane components yields intact X. However, treatment of erythrocyte "ghosts" (which consist of only membranes, produced by disrupting the cells and washing out the hemoglobin) with proteolytic enzymes followed by disruption and extraction yields extensively fragmented X. What do these observations indicate about the location of X in the plasma membrane? Do the properties of X resemble those of an integral or peripheral membrane protein?

12. Membrane Self-Sealing Cellular membranes are self-sealing—if they are punctured or disrupted mechanically, they quickly and automatically reseal. What properties of membranes are responsible for this important feature?

13. Lipid Melting Temperatures Membrane lipids in tissue samples obtained from different parts of the leg of a reindeer have different fatty acid compositions. Membrane lipids from tissue near the hooves contain a larger proportion of unsaturated fatty acids than those from tissue in the upper leg. What is the significance of this observation?

14. Flip-Flop Diffusion The inner face (monolayer) of the human erythrocyte membrane consists predominantly of phosphatidylethanolamine and phosphatidylserine. The outer face consists predominantly of phosphatidylcholine and sphingomyelin. Although the phospholipid components of the membrane can diffuse in the fluid bilayer, this sidedness is preserved at all times. How?

15. Membrane Permeability At pH 7, tryptophan crosses a lipid bilayer about 1,000 times more slowly than does the closely related substance indole:

N
H

Suggest an explanation for this observation.

16. Water Flow through an Aquaporin Each human erythrocyte has about 2×10^5 AQP-1 monomers. If water flows through the plasma membrane at a rate of 5×10^8 H_2O molecules per AQP-1 tetramer per second, and the volume of an erythrocyte is 5×10^{-11} mL, how rapidly could an erythrocyte halve its volume as it encounters the high osmolarity (1 M) in the interstitial fluid of the renal medulla?

17. Labeling the Galactoside Transporter The galactoside transporter of a bacterium, which is highly specific for its substrate lactose, contains a Cys residue that is essential to its transport activity. Covalent reaction of *N*-ethylmaleimide (NEM) with this Cys residue irreversibly inactivates the transporter. A high concentration of lactose in the medium prevents inactivation by NEM, presumably by sterically protecting the Cys residue, which is in or near the lactose binding site. You know nothing else about the transporter protein. Suggest an experiment that might allow you to determine the M_r of the Cys-containing transporter polypeptide.

18. Predicting Membrane Protein Topology from Sequence You have cloned the gene for a human erythrocyte protein, which you suspect is a membrane protein. From the nucleotide sequence of the gene, you know the amino acid sequence. From this sequence alone, how would you evaluate the possibility that the protein is an integral protein? Suppose the protein proves to be an integral protein, either type I or type II. Suggest biochemical or chemical experiments that might allow you to determine which type it is.

19. Intestinal Uptake of Leucine You are studying the uptake of L-leucine by epithelial cells of the mouse intestine. Measurements of the rate of uptake of L-leucine and several of its analogs, with and without Na^+ in the assay buffer, yield the following results. What can you conclude about the properties and mechanism of the leucine transporter? Would you expect L-leucine uptake to be inhibited by ouabain?

Substrate	Uptake in presence of Na^+		Uptake in absence of Na^+	
	V_{max}	K_t (mM)	V_{max}	K_t (mM)
L-Leucine	420	0.24	23	0.24
D-Leucine	310	4.7	5	4.7
L-Valine	225	0.31	19	0.31

20. Effect of an Ionophore on Active Transport Consider the leucine transporter described in Problem 19. Would V_{max} and/or K_t change if you added a Na^+ ionophore to the assay solution containing Na^+? Explain.

21. Surface Density of a Membrane Protein *E. coli* can be induced to make about 10,000 copies of the galactoside transporter (M_r 31,000) per cell. Assume that *E. coli* is a cylinder 1 μm in diameter and 2 μm long. What fraction of the plasma membrane surface is occupied by the galactoside transporter molecules? Explain how you arrived at this conclusion.

Biochemistry on the Internet

22. Membrane Protein Topology The receptor for the hormone epinephrine in animal cells is an integral membrane protein (M_r 64,000) that is believed to have seven membrane-spanning regions.

(a) Show that a protein of this size is capable of spanning the membrane seven times.

(b) Given the amino acid sequence of this protein, how would you predict which regions of the protein form the membrane-spanning helices?

(c) Go to the Protein Data Bank Web site. Use the PDB identifier 1DEP to retrieve the data page for a portion of the β-adrenergic receptor (one type of epinephrine receptor) from turkey. Using Chime to explore the structure, predict where this portion of the receptor is located: within the membrane or at the membrane surface. Explain.

(d) Retrieve the data for a portion of another receptor, the acetylcholine receptor of neurons and myocytes, using the PDB identifier 1A11. As in (c), predict where this portion of the receptor is located and explain your answer.

If you have not used the PDB or Chemscape Chime, you can find instructions at http://www.worthpublishers.com/lehninger.

chapter 13

Biosignaling

The ability of cells to receive and act on signals from beyond the plasma membrane is fundamental to life. Bacterial cells receive constant input from membrane receptors that sample the surrounding medium for pH; osmotic strength; the availability of food, oxygen, and light; and the presence of noxious chemicals, predators, or competitors for food. These signals elicit appropriate responses, such as motion toward food or away from toxic substances or the formation of dormant spores in a nutrient-depleted medium. In multicellular organisms, cells with different functions exchange a wide variety of signals. Plant cells respond to growth hormones and to variations in sunlight. Animal cells exchange information about their correct placement in a developing embryo, the concentrations of ions and glucose in extracellular fluids, and the interdependent metabolic activities taking place in different tissues. The signals in animals may be autocrine (acting on the same cell that produces them), paracrine (acting on a near neighbor), or endocrine (carried in the bloodstream from the producer cell to a distant target cell). In all three cases, the signal is detected by a specific receptor and is converted to a cellular response.

Although the number of biological signals is legion (Table 13–1), as is the variety of biological responses to these signals, organisms use just a few evolutionarily conserved mechanisms to detect extracellular signals and *transduce* them into intracellular changes. In this chapter we examine some examples of the major classes of signaling mechanisms, and we see how they are integrated in specific biological functions such as the transmission of nerve signals, responses to hormones and growth factors, the senses of sight, smell, and taste, and control of the cell cycle. Often, the end result of a signaling pathway is the phosphorylation of a few specific target-cell proteins, which changes their activities and thus the activities of the cell. Throughout our discussion we emphasize the conservation of fundamental mechanisms for the transduction of biological signals and the adaptation of these basic membrane-linked processes to a wide variety of signaling pathways.

table 13–1

Some Signals to Which Cells Respond
Antigens
Cell surface glycoproteins/oligosaccharides
Developmental signals
Extracellular matrix components
Growth factors
Hormones
Light
Mechanical touch
Neurotransmitters
Odorants
Pheromones
Tastants

Molecular Mechanisms of Signal Transduction

Signal transductions are remarkably specific and exquisitely sensitive. **Specificity** is achieved by precise molecular complementarity between the signal and receptor molecules (Fig. 13–1a), mediated by the same kinds of weak (noncovalent) forces that occur in enzyme-substrate and antigen-antibody interactions. In multicellular organisms, specificity is further promoted because the receptors for a given signal, or the intracellular targets of a given signal pathway, are present only in certain cell types.

(a) Specificity
Signal molecule fits binding site on its complementary receptor; other signals do not fit.

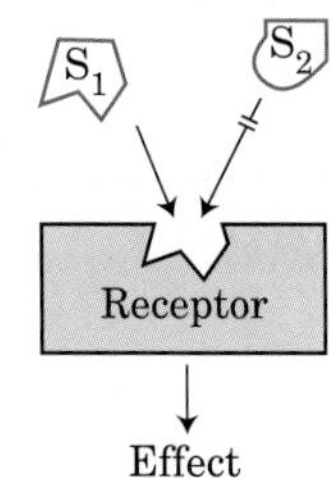

(b) Amplification
When enzymes activate enzymes, the number of affected molecules increases geometrically in an enzyme cascade.

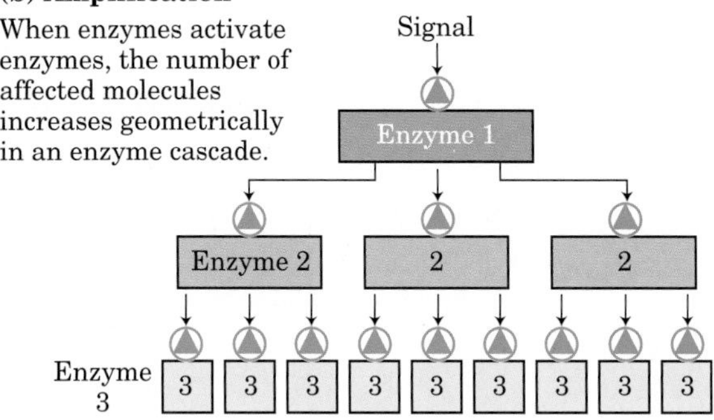

(c) Desensitization/Adaptation
Receptor activation triggers a feedback circuit that shuts off the receptor or removes it from the cell surface.

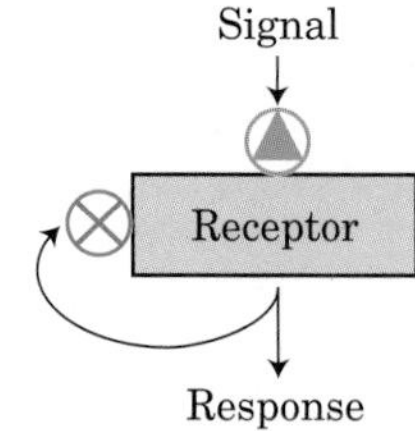

(d) Integration
When two signals have opposite effects on a metabolic characteristic such as the concentration of a second messenger X, or the membrane potential V_m, the regulatory outcome results from the integrated input from both receptors.

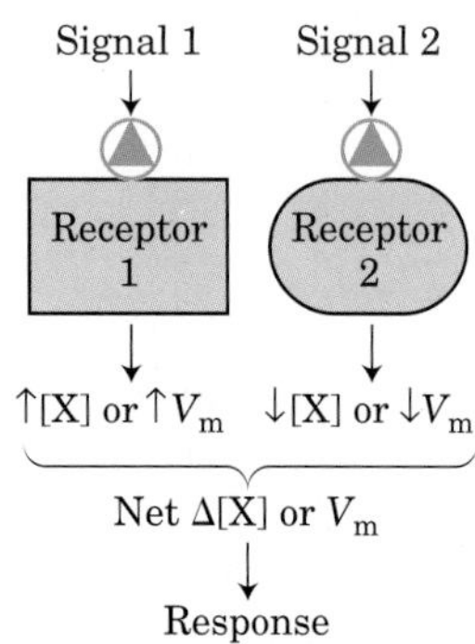

figure 13–1
Four features of signal-transducing systems.

Thyrotropin-releasing hormone, for example, triggers responses in the cells of the anterior pituitary but not in hepatocytes, which lack receptors for this hormone. Epinephrine alters glycogen metabolism in hepatocytes but not in erythrocytes, although in this case both cell types have receptors for the hormone; however, hepatocytes have the glycogen-metabolizing enzyme that is stimulated by epinephrine, but erythrocytes do not.

Three factors account for the extraordinary sensitivity of signal transducers: the high affinity of receptors for signal molecules, cooperativity in the ligand-receptor interaction, and amplification of the signal by enzyme cascades. The affinity between signal (ligand) and receptor can be expressed as the dissociation constant K_d, often 10^{-10} M or smaller, meaning that the receptor can detect picomolar concentrations of a signal molecule. Receptor-ligand interactions can be quantified by Scatchard analysis, which, in the best cases, yields a quantitative measure of affinity (K_d) and the number of ligand-binding sites in a receptor sample (Box 13–1). Cooperativity in receptor-ligand interactions results in large changes in receptor activation with small changes in ligand concentration (recall the effect of cooperativity on oxygen binding to hemoglobin; see Fig. 7–12). **Amplification** by **enzyme cascades** results when an enzyme associated with a signal receptor is activated and, in turn, catalyzes the activation of many molecules of a second enzyme, each of which activates many molecules of a third enzyme, and so on (Fig. 13–1b). Amplifications of several orders of magnitude can be produced in milliseconds by such cascades.

The sensitivity of receptor systems is subject to modification. When a signal is present continuously, **desensitization** of the receptor system results (Fig. 13–1c); when the stimulus falls below a certain threshold, the system again becomes sensitive. Think of what happens to your visual transduction system when you walk from bright sunlight into a darkened room or vice versa.

A final noteworthy feature of signal-transducing systems is **integration** (Fig. 13–1d), the ability of the system to receive multiple signals and produce a unified response appropriate to the needs of the cell or organism. Different signaling pathways converse with each other at several levels, generating a wealth of interactions that maintain homeostasis in the cell and the organism.

box 13–1 Scatchard Analysis Quantifies the Receptor-Ligand Interaction

The cellular actions of a hormone begin when the hormone (ligand), L, binds specifically and tightly to its protein receptor, R, on or in the target cell. Binding is mediated by noncovalent interactions (hydrogen bonds, hydrophobic and electrostatic interactions) between the complementary surfaces of ligand and receptor. Receptor-ligand interaction brings about a conformational change that alters the biological activity of the receptor, which may be an enzyme, an enzyme regulator, an ion channel, or a regulator of gene expression.

Receptor-ligand binding is described by the equation

$$\underset{\text{Receptor}}{\text{R}} + \underset{\text{Ligand}}{\text{L}} \rightleftharpoons \underset{\text{Receptor-ligand complex}}{\text{RL}}$$

This binding, like that of an enzyme to its substrate, is dependent on the concentrations of the interacting components and can be described by an equilibrium constant:

$$\text{R} + \text{L} \underset{k_{-1}}{\overset{k_{+1}}{\rightleftharpoons}} \text{RL}$$

$$K_a = \frac{[\text{RL}]}{[\text{R}][\text{L}]} = \frac{k_{+1}}{k_{-1}} = 1/K_d$$

where K_a is the association constant and K_d is the dissociation constant.

As in the case of enzyme-substrate binding, receptor-ligand binding is saturable. As more ligand is added to a fixed amount of receptor, an increasing fraction of receptor molecules is occupied by ligand (Fig. 1a). A rough measure of the receptor-ligand affinity is given by the concentration of ligand needed to give half-saturation of the receptor. **Scatchard analysis** of receptor-ligand binding allows an estimation of both the dissociation constant K_d and the number of receptor-binding sites in a given preparation. When binding has reached equilibrium, the total number of possible binding sites, B_{max}, is equal to the number of unoccupied sites, represented by [R], plus the number of occupied or ligand-bound sites, [RL]; that is, $B_{max} = [\text{R}] + [\text{RL}]$. The number of unbound sites can be expressed in terms of total sites minus occupied sites: $[\text{R}] = B_{max} - [\text{RL}]$. The equilibrium expression can now be written

$$K_a = \frac{[\text{RL}]}{[\text{L}](B_{max} - [\text{RL}])}$$

Rearranging to obtain the ratio of receptor-bound ligand to free (unbound) ligand, we get

$$\frac{[\text{Bound}]}{[\text{Free}]} = \frac{[\text{RL}]}{[\text{L}]} = K_a(B_{max} - [\text{RL}])$$
$$= \frac{1}{K_d}(B_{max} - [\text{RL}])$$

This slope-intercept form of the equation shows that a plot of [bound ligand]/[free ligand] versus [bound ligand] should give a straight line with a slope of $-K_a$ (or $-1/K_d$) and an intercept on the abscissa of B_{max}, the total number of binding sites (Fig. 1b). Scatchard analysis of a number of different hormone-ligand interactions has yielded K_d values of about 10^{-9} to 10^{-11} M, corresponding to very tight binding of ligand by receptor.

Scatchard analysis is reliable for the simplest cases, but as with the Lineweaver-Burk plot for enzymes, when the receptor is an allosteric protein, the plots show deviations from linearity.

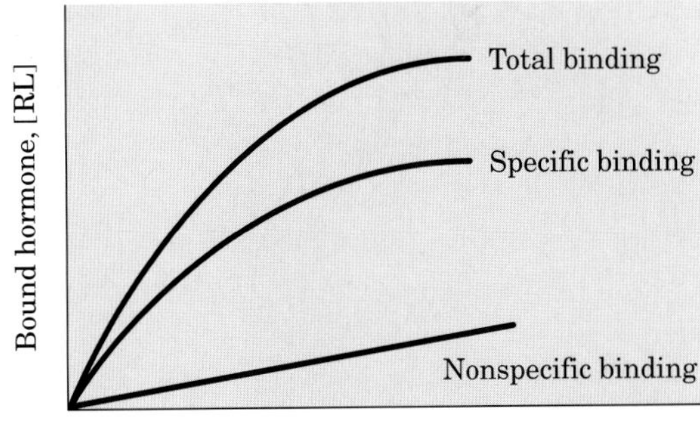

figure 1
Scatchard analysis of the receptor-ligand interaction. A radiolabeled ligand (L)—a hormone, for example—is added at several concentrations to a fixed amount of receptor (R), and the fraction of the hormone bound to receptor is determined by separating the receptor-hormone complex (RL) from free hormone. **(a)** A plot of [RL] versus total hormone added, [L] + [RL], is hyperbolic, rising toward a maximum for [RL] as the receptor sites become saturated. To control for nonsaturable, nonspecific binding sites (eicosanoid hormones bind nonspecifically to the lipid bilayer, for example), in a separate series of binding experiments a large excess of unlabeled hormone is added along with the dilute solution of labeled hormone. The unlabeled molecules compete with the labeled molecules for specific binding to the saturable site on the receptor, but not for the nonspecific binding. The true value for specific binding is obtained by subtracting nonspecific binding from total binding. **(b)** A linear plot of [RL]/[L] versus [RL] gives the K_d and B_{max} for the receptor-hormone complex. Compare these plots with those of V_0 versus [S] (see Fig. 8–12) and $1/V_0$ versus 1/[S] for an enzyme-substrate complex (see Box 8–1).

We will consider the molecular details of several representative signal-transduction systems. The trigger for each system is different, but the general features of signal transduction are common to all: a signal interacts with a receptor; the activated receptor interacts with cellular machinery, producing a second signal or a change in the activity of a cellular protein; the metabolic activity of the target cell undergoes a change; and finally, the transduction event is terminated and the cell returns to its prestimulus state. To illustrate these general features of signaling systems, we provide examples of each of the four basic signaling mechanisms (Fig. 13–2).

The simplest signal transducers are ion channels of the plasma membrane that open and close (hence the term "gating") in response to the binding of chemical ligands or changes in transmembrane potential. The acetylcholine receptor–ion channel is an example of this mechanism. The second basic signaling mechanism involves plasma membrane receptors that are also enzymes (receptor enzymes). When one of these receptors is activated by its extracellular ligand, it catalyzes the production of an intracellular second messenger. This mechanism is exemplified by the insulin receptor. The third transduction mechanism, illustrated by the β-adrenergic receptor system that detects epinephrine (adrenaline), is mediated by plasma membrane receptor proteins that *indirectly* activate (through GTP-binding proteins) enzymes that generate intracellular second messengers. Finally, the nucleus has a large class of receptors that, when bound to their specific ligand (such as the steroid hormone estrogen), alter the rate at which specific genes are transcribed and translated into cellular proteins. Because steroid hormones function through mechanisms intimately related to the regulation of gene expression, we consider them here only briefly and defer a detailed discussion of their action until Chapter 28.

Before proceeding to our discussion of the gated ion channels, we need to consider some basic information on ion channels and membrane potentials.

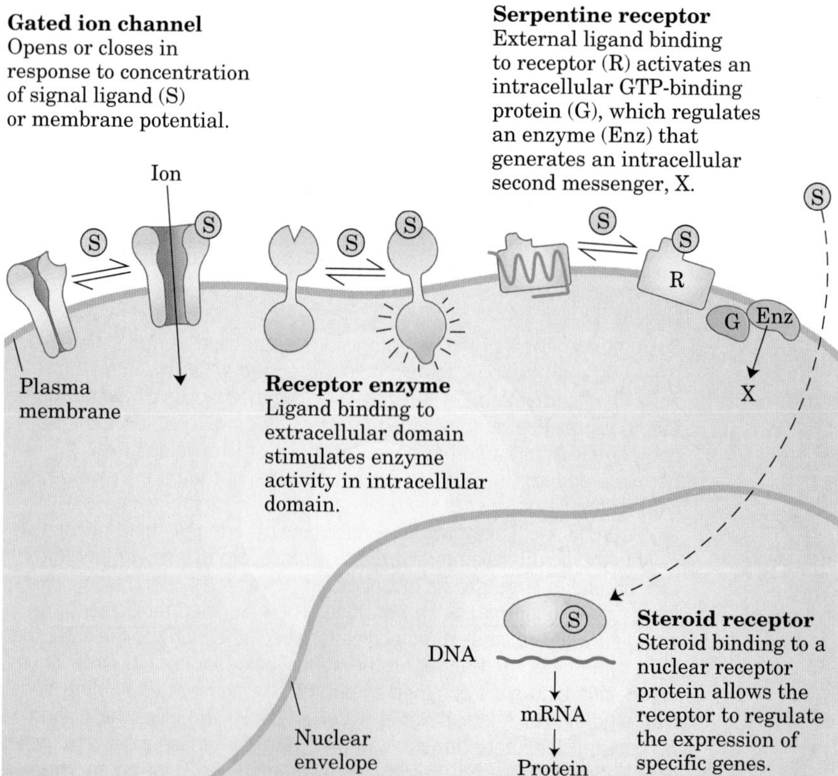

Gated ion channel
Opens or closes in response to concentration of signal ligand (S) or membrane potential.
Ion
S
Plasma membrane
Serpentine receptor
External ligand binding to receptor (R) activates an intracellular GTP-binding protein (G), which regulates an enzyme (Enz) that generates an intracellular second messenger, X.
R
G
Enz
X
Receptor enzyme
Ligand binding to extracellular domain stimulates enzyme activity in intracellular domain.
Steroid receptor
Steroid binding to a nuclear receptor protein allows the receptor to regulate the expression of specific genes.
DNA
mRNA
Protein
Nuclear envelope

figure 13–2
Four general types of signal transducers.

Gated Ion Channels

Ion Channels Underlie Electrical Signaling in Excitable Cells

The excitability of sensory cells, neurons, and myocytes depends on ion channels, signal transducers that provide a regulated path for the movement of inorganic ions such as Na^+, K^+, Ca^{2+}, and Cl^- across the plasma membrane in response to various stimuli. Recall from Chapter 12 that these ion channels are "gated"; they may be open or closed, depending on whether the associated receptor has been activated by the binding of its specific ligand (a neurotransmitter, for example) or by a change in the transmembrane electrical potential, V_m. The Na^+K^+ ATPase creates a charge imbalance across the plasma membrane by carrying 3 Na^+ out of the cell for every 2 K^+ carried in (Fig. 13–3a), making the inside negative relative to the outside. The membrane is now said to be polarized. By convention, V_m is negative when the inside of the cell is negative relative to the outside. For a typical animal cell, $V_m = -60$ to -70 mV.

Because ion channels generally allow passage of either anions or cations but not both, ion flux through a channel causes a redistribution of charge on the two sides of the membrane, changing V_m. Influx of a positively charged ion such as Na^+, or efflux of a negatively charged ion such as Cl^-, depolarizes the membrane and brings V_m closer to zero. Conversely, efflux of the positive ion K^+ hyperpolarizes the membrane and V_m becomes more negative. These ion fluxes through channels are passive, in contrast with the active transport by the Na^+K^+ ATPase.

The direction of spontaneous ion flow across a polarized membrane is dictated by the ion's electrochemical potential across the membrane. The force (ΔG) that causes a given ion to pass spontaneously through an ion

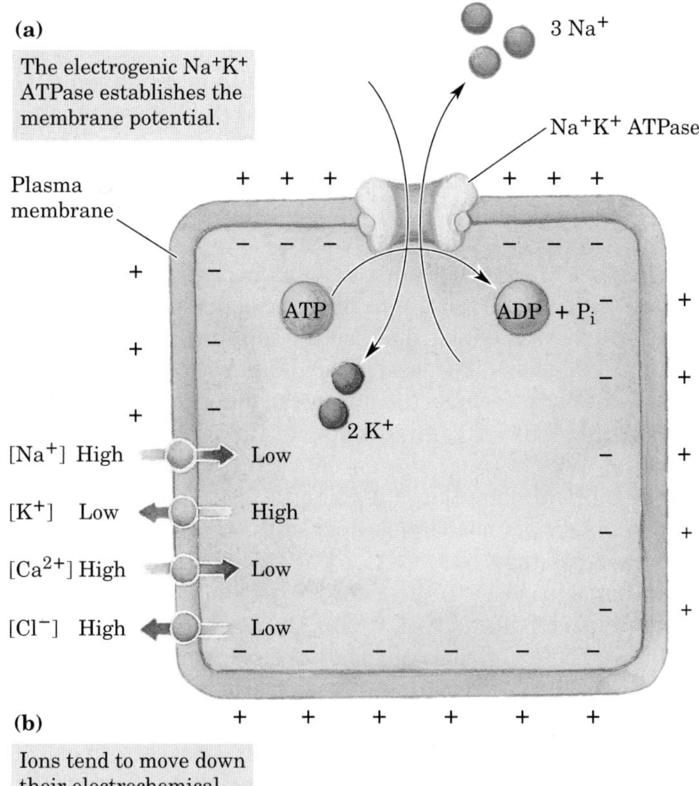

figure 13–3

Transmembrane electrical potential. (a) The electrogenic Na^+K^+ ATPase produces a transmembrane electrical potential of −60 mV (inside negative). **(b)** Blue arrows show the direction in which ions tend to move spontaneously across the plasma membrane in an animal cell, driven by the combination of chemical and electrical gradients. The chemical gradient drives Na^+ and Ca^{2+} inward (producing depolarization) and K^+ outward (producing hyperpolarization). The electrical gradient drives Cl^- outward, against its concentration gradient (producing depolarization).

channel is a function of the ratio of its concentrations on the two sides of the membrane (C_2/C_1) and of the difference in electrical potential ($\Delta\psi$ or V_m):

$$\Delta G = RT \ln (C_{out}/C_{in}) + Z\mathcal{J} V_m \qquad (13\text{–}1)$$

where R is the gas constant, T the absolute temperature, Z the charge on the ion, and $\mathcal{J}$ the Faraday constant. In a typical neuron or myocyte, the concentrations of Na^+, K^+, Ca^{2+}, and Cl^- in the cytosol are very different from those in the extracellular fluid (Table 13–2). Given these concentration differences, the resting V_m of -60 mV, and the relationship shown in Equation 13–1, opening of a Na^+ or Ca^{2+} channel will result in a spontaneous inward flow of Na^+ or Ca^{2+} (and depolarization), whereas opening of a K^+ channel will result in a spontaneous outward flux of K^+ (and hyperpolarization) (Fig. 13–3b).

table 13–2

Ion Concentrations in Cells and Extracellular Fluids (mM)

	K^+		Na^+		Ca^{2+}		Cl^-	
	In	Out	In	Out	In	Out	In	Out
Cell type								
Squid axon	400	20	50	440	≤0.4	10	40–150	560
Frog muscle	124	2.3	10.4	109	<0.1	2.1	1.5	78

A given ionic species continues to flow through a channel only as long as the combination of concentration gradient and electrical potential provides a driving force, according to Equation 13–1. For example, as Na^+ flows down its concentration gradient it depolarizes the membrane. When the membrane potential reaches $+70$ mV, the effect of this membrane potential (to resist further entry of Na^+) exactly equals the effect of the Na^+ concentration gradient (to cause more Na^+ to flow inward). At this equilibrium potential (E), the driving force tending to move an ion (ΔG) is zero, and Equation 13–1 can be rearranged to yield the Nernst equation:

$$E = -(RT/Z\mathcal{J}) \ln (C_{out}/C_{in})$$

The equilibrium potential is different for each ionic species because the concentration gradients differ for each ion.

The number of ions that must flow to change the membrane potential significantly is negligible relative to the large concentrations of Na^+, K^+, and Cl^- in cells and extracellular fluid, so the ion fluxes that occur during signaling in excitable cells have essentially no effect on the concentrations of those ions. However, because the intracellular concentration of Ca^{2+} is generally very low ($\sim 10^{-7}$ M), inward flow of Ca^{2+} can significantly alter the cytosolic [Ca^{2+}].

The membrane potential of a cell at a given time is the result of the types and numbers of ion channels open at that instant. In most cells at rest, more K^+ channels than Na^+, Cl^-, or Ca^{2+} channels are open and thus the resting potential is closer to the E for K^+ (–98 mV) than that for any other ions. When channels for Na^+, Ca^{2+}, or Cl^- open, the membrane potential moves toward the E for that ion. The precisely timed opening and closing of ion channels and the resulting transient changes in membrane potential underlie the electrical signaling by which the nervous system signals the skeletal muscles to contract, the heart to beat, or secretory cells to release their contents. Moreover, many hormones exert their effects by altering the membrane potentials of their target cells. These mechanisms are not limited to complex organisms; ion channels play important roles in the responses of bacteria and other unicellular organisms to environmental signals.

To illustrate the action of ion channels in cell-to-cell signaling, we now describe the mechanisms by which a neuron passes a signal along its length and across a synapse to the next neuron (or to a myocyte) in a cellular circuit, using acetylcholine as the neurotransmitter.

The Nicotinic Acetylcholine Receptor Is a Ligand-Gated Ion Channel

One of the best-understood examples of a ligand-gated receptor channel is the **nicotinic acetylcholine receptor** (see Fig. 12–39). The receptor channel opens in response to the neurotransmitter acetylcholine (and to nicotine, hence the name). This receptor is found in the postsynaptic membrane of neurons at certain synapses and in muscle fibers (myocytes) at neuromuscular junctions.

$$\mathrm{CH_3{-}\overset{+}{N}(CH_3)_2{-}CH_2CH_2O{-}\overset{O}{\overset{\|}{C}}{-}CH_3}$$

Acetylcholine (Ach)

Acetylcholine released by an excited neuron diffuses a few micrometers across the synaptic cleft or neuromuscular junction to the postsynaptic neuron or myocyte, where it interacts with the acetylcholine receptor and triggers electrical excitation (depolarization) of the receiving cell. The acetylcholine receptor is an allosteric protein with two high-affinity binding sites for acetylcholine, about 3.0 nm from the ion gate, on the two α subunits. The binding of acetylcholine causes a change from the closed to the open conformation. The process is positively cooperative: binding of acetylcholine to the first site increases the acetylcholine-binding affinity of the second site. When the presynaptic cell releases a brief pulse of acetylcholine, both sites on the postsynaptic cell receptor are occupied briefly, and the channel opens (Fig. 13–4). Either Na^+ or Ca^{2+} can now pass, and the inward flux of these ions depolarizes the plasma membrane, initiating subsequent events that vary with the type of tissue. In a postsynaptic neuron, depolarization initiates an action potential (see below); at a neuromuscular junction, depolarization of the muscle fiber triggers muscle contraction.

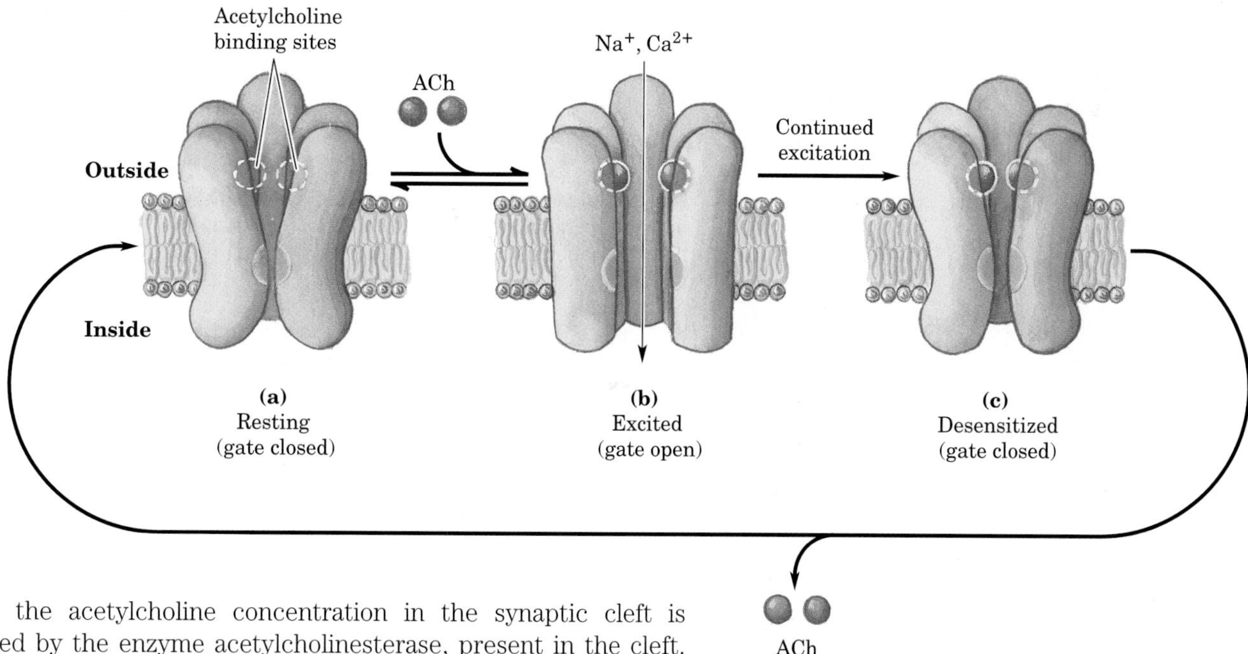

figure 13–4
Three states of the acetylcholine receptor. Brief exposure of the resting (closed) ion channel **(a)** to acetylcholine produces the excited (open) state **(b).** Longer exposure leads to desensitization and channel closure **(c).**

Normally, the acetylcholine concentration in the synaptic cleft is quickly lowered by the enzyme acetylcholinesterase, present in the cleft. When acetylcholine levels remain high for more than a few milliseconds, receptor desensitization occurs (Fig. 13–1c). The receptor channel is converted to a third conformation (Fig. 13–4c), in which the channel is closed and the acetylcholine is very tightly bound. The slow release (in tens of milliseconds) of acetylcholine from its binding sites eventually allows the receptor to return to its resting state—closed and resensitized to acetylcholine levels.

Voltage-Gated Ion Channels Produce Neuronal Action Potentials

Signaling in the nervous system is accomplished by networks of neurons, specialized cells that carry an electrical impulse (action potential) from one end of the cell (the cell body) through an elongated cytoplasmic extension (the axon). The electrical signal triggers release of neurotransmitter molecules at the synapse, carrying the signal to the next cell in the circuit. Three types of voltage-gated ion channels are essential to this signaling mechanism. Along the entire length of the axon are **voltage-gated Na^+ channels** (Fig. 13–5; see also Fig. 12–40), which are closed when the membrane is at rest ($V_m = -60$ mV) but open briefly when the membrane is depolarized locally in response to acetylcholine (or some other neurotransmitter). The depolarization induced by the opening of Na^+ channels causes **voltage-gated K^+ channels** to open and the resulting efflux of K^+ repolarizes the membrane locally. A brief pulse of depolarization traverses the axon as local depolarization triggers the brief opening of neighboring Na^+ channels, then K^+ channels. After each opening of a Na^+ channel, a short refractory period follows during which that channel cannot open again, and thus a unidirectional wave of depolarization sweeps from the nerve cell body toward the end of the axon. The voltage sensitivity of ion channels is due to the presence at critical positions of charged amino acid side chains that interact with the electric field across the membrane. Changes in transmembrane potential produce subtle conformational changes in the channel protein (see Fig. 12–40).

At the distal tip of the axon are **voltage-gated Ca^{2+} channels.** When the wave of depolarization reaches them, these channels open, and Ca^{2+} enters from the extracellular space. Acting as an intracellular second messenger, Ca^{2+} then triggers release of acetylcholine by exocytosis into the synaptic cleft (step ③ in Fig. 13–5). Acetylcholine diffuses to the postsynaptic cell (another neuron or a myocyte), where it binds to acetylcholine receptors and triggers depolarization. Thus the message is passed to the next cell in the circuit.

We see, then, that gated ion channels convey signals in either of two ways: by changing the cytosolic concentration of an ion (such as Ca^{2+}),

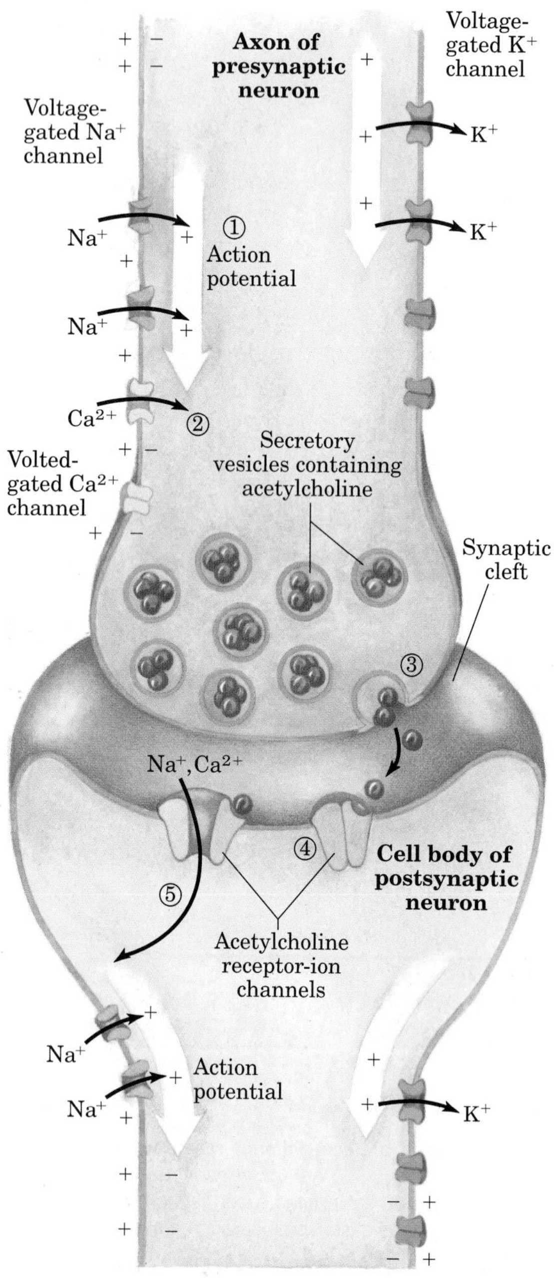

figure 13–5

Role of voltage-gated and ligand-gated ion channels in neural transmission. Initially, the plasma membrane of the presynaptic neuron is polarized (inside negative) through the action of the electrogenic Na^+K^+ ATPase, which pumps 3 Na^+ out for every 2 K^+ pumped into the neuron (see Fig. 13–3). ① A stimulus to this neuron causes an action potential to move along the axon (white arrow), away from the cell body. The opening of one voltage-gated Na^+ channel allows Na^+ entry, and the resulting local depolarization causes the adjacent Na^+ channel to open, and so on. The directionality of movement of the action potential is ensured by the brief refractory period that follows the opening of each voltage-gated Na^+ channel. ② When the wave of depolarization reaches the axon tip, voltage-gated Ca^{2+} channels open, allowing Ca^{2+} entry into the presynaptic neuron. ③ The resulting increase in internal [Ca^{2+}] triggers exocytic release of the neurotransmitter acetylcholine into the synaptic cleft. ④ Acetylcholine binds to a receptor on the postsynaptic neuron, causing its ligand-gated ion channel to open. ⑤ Extracellular Na^+ and Ca^{2+} enter through this channel, depolarizing the postsynaptic cell. The electrical signal has thus passed to the cell body of the postsynaptic neuron and will move along its axon to a third neuron by this same sequence of events.

which then serves as an intracellular second messenger, or by changing V_m and affecting other membrane proteins that are sensitive to V_m. The passage of an electrical signal through one neuron and on to the next illustrates both types of mechanism.

Neurons Have Receptor Channels That Respond to a Variety of Neurotransmitters

Animal cells, especially those of the nervous system, contain a variety of ion channels gated by ligands, voltage, or both. The neurotransmitters 5-hydroxytryptamine (serotonin), glutamate, and glycine can all act through receptor channels that are structurally related to the acetylcholine receptor. Serotonin and glutamate trigger the opening of cation (K^+, Na^+, Ca^{2+}) channels, whereas glycine opens Cl^--specific channels. Cation and anion channels are distinguished by subtle differences in the amino acid residues that line the hydrophilic channel. Cation channels have negatively charged Glu and Asp side chains at crucial positions. When a few of these acidic residues are experimentally replaced with basic residues, the cation channel is converted to an anion channel.

Depending on which ion passes through a channel, the ligand (neurotransmitter) for that channel either depolarizes or hyperpolarizes the target cell. A single neuron normally receives input from several (or many) other neurons, each releasing its own characteristic neurotransmitter with its characteristic depolarizing or hyperpolarizing effect. The target cell's V_m therefore reflects the integrated input (Fig. 13–1d) from multiple neurons. The cell responds with an action potential only if the integrated input adds up to a net depolarization.

The receptor channels for acetylcholine, glycine, glutamate, and γ-aminobutyric acid (GABA) are gated by *extra*cellular ligands. *Intra*cellular second messengers—such as cAMP, cGMP (3′,5′-cyclic GMP, a close analog of cAMP), IP_3 (inositol 1,4,5-trisphosphate), Ca^{2+}, and ATP—regulate ion channels of another class, which, as we shall see later, participate in the sensory transductions of vision, olfaction, and gustation.

Serotonin
(5-Hydroxytryptamine)

Glutamate

Receptor Enzymes

A fundamentally different mechanism of signal transduction is carried out by the receptor enzymes. These proteins have a ligand-binding domain on the extracellular surface of the plasma membrane and an enzyme active site on the cytosolic side, with the two domains connected by a single transmembrane segment. Commonly, the receptor enzyme is a protein kinase that phosphorylates Tyr residues in specific target proteins; the insulin receptor is the prototype for this group. Other receptor enzymes synthesize the intracellular second messenger cGMP in response to extracellular signals. The receptor for atrial natriuretic factor is typical of this type.

The Insulin Receptor Is a Tyrosine-Specific Protein Kinase

Insulin regulates both metabolism and gene expression: the insulin signal passes from the plasma membrane receptor to insulin-sensitive metabolic enzymes and to the nucleus, where it stimulates the transcription of specific genes. The active insulin receptor consists of two identical α chains protruding from the outer face of the plasma membrane and two transmembrane β subunits with their carboxyl termini protruding into the cytosol (Fig. 13–6). The α chains contain the insulin-binding domain, and the intracellular domains of the β chains contain the protein kinase activity that transfers a phosphoryl group from ATP to the hydroxyl group of Tyr residues in specific target proteins. Signaling through the insulin receptor is

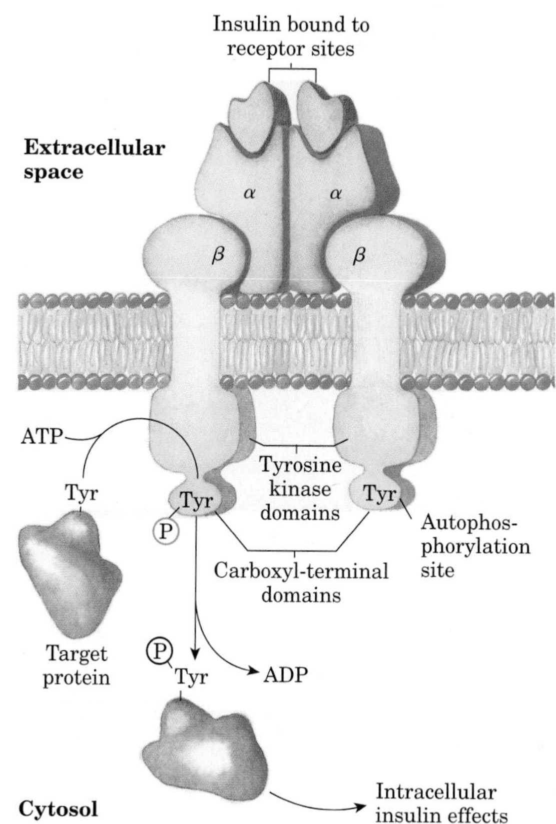

figure 13–6
Insulin receptor. The insulin receptor consists of two α chains on the outer face of the plasma membrane and two β chains that traverse the membrane and protrude from the cytosolic face. Binding of insulin to the α chains triggers a conformational change that allows the autophosphorylation of Tyr residues in the carboxyl-terminal domain of the β subunits. Autophosphorylation further activates the tyrosine kinase domain, which then catalyzes phosphorylation of other target proteins.

illustrated in Figure 13–7. It begins (step ①) when binding of insulin to the α chains activates the tyrosine kinase activity of the β chains, and each $\alpha\beta$ monomer phosphorylates critical Tyr residues near the carboxyl terminus of the β chain of its partner in the dimer. This autophosphorylation opens up the active site, allowing the enzyme to phosphorylate Tyr residues of other target proteins.

One of these target proteins (step ②) is insulin receptor substrate-1 (IRS-1). Once phosphorylated on Tyr residues, IRS-1 becomes the point of nucleation for a complex of proteins (step ③) that carry the message from the insulin receptor to end targets in the cytosol and nucleus, through a long series of intermediate proteins. First, a Ⓟ-Tyr residue in IRS-1 is bound by the **SH2 domain** of the protein Grb2. (SH2 is an abbreviation of Src homology 2; the sequences of SH2 domains are similar to a domain in another protein tyrosine kinase called Src. A number of signaling proteins contain SH2 domains, all of which bind Ⓟ-Tyr residues in a protein partner.) Grb2 in turn recruits another protein, Sos, to the growing complex. When bound to Grb2, Sos catalyzes the replacement of bound GDP by GTP on Ras, one of a family of guanosine nucleotide–binding proteins **(G proteins)** that mediate a wide variety of signal transductions. When GTP is bound, Ras can activate a protein kinase, Raf-1 (step ④), the first of three protein kinases (Raf-1, MEK, and MAPK) that form a cascade in which each kinase activates the next by phosphorylation (step ⑤). The last protein kinase, mitogen-activated protein kinase (MAPK), is activated by phosphorylation of both a Thr and a Tyr residue. When activated, it mediates some of the biological effects of insulin by entering the nucleus and phosphorylating proteins such as Elk1, which modulates the transcription of certain insulin-regulated genes (step ⑥).

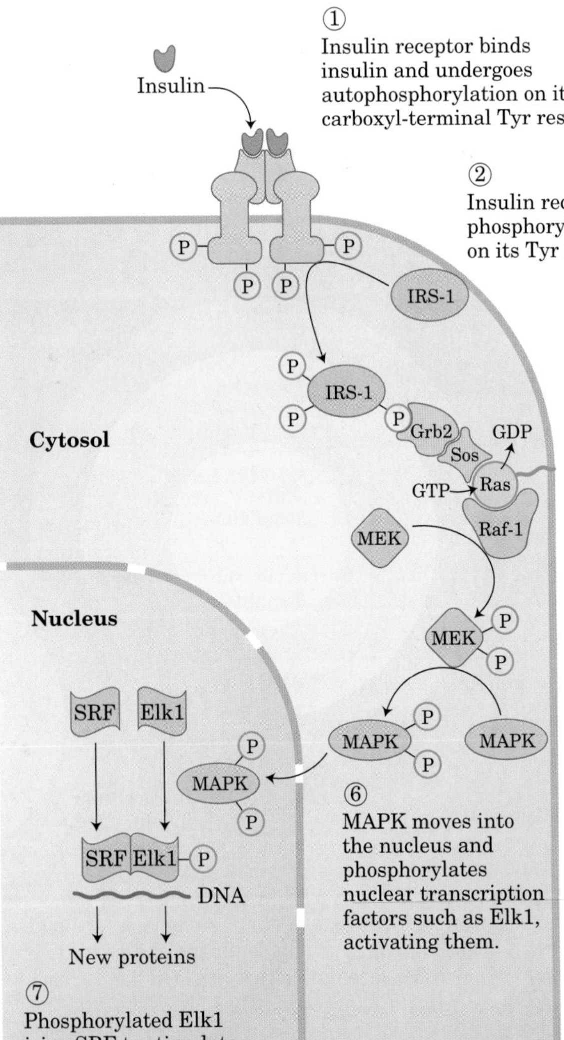
Insulin
①
Insulin receptor binds insulin and undergoes autophosphorylation on its carboxyl-terminal Tyr residues.
②
Insulin receptor phosphorylates IRS-1 on its Tyr residues.
P
IRS-1
Cytosol
Grb2
Sos
GDP
GTP
Ras
Raf-1
MEK
③
SH2 domain of Grb2 binds to Ⓟ-Tyr of IRS-1. Sos binds to Grb2, then to Ras, causing GDP release and GTP binding to Ras.
④
Activated Ras binds and activates Raf-1.
⑤
Raf-1 phosphorylates MEK on two Ser residues, activating it. MEK phosphorylates MAPK on a Thr and a Tyr residue, activating it.
Nucleus
SRF
Elk1
MAPK
DNA
New proteins
⑥
MAPK moves into the nucleus and phosphorylates nuclear transcription factors such as Elk1, activating them.
⑦
Phosphorylated Elk1 joins SRF to stimulate the transcription and translation of a set of genes needed for cell division.

figure 13–7

Regulation of gene expression by insulin. The signaling pathway by which insulin regulates the expression of specific genes involves a cascade of protein kinases, each of which activates the next. The insulin receptor is a Tyr-specific kinase; the other kinases (all shown in blue) phosphorylate Ser or Thr residues. MEK is a dual-specificity kinase, which phosphorylates both a Thr and a Tyr residue in MAPK. MAPK is sometimes called ERK (extracellular regulated kinase); MEK is mitogen-activated, ERK-activating kinase; SRF is serum response factor.

Biochemists now recognize the insulin pathway as but one instance of a more general theme in which hormone signals, via pathways similar to that shown in Figure 13–7, result in phosphorylation of target enzymes by protein kinases. The target of phosphorylation is often another protein kinase, which then phosphorylates a third protein kinase, and so on. The result of a catalyst activating a catalyst that activates a catalyst is a cascade of reactions that amplifies the initial signal by many orders of magnitude (see Fig. 13–1b).

Grb2 is not the only protein activated by association with phosphorylated IRS-1. PI-3 kinase (PI-3K) associates with IRS-1 through the latter's SH2 domain (Fig. 13–8). Thus activated, PI-3K converts the membrane lipid phosphatidylinositol 4,5-bisphosphate (see Fig. 11–13), also called PIP_2, to phosphatidylinositol 3,4,5-trisphosphate (PIP_3), which indirectly activates another kinase, protein kinase B (PKB). When bound to PIP_3, PKB is phosphorylated and activated by yet another protein kinase, PDK1. PKB then phosphorylates Ser or Thr residues on its target proteins, one of which is glycogen synthase kinase 3 (GSK3). In its active, nonphosphorylated form, GSK3 phosphorylates glycogen synthase, inactivating it and thereby slowing glycogen synthesis. When phosphorylated by PKB, GSK3 is inactivated. By thus preventing inactivation of glycogen synthase, the cascade of protein phosphorylations initiated by insulin stimulates glycogen synthesis (Fig. 13–8). PKB is also believed to trigger the movement of glucose transporters (GluT4) from internal vesicles to the plasma membrane, stimulating glucose uptake from the blood (see Box 12–2).

What spurred the evolution of such a complicated regulatory system? This system allows one activated receptor to activate several IRS-1 molecules, amplifying the insulin signal, and it provides for the integration of signals from several receptors, each of which can phosphorylate IRS-1. Furthermore, because IRS-1 can activate any of several proteins that contain SH2 domains, a single receptor acting through IRS-1 can trigger two or more signaling pathways; insulin affects gene expression through the Grb2-Sos-Ras-MAPK pathway and affects glycogen metabolism through the PI-3K–PKB pathway.

The insulin receptor is the prototype for a number of receptor enzymes that have a similar structure and have protein tyrosine kinase activity. The receptors for epidermal growth factor and platelet-derived growth factor, for example, show structural and sequence similarities with the insulin receptor, and both have a protein tyrosine kinase activity that phosphorylates IRS-1. Many of these receptors dimerize upon binding ligand; the insulin receptor is already a dimer before insulin binds.

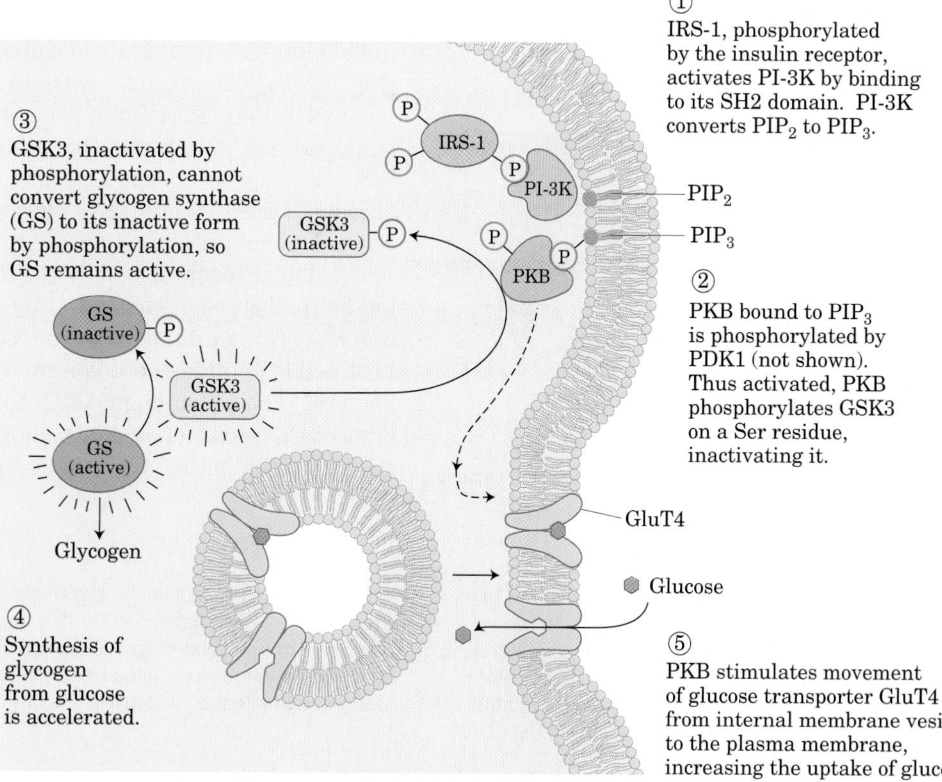
①
IRS-1, phosphorylated by the insulin receptor, activates PI-3K by binding to its SH2 domain. PI-3K converts PIP_2 to PIP_3.
P
IRS-1
P
P
PI-3K
PIP_2
PIP_3
③
GSK3, inactivated by phosphorylation, cannot convert glycogen synthase (GS) to its inactive form by phosphorylation, so GS remains active.
GSK3 (inactive)
P
P
PKB
P
②
PKB bound to PIP_3 is phosphorylated by PDK1 (not shown). Thus activated, PKB phosphorylates GSK3 on a Ser residue, inactivating it.
GS (inactive)
P
GSK3 (active)
GS (active)
Glycogen
GluT4
Glucose
④
Synthesis of glycogen from glucose is accelerated.
⑤
PKB stimulates movement of glucose transporter GluT4 from internal membrane vesicles to the plasma membrane, increasing the uptake of glucose.

figure 13–8
Activation of glycogen synthase by insulin. Transmission of the signal is mediated by PI-3 kinase (PI-3K) and protein kinase B (PKB).

In addition to the many receptors that act as protein tyrosine kinases, a number of receptorlike plasma membrane proteins have protein tyrosine phosphatase activity. Based on the structures of these proteins, we can surmise that their ligands are components of the extracellular matrix or the surfaces of other cells. Although their signaling roles are not yet as well understood as those of the receptor tyrosine kinases, they clearly have the potential to reverse the actions of signals that stimulate these kinases.

Guanylyl Cyclase Is a Receptor Enzyme That Generates the Second Messenger cGMP

Guanylyl cyclase (Fig. 13–9) is another type of receptor enzyme. When activated, it produces **guanosine 3′,5′-cyclic monophosphate (cyclic GMP or cGMP)** from GTP:

$$\text{GTP} \xrightarrow[\text{guanylyl cyclase}]{\text{PP}_i} \text{Guanosine 3}',5'\text{-cyclic monophosphate (cGMP)}$$

Cyclic GMP is a second messenger that carries different messages in different tissues. In the kidney and intestine it triggers changes in ion transport and water retention; in cardiac (smooth) muscle it signals relaxation; in the brain it may be involved in both development and adult brain function.

Guanylyl cyclase in the kidney is activated by the hormone **atrial natriuretic factor (ANF),** which is released by cells in the atrium of the heart when it is stretched by increased blood volume. Carried in the blood to the kidney, ANF activates guanylyl cyclase in cells of the collecting ducts (Fig. 13–9a). The resulting rise in cGMP concentration triggers increased renal excretion of Na^+ and, consequently, of water. Water loss reduces the blood volume, countering the stimulus that initially led to ANF secretion. Vascular smooth muscle also has an ANF receptor–guanylyl cyclase; on binding to this receptor, ANF causes vessel relaxation (vasodilation), which reduces blood pressure.

A similar receptor–guanylyl cyclase in the plasma membrane of intestinal epithelial cells is activated by an intestinal peptide, **guanylin,** which regulates Cl^- secretion in the intestine. This receptor is also the target of a heat-stable peptide endotoxin produced by *E. coli* and other gram-negative bacteria. The elevation in cGMP concentration caused by the endotoxin increases Cl^- secretion and consequently decreases reabsorption of water by the intestinal epithelium, producing diarrhea.

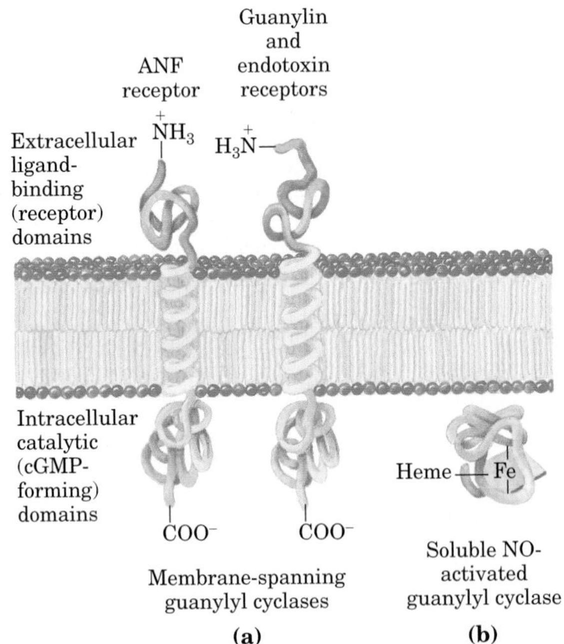

figure 13–9

Two types (isozymes) of guanylyl cyclase involved in signal transduction. (a) The first type exists in two similar membrane-spanning forms that are activated by their extracellular ligands: atrial natriuretic factor (ANF) (receptors in cells of the renal collecting ducts and the smooth muscle of blood vessels) and guanylin (receptors in intestinal epithelial cells). The guanylin receptor is also the target of a bacterial endotoxin that triggers severe diarrhea. **(b)** The second type is a soluble enzyme that is activated by intracellular nitric oxide (NO); this form is found in many tissues, including smooth muscle of the heart and blood vessels.

A distinctly different type of guanylyl cyclase is a cytosolic protein with a tightly associated heme group (Fig. 13–9b). This enzyme is activated by nitric oxide (NO). Nitric oxide is produced from arginine by Ca^{2+}-dependent **NO synthase,** present in many mammalian tissues, and diffuses from its cell of origin into nearby cells. NO is sufficiently nonpolar to cross plasma membranes without a carrier. In the target cell, it binds to the heme group of guanylyl cyclase and activates cGMP production. In the heart, cGMP brings about less forceful contractions by stimulating the ion pump(s) that expel Ca^{2+} from the cytosol.

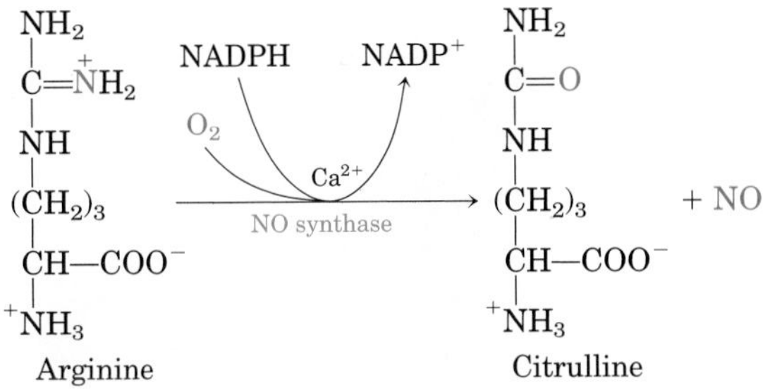

Arginine + NADPH + O_2 → (NO synthase, Ca^{2+}) → Citrulline + NO + $NADP^+$

This NO-induced relaxation of heart muscle is the same response brought about by nitroglycerin tablets and other nitrovasodilators taken to relieve angina, the pain caused by contraction of a heart deprived of O_2 because of blocked coronary arteries. Nitric oxide is unstable and its action is brief; within seconds of its formation, it undergoes oxidation to nitrite or nitrate. Nitrovasodilators produce long-lasting relaxation of cardiac muscle because they break down over several hours, yielding a steady stream of NO.

Most of the actions of cGMP are believed to be mediated by **cGMP-dependent protein kinase,** also called **protein kinase G** or **PKG,** which phosphorylates Ser and Thr residues in target proteins when activated by cGMP. The catalytic and regulatory domains of this enzyme are present in a single polypeptide (M_r ~80,000). Binding of cGMP forces a substratelike part of the regulatory domain out of the substrate-binding site, activating the catalytic domain.

Cyclic GMP has a second mode of action in the vertebrate eye: it causes ion-specific channels to open in the retinal rod and cone cells. We return to this role of cGMP in our discussion of vision.

G Protein–Coupled Receptors and Second Messengers

A third mechanism of signal transduction, distinct from gated ion channels and receptor enzymes, is defined by three essential components: a plasma membrane receptor with seven transmembrane segments, an enzyme in the plasma membrane that generates an intracellular second messenger, and a GTP-binding protein that dissociates from the occupied receptor and binds to the enzyme, activating it. The β-adrenergic receptor, which mediates the effects of epinephrine (Fig. 13–10) on many tissues, is the prototype for this third type of transducing system.

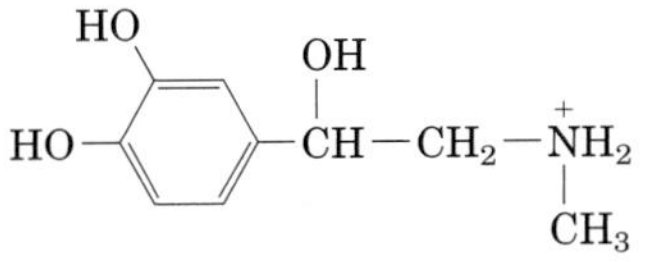

Epinephrine

figure 13–10

Epinephrine. This compound, also called adrenaline, is released from the adrenal gland and regulates energy-yielding metabolism in muscle, liver, and adipose tissue. It also serves as a neurotransmitter in adrenergic neurons.

The β-Adrenergic Receptor System Acts through the Second Messenger cAMP

Epinephrine action begins when the hormone binds to a protein receptor in the plasma membrane of a hormone-sensitive cell. **Adrenergic receptors** ("adrenergic" reflects the alternative name for epinephrine, adrenaline) are of four general types, defined by subtle differences in their affinities and responses to a group of agonists and antagonists. Agonists are structural analogs that bind to a receptor and mimic the effects of its natural ligand; antagonists are analogs that bind without triggering the normal effect and thereby block the effects of agonists. The four types of adrenergic receptors (α_1, α_2, β_1, β_2) are found in different target tissues and mediate different responses to epinephrine. Here we focus on the ***β*-adrenergic receptors** of muscle, liver, and adipose tissue. These receptors mediate changes in fuel metabolism, as described in Chapter 23, including the increased breakdown of glycogen and fat.

The β-adrenergic receptor is an integral protein with seven hydrophobic regions of 20 to 28 residues that "snake" back and forth seven times

Alfred G. Gilman

Martin Rodbell
1925–1998

across the plasma membrane. This protein is a member of a very large family of receptors, all with seven transmembrane helices, that are commonly called **serpentine receptors.** The binding of epinephrine to a site on the receptor deep within the membrane (Fig. 13–11, step ①) apparently promotes a conformational change in the intracellular domain of the receptor that affects its interaction with the second protein in the signal-transduction pathway, a heterotrimeric GTP-binding **stimulatory G protein,** or $\mathbf{G_s}$**,** on the cytosolic side of the plasma membrane. Alfred G. Gilman and Martin Rodbell discovered that when GTP is bound to G_s, it stimulates the production of cAMP by adenylyl cyclase (see p. 451) in the plasma membrane. The function of G_s as a molecular switch resembles that of another class of G proteins typified by Ras, discussed above in the context of the insulin receptor. Structurally, G_s and Ras are quite distinct; G proteins of the Ras type are monomers of about 20 kDa, whereas the G proteins that interact with serpentine receptors are trimers of three different subunits, α (M_r 43,000), β (M_r 37,000), and γ (M_r 7,500–10,000). However, Ras and α subunits of G proteins share many structural features.

When the nucleotide-binding site of G_s (on the α subunit) is occupied by GTP, G_s is active and can activate adenylyl cyclase; with GDP bound to the site, G_s is inactive. Binding of epinephrine enables the receptor to catalyze displacement of bound GDP by GTP, converting G_s to its active form (Fig. 13–11, step ②). As this occurs, the β and γ subunits of G_s dissociate from the α subunit, and $G_{s\alpha}$, with its bound GTP, moves in the plane of the membrane from the receptor to a nearby molecule of adenylyl cyclase (step ③). The $G_{s\alpha}$ is held to the membrane by a covalently attached lipid.

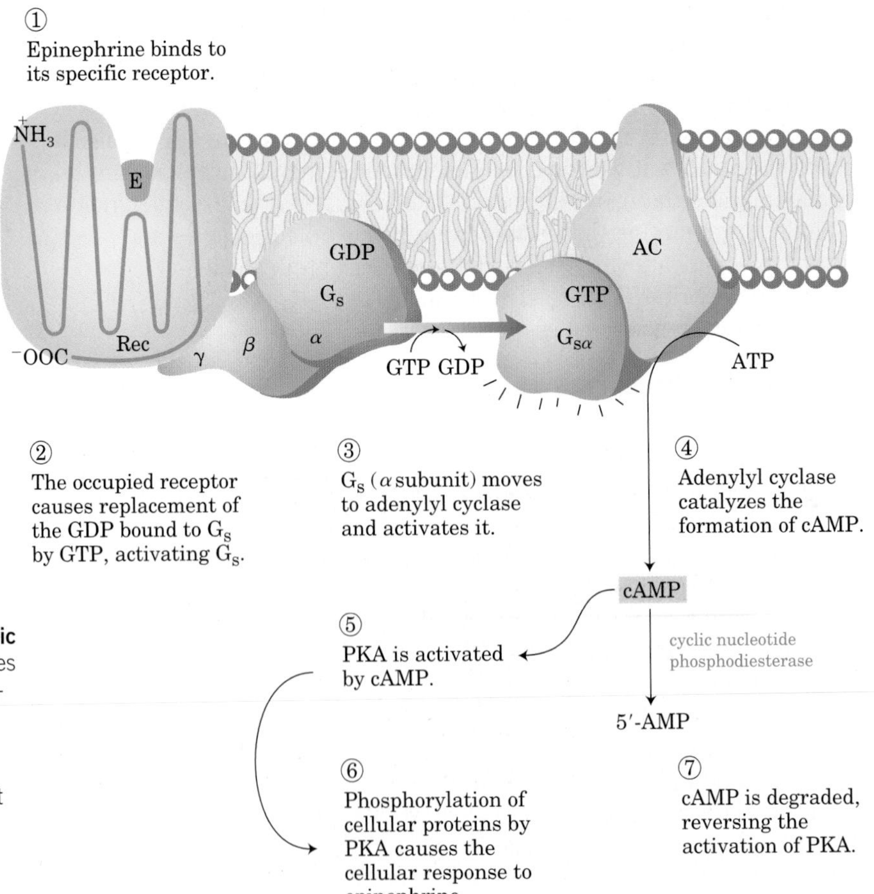
①
Epinephrine binds to its specific receptor.
$\overset{+}{N}H_3$
E
Rec
$^-$OOC
GDP
G_S
γ
β
α
GTP GDP
GTP
$G_{S\alpha}$
AC
ATP
②
The occupied receptor causes replacement of the GDP bound to G_S by GTP, activating G_S.
③
G_S (α subunit) moves to adenylyl cyclase and activates it.
④
Adenylyl cyclase catalyzes the formation of cAMP.
cAMP
⑤
PKA is activated by cAMP.
cyclic nucleotide phosphodiesterase
5′-AMP
⑥
Phosphorylation of cellular proteins by PKA causes the cellular response to epinephrine.
⑦
cAMP is degraded, reversing the activation of PKA.

figure 13–11

Transduction of the epinephrine signal: the β-adrenergic pathway. The seven steps of the mechanism that couples binding of epinephrine (E) to its receptor (Rec) with activation of adenylyl cyclase (AC) are discussed further in the text. The same adenylyl cyclase molecule in the plasma membrane may be regulated by a stimulatory G protein (G_s), as shown, or an inhibitory G protein (G_i, not shown). G_s and G_i are under the influence of different hormones. Hormones that induce GTP binding to G_i cause *inhibition* of adenylyl cyclase, resulting in lower cellular [cAMP].

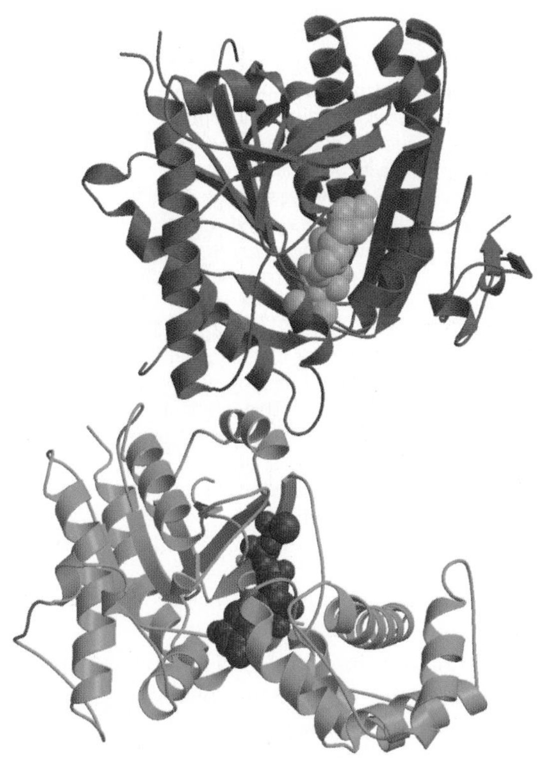

figure 13–12
Interaction of $G_{s\alpha}$ with adenylyl cyclase. The soluble catalytic core of the adenylyl cyclase (blue), severed from its membrane anchor, was cocrystallized with $G_{s\alpha}$ (green) to give this crystal structure. The plant terpene forskolin (yellow) is a drug that strongly stimulates the enzyme, and GTP (red) bound to $G_{s\alpha}$ triggers interaction of $G_{s\alpha}$ with adenylyl cyclase.

Adenylyl cyclase (Fig. 13–12) is an integral protein of the plasma membrane with its active site on the cytosolic face. It catalyzes the synthesis of cAMP from ATP:

$$\text{ATP} \xrightarrow[\text{adenylyl cyclase}]{PP_i} \text{Adenosine 3}',5'\text{-cyclic monophosphate (cAMP)}$$

The association of active $G_{s\alpha}$ with adenylyl cyclase stimulates the enzyme to catalyze cAMP synthesis (Fig. 13–11, step ④), raising the cytosolic [cAMP]. This stimulation by $G_{s\alpha}$ is self-limiting; $G_{s\alpha}$ is a GTPase that turns itself off by converting its bound GTP to GDP (Fig. 13–13). The now inactive $G_{s\alpha}$ dissociates from adenylyl cyclase, rendering the enzyme inactive. After $G_{s\alpha}$ reassociates with the β and γ subunits ($G_{s\beta\gamma}$), G_s is again available to interact with a hormone-bound receptor.

① G_s with GDP bound is turned off; it cannot activate adenylyl cyclase.

② Contact of G_s with hormone-receptor complex causes displacement of bound GDP by GTP.

③ G_s with GTP bound dissociates into α and $\beta\gamma$ subunits. $G_{s\alpha}$-GTP is turned on; it can activate adenylyl cyclase.

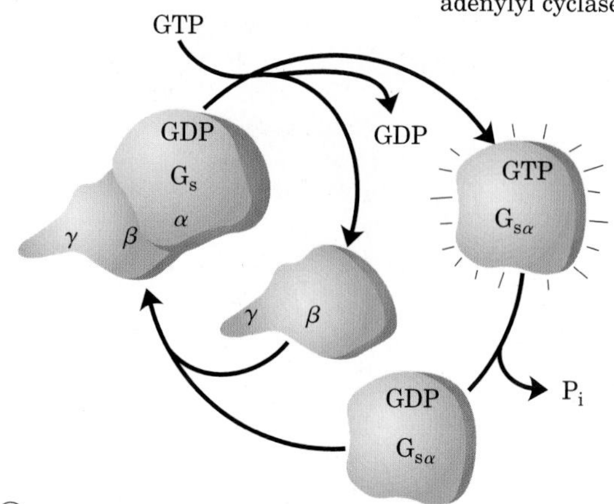

④ GTP bound to $G_{s\alpha}$ is hydrolyzed by the protein's intrinsic GTPase; $G_{s\alpha}$ thereby turns itself off. The inactive α subunit reassociates with the β, γ subunits.

figure 13–13
Self-inactivation of G_s. The steps are further described in the text. The protein's intrinsic GTPase activity, in many cases stimulated by RGS proteins (*r*egulators of *G* protein *s*ignaling), determines how quickly bound GTP is hydrolyzed to GDP, and therefore how long the G protein remains active.

Inactive PKA

Regulatory subunits: empty cAMP sites

Catalytic subunits: substrate-binding sites blocked by autoinhibitory domains of R subunits

R R

C C

4 cAMP 4 cAMP

Regulatory subunits: autoinhibitory domains buried

R R

+

Active PKA

Catalytic subunits: open substrate-binding sites

C C

(a)

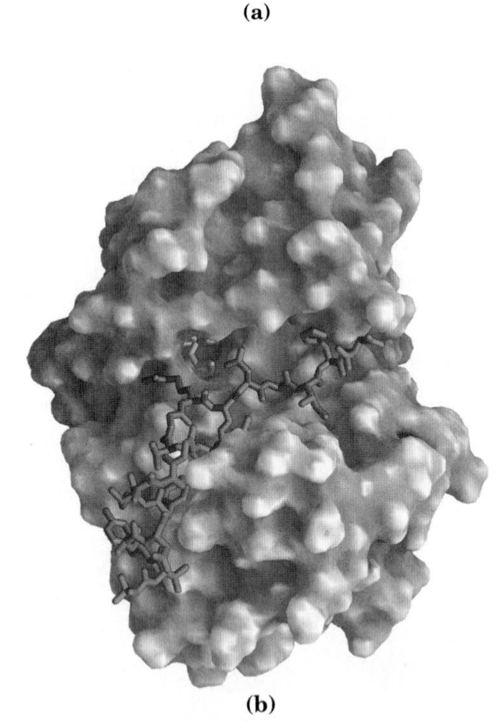

(b)

figure 13–14

Activation of cAMP-dependent protein kinase, PKA. **(a)** In the inactive R_2C_2 tetramer, the autoinhibitory domain of a regulatory (R) subunit occupies the substrate-binding site, inhibiting the activity of the catalytic (C) subunit. Cyclic AMP activates PKA by causing dissociation of the C subunits from the inhibitory R subunits. This allows PKA to phosphorylate a variety of protein substrates (Table 13–3) that contain the PKA consensus sequence (see Table 8–9), including phosphorylase *b* kinase. **(b)** A C subunit of PKA. A potent inhibitor peptide (PKI), which mimics the normal substrates for phosphorylation, is shown here occupying the substrate-binding site. PKI is in red and ATP at the active site is blue. This inhibitor contains the sequence Arg–Arg–Gln–Ala–Ile, which corresponds to the consensus sequence recognized by PKA, except that the Ser residue phosphorylated in a substrate is replaced by an Ala residue in the inhibitor.

One effect of epinephrine is to activate glycogen phosphorylase *b*. This conversion is promoted by the enzyme phosphorylase *b* kinase, which catalyzes the phosphorylation of two specific Ser residues in phosphorylase *b*, converting it to phosphorylase *a* (see Fig. 8–28). Cyclic AMP does not affect phosphorylase *b* kinase directly. Rather, **cAMP-dependent protein kinase,** also called **protein kinase A** or **PKA,** which is allosterically activated by cAMP, catalyzes the phosphorylation of inactive phosphorylase *b* kinase to yield the active form (steps ⑤ and ⑥ in Fig. 13–11).

The inactive form of PKA contains two catalytic subunits (C) and two regulatory subunits (R) (Fig. 13–14a), which are similar in sequence to the catalytic and regulatory domains of PKG (cGMP-dependent protein kinase). The tetrameric R_2C_2 complex is catalytically inactive, because an autoinhibitory domain of each R subunit occupies the substrate-binding site of each C subunit. When cAMP binds to two sites on each R subunit, the R subunits undergo a conformational change and the R_2C_2 complex dissociates to yield two free, catalytically active C subunits. This same basic mechanism—displacement of an autoinhibitory domain—mediates the allosteric activation of many types of protein kinases by their second messengers.

Signal transduction by adenylyl cyclase involves several steps that amplify the original hormone signal (Fig. 13–15). First, the binding of one hormone molecule to one receptor catalytically activates several G_s molecules. Next, by activating a molecule of adenylyl cyclase, each active $G_{s\alpha}$ molecule stimulates the catalytic synthesis of many molecules of cAMP. The second messenger cAMP now activates PKA, each molecule of which catalyzes the phosphorylation of many molecules of the target protein (phosphorylase *b* kinase). This kinase activates glycogen phosphorylase *b*, which leads to the rapid mobilization of glucose from glycogen. The net effect of the cascade is amplification of the hormonal signal by several orders of magnitude, which accounts for the very low concentration of epinephrine (or any other hormone) required for hormone activity.

PKA regulates a number of enzymes in addition to phosphorylase *b* kinase (Table 13–3). Although the proteins regulated by cAMP-dependent phosphorylation have diverse functions, they share a region of sequence similarity around the Ser or Thr residue that undergoes phosphorylation, a sequence that marks them for regulation by PKA. Comparison of the sequences of a number of protein substrates for PKA yields the consensus sequence—the minimum of specific neighboring residues needed to mark a Ser or Thr for phosphorylation (see Table 8–9).

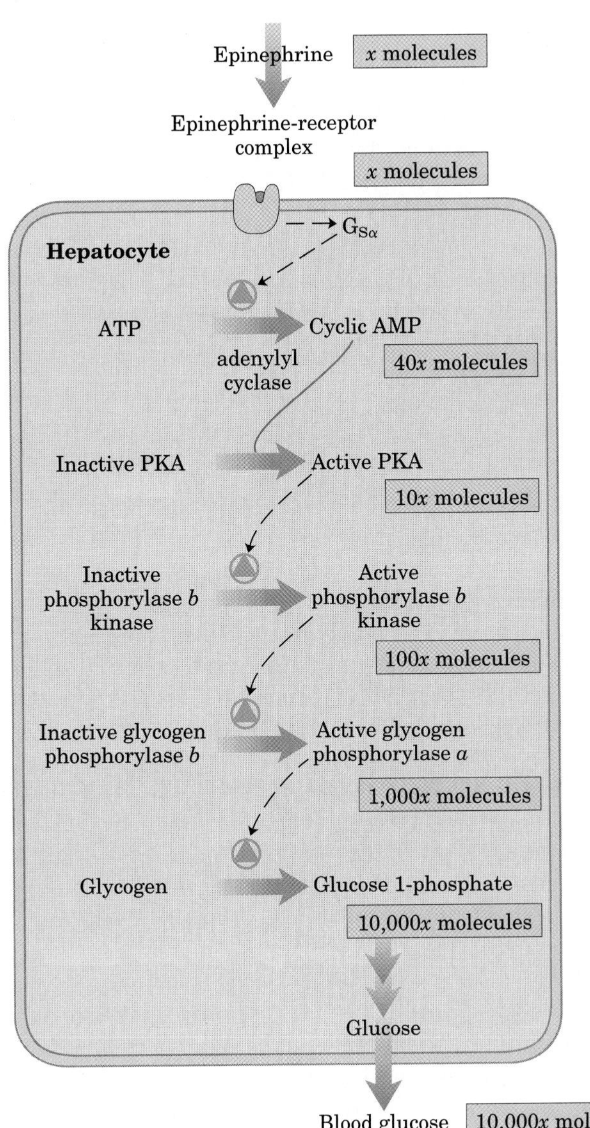

figure 13–15

Epinephrine cascade. Epinephrine triggers a series of reactions in hepatocytes in which catalysts activate catalysts, resulting in great amplification of the signal. Binding of a small number of molecules of epinephrine to specific β-adrenergic receptors on the cell surface activates adenylyl cyclase. To illustrate amplification, we show 40 molecules of cAMP produced by each molecule of adenylyl cyclase, the 40 cAMP molecules activating 10 molecules of PKA, each PKA molecule activating 10 molecules of the next enzyme, and so forth. These amplifications are probably gross underestimates.

table 13–3

Some Enzymes Regulated by cAMP-Dependent Phosphorylation (by PKA)

Enzyme	Sequence phosphorylated*	Pathway
Glycogen synthase	RASCTSSS	Glycogen synthesis
Phosphorylase *b* kinase	α subunit: VEFRRLSI β subunit: RTKRSGSV	Glycogen breakdown
Pyruvate kinase (rat liver)	GVLRRASVAZL	Glycolysis
Pyruvate dehydrogenase complex (type L)	GYLRRASV	Pyruvate to acetyl-CoA
Hormone-sensitive lipase	PMRRSV	Triacylglycerol mobilization and fatty acid oxidation
Phosphofructokinase-2/fructose 2,6-bisphosphatase	LQRRRGSSIPQ	Glycolysis/gluconeogenesis
Tyrosine hydroxylase	FIGRRQSL	Synthesis of L-DOPA, dopamine, norepinephrine, and epinephrine
Histone H1	AKRKASGPPVS	DNA condensation
Histone H2B	KKAKASRKESYSVYVYK	DNA condensation
Cardiac phospholamban (a cardiac pump regulator)	AIRRAST	Regulation of intracellular [Ca^{2+}]
Protein phosphatase-1 inhibitor-1	IRRRRPTP	Regulation of protein dephosphorylation
CREB	ILSRRPSY	cAMP regulation of gene expression
PKA consensus sequence†	XR(R/K)X(S/T)B	

*The phosphorylated S or T residue is shown in red. All residues are given as their one-letter abbreviations (see Table 5–1).

†X is any amino acid; B is any hydrophobic amino acid.

Cyclic AMP, the intracellular second messenger in this system, is short-lived. It is quickly degraded by **cyclic nucleotide phosphodiesterase** to 5′-AMP (Fig. 13–11, step ⑦), which is not active as a second messenger:

Cyclic AMP —(H_2O; cyclic nucleotide phosphodiesterase; ⊗ ← caffeine, theophylline)→ Adenosine 5′-monophosphate (AMP)

The intracellular signal therefore persists only as long as the hormone receptor remains occupied by epinephrine. Methyl xanthines such as caffeine and theophylline (a component of tea) inhibit the phosphodiesterase, increasing the half-life of cAMP and thereby potentiating agents that act by stimulating adenylyl cyclase.

The β-Adrenergic Receptor Is Desensitized by Phosphorylation

As noted earlier, signal-transducing systems undergo desensitization when the signal persists. Desensitization of the β-adrenergic receptor is mediated by a protein kinase that phosphorylates the receptor on the (intracellular) domain that normally interacts with G_s (Fig. 13–16). When the receptor is occupied by epinephrine, **β-adrenergic receptor kinase (βARK)** phosphorylates Ser residues near the carboxyl terminus of the receptor. Normally located in the cytosol, βARK is drawn to the plasma membrane by its association with the $G_{s\beta\gamma}$ subunits and is thus positioned to phosphorylate the receptor. The phosphorylation creates a binding site for the protein **β-arrestin,** and binding of arrestin effectively prevents interaction between the receptor and the G protein. The binding of arrestin also facilitates receptor sequestration, the removal of receptors from the plasma membrane by endocytosis into small intracellular vesicles. Receptors in the endocytic vesicles are dephosphorylated, then returned to the plasma membrane, completing the circuit and resensitizing the system to epinephrine. β-Adrenergic receptor kinase is a member of a family of **G protein–coupled receptor kinases (GRKs),** all of which phosphorylate serpentine receptors on their carboxyl-terminal cytosolic domains and play roles similar to that of βARK in desensitization and resensitization of their receptors.

table 13–4

Some Signals That Use cAMP as Second Messenger
Corticotropin (ACTH)
Corticotropin-releasing hormone (CRH)
Dopamine [D-1, D-2]*
Epinephrine (β-adrenergic)
Follicle-stimulating hormone (FSH)
Glucagon
Histamine [H-2]*
Luteinizing hormone (LH)
Melanocyte-stimulating hormone (MSH)
Odorants (many)
Parathyroid hormone
Prostaglandins E_1, E_2 (PGE_1, PGE_2)
Serotonin [5-HT-1α, 5-HT-2]*
Somatostatin
Tastants (sweet, bitter)
Thyroid-stimulating hormone (TSH)

*Some signals have two or more receptor subtypes (shown in square brackets), which may have different transduction mechanisms. For example, serotonin is detected in some tissues by receptor subtypes 5-HT-1a and 5-HT-1b, which act through adenylyl cyclase and cAMP, and in other tissues by receptor subtype 5-HT-1c, acting through the phospholipase C–IP_3 mechanism (see Table 13–5).

Cyclic AMP Acts as a Second Messenger for a Number of Regulatory Molecules

Epinephrine is only one of many hormones, growth factors, and other regulatory molecules that act by changing the intracellular [cAMP] and thus the activity of PKA (Table 13–4). For example, glucagon binds to its receptors in the plasma membrane of adipocytes, activating (via a G_s protein) adenylyl cyclase. PKA, stimulated by the resulting rise in [cAMP], phosphorylates and activates triacylglycerol lipase, leading to the mobilization of fatty acids. Similarly, the peptide hormone ACTH (adrenocorticotropic hormone, also called corticotropin), produced by the anterior pituitary, binds to specific receptors in the adrenal cortex, activating adenylyl cyclase and raising the intracellular [cAMP]. PKA then phosphorylates and activates several of the enzymes required for the synthesis of cortisol and other steroid hormones.

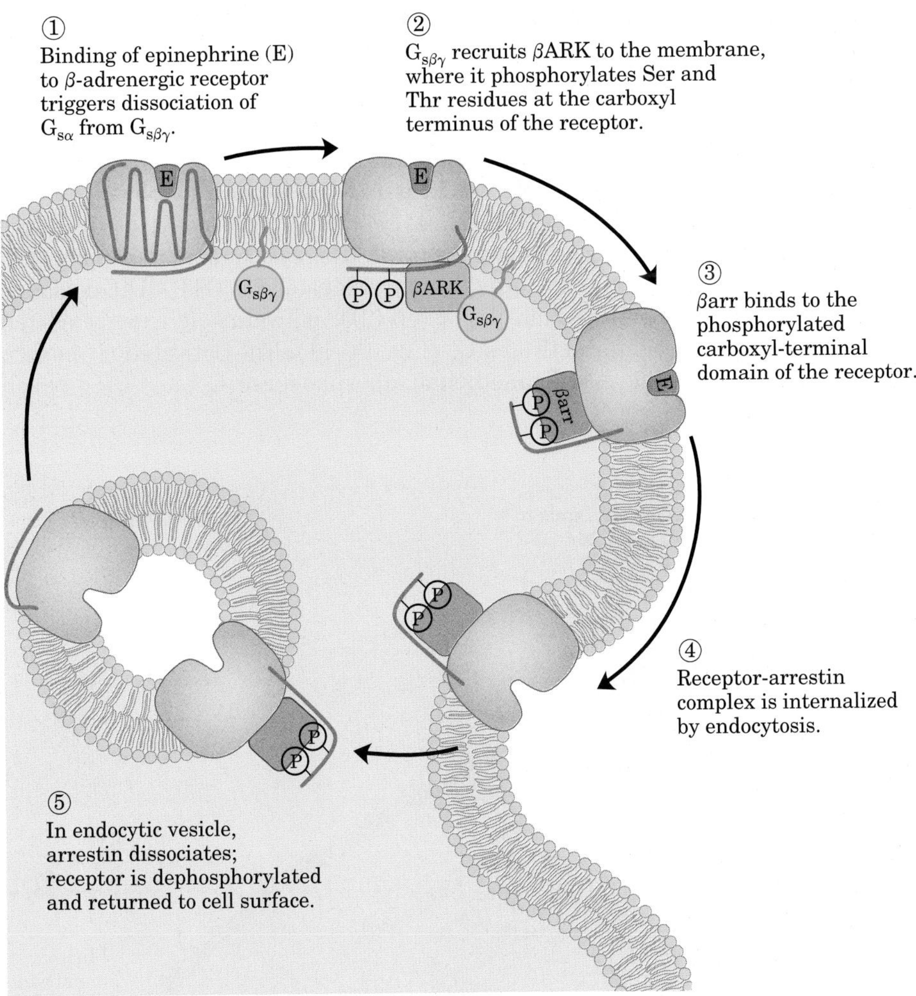

①
Binding of epinephrine (E) to β-adrenergic receptor triggers dissociation of $G_{s\alpha}$ from $G_{s\beta\gamma}$.
②
$G_{s\beta\gamma}$ recruits βARK to the membrane, where it phosphorylates Ser and Thr residues at the carboxyl terminus of the receptor.
③
βarr binds to the phosphorylated carboxyl-terminal domain of the receptor.
④
Receptor-arrestin complex is internalized by endocytosis.
⑤
In endocytic vesicle, arrestin dissociates; receptor is dephosphorylated and returned to cell surface.
E
$G_{s\beta\gamma}$
P
βARK
βarr

figure 13–16
Desensitization of the β-adrenergic receptor in the continued presence of epinephrine. This process is mediated by two proteins: β-adrenergic protein kinase (βARK) and β-arrestin (βarr).

Some hormones act by *inhibiting* adenylyl cyclase, *lowering* cAMP levels, and *suppressing* protein phosphorylation. For example, the binding of somatostatin to its receptor leads to activation of an **inhibitory G protein,** or $\mathbf{G_i}$**,** structurally homologous to G_s, that inhibits adenylyl cyclase and lowers [cAMP]. Somatostatin therefore counterbalances the effects of glucagon. In adipose tissue, prostaglandin E_1 (PGE_1; see Fig. 11–16b) inhibits adenylyl cyclase, thereby lowering [cAMP] and slowing the mobilization of lipid reserves triggered by epinephrine and glucagon. In certain other tissues, PGE_1 stimulates cAMP synthesis because its receptors are coupled to adenylyl cyclase through a stimulatory G protein, G_s. In tissues with α_2-adrenergic receptors, epinephrine lowers [cAMP] because the α_2 receptors are coupled to adenylyl cyclase through an inhibitory G protein, G_i. In short, an extracellular signal such as epinephrine or PGE_1 can have quite different effects on different tissues or cell types, depending on (1) the type of receptor, (2) the type of G protein (G_s or G_i) with which the receptor is coupled, and (3) the set of PKA target enzymes in the cell.

Inositol 1,4,5-trisphosphate

Two Second Messengers Are Derived from Phosphatidylinositols

A second class of serpentine receptors are coupled through a G protein to a plasma membrane **phospholipase C** that is specific for the plasma membrane lipid phosphatidylinositol 4,5-bisphosphate (see Fig. 11–13). This hormone-sensitive enzyme catalyzes the formation of two potent second messengers: **diacylglycerol** and **inositol 1,4,5-trisphosphate,** or $\mathbf{IP_3}$ (not to be confused with PIP_3 (p. 447)). When a hormone of this class (Table 13–5) binds its specific receptor in the plasma membrane (Fig. 13–17, step ①), the receptor-hormone complex catalyzes GTP-GDP exchange on an associated G protein, $\mathbf{G_q}$ (step ②), activating it exactly as the β-adrenergic receptor activates G_s (Fig. 13–11). The activated G_q in turn activates a specific membrane-bound phospholipase C (step ③), which

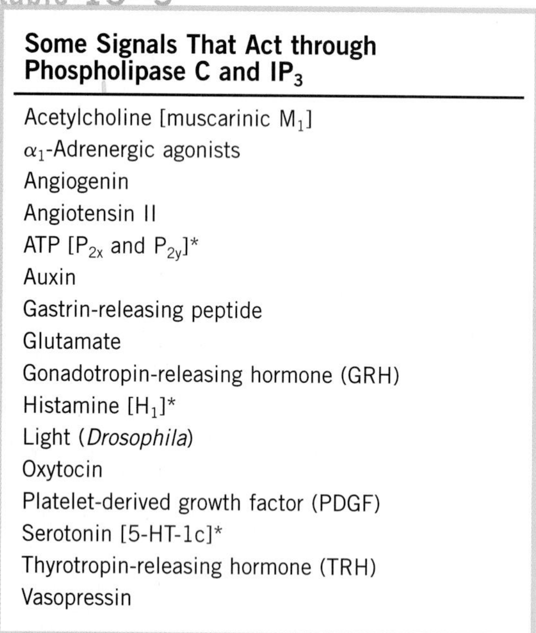

table 13–5

Some Signals That Act through Phospholipase C and IP_3

Acetylcholine [muscarinic M_1]
α_1-Adrenergic agonists
Angiogenin
Angiotensin II
ATP [P_{2x} and P_{2y}]*
Auxin
Gastrin-releasing peptide
Glutamate
Gonadotropin-releasing hormone (GRH)
Histamine [H_1]*
Light (*Drosophila*)
Oxytocin
Platelet-derived growth factor (PDGF)
Serotonin [5-HT-1c]*
Thyrotropin-releasing hormone (TRH)
Vasopressin

*Receptor subtypes are in square brackets; see footnote to Table 13–4.

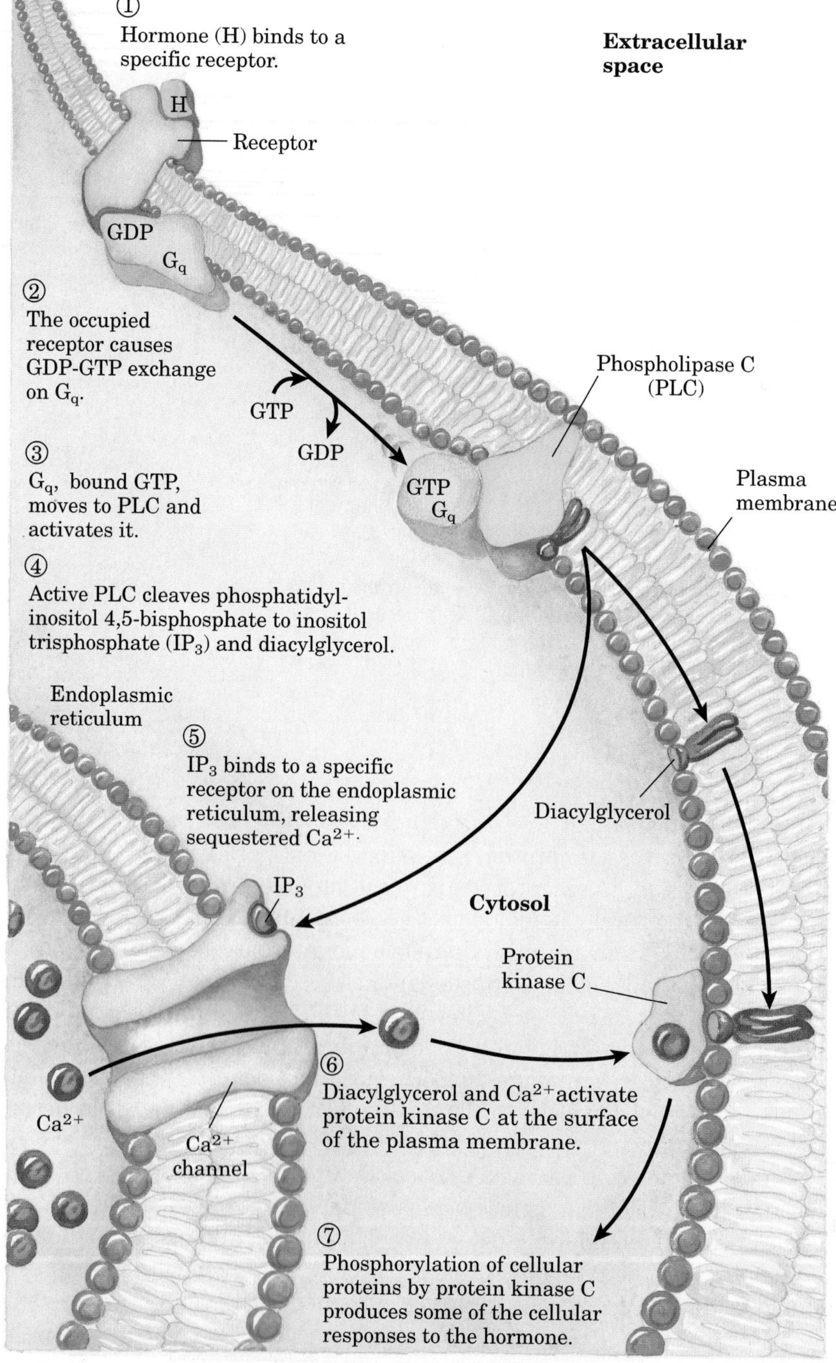
① Hormone (H) binds to a specific receptor.
Extracellular space
H
Receptor
GDP
G_q
② The occupied receptor causes GDP-GTP exchange on G_q.
GTP
GDP
Phospholipase C (PLC)
③ G_q, bound GTP, moves to PLC and activates it.
GTP
G_q
Plasma membrane
④ Active PLC cleaves phosphatidylinositol 4,5-bisphosphate to inositol trisphosphate (IP_3) and diacylglycerol.
Endoplasmic reticulum
⑤ IP_3 binds to a specific receptor on the endoplasmic reticulum, releasing sequestered Ca^{2+}.
Diacylglycerol
IP_3
Cytosol
Protein kinase C
Ca^{2+}
Ca^{2+} channel
⑥ Diacylglycerol and Ca^{2+} activate protein kinase C at the surface of the plasma membrane.
⑦ Phosphorylation of cellular proteins by protein kinase C produces some of the cellular responses to the hormone.

figure 13–17

Hormone-activated phospholipase C and IP_3. Two intracellular second messengers are produced in the hormone-sensitive phosphatidylinositol system: inositol 1,4,5-trisphosphate (IP_3) and diacylglycerol. Both contribute to the activation of protein kinase C. By raising cytosolic [Ca^{2+}], IP_3 also activates other Ca^{2+}-dependent enzymes; thus Ca^{2+} also acts as a second messenger.

catalyzes the production of two second messengers by hydrolysis of phosphatidylinositol 4,5-bisphosphate in the plasma membrane (step ④).

Inositol trisphosphate (IP_3), the water-soluble product of phospholipase C action, diffuses from the plasma membrane to the endoplasmic reticulum, where it binds to specific IP_3 receptors and causes Ca^{2+} channels within the ER to open. Sequestered Ca^{2+} is thus released into the cytosol (step ⑤). The cytosolic [Ca^{2+}] rises sharply to about 10^{-6} M. One effect of elevated [Ca^{2+}] is the activation of **protein kinase C (PKC).** Diacylglycerol cooperates with Ca^{2+} in activating PKC and is therefore also a second messenger (step ⑥). PKC phosphorylates Ser or Thr residues of specific target proteins, changing their catalytic activities (step ⑦). There are a number of isozymes of PKC, each with a characteristic tissue distribution, target protein specificity, and tissue-specific role.

The action of a group of compounds known as **tumor promoters** is attributable to their effects on PKC. The best understood of these, the phorbol esters, are synthetic compounds that are potent activators of PKC. They apparently mimic cellular diacylglycerol as second messengers, but unlike naturally occurring diacylglycerols they are not rapidly metabolized. By continuously activating PKC, these synthetic tumor promoters interfere with the normal regulation of cell growth and division (discussed later in this chapter).

Myristoylphorbol acetate (a phorbol ester)

Calcium Is a Second Messenger in Many Signal Transductions

In many cells that respond to extracellular signals, Ca^{2+} serves as a second messenger that triggers intracellular responses, such as exocytosis in neurons and endocrine cells, contraction in muscle, or cytoskeletal rearrangement during amoeboid movement. Normally, cytosolic [Ca^{2+}] is kept very low ($<10^{-7}$ M) by the action of Ca^{2+} pumps in the ER, mitochondria, and plasma membrane. Hormonal, neural, or other stimuli cause either an influx of Ca^{2+} into the cell through specific Ca^{2+} channels in the plasma membrane or release of sequestered Ca^{2+} from the ER or mitochondria, in either case raising the cytosolic [Ca^{2+}] and triggering a cellular response.

Very commonly, [Ca^{2+}] does not simply rise and then decay, but rather oscillates with a period of a few seconds (Fig. 13–18), even when the extracellular concentration of hormone remains constant. The mechanism underlying [Ca^{2+}] oscillations presumably involves feedback regulation by Ca^{2+} of either the phospholipase that generates IP_3 or the ion channel that regulates Ca^{2+} release from the ER, or both. Whatever the mechanism, the effect is that one kind of signal (hormone concentration, for example) is converted into another (frequency and amplitude of intracellular [Ca^{2+}] "spikes").

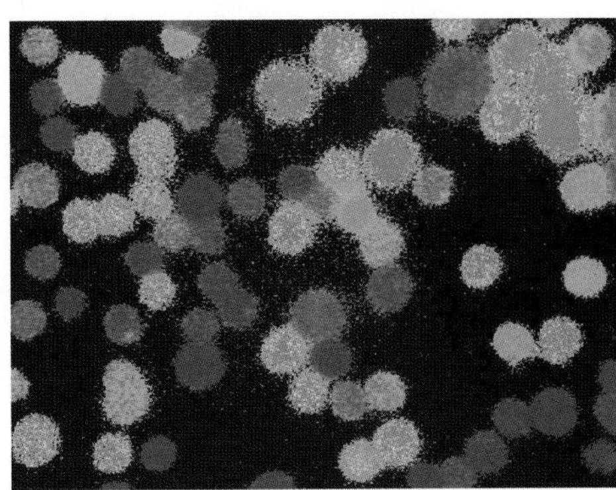

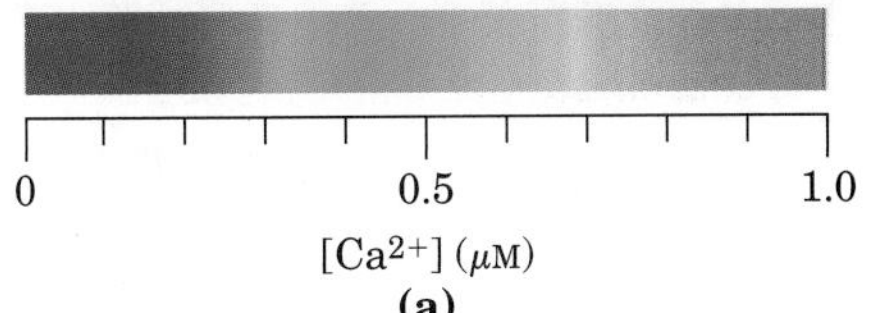

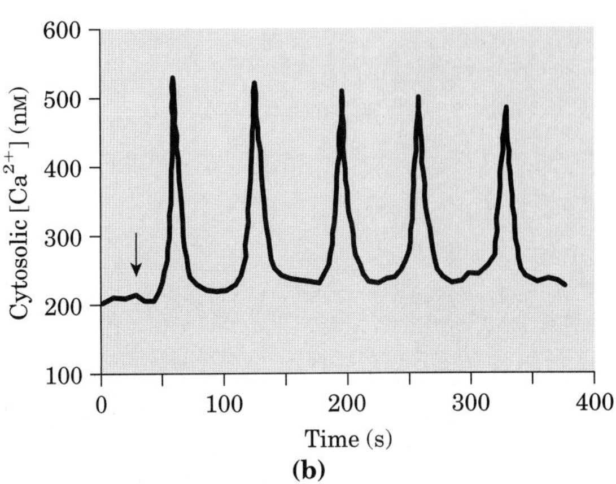

figure 13–18

Triggering of oscillations in intracellular [Ca^{2+}] by extracellular signals. (a) A dye (fura) whose fluorescence changes when it binds Ca^{2+} is allowed to diffuse into cells, and its instantaneous light output is measured by fluorescence microscopy. Fluorescence intensity is represented by color; the color scale relates intensity of color to [Ca^{2+}], allowing determination of the absolute [Ca^{2+}]. In this case, thymocytes have been stimulated with extracellular ATP, which raises their internal [Ca^{2+}]. The cells are heterogeneous in their responses; some have high intracellular [Ca^{2+}] (red), others much lower (blue). **(b)** When such a probe is used to measure [Ca^{2+}] in a single hepatocyte, we observe that the agonist norepinephrine (added at the arrow) causes oscillations of [Ca^{2+}] from 200 to 500 nM. Similar oscillations are induced in other cell types by other extracellular signals.

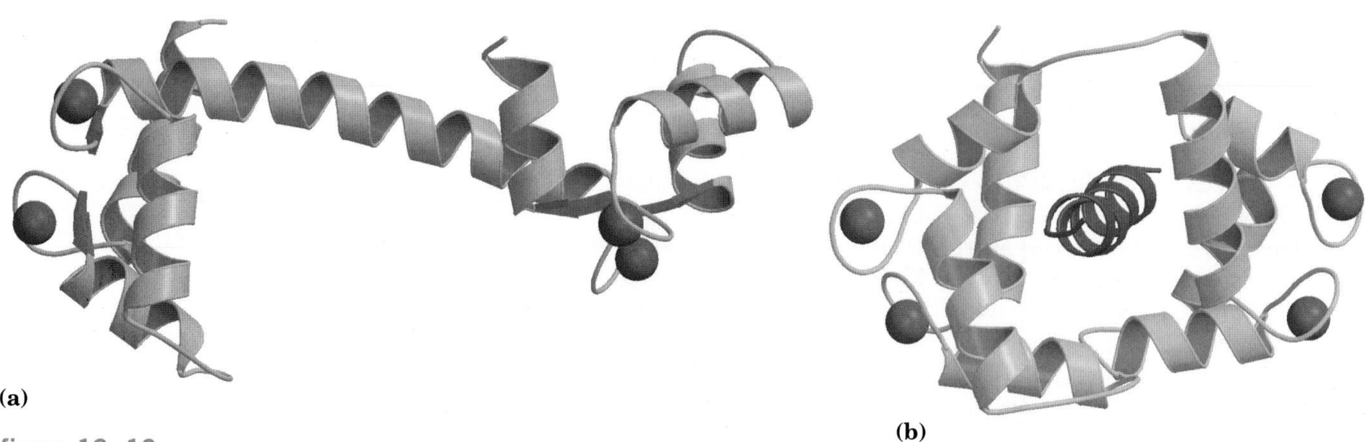

figure 13–19
Calmodulin. The protein mediator of many Ca^{2+}-stimulated enzymatic reactions, calmodulin has four high-affinity Ca^{2+}-binding sites (K_d of about 0.1 to 1 μM). **(a)** A ribbon model of the crystal structure of calmodulin. The four Ca^{2+}-binding sites are occupied by Ca^{2+} (purple). The amino-terminal domain is on the left; the carboxyl-terminal domain on the right. **(b)** Calmodulin associated with a helical domain (red) of one of the many enzymes that it regulates, calmodulin-dependent protein kinase II. Notice that the long central α helix visible in **(a)** has bent back on itself in binding to the helical substrate domain.

Changes in intracellular [Ca^{2+}] are detected by calcium-binding proteins that regulate a variety of Ca^{2+}-dependent enzymes. **Calmodulin (CaM;** M_r 17,000) is an acidic protein with four high-affinity Ca^{2+}-binding sites. When intracellular [Ca^{2+}] rises to about 10^{-6} M (1 μM), the binding of Ca^{2+} to calmodulin drives a conformational change (Fig. 13–19). Calmodulin associates with a variety of proteins and, in its Ca^{2+}-bound state, modulates their activities. Calmodulin is a member of a family of Ca^{2+}-binding proteins that also includes troponin (p. 238), which triggers skeletal muscle contraction in response to increased [Ca^{2+}].

Calmodulin is an integral subunit of **Ca^{2+}/calmodulin-dependent protein kinase (CaM kinase).** When intracellular [Ca^{2+}] increases in response to some stimulus, calmodulin binds Ca^{2+}, undergoes a change in conformation, and activates CaM kinase. The kinase then phosphorylates a number of target enzymes, regulating their activities. Calmodulin is also a regulatory subunit of phosphorylase *b* kinase of muscle, which is activated by Ca^{2+}. Thus Ca^{2+} triggers ATP-requiring muscle contractions while also activating glycogen breakdown, providing fuel for ATP synthesis. Many other enzymes are also known to be modulated by Ca^{2+} through calmodulin (Table 13–6).

table 13–6

Some Proteins Regulated by Ca^{2+} and Calmodulin
Adenylyl cyclase (brain)
Ca^{2+}/calmodulin-dependent protein kinases
Ca^{2+}-dependent Na^+ channel (*Paramecium*)
Ca^{2+} release channel of sarcoplasmic reticulum
Calcineurin (phosphoprotein phosphatase 2B)
cAMP phosphodiesterase
cAMP-gated olfactory channel
cGMP-gated Na^+, Ca^{2+} channels (rod and cone cells)
Myosin light chain kinases
NADH kinase
Nitric oxide synthase
PI-3 kinase
Plasma membrane Ca^{2+} ATPase (Ca^{2+} pump)
RNA helicase (p68)

Sensory Transduction in Vision, Olfaction, and Gustation

The detection of light, smells, and tastes (vision, olfaction, and gustation, respectively) in animals is accomplished by specialized sensory neurons that use signal-transduction mechanisms fundamentally similar to those used to detect hormones, neurotransmitters, and growth factors. An initial sensory signal is amplified greatly by mechanisms that involve gated ion channels and intracellular second messengers; the system adapts to continued stimulation by changing its sensitivity to the stimulus (desensitization); and sensory input from several receptors is integrated before the final signal goes to the brain.

Light Hyperpolarizes Rod and Cone Cells of the Vertebrate Eye

In the vertebrate eye, a beam of light entering through the pupil is focused on a highly organized collection of light-sensitive neurons (Fig. 13–20). The light-sensing cells are of two types: **rods** (about 10^9 per retina), which sense low levels of light but cannot discriminate colors, and **cones** (about 3×10^6 per retina), which are less sensitive but can discriminate colors. Both cell types are long, narrow, specialized sensory neurons with two distinct cellular compartments: an **outer segment,** which contains dozens of

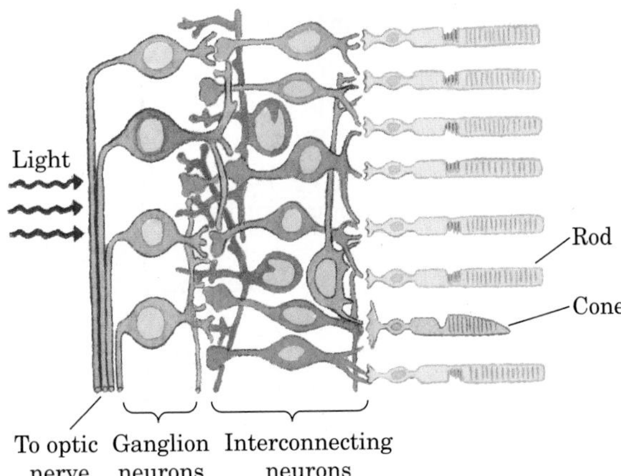

figure 13–20

Light reception in the vertebrate eye. The lens of the eye focuses light on the retina, which is composed of layers of neurons. The primary photosensory neurons are rod cells (orange), which are responsible for high-resolution and night vision, and cone cells of three subtypes (red), which initiate color vision. The rods and cones form synapses with several ranks of interconnecting neurons that convey and integrate the electrical signals. The signals eventually pass from ganglion neurons through the optic nerve to the brain.

membranous disks loaded with the membrane protein rhodopsin, and an **inner segment** containing the nucleus and many mitochondria, which produce the ATP essential to phototransduction.

Like other neurons, rods and cones have a transmembrane electrical potential (V_m), produced by the electrogenic pumping of the Na^+K^+ ATPase in the plasma membrane of the inner segment (Fig. 13–21). Also contributing to the membrane potential is an ion channel in the outer segment that permits passage of either Na^+ or Ca^{2+} and is gated (opened) by cGMP. In the dark, rod cells contain enough cGMP to keep this channel open. The membrane potential is therefore determined by the net difference between the Na^+ and K^+ pumped by the inner segment (which polarizes the membrane) and the influx of Na^+ through the ion channels of the outer segment (which tends to depolarize the membrane).

The essence of signaling in the retinal rod or cone cell is a light-induced decrease in the concentration of cGMP, which causes the cGMP-gated ion channel to close. The plasma membrane then becomes hyperpolarized by the Na^+K^+ ATPase. Rod and cone cells synapse with interconnecting neurons (Fig. 13–20) that carry information about the electrical activity to the ganglion neurons near the inner surface of the retina. The ganglion neurons integrate the output from many rod or cone cells and send the resulting signal through the optic nerve to the visual cortex of the brain.

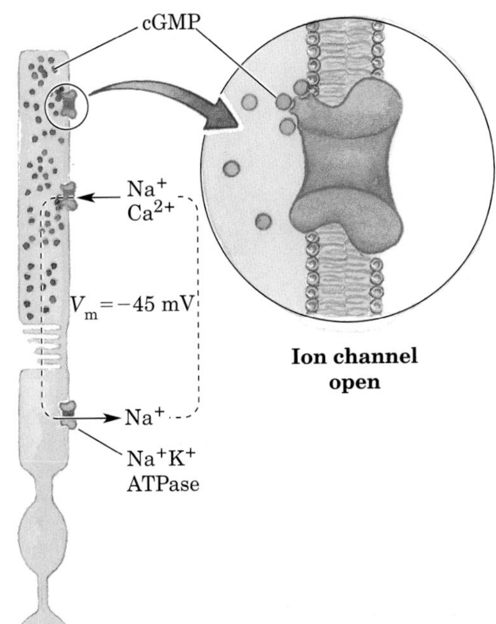

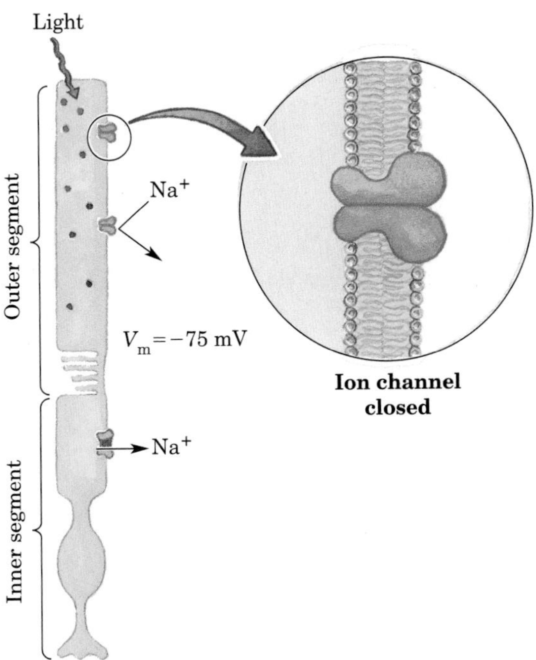

figure 13–21

Light-induced hyperpolarization of rod cells. The rod cell consists of an outer segment, filled with stacks of membranous disks (not shown) containing the photoreceptor rhodopsin, and an inner segment that contains the nucleus and other organelles. Cones have a similar structure. ATP in the inner segment powers the Na^+K^+ ATPase, which creates a transmembrane electrical potential by pumping 3 Na^+ out for every 2 K^+pumped in. The membrane potential is reduced by the flow of Na^+ and Ca^{2+} back into the cell through cGMP-gated cation channels in the plasma membrane of the outer segment. When rhodopsin absorbs light, it triggers degradation of cGMP (green dots) in the outer segment, causing closure of the cation channel. Without cation influx through this channel, the cell becomes hyperpolarized. This electrical signal is passed to the brain through the ranks of neurons shown in Figure 13–20.

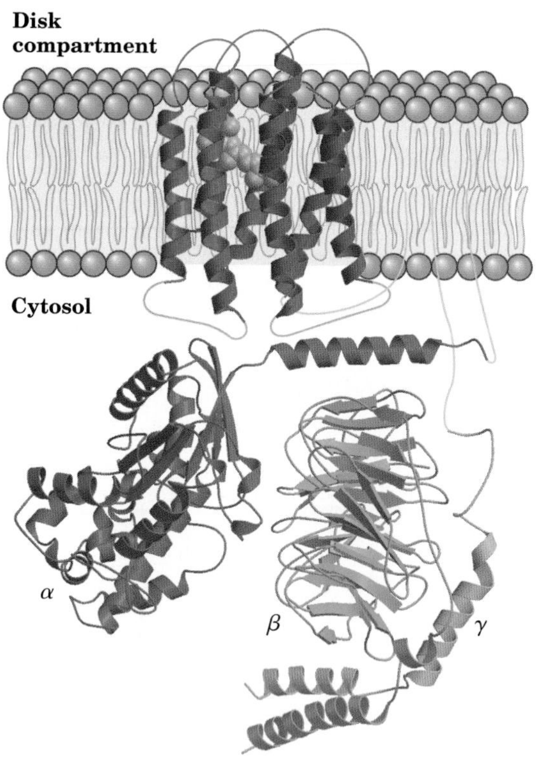

figure 13–22
Likely structure of rhodopsin complexed with the G protein transducin. Rhodopsin (red) has seven transmembrane helices embedded in the disk membranes of rod outer segments and is oriented with its carboxyl terminus on the cytosolic side, its amino terminus inside the disk. The chromophore 11-*cis* retinal (blue), attached through a Schiff base linkage to Lys^{256} on the seventh helix, lies near the center of the bilayer. (This location is similar to that of the epinephrine-binding site in the β-adrenergic receptor.) A number of Ser and Thr residues near the carboxyl terminus are substrates for phosphorylations that are part of the desensitization mechanism for rhodopsin. Cytosolic loops that interact with the G protein transducin are shown in orange; their exact positions are not yet known. The three subunits of transducin (green) are shown in their likely arrangement. Rhodopsin is palmitoylated at its carboxyl terminus, and both the α and γ subunits of transducin have attached lipids that assist in anchoring them to the membrane; these hydrophobic tails are in yellow.

Light Triggers Conformational Changes in the Receptor Rhodopsin

Visual transduction begins when light falls on rhodopsin, many thousands of molecules of which are present in each of the disks of the outer segment. **Rhodopsin** (M_r 40,000) is an integral protein with seven membrane-spanning α helices (Fig. 13–22), the characteristic serpentine architecture. The amino-terminal domain projects into the disk, and the carboxyl-terminal domain faces the cytosol of the outer segment. The light-absorbing pigment (chromophore) **11-*cis*-retinal** is covalently attached to **opsin,** the protein component of rhodopsin, through a Schiff base to a Lys residue. The retinal lies near the middle of the bilayer (Fig. 13–22), oriented with its long axis approximately in the plane of the membrane. When a photon is absorbed by the retinal component of rhodopsin, the energy causes a photochemical change; 11-*cis*-retinal is converted to **all-*trans*-retinal** (see Figs. 3–8b and 11–19). This change in the structure of the chromophore causes conformational changes in the rhodopsin molecule—the first stage in visual transduction.

Excited Rhodopsin Acts through the G Protein Transducin to Reduce the cGMP Concentration

In its excited conformation, rhodopsin is able to interact with a second protein, **transducin,** which hovers nearby on the cytosolic face of the disk membrane (Fig. 13–22). Transducin (T) belongs to the same family of trimeric GTP-binding proteins as G_s and G_i. Although specialized for visual transduction, transducin shares many functional features with G_s and G_i. It can bind either GDP or GTP. In the dark, GDP is bound, all three subunits of the protein (T_α, T_β, and T_γ) remain together, and no signal is sent. When rhodopsin is excited by light, it interacts with transducin, catalyzing the replacement of bound GDP by GTP from the cytosol (Fig. 13–23, steps ① and ②). Transducin then dissociates into T_α and $T_{\beta\gamma}$, and the GTP-bound T_α carries the signal from the excited receptor to the next element in the transduction pathway, cGMP phosphodiesterase (PDE), an enzyme that converts cGMP to 5′-GMP (steps ③ and ④). Note that this is not the same cyclic nucleotide phosphodiesterase that hydrolyzes cAMP to terminate the β-adrenergic response. The cGMP-specific PDE is unique to the visual cells of the retina.

PDE is an integral protein with its active site on the cytosolic side of the disk membrane. In the dark, a tightly bound inhibitory subunit very effectively suppresses PDE activity. When T_α-GTP encounters PDE, the inhibitory subunit is released, and the enzyme's activity immediately increases by several orders of magnitude. Each molecule of active PDE degrades many molecules of cGMP to the biologically inactive 5′-GMP, lowering [cGMP] in the outer segment within a fraction of a second. At the new, lower [cGMP], the cGMP-gated ion channels close, blocking reentry of Na^+ and Ca^{2+} into the outer segment and hyperpolarizing the membrane of the rod or cone cell (step ⑤). By this process, the initial stimulus—a photon—changes the V_m of the cell.

Signal Amplification Occurs in Rod and Cone Cells

Several steps in the visual transduction process result in great amplification of the signal. Each excited rhodopsin molecule activates at least 500 molecules of transducin, each of which can activate a molecule of PDE. PDE has a remarkably high turnover number, each activated molecule hydrolyzing 4,200 molecules of cGMP per second. The binding of cGMP to cGMP-gated ion channels is cooperative (at least three cGMP molecules must be bound to open one channel), and a relatively small change in [cGMP] therefore registers as a large change in ion conductance. The result of these amplifications is exquisite sensitivity to light. Absorption of a single photon closes a

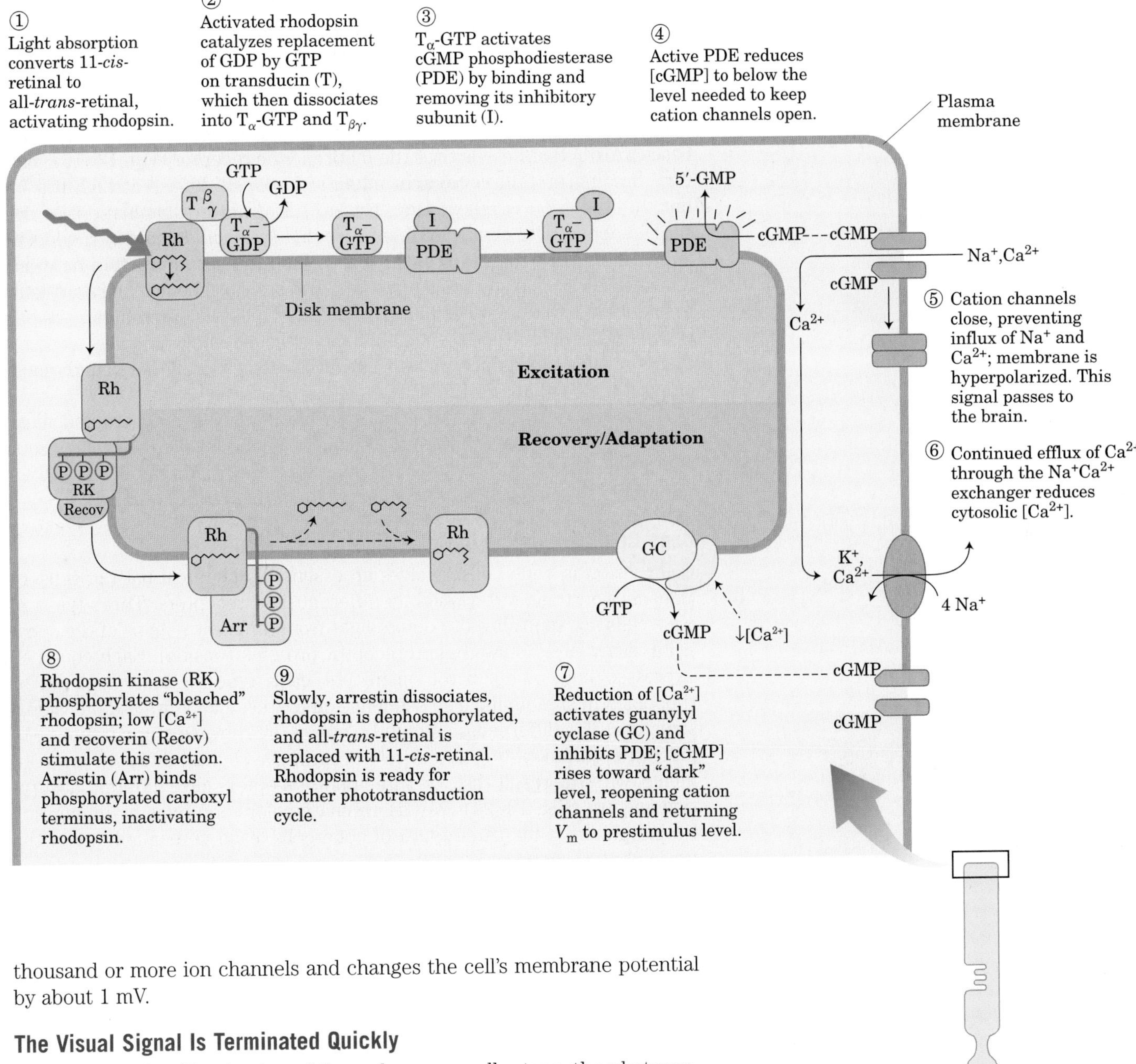

① Light absorption converts 11-cis-retinal to all-trans-retinal, activating rhodopsin.
② Activated rhodopsin catalyzes replacement of GDP by GTP on transducin (T), which then dissociates into Tα-GTP and Tβγ.
③ Tα-GTP activates cGMP phosphodiesterase (PDE) by binding and removing its inhibitory subunit (I).
④ Active PDE reduces [cGMP] to below the level needed to keep cation channels open.
Plasma membrane
GTP
GDP
Rh
PDE
I
5′-GMP
cGMP---cGMP
Na+,Ca2+
cGMP
Ca2+
Disk membrane
Excitation
Recovery/Adaptation
⑤ Cation channels close, preventing influx of Na+ and Ca2+; membrane is hyperpolarized. This signal passes to the brain.
⑥ Continued efflux of Ca2+ through the Na+Ca2+ exchanger reduces cytosolic [Ca2+].
RK
Recov
Arr
GC
K+, Ca2+
4 Na+
GTP
cGMP ↓[Ca2+]
⑧ Rhodopsin kinase (RK) phosphorylates "bleached" rhodopsin; low [Ca2+] and recoverin (Recov) stimulate this reaction. Arrestin (Arr) binds phosphorylated carboxyl terminus, inactivating rhodopsin.
⑨ Slowly, arrestin dissociates, rhodopsin is dephosphorylated, and all-trans-retinal is replaced with 11-cis-retinal. Rhodopsin is ready for another phototransduction cycle.
⑦ Reduction of [Ca2+] activates guanylyl cyclase (GC) and inhibits PDE; [cGMP] rises toward "dark" level, reopening cation channels and returning Vm to prestimulus level.

figure 13–23
Molecular consequences of photon absorption by rhodopsin in the rod outer segment. The top half of the figure (steps ① to ⑤) describes excitation; the bottom (steps ⑥ to ⑨), recovery and adaptation after illumination.

thousand or more ion channels and changes the cell's membrane potential by about 1 mV.

The Visual Signal Is Terminated Quickly

Very shortly after illumination of the rod or cone cells stops, the photosensory system shuts off. The α subunit of transducin (with bound GTP) has intrinsic GTPase activity. Within milliseconds after the decrease in light intensity, GTP is hydrolyzed, and T_α reassociates with $T_{\beta\gamma}$. The inhibitory subunit of PDE, which had been bound to T_α-GTP, is released and reassociates with PDE, inhibiting that enzyme very strongly. To return [cGMP] to its "dark" level, the enzyme guanylyl cyclase converts GTP to cGMP (step ⑦ in Fig. 13–23) in a reaction that is inhibited by high [Ca^{2+}] (>100 nM). During illumination, Ca^{2+} levels drop because the steady-state [Ca^{2+}] in the outer segment is the result of outward pumping of Ca^{2+} through the Na^+Ca^{2+} exchanger of the plasma membrane and inward movement of Ca^{2+} through open cGMP-gated channels. In the dark, this produces a [Ca^{2+}] of about 500 nM—enough to inhibit cGMP synthesis. After brief illumination, Ca^{2+} entry slows and [Ca^{2+}] drops (step ⑥). The inhibition of guanylyl cyclase by Ca^{2+} is relieved, and the enzyme converts GTP to cGMP to return the system to its prestimulus state (step ⑦).

Rhodopsin Is Desensitized by Phosphorylation

Rhodopsin itself also undergoes changes in response to prolonged illumination. The conformational change induced by light absorption exposes several Thr and Ser residues in the carboxyl-terminal domain. These residues are quickly phosphorylated by **rhodopsin kinase** (step ⑧ in Fig. 13–23), which is functionally and structurally homologous to the β-adrenergic kinase (βARK) that desensitizes the β-adrenergic receptor (Fig. 13–16). The Ca^{2+}-binding protein **recoverin** inhibits rhodopsin kinase at high [Ca^{2+}], but the inhibition is relieved when [Ca^{2+}] drops after illumination, as described above. The phosphorylated carboxyl-terminal domain of rhodopsin is bound by the protein **arrestin,** preventing further interaction between activated rhodopsin and transducin. Arrestin is a close homolog of β-arrestin (Fig. 13–16). On a relatively long time scale (seconds to minutes), the all-*trans*-retinal of an excited rhodopsin molecule is removed and replaced by 11-*cis*-retinal, to produce rhodopsin that is ready for another round of excitation (step ⑨ in Fig. 13–23).

Humans cannot synthesize retinal from simpler precursors and must obtain it in the diet in the form of vitamin A (p. 381). Given the role of retinal in the process of vision, it is not surprising that dietary deficiency of vitamin A causes night blindness (poor vision at night or in dim light).

Cone Cells Specialize in Color Vision

Color vision in cone cells involves an essentially identical path of sensory transduction triggered by slightly different light receptors. Three types of cone cells are specialized to detect light from different regions of the spectrum, using three related photoreceptor proteins (opsins). Each cone cell expresses only one kind of opsin, but each type is closely related to rhodopsin in size, amino acid sequence, and presumably three-dimensional structure. The differences among the opsins are, however, great enough to place the chromophore, 11-*cis*-retinal, in three slightly different environments, with the result that the three photoreceptors have different absorption spectra (Fig. 13–24). We discriminate colors and hues by integrating the output from the three types of cone cells, each containing one of the three photoreceptors.

Color blindness, such as the inability to distinguish red from green, is a fairly common, genetically inherited trait in humans. The various types of color blindness result from different opsin mutations. One form is due to

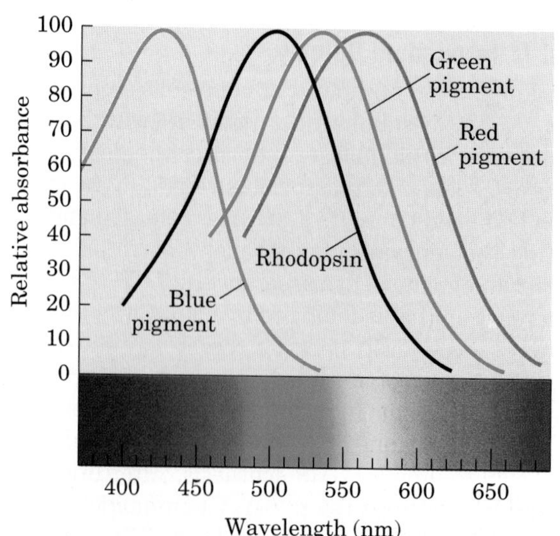

figure 13–24
Absorption spectra of purified rhodopsin and the red, green, and blue receptors of cone cells. The spectra, obtained from individual cone cells isolated from cadavers, peak at about 420, 530, and 560 nm, and the maximum absorption for rhodopsin is at about 500 nm. For reference, the visible spectrum for humans is about 380 to 750 nm.

box 13–2 Color Blindness: John Dalton's Experiment from the Grave

The chemist John Dalton (of atomic theory fame) was color-blind. He thought it probable that the vitreous humor of his eyes (the fluid that fills the eyeball behind the lens) was tinted blue, unlike the colorless fluid found in normal eyes. He proposed that after his death, his eyes should be dissected and the color of the vitreous humor determined. His wish was honored; the day after Dalton's death in July 1844, Joseph Ransome dissected his eyes and found the vitreous humor to be perfectly colorless. Ransome, like many scientists, was reluctant to throw samples away. He placed Dalton's eyes in a jar of preservative (Fig. 1), where they stayed until the mid-1990s, when molecular biologists in England took small samples of the 150-year-old retinas and extracted DNA. Using the known gene sequences for the opsins of the red and green photopigments, they amplified the relevant sequences (using techniques described in Chapter 29) and determined that Dalton had the opsin gene for the red photopigment but lacked the opsin gene for the green photopigment. Dalton was a green$^-$ dichromat. One hundred and fifty years after his death, the experiment he started—by hypothesizing about the cause of his color blindness—was finally finished.

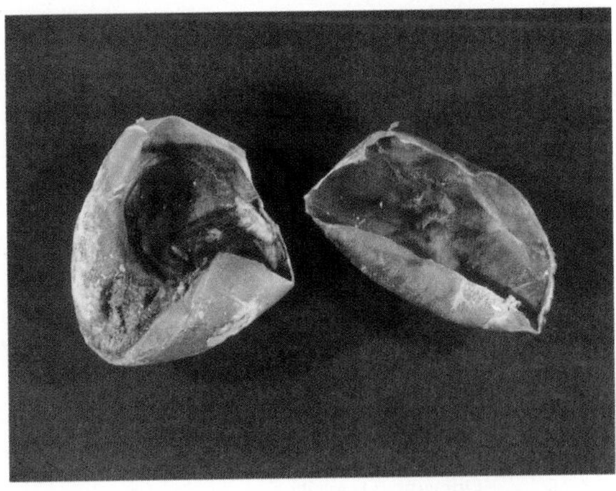

figure 1
Dalton's eyes. From Hunt, D.M., Dulai, K.S., Bowmaker, J.K., & Mollon, J.D. (1995) The chemistry of John Dalton's color blindness. *Science* **267,** 984–988.

loss of the red photoreceptor; affected individuals are called **red$^-$ dichromats** (they see only two primary colors). Others lack the green pigment and are **green$^-$ dichromats.** In some cases, the red and green photoreceptors are present but have a changed amino acid sequence that causes a change in their absorption spectra, resulting in abnormal color vision. Depending on which pigment is altered, such individuals are termed **red-anomalous trichromats** or **green-anomalous trichromats.** Examination of the genes for the visual receptors recently allowed the diagnosis of color blindness in a famous "patient" more than a century after his death (Box 13–2)!

Vertebrate Olfaction and Gustation Use Mechanisms Similar to the Visual System

The sensory cells used to detect odors and tastes have much in common with the rod and cone cells that detect light. Olfactory neurons have a number of long thin cilia extending from one end of the cell into a mucous layer that overlays the cell. These cilia present a large surface area for interaction with olfactory signals. The receptors for olfactory stimuli are ciliary membrane proteins with the familiar serpentine structure of seven transmembrane α helices. The olfactory signal can be any of a large number of volatile compounds for which there are specific receptor proteins. Our ability to discriminate odors stems from hundreds of different olfactory receptors in the tongue and nasal passage and from the brain's ability to integrate input from different types of olfactory receptors to recognize a "hybrid" pattern, extending our range of discrimination far beyond the number of receptors.

The olfactory stimulus arrives at the sensory cells by diffusion through the air. In the mucous layer overlaying the olfactory neurons, the odorant molecule binds directly to an olfactory receptor or to a specific binding protein that carries the odorant to a receptor (Fig. 13–25). Interaction between odorant and receptor triggers a change in receptor conformation that results in the replacement of bound GDP by GTP on a G protein, G_{olf}, analogous to transducin and to G_s of the β-adrenergic system. The activated G_{olf} then activates adenylyl cyclase of the ciliary membrane, which synthesizes cAMP from ATP, raising the local [cAMP]. The cAMP-gated Na^+ channels of the ciliary membrane open, and the influx of Na^+ produces a small depolarization called the **receptor potential.** If enough molecules of odorant encounter receptors, the receptor potential is strong enough to cause the neuron to fire an action potential. This is relayed to the brain in several stages and registers as a specific smell. All of these events occur within 100 to 200 ms.

Some olfactory neurons may use a second transduction mechanism. They have receptors coupled through G proteins to phospholipase C rather than to adenylyl cyclase. Signal reception in these cells triggers production of IP_3 (Fig. 13–17), which opens IP_3-gated Ca^{2+} channels in the ciliary membrane. Influx of Ca^{2+} then depolarizes the ciliary membrane and generates a receptor potential or regulates Ca^{2+}-dependent enzymes in the olfactory pathway.

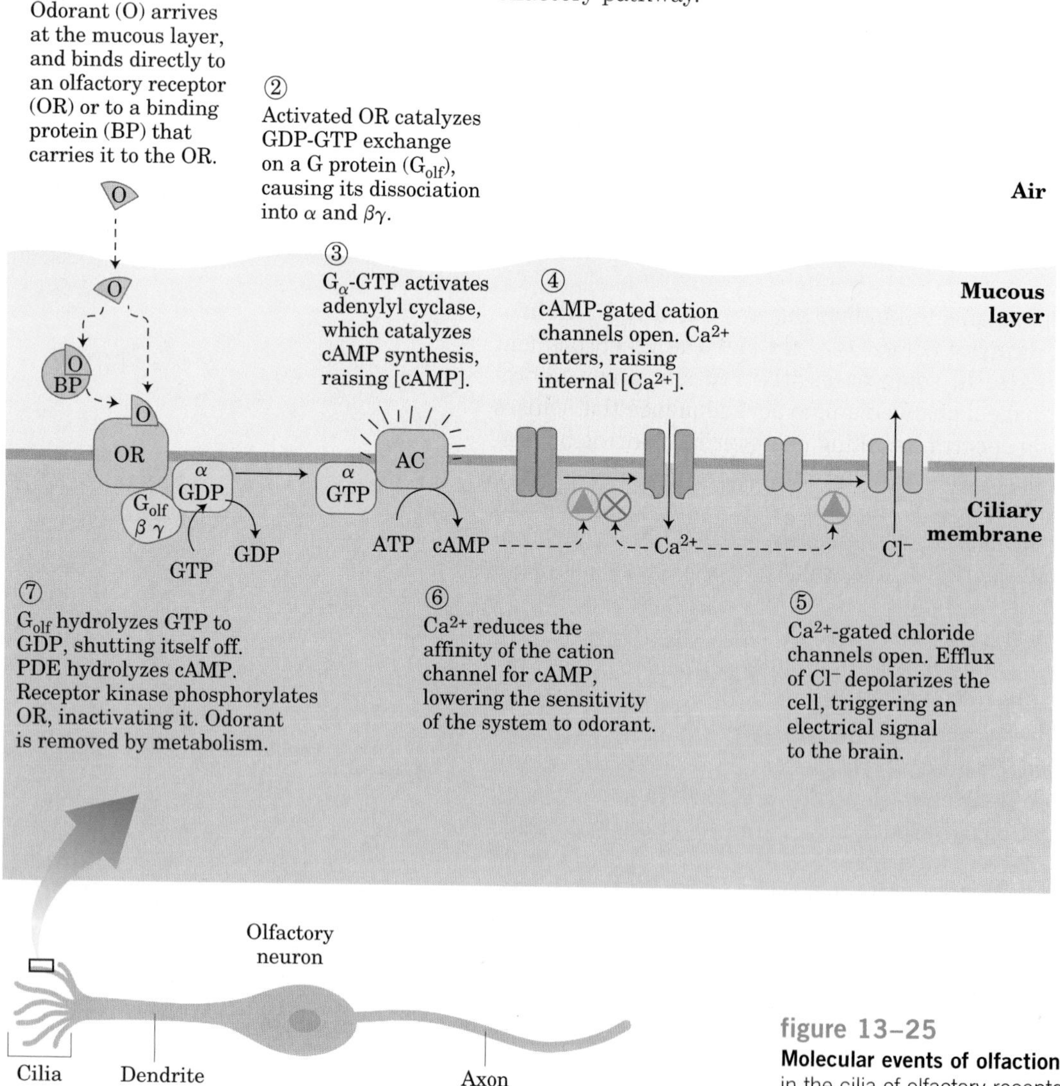

① Odorant (O) arrives at the mucous layer, and binds directly to an olfactory receptor (OR) or to a binding protein (BP) that carries it to the OR.
② Activated OR catalyzes GDP-GTP exchange on a G protein (G_{olf}), causing its dissociation into α and $\beta\gamma$.
③ G_α-GTP activates adenylyl cyclase, which catalyzes cAMP synthesis, raising [cAMP].
④ cAMP-gated cation channels open. Ca^{2+} enters, raising internal [Ca^{2+}].
⑤ Ca^{2+}-gated chloride channels open. Efflux of Cl^- depolarizes the cell, triggering an electrical signal to the brain.
⑥ Ca^{2+} reduces the affinity of the cation channel for cAMP, lowering the sensitivity of the system to odorant.
⑦ G_{olf} hydrolyzes GTP to GDP, shutting itself off. PDE hydrolyzes cAMP. Receptor kinase phosphorylates OR, inactivating it. Odorant is removed by metabolism.
Air
Mucous layer
Ciliary membrane
O
BP
OR
α
GDP
G_{olf}
$\beta\ \gamma$
GTP
GDP
α
GTP
AC
ATP
cAMP
Ca^{2+}
Cl^-
Olfactory neuron
Cilia
Dendrite
Axon

figure 13–25

Molecular events of olfaction. These interactions occur in the cilia of olfactory receptor cells.

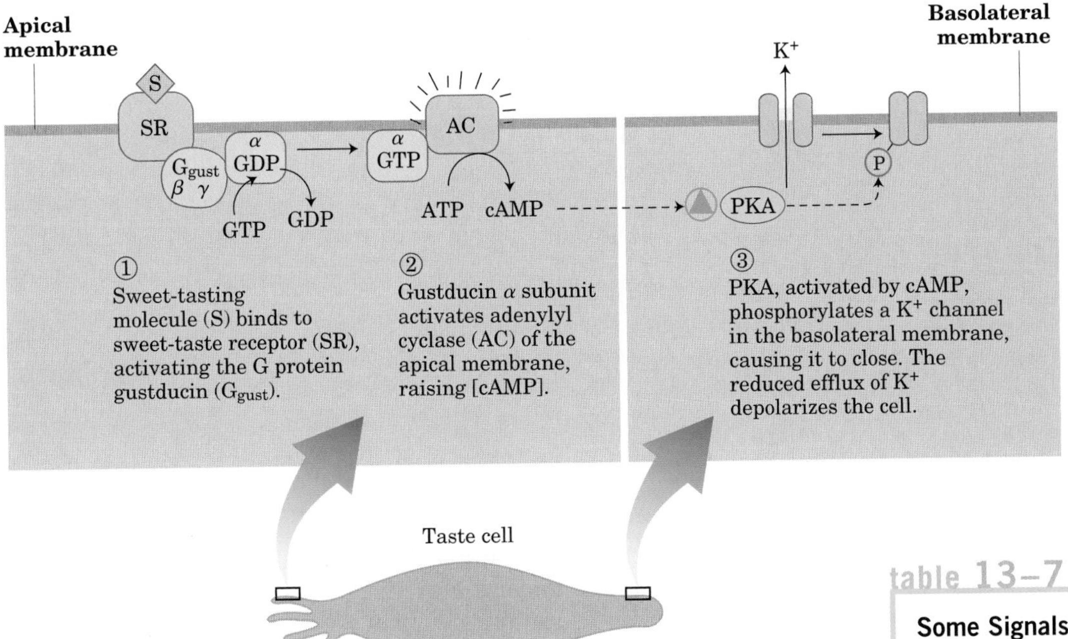
Apical membrane
Basolateral membrane
S
SR
α
Ggust
GDP
β γ
GTP
GDP
α
GTP
AC
ATP cAMP
K+
PKA
P
①
Sweet-tasting molecule (S) binds to sweet-taste receptor (SR), activating the G protein gustducin (Ggust).
②
Gustducin α subunit activates adenylyl cyclase (AC) of the apical membrane, raising [cAMP].
③
PKA, activated by cAMP, phosphorylates a K+ channel in the basolateral membrane, causing it to close. The reduced efflux of K+ depolarizes the cell.
Taste cell

figure 13–26
Transduction mechanism for sweet tastants.

In either type of olfactory neuron, when the stimulus is no longer present, the transducing machinery shuts itself off in several ways. A cAMP phosphodiesterase returns [cAMP] to the prestimulus level. G_{olf} hydrolyzes its bound GTP to GDP, thereby inactivating itself. Phosphorylation of the receptor by a specific kinase prevents its interaction with G_{olf} by a mechanism analogous to that used to desensitize the β-adrenergic receptor and rhodopsin. And lastly, some odorants are enzymatically destroyed by oxidases.

The sense of taste in vertebrates reflects the activity of gustatory neurons clustered in taste buds on the surface of the tongue. In these sensory neurons, serpentine receptors are coupled to the trimeric G protein **gustducin** (very similar to the transducin of rod and cone cells). Sweet-tasting molecules are those that bind receptors in "sweet" taste buds. When the molecule (tastant) binds, gustducin is activated by replacement of bound GDP with GTP and then stimulates cAMP production by adenylyl cyclase. The resulting elevation of [cAMP] activates PKA, which phosphorylates K^+ channels in the plasma membrane, causing them to close. Reduced efflux of K^+ depolarizes the cell (Fig. 13–26). Other taste buds specialize in detecting bitter, sour or salty tastants, using various combinations of second messengers and ion channels in the transduction mechanisms.

G Protein–Coupled Serpentine Receptor Systems Share Several Features

We have now looked at four systems (hormone signaling, vision, olfaction, and gustation) in which membrane receptors are coupled to second messenger–generating enzymes through G proteins. It is clear that signaling mechanisms arose early in evolution; serpentine receptors, heterotrimeric G proteins, and adenylyl cyclase are all found in brewer's yeast. Overall patterns have been conserved, while the introduction of variety has given modern organisms the ability to respond to a wide range of stimuli (Table 13–7). Of the approximately 10^5 genes in the mammalian genome, as many as 10^3 encode serpentine receptors, including hundreds for olfactory stimuli and a number of "orphan receptors" for which the natural ligand is not yet known.

table 13–7

Some Signals Transduced by G Protein–Coupled Serpentine Receptors
Acetylcholine (muscarinic)
Adenosine
Angiotensin
ATP (extracellular)
Bradykinin
Calcitonin
Cannabinoids
Catecholamines
Cholecystokinin
Corticotropin-releasing factor (CRF)
Cyclic AMP (*Dictyostelium discoideum*)
Dopamine
Endothelin
Follicle-stimulating hormone (FSH)
γ-Aminobutyric acid (GABA)
Glucagon
Glutamate (metabotropic)
Growth hormone releasing hormone (GHRH)
Histamine
Leukotrienes
Light
Luteinizing hormone (LH)
Melatonin
Odorants
Opioids
Oxytocin
Platelet-activating factor
Prostaglandins
Secretin
Serotonin
Somatostatin
Tastants
Thyrotropin
Thyrotropin-releasing hormone (TRH)
Vasoactive intestinal peptide
Vasopressin
Yeast mating factors

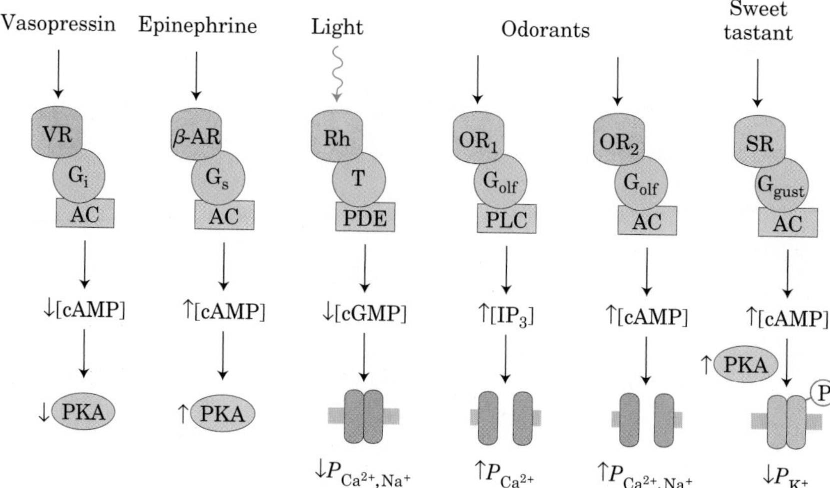

figure 13–27

Common features of signaling systems that detect hormones, light, smells, and tastes. Serpentine receptors provide signal specificity, and their interaction with G proteins provides signal amplification. Heterotrimeric G proteins activate effector enzymes: adenylyl cyclase (AC), phospholipase C (PLC), and phosphodiesterases (PDE) that degrade cAMP or cGMP. Changes in concentrations of the second messengers (cAMP, cGMP, IP_3) result in alterations of enzymatic activities by phosphorylation or alterations in the permeability (P) of surface membranes to Ca^{2+}, Na^+, and K^+. The resulting depolarization or hyperpolarization of the sensory cell (the signal) is passed through relay neurons to sensory centers in the brain. In the best-studied cases, desensitization includes phosphorylation of the receptor and binding of a protein (arrestin) that interrupts receptor–G protein interactions. (VR represents the vasopressin receptor; other receptor and G protein abbreviations are as used in earlier illustrations.)

All well-studied transducing systems that act through heterotrimeric G proteins share certain common features (Fig. 13–27). The receptors have seven transmembrane segments, a domain (generally the loop between transmembrane helices 6 and 7) that interacts with a G protein, and a carboxyl-terminal cytoplasmic domain that undergoes reversible phosphorylation on several Ser or Thr residues. The ligand-binding site (or, in the case of light reception, the light receptor) is buried deep in the membrane and involves residues from several of the transmembrane segments. Ligand binding (or light) induces a conformational change in the receptor, exposing a domain that can interact with a G protein. Heterotrimeric G proteins activate or inhibit effector enzymes (adenylyl cyclase, PDE, or phospholipase C), which change the concentration of a second messenger (cAMP, cGMP, IP_3, or Ca^{2+}). In the hormone-detecting systems, the final output is an activated protein kinase that regulates some cellular process by phosphorylating a protein critical to that process. In sensory neurons, the output is a change in membrane potential and a consequent electrical signal that passes to another neuron in the pathway connecting the sensory cell to the brain.

All of these systems self-inactivate. Bound GTP is converted to GDP by the intrinsic GTPase activity of G proteins, often augmented by GTPase activating proteins (GAPs) or RGS proteins (*r*egulators of *G*-protein *s*ignaling). In some cases, the effector enzymes that are the targets of G protein modulation also serve as GAPs.

Disruption of G-Protein Signaling Causes Disease

Biochemical studies of signal transductions have led to an improved understanding of the pathological effects of toxins produced by the bacteria that cause cholera and pertussis (whooping cough). Both toxins are enzymes that interfere with normal signal transductions in the host animal. **Cholera toxin,** secreted by *Vibrio cholerae* found in contaminated drinking water, catalyzes the transfer of ADP-ribose from NAD^+ to the α subunit of G_s (Fig. 13–28), blocking its GTPase activity and thereby rendering G_s permanently activated. This results in continuous activation of the adenylyl cyclase of intestinal epithelial cells and chronically high [cAMP], which triggers continual secretion of Cl^-, HCO_3^-, and water into the intestinal lumen. The resulting dehydration and electrolyte loss are the major pathologies in cholera. The **pertussis toxin,** produced by *Bordetella pertussis,* catalyzes ADP-ribosylation of G_i, preventing displacement of GDP by GTP and blocking inhibition of adenylyl cyclase by G_i. This defect produces two of the symptoms of whooping cough: hypersensitivity to histamines and lowered blood glucose.

Normal G_s: GTPase activity terminates the signal from receptor to adenylate cyclase.

NAD$^+$

cholera toxin

ADP-ribosylated G_s: GTPase activity is inactivated; G_s constantly activates adenylate cyclase.

ADP-ribose

figure 13–28

Toxins produced by bacteria that cause cholera and whooping cough (pertussis). These toxins are enzymes that catalyze transfer of the ADP-ribose moiety of NAD$^+$ to an Arg residue of G proteins: G_s in the case of cholera (as shown here) and G_i in whooping cough. The G proteins thus modified fail to respond to normal hormonal stimuli. The pathology of both diseases results from defective regulation of adenylyl cyclase and overproduction of cAMP.

Phosphorylation as a Regulatory Mechanism

A common denominator in signal transductions is the eventual regulation of the activity of a protein kinase, an enzyme that phosphorylates a protein. We have seen examples of kinases activated by cGMP, cAMP, insulin, Ca^{2+}/calmodulin, Ca^{2+}/diacylglycerol, and activated Ras, and by phosphorylation by another protein kinase. The number of known protein kinases has grown remarkably since their discovery by Edwin G. Krebs and Edmond H. Fischer in 1959. The yeast genome encodes about 120 different protein kinases, representing 2% of all the genes in that organism. Other eukaryotes probably have at least this many protein kinase genes, each with its specific activator and protein target(s).

Edwin G. Krebs

Edmond H. Fischer

Localization of Protein Kinases and Phosphatases Affects the Specificity for Target Proteins

Several types of protein kinases and phosphatases are localized to specific subcellular regions, held in close proximity to their protein substrates by a targeting protein. For example, the enzymes that catalyze and regulate glycogen synthesis and breakdown in skeletal muscle are tightly associated with glycogen granules. Protein phosphatase 1 (PP-1) is part of that regulatory machinery; it removes phosphates from Ⓟ-Ser and Ⓟ-Thr residues of glycogen synthase and phosphorylase kinase, reversing the regulatory effects of cAMP-dependent protein kinase, PKA. PP-1 has two subunits: the catalytic (C) subunit and the R_G subunit, which binds the C subunit to the glycogen granule, assuring easy access to the substrates that PP-1 dephosphorylates.

Targeting subunits also hold some protein kinases close to their substrates. The R subunit of PKA (Fig. 13–14) associates tightly with a family of targeting proteins called AKAPs (*A kinase anchoring proteins*), each of which is thought to anchor PKA to a specific region of the cell—mitochondria, Golgi apparatus, endoplasmic reticulum, nuclear matrix, or secretory granules—presumably in close proximity to critical protein substrates of PKA.

Such PKA localization, combined with the generation of spatial gradients in [cAMP] by localized adenylyl cyclase and phosphodiesterase activities, may provide a mechanism for the separate regulation of several PKA substrates in the same cell. The various isozymes of protein kinase C are also differentially localized by their association with anchoring proteins called RACKs (*r*eceptors for *a*ctivated *C* *k*inase) or PICKs (*p*roteins that *i*nteract with *C* *k*inase).

Regulation of Transcription by Steroid Hormones

The large group of steroid, retinoic acid (retinoid), and thyroid hormones exert their effects by a mechanism fundamentally different from that of other hormones: they act in the nucleus to alter gene expression. We therefore discuss their mode of action in detail in Chapter 28, along with other mechanisms for regulating gene expression. Here we give a brief overview.

Steroid hormones (estrogen, progesterone, and cortisol, for example), too hydrophobic to dissolve readily in the blood, are carried on specific carrier proteins from their point of release to their target tissues. In target cells, these hormones pass through the plasma membranes by simple diffusion and bind to specific receptor proteins in the nucleus (Fig. 13–29). Hormone binding triggers changes in the conformation of the receptor proteins so that they become capable of interacting with specific regulatory sequences in DNA called **hormone response elements (HREs),** thus altering gene expression (see Fig. 28–33). The bound receptor-hormone complex can either enhance or suppress the expression of specific genes

figure 13–29
General mechanism by which steroid and thyroid hormones, retinoids, and vitamin D regulate gene expression. The details of transcription and protein synthesis are discussed in Chapters 26 and 27.

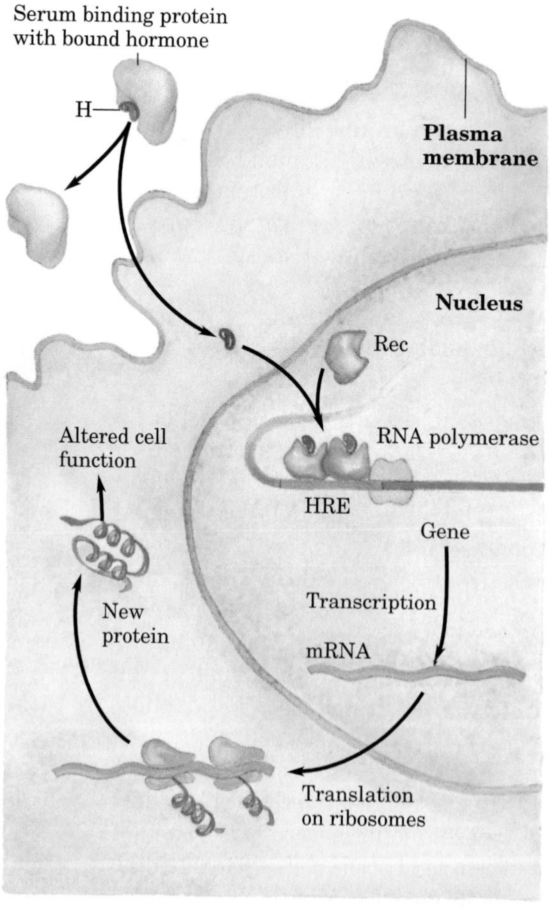

① Hormone (H), carried to the target tissue on serum binding proteins, diffuses across the plasma membrane and binds to its specific receptor protein (Rec) in the nucleus.

② Hormone binding changes the conformation of Rec; it forms homo- or heterodimers with other hormone-receptor complexes and binds to specific regulatory regions called hormone response elements (HREs) in the DNA adjacent to specific genes.

③ Binding regulates transcription of the adjacent gene(s), increasing or decreasing the rate of mRNA formation.

④ Altered levels of the hormone-regulated gene product produce the cellular response to the hormone.

adjacent to HREs. Hours or days are required for these regulators to have their full effect—the time required for the changes in RNA synthesis and subsequent protein synthesis to become evident in altered metabolism.

The specificity of the steroid-receptor interaction is exploited in the use of the drug **tamoxifen** to treat breast cancer. In some types of breast cancer, division of the cancerous cells depends on the continued presence of the hormone estrogen. Tamoxifen competes with estrogen for binding to the estrogen receptor, but the tamoxifen-receptor complex has no effect on gene expression; tamoxifen is an antagonist of estrogen. Consequently, tamoxifen administered after surgery or during chemotherapy for hormone-dependent breast cancer slows or stops the growth of remaining cancerous cells, prolonging the life of the patient. Another steroid analog, the drug **RU486,** is used to terminate early pregnancies. An antagonist of the hormone progesterone, RU486 binds to the progesterone receptor and blocks hormone actions essential to implantation of the fertilized ovum in the uterus.

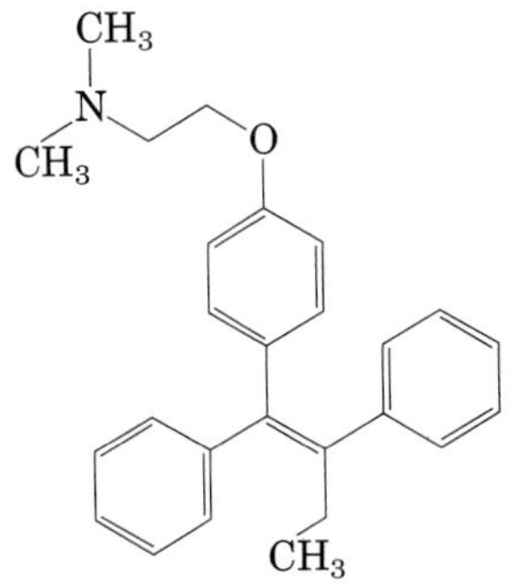

Tamoxifen

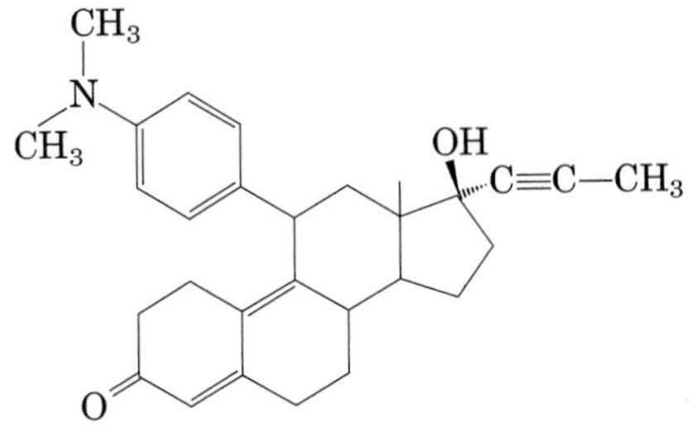

RU486 (mifepristone)

Regulation of the Cell Cycle by Protein Kinases

One of the most dramatic roles for protein phosphorylation is in the regulation of the eukaryotic cell cycle. During embryonic growth and later development, cell division occurs in virtually every tissue. In the adult organism most tissues become quiescent. A cell's "decision" to divide or not is of crucial importance to the organism. When the regulatory mechanisms that limit cell division are defective and cells undergo unregulated division, the result is catastrophic—cancer. Proper cell division requires a precisely ordered sequence of biochemical events that assures every daughter cell a full complement of the molecules required for life. Investigations into the control of cell division in diverse eukaryotic cells have revealed universal regulatory mechanisms. Protein kinases and protein phosphorylation are central to the timing mechanism that determines entry into cell division and assures orderly passage through these events.

The Cell Cycle Has Four Stages

Cell division in eukaryotes occurs in four well-defined stages (Fig. 13–30). In the S (synthesis) phase, the DNA is replicated to produce copies for both daughter cells. In the G2 phase (G indicates the gap between divisions), new proteins are synthesized and the cell approximately doubles in size. In the M phase (mitosis), the maternal nuclear envelope breaks down, matching chromosomes are pulled to opposite poles of the cell, each set of daughter chromosomes is surrounded by a newly formed nuclear envelope, and cytokinesis pinches the cell in half, producing two daughter cells. In embryonic or rapidly proliferating tissue, each daughter cell divides again, but only after a waiting period (G1). In cultured animal cells the entire process takes about 24 hours.

After passing through mitosis and into G1, a cell either continues through another division or ceases to divide, entering a quiescent phase (G0) that may last hours, days, or the lifetime of the cell. Terminally differentiated cells are in the G0 phase. When a cell in G0 begins to divide again, it reenters the division cycle through the G1 phase.

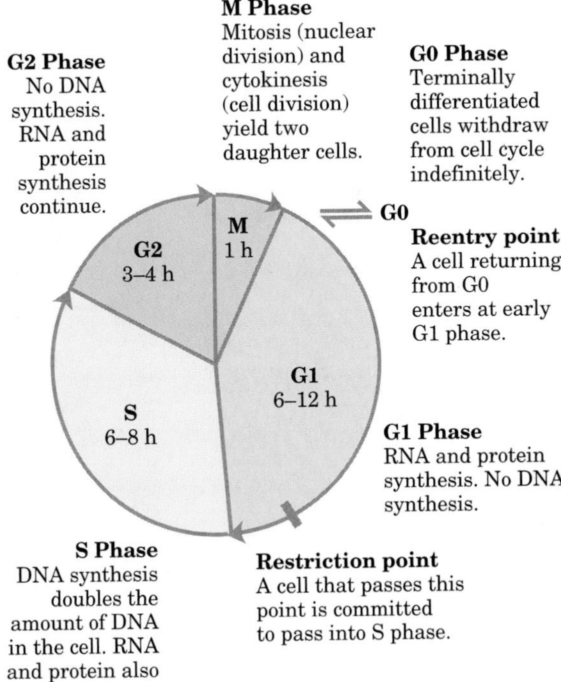
M Phase
Mitosis (nuclear division) and cytokinesis (cell division) yield two daughter cells.
G2 Phase
No DNA synthesis. RNA and protein synthesis continue.
G0 Phase
Terminally differentiated cells withdraw from cell cycle indefinitely.
G0
Reentry point
A cell returning from G0 enters at early G1 phase.
M
1 h
G2
3–4 h
G1
6–12 h
S
6–8 h
G1 Phase
RNA and protein synthesis. No DNA synthesis.
S Phase
DNA synthesis doubles the amount of DNA in the cell. RNA and protein also synthesized.
Restriction point
A cell that passes this point is committed to pass into S phase.

figure 13–30
Eukaryotic cell cycle. The times for the four stages vary, but those shown are typical.

Levels of Cyclin-Dependent Protein Kinases Oscillate

The timing of the cell cycle is controlled by a family of protein kinases with activities that change in response to cellular signals. By phosphorylating specific proteins at precisely timed intervals, these protein kinases orchestrate the metabolic activities of the cell to produce orderly cell division.

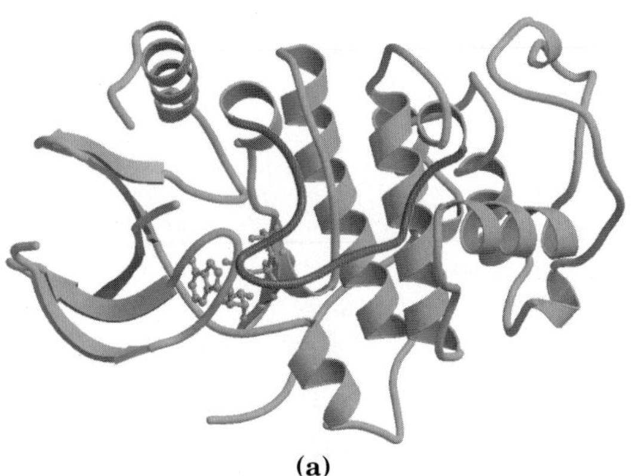

(a)

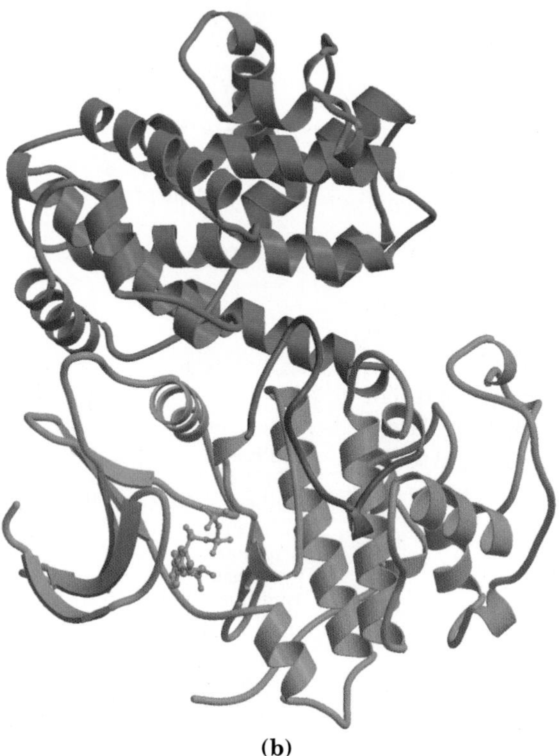

(b)

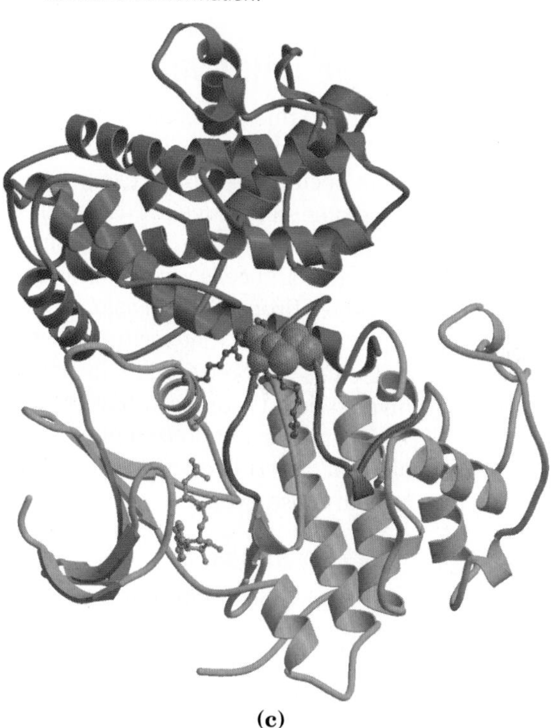

(c)

figure 13–31
Activation of cyclin-dependent protein kinases (CDKs) by cyclin and phosphorylation. CDKs, a family of related enzymes, are active only when associated with cyclins, another protein family. From the crystal structure of CDK2 with and without cyclin, the basis for this activation is apparent. Without cyclin **(a),** CDK2 folds so that one segment, the T loop (red), obstructs the binding site for protein substrates and thus inhibits protein kinase activity. The binding site for ATP (blue) is also near the T loop. The binding of cyclin **(b)** forces conformational changes that move the T loop away from the active site and reorients an amino-terminal helix (green), bringing one of the residues critical to catalysis (Glu^{51}) into the active site. **(c)** Phosphorylation of a Thr residue (brown space-filling structure) in the T-loop produces a negatively charged residue that is stabilized by interaction with three Arg residues (red ball-and-stick structures), holding CDK in its active conformation.

These kinases are heterodimers with a regulatory subunit, **cyclin,** and a catalytic subunit, **cyclin-dependent protein kinase (CDK).** In the absence of cyclin, the catalytic subunit is virtually inactive. When cyclin binds, the catalytic site opens up, a residue essential to catalysis becomes accessible, and the activity of the catalytic subunit increases 10,000-fold (Fig. 13–31). Animal cells have at least ten different cyclins (designated A, B, and so forth) and at least eight cyclin-dependent kinases (CDK1 through CDK8), which act in various combinations at specific points in the cell cycle. Plants also use a family of CDKs to regulate their cell division.

In a population of animal cells undergoing synchronous division, some CDK activities show striking oscillations (Fig. 13–32). These oscillations are the result of four mechanisms for regulating CDK activity: phosphorylation or dephosphorylation of the CDK, controlled degradation of the cyclin subunit, periodic synthesis of CDKs and cyclins, and the action of specific CDK-inhibiting proteins.

Regulation of CDKs by Phosphorylation The activity of a CDK is strikingly affected by phosphorylation and dephosphorylation of two critical residues in the protein (Fig. 13–33a). Phosphorylation of Tyr^{15} near the amino terminus renders CDK2 inactive; the Ⓟ-Tyr residue is in the ATP-binding site of the kinase, and the negatively charged phosphate group blocks the entry of ATP. A specific phosphatase dephosphorylates this Ⓟ-Tyr residue, permitting the binding of ATP. Phosphorylation of Thr^{160} in the "T loop" of CDK, catalyzed by the CDK-activating kinase, forces the T loop out of the substrate-binding cleft, permitting substrate binding and catalytic activity.

One circumstance that triggers this control mechanism is the presence of single-strand breaks in DNA, which leads to arrest of the cell cycle in G2. A specific protein kinase (called Rad3 in yeast), which is activated by single-strand breaks, triggers a cascade leading to the inactivation of the phosphatase that dephosphorylates Tyr^{15} of CDK. The CDK remains inactive and the cell is arrested in G2. The cell will not divide until the DNA is repaired and the effects of the cascade are reversed.

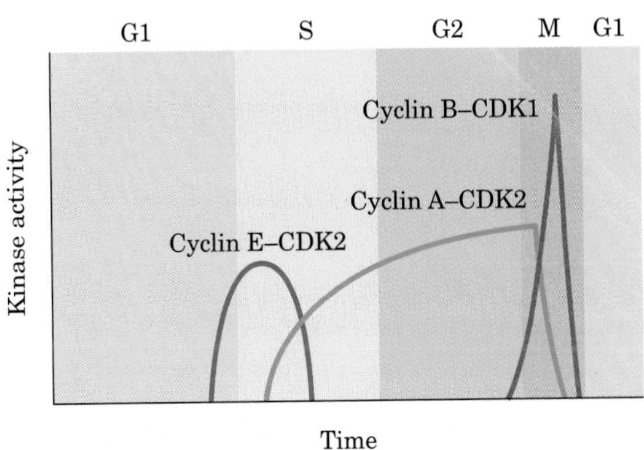

figure 13–32
Variations in the activities of specific CDKs during the cell cycle in animals. The cyclin E–CDK2 activity peaks near the G1 phase–S phase boundary, where the active enzyme triggers synthesis of enzymes required for DNA synthesis (see Fig. 13–35). Cyclin A–CDK2 rises during S phase and G2 phase, then drops sharply in M phase, as cyclin B–CDK1 peaks.

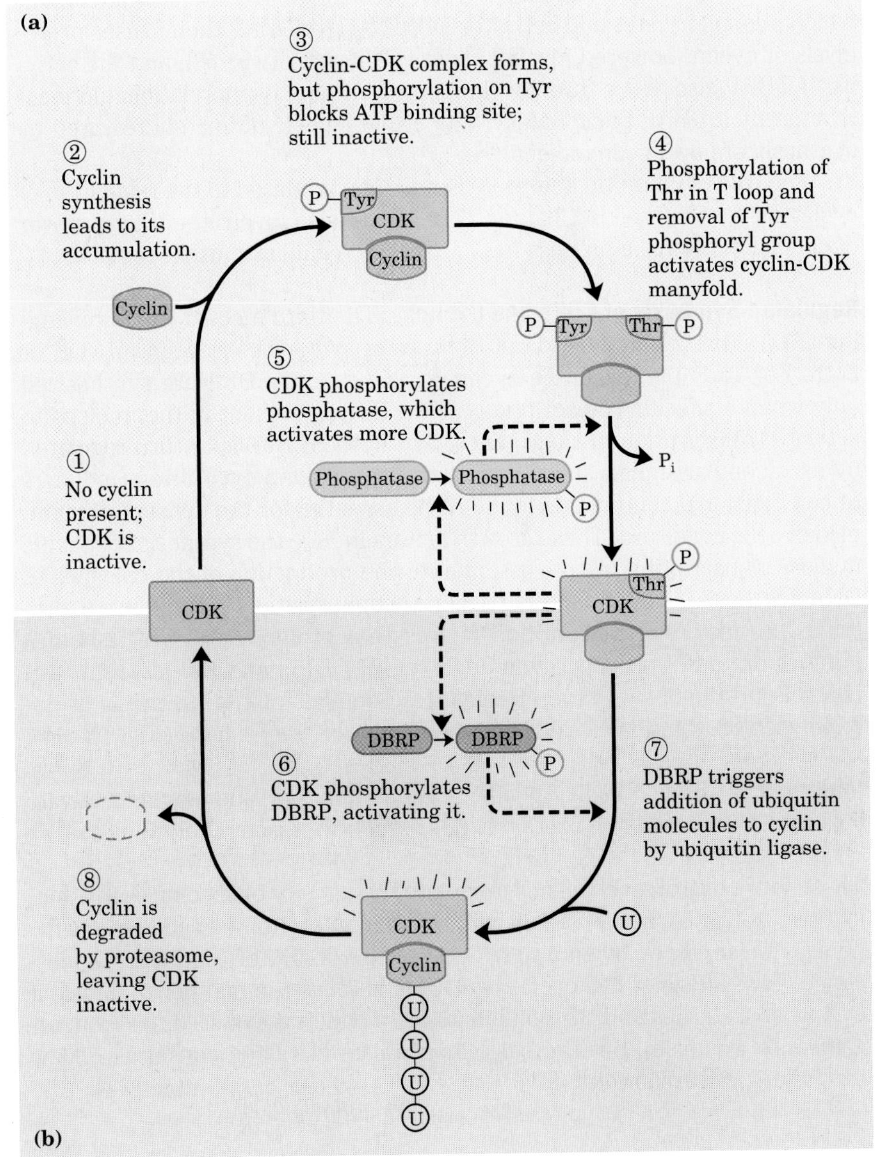

figure 13–33
Regulation of CDK by phosphorylation and proteolysis. **(a)** The cyclin-dependent protein kinase activated at the time of mitosis (the M phase CDK) has a "T loop" that can fold into the substrate-binding site. When a Thr residue in the T loop is phosphorylated, the loop moves out of the substrate-binding site, activating the CDK manyfold. **(b)** The active complex of cyclin and CDK triggers its own inactivation by phosphorylation of DBRP (destruction box recognizing protein). DBRP and ubiquitin ligase then attach several molecules of ubiquitin (U) to cyclin, targeting it for destruction by proteasomes, proteolytic enzyme complexes.

Controlled Degradation of Cyclin Highly specific and precisely timed proteolytic breakdown of mitotic cyclins regulates CDK activity throughout the cell cycle. Progress through mitosis requires first the activation then the destruction of cyclins A and B, which activate the catalytic subunit of the M-phase CDK. These cyclins contain near their amino terminus the sequence Arg–Thr–Ala–Leu–Gly–Asp–Ile–Gly–Asn, the "destruction box," which targets them for degradation. (This usage of "box" derives from the common practice, in diagramming the sequence of a nucleic acid or protein, of enclosing within a box a short sequence of nucleotide or amino acid residues with some specific function. It does not imply any three-dimensional structure.) A protein that recognizes this sequence, called DBRP (destruction box recognizing protein), initiates the process of cyclin degradation by bringing together the cyclin and another protein, **ubiquitin.** Cyclin and activated ubiquitin are covalently joined by the enzyme ubiquitin ligase (Fig. 13–33b). Several more ubiquitin molecules are then appended, providing the signal for a proteolytic enzyme complex, or **proteasome,** to degrade cyclin.

What controls the timing of cyclin breakdown? There is a feedback loop in which increased CDK activity activates cyclin proteolysis. As shown in Figure 13–33, newly synthesized cyclin associates with and activates CDK, which phosphorylates and activates DBRP. Active DBRP then causes proteolysis of cyclin. Lowered [cyclin] causes CDK activity to fall, and the activity of DBRP also drops through slow, constant dephosphorylation and inactivation by a DBRP phosphatase. The cyclin level is ultimately restored by synthesis of new cyclin molecules.

The role of ubiquitin and proteasomes is not limited to the regulation of cyclin; as we shall see in Chapter 27, both are also involved in the turnover of cellular proteins, a process fundamental to cellular housekeeping.

Regulated Synthesis of CDKs and Cyclins The third mechanism for changing CDK activity is regulation of the rate of synthesis of cyclin or CDK or both. For example, cyclin D, cyclin E, CDK2, and CDK4 are synthesized only when a specific transcription factor, E2F, is present in the nucleus to activate transcription of their genes. Synthesis of E2F is in turn regulated by extracellular signals such as **growth factors** and **cytokines** (inducers of cell division), compounds found to be essential for the division of mammalian cells in culture. These growth factors induce the synthesis of specific nuclear transcription factors essential to the production of the enzymes of DNA synthesis. Growth factors trigger phosphorylation of the nuclear proteins Jun and Fos, transcription factors that promote the synthesis of a number of gene products, including cyclins, CDKs, and E2F. E2F in turn controls production of several enzymes essential for the synthesis of deoxynucleotides and DNA, allowing cells to enter the S phase (Fig. 13–34).

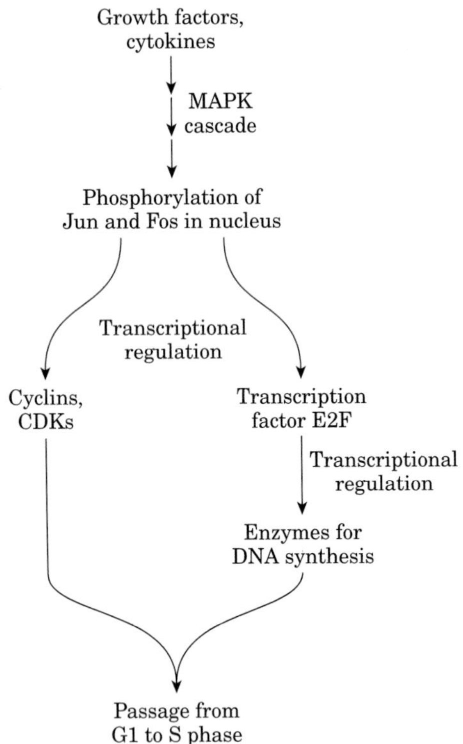

figure 13–34
Regulation of cell division by growth factors. The path from growth factors to cell division leads through the enzyme cascade that activates MAPK (mitogen-activated protein kinase; see Fig. 13–7); phosphorylation of the nuclear transcription factors Jun and Fos; and the activity of the transcription factor E2F, which promotes synthesis of several enzymes essential for DNA synthesis.

Inhibition of CDKs Finally, specific protein inhibitors bind to and inactivate specific CDKs. One such protein is p21, which we discuss below.

These four control mechanisms modulate the activity of specific CDKs that, in turn, control whether a cell will divide, differentiate, become permanently quiescent, or begin a new cycle of division after a period of quiescence. The details of cell cycle regulation, such as the number of different cyclins and kinases and the combinations in which they act, differ from organism to organism, but the basic mechanism has been conserved in the evolution of all eukaryotic cells.

CDKs Regulate Cell Division by Phosphorylating Critical Proteins

We have examined how close control of CDK activity is maintained, but how does the activity of CDK control the cell cycle? The list of target proteins that CDKs are known to act upon continues to grow, and much remains to be learned. But we can see a general pattern behind CDK regulation by inspecting the effect of CDKs on the structures of laminin and myosin and on the activity of retinoblastoma protein.

The structure of the nuclear envelope is maintained in part by highly organized meshworks of intermediate filaments composed of the protein laminin. Breakdown of the nuclear membrane before segregation of the sister chromatids in mitosis is due in part to the phosphorylation of laminin by a CDK, which causes laminin filaments to depolymerize.

A second kinase target is the ATP-driven actomyosin contractile machinery that pinches a dividing cell into two equal parts during cytokinesis. After the division, CDK phosphorylates a small regulatory subunit of myosin, causing dissociation of myosin from actin filaments and inactivating the contractile machinery. Subsequent dephosphorylation allows reassembly of the contractile apparatus for the next round of cytokinesis.

A third and very important CDK substrate is the **retinoblastoma protein, pRb,** which participates in a mechanism that arrests cell division in G1 if DNA damage is detected (Fig. 13–35). Named for the retinal tumor cell line in which it was discovered, pRb functions in all cell types to regulate cell division in response to a variety of stimuli. Unphosphorylated pRb binds the transcription factor E2F; while bound to pRb, E2F cannot promote transcription of a group of genes necessary for DNA synthesis (the genes for DNA polymerase α, ribonucleotide reductase, and other proteins). In this state, the cell cycle cannot proceed from the G1 to the S phase, the step that commits a cell to mitosis and cell division. The pRb-E2F blocking mechanism is relieved when pRb is phosphorylated by cyclin E–CDK2, which occurs in response to a signal for cell division to proceed.

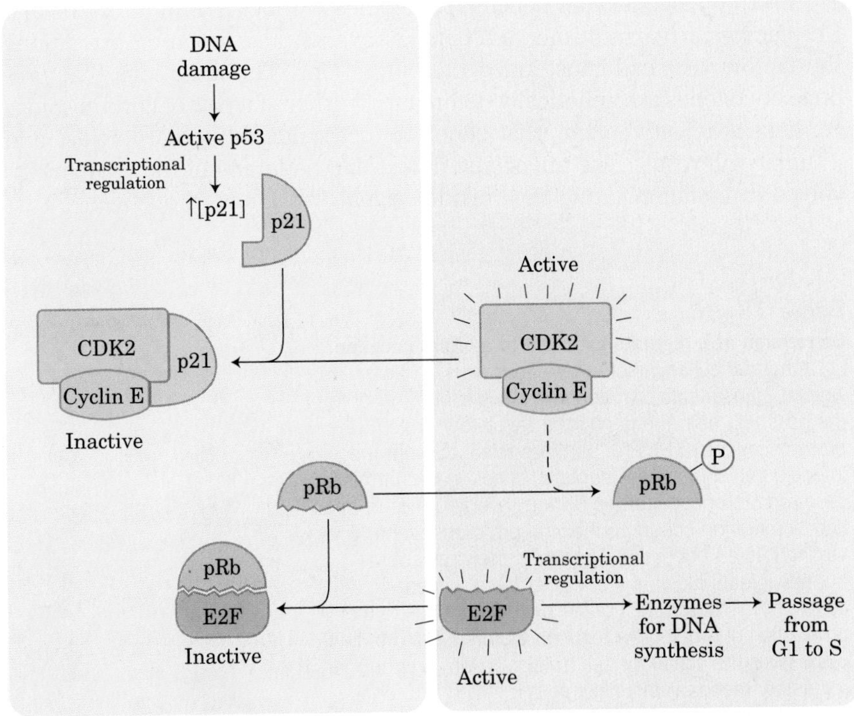

figure 13–35

Regulation of passage from G1 to S by phosphorylation of pRb. When the retinoblastoma protein, pRb, is phosphorylated, it cannot bind and inactivate EF2, a transcription factor that promotes synthesis of enzymes essential to DNA synthesis. If the regulatory protein p53 is activated by ATM and ATR, protein kinases that detect damaged DNA, it stimulates the synthesis of p21. p21 can bind to and inhibit cyclin E–CDK2, preventing phosphorylation of pRb. Unphosphorylated pRb binds and inactivates E2F, blocking passage from G1 to S until the DNA has been repaired.

When damage to DNA, such as a single-strand break, is detected by the protein kinases ATM and ATR, they activate p53 to serve as a transcription factor that stimulates the synthesis of the protein p21. This protein inhibits the protein kinase activity of cyclin E–CDK2. In the presence of p21, pRb remains unphosphorylated and bound to E2F, blocking the activity of this transcription factor, and the cell cycle is arrested in G1. This gives the cell time to repair its DNA before entering the S phase, thereby avoiding the potentially disastrous transfer of a defective genome to one or both daughter cells.

Oncogenes, Tumor Suppressor Genes, and Programmed Cell Death

Tumors and cancer are the result of uncontrolled cell division. Normally, cell division is regulated by a family of extracellular **growth factors,** proteins that cause resting cells to divide and, in some cases, differentiate. Some growth factors stimulate division of only those cells with appropriate receptors; others have a more general effect. Defects in the synthesis, regulation, or recognition of growth factors can lead to cancer.

Oncogenes Are Mutant Forms of the Genes for Proteins that Regulate the Cell Cycle

Oncogenes were originally discovered in tumor-causing viruses, then later found to be closely similar to or derived from genes present in the animal host cells, called proto-oncogenes, which encode growth-regulating proteins. During viral infections, the DNA sequence of a proto-oncogene is sometimes copied by the virus and incorporated into its genome (Fig. 13–36). At some point during the viral infection cycle, the gene can become defective by truncation or mutation. When this viral oncogene is expressed in its host cell during a subsequent infection, the abnormal protein product interferes with normal regulation of cell growth, sometimes resulting in a tumor.

Proto-oncogenes can become oncogenes without a viral intermediary. Chromosomal rearrangements, chemical agents, and radiation are among the factors that can cause oncogenic mutations. The mutations that produce oncogenes are genetically dominant; if either of a pair of chromosomes contains a defective gene, that gene product sends the signal "divide" and a tumor will result. The oncogenic defect can be in any of the proteins involved in communicating the "divide" signal. We know of oncogenes that

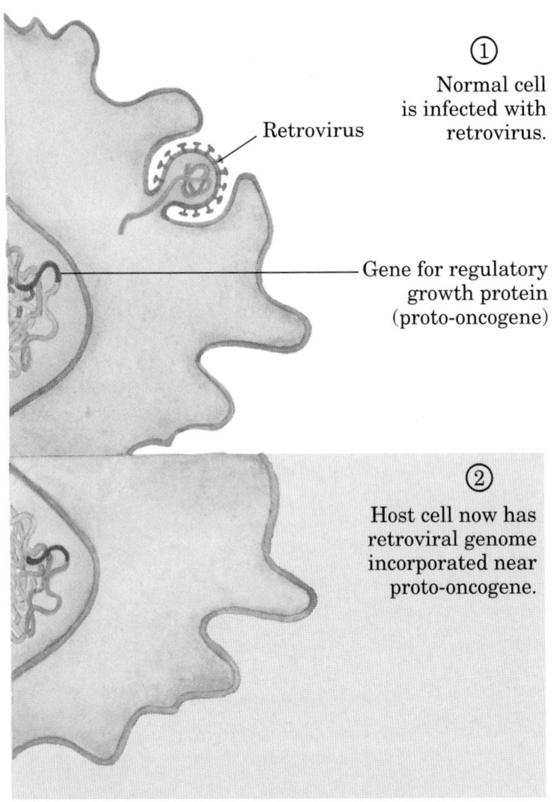
①
Normal cell is infected with retrovirus.
Retrovirus
Gene for regulatory growth protein (proto-oncogene)
②
Host cell now has retroviral genome incorporated near proto-oncogene.

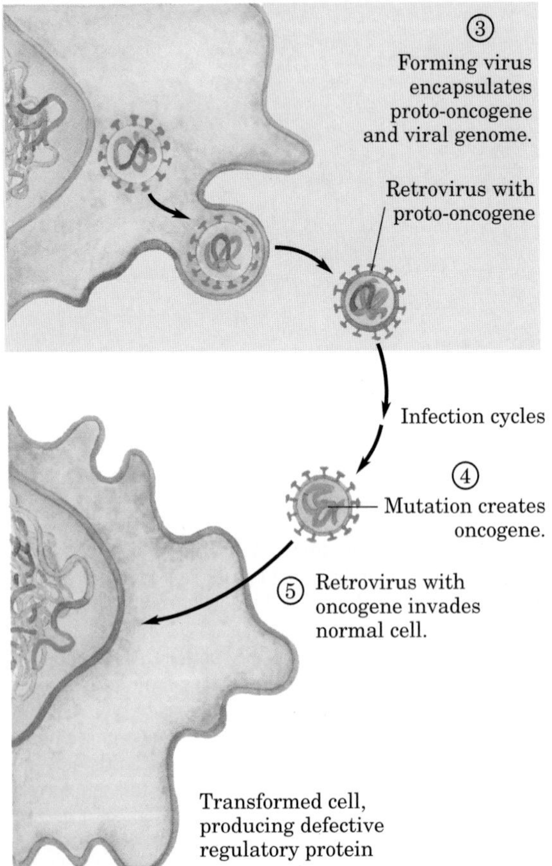
③
Forming virus encapsulates proto-oncogene and viral genome.
Retrovirus with proto-oncogene
Infection cycles
④
Mutation creates oncogene.
⑤ Retrovirus with oncogene invades normal cell.
Transformed cell, producing defective regulatory protein

figure 13–36
Conversion of a regulatory gene into a viral oncogene. ① A normal cell is infected by a retrovirus (Chapter 26), which ② inserts its own genome into the chromosome of the host cell, near the gene for a regulatory protein (the proto-oncogene). ③ Viral particles released from the infected cell sometimes "capture" a host gene, in this case a proto-oncogene. ④ During several cycles of infection, a mutation occurs in the viral proto-oncogene, converting it to an oncogene. ⑤ When the virus subsequently infects a cell, it introduces the oncogene into the cell's DNA. Transcription of the oncogene leads to the production of a defective regulatory protein that continuously gives the signal for cell division, overriding normal regulatory mechanisms. Host cells infected with oncogene-carrying viruses therefore undergo unregulated cell division—they form tumors. Proto-oncogenes can also undergo mutation to oncogenes without the intervention of a retrovirus, as described in the text.

encode secreted proteins, growth factors, transmembrane proteins (receptors), cytoplasmic proteins (G proteins and protein kinases), and the nuclear transcription factors that control the expression of genes essential for cell division (Jun, Fos).

Some oncogenes encode surface receptors with defective or missing signal-binding sites such that the tyrosine kinase activity is unregulated. For example, the protein ErbB is essentially identical to the normal receptor for epidermal growth factor (EGF), except that ErbB lacks the amino-terminal domain that normally binds EGF (Fig. 13–37) and as a result sends the signal to divide whether EGF is present or not. One ErbB variant, the product of the *erb*B2 oncogene (for an explanation of the use of abbreviations in naming genes and their products, see Chapter 25), is commonly associated with cancers of the glandular epithelium in breast, stomach, and ovary.

Mutant forms of the G protein Ras are common in tumor cells. The *ras* oncogene encodes a protein with normal GTP binding but no GTPase activity. The mutant Ras protein is therefore always in its activated (GTP-bound) form, regardless of the signals coming through normal receptors. The result can be unregulated growth. Mutations in *ras* are associated with 30% to 50% of lung and colon carcinomas and more than 90% of pancreatic carcinomas.

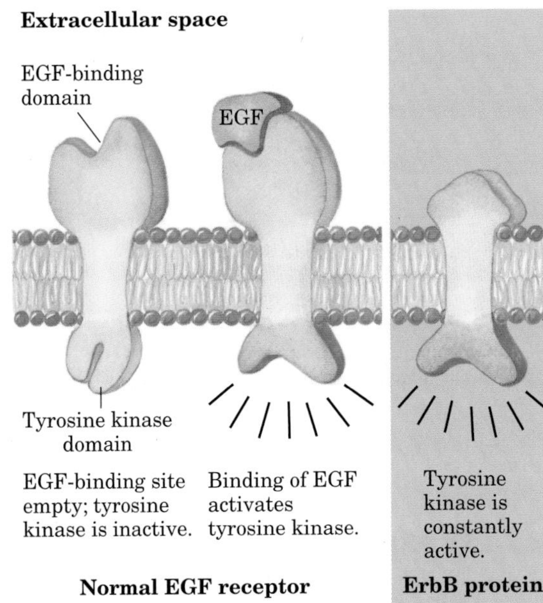
Extracellular space
EGF-binding domain
EGF
Tyrosine kinase domain
EGF-binding site empty; tyrosine kinase is inactive.
Binding of EGF activates tyrosine kinase.
Tyrosine kinase is constantly active.
Normal EGF receptor
ErbB protein

figure 13–37
Oncogene-encoded defective EGF receptor. The product of the *erb*B oncogene (the ErbB protein) is a truncated version of the normal receptor for epidermal growth factor (EGF). Its intracellular domain has the structure normally induced by EGF binding, but the protein lacks the extracellular binding site for EGF. Unregulated by EGF, ErbB continuously signals cell division.

Defects in Tumor Suppressor Genes Remove Normal Restraints on Cell Division

Tumor suppressor genes encode proteins that normally restrain cell division. Mutation in one or more of these genes can lead to tumor formation. Unregulated growth due to defective tumor suppressor genes, unlike that due to oncogenes, is genetically recessive; tumors form only if *both* chromosomes of a pair contain a defective gene. A person who inherits one correct copy and one defective copy will not be diseased, but every cell in that person's body will have one defective copy of the gene. If any of those 10^{12} somatic cells subsequently has a mutation in the one good copy, a tumor may grow from that doubly mutant cell. Mutations in both copies of the genes for pRb, p53, or p21 yield cells in which the normal restraint on cell division is absent, and tumor formation results.

Retinoblastoma is a cancer of the retina that occurs in children who have two defective *Rb* alleles. Very young children who develop retinoblastoma commonly have multiple tumors in both eyes. Each tumor is derived from a single retinal cell that has undergone a mutation in the one good copy of the *Rb* gene. (A fetus with two mutant alleles in every cell is nonviable.) Retinoblastoma patients also have a high incidence of cancers of the lung, prostate, and breast.

A far less likely event is that a person born with two good copies of a gene will have two independent mutations in the *same* gene in the *same* cell, but this does occur. Some individuals develop retinoblastomas later in childhood, usually with only one tumor in only one eye. These individuals were presumably born with two good copies of *Rb* in every cell, but have had mutations in both *Rb* genes in a single retinal cell, leading to a tumor.

Mutations in the gene for p53 also cause tumors; in more than 90% of human cutaneous squamous cell carcinomas (skin cancers) and about 50% of all other human cancers, *p53* is defective. Those very rare individuals who inherit one defective copy of *p53* commonly have the Li-Fraumeni syndrome, in which multiple cancers (of the breast, brain, bone, blood, lung, and skin) occur at high frequency and at an early age. The explanation for multiple tumors in this case is the same as that for *Rb* mutations: an individual born with one defective copy of *p53* in every somatic cell is likely to suffer a second *p53* mutation in more than one cell in his or her lifetime.

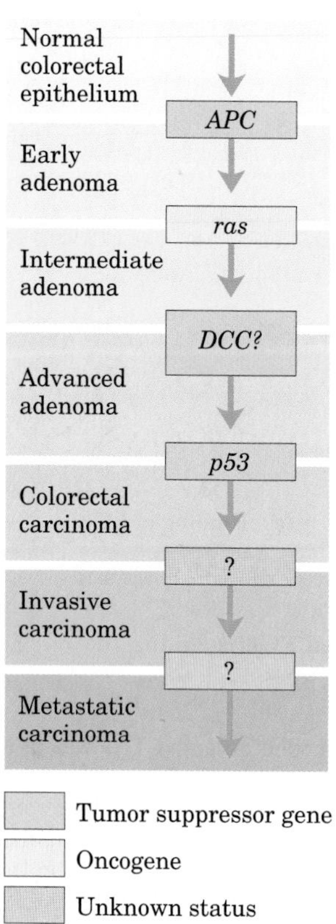

figure 13–38

From normal epithelial cell to colorectal cancer. In the colon, mutations in both copies of the tumor suppressor gene *APC* lead to benign clusters of epithelial cells that multiply too rapidly (early adenoma). If a cell already defective in *APC* suffers a second mutation in the proto-oncogene *ras,* the doubly mutant cell gives rise to an intermediate adenoma, forming a benign polyp of the colon. When one of these cells undergoes further mutations in the tumor suppressor genes *DCC* (probably) and *p53,* progressively more aggressive tumors form. Finally, mutations in genes not yet characterized lead to a malignant tumor and finally to a metastatic tumor that can spread to other tissues. Most malignant tumors are probably the result of a series of mutations such as this.

The effect of mutations in oncogenes and tumor suppressor genes is not all-or-none. In some cancers, perhaps in all, the progression from a normal cell to a malignant tumor involves the accumulation of mutations (sometimes over several decades), none of which alone is responsible for the end effect. For example, the development of colorectal cancer has several recognizable stages, each associated with a mutation (Fig. 13–38). If a normal epithelial cell in the colon undergoes mutation of both copies of the tumor suppressor gene *APC* (adenomatous polyposis coli), it begins to divide faster than normal and produces a clone of itself, a benign polyp (early adenoma). For reasons not yet known, the *APC* mutation results in chromosomal instability; whole regions of a chromosome are lost or rearranged during cell division. This instability can lead to another mutation, commonly in *ras*, that converts the clone into an intermediate adenoma. A third mutation (probably in the tumor suppressor gene *DCC*) leads to a late adenoma. Only when both copies of *p53* become defective does this cell mass become a carcinoma, a malignant, life-threatening cancer. The full sequence therefore requires at least seven genetic "hits": two on each of three tumor suppressor genes (*APC, DCC,* and *p53*) and one on the proto-oncogene *ras*. There are probably several other routes to colorectal cancer as well, but the principle that full malignancy results only from multiple mutations is likely to hold. When a polyp is detected in the early adenoma stage and the cells containing the first mutations are removed surgically, late adenomas and carcinomas will not develop; hence the importance of early detection.

Apoptosis Is Programmed Cell Suicide

Many cells can precisely control the time of their own death by the process of **programmed cell death,** or **apoptosis** (from the Greek for "dropping off," as in leaves dropping in the fall; pronounced app′-a-toe′-sis). In the development of an embryo, for example, some cells must die. Carving fingers from stubby limb buds involves the precisely timed death of cells between developing finger bones. During the development of the nematode *Caenorhabditis elegans* from a fertilized egg, exactly 131 cells (of a total of 1,090 somatic cells in the embryo) must undergo programmed death in order to construct the adult body.

Apoptosis also has roles in processes other than development. When an antibody-producing cell is found to be making antibodies against an antigen normally present in the body, that cell undergoes programmed death in the thymus gland—an essential mechanism for eliminating anti-self antibodies.

The monthly sloughing of cells of the uterine wall (menstruation) is another case of apoptosis mediating normal cell death. Sometimes cell suicide is not programmed but occurs in response to biological circumstances that threaten the rest of the organism. For example, a virus-infected cell that dies before completion of the infection cycle prevents spread of the virus to nearby cells. Severe stresses such as heat, hyperosmolarity, UV light, and gamma irradiation also trigger cell suicide; presumably the organism is better off with these cells dead than alive and aberrant.

The regulatory mechanisms that trigger apoptosis involve some of the same proteins that regulate the cell cycle. The signal for suicide often comes from outside, through a surface receptor. Tumor necrosis factor (TNF), produced by cells of the immune system, interacts with cells through specific TNF receptors. These receptors have TNF-binding sites on the outer face of the plasma membrane and a "death domain" of about 80 amino acid residues that passes the self-destruct signal through the membrane to cytosolic proteins such as TRADD (*T*NF *r*eceptor–*a*ssociated *d*eath *d*omain) (Fig. 13–39). Another receptor, Fas, has a similar death domain that allows it to interact with the cytosolic protein FADD (*F*as-*a*ssociated *d*eath *d*omain), which activates a cytosolic protease called caspase 8. This enzyme belongs to a family of proteases involved in apoptosis; all are synthesized as inactive proenzymes, all have a critical Cys residue at the active site, and all hydrolyze their target proteins on the carboxyl-terminal side of specific Asp residues.

When caspase 8, an "initiator" caspase, is activated by an apoptotic signal carried through FADD, it further self-activates by cleaving its own proenzyme form. Mitochondria are one of the targets of active caspase 8. The protease causes the release of certain proteins found between the inner and outer mitochondrial membranes: cytochrome *c* (Chapter 19) and several "effector" caspases. Cytochrome *c* binds to the proenzyme forms of effector caspases and stimulates their proteolytic activation by caspase 8. The activated effector caspases in turn catalyze wholesale destruction of cellular proteins, which is a major cause of apoptotic cell death. One specific target of caspase action is a caspase-activated DNase.

The monomeric products of protein and DNA degradation (amino acids and nucleotides) are released in a controlled process that allows them to be taken up and reused by neighboring cells. Apoptosis thus allows the organism to eliminate a cell without wasting its components.

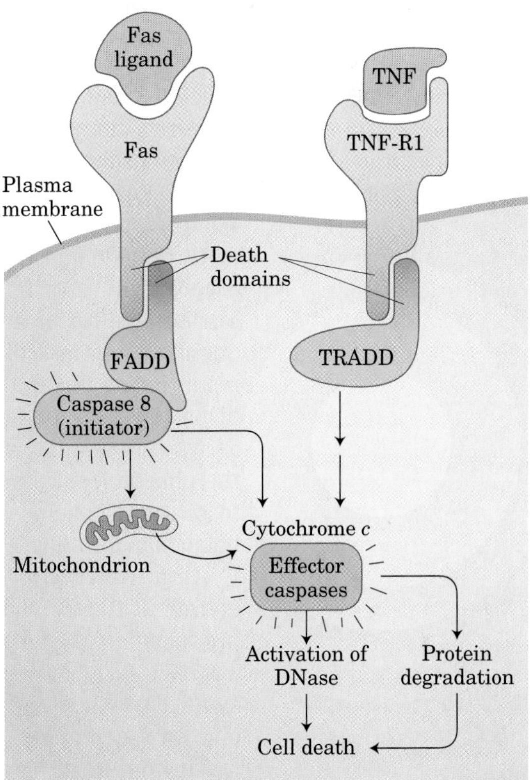

figure 13–39
Initial events of apoptosis. Receptors in the plasma membrane (Fas, TNF-R1) receive signals from outside the cell (the Fas ligand or tumor necrosis factor (TNF), respectively). Receptor activation allows interaction between the "death domain" (an 80-residue sequence) in Fas or TNF-R1 and a similar death domain in the cytosolic proteins FADD or TRADD. FADD activates a cytosolic protease called caspase-8, which proteolytically activates other cellular proteases. TRADD also activates proteases. The resulting proteolysis is a major factor in cell death.

summary

All cells have specific and highly sensitive signal-transducing mechanisms, which have been conserved during evolution. A wide variety of stimuli, including hormones, neurotransmitters, and growth factors, act through specific protein receptors in the plasma membrane. The receptors bind the signal molecule, amplify the signal, integrate it with input from other receptors, and transmit it into the cell. If the signal persists, receptor desensitization reduces or ends the response. Eukaryotic cells have four general signaling mechanisms: (1) gated ion channels in the plasma membrane; (2) enzymes with a catalytic domain inside the cell and an extracellular receptor for the signal; (3) membrane proteins that stimulate target proteins (enzymes or ion channels) by activating G proteins; and (4) nuclear proteins that bind steroids and act as transcription factors.

Ion channels gated by ligands or membrane potential are central to signaling in neurons and other cells. The voltage-gated Na^+ and K^+ channels of neuronal membranes carry the action potential along the axon as a wave of depolarization (Na^+ influx) followed by repolarization (K^+ efflux). The arrival of an action potential triggers neurotransmitter release from the presynaptic cell. The neurotransmitter (acetylcholine, for example) diffuses to the postsynaptic cell, binds to specific receptors in the plasma membrane, and triggers a change in V_m.

The insulin receptor is the prototype of receptor enzymes with tyrosine kinase activity. When insulin binds to the receptor, the enzymatic domain phosphorylates first itself and then other target proteins on Tyr residues, creating a binding site for proteins with SH2 domains. Sequential protein-protein interactions eventually result in the binding of GTP to the Ras protein, displacing GDP and activating Ras. Ras in turn activates a protein kinase cascade that ends with the phosphorylation of target proteins in the cytosol and the nucleus. The result is specific metabolic changes and altered gene expression.

Atrial natriuretic factor binds to a receptor enzyme with guanylyl cyclase activity. The cGMP produced acts as a second messenger, activating cGMP-dependent protein kinase (PKG). This enzyme alters metabolism by phosphorylating specific enzyme targets.

The β-adrenergic receptor typifies the large family of receptors that have seven transmembrane segments and act through trimeric G proteins. On ligand binding, these serpentine receptors catalyze the exchange of GTP for GDP bound to G_s. GTP binding forces dissociation of the α subunit of G_s, and this subunit activates adenylyl cyclase in the plasma membrane. The cAMP produced is an intracellular second messenger that stimulates cAMP-dependent protein kinase (PKA). PKA mediates the effects of epinephrine by phosphorylating key proteins in carbohydrate- and fat-metabolizing pathways. This cascade of events, in which a single molecule of hormone activates a catalyst that in turn activates another catalyst and so on, results in large signal amplification; this is characteristic of all hormone-activated systems. The second messenger is eventually destroyed by cAMP phosphodiesterase, and G_s turns itself off by hydrolysis of its bound GTP to GDP, ending the response. When the epinephrine signal persists, the action of a β-adrenergic receptor-specific protein kinase (βARK) and arrestin temporarily desensitizes the cell to epinephrine. Some receptors stimulate adenylyl cyclase through G_s; others inhibit it through G_i. Thus cellular [cAMP] reflects the integrated input of two (or more) signals.

Some serpentine receptors are coupled to a plasma membrane phospholipase C (PLC) that cleaves PIP_2 to diacylglycerol and IP_3. By opening Ca^{2+} channels in the endoplasmic reticulum, IP_3 raises cytosolic [Ca^{2+}], and Ca^{2+} combines with diacylglycerol to activate protein kinase C (PKC). This enzyme, by phosphorylating specific cellular proteins, produces the cell's response to the extracellular signal. Cellular [Ca^{2+}] regulates a number of enzymes, often through the Ca^{2+}-binding protein calmodulin.

Vision, olfaction, and gustation in vertebrates employ serpentine receptors, which act through trimeric G proteins to change the sensory neuron's V_m. In rod and cone cells of the retina, light activates rhodopsin, which stimulates replacement of GDP by GTP on the G protein transducin. The freed α subunit of transducin activates cGMP phosphodiesterase, which then lowers [cGMP] and thus closes cGMP-dependent ion channels in the neurons' outer segments. The resulting hyperpolarization of the rod or cone cell carries the signal to the next neuron in the pathway, and eventually to the brain. In olfactory neurons, olfactory stimuli, acting through serpentine receptors and G proteins, trigger either an increase in [cAMP] (by activating adenylyl cyclase) or an increase in [Ca^{2+}] (by activating

PLC). These second messengers affect ion channels and thus the V_m. Gustatory neurons have serpentine receptors that respond to tastes by raising (sweet) or lowering (bitter) [cAMP], which in turn changes V_m by gating ion channels.

Steroid hormones enter cells and bind to specific receptor proteins. The hormone-receptor complex binds specific regions of DNA called hormone response elements and regulates the expression of nearby genes.

Many signaling pathways lead eventually to activation of a protein kinase, which alters the activity of its target proteins by adding a phosphoryl group to a Ser, Thr, or Tyr residue. Progression through the cell cycle is regulated by a family of protein kinases that act at specific periods of the cycle, phosphorylating key proteins and modulating their activities. The catalytic subunit of these cyclin-dependent protein kinases (CDKs) is inactive unless associated with the regulatory cyclin subunit. The activity of a cyclin-CDK complex changes during the cell cycle through differential synthesis of CDKs, specific degradation of cyclin, phosphorylation and dephosphorylation of critical residues in CDKs, and binding of inhibitory proteins to specific cyclin-CDKs.

Oncogenes encode defective signaling proteins. By continually giving the signal for cell division, they lead to tumor formation. Oncogenes are genetically dominant and may encode defective growth factors, receptors, G proteins, protein kinases, or nuclear regulators of transcription. Tumor suppressor genes encode regulatory proteins that normally inhibit cell division; mutations in these genes can lead to tumor formation, but are genetically recessive. Cancer is generally the result of an accumulation of mutations in oncogenes and tumor suppressor genes.

Apoptosis (programmed cell death) involves some proteins that regulate the cell cycle and others unique to the cell death pathway. Triggered by external signals such as tumor necrosis factor or by internal mechanisms, apoptosis results in activation of proteases that degrade cellular proteins.

further reading

Receptor–Ion Channels

(See also the further readings on ion channels in Chapter 12, under Solute Transport across Membranes.)

Aidley, D.J. & Stanfield, P.R. (1996) Ion Channels: Molecules in Action. *Cambridge University Press, Cambridge.* **89,** 829–830.

Clear, concise introduction to the physics, chemistry, and molecular biology used in research on ion channels; emphasis is on molecular approaches.

Changeux, J.-P. (1993) Chemical signaling in the brain. *Sci. Am.* **269** (November), 58–62.

Introductory account of the structure and signaling mechanism of the acetylcholine receptor.

Lehmann-Horn, F. & Jurkat-Rott, K. (1999) Voltage-gated ion channels and hereditary disease. *Physiol. Rev.* **79,** 1317–1372.

Advanced review of the structure and function of ion channels, with special emphasis on cases in which ion channel defects produce human diseases.

Receptor Enzymes

Fantl, W.J., Johnson, D.E., & Williams, L.T. (1993) Signaling by receptor tyrosine kinases. *Annu. Rev. Biochem.* **62,** 453–481.

Foster, D.C., Wedel, B.J., Robinson, S.W., & Garbers, D.L. (1999) Mechanisms of regulation and functions of guanylyl cyclases. *Rev. Physiol. Biochem. Pharmacol.* **135,** 1–39.

Advanced review of the structure and function of signal-transducing guanylyl cyclases.

Schaeffer, H.J. & Weber, M.J. (1999) Mitogen-activated protein kinases: specific messages from ubiquitous messengers. *Mol. Cell. Biol.* **19,** 2435–2444.

Intermediate review of MAPKs and the basis for specific signaling through these general signaling proteins.

Seger, R. & Krebs, E.G. (1995) The MAPK signaling cascade. *FASEB J.* **9,** 726–735.

Intermediate review.

Shepherd, P.R., Withers, D.J., & Siddle, K. (1998) Phosphoinositide 3-kinase: the key switch mechanism in insulin signalling. *Biochem. J.* **333,** 471–490.

Intermediate review of the importance of PKB and PI-3K in metabolic regulation by insulin.

Widmann, C., Gibson, S., Jarpe, M.B., & Johnson, G.L. (1999) Mitogen-activated protein kinase: conservation of a three-kinase module from yeast to human. *Physiol. Rev.* **79,** 143–180.

Advanced review of the roles of MAP kinases in diverse organisms, from yeast, slime mold, and nematode to vertebrates and plants.

Serpentine Receptors

Berridge, M.J. (1993) Inositol triphosphate and calcium signalling. *Nature* **361,** 315–325.

Classic description of the IP_3 signaling system.

Hamm, H.E. (1998) The many faces of G protein signaling. *J. Biol. Chem.* **273,** 669–672.

Introduces a series of short reviews on G proteins.

Helmreich, E.J. & Hofmann, K.P. (1996) Structure and function of proteins in G-protein-coupled signal transfer. *Biochim. Biophys. Acta* **1286,** 285–322.

Lee, A.G. (ed.) (1996) *Biomembranes,* Vols 2A and 2B, *Rhodopsins, Receptors, and G-proteins,* JAI Press, Hampton Hill, England.

Twelve excellent articles on serpentine receptors and their transduction mechanisms.

Martin, T.F.J. (1998) Phosphoinositide lipids as signaling molecules. *Annu. Rev. Cell Dev. Biol.* **14,** 231–264.

Discussion of the roles of phosphatidylinositol derivatives in signal transduction, cytoskeletal regulation, and membrane trafficking.

Skiba, N.P. & Hamm, H.E. (1998) How $G_{s\alpha}$ activates adenylyl cyclase. *Nat. Struct. Biol.* **5,** 88–92.

Intermediate review of the mechanism of G-protein action, based on structural studies.

Takahashi, A., Camacho, P., Lechleiter, J.D. & Herman, B. (1999) Measurement of intracellular calcium. *Physiol Rev.* **79,** 1089–1125.

Advanced review of methods for estimating intracellular Ca^{2+} levels in real time.

Thomas, A.P., Bird, G.S.J., Hajnoczky, G., Robb-Gaspers, L.D., & Putney, J.W. Jr. (1996) Spatial and temporal aspects of cellular calcium signaling. *FASEB J.* **10,** 1505–1517.

Vision, Olfaction, and Gustation

Baylor, D. (1996) How photons start vision. *Proc. Natl. Acad. Sci. USA* **93,** 560–565.

One of six short reviews on vision in this journal issue.

Herness, M.S. & Gilbertson, T. (1999) Cellular mechanisms of taste transduction. *Annu. Rev. Physiol.* **61,** 873–900.

Advanced review of gustation.

Mombaerts, P. (1999) Seven-transmembrane proteins as odorant and chemosensory receptors. *Science* **286,** 707–711.

Intermediate review of olfactory transducers in vertebrates, nematodes, and fruit flies.

Nathans, J. (1989) The genes for color vision. *Sci. Am.* **260** (February), 42–49.

Schild, D. & Restrepo, D. (1998) Transduction mechanisms in vertebrate olfactory receptor cells. *Physiol. Rev.* **78,** 429–466.

Advanced review.

Scott, K. & Zuker, C. (1997) Lights out: deactivation of the phototransduction cascade. *Trends Biochem. Sci.* **22,** 350–354.

Protein Kinases and Protein Phosphorylation

(See also the further readings on protein phosphorylation in Chapter 8, under Regulatory Enzymes.)

Faux, M.C. & Scott, J.D. (1996) More on target with protein phosphorylation: conferring specificity by location. *Trends Biochem. Sci.* **21,** 312–315.

Description of the role of targeting proteins in holding protein kinases at their site(s) of action.

Pinna, L.A. & Ruzzene, M. (1996) How do protein kinases recognize their substrates? *Biochim. Biophys. Acta* **1314,** 191–225.

Advanced review of the factors, including consensus sequences, that give protein kinases their specificity.

Roach, P.J. (1991) Multisite and hierarchal protein phosphorylation. *J. Biol. Chem.* **266,** 14,139–14,142.

Looks at the importance of multiple phosphorylation sites in the fine regulation of protein function.

Steroid Hormone Receptors and Action

Carson-Jurica, M.A., Schrader, W.T., & O'Malley, B.W. (1990) Steroid receptor family—structure and functions. *Endocr. Rev.* **11,** 201–220.

Advanced discussion of the structure of nuclear hormone receptors and the mechanisms of their action.

Jordan, V.C. (1998) Designer estrogens. *Sci. Am.* (October), 60–67.

Introductory review of the mechanism of estrogen action and the effects of estrogenlike compounds in medicine.

Cell Cycle and Cancer

Cavenee, W.K. & White, R.L. (1995) The genetic basis of cancer. *Sci. Am.* **272** (March), 72–79.

Fearon, E.R. (1997) Human cancer syndromes—clues to the origin and nature of cancer. *Science* **278,** 1043–1050.

Interemediate review of the role of inherited mutations in the development of cancer.

Herwig, S. & Strauss, M. (1997) The retinoblastoma protein: a master regulator of cell cycle, differentiation and apoptosis. *Eur. J. Biochem.* **246,** 581–601.

Hunt, M. & Hunt, T. (1993) *The Cell Cycle: An Introduction,* W.H. Freeman and Company/Oxford University Press, New York/Oxford.

King, R.W., Deshaies, R.J., Peters, J.M., & Kirschner, M.W. (1996) How proteolysis drives the cell cycle. *Science* **274,** 1652–1659.

Intermediate description of the role of ubiquitin-dependent degradation of cyclins in regulating the cell cycle.

Kinzler, K.W. & Vogelstein, B. (1996) Lessons from hereditary colorectal cancer. *Cell* **87,** 159–170.

Evidence for multistep processes in the development of cancer.

Levine, A.J. (1997) p53, the cellular gatekeeper for growth and division. *Cell* **88,** 323–331.

Intermediate coverage of the function of protein p53 in the normal cell cycle and in cancer.

Morgan, D.O. (1997) Cyclin-dependent kinases: engines, clocks, and microprocessors. *Annu. Rev. Cell Dev. Biol.* **13,** 261–291.

Advanced review.

Weinberg, R.A. (1996) How cancer arises. *Sci. Am.* **275** (September), 62–70.

Apoptosis

Anderson, P. (1997) Kinase cascades regulating entry into apoptosis. *Microbiol. Mol. Biol. Rev.* **61,** 33–46.

Ashkenazi, A. & Dixit, V.M. (1998) Death receptors: signaling and modulation. *Science* **281,** 1305–1308.

This and the papers by Green and Reed and by Thornberry and Lazebnik (below) are in an issue of *Science* devoted to apoptosis.

Duke, R.C., Ojcius, D.M., & Young, J.D.-E. (1996) Cell suicide in health and disease. *Sci. Am.* **275** (December), 80–87.

Green, D.R. & Reed, J.C. (1998) Mitochondria and apoptosis. *Science* **281**, 1309–1312.

Jacobson, M.D., Weil, M., & Raff, M.C. (1997) Programmed cell death in animal development. *Cell* **88,** 347–354.

Thornberry, N.A. & Lazebnik, Y. (1998) Caspases: enemies within. *Science* **281**, 1312–1316.

problems

1. Therapeutic Effects of Albuterol The respiratory symptoms of asthma result from constriction of the bronchi and bronchioles of the lungs due to contraction of the smooth muscle of their walls. This constriction can be reversed by raising the [cAMP] in the smooth muscle. Explain the therapeutic effects of albuterol, a β-adrenergic agonist taken (by inhalation) for asthma. Would you expect this drug to have any side effects? How might one design a better drug that did not have these effects?

2. Amplification of Hormonal Signals Describe all the sources of amplification in the insulin receptor system.

3. Termination of Hormonal Signals Signals carried by hormones must eventually be terminated. Describe several different mechanisms for signal termination.

4. Specificity of a Signal for a Single Cell Type Discuss the validity of the following proposition. A signaling molecule (hormone, growth factor, or neurotransmitter) elicits identical responses in different types of target cells if they contain identical receptors.

5. Resting Membrane Potential A variety of unusual invertebrates, including giant clams, mussels, and polychaete worms live on the fringes of hydrothermal vents on the ocean bottom.

(a) The adductor muscle of a deep-sea giant clam has a resting membrane potential of −95 mV. Given the intracellular and extracellular ionic compositions shown below, would you have predicted this membrane potential? Why or why not?

Ion	Concentration (mM)	
	Intracellular	Extracellular
Na^+	440	50
K^+	20	400
Cl^-	560	21
Ca^{2+}	10	0.4

(b) Assuming that the adductor muscle membrane is permeable to only one of the ions listed above, which ion could determine the V_m?

6. Membrane Potentials in Frog Eggs Fertilization of a frog oocyte by a sperm triggers ionic changes similar to those observed in neurons (during movement of the action potential) and initiates the events that result in cell division and development of the embryo. Oocytes can be stimulated to divide without fertilization by suspending them in 80 mM KCl (normal pond water contains 9 mM KCl).

(a) How does the change in extracellular [KCl] affect the resting membrane potential of the oocyte? (Hint: Assume the oocyte contains 120 mM K^+ and is permeable *only* to K^+.) Assume a temperature of 20 °C.

(b) When the experiment is repeated in Ca^{2+}-free water, elevated [KCl] has no effect. What does this suggest about the mechanism of the KCl effect?

7. Excitation Triggered by Hyperpolarization In most neurons, membrane *depolarization* leads to the opening of voltage-dependent ion channels, generation of an action potential, and ultimately an influx of Ca^{2+}, which causes release of neurotransmitter at the axon terminus. Devise a cellular strategy by which *hyperpolarization* in rod cells could produce excitation of the visual pathway and passage of visual signals to the brain. (Hint: The neuronal signaling pathway in higher organisms consists of a *series* of neurons that relay information to the brain (see Fig. 13–20). The signal released by one neuron can be either excitatory or inhibitory to the following, postsynaptic neuron.)

8. Hormone Experiments in Cell-Free Systems In the 1950s, Earl W. Sutherland, Jr., and his colleagues carried out pioneering experiments to elucidate the mechanism of action of epinephrine and glucagon. Given what you have learned about hormone action in this chapter, interpret each of the experiments described below. Identify substance X and indicate the significance of the results.

(a) Addition of epinephrine to a homogenate of normal liver resulted in an increase in the activity of glycogen phosphorylase. However, if the homogenate was first centrifuged at a high speed and epinephrine or glucagon was added to the clear supernatant fraction that contains phosphorylase, no increase in the phosphorylase activity was observed.

(b) When the particulate fraction from the centrifugation in (a) was treated with epinephrine, substance X was produced. The substance was isolated and purified. Unlike epinephrine, substance X activated glycogen phosphorylase when added to the clear supernatant fraction of the centrifuged homogenate.

(c) Substance X was heat-stable; that is, heat treatment did not affect its capacity to activate phosphorylase. (Hint: Would this be the case if substance X were a protein?) Substance X was nearly identical to a compound obtained when pure ATP was treated with barium hydroxide. (Figure 10–6 will be helpful.)

9. Effect of Cholera Toxin on Adenylyl Cyclase The gram-negative bacterium *Vibrio cholerae* produces a protein, cholera toxin (M_r 90,000), that is responsible for the characteristic symptoms of cholera: extensive loss of body water and Na^+ through continuous, debilitating diarrhea. If body fluids and Na^+ are not replaced, severe dehydration results; untreated, the disease is often fatal. When the cholera toxin gains access to the human intestinal tract it binds tightly to specific sites in the plasma membrane of the epithelial cells lining the small intestine, causing adenylyl cyclase to undergo prolonged activation (hours or days).

(a) What is the effect of cholera toxin on [cAMP] in the intestinal cells?

(b) Based on the information above, suggest how cAMP normally functions in intestinal epithelial cells.

(c) Suggest a possible treatment for cholera.

10. Effect of Dibutyryl cAMP versus cAMP on Intact Cells The physiological effects of epinephrine should in principle be mimicked by addition of cAMP to the target cells. In practice, addition of cAMP to intact target cells elicits only a minimal physiological response. Why? When the structurally related derivative dibutyryl cAMP (shown below) is added to intact cells, the expected physiological response is readily apparent. Explain the basis for the difference in cellular response to these two substances. Dibutyryl cAMP is widely used in studies of cAMP function.

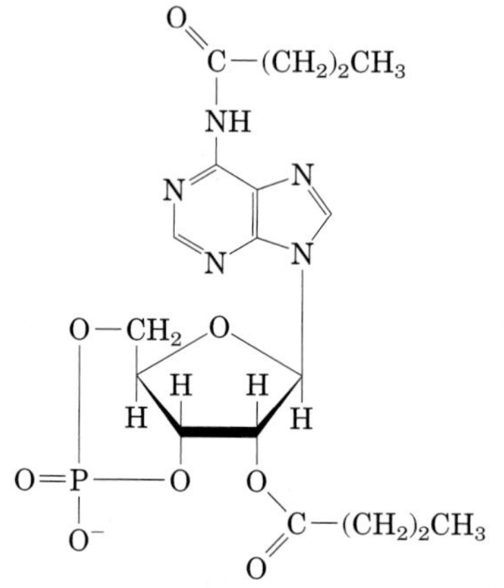

Dibutyryl cAMP
(N^6,$O^{2'}$-Dibutyryl adenosine 3′,5′-cyclic monophosphate)

11. Nonhydrolyzable GTP Analogs Many enzymes can hydrolyze GTP between the β and γ phosphates. The GTP analog β,γ-imidoguanosine 5′-triphosphate Gpp(NH)p, shown below, cannot be hydrolyzed between the β and γ phosphates. Predict the effect of microinjecting Gpp(NH)p into a myocyte on the cell's response to β-adrenergic stimulation.

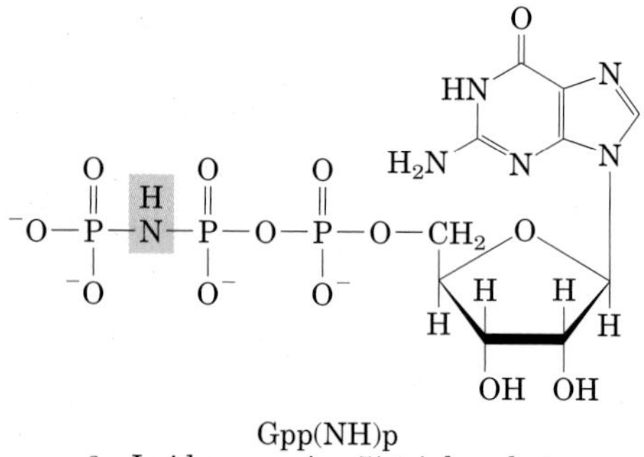

Gpp(NH)p
β,γ-Imidoguanosine 5′-triphosphate

12. G Protein Differences Compare the G proteins G_s, which acts in transducing the signal from β-adrenergic receptors, and Ras. What properties do they share? How do they differ? What is the functional difference between G_s and G_i?

13. EGTA Injection EGTA (ethylene glycol-bis(β-aminoethyl ether)-N,N,N′,N′-tetraacetic acid) is a chelating agent with high affinity and specificity for Ca^{2+}. By microinjecting a cell with an appropriate Ca^{2+}-EDTA solution, an experimenter can prevent cytosolic [Ca^{2+}] from rising above 10^{-7} M. How would EGTA microinjection affect a cell's response to vasopressin (see Table 13–5)? To glucagon?

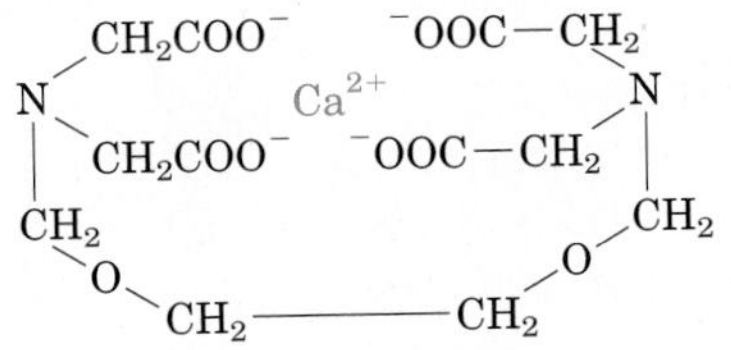

```
    CH2COO⁻        ⁻OOC—CH2
  /         Ca2+           \
N                            N
|  \                       / |
|   CH2COO⁻   ⁻OOC—CH2      |
CH2                         CH2
   \                       /
    O                     O
     \                   /
      CH2 ————————— CH2
```

EGTA (with bound Ca^{2+})
(Ethylene glycol-bis(β-aminoethyl ether)-N,N,N′,N′-tetraacetic acid)

14. Visual Desensitization Oguchi's disease is an inherited form of night blindness. Affected individuals are slow to recover vision after a flash of bright light against a dark background, such as the headlights of a car on the freeway. Suggest what the molecular defect(s) might be in Oguchi's disease. Explain in molecular terms how this defect accounts for the night blindness.

15. Mutations in PKA Explain how mutations in the R or C subunit of cAMP-dependent protein kinase (PKA) might lead to (a) a constantly active PKA or (b) a constantly inactive PKA.

16. Mechanisms for Regulating Protein Kinases Identify eight general types of protein kinases found in eukaryotic cells, and explain what factor is *directly* responsible for activating each type.

17. Mutations in Tumor Suppressor Genes and Oncogenes Explain why mutations in tumor suppressor genes are recessive (both copies of the gene must be defective for the regulation of cell division to be defective) whereas mutations in oncogenes are dominant.

18. Retinoblastoma in Children Explain why some children with retinoblastoma develop multiple tumors of the retina in both eyes, while others have a single tumor in only one eye.

19. Mutations in *ras* How does a mutation in the *ras* gene that leads to a Ras protein with no GTPase activity affect a cell's response to insulin?

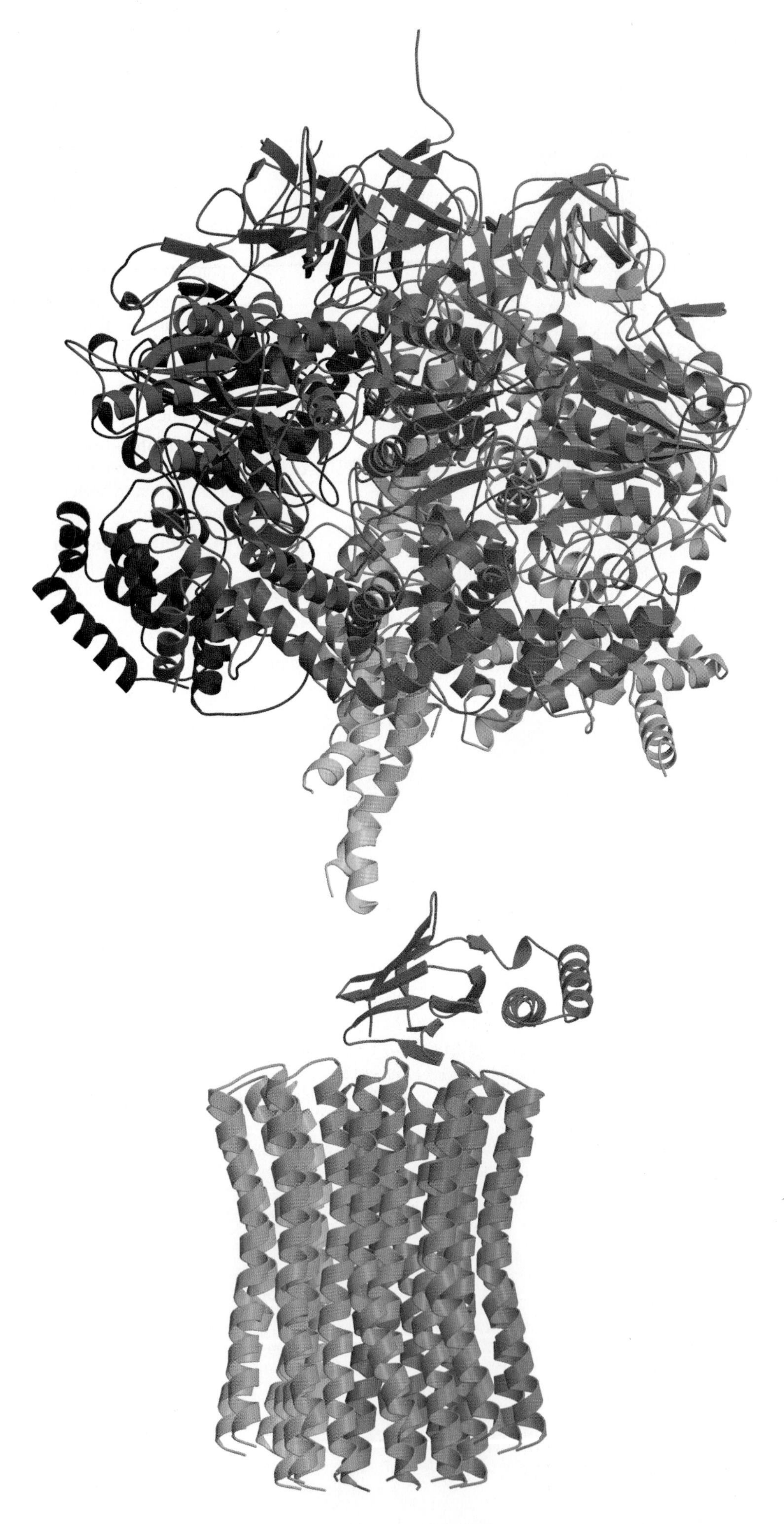

part

III

Bioenergetics and Metabolism

Metabolism is a highly coordinated cellular activity in which many multienzyme systems (metabolic pathways) cooperate to accomplish four functions: (1) obtain chemical energy by capturing solar energy or degrading energy-rich nutrients from the environment; (2) convert nutrient molecules into the cell's own characteristic molecules, including precursors of macromolecules; (3) polymerize monomeric precursors into macromolecules: proteins, nucleic acids, and polysaccharides; and (4) synthesize and degrade biomolecules required in specialized cellular functions, such as membrane lipids, intracellular messengers, and pigments.

Although metabolism embraces hundreds of different enzyme-catalyzed reactions, our major concern will be the central metabolic pathways, which are few in number and remarkably similar in all forms of life. Living organisms can be divided into two large groups according to the chemical form in which they obtain carbon from the environment. **Autotrophs** (such as photosynthetic bacteria and higher plants) can use carbon dioxide from the atmosphere as their sole source of carbon, from which they construct all their carbon-containing biomolecules (see Fig. 2–3). Some autotrophic organisms, such as cyanobacteria, can also use atmospheric nitrogen to generate all their nitrogenous components. **Heterotrophs** cannot use atmospheric carbon dioxide and must obtain carbon from their environment in the form of relatively complex organic molecules such as glucose. Higher animals and most microorganisms are heterotrophic. Autotrophic cells and organisms are relatively self-sufficient, whereas heterotrophic cells and organisms, with their requirements for carbon in more complex forms, must subsist on the products of other cells.

Many autotrophic organisms are photosynthetic and obtain their energy from sunlight, whereas heterotrophic organisms obtain their energy

facing page
ATP synthase, the enzyme responsible for the synthesis of most of the ATP made by aerobic cells and photosynthetic cells. This enzyme is central to chemiosmotic coupling, the universal mechanism by which bacterial, plant, and animal cells convert the energy of substrate oxidation or sunlight into the bonds of ATP. This view does not show several polypeptides that join the F_0 and F_1 subunits.

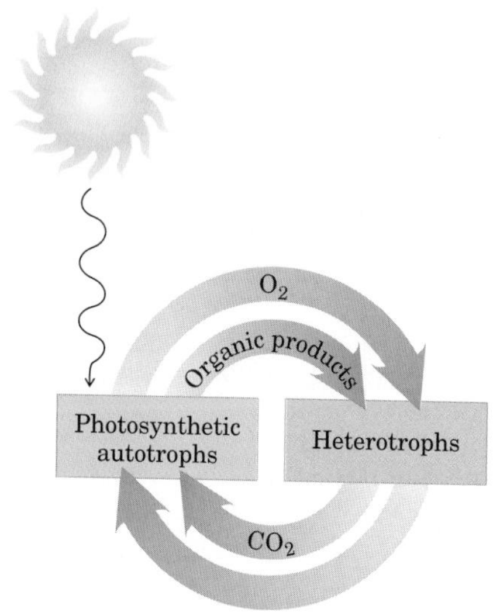

figure 1

Cycling of carbon dioxide and oxygen between the autotrophic (photosynthetic) and heterotrophic domains in the biosphere. The flow of mass through this cycle is enormous; about 4×10^{11} metric tons of carbon are turned over in the biosphere annually.

from the degradation of organic nutrients produced by autotrophs. In our biosphere, autotrophs and heterotrophs live together in a vast, interdependent cycle in which autotrophic organisms use atmospheric carbon dioxide to build their organic biomolecules, some of them generating oxygen from water in the process. Heterotrophs in turn use the organic products of autotrophs as nutrients and return carbon dioxide to the atmosphere. Some of the oxidation reactions that produce carbon dioxide also consume oxygen, converting it to water. Thus carbon, oxygen, and water are constantly cycled between the heterotrophic and autotrophic worlds, with solar energy as the driving force for this massive process (Fig. 1).

All living organisms also require a source of nitrogen, which is necessary for the synthesis of amino acids, nucleotides, and other compounds. Plants can generally use either ammonia or soluble nitrates as their sole source of nitrogen, but vertebrates must obtain nitrogen in the form of amino acids or other organic compounds. Only a few organisms—the cyanobacteria and many species of soil bacteria that live symbiotically on the roots of some plants—are capable of converting ("fixing") atmospheric nitrogen (N_2) into ammonia. Other bacteria (the nitrifying bacteria) oxidize ammonia to nitrites and nitrates, yet others convert nitrate to N_2. Thus, in addition to the global carbon and oxygen cycle, a nitrogen cycle operates in the biosphere in which huge amounts of nitrogen are turned over (Fig. 2). The cycling of carbon, oxygen, and nitrogen, which ultimately involves all species, depends on a proper balance between the activities of the producers (autotrophs) and consumers (heterotrophs) in our biosphere.

These cycles of matter are driven by an enormous flow of energy into and through the biosphere, beginning with the capture of solar energy by photosynthetic organisms and use of this energy to generate energy-rich carbohydrates and other organic nutrients; these nutrients are then used as energy sources by heterotrophic organisms. In metabolic processes, and in all energy transformations, there is a loss of useful energy (free energy) and an inevitable increase in the amount of unusable energy (heat and entropy). In contrast to the cycling of matter, therefore, energy flows one way through the biosphere; organisms cannot regenerate useful energy from energy dissipated as heat and entropy. Carbon, oxygen, and nitrogen recycle continuously, but energy is constantly transformed into unusable forms such as heat.

Metabolism, the sum of all the chemical transformations taking place in a cell or organism, occurs through a series of enzyme-catalyzed reactions that constitute **metabolic pathways.** Each of the consecutive steps in a metabolic pathway brings about a specific, small chemical change, usually the removal, transfer, or addition of a particular atom or functional group. The precursor is converted into a product through a series of metabolic intermediates called **metabolites.** The term **intermediary metabolism** is often applied to the combined activities of all the metabolic pathways that interconvert precursors, metabolites, and products of low molecular weight (generally M_r <1,000).

Catabolism is the degradative phase of metabolism in which organic nutrient molecules (carbohydrates, fats, and proteins) are converted into smaller, simpler end products (e.g., lactic acid, CO_2, NH_3). Catabolic pathways release energy, some of which is conserved in the formation of ATP and reduced electron carriers (NADH, NADPH, and $FADH_2$); the rest is lost as heat. In **anabolism,** also called biosynthesis, small, simple precursors are built up into larger and more complex molecules, including lipids, polysaccharides, proteins, and nucleic acids. Anabolic reactions require an input of energy, generally in the form of the phosphoryl group transfer potential of ATP and the reducing power of NADH, NADPH, and $FADH_2$ (Fig. 3).

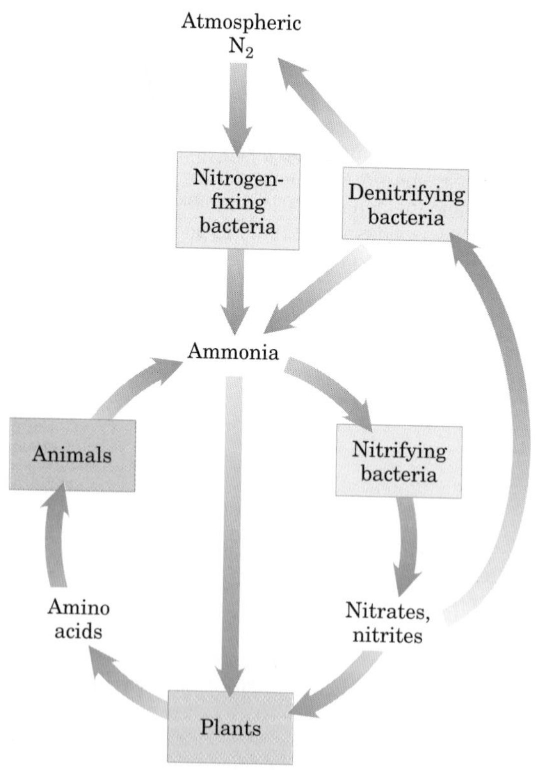

figure 2

Cycling of nitrogen in the biosphere. Gaseous nitrogen (N_2) makes up 80% of the earth's atmosphere.

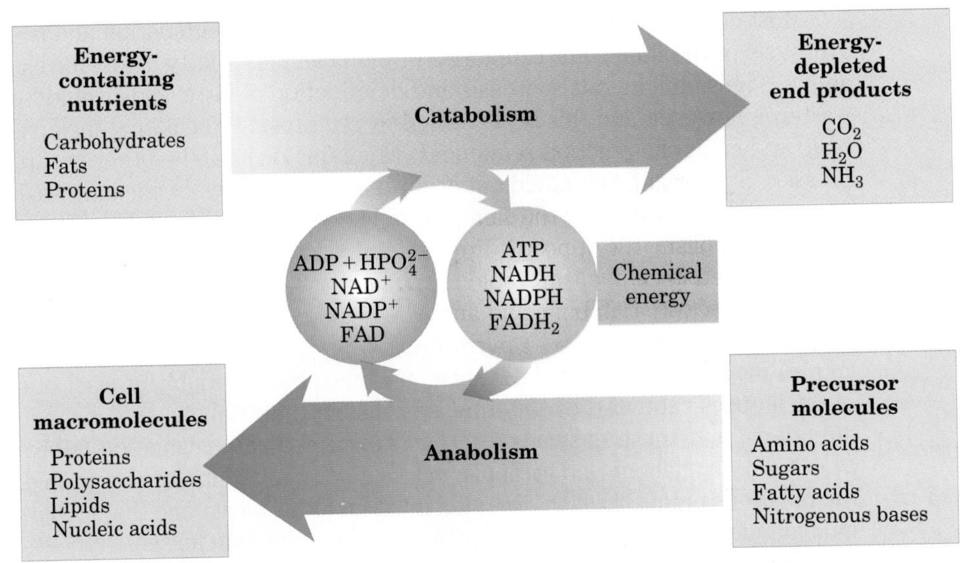

figure 3
Energy relationships between catabolic and anabolic pathways. Catabolic pathways deliver chemical energy in the form of ATP, NADH, NADPH, and $FADH_2$. These energy carriers are used in anabolic pathways to convert small precursor molecules into cell macromolecules.

Some metabolic pathways are linear, and some are branched, yielding multiple useful end products from a single precursor or converting several starting materials into a single product. In general, catabolic pathways are convergent and anabolic pathways divergent (Fig. 4). Some pathways are cyclic: one starting component of the pathway is regenerated in a series of reactions that converts another starting component into a product. We shall see examples of each type of pathway in the following chapters.

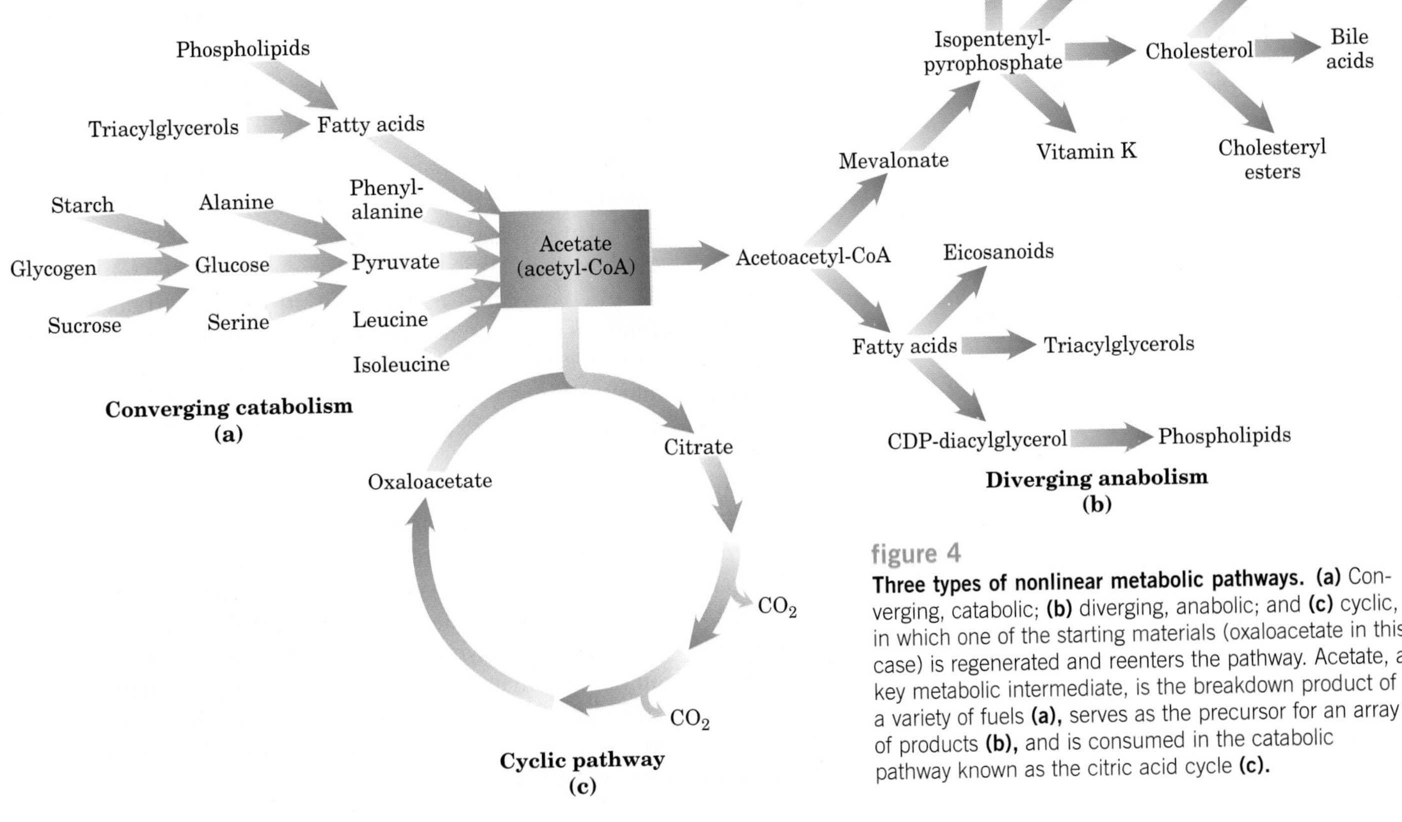

figure 4
Three types of nonlinear metabolic pathways. (a) Converging, catabolic; **(b)** diverging, anabolic; and **(c)** cyclic, in which one of the starting materials (oxaloacetate in this case) is regenerated and reenters the pathway. Acetate, a key metabolic intermediate, is the breakdown product of a variety of fuels **(a),** serves as the precursor for an array of products **(b),** and is consumed in the catabolic pathway known as the citric acid cycle **(c).**

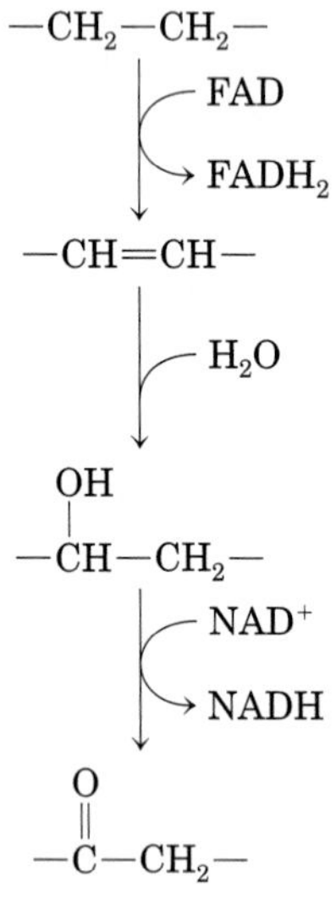

Most cells have the enzymes to carry out both the degradation and the synthesis of the important categories of biomolecules—fatty acids, for example. The simultaneous synthesis and degradation of fatty acids would be wasteful, however, and this is prevented by reciprocally regulating the anabolic and catabolic reaction sequences: when one occurs, the other is suppressed. Such regulation could not occur if anabolic and catabolic pathways were catalyzed by exactly the same set of enzymes, operating in one direction for anabolism, the opposite direction for catabolism: inhibition of an enzyme involved in catabolism would also inhibit the reaction sequence in the anabolic direction. Catabolic and anabolic pathways that connect the same two end points (glucose →→ pyruvate and pyruvate →→ glucose, for example) may employ many of the same enzymes, but invariably at least one of the steps is catalyzed by different enzymes in the catabolic and anabolic directions, and these enzymes are the sites of separate regulation. Moreover, for both anabolic and catabolic pathways to be essentially irreversible, the reactions unique to each direction must include at least one that is thermodynamically very favorable—in other words, a reaction for which the reverse reaction is very unfavorable. As a further contribution to the separate regulation of catabolic and anabolic reaction sequences, paired catabolic and anabolic pathways commonly occur in different cellular compartments: for example, fatty acid catabolism in mitochondria, fatty acid synthesis in the cytosol. The concentrations of intermediates, enzymes, and regulators can be maintained at different levels in these different compartments. Because metabolic pathways are subject to kinetic control by substrate concentration, separate pools of anabolic and catabolic intermediates also contribute to the control of metabolic rates. Devices that separate anabolic and catabolic processes will be of particular interest in our discussions of metabolism.

Metabolic pathways are regulated at several levels, from within the cell and from outside. The most immediate regulation is by the availability of substrate; when the intracellular concentration of an enzyme's substrate is below K_m, the enzyme acts below its V_{max}. A second type of rapid control from within is allosteric regulation (p. 278) by a metabolic intermediate or coenzyme (an amino acid or ATP, for example) that signals the cell's internal metabolic state. When the cell has an amount of an amino acid sufficient to its immediate needs, or when the cellular level of ATP indicates that further fuel consumption is unnecessary at the moment, these signals allosterically inhibit the activity of one or more enzymes in the relevant pathway. In multicellular organisms the metabolic activities of different tissues are regulated and integrated by growth factors and hormones that act from outside the cell. In some cases this regulation occurs virtually instantaneously (sometimes in less than a millisecond) through changes in the levels of intracellular messengers that modify the activity of existing enzyme molecules by allosteric mechanisms or by covalent modification such as phosphorylation. In other cases, the extracellular signal changes the cellular concentration of an enzyme by altering the rate of its synthesis or degradation, so the effect is seen only after many minutes or hours.

The number of metabolic transformations taking place in a typical cell can seem overwhelming to a beginning student. Fortunately, there are recurring patterns in metabolic pathways that make learning them easier. Certain types of reactions occur in many different metabolic pathways but always employ the same coenzyme(s) and the same general mechanism (Fig. 5). For example, succinate is converted to oxaloacetate in the citric acid cycle (see Fig. 16-7) by a three-step sequence that uses the same coenzymes and the same chemical mechanism as the conversion of a fatty acyl derivative to a β-ketoacyl derivative in the oxidation of fatty acids (see Fig. 17-8).

$$-CH_2-CH_2- \xrightarrow{FAD \rightarrow FADH_2} -CH{=}CH- \xrightarrow{H_2O} -CH(OH)-CH_2- \xrightarrow{NAD^+ \rightarrow NADH} -C({=}O)-CH_2-$$

figure 5

Common mechanism to oxidize an alkaline. The three-step process involves first dehydrogenation to introduce a double bond, with FAD as the electron acceptor, then addition of water across the double bond, and finally, oxidation of the alcohol to a ketone, with NAD^+ as the electron acceptor. This same mechanism is employed in the citric acid cycle, oxidation of fatty acids, and amino acid catabolism.

Once you have learned the general mechanism of a reaction, including the role of any necessary coenzyme(s), you will easily recognize the pattern in other metabolic pathways. In the chapters that follow, we usually discuss the general mechanism for reactions involving a coenzyme when we first encounter the coenzyme in its typical role.

In the first half of Part III we consider the major catabolic pathways by which cells obtain energy from the oxidation of various fuels: first, the central pathways for the conversion of hexoses to trioses (Chapter 15) and the oxidation of trioses to carbon dioxide (Chapter 16); then the pathways of fatty acid oxidation (Chapter 17) and amino acid oxidation (Chapter 18). Chapter 19 is the pivotal point of our discussion of metabolism; it concerns chemiosmotic energy coupling, the universal mechanism in which a transmembrane electrochemical potential, produced either by substrate oxidation or by light absorption, drives the synthesis of ATP.

The second half of Part III describes the major anabolic pathways by which cells use the energy in ATP to produce carbohydrates (Chapter 20), lipids (Chapter 21), and amino acids and nucleotides (Chapter 22) from simpler precursors. In Chapter 23 we step back from our detailed look at the metabolic pathways (as they occur in all organisms from *Eschericia coli* to humans) and consider how they are regulated and integrated in mammals by hormonal mechanisms.

We begin our study of intermediary metabolism with an introduction to bioenergetics (Chapter 14). But before we begin, a final word. Try not to forget that the myriad reactions described on these pages take place in, and play crucial roles in, living organisms. Ask for each reaction and for each pathway, "What does this chemical transformation do for the organism? How does this pathway interconnect with the other pathways operating simultaneously in the same cell to produce the energy and products required for cell maintenance and growth? How do the multilayered regulatory mechanisms cooperate to balance metabolic and energy inputs and outputs, achieving the dynamic steady state of life?" Learned with this perspective, metabolism provides fascinating and revealing insights into life, with countless applications in medicine, agriculture, and biotechnology.

chapter

14 Principles of Bioenergetics

Living cells and organisms must perform work to stay alive, to grow, and to reproduce. The ability to harness energy and to channel it into biological work is a fundamental property of all living organisms; it must have been acquired very early in cellular evolution. Modern organisms carry out a remarkable variety of energy transductions, conversions of one form of energy to another. They use the chemical energy in fuels to bring about the synthesis of complex, highly ordered macromolecules from simple precursors. They also convert the chemical energy of fuels into concentration gradients and electrical gradients, into motion and heat, and, in a few organisms such as fireflies and deep-sea fish, into light. Photosynthetic organisms transduce light energy into all these other forms of energy.

The chemical mechanisms that underlie biological energy transductions have fascinated and challenged biologists for centuries. Antoine Lavoisier, before he lost his head in the French Revolution, recognized that animals somehow transform chemical fuels (foods) into heat and that this process of respiration is essential to life. He observed that

Antoine Lavoisier
1743–1794

> . . . in general, respiration is nothing but a slow combustion of carbon and hydrogen, which is entirely similar to that which occurs in a lighted lamp or candle, and that, from this point of view, animals that respire are true combustible bodies that burn and consume themselves. . . . One may say that this analogy between combustion and respiration has not escaped the notice of the poets, or rather the philosophers of antiquity, and which they had expounded and interpreted. This fire stolen from heaven, this torch of Prometheus, does not only represent an ingenious and poetic idea, it is a faithful picture of the operations of nature, at least for animals that breathe; one may therefore say, with the ancients, that the torch of life lights itself at the moment the infant breathes for the first time, and it does not extinguish itself except at death.*

In this century, biochemical studies have revealed much of the chemistry underlying that "torch of life." Biological energy transductions obey the same physical laws that govern all other natural processes. It is therefore essential for a student of biochemistry to understand these laws and how they apply to the flow of energy in the biosphere. In this chapter we first review the laws of thermodynamics and the quantitative relationships among free energy, enthalpy, and entropy. We then describe the special role of ATP in biological energy exchanges. Finally, we consider the importance of oxidation-reduction reactions in living cells, the energetics of electron-transfer reactions, and the electron carriers commonly employed as cofactors of the enzymes that catalyze these reactions.

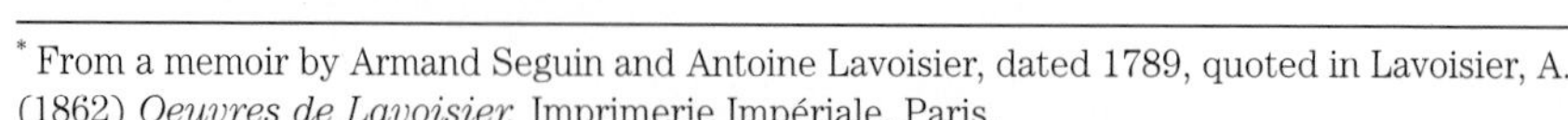

* From a memoir by Armand Seguin and Antoine Lavoisier, dated 1789, quoted in Lavoisier, A. (1862) *Oeuvres de Lavoisier,* Imprimerie Impériale, Paris.

Bioenergetics and Thermodynamics

Bioenergetics is the quantitative study of the energy transductions that occur in living cells and of the nature and function of the chemical processes underlying these transductions. Although many of the principles of thermodynamics have been introduced in earlier chapters and may be familiar to you, a review of the quantitative aspects of these principles is useful here.

Biological Energy Transformations Obey the Laws of Thermodynamics

Many quantitative observations made by physicists and chemists on the interconversion of different forms of energy led, in the nineteenth century, to the formulation of two fundamental laws of thermodynamics. The first law is the principle of the conservation of energy: *for any physical or chemical change, the total amount of energy in the universe remains constant; energy may change form or it may be transported from one region to another, but it cannot be created or destroyed.* The second law of thermodynamics, which can be stated in several forms, says that the universe always tends toward increasing disorder: *in all natural processes, the entropy of the universe increases.*

Living organisms consist of collections of molecules much more highly organized than the surrounding materials from which they are constructed, and organisms maintain and produce order, seemingly oblivious to the second law of thermodynamics. But living organisms do not violate the second law; they operate strictly within it. To discuss the application of the second law to biological systems, we must first define those systems and their surroundings. The reacting system is the collection of matter that is undergoing a particular chemical or physical process; it may be an organism, a cell, or two reacting compounds. The reacting system and its surroundings together constitute the universe. In the laboratory, some chemical or physical processes can be carried out in isolated or closed systems, in which no material or energy is exchanged with the surroundings. Living cells and organisms, however, are open systems, exchanging both material and energy with their surroundings; living systems are never at equilibrium with their surroundings, and the constant transactions between system and surroundings explain how organisms can create order within themselves while operating within the second law of thermodynamics.

"Now, in the *second* law of thermodynamics . . ."

Earlier in this text we defined three thermodynamic quantities that describe the energy changes occurring in a chemical reaction. Gibbs free energy (G) expresses the amount of energy capable of doing work during a reaction at constant temperature and pressure (p. 9). When a reaction proceeds with the release of free energy (i.e., when the system changes so as to possess less free energy), the free-energy change, ΔG, has a negative value and the reaction is said to be exergonic. In endergonic reactions, the system gains free energy and ΔG is positive. Enthalpy, H, is the heat content of the reacting system. It reflects the number and kinds of chemical bonds in the reactants and products. When a chemical reaction releases heat, it is said to be exothermic; the heat content of the products is less than that of the reactants and ΔH has, by convention, a negative value. Reacting systems that take up heat from their surroundings are endothermic and have positive values of ΔH (p. 65). Entropy, S, is a quantitative expression for the randomness or disorder in a system (Box 14–1). When the products of a reaction are less complex and more disordered than the reactants, the reaction is said to proceed with a gain in entropy (p. 72). The units of ΔG and ΔH are joules/mole or calories/mole (recall that 1 cal equals 4.184 J); units of entropy are joules/mole · Kelvin (J/mol · K) (Table 14–1).

table 14–1

Some Physical Constants and Units Used in Thermodynamics

Boltzmann constant, $\mathbf{k} = 1.381 \times 10^{-23}$ J/K
Avogadro's number, $N = 6.022 \times 10^{23}$ mol^{-1}
Faraday constant, $\mathcal{F} = 96{,}480$ J/V · mol
Gas constant, $R = 8.315$ J/mol · K
(= 1.987 cal/mol · K)

Units of ΔG and ΔH are J/mol (or cal/mol)
Units of ΔS are J/mol · K (or cal/mol · K)
1 cal = 4.184 J

Units of absolute temperature, T, are Kelvin, K
25 °C = 298 K
At 25 °C, $RT = 2.479$ kJ/mol
(= 0.592 kcal/mol)

Under the conditions existing in biological systems (including constant temperature and pressure), changes in free energy, enthalpy, and entropy are related to each other quantitatively by the equation

$$\Delta G = \Delta H - T\,\Delta S \tag{14–1}$$

in which ΔG is the change in Gibbs free energy of the reacting system, ΔH is the change in enthalpy of the system, T is the absolute temperature, and ΔS is the change in entropy of the system. By convention, ΔS has a positive sign when entropy increases and ΔH, as noted above, has a negative sign when heat is released by the system to its surroundings. Either of these conditions, which are typical of favorable processes, tend to make ΔG negative. In fact, ΔG of a spontaneously reacting system is always negative.

The second law of thermodynamics states that the entropy *of the universe* increases during all chemical and physical processes, but it does not require that the entropy increase take place *in the reacting system* itself.

box 14–1 Entropy: The Advantages of Being Disorganized

The term entropy, which literally means "a change within," was first used in 1851 by Rudolf Clausius, one of the formulators of the second law of thermodynamics. A rigorous quantitative definition of entropy involves statistical and probability considerations. However, its nature can be illustrated qualitatively by three simple examples, each demonstrating one aspect of entropy. The key descriptors of entropy are *randomness* and *disorder*, manifested in different ways.

Case 1: The Teakettle and the Randomization of Heat We know that steam generated from boiling water can do useful work. But suppose we turn off the burner under a teakettle full of water at 100 °C (the "system") in the kitchen (the "surroundings") and allow the teakettle to cool. As it cools, no work is done, but heat passes from the teakettle to the surroundings, raising the temperature of the surroundings (the kitchen) by an infinitesimally small amount until complete equilibrium is attained. At this point all parts of the teakettle and the kitchen are at precisely the same temperature. The free energy that was once concentrated in the teakettle of hot water at 100 °C, *potentially* capable of doing work, has disappeared. Its equivalent in heat energy is still present in the teakettle + kitchen (i.e., the "universe") but has become completely randomized throughout. This energy is no longer available to do work because there is no temperature differential within the kitchen. Moreover, the increase in entropy of the kitchen (the surroundings) is irreversible. We know from everyday experience that heat never spontaneously passes back from the kitchen into the teakettle to raise the temperature of the water to 100 °C again.

Case 2: The Oxidation of Glucose Entropy is a state or condition not only of energy but of matter. Aerobic (heterotrophic) organisms extract free energy from glucose obtained from their surroundings by oxidizing the glucose with molecular oxygen, also obtained from the surroundings. The end products of this oxidative metabolism, CO_2 and H_2O, are returned to the surroundings. In this process the surroundings undergo an increase in entropy, whereas the organism itself remains in a steady state and undergoes no change in its internal order. Although some entropy arises from the dissipation of heat, entropy also arises from another kind of disorder, illustrated by the equation for the oxidation of glucose:

$$C_6H_{12}O_6 + 6O_2 \longrightarrow 6CO_2 + 6H_2O$$

We can represent this schematically as

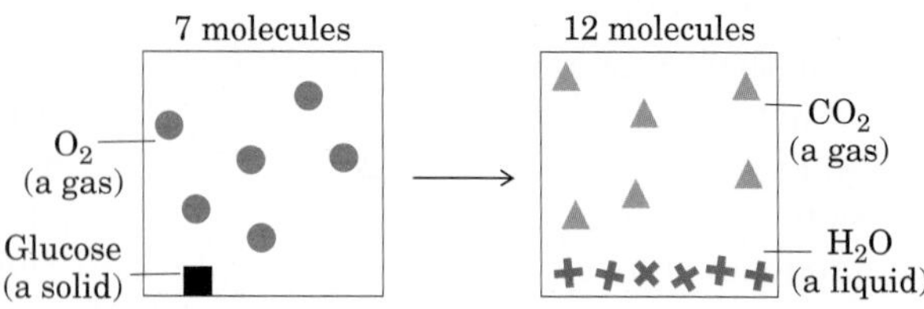

The order produced within cells as they grow and divide is more than compensated for by the disorder they create in their surroundings in the course of growth and division (Box 14–1, case 2). In short, living organisms preserve their internal order by taking from the surroundings free energy in the form of nutrients or sunlight, and returning to their surroundings an equal amount of energy as heat and entropy.

Cells Require Sources of Free Energy

Cells are isothermal systems—they function at essentially constant temperature (and constant pressure). Heat flow is not a source of energy for cells because heat can do work only as it passes to a zone or object at a lower temperature. The energy that cells can and must use is free energy, described by the Gibbs free-energy function G, which allows prediction of the direction of chemical reactions, their exact equilibrium position, and the amount of work they can in theory perform at constant temperature and

The atoms contained in 1 molecule of glucose plus 6 molecules of oxygen, a total of 7 molecules, are more randomly dispersed by the oxidation reaction and are now present in a total of 12 molecules ($6CO_2$ + $6H_2O$).

Whenever a chemical reaction results in an increase in the number of molecules—or when a solid substance is converted into liquid or gaseous products, which allow more freedom of molecular movement than solids—molecular disorder, and thus entropy, increases.

Case 3: Information and Entropy The following short passage from *Julius Caesar,* Act IV, Scene 3, is spoken by Brutus, when he realizes that he must face Mark Antony's army. It is an information-rich nonrandom arrangement of 125 letters of the English alphabet:

> There is a tide in the affairs of men,
> Which, taken at the flood, leads on to fortune;
> Omitted, all the voyage of their life
> Is bound in shallows and in miseries.

In addition to what this passage says overtly, it has many hidden meanings. It not only reflects a complex sequence of events in the play, it also echoes the play's ideas on conflict, ambition, and the demands of leadership. Permeated with Shakespeare's understanding of human nature, it is very rich in information.

However, if the 125 letters making up this quotation were allowed to fall into a completely random, chaotic pattern, as shown in the following box, they would have no meaning whatsoever.

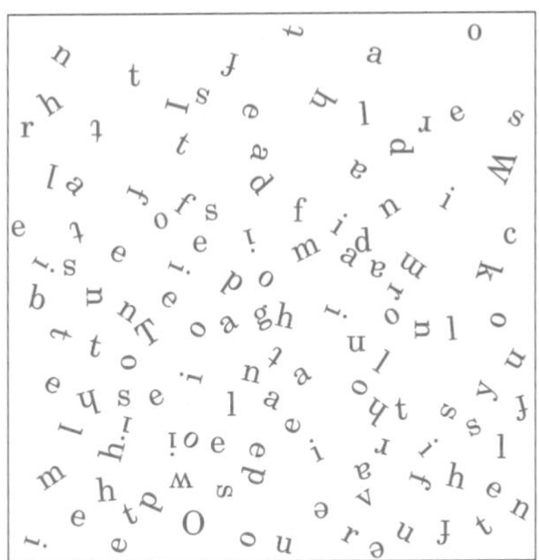

In this form the 125 letters contain little or no information, but they are very rich in entropy. Such considerations have led to the conclusion that information is a form of energy; information has been called "negative entropy." In fact, the branch of mathematics called information theory, which is basic to the programming logic of computers, is closely related to thermodynamic theory. Living organisms are highly ordered, nonrandom structures, immensely rich in information and thus entropy-poor.

pressure. Heterotrophic cells acquire free energy from nutrient molecules, and photosynthetic cells acquire it from absorbed solar radiation. Both kinds of cells transform this free energy into ATP and other energy-rich compounds capable of providing energy for biological work at constant temperature.

The Standard Free-Energy Change Is Directly Related to the Equilibrium Constant

The composition of a reacting system (a mixture of chemical reactants and products) tends to continue changing until equilibrium is reached. At the equilibrium concentration of reactants and products, the rates of the forward and reverse reactions are exactly equal and no further net change occurs in the system. The concentrations of reactants and products *at equilibrium* define the equilibrium constant, K_{eq} (p. 96). In the general reaction $a\text{A} + b\text{B} \rightleftharpoons c\text{C} + d\text{D}$, where a, b, c, and d are the number of molecules of A, B, C, and D participating, the equilibrium constant is given by

$$K_{eq} = \frac{[\text{C}]^c[\text{D}]^d}{[\text{A}]^a[\text{B}]^b} \quad (14\text{–}2)$$

where [A], [B], [C], and [D] are the molar concentrations of the reaction components at the point of equilibrium.

When a reacting system is not at equilibrium, the tendency to move toward equilibrium represents a driving force, the magnitude of which can be expressed as the free-energy change for the reaction, ΔG. Under standard conditions (298 K = 25 °C), when reactants and products are initially present at 1 M concentrations or, for gases, at partial pressures of 101.3 kilopascals (kPa) or 1 atm, the force driving the system toward equilibrium is defined as the standard free-energy change, $\Delta G°$. By this definition, the standard state for reactions that involve hydrogen ions is $[H^+] = 1$ M, or pH 0. Most biochemical reactions occur in well-buffered aqueous solutions near pH 7; both the pH and the concentration of water (55.5 M) are essentially constant. For convenience of calculations, biochemists therefore define a different standard state, in which the concentration of H^+ is 10^{-7} M (pH 7) and that of water is 55.5 M; for reactions that involve Mg^{2+} (including most reactions for which ATP is a substrate), its concentration in solution is commonly taken to be constant at 1 mM. Physical constants based on this biochemical standard state are called **standard transformed constants** and are written with a prime (e.g., $\Delta G'°$ and K'_{eq}) to distinguish them from the untransformed constants used by chemists and physicists. (Notice that the symbol $\Delta G'°$ is a change from the symbol $\Delta G°'$ used in earlier editions of this book and in most other textbooks. The change, recommended by an international committee of chemists and biochemists, is intended to emphasize that the transformed free energy G' is the criterion for equilibrium.) By convention, when H_2O, H^+, or Mg^{2+} are reactants or products, their concentrations are not included in equations such as Equation 14–2 but are instead incorporated into the constants $\Delta G'°$ and K'_{eq}.

Just as K'_{eq} is a physical constant characteristic for each reaction, so too is $\Delta G'°$ a constant. As we noted in Chapter 8 (Eqn 8–3, p. 249), there is a simple relationship between K'_{eq} and $\Delta G'°$:

$$\Delta G'° = -RT \ln K'_{eq}$$

The standard free-energy change of a chemical reaction is simply an alternative mathematical way of expressing its equilibrium constant. Table 14–2 shows the relationship between $\Delta G'°$ and K'_{eq}. If the equilibrium

table 14–2

Relationship between the Equilibrium Constants and Standard Free-Energy Changes of Chemical Reactions

	$\Delta G'°$	
K'_{eq}	(kJ/mol)	(kcal/mol)*
10^3	−17.1	−4.1
10^2	−11.4	−2.7
10^1	−5.7	−1.4
1	0.0	0.0
10^{-1}	5.7	1.4
10^{-2}	11.4	2.7
10^{-3}	17.1	4.1
10^{-4}	22.8	5.5
10^{-5}	28.5	6.8
10^{-6}	34.2	8.2

*Although joules and kilojoules are the standard units of energy and are used throughout this text, biochemists sometimes express $\Delta G'°$ values in kilocalories per mole. We have therefore included values in both kilojoules and kilocalories in this table and in Tables 14–4 and 14–6. To convert kilojoules to kilocalories, divide the number of kilojoules by 4.184.

table 14–3

Relationships among K'_{eq}, $\Delta G'^{\circ}$, and the Direction of Chemical Reactions under Standard Conditions

When K'_{eq} is	$\Delta G'^{\circ}$ is	Starting with 1 M components the reaction
>1.0	Negative	Proceeds forward
1.0	Zero	Is at equilibrium
<1.0	Positive	Proceeds in reverse

constant for a given chemical reaction is 1.0, the standard free-energy change of that reaction is 0.0 (the natural logarithm of 1.0 is zero). If K'_{eq} of a reaction is greater than 1.0, its $\Delta G'^{\circ}$ is negative. If K'_{eq} is less than 1.0, $\Delta G'^{\circ}$ is positive. Because the relationship between $\Delta G'^{\circ}$ and K'_{eq} is exponential, relatively small changes in $\Delta G'^{\circ}$ correspond to large changes in K'_{eq}.

It may be helpful to think of the standard free-energy change in another way. $\Delta G'^{\circ}$ is the difference between the free-energy content of the products and the free-energy content of the reactants under standard conditions. When $\Delta G'^{\circ}$ is negative, the products contain less free energy than the reactants and the reaction will proceed spontaneously under standard conditions; all chemical reactions tend to go in the direction that results in a decrease in the free energy of the system. A positive value of $\Delta G'^{\circ}$ means that the products of the reaction contain more free energy than the reactants and this reaction will tend to go in the reverse direction if we start with 1.0 M concentrations of all components (standard conditions). Table 14–3 summarizes these points.

As an example, let us make a simple calculation of the standard free-energy change of the reaction catalyzed by the enzyme phosphoglucomutase:

$$\text{Glucose 1-phosphate} \rightleftharpoons \text{glucose 6-phosphate}$$

Chemical analysis shows that whether we start with, say, 20 mM glucose 1-phosphate (but no glucose 6-phosphate) or with 20 mM glucose 6-phosphate (but no glucose 1-phosphate), the final equilibrium mixture will contain 1 mM glucose 1-phosphate and 19 mM glucose 6-phosphate at 25 °C and pH 7.0. (Remember that enzymes do not affect the point of equilibrium of a reaction; they merely hasten its attainment.) From these data we can calculate the equilibrium constant:

$$K'_{eq} = \frac{[\text{glucose 6-phosphate}]}{[\text{glucose 1-phosphate}]} = \frac{19\ \text{mM}}{1\ \text{mM}} = 19$$

From this value of K'_{eq} we can calculate the standard free-energy change:

$$\begin{aligned}\Delta G'^{\circ} &= -RT \ln K'_{eq} \\ &= -(8.315\ \text{J/mol}\cdot\text{K})(298\ \text{K})(\ln 19) \\ &= -7{,}296\ \text{J/mol} = -7.3\ \text{kJ/mol}\end{aligned}$$

Because the standard free-energy change is negative, when the reaction starts with 1.0 M glucose 1-phosphate and 1.0 M glucose 6-phosphate, the conversion of glucose 1-phosphate to glucose 6-phosphate proceeds with a loss (release) of free energy. For the reverse reaction (the conversion of glucose 6-phosphate to glucose 1-phosphate), $\Delta G'^{\circ}$ has the same magnitude but the opposite sign.

table 14–4

Standard Free-Energy Changes of Some Chemical Reactions at pH 7.0 and 25 °C (298 K)

Reaction type	$\Delta G'^{\circ}$ (kJ/mol)	$\Delta G'^{\circ}$ (kcal/mol)
Hydrolysis reactions		
Acid anhydrides		
Acetic anhydride + H_2O ⟶ 2 acetate	−91.1	−21.8
ATP + H_2O ⟶ ADP + P_i	−30.5	−7.3
ATP + H_2O ⟶ AMP + PP_i	−45.6	−10.9
PP_i + H_2O ⟶ 2P_i	−19.2	−4.6
UDP-glucose +H_2O ⟶ UMP + glucose 1-phosphate	−43.0	−10.3
Esters		
Ethyl acetate + H_2O ⟶ ethanol + acetate	−19.6	−4.7
Glucose 6-phosphate + H_2O ⟶ glucose + P_i	−13.8	−3.3
Amides and peptides		
Glutamine + H_2O ⟶ glutamate + NH_4^+	−14.2	−3.4
Glycylglycine + H_2O ⟶ 2 glycine	−9.2	−2.2
Glycosides		
Maltose + H_2O ⟶ 2 glucose	−15.5	−3.7
Lactose + H_2O ⟶ glucose + galactose	−15.9	−3.8
Rearrangements		
Glucose 1-phosphate ⟶ glucose 6-phosphate	−7.3	−1.7
Fructose 6-phosphate ⟶ glucose 6-phosphate	−1.7	−0.4
Elimination of water		
Malate ⟶ fumarate + H_2O	3.1	0.8
Oxidations with molecular oxygen		
Glucose + $6O_2$ ⟶ $6CO_2$ + $6H_2O$	−2,840	−686
Palmitate + $23O_2$ ⟶ $16CO_2$ + $16H_2O$	−9,770	−2,338

Table 14–4 gives the standard free-energy changes for some representative chemical reactions. Note that hydrolysis of simple esters, amides, peptides, and glycosides, as well as rearrangements and eliminations, proceed with relatively small standard free-energy changes, whereas hydrolysis of acid anhydrides occurs with relatively large decreases in standard free energy. The complete oxidation of organic compounds such as glucose or palmitate to CO_2 and H_2O, which in cells requires many steps, results in very large decreases in standard free energy. However, standard free-energy changes such as those in Table 14–4 indicate how much free energy is available from a reaction under *standard conditions*. To describe the energy released under the conditions existing in cells, an expression for the *actual* free-energy change is essential.

Actual Free-Energy Changes Depend on Reactant and Product Concentrations

We must be careful to distinguish between two different quantities: the free-energy change, ΔG, and the standard free-energy change, $\Delta G'^{\circ}$. Each chemical reaction has a characteristic standard free-energy change, which may be positive, negative, or zero, depending on the equilibrium constant of the reaction. The standard free-energy change tells us in which direction and how far a given reaction must go to reach equilibrium *when the initial concentration of each component is 1.0 M,* the pH is 7.0, the temperature is 25 °C, and the pressure is 101.3 kPa (1 atm). Thus $\Delta G'^\circ$ is a constant: it has a characteristic, unchanging value for a given reaction. But the *actual* free-energy change, ΔG, is a function of reactant and product concentrations and of the temperature prevailing during the reaction, which will not necessarily match the standard conditions as defined above. Moreover, the ΔG of any reaction proceeding spontaneously toward its equilibrium is always negative, becomes less negative as the reaction proceeds, and is zero at the point of equilibrium, indicating that no more work can be done by the reaction.

ΔG and $\Delta G'^\circ$ for any reaction A + B $\rightleftharpoons$ C + D are related by the equation

$$\Delta G = \Delta G'^\circ + RT \ln \frac{[\mathrm{C}][\mathrm{D}]}{[\mathrm{A}][\mathrm{B}]} \qquad (14\text{–}3)$$

in which the terms in red are those *actually prevailing* in the system under observation. The concentration terms in this equation express the effects commonly called mass action. As an example, let us suppose that the reaction A + B $\rightleftharpoons$ C + D is taking place at the standard conditions of temperature (25 °C) and pressure (101.3 kPa) but that the concentrations of A, B, C, and D are *not* equal and that none of the components is present at the standard concentration of 1.0 M. To determine the actual free-energy change, ΔG, under these nonstandard conditions of concentration as the reaction proceeds from left to right, we simply enter the *actual* concentrations of A, B, C, and D in Equation 14–3; the values of R, T, and $\Delta G'^\circ$ are the standard values. ΔG is negative and approaches zero as the reaction proceeds because the actual concentrations of A and B decrease and the concentrations of C and D increase. Notice that when a reaction is at equilibrium—when there is no force driving the reaction in either direction and ΔG is zero—Equation 14–3 reduces to

$$0 = \Delta G'^\circ = RT \ln \frac{[\mathrm{C}]_{\mathrm{eq}}[\mathrm{D}]_{\mathrm{eq}}}{[\mathrm{A}]_{\mathrm{eq}}[\mathrm{B}]_{\mathrm{eq}}}$$

or

$$\Delta G'^\circ = -RT \ln K'_{\mathrm{eq}}$$

the equation relating the standard free-energy change and equilibrium constant as noted above.

The criterion for spontaneity of a reaction is the value of ΔG, not $\Delta G'^\circ$. A reaction with a positive $\Delta G'^\circ$ can go in the forward direction *if* ΔG *is negative.* This is possible if the term RT ln ([products]/[reactants]) in Equation 14–3 is negative and has a larger absolute value than $\Delta G'^\circ$. For example, the immediate removal of the products of a reaction can keep the ratio [products]/[reactants] well below 1, such that the term RT ln ([products]/[reactants]) has a large, negative value.

$\Delta G'^\circ$ and ΔG are expressions of the *maximum* amount of free energy that a given reaction can *theoretically* deliver—an amount of energy that could be realized only if a perfectly efficient device were available to trap or harness it. Given that no such device is possible (some free energy is always lost to entropy during any process), the amount of work done by the reaction at constant temperature and pressure is always less than the theoretical amount.

Another important point is that some thermodynamically favorable reactions (that is, reactions for which $\Delta G'^\circ$ is large and negative) do not occur at measurable rates. For example, combustion of firewood to CO_2 and H_2O

is very favorable thermodynamically, but firewood remains stable for years because the activation energy (see Fig. 8–3) for the combustion reaction is higher than the energy available at room temperature. If the necessary activation energy is provided (with a lighted match, for example), combustion will begin, converting the wood to the more stable products CO_2 and H_2O and releasing energy as heat and light. The heat released by this exothermic reaction provides the activation energy for combustion of neighboring regions of the firewood; the process is self-perpetuating.

In living cells, reactions that would be extremely slow *if uncatalyzed* are caused to occur, not by supplying additional heat, but by lowering the activation energy with an enzyme. An enzyme provides an alternative reaction pathway with a lower activation energy than the uncatalyzed reaction, so that at room temperature a large fraction of the substrate molecules have enough thermal energy to overcome the activation barrier, and the reaction rate increases dramatically. *The free-energy change for a reaction is independent of the pathway by which the reaction occurs;* it depends only on the nature and concentration of the initial reactants and the final products. Enzymes cannot, therefore, change equilibrium constants; but they can and do increase the *rate* at which a reaction proceeds in the direction dictated by thermodynamics.

Standard Free-Energy Changes Are Additive

In the case of two sequential chemical reactions, $A \rightleftharpoons B$ and $B \rightleftharpoons C$, each reaction has its own equilibrium constant and each has its characteristic standard free-energy change, $\Delta G_1'^{\circ}$ and $\Delta G_2'^{\circ}$. As the two reactions are sequential, B cancels out to give the overall reaction $A \rightleftharpoons C$, which has its own equilibrium constant and thus its own standard free-energy change, $\Delta G_{\text{total}}'^{\circ}$. *The* $\Delta G'^{\circ}$ *values of sequential chemical reactions are additive.* For the overall reaction $A \rightleftharpoons C$, $\Delta G_{\text{total}}'^{\circ}$ is the algebraic sum of the individual standard free-energy changes, $\Delta G_1'^{\circ}$ and $\Delta G_2'^{\circ}$, of the two separate reactions: $\Delta G_{\text{total}}'^{\circ} = \Delta G_1'^{\circ} + \Delta G_2'^{\circ}$. This principle of bioenergetics explains how a thermodynamically unfavorable (endergonic) reaction can be driven in the forward direction by coupling it to a highly exergonic reaction through a common intermediate. For example, the synthesis of glucose 6-phosphate is the first step in the utilization of glucose by many organisms:

$$
\begin{array}{lll}
(1) & A \longrightarrow B & \Delta G_1'^{\circ} \\
(2) & B \longrightarrow C & \Delta G_2'^{\circ} \\
\hline
\textit{Sum:} & A \longrightarrow C & \Delta G_1'^{\circ} + \Delta G_2'^{\circ}
\end{array}
$$

$$\text{Glucose} + P_i \longrightarrow \text{glucose 6-phosphate} + H_2O \qquad \Delta G'^{\circ} = 13.8 \text{ kJ/mol}$$

The positive value of $\Delta G'^{\circ}$ predicts that under standard conditions the reaction will tend not to proceed spontaneously in the direction written. Another cellular reaction, the hydrolysis of ATP to ADP and P_i, is very exergonic:

$$\text{ATP} + H_2O \longrightarrow \text{ADP} + P_i \qquad \Delta G'^{\circ} = -30.5 \text{ kJ/mol}$$

These two reactions share the common intermediates P_i and H_2O and may be expressed as sequential reactions:

$$
\begin{array}{ll}
(1) & \text{Glucose} + P_i \longrightarrow \text{glucose 6-phosphate} + H_2O \\
(2) & \text{ATP} + H_2O \longrightarrow \text{ADP} + P_i \\
\hline
\textit{Sum:} & \text{ATP} + \text{glucose} \longrightarrow \text{ADP} + \text{glucose 6-phosphate}
\end{array}
$$

The overall standard free-energy change is obtained by adding the $\Delta G'^{\circ}$ values for individual reactions:

$$\Delta G'^{\circ} = 13.8 \text{ kJ/mol} + (-30.5 \text{ kJ/mol}) = -16.7 \text{ kJ/mol}$$

The overall reaction is exergonic. In this case, energy stored in the bonds of ATP is used to drive the synthesis of glucose 6-phosphate, even though its formation from glucose and phosphate is endergonic. The *pathway* of glucose 6-phosphate formation by phosphoryl transfer from ATP is different from reactions (1) and (2) above, but the net result is the same as the sum of the two reactions. In thermodynamic calculations, all that matters is the state of the system at the beginning of the process, and its state at the end; the route between the initial and final states is immaterial.

We have said that $\Delta G'^{\circ}$ is a way of expressing the equilibrium constant for a reaction. For reaction (1) above,

$$K'_{eq_1} = \frac{[\text{glucose 6-phosphate}]}{[\text{glucose}][\text{P}_i]} = 3.9 \times 10^{-3}\ \text{M}^{-1}$$

Notice that H_2O is not included in this expression. The equilibrium constant for the hydrolysis of ATP is

$$K'_{eq_2} = \frac{[\text{ADP}][\text{P}_i]}{[\text{ATP}]} = 2 \times 10^{5}\ \text{M}$$

The equilibrium constant for the two coupled reactions is

$$\begin{aligned} K'_{eq_3} &= \frac{[\text{glucose 6-phosphate}][\text{ADP}][\text{P}_i]}{[\text{glucose}][\text{P}_i][\text{ATP}]} \\ &= (K'_{eq_1})(K'_{eq_2}) = (3.9 \times 10^{-3}\ \text{M}^{-1})(2.0 \times 10^{5}\ \text{M}) \\ &= 7.8 \times 10^{2} \end{aligned}$$

This calculation illustrates an important point about equilibrium constants: although the $\Delta G'^{\circ}$ values for two reactions that sum to a third are *additive,* the K'_{eq} for a reaction that is the sum of two reactions is the *product* of their individual K'_{eq} values. Equilibrium constants are *multiplicative.* By coupling ATP hydrolysis to glucose 6-phosphate synthesis, the K'_{eq} for formation of glucose 6-phosphate has been raised by a factor of about 2×10^5.

This common-intermediate strategy is employed by all living cells in the synthesis of metabolic intermediates and cellular components. Obviously, the strategy works only if compounds such as ATP are continuously available. In the following chapters we consider several of the most important cellular pathways for producing ATP.

Phosphoryl Group Transfers and ATP

Having developed some fundamental principles of energy changes in chemical systems, we can now examine the energy cycle in cells and the special role of ATP as the energy currency that links catabolism and anabolism (see Fig. 1–14). Heterotrophic cells obtain free energy in a chemical form by the catabolism of nutrient molecules, and they use that energy to make ATP from ADP and P_i. ATP then donates some of its chemical energy to endergonic processes such as the synthesis of metabolic intermediates and macromolecules from smaller precursors, the transport of substances across membranes against concentration gradients, and mechanical motion. This donation of energy from ATP generally involves the covalent participation of ATP in the reaction that is to be driven, with the eventual result that ATP is converted to ADP and P_i or, in some reactions, to AMP and 2 P_i. We discuss here the chemical basis for the large free-energy changes that accompany hydrolysis of ATP and other high-energy phosphate compounds, and we show that most cases of energy donation by ATP involve group transfer, not simple hydrolysis of ATP. To illustrate the range of energy transductions in which ATP provides the energy, we consider the synthesis of information-rich macromolecules, the transport of solutes across membranes, and motion produced by muscle contraction.

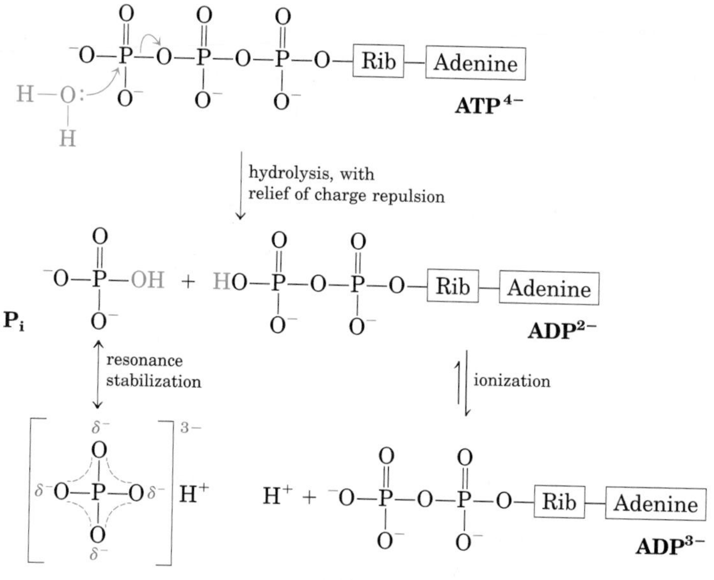

$$ATP^{4-} + H_2O \longrightarrow ADP^{3-} + P_i^{2-} + H^+$$
$$\Delta G'^{\circ} = -30.5 \text{ kJ/mol}$$

figure 14–1

Chemical basis for the large free-energy change associated with ATP hydrolysis. First, hydrolysis, by causing charge separation, relieves electrostatic repulsion among the four negative charges on ATP. Second, inorganic phosphate (P_i) released by hydrolysis is stabilized by formation of a resonance hybrid, in which each of the four P—O bonds has the same degree of double-bond character and the hydrogen ion is not permanently associated with any one of the oxygens. Some degree of resonance stabilization also occurs in phosphates involved in ester or anhydride linkages, but fewer resonance forms are possible than for P_i. Third, ADP^{2-} produced by the hydrolysis immediately ionizes, releasing a proton into a medium of very low [H$^+$] (pH 7). A fourth factor (not shown) that favors ATP hydrolysis is the greater degree of solvation (hydration) of the products P_i and ADP relative to ATP, which further stabilizes the products relative to the reactants.

The Free-Energy Change for ATP Hydrolysis Is Large and Negative

Figure 14–1 summarizes the chemical basis for the relatively large, negative, standard free energy of hydrolysis of ATP. The hydrolytic cleavage of the terminal phosphoric acid anhydride (phosphoanhydride) bond in ATP separates one of the three negatively charged phosphates and thus relieves some of the electrostatic repulsion in ATP; the P_i (HPO_4^{2-}) released is stabilized by the formation of several resonance forms not possible in ATP; and ADP^{2-}, the other direct product of hydrolysis, immediately ionizes, releasing H^+ into a medium of very low [H$^+$] ($\sim10^{-7}$ M). Because the concentrations of the direct products of ATP hydrolysis are far below the concentrations at equilibrium, mass action favors the hydrolysis reaction.

Although the hydrolysis of ATP is highly exergonic ($\Delta G'^{\circ} = -30.5$ kJ/mol), the molecule is kinetically stable at pH 7 because the activation energy for ATP hydrolysis is relatively high. Rapid cleavage of the phosphoanhydride bonds occurs only when catalyzed by an enzyme.

The free-energy change for ATP hydrolysis is -30.5 kJ/mol under standard conditions, but the *actual* free energy of hydrolysis (ΔG) of ATP in living cells is very different: the cellular concentrations of ATP, ADP, and P_i are not identical and are much lower than the 1.0 M of standard conditions (Table 14–5). Furthermore, Mg^{2+} in the cytosol binds to ATP and ADP (Fig. 14–2), and for most enzymatic reactions that involve ATP as phosphoryl group donor, the true substrate is $MgATP^{2-}$. The relevant $\Delta G'^{\circ}$ is therefore that for $MgATP^{2-}$ hydrolysis. Box 14–2 shows how ΔG for ATP hydrolysis in the intact erythrocyte can be calculated from the data in Table 14–5. In intact cells, ΔG for ATP hydrolysis, usually designated ΔG_p, is much more negative than $\Delta G'^{\circ}$, ranging from -50 to -65 kJ/mol. ΔG_p is often called the **phosphorylation potential.** In the following discussions we use the standard free-energy change for ATP hydrolysis, because this allows comparison, on the same basis, with the energetics of other cellular reactions. Remember, however, that in living cells ΔG is the relevant quantity—for ATP hydrolysis and all other reactions—and may be quite different from $\Delta G'^{\circ}$.

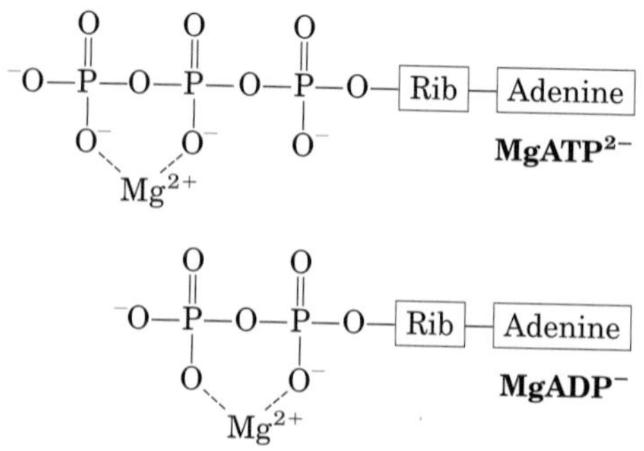

figure 14–2

Mg^{2+} and ATP. Formation of Mg^{2+} complexes partially shields the negative charges and influences the conformation of the phosphate groups in nucleotides such as ATP and ADP.

box 14–2 The Free Energy of Hydrolysis of ATP within Cells: The Real Cost of Doing Metabolic Business

The standard free energy of hydrolysis of ATP is −30.5 kJ/mol. In the cell, however, the concentrations of ATP, ADP, and P_i are not only unequal but much lower than the standard 1 M concentrations (see Table 14–5). Moreover, the cellular pH may differ somewhat from the standard pH of 7.0. Thus the *actual* free energy of hydrolysis of ATP under intracellular conditions (ΔG_p) differs from the standard free-energy change, $\Delta G'^{\circ}$. We can easily calculate ΔG_p. For example, in human erythrocytes the concentrations of ATP, ADP, and P_i are 2.25, 0.25, and 1.65 mM, respectively. Let us assume for simplicity that the pH is 7.0 and the temperature is 25 °C, the standard pH and temperature. The actual free energy of hydrolysis of ATP in the erythrocyte under these conditions is given by the relationship

$$\Delta G_p = \Delta G'^{\circ} + RT \ln \frac{[\text{ADP}][\text{P}_i]}{[\text{ATP}]}$$

Substituting the appropriate values we obtain

$$\begin{aligned}\Delta G_p &= -30{,}500 \text{ J/mol} + (8.315 \text{ J/mol}\cdot\text{K})(298 \text{ K}) \\ &\quad \times \ln \frac{(2.50 \times 10^{-4})(1.65 \times 10^{-3})}{2.25 \times 10^{-3}} \\ &= -30{,}500 \text{ J/mol} + (2{,}480 \text{ J/mol}) \ln (1.83 \times 10^{-4}) \\ &= -30{,}500 \text{ J/mol} - 21{,}300 \text{ J/mol} \\ &= -51{,}800 \text{ J/mol} \\ &= -51.8 \text{ kJ/mol}\end{aligned}$$

Thus ΔG_p, the actual free-energy change for ATP hydrolysis in the intact erythrocyte (−51.8 kJ/mol), is much larger than the standard free-energy change (−30.5 kJ/mol). By the same token, the free energy required to *synthesize* ATP from ADP and P_i under the conditions prevailing in the erythrocyte would be 51.8 kJ/mol.

Because the concentrations of ATP, ADP, and P_i differ from one cell type to another (see Table 14–5), ΔG_p for ATP hydrolysis likewise differs among cells. Moreover, in any given cell, ΔG_p can vary from time to time, depending on the metabolic conditions in the cell and how they influence the concentrations of ATP, ADP, P_i, and H^+ (pH). We can calculate the actual free-energy change for any given metabolic reaction as it occurs in the cell, providing we know the concentrations of all the reactants and products of the reaction and other factors (such as pH, temperature, and concentration of Mg^{2+}) that may affect the $\Delta G'^{\circ}$ and thus the calculated free-energy change, ΔG_p.

To further complicate the issue, the total concentrations of ATP, ADP, P_i, and H^+ may be substantially higher than the free concentrations, which are the thermodynamically relevant values. The difference is due to tight binding of ATP, ADP, and P_i to cellular proteins. For example, the concentration of free ADP in resting muscle has been variously estimated at between 1 and 37 μM. Using the value 25 μM in the calculation outlined above, we get a ΔG_p of −57.5 kJ/mol. The exact value of ΔG_p is perhaps less instructive than the generalization we can make about actual free-energy changes: in vivo, the energy released by ATP hydrolysis is greater than the standard free-energy change, $\Delta G'^{\circ}$.

table 14–5

Adenine Nucleotide, Inorganic Phosphate, and Phosphocreatine Concentrations in Some Cells*

	Concentration (mM)				
	ATP	ADP†	AMP	P_i	PCr
Rat hepatocyte	3.38	1.32	0.29	4.8	0
Rat myocyte	8.05	0.93	0.04	8.05	28
Rat neuron	2.59	0.73	0.06	2.72	4.7
Human erythrocyte	2.25	0.25	0.02	1.65	0
E. coli cell	7.90	1.04	0.82	7.9	0

*For erythrocytes the concentrations are those of the cytosol (human erythrocytes lack a nucleus and mitochondria). In the other types of cells the data are for the entire cell contents, although the cytosol and the mitochondria have very different concentrations of ADP. PCr is phosphocreatine, discussed on p. 502.

†This value reflects total concentration; the true value for free ADP may be much lower (see Box 14–2).

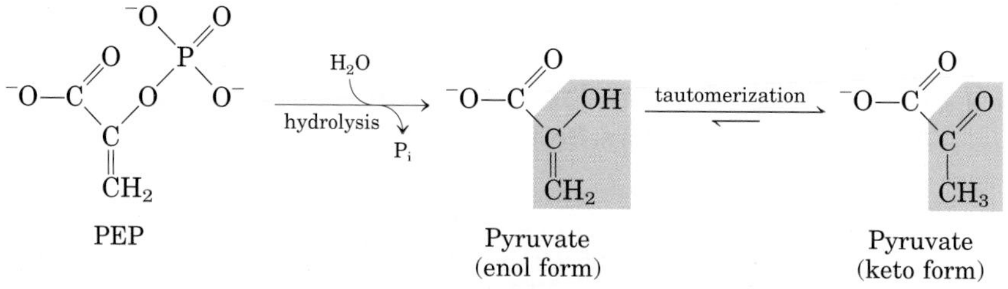

figure 14–3
Hydrolysis of phosphoenolpyruvate (PEP). Catalyzed by pyruvate kinase, this reaction is followed by spontaneous tautomerization of the product, pyruvate. Tautomerization is not possible in PEP, and thus the products of hydrolysis are stabilized relative to the reactants. Resonance stabilization of P_i also occurs, as shown in Figure 14–1.

PEP → (hydrolysis; H_2O in, P_i out) Pyruvate (enol form) ⇌ (tautomerization) Pyruvate (keto form)

$$PEP^{3-} + H_2O \longrightarrow pyruvate^{-} + P_i^{2-}$$
$$\Delta G'^{\circ} = -61.9 \text{ kJ/mol}$$

1,3-Bisphosphoglycerate → (hydrolysis; H_2O in, P_i out) 3-Phosphoglyceric acid ⇌ (ionization; H^+ out) 3-Phosphoglycerate (resonance stabilization)

$$\text{1,3-Bisphosphoglycerate}^{4-} + H_2O \longrightarrow \text{3-phosphoglycerate}^{3-} + P_i^{2-} + H^+$$
$$\Delta G'^{\circ} = -49.3 \text{ kJ/mol}$$

figure 14–4
Hydrolysis of 1,3-bisphosphoglycerate. The direct product of hydrolysis is 3-phosphoglyceric acid, which has an undissociated carboxylic acid group, but dissociation occurs immediately. This ionization and the resonance structures it makes possible stabilize the product relative to the reactants. Resonance stabilization of P_i further contributes to the negative free-energy change.

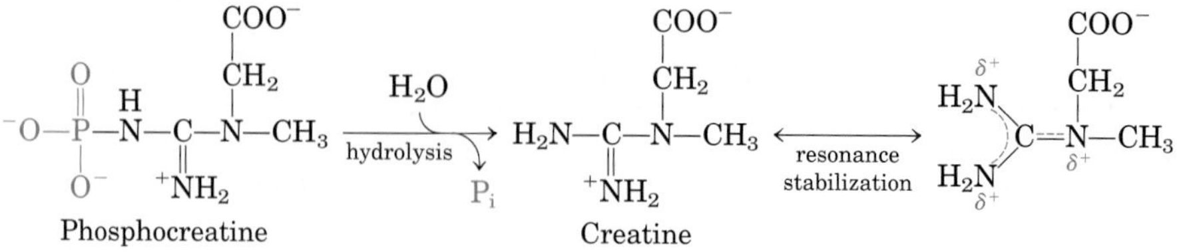

Other Phosphorylated Compounds and Thioesters Also Have Large Free Energies of Hydrolysis

Phosphoenolpyruvate (Fig. 14–3) contains a phosphate ester bond that can undergo hydrolysis to yield the enol form of pyruvate, and this direct product of hydrolysis can immediately tautomerize to the more stable keto form of pyruvate. Because the reactant (phosphoenolpyruvate) has only one form (enol) and the product (pyruvate) has two possible forms, the product is stabilized relative to the reactant. This is the greatest contributing factor to the high standard free energy of hydrolysis of phosphoenolpyruvate: $\Delta G'^{\circ} = -61.9$ kJ/mol.

Another three-carbon compound, 1,3-bisphosphoglycerate (Fig. 14–4), contains an anhydride bond between the carboxyl group at C-1 and phosphoric acid. Hydrolysis of this acyl phosphate is accompanied by a large, negative, standard free-energy change ($\Delta G'^{\circ} = -49.3$ kJ/mol), which can, again, be explained in terms of the structure of reactant and products. When H_2O is added across the anhydride bond of 1,3-bisphosphoglycerate, one of the direct products, 3-phosphoglyceric acid, can immediately lose a proton to give the carboxylate ion, 3-phosphoglycerate, which has two equally probable resonance forms (Fig. 14–4). Removal of the direct product (3-phosphoglyceric acid) and formation of the resonance-stabilized ion favors the forward reaction.

In phosphocreatine (Fig. 14–5), the P—N bond can be hydrolyzed to generate free creatine and P_i. The release of P_i and the resonance stabilization of creatine favor the forward reaction. The standard free-energy change of phosphocreatine hydrolysis is again large, about −43 kJ/mol.

In all these phosphate-releasing reactions, the several resonance forms available to P_i (Fig. 14–1) stabilize this product relative to the reactant,

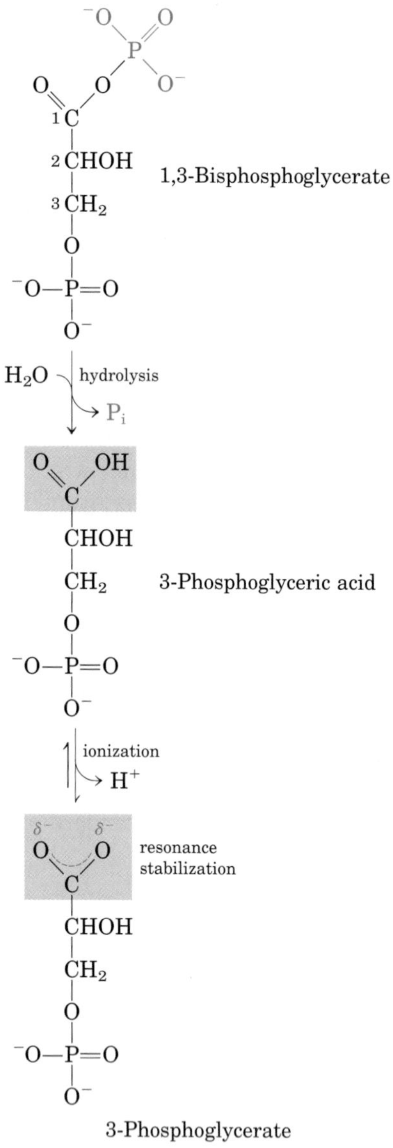

Phosphocreatine → (hydrolysis; H_2O in, P_i out) Creatine ⇌ (resonance stabilization)

$$\text{Phosphocreatine}^{2-} + H_2O \longrightarrow \text{creatine} + P_i^{2-}$$
$$\Delta G'^{\circ} = -43.0 \text{ kJ/mol}$$

figure 14–5
Hydrolysis of phosphocreatine. Breakage of the P—N bond in phosphocreatine produces creatine, which is stabilized by formation of a resonance hybrid. The other product, P_i, is also resonance stabilized.

table 14–6

Standard Free Energies of Hydrolysis of Some Phosphorylated Compounds and Acetyl-CoA (a Thioester)

	$\Delta G'^{\circ}$ (kJ/mol)	$\Delta G'^{\circ}$ (kcal/mol)
Phosphoenolpyruvate	−61.9	−14.8
1,3-bisphosphoglycerate (→ 3-phosphoglycerate + P_i)	−49.3	−11.8
Phosphocreatine	−43.0	−10.3
ADP (→ AMP + P_i)	−32.8	−7.8
ATP (→ ADP + P_i)	−30.5	−7.3
ATP (→ AMP + PP_i)	−45.6	−10.9
AMP (→ adenosine + P_i)	−14.2	−3.4
PP_i (→ $2P_i$)	−19	−4.0
Glucose 1-phosphate	−20.9	−5.0
Fructose 6-phosphate	−15.9	−3.8
Glucose 6-phosphate	−13.8	−3.3
Glycerol 1-phosphate	−9.2	−2.2
Acetyl-CoA	−31.4	−7.5

Source: Data mostly from Jencks, W.P. (1976) in *Handbook of Biochemistry and Molecular Biology,* 3rd edn (Fasman, G.D., ed.), *Physical and Chemical Data,* Vol. I, pp. 296–304, CRC Press, Boca Raton, FL.

contributing to an already negative free-energy change. Table 14–6 lists the standard free energies of hydrolysis for a number of phosphorylated compounds.

Thioesters, in which a sulfur atom replaces the usual oxygen in the ester bond, also have large, negative, standard free energies of hydrolysis. Acetyl-coenzyme A, or acetyl-CoA (Fig. 14–6), is one of many thioesters important in metabolism. The acyl group in these compounds is activated for transacylation, condensation, or oxidation-reduction reactions. Thioesters undergo much less resonance stabilization than do oxygen esters (Fig. 14–7); consequently, the difference in free energy between the reactant and its hydrolysis products, which *are* resonance-stabilized, is greater for thioesters than for comparable oxygen esters. In both cases, hydrolysis of the ester generates a carboxylic acid, which can ionize and assume several resonance forms (Fig. 14–6). Together, these factors result in the large, negative $\Delta G'^{\circ}$ (−31 kJ/mol) for acetyl-CoA hydrolysis.

Acetyl-CoA
H_2O hydrolysis → CoASH
Acetic acid
ionization → H^+
Acetate
resonance stabilization

$$\text{Acetyl-CoA} + H_2O \longrightarrow \text{acetate}^- + \text{CoA} + H^+$$
$$\Delta G'^{\circ} = -32.2 \text{ kJ/mol}$$

figure 14–6

Hydrolysis of acetyl-coenzyme A. Acetyl-CoA is a thioester with a large, negative, standard free energy of hydrolysis. Thioesters contain a sulfur atom in the position occupied by an oxygen atom in oxygen esters. The complete structure of coenzyme A (CoASH) is shown in Figure 10–41.

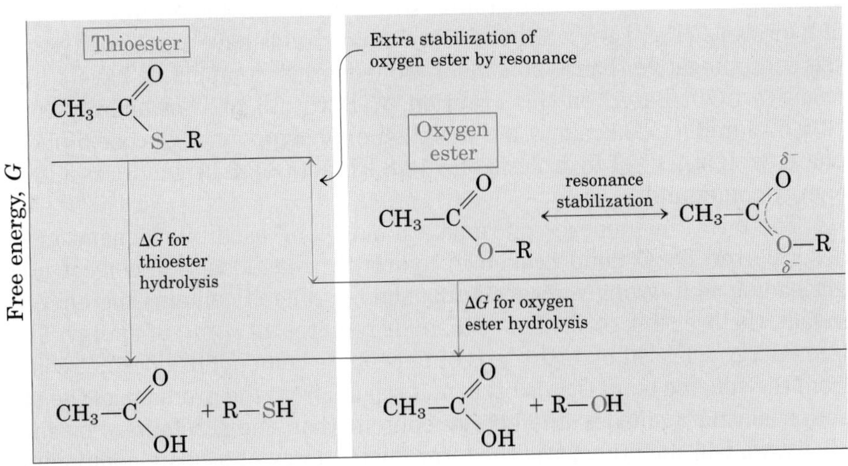

figure 14–7

Free energy of hydrolysis for thioesters and oxygen esters. The *products* of both types of hydrolysis reaction have about the same free-energy content (*G*), but the thioester has a higher free-energy content than the oxygen ester. Orbital overlap between the O and C atoms allows resonance stabilization in oxygen esters, but orbital overlap between S and C atoms is poorer and little resonance stabilization occurs.

To summarize, for hydrolysis reactions with large, negative, standard free-energy changes, the products are more stable than the reactants for one or more of the following reasons: (1) the bond strain in reactants due to electrostatic repulsion is relieved by charge separation, as for ATP (described earlier); (2) the products are stabilized by ionization, as for ATP, acyl phosphates, and thioesters; (3) the products are stabilized by isomerization (tautomerization), as for phosphoenolpyruvate; and/or (4) the products are stabilized by resonance, as for creatine released from phosphocreatine, carboxylate ion released from acyl phosphates and thioesters, and phosphate (P_i) released from anhydride or ester linkages.

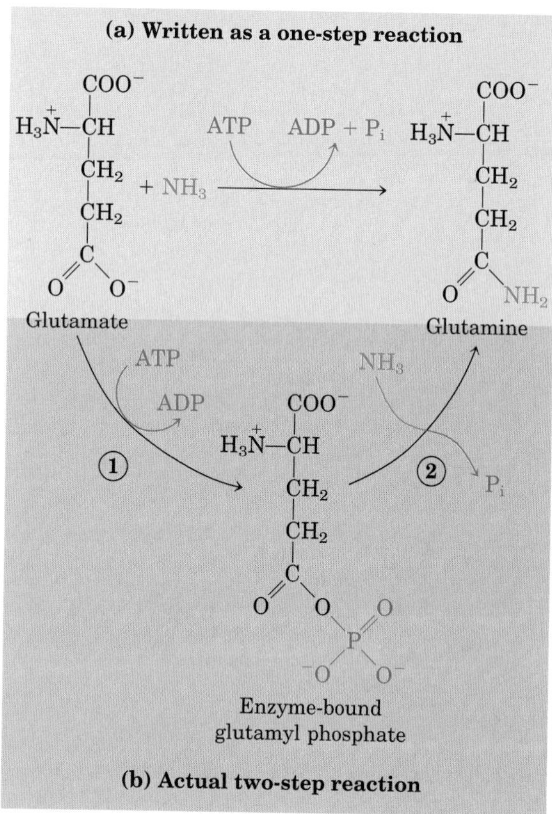

figure 14–8

ATP hydrolysis in two steps. The contribution of ATP to a reaction is often shown as a single step **(a)**, but is almost always a two-step process, such as that shown here for the reaction catalyzed by ATP-dependent glutamine synthetase **(b)**. ① A phosphoryl group is first transferred from ATP to glutamate, then ② the phosphoryl group is displaced by NH_3 and released as P_i.

ATP Provides Energy by Group Transfers, Not by Simple Hydrolysis

Throughout this book you will encounter reactions or processes for which ATP supplies energy, and the contribution of ATP to these reactions is commonly indicated as in Figure 14–8a, with a single arrow showing the conversion of ATP to ADP and P_i, or of ATP to AMP and PP_i (pyrophosphate). When written this way, these reactions of ATP appear to be simple hydrolysis reactions in which water displaces either P_i or PP_i, and one is tempted to say that an ATP-dependent reaction is "driven by the hydrolysis of ATP." This is *not* the case. ATP hydrolysis per se usually accomplishes nothing but the liberation of heat, which cannot drive a chemical process in an isothermal system. Single reaction arrows such as those in Figure 14–8a almost invariably represent two-step processes (Fig. 14–8b) in which part of the ATP molecule, a phosphoryl or pyrophosphoryl group or the adenylate moiety (AMP), is first transferred to a substrate molecule or to an amino acid residue in an enzyme, becoming covalently attached to the substrate or the enzyme and raising its free-energy content. In the second step, the phosphate-containing moiety that was transferred in the first step is displaced, generating P_i, PP_i, or AMP. Thus ATP participates *covalently* in the enzyme-catalyzed reaction to which it contributes free energy.

Some processes *do* involve direct hydrolysis of ATP (or GTP), however. For example, noncovalent binding of ATP (or of GTP), followed by its hydrolysis to ADP (or GDP) and P_i, can provide the energy to cycle some proteins between two conformations, producing mechanical motion. This occurs in muscle contraction and in the movement of enzymes along DNA or of ribosomes along messenger RNA. The energy-dependent reactions catalyzed by helicases, RecA protein, and some topoisomerases (Chapter 25) also involve direct hydrolysis of phosphoanhydride bonds. GTP-binding proteins that act in signaling pathways directly hydrolyze GTP to drive conformational changes that terminate signals triggered by hormones or by other extracellular factors (Chapter 13).

The phosphate compounds found in living organisms can be divided somewhat arbitrarily into two groups, based on their standard free energies of hydrolysis (Fig. 14–9). "High-energy" compounds have a $\Delta G'^{\circ}$ of hydrolysis more negative than -25 kJ/mol; "low-energy" compounds have a less negative $\Delta G'^{\circ}$. Based on this criterion, ATP, with a $\Delta G'^{\circ}$ of hydrolysis of -30.5 kJ/mol (-7.3 kcal/mol), is a high-energy compound; glucose 6-phosphate, with a $\Delta G'^{\circ}$ of hydrolysis of -13.8 kJ/mol (-3.3 kcal/mol), is a low-energy compound.

The term "high-energy phosphate bond," long used by biochemists to describe the P—O bond broken in hydrolysis reactions, is incorrect and misleading as it wrongly suggests that the bond itself contains the energy. In fact, the breaking of all chemical bonds requires an *input* of energy. The free energy released by hydrolysis of phosphate compounds does not come from the specific bond that is broken; it results from the products of the reaction having a smaller free-energy content than the reactants. For simplicity, we will sometimes use the term "high-energy phosphate compound"

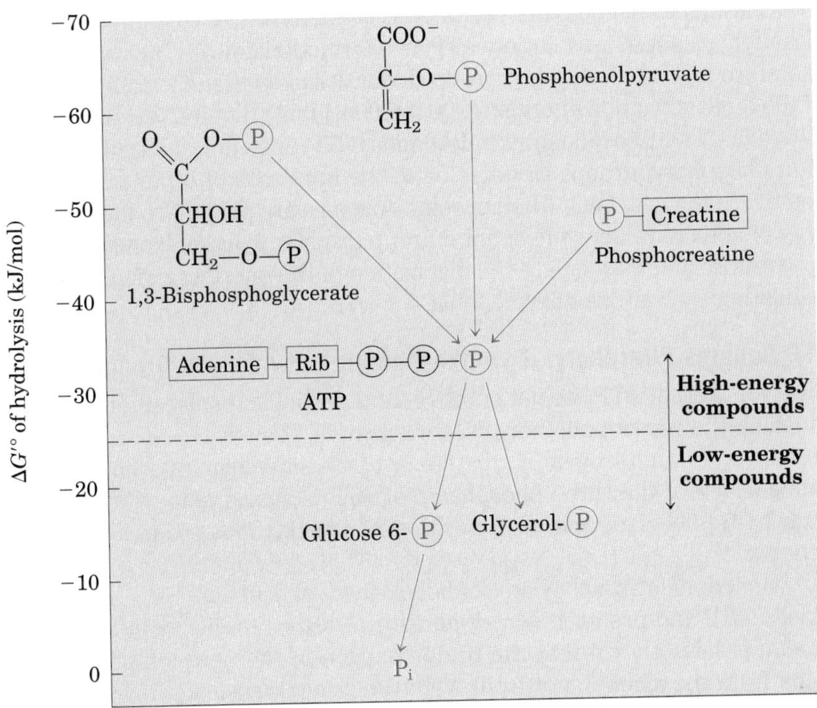

figure 14–9

Ranking of biological phosphate compounds by standard free energies of hydrolysis. This shows the flow of phosphoryl groups, represented by Ⓟ, from high-energy phosphoryl donors via ATP to acceptor molecules (such as glucose and glycerol) to form their low-energy phosphate derivatives. This flow of phosphoryl groups, catalyzed by enzymes called **kinases,** proceeds with an overall loss of free energy under intracellular conditions. Hydrolysis of low-energy phosphate compounds releases P_i, which has an even lower phosphoryl group transfer potential (as defined in the text).

when referring to ATP or other phosphate compounds with a large, negative, standard free energy of hydrolysis.

As is evident from the additivity of free-energy changes of sequential reactions, any phosphorylated compound can be synthesized by coupling the synthesis to the breakdown of another phosphorylated compound with a more negative free energy of hydrolysis. For example, because cleavage of P_i from phosphoenolpyruvate (PEP) releases more energy than is needed to drive the condensation of P_i with ADP, the direct donation of a phosphoryl group from PEP to ADP is thermodynamically feasible:

		$\Delta G'^{\circ}$ (kJ/mol)
(1)	$PEP + H_2O \longrightarrow pyruvate + P_i$	−61.9
(2)	$ADP + P_i \longrightarrow ATP + H_2O$	+30.5
Sum:	$PEP + ADP \longrightarrow pyruvate + ATP$	−31.4

We can therefore describe phosphorylated compounds as having a high or low phosphoryl group transfer potential. The phosphoryl group transfer potential of phosphoenolpyruvate is very high, that of ATP is high, and that of glucose 6-phosphate is low.

Much of catabolism is directed toward the synthesis of high-energy phosphate compounds, but their formation is not an end in itself; they are the means of activating a very wide variety of compounds for further chemical transformation. The transfer of a phosphoryl group to a compound effectively puts free energy into that compound, so that it has more free energy to give up during subsequent metabolic transformations. We described above how the synthesis of glucose 6-phosphate is accomplished by phosphoryl group transfer from ATP. In the next chapter we see how this phosphorylation of glucose activates or "primes" the glucose for catabolic reactions that occur in nearly every living cell. Because of its intermediate position on the scale of group transfer potential, ATP can carry energy from high-energy phosphate compounds produced by catabolism to compounds such as glucose, converting them into more reactive species. ATP thus serves as the universal energy currency in all living cells.

One more chemical feature of ATP is crucial to its role in metabolism: although in aqueous solution ATP is thermodynamically unstable and is therefore a good phosphoryl group donor, it is *kinetically* stable. Because of the huge activation energies (200 to 400 kJ/mol) required for uncatalyzed cleavage of its phosphoanhydride bonds, ATP does not spontaneously donate phosphoryl groups to water or to the hundreds of other potential acceptors in the cell. Only when specific enzymes are present to lower the energy of activation does phosphoryl group transfer from ATP occur. The cell is therefore able to regulate the disposition of the energy carried by ATP by regulating the various enzymes that act on it.

ATP Donates Phosphoryl, Pyrophosphoryl, and Adenylyl Groups

The reactions of ATP are generally SN2 nucleophilic displacements (p. 66), in which the nucleophile may be, for example, the oxygen of an alcohol or carboxylate or a nitrogen of creatine or of the side chain of arginine or histidine. Each of the three phosphates of ATP is susceptible to nucleophilic attack (Fig. 14–10), and each position of attack yields a different type of product.

Nucleophilic attack by an alcohol on the γ phosphate (Fig. 14–10a) displaces ADP and produces a new phosphate ester. Studies with ^{18}O-labeled reactants have shown that the bridge oxygen in the new compound is derived from the alcohol, not from ATP; the group transferred from ATP is a phosphoryl ($—PO_3^{2-}$), not a phosphate ($—OPO_3^{2-}$). Phosphoryl group transfer from ATP to glutamate (Fig. 14–8) or to glucose (p. 275) involves attack at the γ position of the ATP molecule.

Attack at the β phosphate of ATP displaces AMP and transfers a pyrophosphoryl (not pyrophosphate) group to the attacking nucleophile (Fig. 14–10b). For example, the formation of 5′-phosphoribose 1-pyrophosphate, a key intermediate in nucleotide synthesis, occurs as an —OH of the ribose attacks the β phosphate.

Nucleophilic attack at the α position of ATP displaces PP_i and transfers adenylate (5′-AMP) as an adenylyl group (Fig. 14–10c); the reaction is an **adenylylation** (a-den′-i-li-lā′-shun, probably the most ungainly word in

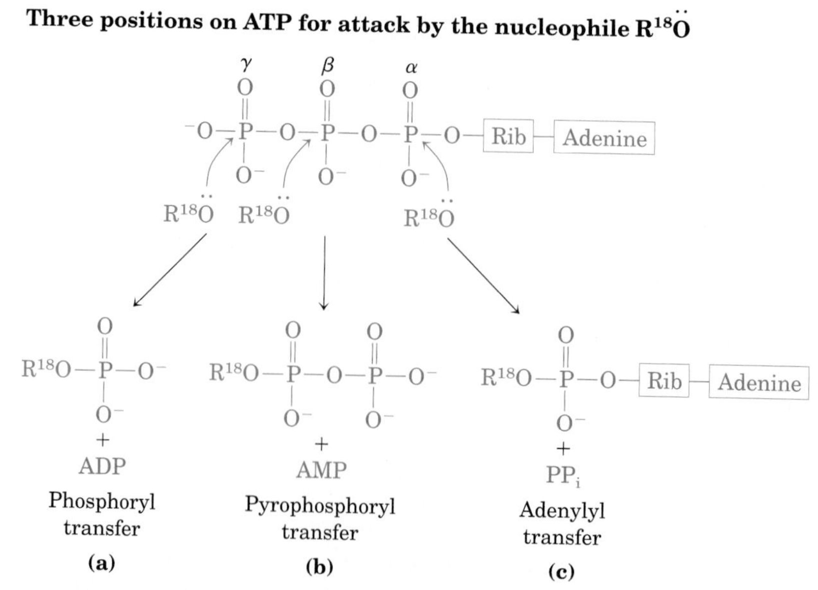

figure 14–10
Nucleophilic displacement reactions of ATP. Any of the three P atoms (α, β, or γ) may serve as the electrophilic target for nucleophilic attack—by the labeled nucleophile $R—^{18}O$:, in this case. The nucleophile may be an alcohol (ROH), a carboxyl group ($RCOO^-$), or a phosphoanhydride (a nucleoside mono- or diphosphate, for example). **(a)** When the oxygen of the nucleophile attacks the γ position, the bridge oxygen of the product is labeled, indicating that the group transferred from ATP is a phosphoryl ($—PO_3^{2-}$), not a phosphate ($—OPO_3^{2-}$). **(b)** Attack on the β position displaces AMP and leads to the transfer of a pyrophosphoryl (not pyrophosphate) group to the nucleophile. **(c)** Attack on the α position displaces PP_i and transfers the adenylyl group to the nucleophile.

Three positions on ATP for attack by the nucleophile $R^{18}\ddot{O}$

γ β α
Rib Adenine
$R^{18}\ddot{O}$ $R^{18}\ddot{O}$ $R^{18}\ddot{O}$
+ ADP
Phosphoryl transfer
(a)
+ AMP
Pyrophosphoryl transfer
(b)
Rib Adenine
+ PP_i
Adenylyl transfer
(c)

the biochemical language). Notice that hydrolysis of the α–β phosphoanhydride bond releases considerably more energy (~46 kJ/mol) than hydrolysis of the β–γ bond (~31 kJ/mol) (Table 14–6). Furthermore, the PP_i formed as a byproduct of the adenylylation is hydrolyzed to two P_i by the ubiquitous enzyme **inorganic pyrophosphatase,** releasing 19 kJ/mol and thereby providing a further energy "push" for the adenylylation reaction. In effect, both phosphoanhydride bonds of ATP are split in the overall reaction. Adenylylation reactions are therefore thermodynamically very favorable. When the energy of ATP is used to drive a particularly unfavorable metabolic reaction, adenylylation is often the mechanism of energy coupling. Fatty acid activation is a good example of this energy-coupling strategy.

The first step in the activation of a fatty acid—either for energy-yielding oxidation (see Fig. 17–5) or for use in the synthesis of more complex lipids (Chapter 21)—is its attachment to the carrier coenzyme A (Fig. 14–11). The direct condensation of a fatty acid with coenzyme A is endergonic, but the formation of fatty acyl–CoA is made exergonic by stepwise removal of *two* phosphoryl groups from ATP. First, adenylate (AMP) is transferred from ATP to the carboxyl group of the fatty acid, forming a mixed anhydride (fatty acyl adenylate) and liberating PP_i. The thiol group of coenzyme A then displaces the adenylate group and forms a thioester with the fatty acid.

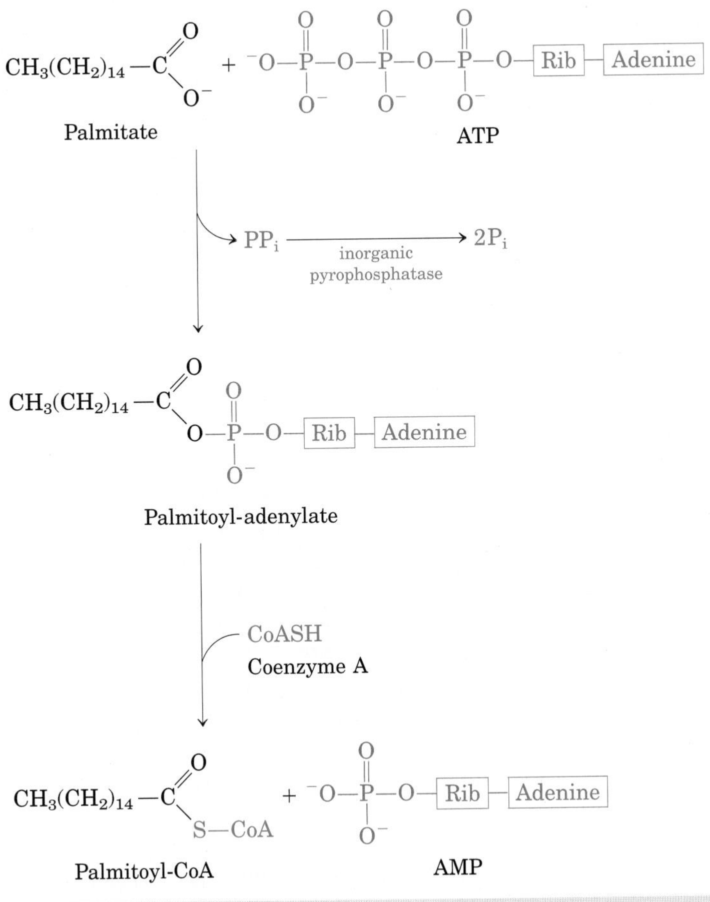

$CH_3(CH_2)_{14}-COO^-$ + $^-O-P(O)(O^-)-O-P(O)(O^-)-O-P(O)(O^-)-O-$ Rib – Adenine

Palmitate ATP

$\rightarrow PP_i \xrightarrow[\text{pyrophosphatase}]{\text{inorganic}} 2P_i$

$CH_3(CH_2)_{14}-C(=O)-O-P(O)(O^-)-O-$ Rib – Adenine

Palmitoyl-adenylate

CoASH
Coenzyme A

$CH_3(CH_2)_{14}-C(=O)-S-CoA$ + $^-O-P(O)(O^-)-O-$ Rib – Adenine

Palmitoyl-CoA AMP

Overall reaction:
Palmitate + ATP + CoASH $\longrightarrow$ palmitoyl-CoA + AMP + $2P_i$
$\Delta G'^{\circ} = -32.5$ kJ/mol

figure 14–11
Adenylylation reaction in activation of a fatty acid. Both phosphoanhydride bonds of ATP are eventually broken in the formation of palmitoyl-coenzyme A. First, ATP donates adenylate (AMP), forming the fatty acyl–adenylate and releasing PP_i, which is hydrolyzed by inorganic pyrophosphatase. The "energized" fatty acyl group is then transferred to coenzyme A (CoASH), with which it forms a thioester bond that conserves some of the energy "invested" from ATP.

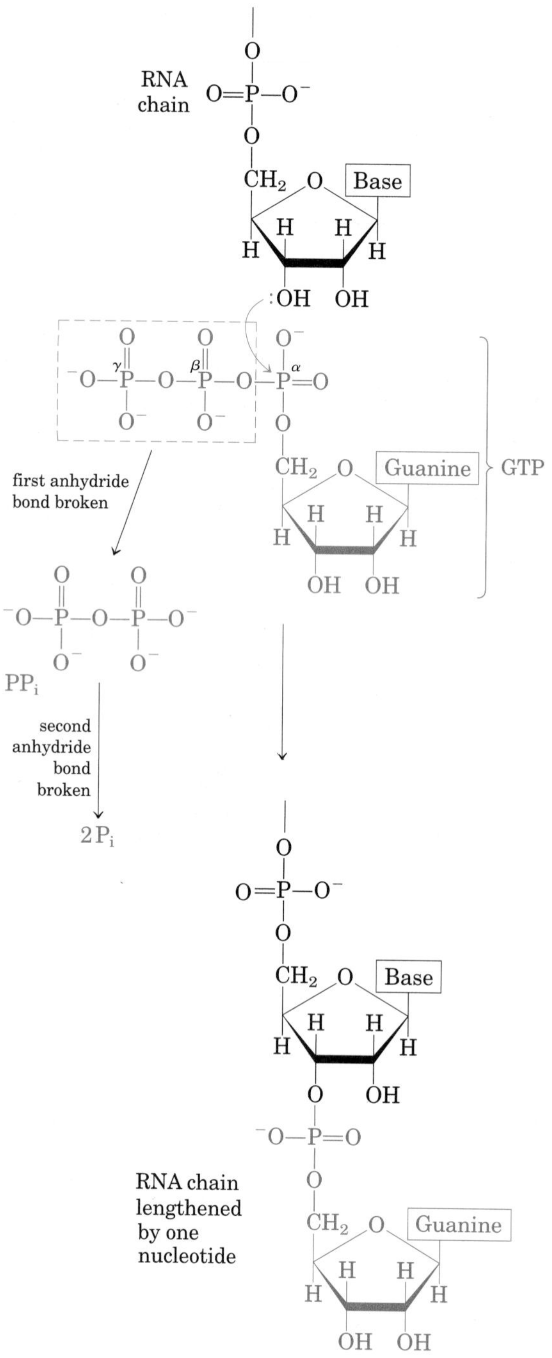

The sum of these two reactions is energetically equivalent to the exergonic hydrolysis of ATP to AMP and PP_i ($\Delta G'^\circ = -45.6$ kJ/mol) and the endergonic formation of fatty acyl–CoA ($\Delta G'^\circ = 31.4$ kJ/mol). The formation of fatty acyl–CoA is made energetically favorable by hydrolysis of the PP_i by inorganic pyrophosphatase. Thus, in the activation of a fatty acid, both of the phosphoanhydride bonds of ATP are broken. The resulting $\Delta G'^\circ$ is the sum of the $\Delta G'^\circ$ values for the breakage of these bonds, or -45.6 kJ/mol + (-19.22) kJ/mol:

$$\text{ATP} + 2\text{H}_2\text{O} \longrightarrow \text{AMP} + 2\text{P}_i \qquad \Delta G'^\circ = -64.8 \text{ kJ/mol}$$

The activation of amino acids before their polymerization into proteins (see Fig. 27–16) is accomplished by an analogous set of reactions in which a transfer RNA molecule takes the place of coenzyme A. An unusual use of the cleavage of ATP to AMP and PP_i occurs in the firefly, which uses ATP as an energy source to produce light flashes (Box 14–3).

Assembly of Informational Macromolecules Requires Energy

When simple precursors are assembled into high molecular weight polymers with defined sequences (DNA, RNA, proteins), as described in detail in Part IV of this book, energy is required both for the condensation of monomeric units and for the creation of *ordered* sequences. The precursors for DNA and RNA synthesis are nucleoside triphosphates, and polymerization is accompanied by cleavage of the phosphoanhydride linkage between the α and β phosphates, with the release of PP_i (Fig. 14–12). The moieties transferred to the growing polymer in these reactions are adenylate (AMP), guanylate (GMP), cytidylate (CMP), or uridylate (UMP) for RNA synthesis, and their deoxy analogs (with TMP in place of UMP) for DNA synthesis. As noted above, the activation of amino acids for protein synthesis involves the donation of adenylate groups from ATP, and we shall see in Chapter 27 that several steps of protein synthesis on the ribosome are also accompanied by GTP hydrolysis. In all of these cases, the exergonic breakdown of a nucleoside triphosphate is coupled to the endergonic process of synthesizing a polymer of a specific sequence.

ATP Energizes Active Transport and Muscle Contraction

ATP can supply the energy for transporting an ion or a molecule across a membrane into another aqueous compartment where its concentration is higher (see Fig. 12–30). Transport processes are major consumers of energy; in human kidney and brain, for example, as much as two-thirds of the energy consumed at rest is used to pump Na^+ and K^+ across plasma membranes via the Na^+K^+ ATPase. The transport of Na^+ and K^+ is driven by cyclic phosphorylation and dephosphorylation of the transporter protein, with ATP as the phosphoryl group donor (see Fig. 12–34). Na^+-dependent phosphorylation of the Na^+K^+ ATPase forces a change in the protein's conformation, and K^+-dependent dephosphorylation favors return to the original conformation. Each cycle in the transport process results in the conversion of ATP to ADP and P_i, and it is the free-energy change of ATP hydrolysis that drives the cyclic changes in protein conformation that result in the electrogenic pumping of Na^+ and K^+. Note that, in this case, ATP interacts covalently by phosphoryl group transfer to the enzyme, not the substrate.

In the contractile system of skeletal muscle cells, myosin and actin are specialized to transduce the chemical energy of ATP into motion (see Fig. 7–33). ATP binds tightly but noncovalently to one conformation of myosin, holding the protein in that conformation. When myosin catalyzes the hydrolysis of its bound ATP, the ADP and P_i dissociate from the protein, allowing it to relax into a second conformation until another molecule of ATP

figure 14–12
Nucleoside triphosphates in RNA synthesis.
With each nucleoside monophosphate added to the growing chain, one PP_i is released and hydrolyzed to two P_i. The hydrolysis of two phosphoanhydride bonds for each nucleotide added provides the energy for forming the bonds in the RNA polymer and for assembling a specific sequence of nucleotides.

box 14–3 Firefly Flashes: Glowing Reports of ATP

Bioluminescence requires considerable amounts of energy. In the firefly, ATP is used in a set of reactions that converts chemical energy into light energy. From many thousands of fireflies collected by children in and around Baltimore, William McElroy and his colleagues at The Johns Hopkins University isolated the principal biochemical components: luciferin, a complex carboxylic acid, and luciferase, an enzyme. The generation of a light flash requires activation of luciferin by an enzymatic reaction involving pyrophosphate cleavage of ATP to form luciferyl adenylate. In the presence of molecular oxygen and luciferase, the luciferin undergoes a multistep oxidative decarboxylation to oxyluciferin. This process is accompanied by emission of light. The color of the light flash differs with the firefly species and appears to be determined by differences in the structure of the luciferase. Luciferin is regenerated from oxyluciferin in a subsequent series of reactions.

In the laboratory, pure firefly luciferin and luciferase are used to measure minute quantities of ATP by the intensity of the light flash produced. As little as a few picomoles (10^{-12} mol) of ATP can be measured in this way. An enlightening extension of the studies in luciferase is the cloning of the luciferase gene into tobacco plants. When watered with a solution containing luciferin, the plants glowed in the dark (see Fig. 29–20).

The firefly, a beetle of the Lampyridae family.

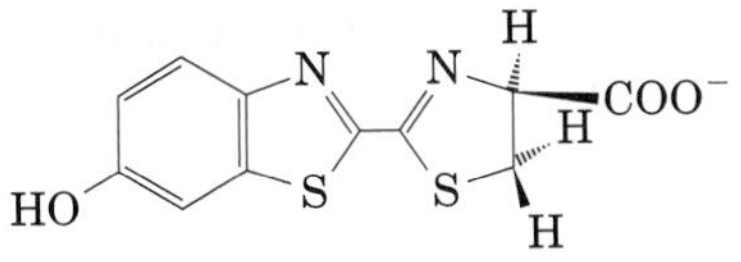

Firefly luciferin

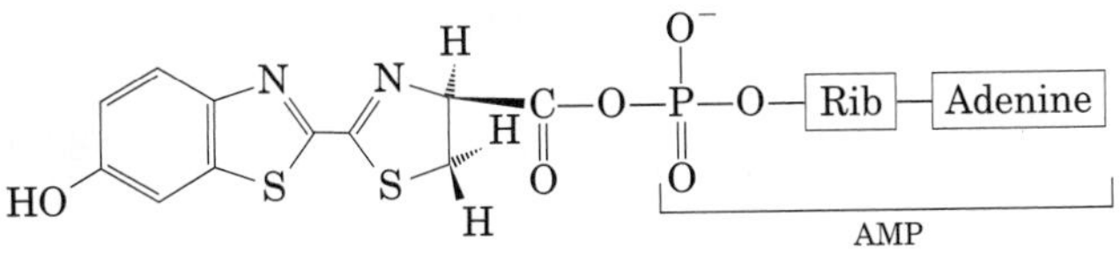

Luciferyl adenylate

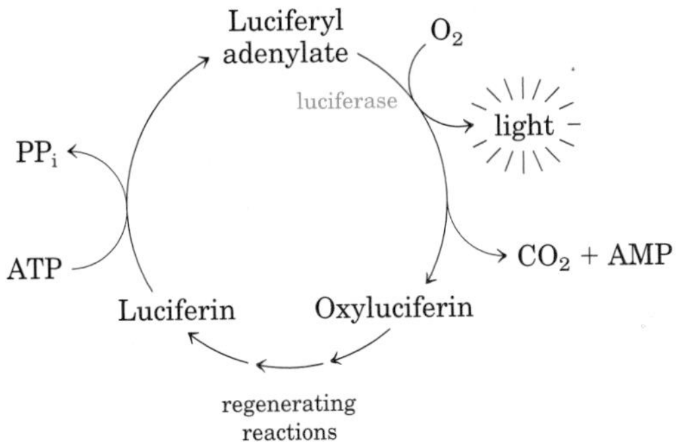

Important components in firefly bioluminescence, and the firefly bioluminescence cycle.

binds. The binding and subsequent hydrolysis of ATP (by myosin ATPase) provide the energy that forces cyclic changes in the conformation of the myosin head. The change in conformation of many individual myosin molecules results in the sliding of myosin fibrils along actin filaments (see Fig. 7–32), which translates into macroscopic contraction of the muscle fiber. This production of mechanical motion at the expense of ATP is, as noted earlier, one of the few cases in which ATP hydrolysis per se, rather than group transfer from ATP, is the source of the chemical energy in a coupled process.

Transphosphorylations between Nucleotides Occur in All Cell Types

Although we have focused on ATP as the cell's energy currency and donor of phosphoryl groups, all other nucleoside triphosphates (GTP, UTP, and CTP) and all the deoxynucleoside triphosphates (dATP, dGTP, dTTP, and dCTP) are energetically equivalent to ATP. The free-energy changes associated with hydrolysis of their phosphoanhydride linkages are very nearly identical with those shown in Table 14–6 for ATP. In preparation for their various biological roles, these other nucleotides are generated and maintained as the nucleoside triphosphate (NTP) forms by phosphoryl group transfer to the corresponding nucleoside diphosphates (NDPs) and monophosphates (NMPs).

ATP is the primary high-energy phosphate compound produced by catabolism, in the processes of glycolysis, oxidative phosphorylation, and, in photosynthetic cells, photophosphorylation. Several enzymes then carry phosphoryl groups from ATP to the other nucleotides. **Nucleoside diphosphate kinase,** found in all cells, catalyzes the reaction

$$\text{ATP} + \text{NDP (or dNDP)} \xrightleftharpoons{\text{Mg}^{2+}} \text{ADP} + \text{NTP (or dNTP)} \qquad \Delta G'^{\circ} \approx 0$$

Although this reaction is fully reversible, the relatively high [ATP]/[ADP] ratio in cells normally drives the reaction to the right, with the net formation of NTPs and dNTPs. The enzyme actually catalyzes a two-step phosphoryl transfer. First, phosphoryl group transfer from ATP to an active-site His residue produces a phosphoenzyme intermediate; then the phosphoryl group is transferred from the Ⓟ-His residue to an NDP acceptor. Because the enzyme is nonspecific for the base in the NDP and works equally well on dNDPs and NDPs, it can synthesize all NTPs and dNTPs, given the corresponding NDPs and a supply of ATP.

When ADP accumulates as a result of phosphoryl group transfers from ATP, such as when muscle is contracting vigorously, the ADP interferes with ATP-dependent contraction. **Adenylate kinase** removes ADP by the reaction

$$2\text{ADP} \xrightleftharpoons{\text{Mg}^{2+}} \text{ATP} + \text{AMP} \qquad \Delta G'^{\circ} \approx 0$$

This reaction is fully reversible, so the enzyme can also convert AMP (produced by pyrophosphoryl or adenylyl group transfer from ATP) into ADP, which can then be phosphorylated to ATP through one of the catabolic pathways. A similar enzyme, guanylate kinase, converts GMP to GDP at the expense of ATP. By pathways such as these, energy conserved in the catabolic production of ATP is used to supply the cell with all required NTPs and dNTPs.

Phosphocreatine (Fig. 14–5), also called creatine phosphate, serves as a ready source of phosphoryl groups for the quick synthesis of ATP from ADP. The phosphocreatine (PCr) concentration in skeletal muscle is approximately 30 mM, nearly ten times the concentration of ATP, and in other tissues such as smooth muscle, brain, and kidney is 5 to 10 mM. The enzyme **creatine kinase** catalyzes the reversible reaction

$$\text{ATP} + \text{PCr} \xrightleftharpoons{\text{Mg}^{2+}} \text{ATP} + \text{Cr} \qquad \Delta G'^{\circ} = -12.5\ \text{kJ/mol}$$

When a sudden demand for energy depletes ATP, the PCr reservoir is used to replenish ATP at a rate considerably faster than ATP can be synthesized by catabolic pathways. When the demand for energy slackens, ATP produced by catabolism is used to replenish the PCr reservoir by reversal of the creatine kinase reaction. Organisms in the lower phyla employ other PCr-like molecules (collectively called **phosphagens**) as phosphoryl reservoirs.

Inorganic Polyphosphate Is a Potential Phosphoryl Group Donor

Inorganic polyphosphate (polyP) is a linear polymer composed of hundreds of P_i residues linked through phosphoanhydride bonds. This polymer, present in cells of all organisms, has about the same phosphoryl group transfer potential as PP_i, but its biological roles remain uncertain. In *Escherichia coli,* polyP accumulates when cells are grown in a medium with excess P_i, and this accumulation confers a survival advantage during periods of nutritional or oxidative stress. The enzyme **polyphosphate kinase** catalyzes the reaction

$$\text{ATP} + \text{polyP}_n \xrightleftharpoons{\text{Mg}^{2+}} \text{ADP} + \text{polyP}_{n+1} \qquad \Delta G'^{\circ} = -20 \text{ kJ/mol}$$

by a mechanism involving an enzyme-bound phosphohistidine intermediate (recall the mechanism of nucleoside diphosphate kinase, described above). Because the reaction is reversible, polyP (like PCr) could serve as a reservoir of phosphoryl groups or even as a phosphoryl group donor analogous to ATP for kinase-catalyzed transfers. The shortest polyphosphate, PP_i ($n = 2$) can serve as the energy source for active transport of H^+ in plant vacuoles. PP_i is also the usual phosphoryl group donor for at least one form of the enzyme phosphofructokinase in plants, a role normally played by ATP in animals and microbes (p. 533). The finding of high concentrations of polyP in volcanic condensates and steam vents suggests that it could have served as an energy source in prebiotic and early cellular evolution.

$$^{-}\text{O}-\overset{\text{O}}{\overset{\|}{\underset{\text{O}^-}{\underset{|}{\text{P}}}}}-\text{O}-\overset{\text{O}}{\overset{\|}{\underset{\text{O}^-}{\underset{|}{\text{P}}}}}-\text{O}-\overset{\text{O}}{\overset{\|}{\underset{\text{O}^-}{\underset{|}{\text{P}}}}}-\text{O}-\overset{\text{O}}{\overset{\|}{\underset{\text{O}^-}{\underset{|}{\text{P}}}}}-\text{O}-\overset{\text{O}}{\overset{\|}{\underset{\text{O}^-}{\underset{|}{\text{P}}}}}-\text{O}-$$

Inorganic polyphosphate (polyP)

Biochemical and Chemical Equations Are Not Identical

Biochemists write metabolic equations in a simplified way, and this is particularly evident for reactions involving ATP. Phosphorylated compounds can exist in several ionization states, and, as we have noted, the different species can bind Mg^{2+}. For example, at pH 7 and 2 mM Mg^{2+}, ATP exists in the forms ATP^{4-}, $HATP^{3-}$, H_2ATP^{2-}, $MgHATP^-$, and Mg_2ATP. In thinking about the biological role of ATP, however, we are not always interested in all of this detail, and so we consider ATP as an entity made up of a sum of species and write its hydrolysis as the biochemical equation

$$\text{ATP} + \text{H}_2\text{O} \longrightarrow \text{ADP} + \text{P}_\text{i}$$

where ATP, ADP, and P_i are sums of species. The corresponding apparent equilibrium constant,

$$K'_{\text{eq}} = [\text{ADP}][\text{P}_\text{i}]/[\text{ATP}]$$

depends on the pH and the concentration of free Mg^{2+}. Note that H^+ and Mg^{2+} do not appear in the biochemical equation because they are held constant. Thus, a biochemical equation does not balance H, Mg, or charge, although it does balance all other elements involved in the reaction (C, N, O, and P in the equation above).

We can write a chemical equation that *does* balance for all elements and for charge. For example, when ATP is hydrolyzed at a pH above 8.5 in the absence of Mg^{2+}, the chemical reaction is represented by

$$\text{ATP}^{4-} + \text{H}_2\text{O} \longrightarrow \text{ADP}^{3-} + \text{HPO}_4^{2-} + \text{H}^+$$

The corresponding equilibrium constant,

$$K'_{\text{eq}} = [\text{ADP}^{3-}][\text{HPO}_4^{2-}][\text{H}^+]/[\text{ATP}^{4-}]$$

depends only on temperature, pressure, and ionic strength.

Both ways of writing a metabolic reaction have value in biochemistry. Chemical equations are needed when we want to account for all atoms and charges in a reaction, as when we are considering the mechanism of a chemical reaction. Biochemical equations are used to determine in which direction a reaction will proceed spontaneously, given a specified pH and [Mg^{2+}], or to calculate the equilibrium constant of such a reaction.

Throughout this book we use biochemical equations, unless the focus is on chemical mechanism, and we use values of $\Delta G'^{\circ}$ and K'_{eq} as determined at pH 7 and 1 mM Mg^{2+}.

Biological Oxidation-Reduction Reactions

The transfer of phosphoryl groups is a central feature of metabolism. Equally important is another kind of transfer, electron transfer in oxidation-reduction reactions. These reactions involve the loss of electrons by one chemical species, which is thereby oxidized, and the gain of electrons by another, which is reduced. The flow of electrons in oxidation-reduction reactions is responsible, directly or indirectly, for all work done by living organisms. In nonphotosynthetic organisms, the sources of electrons are reduced compounds (foods); in photosynthetic organisms, the initial electron donor is a chemical species excited by the absorption of light. The path of electron flow in metabolism is complex. Electrons move from various metabolic intermediates to specialized electron carriers in enzyme-catalyzed reactions. The carriers in turn donate electrons to acceptors with higher electron affinities, with the release of energy. Cells contain a variety of molecular energy transducers, which convert the energy of electron flow into useful work.

We begin our discussion with a description of the general types of metabolic reactions in which electrons are transferred. After considering the theoretical and experimental basis for measuring the energy changes in oxidation reactions in terms of electromotive force, we discuss the relationship between this force, expressed in volts, and the free-energy change, expressed in joules. We conclude by describing the structures and oxidation-reduction chemistry of the most common of the specialized electron carriers, which you will encounter repeatedly in later chapters.

The Flow of Electrons Can Do Biological Work

Every time we use a motor, an electric light or heater, or a spark to ignite gasoline in a car engine, we use the flow of electrons to accomplish work. In the circuit that powers a motor, the source of electrons can be a battery containing two chemical species that differ in affinity for electrons. Electrical wires provide a pathway for electron flow from the chemical species at one pole of the battery, through the motor, to the chemical species at the other pole of the battery. Because the two chemical species differ in their affinity for electrons, electrons flow spontaneously through the circuit, driven by a force proportional to the difference in electron affinity, the **electromotive force (emf).** The electromotive force (typically a few volts) can accomplish work if an appropriate energy transducer—in this case a motor—is placed in the circuit. The motor can be coupled to a variety of mechanical devices to accomplish useful work.

Living cells have an analogous biological "circuit," with a relatively reduced compound such as glucose as the source of electrons. As glucose is enzymatically oxidized, the electrons released flow spontaneously through a series of electron-carrier intermediates to another chemical species, such as O_2. This electron flow is exergonic because O_2 has a higher affinity for electrons than do the electron-carrier intermediates. The resulting electromotive force provides energy to a variety of molecular energy transducers (enzymes and other proteins) that do biological work. In the mitochondrion, for example, membrane-bound enzymes couple electron flow to the production of a transmembrane pH difference, accomplishing osmotic and electrical work. The proton gradient thus formed has potential energy, sometimes called the proton-motive force by analogy with electromotive force. Another enzyme, ATP synthase in the inner mitochondrial membrane, uses the proton-motive force to do chemical work: synthesis of ATP from ADP and P_i as protons flow spontaneously across the membrane. Similarly, membrane-localized enzymes in *E. coli* convert electromotive force to proton-motive force, which is then used to power flagellar motion.

The principles of electrochemistry that govern energy changes in the macroscopic circuit with a motor and battery apply with equal validity to the molecular processes accompanying electron flow in living cells. We turn now to a discussion of those principles.

Oxidation-Reductions Can Be Described as Half-Reactions

Although oxidation and reduction must occur together, it is convenient when describing electron transfers to consider the two halves of an oxidation-reduction reaction separately. For example, the oxidation of ferrous ion by cupric ion,

$$Fe^{2+} + Cu^{2+} \rightleftharpoons Fe^{3+} + Cu^{+}$$

can be described in terms of two half-reactions:

$$(1)\quad Fe^{2+} \rightleftharpoons Fe^{3+} + e^{-}$$
$$(2)\quad Cu^{2+} + e^{-} \rightleftharpoons Cu^{+}$$

The electron-donating molecule in an oxidation-reduction reaction is called the reducing agent or reductant; the electron-accepting molecule is the oxidizing agent or oxidant. A given agent, such as an iron cation existing in the ferrous (Fe^{2+}) or ferric (Fe^{3+}) state, functions as a conjugate reductant-oxidant pair (redox pair), just as an acid and corresponding base function as a conjugate acid-base pair. Recall from Chapter 4 that in acid-base reactions we can write a general equation: proton donor $\rightleftharpoons H^{+}$ + proton acceptor. In redox reactions we can write a similar general equation: electron donor $\rightleftharpoons e^{-}$ + electron acceptor. In the reversible half-reaction (1) above, Fe^{2+} is the electron donor and Fe^{3+} is the electron acceptor; together, Fe^{2+} and Fe^{3+} constitute a **conjugate redox pair.**

The electron transfers in the oxidation-reduction reactions of organic compounds are not fundamentally different from those of inorganic species. In Chapter 9 we considered the oxidation of a reducing sugar (an aldehyde or ketone) by cupric ion (see Fig. 9–10a):

$$R{-}C({=}O){-}H + 4OH^{-} + 2Cu^{2+} \rightleftharpoons R{-}C({=}O){-}OH + Cu_2O + 2H_2O$$

This overall reaction can be expressed as two half-reactions:

$$(1)\quad R{-}C({=}O){-}H + 2OH^{-} \rightleftharpoons R{-}C({=}O){-}OH + 2e^{-} + H_2O$$

$$(2)\quad 2Cu^{2+} + 2e^{-} + 2OH^{-} \rightleftharpoons Cu_2O + H_2O$$

Because two electrons are removed from the aldehyde carbon, the second half-reaction (the one-electron reduction of cupric to cuprous ion) must be doubled to balance the overall equation.

Compound	Structure	Electrons owned
Methane	H:C:H (with H above and below C)	8
Ethane (alkane)	H:C:C:H (with H H above and H H below)	7
Ethene (alkene)	H₂C::CH₂	6
Ethanol (alcohol)	H:C:C:O:H (with H H above and H H below)	5
Acetylene (alkyne)	H:C:::C:H	5
Formaldehyde	H₂C::O	4
Acetaldehyde (aldehyde)	H:C:C(H)::O (with H above and below the first C)	3
Acetone (ketone)	H:C:C:C:H (O above the middle C; H above and below the end carbons)	2
Formic acid (carboxylic acid)	H:C(::O):O:H	2
Carbon monoxide	:C:::O:	2
Acetic acid (carboxylic acid)	H:C:C(::O):O:H (with H above and below the first C)	1
Carbon dioxide	O::C::O	0

figure 14–13

Oxidation states of carbon occurring in the biosphere. The oxidation states are illustrated with some representative compounds. Focus on the red carbon atom and its bonding electrons. When this carbon is bonded to the less electronegative H atom, both bonding electrons (red) are assigned to the carbon. When carbon is bonded to another carbon, bonding electrons are shared equally, so one of the two electrons is assigned to the red carbon. When the red carbon is bonded to the more electronegative O atom, the bonding electrons are assigned to the oxygen. The number to the right of each compound is the number of electrons "owned" by the red carbon, a rough expression of the oxidation state of that carbon. When the red carbon undergoes oxidation (loses electrons), the number gets smaller. Thus the order of increasing oxidation state is alkane < alkene < alcohol < alkyne < aldehyde < ketone < carboxylic acid < carbon dioxide.

Biological Oxidations Often Involve Dehydrogenation

The carbon in living cells exists in a range of oxidation states (Fig. 14–13). When a carbon atom shares an electron pair with another atom (typically H, C, S, N, or O), the sharing is unequal in favor of the more electronegative atom. (Recall from Table 3–2 the order of increasing electronegativity: H < C < S < N < O.) In oversimplified but useful terms, the more electronegative atom "owns" the bonding electrons it shares with another atom. For example, in methane (CH_4), carbon is more electronegative than the four hydrogens bonded to it, and the C atom therefore "owns" all eight bonding electrons (Fig. 14–13). In ethane, the electrons in the C—C bond are shared equally, so each C atom owns only seven of its eight bonding electrons. In ethanol, C-1 is less electronegative than the oxygen to which it is bonded, and the O atom therefore "owns" both electrons of the C—O bond, leaving C-1 with only five bonding electrons. With each formal loss of electrons, the carbon atom has undergone oxidation—even when no oxygen is involved, as in the conversion of an alkane ($CH_2—CH_2$) to an alkene (CH═CH). In this case, oxidation (loss of electrons) is coincident with the loss of hydrogen. In biological systems, oxidation is often synonymous with **dehydrogenation,** and many enzymes that catalyze oxidation reactions are **dehydrogenases.** Notice that the more reduced compounds in Figure 14–13 (top) are richer in hydrogen than in oxygen, whereas the more oxidized compounds (bottom) have more oxygen and less hydrogen.

Not all biological oxidation-reduction reactions involve carbon. For example, in the conversion of molecular nitrogen to ammonia, $6H^+ + 6e^- + N_2 \rightarrow 2NH_3$, the nitrogen atoms are reduced.

Electrons are transferred from one molecule (electron donor) to another (electron acceptor) in one of four different ways:

1. Directly as *electrons.* For example, the Fe^{2+}/Fe^{3+} redox pair can transfer an electron to the Cu^+/Cu^{2+} redox pair:

$$Fe^{2+} + Cu^{2+} \rightleftharpoons Fe^{3+} + Cu^+$$

2. As *hydrogen atoms.* Recall that a hydrogen atom consists of a proton (H^+) and a single electron (e^-). In this case we can write the general equation

$$AH_2 \rightleftharpoons A + 2e^- + 2H^+$$

where AH_2 is the hydrogen/electron donor. (Do not mistake the above reaction for an acid dissociation; the H^+ arises from the removal of a hydrogen atom ($H^+ + e^-$).) AH_2 and A together constitute a conjugate redox pair (A/AH_2), which can reduce another compound B (or redox pair, B/BH_2) by transfer of hydrogen atoms:

$$AH_2 + B \rightleftharpoons A + BH_2$$

3. As a *hydride ion* (:H^-), which has two electrons. This occurs in the case of NAD-linked dehydrogenases, described below.

4. Through direct *combination with oxygen.* In this case, oxygen combines with an organic reductant and is covalently incorporated in the product, as in the oxidation of a hydrocarbon to an alcohol:

$$R{-}CH_3 + \tfrac{1}{2}O_2 \longrightarrow R{-}CH_2{-}OH$$

The hydrocarbon is the electron donor and the oxygen atom is the electron acceptor.

All four types of electron transfer occur in cells. The neutral term **reducing equivalent** is commonly used to designate a single electron equivalent participating in an oxidation-reduction reaction, no matter whether this equivalent is an electron per se, a hydrogen atom, or a hydride ion, or whether the electron transfer takes place in a reaction with oxygen to yield an oxygenated product. Because biological fuel molecules are usually enzymatically dehydrogenated to lose *two* reducing equivalents at a time, and because each oxygen atom can accept two reducing equivalents, biochemists by convention regard the unit of biological oxidations as two reducing equivalents passing from substrate to oxygen.

Reduction Potentials Measure Affinity for Electrons

When two conjugate redox pairs are together in solution, electron transfer from the electron donor of one pair to the electron acceptor of the other may occur spontaneously. The tendency for such a reaction depends on the relative affinity of the electron acceptor of each redox pair for electrons. The **standard reduction potential, $E°$**, a measure (in volts) of this affinity, can be determined in an experiment such as that described in Figure 14–14. Electrochemists have chosen as a standard of reference the half-reaction

$$H^+ + e^- \longrightarrow \tfrac{1}{2}H_2$$

The electrode at which this half-reaction occurs (called a half-cell) is arbitrarily assigned a standard reduction potential of 0.00 V. When this hydrogen electrode is connected through an external circuit to another half-cell in which an oxidized species and its corresponding reduced species are present at standard concentrations (each solute at 1 M, each gas at 101.3 kPa or 1 atm), electrons tend to flow through the external circuit from the half-cell of lower standard reduction potential to the half-cell of higher standard reduction potential. By convention, the half-cell with the stronger tendency to acquire electrons is assigned a positive value of $E°$.

The reduction potential of a half-cell depends not only on the chemical species present but also on their activities, approximated by their concentrations. About a century ago, Walther Nernst derived an equation that relates standard reduction potential ($E°$) to the reduction potential (E) at any concentration of oxidized and reduced species in the cell:

$$E = E° + \frac{RT}{n\mathcal{J}} \ln \frac{[\text{electron acceptor}]}{[\text{electron donor}]} \quad (14\text{–}4)$$

where R and T have their usual meanings, n is the number of electrons transferred per molecule, and $\mathcal{J}$ is the Faraday constant (Table 14–1). At 298 K (25 °C), this expression reduces to

$$E = E° + \frac{0.026\text{ V}}{n} \ln \frac{[\text{electron acceptor}]}{[\text{electron donor}]} \quad (14\text{–}5)$$

Many half-reactions of interest to biochemists involve protons. As in the definition of $\Delta G'°$, biochemists define the standard state for oxidation-reduction reactions as pH 7 and express reduction potential as $E'°$, the

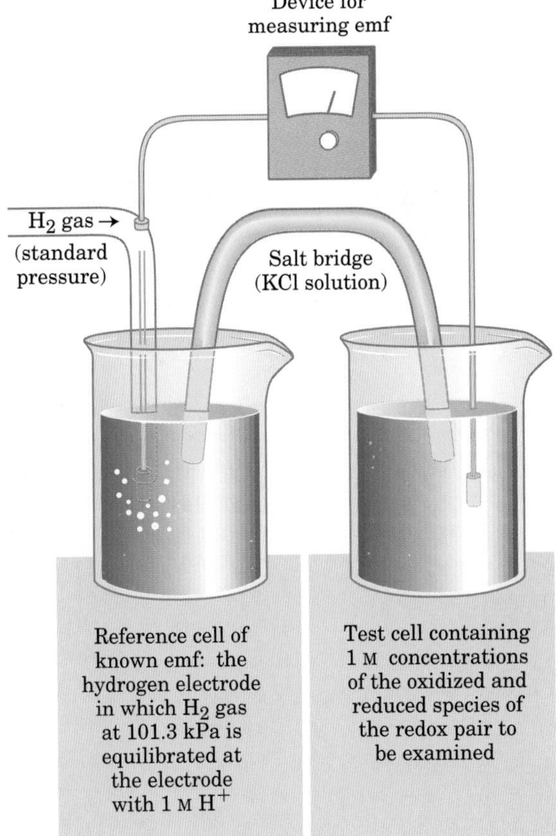

figure 14–14
Measurement of the standard reduction potential ($E'°$) of a redox pair. Electrons flow from the test electrode to the reference electrode, or vice versa. The ultimate reference half-cell is the hydrogen electrode, as shown here. The electromotive force (emf) of this electrode is designated 0.00 V. At pH 7, $E'°$ for the hydrogen electrode is −0.414 V. The direction of electron flow depends on the relative electron "pressure" or potential of the two cells. A salt bridge containing a saturated KCl solution provides a path for counter-ion movement between the test cell and the reference cell. From the observed emf and the known emf of the reference cell, the emf of the test cell containing the redox pair is obtained. The cell that gains electrons has, by convention, the more positive reduction potential.

table 14–7

Standard Reduction Potentials of Some Biologically Important Half-Reactions, at 25 °C and pH 7

Half-reaction	E'° (V)
$\frac{1}{2}O_2 + 2H^+ + 2e^- \longrightarrow H_2O$	0.816
$Fe^{3+} + e^- \longrightarrow Fe^{2+}$	0.771
$NO_3^- + 2H^+ + 2e^- \longrightarrow NO_2^- + H_2O$	0.421
Cytochrome *f* (Fe^{3+}) + e^- ⟶ cytochrome *f* (Fe^{2+})	0.365
$Fe(CN)_6^{3-}$ (ferricyanide) + e^- ⟶ $Fe(CN)_6^{4-}$	0.36
Cytochrome a_3 (Fe^{3+}) + e^- ⟶ cytochrome a_3(Fe^{2+})	0.35
$O_2 + 2H^+ + 2e^- \longrightarrow H_2O_2$	0.295
Cytochrome *a* (Fe^{3+}) + e^- ⟶ cytochrome *a* (Fe^{2+})	0.29
Cytochrome *c* (Fe^{3+}) + e^- ⟶ cytochrome *c* (Fe^{2+})	0.254
Cytochrome c_1 (Fe^{3+}) + e^- ⟶ cytochrome c_1 (Fe^{2+})	0.22
Cytochrome *b* (Fe^{3+}) + e^- ⟶ cytochrome *b* (Fe^{2+})	0.077
Ubiquinone + $2H^+$ + $2e^-$ ⟶ ubiquinol + H_2	0.045
Fumarate^{2-} + $2H^+$ + $2e^-$ ⟶ succinate^{2-}	0.031
$2H^+ + 2e^- \longrightarrow H_2$ (at standard conditions, pH 0)	0.000
Crotonyl-CoA + $2H^+$ + $2e^-$ ⟶ butyryl-CoA	−0.015
Oxaloacetate^{2-} + $2H^+$ + $2e^-$ ⟶ malate^{2-}	−0.166
Pyruvate$^-$ + $2H^+$ + $2e^-$ ⟶ lactate$^-$	−0.185
Acetaldehyde + $2H^+$ + $2e^-$ ⟶ ethanol	−0.197
FAD + $2H^+$ + $2e^-$ ⟶ $FADH_2$	−0.219*
Glutathione + $2H^+$ + $2e^-$ ⟶ 2 reduced glutathione	−0.23
$S + 2H^+ + 2e^- \longrightarrow H_2S$	−0.243
Lipoic acid + $2H^+$ + $2e^-$ ⟶ dihydrolipoic acid	−0.29
NAD^+ + H^+ + $2e^-$ ⟶ NADH	−0.320
$NADP^+$ + H^+ + $2e^-$ ⟶ NADPH	−0.324
Acetoacetate + $2H^+$ + $2e^-$ ⟶ β-hydroxybutyrate	−0.346
α-Ketoglutarate + CO_2 + $2H^+$ + $2e^-$ ⟶ isocitrate	−0.38
$2H^+ + 2e^- \longrightarrow H_2$ (at pH 7)	−0.414
Ferredoxin (Fe^{3+}) + e^- ⟶ ferredoxin (Fe^{2+})	−0.432

Data mostly from Loach, P.A. (1976) In *Handbook of Biochemistry and Molecular Biology,* 3rd edn (Fasman, G.D., ed.), *Physical and Chemical Data,* Vol. I, pp. 122–130, CRC Press, Boca Raton, FL.

*This is the value for free FAD; FAD bound to a specific flavoprotein (for example succinate dehydrogenase) has a different E'°.

standard reduction potential at pH 7. The standard reduction potentials given in Table 14–7 and used throughout this book are values for E'° and are therefore only valid for systems at neutral pH. Each value represents the potential difference when the conjugate redox pair, at 1 M concentrations and pH 7, is connected with the standard (pH 0) hydrogen electrode. Notice in Table 14–7 that when the conjugate pair $2H^+/H_2$ at pH 7 is connected with the standard hydrogen electrode (pH 0), electrons tend to flow from the pH 7 cell to the standard (pH 0) cell; the measured E'° for the $2H^+/H_2$ pair is −0.414 V.

Standard Reduction Potentials Can Be Used to Calculate the Free-Energy Change

The usefulness of reduction potentials stems from the fact that when *E* values have been determined for any two half-cells, relative to the standard hydrogen electrode, their reduction potentials relative to each other are also known. We can then predict the direction in which electrons will tend to flow when the two half-cells are connected through an external circuit or when components of both half-cells are present in the same solution. Electrons tend to flow to the half-cell with the more positive E, and the strength of that tendency is proportional to the difference in reduction potentials, ΔE.

The energy made available by this spontaneous electron flow (the free-energy change for the oxidation-reduction reaction) is proportional to ΔE:

$$\Delta G = -n\mathcal{F}\,\Delta E, \quad \text{or} \quad \Delta G'^{\circ} = -n\mathcal{F}\,\Delta E'^{\circ} \qquad (14\text{–}6)$$

Here n represents the number of electrons transferred in the reaction. With this equation we can calculate the free-energy change for any oxidation-reduction reaction from the values of E'° in a table of reduction potentials (Table 14–7) and the concentrations of the species participating in the reaction.

Consider the reaction in which acetaldehyde is reduced by the biological electron carrier NADH:

$$\text{Acetaldehyde} + \text{NADH} + \text{H}^+ \longrightarrow \text{ethanol} + \text{NAD}^+$$

The relevant half-reactions and their E'° values are:

$$\begin{aligned}
&(1)\ \ \text{Acetaldehyde} + 2\text{H}^+ + 2e^- \longrightarrow \text{ethanol} && E'^{\circ} = -0.197\ \text{V}\\
&(2)\ \ \text{NAD}^+ + 2\text{H}^+ + 2e^- \longrightarrow \text{NADH} + \text{H}^+ && E'^{\circ} = -0.320\ \text{V}
\end{aligned}$$

By convention, $\Delta E'^{\circ}$ is expressed as E'° of the electron acceptor minus E'° of the electron donor. Because acetaldehyde is accepting electrons from NADH in our example, $\Delta E'^{\circ} = -0.197\ \text{V} - (-0.320\ \text{V}) = 0.123\ \text{V}$, and n is 2. Therefore,

$$\Delta G'^{\circ} = -n\mathcal{F}\,\Delta E'^{\circ} = -2(96.5\ \text{kJ/V}\cdot\text{mol})(0.123\ \text{V}) = -23.7\ \text{kJ/mol}$$

This is the free-energy change for the oxidation-reduction reaction at pH 7 when acetaldehyde, ethanol, NAD^+, and NADH are all present at 1 M concentrations. If, instead, acetaldehyde and NADH were present at 1 M but ethanol and NAD^+ were present at 0.1 M, the value for ΔG would be calculated as follows. First, the values of E for both reductants are determined (Eqn 14–4):

$$\begin{aligned}
E_{\text{acetaldehyde}} &= E'^{\circ} + \frac{RT}{n\mathcal{F}}\ln\frac{[\text{acetaldehyde}]}{[\text{ethanol}]}\\
&= -0.197\ \text{V} + \frac{0.026\ \text{V}}{2}\ln\frac{1.0}{0.1} = -0.167\ \text{V}\\
E_{\text{NADH}} &= E'^{\circ} + \frac{RT}{n\mathcal{F}}\ln\frac{[\text{NAD}^+]}{[\text{NADH}]}\\
&= -0.320\ \text{V} + \frac{0.026\ \text{V}}{2}\ln\frac{0.1}{1.0} = -0.350\ \text{V}
\end{aligned}$$

Then ΔE is used to calculate ΔG (Eqn 14–5):

$$\begin{aligned}
\Delta E &= -0.167\ \text{V} - (-0.350)\ \text{V} = 0.183\ \text{V}\\
\Delta G &= -n\mathcal{F}\,\Delta E\\
&= -2(96.5\ \text{kJ/Vmol})(0.183\ \text{V})\\
&= -35.3\ \text{kJ/mol}
\end{aligned}$$

It is thus possible to calculate the free-energy change for any biological redox reaction at any concentrations of the redox pairs.

Cellular Oxidation of Glucose to Carbon Dioxide Requires Specialized Electron Carriers

The principles of oxidation-reduction energetics described above apply to the many metabolic reactions that involve electron transfers. For example, in many organisms, the oxidation of glucose supplies energy for the production of ATP. The complete oxidation of glucose:

$$C_6H_{12}O_6 + 6O_2 \longrightarrow 6CO_2 + 6H_2O$$

has a $\Delta G'^{\circ}$ of $-2{,}840$ kJ/mol. This is a much larger release of free energy than is required for ATP synthesis (50 to 60 kJ/mol; see Box 14–2). Cells do not convert glucose to CO_2 in a single, high-energy-releasing reaction, but rather in a series of controlled reactions, some of which are oxidations. The free energy released in these oxidation steps is of the same order of magnitude as that required for ATP synthesis from ADP, with some energy to spare. Electrons removed in these oxidation steps are transferred to coenzymes specialized for carrying electrons, such as NAD^+ and FAD (described below).

A Few Types of Coenzymes and Proteins Serve as Universal Electron Carriers

The multitude of enzymes that catalyze cellular oxidations channel electrons from their hundreds of different substrates into just a few types of universal electron carriers. The reduction of these carriers in catabolic processes results in the conservation of free energy released by substrate oxidation. NAD^+, $NADP^+$, FMN, and FAD are water-soluble coenzymes that undergo reversible oxidation and reduction in many of the electron-transfer reactions of metabolism. The nucleotides NAD^+ and $NADP^+$ move readily from one enzyme to another; the flavin nucleotides FMN and FAD are usually very tightly bound to the enzymes, called flavoproteins, for which they serve as prosthetic groups. Lipid-soluble quinones such as ubiquinone and plastoquinone act as electron carriers and proton donors in the nonaqueous environment of membranes. Iron-sulfur proteins and cytochromes, which have tightly bound prosthetic groups that undergo reversible oxidation and reduction, also serve as electron carriers in many oxidation-reduction reactions. Some of these proteins are water-soluble, but others are peripheral or integral membrane proteins (see Fig. 12–11).

We conclude this chapter by describing some chemical features of nucleotide coenzymes and some of the enzymes (dehydrogenases and flavoproteins) that use them. The oxidation-reduction chemistry of quinones, iron-sulfur proteins, and cytochromes is discussed in Chapter 19.

NADH and NADPH Act with Dehydrogenases as Soluble Electron Carriers

Nicotinamide adenine dinucleotide (NAD^+ in its oxidized form) and its close analog nicotinamide adenine dinucleotide phosphate ($NADP^+$) are composed of two nucleotides joined through their phosphate groups by a phosphoanhydride bond (Fig. 14–15). Because the nicotinamide ring resembles pyridine, these compounds are sometimes called pyridine nucleotides. The vitamin niacin is the source of the nicotinamide moiety in nicotinamide nucleotides.

Both coenzymes undergo reversible reduction of the nicotinamide ring (Fig. 14–15). As a substrate molecule undergoes oxidation (dehydrogenation), giving up two hydrogen atoms, the oxidized form of the nucleotide (NAD^+ or $NADP^+$) accepts a hydride ion ($:H^-$, the equivalent of a proton and two electrons) and is transformed into the reduced form (NADH or NADPH). The second proton removed from the substrate is released to the aqueous solvent. The half-reaction for each type of nucleotide is therefore

$$NAD^+ + 2e^- + 2H^+ \longrightarrow NADH + H^+$$
$$NADP^+ + 2e^- + 2H^+ \longrightarrow NADPH + H^+$$

Reduction of NAD^+ or $NADP^+$ converts the benzenoid ring of the nicotinamide moiety (with a fixed positive charge on the ring nitrogen) to the quinonoid form (with no charge on the nitrogen). Note that the reduced nucleotides absorb light at 340 nm; the oxidized forms do not (Fig. 14–15b). The plus sign in the abbreviations NAD$^+$ and NADP$^+$ does *not* indicate the net charge on these molecules (they are both negatively charged), but rather that the nicotinamide ring is in its oxidized form, with a positive charge on the nitrogen atom. In the abbreviations NADH and NADPH, the "H" denotes the added hydride ion. To refer to these nucleotides without specifying their oxidation state, we use NAD and NADP.

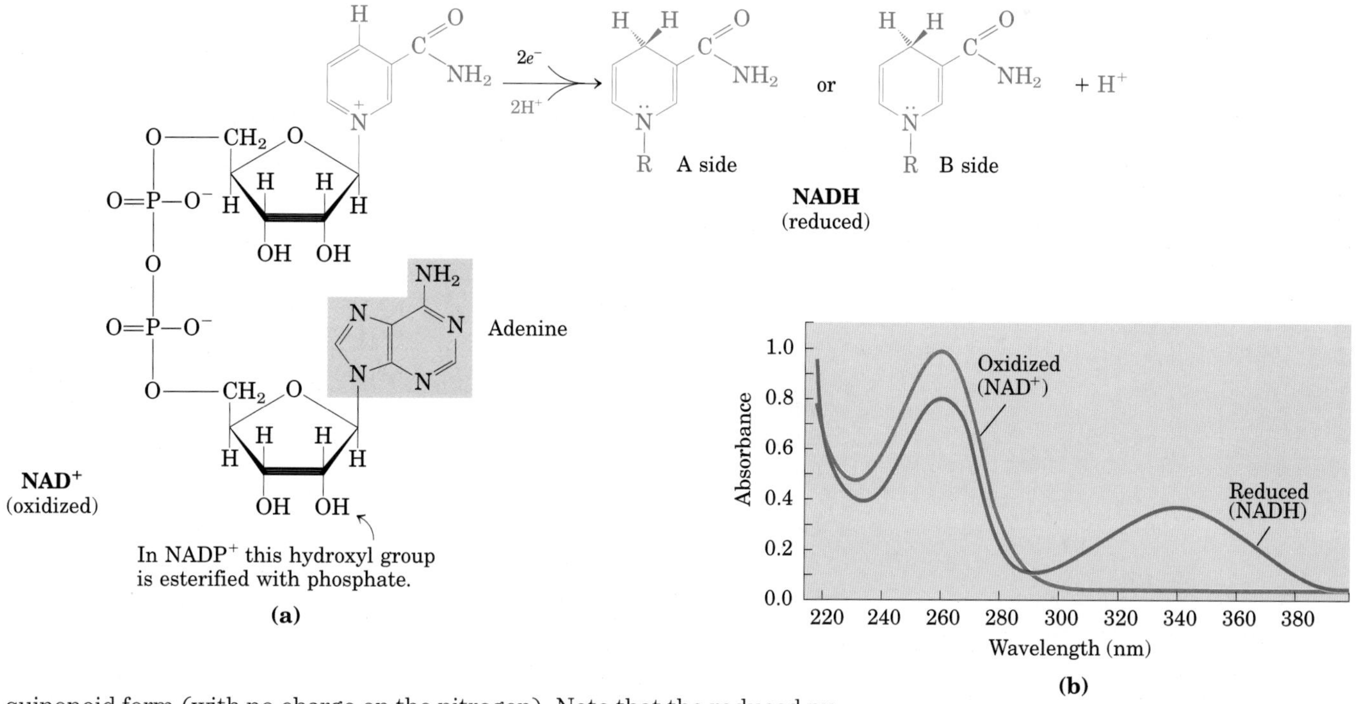

The total concentration of NAD$^+$ + NADH in most tissues is about 10^{-5} M; that of NADP$^+$ + NADPH is about 10 times lower. In many cells and tissues, the ratio of NAD$^+$ (oxidized) to NADH (reduced) is high, favoring hydride transfer from a substrate *to* NAD$^+$ to form NADH. By contrast, NADPH (reduced) is generally present in greater amounts than its oxidized form, NADP$^+$, favoring hydride transfer *from* NADPH to a substrate. This reflects the specialized metabolic roles of the two coenzymes: NAD$^+$ generally functions in oxidations—usually as part of a catabolic reaction; and NADPH is the usual coenzyme in reductions—nearly always as part of an anabolic reaction. A few enzymes can use either coenzyme, but most show a strong preference for one over the other. This functional specialization allows a cell to maintain two distinct pools of electron carriers, with two distinct functions, in the same cellular compartment.

More than 200 enzymes are known to catalyze reactions in which NAD$^+$ (or NADP$^+$) accepts a hydride ion from a reduced substrate, or NADPH (or NADH) donates a hydride ion to an oxidized substrate. The general reactions are

$$AH_2 + NAD^+ \longrightarrow A + NADH + H^+$$
$$A + NADPH + H^+ \longrightarrow AH_2 + NADP^+$$

where AH_2 is the reduced substrate and A the oxidized substrate. The general name for an enzyme of this type is **oxidoreductase;** they are also commonly called **dehydrogenases.** For example, alcohol dehydrogenase catalyzes the first step in the catabolism of ethanol, in which ethanol is oxidized to acetaldehyde:

$$\underset{\text{Ethanol}}{CH_3CH_2OH} + NAD^+ \longrightarrow \underset{\text{Acetaldehyde}}{CH_3CHO} + NADH + H^+$$

Notice that one of the carbon atoms in ethanol has lost a hydrogen; the compound has been oxidized from an alcohol to an aldehyde (Fig. 14–13).

figure 14–15

NAD and NADP. (a) Nicotinamide adenine dinucleotide (NAD$^+$) and its phosphorylated analog NADP$^+$ undergo reduction to NADH and NADPH, accepting a hydride ion (two electrons and one proton) from an oxidizable substrate. The hydride ion is added to either the front (the A side) or the back (the B side) of the planar nicotinamide ring (see Table 14–8). **(b)** The UV absorption spectra of NAD$^+$ and NADH. Reduction of the nicotinamide ring produces a new, broad absorption band with a maximum at 340 nm. The production of NADH during an enzyme-catalyzed reaction can be conveniently followed by observing the appearance of the absorbance at 340 nm; ϵ_{340} = 6,200 $M^{-1}cm^{-1}$.

table 14–8

Stereospecificity of Dehydrogenases That Employ NAD^+ or $NADP^+$ as Coenzymes

Enzyme	Coenzyme	Stereochemical specificity for nicotinamide ring (A or B)
Isocitrate dehydrogenase	NAD^+	A
α-Ketoglutarate dehydrogenase	NAD^+	B
Glucose 6-phosphate dehydrogenase	$NADP^+$	B
Malate dehydrogenase	NAD^+	A
Glutamate dehydrogenase	NAD^+ or $NADP^+$	B
Glyceraldehyde 3-phosphate dehydrogenase	NAD^+	B
Lactate dehydrogenase	NAD^+	A
Alcohol dehydrogenase	NAD^+	A

When NAD^+ or $NADP^+$ is reduced, the hydride ion could in principle be transferred to either side of the nicotinamide ring: the front (A side) or the back (B side) as represented in Figure 14–15. Studies with isotopically labeled substrates have shown that a given enzyme catalyzes either an A-type or B-type transfer, but not both. For example, yeast alcohol dehydrogenase and lactate dehydrogenase of vertebrate heart transfer a hydride ion to (or remove a hydride ion from) the A side of the nicotinamide ring; they are classed as type A dehydrogenases to distinguish them from another group of enzymes that transfer a hydride ion to (or remove a hydride ion from) the B side of the nicotinamide ring (Table 14–8).

The association between a dehydrogenase and NAD or NADP is relatively loose; the coenzyme readily diffuses from one enzyme to another, acting as a water-soluble carrier of electrons from one metabolite to another. For example, in the production of alcohol during fermentation of glucose by yeast cells, a hydride ion is removed from glyceraldehyde 3-phosphate by one enzyme (glyceraldehyde 3-phosphate dehydrogenase, a type B enzyme) and transferred to NAD^+. The NADH produced then leaves the enzyme surface and diffuses to another enzyme (alcohol dehydrogenase, a type A enzyme), which transfers a hydride ion to acetaldehyde, producing ethanol:

$$\text{(1)}\quad \text{Glyceraldehyde 3-phosphate} + NAD^+ \longrightarrow \text{3-phosphoglycerate} + NADH + H^+$$

$$\text{(2)}\quad \text{Acetaldehyde} + NADH + H^+ \longrightarrow \text{ethanol} + NAD^+$$

$$\textit{Sum:}\quad \text{Glyceraldehyde 3-phosphate} + \text{acetaldehyde} \longrightarrow \text{3-phosphoglycerate} + \text{ethanol}$$

Notice that in the overall reaction there is no net production or consumption of NAD^+ or NADH; the coenzymes function catalytically and are recycled repeatedly without a net change in the concentration of NAD^+ + NADH.

table 14–9

Some Enzymes (Flavoproteins) That Employ Flavin Nucleotide Coenzymes

Enzyme	Flavin nucleotide
Fatty acyl–CoA dehydrogenase	FAD
Dihydrolipoyl dehydrogenase	FAD
Succinate dehydrogenase	FAD
Glycerol 3-phosphate dehydrogenase	FAD
Thioredoxin reductase	FAD
NADH dehydrogenase (Complex I)	FMN
Glycolate dehydrogenase	FMN

Flavin Nucleotides Are Tightly Bound in Flavoproteins

Flavoproteins (Table 14–9) are enzymes that catalyze oxidation-reduction reactions using either flavin mononucleotide (FMN) or flavin adenine dinucleotide (FAD) as coenzyme (Fig. 14–16). These coenzymes are derived from the vitamin riboflavin. The fused ring structure of flavin nucleotides (the isoalloxazine ring) undergoes reversible reduction, accepting either

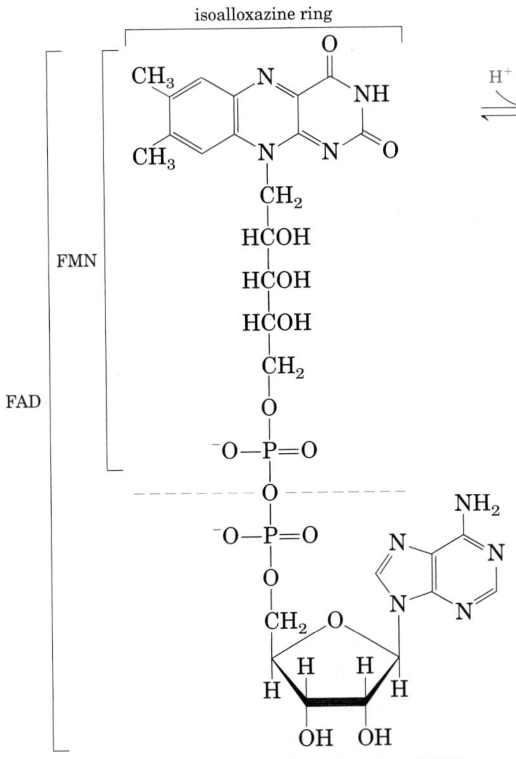

Flavin adenine dinucleotide (FAD) and flavin mononucleotide (FMN)

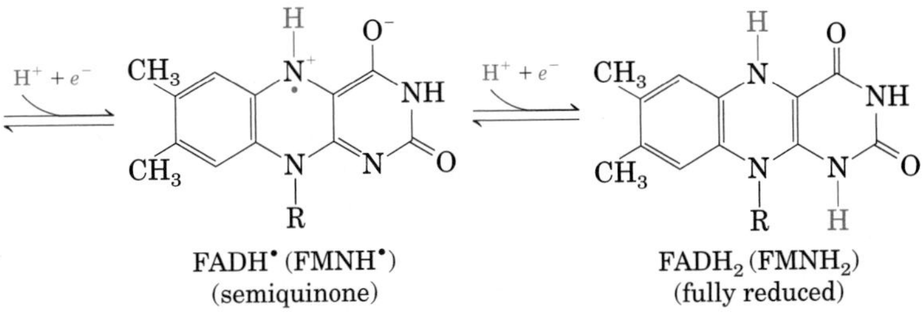

figure 14–16
Structures of oxidized and reduced FAD and FMN. FMN consists of the structure above the dashed line shown on the oxidized (FAD) structure. The flavin nucleotides accept two hydrogen atoms (two electrons and two protons), both of which appear in the flavin ring system. When FAD or FMN accepts only one hydrogen atom, the semiquinone, a stable free radical, forms.

one or two electrons in the form of one or two hydrogen atoms (each atom an electron plus a proton) from a reduced substrate. The fully reduced forms are abbreviated $FADH_2$ and $FMNH_2$. When a fully oxidized flavin nucleotide accepts only one electron (one hydrogen atom), the semiquinone form of the isoalloxazine ring is produced, abbreviated $FADH^{\bullet}$ and $FMNH^{\bullet}$. Because flavoproteins can participate in either one- or two-electron transfers, this class of proteins is involved in a greater diversity of reactions than the pyridine nucleotide–linked dehydrogenases.

Like the nicotinamide coenzymes (Fig. 14–15), the flavin nucleotides undergo a shift in a major absorption band on reduction. Oxidized flavoproteins generally have an absorption maximum near 570 nm; when reduced, the absorption maximum shifts to about 450 nm. This change can be used to assay reactions involving a flavoprotein.

The flavin nucleotide in most flavoproteins is bound rather tightly to the protein, and in some enzymes, such as succinate dehydrogenase, it is bound covalently. Such tightly bound coenzymes are properly called prosthetic groups. They do not transfer electrons by diffusing from one enzyme to another; rather, they provide a means by which the flavoprotein can temporarily hold electrons while it catalyzes electron transfer from a reduced substrate to an electron acceptor. One important feature of the flavoproteins is the variability in the standard reduction potential (E'°) of the bound flavin nucleotide. Tight association between the enzyme and prosthetic group confers on the flavin ring a reduction potential typical of that particular flavoprotein, sometimes quite different from that of the free flavin nucleotide. FAD bound to succinate dehydrogenase, for example, has an E'° close to 0.0 V, compared with −0.219 V for free FAD. Flavoproteins are often very complex; some have, in addition to a flavin nucleotide, tightly bound inorganic ions (iron or molybdenum, for example) capable of participating in electron transfers.

summary

Living cells constantly perform work and thus require energy for the maintenance of highly organized structures, for the synthesis of cellular components, for movement, for the generation of electric currents, for the production of light, and for many other processes. Bioenergetics is the quantitative study of energy relationships and energy conversions in biological systems. Biological energy transformations obey the laws of thermodynamics. All chemical reactions are influenced by two forces: the tendency to achieve the most stable bonding state (for which enthalpy, H, is a useful expression) and the tendency to achieve the highest degree of randomness, expressed as entropy, S. The net driving force in a reaction is ΔG, the free-energy change, which represents the net effect of these two factors: $\Delta G = \Delta H - T\,\Delta S$. Cells require sources of free energy to perform work.

The standard transformed free-energy change, $\Delta G'^\circ$, is a physical constant characteristic for a given reaction and can be calculated from the equilibrium constant for the reaction: $\Delta G'^\circ = -RT \ln K'_{eq}$. The actual free-energy change, ΔG, is a variable, which depends on $\Delta G'^\circ$ and on the concentrations of reactants and products: $\Delta G = \Delta G'^\circ + RT \ln$ ([products]/[reactants]). When ΔG is large and negative, the reaction tends to go in the forward direction; when it is large and positive, the reaction tends to go in the reverse direction; and when $\Delta G = 0$, the system is at equilibrium. The free-energy change for a reaction is independent of the pathway by which the reaction occurs. Free-energy changes are additive; the net chemical reaction that results from the successive occurrence of reactions sharing a common intermediate has an overall free-energy change that is the sum of the ΔG values for the individual reactions.

ATP is the chemical link between catabolism and anabolism. It constitutes the energy currency of the living cell. Its exergonic conversion to ADP and P_i, or to AMP and PP_i, is coupled to a large number of endergonic reactions and processes. In general, it is not ATP hydrolysis, but the transfer of a phosphoryl, pyrophosphoryl, or adenylyl group from ATP to a substrate or enzyme molecule that couples the energy of ATP breakdown to endergonic transformations of substrates. By these group transfer reactions, ATP provides the energy for anabolic reactions, including the synthesis of informational molecules, and for the transport of molecules and ions across membranes against concentration gradients and electrical potential gradients. Muscle contraction is one of several exceptions to this generalization; the conformational changes that produce muscle contraction are driven by ATP hydrolysis directly.

Cells contain other metabolites with large, negative, free energies of hydrolysis, including phosphoenolpyruvate, 1,3-bisphosphoglycerate, and phosphocreatine. These high-energy compounds, like ATP, have a high phosphoryl group transfer potential; they are good donors of the phosphoryl group. Thioesters also have high free energies of hydrolysis.

Biological oxidation-reduction reactions can be described in terms of two half-reactions, each with a characteristic standard reduction potential, E'°. When two electrochemical half-cells, each containing the components of a half-reaction, are connected, electrons tend to flow to the half-cell with the higher reduction potential. The strength of this tendency is proportional to the difference between the two reduction potentials (ΔE) and is a function of the concentrations of oxidized and reduced species. The standard free-energy change for an oxidation-reduction reaction is directly proportional to the difference in standard reduction potentials of the two half-cells: $\Delta G'^\circ = -n\mathcal{F}\,\Delta E'^\circ$.

Many biological oxidation reactions are dehydrogenations in which one or two hydrogen atoms (electron and proton) are transferred from a substrate to a hydrogen acceptor. Oxidation-reduction reactions in cells involve specialized electron carriers. NAD and NADP are the freely diffusible coenzymes of many dehydrogenases. Both NAD^+ and $NADP^+$ accept two electrons and one proton. FAD and FMN, the flavin nucleotides, serve as tightly bound prosthetic groups of flavoproteins. They can accept either one or two electrons. In many organisms, a central energy-conserving process is the stepwise oxidation of glucose to CO_2, in which some of the energy of oxidation is conserved in ATP as electrons are passed to O_2.

further reading

Bioenergetics and Thermodynamics

Atkins, P.W. (1984) *The Second Law,* Scientific American Books, Inc., New York.

A well-illustrated and elementary discussion of the second law and its implications.

Becker, W.M. (1977) *Energy and the Living Cell: An Introduction to Bioenergetics,* J.B. Lippincott Company, Philadelphia.

A clear introductory account of cellular metabolism, in terms of energetics.

Bergethon, P.R. (1998) *The Physical Basis of Biochemistry,* Springer Verlag, New York.

Chapters 11 through 13 of this book, and the books by Tinoco et al. and van Holde et al. (below), are excellent general references for physical biochemistry, with good discussions of the applications of thermodynamics to biochemistry.

Edsall, J.T. & Gutfreund, H. (1983) *Biothermodynamics: The Study of Biochemical Processes at Equilibrium,* John Wiley & Sons, Inc., New York.

Harold, F.M. (1986) *The Vital Force: A Study of Bioenergetics,* W.H. Freeman and Company, New York.

A beautifully clear discussion of thermodynamics in biological processes.

Harris, D.A. (1995) *Bioenergetics at a Glance,* Blackwell Science, Oxford.

A short, clearly written account of cellular energetics, including introductory chapters on thermodynamics.

Morowitz, H.J. (1978) *Foundations of Bioenergetics,* Academic Press, Inc., New York.

Clear, rigorous description of thermodynamics in biology. Out of print.

Tinoco, I., Jr., Sauer, K., & Wang, J.C. (1996) *Physical Chemistry: Principles and Applications in Biological Sciences,* 3rd edn, Prentice-Hall, Inc., Upper Saddle River, NJ.

Chapters 2 through 5 cover thermodynamics.

van Holde, K.E., Johnson, W.C., & Ho, P.S. (1998) *Principles of Physical Biochemistry,* Prentice-Hall, Inc., Upper Saddle River, NJ.

Chapters 2 and 3 are especially relevant.

Phosphoryl Group Transfers and ATP

Alberty, R.A. (1994) Biochemical thermodynamics. *Biochim. Biophys. Acta* **1207,** 1–11.

Explains the distinction between biochemical and chemical equations, and the calculation and meaning of transformed thermodynamic properties for ATP and other phosphorylated compounds.

Bridger, W.A. & Henderson, J.F. (1983) *Cell ATP,* John Wiley & Sons, Inc., New York.

The chemistry of ATP, its role in metabolic regulation, and its catabolic and anabolic roles.

Frey, P.A. & Arabshahi, A. (1995) Standard free-energy change for the hydrolysis of the α–β-phosphoanhydride bridge in ATP. *Biochemistry* **34,** 11,307–11,310.

Hanson, R.W. (1989) The role of ATP in metabolism. *Biochem. Educ.* **17,** 86–92.

Excellent summary of the chemistry and biology of ATP.

Lipmann, F. (1941) Metabolic generation and utilization of phosphate bond energy. *Adv. Enzymol.* **11,** 99–162.

The classic description of the role of high-energy phosphate compounds in biology.

Pullman, B. & Pullman, A. (1960) Electronic structure of energy-rich phosphates. *Radiat. Res.* Suppl. 2, pp. 160–181.

An advanced discussion of the chemistry of ATP and other "energy-rich" compounds.

Veech, R.L., Lawson, J.W.R., Cornell, N.W., & Krebs, H.A. (1979) Cytosolic phosphorylation potential. *J. Biol. Chem.* **254,** 6538–6547.

Experimental determination of ATP, ADP, and P_i concentrations in brain, muscle, and liver, and a discussion of the problems in determining the real free-energy change for ATP synthesis in cells.

Westheimer, F.H. (1987) Why nature chose phosphates. *Science* **235,** 1173–1178.

A chemist's description of the unique suitability of phosphate esters and anhydrides for metabolic transformations.

Biological Oxidation-Reduction Reactions

Dolphin, D., Avramovic, O., & Poulson, R. (eds) (1987) *Pyridine Nucleotide Coenzymes: Chemical, Biochemical, and Medical Aspects,* John Wiley & Sons, Inc., New York.

An excellent two-volume collection of authoritative reviews. Among the most useful are the chapters by Kaplan, Westheimer, Veech, and Ohno and Ushio.

problems

1. Entropy Changes during Egg Development Consider a system consisting of an egg in an incubator. The white and yolk of the egg contain proteins, carbohydrates, and lipids. If fertilized, the egg is transformed from a single cell to a complex organism. Discuss this irreversible process in terms of the entropy changes in the system, surroundings, and universe. Be sure that you first clearly define the system and surroundings.

2. Calculation of $\Delta G'^\circ$ from Equilibrium Constants Calculate the standard free-energy changes of the following metabolically important enzyme-catalyzed reactions at 25 °C and pH 7.0 from the equilibrium constants given.

(a) $\text{Glutamate} + \text{oxaloacetate} \xrightleftharpoons{\text{aspartate aminotransferase}} \text{aspartate} + \alpha\text{-ketoglutarate} \quad K'_{eq} = 6.8$

(b) $\text{Dihydroxyacetone phosphate} \xrightleftharpoons{\text{triose phosphate isomerase}} \text{glyceraldehyde 3-phosphate} \quad K'_{eq} = 0.0475$

(c) $\text{Fructose 6-phosphate} + \text{ATP} \xrightleftharpoons{\text{phosphofructokinase}} \text{fructose 1,6-bisphosphate} + \text{ADP} \quad K'_{eq} = 254$

3. Calculation of Equilibrium Constants from $\Delta G'^\circ$ Calculate the equilibrium constants K'_{eq} for each of the following reactions at pH 7.0 and 25 °C, using the $\Delta G'^\circ$ values of Table 14–4:

(a) $\text{Glucose 6-phosphate} + H_2O \xrightleftharpoons{\text{glucose 6-phosphatase}} \text{glucose} + P_i$

(b) $\text{Lactose} + H_2O \xrightleftharpoons{\beta\text{-galactosidase}} \text{glucose} + \text{galactose}$

(c) $\text{Malate} \xrightleftharpoons{\text{fumarase}} \text{fumarate} + H_2O$

4. Experimental Determination of K'_{eq} and $\Delta G'^\circ$ If a 0.1 M solution of glucose 1-phosphate is incubated with a catalytic amount of phosphoglucomutase, the glucose 1-phosphate is transformed to glucose 6-phosphate. At equilibrium, the concentrations of the reaction components are

$$\underset{4.5 \times 10^{-3}\ \text{M}}{\text{Glucose 1-phosphate}} \rightleftharpoons \underset{9.6 \times 10^{-2}\ \text{M}}{\text{glucose 6-phosphate}}$$

Calculate K'_{eq} and $\Delta G'^\circ$ for this reaction at 25 °C.

5. Experimental Determination of $\Delta G'^\circ$ for ATP Hydrolysis A direct measurement of the standard free-energy change associated with the hydrolysis of ATP is technically demanding because the minute amount of ATP remaining at equilibrium is difficult to measure accurately. The value of $\Delta G'^\circ$ can be calculated indirectly, however, from the equilibrium constants of two other enzymatic reactions having less favorable equilibrium constants:

$$\text{Glucose 6-phosphate} + H_2O \longrightarrow \text{glucose} + P_i \quad K'_{eq} = 270$$

$$\text{ATP} + \text{glucose} \longrightarrow \text{ADP} + \text{glucose 6-phosphate} \quad K'_{eq} = 890$$

Using this information, calculate the standard free energy of hydrolysis of ATP at 25 °C.

6. Difference between $\Delta G'^\circ$ and ΔG Consider the following interconversion, which occurs in glycolysis (Chapter 15):

$$\text{Fructose 6-phosphate} \rightleftharpoons \text{glucose 6-phosphate} \quad K'_{eq} = 1.97$$

(a) What is $\Delta G'^\circ$ for the reaction (assuming that the temperature is 25 °C)?

(b) If the concentration of fructose 6-phosphate is adjusted to 1.5 M and that of glucose 6-phosphate is adjusted to 0.5 M, what is ΔG?

(c) Why are $\Delta G'^\circ$ and ΔG different?

7. Dependence of ΔG on pH The free energy released by the hydrolysis of ATP under standard conditions at pH 7.0 is −30.5 kJ/mol. If ATP is hydrolyzed under standard conditions but at pH 5.0, is more or less free energy released? Why?

8. The $\Delta G'^\circ$ for Coupled Reactions Glucose 1-phosphate is converted into fructose 6-phosphate in two successive reactions:

$$\text{Glucose 1-phosphate} \longrightarrow \text{glucose 6-phosphate}$$
$$\text{Glucose 6-phosphate} \longrightarrow \text{fructose 6-phosphate}$$

Using the $\Delta G'^\circ$ values in Table 14–4, calculate the equilibrium constant, K'_{eq}, for the sum of the two reactions at 25 °C:

$$\text{Glucose 1-phosphate} \longrightarrow \text{fructose 6-phosphate}$$

9. Strategy for Overcoming an Unfavorable Reaction: ATP-Dependent Chemical Coupling The phosphorylation of glucose to glucose 6-phosphate is the initial step in the catabolism of glucose. The direct phosphorylation of glucose by P_i is described by the equation

$$\text{Glucose} + P_i \longrightarrow \text{glucose 6-phosphate} + H_2O \quad \Delta G'^\circ = 13.8 \text{ kJ/mol}$$

(a) Calculate the equilibrium constant for the above reaction. In the rat hepatocyte the physiological concentrations of glucose and P_i are maintained at approximately 4.8 mM. What is the equilibrium concentration of glucose 6-phosphate obtained by the direct phosphorylation of glucose by P_i? Does this reaction represent a reasonable metabolic step for the catabolism of glucose? Explain.

(b) In principle, at least, one way to increase the concentration of glucose 6-phosphate is to drive the equilibrium reaction to the right by increasing the intracellular concentrations of glucose and P_i. Assuming a fixed concentration of P_i at 4.8 mM, how high would the intracellular concentration of glucose have to be to give an equilibrium concentration of glucose 6-phosphate of 250 μM (normal physiological concentration)? Would this route be physiologically reasonable, given that the maximum solubility of glucose is less than 1 M?

(c) The phosphorylation of glucose in the cell is coupled to the hydrolysis of ATP; that is, part of the free energy of ATP hydrolysis is utilized to effect the endergonic phosphorylation of glucose:

(1) Glucose + P_i $\longrightarrow$ glucose 6-phosphate + H_2O
$\Delta G'^\circ = 13.8$ kJ/mol

(2) ATP + H_2O $\longrightarrow$ ADP + P_i
$\Delta G'^\circ = -30.5$ kJ/mol

Sum: Glucose + ATP $\longrightarrow$ glucose 6-phosphate + ADP

Calculate K'_{eq} for the overall reaction. For the ATP-dependent phosphorylation of glucose, what concentration of glucose is needed to achieve a 250 μM intracellular concentration of glucose 6-phosphate when the concentrations of ATP and ADP are 3.38 and 1.32 mM, respectively? Does this coupling process provide a feasible route, at least in principle, for the phosphorylation of glucose in the cell? Explain.

(d) Although coupling ATP hydrolysis to glucose phosphorylation makes thermodynamic sense, how this coupling is to take place has not been specified. Given that coupling requires a common intermediate, one conceivable route is to use ATP hydrolysis to raise the intracellular concentration of P_i and thus drive the unfavorable phosphorylation of glucose by P_i. Is this a reasonable route? (Think about the solubility products of metabolic intermediates.)

(e) The ATP-coupled phosphorylation of glucose is catalyzed in hepatocytes by the enzyme glucokinase. This enzyme binds ATP and glucose to form a glucose-ATP-enzyme complex, and the phosphoryl group is transferred directly from ATP to glucose. Explain the advantages of this route.

10. Calculations of $\Delta G'^\circ$ for ATP-Coupled Reactions From data in Table 14–6 calculate the $\Delta G'^\circ$ value for the reactions

(a) Phosphocreatine + ADP $\longrightarrow$ creatine + ATP

(b) ATP + fructose $\longrightarrow$ ADP + fructose 6-phosphate

11. Coupling ATP Cleavage to an Unfavorable Reaction To explore the consequences of coupling ATP hydrolysis under physiological conditions to a thermodynamically unfavorable biochemical reaction, consider the hypothetical transformation X $\rightarrow$ Y, for which $\Delta G'^\circ = 20$ kJ/mol.

(a) What is the ratio [Y]/[X] at equilibrium?

(b) Suppose X and Y participate in a sequence of reactions during which ATP is hydrolyzed to ADP and P_i. The overall reaction is

$$X + ATP + H_2O \longrightarrow Y + ADP + P_i$$

Calculate [Y]/[X] for this reaction at equilibrium. Assume that the concentrations of ATP, ADP, and P_i are all 1 M when the reaction is at equilibrium.

(c) We know that [ATP], [ADP], and [P_i] are *not* 1 M under physiological conditions. Calculate [Y]/[X] for the ATP-coupled reaction when the values of [ATP], [ADP], and [P_i] are those found in rat myocytes (Table 14–5).

12. Calculations of ΔG at Physiological Concentrations Calculate the physiological ΔG (not $\Delta G'^\circ$) for the reaction

Phosphocreatine + ADP $\longrightarrow$ creatine + ATP

at 25 °C as it occurs in the cytosol of neurons, in which phosphocreatine is present at 4.7 mM, creatine at 1.0 mM, ADP at 0.73 mM, and ATP at 2.6 mM.

13. Free Energy Required for ATP Synthesis under Physiological Conditions In the cytosol of rat hepatocytes, the mass-action ratio is

$$\frac{[ATP]}{[ADP][P_i]} = 5.33 \times 10^2\ \text{M}^{-1}$$

Calculate the free energy required to synthesize ATP in a rat hepatocyte.

14. Daily ATP Utilization by Human Adults

(a) A total of 30.5 kJ/mol of free energy is needed to synthesize ATP from ADP and P_i when the reactants and products are at 1 M concentration (standard state). Because the actual physiological concentrations of ATP, ADP, and P_i are not 1 M, the free energy required to synthesize ATP under physiological conditions is different from $\Delta G'^\circ$. Calculate the free energy required to synthesize ATP in the human hepatocyte when the physiological concentrations of ATP, ADP, and P_i are 3.5, 1.50, and 5.0 mM, respectively.

(b) A 68 kg (150 lb) adult requires a caloric intake of 2,000 kcal (8,360 kJ) of food per day (24 h). The food is metabolized and the free energy is used to synthesize ATP, which then provides energy for the body's daily chemical and mechanical work. Assuming that the efficiency of converting food energy into ATP is 50%, calculate the weight of ATP used by a human adult in 24 h. What percentage of the body weight does this represent?

(c) Although adults synthesize large amounts of ATP daily, their body weight, structure, and composition do not change significantly during this period. Explain this apparent contradiction.

15. Rates of Turnover of γ and β Phosphates of ATP If a small amount of ATP labeled with radioactive phosphorus in the terminal position, [γ-^{32}P]ATP, is added to a yeast extract, about half of the ^{32}P activity is found in P_i within a few minutes, but the concentration of ATP remains unchanged. Explain. If the same experiment is carried out using ATP labeled with ^{32}P in the central position, [β-^{32}P]ATP, the ^{32}P does not appear in P_i within such a short time. Why?

16. Cleavage of ATP to AMP and PP_i during Metabolism The synthesis of the activated form of acetate (acetyl-CoA) is carried out in an ATP-dependent process:

$$\text{Acetate} + \text{CoA} + \text{ATP} \longrightarrow \text{acetyl-CoA} + \text{AMP} + \text{PP}_i$$

(a) The $\Delta G'^\circ$ for the hydrolysis of acetyl-CoA to acetate and CoA is -32.2 kJ/mol and that for hydrolysis of ATP to AMP and PP_i is -30.5 kJ/mol. Calculate $\Delta G'^\circ$ for the ATP-dependent synthesis of acetyl-CoA.

(b) Almost all cells contain the enzyme inorganic pyrophosphatase, which catalyzes the hydrolysis of PP_i to P_i. What effect does the presence of this enzyme have on the synthesis of acetyl-CoA? Explain.

17. Energy for H^+ Pumping The parietal cells of the stomach lining contain membrane "pumps" that transport hydrogen ions from the cytosol of these cells (pH 7.0) into the stomach, contributing to the acidity of gastric juice (pH 1.0). Calculate the free energy required to transport 1 mol of hydrogen ions through these pumps. (Hint: See Chapter 13.) Assume a temperature of 25 °C.

18. Standard Reduction Potentials The standard reduction potential, E'°, of any redox pair is defined for the half-cell reaction:

$$\text{Oxidizing agent} + n \text{ electrons} \longrightarrow \text{reducing agent}$$

The E'° values for the NAD^+/NADH and pyruvate/lactate conjugate redox pairs are -0.32 and -0.19 V, respectively.

(a) Which conjugate pair has the greater tendency to lose electrons? Explain.

(b) Which is the stronger oxidizing agent? Explain.

(c) Beginning with 1 M concentrations of each reactant and product at pH 7, in which direction will the following reaction proceed?

$$\text{Pyruvate} + \text{NADH} + \text{H}^+ \rightleftharpoons \text{lactate} + \text{NAD}^+$$

(d) What is the standard free-energy change ($\Delta G'^\circ$) at 25 °C for the conversion of pyruvate to lactate?

(e) What is the equilibrium constant (K'_{eq}) for this reaction?

19. Energy Span of the Respiratory Chain Electron transfer in the mitochondrial respiratory chain may be represented by the net reaction equation

$$\text{NADH} + \text{H}^+ + \tfrac{1}{2}\text{O}_2 \rightleftharpoons \text{H}_2\text{O} + \text{NAD}^+$$

(a) Calculate the value of $\Delta E'^\circ$ for the net reaction of mitochondrial electron transfer.

(b) Calculate $\Delta G'^\circ$ for this reaction.

(c) How many ATP molecules can *theoretically* be generated by this reaction if the free energy of ATP synthesis under cellular conditions is 52 kJ/mol?

20. Dependence of Electromotive Force on Concentrations Calculate the electromotive force (in volts) registered by an electrode immersed in a solution containing the following mixtures of NAD^+ and NADH at pH 7.0 and 25 °C, with reference to a half-cell of E'° 0.00 V.

(a) 1.0 mM NAD^+ and 10 mM NADH

(b) 1.0 mM NAD^+ and 1.0 mM NADH

(c) 10 mM NAD^+ and 1.0 mM NADH

21. Electron Affinity of Compounds List the following substances in order of increasing tendency to accept electrons: (a) α-ketoglutarate + CO_2 (yielding isocitrate); (b) oxaloacetate; (c) O_2; (d) $NADP^+$.

22. Direction of Oxidation-Reduction Reactions Which of the following reactions would you expect to proceed in the direction shown under standard conditions, assuming that the appropriate enzymes are present to catalyze them?

(a) Malate + NAD^+ $\longrightarrow$ oxaloacetate + NADH + H^+

(b) Acetoacetate + NADH + H^+ $\longrightarrow$ β-hydroxybutyrate + NAD^+

(c) Pyruvate + NADH + H^+ $\longrightarrow$ lactate + NAD^+

(d) Pyruvate + β-hydroxybutyrate $\longrightarrow$ lactate + acetoacetate

(e) Malate + pyruvate $\longrightarrow$ oxaloacetate + lactate

(f) Acetaldehyde + succinate $\longrightarrow$ ethanol + fumarate

chapter 15

Glycolysis and the Catabolism of Hexoses

D-Glucose is the major fuel of most organisms and occupies a central position in metabolism. It is relatively rich in potential energy; the complete oxidation of glucose to carbon dioxide and water proceeds with a standard free-energy change of −2,840 kJ/mol. By storing glucose as a high molecular weight polymer such as starch or glycogen, a cell can stockpile large quantities of hexose units while maintaining a relatively low cytosolic osmolarity. When energy demands suddenly increase, glucose can be released quickly from these intracellular storage polymers and used to produce ATP either aerobically or anaerobically.

Glucose is not only an excellent fuel, it is also a remarkably versatile precursor, capable of supplying a huge array of metabolic intermediates for biosynthetic reactions. A bacterium such as *Escherichia coli* can obtain from glucose the carbon skeletons for every amino acid, nucleotide, coenzyme, fatty acid, or other metabolic intermediate needed for growth. A comprehensive study of the metabolic fates of glucose would encompass hundreds or thousands of transformations. In higher plants and animals, glucose has three major fates: it may be stored (as a polysaccharide or as sucrose), oxidized to a three-carbon compound (pyruvate) via glycolysis, or oxidized to pentoses via the pentose phosphate (phosphogluconate) pathway (Fig. 15–1).

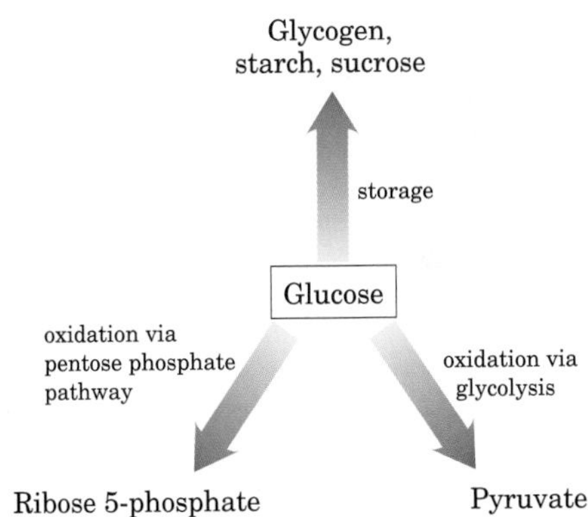

figure 15–1
Major pathways of glucose utilization in cells of higher plants and animals. Although not the only possible fates for glucose, these three pathways are the most significant in terms of the amount of glucose that flows through them in most cells.

This chapter describes the individual reactions of glycolysis and the enzymes that catalyze them; fermentation, the operation of glycolysis under anaerobic conditions; and the pathways that produce the starting material for glycolysis from nonglucose hexoses, disaccharides, and polysaccharides. Using glycolysis as an example of a metabolic pathway under tight regulation, we discuss the general principles of metabolic control. We conclude with a brief description of the catabolic pathway that leads to pentoses.

Glycolysis

In **glycolysis** (from the Greek *glykys,* meaning "sweet," and *lysis,* meaning "splitting") a molecule of glucose is degraded in a series of enzyme-catalyzed reactions to yield two molecules of the three-carbon compound pyruvate. During the sequential reactions of glycolysis, some of the free energy released from glucose is conserved in the form of ATP and NADH. Glycolysis was the first metabolic pathway to be elucidated and is probably the best understood. From the discovery by Eduard Büchner (in 1897) of fermentation in broken extracts of yeast cells until the clear recognition by Fritz Lipmann and Herman Kalckar (in 1941) of the metabolic role of high-energy compounds such as ATP, the reactions of glycolysis in extracts of yeast and muscle were central to biochemical research. The development of

Fritz Lipmann
1899–1986

Herman Kalckar
1908–1991

methods of enzyme purification, the discovery and recognition of the importance of coenzymes such as NAD, and the discovery of the pivotal metabolic role of ATP and other phosphorylated compounds all came out of studies of glycolysis. The glycolytic enzymes of many species have long since been purified and thoroughly studied.

Glycolysis is an almost universal central pathway of glucose catabolism, the pathway with the largest flux of carbon in most cells. The glycolytic breakdown of glucose is the sole source of metabolic energy in some mammalian tissues and cell types (erythrocytes, renal medulla, brain, and sperm, for example). Some plant tissues that are modified to store starch (such as potato tubers) and some aquatic plants (watercress, for example) derive most of their energy from glycolysis; many anaerobic microorganisms are entirely dependent on glycolysis.

Fermentation is a general term for the *anaerobic* degradation of glucose or other organic nutrients to obtain energy, conserved as ATP. Because living organisms first arose in an atmosphere without oxygen, anaerobic breakdown of glucose is probably the most ancient biological mechanism for obtaining energy from organic fuel molecules. In the course of evolution, the chemistry of this reaction sequence has been completely conserved; the glycolytic enzymes of vertebrates are closely similar, in amino acid sequence and three-dimensional structure, to their homologs in yeast and spinach. Glycolysis differs among species only in the details of its regulation and in the subsequent metabolic fate of the pyruvate formed. The thermodynamic principles and the types of regulatory mechanisms in glycolysis are common to all pathways of cell metabolism. A study of glycolysis can therefore serve as a model for many aspects of the pathways discussed later in this book.

Before examining each step of the pathway in some detail, we will take a look at glycolysis as a whole.

An Overview: Glycolysis Has Two Phases

The breakdown of the six-carbon glucose into two molecules of the three-carbon pyruvate occurs in ten steps, the first five of which constitute the *preparatory phase* (Fig. 15–2a). In these reactions, glucose is first phosphorylated at the hydroxyl group on C-6 (step ①). The D-glucose 6-phosphate thus formed is converted to D-fructose 6-phosphate (step ②), which is again phosphorylated, this time at C-1, to yield D-fructose 1,6-bisphosphate (step ③). For both phosphorylations, ATP is the phosphoryl group donor. As all sugar derivatives in the glycolytic pathway are the D isomers, we will omit the D designation except when emphasizing stereochemistry.

Fructose 1,6-bisphosphate is next split to yield two three-carbon molecules, dihydroxyacetone phosphate and glyceraldehyde 3-phosphate (step ④); this is the "lysis" step that gives the pathway its name. The dihydroxyacetone phosphate is isomerized to a second molecule of glyceraldehyde 3-phosphate (step ⑤), ending the first phase of glycolysis. Note that two molecules of ATP must be invested to activate the glucose molecule for its cleavage into two three-carbon pieces; later there will be a good return on this investment. To sum up: in the preparatory phase of glycolysis the energy of ATP is invested, raising the free-energy content of the intermediates, and the carbon chains of all the metabolized hexoses are converted into a common product, glyceraldehyde 3-phosphate.

The energy gain comes in the *payoff phase* of glycolysis (Fig. 15–2b). Each molecule of glyceraldehyde 3-phosphate is oxidized and phosphorylated by inorganic phosphate (*not* by ATP) to form 1,3-bisphosphoglycerate (step ⑥). Energy is then released as the two molecules of 1,3-bisphosphoglycerate are converted to two molecules of pyruvate (steps ⑦ through ⑩).

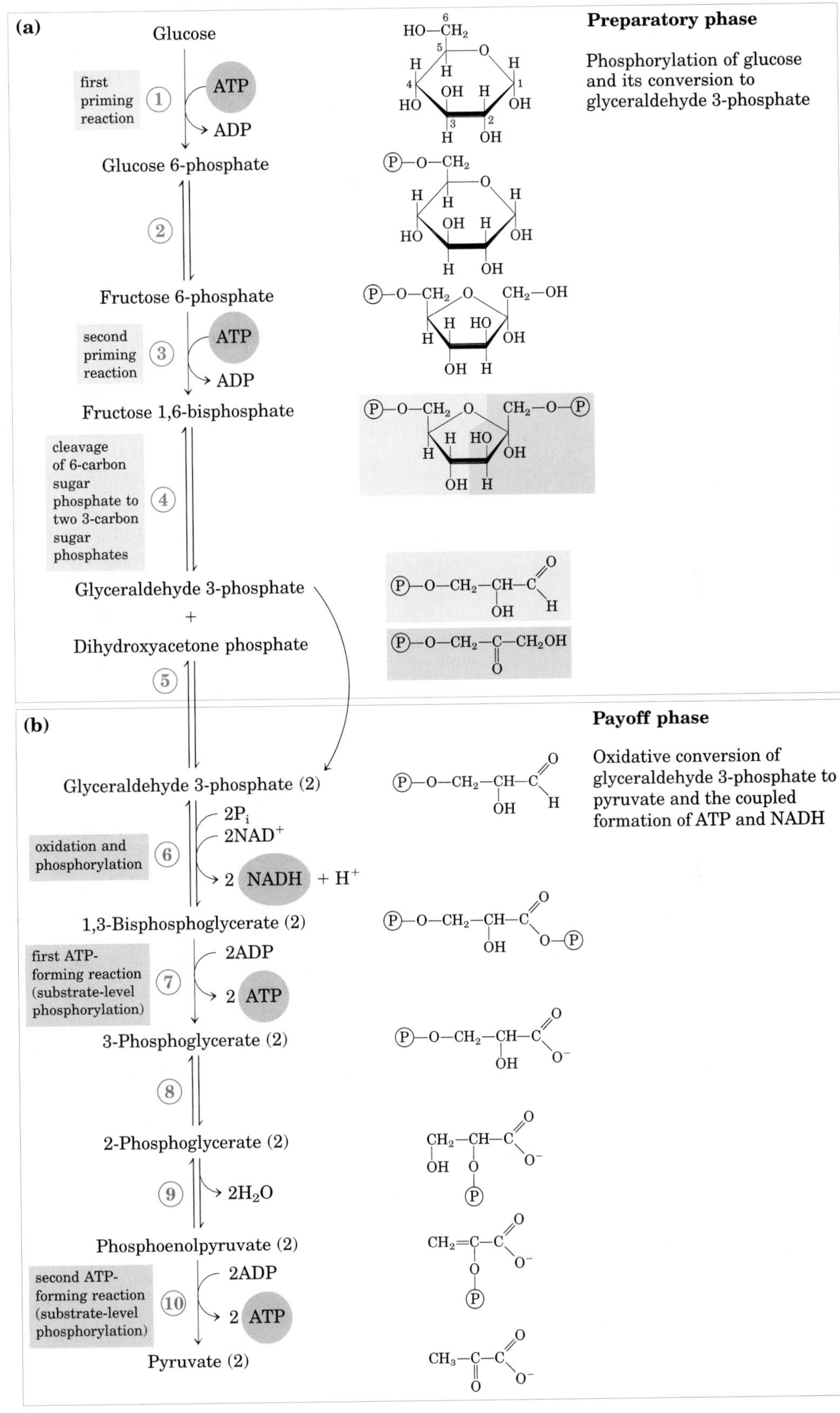

figure 15–2

The two phases of glycolysis. For each molecule of glucose that passes through the preparatory phase **(a),** two molecules of glyceraldehyde 3-phosphate are formed; both pass through the payoff phase **(b).** Pyruvate is the end product of the second phase of glycolysis. For each glucose molecule, two ATP are consumed in the preparatory phase and four ATP are produced in the payoff phase, giving a net yield of two ATP per molecule of glucose converted to pyruvate. The number beside each reaction step corresponds to its numbered heading in the text discussion. Keep in mind that each phosphoryl group, represented here as Ⓟ, has two negative charges ($—PO_3^{2-}$).

Much of this energy is conserved by the coupled phosphorylation of four molecules of ADP to ATP. The net yield is two molecules of ATP per molecule of glucose used, because two molecules of ATP were invested in the preparatory phase. Energy is also conserved in the payoff phase in the formation of two molecules of NADH per molecule of glucose.

In the sequential reactions of glycolysis, three types of chemical transformations are particularly noteworthy: (1) degradation of the carbon skeleton of glucose to yield pyruvate, (2) phosphorylation of ADP to ATP by high-energy phosphate compounds formed during glycolysis, and (3) transfer of a hydride ion with its electrons to NAD^+, forming NADH. The fate of the pyruvate depends on the cell type and the metabolic circumstances.

Fates of Pyruvate Barring some interesting variations in the bacterial realm, the pyruvate formed by glycolysis is further metabolized via one of three catabolic routes. In aerobic organisms or tissues, under aerobic conditions, glycolysis is only the first stage in the complete degradation of glucose (Fig. 15–3). Pyruvate is oxidized, with loss of its carboxyl group as CO_2, to yield the acetyl group of acetyl-coenzyme A; the acetyl group is then oxidized completely to CO_2 by the citric acid cycle (Chapter 16). The electrons from these oxidations are passed to O_2 through a chain of carriers in the mitochondrion, forming H_2O. The energy from the electron transfer reactions drives the synthesis of ATP in the mitochondrion (Chapter 19).

The second route for pyruvate is its reduction to lactate via **lactic acid fermentation.** When vigorously contracting skeletal muscle must function under low-oxygen conditions (**hypoxia**), NADH cannot be reoxidized to NAD^+, and NAD^+ is required as an electron acceptor for the further oxidation of pyruvate. Under these conditions pyruvate is reduced to lactate, accepting electrons from NADH and thereby regenerating the NAD^+ necessary for glycolysis to continue. Certain tissues and cell types (retina, brain, erythrocytes) convert glucose to lactate even under aerobic conditions, and lactate is also the product of glycolysis under anaerobic conditions in some microorganisms (Fig. 15–3).

The third major route of pyruvate catabolism leads to ethanol. In some plant tissues and in certain invertebrates, protists, and microorganisms such as brewer's yeast, pyruvate is converted under hypoxic or anaerobic conditions into ethanol and CO_2, a process called **alcohol** (or **ethanol**) **fermentation** (Fig. 15–3).

The focus of this chapter is catabolism, but pyruvate has anabolic fates as well. It can, for example, provide the carbon skeleton for the synthesis of the amino acid alanine. We return to these anabolic reactions of pyruvate in later chapters.

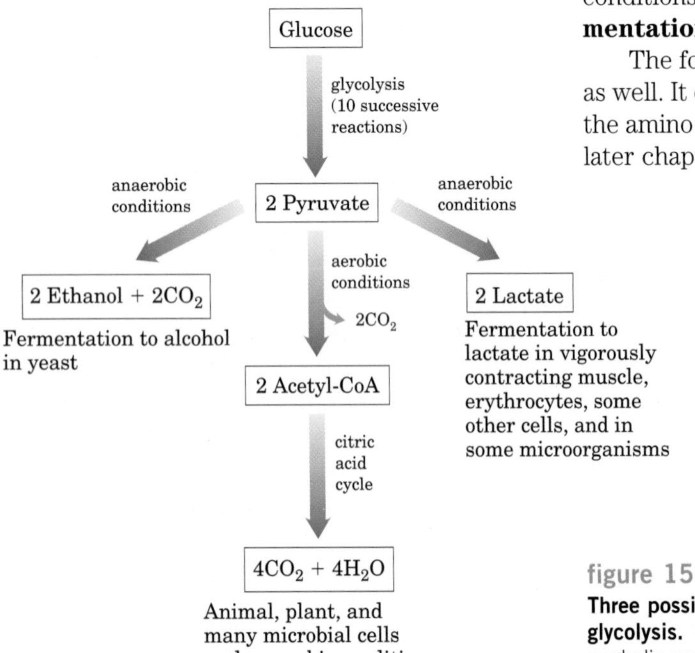

figure 15–3
Three possible catabolic fates of the pyruvate formed in glycolysis. Pyruvate also serves as a precursor in many anabolic reactions, not shown here.

ATP Formation Coupled to Glycolysis During glycolysis some of the energy of the glucose molecule is conserved in ATP, while much remains in the product, pyruvate. The overall equation for glycolysis is

$$\text{Glucose} + 2\text{NAD}^+ + 2\text{ADP} + 2\text{P}_i \longrightarrow 2\ \text{pyruvate} + 2\text{NADH} + 2\text{H}^+ + 2\text{ATP} + 2\text{H}_2\text{O} \quad (15\text{–}1)$$

For each molecule of glucose degraded to pyruvate, two molecules of ATP are generated from ADP and P_i. We can now resolve the equation of glycolysis into two processes: (1) the conversion of glucose to pyruvate, which is exergonic:

$$\text{Glucose} + 2\text{NAD}^+ \longrightarrow 2\ \text{pyruvate} + 2\text{NADH} + 2\text{H}^+ \quad (15\text{–}2)$$
$$\Delta G_1{'}^{\circ} = -146\ \text{kJ/mol}$$

and (2) the formation of ATP from ADP and P_i, which is endergonic:

$$2\ \text{ADP} + 2\text{P}_i \longrightarrow 2\text{ATP} + 2\text{H}_2\text{O} \quad (15\text{–}3)$$
$$\Delta G_2'^{\circ} = 2(30.5\ \text{kJ/mol}) = 61\ \text{kJ/mol}$$

The sum of Equations 15–2 and 15–3 gives the overall standard free-energy change of glycolysis, $\Delta G_s'^{\circ}$:

$$\Delta G_s'^{\circ} = \Delta G_1'^{\circ} + \Delta G_2'^{\circ} = -146\ \text{kJ/mol} + 61\ \text{kJ/mol} = -85\ \text{kJ/mol}$$

Under standard conditions and in the cell, glycolysis is an essentially irreversible process, driven to completion by a large net decrease in free energy. At the actual intracellular concentrations of ATP, ADP, and P_i (see Box 14–2) and of glucose and pyruvate, the energy released in glycolysis (with pyruvate as the end product) is recovered as ATP with an efficiency of over 60%.

Energy Remaining in Pyruvate Glycolysis releases only a small fraction of the total available energy of the glucose molecule. When glucose is oxidized completely to CO_2 and H_2O, the total standard free-energy change is −2,840 kJ/mol. Glycolytic degradation of glucose to two molecules of pyruvate ($\Delta G'^{\circ} = -146$ kJ/mol) therefore yields only (146/2,840)100 = 5.2% of the total energy that can be released from glucose by complete oxidation. The two molecules of pyruvate formed by glycolysis still contain most of the chemical potential energy of the glucose molecule, energy that can be extracted by oxidative reactions in the citric acid cycle (Chapter 16) and oxidative phosphorylation (Chapter 19).

Importance of Phosphorylated Intermediates Each of the nine glycolytic intermediates between glucose and pyruvate is phosphorylated (Fig. 15–2). The phosphoryl groups appear to have three functions.

1. They are ionized at pH 7, giving each glycolytic intermediate a net negative charge. Because the plasma membrane is impermeable to charged molecules, the phosphorylated intermediates cannot diffuse out of the cell. After the initial phosphorylation, no further energy is necessary to retain phosphorylated intermediates in the cell, despite the large difference in their intracellular and extracellular concentrations.
2. Phosphoryl groups are essential components in the enzymatic conservation of metabolic energy. Energy released in the breakage of phosphoanhydride bonds (such as those in ATP) is partially conserved in the formation of phosphate esters such as glucose 6-phosphate. High-energy phosphate compounds formed in glycolysis (1,3-bisphosphoglycerate and phosphoenolpyruvate) donate phosphoryl groups to ADP to form ATP.

3. Binding energy resulting from the binding of phosphate groups to the active sites of enzymes lowers the activation energy and increases the specificity of the enzymatic reactions (see p. 251). The phosphate groups of ADP, ATP, and the glycolytic intermediates form complexes with Mg^{2+}, and the substrate binding sites of many glycolytic enzymes are specific for these Mg^{2+} complexes. Most glycolytic enzymes require Mg^{2+} for activity.

Arthur Harden
1865–1940

William Young
1878–1942

The Preparatory Phase of Glycolysis Requires ATP

In the preparatory phase of glycolysis, two molecules of ATP are invested and the hexose chain is cleaved into two triose phosphates. The realization that *phosphorylated* hexoses were intermediates in glycolysis came slowly and serendipitously. In 1906, Arthur Harden and William Young tested their hypothesis that inhibitors of proteolytic enzymes would stabilize the glucose-fermenting enzymes in yeast extract. They added blood serum (known to contain inhibitors of proteolytic enzymes) to yeast extracts and observed the predicted stimulation of glucose metabolism. However, in a control experiment intended to show that boiling the serum destroyed the stimulatory activity, they discovered that boiled serum was just as effective at stimulating glycolysis. Careful examination and testing of the contents of the boiled serum revealed that inorganic phosphate was responsible for the stimulation. Harden and Young soon discovered that glucose added to their yeast extract was converted into a hexose bisphosphate (the "Harden-Young ester," eventually identified as fructose 1,6-bisphosphate). This was the beginning of a long series of investigations on the role of organic esters of phosphate in biochemistry, which has led to our current understanding of the central role of phosphoryl group transfer in biology.

① Phosphorylation of Glucose In the first step of glycolysis, glucose is activated for subsequent reactions by its phosphorylation at C-6 to yield **glucose 6-phosphate,** with ATP as the phosphoryl donor:

$$\text{Glucose} + \text{ATP} \xrightarrow[\text{hexokinase}]{Mg^{2+}} \text{Glucose 6-phosphate} + \text{ADP}$$

$$\Delta G'^{\circ} = -16.7 \text{ kJ/mol}$$

This reaction, which is irreversible under intracellular conditions, is catalyzed by **hexokinase.** Recall that kinases are enzymes that catalyze the transfer of the terminal phosphoryl group from ATP to some acceptor nucleophile (see Fig. 14–10). Kinases are a subclass of transferases (see Table 8–3). The acceptor in the case of hexokinase is a hexose, normally D-glucose, although hexokinase also catalyzes the phosphorylation of other common hexoses, such as D-fructose and D-mannose.

Hexokinase, like many other kinases, requires Mg^{2+} for its activity, because the true substrate of the enzyme is not ATP^{4-} but the $MgATP^{2-}$ complex (see Fig. 14–2). Detailed studies of yeast hexokinase have shown that the enzyme undergoes a profound change in shape, an induced fit, when it binds the hexose molecule (see Fig. 8–21). Hexokinase is present in all cells of all organisms. Hepatocytes also contain a form of hexokinase called hexokinase D or glucokinase, which is more specific for glucose and differs from other forms of hexokinase in kinetic and regulatory properties (p. 555).

Like the other nine enzymes of glycolysis, hexokinase is a soluble, cytosolic protein, although, as we note later, there may be organized complexes of several glycolytic enzymes (see Fig. 15–8).

② Conversion of Glucose 6-Phosphate to Fructose 6-Phosphate The enzyme **phosphohexose isomerase (phosphoglucose isomerase)** catalyzes the reversible isomerization of glucose 6-phosphate, an aldose, to **fructose 6-phosphate,** a ketose:

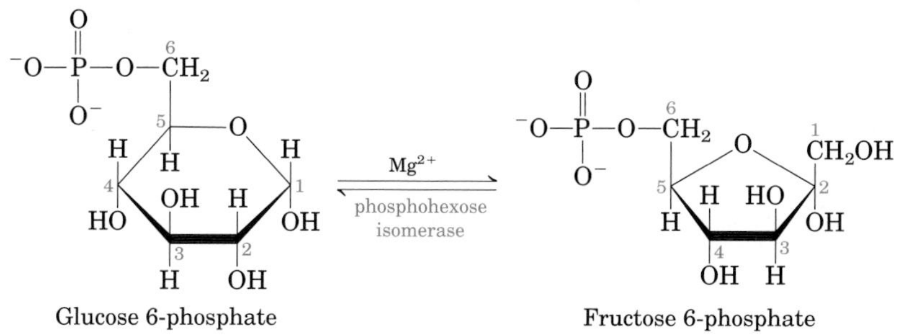

Glucose 6-phosphate $\underset{\text{phosphohexose isomerase}}{\overset{Mg^{2+}}{\rightleftharpoons}}$ Fructose 6-phosphate

$\Delta G'^{\circ} = 1.7$ kJ/mol

This reaction proceeds readily in either direction, as predicted from the relatively small change in standard free energy. Phosphohexose isomerase requires Mg^{2+} and is specific for glucose 6-phosphate and fructose 6-phosphate.

③ Phosphorylation of Fructose 6-Phosphate to Fructose 1,6-Bisphosphate In the second of the two priming reactions of glycolysis, **phosphofructokinase-1** catalyzes the transfer of a phosphoryl group from ATP to fructose 6-phosphate to yield **fructose 1,6-bisphosphate:**

Fructose 6-phosphate + ATP $\xrightarrow[\text{phosphofructokinase-1}]{Mg^{2+}}$ Fructose 1,6-bisphosphate + ADP

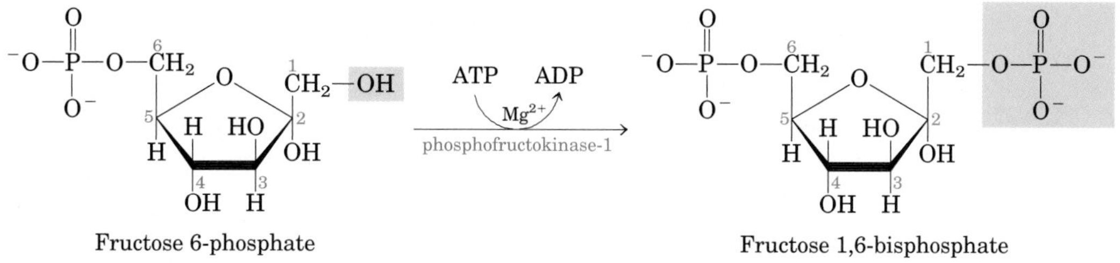

$\Delta G'^{\circ} = -14.2$ kJ/mol

The reaction is essentially irreversible under cellular conditions. (This enzyme is called phosphofructokinase-1 (PFK-1) to distinguish it from a second enzyme (PFK-2) that catalyzes the formation of fructose 2,6-bisphosphate from fructose 6-phosphate; see Fig. 20–8.)

Some bacteria and protists and perhaps all plants have a phosphofructokinase that uses pyrophosphate (PP_i), not ATP, as the phosphoryl group donor in the synthesis of fructose 1,6-bisphosphate:

$$\text{Fructose 6-phosphate} + PP_i \xrightarrow{Mg^{2+}} \text{fructose 1,6-bisphosphate} + P_i \qquad \Delta G'^{\circ} = -14 \text{ kJ/mol}$$

Phosphofructokinase-1 is a regulatory enzyme (Chapter 8), one of the most complex known. It is the major point of regulation in glycolysis. The activity of PFK-1 is increased whenever the cell's ATP supply is depleted or when the ATP breakdown products—ADP and AMP, particularly the latter—are in excess. The enzyme is inhibited whenever the cell has ample ATP and is well supplied by other fuels such as fatty acids. In some organisms, fructose 2,6-bisphosphate (not to be confused with the PFK-1 reaction product, fructose 1,6-bisphosphate) is a potent allosteric activator of phosphofructokinase-1. The regulation of this step in glycolysis is discussed in greater detail later in the chapter.

④ Cleavage of Fructose 1,6-Bisphosphate The enzyme **fructose 1,6-bisphosphate aldolase,** often called simply **aldolase,** catalyzes a reversible aldol condensation. Fructose 1,6-bisphosphate is cleaved to yield two different triose phosphates, **glyceraldehyde 3-phosphate,** an aldose, and **dihydroxyacetone phosphate,** a ketose:

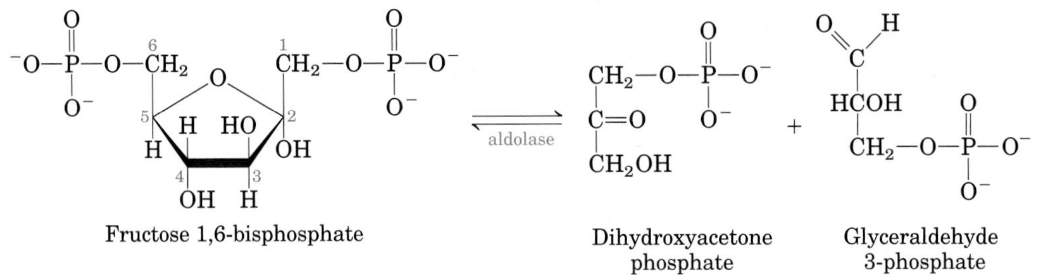

Fructose 1,6-bisphosphate ⇌ (aldolase) Dihydroxyacetone phosphate + Glyceraldehyde 3-phosphate

$\Delta G'^{\circ} = 23.8$ kJ/mol

The aldolase of vertebrate animal tissues does not require a divalent cation, but in many microorganisms aldolase is a Zn^{2+}-containing enzyme. Although the aldolase reaction has a strongly positive standard free-energy change in the direction of fructose 1,6-bisphosphate cleavage, in cells it can proceed readily in either direction. During glycolysis the reaction products (two triose phosphates) are removed quickly by the next two steps, pulling the reaction in the direction of cleavage.

⑤ Interconversion of the Triose Phosphates Only one of the two triose phosphates formed by aldolase—glyceraldehyde 3-phosphate—can be directly degraded in the subsequent steps of glycolysis. The other product, dihydroxyacetone phosphate, is rapidly and reversibly converted to glyceraldehyde 3-phosphate by the fifth enzyme of the glycolytic sequence, **triose phosphate isomerase:**

Dihydroxyacetone phosphate ⇌ (triose phosphate isomerase) Glyceraldehyde 3-phosphate

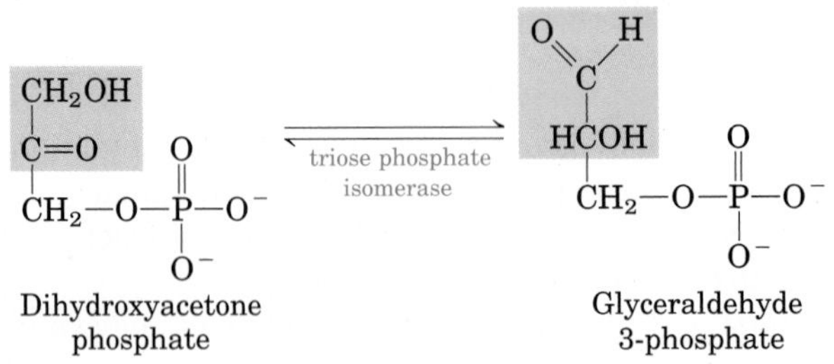

$\Delta G'^{\circ} = 7.5$ kJ/mol

By this reaction C-1, C-2, and C-3 of the starting glucose now become chemically indistinguishable from C-6, C-5, and C-4, respectively (Fig. 15–4).

This reaction completes the preparatory phase of glycolysis. The hexose molecule has been phosphorylated at C-1 and C-6 and then cleaved to form two molecules of glyceraldehyde 3-phosphate. Other hexoses, such as D-fructose, D-mannose, and D-galactose, can also be converted into glyceraldehyde 3-phosphate, as we shall see later.

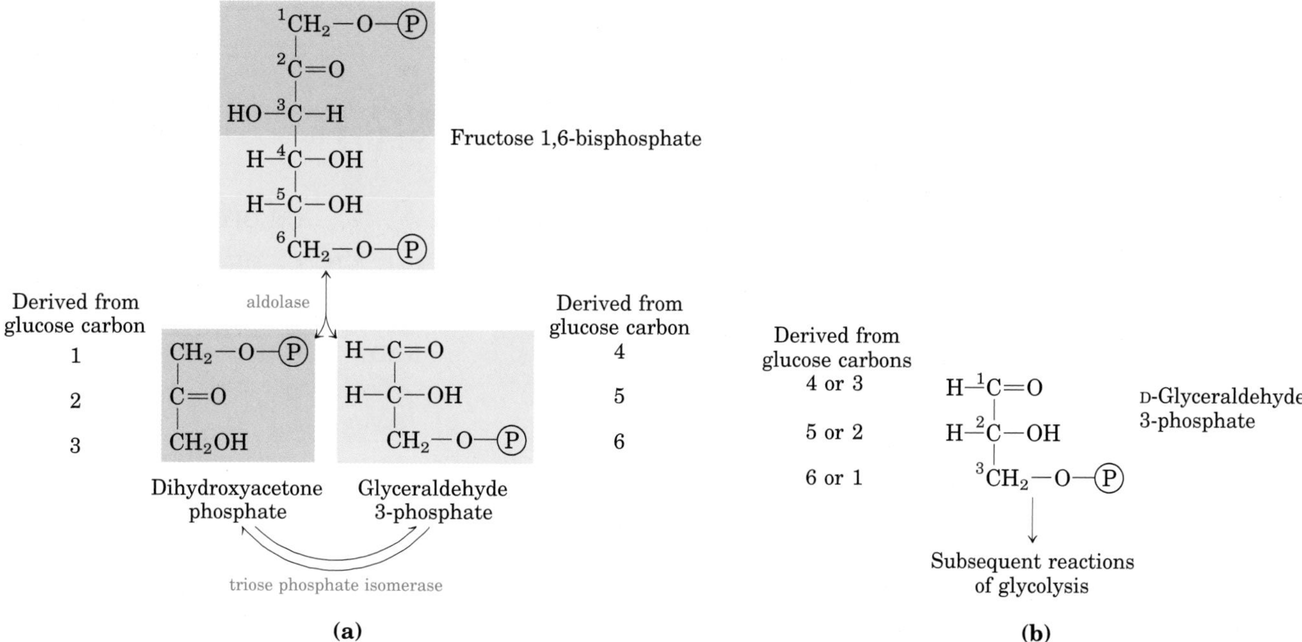

figure 15–4

Fate of the carbon atoms of glucose in the formation of glyceraldehyde 3-phosphate. (a) The origin of the carbons in the two three-carbon products of the aldolase and triose phosphate isomerase reactions. The end product of the two reactions is two molecules of glyceraldehyde 3-phosphate. Each carbon of glyceraldehyde 3-phosphate is derived from either of two specific carbons of glucose **(b).** Note that the numbering of the carbon atoms of glyceraldehyde 3-phosphate differs from that of the glucose from which it is derived. In glyceraldehyde 3-phosphate, the most complex functional group (the carbonyl) is specified as C-1. This numbering change is important for interpreting experiments with glucose in which a single carbon is labeled with a radioisotope. (See Problems 3 and 5 at the end of this chapter.)

The Payoff Phase of Glycolysis Produces ATP and NADH

The payoff phase of glycolysis (Fig. 15–2b) includes the energy-conserving phosphorylation steps in which some of the free energy of the glucose molecule is conserved in the form of ATP. Remember that one molecule of glucose yields two molecules of glyceraldehyde 3-phosphate; both halves of the glucose molecule follow the same pathway in the second phase of glycolysis. The conversion of two molecules of glyceraldehyde 3-phosphate to two molecules of pyruvate is accompanied by the formation of four molecules of ATP from ADP. However, the net yield of ATP per molecule of glucose degraded is only two, because two ATP were invested in the preparatory phase of glycolysis to phosphorylate the two ends of the hexose molecule.

⑥ Oxidation of Glyceraldehyde 3-Phosphate to 1,3-Bisphosphoglycerate

The first step in the payoff phase is the oxidation of glyceraldehyde 3-phosphate to **1,3-bisphosphoglycerate,** catalyzed by **glyceraldehyde 3-phosphate dehydrogenase:**

$$\underset{\text{Glyceraldehyde 3-phosphate}}{\mathrm{HC(=O)-HCOH-CH_2OPO_3^{2-}}} + \underset{\text{Inorganic phosphate}}{\mathrm{HO-P(=O)(O^-)-O^-}} \xrightleftharpoons[\text{glyceraldehyde 3-phosphate dehydrogenase}]{\mathrm{NAD^+}\ \ \mathrm{NADH + H^+}} \underset{\text{1,3-Bisphosphoglycerate}}{\mathrm{O{=}C(O-P(=O)(O^-)-O^-)-HCOH-CH_2OPO_3^{2-}}}$$

$$\Delta G'^{\circ} = 6.3\ \text{kJ/mol}$$

This is the first of the two energy-conserving reactions of glycolysis that eventually lead to the formation of ATP. The aldehyde group of glyceraldehyde 3-phosphate is dehydrogenated, not to a free carboxyl group but to a carboxylic acid anhydride with phosphoric acid. This type of anhydride,

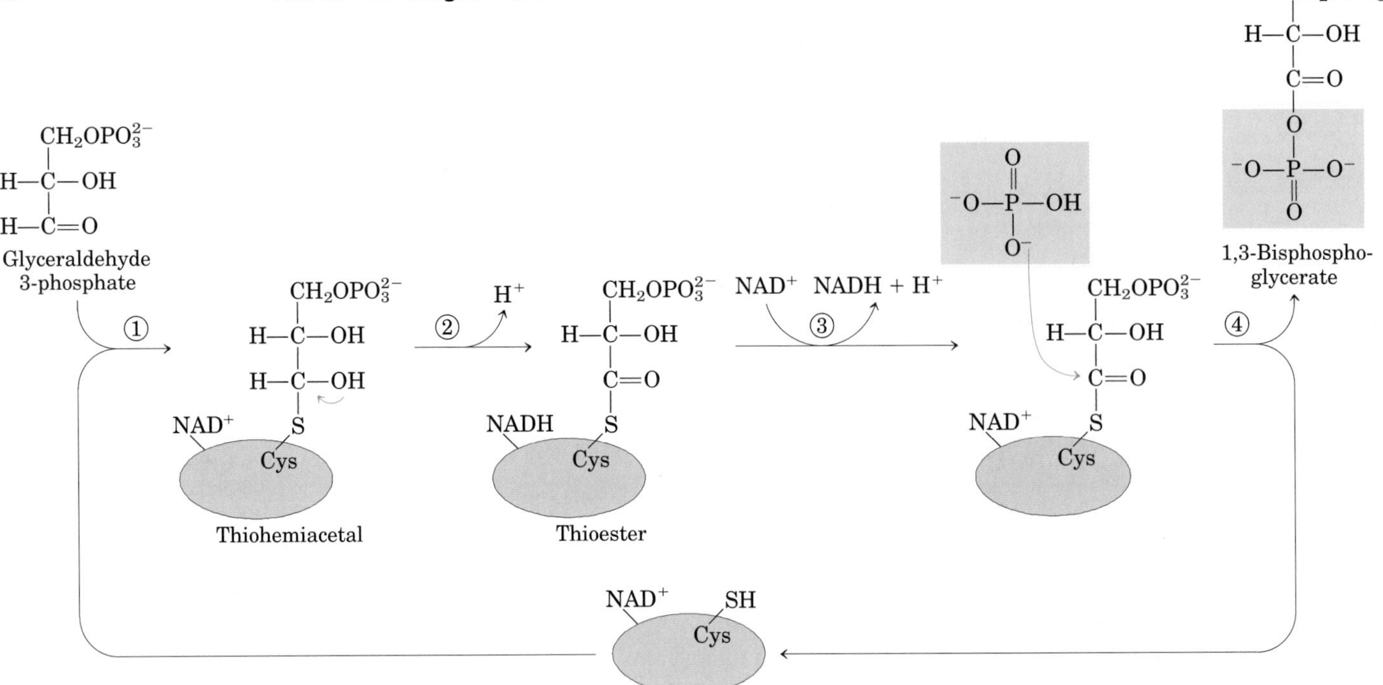

figure 15–5
Glyceraldehyde 3-phosphate dehydrogenase reaction: a more detailed representation. In step ①, a covalent thiohemiacetal linkage forms between the substrate and the —SH group of a Cys residue in the enzyme's active site. This enzyme-substrate intermediate is oxidized by NAD^+ (step ②), also bound to the active site, to form a covalent acyl-enzyme intermediate, a thioester. The enzyme-bound NADH is reoxidized by free NAD^+ (step ③). The bond between the acyl group and the thiol group of the enzyme has a very high standard free energy of hydrolysis. In step ④, this bond undergoes phosphorolysis (attack by P_i), releasing the free enzyme and an acyl phosphate, 1,3-bisphosphoglycerate. Formation of this product conserves much of the free energy liberated during oxidation of the aldehyde group of glyceraldehyde 3-phosphate.

called an **acyl phosphate,** has a very high standard free energy of hydrolysis ($\Delta G'^{\circ} = -49.3$ kJ/mol; see Fig. 14–4, Table 14–6). Much of the free energy of oxidation of the aldehyde group of glyceraldehyde 3-phosphate is conserved by formation of the acyl phosphate group at C-1 of 1,3-bisphosphoglycerate.

The acceptor of hydrogen in the glyceraldehyde 3-phosphate dehydrogenase reaction is NAD^+ (see Fig. 14–15). The reduction of NAD^+ proceeds by the enzymatic transfer of a hydride ion (:H^-) from the aldehyde group of glyceraldehyde 3-phosphate to the nicotinamide ring of NAD^+, yielding the reduced coenzyme NADH. The other hydrogen atom of the substrate molecule is released to solution as H^+.

Glyceraldehyde 3-phosphate is covalently bound to the enzyme during the reaction catalyzed by glyceraldehyde 3-phosphate dehydrogenase (Fig. 15–5). The aldehyde group of glyceraldehyde 3-phosphate reacts with the —SH group of an essential Cys residue in the active site, in a reaction analogous to the formation of a hemiacetal (see Fig. 9–5), in this case producing a *thio*hemiacetal. The discovery that glyceraldehyde 3-phosphate dehydrogenase is inhibited by iodoacetate was important in the history of research on glycolysis; the addition of this enzyme inhibitor to crude extracts of yeast or muscle caused the accumulation of the hexose phosphates produced in glycolysis, allowing them to be isolated and identified.

Iodoacetate is a potent inhibitor of glyceraldehyde 3-phosphate dehydrogenase because it forms a covalent derivative of the essential —SH group of the active site, rendering the enzyme inactive.

NAD^+ SH Cys + ICH_2—COO^- $\xrightarrow{HI}$ NAD^+ Cys—S—CH_2—COO^-

Iodoacetate Inactive enzyme

Because cells contain limited amounts of NAD^+, glycolysis would soon come to a halt for lack of NAD^+ if the NADH formed in this step of glycolysis were not continuously reoxidized. The reactions in which NAD^+ is regenerated anaerobically are described in detail later, in connection with the alternative fates of pyruvate.

⑦ Phosphoryl Transfer from 1,3-Bisphosphoglycerate to ADP The enzyme **phosphoglycerate kinase** transfers the high-energy phosphoryl group from the carboxyl group of 1,3-bisphosphoglycerate to ADP, forming ATP and **3-phosphoglycerate:**

1,3-Bisphosphoglycerate + ADP $\xrightleftharpoons[\text{phosphoglycerate kinase}]{Mg^{2+}}$ 3-Phosphoglycerate + ATP

$$\Delta G'^{\circ} = -18.5 \text{ kJ/mol}$$

Notice that phosphoglycerate kinase is named for the reverse reaction. Like all enzymes, it catalyzes the reaction in both directions. This enzyme acts in the direction suggested by its name during photosynthetic CO_2 fixation (see Fig. 20–25).

Steps ⑥ and ⑦ together constitute an energy-coupling process in which 1,3-bisphosphoglycerate is the common intermediate; it is formed in the first reaction (which would be endergonic in isolation), and its acyl phosphate group is transferred to ADP in the second reaction (which is strongly exergonic). The sum of these two reactions is

Glyceraldehyde 3-phosphate + ADP + P_i + NAD^+ $\rightleftharpoons$
3-phosphoglycerate + ATP + NADH + H^+
$$\Delta G'^{\circ} = -12.5 \text{ kJ/mol}$$

Thus the overall reaction is exergonic.

Recall from Chapter 14 that the actual free-energy change, ΔG, is determined by the standard free-energy change, $\Delta G'^{\circ}$, and the mass-action ratio, which is the ratio [products]/[reactants] (see Eqn 14–3). For step ⑥

$$\Delta G = \Delta G'^{\circ} + RT \ln \frac{[\text{1,3-bisphosphoglycerate}][\text{NADH}]}{[\text{glyceraldehyde 3-phosphate}][\text{P}_i][\text{NAD}^+]}$$

Notice that $[H^+]$ is not included in the mass-action ratio. In biochemical calculations, $[H^+]$ is assumed to be a constant (10^{-7} M), and this constant is included in the definition of $\Delta G'^{\circ}$ (see p. 494).

Step ⑦, by consuming the 1,3-bisphosphoglycerate produced in step ⑥, reduces [1,3-bisphosphoglycerate] and thereby reduces the mass-action ratio for the overall energy-coupling process. When this ratio is less than 1.0, its natural logarithm has a negative sign. If the mass-action ratio is very small, the contribution of the logarithmic term can make ΔG strongly negative. This is simply another way of showing that the two reactions, steps ⑥ and ⑦, are coupled through a common intermediate.

The outcome of these coupled reactions, both reversible under cellular conditions, is that the energy released on oxidation of an aldehyde to a carboxylate group is conserved by the coupled formation of ATP from ADP and P_i. The formation of ATP by phosphoryl group transfer from a substrate such as 1,3-bisphosphoglycerate is referred to as a **substrate-level phosphorylation** to distinguish this mechanism from **respiration-linked phosphorylation.** Substrate-level phosphorylations involve soluble enzymes and chemical intermediates (1,3-bisphosphoglycerate in this case). Respiration-linked phosphorylations, on the other hand, involve membrane-bound enzymes and transmembrane gradients of protons (Chapter 19).

⑧ Conversion of 3-Phosphoglycerate to 2-Phosphoglycerate The enzyme **phosphoglycerate mutase** catalyzes a reversible shift of the phosphoryl group between C-2 and C-3 of glycerate; Mg^{2+} is essential for this reaction:

$$\text{3-Phosphoglycerate} \underset{\text{phosphoglycerate mutase}}{\overset{Mg^{2+}}{\rightleftharpoons}} \text{2-Phosphoglycerate}$$

3-Phosphoglycerate 2-Phosphoglycerate

$$\Delta G'^{\circ} = 4.4 \text{ kJ/mol}$$

The reaction occurs in two steps (Fig. 15–6). A phosphoryl group initially attached to a His residue of the mutase is transferred to the hydroxyl group at C-2 of 3-phosphoglycerate, forming 2,3-bisphosphoglycerate (BPG). The phosphoryl group at C-3 of 2,3-bisphosphoglycerate is then transferred to the same His residue, producing 2-phosphoglycerate and regenerating the phosphorylated enzyme. Phosphoglycerate mutase is initially phosphorylated by phosphoryl transfer from BPG, which thus functions as a coenzyme; it is required in small quantities to initiate the catalytic cycle and is continuously regenerated by that cycle. Although in most cells BPG is present in only trace amounts, it is a major component (~5 mM) of erythrocytes, where it regulates the affinity of hemoglobin for oxygen (see Fig. 7–16).

The enzyme phosphoglucomutase (p. 548) employs essentially the same mechanism as phosphoglycerate mutase. Phosphoglucomutase converts glucose 1-phosphate into glucose 6-phosphate, with glucose 1,6-bisphosphate serving as the coenzyme. The general name **mutase** is given to enzymes that catalyze the transfer of a functional group from one position to another in the same molecule. Mutases are a subclass of **isomerases,** enzymes that interconvert stereoisomers or structural or positional isomers (see Table 8–3).

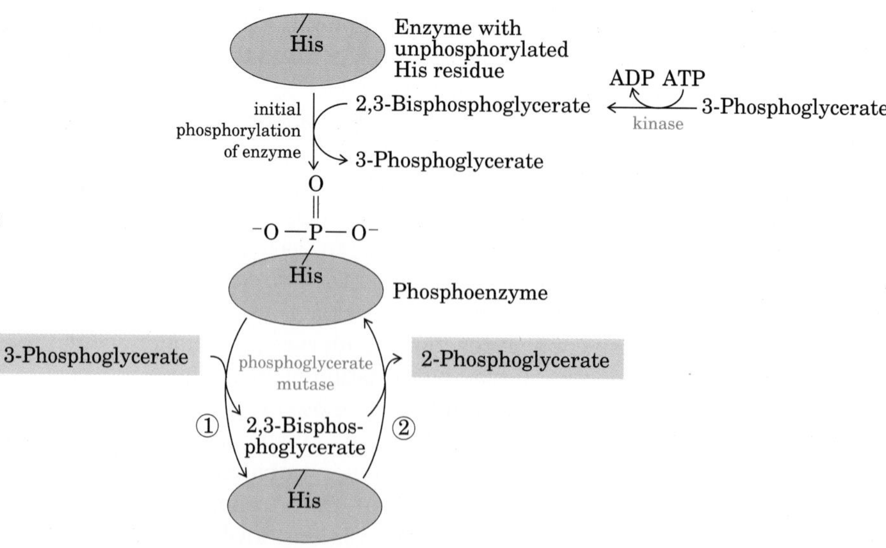

figure 15–6
Mechanism of the phosphoglycerate mutase reaction. The enzyme is initially phosphorylated on a His residue by transfer of a phosphoryl group from 2,3-bisphosphoglycerate (BPG). In step ① of the catalytic reaction, the phosphoenzyme transfers its phosphoryl group to 3-phosphoglycerate, forming BPG. In step ② the phosphoryl group at C-3 of BPG is transferred to the same His residue on the enzyme, producing 2-phosphoglycerate and regenerating the phosphoenzyme. The BPG required initially to phosphorylate the enzyme is formed from 3-phosphoglycerate by a specific ATP-dependent kinase; it is then regenerated in step ① of each catalytic cycle.

⑨ Dehydration of 2-Phosphoglycerate to Phosphoenolpyruvate In the second glycolytic reaction that generates a compound with high phosphoryl group transfer potential, **enolase** promotes reversible removal of a molecule of water from 2-phosphoglycerate to yield **phosphoenolpyruvate:**

2-Phosphoglycerate ⇌ (enolase; H_2O) Phosphoenolpyruvate

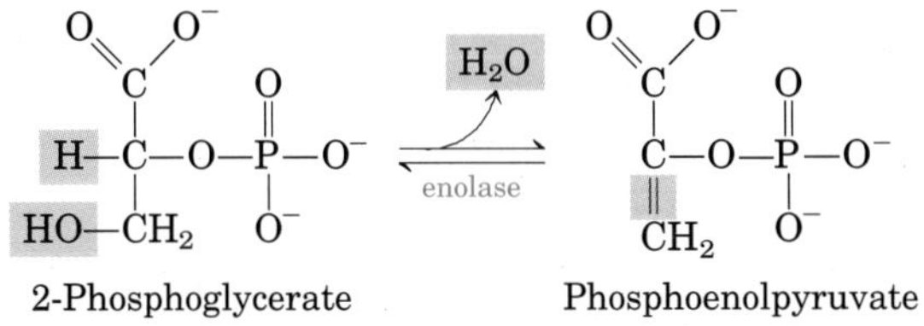

$$\Delta G'^\circ = 7.5 \text{ kJ/mol}$$

Despite the relatively small standard free-energy change of this reaction, there is a very large difference in the standard free energy of hydrolysis of the phosphate groups of the reactant and product: −17.6 kJ/mol for 2-phosphoglycerate (a low-energy phosphate ester) and −61.9 kJ/mol for phosphoenolpyruvate (a super high-energy phosphate compound) (see Fig. 14–3, Table 14–6). Although 2-phosphoglycerate and phosphoenolpyruvate contain nearly the same *total* amount of energy, the loss of the water molecule from 2-phosphoglycerate causes a redistribution of energy within the molecule, greatly increasing the standard free energy of hydrolysis of the phosphate group.

⑩ Transfer of the Phosphoryl Group from Phosphoenolpyruvate to ADP The last step in glycolysis is the transfer of the phosphoryl group from phosphoenolpyruvate to ADP, catalyzed by **pyruvate kinase,** which requires K^+ and either Mg^{2+} or Mn^{2+}:

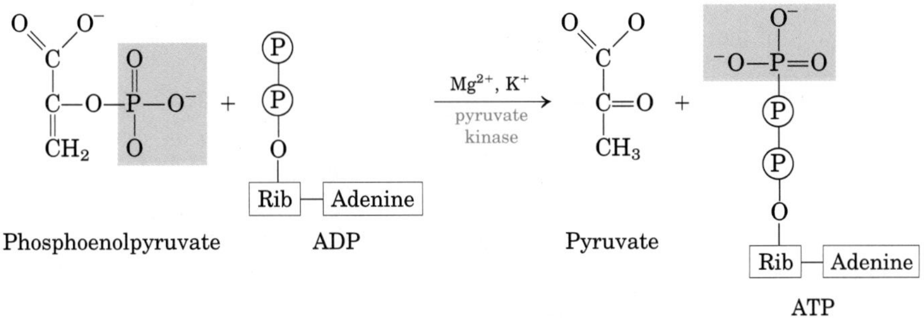

Phosphoenolpyruvate + ADP → (Mg^{2+}, K^+; pyruvate kinase) Pyruvate + ATP

$$\Delta G'^\circ = -31.4 \text{ kJ/mol}$$

In this substrate-level phosphorylation, the product **pyruvate** first appears in its enol form, then tautomerizes rapidly and nonenzymatically to its keto form, which predominates at pH 7. The overall reaction has a large, negative standard free-energy change, due in large part to the spontaneous conversion of the enol form of pyruvate to the keto form (see Fig. 14–3). The $\Delta G'^\circ$ of phosphoenolpyruvate hydrolysis is −61.9 kJ/mol; about half of this energy is conserved in the formation of the phosphoanhydride bond of ATP ($\Delta G'^\circ$ = −30.5 kJ/mol) and the rest (−31.4 kJ/mol) constitutes a large driving force pushing the reaction toward ATP synthesis. The pyruvate kinase reaction is essentially irreversible under intracellular conditions and is an important site of regulation, as described later.

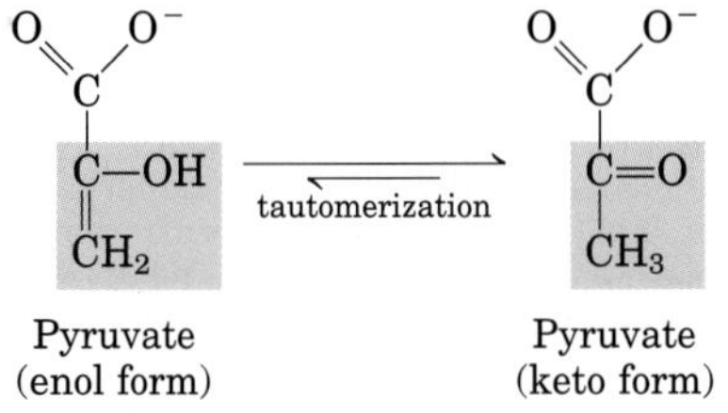

Pyruvate (enol form) ⇌ (tautomerization) Pyruvate (keto form)

The Overall Balance Sheet Shows a Net Gain of ATP

We can now construct a balance sheet for glycolysis to account for (1) the fate of the carbon skeleton of glucose, (2) the input of P_i and ADP and the output of ATP, and (3) the pathway of electrons in the oxidation-reduction reactions. The left-hand side of the following equation shows all the inputs of ATP, NAD^+, ADP, and P_i (consult Fig. 15–2), and the right-hand side shows all the outputs (keep in mind that each molecule of glucose yields two molecules of glyceraldehyde 3-phosphate):

$$\text{Glucose} + 2\text{ATP} + 2\text{NAD}^+ + 4\text{ADP} + 2\text{P}_i \longrightarrow 2\text{ pyruvate} + 2\text{ADP} + 2\text{NADH} + 2\text{H}^+ + 4\text{ATP} + 2\text{H}_2\text{O}$$

Canceling out common terms on both sides of the equation gives the overall equation for glycolysis under aerobic conditions:

$$\text{Glucose} + 2\text{NAD}^+ + 2\text{ADP} + 2\text{P}_i \longrightarrow 2\text{ pyruvate} + 2\text{NADH} + 2\text{H}^+ + 2\text{ATP} + 2\text{H}_2\text{O}$$

The two molecules of NADH formed by glycolysis in the cytosol are, under aerobic conditions, reoxidized to NAD^+ by transfer of their electrons to the respiratory chain, which in eukaryotic cells is located in the mitochondria. The respiratory chain passes these electrons to their ultimate destination, O_2:

$$2\text{NADH} + 2\text{H}^+ + \text{O}_2 \longrightarrow 2\text{NAD}^+ + 2\text{H}_2\text{O}$$

Electron transfer from NADH to O_2 in mitochondria provides the energy for synthesis of ATP by respiration-linked phosphorylation (Chapter 19).

In the overall glycolytic process, one molecule of glucose is converted to two molecules of pyruvate (the pathway of carbon). Two molecules of ADP and two of P_i are converted to two molecules of ATP (the pathway of phosphoryl groups). Four electrons, as two hydride ions, are transferred from two molecules of glyceraldehyde 3-phosphate to two of NAD^+ (the pathway of electrons).

Intermediates Are Channeled between Glycolytic Enzymes

Although the enzymes of glycolysis are usually described as soluble components of the cytosol, growing evidence suggests that within the cell these enzymes exist as multienzyme complexes. The classic approach of enzymology—purification of individual proteins from extracts of broken cells—was applied with great success to the enzymes of glycolysis. However, the first casualty of cell breakage is higher-level organization within a cell—the noncovalent, reversible interaction of one protein with another, or of an enzyme with some structural component such as a membrane, microtubule, or microfilament. When cells are broken open, their contents, including enzymes, are diluted 100- or 1,000-fold (Fig. 15–7).

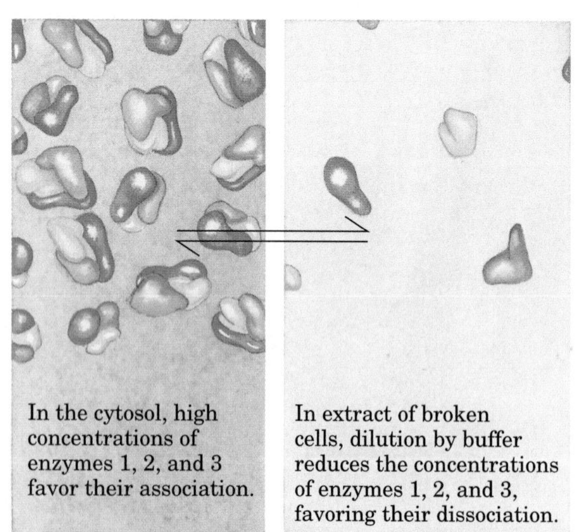

In the cytosol, high concentrations of enzymes 1, 2, and 3 favor their association.
In extract of broken cells, dilution by buffer reduces the concentrations of enzymes 1, 2, and 3, favoring their dissociation.

figure 15–7
Dilution of a solution containing a noncovalent protein complex favors dissociation of the complex into its constituents.

When the purified enzymes of glycolysis are combined in vitro at relatively high concentrations, they form specific, functional aggregates, which may reflect their true state inside cells. Several types of evidence suggest that, in cells, such complexes ensure efficient passage of the product of one enzyme reaction to the next enzyme in the pathway. Kinetic evidence for the **channeling** of 1,3-bisphosphoglycerate from glyceraldehyde 3-phosphate dehydrogenase to phosphoglycerate kinase without its entering solution (Fig. 15–8) is corroborated by physical evidence that these two enzymes form stable, noncovalent complexes. There is similar evidence for channeling of intermediates between other glycolytic enzymes, such as glyceraldeyde 3-phosphate from aldolase to glyceraldehyde 3-phosphate dehydrogenase.

Furthermore, some glycolytic enzymes form specific noncovalent complexes with structural components of the cell, which may organize reaction sequences and assure efficient transfer of intermediates between cellular compartments. Phosphofructokinase-1 and aldolase, for example, bind to actin microfilaments, bringing the two enzymes into close association and altering the catalytic properties of aldolase (both K_m and V_{max} increase). Hexokinase binds specifically to the outer mitochondrial membrane, perhaps allowing ATP produced within the mitochondrion to move directly to the catalytic site of hexokinase without entering, and being diluted by, the cytosol. There is strong evidence for substrate channeling through multienzyme complexes in other metabolic pathways, and many enzymes now thought of as "soluble" probably function in the cell as highly organized complexes that channel intermediates.

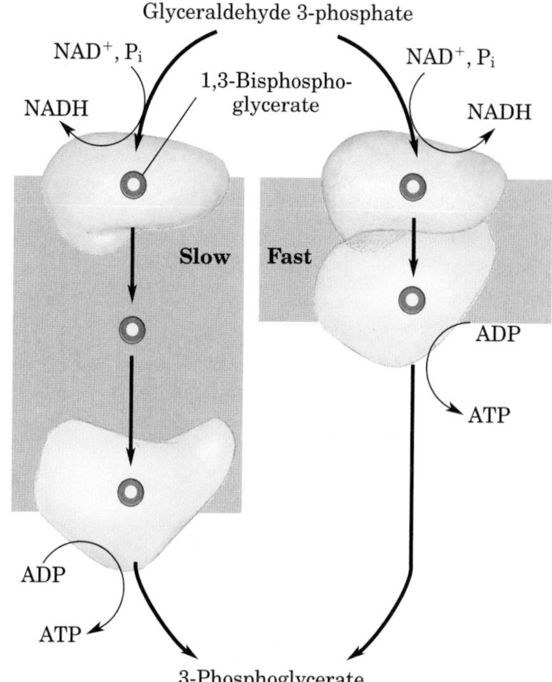
Glyceraldehyde 3-phosphate
NAD+, Pi
NADH
1,3-Bisphospho-
glycerate
NAD+, Pi
NADH
Slow
Fast
ADP
ATP
ADP
ATP
3-Phosphoglycerate
Sequential action of two separate enzymes: the product of the first enzyme (1,3-bisphosphoglycerate) diffuses to the second enzyme.
Substrate channeling through a functional complex of two enzymes: the intermediate (1,3-bisphosphoglycerate) is never released to the solvent.

figure 15–8

Channeling of a substrate between two enzymes in the glycolytic pathway. When glyceraldehyde 3-phosphate dehydrogenase (blue) and 3-phosphoglycerate kinase (yellow) are combined in vitro, the rate at which they catalyze the two-step conversion of glyceraldehyde 3-phosphate to 3-phosphoglycerate (see Figs 15–5, 15–6) exceeds the rate at which the first step is catalyzed in the presence of the first enzyme only. Apparently the direct transfer of 1,3-bisphosphoglycerate from the dehydrogenase to the kinase (right) occurs faster than the dissociation of 1,3-bisphosphoglycerate from the dehydrogenase into the surrounding medium (left). Physical studies show that the two enzymes can form a stable complex, as is required for substrate channeling between them.

Glycolysis Is under Tight Regulation

During his studies on the fermentation of glucose by yeast, Louis Pasteur discovered that both the rate and the total amount of glucose consumption were many times greater under anaerobic than aerobic conditions. Later studies of muscle showed the same large difference in the rates of anaerobic and aerobic glycolysis. The biochemical basis of this "Pasteur effect" is now clear. The ATP yield from glycolysis under anaerobic conditions (2 ATP per molecule of glucose) is much smaller than that from the complete oxidation of glucose to CO_2 under aerobic conditions (30 or 32 ATP per glucose; see Table 19–5). About 18 times as much glucose must therefore be consumed anaerobically as aerobically to yield the same amount of ATP.

The flux of glucose through the glycolytic pathway is regulated to achieve constant ATP levels (as well as adequate supplies of glycolytic intermediates that serve biosynthetic roles). The required adjustment in the rate of glycolysis is achieved by the allosteric regulation of two glycolytic enzymes—phosphofructokinase-1 and pyruvate kinase—by second-to-second fluctuations in the concentration of key metabolites that reflect the cellular balance between ATP production and consumption. We return to a more detailed discussion of the regulation of glycolysis later in the chapter.

Glucose Catabolism Is Deranged in Cancerous Tissue

Glucose uptake and glycolysis proceed about ten times faster in most solid tumors than in noncancerous tissues. Tumor cells commonly experience hypoxia (limited oxygen supply) because they initially lack an extensive capillary network to supply the tumor with oxygen. As a result, cancer cells more than 100 to 200 μm from the nearest capillaries depend on anaerobic glycolysis for much of their ATP production. They take up more glucose than normal cells, converting it to pyruvate and then to lactate as they recycle NADH. The high glycolytic rate may also result in part from smaller numbers of mitochondria in tumor cells; less ATP made by mitochondrial oxidative phosphorylation means more needed from glycolysis. In addition, some tumor cells overproduce several glycolytic enzymes, including an isozyme of hexokinase that associates with the cytosolic face of the mitochondrial inner membrane and is insensitive to feedback inhibition by glucose 6-phosphate (p. 555). This enzyme may monopolize the ATP produced in mitochondria, using it to convert glucose to glucose 6-phosphate and committing the cell to continued glycolysis. The protein products of oncogenes and tumor supressor genes such as *p53* and *ras* (discussed in Chapter 13) are involved in the increased production of glycolytic enzymes in tumor cells.

Fates of Pyruvate under Aerobic and Anaerobic Conditions

Pyruvate represents an important junction in carbohydrate catabolism (Fig. 15–3). Under aerobic conditions pyruvate is oxidized to acetate, which enters the citric acid cycle and is oxidized to CO_2 and H_2O, and NADH formed by the dehydrogenation of glyceraldehyde 3-phosphate is reoxidized to NAD^+ by passage of its electrons to O_2 in mitochondrial respiration. However, under hypoxic conditions such as in very active skeletal muscle, in submerged plant parts, or in lactic acid bacteria, NADH generated by glycolysis cannot be reoxidized by O_2. Failure to regenerate NAD^+ would leave the cell with no electron acceptor for the oxidation of glyceraldehyde 3-phosphate, and the energy-yielding reactions of glycolysis would stop. NAD^+ must therefore be regenerated in some other way.

The earliest cells to arise during evolution lived in an atmosphere almost devoid of oxygen and had to develop strategies for deriving energy from fuel molecules under anaerobic conditions. Most modern organisms have retained the ability to continually regenerate NAD^+ during anaerobic glycolysis by transferring electrons from NADH to form a reduced end product such as lactate or ethanol.

Pyruvate Is the Terminal Electron Acceptor in Lactic Acid Fermentation

Pyruvate

lactate dehydrogenase: NADH + H^+ → NAD^+

Lactate

$\Delta G'^\circ = -25.1$ kJ/mol

When animal tissues cannot be supplied with sufficient oxygen to support aerobic oxidation of the pyruvate and NADH produced in glycolysis, NAD^+ is regenerated from NADH by the reduction of pyruvate to **lactate.** As mentioned earlier, some tissues and cell types (such as erythrocytes) produce lactate from glucose even under aerobic conditions. The reduction of pyruvate is catalyzed by **lactate dehydrogenase,** which forms the L isomer of lactate at pH 7. The overall equilibrium of this reaction strongly favors lactate formation, as shown by the large negative standard free-energy change.

In glycolysis, dehydrogenation of the two molecules of glyceraldehyde 3-phosphate derived from each molecule of glucose converts two molecules of NAD^+ to two of NADH. Because the reduction of two molecules of pyruvate to two of lactate regenerates two molecules of NAD^+, the overall process is balanced and can continue indefinitely.

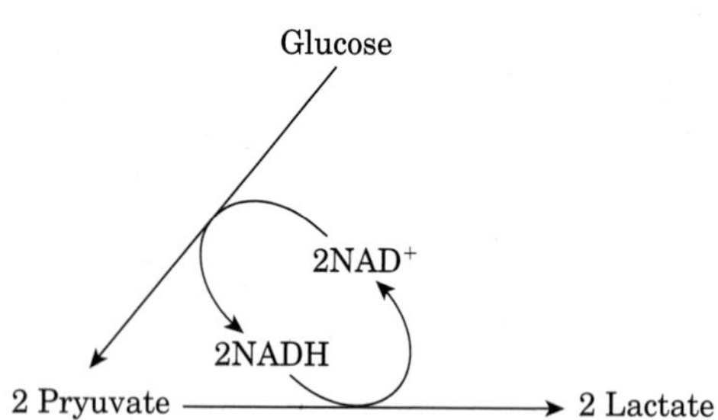

Although conversion of glucose to lactate includes two oxidation-reduction steps, there is no net change in the oxidation state of carbon; in glucose ($C_6H_{12}O_6$) and lactic acid ($C_3H_6O_3$), the H : C ratio is the same. Nevertheless, some of the energy of the glucose molecule has been extracted by its conversion to lactate, enough to give a net yield of two molecules of ATP for every glucose molecule consumed. The lactate formed by active skeletal muscles can be recycled; it is carried in the blood to the liver where it is converted to glucose during the recovery from strenuous muscular activity. When lactate is produced in large quantities during vigorous muscle contraction (during a sprint, for example), the acidification that results from ionization of lactic acid in muscle and blood causes pain and limits the period of vigorous activity. The best-conditioned athletes can sprint for no more than a minute (Box 15–1).

Many microorganisms ferment glucose and other hexoses to lactate. Certain lactobacilli and streptococci, for example, ferment the lactose in milk to lactic acid. The dissociation of lactic acid to lactate and H^+ in the fermentation mixture lowers the pH, denaturing casein and other milk proteins and causing them to precipitate. Under the correct, controlled conditions, the resultant curdling produces cheese or yogurt, depending on the microorganism.

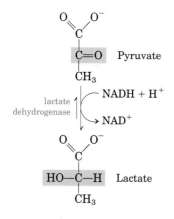

box 15–1 Glycolysis at Limiting Concentrations of Oxygen: Athletes, Alligators, and Coelacanths

Most vertebrates are essentially aerobic organisms; they convert glucose to pyruvate by glycolysis, then oxidize the pyruvate completely to CO_2 and H_2O using molecular oxygen. Anaerobic catabolism of glucose to lactate occurs during short bursts of extreme muscular activity, for example in a 100 m sprint, during which oxygen cannot be carried to the muscles fast enough to oxidize pyruvate. Instead, the muscles use their stored glucose (glycogen) as fuel to generate ATP by fermentation, with lactate as the end product. In a sprint, lactate in the blood builds up to high concentrations. It is slowly converted back to glucose by gluconeogenesis in the liver (Chapter 19) in the subsequent rest or recovery period, during which oxygen is consumed at a gradually diminishing rate until the breathing rate returns to normal. The excess oxygen consumed in the recovery period represents a repayment of the oxygen debt. This is the amount of oxygen required to supply ATP for gluconeogenesis during recovery respiration, in order to regenerate the glycogen "borrowed" from liver and muscle to carry out intense muscular activity in the sprint. The cycle of reactions that includes glucose conversion to lactate in muscle and lactate conversion to glucose in liver is called the Cori cycle, for Carl and Gerty Cori, whose studies in the 1930s and 1940s clarified the pathway and its role.

The circulatory systems of most small vertebrates can carry oxygen to their muscles fast enough to avoid having to use muscle glycogen anaerobically. For example, migrating birds often fly great distances at high speeds without rest and without incurring an oxygen debt. Many running animals of moderate size also maintain an essentially aerobic metabolism in their skeletal muscle. However, the circulatory systems of larger animals, including humans, cannot completely sustain aerobic metabolism in skeletal muscles during long bursts of muscular activity. These animals generally are slow-moving under normal circumstances and engage in intense muscular activity only in the gravest emergencies, because such bursts of activity require long recovery periods to repay the oxygen debt.

Alligators and crocodiles, for example, are normally sluggish and torpid. Yet when provoked these animals are capable of lightning-fast charges and dangerous lashings of their powerful tails. Such intense bursts of activity are short and must be followed by long periods of recovery. The fast emergency movements require lactic acid fermentation to generate ATP in skeletal muscles. The stores of muscle glycogen are rapidly expended in intense muscular activity, and lactate reaches very high concentrations in muscles and extracellular fluid. Whereas a trained athlete can recover from a 100 m sprint in 30 min or less, an alligator may require many hours of rest and extra oxygen consumption to clear the excess lactate from its blood and regenerate muscle glycogen after a burst of activity.

Other large animals, such as the elephant and rhinoceros, have similar metabolic problems, as do diving mammals such as whales and seals. Dinosaurs and other huge, now-extinct animals probably had to depend on lactate fermentation to supply energy for muscular activity, followed by very long recovery periods during which they were vulnerable to attack by smaller predators better able to use oxygen and thus better adapted to continuous, sustained muscular activity.

Deep-sea explorations have revealed many species of marine life at great ocean depths, where the oxygen concentration is near zero. For example, the primitive coelacanth, a large fish recovered from depths of 4,000 m or more off the coast of South Africa, has an essentially anaerobic metabolism in virtually all its tissues. It converts carbohydrates to lactate and other products, most of which must be excreted. Some marine vertebrates ferment glucose to ethanol and CO_2 in order to generate ATP.

box 15–2 Brewing Beer

Beer is made by alcohol fermentation of the carbohydrates in cereal grains (seeds) such as barley by yeast glycolytic enzymes. The carbohydrates, largely polysaccharides, must first be degraded to disaccharides and monosaccharides. In a process called malting, the barley seeds are allowed to germinate until they form the hydrolytic enzymes required to break down their polysaccharides, at which point germination is stopped by controlled heating. The product is malt, which contains enzymes that catalyze the hydrolysis of the β linkages of cellulose and other cell-wall polysaccharides of the barley husks, and enzymes such as α-amylase and maltase.

The brewer next prepares the wort, the nutrient medium required for fermentation by yeast cells. The malt is mixed with water and then mashed or crushed. This allows the enzymes formed in the malting process to act on the cereal polysaccharides to form maltose, glucose, and other simple sugars, which are soluble in the aqueous medium. The remaining cell matter is then separated, and the liquid wort is boiled with hops to give flavor. The wort is cooled and then aerated.

Now the yeast cells are added. In the aerobic wort the yeast grows and reproduces very rapidly, using energy obtained from available sugars. No alcohol is formed because the yeast, amply supplied with oxygen, oxidizes the pyruvate formed by glycolysis to CO_2 and H_2O via the citric acid cycle. When all the dissolved oxygen in the vat of wort has been consumed, the yeast cells switch to anaerobic metabolism, and from this point they ferment the sugars into ethanol and CO_2. The fermentation process is controlled in part by the concentration of the ethanol formed, by the pH, and by the amount of remaining sugar. After fermentation has been stopped, the cells are removed and the "raw" beer is ready for final processing.

In the final steps of brewing, the amount of foam or head on the beer, which results from dissolved proteins, is adjusted. Normally this is controlled by proteolytic enzymes that arise in the malting process. If these enzymes act on the proteins too long, the beer will have very little head and will be flat; if they do not act long enough, the beer will not be clear when it is cold. Sometimes proteolytic enzymes from other sources are added to control the head.

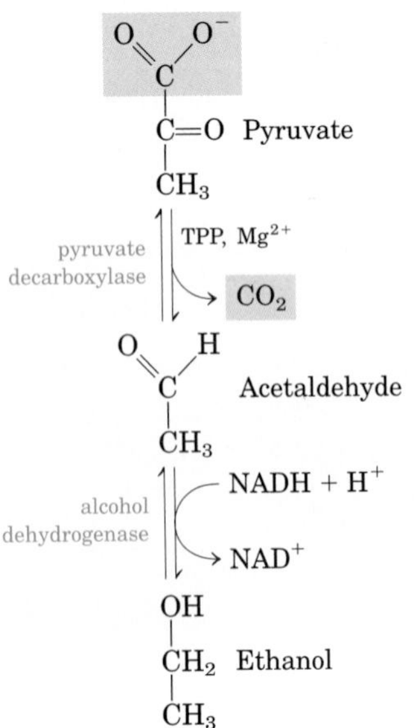

Ethanol Is the Reduced Product in Alcohol Fermentation

Yeast and other microorganisms ferment glucose to ethanol and CO_2, rather than to lactate. Glucose is converted to pyruvate by glycolysis, and the pyruvate is converted to ethanol and CO_2 in a two-step process. In the first step, pyruvate is decarboxylated in an irreversible reaction catalyzed by pyruvate decarboxylase. This reaction is a simple decarboxylation and does not involve the net oxidation of pyruvate. Pyruvate decarboxylase requires Mg^{2+} and has a tightly bound coenzyme, thiamine pyrophosphate, discussed in more detail below. In the second step, acetaldehyde is reduced to ethanol through the action of alcohol dehydrogenase, with NADH derived from glyceraldehyde 3-phosphate dehydrogenation furnishing the reducing power. Ethanol and CO_2 are thus the end products of alcohol fermentation, and the overall equation is

$$\text{Glucose} + 2\text{ADP} + 2\text{P}_i \longrightarrow 2\text{ ethanol} + 2CO_2 + 2\text{ATP} + 2H_2O$$

As in lactic acid fermentation, there is no net change in the ratio of hydrogen to carbon atoms when glucose (H:C ratio = 12/6 = 2) is fermented to two ethanol and two CO_2 (combined H:C ratio = 12/6 = 2). In all fermentations, the H:C ratio of the reactants and products remains the same.

Pyruvate decarboxylase is characteristically present in brewer's and baker's yeast and in all other organisms in which alcohol fermentation occurs, including some plants. The CO_2 produced by pyruvate decarboxylation in brewer's yeast is responsible for the characteristic carbonation of champagne. The ancient art of brewing beer involves a number of enzymatic processes in addition to the reactions of alcohol fermentation (Box 15–2). In baking, CO_2 released by pyruvate decarboxylase when yeast is mixed with a fermentable sugar causes dough to rise. The enzyme is absent in vertebrate tissues and in other organisms that carry out lactic acid fermentation.

Alcohol dehydrogenase is present in many organisms that metabolize alcohol, including humans. In human liver it catalyzes the *oxidation* of ethanol, either ingested or produced by intestinal microorganisms, with the concomitant reduction of NAD^+ to NADH.

Thiamine Pyrophosphate Carries "Active Aldehyde" Groups

The pyruvate decarboxylase reaction is our first encounter with **thiamine pyrophosphate (TPP)** (Fig. 15–9), a coenzyme derived from vitamin B_1. Lack of vitamin B_1 in the human diet leads to the condition known as beriberi, characterized by an accumulation of body fluids (swelling), pain, paralysis, and ultimately death.

Thiamine pyrophosphate plays an important role in the cleavage of bonds adjacent to a carbonyl group, such as the decarboxylation of α-keto acids, and in chemical rearrangements in which an activated aldehyde group is transferred from one carbon atom to another (Table 15–1). The functional part of TPP, the thiazolium ring, has a relatively acidic proton at C-2. Loss of this proton produces a carbanion that is the active species in TPP-dependent reactions (Fig. 15–9). The carbanion readily adds to carbonyl groups, and the thiazolium ring is thereby positioned to act as an "electron sink" that greatly facilitates reactions such as the decarboxylation catalyzed by pyruvate decarboxylase.

figure 15–9

Thiamine pyrophosphate (TPP) and its role in pyruvate decarboxylation. TPP is the coenzyme form of vitamin B_1 (thiamine). The reactive carbon atom in the thiazolium ring of TPP **(a)** is shown in red. In the reaction catalyzed by pyruvate decarboxylase, two of the three carbons of pyruvate are carried transiently on TPP in the form of a hydroxyethyl, or "active acetaldehyde," group **(b),** which is subsequently released as acetaldehyde. **(c)** After cleavage of a carbon–carbon bond, one product often has a free electron pair or carbanion, and the strong tendency of the carbanion to form a new bond generally renders this product unstable. The thiazolium ring of TPP stabilizes carbanion intermediates by providing an electrophilic (electron-deficient) structure into which the carbanion electrons can be delocalized by resonance. Structures with this property, often called "electron sinks," play a role in many biochemical reactions. This principle is illustrated here for the reaction catalyzed by pyruvate decarboxylase. In step ① the TPP carbanion acts as a nucleophile, attacking the carbonyl group of pyruvate. In step ②, decarboxylation produces a carbanion that is stabilized by the thiazolium ring. After protonation to form hydroxyethyl TPP (step ③), acetaldehyde is released (step ④).

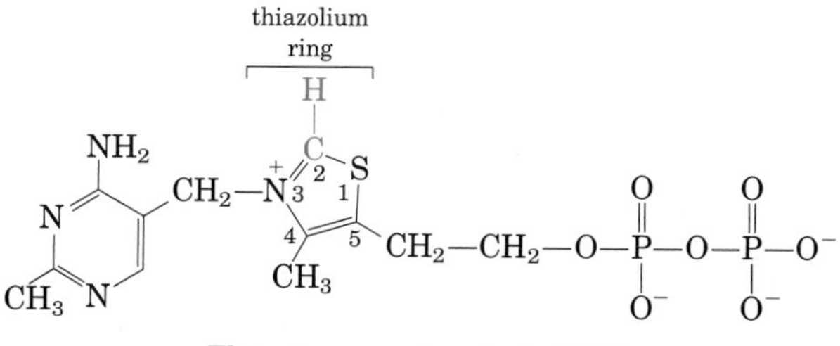

thiazolium ring

Thiamine pyrophosphate (TPP)

(a)

active aldehyde

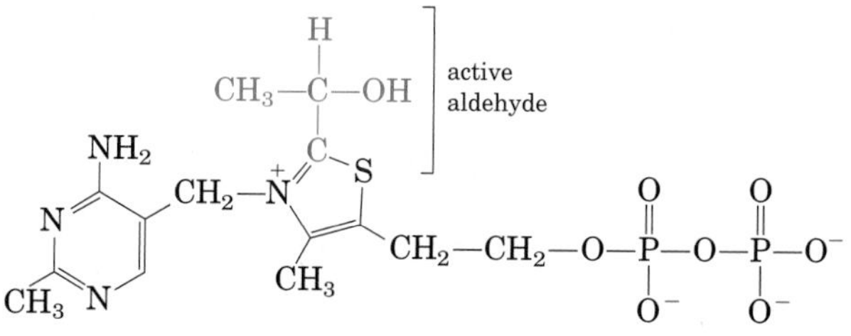

Hydroxyethyl thiamine pyrophosphate

(b)

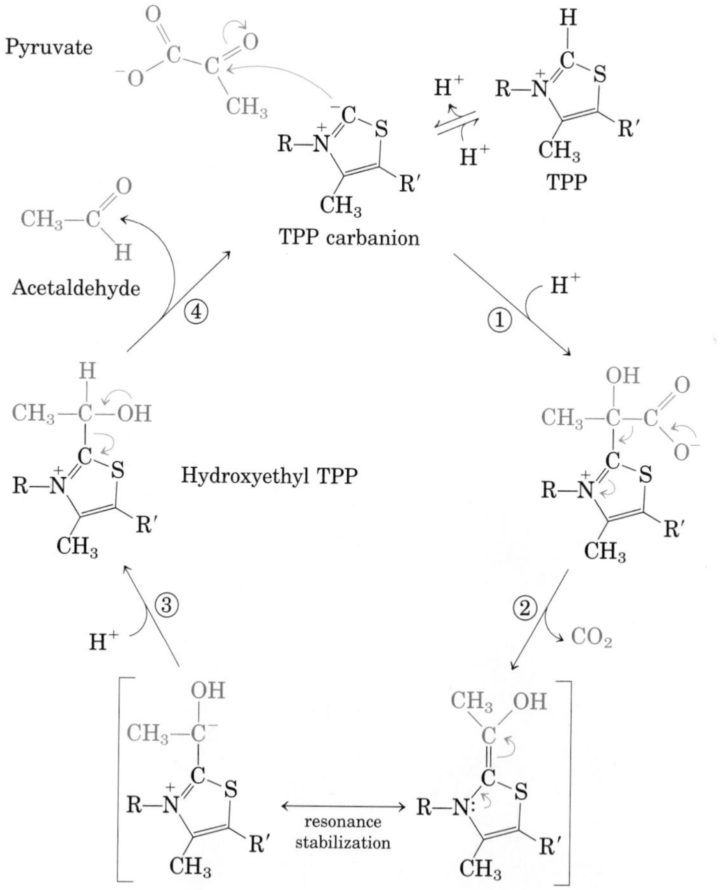

Pyruvate

TPP

TPP carbanion

Acetaldehyde

Hydroxyethyl TPP

resonance stabilization

(c)

table 15–1

Some TPP-Dependent Reactions

Enzyme	Pathway	Bond cleaved	Bond formed
Pyruvate decarboxylase	Alcohol fermentation	$R^1{-}C(=O){-}C(=O)O^-$	$R^1{-}C(=O)H$
Pyruvate dehydrogenase α-Ketoglutarate dehydrogenase	Synthesis of acetyl-CoA Citric acid cycle	$R^2{-}C(=O){-}C(=O)O^-$	$R^2{-}C(=O){-}S\text{-}CoA$
Transketolase	Carbon-fixation reactions of photosynthesis	$R^3{-}C(=O){-}C(OH)(H){-}R^4$	$R^3{-}C(=O){-}C(OH)(H){-}R^5$
Acetolactate synthetase	Valine, leucine biosynthesis	$R^6{-}C(=O){-}C(=O)O^-$	$R^6{-}C(=O){-}C(OH)(C(=O)O^-){-}$

Microbial Fermentations Yield Other End Products of Commercial Value

Lactate and ethanol are common products of microbial fermentations, but they are by no means the only possible ones. In 1910 Chaim Weizmann (later to become the first president of Israel) discovered that the bacterium *Clostridium acetobutyricum* ferments starch to butanol and acetone. This discovery opened the field of industrial fermentations, in which some readily available material rich in carbohydrate (corn starch or molasses, for example) is supplied to a pure culture of a specific microorganism, which ferments it into a product of greater value. The methanol used to make "gasohol" is produced by microbial fermentation, as are formic, acetic, propionic, butyric, and succinic acids, and glycerol, ethanol, isopropanol, butanol, and butanediol. These fermentations are generally carried out in huge, closed vats in which temperature and access to air are adjusted to favor the multiplication of the desired microorganism and to exclude contaminating organisms (Fig. 15–10). The beauty of industrial fermentations is that complicated, multistep chemical transformations are carried out in high yields and with few side products by chemical factories that reproduce themselves—microbial cells. In some cases it is possible to immobilize the cells in an inert support, to pass the starting material continuously through a bed of immobilized cells, and to collect the desired product in the effluent—an engineer's dream!

figure 15–10

Industrial-scale fermentation. Microorganisms are cultured in a sterilizable vessel containing thousands of liters of growth medium—an inexpensive source of both carbon and energy—under carefully controlled conditions, including low oxygen concentration and constant temperature. After centrifugal separation of the cells from the growth medium, the valuable products of the fermentation are recovered from the cells or from the supernatant fluid.

Feeder Pathways for Glycolysis

Many carbohydrates besides glucose meet their catabolic fate in the glycolytic pathway, after being transformed to one of the glycolytic intermediates. The most significant are the storage polysaccharides glycogen and starch, the disaccharides maltose, lactose, trehalose, and sucrose, and the monosaccharides fructose, mannose, and galactose (Fig. 15–11).

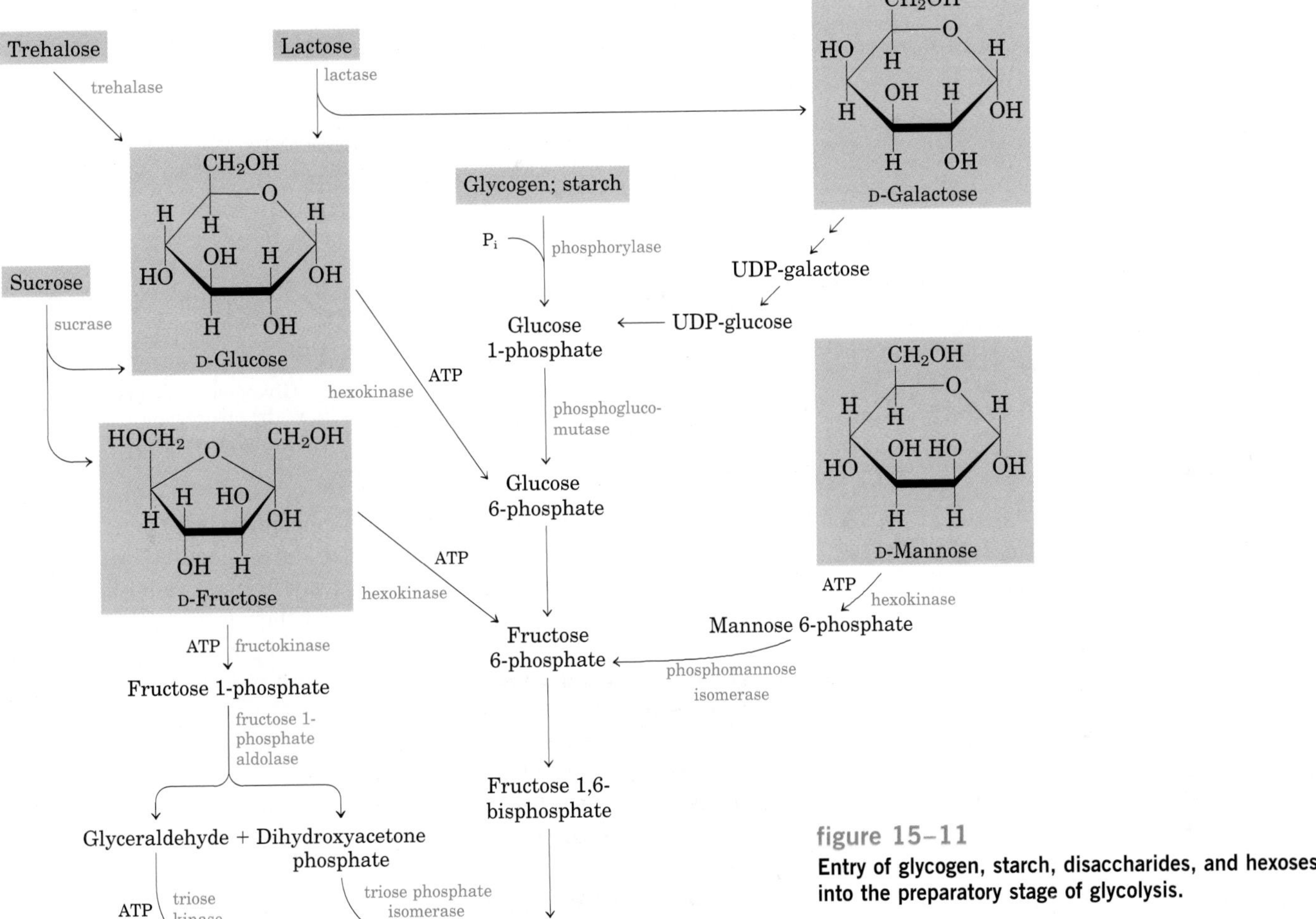

figure 15–11
Entry of glycogen, starch, disaccharides, and hexoses into the preparatory stage of glycolysis.

Glycogen and Starch Are Degraded by Phosphorolysis

The glucose units of the outer branches of glycogen and starch enter the glycolytic pathway through the sequential action of two enzymes: glycogen phosphorylase, or the similar starch phosphorylase in plants, and phosphoglucomutase. Glycogen phosphorylase catalyzes the reaction in which an (α1→4) glycosidic linkage joining two glucose residues in glycogen undergoes attack by inorganic phosphate, removing the terminal glucose residue as α-D-glucose 1-phosphate (Fig. 15–12). This *phosphorolysis* reaction is different from the *hydrolysis* of glycosidic bonds by amylase during intestinal degradation of glycogen or starch. In phosphorolysis, some of the energy of the glycosidic bond is preserved in the formation of the phosphate ester, glucose 1-phosphate.

Pyridoxal phosphate is an essential cofactor in the glycogen phosphorylase reaction; its phosphate group acts as a general acid catalyst, promoting attack by P_i on the glycosidic bond. (A quite different role of pyridoxal phosphate as a cofactor in amino acid metabolism is described in detail in Fig. 18–6.)

Pyridoxal phosphate

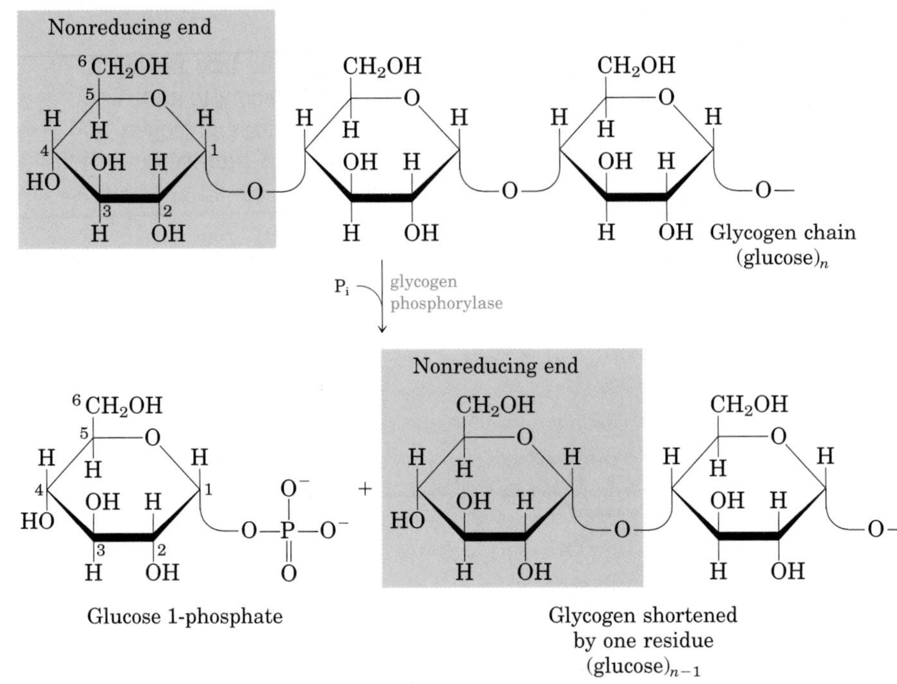

figure 15–12

Removal of a terminal glucose residue from the nonreducing end of a glycogen chain by glycogen phosphorylase. This process is repetitive; the enzyme removes successive glucose residues until it reaches the fourth glucose unit from a branch point (see Fig. 15–13). Amylopectin is degraded in a similar fashion by starch phosphorylase.

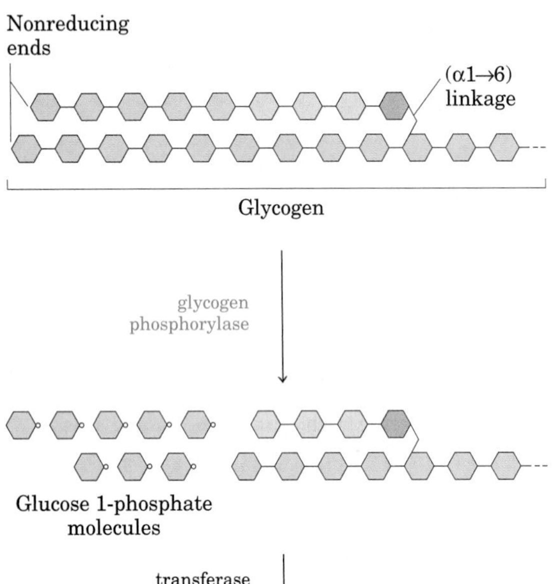

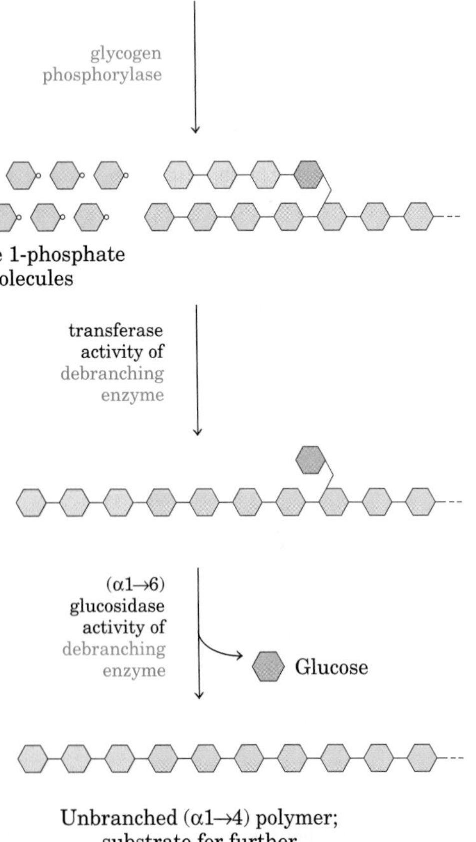

figure 15–13

Glycogen breakdown near an (α1→6) branch point. Following the sequential removal of terminal glucose residues by glycogen phosphorylase (see Fig. 15–12), glucose residues near a branch are removed in a two-step process that requires a bifunctional "debranching enzyme." First, the transferase activity of the enzyme shifts a block of three glucose residues from the branch to a nearby nonreducing end, to which they are reattached in (α1→4) linkage. The single glucose residue remaining at the branch point, in (α1→6) linkage, is then released as free glucose by the debranching enzyme's (α1→6) glucosidase activity. The glucose residues are shown here in shorthand form.

Glycogen phosphorylase (or starch phosphorylase) acts repetitively on the nonreducing ends of glycogen (or amylopectin) branches until it reaches a point four glucose residues away from an (α1→6) branch point (see Fig. 9–15), where its action stops. Further degradation can occur only after the **debranching enzyme,** formerly known as **oligo (α1→6) to (α1→4) glucantransferase,** catalyzes two successive reactions that remove branches (Fig. 15–13).

Glucose 1-phosphate, the end product of the glycogen and starch phosphorylase reactions, is converted to glucose 6-phosphate by **phosphoglucomutase,** which catalyzes the reversible reaction

$$\text{Glucose 1-phosphate} \rightleftharpoons \text{glucose 6-phosphate}$$

Phosphoglucomutase requires as a cofactor **glucose 1,6-bisphosphate,** which has a role analogous to that of 2,3-bisphosphoglycerate in the phosphoglycerate mutase reaction (Fig. 15–6). In phosphoglucomutase, however, the hydroxyl group of a Ser residue (rather than a His residue as in phosphoglycerate mutase) undergoes phosphorylation and dephosphorylation in the catalytic cycle.

Other Monosaccharides Enter the Glycolytic Pathway at Several Points

In most organisms, hexoses other than glucose can undergo glycolysis after conversion to a phosphorylated derivative. D-Fructose, present in free form in many fruits and formed by hydrolysis of sucrose in the small intestine of vertebrates, is phosphorylated by hexokinase:

$$\text{Fructose} + \text{ATP} \xrightarrow{Mg^{2+}} \text{fructose 6-phosphate} + \text{ADP}$$

This is a major pathway of fructose entry into glycolysis in the muscles and kidney. In the liver, however, fructose enters by a different pathway. The liver enzyme **fructokinase** catalyzes the phosphorylation of fructose at C-1 rather than C-6:

$$\text{Fructose} + \text{ATP} \xrightarrow{Mg^{2+}} \text{fructose 1-phosphate} + \text{ADP}$$

The fructose 1-phosphate is then cleaved to glyceraldehyde and dihydroxyacetone phosphate by **fructose 1-phosphate aldolase:**

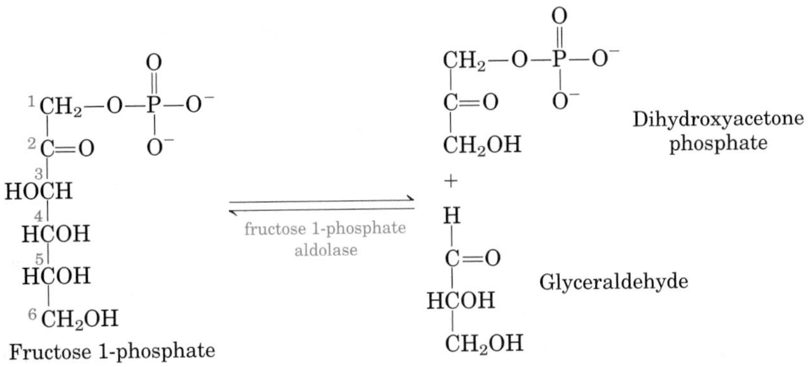

Dihydroxyacetone phosphate is converted to glyceraldehyde 3-phosphate by the glycolytic enzyme triose phosphate isomerase. Glyceraldehyde is phosphorylated by ATP and **triose kinase** to glyceraldehyde 3-phosphate:

$$\text{Glyceraldehyde} + \text{ATP} \xrightarrow{Mg^{2+}} \text{glyceraldehyde 3-phosphate} + \text{ADP}$$

Thus both products of fructose 1-phosphate hydrolysis enter the glycolytic pathway as glyceraldehyde 3-phosphate.

D-Galactose, a product of hydrolysis of the disaccharide lactose (milk sugar), is first phosphorylated at C-1 at the expense of ATP by the enzyme **galactokinase:**

$$\text{Galactose} + \text{ATP} \xrightarrow{Mg^{2+}} \text{galactose 1-phosphate} + \text{ADP}$$

The galactose 1-phosphate is then converted to its epimer at C-4, glucose 1-phosphate, by a set of reactions in which **uridine diphosphate** (UDP) functions as a coenzyme-like carrier of hexose groups (Fig. 15–14).

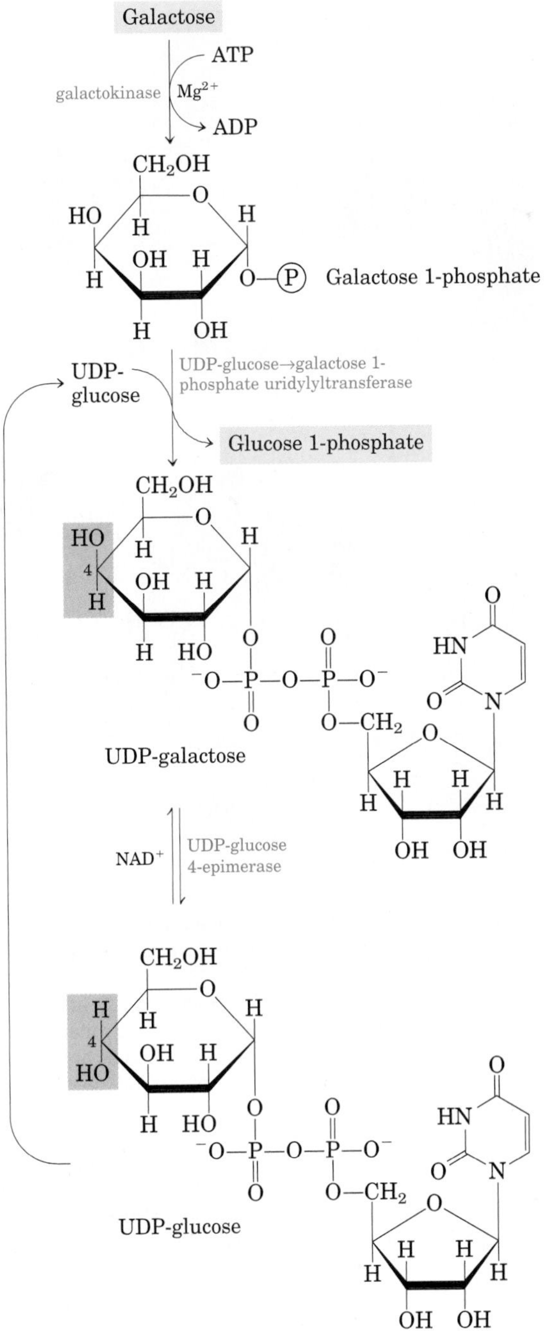

figure 15–14
Conversion of D-galactose 1-phosphate to D-glucose 1-phosphate. The conversion proceeds through a sugar-nucleotide derivative, UDP-galactose, which is formed when galactose 1-phosphate displaces glucose 1-phosphate from UDP-glucose. UDP-galactose is then converted by UDP-glucose 4-epimerase to UDP-glucose. The UDP-glucose is recycled through another round of the same reaction. The net effect of this cycle is the conversion of galactose 1-phosphate to glucose 1-phosphate; there is no net production or consumption of UDP-galactose or UDP-glucose.

Several human genetic diseases result in disordered galactose metabolism. In the most common form of **galactosemia,** the enzyme UDP-glucose $\rightarrow$ galactose 1-phosphate uridylyltransferase is genetically defective, preventing the overall conversion of galactose to glucose. Other forms of galactosemia result when either galactokinase or UDP-glucose 4-epimerase is genetically defective.

D-Mannose, released in the digestion of various polysaccharides and glycoproteins of foods, can be phosphorylated at C-6 by hexokinase:

$$\text{Mannose} + \text{ATP} \xrightarrow{Mg^{2+}} \text{mannose 6-phosphate} + \text{ADP}$$

Mannose 6-phosphate is isomerized by **phosphomannose isomerase** to yield fructose 6-phosphate, an intermediate of glycolysis.

Dietary Polysaccharides and Disaccharides Are Hydrolyzed to Monosaccharides

For most humans, starch is the major source of carbohydrates in the diet. Digestion begins in the mouth, where salivary **α-amylase** hydrolyzes the internal glycosidic linkages of starch, producing short polysaccharide fragments or oligosaccharides. In the stomach, salivary α-amylase is inactivated by the low pH, but a second form of α-amylase, secreted by the pancreas into the small intestine, continues the breakdown process. Pancreatic α-amylase yields mainly maltose (the disaccharide of $\alpha(1\rightarrow4)$ glucose) and oligosaccharides called dextrins, fragments of amylopectin containing $\alpha(1\rightarrow6)$ branch points. Maltose and dextrins are degraded by enzymes of the intestinal brush border (the fingerlike microvilli of intestinal epithelial cells, which greatly increase the area of the intestinal surface). Dietary glycogen has essentially the same structure as starch, and its digestion proceeds by the same pathway.

Disaccharides must be hydrolyzed to monosaccharides before entering cells. Intestinal disaccharides and dextrins are hydrolyzed by enzymes attached to the outer surface of the intestinal epithelial cells:

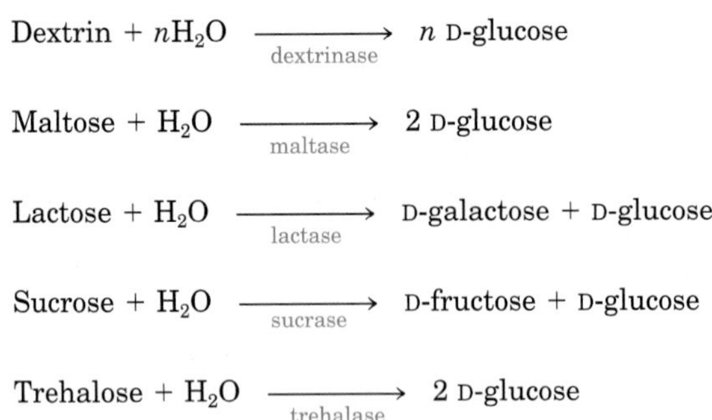

$$\text{Dextrin} + n\text{H}_2\text{O} \xrightarrow[\text{dextrinase}]{} n\ \text{D-glucose}$$

$$\text{Maltose} + \text{H}_2\text{O} \xrightarrow[\text{maltase}]{} 2\ \text{D-glucose}$$

$$\text{Lactose} + \text{H}_2\text{O} \xrightarrow[\text{lactase}]{} \text{D-galactose} + \text{D-glucose}$$

$$\text{Sucrose} + \text{H}_2\text{O} \xrightarrow[\text{sucrase}]{} \text{D-fructose} + \text{D-glucose}$$

$$\text{Trehalose} + \text{H}_2\text{O} \xrightarrow[\text{trehalase}]{} 2\ \text{D-glucose}$$

The monosaccharides so formed are transported into the epithelial cells, then pass into the blood to be carried to various tissues where they are phosphorylated and funneled into the glycolytic sequence.

Lactose intolerance, common among adults of most human populations except those originating in Northern Europe and some parts of Africa, is due to the disappearance after childhood of most or all of the lactase activity of the intestinal cells (Fig. 15–15). Lactose cannot be completely digested and absorbed in the small intestine, and in the large intestine the lactose is converted by bacteria into toxic products that cause abdominal cramps and diarrhea. In most parts of the world where lactose intolerance is prevalent, milk is not used as a food by adults, although milk products

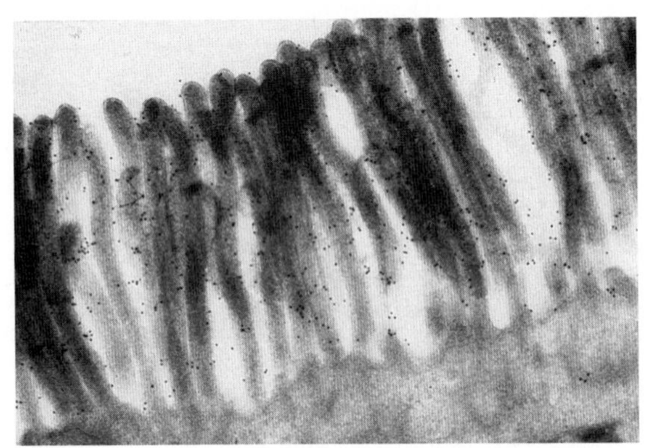

(a)

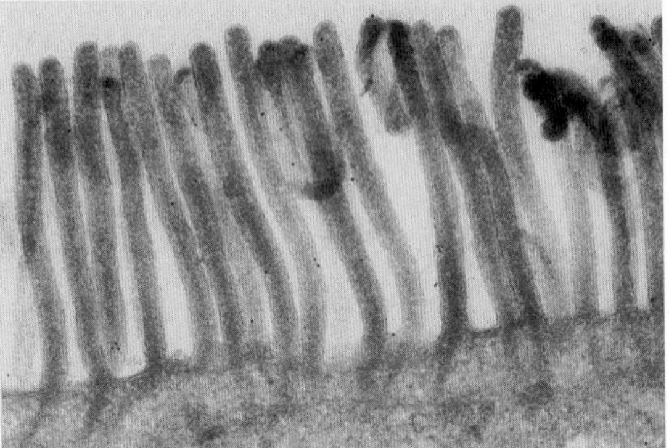

(b)

figure 15–15

Lactase activity of intestinal epithelial cells. Although all children have intestinal lactase, most people cease to produce the enzyme by adulthood and therefore are lactose intolerant. Lactase can be detected by treating a thin section of intestinal tissue with an antibody that specifically binds to the enzyme. The antibodies are made visible in the electron microscope by attaching to them colloidal particles of gold, which appear as black (electron-dense) dots in electron micrographs. **(a)** Lactase-specific antibodies heavily label the intestinal microvilli of a lactose-tolerant adult. **(b)** Intestinal microvilli from a lactose-intolerant adult show little or no labeling with the same antibody.

predigested with lactase are commercially available in some countries. In certain human disorders, several or all of the intestinal disaccharidases are missing. In these cases, the digestive disturbances triggered by dietary disaccharides can sometimes be minimized by a controlled diet.

Regulation of Carbohydrate Catabolism

Carbohydrate catabolism provides both ATP and a variety of precursors for biosynthetic processes. It is crucial for a cell to maintain the concentration of ATP at a nearly constant level—regardless of which fuel is used to produce the ATP and the rate at which ATP is consumed. Similarly, biosynthetic precursors derived from carbohydrate catabolism (for example, fructose 6-phosphate for glucosamine synthesis, or 3-phosphoglycerate for amino acid synthesis) must be provided in adequate amounts. An organism that undergoes a change in circumstances, such as increased muscular activity, decreased availability of oxygen, or decreased dietary intake of carbohydrate, must adjust the rate of flux through its catabolic pathways and possibly the source of fuel as well, perhaps turning to fuel reserves mobilized only in time of need. These changes in catabolic patterns are accomplished by the regulation of key enzymes in the catabolic pathways. In glycolysis in muscle and liver tissue, four enzymes play a regulatory role: glycogen phosphorylase, hexokinase, phosphofructokinase-1, and pyruvate kinase. Our discussion of the regulation of glycolysis necessarily involves some details of the reciprocally regulated process of glucose synthesis (gluconeogenesis), which is more fully discussed in Chapter 20.

Before describing the regulation of glucose catabolism, we consider some general principles that apply to the regulation of all biochemical pathways.

Regulatory Enzymes Act as Metabolic Valves

Although not at equilibrium with their surroundings, adult organisms generally exist in a steady state. By managing a constant influx of energy and nutrients and a constant release of waste products, the organism maintains a constant composition. When the steady state is disturbed by some change in external circumstances or energy supply, the temporarily altered fluxes through individual metabolic pathways trigger regulatory mechanisms intrinsic to each pathway. The net effect of all these adjustments is to return the organism to the steady state—to achieve **homeostasis.** ATP has the central role in cellular activities, and under selective pressure organisms have developed a highly integrated network of catabolic enzymes with regulatory properties that ensure a high steady-state concentration of ATP (high relative to the breakdown products ADP and AMP).

The flux through a biochemical pathway depends on the activities of the enzymes that catalyze each reaction. For some steps in a pathway such as glycolysis, the reaction is essentially at equilibrium within the cell; the activity of the enzyme is sufficiently high that the substrate equilibrates with product as fast as the substrate is supplied. The flux through this step is essentially *substrate-limited*—determined by the instantaneous concentration of the substrate.

Other cellular reactions are far from equilibrium. The equilibrium constant (K'_{eq}) for the phosphofructokinase-1 reaction is about 250, but the mass-action ratio [fructose 1,6-bisphosphate][ADP]/[fructose 6-phosphate][ATP] in a typical cell in the steady state is about 0.04. (The intracellular concen-

table 15–2

Cytosolic Concentrations of Enzymes and Intermediates of the Glycolytic Pathway in Skeletal Muscle

Enzyme	Concentration (μM)	Intermediate	Concentration (μM)
Aldolase	810	Glucose 6-phosphate	3,900
Triose phosphate isomerase	220	Fructose 6-phosphate	1,500
Glyceraldehyde 3-phosphate dehydrogenase	1,400	Fructose 1,6-bisphosphate	80
Phosphoglycerate kinase	130	Dihydroxyacetone phosphate	160
Phosphoglycerate mutase	240	Glyceraldehyde 3-phosphate	80
Enolase	540	1,3-Bisphosphoglycerate	50
Pyruvate kinase	170	3-Phosphoglycerate	200
Lactate dehydrogenase	300	2-Phosphoglycerate	20
Phosphoglucomutase	32	Phosphenolpyruvate	65
		Pyruvate	380
		Lactate	3,700
		ATP	8,000
		ADP	600
		P_i	8,000
		NAD^+	540
		NADH	50

Source: From Srivastava, D.K. & Bernhard, S.A. (1987) Biophysical chemistry of metabolic reaction sequences in concentrated solution and in the cell. *Annu. Rev. Biophys. Biophys. Chem.* **16,** 175–204.

trations of some glycolytic enzymes and intermediates are given in Table 15–2.) The reaction is so far from equilibrium because the rate of conversion of fructose 6-phosphate to fructose 1,6-bisphosphate is limited by the activity of PFK-1. Increased production of fructose 6-phosphate by the preceding enzymes in the glycolytic pathway does not increase the flux through this step, but instead leads to the accumulation of fructose 6-phosphate. Thus PFK-1 functions as a valve, regulating the flow of carbon through glycolysis; increasing the activity of this enzyme (by allosteric activation, for example) increases the overall flux through the pathway. Metabolite flux through this pathway is determined not by mass action (by substrate and product concentrations) but by the extent of "opening" of this enzymatic valve.

Every metabolic pathway has at least one reaction that, in the cell, is far from equilibrium because of the relatively low activity of the enzyme that catalyzes it (Fig. 15–16). The rate of this reaction is not limited by substrate availability but by the activity of the enzyme. The reaction is said to be *enzyme-limited,* and because its rate limits the rate of the entire reaction sequence, the step is the *rate-limiting step* in the pathway. In general, these

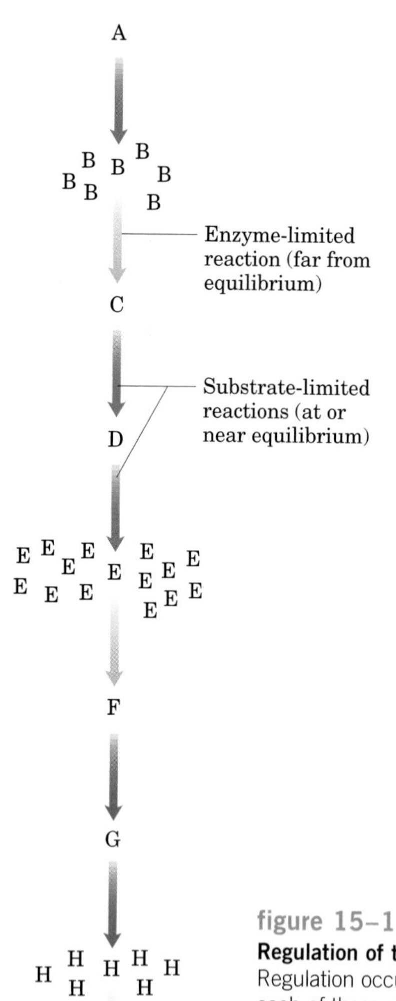

figure 15–16
Regulation of the flux through a multistep pathway. Regulation occurs at steps that are enzyme-limited. At each of these steps (orange arrows), which are generally exergonic, the substrate is not in equilibrium with the product because the reaction is relatively slow; the substrate tends to accumulate, just as river water accumulates behind a dam. In the substrate-limited reactions (blue arrows), the substrate and product are essentially at their equilibrium concentrations. In the steady state, all reactions in the sequence occur at the same rate, which is determined by the rate-limiting step.

rate-limiting steps are very exergonic reactions, essentially irreversible under cellular conditions. *The enzymes that catalyze these exergonic, rate-limiting steps are commonly the targets of metabolic regulation.* Very rapid allosteric enzyme regulation is a primary level of control in cells. In multicellular organisms, the concentrations of allosteric regulators may themselves be under slower hormonal control. Hormone action alters the activities of key enzymes, often within seconds or minutes, allowing the metabolic activities of different tissues and organs to be coordinated (Chapter 23). When external circumstances change over a longer time, as when a human's diet shifts from primarily fat to primarily carbohydrate, flux through specific pathways can be changed by adjusting the relative rates of synthesis and degradation of the regulatory enzymes themselves, the slowest level of control (Chapters 27 and 28).

Many regulatory enzymes are situated at critical branch points in metabolism, their activities determining the allocation of a metabolite to each of the several pathways through which it might pass. For example, glucose 6-phosphate can be metabolized either by glycolysis or by the pentose phosphate pathway (described later in this chapter). The first enzyme unique to each of these pathways (PFK-1 and glucose 6-phosphate dehydrogenase, respectively) catalyzes the "committed" step for its pathway. Both are regulatory enzymes subject to control by a variety of allosteric regulators that signal the need for the products of each pathway.

Cells commonly have the enzymatic capacity to carry out both the catabolism of a complex molecule into a simpler product and the anabolic conversion of that product back into the starting molecule. Glycolysis degrades glucose to pyruvate; gluconeogenesis converts pyruvate to glucose. Paired catabolic and anabolic pathways often employ many of the same enzymes—those that catalyze readily reversible reactions. Phosphoglycerate mutase, for example, acts in both glycolysis and gluconeogenesis. However, paired pathways invariably have at least one reaction in the catabolic direction that differs from the corresponding step in the anabolic direction and is catalyzed by a different enzyme. These distinctive enzymes are the points of regulation of the two opposing pathways. The reactions catalyzed by these path-specific enzymes are generally exergonic reactions, irreversible under cellular conditions and out of equilibrium in the steady state; they are enzyme-limited, not substrate-limited. Having one or more separate enzymes for catabolic and anabolic pathways allows separate regulation of the flux in each direction, avoiding the wasteful "futile cycling" that would result if the breakdown and energy-consuming resynthesis of a compound were allowed to proceed simultaneously. The regulatory enzymes that control the rate of breakdown of carbohydrates via glycolysis illustrate these general principles of metabolic regulation.

Glycolysis and Gluconeogenesis Are Coordinately Regulated

Most organisms can synthesize glucose from simpler precursors such as pyruvate or lactate. In mammals this process, called **gluconeogenesis,** occurs primarily in the liver, and its role is to provide glucose for export to other tissues when other sources of glucose are exhausted. Gluconeogenesis employs most of the same enzymes that act in glycolysis, but it is not simply the reversal of glycolysis. Seven of the glycolytic reactions are freely reversible, and the enzymes that catalyze these reactions also function in gluconeogenesis. Three reactions of glycolysis are so exergonic as to be essentially irreversible: those catalyzed by hexokinase, PFK-1, and pyruvate kinase. Gluconeogenesis uses detours around each of these irreversible steps; for example, the conversion of fructose 1,6-bisphosphate to fructose 6-phosphate is catalyzed by **fructose 1,6-bisphosphatase (FBPase-1)** (Fig. 15–17).

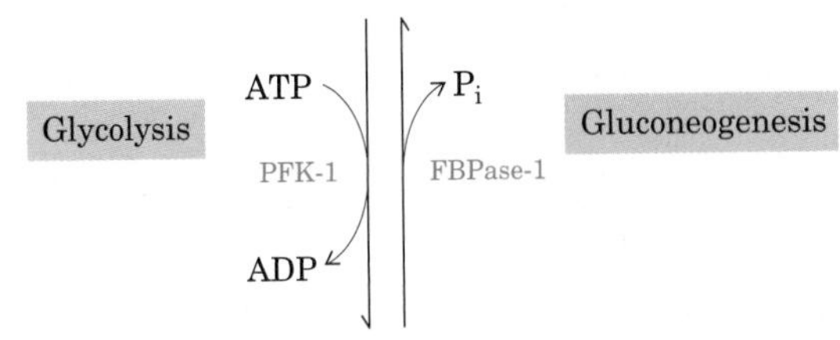

figure 15–17
The reaction of gluconeogenesis that bypasses the irreversible phosphofructokinase-1 reaction of glycolysis. Conversion of fructose 1,6-bisphosphate to fructose 6-phosphate is catalyzed by fructose 1,6-bisphosphatase (called FBPase-1 to distinguish it from a similar enzyme that acts in gluconeogenesis; see Table 20–2).

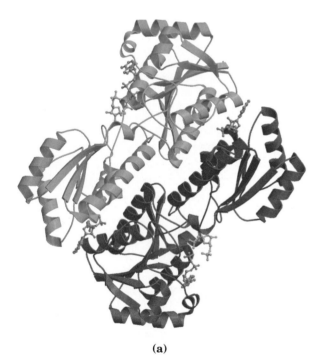

To prevent a wasteful cycle in which glucose is simultaneously degraded by glycolysis and resynthesized by gluconeogenesis, the enzymes unique to each pathway are reciprocally regulated by common allosteric effectors. Fructose 2,6-bisphosphate, a potent activator of liver PFK-1 and therefore of glycolysis, is an inhibitor of FBPase-1 and therefore of gluconeogenesis.

Fructose 2,6-bisphosphate

Glucagon, a hormone released by the pancreas to signal low blood sugar, lowers the level of fructose 2,6-bisphosphate in liver, slowing the consumption of glucose by glycolysis and stimulating the production of glucose by gluconeogenesis; the liver releases the glucose into the blood. We will return to a more complete discussion of this coordinate regulation in Chapter 20.

Fructose 2,6-bisphosphate is found in all animals, in fungi, and in some plants, but not in bacteria. It stimulates PFK-1 activity in all animals and in yeast. In plants, fructose 2,6-bisphosphate regulates carbohydrate metabolism by activating the PP_i-dependent phosphofructokinase that is responsible for fructose 1,6-bisphosphate formation in glycolysis (p. 533), but it does not activate the ATP-dependent PFK-1 of plants. Plant PFK-1 is, however, strongly inhibited by phosphoenolpyruvate, a glycolytic intermediate downstream from fructose 1,6-bisphosphate.

Phosphofructokinase-1 Is under Complex Allosteric Regulation

Glucose 6-phosphate can flow either into glycolysis or through alternative oxidative pathways described later in this chapter. The irreversible reaction catalyzed by PFK-1 is the step that commits a cell to channeling glucose into glycolysis. In addition to its substrate-binding sites, this complex enzyme has several regulatory sites at which allosteric activators or inhibitors bind.

ATP is not only a substrate for PFK-1 but also an end product of the glycolytic pathway. When high ATP levels in the cell signal that ATP is being produced faster than it is being consumed, ATP inhibits PFK-1 by binding to an allosteric site and lowering the affinity of the enzyme for fructose 6-phosphate (Fig. 15–18). ADP and AMP, which increase in concentration as consumption of ATP outpaces production, act allosterically to relieve this

(a)

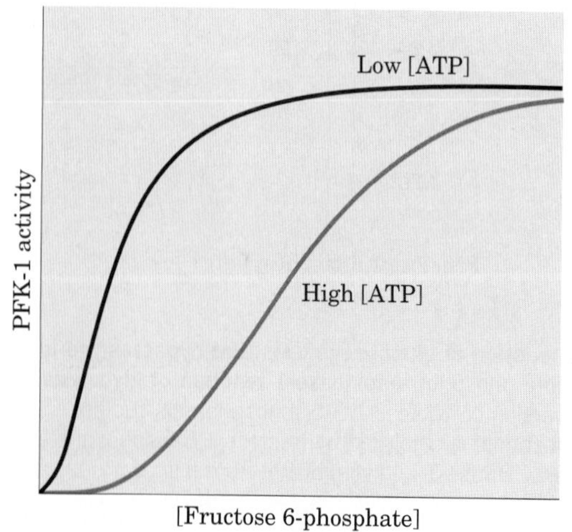

(b)

figure 15–18

Phosphofructokinase-1 and its regulation. (a) A ribbon diagram of *E. coli* PFK-1, showing two of its four identical subunits. Each subunit has its own catalytic site where ADP (blue) and fructose 1,6-bisphosphate (yellow) are almost in contact, and its own binding sites for the allosteric regulator ADP (blue), located at the interface between subunits. **(b)** Allosteric regulation of muscle PFK-1 by ATP, shown by a substrate-activity curve. At low [ATP] the $K_{0.5}$ (p. 281) for fructose 6-phosphate is relatively low, enabling the enzyme to function at a high rate at relatively low concentrations of fructose 6-phosphate. When [ATP] is high, $K_{0.5}$ for fructose 6-phosphate is greatly increased, as indicated by the sigmoid relationship between substrate concentration and enzyme activity. **(c)** A summary of the regulators affecting PFK-1 activity. The symbols here are those used throughout the book: ⊗ for inhibition and ⊕ for activation.

ATP AMP, ADP
Fructose 6-phosphate + ATP → Fructose 1,6-bisphosphate + ADP
citrate fructose 2,6-bisphosphate

(c)

inhibition by ATP. These effects combine to produce higher enzyme activity when fructose 6-phosphate, ADP, or AMP accumulates and to lower activity when ATP accumulates.

Citrate (the ionized form of citric acid), a key intermediate in the aerobic oxidation of pyruvate (Chapter 16), also serves as an allosteric regulator of PFK-1; high citrate concentration increases the inhibitory effect of ATP, further reducing the flow of glucose through glycolysis. In this case, as in several others to be encountered later, citrate serves as an intracellular signal that the cell is meeting its current needs for energy-yielding metabolism and for biosynthetic intermediates.

The most significant allosteric regulator of PFK-1 is fructose 2,6-bisphosphate, which, as noted earlier, strongly activates the enzyme.

Hexokinase Is Allosterically Inhibited by Its Reaction Product

Hexokinase, which catalyzes the entry of free glucose into the glycolytic pathway, is another regulatory enzyme. The hexokinase of myocytes has a high affinity for glucose—it is half-saturated at about 0.1 mM. Because glucose entering myocytes from the blood (in which the glucose concentration is 4 to 5 mM) produces an intracellular glucose concentration high enough to saturate hexokinase, the enzyme normally acts at its maximal rate. Muscle hexokinase is allosterically inhibited by its product, glucose 6-phosphate. Whenever the cellular concentration of glucose 6-phosphate rises above its normal level, hexokinase is temporarily and reversibly inhibited, bringing the rate of glucose 6-phosphate formation into balance with the rate of its utilization and reestablishing the steady state.

Mammals have several forms of hexokinase, all of which catalyze the conversion of glucose to glucose 6-phosphate. Different proteins that catalyze the same reaction are called **isozymes** (Box 15–3). The different isozymes of liver and muscle reflect the different roles of these organs in carbohydrate metabolism: muscle consumes glucose, using it for energy production, whereas liver produces and distributes glucose for other tissues. The predominant hexokinase isozyme of liver is hexokinase D, also called **glucokinase,** which differs in two important respects from the hexokinase isozymes of muscle.

First, the glucose concentration at which glucokinase is half-saturated (about 10 mM) is higher than the usual concentration of glucose in the blood. Because an efficient glucose transporter in hepatocytes maintains a glucose concentration close to that in the blood, this property of glucokinase allows its direct regulation by the level of blood glucose. When the blood glucose concentration is high, as it is after a meal rich in carbohydrates, excess blood glucose is transported into hepatocytes, where glucokinase converts it to glucose 6-phosphate.

Second, glucokinase is inhibited not by its reaction product glucose 6-phosphate but by fructose 6-phosphate, which is always in equilibrium with glucose 6-phosphate through the action of phosphoglucose isomerase. Partial inhibition of glucokinase by fructose 6-phosphate is mediated by an additional protein, the **glucokinase regulatory protein.** This protein also has an affinity for fructose 1-phosphate, which competes with fructose 6-phosphate and cancels its inhibitory effect on glucokinase. Because fructose 1-phosphate is present in liver only when fructose is present in the blood, this property of the glucokinase regulatory protein explains the observation that ingested fructose stimulates the phosphorylation of glucose in the liver.

Cells of the pancreas known as β cells, which are responsible for the release of the hormone insulin when blood glucose levels rise, also contain glucokinase and the glucokinase regulatory protein.

box 15–3 Isozymes: Different Proteins, Same Reaction

The several forms of hexokinase found in mammalian tissues are but one example of a common situation in which the same reaction is catalyzed by two or more different molecular forms of an enzyme. These multiple forms, called isozymes or isoenzymes, may occur in the same species, in the same tissue, or even in the same cell. The different forms of the enzyme generally differ in kinetic or regulatory properties, in the cofactor they use (NADH or NADPH for dehydrogenase isozymes, for example), or in their subcellular distribution (soluble or membrane-bound). Isozymes may have similar, but not identical, amino acid sequences, and in many cases they clearly share a common evolutionary origin.

One of the first enzymes found to have isozymes was lactate dehydrogenase (LDH) (p. 542), which in vertebrate tissues exists as at least five different isozymes separable by electrophoresis. All LDH isozymes contain four polypeptide chains (each of M_r 33,500), each type containing a different ratio of two kinds of polypeptides. The A chain (also designated M for muscle) and the B chain (also designated H for heart) are encoded by two different genes. In skeletal muscle the predominant isozyme contains four A chains, and in heart the predominant isozyme contains four B chains. LDH isozymes in other tissues are a mixture of the five possible forms, which may be designated A_4, A_3B, A_2B_2, AB_3, and B_4. The different LDH isozymes have significantly different values of V_{max} and K_m, particularly for pyruvate. The properties of LDH isozyme A_4 favor rapid reduction of very low concentrations of pyruvate to lactate in skeletal muscle, whereas those of isozyme B_4 favor rapid oxidation of lactate to pyruvate in the heart.

The distribution of different isozymes of a given enzyme reflects at least four factors:

1. *Different metabolic patterns in different organs.* For glycogen phosphorylase, the isozymes in skeletal muscle and liver have different regulatory properties, reflecting the different roles of glycogen breakdown in these two tissues.
2. *Different locations and metabolic roles for isozymes in the same cell.* The isocitrate dehydrogenase isozymes of the cytosol and the mitochondrion are an example (Chapter 16).
3. *Different stages of development in embryonic or fetal tissues and in adult tissues.* For example, the fetal liver has a characteristic isozyme distribution of LDH, which changes as the organ develops into its adult form. Some enzymes of glucose catabolism in malignant (cancer) cells occur as their fetal, not adult, isozymes.
4. *Different responses of isozymes to alosteric modulators.* This difference is useful in fine-tuning metabolic rates. Hexokinase D (glucokinase) of liver and the hexokinase isozymes of other tissues differ in their sensitivity to inhibition by glucose 6-phosphate.

Pyruvate Kinase Is Inhibited by ATP

At least three isozymes of pyruvate kinase are found in vertebrates, differing somewhat in their tissue distribution and their response to modulators. High concentrations of ATP allosterically inhibit pyruvate kinase by decreasing its affinity for the substrate phosphoenolpyruvate (PEP). Under normal conditions, the level of PEP in cells is not high enough to saturate the enzyme, and the reaction rate accordingly is low.

Pyruvate kinase is also inhibited by acetyl-CoA and by long-chain fatty acids, both of which are important fuels for the citric acid cycle, a major pathway in the production of ATP. (Recall that acetyl-CoA (acetate) is produced by the catabolism of fats and amino acids, as well as by glucose catabolism; see Fig. 4, p. 487.) When these other fuels are available, an active citric acid cycle produces plenty of ATP, and concomitant inhibition of pyruvate kinase slows the rate of flux through glycolysis. When the ATP concentration falls, the affinity of pyruvate kinase for PEP increases, enabling the enzyme to catalyze the formation of ATP via substrate-level phosphorylation even though the PEP concentration is relatively low. The result is maintenance of the steady-state concentration of ATP.

Glycogen Phosphorylase Is Regulated Allosterically and Hormonally

In muscle, the end served by glycolysis is ATP production, and the rate of glycolysis increases as muscle works harder, demanding more ATP. The liver has a different role: to maintain a constant level of blood glucose by producing and exporting glucose when the tissues demand it, and importing and storing glucose when provided in excess in the diet. As we have seen, the mobilization of stored glycogen is brought about by glycogen phosphorylase, which degrades glycogen to glucose 1-phosphate (Fig. 15–12). Glycogen phosphorylase provides an especially instructive example of enzyme regulation. It was the first known example of an allosterically regulated enzyme and the first enzyme shown to be controlled by reversible phosphorylation. It was also one of the first allosteric enzymes for which the detailed three-dimensional structures of the active and inactive forms were revealed by x-ray crystallographic studies.

The glycogen phosphorylase isozyme of skeletal muscle exists in two interconvertible forms: **phosphorylase *a*,** which is catalytically active, and **phosphorylase *b*,** which is less active. Phosphorylase *b* predominates in resting muscle, but during vigorous muscular activity the hormone epinephrine triggers phosphorylation of a specific Ser residue in phosphorylase *b*, converting it to phosphorylase *a*. The mechanistic details of phosphorylase activation by phosphorylation are presented on pages 282–284, where glycogen phosphorylase served as an example of this type of control. The activated form of glycogen phosphorylase speeds glycogen breakdown, supplying glucose 1-phosphate for production of the ATP required to power muscle contraction. The enzyme (phosphorylase *b* kinase) responsible for activating phosphorylase by transferring a phosphoryl group to its Ser residue is itself activated by epinephrine through a series of steps shown in Figure 13–15. When the muscle returns to rest, a second enzyme (phosphorylase *a* phosphatase) removes the phosphoryl groups from phosphorylase *a*, converting it to the inactive form phosphorylase *b*.

Superimposed on the regulation of phosphorylase by covalent modification are two allosteric control mechanisms. Ca^{2+}, the signal for muscle contraction, binds to and activates phosphorylase *b* kinase, promoting conversion of inactive phosphorylase *b* to the active *a* form. AMP, which accumulates in vigorously contracting muscle, binds to and activates phosphorylase. When ATP levels are adequate, ATP blocks the allosteric site to which AMP binds, inactivating phosphorylase.

In the liver, glycogen phosphorylase is regulated by the hormone glucagon and by allosteric mechanisms. When the blood glucose level is too low, glucagon activates phosphorylase *b* kinase, which converts inactive phosphorylase *b* to its active *a* form, initiating the release of glucose into the blood. When blood glucose levels return to normal, glucose enters hepatocytes and binds to an allosteric site on phosphorylase *a* (Fig. 15–19). This produces a conformational change that exposes the phosphorylated Ser residues to phosphorylase *a* phosphatase, which removes the phosphoryl groups and inactivates phosphorylase. The details of these complex regulatory processes are discussed in Chapter 23, where we compare the regulation of glycogen phosphorylase with that of its counterpart in glycogen synthesis, glycogen synthase. These two enzymes are under finely tuned and reciprocal regulation.

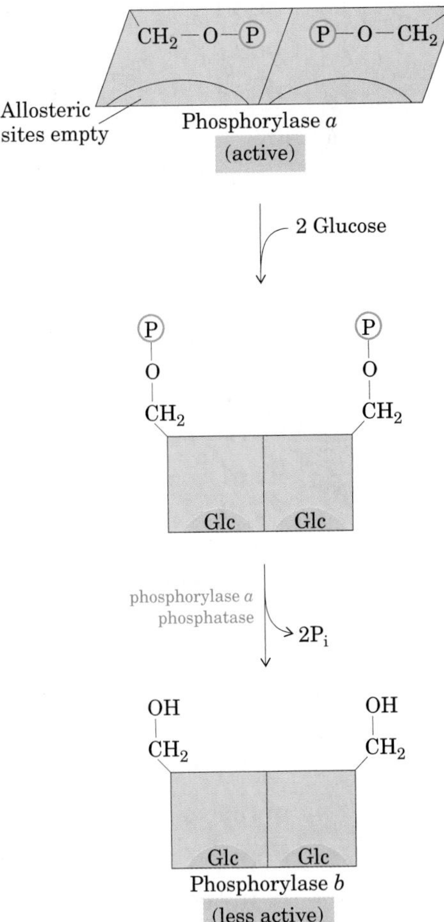

figure 15–19
Glycogen phosphorylase of liver as a glucose sensor. Glucose binding to an allosteric site of phosphorylase *a* induces a conformational change that exposes the phosphorylated Ser residues to the action of phosphorylase *a* phosphatase. This phosphatase converts phosphorylase *a* to phosphorylase *b*, reducing the activity of phosphorylase in response to high blood glucose.

The Pentose Phosphate Pathway of Glucose Oxidation

In most animal tissues, the major fate of glucose is glycolytic breakdown to pyruvate, most of which is oxidized via the citric acid cycle—the main function being formation of ATP. Glucose does have other catabolic fates, however, which lead to specialized products needed by the cell. Of particular importance is the synthesis of pentose phosphates.

The **pentose phosphate pathway,** also called the **phosphogluconate pathway,** produces NADPH and ribose 5-phosphate. Recall from Chapter 14 (p. 519) that NADPH carries chemical energy in the form of reducing power and is used almost universally as the reductant in anabolic pathways. In mammals this role of NADPH—and thus the activity of the pentose phosphate pathway—is especially prominent in tissues actively synthesizing fatty acids and steroids, such as the mammary gland, adrenal cortex, liver, and adipose tissue. These tissues use NADPH to reduce the double bonds and carbonyl groups of intermediates in the synthetic process (Chapter 21). Tissues less active in synthesizing fatty acids, such as skeletal muscle, are virtually lacking in the pentose phosphate pathway.

A second function of the pentose phosphate pathway is to generate pentoses, particularly D-ribose, necessary for the biosynthesis of nucleic acids (Chapter 22). Nucleic acid biosynthesis occurs at high rates in growing and regenerating tissues and in tumors.

The first reaction of the pentose phosphate pathway (Fig. 15–20) is the dehydrogenation of glucose 6-phosphate by **glucose 6-phosphate dehydrogenase** to form 6-phosphoglucono-δ-lactone, an intramolecular ester. $NADP^+$ is the electron acceptor, and the overall equilibrium lies far in the direction of NADPH formation. The lactone is hydrolyzed to the free acid 6-phosphogluconate by a specific **lactonase,** then 6-phosphogluconate undergoes dehydrogenation and decarboxylation by **6-phosphogluconate dehydrogenase** to form the ketopentose D-ribulose 5-phosphate. This reaction generates a second molecule of NADPH. **Phosphopentose isomerase** converts ribulose 5-phosphate to its aldose isomer, D-ribose 5-phosphate. In some tissues, the pentose phosphate pathway ends at this point, and its overall equation is

$$\text{Glucose 6-phosphate} + 2NADP^+ + H_2O \longrightarrow \text{ribose 5-phosphate} + CO_2 + 2NADPH + 2H^+$$

The net result is the production of NADPH, a reductant for biosynthetic reactions, and ribose 5-phosphate, a precursor for nucleotide synthesis.

In tissues that require primarily NADPH rather than ribose 5-phosphate, pentose phosphates are recycled into glucose 6-phosphate in a series of reactions (Fig. 15–21) to be examined in more detail in Chapter 20. First, ribulose 5-phosphate is epimerized to xylulose 5-phosphate. Then, in a series of rearrangements of the carbon skeletons, six five-carbon sugar phosphates are converted to five six-carbon sugar phosphates, completing the cycle and allowing continued oxidation of glucose 6-phosphate with production of NADPH. In erythrocytes, the NADPH produced by this pathway is essential in protecting the cells from oxidative damage. A genetic defect in glucose 6-phosphate dehydrogenase can have serious consequences (Box 15–4).

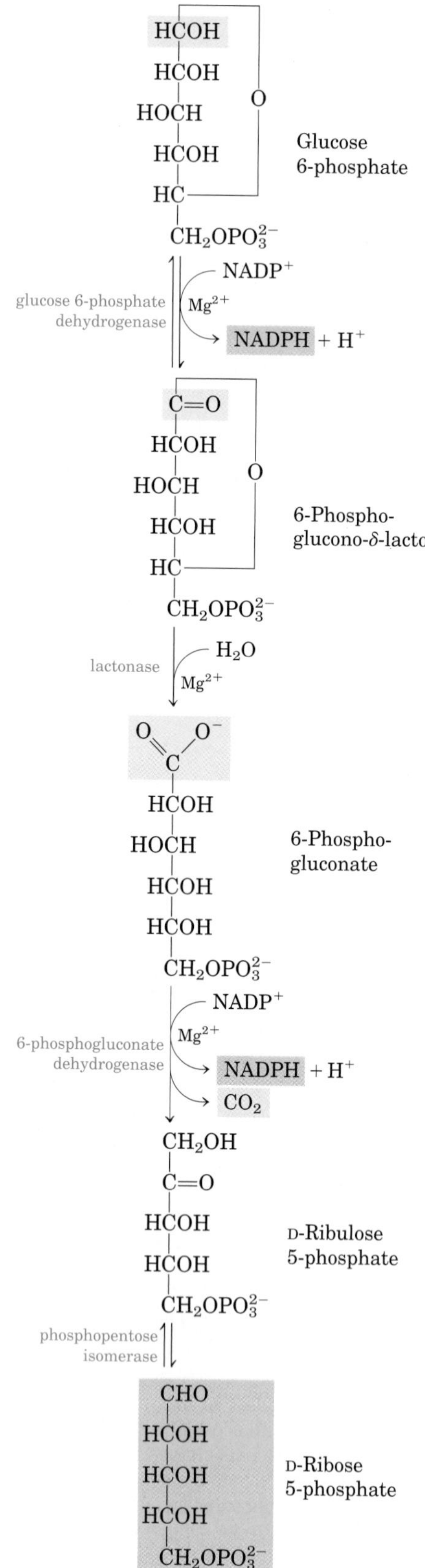

figure 15–20
Oxidative reactions of the pentose phosphate pathway. The end products are D-ribose 5-phosphate and NADPH.

In the nonoxidative part of the pentose phosphate pathway (Fig. 15–21a), transketolase, a TPP-dependent enzyme, catalyzes the transfer of a two-carbon fragment (C-1 and C-2) of xylulose 5-phosphate to ribose 5-phosphate, forming the seven-carbon product sedoheptulose 7-phosphate. The remaining three-carbon fragment of xylulose is glyceraldehyde 3-phosphate (see Fig. 20–27.) Transaldolase then catalyzes a reaction similar to the aldolase reaction of glycolysis: a three-carbon fragment is removed from sedoheptulose 7-phosphate and condensed with glyceraldehyde 3-phosphate, forming fructose 6-phosphate. The remaining four-carbon fragment of sedoheptulose is erythrose 4-phosphate. Now transketolase acts again, forming fructose 6-phosphate and glyceraldehyde 3-phosphate from erythrose 4-phosphate and xylulose 5-phosphate. Two molecules of glyceraldehyde 3-phosphate formed by two iterations of these reactions can be converted into a molecule of fructose 1,6-bisphosphate (Fig. 15–21b). The cycle is then complete: six pentose phosphates have been converted to five hexose phosphates.

All reactions of the nonoxidative part of the pentose phosphate pathway are readily reversible and thus also provide a means of converting hexose phosphates to pentose phosphates. As we shall see in Chapter 20, this is essential in the photosynthetic fixation of CO_2 by plants.

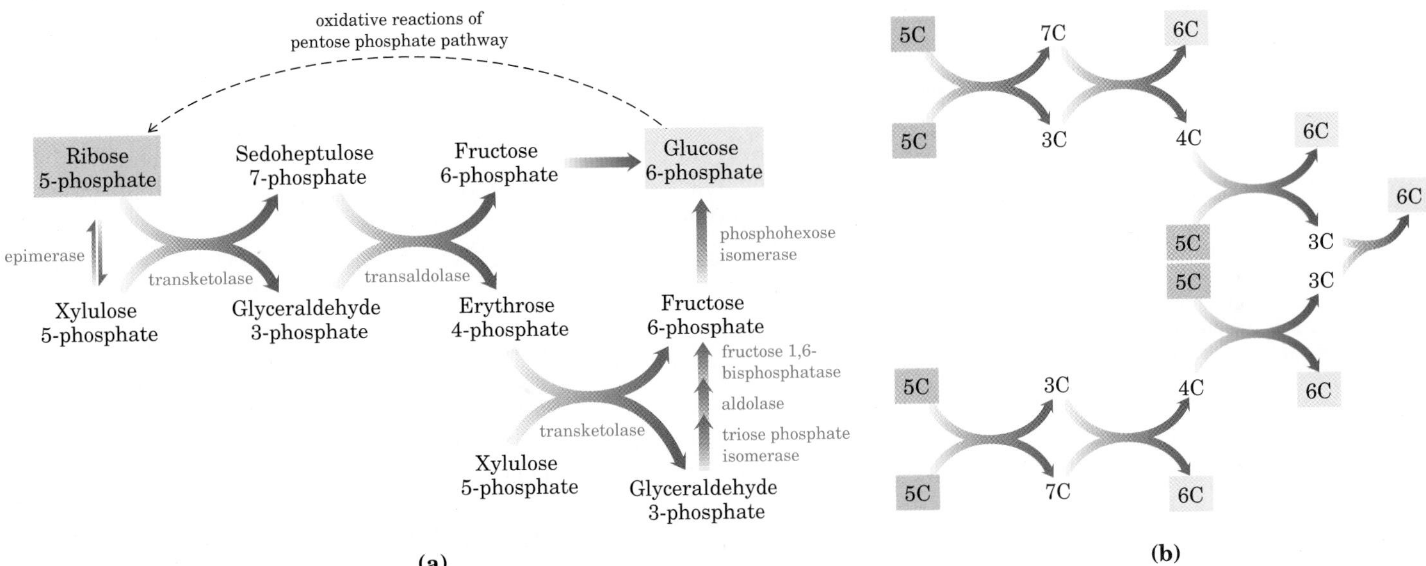

figure 15–21
Nonoxidative reactions of the pentose phosphate pathway. These reactions **(a)** convert pentose phosphates back to hexose phosphates, allowing the oxidative reactions (see Fig. 15–20) to continue. The enzymes transketolase and transaldolase (discussed in more detail in Chapter 20) are specific to this pathway; the other enzymes also serve in the glycolytic or gluconeogenic pathways. **(b)** A schematic diagram showing the pathway from six pentoses (5C) to five hexoses (6C). Note that this involves two sets of the interconversions shown in **(a).** Every reaction shown here is reversible; unidirectional arrows are used only to make clear the direction of the reactions during continuous oxidation of glucose 6-phosphate. In the light-independent reactions of photosynthesis, the direction of these reactions is reversed (see Fig. 20–26).

box 15–4 Glucose 6-Phosphate Dehydrogenase Deficiency: Why Pythagoras Wouldn't Eat Falafel

Fava beans, a traditional ingredient of falafel, have been an important food source in the Mediterranean and Middle East since antiquity. The Greek philosopher and mathematicia Pythagoras prohibited his followers from dining on fava beans perhaps because they make many people sick with a condition called favism, which can be fatal. In favism, erythrocytes begin to lyse 24 to 48 hours after ingestion of the beans, releasing free hemoglobin into the blood. Jaundice and sometimes kidney failure can result. Similar symptoms can occur with ingestion of the antimalarial drug primaquine or sulfa antibiotics or following exposure to certain herbicides. These symptoms have a genetic basis: glucose 6-phosphate dehydrogenase (G6PD) deficiency, which affects about 400 million people. Most G6PD-deficient individuals are asymptomatic; only in combination with certain environmental factors do the clinical manifestations occur.

G6PD catalyzes the first step in the pentose phosphate pathway (see Fig. 15–20), which produces NADPH. This reductant, essential in many biosynthetic pathways, also protects cells from oxidative damage by hydrogen peroxide (H_2O_2) and superoxide free radicals, highly reactive oxidants generated as metabolic byproducts and through the actions of drugs such as primaquine and natural products such as divicine—the toxic ingredient of fava beans. During normal detoxification, H_2O_2 is converted to H_2O by reduced glutathione and glutathione peroxidase, and the oxidized glutathione is converted back to the reduced form by glutathione reductase and NADPH (Fig. 1). H_2O_2 is also broken down to H_2O and O_2 by catalase, which also requires NADPH. In G6PD-deficient individuals, the NADPH production is diminished and detoxification of H_2O_2 is inhibited. Cellular damage results: lipid peroxidation leading to erythrocyte membrane breakdown, and protein and DNA oxidation.

The geographical distribution of G6PD deficiency is instructive. Frequencies as high as 25% occur in tropical Africa, parts of the Middle East, and Southeast Asia, areas where malaria is most prevalent. In addition to such epidemiological observations, in vitro studies show that growth of one malaria parasite, *Plasmodium falciparum*, is inhibited in G6PD-deficient erythrocytes. The parasite is very sensitive to oxidative damage and is killed by a level of oxidative stress that is tolerable to a G6PD-deficient human host. Because the advantage of resistance to malaria balances the disadvantage of lowered resistance to oxidative damage, natural selection sustains the G6PD-deficient genotype in human populations where malaria is prevalent. Only under overwhelming oxidative stress, caused by drugs, herbicides, or divicine, does G6PD deficiency cause serious medical problems.

An antimalarial drug such as primaquine is believed to act by causing oxidative stress to the parasite. It is ironic that antimalarial drugs can cause illness through the same biochemical mechanism that provides resistance to malaria. Divicine also acts as an antimalarial drug, and ingestion of fava beans may protect against malaria. By refusing to eat falafel, many Pythagoreans with normal G6PD activity may have unwittingly increased their risk of malaria!

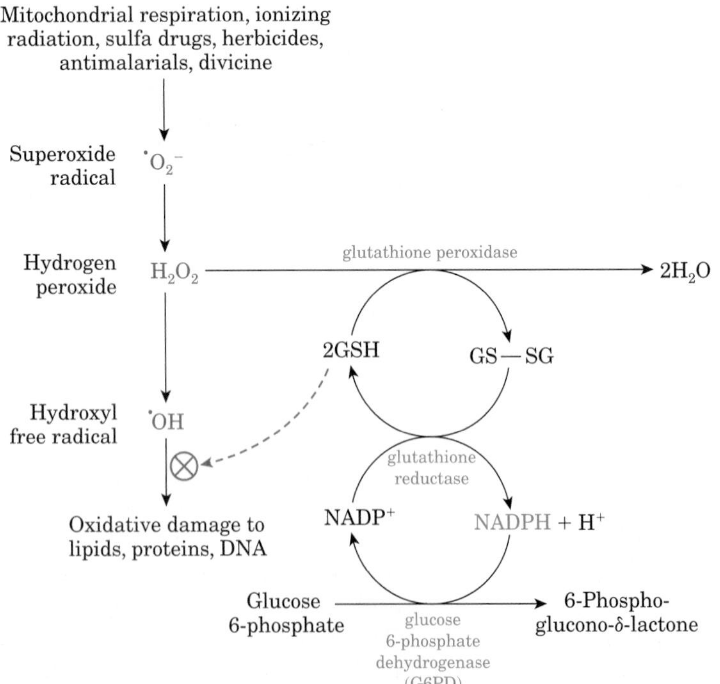

figure 1
Role of NADPH and glutathione in protecting cells against highly reactive oxygen derivatives. Reduced glutathione (GSH) protects the cell by destroying hydrogen peroxide and hydroxyl free radicals. Regeneration of GSH from its oxidized form (GS-SG) requires the NADPH produced in the glucose 6-phosphate dehydrogenase reaction.

summary

Glycolysis is a universal metabolic pathway for the catabolism of glucose to pyruvate accompanied by the formation of ATP. The process is catalyzed by ten cytosolic enzymes, and all intermediates after glucose are phosphorylated compounds. In the preparatory phase of glycolysis, ATP is invested to convert glucose to fructose 1,6-bisphosphate, then the carbon–carbon bond between C-3 and C-4 is broken to yield two molecules of triose phosphate. In the payoff phase, each of the two molecules of glyceraldehyde 3-phosphate derived from glucose undergoes oxidation at C-1; the energy of this oxidation reaction is conserved in the formation of NADH and an acyl phosphate bond of 1,3-bisphosphoglycerate. This intermediate has a high phosphoryl group transfer potential, and in a substrate-level phosphorylation the phosphoryl group is transferred to ADP, forming ATP and 3-phosphoglycerate. Rearrangement of the atoms in 3-phosphoglycerate with the loss of H_2O gives rise to phosphoenolpyruvate, another compound with a high phosphoryl group transfer potential. A second substrate-level phosphorylation follows. The pyruvate formed in this reaction is the end product of the payoff phase. The overall equation for glycolysis is

$$\text{Glucose} + 2\text{NAD}^+ + 2\text{ADP} + 2\text{P}_i \longrightarrow 2\ \text{pyruvate} + 2\text{NADH} + 2\text{H}^+ + 2\text{ATP} + 2\text{H}_2\text{O}$$

There is a net gain of two ATP per molecule of glucose.

The NADH formed in glycolysis must be recycled to regenerate NAD^+, which is required as electron acceptor in the first step of the payoff phase. Under aerobic conditions, electrons pass from NADH to O_2 in mitochondrial respiration. Under anaerobic or hypoxic conditions, many organisms regenerate NAD^+ by transferring electrons from NADH to pyruvate, forming lactate. Other organisms, such as yeast, regenerate NAD^+ by reducing pyruvate to ethanol and CO_2. In these anaerobic processes, called fermentations, no *net* oxidation or reduction of the carbons of glucose occurs.

Glycogen and starch, polymeric storage forms of glucose, enter glycolysis in a two-step process that begins with phosphorolytic cleavage of a glucose residue from an end of the polymer, forming glucose 1-phosphate. This is catalyzed by glycogen or starch phosphorylase. Phosphoglucomutase then converts the glucose 1-phosphate to glucose 6-phosphate (the first intermediate in glycolysis). Ingested polysaccharides and disaccharides are converted to monosaccharides by intestinal hydrolytic enzymes then taken up by intestinal cells and transported to the liver or other tissues.

A variety of D-hexoses, including fructose, mannose, and galactose, can be funneled into glycolysis. Each is first phosphorylated, then converted to either glucose 6-phosphate or fructose 6-phosphate (the second intermediate in glycolysis). Conversion of galactose 1-phosphate to glucose 1-phosphate involves two nucleotide derivatives: UDP-galactose and UDP-glucose. Genetic defects in the enzymes that catalyze conversion of galactose to glucose 1-phosphate result in galactosemia, a serious human disease.

In a metabolically active cell in a steady state, intermediates are formed and consumed at equal rates. Of paramount importance to the cell is the maintenance of an adequate steady-state concentration of ATP. When a perturbation alters the rate of formation or consumption of ATP (or other metabolite), compensating changes in enzyme activities bring the system back to the steady state. These changes in enzyme activity are achieved by allosteric regulation or covalent modification (such as phosphorylation), often triggered by hormonal signals.

In multistep processes such as glycolysis, certain of the reactions are essentially at equilibrium in the steady state; the rates of these substrate-limited reactions rise and fall with substrate concentration. Other reactions are out of equilibrium; their rates are too slow to produce instant equilibration of substrate and product. These enzyme-limited reactions are often highly exergonic and therefore irreversible, and the enzymes that catalyze them are commonly the points at which flux through the pathway is regulated. In the glycolytic pathway from glycogen to pyruvate, the regulated enzymes include glycogen phosphorylase, hexokinase, phosphofructokinase-1, and pyruvate kinase.

Gluconeogenesis is a multistep process in which pyruvate is converted to glucose. Some of the steps in gluconeogenesis are catalyzed by the same enzymes used in glycolysis; these are the substrate-limited (near-equilibrium) reactions in both processes. In gluconeogenesis, the essentially irreversible steps in glycolysis are bypassed by reactions catalyzed by different enzymes. To prevent futile cycling of glucose, the enzyme-limited reactions of glycolysis and gluconeogenesis are under reciprocal allosteric control: when the glycolytic reactions are stimulated, the gluconeogenic reactions are inhibited, and vice versa. Fructose 2,6-bisphosphate is an allosteric activator of PFK-1 (glycolysis) and an allosteric inhibitor of fructose 1,6-bisphosphatase (gluconeogenesis). Glucagon triggers a series of enzymatic changes that reduce the concentration of fructose 2,6-bisphosphate, with a consequent slowing of glycolysis and stimulation of gluconeogenesis.

Hexokinase is inhibited by high concentrations of glucose 6-phosphate; thus, when the product of the hexokinase reaction accumulates, its rate of production is lowered. Pyruvate kinase is likewise allosterically inhibited by one of its products, ATP.

Glycogen phosphorylase, which mobilizes glucose from glycogen stores, is activated by phosphorylation catalyzed by phosphorylase kinase, which is itself activated by events triggered by epinephrine. Removal of the phosphoryl groups by a specific phosphatase inactivates the enzyme. Glycogen phosphorylase isozymes in muscle and liver differ in their mode of regulation, reflecting the different roles of muscle and liver in glucose metabolism. The activity of liver glycogen phosphorylase is further regulated allosterically by glucose, allowing the liver to buffer changes in blood glucose concentration.

The pentose phosphate pathway is an alternative oxidative route for glucose. It results in oxidation and decarboxylation at the C-1 position of glucose, producing NADPH and pentose phosphates. NADPH provides reducing power for biosynthetic reactions, and pentose phosphates are essential precursors for nucleotide and nucleic acid synthesis.

further reading

General

Dennis, D.T. (1987) *The Biochemistry of Energy Utilization in Plants,* Chapman & Hall, New York.

A short presentation of energy-conserving reactions in higher plants, written at about the level of this textbook. It includes a very good chapter (Chapter 6) on glycolysis and the pentose phosphate pathway.

Fell, D. (1997) *Understanding the Control of Metabolism,* Portland Press, Miami.

A very appealing introduction to the basic theories of metabolism and metabolic regulation. Includes controversial topics not covered in this chapter.

Fruton, J.S. (1972) *Molecules and Life: Historical Essays on the Interplay of Chemistry and Biology,* Wiley-Interscience, New York.

This text includes a detailed historical account of research on glycolysis.

Suarez, R.K., Staples, J.F., Lighton, J.R., & West, T.G. (1997) Relationships between enzymatic flux capacities and metabolic flux rates: nonequilibrium reactions in muscle glycolysis. *Proc. Natl. Acad. Sci. USA* **94,** 7065–7069.

Glycolysis

Phillips, D., Blake, C.C.F., & Watson, H.C. (eds) (1981) The Enzymes of Glycolysis: Structure, Activity and Evolution. *Philos. Trans. R. Soc. Lond. [Biol.]* **293,** 1–214.

A collection of excellent reviews on the enzymes of glycolysis, written at a level challenging but comprehensible to a beginning student of biochemistry.

Physiological Significance of Metabolite Channeling. (1991) *J. Theor. Biol.* **152,** 1–140.

A special issue containing 29 short papers on all aspects of metabolite channeling.

Plaxton, W.C. (1996) The organization and regulation of plant glycolysis. *Annu. Rev. Plant Physiol. Plant Mol. Biol.* **47,** 185–214.

Very helpful review of the subcellular localization of glycolytic enzymes and the regulation of glycolysis in plants.

Srivastava, D.K. & Bernhard, S.A. (1987) Biophysical chemistry of metabolic reaction sequences in concentrated enzyme solution and in the cell. *Annu. Rev. Biophys. Biophys. Chem.* **16,** 175–204.

Evidence for protein-protein interactions and substrate channeling in concentrated protein solutions, with examples from the glycolytic pathway.

Van Beers, E.H., Buller, H.A., Grand, R.J., Einerhand, A.W.C., & Dekker, J. (1995) Intestinal brush border glycohydrolases: structure, function, and development. *Crit. Rev. Biochem. Mol. Biol.* **30,** 197–262.

Regulation of Carbohydrate Metabolism

Barford, D., Hu, S.-H., & Johnson, L.N. (1991) Structural mechanism for glycogen phosphorylase control by phosphorylation and AMP. *J. Mol. Biol.* **218,** 233–260.

Clear discussion of the regulatory changes in the structure of glycogen phosphorylase, based on the structures (from x-ray diffraction studies) of the active and inactive forms of the enzyme.

Dang, C.V. & Semenza, G.L. (1999) Oncogenic alterations of metabolism. *Trends Biochem. Sci.* **24,** 68–72.

Depre, C., Rider, M.H., & Hue, L. (1998) Mechanisms of control of heart glycolysis. *Eur. J. Biochem.* **258,** 277–290.

Heinrich, R., Melendez-Hevia, E., Montero, F., Nuno, J.C., Stephani, A., & Waddell, T.G. (1999) The structural design of glycolysis: an evolutionary approach. *Biochem. Soc. Trans.* **27,** 261–264.

Hudson, J.W., Golding, G.B., & Crerar, M.M. (1993) Evolution of allosteric control in glycogen phosphorylase. *J. Mol. Biol.* **234,** 700–721.

Hue, L. & Rider, M.H. (1987) Role of fructose 2,6-bisphosphate in the control of glycolysis in mammalian tissues. *Biochem. J.* **245,** 313–324.

Mathupala, S.P., Rempel, A., & Pedersen, P.L. (1997) Aberrant glycolytic metabolism of cancer cells: a remarkable coordination of genetic, transcriptional, post-translational, and mutational events that leads to a critical role for type II hexokinase. *J. Bioenerg. Biomembr.* **29,** 339–343.

Ochs, R.S., Hanson, R.W., & Hall, J. (eds) (1985) *Metabolic Regulation,* Elsevier Science Publishing Co., Inc., New York.

A collection of short essays first published in *Trends in Biochemical Sciences,* better known as *TIBS.*

Pilkis, S.J. & Granner, D.K. (1992) Molecular physiology of the regulation of hepatic gluconeogenesis and glycolysis. *Annu. Rev. Physiol.* **54,** 885–909.

Schirmer, T. & Evans, P.R. (1990) Structural basis of the allosteric behavior of phosphofructokinase. *Nature* **343,** 140–145.

Alternative Pathways of Glucose Oxidation

Chayen, J., Howat, D.W., & Bitensky, L. (1986) Cellular biochemistry of glucose 6-phosphate and 6-phosphogluconate dehydrogenase activities. *Cell Biochem. Funct.* **4,** 249–253.

Luzzato, L. & Mehta, A. (1995) Glucose 6-phosphate dehydrogenase deficiency. In *The Metabolic and Molecular Bases of Inherited Disease, 7th edn* (Scriver, C.R., Beaudet, A.L., Sly, W.S., & Valle, D., eds), pp. 3367–3398, McGraw-Hill Book Company, New York.

This three-volume treatise is filled with fascinating information about the clinical and biochemical features of hundreds of inherited diseases of metabolism.

Tephyl, T.R. & Burchell, B. (1990) UDP-glucuronosyl transferases: a family of detoxifying enzymes. *Trends Pharmacol. Sci.* **11,** 276–279.

Wood, T. (1986) Physiological functions of the pentose phosphate pathway. *Cell Biochem. Funct.* **4,** 241–247.

problems

1. Equation for the Preparatory Phase of Glycolysis Write balanced biochemical equations for all the reactions in the catabolism of D-glucose to two molecules of D-glyceraldehyde 3-phosphate (the preparatory phase of glycolysis), including the standard free-energy change for each reaction. Then write the overall or net equation for the preparatory phase of glycolysis, with the net standard free-energy change.

2. The Payoff Phase of Glycolysis in Skeletal Muscle In working skeletal muscle under anaerobic conditions, glyceraldehyde 3-phosphate is converted to pyruvate (the payoff phase of glycolysis), and the pyruvate is reduced to lactate. Write balanced biochemical equations for all the reactions in this process, with the standard free-energy change for each reaction. Then write the overall or net equation for the payoff phase of glycolysis (with lactate as the end product), including the net standard free-energy change.

3. Pathway of Atoms in Fermentation A "pulse-chase" experiment using ^{14}C-labeled carbon sources is carried out on a yeast extract maintained under strictly anaerobic conditions to produce ethanol. The experiment consists of incubating a small amount of ^{14}C-labeled substrate (the pulse) with the yeast extract just long enough for each intermediate in the fermentation pathway to become labeled. The label is then "chased" through the pathway by the addition of excess unlabeled glucose. The chase effectively prevents any further entry of labeled glucose into the pathway.

(a) If [1-^{14}C]glucose (glucose labeled at C-1 with ^{14}C) is used as a substrate, what is the location of ^{14}C in the product ethanol? Explain.

(b) Where would ^{14}C have to be located in the starting glucose to ensure that all the ^{14}C activity is liberated as $^{14}CO_2$ during fermentation to ethanol? Explain.

4. Fermentation to Produce Soy Sauce Soy sauce is prepared by fermenting a salted mixture of soybeans and wheat with several microorganisms, including yeast, over a period of 8 to 12 months. The resulting sauce (after solids are removed) is rich in lactate and ethanol. How are these two compounds produced? To prevent the soy sauce from having a strong vinegar taste (vinegar is dilute acetic acid), oxygen must be kept out of the fermentation tank. Why?

5. Equivalence of Triose Phosphates ^{14}C-Labeled glyceraldehyde 3-phosphate was added to a yeast extract. After a short time, fructose 1,6-bisphosphate labeled with ^{14}C at C-3 and C-4 was isolated. What was the location of the ^{14}C label in the starting glyceraldehyde 3-phosphate? Where did the second ^{14}C label in fructose 1,6-bisphosphate come from? Explain.

6. Glycolysis Shortcut Suppose you discovered a mutant yeast whose glycolytic pathway was shorter because of the presence of a new enzyme catalyzing the reaction:

$$\text{Glyceraldehyde 3-phosphate} + H_2 \xrightarrow[\text{3-phosphoglycerate}]{NAD^+ \curvearrowright NADH + H^+}$$

Would shortening the glycolytic pathway in this way benefit the cell? Explain.

7. Role of Lactate Dehydrogenase During strenuous activity, the demand for ATP in muscle tissue is vastly increased. In rabbit leg muscle or turkey flight muscle, the ATP is produced almost exclusively by lactic acid fermentation. ATP is formed in the payoff phase of glycolysis by two reactions, promoted by phosphoglycerate kinase and pyruvate kinase. Suppose skeletal muscle were devoid of lactate dehydrogenase. Could it carry out strenuous physical activity; that is, could it generate ATP at a high rate by glycolysis? Explain.

8. Efficiency of ATP Production in Muscle The transformation of glucose to lactate in myocytes releases only about 7% of the free energy released when glucose is completely oxidized to CO_2 and H_2O. Does this mean that anaerobic glycolysis in muscle is a wasteful use of glucose? Explain.

9. Free-Energy Change for Triose Phosphate Oxidation The oxidation of glyceraldehyde 3-phosphate to 1,3-bisphosphoglycerate, catalyzed by glyceraldehyde 3-phosphate dehydrogenase, proceeds with an unfavorable equilibrium constant ($K'_{eq} = 0.08$; $\Delta G'^{\circ} = 6.3$ kJ/mol), yet the flow through this point in the glycolytic pathway proceeds smoothly. How does the cell overcome the unfavorable equilibrium?

10. Are All Metabolic Reactions at Equilibrium?

(a) Phosphoenolpyruvate (PEP) is one of the two phosphoryl group donors in the synthesis of ATP during glycolysis. In human erythrocytes, the steady-state concentration of ATP is 2.24 mM, that of ADP is 0.25 mM, and that of pyruvate is 0.051 mM. Calculate the concentration of PEP at 25 °C, assuming that the pyruvate kinase reaction (see Fig. 14–3) is at equilibrium in the cell.

(b) The physiological concentration of PEP in human erythrocytes is 0.023 mM. Compare this with the value obtained in (a). Explain the significance of this difference.

11. Arsenate Poisoning Arsenate is structurally and chemically similar to inorganic phosphate (P_i), and many enzymes that require phosphate will also use arsenate. Organic compounds of arsenate are less stable than analogous phosphate compounds, however. For example, acyl *arsenates* decompose rapidly by hydrolysis:

$$R{-}\overset{\overset{\displaystyle O}{\|}}{C}{-}O{-}\underset{\underset{\displaystyle O^-}{|}}{\overset{\overset{\displaystyle O}{\|}}{As}}{-}O^- + H_2O \longrightarrow R{-}\overset{\overset{\displaystyle O}{\|}}{C}{-}O^- + HO{-}\underset{\underset{\displaystyle O^-}{|}}{\overset{\overset{\displaystyle O}{\|}}{As}}{-}O^- + H^+$$

On the other hand, acyl *phosphates,* such as 1,3-bisphosphoglycerate, are more stable and undergo further enzyme-catalyzed transformation in cells.

(a) Predict the effect on the net reaction catalyzed by glyceraldehyde 3-phosphate dehydrogenase if phosphate were replaced by arsenate.

(b) What would be the consequence to an organism if arsenate were substituted for phosphate? Arsenate is very toxic to most organisms. Explain why.

12. Requirement for Phosphate in Alcohol Fermentation In 1906 Harden and Young, in a series of classic studies on the fermentation of glucose to ethanol and CO_2 by extracts of brewer's yeast, made the following observations. (1) Inorganic phosphate was essential to fermentation; when the supply of phosphate was exhausted, fermentation ceased before all the glucose was used. (2) During fermentation under these conditions, ethanol, CO_2, and a bisphosphohexose accumulated. (3) When arsenate was substituted for phosphate, no bisphosphohexose accumulated, but the fermentation proceeded until all the glucose was converted to ethanol and CO_2.

(a) Why did fermentation cease when the supply of phosphate was exhausted?

(b) Why did ethanol and CO_2 accumulate? Was the conversion of pyruvate to ethanol and CO_2 essential? Why? Identify the bisphosphohexose that accumulated. Why did it accumulate?

(c) Why did the substitution of arsenate for phosphate prevent the accumulation of the bisphosphohexose yet allow fermentation to ethanol and CO_2 to go to completion? (See Problem 11.)

13. Role of the Vitamin Niacin Adults engaged in strenuous physical activity require an intake of about 160 g of carbohydrate daily but only about 20 mg of niacin for optimal nutrition. Given the role of niacin in glycolysis, how do you explain the observation?

14. Cellular Glucose Concentration The concentration of glucose in human blood plasma is maintained at about 5 mM. The concentration of free glucose inside a myocyte is much lower. Why is the concentration so low in the cell? What happens to glucose after entry into the cell? Glucose is administered intravenously as a food source in certain clinical situations. Given that the transformation of glucose to glucose 6-phosphate consumes ATP, why not administer intravenous glucose 6-phosphate instead?

15. Metabolism of Glycerol Glycerol obtained from the breakdown of fat is metabolized by conversion to dihydroxyacetone phosphate, a glycolytic intermediate, in two enzyme-catalyzed reactions. Propose a reaction sequence for glycerol metabolism. On which known enzyme-catalyzed reactions is your proposal based? Write the net equation for the conversion of glycerol to pyruvate according to your scheme.

$$\begin{array}{c} \mathrm{OH} \\ | \\ \mathrm{HOCH_2{-}C{-}CH_2OH} \\ | \\ \mathrm{H} \end{array}$$

Glycerol

16. Measurement of Intracellular Metabolite Concentrations Measuring the concentrations of metabolic intermediates in a living cell presents great experimental difficulties. Enzymes catalyze metabolic interconversions very rapidly, so a common problem associated with these measurements is that they reflect not the physiological concentrations of metabolites but the equilibrium concentrations in the cell extracts. A reliable experimental technique requires all enzyme-catalyzed reactions to be instantaneously stopped in the *intact* tissue so that intermediates do not undergo further change. This objective is accomplished by rapidly compressing the tissue between large aluminum plates cooled with liquid nitrogen (−190 °C), a process called **freeze-clamping.** After freezing, the tissue is powdered and the enzymes are inactivated by precipitation with perchloric acid. The precipitate is removed by centrifugation, and the clear supernatant extract is analyzed for metabolites. To calculate intracellular concentrations, the intracellular volume is determined from the total water content of the tissue and a measurement of the extracellular volume.

The intracellular concentrations of the substrates and products of the phosphofructokinase-1 reaction in rat heart tissue are given in the table below.

Metabolite	Concentration (mM)*
Fructose 6-phosphate	0.087
Fructose 1,6-bisphosphate	0.022
ATP	11.4
ADP	1.32

Source: From Williamson, J.R. (1965) Glycolytic control mechanisms I. Inhibition of glycolysis by acetate and pyruvate in the isolated, perfused rat heart. *J. Biol. Chem.* **240,** 2308–2321.

*Calculated as μmol/mL of intracellular water.

(a) Calculate the mass-action ratio, [fructose 1,6-bisphosphate][ADP]/[fructose 6-phosphate][ATP], for the PFK-1 reaction under physiological conditions.

(b) Given that $\Delta G'^\circ$ for the PFK-1 reaction is −14.2 kJ/mol, calculate the equilibrium constant for the reaction.

(c) Compare the values of the mass-action ratio and K'_{eq}. Is the physiological reaction at equilibrium? Explain. What does this experiment tell you about the role of PFK-1 as a regulatory enzyme?

17. Effect of O_2 Supply on Glycolytic Rates The regulated steps of glycolysis in intact cells can be identified by studying the catabolism of glucose in whole tissues or organs. For example, glucose consumption by heart muscle can be measured by artificially circulating blood through an isolated intact heart and measuring the blood glucose before and after the blood passes through the heart. If the circulating blood is deoxygenated, heart muscle consumes glucose at a steady rate. When oxygen is added to the blood, the rate of glucose consumption drops dramatically, then is maintained at the new, lower rate. Why?

18. Regulation of Phosphofructokinase-1 The effect of ATP on the allosteric enzyme PFK-1 is shown below. For a given concentration of fructose 6-phosphate, PFK-1 activity increases with increasing [ATP], but a point is reached beyond which increasing [ATP] inhibits the enzyme.

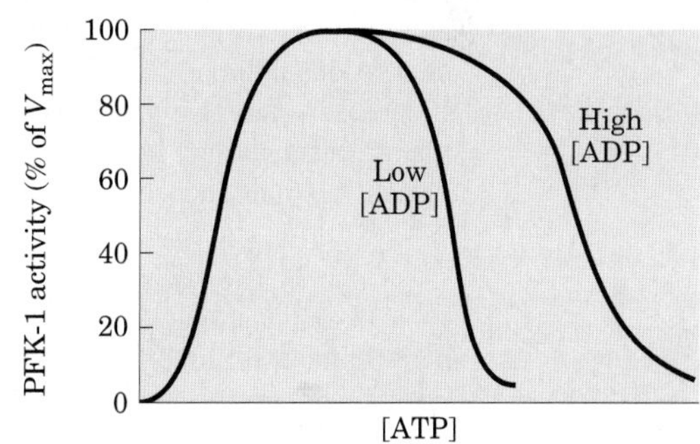

(a) Explain how ATP can be both a substrate and an inhibitor of PFK-1. How is the enzyme regulated by ATP?

(b) How does [ATP] regulate glycolysis?

(c) Explain why the inhibition of PFK-1 by ATP is diminished when [ADP] is high (see graph).

19. Enzyme Activity and Physiological Function The V_{max} of glycogen phosphorylase from skeletal muscle is much larger than the V_{max} of the same enzyme from liver.

(a) What is the physiological function of glycogen phosphorylase in skeletal muscle? In liver?

(b) Why does the V_{max} of the muscle enzyme need to be larger than that of the liver enzyme?

20. Glycogen Phosphorylase Equilibrium Glycogen phosphorylase catalyzes the removal of glucose from glycogen. The $\Delta G'^\circ$ for this reaction is 3.1 kJ/mol.

(a) Calculate the ratio of [P_i] to [glucose 1-phosphate] when the reaction is at equilibrium. (Hint: The removal of glucose units from glycogen does not change the glycogen concentration.)

(b) The measured ratio of [P_i] to [glucose 1-phosphate] in myocytes under physiological conditions is more than 100 to 1. Why are the equilibrium and physiological ratios different? What is the possible significance of this difference?

21. Regulation of Glycogen Phosphorylase In muscle tissue, the rate of conversion of glycogen to glucose 6-phosphate is determined by the ratio of phosphorylase *a* (active) to phosphorylase *b* (less active). What happens to the rate of glycogen breakdown if a muscle preparation containing glycogen phosphorylase is treated with (a) phosphorylase kinase and ATP; (b) phosphorylase phosphatase; (c) epinephrine?

22. Enzyme Defects in Carbohydrate Metabolism Summaries of three clinical case studies follow. For each case determine which enzyme is defective and designate the appropriate treatment, from the lists provided. Justify your choices. Answer the questions contained in each case study.

Case A The patient develops vomiting and diarrhea shortly after milk ingestion. A lactose tolerance test is administered. The patient ingests a standard amount of lactose, and the glucose and galactose concentrations in blood plasma are measured at intervals. In lactose-tolerant individuals the levels increase to a maximum in about 1 h, then decline. Explain why. The patient's blood glucose and galactose levels do not increase during the test. Explain why.

Case B The patient develops vomiting and diarrhea after ingestion of milk. His blood is found to have a low concentration of glucose but a much higher than normal concentration of reducing sugars. A urine test reveals galactose in the urine. Why is the level of reducing sugar in the blood high? Why does galactose appear in the urine?

Case C The patient is lethargic and her liver is enlarged. A liver biopsy shows large amounts of excess glycogen. She also has lower than normal blood glucose. Why does the patient have low blood glucose?

Defective Enzyme
(a) Muscle phosphofructokinase-1
(b) Phosphomannose isomerase
(c) Galactose 1-phosphate uridylyltransferase
(d) Liver glycogen phosphorylase
(e) Triose kinase
(f) Lactase of intestinal epithelial cells
(g) Maltase of intestinal epithelial cells

Treatment
1. Frequent regular feedings
2. Fat-free diet
3. Low-lactose diet
4. Large doses of niacin (the precursor of NAD^+)

23. Severity of Clinical Symptoms Due to Enzyme Deficiency The clinical symptoms of the two forms of galactosemia—deficiency of either galactokinase or UDP glucose → galactose 1-phosphate uridylyltransferase—show radically different severity. Although both types produce gastric discomfort after milk ingestion, deficiency of the latter enzyme also leads to liver, kidney, spleen, and brain dysfunction and eventual death. What products accumulate in the blood and tissues with each type of enzyme deficiency? Estimate the relative toxicities of these products from the above information.

chapter

16

The Citric Acid Cycle

As we saw in Chapter 15, some cells obtain energy (ATP) by fermentation, breaking down glucose in the absence of oxygen. For most eukaryotic cells and many bacteria, living under aerobic conditions and oxidizing their organic fuels to carbon dioxide and water, glycolysis is but the first stage in the complete oxidation of glucose. The pyruvate produced by glycolysis, rather than being reduced to lactate, ethanol, or some other fermentation product, is further oxidized to H_2O and CO_2. This aerobic phase of catabolism is called **respiration.** In the broader physiological or macroscopic sense, respiration refers to a multicellular organism's uptake of O_2 and release of CO_2. Biochemists and cell biologists use the term in a narrower sense to refer to the molecular processes by which *cells* consume O_2 and produce CO_2— processes more precisely termed **cellular respiration.**

Cellular respiration occurs in three major stages (Fig. 16–1). In the first, organic fuel molecules—glucose, fatty acids, and some amino acids—are oxidized to yield two-carbon fragments in the form of the acetyl group of acetyl-coenzyme A (acetyl-CoA). In the second stage, the acetyl groups are fed into the citric acid cycle, which enzymatically oxidizes them to CO_2. The energy released by oxidation is conserved in the reduced electron carriers NADH and $FADH_2$. In the third stage of respiration, these reduced coenzymes are themselves oxidized, giving up protons (H^+) and electrons. The electrons are transferred to O_2—the final electron acceptor—via a chain of electron-carrying molecules known as the respiratory chain. In the course of electron transfer, the large amount of energy released is conserved in the form of ATP by a process called oxidative phosphorylation. Respiration is more complex than glycolysis and is believed to have evolved much later, after the appearance of cyanobacteria. The metabolic activities of cyanobacteria account for the rise of oxygen in the earth's atmosphere, a dramatic turning point in evolutionary history.

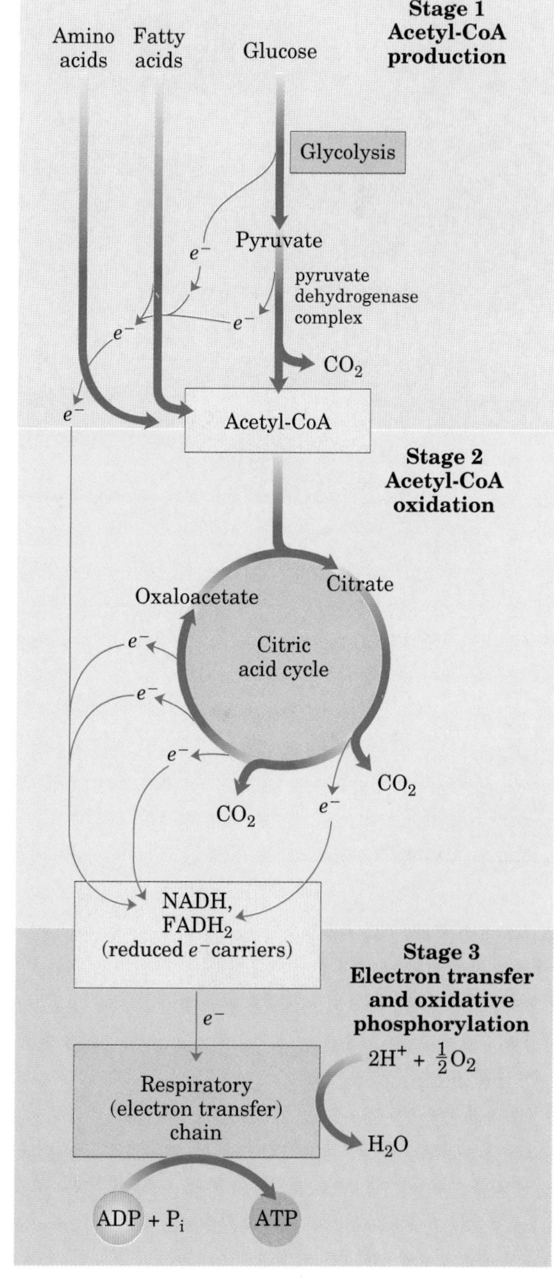

figure 16–1

Catabolism of proteins, fats, and carbohydrates in the three stages of cellular respiration. Stage 1: Oxidation of fatty acids, glucose, and some amino acids yields acetyl-CoA. Stage 2: Oxidation of acetyl groups in the citric acid cycle includes four steps in which electrons are abstracted. Stage 3: Electrons carried by NADH and $FADH_2$ are funneled into a chain of mitochondrial (or, in bacteria, plasma membrane–bound) electron carriers—the respiratory chain—ultimately reducing O_2 to H_2O. This electron flow drives the production of ATP.

Hans Krebs
1900–1981

This chapter begins with the conversion of pyruvate to acetyl groups, followed by entry of those groups into the **citric acid cycle,** also called the **tricarboxylic acid cycle** or the **Krebs cycle** (after its discoverer, Sir Hans Krebs). We then examine the cycle reactions and the enzymes that catalyze them. Because intermediates of the citric acid cycle are also siphoned off as biosynthetic precursors, we go on to consider some ways in which these intermediates are replenished. The citric acid cycle is a hub in metabolism, with degradative pathways leading in and anabolic pathways leading out, and it is closely regulated in coordination with other pathways. The chapter ends with a description of the glyoxylate pathway, a metabolic sequence in some organisms that employs several of the same enzymes and reactions used in the citric acid cycle, bringing about the net synthesis of glucose from stored triacylglycerols.

Production of Acetate

In aerobic organisms, glucose and other sugars, fatty acids, and most amino acids are ultimately oxidized to CO_2 and H_2O via the citric acid cycle and the respiratory chain. Before entering the citric acid cycle, the carbon skeletons of sugars and fatty acids are degraded to the acetyl group of acetyl-CoA, the form in which the cycle accepts most of its fuel input. Many amino acid carbons also enter the cycle this way, although several amino acids are degraded to other cycle intermediates. Here we focus on how pyruvate, derived from glucose and other sugars by glycolysis, is oxidized to acetyl-CoA and CO_2 by the **pyruvate dehydrogenase complex,** a cluster of three enzymes located in the mitochondria of eukaryotic cells and in the cytosol of prokaryotes.

A careful examination of this enzyme complex is rewarding in several respects. The pyruvate dehydrogenase complex is a classic, very well studied example of a multienzyme complex in which a series of chemical intermediates remain bound to the surface of the enzyme molecules as the substrate is transformed into the final product. Five cofactors derived from vitamins participate in the reaction mechanism. The regulation of this enzyme complex also illustrates how a combination of covalent modification and allosteric regulation results in precisely regulated flux through a metabolic step. Finally, the pyruvate dehydrogenase complex is the prototype for two other important enzyme complexes: α-ketoglutarate dehydrogenase of the citric acid cycle and the branched-chain α-ketoacid dehydrogenase of the oxidative pathways of several amino acids (see Fig. 18–27). The remarkable similarity in the protein structure, cofactor requirements, and reaction mechanisms of these three complexes doubtless reflects a common evolutionary origin.

Pyruvate Is Oxidized to Acetyl-CoA and CO_2

The overall reaction catalyzed by the pyruvate dehydrogenase complex is an **oxidative decarboxylation,** an irreversible oxidation process in which the carboxyl group is removed from pyruvate as a molecule of CO_2 and the two remaining carbons become the acetyl group of acetyl-CoA (Fig. 16–2). The NADH formed in this reaction gives up a hydride ion ($:H^-$) to the respiratory chain (Fig. 16–1), which carries the two electrons to oxygen or, in anaerobic microorganisms, to an alternative electron acceptor such as nitrate or sulfate. The transfer of electrons from NADH to oxygen ultimately generates 2.5 molecules of ATP per pair of electrons.

The irreversibility of the pyruvate dehydrogenase reaction has been proved by isotopic labeling experiments: the complex cannot reattach radioactively labeled CO_2 to acetyl-CoA to yield carboxyl-labeled pyruvate.

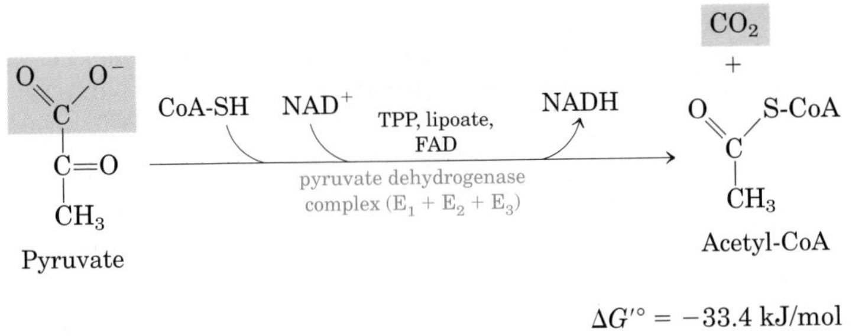

figure 16–2
Overall reaction catalyzed by the pyruvate dehydrogenase complex. The five coenzymes participating in this reaction, and the three enzymes that make up the enzyme complex, are discussed in the text.

The Pyruvate Dehydrogenase Complex Requires Five Coenzymes

The combined dehydrogenation and decarboxylation of pyruvate to the acetyl group of acetyl-CoA (Fig. 16–2) requires the sequential action of three different enzymes and five different coenzymes or prosthetic groups—thiamine pyrophosphate (TPP), flavin adenine dinucleotide (FAD), coenzyme A (CoA), nicotinamide adenine dinucleotide (NAD), and lipoate. Four different vitamins required in human nutrition are vital components of this system: thiamine (in TPP), riboflavin (in FAD), niacin (in NAD), and pantothenate (in CoA).

We have already described the roles of FAD and NAD as electron carriers (Chapter 14), and we have encountered TPP as the coenzyme of pyruvate decarboxylase (see Fig. 15–9). Coenzyme A contains pantothenate (Fig. 16–3) and has a reactive thiol (—SH) group that is critical to its role as an acyl carrier in a number of metabolic reactions. Acyl groups are covalently linked to the thiol group, forming **thioesters.** Because of their relatively high standard free energies of hydrolysis (see Figs. 14–6, 14–7), thioesters have a high acyl group transfer potential, allowing them to donate their acyl groups to a variety of acceptor molecules. The acyl group attached to coenzyme A may thus be thought of as "activated" for group transfer.

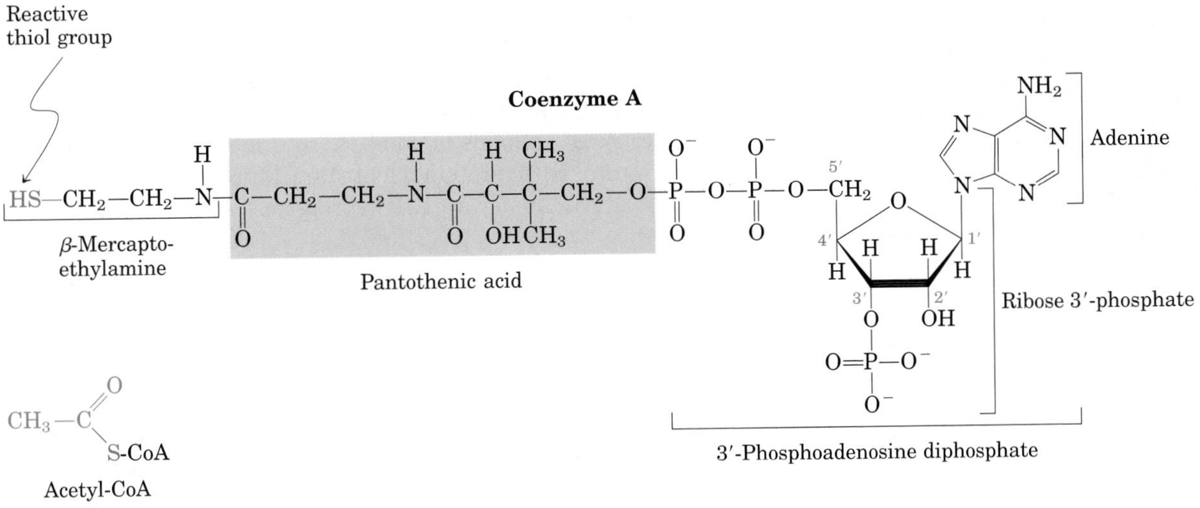

figure 16–3
Coenzyme A (CoA). A hydroxyl group of pantothenic acid is joined to a modified ADP moiety by a phosphate ester bond, and its carboxyl group is attached to β-mercaptoethylamine in amide linkage. The hydroxyl group at the 3′ position of the ADP moiety has a phosphoryl group not present in ADP itself. The —SH group of the mercaptoethylamine moiety forms a thioester with acetate in acetyl-coenzyme A (acetyl-CoA) (lower left).

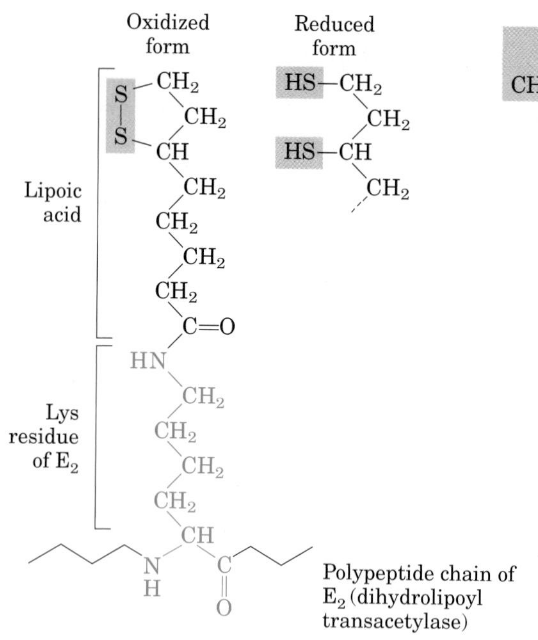

figure 16–4
Lipoic acid (lipoate) in amide linkage with the side chain of a Lys residue. The lipoyllysyl moiety is the prosthetic group of dihydrolipoyl transacetylase (E_2 of the pyruvate dehydrogenase complex). The lipoyl group occurs in oxidized (disulfide) and reduced (dithiol) forms and acts as a carrier of both hydrogen and an acetyl (or other acyl) group.

The fifth cofactor for the pyruvate dehydrogenase reaction, **lipoate** (Fig. 16–4), has two thiol groups that can undergo reversible oxidation to a disulfide bond (—S—S—), similar to that between two Cys residues in a protein. Because of its capacity to undergo oxidation-reduction reactions, lipoate can serve both as an electron carrier and as an acyl carrier, as we shall see.

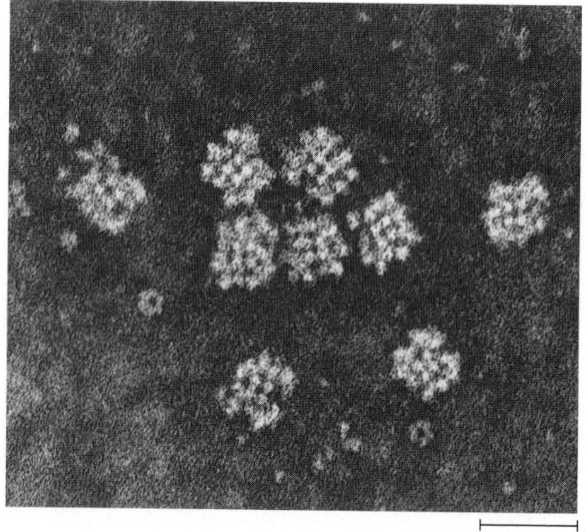

figure 16–5
Electron micrograph of pyruvate dehydrogenase complexes isolated from *E. coli*, showing the subunit structure.

The Pyruvate Dehydrogenase Complex Consists of Three Distinct Enzymes

The pyruvate dehydrogenase complex contains three enzymes—**pyruvate dehydrogenase** (E_1), **dihydrolipoyl transacetylase** (E_2), and **dihydrolipoyl dehydrogenase** (E_3)—each present in multiple copies. The number of copies of each enzyme and therefore the size of the complex varies among organisms. The pyruvate dehydrogenase complex isolated from *Escherichia coli* (M_r >4.5 × 10^6) is about 45 nm in diameter, more than five times the size of an entire *E. coli* ribosome, and can be visualized with the electron microscope (Fig. 16–5). The "core" of the cluster, to which the other enzymes are attached, is dihydrolipoyl transacetylase (E_2). In the *E. coli* complex, 24 copies of this polypeptide chain, each containing three molecules of covalently bound lipoate, constitute the core. The complex from mammals has 60 copies of E_2 and six of a related protein, protein X, which also contains covalently bound lipoate. The attachment of lipoate to the end of Lys side chains in E_2 produces long, flexible arms that can carry acetyl groups from one active site to another in the pyruvate dehydrogenase complex. Bound to this core of E_2 molecules in the *E. coli* complex are 24 copies of pyruvate dehydrogenase (E_1) and 12 copies of dihydrolipoyl dehydrogenase (E_3). E_1 contains bound TPP, and E_3 contains bound FAD. Two regulatory proteins, a protein kinase and a phosphoprotein phosphatase, are also part of the pyruvate dehydrogenase complex. Their role is discussed below.

Intermediates Remain Bound to the Enzyme Surface

Figure 16–6 shows schematically how the pyruvate dehydrogenase complex carries out the five consecutive reactions in the decarboxylation and dehydrogenation of pyruvate. Step ① is essentially identical to the reaction catalyzed by pyruvate decarboxylase (see Fig. 15–9c); C-1 of pyruvate is released as CO_2, and C-2, which in pyruvate has the oxidation state of an aldehyde, is attached to TPP as a hydroxyethyl group. In step ② this group is oxidized to a carboxylic acid (acetate). The two electrons removed in the oxidation reaction reduce the —S—S— of a lipoyl group on E_2 to two thiol (—SH) groups. The acetate produced in this oxidation-reduction reaction is first esterified to one of the lipoyl —SH groups, then transesterified to CoA to form acetyl-CoA (step ③). Thus the energy of oxidation drives the formation of a high-energy thioester of acetate. The remaining reactions catalyzed by the pyruvate dehydrogenase complex (steps ④ and ⑤) are electron transfers necessary to regenerate the oxidized (disulfide) form of the lipoyl group of E_2 to prepare the enzyme complex for another round of

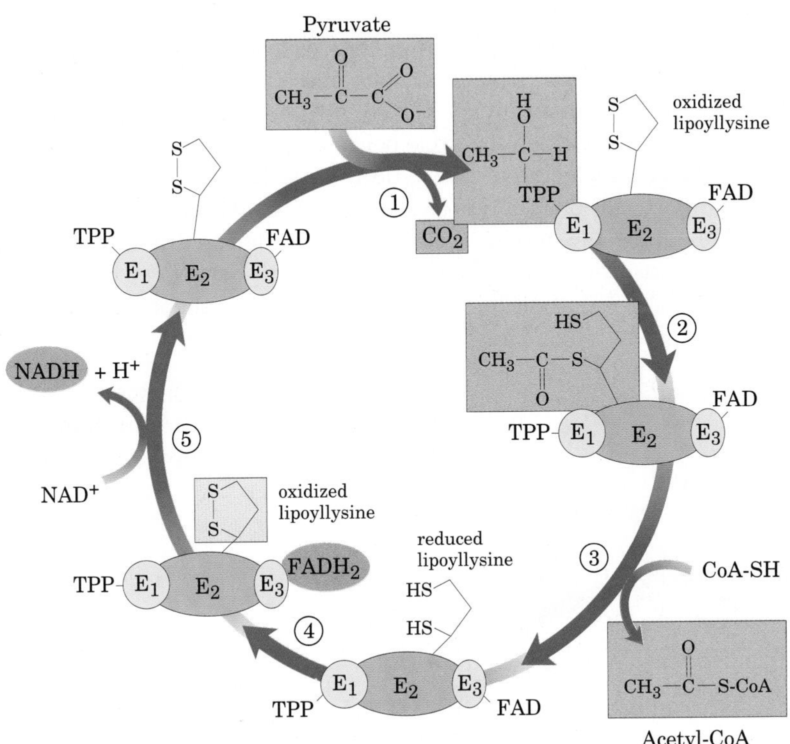

E_1 pyruvate dehydrogenase
E_2 dihydrolipoyl transacetylase
E_3 dihydrolipoyl dehydrogenase

figure 16–6
Oxidative decarboxylation of pyruvate to acetyl-CoA by the pyruvate dehydrogenase complex. The fate of pyruvate is traced in pink. In step ① pyruvate reacts with the bound thiamine pyrophosphate (TPP) of pyruvate dehydrogenase (E_1), undergoing decarboxylation to the hydroxyethyl derivative. Pyruvate dehydrogenase also carries out step ②, the transfer of two electrons and the acetyl group from TPP to the oxidized form of the lipoyllysyl group of the core enzyme, dihydrolipoyl transacetylase (E_2), to form the acetyl thioester of the reduced lipoyl group. Step ③ is a transesterification in which the —SH group of CoA replaces the —SH group of E_2 to yield acetyl-CoA and the fully reduced (dithiol) form of the lipoyl group. In step ④ dihydrolipoyl dehydrogenase (E_3) promotes transfer of two hydrogen atoms from the reduced lipoyl groups of E_2 to the FAD prosthetic group of E_3, restoring the oxidized form of the lipoyllysyl group of E_2 (shaded yellow). In step ⑤ the reduced $FADH_2$ of E_3 transfers a hydride ion to NAD^+, forming NADH. The enzyme complex is now ready for another catalytic cycle.

oxidation. The electrons removed from the hydroxyethyl group derived from pyruvate pass through FAD to NADH.

Central to the mechanism of the pyruvate dehydrogenase complex are the swinging lipoyllysyl arms of E_2, which accept from E_1 the two electrons and the acetyl group derived from pyruvate, passing them to E_3. All these enzymes and coenzymes are clustered, allowing the intermediates to react quickly without diffusing away from the surface of the enzyme complex. The five-reaction sequence shown in Figure 16–6 is an example of substrate channeling (see Fig. 15–8).

As one might predict, mutations in the genes for the subunits of the pyruvate dehydrogenase complex, or a dietary thiamine deficiency, can have severe consequences. Thiamine-deficient animals are unable to oxidize pyruvate normally. This is of particular importance to the brain, which usually obtains all its energy from the aerobic oxidation of glucose in a pathway that necessarily includes the oxidation of pyruvate. Beriberi, a disease that results from thiamine deficiency, is characterized by loss of neural function. This disease occurs primarily in populations that rely on a diet consisting mainly of white (polished) rice, which lacks the hulls in which most of the thiamine of rice is found. People who habitually consume large amounts of alcohol can also develop thiamine deficiency because much of their dietary intake consists of the vitamin-free "empty calories" of distilled spirits. An elevated level of pyruvate in the blood is often an indicator of defects in pyruvate oxidation due to one of these causes.

Reactions of the Citric Acid Cycle

We are now ready to trace the process by which acetyl-CoA undergoes oxidation. This chemical transformation is carried out by the citric acid cycle, the first *cyclic* pathway we have encountered (Fig. 16–7). To begin a turn of the cycle, acetyl-CoA donates its acetyl group to the four-carbon compound oxaloacetate to form the six-carbon citrate. Citrate is then transformed into isocitrate, also a six-carbon molecule, which is dehydrogenated with loss of CO_2 to yield the five-carbon compound α-ketoglutarate (also called oxoglutarate). α-Ketoglutarate undergoes loss of CO_2 and ultimately yields the four-carbon compound succinate and a second molecule of CO_2. Succinate is then enzymatically converted in three steps into the four-carbon oxaloacetate, with which the cycle began; oxaloacetate is then ready to react with another molecule of acetyl-CoA. In each turn of the cycle, one acetyl group (two carbons) enters as acetyl-CoA and two molecules of CO_2 leave, and one molecule of oxaloacetate is used to form citrate, but after a series of reactions, the oxaloacetate is regenerated. No net removal of oxaloacetate occurs; one molecule of oxaloacetate can theoretically bring

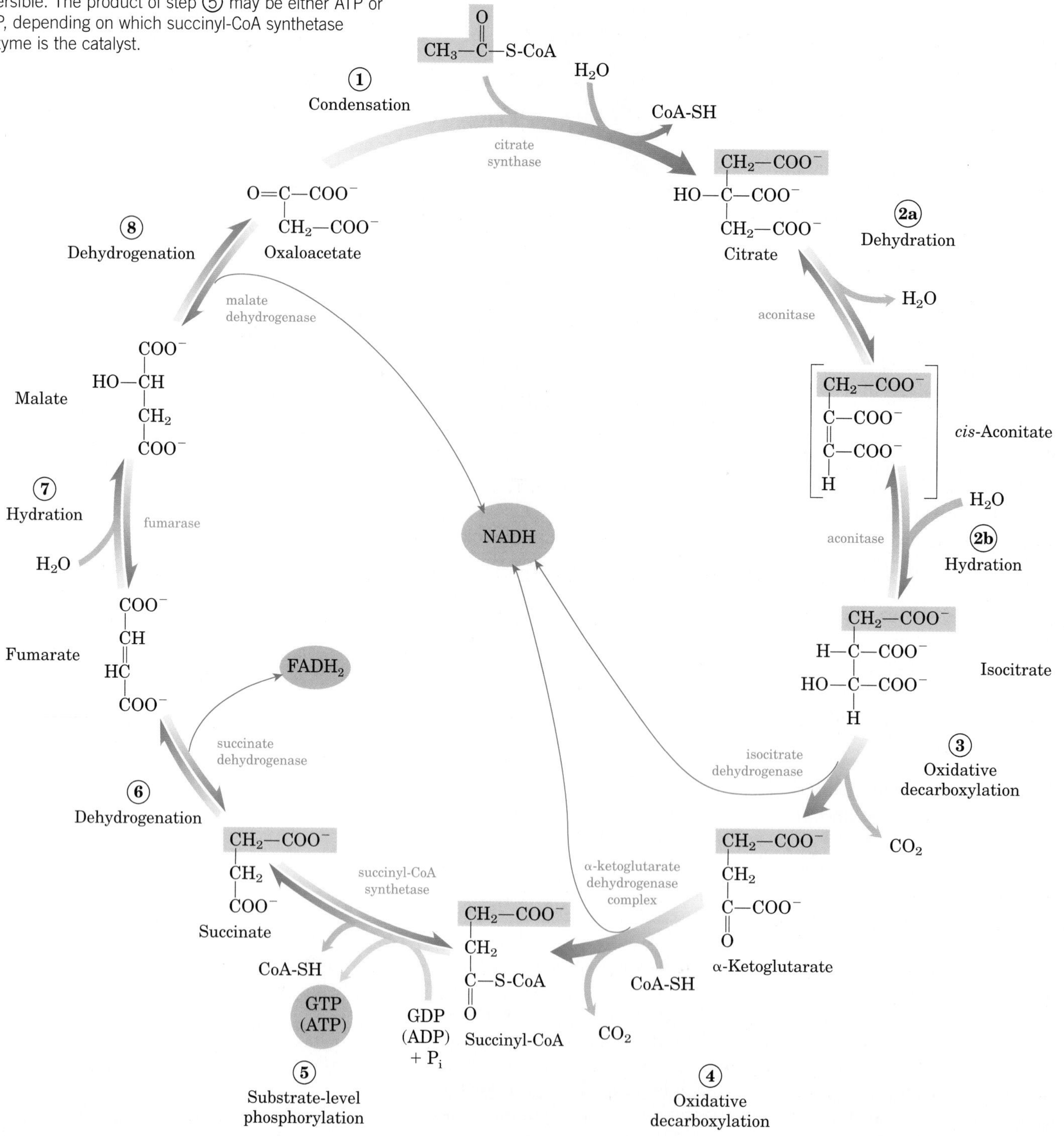

figure 16–7

Reactions of the citric acid cycle. The carbon atoms shaded in red are those derived from the acetate of acetyl-CoA in the first turn of the cycle; these are *not* the carbons released as CO_2 in the first turn. Note that in fumarate, the two-carbon group derived from acetate can no longer be specifically denoted; because succinate and fumarate are symmetric molecules, C-1 and C-2 are indistinguishable from C-4 and C-3. The number beside each reaction step corresponds to a numbered heading in the text. The red arrows show where energy is conserved by electron transfer to FAD or NAD^+, forming $FADH_2$ or NADH + H^+. Steps ①, ③, and ④ are essentially irreversible in the cell; all other steps are reversible. The product of step ⑤ may be either ATP or GTP, depending on which succinyl-CoA synthetase isozyme is the catalyst.

about oxidation of an infinite number of acetyl groups, and in fact, oxaloacetate is present in cells in very low concentrations. Four of the eight steps in this process are oxidations, in which the energy of oxidation is very efficiently conserved in the form of the reduced coenzymes NADH and $FADH_2$.

As noted earlier, although the citric acid cycle is central to energy-yielding metabolism its role is not limited to energy conservation. Four- and five-carbon intermediates of the cycle serve as precursors for a wide variety of products. To replace intermediates removed for this purpose, cells employ anaplerotic (replenishing) reactions, which are described below.

Eugene Kennedy and Albert Lehninger showed in 1948 that in eukaryotes the entire set of reactions of the citric acid cycle takes place in mitochondria. Isolated mitochondria were found to contain not only all the enzymes and coenzymes required for the citric acid cycle, but also all the enzymes and proteins necessary for the last stage of respiration—electron transfer and ATP synthesis by oxidative phosphorylation. As we shall see in later chapters, mitochondria also contain the enzymes for the oxidation of fatty acids and some amino acids to acetyl-CoA, and the oxidative degradation of other amino acids to α-ketoglutarate, succinyl-CoA, or oxaloacetate. Thus in nonphotosynthetic eukaryotes, the mitochondrion is the site of most energy-yielding oxidative reactions and of the coupled synthesis of ATP. In photosynthetic eukaryotes, mitochondria are the major site of ATP production in the dark, but in daylight chloroplasts produce most of the organism's ATP. In most prokaryotes, the enzymes of the citric acid cycle are in the cytosol, and the plasma membrane plays a role analogous to that of the inner mitochondrial membrane in ATP synthesis (Chapter 19).

The Citric Acid Cycle Has Eight Steps

In examining the eight successive reaction steps of the citric acid cycle, we place special emphasis on the chemical transformations taking place as citrate formed from acetyl-CoA and oxaloacetate is oxidized to yield CO_2 and the energy of this oxidation is conserved in the reduced coenzymes NADH and $FADH_2$.

① Formation of Citrate The first reaction of the cycle is the condensation of acetyl-CoA with **oxaloacetate** to form **citrate,** catalyzed by **citrate synthase:**

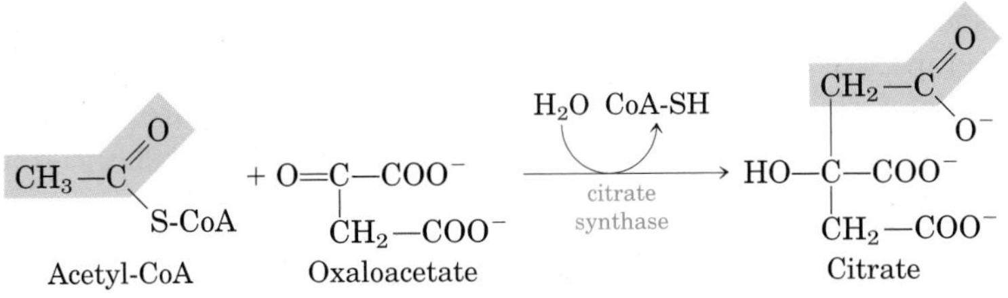

$$CH_3{-}C(=O){-}S\text{-}CoA + O{=}C(COO^-){-}CH_2{-}COO^- \xrightarrow[\text{citrate synthase}]{H_2O \quad CoA\text{-}SH} HO{-}C(COO^-)(CH_2{-}COO^-){-}CH_2{-}C(=O)O^-$$

Acetyl-CoA   Oxaloacetate   Citrate

$$\Delta G'^{\circ} = -32.2 \text{ kJ/mol}$$

In this reaction the methyl carbon of the acetyl group is joined to the carbonyl group (C-2) of oxaloacetate. Citroyl-CoA is a transient intermediate formed on the active site of the enzyme. It rapidly undergoes hydrolysis to free CoA and citrate, which are released from the active site. The hydrolysis of this high-energy thioester intermediate makes the forward reaction highly exergonic. The large, negative standard free-energy change of the citrate synthase reaction is essential to the operation of the cycle because, as noted earlier, the concentration of oxaloacetate is normally very low. The CoA liberated in this reaction is recycled to participate in the oxidative decarboxylation of another molecule of pyruvate by the pyruvate dehydrogenase complex.

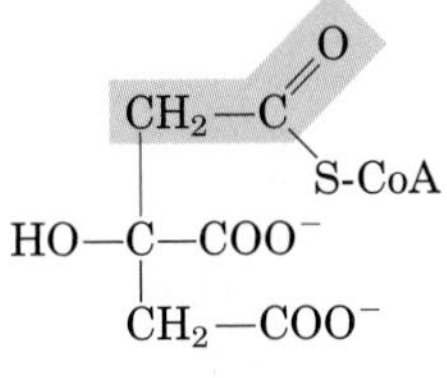

$$HO{-}C(COO^-)(CH_2{-}COO^-){-}CH_2{-}C(=O){-}S\text{-}CoA$$

Citroyl-CoA

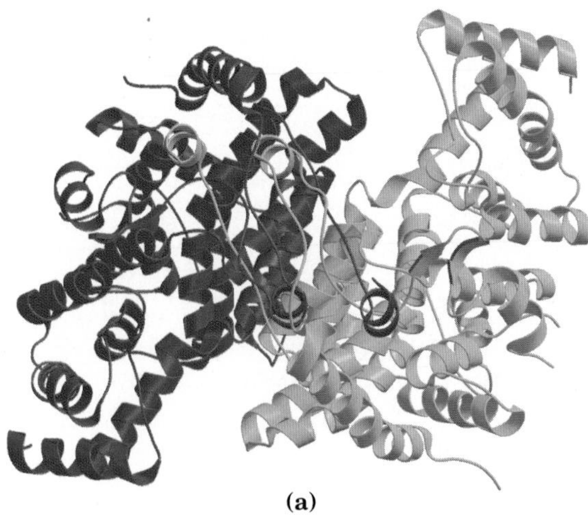

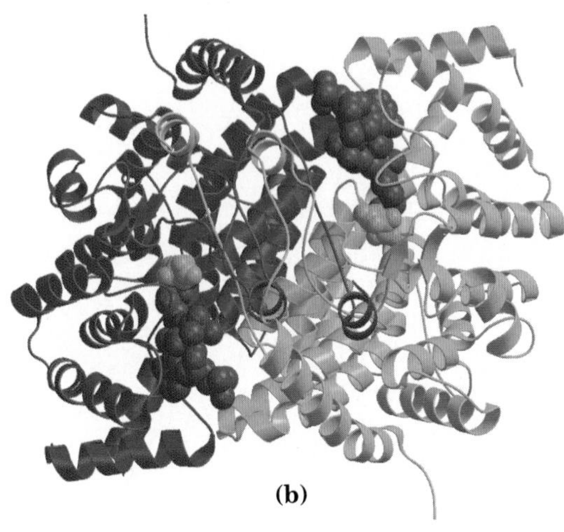

figure 16–8
Citrate synthase. The enzyme undergoes a large conformational change on binding oxaloacetate: **(a)** open form of enzyme alone; **(b)** closed form with oxaloacetate (yellow) and a stable analog of acetyl-CoA (carboxymethyl-CoA; red) bound.

Citrate synthase from mitochondria has been crystallized and visualized by x-ray crystallography in the presence and absence of its substrates and inhibitors (Fig. 16–8). Oxaloacetate, the first substrate to bind to the enzyme, induces a large conformational change, creating a binding site for the second substrate, acetyl-CoA. When citroyl-CoA forms on the enzyme surface, another conformational change brings the side chain of a crucial Asp residue into position to cleave the thioester. This induced fit of the enzyme first to its substrate and then to its intermediate decreases the likelihood of premature and unproductive cleavage of the thioester bond of acetyl-CoA.

② Formation of Isocitrate via *cis*-Aconitate The enzyme **aconitase** (more formally, **aconitate hydratase**) catalyzes the reversible transformation of citrate to **isocitrate,** through the intermediary formation of the tricarboxylic acid ***cis*-aconitate,** which normally does not dissociate from the active site. Aconitase can promote the reversible addition of H_2O to the double bond of enzyme-bound *cis*-aconitate in two different ways, one leading to citrate and the other to isocitrate:

$$\underset{\text{Citrate}}{\begin{array}{c}CH_2{-}COO^-\\ HO{-}C{-}COO^-\\ H{-}C{-}COO^-\\ H\end{array}} \underset{\text{aconitase}}{\overset{H_2O}{\rightleftharpoons}} \left[\underset{cis\text{-Aconitate}}{\begin{array}{c}CH_2{-}COO^-\\ C{-}COO^-\\ \| \\ C{-}COO^-\\ H\end{array}}\right] \underset{\text{aconitase}}{\overset{H_2O}{\rightleftharpoons}} \underset{\text{Isocitrate}}{\begin{array}{c}CH_2{-}COO^-\\ H{-}C{-}COO^-\\ HO{-}C{-}COO^-\\ H\end{array}}$$

$\Delta G'^{\circ} = 13.3\ \text{kJ/mol}$

Although the equilibrium mixture at pH 7.4 and 25 °C contains less than 10% isocitrate, in the cell the reaction is pulled to the right because isocitrate is rapidly consumed in the next step of the cycle, lowering its steady-state concentration. Aconitase contains an **iron-sulfur center** (Fig. 16–9), which acts both in the binding of the substrate at the active site and in the catalytic addition or removal of H_2O.

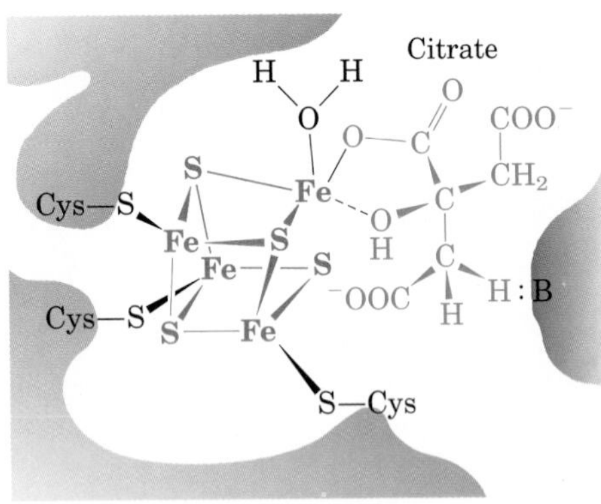

figure 16–9
Iron-sulfur center in aconitase. The iron-sulfur center is in red, the citrate molecule in blue. Three Cys residues of the enzyme bind three iron atoms; the fourth iron is bound to one of the carboxyl groups of citrate. A basic residue, **:** B, on the enzyme helps to position the citrate in the active site. The iron-sulfur center acts in both substrate binding and catalysis. The general properties of iron-sulfur proteins are discussed in Chapter 19 (see Fig. 19–5).

③ Oxidation of Isocitrate to α-Ketoglutarate and CO_2 In the next step, **isocitrate dehydrogenase** catalyzes oxidative decarboxylation of isocitrate to form **α-ketoglutarate:**

$$\underset{\text{Isocitrate}}{\begin{array}{c}CH_2{-}COO^-\\ H{-}C{-}COO^-\\ HO{-}C{-}COO^-\\ H\end{array}} \xrightarrow[\text{isocitrate dehydrogenase}]{NAD(P)^+ \quad NAD(P)H + H^+} \underset{\alpha\text{-Ketoglutarate}}{\begin{array}{c}CH_2{-}COO^-\\ CH_2\\ C{-}COO^-\\ \| \\ O\end{array}} + CO_2$$

$\Delta G'^{\circ} = -20.9\ \text{kJ/mol}$

There are two different forms of isocitrate dehydrogenase, one requiring NAD^+ as electron acceptor and the other requiring $NADP^+$. The overall reactions are otherwise identical. In eukaryotic cells, the NAD-dependent enzyme occurs in the mitochondrial matrix and serves in the citric acid cycle. The NADP-dependent enzyme, found in both the mitochondrial matrix and the cytosol, may have as its main function the generation of NADPH, which is essential for reductive anabolic reactions.

④ Oxidation of α-Ketoglutarate to Succinyl-CoA and CO_2 The next step is another oxidative decarboxylation, in which α-ketoglutarate is converted to **succinyl-CoA** and CO_2 by the action of the **α-ketoglutarate dehydrogenase complex;** NAD^+ serves as electron acceptor and CoA as the carrier of the succinyl group. The energy of oxidation of α-ketoglutarate is conserved in the formation of the thioester bond of succinyl-CoA:

$$\underset{\alpha\text{-Ketoglutarate}}{\begin{array}{l}CH_2{-}COO^-\\ |\\ CH_2\\ |\\ C{-}COO^-\\ \|\\ O\end{array}} \xrightarrow[\alpha\text{-ketoglutarate dehydrogenase complex}]{\text{CoA-SH}\quad NAD^+\quad \to\quad NADH} \underset{\text{Succinyl-CoA}}{\begin{array}{l}CH_2{-}COO^-\\ |\\ CH_2\\ |\\ C{-}S\text{-CoA}\\ \|\\ O\end{array}} + CO_2$$

$$\Delta G'^\circ = -33.5 \text{ kJ/mol}$$

This reaction is virtually identical to the pyruvate dehydrogenase reaction discussed above, and the α-ketoglutarate dehydrogenase complex closely resembles the pyruvate dehydrogenase complex in both structure and function. It includes three enzymes, homologous to E_1, E_2, and E_3 of the pyruvate dehydrogenase complex, as well as enzyme-bound TPP, bound lipoate, FAD, NAD, and coenzyme A. Although the E_1 components of the two complexes are structurally similar, their amino acid sequences differ and they have different binding specificities: E_1 of the pyruvate dehydrogenase complex binds pyruvate, and E_1 of the α-ketoglutarate dehydrogenase complex binds α-ketoglutarate. The E_2 components of the two complexes are also very similar, both having covalently bound lipoyl moieties. The subunits of E_3 for the two complexes are virtually identical. The proteins of these two multienzyme complexes most likely share a common evolutionary origin.

⑤ Conversion of Succinyl-CoA to Succinate Succinyl-CoA, like acetyl-CoA, has a thioester bond with a strongly negative standard free energy of hydrolysis ($\Delta G'^\circ \approx -36$ kJ/mol). In the next step of the citric acid cycle, energy released in the breakage of this bond is used to drive the synthesis of a phosphoanhydride bond in either GTP or ATP, with a net $\Delta G'^\circ$ of only -2.9 kJ/mol. **Succinate** is formed in the process:

$$\underset{\text{Succinyl-CoA}}{\begin{array}{l}CH_2{-}COO^-\\ |\\ CH_2\\ |\\ C{-}S\text{-CoA}\\ \|\\ O\end{array}} \underset{\text{succinyl-CoA synthetase}}{\overset{GDP + P_i\quad \to\quad GTP\quad \text{CoA-SH}}{\rightleftharpoons}} \underset{\text{Succinate}}{\begin{array}{l}COO^-\\ |\\ CH_2\\ |\\ CH_2\\ |\\ COO^-\end{array}}$$

$$\Delta G'^\circ = -2.9 \text{ kJ/mol}$$

The enzyme that catalyzes this reversible reaction is called **succinyl-CoA synthetase** or **succinic thiokinase;** both names indicate the participation of a nucleoside triphosphate in the reaction (Box 16–1).

This energy-conserving reaction involves an intermediate step in which the enzyme molecule itself becomes phosphorylated at a His residue in the active site (Fig. 16–10a). This phosphoryl group, which has a high group transfer potential, is transferred to ADP (or GDP) to form ATP (or GTP). Animal cells have two isozymes, one specific for ADP, and the other for GDP. Succinyl-CoA synthetase has two subunits, α (M_r 32,000), which has the

box 16–1 Synthases and Synthetases; Ligases and Lyases; Kinases, Phosphatases, and Phosphorylases: Yes, the Names Are Confusing!

Citrate synthase is one of many enzymes that catalyze condensation reactions, yielding a product more chemically complex than its precursors. **Synthases** catalyze condensation reactions in which no nucleoside triphosphate (ATP, GTP, etc.) is required as an energy source. **Synthetases** catalyze condensations that *do* use ATP or another nucleoside triphosphate as a source of energy for the synthetic reaction. Succinyl-CoA synthetase is such an enzyme. **Ligases** (from the Latin *ligare,* "to tie together") are enzymes that catalyze condensation reactions in which two atoms are joined using the energy of ATP or another energy source. (Thus synthetases are ligases.) DNA ligase, for example, closes breaks in DNA molecules, using energy supplied by either ATP or NAD^+; it is widely used in joining DNA pieces for genetic engineering. Ligases are not to be confused with **lyases,** enzymes that catalyze cleavages (or, in the reverse directions, additions) in which electronic rearrangements occur. The pyruvate dehydrogenase complex, which oxidatively cleaves CO_2 from pyruvate, is a member of the large class of lyases.

The name **kinase** is applied to enzymes that transfer a phosphoryl group from a nucleoside triphosphate such as ATP to an acceptor molecule—a sugar (as in hexokinase and glucokinase), a protein (as in glycogen phosphorylase kinase), another nucleotide (as in nucleoside diphosphate kinase), or a metabolic intermediate such as oxaloacetate (as in PEP carboxykinase). The reaction catalyzed by a kinase is a *phosphorylation.* On the other hand, *phosphorolysis* is a displacement reaction in which phosphate is the attacking species and becomes covalently attached at the point of bond breakage. Such reactions are catalyzed by **phosphorylases.** Glycogen phosphorylase, for example, catalyzes the phosphorolysis of glycogen, producing glucose 1-phosphate. *Dephosphorylation,* the removal of a phosphoryl group from a phosphate ester, is catalyzed by **phosphatases,** with water as the attacking species. Fructose bisphosphatase-1 converts fructose 1,6-bisphosphate to fructose 6-phosphate in gluconeogenesis, and phosphorylase *a* phosphatase removes phosphoryl groups from phosphoserine in phosphorylated glycogen phosphorylase. Whew!

Unfortunately, these descriptions of enzyme types overlap, and many enzymes are commonly called by two or more names. Succinyl-CoA synthetase, for example, is also called succinate thiokinase; the enzyme is both a synthetase in the citric acid cycle and a kinase when acting in the direction of succinyl-CoA synthesis. This raises another source of confusion in the naming of enzymes. An enzyme may have been discovered by the use of an assay in which, say, A is converted to B. The enzyme is then named for that reaction. Later work may show, however, that in the cell, the enzyme functions primarily in converting B to A. Commonly, the first name continues to be used, although the metabolic role of the enzyme would be better described by naming it for the reverse reaction. The glycolytic enzyme pyruvate kinase illustrates this situation (p. 539). To a beginner in biochemistry, this duplication in nomenclature can be bewildering. International committees have made heroic efforts to systematize the nomenclature of enzymes (see Table 8–3 for a brief summary of the system), but some systematic names have proved too long and cumbersome and are not frequently used in biochemical conversation.

We have tried throughout this book to use the enzyme name most commonly used by working biochemists and to point out cases in which an enzyme has more than one widely used name. For current information on enzyme nomenclature, the Lehninger Web site

(http://www.worthpublishers.com/lehninger)

has a link to the recommendations of The Nomenclature Committee of the International Union of Biochemistry and Molecular Biology.

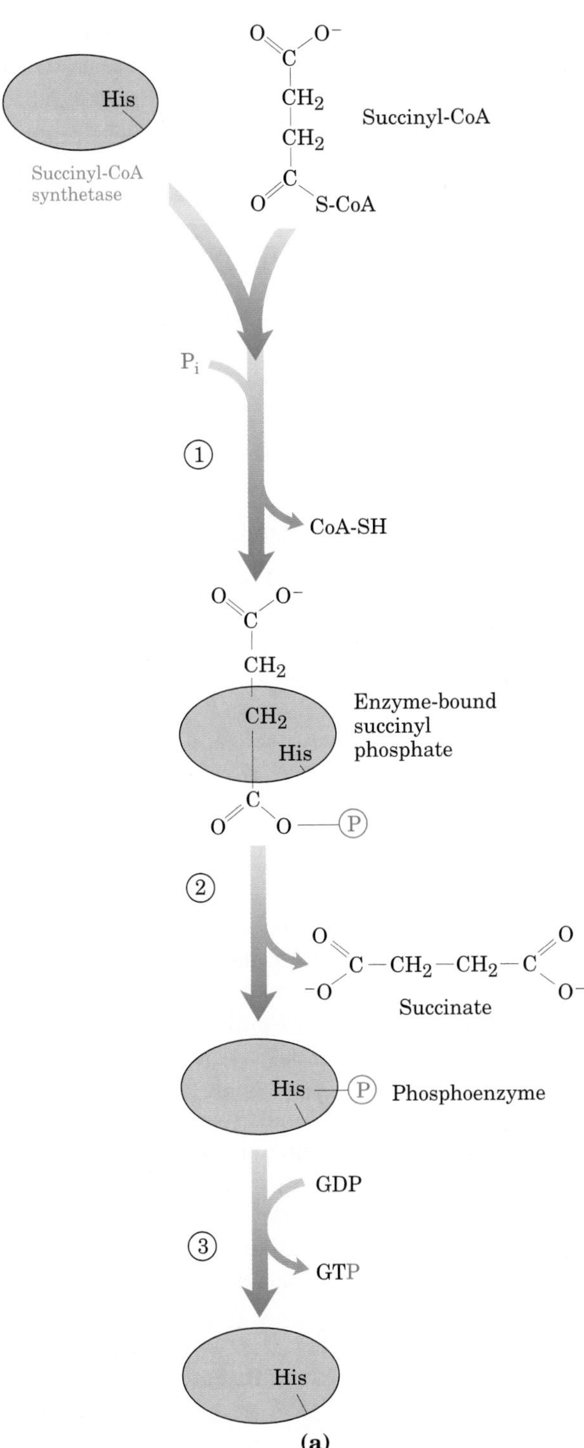

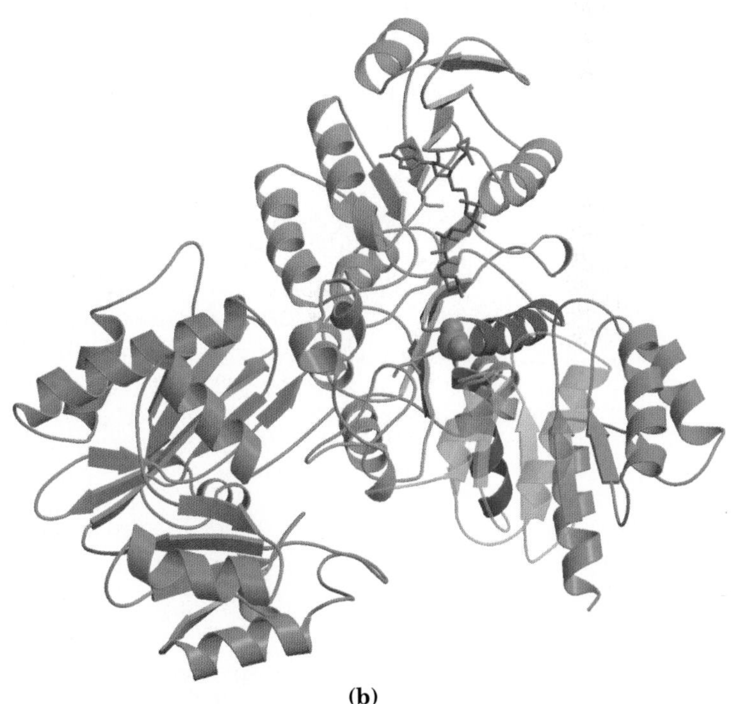

figure 16–10

The succinyl-CoA synthetase reaction. (a) In step ① a phosphoryl group replaces CoA in succinyl-CoA bound to the enzyme, forming a high-energy acyl phosphate. In step ② the succinyl phosphate donates its phosphoryl group to a His residue on the enzyme, forming a high-energy phosphohistidyl enzyme. In step ③ the phosphoryl group is transferred from the His residue to the terminal phosphate of GDP (or ADP), forming GTP (or ATP). **(b)** Succinyl-CoA synthetase from *E. coli*. The bacterial and mammalian enzymes have similar amino acid sequences and presumably have very similar three-dimensional structures. The active site includes part of both the α (blue) and β (brown) subunits. The power helices (dark blue, dark brown) place the partial positive charges of the helix dipole near the phosphate group (orange) on His^{246} of the α chain, stabilizing the phosphoenzyme intermediate. Coenzyme A is shown here as a red stick structure.

phospho-His residue (His^{246}) and the binding site for CoA, and β (M_r 42,000), which confers specificity for either ATP or GTP. The active site is at the interface between subunits. The crystallographic structure of succinyl-CoA synthetase reveals two "power helices" (one from each subunit) oriented so that their electric dipoles put partial positive charges close to the negatively charged phospho-His (Fig. 16–10b), stabilizing the phosphoenzyme intermediate. (Recall the similar role of helix dipoles in stabilizing K^+ ions in the K^+ channel (see Fig. 12–38).)

The formation of ATP (or GTP) at the expense of the energy released by the oxidative decarboxylation of α-ketoglutarate is a substrate-level phosphorylation, like the synthesis of ATP in the glycolytic reactions catalyzed by glyceraldehyde 3-phosphate dehydrogenase and pyruvate kinase (see Fig. 15–2).

The GTP formed by succinyl-CoA synthetase can donate its terminal phosphoryl group to ADP to form ATP, in a reversible reaction catalyzed by **nucleoside diphosphate kinase** (p. 510).

$$\text{GTP} + \text{ADP} \xrightleftharpoons{Mg^{2+}} \text{GDP} + \text{ATP} \qquad \Delta G'^{\circ} = 0 \text{ kJ/mol}$$

Thus the net result of the activity of either isozyme of succinyl-CoA synthetase is the conservation of energy as ATP. There is no change in free energy for the nucleoside diphosphate kinase reaction; ATP and GTP are energetically equivalent.

⑥ Oxidation of Succinate to Fumarate The succinate formed from succinyl-CoA is oxidized to **fumarate** by the flavoprotein **succinate dehydrogenase:**

$$\underset{\text{Succinate}}{{}^-OOC{-}CH_2{-}CH_2{-}COO^-} \xrightleftharpoons[\text{succinate dehydrogenase}]{\text{FAD} \;\rightarrow\; \text{FADH}_2} \underset{\text{Fumarate}}{{}^-OOC{-}CH{=}CH{-}COO^-}$$

$$\Delta G'^{\circ} = 0 \text{ kJ/mol}$$

In eukaryotes, succinate dehydrogenase is tightly bound to the inner mitochondrial membrane; in prokaryotes, to the plasma membrane. It is the only enzyme of the citric acid cycle that is membrane-bound. The enzyme from beef heart mitochondria contains three different iron-sulfur clusters and one molecule of covalently bound FAD. Electrons pass from succinate through the FAD and iron-sulfur centers before entering the chain of electron carriers in the mitochondrial inner membrane (or the plasma membrane in bacteria). Electron flow from succinate through these carriers to the final electron acceptor, O_2, is coupled to the synthesis of about 1.5 ATP molecules per pair of electrons (respiration-linked phosphorylation). Malonate, an analog of succinate, is a strong competitive inhibitor of succinate dehydrogenase and therefore blocks the activity of the citric acid cycle.

$$\underset{\text{Malonate}}{{}^-OOC{-}CH_2{-}COO^-} \qquad \underset{\text{Succinate}}{{}^-OOC{-}CH_2{-}CH_2{-}COO^-}$$

⑦ Hydration of Fumarate to Malate The reversible hydration of fumarate to **L-malate** is catalyzed by **fumarase (fumarate hydratase):**

$$\underset{\text{Fumarate}}{{}^-OOC{-}CH{=}CH{-}COO^-} + H_2O \xrightleftharpoons[\text{fumarase}]{} \underset{\text{L-Malate}}{{}^-OOC{-}CH(OH){-}CH_2{-}COO^-}$$

$$\Delta G'^{\circ} = -3.8 \text{ kJ/mol}$$

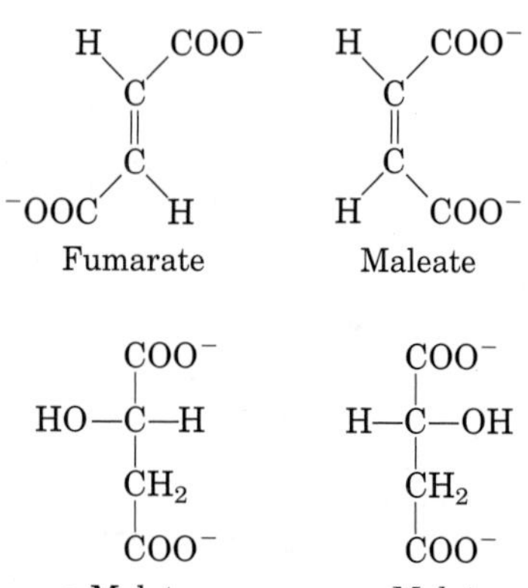

This enzyme is highly stereospecific; it catalyzes hydration of the trans double bond of fumarate but not the cis double bond of maleate (the cis isomer of fumarate). In the reverse direction (from L-malate to fumarate), fumarase is equally stereospecific: D-malate is not a substrate.

⑧ Oxidation of Malate to Oxaloacetate In the last reaction of the citric acid cycle, NAD-linked **L-malate dehydrogenase** catalyzes the oxidation of L-malate to oxaloacetate:

$$\text{L-Malate} + NAD^+ \underset{\text{malate dehydrogenase}}{\rightleftharpoons} \text{Oxaloacetate} + NADH + H^+$$

$$\Delta G'^{\circ} = 29.7 \text{ kJ/mol}$$

The equilibrium of this reaction lies far to the left under standard thermodynamic conditions, but in intact cells oxaloacetate is continually removed by the highly exergonic citrate synthase reaction (step ①). This keeps the concentration of oxaloacetate in the cell extremely low ($<10^{-6}$ M), pulling the malate dehydrogenase reaction toward the formation of oxaloacetate.

Although the individual reactions of the citric acid cycle were initially worked out in vitro using minced muscle tissue, the pathway and its regulation have also been studied extensively in vivo. By using radioactively labeled precursors such as [^{14}C]pyruvate and [^{14}C]acetate, we can trace the fate of individual carbon atoms through the citric acid cycle. Some of the earliest experiments with isotopes produced an unexpected result, however, which aroused considerable controversy about the pathway and mechanism of the citric acid cycle. In fact, these experiments at first seemed to show that citrate was not the first tricarboxylic acid to be formed. Box 16–2 gives some details of this episode in the history of citric acid cycle research. Metabolic flux through the cycle can now be monitored in living tissue by using ^{13}C-labeled precursors and whole-tissue NMR spectroscopy. Because the NMR signal is unique to the compound containing the ^{13}C, we can trace the movement of precursor carbons into each cycle intermediate and into compounds derived from these intermediates. This technique has great promise for studies of regulation of the citric acid cycle and its interconnections with other metabolic pathways such as glycolysis.

The Energy of Oxidations in the Cycle Is Efficiently Conserved

We have now covered one complete turn of the citric acid cycle (Fig. 16–11). A two-carbon acetyl group entered the cycle by combining with oxaloacetate. Two carbon atoms emerged from the cycle as CO_2 from the oxidation of isocitrate and α-ketoglutarate. The energy released by these oxidations was conserved in the reduction of three NAD^+ and one FAD and the production of one ATP or GTP. At the end of the cycle a molecule of oxaloacetate was regenerated. Note that the two carbon atoms appearing as CO_2 are not the same two carbons that entered in the form of the acetyl group; additional turns around the cycle are required to release these carbons as CO_2 (Fig. 16–7).

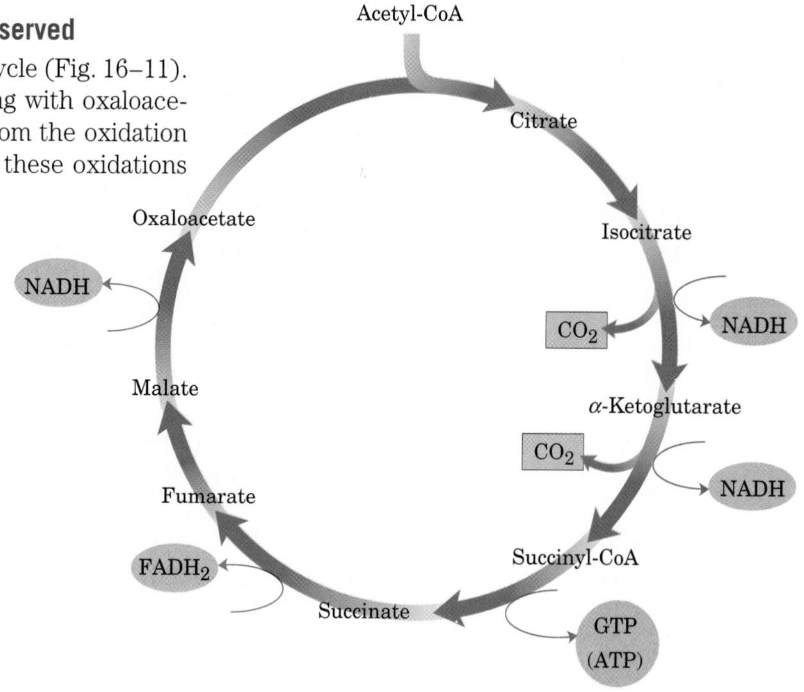

figure 16–11
Products of one turn of the citric acid cycle. Three NADH, one $FADH_2$, one GTP (or ATP), and two CO_2 are released in oxidative decarboxylation reactions. Here and in several following figures, all cycle reactions are shown in one direction only, but keep in mind that most of the reactions are reversible, as shown in Figure 16–7.

box 16–2 Citrate: A Symmetrical Molecule That Reacts Asymmetrically

When compounds enriched in the heavy-carbon isotope ^{13}C and the radioactive carbon isotopes ^{11}C and ^{14}C became available about fifty years ago, they were soon put to use to trace the pathway of carbon atoms through the citric acid cycle. One such experiment initiated the controversy over the role of citrate. Acetate labeled in the carboxyl group (designated [1-^{14}C]acetate) was incubated aerobically with an animal tissue preparation. Acetate is enzymatically converted to acetyl-CoA in animal tissues, and the pathway of the labeled carboxyl carbon of the acetyl group in the cycle reactions could thus be traced. α-Ketoglutarate was isolated from the tissue after incubation, then degraded by known chemical reactions to establish the position(s) of the isotopic carbon.

Condensation of unlabeled oxaloacetate with carboxyl-labeled acetate would be expected to produce citrate labeled in one of the two primary carboxyl groups. Citrate is a symmetric molecule, its two terminal carboxyl groups being chemically indistinguishable. Therefore, half of the labeled citrate molecules were expected to yield α-ketoglutarate labeled in the α-carboxyl group and the other half to yield α-ketoglutarate labeled in the γ-carboxyl group; that is, the α-ketoglutarate isolated was expected to be a mixture of the two types of labeled molecules (Fig. 1).

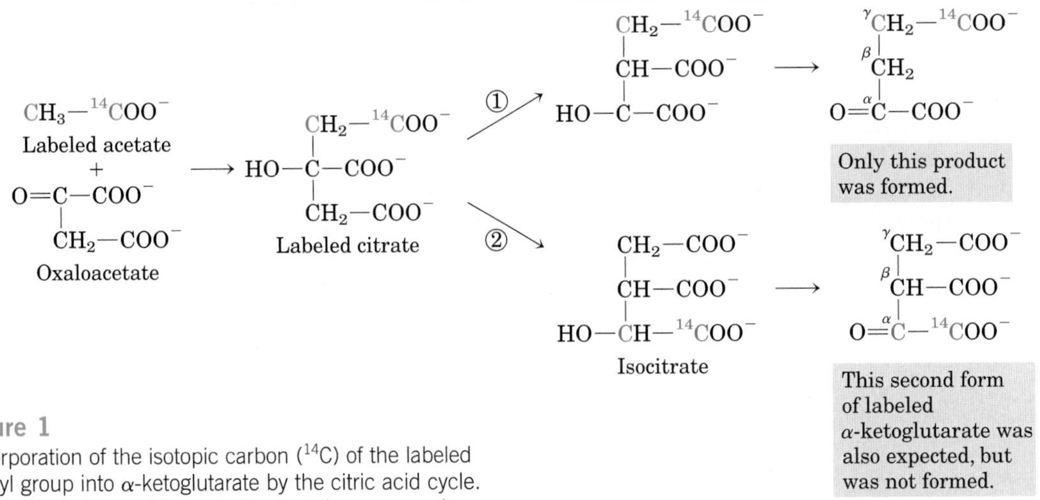

figure 1
Incorporation of the isotopic carbon (^{14}C) of the labeled acetyl group into α-ketoglutarate by the citric acid cycle. The carbon atoms of the entering acetyl group are shown in red.

Although the citric acid cycle itself directly generates only one ATP per turn (in the conversion of succinyl-CoA to succinate), the four oxidation steps in the cycle provide a large flow of electrons into the respiratory chain via NADH and $FADH_2$ and thus lead to formation of a large number of ATP molecules during oxidative phosphorylation.

We saw in Chapter 15 that the energy yield from the production of two molecules of pyruvate from one molecule of glucose in glycolysis is 2 ATP and 2 NADH. In oxidative phosphorylation (Chapter 19), passage of two electrons from NADH to O_2 drives the formation of about 2.5 ATP, and passage of two electrons from $FADH_2$ to O_2 yields about 1.5 ATP. This stoichiometry allows us to calculate the overall yield of ATP from the complete oxidation of glucose. When both pyruvate molecules are oxidized to 6 CO_2 via the pyruvate dehydrogenase complex and the citric acid cycle, and the

Contrary to this expectation, the labeled α-ketoglutarate isolated from the tissue suspension contained ^{14}C only in the γ-carboxyl group (Fig. 1, pathway ①). The investigators concluded that citrate (or any other symmetric molecule) could not be an intermediate in the pathway from acetate to α-ketoglutarate. Rather, an asymmetric tricarboxylic acid, presumably *cis*-aconitate or isocitrate, must be the first product formed from condensation of acetate and oxaloacetate.

In 1948, however, Alexander Ogston pointed out that although citrate has no chiral center (see Fig. 3–9), it has the *potential* to react asymmetrically if an enzyme with which it interacts has an active site that is asymmetric. He suggested that the active site of aconitase may have three points to which the citrate must be bound and that the citrate must undergo a specific three-point attachment to these binding points. As seen in Figure 2, the binding of citrate to three such points can happen in only one way, and this would account for the formation of only one type of labeled α-ketoglutarate. Organic molecules such as citrate that have no chiral center but are potentially capable of reacting asymmetrically with an asymmetric active site are now called **prochiral** molecules.

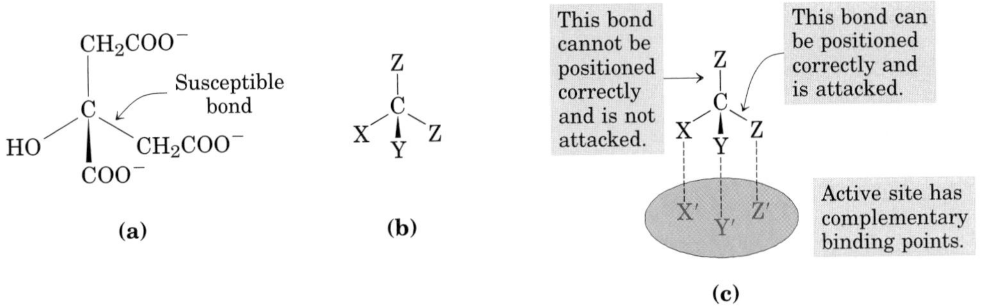

figure 2
The prochiral nature of citrate. **(a)** Structure of citrate; **(b)** schematic representation: X = —OH; Y = $—COO^-$; Z = $—CH_2COO^-$. **(c)** Correct complementary fit of citrate to the binding site of aconitase. There is only one way in which the three specified groups of citrate can fit on the three points of the binding site. Thus only one of the two $—CH_2COO^-$ groups is bound by aconitase.

electrons are transferred to O_2 via oxidative phosphorylation, as many as 32 ATP are obtained per glucose (Table 16–1). In round numbers, this represents the conservation of 32 × 30.5 kJ/mol = 976 kJ/mol, or 34% of the theoretical maximum of 2,840 kJ/mol available from the complete oxidation of glucose. These calculations employ the standard free-energy changes; when corrected for the actual free energy required to form ATP within cells (Box 14–2), the calculated efficiency of the process is closer to 65%.

Why Is the Oxidation of Acetate So Complicated?

The eight-step, cyclic process for oxidation of simple two-carbon acetyl groups to CO_2 may seem unnecessarily cumbersome and not in keeping with the biological principle of maximum economy. The role of the citric acid cycle is not confined to the oxidation of acetate, however. This path-

table 16–1

Stoichiometry of Coenzyme Reduction and ATP Formation in the Aerobic Oxidation of Glucose via Glycolysis, the Pyruvate Dehydrogenase Reaction, the Citric Acid Cycle, and Oxidative Phosphorylation

Reaction	Number of ATP or reduced coenzymes directly formed	Number of ATP ultimately formed*
Glucose ⟶ glucose 6-phosphate	−1 ATP	−1
Fructose 6-phosphate ⟶ fructose 1,6-bisphosphate	−1 ATP	−1
2 Glyceraldehyde 3-phosphate ⟶ 2 1,3-bisphosphoglycerate	2 NADH	3–5
2 1,3-Bisphosphoglycerate ⟶ 2 3-phosphoglycerate	2 ATP	2
2 Phosphoenolpyruvate ⟶ 2 pyruvate	2 ATP	2
2 Pyruvate ⟶ 2 acetyl-CoA	2 NADH	5
2 Isocitrate ⟶ 2 α-ketoglutarate	2 NADH	5
2 α-Ketoglutarate ⟶ 2 succinyl-CoA	2 NADH	5
2 Succinyl-CoA ⟶ 2 succinate	2 ATP (or 2 GTP)	2
2 Succinate ⟶ 2 fumarate	2 $FADH_2$	3
2 Malate ⟶ 2 oxaloacetate	2 NADH	5
Total		30–32

*This is calculated as 2.5 ATP per NADH and 1.5 ATP per $FADH_2$. A negative value indicates consumption.

way is the hub of intermediary metabolism. Four- and five-carbon end products of many catabolic processes feed into the cycle to serve as fuels. Oxaloacetate and α-ketoglutarate, for example, are produced from aspartate and glutamate, respectively, when proteins are degraded. Under some metabolic circumstances, intermediates are drawn out of the cycle to be used as precursors in a variety of biosynthetic pathways.

The citric acid cycle, like all other metabolic pathways, is the product of evolution, and much of this evolution occurred before the advent of aerobic organisms. It does not necessarily represent the *shortest* pathway from acetate to CO_2, but it is the pathway that has, over time, conferred the greatest selective advantage. Early anaerobes very probably used some of the reactions of the citric acid cycle in linear biosynthetic processes. In fact, some modern anaerobic microorganisms use an incomplete citric acid cycle as a source, not of energy, but of biosynthetic precursors (Fig. 16–12). These organisms use the first three reactions of the cycle to make α-ketoglutarate but, lacking α-ketoglutarate dehydrogenase, they cannot carry out the complete set of citric acid cycle reactions. They do have the four enzymes that catalyze the reversible conversion of oxaloacetate to succinyl-CoA and can produce malate, fumarate, succinate, and succinyl-CoA from oxaloacetate in a reversal of the "normal" (oxidative) direction of flow through the cycle. This is a fermentation, with the NADH produced by isocitrate oxidation recycled by reduction of oxaloacetate to succinate.

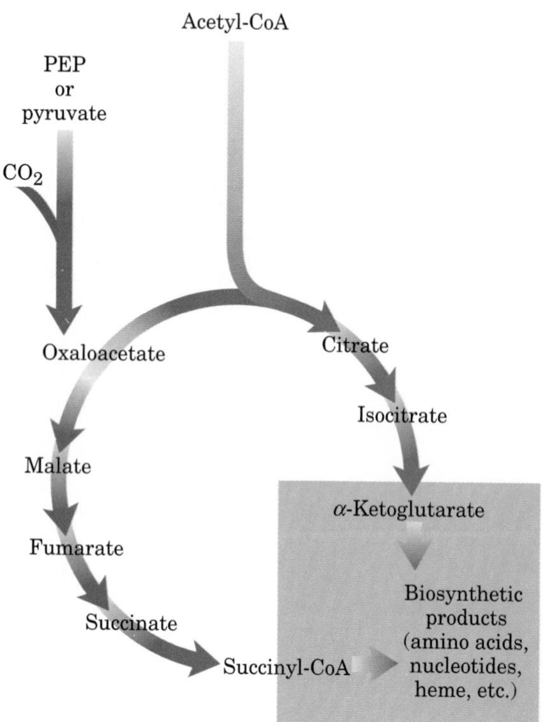

figure 16–12
Biosynthetic precursors produced by an incomplete citric acid cycle in anaerobic bacteria. These anaerobes lack α-ketoglutarate dehydrogenase and therefore cannot carry out the complete citric acid cycle. α-Ketoglutarate and succinyl-CoA serve as precursors in a variety of biosynthetic pathways. See Figure 16–11 for the "normal" direction of these reactions in the citric acid cycle.

With the evolution of cyanobacteria that produced O_2 from water, the earth's atmosphere became aerobic and organisms were under selective pressure to develop aerobic metabolism, which, as we have seen, is much more efficient than anaerobic fermentation.

Citric Acid Cycle Components Are Important Biosynthetic Intermediates

In aerobic organisms, the citric acid cycle is an **amphibolic pathway,** one that serves in both catabolic and anabolic processes. Besides its role in the oxidative catabolism of carbohydrates, fatty acids, and amino acids, the cycle provides precursors for many biosynthetic pathways (Fig. 16–13), through reactions that served the same purpose in anaerobic ancestors. α-Ketoglutarate and oxaloacetate can, for example, serve as precursors of the amino acids aspartate and glutamate by simple transamination (Chapter 22). Through aspartate and glutamate, the carbons of oxaloacetate and α-ketoglutarate are then used to build other amino acids as well as purine and pyrimidine nucleotides. Oxaloacetate is converted to glucose in gluconeogenesis (see Fig. 20–2). Succinyl-CoA is a central intermediate in the synthesis of the porphyrin ring of heme groups, which serve as oxygen carriers (in hemoglobin and myoglobin) and electron carriers (in cytochromes). Citrate is released by some commercially exploited microorganisms (Box 16–3).

box 16–3 Citrate Synthase, Soda Pop, and the World Food Supply

Citrate has a number of important industrial applications. A quick examination of the ingredients in most soft drinks reveals the common use of citric acid to provide a tart or fruity flavor. Citric acid is also used as a plasticizer and foam inhibitor in the manufacture of certain resins, as a mordant to brighten colors, and as an antioxidant to preserve the flavors of foods. Citric acid is produced industrially by growing the fungus *Aspergillus niger* in the presence of an inexpensive sugar source, usually beet molasses. Culture conditions are designed to inhibit the reactions of the citric acid cycle such that citrate accumulates.

On a grander scale, citric acid may play a spectacular role in the alleviation of world hunger. With its three negatively charged carboxyl groups, citrate is a good chelator of metal ions, and some plants exploit this property by releasing citrate into the soil, where it binds metal ions and prevents their absorption by the plant. Of particular importance is the aluminum ion (Al^{3+}), which is toxic to many plants and causes decreased crop yields on 30% to 40% of the world's arable land. While aluminum is the most abundant metal in the earth's crust, it occurs mostly in chemical compounds, such as $Al(OH)_3$, that are biologically inert. However, when soil pH is less than 5, Al^{3+} becomes soluble and thus can be absorbed by plant roots. Acidic soil and Al^{3+} toxicity are most prevalent in the tropics, where maize yields can be depressed as much as 80%. In Mexico, Al^{3+} toxicity limits papaya production to 20,000 hectares instead of the 3 million hectares that could theoretically be cultivated. One solution would be to raise soil pH with lime, but this is economically and environmentally unsound. An alternative would be to breed Al^{3+}-resistant plants. Naturally resistant plants do exist, and provide the means for a third solution: transferring resistance to crop plants by genetic engineering.

A group of researchers in Mexico have genetically engineered tobacco and papaya plants to express elevated levels of bacterial citrate synthase. These plants secrete five to six times their normal amount of Al^{3+}-chelating citric acid and grow at levels of Al^{3+} ten times those at which control plants can grow. This level of resistance would allow Mexico to grow papaya on the 3 million hectares of land currently rendered unsuitable by Al^{3+}.

Given projected levels of population growth, world food production must more than triple in fifty years to adequately feed 9.6 billion people. A long-term solution may turn on increasing crop productivity on the arable land affected by aluminum toxicity, and citric acid may play an important role in achieving this goal.

Anaplerotic Reactions Replenish Citric Acid Cycle Intermediates

As intermediates of the citric acid cycle are removed to serve as biosynthetic precursors, they are replenished by **anaplerotic reactions** (Fig. 16–13; Table 16–2). Under normal circumstances, the reactions by which cycle intermediates are siphoned off into other pathways and those by which they are replenished are in dynamic balance. Thus, concentrations of the citric acid cycle intermediates remain almost constant.

Table 16–2 shows the most common anaplerotic reactions, all of which, in various tissues and organisms, convert either pyruvate or phosphoenolpyruvate to oxaloacetate or malate. The most important anaplerotic reaction in mammalian liver and kidney is the reversible carboxylation of pyruvate by CO_2 to form oxaloacetate, catalyzed by **pyruvate carboxylase.** When the citric acid cycle is deficient in oxaloacetate or any other intermediates, pyruvate is carboxylated to produce more oxaloacetate. The enzymatic addition of a carboxyl group to pyruvate requires energy, which is supplied by ATP—the free energy required to attach a carboxyl group to pyruvate is about equal to the free energy available from ATP.

Pyruvate carboxylase is a regulatory enzyme and is virtually inactive in the absence of acetyl-CoA, its positive allosteric modulator. Whenever acetyl-CoA, the fuel for the citric acid cycle, is present in excess, it stimulates the pyruvate carboxylase reaction to produce more oxaloacetate, enabling the cycle to use more acetyl-CoA in the citrate synthase reaction.

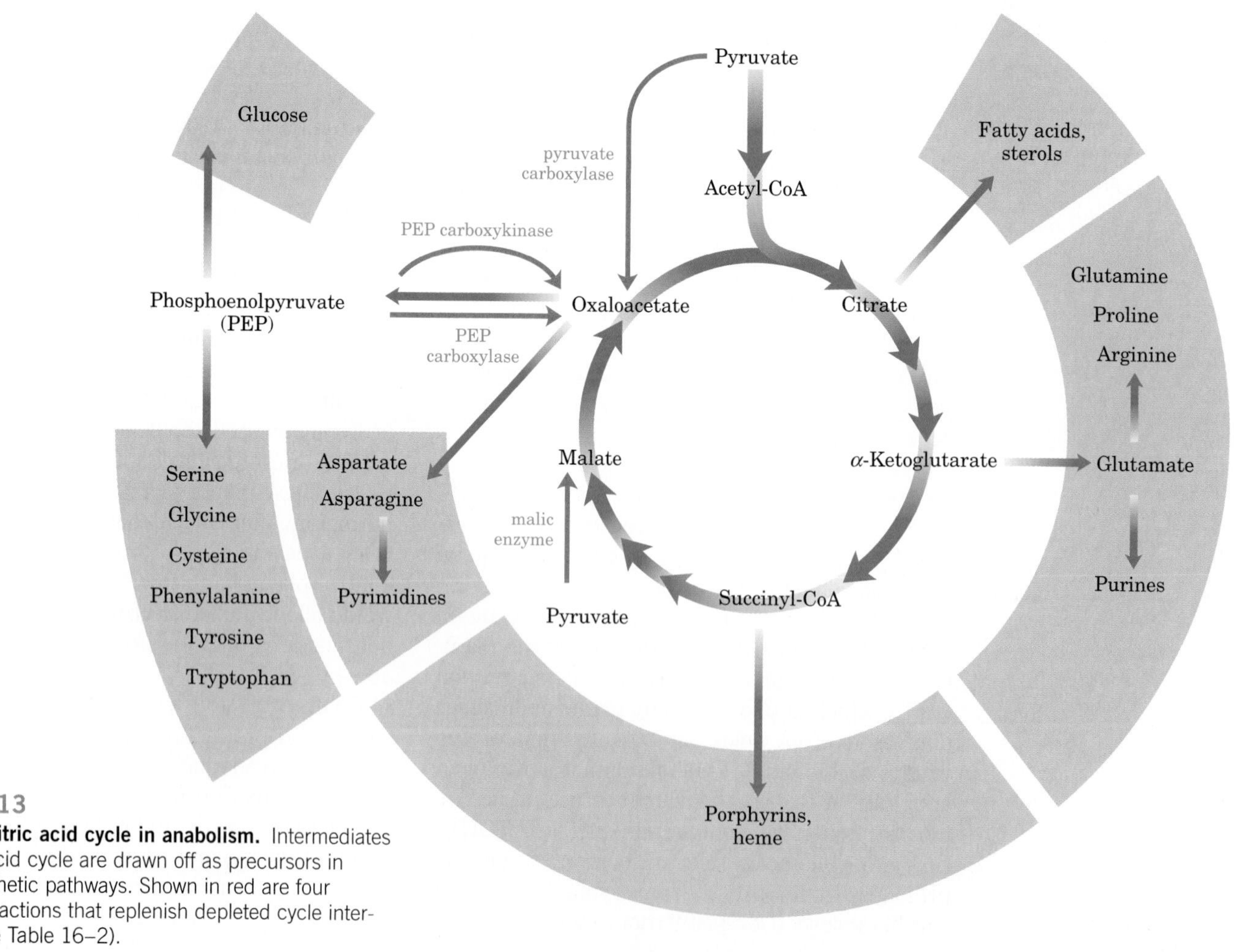

figure 16–13
Role of the citric acid cycle in anabolism. Intermediates of the citric acid cycle are drawn off as precursors in many biosynthetic pathways. Shown in red are four anaplerotic reactions that replenish depleted cycle intermediates (see Table 16–2).

table 16–2

Anaplerotic Reactions

Reaction	Tissue(s)/organism(s)
Pyruvate + HCO_3^- + ATP $\xrightarrow{\text{pyruvate carboxylase}}$ oxaloacetate + ADP + P_i	Liver, kidney
Phosphoenolpyruvate + CO_2 + GDP $\xrightleftharpoons{\text{PEP carboxykinase}}$ oxaloacetate + GTP	Heart, skeletal muscle
Phosphoenolpyruvate + HCO_3^- $\xrightarrow{\text{PEP carboxylase}}$ oxaloacetate + P_i	Higher plants, yeast, bacteria
Pyruvate + HCO_3^- + NAD(P)H $\xrightleftharpoons{\text{malic enzyme}}$ malate + $NAD(P)^+$	Widely distributed in eukaryotes and prokaryotes

The other anaplerotic reactions shown in Table 16–2 are also regulated to keep the level of intermediates high enough to support the activity of the citric acid cycle. Phosphoenolpyruvate carboxylase, for example, is activated by the glycolytic intermediate fructose 1,6-bisphosphate, which accumulates when the citric acid cycle operates too slowly to process the pyruvate generated by glycolysis.

Biotin in Pyruvate Carboxylase Carries CO_2 Groups

The pyruvate carboxylase reaction requires the vitamin **biotin** (Fig. 16–14), which is the prosthetic group of the enzyme. Biotin plays a key role in many carboxylation reactions. It is a specialized carrier of one-carbon groups in their most oxidized form: CO_2. (The transfer of one-carbon groups in more reduced forms is mediated by other cofactors, notably tetrahydrofolate and *S*-adenosylmethionine, as described in Chapter 18.) Carboxyl groups are activated in a reaction that splits ATP and joins CO_2 to enzyme-bound biotin. This "activated" CO_2 is then passed to an acceptor (pyruvate in this case) in a carboxylation reaction.

ATP

Bicarbonate

Biotinyl-enzyme

① ADP + P_i

Carboxybiotinyl-enzyme

Pyruvate enolate

②

Oxaloacetate

Biotinyl-enzyme

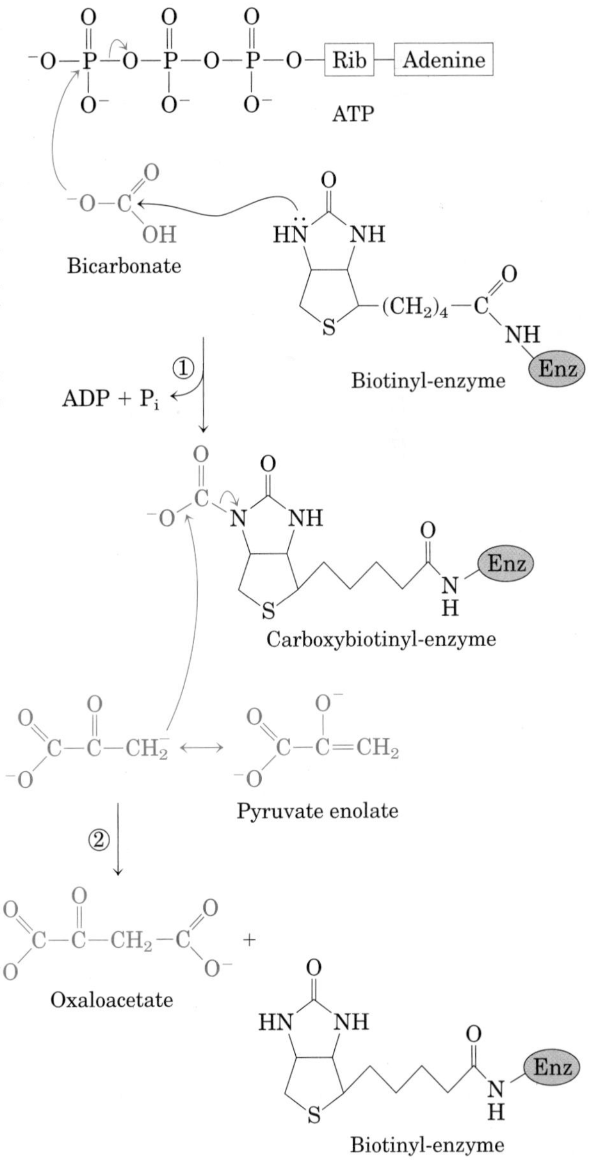

figure 16–14

Role of biotin in the pyruvate carboxylase reaction. Biotin is attached to the enzyme through an amide linkage to the ϵ-amino group of a Lys residue, forming biotinyl-enzyme. The enzyme catalyzes a two-step process. In step ①, a nitrogen atom of biotin makes a nucleophilic attack on bicarbonate ion (red), the predominant form of CO_2 at pH 7. Simultaneously, an oxygen atom of bicarbonate attacks the terminal phosphate of ATP, displacing P_i and forming a transient carboxyphosphate (not shown). This step adds the energy of ATP hydrolysis to the overall free-energy change for the reaction. The rest of the atoms of bicarbonate are now joined to biotin as "activated CO_2," carboxybiotinyl-enzyme. The carboxybiotinyl group is at the end of a long, flexible arm (the side chain of Lys and the methylene groups of biotin), which swings from the CO_2-activation site to the pyruvate carboxylation site. The covalent tether assures that the activated intermediate does not leave the enzyme surface. In step ②, pyruvate in its ionized enol (enolate) form (blue) makes a nucleophilic attack on the activated CO_2, displacing biotinyl-enzyme and forming oxaloacetate. Similar mechanisms occur in other biotin-dependent carboxylation reactions, such as propionyl-CoA carboxylase (see Fig. 17–11) and acetyl-CoA carboxylase (see Fig. 21–1).

Pyruvate carboxylase is composed of four identical subunits, each containing a molecule of biotin covalently attached through an amide linkage to the ϵ-amino group of a specific Lys residue in the enzyme active site (Fig. 16–14). Carboxylation of pyruvate proceeds in two steps: first, a carboxyl group derived from HCO_3^- is attached to biotin, then the carboxyl group is transferred to pyruvate to form oxaloacetate. These two steps occur at separate active sites; the long flexible arm of biotin permits the transfer of activated carboxyl groups from the first active site to the second, functioning much like the long lipoyllysyl arm of E_2 in the pyruvate dehydrogenase complex.

Biotin is a vitamin required in the human diet; it is abundant in many foods and is synthesized by intestinal bacteria. Biotin deficiency is rare, generally occurring only when large quantities of raw eggs are consumed. Egg whites contain a large amount of the protein **avidin** (M_r 70,000), which binds very tightly to biotin and prevents its absorption in the intestine. The avidin of egg whites may be a defense mechanism for the potential chick embryo, inhibiting the growth of bacteria. When eggs are cooked, avidin is denatured (and thereby inactivated) along with all other egg white proteins.

Regulation of the Citric Acid Cycle

Regulation of key enzymes in metabolic pathways, by allosteric effectors and covalent modification, assures production of intermediates and products at the rates required to keep the cell in a stable steady state and avoids wasteful overproduction of intermediates (see pp. 551–553). The flow of carbon atoms from pyruvate into and through the citric acid cycle is under tight regulation at two levels: the conversion of pyruvate to acetyl-CoA, the starting material for the cycle (the pyruvate dehydrogenase complex reaction) and the entry of acetyl-CoA into the cycle (the citrate synthase reaction). Because pyruvate is not the sole source of acetyl-CoA—most cells also produce acetyl-CoA by the oxidation of fatty acids and certain amino acids—the availability of intermediates from these other pathways is also important in the regulation of pyruvate oxidation and of the citric acid cycle. The cycle is also regulated at the isocitrate dehydrogenase and α-ketoglutarate dehydrogenase reactions.

Production of Acetyl-CoA by the Pyruvate Dehydrogenase Complex Is Regulated by Allosteric and Covalent Mechanisms

The pyruvate dehydrogenase complex of vertebrates is strongly inhibited by ATP and by acetyl-CoA and NADH, the products of the reaction catalyzed by the complex (Fig. 16–15). The allosteric inhibition of pyruvate oxidation is greatly enhanced when long-chain fatty acids are available. AMP, CoA, and NAD^+, all of which accumulate when too little acetate flows into the citric acid cycle, allosterically activate the pyruvate dehydrogenase complex. Thus this enzyme activity is turned off when ample fuel is available in the form of fatty acids and acetyl-CoA and when the cell's [ATP]/[ADP] and [NADH]/[NAD^+] ratios are high, and it is turned on again when energy demands are high and greater flux of acetyl-CoA into the citric acid cycle is required.

In vertebrates, these allosteric regulatory mechanisms are complemented by a second level of regulation, covalent protein modification. The pyruvate dehydrogenase complex is inhibited by reversible phosphorylation of a specific Ser residue on one of the two subunits of E_1. As noted earlier, in addition to the enzymes E_1, E_2, and E_3, the vertebrate complex contains two regulatory proteins whose sole purpose is to regulate the activity of the

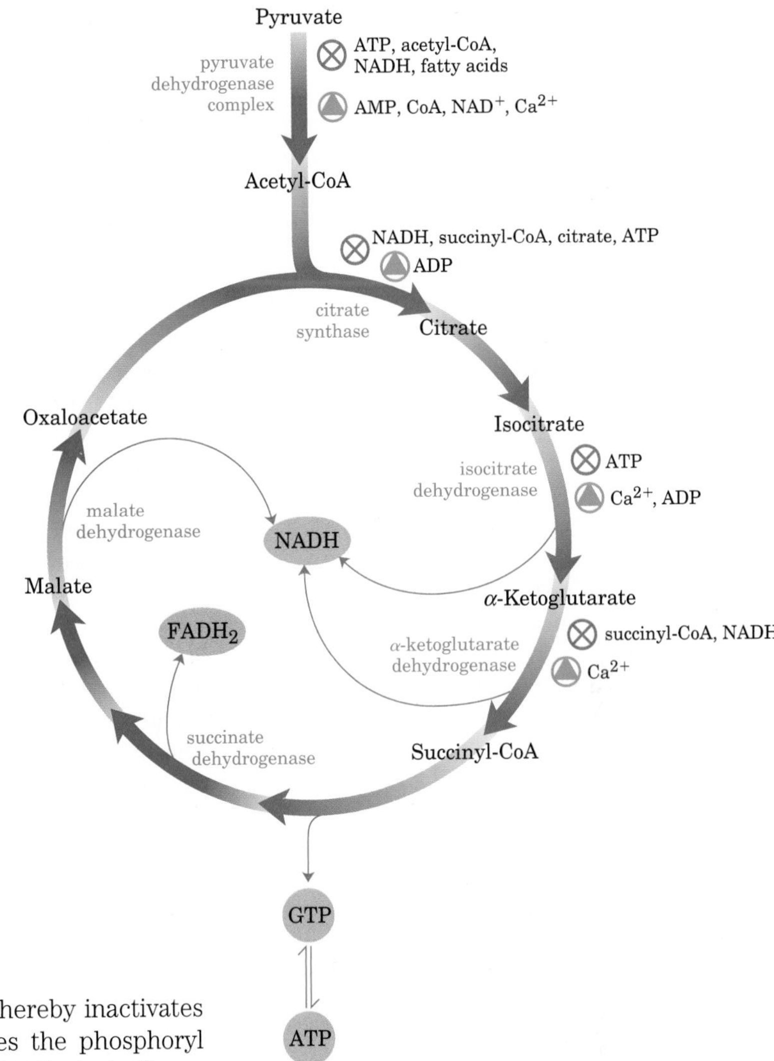

figure 16–15
Regulation of metabolite flow from pyruvate through the citric acid cycle. The pyruvate dehydrogenase complex is allosterically inhibited when the ratios of [ATP]/[ADP], [NADH]/[NAD^+], and [acetyl-CoA]/[CoA] are high, indicating an energy-sufficient metabolic state. When these ratios decrease, allosteric activation of pyruvate oxidation results. The rate of flow through the citric acid cycle can be limited by the availability of the citrate synthase substrates, oxaloacetate and acetyl-CoA, or of NAD^+ (depleted by its conversion to NADH), which slows the three NAD-dependent oxidation steps. Feedback inhibition by succinyl-CoA, citrate, and ATP also slows the cycle by inhibiting early steps. In muscle tissue, Ca^{2+} signals contraction and, as shown here, stimulates energy-yielding metabolism to replace the ATP consumed by contraction.

complex. A specific protein kinase phosphorylates and thereby inactivates E_1, and a specific phosphoprotein phosphatase removes the phosphoryl group by hydrolysis and thereby activates E_1. The kinase is allosterically activated by ATP: when [ATP] is high (reflecting a sufficient supply of energy), the pyruvate dehydrogenase complex is inactivated by phosphorylation of E_1. When [ATP] declines, kinase activity decreases and phosphatase action removes the phosphoryl groups from E_1, activating the complex.

The pyruvate dehydrogenase complex of plants, located in the mitochondrial matrix and in plastids, is strongly inhibited by NADH, which may be its primary regulator. There is also evidence for inactivation of the plant mitochondrial enzyme by reversible phosphorylation. The pyruvate dehydrogenase complex of *E. coli* is under allosteric regulation similar to that of the vertebrate enzyme, but does not seem to be regulated by phosphorylation.

The Citric Acid Cycle Is Regulated at Its Three Exergonic Steps

The flow of metabolites through the citric acid cycle is under stringent regulation. Three factors govern the rate of flux through the cycle: substrate availability, inhibition by accumulating products, and allosteric feedback inhibition of the enzymes that catalyze early steps in the cycle.

Each of the three strongly exergonic steps in the cycle—those catalyzed by citrate synthase, isocitrate dehydrogenase, and α-ketoglutarate dehydrogenase (Fig. 16–15)—can become the rate-limiting step under some circumstances. The availability of the substrates for citrate synthase (acetyl-CoA and oxaloacetate) varies with the metabolic state of the cell and sometimes limits the rate of citrate formation. NADH, a product of isocitrate and α-ketoglutarate oxidation, accumulates under some conditions, and at high [NADH]/[NAD^+] both dehydrogenase reactions are severely inhibited by mass action. Similarly, in the cell, the malate dehydrogenase reaction is essentially at equilibrium (i.e., it is substrate-limited; see Fig. 15–16), and when [NADH]/[NAD^+] is high the concentration of oxaloacetate is low, slowing the first step in the cycle. Product accumulation inhibits all three limiting steps of the cycle: succinyl-CoA inhibits α-ketoglutarate dehydrogenase (and also citrate synthase); citrate blocks citrate synthase; and the end product, ATP, inhibits both citrate synthase and isocitrate dehydrogenase. The inhibition of citrate synthase by ATP is relieved by ADP, an allosteric activator of this enzyme. In vertebrate muscle, Ca^{2+}, the signal for contraction and a concomitant increase in demand for ATP, activates both isocitrate dehydrogenase and α-ketoglutarate dehydrogenase, as well as the pyruvate dehydrogenase complex. In short, the concentrations of substrates and intermediates in the citric acid cycle set the flux through this pathway at a rate that provides optimal concentrations of ATP and NADH.

Under normal conditions the rates of glycolysis and of the citric acid cycle are integrated so that only as much glucose is metabolized to pyruvate as is needed to supply the citric acid cycle with its fuel, the acetyl groups of acetyl-CoA. Pyruvate, lactate, and acetyl-CoA are normally maintained at steady-state concentrations. The rate of glycolysis is matched to the rate of the citric acid cycle not only through its inhibition by high levels of ATP and NADH, which are common components of both the glycolytic and respiratory stages of glucose oxidation, but also by the concentration of citrate. Citrate, the product of the first step of the citric acid cycle, is as an important allosteric inhibitor of phosphofructokinase-1 in the glycolytic pathway (p. 555).

The Glyoxylate Cycle

Vertebrates cannot convert fatty acids, or the acetate derived from them, to carbohydrates. Conversion of phosphoenolpyruvate to pyruvate (p. 539) and of pyruvate to acetyl-CoA (Fig. 16–2) are so exergonic as to be essentially irreversible. If a cell cannot convert acetate into phosphoenolpyruvate, acetate cannot serve as the starting material for the gluconeogenic pathway, which leads from phosphoenolpyruvate to glucose (see Fig. 20–2). Without this capacity, a cell or organism is unable to convert fuels or metabolites that are degraded to acetate (fatty acids and certain amino acids) into carbohydrates.

As we saw in our discussion of anaplerotic reactions (Table 16–2), phosphoenolpyruvate can be synthesized from oxaloacetate in the reversible reaction catalyzed by PEP carboxykinase:

$$\text{Oxaloacetate} + \text{GTP} \rightleftharpoons \text{phosphoenolpyruvate} + CO_2 + \text{GDP}$$

Because carbon atoms of acetate molecules that enter the citric acid cycle appear eight steps later in oxaloacetate, it might seem that this pathway could generate oxaloacetate from acetate and thus generate phosphoenolpyruvate for gluconeogenesis. However, an examination of the stoichiometry of the citric acid cycle shows that there is no *net* conversion of acetate to oxaloacetate; in vertebrates, for every two carbons that enter the cycle as acetyl-CoA, two leave as CO_2. In many organisms other than vertebrates, the **glyoxylate cycle** serves as a mechanism for converting acetate to carbohydrate.

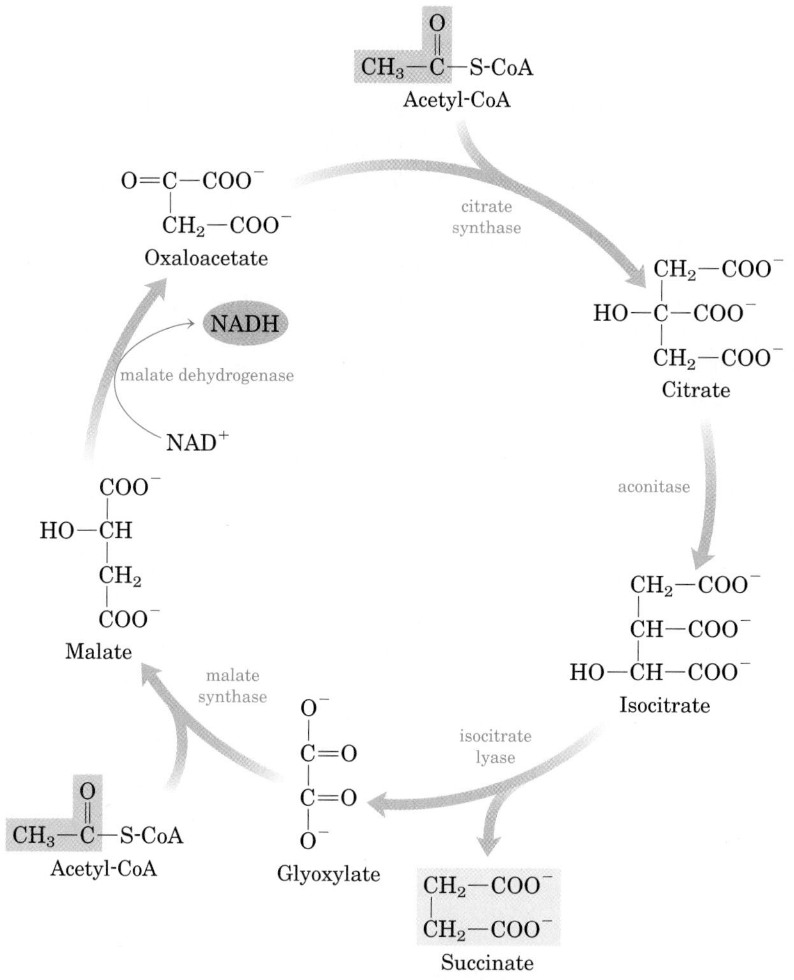

figure 16–16
Glyoxylate cycle. Citrate synthase, aconitase, and malate dehydrogenase are isozymes of the citric acid cycle enzymes; isocitrate lyase and malate synthase are unique to the glyoxylate cycle. Notice that two acetyl groups (red) enter the cycle and four carbons leave as succinate (blue). The glyoxylate cycle was elucidated by Hans Kornberg and Neil Madsen in the laboratory of Hans Krebs.

The Glyoxylate Cycle Produces Four-Carbon Compounds from Acetate

In plants, certain invertebrates, and some microorganisms such as *E. coli* and yeast, acetate can serve both as an energy-rich fuel and as a source of phosphoenolpyruvate for carbohydrate synthesis. In these organisms, enzymes of the **glyoxylate cycle** catalyze the net conversion of acetate to succinate or other four-carbon intermediates of the citric acid cycle:

$$2\ \text{Acetyl-CoA} + NAD^+ + 2H_2O \longrightarrow \text{succinate} + 2\text{CoA} + NADH + H^+$$

In the glyoxylate cycle, acetyl-CoA condenses with oxaloacetate to form citrate and citrate is converted to isocitrate, exactly as in the citric acid cycle. The next step, however, is not the breakdown of isocitrate by isocitrate dehydrogenase but the cleavage of isocitrate by **isocitrate lyase,** forming succinate and **glyoxylate.** The glyoxylate then condenses with a second molecule of acetyl-CoA to yield malate in a reaction catalyzed by **malate synthase.** The malate is subsequently oxidized to oxaloacetate, which can condense with another molecule of acetyl-CoA to start another turn of the cycle (Fig. 16–16). Each turn of the glyoxylate cycle consumes two molecules of acetyl-CoA and produces one molecule of succinate, available for biosynthetic purposes. The succinate may be converted through fumarate and malate into oxaloacetate, which can then be converted to phosphoenolpyruvate by PEP carboxykinase, and thus to glucose by gluconeogenesis.

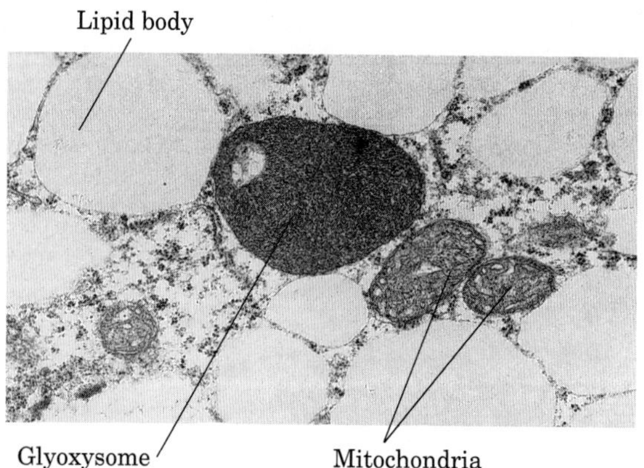

figure 16–17
Electron micrograph of a germinating cucumber seed, showing a glyoxysome, mitochondria, and surrounding lipid bodies.

In plants, the enzymes of the glyoxylate cycle are sequestered in membrane-bounded organelles called glyoxysomes (Fig. 16–17). Those enzymes common to the citric acid and glyoxylate cycles have two isozymes, one specific to mitochondria, the other to glyoxysomes. Glyoxysomes are not present in all plant tissues at all times. They develop in lipid-rich seeds during germination, before the developing plants acquire the ability to make glucose by photosynthesis. In addition to glyoxylate cycle enzymes, glyoxysomes also contain all the enzymes needed for the degradation of the fatty acids stored in seed oils (see Fig. 17–12). Acetyl-CoA formed from lipid breakdown is converted to succinate via the glyoxylate cycle and the succinate is exported to mitochondria, where citric acid cycle enzymes transform it to malate. A cytosolic isozyme of malate dehydrogenase oxidizes malate to oxaloacetate, a precursor for gluconeogenesis. Germinating seeds can therefore convert the carbon of stored lipids into glucose.

Vertebrate animals do not have the enzymes specific to the glyoxylate cycle (isocitrate lyase and malate synthase) and therefore cannot bring about the net synthesis of glucose from lipids.

The Citric Acid and Glyoxylate Cycles Are Coordinately Regulated

In germinating seeds, the enzymatic transformations of dicarboxylic and tricarboxylic acids occur in three intracellular compartments: mitochondria, glyoxysomes, and the cytosol. There is a continuous interchange of intermediates among these compartments (Fig. 16–18).

figure 16–18
Relationship between the glyoxylate and citric acid cycles. The reactions of the glyoxylate cycle (in glyoxysomes) proceed simultaneously with, and mesh with, those of the citric acid cycle (in mitochondria), as intermediates pass through the cytosol between these compartments. The conversion of succinate to oxaloacetate is catalyzed by citric acid cycle enzymes. The oxidation of fatty acids to acetyl-CoA is described in Chapter 17; the synthesis of hexoses from oxaloacetate in Chapter 20.

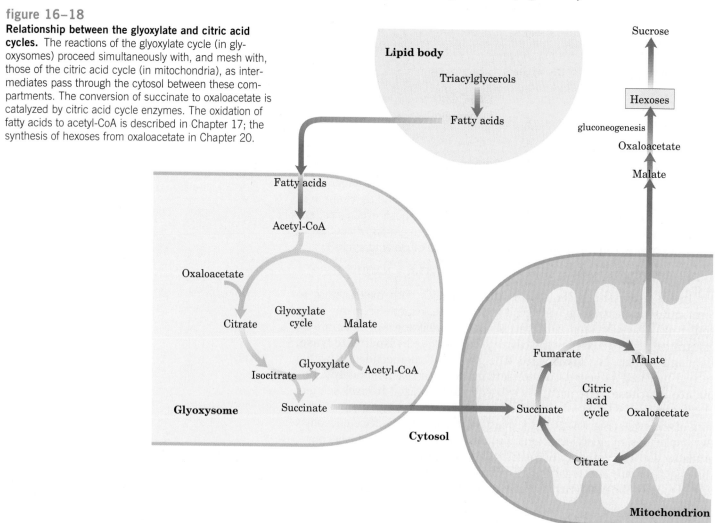

Lipid body
Triacylglycerols
Fatty acids
Sucrose
Hexoses
gluconeogenesis
Oxaloacetate
Malate
Fatty acids
Acetyl-CoA
Oxaloacetate
Citrate
Glyoxylate cycle
Malate
Glyoxylate
Acetyl-CoA
Isocitrate
Succinate
Glyoxysome
Cytosol
Fumarate
Malate
Citric acid cycle
Succinate
Oxaloacetate
Citrate
Mitochondrion

The carbon skeleton of oxaloacetate from the citric acid cycle (in the mitochondrion) is carried to the glyoxysome in the form of aspartate. Aspartate is converted to oxaloacetate, which condenses with acetyl-CoA derived from fatty acid breakdown. The citrate thus formed is converted to isocitrate by aconitase, then split into glyoxylate and succinate by isocitrate lyase. The succinate returns to the mitochondrion, where it reenters the citric acid cycle and is transformed into malate, which enters the cytosol and is oxidized (by cytosolic malate dehydrogenase) to oxaloacetate. Oxaloacetate is converted via gluconeogenesis into hexoses and sucrose, which can be transported to the growing roots and shoot. Four distinct pathways participate in these conversions: fatty acid breakdown to acetyl-CoA (in glyoxysomes), the glyoxylate cycle (in glyoxysomes), the citric acid cycle (in mitochondria), and gluconeogenesis (in the cytosol).

The sharing of common intermediates requires that these pathways be coordinately regulated. Isocitrate is a crucial intermediate, at the branch point between the glyoxylate and citric acid cycles (Fig. 16–19). Isocitrate dehydrogenase is regulated by covalent modification: a specific protein kinase phosphorylates and thereby inactivates the dehydrogenase. This inactivation shunts isocitrate to the glyoxylate cycle, where it begins the synthetic route toward glucose. A phosphoprotein phosphatase removes the phosphoryl group from isocitrate dehydrogenase, reactivating the enzyme and sending more isocitrate through the energy-yielding citric acid cycle. The regulatory protein kinase and phosphoprotein phosphatase are separate enzymatic activities on a single polypeptide.

Some bacteria, including *E. coli,* have the full complement of enzymes for the glyoxylate and citric acid cycles in the cytosol and can therefore grow on acetate as their sole source of carbon and energy. The phosphatase that activates isocitrate dehydrogenase is stimulated by intermediates of the citric acid cycle and glycolysis and by indicators of reduced cellular energy supply (Fig. 16–19). The same metabolites *inhibit* the protein kinase activity of the bifunctional enzyme. Thus, the accumulation of intermediates of the central energy-yielding pathways—indicating energy depletion—results in the activation of isocitrate dehydrogenase. When the concentration of these regulators falls, signaling a sufficient flux through the energy-yielding citric acid cycle, isocitrate dehydrogenase is inactivated by the protein kinase.

The same intermediates of the glycolytic and citric acid cycles that activate isocitrate dehydrogenase are allosteric inhibitors of isocitrate lyase. When energy-yielding metabolism is sufficiently fast to keep the concentrations of glycolytic and citric acid cycle intermediates low, isocitrate dehydrogenase is inactivated, the inhibition of isocitrate lyase is relieved, and isocitrate flows into the glyoxylate pathway, to be used in the biosynthesis of carbohydrates, amino acids, and other cellular components.

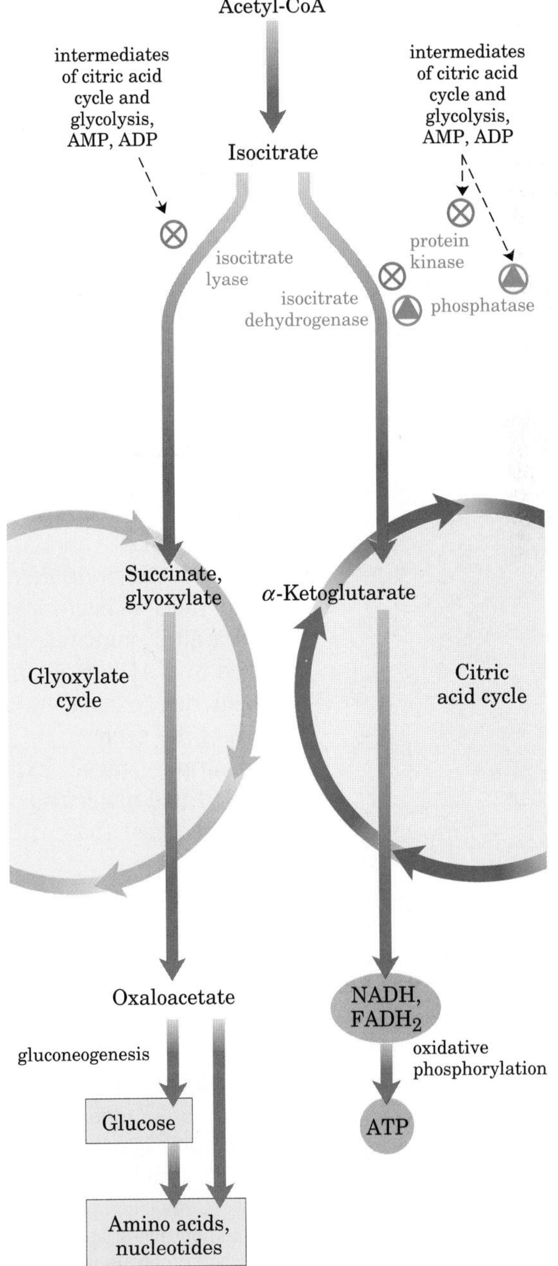

figure 16–19
Regulation of isocitrate dehydrogenase activity that determines partitioning of isocitrate between the glyoxylate and citric acid cycles. When isocitrate dehydrogenase is inactivated by phosphorylation (by a specific protein kinase), isocitrate is directed into biosynthetic reactions via the glyoxylate cycle. When the enzyme is activated by dephosphorylation (by a specific phosphatase), isocitrate enters the citric acid cycle and ATP is produced.

summary

Cellular respiration occurs in three stages: (1) the oxidative formation of acetyl-CoA from pyruvate, fatty acids, and some amino acids; (2) the degradation of acetyl residues by the citric acid cycle to yield CO_2 and reduced coenzymes; and (3) the stepwise transfer of electrons from the reduced coenzymes to molecular oxygen, coupled to the phosphorylation of ADP to ATP. The oxidative catabolism of glucose yields much more energy than the anaerobic fermentation pathways.

Pyruvate, the end product of glycolysis, undergoes dehydrogenation and decarboxylation by the pyruvate dehydrogenase complex—a huge complex containing three sequentially acting enzymes and requiring five coenzymes—to yield acetyl-CoA and CO_2. Acetyl-CoA enters the citric acid cycle (in the mitochondria of eukaryotes, the cytosol of prokaryotes) as citrate synthase catalyzes its condensation with oxaloacetate to form citrate. Aconitase catalyzes the reversible formation of isocitrate from citrate; isocitrate is then oxidized to α-ketoglutarate by isocitrate dehydrogenase in a reaction that also yields CO_2. The α-ketoglutarate undergoes another dehydrogenation and decarboxylation to succinyl-CoA and CO_2. Succinyl-CoA reacts with ADP (or GDP) and P_i to form free succinate and ATP (or GTP), in a substrate-level phosphorylation. Succinate is then oxidized to fumarate by succinate dehydrogenase, an FAD-linked enzyme (in the inner mitochondrial membrane, the plasma membrane in bacteria). Fumarate is reversibly hydrated by fumarase to L-malate, which is oxidized by NAD-linked L-malate dehydrogenase to regenerate oxaloacetate, available for another turn of the cycle.

Isotopic tracer experiments with carbon-labeled fuel molecules or intermediates have established that the citric acid cycle is the major pathway of carbohydrate oxidation in aerobic cells. The pyruvate dehydrogenase complex of vertebrates is inhibited by the allosteric effectors NADH, ATP, and acetyl-CoA, and by reversible phosphorylation catalyzed by a protein kinase and phosphatase that are part of the enzyme complex. The overall rate of the citric acid cycle is controlled by the rate of conversion of pyruvate to acetyl-CoA and by the flux through citrate synthase, isocitrate dehydrogenase, and α-ketoglutarate dehydrogenase. These fluxes are largely determined by the concentrations of substrates and products; the end products ATP and NADH are inhibitory.

Citric acid cycle intermediates are also used as precursors in the biosynthesis of amino acids and other biomolecules. The cycle intermediates are replenished by anaplerotic reactions catalyzed by pyruvate carboxylase, PEP carboxykinase, PEP carboxylase, and malic enzyme. In the germinating seeds of some plants, and in certain microorganisms that can live on acetate as sole carbon source for the synthesis of carbohydrate, the glyoxylate cycle comes into play. This pathway (in glyoxysomes) involves several citric acid cycle enzymes and two additional enzymes: isocitrate lyase and malate synthase. Because the glyoxylate cycle bypasses the two decarboxylation steps of the citric acid cycle, it makes possible the *net* formation of succinate, oxaloacetate, and other cycle intermediates from acetyl-CoA. Oxaloacetate thus formed can be used to synthesize glucose via gluconeogenesis. The partitioning of isocitrate between the citric acid cycle and the glyoxylate cycle is controlled at the level of isocitrate dehydrogenase. This enzyme is subject to regulation by reversible phosphorylation. Vertebrates lack the glyoxylate cycle and cannot synthesize glucose from acetate.

further reading

General

Holmes, F.L. (1990, 1993) *Hans Krebs, Volume 1, Formation of a Scientific Life, 1900–1933*; *Volume 2, Architect of Intermediary Metabolism, 1933–1937*, Oxford University Press, Oxford.

A scientific and personal biography of Krebs by an eminent historian of science, with a thorough description of the work that revealed the urea and citric acid cycles.

Kay, J. & Weitzman, P.D.J. (eds) (1987) *Krebs' Citric Acid Cycle: Half a Century and Still Turning*, Biochemical Society Symposium 54, The Biochemical Society, London.

A multiauthor book on the citric acid cycle, including molecular genetics, regulatory mechanisms, variations on the cycle in microorganisms from unusual ecological niches, and evolution of the pathway.

Especially relevant are the chapters by H. Gest (Evolutionary roots of the citric acid cycle in prokaryotes), W.H. Holms (Control of flux through the citric acid cycle and the glyoxylate bypass in *Escherichia coli),* and R.N. Perham et al. (α-Keto acid dehydrogenase complexes).

Pyruvate Dehydrogenase Complex

Behal, R.H., Buxton, D.B., Robertson, J.G., & Olson, M.S. (1993) Regulation of the pyruvate dehydrogenase multienzyme complex. *Annu. Rev. Nutr.* **13,** 497–520.

de Kok, A., Hangeveld, A.F., Martin, A., & Westphal, A.H. (1998) The pyruvate dehydrogenase multi-enzyme complex from Gram-negative bacteria. *Biochim. Biophys. Acta* **1385,** 353–366.

Roche, T.E. & Patel, M.S. (eds) (1990) Alpha-Keto Acid Dehydrogenase Complexes: Organization, Regulation, and Biomedical Ramifications. *Ann. N.Y. Acad. Sci.* **573.**

This volume contains about 60 papers covering all aspects of the enzyme group that includes the pyruvate dehydrogenase and α-ketoglutarate dehydrogenase complexes.

Citric Acid Cycle Enzymes

Goward C.R. & Nicholls D.J. (1994) Malate dehydrogenase: a model for structure, evolution, and catalysis. *Protein Sci.* **3,** 1883–1888.

A good, short review.

Hagerhall, C. (1997) Succinate: quinone oxidoreductases. Variations on a conserved theme. *Biochim. Biophys. Acta* **1320,** 107–141.

A review of the structure and function of succinate dehydrogenases.

Knowles, J. (1989) The mechanism of biotin-dependent enzymes. *Annu. Rev. Biochem.* **58,** 195–221.

Musrati, R.A., Kollarova, M., Mernik, N., & Mikulasova, D. (1998) Malate dehydrogenase: distribution, function and properties. *Gen. Physiol. Biophys.* **17,** 193–210.

A short review.

Nishimura, J.S. (1986) Succinyl-CoA synthetase structure-function relationships and other considerations. *Adv. Enzymol. Relat. Areas Mol. Biol.* **58,** 141–72.

Remington, S.J. (1992) Structure and mechanism of citrate synthase. *Curr. Top. Cell. Regul.* **33,** 209–229.

A thorough review of this enzyme.

Singer, T.P. & Johnson, M.K. (1985) The prosthetic groups of succinate dehydrogenase: 30 years from discovery to identification. *FEBS Lett.* **190,** 189–198.

A description of the structure and role of the iron-sulfur centers in this enzyme.

Velot, C., Mixon, M.B., Teige, M., & Srere, P.A. (1997) Model of a quinary structure between Krebs TCA cycle enzymes: a model for the metabolon. *Biochemistry* **36,** 14,271–14,276.

An interesting suggestion for how the citric acid cycle enzymes may interact to channel intermediates in the path.

Weigand, G. & Remington, S.J. (1986) Citrate synthase: structure, control, and mechanism. *Annu. Rev. Biophys. Biophys. Chem.* **15,** 97–117.

Wolodko, W.T., Fraser, M.E., James, M.N.G., & Bridger, W.A. (1994) The crystal structure of succinyl-CoA synthetase from *Escherichia coli* at 2.5-Å resolution. *J. Biol. Chem.* **269,** 10,883–10,890.

Regulation of the Citric Acid Cycle

Hansford, R.G. (1980) Control of mitochondrial substrate oxidation. *Curr. Top. Bioenerget.* **10,** 217–278.

A detailed review of the regulation of the citric acid cycle.

Kaplan, N.O. (1985) The role of pyridine nucleotides in regulating cellular metabolism. *Curr. Top. Cell. Regul.* **26,** 371–381.

An excellent general discussion of the importance of the [NADH]/[NAD^+] ratio in cellular regulation.

Reed, L.J., Damuni, Z., & Merryfield, M.L. (1985) Regulation of mammalian pyruvate and branched-chain α-keto acid dehydrogenase complexes by phosphorylation-dephosphorylation. *Curr. Top. Cell. Regul.* **27,** 41–49.

Glyoxylate Cycle

Holms, W.H. (1986) The central metabolic pathways of *Escherichia coli:* relationship between flux and control at a branch point, efficiency of conversion to biomass, and excretion of acetate. *Curr. Top. Cell. Regul.* **28,** 69–106.

problems

1. Balance Sheet for the Citric Acid Cycle The citric acid cycle has eight enzymes: citrate synthase, aconitase, isocitrate dehydrogenase, α-ketoglutarate dehydrogenase, succinyl-CoA synthetase, succinate dehydrogenase, fumarase, and malate dehydrogenase.

(a) Write a balanced equation for the reaction catalyzed by each enzyme.

(b) Name the cofactor(s) required by each enzyme reaction.

(c) For each enzyme determine which of the following describes the type of reaction(s) catalyzed: condensation (carbon–carbon bond formation); dehydration (loss of water); hydration (addition of water); decarboxylation (loss of CO_2); oxidation-reduction; substrate-level phosphorylation; isomerization.

(d) Write a balanced net equation for the catabolism of acetyl-CoA to CO_2.

2. Recognizing Oxidation and Reduction Reactions One biochemical strategy of many living organisms is the stepwise oxidation of organic compounds to CO_2 and H_2O and the conservation of a major part of the energy thus produced in the form of ATP. It is important to be able to recognize oxidation-reduction processes in metabolism. Reduction of an organic molecule results from the hydrogenation of a double bond (Eqn 1 below) or of a single bond with accompanying cleavage (Eqn 2). Conversely, oxidation results from dehydrogenation. In biochemical redox reactions, the coenzymes NAD and FAD dehydrogenate/hydrogenate organic molecules in the presence of the proper enzymes.

$$\underset{\text{Acetaldehyde}}{CH_3\text{—}C(\text{=O})\text{—}H} + H\text{—}H \underset{\text{oxidation}}{\overset{\text{reduction}}{\rightleftharpoons}} \left[CH_3\text{—}C(\text{=O})(H)\cdots H\text{—}H\right] \underset{\text{oxidation}}{\overset{\text{reduction}}{\rightleftharpoons}} \underset{\text{Ethanol}}{CH_3\text{—}C(H)(H)\text{—}O\text{—}H} \quad (1)$$

$$\underset{\text{Acetate}}{CH_3\text{—}C(\text{=O})\text{—}O^-} + H^+ + H\text{—}H \underset{\text{oxidation}}{\overset{\text{reduction}}{\rightleftharpoons}} \left[CH_3\text{—}C(\text{=O})\text{—}O^- \;\; H\text{—}H \;\; H^+\right] \underset{\text{oxidation}}{\overset{\text{reduction}}{\rightleftharpoons}} \underset{\text{Acetaldehyde}}{CH_3\text{—}C(\text{=O})\text{—}H} + H\text{—}O\text{—}H \quad (2)$$

For each of the metabolic transformations in (a) through (h), determine whether oxidation or reduction has occurred. Balance each transformation by inserting H—H and, where necessary, H_2O.

(a) $\underset{\text{Methanol}}{CH_3\text{—}OH} \longrightarrow \underset{\text{Formaldehyde}}{H\text{—}C(\text{=O})\text{—}H}$

(b) $\underset{\text{Formaldehyde}}{H\text{—}C(\text{=O})\text{—}H} \longrightarrow \underset{\text{Formate}}{H\text{—}C(\text{=O})\text{—}O^-} + H^+$

(c) $\underset{\text{Carbon dioxide}}{O\text{=}C\text{=}O} \longrightarrow \underset{\text{Formate}}{H\text{—}C(\text{=O})\text{—}O^-} + H^+$

(d) $\underset{\text{Glycerate}}{CH_2(OH)\text{—}CH(OH)\text{—}C(\text{=O})\text{—}O^-} + H^+ \longrightarrow \underset{\text{Glyceraldehyde}}{CH_2(OH)\text{—}CH(OH)\text{—}C(\text{=O})\text{—}H}$

(e) $\underset{\text{Glycerol}}{CH_2(OH)\text{—}CH(OH)\text{—}CH_2(OH)} \longrightarrow \underset{\text{Dihydroxyacetone}}{CH_2(OH)\text{—}C(\text{=O})\text{—}CH_2(OH)}$

(f) $\underset{\text{Toluene}}{C_6H_5\text{—}CH_3} \longrightarrow \underset{\text{Benzoate}}{C_6H_5\text{—}C(\text{=O})\text{—}O^-} + H^+$

(g) $\underset{\text{Succinate}}{^-O\text{—}C(\text{=O})\text{—}CH_2\text{—}CH_2\text{—}C(\text{=O})\text{—}O^-} \longrightarrow \underset{\text{Fumarate}}{^-O\text{—}C(\text{=O})\text{—}CH\text{=}CH\text{—}C(\text{=O})\text{—}O^-}$

(h) $\underset{\text{Pyruvate}}{CH_3\text{—}C(\text{=O})\text{—}C(\text{=O})\text{—}O^-} \longrightarrow \underset{\text{Acetate}}{CH_3\text{—}C(\text{=O})\text{—}O^-} + CO_2$

3. Relationship between Energy Release and the Oxidation State of Carbon A eukaryotic cell can use glucose ($C_6H_{12}O_6$) and hexanoic acid ($C_6H_{14}O_2$) as fuels for cellular respiration. On the basis of their structural formulas, which substance releases more energy on complete combustion to CO_2 and H_2O?

4. Nicotinamide Coenzymes as Reversible Redox Carriers The nicotinamide coenzymes (see Fig. 14–15) can undergo reversible oxidation-reduction reactions with specific substrates in the presence of the appropriate dehydrogenase. In these reactions, NADH + H^+ serves as the hydrogen source, as described in Problem 2. Whenever the coenzyme is oxidized, a substrate must be simultaneously reduced:

$$\underset{\text{Oxidized}}{\text{Substrate}} + \underset{\text{Reduced}}{\text{NADH} + \text{H}^+} \rightleftharpoons \underset{\text{Reduced}}{\text{product}} + \underset{\text{Oxidized}}{\text{NAD}^+}$$

For each of the reactions in (a) through (f), determine whether the substrate has been oxidized or reduced or is unchanged in oxidation state (see Problem 2). If a redox change has occurred, balance the reaction with the necessary amount of NAD^+, NADH, H^+, and H_2O. The objective is to recognize when a redox coenzyme is necessary in a metabolic reaction.

(a) $CH_3CH_2OH \longrightarrow CH_3{-}CHO$
Ethanol → Acetaldehyde

(b) $^{2-}O_3PO{-}CH_2{-}CH(OH){-}C(=O)OPO_3^{2-} \longrightarrow {}^{2-}O_3PO{-}CH_2{-}CH(OH){-}CHO + HPO_4^{2-}$
1,3-Bisphosphoglycerate → Glyceraldehyde 3-phosphate

(c) $CH_3{-}C(=O){-}COO^- \longrightarrow CH_3{-}CHO + CO_2$
Pyruvate → Acetaldehyde

(d) $CH_3{-}C(=O){-}COO^- \longrightarrow CH_3{-}COO^- + CO_2$
Pyruvate → Acetate

(e) $^-OOC{-}CH_2{-}C(=O){-}COO^- \longrightarrow {}^-OOC{-}CH_2{-}CH(OH){-}COO^-$
Oxaloacetate → Malate

(f) $CH_3{-}C(=O){-}CH_2{-}COO^- + H^+ \longrightarrow CH_3{-}C(=O){-}CH_3 + CO_2$
Acetoacetate → Acetone

5. Stimulation of Oxygen Consumption by Oxaloacetate and Malate In the early 1930s, Albert Szent-Györgyi reported the interesting observation that the addition of small amounts of oxaloacetate or malate to suspensions of minced pigeon-breast muscle stimulated the oxygen consumption of the preparation. Surprisingly, the amount of oxygen consumed was about seven times more than the amount necessary for complete oxidation (to CO_2 and H_2O) of the added oxaloacetate or malate. Why did the addition of oxaloacetate or malate stimulate oxygen consumption? Why was the amount of oxygen consumed so much greater than the amount necessary to completely oxidize the added oxaloacetate or malate?

6. Formation of Oxaloacetate in a Mitochondrion In the last reaction of the citric acid cycle, malate is dehydrogenated to regenerate the oxaloacetate necessary for the entry of acetyl-CoA into the cycle:

$$\text{L-Malate} + \text{NAD}^+ \longrightarrow \text{oxaloacetate} + \text{NADH} + \text{H}^+ \qquad \Delta G'^\circ = 30 \text{ kJ/mol}$$

(a) Calculate the equilibrium constant for this reaction at 25 °C.

(b) Because $\Delta G'^\circ$ assumes a standard pH of 7, the equilibrium constant calculated in (a) corresponds to

$$K'_{eq} = \frac{[\text{oxaloacetate}][\text{NADH}]}{[\text{L-malate}][\text{NAD}^+]}$$

The measured concentration of L-malate in rat liver mitochondria is about 0.20 mM when $[NAD^+]/[NADH]$ is 10. Calculate the concentration of oxaloacetate at pH 7 in these mitochondria.

(c) Rat liver mitochondria are roughly spherical, with a diameter of about 2 μm. To appreciate the magnitude of the mitochondrial oxaloacetate concentration, calculate the number of oxaloacetate molecules in a single rat liver mitochondrion.

7. Energy Yield from the Citric Acid Cycle The reaction catalyzed by succinyl-CoA synthetase produces the high-energy compound GTP. How is the free energy contained in GTP incorporated into the cellular ATP pool?

8. Respiration Studies in Isolated Mitochondria Cellular respiration can be studied in isolated mitochondria by measuring oxygen consumption under different conditions. If 0.01 M sodium malonate is added to actively respiring mitochondria that are using pyruvate as fuel source, respiration soon stops and a metabolic intermediate accumulates.

(a) What is the structure of this intermediate?

(b) Explain why it accumulates.

(c) Explain why oxygen consumption stops.

(d) Aside from removing the malonate, how can this inhibition of respiration be overcome? Explain.

9. Labeling Studies in Isolated Mitochondria The metabolic pathways of organic compounds have often been delineated by using a radioactively labeled substrate and following the fate of the label.

(a) How can you determine whether glucose added to a suspension of isolated mitochondria is metabolized to CO_2 and H_2O?

(b) Suppose you add a brief pulse of [3-^{14}C]pyruvate (labeled in the methyl position) to the mitochondria. After one turn of the citric acid cycle, what is the location of the ^{14}C in the oxaloacetate? Explain by tracing the ^{14}C label through the pathway. How many turns of the cycle are required to release all of the [3-^{14}C]pyruvate as CO_2?

10. [1-^{14}C]Glucose Catabolism An actively respiring bacterial culture is briefly incubated with [1-^{14}C] glucose, and the glycolytic and citric acid cycle intermediates are isolated. Where is the ^{14}C in each of the intermediates listed below? Consider only the initial incorporation of ^{14}C, in the first pass of labeled glucose through the pathways.

(a) Fructose 1,6-bisphosphate

(b) Glyceraldehyde 3-phosphate

(c) Phosphoenolpyruvate

(d) Acetyl-CoA

(e) Citrate

(f) α-Ketoglutarate

(g) Oxaloacetate

11. Role of the Vitamin Thiamin People with beriberi, a disease caused by thiamin deficiency, have elevated levels of blood pyruvate and α-ketoglutarate, especially after consuming a meal rich in glucose. How are these effects related to thiamin deficiency?

12. Synthesis of Oxaloacetate by the Citric Acid Cycle Oxaloacetate is formed in the last step of the citric acid cycle by the NAD$^+$-dependent oxidation of L-malate. Can a net synthesis of oxaloacetate from acetyl-CoA occur using only the enzymes and cofactors of the citric acid cycle, without depleting the intermediates of the cycle? Explain. How is oxaloacetate that is lost from the cycle (to biosynthetic reactions) replenished?

13. Mode of Action of the Rodenticide Fluoroacetate Fluoroacetate, prepared commercially for rodent control, is also produced by a South African plant. After entering a cell, fluoroacetate is converted to fluoroacetyl-CoA in a reaction catalyzed by the enzyme acetate thiokinase:

$$\text{F—CH}_2\text{COO}^- + \text{CoA-SH} + \text{ATP} \longrightarrow \text{F—CH}_2\overset{\text{O}}{\overset{\|}{\text{C}}}\text{—S-CoA} + \text{AMP} + \text{PP}_i$$

The toxic effect of fluoroacetate was studied in an experiment using intact isolated rat heart. After the heart was perfused with 0.22 mM fluoroacetate, the measured rate of glucose uptake and glycolysis decreased, and glucose 6-phosphate and fructose 6-phosphate accumulated. Examination of the citric acid cycle intermediates revealed that their concentrations were below normal, except for citrate, with a concentration 10 times higher than normal.

(a) Where did the block in the citric acid cycle occur? What caused citrate to accumulate and the other cycle intermediates to be depleted?

(b) Fluoroacetyl-CoA is enzymatically transformed in the citric acid cycle. What is the structure of the end product of fluoroacetate metabolism? Why does it block the citric acid cycle? How might the inhibition be overcome?

(c) In the heart perfusion experiments, why did glucose uptake and glycolysis decrease? Why did hexose monophosphates accumulate?

(d) Why is fluoroacetate poisoning fatal?

14. Synthesis of L-Malate in Wine Making The tartness of some wines is due to high concentrations of L-malate. Write a sequence of reactions showing how yeast cells synthesize L-malate from glucose under anaerobic conditions in the presence of dissolved CO_2 (HCO_3^-). Note that the overall reaction for this fermentation cannot involve the consumption of nicotinamide coenzymes or citric acid cycle intermediates.

15. Net Synthesis of α-Ketoglutarate α-Ketoglutarate plays a central role in the biosynthesis of several amino acids. Write a sequence of enzymatic reactions that could result in the net synthesis of α-ketoglutarate from pyruvate. Your proposed sequence must not involve the net consumption of other citric acid cycle intermediates. Write an equation for the overall reaction and identify the source of each reactant.

16. Regulation of Pyruvate Dehydrogenase In animal tissue, the rate of conversion of pyruvate to acetyl-CoA is regulated by the ratio of active, phosphorylated pyruvate dehydrogenase to inactive, unphosphorylated pyruvate dehydrogenase phosphate. Determine what happens to the rate of this reaction when a preparation of rabbit muscle mitochondria containing pyruvate dehydrogenase is treated with (a) pyruvate dehydrogenase kinase, ATP, and NADH; (b) pyruvate dehydrogenase phosphatase and Ca^{2+}; (c) malonate.

17. Commercial Synthesis of Citric Acid Citric acid is used as a flavoring agent in soft drinks, fruit juices, and numerous other foods. Worldwide, the market for citric acid is hundreds of millions of dollars per year. Commercial production uses the mold

Aspergillus niger, acting on sucrose under carefully controlled conditions.

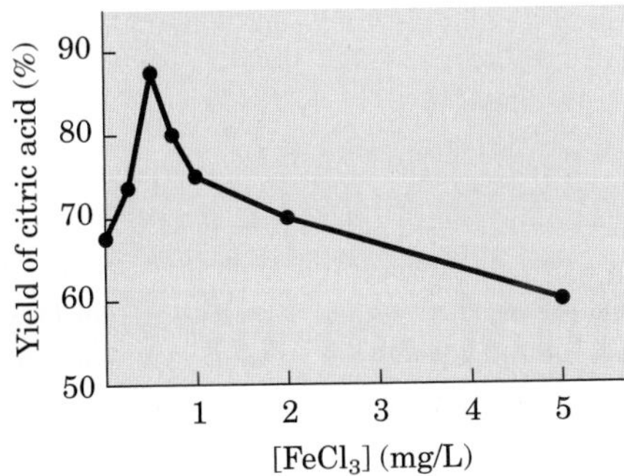

(a) The yield of citric acid is strongly dependent on the concentration of $FeCl_3$ in the culture medium, as indicated in the graph. Why does the yield decrease when the concentration of Fe^{3+} is above or below the optimal value of 0.5 mg/L?

(b) Write the sequence of reactions by which *A. niger* synthesizes citric acid from sucrose. Write an equation for the overall reaction.

(c) Does the commercial process require the culture medium to be aerated—that is, is this a fermentation or an aerobic process? Explain.

18. Regulation of Citrate Synthase In the presence of saturating amounts of oxaloacetate, the activity of citrate synthase from pig heart tissue shows a sigmoid dependence on the concentration of acetyl-CoA, as shown in the graph. When succinyl-CoA is added, the curve shifts to the right and the sigmoid dependence is more pronounced.

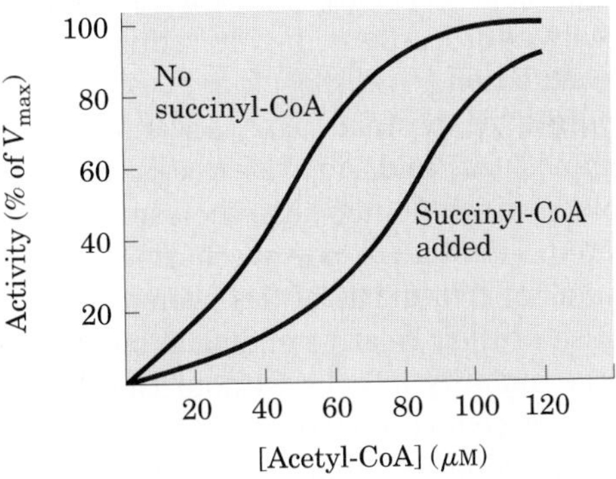

On the basis of these observations, suggest how succinyl-CoA regulates the activity of citrate synthase. (Hint: See Figure 8–26.) Why is succinyl-CoA an appropriate signal for regulation of the citric acid cycle? How does the regulation of citrate synthase control the rate of cellular respiration in pig heart tissue?

19. Regulation of Pyruvate Carboxylase The carboxylation of pyruvate by pyruvate carboxylase occurs at a very low rate unless acetyl-CoA, a positive allosteric modulator, is present. If you have just eaten a meal rich in fatty acids (triacylglycerols) but low in carbohydrates (glucose), how does this regulatory property shut down the oxidation of glucose to CO_2 and H_2O but increase the oxidation of acetyl-CoA derived from fatty acids?

20. Relationship between Respiration and the Citric Acid Cycle Although oxygen does not participate directly in the citric acid cycle, the cycle operates only when O_2 is present. Why?

21. Regulation of Metabolism in Rabbit Muscle The intracellular use of glucose and glycogen is tightly regulated at four points. To compare the regulation of glycolysis when oxygen is plentiful and when it is depleted, consider the utilization of glucose and glycogen by rabbit leg muscle in two physiological settings: a resting rabbit, with low ATP demands, and a rabbit that sights its mortal enemy, the coyote, and dashes for shelter. For each setting, determine the relative levels (high, intermediate, or low) of AMP, ATP, citrate, and acetyl-CoA and how these levels affect the flow of metabolites through glycolysis by regulating specific enzymes. (In periods of stress, rabbit leg muscle produces much of its ATP by anaerobic glycolysis (lactic acid fermentation) and very little by oxidation of acetyl-CoA derived from fat breakdown.)

22. Regulation of Metabolism in Migrating Birds Unlike the rabbit with its short dash, migratory birds require energy for extended periods of time. For example, ducks generally fly several thousand miles during their annual migration. The flight muscles of migratory birds have a high oxidative capacity and obtain the necessary ATP through the oxidation of acetyl-CoA (obtained from fats) via the citric acid cycle. Compare the regulation of muscle glycolysis during short-term intense activity, as in the fleeing rabbit, and during extended activity, as in the migrating duck. Why must the regulation in these two settings be different?

23. Thermodynamics of Citrate Synthase Reaction in Cells Citrate is formed by the condensation of acetyl-CoA with oxaloacetate, catalyzed by citrate synthase:

$$\text{Oxaloacetate} + \text{acetyl-CoA} + H_2O \longrightarrow \text{Citrate} + \text{CoA} + H^+$$

In rat heart mitochondria at pH 7.0 and 25 °C, the concentrations of reactants and products are: oxaloacetate, 1 μM; acetyl-CoA, 1 μM; citrate, 220 μM; and CoA, 65 μM. Given that the standard free-energy change for the citrate synthase reaction is −32.2 kJ/mol, what is the direction of metabolite flow through the citrate synthase reaction in cells of the rat heart? Explain.

24. Reactions of the Pyruvate Dehydrogenase Complex Two of the steps in the oxidative decarboxylation of pyruvate (steps ④ and ⑤ in Fig. 16–6) do not involve any of the three carbons of pyruvate yet are essential to the operation of the pyruvate dehydrogenase complex. Explain.

chapter

17 Oxidation of Fatty Acids

The oxidation of long-chain fatty acids to acetyl-CoA is a central energy-yielding pathway in animals, many protists, and some bacteria. The electrons removed from fatty acids in this process pass through the respiratory chain, driving ATP synthesis; acetyl-CoA produced from the fatty acids may be completely oxidized to CO_2 in the citric acid cycle, resulting in further energy conservation. In some organisms and in some tissues, the acetyl-CoA has alternative fates. In liver, acetyl-CoA may be converted to ketone bodies—water-soluble fuels exported to the brain and other tissues when glucose is not available. In higher plants, acetyl-CoA serves primarily as a biosynthetic precursor, only secondarily as fuel. Although the biological role of fatty acid oxidation differs from organism to organism, the mechanism is essentially the same. The repetitive four-step process, called ***β* oxidation,** by which fatty acids are converted into acetyl-CoA is the main topic of this chapter.

In Chapter 11 we described the properties of triacylglycerols (also called triglycerides or neutral fats) that make them especially suitable as storage fuels. The long alkyl chains of their constituent fatty acids are essentially hydrocarbons, highly reduced structures with an energy of complete oxidation (~38 kJ/g) more than twice that for the same weight of carbohydrate or protein. This advantage is compounded by the extreme insolubility of lipids in water; cellular triacylglycerols aggregate in lipid droplets, which do not raise the osmolarity of the cytosol, and they are unsolvated. (In storage polysaccharides, by contrast, water of solvation can account for two-thirds of the overall weight of the stored molecules.) And because of their relative chemical inertness, triacylglycerols can be stored in large quantity in cells without the risk of undesired chemical reactions with other cellular constituents.

The properties that make triacylglycerols good storage compounds, however, present problems in their role as fuels. Because they are insoluble in water, ingested triacylglycerols must be emulsified before they can be digested by water-soluble enzymes in the intestine, and triacylglycerols absorbed in the intestine or mobilized from storage tissues must be carried in

Success in the marathon requires training, strategy, willpower, and a constant supply of energy to power muscles. Long-distance runners get much of the energy they need by oxidizing stored fats in pathways described in this chapter.

the blood bound to proteins that counteract their insolubility. To overcome the relative stability of the C—C bonds in a fatty acid, the carboxyl group at C-1 is activated by attachment to coenzyme A, which allows stepwise oxidation of the fatty acyl group at the C-3 or β position—hence the name β oxidation.

We begin this chapter with a brief discussion of the sources of fatty acids and the routes by which they travel to the site of their oxidation, with special emphasis on the process in vertebrates. The chemical steps of fatty acid oxidation in mitochondria are then described. There are three stages in the complete oxidation of fatty acids to CO_2 and H_2O: the oxidation of long-chain fatty acids to two-carbon fragments, in the form of acetyl-CoA (β oxidation); the oxidation of acetyl-CoA to CO_2 in the citric acid cycle (Chapter 16); and the transfer of electrons from reduced electron carriers to the mitochondrial respiratory chain (Chapter 19). In this chapter we focus on the first of these stages. We consider the simple case in which a fully saturated fatty acid with an even number of carbon atoms is degraded to acetyl-CoA, then we look briefly at the extra transformations necessary for the degradation of unsaturated fatty acids and fatty acids with an odd number of carbons. Finally, we discuss variations on the β-oxidation theme in specialized organelles—peroxisomes and glyoxysomes. The chapter concludes with a description of an alternative fate for the acetyl-CoA formed by β oxidation in vertebrates: the production of ketone bodies in the liver.

Digestion, Mobilization, and Transport of Fats

Cells can obtain fatty acid fuels from three sources: fats consumed in the diet, fats stored in cells as lipid droplets, and fats synthesized in one organ for export to another. Some organisms use all three sources under various circumstances, others use one or two. Vertebrates, for example, obtain fats in the diet, mobilize fats stored in specialized tissue (adipose tissue, consisting of cells called adipocytes), and convert excess dietary carbohydrates to fats in the liver for export to other tissues. On average, 40% or more of the daily energy requirement of humans in highly industrialized countries is supplied by dietary triacylglycerols (although most nutritional guidelines recommend that no more than 30% of the daily caloric intake be from fats). Triacylglycerols provide more than half the energy requirements of some organs, particularly the liver, heart, and resting skeletal muscle. Stored triacylglycerols are virtually the sole source of energy in hibernating animals and migrating birds. Protists obtain fats by consuming organisms lower in the food chain, and some also store fats as cytosolic lipid droplets. Higher plants mobilize fats stored in seeds during germination, but do not otherwise depend on fats for energy.

Dietary Fats Are Absorbed in the Small Intestine

Before ingested triacylglycerols can be absorbed through the intestinal wall, they must be converted from insoluble macroscopic fat particles to finely dispersed microscopic micelles. Bile salts such as taurocholic acid are synthesized from cholesterol in the liver, stored in the gallbladder, and released into the small intestine after ingestion of a fatty meal. These amphipathic compounds act as biological detergents, converting dietary fats into mixed micelles of bile salts and triacylglycerols (Fig. 17–1, step ①). Micelle formation enormously increases the fraction of lipid molecules accessible to the action of water-soluble lipases in the intestine, and lipase action converts triacylglycerols to monoacylglycerols (monoglycerides) and diacylglycerols (diglycerides), free fatty acids, and glycerol (step ②). These

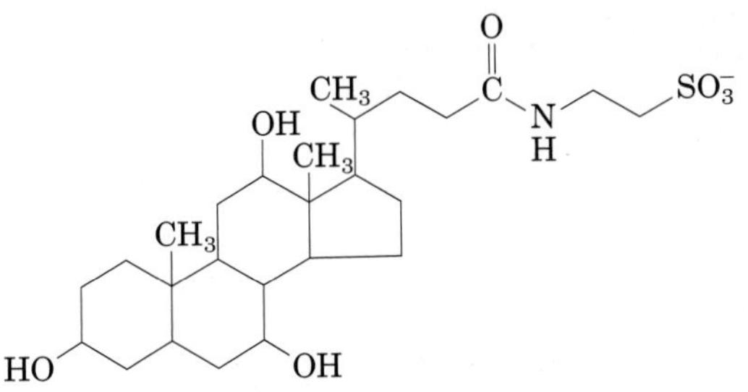

Taurocholic acid

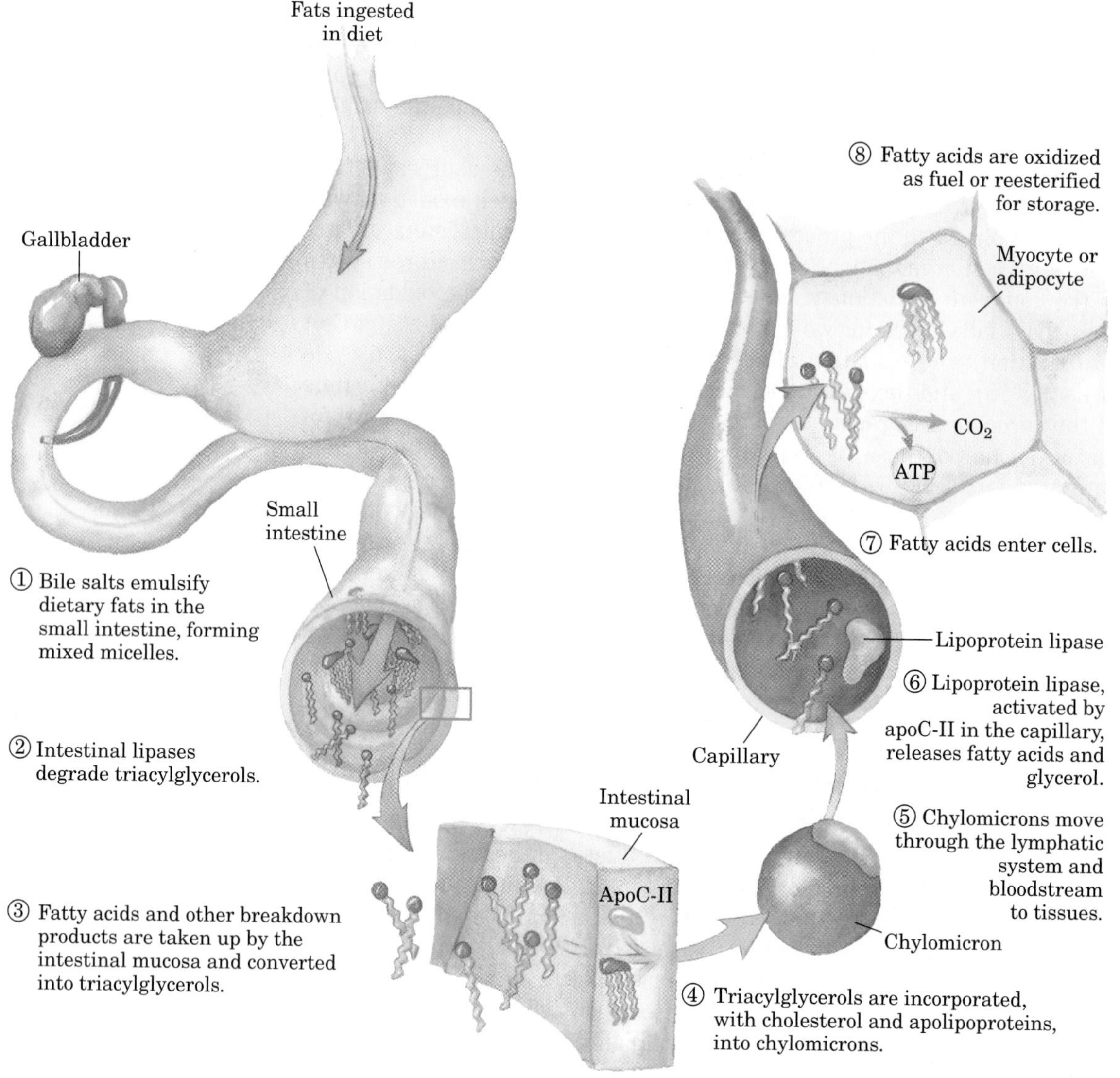
Fats ingested in diet
Gallbladder
Small intestine
① Bile salts emulsify dietary fats in the small intestine, forming mixed micelles.
② Intestinal lipases degrade triacylglycerols.
③ Fatty acids and other breakdown products are taken up by the intestinal mucosa and converted into triacylglycerols.
Intestinal mucosa
ApoC-II
④ Triacylglycerols are incorporated, with cholesterol and apolipoproteins, into chylomicrons.
Chylomicron
⑤ Chylomicrons move through the lymphatic system and bloodstream to tissues.
Capillary
Lipoprotein lipase
⑥ Lipoprotein lipase, activated by apoC-II in the capillary, releases fatty acids and glycerol.
⑦ Fatty acids enter cells.
⑧ Fatty acids are oxidized as fuel or reesterified for storage.
Myocyte or adipocyte
CO₂
ATP

figure 17–1

Processing of dietary lipids in vertebrates. Digestion and absorption of dietary lipids occur in the small intestine, and the fatty acids released from triacylgycerols are packaged and delivered to muscle and adipose tissues. The eight steps are discussed in the text.

products of lipase action diffuse into the epithelial cells lining the intestinal surface (the intestinal mucosa) (step ③), where they are reconverted to triacylglycerols and packaged with dietary cholesterol and specific proteins into lipoprotein aggregates called **chylomicrons** (Fig. 17–2; see also Fig. 17–1, step ④).

Apolipoproteins are lipid-binding proteins in the blood, responsible for the transport of triacylglycerols, phospholipids, cholesterol, and cholesteryl esters between organs. Apolipoproteins ("apo" designates the protein in its lipid-free form) combine with lipids to form several classes of lipoprotein particles, spherical aggregates with hydrophobic lipids at the core and hydrophilic protein side chains and lipid head groups at the surface. Various combinations of lipid and protein produce particles of different densities, ranging from chylomicrons and very low-density lipoproteins (VLDL) to very high-density lipoproteins (VHDL), which may be separated by ultracentrifugation. The structures and roles of these lipoprotein particles in lipid transport are detailed in Chapter 21.

The protein moieties of lipoproteins are recognized by receptors on cell surfaces. In lipid uptake from the intestine, chylomicrons, which contain apolipoprotein C-II (apoC-II), move from the intestinal mucosa into the lymphatic system, from which they enter the blood and are carried to

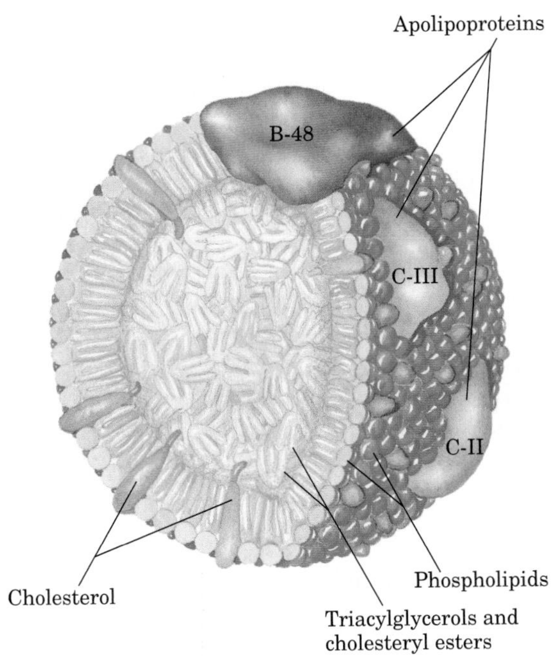

figure 17–2

Molecular structure of a chylomicron. The surface is a layer of phospholipids, with head groups facing the aqueous phase. Triacylglycerols sequestered in the interior (yellow) make up more than 80% of the mass. Several apolipoproteins that protrude from the surface (B-48, C-III, C-II) act as signals in the uptake and metabolism of chylomicron contents. The diameter of chylomicrons ranges from about 100 nm to about 500 nm.

muscle and adipose tissue (Fig. 17–1, step ⑤). In the capillaries of these tissues, the extracellular enzyme **lipoprotein lipase,** activated by apoC-II, hydrolyzes triacylglycerols to fatty acids and glycerol (step ⑥), which are taken up by cells in the target tissues (step ⑦). In muscle, the fatty acids are oxidized for energy; in adipose tissue, they are reesterified for storage as triacylglycerols (step ⑧).

The remnants of chylomicrons, depleted of most of their triacylglycerols but still containing cholesterol and apolipoproteins, travel in the blood to the liver, where they are taken up by endocytosis, mediated by receptors for their apolipoproteins. Triacylglycerols that enter the liver by this route may be oxidized to provide energy or to provide precursors for the synthesis of ketone bodies, as described later in this chapter. When the diet contains more fatty acids than are needed immediately for fuel or as precursors, the liver converts them to triacylglycerols, which are packaged with specific apolipoproteins into VLDLs. The VLDLs are transported in the blood to adipose tissues, where the triacylglycerols are removed and stored in lipid droplets within adipocytes.

Hormones Trigger Mobilization of Stored Triacylglycerols

When hormones signal the need for metabolic energy, triacylglycerols stored in adipose tissue are mobilized (brought out of storage) and transported to tissues (skeletal muscle, heart, and renal cortex) in which fatty acids can be oxidized for energy production. The hormones epinephrine and glucagon, secreted in response to low blood glucose levels, activate the enzyme adenylyl cyclase in the adipocyte plasma membrane (Fig. 17–3), which produces an intracellular second messenger, cyclic AMP (cAMP). A cAMP-dependent protein kinase phosphorylates and thereby activates **hormone-sensitive triacylglycerol lipase,** which catalyzes hydrolysis of the ester linkages of triacylglycerols. The fatty acids thus released pass from the adipocyte into the blood, where they bind to the blood protein **serum albumin.** This protein (M_r 62,000), which makes up about half of the total serum protein, noncovalently binds as many as 10 fatty acids per protein monomer. Bound to this soluble protein, the otherwise insoluble fatty acids are carried to tissues such as skeletal muscle, heart, and renal cortex. Here, fatty acids dissociate from albumin and are transported into cells to serve as fuel.

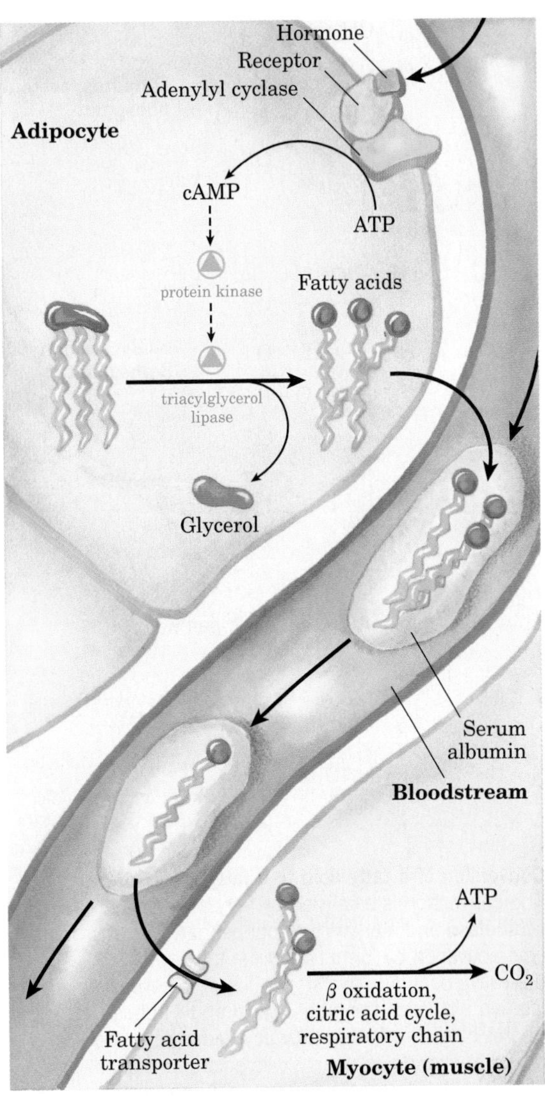

figure 17–3

Mobilization of triacylglycerols stored in adipose tissue. Low levels of glucose in the blood trigger the mobilization of triacylglycerols through the action of epinephrine and glucagon on the adipocyte adenylyl cyclase. The subsequent steps in mobilization are described in the text.

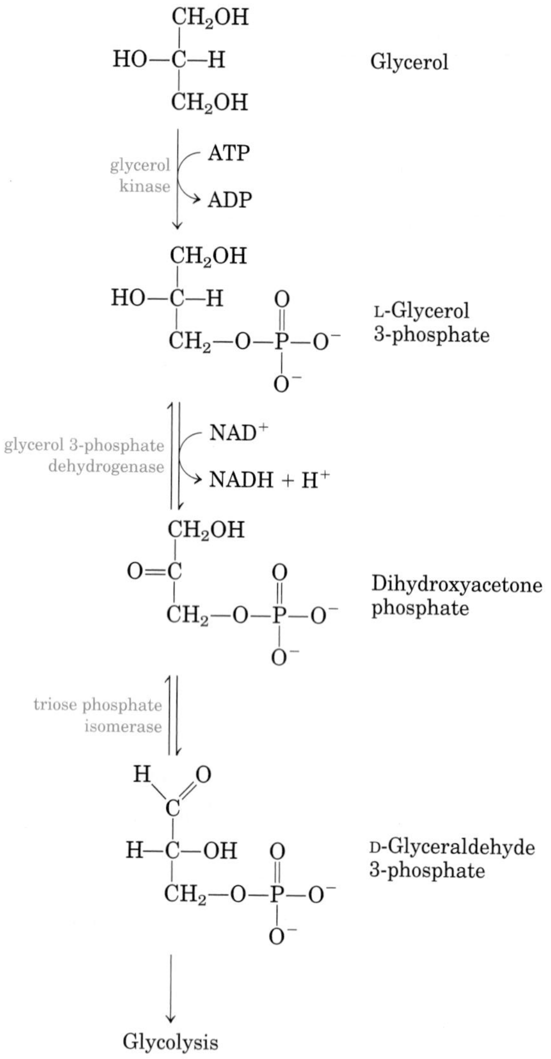

About 95% of the biologically available energy of triacylglycerols resides in their three long-chain fatty acids; only 5% is contributed by the glycerol moiety. The glycerol released by lipase action is phosphorylated by **glycerol kinase** (Fig. 17–4), and the resulting glycerol 3-phosphate is oxidized to dihydroxyacetone phosphate. The glycolytic enzyme triose phosphate isomerase converts this compound to glyceraldehyde 3-phosphate, which is oxidized via glycolysis.

Fatty Acids Are Activated and Transported into Mitochondria

The enzymes of fatty acid oxidation in animal cells are located in the mitochondrial matrix, as demonstrated in 1948 by Eugene P. Kennedy and Albert Lehninger. The free fatty acids that enter the cytosol from the blood cannot pass directly through the mitochondrial membranes, but must first undergo a series of three enzymatic reactions. The first is catalyzed by a family of isozymes present in the outer mitochondrial membrane, the **acyl-CoA synthetases,** which promote the general reaction

$$\text{Fatty acid} + \text{CoA} + \text{ATP} \rightleftharpoons \text{fatty acyl–CoA} + \text{AMP} + \text{PP}_i$$

The different acyl-CoA synthetase isozymes are specific for fatty acids having short, intermediate, or long carbon chains. Acyl-CoA synthetases catalyze the formation of a thioester linkage between the fatty acid carboxyl group and the thiol group of coenzyme A to yield a **fatty acyl–CoA,** coupled to the cleavage of ATP to AMP and PP_i. (Recall the description of this reaction in Chapter 14, to illustrate how the free energy released by cleavage of phosphoanhydride bonds in ATP could be coupled to the formation of a high-energy compound; see Fig. 14–11.) The reaction occurs in two steps and involves a fatty acyl–adenylate intermediate (Fig. 17–5).

Fatty acyl–CoAs, like acetyl-CoA, are high-energy compounds; their hydrolysis to free fatty acid and CoA has a large, negative standard free-energy change ($\Delta G'^\circ \approx -31$ kJ/mol). The formation of a fatty acyl–CoA is

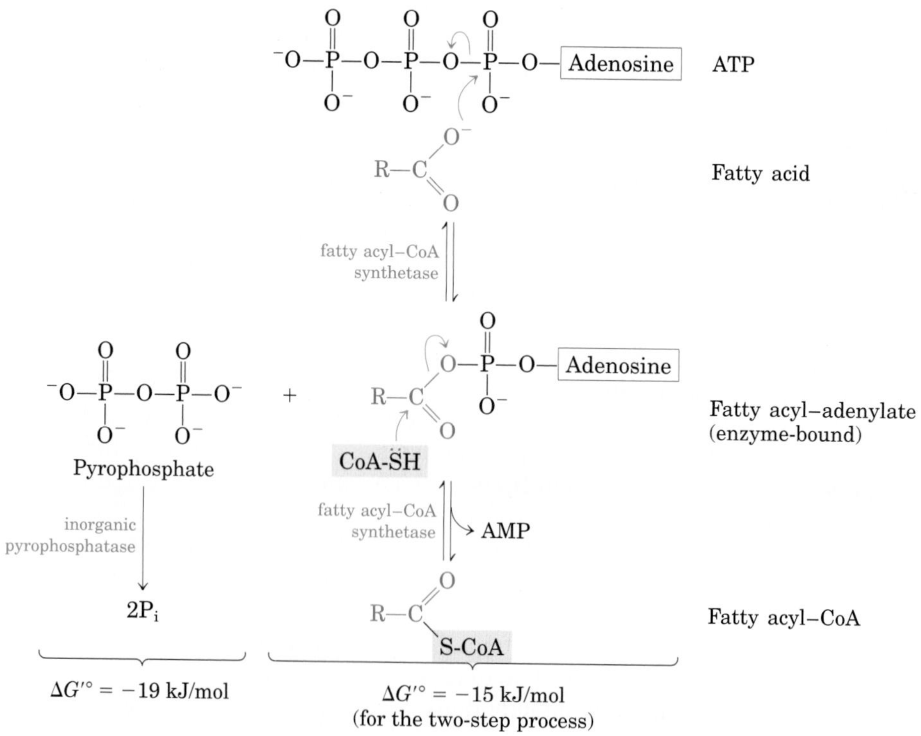

Glycerol

glycerol kinase; ATP → ADP

L-Glycerol 3-phosphate

glycerol 3-phosphate dehydrogenase; NAD^+ → NADH + H^+

Dihydroxyacetone phosphate

triose phosphate isomerase

D-Glyceraldehyde 3-phosphate

Glycolysis

figure 17–4
Entry of glycerol into the glycolytic pathway.

ATP

Fatty acid

fatty acyl–CoA synthetase

Pyrophosphate + Fatty acyl–adenylate (enzyme-bound)

CoA-SH

inorganic pyrophosphatase → $2P_i$

fatty acyl–CoA synthetase → AMP

Fatty acyl–CoA

$\Delta G'^\circ = -19$ kJ/mol

$\Delta G'^\circ = -15$ kJ/mol (for the two-step process)

figure 17–5
Conversion of a fatty acid to a fatty acyl–CoA. The conversion is catalyzed by fatty acyl–CoA synthetase and inorganic pyrophosphatase. Fatty acid activation by formation of the fatty acyl–CoA derivative occurs in two steps. First, the carboxylate ion displaces the outer two (β and γ) phosphates of ATP to form a fatty acyl–adenylate, the mixed anhydride of a carboxylic acid and a phosphoric acid. The other product is PP_i, an excellent leaving group that is immediately hydrolyzed to two P_i, pulling the reaction in the forward direction. The thiol group of coenzyme A carries out nucleophilic attack on the enzyme-bound mixed anhydride, displacing AMP and forming the thioester fatty acyl–CoA. The overall reaction is highly exergonic.

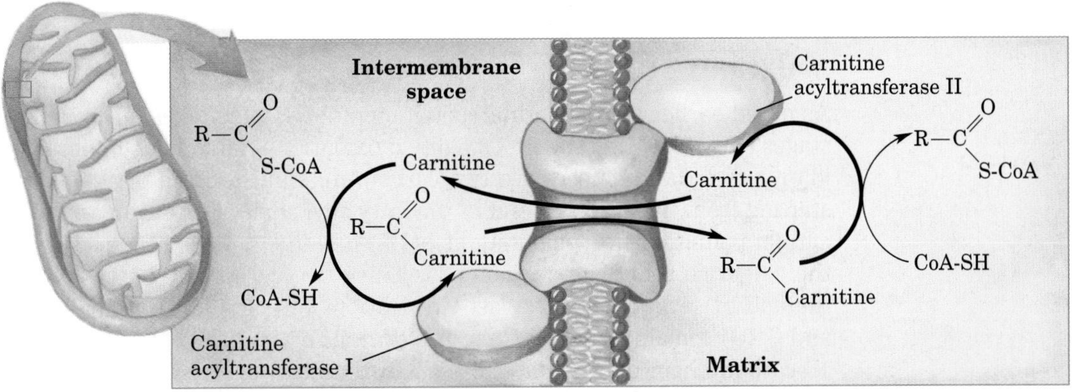

figure 17–6

Fatty acid entry into mitochondria via the acyl-carnitine/carnitine transporter. After fatty acyl–carnitine is formed at the outer surface of the inner mitochondrial membrane, it moves into the matrix by facilitated diffusion through the transporter. In the matrix, the acyl group is transferred to mitochondrial coenzyme A, freeing carnitine to return to the intermembrane space through the same transporter. The enzymes acyltransferase I and II are bound to the outer and inner surfaces, respectively, of the inner membrane. Acyltransferase I is inhibited by malonyl-CoA, the first intermediate in fatty acid synthesis (see Fig. 21–1). This inhibition prevents the simultaneous synthesis and degradation of fatty acids.

made more favorable by the hydrolysis of *two* high-energy bonds in ATP; the pyrophosphate formed in the activation reaction is immediately hydrolyzed by inorganic pyrophosphatase (left side of Fig. 17–5), which pulls the preceding activation reaction in the direction of fatty acyl–CoA formation. The overall reaction is

$$\text{Fatty acid} + \text{CoA} + \text{ATP} \longrightarrow \text{fatty acyl–CoA} + \text{AMP} + 2\text{P}_i \qquad (17\text{–}1)$$
$$\Delta G'^{\circ} = -34 \text{ kJ/mol}$$

Fatty acyl–CoA esters formed in the outer mitochondrial membrane do not cross the inner mitochondrial membrane intact. Instead, the fatty acyl group is transiently attached to the hydroxyl group of **carnitine** to form fatty acyl–carnitine. This transesterification is catalyzed by **carnitine acyltransferase I,** on the outer face of the inner membrane. The fatty acyl–carnitine ester then enters the matrix by facilitated diffusion through the **acyl-carnitine/carnitine transporter** of the inner mitochondrial membrane (Fig. 17–6).

$$CH_3—N^+(CH_3)_2—CH_2—CH(OH)—CH_2—COO^-$$

Carnitine

In the third and final step of the entry process, the fatty acyl group is enzymatically transferred from carnitine to intramitochondrial coenzyme A by **carnitine acyltransferase II.** This isozyme, located on the inner face of the inner mitochondrial membrane, regenerates fatty acyl–CoA and releases it, along with free carnitine, into the matrix (Fig. 17–6). Carnitine reenters the space between the inner and outer mitochondrial membranes via the acyl-carnitine/carnitine transporter.

This three-step process for transferring fatty acids into the mitochondrion—esterification to CoA, transesterification to carnitine followed by transport, and transesterification back to CoA—links two separate pools of coenzyme A, one in the cytosol, the other in mitochondria. These pools have different functions. Coenzyme A in the mitochondrial matrix is largely used in oxidative degradation of pyruvate, fatty acids, and some amino acids, whereas cytosolic coenzyme A is used in the biosynthesis of fatty acids (see Fig. 21–11).

The carnitine-mediated entry process is the rate-limiting step for oxidation of fatty acids in mitochondria and, as discussed later in this chapter, is a regulation point. Once inside the mitochondrion, the fatty acyl–CoA is quickly acted upon by a set of enzymes in the matrix.

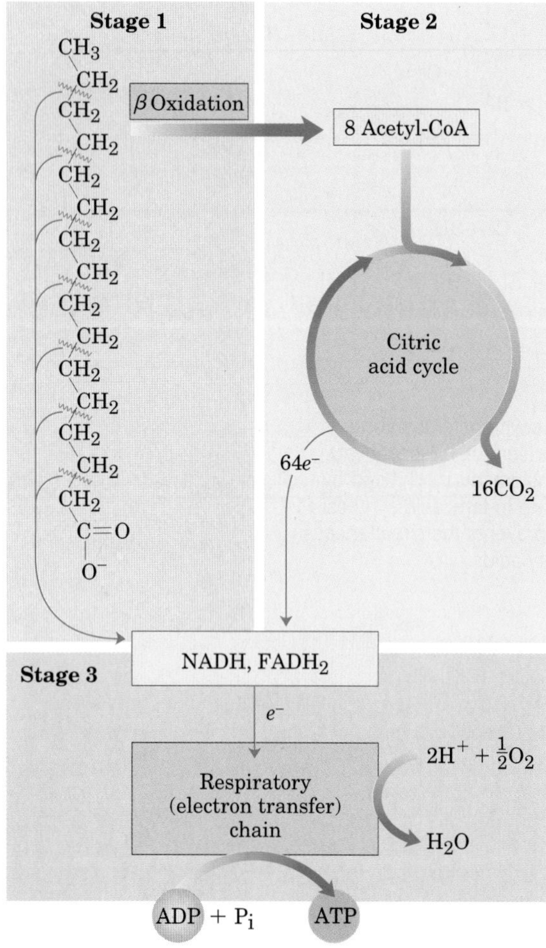

figure 17–7

Stages of fatty acid oxidation. Stage 1: A long-chain fatty acid is oxidized to yield acetyl residues in the form of acetyl-CoA. This process is called β oxidation. Stage 2: The acetyl groups are oxidized to CO_2 via the citric acid cycle. Stage 3: Electrons derived from the oxidations of stages 1 and 2 pass to O_2 via the mitochondrial respiratory chain, providing the energy for ATP synthesis by oxidative phosphorylation.

β Oxidation

As noted earlier, mitochondrial oxidation of fatty acids takes place in three stages (Fig. 17–7). In the first stage—β oxidation—fatty acids undergo oxidative removal of successive two-carbon units in the form of acetyl-CoA, starting from the carboxyl end of the fatty acyl chain. For example, the 16-carbon palmitic acid (palmitate at pH 7) undergoes seven passes through the oxidative sequence, in each pass losing two carbons as acetyl-CoA. At the end of seven cycles the last two carbons of palmitate (originally C-15 and C-16) remain as acetyl-CoA. The overall result is the conversion of the 16-carbon chain of palmitate to eight two-carbon acetyl groups of acetyl-CoA molecules. Formation of each acetyl-CoA requires removal of four hydrogen atoms (two pairs of electrons and four H^+) from the fatty acyl moiety by dehydrogenases.

In the second stage of fatty acid oxidation, the acetyl groups of acetyl-CoA are oxidized to CO_2 in the citric acid cycle, which also takes place in the mitochondrial matrix. Acetyl-CoA derived from fatty acids thus enters a final common pathway of oxidation along with acetyl-CoA derived from glucose via glycolysis and pyruvate oxidation (see Fig. 16–1).

The first two stages of fatty acid oxidation produce the reduced electron carriers NADH and $FADH_2$, which in the third stage donate electrons to the mitochondrial respiratory chain, through which the electrons pass to oxygen with the concomitant phosphorylation of ADP to ATP (Fig. 17–7). Thus energy released by fatty acid oxidation is conserved as ATP.

We now take a closer look at the first stage of fatty acid oxidation, beginning with the simple case of a saturated fatty acyl chain with an even number of carbons, then turning to the slightly more complicated cases of unsaturated and odd-number chains. We then consider the regulation of fatty acid oxidation, and finally the β-oxidative processes as they occur in organelles other than mitochondria.

The β Oxidation of Saturated Fatty Acids Has Four Basic Steps

Four enzyme-catalyzed reactions make up the first stage of fatty acid oxidation (Fig. 17–8a). First, dehydrogenation of fatty acyl–CoA produces a double bond between the α and β carbon atoms (C-2 and C-3), yielding a ***trans*-Δ^2-enoyl-CoA** (the symbol Δ^2 designates the position of the double bond; you may want to review fatty acid nomenclature, p. 363.) The new double bond has the trans configuration, whereas the double bonds in naturally occurring unsaturated fatty acids are normally in the cis configuration. We consider the significance of this difference later.

This first step is catalyzed by three isozymes of **acyl-CoA dehydrogenase.** Long-chain acyl-CoA dehydrogenase (LCAD) acts on fatty acids of 12 to 18 carbons, and the isozymes specific for medium-chain (MCAD) and short-chain (SCAD) fatty acids act on fatty acids of 4 to 14 and 4 to 8 carbons, respectively. All three isozymes have FAD (see Fig. 14–16) as a prosthetic group. The electrons removed from the fatty acyl–CoA are transferred to FAD, and the reduced form of the dehydrogenase immediately donates its electrons to an electron carrier of the mitochondrial respiratory chain, the **electron-transferring flavoprotein (ETF)** (see Fig. 19–8). The oxidation catalyzed by acyl-CoA dehydrogenase is analogous to succinate dehydrogenation in the citric acid cycle (p. 578); in both reactions the enzyme is bound to the inner membrane, a double bond is introduced into a carboxylic acid between the α and β carbons, FAD is the electron acceptor, and electrons from the reaction ultimately enter the respiratory chain and pass to O_2, with the concomitant synthesis of about 1.5 ATP molecules per electron pair.

In the second step of the β-oxidation cycle (Fig. 17–8a), water is added to the double bond of the *trans*-Δ^2-enoyl-CoA to form the L stereoisomer of

β-hydroxyacyl-CoA (also designated **3-hydroxyacyl-CoA**). This reaction, catalyzed by **enoyl-CoA hydratase,** is formally analogous to the fumarase reaction in the citric acid cycle, in which H_2O adds across an α–β double bond (p. 578).

In the third step, the L-β-hydroxyacyl-CoA is dehydrogenated to form **β-ketoacyl-CoA** by the action of **β-hydroxyacyl-CoA dehydrogenase;** NAD^+ is the electron acceptor. This enzyme is absolutely specific for the L stereoisomer of hydroxyacyl-CoA. The NADH formed in the reaction donates its electrons to **NADH dehydrogenase,** an electron carrier of the respiratory chain, and ATP is formed from ADP as the electrons pass along the chain to O_2. The reaction catalyzed by β-hydroxyacyl-CoA dehydrogenase is closely analogous to the malate dehydrogenase reaction of the citric acid cycle (p. 579).

The fourth and last step of the β-oxidation cycle is catalyzed by **acyl-CoA acetyltransferase,** more commonly called **thiolase,** which promotes reaction of β-ketoacyl-CoA with a molecule of free coenzyme A to split off the carboxyl-terminal two-carbon fragment of the original fatty acid as acetyl-CoA. The other product is the coenzyme A thioester of the fatty acid, now shortened by two carbon atoms (Fig. 17–8a). This reaction is called thiolysis, by analogy with the process of hydrolysis, because the β-ketoacyl-CoA is cleaved by reaction with the thiol group of coenzyme A.

As noted earlier, the single bond between methylene (—CH_2—) groups in fatty acids is relatively stable. The β-oxidation sequence is an elegant mechanism for breaking these bonds. The first three reactions of β oxidation create a much less stable, more easily broken, C—C bond, in which the α carbon (C-2) is bonded to *two* carbonyl carbons (the β-ketoacyl-CoA intermediate). The ketone function on the β carbon (C-3) makes it a good target for nucleophilic attack by —SH of coenzyme A, catalyzed by thiolase. The acidity of the α hydrogen and the resonance stabilization of the carbanion generated by the departure of this hydrogen make the terminal —CH_2—CO—S-CoA a good leaving group, facilitating breakage of the α–β bond.

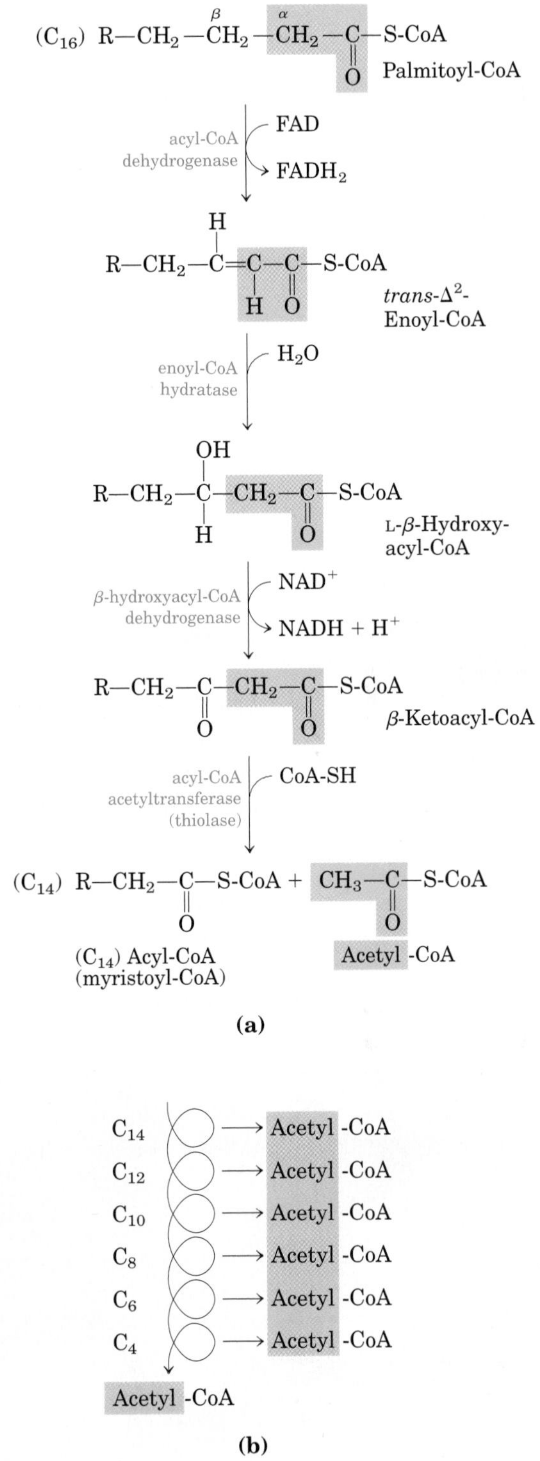

figure 17–8
The β-oxidation pathway. (a) In each pass through this four-step sequence, one acetyl residue (shaded in red) is removed in the form of acetyl-CoA from the carboxyl end of the fatty acyl chain—in this example palmitate (C_{16}), which enters as palmitoyl-CoA. **(b)** Six more passes through the pathway yield seven more molecules of acetyl-CoA, the seventh arising from the last two carbon atoms of the 16-carbon chain. Eight molecules of acetyl-CoA are formed in all.

The Four Steps Are Repeated to Yield Acetyl-CoA and ATP

In one pass through the β-oxidation sequence, one molecule of acetyl-CoA, two pairs of electrons, and four protons (H^+) are removed from the long-chain fatty acyl–CoA, shortening it by two carbon atoms. The equation for one pass, beginning with the coenzyme A ester of our example, palmitate, is

$$\text{Palmitoyl-CoA} + \text{CoA} + \text{FAD} + \text{NAD}^+ + \text{H}_2\text{O} \longrightarrow \text{myristoyl-CoA} + \text{acetyl-CoA} + \text{FADH}_2 + \text{NADH} + \text{H}^+ \quad (17\text{–}2)$$

Following removal of one acetyl-CoA unit from palmitoyl-CoA, the coenzyme A thioester of the shortened fatty acid (now the 14-carbon myristate) remains. The myristoyl-CoA can now go through another set of four β-oxidation reactions, exactly analogous to the first, to yield a second molecule of acetyl-CoA and lauroyl-CoA, the coenzyme A thioester of the 12-carbon laurate. Altogether, seven passes through the β-oxidation sequence are required to oxidize one molecule of palmitoyl-CoA to eight molecules of acetyl-CoA (Fig. 17–8b). The overall equation is

$$\text{Palmitoyl-CoA} + 7\text{CoA} + 7\text{FAD} + 7\text{NAD}^+ + 7\text{H}_2\text{O} \longrightarrow 8\ \text{acetyl-CoA} + 7\text{FADH}_2 + 7\text{NADH} + 7\text{H}^+ \quad (17\text{–}3)$$

Each molecule of $FADH_2$ formed during oxidation of the fatty acid donates a pair of electrons to ETF of the respiratory chain, and about 1.5 molecules of ATP are generated during the ensuing transfer of each electron pair to O_2. Similarly, each molecule of NADH formed delivers a pair of electrons to the mitochondrial NADH dehydrogenase, and the subsequent transfer of each pair of electrons to O_2 results in formation of about 2.5 molecules of ATP. Thus four molecules of ATP are formed for each two-carbon unit removed in one pass through the sequence. Note that water is also produced in this process. Transfer of electrons from NADH or $FADH_2$ to O_2 yields one H_2O per electron pair. Reduction of O_2 by NADH also consumes one H^+ per NADH molecule: $NADH + H^+ + \frac{1}{2}O_2 \longrightarrow NAD^+ + H_2O$. In hibernating animals, fatty acid oxidation provides metabolic energy, heat, and water—all essential for survival of an animal that neither eats nor drinks for long periods (Box 17–1). Camels obtain water to supplement the meager supply available in their natural environment by oxidation of fats stored in their hump.

The overall equation for the oxidation of palmitoyl-CoA to eight molecules of acetyl-CoA, including the electron transfers and oxidative phosphorylations, is

$$\text{Palmitoyl-CoA} + 7\text{CoA} + 7O_2 + 28P_i + 28\text{ADP} \longrightarrow 8\text{ acetyl-CoA} + 28\text{ATP} + 7H_2O \quad (17\text{–}4)$$

box 17–1 Fat Bears Carry Out β Oxidation in Their Sleep

Many animals depend on fat stores for energy during hibernation, during migratory periods, and in other situations involving radical metabolic adjustments. One of the most pronounced adjustments of fat metabolism occurs in hibernating grizzly bears. These animals remain in a continuous state of dormancy for periods as long as seven months. Unlike most hibernating species, the bear maintains a body temperature of between 32 and 35 °C, close to the normal (nonhibernating) level. Although expending about 25,000 kJ/day (6,000 kcal/day), the bear does not eat, drink, urinate, or defecate for months at a time.

A grizzly bear prepares its hibernation nest, near the McNeil River in Canada.

Experimental studies have shown that hibernating grizzly bears use body fat as their sole fuel. Fat oxidation yields sufficient energy for maintenance of body temperature, active synthesis of amino acids and proteins, and other energy-requiring activities, such as membrane transport. Fat oxidation also releases large amounts of water, as described in the text, which replenishes water lost in breathing. The glycerol released by degradation of triacylglycerols is converted into blood glucose by gluconeogenesis. Urea formed during breakdown of amino acids is reabsorbed in the kidneys and recycled, the amino groups reused to make new amino acids for maintaining body proteins.

Bears store an enormous amount of body fat in preparation for their long sleep. An adult grizzly consumes about 38,000 kJ/day during the late spring and summer, but as winter approaches it feeds 20 hours a day, consuming up to 84,000 kJ daily. This change in feeding is a response to a seasonal change in hormone secretion. Large amounts of triacylglycerols are formed from the huge intake of carbohydrates during the fattening-up period. Other hibernating species, including the tiny dormouse, also accumulate large amounts of body fat. The camel has a huge store of triacylglycerols in its hump, a metabolic source of both energy and water under desert conditions.

Acetyl-CoA Can Be Further Oxidized in the Citric Acid Cycle

The acetyl-CoA produced from the oxidation of fatty acids can be oxidized to CO_2 and H_2O by the citric acid cycle. The following equation represents the balance sheet for the second stage in the oxidation of palmitoyl-CoA, together with the coupled phosphorylations of the third stage:

$$8\ \text{Acetyl-CoA} + 16O_2 + 80P_i + 80\text{ADP} \longrightarrow 8\text{CoA} + 80\text{ATP} + 16H_2O + 16CO_2 \qquad (17\text{–}5)$$

Combining Equations 17–4 and 17–5, we obtain the overall equation for the complete oxidation of palmitoyl-CoA to carbon dioxide and water:

$$\text{Palmitoyl-CoA} + 23O_2 + 108P_i + 108\text{ADP} \longrightarrow \text{CoA} + 108\text{ATP} + 16CO_2 + 23H_2O \qquad (17\text{–}6)$$

Table 17–1 summarizes the yields of NADH, $FADH_2$, and ATP in the successive steps of palmitoyl-CoA oxidation. Because the activation of palmitate to palmitoyl-CoA breaks both phosphoanhydride bonds in ATP (see p. 602 and Fig. 14–11), the energetic cost of activating a fatty acid is equivalent to two ATPs, and the net gain per molecule of palmitate is 106 ATP. The standard free-energy change for the oxidation of palmitate to CO_2 and H_2O is about 9,800 kJ/mol. Under standard conditions, 106 × 30.5 kJ/mol = 3,230 kJ/mol (about 33% of the theoretical maximum) is recovered as the phosphate bond energy of ATP. However, when the free-energy changes are calculated from actual concentrations of reactants and products under intracellular conditions (see Box 14–2), the free-energy recovery is over 60%; the energy conservation is remarkably efficient.

table 17–1

Yield of ATP during Oxidation of One Molecule of Palmitoyl-CoA to CO_2 and H_2O

Enzyme catalyzing the oxidation step	Number of NADH or $FADH_2$ formed	Number of ATP ultimately formed*
Acyl-CoA dehydrogenase	7 $FADH_2$	10.5
β-Hydroxyacyl-CoA dehydrogenase	7 NADH	17.5
Isocitrate dehydrogenase	8 NADH	20
α-Ketoglutarate dehydrogenase	8 NADH	20
Succinyl-CoA synthetase		8†
Succinate dehydrogenase	8 $FADH_2$	12
Malate dehydrogenase	8 NADH	20
Total		108

*These calculations assume that mitochondrial oxidative phosphorylation produces 1.5 ATP per $FADH_2$ oxidized and 2.5 ATP per NADH oxidized.

†GTP produced directly in this step yields ATP in the reaction catalyzed by nucleoside diphosphate kinase (p. 578).

Oxidation of Unsaturated Fatty Acids Requires Two Additional Reactions

The fatty acid oxidation sequence just described is typical when the incoming fatty acid is saturated (i.e., has only single bonds in its carbon chain). However, most of the fatty acids in the triacylglycerols and phospholipids of animals and plants are unsaturated, having one or more double bonds. These bonds are in the cis configuration and cannot be acted upon by enoyl-CoA hydratase, the enzyme catalyzing the addition of H_2O to the trans double bond of the Δ^2-enoyl-CoA generated during β oxidation. However, by the action of two auxiliary enzymes, the fatty acid oxidation sequence described above can also break down the common unsaturated

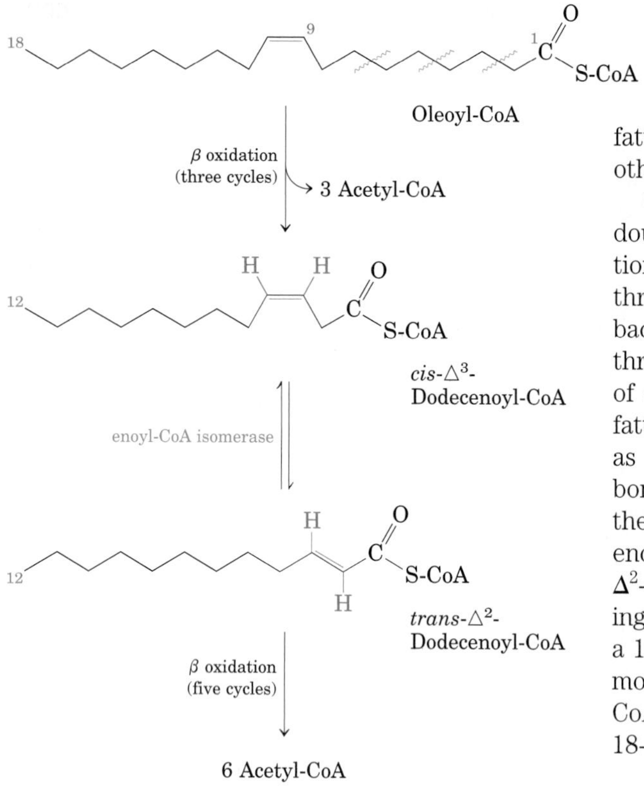

18 9 1 O C S-CoA

Oleoyl-CoA

β oxidation (three cycles) → 3 Acetyl-CoA

H H O
12 C S-CoA

cis-Δ^3-
Dodecenoyl-CoA

enoyl-CoA isomerase

H O
C
12 S-CoA
H

trans-Δ^2-
Dodecenoyl-CoA

β oxidation (five cycles)

6 Acetyl-CoA

figure 17–9

Oxidation of a monounsaturated fatty acid. Oleic acid, as oleoyl-CoA (Δ^9), is the example used here. Oxidation requires an additional enzyme, enoyl-CoA isomerase, to reposition the double bond, converting the cis isomer to a trans isomer, a normal intermediate in β oxidation.

fatty acids. The actions of these two enzymes, one an isomerase and the other a reductase, are illustrated here by two examples.

Oleate is an abundant 18-carbon monounsaturated fatty acid with a cis double bond between C-9 and C-10 (denoted Δ^9). In the first step of oxidation, oleate is converted to oleoyl-CoA (Fig. 17–9), which is transported through the mitochondrial membrane as oleoyl-carnitine and converted back to oleoyl-CoA in the matrix (Fig. 17–6). Oleoyl-CoA then undergoes three passes through the fatty acid oxidation cycle to yield three molecules of acetyl-CoA and the coenzyme A ester of a Δ^3, 12-carbon unsaturated fatty acid, *cis*-Δ^3-dodecenoyl-CoA (Fig. 17–9). This product cannot serve as a substrate for enoyl-CoA hydratase, which acts only on trans double bonds. However, the auxiliary enzyme **enoyl-CoA isomerase** isomerizes the *cis*-Δ^3-enoyl-CoA to the *trans*-Δ^2-enoyl-CoA, which is converted by enoyl-CoA hydratase into the corresponding L-β-hydroxyacyl-CoA (*trans*-Δ^2-dodecenoyl-CoA). This intermediate is now acted upon by the remaining enzymes of β oxidation to yield acetyl-CoA and the coenzyme A ester of a 10-carbon saturated fatty acid, decanoyl-CoA. The latter undergoes four more passes through the pathway to yield five more molecules of acetyl-CoA. Altogether, nine acetyl-CoAs are produced from one molecule of the 18-carbon oleate.

The other auxiliary enzyme (a reductase) is required for oxidation of polyunsaturated fatty acids—for example, the 18-carbon linoleate, which has a *cis*-Δ^9,*cis*-Δ^{12} configuration (Fig. 17–10). Linoleoyl-CoA undergoes three passes through the standard β-oxidation sequence to yield three molecules of acetyl-CoA and the coenzyme A ester of a 12-carbon unsaturated fatty acid with a *cis*-Δ^3,*cis*-Δ^6 configuration. This intermediate cannot be used by the enzymes of the β-oxidation pathway; its double bonds are in the wrong position and have the wrong configuration (cis, not trans). However, the combined action of enoyl-CoA isomerase and **2,4-dienoyl-CoA reductase** (Fig. 17–10) allows reentry of this intermediate into the normal β-oxidation pathway and its degradation to six acetyl-CoAs. The overall result is conversion of linoleate to nine molecules of acetyl-CoA.

Complete Oxidation of Odd-Number Fatty Acids Requires Three Extra Reactions

Although most naturally occurring lipids contain fatty acids with an even number of carbon atoms, fatty acids with an odd number of carbons are common in the lipids of many plants and some marine organisms. Small quantities of the three-carbon **propionate** are added as a mold inhibitor to some breads and cereals, and thus propionate enters the human diet. Moreover, cattle and other ruminant animals form large amounts of propionate during fermentation of carbohydrates in the rumen. The propionate so formed is absorbed into the blood and oxidized by the liver and other tissues.

$CH_3—CH_2—COO^-$
Propionate

Long-chain odd-number fatty acids are oxidized by the same pathway as the even-number acids, beginning at the carboxyl end of the chain. However, the substrate for the last pass through the β-oxidation sequence is a fatty acyl–CoA in which the fatty acid has five carbon atoms. When this is oxidized and cleaved, the products are acetyl-CoA and **propionyl-CoA.** The acetyl-CoA can be oxidized in the citric acid cycle, of course, but propionyl-CoA enters a different enzymatic pathway involving three enzymes.

Propionyl-CoA is carboxylated to form the D stereoisomer of **methylmalonyl-CoA** (Fig. 17–11) by **propionyl-CoA carboxylase,** which contains the cofactor biotin. In this enzymatic reaction, as in the pyruvate carboxylase reaction (see Fig. 16–14), CO_2 (or its hydrated ion, HCO_3^-) is

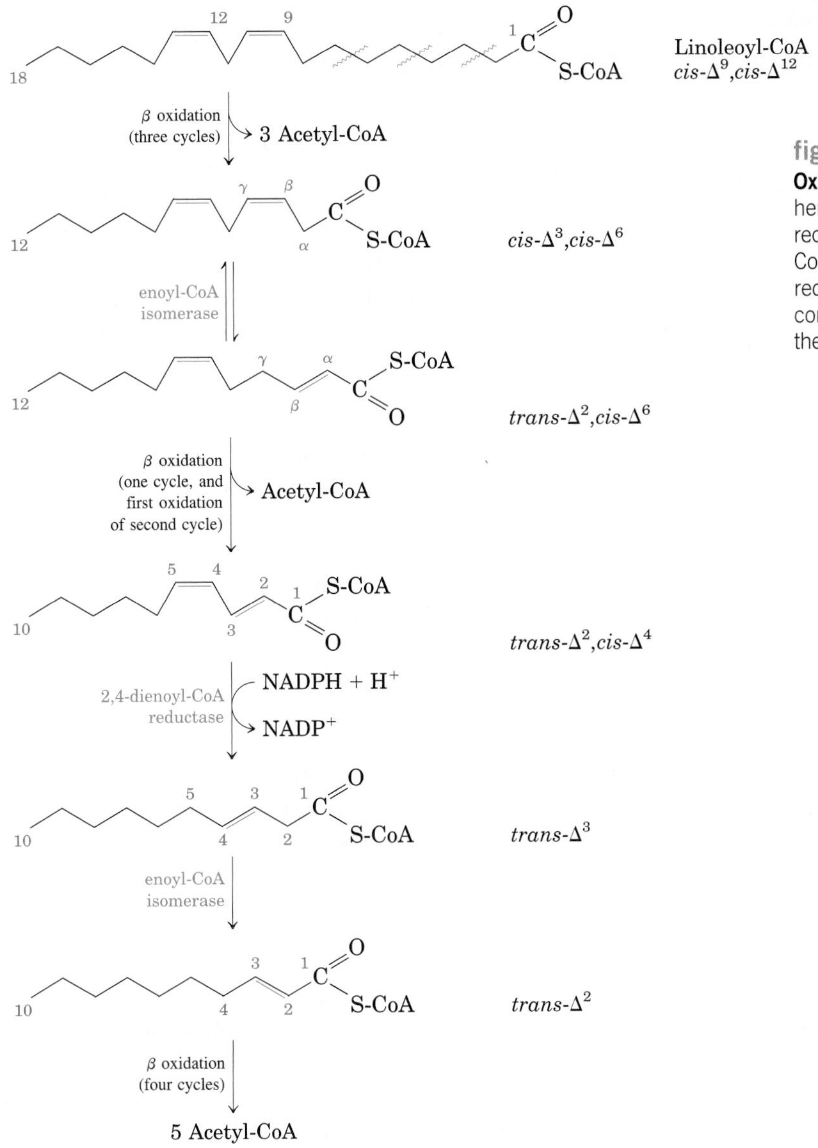

figure 17–10
Oxidation of a polyunsaturated fatty acid. The example here is linoleic acid, as linoleoyl-CoA ($\Delta^{9,12}$). Oxidation requires a second auxiliary enzyme in addition to enoyl-CoA isomerase: NADPH-dependent 2,4-dienoyl-CoA reductase. The combined action of these two enzymes converts a *trans*-Δ^2,*cis*-Δ^4-dienoyl-CoA intermediate into the *trans*-Δ^2-enoyl-CoA substrate necessary for β oxidation.

activated by attachment to biotin before its transfer to the substrate, in this case the propionate moiety. Formation of the carboxybiotin intermediate requires energy, which is provided by the cleavage of ATP to AMP and PP_i. The D-methylmalonyl-CoA thus formed is enzymatically epimerized to its L stereoisomer by **methylmalonyl-CoA epimerase** (Fig. 17–11). The L-methylmalonyl-CoA then undergoes an intramolecular rearrangement to form succinyl-CoA, which can enter the citric acid cycle. This rearrangement is catalyzed by **methylmalonyl-CoA mutase,** which requires as its coenzyme **deoxyadenosyl-cobalamin,** or **coenzyme B_{12},** derived from vitamin B_{12} (cobalamin). Box 17–2 describes the role of coenzyme B_{12} in this remarkable exchange reaction.

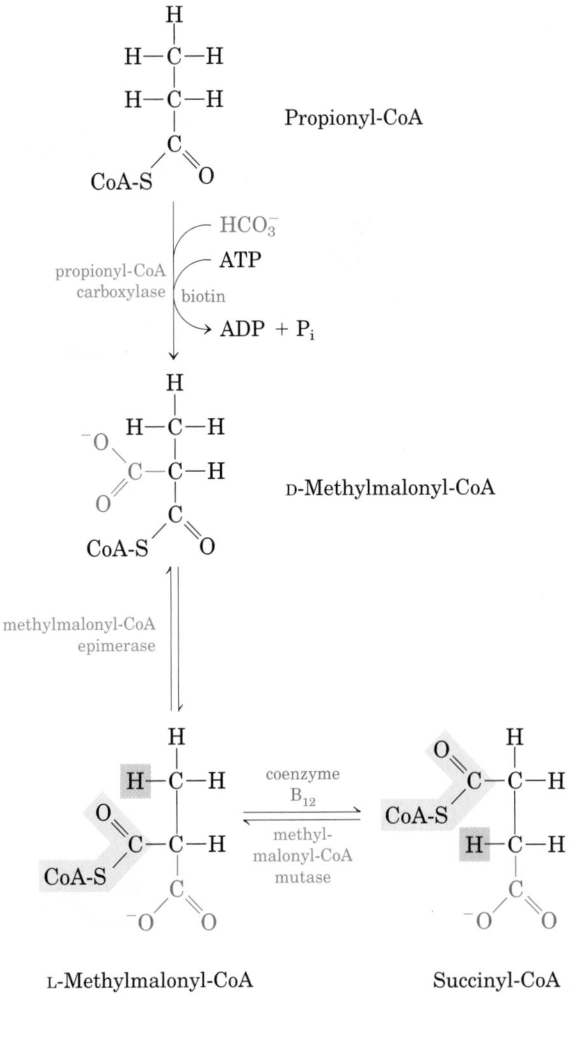

figure 17–11
Oxidation of propionyl-CoA produced by β oxidation of odd-number fatty acids. The sequence involves the carboxylation of propionyl-CoA to D-methylmalonyl-CoA and conversion of the latter to succinyl-CoA. This conversion requires epimerization of D- to L-methylmalonyl-CoA, followed by a remarkable reaction in which substituents on adjacent carbon atoms exchange positions (see Box 17–2).

Coenzyme B_{12}: A Radical Solution to a Perplexing Problem

figure 1

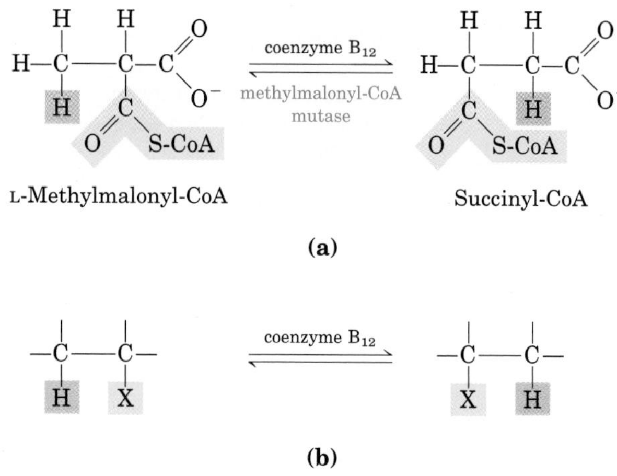

L-Methylmalonyl-CoA ⇌ (coenzyme B_{12}; methylmalonyl-CoA mutase) Succinyl-CoA

(a)

coenzyme B_{12}

(b)

figure 2

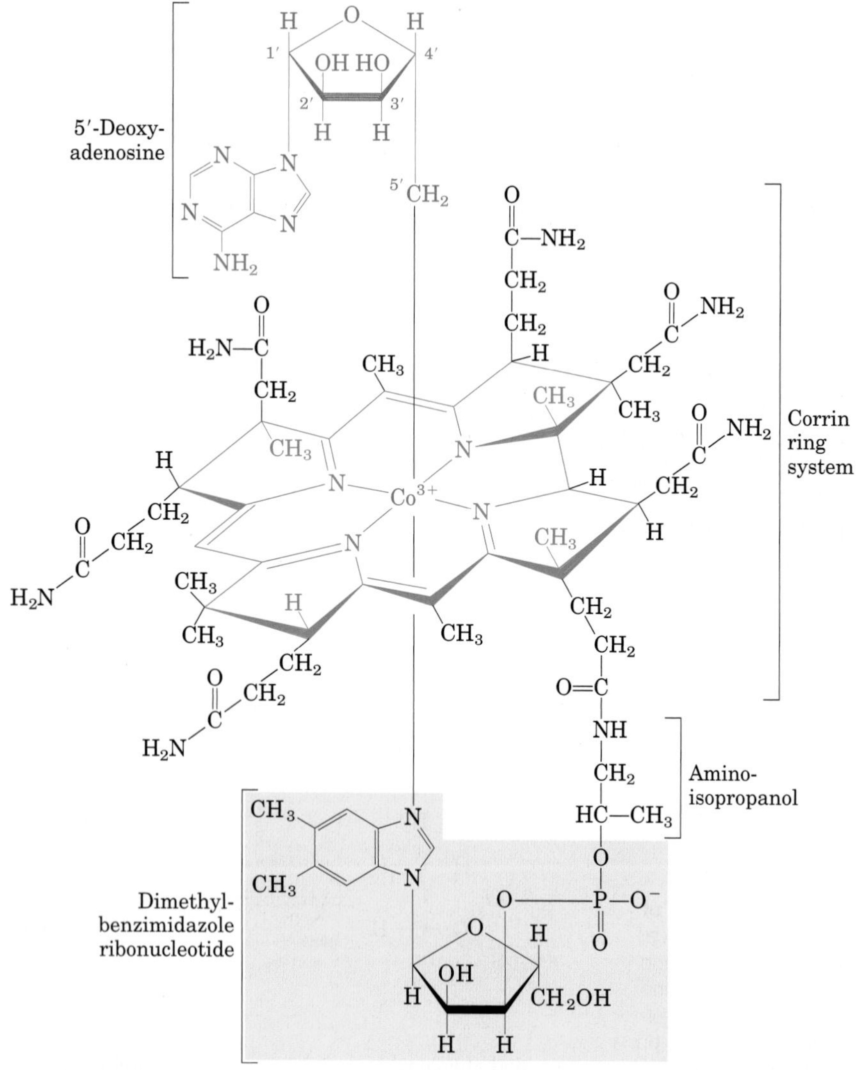

In the methylmalonyl-CoA mutase reaction (see Fig. 17–11), the group —CO—S-CoA at C-2 of the original propionate exchanges position with a hydrogen atom at C-3 of the original propionate (Fig. 1a). Coenzyme B_{12} is the cofactor for this reaction, as it is for almost all enzymes that catalyze reactions of this general type (Fig. 1b). These coenzyme B_{12}–dependent reactions are among the very few enzymatic reactions in biology in which there is an exchange of an alkyl or substituted alkyl group (X) with a hydrogen atom on an adjacent carbon, *with no mixing of the transferred hydrogen atom with the hydrogen of the solvent,* H_2O. How can the hydrogen atom move between two carbons without mixing with the enormous excess of hydrogen atoms in the solvent?

Coenzyme B_{12} is the cofactor form of vitamin B_{12}, which is unique among all the vitamins in that it contains not only a complex organic molecule but an essential trace element, cobalt. The complex **corrin ring system** of vitamin B_{12} (colored blue in Fig. 2), to which cobalt (as Co^{3+}) is coordinated, is chemically related to the porphyrin ring system of heme and heme proteins (see Fig. 7–1). A fifth coordination position of cobalt is filled by the nucleotide dimethylbenzimidazole ribonucleotide (shaded yellow), bound covalently by its 3′-phosphate group to a side chain of the corrin ring, through aminoisopropanol.

Vitamin B_{12} as usually isolated is called **cyanocobalamin** because it contains a cyano group (picked up during purification) attached to cobalt in the sixth coordination position. In **5′-deoxyadenosylcobalamin,** the cofactor for methylmalonyl-CoA mutase, the cyano group is replaced by the **5′-deoxyadenosyl** group (red in Fig. 2), covalently bound through C-5′ to the cobalt. The three-dimensional structure of the cofactor was determined by Dorothy Crowfoot Hodgkin in 1956, using x-ray crystallography.

The formation of this complex cofactor (Fig. 3) occurs in one of only two known reactions in which triphosphate is cleaved from ATP; the other reaction is the formation of *S*-adenosylmethionine from ATP and methionine (see Fig. 18–17).

The key to understanding how coenzyme B_{12} catalyzes hydrogen exchange lies in the properties of the covalent bond between cobalt and C-5′ of the deoxyadenosyl group (Fig. 2). This is a relatively weak bond; its bond dissociation energy is about 110 kJ/mol, compared with 348 kJ/mol for a typical C—C bond or 414 kJ/mol for a C—H bond. Merely illuminating the compound with visible light is enough to break this bond. (This

figure 3

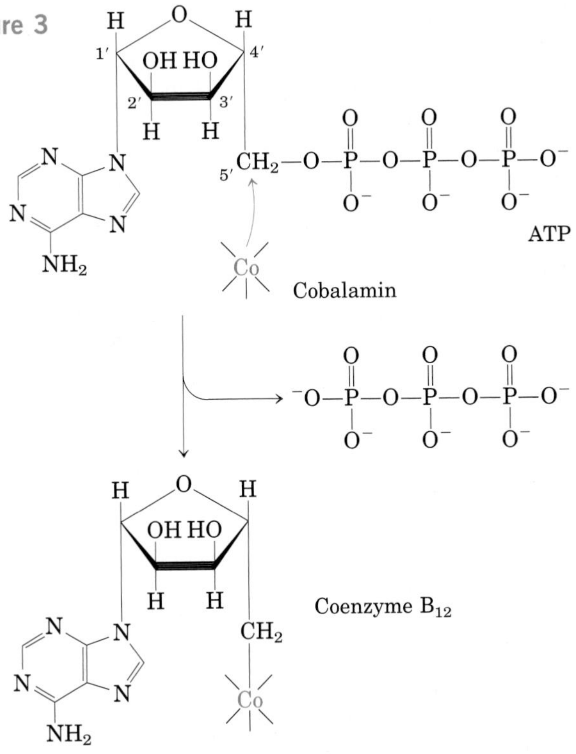

ATP

Cobalamin

Coenzyme B_{12}

figure 4

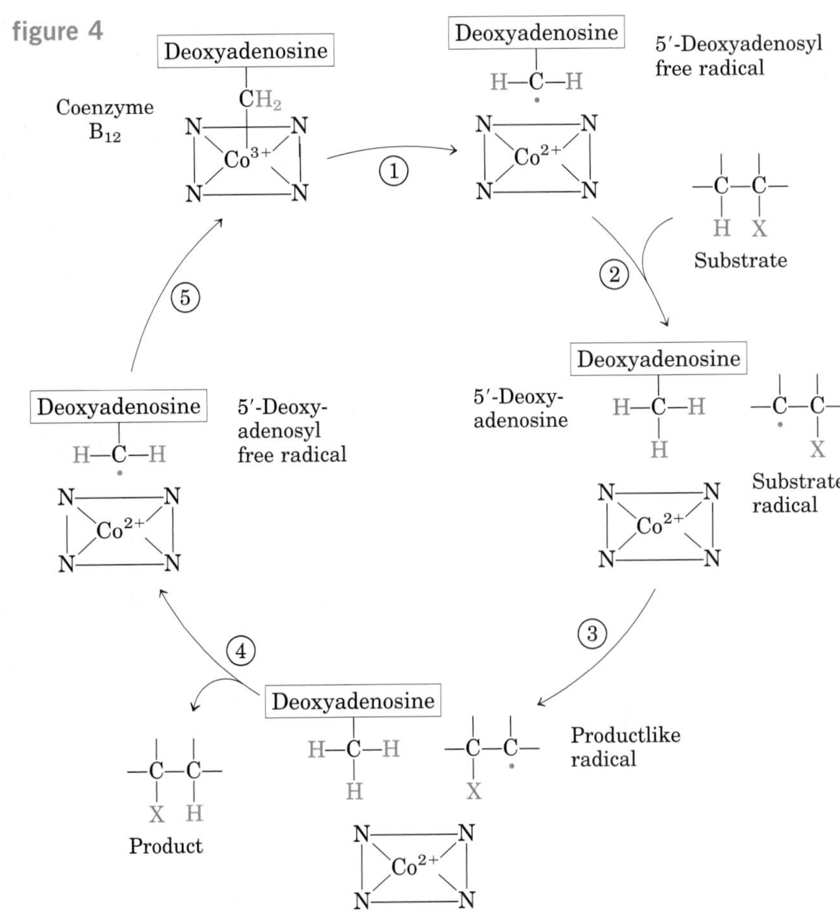

extreme photolability probably accounts for the fact that plants do not contain vitamin B_{12}.) Dissociation produces a 5′-deoxyadenosyl radical and the Co^{2+} form of the vitamin. The chemical function of 5′-deoxyadenosylcobalamin is to generate free radicals in this way, thus initiating a series of transformations such as that illustrated in Figure 4—a postulated mechanism for the reaction catalyzed by methylmalonyl-CoA mutase and a number of other coenzyme B_{12}–dependent transformations.

The enzyme first breaks the Co—C bond in the cofactor, leaving the coenzyme in its Co^{2+} form and producing the 5′-deoxyadenosyl free radical (step ①). This radical now abstracts a hydrogen atom from the substrate, converting the substrate to a radical and producing 5′-deoxyadenosine (step ②). Rearrangement of the substrate radical (step ③) yields another radical, in which the migrating group X (—CO—S-CoA for methylmalonyl-CoA mutase) has moved to the adjacent carbon to form a radical that has the carbon skeleton of the eventual product (a four-carbon straight chain). The hydrogen atom initially abstracted from the substrate is now part of the $—CH_3$ group of 5′-deoxyadenosine; one of the hydrogens from this same $—CH_3$ group (it can be the same one originally abstracted) is returned to the productlike radical, generating the product and regenerating the deoxyadenosyl free radical (step ④). Finally, the bond re-forms between cobalt and the $—\dot{C}H_2$ group of the deoxyadenosyl radical (step ⑤), destroying the free radical and regenerating the cofactor in its Co^{3+} form, ready to undergo another catalytic cycle.

In this postulated mechanism, the migrating hydrogen atom never exists as a free species and is thus never free to exchange with the hydrogen of surrounding water molecules.

Vitamin B_{12} deficiency results in serious disease. Vitamin B_{12} is not made by plants or animals and can be synthesized only by a few species of microorganisms. It is required by healthy people in only minute amounts, about 3 μg/day. The severe disease **pernicious anemia** results from failure to absorb vitamin B_{12} efficiently from the intestine, where it is synthesized by intestinal bacteria or obtained from digestion of meat. Individuals with this disease do not produce sufficient amounts of **intrinsic factor,** a glycoprotein essential to vitamin B_{12} absorption. The pathology in pernicious anemia includes reduced production of erythrocytes, reduced levels of hemoglobin, and severe, progressive impairment of the central nervous system. Administration of large doses of vitamin B_{12} alleviates these symptoms in at least some cases.

Dorothy Crowfoot Hodgkin
1910–1994

Fatty Acid Oxidation Is Tightly Regulated

Oxidation of fatty acids consumes a precious fuel, and it is regulated so as to occur only when the need for energy requires it. In the liver, fatty acyl–CoA formed in the cytosol has two major pathways open to it: (1) β oxidation by enzymes in the mitochondria or (2) conversion into triacylglycerols and phospholipids by enzymes in the cytosol. The pathway taken depends on the rate of transfer of long-chain fatty acyl–CoA into the mitochondria. The three-step process by which fatty acyl groups are carried from cytosolic fatty acyl–CoA into the mitochondrial matrix (Fig. 17–6) is rate-limiting for fatty acid oxidation and is an important point of regulation. Once fatty acyl groups have entered the mitochondria, they are committed to oxidation to acetyl-CoA.

$$^{-}OOC{-}CH_2{-}\overset{\overset{\displaystyle O}{\|}}{C}{-}S\text{-}CoA$$

Malonyl-CoA

Malonyl-CoA, the first intermediate in the cytosolic biosynthesis of long-chain fatty acids from acetyl-CoA (see Fig. 21–1), increases in concentration whenever the animal is well supplied with carbohydrate; excess glucose that cannot be oxidized or stored as glycogen is converted in the cytosol into fatty acids for storage as triacylglycerol. The inhibition of carnitine acyltransferase I by malonyl-CoA assures that the oxidation of fatty acids is inhibited whenever the liver is amply supplied with glucose as fuel and is actively making triacylglycerols from excess glucose.

Two of the enzymes of β oxidation are also regulated by metabolites that signal energy sufficiency. When the [NADH]/[NAD^+] ratio is high, β-hydroxyacyl-CoA dehydrogenase is inhibited; in addition, high concentrations of acetyl-CoA inhibit thiolase.

Peroxisomes Also Carry Out β Oxidation

Although the major site of fatty acid oxidation in animal cells is the mitochondrial matrix, other compartments in certain cells also contain enzymes capable of oxidizing fatty acids to acetyl-CoA, by a pathway similar to, but not identical with, that in mitochondria. Peroxisomes are membrane-enclosed cellular compartments (p. 34) of animal and plants cells in which fatty acid oxidation produces H_2O_2, which is then enzymatically destroyed. As in mitochondria, the intermediates for β oxidation of fatty acids in peroxisomes are coenzyme A derivatives, and the process consists of four steps (Fig. 17–12): (1) dehydrogenation; (2) addition of water to the resulting double bond; (3) oxidation of the β-hydroxyacyl-CoA to a ketone, and (4) thiolytic cleavage by coenzyme A. (The identical reactions also occur in glyoxysomes, as discussed below.) The difference between the peroxisomal and mitochondrial pathways is in the first step. In peroxisomes, the flavoprotein dehydrogenase that introduces the double bond passes electrons directly to O_2, producing H_2O_2 (Fig. 17–12). This strong and potentially damaging oxidant is immediately cleaved to H_2O and O_2 by **catalase.** Recall that in mitochondria, the electrons removed in the first oxidation step pass through the respiratory chain to O_2 to produce H_2O, and this process is accompanied by ATP synthesis. In peroxisomes, the energy released in the first oxidative step of fatty acid breakdown is not conserved as ATP, but is dissipated as heat.

High concentrations of fats in the diet of a mammal result in increased synthesis of the enzymes of peroxisomal β oxidation in the liver. Liver peroxisomes do not contain the enzymes of the citric acid cycle and cannot catalyze the oxidation of acetyl-CoA to CO_2. Instead, the acetate produced by fatty acid oxidation is exported from peroxisomes; presumably some of this can enter mitochondria to be oxidized.

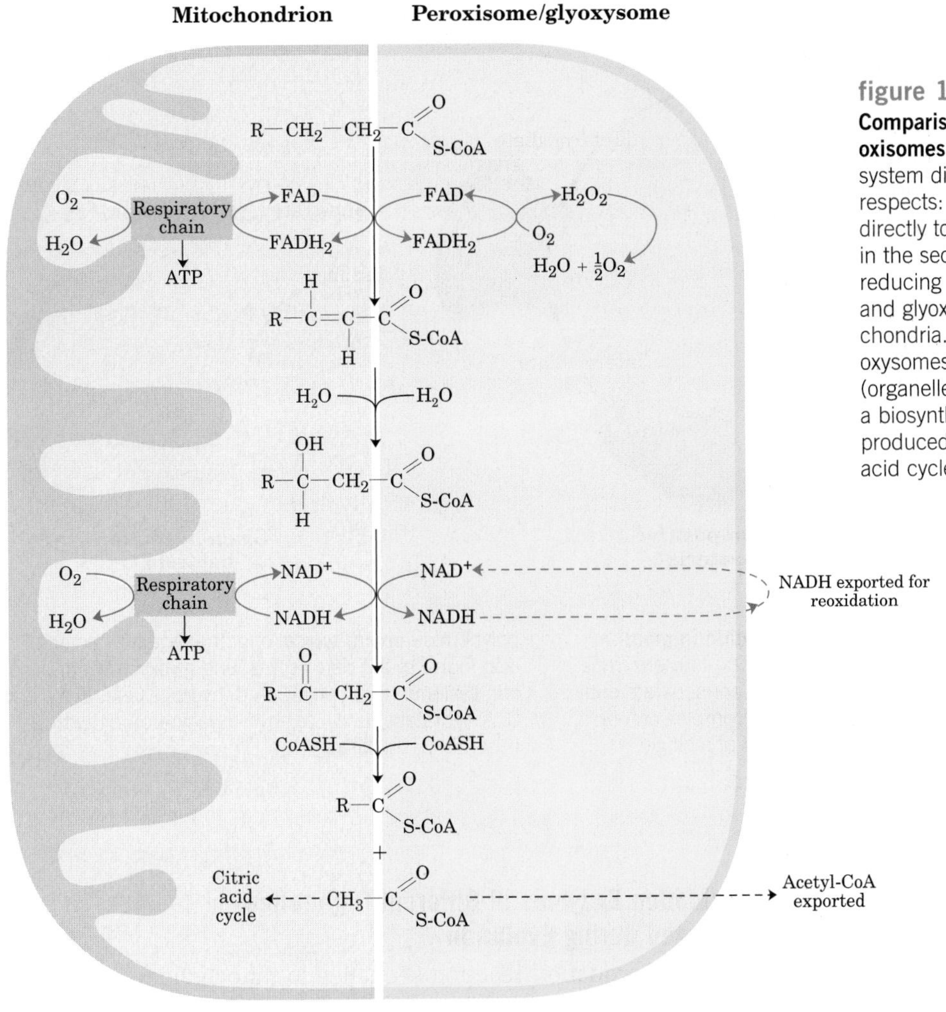

figure 17–12

Comparison of β oxidation in mitochondria and in peroxisomes and glyoxysomes. The peroxisomal/glyoxysomal system differs from the mitochondrial system in two respects: (1) in the first oxidative step electrons pass directly to O_2, generating H_2O_2, and (2) the NADH formed in the second oxidative step cannot be reoxidized, so reducing equivalents are exported from the peroxisome and glyoxysome to the cytosol, eventually entering mitochondria. Acetyl-CoA produced by peroxisomes and glyoxysomes is also exported; the acetate from glyoxysomes (organelles found only in germinating seeds) serves as a biosynthetic precursor (see Fig. 17–13). Acetyl-CoA produced in mitochondria is further oxidized in the citric acid cycle.

Plant Peroxisomes and Glyoxysomes Use Acetyl-CoA from β Oxidation as a Biosynthetic Precursor

Fatty acid oxidation in plants occurs not in mitochondria but in the peroxisomes of leaf tissue and the glyoxysomes of germinating seeds. Plant peroxisomes and glyoxysomes are similar in structure and function. Glyoxysomes occur only during seed germination and may be considered specialized peroxisomes.

The biological role of β oxidation in plant peroxisomes and glyoxysomes is to provide biosynthetic precursors from stored lipids. The β-oxidation pathway is not an important source of metabolic energy in plants; in fact, plant mitochondria do not contain the enzymes of β oxidation. During seed germination, stored triacylglycerols are converted into glucose, sucrose, and a wide variety of essential metabolites (Fig. 17–13). Fatty acids released from the triacylglycerols are activated to their coenzyme A derivatives and oxidized in glyoxysomes by the same four-step process that occurs in peroxisomes (Fig. 17–12). The acetyl-CoA produced is converted via the glyoxylate cycle (see Fig. 16–16) to four-carbon precursors for gluconeogenesis (see Fig. 20–2). Glyoxysomes, like peroxisomes, contain high concentrations of catalase, which converts the H_2O_2 produced by β oxidation to H_2O and O_2.

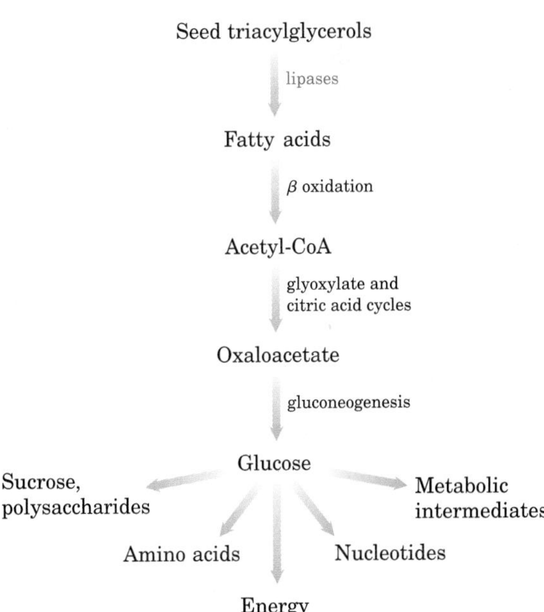

figure 17–13

Role of β oxidation in the conversion of stored triacylglycerols to glucose in germinating seeds.

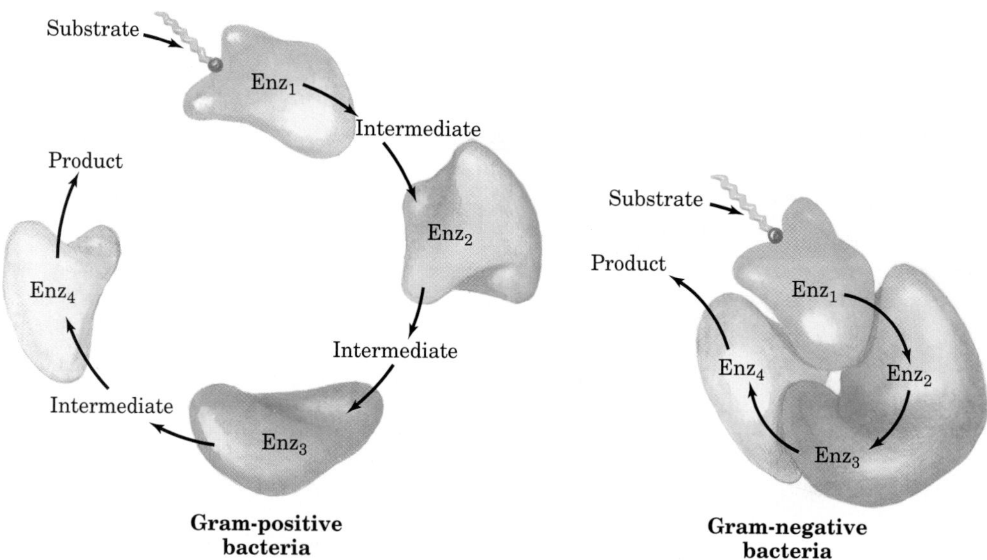

figure 17–14

Comparison of the enzymes of β oxidation in gram-positive and gram-negative bacteria. The four enzymes of β oxidation in gram-positive bacteria are separate entities, as are those of mitochondria. The complex of four enzyme activities (two of which are part of a single polypeptide chain) typical of gram-negative bacteria is also found in the peroxisomal and glyoxysomal β-oxidation systems. Enz$_1$, acyl-CoA dehydrogenase; Enz$_2$, enoyl-CoA hydratase; Enz$_3$, L-β-hydroxyacyl-CoA dehydrogenase; Enz$_4$, thiolase.

The β-Oxidation Enzymes of Different Organelles Have Diverged during Evolution

Although the β-oxidation reaction sequence in mitochondria is essentially the same as that in peroxisomes and glyoxysomes, the enzymes (isozymes) differ significantly between the two types of organelles. The differences apparently reflect an evolutionary divergence that occurred very early, with the separation of gram-positive and gram-negative bacteria (see Fig. 2–5). In mitochondria, the four β-oxidation enzymes are separate, soluble proteins, similar in structure to the analogous enzymes of gram-positive bacteria. In contrast, the β-oxidation enzymes of peroxisomes and glyoxysomes form a complex of proteins, at least one of which contains two enzymatic activities in a single polypeptide chain (Fig. 17–14). Enoyl-CoA hydratase and L-β-hydroxyacyl-CoA dehydrogenase activities both reside in a single, monomeric protein (M_r ~150,000), closely similar to the bifunctional protein of the gram-negative bacterium *Escherichia coli.* The evolutionary selective value of retaining both types of β-oxidation system in the same organism is not yet apparent.

Omega Oxidation Occurs in the Endoplasmic Reticulum

Although mitochondrial β oxidation, in which enzymes act at the carboxyl end of a fatty acid, is by far the most important catabolic fate for fatty acids, there is another pathway in some organisms, including vertebrates, that involves oxidation of the omega (ω) carbon—the carbon most distant from the carboxyl group. The enzymes unique to this pathway are located (in vertebrates) in the endoplasmic reticulum of liver and kidney, and the preferred substrates are fatty acids of 10 or 12 carbon atoms. The first step introduces a hydroxyl group onto the ω carbon (Fig. 17–15). The oxygen for this group comes from molecular oxygen (O_2) in a complex reaction that involves cytochrome P450 and the electron donor NADPH. Reactions of this type are catalyzed by **mixed-function oxidases,** described in more detail in Box 21–1.

Two more enzymes now act on the ω carbon: **alcohol dehydrogenase** oxidizes the hydroxyl group to an aldehyde, and **aldehyde dehydrogenase** oxidizes the aldehyde group to a carboxylic acid, producing a fatty acid with a carboxyl group at each end. Now, either end can be attached to coenzyme A, and the molecule can enter the mitochondrion and undergo β oxidation by the normal route. In each pass through the β-oxidation pathway, the "double-ended" fatty acid yields dicarboxylic acids such as succinic and adipic acids (Fig. 17–15).

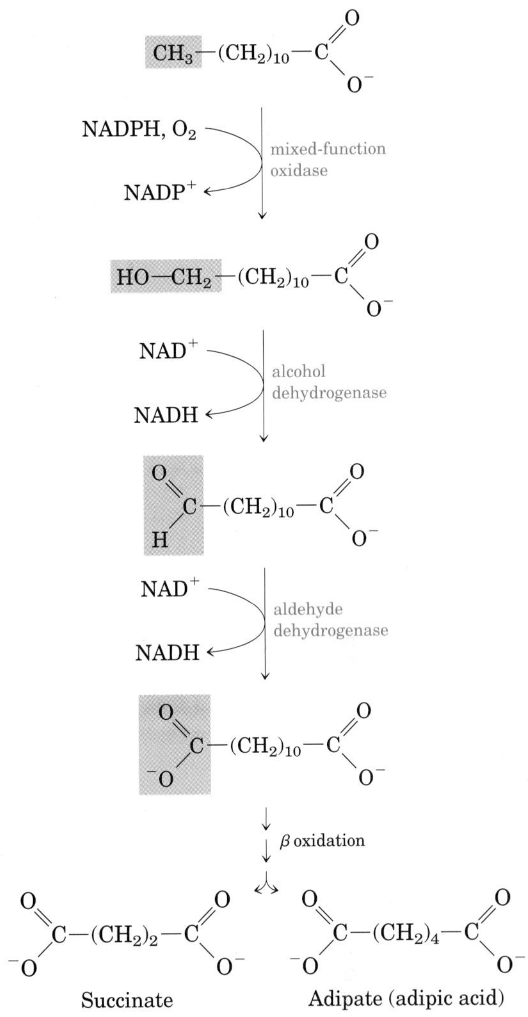

Genetic Defects in Fatty Acyl–CoA Dehydrogenases Cause Serious Disease

Stored triacylglycerols are typically the chief source of energy for muscle contraction, and the inability to oxidize fatty acids from triacylglycerols therefore has serious consequences for health. The most common human genetic defect in fatty acid catabolism in U.S. and northern European populations is due to a mutation in the gene encoding the **medium-chain acyl-CoA dehydrogenase (MCAD).** Among northern Europeans, the frequency of carriers (individuals with this recessive mutation on one of the two homologous chromosomes) is about 1 in 40. Individuals with two copies of the mutant MCAD allele are unable to oxidize fatty acids of 6 to 12 carbons. They have recurring episodes of a syndrome that includes fat accumulation in the liver, high blood levels of octanoic acid, low blood glucose (hypoglycemia), sleepiness, vomiting, and coma. The pattern of organic acids in the urine helps in the diagnosis of this disease. The urine commonly contains high levels of C-6 to C-10 dicarboxylic acids (produced by ω oxidation), but the high levels of urinary ketone bodies (see below) commonly associated with hypoglycemia are absent. Although individuals may have no symptoms between episodes, the episodes are very serious; mortality from this disease is 25% to 60% in early childhood. If the genetic defect is detected shortly after birth, the affected individual can be given a low-fat, high-carbohydrate diet. Long intervals between meals must be avoided to prevent the body from turning to its fat reserves for energy. With early detection and careful management of the diet, the prognosis for these individuals is good.

$CH_3-(CH_2)_{10}-COO^-$

NADPH, O_2 → $NADP^+$ (mixed-function oxidase)

$HO-CH_2-(CH_2)_{10}-COO^-$

NAD^+ → NADH (alcohol dehydrogenase)

$OHC-(CH_2)_{10}-COO^-$

NAD^+ → NADH (aldehyde dehydrogenase)

$^-OOC-(CH_2)_{10}-COO^-$

β oxidation

$^-OOC-(CH_2)_2-COO^-$ Succinate

$^-OOC-(CH_2)_4-COO^-$ Adipate (adipic acid)

figure 17–15
Omega oxidation in the endoplasmic reticulum. This alternative to β oxidation begins with oxidation of the carbon most distant from the α carbon—the omega (ω) carbon. The substrate is usually a medium-chain fatty acid. This pathway is generally not the major route for oxidative catabolism of fatty acids.

Ketone Bodies

In human beings and most other mammals, acetyl-CoA formed in the liver during oxidation of fatty acids can enter the citric acid cycle (stage 2 of Fig. 17–7) or can be converted to the "ketone bodies" **acetone, acetoacetate,** and **D-β-hydroxybutyrate** for export to other tissues. (The term "bodies" is a historical artifact; the term is occasionally applied to insoluble particles, but these compounds are quite soluble in blood and urine.) Acetone, produced in smaller quantities than the other ketone bodies, is exhaled. Acetoacetate and D-β-hydroxybutyrate are transported by the blood to tissues other than the liver (the extrahepatic tissues), where they are oxidized in the citric acid cycle to provide much of the energy required by tissues such as skeletal and heart muscle and the renal cortex. The brain, which preferentially uses glucose as fuel, can adapt to the use of acetoacetate or D-β-hydroxybutyrate under starvation conditions, when glucose is unavailable. The production and export of ketone bodies from the liver to extrahepatic tissues allows continued oxidation of fatty acids in the liver when acetyl-CoA is not being oxidized in the citric acid cycle.

$CH_3-CO-CH_3$ Acetone

$CH_3-CO-CH_2-COO^-$ Acetoacetate

$CH_3-CH(OH)-CH_2-COO^-$ D-β-Hydroxybutyrate

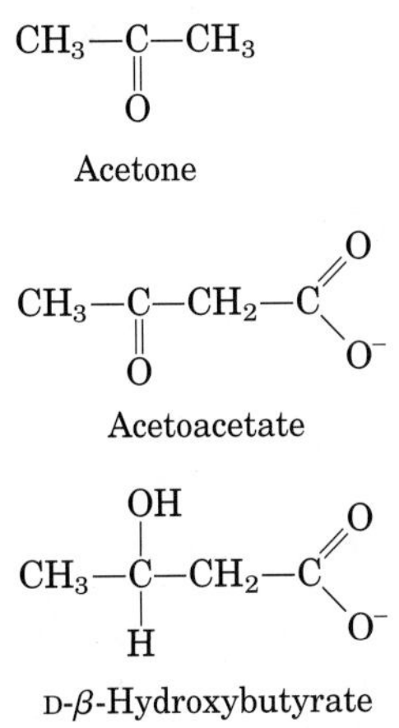

Ketone Bodies Formed in the Liver Are Exported to Other Organs

The first step in the formation of acetoacetate in the liver (Fig. 17–16) is the enzymatic condensation of two molecules of acetyl-CoA, catalyzed by thiolase; this is simply the reversal of the last step of β oxidation. The acetoacetyl-CoA then condenses with acetyl-CoA to form **β-hydroxy-β-methylglutaryl-CoA** (HMG-CoA), which is cleaved to free acetoacetate and acetyl-CoA.

The free acetoacetate so produced is reversibly reduced by D-β-hydroxybutyrate dehydrogenase, a mitochondrial enzyme, to D-β-hydroxybutyrate (Fig. 17–16). This enzyme is specific for the D stereoisomer; it does not act on L-β-hydroxyacyl-CoAs and is not to be confused with L-β-hydroxyacyl-CoA dehydrogenase of the β-oxidation pathway. In healthy people, acetone is formed in very small amounts from acetoacetate by the loss of a carboxyl group. Acetoacetate is easily decarboxylated, either spontaneously or by the action of **acetoacetate decarboxylase** (Fig. 17–16). Because individuals with untreated diabetes produce large quantities of acetoacetate, their blood contains significant amounts of acetone, which is toxic. Acetone is volatile and imparts a characteristic odor to the breath, which is sometimes useful in diagnosing diabetes.

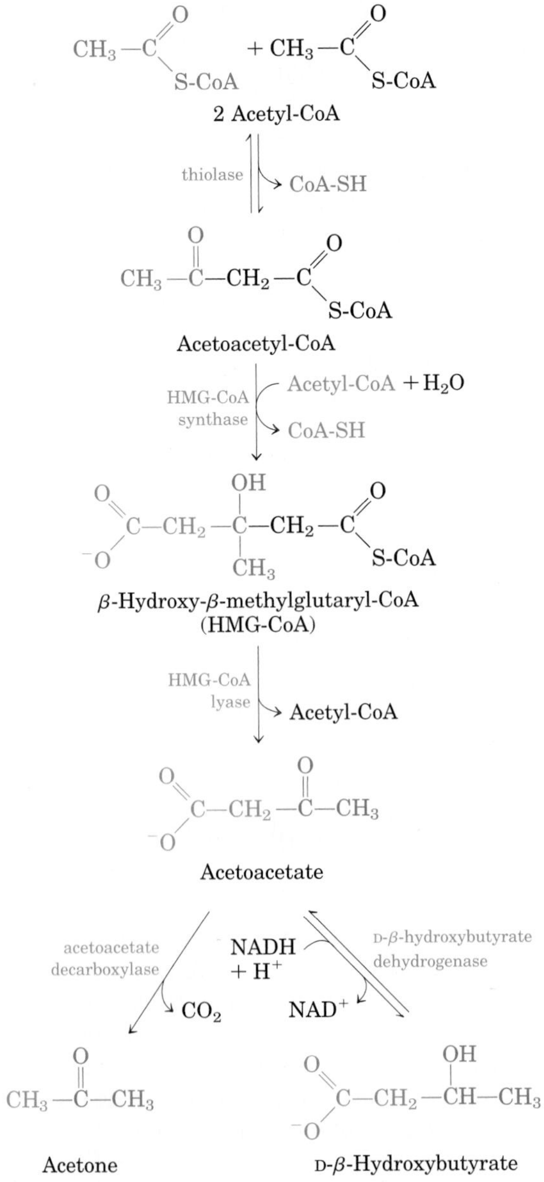

$CH_3-C(=O)-S\text{-}CoA + CH_3-C(=O)-S\text{-}CoA$

2 Acetyl-CoA

thiolase ⇅ CoA-SH

$CH_3-C(=O)-CH_2-C(=O)-S\text{-}CoA$

Acetoacetyl-CoA

HMG-CoA synthase ↓ Acetyl-CoA + H_2O → CoA-SH

$^{-}O(O=)C-CH_2-C(OH)(CH_3)-CH_2-C(=O)-S\text{-}CoA$

β-Hydroxy-β-methylglutaryl-CoA (HMG-CoA)

HMG-CoA lyase ↓ Acetyl-CoA

$^{-}O(O=)C-CH_2-C(=O)-CH_3$

Acetoacetate

acetoacetate decarboxylase ↙ CO_2 | NADH + H^+ → NAD^+ D-β-hydroxybutyrate dehydrogenase

$CH_3-C(=O)-CH_3$

Acetone

$^{-}O(O=)C-CH_2-CH(OH)-CH_3$

D-β-Hydroxybutyrate

figure 17–16

Formation of ketone bodies from acetyl-CoA. Well-nourished individuals produce ketone bodies at a relatively low rate. When acetyl-CoA accumulates (as in starvation or untreated diabetes, for example), thiolase catalyzes the condensation of two acetyl-CoA molecules to acetoacetyl-CoA, the parent compound of the three ketone bodies. The reactions of ketone body formation occur in the matrix of liver mitochondria. The six-carbon compound β-hydroxy-β-methylglutaryl-CoA (HMG-CoA) is also an intermediate of sterol biosynthesis, but the enzyme that forms HMG-CoA in that pathway is cytosolic. HMG-CoA lyase is present only in the mitochondrial matrix.

Extrahepatic Tissues Use Ketone Bodies as Fuels

In extrahepatic tissues, D-β-hydroxybutyrate is oxidized to acetoacetate by D-β-hydroxybutyrate dehydrogenase (Fig. 17–17). Acetoacetate is activated to its coenzyme A ester by transfer of CoA from succinyl-CoA, an intermediate of the citric acid cycle (see Fig. 16–7), in a reaction catalyzed by **β-ketoacyl-CoA transferase.** The acetoacetyl-CoA is then cleaved by thiolase to yield two acetyl-CoAs, which enter the citric acid cycle.

Ketone Bodies Are Overproduced in Diabetes and during Starvation

The production and export of ketone bodies from the liver allows continued oxidation of fatty acids with only minimal oxidation of acetyl-CoA (Fig. 17–18). When, for example, intermediates of the citric acid cycle are being used for glucose synthesis by gluconeogenesis, oxidation of citric acid cycle intermediates slows, and so does acetyl-CoA oxidation. Moreover, the liver contains a limited amount of coenzyme A, and when most of it is tied up in acetyl-CoA, β oxidation slows for want of the free coenzyme. The production and export of ketone bodies frees coenzyme A, allowing continued fatty acid oxidation.

Severe starvation or untreated diabetes mellitus leads to overproduction of ketone bodies, with several associated medical problems. During starvation, gluconeogenesis depletes citric acid cycle intermediates, diverting acetyl-CoA to ketone body production (Fig. 17–18). In untreated diabetes, when the insulin level is insufficient, extrahepatic tissues cannot take up glucose efficiently from the blood to use as a fuel or to convert to fats for storage. Under these conditions malonyl-CoA (the starting material for fatty acid synthesis) is not formed and thus is not present to inhibit carnitine acyltransferase I. Fatty acids therefore enter the mitochondria to be degraded to acetyl-CoA, which cannot pass through the citric acid cycle

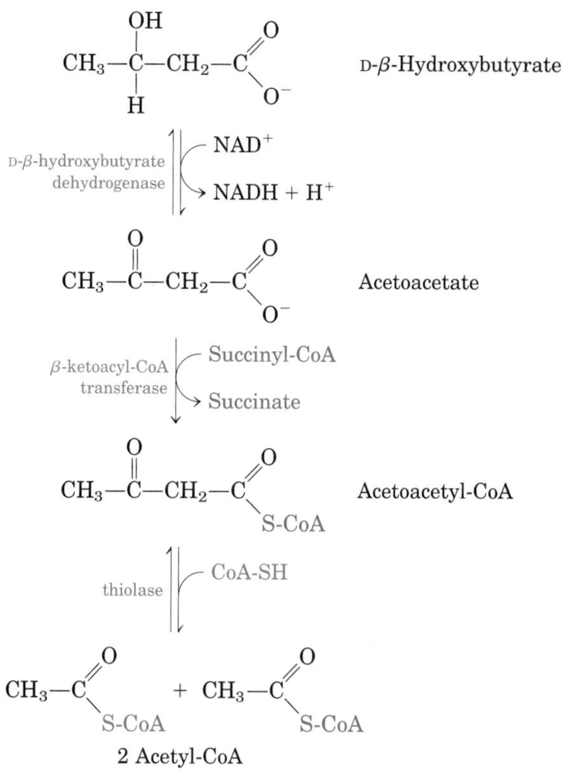

$CH_3—CH(OH)—CH_2—COO^-$ D-β-Hydroxybutyrate

D-β-hydroxybutyrate dehydrogenase (NAD^+ → $NADH + H^+$)

$CH_3—CO—CH_2—COO^-$ Acetoacetate

β-ketoacyl-CoA transferase (Succinyl-CoA → Succinate)

$CH_3—CO—CH_2—CO—S\text{-}CoA$ Acetoacetyl-CoA

thiolase (CoA-SH)

$CH_3—CO—S\text{-}CoA$ + $CH_3—CO—S\text{-}CoA$

2 Acetyl-CoA

figure 17–17

β-Hydroxybutyrate as a fuel. D-β-Hydroxybutyrate synthesized in the liver passes into the blood and thus to other tissues, where it is converted in three steps to acetyl-CoA. It is first oxidized to acetoacetate, which is activated with coenzyme A donated from succinyl-CoA, then split by thiolase. The acetyl-CoA thus formed is used for energy production.

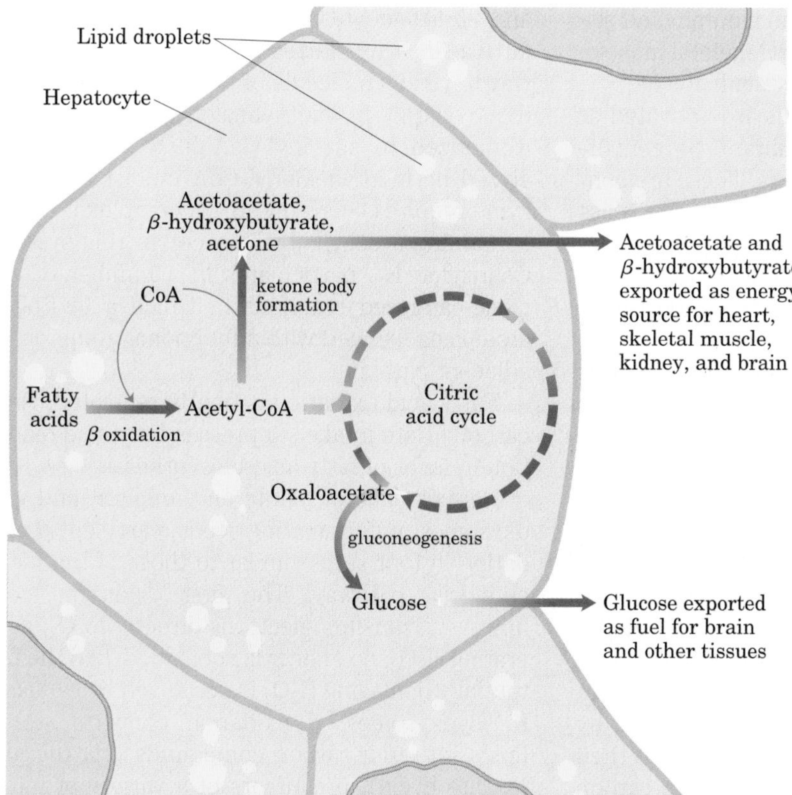

figure 17–18

Ketone body formation and export from the liver. Conditions that promote gluconeogenesis (untreated diabetes, severe dieting, fasting) slow the citric acid cycle (by drawing off oxaloacetate) and enhance the conversion of acetyl-CoA to acetoacetate. The released coenzyme A allows continued β oxidation of fatty acids.

table 17–2

Ketone Body Accumulation in Diabetic Ketosis

	Urinary excretion (mg/24 h)	Blood concentration (mg/100 mL)
Normal	≤125	<3
Extreme ketosis (untreated diabetes)	5,000	90

because cycle intermediates have been drawn off for use as substrates in gluconeogenesis (Chapter 20). The accumulation of acetyl-CoA accelerates the formation of ketone bodies beyond the capacity of extrahepatic tissues to oxidize them. The increased blood levels of acetoacetate and D-β-hydroxybutyrate lower the blood pH, causing the condition known as **acidosis.** Extreme acidosis can lead to coma and in some cases death. Ketone bodies in the blood and urine of untreated diabetics can reach extraordinary levels (Table 17–2), a condition called **ketosis.** In individuals on very low-calorie diets, using the fats stored in adipose tissue as their major energy source, levels of ketone bodies in the blood and urine must be monitored to avoid the dangers of acidosis and ketosis (ketoacidosis).

summary

The fatty acid components of triacylglycerols furnish a large fraction of the oxidative energy in animals. Dietary triacylglycerols are emulsified in the small intestine by bile salts, hydrolyzed by intestinal lipases, absorbed by intestinal epithelial cells, reconverted into triacylglycerols, then formed into chylomicrons by combination with specific apolipoproteins. Chylomicrons deliver triacylglycerols to tissues, where lipoprotein lipase releases free fatty acids for entry into cells. Triacylglycerols stored in adipose tissue are mobilized by the action of hormones through a hormone-sensitive triacylglycerol lipase. The released fatty acids bind to serum albumin and are carried in the blood to the heart, skeletal muscle, and other tissues that use fatty acids for fuel.

Once inside cells, fatty acids are activated at the outer mitochondrial membrane by conversion to fatty acyl–CoA thioesters. These are converted into fatty acyl–carnitine esters, pass through a specific transporter in the inner mitochondrial membrane, are reconverted to fatty acyl–CoA esters in the matrix, then enter the β-oxidation pathway.

In the first stage of β oxidation, four reactions are required to remove each acetyl-CoA unit from the carboxyl end of a saturated fatty acyl–CoA: (1) dehydrogenation of the α and β carbons (C-2 and C-3) by FAD-linked acyl-CoA dehydrogenases, (2) hydration of the resulting *trans*-Δ^2 double bond by enoyl-CoA hydratase, (3) dehydrogenation of the resulting L-β-hydroxyacyl-CoA by NAD-linked β-hydroxyacyl-CoA dehydrogenase, and (4) CoA-requiring cleavage of the resulting β-ketoacyl-CoA by thiolase to form acetyl-CoA and a fatty acyl–CoA shortened by two carbons. The shortened fatty acyl–CoA then reenters the sequence. Overall, the 16-carbon palmitate, for example, yields eight molecules of acetyl-CoA. In the second stage of fatty acid oxidation the acetyl-CoA is oxidized to CO_2 in the citric acid cycle. A large fraction of the theoretical yield of free energy from fatty acid oxidation is recovered as ATP by oxidative phosphorylation in the final stage of the oxidative pathway. Genetic defects in the medium-chain acyl-CoA dehydrogenase (MCAD), common in northern European and U.S. populations, result in serious human disease.

Oxidation of unsaturated fatty acids requires two additional enzymes: enoyl-CoA isomerase and 2,4-dienoyl-CoA reductase. Odd-number fatty acids are oxidized by the β-oxidation pathway to yield a molecule of propionyl-CoA. This is carboxylated to methylmalonyl-CoA, which is isomerized to succinyl-CoA in a reaction catalyzed by methylmalonyl-CoA mutase. This enzyme requires coenzyme B_{12}, a complex cofactor containing a cobalt ion in a corrin ring system. Coenzyme B_{12} participates in a number of enzyme-catalyzed reactions in which a hydrogen atom is exchanged with a functional group on an adjacent carbon.

Fatty acid oxidation is tightly regulated. High carbohydrate intake suppresses fatty acid oxidation in favor of fatty acid biosynthesis.

Peroxisomes of plants and animals, and glyoxysomes of germinating seeds, carry out β oxidation in four steps similar to those of the mitochondrial pathway. The first oxidation step, however, transfers electrons directly to O_2, generating H_2O_2; no energy is conserved, and the potentially damaging H_2O_2 is destroyed by catalase. In glyoxysomes, β oxidation converts stored lipids into four-carbon compounds (via the glyoxylate cycle), precursors of a variety of intermediates and products required during seed germination.

Omega oxidation of fatty acids in the endoplasmic reticulum produces dicarboxylic fatty acids, with carboxyl groups at each end of a fatty acyl chain. The intermediates can undergo β oxidation at either end, yielding short dicarboxylic acids such as succinate.

The ketone bodies—acetone, acetoacetate, and D-β-hydroxybutyrate—are formed in the liver. The latter two compounds travel in the blood to other tissues, where they serve as fuel molecules; they are oxidized to acetyl-CoA and enter the citric acid cycle. Overproduction of ketone bodies in uncontrolled diabetes or severe starvation can lead to acidosis or ketosis.

further reading

General

Boyer, P.D. (1983) *The Enzymes,* 3rd edn, Vol. 16: *Lipid Enzymology,* Academic Press, Inc., San Diego, CA.

Gurr, M.I. & Harwood, J.L. (1991) *Lipid Biochemistry: An Introduction,* 4th edn, Chapman & Hall, London.

Langin, D., Holm, C., & Lafontan, M. (1996) Adipocyte hormone-sensitive lipase: a major regulator of lipid metabolism. *Proc. Nutr. Soc.* **55,** 93–109.

Ramsay, T.G. (1996) Fat cells. *Endocrinol. Metab. Clin. N. Am.* **25,** 847–870.

A review of all aspects of fat storage and mobilization in adipocytes.

Wang, C.S., Hartsuck, J., & McConathy, W.J. (1992) Structure and functional properties of lipoprotein lipase. *Biochim. Biophys. Acta* **1123,** 1–17.

Advanced-level discussion of the enzyme that releases fatty acids from lipoproteins in the capillaries of muscle and adipose tissue.

Mitochondrial β Oxidation

Bannerjee, R. (1997) The yin-yang of cobalamin biochemistry. *Chem. Biol.* **4,** 175–186.

A review of the biochemistry of coenzyme B_{12} reactions, including the methylmalonyl-CoA mutase reaction.

Eaton, S., Bartlett, K., & Pourfarzam, M. (1996) Mammalian mitochondrial β-oxidation. *Biochem. J.* **320,** 345–357.

A review of the enzymology of β oxidation, inherited defects in this pathway, and regulation of the process in mitochondria.

Harwood, J.L. (1988) Fatty acid metabolism. *Annu. Rev. Plant Physiol. Plant Mol. Biol.* **39,** 101–138.

Jeukendrup, A.E., Saris, W.H., & Wagenmakers, A.J. (1998) Fat metabolism during exercise: a review. Part III: effects of nutritional interventions. *Int. J. Sports Med.* **19,** 371–379.

This paper is one of a series that reviews the factors that influence fat mobilization and utilization during exercise.

Kerner, J. & Hoppel, C. (1998) Genetic disorders of carnitine metabolism and their nutritional management. *Annu. Rev. Nutr.* **18,** 179–206.

Kunau, W.H., Dommes, V., & Schulz, H. (1995) β-Oxidation of fatty acids in mitochondria, peroxisomes, and bacteria: a century of continued progress. *Prog. Lipid Res.* **34,** 267–342.

A good historical account and a useful comparison of β oxidation in different systems.

Rinaldo, P., Raymond, K., al-Odaib, A., & Bennett, M.J. (1998) Clinical and biochemical features of fatty acid oxidation disorders. *Curr. Opin. Pediatr.* **10,** 615–621.

Review of metabolic defects in fat oxidation, including MCAD mutations.

Sherratt, H.S. (1994) Introduction: the regulation of fatty acid oxidation in cells. *Biochem. Soc. Trans.* **22,** 421–422.

Introduction to reviews (in this journal issue) of various aspects of fatty acid oxidation and its regulation.

Thorpe, C. & Kim, J.J. (1995) Structure and mechanism of action of the acyl-CoA dehydrogenases. *FASEB J.* **9,** 718–725.

Short, clear description of the three-dimensional structure and catalytic mechanism of these enzymes.

Peroxisomal β Oxidation

Hashimoto, T. (1996) Peroxisomal β-oxidation: enzymology and molecular biology. *Ann. N. Y. Acad. Sci.* **804,** 86–98.

Mannaerts, G.P. & van Veldhoven, P.P. (1996) Functions and organization of peroxisomal β-oxidation. *Ann. N. Y. Acad. Sci.* **804,** 99–115.

Ketone Bodies

Foster, D.W. & McGarry, J.D. (1983) The metabolic derangements and treatment of diabetic ketoacidosis. *N. Engl. J. Med.* **309,** 159–169.

McGarry, J.D. & Foster, D.W. (1980) Regulation of hepatic fatty acid oxidation and ketone body production. *Annu. Rev. Biochem.* **49,** 395–420.

Robinson, A.M. & Williamson, D.H. (1980) Physiological roles of ketone bodies as substrates and signals in mammalian tissues. *Physiol. Rev.* **60,** 143–187.

problems

1. Energy in Triacylglycerols On a per-carbon basis, where does the largest amount of biologically available energy in triacylglycerols reside: in the fatty acid portions or the glycerol portion? Indicate how knowledge of the chemical structure of triacylglycerols provides the answer.

2. Fuel Reserves in Adipose Tissue Triacylglycerols, with their hydrocarbon-like fatty acids, have the highest energy content of the major nutrients.

(a) If 15% of the body mass of a 70 kg adult consists of triacylglycerols, calculate the total available fuel reserve, in both kilojoules and kilocalories, in the form of triacylglycerols. Recall that 1.00 kcal = 4.18 kJ.

(b) If the basal energy requirement is approximately 8,400 kJ/day (2,000 kcal/day), how long could this person survive if the oxidation of fatty acids stored as triacylglycerols were the only source of energy?

(c) What would be the weight loss in pounds per day under such starvation conditions (1 lb = 0.454 kg)?

3. Common Reaction Steps in the Fatty Acid Oxidation Cycle and Citric Acid Cycle Cells often use the same enzyme reaction pattern for analogous metabolic conversions. For example, the steps in the oxidation of pyruvate to acetyl-CoA and α-ketoglutarate to succinyl-CoA, although catalyzed by different enzymes, are very similar. The first stage of fatty acid oxidation follows a reaction sequence closely resembling a sequence in the citric acid cycle. Use equations to show the analogous reaction sequences in the two pathways.

4. Chemistry of the Acyl-CoA Synthetase Reaction Fatty acids are converted to their coenzyme A esters in a reversible reaction catalyzed by acyl-CoA synthetase:

$$\text{R—COO}^- + \text{ATP} + \text{CoA} \rightleftharpoons \text{R—}\overset{\overset{\text{O}}{\|}}{\text{C}}\text{—CoA} + \text{AMP} + \text{PP}_i$$

(a) The enzyme-bound intermediate in this reaction has been identified as the mixed anhydride of the fatty acid and adenosine monophosphate (AMP), acyl-AMP:

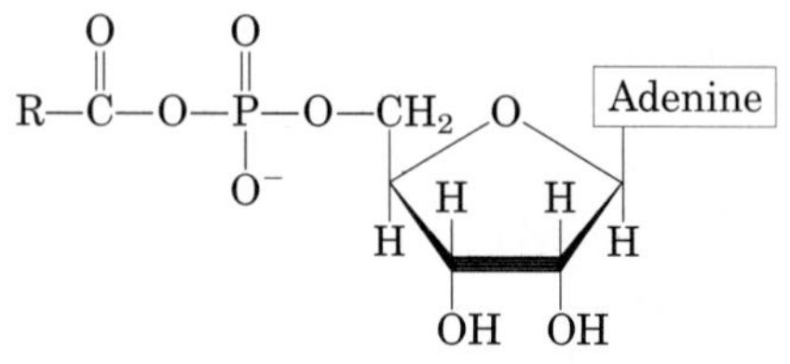

$$\text{R—}\overset{\overset{\text{O}}{\|}}{\text{C}}\text{—O—}\underset{\underset{\text{O}^-}{|}}{\overset{\overset{\text{O}}{\|}}{\text{P}}}\text{—O—CH}_2 \text{ (ribose: O, H, H, H, H, OH, OH) — Adenine}$$

Write two equations corresponding to the two steps of the reaction catalyzed by acyl-CoA synthetase.

(b) The reaction in the above equation is readily reversible, with an equilibrium constant near 1. How can this reaction be made to favor formation of fatty acyl–CoA?

5. Oxidation of Tritiated Palmitate Palmitate uniformly labeled with tritium (^{3}H) to a specific activity of 2.48×10^8 counts per minute (cpm) per micromole of palmitate is added to a mitochondrial preparation that oxidizes it to acetyl-CoA. The acetyl-CoA is isolated and hydrolyzed to acetate. The specific activity of the isolated acetate is 1.00×10^7 cpm/μmol. Is this result consistent with the β-oxidation pathway? Explain. What is the final fate of the removed tritium?

6. Compartmentation in β Oxidation Free palmitate is activated to its coenzyme A derivative (palmitoyl-CoA) in the cytosol before it can be oxidized in the mitochondrion. If palmitate and [^{14}C]coenzyme A are added to a liver homogenate, palmitoyl-CoA isolated from the cytosolic fraction is radioactive, but that isolated from the mitochondrial fraction is not. Explain.

7. Comparative Biochemistry: Energy-Generating Pathways in Birds One indication of the relative importance of various ATP-producing pathways is the V_{max} of certain enzymes of these pathways. The V_{max} of several enzymes from the pectoral muscles (chest muscles used for flying) of pigeon and pheasant are listed below.

Enzyme	V_{max} (μmol substrate/min/g tissue)	
	Pigeon	Pheasant
Hexokinase	3.0	2.3
Glycogen phosphorylase	18.0	120.0
Phosphofructokinase-1	24.0	143.0
Citrate synthase	100.0	15.0
Triacylglycerol lipase	0.07	0.01

(a) Discuss the relative importance of glycogen metabolism and fat metabolism in generating ATP in the pectoral muscles of these birds.

(b) Compare oxygen consumption in the two birds.

(c) Judging from the data in the table, which bird is the long-distance flyer? Justify your answer.

(d) Why were these particular enzymes selected for comparison? Would the activities of triose phosphate isomerase and malate dehydrogenase be equally good bases for comparison? Explain.

8. Effect of Carnitine Deficiency An individual developed a condition characterized by progressive muscular weakness and aching muscle cramps. The symptoms were aggravated by fasting, exercise, and a high-fat diet. The homogenate of a skeletal muscle specimen from the patient oxidized added oleate more slowly than did control homogenates of muscle specimens from healthy individuals. When carnitine was added to the patient's muscle homogenate, the rate of oleate oxidation equaled that in the control homogenates. The patient was diagnosed as having a carnitine deficiency.

(a) Why did added carnitine increase the rate of oleate oxidation in the patient's muscle homogenate?

(b) Why were the patient's symptoms aggravated by fasting, exercise, and a high-fat diet?

(c) Suggest two possible reasons for the deficiency of muscle carnitine in this individual.

9. Fatty Acids as a Source of Water Contrary to legend, camels do not store water in their humps, which actually consist of large fat deposits. How can these fat deposits serve as a source of water? Calculate the amount of water (in liters) that a camel can produce from 1 kg of fat. Assume for simplicity that the fat consists entirely of tripalmitoylglycerol.

10. Petroleum as a Microbial Food Source Some microorganisms of the genera *Nocardia* and *Pseudomonas* can grow in an environment where hydrocarbons are the only food source. These bacteria oxidize straight-chain aliphatic hydrocarbons, such as octane, to their corresponding carboxylic acids:

$$CH_3(CH_2)_6CH_3 + NAD^+ + O_2 \rightleftharpoons CH_3(CH_2)_6COOH + NADH + H^+$$

How can these bacteria be used to clean up oil spills? What are some of the limiting factors to the efficiency of this process?

11. Metabolism of a Straight-Chain Phenylated Fatty Acid A crystalline metabolite was isolated from the urine of a rabbit that had been fed a straight-chain fatty acid containing a terminal phenyl group:

$$C_6H_5—CH_2—(CH_2)_n—COO^-$$

A 302 mg sample of the metabolite in aqueous solution was completely neutralized by adding 22.2 mL of 0.1 M NaOH.

(a) What is the probable molecular weight and structure of the metabolite?

(b) Did the straight-chain fatty acid contain an even or an odd number of methylene ($—CH_2—$) groups (i.e., is n even or odd)? Explain.

12. Fatty Acid Oxidation in Diabetes When the acetyl-CoA produced during β oxidation in the liver exceeds the capacity of the citric acid cycle, the excess acetyl-CoA forms ketone bodies—acetone, acetoacetate, and D-β-hydroxybutyrate. This occurs in severe diabetes: because the tissues cannot use glucose, they oxidize large amounts of fatty acids instead. Although acetyl-CoA is not toxic, the mitochondrion must divert the acetyl-CoA to ketone bodies. What problem would arise if acetyl-CoA were not converted to ketone bodies? How does this diversion solve the problem?

13. Consequences of a High-Fat Diet with No Carbohydrates Suppose you had to subsist on a diet of whale blubber and seal blubber with little or no carbohydrate.

(a) What would be the effect of carbohydrate deprivation on the utilization of fats for energy?

(b) If your diet were totally devoid of carbohydrate, would it be better to consume odd- or even-numbered fatty acids? Explain.

14. Metabolic Consequences of Ingesting ω-Fluorooleate The shrub *Dichapetalum toxicanium*, which grows in Sierra Leone, produces ω-fluorooleate, which is highly toxic to warm-blooded animals.

$$F—CH_2—(CH_2)_7—\overset{H}{\overset{|}{C}}=\overset{H}{\overset{|}{C}}—(CH_2)_7—COO^-$$

ω-Fluorooleate

This substance has been used as an arrow poison, and powdered fruit from the plant is sometimes used as a rat poison (hence the plant's common name, ratsbane). Why is this substance so toxic? (Hint: Review Chapter 16, Problem 13).

15. Role of FAD as Electron Acceptor Acyl-CoA dehydrogenase uses enzyme-bound FAD as a prosthetic group to dehydrogenate the α and β carbons of fatty acyl–CoA. What is the advantage of using FAD as an electron acceptor rather than NAD^+? Explain in terms of the standard reduction potentials for the Enz-FAD/$FADH_2$ ($E'° = -0.219$ V) and NAD^+/NADH ($E'° = -0.320$ V) half-reactions.

16. β Oxidation of Arachidic Acid How many turns of the fatty acid oxidation cycle are required for complete oxidation of arachidic acid (see Table 11–1) to acetyl-CoA?

17. Fate of Labeled Propionate If [3-^{14}C]propionate (^{14}C in the methyl group) is added to a liver homogenate, ^{14}C-labeled oxaloacetate is rapidly produced. Draw a flow chart for the pathway by which propionate is transformed to oxaloacetate, and indicate the location of the ^{14}C in oxaloacetate.

18. Sources of H_2O Produced in β Oxidation The complete oxidation of palmitoyl-CoA to carbon dioxide and water is represented by the overall equation

$$\text{Palmitoyl-CoA} + 23O_2 + 108P_i + 108\text{ADP} \longrightarrow \text{CoA} + 16CO_2 + 108\text{ATP} + 23H_2O$$

Water is also produced in the reaction

$$\text{ADP} + P_i \longrightarrow \text{ATP} + H_2O$$

Why is this water not included as a net product of the reaction?

19. Biological Importance of Cobalt In cattle, deer, sheep, and other ruminant animals, large amounts of propionate are produced in the rumen through the bacterial fermentation of ingested plant matter. Propionate is the principal source of glucose for these animals via the route propionate $\longrightarrow$ oxaloacetate $\longrightarrow$ glucose. In some areas of the world, notably Australia, ruminant animals sometimes show symptoms of anemia with concomitant loss of appetite and retarded growth, resulting from an inability to transform propionate to oxaloacetate. This condition is due to a cobalt deficiency caused by very low cobalt levels in the soil and thus in plant matter. Explain.

20. Fat Loss during Hibernation Bears expend about 25×10^6 J/day during periods of hibernation, which may last as long as seven months. The energy required to sustain life is obtained from fatty acid oxidation. How much weight loss (in kilograms) has occurred after seven months? How might ketosis be minimized during hibernation?

chapter

18

Amino Acid Oxidation and the Production of Urea

We now turn our attention to the amino acids, the final class of biomolecules that, through their oxidative degradation, make a significant contribution to the generation of metabolic energy. The fraction of metabolic energy obtained from amino acids, whether they are derived from dietary protein or from tissue protein, varies greatly with the type of organism and with metabolic conditions. Carnivores (immediately following a meal) can obtain up to 90% of their energy requirements from amino acid oxidation, whereas herbivores may fill only a small fraction of their energy needs by this route. Most microorganisms can scavenge amino acids from their environment and use them as fuel when required by metabolic conditions. Plants, however, rarely if ever oxidize amino acids to provide energy; the carbohydrate produced from CO_2 and H_2O in photosynthesis is used almost exclusively as their energy source. Amino acid concentrations in plant tissues are carefully regulated to just meet the requirements for biosynthesis of proteins, nucleic acids, and other molecules needed to support growth. Amino acid catabolism does occur in plants, but its purpose is to produce metabolites for other biosynthetic pathways.

In animals, amino acids undergo oxidative degradation in three different metabolic circumstances.

1. During the normal synthesis and degradation of cellular proteins (protein turnover; Chapter 27), some amino acids released during protein breakdown undergo oxidative degradation if they are not needed for new protein synthesis.
2. When a diet is rich in protein and ingested amino acids exceed the body's needs for protein synthesis, the surplus is catabolized; amino acids cannot be stored.
3. During starvation or in diabetes mellitus, when carbohydrates are either unavailable or not properly utilized, cellular proteins are used as fuel.

Under all of these metabolic conditions, amino acids lose their amino groups to form α-keto acids, the "carbon skeletons" of amino acids. The α-keto acids undergo oxidation to CO_2 and H_2O or, often more importantly, provide three- and four-carbon units that can be converted by gluconeogenesis into glucose, the fuel for brain, skeletal muscle, and other tissues.

The pathways of amino acid catabolism are quite similar in most organisms. The focus of this chapter is on the pathways in vertebrates, because these have received the most research attention. As in carbohydrate and fatty acid catabolism, the processes of amino acid degradation converge on the central catabolic pathways, with the carbon skeletons of most amino acids finding their way to the citric acid cycle. In some cases the reaction

A view on San Lorenzo Island, one of the guano islands off the coast of Peru. Hundreds of thousands of "gooney birds," black-footed albatrosses, nest on these islands. Birds excrete excess amino groups as uric acid, which is the major component of guano. Over the centuries, enormous clifflike deposits of guano have built up. Guano is a valuable fertilizer because of its nitrogen content.

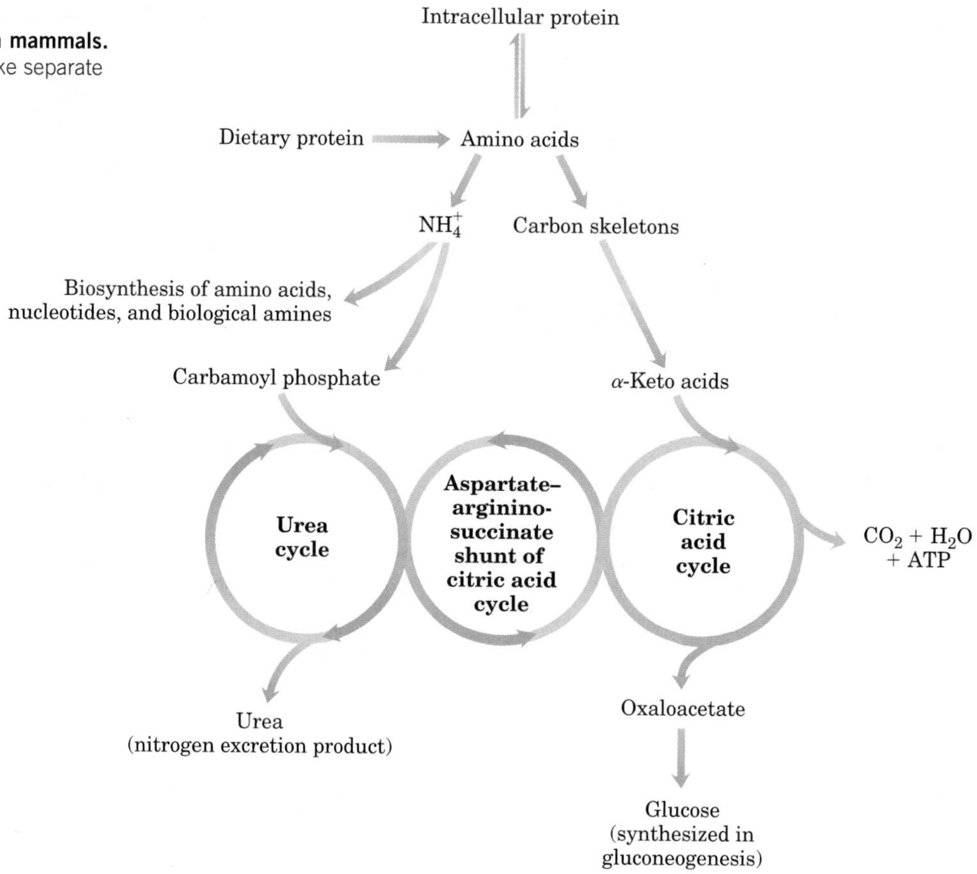

figure 18–1
Overview of the catabolism of amino acids in mammals. The amino groups and the carbon skeleton take separate but interconnected pathways.

pathways of amino acid breakdown closely parallel steps in the catabolism of fatty acids (Chapter 17).

One important feature distinguishes amino acid degradation from other catabolic processes described to this point: every amino acid contains an amino group, and the pathways for amino acid degradation therefore include a key step in which the α-amino group is separated from the carbon skeleton and shunted into the pathways of amino group metabolism (Fig. 18–1). We deal first with amino group metabolism and nitrogen excretion, then with the fate of the carbon skeletons derived from the amino acids; along the way we see how the pathways are interconnected.

Metabolic Fates of Amino Groups

Nitrogen, N_2, is abundant in the atmosphere but is too inert for use in most biochemical processes. Because only a few microorganisms can convert N_2 to biologically useful forms such as NH_3 (described in Chapter 22), amino groups are carefully husbanded in biological systems.

An overview of the catabolism of ammonia and amino groups in vertebrates is provided in Figure 18–2a. Amino acids derived from dietary protein are the source of most amino groups. Most amino acids are metabolized in the liver. Some of the ammonia generated in this process is recycled and used in a variety of biosynthetic pathways; the excess is either excreted directly or converted to urea or uric acid for excretion, depending on the organism (Fig. 18–2b). Excess ammonia generated in other (extrahepatic) tissues travels to the liver (in the form of amino groups, as described below) for conversion to the form in which it is excreted.

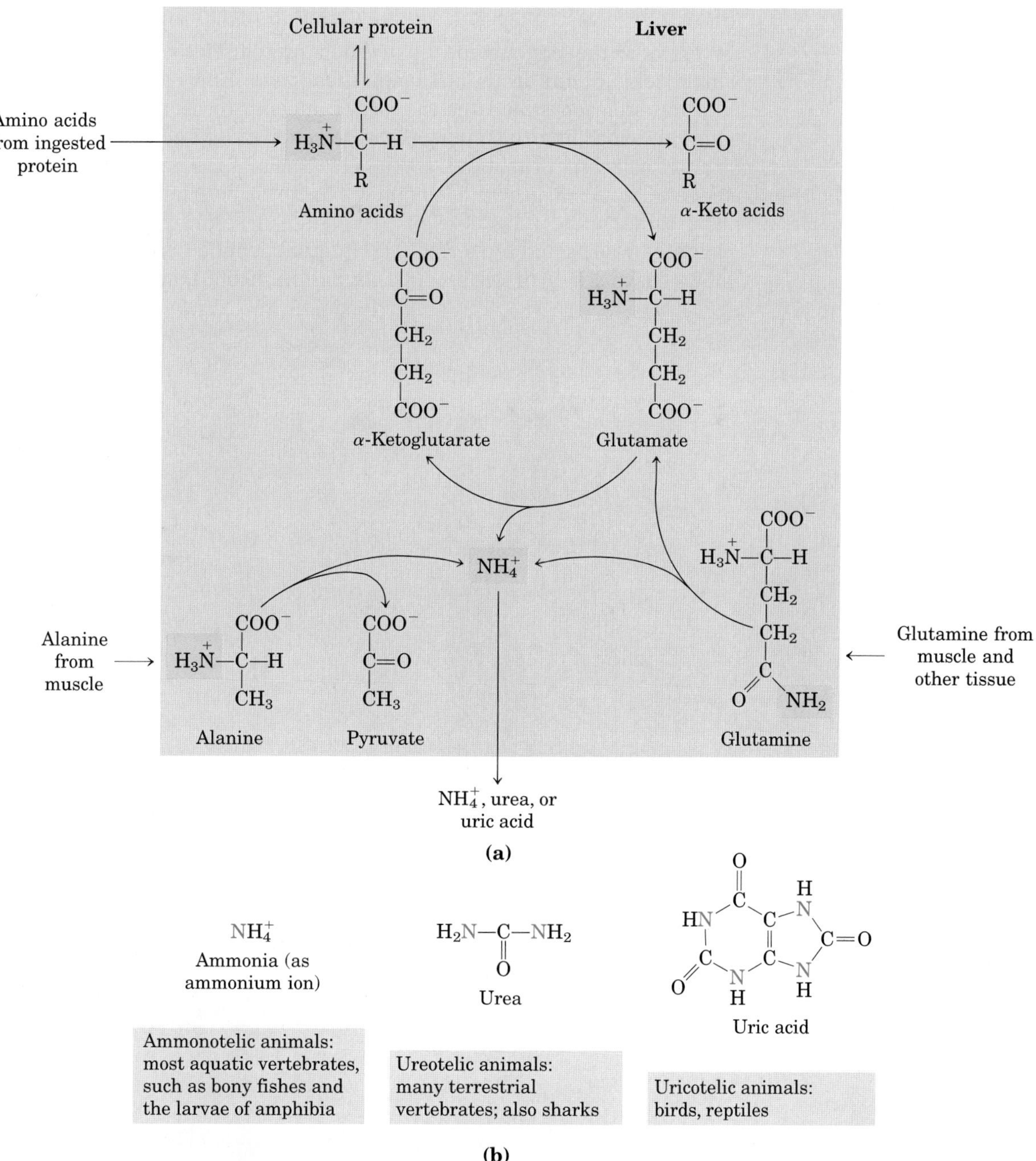

figure 18–2

Amino group catabolism. (a) Overview of catabolism of amino groups (shaded) in vertebrate liver. **(b)** Excretory forms of nitrogen. Excess NH_4^+ is excreted as ammonia (microbes, bony fishes), urea (most terrestrial vertebrates), or uric acid (birds and terrestrial reptiles). Notice that the carbon atoms of urea and uric acid are highly oxidized; the organism discards carbon only after extracting most of its available energy of oxidation.

Glutamate and glutamine play especially critical roles in nitrogen metabolism. In the cytosol of hepatocytes, amino groups from most amino acids are transferred to α-ketoglutarate to form glutamate. Glutamate is then transported into mitochondria, where the amino group is removed to form NH_4^+. Excess ammonia generated in most other tissues is converted to the amide nitrogen of glutamine, which passes to the liver, then into liver mitochondria. In most tissues, glutamine or glutamate or both are present in higher concentrations than other amino acids.

In muscle, excess amino groups are generally transferred to pyruvate to form alanine, another important molecule in the transport of amino groups to the liver.

We turn now to a discussion of the breakdown of dietary proteins, followed by a general description of the metabolic fates of amino groups.

Dietary Protein Is Enzymatically Degraded to Amino Acids

In humans, the degradation of ingested proteins into their constituent amino acids occurs in the gastrointestinal tract. Entry of dietary protein into the stomach stimulates the gastric mucosa to secrete the hormone **gastrin,** which in turn stimulates the secretion of hydrochloric acid by the parietal cells and pepsinogen by the chief cells of the gastric glands (Fig. 18–3a). The acidic gastric juice (pH 1.0 to 2.5) is both an antiseptic, killing most bacteria and other foreign cells, and a denaturing agent, unfolding globular proteins and rendering their internal peptide bonds more accessible to enzymatic hydrolysis. **Pepsinogen** (M_r 40,000), an inactive precur-

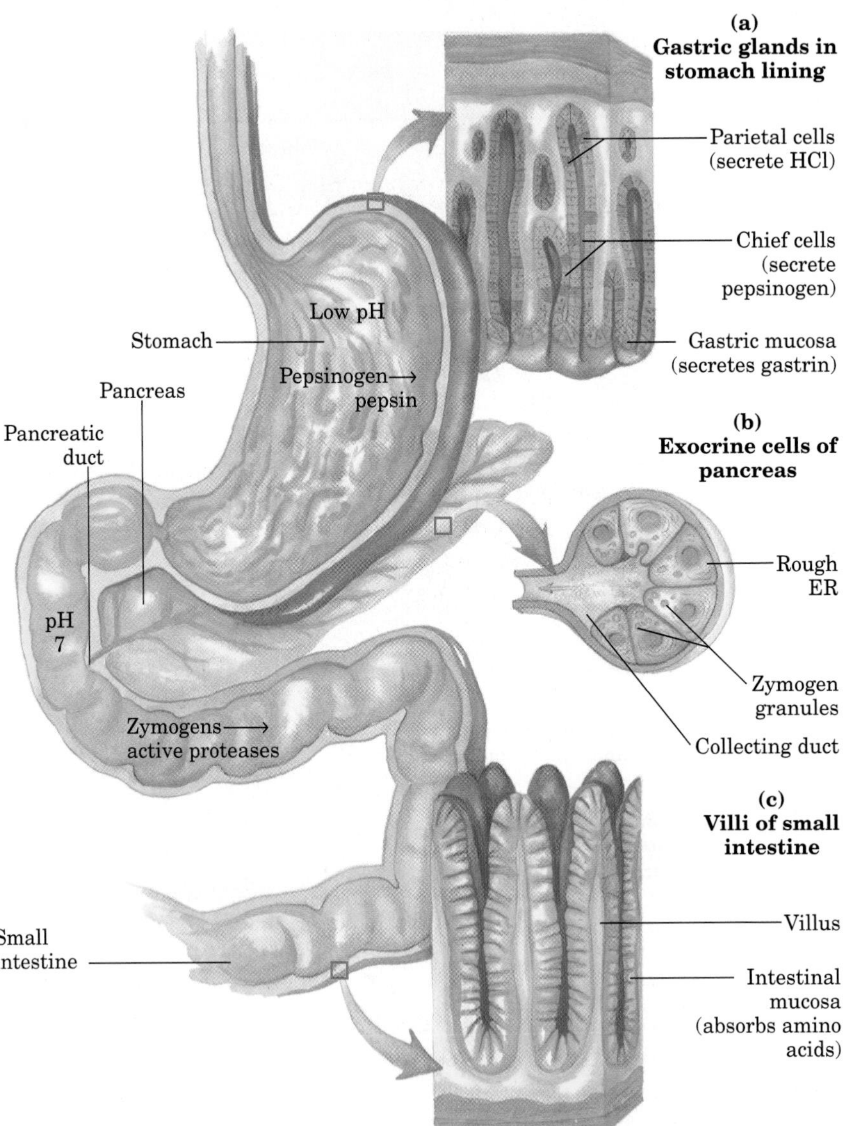

figure 18–3

Portion of the human digestive (gastrointestinal) tract. **(a)** Gastric glands in the stomach lining. The parietal cells and chief cells secrete their products in response to the hormone gastrin. Pepsin begins the process of protein degradation in the stomach. **(b)** Exocrine cells of the pancreas. The cytoplasm of these cells is completely filled with rough endoplasmic reticulum, the site of synthesis of the zymogens of many digestive enzymes. The zymogens are concentrated in membrane-enclosed transport particles called zymogen granules. When an exocrine cell is stimulated, its plasma membrane fuses with the zymogen granule membrane and zymogens are released into the lumen of the collecting duct by exocytosis. The collecting ducts ultimately lead to the pancreatic duct and thence to the small intestine. **(c)** Villi of the small intestine. Amino acids are absorbed through the epithelial cell layer (intestinal mucosa) and enter the capillaries. Recall that the products of lipid hydrolysis in the small intestine enter the lymphatic system following absorption by the intestinal mucosa (see Fig. 17–1).

sor or zymogen (see Fig. 8–31), is converted to active pepsin by the enzymatic action of pepsin itself. In this process, 42 amino acid residues are removed from the amino-terminal end of the polypeptide chain and the portion that remains is **pepsin** (M_r 33,000). In the stomach, pepsin hydrolyzes ingested proteins at peptide bonds on the amino-terminal side of the aromatic amino acid residues Phe, Trp, and Tyr (see Table 5–7), cleaving long polypeptide chains into a mixture of smaller peptides.

As the acidic stomach contents pass into the small intestine, the low pH triggers secretion of the hormone **secretin** into the blood. Secretin stimulates the pancreas to secrete bicarbonate into the small intestine to neutralize the gastric HCl, abruptly increasing the pH to about 7. (All pancreatic secretions pass into the small intestine through the pancreatic duct.) The digestion of proteins now continues in the small intestine. Arrival of amino acids in the upper part of the intestine (duodenum) causes release into the blood of the hormone **cholecystokinin,** which stimulates secretion of several pancreatic enzymes with activity optima at pH 7 to 8. **Trypsinogen, chymotrypsinogen,** and **procarboxypeptidases A** and **B,** the zymogens of **trypsin, chymotrypsin,** and **carboxypeptidases A** and **B,** are synthesized and secreted by the exocrine cells of the pancreas (Fig. 18–3b). Trypsinogen is converted into its active form, trypsin, by **enteropeptidase,** a proteolytic enzyme secreted by intestinal cells. Free trypsin then catalyzes the conversion of additional trypsinogen into trypsin (see Fig. 8–31). Trypsin also activates chymotrypsinogen, the procarboxypeptidases, and proelastase.

Synthesis of the digestive enzymes as inactive precursors protects the exocrine cells from destructive proteolytic attack. The pancreas further protects itself against self-digestion by making a specific inhibitor, a protein called **pancreatic trypsin inhibitor** (p. 286), that effectively prevents premature production of active proteolytic enzymes within the pancreatic cells.

Trypsin and chymotrypsin further hydrolyze the peptides produced by pepsin in the stomach. This stage of protein digestion is accomplished very efficiently because pepsin, trypsin, and chymotrypsin have different amino acid specificities. Trypsin hydrolyzes peptide bonds with carbonyl groups contributed by Lys and Arg residues, and chymotrypsin hydrolyzes peptide bonds on the carboxyl-terminal side of Phe, Trp, and Tyr residues (see Table 5–7).

Degradation of the short peptides in the small intestine is now completed by other intestinal peptidases. These include carboxypeptidases A and B, both zinc-containing enzymes, which remove successive carboxyl-terminal residues from peptides, and an **aminopeptidase** that hydrolyzes successive amino-terminal residues from short peptides. The sequential action of these proteolytic enzymes and peptidases yields a mixture of free amino acids that can be transported into the epithelial cells lining the small intestine (Fig. 18–3c). The free amino acids enter the blood capillaries in the villi and travel to the liver.

In humans, most globular proteins from animal sources are almost completely hydrolyzed into amino acids in the gastrointestinal tract, but some fibrous proteins, such as keratin, are only partly digested. In addition, the protein content of some plant foods is protected against breakdown by indigestible cellulose husks.

Acute pancreatitis is a disease caused by obstruction of the normal pathway by which pancreatic secretions enter the intestine. The zymogens of the proteolytic enzymes are converted into their catalytically active forms prematurely, *inside* the pancreatic cells, and attack the pancreatic tissue itself. This causes excruciating pain and damage to the organ that can prove fatal.

Pyridoxal Phosphate Participates in the Transfer of α-Amino Groups to α-Ketoglutarate

The first step in the catabolism of most L-amino acids, once they have reached the liver, is removal of the α-amino groups, promoted by enzymes called **aminotransferases** or **transaminases.** In these **transamination** reactions, the α-amino group is transferred to the α-carbon atom of α-ketoglutarate, leaving behind the corresponding α-keto acid analog of the amino acid (Fig. 18–4). There is no net deamination (i.e., loss of amino groups) in these reactions because the α-ketoglutarate becomes aminated as the α-amino acid is deaminated. The effect of transamination reactions is to collect the amino groups from many different amino acids in the form of L-glutamate. The glutamate then functions as an amino group donor for biosynthetic pathways or excretion pathways that lead to the elimination of nitrogenous waste products.

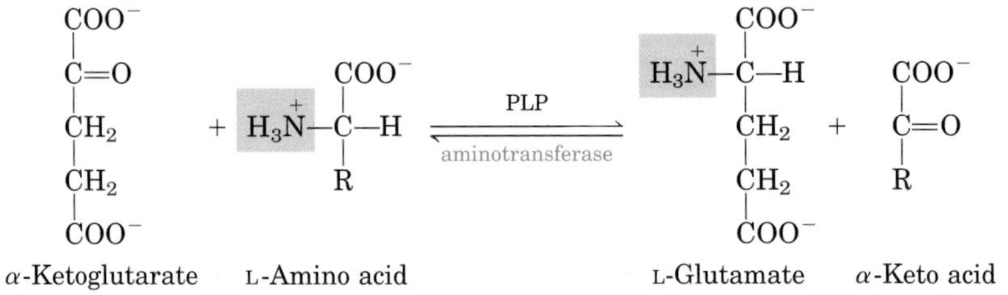

$$\underset{\alpha\text{-Ketoglutarate}}{\begin{array}{c}COO^-\\|\\C{=}O\\|\\CH_2\\|\\CH_2\\|\\COO^-\end{array}} + \underset{\text{L-Amino acid}}{H_3\overset{+}{N}{-}\begin{array}{c}COO^-\\|\\C{-}H\\|\\R\end{array}} \underset{\text{aminotransferase}}{\overset{\text{PLP}}{\rightleftharpoons}} \underset{\text{L-Glutamate}}{\begin{array}{c}COO^-\\|\\H_3\overset{+}{N}{-}C{-}H\\|\\CH_2\\|\\CH_2\\|\\COO^-\end{array}} + \underset{\alpha\text{-Keto acid}}{\begin{array}{c}COO^-\\|\\C{=}O\\|\\R\end{array}}$$

figure 18–4

Enzyme-catalyzed transaminations. In many aminotransferase reactions, α-ketoglutarate is the amino group acceptor. All aminotransferases have pyridoxal phosphate (PLP) as cofactor.

Cells contain a number of different aminotransferases. Many are specific for α-ketoglutarate as the amino group acceptor, but differ in their specificity for the L-amino acid. The enzymes are named for the amino group donor (e.g., alanine aminotransferase, aspartate aminotransferase). The reactions catalyzed by aminotransferases are freely reversible, having an equilibrium constant of about 1.0 ($\Delta G'^{\circ} \approx 0$ kJ/mol).

All aminotransferases have the same prosthetic group and the same reaction mechanism. The prosthetic group is **pyridoxal phosphate (PLP),** the coenzyme form of pyridoxine or vitamin B_6. We encountered pyridoxal phosphate briefly in Chapter 15 (p. 547) as a coenzyme in the glycogen phosphorylase reaction, but its role in that reaction is not representative of its usual coenzyme function. Its primary role in cells is in the metabolism of molecules with amino groups.

Pyridoxal phosphate functions as an intermediate carrier of amino groups at the active site of aminotransferases. It undergoes reversible transformations between its aldehyde form, pyridoxal phosphate, which can accept an amino group, and its aminated form, pyridoxamine phosphate, which can donate its amino group to an α-keto acid (Fig. 18–5a). Pyridoxal phosphate is generally covalently bound to the enzyme's active site through an aldimine (Schiff-base) linkage to the ϵ-amino group of a Lys residue (Fig. 18–5b, d).

Pyridoxal phosphate is involved in a variety of reactions at the α, β, and γ carbons (C-2 to C-4) of amino acids. Reactions at the α carbon (Fig. 18–6) include racemizations (interconverting L- and D-amino acids) and decarboxylations, as well as transaminations. Pyridoxal phosphate plays the same chemical role in each of these reactions. A bond to the α carbon is broken, removing either a proton or a carboxyl group and leaving behind a free electron pair on the α carbon. This carbanion intermediate is very unstable; pyridoxal phosphate stabilizes it by providing a highly conjugated structure (an electron sink) that permits delocalization of the negative charge.

Pyridoxal phosphate (PLP)

Pyridoxamine phosphate

(a)

Enz —Lys—NH_2

H_2O

Enz —Lys—$\overset{+}{N}$=C—

Schiff base

(b)

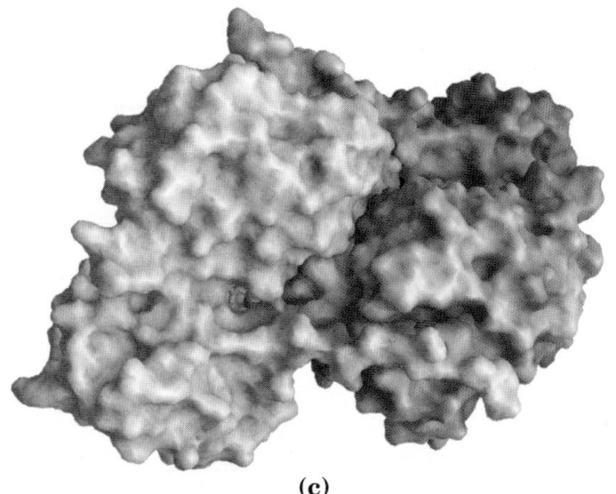

(c)

figure 18–5

Pyridoxal phosphate, the prosthetic group of aminotransferases. (a) Pyridoxal phosphate (PLP) and its aminated form, pyridoxamine phosphate, are the tightly bound coenzymes of aminotransferases. The functional groups are shaded in red. **(b)** Pyridoxal phosphate is bound to the enzyme through noncovalent interactions and a Schiff-base linkage to a Lys residue at the active site. **(c)** PLP (red) is shown bound to one of the two active sites of the dimeric enzyme aspartate aminotransferase, a typical aminotransferase. **(d)** A close-up view of the active site, with PLP (red, with yellow phosphorus) in aldimine linkage with the side chain of Lys^{258} (purple). **(e)** Another close-up view of the active site, with PLP linked to the substrate analog 2-methylaspartate (green) via a Schiff base.

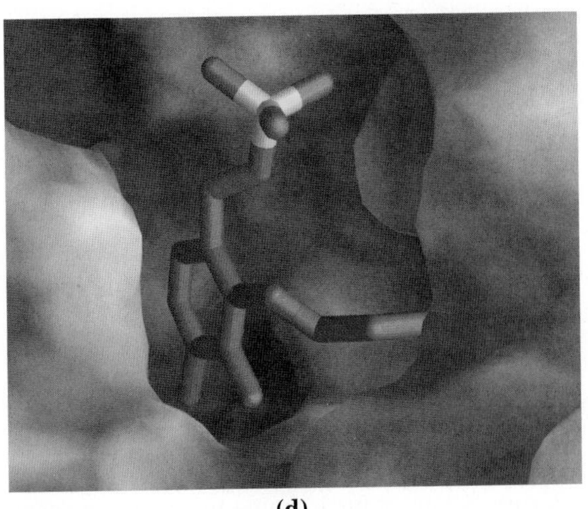

(d)

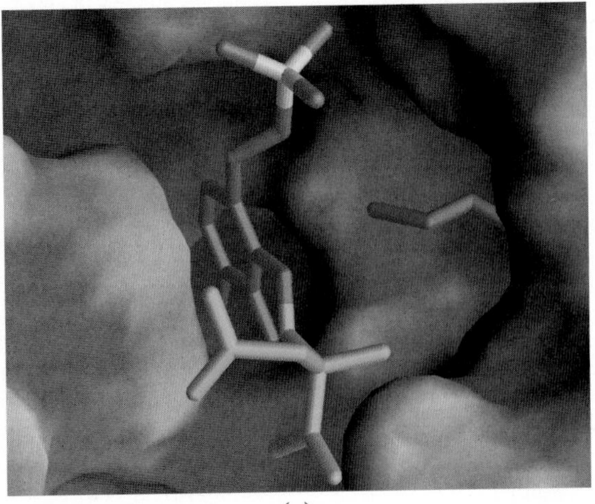

(e)

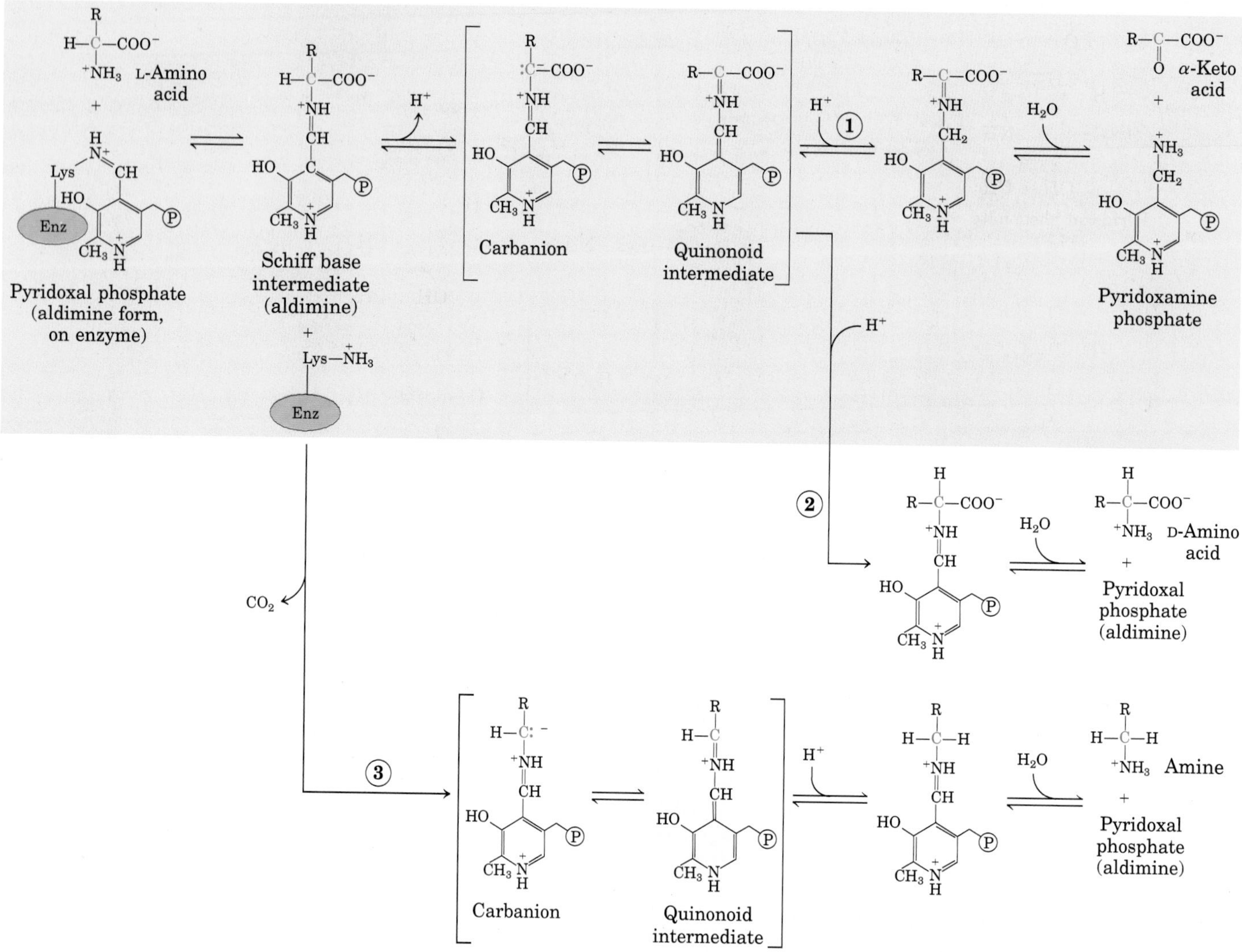

figure 18–6

Some amino acid transformations at the α carbon facilitated by pyridoxal phosphate. Pyridoxal phosphate is generally bonded to the enzyme through a Schiff base (see Fig. 18–5b, d). Reactions begin with formation of a new Schiff base (aldimine) between the α-amino group of the amino acid and PLP, which substitutes for the enzyme-PLP linkage. Three alternative fates, each involving formation of a carbanion, are shown: ① transamination, ② racemization, or ③ decarboxylation. The Schiff base formed between PLP and the amino acid is in conjugation with the pyridine ring, an electron sink that permits delocalization of the negative charge on the carbanion (shown in brackets). A quinonoid intermediate is involved in all three types of reactions. The transamination route is especially important in the pathways described in this chapter. The highlighted transamination pathway (shown left to right) represents only part of the overall reaction catalyzed by aminotransferases. To complete the process, a second α-keto acid replaces the one that is released and is converted to an amino acid in a reversal of the reaction steps (right to left). Pyridoxal phosphate is also involved in certain reactions at the β and γ carbons of some amino acids (not shown).

Aminotransferases (Fig. 18–5c,d,e) are classic examples of enzymes catalyzing bimolecular Ping-Pong reactions (see Fig. 8–13b), in which the first substrate reacts, then the product must leave the active site before the second substrate can bind. Thus the incoming amino acid binds to the active site, donates its amino group to pyridoxal phosphate, and departs in the form of an α-keto acid. The incoming α-keto acid then binds, accepts the amino group from pyridoxamine phosphate, and departs in the form of an amino acid.

Measurement of the alanine aminotransferase and aspartate aminotransferase levels in blood serum is important in some medical diagnoses (Box 18–1).

box 18–1 Assays for Tissue Damage

Analysis of some enzyme activities in blood serum gives valuable diagnostic information for a number of disease conditions.

Alanine aminotransferase (ALT; also called glutamate-pyruvate transaminase, GPT) and aspartate aminotransferase (AST; also called glutamate-oxaloacetate transaminase, GOT) are important in the diagnosis of heart and liver damage caused by heart attack, drug toxicity, or infection. After a heart attack, a variety of enzymes, including these aminotransferases, leak from the injured heart cells into the bloodstream. Measurements of the blood serum concentration of the two aminotransferases by the SGPT and SGOT tests (S for serum), and of another enzyme, **creatine kinase,** by the SCK test, can provide information about the severity of the damage. Creatine kinase is the first heart enzyme to appear in the blood after a heart attack; it also disappears quickly from the blood. GOT is the next to appear, and GPT follows later. Lactate dehydrogenase also leaks from injured or anaerobic heart muscle.

The SGOT and SGPT tests are also important in industrial medicine, to determine whether people exposed to carbon tetrachloride, chloroform, or other industrial solvents have suffered liver damage. Liver degeneration caused by these solvents is accompanied by leakage of various enzymes from injured hepatocytes into the blood. Aminotransferases are most useful in the monitoring of people exposed to these chemicals because they are very active in liver and their activity can be detected in very small amounts.

Glutamate Releases Ammonia in the Liver

As we have seen, the amino groups from many of the α-amino acids are collected in the liver in the form of the amino group of L-glutamate. How are the amino groups removed from glutamate molecules to prepare them for excretion?

In hepatocytes, glutamate is transported from the cytosol into the mitochondria, where it undergoes **oxidative deamination** catalyzed by **L-glutamate dehydrogenase** (M_r 330,000). In mammals, this enzyme is present in the mitochondrial matrix and can use either NAD^+ or $NADP^+$ as acceptor of the reducing equivalents (Fig. 18–7). The combined action of an aminotransferase and glutamate dehydrogenase is referred to as **transdeamination.** A few amino acids bypass the transdeamination pathway and undergo direct oxidative deamination. The fate of the NH_4^+ produced by any of these deamination processes is discussed in detail later in this chapter. The α-ketoglutarate formed from glutamate deamination can be used in the citric acid cycle and for glucose synthesis.

Glutamate dehydrogenase operates at an important intersection of carbon and nitrogen metabolism. An allosteric enzyme with six identical subunits, its activity is influenced by a complicated array of allosteric modulators. The best-studied of these are the positive modulator ADP and the negative modulator GTP (Fig. 18–7). The metabolic rationale for this

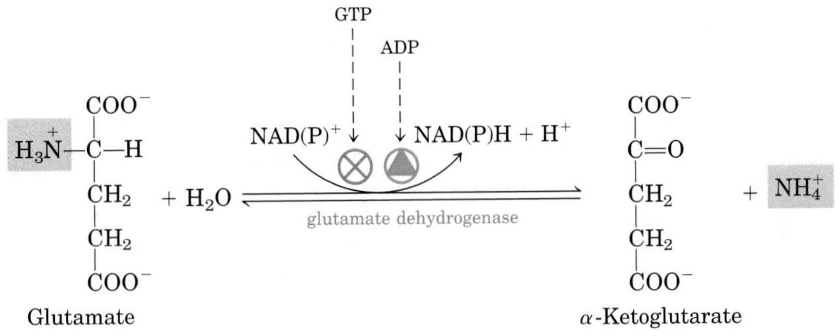

GTP
ADP
$NAD(P)^+$
$NAD(P)H + H^+$
Glutamate (COO^-, $H_3\overset{+}{N}$—C—H, CH_2, CH_2, COO^-) + H_2O ⇌ (glutamate dehydrogenase) α-Ketoglutarate (COO^-, C=O, CH_2, CH_2, COO^-) + NH_4^+

figure 18–7
Reaction catalyzed by glutamate dehydrogenase. The glutamate dehydrogenase of mammalian liver has the unusual capacity to use either NAD^+ or $NADP^+$ as cofactor. The glutamate dehydrogenases of plants and microorganisms are generally specific for one or the other. The mammalian enzyme is allosterically regulated by GTP and ADP.

regulatory pattern has not been elucidated in detail. Mutations that alter the allosteric binding site for GTP or otherwise cause permanent activation of glutamate dehydrogenase lead to a human genetic disorder called hyperinsulinism-hyperammonemia syndrome, characterized by elevated levels of ammonia in the bloodstream and accompanying hypoglycemia.

Glutamine Transports Ammonia in the Bloodstream

Ammonia is quite toxic to animal tissues (we examine some possible reasons for this toxicity later). In most animals excess ammonia is converted into a nontoxic compound before export from extrahepatic tissues into the blood and transport to the liver or kidneys. Glutamate, critical to intracellular amino group metabolism, is supplanted by L-glutamine for this transport function. In many tissues, including the brain, some ammonia is generated by processes such as nucleotide degradation. The free ammonia is combined with glutamate to yield glutamine by the action of **glutamine synthetase.** This reaction requires ATP and occurs in two steps. First, glutamate and ATP react to form ADP and a γ-glutamyl phosphate intermediate that reacts with ammonia to produce glutamine and inorganic phosphate. Glutamine is a nontoxic transport form of ammonia; it is normally present in blood in much higher concentrations than other amino acids. Glutamine also serves as a source of amino groups in a variety of biosynthetic reactions. Glutamine synthetase is found in all organisms, always playing a central metabolic role. In microorganisms, the enzyme serves as an essential portal for the entry of fixed nitrogen into biological systems. (The roles of glutamine and glutamine synthetase in metabolism are further discussed in Chapter 22.)

$^-OOC—CH_2—CH_2—CH(\overset{+}{N}H_3)—COO^-$ L-Glutamate

glutamine synthetase: ATP → ADP

$^-O—P(=O)(O^-)—O—C(=O)—CH_2—CH_2—CH(\overset{+}{N}H_3)—COO^-$ γ-Glutamyl phosphate

glutamine synthetase: NH_4^+ → P_i

$H_2N—C(=O)—CH_2—CH_2—CH(\overset{+}{N}H_3)—COO^-$ L-Glutamine

glutaminase (liver mitochondria): H_2O → NH_4^+ ⟶ ⟶ Urea

$^-O—C(=O)—CH_2—CH_2—CH(\overset{+}{N}H_3)—COO^-$ L-Glutamate

In most terrestrial animals, glutamine in excess of that required for biosynthesis is transported in the blood to the liver and kidneys for processing. The amide nitrogen is released as ammonia only in liver and kidney mitochondria, where the enzyme **glutaminase** converts glutamine to glutamate and NH_4^+. In the liver, this reaction provides another source of ammonia to be disposed of by urea synthesis. The glutamate is further processed in the liver by glutamate dehydrogenase, releasing more ammonia and producing carbon skeletons for metabolic fuel.

The kidneys extract little glutamine from the blood under normal conditions, but in metabolic acidosis they increase glutamine processing. The NH_4^+ thus produced by glutaminase and glutamate dehydrogenase in the kidneys is not released into the bloodstream or converted to urea, but is excreted directly into the urine. In the kidney, the NH_4^+ forms salts with metabolic acids, facilitating their removal in the urine. Decarboxylation of α-ketoglutarate in the citric acid cycle also provides bicarbonate that can serve as a buffer in blood plasma. Taken together, these effects of glutamine metabolism in the kidney tend to counteract acidosis.

Alanine Transports Ammonia from Muscles to the Liver

Alanine also plays a special role in transporting amino groups to the liver in a nontoxic form, via a pathway called the **glucose-alanine cycle** (Fig. 18–8). In muscle and certain other tissues that degrade amino acids for fuel, amino groups are collected in the form of glutamate by transamination (Fig. 18–2). Glutamate can be converted to glutamine for transport to the liver or it can transfer its α-amino group to pyruvate, a readily available product of muscle glycolysis, by the action of **alanine aminotransferase** (Fig. 18–8). The alanine so formed passes into the blood and travels to the liver. In the cytosol of hepatocytes, alanine aminotransferase transfers the amino group from alanine to α-ketoglutarate, forming pyruvate and gluta-

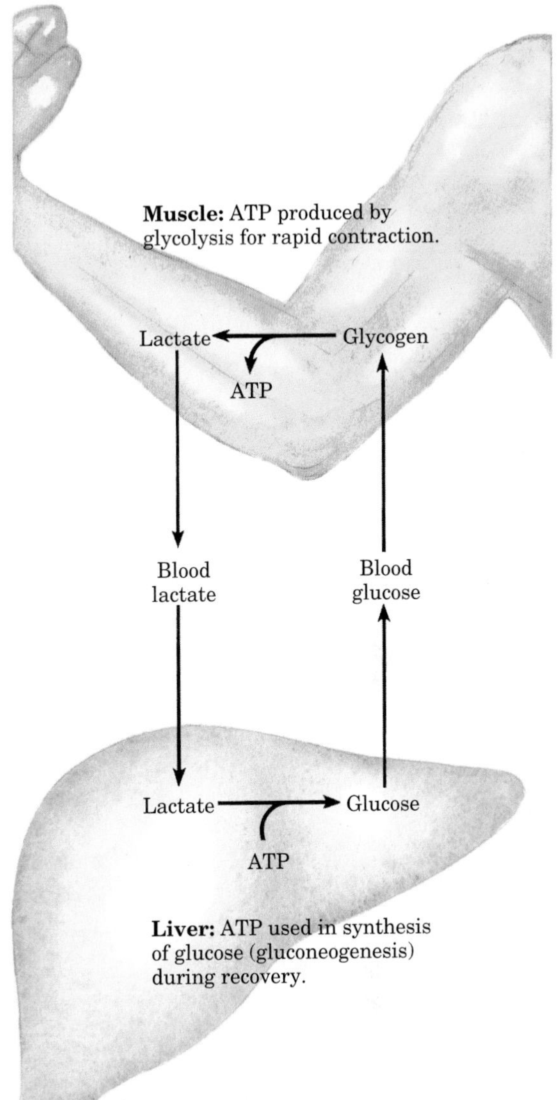
Muscle: ATP produced by glycolysis for rapid contraction.
Lactate
Glycogen
ATP
Blood lactate
Blood glucose
Lactate
Glucose
ATP
Liver: ATP used in synthesis of glucose (gluconeogenesis) during recovery.

figure 18–8
Glucose-alanine cycle. Alanine serves as a carrier of ammonia and of the carbon skeleton of pyruvate from muscle to liver. The ammonia is excreted and the pyruvate is used to produce glucose, which is returned to the muscle.

mate. Glutamate can then enter mitochondria, where the glutamate dehydrogenase reaction releases NH_4^+ (Fig. 18–7), or can undergo transamination with oxaloacetate to form aspartate, another nitrogen donor in urea synthesis.

The use of alanine to transport ammonia from skeletal muscles to the liver is another example of the intrinsic economy of living organisms. Vigorously contracting skeletal muscles operate anaerobically, producing pyruvate and lactate from glycolysis as well as ammonia from protein breakdown. These products must find their way to the liver, where pyruvate and lactate are incorporated into glucose, which is returned to the muscles, and ammonia is converted into urea for excretion. The glucose-alanine cycle, in concert with the Cori cycle (see Box 15–1 and Fig. 23–6), accomplishes this transaction. The energetic burden of gluconeogenesis is thus imposed on the liver rather than the muscle, and all available ATP in muscle is devoted to muscle contraction.

Ammonia Is Toxic to Animals

The catabolic production of ammonia poses a serious biochemical problem because ammonia is very toxic. The molecular basis for this toxicity is not entirely understood. The terminal stages of ammonia intoxication in humans are characterized by onset of a comatose state and other effects on the brain, so research and speculation on ammonia toxicity have focused on this tissue. Speculation centers on a potential depletion of ATP in brain cells.

Ridding the cytosol of excess ammonia requires reductive amination of α-ketoglutarate to glutamate by glutamate dehydrogenase (the reverse of the reaction described earlier; see Fig. 18–7) and conversion of glutamate to glutamine by glutamine synthetase. Both enzymes are present at high levels in the brain, although the glutamine synthetase reaction is probably the more important pathway for removal of ammonia. A shift in the equilibrium of the glutamate dehydrogenase reaction toward glutamate production could limit the availability of α-ketoglutarate for the citric acid cycle, and the glutamine synthetase reaction depletes ATP. Overall, toxic concentrations of NH_4^+ may interfere with the very high levels of ATP production required to maintain brain function.

Depletion of glutamate in the glutamine synthetase reaction may have additional effects on the brain. Glutamate and its derivative, γ-aminobutyrate (GABA; Fig. 22–27), are important neurotransmitters; the sensitivity of the brain to ammonia may reflect a depletion of neurotransmitters as well as changes in cellular ATP metabolism.

As we close this discussion of amino group metabolism, note that we have described several processes that deposit excess ammonia in the mitochondria of hepatocytes (Fig. 18–2). We now look at the fate of that ammonia.

Nitrogen Excretion and the Urea Cycle

If not reused for the synthesis of new amino acids or other nitrogenous products, amino groups are channeled into a single excretory end product (Fig. 18–9). Most aquatic species, such as the bony fishes, excrete amino nitrogen as ammonia and are thus called **ammonotelic** animals; most terrestrial animals are **ureotelic,** excreting amino nitrogen in the form of urea; birds and reptiles are **uricotelic,** excreting amino nitrogen as uric acid. (The pathway for synthesis of uric acid is described in Figure 22–43.) Plants recycle virtually all amino groups, and nitrogen excretion occurs only under very unusual circumstances.

In ureotelic organisms, the ammonia deposited in the mitochondria of hepatocytes is converted to urea in the **urea cycle.** This pathway was

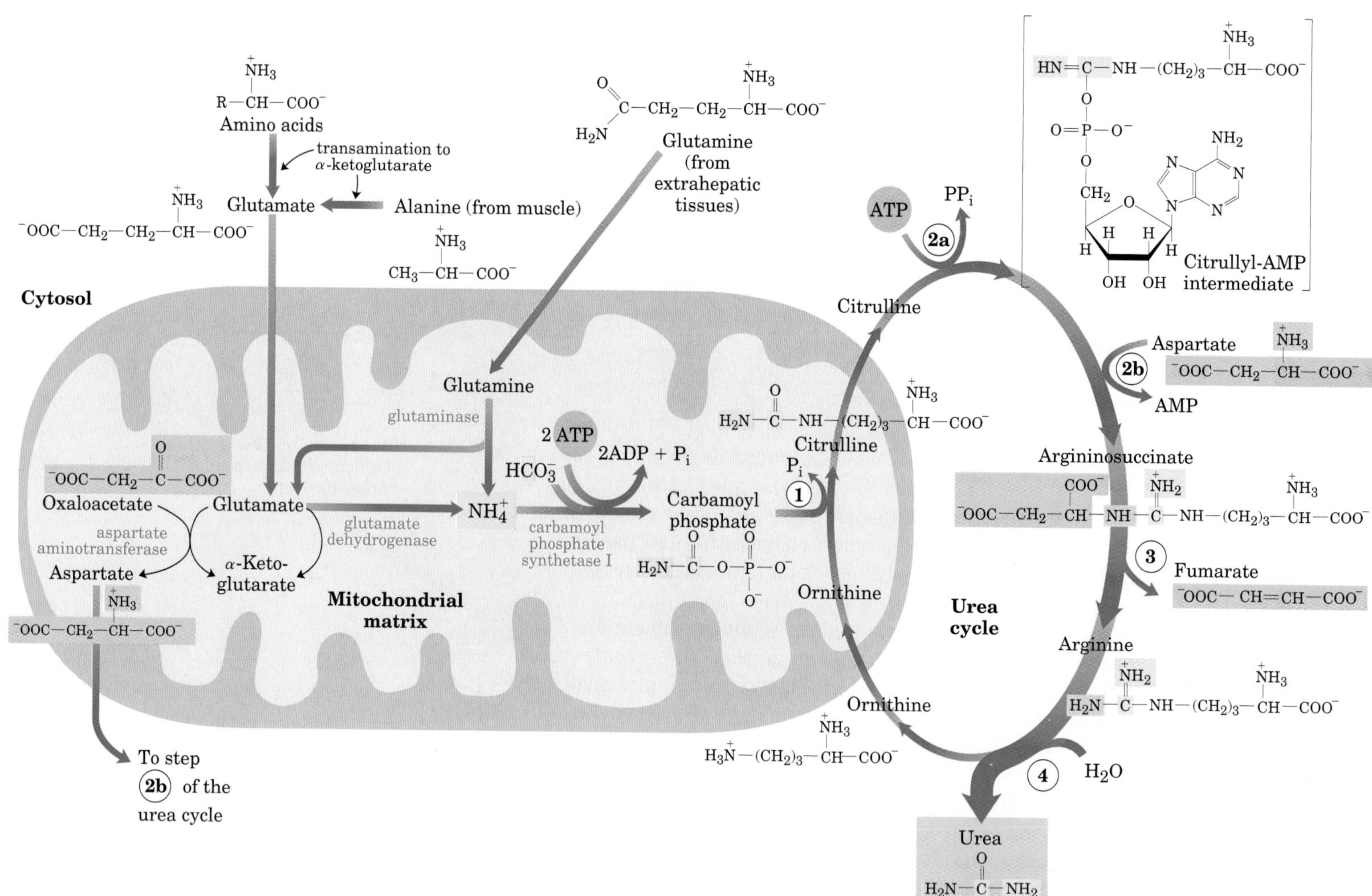

figure 18–9

Urea cycle and reactions that feed amino groups into the cycle. The enzymes catalyzing these reactions (named in the text) are distributed between the mitochondrial matrix and the cytosol. One amino group enters the urea cycle as carbamoyl phosphate (step ①), formed in the matrix; the other (entering at step ②b) enters as aspartate, formed in the matrix by transamination of oxaloacetate and glutamate catalyzed by aspartate aminotransferase. The urea cycle itself consists of four steps: ① Formation of citrulline from ornithine and carbamoyl phosphate; the citrulline passes into the cytosol. ② Formation of argininosuccinate through a citrullyl-AMP intermediate. ③ Formation of arginine from argininosuccinate; this reaction releases fumarate, which enters the mitochondrial citric acid cycle. ④ Formation of urea. The arginase reaction also regenerates the starting compound, ornithine. The pathways by which NH_4^+ arrives in the mitochondrial matrix of hepatocytes are discussed earlier in the chapter.

discovered in 1932 by Hans Krebs (who later also discovered the citric acid cycle) and a medical student associate, Kurt Henseleit. Urea production occurs almost exclusively in the liver and is the fate of most of the ammonia channeled there. The urea passes into the bloodstream and thus to the kidneys and is excreted into the urine. The production of urea now becomes the focus of our discussion.

Urea Is Produced from Ammonia in Five Enzymatic Steps

The urea cycle begins inside liver mitochondria, but three of the subsequent steps occur in the cytosol; the cycle thus spans two cellular compartments (Fig. 18–9). The first amino group to enter the urea cycle is derived from ammonia in the mitochondrial matrix, arising by the multiple pathways described above. The liver also receives some ammonia via the portal vein from the intestine, where it is produced by bacterial oxidation of amino acids. Whatever its source, the NH_4^+ generated in liver mitochondria is immediately used, together with CO_2 (as HCO_3^-) produced by mitochondrial respiration, to form carbamoyl phosphate in the matrix (Fig. 18–10; see also Fig. 18–9). This ATP-dependent reaction is catalyzed by **carbamoyl phosphate synthetase I,** a regulatory enzyme (see below). The mitochondrial form of the enzyme is distinct from the cytosolic (II) form, which has a separate function in pyrimidine biosynthesis (Chapter 22).

The carbamoyl phosphate, which may be regarded as an activated carbamoyl group donor, now enters the urea cycle. The cycle has four enzymatic steps. First, carbamoyl phosphate donates its carbamoyl group to ornithine to form citrulline, with the release of P_i (Fig. 18–9, step ①). Ornithine thus plays a role resembling that of oxaloacetate in the citric acid cycle, accepting material at each turn of the cycle. The reaction is catalyzed by **ornithine transcarbamoylase,** and the citrulline that results passes from the mitochondrion to the cytosol.

The second amino group is introduced from aspartate (generated in mitochondria by transamination and transported into the cytosol) by a condensation reaction between the amino group of aspartate and the ureido (carbonyl) group of citrulline, forming argininosuccinate (step ②). This cytosolic reaction, catalyzed by **argininosuccinate synthetase,** requires ATP and proceeds through a citrullyl-AMP intermediate. The argininosuccinate is then reversibly cleaved by **argininosuccinate lyase** (step ③) to form free arginine and fumarate, the latter entering mitochondria to join the pool of citric acid cycle intermediates. In the last reaction of the urea cycle (step ④), the cytosolic enzyme **arginase** cleaves arginine to yield **urea** and ornithine. Ornithine is transported into the mitochondrion to initiate another round of the urea cycle.

As we noted in Chapter 15, the enzymes of many metabolic pathways are clustered (p. 541), the product of one enzyme reaction being channeled directly to the next enzyme in the pathway. In the urea cycle, the mitochondrial and cytosolic enzymes appear to be clustered in this way. The citrulline transported out of the mitochondrion is not diluted into the general pool of metabolites in the cytosol but is passed directly to the active site of argininosuccinate synthetase. This channeling between enzymes continues for argininosuccinate, arginine, and ornithine. Only urea is released into the general cytosolic pool of metabolites.

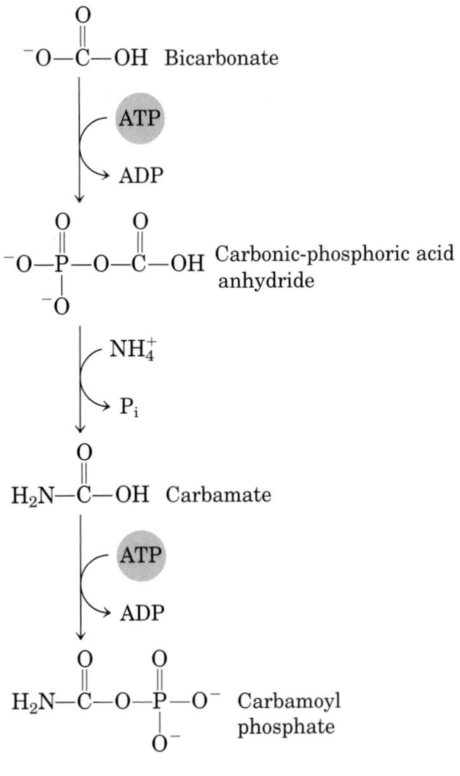

$^-O{-}C(=O){-}OH$ Bicarbonate

↓ ATP → ADP

$^-O{-}P(=O)(O^-){-}O{-}C(=O){-}OH$ Carbonic-phosphoric acid anhydride

↓ NH_4^+ → P_i

$H_2N{-}C(=O){-}OH$ Carbamate

↓ ATP → ADP

$H_2N{-}C(=O){-}O{-}P(=O)(O^-){-}O^-$ Carbamoyl phosphate

figure 18–10

Reaction catalyzed by carbamoyl phosphate synthetase I. The terminal phosphate groups of two molecules of ATP are used to form one molecule of carbamoyl phosphate. In other words, this reaction has two activation steps.

The Citric Acid and Urea Cycles Can Be Linked

Because the fumarate produced in the argininosuccinate lyase reaction is also an intermediate of the citric acid cycle, the cycles are, in principle, interconnected—in a process dubbed the "Krebs bicycle" (Fig. 18–11). However, each cycle can operate independently and communication between them depends on the transport of key intermediates between the mitochondrion and cytosol. Several enzymes of the citric acid cycle, including fumarase (fumarate hydratase) and malate dehydrogenase (pp. 578–579), also occur as isozymes in the cytosol. The fumarate generated in cytosolic arginine synthesis can therefore be converted to malate and then to oxaloacetate in the cytosol, and these intermediates can be further metabolized in the cytosol or transported into mitochondria for use in the citric acid cycle.

Aspartate formed in mitochondria by transamination between oxaloacetate and glutamate can be transported to the cytosol, where it serves as nitrogen donor in the urea cycle reaction catalyzed by argininosuccinate synthetase. These reactions, making up the **aspartate-argininosuccinate shunt,** provide metabolic links between the separate pathways by which the amino groups and carbon skeletons of amino acids are processed.

The Activity of the Urea Cycle Is Regulated at Two Levels

The flux of nitrogen through the urea cycle varies with an organisms's diet. When the diet is primarily protein, the carbon skeletons of amino acids are used for fuel, producing much urea from the excess amino groups. During prolonged starvation, when breakdown of muscle protein begins to supply much of the organism's metabolic energy, urea production also increases substantially.

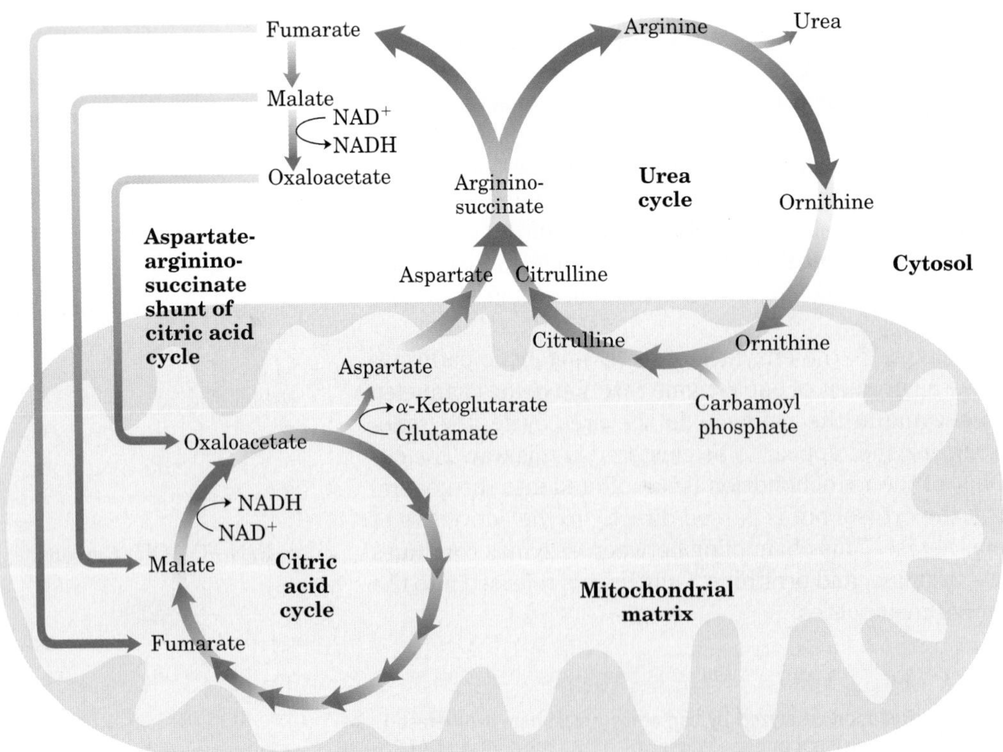

figure 18–11
Links between the urea cycle and citric acid cycle. The interconnected pathways at the top of the figure have been called the "Krebs bicycle." When the complete citric acid cycle is included, the name "Krebs tricycle" is also used. The cycle linking the citric acid and urea cycles is called the aspartate-argininosuccinate shunt. These pathways effectively link the fates of the amino groups and the carbon skeletons of amino acids. The interconnections are even more elaborate than the arrows suggest. For example, some citric acid cycle enzymes, such as fumarase and malate dehydrogenase, have both cytosolic and mitochondrial isozymes. Fumarate produced in the cytosol—whether by the urea cycle, purine biosynthesis, or other processes—can be converted to cytosolic malate and oxaloacetate, which also can be used in the cytosol. Oxaloacetate is a precursor for glucose (see Fig. 20–5) and for some amino acids (see Fig. 22–9). Alternatively, cytosolic malate and oxaloacetate can be transported into mitochondria and used in the citric acid cycle. Malate transport into the mitochondrion is particularly well-characterized and is part of the malate-aspartate shuttle described in Figure 19–26.

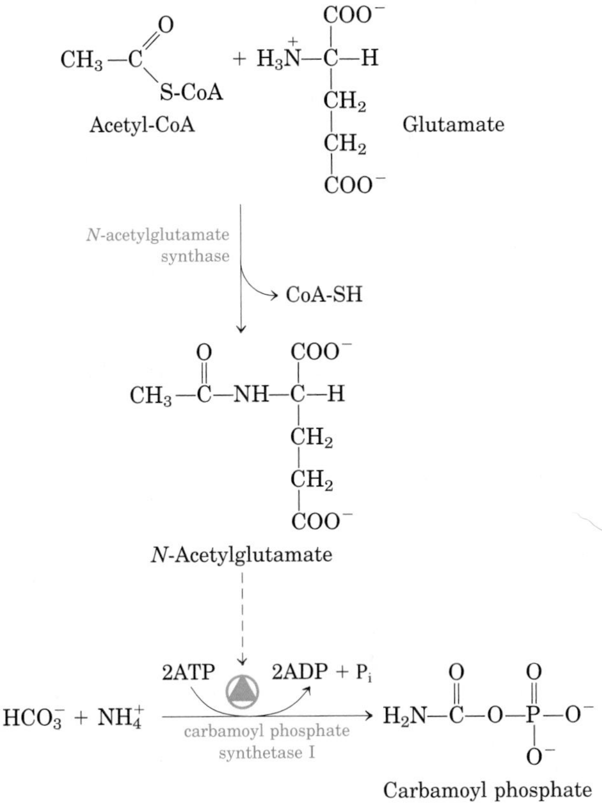

figure 18–12
Synthesis of *N*-acetylglutamate and its activation of carbamoyl phosphate synthetase I.

These changes in demand for urea cycle activity are met in the long term by regulation of the rates of synthesis of the four urea cycle enzymes and carbamoyl phosphate synthetase I in the liver. All five enzymes are synthesized at higher rates in starving animals and in animals on very high-protein diets than in well-fed animals eating primarily carbohydrates and fats. Animals on protein-free diets produce lower levels of urea cycle enzymes.

On a shorter time scale, allosteric regulation of at least one key enzyme adjusts the flux through the urea cycle. The first enzyme in the pathway, carbamoyl phosphate synthetase I, is allosterically activated by ***N*-acetylglutamate,** which is synthesized from acetyl-CoA and glutamate by ***N*-acetylglutamate synthase** (Fig. 18–12). This enzyme catalyzes the first step in the de novo synthesis of arginine from glutamate in plants and microorganisms (see Fig. 22–10). Mammals, however, have *N*-acetylglutamate synthase activity in the liver but lack the other enzymes needed to convert glutamate to arginine. Thus, the use of *N*-acetylglutamate to activate a step in the mammalian urea cycle is enigmatic.

Pathway Interconnections Reduce the Energetic Cost of Urea Synthesis

If the urea cycle is considered in isolation, the synthesis of one molecule of urea requires four high-energy phosphate groups. Two ATPs are required to make carbamoyl phosphate, and one ATP to make argininosuccinate—the latter ATP undergoing a pyrophosphate cleavage to AMP and PP_i, which is hydrolyzed to two P_i. The overall equation of the urea cycle is

$$2NH_4^+ + HCO_3^- + 3ATP^{4-} + H_2O \longrightarrow \text{urea} + 2ADP^{3-} + 4P_i^{2-} + AMP^{2-} + 5H^+$$

However, the urea cycle also causes a net conversion of oxaloacetate to fumarate (via aspartate), and the regeneration of oxaloacetate (Fig. 18–11) produces NADH in the malate dehydrogenase reaction. Each NADH molecule can generate up to 2.5 ATPs during mitochondrial respiration (Chapter 19), greatly reducing the overall energetic cost of urea synthesis.

Genetic Defects in the Urea Cycle Can Be Life-Threatening

People with genetic defects in any enzyme involved in urea formation cannot tolerate protein-rich diets. Amino acids ingested in excess of the minimum daily requirements for protein synthesis are deaminated in the liver, producing free ammonia that cannot be converted to urea and exported into the bloodstream and, as we have seen, ammonia is highly toxic. Humans, however, cannot live on a protein-free diet. We are incapable of synthesizing half of the 20 standard amino acids, and these **essential amino acids** (Table 18–1) must be provided in the diet.

A variety of treatments are used for individuals with urea cycle defects. Careful administration of the aromatic acids benzoate or phenylacetate in the diet can help lower the levels of ammonia in the bloodstream (Fig. 18–13). Benzoate is converted to benzoyl-CoA, which combines with glycine to form hippurate. The glycine used up in this reaction must be regenerated, and ammonia is thus taken up in the glycine synthase reaction (Chapter 22). Phenylacetate combines with glutamine to form phenylacetylglutamine, and further synthesis of glutamine by glutamine synthetase (Eqn 22–1)

table 18–1

Nonessential and Essential Amino Acids for Humans and the Albino Rat

Nonessential	Essential
Alanine	Arginine*
Asparagine	Histidine
Aspartate	Isoleucine
Cysteine	Leucine
Glutamate	Lysine
Glutamine	Methionine
Glycine	Phenylalanine
Proline	Threonine
Serine	Tryptophan
Tyrosine	Valine

*Essential in young, growing animals but not in adults.

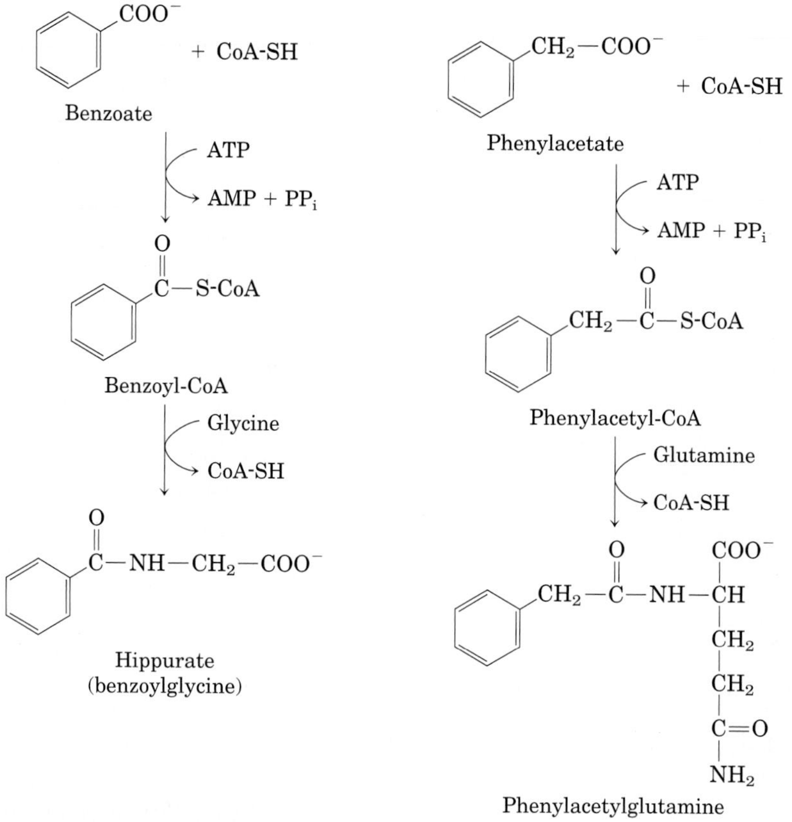

figure 18–13

Treatment for deficiencies in urea cycle enzymes. The aromatic acids benzoate and phenylacetate administered in the diet combine with glycine and glutamine, respectively, and the products are excreted in the urine. Subsequent synthesis of glycine and glutamine to replenish the pool of these intermediates removes ammonia from the bloodstream.

Benzoate + CoA-SH → (ATP → AMP + PP_i) → Benzoyl-CoA → (Glycine → CoA-SH) → Hippurate (benzoylglycine)

Phenylacetate + CoA-SH → (ATP → AMP + PP_i) → Phenylacetyl-CoA → (Glutamine → CoA-SH) → Phenylacetylglutamine

helps remove ammonia. Both hippurate and phenylacetylglutamine are nontoxic compounds that are excreted in the urine. Although the pathways shown in Figure 18–13 make only minor contributions to normal metabolism, they become prominent when aromatic acids are ingested.

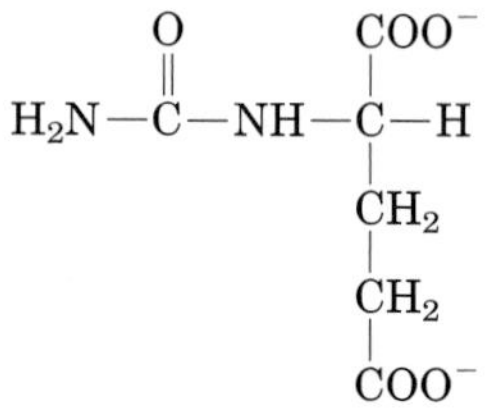

Other therapies are more specific to a particular enzyme deficiency. Deficiency of *N*-acetylglutamate synthase results in the absence of the normal activator of carbamoyl phosphate synthetase I (Fig. 18–12). This condition can be treated by administering carbamoyl glutamate, an analog of *N*-acetylglutamate that is effective in activating carbamoyl phosphate synthetase I. Supplementing the diet with arginine is useful in treating deficiencies of ornithine transcarbamoylase, argininosuccinate synthetase, and argininosuccinate lyase. Many of these treatments must be accompanied by strict dietary control and supplements of essential amino acids. In the rare cases of arginase deficiency, arginine, the substrate of the defective enzyme, must be excluded from the diet.

Carbamoyl glutamate

Natural Habitat Determines the Pathway for Nitrogen Excretion

Urea synthesis is not the only, or even the most common, pathway for excreting ammonia. The basis for the particular molecular form in which amino groups are excreted lies in an organism's anatomy and physiology in relation to its natural habitat. Bacteria and free-living protozoa simply release ammonia to their aqueous environment, where it is diluted and thus made harmless. In the bony fishes (ammonotelic animals), the liver is the primary site of amino acid catabolism. The NH_4^+ produced by transdeamination is simply released from the liver into the bloodstream for transport to the gills, and is rapidly cleared from the blood as water passes through the gills. The bony fishes thus do not require a complex urinary system.

In birds and reptiles, availability of water for the excretory process is an especially important consideration. Excretion of urea in urine requires the simultaneous excretion of a relatively large volume of water; birds in flight would be impeded by the weight, and reptiles in arid environments cannot spare the water. These animals convert amino nitrogen into purines, which are catabolized to uric acid, a relatively insoluble compound excreted with the feces as a semisolid mass of uric acid crystals (see p. 623). To gain the advantage of excreting amino nitrogen in this form, birds and reptiles carry out considerable metabolic work; the pathway from amino acid amino groups to purines to uric acid is a complex energy-requiring process (Chapter 22).

Pathways of Amino Acid Degradation

The pathways of amino acid catabolism, taken together, normally account for only 10% to 15% of the human body's energy production; these pathways are not nearly as active as glycolysis and fatty acid oxidation. Flux through these pathways also varies greatly, depending on the balance between requirements for biosynthetic processes and the availability of a particular amino acid. The 20 catabolic pathways converge to form only five products, all of which enter the citric acid cycle (Fig. 18–14). From here the carbon skeletons are diverted to gluconeogenesis or ketogenesis or are completely oxidized to CO_2 and H_2O.

All or part of the carbon skeletons of ten amino acids are ultimately broken down to acetyl-CoA. Five amino acids are converted to α-ketoglutarate, four to succinyl-CoA, two to fumarate, and two to oxaloacetate. We summarize the individual pathways for the 20 amino acids in flow diagrams, each leading to a specific point of entry into the citric acid cycle. In these diagrams the carbon atoms that enter the citric acid cycle are shown in color. Note that some amino acids appear more than once, reflecting different fates for different parts of their carbon skeletons. Rather than examining every step of every pathway in amino acid catabolism, we single out for special discussion some enzymatic reactions that are particularly noteworthy for their mechanisms or their medical significance.

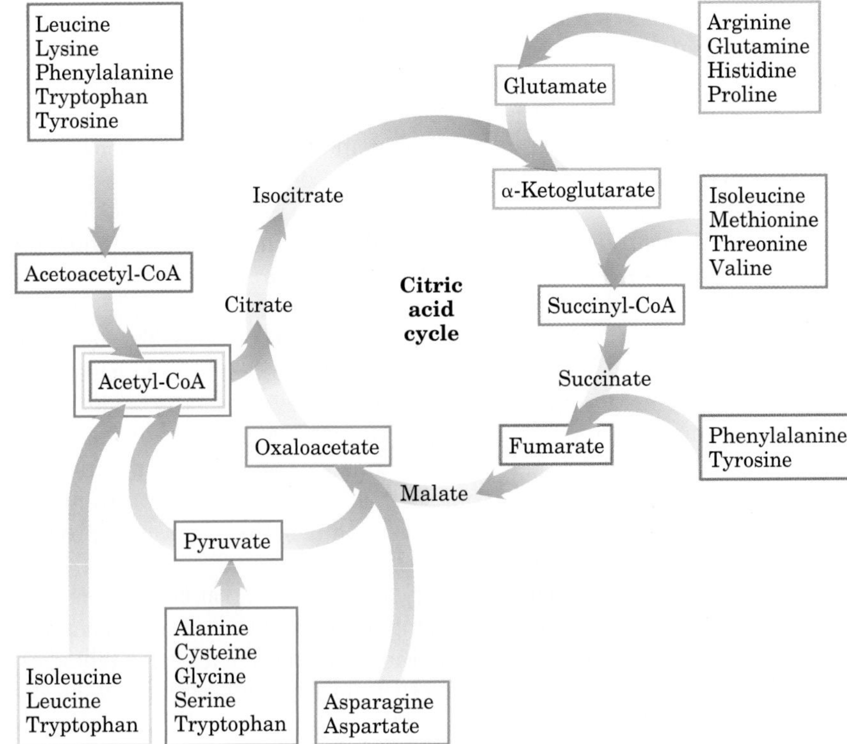

figure 18–14
Summary of the points of entry of the standard amino acids into the citric acid cycle. (The boxes around the amino acids are color-matched to the end products of the catabolic pathways. This color scheme is used in figures throughout the rest of the chapter.) Some amino acids are listed more than once because they yield more than one end product. The figure shows the major catabolic pathways in vertebrate animals, but there are minor variations among vertebrate species. Threonine, for instance, is degraded to acetyl-CoA via pyruvate in some organisms (see Fig. 18–18).

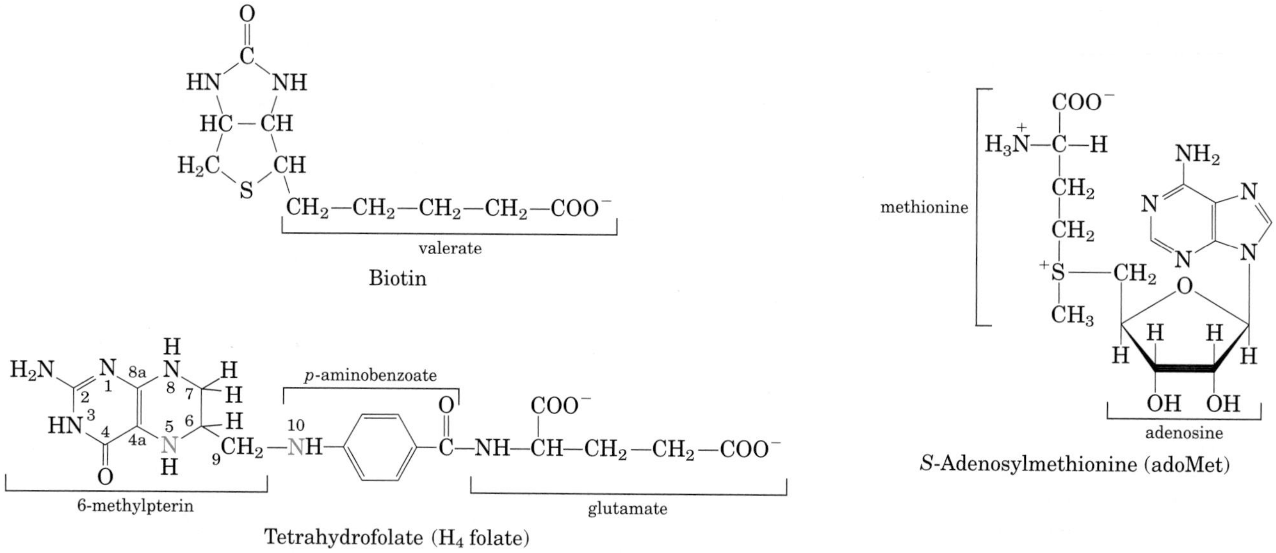

figure 18–15

Some enzyme cofactors important in one-carbon transfer reactions. The nitrogen atoms to which one-carbon groups are attached in tetrahydrofolate are shown in blue.

Several Enzyme Cofactors Play Important Roles in Amino Acid Catabolism

A variety of interesting chemical rearrangements occur in the catabolic pathways of amino acids. It is useful to begin our study of these pathways by noting the classes of reactions that recur and introducing their enzyme cofactors. We have already considered one important class: transamination reactions requiring pyridoxal phosphate. Another common type of reaction in amino acid catabolism is one-carbon transfers, which usually involve one of three cofactors: biotin, tetrahydrofolate, or *S*-adenosylmethionine (Fig. 18–15). These cofactors transfer one-carbon groups in different oxidation states: biotin transfers carbon in its most oxidized state, CO_2 (see Fig. 16–14); tetrahydrofolate transfers one-carbon groups in intermediate oxidation states and sometimes as methyl groups; and *S*-adenosylmethionine transfers methyl groups, the most reduced state of carbon. The latter two cofactors are especially important in amino acid and nucleotide metabolism.

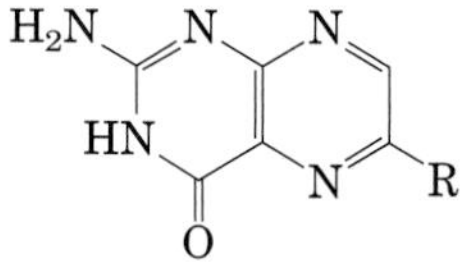

Tetrahydrofolate (H_4 folate) consists of substituted pterin (6-methylpterin), *p*-aminobenzoate, and glutamate moieties (Fig. 18–15). This cofactor is synthesized in bacteria. The oxidized form, folate, is a vitamin for mammals and is converted in two steps to tetrahydrofolate by the enzyme dihydrofolate reductase. The one-carbon group undergoing transfer, in any of three oxidation states, is bonded to N-5 or N-10 or both. The most reduced form of the cofactor carries a methyl group, a more oxidized form carries a methylene group, and the most oxidized forms carry a methenyl, formyl, or formimino group (Fig. 18–16). Most forms of tetrahydrofolate are interconvertible and serve as donors of one-carbon units in a variety of metabolic reactions. The major source of one-carbon units for tetrahydrofolate is the carbon removed in the conversion of serine to glycine, producing N^5,N^{10}-methylenetetrahydrofolate.

Although tetrahydrofolate can carry a methyl group at N-5, this methyl group's transfer potential is insufficient for most biosynthetic reactions. ***S*-Adenosylmethionine** (adoMet) is the preferred cofactor for biological methyl group transfers. It is synthesized from ATP and methionine by the

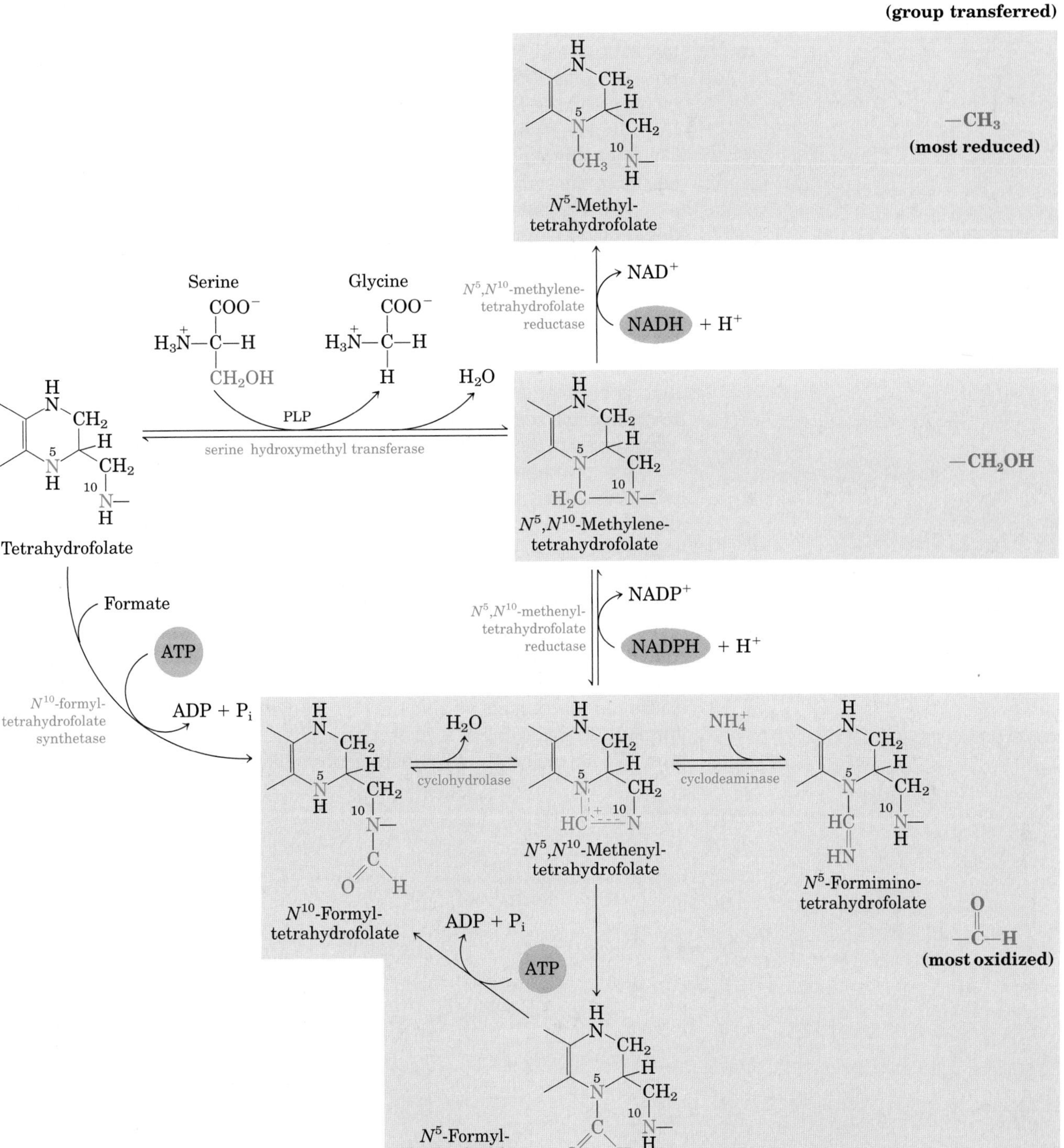

figure 18–16

Conversions of one-carbon units on tetrahydrofolate. The different molecular species are grouped according to oxidation state, with the most reduced at the top and most oxidized at the bottom. All species within a single shaded box are at the same oxidation state. The conversion of N^5, N^{10}-methylene tetrahydrofolate to N^5-methyltetrahydrofolate is effectively irreversible. The enzymatic transfer of formyl groups, as in purine synthesis (see Fig. 22–30) and in the formation of formylmethionine in prokaryotes (Chapter 27), generally uses N^{10}-formyltetrahydrofolate rather than N^5-formyltetrahydrofolate. The latter species is significantly more stable and therefore a weaker donor of formyl groups. The equilibria of the reactions that interconnect these molecular species favor formation of N^5-formyltetrahydrofolate. Conversion of N^5- to N^{10}-formyltetrahydrofolate requires ATP because of an otherwise unfavorable equilibrium. Note that N^5-formiminotetrahydrofolate is derived from histidine in a pathway shown in Figure 18–25.

action of **methionine adenosyl transferase** (Fig. 18–17, step ①). This reaction is unusual in that the nucleophilic sulfur atom of methionine attacks the 5′ carbon of the ribose moiety of ATP rather than one of the phosphorus atoms. Triphosphate is released and is cleaved to P_i and PP_i on the enzyme and the PP_i is later cleaved by inorganic pyrophosphatase, so that three bonds, including two high-energy bonds, are broken in this reaction. The only other known reaction in which triphosphate is displaced from ATP occurs in the synthesis of coenzyme B_{12} (see Box 17–2, Fig. 3).

S-Adenosylmethionine is a potent alkylating agent by virtue of its destabilizing sulfonium ion. The methyl group is subject to attack by nucleophiles and is about 1,000 times more reactive than the methyl group of N^5-methyltetrahydrofolate.

Transfer of the methyl group from *S*-adenosylmethionine to an acceptor yields ***S*-adenosylhomocysteine** (Fig. 18–17, step ②), which is subsequently broken down to homocysteine and adenosine (step ③). Methionine is regenerated by transfer of a methyl group to homocysteine in a reaction catalyzed by **methionine synthase** (step ④), and methionine is reconverted to *S*-adenosylmethionine to complete an activated-methyl cycle.

One form of methionine synthase common in bacteria uses N^5-methyltetrahydrofolate as a methyl donor (as shown in Fig. 18–17). Another form of the enzyme present in bacteria and mammals uses either N^5-methyltetrahydrofolate or methylcobalamin derived from coenzyme B_{12}. This reaction and the rearrangement of L-methylmalonyl-CoA to succinyl-CoA (see Box 17–2, Fig. 1a) are the only known coenzyme B_{12}–dependent reactions in mammals. In cases of vitamin B_{12} deficiency, some symptoms can be alleviated by the administration not only of vitamin B_{12} but of folate. The methyl group of methylcobalamin is derived from N^5-methyltetrahydrofolate. Because the reaction converting the N^5,N^{10}-methylene form to the N^5-methyl form of tetrahydrofolate is irreversible (Fig. 18–16), if coenzyme B_{12} is not available for the synthesis of methylcobalamin, then no acceptor is available

figure 18–17

Synthesis of methionine and *S*-adenosylmethionine in an activated-methyl cycle. The steps are described in the text. In some organisms, the methyl group donor in the methionine synthase reaction (step ④) is methylcobalamin. *S*-Adenosylmethionine, which has a positively charged sulfur (and is thus a sulfonium ion), is a powerful methylating agent in a number of biosynthetic reactions. The methyl group acceptor (step ②) is designated R.

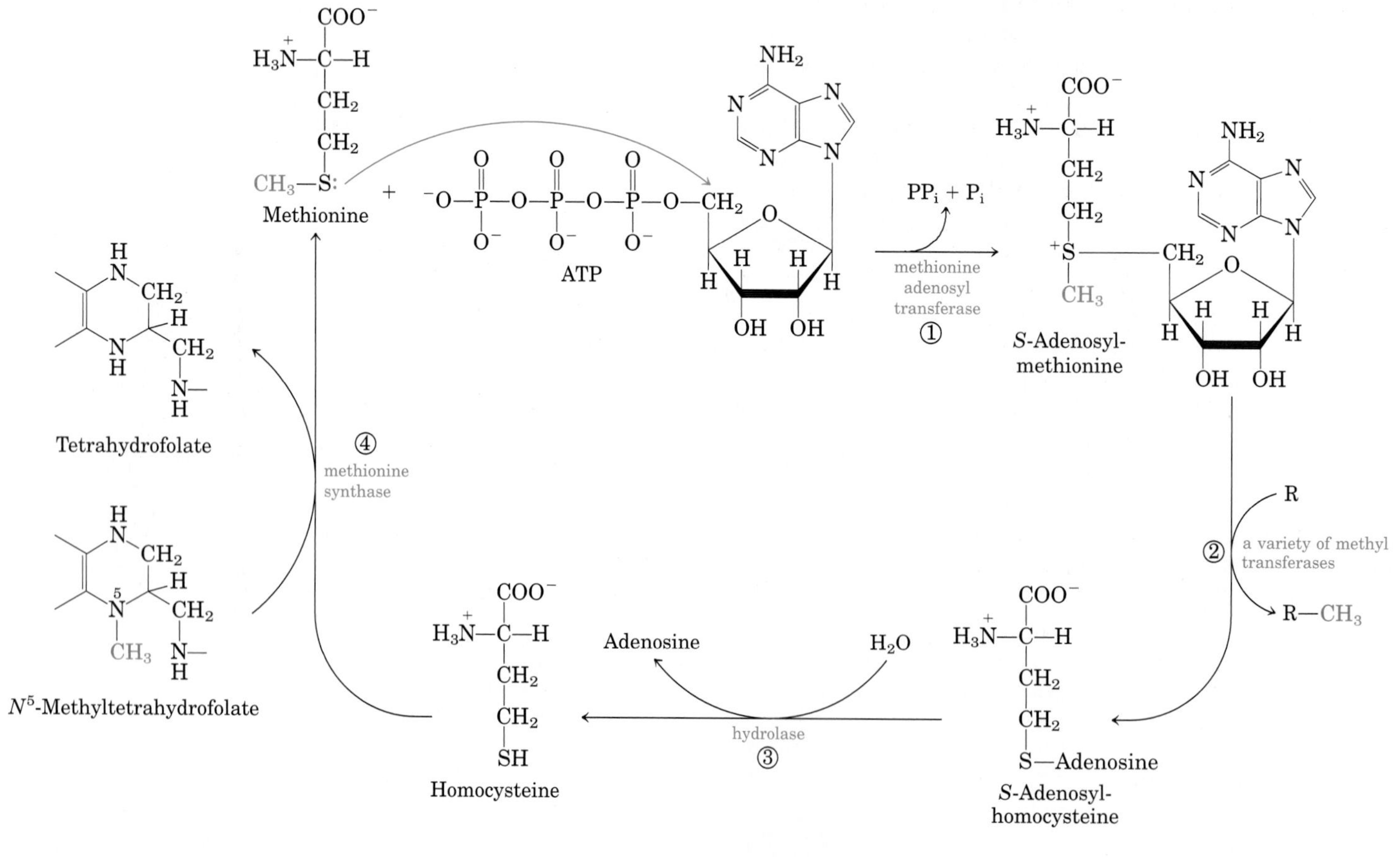

for the methyl group of N^5-methyltetrahydrofolate and metabolic folates become trapped in the N^5-methyl form. This sequestering of folates in one form may be the cause of some symptoms of the vitamin B_{12} deficiency disease pernicious anemia. However, we do not know whether this is the only effect of insufficient vitamin B_{12}.

Tetrahydrobiopterin, another cofactor of amino acid catabolism, is similar to the pterin moiety of tetrahydrofolate, but it is not involved in one-carbon transfers; instead it participates in oxidation reactions. We consider its mode of action when we discuss phenylalanine degradation.

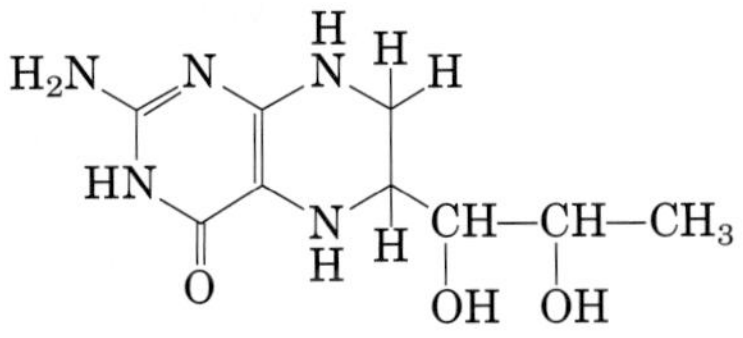

Tetrahydrobiopterin

Ten Amino Acids Are Degraded to Acetyl-CoA

The carbon skeletons of ten amino acids yield acetyl-CoA, which enters the citric acid cycle directly (Fig. 18–14) or is used for fatty acid synthesis. Five of the ten are degraded to acetyl-CoA via pyruvate, which can be used to synthesize glucose. The other five are converted directly to acetyl-CoA and/or to acetoacetyl-CoA, which is then cleaved to acetyl-CoA.

The five amino acids entering via pyruvate are **alanine, tryptophan, cysteine, serine,** and **glycine** (Fig. 18–18). In some organisms **threonine** is also degraded to acetyl-CoA; in humans it is degraded to succinyl-CoA, as described later. Alanine yields pyruvate directly on transamination with α-ketoglutarate, and the side chain of tryptophan is cleaved to yield alanine and thus pyruvate. Cysteine is converted to pyruvate in two steps; one removes the sulfur atom, the other is a transamination. Serine is converted to pyruvate by serine dehydratase. Both the β-hydroxyl and the α-amino groups of serine are removed in this single pyridoxal phosphate–dependent reaction (an analogous reaction with threonine is shown in Fig. 18–26).

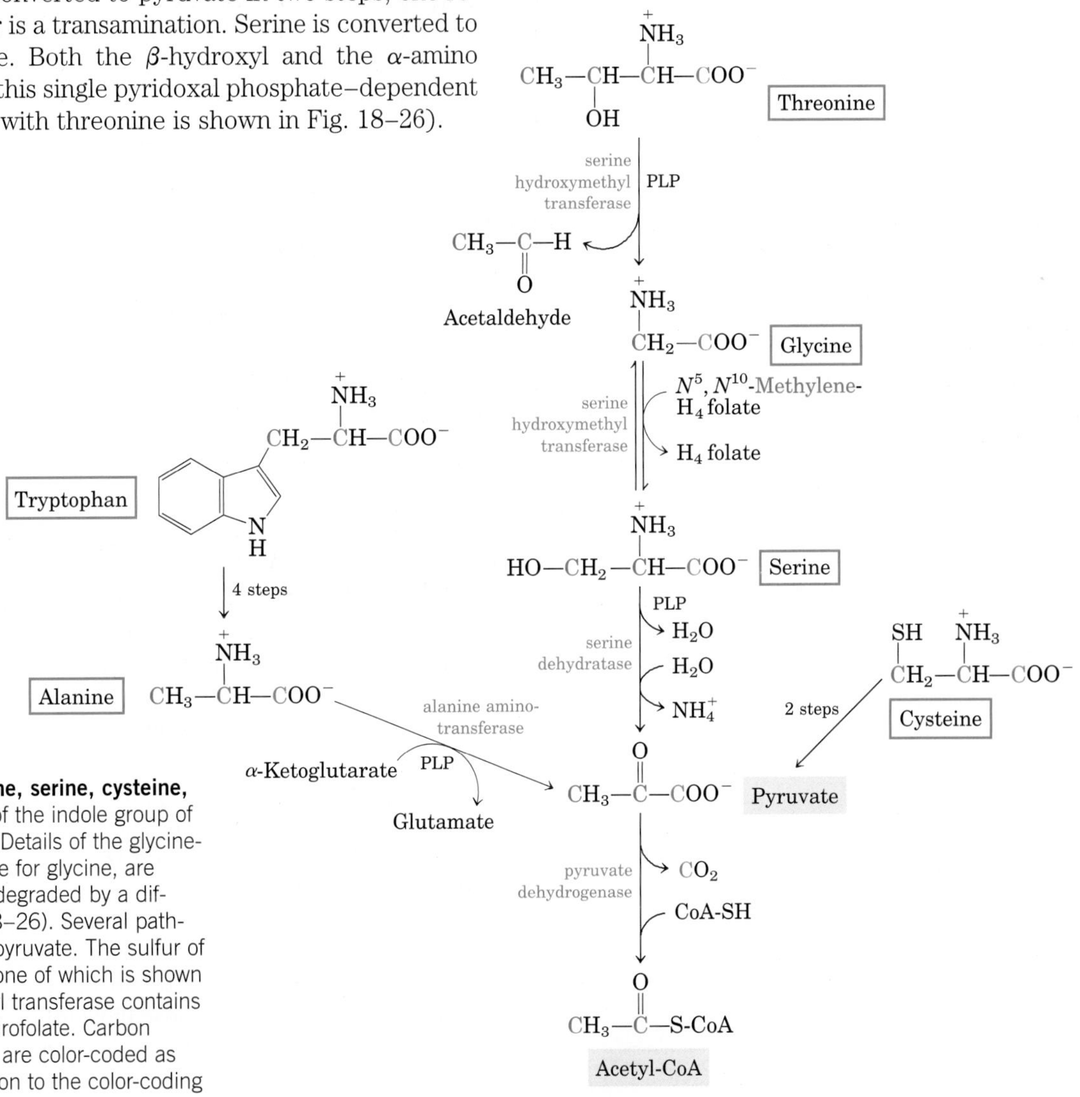

Threonine
serine hydroxymethyl transferase
PLP
Acetaldehyde
Glycine
serine hydroxymethyl transferase
N^5,N^{10}-Methylene-H_4 folate
H_4 folate
Tryptophan
Serine
4 steps
PLP
serine dehydratase
H_2O
H_2O
NH_4^+
Cysteine
Alanine
alanine aminotransferase
2 steps
α-Ketoglutarate
PLP
Glutamate
Pyruvate
pyruvate dehydrogenase
CO_2
CoA-SH
Acetyl-CoA

figure 18–18
Catabolic pathways for alanine, glycine, serine, cysteine, tryptophan, and threonine. The fate of the indole group of tryptophan is shown in Figure 18–20. Details of the glycine-to-serine conversion, and a second fate for glycine, are shown in Figure 18–19. Threonine is degraded by a different pathway in humans (see Fig. 18–26). Several pathways for cysteine degradation lead to pyruvate. The sulfur of cysteine has several alternative fates, one of which is shown in Figure 22–15. Serine hydroxymethyl transferase contains both pyridoxal phosphate and tetrahydrofolate. Carbon atoms here and in subsequent figures are color-coded as necessary to trace their fates, in addition to the color-coding for pathways described in Figure 18–14.

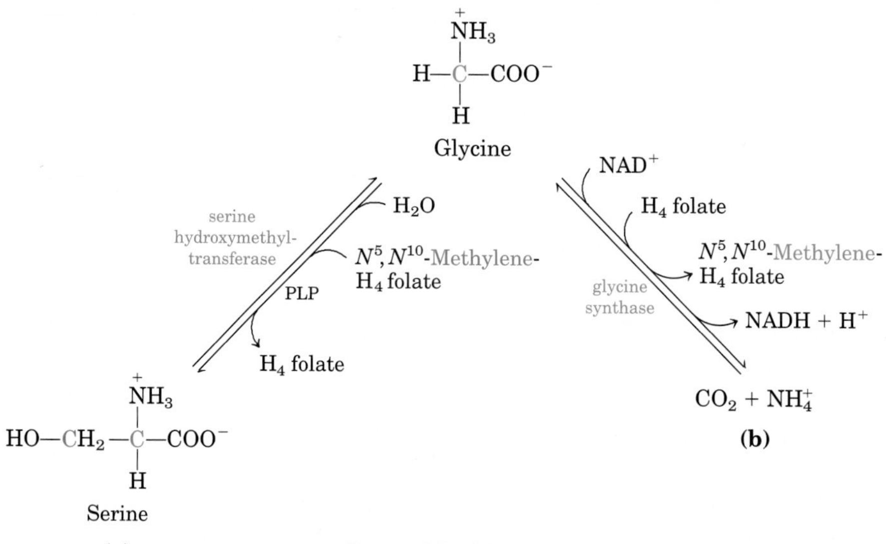

figure 18–19

Two metabolic fates of glycine. (a) Conversion to serine and **(b)** breakdown to CO_2 and NH_4^+. Tetrahydrofolate carries one-carbon units in both of these reactions (see Figs 18–16, 18–17).

Glycine has two pathways. In a major bacterial pathway, glycine is converted to serine by enzymatic addition of a hydroxymethyl group (Fig. 18–19a). This reaction, catalyzed by **serine hydroxymethyl transferase,** requires the coenzymes tetrahydrofolate and pyridoxal phosphate. In the second pathway, which predominates in animals, glycine undergoes oxidative cleavage into CO_2, NH_4^+, and a methylene group (—CH_2—) (Fig. 18–19b). This readily reversible reaction, catalyzed by **glycine synthase,** also requires tetrahydrofolate, which accepts the methylene group. In this oxidative cleavage pathway the two carbon atoms of glycine do not enter the citric acid cycle. One carbon is lost as CO_2 and the other becomes the methylene group of N^5,N^{10}-methylenetetrahydrofolate (Fig. 18–16), a one-carbon group donor in certain biosynthetic pathways.

Portions of the carbon skeletons of six amino acids—**tryptophan, lysine, phenylalanine, tyrosine, leucine,** and **isoleucine**—yield acetyl-CoA or acetoacetyl-CoA or both; the latter is then converted to acetyl-CoA (Fig. 18–20). Some of the final steps in the degradative pathways for leucine, lysine, and tryptophan resemble steps in the oxidation of fatty acids. The degradative pathways of two of these six amino acids deserve special mention.

Tryptophan breakdown is the most complex of all the pathways of amino acid catabolism in animal tissues; portions of tryptophan (six of its total carbons) yield acetyl-CoA by two different pathways, one via pyruvate and one via acetoacetyl-CoA. Some of the intermediates in tryptophan catabolism are precursors for the synthesis of other biomolecules (Fig. 18–21), including nicotinate, a precursor of NAD and NADP in animals; serotonin, a neurotransmitter in vertebrates; and indoleacetate, a growth factor in plants. Some of these biosynthetic pathways are described in more detail in Chapter 22 (see Figs 22–26 and 22–27).

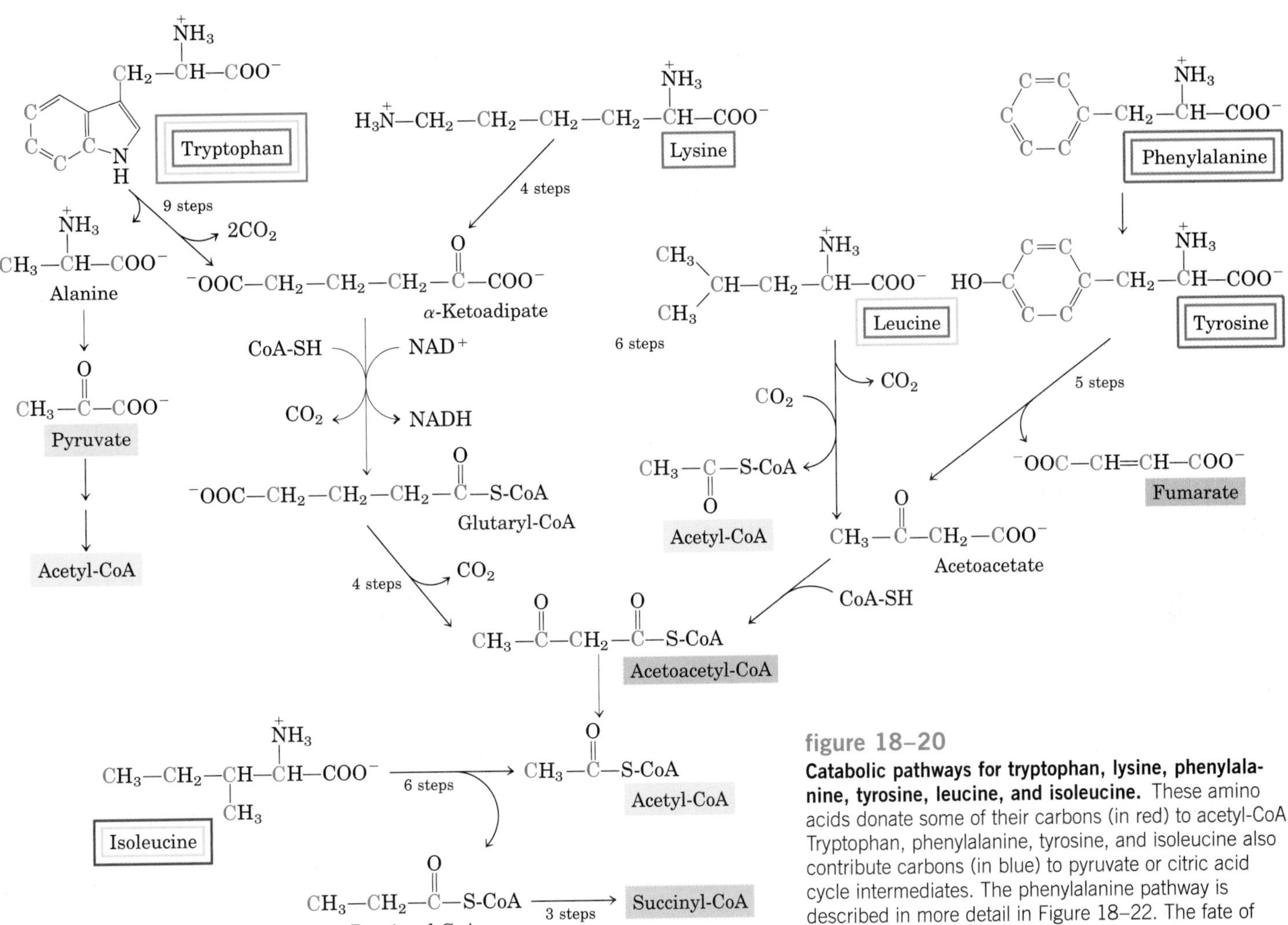

figure 18–20

Catabolic pathways for tryptophan, lysine, phenylalanine, tyrosine, leucine, and isoleucine. These amino acids donate some of their carbons (in red) to acetyl-CoA. Tryptophan, phenylalanine, tyrosine, and isoleucine also contribute carbons (in blue) to pyruvate or citric acid cycle intermediates. The phenylalanine pathway is described in more detail in Figure 18–22. The fate of nitrogen atoms is not traced in this scheme; in most cases they are transferred to α-ketoglutarate to form glutamate.

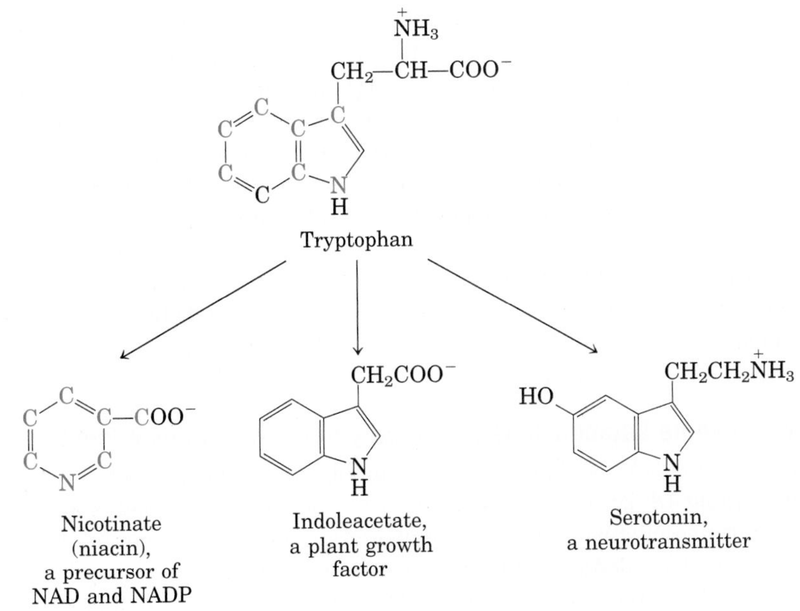

Tryptophan

Nicotinate (niacin), a precursor of NAD and NADP

Indoleacetate, a plant growth factor

Serotonin, a neurotransmitter

figure 18–21

Tryptophan as precursor. The aromatic rings of tryptophan give rise to nicotinate, indoleacetate, and serotonin. Colored atoms are used to trace the source of the ring atoms in nicotinate.

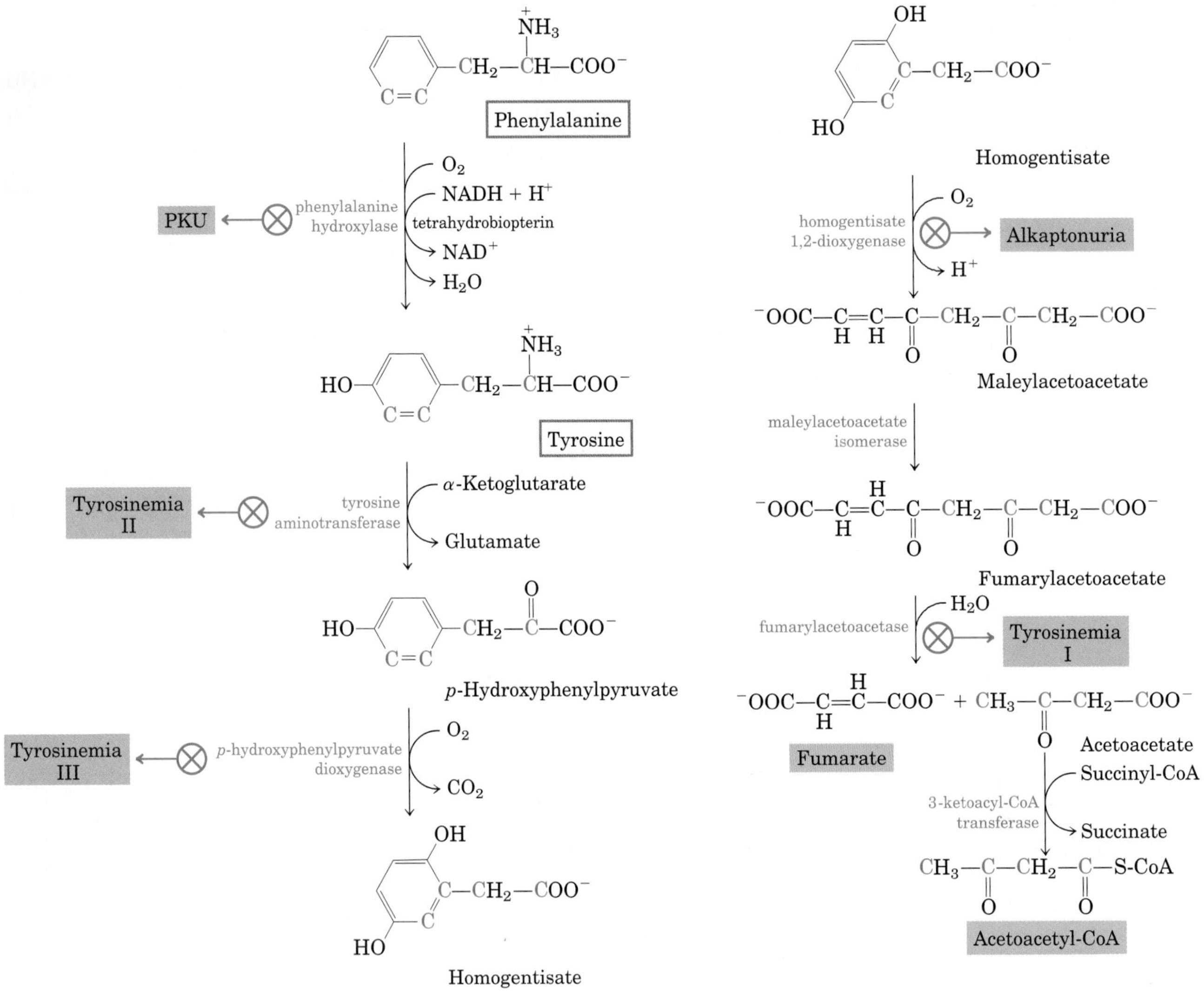

Phenylalanine

PKU ← phenylalanine hydroxylase (O_2, NADH + H^+, tetrahydrobiopterin → NAD^+, H_2O)

Tyrosine

Tyrosinemia II ← tyrosine aminotransferase (α-Ketoglutarate → Glutamate)

p-Hydroxyphenylpyruvate

Tyrosinemia III ← *p*-hydroxyphenylpyruvate dioxygenase (O_2 → CO_2)

Homogentisate

homogentisate 1,2-dioxygenase (O_2 → H^+) → Alkaptonuria

Maleylacetoacetate

maleylacetoacetate isomerase

Fumarylacetoacetate

fumarylacetoacetase (H_2O) → Tyrosinemia I

Fumarate + Acetoacetate

3-ketoacyl-CoA transferase (Succinyl-CoA → Succinate)

Acetoacetyl-CoA

figure 18–22

Catabolic pathways for phenylalanine and tyrosine. These amino acids are normally converted to acetoacetyl-CoA and fumarate in humans. Genetic defects in many of these enzymes are known to cause inheritable human diseases (shaded in red).

The breakdown of phenylalanine is noteworthy because genetic defects in the enzymes of this pathway lead to several inheritable human diseases (Fig. 18–22), as discussed below. Phenylalanine and its oxidation product tyrosine (both with nine carbons) are degraded into two fragments, both of which can enter the citric acid cycle: four of the nine carbon atoms yield free acetoacetate, which is converted to acetoacetyl-CoA and thus acetyl-CoA, and a second four-carbon fragment is recovered as fumarate. Eight of the nine carbons of these two amino acids thus enter the citric acid cycle; the remaining carbon is lost as CO_2. Phenylalanine, after its hydroxylation to tyrosine, is also the precursor of dopamine, a neurotransmitter, and of norepinephrine and epinephrine, hormones secreted by the adrenal medulla (see Fig. 22–27). Melanin, the black pigment of skin and hair, is also derived from tyrosine.

Phenylalanine Catabolism Is Genetically Defective in Some People

Many genetic defects of amino acid metabolism have been identified in humans (Table 18–2). In most such diseases specific intermediates accumulate, causing defective neural development and mental retardation. For example, a genetic defect in **phenylalanine hydroxylase,** the first enzyme in the catabolic pathway for phenylalanine (Fig. 18–22), is responsible for

table 18–2

Some Human Genetic Disorders Affecting Amino Acid Catabolism

Medical condition	Approximate incidence (per 100,000 births)	Defective process	Defective enzyme	Symptoms and effects
Albinism	3	Melanin synthesis from tyrosine	Tyrosine 3-monooxygenase (tyrosinase)	Lack of pigmentation; white hair, pink skin
Alkaptonuria	0.4	Tyrosine degradation	Homogentisate 1,2-dioxygenase	Dark pigment in urine; late-developing arthritis
Argininemia	<0.5	Urea synthesis	Arginase	Mental retardation
Argininosuccinic acidemia	1.5	Urea synthesis	Argininosuccinate lyase	Vomiting, convulsions
Carbamoyl phosphate synthetase I deficiency	>0.5	Urea synthesis	Carbamoyl phosphate synthetase I	Lethargy, convulsions, early death
Homocystinuria	0.5	Methionine degradation	Cystathionine β-synthase	Faulty bone development, mental retardation
Maple syrup urine disease (branched-chain ketoaciduria)	0.4	Isoleucine, leucine, and valine degradation	Branched-chain α-keto acid dehydrogenase complex	Vomiting, convulsions, mental retardation, early death
Methylmalonic acidemia	<0.5	Conversion of propionyl-CoA to succinyl-CoA	Methylmalonyl-CoA mutase	Vomiting, convulsions, mental retardation, early death
Phenylketonuria	8	Conversion of phenylalanine to tyrosine	Phenylalanine hydroxylase	Neonatal vomiting; mental retardation

the disease **phenylketonuria (PKU),** the most common cause of elevated levels of phenylalanine (hyperphenylalaninemia).

Phenylalanine hydroxylase (also called phenylalanine-4-monooxygenase) is one of a general class of enzymes called **mixed-function oxidases** (see Box 21–1), all of which catalyze simultaneous hydroxylation of a substrate by an oxygen atom of O_2 and reduction of the other oxygen atom to H_2O. Phenylalanine hydroxylase requires the cofactor tetrahydrobiopterin, which carries electrons from NADH to O_2 and becomes oxidized to dihydrobiopterin in the process (Fig. 18–23). It is subsequently reduced by the enzyme **dihydrobiopterin reductase** in a reaction that requires NADH.

figure 18–23
Role of tetrahydrobiopterin in the phenylalanine hydroxylase reaction. Note that NADH is required to restore the reduced form of the coenzyme.

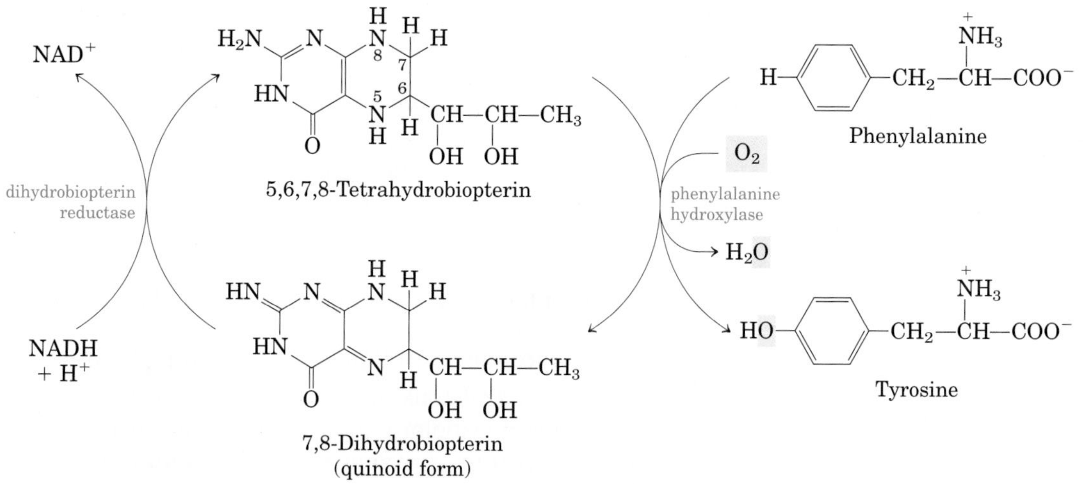

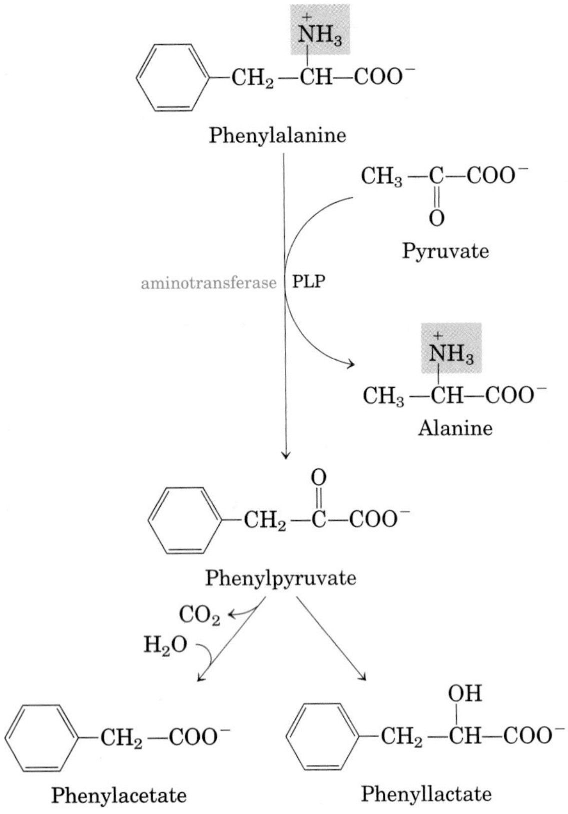

figure 18–24
Alternative pathways for catabolism of phenylalanine in phenylketonuria. Phenylpyruvate accumulates in the tissues, blood, and urine. Phenylacetate and phenyllactate can also be found in the urine.

In individuals with PKU, a secondary pathway of phenylalanine metabolism, normally little used, comes into play in which phenylalanine undergoes transamination with pyruvate to yield **phenylpyruvate** (Fig. 18–24). Phenylalanine and phenylpyruvate accumulate in the blood and tissues and are excreted in the urine: hence the name "phenylketonuria." Much of the phenylpyruvate, rather than being excreted as such, is either decarboxylated to phenylacetate or reduced to phenyllactate. Phenylacetate imparts a characteristic odor to the urine, which nurses have traditionally used to detect PKU in infants. The accumulation of phenylalanine or its metabolites in early life impairs normal development of the brain, causing severe mental retardation. This may be caused by excess phenylalanine competing with other amino acids for transport across the blood-brain barrier, resulting in a deficit of required metabolites.

Phenylketonuria was among the first inheritable metabolic defects discovered in humans. When this condition is recognized early in infancy, mental retardation can largely be prevented by rigid dietary control. The diet must supply only enough phenylalanine and tyrosine to meet the needs for protein synthesis. Consumption of protein-rich foods must be curtailed. Natural proteins, such as casein of milk, must first be hydrolyzed and much of the phenylalanine removed to provide an appropriate diet, at least through childhood. Because the artificial sweetener aspartame is a dipeptide of aspartate and the methyl ester of phenylalanine (see Fig. 3–14), foods sweetened with aspartame bear warnings addressed to individuals on phenylalanine-controlled diets.

Phenylketonuria can also be caused by a defect in the enzyme that catalyzes the regeneration of tetrahydrobiopterin (Fig. 18–23). The treatment in this case is more complex than restricting the intake of phenylalanine and tyrosine. Tetrahydrobiopterin is also required for the formation of L-3,4-dihydroxyphenylalanine (L-dopa) and 5-hydroxytryptophan—precursors of the neurotransmitters norepinephrine and serotonin, respectively—and in phenylketonuria of this type, these precursors must be supplied in the diet. Supplementing the diet with tetrahydrobiopterin itself is ineffective because it is unstable and does not cross the blood-brain barrier.

Screening newborns for genetic diseases can be highly cost-effective, especially in the case of PKU. The tests (no longer relying on urine odor) are relatively inexpensive, and the detection and early treatment of PKU in infants (eight to ten cases per 100,000 newborns) saves millions of dollars in later health care costs each year. More importantly, the emotional trauma avoided by these simple tests is inestimable.

Another inheritable disease of phenylalanine catabolism is **alkaptonuria,** in which the defective enzyme is **homogentisate dioxygenase.** Less serious than PKU, this condition produces few ill effects, although large amounts of homogentisate are excreted, and its oxidation turns the urine black. Individuals with alkaptonuria are also prone to develop a form of arthritis. Alkaptonuria is of considerable historical interest. Archibald Garrod in the early 1900s discovered that this condition is inherited, and he traced the cause to the absence of a single enzyme. Garrod was the first to make a connection between an inheritable trait and an enzyme, a major advance on the path that ultimately led to our current understanding of genes and the information pathways described in Part IV of this book.

Five Amino Acids Are Converted to α-Ketoglutarate

The carbon skeletons of five amino acids (proline, glutamate, glutamine, arginine, and histidine) enter the citric acid cycle as α-ketoglutarate (Fig. 18–25). **Proline, glutamate,** and **glutamine** have five-carbon skeletons. The cyclic structure of proline is opened by oxidation of the carbon most

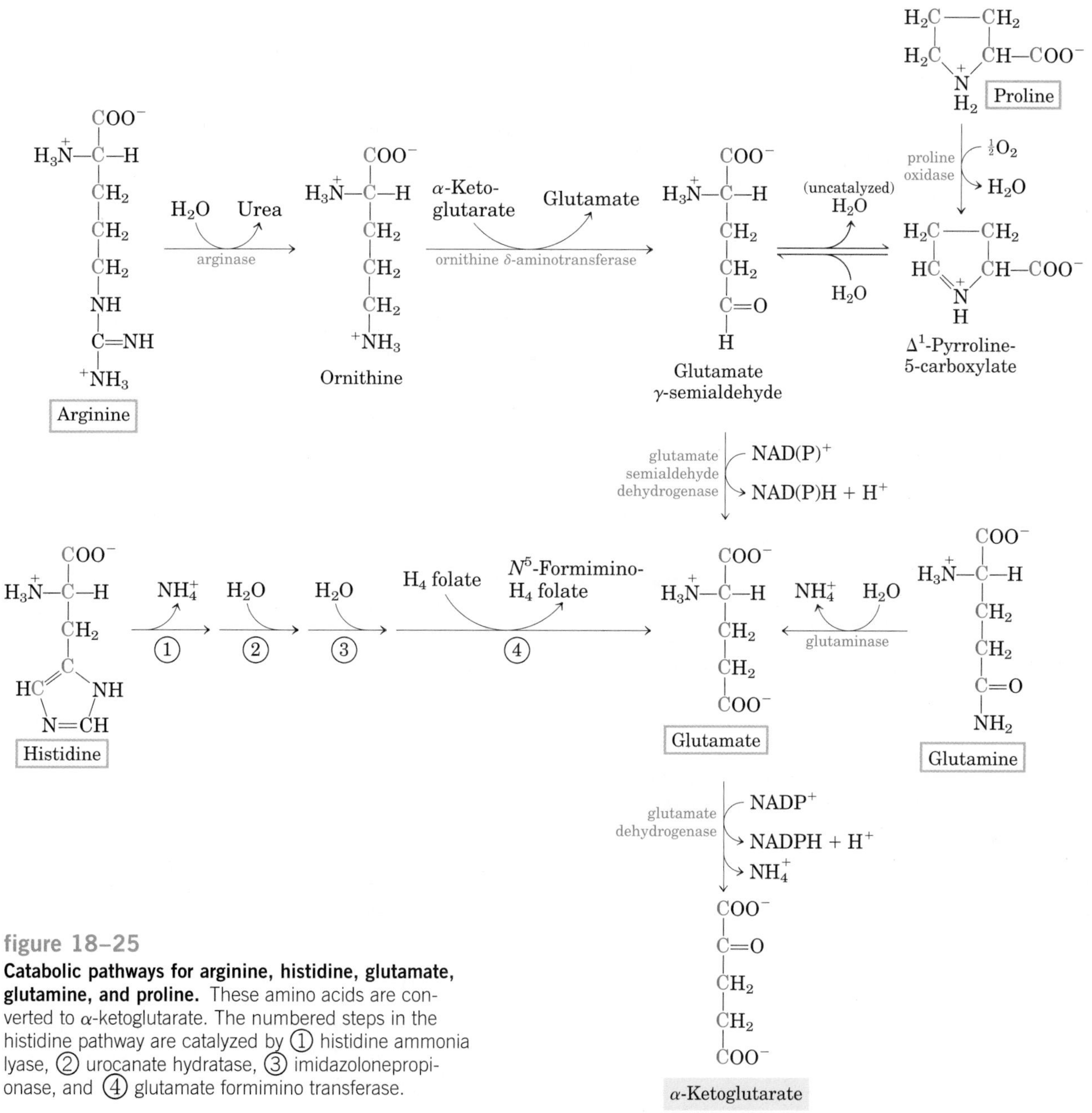

figure 18–25

Catabolic pathways for arginine, histidine, glutamate, glutamine, and proline. These amino acids are converted to α-ketoglutarate. The numbered steps in the histidine pathway are catalyzed by ① histidine ammonia lyase, ② urocanate hydratase, ③ imidazolonepropionase, and ④ glutamate formimino transferase.

distant from the carboxyl group to create a Schiff base, then hydrolysis of the Schiff base to a linear semialdehyde, glutamate γ-semialdehyde. This is further oxidized at the same carbon to produce glutamate. The action of glutaminase, or any of several enzyme reactions in which glutamine donates its amide nitrogen to an acceptor, converts glutamine to glutamate. Transamination or deamination of glutamate produces α-ketoglutarate.

Arginine and **histidine** contain five adjacent carbons and a sixth carbon attached through a nitrogen atom. The catabolic conversion of these amino acids to glutamate is therefore slightly more complex than the path from proline or glutamine (Fig. 18–25). Arginine is converted to the five-carbon skeleton of ornithine in the urea cycle (see Fig. 18–9), and the ornithine is transaminated to glutamate γ-semialdehyde. Conversion of histidine to the five-carbon glutamate occurs in a multistep pathway; the extra carbon is removed in a step that uses tetrahydrofolate as cofactor.

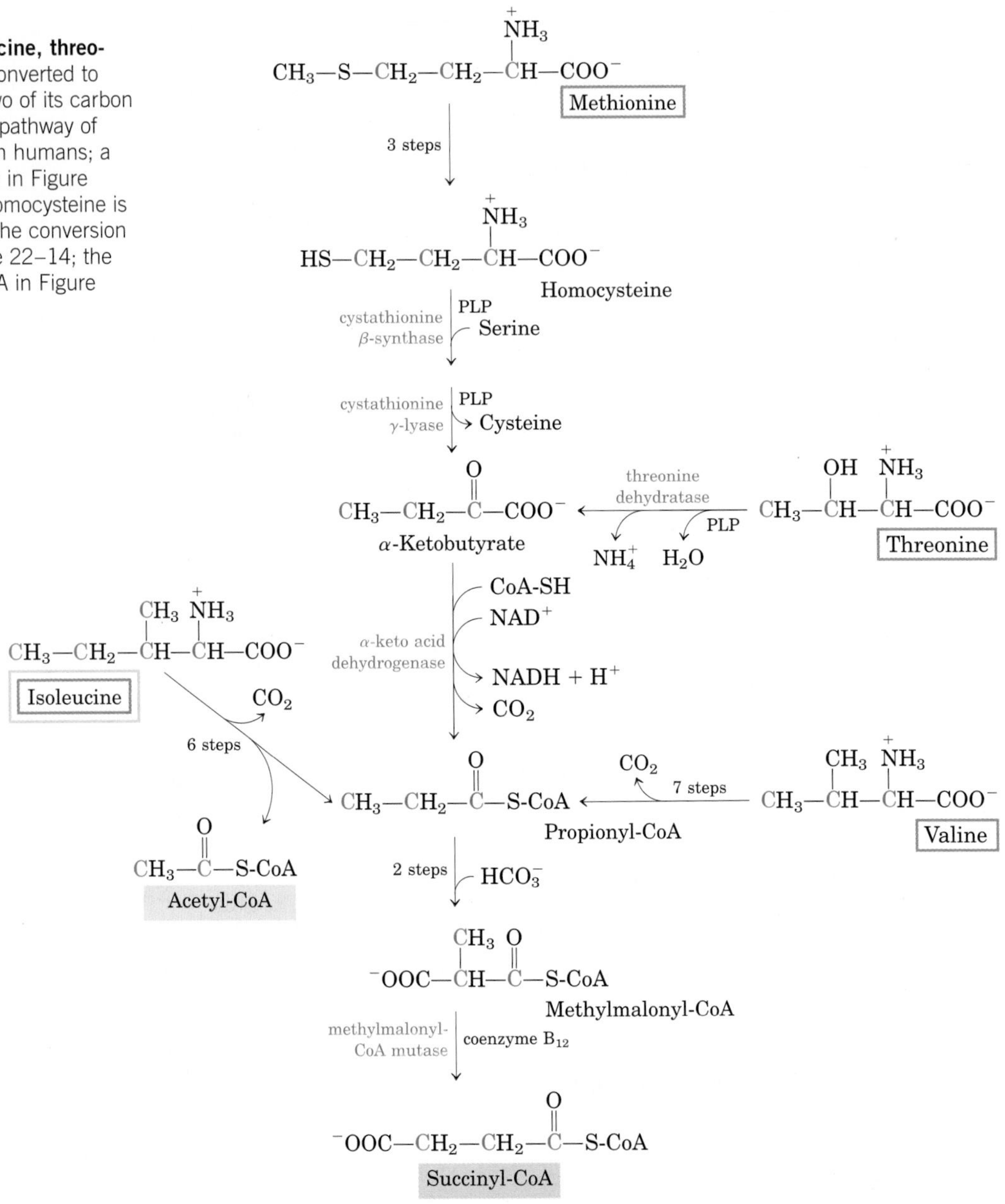

figure 18–26

Catabolic pathways for methionine, isoleucine, threonine, and valine. These amino acids are converted to succinyl-CoA; isoleucine also contributes two of its carbon atoms to acetyl-CoA (see Fig. 18–20). The pathway of threonine degradation shown here occurs in humans; a pathway found in other organisms is shown in Figure 18–18. The pathway from methionine to homocysteine is described in more detail in Figure 18–17; the conversion of homocysteine to α-ketobutyrate in Figure 22–14; the conversion of propionyl-CoA to succinyl-CoA in Figure 17–11.

$CH_3{-}S{-}CH_2{-}CH_2{-}CH(\overset{+}{N}H_3){-}COO^-$ Methionine

3 steps

$HS{-}CH_2{-}CH_2{-}CH(\overset{+}{N}H_3){-}COO^-$ Homocysteine

cystathionine β-synthase; PLP; Serine

cystathionine γ-lyase; PLP; → Cysteine

$CH_3{-}CH_2{-}C(=O){-}COO^-$ α-Ketobutyrate ← threonine dehydratase; PLP; NH_4^+, H_2O ← $CH_3{-}CH(OH){-}CH(\overset{+}{N}H_3){-}COO^-$ Threonine

α-keto acid dehydrogenase; CoA-SH, NAD^+ → $NADH + H^+$, CO_2

$CH_3{-}CH_2{-}CH(CH_3){-}CH(\overset{+}{N}H_3){-}COO^-$ Isoleucine → 6 steps (CO_2) → $CH_3{-}C(=O){-}S\text{-}CoA$ Acetyl-CoA

$CH_3{-}CH_2{-}C(=O){-}S\text{-}CoA$ Propionyl-CoA ← 7 steps (CO_2) ← $CH_3{-}CH(CH_3){-}CH(\overset{+}{N}H_3){-}COO^-$ Valine

2 steps; HCO_3^-

$^-OOC{-}CH(CH_3){-}C(=O){-}S\text{-}CoA$ Methylmalonyl-CoA

methylmalonyl-CoA mutase; coenzyme B_{12}

$^-OOC{-}CH_2{-}CH_2{-}C(=O){-}S\text{-}CoA$ Succinyl-CoA

Four Amino Acids Are Converted to Succinyl-CoA

The carbon skeletons of methionine, isoleucine, threonine, and valine are degraded by pathways that yield succinyl-CoA (Fig. 18–26), an intermediate of the citric acid cycle. **Methionine** donates its methyl group to one of several possible acceptors through *S*-adenosylmethionine, and three of its four remaining carbon atoms are converted to the propionate of propionyl-CoA, a precursor of succinyl-CoA. **Isoleucine** undergoes transamination, followed by oxidative decarboxylation of the resulting α-keto acid. The remaining five-carbon skeleton is further oxidized to acetyl-CoA and propionyl-CoA. **Valine** undergoes transamination and decarboxylation, then a series of oxidation reactions that convert the remaining four carbons to propionyl-CoA. Some parts of the valine and isoleucine degradative pathways closely parallel steps in fatty acid degradation (see Fig. 17–8). In human tissues, **threonine** is also converted to propionyl-CoA.

The propionyl-CoA derived from these three amino acids is converted to succinyl-CoA by a pathway described in Chapter 17: carboxylation to methylmalonyl-CoA, epimerization of the methylmalonyl-CoA, and conversion to succinyl-CoA by the coenzyme B_{12}–dependent methylmalonyl-CoA mutase (see Fig. 17–11). In the rare genetic disease known as methylmalonic acidemia, methylmalonyl-CoA mutase is lacking—with serious metabolic consequences (Table 18–2; Box 18–2).

Branched-Chain Amino Acids Are Not Degraded in the Liver

Although much of the catabolism of amino acids takes place in the liver, the three amino acids with branched side chains (leucine, isoleucine, and valine) are oxidized as fuels primarily in muscle, adipose, kidney, and brain tissue. These extrahepatic tissues contain an aminotransferase, absent in liver, that acts on all three branched-chain amino acids to produce the corresponding α-keto acids (Fig. 18–27). The **branched-chain α-keto acid dehydrogenase complex** then catalyzes oxidative decarboxylation of all three α-keto acids, in each case releasing the carboxyl group as CO_2 and producing the acyl-CoA derivative. This reaction is formally analogous to two other oxidative decarboxylations encountered in Chapter 16: oxidation of pyruvate to acetyl-CoA by the pyruvate dehydrogenase complex (see Fig. 16–6) and oxidation of α-ketoglutarate to succinyl-CoA by the α-ketoglutarate dehydrogenase complex (p. 575). In fact, all three enzyme complexes are similar in structure and share essentially the same reaction mechanism. Five cofactors (thiamine pyrophosphate, FAD, NAD, lipoate, and coenzyme A) participate, and the three proteins in each complex catalyze homologous reactions. This is clearly a case in which enzymatic machinery that evolved to catalyze one reaction was "borrowed" by gene duplication and further evolved to catalyze similar reactions in other pathways.

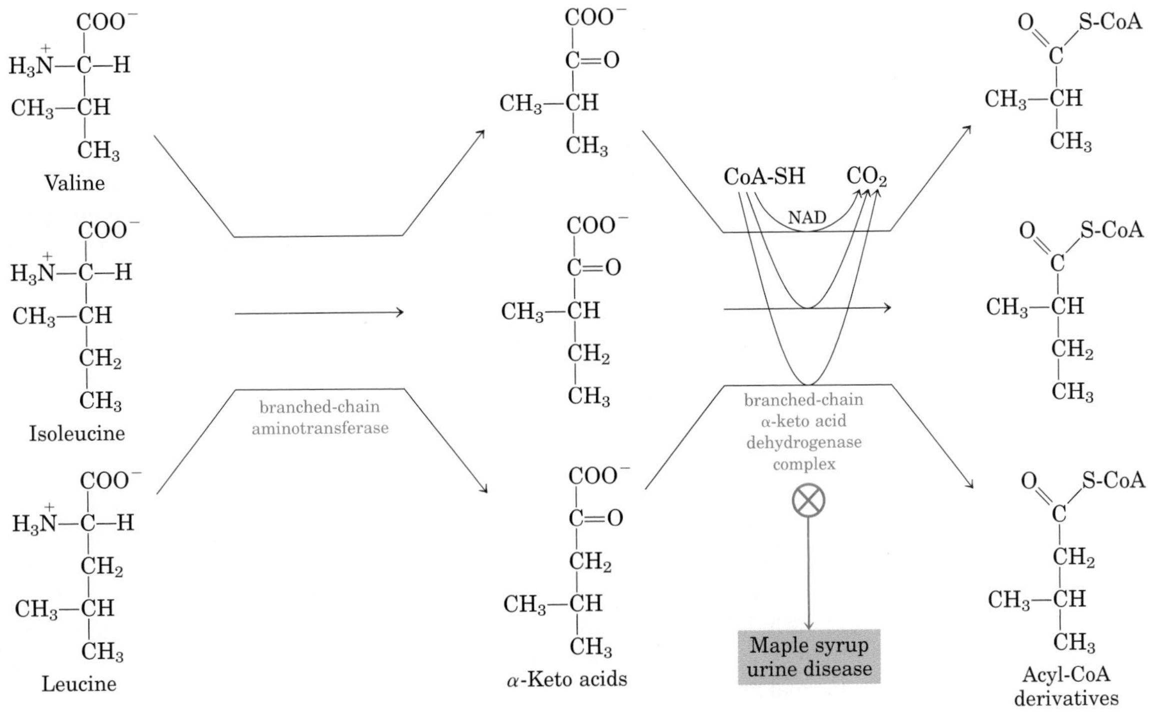

figure 18–27

Catabolic pathways for the three branched-chain amino acids: valine, isoleucine, and leucine. The three pathways, which occur in extrahepatic tissues, share the first two enzymes. The branched-chain α-keto acid dehydrogenase complex is analogous to the pyruvate and α-ketoglutarate dehydrogenase complexes and requires the same five cofactors (some not shown here). This enzyme is defective in people with maple syrup urine disease.

box 18–2 Scientific Sleuths Solve a Murder Mystery

Truth can sometimes be stranger than fiction—or at least as strange as a made-for-TV movie. Take, for example, the case of Patricia Stallings. Convicted of the murder of her infant son, she was sentenced to life in prison—but was later found innocent, thanks to the medical sleuthing of three persistent researchers.

The story began in the summer of 1989 when Stallings brought her three-month-old son, Ryan, to the emergency room of Cardinal Glennon Children's Hospital in St. Louis. The child had labored breathing, uncontrollable vomiting, and gastric distress. According to the attending physician, a toxicologist, the child's symptoms indicated that he had been poisoned with ethylene glycol, an ingredient of antifreeze, a conclusion apparently confirmed by analysis at a commercial lab.

After he recovered, the child was placed in a foster home, and Stallings and her husband, David, were allowed to see him in supervised visits. But when the infant became ill, and subsequently died, after a visit in which Stallings had been briefly left alone with him, she was charged with first-degree murder and held without bail. At the time, the evidence seemed compelling as both the commercial lab and the hospital lab found large amounts of ethylene glycol in the boy's blood and traces of it in a bottle of milk Stallings had fed her son during the visit.

But without knowing it, Stallings had performed a brilliant experiment. While in custody, she learned she was pregnant; she subsequently gave birth to another son, David Stallings Jr., in February 1990. He was placed immediately in a foster home, but within two weeks he started having symptoms similar to Ryan's. David was eventually diagnosed with a rare metabolic disorder called methylmalonic acidemia (MMA). A recessive genetic disorder of amino acid metabolism, MMA affects about 1 in 48,000 newborns and presents symptoms almost identical with those caused by ethylene glycol poisoning.

Stallings couldn't possibly have poisoned her second son, but the Missouri state prosecutor's office was not impressed by the new developments and pressed forward with her trial anyway. The court wouldn't allow the MMA diagnosis of the second child to be introduced as evidence, and in January 1991 Patricia Stallings was convicted of assault with a deadly weapon and sentenced to life in prison.

Fortunately for Stallings, however, William Sly, chairman of the Department of Biochemistry

Experiments with rats have shown that the branched-chain α-keto acid dehydrogenase complex is regulated by covalent modification in response to the content of branched-chain amino acids in the diet. With little or no excess dietary intake of branched-chain amino acids, the enzyme complex is phosphorylated and thereby inactivated by a protein kinase. Addition of excess branched-chain amino acids to the diet results in dephosphorylation and consequent activation of the enzyme. Recall that the pyruvate dehydrogenase complex is subject to similar regulation by phosphorylation and dephosphorylation (p. 586).

There is a relatively rare genetic disease in which the three branched-chain α-keto acids (as well as their precursor amino acids, especially leucine) accumulate in the blood and "spill over" into the urine. This condition, called **maple syrup urine disease** because of the characteristic odor imparted to the urine by the α-keto acids, results from a defective branched-chain α-ketoacid dehydrogenase complex. Untreated, the disease results in abnormal development of the brain, mental retardation, and death in early infancy. Treatment entails rigid control of the diet, limiting the intake of valine, isoleucine, and leucine to the minimum required to permit normal growth.

and Molecular Biology at St. Louis University, and James Shoemaker, head of a metabolic screening lab at the university, got interested in her case when they heard about it from a television broadcast. Shoemaker performed his own analysis of Ryan's blood and didn't detect ethylene glycol. He and Sly then contacted Piero Rinaldo, a metabolic disease expert at Yale University School of Medicine whose lab is equipped to diagnose MMA from blood samples.

When Rinaldo analyzed Ryan's blood serum, he found high concentrations of methylmalonic acid, a breakdown product of the branched-chain amino acids isoleucine and valine, which accumulates in MMA patients because the enzyme that should convert it to the next product in the metabolic pathway is defective. And particularly telling, he says, the child's blood and urine contained massive amounts of ketones, another metabolic consequence of the disease. Like Shoemaker, he did not find any ethylene glycol in a sample of the baby's bodily fluids. The bottle couldn't be tested, since it had mysteriously disappeared. Rinaldo's analyses convinced him that Ryan had died from MMA, but how to account for the results from two labs, indicating that the boy had ethylene glycol in his blood? Could they both be wrong?

When Rinaldo obtained the lab reports, what he saw was, he says, "scary." One lab said that Ryan Stallings' blood contained ethylene glycol, even though the blood sample analysis did not match the lab's own profile for a known sample containing ethylene glycol. "This was not just a matter of questionable interpretation. The quality of their analysis was unacceptable," Rinaldo says. And the second laboratory? According to Rinaldo, that lab detected an abnormal component in Ryan's blood and just "assumed it was ethylene glycol." Samples from the bottle had produced nothing unusual, says Rinaldo, yet the lab claimed evidence of ethylene glycol in that, too.

Rinaldo presented his findings to the case's prosecutor, George McElroy, who called a press conference the very next day. "I no longer believe the laboratory data," he told reporters. Having concluded that Ryan Stallings had died of MMA after all, McElroy dismissed all charges against Patricia Stallings on September 20, 1991.

Asparagine and Aspartate Are Degraded to Oxaloacetate

The carbon skeletons of **asparagine** and **aspartate** ultimately enter the citric acid cycle as oxaloacetate. The enzyme **asparaginase** catalyzes the hydrolysis of asparagine to aspartate, which undergoes transamination with α-ketoglutarate to yield glutamate and oxaloacetate (Fig. 18–28).

We have now seen how the 20 standard amino acids, after losing their nitrogen atoms, are degraded by dehydrogenation, decarboxylation, and other reactions to yield portions of their carbon backbones in the form of five central metabolites that can enter the citric acid cycle. Those portions degraded to acetyl-CoA are completely oxidized to carbon dioxide and water, with generation of ATP by oxidative phosphorylation. We close with a short discussion of some alternative fates of the carbon skeletons of amino acids.

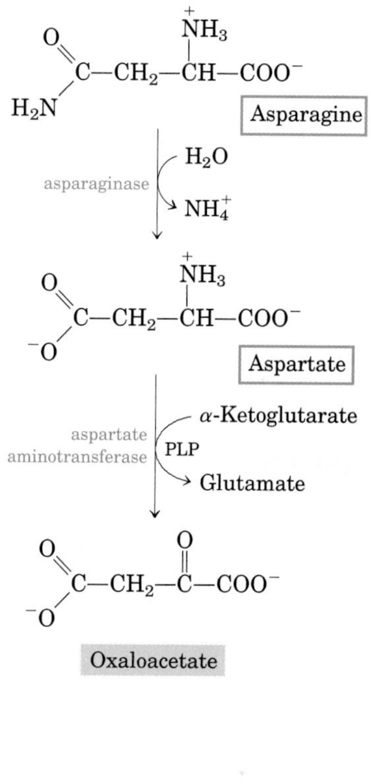

$H_2N-C(=O)-CH_2-CH(\overset{+}{N}H_3)-COO^-$ Asparagine

asparaginase: $H_2O \rightarrow NH_4^+$

$^-O-C(=O)-CH_2-CH(\overset{+}{N}H_3)-COO^-$ Aspartate

aspartate aminotransferase, PLP: α-Ketoglutarate $\rightarrow$ Glutamate

$^-O-C(=O)-CH_2-C(=O)-COO^-$ Oxaloacetate

figure 18–28

Catabolic pathway for asparagine and aspartate. Both amino acids are converted to oxaloacetate.

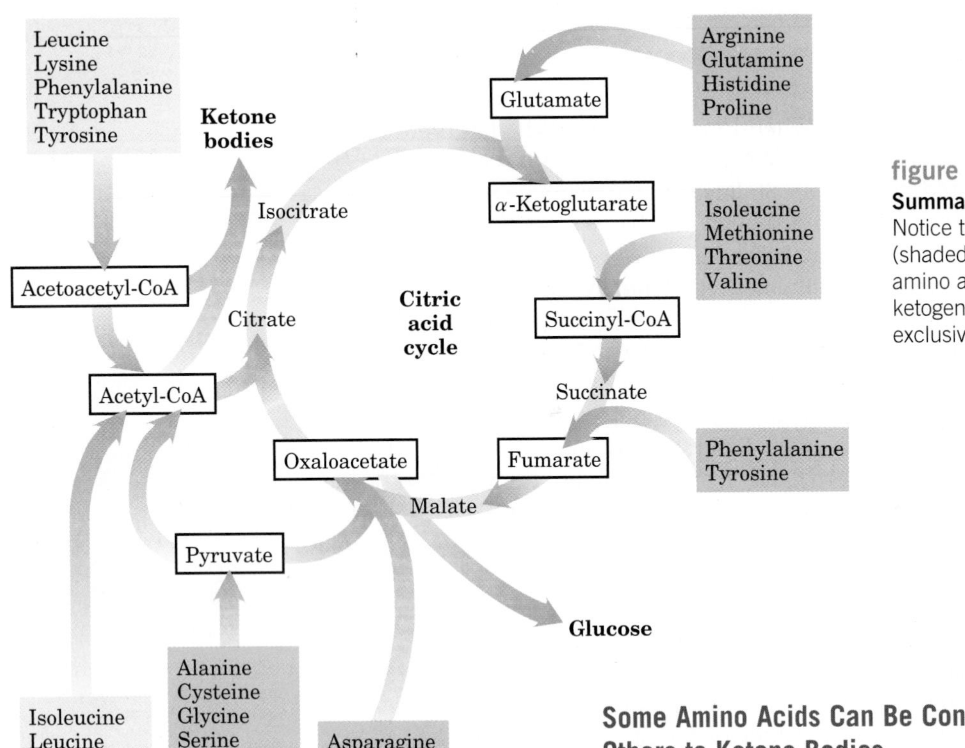

figure 18–29
Summary of the glucogenic and ketogenic amino acids. Notice that four of the amino acids are both glucogenic (shaded red) and ketogenic (shaded blue). The five amino acids degraded to pyruvate are also potentially ketogenic. Only two amino acids, leucine and lysine, are exclusively ketogenic.

Some Amino Acids Can Be Converted to Glucose, Others to Ketone Bodies

The six amino acids that are degraded to acetoacetyl-CoA and/or acetyl-CoA—tryptophan, phenylalanine, tyrosine, isoleucine, leucine, and lysine—can yield ketone bodies in the liver, where acetoacetyl-CoA is converted to acetoacetate and β-hydroxybutyrate (see Fig. 17–16). These are the **ketogenic** amino acids (Fig. 18–29). Their ability to form ketone bodies is particularly evident in untreated diabetes mellitus, in which the liver produces large amounts of ketone bodies from both fatty acids and the ketogenic amino acids.

The amino acids that are degraded to pyruvate, α-ketoglutarate, succinyl-CoA, fumarate, and/or oxaloacetate can be converted into glucose and glycogen by pathways described in Chapter 20. They are the **glucogenic** amino acids. The division between ketogenic and glucogenic amino acids is not sharp; four amino acids—tryptophan, phenylalanine, tyrosine, and isoleucine—are both ketogenic and glucogenic. Catabolism of amino acids is particularly critical to the survival of animals with high-protein diets or during starvation. Leucine is an exclusively ketogenic amino acid that is very common in proteins. Its degradation makes a substantial contribution to ketosis under starvation conditions.

summary

Humans derive a small fraction of their oxidative energy from the catabolism of amino acids. Amino acids are derived from the normal breakdown (recycling) of cellular proteins, degradation of ingested proteins, or breakdown of body proteins in lieu of other fuel sources during starvation or in untreated diabetes mellitus. Ingested proteins are degraded in the stomach and small intestine by proteases. Most proteases are initially synthesized as inactive zymogens.

An early step in the catabolism of amino acids is the separation of the amino group from the carbon skeleton. In most cases, the amino group is transferred to α-ketoglutarate to form glutamate. This transamination reaction requires the coenzyme pyridoxal phosphate. Glutamate is transported to liver mitochondria, where glutamate dehydrogenase liberates the amino group as ammonia (NH_4^+). Ammonia formed in other tissues is transported to the liver as the amide nitrogen of glutamine or, in transport from skeletal muscle, as the amino group of alanine. The pyruvate produced by deamination of alanine in the liver is converted to glucose, which is transported back to muscle as part of the glucose-alanine cycle.

Ammonia is highly toxic to animal tissues. In the urea cycle, ornithine combines with ammonia, in the form of carbamoyl phosphate, to form citrulline. A second amino group is transferred to citrulline from aspartate to form arginine—the immediate precursor of urea. Arginase catalyzes hydrolysis of arginine to urea and ornithine; thus ornithine is regenerated in each turn of the cycle. The urea cycle results in a net conversion of oxaloacetate to fumarate, both of which are intermediates in the citric acid cycle. The two cycles are thus interconnected. The activity of the urea cycle is regulated at the level of enzyme synthesis and by allosteric regulation of the enzyme that forms carbamoyl phosphate.

After removal of their amino groups, the carbon skeletons of amino acids undergo oxidation to compounds that can enter the citric acid cycle for oxidation to CO_2 and H_2O. The reactions of these pathways require a number of cofactors, including tetrahydrofolate and *S*-adenosylmethionine in one-carbon transfer reactions and tetrahydrobiopterin in the oxidation of phenylalanine by phenylalanine hydroxylase. The carbon skeletons of amino acids enter the citric acid cycle through five intermediates: acetyl-CoA, α-ketoglutarate, succinyl-CoA, fumarate, and oxaloacetate. The amino acids producing acetyl-CoA are divided into two groups: alanine, cysteine, glycine, tryptophan, and serine yield acetyl-CoA via pyruvate; leucine, lysine, phenylalanine, tyrosine, and tryptophan yield acetyl-CoA via acetoacetyl-CoA. Isoleucine, leucine, and tryptophan also form acetyl-CoA directly. Proline, histidine, arginine, glutamine, and glutamate produce α-ketoglutarate; threonine, methionine, isoleucine, and valine produce succinyl-CoA; four carbon atoms of phenylalanine and tyrosine give rise to fumarate; and asparagine and aspartate produce oxaloacetate. The branched-chain amino acids (leucine, isoleucine, and valine), unlike the other amino acids, are degraded only in extrahepatic tissues. A number of serious human diseases can be traced to genetic defects in the enzymes of amino acid catabolism.

Depending on their degradative end-product, some amino acids can be converted to ketone bodies; some can be converted to glucose; some to both. These metabolic pathways integrate amino acid degradation into intermediary metabolism and can be critical to survival under conditions in which amino acids are a significant source of metabolic energy.

further reading

General

Bender, D.A. (1985) *Amino Acid Metabolism,* 2nd edn, Wiley-Interscience, Inc., New York.

Campbell, J.W. (1991) Excretory nitrogen metabolism. In *Environmental and Metabolic Animal Physiology,* 4th edn (Prosser, C.L., ed.), pp. 277–324, John Wiley & Sons, Inc., New York.

Coomes, M.W. (1997) Amino acid metabolism. In *Textbook of Biochemistry with Clinical Correlations,* 4th edn (Devlin, T.M., ed.), pp. 445–488, Wiley-Liss, New York.

Hayashi, H. (1995) Pyridoxal enzymes: mechanistic diversity and uniformity. *J. Biochem.* **118,** 463–473.

Mazelis, M. (1980) Amino acid catabolism. In *The Biochemistry of Plants: A Comprehensive Treatise* (Stumpf, P.K. & Conn, E.E., eds), Vol. 5: *Amino Acids and Derivatives* (Miflin, B.J., ed.), pp. 541–567, Academic Press, Inc., New York.

A discussion of the various fates of amino acids in plants.

Powers-Lee, S.G. & Meister, A. (1988) Urea synthesis and ammonia metabolism. In *The Liver: Biology and Pathobiology,* 2nd edn (Arias, I.M., Jakoby, W.B., Popper, H., Schachter, D., & Shafritz, D.A., eds), pp. 317–329, Raven Press, New York.

Walsh, C. (1979) *Enzymatic Reaction Mechanisms,* W.H. Freeman and Company, San Francisco.

A good source for in-depth discussion of the classes of enzymatic reaction mechanisms described in the chapter.

Amino Acid Metabolism

Christen, P. & Metzler, D.E. (1985) *Transaminases,* Wiley-Interscience, Inc., New York.

Curthoys, N.P. & Watford, M. (1995) Regulation of glutaminase activity and glutamine metabolism. *Annu. Rev. Nutr.* **15,** 133–159.

Fitzpatrick, P.F. (1999) Tetrahydropterin-dependent amino acid hydroxylases. *Annu. Rev. Biochem.* **68,** 355–382.

The Urea Cycle

Holmes, F.L. (1980) Hans Krebs and the discovery of the ornithine cycle. *Fed. Proc.* **39,** 216–225.

A medical historian reconstructs the events leading to the discovery of the urea cycle.

Kirsch, J.F., Eichele, G., Ford, G.C., Vincent, M.G., Jansonius, J.N., Gehring, H., & Christen, P. (1984) Mechanism of action of aspartate aminotransferase proposed on the basis of its spatial structure. *J. Mol. Biol.* **174,** 497–525.

Disorders of Amino Acid Degradation

Ledley, F.D., Levy, H.L., & Woo, S.L.C. (1986) Molecular analysis of the inheritance of phenylketonuria and mild hyperphenylalaninemia in families with both disorders. *N. Engl. J. Med.* **314,** 1276–1280.

Nyhan, W.L. (1984) *Abnormalities in Amino Acid Metabolism in Clinical Medicine,* Appleton-Century-Crofts, Norwalk, CT.

Scriver, C.R., Beaudet, A.L., Sly, W.S., & Valle, D. (eds) (1995) *The Metabolic and Molecular Bases of Inherited Disease,* 7th edn, Part 5: *Amino Acids,* McGraw-Hill Book Company, Inc., New York.

Scriver, C.R., Kaufman, S., & Woo, S.L.C. (1988) Mendelian hyperphenylalaninemia. *Annu. Rev. Genet.* **22,** 301–321.

problems

1. Products of Amino Acid Transamination Name and draw the structure of the α-keto acid resulting when the following amino acids undergo transamination with α-ketoglutarate: (a) aspartate, (b) glutamate, (c) alanine, (d) phenylalanine.

2. Measurement of Alanine Aminotransferase Activity The activity (reaction rate) of alanine aminotransferase is usually measured by including an excess of pure lactate dehydrogenase and NADH in the reaction system. The rate of alanine disappearance is equal to the rate of NADH disappearance measured spectrophotometrically. Explain how this assay works.

3. Distribution of Amino Nitrogen If your diet is rich in alanine but deficient in aspartate, will you show signs of aspartate deficiency? Explain.

4. A Genetic Defect in Amino Acid Metabolism: A Case History A two-year-old child was taken to the hospital. His mother said that he vomited frequently, especially after feedings. The child's weight and physical development were below normal. His hair, although dark, contained patches of white. A urine sample treated with ferric chloride ($FeCl_3$) gave a green color characteristic of the presence of phenylpyruvate. Quantitative analysis of urine samples gave the results shown in the table.

Substance	Concentration (mM)	
	Patient's urine	Normal urine
Phenylalanine	7.0	0.01
Phenylpyruvate	4.8	0
Phenyllactate	10.3	0

(a) Suggest which enzyme might be deficient in this child. Propose a treatment.

(b) Why does phenylalanine appear in the urine in large amounts?

(c) What is the source of phenylpyruvate and phenyllactate? Why does this pathway (normally not functional) come into play when the concentration of phenylalanine rises?

(d) Why does the boy's hair contain patches of white?

5. Role of Cobalamin in Amino Acid Catabolism Pernicious anemia is caused by impaired absorption of vitamin B_{12}. What is the effect of this impairment on the catabolism of amino acids? Are all amino acids equally affected ? (Hint: See Box 17–2.)

6. Lactate versus Alanine as Metabolic Fuel: The Cost of Nitrogen Removal The three carbons in lactate and alanine have identical oxidation states, and animals can use either carbon source as a metabolic fuel. Compare the net ATP yield (moles of ATP per mole of substrate) for the complete oxidation (to CO_2 and H_2O) of lactate versus alanine when the cost of nitrogen excretion as urea is included.

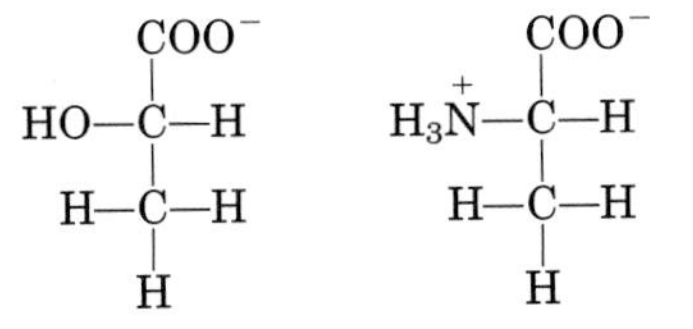

Lactate Alanine

7. Pathway of Carbon and Nitrogen in Glutamate Metabolism When [2-^{14}C,^{15}N]glutamate undergoes oxidative degradation in the liver of a rat, in which atoms of the following metabolites will each isotope be found? (a) urea, (b) succinate, (c) arginine, (d) citrulline, (e) ornithine, (f) aspartate.

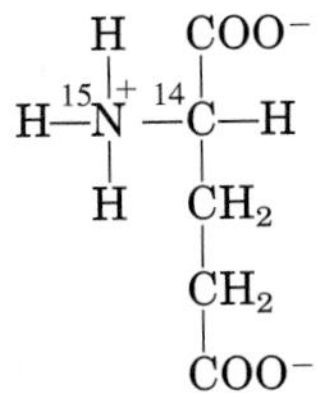

Glutamate

8. Chemical Strategy of Isoleucine Catabolism Isoleucine is degraded in six steps to propionyl-CoA and acetyl-CoA:

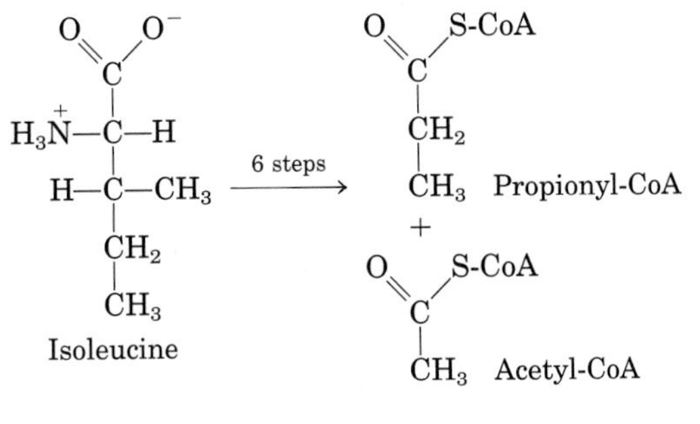

Isoleucine —6 steps→ Propionyl-CoA + Acetyl-CoA

(a) The chemical process of isoleucine degradation includes strategies analogous to those used in the citric acid cycle and the β oxidation of fatty acids. The intermediates of isoleucine degradation (I to V) shown below are not in the proper order. Use your knowledge and understanding of the citric acid cycle and β-oxidation pathway to arrange the intermediates into the proper metabolic sequence for isoleucine degradation.

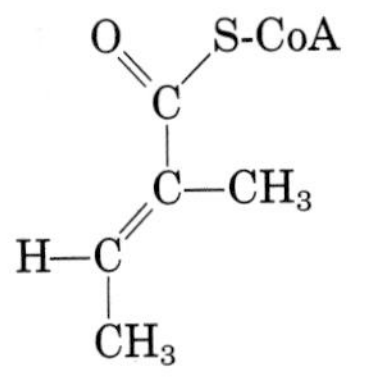

I 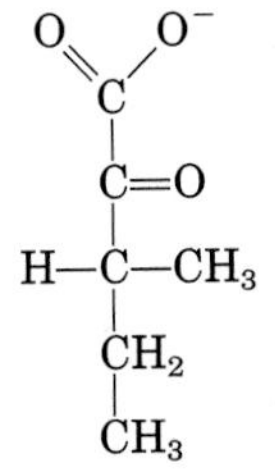 II

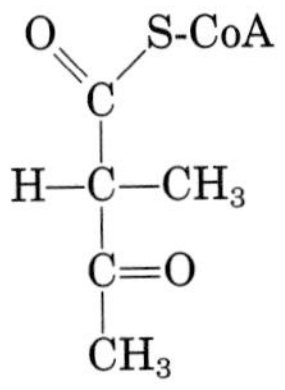

III 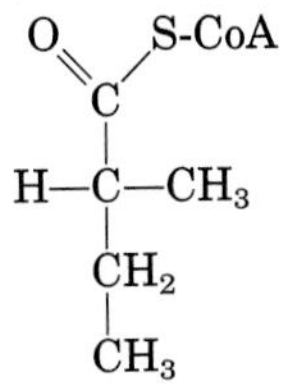 IV

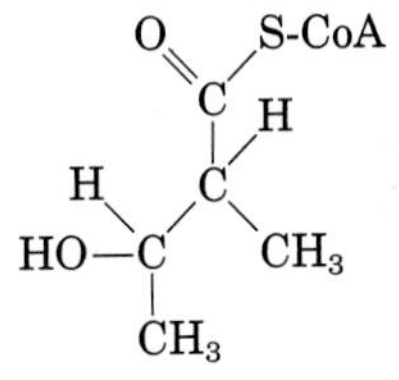

V

(b) For each step you propose, describe the chemical process, provide an analogous example from the citric acid cycle or β-oxidation pathway (where possible), and indicate any necessary cofactors.

9. Ammonia Intoxication Resulting from an Arginine-Deficient Diet In a study conducted some years ago, cats were fasted overnight then given a single meal complete in all amino acids except arginine. Within 2 hours, blood ammonia levels increased from a normal level of 18 μg/L to 140 μg/L, and the cats showed the clinical symptoms of ammonia toxicity. A control group fed a complete amino acid diet or an amino acid diet in which arginine was replaced by ornithine showed no unusual clinical symptoms.

(a) What was the role of fasting in the experiment?

(b) What caused the ammonia levels to rise in the experimental group? Why did the absence of arginine lead to ammonia toxicity? Is arginine an essential amino acid in cats? Why or why not?

(c) Why can ornithine be substituted for arginine?

10. Oxidation of Glutamate Write a series of balanced equations, and an overall equation for the net reaction, describing the oxidation of 2 moles of glutamate to 2 moles of α-ketoglutarate and 1 mole of urea.

11. Role of Pyridoxal Phosphate in Glycine Metabolism The enzyme serine hydroxymethyltransferase (Fig. 18–19) requires pyridoxal phosphate as cofactor. Propose a mechanism for the reaction that explains this requirement. (Hint: See Figure 18–6.)

12. Parallel Pathways for Amino Acid and Fatty Acid Degradation The carbon skeleton of leucine is degraded by a series of reactions closely analogous to those of the citric acid cycle and β oxidation. For each reaction, (a) through (f), indicate its type, provide an analogous example from the citric acid cycle or β-oxidation pathway (where possible), and note any necessary cofactors.

$$CH_3{-}CH(CH_3){-}CH_2{-}CH(\overset{+}{N}H_3){-}COO^-$$

Leucine

(a) ↓

$$CH_3{-}CH(CH_3){-}CH_2{-}C({=}O){-}COO^-$$

α-Ketoisocaproate

(b) ↓ CoA-SH → CO_2

$$CH_3{-}CH(CH_3){-}CH_2{-}C({=}O){-}S\text{-}CoA$$

Isovaleryl-CoA

(c) ↓

$$CH_3{-}C(H_3C){=}C(H){-}C({=}O){-}S\text{-}CoA$$

β-Methylcrotonyl-CoA

(d) ↓ HCO_3^-

$$^-OOC{-}CH_2{-}C(H_3C){=}C(H){-}C({=}O){-}S\text{-}CoA$$

β-Methylglutaconyl-CoA

(e) ↓ H_2O

$$^-OOC{-}CH_2{-}C(OH)(CH_3){-}CH_2{-}C({=}O){-}S\text{-}CoA$$

β-Hydroxy-β-methylglutaryl-CoA

(f) ↓

$$^-OOC{-}CH_2{-}C({=}O){-}CH_3 \quad + \quad CH_3{-}C({=}O){-}S\text{-}CoA$$

Acetoacetate + Acetyl-CoA

13. Transamination and the Urea Cycle Aspartate aminotransferase has the highest activity of all the mammalian liver aminotransferases. Why?

14. The Case against the Liquid Protein Diet A weight-reducing diet heavily promoted some years ago required the daily intake of "liquid protein" (soup of hydrolyzed gelatin), water, and an assortment of vitamins. All other food and drink were to be avoided. People on this diet typically lost 10 to 14 lb in the first week.

(a) Opponents argued that the weight loss was almost entirely due to water loss and would be regained almost immediately when a normal diet was resumed. What is the biochemical basis for this argument?

(b) A number of people on this diet died. What are some of the dangers inherent in the diet and how can they lead to death?

15. Alanine and Glutamine in the Blood Normal human blood plasma contains all the amino acids required for the synthesis of body proteins, but not in equal concentrations. Alanine and glutamine are present in much higher concentrations than any other amino acids. Suggest why.

chapter

19

Oxidative Phosphorylation and Photophosphorylation

Oxidative phosphorylation is the culmination of energy-yielding metabolism in aerobic organisms. All oxidative steps in the degradation of carbohydrates, fats, and amino acids converge at this final stage of cellular respiration, in which the energy of oxidation drives the synthesis of ATP. Photophosphorylation is the means by which photosynthetic organisms capture the energy of sunlight—the ultimate source of energy in the biosphere—and use it to make ATP. Together, oxidative phosphorylation and photophosphorylation account for most of the ATP synthesized by most organisms most of the time.

In eukaryotes, oxidative phosphorylation occurs in mitochondria, photophosphorylation in chloroplasts. Oxidative phosphorylation involves the *reduction* of O_2 to H_2O with electrons donated by NADH and $FADH_2$, and occurs equally well in light or darkness. Photophosphorylation involves the *oxidation* of H_2O to O_2, with $NADP^+$ as ultimate electron acceptor; it is absolutely dependent on the energy of light. Despite their differences, these two highly efficient energy-converting processes have fundamentally similar mechanisms.

Our current understanding of ATP synthesis in mitochondria and chloroplasts is based on the hypothesis, introduced by Peter Mitchell in 1961, that transmembrane differences in proton concentration are the reservoir for the energy extracted from biological oxidation reactions. This **chemiosmotic theory** has been accepted as one of the great unifying principles of twentieth century biology. It provides insight into the processes of oxidative phosphorylation and photophosphorylation, and into such apparently disparate energy transductions as active transport across membranes and the motion of bacterial flagella.

Oxidative phosphorylation and photophosphorylation are mechanistically similar in three respects. (1) Both processes involve the flow of electrons through a chain of membrane-bound carriers. (2) The free energy made available by this "downhill" (exergonic) electron flow is coupled to the "uphill" transport of protons across a proton-impermeable membrane, conserving the free energy of fuel oxidation as a transmembrane electrochemical potential (p. 408). (3) The transmembrane flow of protons down their concentration gradient through specific protein channels provides the free energy for synthesis of ATP, catalyzed by a membrane protein complex (ATP synthase) that couples proton flow to phosphorylation of ADP.

We begin this chapter with oxidative phosphorylation. We first describe the components of the electron-transfer chain, their organization into large functional complexes in the mitochondrial inner membrane, the path of electron flow through them, and the proton movements that accompany this flow. Our next topic is the remarkable enzyme complex that, by "rotational catalysis," captures the energy of proton flow in ATP. We then consider the regulatory mechanisms that coordinate oxidative phosphorylation with the many catabolic pathways by which fuels are oxidized. With this understanding of mitochondrial oxidative phosphorylation, we turn to photophosphorylation, beginning with the absorption of light by photosynthetic pigments, followed by the light-driven flow of electrons from H_2O to $NADP^+$ and the molecular basis for coupling electron and proton flow. Finally, we look at the similarities of structure and mechanism between the ATP synthases of chloroplasts and mitochondria, and at the evolutionary basis for this conservation of mechanism.

Albert L. Lehninger
1917–1986

Oxidative Phosphorylation

Electron-Transfer Reactions in Mitochondria

The discovery in 1948 by Eugene Kennedy and Albert Lehninger that mitochondria are the site of oxidative phosphorylation in eukaryotes marked the beginning of the modern phase of studies in biological energy transductions. Mitochondria, like gram-negative bacteria, have two membranes (Fig. 19–1). The outer mitochondrial membrane is readily permeable to small molecules (M_r <5000) and ions, which move freely through transmembrane channels formed by a family of integral membrane proteins called porins. The inner membrane is impermeable to most small molecules and ions, including protons (H^+); the only species that cross the inner membrane are those for which there are specific transporters. The inner membrane bears the components of the respiratory chain and the ATP synthase.

The mitochondrial matrix enclosed by the inner membrane contains the pyruvate dehydrogenase complex and the enzymes of the citric acid cycle, the fatty acid β-oxidation pathway, and the pathways of amino acid oxidation—all the pathways of fuel oxidation except glycolysis, which takes place in the cytosol. The selectively permeable inner membrane segregates the intermediates and enzymes of cytosolic metabolic pathways from those of metabolic processes occurring in the matrix. However, specific transporters carry pyruvate, fatty acids, and amino acids or their α-keto derivatives into the matrix for access to the machinery of the citric acid cycle. Similarly, ADP and P_i are specifically transported into the matrix as newly synthesized ATP is transported out.

figure 19–1

Biochemical anatomy of a mitochondrion. The convolutions (cristae) of the inner membrane provide a very large surface area. The inner membrane of a single liver mitochondrion may have over 10,000 sets of electron-transfer systems (respiratory chains) and ATP synthase molecules, distributed over the membrane surface. Heart mitochondria, which have more profuse cristae and thus a much larger area of inner membrane, contain over three times as many sets of electron-transfer systems as liver mitochondria. The mitochondrial pool of coenzymes and intermediates is functionally separate from the cytosolic pool. The mitochondria of invertebrates, plants, and microbial eukaryotes are similar to those shown here, but with much variation in size, shape, and degree of convolution of the inner membrane. See Chapter 2 (pp. 36–37) for other details of mitochondrial structure.

Electrons Are Funneled to Universal Electron Acceptors

Oxidative phosphorylation begins with the entry of electrons into the respiratory chain. Most of these electrons arise from the action of dehydrogenases that collect electrons from catabolic pathways and funnel them into universal electron acceptors—nicotinamide nucleotides (NAD^+ or $NADP^+$) or flavin nucleotides (FMN or FAD).

Nicotinamide nucleotide–linked dehydrogenases catalyze reversible reactions of the following general types:

$$\text{Reduced substrate} + NAD^+ \rightleftharpoons \text{oxidized substrate} + NADH + H^+$$

$$\text{Reduced substrate} + NADP^+ \rightleftharpoons \text{oxidized substrate} + NADPH + H^+$$

Most dehydrogenases that act in catabolism are specific for NAD^+ as electron acceptor (Table 19–1). Some are in the cytosol, others are in mitochondria, and still others have mitochondrial and cytosolic isozymes.

NAD-linked dehydrogenases remove two hydrogen atoms from their substrates. One of these is transferred as a hydride ion ($:H^-$) to NAD^+; the other is released as H^+ in the medium (see Fig. 14–15). NAD^+ can also collect reducing equivalents from substrates acted upon by NADP-linked dehydrogenases. This is made possible by **nicotinamide nucleotide transhydrogenase,** which catalyzes the reaction

$$NADPH + NAD^+ \rightleftharpoons NADP^+ + NADH$$

NADH and NADPH are water-soluble electron carriers that associate *reversibly* with dehydrogenases. NADH carries electrons from catabolic reactions to their point of entry into the respiratory chain, the NADH dehydrogenase complex described below. NADPH generally supplies electrons to anabolic reactions. Neither NADH nor NADPH can cross the inner mitochondrial membrane, but the electrons they carry can be shuttled across indirectly, as we shall see.

table 19–1

Some Important Reactions Catalyzed by NAD(P)H-Linked Dehydrogenases

Reaction*	Location†
NAD-linked	
α-Ketoglutarate + CoA + $NAD^+ \rightleftharpoons$ succinyl-CoA + CO_2 + NADH + H^+	M
L-Malate + $NAD^+ \rightleftharpoons$ oxaloacetate + NADH + H^+	M and C
Pyruvate + CoA + $NAD^+ \rightleftharpoons$ acetyl-CoA + CO_2 + NADH + H^+	M
Glyceraldehyde 3-phosphate + P_i + $NAD^+ \rightleftharpoons$ 1,3-bisphosphoglycerate + NADH + H^+	C
Lactate + $NAD^+ \rightleftharpoons$ pyruvate + NADH + H^+	C
β-Hydroxyacyl-CoA + $NAD^+ \rightleftharpoons$ β-ketoacyl-CoA + NADH + H^+	M
NADP-linked	
Glucose 6-phosphate + $NADP^+ \rightleftharpoons$ 6-phosphogluconate + NADPH + H^+	C
NAD- or NADP-linked	
L-Glutamate + H_2O + $NAD(P)^+ \rightleftharpoons$ α-ketoglutarate + NH_4^+ + NAD(P)H	M
Isocitrate + $NAD(P)^+ \rightleftharpoons$ α-ketoglutarate + CO_2 + NAD(P)H + H^+	M and C

*These reactions and their enzymes are discussed in Chapters 15 through 18.

†M designates mitochondria; C, cytosol.

Flavoproteins contain a very tightly, sometimes covalently, bound flavin nucleotide, either FMN or FAD (see Fig. 14–16). The oxidized flavin nucleotide can accept either one electron (yielding the semiquinone form) or two (yielding $FADH_2$ or $FMNH_2$). Electron transfer occurs because the flavoprotein has a higher reduction potential than the compound oxidized. The standard reduction potential of a flavin nucleotide, unlike that of NAD or NADP, depends on the protein with which it is associated. Local interactions with functional groups in the protein distort the electron orbitals in the flavin ring, changing the relative stabilities of oxidized and reduced forms. The relevant standard reduction potential is therefore that of the particular flavoprotein, not that of isolated FAD or FMN. The flavin nucleotide should be considered part of the flavoprotein's active site, not as a reactant or product in the electron-transfer reaction. Because flavoproteins can participate in either one- or two-electron transfers, they can serve as intermediates between reactions in which two electrons are donated (as in dehydrogenations) and those in which only one electron is accepted (as in the reduction of a quinone to a hydroquinone, described below).

Electrons Pass through a Series of Membrane-Bound Carriers

The mitochondrial respiratory chain consists of a series of sequentially acting electron carriers, most of which are integral proteins with prosthetic groups capable of accepting and donating either one or two electrons. Three types of electron transfers occur in oxidative phosphorylation: (1) direct transfer of electrons, as in the reduction of Fe^{3+} to Fe^{2+}; (2) transfer as a hydrogen atom ($H^+ + e^-$); and (3) transfer as a hydride ion ($:H^-$), which bears two electrons. The term **reducing equivalent** is used to designate a single electron equivalent transferred in an oxidation-reduction reaction.

In addition to NAD and flavoproteins, three other types of electron-carrying molecules function in the respiratory chain: a hydrophobic quinone (ubiquinone) and two different types of iron-containing proteins (cytochromes and iron-sulfur proteins).

Ubiquinone (also called **coenzyme Q,** or simply **Q**) is a lipid-soluble benzoquinone with a long isoprenoid side chain (Fig. 19–2). The closely related compounds plastoquinone (of plant chloroplasts) and menaquinone (of bacteria) play roles analogous to that of ubiquinone, carrying electrons in membrane-associated electron-transfer chains. Ubiquinone can accept one electron to become the semiquinone radical ($QH^•$) or two electrons to form ubiquinol (QH_2) (Fig. 19–2) and, like flavoprotein carriers, it can act at the junction between a two-electron donor and a one-electron acceptor. Because ubiquinone is both small and hydrophobic, it is freely diffusible within the lipid bilayer of the inner mitochondrial membrane and can shuttle reducing equivalents between other, less mobile electron carriers in the membrane. And because it carries both electrons and protons, it plays a central role in coupling electron flow to proton movement.

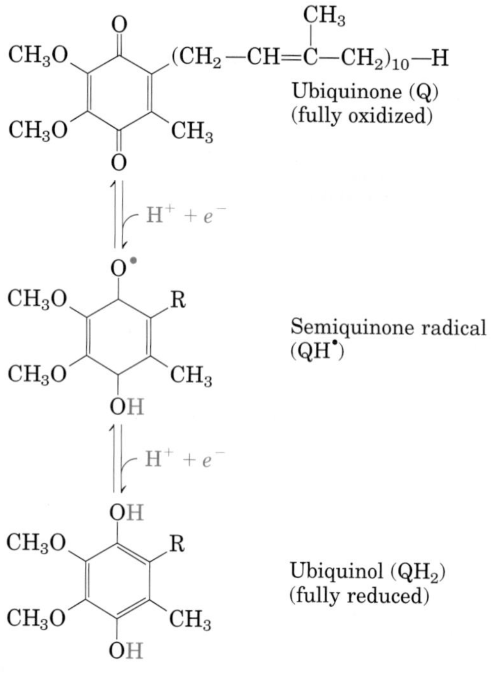

CH_3O, CH_3O, CH_3, $(CH_2{-}CH{=}C(CH_3){-}CH_2)_{10}{-}H$
Ubiquinone (Q)
(fully oxidized)

$H^+ + e^-$

Semiquinone radical
($QH^•$)

$H^+ + e^-$

Ubiquinol (QH_2)
(fully reduced)

figure 19–2
Ubiquinone (Q, or coenzyme Q). Complete reduction of ubiquinone requires two electrons and two protons, and occurs in two steps through the semiquinone radical intermediate.

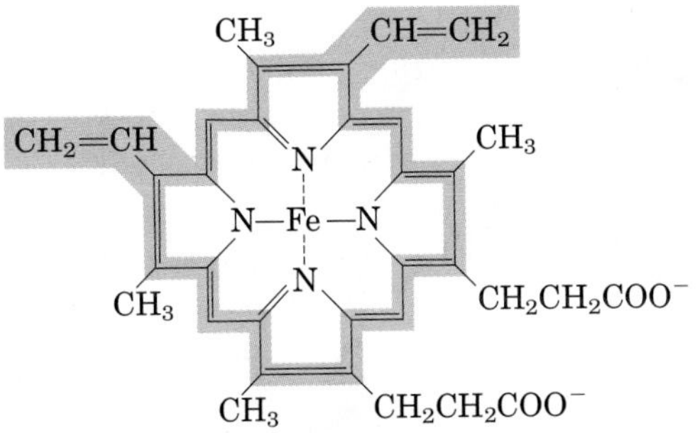

Iron protoporphyrin IX
(in *b*-type cytochromes)

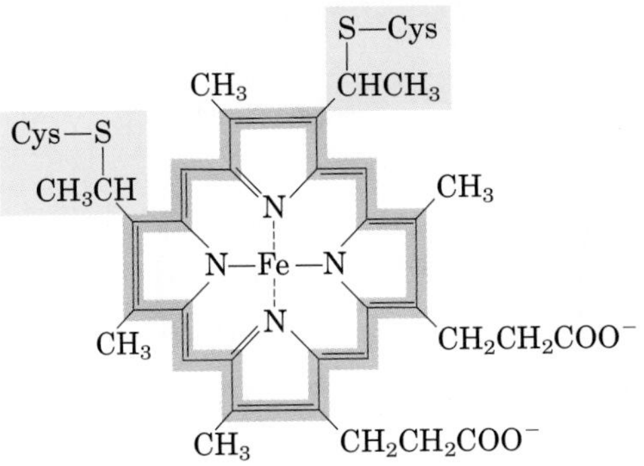

Heme C
(in *c*-type cytochromes)

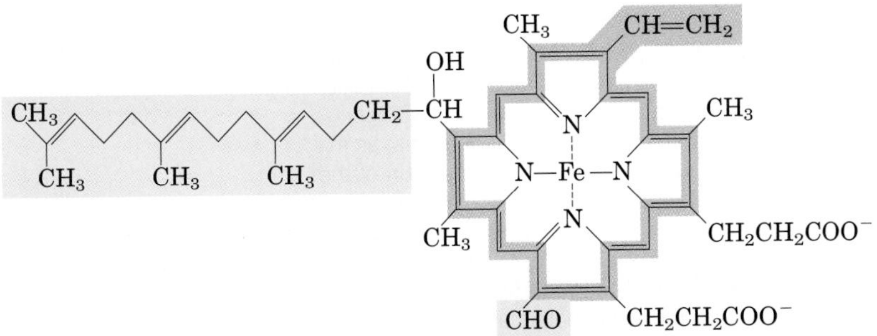

Heme A
(in *a*-type cytochromes)

figure 19–3
Prosthetic groups of cytochromes. Each consists of four five-membered, nitrogen-containing rings in a cyclic structure called a porphyrin. The four nitrogen atoms are coordinated with a central Fe ion, either Fe^{2+} or Fe^{3+}. Iron protoporphyrin IX is found in *b*-type cytochromes and in hemoglobin and myoglobin (see Fig. 6–17). Heme C is covalently bound to the protein of cytochrome *c* through thioether bonds to two Cys residues. Heme A, found in the *a*-type cytochromes, has a long isoprenoid tail attached to one of the five-membered rings. The conjugated double-bond system (shaded in red) of the porphyrin ring accounts for the absorption of visible light by these hemes.

The **cytochromes** are proteins with characteristic strong absorption of visible light, due to their iron-containing heme prosthetic groups (Fig. 19–3). Mitochondria contain three classes of cytochromes, designated *a*, *b*, and *c*, distinguished by differences in their light-absorption spectra. Each type of cytochrome in its reduced (Fe^{2+}) state has three absorption bands in the visible range (Fig. 19–4). The longest-wavelength band is near 600 nm in type *a* cytochromes, near 560 nm in type *b*, and near 550 nm in type *c*. To distinguish among closely related cytochromes of one type, the exact absorption maximum is sometimes used in the names, as in cytochrome b_{562}.

The heme cofactors of *a* and *b* cytochromes are tightly, but not covalently, bound to their associated proteins; the hemes of *c*-type cytochromes are covalently attached through Cys residues (Fig. 19–3). As with the flavoproteins, the standard reduction potential of the heme iron atom of a cytochrome depends on its interaction with protein side chains and is therefore different for each cytochrome. The cytochromes of type *a* and *b* and some of type *c* are integral proteins of the inner mitochondrial membrane. One striking exception is cytochrome *c* of mitochondria, a soluble protein that associates through electrostatic interactions with the outer surface of the inner membrane. We encountered cytochrome *c* in earlier discussions of protein evolution (see Box 5–2) and protein structure (see Fig. 6–18a).

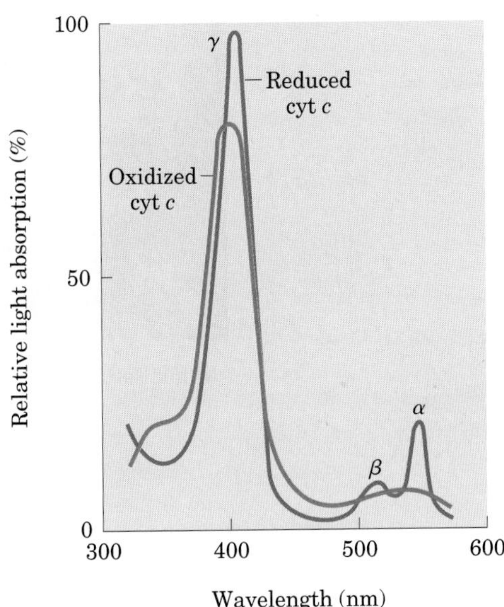

figure 19–4
Absorption spectra of cytochrome *c* in its oxidized (red) and reduced (blue) forms. Also labeled are the characteristic α, β, and γ bands of the reduced form.

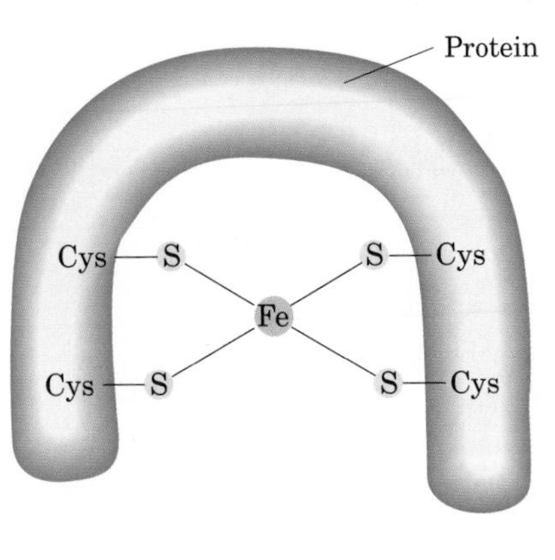

(a)

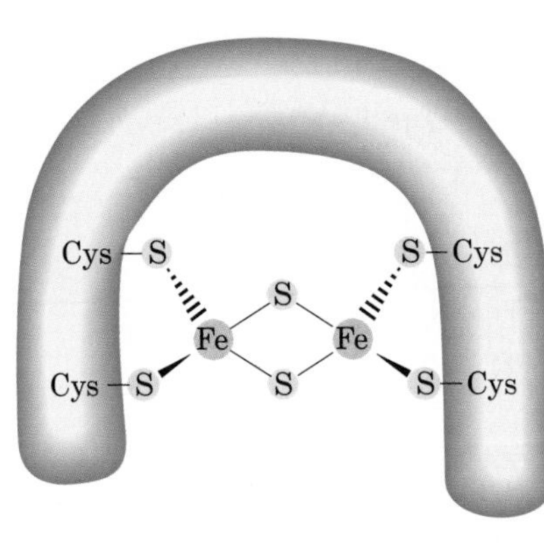

(b)

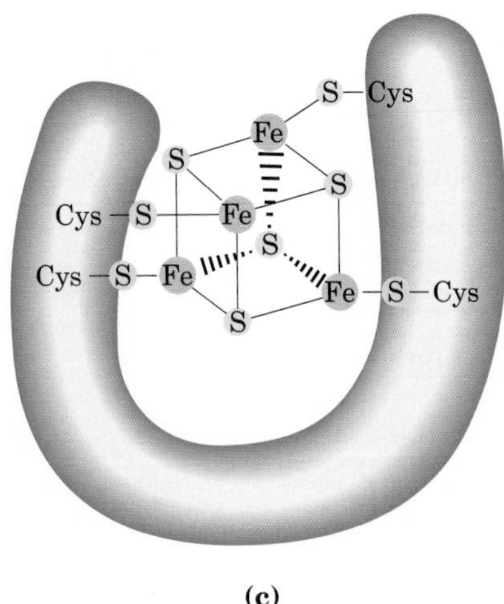

(c)

(d)

figure 19–5

Iron-sulfur centers. The Fe-S centers of iron-sulfur proteins may be as simple as **(a),** with a single Fe ion surrounded by the S atoms of four Cys residues. Other centers include both inorganic and Cys S atoms, as in 2Fe-2S **(b)** or 4Fe-4S **(c)** centers. The ferredoxin of the cyanobacterium *Anabaena* 7120 **(d)** has one 2Fe-2S center. (Note that only the inorganic S atoms are counted in these designations. For example, in the 2Fe-2S center **(b),** each Fe ion is actually surrounded by four S atoms.) The exact standard reduction potential of the iron in these centers depends on the type of center and its interaction with the associated protein.

Helmut Beinert

In **iron-sulfur proteins,** first discovered by Helmut Beinert, the iron is present not in heme but in association with inorganic sulfur atoms or with the sulfur atoms of Cys residues in the protein, or both. These iron-sulfur (Fe-S) centers range from simple structures with a single Fe atom coordinated to four Cys —SH groups to more complex Fe-S centers with two or four Fe atoms (Fig. 19–5). **Rieske iron-sulfur proteins** (named after their discoverer) are a variation on this theme, in which one Fe atom is coordinated to two His residues rather than two Cys residues. All iron-sulfur proteins participate in one-electron transfers in which one iron atom of the iron-sulfur cluster is oxidized or reduced. At least eight Fe-S proteins function in mitochondrial electron transfer. The reduction potential of Fe-S proteins varies from -0.65 V to $+0.45$ V, depending on the microenvironment of the iron within the protein.

In the overall reaction catalyzed by the mitochondrial respiratory chain, electrons move from NADH, succinate, or some other primary electron donor through flavoproteins, ubiquinone, iron-sulfur proteins, and cytochromes, and finally to O_2. A look at the methods used to determine the sequence in which the carriers act is instructive, as the same general approaches have been used to study other electron-transfer chains, such as those of chloroplasts.

First, the standard reduction potentials of the individual electron carriers have been determined experimentally (Table 19–2). One expects the carriers to function in order of increasing reduction potential, because electrons tend to flow spontaneously from carriers of lower $E'°$ to carriers of higher $E'°$. The order of carriers deduced by this method is NADH $\rightarrow$ Q $\rightarrow$ cytochrome $b \rightarrow$ cytochrome $c_1 \rightarrow$ cytochrome $c \rightarrow$ cytochrome $a \rightarrow$ cytochrome $a_3 \rightarrow O_2$. Note, however, that the order of standard reduction

table 19–2

Standard Reduction Potentials of Respiratory Chain and Related Electron Carriers

Redox reaction (half-reaction)	E'° (V)
$2H^+ + 2e^- \longrightarrow H_2$	−0.414
$NAD^+ + H^+ + 2e^- \longrightarrow NADH$	−0.320
$NADP^+ + H^+ + 2e^- \longrightarrow NADPH$	−0.324
NADH dehydrogenase (FMN) + $2H^+ + 2e^- \longrightarrow$ NADH dehydrogenase ($FMNH_2$)	−0.30
Ubiquinone + $2H^+ + 2e^- \longrightarrow$ ubiquinol	0.045
Cytochrome *b* (Fe^{3+}) + $e^- \longrightarrow$ cytochrome *b* (Fe^{2+})	0.077
Cytochrome c_1 (Fe^{3+}) + $e^- \longrightarrow$ cytochrome c_1 (Fe^{2+})	0.22
Cytochrome *c* (Fe^{3+}) + $e^- \longrightarrow$ cytochrome *c* (Fe^{2+})	0.254
Cytochrome *a* (Fe^{3+}) + $e^- \longrightarrow$ cytochrome *a* (Fe^{2+})	0.29
Cytochrome a_3 (Fe^{3+}) + $e^- \longrightarrow$ cytochrome a_3 (Fe^{2+})	0.55
$\frac{1}{2}O_2 + 2H^+ + 2e^- \longrightarrow H_2O$	0.816

potentials is not necessarily the same as the order of actual reduction potentials under cellular conditions, which depend on the concentration of reduced and oxidized forms (p. 517). A second method for determining the sequence of electron carriers involves reducing the entire chain of carriers experimentally by providing an electron source but no electron acceptor (no O_2). When O_2 is suddenly introduced into the system, the rate at which each electron carrier becomes oxidized (measured spectroscopically) shows the order in which the carriers function. The carrier nearest O_2 (at the end of the chain) gives up its electrons first, the second carrier from the end is oxidized next, and so on. Such experiments have confirmed the sequence deduced from standard reduction potentials.

In a final confirmation, agents that inhibit the flow of electrons through the chain have been used in combination with measurement of the degree of oxidation of each carrier. In the presence of O_2 and an electron donor, carriers that function before the inhibited step become fully reduced, and those that function after the block are completely oxidized (Fig. 19–6). By using several inhibitors that block different steps in the chain, the entire sequence has been deduced; it is the same as predicted from the first two approaches.

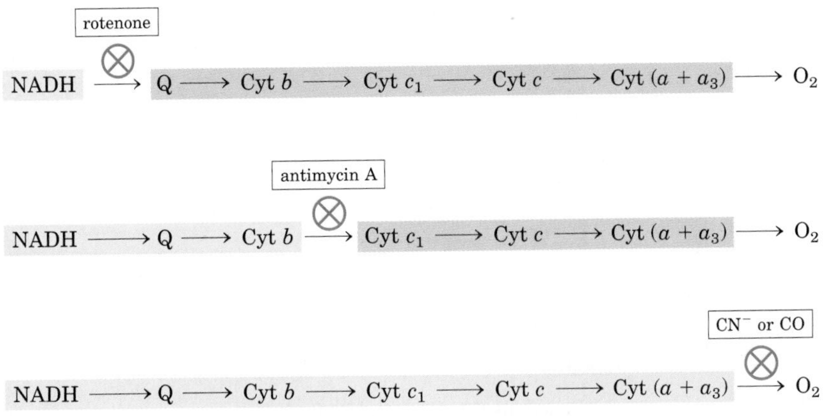

rotenone

NADH ⊗ ⟶ Q ⟶ Cyt *b* ⟶ Cyt c_1 ⟶ Cyt *c* ⟶ Cyt ($a + a_3$) ⟶ O_2

antimycin A

NADH ⟶ Q ⟶ Cyt *b* ⊗ ⟶ Cyt c_1 ⟶ Cyt *c* ⟶ Cyt ($a + a_3$) ⟶ O_2

CN^- or CO

NADH ⟶ Q ⟶ Cyt *b* ⟶ Cyt c_1 ⟶ Cyt *c* ⟶ Cyt ($a + a_3$) ⊗ ⟶ O_2

figure 19–6
Method for determining the sequence of electron carriers. This method measures the effects of inhibitors of electron transfer on the oxidation state of each carrier. In the presence of an electron donor and O_2, each inhibitor causes a characteristic pattern of oxidized/reduced carriers: those before the block become reduced (blue), and those after the block become oxidized (red).

Electron Carriers Function in Multienzyme Complexes

The electron carriers of the respiratory chain are organized into membrane-embedded supramolecular complexes that can be physically separated. Gentle treatment of the inner mitochondrial membrane with detergents allows the resolution of four unique electron-carrier complexes, each capable of catalyzing electron transfer through a portion of the chain (Fig. 19–7; Table 19–3). Complexes I and II catalyze electron transfer to ubiquinone from two different electron donors: NADH (Complex I) and succinate (Complex II). Complex III carries electrons from ubiquinone to cytochrome *c*, and Complex IV completes the sequence by transferring electrons from cytochrome *c* to O_2.

We now look in more detail at the structure and function of each complex of the mitochondrial respiratory chain.

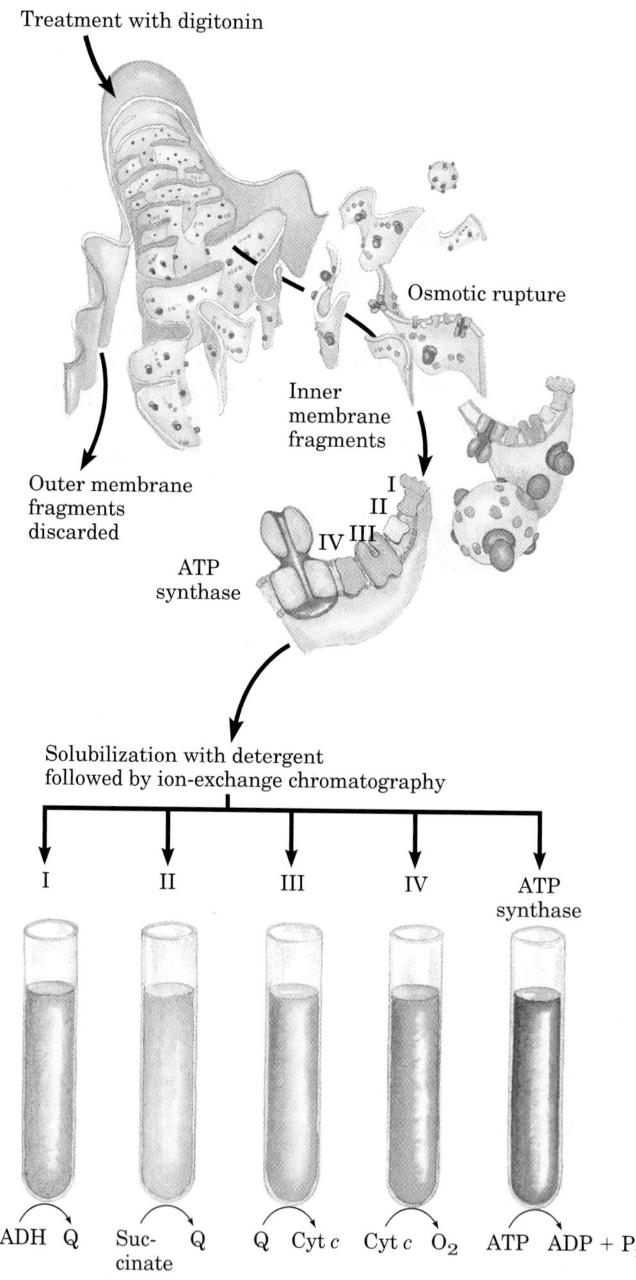

figure 19–7
Separation of functional complexes of the respiratory chain. The outer mitochondrial membrane is first removed by treatment with the detergent digitonin. Fragments of inner membrane are then obtained by osmotic rupture of the mitochondria, and the fragments are gently dissolved in a second detergent. The resulting mixture of inner membrane proteins is resolved by ion-exchange chromatography into different complexes (I through IV) of the respiratory chain, each with its unique protein composition (see Table 19–3), and the enzyme ATP synthase (sometimes called Complex V). The isolated Complexes I through IV catalyze transfers between donors (NADH and succinate), intermediate carriers (Q and cytochrome *c*), and O_2, as shown. In vitro, ATP synthase has only ATP-hydrolyzing (ATPase), not ATP-synthesizing, activity.

table 19–3

Protein Components of the Mitochondrial Electron-Transfer Chain

Enzyme complex	Mass (kDa)	Number of subunits*	Prosthetic group(s)
I NADH dehydrogenase	850	42 (14)	FMN, Fe-S
II Succinate dehydrogenase	140	5	FAD, Fe-S
III Ubiquinone: cytochrome *c* oxidoreductase	250	11	Hemes, Fe-S
Cytochrome $c^{\dagger}$	13	1	Heme
IV Cytochrome oxidase	160	13 (3–4)	Hemes; Cu_A, Cu_B

*Numbers of subunits in the bacterial equivalents in parentheses.

†Cytochrome *c* is not part of an enzyme complex; it moves between Complexes III and IV as a freely soluble protein.

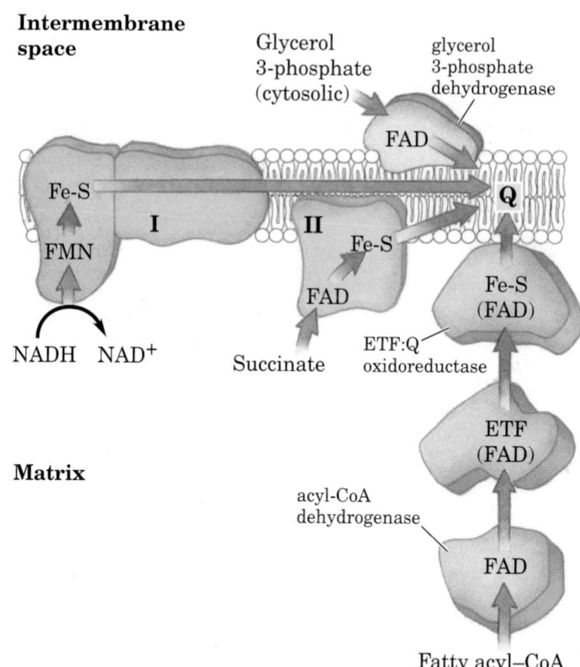

figure 19–8

Path of electrons from NADH, succinate, fatty acyl–CoA, and glycerol 3-phosphate to ubiquinone. Electrons from NADH pass through a flavoprotein to a series of iron-sulfur proteins (in Complex I) and then to Q. Electrons from succinate pass through a flavoprotein and several Fe-S centers (in Complex II) on the way to Q. Glycerol 3-phosphate donates electrons to a flavoprotein (glycerol 3-phosphate dehydrogenase) on the outer face of the inner mitochondrial membrane, from which they pass to Q. Acyl-CoA dehydrogenase (the first enzyme of β oxidation) transfers electrons to electron-transferring flavoprotein (ETF), from which they pass via ETF: ubiquinone oxidoreductase to Q.

Complex I: NADH to Ubiquinone Figure 19–8 illustrates the relationship between Complexes I and II and ubiquinone. **Complex I,** also called **NADH:ubiquinone oxidoreductase,** is a large enzyme composed of 42 different polypeptide chains, including an FMN-containing flavoprotein and at least six iron-sulfur centers (Table 19–3). High-resolution electron microscopy shows Complex I to be L-shaped, with one arm of the L in the membrane and the other extending into the matrix. As shown in Figure 19–9, Complex I catalyzes two simultaneous and obligately coupled processes: (1) the exergonic transfer to ubiquinone of a hydride ion from NADH and a proton from the matrix:

$$NADH + H^+ + Q \longrightarrow NAD^+ + QH_2 \quad (19\text{–}1)$$

and (2) the endergonic transfer of four protons from the matrix to the intermembrane space. Complex I is therefore a proton pump driven by the energy of electron transfer, and the reaction it catalyzes is **vectorial:** it moves protons in a specific direction from one location (the matrix, which becomes negatively charged with the departure of protons) to another (the intermembrane space, which becomes positively charged). To emphasize

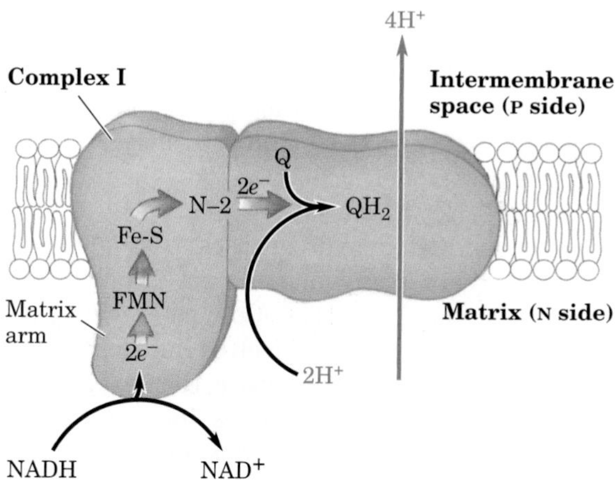

figure 19–9

NADH:ubiquinone oxidoreductase (Complex I). Complex I catalyzes the transfer of a hydride ion from NADH to FMN, from which two electrons pass through a series of Fe-S centers to the iron-sulfur protein N-2 in the matrix arm of the complex. Electron transfer from N-2 to ubiquinone on the membrane arm forms QH_2, which diffuses into the lipid bilayer. It also drives the expulsion from the matrix of four protons per pair of electrons. The detailed mechanism that couples electron and proton transfer in Complex I is not yet known, but probably involves a Q cycle similar to that in Complex III in which QH_2 participates twice per electron pair (see Fig. 19–11). This proton flux produces an electrochemical potential across the inner mitochondrial membrane (N side negative, P side positive), which conserves some of the energy released by the electron-transfer reactions.This electrochemical potential drives ATP synthesis.

table 19–4

Some Agents That Interfere with Oxidative Phosphorylation or Photophosphorylation

Type of interference	Compound*	Target/mode of action
Inhibition of electron transfer	Cyanide Carbon monoxide	Inhibit cytochrome oxidase
	Antimycin A	Blocks electron transfer from cytochrome *b* to cytochrome c_1
	Myxothiazol Rotenone Amytal Piericidin A	Prevent electron transfer from Fe-S center to ubiquinone
	DCMU	Competes with Q_B for binding site in PSII
Inhibition of ATP synthase	Aurovertin	Inhibits F_1
	Oligomycin Venturicidin	Inhibit F_o and CF_o
	DCCD	Blocks proton flow through F_o and CF_o
Uncoupling of phosphorylation from electron transfer	FCCP DNP	Hydrophobic proton carriers
	Valinomycin	K^+ ionophore
	Thermogenin	Forms proton-conducting pores in inner membrane of brown fat mitochondria
Inhibition of ATP-ADP exchange	Atractyloside	Inhibits adenine nucleotide translocase

*DCMU is 3-(3,4-dichlorophenyl)-1,1-dimethylurea; DCCD, dicyclohexylcarbodiimide; FCCP, cyanide-*p*-trifluoromethoxyphenylhydrazone; DNP, 2,4-dinitrophenol.

the vectorial nature of the process, the overall reaction is often written with subscripts that indicate the locations of the protons: P for the positive side of the inner membrane (the intermembrane space), N for the negative side (the matrix):

$$\text{NADH} + 5\text{H}^+_\text{N} + \text{Q} \longrightarrow \text{NAD}^+ + \text{QH}_2 + 4\text{H}^+_\text{P} \qquad (19\text{–}2)$$

Amytal (a barbiturate drug), rotenone (a plant product commonly used as an insecticide), and piericidin A (an antibiotic) inhibit electron flow from the Fe-S centers of Complex I to ubiquinone (Table 19–4) and therefore block the overall process of oxidative phosphorylation.

Ubiquinol (QH_2, the fully reduced form; Fig. 19–2) diffuses in the mitochondrial inner membrane from Complex I to Complex III, where it is oxidized to Q in a process that also involves outward movement of H^+.

Complex II: Succinate to Ubiquinone We encountered **Complex II** in Chapter 16 as **succinate dehydrogenase,** the only membrane-bound enzyme in the citric acid cycle (p. 578). Although smaller and simpler than Complex I, it contains two types of prosthetic groups and at least four different proteins (Table 19–3). One protein has a covalently bound FAD and an Fe-S center with four Fe atoms; a second iron-sulfur protein is also present (see frontispiece). Electrons pass from succinate to FAD, then through the Fe-S centers to ubiquinone (Fig. 19–8).

Other substrates for mitochondrial dehydrogenases pass electrons into the respiratory chain at the level of ubiquinone, but not through Complex II. The first step in the β oxidation of fatty acyl–CoA, catalyzed by the flavoprotein **acyl-CoA dehydrogenase** (p. 604), involves transfer of electrons from the substrate to the FAD of the dehydrogenase, then to electron-transferring flavoprotein (ETF), which in turn passes its electrons to **ETF:ubiquinone oxidoreductase** (Fig. 19–8). This enzyme passes electrons into the respiratory chain by reducing ubiquinone. Glycerol 3-phosphate, formed either from glycerol released by triacylglycerol breakdown or by the reduction of dihydroxyacetone phosphate from glycolysis, is oxidized by **glycerol 3-phosphate dehydrogenase** (see Fig. 17–4). This enzyme is a flavoprotein located on the outer face of the inner mitochondrial membrane, and like succinate dehydrogenase and acyl-CoA dehydrogenase it channels electrons into the respiratory chain by reducing ubiquinone (Fig. 19–8). The important role of glycerol 3-phosphate dehydrogenase in shuttling reducing equivalents from cytosolic NADH into the mitochondrial matrix is described later (see Fig. 19–27). The effect of each of these electron-transferring enzymes is to contribute to the pool of reduced ubiquinone. QH_2 from all these reactions is reoxidized by Complex III, the next component in the mitochondrial electron-transfer chain.

Complex III: Ubiquinone to Cytochrome *c* The next respiratory complex, **Complex III**—also called **cytochrome *bc*$_1$ complex** or **ubiquinone:cytochrome *c* oxidoreductase**—couples the transfer of electrons from ubiquinol (QH_2) to cytochrome *c* with the vectorial transport of protons from the matrix to the intermembrane space. The determinations of the complete structure of this huge complex (Fig. 19–10) and of Complex IV (below) by x-ray crystallography in 1995–1998 were landmarks in the study of mitochondrial electron transfer, providing the structural framework to integrate the many biochemical observations on the function of the complexes.

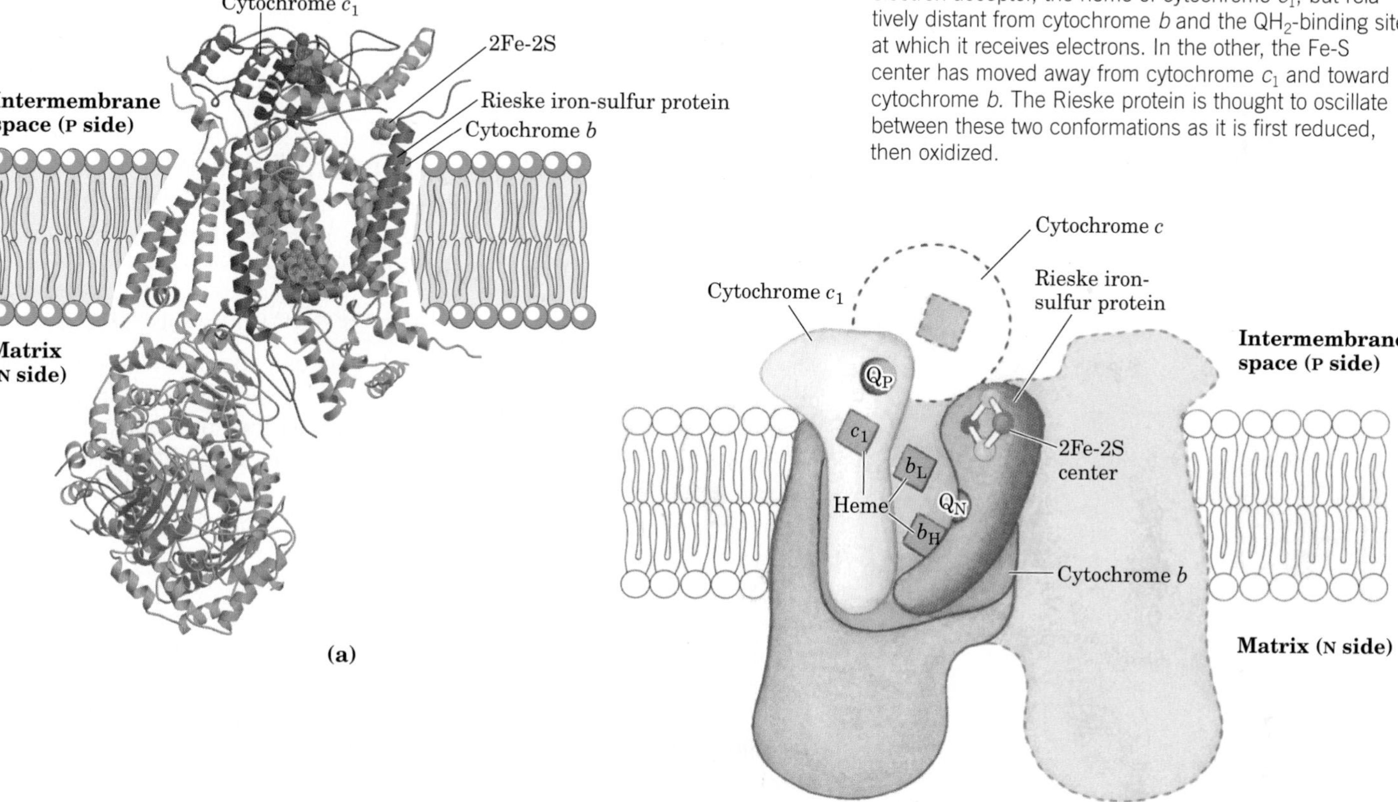

figure 19–10

Cytochrome *bc*$_1$ complex (Complex III). The complex is a dimer of identical monomers, each with 11 different subunits. **(a)** Structure of a monomer. The functional core is three subunits: cytochrome *b* (green) with its two hemes (b_H and b_L; light red), the Rieske iron-sulfur protein (purple) with its 2Fe-2S center (yellow); and cytochrome c_1 (blue) with its heme (red). **(b)** The dimeric functional unit. Cytochrome c_1 and the Rieske iron-sulfur protein project from the P surface, and can interact with cytochrome *c* (shown here, but not part of the functional complex) in the intermembrane space. The complex has two distinct binding sites for ubiquinone, Q_N and Q_P, which correspond to the sites of inhibition by two drugs that block oxidative phosphorylation. Antimycin A, which blocks electron flow from heme b_H to Q, binds at Q_N, close to heme b_H on the N (matrix) side of the membrane. Myxothiazol, which prevents electron flow from QH_2 to the Rieske iron-sulfur protein, binds at Q_P, near the 2Fe-2S center and heme b_L on the P side of the membrane. The dimeric structure is essential to the function of Complex III. The interface between monomers forms two pockets, each containing a Q_P site from one monomer and a Q_N site from the other. The movement of ubiquinone intermediates occurs within these sheltered pockets.

Complex III crystallizes in two distinct conformations (not shown). In one, the Rieske Fe-S center is close to its electron acceptor, the heme of cytochrome c_1, but relatively distant from cytochrome *b* and the QH_2-binding site at which it receives electrons. In the other, the Fe-S center has moved away from cytochrome c_1 and toward cytochrome *b*. The Rieske protein is thought to oscillate between these two conformations as it is first reduced, then oxidized.

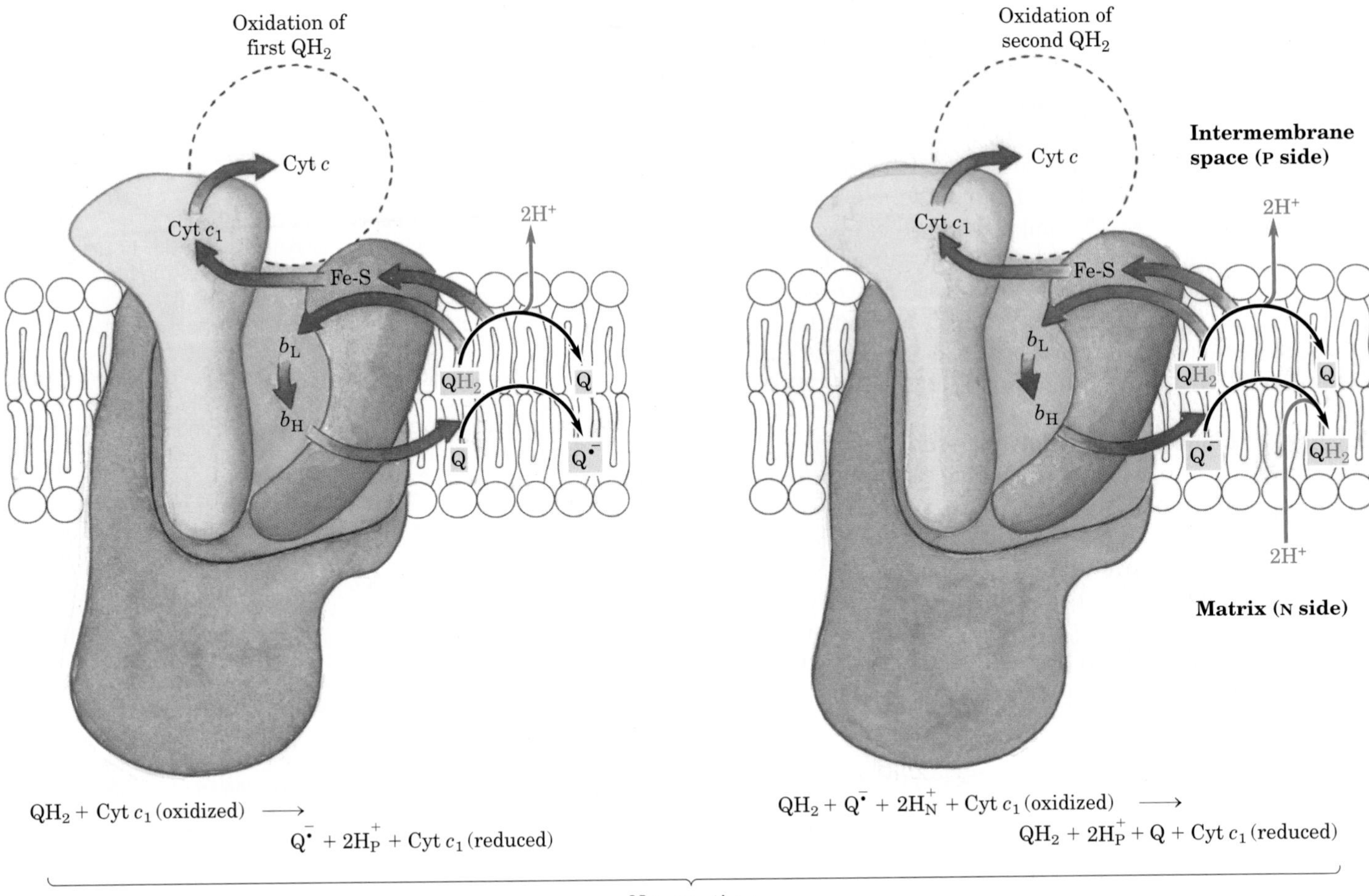

figure 19–11
The Q cycle. The path of electrons through Complex III is shown by blue arrows. On the P side of the membrane, two molecules of QH_2 are oxidized to Q at the Q_P site, releasing four protons into the intermembrane space. Each QH_2 donates one electron (via the Rieske Fe-S center) to cytochrome c_1, and one electron (via cytochrome *b*) to a molecule of Q at the Q_N site, reducing it in two steps to QH_2. This reduction also uses two protons taken up from the matrix.

Based on the structure of Complex III and detailed biochemical studies of the redox reactions, a reasonable model has been proposed for the passage of electrons and protons through the complex. The net equation for the redox reactions of this **Q cycle** (Fig. 19–11) is

$$QH_2 + 2\ \text{cyt}\ c_{1(\text{oxidized})} + 2H^+_N \longrightarrow Q + 2\ \text{cyt}\ c_{1(\text{reduced})} + 4H^+_P \qquad (19\text{–}3)$$

The Q cycle accommodates the switch between the two-electron carrier ubiquinone and the one-electron carriers—cytochromes b_{562}, b_{566}, c_1, and *c*—and explains the measured stoichiometry of four protons translocated per pair of electrons passing through the complex to cytochrome *c*. Although the path of electrons through this segment of the respiratory chain is complicated, the net effect of the transfer is simple: QH_2 is oxidized to Q and two molecules of cytochrome *c* are reduced.

Cytochrome *c* (see Fig. 6–18) is a soluble protein of the intermembrane space. After its single heme accepts an electron from Complex III, cytochrome *c* moves to Complex IV to donate the electron to a binuclear copper center in that enzyme.

Complex IV: Cytochrome *c* to O_2 In the final step of the respiratory chain, **Complex IV,** also called **cytochrome oxidase,** carries electrons from cytochrome *c* to molecular oxygen, reducing it to H_2O. Complex IV is a large enzyme (13 subunits; M_r 204,000) of the inner mitochondrial membrane. Bacteria contain a much simpler form, with only three or four subunits, but

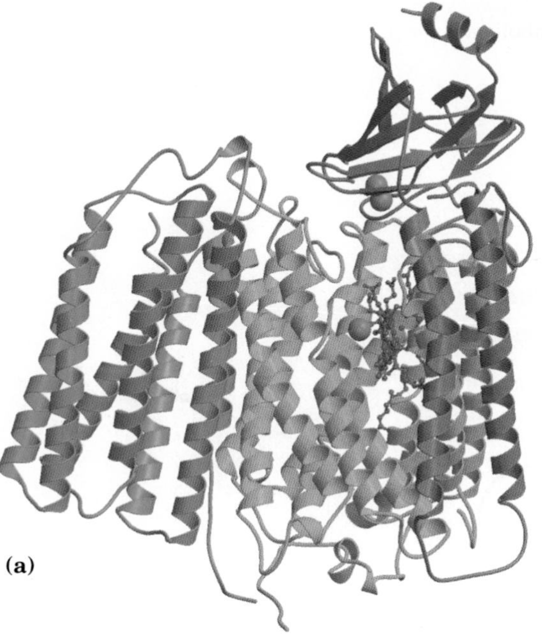

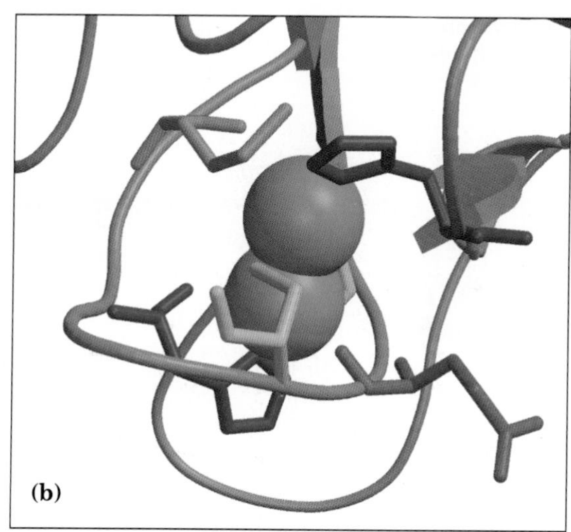

figure 19–12

Critical subunits of cytochrome oxidase (Complex IV). The bovine complex is shown here. **(a)** The core of Complex IV has three subunits. Subunit I (yellow) has two heme groups, *a* and a_3 (red), and a copper ion, Cu_B (green sphere). Heme a_3 and Cu_B form a binuclear Fe-Cu center. Subunit II (blue) contains two Cu ions (green spheres) complexed with the —SH groups of two Cys residues in a binuclear center, Cu_A, that resembles the 2Fe-2S centers of iron-sulfur proteins. This binuclear center and the cytochrome *c*–binding site are located in a domain of subunit II that protrudes from the P side of the inner membrane (into the intermembrane space). Subunit III (light green) is apparently essential for Complex IV function, but its role is not well understood. **(b)** The binuclear center of Cu_A. The Cu ions (green spheres) share electrons equally. When the center is reduced they have the formal charges $Cu^{1+}Cu^{1+}$; when oxidized, $Cu^{1.5+}Cu^{1.5+}$. Ligands around the Cu ions include two His (dark blue), two Cys (yellow), an Asp (red), and Met (orange).

still capable of catalyzing both electron transfer and proton pumping. Comparison of the mitochondrial and bacterial complexes suggests that three subunits are critical to the function (Fig. 19–12).

Mitochondrial subunit II contains two Cu ions complexed with the —SH groups of two Cys residues in a binuclear center (called Cu_A)(Fig. 19–12b) that resembles the 2Fe-2S centers of iron-sulfur proteins. Subunit I contains two heme groups, designated a and a_3, and another copper ion (Cu_B). Heme a_3 and Cu_B form a second binuclear center that accepts electrons from heme a and transfers them to O_2 bound to heme a_3.

Electron transfer through Complex IV is from cytochrome c to the Cu_A center, to heme a, to the heme a_3–Cu_B center, and finally to O_2 (Fig. 19–13). For every four electrons passing through this complex, the enzyme consumes four "substrate" H^+ from the matrix (N side) in converting O_2 to 2 H_2O. It also uses the energy of this redox reaction to pump one proton outward into the intermembrane space (P side) for each electron that passes through, adding to the electrochemical potential produced by redox-driven proton transport through Complexes I and III. The overall reaction catalyzed by Complex IV is

$$4 \text{ cyt } c_{(\text{reduced})} + 8\text{H}^+_\text{N} + \text{O}_2 \longrightarrow 4 \text{ cyt } c_{(\text{oxidized})} + 4\text{H}^+_\text{P} + 2\text{H}_2\text{O} \qquad (19\text{–}4)$$

This four-electron reduction of O_2 involves redox centers that carry only one electron at a time, and it must occur without the release of incompletely reduced intermediates such as hydrogen peroxide or hydroxyl free radicals—very reactive species that would damage cellular components. The intermediates remain tightly bound to the complex until completely converted to water.

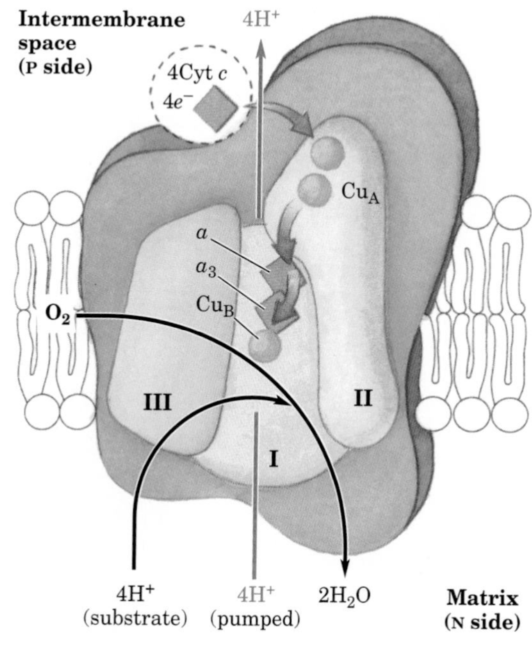

figure 19–13

Path of electrons through Complex IV. The three proteins critical to electron flow are I, II, and III. The lighter outline includes the other ten proteins in the complex. Electron transfer through Complex IV begins when two molecules of reduced cytochrome *c* each donate an electron to the binuclear center Cu_A. From here electrons pass through heme *a* to the Fe-Cu center (cytochrome a_3 and Cu_B). Oxygen now binds to heme a_3 and is reduced to its peroxy derivative (O_2^{2-}) by two electrons from the Fe-Cu center. Delivery of two more electrons from cytochrome *c* converts the O_2^{2-} to two molecules of water, consuming four "substrate" protons from the matrix. At the same time, four more protons are pumped from the matrix by an as yet unknown mechanism.

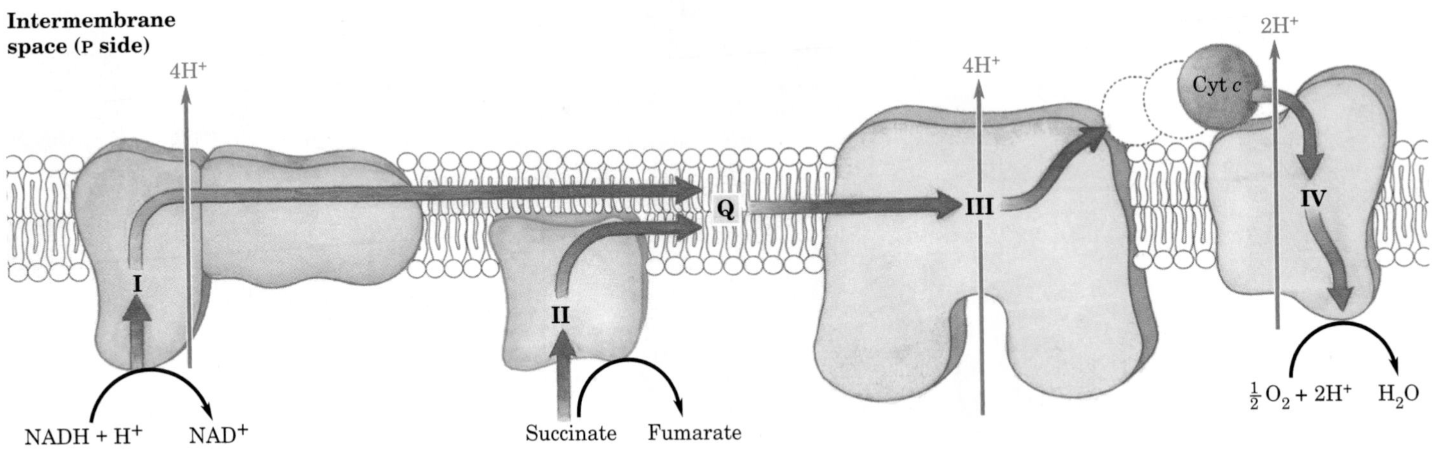

figure 19–14

Summary of the flow of electrons and protons through the four complexes of the respiratory chain. Electrons reach Q through Complexes I and II. QH_2 serves as a mobile carrier of electrons and protons. It passes electrons to Complex III, which passes them to another mobile connecting link, cytochrome *c*. Complex IV then transfers electrons from reduced cytochrome *c* to O_2. Electron flow through Complexes I, III, and IV is accompanied by proton flow from the matrix to the intermembrane space. Recall that electrons from β oxidation of fatty acids can also enter the respiratory chain through Q (see Fig. 19–8).

The Energy of Electron Transfer Is Efficiently Conserved in a Proton Gradient

The transfer of two electrons from NADH through the respiratory chain to molecular oxygen can be written as

$$\text{NADH} + \text{H}^+ + \tfrac{1}{2}\text{O}_2 \longrightarrow \text{NAD}^+ + \text{H}_2\text{O} \qquad (19\text{–}5)$$

This net reaction is highly exergonic. For the redox pair $NAD^+/NADH$, E'° is -0.320 V, and for the pair O_2/H_2O, E'° is 0.816 V. The $\Delta E'^\circ$ for this reaction is therefore 1.14 V, and the standard free-energy change (Eqn 14–6, p. 517) is

$$\begin{aligned}\Delta G'^\circ &= -n\mathcal{F}\Delta E'^\circ \\ &= -2(96.5\ \text{kJ/V}\cdot\text{mol})(1.14\ \text{V}) \\ &= -220\ \text{kJ/mol (of NADH)}\end{aligned} \qquad (19\text{–}6)$$

This *standard* free-energy change is based on the assumption of equal concentrations (1 M) of NADH and NAD^+. In actively respiring mitochondria, the actions of many dehydrogenases keep the actual $NADH/NAD^+$ ratio well above unity, and the real free-energy change for the reaction shown in Equation 19–5 is therefore substantially greater (more negative) than -220 kJ/mol. A similar calculation for the oxidation of succinate shows that electron transfer from succinate (E'° for fumarate/succinate = 0.031 V) to O_2 has a smaller, but still negative, standard free-energy change of about -150 kJ/mol.

Much of this energy is used to pump protons out of the matrix. For each pair of electrons transferred to O_2, four protons are pumped out by Complex I, four by Complex III, and two by Complex IV (Fig. 19–14). The *vectorial* equation for the process is therefore

$$\text{NADH} + 11\text{H}^+_\text{N} + \tfrac{1}{2}\text{O}_2 \longrightarrow \text{NAD}^+ + 10\text{H}^+_\text{P} + \text{H}_2\text{O} \qquad (19\text{–}7)$$

The electrochemical energy inherent in this difference in proton concentration and separation of charge represents a temporary conservation of much of the energy of electron transfer. The energy stored in such a gradient, termed the **proton-motive force,** has two components: (1) the chemical potential energy due to the difference in concentration of a chemical species (H^+) in the two regions separated by the membrane, and (2) the electrical potential energy that results from the separation of charge when a proton moves across the membrane without a counterion (Fig. 19–15).

As we showed in Chapter 12, the free-energy change for the creation of an electrochemical gradient by an ion pump is

$$\Delta G = RT \ln (C_2/C_1) + Z\mathcal{J}\Delta\psi \qquad (19\text{–}8)$$

where C_2/C_1 is the concentration ratio for the ion that moves, Z is the absolute value of its electrical charge (1 for a proton), and $\Delta\psi$ is the transmembrane difference in electrical potential, measured in volts.

For protons at 25 °C,

$$\ln (C_2/C_1) = 2.3(\log [H^+]_P - \log [H^+]_N) = 2.3(pH_N - pH_P) = 2.3\ \Delta pH$$

and Equation 19–8 reduces to

$$\begin{aligned}\Delta G &= 2.3RT\ \Delta pH + \mathcal{J}\Delta\psi \\ &= (5.70\ kJ/mol)\Delta pH + (96.5\ kJ/V \cdot mol)\Delta\psi\end{aligned} \qquad (19\text{–}9)$$

In actively respiring mitochondria, the measured $\Delta\psi$ is 0.15 V–0.2 V and the pH of the matrix is about 0.75 units more alkaline than that of the intermembrane space, so the calculated free-energy change for pumping protons outward is about +20 kJ/mol (of H^+), most of which is contributed by the electrical portion of the electrochemical potential. Because the transfer of two electrons from NADH to O_2 is accompanied by the outward pumping of 10 H^+ (Eqn 19–7), roughly 200 kJ of the 220 kJ released by oxidation of a mole of NADH are conserved in the proton gradient.

When protons flow spontaneously *down* their electrochemical gradient, energy is made available to do work. In mitochondria, chloroplasts, and aerobic bacteria, the electrochemical energy in the proton gradient drives the synthesis of ATP from ADP and P_i. We return to the energetics and stoichiometry of ATP synthesis driven by the electrochemical potential in the proton gradient later in this chapter.

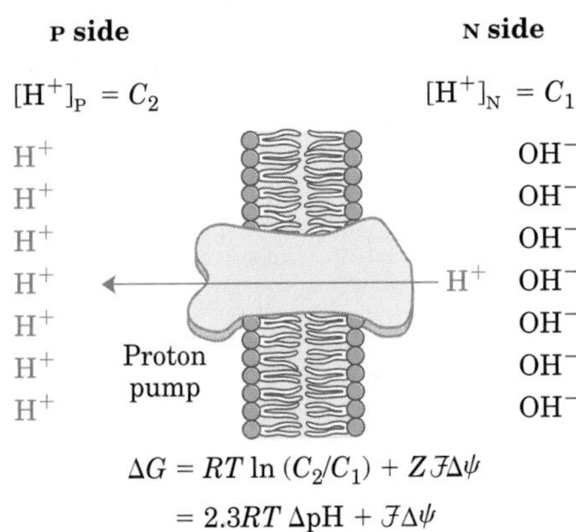

figure 19–15

Proton-motive force. The inner mitochondrial membrane separates two compartments of different $[H^+]$, resulting in differences in chemical concentration (ΔpH) and charge distribution ($\Delta\psi$), across the membrane. The net effect is the proton-motive force (ΔG), which can be calculated as shown here. This is explained more fully in the text.

Plant Mitochondria Have Alternative Mechanisms for Oxidizing NADH

Plant mitochondria supply ATP during periods of low illumination or darkness by mechanisms entirely analogous to those used by nonphotosynthetic organisms. In the light, the principal source of mitochondrial NADH is a reaction in which glycine produced by photorespiration is converted to serine (see Fig. 20–39):

$$2\ \text{Glycine} + NAD^+ \longrightarrow \text{serine} + NADH + H^+ + CO_2 + NH_4^+$$

For reasons discussed in Chapter 20, plants must carry out this reaction even when they do not need NADH for ATP production. To regenerate NAD^+ from unneeded NADH, plant mitochondria transfer electrons from NADH directly to ubiquinone and from ubiquinone directly to O_2, bypassing Complexes III and IV and their proton pumps. The energy in NADH is dissipated as heat, which can sometimes be of value to a plant (Box 19–1). Unlike cytochrome oxidase (Complex IV), the alternative QH_2 oxidase is not inhibited by cyanide. Cyanide-resistant NADH oxidation is therefore the hallmark of this unique plant electron-transfer pathway.

box 19–1 Alternative Respiratory Pathways and Hot, Stinking Plants

Many flowering plants attract insect pollinators by releasing odorant molecules that mimic an insect's natural food sources or potential egg-laying sites. Plants pollinated by flies or beetles that normally feed on or lay their eggs in dung or carrion sometimes use foul-smelling compounds to attract these insects.

One family of stinking plants is the Araceae, which includes philodendrons, arum lilies, and skunk cabbages. These plants have tiny flowers densely packed on an erect structure called a spadix, surrounded by a modified leaf called a spathe. The spadix releases odors of rotting flesh or dung. Before pollination the spadix also warms, in some species to as much as 20 to 40 °C above the ambient temperature. Heat production (thermogenesis) helps evaporate odorant molecules for better dispersal, and because rotting flesh and dung are usually warm from the hyperactive metabolism of scavenging microbes, the heat itself might also attract insects. In the case of the eastern skunk cabbage (Fig. 1), which flowers in late winter or early spring when snow still covers the ground, thermogenesis allows the spadix to grow up through the snow.

How does a skunk cabbage heat its spadix? Although the mitochondria of plants, fungi, and unicellular eukaryotes have electron-transfer systems that are essentially the same as those in animals, they also have an alternative respiratory pathway. In this pathway, a cyanide-resistant QH_2 oxidase transfers electrons from the ubiquinone pool directly to oxygen, bypassing the two proton-translocating steps of Complexes III and IV (Fig. 2). Energy that might have been conserved as ATP is instead released as heat. Plant mitochondria also have an alternative NADH dehydrogenase, insensitive to the Complex I inhibitor rotenone (see Table 19–4), that transfers electrons from NADH in the matrix directly to ubiquinone, bypassing Complex I and its associated proton pumping. And plant mitochondria have yet another NADH dehydrogenase, on the external face of the inner membrane, that faces the intermembrane space and transfers electrons from NADPH or NADH to ubiquinone, again bypassing Complex I. Thus, when electrons enter the alternative respiratory pathway through the rotenone-insensitive NADH dehydrogenase, the external NADH dehydrogenase, or succinate dehydrogenase (Complex II) and pass to O_2 via the cyanide-resistant alternative oxidase, energy is not conserved as ATP but is released as heat. A skunk cabbage can use the heat to melt snow, produce a foul stench, or attract beetles or flies.

figure 1 Eastern skunk cabbage.

figure 2 Electron carriers of the inner membrane of plant mitochondria. Electrons can flow through Complexes I, III, and IV, as in animal mitochondria, or through plant-specific alternative carriers by the paths shown with blue arrows.

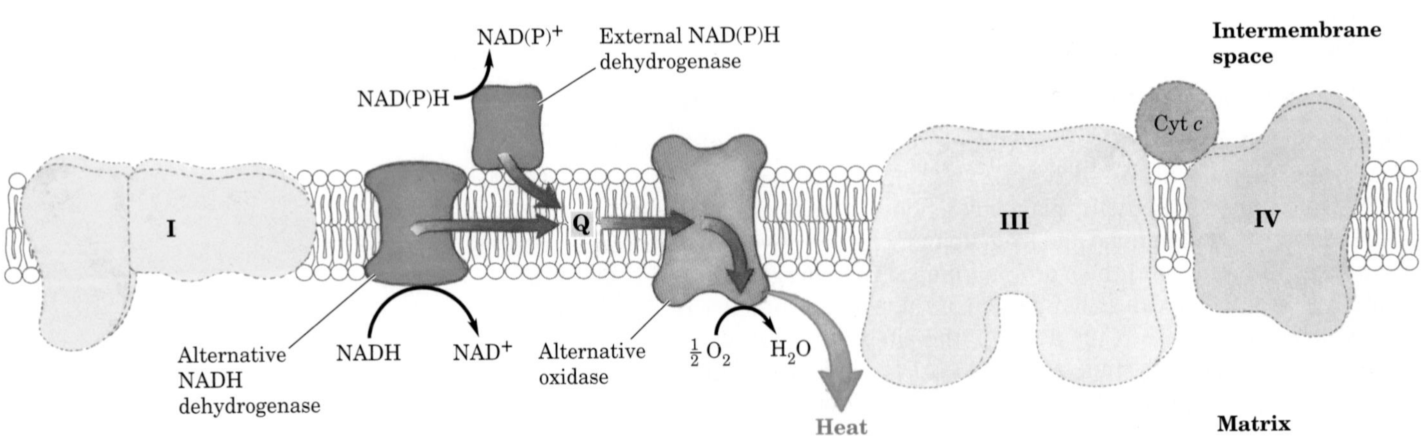

ATP Synthesis

How is a concentration gradient of protons transformed into ATP? We have seen that electron transfer releases, and the proton-motive force conserves, more than enough free energy (about 200 kJ) per "mole" of electron pairs to drive the formation of a mole of ATP, which requires about 50 kJ (see Box 14–2). Mitochondrial oxidative phosphorylation therefore poses no thermodynamic problem. We now consider the chemical mechanism that couples proton flux with phosphorylation.

Peter Mitchell
1920–1992

The **chemiosmotic model** proposed by Peter Mitchell is the paradigm for this mechanism. According to the model (Fig. 19–16), the electrochemical energy inherent in the difference in proton concentration and separation of charge across the mitochondrial inner membrane, the proton-motive force, drives the synthesis of ATP as protons flow passively back into the matrix through a proton pore associated with **ATP synthase.** To emphasize this crucial role of the proton-motive force, the equation for ATP synthesis is sometimes written

$$ADP + P_i + nH^+_P \rightarrow ATP + H_2O + nH^+_N \qquad (19\text{–}10)$$

The operational definition of "coupling" is shown in Figure 19–17. When isolated mitochondria are suspended in a buffer containing ADP, P_i, and an oxidizable substrate such as succinate, three easily measured processes occur: (1) the substrate is oxidized (succinate yields fumarate), (2) O_2 is consumed, and (3) ATP is synthesized. Oxygen consumption and ATP synthesis are dependent on the presence of an oxidizable substrate (succinate in this case) as well as ADP and P_i.

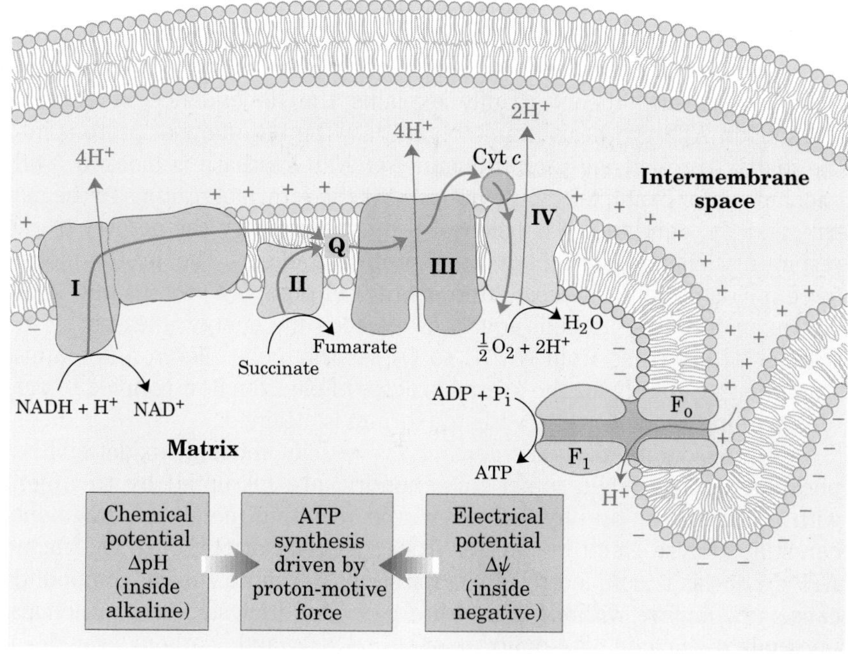

figure 19–16

Chemiosmotic model. In this simple representation of the chemiosmotic theory applied to mitochondria, electrons from NADH and other oxidizable substrates pass through a chain of carriers arranged asymmetrically in the inner membrane. Electron flow is accompanied by proton transfer across the membrane, producing both a chemical gradient (ΔpH) and an electrical gradient (Δψ). The inner mitochondrial membrane is impermeable to protons; protons can reenter the matrix only through proton-specific channels (F_o). The proton-motive force that drives protons back into the matrix provides the energy for ATP synthesis, catalyzed by the F_1 complex associated with F_o.

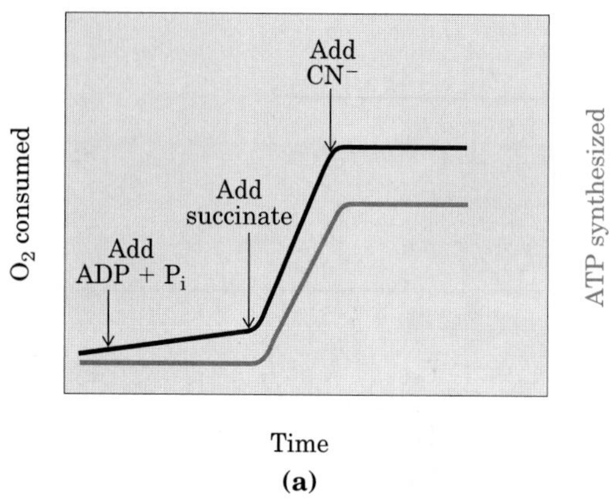

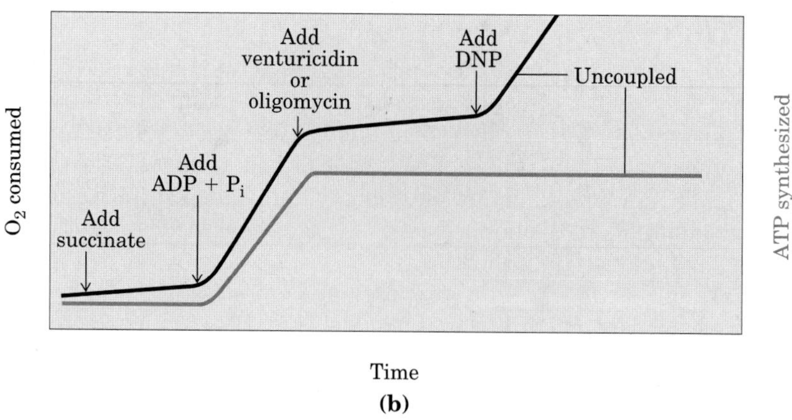

figure 19–17

Coupling of electron transfer and ATP synthesis in mitochondria. In experiments to demonstrate coupling, mitochondria are suspended in a buffered medium and an O_2 electrode is used to monitor O_2 consumption. At intervals, samples are removed and assayed for the presence of ATP. **(a)** Addition of ADP and P_i alone results in little or no increase in either respiration (O_2 consumption; black) or ATP synthesis (red). When succinate is added, respiration begins immediately and ATP is synthesized. Addition of cyanide (CN^-), which blocks electron transfer between cytochrome oxidase and O_2, inhibits both respiration and ATP synthesis. **(b)** Mitochondria provided with succinate respire and synthesize ATP only when ADP and P_i are added. Subsequent addition of venturicidin or oligomycin, inhibitors of ATP synthase, blocks both ATP synthesis and respiration. Dinitrophenol (DNP) is an uncoupler, allowing respiration to continue without ATP synthesis.

Because the energy of substrate oxidation drives ATP synthesis in mitochondria, it is not unexpected that inhibitors of the passage of electrons to O_2 (e.g., cyanide, carbon monoxide, and antimycin A) block ATP synthesis (Fig. 19–17a). More surprising is the finding that the converse is also true: inhibition of ATP synthesis blocks electron transfer in intact mitochondria. This obligatory coupling can be demonstrated in isolated mitochondria by providing O_2 and oxidizable substrates, but not ADP (Fig. 19–17b). Under these conditions, no ATP synthesis can occur and electron transfer to O_2 does not proceed. Coupling of oxidation and phosphorylation can also be demonstrated using oligomycin or venturicidin, toxic antibiotics that bind to the ATP synthase in mitochondria. These compounds are potent inhibitors of both ATP synthesis *and* the transfer of electrons through the chain of carriers to O_2 (Fig. 19–17b). Because oligomycin is known not to interact directly with the electron carriers but only with ATP synthase, it follows that electron transfer and ATP synthesis are obligately coupled; neither reaction occurs without the other.

Chemiosmotic theory readily explains the dependence of electron transfer on ATP synthesis in mitochondria. When the flow of protons into the matrix through the proton channel of ATP synthase is blocked (with oligomycin, for example), no path exists for the return of protons to the matrix, and the continued extrusion of protons driven by the activity of the respiratory chain generates a large proton gradient. The proton-motive force builds up until the cost (free energy) of pumping protons out of the matrix against this gradient equals or exceeds the energy released by the transfer of electrons from NADH to O_2. At this point electron flow must stop; the free energy for the overall process of electron flow coupled to proton pumping becomes zero, and equilibrium is attained.

Certain conditions and reagents, however, can uncouple oxidation from phosphorylation. When intact mitochondria are disrupted by treatment with detergent or by physical shear, the resulting membrane fragments can still catalyze electron transfer from succinate or NADH to O_2, but no ATP synthesis is coupled to this respiration. Certain chemical compounds cause uncoupling without disrupting mitochondrial structure. Chemical uncouplers include 2,4-dinitrophenol (DNP) and carbonylcyanide-*p*-trifluoromethoxyphenylhydrazone (FCCP) (Table 19–4; Fig. 19–18), weak acids with hydrophobic properties. The hydrophobicity of these compounds allows them to diffuse readily across mitochondrial membranes. After entering the mitochondrial matrix in the protonated form, they can release a proton, thus dissipating the proton gradient. Ionophores such as valinomycin (see Fig. 12–37; Table 19–4) allow inorganic ions to pass easily

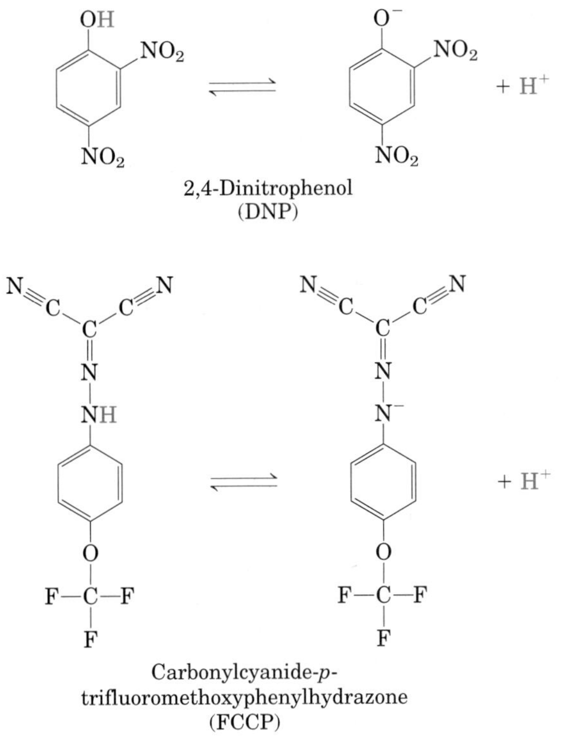

2,4-Dinitrophenol (DNP) ⇌ + H^+

Carbonylcyanide-*p*-trifluoromethoxyphenylhydrazone (FCCP) ⇌ + H^+

figure 19–18
Two chemical uncouplers of oxidative phosphorylation. Both DNP and FCCP have a dissociable proton and are very hydrophobic. They carry protons across the inner mitochondrial membrane, dissipating the proton gradient. Both also uncouple photophosphorylation (p. 709).

through membranes. Ionophores uncouple electron transfer from oxidative phosphorylation by dissipating the electrical contribution to the electrochemical gradient across the mitochondrial membrane.

If the role of electron transfer in mitochondrial ATP synthesis is simply to pump protons to create the electrochemical potential of the proton-motive force, an artificially created proton gradient should be able to replace electron transfer in driving ATP synthesis. This prediction of the chemiosmotic model has been experimentally tested and confirmed (Fig. 19–19). Mitochondria manipulated so as to impose a difference of proton concentration and a separation of charge across the inner membrane synthesize ATP *in the absence of an oxidizable substrate;* the proton-motive force alone suffices to drive ATP synthesis.

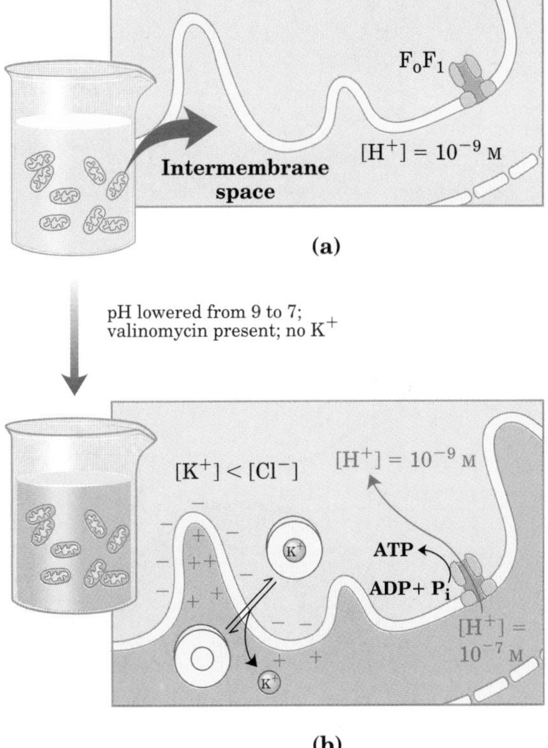

figure 19–19
Evidence for the role of a proton gradient in ATP synthesis. An artificially imposed electrochemical gradient can drive ATP synthesis in the absence of an oxidizable substrate as electron donor. In this two-step experiment, isolated mitochondria are first incubated in a pH 9 buffer containing 0.1 M KCl **(a).** Slow leakage of buffer and KCl into the mitochondria eventually brings the matrix into equilibrium with the surrounding medium. No oxidizable substrates are present. **(b)** Mitochondria are now separated from the pH 9 buffer and resuspended in pH 7 buffer containing valinomycin but no KCl. The change in buffer creates a difference of two pH units across the inner membrane of the mitochondria. The outward flow of K^+, carried down its concentration gradient without a counterion by valinomycin, creates a charge imbalance across the membrane (matrix negative). The sum of the chemical potential provided by the pH difference and the electrical potential provided by the separation of charges is a proton-motive force large enough to support ATP synthesis in the absence of an oxidizable substrate.

Efraim Racker
1913–1991

ATP Synthase Has Two Functional Domains, F_o and F_1

Mitochondrial **ATP synthase** is an F-type ATPase (see Fig. 12–31c; Table 12–4) similar in structure and mechanism to the ATP synthases of chloroplasts and eubacteria. This large enzyme complex of the inner mitochondrial membrane catalyzes the formation of ATP from ADP and P_i, accompanied by the flow of protons from the P to the N side of the membrane (Eqn 19–10). ATP synthase, also called Complex V, has two distinct components: F_1, a peripheral membrane protein, and F_o, which is integral to the membrane. The *o* in F_o stands for oligomycin-sensitive. F_1 was the first factor identified as essential for oxidative phosphorylation. It was identified and purified by Efraim Racker and his colleagues in the early 1960s. In the laboratory, small membrane vesicles formed from inner mitochondrial membranes carry out ATP synthesis coupled to electron transfer. When F_1 is gently extracted from these vesicles, the "stripped" vesicles still contain intact respiratory chains and the F_o portion of ATP synthase. The vesicles can catalyze electron transfer from NADH to O_2 but cannot produce a proton gradient: F_o has a proton pore through which protons leak as fast as they are pumped by electron transfer, and without a proton gradient the F_1-depleted vesicles cannot make ATP. Isolated F_1 catalyzes ATP hydrolysis (the reversal of synthesis) and was therefore originally called **F_1ATPase.** When purified F_1 is added back to the depleted vesicles, it reassociates with F_o, plugging its proton pore and restoring the membrane's capacity to couple electron transfer and ATP synthesis.

ATP Is Stabilized Relative to ADP on the Surface of F_1

Isotope exchange experiments with purified F_1 reveal a remarkable fact about the enzyme's catalytic mechanism: on the enzyme's surface, the reaction ADP + P_i $\rightleftharpoons$ ATP + H_2O is readily reversible—the free-energy change for ATP synthesis is close to zero! When ATP is hydrolyzed by F_1 (the reverse of ATP synthesis) in water labeled with ^{18}O, the P_i that is formed contains an ^{18}O atom. Careful measurement of the ^{18}O content of P_i formed in vitro by F_1-catalyzed hydrolysis of ATP reveals that the P_i contains not one, but three or four ^{18}O atoms (Fig. 19–20). This indicates that the terminal pyrophosphate bond in ATP is cleaved and re-formed repeatedly before P_i leaves the enzyme surface. With the P_i free to tumble in its binding site, each hydrolysis inserts ^{18}O randomly at one of the four positions in P_i. This exchange reaction occurs in unenergized F_oF_1 complexes (with no proton gradient) and with isolated F_1; it does not require the input of energy.

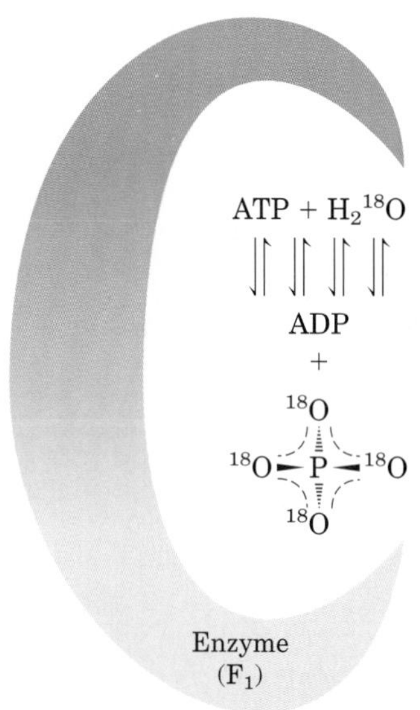

figure 19–20
Catalytic mechanism of F_1: ^{18}O exchange experiment. F_1 solubilized from mitochondrial membranes is incubated with ATP in the presence of ^{18}O-labeled water. At intervals, a sample of the solution is withdrawn and analyzed for the incorporation of ^{18}O into the P_i produced from ATP hydrolysis. In minutes, the P_i contains three or four ^{18}O atoms, indicating that both ATP hydrolysis and ATP synthesis have occurred several times during the incubation.

Kinetic studies of the initial rates of ATP synthesis and hydrolysis confirm the conclusion that $\Delta G'^{\circ}$ for ATP synthesis on the enzyme is near zero. From the measured rates of hydrolysis ($k_1 = 10\ \text{s}^{-1}$) and synthesis ($k_{-1} = 24\ \text{s}^{-1}$), the calculated equilibrium constant for the reaction

$$\text{Enz-ATP} \underset{k_{-1}}{\overset{k_1}{\rightleftharpoons}} \text{Enz-(ADP + P}_i\text{)}$$

is

$$K_{eq} = \frac{k_{-1}}{k_1} = \frac{24\ \text{s}^{-1}}{10\ \text{s}^{-1}} = 2.4$$

From this K_{eq}, the calculated apparent $\Delta G'^{\circ}$ is close to zero. This contrasts with K_{eq} of about 10^5 ($\Delta G'^{\circ} = -30.5$ kJ/mol) for the hydrolysis of ATP free in solution (not on the enzyme surface).

What accounts for the huge difference? ATP synthase stabilizes ATP relative to ADP + P_i by binding ATP more tightly, releasing enough energy to counterbalance the cost of making ATP. Careful measurements of the binding constants show that F_oF_1 binds ATP with very high affinity ($K_d \leq 10^{-12}$ M) and ADP with much lower affinity ($K_d \approx 10^{-5}$ M). This difference in K_d corresponds to a difference of about 40 kJ/mol in binding energy, and this binding energy drives the equilibrium toward formation of the product ATP.

The Proton Gradient Drives the Release of ATP from the Enzyme Surface

Although ATP synthase equilibrates ATP with ADP + P_i, newly synthesized ATP does not leave the surface of the enzyme in the absence of a proton gradient. It is the proton gradient that causes the enzyme to release the ATP formed on its surface. The reaction coordinate diagram of the process (Fig. 19–21) illustrates the difference between the mechanism of ATP synthase and that of many other enzymes that catalyze endergonic reactions.

For the continued synthesis of ATP, the enzyme must cycle between a form that binds ATP very tightly and a form that releases ATP. Chemical and crystallographic studies of the ATP synthase have revealed the structural basis for this alternation in function.

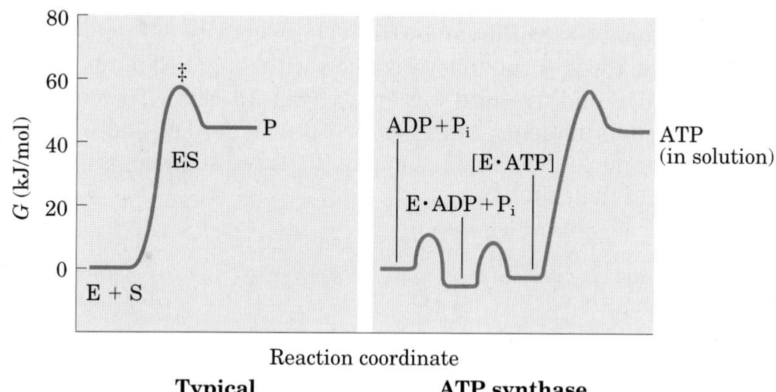

figure 19–21

Reaction coordinate diagrams for ATP synthase and a typical enzyme. In a typical enzyme-catalyzed reaction (left), reaching the transition state (‡) between substrate and product is the major energy barrier to overcome. In the reaction catalyzed by ATP synthase (right), release of ATP from the enzyme, not formation of ATP, is the major energy barrier. Although the free-energy change for the formation of ATP from ADP and P_i in aqueous solution is large and positive, the very tight binding of ATP to the enzyme provides sufficient binding energy to bring the free energy of the enzyme-bound ATP close to that of ADP + P_i. On the enzyme surface, the reaction is therefore readily reversible; the equilibrium constant is near 1. The free energy required for the release of ATP is provided by the proton-motive force.

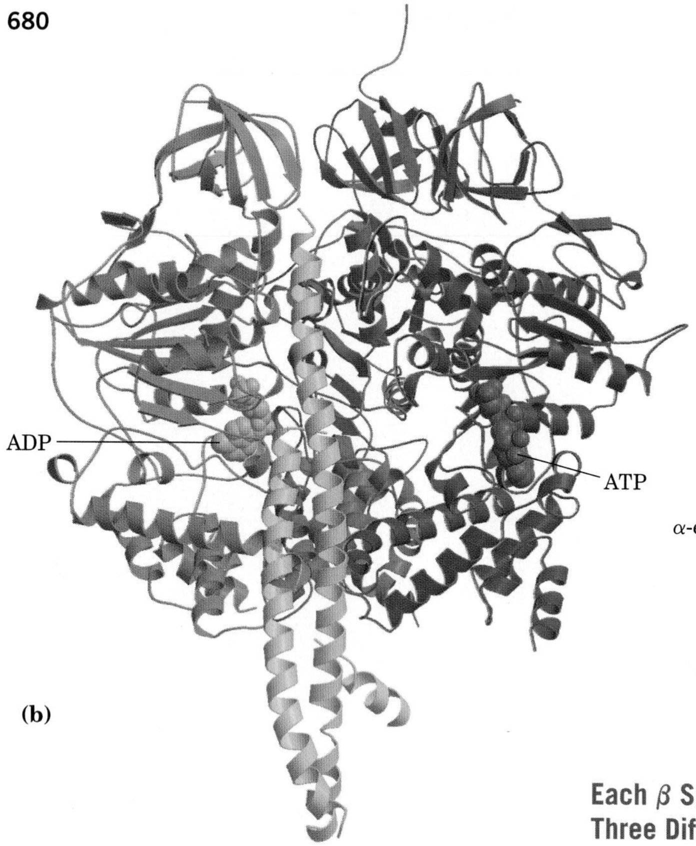

(b)

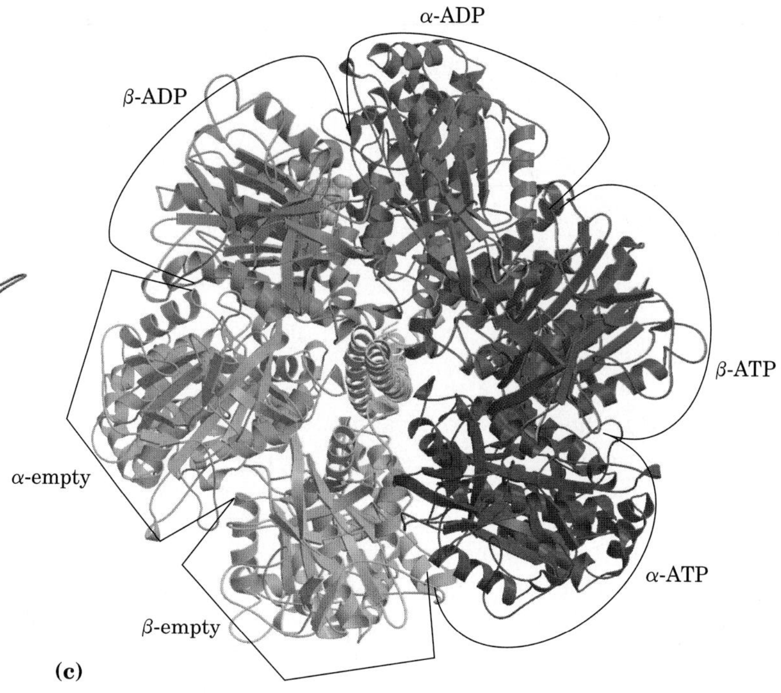

(c)

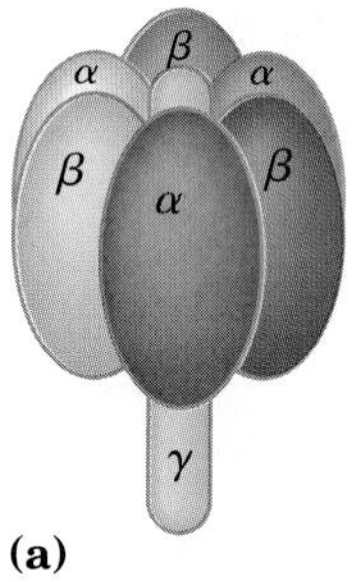

(a)

John E. Walker

figure 19–22

Mitochondrial ATP synthase complex. (a) The structure of the F_1 complex, deduced from crystallographic and biochemical studies. In F_1, three α and three β subunits are arranged like the segments of an orange, with alternating α (shades of gray) and β (shades of purple) subunits about a central shaft, the γ subunit (green). **(b)** The crystal structure of F_1, viewed from the side. Two α subunits and one β subunit have been deleted to reveal the central shaft (γ subunit) and the binding sites for ATP (red) and ADP (yellow) on β subunits. The δ and ϵ subunits are not shown here. **(c)** F_1 viewed from above (that is, from the N side of the membrane), showing the three β and three α subunits and the central shaft (γ subunit). On each β subunit near its interface with the neighboring α subunit there is a nucleotide binding site critical to the catalytic activity. The single γ subunit associates primarily with one of the three $\alpha\beta$ pairs, forcing each of the three β subunits into slightly different conformations, with different nucleotide binding sites. In the crystalline enzyme, one subunit (β-ADP) has ADP (yellow) in its binding site, the next (β-ATP) has ATP (red), and the third (β-empty) has no bound nucleotide.

Each β Subunit of ATP Synthase Can Assume Three Different Conformations

Mitochondrial F_1 has nine subunits of five different types, with the composition $\alpha_3\beta_3\gamma\delta\epsilon$. Each of the three β subunits has one catalytic site for ATP synthesis. The crystallographic determination of the F_1 structure by John E. Walker and colleagues revealed structural details very helpful in explaining the catalytic mechanism of the enzyme. The knoblike portion of F_1 is a flattened sphere 8 nm high and 10 nm across consisting of alternating α and β subunits arranged like the sections of an orange (Fig. 19–22a, b, c). The polypeptides that make up the stalk in the F_1 crystal structure are asymmetrically arranged, with one domain of the single γ subunit making up a central shaft that passes through F_1, and another domain of γ associated primarily with one of the three β subunits, designated β-empty (Fig. 19–22c). Although the amino acid sequences of the three β subunits are identical, their conformations differ, in part because of the association of the γ subunit with just one of the three. The structures of the δ and ϵ subunits are not revealed in these crystallographic studies.

The conformational differences among β subunits extends to differences in their ATP/ADP-binding sites. When researchers crystallized the protein in the presence of ADP and App(NH)p, a close structural analog of ATP that cannot be hydrolyzed by the ATPase activity of F_1, the binding site of one of the three β subunits was filled with App(NH)p, the second was filled with ADP, and the third was empty (Fig. 19–22c). The corresponding β subunit conformations are designated β-ATP, β-ADP, and β-empty. This difference in nucleotide binding among the three subunits is critical to the mechanism of the complex.

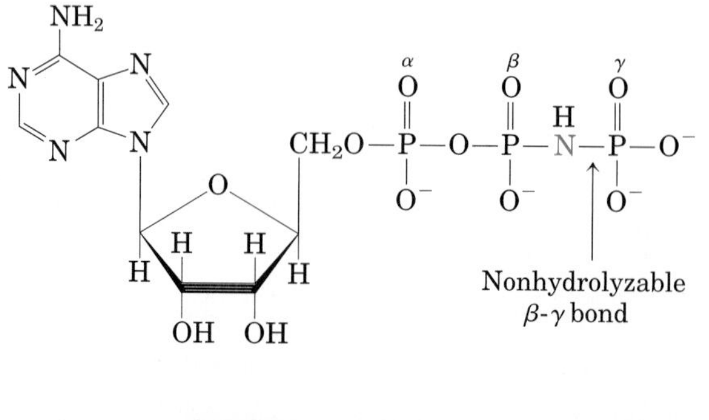

App(NH)p
(β,γ-imidoadenosine
5′-triphosphate)

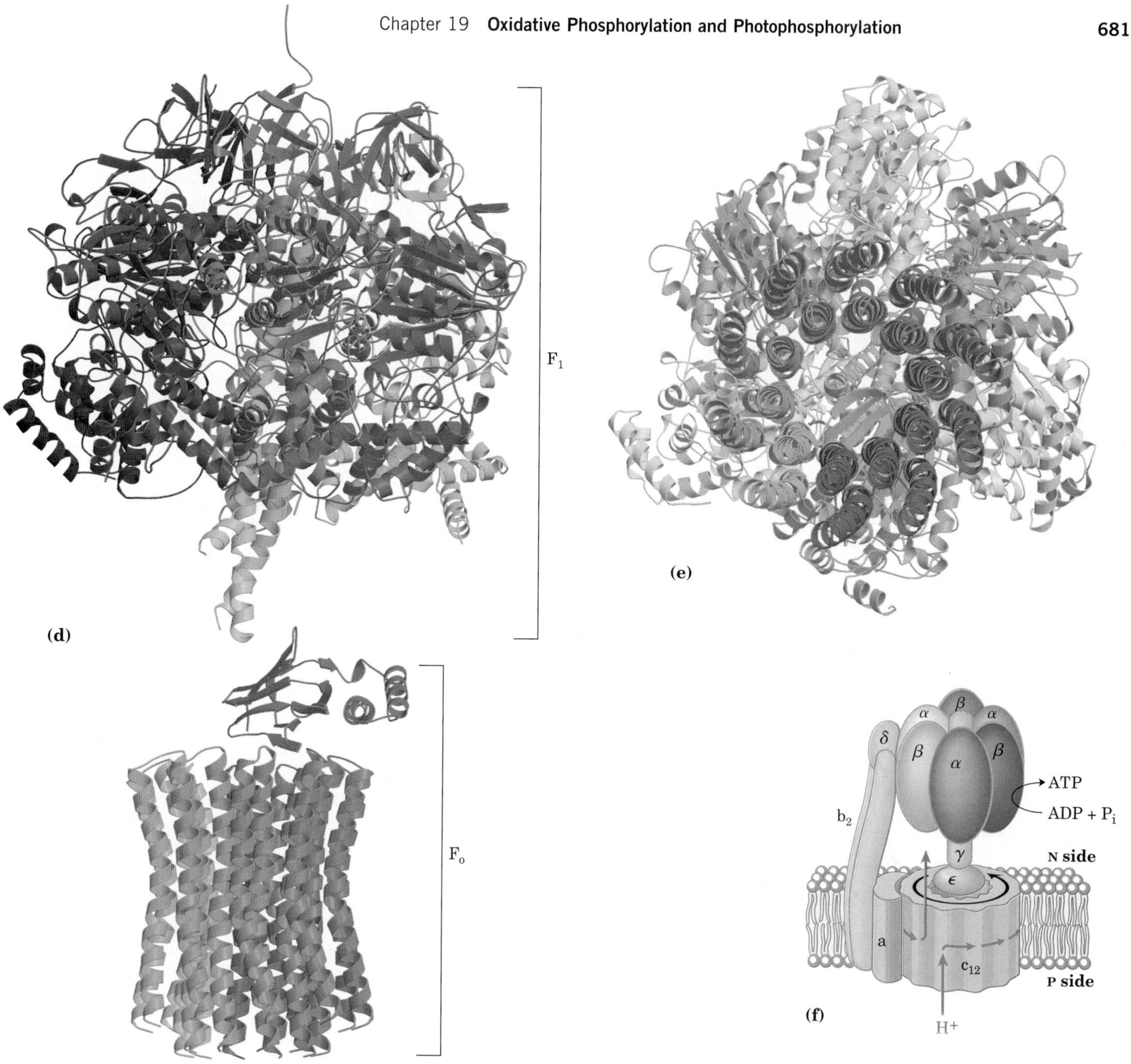

figure 19–22 ***(continued)***
(d) A side view of the F_oF_1 structure. This is a composite, in which the crystallographic coordinates of bovine mitochondrial F_1 (shades of purple and gray) have been combined with those of yeast mitochondrial F_o (shades of yellow and orange). Subunits a, b, δ, and ε were not part of the crystal structure shown here. **(e)** The yeast F_oF_1 structure, viewed end-on in the direction P side to N side. The major structures visible in this cross section are the two transmembrane helices of each of ten c subunits arranged in concentric circles. **(f)** The structure of the F_oF_1 complex, deduced from biochemical and crystallographic studies. The two b subunits of F_o associate firmly with the α and β subunits of F_1, holding them fixed relative to the membrane. In F_o, the membrane-embedded cylinder of c subunits is attached to the shaft made up of F_1 subunits γ and ε. As protons flow through the membrane from the P side to the N side via F_o, the cylinder and shaft rotate, and the β subunits of F_1 change conformation as the γ subunit associates with each in turn.

The F_o complex making up the proton pore is composed of three subunits, a, b, and c, in the proportion ab_2c_{10-12}. Subunit c is a small (M_r 8,000), very hydrophobic polypeptide, consisting almost entirely of two transmembrane helices, with a small loop extending from the matrix side of the membrane. The crystal structure of F_oF_1 from yeast, solved in 1999, shows the arrangement of the c subunits. There are 10 c subunits in yeast, each with two transmembrane helices roughly perpendicular to the plane of the membrane and arranged in two concentric circles (Fig. 19–22d, e). The inner circle is made up of the amino-terminal helices of each c subunit; the outer circle, about 55 Å in diameter, is made up of the carboxyl-terminal helices. The ε and γ subunits of F_1 form a leg-and-foot that projects from the bottom (membrane) side of F_1 and stands firmly on the ring of c subunits. A schematic drawing (Fig. 19–22f) combines the structural information from studies of bovine F_1 and yeast F_oF_1.

Paul Boyer

Rotational Catalysis Is Key to the Binding-Change Mechanism for ATP Synthesis

On the basis of detailed kinetic and binding studies of the reactions catalyzed by F_oF_1, Paul Boyer proposed a mechanism in which the three active sites of F_1 take turns catalyzing ATP synthesis (Fig. 19–23). A given β subunit starts in the β-ADP conformation, which binds ADP and P_i from the surrounding medium. The subunit now changes conformation, assuming the β-ATP form that tightly binds and stabilizes ATP, bringing about the ready equilibration of ADP + P_i with ATP on the enzyme surface. Finally, the subunit changes to the β-empty conformation, which has very low affinity for ATP, and the newly synthesized ATP leaves the enzyme surface. Another round of catalysis begins when this subunit again assumes the β-ADP form and binds ADP and P_i.

The conformational changes central to this mechanism are driven by the passage of protons through the F_o portion of ATP synthase. The streaming of protons through the F_o "pore" causes the cylinder of c subunits and the attached γ subunit to rotate about the long axis of γ, which is perpendicular to the plane of the membrane. The γ subunit passes through the center of the $\alpha_3\beta_3$ spheroid, which is held stationary relative to the membrane surface by the b_2 and δ subunits (Fig. 19–22f). With each rotation of 120°, γ comes into contact with a different β subunit, and the contact forces that β subunit into the β-empty conformation.

The three β subunits interact in such a way that when one assumes the β-empty conformation, its neighbor to one side *must* assume the β-ADP form, and the other neighbor the β-ATP form. Thus one complete rotation of the γ subunit causes each β subunit to cycle through all three of its possible conformations, and for each rotation, three ATP are synthesized and released from the enzyme's surface.

One strong prediction of this binding-change model is that the γ subunit should rotate in one direction when F_oF_1 is synthesizing ATP and in the opposite direction when the enzyme is hydrolyzing ATP. This prediction was confirmed in elegant experiments in the laboratories of Masamitsu Yoshida and Kazuhiko Kinosita, Jr. The rotation of γ in a single F_1 molecule

figure 19–23
Binding-change model for ATP synthase. The F_1 complex has three nonequivalent adenine nucleotide–binding sites, one for each pair of α and β subunits. At any given moment, one of these sites is in the β-ATP conformation (which binds ATP tightly), a second is in the β-ADP (loose-binding) conformation, and a third is in the β-empty (very loose-binding) conformation. The proton-motive force causes rotation of the central shaft—the γ subunit, shown as a green arrowhead—which comes into contact with each $\alpha\beta$ subunit pair in succession. This produces a cooperative conformational change in which the β-ATP site is converted to the β-empty conformation, and ATP dissociates; the β-ADP site is converted to the β-ATP conformation, which promotes condensation of bound ADP + P_i to form ATP; and the β-empty site becomes a β-ADP site, which loosely binds ADP and P_i entering from the solvent. The model, based on experimental findings, requires that at least two of the three catalytic sites alternate in activity; ATP cannot be released from one site unless and until ADP and P_i are bound at the other.

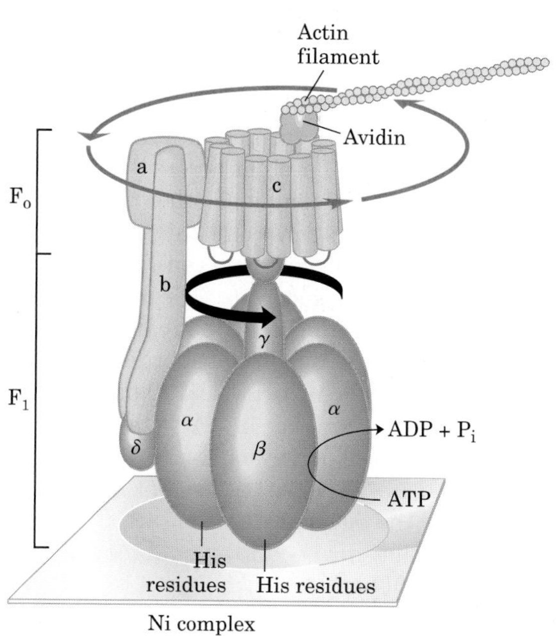

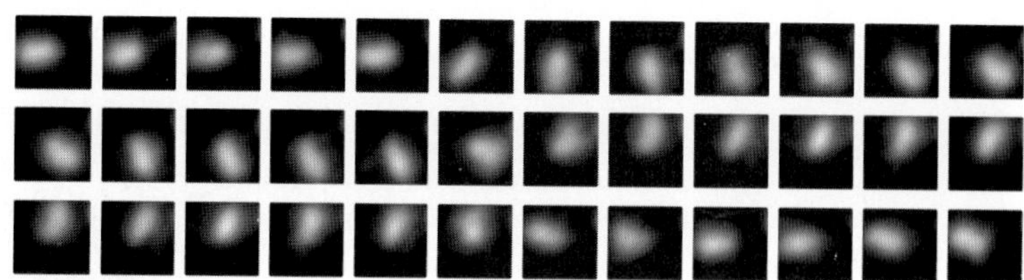

figure 19–24

Rotation of F_o and γ experimentally demonstrated. F_1 genetically engineered to contain a run of His residues adheres tightly to a microscope slide coated with a Ni complex; biotin is covalently attached to a c subunit. The protein avidin, which binds biotin very tightly, is covalently attached to long filaments of actin labeled with a fluorescent probe. Biotin-avidin binding now attaches the actin filaments to the c subunit. When ATP is provided as a substrate for the ATPase activity of F_1, the labeled filament is seen to rotate continuously in one direction, proving that the F_o cylinder of c subunits rotates. In another experiment (not shown), attachment of an actin filament directly to the γ subunit showed that it, too, rotates; presumably the cylinder and shaft move as one unit.

was detected microscopically by attaching a long, thin, fluorescent actin polymer to γ and watching it move relative to $\alpha_3\beta_3$ immobilized on a microscope slide, as ATP was hydrolyzed. When the entire F_oF_1 complex (not just F_1) was used in a similar experiment, the entire ring of c subunits rotated with γ (Fig. 19–24). The "shaft" rotated in the predicted direction through 360°. The rotation was not smooth, but occurred in three discrete steps of 120°. As calculated from the known rate of ATP hydrolysis by one F_1 molecule and the frictional drag on the long actin polymer, the efficiency of this mechanism in converting chemical energy into motion is close to 100%. It is, in Boyer's words, "a splendid molecular machine!"

Chemiosmotic Coupling Allows Nonintegral Stoichiometries of O_2 Consumption and ATP Synthesis

Before the general acceptance of the chemiosmotic model for oxidative phosphorylation, it was assumed that the overall reaction equation would take the following form:

$$\text{ADP} + \text{P}_i + x(\tfrac{1}{2})\text{O}_2 + x\text{H}^+ + x\text{NADH} \longrightarrow \text{ATP} + x\text{H}_2\text{O} + x\text{NAD}^+ \quad (19\text{–}11)$$

with the value of x—sometimes called the **P/O ratio** or the **P/2*e* ratio**—always an integer. When intact mitochondria are suspended in solution with an oxidizable substrate such as succinate or NADH and provided with O_2, ATP synthesis is readily measurable, as is the decrease in O_2. Measurement of P/O, however, is complicated by the fact that intact mitochondria consume ATP in many reactions occurring in the matrix, and they consume O_2 for purposes other than oxidative phosphorylation. Most experiments yielded P/O (ATP to $\frac{1}{2}O_2$) ratios of more than two when NADH was the electron donor, and more than one when succinate was the donor. Given the assumption that P/O should have an integral value, most experimenters agreed that the P/O ratios must be 3 for NADH and 2 for succinate, and for years those values have been found in research papers and textbooks.

With the introduction of the chemiosmotic paradigm for coupling ATP synthesis to electron transfer, there was no theoretical requirement for P/O to be integral. The relevant questions about stoichiometry became: how many protons are pumped outward by electron transfer from one NADH to O_2, and how many protons must flow inward through the F_oF_1 complex to drive the synthesis of one ATP? The measurement of proton fluxes is technically complicated; one has to account for the buffering capacity of mitochondria, the nonproductive leakage of protons across the inner membrane, and the use of the proton gradient for functions other than ATP synthesis, such as driving transport of substrates across the mitochondrial membrane (described below). The consensus values for protons pumped out per pair of electrons are 10 for NADH and 6 for succinate. The most widely accepted experimental value for number of protons required to drive the synthesis of an ATP molecule is 4, of which one is used in transporting P_i, ATP, and ADP across the mitochondrial membrane (see below). If 10 protons are pumped out per NADH and 4 must flow in to produce one ATP, the proton-based P/O ratio is 2.5 for NADH as the electron donor and 1.5 (6/4) for succinate. We use the P/O values of 2.5 and 1.5 throughout this book, but the values 3.0 and 2.0 are still often given in the biochemical literature. The final word on proton stoichiometry will probably not be written until the full details of the F_1F_o reaction mechanism are known.

The Proton-Motive Force Energizes Active Transport

Although the primary role of the proton gradient in mitochondria is to furnish energy for the synthesis of ATP, the proton-motive force also drives several transport processes essential to oxidative phosphorylation. The inner mitochondrial membrane is generally impermeable to charged species, but two specific systems transport ADP and P_i into the matrix and ATP out to the cytosol (Fig. 19–25).

The **adenine nucleotide translocase,** integral to the inner membrane, binds ADP^{3-} in the intermembrane space and transports it into the matrix in exchange for an ATP^{4-} molecule simultaneously transported outward (see Fig. 14–1 for the ionic forms of ATP and ADP). Because this antiporter moves four negative charges out for three moved in, its activity is favored by the transmembrane electrochemical gradient, which gives the matrix a net negative charge; the proton-motive force drives ATP-ADP exchange. Adenine nucleotide translocase is specifically inhibited by atractyloside, a toxic glycoside formed by a species of thistle. If the transport of ADP into and ATP out of mitochondria is inhibited, cytosolic ATP cannot be regenerated from ADP, explaining the toxicity of atractyloside (Table 19–4).

A second membrane transport system essential to oxidative phosphorylation is the **phosphate translocase,** which promotes symport of one $H_2PO_4^-$ and one H^+ into the matrix. This transport process, too, is favored by the transmembrane proton gradient (Fig. 19–25). Notice that it requires one proton to move from the P to the N side of the inner membrane, consuming some of the energy of electron transfer.

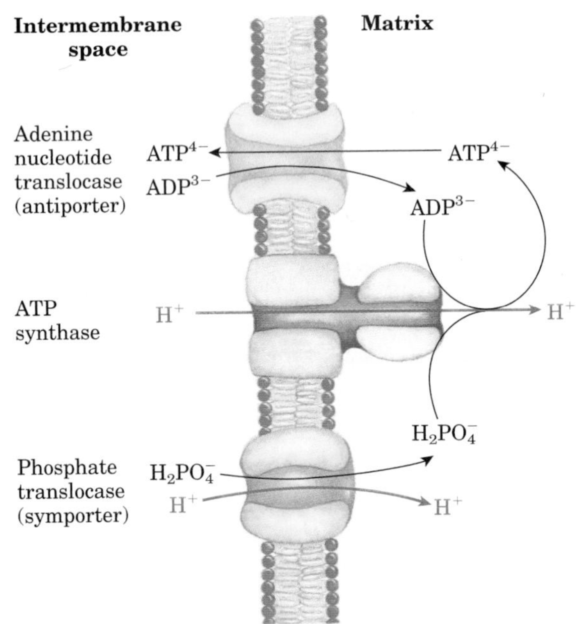

figure 19–25
Adenine nucleotide and phosphate translocases. Transport systems of the mitochondrial inner membrane carry ADP and P_i into the matrix and newly synthesized ATP into the cytosol. The adenine nucleotide translocase is an antiporter; the same protein moves ADP into the matrix and ATP out. The effect of replacing ATP^{4-} with ADP^{3-} is the net efflux of one negative charge, which is favored by the charge difference across the inner membrane (outside positive). At pH 7, P_i is present as both HPO_4^{2-} and $H_2PO_4^-$; the phosphate translocase is specific for $H_2PO_4^-$. There is no net flow of charge during symport of $H_2PO_4^-$ and H^+, but the relatively low proton concentration in the matrix favors the inward movement of H^+. Thus the proton-motive force is responsible both for providing the energy for ATP synthesis and for transporting substrates (ADP and P_i) in and product (ATP) out of the mitochondrial matrix.

Shuttle Systems Are Required for Mitochondrial Oxidation of Cytosolic NADH

The NADH dehydrogenase of the inner mitochondrial membrane of animal cells can accept electrons only from NADH in the matrix. Given that the inner membrane is not permeable to NADH, how can the NADH generated by glycolysis in the cytosol be reoxidized to NAD^+ by O_2 via the respiratory chain? Special shuttle systems carry reducing equivalents from cytosolic NADH into mitochondria by an indirect route. The most active NADH shuttle, which functions in liver, kidney, and heart mitochondria, is the **malate-aspartate shuttle** (Fig. 19–26). The reducing equivalents of cytosolic NADH are first transferred to cytosolic oxaloacetate to yield malate, catalyzed by cytosolic malate dehydrogenase. The malate thus formed passes through the inner membrane via the malate–α-ketoglutarate transporter. Within the matrix the reducing equivalents are passed to NAD^+ by the action of matrix malate dehydrogenase, forming NADH; this NADH can pass electrons directly to the respiratory chain. About 2.5 molecules of ATP are generated as this pair of electrons passes to O_2. Cytosolic oxaloacetate must be regenerated by transamination reactions and the activity of membrane transporters to start another cycle of the shuttle.

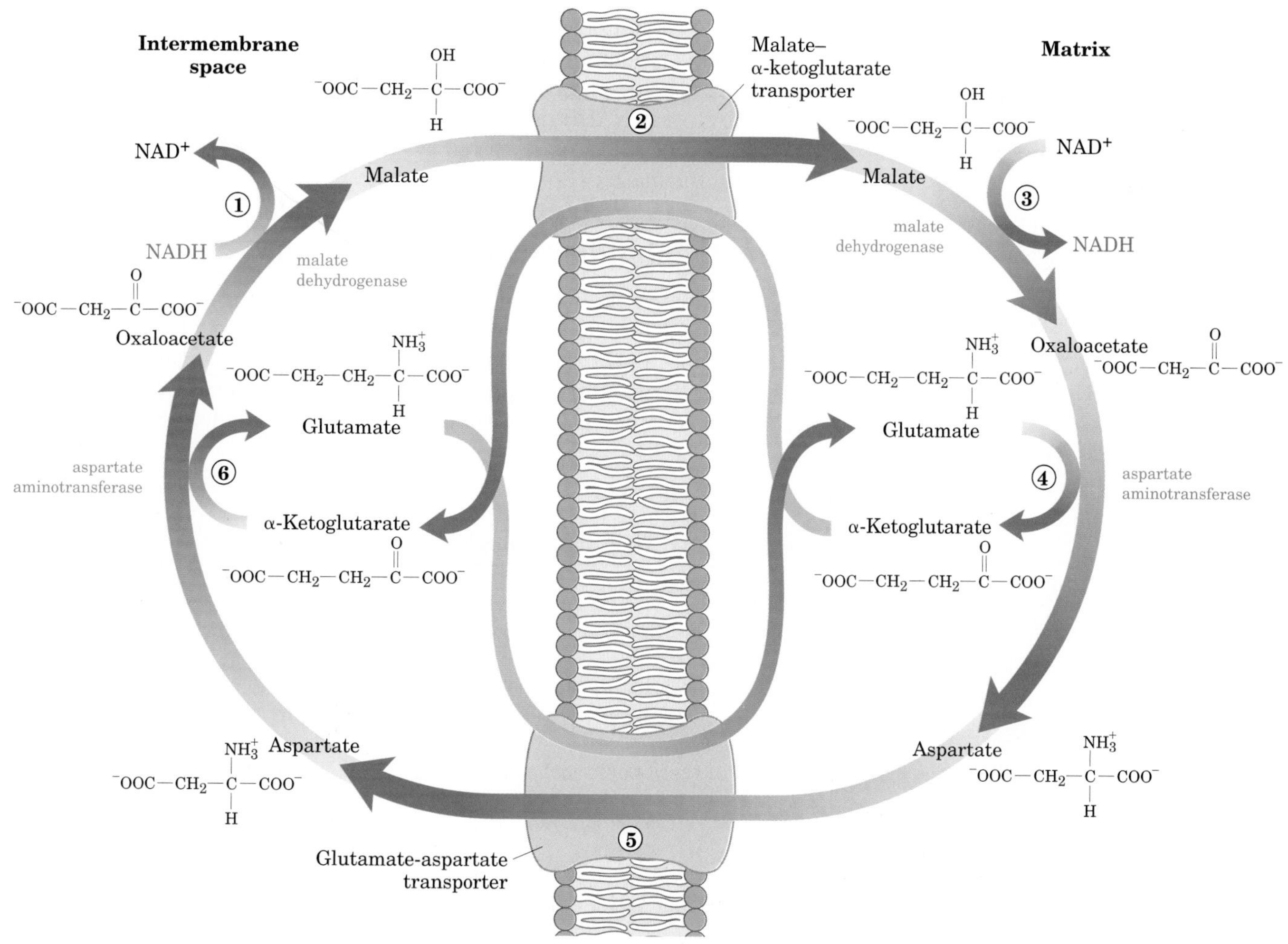

figure 19–26

Malate-aspartate shuttle. This shuttle for transporting reducing equivalents from cytosolic NADH into the mitochondrial matrix is used in liver, kidney, and heart. ① NADH in the cytosol (intermembrane space) passes two reducing equivalents to oxaloacetate, producing malate. ② Malate is transported across the inner membrane by the malate–α-ketoglutarate transporter. ③ In the matrix, malate passes two reducing equivalents to NAD^+, and the resulting NADH is oxidized by the respiratory chain. The oxaloacetate formed from malate cannot pass directly into the cytosol. It is first transaminated to aspartate ④, which can leave via the glutamate-aspartate transporter ⑤. Oxaloacetate is regenerated in the cytosol ⑥, completing the cycle.

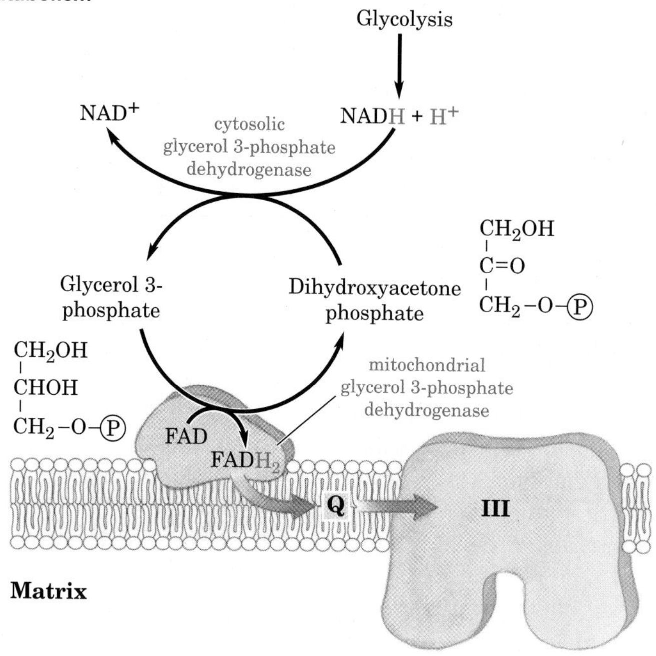

figure 19–27
Glycerol 3-phosphate shuttle. This alternative means of moving reducing equivalents from the cytosol to the mitochondrial matrix operates in skeletal muscle and the brain. In the cytosol, dihydroxyacetone phosphate accepts two reducing equivalents from NADH in a reaction catalyzed by cytosolic glycerol 3-phosphate dehydrogenase. An isozyme of glycerol 3-phosphate dehydrogenase bound to the outer face of the inner membrane then transfers two reducing equivalents from glycerol 3-phosphate in the intermembrane space to ubiquinone. Note that this shuttle does not involve membrane transport systems.

Skeletal muscle and brain use a different NADH shuttle, the **glycerol 3-phosphate shuttle** (Fig. 19–27). It differs from the malate-aspartate shuttle in that it delivers the reducing equivalents from NADH (via ubiquinone) into Complex III, not Complex I (Fig. 19–8), providing only enough energy to synthesize 1.5 ATP molecules per pair of electrons.

The mitochondria of higher plants have an *externally* oriented NADH dehydrogenase that can transfer electrons directly from cytosolic NADH into the respiratory chain at the level of ubiquinone. Because this pathway bypasses NADH dehydrogenase and the associated proton movement, the yield of ATP from cytosolic NADH is less than that from NADH generated in the matrix (Box 19–1).

Regulation of Oxidative Phosphorylation

Oxidative phosphorylation produces most of the ATP made in aerobic cells. Complete oxidation of a molecule of glucose to CO_2 yields 30 or 32 ATP (Table 19–5). By comparison, glycolysis under anaerobic conditions (lactate fermentation) yields only two ATP per glucose. Clearly, the evolution of oxidative phosphorylation provided a tremendous increase in the energy

table 19–5

ATP Yield from Complete Oxidation of Glucose

Process	Direct product	Final ATP
Glycolysis	2 NADH (cytosolic)	3 or 5*
	2 ATP	2
Pyruvate oxidation (two per glucose)	2 NADH (mitochondrial matrix)	5
Acetyl-CoA oxidation in citric acid cycle (two per glucose)	6 NADH (mitochondrial matrix)	15
	2 $FADH_2$	3
	2 ATP or 2 GTP	2
Total yield per glucose		30 or 32

*The number depends on which shuttle system transfers reducing equivalents into mitochondria.

efficiency of catabolism. Complete oxidation to CO_2 of the coenzyme A derivative of the 16-carbon saturated fatty acid palmitate, which also occurs in the mitochondrial matrix, yields 108 ATP per palmitoyl-CoA (see Table 17–1). A similar calculation can be made for the ATP yield from oxidation of each of the amino acids (Chapter 18). Aerobic oxidative pathways that result in electron transfer to O_2 accompanied by oxidative phosphorylation therefore account for the vast majority of the ATP produced in catabolism. Thus the regulation of ATP production by oxidative phosphorylation to match the cell's fluctuating needs for ATP is absolutely essential.

Oxidative Phosphorylation Is Regulated by Cellular Energy Needs

The rate of respiration (O_2 consumption) in mitochondria is tightly regulated; it is generally limited by the availability of ADP as a substrate for phosphorylation. Dependence of the rate of O_2 consumption on the availability of the P_i acceptor ADP (see Fig. 19–17b), called **acceptor control** of respiration, can be dramatic. In some animal tissues, the **acceptor control ratio,** the ratio of the maximal rate of ADP-induced O_2 consumption to the basal rate in the absence of ADP, is at least 10.

The intracellular concentration of ADP is one measure of the energy status of cells. Another, related measure is the **mass-action ratio** of the ATP-ADP system: [ATP]/([ADP][P_i]). Normally this ratio is very high, so that the ATP-ADP system is almost fully phosphorylated. When the rate of some energy-requiring process (protein synthesis, for example) increases, the rate of breakdown of ATP to ADP and P_i increases, lowering the mass-action ratio. With more ADP available for oxidative phosphorylation, the rate of respiration increases, causing regeneration of ATP. This continues until the mass-action ratio returns to its normal high level, at which point respiration slows again. The rate of oxidation of cellular fuels is regulated with such sensitivity and precision that the ratio [ATP]/([ADP][P_i]) fluctuates only slightly in most tissues, even during extreme variations in energy demand. In short, ATP is formed only as fast as it is used in energy-requiring cellular activities.

Uncoupled Mitochondria in Brown Fat Produce Heat

There is a remarkable and instructive exception to the general rule that respiration slows when the ATP supply is adequate. Most newborn mammals, including humans, have a type of adipose tissue called **brown fat** in which fuel oxidation serves not to produce ATP, but to generate heat to keep the newborn warm. This specialized adipose tissue is brown because of the presence of large numbers of mitochondria and thus large amounts of cytochromes, whose heme groups are strong absorbers of visible light.

The mitochondria of brown fat are like those of other mammalian cells in all respects, except that they have a unique protein in their inner membrane. **Thermogenin,** also called the **uncoupling protein** (Table 19–4), provides a path for protons to return to the matrix without passing through the F_oF_1 complex (Fig. 19–28). As a result of this short-circuiting of protons, the energy of oxidation is not conserved by ATP formation but is dissipated as heat, which contributes to maintaining the body temperature of the newborn. Hibernating animals also depend on uncoupled mitochondria of brown fat to generate heat during their long dormancy (see Box 17–1).

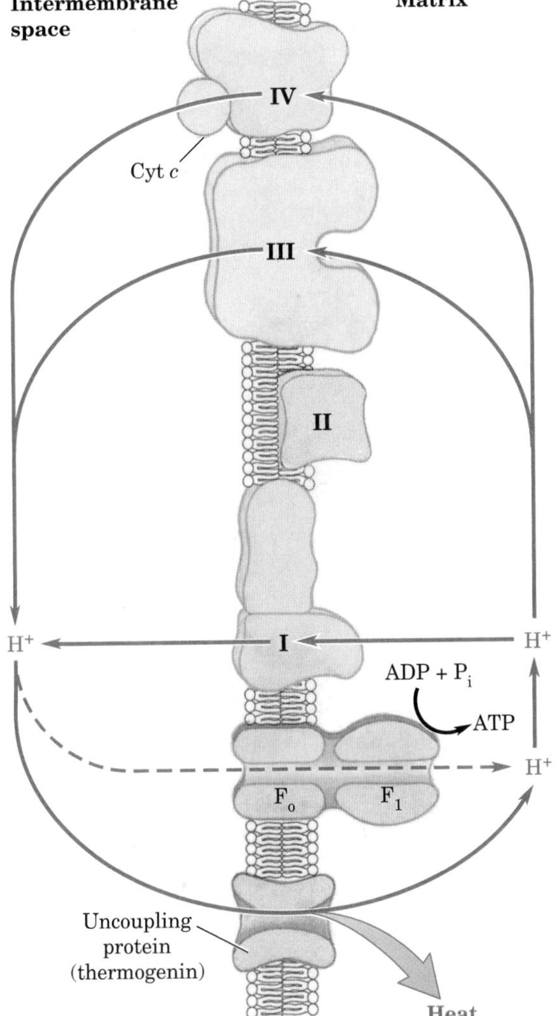

figure 19–28
Heat generation by uncoupled mitochondria. The uncoupling protein (thermogenin) of brown fat mitochondria, by providing an alternative route for protons to reenter the mitochondrial matrix, causes the energy conserved by proton pumping to be dissipated as heat.

figure 19–29

Regulation of the ATP-producing pathways. This diagram shows the interlocking regulation of glycolysis, pyruvate oxidation, the citric acid cycle, and oxidative phosphorylation by the relative concentrations of ATP, ADP, and AMP, and by NADH. High [ATP] (or low [ADP] and [AMP]) produces low rates of glycolysis, pyruvate oxidation, acetate oxidation via the citric acid cycle, and oxidative phosphorylation. All four pathways are accelerated when the use of ATP and the formation of ADP, AMP, and P_i increase. Interlocking of glycolysis and the citric acid cycle by citrate, which inhibits glycolysis, supplements the action of the adenine nucleotide system. In addition, increased levels of NADH and acetyl-CoA also inhibit the oxidation of pyruvate to acetyl-CoA, and high [NADH]/[NAD^+] ratios inhibit the dehydrogenase reactions of the citric acid cycle (see Fig. 16–15).

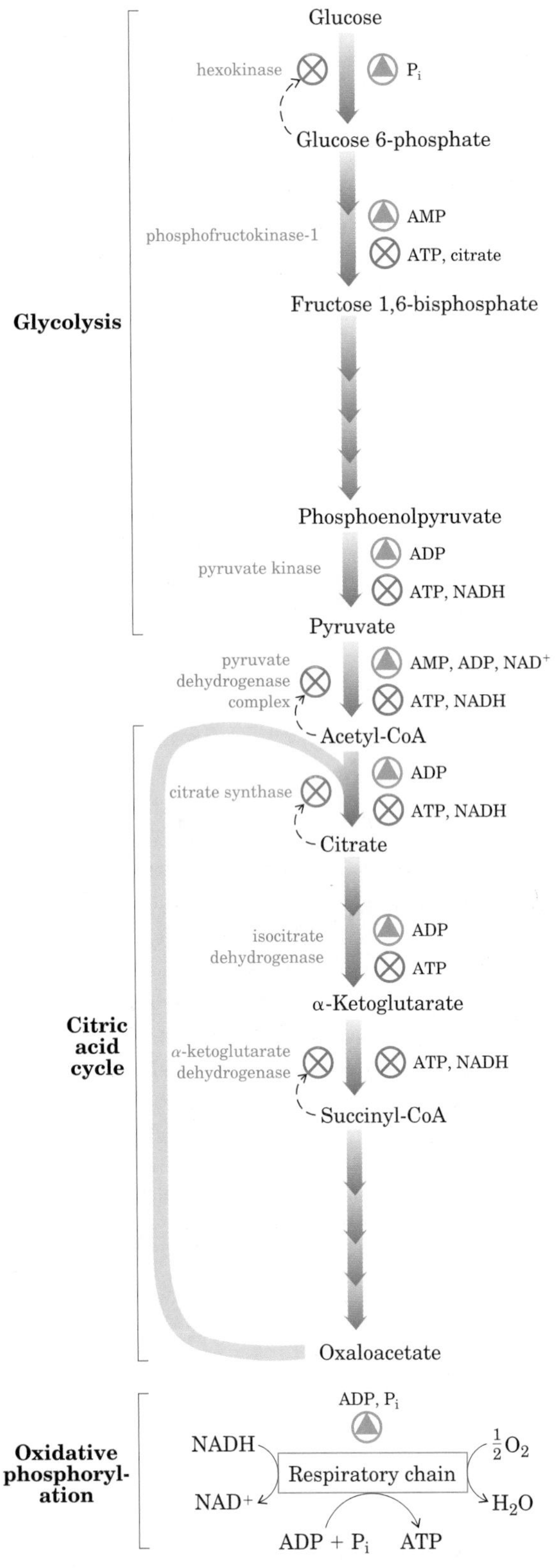

ATP-Producing Pathways Are Coordinately Regulated

The major catabolic pathways have interlocking and concerted regulatory mechanisms that allow them to function together in an economical and self-regulating manner to produce ATP and biosynthetic precursors. The relative concentrations of ATP and ADP control not only the rates of electron transfer and oxidative phosphorylation but also the rates of the citric acid cycle, pyruvate oxidation, and glycolysis (Fig. 19–29). Whenever ATP consumption increases, the rate of electron transfer and oxidative phosphorylation increases. Simultaneously, the rate of pyruvate oxidation via the citric acid cycle increases, thus increasing the flow of electrons into the respiratory chain. These events can in turn evoke an increase in the rate of glycolysis, increasing the rate of pyruvate formation. When conversion of ADP to ATP lowers the ADP concentration, acceptor control slows electron transfer and thus oxidative phosphorylation. Glycolysis and the citric acid cycle also slow, because ATP is an allosteric inhibitor of the glycolytic enzyme phosphofructokinase-1 (see Fig. 15–18) and of pyruvate dehydrogenase (see Fig. 16–15).

Phosphofructokinase-1 is inhibited not only by ATP but by citrate, the first intermediate of the citric acid cycle. When the cycle is "idling," citrate accumulates within mitochondria, then spills into the cytosol. When the concentrations of both ATP and citrate are elevated they produce a concerted allosteric inhibition of phosphofructokinase-1 that is greater than the sum of their individual effects, slowing glycolysis.

Mutations in Mitochondrial Genes Cause Human Disease

Mitochondria contain their own genome, a circular, double-stranded DNA molecule. The human mitochondrial chromosome (Fig. 19–30) contains 37 genes (16,569 base pairs), including 13 that encode subunits of proteins of the respiratory chain (Table 19–6); the remaining genes code for rRNA and tRNA molecules essential to the protein-synthesizing machinery of mitochondria. Many mitochondrial proteins are encoded by nuclear genes, synthesized on cytoplasmic ribosomes, then imported posttranslationally and assembled within mitochondria (see Fig. 27–39).

A growing number of human diseases can be attributed to mutations in mitochondrial genes. These diseases are invariably inherited from the mother, because all the mitochondria of a developing embryo are derived from the mother's egg. The rare disease **Leber's hereditary optic neuropathy** (LHON) affects the central nervous system, including the optic nerves, causing bilateral loss of vision in early adulthood. A single base change in the mitochondrial gene *ND4* (Fig. 19–30a) changes an Arg residue to a His residue in a polypeptide of Complex I, and the result is mitochondria partially defective in electron transfer from NADH to ubiquinone. Although these mitochondria can produce some ATP by electron transfer from succinate, they apparently cannot supply sufficient ATP to support the very active metabolism of neurons. One result is damage to the optic nerve, leading to blindness. A single base change in the mitochondrial gene for cytochrome *b*, a component of Complex III, also produces LHON,

figure 19–30

Mitochondrial genes and mutations. (a) Map of human mitochondrial DNA, showing the genes that encode proteins of Complex I, the NADH dehydrogenase (*ND1* to *ND6*); the cytochrome *b* of Complex III (*Cyt b*); the subunits of cytochrome oxidase (Complex IV) (*COI* to *COIII*); and two subunits of ATP synthase (*ATPase6* and *ATPase8*). The colors of the genes correspond to those of the complexes shown in Figure 19–7. Also included here are the genes for ribosomal RNAs (*rRNA*) and for a number of mitochondrion-specific transfer RNAs; tRNA specificity is indicated by the one-letter codes for amino acids. Arrows indicate the positions of mutations that cause Leber's hereditary optic neuropathy (LHON) and myoclonic epilepsy and ragged-red fiber disease (MERRF). Numbers in parentheses indicate the position of the altered nucleotides (nucleotide 1 is at the top of the circle). **(b)** Electron micrograph of an abnormal mitochondrion from the muscle of an individual with MERRF, showing the paracrystalline protein inclusions sometimes present in the mutant mitochondria.

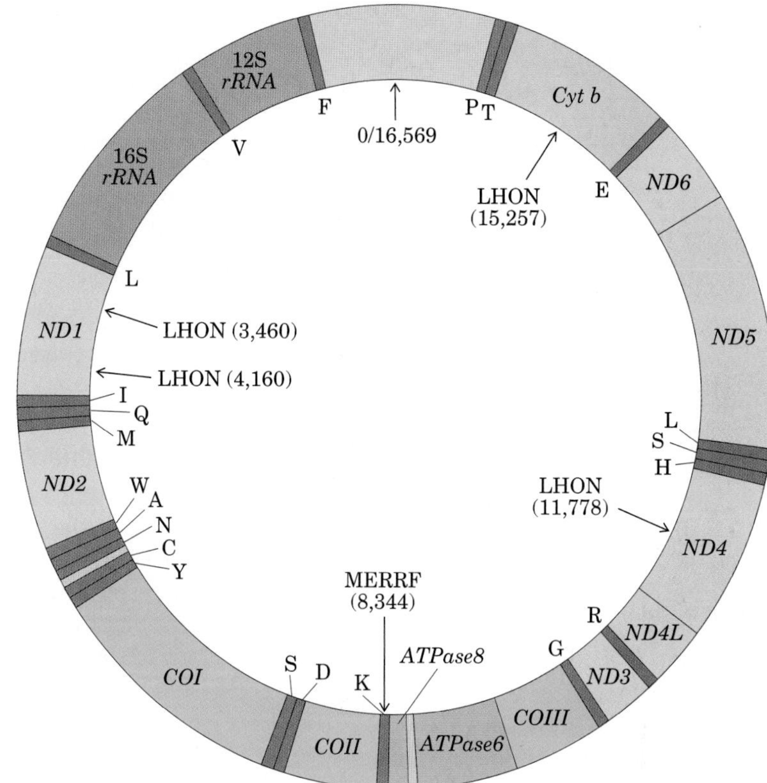

- Complex I
- Complex III
- Complex IV
- ATP synthase
- Transfer RNA
- Ribosomal RNA
- Control region of DNA

(a)

demonstrating that the pathology results from a general reduction of mitochondrial function, not specifically from a defect in electron transfer through Complex I.

Myoclonic epilepsy and ragged-red fiber disease (MERRF) is caused by a mutation in the mitochondrial gene that encodes a tRNA specific for leucine (leucyl-tRNA). This disease, characterized by uncontrollable muscular jerking, apparently results from defective production of several of the proteins synthesized using mitochondrial tRNAs. Skeletal muscle fibers of individuals with MERRF have abnormally shaped mitochondria that sometimes contain paracrystalline structures (Fig. 19–30b). Mutations in the mitochondrial leucyl-tRNA gene are one of the causes of adult-onset diabetes mellitus. Other mutations in mitochondrial genes are believed to be responsible for the progressive muscular weakness that characterizes mitochondrial myopathy, and for enlargement and deterioration of the heart muscle (hypertrophic cardiomyopathy). According to one hypothesis on the progressive changes that accompany aging, the accumulation of mutations in mitochondrial DNA during a lifetime of exposure to DNA-damaging agents results in mitochondria that cannot supply sufficient ATP for normal cellular function.

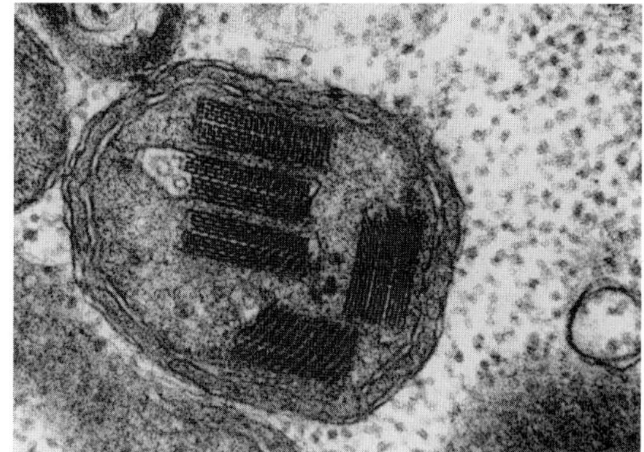

(b)

table 19–6

Respiratory Proteins Encoded by the Human Mitochondrial Chromosome

Complex	Total number of subunits	Number of subunits encoded by mitochondrial DNA
I NADH dehydrogenase	>25	7
II Succinate dehydrogenase	4	0
III Ubiquinone:cytochrome *c* oxidoreductase	9	1
IV Cytochrome oxidase	13	3
V ATP synthase	12	2

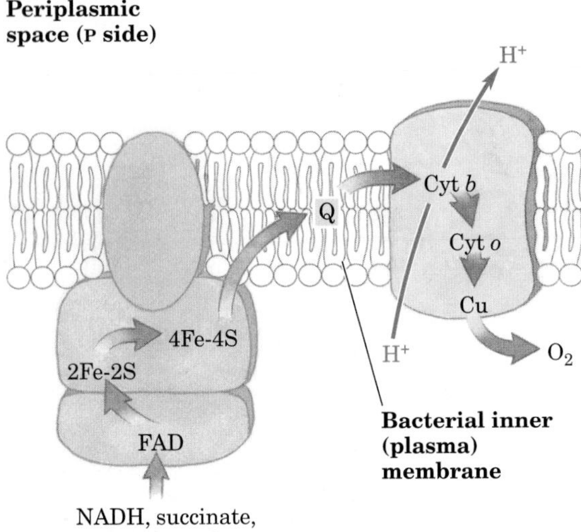

Cytosol (N side)

$n = 7–9$

Menaquinone

figure 19–31

Bacterial respiratory chain. Shown here are the respiratory carriers of the inner membrane of *E. coli.* Eubacteria contain a minimal form of Complex I, containing all the prosthetic groups normally associated with the mitochondrial complex, but only 14 polypeptides. This plasma membrane complex carries NADH to ubiquinone or menaquinone, the bacterial equivalent of ubiquinone, while pumping protons outward and creating an electrochemical potential that drives ATP synthesis.

Mitochondria Probably Evolved from Endosymbiotic Bacteria

The existence of mitochondrial DNA, ribosomes, and tRNAs supports the theory of the endosymbiotic origin of mitochondria (see Fig. 2–15). This theory holds that the first organisms capable of aerobic metabolism, including respiration-linked ATP production, were prokaryotes. Primitive eukaryotes that lived anaerobically (by fermentation) acquired the ability to carry out oxidative phosphorylation when they established a symbiotic relationship with bacteria living in their cytosol. After much evolution and the movement of many bacterial genes into the nucleus of the "host" eukaryote, the endosymbiotic bacteria eventually became mitochondria.

This theory presumes that early free-living prokaryotes had the enzymatic machinery for oxidative phosphorylation. It predicts that their modern prokaryotic descendants have respiratory chains closely similar to those of modern eukaryotes. They do. Aerobic bacteria carry out NAD-linked electron transfer from substrates to O_2, coupled to the phosphorylation of cytosolic ADP. The dehydrogenases are located in the bacterial cytosol and the respiratory chain in the plasma membrane. The electron carriers are similar to some mitochondrial electron carriers (Fig. 19–31). They translocate protons outward across the plasma membrane as electrons are transferred to O_2. Bacteria such as *Escherichia coli* have F_oF_1 complexes in their plasma membranes; the F_1 portion protrudes into the cytosol and catalyzes ATP synthesis from ADP and P_i as protons flow back into the cell through the proton channel of F_o.

The respiration-linked extrusion of protons across the bacterial plasma membrane also provides the driving force for other processes. Certain bacterial transport systems bring about uptake of extracellular nutrients (lactose, for example) against a concentration gradient, in symport with protons (see Fig. 12–35). And the rotary motion of bacterial flagella is provided by "proton turbines," molecular rotary motors driven not by ATP but directly by the transmembrane electrochemical potential generated by respiration-linked proton pumping (Fig. 19–32). It appears likely that the chemiosmotic mechanism evolved early, before the emergence of eukaryotes.

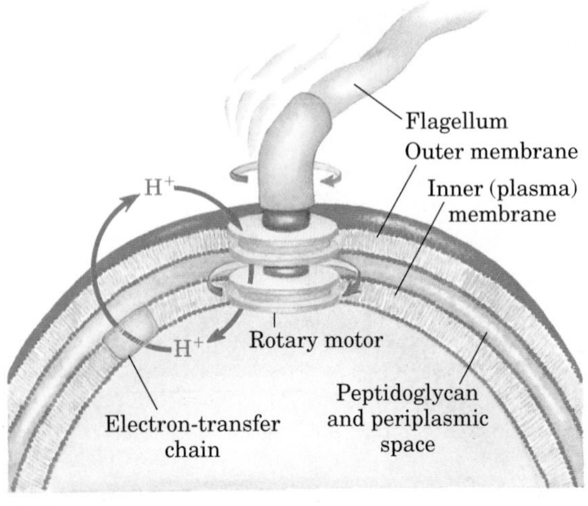

figure 19–32

Rotation of bacterial flagella by proton-motive force. The shaft and rings at the base of the flagellum make up a rotary motor that has been called a "proton turbine." Protons ejected by electron transfer flow back into the cell through the turbine, causing rotation of the shaft of the flagellum. This motion differs fundamentally from the motion of muscle and of eukaryotic flagella and cilia, for which ATP hydrolysis is the energy source.

Photosynthesis: Harvesting Light Energy

We now turn to another reaction sequence in which the flow of electrons is coupled to the synthesis of ATP: light-driven phosphorylation. The capture of solar energy by photosynthetic organisms and its conversion into the chemical energy of reduced organic compounds is the ultimate source of nearly all biological energy. Photosynthetic and heterotrophic organisms live in a balanced steady state in the biosphere (Fig. 19–33). Photosynthetic organisms trap solar energy and form ATP and NADPH, which they use as energy sources to make carbohydrates and other organic compounds from CO_2 and H_2O; simultaneously, they release O_2 into the atmosphere. Aerobic heterotrophs (humans, for example) use the O_2 so formed to degrade the energy-rich organic products of photosynthesis to CO_2 and H_2O, generating ATP for their own activities. The CO_2 formed by respiration in heterotrophs returns to the atmosphere, to be used again by photosynthetic organisms. Solar energy thus provides the driving force for the continuous cycling of CO_2 and O_2 through the biosphere and provides the reduced substrates (fuels), such as glucose, on which nonphotosynthetic organisms depend.

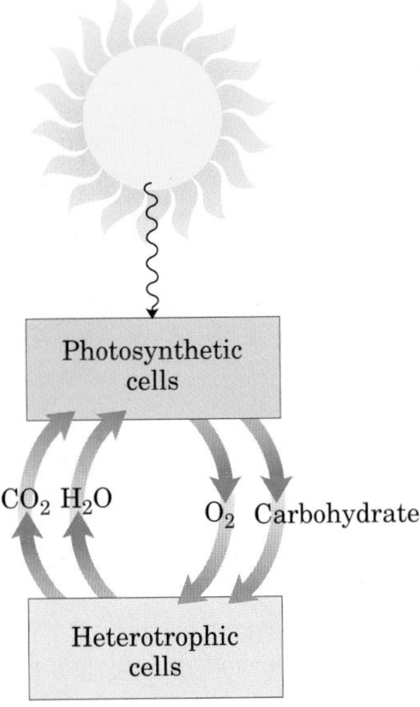

figure 19–33
Solar energy is the ultimate source of all biological energy. Photosynthetic organisms use the energy of sunlight to manufacture glucose and other organic products, which heterotrophic cells use as energy and carbon sources.

Photosynthesis occurs in a variety of bacteria and in unicellular eukaryotes (algae) as well as in higher plants. Although the process in these organisms differs in detail, the underlying mechanisms are remarkably similar, and much of our understanding of photosynthesis in higher plants is derived from studies of simpler organisms.The overall equation for photosynthesis in higher plants describes an oxidation-reduction reaction in which H_2O donates electrons (as hydrogen) for the reduction of CO_2 to carbohydrate (CH_2O):

$$CO_2 + H_2O \xrightarrow{\text{light}} O_2 + (CH_2O)$$

General Features of Photophosphorylation

Unlike NADH (the major electron donor in oxidative phosphorylation), H_2O is a poor donor of electrons; its standard reduction potential is +0.82 V, compared with −0.32 V for NADH. Photophosphorylation differs from oxidative phosphorylation in requiring the input of energy in the form of light to *create* a good electron donor. In photophosphorylation, electrons flow through a series of membrane-bound carriers including cytochromes, quinones, and iron-sulfur proteins, while protons are pumped across a membrane to create an electrochemical potential. Electron transfer and proton pumping are catalyzed by membrane complexes homologous in structure and function to Complex III of mitochondria. The electrochemical potential they produce is the driving force for ATP synthesis from ADP and P_i, catalyzed by a membrane-bound ATP synthase complex closely similar to that of oxidative phosphorylation.

Photosynthesis in higher plants encompasses two processes: the **light-dependent reactions,** or **light reactions,** which occur only when plants are illuminated, and the **carbon-assimilation** or **carbon-fixation reactions,** sometimes misleadingly called the dark reactions, which are driven by products of the light reactions (Fig. 19–34). In the light reactions, chlorophyll and other pigments of photosynthetic cells absorb light energy and conserve it as ATP and NADPH; simultaneously, O_2 is evolved. In the carbon-assimilation reactions, ATP and NADPH are used to reduce CO_2 to form triose phosphates, starch, and sucrose, and other products derived from them. In this chapter we are concerned only with the light-dependent reactions that lead to the synthesis of ATP and NADPH. The reduction of CO_2 is described in Chapter 20 with other carbohydrate-synthesizing pathways.

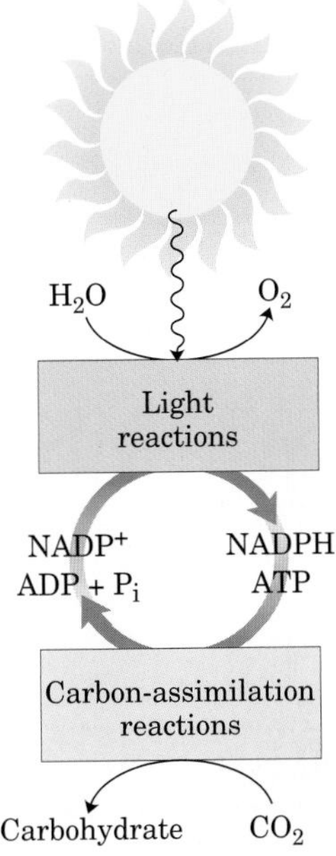

figure 19–34
The light reactions of photosynthesis generate energy-rich NADPH and ATP at the expense of solar energy. These products are used in the carbon-assimilation reactions, which occur in light or darkness, to reduce CO_2 to form trioses and more complex compounds (such as glucose) derived from trioses.

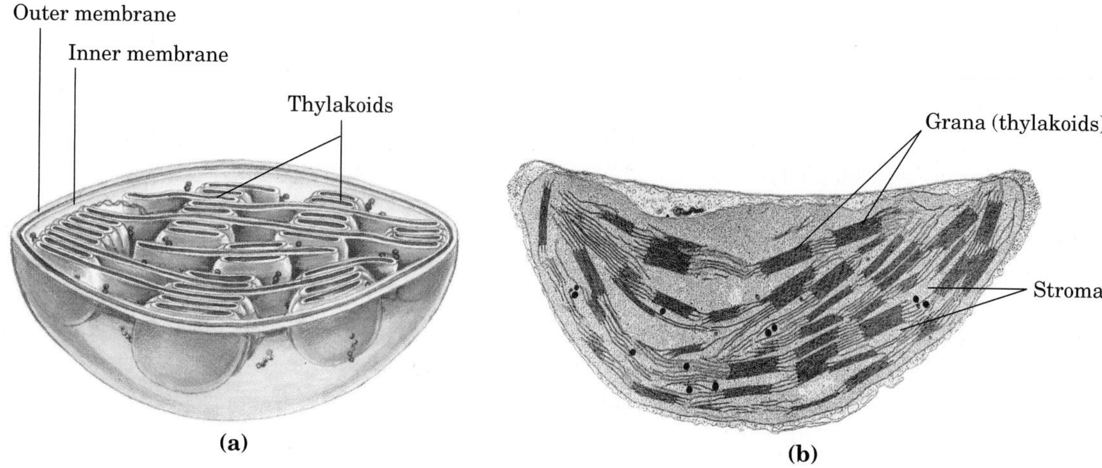

figure 19–35
Chloroplast. (a) Schematic diagram. **(b)** Electron micrograph at high magnification showing grana, stacks of thylakoid membranes.

Photosynthesis in Higher Plants Takes Place in Chloroplasts

In photosynthetic eukaryotic cells, both the light-dependent and the carbon-assimilation reactions take place in the chloroplasts (Fig. 19–35), membrane-bounded intracellular organelles that are variable in shape and generally a few microns in diameter (see Fig. 2–7b and p. 37). Like mitochondria, they are surrounded by two membranes, an outer membrane that is permeable to small molecules and ions, and an inner membrane that encloses the internal compartment. This compartment contains many flattened, membrane-surrounded vesicles or sacs called **thylakoids,** usually arranged in stacks called **grana** (Fig. 19–35b). Embedded in the thylakoid membranes (commonly called **lamellae**) are the photosynthetic pigments and the enzyme complexes that carry out the light reactions and ATP synthesis. The **stroma** (the aqueous phase enclosed by the inner membrane) contains most of the enzymes required for the carbon-assimilation reactions.

Light Drives Electron Flow in Chloroplasts

In 1937, Robert Hill found that when leaf extracts containing chloroplasts were illuminated, they (1) evolved O_2 and (2) reduced a nonbiological electron acceptor added to the medium, according to the **Hill reaction:**

$$2H_2O + 2A \xrightarrow{\text{light}} 2AH_2 + O_2$$

where A is the artificial electron acceptor or **Hill reagent.** One Hill reagent, the dye 2,6-dichlorophenolindophenol, is blue when oxidized (A) and colorless when reduced (AH_2). When the leaf extract supplemented with the dye was illuminated, the blue dye became colorless and O_2 was evolved. In the dark, neither O_2 evolution nor dye reduction took place. This was the first evidence that absorbed light energy causes electrons to flow from H_2O to an electron acceptor. Moreover, Hill found that CO_2 was neither required nor reduced to a stable form under these conditions; O_2 evolution could be dissociated from CO_2 reduction. Several years later Severo Ochoa showed that $NADP^+$ is the biological electron acceptor in chloroplasts, according to the equation

Oxidized form (blue) → Reduced form (colorless)

Dichlorophenolindophenol

$$2H_2O + 2NADP^+ \xrightarrow{\text{light}} 2NADPH + 2H^+ + O_2$$

To understand this photochemical process, we must first consider the effects of light absorption on molecular structure.

Light Absorption

Visible light is electromagnetic radiation of wavelengths 400 to 700 nm, a small part of the electromagnetic spectrum (Fig. 19–36), ranging from violet to red. The energy of a single **photon** (a quantum of light) is greater at the violet end of the spectrum than at the red end; shorter wavelength (and higher frequency) corresponds to higher energy. The energy in a "mole" of photons (one einstein; 6×10^{23} photons) of visible light is 170 to 300 kJ, almost an order of magnitude greater than the 30 to 50 kJ required to synthesize a mole of ATP from ADP and P_i.

When a photon is absorbed, an electron in the absorbing molecule (chromophore) is lifted to a higher energy level. This is an all-or-nothing event; to be absorbed, the photon must contain a quantity of energy (a **quantum**) that exactly matches the energy of the electronic transition. A molecule that has absorbed a photon is in an **excited state,** which is generally unstable. An electron lifted into a higher-energy orbital usually returns rapidly to its normal lower-energy orbital; the excited molecule decays to the stable **ground state,** giving up the absorbed quantum as light or heat or using it to do chemical work. Light emission accompanying decay of excited molecules (called **fluorescence**) is always at a longer wavelength (lower energy) than that of the absorbed light. An alternative mode of decay important in photosynthesis involves direct transfer of excitation energy from an excited molecule to a neighboring molecule.

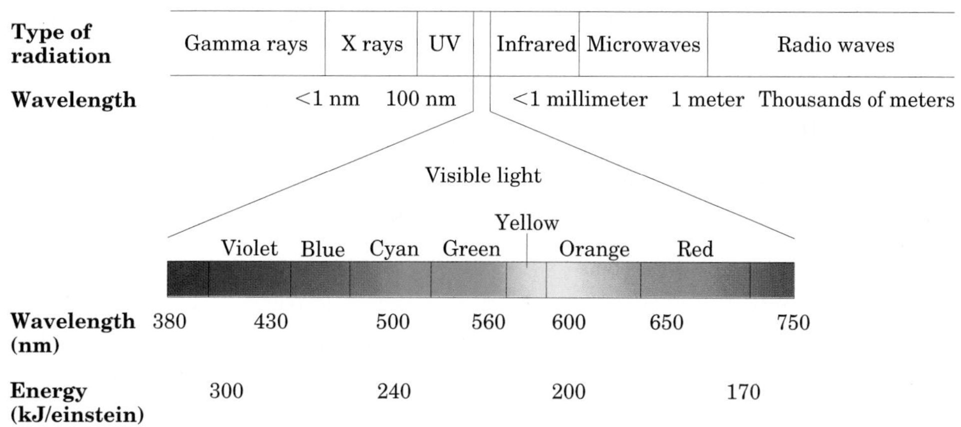

figure 19–36
The spectrum of electromagnetic radiation, and the energy of photons in the visible range of the spectrum. One einstein is 6×10^{23} photons.

Chlorophylls Absorb Light Energy for Photosynthesis

The most important light-absorbing pigments in the thylakoid membranes are the **chlorophylls,** green pigments with polycyclic, planar structures resembling the protoporphyrin of hemoglobin (see Fig. 7–1), except that Mg^{2+}, not Fe^{2+}, occupies the central position (Fig. 19–37). All chlorophylls have a long **phytol** side chain, esterified to a carboxyl-group substituent in ring IV. The four inward-oriented nitrogen atoms of chlorophyll are coordinated with the Mg^{2+}.

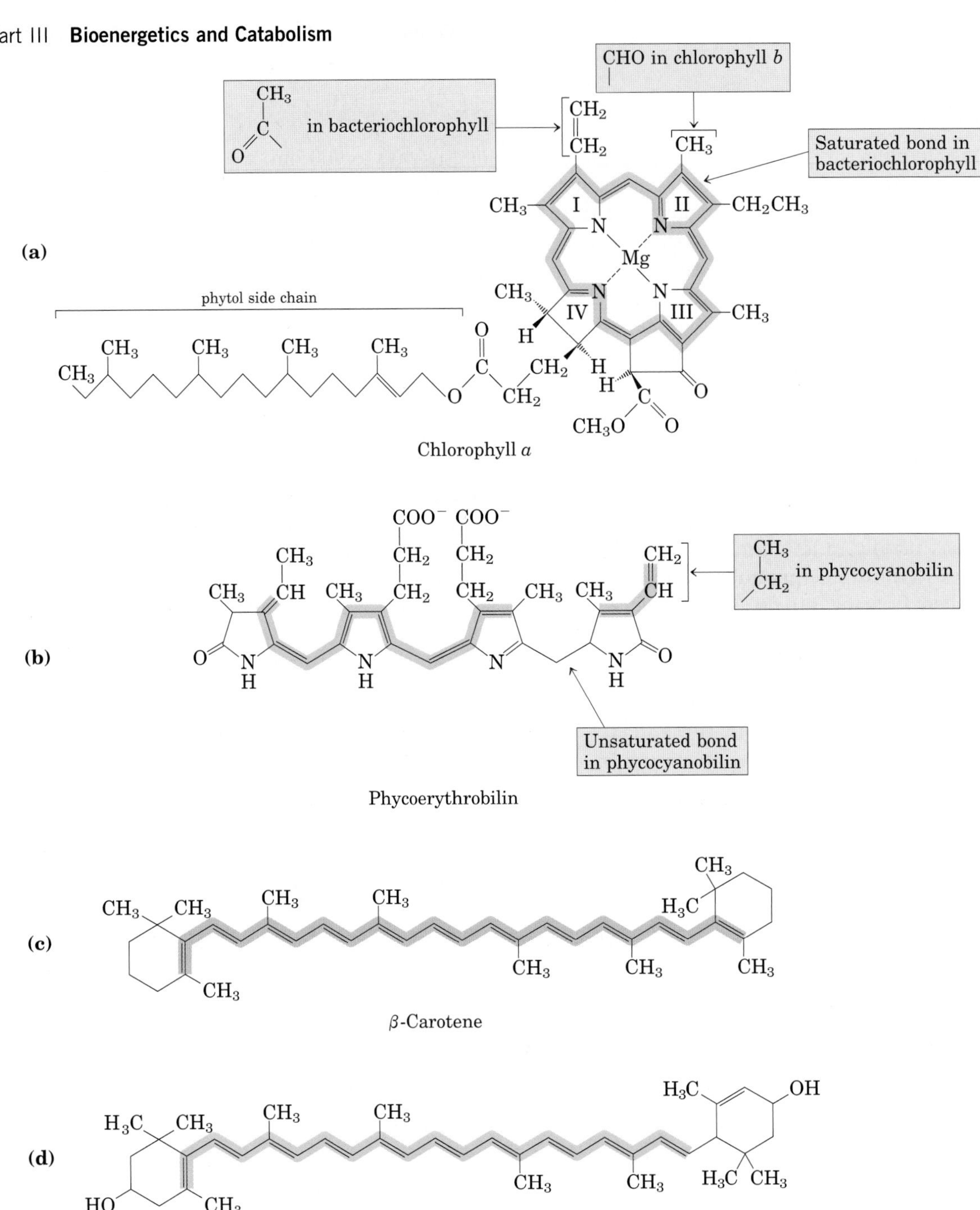

figure 19–37

Primary and secondary photopigments. (a) Chlorophylls *a* and *b* and bacteriochlorophyll are the primary gatherers of light energy. **(b)** Phycoerythrobilin and phycocyanobilin (phycobilins) are the antenna pigments in cyanobacteria and red algae. **(c)** β-Carotene (a carotenoid) and **(d)** lutein (also called xanthophyll) are accessory pigments in plants. The areas shaded pink are the conjugated systems (alternating single and double bonds) that largely account for the absorption of visible light.

The heterocyclic five-ring system that surrounds the Mg^{2+} has an extended polyene structure, with alternating single and double bonds. Such polyenes characteristically show strong absorption in the visible region of the spectrum (Fig. 19–38); the chlorophylls have unusually high molar extinction coefficients (see Box 5–1) and are therefore particularly well-suited for absorbing visible light during photosynthesis.

Chloroplasts of higher plants always contain both chlorophyll *a* and chlorophyll *b* (Fig. 19–37a). Although both are green, their absorption spectra are sufficiently different (Fig. 19–38) to allow them to complement each other's range of light absorption in the visible region. Most higher plants contain about twice as much chlorophyll *a* as chlorophyll *b*. The pigments in algae and photosynthetic bacteria include chlorophylls that differ only slightly from the plant pigments.

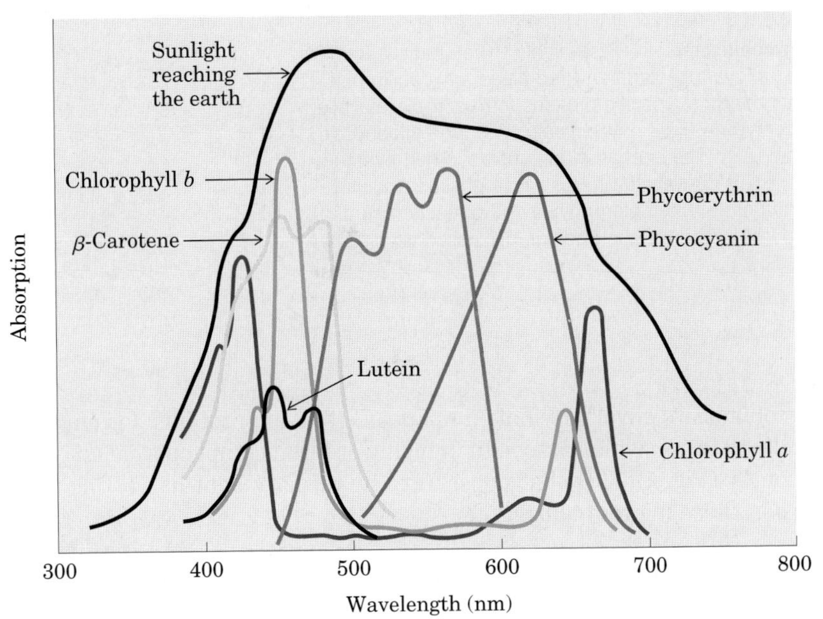

figure 19–38

Absorption of visible light by photopigments. Plants are green because their pigments absorb light from the red and blue regions of the spectrum, leaving primarily green light to be reflected or transmitted. Compare the absorption spectra of the pigments with the spectrum of sunlight reaching the earth's surface; the combination of chlorophylls (*a* and *b*) and accessory pigments enables plants to harvest most of the energy available in sunlight.

The relative amounts of chlorophylls and accessory pigments are characteristic for different plant species. Variation in the proportions of these pigments is responsible for the range of colors of photosynthetic organisms, from the deep blue-green of spruce needles, to the greener green of maple leaves, to the red, brown, or purple color of some species of multicellular algae and the leaves of some decorative plants.

Chlorophyll is always associated with specific binding proteins, forming **light-harvesting complexes (LHCs)** in which chlorophyll molecules are fixed in relation to each other, to other protein complexes, and to the membrane. The detailed structure of one light-harvesting complex is known from x-ray crystallography (Fig. 19–39). It contains seven molecules of chlorophyll *a*, five of chlorophyll *b*, and two of the accessory pigment lutein (see below).

Cyanobacteria and red algae employ **phycobilins** such as phycoerythrobilin and phycocyanobilin (Fig. 19–37b) as their light-harvesting pigments. These open-chain tetrapyrroles have the extended polyene system found in chlorophylls, but not their cyclic structure or central Mg^{2+}.

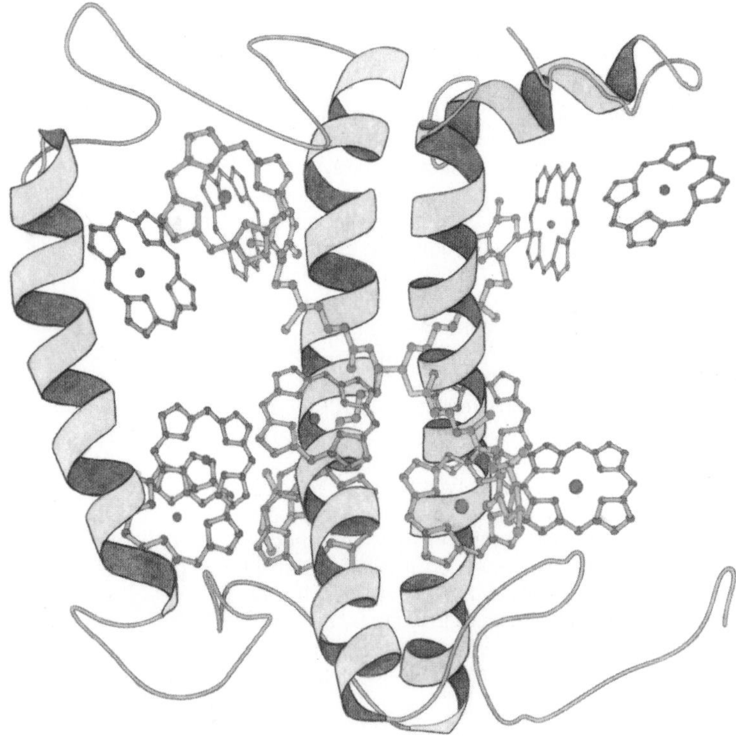

figure 19–39

A light-harvesting complex, LHCII. The functional unit is an LHC trimer, with 36 chlorophyll and six lutein molecules. Shown here is a monomer viewed in the plane of the membrane. There are three transmembrane α-helical segments, seven chlorophyll *a* molecules (green), five chlorophyll *b* molecules (red), and two molecules of the accessory pigment lutein (yellow), which form an internal cross-brace.

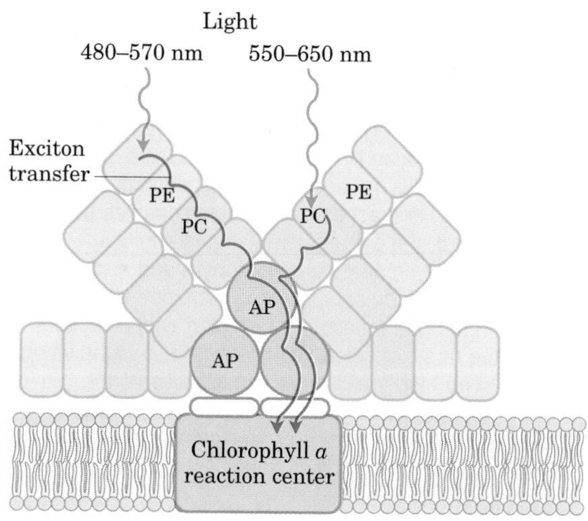

figure 19–40

A phycobilisome. In these highly structured assemblies found in cyanobacteria and red algae, phycobilin pigments bound to specific proteins form complexes called phycoerythrin (PE), phycocyanin (PC), and allophycocyanin (AP). The energy of photons absorbed by either PE or PC is conveyed through AP (a phycocyanobilin-binding protein) to chlorophyll *a* of the reaction center by a process called exciton transfer, discussed on p. 699.

Phycobilins are covalently linked to specific binding proteins, forming **phycobiliproteins,** which associate in highly ordered complexes called phycobilisomes (Fig. 19–40) that constitute the primary light-harvesting structures in these microorganisms.

Accessory Pigments Extend the Range of Light Absorption

In addition to chlorophylls, the thylakoid membranes contain secondary light-absorbing pigments **(accessory pigments),** the carotenoids. **Carotenoids** may be yellow, red, or purple. The most important are **β-carotene** (Fig. 19–37c), which is a red-orange isoprenoid, and the yellow carotenoid **lutein** (Fig. 19–37d). The carotenoid pigments absorb light at wavelengths not absorbed by the chlorophylls (Fig. 19–38) and thus are supplementary light receptors.

Experimental determination of the effectiveness of light of different colors in promoting photosynthesis yields an **action spectrum** (Fig. 19–41), often useful in identifying the pigment primarily responsible for a biological effect of light. By capturing light in a region of the spectrum not used by other organisms, a photosynthetic organism can claim a unique ecological niche. For example, the phycobilins in red algae and cyanobacteria absorb

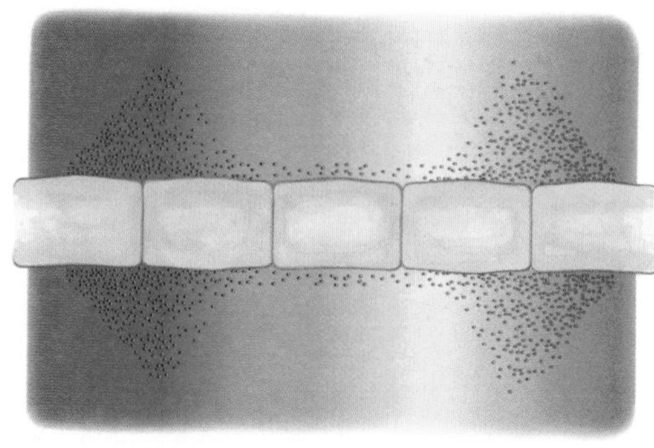
(a)

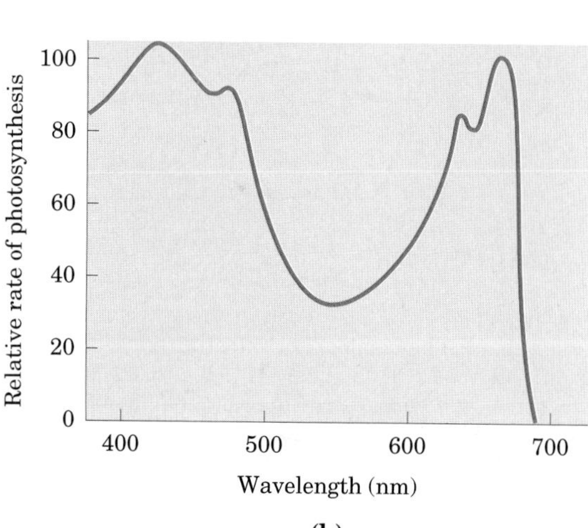

(b)

figure 19–41

Two ways to determine the action spectrum for photosynthesis. (a) The results of a classic experiment by T.W. Englemann in 1882 to determine what wavelength of light was most effective in supporting photosynthesis. Englemann placed cells of a filamentous photosynthetic alga on a microscope stage and illuminated them with light from a prism, so that cells in one part of the filament received mainly blue light, another part yellow, another red. To determine which cells carried out photosynthesis most actively, bacteria known to migrate toward regions of high O_2 concentration were also placed on the microscope slide. After a period of illumination, the distribution of bacteria showed highest O_2 levels (produced by photosynthesis) in the regions illuminated with violet and red light. **(b)** A similar experiment using modern techniques (an oxygen electrode) for the measurement of O_2 production yields the same result. An action spectrum describes the relative rate of photosynthesis for illumination with a constant number of photons of different wavelengths. An action spectrum is useful because it suggests, by comparison with absorption spectra (such as those in Fig. 19–38), which pigments can channel energy into photosynthesis.

light in the range 520 to 630 nm (Fig. 19–38), allowing them to occupy niches where light of lower or higher wavelength has been filtered out by the pigments of other organisms living in the water above them, or by the water itself.

Chlorophyll Funnels Absorbed Energy to Reaction Centers by Exciton Transfer

The light-absorbing pigments of thylakoid or bacterial membranes are arranged in functional arrays called **photosystems.** In spinach chloroplasts, for example, each photosystem contains about 200 molecules of chlorophylls and about 50 molecules of carotenoids. All the pigment molecules in a photosystem can absorb photons, but only a few chlorophyll molecules associated with the **photochemical reaction center** are specialized to transduce light into chemical energy. The other pigment molecules in a photosystem are called **light-harvesting** or **antenna molecules.** They absorb light energy and transmit it rapidly and efficiently to the reaction center (Fig. 19–42).

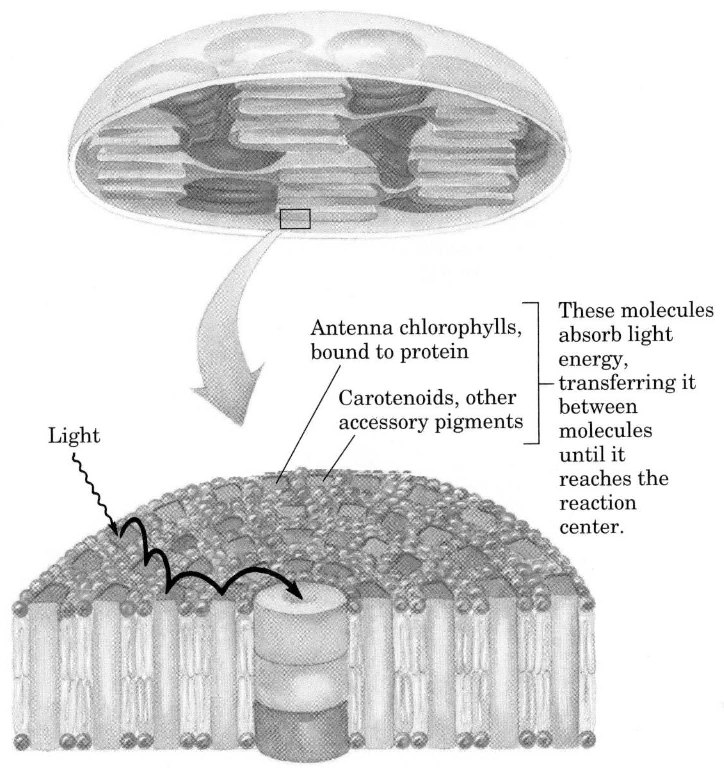
Antenna chlorophylls, bound to protein
Carotenoids, other accessory pigments
These molecules absorb light energy, transferring it between molecules until it reaches the reaction center.
Light
Reaction center
Photochemical reaction here converts the energy of a photon into a separation of charge, initiating electron flow.

figure 19–42
Organization of photosystems in the thylakoid membrane. Photosystems are tightly packed in the thylakoid membrane, with several hundred antenna chlorophylls and accessory pigments surrounding a photoreaction center. Absorption of a photon by any of the antenna chlorophylls leads to excitation of the reaction center by exciton transfer (black arrows). Also embedded in the thylakoid membrane are the cytochrome b_6f complex and ATP synthase (see Fig. 19–47).

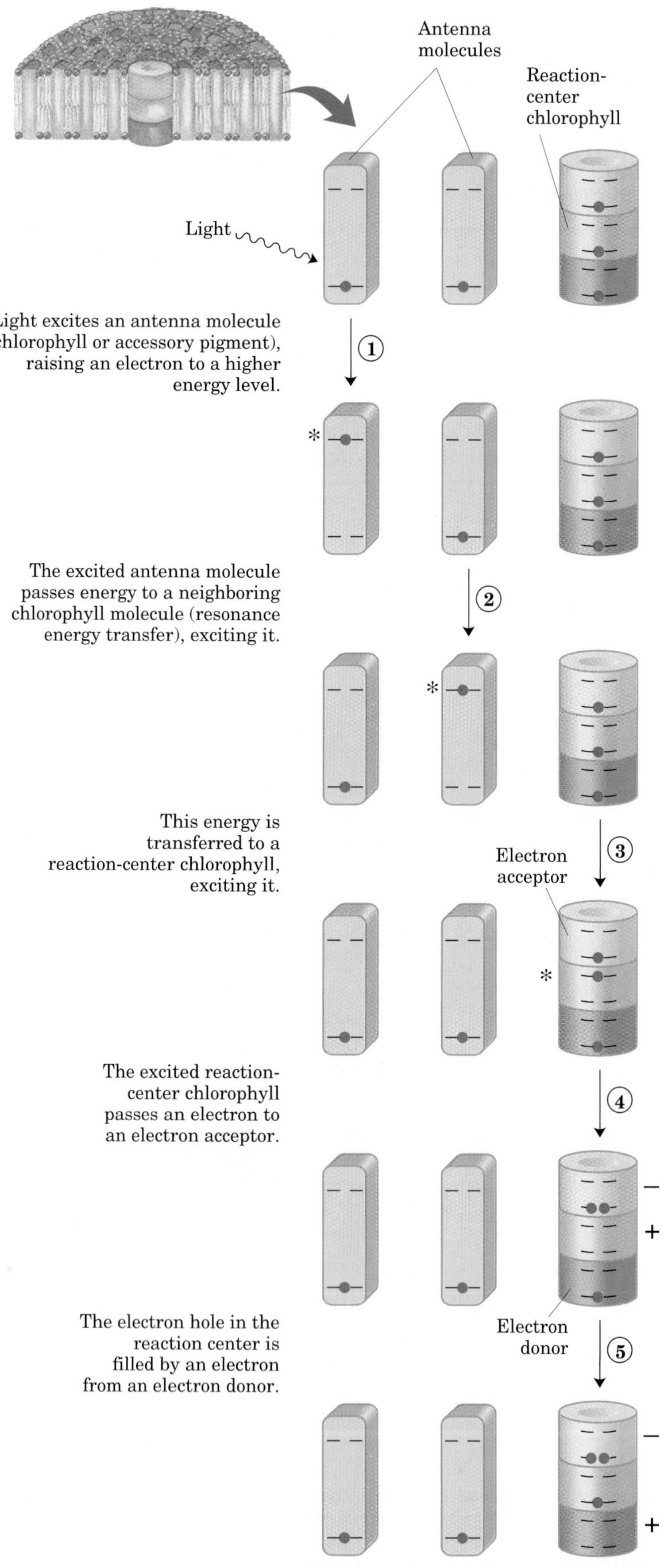
Antenna molecules
Reaction-center chlorophyll
Light
Light excites an antenna molecule (chlorophyll or accessory pigment), raising an electron to a higher energy level.
1
*
The excited antenna molecule passes energy to a neighboring chlorophyll molecule (resonance energy transfer), exciting it.
2
*
This energy is transferred to a reaction-center chlorophyll, exciting it.
Electron acceptor
3
*
The excited reaction-center chlorophyll passes an electron to an electron acceptor.
4
–
+
The electron hole in the reaction center is filled by an electron from an electron donor.
Electron donor
5
–
+
The absorption of a photon has caused separation of charge in the reaction center.

figure 19–43

Exciton transfer. This generalized scheme shows the conversion of energy from an absorbed photon into separation of charges at the reaction center. The steps are further described in the text. Note that step ① may be repeated a number of times between successive antenna molecules until a reaction-center chlorophyll is reached. The asterisk (*) represents the excited state of an antenna molecule.

The chlorophyll molecules in light-harvesting complexes have light-absorption properties that are subtly different from those of free chlorophyll. When isolated chlorophyll molecules in vitro are excited by light, the absorbed energy is quickly released as fluorescence and heat, but when chlorophyll in intact leaves is excited by visible light (Fig. 19–43, step ①), very little fluorescence is observed. Instead, the excited antenna chlorophyll transfers energy directly to a neighboring chlorophyll molecule, which becomes excited as the first molecule returns to its ground state (step ②). This transfer of energy, also called **exciton transfer,** extends to a third, fourth, or subsequent neighbor, until one of a special pair of chlorophyll *a* molecules at the photochemical reaction center is excited (step ③). In this excited chlorophyll molecule, an electron is promoted to a higher-energy orbital. This electron then passes to a nearby electron acceptor that is part of the electron-transfer chain, leaving the reaction-center chlorophyll with an empty orbital (an "electron hole") (step ④). The electron acceptor acquires a negative charge in this transaction. The electron lost by the reaction-center chlorophyll is replaced by an electron from a neighboring electron-donor molecule (step ⑤), which thereby becomes positively charged. In this way, *excitation by light causes electric charge separation and initiates an oxidation-reduction chain.*

The Central Photochemical Event: Light-Driven Electron Flow

Light-driven electron transfer in plant chloroplasts during photosynthesis is accomplished by multienzyme systems in the thylakoid membrane. Our current picture of photosynthetic mechanisms is a composite, drawn from studies of plant chloroplasts and a variety of bacteria and algae. Determination of the molecular structures of bacterial photosynthetic complexes (by x-ray crystallography) has given us a much improved understanding of the molecular events in photosynthesis.

Bacteria Have One of Two Types of Single Photochemical Reaction Centers

One major insight from studies of photosynthetic bacteria came in 1952 when Louis Duysens found that illumination of the photosynthetic membranes of the purple bacterium *Rhodospirillum rubrum* with a pulse of light of a specific wavelength (870 nm) caused a temporary decrease in the absorption of light at that wavelength; a pigment was "bleached" by 870 nm light. Later studies by Bessel Kok and Horst Witt showed similar bleaching of plant chloroplast pigments by light of 680 and 700 nm. Furthermore, addition of the (nonbiological) electron acceptor $[Fe(CN)_6]^{3-}$ caused bleaching at these wavelengths *without illumination.* Bleaching of the pigments was therefore due to the loss of an electron from a **photochemical reaction center.** These pigments were named for the wavelength of maximum bleaching: P870, P680, and P700.

Photosynthetic bacteria have relatively simple phototransduction machinery, with one of two general types of reaction center. One type (found in purple bacteria) passes electrons through **pheophytin** (chlorophyll lacking the central Mg^{2+} ion) to a quinone. The other (in green sulfur bacteria) passes electrons through a quinone to an iron-sulfur center. Cyanobacteria and higher plants have two photosystems (PSI, PSII), one of each type, acting in tandem. Biochemical and biophysical studies of the phototransduction machinery of bacteria have revealed many of the molecular details of reaction centers. These bacterial systems therefore serve as prototypes for the more complex phototransduction systems of higher plants.

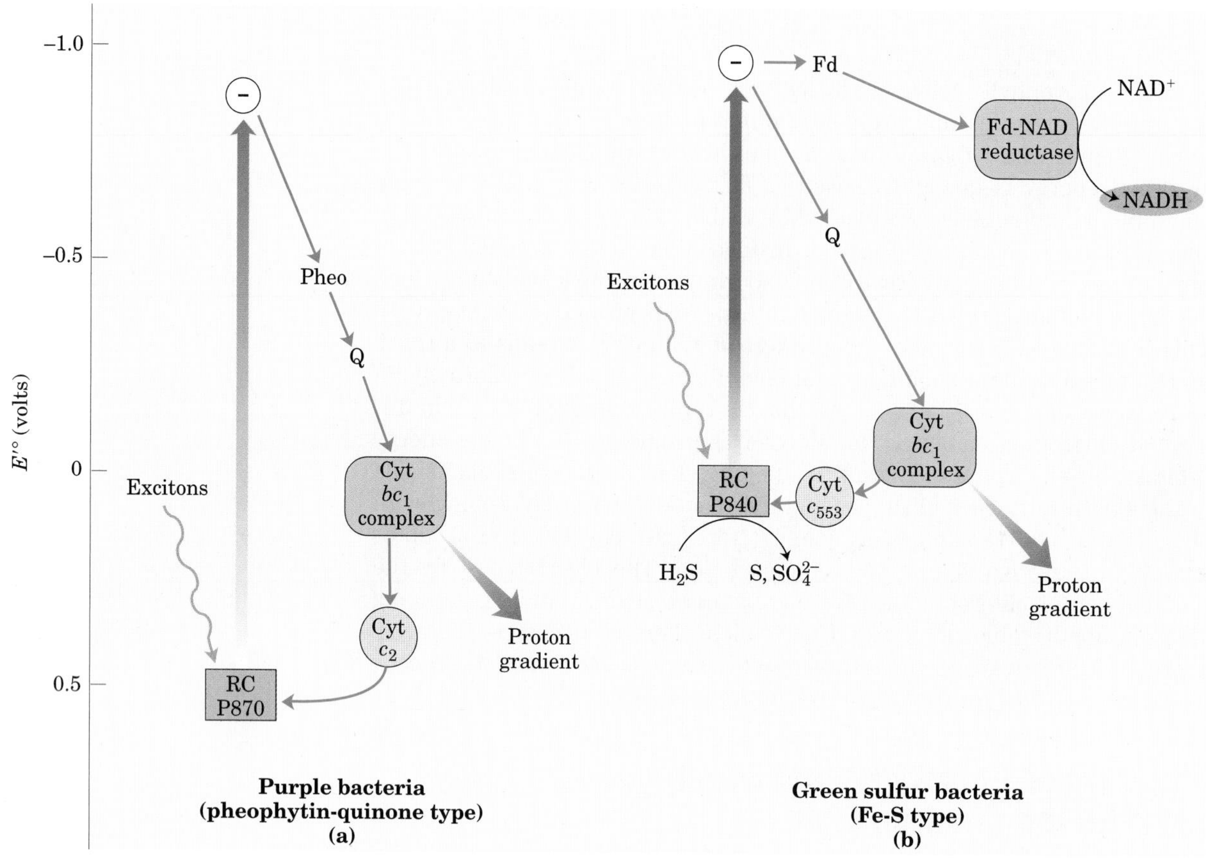

figure 19–44

Functional modules of photosynthetic machinery in purple bacteria and green sulfur bacteria. (a) In the purple bacteria, light energy drives electrons from the reaction center P870 through pheophytin (Pheo), a quinone (Q), and the cytochrome bc_1 complex, then through cytochrome c_2 and back to the reaction center. Electron flow through the cytochrome bc_1 complex causes proton pumping, creating an electrochemical potential that powers ATP synthesis. **(b)** Green sulfur bacteria have two routes for electrons driven by excitation of P840. A cyclic route passes through a quinone (Q) to the cytochrome bc_1 complex and back to the reaction center via cytochrome *c*. A noncyclic route passes from the reaction center through the iron-sulfur protein ferredoxin (Fd), then to NAD^+ in a reaction catalyzed by ferredoxin-NAD reductase.

The Pheophytin-Quinone Reaction Center (Type II Reaction Center) The photosynthetic machinery in purple bacteria consists of three basic modules (Fig. 19–44a): a single reaction center (P870), a cytochrome bc_1 electron-transfer complex similar to Complex III of the mitochondrial electron-transfer chain, and an ATP synthase, also similar to that of mitochondria. Illumination drives electrons through pheophytin and a quinone to the cytochrome bc_1 complex; after passing through the complex, electrons flow through cytochrome c_2 back to the reaction center, restoring its preillumination state. This light-driven cyclic flow of electrons provides the energy for proton pumping by the cytochrome bc_1 complex. Powered by the resulting proton gradient, the ATP synthase module produces ATP, exactly as in mitochondria.

The three-dimensional structures of the reaction centers of purple bacteria (*Rhodopseudomonas viridis* and *Rhodobacter sphaeroides*), deduced from x-ray crystallography, shed light on how phototransduction takes place in a pheophytin-quinone reaction center. The *R. viridis* reaction center (Fig. 19–45a) is a large protein complex containing four polypeptide subunits and 13 cofactors: two pairs of bacterial chlorophylls, a pair of pheophytins, two quinones, a nonheme iron, and four hemes in the associated *c*-type cytochrome.

The extremely rapid sequence of electron transfers shown in Figure 19–45b has been deduced from physical studies of the bacterial pheophytin-quinone centers, using brief flashes of light to trigger phototransduction and a variety of spectroscopic techniques to follow the flow of electrons through several carriers. A pair of bacteriochlorophylls—the "special pair," designated $(Chl)_2$—is the site of the initial photochemistry in the bacterial reaction center. Energy from a photon absorbed by one of the many antenna chlorophyll molecules surrounding the reaction center reaches $(Chl)_2$ by exciton transfer. When these two chlorophyll molecules—so close that their bonding orbitals overlap—absorb an exciton, one of them gives up a loosely bound electron, which passes through a neighboring chloro-

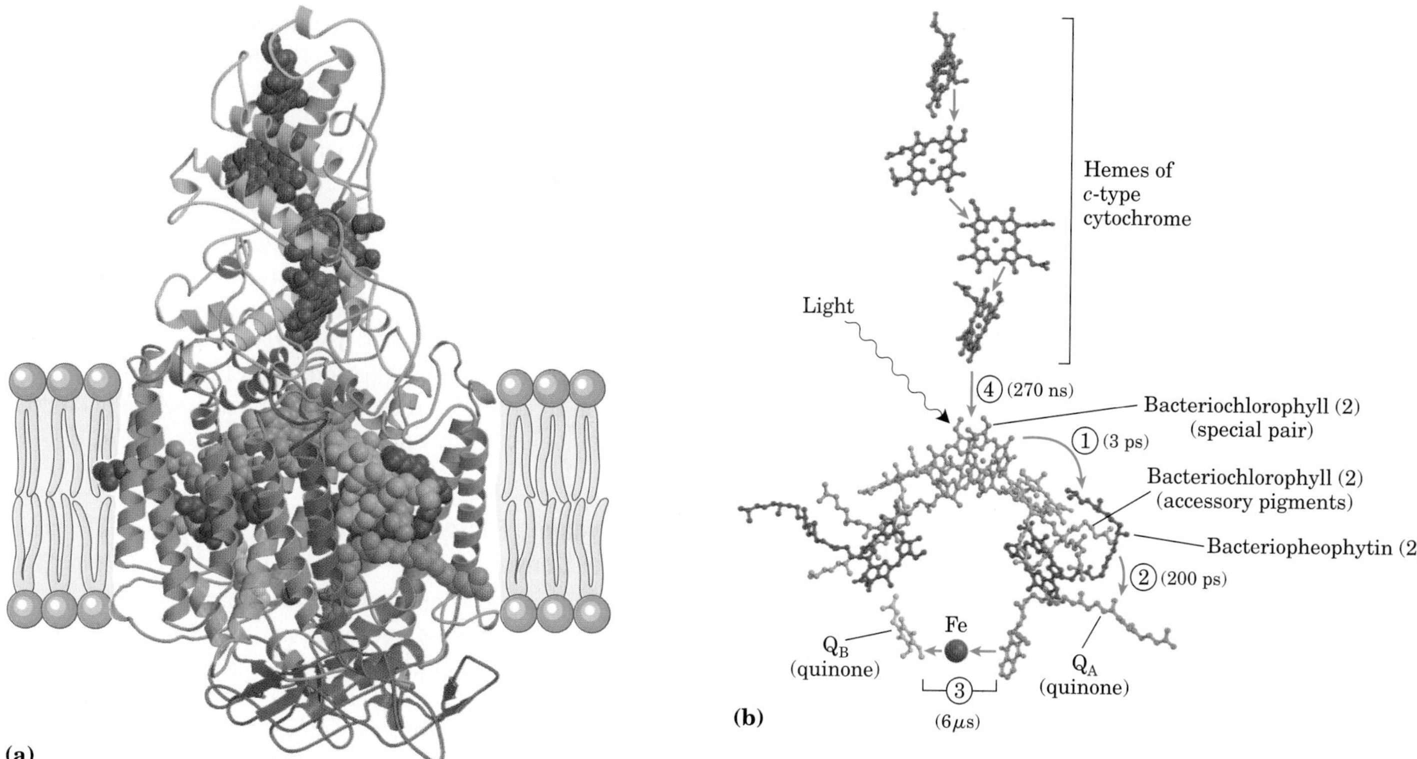

figure 19–45

Photoreaction center of the purple bacterium *Rhodopseudomonas viridis.* (a) The system has four components: three subunits, H, M, and L (brown, blue, and gray, respectively), with a total of 11 transmembrane helical segments, and a fourth protein, cytochrome *c* (yellow), associated with the complex at the membrane surface. Subunits L and M are paired transmembrane proteins that together form a cylindrical structure with roughly bilateral symmetry about its long axis. Shown as ball-and-stick structures are the prosthetic groups that participate in the photochemical events. Bound to the L and M chains are two pairs of bacteriochlorophyll molecules (green); one of the pairs (the "special pair," dark green) is the site of the first photochemical changes after light absorption. Also incorporated are a pair of pheophytin *a* (Pheo *a*) molecules (light blue); two quinones, menaquinone (Q_A) and ubiquinone (Q_B) (yellow), also arranged with bilateral symmetry; and a single nonheme Fe (red) located approximately on the axis of symmetry between the quinones. Shown at the top of the figure are four heme groups (red) associated with the *c*-type cytochrome of the reaction center. The reaction center of another purple bacterium, *Rhodobacter sphaeroides,* is very similar, except that cytochrome *c* is not part of the crystalline complex. **(b)** The sequence of events following excitation of the special pair of bacteriochlorophylls and the time scale of the electron transfers (in parentheses; ps = picoseconds). The excited special pair passes an electron to pheophytin ①, from which the electron moves rapidly to the tightly bound menaquinone, Q_A ②. This quinone passes electrons much more slowly to the diffusible ubiquinone, Q_B, through the nonheme Fe ③. Meanwhile, the "electron hole" in the special pair is filled by an electron from a heme of the cytochrome *c* ④.

phyll monomer to pheophytin (Pheo). This produces two radicals, one positively charged (the special pair of chlorophylls) and one negatively charged (the pheophytin):

$$(\text{Chl})_2 + 1 \text{ exciton} \longrightarrow (\text{Chl})_2^* \qquad \text{(excitation)}$$

$$(\text{Chl})_2^* + \text{Pheo} \longrightarrow (\text{Chl})_2^{\bullet +} + \text{Pheo}^{\bullet -} \qquad \text{(charge separation)}$$

The pheophytin radical now passes its electron to a tightly bound molecule of quinone (Q_A), converting it to a semiquinone radical, which immediately donates its extra electron to a second, loosely bound quinone (Q_B). Two such electron transfers convert Q_B to its fully reduced form, Q_BH_2, which is free to diffuse in the membrane bilayer, away from the reaction center:

$$2\text{Pheo}^{\bullet -} + 2\text{H}^+ + \text{Q}_\text{B} \longrightarrow 2\text{ Pheo} + \text{Q}_\text{B}\text{H}_2 \qquad \text{(quinone reduction)}$$

The hydroquinone (Q_BH_2), carrying in its chemical bonds some of the energy of the photons that originally excited P870, enters the pool of reduced quinone (QH_2) dissolved in the membrane and moves through the lipid phase of the bilayer to the cytochrome bc_1 complex.

Like the homologous complex III in mitochondria, the cytochrome bc_1 complex of purple bacteria carries electrons from a quinol donor (QH_2) to an electron acceptor, using the energy of electron transfer to pump protons across the membrane, producing a proton-motive force. The path of electron flow through this complex is believed to be very similar to that through mitochondrial Complex III, involving a Q cycle (Fig. 19–11) in which protons are consumed on one side of the membrane and released on the other. The ultimate electron acceptor in purple bacteria is the electron-depleted form of P870, $(\text{Chl})_2^{\bullet +}$ (Fig. 19–44a). Electrons move from the cytochrome bc_1 complex to P870 via a soluble *c*-type cytochrome, cytochrome c_2. The electron-transfer process completes the cycle, returning the reaction center to its unbleached state, ready to absorb another exciton from antenna chlorophyll.

The Fe-S Reaction Center (Type I Reaction Center) Photosynthesis in green sulfur bacteria involves the same three modules as in purple bacteria, but the process differs in several respects and involves additional enzymatic reactions (Fig. 19–44b). Excitation causes an electron to move from the reaction center to the cytochrome bc_1 complex via a quinone carrier. Electron transfer through this complex powers proton transport and creates the proton-motive force used for ATP synthesis, just as in purple bacteria and in mitochondria. However, in contrast to the cyclic flow of electrons in purple bacteria, some electrons flow from the reaction center to an iron-sulfur protein, **ferredoxin,** which then passes electrons via ferredoxin-NAD reductase to NAD^+, producing NADH. The electrons taken from the reaction center to reduce NAD^+ are replaced by the oxidation of H_2S to HSO_4^{2-} (Fig. 19–44b), in the reaction that defines the green sulfur bacteria. This oxidation of H_2S by bacteria is chemically analogous to the oxidation of H_2O by higher plants.

Kinetic and Thermodynamic Factors Prevent Energy Dissipation by Internal Conversion

The complex construction of reaction centers is the product of evolutionary selection for efficiency in the photosynthetic process. The excited state $(\text{Chl})_2^*$ could in principle decay to its ground state by internal conversion, a very rapid process (10 picoseconds; ps $= 10^{-12}$ s) in which the energy of the absorbed photon is converted to heat (molecular motion). Reaction centers are constructed to prevent the inefficiency that would result from internal conversion. The proteins of the reaction center hold the bacteriochlorophylls, bacteriopheophytins, and quinones in a fixed orientation relative to each other. The photochemical reactions among these components therefore take place in a virtually solid state. This accounts for the high efficiency and rapidity of the reactions; nothing is left to chance collision or depends on random diffusion. Exciton transfer from antenna chlorophyll to the special pair of the reaction center is accomplished in less than 100 ps with >90% efficiency. Within 3 ps of the excitation of P870, pheophytin has received an electron and become a negatively charged radical; less than 200 ps later, the electron has reached the quinone Q_B (Fig. 19–45b). The electron-transfer reactions not only are fast, but are thermodynamically "downhill"; the excited special pair $(\text{Chl})_2^*$ is a very good electron donor ($E'^\circ \approx -1$ V), and each successive electron transfer is to an acceptor of substantially less negative E'°. The standard free-energy change for the process is therefore negative and large; recall from Chapter 14 that $\Delta G'^\circ = -n\mathcal{J}\Delta E'^\circ$, where $\Delta E'^\circ$ is the difference between the standard reduction potentials of the two half-reactions:

$$(1)\quad (\text{Chl})_2^* \longrightarrow (\text{Chl})_2^{\cdot +} + e^- \qquad E'^\circ \approx -1\text{ V}$$

$$(2)\quad \text{Q} + 2\text{H}^+ + 2e^- \longrightarrow \text{QH}_2 \qquad E'^\circ = -0.045\text{ V}$$

Thus $\Delta E'^\circ = -0.045\text{ V} - (-1\text{ V}) \approx 0.95$ V, and

$$\Delta G'^\circ = -2(96.5\text{ kJ/V}\cdot\text{mol})(0.95\text{ V}) \approx -180\text{ kJ/mol}$$

The combination of fast kinetics and favorable thermodynamics makes the process virtually irreversible and highly efficient. The overall energy yield for the process (the percentage of the photon's energy conserved in QH_2) is >30%, with the remainder of the energy dissipated as heat.

In Higher Plants, Two Reaction Centers Act in Tandem

The photosynthetic apparatus of modern cyanobacteria, algae, and higher plants is more complex than the one-center bacterial systems, and it appears to have evolved through the combination of two simpler bacterial photo-centers. The thylakoid membranes of chloroplasts have two different kinds of photosystems, each with its own type of photochemical reaction center and set of antenna molecules. The two systems have distinct and complementary functions (Fig. 19–46). **Photosystem II (PSII)** is a pheophytin-quinone type of system (like the single photosystem of purple bacteria) containing roughly equal amounts of chlorophylls *a* and *b*. Excitation of its reaction center P680 drives electrons through the cytochrome b_6f complex with concomitant movement of protons across the thylakoid membrane. **Photosystem I (PSI)** is of the ferredoxin type, structurally and functionally related to the reaction center of green sulfur bacteria. It has a reaction center designated P700 and a high ratio of chlorophyll *a* to chlorophyll *b*. Excited P700 passes electrons to the Fe-S protein ferredoxin, then to $NADP^+$, producing NADPH. The thylakoid membranes of a single spinach chloroplast have many hundreds of each kind of photosystem.

These two reaction centers in plants act in tandem to catalyze the light-driven movement of electrons from H_2O to $NADP^+$ (Fig. 19–46). Electrons are carried between the two photosystems by the soluble protein plastocyanin, a one-electron carrier functionally similar to cytochrome *c* of mitochondria. To replace the electrons that move from PSII through PSI to $NADP^+$, cyanobacteria and plants oxidize H_2O (as green sulfur bacteria oxidize H_2S), producing O_2 (Fig. 19–46, bottom left). This process is called **oxygenic photosynthesis** to distinguish it from the anoxygenic photosynthesis of purple and green sulfur bacteria. All O_2-evolving photosynthetic cells—those of higher plants, algae, and cyanobacteria—contain both PSI

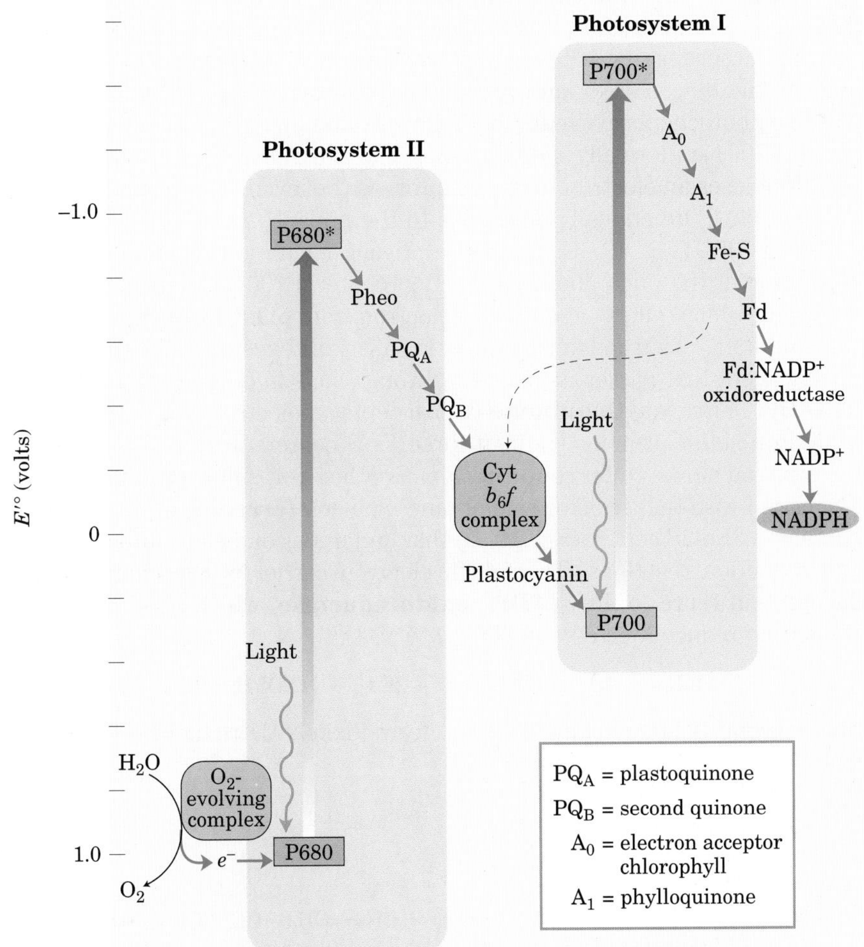

figure 19–46
Integration of photosystems I and II in chloroplasts. This "Z scheme" shows the pathway of electron transfer from H_2O (lower left) to $NADP^+$ (far right) in noncyclic photosynthesis. The position on the vertical scale of each electron carrier reflects its standard reduction potential. To raise the energy of electrons derived from H_2O to the energy level required to reduce $NADP^+$ to NADPH, each electron must be "lifted" twice (heavy arrows) by photons absorbed in PSI and PSII. One photon is required per electron in each photosystem. After excitation, the high-energy electrons flow "downhill" through the carrier chains shown. Protons move across the thylakoid membrane during the water-splitting reaction and during electron transfer through the cytochrome b_6f complex, producing the proton gradient that is central to ATP formation. The dashed arrow is the path of cyclic electron transfer (discussed later in the text), in which only PSI is involved; electrons return via the cyclic pathway to PSI, instead of reducing $NADP^+$ to NADPH.

and PSII; organisms with only one photosystem do not evolve O_2. The diagram in Figure 19–46, often called the **Z scheme** because of its overall form, outlines the pathway of electron flow between the two photosystems and the energy relationships in the light reactions. The Z scheme thus describes the complete route by which electrons flow from H_2O to $NADP^+$ according to the equation

$$2H_2O + 2NADP^+ + 8\text{ photons} \longrightarrow O_2 + 2NADPH + 2H^+$$

For every two photons absorbed (one by each photosystem), one electron is transferred from H_2O to $NADP^+$. To form one molecule of O_2, which requires transfer of four electrons from two H_2O to two $NADP^+$, a total of eight photons must be absorbed, four by each photosystem.

The mechanistic details of the photochemical reactions in PSII and PSI are essentially similar to those of the two bacterial photosystems. Excitation of P680 (in PSII) produces P680*, an excellent electron donor, which within picoseconds transfers an electron to **pheophytin,** giving it a negative charge ($Pheo^-$) (Fig. 19–46, left side). With the loss of its electron, P680* is transformed into a radical cation, P680$^+$. $Pheo^-$ very rapidly passes its extra electron to a protein-bound **plastoquinone,** PQ_A, which in turn passes its electron to another, more loosely bound plastoquinone, PQ_B. When PQ_B has acquired two electrons in two such transfers from PQ_A and two protons from the solvent water, it is in its fully reduced quinol form, PQ_BH_2. The overall reaction initiated by light in PSII is

$$4\text{ P680} + 4H^+ + 2PQ_B + 4\text{ photons} \longrightarrow 4\text{ P680}^+ + 2PQ_BH_2 \qquad (19\text{–}12)$$

Eventually, the electrons in PQ_BH_2 pass through the cytochrome b_6f complex. The binding site for plastoquinone is the point of action of many commercial herbicides that kill plants by blocking electron transfer through the cytochrome b_6f complex and preventing photosynthetic ATP production.

The photochemical events that follow excitation of PSI (at the reaction center P700) are formally similar to those in PSII. The excited reaction center P700* loses an electron to an acceptor, A_0 (believed to be a special form of chlorophyll, functionally analogous to the pheophytin of PSII), creating A_0^- and P700$^+$ (Fig. 19–46, right side); again, excitation results in charge separation at the photochemical reaction center. P700$^+$ is a strong oxidizing agent, which quickly acquires an electron from **plastocyanin,** a soluble Cu-containing electron-transfer protein. A_0^- is an exceptionally strong reducing agent that passes its electron through a chain of carriers that leads to $NADP^+$. First, **phylloquinone** (A_1) accepts an electron and passes it on to an iron-sulfur protein (through three Fe-S centers in PSI). From here, the electron moves to **ferredoxin** (Fd), another iron-sulfur protein loosely associated with the thylakoid membrane. Spinach **ferredoxin** (M_r 10,700) contains a 2Fe-2S center (Fig. 19–5) that undergoes one-electron oxidation and reduction reactions. The fourth electron carrier in the chain is the flavoprotein **ferredoxin: NADP$^+$ oxidoreductase,** which transfers electrons from reduced ferredoxin (Fd_{red}) to $NADP^+$:

$$2Fd_{red} + 2H^+ + NADP^+ \longrightarrow 2Fd_{ox} + NADPH + H^+$$

This enzyme is homologous with the ferredoxin-NAD reductase of green sulfur bacteria (Fig. 19–44b).

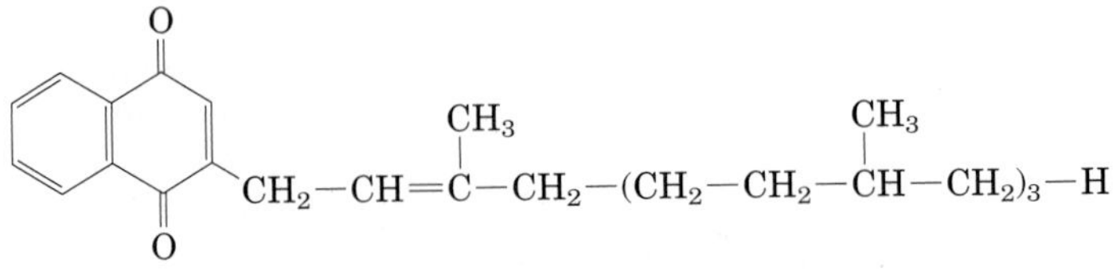

$$\text{(naphthoquinone ring)}-CH_2-CH=\overset{\overset{CH_3}{|}}{C}-CH_2-(CH_2-CH_2-\overset{\overset{CH_3}{|}}{CH}-CH_2)_3-H$$

Phylloquinone

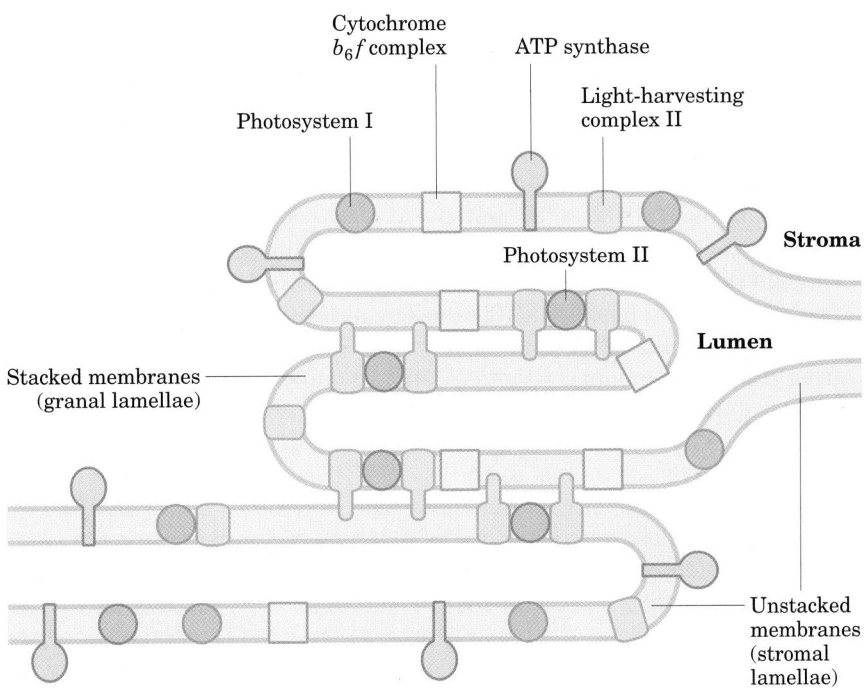

figure 19–47
Localization of PSI and PSII in thylakoid membranes. The light-harvesting complex LHCII and ATP synthase are located in both stacked (granal lamellae) and unstacked (stromal lamellae) regions of the thylakoid membrane, and have ready access to ADP and $NADP^+$ in the stroma. Photosystem II (PSII) is found almost exclusively in the stacked regions, and photosystem I (PSI) almost exclusively in unstacked regions exposed to the stroma. LHCII is the "adhesive" that holds stacked lamellae together (see Fig. 19–48).

Spatial Separation of Photosystems I and II Prevents Exciton Larceny

The energy required to excite PSI (P700) is less than that needed to excite PSII (P680) (shorter wavelength corresponds to higher energy). If PSI and PSII were physically contiguous, excitons originating in the antenna system of PSII would migrate to the reaction center of PSI, leaving PSII chronically underexcited and interfering with the operation of the two-center system. This exciton larceny is prevented by separation of PSI and PSII in the thylakoid membrane (Fig. 19–47). PSII is located almost exclusively in the tightly apposed membrane stacks of thylakoid grana (granal lamellae); its associated light-harvesting complex (LHCII) mediates the tight association of adjacent membranes in the grana. PSI and the ATP synthase complex are located almost exclusively in the unstacked thylakoid membranes (the stromal lamellae), where both have access to the contents of the stroma, including ADP and $NADP^+$. The cytochrome b_6f complex is present throughout the thylakoid membrane.

The association of LHCII with PSII is regulated by light intensity and wavelength. In bright sunlight (with a large component of blue light), PSII absorbs more light that PSI and produces reduced plastoquinone (plastoquinol, PQH_2) faster than PSI can oxidize it. The resulting accumulation of PQH_2 activates a protein kinase that phosphorylates a Thr residue on LHCII (Fig. 19–48). Phosphorylation weakens the interaction of LHCII with PSII, and some LHCII dissociates and moves to the stromal lamellae; here it captures photons for PSI, speeding the oxidation of PQH_2 and reversing the imbalance between electron flow in PSI and PSII. In less intense light with more red (in the shade), PSI oxidizes PQH_2 faster than PSII can make it, and the resulting increase in PQ concentration triggers dephosphorylation of LHCII, reversing the effect of phosphorylation.

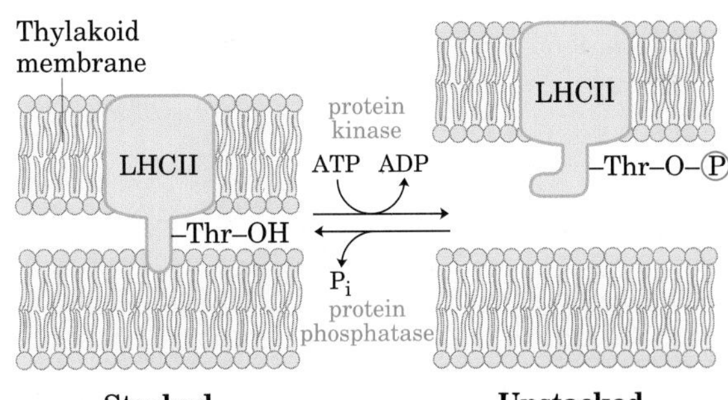

figure 19–48
Modulation of granal stacking equalizes electron flow in PSI and PSII. A hydrophobic domain of the light-harvesting complex LHCII in one thylakoid lamella inserts into the neighboring lamella and holds the two membranes in close apposition (granal lamellae). Accumulation of plastoquinol stimulates a protein kinase that phosphorylates a Thr residue in the hydrophobic domain of LHCII, which reduces its affinity for the neighboring thylakoid membrane, converting granal lamellae into stromal lamellae. A specific protein phosphatase reverses this regulatory phosphorylation when the concentration of PQ increases.

The Cytochrome b_6f Complex Links Photosystems II and I

Electrons temporarily stored in plastoquinol as a result of the excitation of P680 in PSII are carried to P700 of PSI via the cytochrome b_6f complex and the soluble protein plastocyanin (Fig. 19–46, center). Like Complex III of mitochondria, the cytochrome b_6f complex (Fig. 19–49) contains a *b*-type cytochrome with two heme groups (designated b_H and b_L), a Rieske iron-sulfur protein (M_r 20,000), and the *c*-type cytochrome c_{552}, commonly called cytochrome *f* (for the Latin *frons*, meaning "leaf"). Electrons flow through the cytochrome b_6f complex from PQ_BH_2 to cytochrome *f*, then to plastocyanin, and finally to P700, thereby reducing it.

Like Complex III of mitochondria, cytochrome b_6f conveys electrons from a reduced quinone—a mobile, lipid-soluble carrier of two electrons (Q in mitochondria, PQ_B in chloroplasts)—to a water-soluble protein that carries one electron (cytochrome *c* in mitochondria, plastocyanin in chloroplasts). As in mitochondria, the function of this complex involves a Q cycle (Fig. 19–11) in which electrons pass, one at a time, from PQ_BH_2 to cytochrome b_6. This cycle results in the pumping of protons across the membrane; in chloroplasts, the direction of proton movement is from the stromal compartment to the thylakoid lumen, up to four protons moving for each pair of electrons. The result is production of a proton gradient across the thylakoid membrane as electrons pass from PSII to PSI. Because the volume of the flattened thylakoid lumen is small, the influx of a small number of protons has a relatively large effect on lumenal pH. The measured difference in pH between the stroma (pH 8) and the thylakoid lumen (pH 5.0) represents a 1,000-fold difference in proton concentration—a powerful driving force for ATP synthesis.

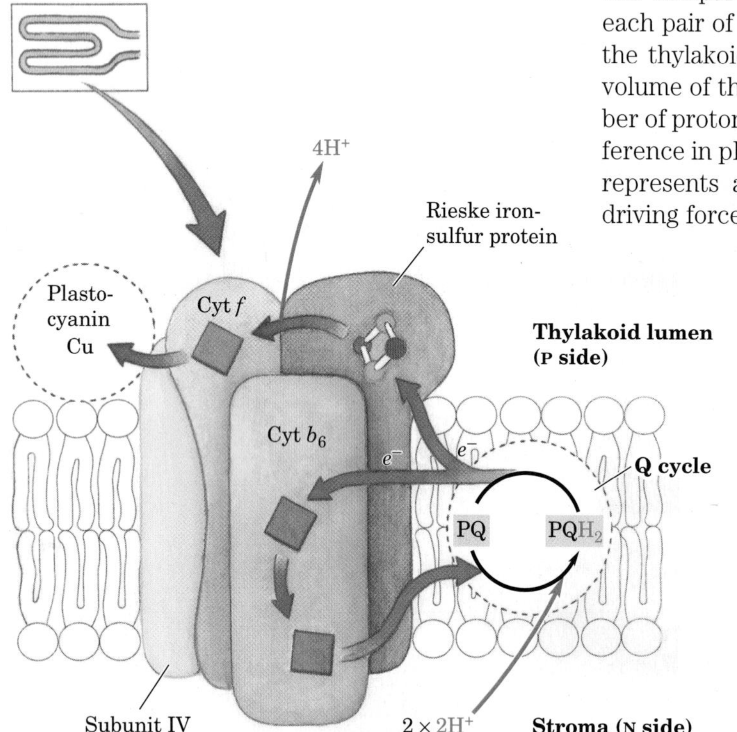

figure 19–49
Electron and proton flow through the cytochrome b_6f complex. Plastoquinol (PQH_2) formed in PSII is oxidized by the cytochrome b_6f in a series of steps like those of the Q cycle in the cytochrome bc_1 complex (Complex III) of mitochondria (see Fig. 19–11). One electron passes to the Fe-S center of the Rieske protein (purple), the other to the hemes of cytochrome b_6 (green). The net effect is passage of electrons from PQH_2 to the soluble protein plastocyanin, which carries them to PSI.

Cyanobacteria Use the Cytochrome b_6f Complex and Cytochrome *c* in Both Oxidative Phosphorylation and Photophosphorylation

Cyanobacteria can synthesize ATP by either oxidative phosphorylation or photophosphorylation, although they have neither mitochondria nor chloroplasts. The enzymatic machinery for both processes is in a highly convoluted plasma membrane (see Fig. 2–5). Two protein components function in both processes (Fig. 19–50). The proton-pumping cytochrome b_6f complex carries electrons from plastoquinone to cytochrome *c* in photosynthesis, and also carries electrons from ubiquinone to cytochrome *c* in oxidative phosphorylation—the role played by cytochrome bc_1 in higher plants. Cytochrome *c*, which in mitochondria carries electrons from Complex III to

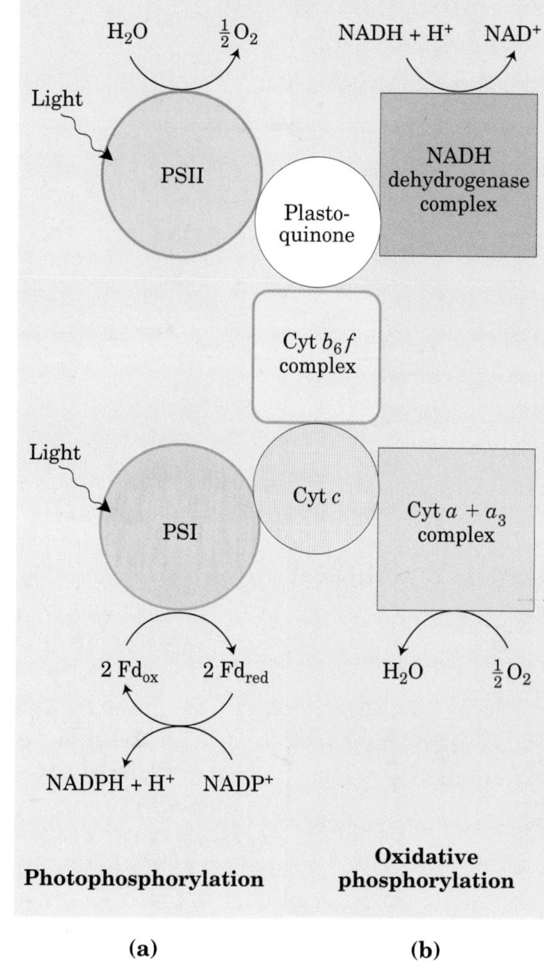

figure 19–50

Dual roles of cytochrome b_6f and cytochrome *c* in cyanobacteria. These organisms use cytochrome b_6f, cytochrome *c,* and plastoquinone for both oxidative phosphorylation and photophosphorylation. **(a)** In photophosphorylation, electrons flow from water to $NADP^+$. **(b)** In oxidative phosphorylation, electrons flow from NADH to O_2. Both processes are accompanied by proton movement across the membrane, accomplished by a Q cycle.

Complex IV, performs the same function in cyanobacteria, but also carries electrons from the cytochrome b_6f complex to PSI—a role carried out in higher plants by plastocyanin. We therefore see the functional homology between the cyanobacterial cytochrome b_6f complex and the plant cytochrome bc_1 complex, and between cyanobacterial cytochrome *c* and plant plastocyanin.

Water Is Split by the Oxygen-Evolving Complex

The ultimate source of the electrons passed to NADPH in plant (oxygenic) photosynthesis is water. Having given up an electron to pheophytin, $P680^+$ (of PSII) must acquire an electron to return to its ground state in preparation for capture of another photon. In principle, the required electron might come from any number of organic or inorganic compounds. Photosynthetic bacteria use a variety of electron donors for this purpose—acetate, succinate, malate, or sulfide—depending on what is available in a particular ecological niche. About 3 billion years ago, evolution of primitive photosynthetic bacteria (the progenitors of the modern cyanobacteria) produced a photosystem capable of taking electrons from a donor that is always available—water. In this process two water molecules are split, yielding four electrons, four protons, and molecular oxygen:

$$2H_2O \longrightarrow 4H^+ + 4e^- + O_2$$

A single photon of visible light does not have enough energy to break the bonds in water; four photons are required in this photolytic cleavage reaction.

The four electrons abstracted from water do not pass directly to $P680^+$, which can accept only one electron at a time. Instead, a remarkable molecular device, the **oxygen-evolving complex** (also called the **water-splitting complex),** passes four electrons *one at a time* to $P680^+$ (Fig. 19–51). The

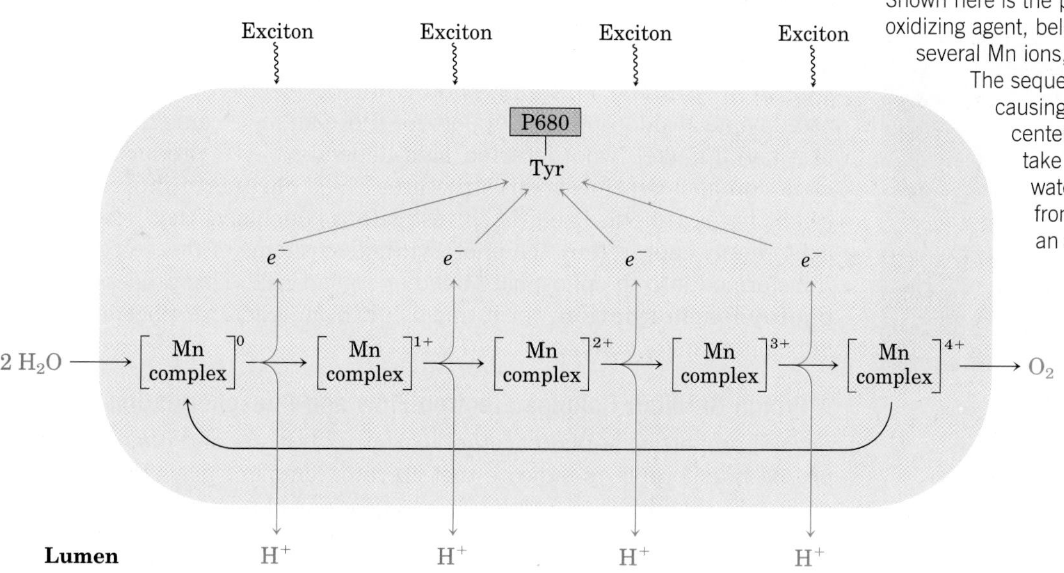

figure 19–51

Water-splitting activity of the oxygen-evolving complex. Shown here is the process that produces a four-electron oxidizing agent, believed to be a multinuclear center with several Mn ions, in the water-splitting complex of PSII. The sequential absorption of four photons, each causing the loss of one electron from the Mn center, produces an oxidizing agent that can take four electrons from two molecules of water, producing O_2. The electrons lost from the Mn center pass one at a time to an oxidized Tyr residue in a PSII protein.

immediate electron donor to $P680^+$ is a Tyr residue (often designated Z or Tyr_Z) in protein subunit D1 of the PSII reaction center. The Tyr residue loses both a proton and an electron, generating the electrically neutral Tyr free radical, $Tyr^\bullet$

$$4\ P680^+ + 4\ Tyr \longrightarrow 4\ P680 + 4\ Tyr^\bullet \qquad (19\text{–}13)$$

The Tyr radical regains its missing electron and proton by oxidizing a cluster of four manganese ions in the water-splitting complex. With each single-electron transfer, the Mn cluster becomes more oxidized; four single-electron transfers, each corresponding to the absorption of one photon, produce a charge of +4 on the Mn complex (Fig. 19–51):

$$4\ Tyr^\bullet + [\text{Mn complex}]^0 \longrightarrow 4\ Tyr + [\text{Mn complex}]^{4+} \qquad (19\text{–}14)$$

In this state, the Mn complex can take four electrons from a pair of water molecules, releasing 4 H^+ and O_2:

$$[\text{Mn complex}]^{4+} + 2H_2O \longrightarrow [\text{Mn complex}]^0 + 4H^+ + O_2 \qquad (19\text{–}15)$$

Because the four protons produced in this reaction are released into the thylakoid lumen, the oxygen-evolving complex acts as a proton pump, driven by electron transfer. The sum of Equations 19–12 through 19–15 is

$$2H_2O + 2PQ_B + 4\ \text{photons} \longrightarrow O_2 + 2PQ_BH_2 \qquad (19\text{–}16)$$

The water-splitting activity associated with the PSII reaction center has proved exceptionally difficult to purify. A peripheral membrane protein (M_r 33,000) on the lumenal side of the thylakoid membrane is believed to stabilize the Mn complex. The detailed structure of the Mn cluster is not yet known. Manganese can exist in stable oxidation states from +2 to +7, so a cluster of four Mn ions can certainly donate or accept four electrons. The mechanism shown in Figure 19–51 is consistent with the experimental facts, but until the exact chemical structures of all the intermediates of the Mn cluster are known, the detailed mechanism remains elusive.

ATP Synthesis By Photophosphorylation

The combined activities of the two plant photosystems move electrons from water to $NADP^+$, conserving some of the energy of absorbed light as NADPH (Fig. 19–46). Simultaneously, protons are pumped across the thylakoid membrane and energy is conserved as an electrochemical potential. We turn now to the process by which this proton gradient drives the synthesis of ATP, the other energy-conserving product of the light-dependent reactions.

In 1954 Daniel Arnon and his colleagues discovered that ATP is generated from ADP and P_i during photosynthetic electron transfer in illuminated spinach chloroplasts. Support for these findings came from the work of Albert Frenkel, who detected light-dependent ATP production in pigment-containing membranous structures called **chromatophores,** derived from photosynthetic bacteria. Investigators concluded that some of the light energy captured by the photosynthetic systems of these organisms is transformed into the phosphate bond energy of ATP. This process is called **photophosphorylation,** to distinguish it from oxidative phosphorylation in respiring mitochondria.

Daniel Arnon

A Proton Gradient Couples Electron Flow and Phosphorylation

Several properties of photosynthetic electron transfer and photophosphorylation in chloroplasts indicate that a proton gradient plays the same role as in mitochondrial oxidative phosphorylation. (1) The reaction centers, electron carriers, and ATP-forming enzymes are located in a proton-impermeant membrane—the thylakoid membrane—which must be intact to support photophosphorylation. (2) Photophosphorylation can be uncoupled from electron flow by reagents that promote the passage of protons through the thylakoid membrane. (3) Photophosphorylation can be blocked by venturicidin and similar agents that inhibit the formation of ATP from ADP and P_i by the mitochondrial ATP synthase (Table 19–4). (4) ATP synthesis is catalyzed by F_oF_1 complexes, located on the outer surface of the thylakoid membranes, that are very similar in structure and function to the F_oF_1 complexes of mitochondria.

Electron-transferring molecules in the chain of carriers connecting PSII and PSI are oriented asymmetrically in the thylakoid membrane, so that photoinduced electron flow results in the net movement of protons across the membrane, from the stromal side to the thylakoid lumen (Fig. 19–52).

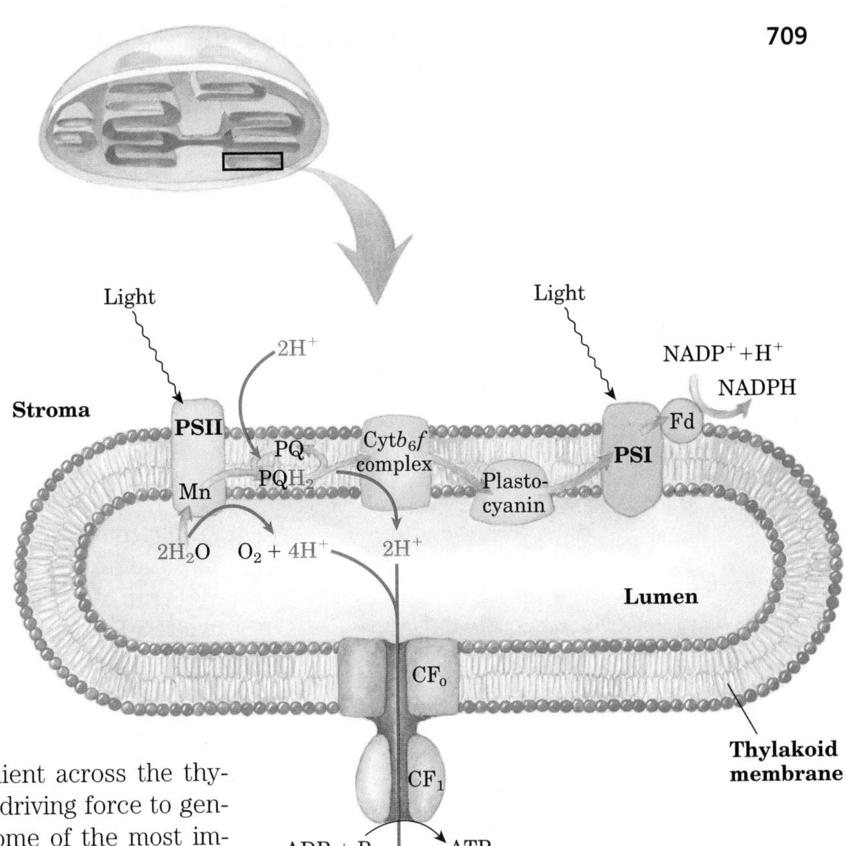

figure 19–52

Proton and electron circuits in thylakoids. Electrons (blue arrows) move from H_2O through PSII, the intermediate chain of carriers, PSI, and finally to $NADP^+$. Protons (red arrows) are pumped into the thylakoid lumen by the flow of electrons through the chain of carriers between PSII and PSI, and reenter the stroma through proton channels formed by the F_o portion of the ATP synthase, designated CF_o in the chloroplast enzyme. The F_1 subunit, CF_1, catalyzes synthesis of ATP.

In 1966 André Jagendorf showed that a pH gradient across the thylakoid membrane (alkaline outside) could furnish the driving force to generate ATP. Jagendorf's early observations provided some of the most important experimental evidence in support of Mitchell's chemiosmotic hypothesis. He incubated chloroplasts, in the dark, in a pH 4 buffer; the buffer slowly penetrated into the inner compartment of the thylakoids, lowering their internal pH. He added ADP and P_i to the dark suspension of chloroplasts and then suddenly raised the pH of the outer medium to 8, momentarily creating a large pH gradient across the membrane. As protons moved out of the thylakoids into the medium, ATP was generated from ADP and P_i. Because the formation of ATP occurred in the dark (with no input of energy from light), this experiment showed that a pH gradient across the membrane is a high-energy state that can, as in mitochondrial oxidative phosphorylation, mediate the transduction of energy from electron transfer into the chemical energy of ATP.

André Jagendorf

The Approximate Stoichiometry of Photophosphorylation Has Been Established

As electrons move from water to $NADP^+$, about 12 H^+ move from the stroma to the thylakoid lumen per four electrons passed (that is, per O_2 formed). Four of these protons are moved by the oxygen-evolving complex, and up to eight by the cytochrome b_6f complex. The measurable result is a 1,000-fold difference in proton concentration across the thylakoid membrane ($\Delta pH = 3$). Recall that the energy stored in a proton gradient (the electrochemical potential) has two components: a proton concentration difference (ΔpH) and an electrical potential ($\Delta\psi$) due to charge separation. In chloroplasts, ΔpH is the dominant component; counterion movement apparently dissipates most of the electrical potential. In illuminated chloroplasts, the energy stored in the proton gradient per mole of protons is

$$\Delta G = RT \ln \Delta pH + Z\mathcal{J}\Delta\psi = -17 \text{ kJ/mol}$$

so the movement of 12 mol of protons across the thylakoid membrane represents conservation of about 200 kJ of energy—enough energy to drive the synthesis of several moles of ATP ($\Delta G'^\circ = 30.5$ kJ/mol). Experimental measurements yield values for ATP per O_2 of about 3.

At least eight photons must be absorbed to drive four electrons from H_2O to NADPH (one photon per electron at each reaction center). The energy in eight photons of visible light is more than enough for the synthesis of three molecules of ATP.

ATP synthesis is not the only energy-conserving reaction of photosynthesis; the NADPH formed in the final electron transfer is also energetically rich, like its close analog NADH. The overall equation for noncyclic (see below) photophosphorylation is

$$2H_2O + 8 \text{ photons} + 2NADP^+ + {\sim}3ADP + {\sim}3P_i \longrightarrow O_2 + {\sim}3ATP + 2NADPH \qquad (19\text{–}17)$$

Cyclic Electron Flow Produces ATP but Not NADPH or O_2

An alternative path of light-induced electron flow allows chloroplasts to vary the ratio of NADPH and ATP formed in the light; this is called **cyclic electron flow** to differentiate it from the normally unidirectional or **noncyclic electron flow** from H_2O to $NADP^+$, as discussed thus far. Cyclic electron flow (Fig. 19–46) involves only PSI. Electrons passed from P700 to ferredoxin do not continue to $NADP^+$, but move back through the cytochrome b_6f complex to plastocyanin. The path of electrons matches that in green sulfur bacteria (Fig. 19–44b). Plastocyanin donates electrons to P700, which transfers them to ferredoxin when illuminated. Thus in the light, PSI can cause electrons to cycle continuously out of the reaction center of PSI and back into it, each electron propelled around the cycle by the energy yielded by absorption of one photon. Cyclic electron flow is not accompanied by net formation of NADPH or evolution of O_2. However, it is accompanied by proton pumping by the cytochrome b_6f complex and by phosphorylation of ADP to ATP, referred to as **cyclic photophosphorylation.** The overall equation for cyclic electron flow and photophosphorylation is simply

$$ADP + P_i \xrightarrow{\text{light}} ATP + H_2O$$

By regulating the partitioning of electrons between $NADP^+$ reduction and cyclic photophosphorylation, a plant adjusts the ratio of ATP and NADPH produced in the light-dependent reactions to match its needs for these products in the carbon-assimilation reactions. As we shall see in Chapter 20, the carbon-assimilation reactions require ATP and NADPH in the ratio 3:2.

The ATP Synthase of Chloroplasts Is Like That of Mitochondria

The enzyme responsible for ATP synthesis in chloroplasts is a large complex with two functional components, CF_o and CF_1 (the C denoting its origin in chloroplasts). CF_o is a transmembrane proton pore composed of several integral membrane proteins and is homologous with mitochondrial F_o. CF_1 is a peripheral membrane protein complex very similar in subunit composition, structure, and function to mitochondrial F_1.

Electron microscopy of sectioned chloroplasts shows ATP synthase complexes as knoblike projections on the *outside* (stromal or N) surface of thylakoid membranes; these complexes correspond to the ATP synthase complexes seen to project on the *inside* (matrix or N) surface of the inner mitochondrial membrane. Thus both the orientation of the ATP synthase and the direction of proton pumping in chloroplasts are opposite to those in mitochondria. In both cases, the F_1 portion of ATP synthase is located on the more alkaline side (N) of the membrane through which protons flow down their concentration gradient; the direction of proton flow relative to F_1 is the same in both cases: P to N (Fig. 19–53).

The mechanism of chloroplast ATP synthase is also believed to be essentially identical to that of its mitochondrial analog; ADP and P_i readily

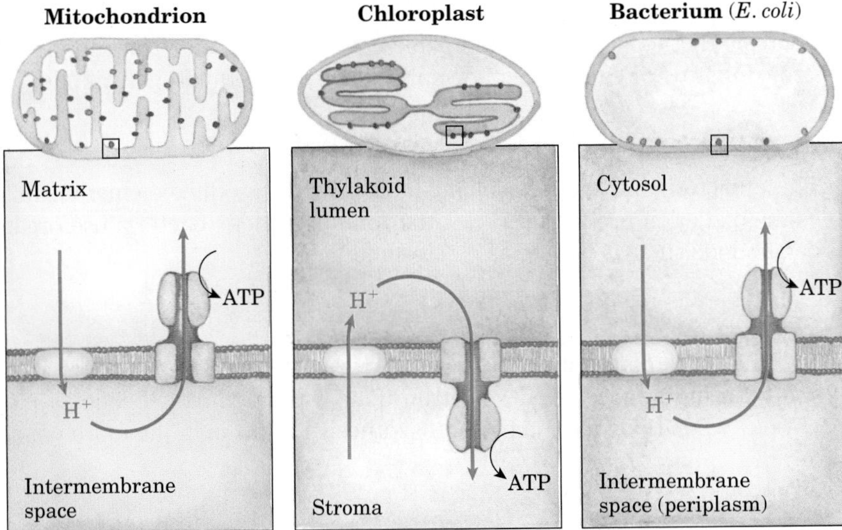

figure 19–53
Comparison of the topology of proton movement and ATP synthase orientation in the membranes of mitochondria, chloroplasts, and the bacterium *E. coli.* In each case, orientation of the proton gradient relative to the ATP synthase is the same.

condense to form ATP on the enzyme surface, and the release of this enzyme-bound ATP requires a proton-motive force. Rotational catalysis sequentially engages each of the three β subunits of the ATP synthase in ATP synthesis, ATP release, and ADP + P_i binding (Figs. 19–23, 19–24).

Chloroplasts Probably Evolved from Endosymbiotic Bacteria

Like mitochondria, chloroplasts contain their own DNA and protein-synthesizing machinery. Some of the polypeptides of chloroplast proteins are encoded by chloroplast genes and synthesized in the chloroplast; others are encoded by nuclear genes, synthesized outside the chloroplast, and imported (see Fig. 27–39b). When plant cells grow and divide, chloroplasts give rise to new chloroplasts by division, during which their DNA is replicated and divided between daughter chloroplasts. The machinery and mechanism for light capture, electron flow, and ATP synthesis in photosynthetic bacteria are similar in many respects to those in the chloroplasts of higher plants. These observations led to the now widely accepted hypothesis that the evolutionary progenitors of modern plant cells were primitive eukaryotes that engulfed photosynthetic bacteria and established stable endosymbiotic relationships with them (see Fig. 2–15). By several criteria, plant PSII may have been derived from the reaction centers of purple bacteria, whereas PSI may have been derived from green bacteria.

Prochlorophytes, photosynthetic bacteria that contain both chlorophyll *a* and chlorophyll *b* but not phycobilins, are the likely progenitors of the chloroplasts of modern higher plants. Comparisons of the DNA sequences in genes that encode photosystem proteins in higher plants and in prochlorophytes support this evolutionary relationship. Cyanobacteria use both chlorophyll *a* and phycobilins to harvest light and are probably the progenitors of the chloroplasts of modern red algae, which also use chlorophyll *a* and phycobilins.

Diverse Photosynthetic Organisms Use Hydrogen Donors Other Than Water

At least half of the photosynthetic activity on Earth occurs in microorganisms—algae, other photosynthetic eukaryotes, and photosynthetic bacteria. Cyanobacteria have PSII and PSI in tandem, and the PSII has an associated H_2O-splitting activity resembling that of plants. However, the other groups of photosynthetic bacteria have single reaction centers and do not split H_2O or produce O_2. Many are obligate anaerobes and cannot tolerate

O_2; they must use some compound other than H_2O as electron donor. Some photosynthetic bacteria use inorganic compounds as electron (and hydrogen) donors. For example, green sulfur bacteria use hydrogen sulfide:

$$2H_2S + CO_2 \xrightarrow{\text{light}} (CH_2O) + H_2O + 2S$$

These bacteria, instead of evolving molecular O_2, produce elemental sulfur as the oxidation product of H_2S. Other photosynthetic bacteria use organic compounds such as lactate as electron donors:

$$2 \text{ Lactate} + CO_2 \xrightarrow{\text{light}} (CH_2O) + H_2O + 2 \text{ pyruvate}$$

The fundamental similarity of photosynthesis in plants and bacteria, despite the differences in the electron donors they employ, becomes more obvious when the equation of photosynthesis is written in the more general form

$$2H_2D + CO_2 \xrightarrow{\text{light}} (CH_2O) + H_2O + 2D$$

in which H_2D is an electron (and hydrogen) donor and D is its oxidized form. H_2D may be water, hydrogen sulfide, lactate, or some other organic compound, depending on the species. It is very likely that the bacteria that first developed photosynthetic ability used H_2S as their electron source, and that only after the later development of oxygenic photosynthesis (about 2.3 billion years ago) did oxygen become a significant proportion of the earth's atmosphere. With that development, the evolution of electron-transfer systems that used O_2 as their ultimate electron acceptor became possible, leading to the highly efficient energy extraction of oxidative phosphorylation.

In Halophilic Bacteria, a Single Protein Absorbs Light and Pumps Protons to Drive ATP Synthesis

The halophilic ("salt-loving") bacterium *Halobacterium salinarum*, an archaebacterium derived from very ancient evolutionary progenitors, traps the energy of sunlight in a process very different from the photosynthetic mechanisms we have described. This bacterium lives only in brine ponds and salt lakes (Great Salt Lake and the Dead Sea, for example), where the high salt concentration—which can exceed 4 M—results from water loss by evaporation; indeed, halobacteria cannot live in NaCl concentrations lower than 3 M. These organisms are aerobes and normally use O_2 to oxidize organic fuel molecules. However, the solubility of O_2 is so low in brine ponds that sometimes oxidative metabolism must be supplemented by sunlight as an alternative source of energy.

The plasma membrane of *H. salinarum* contains patches of the light-absorbing pigment **bacteriorhodopsin,** which contains retinal (the aldehyde derivative of vitamin A; see Fig. 11–19) as a prosthetic group. When the cells are illuminated, all-*trans*-retinal bound to the bacteriorhodopsin absorbs a photon and undergoes photoisomerization to 13-*cis*-retinal. The restoration of all-*trans*-retinal is accompanied by the outward movement of protons through the plasma membrane. Bacteriorhodopsin, with only 247 amino acid residues, is the simplest light-driven proton pump known. The difference in the three-dimensional structure of bacteriorhodopsin in the dark and after illumination (Fig. 19–54a) suggests a pathway by which a concerted series of proton "hops" could effectively move a proton across the membrane as the conformational change occurs. The chromophore retinal is bound through a Schiff-base linkage to the ϵ-amino group of a Lys residue. In the dark the N of this Schiff base is protonated, but photoisomerization of retinal lowers the pK_a of this group and it releases its proton to a nearby Asp residue, triggering a series of proton hops that ultimately result in the release of a proton at the outer surface of the membrane (Fig. 19–54b).

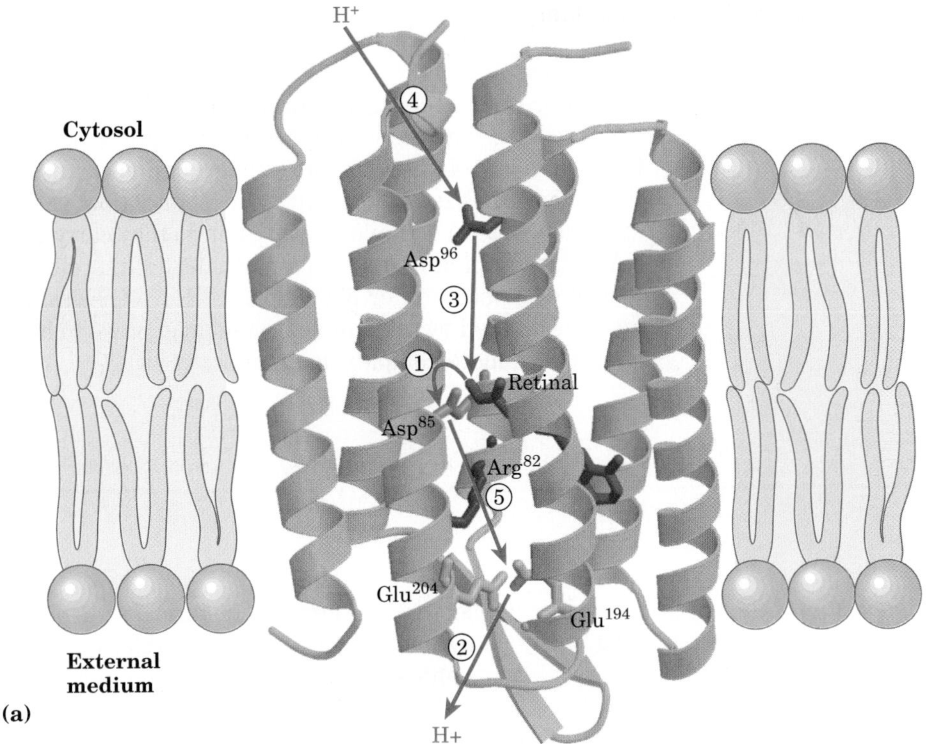

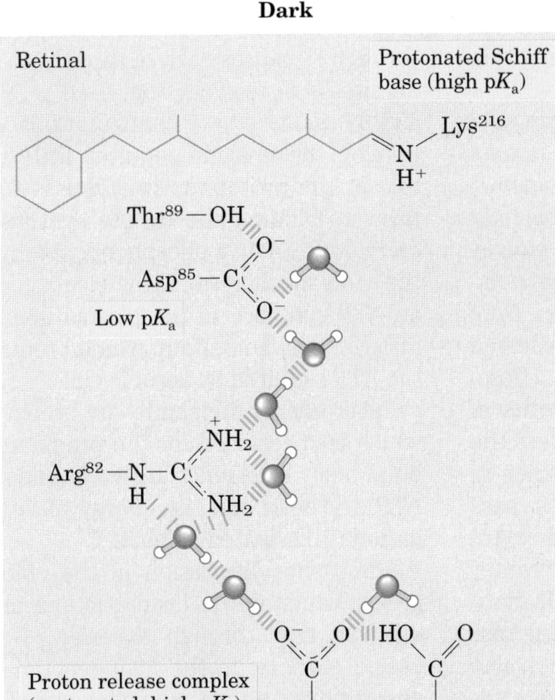

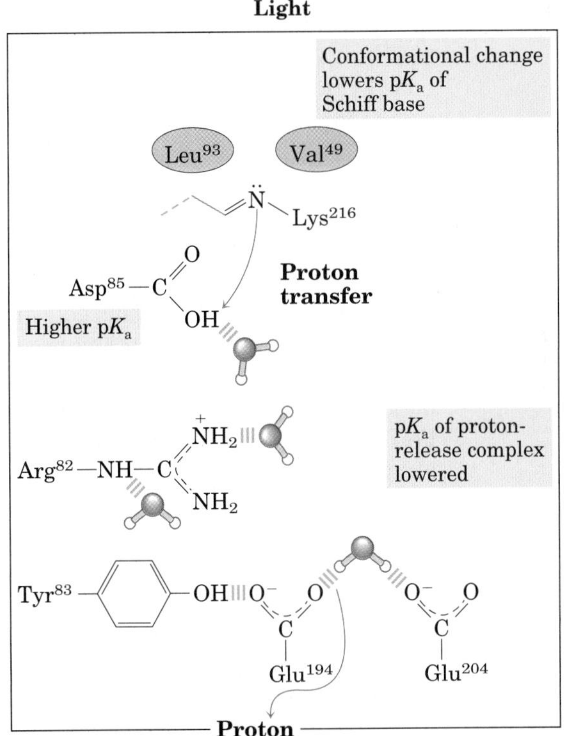

figure 19–54

Light-driven proton pumping by bacteriorhodopsin. **(a)** Bacteriorhodopsin (M_r 26,000) has seven membrane-spanning α helices; the chromophore all-*trans*-retinal (purple) is covalently attached via a Schiff base to the ϵ-amino group of a Lys residue deep in the membrane interior. Running through the protein are a series of Asp and Glu residues and a series of closely associated water molecules that together provide the transmembrane path for protons (red arrows). Steps ① through ⑤ indicate proton movements, described below. **(b)** In the dark (left panel), the Schiff base is protonated. Illumination (right panel) photoisomerizes the retinal, forcing subtle conformational changes in the protein that alter the distance between the Schiff base and its neighboring amino acid residues. Interaction with these neighbors lowers the pK_a of the protonated Schiff base, and the base gives up its proton to a nearby carboxyl group on Asp^{85} (step ① in **(a)**). This initiates a series of concerted proton hops between water molecules (see Fig. 4–12) in the interior of the protein, which ends with step ②, the release of a proton that was shared by Glu^{194} and Glu^{204} near the extracellular surface. The Schiff base reacquires a proton from Asp^{96} (③), which in turn takes up a proton from the cytosol (④). Finally, Asp^{85} gives up its proton, leading to the reprotonation of the Glu^{204}–Glu^{194} pair (⑤). The system is now ready for another round of proton pumping.

The resulting electrochemical potential drives protons back into the cell through a membrane ATP synthase complex very similar to that of mitochondria and chloroplasts. Thus halobacteria can use light to supplement the ATP synthesized by oxidative phosphorylation when O_2 is limited. However, halobacteria do not evolve O_2, nor do they carry out photoreduction of $NADP^+$; their phototransducing machinery is therefore much simpler than that of cyanobacteria or higher plants. Nevertheless, the proton-pumping mechanism used by this simple protein may prove to be prototypical for the many other, more complex, ion pumps.

summary

Chemiosmotic theory provides the intellectual framework for understanding many biological energy transductions, including oxidative phosphorylation and photophosphorylation. The mechanism of energy coupling is similar in both cases: the energy of electron flow is conserved by the concomitant pumping of protons across the membrane, producing an electrochemical gradient, the proton-motive force.

In mitochondria, H atoms removed from substrates by NAD-linked dehydrogenases donate electrons to the respiratory (electron-transfer) chain, which transfers the electrons to molecular O_2, reducing it to H_2O. Shuttle systems convey reducing equivalents from cytosolic NADH to mitochondrial NADH. Reducing equivalents from all NAD-linked dehydrogenations are transferred to mitochondrial NADH dehydrogenase (Complex I). They are then passed through a series of Fe-S centers to ubiquinone, which transfers the electrons to cytochrome *b*, the first carrier in Complex III. In this complex, electrons pass through two *b*-type cytochromes and cytochrome c_1 to an Fe-S center. The Fe-S center passes electrons, one at a time, through cytochrome *c* and into Complex IV, cytochrome oxidase. This copper-containing enzyme, which also contains cytochromes *a* and a_3, accumulates electrons then passes them to O_2, reducing it to H_2O.

Some electrons enter this chain of carriers through alternative paths. Succinate, for example, is oxidized by succinate dehydrogenase (Complex II), which contains a flavoprotein (with FAD) that passes electrons through several Fe-S centers to ubiquinone. Electrons derived from the oxidation of fatty acids pass to ubiquinone via the electron-transferring flavoprotein (ETF).

The flow of electrons through Complexes I, III, and IV results in the pumping of protons across the mitochondrial inner membrane, making the matrix alkaline relative to the intermembrane space. This proton gradient provides the energy (proton-motive force) for ATP synthesis from ADP and P_i by the inner-membrane ATP synthase (F_oF_1 complex). This enzyme carries out "rotational catalysis" in which the flow of protons through F_o causes each of three nucleotide-binding sites in F_1 to cycle from (ADP + P_i)-bound to ATP-bound to empty configurations. ATP formation on the enzyme requires little energy; the role of the proton-motive force is to push ATP from its binding site on the synthase. Bacteria carry out oxidative phosphorylation by essentially the same mechanism, using electron carriers and an ATP synthase in the plasma membrane. Oxidative phosphorylation, which produces most of the ATP required by aerobic cells, is regulated by cellular energy demands. In brown fat tissue, which is specialized for the production of metabolic heat, electron transfer is uncoupled from ATP synthesis and the energy of fatty acid oxidation is dissipated as heat.

Photophosphorylation in the chloroplasts of green plants and in cyanobacteria also involves electron flow through a series of membrane-bound carriers. In the light reactions of plants, absorption of a photon excites chlorophyll molecules and other (accessory) pigments, which funnel the energy into reaction centers in the thylakoid membranes. In the reaction centers, photoexcitation results in a charge separation that produces a strong electron donor (reducing agent) and a strong electron acceptor. Bacteria have a single reaction center; in purple bacteria, it is of the pheophytin-quinone type, and in green sulfur bacteria, the Fe-S type. Structural studies of the reaction center of a purple bacterium have provided information about light-driven electron flow from an excited special pair of chlorophyll molecules, through pheophytin, to quinones. Electrons then pass from quinones through the cytochrome bc_1 complex, and back to the photoreaction center. An alternative path, in green sulfur bacteria, sends electrons from reduced quinones to NAD^+. Electron flow through the cytochrome bc_1 complex drives protons across the plasma membrane, creating a proton-motive force that provides the energy for ATP synthesis by an ATP synthase similar to that of mitochondria.

Cyanobacteria and plants have two different photoreaction centers, arranged in tandem. Plant photosystem I passes electrons from its excited reaction center, P700, through a series of carriers to ferredoxin, which then reduces $NADP^+$ to NADPH. The reaction center of photosystem II, P680, passes electrons to plastoquinone, and the electrons lost from P680 are replaced by electrons from H_2O (electron donors other than H_2O are used in other organisms). This light-driven splitting of H_2O is catalyzed by a Mn-containing protein complex; O_2 is produced. The reduced plastoquinone carries electrons to the cytochrome b_6f complex; from here they pass to plastocyanin, and then to P700 to replace those lost during its photoexcitation. Both the water-splitting and the electron flow through the cytochrome b_6f complex are accompanied by proton pumping across the thylakoid membrane, and the proton-motive force thus created drives ATP synthesis by a CF_oCF_1 complex closely similar to the mitochondrial F_oF_1 complex. This flow of electrons through the photosystems thus produces both NADPH and ATP, in the ratio of about 2:3. A second type of electron flow (cyclic flow) produces ATP only and allows variability in the proportions of NADPH and ATP formed. The localization of PSI and PSII between the granal and stromal lamellae is variable and is indirectly controlled by light intensity, optimizing the distribution of excitons between PSI and PSII for efficient energy capture.

Both mitochondria and chloroplasts contain their own genomes and are believed to have originated from prokaryotic endosymbionts of early eukaryotic cells. Oxidative phosphorylation in aerobic bacteria and photophosphorylation in photosynthetic bacteria are closely similar, in machinery and mechanism, to the homologous processes in mitochondria and chloroplasts. Cyanobacteria use several enzyme complexes for both oxidative phosphorylation and photophosphorylation, illustrating the similarity between the two processes. Bacteriorhodopsin is the best understood and simplest light-driven proton pump and photosynthetic system.

further reading

History and General Background

Arnon, D.I. (1984) The discovery of photosynthetic phosphorylation. *Trends Biochem. Sci.* **9,** 258–262.

Beinert, H. (1995) These are the moments when we live! From Thunberg tubes and manometry to phone, fax and FedEx. In *Selected Topics in the History of Biochemistry: Personal Recollections,* IV, Comprehensive Biochemistry, Vol. 38, Elsevier Science Publishing Co., Inc., New York.

An engaging personal account of the exciting period in which the biochemistry of respiratory electron transfer was worked out.

Blankenship, R.E., Madigan, M.T., & Bauer, C.E. (eds) (1995) *Anoxygenic Photosynthetic Bacteria,* Advances in Photosynthesis, Vol. 2, Kluwer Academic Publishers, Dordrecht, The Netherlands.

The 60 chapters cover every aspect of photosynthesis in bacteria.

Gray, M.W., Berger, G., & Lang, B.F. (1999) Mitochondrial evolution. *Science* **283,** 1476–1481.

Compact review of the endosymbiotic origin hypothesis and the evidence for and against it.

Harold, F.M. (1986) *The Vital Force: A Study in Bioenergetics,* W.H. Freeman and Company, New York.

A very readable synthesis of the principles of bioenergetics and their application to energy transductions.

Harris, D.A. (1995) *Bioenergetics at a Glance,* Blackwell Science, London.

A short (100 pages), clearly written, well-illustrated summary of energy transductions in mitochondria, chloroplasts, and bacteria.

Heldt, H.-W. (1997) *Plant Biochemistry and Molecular Biology,* Oxford University Press, Oxford.

A textbook of plant biochemistry with excellent discussions of photophosphorylation.

Keilin, D. (1966) *The History of Cell Respiration and Cytochrome,* Cambridge University Press, London.

An authoritative and absorbing account of the discovery of cytochromes and their roles in respiration, written by the man who discovered cytochromes.

Mitchell, P. (1979) Keilin's respiratory chain concept and its chemiosmotic consequences. *Science* **206,** 1148–1159.

Mitchell's Nobel lecture, outlining the evolution of the chemiosmotic hypothesis.

Ort, D.R., Yocum, C.F., & Heichel, I.F. (eds) (1996) *Oxygenic Photosynthesis: The Light Reactions,* Advances in Photosynthesis, Vol. 4, Kluwer Academic Publishers, Dordrecht, The Netherlands.

A comprehensive, advanced treatise.

Raghavendra, A.S. (ed.) (1998) *Photosynthesis: A Comprehensive Treatise*, Cambridge University Press, Cambridge.

An advanced treatment of all aspects of photosynthesis.

Skulachev, V.P. (1992) The laws of cell energetics. *Eur. J. Biochem.* **208,** 203–209.

Discussion of the interconvertibility of ATP and ion gradients.

Slater, E.C. (1987) The mechanism of the conservation of energy of biological oxidations. *Eur. J. Biochem.* **166,** 489–504.

A clear and critical account of the evolution of the chemiosmotic model.

Oxidative Phosphorylation

Respiratory Electron Flow

Babcock, G.T. & Wikström, M. (1992) Oxygen activation and the conservation of energy in cell respiration. *Nature* **356,** 301–309.

Advanced discussion of the reduction of water and pumping of protons by cytochrome oxidase.

Brandt, U. (1997) Proton-translocation by membrane-bound NADH:ubiquinone-oxidoreductase (complex I) through redox-gated ligand conduction. *Biochim. Biophys. Acta* **1318,** 79–91.

Advanced discussion of models for electron movement through Complex I.

Brandt, U. & Trumpower, B. (1994) The protonmotive Q cycle in mitochondria and bacteria. *Crit. Rev. Biochem. Mol. Biol.* **29,** 165–197.

Crofts, A.R. & Berry, E.A. (1998) Structure and function of the cytochrome bc_1 complex of mitochondria and photosynthetic bacteria. *Curr. Opin. Struct. Biol.* **8,** 501–509.

Douce, R. & Neuburger, M. (1989) The uniqueness of plant mitochondria. *Annu. Rev. Plant Physiol. Plant Mol. Biol.* **40,** 371–414.

A focus on the features of plant mitochondria that distinguish them from mitochondria of animal cells.

Michel, H., Behr, J., Harrenga, A., & Kannt, A. (1998) Cytochrome *c* oxidase: structure and spectroscopy. *Annu. Rev. Biophys. Biomol. Struct.* **27,** 329–356.

Advanced review of Complex IV structure and function.

Rottenberg, H. (1998) The generation of proton electrochemical potential gradient by cytochrome *c* oxidase. *Biochim. Biophys. Acta* **1364,** 1–16.

Soole, K.L. & Menz, R.I. (1995) Functional molecular aspects of the NADH dehydrogenases of plant mitochondria. *J. Bioenerg. Biomembr.* **27,** 397–406.

Tsukihara, T., Aoyama, H., Yamashita, E., Tomizaki, T., Yamaguchi, H., Shinzawa-Itoh, K., Nakashima, R., Yaono, R., & Yoshikawa, S. (1996) The whole structure of the 13-subunit oxidized cytochrome *c* oxidase at 2.8 Å. *Science* **272,** 1136–1144.

The solution by x-ray crystallography of the structure of this huge membrane protein.

Vanlerberghe, G.C. (1997) Alternative oxidase: from gene to function. *Annu. Rev. Plant Physiol. Plant Mol. Biol.* **48,** 703–734.

Advanced review of the alternative oxidase of plant mitochondria.

Xia, D., Yu, C.-A., Kim, H., Xia, J.-Z., Kachurin, A.M., Zhang, L., Yu, L., & Deisenhofer, J. (1997) Crystal structure of the cytochrome bc_1 complex from bovine heart mitochondria. *Science* **277,** 60–66.

Crystallographic structure of Complex III.

Coupling ATP Synthesis to Respiratory Electron Flow

Abrahams, J.P., Leslie, A.G.W., Lutter, R., & Walker, J.E. (1994) The structure of F_1-ATPase from bovine heart mitochondria determined at 2.8 Å resolution. *Nature* **370,** 621–628.

Bianchet, M.A., Hullihen, J., Pedersen, P.L., & Amzel, L.M. (1998) The 2.80 Å structure of rat liver F_1-ATPase: configuration of a critical intermediate in ATP synthesis-hydrolysis. *Proc. Natl. Acad. Sci. USA* **95,** 11,065–11,070.

Research paper that provided important structural detail in support of the catalytic mechanism.

Boyer, P.D. (1997) The ATP synthase—a splendid molecular machine. *Annu. Rev. Biochem.* **66,** 717–749.

An account of the historical development and current state of the binding-change model, written by its principal architect.

Hinkle, P.C., Kumar, M.A., Resetar, A., & Harris, D.L. (1991) Mechanistic stoichiometry of mitochondrial oxidative phosphorylation. *Biochemistry* **30,** 3576–3582.

A careful analysis of experimental results and theoretical considerations on the question of nonintegral P/O ratios.

Junge, W., Lill, H., & Engelbrecht, S. (1997) ATP synthase: an electrochemical transducer with rotatory mechanics. *Trends Biochem. Sci.* **22,** 420–424.

A short, clear description of the evidence for the rotary motion of ATP synthase.

Khan, S. (1997) Rotary chemiosmotic machines. *Biochim. Biophys. Acta* **1322,** 86–105.

Detailed review of the structures that underlie proton-driven rotary motion of ATP synthase and bacterial flagella.

Kinosita, K., Jr., Yasuda, R., Noji, H., Ishiwata, S., & Yoshida, M. (1998) F_1ATPase: a rotary motor made of a single molecule. *Cell* **93,** 21–24.

A short review of the evidence for rotation of ATP synthase.

Sambongi, Y., Iko, Y., Tanabe, M., Omote, H., Iwamoto-Kihara, A., Ueda, I., Yanagida, T., Wada, Y., & Futai, M. (1999) Mechanical rotation of the c subunit oligomer in ATP synthase (F_oF_1): direct observation. *Science* **286,** 1722–1724.

The experimental evidence for rotation of the entire cylinder of c subunits in F_oF_1.

Stock, D., Leslie, A.G.W., & Walker, J.E. (1999) Molecular architecture of the rotary motor in ATP synthase. *Science* **286,** 1700–1705.

The first crystallographic view of the F_o subunit, in the yeast F_oF_1. See also the editorial comment of R.H. Fillingame in the same issue of *Science.*

Weber, J. & Senior, A.E. (1997) Catalytic mechanism of F_1-ATPase. *Biochim. Biophys. Acta* **1319,** 19–58.

An advanced review of kinetic, structural, and biochemical evidence for the ATP synthase mechanism.

Yasuda, R., Noji, H., Kinosita, Jr., K., & Yoshida, M. (1998) F_1-ATPase is a highly efficient molecular motor that rotates with discrete 120° steps. *Cell* **93,** 1117–1124.

Graphical demonstration of the rotation of ATP synthase.

Regulation of Mitochondrial Oxidative Phosphorylation

Brand, M.D. & Murphy, M.P. (1987) Control of electron flux through the respiratory chain in mitochondria and cells. *Biol. Rev. Camb. Philos. Soc.* **62,** 141–193.

An advanced description of respiratory control.

Harris, D.A. & Das, A.M. (1991) Control of mitochondrial ATP synthesis in the heart. *Biochem. J.* **280,** 561–573.

Advanced discussion of the regulation of ATP synthase by Ca^{2+} and other factors.

Klingenberg, M. & Huang, S.-G. (1999) Structure and function of the uncoupling protein from brown adipose tissue. *Biochim. Biophys. Acta* **1415,** 271–296.

Mitochondrial Diseases

Schapira, A.H.V. (1999) Mitochondrial disorders. *Biochim. Biophys. Acta* **1410,** 99–102.

A short editorial introduction to an entire volume (eight reviews) on the role of mitochondria in human diseases.

Wallace, D.C. (1997) Mitochondrial DNA in aging and disease. *Sci. Am.* **277** (August), 40–47.

Wallace, D.C. (1999) Mitochondrial disease in man and mouse. *Science* **283,** 1482–1487.

Photosynthesis

Harvesting Light Energy

Cogdell, R.J., Isaacs, N.W., Howard, T.D., McLuskey, K., Fraser, N.J., & Prince, S.M. (1999) How photosynthetic bacteria harvest solar energy. *J. Bacteriol.* **181,** 3869–3879.

A short, intermediate level review of the structure and function of the light-harvesting complex of the purple bacteria and exciton flow to the reaction center.

Green, B.R., Pichersky, E., & Kloppstech, K. (1991) Chlorophyll *a/b*-binding proteins: an extended family. *Trends Biochem. Sci.* **16,** 181–186.

An intermediate-level description of the proteins that orient chlorophyll molecules in chloroplasts.

Zuber, H. (1986) Structure of light-harvesting antenna complexes of photosynthetic bacteria, cyanobacteria and red algae. *Trends Biochem. Sci.* **11,** 414–419.

Light-Driven Electron Flow

Cogdell, R.J., Isaacs, N.W., Howard, T.D., McLuskey, K., Fraser, N.J., & Prince, S.M. (1999) How photosynthetic bacteria harvest solar energy. *J. Bacteriol.* **181,** 3869–3879.

Review of recent developments in the chemistry and biology of bacterial light-harvesting and the photoreaction center.

Deisenhofer, J. & Michel, H. (1991) Structures of bacterial photosynthetic reaction centers. *Annu. Rev. Cell Biol.* **7,** 1–23.

The structure of the reaction center of purple bacteria, and implications for the function of bacterial and plant reaction centers.

Huber, R. (1990) A structural basis of light energy and electron transfer in biology. *Eur. J. Biochem.* **187,** 283–305.

Huber's Nobel lecture, describing the physics and chemistry of phototransductions; an exceptionally clear and well-illustrated discussion, based on crystallographic studies of reaction centers.

Jagendorf, A.T. (1967) Acid-base transitions and phosphorylation by chloroplasts. *Fed. Proc.* **26,** 1361–1369.

The classic experiment establishing the ability of a proton gradient to drive ATP synthesis in the dark.

Kerfeld, C.A. & Krogmann, D.W. (1998) Photosynthetic cytochromes *c* in cyanobacteria, algae, and plants. *Annu. Rev. Plant Physiol. Plant Mol. Biol.* **49,** 397–425.

Water-Splitting Complex

Rögner, M., Boekema, E.J., & Barber, J. (1996) How does photosystem 2 split water? The structural basis of efficient energy conversion. *Trends Biochem. Sci.* **21,** 44–49.

Szalai, V.A. & Brudvig, G.W. (1998) How plants produce dioxygen. *Am. Sci.* **86,** 542–551.

A well-illustrated introduction to the oxygen-evolving complex of plants.

Bacteriorhodopsin

Luecke, H., Schobert, B., Richter, H.-T., Cartailler, J.-P., & Lanyi, J.K. (1999) Structural changes in bacteriorhodopsin during ion transport at 2 angstrom resolution. *Science* **286,** 255–264.

This paper, and an editorial comment in the same *Science* issue, describe the model for H^+ translocation by proton hopping.

problems

1. Oxidation-Reduction Reactions The NADH dehydrogenase complex of the mitochondrial respiratory chain promotes the following series of oxidation-reduction reactions, in which Fe^{3+} and Fe^{2+} represent the iron in iron-sulfur centers, Q is ubiquinone, QH_2 is ubiquinol, and E is the enzyme:

(1) $\text{NADH} + \text{H}^+ + \text{E-FMN} \rightarrow \text{NAD}^+ + \text{E-FMNH}_2$

(2) $\text{E-FMNH}_2 + 2\text{Fe}^{3+} \rightarrow \text{E-FMN} + 2\text{Fe}^{2+} + 2\text{H}^+$

(3) $2\text{Fe}^{2+} + 2\text{H}^+ + \text{Q} \rightarrow 2\text{Fe}^{3+} + \text{QH}_2$

Sum: $\text{NADH} + \text{H}^+ + \text{Q} \rightarrow \text{NAD}^+ + \text{QH}_2$

For each of the three reactions catalyzed by the NADH dehydrogenase complex, identify (a) the electron donor, (b) the electron acceptor, (c) the conjugate redox pair, (d) the reducing agent, and (e) the oxidizing agent.

2. All Parts of Ubiquinone Have a Function In electron transfer, only the quinone portion of ubiquinone undergoes oxidation-reduction; the isoprenoid side chain remains unchanged. What is the function of this chain?

3. Use of FAD Rather Than NAD$^+$ in Succinate Oxidation All the dehydrogenases of glycolysis and the citric acid cycle use NAD$^+$ (E'° for NAD$^+$/NADH is -0.32 V) as electron acceptor except succinate dehydrogenase, which uses covalently bound FAD (E'° for FAD/FADH$_2$ in this enzyme is 0.05 V). Suggest why FAD is a more appropriate electron acceptor than NAD$^+$ in the dehydrogenation of succinate, based on the E'° values of fumarate/succinate ($E'^\circ = 0.03$), NAD$^+$/NADH, and the succinate dehydrogenase FAD/FADH$_2$.

4. Degree of Reduction of Electron Carriers in the Respiratory Chain The degree of reduction of each carrier in the respiratory chain is determined by conditions in the mitochondrion. For example, when NADH and O_2 are abundant, the steady-state degree of reduction of the carriers decreases as electrons pass from the substrate to O_2. When electron transfer is blocked, the carriers before the block become more reduced and those beyond the block become more oxidized (see Fig. 19–6). For each of the conditions below, predict the state of oxidation of ubiquinone and cytochromes b, c_1, c, and $a + a_3$.

(a) Abundant NADH and O_2, but cyanide added

(b) Abundant NADH, but O_2 exhausted

(c) Abundant O_2, but NADH exhausted

(d) Abundant NADH and O_2

5. Effect of Rotenone and Antimycin A on Electron Transfer Rotenone, a toxic natural product from plants, strongly inhibits NADH dehydrogenase of insect and fish mitochondria. Antimycin A, a toxic antibiotic, strongly inhibits the oxidation of ubiquinol.

(a) Explain why rotenone ingestion is lethal to some insect and fish species.

(b) Explain why antimycin A is a poison.

(c) Assuming that rotenone and antimycin A are equally effective in blocking their respective sites in the electron-transfer chain, which would be a more potent poison? Explain.

6. Uncouplers of Oxidative Phosphorylation In normal mitochondria the rate of electron transfer is tightly coupled to the demand for ATP. When the rate of use of ATP is relatively low, the rate of electron transfer is low; when demand for ATP increases, electron-transfer rate increases. Under these conditions of tight coupling, the number of ATP molecules produced per atom of oxygen consumed when NADH is the electron donor—the P/O ratio—is about 2.5.

(a) Predict the effect of a relatively low and a relatively high concentration of uncoupling agent on the rate of electron transfer and the P/O ratio.

(b) Ingestion of uncouplers causes profuse sweating and an increase in body temperature. Explain this phenomenon in molecular terms. What happens to the P/O ratio in the presence of uncouplers?

(c) The uncoupler 2,4-dinitrophenol was once prescribed as a weight-reducing drug. How could this agent, in principle, serve as a weight-reducing aid? Uncoupling agents are no longer prescribed because some deaths occurred following their use. Why might the ingestion of uncouplers lead to death?

7. Effects of Valinomycin on Oxidative Phosphorylation When the antibiotic valinomycin is added to actively respiring mitochondria, several things happen: the yield of ATP decreases, the rate of O_2 consumption increases, heat is released, and the pH gradient across the inner mitochondrial membrane increases. Does valinomycin act as an uncoupler or an inhibitor of oxidative phosphorylation? Explain the experimental observations in terms of the antibiotic's ability to transfer K^+ ions across the inner mitochondrial membrane.

8. Mode of Action of Dicyclohexylcarbodiimide (DCCD) When DCCD is added to a suspension of tightly coupled, actively respiring mitochondria, the rate of electron transfer (measured by O_2 consumption) and the rate of ATP production dramatically decrease. If a solution of 2,4-dinitrophenol is now added to the preparation, O_2 consumption returns to normal but ATP production remains inhibited.

(a) What process in electron transfer or oxidative phosphorylation is affected by DCCD?

(b) Why does DCCD affect the O_2 consumption of mitochondria? Explain the effect of 2,4-dinitrophenol on the inhibited mitochondrial preparation.

(c) Which of the following inhibitors does DCCD most resemble in its action: antimycin A, rotenone, or oligomycin?

9. Compartmentalization of Citric Acid Cycle Components Isocitrate dehydrogenase is found only in the mitochondrion, but malate dehydrogenase is found in both the cytosol and mitochondrion. What is the role of cytosolic malate dehydrogenase?

10. The Malate–α-Ketoglutarate Transport System The transport system that conveys malate and α-ketoglutarate across the inner mitochondrial membrane (Fig. 19–26) is inhibited by *n*-butylmalonate. Suppose *n*-butylmalonate is added to an aerobic suspension of kidney cells using glucose exclusively as fuel. Predict the effect of this inhibitor on (a) glycolysis, (b) oxygen consumption, (c) lactate formation, and (d) ATP synthesis.

11. Cellular ADP Concentration Controls ATP Formation Although both ADP and P_i are required for the synthesis of ATP, the rate of synthesis depends mainly on the concentration of ADP, not P_i. Why?

12. The Pasteur Effect When O_2 is added to an anaerobic suspension of cells consuming glucose at a high rate, the rate of glucose consumption declines dramatically as the O_2 is used up, and accumulation of lactate ceases. This effect, first observed by Louis Pasteur in the 1860s, is characteristic of most cells capable of both aerobic and anaerobic glucose catabolism.

(a) Why does the accumulation of lactate cease after O_2 is added?

(b) Why does the presence of O_2 decrease the rate of glucose consumption?

(c) How does the onset of O_2 consumption slow down the rate of glucose consumption? Explain in terms of specific enzymes.

13. Respiration-Deficient Yeast Mutants and Ethanol Production Respiration-deficient yeast mutants (p^-; "petites") can be produced from wild-type parents by treatment with mutagenic agents. The mutants lack cytochrome oxidase, a deficit that markedly affects their metabolic behavior. One striking effect is that fermentation is not suppressed by O_2—that is, the mutants lack the Pasteur effect (see Problem 14). Some companies are very interested in using these mutants to ferment wood chips to ethanol for energy use. Explain the advantages of using these mutants rather than wild-type yeast for large-scale ethanol production. Why does the absence of cytochrome oxidase eliminate the Pasteur effect?

14. How Many Protons in a Mitochondrion? Electron transfer translocates protons from the mitochondrial matrix to the external medium, establishing a pH gradient across the inner membrane (outside more acidic than inside). The tendency of protons to diffuse back into the matrix is the driving force for ATP synthesis by ATP synthase. During oxidative phosphorylation by a suspension of mitochondria in a medium of pH 7.4, the pH of the matrix has been measured as 7.7.

(a) Calculate $[H^+]$ in the external medium and in the matrix under these conditions.

(b) What is the outside: inside ratio of $[H^+]$? Comment on the energy inherent in this concentration. (Hint: See p. 416, Eqn 12–3.)

(c) Calculate the number of protons in a respiring liver mitochondrion, assuming its inner matrix compartment is a sphere of diameter 1.5 μm.

(d) From these data, is the pH gradient alone sufficient to generate ATP?

(e) If not, suggest how the necessary energy for synthesis of ATP arises.

15. Rate of ATP Turnover in Rat Heart Muscle Rat heart muscle operating aerobically fills more than 90% of its ATP needs by oxidative phosphorylation. Each gram of tissue consumes O_2 at the rate of 10 μmol/min, with glucose as the fuel source.

(a) Calculate the rate at which the heart muscle consumes glucose and produces ATP.

(b) For a steady-state concentration of ATP of 5 μmol/g of heart muscle tissue, calculate the time required (in seconds) to completely turn over the cellular pool of ATP. What does this result indicate about the need for tight regulation of ATP production? (Note: Concentrations are expressed as micromoles per gram of muscle tissue because the tissue is mostly water.)

16. Rate of ATP Breakdown in Flight Muscle ATP production in the flight muscle of the fly *Lucilia sericata* results almost exclusively from oxidative phosphorylation. During flight, 187 ml of O_2/h · g of body weight is needed to maintain an ATP concentration of 7 μmol/g of flight muscle. Assuming that flight muscle makes up 20% of the weight of the fly, calculate the rate at which the flight-muscle ATP pool turns over. How long would the reservoir of ATP last in the absence of oxidative phosphorylation? Assume that reducing equivalents are transferred by the glycerol 3-phosphate shuttle and that O_2 is at 25 °C and 101.3 kPa (1 atm). (Note: Concentrations are expressed in micromoles per gram of flight muscle.)

17. Transmembrane Movement of Reducing Equivalents Under aerobic conditions, extramitochondrial NADH must be oxidized by the mitochondrial electron-transfer chain. Consider a preparation of rat hepatocytes containing mitochondria and all the cytosolic enzymes. If [4-^{3}H]NADH is introduced, radioactivity soon appears in the mitochondrial matrix. However, if [7-^{14}C]NADH is introduced, no radioactivity appears in the matrix. What do these observations reveal about the oxidation of extramitochondrial NADH by the electron-transfer chain?

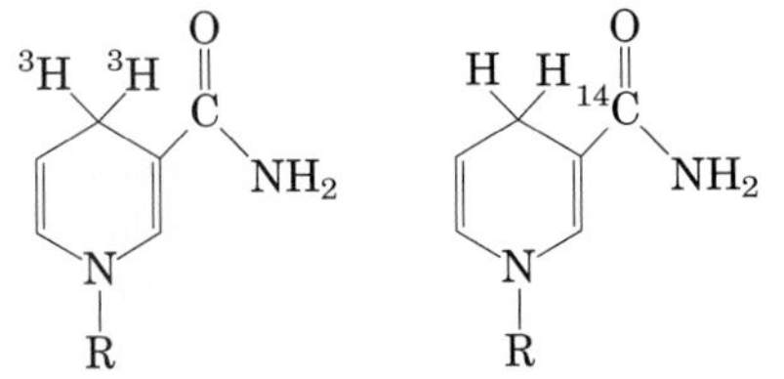

[4-^{3}H]NADH [7-^{14}C]NADH

18. NAD Pools and Dehydrogenase Activities Although both pyruvate dehydrogenase and glyceraldehyde 3-phosphate dehydrogenase use NAD^+ as their electron acceptor, the two enzymes do not compete for the same cellular NAD pool. Why?

19. Photochemical Efficiency of Light at Different Wavelengths The rate of photosynthesis, measured by O_2 production, is higher when a green plant is illuminated with light of wavelength 680 nm than with light of 700 nm. However, illumination by a combination of light of 680 nm and 700 nm gives a higher rate of photosynthesis than light of either wavelength alone. Explain.

20. Balance Sheet for Photosynthesis In 1804 Theodore de Saussure observed that the total weights of oxygen and dry organic matter produced by plants is greater than the weight of carbon dioxide consumed during photosynthesis. Where does the extra weight come from?

21. Role of H_2S in Some Photosynthetic Bacteria Illuminated purple sulfur bacteria carry out photosynthesis in the presence of H_2O and $^{14}CO_2$, but only if H_2S is added and O_2 is absent. During the course of photosynthesis, measured by formation of [^{14}C]carbohydrate, H_2S is converted to elemental sulfur, but no O_2 is evolved. What is the role of the conversion of H_2S to sulfur? Why is no O_2 evolved?

22. Boosting the Reducing Power of Photosystem I by Light Absorption When photosystem I absorbs red light at 700 nm, the standard reduction potential of P700 changes from 0.4 V to about −1.2 V. What fraction of the absorbed light is trapped in the form of reducing power?

23. Limited ATP Synthesis in the Dark Spinach chloroplasts are illuminated in the absence of ADP and P_i, then the light is turned off and ADP and P_i are added. ATP is then synthesized for a short time in the dark. Explain this finding.

24. Mode of Action of the Herbicide DCMU When chloroplasts are treated with 3-(3,4-dichlorophenyl)-1,1-dimethylurea (DCMU, or Diuron), a potent herbicide, O_2 evolution and photophosphorylation cease. Oxygen evolution, but not photophosphorylation, can be restored by addition of an external electron acceptor, or Hill reagent. How does DCMU act as a weed killer? Suggest a location for the inhibitory action of this herbicide in the scheme shown in Figure 19–46. Explain.

25. Bioenergetics of Photophosphorylation The steady-state concentrations of ATP, ADP, and P_i in isolated spinach chloroplasts under full illumination at pH 7.0 are 120, 6, and 700 μm, respectively.

(a) What is the free-energy requirement for the synthesis of 1 mol of ATP under these conditions?

(b) The energy for ATP synthesis is furnished by light-induced electron transfer in the chloroplasts. What is the minimum voltage drop necessary (during transfer of a pair of electrons) to synthesize ATP under these conditions? (You may need to refer to p. 517, Eqn 14–6.)

26. Light Energy for a Redox Reaction Suppose you have isolated a new photosynthetic microorganism that oxidizes H_2S and passes the electrons to NAD^+. What wavelength of light would provide enough energy for H_2S to reduce NAD^+ under standard conditions? Assume 100% efficiency in the photochemical event, and use E'° of −230 mV for H_2S and −320 mV for NAD^+. See Figure 19–36 for energy equivalents of wavelengths of light.

27. Equilibrium Constant for Water-Splitting Reactions The coenzyme $NADP^+$ is the terminal electron acceptor in chloroplasts, according to the reaction

$$2H_2O + 2NADP^+ \longrightarrow 2NADPH + 2H^+ + O_2$$

Use the information in Table 19–2 to calculate the equilibrium constant for this reaction at 25 °C. (The relationship between K'_{eq} and $\Delta G'^\circ$ is discussed on p. 494.) How can the chloroplast overcome this unfavorable equilibrium?

28. Energetics of Phototransduction During photosynthesis, eight photons must be absorbed (four by each photosystem) for every O_2 molecule produced:

$$2H_2O + 2NADP^+ + 8 \text{ photons} \longrightarrow 2NADPH + 2H^+ + O_2$$

Assuming that these photons have a wavelength of 700 nm (red) and that the absorption and use of light energy are 100% efficient, calculate the free-energy change for the process.

29. Electron Transfer to a Hill Reagent Isolated spinach chloroplasts evolve O_2 when illuminated in the presence of potassium ferricyanide (the Hill reagent), according to the equation

$$2H_2O + 4Fe^{3+} \longrightarrow O_2 + 4H^+ + 4Fe^{2+}$$

where Fe^{3+} represents ferricyanide and Fe^{2+}, ferrocyanide. Is NADPH produced in this process? Explain.

30. How Often Does a Chlorophyll Molecule Absorb a Photon? The amount of chlorophyll *a* (M_r 892) in a spinach leaf is about 20 μg/cm^2 of leaf. In noonday sunlight (average energy 5.4 J/cm^2 · min), the leaf absorbs about 50% of the radiation. How often does a single chlorophyll molecule absorb a photon? Given that the average lifetime of an excited chlorophyll molecule in vivo is 1 ns, what fraction of chlorophyll molecules is excited at any one time?

31. Effect of Monochromatic Light on Electron Flow The extent to which an electron carrier is oxidized or reduced during photosynthetic electron transfer can sometimes be observed directly with a spectrophotometer. When chloroplasts are illuminated with 700 nm light, cytochrome *f*, plastocyanin, and plastoquinone are oxidized. When chloroplasts are illuminated with 680 nm light, however, these electron carriers are reduced. Explain.

32. Function of Cyclic Photophosphorylation When the [NADPH]/[NADP$^+$] ratio in chloroplasts is high, photophosphorylation is predominantly cyclic (see Fig. 19–46). Is O_2 evolved during cyclic photophosphorylation? Is NADPH produced? Explain. What is the main function of cyclic photophosphorylation?

chapter

20 Carbohydrate Biosynthesis

Sugar cane carries out photosynthesis, in which sunlight drives the reduction of CO_2 to carbohydrates, which are stored in the cane as sucrose. Photosynthetic production of carbohydrates provides reduced substrates, which other organisms can consume as fuels (or confections!).

We have now reached a turning point in our study of cellular metabolism. Thus far in Part III we have described how the major foodstuffs—carbohydrates, fatty acids, and amino acids—are degraded via converging *catabolic* pathways to enter the citric acid cycle and yield their electrons to the respiratory chain, with this exergonic flow of electrons to oxygen coupled to the endergonic synthesis of ATP. We now turn to *anabolic* pathways, which use chemical energy in the form of ATP and NADH or NADPH to synthesize cellular components from simple precursor molecules. Anabolic pathways are generally reductive rather than oxidative. Catabolism and anabolism proceed simultaneously in a dynamic steady state, so that the energy-yielding degradation of cellular components is counterbalanced by biosynthetic processes, which create and maintain the intricate orderliness of living cells.

Several organizing principles of biosynthesis deserve emphasis at the outset. First, molecules are synthesized and degraded by different pathways and, though the two opposing (anabolic and catabolic) pathways may share many reversible reactions, each has at least one enzymatic step unique to that pathway and essentially irreversible. If the reactions of catabolism and anabolism were catalyzed by the same set of enzymes acting reversibly, the direction of carbon flow through these pathways would be dictated exclusively by mass action (p. 494), not by the cell's changing needs for energy, precursors, or macromolecules.

Second, corresponding anabolic and catabolic pathways are controlled at one or more of the reactions unique to each pathway. Opposing pathways are regulated in a coordinated, reciprocal manner, so that stimulation of the biosynthetic pathway is accompanied by inhibition of the corresponding degradative pathway, and vice versa. Like all complex pathways, a biosynthetic pathway is usually regulated at an early exergonic step that commits intermediates to that pathway. By regulating an early step, the cell avoids wasting precursors to make unneeded intermediates; intrinsic economy prevails in the molecular logic of living cells.

Third, energy-requiring biosynthetic processes are coupled to the energy-yielding breakdown of ATP in such a way that the overall process is essentially irreversible in vivo. Thus the total amount of ATP and NAD(P)H energy used in a given biosynthetic pathway always exceeds the minimum amount of free energy required to convert the precursor into the biosynthetic product. As a result, the biosynthetic process is thermodynamically favorable—the free-energy change for the overall process remains negative even when the concentrations of precursors are relatively low.

This chapter provides many opportunities for elaboration of the three principles outlined above, as we describe pathways for carbohydrate biosynthesis. The chapter is divided into four parts. First we consider gluconeogenesis, the pathway for synthesis of glucose from simpler precursors. We then describe how glucose is converted into a variety of disaccharides and polysaccharides: glycogen in animals and many microorganisms, starch and sucrose in plants. At this point the focus shifts entirely to plants. The third topic is the incorporation of CO_2 into more complex molecules (CO_2 assimilation), the ultimate source of reduced carbon compounds for all organisms. The chapter ends with a discussion of the regulation of carbohydrate metabolism in plants. The overall regulation of carbohydrate metabolism in mammals is covered within a broader discussion of hormonal regulation in Chapter 23.

Gluconeogenesis

We begin our survey of biosynthetic processes with the central pathway that leads to the formation of carbohydrates from noncarbohydrate precursors (Fig. 20–1). Our focus is primarily on animal tissues. The biosynthesis of glucose is an absolute necessity in all mammals, because the brain and nervous system, as well as the erythrocytes, testes, renal medulla, and embryonic tissues, require glucose from the blood as their sole or major fuel source. The human brain alone requires over 120 g of glucose each day.

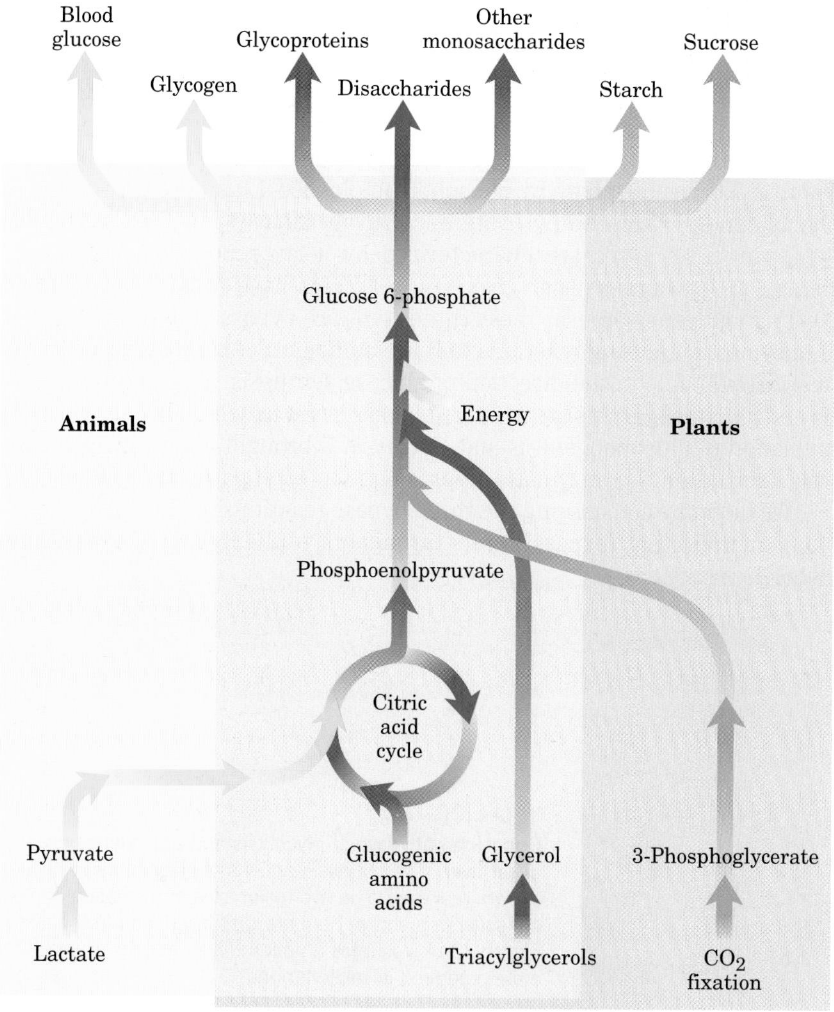

figure 20–1
Carbohydrate synthesis from simple precursors. The pathway from phosphoenolpyruvate to glucose 6-phosphate is common to the biosynthetic conversion of many different precursors to carbohydrates in animals and plants. Plants and photosynthetic bacteria are uniquely able to convert CO_2 into carbohydrates.

The formation of glucose from nonhexose precursors is called **gluconeogenesis** ("formation of new sugar"). Gluconeogenesis occurs in all animals, plants, fungi, and microorganisms. The reactions are essentially the same in every case. The important precursors of glucose in animals are lactate, pyruvate, glycerol, and certain amino acids (Fig. 20–1). In higher animals gluconeogenesis occurs largely in the liver and to a much smaller extent in renal cortex. The glucose produced passes into the blood to supply other tissues. In plant seedlings, stored fats and proteins are converted to the disaccharide sucrose for transport throughout the developing plant. Glucose and its derivatives are precursors in the synthesis of plant cell walls, nucleotides and coenzymes, and a variety of other essential metabolites. Many microorganisms can grow on simple organic compounds such as acetate, lactate, and propionate, which they convert to glucose by gluconeogenesis.

Although the reactions of gluconeogenesis are the same in all organisms, the metabolic context and the regulation of the pathway differ from one species to another and from tissue to tissue. In this chapter we focus first on gluconeogenesis as it occurs in the mammalian liver. We then consider its role and regulation in plants.

Just as the glycolytic conversion of glucose to pyruvate is a central pathway of carbohydrate catabolism, the conversion of pyruvate to glucose is a central pathway in gluconeogenesis. In animals, both pathways occur largely in the cytosol, necessitating their reciprocal and coordinated regulation. These pathways are not identical, although they share several steps (Fig. 20–2); seven of the ten enzymatic reactions of gluconeogenesis are the reverse of glycolytic reactions (discussed in Chapter 15).

Three reactions of glycolysis are essentially irreversible in vivo and cannot be used in gluconeogenesis: the conversion of glucose to glucose 6-phosphate by hexokinase, the phosphorylation of fructose 6-phosphate to fructose 1,6-bisphosphate by phosphofructokinase-1, and the conversion of phosphoenolpyruvate to pyruvate by pyruvate kinase (Fig. 20–2). In cells, these three reactions are characterized by a large negative free-energy change, ΔG, whereas other glycolytic reactions have a ΔG near 0 (Table 20–1). In gluconeogenesis, these three steps are bypassed by a separate set of enzymes, catalyzing reactions that are sufficiently exergonic to be effectively irreversible in the direction of glucose synthesis. Thus, both glycolysis and gluconeogenesis are irreversible processes in cells. The independent regulation of gluconeogenesis and glycolysis is brought about through controls exerted on the enzymatic steps unique to each pathway.

We begin by considering the three bypass reactions of gluconeogenesis. (Keep in mind that "bypass" refers throughout to the bypass of irreversible glycolytic reactions.)

figure 20–2

Opposing pathways of glycolysis and gluconeogenesis in rat liver. The bypass reactions of gluconeogenesis are shown in orange. The two major sites of regulation of gluconeogenesis shown here are discussed later in the text. Figure 20–4 illustrates an alternative route for oxaloacetate produced in mitochondria.

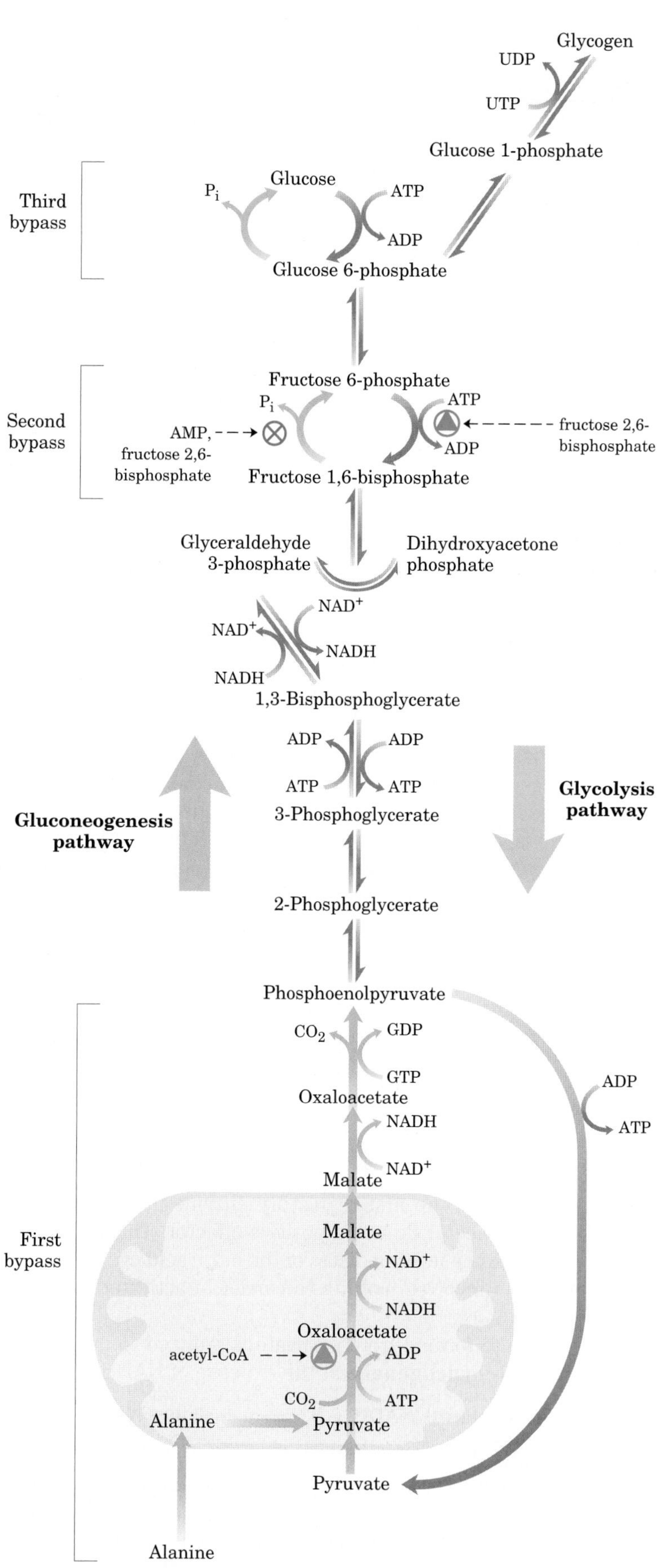
Glycogen
UDP
UTP
Glucose 1-phosphate
Third bypass
Pi
Glucose
ATP
ADP
Glucose 6-phosphate
Second bypass
Fructose 6-phosphate
Pi
ATP
AMP,
fructose 2,6-bisphosphate
fructose 2,6-bisphosphate
ADP
Fructose 1,6-bisphosphate
Glyceraldehyde 3-phosphate
Dihydroxyacetone phosphate
NAD+
NAD+
NADH
NADH
1,3-Bisphosphoglycerate
ADP
ADP
ATP
ATP
3-Phosphoglycerate
Gluconeogenesis pathway
Glycolysis pathway
2-Phosphoglycerate
Phosphoenolpyruvate
CO2
GDP
GTP
Oxaloacetate
NADH
NAD+
Malate
ADP
ATP
First bypass
Malate
NAD+
NADH
Oxaloacetate
acetyl-CoA
ADP
CO2
ATP
Alanine
Pyruvate
Pyruvate
Alanine

table 20–1

Free-Energy Changes of Glycolytic Reactions in Erythrocytes*

Glycolytic reaction step	$\Delta G'^{\circ}$ (kJ/mol)	ΔG (kJ/mol)
① Glucose + ATP ⟶ glucose 6-phosphate + ADP + H^+	−16.7	−33.4
② Glucose 6-phosphate ⇌ fructose 6-phosphate	1.7	−2.5
③ Fructose 6-phosphate + ATP ⟶ fructose 1,6-bisphosphate + ADP + H^+	−14.2	−22.2
④ Fructose 1,6-bisphosphate ⇌ dihydroxyacetone phosphate + glyceraldehyde 3-phosphate	23.8	−1.25
⑤ Dihydroxyacetone phosphate ⇌ glyceraldehyde 3-phosphate	7.5	2.5
⑥ Glyceraldehyde 3-phosphate + P_i + NAD^+ ⇌ 1,3-bisphosphoglycerate + NADH + H^+	6.3	−1.7
⑦ 1,3-Bisphosphoglycerate + ADP ⇌ 3-phosphoglycerate + ATP	−18.8	1.25
⑧ 3-Phosphoglycerate ⇌ 2-phosphoglycerate	4.4	0.8
⑨ 2-Phosphoglycerate ⇌ phosphoenolpyruvate + H_2O	7.5	−3.3
⑩ Phosphoenolpyruvate + ADP + H^+ ⟶ pyruvate + ATP	−31.4	−16.7

*$\Delta G'^{\circ}$ is the standard free-energy change, as defined in Chapter 14 (see p. 494). At pH 7.0, ΔG is the free-energy change calculated from the actual concentrations of glycolytic intermediates present under physiological conditions in erythrocytes. The glycolytic reactions bypassed in gluconeogenesis are shown in red.

Conversion of Pyruvate to Phosphoenolpyruvate Requires Two Exergonic Reactions

The first of the bypass reactions in gluconeogenesis is the conversion of pyruvate to phosphoenolpyruvate. This reaction cannot occur by reversal of the pyruvate kinase reaction of glycolysis (p. 539), which has a large, negative standard free-energy change and is irreversible under the conditions prevailing in intact cells (Table 20–1, step ⑩). Instead, the phosphorylation of pyruvate is achieved by a roundabout sequence of reactions that in mammals and some other organisms requires the cooperation of enzymes in both the cytosol and mitochondria. As we will see, the pathway shown in Figure 20–2 and described in more detail below is one of two paths from pyruvate to phosphoenolpyruvate; it is the predominant path when pyruvate or alanine is the glucogenic precursor. A second pathway, described later, predominates when lactate is the glucogenic precursor.

Pyruvate is first transported from the cytosol into mitochondria or generated from alanine within mitochondria by transamination. Then **pyruvate carboxylase,** a mitochondrial enzyme that requires biotin, converts the pyruvate to oxaloacetate (Fig. 20–3):

$$\text{Pyruvate} + HCO_3^- + \text{ATP} \longrightarrow \text{oxaloacetate} + \text{ADP} + P_i \qquad (20\text{–}1)$$

Pyruvate carboxylase is the first regulatory enzyme in the gluconeogenic pathway; it requires acetyl-CoA as a positive effector. This reaction is also anaplerotic, replenishing intermediates of the citric acid cycle (p. 584). The reaction mechanism involves biotin as a carrier of activated HCO_3^-, as described in Figure 16–14.

The oxaloacetate formed from pyruvate is reduced to malate by mitochondrial **malate dehydrogenase** at the expense of NADH:

$$\text{Oxaloacetate} + \text{NADH} + H^+ \rightleftharpoons \text{L-malate} + NAD^+ \qquad (20\text{–}2)$$

Recall from Chapter 16 (p. 579) that this reaction, judged by its $\Delta G'^{\circ}$, is highly exergonic. However, under physiological conditions the reaction has a ΔG of about 0 and is thus readily reversible. Mitochondrial malate dehydrogenase therefore functions in both gluconeogenesis and the citric acid cycle, even though the overall flow of metabolites in the two processes is in opposite directions.

Malate leaves the mitochondrion through the malate–α-ketoglutarate transporter of the inner mitochondrial membrane (see Fig. 19–26), and in the cytosol it is reoxidized to oxaloacetate, with the production of cytosolic NADH:

$$\text{Malate} + \text{NAD}^+ \longrightarrow \text{oxaloacetate} + \text{NADH} + \text{H}^+ \qquad (20\text{–}3)$$

The oxaloacetate is then converted to phosphoenolpyruvate (PEP) by **phosphoenolpyruvate carboxykinase.** This Mg^{2+}-dependent reaction requires GTP as the phosphate donor (Fig. 20–3):

$$\text{Oxaloacetate} + \text{GTP} \rightleftharpoons \text{phosphoenolpyruvate} + CO_2 + \text{GDP} \qquad (20\text{–}4)$$

The reaction is reversible under intracellular conditions; the formation of one high-energy phosphate compound (PEP) is balanced by the hydrolysis of another (GTP).

The overall equation for this set of bypass reactions, the sum of Equations 20–1 through 20–4, is

$$\text{Pyruvate} + \text{ATP} + \text{GTP} + HCO_3^- \longrightarrow \text{phosphoenolpyruvate} + \text{ADP} + \text{GDP} + P_i + CO_2 \qquad (20\text{–}5)$$
$$\Delta G'^\circ = 0.9 \text{ kJ/mol}$$

Two high-energy phosphate equivalents (one from ATP, one from GTP), each yielding about 50 kJ/mol under cellular conditions, must be expended to phosphorylate one molecule of pyruvate to PEP. In contrast, when PEP is converted to pyruvate during glycolysis, only one ATP is generated from ADP. Although the standard free-energy change ($\Delta G'^\circ$) of the two-step path from pyruvate to PEP is 0.9 kJ/mol, the actual free-energy change (ΔG), calculated from measured cellular concentrations of intermediates, is very strongly negative (−25 kJ/mol); this results from the ready consumption of PEP in other reactions, such that its concentration remains relatively low. The reaction is thus effectively irreversible in the cell.

Note that the CO_2 added to pyruvate in the pyruvate carboxylase step is the same molecule that is lost in the PEP carboxykinase reaction (Fig. 20–3). This carboxylation-decarboxylation sequence represents a way of "activating" pyruvate in that the decarboxylation of oxaloacetate facilitates PEP formation. In Chapter 21 we will see that a similar carboxylation-decarboxylation sequence is used to activate acetyl-CoA for fatty acid biosynthesis (see Fig. 21–1).

There is a logic to the path of these reactions through the mitochondria. The [NADH]/[NAD$^+$] ratio in the cytosol is 8×10^{-4}, about 10^5 times lower than in mitochondria. Because cytosolic NADH is consumed in gluconeogenesis (in the conversion of 1,3-bisphosphoglycerate to glyceraldehyde 3-phosphate; Fig. 20–2), glucose biosynthesis cannot proceed unless NADH is available. The transport of malate from the mitochondrion to the cytosol and its reconversion there to oxaloacetate effectively moves reducing equivalents to the cytosol, where they are scarce. This path from pyruvate to PEP therefore provides an important balance between NADH produced and consumed in the cytosol during gluconeogenesis.

A second (and shorter) pyruvate → PEP bypass predominates when lactate is the glucogenic precursor (Fig. 20–4). This pathway makes use of lactate produced by glycolysis in erythrocytes or anaerobic muscle, for example, and it is particularly important in large vertebrates after vigorous exercise (see Box 15–1). The conversion of lactate to pyruvate in the hepatocyte cytosol yields NADH, and the export of reducing equivalents (as malate) from mitochondria is therefore unnecessary. After the pyruvate produced by the lactate dehydrogenase reaction is transported into the mitochondrion, it is converted to oxaloacetate by pyruvate carboxylase as described above. This oxaloacetate, however, is converted directly to PEP by

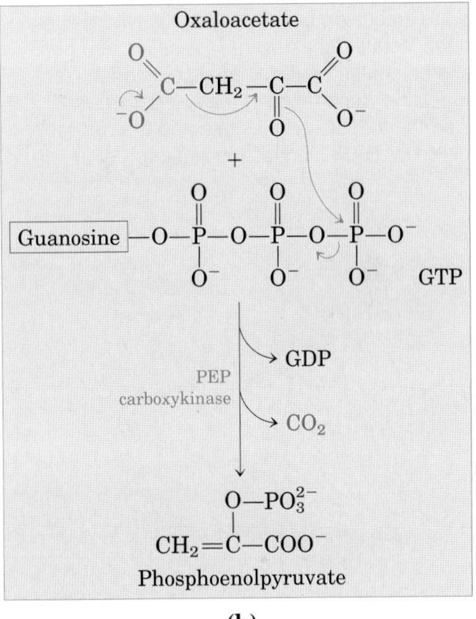

figure 20–3
Synthesis of phosphoenolpyruvate from pyruvate. **(a)** In mitochondria, pyruvate is converted to oxaloacetate in a biotin-requiring reaction catalyzed by pyruvate carboxylase. **(b)** In the cytosol, oxaloacetate is converted to phosphoenolpyruvate by PEP carboxykinase. The CO_2 incorporated in the pyruvate carboxylase reaction is lost here as CO_2. The decarboxylation leads to a rearrangement of electrons that facilitates attack of the carbonyl oxygen of the pyruvate moiety on the γ phosphate of GTP.

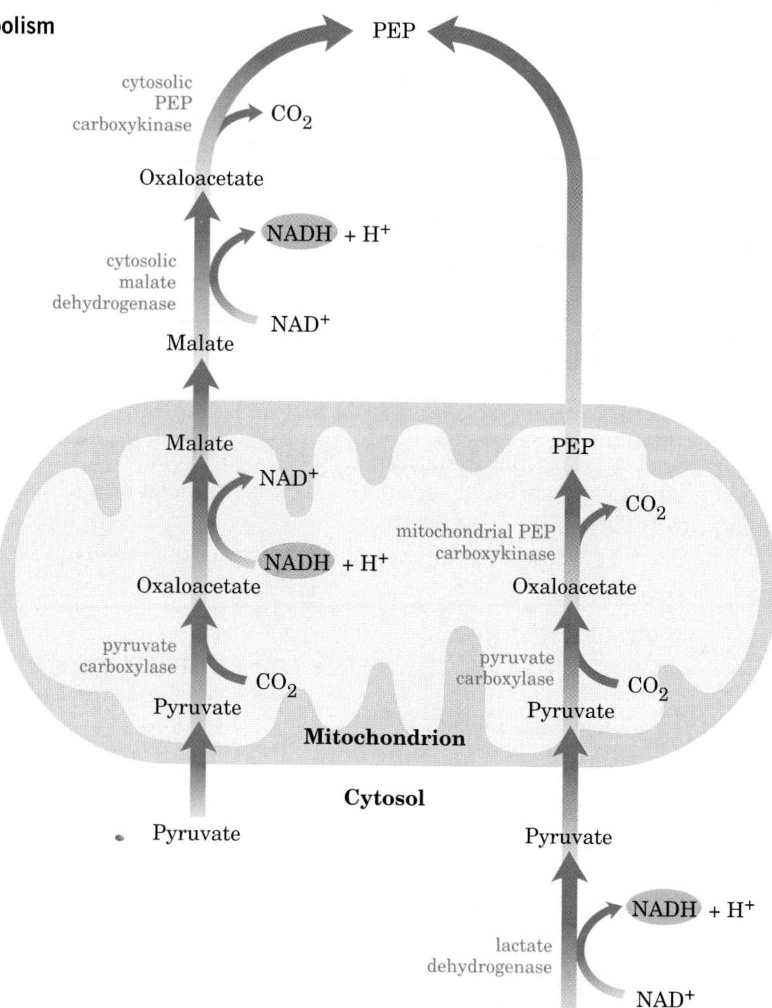

figure 20–4

Alternative paths from pyruvate to phosphoenolpyruvate. The path that predominates depends on the glucogenic precursor (lactate or pyruvate). The path on the right (a step shorter than the path on the left) predominates when lactate is the precursor, because cytosolic NADH is generated in the lactate dehydrogenase reaction and does not have to be shuttled out of the mitochondrion (see text). The relative importance of the two pathways depends on the availability of lactate and the cytosolic requirements for NADH in gluconeogenesis.

a mitochondrial isozyme of PEP carboxykinase, and the PEP is transported out of the mitochondrion to continue on the gluconeogenic path. The mitochondrial and cytosolic isozymes of PEP carboxykinase are encoded by separate nuclear genes, providing an example of two distinct enzymes catalyzing the same reaction but having different cellular locations and metabolic roles (recall the description of isozymes in Box 15–3).

Conversion of Fructose 1,6-Bisphosphate to Fructose 6-Phosphate Is the Second Bypass

The second glycolytic reaction that cannot participate in gluconeogenesis is the phosphorylation of fructose 6-phosphate by phosphofructokinase-1 (Table 20–1, step ③). Because this reaction is highly exergonic and therefore irreversible in intact cells, the generation of fructose 6-phosphate from fructose 1,6-bisphosphate (Fig. 20–2) is catalyzed by a different enzyme, Mg^{2+}-dependent **fructose 1,6-bisphosphatase,** which promotes the essentially irreversible *hydrolysis* of the C-1 phosphate (*not* phosphoryl group transfer to ADP):

$$\text{Fructose 1,6-bisphosphate} + H_2O \longrightarrow \text{fructose 6-phosphate} + P_i \qquad \Delta G'^{\circ} = -16.3 \text{ kJ/mol}$$

Conversion of Glucose 6-Phosphate to Free Glucose Is the Third Bypass

The third bypass is the final reaction of gluconeogenesis, the dephosphorylation of glucose 6-phosphate to yield free glucose (Fig. 20–2). Reversal of the hexokinase reaction (see p. 532) would require phosphoryl group transfer from glucose 6-phosphate to ADP, forming ATP, an energetically unfavorable reaction (Table 20–1, step ①). The reaction catalyzed by **glucose 6-phosphatase** does not require synthesis of ATP, but is a simple hydrolysis of a phosphate ester:

$$\text{Glucose 6-phosphate} + H_2O \longrightarrow \text{glucose} + P_i \qquad \Delta G'^\circ = -13.8 \text{ kJ/mol}$$

This Mg^{2+}-activated enzyme is found on the lumenal side of the endoplasmic reticulum of hepatocytes and renal cells. The enzyme is not present in muscle or in the brain, and gluconeogenesis does not occur in these tissues. Instead, glucose produced by gluconeogenesis in the liver or kidney or ingested in the diet is delivered to brain and muscle through the bloodstream.

Gluconeogenesis Is Expensive

The sum of the biosynthetic reactions leading from pyruvate to free blood glucose (Table 20–2) is

$$2 \text{ Pyruvate} + 4\text{ATP} + 2\text{GTP} + 2\text{NADH} + 4H_2O \longrightarrow \text{glucose} + 4\text{ADP} + 2\text{GDP} + 6P_i + 2\text{NAD}^+ + 2H^+ \qquad (20\text{–}6)$$

For each molecule of glucose formed from pyruvate, six high-energy phosphate groups are required, four from ATP and two from GTP. In addition, two molecules of NADH are required for the reduction of two molecules of 1,3-bisphosphoglycerate. Equation 20–6 is clearly not simply the reverse of the equation for the conversion of glucose to pyruvate by glycolysis, which requires only two molecules of ATP:

$$\text{Glucose} + 2\text{ADP} + 2P_i + 2\text{NAD}^+ \longrightarrow 2 \text{ pyruvate} + 2\text{ATP} + 2\text{NADH} + 2H^+ + 2H_2O$$

Thus the synthesis of glucose from pyruvate is a relatively expensive process. Much of this high energy cost is necessary to ensure the irreversibility of gluconeogenesis. Under intracellular conditions, the overall free-energy change of glycolysis is at least −63 kJ/mol. Under the same conditions the overall ΔG of gluconeogenesis is −16 kJ/mol. Thus both glycolysis and gluconeogenesis are essentially irreversible processes in cells.

table 20–2

Sequential Reactions in Gluconeogenesis Starting from Pyruvate*

Reaction	
Pyruvate + HCO_3^- + ATP ⟶ oxaloacetate + ADP + P_i + H^+	×2
Oxaloacetate + GTP ⇌ phosphoenolpyruvate + CO_2 + GDP	×2
Phosphoenolpyruvate + H_2O ⇌ 2-phosphoglycerate	×2
2-Phosphoglycerate ⇌ 3-phosphoglycerate	×2
3-Phosphoglycerate + ATP ⇌ 1,3-bisphosphoglycerate + ADP + H^+	×2
1,3-Bisphosphoglycerate + NADH + H^+ ⇌ glyceraldehyde 3-phosphate + NAD^+ + P_i	×2
Glyceraldehyde 3-phosphate ⇌ dihydroxyacetone phosphate	
Glyceraldehyde 3-phosphate + dihydroxyacetone phosphate ⇌ fructose 1,6-bisphosphate	
Fructose 1,6-bisphosphate + H_2O ⟶ fructose 6-phosphate + P_i	
Fructose 6-phosphate ⇌ glucose 6-phosphate	
Glucose 6-phosphate + H_2O ⟶ glucose + P_i	
Sum: 2 Pyruvate + 4ATP + 2GTP + 2NADH + $4H_2O$ ⟶ glucose + 4ADP + 2GDP + $6P_i$ + $2NAD^+$ + $2H^+$	

*The bypass reactions are in red; all other reactions are reversible steps of glycolysis. The figures at the right indicate that the reaction is to be counted twice, because two three-carbon precursors are required to make a molecule of glucose. Note that the reactions required to replace the cytosolic NADH consumed in the glyceraldehyde 3-phosphate dehydrogenase reaction (the conversion of lactate to pyruvate in the cytosol or the transport of reducing equivalents from mitochondria to the cytosol in the form of malate) are not considered in this summary.

table 20–3

Glucogenic Amino Acids, Grouped by Site of Entry*

Pyruvate	Succinyl-CoA
Alanine	Isoleucine†
Cysteine	Methionine
Glycine	Threonine
Serine	Valine
Tryptophan†	
α-Ketoglutarate	**Fumarate**
Arginine	Phenylalanine†
Glutamate	Tyrosine†
Glutamine	**Oxaloacetate**
Histidine	Asparagine
Proline	Aspartate

*These amino acids are precursors of blood glucose or liver glycogen because they can be converted to pyruvate or citric acid cycle intermediates. Only leucine and lysine are unable to furnish carbon for net glucose synthesis.

†These amino acids are also ketogenic (see Fig. 18–19).

Citric Acid Cycle Intermediates and Many Amino Acids Are Glucogenic

The biosynthetic pathway to glucose described above allows the net synthesis of glucose not only from pyruvate but also from the citric acid cycle intermediates: citrate, isocitrate, α-ketoglutarate, succinyl-CoA, succinate, fumarate, and malate, all of which can undergo oxidation to oxaloacetate.

In Chapter 18 we showed that some or all of the carbon atoms of many of the amino acids derived from proteins are ultimately converted to pyruvate or to intermediates of the citric acid cycle. Such amino acids can therefore undergo net conversion to glucose and are said to be **glucogenic** (Table 20–3). Alanine and glutamine, the principal molecules used to transport amino groups from extrahepatic tissues to the liver, are particularly important glucogenic amino acids. After removal of their amino groups in liver mitochondria, the carbon skeletons remaining (pyruvate and α-ketoglutarate, respectively) are readily funneled into gluconeogenesis.

In contrast, no net conversion of fatty acids to glucose occurs in mammals. Even-number fatty acids yield only acetyl-CoA on oxidative cleavage, and mammals cannot use acetyl-CoA as a precursor of glucose: the pyruvate dehydrogenase reaction is irreversible in cells and no other pathway exists by which acetyl-CoA can be converted to pyruvate. For every two carbons of acetate (acetyl-CoA) that enter the citric acid cycle, two carbons are lost as CO_2 (Fig. 20–5); no net conversion of acetate to oxaloacetate or pyruvate is possible. Acetate can be converted to oxaloacetate by organisms having the glyoxylate cycle, a pathway introduced in Chapter 16 (see Fig. 16–16) and revisited later in this chapter, but this pathway is absent in animals. Although fatty acids cannot supply carbon atoms for gluconeogenesis in animals, they do provide energy for glucose synthesis.

Futile Cycles in Carbohydrate Metabolism Consume ATP

At each of the three points where glycolytic reactions are bypassed by alternative, gluconeogenic reactions, simultaneous operation of both pathways would be wasteful. For example, phosphofructokinase-1 and fructose 1,6-bisphosphatase catalyze opposing reactions:

$$\text{ATP} + \text{fructose 6-phosphate} \longrightarrow \text{ADP} + \text{fructose 1,6-bisphosphate}$$

$$\text{Fructose 1,6-bisphosphate} + H_2O \longrightarrow \text{fructose 6-phosphate} + P_i$$

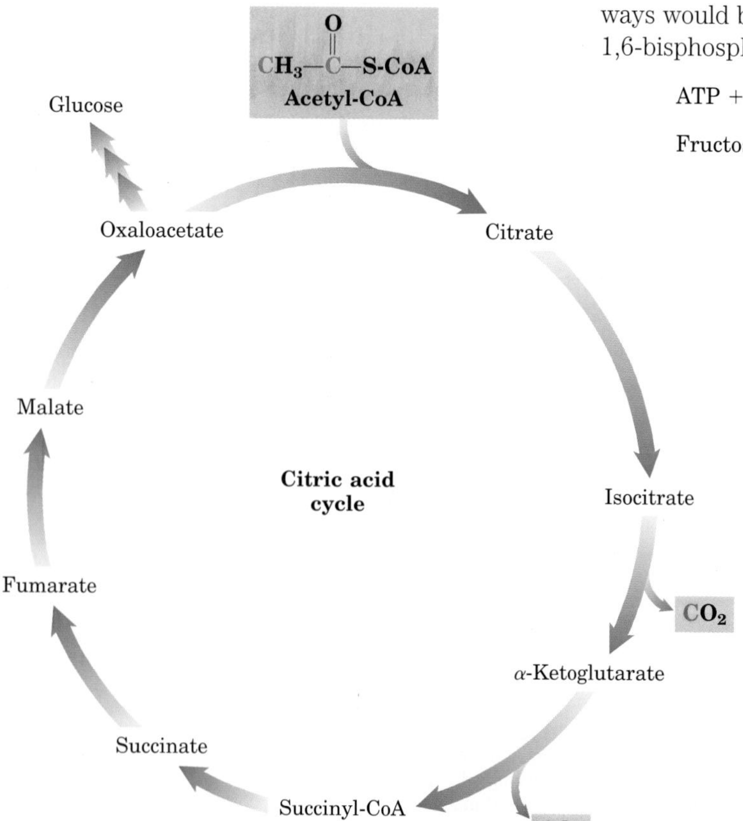

figure 20–5
Even-number fatty acids cannot be a source of carbon for the net synthesis of glucose in animals and microorganisms. Fatty acids are catabolized to acetyl-CoA, and for every two carbons entering the citric acid cycle as acetyl-CoA, two are lost as CO_2, so there is no net production of oxaloacetate to support glucose biosynthesis. However, fatty acid oxidation does provide extraordinary amounts of energy in the form of NADH, ATP, and GTP (Chapter 17) for gluconeogenesis. Amino acids degraded to acetyl-CoA are also not glucogenic (see Fig. 18–29), for the reason illustrated here.

The sum of these two reactions is

$$ATP + H_2O \longrightarrow ADP + P_i + heat$$

which represents a waste of ATP—hydrolysis of ATP without any net metabolic work being done. Clearly, if these two reactions were allowed to proceed simultaneously at a high rate in the same cell, a large amount of chemical energy would be dissipated as heat. This uneconomical process is called a **futile cycle.** Similar futile cycles could in principle occur with the other two sets of bypass reactions.

Under normal circumstances futile cycling does not occur at a significant rate, because it is prevented by reciprocal regulatory mechanisms (as discussed below). However, futile cycling sometimes serves physiologically to generate heat that raises or maintains body temperature. For example, in cold weather bumblebees cannot fly until they have warmed their muscles to about 30 °C by futile cycling of fructose 6-phosphate and fructose 1,6-bisphosphate with the consequent heat-generating hydrolysis of ATP.

Gluconeogenesis and Glycolysis Are Reciprocally Regulated

To meet metabolic needs and ensure that futile cycling does not occur under normal circumstances, gluconeogenesis and glycolysis are regulated coordinately and reciprocally. The first control point determines the fate of pyruvate in the mitochondrion. Pyruvate can be converted either to acetyl-CoA (by the pyruvate dehydrogenase complex; Chapter 16) to fuel the citric acid cycle, or to oxaloacetate (by pyruvate carboxylase) to start the process of gluconeogenesis (Fig. 20–6). Acetyl-CoA is a positive allosteric modulator of pyruvate carboxylase and a negative modulator of pyruvate dehydrogenase, through stimulation of a protein kinase that inactivates the dehydrogenase. When the cell's energetic needs are being met, oxidative phosphorylation slows, NADH accumulates and inhibits the citric acid cycle, and acetyl-CoA accumulates. The increased concentration of acetyl-CoA inhibits the pyruvate dehydrogenase complex, slowing the formation of acetyl-CoA from pyruvate, and stimulates gluconeogenesis by activating pyruvate carboxylase, allowing excess pyruvate to be converted to glucose.

The second control point in gluconeogenesis is the reaction catalyzed by fructose 1,6-bisphosphatase, which is strongly inhibited by AMP. The corresponding glycolytic enzyme, phosphofructokinase-1, is stimulated by AMP and ADP but inhibited by citrate and ATP. Thus these opposing steps in the two pathways are regulated in a coordinated and reciprocal manner. In general, then, when sufficient concentrations of acetyl-CoA or citrate (the product of acetyl-CoA condensation with oxaloacetate) are present, or when a high proportion of the cell's adenylate is in the form of ATP, gluconeogenesis is favored.

The special role of liver in maintaining a constant blood glucose level requires additional regulatory mechanisms to coordinate glucose production and consumption. When the blood glucose level decreases, the hormone **glucagon** signals the liver to produce and release more glucose. One source of glucose is glycogen stored in the liver; another source is gluconeogenesis.

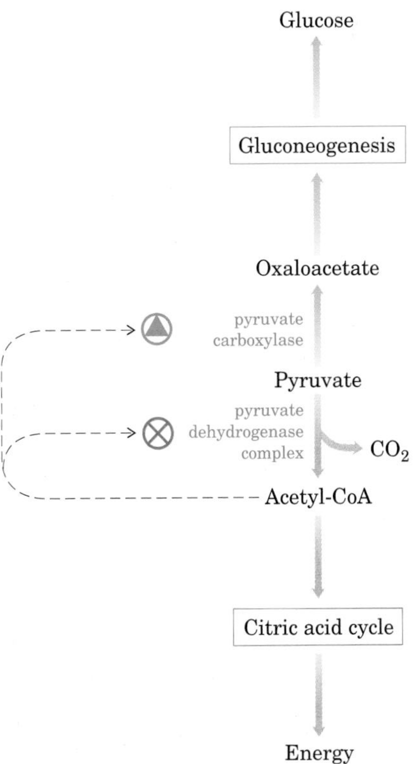

figure 20–6
Two alternative fates for pyruvate. Pyruvate can be converted to glucose and glycogen via gluconeogenesis or oxidized to acetyl-CoA for energy production. The first enzyme in each path is regulated allosterically; acetyl-CoA stimulates pyruvate carboxylase and inhibits the pyruvate dehydrogenase complex.

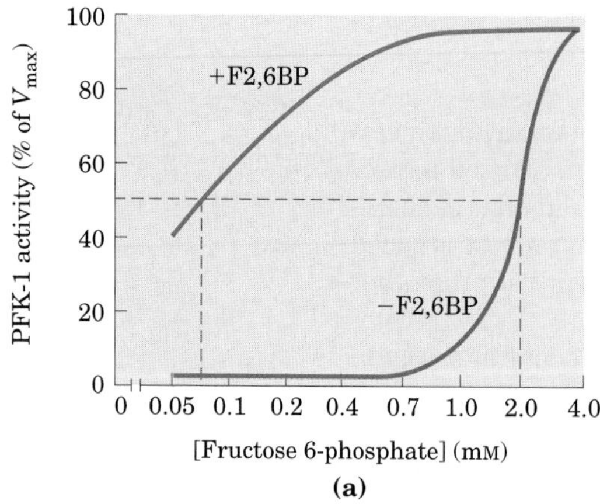

(a)

FBPase-1 activity (% of V_{max})

[Fructose 1,6-bisphosphate] (μM)

(b)

figure 20–7

Role of fructose 2,6-bisphosphate in regulation of glycolysis and gluconeogenesis. Fructose 2,6-bisphosphate has opposite effects on the enzymatic activities of phosphofructokinase-1 (PFK-1, a glycolytic enzyme) and fructose 1,6-bisphosphatase (FBPase-1, an enzyme of gluconeogenesis). **(a)** PFK-1 activity in the absence of fructose 2,6-bisphosphate (F2,6BP) (blue curve) is half-maximal when the concentration of fructose 6-phosphate is 2 mM (that is, $K_{0.5}$ = 2 mM; recall from Chapter 8 (p. 280) that $K_{0.5}$ or K_m is equivalent to the substrate concentration at which half-maximal enzyme activity occurs). When 0.13 μM fructose 2,6-bisphosphate is present (red curve), the $K_{0.5}$ for fructose 6-phosphate is only 0.08 mM. Thus fructose 2,6-bisphosphate activates PFK-1 by increasing its apparent affinity (p. 262) for fructose 6-phosphate. **(b)** FBPase-1 activity is inhibited by as little as 1 μM fructose 2,6-bisphosphate and is strongly inhibited by 25 μM. In the absence of this inhibitor (blue curve) the $K_{0.5}$ for fructose 1,6-bisphosphate is 5 μM, but in the presence of 25 μM fructose 2,6-bisphosphate (red curve) the $K_{0.5}$ is >70 μM. Fructose 2,6-bisphosphate also makes FBPase-1 more sensitive to inhibition by another allosteric regulator, AMP.

Fructose 2,6-bisphosphate

The hormonal regulation of glycolysis and gluconeogenesis in liver is mediated by **fructose 2,6-bisphosphate,** an allosteric effector for the enzymes phosphofructokinase-1 (PFK-1; see Fig. 15–18) and fructose 1,6-bisphosphatase (FBPase-1) (Fig. 20–7). When fructose 2,6-bisphosphate binds to its allosteric site on PFK-1, it increases that enzyme's affinity for its substrate fructose 6-phosphate and reduces its affinity for the allosteric inhibitors ATP and citrate. Fructose 2,6-bisphosphate therefore activates PFK-1 and stimulates glycolysis in liver. At the same time, fructose 2,6-bisphosphate *inhibits* FBPase-1, thereby slowing gluconeogenesis.

Although structurally related to fructose 1,6-bisphosphate, fructose 2,6-bisphosphate is *not* an intermediate in gluconeogenesis or glycolysis; it is a *regulator* whose cellular level reflects the level of glucagon in the blood, which rises when blood glucose falls.

The cellular concentration of fructose 2,6-bisphosphate is set by the relative rates of its formation and breakdown (Fig. 20–8). Fructose 2,6-bisphosphate is formed by phosphorylation of fructose 6-phosphate, catalyzed by **phosphofructokinase-2** (PFK-2), and is broken down by **fructose 2,6-bisphosphatase** (FBPase-2). (Note that these enzymes are distinct from PFK-1 and FBPase-1, which catalyze the formation and breakdown, respectively, of fructose 1,6-bisphosphate.) PFK-2 and FBPase-2 are two distinct enzymatic activities of a single, bifunctional protein. The balance of these two activities in the liver, which determines the cellular level of fructose 2,6-bisphosphate, is regulated by glucagon (Fig. 20–8). We saw in Chapter 13 (see p. 454) that glucagon stimulates adenylyl cyclase through a G protein. Adenylyl cyclase synthesizes 3′,5′-cyclic AMP (cAMP) from ATP, then cAMP stimulates a cAMP-dependent protein kinase, which transfers a phosphoryl

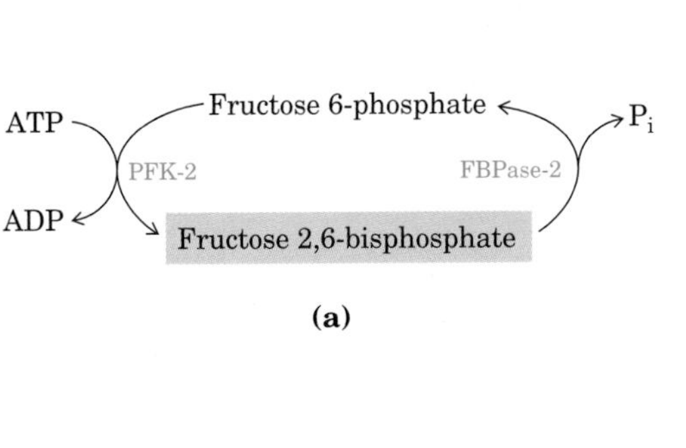

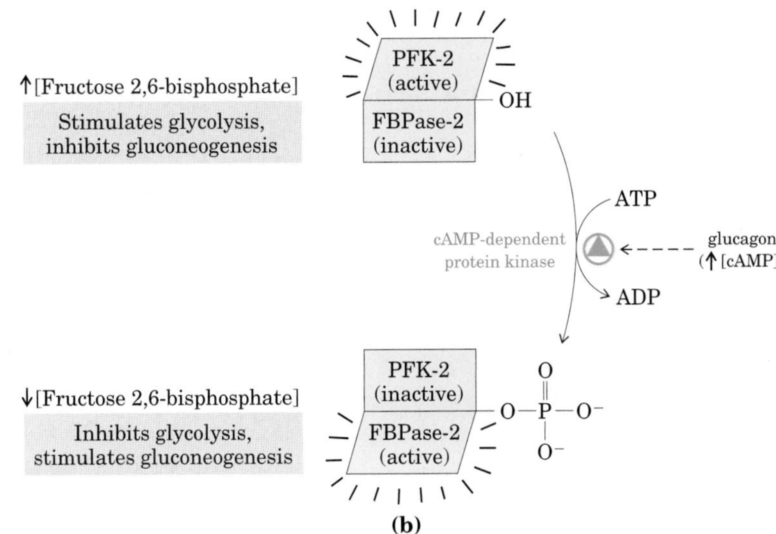

figure 20–8

Regulation of fructose 2,6-bisphosphate level. (a) The cellular concentration of the regulator fructose 2,6-bisphosphate is determined by the rates of its synthesis by phosphofructokinase-2 (PFK-2) and breakdown by fructose 2,6-bisphosphatase (FBPase-2). **(b)** Both of these enzymes are part of the same polypeptide chain, and both are regulated, in a reciprocal fashion, by glucagon. Here and elsewhere, arrows are used to indicate increasing (↑) and decreasing (↓) levels of metabolites.

group from ATP to the bifunctional protein PFK-2/FBPase-2. Phosphorylation of this protein enhances its FBPase-2 activity and inhibits its PFK-2 activity. Glucagon thereby lowers the cellular level of fructose 2,6-bisphosphate, inhibiting glycolysis and stimulating gluconeogenesis. The resulting production of more glucose enables the liver to replenish blood glucose in response to glucagon.

Gluconeogenesis Converts Fats and Proteins to Glucose in Germinating Seeds

Many plants store lipids and proteins in their seeds, to be used as sources of energy and biosynthetic precursors during germination, before photosynthetic mechanisms can supply both. Active gluconeogenesis occurs in germinating seeds, providing glucose for the synthesis of sucrose (the transport form of carbon in plants), polysaccharides, and many metabolites derived from hexoses. In plant seedlings, sucrose provides much of the chemical energy needed for initial growth.

Unlike animals, plants and some microorganisms can convert acetyl-CoA derived from fatty acid oxidation into glucose. Some of the enzymes essential to this conversion are sequestered in glyoxysomes, organelles in which fatty acids are oxidized to acetyl-CoA by glyoxysome-specific isozymes of the β-oxidation enzymes (see Fig. 16–18). The physical separation of these glyoxylate cycle and β-oxidation enzymes from the mitochondrial citric acid cycle enzymes prevents the further oxidation of acetyl-CoA to CO_2. Instead, the acetyl-CoA is converted to succinate in the glyoxylate cycle. The succinate passes into the mitochondrial matrix, where it is converted by citric acid cycle enzymes to oxaloacetate, which passes into the cytosol. Cytosolic oxaloacetate is converted by gluconeogenesis to fructose 6-phosphate, the precursor of sucrose. Thus, the integration of reaction sequences in three subcellular compartments is required for the production of fructose 6-phosphate or sucrose from stored lipids. Because only

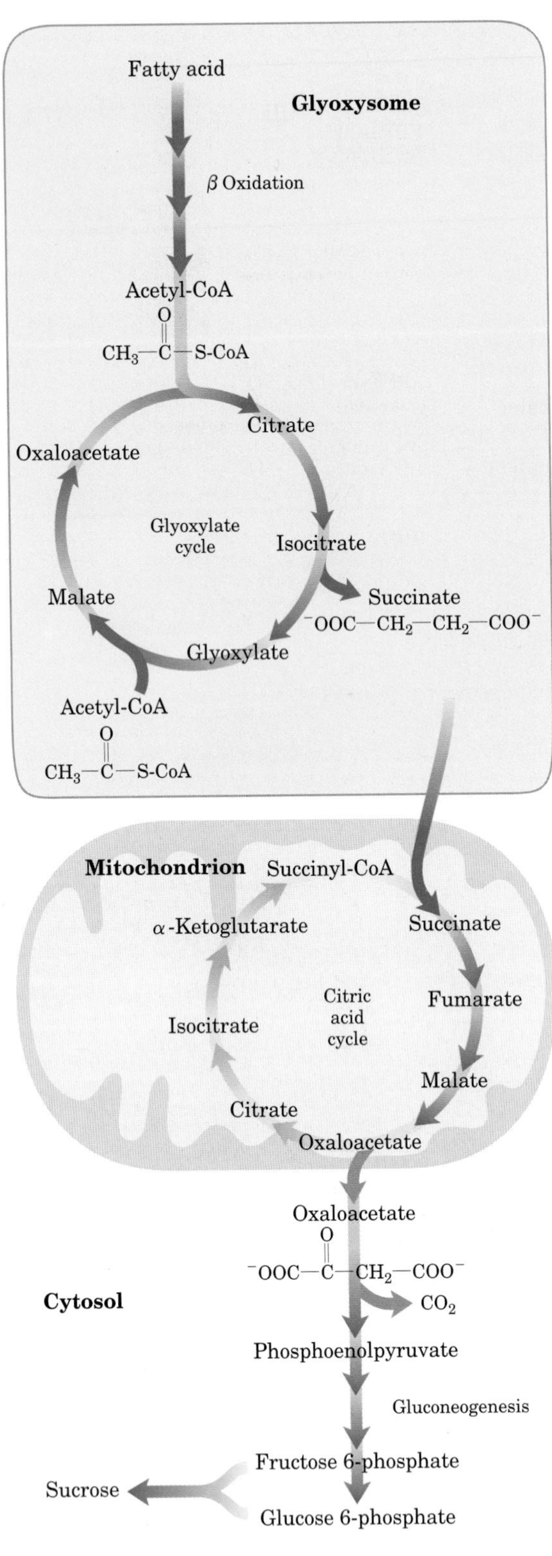

figure 20–9

Conversion of stored fatty acids to sucrose in germinating seeds. This pathway begins in glyoxysomes. Succinate is produced and exported to mitochondria, where it is converted to oxaloacetate by enzymes of the citric acid cycle. Oxaloacetate enters the cytosol and serves as the starting material for gluconeogenesis and the synthesis of sucrose, the transport form of carbon in plants.

three of the four carbons in each molecule of oxaloacetate are converted to hexose in the cytosol, about 75% of the carbon in the fatty acids of the stored seed lipids is converted to carbohydrate by the combined pathways of Figure 20–9. The other 25% is lost as CO_2 in the conversion of oxaloacetate to phosphoenolpyruvate. Hydrolysis of storage triacylglycerols also produces glycerol 3-phosphate, which can enter the gluconeogenic pathway after its oxidation to dihydroxyacetone phosphate (Fig. 20–10).

Glucogenic amino acids (Table 20–3) derived from the breakdown of stored seed proteins also yield precursors for gluconeogenesis, following transamination and oxidation to succinyl-CoA, pyruvate, oxaloacetate, fumarate, and α-ketoglutarate (Chapter 18)—all good starting materials for gluconeogenesis.

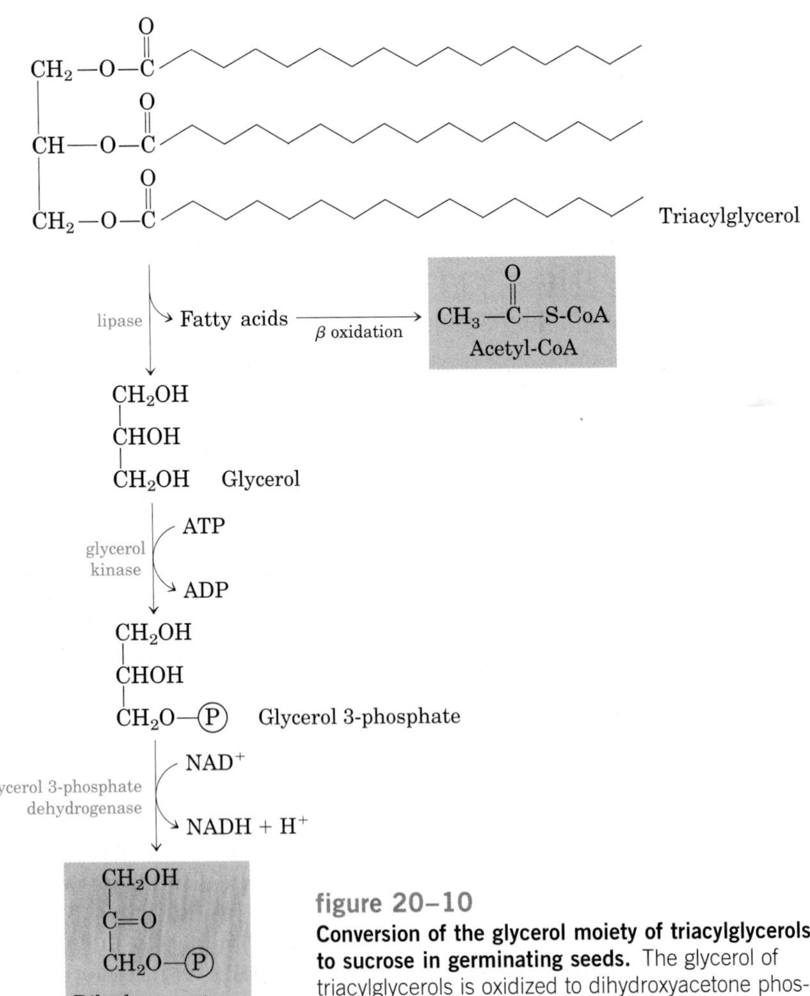

figure 20–10

Conversion of the glycerol moiety of triacylglycerols to sucrose in germinating seeds. The glycerol of triacylglycerols is oxidized to dihydroxyacetone phosphate, which enters the gluconeogenic pathway at the triose phosphate isomerase reaction.

Biosynthesis of Glycogen, Starch, Sucrose, and Other Carbohydrates

In a wide range of organisms, excess glucose is converted to polymeric forms for storage and to disaccharides for transport. The principal storage forms of glucose are glycogen in vertebrates and many microorganisms, and starch in plants. In vertebrates, glucose itself is generally transported in the blood, but the transport form in plants is sucrose, and in insects is trehalose.

Many of the reactions in which hexoses are transformed or polymerized involve **sugar nucleotides,** compounds in which the anomeric carbon of a sugar is activated by attachment to a nucleotide through a phosphodiester linkage. Sugar nucleotides are the substrates for polymerization of monosaccharides into disaccharides, glycogen, starch, cellulose, and more complex extracellular polysaccharides. They are also key intermediates in the production of aminohexoses and deoxyhexoses found in some of these polysaccharides, and in the synthesis of vitamin C (L-ascorbic acid); a sugar nucleotide also is essential in some detoxification reactions in vertebrates. The role of sugar nucleotides in the biosynthesis of glycogen and many other carbohydrate derivatives was first discovered by the Argentine biochemist Luis Leloir, who identified the role of **UDP-glucose.**

Luis Leloir
1906–1988

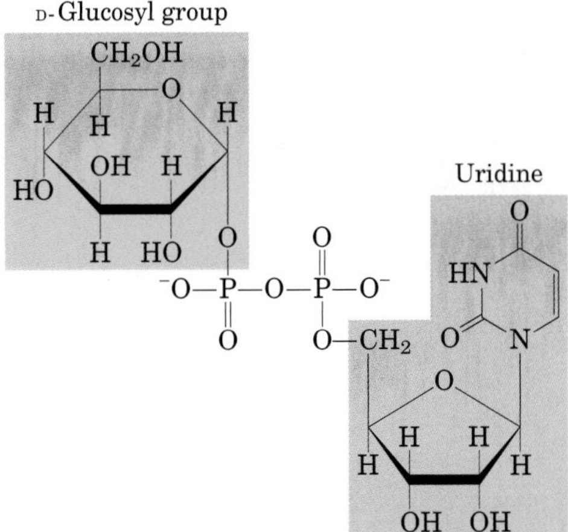

UDP-glucose

The suitability of sugar nucleotides for biosynthetic reactions stems from several properties:

1. Their formation is metabolically irreversible, contributing to the irreversibility of the synthetic pathways in which they are intermediates. The condensation of a nucleoside triphosphate with a hexose 1-phosphate to form a sugar nucleotide has a free-energy change near zero in the cell, but the reaction releases PP_i, which is hydrolyzed by inorganic pyrophosphatase (Fig. 20–11). The large, negative free-energy change of PP_i hydrolysis drives the synthetic reaction, a common strategy in biological polymerization reactions.

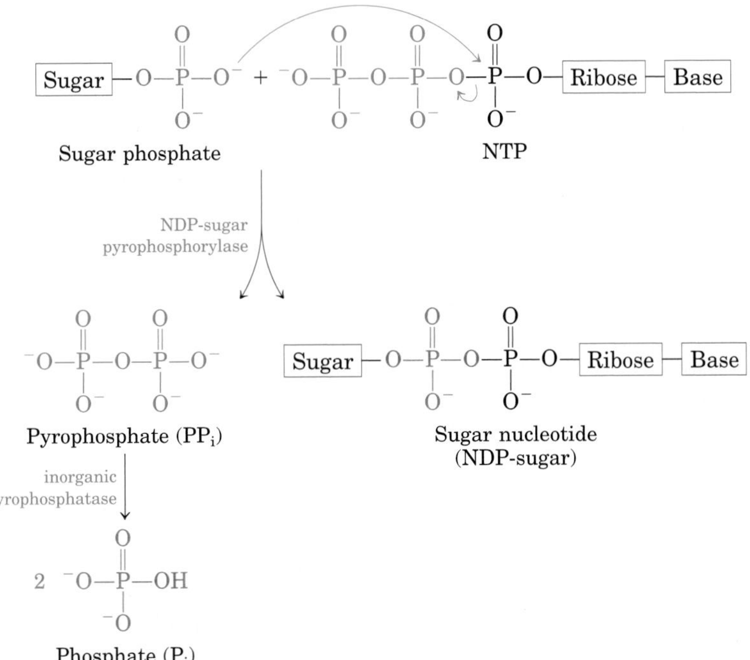

Net reaction: Sugar phosphate + NTP ⟶ NDP-sugar + $2P_i$

figure 20–11

Formation of a sugar nucleotide. A condensation reaction occurs between a nucleoside triphosphate (NTP) and a sugar phosphate. The negatively charged oxygen on the sugar phosphate serves as a nucleophile, attacking the α phosphate of the nucleoside triphosphate and displacing pyrophosphate. The reaction is pulled in the forward direction by the hydrolysis of PP_i by inorganic pyrophosphatase.

2. Although the chemical transformations of sugar nucleotides do not involve the atoms of the nucleotide itself, the nucleotide moiety has many groups that can undergo noncovalent interactions with enzymes; the additional free energy of binding can contribute significantly to catalytic activity (Chapter 8; see also p. 356).
3. Like phosphate, the nucleotidyl group (UMP or AMP, for example) is an excellent leaving group, facilitating nucleophilic attack by activating the sugar carbon to which it is attached.
4. By "tagging" some hexoses with nucleotidyl groups, cells can set them aside for one purpose (glycogen synthesis, for example) in a pool separate from hexose phosphates earmarked for another purpose (such as glycolysis).

UDP-Glucose Is the Substrate for Glycogen Synthesis

Glycogen synthesis occurs in virtually all animal tissues but is especially prominent in the liver and skeletal muscles. In the liver, glycogen serves as a reservoir of glucose, readily converted into blood glucose for distribution to other tissues; in muscle, glucose produced from the breakdown of glycogen is metabolized via glycolysis to provide ATP energy for muscle contraction.

The starting point for synthesis of glycogen is **glucose 6-phosphate.** This can be derived from free glucose in a reaction catalyzed by **glucokinase** (in liver) or **hexokinase** (in muscle):

$$\text{D-Glucose} + \text{ATP} \longrightarrow \text{D-glucose 6-phosphate} + \text{ADP}$$

However, some ingested glucose takes a more roundabout path to glycogen. It is first taken up by erythrocytes and converted to lactate glycolytically; the lactate is then taken up by the liver and converted to glucose 6-phosphate by gluconeogenesis.

To initiate glycogen synthesis, the glucose 6-phosphate is converted to **glucose 1-phosphate** by **phosphoglucomutase:**

$$\text{Glucose 6-phosphate} \rightleftharpoons \text{glucose 1-phosphate}$$

Formation of UDP-glucose by the action of **UDP-glucose pyrophosphorylase** is a key reaction in glycogen biosynthesis:

$$\text{Glucose 1-phosphate} + \text{UTP} \longrightarrow \text{UDP-glucose} + \text{PP}_i$$

(Notice that this enzyme is named for the reverse reaction; recall our warning in Box 16–1 about nomenclature issues such as this.) In the cell, this reaction proceeds in the direction of UDP-glucose formation because pyrophosphate is rapidly hydrolyzed to inorganic phosphate by inorganic pyrophosphatase ($\Delta G'^{\circ} = -25$ kJ/mol) (Fig. 20–11).

UDP-glucose is the immediate donor of glucose residues in the reaction catalyzed by **glycogen synthase,** which promotes the transfer of the glucose residue from UDP-glucose to a nonreducing end of a branched glycogen molecule (Fig. 20–12). The overall equilibrium of the path from glucose 6-phosphate to lengthened glycogen greatly favors synthesis of glycogen. Glycogen synthase requires as a primer an (α1→4) polyglucose chain or branch having at least eight glucose residues.

Glycogen synthase cannot make the (α1→6) bonds found at the branch points of glycogen (see Fig. 9–15 for similar branch points in starch); these are formed by a glycogen-branching enzyme called **amylo (1→4) to (1→6) transglycosylase** or **glycosyl-(4→6)-transferase.** Glycosyl-(4→6)-transferase catalyzes transfer of a terminal fragment of six or seven glucose residues from the nonreducing end of a glycogen branch having at

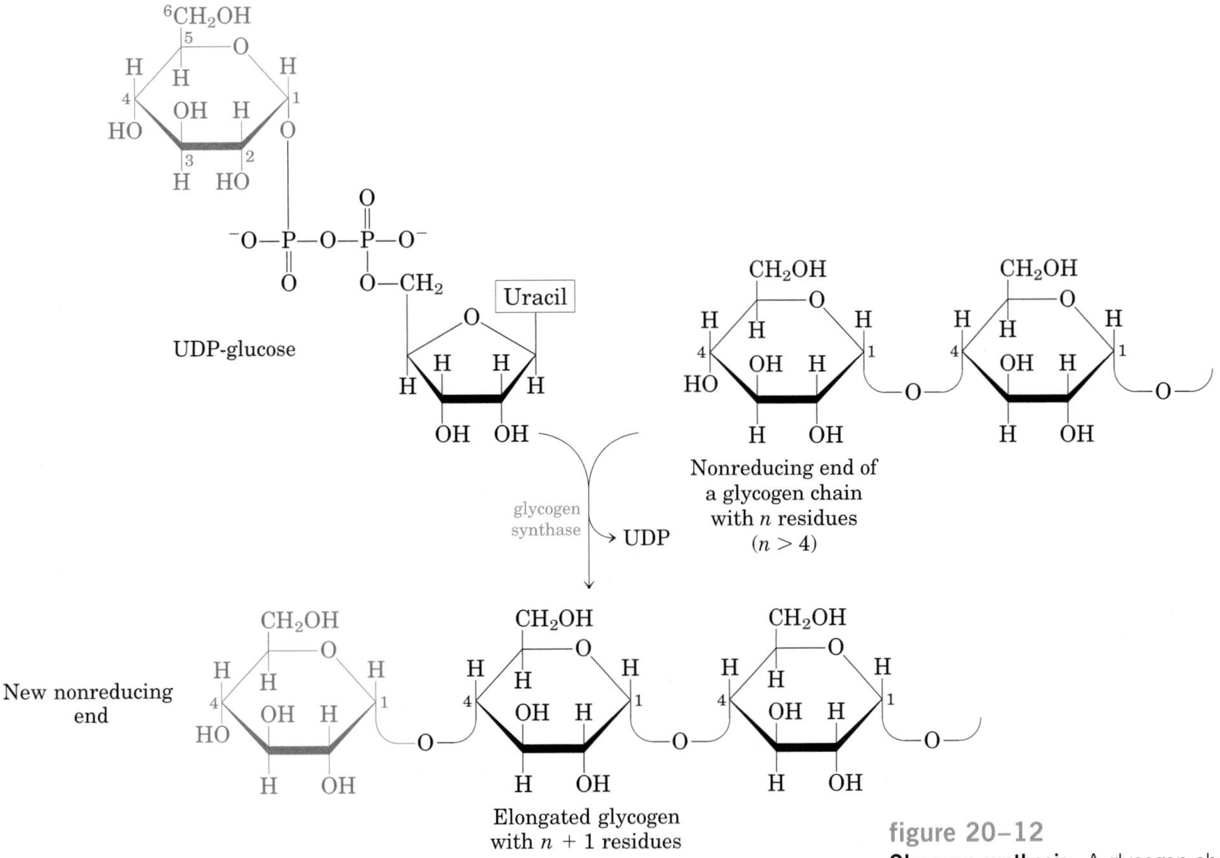

UDP-glucose

Uracil

glycogen synthase → UDP

Nonreducing end of a glycogen chain with n residues ($n > 4$)

New nonreducing end

Elongated glycogen with $n + 1$ residues

figure 20–12

Glycogen synthesis. A glycogen chain is elongated by glycogen synthase. The enzyme transfers the glucose residue of UDP-glucose to the nonreducing end of a glycogen branch (see Fig. 9–15) to make a new (α1→4) linkage.

least 11 residues to the C-6 hydroxyl group of a glucose residue at a more interior position of the same or another glycogen chain, thus creating a new branch (Fig. 20–13). Further glucose residues may be added to the new branch by glycogen synthase. The biological effect of branching is to make the glycogen molecule more soluble and to increase the number of nonreducing ends. This increases the number of sites accessible to glycogen phosphorylase and glycogen synthase, both of which act only at nonreducing ends.

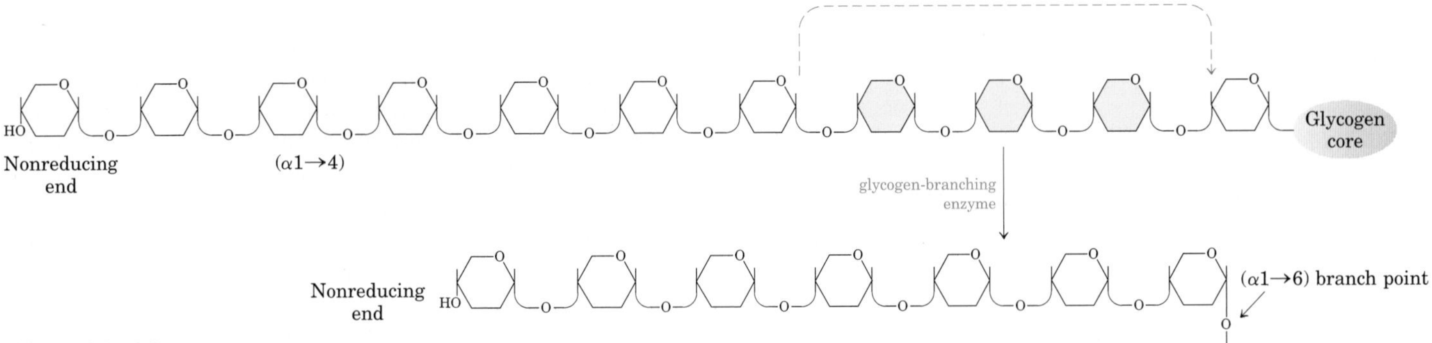

Nonreducing end

(α1→4)

Glycogen core

glycogen-branching enzyme

Nonreducing end

(α1→6) branch point

Nonreducing end

Glycogen core

figure 20–13

Branch synthesis in glycogen. The glycogen-branching enzyme, glycosyl-(4→6)-transferase (also called amylo (1→4) to (1→6) transglycosylase), forms a new branch point during glycogen synthesis.

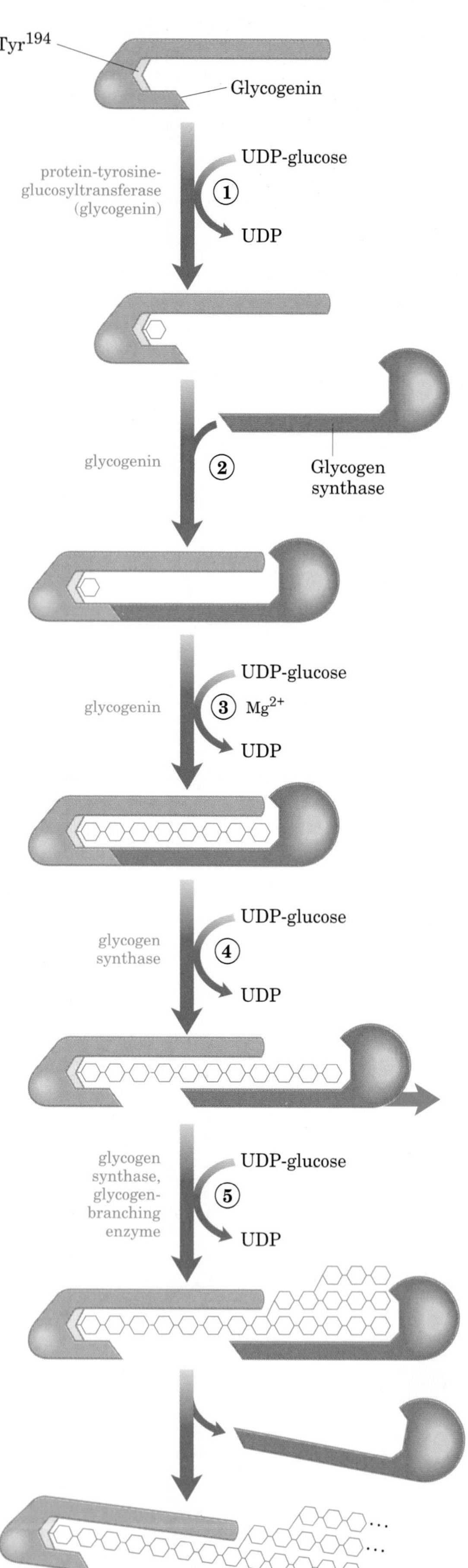

figure 20–14

Initiation of a glycogen particle by glycogenin. Steps ① through ⑤ are described in the text. Glycogenin is found within mature glycogen particles, still covalently attached to the reducing end of the molecule.

Given that glycogen synthase requires a primer, how is a new glycogen molecule initiated? The answer lies in an intriguing protein called **glycogenin** (M_r 37,284), which is both the primer on which new chains are assembled and the enzyme that catalyzes their assembly (Fig. 20–14). The first step in the synthesis of a new glycogen molecule is the covalent attachment of a glucose residue to Tyr^{194} of glycogenin, catalyzed by the protein's glucosyltransferase activity (step ①). Glycogenin then forms a tight complex with glycogen synthase (step ②), and the next few steps occur within this complex. The nascent chain is extended by the sequential addition of up to seven more glucose residues (step ③), each derived from UDP-glucose. The reactions are autocatalytic, mediated by the glucosyltransferase of glycogenin. At this point, glycogen synthase takes over, extending the glycogen chain and ultimately dissociating from glycogenin (step ④). The combined action of glycogen synthase and the branching enzyme (step ⑤) completes the glycogen particle. Glycogenin remains buried within the particle, covalently attached to the single reducing end of the glycogen molecule.

Glycogen Synthase and Glycogen Phosphorylase Are Reciprocally Regulated

Glycogen synthase exists in phosphorylated and dephosphorylated forms (Fig. 20–15). Its active form, **glycogen synthase *a*,** is unphosphorylated. When it is phosphorylated at several Ser hydroxyl groups by specific protein kinases (see Fig. 8–28), glycogen synthase *a* is converted to the less active **glycogen synthase *b*.** Conversion of glycogen synthase *b* back into the active form is promoted by a **phosphoprotein phosphatase,** which removes the phosphate groups from the Ser residues.

In Chapter 15 we saw that the breakdown of glycogen is regulated by both covalent and allosteric modulation of glycogen phosphorylase (see Fig. 15–19). Phosphorylase *a*, the active form, which is phosphorylated at Ser^{14} on both of its subunits, is dephosphorylated by phosphorylase *a* phosphatase to yield phosphorylase *b*, the relatively inactive form, which can be stimulated by AMP, its allosteric modulator. Phosphorylase *b* kinase can convert phosphorylase *b* to the active phosphorylase *a* by phosphorylating Ser^{14}. Glycogen phosphorylase and glycogen synthase are therefore reciprocally regulated by this phosphorylation-dephosphorylation cycle; when one enzyme is stimulated, the other is inhibited (Fig. 20–15). It appears that these two enzymes are never fully active simultaneously.

In Chapter 13 we examined in detail the nature of signaling pathways, and our examples included hormonal control of glycogen synthase and glycogen phosphorylase. The balance between glycogen synthesis and breakdown in liver is controlled by the hormones glucagon and insulin.

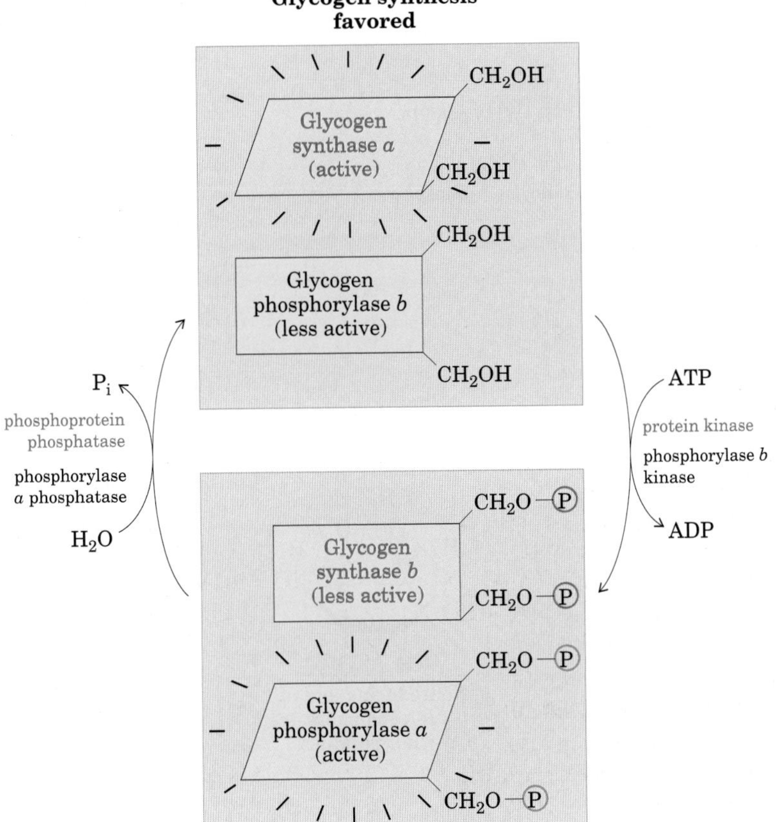

figure 20–15
Reciprocal regulation of glycogen synthase and glycogen phosphorylase. Regulation occurs by a phosphorylation-dephosphorylation cycle. In both enzymes the sites of phosphorylation are Ser residues (represented here by —CH_2OH). The enzymes of glycogen synthesis are shown in red; those of glycogen breakdown in black.

Each of these hormones, acting through specific plasma membrane receptors, triggers a chain of events that leads to changes in the phosphorylation of target proteins (see Figs 13–7, 13–8), including glycogen synthase and glycogen phosphorylase, setting their levels of activity. These hormones also regulate the concentration of fructose 2,6-bisphosphate and thereby the balance between gluconeogenesis and glycolysis. **Epinephrine** has effects similar to those of glucagon, but its target is primarily muscle whereas glucagon's primary action is on liver. In Chapter 23 we describe how hormones and other signals integrate metabolism in the tissues of mammals.

ADP-Glucose Is the Substrate for Starch Synthesis in Plants and Glycogen Synthesis in Bacteria

Starch, like glycogen, is a high molecular weight polymer of D-glucose in ($\alpha1\rightarrow4$) linkage. It is synthesized in chloroplasts as one of the stable end products of photosynthesis, and is also made in other organelles and tissues (in the amyloplasts of the colorless parts of plants—seeds, roots, and tubers).

The mechanism of hexose polymerization in starch synthesis is similar to that in glycogen synthesis. An activated nucleotide sugar, in this case **ADP-glucose,** is formed by condensation of glucose 1-phosphate with ATP.

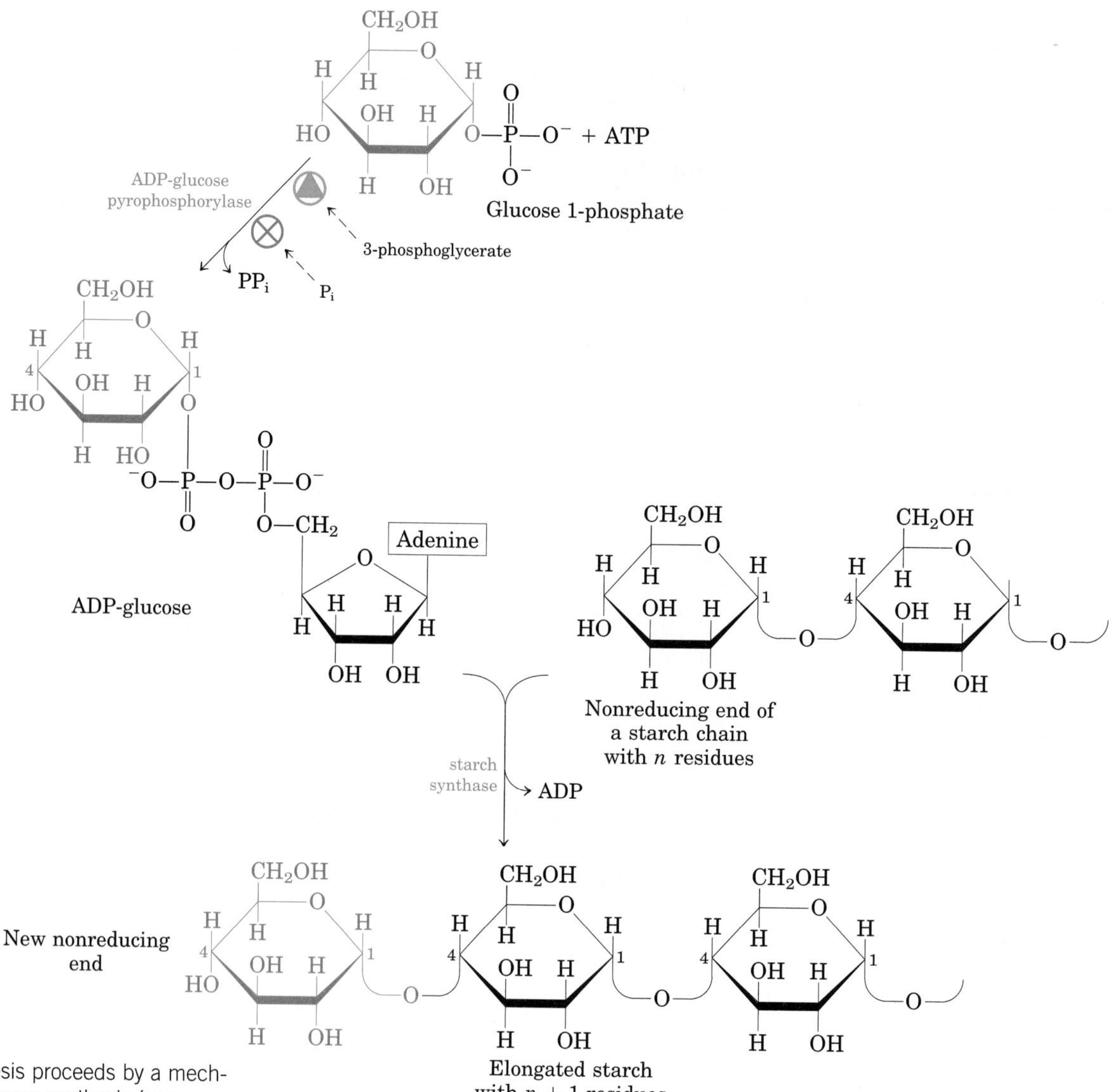

figure 20–16
Starch synthesis. Starch synthesis proceeds by a mechanism analogous to that for glycogen synthesis (see Fig. 20–12), except that the activated substrate is ADP-glucose instead of UDP-glucose. Glucose is transferred to the nonreducing end of an existing starch molecule, in (α1→4) linkage. ADP-glucose pyrophosphorylase is a regulatory enzyme, as discussed later in this chapter.

Starch synthase then transfers glucose residues from ADP-glucose to the nonreducing end of preexisting starch molecules that act as primers (Fig. 20–16). The reaction involves displacement of the ADP of ADP-glucose by the attacking C-4 hydroxyl of the primer, forming the characteristic (α1→4) linkages of starch. The amylose of starch is unbranched, but amylopectin has numerous (α1→6)-linked branches, though fewer than glycogen (see Fig. 9–15). Chloroplasts contain a branching enzyme, similar to glycogen-branching enzyme (Fig. 20–13), that introduces the (α1→6) branches of amylopectin.

With the hydrolysis by inorganic pyrophosphatase of the PP_i produced during ADP-glucose synthesis, the overall reaction for starch formation from glucose 1-phosphate is

$$\text{Starch}_n + \text{glucose 1-phosphate} + \text{ATP} \longrightarrow \text{starch}_{n+1} + \text{ADP} + 2\text{P}_i$$
$$\Delta G'^\circ = -50 \text{ kJ/mol}$$

Starch synthesis is regulated at the level of ADP-glucose formation (Fig. 20–16), as discussed later in this chapter.

Bacteria store carbohydrates in the form of glycogen, which they synthesize in a reaction analogous to that catalyzed by glycogen synthase in animals. Bacteria, however, use ADP-glucose rather than UDP-glucose as the activated form of glucose.

UDP-Glucose Is the Substrate for Sucrose Synthesis in Plants

Most of the triose phosphates generated by CO_2 fixation in plants are converted to sucrose (Fig. 20–17) or starch. In the course of evolution, sucrose may have been selected as the transport form of carbon because of its unusual linkage between the anomeric C-1 of glucose and the anomeric C-2 of fructose. This bond is not hydrolyzed by amylases or other common carbohydrate-cleaving enzymes, and the unavailability of the anomeric carbons prevents sucrose from reacting nonenzymatically (as does glucose) with amino acids and proteins.

Sucrose is synthesized in the cytosol, beginning with dihydroxyacetone phosphate and glyceraldehyde 3-phosphate exported from the chloroplast. After condensation to fructose 1,6-bisphosphate (catalyzed by aldolase), hydrolysis by fructose 1,6-bisphosphatase yields fructose 6-phosphate. **Sucrose 6-phosphate synthase** then catalyzes the reaction of fructose 6-phosphate with UDP-glucose to form **sucrose 6-phosphate.** Finally, **sucrose 6-phosphate phosphatase** removes the phosphate group, making sucrose available for export to other plant tissues. Sucrose synthesis is regulated and closely coordinated with starch synthesis, as we shall see.

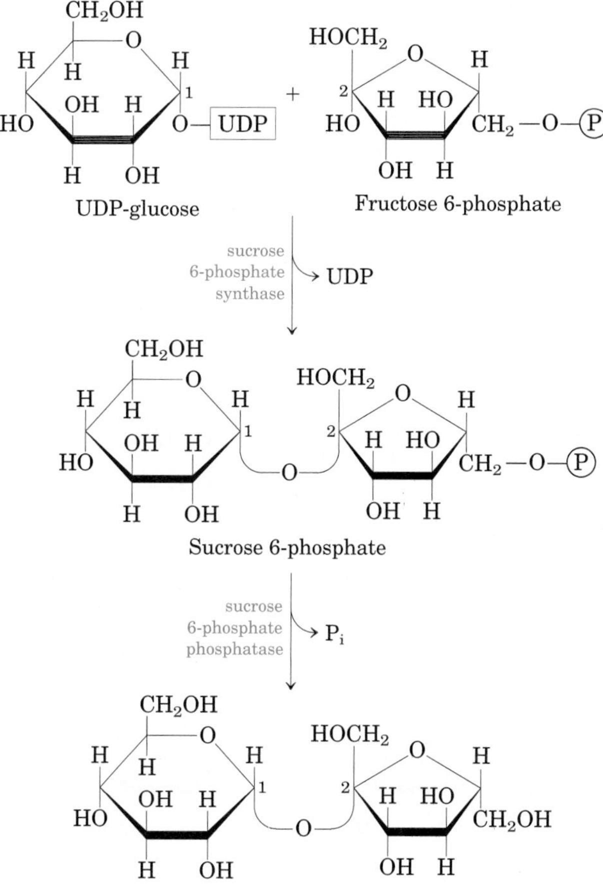

figure 20–17
Sucrose synthesis. Sucrose is synthesized from UDP-glucose and fructose 6-phosphate, which are synthesized from triose phosphates in the plant cell cytosol by pathways shown in Figures 20–11 and 20–25. The sucrose 6-phosphate synthase of most plant species is allosterically regulated by glucose 6-phosphate and P_i.

Lactose Synthesis Is Regulated in a Unique Way

Lactose is synthesized in lactating mammary gland by a mechanism similar to that for glycogen synthesis. However, the regulation of lactose synthesis has an unusual twist. Under certain circumstances, an enzyme normally specific for a different substrate and a different metabolic role changes to a form that catalyzes the synthesis of lactose. Most vertebrate tissues contain the enzyme **galactosyltransferase** (Fig. 20–18a), which contributes to the biosynthesis of glycoproteins by promoting the transfer of the activated galactose residue of UDP-galactose to the monosaccharide *N*-acetylglucosamine already attached to protein:

$$\text{UDP-D-galactose} + N\text{-acetyl-D-glucosamine} \longrightarrow \text{D-galactosyl-}N\text{-acetyl-D-glucosamine} + \text{UDP}$$

This reaction has no role in lactose synthesis, and galactosyltransferase is only slightly active with glucose as the galactosyl group acceptor. Immediately after a female mammal gives birth, however, the specificity of galactosyltransferase in the mammary gland changes: it now transfers the galactosyl group of UDP-galactose to glucose at a very high rate, forming lactose:

$$\text{UDP-D-galactose} + \text{D-glucose} \longrightarrow \text{D-lactose} + \text{UDP}$$

This "new" enzyme is called **lactose synthase** (Fig. 20–18b). The change in specificity of galactosyltransferase is caused by its association with newly (postpartum) synthesized **α-lactalbumin** (M_r 13,500), a milk protein whose function was long unknown. Its synthesis in the mammary gland, which is regulated by the hormones promoting lactation, leads to formation of an **α-lactalbumin–galactosyltransferase** complex—that is, lactose synthase.

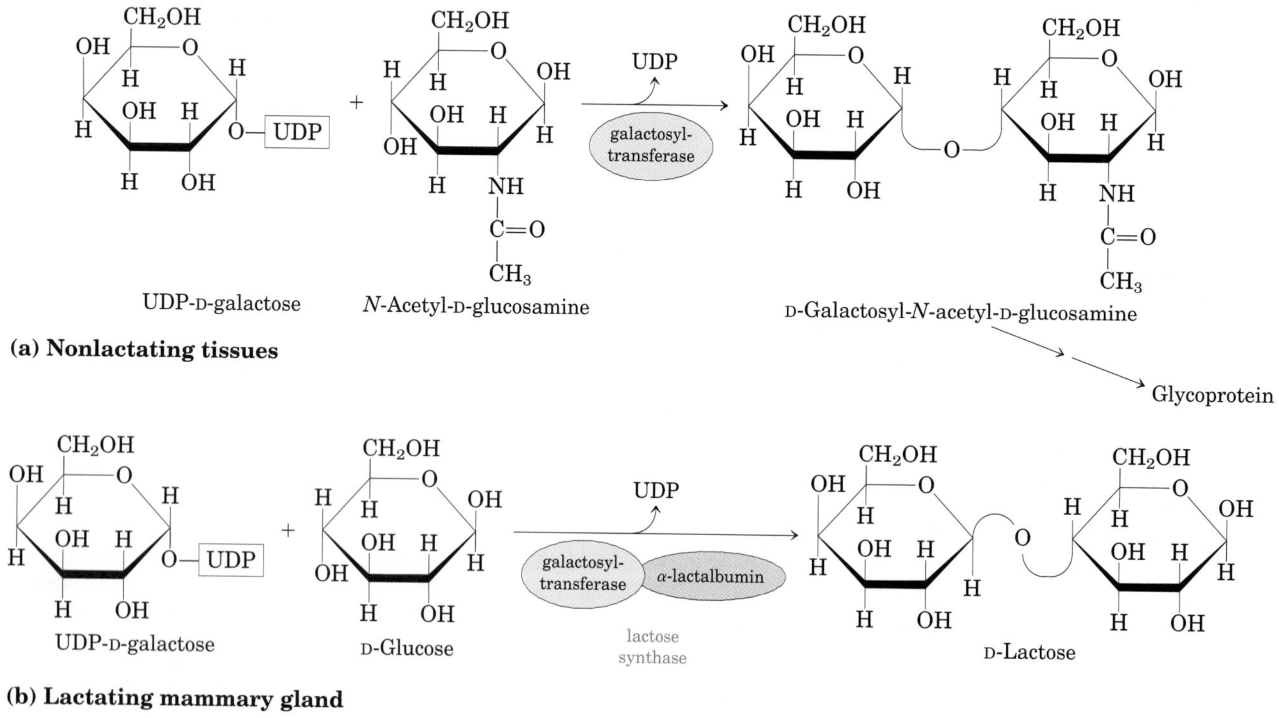

figure 20–18
Lactose synthesis. Two distinct reactions are catalyzed by galactosyltransferase, depending on whether the protein α-lactalbumin, produced only in lactating mammary gland, is present. **(a)** The reaction in nonlactating tissues; **(b)** the reaction in lactating mammary gland.

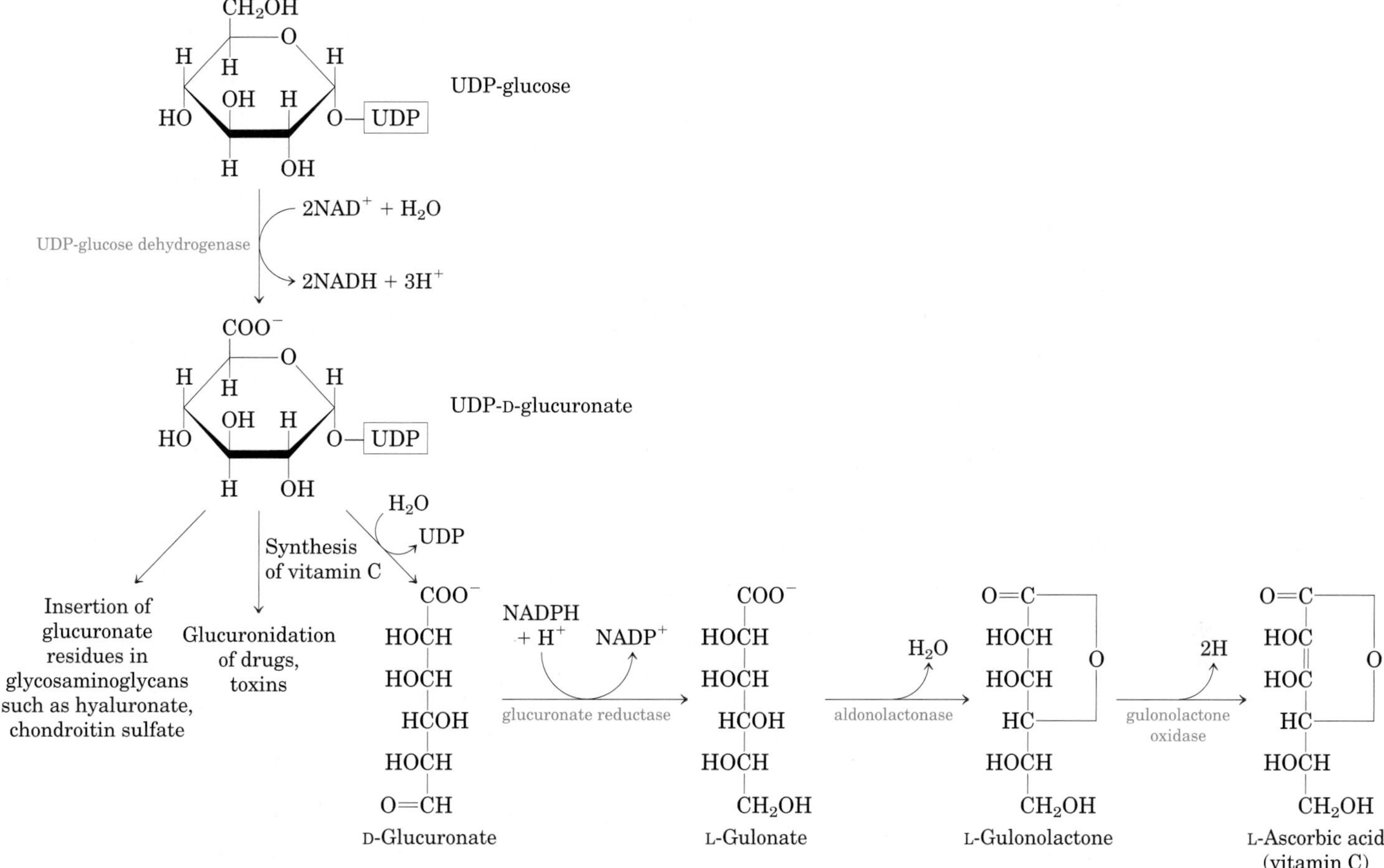

figure 20–19
Synthesis of glucuronate and vitamin C. UDP-glucuronate, formed from glucose as shown here, is the precursor of glucuronate (GlcA) residues in glycosaminoglycans (see Fig. 9–20) and of vitamin C (ascorbic acid), and participates in detoxification reactions.

UDP-Glucose Is an Intermediate in the Formation of Glucuronate and Vitamin C

In many organisms, glucose, through UDP-glucose, gives rise to **D-glucuronate,** a component of glycosaminoglycans and an essential participant in certain detoxification processes, and L-ascorbic acid or **vitamin C.** The glucose portion of UDP-glucose is oxidized to yield **UDP-glucuronate** by UDP-glucose dehydrogenase (Fig. 20–19). UDP-glucuronate is the precursor of the glucuronate residues of such acidic polysaccharides as the glycosaminoglycans hyaluronate and chondroitin sulfate (see Fig. 9–20). This UDP derivative is also involved in the detoxification and excretion of foreign organic compounds.

UDP-glucuronate is the glucuronosyl donor used by a family of detoxifying enzymes that act on a variety of relatively nonpolar drugs, environmental toxins, and carcinogens. The conjugation of these compounds with glucuronate, a process called **glucuronidation,** converts them to polar derivatives that are more easily cleared from the blood by the kidneys and excreted in the urine. For example, the sedative drug phenobarbital,

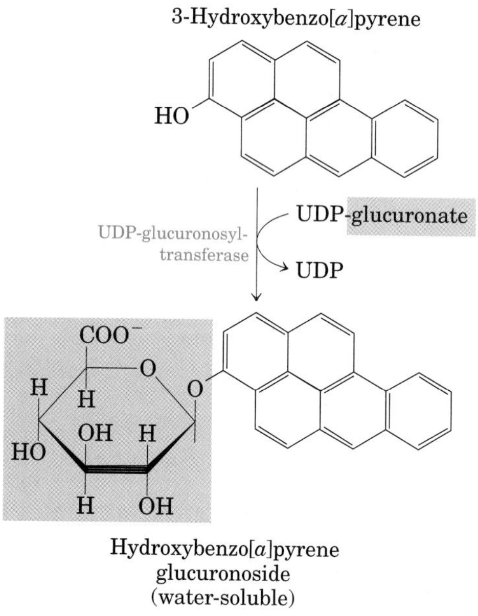

figure 20–20
Role of UDP-glucuronate in detoxification. Shown here is the detoxification of 3-hydroxybenzo[*a*]pyrene, a toxic component of tobacco smoke. Glucuronidation by transfer of glucuronate from UDP-glucuronate converts the nonpolar toxin to a polar compound more easily removed from the blood by the kidneys.

the anti-HIV drug AZT, and the hydroxylated form of the carcinogen benzo[*a*]pyrene (3-hydroxybenzo[*a*]pyrene) all undergo glucuronidation catalyzed by UDP-glucuronosyltransferases in the human liver (Fig. 20–20). Chronic exposure to the drug or toxin induces increased synthesis of the enzyme specific for that compound, increasing tolerance for the drug or resistance to the toxin.

D-Glucuronate, formed by hydrolysis of UDP-D-glucuronate, is the precursor of L-ascorbic acid (Fig. 20–19). In this pathway, D-glucuronate is reduced to the sugar acid L-gulonate, which is converted to its lactone. L-Gulonolactone then undergoes dehydrogenation by the flavoprotein **gulonolactone oxidase** to yield L-ascorbic acid (vitamin C). Some animal species, including humans, guinea pigs, monkeys, some birds, and some fish, lack the enzyme gulonolactone oxidase and are unable to synthesize ascorbic acid, thus requiring it in the diet. Citrus fruits and tomatoes are especially rich sources of vitamin C. Humans who do not obtain enough vitamin C develop the serious disease scurvy, in which the synthesis of collagen-containing connective tissue is defective. The symptoms of scurvy include swollen and bleeding gums with loosened teeth; stiffness and soreness of joints; bleeding under the skin; and slow wound healing. For centuries the disease was very common among sailors on long sea voyages, during which no fresh fruit was available. In 1753 the Scottish naval surgeon James Lind showed that scurvy could be prevented and cured by ingestion of citrus juice. In 1932 the antiscurvy compound vitamin C was isolated from lemon juice and named ascorbic acid (from the Latin *scorbutus*, meaning "scurvy").

Sugar Nucleotides Are Precursors in Bacterial Cell Wall Synthesis

The peptidoglycan that gives bacterial envelopes their strength and rigidity is an alternating linear copolymer of *N*-acetylglucosamine (GlcNAc) and *N*-acetylmuramic acid (Mur2Ac), linked by ($\beta 1 \rightarrow 4$) glycosidic bonds and cross-linked by short peptides attached to the Mur2Ac (see Fig. 9–19). During assembly of the polysaccharide backbone of this complex macromolecule, both GlcNAc and Mur2Ac are activated by attachment of a uridine nucleotide at their anomeric carbons. First, GlcNAc 1-phosphate condenses with UTP to form UDP-GlcNAc (Fig. 20–21, step ①), which reacts with phosphoenolpyruvate to form UDP-Mur2Ac (step ②); five amino acids are then added (step ③). The Mur2Ac-pentapeptide moiety is transferred from the uridine nucleotide to a long-chain isoprenoid alcohol (dolichol; see Fig. 11–20f) (step ④), and a GlcNAc residue is donated by UDP-GlcNAc (step ⑤). In many bacteria, five glycines are added in peptide linkage to the amino group of the Lys residue of the pentapeptide (step ⑥). Finally, this disaccharide decapeptide is added to the nonreducing end of an existing peptidoglycan molecule (step ⑦). A transpeptidation reaction cross-links adjacent polysaccharide chains (step ⑧), contributing to a huge, strong macromolecular wall around the bacterial cell. Many of the most effective antibiotics in use today act by inhibiting reactions in the synthesis of the peptidoglycan (Box 20–1). Without the support of this peptidoglycan layer, the bacterial cell is fragile and prone to osmotic lysis.

Many other oligosaccharides and polysaccharides are synthesized by similar routes in which sugars are activated for subsequent reactions by attachment to nucleotides. In the glycosylation of proteins, for example (see Fig. 27–36), the precursors of the carbohydrate moieties are sugar nucleotides.

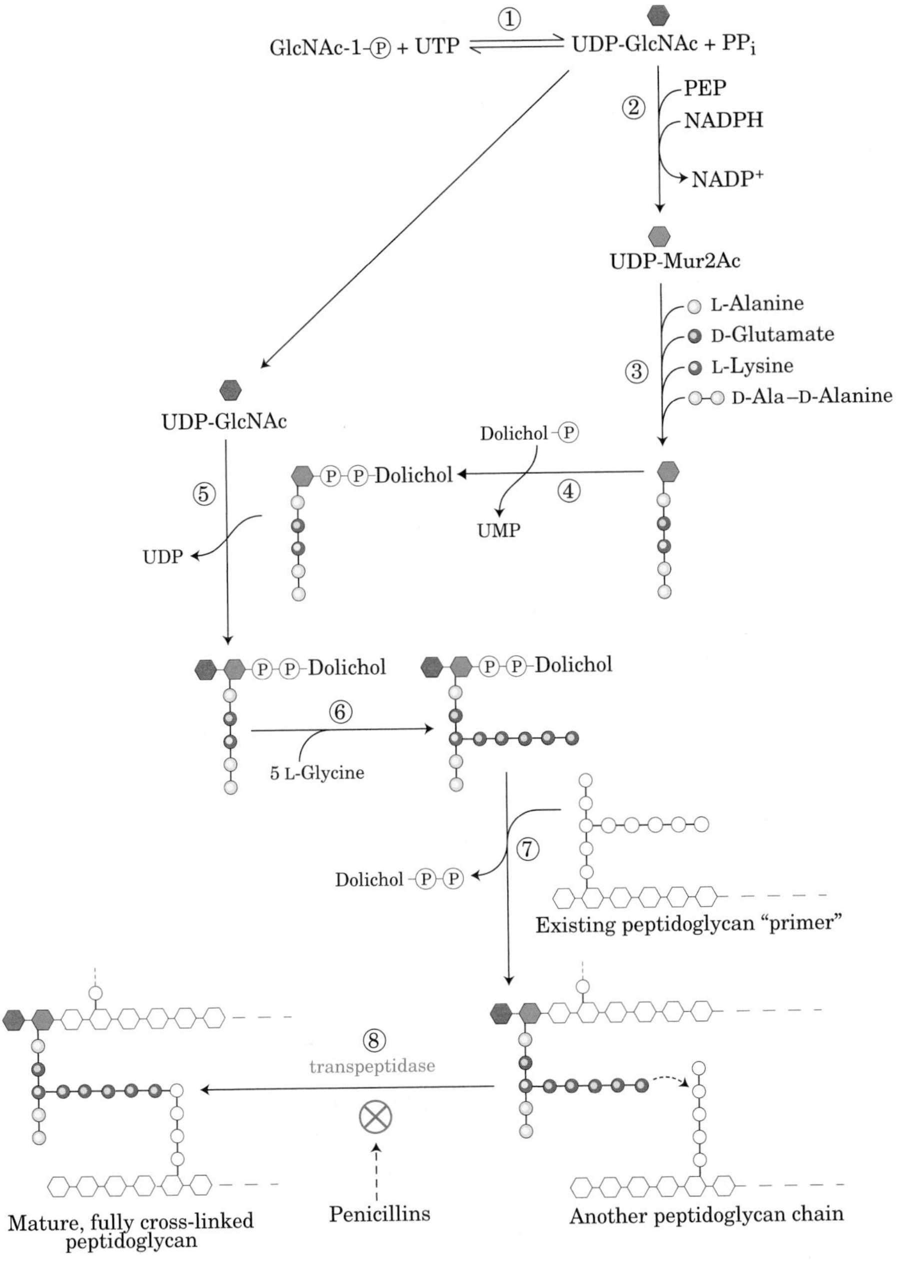

figure 20–21

Synthesis of bacterial peptidoglycan. In the early steps of this pathway (① through ④), *N*-acetylglucosamine (GlcNAc) and *N*-acetylmuramic acid (Mur2Ac) are activated by attachment of a uridine nucleotide (UDP) to their anomeric carbons and, in the case of Mur2Ac, of a long-chain isoprenyl alcohol (dolichol) through a phosphodiester bond. These activating groups participate in the formation of glycosidic linkages; they serve as excellent leaving groups. After assembly (⑤, ⑥) of a disaccharide with a peptide side chain (10 amino acid residues), this precursor is transferred (⑦) to the nonreducing end of an existing peptidoglycan chain, which serves as a primer for the polymerization reaction. Finally, a transpeptidation reaction (⑧) occurs between the peptide side chains on two different peptidoglycan molecules. A Gly residue at the end of one peptide chain displaces a terminal D-Ala in the other peptide chain, forming a cross-link. This transpeptidation reaction is inhibited by the penicillins, which kill bacteria by weakening their cell walls.

box 20–1 Penicillin and β-Lactamase: The Magic Bullet versus the Bulletproof Vest

Because peptidoglycans are unique to bacterial cell walls, with no homologous structures known in mammals, the enzymes responsible for their synthesis are ideal targets for antibiotic action. Antibiotics that hit specific bacterial targets are sometimes called "magic bullets." Penicillin and its many synthetic analogs have been used to treat bacterial infections since these drugs came into wide application in the Second World War.

Penicillins and related antibiotics contain the β-lactam ring (Fig. 1), variously modified. All penicillins have a thiazolidine ring attached to the β-lactam, but they differ in the substituent at position six, which accounts for the different pharmacological properties of the various penicillins. For example, penicillin V is acid stable and can be administered orally, but methicillin is acid labile and must be given intravenously or intramuscularly. However, methicillin resists breakdown by bacterial enzymes (β-lactamases) whereas many other penicillins do not. The β-lactams have many of the properties that make a good drug. First, they target a metabolic pathway present in bacteria but not in people. Second, they have half-lives in the body long enough to be clinically useful. Third, they reach therapeutic concentrations in most, if not all, tissues and organs. Finally, they are effective against a broad range of bacterial species.

Penicillins block formation of the peptide cross-links in peptidoglycans, acting as mechanism-based (suicide) inhibitors. The normal catalytic mechanism of the target enzyme activates the inhibitor, which then covalently modifies a critical residue in the active site. Transpeptidases employ a reaction mechanism (involving Ser residues) similar to that of chymotrypsin (see Fig. 8–19); the reaction activates β-lactams such as penicillin, which in turn inactivate the transpeptidases. After penicillin enters the transpeptidase active site, the proton on the hydroxyl

figure 1

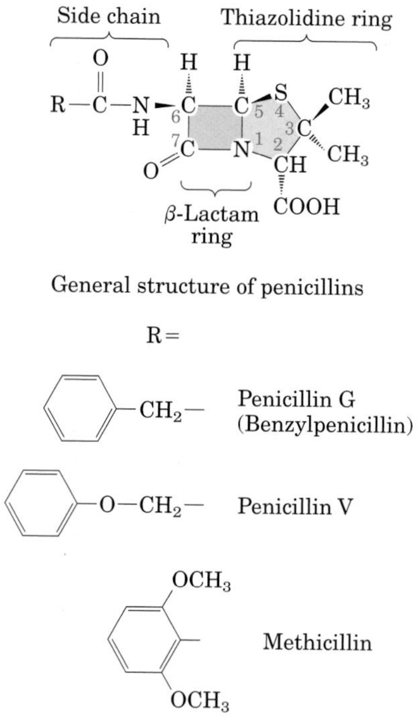

Side chain

Thiazolidine ring

β-Lactam ring

General structure of penicillins

R=

CH_2— Penicillin G (Benzylpenicillin)

O—CH_2— Penicillin V

OCH_3 / OCH_3 Methicillin

Photosynthetic Carbohydrate Synthesis

The synthesis of carbohydrates in animal cells always employs precursors having at least three carbons, all of which are less oxidized than the carbon in CO_2. Photosynthetic organisms, by contrast, can synthesize carbohydrates from CO_2 and water, reducing CO_2 at the expense of energy furnished by ATP and NADPH generated by photosynthetic electron transfer. This represents a fundamental difference between autotrophic (phototrophic or chemotrophic) and heterotrophic organisms. Autotrophs such as plants can use CO_2 as the sole source of all the carbon atoms required for the biosynthesis of cellulose and starch, lipids and proteins, and the many other organic components of plant cells. By contrast, heterotrophs in general cannot bring about the net reduction of CO_2 to form "new" glucose. Carbon dioxide can be incorporated into organic compounds by animal tissues, as in the pyruvate carboxylase reaction of gluconeogenesis, but the CO_2 is lost in a subsequent reaction step (Fig. 20–3). Similarly, the CO_2

group of an active site Ser residue is abstracted to the nitrogen of the β-lactam ring, and the activated oxygen of the Ser hydroxyl attacks the carbonyl carbon at position 7 of the β-lactam, opening the ring and forming a stable penicilloyl-enzyme derivative that inactivates the enzyme (Fig. 2a).

Widespread use of antibiotics has driven the selection and evolution of antibiotic resistance in many pathogenic bacteria. The most important mechanism of resistance is inactivation of the antibiotic by enzymatic hydrolysis of the lactam ring, catalyzed by bacterial β-lactamases, which provide bacteria with a bulletproof vest (Fig. 2b). β-Lactamase forms a temporary covalent adduct with the carboxyl group of the opened β-lactam ring, which is immediately hydrolyzed, regenerating active enzyme. One approach to circumventing antibiotic resistance of this type is to synthesize penicillin analogs, such as methicillin, that are poor substrates for β-lactamases. Another approach is to administer along with antibiotics a β-lactamase inhibitor such as clavulanate or sulbactam.

Antibiotic resistance is a significant threat to public health. Some bacterial infections are now essentially untreatable with antibiotics. By the early 1990s, 20% to 40% of *Staphylococcus aureus* (the causative agent of "staph" infections) was resistant to methicillin and 32% of *Neisseria gonorrhoeae* (the causative agent of gonorrhea) was resistant to penicillin. By 1986, 32% of *Shigella* (a pathogen responsible for severe forms of dysentery, some with a lethality of up to 15%) was resistant to ampicillin. Significantly, many of these pathogens are also resistant to many other antibiotics. In the future, we will need to develop new drugs that circumvent the resistance mechanisms that have evolved in bacteria or that act on different bacterial targets.

figure 2

Penicillin

Enz—Ser—ÖH

(a) (b)

Trans-peptidase —Ser—O

Stably derivatized, inactive transpeptidase

β-Lactamase —Ser—O

H_2O → β-Lactamase

Inactive penicillin

taken up by acetyl-CoA carboxylase during fatty acid synthesis (see Fig. 21–1) or by carbamoyl phosphate synthetase I during urea formation (see Fig. 18–9) is lost in later steps.

Green plants contain in their chloroplasts unique enzymatic machinery that catalyzes the conversion of CO_2 to simple (reduced) organic compounds, a process called **CO_2 assimilation.** This process has also been called **CO_2 fixation** or **carbon fixation,** but here we reserve these terms for the specific reaction in which CO_2 is incorporated (fixed) into 3-phosphoglycerate. This simple product of photosynthesis is the precursor of more complex biomolecules, including sugars, polysaccharides, and the metabolites derived from them, synthesized in metabolic pathways similar to those of animal tissues. Carbon dioxide is assimilated via a cyclic pathway in which key intermediates are constantly regenerated. The pathway was elucidated in the early 1950s by Melvin Calvin and coworkers and is often called the **Calvin cycle.**

Melvin Calvin
1911–1997

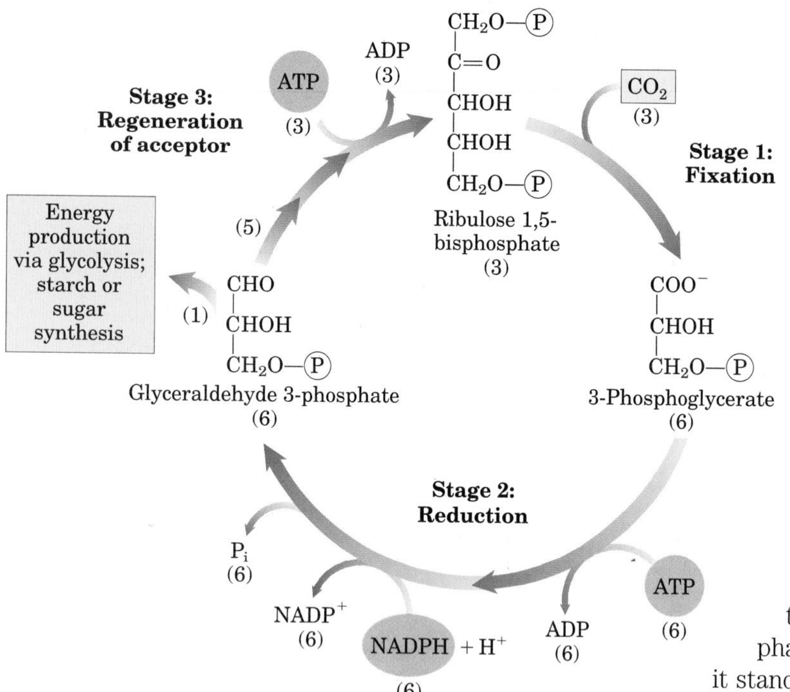

figure 20–22

The three stages of CO_2 assimilation in photosynthetic organisms. Stoichiometries of three key intermediates are shown (numbers in parentheses) so that the fate of carbon atoms entering and leaving the cycle is apparent. As shown here, three CO_2 are fixed to permit the net synthesis of one molecule of glyceraldehyde 3-phosphate. This cycle is called the photosynthetic carbon reduction cycle or the Calvin cycle.

Carbon Dioxide Assimilation Occurs in Three Stages

The first stage in the assimilation of CO_2 into biomolecules (Fig. 20–22) is the carbon fixation reaction: condensation of CO_2 with a five-carbon acceptor, **ribulose 1,5-bisphosphate,** to form two molecules of 3-phosphoglycerate. (Note that Figure 20–22 shows the number of molecules reacting to form one molecule of triose—this takes three molecules of CO_2.) In the second stage, the 3-phosphoglycerate is reduced to triose phosphates. Thus three molecules of CO_2 are fixed to three molecules of ribulose 1,5-bisphosphate to form six molecules of glyceraldehyde 3-phosphate (18 carbons) in equilibrium with dihydroxyacetone phosphate. In the third stage, five of the six molecules of triose phosphate (15 carbons) are used to regenerate three molecules of ribulose 1,5-bisphosphate (15 carbons), the starting material. Thus the overall process is cyclical, with the continuous conversion of CO_2 into triose and hexose phosphates. Fructose 6-phosphate is a key intermediate in stage 3; it stands at a branch point, leading either to regeneration of ribulose 1,5-bisphosphate or to synthesis of starch. The pathway from hexose phosphate to pentose bisphosphate involves many of the same reactions used in animal cells for the conversion of pentose phosphates to hexose phosphates during the oxidative reactions of the **pentose phosphate pathway,** an alternative route for glucose oxidation (see Fig. 15–20). In the photosynthetic assimilation of CO_2, this pathway operates in the opposite direction, converting hexose phosphates to pentose phosphates. This reductive pentose phosphate cycle uses the same enzymes as the oxidative pathway, and several more that make the reductive cycle irreversible.

Stage 1: Fixation of CO_2 into 3-Phosphoglycerate An important clue to the nature of the CO_2 assimilation mechanisms in photosynthetic organisms came in the late 1940s. Calvin and his associates illuminated a suspension of green algae in the presence of radioactive carbon dioxide ($^{14}CO_2$) for just a few seconds, then quickly killed the cells, extracted their contents, and with the help of chromatographic methods searched for the metabolites in which the labeled carbon first appeared. The first compound that became labeled was **3-phosphoglycerate,** with the ^{14}C predominantly located in the carboxyl carbon atom. This atom is not rapidly labeled in animal tissues in the presence of radioactive $^{14}CO_2$. These experiments strongly suggested that 3-phosphoglycerate is an early intermediate in photosynthesis. The enzyme in chloroplasts that catalyzes incorporation of CO_2 into an organic form is **ribulose 1,5-bisphosphate carboxylase** or **RuBP carboxylase/oxygenase,** often called **rubisco** for short. (The enzyme's oxygenase activity is discussed later in the chapter.) As a carboxylase, rubisco catalyzes the covalent attachment of CO_2 to the five-carbon sugar ribulose 1,5-bisphosphate and the cleavage of the unstable six-carbon intermediate to form two molecules of 3-phosphoglycerate, one of which bears the carbon introduced as CO_2 in its carboxyl group (Fig. 20–23).

Plant rubisco, the key enzyme in the production of biomass from CO_2, has a complex structure (Fig. 20–24a, b), with eight large subunits (each of M_r 53,000), each containing an active site, and eight small subunits (each of M_r 14,000), whose function is not well understood. The rubisco of photosynthetic bacteria is quite different in structure, having only two subunits, which in many respects resemble the large subunits of the plant enzyme (Fig. 20–24c). The plant enzyme is located in the chloroplast stroma, where it makes up about 50% of the total chloroplast protein. Rubisco is the most abundant enzyme in the biosphere.

Ribulose 1,5-bisphosphate ⇌ Enediol intermediate ⇌ (CO_2, H_2O) 2′-Carboxy-3-keto-D-arabinitol 1,5-bisphosphate (β-keto acid intermediate) ⇌ (H^+) Hydrated intermediate ⇌ Carbanion + 3-Phosphoglycerate → (H^+) 3-Phosphoglycerate

Six-carbon reaction intermediates bound to ribulose 1,5-bisphosphate carboxylase (rubisco)

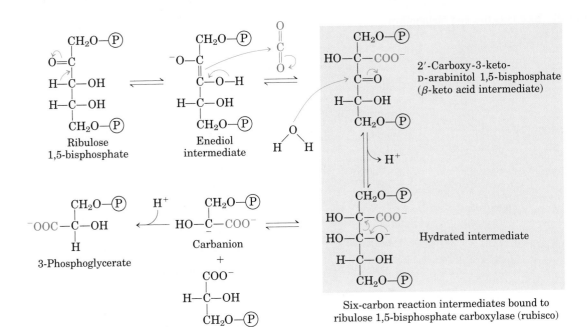

figure 20–23

First stage of CO_2 assimilation. The CO_2 fixation reaction is catalyzed by ribulose 1,5-bisphosphate carboxylase (rubisco). The carboxylated reaction intermediate is believed to be the enzyme-bound six-carbon β-keto acid shown here, which is hydrolyzed and released from the enzyme surface as two identical three-carbon products, one of which contains the carbon atom from CO_2 (red).

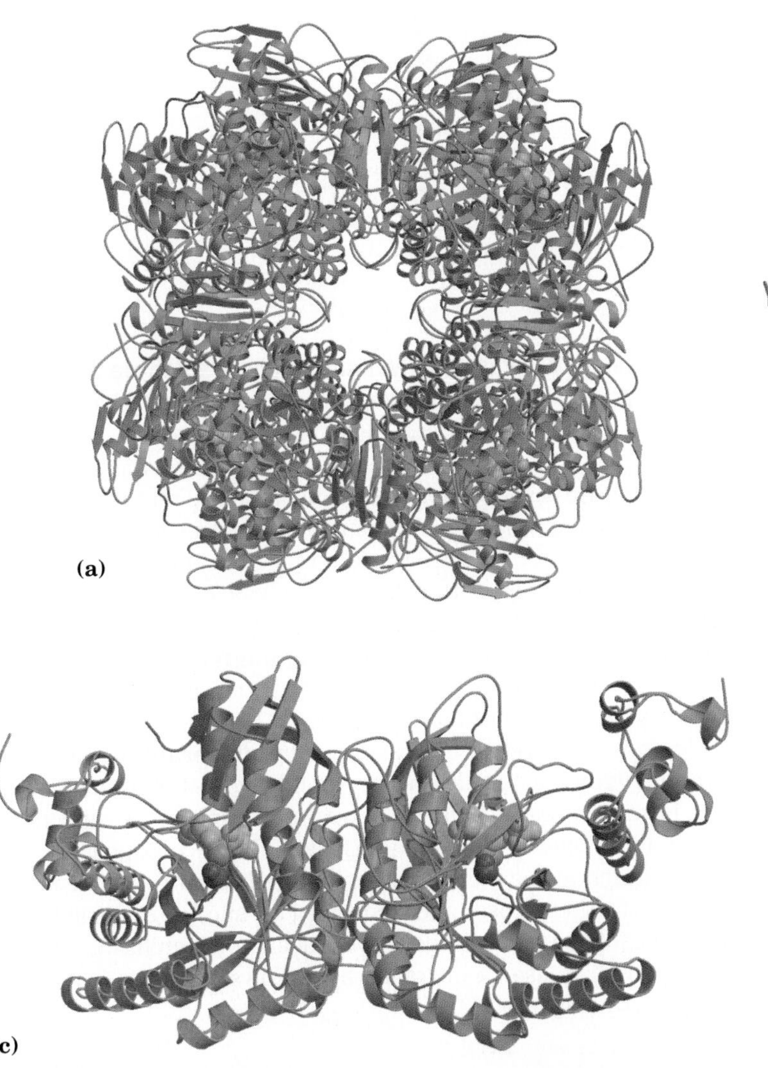

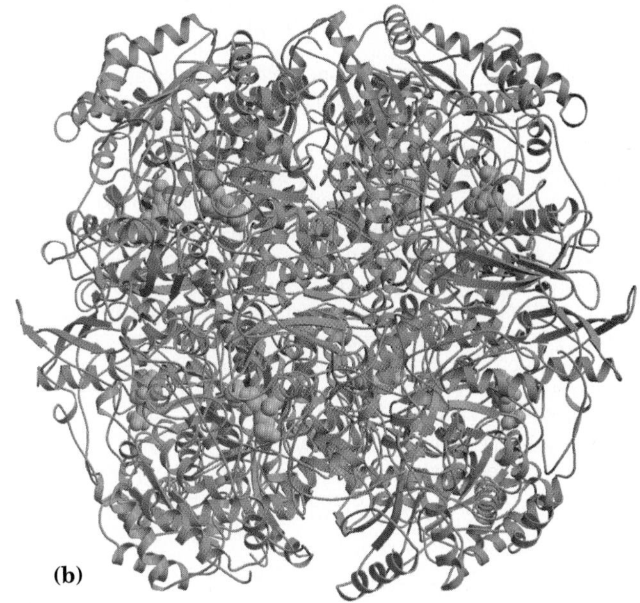

figure 20–24

Structure of ribulose 1,5-bisphosphate carboxylase (rubisco). (a) Top view and **(b)** side view of a ribbon model of rubisco from spinach, based on x-ray diffraction analysis of the crystalline enzyme. There are eight large subunits (shown in blue) and eight small ones (gray), tightly packed into a structure of M_r 550,000. Rubisco is present at a concentration of about 250 mg/mL in the chloroplast stroma, corresponding to an extraordinarily high concentration of active sites (~4 mM). Amino acid residues of the active site are shown in yellow. **(c)** Ribbon model of rubisco from the bacterium *Rhodospirillum rubrum.* The subunits are in gray and blue. A Lys residue at the active site, which is carboxylated to form a carbamate in the active enzyme (see Fig. 20–33), is in red. The substrate, ribulose 1,5-bisphosphate, is yellow.

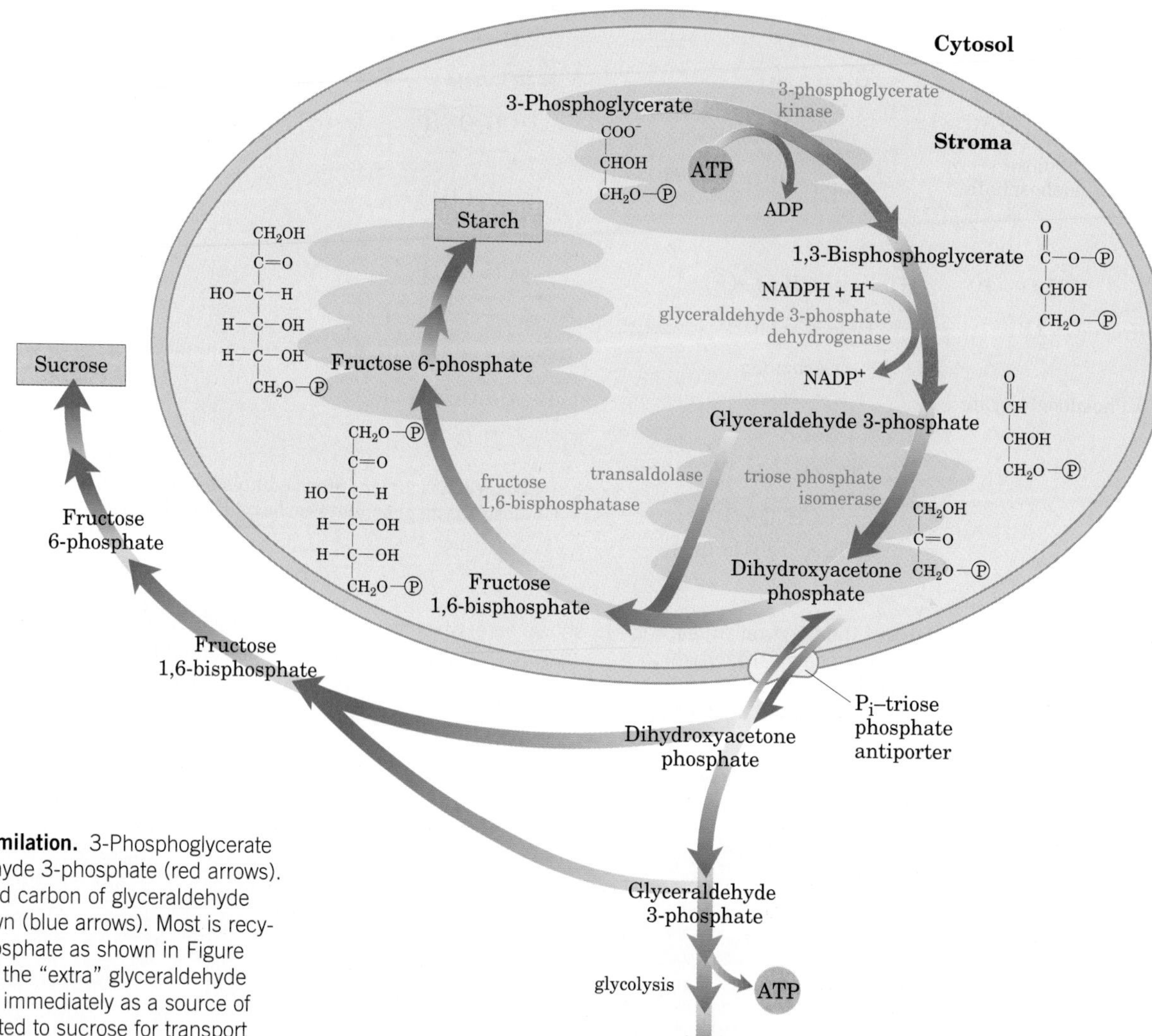

figure 20–25

Second stage of CO_2 assimilation. 3-Phosphoglycerate is converted to glyceraldehyde 3-phosphate (red arrows). Alternative fates of the fixed carbon of glyceraldehyde 3-phosphate are also shown (blue arrows). Most is recycled to ribulose 1,5-bisphosphate as shown in Figure 20–26. A small fraction of the "extra" glyceraldehyde 3-phosphate may be used immediately as a source of energy, but most is converted to sucrose for transport or is stored as starch for future use. If glyceraldehyde 3-phosphate is needed for starch synthesis, it condenses with dihydroxyacetone phosphate in the stroma to form fructose 1,6-bisphosphate, a precursor of starch. In other situations it is converted to dihydroxyacetone phosphate, which leaves the chloroplast via a specific transporter (see Fig. 20–31). In the cytosol, dihydroxyacetone phosphate can be degraded glycolytically to provide energy or can be used to form fructose 6-phosphate and hence sucrose.

Stage 2: Conversion of 3-Phosphoglycerate to Glyceraldehyde 3-Phosphate

The 3-phosphoglycerate formed in stage 1 is converted to glyceraldehyde 3-phosphate in two steps that are essentially the reversal of the corresponding steps in glycolysis, with one exception: the nucleotide cofactor for the reduction of 1,3-bisphosphoglycerate is NADPH rather than NADH (Fig. 20–25). The chloroplast stroma contains all of the glycolytic enzymes except phosphoglycerate mutase. These stromal enzymes are isozymes of those found in the cytosol; both sets of enzymes catalyze the same reactions, but they are the products of different genes.

In the first step of the sequence, **3-phosphoglycerate kinase** in the stroma catalyzes the transfer of a phosphoryl group from ATP to 3-phosphoglycerate, yielding 1,3-bisphosphoglycerate (Fig. 20–25). Next, NADPH donates electrons in a reduction catalyzed by **glyceraldehyde 3-phosphate dehydrogenase,** producing glyceraldehyde 3-phosphate. Triose phosphate isomerase interconverts the triose phosphates glyceraldehyde 3-phosphate and dihydroxyacetone phosphate. Triose phosphate is either converted to starch in the chloroplast and stored for later use or immediately exported to the cytosol and converted to sucrose for transport to growing regions of the plant. In developing leaves, a significant portion of triose phosphate may also be degraded by glycolysis to provide additional energy for growth (Fig. 20–25).

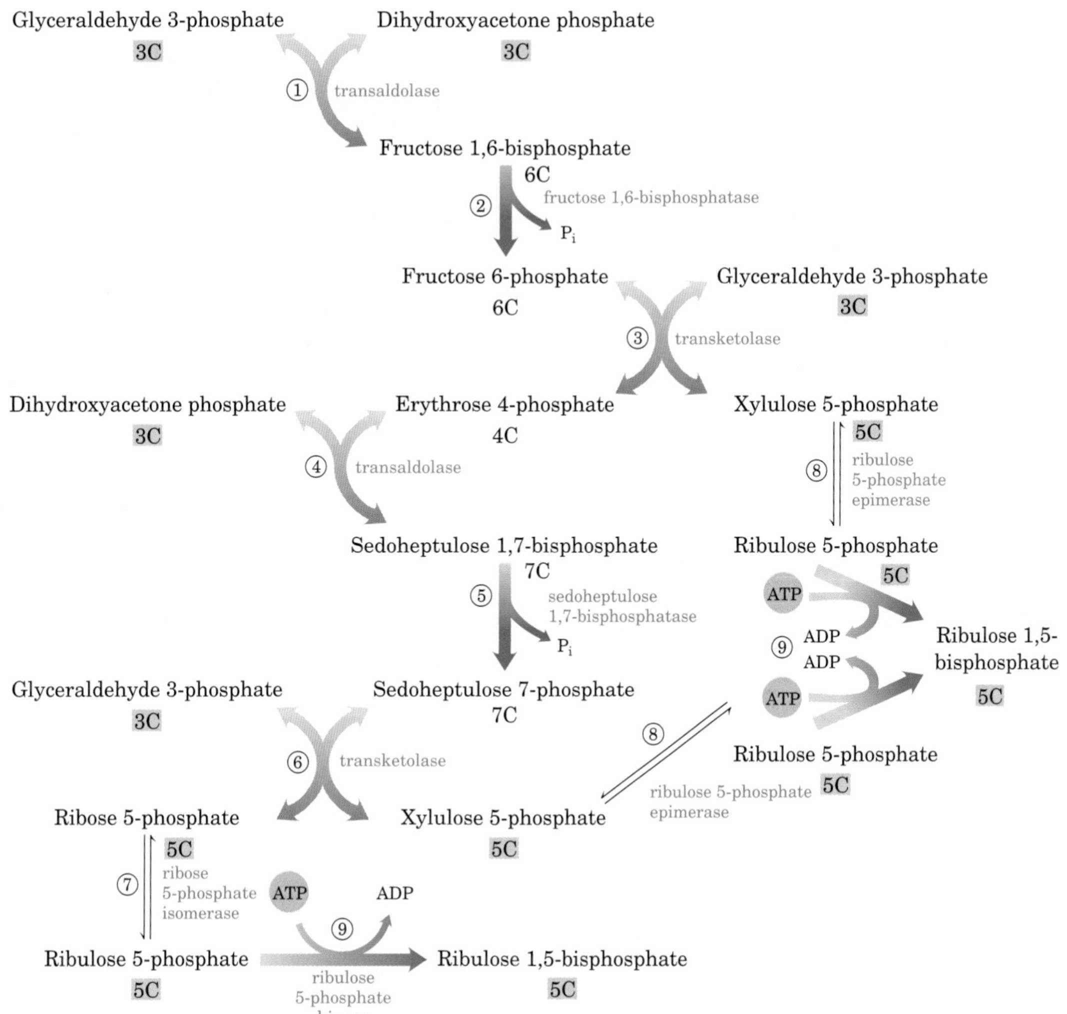

figure 20–26

Third stage of CO_2 assimilation. This schematic diagram shows the interconversions of triose phosphates (3C compounds; green screen) and pentose phosphates (5C compounds; red screen). The starting materials are glyceraldehyde 3-phosphate and dihydroxyacetone phosphate. Reactions catalyzed by transaldolase (① and ④) and transketolase (③ and ⑥) produce pentose phosphates that are converted to ribulose 1,5-bisphosphate—ribose 5-phosphate by ribose 5-phosphate isomerase (⑦) and xylulose 5-phosphate by ribulose 5-phosphate epimerase (⑧). In step ⑨, ribulose 5-phosphate is phosphorylated, regenerating ribulose 1,5-bisphosphate. The steps with blue arrows are exergonic and make the whole process irreversible: steps ② (fructose 1,6-bisphosphatase), ⑤ (sedoheptulose bisphosphatase), and ⑨ (ribulose 5-phosphate kinase).

Stage 3: Regeneration of Ribulose 1,5-Bisphosphate from Triose Phosphates As we have seen, the first reaction in the assimilation of CO_2 into triose phosphates consumes ribulose 1,5-bisphosphate. For continuous flow of CO_2 into carbohydrate, ribulose 1,5-bisphosphate must be constantly regenerated. This is accomplished in a series of reactions that, together with stages 1 and 2 discussed above, form a cyclic pathway (Fig. 20–26) in which the product of the first reaction (3-phosphoglycerate) passes through a series of transformations that regenerate ribulose 1,5-bisphosphate. This process involves rearrangements of the carbon skeletons of glyceraldehyde 3-phosphate and dihydroxyacetone phosphate. The intermediates in the pathway include three-, four-, five-, six-, and seven-carbon sugars. In the following discussion, all step numbers refer to Figure 20–26.

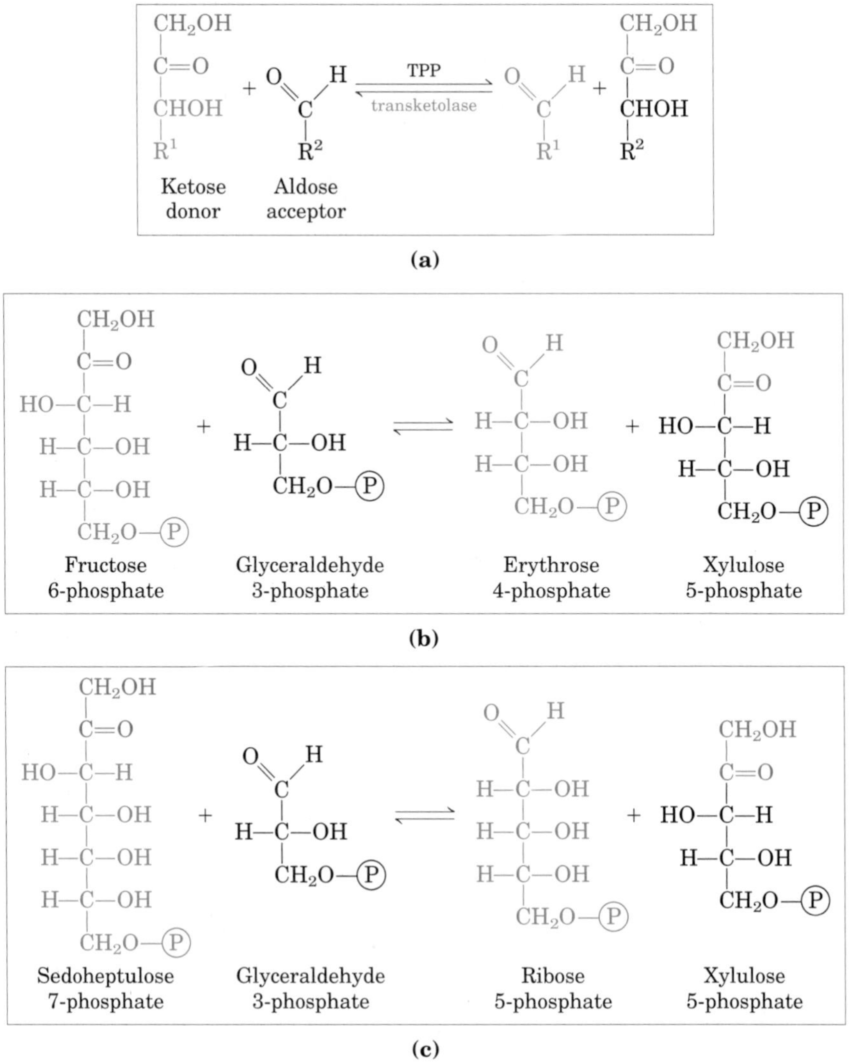

$$\begin{array}{c} CH_2OH \\ | \\ C{=}O \\ | \\ CHOH \\ | \\ R^1 \end{array} + \begin{array}{c} O \quad H \\ \backslash\!\!\backslash \; / \\ C \\ | \\ R^2 \end{array} \xrightleftharpoons[\text{transketolase}]{\text{TPP}} \begin{array}{c} O \quad H \\ \backslash\!\!\backslash \; / \\ C \\ | \\ R^1 \end{array} + \begin{array}{c} CH_2OH \\ | \\ C{=}O \\ | \\ CHOH \\ | \\ R^2 \end{array}$$

Ketose donor · Aldose acceptor

(a)

Fructose 6-phosphate + Glyceraldehyde 3-phosphate ⇌ Erythrose 4-phosphate + Xylulose 5-phosphate

(b)

Sedoheptulose 7-phosphate + Glyceraldehyde 3-phosphate ⇌ Ribose 5-phosphate + Xylulose 5-phosphate

(c)

figure 20–27

Transketolase-catalyzed reactions of the Calvin cycle. **(a)** The general reaction catalyzed by transketolase is the transfer of a two-carbon group, carried temporarily on enzyme-bound TPP, from a ketose donor to an aldose acceptor. **(b)** Conversion of a hexose and a triose to a four-carbon sugar and a pentose (step ③ of Fig. 20–26). **(c)** Conversion of seven-carbon and three-carbon sugars to two pentoses (step ⑥ of Fig. 20–26).

Steps ① and ④ are promoted by transaldolase, which catalyzes the reversible condensation of glyceraldehyde 3-phosphate with dihydroxyacetone phosphate (step ①) as in glycolysis and of erythrose 4-phosphate with dihydroxyacetone phosphate (step ④), yielding the seven-carbon **sedoheptulose 1,7-bisphosphate.** Steps ③ and ⑥ are catalyzed by **transketolase,** which contains thiamine pyrophosphate (TPP) as its prosthetic group (see Fig. 15–9a) and requires Mg^{2+}. Transketolase catalyzes the reversible transfer of a ketol group (CH_2OH—CO—) from a ketose phosphate donor, fructose 6-phosphate in step ③, to an aldose phosphate acceptor, glyceraldehyde 3-phosphate (Fig. 20–27a, b). The same basic reaction converts sedoheptulose 7-phosphate and glyceraldehyde 3-phosphate to two pentose phosphates in step ⑥ (Fig. 20–27c). Figure 20–28 shows how a two-carbon fragment is temporarily carried on TPP and condensed with the three carbons of glyceraldehyde 3-phosphate in step ⑥.

The pentose phosphates formed in the transketolase reactions—ribose 5-phosphate and xylulose 5-phosphate—are converted to **ribulose 5-phosphate** (steps ⑦ and ⑧), which in the final step of the cycle (step ⑨) is phosphorylated to ribulose 1,5-bisphosphate by ribulose 5-phosphate kinase (Fig. 20–29). This is the third very exergonic reaction of the pathway.

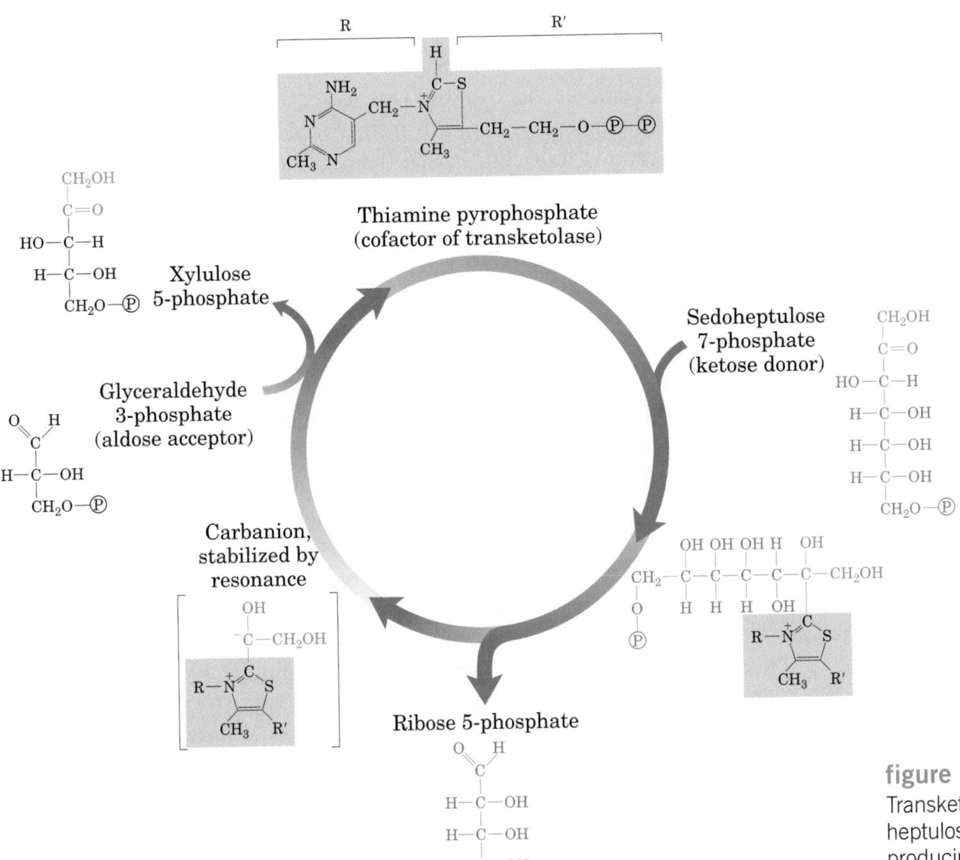

figure 20–28
Transketolase transfers a two-carbon group from sedoheptulose 7-phosphate to glyceraldehyde 3-phosphate, producing two pentose phosphates (step ⑥ in Fig. 20–26). Thiamine pyrophosphate serves as a temporary carrier of the two-carbon unit and as an electron sink (see Fig. 15–9) to facilitate the reactions.

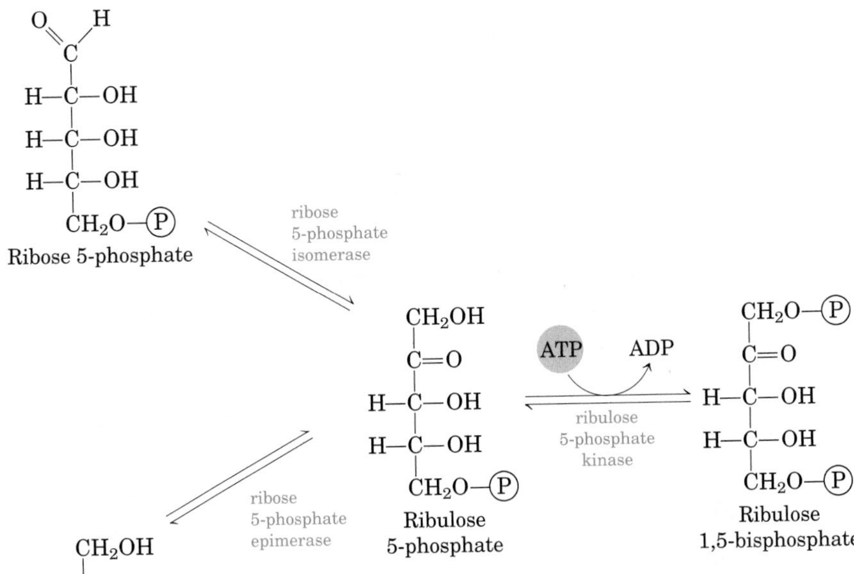

figure 20–29
Regeneration of ribulose 1,5-bisphosphate. The starting material for the Calvin cycle, ribulose 1,5-bisphosphate, is regenerated from two pentose phosphates produced in the cycle. This pathway involves the action of an isomerase and an epimerase, then phosphorylation by a kinase, with ATP as phosphate group donor (steps ⑦, ⑧, and ⑨ of Fig. 20–26).

Each Triose Phosphate Synthesized from CO_2 Costs Six NADPH and Nine ATP

The net result of the Calvin cycle is the conversion of three molecules of CO_2 and one molecule of phosphate into a molecule of triose phosphate. The stoichiometry of the overall path from CO_2 to triose phosphate, with the regeneration of ribulose 1,5-bisphosphate, is shown in Figure 20–30. Three molecules of ribulose 1,5-bisphosphate (a total of 15 carbons) condense with three CO_2 (three carbons) to form six molecules of 3-phosphoglycerate (18 carbons). These six molecules of 3-phosphoglycerate are reduced to six molecules of glyceraldehyde 3-phosphate (which is in equilibrium with dihydroxyacetone phosphate), with the expenditure of six ATP (in the synthesis of 1,3-bisphosphoglycerate) and six NADPH (in the reduction of 1,3-bisphosphoglycerate to glyceraldehyde 3-phosphate). *One molecule of glyceraldehyde 3-phosphate is the net product of the carbon assimilation pathway.* The other five triose phosphate molecules (15 carbons) are rearranged in steps ① to ⑨ of Figure 20–26 to form three molecules of ribulose 1,5-bisphosphate (15 carbons). The last step in this conversion requires one ATP per ribulose 1,5-bisphosphate, or a total of three ATP. Thus, for every molecule of triose phosphate produced by photosynthetic CO_2 assimilation, six NADPH and nine ATP are required.

The source of NADPH and ATP for triose phosphate synthesis is the light-driven reactions of photosynthesis (Chapter 19), which produce NADPH and ATP in about the same ratio (2:3) at which they are consumed in the Calvin cycle. Of the nine ATP molecules converted to ADP and phosphate in the generation of a molecule of triose phosphate, eight of the phosphates are released as P_i and combined with eight ADP to regenerate ATP. The ninth phosphate is incorporated into the triose phosphate itself. To convert the ninth ADP to ATP, a molecule of P_i must be imported from the cytosol, as we shall see.

In the dark, the production of ATP and NADPH by photophosphorylation ceases, and the incorporation of CO_2 into triose phosphate (by the so-called "dark reactions") also stops. The "dark reactions" of photosynthesis

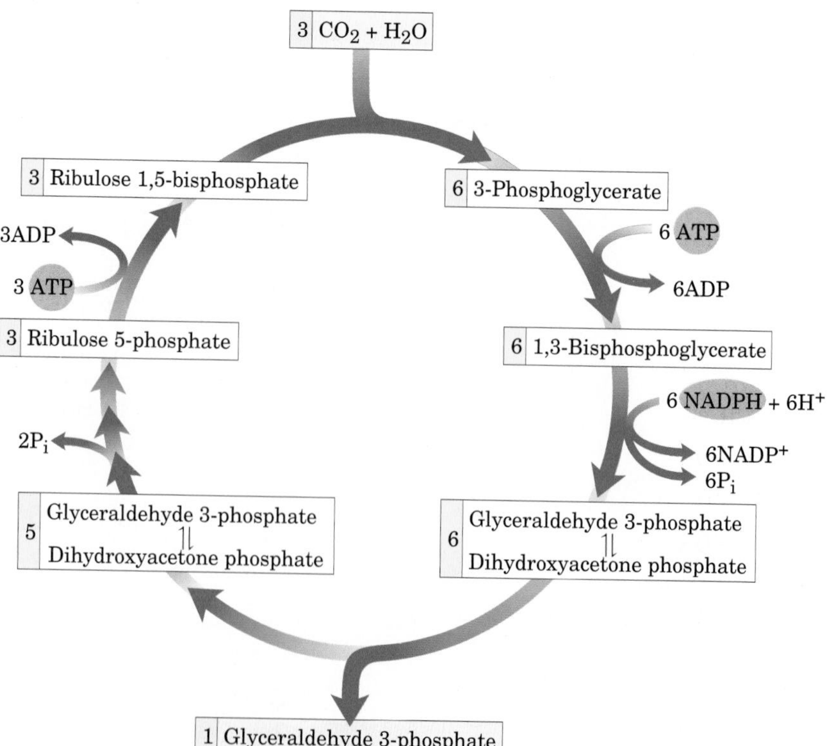

figure 20–30
Stoichiometry of CO_2 assimilation in the Calvin cycle. For every three CO_2 molecules fixed, one molecule of triose phosphate (glyceraldehyde 3-phosphate) is produced and nine ATP and six NADPH are consumed.

were so named to distinguish them from the *primary* light-driven reactions of electron transfer to $NADP^+$ and synthesis of ATP, described in Chapter 19. They do not, in fact, occur at significant rates in the dark in photosynthetic organisms (although they do in chemotrophic organisms) and are thus more appropriately called the **carbon assimilation reactions.** We describe later in this chapter the regulatory mechanisms that turn on carbon assimilation in the light and turn it off in the dark.

Significant amounts of the fixed carbon generated in the chloroplast are also stored there. The chloroplast stroma contains all the enzymes necessary to convert the triose phosphates produced by CO_2 assimilation (glyceraldehyde 3-phosphate and dihydroxyacetone phosphate) into starch, which is stored in the chloroplast as insoluble granules. Aldolase condenses the trioses to fructose 1,6-bisphosphate, fructose 1,6-bisphosphatase produces fructose 6-phosphate, phosphohexose isomerase yields glucose 6-phosphate, and phosphoglucomutase produces glucose 1-phosphate, the starting material for starch synthesis.

All the reactions of the Calvin cycle except those catalyzed by rubisco, sedoheptulose 1,7-bisphosphatase, and ribulose 5-phosphate kinase also take place in animal tissues. For lack of these three enzymes (and the abundant ATP and NADPH provided by the light reactions of photosynthesis), animals cannot carry out net conversion of CO_2 into glucose.

A Transport System Exports Triose Phosphates from the Chloroplast and Imports Phosphate

The inner chloroplast membrane is impermeable to most phosphorylated compounds, including fructose 6-phosphate, glucose 6-phosphate, and fructose 1,6-bisphosphate. It does, however, have a specific transporter (antiporter) that catalyzes the one-for-one exchange of P_i with triose phosphates, either dihydroxyacetone phosphate or 3-phosphoglycerate (Fig. 20–31; see also Fig. 20–25). This antiporter simultaneously moves triose phosphate out of the chloroplast and P_i into the chloroplast, where it is used in photophosphorylation.

Without this antiport system, CO_2 assimilation would be sharply reduced. The net transport of triose phosphates out of the chloroplast serves the important function of removing the triose phosphate products of carbon assimilation. In the cytosol, the triose phosphates are converted to sucrose

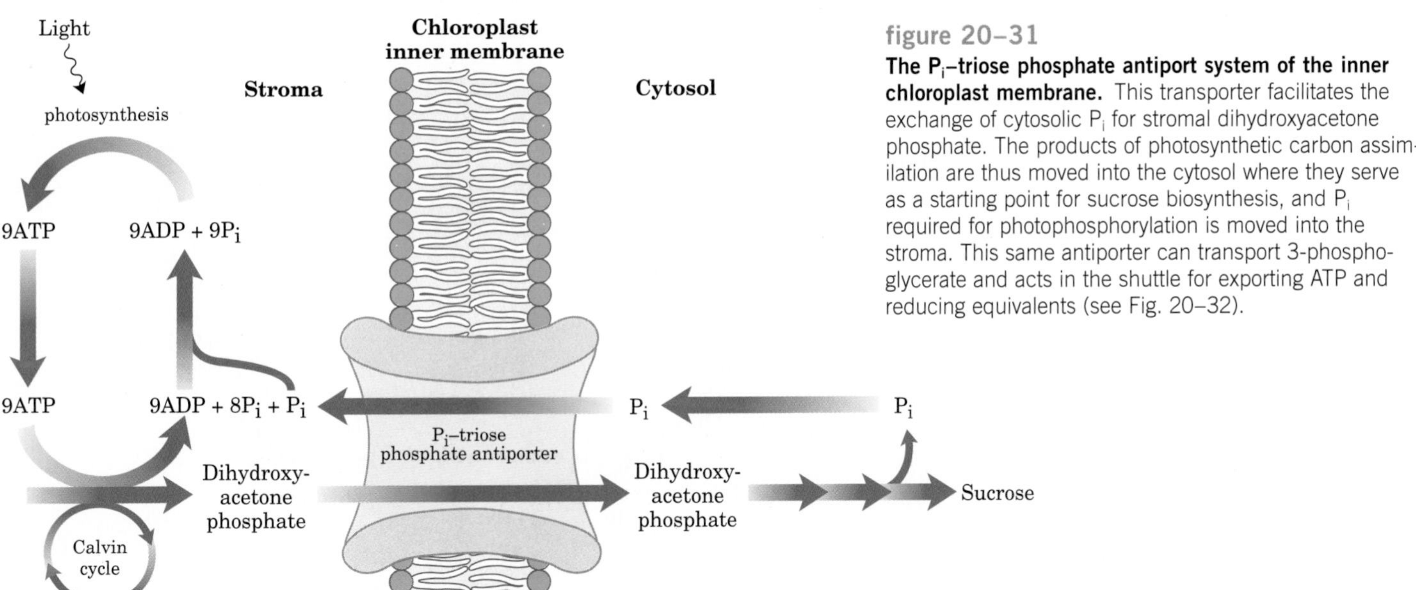

figure 20–31
The P_i–triose phosphate antiport system of the inner chloroplast membrane. This transporter facilitates the exchange of cytosolic P_i for stromal dihydroxyacetone phosphate. The products of photosynthetic carbon assimilation are thus moved into the cytosol where they serve as a starting point for sucrose biosynthesis, and P_i required for photophosphorylation is moved into the stroma. This same antiporter can transport 3-phosphoglycerate and acts in the shuttle for exporting ATP and reducing equivalents (see Fig. 20–32).

by the pathways illustrated in Figures 20–25 and 20–17. Sucrose synthesis in the cytosol and starch synthesis in the chloroplast are the major pathways by which the excess triose phosphates are "harvested." The steps of sucrose synthesis release four P_i molecules from the four triose phosphates required to make sucrose. This P_i is transported back into the chloroplast and used in the synthesis of ATP, effectively replacing the molecule of P_i that is used to generate triose phosphate, as described above. For every molecule of triose phosphate removed from the chloroplast, one P_i is transported into the chloroplast. If this exchange were blocked, triose phosphate synthesis would quickly deplete the available P_i in the chloroplast and thus suppress CO_2 assimilation into starch.

The P_i–triose phosphate antiport system serves one additional function. ATP and reducing power are needed in the cytosol for a variety of synthetic and energy-requiring reactions. These requirements are met to an as yet undetermined degree by mitochondria. A second potential source of energy is the ATP and NADPH generated in the chloroplast stroma during the light reactions of photosynthesis; however, ATP and NADPH do not cross the chloroplast membrane. The P_i–triose phosphate antiport system has the indirect effect of moving ATP and reducing equivalents across the chloroplast membrane (Fig. 20–32). Dihydroxyacetone phosphate formed in the stroma by CO_2 assimilation is transported to the cytosol, where it is converted by glycolytic enzymes to 3-phosphoglycerate, generating ATP and NADH. 3-Phosphoglycerate reenters the chloroplast, completing the cycle. The net effect is transport of NADPH/NADH and ATP from the chloroplast to the cytosol.

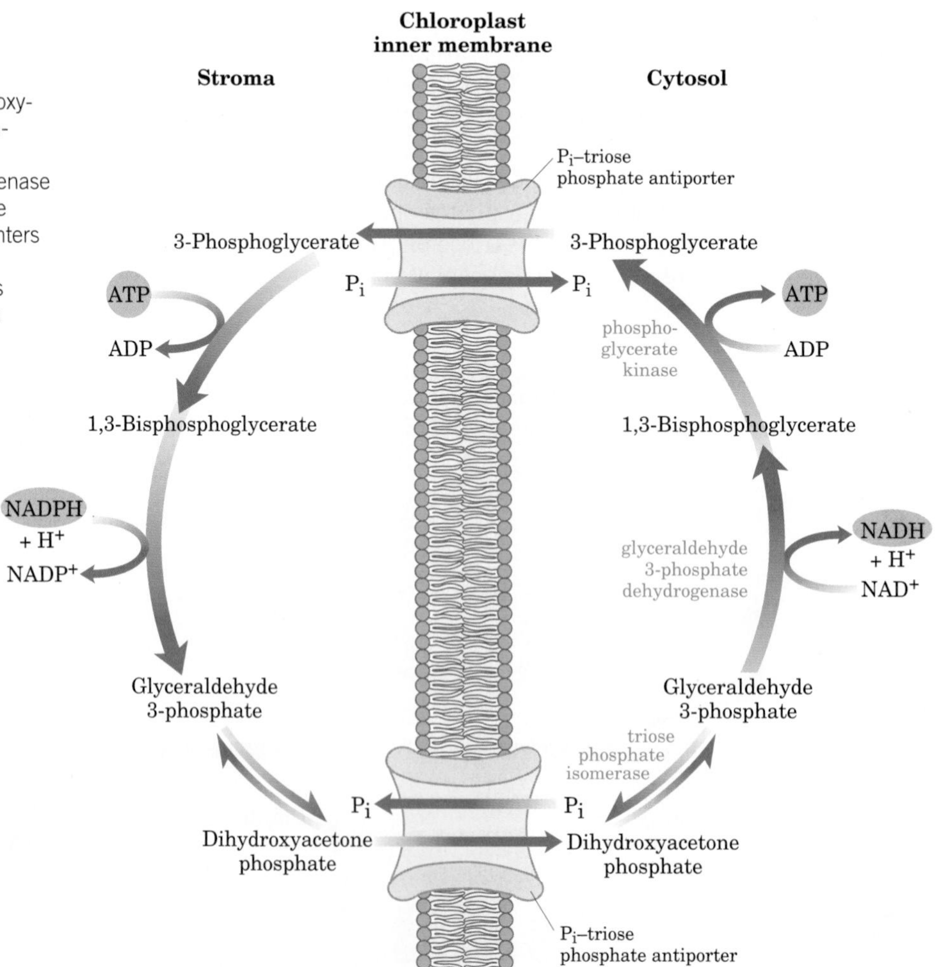

figure 20–32

Role of the P_i–triose phosphate antiporter in the transport of ATP and reducing equivalents. Dihydroxyacetone phosphate leaves the chloroplast and is converted to glyceraldehyde 3-phosphate in the cytosol. The cytosolic glyceraldehyde 3-phosphate dehydrogenase and phosphoglycerate kinase reactions then produce NADH, ATP, and 3-phosphoglycerate. The latter reenters the chloroplast and is reduced to dihydroxyacetone phosphate, completing a cycle that effectively moves ATP and reducing equivalents (NADPH/NADH) from chloroplast to cytosol.

Regulation of Carbohydrate Metabolism in Plants

Carbohydrate metabolism is more complex in plant cells than in animal cells or nonphotosynthetic microorganisms. In addition to the universal pathways of glycolysis and gluconeogenesis, plants have the unique reaction sequences for reduction of CO_2 to triose phosphates and the associated reductive pentose phosphate pathway—all of which must be coordinately regulated to avoid futile cycling and to ensure proper allocation of carbon to energy production and synthesis of starch and sucrose. Key enzymes are regulated by one or more of the following mechanisms:

1. Reduction of disulfide bonds by electrons flowing from photosystem I.
2. Changes in pH and Mg^{2+} concentration that result from illumination.
3. Conventional allosteric regulation by one or more metabolic intermediates.
4. Covalent modification.

Rubisco Is Subject to Both Positive and Negative Regulation

As the site where photosynthetic CO_2 fixation is initiated, rubisco is a prime target for regulation. One type of regulation involves the carbamylation of a Lys residue (Fig. 20–33a, b). At high CO_2 levels this occurs nonenzymatically. However, the substrate for rubisco, ribulose 1,5-bisphosphate, inhibits carbamylation and this inhibition is almost complete at physiological CO_2 concentrations. An enzyme called **rubisco activase** overcomes the inhibition by promoting an ATP-dependent release of ribulose 1,5-bisphosphate, exposing the Lys amino group to carbamylation by CO_2, which activates the enzyme.

The carbamylated form of rubisco is no longer inhibited by ribulose 1,5-bisphosphate, but 2-carboxyarabinitol 1-phosphate, a naturally occurring transition-state analog (see Box 8–3) with a structure similar to that of the β-keto acid intermediate of the rubisco reaction (Figs. 20–23, 20–33c), remains a potent inhibitor. This compound, synthesized in the dark in some plants, is sometimes called the "nocturnal inhibitor." It is broken down when light returns, permitting reactivation of rubisco.

figure 20–33

Regulation of rubisco. (a) Activation of rubisco by formation of a carbamate derivative of a Lys residue at the active site, catalyzed by rubisco activase. **(b)** The active site of the rubisco of *Rhodospirillum rubrum.* The substrate-binding site is occupied here by an inhibitor, 2-carboxy-D-arabinitol 1,5-bisphosphate (yellow), that was cocrystallized with the enzyme. Two acidic amino acid side chains that interact with the bound inhibitor are shown in blue. The carbamylated Lys residue is red, and the green sphere is a Mg^{2+} ion at the active site. **(c)** The naturally occurring transition-state analog 2-carboxyarabinitol 1-phosphate, compared here with the β-keto acid intermediate (see Fig. 20–23) of the rubisco reaction. This analog is quite similar to the inhibitor bound to the enzyme in **(b)**.

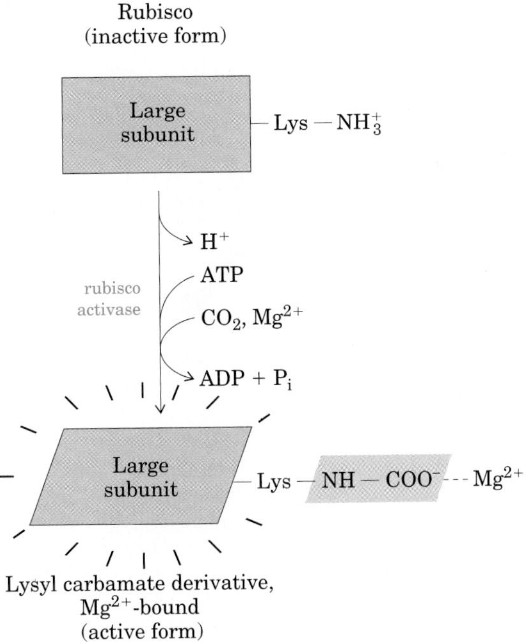

(a)

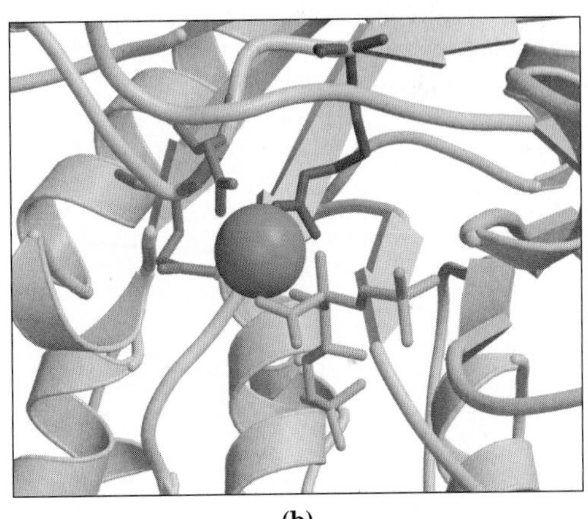

(b)

$$\begin{array}{c} CH_2O-PO_3^{2-} \\ | \\ HO-C-COO^- \\ | \\ H-C-OH \\ | \\ H-C-OH \\ | \\ CH_2-OH \end{array} \qquad \begin{array}{c} CH_2O-PO_3^{2-} \\ | \\ HO-C-COO^- \\ | \\ C=O \\ | \\ H-C-OH \\ | \\ CH_2O-PO_3^{2-} \end{array}$$

2-Carboxyarabinitol 1-phosphate — β-Keto acid intermediate

(c)

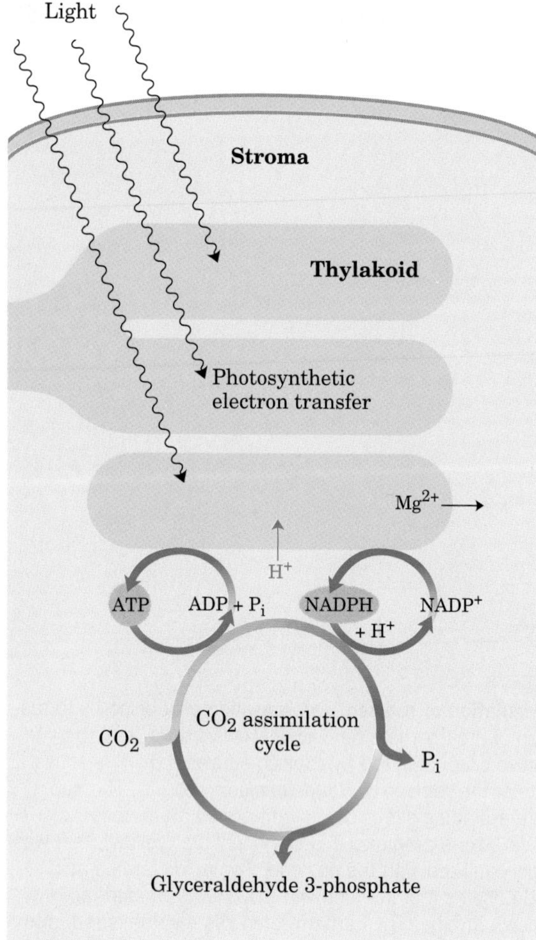

figure 20–34
ATP and NADPH produced by the light reactions are essential substrates for the reduction of CO_2. The photosynthetic reactions that produce ATP and NADPH are accompanied by movement of protons (red) from the stroma into the thylakoid, creating alkaline conditions in the stroma. Magnesium ions pass from the thylakoid into the stroma, increasing the stromal [Mg^{2+}].

Certain Enzymes of the Calvin Cycle Are Indirectly Activated by Light

The reductive assimilation of CO_2 requires ATP and NADPH, and their stromal concentrations increase when chloroplasts are illuminated (Fig. 20–34). The light-induced transport of protons across the thylakoid membrane (Chapter 19) also increases the stromal pH and is accompanied by a flow of Mg^{2+} out of the thylakoid compartment into the stroma. Several stromal enzymes have evolved to take advantage of these light-dependent conditions that signal the availability of ATP and NADPH: these enzymes have pH or [Mg^{2+}] optima that are better suited to alkaline conditions and high [Mg^{2+}]. Activation of rubisco by formation of the lysyl carbamate is faster at alkaline pH, and high stromal [Mg^{2+}] favors formation of the enzyme's active Mg^{2+} complex. Fructose 1,6-bisphosphatase requires Mg^{2+} and is very dependent on pH (Fig. 20–35). Its activity increases more than 100-fold when the pH and [Mg^{2+}] rise during chloroplast illumination.

Four essential Calvin cycle enzymes are subject to a special type of regulation by light. Ribulose 5-phosphate kinase, fructose 1,6-bisphosphatase, sedoheptulose 1,7-bisphosphatase, and glyceraldehyde 3-phosphate dehydrogenase are activated by light-driven reduction of disulfide bonds between two Cys residues critical to their catalytic activities. When these Cys residues are disulfide-bonded (oxidized), the enzymes are inactive; this is the normal situation in the dark. With illumination, electrons flow from photosystem I to ferredoxin (see Fig. 19–46), which passes electrons to a small, soluble, disulfide-containing protein called **thioredoxin** (Fig. 20–36). Thioredoxin donates electrons for the reduction of the disulfide bonds of the light-activated enzymes, and this reductive cleavage is accompanied by conformational changes that increase enzyme activities. Thioredoxin is reactivated in a disulfide-exchange reaction catalyzed by **thioredoxin reductase.** Outside the chloroplast, thioredoxins participate in a variety of other redox processes (see Fig. 22–37). At nightfall, the Cys residues are reoxidized to their disulfide forms, the four enzymes are inactivated, and ATP is not expended in CO_2 assimilation. Instead, starch synthesized and stored during the daytime is degraded to fuel glycolysis at night.

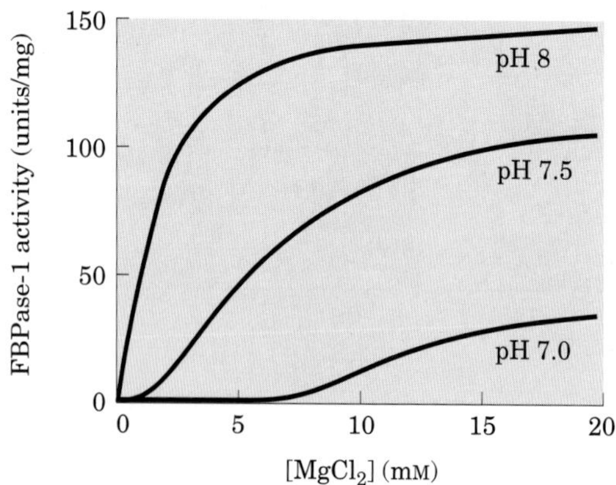

figure 20–35
Activation of chloroplast fructose 1,6-bisphosphatase. Reduced fructose 1,6-bisphosphatase (FBPase-1) is activated by light and by the combination of high pH and high [Mg^{2+}] in the stroma, both of which are produced by illumination.

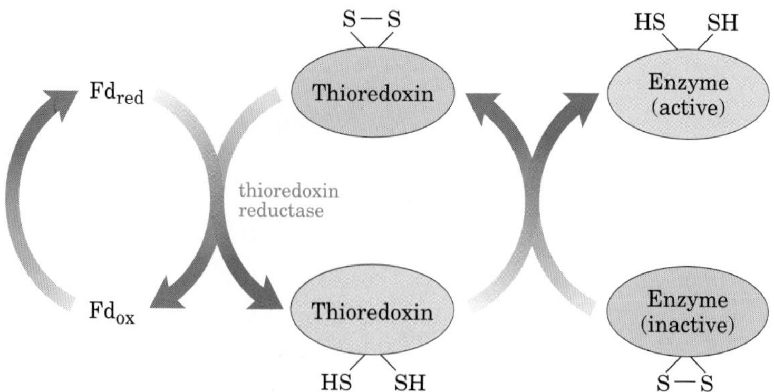

figure 20–36
Thioredoxin. Light activation of several enzymes of the Calvin cycle is mediated by thioredoxin, a small, disulfide-containing protein. In the light, thioredoxin is reduced by electrons from ferredoxin (Fd) (blue arrows), then thioredoxin reduces disulfide bonds of sedoheptulose 1,7-bisphosphatase, fructose 1,6-bisphosphatase, ribulose 5-phosphate kinase, and glyceraldehye 3-phosphate dehydrogenase, thereby activating these enzymes.

The Use of Triose Phosphates for Sucrose and Starch Synthesis Is Tightly Regulated in Plants

Triose phosphates produced by the Calvin cycle in bright sunlight may be temporarily stored in the chloroplast as starch, or converted to sucrose and exported to nonphotosynthetic parts of the plant, or both. The balance between these two processes is tightly regulated, and both must be coordinated with the rate of carbon fixation. Five-sixths of the triose phosphate formed in the Calvin cycle must be recycled to ribulose 1,5-bisphosphate; if more than one-sixth of the triose phosphate is drawn out of the cycle to make sucrose and starch, the cycle will slow or stop. On the other hand, insufficient conversion of triose phosphate to starch or sucrose would tie up phosphate, leaving a chloroplast deficient in P_i, which is also essential for operation of the Calvin cycle.

Fructose 1,6-bisphosphatase (FBPase-1) and the enzyme that effectively reverses its action, PP_i-dependent phosphofructokinase (PP-PFK-1; see p. 533), control the flow of triose phosphates into sucrose and are therefore critical points for regulating the fate of triose phosphates produced by photosynthesis. Both enzymes are regulated by fructose 2,6-bisphosphate, which inhibits FBPase-1 and stimulates PP-PFK-1. In higher plants, the concentration of fructose 2,6-bisphosphate varies inversely with the rate of photosynthesis (Fig. 20–37). Phosphofructokinase-2, responsible for fructose 2,6-bisphosphate synthesis, is inhibited by dihydroxyacetone phosphate or 3-phosphoglycerate and stimulated by fructose 6-phosphate and P_i. During active photosynthesis, dihydroxyacetone phosphate is produced and P_i is consumed, resulting in inhibition of PFK-2 and lowered concentrations of fructose 2,6-bisphosphate. This favors greater flux of triose phosphate into fructose 6-phosphate formation and sucrose synthesis. With this regulatory system, sucrose synthesis occurs only when the level of triose phosphate produced by the Calvin cycle exceeds that needed to maintain the operation of the cycle.

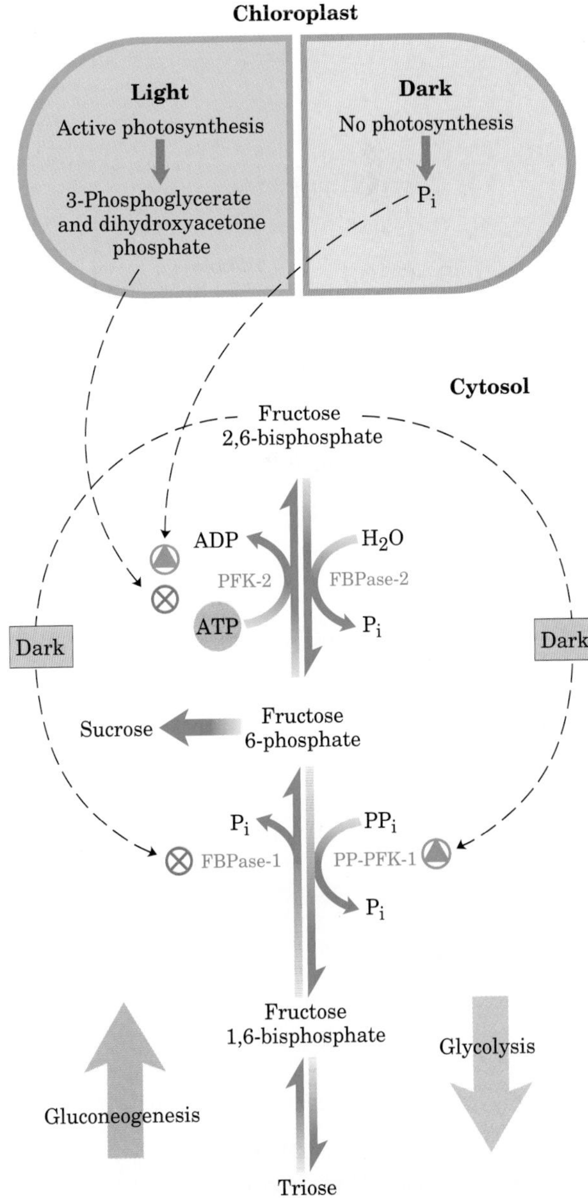

figure 20–37

Fructose 2,6-bisphosphate as regulator of sucrose synthesis. The concentration of the allosteric regulator fructose 2,6-bisphosphate (FBPase-2) in plant cells is regulated by the products of photosynthetic carbon assimilation and by P_i. Dihydroxyacetone phosphate and 3-phosphoglycerate produced by CO_2 assimilation inhibit phosphofructokinase-2 (PFK-2), the enzyme that synthesizes the regulator; P_i stimulates PFK-2. The concentration of the regulator is therefore inversely proportional to the rate of photosynthesis. In the dark, the concentration of fructose 2,6-bisphosphate increases and stimulates the glycolytic enzyme PP_i-dependent phosphofructokinase-1 (PP-PFK-1), while inhibiting the gluconeogenic enzyme fructose 1,6-bisphosphatase (FBPase-1). When photosynthesis is active (in the light), the concentration of the regulator drops and the synthesis of fructose 6-phosphate and sucrose is favored.

Sucrose synthesis is also regulated at the level of sucrose 6-phosphate synthase, which is activated by glucose 6-phosphate and inhibited by P_i. During active photosynthesis, triose phosphates are converted to fructose 6-phosphate, which is rapidly equilibrated with glucose 6-phosphate by phosphohexose isomerase. Because the equilibrium lies far toward glucose 6-phosphate, as soon as fructose 6-phosphate accumulates, the level of glucose 6-phosphate rises and sucrose synthesis is stimulated. Sucrose 6-phosphate synthase is also regulated by covalent modification (phosphorylation) in response to light-dark signals. In the dark, the enzyme is phosphorylated on a Ser residue, producing the less active form. The protein phosphatase that removes the phosphate, activating the enzyme, is itself activated by light; the mechanism is not yet clear.

The key regulatory enzyme in starch synthesis is ADP-glucose pyrophosphorylase (Fig. 20–16), which is activated by 3-phosphoglycerate and inhibited by P_i. When sucrose synthesis slows, 3-phosphoglycerate formed by CO_2 fixation accumulates, activating this enzyme and stimulating the synthesis of starch.

Photorespiration Results from Rubisco's Oxygenase Activity

Rubisco is not absolutely specific for CO_2 as a substrate. Oxygen competes with CO_2 at the active site, and rubisco catalyzes the condensation of O_2 with ribulose 1,5-bisphosphate to form 3-phosphoglycerate and **phosphoglycolate** (Fig. 20–38), a metabolically useless product. This is the oxygenase activity referred to in the full name of the enzyme: RuBP carboxylase/oxygenase. It results in no fixation of carbon and appears to be a net liability to the cell in which it occurs; salvaging the carbons from phosphoglycolate (by the pathway outlined below) uses cellular energy. The condensation of O_2 with ribulose 1,5-bisphosphate occurs concurrently with CO_2 fixation, with the latter predominating by a factor of about three.

Apparently the evolution of rubisco produced an active site not able to discriminate well between CO_2 and O_2, perhaps because much of this evolution occurred before O_2 was an important component of the atmosphere. The K_m for CO_2 is about 9 μM, and for O_2 is about 350 μM. (A solution in equilibrium with air at room temperature contains about 11 μM CO_2 and 250 μM O_2.) The modern atmosphere contains about 20% O_2 and only 0.04% CO_2, proportions that allow O_2 "fixation" by rubisco to constitute a significant waste of energy. As CO_2 is consumed in the assimilation reactions, the proportion of O_2 in the air spaces of the leaf increases. In addition, the affinity of rubisco for CO_2 decreases with increasing temperature, exacerbating the tendency of the enzyme to catalyze the wasteful oxygenase reaction.

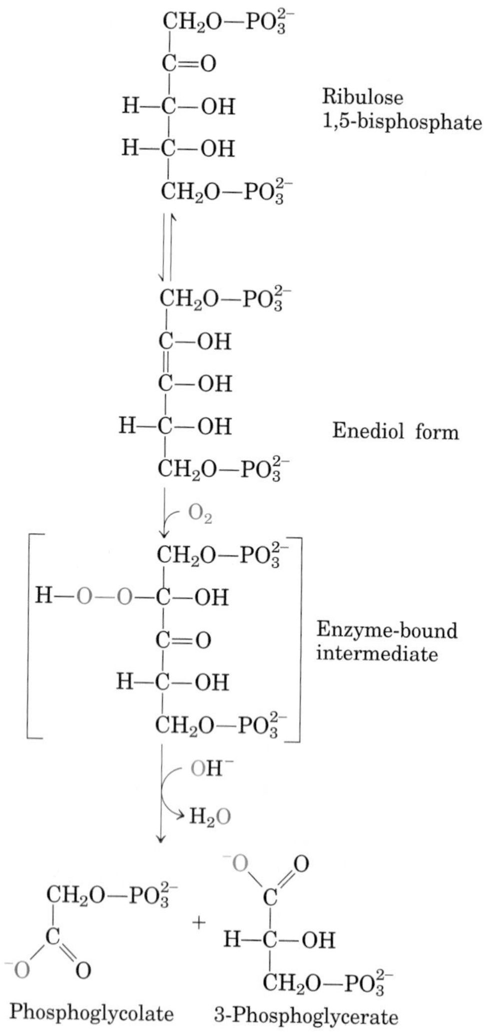

Ribulose 1,5-bisphosphate

Enediol form

Enzyme-bound intermediate

Phosphoglycolate

3-Phosphoglycerate

figure 20–38

Oxygenase activity of rubisco. Rubisco can incorporate O_2 rather than CO_2 into ribulose 1,5-bisphosphate. The unstable intermediate thus formed splits into phosphoglycolate, which is recycled as described in Figure 20–39, and 3-phosphoglycerate, which can reenter the Calvin cycle.

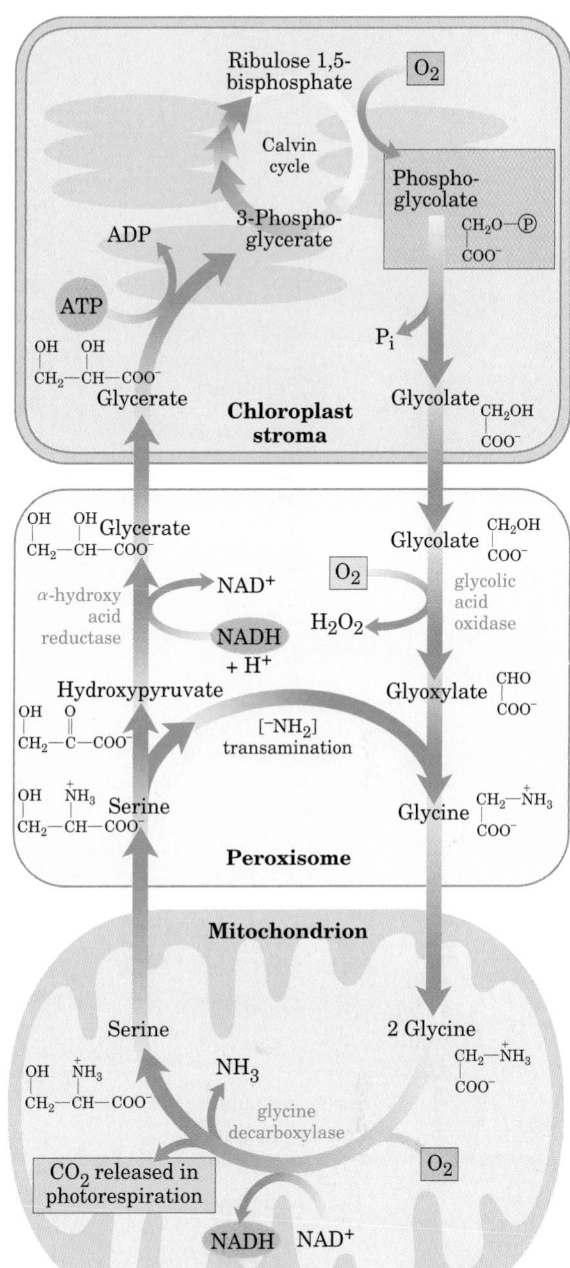

figure 20–39

Glycolate pathway. This pathway, by which phosphoglycolate (shaded red) is salvaged by conversion to serine and thus 3-phosphoglycerate, involves three cellular compartments. Glycolate formed by dephosphorylation of phosphoglycolate in chloroplasts is transaminated to glycine in peroxisomes. In mitochondria, two glycine molecules condense to form serine and the CO_2 released during photorespiration (shaded green). This reaction is catalyzed by glycine decarboxylase, an enzyme present at very high levels in the mitochondria of C_3 plants (see text). The serine is converted to hydroxypyruvate and then to glycerate in peroxisomes, and glycerate reenters the chloroplasts to be phosphorylated, rejoining the Calvin cycle. Oxygen is consumed at three steps during photorespiration (shaded blue).

The pathway that salvages the carbons from phosphoglycolate, the glycolate pathway (Fig. 20–39), involves the conversion of two molecules of phosphoglycolate to a molecule of serine (three carbons) and a molecule of CO_2. The oxygenase activity of rubisco combined with the salvage pathway consumes O_2 and produces CO_2—hence the name **photorespiration.** Unlike mitochondrial respiration, this process does not conserve energy. Photorespiration may inhibit net biomass formation as much as 50%. This has led to some adaptations in the process by which carbon assimilation takes place, particularly in plants that have evolved in warm climates.

Some Plants Have a Mechanism to Minimize Photorespiration

In many plants that grow in the tropics (and in temperate-zone crop plants native to the tropics, such as maize, sugarcane, and sorghum) a mechanism has evolved to circumvent the problem of wasteful photorespiration. The step in which CO_2 is incorporated into a three-carbon product, 3-phosphoglycerate, is preceded by several steps, one of which is a temporary fixation of CO_2 into a compound with four carbon atoms. These plants are referred to as **C_4 plants,** and the assimilation process as C_4 metabolism or the C_4 pathway. Plants in which the *first step* in carbon assimilation is reaction of CO_2 with ribulose 1,5-bisphosphate to form 3-phosphoglycerate—as we have described thus far—are called **C_3 plants.**

C_4 plants, which typically grow at high light intensity and high temperatures, have several important characteristics: high photosynthetic rates, high growth rates, low photorespiration rates, low rates of water loss, and an unusual leaf structure. Photosynthesis in the leaves of C_4 plants involves two cell types: mesophyll and bundle-sheath cells (Fig. 20–40a). There are three known patterns of C_4 metabolism. Here we describe the best-understood pathway, worked out in the 1960s by two plant biochemists, Marshall Hatch and Rodger Slack (Fig. 20–40b).

Mesophyll cell
Plasmodesmata
Bundle-sheath cell

(a)

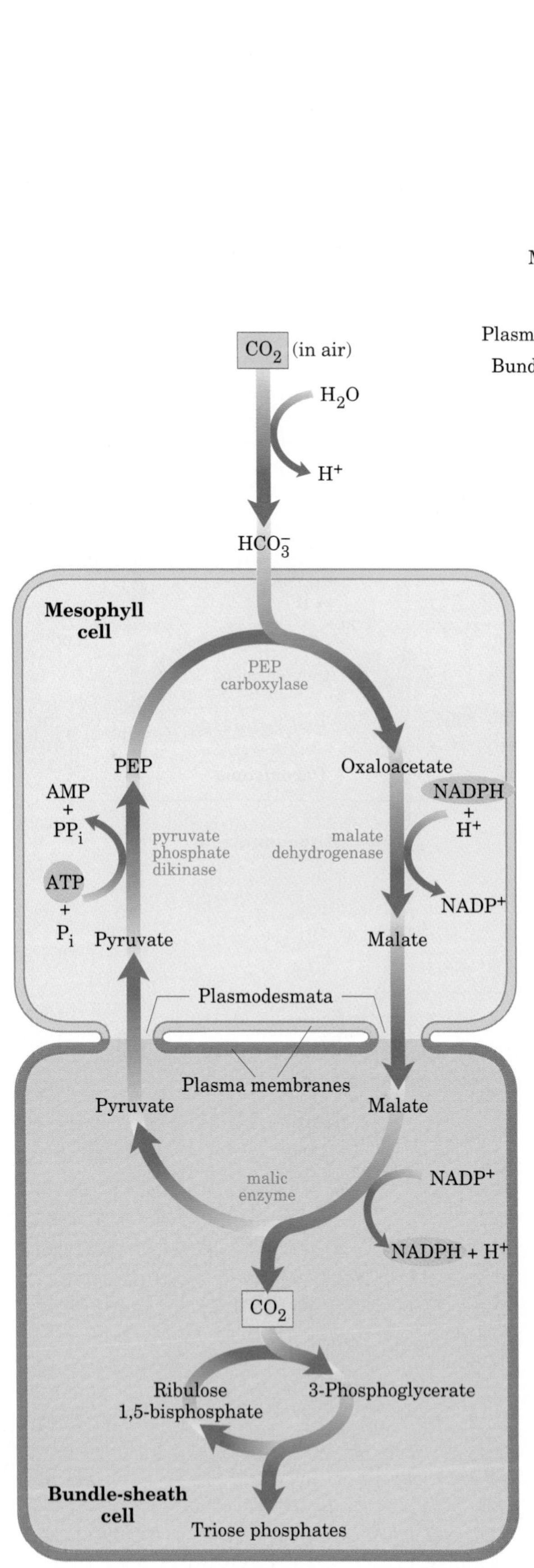

figure 20–40

Carbon assimilation in C_4 plants. The C_4 pathway, involving mesophyll cells and bundle-sheath cells, predominates in plants of tropical origin. **(a)** An electron micrograph showing chloroplasts of adjacent mesophyll and bundle-sheath cells. The bundle-sheath cell contains starch granules. Plasmodesmata connecting the two cells are visible. **(b)** The Hatch-Slack pathway of CO_2 assimilation, through a four-carbon intermediate.

In plants of tropical origin the first intermediate into which $^{14}CO_2$ is fixed is not 3-phosphoglycerate but oxaloacetate, a four-carbon compound. This reaction, which occurs in the cytosol of leaf mesophyll cells, is catalyzed by **phosphoenolpyruvate carboxylase.** The oxaloacetate thus formed is either reduced to malate at the expense of NADPH (as shown in Fig. 20–40b) or converted to aspartate by transamination:

$$\text{Oxaloacetate} + \alpha\text{-amino acid} \longrightarrow \text{L-aspartate} + \alpha\text{-keto acid}$$

The malate or aspartate formed in the mesophyll cells then passes into the neighboring bundle-sheath cells through plasmodesmata (see Fig. 2–22). In the bundle-sheath cells the malate is oxidized and decarboxylated to yield pyruvate and CO_2 by the action of **malic enzyme,** reducing $NADP^+$. In plants that use aspartate as the CO_2 carrier, aspartate arriving in bundle-sheath cells is transaminated to form oxaloacetate and reduced to malate, then the CO_2 is released by malic enzyme or PEP carboxykinase. The free CO_2 released in the bundle-sheath cells is the same CO_2 molecule that was originally fixed into oxaloacetate in the mesophyll cells. This CO_2 is now fixed again, this time by rubisco, in exactly the same reaction that occurs in C_3 plants: incorporation of CO_2 into C-1 of 3-phosphoglycerate.

The pyruvate formed by decarboxylation of malate in bundle-sheath cells is transferred back to the mesophyll cells, where it is converted to PEP by an unusual enzymatic reaction catalyzed by **pyruvate phosphate dikinase** (Fig. 20–40b). This enzyme is called a dikinase because two different molecules are simultaneously phosphorylated by one molecule of

ATP: pyruvate to PEP, and phosphate to pyrophosphate. The pyrophosphate is subsequently hydrolyzed to phosphate, so two high-energy phosphate groups of ATP are used in regenerating PEP. The PEP is now ready to receive another molecule of CO_2 in the mesophyll cell.

The PEP carboxylase of mesophyll cells has a high affinity for HCO_3^- and can fix CO_2 more efficiently than can rubisco, and unlike rubisco it does not use O_2 as an alternative substrate, so there is no competition between CO_2 and O_2 with this enzyme. The PEP carboxylase reaction serves to fix and concentrate CO_2 in the form of malate. Release of CO_2 from malate in the bundle-sheath cells yields a sufficiently high local concentration of CO_2 for rubisco to function near its maximal rate. In addition, the bundle-sheath cells are farther away from the leaf surface than are mesophyll cells and thus have a lower O_2 concentration. These factors minimize the oxygenase activity of rubisco in C_4 plants.

Once CO_2 is fixed into 3-phosphoglycerate in the bundle-sheath cells, all the other reactions of the Calvin cycle take place exactly as described earlier. Thus in C_4 plants, mesophyll cells carry out CO_2 assimilation by the C_4 pathway and bundle-sheath cells synthesize starch and sucrose by the C_3 pathway.

The pathway of CO_2 assimilation has a greater energy cost in C_4 plants than in C_3 plants. For each molecule of CO_2 assimilated in the C_4 pathway, a molecule of PEP must be regenerated at the expense of two high-energy phosphate groups of ATP. Thus C_4 plants need five ATP molecules to assimilate one molecule of CO_2, whereas C_3 plants need only three (nine per triose phosphate). As the temperature increases (and the affinity of rubisco for CO_2 decreases, as noted above), a point is reached (about 28 to 30 °C) at which the gain in efficiency from the elimination of photorespiration more than compensates for this energetic cost. C_4 plants (crabgrass, for example) outgrow most C_3 plants during the summer, as any experienced gardener can attest.

summary

Gluconeogenesis is the formation of carbohydrate from noncarbohydrate precursors, the most important of which are pyruvate, lactate, and alanine. In vertebrates, gluconeogenesis in the liver and kidney provides glucose for use by the brain, muscle, and erythrocytes. Like all biosynthetic pathways, gluconeogenesis proceeds by an enzymatic route that differs from the corresponding catabolic pathway, is independently regulated, and requires ATP. The biosynthetic pathway from pyruvate to glucose occurs in all organisms. It uses seven of the glycolytic enzymes, which function reversibly. Three irreversible steps in the glycolytic pathway are bypassed by reactions catalyzed by gluconeogenic enzymes: (1) conversion of pyruvate to phosphoenolpyruvate via oxaloacetate; (2) dephosphorylation of fructose 1,6-bisphosphate by fructose 1,6-bisphosphatase; and (3) dephosphorylation of glucose 6-phosphate by glucose 6-phosphatase. The path from pyruvate to phosphoenolpyruvate depends on whether lactate or pyruvate itself serves as the glucogenic precursor. Formation of one molecule of glucose from pyruvate requires 4 ATP and 2 GTP. Three carbon atoms of each citric acid cycle intermediate and some or all carbons of many amino acids are convertible to glucose.

Gluconeogenesis in the liver is regulated at two major points: (1) carboxylation of pyruvate by pyruvate carboxylase, which is stimulated by acetyl-CoA, and (2) dephosphorylation of fructose 1,6-bisphosphate by fructose 1,6-bisphosphatase, which is inhibited by fructose 2,6-bisphosphate and AMP and stimulated by citrate. Fructose 2,6-bisphosphate also stimulates phosphofructokinase-1 and is crucial to the balance between gluconeogenesis and glycolysis. Fructose 2,6-bisphosphate levels are hormonally regulated in animals. Reciprocal regulation of gluconeogenesis and glycolysis prevents futile cycling with its accompanying loss of ATP energy.

Unlike animals, plants can convert acetyl-CoA derived from fatty acids into glucose, through the reactions of the glyoxylate cycle and gluconeogenesis, compartmented among glyoxysomes, mitochondria, and the cytosol.

Glycogen synthesis requires conversion of glucose 1-phosphate to UDP-glucose, a sugar nucleotide. In general, sugar phosphates are activated and earmarked for a particular synthetic path by ester linkage of a nucleoside diphosphate to the anomeric carbon of the sugar. Glycogen synthase adds glucose units from UDP-glucose to the nonreducing end of a growing glycogen chain, forming ($\alpha1\rightarrow4$) links; glycosyl-($4\rightarrow6$)-transferase adds ($\alpha1\rightarrow6$) branch points. Initiation of glycogen synthesis requires glycogenin. The synthesis and breakdown of glycogen are reciprocally regulated by hormone-dependent phosphorylation of glycogen synthase (inactivation) and glycogen phosphorylase (activation).

Lactose synthesis in the lactating mammary gland is brought about by the α-lactalbumin–galactosyltransferase (lactose synthase) enzyme complex, using UDP-galactose and glucose as substrates. The α-lactalbumin serves as a specificity-modifying subunit, whose formation is regulated by hormones promoting lactation.

In plants, triose phosphates can be condensed to hexose phosphates and polymerized to starch for temporary storage within chloroplasts. Starch synthase catalyzes the addition of glucose units from ADP-glucose to starch by a mechanism similar to that of glycogen synthase. Alternatively, triose phosphates pass into the cytosol and serve as precursors for sucrose. Sucrose 6-phosphate synthase, which condenses UDP-glucose with fructose 6-phosphate, is inhibited when sucrose accumulates in the cytosol. Starch synthesis is stimulated by sucrose accumulation.

Plant photosynthesis takes place in chloroplasts. In the CO_2-assimilating reactions (the Calvin cycle), ATP and NADPH are used to reduce CO_2 to triose phosphates. These reactions occur in three stages: the fixation reaction itself, catalyzed by ribulose 1,5-bisphosphate carboxylase/oxygenase (rubisco); reduction of the resulting 3-phosphoglycerate to glyceraldehyde 3-phosphate, which can be used in the synthesis of hexoses, sucrose, and starch, or in glycolysis; and regeneration of ribulose 1,5-bisphosphate from triose phosphates.

Rubisco condenses CO_2 with ribulose 1,5-bisphosphate, then hydrolyzes the resulting hexose into two molecules of 3-phosphoglycerate. Stromal isozymes of the glycolytic enzymes catalyze reduction of 3-phosphoglycerate to glyceraldehyde 3-phosphate; each molecule reduced requires one ATP and one NADPH. Finally, stromal enzymes including transketolase and transaldolase rearrange the carbon skeletons of triose phosphates, generating intermediates of three, four, five, six, and seven carbons and eventually yielding pentose phosphates. The pentose phosphates are converted to ribulose 5-phosphate, then phosphorylated to ribulose 1,5-bisphosphate to complete the Calvin cycle. The cost of fixing three CO_2 into triose phosphate is 9 ATP and 6 NADPH.

An antiport system in the inner chloroplast membrane exchanges P_i in the cytosol for 3-phosphoglycerate or dihydroxyacetone phosphate produced by CO_2 assimilation in the stroma. Dihydroxyacetone phosphate oxidation in the cytosol generates ATP and NADH, thus moving ATP and reducing equivalents from the chloroplast to the cytosol.

Rubisco is regulated by covalent modification and by a natural transition-state analog. Other enzymes of the Calvin cycle are activated indirectly by light and are inactive in the dark so that hexose synthesis does not compete with glycolysis, which is required to provide energy in the dark. Gluconeogenesis and glycolysis are regulated in plants by fructose 2,6-bisphosphate, the level of which varies inversely with the rate of photosynthesis: as the photosynthetic rate increases, fructose 2,6-bisphosphate levels fall and gluconeogenesis is activated.

Photorespiration wastes photosynthetic energy in C_3 plants by forming and oxidizing phosphoglycolate, a product of the oxygenation of ribulose 1,5-bisphosphate by rubisco. C_4 plants have a pathway that minimizes photorespiration: CO_2 is fixed in mesophyll cells into a four-carbon compound, which passes into bundle-sheath cells and releases CO_2 in high concentrations. This CO_2 is fixed by rubisco, and the remaining reactions of the Calvin cycle occur as in C_3 plants.

further reading

Gluconeogenesis

Gleeson, T. (1996) Post-exercise lactate metabolism: a comparative review of sites, pathways, and regulation. *Annu. Rev. Physiol.* **58,** 565–581.

Hers, H.G. & Hue, L. (1983) Gluconeogenesis and related aspects of glycolysis. *Annu. Rev. Biochem.* **52,** 617–653.

Jitrapakdee, S. & Wallace, J.C. (1999) Structure, function, and regulation of pyruvate carboxylase. *Biochem. J.* **340,** 1–16.

Review of a key gluconeogenic enzyme, including an account of some genetic defects in this enzyme.

Jungerman, K. & Kietzmann, T. (1996) Zonation of parenchymal and nonparenchymal metabolism in liver. *Annu. Rev. Nutr.* **16,** 179–203.

Some additional information on liver function.

Matte, A., Tari, L.W., Goldie, H., & Delbaere, L.T.J. (1997) Structure and mechanism of phosphoenolpyruvate carboxykinase. *J. Biol. Chem.* **272,** 8105–8108.

Nordlie, R.C., Foster, J.D., & Lange, A.J. (1999) Regulation of glucose production by the liver. *Annu. Rev. Nutr.* **19,** 379–406.

Review with a special focus on short- and long-term regulation of glucose 6-phosphatase and glucokinase.

Pilkis, S.J., El-Maghrabi, M.R., & Claus, T.H. (1988) Hormonal regulation of hepatic gluconeogenesis and glycolysis. *Annu. Rev. Biochem.* **57,** 755–783.

Polysaccharide Synthesis

Ball, S., Guan, H.-P., James, M., Myers, A., Keeling, P., Mouille, G., Buléon, A., Colonna, P., & Preiss, J. (1996) From glycogen to amylopectin: a model for the biogenesis of the plant starch granule. *Cell* **86,** 349–352.

Beck, E. & Ziegler, P. (1989) Biosynthesis and degradation of starch in higher plants. *Annu. Rev. Plant Physiol. Plant Mol. Biol.* **40,** 95–117.

Leloir, L.F. (1971) Two decades of research on the biosynthesis of saccharides. *Science* **172,** 1299–1303.

Leloir's Nobel address, including a discussion of the role of sugar nucleotides in metabolism.

Nuttall, F.Q., Gilboe, D.P., Gannon, M.C., Niewoehner, C.B., & Tan, A.W.H. (1988) Regulation of glycogen synthesis in the liver. *Am. J. Med.* **85** (Suppl. 5A), 77–85.

Preiss, J. & Romeo, T. (1994) Molecular biology and regulatory aspects of glycogen biosynthesis in bacteria. *Prog. Nucl. Acid Res. Mol. Biol.* **47,** 299–329.

Roach, P. & Skurat, A. (1997) Self-glucosylating initiator proteins and their role in glycogen biosynthesis. *Prog. Nucl. Acid Res. Mol. Biol.* **57,** 289–316.

Carbon Dioxide Assimilation

Andersson, I., Knight, S., Schneider, G., Lindqvist, Y., Lundqvist, T., Brändén, C.-I., & Lorimer, G.H. (1989) Crystal structure of the active site of ribulose-bisphosphate carboxylase. *Nature* **337,** 229–234.

Cleland, W.W., Andrews, T.J., Gutteridge, S., Hartman, F.C., & Lorimer, G.H. (1998) Mechanism of rubisco—the carbamate as general base. *Chem. Rev.* **98,** 549–561.

Review with a special focus on the carbamate at the active site.

Douce, R. & Neuberger, M. (1999) Biochemical dissection of photorespiration. *Curr. Opin. Plant Biol.* **2,** 214–222.

Flügge, U.-I. (1999) Phosphate translocaters in plastids. *Annu. Rev. Plant Physiol. Plant Mol. Biol.* **50,** 27–45.

Review of the transporters that carry P_i and various sugar phosphates across plastid membranes.

Hartman, F.C. & Harpel, M.R. (1994) Structure, function, regulation and assembly of D-ribulose-1,5-bisphosphate carboxylase/oxygenase. *Annu. Rev. Biochem.* **63,** 197–234.

Heldt, H.-W. (1997) *Plant Biochemistry and Molecular Biology,* Oxford University Press, New York.

Includes good chapters on CO_2 fixation and assimilation, and synthesis of starch and sucrose.

Horecker, B.L. (1976) Unravelling the pentose phosphate pathway. In *Reflections on Biochemistry* (Kornberg, A., Cornudella, L., Horecker, B.L., & Oro, J., eds), pp. 65–72, Pergamon Press, Inc., Oxford.

Huber, S.C. & Huber, J.L. (1996) Role and regulation of sucrose-phosphate synthase in higher plants. *Annu. Rev. Plant Physiol. Plant Mol. Biol.* **47,** 431–444.

Short review of factors that regulate this critical enzyme

Portis, A.R., Jr. (1990) Rubisco activase. *Biochim. Biophys. Acta* **1015,** 15–28.

Schneider, G., Lindqvist, Y., Brändén, C.-I., & Lorimer, G. (1986) Three-dimensional structure of ribulose-1,5-bisphosphate carboxylase/oxygenase from *Rhodospirillum rubrum* at 2.9 Å resolution. *EMBO J.* **5,** 3409–3415.

Smith, A.M., Denyer, K., & Martin, C. (1997) The synthesis of the starch granule. *Annu. Rev. Plant Physiol. Plant Mol. Biol.* **48,** 67–87.

Review of the role of ADP-glucose pyrophosphorylase in the synthesis of amylose and amylopectin in starch granules.

Stitt, M. (1990) Fructose 2,6-bisphosphate as a regulatory molecule in plants. *Annu. Rev. Plant Physiol. Plant Mol. Biol.* **41,** 153–185.

Stitt, M. (1995) Regulation of metabolism in transgenic plants. *Annu. Rev. Plant Physiol. Plant Mol. Biol.* **46,** 341–368.

Review of the genetic approaches to defining points of regulation in vivo.

Tabita, F.R. (1999) Microbial ribulose 1,5-bisphosphate carboxylase/oxygenase: a different perspective. *Photosynth. Res.* **60,** 1–28.

Discussion of biochemical and genetic studies of the microbial rubisco, and comparison with the enzyme from plants.

Wolosiuk, R., Ballicora, M., & Hagelin, K. (1993) The reductive pentose phosphate cycle for photosynthetic CO_2 assimilation: enzyme modulation. *FASEB J.* **7,** 622–637.

Wood, T. (1985) *The Pentose Phosphate Pathway,* Academic Press, Inc., Orlando, FL.

Woodrow, I.E. & Berry, J.A. (1988) Enzymatic regulation of photosynthetic CO_2 fixation in C_3 plants. *Annu. Rev. Plant Physiol. Plant Mol. Biol.* **39,** 533–594.

problems

1. Role of Oxidative Phosphorylation in Gluconeogenesis Is it possible to obtain a net synthesis of glucose from pyruvate if the citric acid cycle and oxidative phosphorylation are totally inhibited?

2. Pathway of Atoms in Gluconeogenesis A liver extract capable of carrying out all the normal metabolic reactions of the liver is briefly incubated in separate experiments with the following ^{14}C-labeled precursors:

(a) [^{14}C]Bicarbonate, $HO{-}^{14}C(=O)O^-$

(b) [1-^{14}C]Pyruvate, $CH_3{-}C(=O){-}^{14}COO^-$

Trace the pathway of each precursor through gluconeogenesis. Indicate the location of ^{14}C in all intermediates and in the product, glucose.

3. Pathway of CO_2 in Gluconeogenesis In the first bypass step of gluconeogenesis, the conversion of pyruvate to phosphoenolpyruvate, pyruvate is carboxylated by pyruvate carboxylase to oxaloacetate, which is subsequently decarboxylated by PEP carboxykinase to yield phosphoenolpyruvate. The observation that the addition of CO_2 is directly followed by the loss of CO_2 suggests that ^{14}C of $^{14}CO_2$ would not be incorporated into PEP, glucose, or any intermediates in gluconeogenesis. However, when a rat liver preparation synthesizes glucose in the presence of $^{14}CO_2$, ^{14}C slowly appears in PEP and eventually at C-3 and C-4 of glucose. How does the ^{14}C label get into PEP and glucose? (Hint: During gluconeogenesis in the presence of $^{14}CO_2$, several of the four-carbon citric acid cycle intermediates also become labeled.)

4. Energy Cost of a Cycle of Glycolysis and Gluconeogenesis What is the cost (in ATP equivalents) of transforming glucose to pyruvate via glycolysis and back again to glucose via gluconeogenesis?

5. Regulation of Glycolysis and Gluconeogenesis Because both glycolysis and gluconeogenesis are irreversible, there is no thermodynamic barrier to their simultaneous operation. What would the result be if both pathways were operating at the same time and the same rate? What prevents such simultaneous operation in cells? What considerations govern which pathway is operating at any time?

6. Regulation of Fructose 1,6-Bisphosphatase and Phosphofructokinase-1 What are the effects of increasing concentrations of ATP and AMP on the catalytic activities of fructose 1,6-bisphosphatase and phosphofructokinase-1? What are the consequences of these effects on the relative flow of metabolites through gluconeogenesis and glycolysis?

7. Glucogenic Substrates A common procedure for determining the effectiveness of compounds as precursors of glucose in mammals is to starve the animal until the liver glycogen stores are depleted and then administer the compound in question. A substrate that leads to a *net* increase in liver glycogen is termed glucogenic because it must first be converted to glucose 6-phosphate. Show by means of known enzymatic reactions which of the following substances are glucogenic:

(a) $^-OOC_2—CH_2—CH_2—COO^-$
Succinate

(b) $CH_2(OH)—C(H)(OH)—CH_2(OH)$
Glycerol

(c) $CH_3—C(=O)—S\text{-}CoA$
Acetyl-CoA

(d) $CH_3—C(=O)—COO^-$
Pyruvate

(e) $CH_3—CH_2—CH_2—COO^-$
Butyrate

8. Blood Lactate Levels during Vigorous Exercise The concentration of lactate in blood plasma before, during, and after a 400 m sprint are shown in the graph.

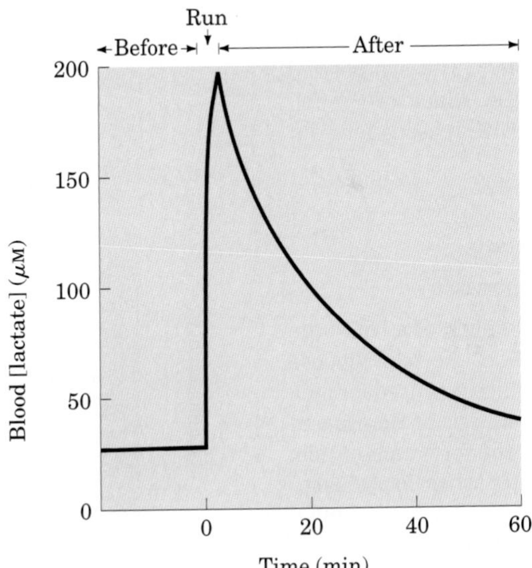

(a) What causes the rapid rise in lactate concentration?

(b) What causes the decline in lactate concentration after completion of the sprint? Why does the decline occur more slowly than the increase?

(c) Why is the concentration of lactate not zero during the resting state?

9. Relationship between Fructose 1,6-Bisphosphatase and Blood Lactate Levels. A congenital defect in the liver enzyme fructose 1,6-bisphosphatase results in abnormally high levels of lactate in the blood plasma. Explain.

10. Ethanol Affects Blood Glucose Levels The consumption of alcohol (ethanol), especially after periods of strenuous activity or after not eating for several hours, results in a deficiency of glucose in the blood, a condition known as hypoglycemia. The first step in the metabolism of ethanol by the liver is oxidation to acetaldehyde, catalyzed by liver alcohol dehydrogenase:

$$CH_3CH_2OH + NAD^+ \longrightarrow CH_3CHO + NADH + H^+$$

Explain how this reaction inhibits the transformation of lactate to pyruvate. Why does this lead to hypoglycemia?

11. Effect of Phloridzin on Carbohydrate Metabolism Phloridzin, a toxic glycoside from the bark of the pear tree, blocks the normal reabsorption of glucose from the kidney tubule, thus causing blood glucose to be almost completely excreted in the urine. Rats fed phloridzin and sodium succinate excreted about 0.5 moles of glucose (made by gluconeogenesis) for every mole of sodium succinate ingested. How is the succinate transformed to glucose? Explain the stoichiometry.

Phloridzin

12. Excess O_2 Uptake during Gluconeogenesis Lactate absorbed by the liver is converted to glucose, with the input of 6 mol of ATP for every mole of glucose produced. The extent of this process in a rat liver preparation can be monitored by administering [^{14}C]lactate and measuring the amount of [^{14}C]glucose produced. Because the stoichiometry of O_2 consumption and ATP production is known (Chapter 19), we can predict the extra O_2 consumption above the normal rate when a given amount of lactate is administered. However, when the extra O_2 used in the synthesis of glucose from lactate is actually measured, it is always higher than predicted by known stoichiometric relationships. Suggest a possible explanation for this observation.

13. At What Point Is Glycogen Synthesis Regulated? Explain how the following observations identify the point of regulation in the synthesis of glycogen in skeletal muscle.

(a) The measured activity of glycogen synthase in resting muscle, expressed in micromoles of UDP-glucose used per gram per minute, is lower than the activity of phosphoglucomutase or UDP-glucose pyrophosphorylase, each measured in terms of micromoles of substrate transformed per gram per minute.

(b) Stimulation of glycogen synthesis leads to a small decrease in the concentrations of glucose 6-phosphate and glucose 1-phosphate, a large decrease in the concentration of UDP-glucose, but a substantial increase in the concentration of UDP.

14. What Is the Cost of Storing Glucose as Glycogen? Write the sequence of steps and the net reaction required to calculate the cost in ATP molecules of converting a molecule of cytosolic glucose 6-phosphate to glycogen and back to glucose 6-phosphate. What fraction of the maximum number of ATP molecules available from complete catabolism of glucose 6-phosphate to CO_2 and H_2O does this cost represent?

15. Identification of a Defective Enzyme in Carbohydrate Metabolism A sample of liver tissue was obtained post mortem from the body of a patient believed to be genetically deficient in one of the enzymes of carbohydrate metabolism. A homogenate of the liver sample had the following characteristics: (1) it degraded glycogen to glucose 6-phosphate, (2) it did not synthesize glycogen from any sugar nor did it use galactose as an energy source, and (3) it synthesized glucose 6-phosphate from lactate. Which of the following enzymes was deficient? Give reasons for your choice.

(a) Glycogen phosphorylase

(b) Fructose 1,6-bisphosphatase

(c) UDP-glucose pyrophosphorylase

16. Ketosis in Sheep After the birth of a lamb, the udder of a ewe uses almost 80% of the total glucose synthesized by the animal. The glucose is used for milk production, principally in the synthesis of lactose and of glycerol 3-phosphate used in the formation of milk triacylglycerols. During the winter when food quality is poor, milk production decreases and the ewes sometimes develop ketosis: increased levels of plasma ketone bodies. Why do these changes occur? A standard treatment for sheep ketosis is the administration of large doses of propionate (readily converted to succinyl-CoA in ruminants). How does this treatment work?

17. Phases of Photosynthesis When a suspension of green algae is illuminated in the absence of CO_2 and then incubated with $^{14}CO_2$ in the dark, $^{14}CO_2$ is converted to [^{14}C]glucose for a brief time. What is the significance of this observation with regard to the CO_2 assimilation process, and how is it related to the light reactions of photosynthesis? Why does the conversion of $^{14}CO_2$ to [^{14}C]glucose stop after a brief time?

18. Identification of Key Intermediates in CO_2 Assimilation Calvin and his colleagues used the unicellular green alga *Chlorella* to study the carbon assimilation reactions of photosynthesis. They incubated $^{14}CO_2$ with illuminated suspensions of algae and followed the time course of appearance of ^{14}C in two compounds, X and Y, under two sets of conditions. Suggest the identities of X and Y based on your understanding of the Calvin cycle.

(a) Illuminated *Chlorella* were grown with unlabeled CO_2, then the lights were turned off and $^{14}CO_2$ was added (vertical dashed line in graph **a**). Under these conditions, X was the first compound to become labeled with ^{14}C; Y was unlabeled.

(b) Illuminated *Chlorella* cells were grown with $^{14}CO_2$. Illumination was continued until all the $^{14}CO_2$ had disappeared (vertical dashed line in graph **b**). Under these conditions, X became labeled quickly but lost its radioactivity with time, whereas Y became more radioactive with time.

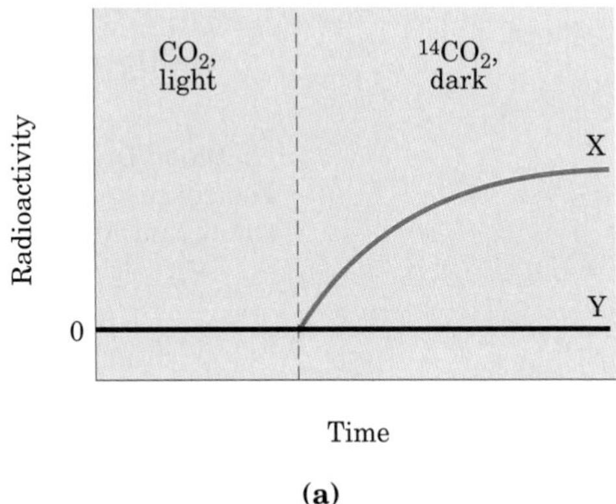

(a)

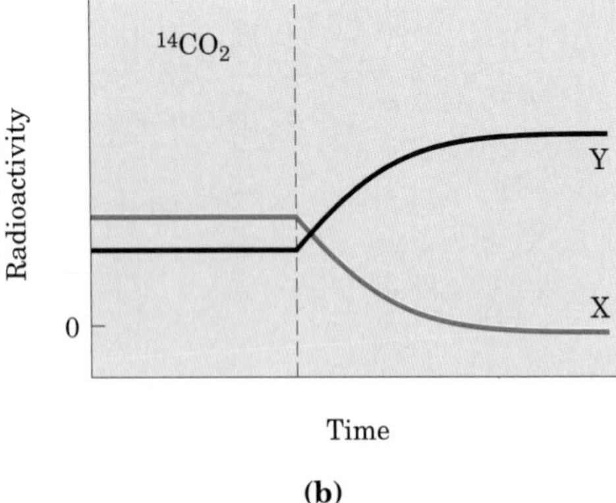

(b)

19. Pathway of CO_2 Assimilation in Maize If a maize (corn) plant is illuminated in the presence of $^{14}CO_2$, after about 1 s more than 90% of all the radioactivity incorporated in the leaves is found at C-4 of malate, aspartate, and oxaloacetate. Only after 60 s does ^{14}C appear at C-1 of 3-phosphoglycerate. Explain.

20. Chemistry of Malic Enzyme: Variation on a Theme Malic enzyme, found in the bundle-sheath cells of C_4 plants, carries out a reaction that has a counterpart in the citric acid cycle. What is the analogous reaction? Explain.

21. Sucrose and Dental Caries The most prevalent infection in humans worldwide is dental caries, which stems from the colonization and destruction of tooth enamel by a variety of acidifying microorganisms. These organisms synthesize and live within a water-insoluble network of dextrans, called dental plaque, composed of (α1→6)-linked polymers of glucose with many (α1→3) branch points. Polymerization of dextran requires dietary sucrose, and the reaction is catalyzed by a bacterial enzyme, dextran-sucrose glucosyltransferase.

(a) Write the overall reaction for dextran polymerization.

(b) In addition to providing a substrate for the formation of dental plaque, how does dietary sucrose also provide oral bacteria with an abundant source of metabolic energy?

22. Regulation of Carbohydrate Synthesis in Plants Sucrose synthesis occurs in the cytosol and starch synthesis in the chloroplast stroma, yet the two processes are intricately balanced.

(a) What factors shift the reactions in favor of starch synthesis?

(b) What factors shift the reactions to favor sucrose synthesis?

(c) Given that these two synthetic pathways occur in separate cellular compartments, what enables the two processes to influence each other?

chapter

21 Lipid Biosynthesis

Lipids play a variety of cellular roles, some only recently recognized. They are the principal form of stored energy in most organisms, as well as major constituents of cell membranes. Specialized lipids serve as pigments (retinal, carotene), cofactors (vitamin K), detergents (bile salts), transporters (dolichols), hormones (vitamin D derivatives, sex hormones), extracellular and intracellular messengers (eicosanoids and derivatives of phosphatidylinositol), and anchors for membrane proteins (covalently attached fatty acids, prenyl groups, and phosphatidylinositol). The ability to synthesize a variety of lipids is essential to all organisms. This chapter describes biosynthetic pathways for some of the principal lipids present in most cells, illustrating the strategies employed in assembling these water-insoluble products from water-soluble precursors such as acetate. Like other biosynthetic pathways, these reaction sequences are endergonic and reductive. They use ATP as a source of metabolic energy and a reduced electron carrier (usually NADPH) as a reductant.

We first describe the biosynthesis of fatty acids, the major components of both triacylglycerols and phospholipids. Then we examine the assembly of fatty acids into triacylglycerols and into the simpler types of membrane phospholipids. Finally, we consider the synthesis of cholesterol, a component of some membranes and the precursor of steroid products such as the bile acids, sex hormones, and adrenal cortical hormones.

Biosynthesis of Fatty Acids and Eicosanoids

When fatty acid oxidation was found to occur by oxidative removal of successive two-carbon (acetyl-CoA) units (see Fig. 17–8), biochemists thought that the biosynthesis of fatty acids might proceed by simple reversal of the same enzymatic steps used in their oxidation. However, fatty acid biosynthesis and breakdown occur by different pathways, are catalyzed by different sets of enzymes, and take place in different parts of the cell. Moreover, a three-carbon intermediate, **malonyl-CoA,** participates in the biosynthesis of fatty acids but not in their breakdown.

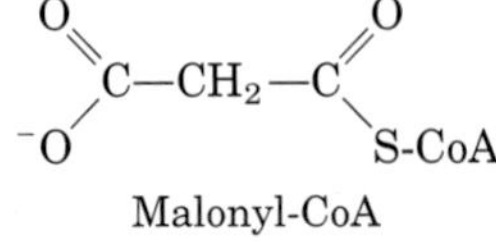

Malonyl-CoA

We focus first on the pathway of fatty acid synthesis, then turn our attention to regulation of the pathway and to the biosynthesis of long-chain fatty acids, unsaturated fatty acids, and their eicosanoid derivatives.

Malonyl-CoA Is Formed from Acetyl-CoA and Bicarbonate

The irreversible formation of malonyl-CoA from acetyl-CoA is catalyzed by **acetyl-CoA carboxylase.** Acetyl-CoA carboxylase from bacteria has three separate polypeptide subunits (Fig. 21–1); in animal cells, all three activities are part of a single multifunctional polypeptide. Plant cells contain

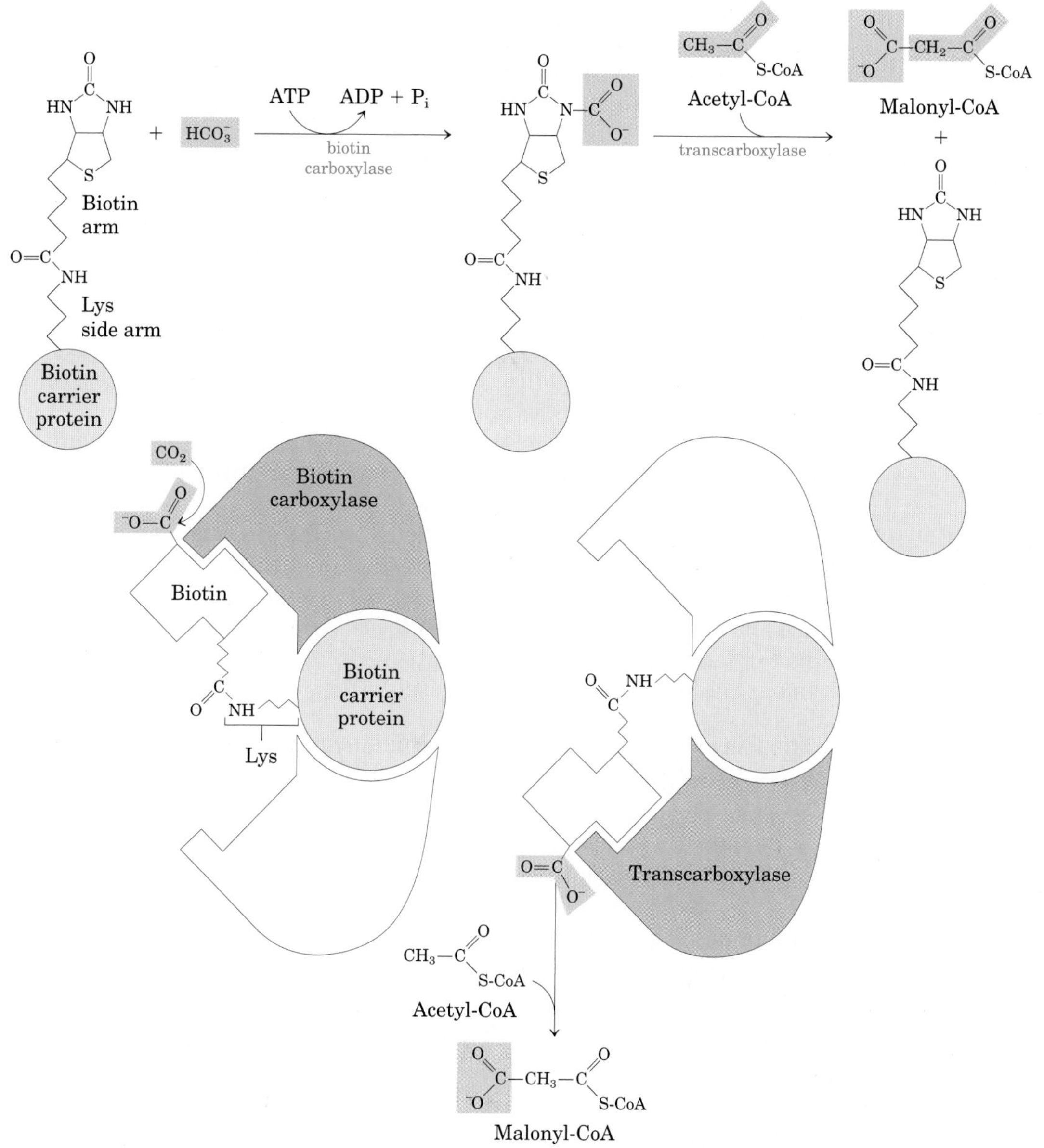

figure 21–1

The acetyl-CoA carboxylase reaction. Acetyl-CoA carboxylase has three functional regions: biotin carrier protein (gray); biotin carboxylase, which activates CO_2 by attaching it to a nitrogen in the biotin ring in an ATP-dependent reaction (see Fig. 16–14); and transcarboxylase, which transfers activated CO_2 (shaded green) from biotin to acetyl-CoA, producing malonyl-CoA. The long, flexible biotin arm carries the activated CO_2 from the biotin carboxylase region to the transcarboxylase active site, as shown in the diagrams below the reaction arrows. The active enzyme in each step is shaded in blue.

both forms of acetyl-CoA carboxylase. In all cases, acetyl-CoA carboxylase contains a biotin prosthetic group covalently bound in amide linkage to the ε-amino group of a Lys residue on one of the three domains of the enzyme molecule. The two-step reaction is very similar to other biotin-dependent carboxylation reactions, such as those catalyzed by pyruvate carboxylase (see Fig. 16–14) and propionyl-CoA carboxylase (see Fig. 17–11). The carboxyl group, derived from bicarbonate (HCO_3^-), is first transferred to biotin in an ATP-dependent reaction. The biotinyl group serves as a temporary carrier of CO_2, transferring it to acetyl-CoA in the second step to yield malonyl-CoA.

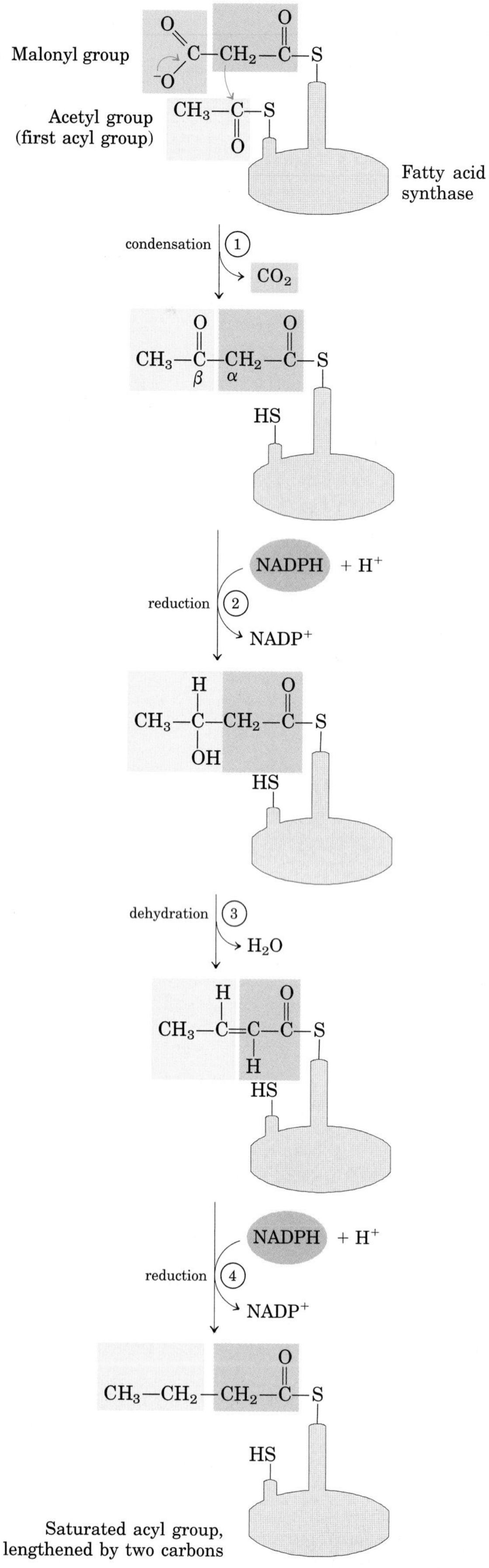

Fatty Acids Are Synthesized by a Repeating Reaction Sequence

The long carbon chains of fatty acids are assembled in a repeating four-step sequence (Fig. 21–2). Saturated acyl groups produced by this set of reactions become the substrate in subsequent condensation with an activated malonyl group. With each passage through the cycle, the fatty acyl chain is extended by two carbons. When the chain length reaches 16 carbons, the product (palmitate, 16:0; see Table 11–1) leaves the cycle. The methyl and carboxyl carbon atoms of the acetyl group become C-16 and C-15, respectively, of the palmitate (Fig. 21–3); the rest of the carbon atoms are derived from acetyl-CoA via malonyl-CoA.

Both the electron carrier cofactor and the activating groups in the reductive anabolic sequence are different from those that act in the oxidative catabolic process. Recall that in β oxidation, NAD^+ and FAD serve as electron acceptors, and the activating group is the thiol (—SH) group of coenzyme A (see Fig. 17–8). By contrast, the reducing agent in the synthetic sequence is NADPH, and the activating groups are two different enzyme-bound —SH groups, described below.

All of the reactions in the synthetic process are catalyzed by a multienzyme complex, **fatty acid synthase.** Although the details of enzyme structure differ in prokaryotes such as *Escherichia coli* and in eukaryotes, the four-step process of fatty acid synthesis is the same in all organisms. We first describe the process as it occurs in *E. coli,* then consider differences in enzyme structure in other organisms.

The Fatty Acid Synthase Complex Has Seven Different Active Sites

The core of the fatty acid synthase system from *E. coli* consists of seven separate polypeptides, and at least three other proteins act at some stage of the process (Table 21–1). The proteins act together to catalyze the formation of fatty acids from acetyl-CoA and malonyl-CoA. Throughout the process, the intermediates remain covalently attached to one of two thiol groups of the complex. One point of attachment is the —SH group of a Cys residue in one of the seven proteins (β-ketoacyl-ACP synthase, described below); the other is the —SH group of acyl carrier protein, with which the acyl intermediates of fatty acid synthesis form a thioester.

figure 21–2

A four-step sequence lengthens a growing fatty acyl chain by two carbons. Each malonyl group and acetyl (or longer acyl) group is activated by a thioester that links it to fatty acid synthase, a multienzyme complex described later in the text. (1) The first step is the condensation of an activated acyl group (an acetyl group is the first acyl group) and two carbons derived from malonyl-CoA, with the elimination of CO_2 from the malonyl group; the net effect is extension of the acyl chain by two carbons. The β-keto product of this condensation is then reduced in three more steps nearly identical to the reactions of β oxidation, but in the reverse sequence: (2) the β-keto group is reduced to an alcohol, (3) the elimination of H_2O creates a double bond, and (4) the double bond is reduced to form the corresponding saturated fatty acyl group.

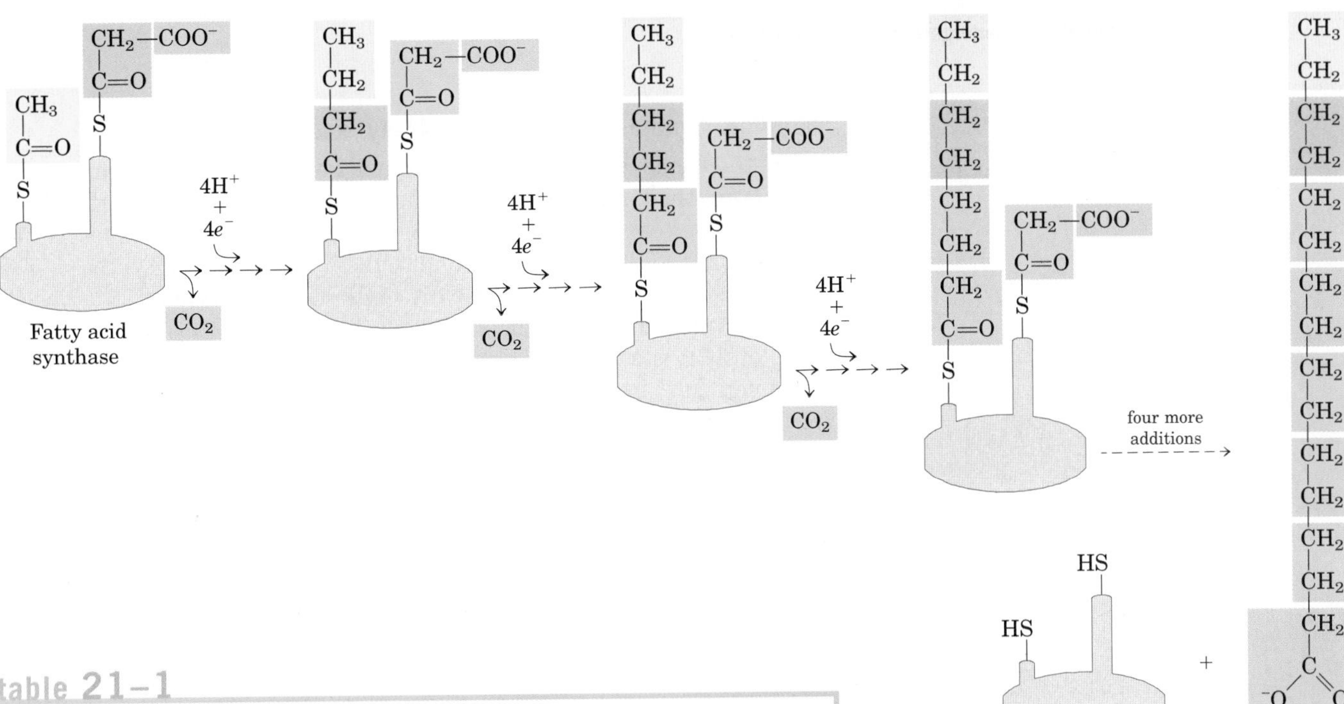

figure 21–3
The overall process of palmitate synthesis. The fatty acyl chain grows by two-carbon units donated by activated malonate, with loss of CO_2 at each step. The initial acetyl group is shaded yellow; C-1 and C-2 of malonate are shaded red; and the carbon released as CO_2 is shaded green. After each two-carbon addition, reductions convert the growing chain to a saturated fatty acid of four, then six, then eight carbons, and so on. The final product is palmitate (16:0).

table 21–1

Proteins of the Fatty Acid Synthase Complex of *E. coli*

Protein	Role
Acyl carrier protein (ACP)	Carries acyl groups in thioester linkage
Acetyl-CoA–ACP transacetylase (AT)	Transfers acyl group from CoA to Cys residue of KS
β-Ketoacyl-ACP synthase (KS)	Condenses acyl and malonyl groups (there are at least three isozymes of KS)
Malonyl-CoA–ACP transferase (MT)	Transfers malonyl group from CoA to ACP
β-Ketoacyl-ACP reductase (KR)	Reduces β-keto group to β-hydroxy group
β-Hydroxyacyl-ACP dehydratase (HD)	Removes H_2O from β-hydroxyacyl-ACP, creating double bond
Enoyl-ACP reductase (ER)	Reduces double bond, forming saturated acyl-ACP

Acyl carrier protein (ACP) of *E. coli* (Table 21–1) is a small protein (M_r 8,860) containing the prosthetic group **4′-phosphopantetheine** (Fig. 21–4; compare to the panthothenic acid and β-mercaptoethylamine moiety of coenzyme A, Fig. 10–41). Hydrolysis of the thioester that links ACP to the fatty acyl group is highly exergonic, and the energy released when this bond is broken helps to make the first reaction in fatty acid synthesis (condensation) thermodynamically favorable. The 4′-phosphopantetheine prosthetic group of ACP is believed to serve as a flexible arm, tethering the growing fatty acyl chain to the surface of the fatty acid synthase complex and carrying the reaction intermediates from one enzyme active site to the next.

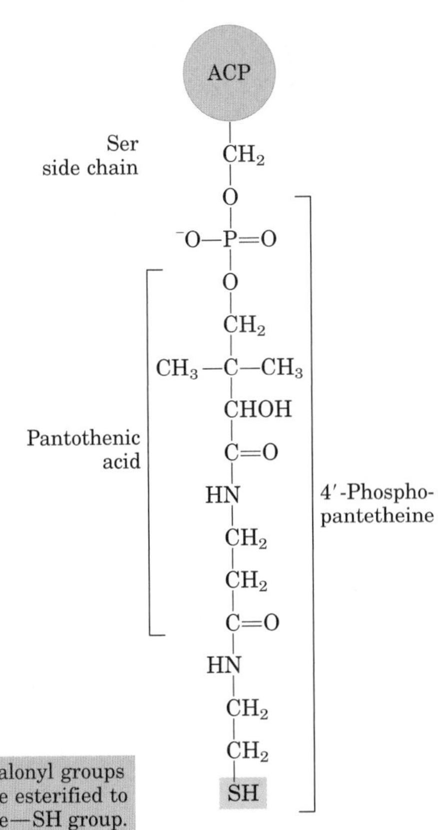

figure 21–4
Acyl carrier protein (ACP). The prosthetic group is 4′-phosphopantetheine, which is covalently attached to the hydroxyl group of a Ser residue in ACP. Phosphopantetheine contains the B vitamin pantothenate, also found in the coenzyme A molecule. Its —SH group is the site of entry of malonyl groups during fatty acid synthesis.

Fatty Acid Synthase Receives the Acetyl and Malonyl Groups

Before the condensation reactions that build up the fatty acid chain can begin, the two thiol groups on the enzyme complex must be charged with the correct acyl groups (top of Fig. 21–5). First, the acetyl group of acetyl-CoA is transferred to the Cys —SH group of the β-ketoacyl-ACP synthase. This reaction is catalyzed by **acetyl-CoA–ACP transacetylase.** The second reaction, transfer of the malonyl group from malonyl-CoA to the —SH group of ACP, is catalyzed by **malonyl-CoA–ACP transferase,** also part of the complex. In the charged synthase complex, the acetyl and malonyl groups are very close to each other and are activated for the chain-lengthening process, which consists of the four steps outlined earlier. These steps, shown in Figure 21–5, are now considered in some detail.

Step ① Condensation The first step in the formation of a fatty acid chain is condensation of the activated acetyl and malonyl groups to form **acetoacetyl-ACP,** an acetoacetyl group bound to ACP through the phosphopantetheine —SH group; simultaneously, a molecule of CO_2 is produced (Fig. 21–5). In this reaction, catalyzed by **β-ketoacyl-ACP synthase,** the acetyl group is transferred from the Cys —SH group of this enzyme to the malonyl group on the —SH of ACP, becoming the methyl-terminal two-carbon unit of the new acetoacetyl group.

The carbon atom in the CO_2 formed in this reaction is the same carbon atom that was originally introduced into malonyl-CoA from HCO_3^- by the acetyl-CoA carboxylase reaction (Fig. 21–1). Thus CO_2 is only transiently in covalent linkage during fatty acid biosynthesis; it is removed as each two-carbon unit is inserted.

Why do cells go to the trouble of adding CO_2 to make a malonyl group from an acetyl group, only to lose CO_2 again during the formation of acetoacetate? Remember that in the β oxidation of fatty acids (see Fig. 17–8), cleavage of the bond between two acyl groups (the cleavage of an acetyl unit from the acyl chain) is highly exergonic, so the simple condensation of two acyl groups (of two acetyl-CoA molecules, for example) is endergonic. The condensation reaction is made thermodynamically favorable by the involvement of activated malonyl groups rather than acetyl groups. The methylene carbon (C-2) of the malonyl group, sandwiched between carbonyl and carboxyl carbons, is an especially good nucleophile. In the condensation step (step ① in Fig. 21–5), decarboxylation of the malonyl group facilitates the nucleophilic attack of this methylene carbon on the thioester linking the acetyl group to β-ketoacyl-ACP synthase, displacing the enzyme's —SH group. Coupling the condensation to the decarboxylation of the malonyl group renders the overall process highly exergonic. Recall that a similar carboxylation-decarboxylation sequence facilitates the formation of phosphoenolpyruvate from pyruvate in gluconeogenesis (see Fig. 20–3).

By using activated malonyl groups in the synthesis of fatty acids and activated acetate in their degradation, the cell manages to make both processes favorable, although one is effectively the reversal of the other. The extra energy required to make fatty acid synthesis favorable is provided by the ATP used to synthesize malonyl-CoA from acetyl-CoA and HCO_3^- (Fig. 21–1).

Step ② Reduction of the Carbonyl Group The acetoacetyl-ACP formed in the condensation (step ① in Fig. 21–5) next undergoes reduction of the carbonyl group at C-3 to form D-β-hydroxybutyryl-ACP (step ②). This reaction is catalyzed by **β-ketoacyl-ACP reductase,** and the electron donor is NADPH. Notice that the D-β-hydroxybutyryl group does not have the same stereoisomeric form as the L-β-hydroxyacyl intermediate in fatty acid oxidation (see Fig. 17–8).

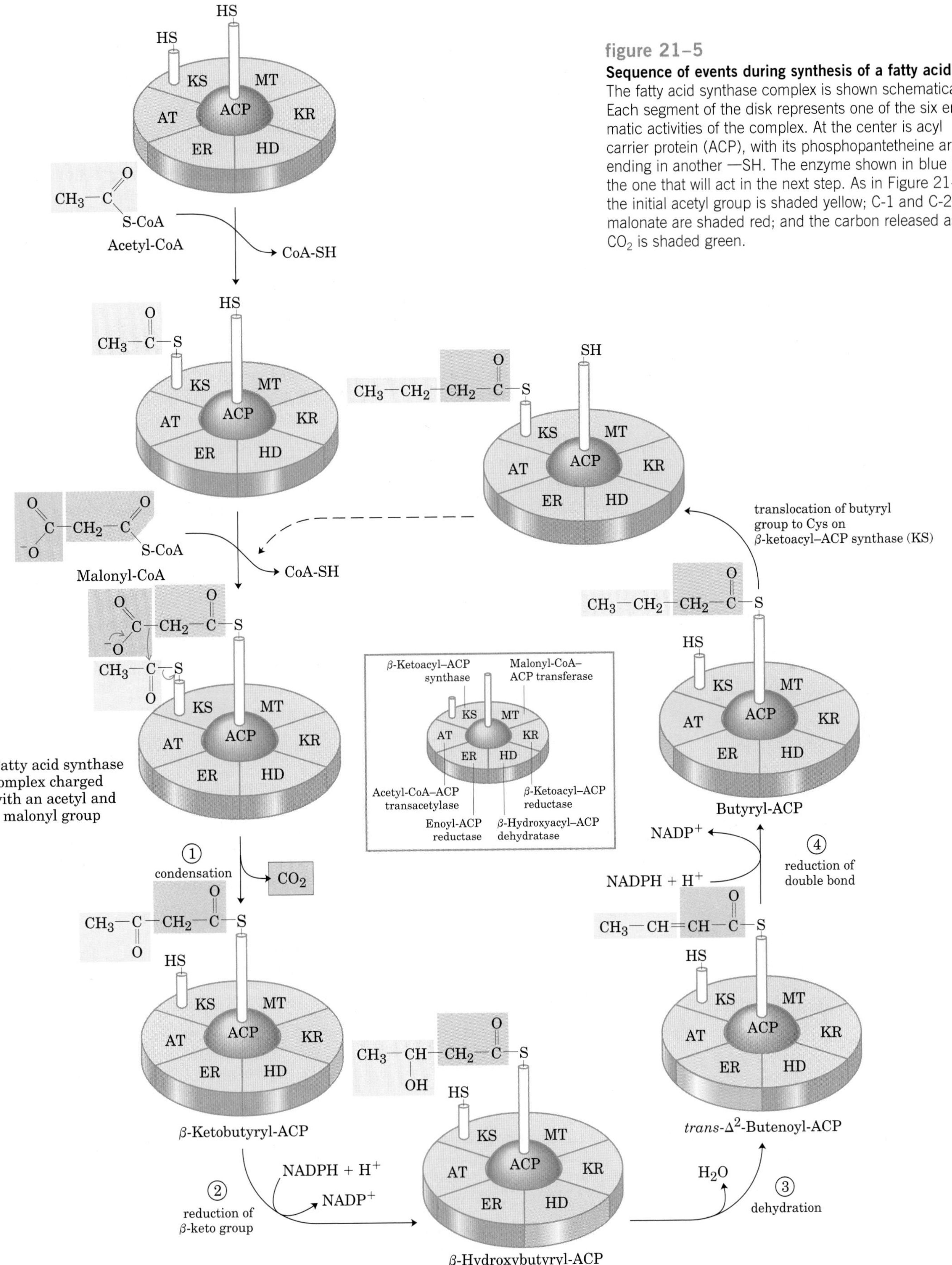

figure 21–5

Sequence of events during synthesis of a fatty acid. The fatty acid synthase complex is shown schematically. Each segment of the disk represents one of the six enzymatic activities of the complex. At the center is acyl carrier protein (ACP), with its phosphopantetheine arm ending in another —SH. The enzyme shown in blue is the one that will act in the next step. As in Figure 21–3, the initial acetyl group is shaded yellow; C-1 and C-2 of malonate are shaded red; and the carbon released as CO_2 is shaded green.

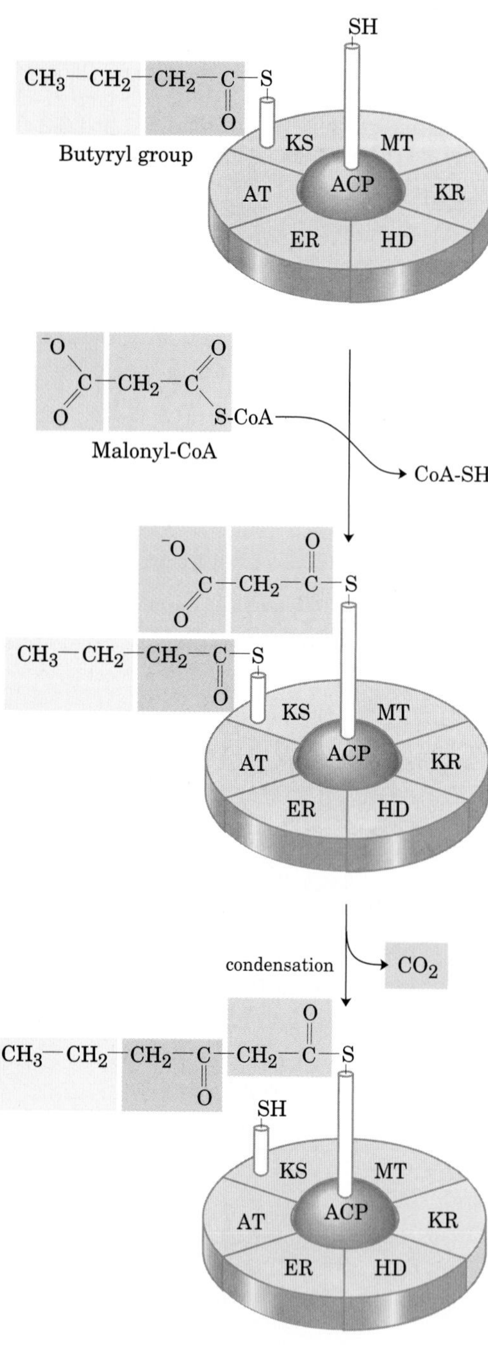

figure 21–6

Beginning of the second round of the fatty acid synthesis cycle. The butyryl group is on the Cys —SH group. The incoming malonyl group is first attached to the phosphopantetheine —SH group. Then, in the condensation step, the entire butyryl group on the Cys —SH is exchanged for the carboxyl group of the malonyl residue, which is lost as CO_2 (green). This step is analogous to step ① in Figure 21–5. The product, a six-carbon β-ketoacyl group, now contains four carbons derived from malonyl-CoA and two derived from the acetyl-CoA that started the reaction. The β-ketoacyl group then undergoes steps ② through ④, as in Figure 21–5.

Step ③ Dehydration In the third step, the elements of water are removed from C-2 and C-3 of D-β-hydroxybutyryl-ACP to yield a double bond in the product, ***trans*-Δ^2-butenoyl-ACP.** The enzyme that catalyzes this dehydration is **β-hydroxyacyl-ACP dehydratase.**

Step ④ Reduction of the Double Bond Finally, the double bond of *trans*-Δ^2-butenoyl-ACP is reduced (saturated) to form **butyryl-ACP** by the action of **enoyl-ACP reductase;** again, NADPH is the electron donor.

The Fatty Acid Synthase Reactions Are Repeated to Form Palmitate

The production of the four-carbon, saturated fatty acyl–ACP completes one pass through the fatty acid synthase complex. The butyryl group is now transferred from the phosphopantetheine —SH group of ACP to the Cys —SH group of β-ketoacyl-ACP synthase, which initially bore the acetyl group (Fig. 21–5). To start the next cycle of four reactions that lengthens the chain by two more carbons, another malonyl group is linked to the now unoccupied phosphopantetheine —SH group of ACP (Fig. 21–6). Condensation occurs as the butyryl group, acting like the acetyl group in the first cycle, is linked to two carbons of the malonyl-ACP group with concurrent loss of CO_2. The product of this condensation is a six-carbon acyl group, covalently bound to the phosphopantetheine —SH group. Its β-keto group is reduced in the next three steps of the synthase cycle to yield the six-carbon saturated acyl group—exactly as in the first round of reactions.

Seven cycles of condensation and reduction produce the 16-carbon saturated palmitoyl group, still bound to ACP. For reasons not well understood, chain elongation generally stops at this point, and free palmitate is released from the ACP molecule by a hydrolytic activity in the synthase complex. Small amounts of longer fatty acids such as stearate (18:0) are also formed. In certain plants (coconut and palm, for example) chain termination occurs earlier; up to 90% of the fatty acids in the oils of these plants are between 8 and 14 carbons long.

The overall reaction for the synthesis of palmitate from acetyl-CoA can be considered in two parts. First, the formation of seven malonyl-CoA molecules:

$$\text{7 Acetyl-CoA} + 7CO_2 + \text{7ATP} \longrightarrow \text{7 malonyl-CoA} + \text{7ADP} + 7P_i \quad (21\text{–}1)$$

then seven cycles of condensation and reduction:

$$\text{Acetyl-CoA} + \text{7 malonyl-CoA} + \text{14NADPH} + 14H^+ \longrightarrow \text{palmitate} + 7CO_2 + \text{8CoA} + \text{14NADP}^+ + 6H_2O \quad (21\text{–}2)$$

The overall process (the sum of Eqns 21–1 and 21–2) is:

$$\text{8 Acetyl-CoA} + \text{7ATP} + \text{14NADPH} + 14H^+ \longrightarrow \text{palmitate} + \text{8CoA} + 6H_2O + \text{7ADP} + 7P_i + \text{14NADP}^+ \quad (21\text{–}3)$$

The biosynthesis of fatty acids such as palmitate thus requires acetyl-CoA and the input of chemical energy in two forms: the group transfer potential of ATP and the reducing power of NADPH. The ATP is required to attach CO_2 to acetyl-CoA to make malonyl-CoA; the NADPH is required to reduce the double bonds. We shall return to the sources of acetyl-CoA and NADPH soon, but let us first consider the structure of the remarkable enzyme complex that catalyzes the synthesis of fatty acids.

The Fatty Acid Synthase of Some Organisms Is Composed of Multifunctional Proteins

The seven active sites for fatty acid synthesis (six enzymes and ACP) reside in seven separate polypeptides in the fatty acid synthase of *E. coli* and some plants (Fig. 21–7). In these complexes, each enzyme is positioned with its active site near that of the preceding and succeeding enzymes of the sequence. The flexible pantetheine arm of ACP can reach all of the active sites, and it carries the growing fatty acyl chain from one site to the next; the intermediates are not released from the enzyme complex until the finished product is obtained. As we have seen in earlier chapters, this channeling of intermediates from one active site to the next increases the efficiency of the overall process.

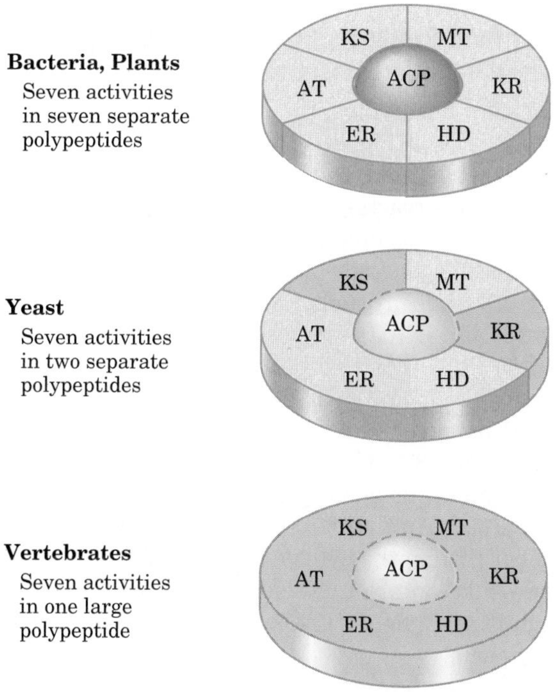

figure 21–7
Structure of fatty acid synthases. The fatty acid synthase from bacteria and plants is a complex of at least seven different polypeptides. In yeast, all seven activities reside in only two polypeptides, and in vertebrates, in a single large polypeptide.

The fatty acid synthases of yeast and of vertebrates are also multienzyme complexes, but their integration is even more complete than in *E. coli* and plants. In yeast, the seven distinct active sites reside in two large, multifunctional polypeptides, with three activities on the α subunit and four on the β subunit; in vertebrates, a single large polypeptide (M_r 240,000) contains all seven enzymatic activities as well as a hydrolytic activity that cleaves the fatty acid from the ACP-like part of the enzyme complex. The enzyme from vertebrates functions as a dimer (M_r 480,000) in which the two identical subunits lie head-to-tail, forming two active sites at their interface. Extension of the growing acyl chain from one subunit is catalyzed by sites on the other subunit.

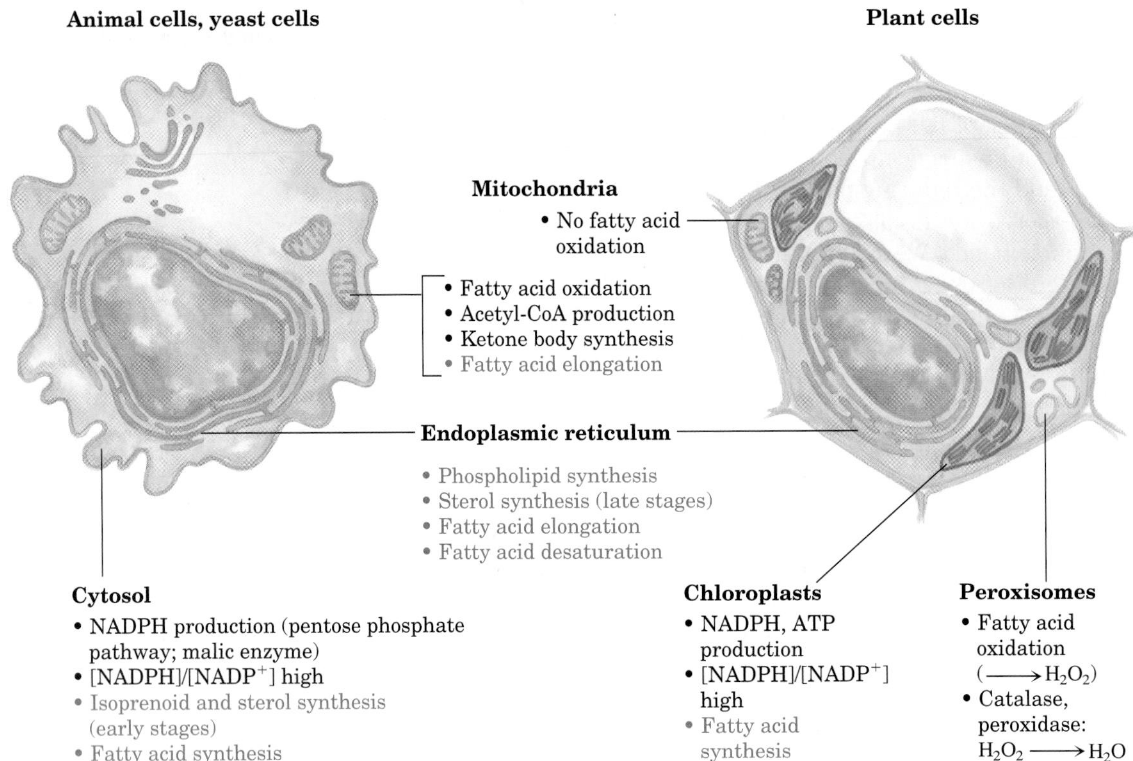

figure 21–8

Subcellular localization of lipid metabolism. Yeast and vertebrate animal cells differ from higher plant cells in the compartmentation of lipid metabolism. Fatty acid synthesis takes place in the compartment in which NADPH is available for reductive synthesis (i.e., where the [NADPH]/[NADP$^+$] ratio is high). Processes highlighted with red type are covered in this chapter.

Fatty Acid Synthesis Occurs in the Cytosol of Many Organisms but in the Chloroplasts of Plants

In higher eukaryotes, the fatty acid synthase complex is found exclusively in the cytosol (Fig. 21–8), as are the biosynthetic enzymes for nucleotides, amino acids, and glucose. This location segregates synthetic processes from degradative reactions, many of which take place in the mitochondrial matrix. There is a corresponding segregation of electron-carrying cofactors for anabolism (generally a reductive process) and those for catabolism (generally oxidative). Usually, NADPH is the electron carrier for anabolic reactions, and NAD$^+$ serves in catabolic reactions. In hepatocytes, the [NADPH]/[NADP$^+$] ratio is very high (about 75) in the cytosol, furnishing a strongly reducing environment for the reductive synthesis of fatty acids and other biomolecules. The cytosolic [NADH]/[NAD$^+$] ratio is much smaller (only about 8×10^{-4}), so the NAD$^+$-dependent oxidative catabolism of glucose can occur in the same compartment, at the same time, as fatty acid synthesis. The [NADH]/[NAD$^+$] ratio within the mitochondrion is much higher than in the cytosol because of the flow of electrons into NAD$^+$ from the oxidation of fatty acids, amino acids, pyruvate, and acetyl-CoA. This high [NADH]/[NAD$^+$] ratio favors the reduction of oxygen via the respiratory chain.

In hepatocytes and adipocytes, cytosolic NADPH is largely generated by the pentose phosphate pathway (see Fig. 15–20) and by **malic enzyme** (Fig. 21–9a). The NADP-linked malic enzyme that operates in the carbon-assimilation pathway of C_4 plants (see Fig. 20–40) is unrelated in function. The pyruvate produced in the reaction shown in Figure 21–9a reenters the mitochondrion. In hepatocytes and in the mammary gland of lactating animals, the NADPH required for fatty acid biosynthesis is supplied primarily by the pentose phosphate pathway (Fig. 21–9b).

figure 21–9

Production of NADPH. Two routes to NADPH, catalyzed by **(a)** malic enzyme and **(b)** the pentose phosphate pathway.

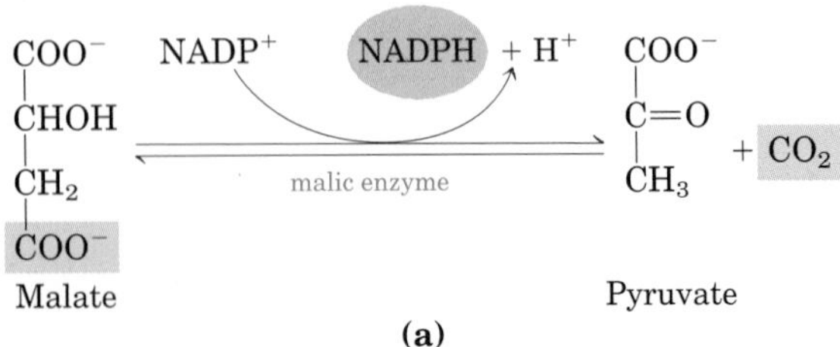

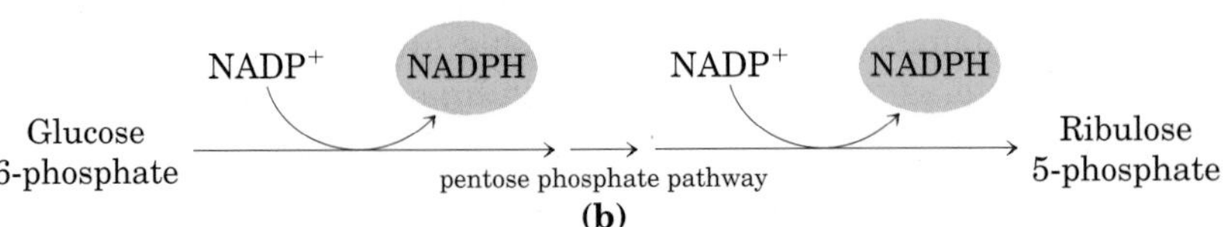

In the photosynthetic cells of plants, fatty acid synthesis occurs not in the cytosol, but in the chloroplast stroma (Fig. 21–8). This makes sense as NADPH is produced in chloroplasts by the light reactions of photosynthesis (Fig. 21–10). Again, the resulting high [NADPH]/[$NADP^+$] ratio provides the reducing environment that favors reductive anabolic processes such as fatty acid synthesis.

$$H_2O + NADP^+ \xrightarrow{\text{light}} \tfrac{1}{2}O_2 + NADPH + H^+$$

figure 21–10
Production of NADPH by photosynthesis.

Acetate Is Shuttled out of Mitochondria as Citrate

In nonphotosynthetic eukaryotes, nearly all the acetyl-CoA used in fatty acid synthesis is formed in mitochondria from pyruvate oxidation and from the catabolism of the carbon skeletons of amino acids. Acetyl-CoA arising from the oxidation of fatty acids does not represent a significant source of acetyl-CoA for fatty acid biosynthesis in animals because the two pathways are regulated reciprocally, as described below.

The mitochondrial inner membrane is impermeable to acetyl-CoA, so an indirect shuttle transfers acetyl group equivalents across the inner membrane (Fig. 21–11). Intramitochondrial acetyl-CoA first reacts with oxaloacetate to form citrate, in the citric acid cycle reaction catalyzed by **citrate synthase** (see Fig. 16–7). Citrate then passes into the cytosol through

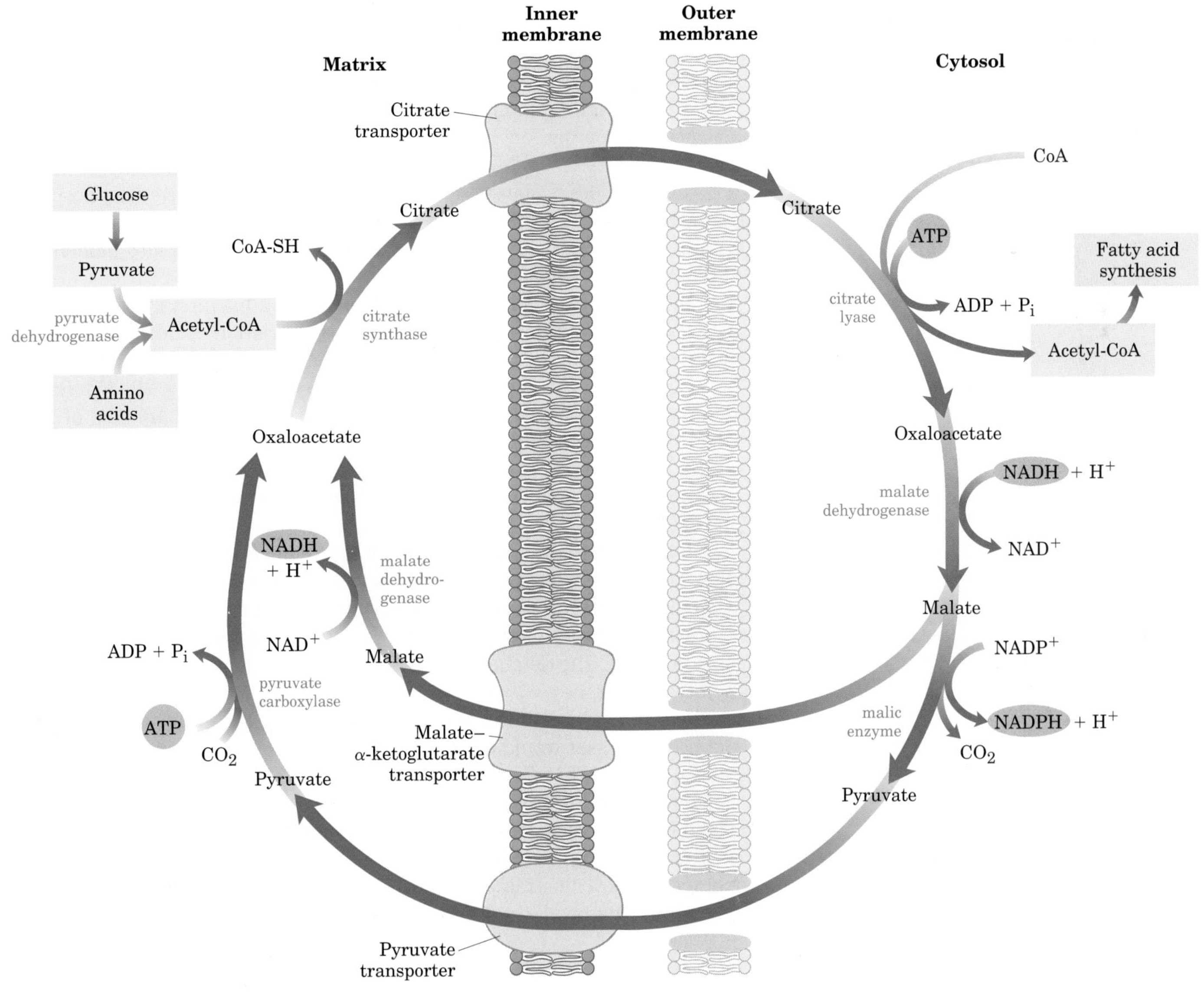

figure 21–11
Shuttle for transfer of acetyl groups from mitochondria to the cytosol. The outer mitochondrial membrane is freely permeable to all of these compounds. Pyruvate derived from amino acid catabolism in the mitochondrial matrix, or from glucose by glycolysis in the cytosol, is converted to acetyl-CoA in the matrix. Acetyl groups pass out of the mitochondrion as citrate; in the cytosol they are delivered as acetyl-CoA for fatty acid synthesis. Oxaloacetate is reduced to malate, which returns to the mitochondrial matrix and is converted to oxaloacetate. An alternative fate for cytosolic malate is oxidation by malic enzyme to generate cytosolic NADPH; the pyruvate produced returns to the mitochondrial matrix.

the mitochondrial inner membrane on the **citrate transporter.** In the cytosol, citrate cleavage by **citrate lyase** regenerates acetyl-CoA in an ATP-dependent reaction. Oxaloacetate cannot return to the mitochondrial matrix directly as there is no oxaloacetate transporter. Instead, oxaloacetate is reduced by cytosolic malate dehydrogenase to malate, which returns to the mitochondrial matrix on the malate–α-ketoglutarate transporter in exchange for citrate. There it is reoxidized to oxaloacetate to complete the shuttle. Alternatively, the malate produced in the cytosol is used to generate cytosolic NADPH through the activity of malic enzyme (Fig. 21–9a).

Fatty Acid Biosynthesis Is Tightly Regulated

When a cell or organism has more than enough metabolic fuel available to meet its energy needs, the excess is generally converted to fatty acids and stored as lipids such as triacylglycerol. The reaction catalyzed by acetyl-CoA carboxylase is the rate-limiting step in the biosynthesis of fatty acids, and this enzyme is an important site of regulation. In vertebrates, palmitoyl-CoA, the principal product of fatty acid synthesis, acts as a feedback inhibitor of the enzyme, and citrate is an allosteric activator (Fig. 21–12a) increasing V_{max}. Citrate plays a central role in diverting cellular metabolism from the consumption (oxidation) of metabolic fuel to its storage as fatty acids. When there is an increase in the concentrations of mitochondrial acetyl-CoA and ATP, citrate is transported out of mitochondria; it then becomes both the precursor of cytosolic acetyl-CoA and an allosteric signal for the activation of acetyl-CoA carboxylase. At the same time, citrate inhibits the activity of phosphofructokinase-1 (Fig. 15–18), reducing the flow of carbon through glycolysis.

Acetyl-CoA carboxylase is also regulated by covalent modification. Phosphorylation triggered by the hormones glucagon and epinephrine inactivates it, thereby slowing fatty acid synthesis. In its active (dephosphorylated) form, acetyl-CoA carboxylase polymerizes into long filaments (Fig. 21–12b); phosphorylation is accompanied by dissociation into monomeric subunits and loss of activity.

The acetyl-CoA carboxylase from plants and bacteria is not regulated by citrate or by a phosphorylation-dephosphorylation cycle. The plant enzyme is activated by an increase in stromal pH and Mg^{2+} concentration, both of which occur upon illumination of the plant (see Fig. 20–35). Bacteria do not use triacylglycerols as energy stores. The primary role of fatty acid synthesis in *E. coli* is to provide precursors for membrane lipids; the regulation of this process is complex, involving guanine nucleotides such as ppGpp that coordinate cell growth with membrane formation (see Figs 10–42 and 28–26).

In addition to the moment-by-moment regulation of enzymatic activity, these pathways are regulated at the level of gene expression. For example, when animals ingest an excess of certain polyunsaturated fatty acids, the expression of genes encoding a wide range of lipogenic enzymes in the liver is suppressed. The detailed mechanism by which these genes are regulated is not yet clear.

If fatty acid synthesis and β oxidation were to occur simultaneously, the two processes would constitute a futile cycle, wasting energy. We noted earlier (p. 612) that β oxidation is blocked by malonyl-CoA, which inhibits carnitine acyltransferase I. Thus during fatty acid synthesis, the production of the first intermediate, malonyl-CoA, shuts down β oxidation at the level of a transport system in the mitochondrial inner membrane. This control mechanism illustrates another advantage of segregating synthetic and degradative pathways in different cellular compartments.

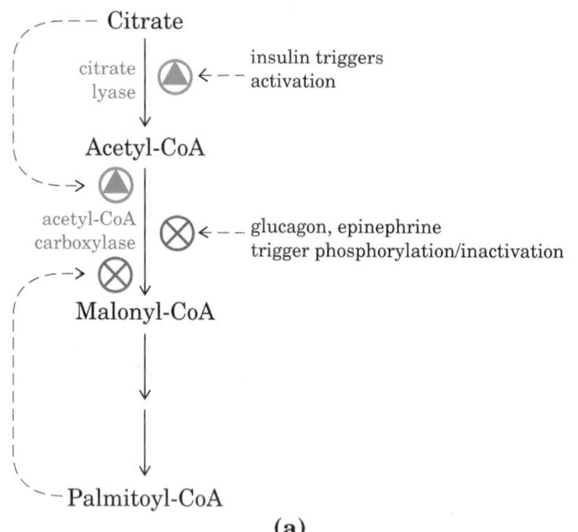

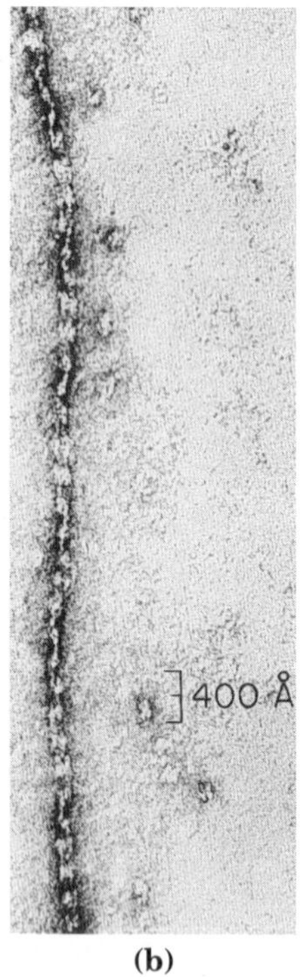

figure 21–12
Regulation of fatty acid synthesis. (a) In the cells of vertebrates, both allosteric regulation and hormone-dependent covalent modification influence the flow of precursors into malonyl-CoA. In plants, acetyl-CoA carboxylase is activated by the changes in [Mg^{2+}] and pH that accompany illumination (not shown here). **(b)** Filaments of acetyl-CoA carboxylase (the active, dephosphorylated form) as seen with the electron microscope.

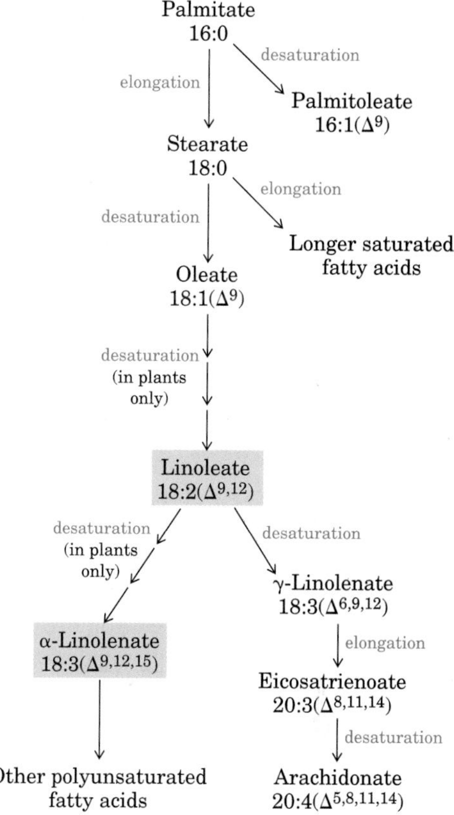

figure 21–13

Routes of synthesis of other fatty acids. Palmitate is the precursor of stearate and longer-chain saturated fatty acids, as well as the monounsaturated acids palmitoleate and oleate. Mammals cannot convert oleate into linoleate or α-linolenate (shaded red), which are therefore required in the diet as essential fatty acids. Conversion of linoleate into other polyunsaturated fatty acids and eicosanoids is outlined. Unsaturated fatty acids are symbolized by indicating the number of carbons and the number and position of the double bonds, as in Table 11–1.

Long-Chain Saturated Fatty Acids Are Synthesized from Palmitate

Palmitate, the principal product of the fatty acid synthase system in animal cells, is the precursor of other long-chain fatty acids (Fig. 21–13). It may be lengthened to form stearate (18:0) or even longer saturated fatty acids by further additions of acetyl groups, through the action of **fatty acid elongation systems** present in the smooth endoplasmic reticulum and the mitochondria. The more active elongation system of the endoplasmic reticulum extends the 16-carbon chain of palmitoyl-CoA by two carbons, forming stearoyl-CoA. Although different enzyme systems are involved, and coenzyme A rather than ACP is the acyl carrier in the reaction, the mechanism of elongation in the endoplasmic reticulum is otherwise identical to that in palmitate synthesis: donation of two carbons by malonyl-CoA, followed by reduction, dehydration, and reduction to the saturated 18-carbon product, stearoyl-CoA.

Some Fatty Acids Are Desaturated

Palmitate and stearate serve as precursors of the two most common monounsaturated fatty acids of animal tissues: palmitoleate, 16:1(Δ^9), and oleate, 18:1(Δ^9) (Fig. 21–13). Each of these fatty acids has a single cis double bond in the Δ^9 position (between C-9 and C-10). The double bond is introduced into the fatty acid chain by an oxidative reaction catalyzed by **fatty acyl–CoA desaturase** (Fig. 21–14), a **mixed-function oxidase** (Box 21–1). Two different substrates, the fatty acid and NADPH, simulta-

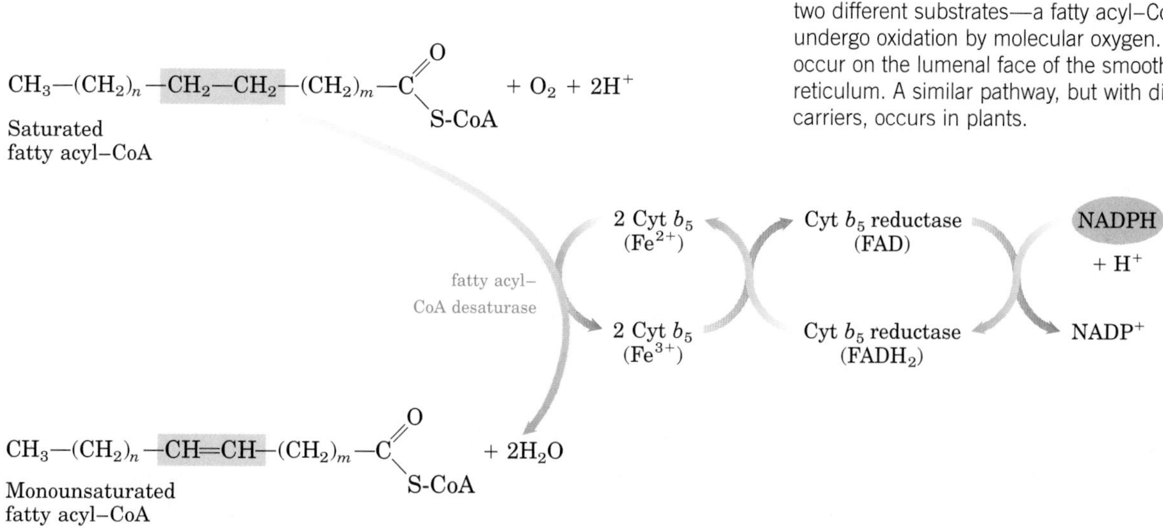

figure 21–14

Electron transfer in the desaturation of fatty acids in vertebrates. Blue arrows show the path of electrons as two different substrates—a fatty acyl–CoA and NADPH—undergo oxidation by molecular oxygen. These reactions occur on the lumenal face of the smooth endoplasmic reticulum. A similar pathway, but with different electron carriers, occurs in plants.

box 21–1 Mixed-Function Oxidases, Oxygenases, and Cytochrome P-450

In this chapter we encounter several enzymes that carry out oxidation-reduction reactions in which molecular oxygen is a participant. The reaction that introduces a double bond into a fatty acyl chain (Fig. 21–14) is such a reaction.

The nomenclature for enzymes that catalyze reactions of this general type is often confusing to students, as is the mechanism. **Oxidase** is the general name for enzymes that catalyze oxidations in which molecular oxygen is the electron acceptor but oxygen atoms do not appear in the oxidized product (however, there is an exception to this "rule," as we shall see!). The enzyme that creates a double bond in fatty acyl–CoA during the oxidation of fatty acids in peroxisomes (see Fig. 17–12) is an oxidase of this type, as is the cytochrome oxidase of the mitochondrial electron transfer chain (see Fig. 19–13). In the first case, the transfer of two electrons to H_2O produces hydrogen peroxide, H_2O_2; in the second, two electrons reduce $\frac{1}{2}O_2$ to H_2O. Many, but not all, oxidases are flavoproteins.

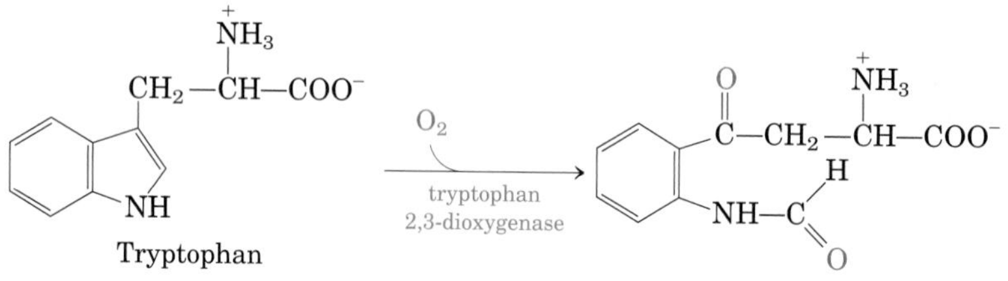

Oxygenases catalyze oxidative reactions in which oxygen atoms *are* directly incorporated into the substrate molecule, forming a new hydroxyl or carboxyl group, for example. **Dioxygenases** catalyze reactions in which both of the oxygen atoms of O_2 are incorporated into the organic substrate molecule. An example of a dioxygenase is tryptophan 2,3-dioxygenase, which catalyzes the opening of the five-membered ring of tryptophan in the catabolism of this amino acid: When this reaction occurs in the presence of $^{18}O_2$, the isotopic oxygen atoms are found in the two carbonyl groups of the product (shown in red).

Tryptophan $\xrightarrow[\text{tryptophan 2,3-dioxygenase}]{O_2}$ *N*-Formylkynurenine

Monooxygenases, which are more abundant and more complex in their action, catalyze reactions in which only one of the two oxygen atoms of O_2 is incorporated into the organic substrate, the other being reduced to H_2O. Monooxygenases require two substrates to serve as reductants of the two oxygen atoms of O_2. The main substrate accepts one of the two oxygen atoms, and a cosubstrate furnishes hydrogen atoms to reduce the other oxygen atom to H_2O. The general reaction equation for monooxygenases is

$$AH + BH_2 + O{-}O \longrightarrow A{-}OH + B + H_2O$$

where AH is the main substrate and BH_2 the cosubstrate. Because most monooxygenases catalyze reactions in which the main substrate becomes hydroxylated, they are also called **hydroxylases.** They are also sometimes called **mixed-function oxidases** or **mixed-function oxygenases,** to indicate that they oxidize two different substrates simultaneously. (Note here the use of "oxidase"—a deviation from the general meaning of this term.)

neously undergo two-electron oxidations. The path of electron flow includes a cytochrome (cytochrome b_5) and a flavoprotein (cytochrome b_5 reductase), both of which, like fatty acyl–CoA desaturase, are present in the smooth endoplasmic reticulum. In plants, oleate is produced by a stearoyl-ACP desaturase in the chloroplast stroma that uses reduced ferredoxin as the electron donor.

Mammalian hepatocytes can readily introduce double bonds at the Δ^9 position of fatty acids but cannot introduce additional double bonds in the fatty acid chain between C-10 and the methyl-terminal end. Linoleate, 18:2($\Delta^{9,12}$), and α-linolenate, 18:3($\Delta^{9,12,15}$), cannot be synthesized by mam-

There are different classes of monooxygenases, depending upon the nature of the cosubstrate. Some use reduced flavin nucleotides ($FMNH_2$ or $FADH_2$), others use NADH or NADPH, and still others use α-ketoglutarate as the cosubstrate. The enzyme that hydroxylates the phenyl ring of phenylalanine to give tyrosine is a monooxygenase for which tetrahydrobiopterin serves as cosubstrate (see Fig. 18–22). This is the enzyme that is defective in the human genetic disease phenylketonuria.

The most numerous and most complex monooxygenation reactions are those employing a type of heme protein called **cytochrome P-450.** This type of cytochrome is usually present in the smooth endoplasmic reticulum rather than the mitochondria. Like mitochondrial cytochrome oxidase, cytochrome P-450 can react with O_2 and binds carbon monoxide, but it can be differentiated from cytochrome oxidase because the carbon monoxide complex of its reduced form absorbs light strongly at 450 nm; thus the name P-450.

Cytochrome P-450 catalyzes hydroxylation reactions in which an organic substrate RH is hydroxylated to R—OH, incorporating one oxygen atom of O_2; the other oxygen atom is reduced to H_2O by reducing equivalents furnished by NADH or NADPH, but usually passed to P-450 by an iron-sulfur protein. Figure 1 shows a simplified outline of the action of cytochrome P-450, which has intermediate steps not yet fully understood.

Cytochrome P-450 is actually a family of closely similar proteins; several hundred members of this protein family are known, each with a different substrate specificity. In the adrenal cortex, for example, a specific cytochrome P-450 participates in the hydroxylation of steroids to yield the adrenocortical hormones (Fig. 21–44). Cytochrome P-450 is also important in the hydroxylation of many different drugs, such as barbiturates and other xenobiotics (substances that are foreign to the body), particularly if they are hydrophobic and relatively insoluble. The environmental carcinogen benzo[*a*]pyrene (found in cigarette smoke) undergoes cytochrome P-450–dependent hydroxylation during detoxification. Hydroxylation of xenobiotics makes them more soluble in water and allows their excretion in the urine. Unfortunately, hydroxylation of some compounds converts them into toxic substances, subverting the detoxification system.

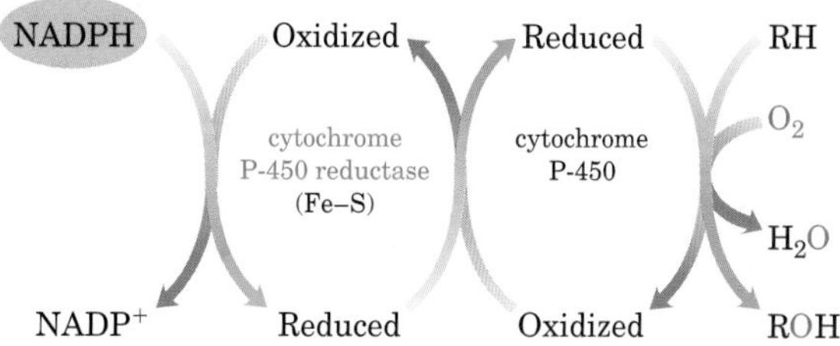

figure 1

Reactions described in this chapter that are catalyzed by mixed-function oxidases are those involved in fatty acyl–CoA desaturation (Fig. 21–14); leukotriene synthesis (Fig. 21–17), plasmalogen synthesis (Fig. 21–28), conversion of squalene to cholesterol (Fig. 21–35), and steroid hormone synthesis (Fig. 21–44).

mals, but plants can synthesize both. The plant desaturases that introduce double bonds at Δ^{12} and Δ^{15} positions are located in the endoplasmic reticulum and the chloroplast. The endoplasmic reticulum enzymes act not on free fatty acids but on a phospholipid, phosphatidylcholine, containing at least one oleate linked to glycerol (Fig. 21–15). Both plants and bacteria must synthesize polyunsaturated fatty acids to ensure membrane fluidity at reduced temperatures.

Because they are necessary precursors for the synthesis of other products, linoleate and linolenate are **essential fatty acids** for mammals; they must be obtained from dietary plant material. Once ingested, linoleate may

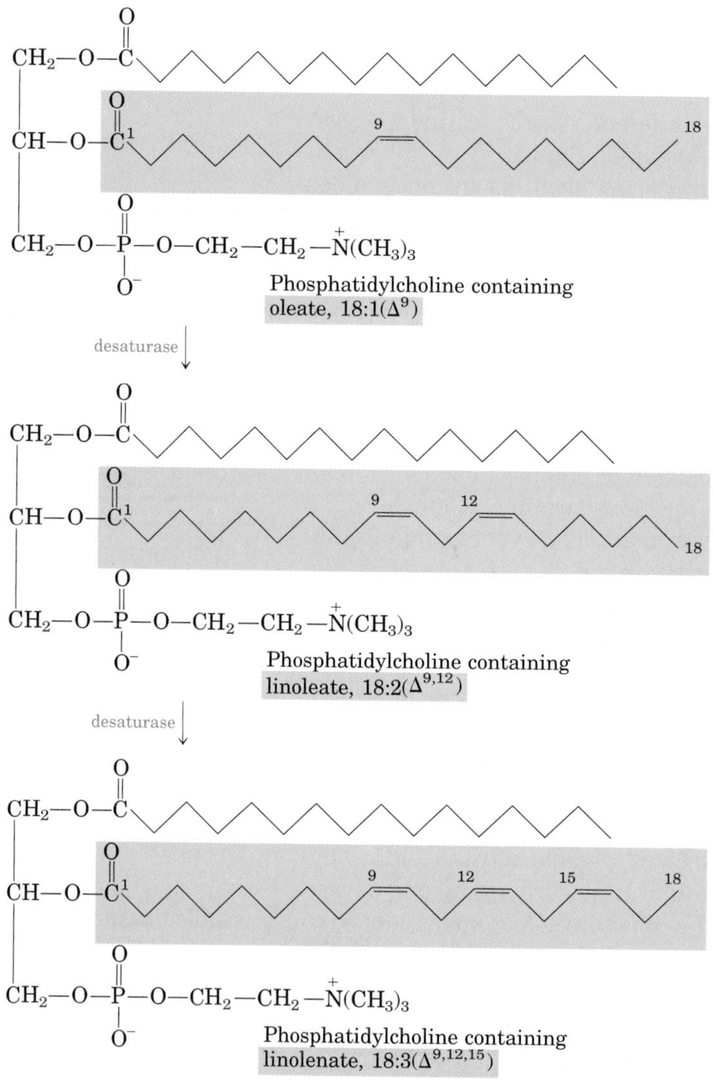

Phosphatidylcholine containing oleate, $18{:}1(\Delta^9)$

desaturase

Phosphatidylcholine containing linoleate, $18{:}2(\Delta^{9,12})$

desaturase

Phosphatidylcholine containing linolenate, $18{:}3(\Delta^{9,12,15})$

figure 21–15
Action of plant desaturases. Desaturases in plants oxidize phosphatidylcholine-bound oleate, producing polyunsaturated fatty acids. Some of the products are released from phosphatidylcholine by hydrolysis.

be converted into certain other polyunsaturated acids, particularly γ-linolenate, eicosatrienoate, and eicosatetraenoate (arachidonate), which can be made only from linoleate (Fig. 21–13). Arachidonate, $20{:}4(\Delta^{5,8,11,14})$, is an essential precursor of regulatory lipids, the eicosanoids. The 20-carbon fatty acids are synthesized from linoleate (and linolenate) by fatty acid elongation reactions analogous to those described on p. 781.

Eicosanoids Are Formed from 20-Carbon Polyunsaturated Fatty Acids

Eicosanoids are a family of very potent biological signaling molecules that act as short-range messengers, affecting tissues near the cells that produce them. In response to hormonal or other stimuli, phospholipase A_2, present in most types of mammalian cells, attacks membrane phospholipids, releasing arachidonate, $20{:}4(\Delta^{5,8,11,14})$ from the middle carbon of glycerol. Enzymes of the smooth endoplasmic reticulum then convert arachidonate into **prostaglandins,** beginning with the formation of PGH_2, the immediate precursor of many other prostaglandins and thromboxanes (Fig. 21–16a). The two reactions that lead to PGH_2 are catalyzed by a bifunctional enzyme, **cyclooxygenase (COX),** also called **prostaglandin H_2 synthase.** In the first of two steps, the cyclooxygenase activity introduces molecular oxygen to convert arachidonate into prostaglandin G_2 (PGG_2). The second step,

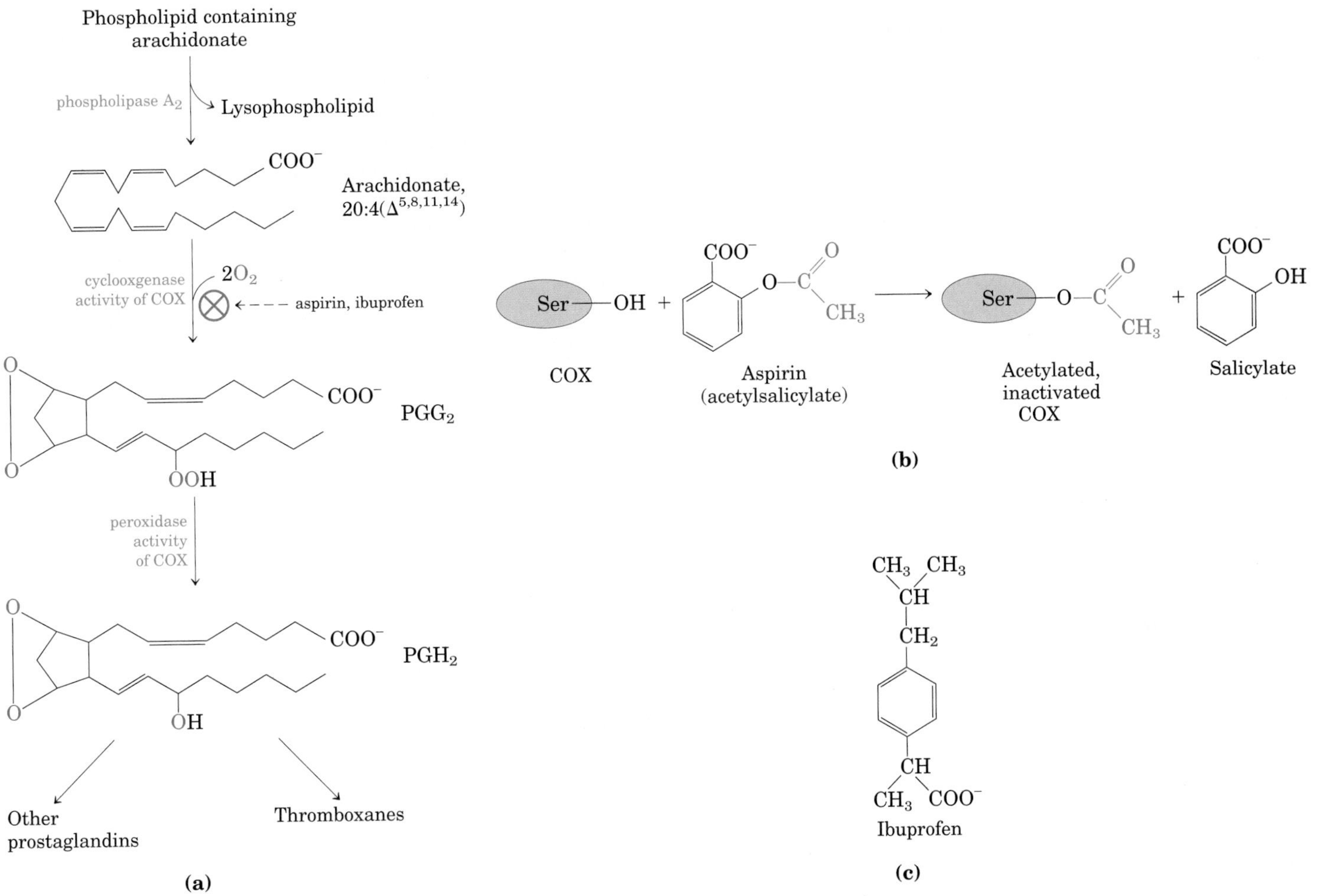

figure 21–16

The "cyclic" pathway from arachidonate to prostaglandins and thromboxanes. (a) After arachidonate is released by the action of phospholipase A_2, the cyclooxygenase and peroxidase activities of COX (also called prostaglandin H_2 synthase) catalyze the production of PGH_2, the precursor of other prostaglandins and thromboxanes. **(b)** Aspirin inhibits the first reaction by acetylating an essential Ser residue on the enzyme. **(c)** Ibuprofen inhibits the same step, probably by mimicking the structure of the substrate or an intermediate in the reaction.

catalyzed by the peroxidase activity of the bifunctional enzyme, converts PGG_2 into PGH_2. Aspirin (acetylsalicylate; Fig. 21–16b) irreversibly inactivates the cyclooxygenase activity by acetylating a Ser residue and blocking the enzyme's active site, thus inhibiting the synthesis of prostaglandins and thromboxanes. Ibuprofen, a widely used nonsteroidal antiinflammatory drug (NSAID; Fig. 21–16c), inhibits the same enzyme. The recent discovery that there are two isozymes of COX has led to the hope that more precisely targeted NSAIDs with fewer undesirable side effects may be developed (Box 21–2).

Thromboxane synthase present in blood platelets (thrombocytes) converts PGH_2 into thromboxane A_2, from which other **thromboxanes** are derived (Fig. 21–16a). Thromboxanes induce constriction of blood vessels and platelet aggregation, early steps in blood clotting. Low doses of aspirin, taken regularly, reduce the probability of heart attacks and strokes by reducing thromboxane production.

box 21–2 Cyclooxygenase Isozymes and the Search for a Better Aspirin: Relief is in (the Active) Site

Each year, several *thousand tons* of aspirin (acetylsalicylate) are consumed around the world for the relief of headaches, sore muscles, inflamed joints, and fever. Because aspirin inhibits platelet aggregation and blood clotting, it is also used at low doses to treat patients at risk of heart attacks. The medicinal properties of the compounds knowns as the salicylates, including aspirin, were first described by western science in 1763, when Edmund Stone of England noted that bark of the willow tree *Salix alba* was effective against fevers, aches, and pains. By the 1830s, German chemists had purified the active components from willow and from another plant rich in salicylates, the meadowsweet, *Spiraea ulmaria.* By the turn of the century, aspirin (from *a* for acetyl and *spir* from *Spirsaüre,* the German term for the acid prepared from *Spiraea*) was being used widely. Aspirin is one of a number of *n*onsteroidal *a*nti-*i*nflammatory *d*rugs, (NSAIDs); others include acetaminophen, ibuprofen, and naproxen (Figure 1), all now sold over the counter. Unfortunately aspirin can have serious side effects, including stomach bleeding, kidney failure and, in extreme cases, death. New NSAIDs with the beneficial effects of aspirin but without its side effects would be medically valuable.

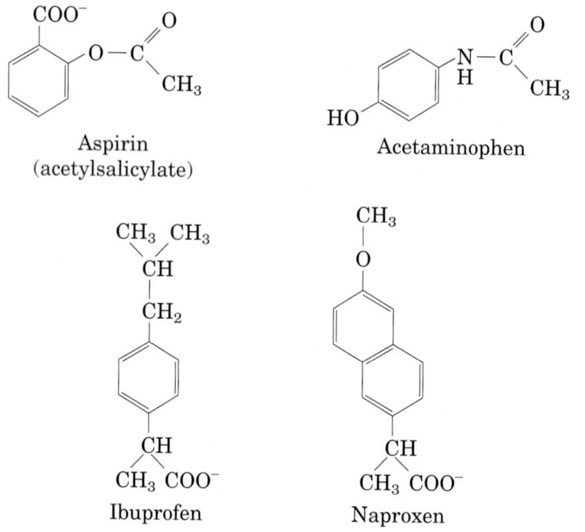

Aspirin (acetylsalicylate)

Acetaminophen

Ibuprofen

Naproxen

figure 1

Aspirin and other NSAIDs inhibit the cyclooxygenase activity of prostaglandin H_2 synthase (also called COX, for cyclooxygenase), which adds molecular oxygen to arachidonate to initiate prostaglandin synthesis (see Fig. 21–16a). Prostaglandins regulate many physiological processes including platelet aggregation, uterine contraction, pain, inflammation, and secretion of mucins that protect the gastric mucosa from acid and proteolytic enzymes in the stomach. The stomach irritation that is a common side effect of aspirin use results from its interference with the secretion of gastric mucin.

In mammals there are two isozymes of prostaglandin H_2 synthase, COX-1 and COX-2. These have different functions but closely similar amino acid sequences (60% to 65% sequence identity) and similar reaction mechanisms at both of their catalytic centers. COX-1 is responsible for the synthesis of the prostaglandins that regulate the secretion of gastric mucin, and COX-2 for the prostaglandins that mediate inflammation, pain, and fever. Aspirin inhibits both isozymes about equally, so a dose sufficient to reduce inflammation also risks stomach irritation. Much research is aimed at developing new NSAIDs that inhibit COX-2 specifically, and several such drugs are becoming available.

The development of COX-2-specific inhibitors has been helped immensely by knowledge of the detailed three-dimensional structure of COX-1 (Figure 2). The protein is a homodimer. Each monomer (M_r 70,000) has an amphipathic domain (orange) that penetrates but does not span

Thromboxanes, like prostaglandins, contain a ring of five or six atoms; the pathway that leads from arachidonate to these two classes of compounds is sometimes called the "cyclic" pathway, to distinguish it from the "linear" pathway that leads from arachidonate to the **leukotrienes,** which are linear (Fig. 21–17). Leukotriene synthesis begins with the action of several lipoxygenases that catalyze the incorporation of molecular oxygen into arachidonate. These enzymes, found in leukocytes and in heart, brain, lung,

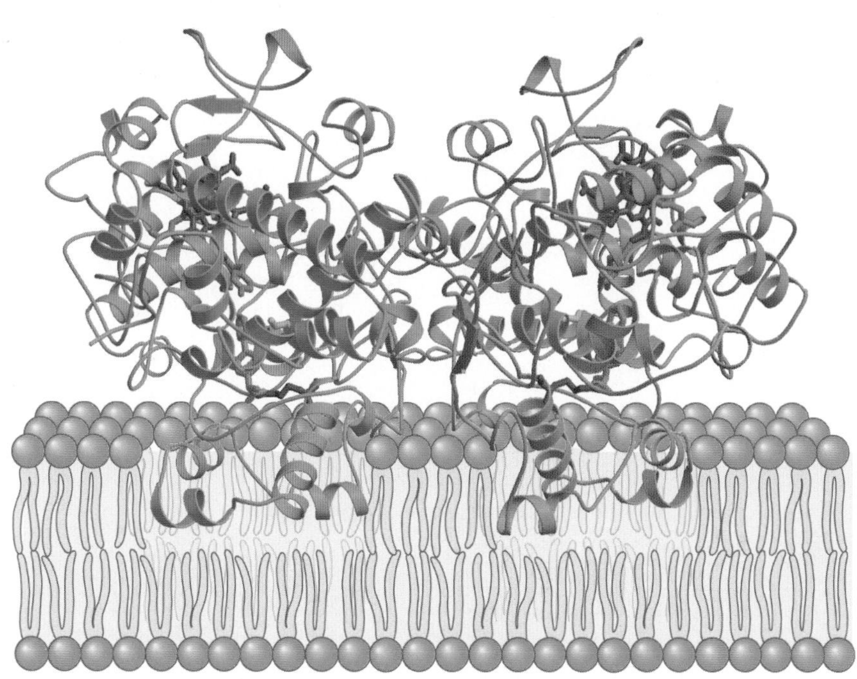

figure 2
Structure of COX-1, determined by x-ray crystallography. Two identical monomers (gray and blue) each have three domains: a membrane anchor consisting of four amphipathic helices, a second domain that somewhat resembles a domain of the epidermal growth factor, and the catalytic domain, which contains the cyclooxygenase and peroxidase activities, as well as the hydrophobic channel in which the substrate (arachidonate) binds. The heme that is part of the peroxidase active site is in red, and a key residue in the cyclooxygenase site, Tyr^{385}, is in turquoise. Other catalytically important residues include Arg^{120} (dark blue), His^{388} (green), and Ser^{530} (yellow). An inhibitory nonsteroidal antiinflammatory drug (flurbiprofin; orange) clearly blocks access to the substrate binding site.

the endoplasmic reticulum membrane; this anchors the enzyme on the lumenal side of the ER (a very unusual topology—generally the hydrophobic regions of integral membrane proteins span the entire bilayer). Both catalytic sites are on the globular domain (blue) protruding into the ER lumen.

COX-1 and COX-2 have virtually identical tertiary and quaternary structures, but they differ subtly in a long thin hydrophobic channel from the membrane interior to the lumenal surface. The channel includes both of the catalytic sites and is presumed to be the binding site for the hydrophobic substrate, arachidonate. By crystallizing COX-1 in the presence of an NSAID, flurbiprofen, it was possible to locate the NSAID-binding site. The bound drug blocks the hydrophobic channel and prevents arachidonate entry. The subtle differences between the channels of COX-1 and COX-2 may allow the design of NSAIDs that perfectly fit one channel but not the other, and therefore inhibit one enzyme more effectively than the other. Some potential drugs do discriminate between COX-1 and COX-2 by a factor of 1,000 in vitro. The use of precise structural information about an enzyme's active site is a powerful tool in the development of better, more specific drugs.

F
CH
CH_3 COO^-
Flurbiprofen

and spleen, are mixed-function oxidases that use cytochrome P-450 (Box 21–1). The various leukotrienes differ in the position of the peroxide group that is introduced by these lipoxygenases. This linear pathway from arachidonate, unlike the cyclic pathway, is not inhibited by aspirin or other NSAIDs.

Plants also derive important signaling molecules from fatty acids. As in animals, a key step in the initiation of signaling involves activation of a

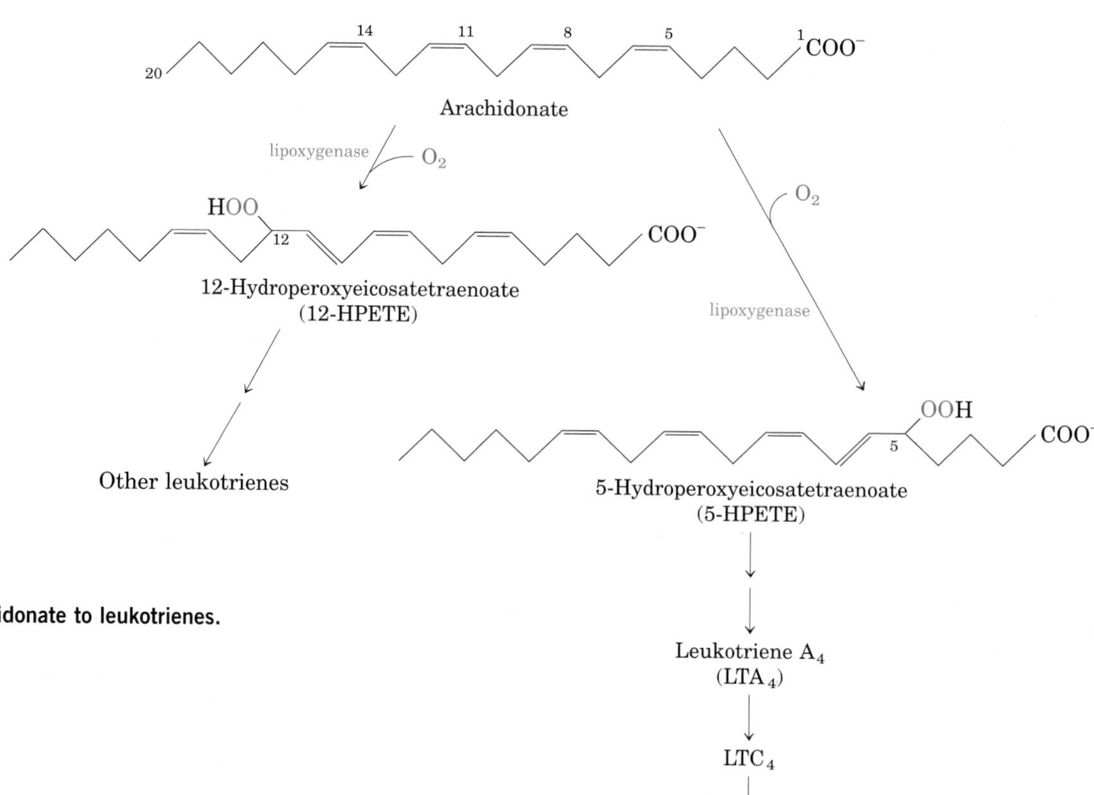

figure 21–17
The "linear" pathway from arachidonate to leukotrienes.

Jasmonate

specific phospholipase. In plants the fatty acid substrate that is released is linolenate. A lipoxygenase then catalyzes the first step in a pathway that converts linolenate (Fig. 21–13) to jasmonate. Specific signaling roles for jasmonate have been identified in insect defense, resistance to fungal pathogens, and in pollen maturation. Jasmonate also affects seed germination, root growth, and fruit and seed development.

Biosynthesis of Triacylglycerols

Most of the fatty acids synthesized or ingested by an organism have one of two fates: incorporation into triacylglycerols for the storage of metabolic energy or incorporation into the phospholipid components of membranes. The partitioning between these alternative fates depends on the needs of the organism. During rapid growth, synthesis of new membranes requires membrane phospholipid synthesis; organisms that have a plentiful supply of food but are not actively growing shunt most of their fatty acids into storage fats. Both pathways begin at the same point: the formation of fatty acyl esters of glycerol. We begin by examining the route to triacylglycerols and its regulation.

Triacylglycerols and Glycerophospholipids Are Synthesized from the Same Precursors

Animals can synthesize and store large quantities of triacylglycerols, to be used later as fuel (see Box 17–1). Humans can store only a few hundred grams of glycogen in liver and muscle cells, barely enough to supply the body's energy needs for 12 hours. In contrast, the total amount of stored triacylglycerol in a 70-kg man of average build is about 15 kg, enough to support basal energy needs for as long as 12 weeks (see Table 23–4). Triacylglycerols have the highest energy content of stored nutrients—more than ~38 kJ/g. Whenever carbohydrate is ingested in excess of the capacity to store glycogen, it is converted into triacylglycerols and stored in adipose tissue. Plants also manufacture triacylglycerols as an energy-rich fuel that is mainly stored in fruits, nuts, and seeds.

Triacylglycerols and glycerophospholipids such as phosphatidylethanolamine share two precursors (fatty acyl–CoAs and L-glycerol 3-phosphate) and several biosynthetic steps in animal tissues. Glycerol 3-phosphate can be formed in two ways (Fig. 21–18). It can arise from dihydroxyacetone phosphate generated during glycolysis by the action of the cytosolic NAD-linked **glycerol 3-phosphate dehydrogenase;** in liver and kidney it is also formed from glycerol by the action of **glycerol kinase.** The other precursors of triacylglycerols are fatty acyl–CoAs, formed from fatty acids by **acyl-CoA synthetases** (Fig. 21–18), the same enzymes responsible for the activation of fatty acids for β oxidation (see Fig. 17–5).

The first stage in the biosynthesis of triacylglycerols is the acylation of the two free hydroxyl groups of L-glycerol 3-phosphate by two molecules of fatty acyl–CoA to yield **diacylglycerol 3-phosphate,** more commonly called **phosphatidic acid** or phosphatidate (Fig. 21–18). Phosphatidic acid

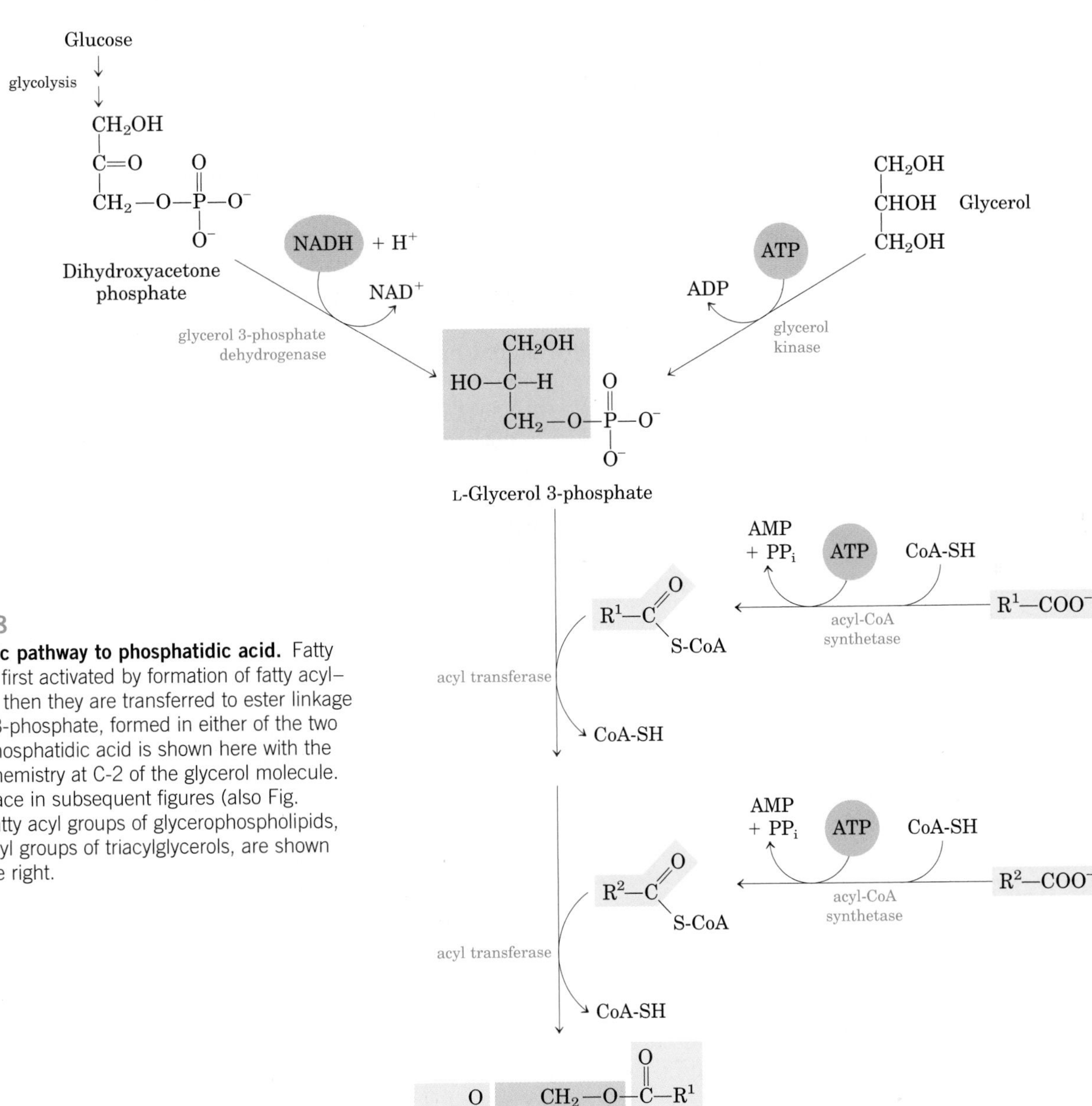

figure 21–18

The biosynthetic pathway to phosphatidic acid. Fatty acyl groups are first activated by formation of fatty acyl–CoA molecules, then they are transferred to ester linkage with L-glycerol 3-phosphate, formed in either of the two ways shown. Phosphatidic acid is shown here with the correct stereochemistry at C-2 of the glycerol molecule. To conserve space in subsequent figures (also Fig. 21–15), both fatty acyl groups of glycerophospholipids, and all three acyl groups of triacylglycerols, are shown projecting to the right.

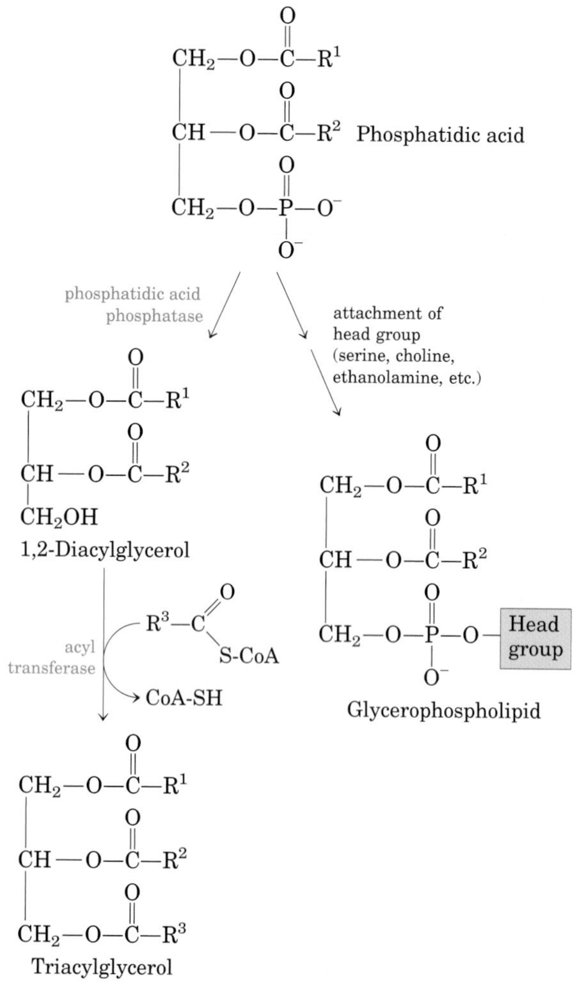

$CH_2—O—C(=O)—R^1$
$CH—O—C(=O)—R^2$ Phosphatidic acid
$CH_2—O—P(=O)(O^-)—O^-$

phosphatidic acid phosphatase

attachment of head group (serine, choline, ethanolamine, etc.)

$CH_2—O—C(=O)—R^1$
$CH—O—C(=O)—R^2$
CH_2OH
1,2-Diacylglycerol

acyl transferase
$R^3—C(=O)—S\text{-}CoA$
→ CoA-SH

$CH_2—O—C(=O)—R^1$
$CH—O—C(=O)—R^2$
$CH_2—O—C(=O)—R^3$
Triacylglycerol

$CH_2—O—C(=O)—R^1$
$CH—O—C(=O)—R^2$
$CH_2—O—P(=O)(O^-)—O—$ Head group
Glycerophospholipid

figure 21–19
Phosphatidic acid in lipid biosynthesis. Phosphatidic acid is the precursor of both triacylglycerols and glycerophospholipids. The mechanisms for head group attachment in phospholipid synthesis are described later.

occurs in only trace amounts in cells, but is a central intermediate in lipid biosynthesis; it can be converted either to a triacylglycerol or to a glycerophospholipid. In the pathway to triacylglycerols, phosphatidic acid is hydrolyzed by **phosphatidic acid phosphatase** to form a 1,2-diacylglycerol (Fig. 21–19). Diacylglycerols are then converted into triacylglycerols by transesterification with a third fatty acyl–CoA.

Triacylglycerol Biosynthesis in Animals Is Regulated by Hormones

In humans, the amount of body fat stays relatively constant over long periods, although there may be minor short-term changes as caloric intake fluctuates. Carbohydrate, fat, or protein consumed in excess of energy needs, is stored in the form of triacylglycerols that can be drawn upon for energy, enabling the body to withstand periods of fasting.

Biosynthesis and degradation of triacylglycerols are regulated reciprocally, with the favored path depending on the metabolic resources and requirements of the moment. The rate of triacylglycerol biosynthesis is profoundly altered by the action of several hormones. Insulin, for example, promotes the conversion of carbohydrate into triacylglycerols (Fig. 21–20). People with severe diabetes mellitus, due to failure of insulin secretion or action, are not only unable to use glucose properly but also fail to synthesize fatty acids from carbohydrates or amino acids. They show increased rates of fat oxidation and ketone body formation (Chapter 17) and therefore lose weight. Triacylglycerol metabolism is also influenced by glucagon (Chapter 23) and by both pituitary growth hormone and adrenal cortical hormones.

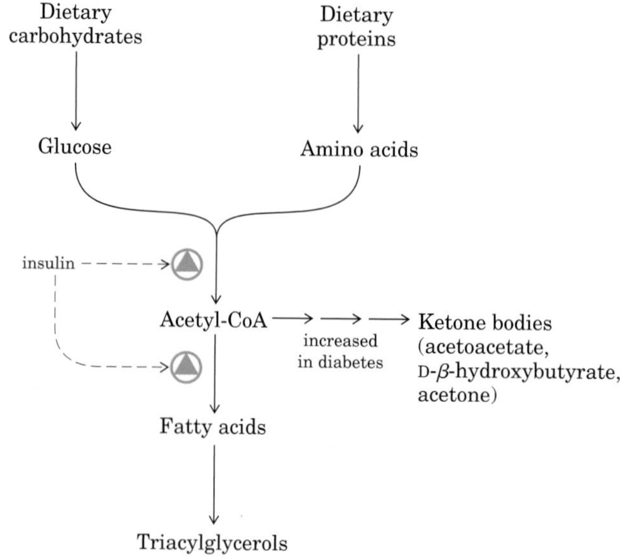

figure 21–20
Regulation of triacylglycerol synthesis by insulin. Insulin stimulates conversion of dietary carbohydrates and proteins into fat. Individuals with untreated diabetes mellitus lack insulin; this results in diminished fatty acid synthesis, and acetyl-CoA arising from catabolism of carbohydrates and proteins is shunted instead to ketone body production. People in severe ketosis smell of acetone, so the condition is sometimes mistaken for drunkenness (see p. 883).

Biosynthesis of Membrane Phospholipids

In Chapter 11 we introduced two major classes of membrane phospholipids: glycerophospholipids and sphingolipids. Many different phospholipid species can be constructed by combining various fatty acids and polar head groups with the glycerol or sphingosine backbones (see Figs. 11–8, 11–10). The many end products of phospholipid biosynthesis are synthesized according to a few basic patterns, which are illustrated in this section. In general, the assembly of phospholipids from simple precursors requires: (1) synthesis of the backbone molecule (glycerol or sphingosine); (2) attachment of fatty acid(s) to the backbone in ester or amide linkage; (3) addition of a hydrophilic head group joined to the backbone through a phosphodiester linkage; and, in some cases, (4) alteration or exchange of the head group to yield the final phospholipid product.

In eukaryotic cells, phospholipid synthesis occurs primarily at the surface of the smooth endoplasmic reticulum and the inner membrane of the mitochondria. Some newly synthesized phospholipids remain there, but most are destined for other cellular locations. The process by which water-insoluble phospholipids move from the site of their synthesis to the point of their eventual function is not fully understood, but we will conclude by discussing some mechanisms that have emerged in recent years.

There Are Two Strategies for Attaching Head Groups

The first steps of glycerophospholipid synthesis are shared with the pathway to triacylglycerols (Fig. 21–19): two fatty acyl groups are esterified to C-1 and C-2 of L-glycerol 3-phosphate to form phosphatidic acid. Commonly but not invariably, the fatty acid at C-1 is saturated and that at C-2 is unsaturated. A second route to phosphatidic acid is the phosphorylation of a diacylglycerol by a specific kinase.

The polar head group of glycerophospholipids is attached through a phosphodiester bond, in which each of two alcoholic hydroxyls (one on the polar head group and one on C-3 of glycerol) forms an ester with phosphoric acid (Fig. 21–21). In the biosynthetic process, one of the hydroxyls is first activated by attachment of a nucleotide, cytidine diphosphate

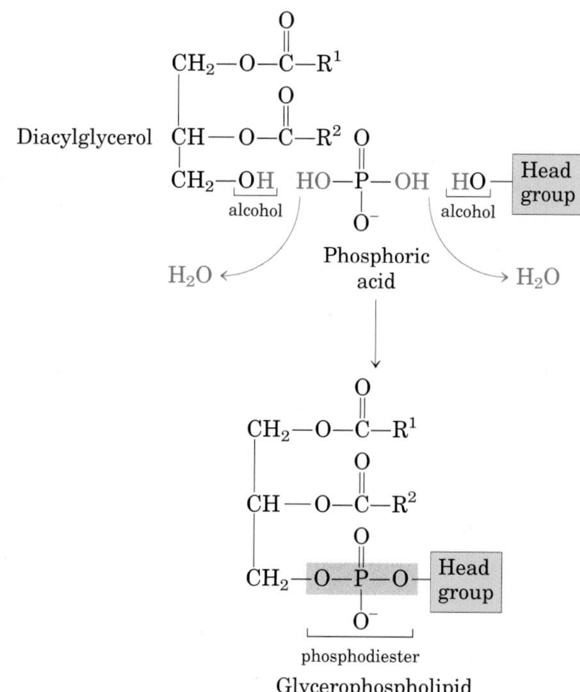

figure 21–21
The phospholipid head group is attached to a diacylglycerol by a phosphodiester bond, formed when phosphoric acid condenses with two alcohols, eliminating two molecules of H_2O.

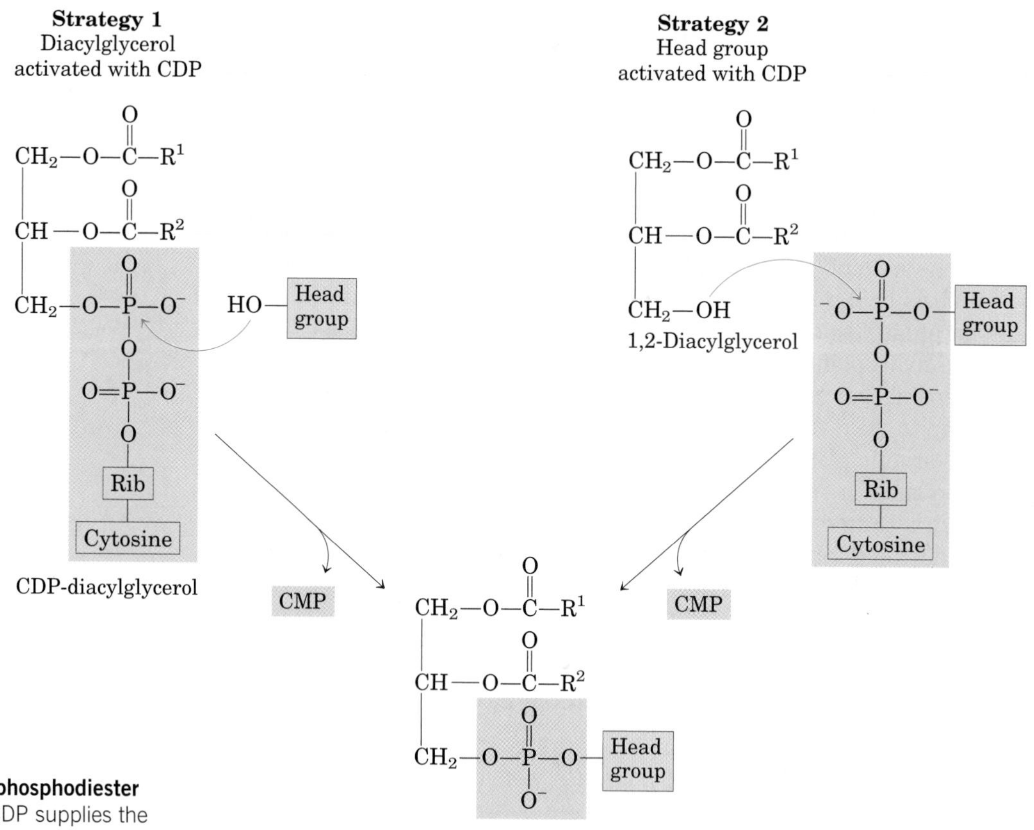

figure 21–22
Two general strategies for forming the phosphodiester bond of phospholipids. In both cases CDP supplies the phosphate group of the phosphodiester bond.

Eugene P. Kennedy

(CDP). Cytidine monophosphate (CMP) is then displaced in a nucleophilic attack by the other hydroxyl (Fig. 21–22). The CDP is attached either to the diacylglycerol, forming the activated phosphatidic acid **CDP-diacylglycerol** (strategy 1), or to the hydroxyl of the head group (strategy 2). Eukaryotic cells employ both strategies whereas prokaryotic cells use only the first. The central importance of cytidine nucleotides in lipid biosynthesis was discovered by Eugene P. Kennedy in the early 1960s.

Phospholipid Synthesis in *E. coli* Employs CDP-Diacylglycerol

The first strategy for head group attachment is illustrated by the synthesis of phosphatidylserine, phosphatidylethanolamine, and phosphatidylglycerol in *E. coli.* The diacylglycerol is activated by condensation of phosphatidic acid with cytidine triphosphate (CTP) to form CDP-diacylglycerol, with the elimination of pyrophosphate (Fig. 21–23). Displacement of CMP through nucleophilic attack by the hydroxyl group of serine or by the C-1 hydroxyl of glycerol 3-phosphate yields **phosphatidylserine** or phosphatidylglycerol 3-phosphate, respectively. The latter is processed further by cleavage of the phosphate monoester (with release of P_i) to yield **phosphatidylglycerol.**

Phosphatidylserine and phosphatidylglycerol can both serve as precursors of other membrane lipids in bacteria (Fig. 21–23). Decarboxylation of the serine moiety in phosphatidylserine by phosphatidylserine decarboxylase yields **phosphatidylethanolamine.** In *E. coli,* condensation of two molecules of phosphatidylglycerol, with the elimination of one glycerol, yields **cardiolipin,** in which two diacylglycerols are joined through a common head group.

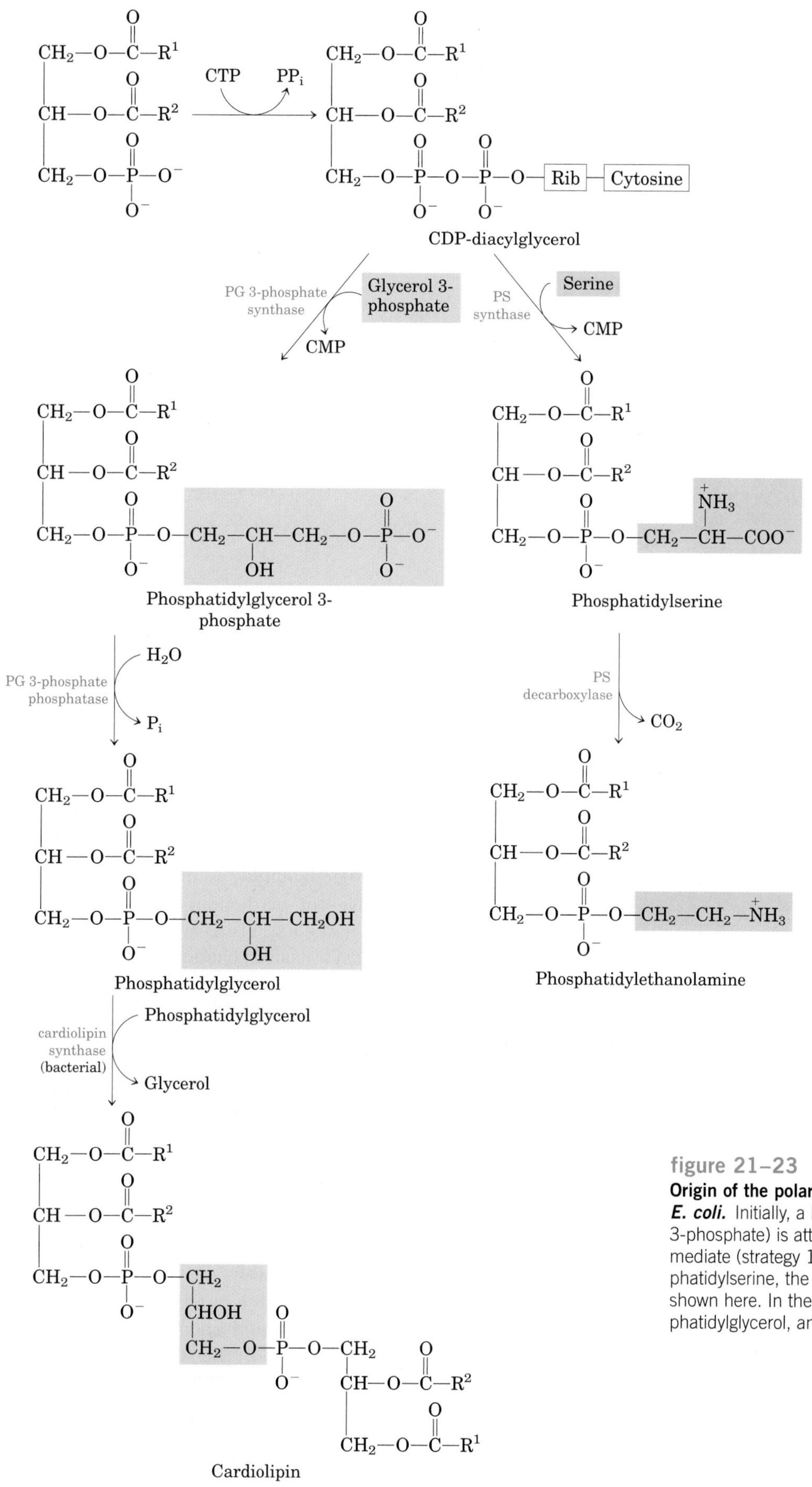

figure 21–23

Origin of the polar head groups of phospholipids in *E. coli.* Initially, a head group (either serine or glycerol 3-phosphate) is attached via a CDP-diacylglycerol intermediate (strategy 1). For phospholipids other than phosphatidylserine, the head group is further modified, as shown here. In the enzyme names, PG represents phosphatidylglycerol, and PS, phosphatidylserine.

figure 21–24

Synthesis of cardiolipin and phosphatidylinositol in eukaryotes. These glycerophospholipids are synthesized using strategy 1 in Figure 21–22. Phosphatidylglycerol is synthesized as in bacteria (see Fig. 21–23). PI represents phosphatidylinositol.

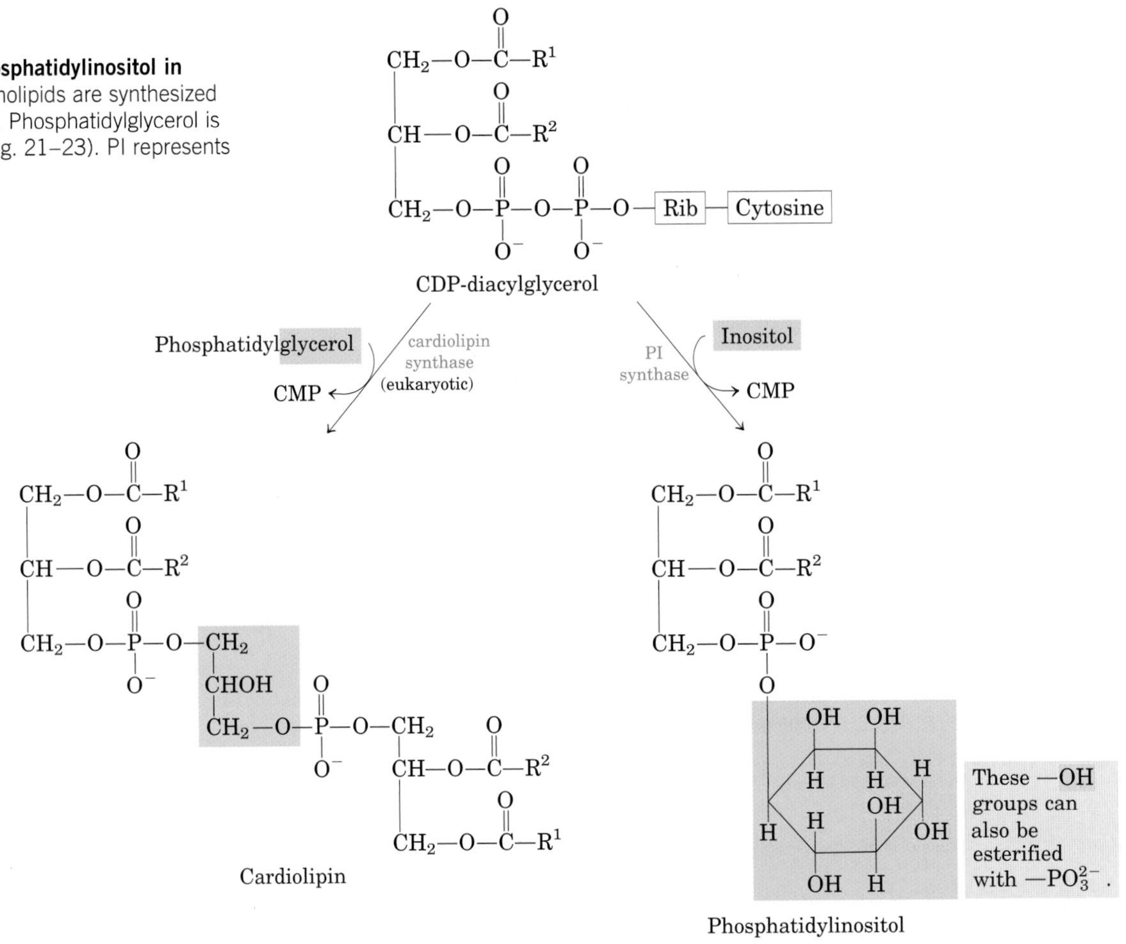

Eukaryotes Synthesize Anionic Phospholipids from CDP-Diacylglycerol

In eukaryotes, phosphatidylglycerol, cardiolipin, and the phosphatidylinositols (all anionic phospholipids; see Fig. 11–8) are synthesized by the same strategy used for phospholipid synthesis in bacteria. Phosphatidylglycerol is made exactly as in bacteria. Cardiolipin synthesis in eukaryotes differs slightly: phosphatidylglycerol condenses with CDP-diacylglycerol (Fig. 21–24), not another molecule of phosphatidylglycerol as in *E. coli* (Fig. 21–23).

Phosphatidylinositol is synthesized by condensation of CDP-diacylglycerol with inositol (Fig. 21–24). Specific **phosphatidylinositol kinases** then convert phosphatidylinositol into its phosphorylated derivatives (see Fig. 11–15). Phosphatidylinositol and its phosphorylated products in the plasma membrane play a central role in signal transduction in eukaryotes, as described in Figures 13–8 and 13–17.

Eukaryotic Pathways to Phosphatidylserine, Phosphatidylethanolamine, and Phosphatidylcholine Are Interrelated

In yeast as in bacteria, phosphatidylserine can be produced by condensation of CDP-diacylglycerol and serine, and phosphatidylethanolamine can be synthesized from phosphatidylserine in the reaction catalyzed by phosphatidylserine decarboxylase (Fig. 21–25). In mammalian cells, an alternative route to phosphatidylserine is a head group exchange reaction, in which free serine displaces ethanolamine. Phosphatidylethanolamine may also be converted to **phosphatidylcholine** (lecithin) by the addition of

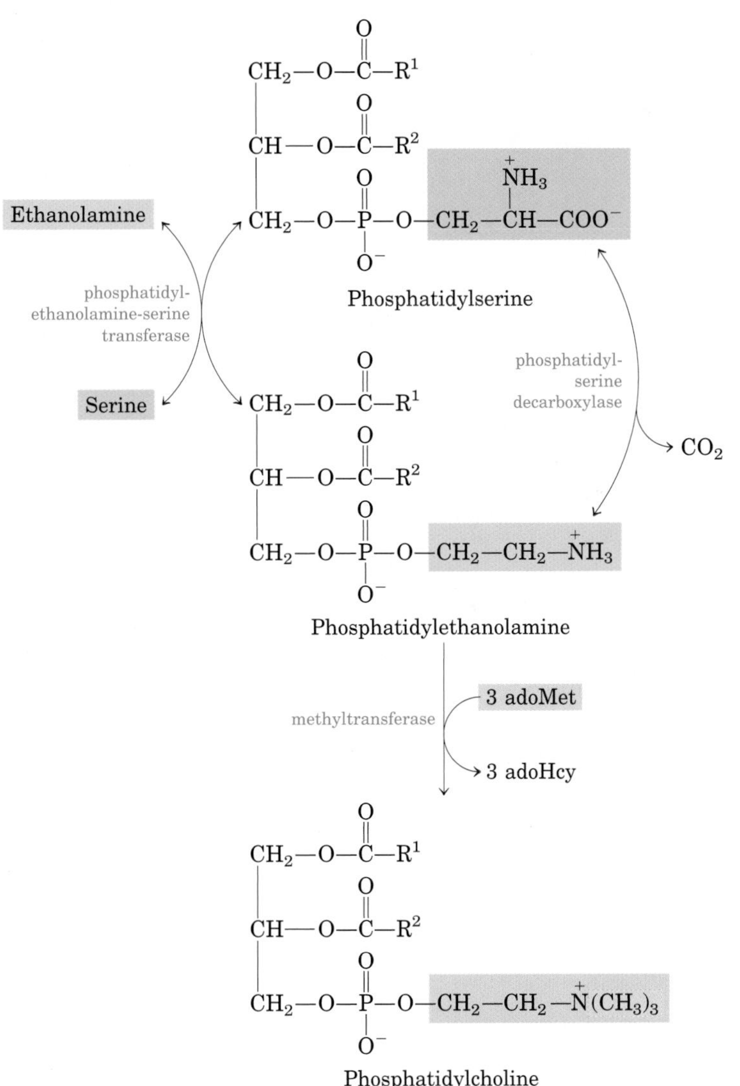

figure 21–25

The "salvage" pathway from phosphatidylserine to phosphatidylethanolamine and phosphatidylcholine in yeast. Phosphatidylserine and phosphatidylethanolamine are interconverted by a reversible head group exchange reaction. In mammals, phosphatidylserine is derived from phosphatidylethanolamine by a reversal of this reaction. adoMet is *S*-adenosylmethionine (see Fig. 18–17); adoHcy represent *S*-adenosylhomocysteine.

three methyl groups to its amino group. *S*-adenosylmethionine is the methyl group donor (see Fig. 18–17) for all three methylation reactions.

In mammals, phosphatidylserine is not synthesized from CDP-diacylglycerol; instead, it is derived from phosphatidylethanolamine via the head group exchange reaction (Fig. 21–25). Synthesis of phosphatidylethanolamine and phosphatidylcholine in mammals occurs by strategy 2 of Figure 21–22: phosphorylation and activation of the head group, followed by condensation with diacylglycerol. For example, choline is reused ("salvaged") by being phosphorylated, then converted into CDP-choline by condensation with CTP. A diacylglycerol displaces CMP from CDP-choline, producing phosphatidylcholine (Fig. 21–26). An analogous salvage pathway converts ethanolamine obtained in the diet into phosphatidylethanolamine. In the liver, phosphatidylcholine is also produced by methylation of phosphatidylethanolamine using *S*-adenosylmethionine (as described above) but in all other tissues, phosphatidylcholine is produced only by condensation of diacylglycerol and CDP-choline. The pathways to phosphatidylcholine and phosphatidylethanolamine in various organisms are summarized in Figure 21–27.

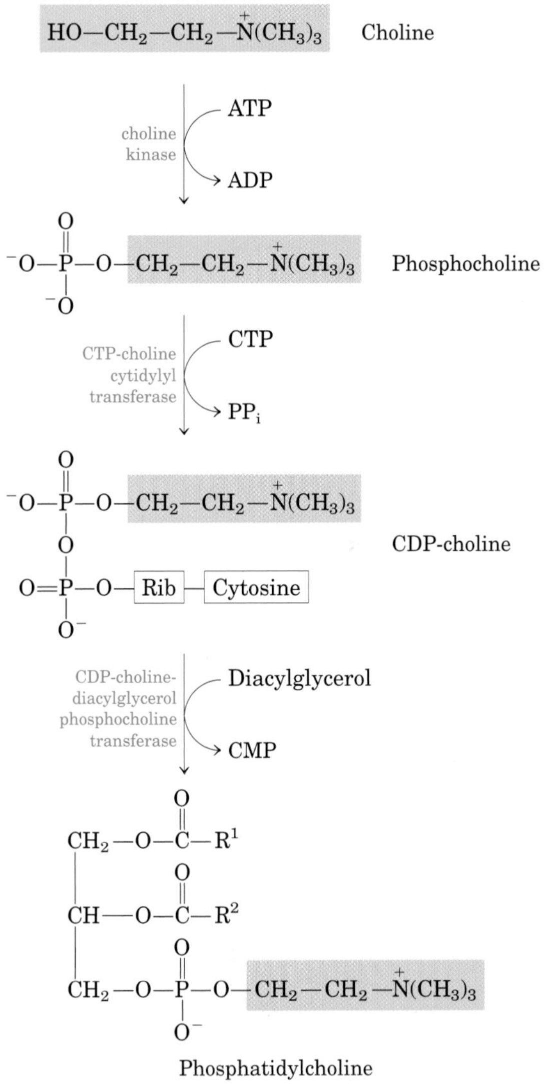

$HO{-}CH_2{-}CH_2{-}\overset{+}{N}(CH_3)_3$ Choline

choline kinase: ATP → ADP

$^{-}O{-}P(=O)(O^{-}){-}O{-}CH_2{-}CH_2{-}\overset{+}{N}(CH_3)_3$ Phosphocholine

CTP-choline cytidylyl transferase: CTP → PP_i

CDP-choline (Rib — Cytosine)

CDP-choline-diacylglycerol phosphocholine transferase: Diacylglycerol → CMP

Phosphatidylcholine

figure 21–26

The pathway for phosphatidylcholine synthesis from choline in mammals. The same strategy (strategy 2 in Figure 21–22) is used for salvaging ethanolamine in phosphatidylethanolamine synthesis.

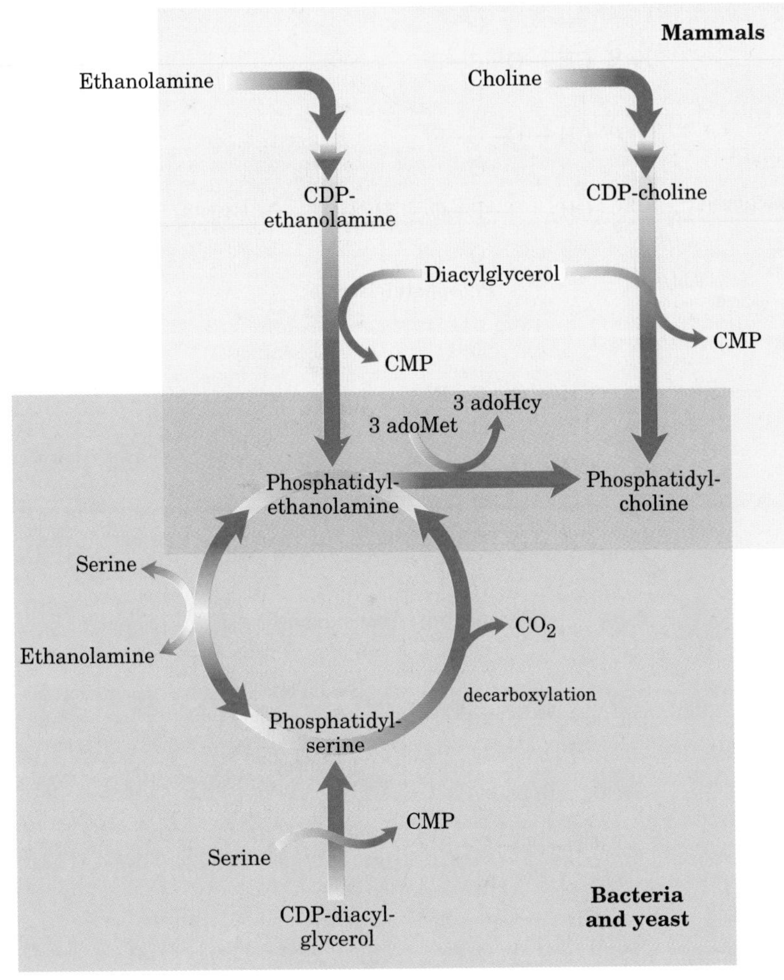

figure 21–27

Summary of the pathways to phosphatidylcholine and phosphatidylethanolamine. Conversion of phosphatidylethanolamine to phosphatidylcholine in mammals occurs only in the liver.

Although the role of lipid composition in membrane function is not entirely understood, dramatic effects of changes can be observed. Fruit flies have been isolated with mutations in the gene that encodes ethanolamine kinase (analogous to choline kinase, Fig. 21–26). Lack of this enzyme eliminates one pathway for phosphatidylethanolamine synthesis, thereby reducing its content in cellular membranes. Flies with the genotype *easily shocked* carry this mutation and exhibit transient paralysis following an electrical stimulation or mechanical shock.

Plasmalogen Synthesis Requires Formation of an Ether-Linked Fatty Alcohol

The biosynthetic pathway to ether lipids, including **plasmalogens** and the **platelet-activating factor** (see Fig. 11–9), involves the displacement of an esterified fatty acyl group by a long-chain alcohol to form the ether linkage (Fig. 21–28). Head group attachment follows, by mechanisms essentially like those for the common ester-linked phospholipids. Finally, the characteristic double bond of plasmalogens (shaded blue in Fig. 21–28) is introduced by the action of a mixed-function oxidase similar to that responsible for desaturation of fatty acids (Fig. 21–14). The peroxisome is a major site of plasmalogen synthesis.

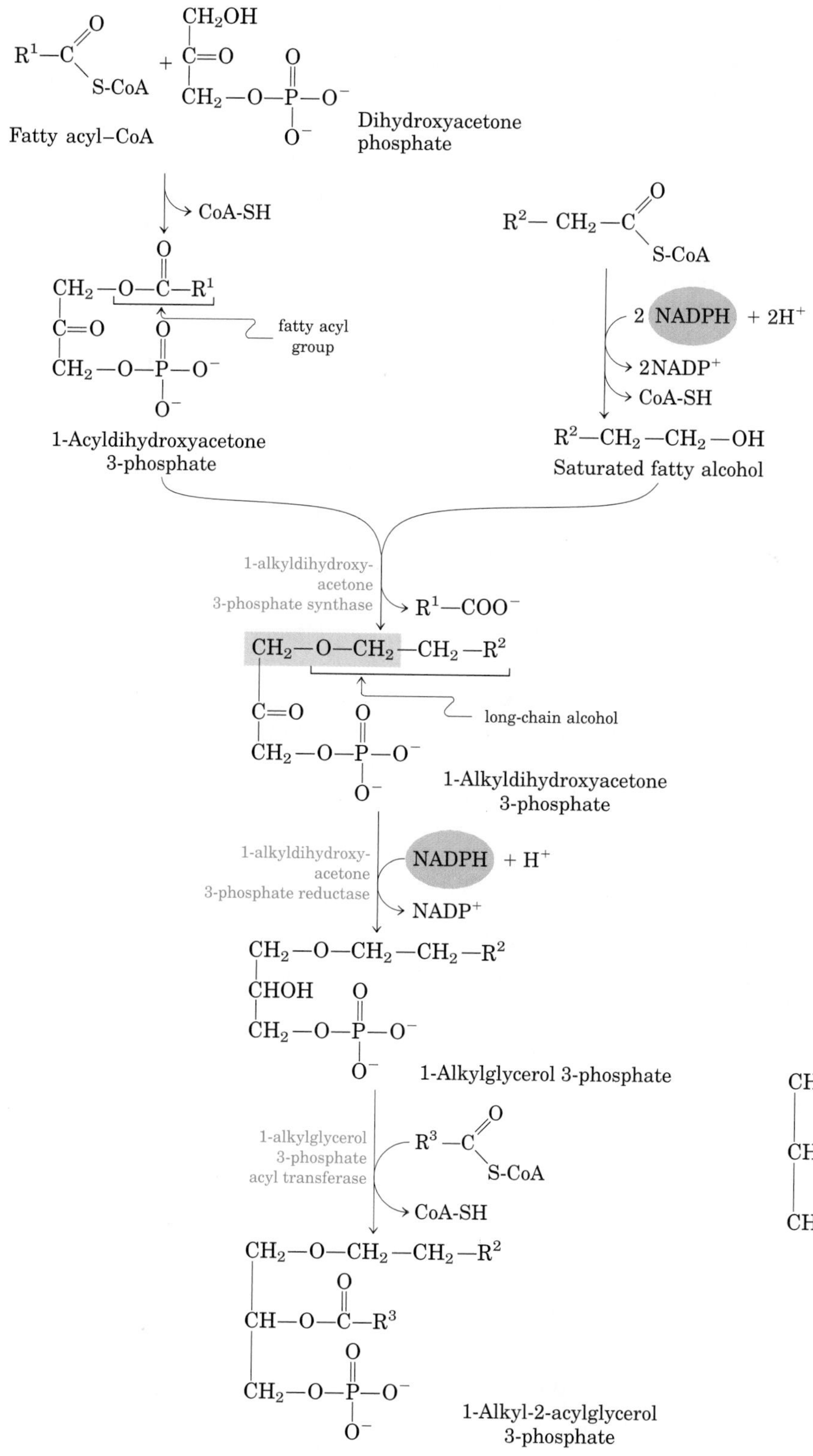

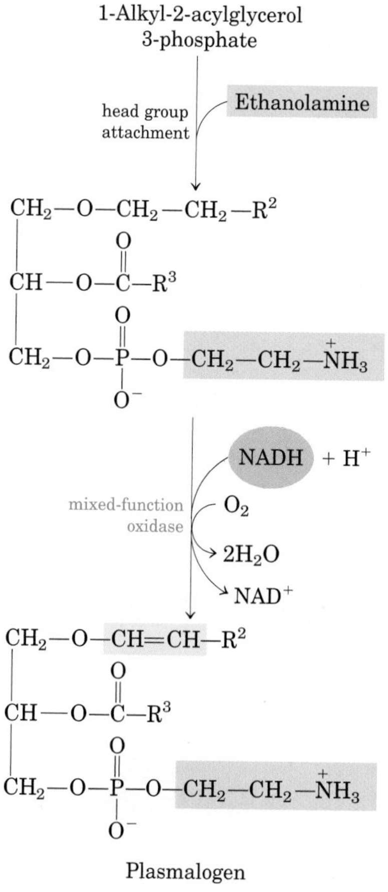

Fatty acyl–CoA + Dihydroxyacetone phosphate

→ CoA-SH

1-Acyldihydroxyacetone 3-phosphate (fatty acyl group)

$R^2-CH_2-C(=O)-S\text{-}CoA$ + 2 NADPH + 2H⁺ → 2NADP⁺ + CoA-SH

$R^2-CH_2-CH_2-OH$

Saturated fatty alcohol

1-alkyldihydroxyacetone 3-phosphate synthase → R^1-COO^-

$CH_2-O-CH_2-CH_2-R^2$ (long-chain alcohol)

1-Alkyldihydroxyacetone 3-phosphate

1-alkyldihydroxyacetone 3-phosphate reductase: NADPH + H^+ → $NADP^+$

1-Alkylglycerol 3-phosphate

1-alkylglycerol 3-phosphate acyl transferase: $R^3-C(=O)-S\text{-}CoA$ → CoA-SH

1-Alkyl-2-acylglycerol 3-phosphate

1-Alkyl-2-acylglycerol 3-phosphate

head group attachment: Ethanolamine

$CH_2-O-CH_2-CH_2-R^2$ / $CH-O-C(=O)-R^3$ / $CH_2-O-P(=O)(O^-)-O-CH_2-CH_2-\overset{+}{N}H_3$

mixed-function oxidase: NADH + H^+, O_2 → $2H_2O$, NAD^+

$CH_2-O-CH{=}CH-R^2$ / $CH-O-C(=O)-R^3$ / $CH_2-O-P(=O)(O^-)-O-CH_2-CH_2-\overset{+}{N}H_3$

Plasmalogen

figure 21–28

Synthesis of ether lipids and plasmalogens. The newly formed ether linkage is shaded red. The intermediate 1-alkyl-2-acylglycerol 3-phosphate is the ether analog of phosphatidic acid. Mechanisms for attaching head groups to ether lipids are essentially the same as for their ester-linked analogs. The characteristic double bond of plasmalogens (shaded blue) is introduced in a final step by a mixed-function oxidase system similar to that shown in Figure 21–14.

Sphingolipid and Glycerophospholipid Synthesis Share Precursors and Some Mechanisms

The biosynthesis of sphingolipids occurs in four stages: (1) synthesis of the 18-carbon amine **sphinganine** from palmitoyl-CoA and serine; (2) attachment of a fatty acid in amide linkage to yield ***N*-acylsphinganine;** (3) desaturation of the sphinganine moiety to form ***N*-acylsphingosine** (ceramide); and (4) attachment of a head group to produce a sphingolipid such as a **cerebroside** or **sphingomyelin** (Fig. 21–29). The pathway shares several features with the pathways leading to glycerophospholipids: NADPH provides reducing power and fatty acids enter as their activated CoA derivatives. In cerebroside formation, sugars enter as their activated nucleotide derivatives. Head group attachment in sphingolipid synthesis has several novel aspects. Phosphatidylcholine, rather than CDP-choline, serves as the donor of phosphocholine in the synthesis of sphingomyelin (Fig. 21–29). In glycolipids, the cerebrosides and **gangliosides** (see Fig. 11–10), the head group is a sugar attached directly to the C-1 hydroxyl of sphingosine in glycosidic linkage, rather than through a phosphodiester bond; the sugar donor is a UDP-sugar (UDP-glucose or UDP-galactose).

Polar Lipids Are Targeted to Specific Cell Membranes

After they are synthesized on the smooth endoplasmic reticulum, the polar lipids, including the glycerophospholipids, sphingolipids, and glycolipids, are inserted into different cell membranes in different proportions by mechanisms that are not yet understood. Membrane lipids are insoluble in water, so they cannot simply diffuse from their point of synthesis (the endoplasmic reticulum) to their point of insertion. Instead, they are delivered in membrane vesicles that bud from the Golgi complex and then move to and fuse with the target membrane (see Figs 2–9, 12–20). Cytosolic proteins also bind phospholipids and sterols and transport them between cell membranes and between faces of lipid bilayers. These mechanisms contribute to the establishment of the characteristic lipid compositions of organelle membranes (see Fig. 12–2).

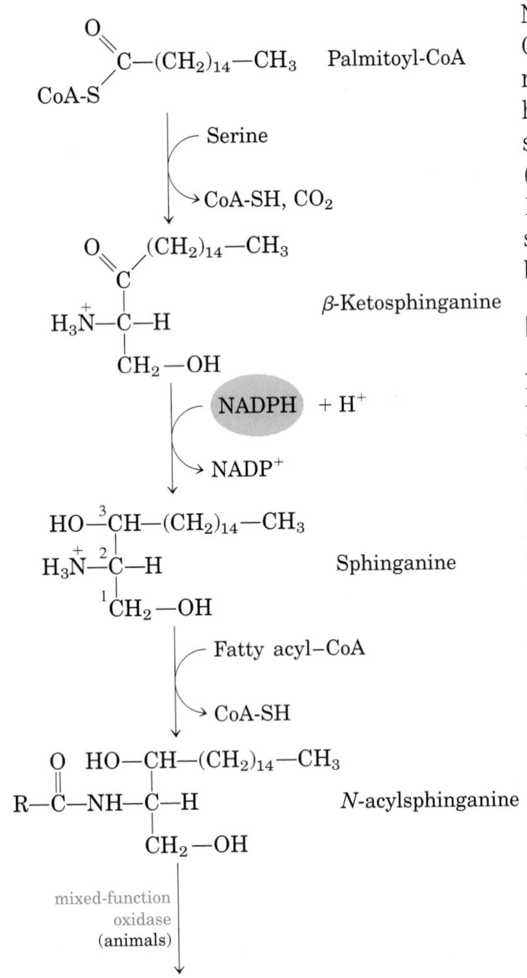

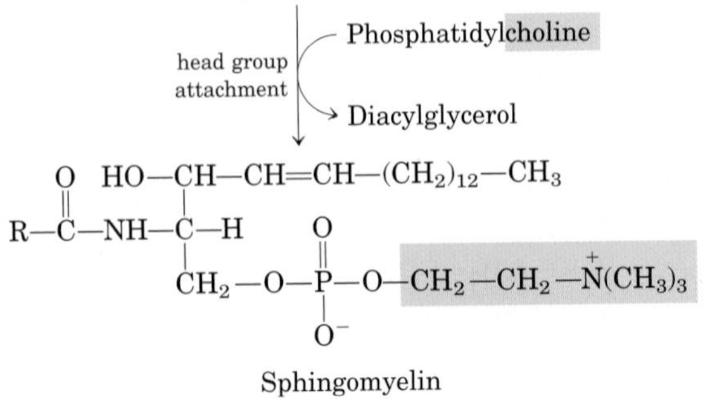

Palmitoyl-CoA: $CoA\text{-}S\text{-}C(=O)\text{-}(CH_2)_{14}\text{-}CH_3$

↓ Serine → $CoA\text{-}SH$, CO_2

β-Ketosphinganine: $O=C[(CH_2)_{14}\text{-}CH_3]\text{-}CH(\overset{+}{N}H_3)\text{-}CH_2\text{-}OH$

↓ NADPH + H^+ → $NADP^+$

Sphinganine: $HO\text{-}{}^3CH[(CH_2)_{14}\text{-}CH_3]\text{-}{}^2CH(\overset{+}{N}H_3)\text{-}{}^1CH_2\text{-}OH$

↓ Fatty acyl–CoA → CoA-SH

N-acylsphinganine: $R\text{-}C(=O)\text{-}NH\text{-}CH[CH(OH)\text{-}(CH_2)_{14}\text{-}CH_3]\text{-}CH_2\text{-}OH$

↓ mixed-function oxidase (animals)

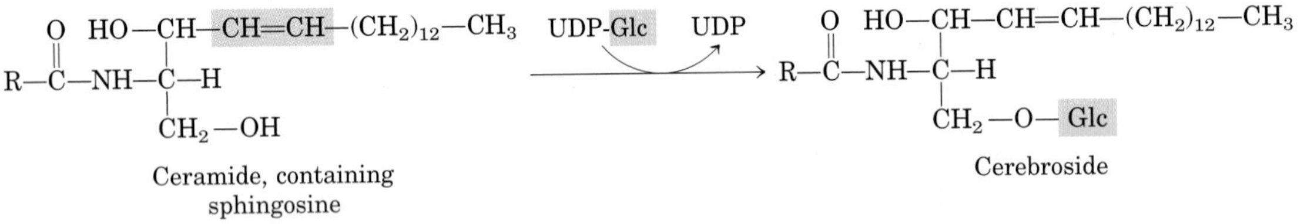

Ceramide, containing sphingosine: $R\text{-}C(=O)\text{-}NH\text{-}CH[CH(OH)\text{-}CH{=}CH\text{-}(CH_2)_{12}\text{-}CH_3]\text{-}CH_2\text{-}OH$

→ UDP-Glc → UDP →

Cerebroside: $R\text{-}C(=O)\text{-}NH\text{-}CH[CH(OH)\text{-}CH{=}CH\text{-}(CH_2)_{12}\text{-}CH_3]\text{-}CH_2\text{-}O\text{-}Glc$

↓ head group attachment: Phosphatidylcholine → Diacylglycerol

Sphingomyelin: $R\text{-}C(=O)\text{-}NH\text{-}CH[CH(OH)\text{-}CH{=}CH\text{-}(CH_2)_{12}\text{-}CH_3]\text{-}CH_2\text{-}O\text{-}P(=O)(O^-)\text{-}O\text{-}CH_2\text{-}CH_2\text{-}\overset{+}{N}(CH_3)_3$

figure 21–29
Biosynthesis of sphingolipids. Condensation of palmitoyl-CoA and serine followed by reduction with NADPH yields sphinganine, which is then acylated to form *N*-acylsphinganine (a ceramide). In animals, a double bond (shaded in red) is then created by a mixed-function oxidase, before the final addition of a head group: phosphatidylcholine, to form sphingomyelin; or glucose, to form a cerebroside.

Biosynthesis of Cholesterol, Steroids, and Isoprenoids

Cholesterol is doubtless the most publicized lipid, because of the strong correlation between high levels of cholesterol in the blood and the incidence of human cardiovascular diseases. Less well advertised is cholesterol's crucial role in the structure of many membranes and as a precursor of steroid hormones and bile acids. Cholesterol is an essential molecule in many animals, including humans, but is not required in the mammalian diet because all cells can synthesize it from simple precursors.

Although the structure of this 27-carbon compound suggests a complex biosynthetic pathway, all of its carbon atoms are provided by a single precursor—acetate (Fig. 21–30). Study of the pathway to cholesterol has led to an understanding of the transport of cholesterol and other lipids between organs, of the process by which cholesterol enters cells (receptor-mediated endocytosis), of the means by which intracellular cholesterol production is influenced by dietary cholesterol, and of how failure to regulate cholesterol production affects health. The **isoprene** units that are key intermediates in the pathway from acetate to cholesterol are precursors to many other natural lipids, and the mechanisms by which isoprene units are polymerized are similar in all of these pathways.

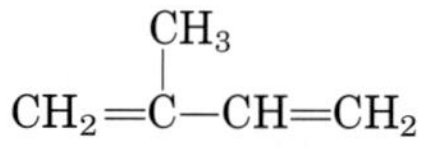

$$CH_2{=}C(CH_3){-}CH{=}CH_2$$

Isoprene

We begin with an account of the major steps in the biosynthesis of cholesterol from acetate, then discuss the transport of cholesterol in the blood, its uptake by cells, and the regulation of cholesterol synthesis in normal individuals and in those with defects in cholesterol uptake or transport. We then consider other cellular components derived from cholesterol, such as bile acids and steroid hormones. Finally, the biosynthetic pathways to some of the many compounds derived from isoprene units, which share early steps with the pathway to cholesterol, are outlined to illustrate the extraordinary versatility of isoprenoid condensations in biosynthesis.

Cholesterol Is Made from Acetyl-CoA in Four Stages

Cholesterol, like long-chain fatty acids, is made from acetyl-CoA, but the assembly plan is quite different in the two cases. In early experiments, animals were fed acetate labeled with ^{14}C in either the methyl carbon or the carboxyl carbon. The pattern of labeling in the cholesterol isolated from the two groups of animals (Fig. 21–30) provided the blueprint for working out the enzymatic steps in cholesterol biosynthesis.

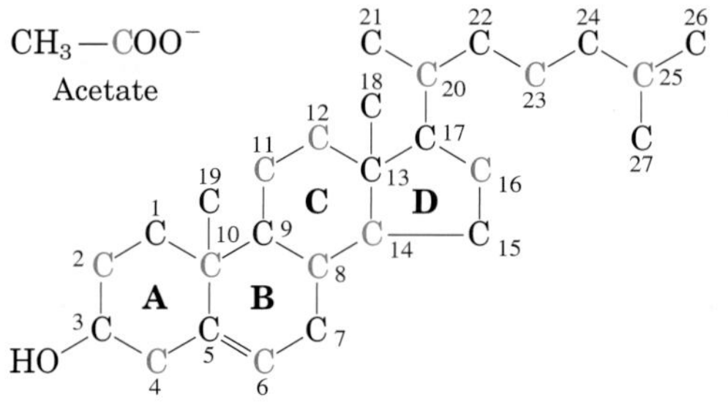

$CH_3{-}COO^-$

Acetate

Cholesterol

figure 21–30

The origin of the carbon atoms of cholesterol. This can be deduced from tracer experiments with acetate labeled in the methyl carbon (black) or the carboxyl carbon (red). The individual rings in cholesterol's fused-ring system are designated A through D.

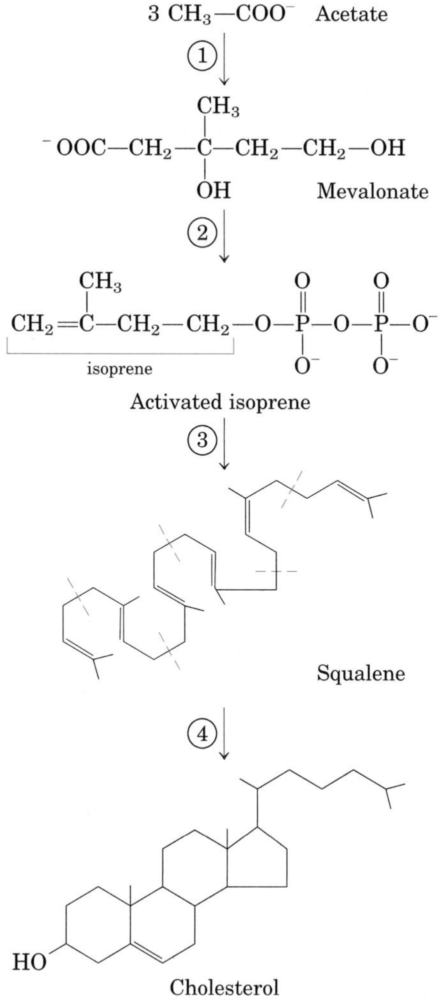

figure 21–31

A summary of cholesterol biosynthesis. The four stages are discussed in the text. Isoprene units in squalene are set off by red dashed lines.

3 $CH_3—COO^-$ Acetate

① ↓

$^-OOC—CH_2—C(CH_3)(OH)—CH_2—CH_2—OH$ Mevalonate

② ↓

$CH_2{=}C(CH_3)—CH_2—CH_2—O—P(O)(O^-)—O—P(O)(O^-)—O^-$

isoprene

Activated isoprene

③ ↓

Squalene

④ ↓

HO

Cholesterol

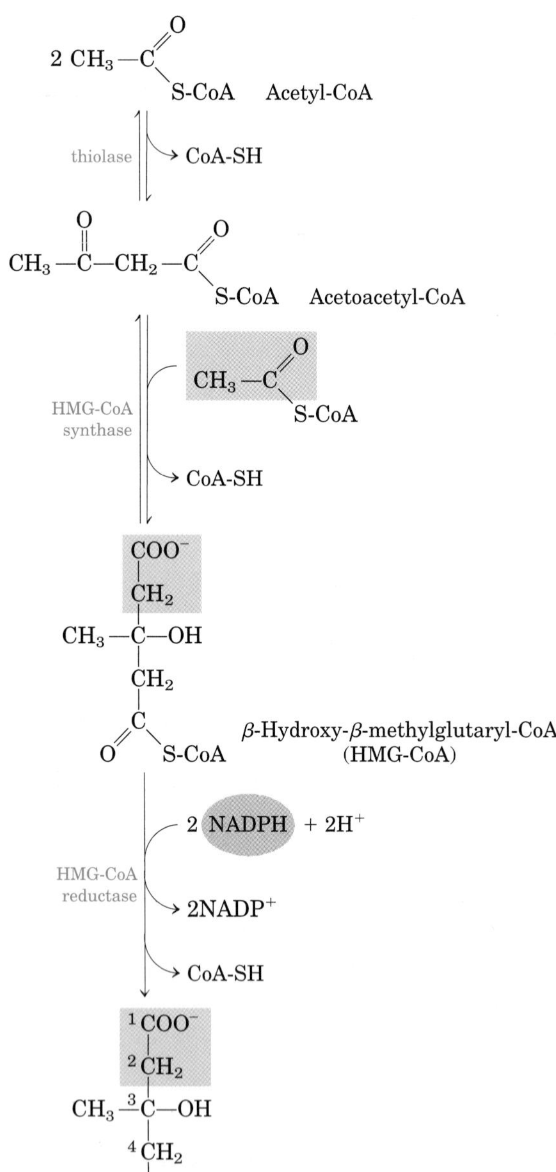

figure 21–32

Formation of mevalonate from acetyl-CoA. The origin of C-1 and C-2 of mevalonate from acetyl-CoA is shown in red.

The process occurs in four stages (Fig. 21–31). In stage ①, the three acetate units condense to form a six-carbon intermediate, mevalonate. Stage ② involves the conversion of mevalonate into activated isoprene units, and stage ③ the polymerization of six 5-carbon isoprene units to form the 30-carbon linear structure of squalene. Finally, in stage ④, the cyclization of squalene forms the four rings of the steroid nucleus, and a further series of changes (oxidations, removal or migration of methyl groups) leads to the final product, cholesterol.

Stage ① Synthesis of Mevalonate from Acetate The first stage in cholesterol biosynthesis leads to the intermediate **mevalonate** (Fig. 21–32). Two molecules of acetyl-CoA condense, forming acetoacetyl-CoA, which condenses with a third molecule of acetyl-CoA to yield the six-carbon compound ***β*-hydroxy-*β*-methylglutaryl-CoA (HMG-CoA).** These first two reactions, catalyzed by **thiolase** and **HMG-CoA synthase,** respectively, are reversible and do not commit the cell to the synthesis of cholesterol or other isoprenoid compounds. The cytosolic HMG-CoA synthase in this pathway is distinct from the mitochondrial isozyme that catalyzes HMG-CoA synthesis in ketone body formation (see Fig. 17–16).

The third reaction is the committed step: the reduction of HMG-CoA to mevalonate, for which two molecules of NADPH each donate two electrons. **HMG-CoA reductase,** an integral membrane protein of the smooth endoplasmic reticulum, is the major point of regulation on the pathway to cholesterol, as we shall see.

Stage ② Conversion of Mevalonate to Two Activated Isoprenes In the next stage of cholesterol synthesis, three phosphate groups are transferred from three ATP molecules to mevalonate (Fig. 21–33). The phosphate attached to the C-3 hydroxyl group of mevalonate in the intermediate 3-phospho-5-pyrophosphomevalonate is a good leaving group; in the next step this phosphate and the nearby carboxyl group both leave, producing a double bond in the five-carbon product, **Δ^3-isopentenyl pyrophosphate.** This is the first of the two activated isoprenes central to cholesterol formation. Isomerization of Δ^3-isopentenyl pyrophosphate yields the second activated isoprene, **dimethylallyl pyrophosphate** (Fig. 21–33). Synthesis of isopentenyl pyrophosphate in the cytoplasm of plant cells follows the pathway described here. However in many bacteria and the chloroplasts of plant cells, a mevalonate-independent pathway is used. This alternative pathway does not occur in animals, so it is an attractive target for the development of new antibiotics.

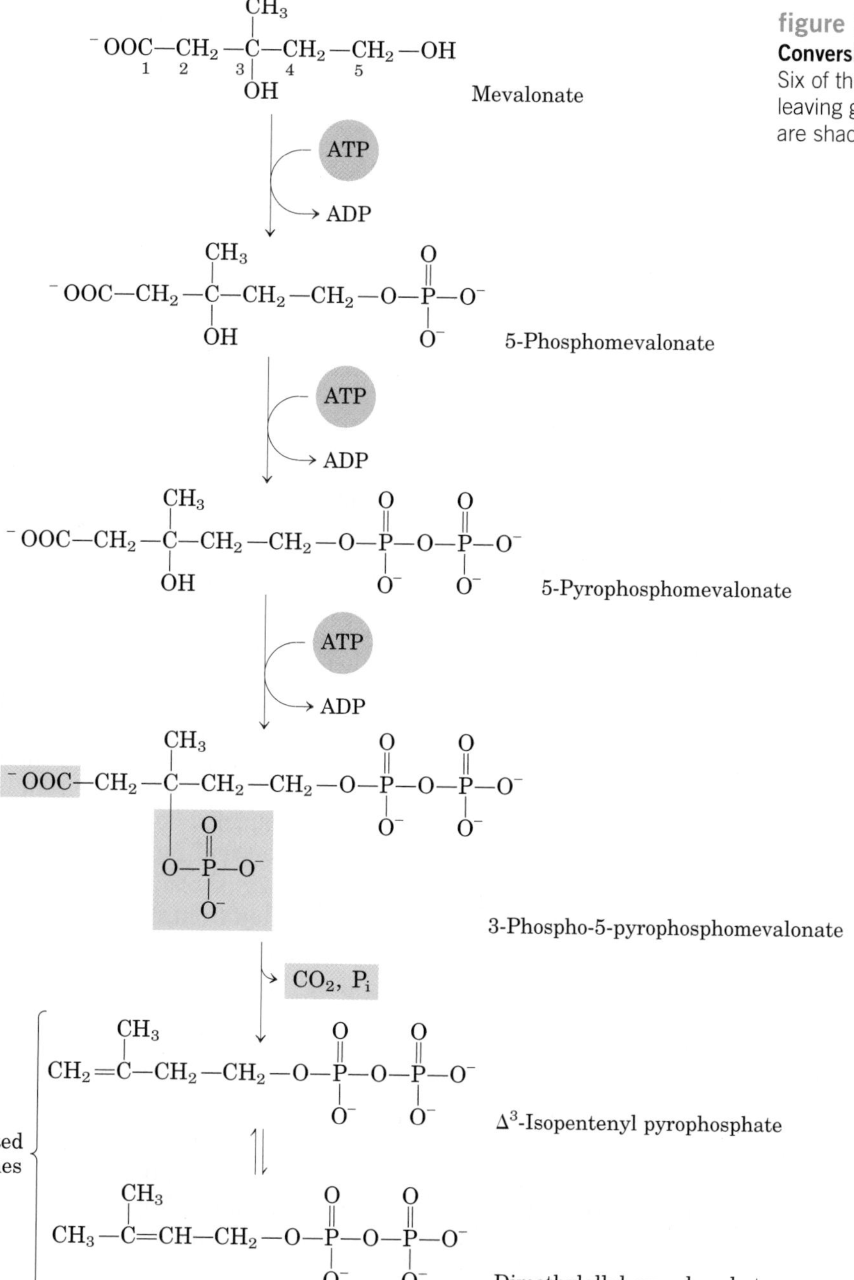

figure 21–33
Conversion of mevalonate into activated isoprene units. Six of these units will combine to form squalene. The leaving groups of 3-phospho-5-pyrophosphomevalonate are shaded in red.

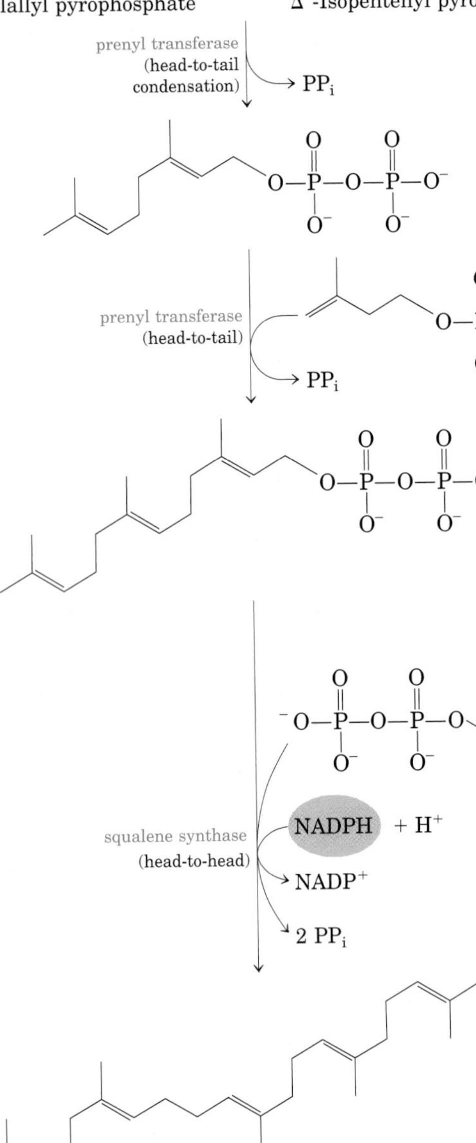

figure 21–34

Formation of squalene. This 30-carbon structure arises through successive condensations of activated isoprene (five-carbon) units.

Stage ③ Condensation of Six Activated Isoprene Units to Form Squalene Isopentenyl pyrophosphate and dimethylallyl pyrophosphate now undergo a head-to-tail condensation, in which one pyrophosphate group is displaced and a 10-carbon chain, **geranyl pyrophosphate,** is formed (Fig. 21–34). (The "head" is the end to which pyrophosphate is joined.) Geranyl pyrophosphate undergoes another head-to-tail condensation with isopentenyl pyrophosphate, yielding the 15-carbon intermediate **farnesyl pyrophosphate.** Finally, two molecules of farnesyl pyrophosphate join head to head, with the elimination of both pyrophosphate groups, forming **squalene** (Fig. 21–34). The common names of these compounds derive from the sources from which they were first isolated. Geraniol, a component of rose oil, has the smell of geraniums, and farnesol is a scent found in the flowers of the Farnese acacia tree. Many natural scents of plant origin are synthesized from isoprene units. Squalene, first isolated from the liver of sharks (genus *Squalus*), has 30 carbons, 24 in the main chain and 6 in the form of methyl group branches.

Stage ④ Conversion of Squalene to the Four-Ring Steroid Nucleus When the squalene molecule is represented as in Figure 21–35, the relationship of its linear structure to the cyclic structure of the sterols is apparent. All of the sterols have four fused rings (the steroid nucleus) and all are alcohols, with a hydroxyl group at C-3; thus the name "sterol." The action of **squalene monooxygenase** adds one oxygen atom from O_2 to the end of the squalene chain, forming an epoxide. This enzyme is another mixed-function oxidase (Box 21–1); NADPH reduces the other oxygen atom of O_2 to H_2O.

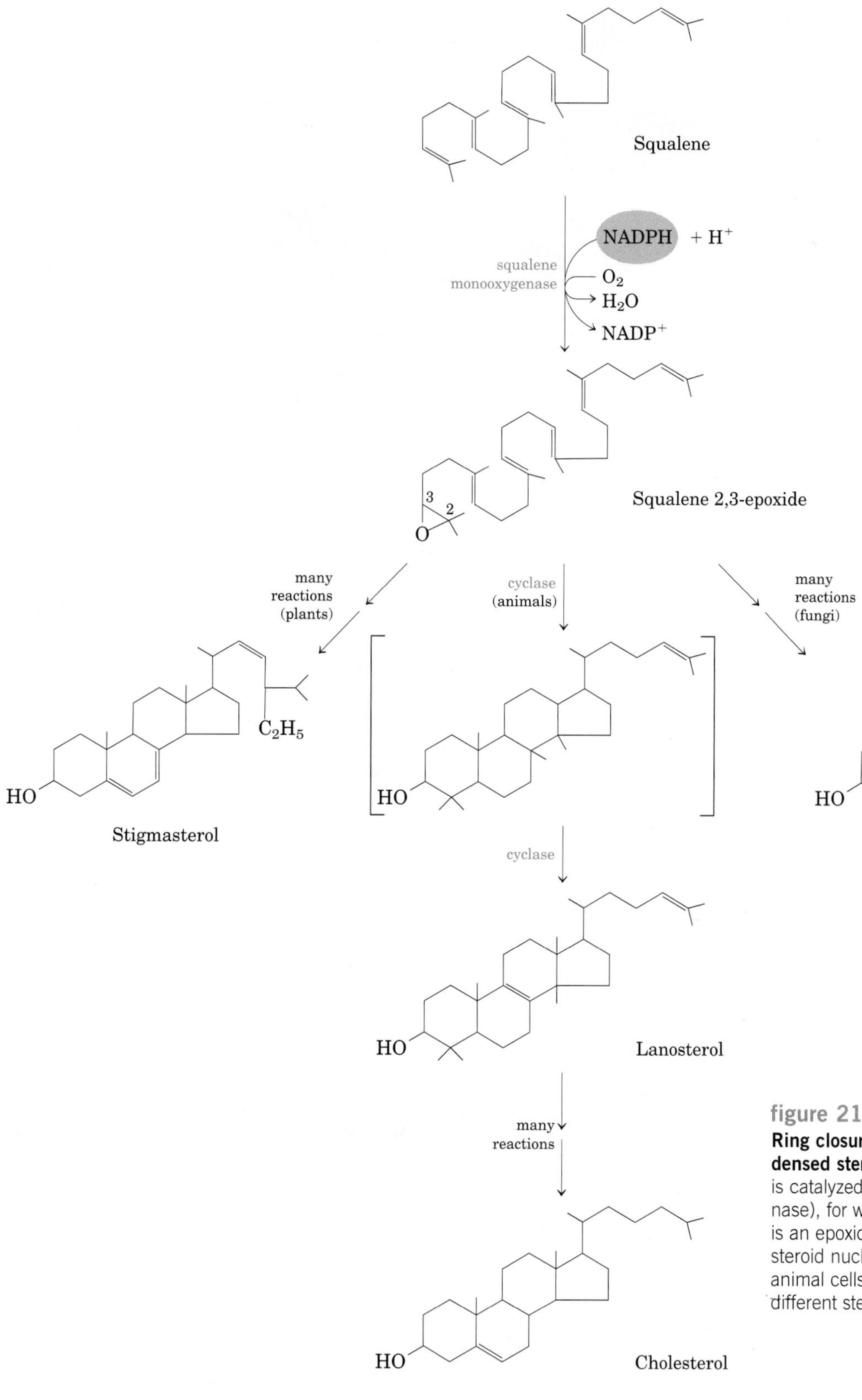

Squalene

squalene monooxygenase

NADPH + H^+

O_2

H_2O

$NADP^+$

Squalene 2,3-epoxide

many reactions (plants)

cyclase (animals)

many reactions (fungi)

C_2H_5

HO

Stigmasterol

HO

HO

Ergosterol

cyclase

HO

Lanosterol

many reactions

HO

Cholesterol

figure 21–35

Ring closure converts linear squalene into the condensed steroid nucleus. The first step in this sequence is catalyzed by a mixed-function oxidase (a monooxygenase), for which the cosubstrate is NADPH. The product is an epoxide, which in the next step is cyclized to the steroid nucleus. The final product of these reactions in animal cells is cholesterol; in other organisms, slightly different sterols are produced.

The double bonds of the product, **squalene 2,3-epoxide,** are positioned so that a remarkable concerted reaction can convert the linear squalene epoxide into a cyclic structure. In animal cells, this cyclization results in the formation of **lanosterol,** which contains the four rings characteristic of the steroid nucleus. Lanosterol is finally converted into cholesterol in a series of

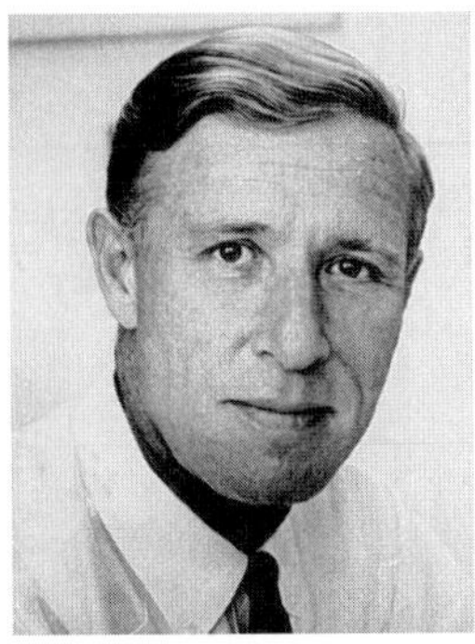

Konrad Bloch

Feodor Lynen
1911–1979

John Cornforth

George Popják

about 20 reactions that include the migration of some methyl groups and the removal of others. Elucidation of this extraordinary biosynthetic pathway, one of the most complex known, was accomplished by Konrad Bloch, Feodor Lynen, John Cornforth, and George Popják in the late 1950s.

Cholesterol is the sterol characteristic of animal cells; plants, fungi, and protists make other closely related sterols instead of cholesterol, using the same synthetic pathway as far as squalene 2,3-epoxide. At this point the synthetic pathways diverge slightly, yielding other sterols, such as stigmasterol in many plants and ergosterol in fungi (Fig. 21–35).

Cholesterol Has Several Fates

Much of the cholesterol synthesis in vertebrates takes place in the liver. A small fraction of the cholesterol made there is incorporated into the membranes of hepatocytes, but most of it is exported in one of three forms: biliary cholesterol, bile acids, or cholesteryl esters. **Bile acids** and their salts are relatively hydrophilic cholesterol derivatives that are synthesized in the liver and aid in lipid digestion (see Fig. 17–1). **Cholesteryl esters** are formed in the liver through the action of **acyl-CoA–cholesterol acyl transferase (ACAT).** This enzyme catalyzes the transfer of a fatty acid from coenzyme A to the hydroxyl group of cholesterol (Fig. 21–36), converting the cholesterol into a more hydrophobic form. Cholesteryl esters are transported in secreted lipoprotein particles to other tissues that use cholesterol, or are stored in the liver.

All growing animal tissues need cholesterol for membrane synthesis, and some organs (adrenal gland and gonads, for example) use cholesterol as a precursor for steroid hormone production (discussed later). Cholesterol is also a precursor of vitamin D (see Fig. 11–18a).

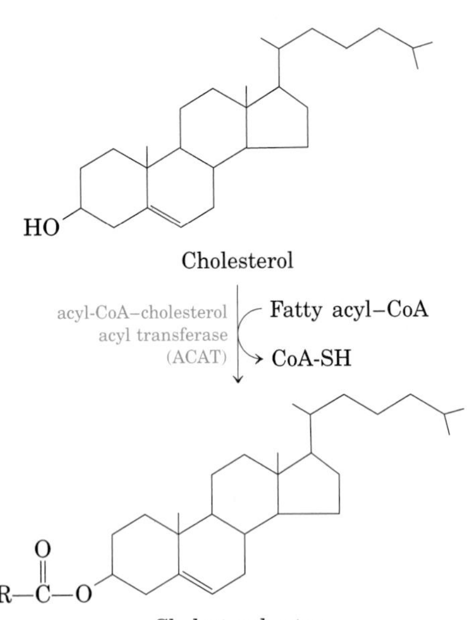

figure 21–36
Synthesis of cholesteryl esters. Esterification converts cholesterol into an even more hydrophobic form for storage and transport.

Cholesterol and Other Lipids Are Carried on Plasma Lipoproteins

Cholesterol and cholesteryl esters, like triacylglycerols and phospholipids, are essentially insoluble in water. These lipids must, however, be moved from the tissue of origin to the tissues in which they will be stored or consumed. They are carried in the blood plasma from one tissue to another as **plasma lipoproteins,** macromolecular complexes of specific carrier proteins called **apolipoproteins** with various combinations of phospholipids, cholesterol, cholesteryl esters, and triacylglycerols.

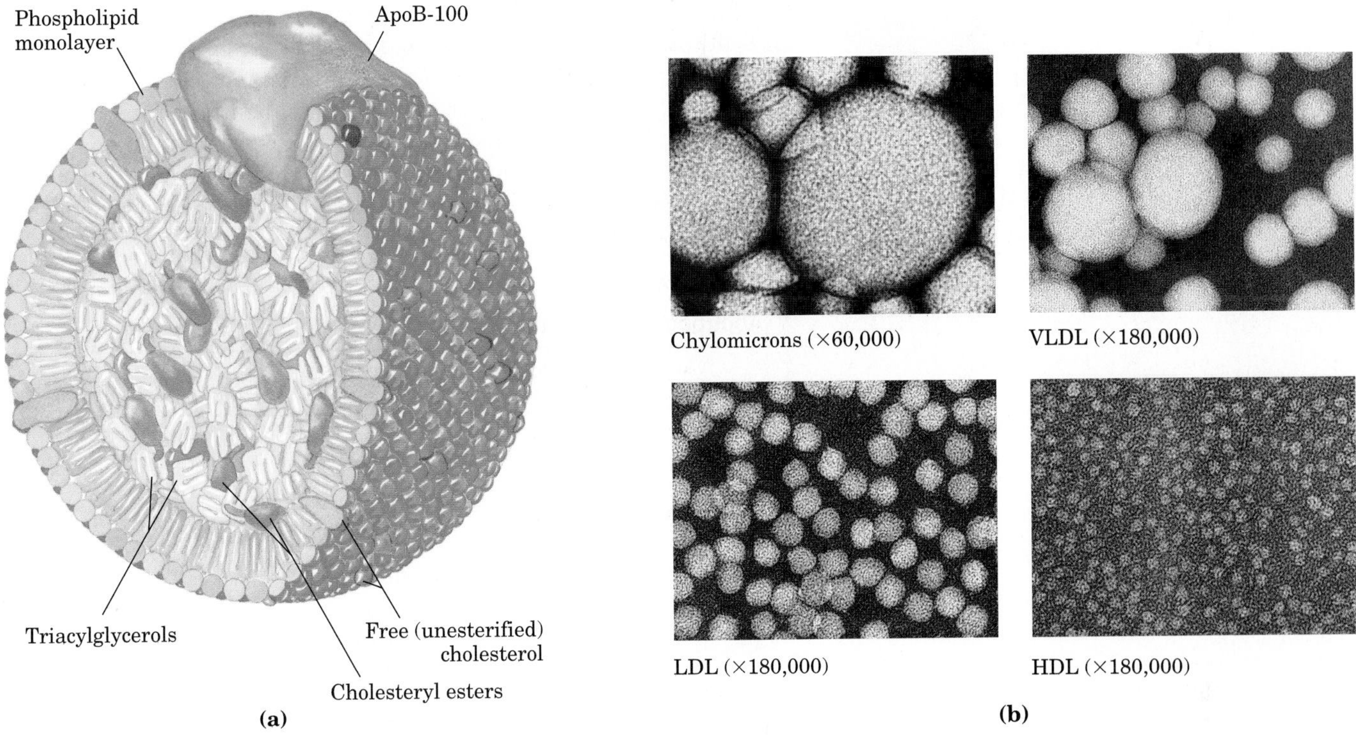

figure 21–37
(a) Structure of a low-density lipoprotein (LDL). Apolipoprotein B-100 (apoB-100) is one of the largest single polypeptide chains known, with 4,636 amino acid residues (M_r 513,000). **(b)** Four classes of lipoproteins visualized in the electron microscope after negative staining. Top left, chylomicrons (50–200 nm in diameter); top right, VLDL (28–70 nm); bottom left, LDL (20–25 nm); bottom right, HDL (8–11 nm). For properties of lipoproteins, see Table 21–2.

Apolipoproteins ("apo" designates the protein in its lipid-free form) combine with lipids to form several classes of lipoprotein particles, spherical complexes with hydrophobic lipids in the core and hydrophilic amino acid side chains at the surface (Fig. 21–37a). Different combinations of lipids and proteins produce particles of different densities, ranging from chylomicrons to high-density lipoproteins (HDL). These may be separated by ultracentrifugation (Table 21–2) and visualized by electron microscopy (Fig. 21–37b).

table 21–2

Major Classes of Human Plasma Lipoproteins: Some Properties

		Composition (wt %)				
Lipoprotein	**Density (g/mL)**	**Protein**	**Phospholipids**	**Free cholesterol**	**Cholesteryl esters**	**Triacylglycerols**
Chylomicrons	<1.006	2	9	1	3	85
VLDL	0.95–1.006	10	18	7	12	50
LDL	1.006–1.063	23	20	8	37	10
HDL	1.063–1.210	55	24	2	15	4

Source: Modified from Kritchevsky, D. (1986) Atherosclerosis and nutrition. *Nutr. Int.* **2,** 290–297.

Each class of lipoprotein has a specific function, determined by its point of synthesis, lipid composition, and apolipoprotein content. At least nine different apolipoproteins are found in the lipoproteins of human plasma (Table 21–3). These can be distinguished by their size, their reactions with specific antibodies, and their characteristic distribution in the lipoprotein classes. These protein components act as signals, targeting lipoproteins to specific tissues or activating enzymes that act on the lipoproteins.

table 21–3

Apolipoproteins of the Human Plasma Lipoproteins

Apolipoprotein	Molecular weight	Lipoprotein association	Function (if known)
ApoA-I	28,331	HDL	Activates LCAT; interacts with ABC transporter
ApoA-II	17,380	HDL	
ApoA-IV	44,000	Chylomicrons, HDL	
ApoB-48	240,000	Chylomicrons	
ApoB-100	513,000	VLDL, LDL	Binds to LDL receptor
ApoC-I	7,000	VLDL, HDL	
ApoC-II	8,837	Chylomicrons, VLDL, HDL	Activates lipoprotein lipase
ApoC-III	8,751	Chylomicrons, VLDL, HDL	Inhibits lipoprotein lipase
ApoD	32,500	HDL	
ApoE	34,145	Chylomicrons, VLDL, HDL	Triggers clearance of VLDL and chylomicron remnants

Source: Modified from Vance, D.E. & Vance, J.E. (eds) (1985) *Biochemistry of Lipids and Membranes.* The Benjamin/Cummings Publishing Company, Menlo Park, CA.

Chylomicrons, discussed previously in connection with the movement of dietary triacylglycerols from the intestine to other tissues, are the largest of the lipoproteins and the least dense, containing a high proportion of triacylglycerols (see Fig. 17–2). Chylomicrons are synthesized in the endoplasmic reticulum of epithelial cells that line the small intestine, then move through the lymphatic system and enter the bloodstream through the left subclavian vein. The apolipoproteins of chylomicrons include apoB-48 (unique to this class of lipoproteins), apoE, and apoC-II (Table 21–3). ApoC-II activates lipoprotein lipase in the capillaries of adipose, heart, skeletal muscle, and lactating mammary tissues, allowing the release of free fatty acids to these tissues. Chylomicrons thus carry dietary fatty acids to tissues where they will be consumed or stored as fuel (Fig. 21–38). The remnants of chylomicrons (depleted of most of their triacylglycerols but still containing cholesterol, apoE, and apoB-48) move through the bloodstream to the liver. Receptors in the liver bind to the apoE in the chylomicron remnants and mediate their uptake by endocytosis. In the liver, they release cholesterol, are degraded in lysosomes.

When the diet contains more fatty acids than are needed immediately as fuel, they are converted into triacylglycerols in the liver and packaged with specific apolipoproteins into **very-low-density lipoprotein, VLDL.** Excess carbohydrate in the diet can also be converted into triacylglycerols in the liver and exported as VLDLs (Fig. 21–38a). In addition to triacylglycerols, VLDLs contain some cholesterol and cholesteryl esters, as well as

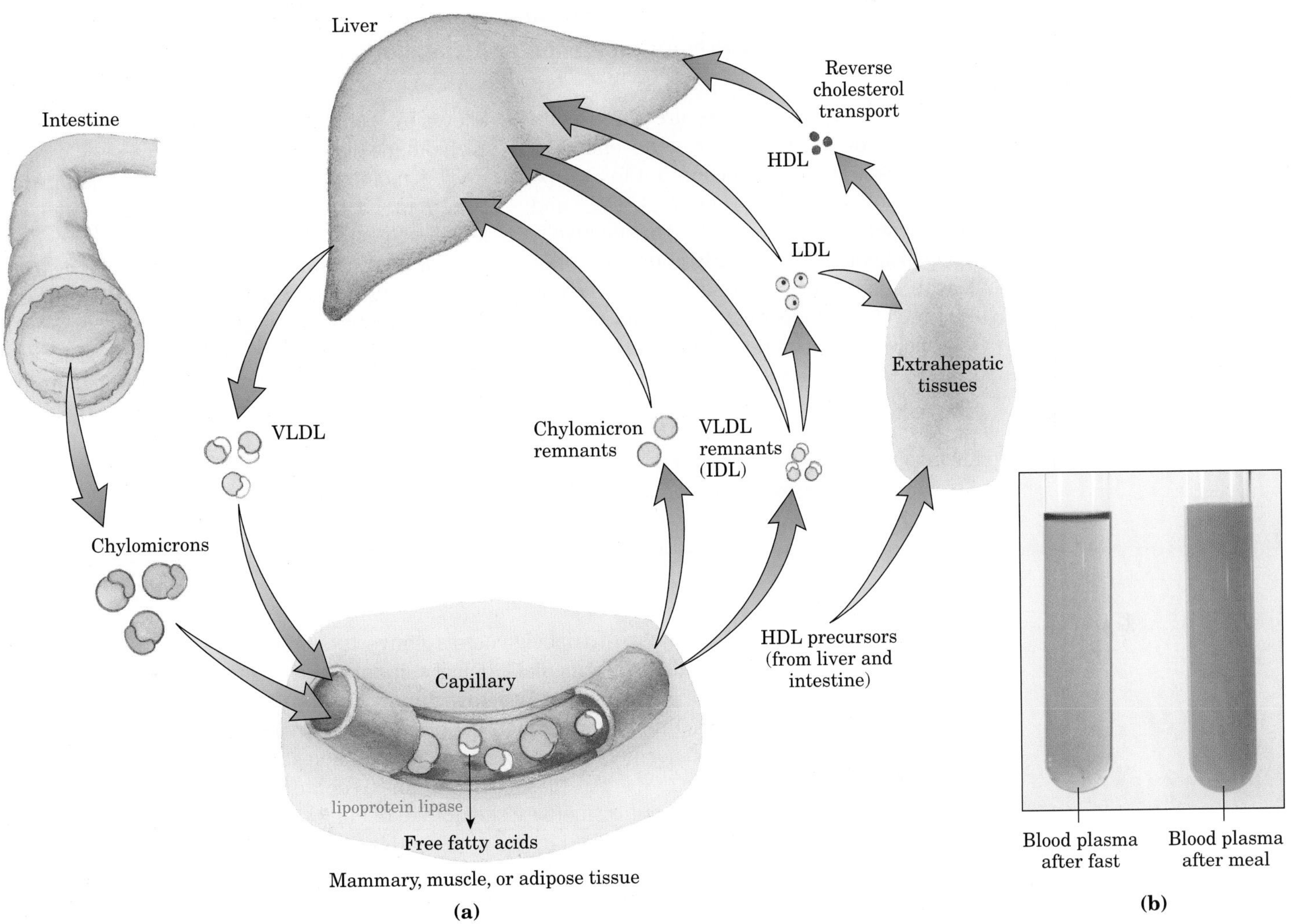

figure 21–38

Lipoproteins and lipid transport. (a) Lipids are transported in the bloodstream as lipoproteins, which occur in several variants that have different functions, different protein and lipid compositions (see Tables 21–2 and 21–3), and, therefore, different densities. Dietary lipids are packaged into chylomicrons; much of their triacylglycerol content is released by lipoprotein lipase to adipose and muscle tissues during transport through capillaries. Chylomicron remnants (containing largely protein and cholesterol) are taken up by the liver. Endogenous lipids and cholesterol from the liver are delivered to adipose and muscle tissue by VLDL. Extraction of lipid from VLDL (along with the loss of some apolipoproteins) gradually converts some of it to LDL, which delivers cholesterol to extrahepatic tissues or is taken up again by the liver. The liver takes up LDL, VLDL remnants, and chylomicron remnants by receptor-mediated endocytosis. Excess cholesterol in extrahepatic tissues is transported back to the liver as HDL. In the liver, some cholesterol is converted to bile salts. **(b)** Blood plasma samples collected after a fast (left) and after a high-fat meal (right). Chylomicrons produced after a fatty meal give the plasma a milky appearance.

apoB-100, apoC-I, apoC-II, apoC-III, and apo-E (Table 21–3). These lipoproteins are transported in the blood from the liver to muscle and adipose tissue, where activation of lipoprotein lipase by apoC-II causes the release of free fatty acids from the triacylglycerols of the VLDL. Adipocytes take up these fatty acids, resynthesize triacylglycerols from them, and store the products in intracellular lipid droplets, whereas myocytes primarily oxidize fatty acids to supply energy. Most VLDL remnants are removed from the circulation by hepatocytes. As for chylomicrons, the uptake is receptor-mediated and depends on the presence of apoE in the VLDL remnants (Box 21–3).

box 21–3 Apolipoprotein E Alleles Predict Incidence of Alzheimer's Disease

In the human population there are three common variants, or alleles, of the gene encoding apoE. The most common, accounting for ~78% of human apoE alleles, is *APOE3; APOE4* and *APOE2* account for 15% and 7%, respectively. The *APOE4* allele is particularly common in humans with Alzheimer's disease and the link is highly predictive. Individuals who inherit the *APOE4* allele have an increased risk of late-onset Alzheimer's disease. Those who are homozygous for the *APOE4* allele have a 16-fold increased risk of developing the disease; for those who do, the mean age of onset is just under 70 years. For people who inherit two copies of the *APOE3* allele, by contrast, the mean age of onset of Alzheimer's disease exceeds 90 years.

The molecular basis for the association between apoE4 and Alzheimer's disease is not yet known. Speculation has focused on a possible role for apoE in stabilizing the cytoskeletal structure of neurons. The apoE2 and apoE3 proteins bind to a number of proteins associated with neuronal microtubules, whereas ApoE4 does not. This may accelerate the death of neurons. Whatever the mechanism proves to be, these observations promise to expand our understanding of the biological functions of apolipoproteins.

The loss of triacylglycerols converts some VLDL to VLDL remnants (also called intermediate density lipoproteins, IDL), and with further removal of tricylglycerol to **low-density lipoprotein, LDL** (Table 21–2). Very rich in cholesterol and cholesteryl esters and containing apoB-100 as their major apolipoprotein, LDLs carry cholesterol to extrahepatic tissues that have specific plasma membrane receptors that recognize apoB-100. These receptors mediate the uptake of cholesterol and cholesteryl esters in a process described in the next section.

The fourth major lipoprotein type, **high-density lipoprotein, HDL,** begins in the liver and small intestine as small, protein-rich particles that contain relatively little cholesterol and no cholesteryl esters (Fig. 21–38). HDLs contain apoA-I, apoC-I, apoC-II, and other apolipoproteins (Table 21–3), as well as the enzyme **lecithin-cholesterol acyl transferase** (LCAT), which catalyzes the formation of cholesteryl esters from lecithin (phosphatidylcholine) and cholesterol (Fig. 21–39). LCAT on the surface of nascent (newly forming) HDL particles converts the cholesterol and phosphatidylcholine of chylomicron and VLDL remnants to cholesteryl esters, which begin to form a core, transforming the disk-shaped nascent HDL to a mature, spherical HDL particle. This cholesterol-rich lipoprotein then returns to the liver, where the cholesterol is unloaded; some of this cholesterol is converted into bile salts.

HDL may be taken up in the liver by receptor-mediated endocytosis, but at least some of the cholesterol in HDL is delivered to other tissues by a novel mechanism. HDL can bind to plasma membrane receptor proteins called SR-BI in hepatic and steroidogenic tissues such as the adrenal gland. These receptors mediate not endocytosis, but a partial and selective transfer of cholesterol and other lipids in HDL to the cell. Depleted HDL then dissociates to recirculate in the bloodstream and extract more lipids from chylomicron and VLDL remnants. Depleted HDL can also pick up cholesterol stored in extrahepatic tissues and carry it to the liver, in **reverse cholesterol transport** pathways. In one reverse transport path, interaction of nascent HDL with SR-BI receptors in cholesterol-rich cells triggers passive movement of cholesterol from the cell surface into HDL, which then carries it back to the liver. In a second pathway, apoA-I in depleted HDL interacts

Phosphatidylcholine (lecithin) + Cholesterol

lecithin-cholesterol acyl transferase (LCAT)

Lysolecithin + Cholesteryl ester

figure 21–39

Reaction catalyzed by lecithin-cholesterol acyl transferase (LCAT). This enzyme is present on the surface of HDL and is stimulated by the HDL component apoA-I. Cholesteryl esters accumulate within nascent HDLs, converting them to mature HDLs.

with an active transporter, the ABC1 protein, in a cholesterol-rich cell. The apoA-I (and presumably the HDL) is taken up by endocytosis, then resecreted with a load of cholesterol, which it transports to the liver.

The ABC1 protein is a member of a large family of multidrug transporters (see Table 12–4), sometimes called ABC transporters because they all have *A*TP-*b*inding *c*assettes; they also have two transmembrane domains with six transmembrane helices. These proteins actively transport a variety of ions, amino acids, vitamins, steroid hormones, and bile salts across plasma membranes. The CFTR protein defective in cystic fibrosis (see Box 12–4) is another member of this ABC family of multidrug transporters.

Cholesteryl Esters Enter Cells by Receptor-Mediated Endocytosis

Each LDL particle in the bloodstream contains apoB-100, which is recognized by specific surface receptor proteins, **LDL receptors,** on cells that need to take up cholesterol. The binding of LDL to an LDL receptor initiates endocytosis (see Fig. 2–9), which brings the LDL and its receptor into the cell within an endosome (Fig. 21–40). This endosome eventually fuses with a lysosome, which contains enzymes that hydrolyze the cholesteryl esters, releasing cholesterol and fatty acid into the cytosol. The apoB-100 of LDL is also degraded to amino acids that are released to the cytosol, but the LDL receptor escapes degradation and is returned to the cell surface, to function again in LDL uptake. ApoB-100 is also present in VLDL, but its receptor-binding domain is not available for binding to the LDL receptor. The conversion of VLDL to LDL exposes the receptor-binding domain of apoB-100. This pathway for the transport of cholesterol in blood and its **receptor-mediated endocytosis** by target tissues was elucidated by Michael Brown and Joseph Goldstein.

Michael Brown and Joseph Goldstein

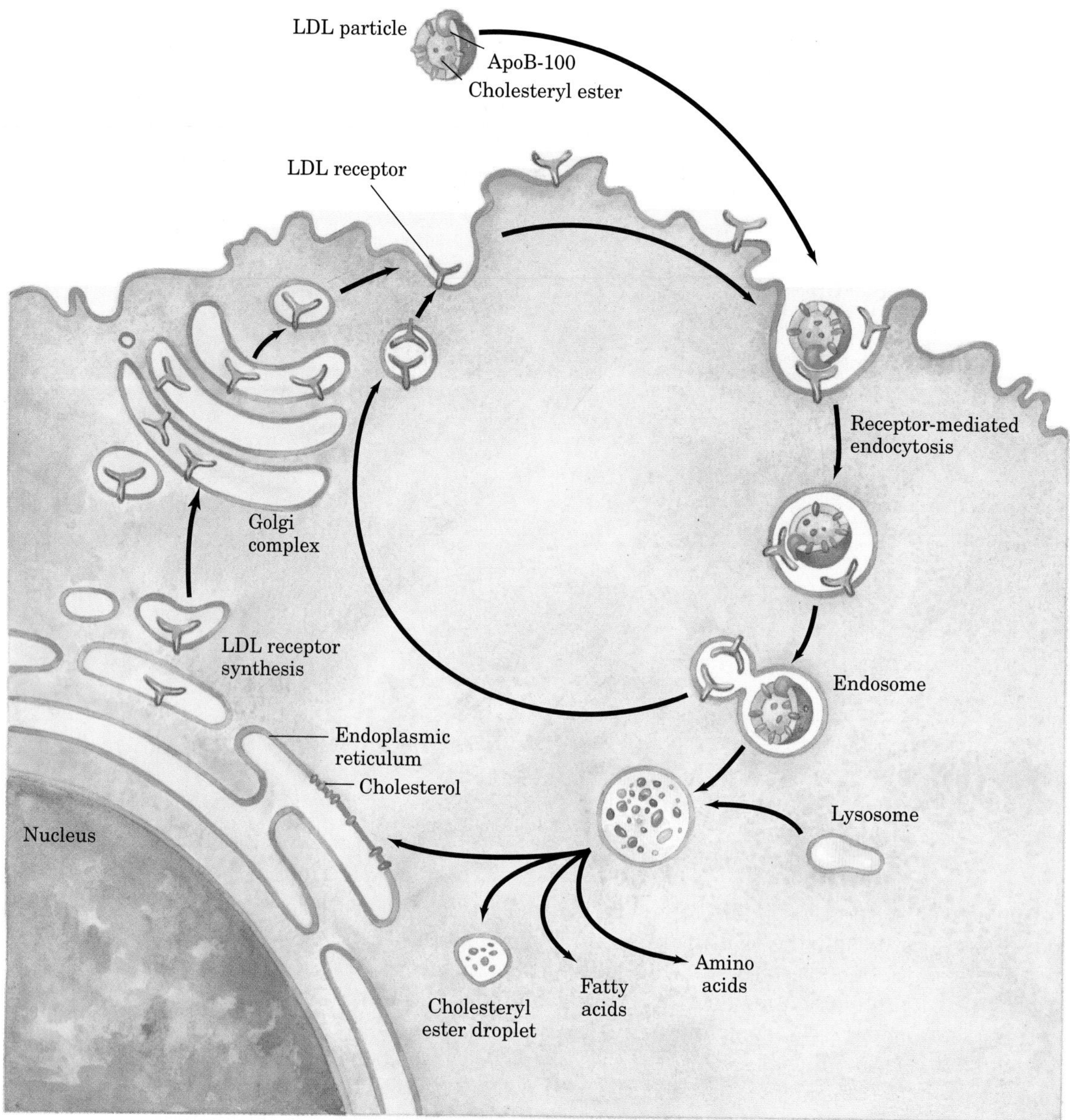

figure 21–40
Uptake of cholesterol by receptor-mediated endocytosis. Endocytosis is also described in Chapter 2 (p. 31).

Cholesterol that enters cells by this path may be incorporated into membranes or may be reesterified by ACAT (Fig. 21–36) for storage within cytosolic lipid droplets. Accumulation of excess intracellular cholesterol is prevented by reducing the rate of cholesterol synthesis when sufficient cholesterol is available from LDL in the blood.

The LDL receptor also binds to apoE and plays a significant role in the hepatic uptake of chylomicrons and VLDL remnants. However, if LDL receptors are unavailable (as, for example, in a mouse strain that lacks the gene for the LDL receptor), VLDL remnants and chylomicrons are still taken up by the liver even though LDL is not. This indicates that a back-up system exists for receptor-mediated endocytosis of VLDL remnants and chylomicrons. One back-up receptor is *l*ipoprotein *r*eceptor-related *p*rotein (LRP), which binds to apoE as well as to a number of other ligands.

Cholesterol Biosynthesis Is Regulated by Several Factors

Cholesterol synthesis is a complex and energy-expensive process, so it is clearly advantageous to an organism to be able to regulate the synthesis of cholesterol to complement dietary intake. In mammals, cholesterol production is regulated by intracellular cholesterol concentration and by the hormones glucagon and insulin. The rate-limiting step in the pathway to cholesterol is the conversion of β-hydroxy-β-methylglutaryl-CoA (HMG-CoA) into mevalonate (Fig. 21–32). The enzyme that catalyzes this reaction, HMG-CoA reductase, is a complex regulatory enzyme whose activity is modulated over a 100-fold range. High levels of an unidentified sterol (perhaps cholesterol, or a cholesterol derivative) promotes rapid degradation of the enzyme (Fig. 21–41) and inhibits transcription of its gene. HMG-CoA reductase is also hormonally regulated. The enzyme exists in phosphorylated (inactive) and unphosphorylated (active) forms. Glucagon stimulates phosphorylation (inactivation), and insulin promotes dephosphorylation, activating the enzyme and favoring cholesterol synthesis.

High intracellular concentrations of cholesterol activate ACAT (Fig. 21–41), which increases esterification of cholesterol for storage. Finally, high cellular cholesterol diminishes transcription of the gene that encodes the LDL receptor, reducing production of the receptor and thus the uptake of cholesterol from the blood.

Unregulated cholesterol production can lead to serious disease. When the sum of the cholesterol synthesized and obtained in the diet exceeds the amount required for the synthesis of membranes, bile salts, and steroids, pathological accumulations of cholesterol in blood vessels (atherosclerotic plaques) can develop in humans, resulting in obstruction of blood vessels **(atherosclerosis).** Heart failure due to occluded coronary arteries is a major cause of death in industrialized societies. Atherosclerosis is linked to high levels of cholesterol in the blood, and particularly to high levels of LDL-bound cholesterol; there is a *negative* correlation between HDL levels and arterial disease.

In familial hypercholesterolemia, a human genetic disorder, blood levels of cholesterol are extremely high and afflicted individuals develop severe atherosclerosis in childhood. The LDL receptor is defective in these individuals, and receptor-mediated uptake of cholesterol carried by LDL does not occur. Consequently, cholesterol is not cleared from the blood; it accumulates and contributes to the formation of atherosclerotic plaques. Endogenous cholesterol synthesis continues despite the excessive cholesterol in the blood, because extracellular cholesterol cannot enter the cell to regulate intracellular synthesis (Fig. 21–41). Two products derived from fungi, **lovastatin** and **compactin,** are used to treat patients with familial hypercholesterolemia. Both compounds, and several synthetic analogs, resemble mevalonate (Fig. 21–42). Both are competitive inhibitors of HMG-CoA reductase and thus inhibit cholesterol synthesis. Lovastatin treatment lowers serum cholesterol by as much as 30% in individuals whose cells carry one defective copy of the gene for the LDL receptor. When combined with an edible resin that binds bile acids and prevents their reabsorption from the intestine, the drug is even more effective.

HDL levels are very low in humans with familial HDL deficiency, and almost undetectable in Tangier disease. Both of these genetic disorders are the result of mutations in the ABC1 protein mentioned above. Cholesterol-depleted HDL cannot take on cholesterol from cells that lack this protein, and cholesterol-poor HDL is rapidly removed from the blood and destroyed. Both FHA and Tangier disease are very rare (fewer than 100 families with Tangier disease are known in the world), but they establish a role for ABC1 protein in the regulation of plasma HDL levels. Because low plasma HDL levels correlate with high incidence of coronary artery disease, the ABC1 protein may prove a useful target for drugs to control HDL levels.

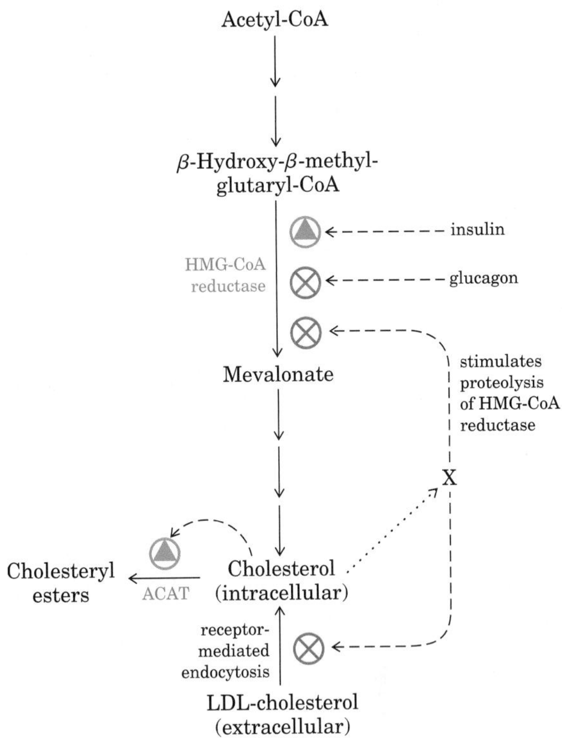

figure 21–41
Regulation of cholesterol biosynthesis balances synthesis with dietary uptake. Glucagon promotes phosphorylation of HMG-CoA reductase, insulin promotes dephosphorylation. X represents unidentified metabolites of cholesterol that stimulate proteolysis of HMG-CoA reductase.

Mevalonate

$R_1 = H$	$R_2 = H$	Compactin
$R_1 = CH_3$	$R_2 = CH_3$	Simvastatin (Zocov)
$R_1 = H$	$R_2 = OH$	Pravastatin (Pravacol)
$R_1 = H$	$R_2 = CH_3$	Lovastatin (Mevacor)

figure 21–42
Inhibitors of HMG-CoA synthase. A comparison of the structures of mevalonate and four compounds that inhibit HMG-CoA synthase.

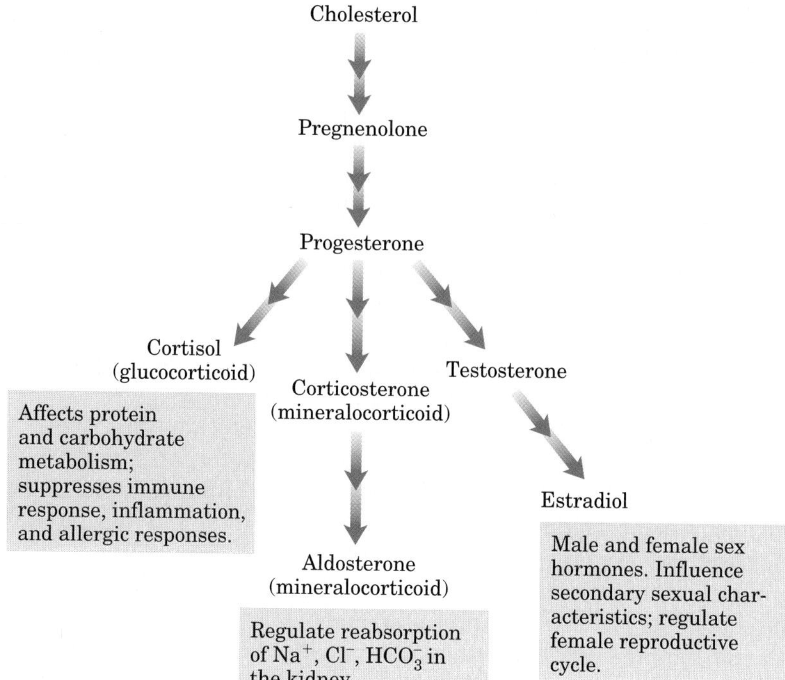

figure 21–43
Some steroid hormones derived from cholesterol. The structures of some of these compounds are shown in Figure 11–17.

Steroid Hormones Are Formed by Side Chain Cleavage and Oxidation of Cholesterol

CH_3
C=O
O
Progesterone

All steroid hormones in humans are derived from cholesterol (Fig. 21–43). Two classes of steroid hormones are synthesized in the cortex of the adrenal gland: **mineralocorticoids,** which control the reabsorption of inorganic ions (Na^+, Cl^-, and HCO_3^-) by the kidney; and **glucocorticoids,** which help regulate gluconeogenesis and reduce the inflammatory response. Sex hormones are produced in male and female gonads and the placenta. They include **progesterone,** which regulates the female reproductive cycle, as well as **androgens** (e.g., testosterone) and **estrogens** (e.g., estradiol), which influence the development of secondary sexual characteristics in males and females, respectively. Steroid hormones are effective at very low concentrations and are therefore synthesized in relatively small quantities. In comparison with the bile salts, their production consumes relatively little cholesterol.

Synthesis of steroid hormones requires removal of some or all of the carbons in the "side chain" that projects from C-17 of the D ring of cholesterol. Side chain removal takes place in the mitochondria of tissues that make steroid hormones. It involves the hydroxylation of two adjacent carbons in the side chain (C-20 and C-22) followed by cleavage of the bond between them (Fig. 21–44). Formation of the individual hormones also involves the introduction of oxygen atoms. All of the hydroxylation and oxygenation reactions in steroid biosynthesis are catalyzed by mixed-function oxidases (Box 21–1) that use NADPH, O_2, and mitochondrial cytochrome P-450.

Intermediates in Cholesterol Biosynthesis Have Many Alternative Fates

In addition to its role as an intermediate in cholesterol biosynthesis, isopentenyl pyrophosphate is the activated precursor of a huge array of biomolecules with diverse biological roles (Fig. 21–45). They include vitamins A, E, and K; plant pigments such as carotene and the phytol chain of chlorophyll;

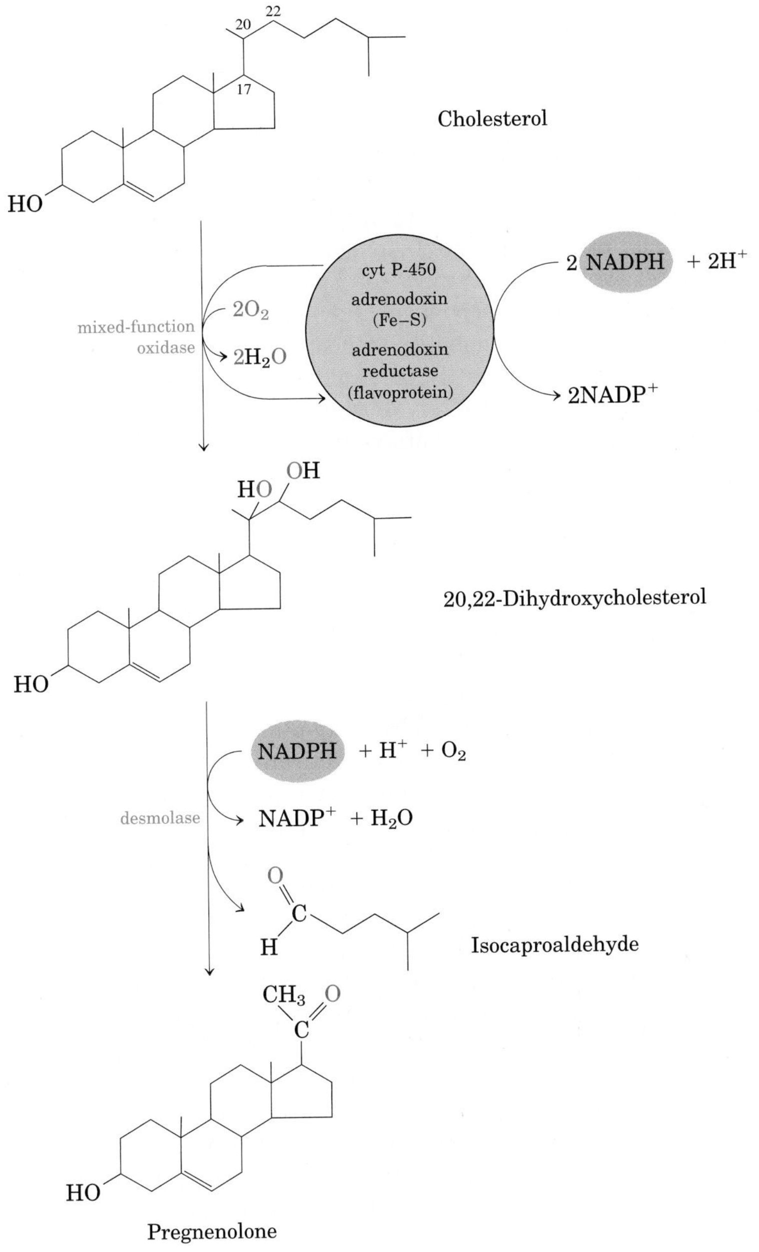

Cholesterol

mixed-function oxidase

$2O_2$ → $2H_2O$

cyt P-450, adrenodoxin (Fe–S), adrenodoxin reductase (flavoprotein)

2 NADPH + $2H^+$ → $2NADP^+$

20,22-Dihydroxycholesterol

desmolase

NADPH + H^+ + O_2 → $NADP^+$ + H_2O

Isocaproaldehyde

Pregnenolone

figure 21–44
Side chain cleavage in the synthesis of steroid hormones. Cytochrome P-450 acts as electron carrier in this mixed-function oxidase system that oxidizes adjacent carbons. The process also requires the electron-transferring proteins adrenodoxin and adrenodoxin reductase. This system for cleaving side chains is found in mitochondria of the adrenal cortex, where active steroid production occurs. Pregnenolone is the precursor of all other steroid hormones (see Fig. 21–43).

natural rubber; many essential oils (such as the fragrant principles of lemon oil, eucalyptus, and musk); insect juvenile hormone, which controls metamorphosis; dolichols, which serve as lipid-soluble carriers in complex polysaccharide synthesis; and ubiquinone and plastoquinone, electron carriers in mitochondria and chloroplasts. Collectively, these molecules are called isoprenoids. More than 20,000 different isoprenoid molecules have been discovered in nature, and hundreds of new ones are reported each year.

Prenylation (covalent attachment of an isoprenoid; see Fig. 27–32) is a common mechanism by which proteins are anchored to the inner surface of mammalian cell membranes (see Fig. 12–13). In some of these proteins the attached lipid is the 15-carbon farnesyl group; others have the 20-carbon geranylgeranyl group. Different enzymes attach the two types of lipids; it is possible that prenylation reactions target proteins to different membranes, depending upon which lipid is attached. Protein prenylation is another important role for the isoprene derivatives of the pathway to cholesterol.

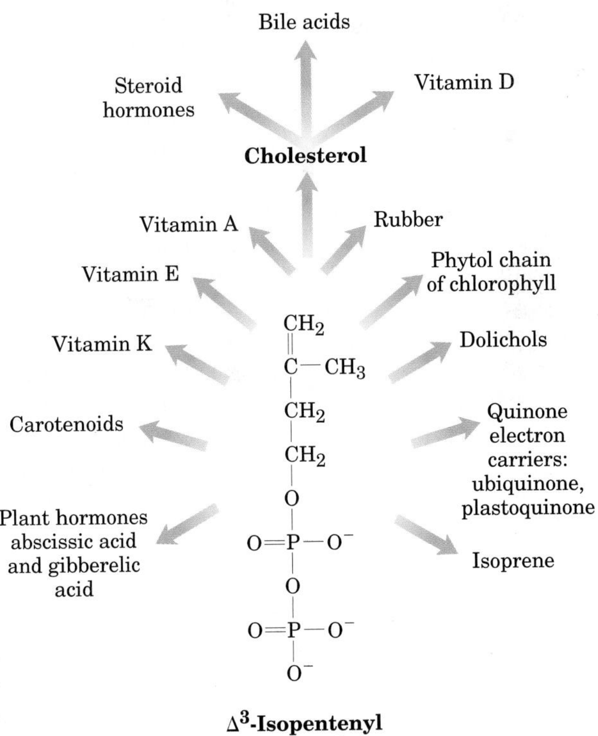

figure 21–45
An overview of isoprenoid biosynthesis. The structures of most of the end products shown here are given in Chapter 11.

summary

Long-chain saturated fatty acids are synthesized from acetyl-CoA by a cytosolic complex of six enzyme activities plus acyl carrier protein (ACP), which contains phosphopantetheine as its prosthetic group. The fatty acid synthase, which in some organisms consists of multifunctional polypeptides, contains two types of —SH groups (one furnished by the phosphopantetheine of ACP and the other by a Cys residue of the enzyme β-ketoacyl–ACP synthase) that function as carriers of the fatty acyl intermediates. Malonyl-ACP, formed from acetyl-CoA (shuttled out of mitochondria) and CO_2, condenses with an acetyl bound to the Cys —SH to yield acetoacetyl-ACP with release of CO_2. Reduction to the D-β-hydroxy derivative and its dehydration to the *trans*-Δ^2-unsaturated acyl-ACP is followed by reduction to butyryl-ACP. For both reduction steps, NADPH is the electron donor. Six more molecules of malonyl-ACP react successively at the carboxyl end of the growing fatty acid chain to form palmitoyl-ACP—the end product of the fatty acid synthase reaction. Free palmitate is released by hydrolysis. Fatty acid synthesis is regulated at the level of malonyl-CoA formation.

Palmitate may be elongated to yield the 18-carbon stearate. Palmitate and stearate in turn can be desaturated to yield palmitoleate and oleate, respectively, by the action of mixed-function oxidases. Mammals cannot make linoleate and must obtain it from plant sources; they convert exogenous linoleate to arachidonate, the parent compound of a family of very potent hormonelike eicosanoids (prostaglandins, thromboxanes, and leukotrienes).

Triacylglycerols are formed by reaction of two molecules of fatty acyl–CoA with glycerol 3-phosphate to form phosphatidic acid; this is dephosphorylated to a diacylglycerol and then acylated by a third molecule of fatty acyl–CoA to yield a triacylglycerol. This process is hormonally regulated. Triacylglycerols are carried in the blood in chylomicrons. Diacylglycerols are the major precursors of glycerophospholipids. In bacteria, phosphatidylserine is formed by the condensation of serine with CDP-diacylglycerol; decarboxylation of phosphatidylserine produces phosphatidylethanolamine. Phosphatidylglycerol is formed by condensation of CDP-diacylglycerol with glycerol 3-phosphate followed by removal of the phosphate in monoester linkage. Yeasts use similar pathways in the synthesis of phosphatidylserine, phosphatidylethanolamine, and phosphatidylglycerol; phosphatidylcholine is formed by methylation of phosphatidylethanolamine. Mammalian cells have somewhat different pathways for synthesizing phosphatidylcholine and phosphatidylethanolamine. The head group alcohol (choline or ethanolamine) is activated as the CDP derivative, then condensed with diacylglycerol. Phosphatidylserine is derived only from phosphatidylethanolamine. Synthesis of plasmalogens involves formation of their characteristic double bond by a mixed-function oxidase. The head groups of sphingolipids are attached by unique mechanisms. Phospholipids are moved to their intracellular destinations by transport vesicles or specific proteins.

Cholesterol is formed from acetyl-CoA in a complex series of reactions through the intermediates β-hydroxy-β-methylglutaryl-CoA, mevalonate, and two activated isoprenes (dimethylallyl pyrophosphate and isopentenyl pyrophosphate). Condensation of isoprene units produces the noncyclic squalene, which is cyclized to yield the steroid ring system and side chain. Cholesterol synthesis is inhibited by elevated intracellular cholesterol. Cholesterol and cholesteryl esters are carried in the blood as plasma lipoproteins. Very low-density lipoprotein (VLDL) carries cholesterol, cholesteryl esters, and triacylglycerols from the liver to other tissues, where the triacylglycerols are degraded by lipoprotein lipase, converting VLDL to low-density lipoprotein (LDL). LDL, rich in cholesterol and its esters, is taken up by receptor-mediated endocytosis, in which the apolipoprotein B-100 of LDL is recognized by receptors in the plasma membrane. High-density lipoprotein (HDL) removes cholesterol from the blood, carrying it to the liver. Dietary conditions or genetic defects in cholesterol metabolism may lead to atherosclerosis and heart disease.

The steroid hormones (glucocorticoids, mineralocorticoids, and sex hormones) are produced from cholesterol by alteration of the side chain and introduction of oxygen atoms into the steroid ring system. In addition to cholesterol, a very wide variety of isoprenoid compounds are derived from mevalonate through condensations of isopentenyl pyrophosphate and dimethylallyl pyrophosphate. Prenylation of certain proteins targets them for association with cell membranes and is essential for their biological activity.

further reading

The general references in Chapters 11 and 17 will also be useful.

General

Bell, S.J., Bradley, D., Forse, R.A., & Bistrian, B.R. (1997) The new dietary fats in health and disease. *J. Am. Dietetic Assoc.* **97,** 280–286.

Gotto, A.M., Jr. (ed.) (1987) *Plasma Lipoproteins,* New Comprehensive Biochemistry, Vol. 14 (Neuberger, A. & van Deenen, L.L.M., series eds), Elsevier Biomedical Press, Amsterdam.

Twelve reviews covering the structure, synthesis, and metabolism of lipoproteins, regulation of cholesterol synthesis, and the enzymes LCAT and lipoprotein lipase.

Hajjar, D.P. & Nicholson, A.C. (1995) Atherosclerosis. *Am. Sci.* **83,** 460–467.

A good description of the molecular basis of the disease and prospects for therapy.

Hawthorne, J.N. & Ansell, G.B. (eds) (1982) *Phospholipids,* New Comprehensive Biochemistry, Vol. 4 (Neuberger, A. & van Deenen, L.L.M., series eds), Elsevier Biomedical Press, Amsterdam.

Excellent reviews of biosynthetic pathways to glycerophospholipids and sphingolipids, phospholipid transfer proteins, and bilayer assembly.

Ohlrogge, J. & Browse, J. (1995) Lipid biosynthesis. *Plant Cell* **7,** 957–970.

A good summary of pathways for lipid biosynthesis in plants.

Vance, D.E. & Vance, J.E. (eds) (1996) *Biochemistry of Lipids, Lipoproteins and Membranes,* New Comprehensive Biochemistry, Vol. 31, Elsevier Science Publishing Co., Inc., New York.

Excellent reviews of lipid structure, biosynthesis, and function.

Biosynthesis of Fatty Acids and Eicosanoids

Capdevila, J.H., Falck, J.R., & Estabrook, R.W. (1992) Cytochrome P450 and the arachidonate cascade. *FASEB J.* **6,** 731–736.

This issue of the *FASEB J.* contains 20 articles on the structure and function of various cytochrome P-450s.

Creelman, R.A. & Mullet, J.E. (1997) Biosynthesis and action of jasmonates in plants. *Annu. Rev. Plant Physiol. Plant Mol. Biol.* **48,** 355–381.

DeWitt, D.L. (1999) Cox-2-selective inhibitors: the new super aspirins. *Molec. Pharmacol.* **55,** 625–631.

A short, clear review of the topic discussed in Box 21–2.

Drazen, J.M., Israel, E., & O'Byrne, P.M. (1999) Drug therapy: treatment of asthma with drugs modifying the leukotriene pathway. *New Engl. J. Med.* **340,** 197–206.

Kim, K.H. (1997) Regulation of mammalian acetyl-coenzyme A carboxylase. *Annu. Rev. Nutr.* **17,** 77–99.

Lands, W.E.M. (1991) Biosynthesis of prostaglandins. *Annu. Rev. Nutr.* **11,** 41–60.

Discussion of the nutritional requirement for unsaturated fatty acids and recent biochemical work on pathways from arachidonate to prostaglandins; advanced level.

Slabas, A.R., Brown, A., Sinden, B.S., Swinhoe, R., Simon, J.W., Ashton, A.R., Whitfeld, P.R., & Elborough, K.M. (1994) Pivotal reactions in fatty acid synthesis. *Prog. Lipid Res.* **33,** 39–46.

Smith, S. (1994) The animal fatty acid synthase: one gene, one polypeptide, seven enzymes. *FASEB J.* **8,** 1248–1259.

Smith, W.L., Garavito, R.M., & DeWitt, D.L. (1996) Prostaglandin endoperoxide H synthases (cyclooxygenases)-1 and -2. *J. Biol. Chem.* **271,** 33,157–33,160.

A concise review of the properties and roles of COX-1 and COX-2.

Biosynthesis of Membrane Phospholipids

Bishop, W.R. & Bell, R.M. (1988) Assembly of phospholipids into cellular membranes: biosynthesis, transmembrane movement and intracellular translocation. *Annu. Rev. Cell Biol.* **4,** 579–610.

Advanced review of the enzymology and cell biology of phospholipid synthesis and targeting.

Dowhan, W. (1997) Molecular basis for membrane phospholipid diversity: why are there so many lipids? *Annu. Rev. Biochem.* **66,** 199–232.

Kennedy, E.P. (1962) The metabolism and function of complex lipids. *Harvey Lectures* **57,** 143–171.

A classic description of the role of cytidine nucleotides in phospholipid synthesis.

Pavlidis, P., Ramaswami, M., & Tanouye, M.A. (1994) The Drosophila *easily shocked* gene: a mutation in a phospholipid synthetic pathway causes seizure, neuronal failure, and paralysis. *Cell* **79,** 23–33.

Fascinating effects of changing the composition of membrane lipids in fruit flies.

Raetz, C.R.H. & Dowhan, W. (1990) Biosynthesis and function of phospholipids in *Escherichia coli.* *J. Biol. Chem.* **265,** 1235–1238.

A brief review of bacterial biosynthesis of phospholipids and lipopolysaccharides.

Biosynthesis of Cholesterol, Steroids, and Isoprenoids

Bittman, R. (ed) (1997) *Subcellular Biochemistry,* Cholesterol: Its Functions and Metabolism in Biology and Medicine, Vol. 28, Plenum Press, New York.

Bloch, K. (1965) The biological synthesis of cholesterol. *Science* **150,** 19–28.

The author's Nobel address; a classic description of cholesterol synthesis in animals.

Chang, T.Y., Chang, C.C.Y., & Cheng, D. (1997) Acyl-coenzyme A: cholesterol acyltransferase. *Annu. Rev. Biochem.* **66,** 613–638.

Edwards, P.A. & Ericsson, J. (1999) Sterols and isoprenoids: signaling molecules derived from the cholesterol biosynthetic pathway. *Annu. Rev. Biochem.* **68,** 157–185.

Goldstein, J.L. & Brown, M.S. (1990) Regulation of the mevalonate pathway. *Nature* **343,** 425–430.

The allosteric and covalent regulation of the enzymes of the mevalonate pathway; includes a short discussion of the prenylation of Ras and other proteins.

Knopp, R.H. (1999) Drug therapy: drug treatment of lipid disorders. *New Engl. J. Med.* **341,** 498–511.

Review of the use of HMG-CoA inhibitors and bile-acid–binding resins to reduce serum cholesterol.

Krieger, M. (1999) Charting the fate of the "good cholesterol": identification and characterization of the high-density lipoprotein receptor SR-BI. *Annu. Rev. Biochem.* **68,** 523–558.

Lawrence, C.M., Rodwell, V.W., & Stauffacher, C.V. (1995) Crystal structure of *Pseudomonas mevalonii* HMG-CoA reductase at 3.0 angstrom resolution. *Science* **268,** 1758–1762.

McGarvey, D.J. & Croteau, R. (1995) Terpenoid metabolism. *Plant Cell* **7,** 1015–1026.

A description of the amazing diversity of isoprenoids in plants.

Olson, R.E. (1998) Discovery of the lipoproteins, their role in fat transport and their significance as risk factors. *J. Nutr.* **128** (2 suppl.), 439S–443S.

Brief, clear historical background to lipoprotein function.

Strittmatter, W.J. & Roses, A.D. (1995) Apolipoprotein E and Alzheimer disease. *Proc. Natl. Acad. Sci. USA* **92,** 4725–4727.

Young, S.G. & Fielding, C.J. (1999) The ABCs of cholesterol efflux. *Nature Genet.* **22,** 316–318.

A brief review of three papers in the same issue of this journal, establishing that mutations in ABC1 cause Tangier disease and familial HDL deficiency.

problems

1. Pathway of Carbon in Fatty Acid Synthesis Using your knowledge of fatty acid biosynthesis, provide an explanation for the following experimental observations:

(a) The addition of uniformly labeled [^{14}C]acetyl-CoA to a soluble liver fraction yields palmitate uniformly labeled with ^{14}C.

(b) However, the addition of a *trace* of uniformly labeled [^{14}C]acetyl-CoA in the presence of an excess of unlabeled malonyl-CoA to a soluble liver fraction yields palmitate labeled with ^{14}C only in C-15 and C-16.

2. Synthesis of Fatty Acids from Glucose After a person has ingested large amounts of sucrose, the glucose and fructose that exceed caloric requirements are transformed to fatty acids for triacylglycerol synthesis. This fatty acid synthesis consumes acetyl-CoA, ATP, and NADPH. How are these substances produced from glucose?

3. Net Equation of Fatty Acid Synthesis Write the net equation for the biosynthesis of palmitate in rat liver, starting from mitochondrial acetyl-CoA and cytosolic NADPH, ATP, and CO_2.

4. Pathway of Hydrogen in Fatty Acid Synthesis Consider a preparation that contains all the enzymes and cofactors necessary for fatty acid biosynthesis from added acetyl-CoA and malonyl-CoA.

(a) If [2-^{2}H]acetyl-CoA (labeled with deuterium, the heavy isotope of hydrogen):

$$^2H-C(^2H)(^2H)-C(=O)-S\text{-}CoA$$

and an excess of unlabeled malonyl-CoA are added as substrates, how many deuterium atoms are incorporated into every molecule of palmitate? What are their locations? Explain.

(b) If unlabeled acetyl-CoA and [2-^{2}H]malonyl-CoA:

$$^-OOC-C(^2H)(^2H)-C(=O)-S\text{-}CoA$$

are added as substrates, how many deuterium atoms are incorporated into every molecule of palmitate? What are their locations? Explain.

5. Energetics of β-Ketoacyl–ACP Synthase In the condensation reaction catalyzed by β-ketoacyl–ACP synthase (Fig. 21–5), a four-carbon unit is synthesized by the combination of a two-carbon unit and a three-carbon unit, with the release of CO_2. What is the thermodynamic advantage of this process over one that simply combines two two-carbon units?

6. Modulation of Acetyl-CoA Carboxylase Acetyl-CoA carboxylase is the principal regulation point in the biosynthesis of fatty acids. Some of the properties of the enzyme are described below:

(a) The addition of citrate or isocitrate raises the V_{max} of the enzyme by as much as a factor of 10.

(b) The enzyme exists in two interconvertible forms that differ markedly in their activities:

$$\text{Protomer (inactive)} \rightleftharpoons \text{filamentous polymer (active)}$$

Citrate and isocitrate bind preferentially to the filamentous form, and palmitoyl-CoA binds preferentially to the protomer.

Explain how these properties are consistent with the regulatory role of acetyl-CoA carboxylase in the biosynthesis of fatty acids.

7. Shuttling of Acetyl Groups across the Inner Mitochondrial Membrane The acetyl group of acetyl-CoA, produced by the oxidative decarboxylation of pyruvate in the mitochondrion, is transferred to the cytosol by the acetyl group shuttle outlined in Figure 21–11.

(a) Write the overall equation for the transfer of one acetyl group from the mitochondrion to the cytosol.

(b) What is the cost of this process in ATPs per acetyl group?

(c) In Chapter 17 we encountered an acyl group shuttle in the transfer of fatty acyl–CoA from the cytosol to the mitochondrion in preparation for β oxidation (see Fig. 17–6). One result of that shuttle was separation of the mitochondrial and cytosolic pools of CoA. Does the acetyl group shuttle also accomplish this?

8. Oxygen Requirement for Desaturases The biosynthesis of palmitoleate (Fig. 21–13), a common unsaturated fatty acid with a cis double bond in the Δ^9 position, uses palmitate as a precursor. Can this be carried out under strictly anaerobic conditions? Explain.

9. Energy Cost of Triacylglycerol Synthesis Use a net equation for the biosynthesis of tripalmitoylglycerol (tripalmitin) from glycerol and palmitate to show how many ATPs are required per molecule of tripalmitin formed.

10. Turnover of Triacylglycerols in Adipose Tissue When [^{14}C]glucose is added to the balanced diet of adult rats, there is no increase in the total amount of stored triacylglycerols, but the triacylglycerols become labeled with ^{14}C. Explain.

11. Energy Cost of Phosphatidylcholine Synthesis Write the sequence of steps and the net reaction for the biosynthesis of phosphatidylcholine by the salvage pathway from oleate, palmitate, dihydroxyacetone phosphate, and choline. Starting from these precursors, what is the cost (in number of ATPs) of the synthesis of phosphatidylcholine by the salvage pathway?

12. Salvage Pathway for Synthesis of Phosphatidylcholine A young rat maintained on a diet deficient in methionine fails to thrive unless choline is included in the diet. Explain.

13. Synthesis of Isopentenyl Pyrophosphate If 2-[^{14}C]acetyl-CoA is added to a rat liver homogenate that is synthesizing cholesterol, where will the ^{14}C label appear in Δ^3-isopentenyl pyrophosphate, the activated form of an isoprene unit?

14. Activated Donors in Lipid Synthesis In the biosynthesis of complex lipids, components are assembled by transfer of the appropriate group from an activated donor. For example, the activated donor of acetyl groups is acetyl-CoA. For each of the following groups, give the form of the activated donor: (a) phosphate; (b) D-glucosyl; (c) phosphoethanolamine; (d) D-galactosyl; (e) fatty acyl; (f) methyl; (g) the two-carbon group in fatty acid biosynthesis; (h) Δ^3-isopentenyl.

15. Importance of Fats in the Diet When young rats are placed on a totally fat-free diet, they grow poorly, develop a scaly dermatitis, lose hair, and soon die—symptoms that can be prevented if linoleate or plant material is included in the diet. What makes linoleate an essential fatty acid? Why can plant material be substituted?

16. Regulation of Cholesterol Biosynthesis Cholesterol in humans can be obtained from the diet or synthesized de novo. An adult human on a low-cholesterol diet typically synthesizes 600 mg of cholesterol per day in the liver. If the amount of cholesterol in the diet is large, de novo synthesis of cholesterol is drastically reduced. How is this regulation brought about?

17. ApoE and Atherosclerosis A strain of laboratory mice that does not express the gene for apoE has elevated levels of LDL. They develop atherosclerosis when fed a normal diet. How could a lack of apoE cause this observed increase in LDL levels?

chapter

22 Biosynthesis of Amino Acids, Nucleotides, and Related Molecules

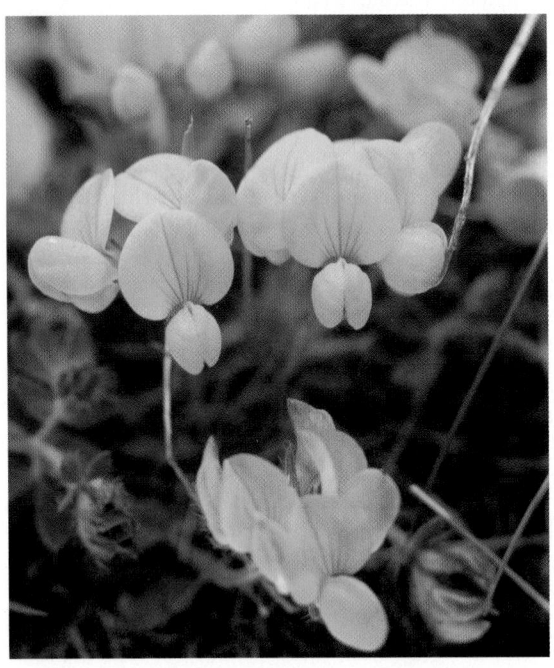

Bird's-foot trefoil in bloom. The bacteria in the root nodules of this and other legumes convert atmospheric nitrogen into the reduced forms required by all living things.

Nitrogen ranks behind only carbon, hydrogen, and oxygen in its contribution to the mass of living systems. Most of this nitrogen is bound up in amino acids and nucleotides. In this chapter we address all aspects of the metabolism of these nitrogen-containing compounds except amino acid catabolism, which is covered in Chapter 18.

Discussing the biosynthetic pathways for amino acids and nucleotides together is a sound approach not only because both classes of molecules contain nitrogen (which arises from common biological sources), but because the two sets of pathways are extensively intertwined, with several key intermediates in common. Certain amino acids or parts of amino acids are incorporated into the structure of purines and pyrimidines, and in one case part of a purine ring is incorporated into an amino acid (histidine). The two sets of pathways also share much common chemistry, in particular a preponderance of reactions involving transfer of nitrogen or one-carbon groups.

The pathways described in the following pages can be intimidating to the beginning biochemistry student. Their complexity arises not so much from the chemistry itself, which in many cases is well understood, but from the sheer number of steps and variety of intermediates. These pathways are best approached by maintaining a focus on metabolic principles we have already discussed, key intermediates and precursors, and common classes of reactions. Even a cursory look at the chemistry can be rewarding, for some of the most unusual chemical transformations in biological systems occur in these pathways; for instance, we find prominent examples of the rare biological use of the metals molybdenum, selenium, and vanadium. The effort also offers a practical dividend, especially for students of human or veterinary medicine. Many genetic diseases of humans and animals have been traced to an absence of one or more enzymes in these pathways. Many pharmaceuticals used to combat infectious diseases are inhibitors of enzymes in these pathways, as are many of the most important agents in cancer chemotherapy.

Regulation is crucial in the biosynthesis of the nitrogen-containing compounds. Because each amino acid and each nucleotide is required in relatively small amounts, the metabolic flow through most of these pathways is not nearly as great as the biosynthetic flow leading to carbohydrate or fat in animal tissues. And because the different amino acids and nucleotides must be made in the correct ratios and at the right time for protein and nucleic acid synthesis, their biosynthetic pathways must be accurately regulated and coordinated with each other. The levels of amino acids and nucleotides, which are charged molecules, must also be regulated to maintain cellular osmotic balance. As discussed in earlier chapters, pathways can be regulated by changes in either the activity or the amounts of specific enzymes. The pathways presented in this chapter provide some of the best-understood examples of the regulation of enzyme activity. Regulation of the amounts of different enzymes in a cell (that is, of their synthesis and degradation) is a topic covered in Chapter 28.

Overview of Nitrogen Metabolism

The biosynthetic pathways leading to amino acids and nucleotides share a requirement for nitrogen. Because soluble, biologically useful nitrogen compounds are generally scarce in natural environments, most organisms maintain strict economy in their use of ammonia, amino acids, and nucleotides. Indeed, as we will see, free amino acids, purines, and pyrimidines formed during metabolic turnover of proteins and nucleic acids are often salvaged and reused. We first examine the pathways by which nitrogen from the environment is introduced into biological systems.

The Nitrogen Cycle Maintains a Pool of Biologically Available Nitrogen

The most important source of nitrogen is air, which is four-fifths molecular nitrogen (N_2). However, relatively few species can convert atmospheric nitrogen into forms useful to living organisms. The metabolic processes of different species function interdependently to salvage and reuse biologically available nitrogen in a vast **nitrogen cycle** (Fig. 22–1). The first step in the cycle is **fixation** (reduction) of atmospheric nitrogen by nitrogen-fixing bacteria to yield ammonia (NH_3 or NH_4^+). Although ammonia can be used by most living organisms, soil bacteria that derive their energy by oxidizing ammonia to nitrite (NO_2^-) and ultimately nitrate (NO_3^-) are so abundant and active that nearly all ammonia reaching the soil is oxidized to nitrate. This process is known as **nitrification.** Plants and many bacteria can take up and readily reduce nitrate and nitrite through the action of nitrate and nitrite reductases. The ammonia so formed is incorporated into amino acids by plants. Animals then use plants as a source of amino acids, both

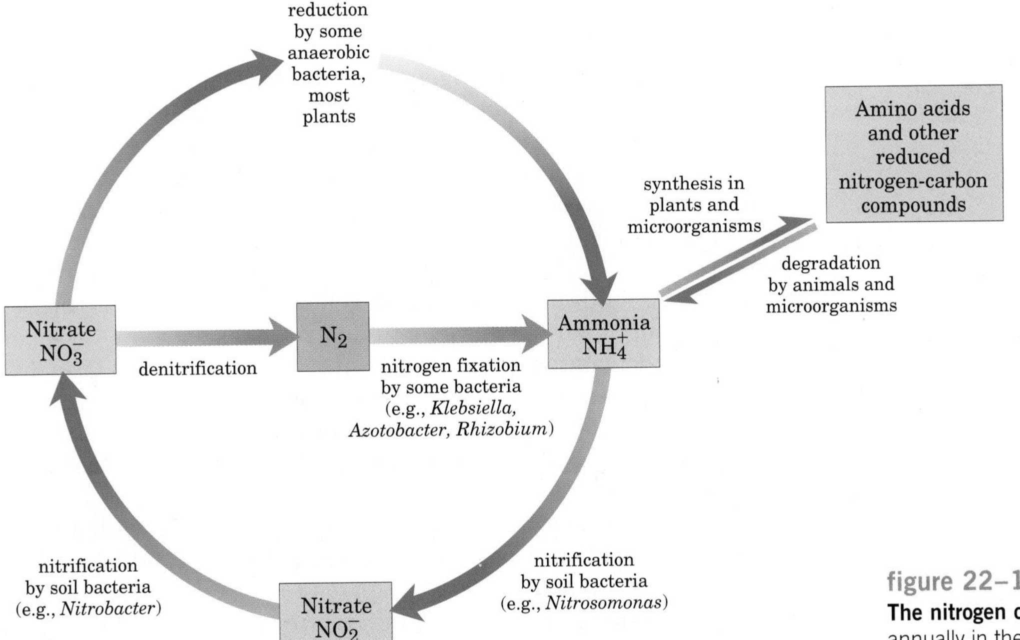

figure 22–1
The nitrogen cycle. The total amount of nitrogen fixed annually in the biosphere exceeds 10^{11} kg.

nonessential and essential, to build their proteins. When organisms die, microbial degradation of their proteins returns ammonia to the soil, where nitrifying bacteria again convert it into nitrite and nitrate. A balance is maintained between fixed nitrogen and atmospheric nitrogen by bacteria that convert nitrate to N_2 under anaerobic conditions, a process called **denitrification** (Fig. 22–1). These soil bacteria use NO_3^- rather than O_2 as the ultimate electron acceptor in a series of reactions that (like oxidative phosphorylation) generates a transmembrane proton gradient, which is used to synthesize ATP.

Now let us examine the process of nitrogen fixation, the first step in the nitrogen cycle.

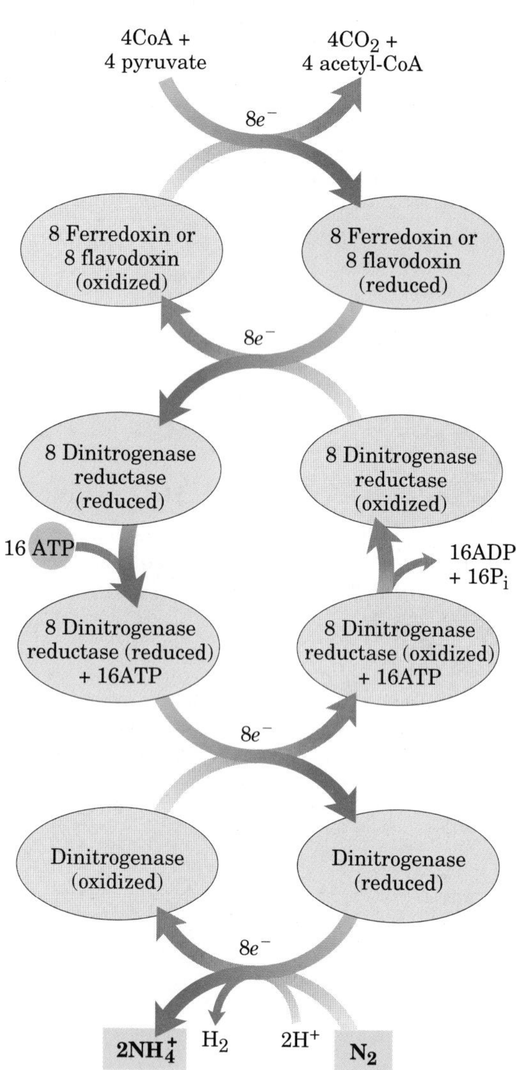

figure 22–2
Nitrogen fixation by the nitrogenase complex. Electrons are transferred from pyruvate to dinitrogenase via ferredoxin (or flavodoxin) and dinitrogenase reductase. Dinitrogenase is reduced one electron at a time by dinitrogenase reductase, with at least six electrons required to fix one molecule of N_2. An additional two electrons are used to reduce 2 H^+ to H_2 in a process that obligatorily accompanies nitrogen fixation in anaerobes, making a total of eight electrons required per N_2 molecule. The subunit structures and metal cofactors of the dinitrogenase reductase and dinitrogenase proteins are described in the text and Figure 22–3.

Nitrogen Is Fixed by Enzymes of the Nitrogenase Complex

Only certain prokaryotes can fix atmospheric nitrogen. These include the cyanobacteria of soils and fresh and salt waters, other kinds of free-living soil bacteria such as *Azotobacter* species, and the nitrogen-fixing bacteria that live as **symbionts** in the root nodules of leguminous plants. The first important product of nitrogen fixation is ammonia, which can be used by other organisms either directly or after its conversion to other soluble compounds such as nitrites, nitrates, or amino acids.

The reduction of nitrogen to ammonia is an exergonic reaction:

$$N_2 + 3H_2 \longrightarrow 2NH_3 \qquad \Delta G'^{\circ} = -33.5 \text{ kJ/mol}$$

The N≡N triple bond, however, is very stable, with a bond energy of 942 kJ/mol. Nitrogen fixation therefore has an extremely high activation energy, and atmospheric nitrogen is almost chemically inert under normal conditions. Ammonia is produced industrially by the Haber process (named for its inventor Fritz Haber), which requires temperatures of 400 to 500 °C and N_2 and H_2 at pressures of tens of thousands of kilopascals (several hundred atmospheres) to provide the necessary activation energy. Biological nitrogen fixation, however, must occur at 0.8 atm of nitrogen and at biological temperatures. The high activation barrier is overcome, at least in part, by the binding and hydrolysis of ATP. The overall reaction can be written

$$N_2 + 10H^+ + 8e^- + 16ATP \longrightarrow 2NH_4^+ + 16ADP + 16P_i + H_2$$

Biological nitrogen fixation is carried out by a highly conserved complex of proteins called the **nitrogenase complex** (Fig. 22–2), the key components of which are **dinitrogenase reductase** and **dinitrogenase** (Fig. 22–3). Dinitrogenase reductase (M_r 60,000) is a dimer of two identical subunits. It contains a single 4Fe-4S redox center (see Fig. 19–5) and can be oxidized and reduced by one electron. It also has two binding sites for ATP/ADP. Dinitrogenase (M_r 240,000), a tetramer with two copies of two different subunits, contains both iron and molybdenum; its redox centers have a total of 2 Mo, 32 Fe, and 30 S per tetramer. About half of the Fe and S is present as two bridged pairs of 4Fe-4S centers called P clusters; the remainder is present as part of a novel iron-molybdenum cofactor. A form of nitrogenase that contains vanadium rather than molybdenum has been discovered, and some bacterial species can produce both types of nitrogenase systems. The vanadium enzyme may be the primary nitrogen-fixation system under some environmental conditions, but it has not yet been well characterized.

Nitrogen fixation is carried out by a highly reduced form of dinitrogenase and requires eight electrons: six for the reduction of N_2 and two to produce one molecule of H_2 as an obligate part of the reaction mechanism. Dinitrogenase is reduced by the transfer of electrons from dinitrogenase reductase (Fig. 22–2). Dinitrogenase has two binding sites for the reductase,

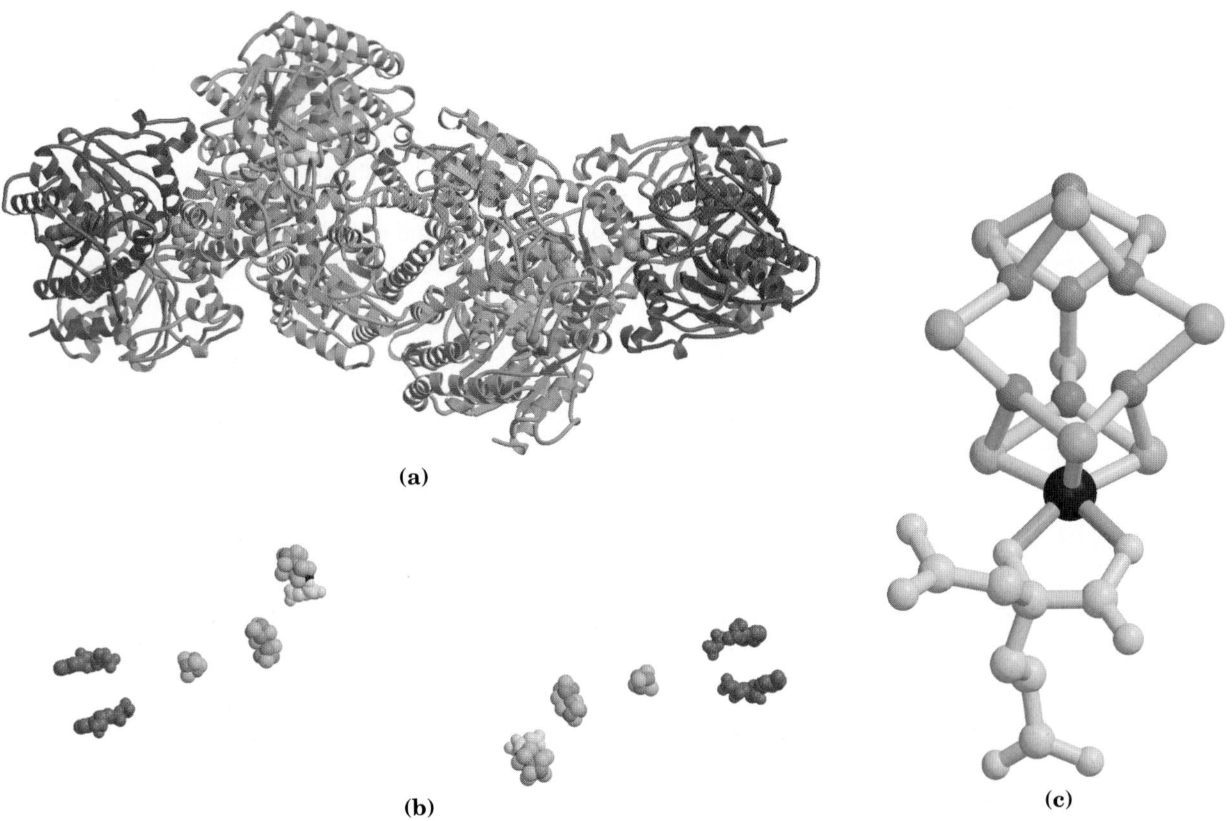

figure 22–3
Enzymes and cofactors of the nitrogenase complex. **(a)** In this ribbon diagram, the dinitrogenase subunits are shown in gray and pink, the dinitrogenase reductase subunits in blue and green. Bound ADP is shown in red. Note the 4Fe-4S complex (Fe and S atoms orange and yellow, respectively) and the Fe-Mo cofactor (Mo black, homocitrate light gray). The P clusters (bridged pairs of 4Fe-4S complexes) are also shown. **(b)** The dinitrogenase complex cofactors without the protein. The colors are as in **(a)**. **(c)** The Fe-Mo cofactor contains 1 Mo (black), 7 Fe (orange), 9 S (yellow), and one molecule of homocitrate (gray).

and the required eight electrons are transferred to dinitrogenase one at a time, with the reduced reductase binding to the dinitrogenase and the oxidized reductase dissociating from it in a repeating cycle. This cycle requires the hydrolysis of ATP by the reductase. The immediate source of electrons to reduce dinitrogenase reductase varies, with reduced ferredoxin (p. 702; see also Fig. 19–5), reduced flavodoxin, and perhaps other sources playing a role. In at least one instance, the ultimate source of electrons to reduce ferredoxin is pyruvate (Fig. 22–2).

The role of ATP in this process is interesting in that it appears to be catalytic rather than thermodynamic. Remember that ATP can contribute not only chemical energy, through the hydrolysis of one or more of its phosphoanhydride bonds, but also binding energy (pp. 251, 356) through noncovalent interactions that lower the activation energy. In the reaction carried out by dinitrogenase reductase, both ATP binding and ATP hydrolysis bring about protein conformational changes that evidently help overcome the high activation energy of nitrogen fixation. ATP binding to the reductase shifts the reduction potential (E'°) of this protein from −250 to −400 mV, an enhancement of its reducing power that is required to transfer electrons to dinitrogenase. Two ATP molecules are then hydrolyzed just before the actual transfer of each electron from dinitrogenase reductase to dinitrogenase.

Another important characteristic of the nitrogenase complex is an extreme lability when oxygen is present. The reductase is inactivated in air, with a half-life of 30 s; dinitrogenase has a half-life of 10 min in air. Free-living bacteria that fix nitrogen cope with this problem in a variety of ways. Some live only anaerobically or repress nitrogenase synthesis when oxygen is present. Some aerobic species, such as *Azotobacter vinelandii*, partially uncouple electron transport from ATP synthesis so that oxygen is burned off as rapidly as it enters the cell (see Box 19–1). When fixing nitrogen, cultures of these bacteria actually warm up as a result of their efforts to rid themselves of oxygen. Some filamentous nitrogen-fixing cyanobacteria use yet another approach: one of every nine cells differentiates into a heterocyst, specialized for nitrogen fixation, with cell walls thick enough to prevent oxygen from entering.

The symbiotic relationship between leguminous plants and the nitrogen-fixing bacteria in their root nodules (Fig. 22–4) takes care of both the energy requirements of the reaction and the oxygen lability of the enzymes. The energy required for nitrogen fixation was probably the evolutionary driving force for this association of plants with bacteria. The bacteria in root nodules have access to a large reservoir of energy in the form of abundant carbohydrate and citric acid cycle intermediates made available by the plant. This may allow the bacteria to fix hundreds of times more nitrogen than do their free-living cousins under conditions generally encountered in soils. To solve the oxygen-toxicity problem, the bacteria in root nodules are bathed in a solution of an oxygen-binding heme protein called **leghemoglobin,** produced by the plant (although the heme may be contributed by the bacteria). Leghemoglobin binds all available oxygen so that it cannot interfere with nitrogen fixation, and it efficiently delivers the oxygen to the bacterial electron-transfer system. The efficiency of the symbiosis between plants and bacteria is evident in the enrichment of soil nitrogen brought about by leguminous plants. This enrichment is the basis of the crop rotation methods used by many farmers, in which plantings of nonleguminous plants (such as maize) that extract fixed nitrogen from the soil are alternated every few years with plantings of legumes such as alfalfa, peas, or clover.

Nitrogen fixation is the subject of intense study because of its immense practical importance. Producing ammonia industrially for use in fertilizers requires a large and expensive input of energy, spurring a drive to develop recombinant or transgenic organisms that can fix nitrogen. Recombinant DNA techniques are being used to transfer the DNA that encodes the enzymes of nitrogen fixation into non–nitrogen-fixing bacteria and plants (Chapter 29). Success in these efforts will depend on overcoming the problem of oxygen toxicity in any cell producing nitrogenase.

figure 22–4

Nitrogen-fixing nodules. (a) Root nodules of bird's-foot trefoil, a legume. **(b)** Artificially colorized electron micrograph of a thin section through a pea root nodule. Symbiotic nitrogen-fixing bacteria, or bacteroids (red), live inside the nodule cells, surrounded by the peribacteroid membrane (blue). Bacteroids produce the nitrogenase complex that converts atmospheric nitrogen (N_2) to ammonium (NH_4^+); without the bacteroids, the plant is unable to utilize N_2. The infected root cells provide some factors essential for nitrogen fixation, including leghemoglobin. This heme protein has a very high binding affinity for oxygen, which strongly inhibits nitrogenase. (The cell nucleus is shown in yellow/green. Not visible in this micrograph are other organelles of the infected root cell that are normally found in plant cells.)

(a)

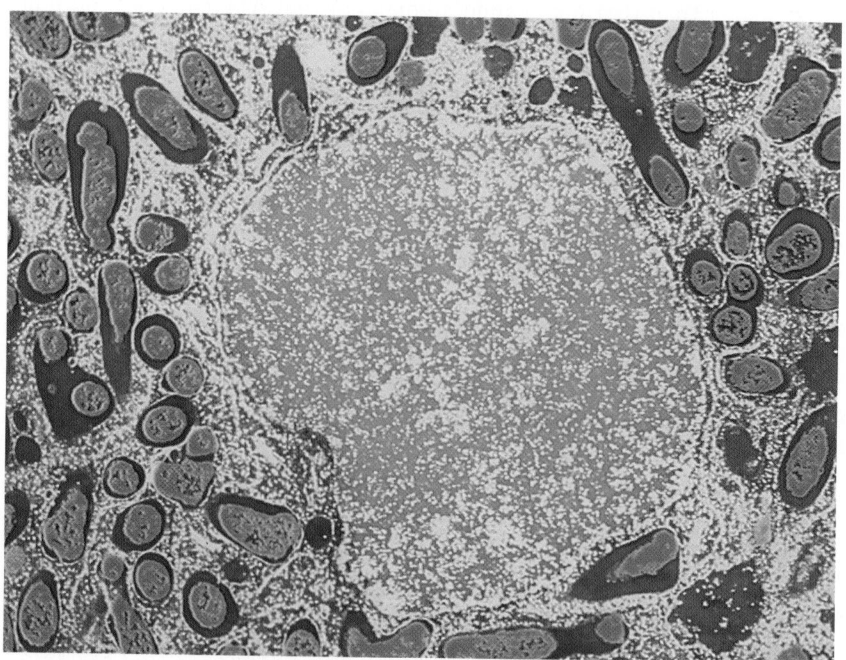

(b)

Ammonia Is Incorporated into Biomolecules through Glutamate and Glutamine

Reduced nitrogen in the form of NH_4^+ is assimilated into amino acids and then into other nitrogen-containing biomolecules. Two amino acids, **glutamate** and **glutamine,** provide the critical entry point. Recall that these same two amino acids play central roles in the catabolism of ammonia and amino groups in amino acid oxidation (Chapter 18). Glutamate is the source of amino groups for most other amino acids, through transamination reactions (the reverse reaction in Fig. 18–4). The amide nitrogen of glutamine is a source of amino groups in a wide range of biosynthetic processes. In most types of cells, and in extracellular fluids in higher organisms, one or both of these amino acids are present at higher concentrations—sometimes an order of magnitude or more higher—than other amino acids. An *E. coli* cell requires so much glutamate that this amino acid is one of the primary solutes in the cytosol. Its concentration is regulated not only in response to the cell's nitrogen requirements, but also to maintain an osmotic balance between the cytosol and the external medium.

The biosynthetic pathways to glutamate and glutamine are simple, and all or some of the steps occur in most organisms. The most important pathway for the assimilation of NH_4^+ into glutamate requires two reactions. First, **glutamine synthetase** catalyzes the reaction of glutamate and NH_4^+ to yield glutamine. Recall that this reaction takes place in two steps, with enzyme-bound γ-glutamyl phosphate as an intermediate (p. 632):

(1) Glutamate + ATP $\longrightarrow$ γ-glutamyl phosphate + ADP
(2) γ-Glutamyl phosphate + NH_4^+ $\longrightarrow$ glutamine + P_i + H^+
Sum: Glutamate + NH_4^+ + ATP $\longrightarrow$ glutamine + ADP + P_i + H^+ (22–1)

Glutamine synthetase is found in all organisms. In addition to its importance for NH_4^+ assimilation in bacteria, it has a central role in amino acid metabolism in mammals, converting toxic free NH_4^+ into glutamine for transport in the blood (Chapter 18).

In bacteria and plants, glutamate is produced from glutamine in a reaction catalyzed by **glutamate synthase.** α-Ketoglutarate, an intermediate of the citric acid cycle, undergoes reductive amination with glutamine as nitrogen donor:

$$\alpha\text{-Ketoglutarate} + \text{glutamine} + \text{NADPH} + H^+ \longrightarrow 2\ \text{glutamate} + \text{NADP}^+ \quad (22\text{–}2)$$

The net reaction of glutamine synthetase and glutamate synthase (Eqns 22–1, 22–2) is

$$\alpha\text{-Ketoglutarate} + NH_4^+ + \text{NADPH} + \text{ATP} \longrightarrow \text{L-glutamate} + \text{NADP}^+ + \text{ADP} + P_i$$

Glutamate synthase is not present in animals; instead, glutamate is maintained at high levels by processes such as the transamination of α-ketoglutarate during amino acid catabolism.

Glutamate can also be formed in yet another, albeit minor, pathway. The reaction of α-ketoglutarate and NH_4^+ to form glutamate in one step is catalyzed by L-glutamate dehydrogenase, an enzyme present in all organisms. Reducing power is furnished by NADPH:

$$\alpha\text{-Ketoglutarate} + NH_4^+ + \text{NADPH} \longrightarrow \text{L-glutamate} + \text{NADP}^+ + H_2O$$

We encountered this reaction in the catabolism of amino acids (Chapter 18). In eukaryotic cells, L-glutamate dehydrogenase is located in the mitochondrial matrix. The reaction equilibrium favors reactants, and the K_m for NH_4^+ (~1 mM) is so high that the reaction probably makes only a modest contribution to NH_4^+ assimilation into amino acids and other metabolites. (Recall that the glutamate dehydrogenase reaction, in reverse (see Fig. 18–9), is a primary source of NH_4^+ destined for the urea cycle.) Soil bacteria and plants generally rely on the two-enzyme pathway outlined above (Eqns 22–1, 22–2). Concentrations of NH_4^+ high enough for the glutamate dehydrogenase reaction to make a significant contribution to glutamate levels generally occur only when NH_3 is added to the soil or when organisms are grown in a laboratory in the presence of high NH_3 concentrations.

Glutamine Synthetase Is a Primary Regulatory Point in Nitrogen Metabolism

The activity of glutamine synthetase is regulated in virtually all organisms—not surprising given its central metabolic role as an entry point for reduced nitrogen. In enteric bacteria such as *Escherichia coli,* the regulation is unusually complex. The enzyme has 12 identical subunits of M_r 50,000 (Fig. 22–5) and is regulated both allosterically and by covalent modification. Alanine, glycine, and at least six end products of glutamine metabolism are allosteric inhibitors of the enzyme (Fig. 22–6). Each inhibitor alone produces only partial inhibition. The effects of multiple inhibitors, however, are more than additive, and all eight together virtually shut down the enzyme. This control mechanism provides continuous adjustment of glutamine levels to match immediate metabolic requirements.

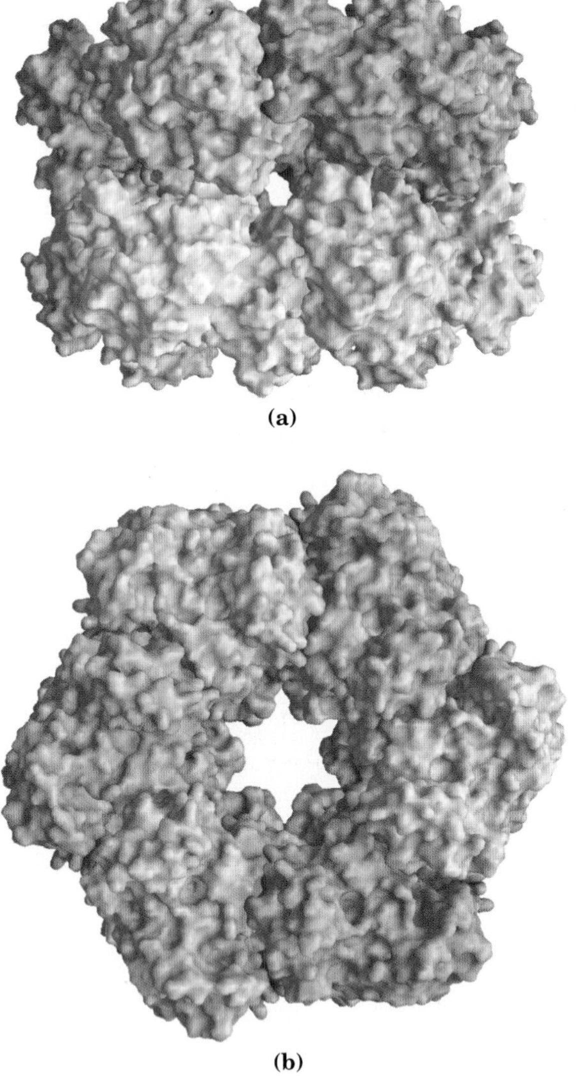

figure 22–5
Subunit structure of glutamine synthetase as determined by x-ray diffraction. (a) Side view. The 12 subunits are identical; they are differently colored to illustrate packing and placement. **(b)** Top view, showing active sites (green).

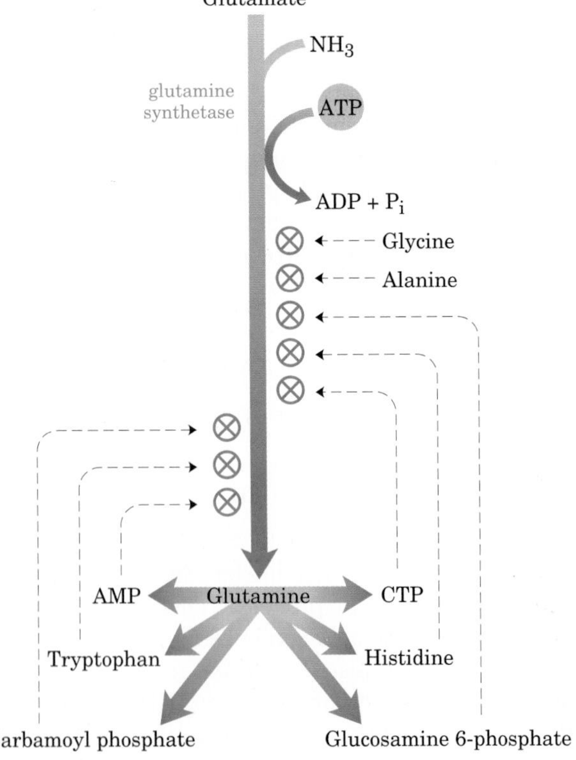

figure 22–6
Allosteric regulation of glutamine synthetase. The enzyme undergoes cumulative regulation by six end products of glutamine metabolism. Alanine and glycine probably serve as indicators of the general status of amino acid metabolism in the cell.

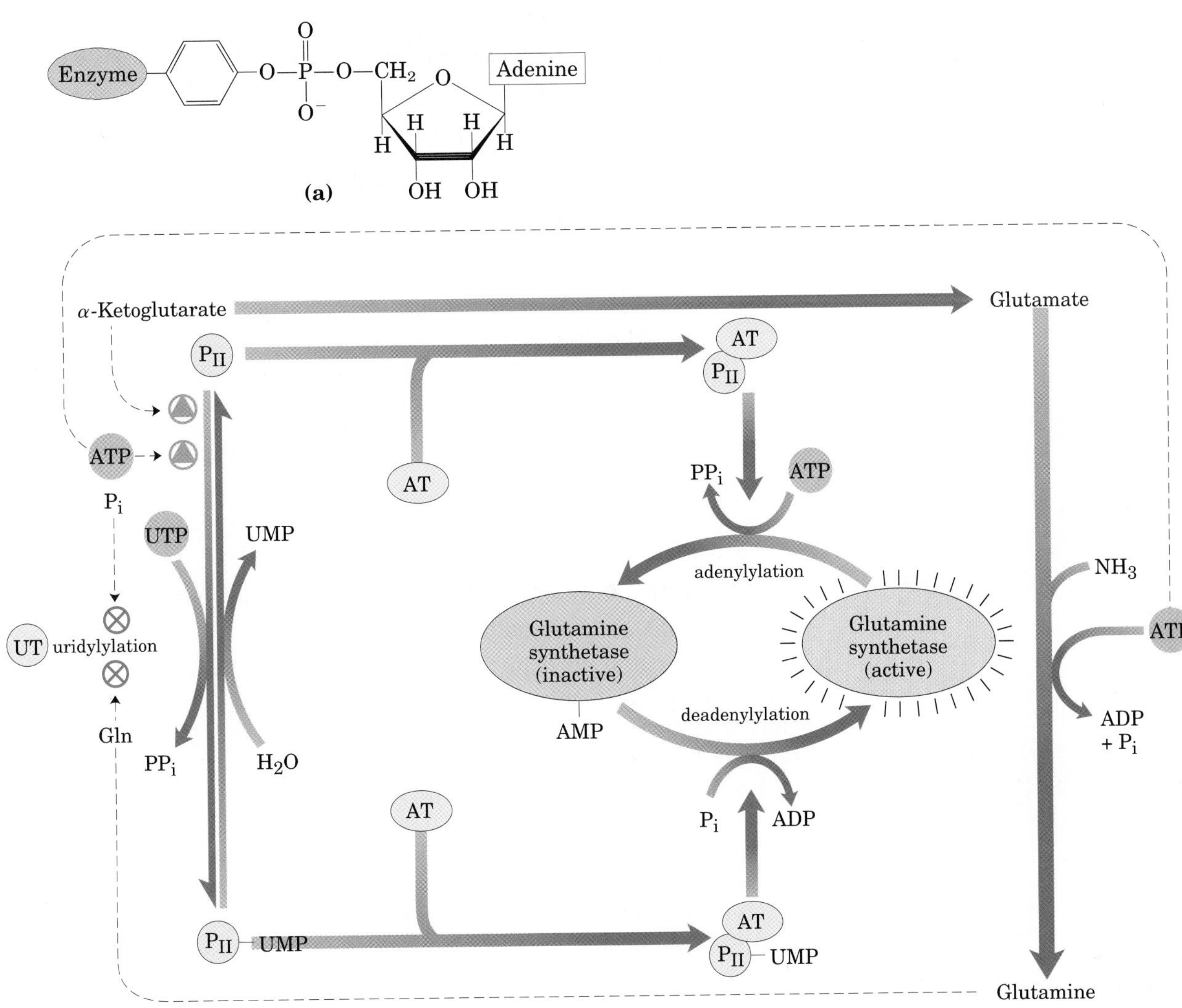

figure 22–7

Second level of regulation of glutamine synthetase: covalent modifications. (a) An adenylylated Tyr residue. **(b)** Cascade leading to adenylylation (inactivation) of glutamine synthetase. AT represents adenylyltransferase; UT, uridylyltransferase. Details of this cascade are discussed in the text.

Superimposed on the allosteric regulation is inhibition by adenylylation of (addition of AMP to) Tyr^{397}, located near the enzyme's active site (Fig. 22–7). This covalent modification increases sensitivity to the allosteric inhibitors, and activity decreases as more subunits are adenylylated. Both adenylylation and deadenylylation are promoted by **adenylyltransferase,** part of a complex enzymatic cascade that responds to levels of glutamine, α-ketoglutarate, ATP, and P_i. The activity of adenylyltransferase is modulated by binding to a regulatory protein called P_{II}, and the activity of P_{II}, in turn, is regulated by covalent modification (uridylylation), again at a Tyr residue. The adenylyltransferase complex with uridylylated P_{II} (P_{II}-UMP) stimulates deadenylylation, whereas the same complex with deuridylylated P_{II} stimulates adenylylation of glutamine synthetase. Both uridylylation and deuridylylation of P_{II} are brought about by a single enzyme, **uridylyltransferase.** Uridylylation is inhibited by binding of glutamine and P_i to uridylyltransferase and is stimulated by binding of α-ketoglutarate and ATP to P_{II}.

The net result of this complex mechanism is a decrease in glutamine synthetase activity when glutamine levels are high, and an increase in activity when glutamine levels are low and α-ketoglutarate and ATP (substrates for the synthetase reaction) are available.

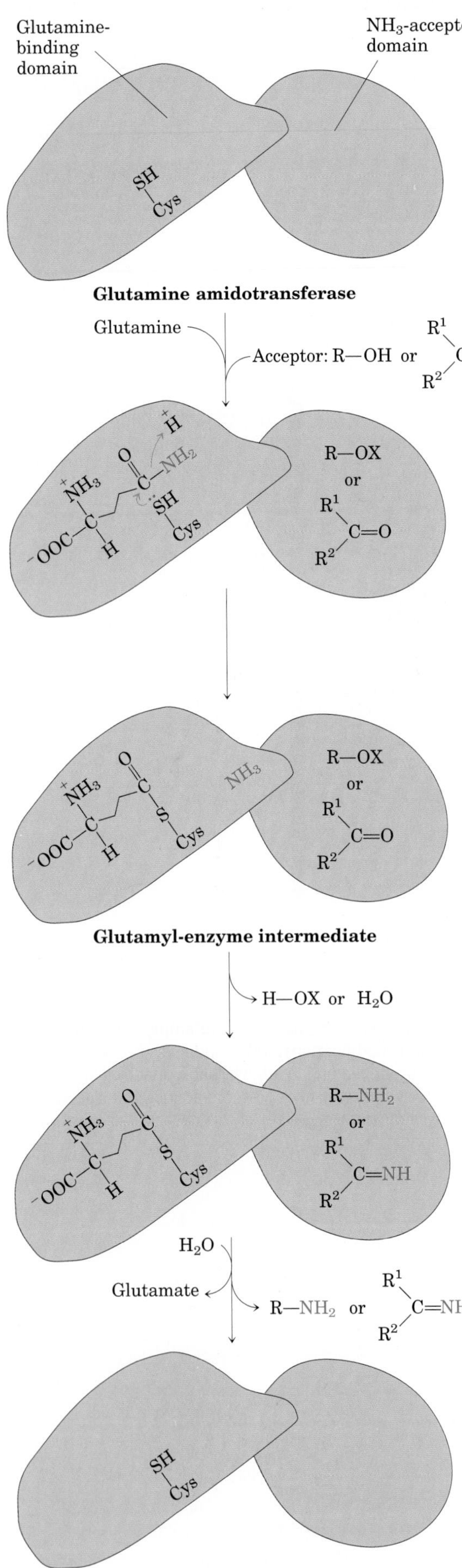

figure 22–8

Proposed mechanism for glutamine amidotransferases. Each enzyme has two domains. The glutamine-binding domain contains a number of structural elements conserved among many of these enzymes, including a Cys residue required for activity. The NH_3-acceptor (second substrate) domain varies. The γ-amido nitrogen of glutamine (red) is released as NH_3 in a reaction that probably involves a covalent glutamyl-enzyme intermediate. Two types of amino acceptors are shown. X represents an activating group, typically a phosphoryl group derived from ATP, that facilitates displacement of a hydroxyl group from R—OH by NH_3.

Several Classes of Reactions Play Special Roles in the Biosynthesis of Amino Acids and Nucleotides

The pathways described in this chapter include a variety of interesting chemical rearrangements. Several of these recur and deserve special note before we progress to the pathways themselves. These are (1) transamination reactions and other rearrangements promoted by enzymes containing pyridoxal phosphate, (2) transfer of one-carbon groups using either tetrahydrofolate or *S*-adenosylmethionine as cofactor, and (3) transfer of amino groups derived from the amide nitrogen of glutamine.

Pyridoxal phosphate (PLP), tetrahydrofolate (H_4 folate), and *S*-adenosylmethionine (adoMet) were described in some detail in Chapter 18 (see Figs 18–6, 18–16, and 18–17). Here we focus on amino group transfer involving the amide nitrogen of glutamine.

There are over a dozen known biosynthetic reactions in which glutamine is the major physiological source of amino groups, and most of these occur in the pathways outlined in this chapter. As a class, the enzymes catalyzing these reactions are called **glutamine amidotransferases,** and all have two structural domains: one binding glutamine, the other binding the second substrate, which serves as amino group acceptor (Fig. 22–8). A conserved Cys residue in the glutamine-binding domain is believed to act as a nucleophile, cleaving the amide bond of glutamine and forming a covalent glutamyl-enzyme intermediate. The NH_3 produced in this reaction remains at the active site and reacts with the second substrate to form the aminated product. The covalent intermediate is hydrolyzed to the free enzyme and glutamate. If the second substrate must be activated, ATP is most often used to generate an acyl phosphate intermediate (R—OX in Fig. 22–8). Glutaminase acts in a similar fashion but uses H_2O as the second substrate, yielding NH_4^+ and glutamate (p. 632).

Biosynthesis of Amino Acids

All amino acids are derived from intermediates in glycolysis, the citric acid cycle, or the pentose phosphate pathway (Fig. 22–9). Nitrogen enters these pathways by way of glutamate and glutamine. Some pathways are simple, others are not. Ten of the amino acids are only one or several steps removed from the common metabolite from which they are derived. The biosynthetic pathways for others, such as the aromatic amino acids, are more complex.

Organisms vary greatly in their ability to synthesize the 20 standard amino acids. Whereas most bacteria and plants can synthesize all 20, mammals can synthesize only about half of them—generally those with simple pathways. These are the **nonessential amino acids,** not needed in the diet (see Table 18–1). The remainder, the **essential amino acids,** must be obtained from food. Unless otherwise indicated, the pathways for the 20 standard amino acids presented below are those operative in bacteria.

A useful way to organize these biosynthetic pathways is to group them into six families corresponding to their metabolic precursors (Table 22–1), and we use this approach to structure the detailed descriptions that follow.

In addition to these six precursors, there is a notable intermediate in several pathways of amino acid and nucleotide synthesis: **5-phosphoribosyl-1-pyrophosphate** (PRPP).

5-Phosphoribosyl-1-pyrophosphate (PRPP)

PRPP is synthesized from ribose 5-phosphate derived from the pentose phosphate pathway (see Fig. 15–20), in a reaction catalyzed by **ribose phosphate pyrophosphokinase:**

$$\text{Ribose 5-phosphate} + \text{ATP} \longrightarrow \text{5-phosphoribosyl-1-pyrophosphate} + \text{AMP}$$

This enzyme is allosterically regulated by many of the biomolecules for which PRPP is a precursor.

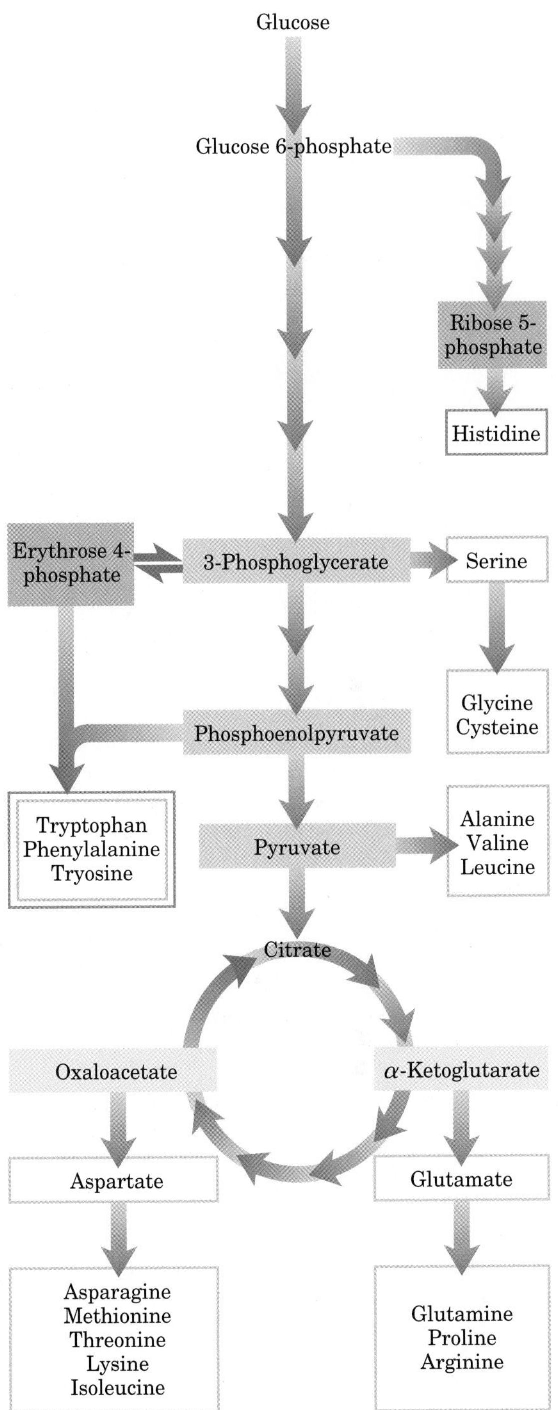

figure 22–9
Overview of amino acid biosynthesis. Precursors from glycolysis (red), the citric acid cycle (blue), and the pentose phosphate pathway (purple) are shaded, and the amino acids derived from them are boxed in the corresponding colors. The same device—the color-matching of precursors and their pathway end products—is used in illustrations of the individual pathways (Figs 22–10 to 22–20).

table 22–1

Amino Acid Biosynthetic Families, Grouped by Metabolic Precursor

α-Ketoglutarate	**Pyruvate**
Glutamate	Alanine
Glutamine	Valine†
Proline	Leucine†
Arginine*	
	Phosphoenolpyruvate and erythrose 4-phosphate
3-Phosphoglycerate	Tryptophan†
Serine	Phenylalanine†
Glycine	Tyrosine‡
Cysteine	
Oxaloacetate	**Ribose 5-phosphate**
Aspartate	Histidine†
Asparagine	
Methionine†	
Threonine†	
Lysine†	
Isoleucine†	

*Essential in young animals.
†Essential amino acids.
‡Derived from phenylalanine in mammals.

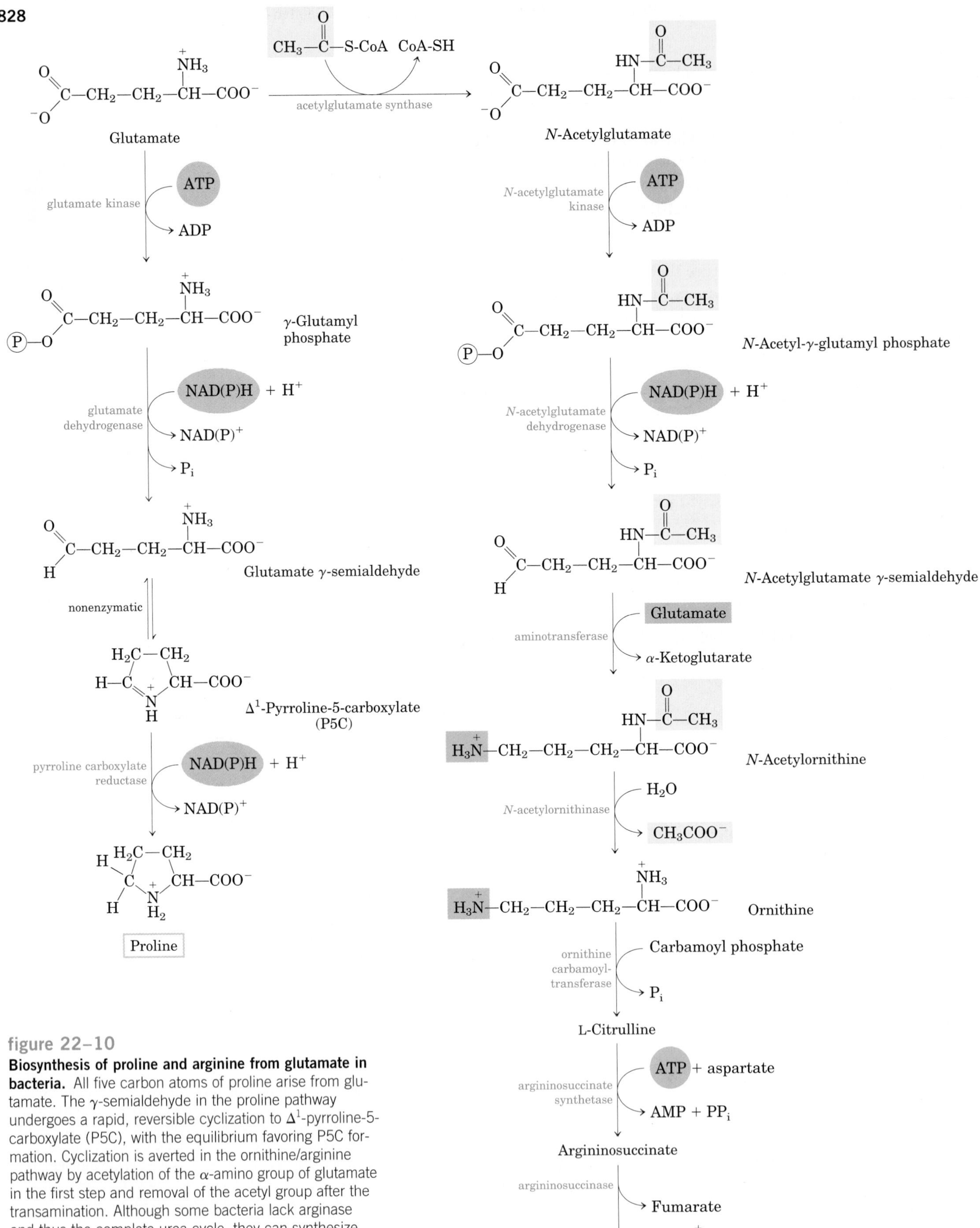

figure 22–10

Biosynthesis of proline and arginine from glutamate in bacteria. All five carbon atoms of proline arise from glutamate. The γ-semialdehyde in the proline pathway undergoes a rapid, reversible cyclization to Δ^1-pyrroline-5-carboxylate (P5C), with the equilibrium favoring P5C formation. Cyclization is averted in the ornithine/arginine pathway by acetylation of the α-amino group of glutamate in the first step and removal of the acetyl group after the transamination. Although some bacteria lack arginase and thus the complete urea cycle, they can synthesize arginine from ornithine in steps that parallel the mammalian urea cycle, with citrulline and argininosuccinate as intermediates (see Fig. 18–9).

α-Ketoglutarate Gives Rise to Glutamate, Glutamine, Proline, and Arginine

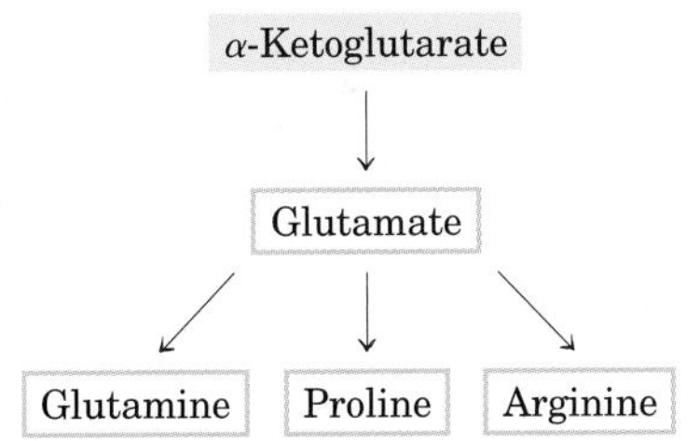

We have already described the biosynthesis of **glutamate** and **glutamine.** **Proline** is a cyclized derivative of glutamate (Fig. 22–10). In the first step of proline synthesis, ATP reacts with the γ-carboxyl group of glutamate to form an acyl phosphate, which is reduced by NADPH to glutamate γ-semialdehyde. This intermediate undergoes rapid spontaneous cyclization and is then reduced further to yield proline.

Arginine is synthesized from glutamate via ornithine and the urea cycle in animals (Chapter 18). In principle, ornithine could also be synthesized from glutamate γ-semialdehyde by transamination, but the spontaneous cyclization of the semialdehyde in the proline pathway precludes a sufficient supply of this intermediate for ornithine synthesis. Bacteria have a de novo biosynthetic pathway for ornithine (and thus arginine) that parallels some steps of the proline pathway but includes two additional steps that avoid the problem of the spontaneous cyclization of glutamate γ-semialdehyde (Fig. 22–10). In the first step, the α-amino group of glutamate is blocked by an acetylation requiring acetyl-CoA; then, after the transamination step, the acetyl group is removed to yield ornithine.

The pathways to proline and arginine are somewhat different in mammals. Proline can be synthesized by the pathway shown in Figure 22–10, but it is also formed from arginine obtained from dietary or tissue protein. Arginase, a urea cycle enzyme, converts arginine to ornithine and urea (see Fig. 18–25). The ornithine is converted to glutamate γ-semialdehyde by the enzyme **ornithine δ-aminotransferase** (Fig. 22–11). The semialdehyde cyclizes to Δ^1-pyrroline-5-carboxylate, which is then converted to proline (Fig. 22–10). The pathway for arginine synthesis shown in Figure 22–10 is absent in mammals. When arginine from dietary intake or protein turnover is insufficient for protein synthesis, the ornithine δ-aminotransferase reaction operates in the direction of ornithine formation. Ornithine is then converted to citrulline and arginine in the urea cycle.

Ornithine ⇌ (α-Ketoglutarate → Glutamate; ornithine δ-aminotransferase) [Glutamate γ-semialdehyde ⇌ Δ^1-Pyrroline-5-carboxylate (P5C)]

figure 22–11
Ornithine δ-aminotransferase reaction: a step in the mammalian pathway to proline. This enzyme is found in the mitochondrial matrix of most tissues. Although the equilibrium favors P5C formation, the reverse reaction is the only mammalian pathway for synthesis of ornithine (and thus arginine) when arginine levels are insufficient for protein synthesis.

Serine, Glycine, and Cysteine Are Derived from 3-Phosphoglycerate

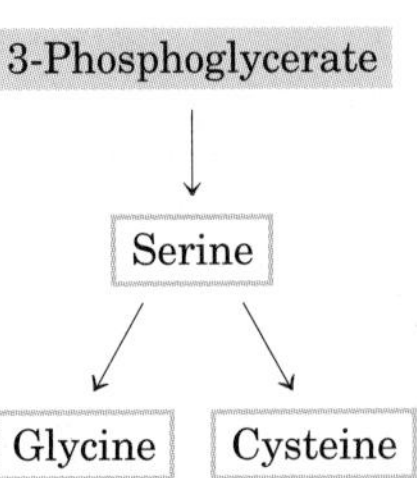

The major pathway for the formation of **serine** is the same in all organisms (Fig. 22–12). In the first step, the hydroxyl group of 3-phosphoglycerate is oxidized by a dehydrogenase (using NAD^+) to yield 3-phosphohydroxypyruvate. Transamination from glutamate yields 3-phosphoserine, which is hydrolyzed to free serine by phosphoserine phosphatase.

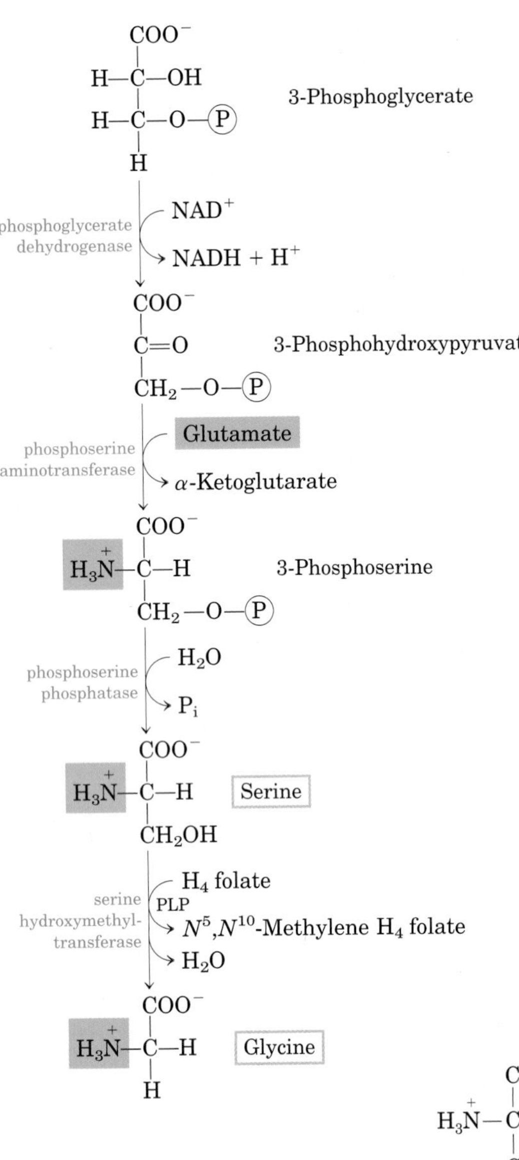

figure 22–12

Biosynthesis of serine from 3-phosphoglycerate and of glycine from serine in all organisms. Glycine is also made from CO_2 and NH_4^+ by the action of glycine synthase, with N^5,N^{10}-methylenetetrahydrofolate as methyl group donor (see text).

Serine (three carbons) is the precursor of **glycine** (two carbons) through removal of a carbon atom by **serine hydroxymethyltransferase** (Fig. 22–12). Tetrahydrofolate accepts the β carbon (C-3) of serine, which forms a methylene bridge between N-5 and N-10 to yield N^5,N^{10}-methylenetetrahydrofolate (see Fig. 18–16). The overall reaction, which is reversible, also requires pyridoxal phosphate.

In the liver of vertebrates, glycine can be made by another route: the reverse of the reaction shown in Figure 18–19b, catalyzed by **glycine synthase:**

$$CO_2 + NH_4^+ + NADH + H^+ + N^5,N^{10}\text{-methylenetetrahydrofolate} \longrightarrow \text{glycine} + NAD^+ + \text{tetrahydrofolate}$$

Plants and bacteria produce the reduced sulfur required for the synthesis of **cysteine** (and methionine, described later) from environmental sulfates; the pathway is shown in Figure 22–13. Sulfate is activated in two steps to produce 3-phosphoadenosine 5′-phosphosulfate (PAPS), which undergoes an eight-electron reduction to sulfide. The sulfide is then used in formation of cysteine in a two-step pathway from serine. In mammals, cysteine is made from two amino acids: methionine furnishes the sulfur atom

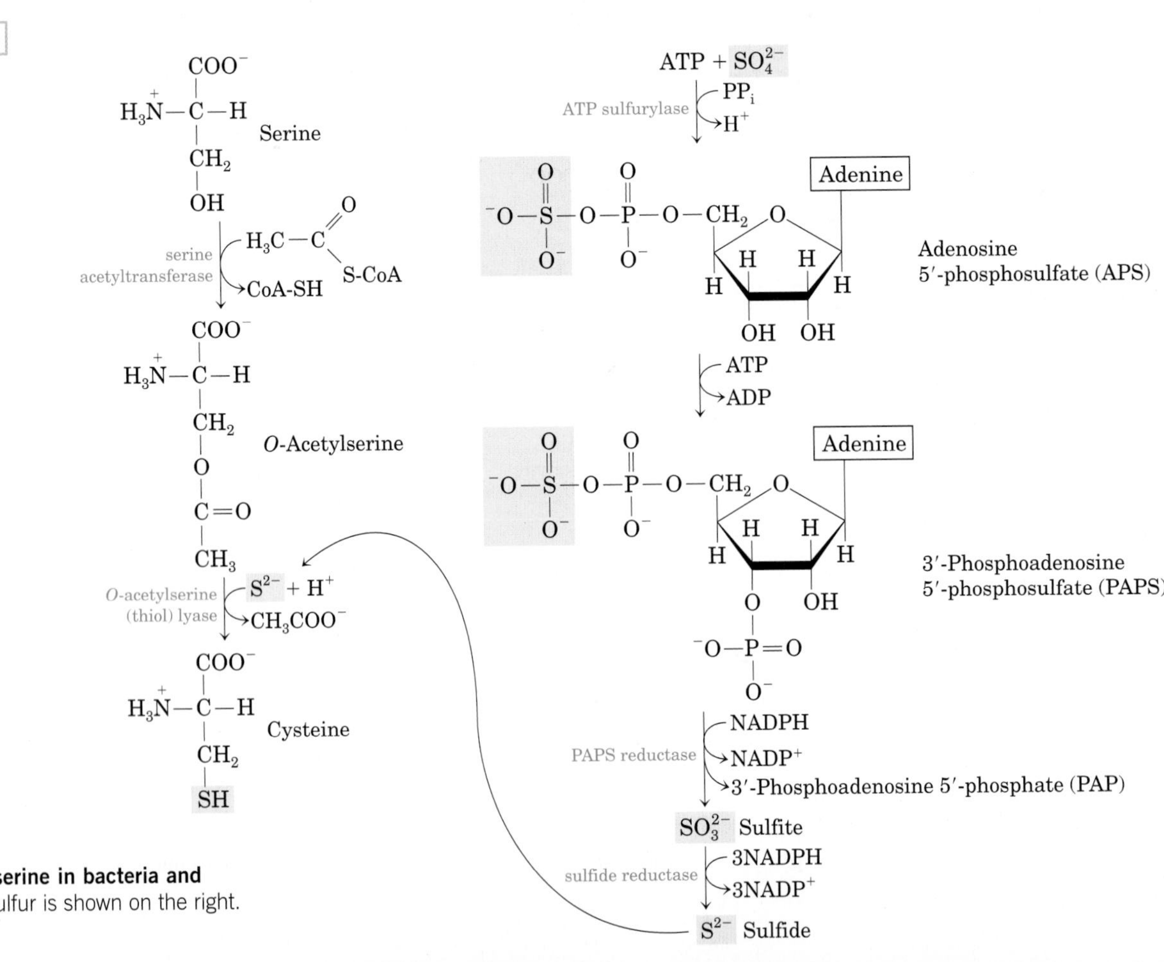

figure 22–13

Biosynthesis of cysteine from serine in bacteria and plants. The origin of reduced sulfur is shown on the right.

$$^{-}OOC{-}\underset{\underset{^{+}NH_3}{|}}{CH}{-}CH_2{-}CH_2{-}SH \;+\; HOCH_2{-}\overset{\overset{^{+}NH_3}{|}}{CH}{-}COO^{-}$$

Homocysteine Serine

cystathionine β-synthase (PLP; → H_2O)

$$^{-}OOC{-}\underset{\underset{^{+}NH_3}{|}}{CH}{-}CH_2{-}CH_2{-}S{-}CH_2{-}\overset{\overset{^{+}NH_3}{|}}{CH}{-}COO^{-}$$

Cystathionine

cystathionine γ-lyase (H_2O; PLP; → NH_4^+)

$$^{-}OOC{-}\underset{\underset{O}{\|}}{C}{-}CH_2{-}CH_3 \;+\; HS{-}CH_2{-}\overset{\overset{^{+}NH_3}{|}}{CH}{-}COO^{-}$$

α-Ketobutyrate Cysteine

figure 22–14

Biosynthesis of cysteine from homocysteine and serine in mammals. The homocysteine is formed from methionine as described in the text.

and serine furnishes the carbon skeleton. Methionine is first converted to *S*-adenosylmethionine (see Fig. 18–17), which can lose its methyl group to any of a number of acceptors to form *S*-adenosylhomocysteine. This demethylated product is hydrolyzed to free homocysteine, which undergoes a reaction with serine, catalyzed by **cystathionine β-synthase,** to yield cystathionine (Fig. 22–14). Finally, **cystathionine γ-lyase,** a PLP-requiring enzyme, catalyzes removal of ammonia and cleavage of cystathionine to yield free cysteine.

Three Nonessential and Six Essential Amino Acids Are Synthesized from Oxaloacetate and Pyruvate

Alanine and **aspartate** are synthesized from pyruvate and oxaloacetate, respectively, by transamination from glutamate. **Asparagine** is synthesized by amidation of aspartate, with glutamine donating the NH_4^+. These are nonessential amino acids and their simple biosynthetic pathways occur in all organisms.

Methionine, threonine, lysine, isoleucine, valine, and leucine are essential amino acids. Their biosynthetic pathways are complex and interconnected. In some cases, the pathways in bacteria, fungi, and plants differ significantly. The bacterial pathways are outlined in Figure 22–15.

Aspartate gives rise to **methionine, threonine,** and **lysine.** Branch points occur at aspartate β-semialdehyde, an intermediate in all three pathways, and at homoserine, a precursor of threonine and methionine. Threonine, in turn, is one of the precursors of isoleucine. The **valine** and **isoleucine** pathways share four enzymes (Fig. 22–15, steps ⑱ to ㉑). Pyruvate gives rise to valine and isoleucine in pathways that begin with condensation of two carbons of pyruvate (in the form of hydroxyethyl thiamine pyrophosphate; see Fig. 15–9) with another molecule of pyruvate (valine path) or with α-ketobutyrate (isoleucine path). The α-ketobutyrate is derived from threonine in a reaction that requires pyridoxal phosphate (Fig. 22–15, step ⑰). An intermediate in the valine pathway, α-ketoisovalerate, is the starting point for a four-step branch pathway leading to **leucine** (steps ㉒ to ㉕).

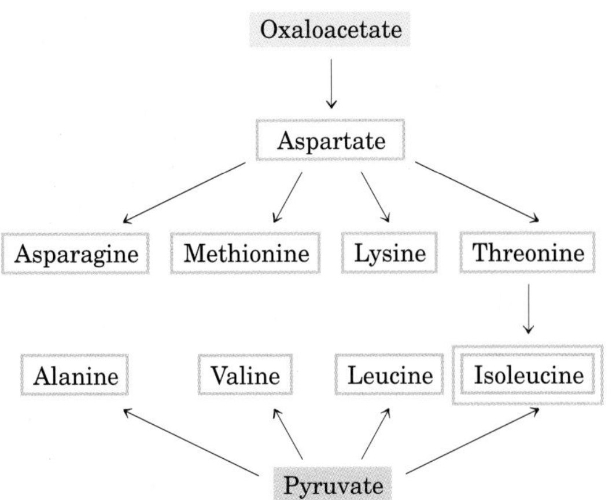

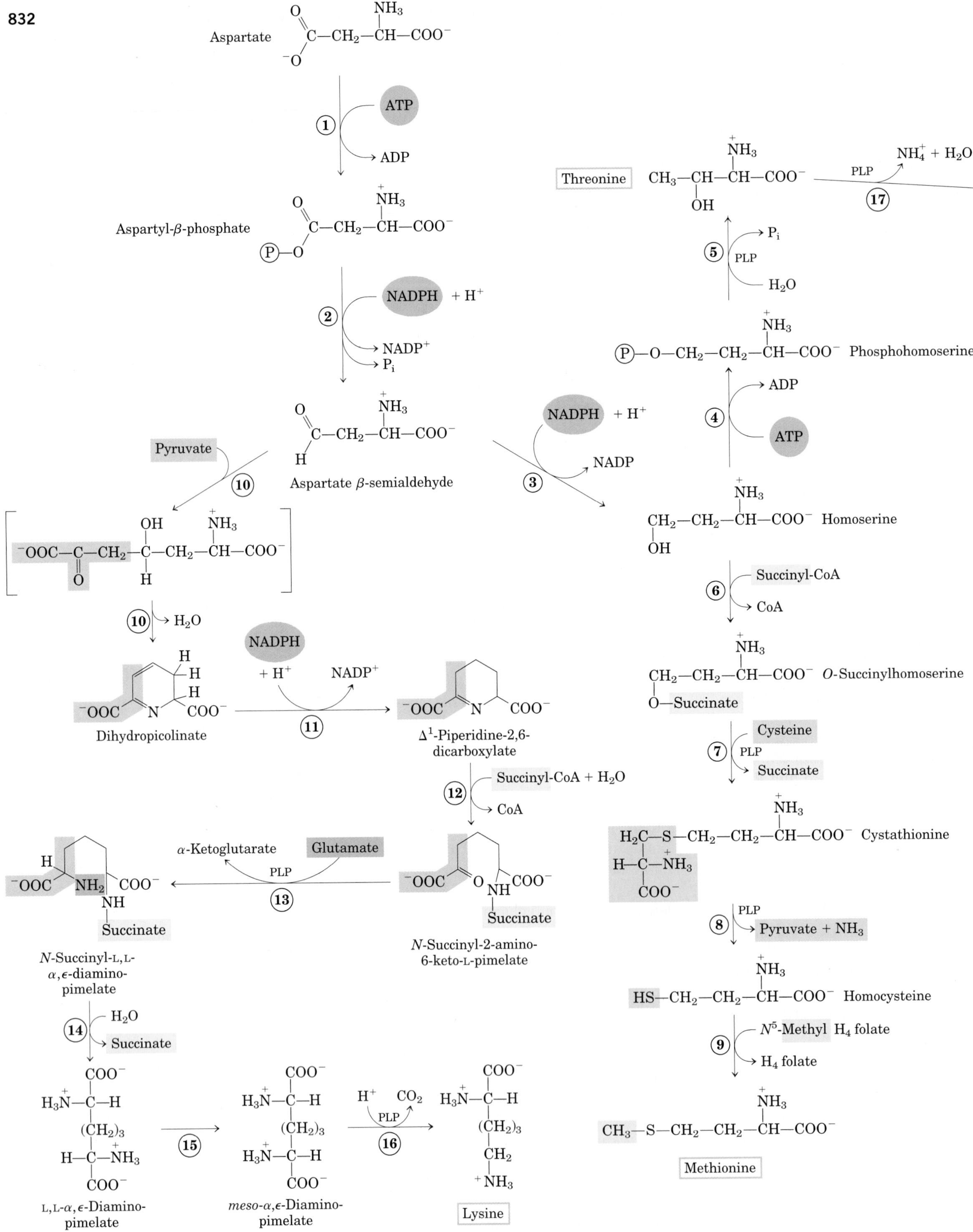

Aspartate
ATP
ADP
Aspartyl-β-phosphate
NADPH + H⁺
NADP⁺
Pᵢ
Aspartate β-semialdehyde
Pyruvate
NADPH + H⁺
NADP
Threonine
PLP
NH₄⁺ + H₂O
Pᵢ
PLP
H₂O
Phosphohomoserine
ADP
ATP
Homoserine
Succinyl-CoA
CoA
O-Succinylhomoserine
Succinate
Cysteine
PLP
Succinate
Cystathionine
H₂O
NADPH
+ H⁺
NADP⁺
Dihydropicolinate
Δ¹-Piperidine-2,6-dicarboxylate
Succinyl-CoA + H₂O
CoA
α-Ketoglutarate
Glutamate
PLP
Succinate
N-Succinyl-L,L-α,ε-diamino-pimelate
N-Succinyl-2-amino-6-keto-L-pimelate
PLP
Pyruvate + NH₃
Homocysteine
N⁵-Methyl H₄ folate
H₄ folate
H₂O
Succinate
Methionine
H⁺ CO₂
PLP
L,L-α,ε-Diamino-pimelate
meso-α,ε-Diamino-pimelate
Lysine

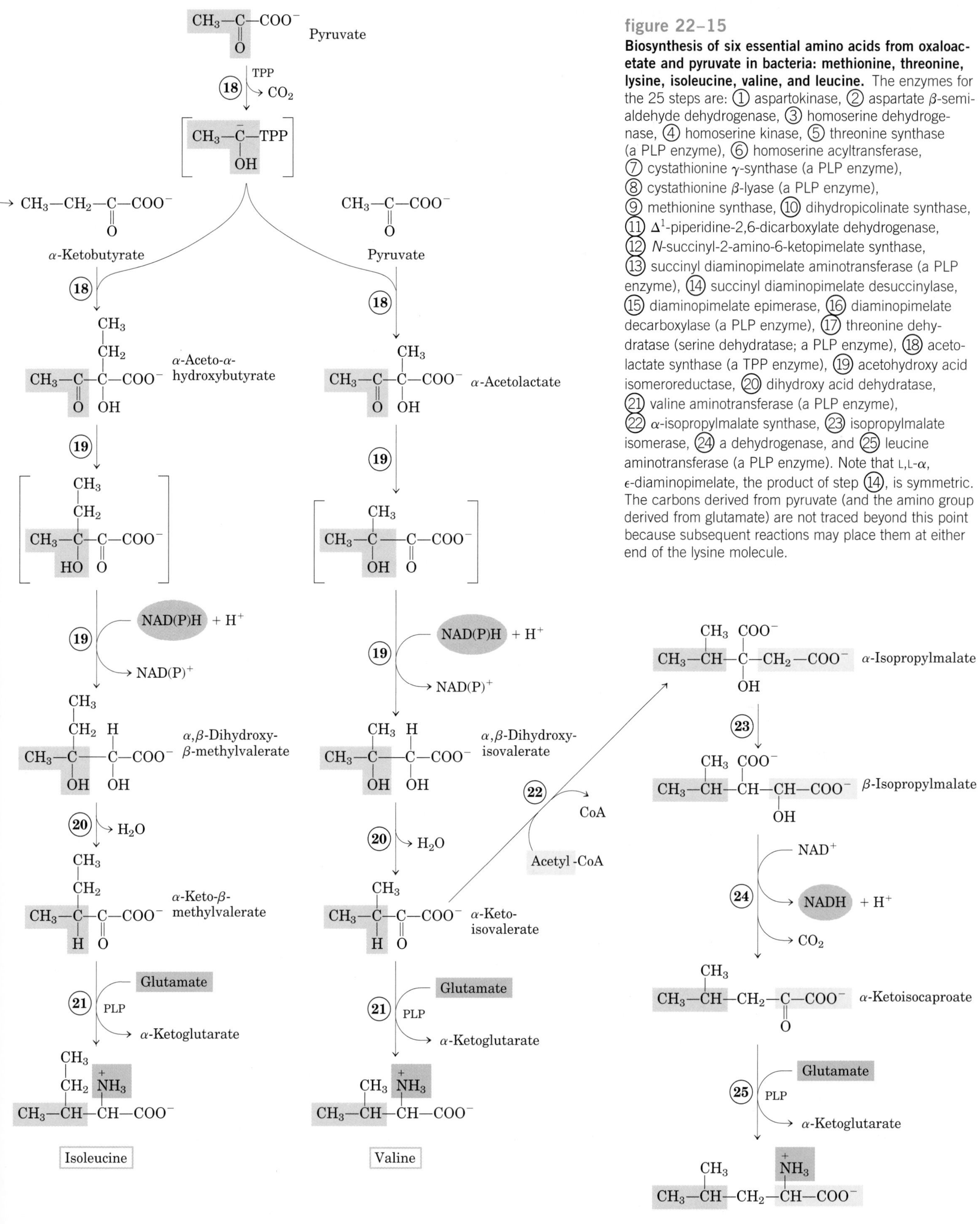

figure 22–15

Biosynthesis of six essential amino acids from oxaloacetate and pyruvate in bacteria: methionine, threonine, lysine, isoleucine, valine, and leucine. The enzymes for the 25 steps are: (1) aspartokinase, (2) aspartate β-semialdehyde dehydrogenase, (3) homoserine dehydrogenase, (4) homoserine kinase, (5) threonine synthase (a PLP enzyme), (6) homoserine acyltransferase, (7) cystathionine γ-synthase (a PLP enzyme), (8) cystathionine β-lyase (a PLP enzyme), (9) methionine synthase, (10) dihydropicolinate synthase, (11) Δ^1-piperidine-2,6-dicarboxylate dehydrogenase, (12) *N*-succinyl-2-amino-6-ketopimelate synthase, (13) succinyl diaminopimelate aminotransferase (a PLP enzyme), (14) succinyl diaminopimelate desuccinylase, (15) diaminopimelate epimerase, (16) diaminopimelate decarboxylase (a PLP enzyme), (17) threonine dehydratase (serine dehydratase; a PLP enzyme), (18) acetolactate synthase (a TPP enzyme), (19) acetohydroxy acid isomeroreductase, (20) dihydroxy acid dehydratase, (21) valine aminotransferase (a PLP enzyme), (22) α-isopropylmalate synthase, (23) isopropylmalate isomerase, (24) a dehydrogenase, and (25) leucine aminotransferase (a PLP enzyme). Note that L,L-α,ε-diaminopimelate, the product of step (14), is symmetric. The carbons derived from pyruvate (and the amino group derived from glutamate) are not traced beyond this point because subsequent reactions may place them at either end of the lysine molecule.

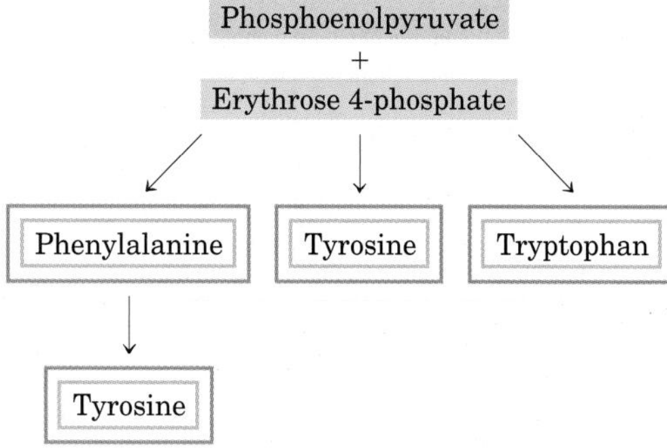

Chorismate Is a Key Intermediate in the Synthesis of Tryptophan, Phenylalanine, and Tyrosine

Aromatic rings are not readily available in the environment, even though the benzene ring is very stable. The branched pathway to tryptophan, phenylalanine, and tyrosine, occurring in bacteria and plants, is the main biological route of aromatic ring formation. It proceeds through ring closure of an aliphatic precursor followed by stepwise addition of double bonds. The first four steps produce shikimate, a seven-carbon molecule derived from erythrose 4-phosphate and phosphoenolpyruvate (Fig. 22–16). Shikimate is converted to chorismate in three steps that include the addition of three more carbons from another molecule of phosphoenolpyruvate. Chorismate is the first branch point of the pathway, with one branch leading to tryptophan, the other to phenylalanine and tyrosine.

figure 22–16

Biosynthesis of chorismate, an intermediate in the synthesis of aromatic amino acids in bacteria and plants. All carbons are derived from either erythrose 4-phosphate (purple) or phosphoenolpyruvate (red). The enzymes are: ① 2-keto-3-deoxy-D-arabinoheptulosonate 7-phosphate synthase, ② dehydroquinate synthase, ③ 3-dehydroquinate dehydratase, ④ shikimate dehydrogenase, ⑤ shikimate kinase, ⑥ 5-enolpyruvylshikimate 3-phosphate synthase, and ⑦ chorismate synthase. Note that the NAD^+ required as a cofactor in step ② is released unchanged; it may be transiently reduced to NADH during the reaction, with formation of an oxidized reaction intermediate. Step ⑥ is inhibited by glyphosate (the active ingredient in the widely used herbicide Roundup), a competitive inhibitor of the enzyme. The herbicide is relatively nontoxic to mammals, which lack this pathway. The chemical names quinate, shikimate, and chorismate are derived from the names of plants in which these intermediates have been found to accumulate.

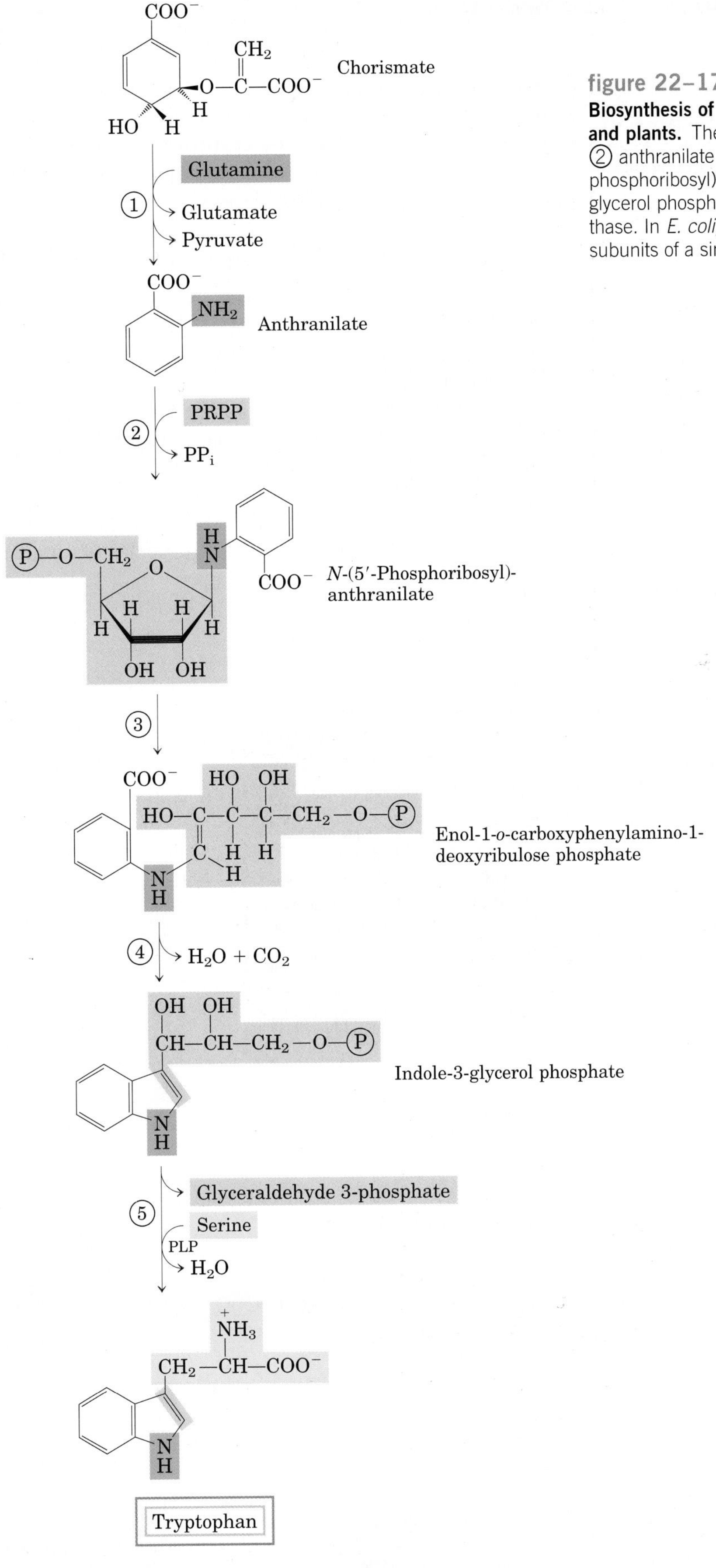

figure 22–17

Biosynthesis of tryptophan from chorismate in bacteria and plants. The enzymes are: ① anthranilate synthase, ② anthranilate phosphoribosyltransferase, ③ *N*-(5′-phosphoribosyl)-anthranilate isomerase, ④ indole-3-glycerol phosphate synthase, and ⑤ tryptophan synthase. In *E. coli,* enzymes catalyzing steps ① and ② are subunits of a single complex, anthranilate synthase.

In the **tryptophan** branch (Fig. 22–17), chorismate is converted to anthranilate in a reaction in which glutamine donates the nitrogen that will become part of the indole ring. Anthranilate then condenses with PRPP. The indole ring of tryptophan is derived from the ring carbons and amino group of anthranilate plus two carbons derived from PRPP. The final reaction in the sequence is catalyzed by **tryptophan synthase.** This enzyme has an $\alpha_2\beta_2$ subunit structure and can be dissociated into two α subunits and a β_2 subunit that catalyze different parts of the overall reaction:

$$\text{Indole-3-glycerol phosphate} \xrightarrow[\alpha \text{ subunit}]{} \text{indole} + \text{glyceraldehyde 3-phosphate}$$

$$\text{Indole} + \text{serine} \xrightarrow[\beta_2 \text{ subunit}]{} \text{tryptophan} + H_2O$$

The second part of the reaction requires pyridoxal phosphate (Fig. 22–18). Indole is not released by the enzyme, but instead moves through a solvent-filled cavity from the α-subunit active site to the β-subunit active site, where it condenses with a Schiff base intermediate derived from serine and PLP. This intermediate channeling may be a feature of the entire pathway from chorismate to tryptophan. Enzyme active sites catalyzing different steps (sometimes not sequential steps) of the pathway to tryptophan are found on single polypeptides in some species of fungi and bacteria, but are

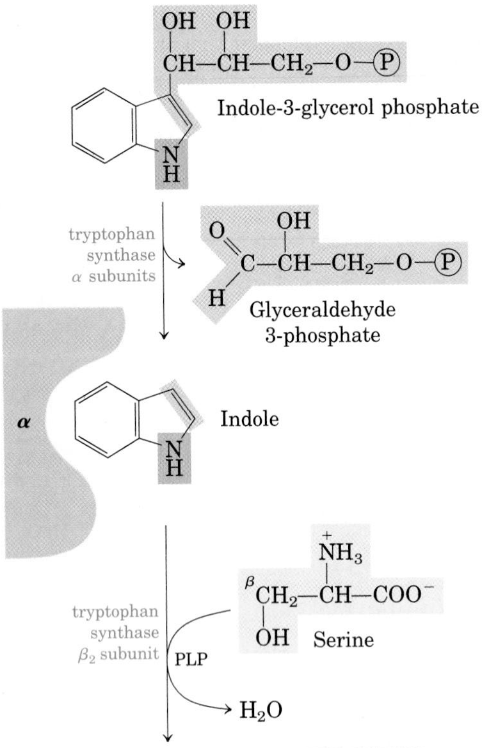

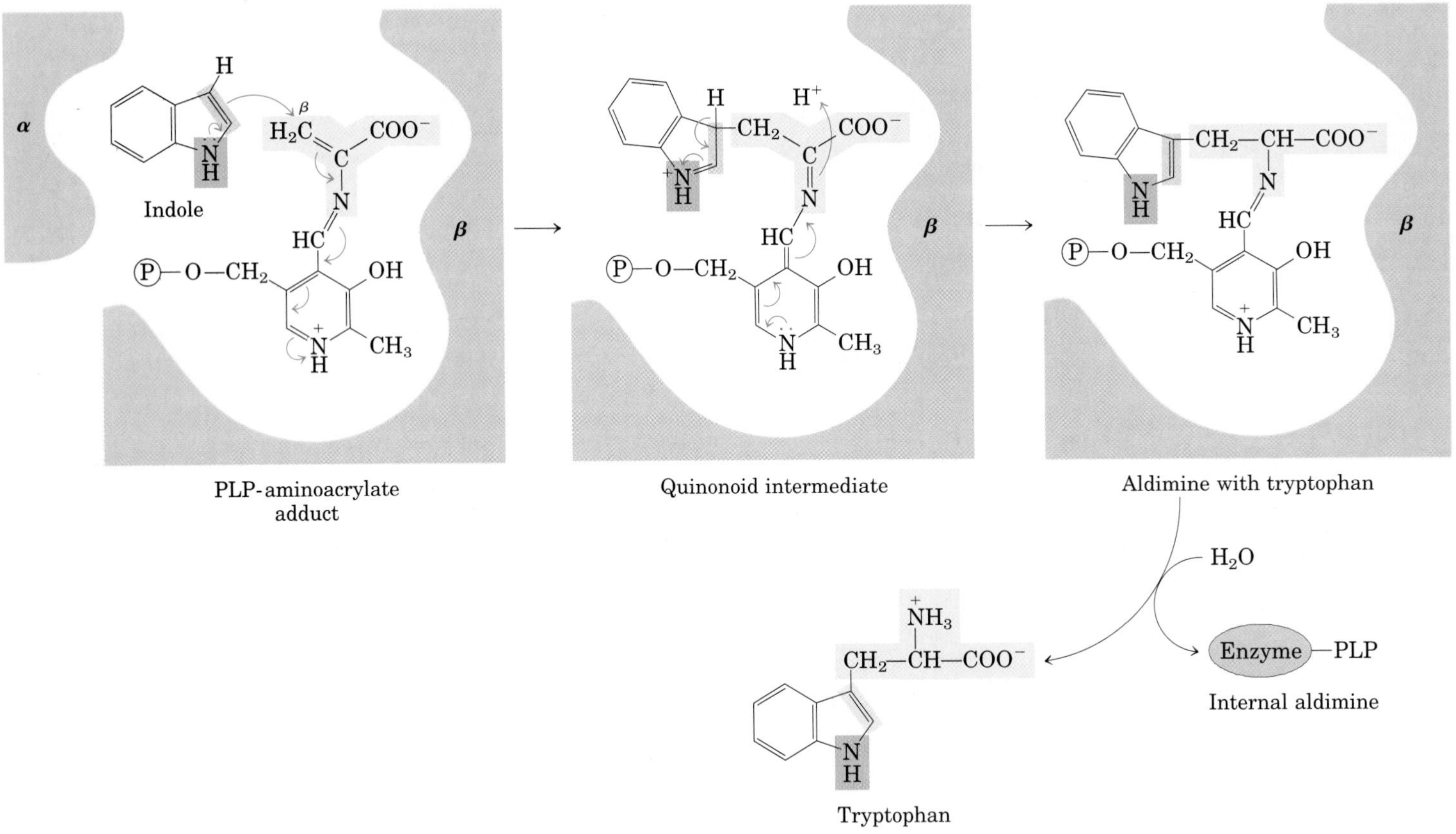

figure 22–18

Tryptophan synthase reaction. This enzyme catalyzes a multistep reaction with different types of chemical rearrangements. First an aldol cleavage produces indole and glyceraldehyde 3-phosphate; this reaction does not require PLP. Next, dehydration of serine forms a PLP-aminoacrylate intermediate; this condenses with indole and the product is hydrolyzed to release tryptophan. These PLP-facilitated transformations occur at the β carbon of the amino acid, as opposed to the α-carbon reactions described in Figure 18–6. The β carbon of serine is attached to the indole ring system.

separate proteins in others. In addition, the activity of some of these enzymes requires a noncovalent association with other enzymes of the pathway. These observations suggest that all are components of a large multienzyme complex in both prokaryotes and eukaryotes. Although such complexes are generally not preserved intact when the enzymes are isolated using traditional biochemical methods, evidence for the existence of multienzyme complexes is accumulating for this and a number of other metabolic pathways (see p. 540).

Phenylalanine and **tyrosine** are synthesized from chorismate in plants and bacteria in pathways much less complex than the tryptophan pathway. The common intermediate is prephenate (Fig. 22–19). The final step in both cases is transamination with glutamate.

$^{-}OOC-CH_2-NH-CH_2-PO_3^{2-}$
Glyphosate

Animals can produce tyrosine directly from phenylalanine through hydroxylation at C-4 of the phenyl group by **phenylalanine hydroxylase;** this enzyme also participates in the degradation of phenylalanine (see Figs 18–22, 18–23). Tyrosine is considered a nonessential amino acid only because it can be synthesized from the essential amino acid phenylalanine.

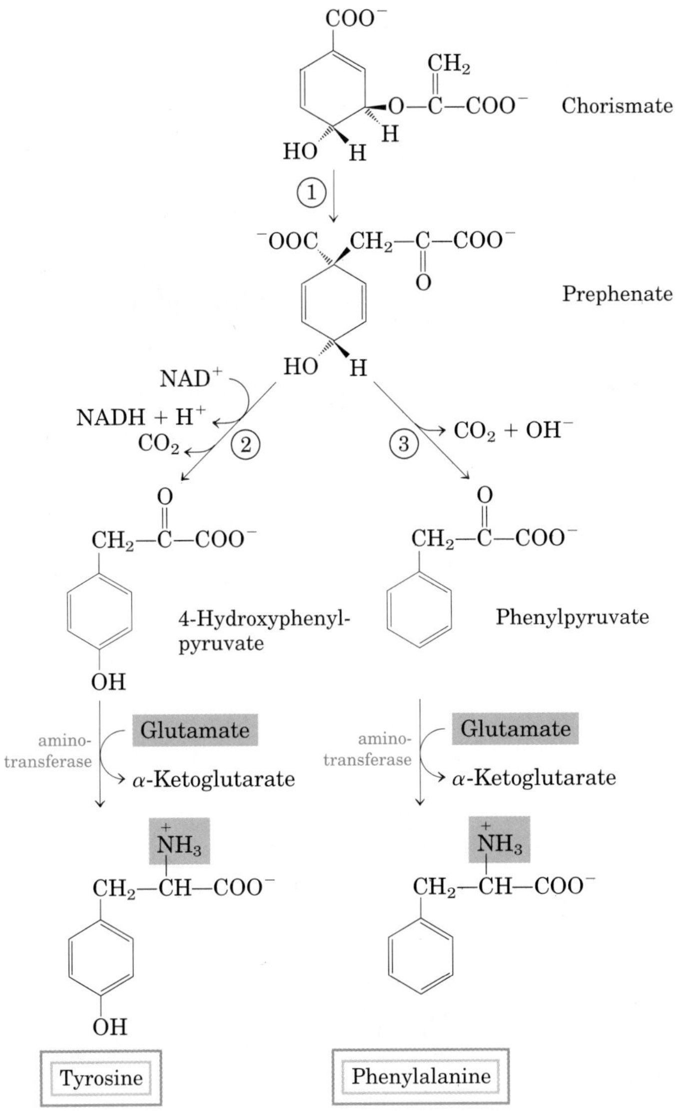

Chorismate

① ↓

Prephenate

NAD^+ → $NADH + H^+$, CO_2 ②

③ $CO_2 + OH^-$

4-Hydroxyphenylpyruvate

Phenylpyruvate

aminotransferase: Glutamate → α-Ketoglutarate

aminotransferase: Glutamate → α-Ketoglutarate

Tyrosine

Phenylalanine

figure 22–19
Biosynthesis of phenylalanine and tyrosine from chorismate in bacteria and plants. The enzymes are: ① chorismate mutase, ② prephenate dehydrogenase, and ③ prephenate dehydratase. Conversion of chorismate to prephenate is a rare biological example of a Claisen rearrangement.

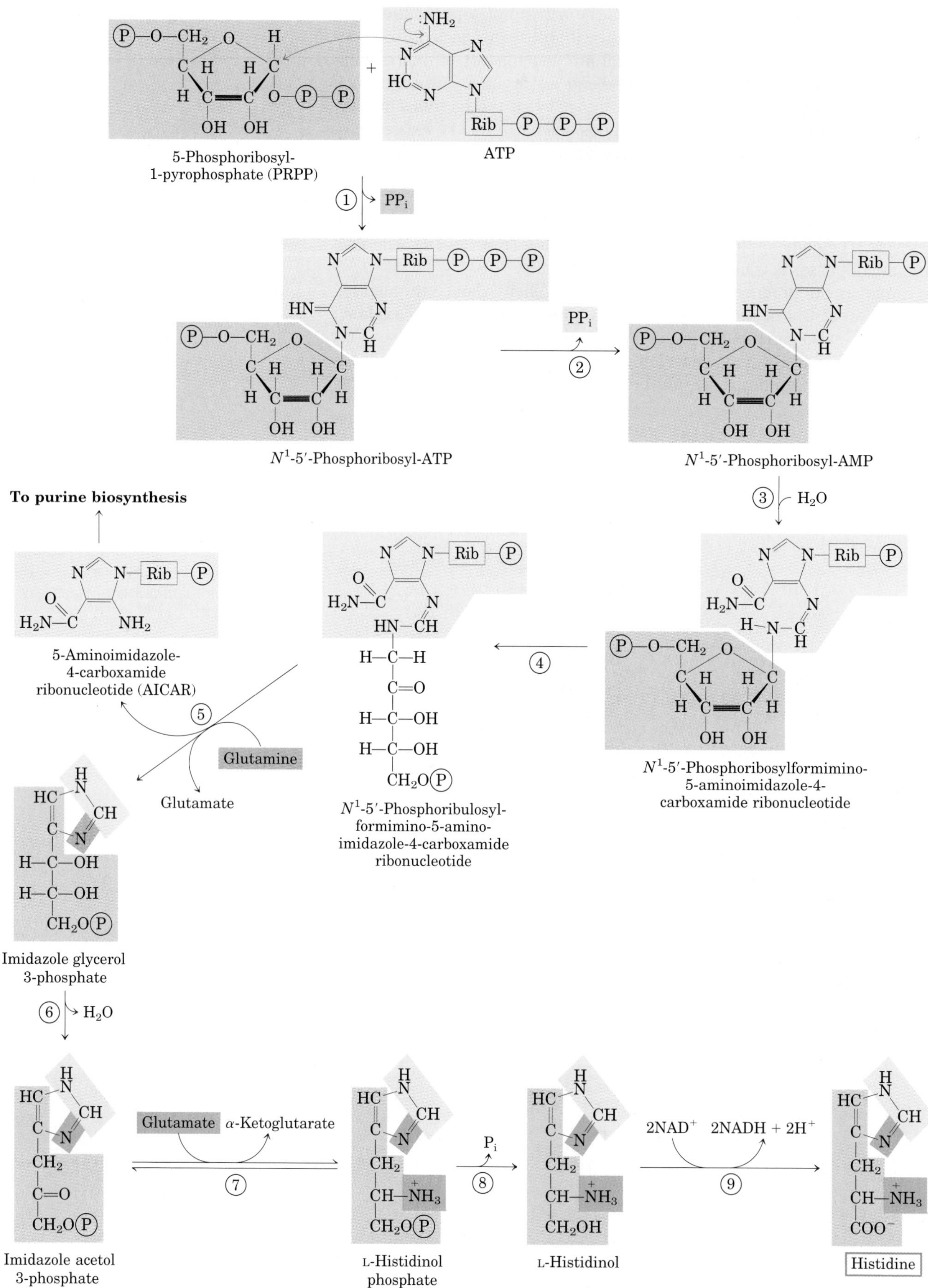
5-Phosphoribosyl-1-pyrophosphate (PRPP)
ATP
PP$_i$
N^1-5′-Phosphoribosyl-ATP
PP$_i$
N^1-5′-Phosphoribosyl-AMP
H_2O
To purine biosynthesis
5-Aminoimidazole-4-carboxamide ribonucleotide (AICAR)
N^1-5′-Phosphoribulosyl-formimino-5-amino-imidazole-4-carboxamide ribonucleotide
N^1-5′-Phosphoribosylformimino-5-aminoimidazole-4-carboxamide ribonucleotide
Glutamine
Glutamate
Imidazole glycerol 3-phosphate
H_2O
Glutamate
α-Ketoglutarate
P$_i$
2NAD$^+$
2NADH + 2H$^+$
Imidazole acetol 3-phosphate
L-Histidinol phosphate
L-Histidinol
Histidine

figure 22–20

Biosynthesis of histidine in bacteria and plants. Atoms derived from PRPP and ATP are shaded red and blue, respectively. Two of the histidine nitrogens are derived from glutamine and glutamate (green). The enzymes are: ① ATP phosphoribosyl transferase, ② pyrophosphohydrolase, ③ phosphoribosyl-AMP cyclohydrolase, ④ phosphoribosylformimino-5-aminoimidazole-4-carboxamide ribonucleotide isomerase, ⑤ glutamine amidotransferase, ⑥ imidazole glycerol 3-phosphate dehydratase, ⑦ L-histidinol phosphate aminotransferase, ⑧ histidinol phosphate phosphatase, and ⑨ histidinol dehydrogenase. Note that the derivative of ATP remaining after step ⑤ (AICAR) is an intermediate in purine biosynthesis (see Fig. 22–31), so ATP is rapidly regenerated.

Histidine Biosynthesis Uses Precursors of Purine Biosynthesis

The pathway to **histidine** in all plants and bacteria differs in several respects from other amino acid biosynthetic pathways. Histidine is derived from three precursors (Fig. 22–20): PRPP contributes five carbons, the purine ring of ATP contributes a nitrogen and a carbon, and glutamine supplies the second ring nitrogen. The key steps are condensation of ATP and PRPP, in which N-1 of the purine ring is linked to the activated C-1 of the ribose of PRPP (step ① in Fig. 22–20); purine ring opening that ultimately leaves N-1 and C-2 of adenine linked to the ribose (step ③); and formation of the imidazole ring, a reaction in which glutamine donates a nitrogen (step ⑤). The use of ATP as a metabolite rather than a high-energy cofactor is unusual, but not wasteful because it dovetails with the purine biosynthetic pathway. The remnant of ATP that is released after the transfer of N-1 and C-2 is 5-aminoimidazole-4-carboxamide ribonucleotide (AICAR), an intermediate of purine biosynthesis (see Fig. 22–31) that is rapidly recycled to ATP.

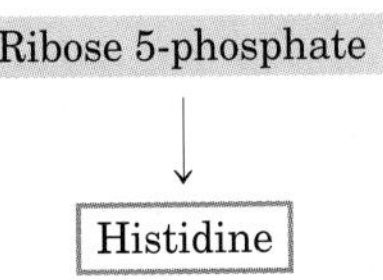

Amino Acid Biosynthesis Is under Allosteric Regulation

The most responsive manner in which amino acid synthesis is controlled is through feedback inhibition of the first reaction in a sequence by the end product of the pathway. This first reaction is usually irreversible and is catalyzed by an allosteric enzyme. As an example, Figure 22–21 shows the allosteric regulation of isoleucine synthesis from threonine (Fig. 22–15). The end product, isoleucine, is an allosteric inhibitor of the first reaction in the sequence. In bacteria, such allosteric modulation of amino acid synthesis is responsive on a minute-to-minute basis.

Allosteric regulation can be considerably more complex. An example is the remarkable set of allosteric controls exerted on glutamine synthetase of *E. coli* (Fig. 22–6). Six products derived from glutamine serve as negative feedback modulators of the enzyme, and the overall effects of these and other modulators are more than additive. Such regulation is called **concerted inhibition.**

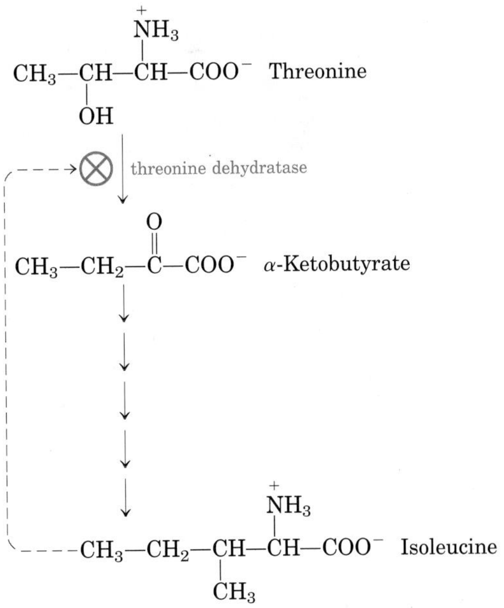

Because the 20 standard amino acids must be made in the correct proportions for protein synthesis, cells have developed ways not only of controlling the rate of synthesis of individual amino acids but also of coordinating their formation. Such coordination is especially well developed in

$CH_3—CH(OH)—CH(\overset{+}{N}H_3)—COO^-$ Threonine

threonine dehydratase

$CH_3—CH_2—C(=O)—COO^-$ α-Ketobutyrate

$CH_3—CH_2—CH(CH_3)—CH(\overset{+}{N}H_3)—COO^-$ Isoleucine

figure 22–21

Allosteric regulation of isoleucine biosynthesis. The first reaction in the pathway from threonine to isoleucine is inhibited by the end product, isoleucine. This was one of the first examples of allosteric feedback inhibition to be discovered. The steps from α-ketobutyrate to isoleucine correspond to steps ⑱ through ㉑ in Figure 22–15 (five steps because ⑲ is a two-step reaction).

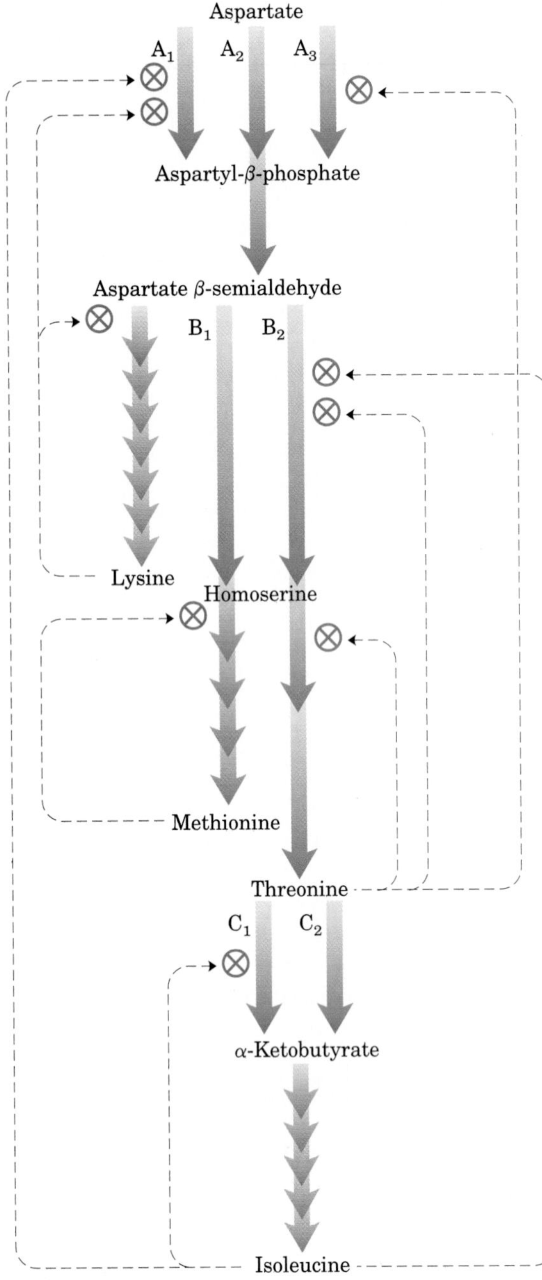

figure 22–22

Interlocking regulatory mechanisms in the biosynthesis of several amino acids derived from aspartate in *E. coli.* Three enzymes (A, B, C) have either two or three isozyme forms, indicated by numerical subscripts. In each case, one isozyme (A_2, B_1, and C_2) has no allosteric regulation; these isozymes are regulated by changes in the amounts synthesized at the genetic level (Chapter 28). Synthesis of isozymes A_2 and B_1 is repressed when methionine levels are high, and synthesis of isozyme C_2 is repressed when isoleucine levels are high. Enzyme A is aspartokinase; B, homoserine dehydrogenase; C, threonine dehydratase.

fast-growing bacterial cells. Figure 22–22 shows how *E. coli* cells coordinate the synthesis of lysine, methionine, threonine, and isoleucine, all made from aspartate. Several important types of inhibition patterns are evident. The step from aspartate to aspartyl-β-phosphate is catalyzed by three isozymes, each independently controlled by different modulators. This **enzyme multiplicity** prevents one biosynthetic end product from shutting down key steps in a pathway when other products of the same pathway are required. The steps from aspartate β-semialdehyde to homoserine and from threonine to α-ketobutyrate (Fig. 22–15) are also catalyzed by dual, independently controlled isozymes. One isozyme for the conversion of aspartate to aspartyl-β-phosphate is allosterically inhibited by two different modulators, lysine and isoleucine, whose action is more than additive—another example of concerted inhibition. The sequence from aspartate to isoleucine shows multiple, overlapping negative feedback inhibition; for example, isoleucine inhibits the conversion of threonine to α-ketobutyrate (as described above), and threonine inhibits its own formation at three points: from homoserine, from aspartate β-semialdehyde, and from aspartate (steps ④, ③, and ① in Fig. 22–15). This overall regulatory mechanism is called **sequential feedback inhibition.**

Molecules Derived from Amino Acids

In addition to their role as the building blocks of proteins, amino acids are precursors of many specialized biomolecules, including hormones, coenzymes, nucleotides, alkaloids, cell wall polymers, porphyrins, antibiotics, pigments, and neurotransmitters. We describe here the pathways to a number of these amino acid derivatives.

Glycine Is a Precursor of Porphyrins

The biosynthesis of **porphyrins,** for which glycine is a major precursor, is our first example because of the central importance of the porphyrin nucleus in heme proteins such as hemoglobin and the cytochromes. The porphyrins are constructed from four molecules of the monopyrrole derivative **porphobilinogen** (Fig. 22–23). In the first step, glycine reacts with succinyl-CoA to yield α-amino-β-ketoadipate, which is then decarboxylated to δ-aminolevulinate. Two molecules of δ-aminolevulinate condense to form porphobilinogen, and four molecules of porphobilinogen come together to form **protoporphyrin,** through a series of complex enzymatic reactions. The iron atom is incorporated after the protoporphyrin has been assembled, in a step catalyzed by ferrochelatase. Porphyrin biosynthesis is regulated by the concentration of the heme product, which can serve as a feedback inhibitor of early steps in the synthetic pathway. Genetic defects in the biosynthesis of porphyrins can lead to the accumulation of pathway intermediates, causing a variety of human diseases known collectively as **porphyrias** (Box 22–1).

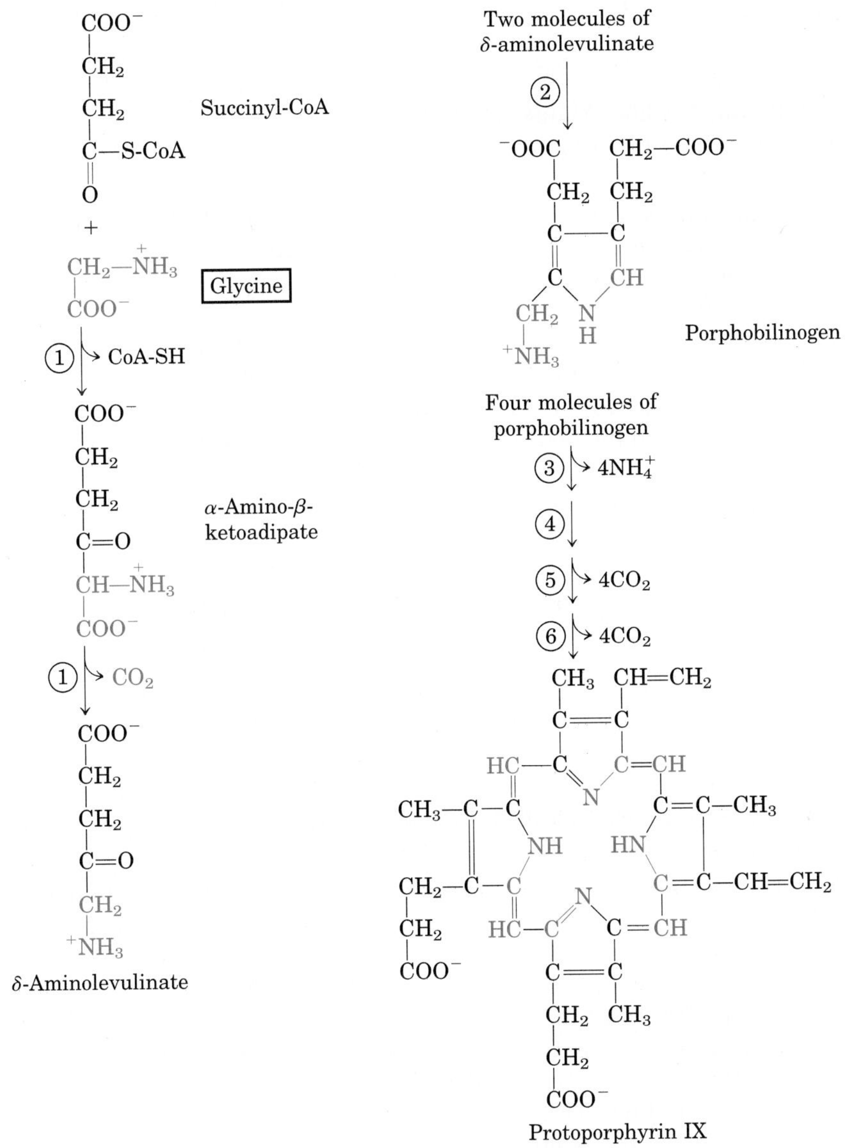

figure 22–23
Biosynthesis of protoporphyrin IX, the porphyrin of hemoglobin and myoglobin in mammals. The atoms furnished by glycine are shown in red. The remaining carbon atoms are derived from the succinyl group of succinyl-CoA. The enzymes are: ① δ-aminolevulinate synthase, ② porphobilinogen synthase, ③ uroporphyrinogen synthase, ④ uroporphyrinogen III cosynthase, ⑤ uroporphyrinogen decarboxylase, and ⑥ coproporphyrinogen oxidase. The Fe^{2+} needed to complete heme (see Fig. 7–1) is added in a step catalyzed by ferrochelatase (not shown). In bacteria and plants, glutamate is the precursor of δ-amino-levulinate.

box 22–1 Biochemistry of Kings and Vampires

Porphyrias are a group of genetic diseases in which defects in certain enzymes of the biosynthetic pathway from glycine to porphyrins lead to an accumulation of specific porphyrin precursors in erythrocytes, body fluids, and the liver. The most common form is acute intermittent porphyria, caused by a deficiency in porphobilinogen deaminase (see Fig. 22–23). Most affected individuals are heterozygotes, usually asymptomatic because the single copy of the normal gene provides a sufficient level of enzyme function. However, poorly understood nutritional or environmental factors can cause a buildup of δ-aminolevulinate and porphobilinogen in these individuals, leading to attacks of acute abdominal pain and neurological dysfunction. King George III, British monarch during the American Revolution, suffered several episodes of apparent madness that tarnished the record of this otherwise accomplished man. The symptoms of his condition suggest that George III suffered from acute intermittent porphyria.

One of the rarer porphyrias, which affects mainly erythrocytes, results in an accumulation of uroporphyrinogen I, an abnormal isomer of a protoporphyrin precursor. This compound stains the urine red, causes the teeth to fluoresce strongly in ultraviolet light, and makes the skin abnormally sensitive to sunlight. Affected individuals are often anemic because insufficient heme is synthesized. This genetic condition may have given rise to the vampire myths of folk legend.

The symptoms of most porphyrias are now readily controlled with dietary changes or the administration of heme or heme derivatives.

Degradation of Heme Yields Bile Pigments

The iron-porphyrin or heme group of hemoglobin, released from dying erythrocytes in the spleen, is degraded to yield free Fe^{3+} and ultimately **bilirubin,** a linear (open) tetrapyrrole derivative. Bilirubin binds to serum albumin and travels to the liver, where it is transformed into the bile pigment bilirubin diglucuronide. This product is sufficiently water-soluble to be secreted with other components of bile into the small intestine. Impaired liver function or blocked bile secretion causes bilirubin to leak into the blood, resulting in a yellowing of the skin and eyeballs, a condition called jaundice. Determination of bilirubin concentration in the blood in cases of jaundice is useful in the diagnosis of underlying liver disease.

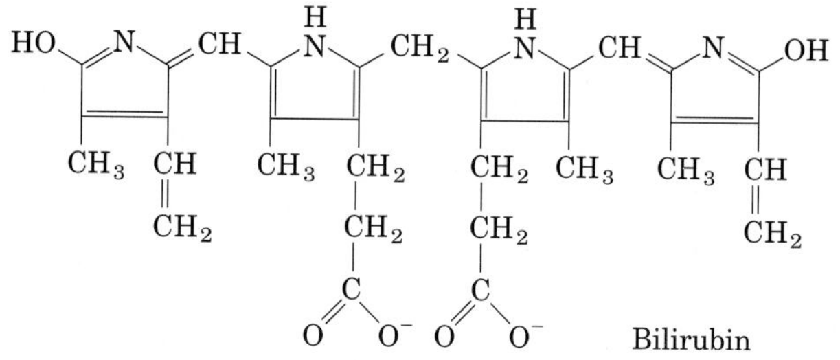

Amino Acids Are Required for the Biosynthesis of Creatine and Glutathione

Phosphocreatine, derived from **creatine,** is an important energy buffer in skeletal muscle (p. 510). Creatine is synthesized from glycine and arginine (Fig. 22–24), with methionine, in the form of *S*-adenosylmethionine, as methyl group donor.

Glutathione (GSH), present in plants, animals, and some bacteria, often at high levels, can be thought of as a redox buffer. It is derived from glycine, glutamate, and cysteine (Fig. 22–25). The γ-carboxyl group of glutamate is activated by ATP to form an acyl phosphate intermediate, which is then attacked by the α-amino group of cysteine. A second condensation reaction follows, with the α-carboxyl group of cysteine activated to an acyl phosphate to permit reaction with glycine.

Glutathione probably helps maintain the sulfhydryl groups of proteins in the reduced state and the iron of heme in the ferrous (Fe^{2+}) state, and it serves as a reducing agent for glutaredoxin in deoxyribonucleotide synthesis (see Fig. 22–37). Its redox function is also used to remove toxic peroxides formed in the normal course of growth and metabolism under aerobic conditions:

$$2GSH + R{-}O{-}O{-}H \longrightarrow GSSG + H_2O + R{-}OH$$

The oxidized form of glutathione (GSSG) contains two glutathione molecules linked by a disulfide bond (Fig. 22–25). This reaction is catalyzed by **glutathione peroxidase,** a remarkable enzyme in that it contains a covalently bound selenium (Se) atom in the form of selenocysteine (see Fig. 5–8a), essential for its activity.

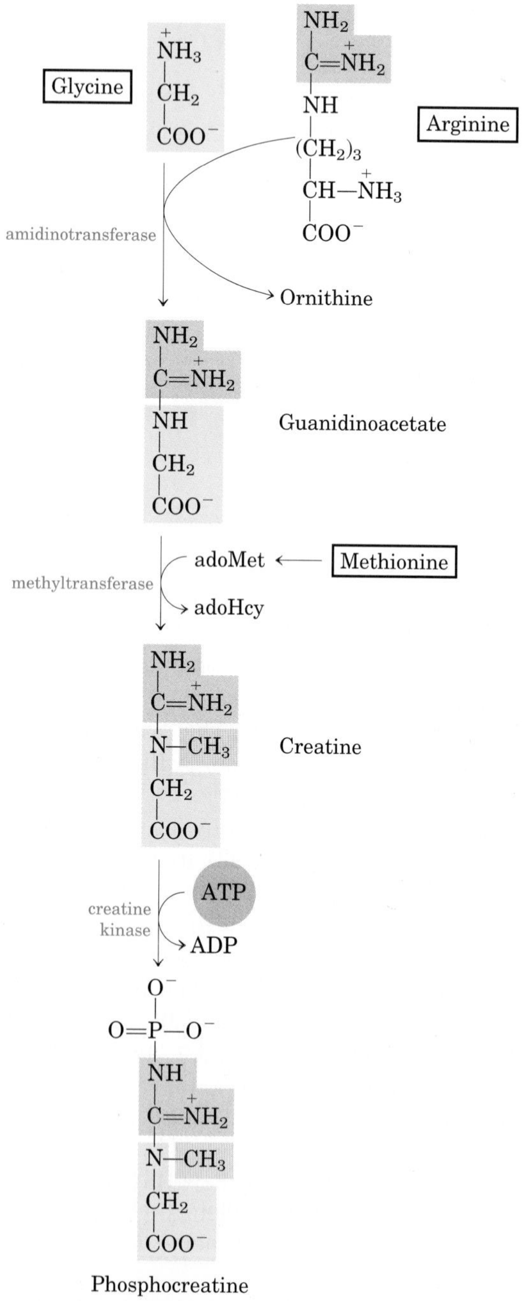

figure 22–24
Biosynthesis of creatine and phosphocreatine. Creatine is made from three amino acids: glycine, arginine, and methionine. This pathway shows the versatility of amino acids as precursors of other nitrogenous biomolecules.

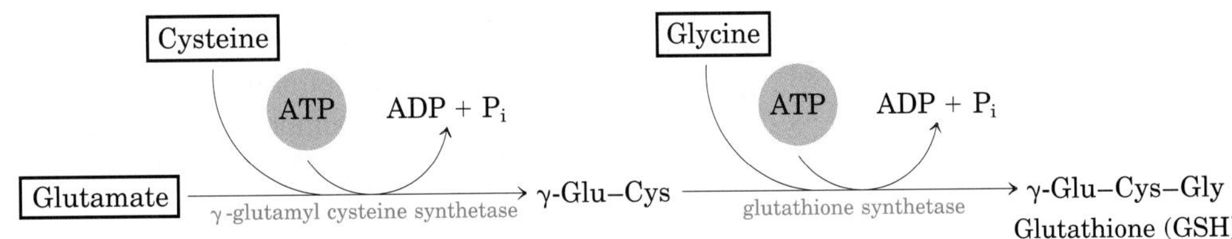

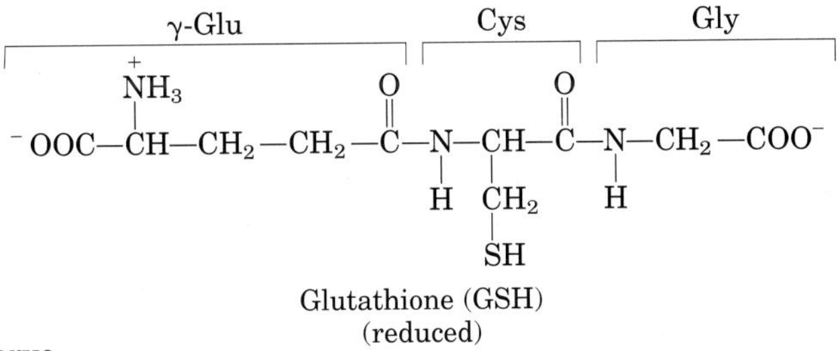

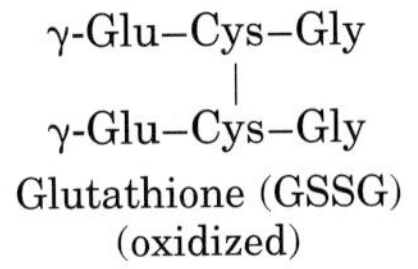

figure 22–25
Biosynthesis and structure of glutathione. The oxidized and reduced forms of glutathione are shown.

Glutathione (GSH)
(reduced)

γ-Glu–Cys–Gly
γ-Glu–Cys–Gly
Glutathione (GSSG)
(oxidized)

D-Amino Acids Are Found Primarily in Bacteria

Although D-amino acids do not generally occur in proteins, they do serve some special functions in the structure of bacterial cell walls and peptide antibiotics. The peptidoglycans (see Fig. 9–19) of bacteria contain both D-alanine and D-glutamate. D-Amino acids arise directly from the L isomers by the action of amino acid racemases, which have pyridoxal phosphate as cofactor (see Fig. 18–6). Amino acid racemization is uniquely important to bacterial metabolism, and enzymes such as alanine racemase are prime targets for pharmaceutical agents. One such agent, **L-fluoroalanine,** is being tested as an antibacterial drug. Another, **cycloserine,** is used to treat tuberculosis. Because these inhibitors also affect some PLP-requiring enzymes of humans, they have potentially undesirable side effects.

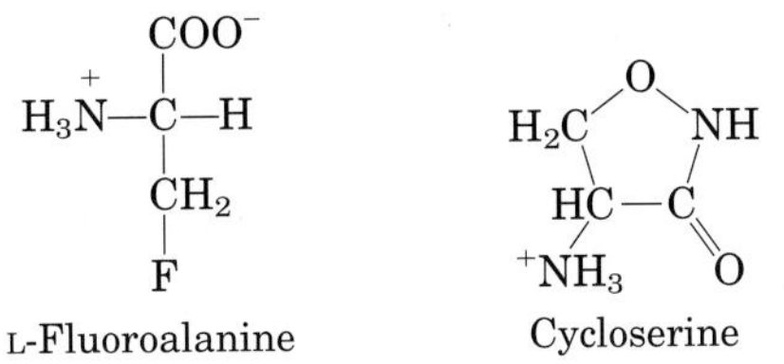

L-Fluoroalanine

Cycloserine

Aromatic Amino Acids Are Precursors of Many Plant Substances

Phenylalanine, tyrosine, and tryptophan are converted to a variety of important compounds in plants. The rigid polymer **lignin** is derived from phenylalanine and tyrosine. It is second only to cellulose in abundance in plant tissues. The structure of the lignin polymer is complex and not well understood. Phenylalanine and tyrosine also give rise to many commercially significant natural products, including the tannins that inhibit oxidation in wines; alkaloids such as morphine, which have potent physiological effects; and the flavor components of products such as cinnamon oil, nutmeg, cloves, vanilla, and cayenne pepper.

Tryptophan is the precursor of the plant growth hormone indole-3-acetate, or **auxin** (Fig. 22–26), which has been implicated in the regulation of a wide range of biological processes in plants.

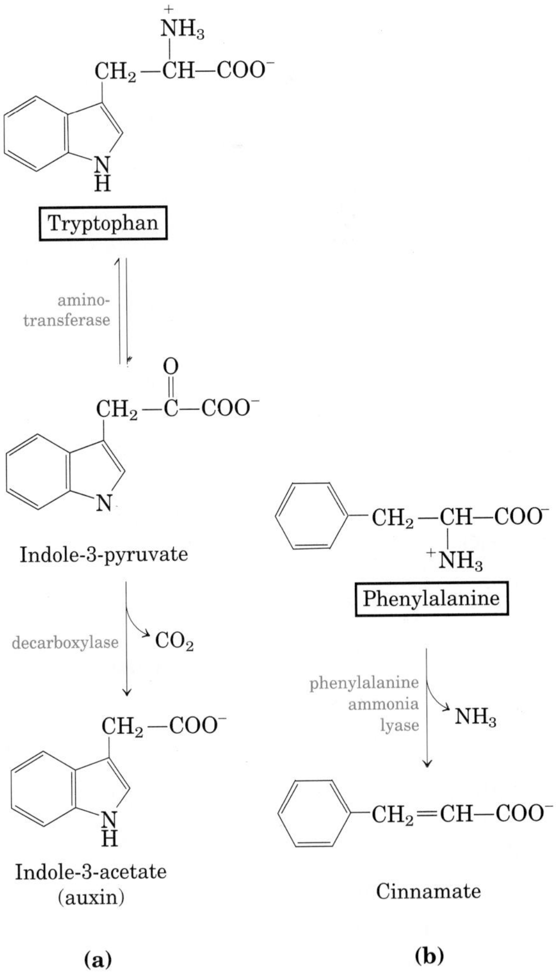

Tryptophan

amino-transferase

Indole-3-pyruvate

decarboxylase CO_2

Indole-3-acetate
(auxin)

(a)

Phenylalanine

phenylalanine ammonia lyase NH_3

Cinnamate

(b)

figure 22–26
Biosynthesis of two plant substances from amino acids. **(a)** Indole-3-acetate (auxin) and **(b)** cinnamate (cinnamon flavor).

Amino Acids Are Converted to Biological Amines by Decarboxylation

Many important neurotransmitters are primary or secondary amines derived from amino acids in simple pathways. In addition, some polyamines that form complexes with DNA are derived from the amino acid ornithine, a component of the urea cycle. A common denominator of many of these pathways is amino acid decarboxylation, another PLP-requiring reaction (see Fig. 18–6).

The synthesis of some neurotransmitters is illustrated in Figure 22–27. Tyrosine gives rise to a family of catecholamines that includes **dopamine, norepinephrine,** and **epinephrine.** Levels of catecholamines are correlated with (among other things) changes in blood pressure. The neurological disorder Parkinson's disease is associated with an underproduction of

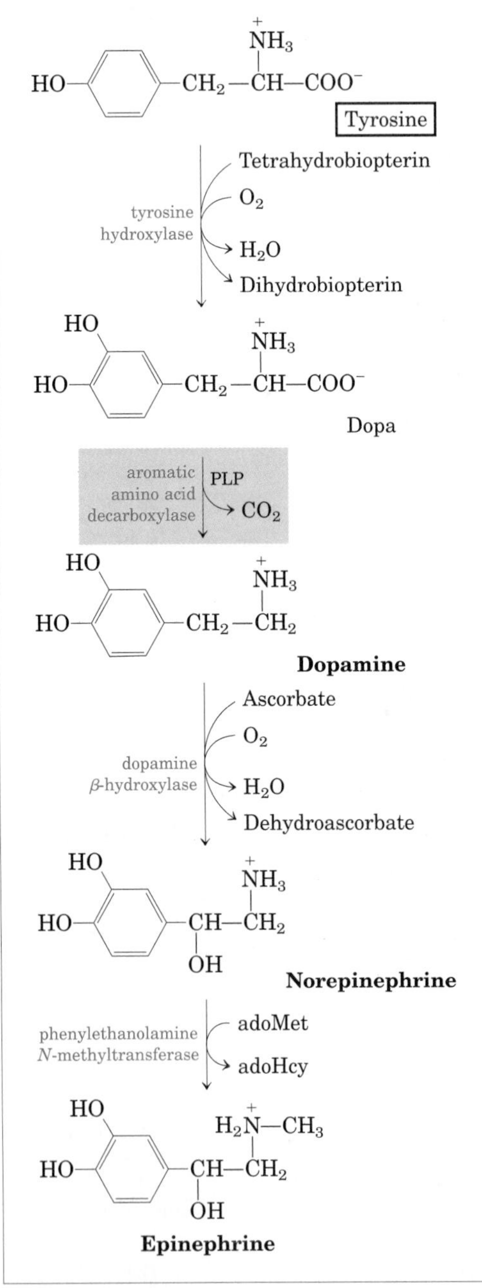

Glutamate
glutamate decarboxylase
PLP
CO_2
γ-Aminobutyrate (GABA)

Histidine
histidine decarboxylase
PLP
CO_2
Histamine

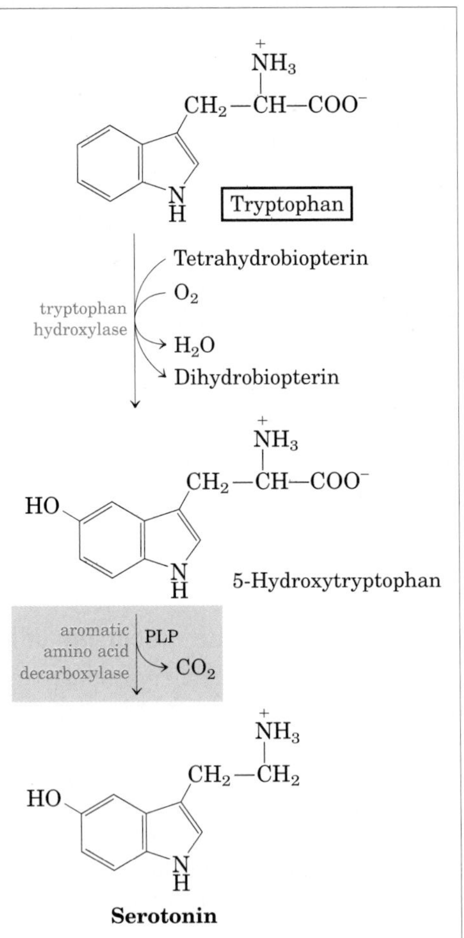

figure 22–27
Biosynthesis of some neurotransmitters from amino acids. The key step is the same in each case: a PLP-dependent decarboxylation (shaded in red).

dopamine, and it has been treated by administering L-dopa. Overproduction of dopamine in the brain may be linked with psychological disorders such as schizophrenia.

Glutamate decarboxylation gives rise to **γ-aminobutyrate (GABA),** an inhibitory neurotransmitter. Its underproduction is associated with epileptic seizures. GABA analogs are used in the treatment of epilepsy and hypertension. Levels of GABA can also be increased by administering inhibitors of the GABA-degrading enzyme GABA aminotransferase.

Another important neurotransmitter, **serotonin,** is derived from tryptophan in a two-step pathway.

Histidine is decarboxylated to form **histamine,** a powerful vasodilator in animal tissues. Histamine is released in large amounts as part of the allergic response, and it also stimulates acid secretion in the stomach. A growing array of pharmaceutical agents are being designed to interfere with either the synthesis or the action of histamine. A prominent example is the histamine receptor antagonist **cimetidine** (Tagamet), a structural analog of histamine. It promotes healing of duodenal ulcers by inhibiting secretion of gastric acid.

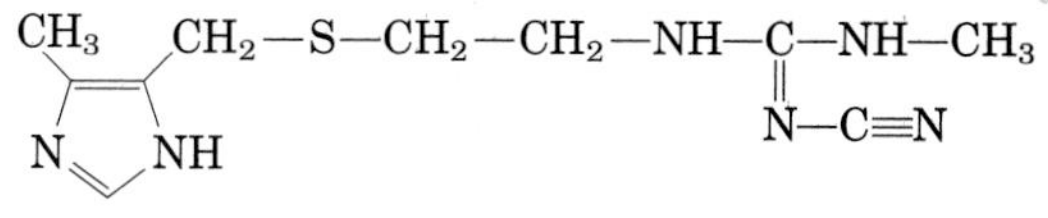

Cimetidine
(Tagamet)

Polyamines such as **spermine** and **spermidine,** used in DNA packaging, are derived from methionine and ornithine by the pathway shown in Figure 22–28. The first step is decarboxylation of ornithine, a precursor of arginine (Fig. 22–10). **Ornithine decarboxylase,** a PLP-requiring enzyme, is the target of several powerful inhibitors used as pharmaceutical agents (Box 22–2).

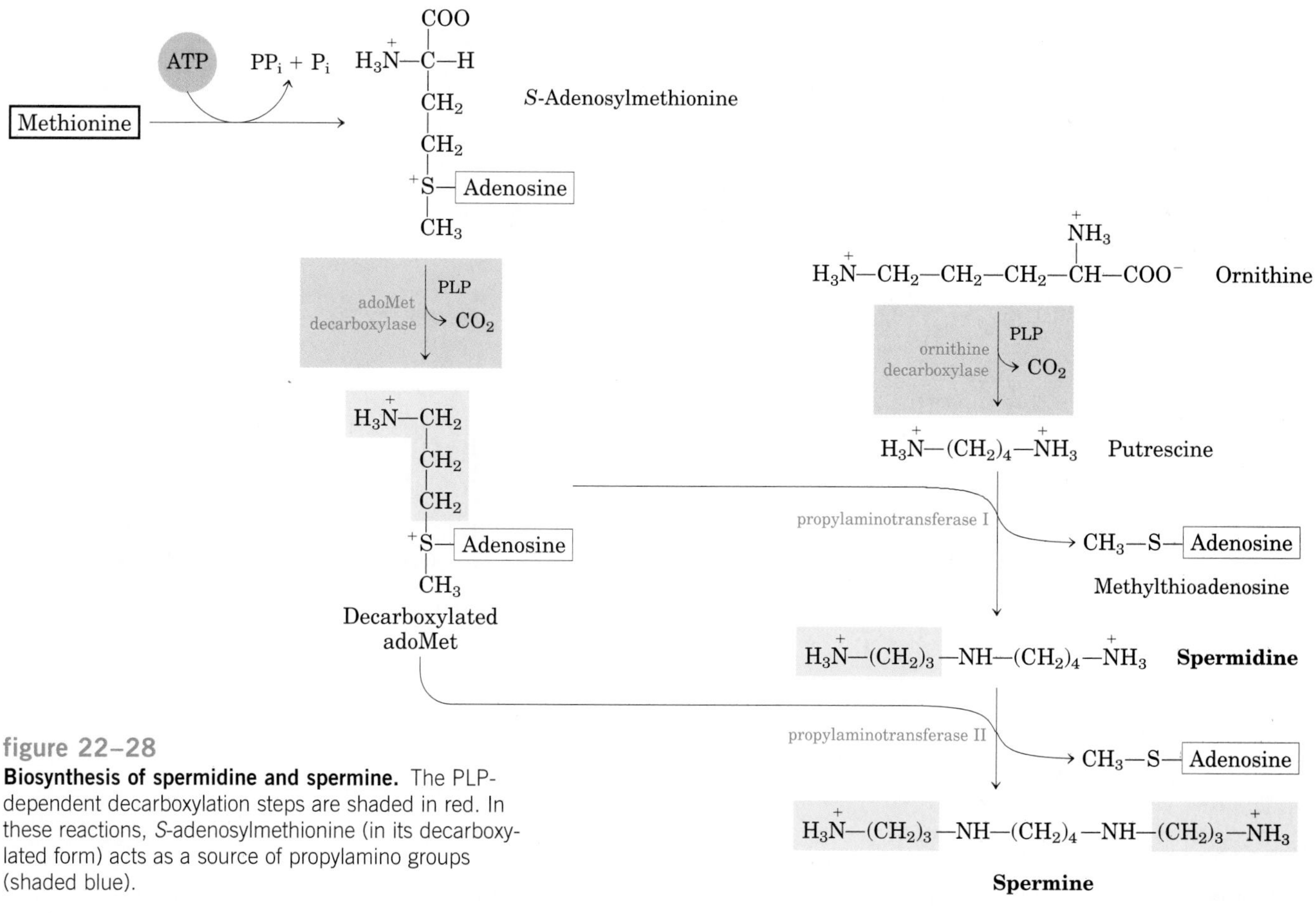

figure 22–28
Biosynthesis of spermidine and spermine. The PLP-dependent decarboxylation steps are shaded in red. In these reactions, *S*-adenosylmethionine (in its decarboxylated form) acts as a source of propylamino groups (shaded blue).

box 22–2 Curing African Sleeping Sickness with a Biochemical Trojan Horse

African sleeping sickness, also called African trypanosomiasis, is caused by protists (single-celled eukaryotes) called trypanosomes (Fig. 1). This disease (and related trypanosome-caused diseases) is medically and economically significant in many developing nations. Until recently, the disease was virtually incurable. Vaccines are ineffective because the parasite has a novel mechanism to evade the host immune system. The cell coat is covered with a single protein, to which the immune system responds. Every so often, however, by a process of genetic recombination (see Table 28–1), a few cells switch to a new protein coat not recognized by the immune system. This process of "changing coats" can occur hundreds of times. The result is a chronic cyclic infection: development of a fever, which subsides as the immune system beats back the first infection; cells with changed coats then become the seed for a second infection, and the fever reappears. This cycle can repeat for weeks, and the weakened person eventually dies.

Some modern approaches to treating African sleeping sickness have been based on an understanding of enzymology and metabolism. In at least one case, this involves pharmaceutical agents designed as mechanism-based enzyme inactivators (suicide inactivators; see p. 269). A vulnerable point in the metabolism of trypanosomes is the pathway for polyamine biosynthesis. The polyamines spermine and spermidine, used in DNA packaging, are required in large amounts in rapidly dividing cells. The first step in their synthesis is catalyzed by ornithine decarboxylase, a PLP-requiring enzyme (see Fig. 22–28). In mammalian cells, ornithine decarboxylase undergoes rapid turnover—that is, enzyme degradation and synthesis of new enzyme occur continuously. For

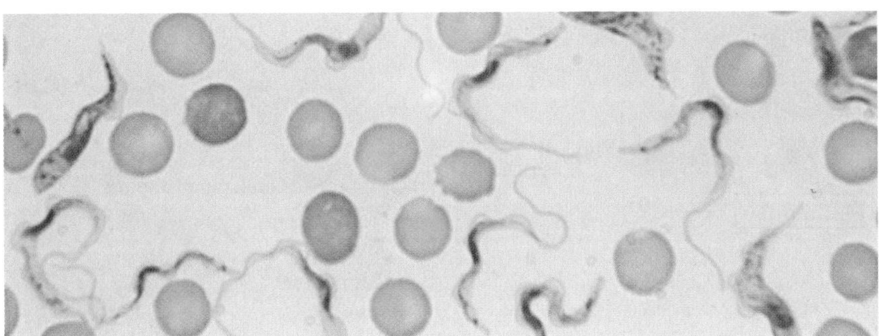

figure 1
Trypanosoma brucei rhodesiense, the causative agent of African sleeping sickness.

Ornithine

$H_2N{-}(CH_2)_3{-}CH(NH_2){-}COO^-$

Pyridoxal phosphate

H_2O

Schiff base

CO_2

Putrescine

H_2O

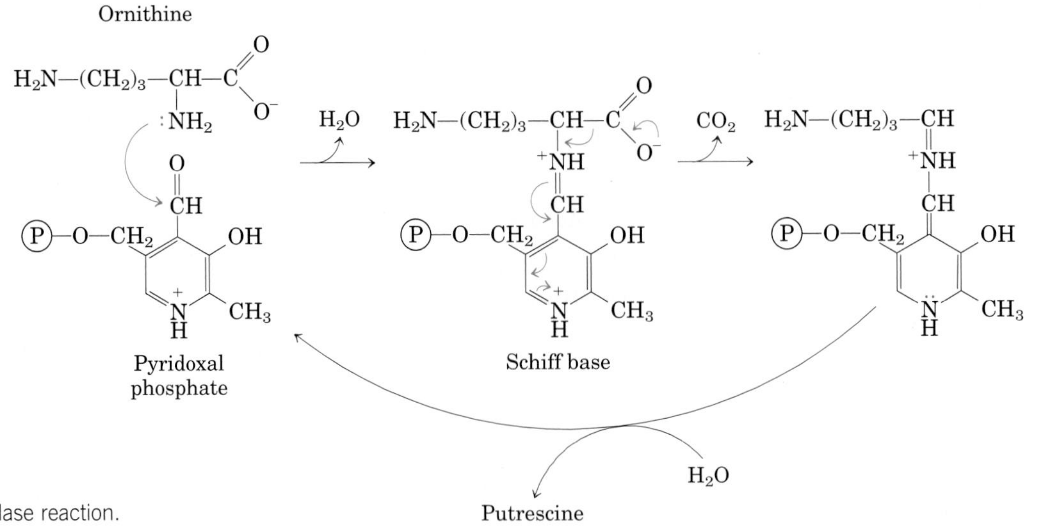

figure 2
Mechanism of ornithine decarboxylase reaction.

reasons not well understood, the trypanosome enzyme is stable, not readily replaced by newly synthesized enzyme. An inhibitor of ornithine decarboxylase that binds permanently to the enzyme would thus have little effect on mammalian cells, which could rapidly replace inactivated enzyme, but would adversely affect the parasite.

The first few steps of the normal reaction catalyzed by ornithine decarboxylase are shown in Figure 2. The flow of electron pairs is denoted by blue arrows. Once CO_2 is released, the electron movement is reversed and putrescine is produced (see Fig. 22–28). Based on this mechanism, several suicide inactivators have been designed for this enzyme, one of which is difluoromethylornithine (DFMO). DFMO is relatively inert in solution. When it binds to ornithine decarboxylase, however, the enzyme is quickly inactivated (Fig. 3). The inhibitor provides an alternative electron sink in the form of two strategically placed fluorine atoms, which are excellent leaving groups. Instead of electrons moving into the ring structure of PLP, the reaction results in displacement of a fluorine atom. Nucleophilic amino acid side chains (represented by B:) at the enzyme's active site then form a covalent complex with the highly reactive PLP-inhibitor adduct in an essentially irreversible reaction. In this way the inhibitor makes use of the enzyme's own reaction mechanisms to kill it. DFMO has proven highly effective against African sleeping sickness in clinical trials.

Approaches such as this show great promise for treating a wide range of diseases. The design of drugs based on enzyme mechanism and structure is replacing the more traditional trial-and-error methods of developing pharmaceuticals.

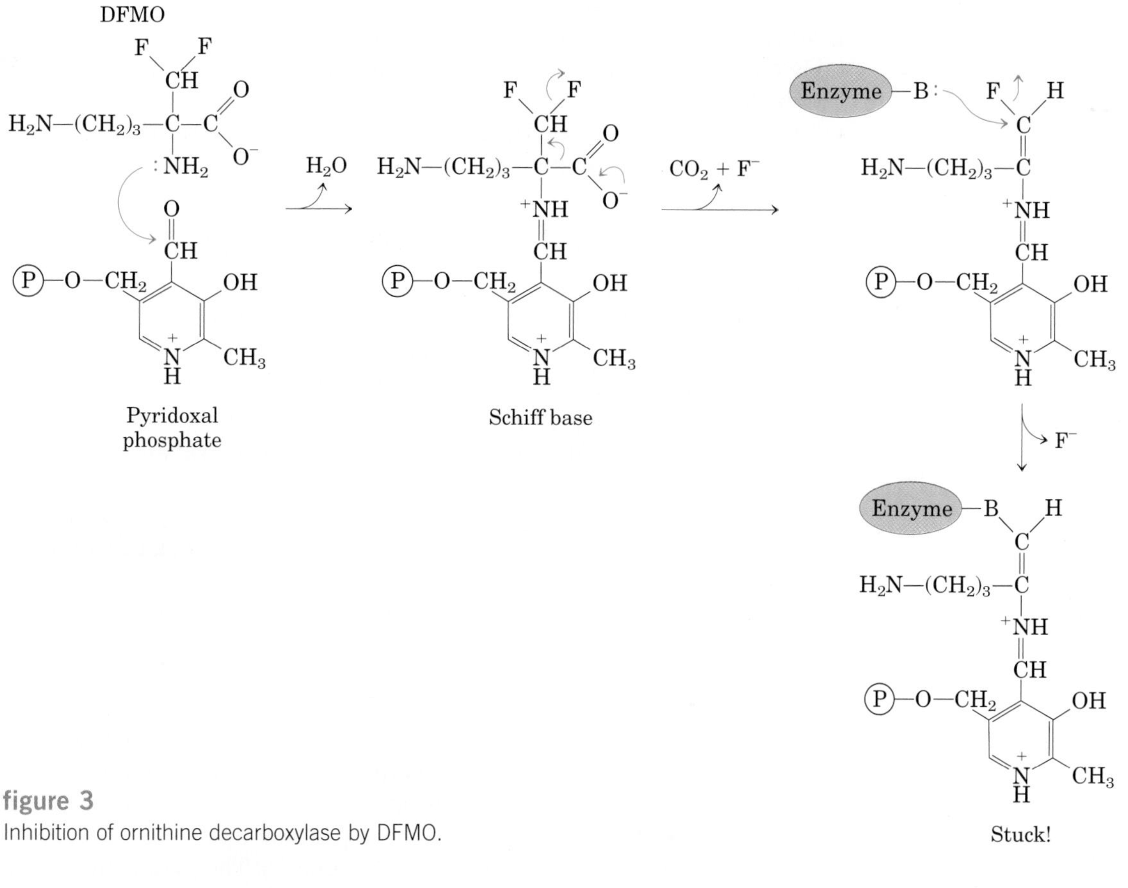

figure 3
Inhibition of ornithine decarboxylase by DFMO.

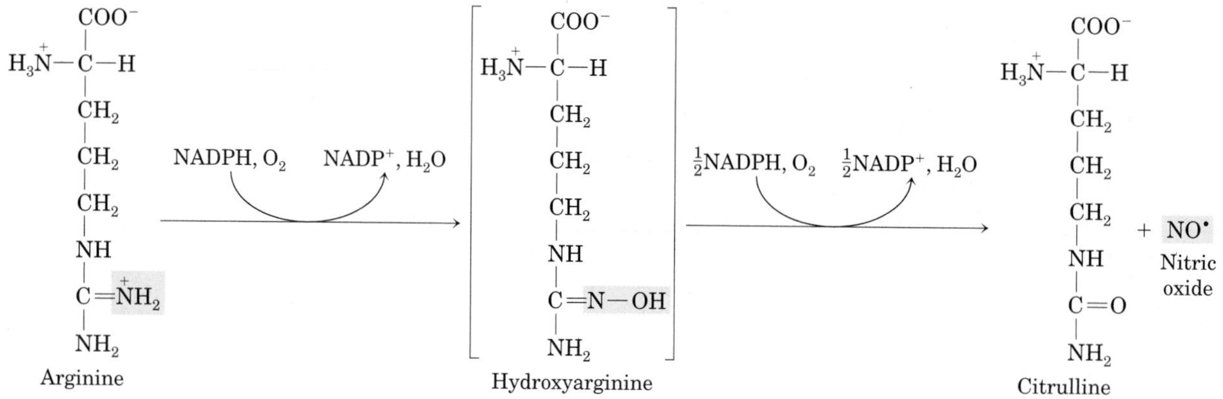

$$\text{Arginine} \xrightarrow[\text{NADPH, O}_2 \;\to\; \text{NADP}^+\text{, H}_2\text{O}]{} [\text{Hydroxyarginine}] \xrightarrow[\frac{1}{2}\text{NADPH, O}_2 \;\to\; \frac{1}{2}\text{NADP}^+\text{, H}_2\text{O}]{} \text{Citrulline} + \text{NO}^{\bullet}$$

Arginine: COO^- / $H_3\overset{+}{N}-C-H$ / CH_2 / CH_2 / CH_2 / NH / $C{=}\overset{+}{N}H_2$ / NH_2

Hydroxyarginine: COO^- / $H_3\overset{+}{N}-C-H$ / CH_2 / CH_2 / CH_2 / NH / $C{=}N{-}OH$ / NH_2

Citrulline: COO^- / $H_3\overset{+}{N}-C-H$ / CH_2 / CH_2 / CH_2 / NH / $C{=}O$ / NH_2

$NO^{\bullet}$ Nitric oxide

figure 22–29
Biosynthesis of nitric oxide. Both steps are catalyzed by nitric oxide synthase. The nitrogen of the NO is derived from the guanidino group of arginine.

Arginine Is the Precursor for Biological Synthesis of Nitric Oxide

A surprise finding in the mid-1980s was the role of nitric oxide (NO), previously known mainly as a component of smog, as an important biological messenger. This simple gaseous substance diffuses readily through membranes, although its high reactivity limits its diffusion to about a 1 mm radius from the site of synthesis. In humans NO plays a role in a range of physiological processes including neurotransmission, blood clotting, and the control of blood pressure. Its mode of action is described in Chapter 13 (see p. 449).

Nitric oxide is synthesized from arginine in an NADPH-dependent reaction catalyzed by nitric oxide synthase, a dimeric enzyme structurally related to NADPH cytochrome P-450 reductase (see Box 21–1). The reaction is a five-electron oxidation (Fig. 22–29). Each subunit of the enzyme contains one bound molecule of each of four different cofactors: FMN, FAD, tetrahydrobiopterin, and Fe^{3+} heme. NO is an unstable molecule and cannot be stored. Its synthesis is stimulated by interaction of nitric oxide synthase with Ca^{2+}-calmodulin (see Fig. 13–19).

Biosynthesis and Degradation of Nucleotides

As discussed in Chapter 10, nucleotides play a variety of important roles in all cells. They are the precursors of DNA and RNA. They are essential carriers of chemical energy—a role primarily of ATP and to some extent GTP. They are components of the cofactors NAD, FAD, *S*-adenosylmethionine, and coenzyme A, as well as of activated biosynthetic intermediates such as UDP-glucose and CDP-diacylglycerol. Some, such as cAMP and cGMP, are also cellular second messengers.

Two types of pathways lead to nucleotides: the **de novo pathways** and the **salvage pathways.** De novo synthesis of nucleotides begins with their metabolic precursors: amino acids, ribose 5-phosphate, CO_2, and NH_3. Salvage pathways recycle the free bases and nucleosides released from nucleic acid breakdown. Both types of pathways are important in cellular metabolism.

The de novo pathways for purine and pyrimidine biosynthesis appear to be identical in nearly all living organisms. Notably, the free bases guanine, adenine, thymine, cytidine, and uracil are *not* intermediates in these pathways; that is, the bases are not synthesized and then attached to ribose, as might be expected. The purine ring structure is built up one or a few atoms at a time, attached to ribose throughout the process. The pyrimidine ring is synthesized as **orotate,** attached to ribose phosphate, and then converted to the common pyrimidine nucleotides used in nucleic acid synthesis. Although the free bases are not intermediates in the de novo pathways, they are intermediates in some of the salvage pathways.

Several important precursors are shared by the de novo pathways for pyrimidines and purines. Phosphoribosyl pyrophosphate (PRPP) is important in both, and here the structure of ribose is retained in the product nucleotide, in contrast to its fate in the tryptophan and histidine biosynthetic pathways discussed earlier. An amino acid is an important precursor in each type of pathway: glycine for purines and aspartate for pyrimidines. Glutamine again is the most important source of amino groups—in five different steps in the de novo pathways. Aspartate is also used twice as the source of an amino group in the purine pathways.

Two other features deserve mention. First, there is evidence, especially in the de novo purine pathway, that the enzymes are present as large, multienzyme complexes in the cell, a recurring theme in our discussion of metabolism. Second, the cellular pools of nucleotides (other than ATP) are quite small, perhaps 1% or less of the amounts required to synthesize the cell's DNA. Therefore, nucleotide synthesis must continue during nucleic acid synthesis and in some cases may limit the rates of DNA replication and transcription. Because of the importance of these processes in dividing cells, agents that inhibit nucleotide synthesis have become particularly important to modern medicine.

We examine here the biosynthetic pathways of purine and pyrimidine nucleotides and their regulation, the formation of the deoxynucleotides, and the degradation of purines and pyrimidines to uric acid and urea. We end with a discussion of chemotherapeutic agents that affect nucleotide synthesis.

De Novo Purine Nucleotide Synthesis Begins with PRPP

The two parent purine nucleotides of nucleic acids are adenosine 5′-monophosphate (AMP; adenylate) and guanosine 5′-monophosphate (GMP; guanylate), containing the purine bases adenine and guanine. Figure 22–30 shows the origin of the carbon and nitrogen atoms of the purine ring system, as determined by John Buchanan using isotopic tracer experiments in birds. The detailed pathway of purine biosynthesis was worked out primarily by Buchanan and G. Robert Greenberg in the 1950s. In the first committed step of the pathway, an amino group donated by glutamine is attached at C-1 of PRPP (Fig. 22–31). The resulting **5-phosphoribosylamine** is highly unstable, with a half-life of 30 s at pH 7.5. The purine ring is subsequently built up on this structure.

John Buchanan

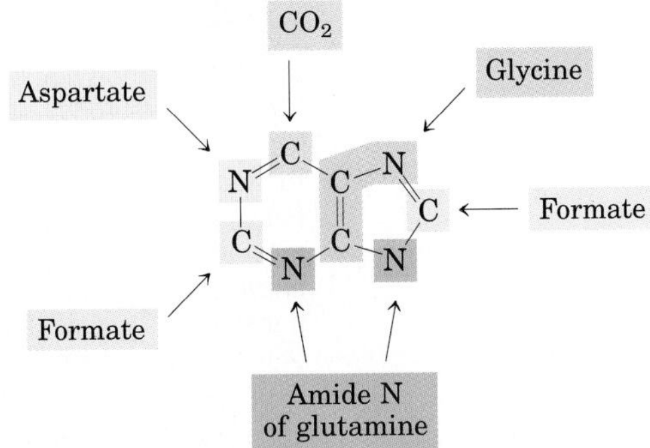

figure 22–30
Origin of the ring atoms of purines. This information was obtained from isotopic experiments with ^{14}C- or ^{15}N-labeled precursors. Formate is supplied in the form of N^{10}-formyltetrahydrofolate.

The second step is the addition of three atoms from glycine (Fig. 22–31, step ②). An ATP is consumed to activate the glycine carboxyl group (in the form of an acyl phosphate) for this condensation reaction. The added glycine amino group is then formylated by N^{10}-formyltetrahydrofolate (step ③), and a nitrogen is contributed by glutamine (step ④), before dehydration and ring closure yield the five-membered imidazole ring of the purine nucleus, as 5-aminoimidazole ribonucleotide (step ⑤).

At this point, three of the six atoms needed for the second ring in the purine structure are in place. To complete the process, a carboxyl group is first added (step ⑥). This carboxylation is unusual in that it does not require biotin, but instead uses the bicarbonate generally present in aqueous solutions. A rearrangement transfers the carboxylate from the exocyclic amino group to position 4 of the imidazole ring (step ⑦). Aspartate then donates its amino group in two steps (⑧ and ⑨): formation of an amide bond is followed by elimination of the carbon skeleton of aspartate (as fumarate). Recall that aspartate plays an analogous role in two steps of the urea cycle (see Fig. 18–9). The final carbon is contributed by N^{10}-formyltetrahydrofolate (step ⑩), and a second ring closure takes place to yield the second fused ring of the purine nucleus (step ⑪). The first intermediate to have a complete purine ring is **inosinate** (IMP).

As in the tryptophan and histidine biosynthetic pathways, the enzymes of IMP synthesis appear to be organized as large multienzyme complexes in the cell. Once again, evidence comes from the existence of single polypeptides with several functions, some of which catalyze nonsequential steps in the pathway. In eukaryotic cells ranging from yeast to fruit flies to chickens, steps ①, ③, and ⑤ in Figure 22–31 are catalyzed by a multifunctional protein. Additional multifunctional proteins catalyze steps ⑦ and ⑧ and steps ⑩ and ⑪. In bacteria, these activities are found on separate proteins, but a large noncovalent complex may exist in these cells. The channeling of reaction intermediates from one enzyme to the next permitted by these complexes is probably especially important for unstable intermediates such as 5-phosphoribosylamine.

figure 22–31

De novo synthesis of purine nucleotides: construction of the purine ring of inosinate (IMP). Each addition to the purine ring is shaded to match Figure 22–30. After step ②, R symbolizes the 5-phospho-D-ribosyl group on which the purine ring is built. Formation of 5-phosphoribosylamine (step ①) is the first committed step in purine synthesis. Note that the product of step ⑨, AICAR, is the remnant of ATP released during histidine biosynthesis (see Fig. 22–20, step ⑤). Abbreviations are given for most intermediates to simplify the naming of the pathway enzymes. The enzymes are: ① glutamine-PRPP amidotransferase, ② GAR synthetase, ③ GAR transformylase, ④ FGAR amidotransferase, ⑤ FGAM cyclase (AIR synthetase), ⑥ N^5-CAIR synthetase, ⑦ N^5-CAIR mutase, ⑧ SAICAR synthetase, ⑨ SAICAR lyase, ⑩ AICAR transformylase, and ⑪ IMP synthase.

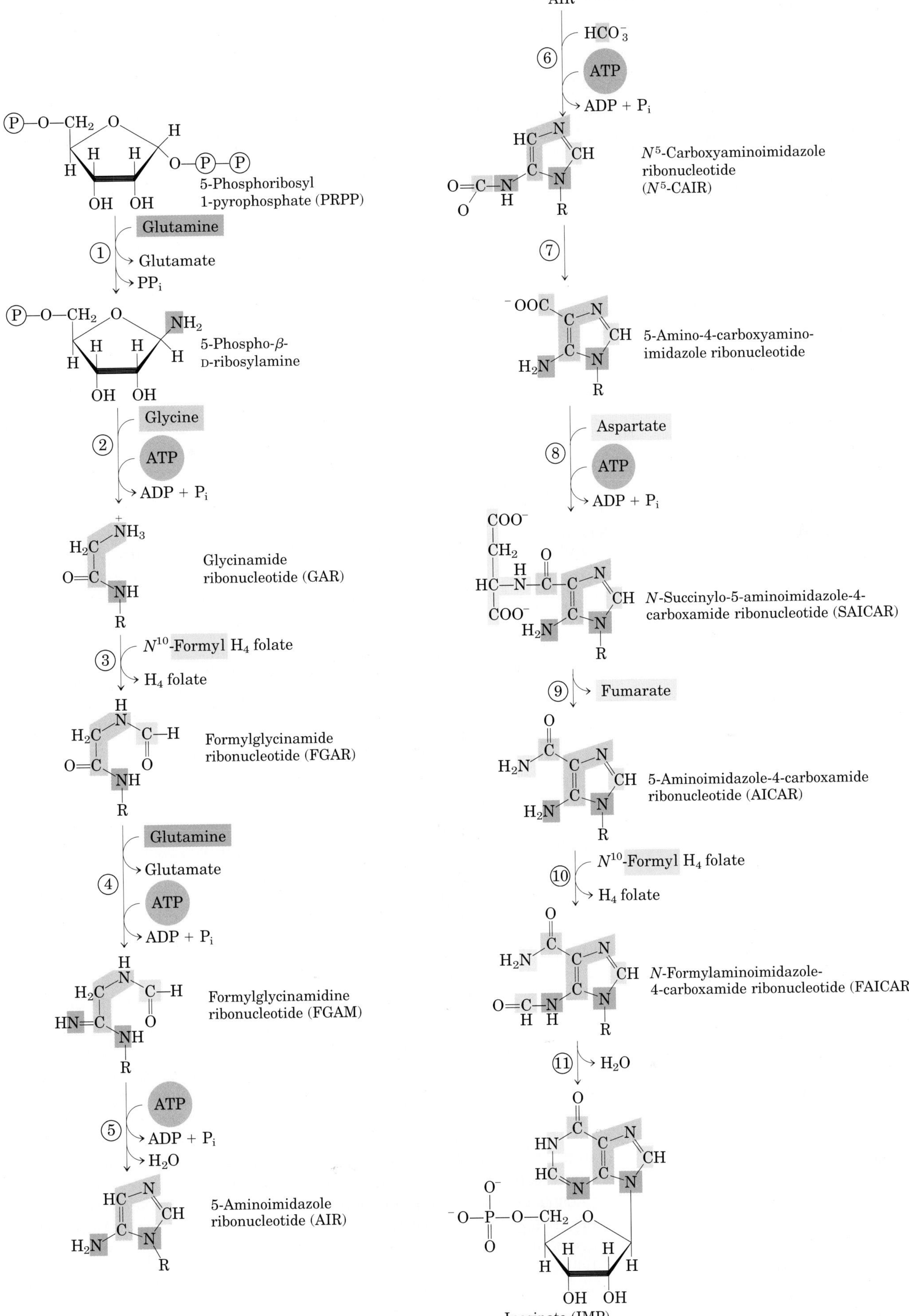
5-Phosphoribosyl 1-pyrophosphate (PRPP)
Glutamine
Glutamate
PPi
5-Phospho-β-D-ribosylamine
Glycine
ATP
ADP + Pi
Glycinamide ribonucleotide (GAR)
N10-Formyl H4 folate
H4 folate
Formylglycinamide ribonucleotide (FGAR)
Glutamine
Glutamate
ATP
ADP + Pi
Formylglycinamidine ribonucleotide (FGAM)
ATP
ADP + Pi
H2O
5-Aminoimidazole ribonucleotide (AIR)
AIR
HCO3-
ATP
ADP + Pi
N5-Carboxyaminoimidazole ribonucleotide (N5-CAIR)
5-Amino-4-carboxyamino-imidazole ribonucleotide
Aspartate
ATP
ADP + Pi
N-Succinylo-5-aminoimidazole-4-carboxamide ribonucleotide (SAICAR)
Fumarate
5-Aminoimidazole-4-carboxamide ribonucleotide (AICAR)
N10-Formyl H4 folate
H4 folate
N-Formylaminoimidazole-4-carboxamide ribonucleotide (FAICAR)
H2O
Inosinate (IMP)

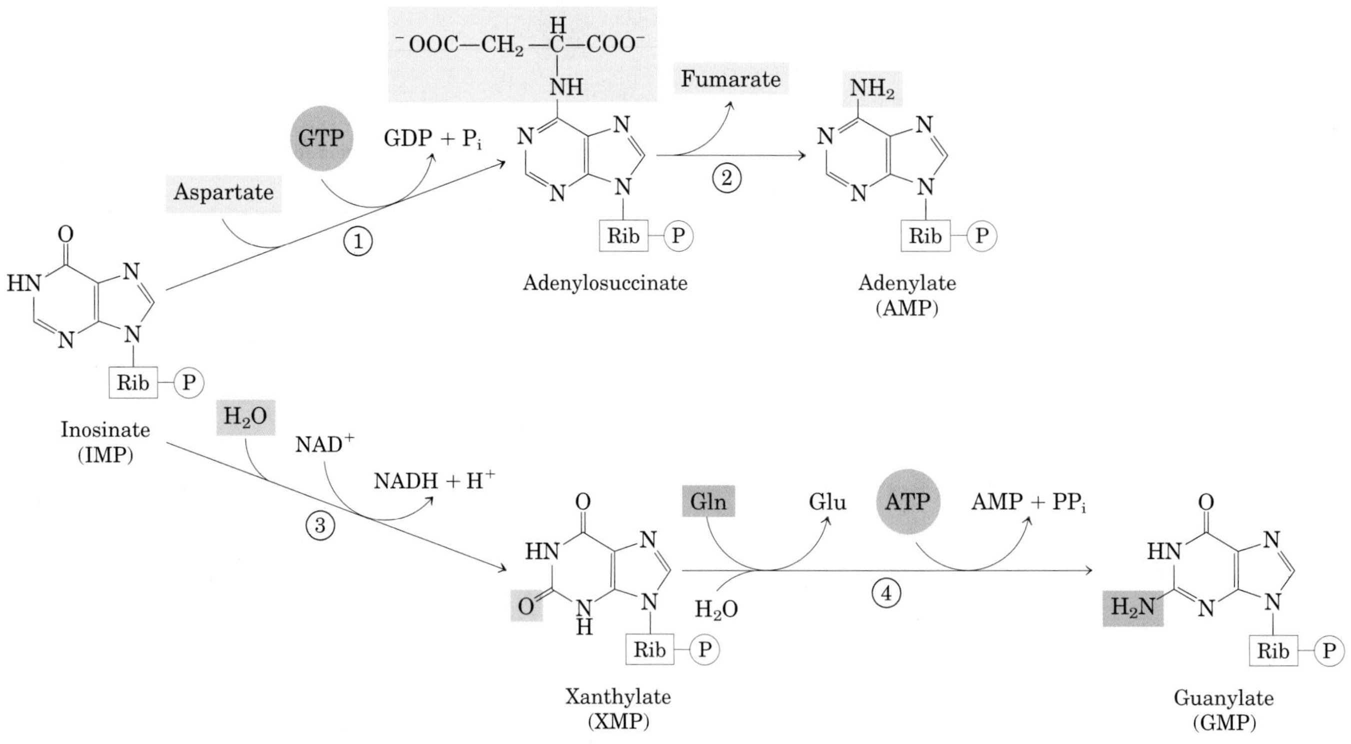

figure 22–32

Biosynthesis of AMP and GMP from IMP. The enzymes are: ① adenylosuccinate synthetase, ② adenylosuccinate lyase, ③ IMP dehydrogenase, and ④ XMP-glutamine amidotransferase.

Conversion of inosinate to adenylate (Fig. 22–32) requires the insertion of an amino group derived from aspartate; this takes place in two reactions similar to those used to introduce N-1 of the purine ring (Fig. 22–31, steps ⑧ and ⑨). A key difference is that GTP rather than ATP is the source of the high-energy phosphate in synthesizing adenylosuccinate. Guanylate is formed by the NAD^+-requiring oxidation of inosinate at C-2, followed by addition of an amino group derived from glutamine. ATP is cleaved to AMP and PP_i in the final step (Fig. 22–32).

Purine Nucleotide Biosynthesis Is Regulated by Feedback Inhibition

Three major feedback mechanisms cooperate in regulating the overall rate of de novo purine nucleotide synthesis and the relative rates of formation of the two end products, adenylate and guanylate (Fig. 22–33). The first mechanism is exerted on the first reaction that is unique to purine synthesis—transfer of an amino group to PRPP to form 5-phosphoribosylamine. This reaction is catalyzed by the allosteric enzyme glutamine-PRPP amidotransferase, which is inhibited by the end products IMP, AMP, and GMP. These same nucleotides also inhibit the synthesis of PRPP from ribose phosphate by ribose phosphate pyrophosphokinase. AMP and GMP act synergistically in this concerted inhibition. Thus, whenever either AMP or GMP accumulates to excess, the first step in its biosynthesis from PRPP is partially inhibited.

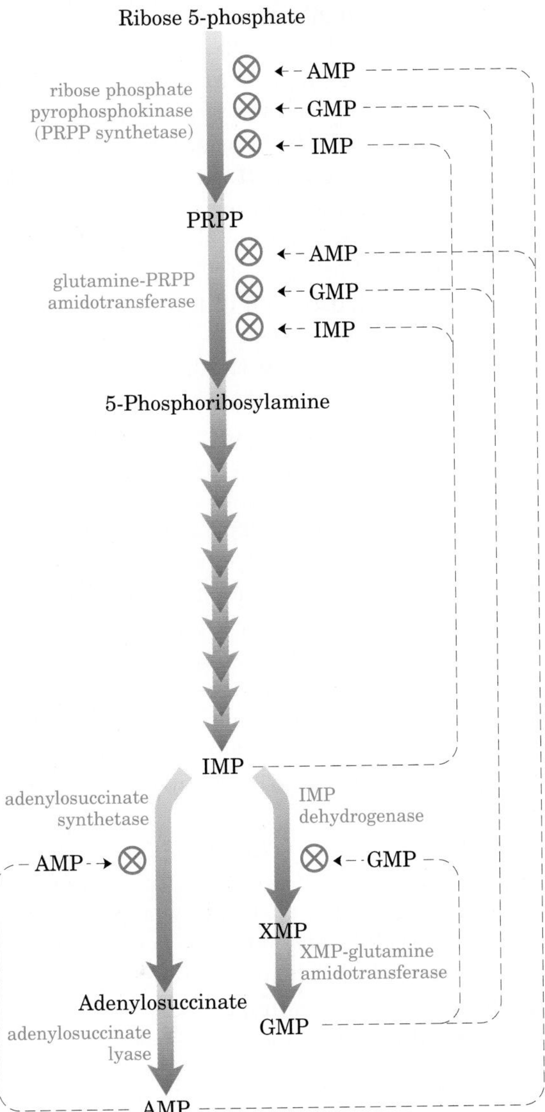

figure 22–33
Regulatory mechanisms in the biosynthesis of adenine and guanine nucleotides in *E. coli*. Regulation of these pathways differs in other organisms.

In the second control mechanism, exerted at a later stage, an excess of GMP in the cell inhibits formation of xanthylate from inosinate by IMP dehydrogenase, without affecting the formation of AMP (Fig. 22–33). Conversely, an accumulation of adenylate inhibits formation of adenylosuccinate by adenylosuccinate synthetase, without affecting the biosynthesis of GMP. In the third mechanism, GTP is required in the conversion of IMP to AMP, whereas ATP is required for conversion of IMP to GMP (Fig. 22–32), a reciprocal arrangement that tends to balance the synthesis of the two ribonucleotides.

Pyrimidine Nucleotides Are Made from Aspartate, PRPP, and Carbamoyl Phosphate

The common pyrimidine ribonucleotides are cytidine 5′-monophosphate (CMP; cytidylate) and uridine 5′-monophosphate (UMP; uridylate), which contain the pyrimidines cytosine and uracil. De novo pyrimidine nucleotide biosynthesis (Fig. 22–34) proceeds in a somewhat different manner from purine nucleotide synthesis; the six-membered pyrimidine ring is made first and then attached to ribose 5-phosphate. Required in this process is car-

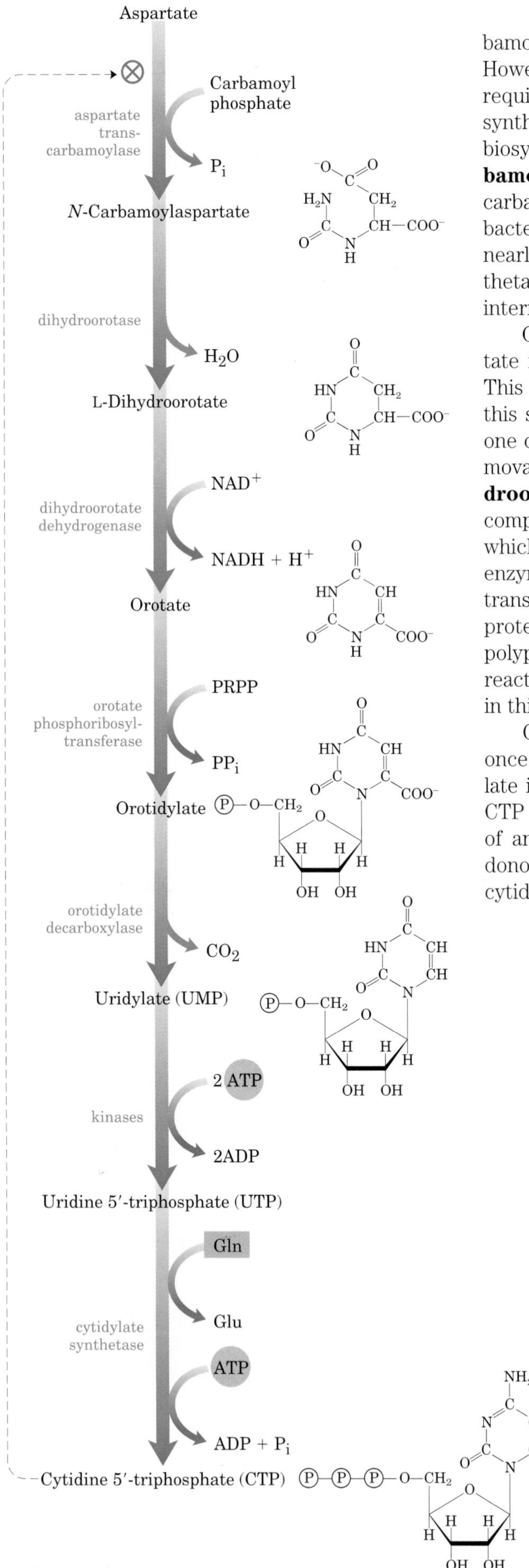

bamoyl phosphate, also an intermediate in the urea cycle (see Fig. 18–9). However, as we noted in Chapter 18, in animals the carbamoyl phosphate required in urea synthesis is made in mitochondria by carbamoyl phosphate synthetase I, whereas the carbamoyl phosphate required in pyrimidine biosynthesis is made in the cytosol by a different form of the enzyme, **carbamoyl phosphate synthetase II.** In bacteria, a single enzyme supplies carbamoyl phosphate for the synthesis of arginine and pyrimidines. The bacterial enzyme has three separate active sites, spaced along a channel nearly 100 Å in length (Fig. 22–35). Bacterial carbamoyl phosphate synthetase provides a vivid illustration of the channeling of unstable reaction intermediates between active sites.

Carbamoyl phosphate reacts with aspartate to yield *N*-carbamoylaspartate in the first committed step of pyrimidine biosynthesis (Fig. 22–34). This reaction is catalyzed by **aspartate transcarbamoylase.** In bacteria, this step is highly regulated, and bacterial aspartate transcarbamoylase is one of the most thoroughly studied allosteric enzymes (see below). By removal of water from *N*-carbamoylaspartate, a reaction catalyzed by **dihydroorotase,** the pyrimidine ring is closed to form L-dihydroorotate. This compound is oxidized to the pyrimidine derivative orotate, a reaction in which NAD^+ is the ultimate electron acceptor. In eukaryotes, the first three enzymes in this pathway—carbamoyl phosphate synthetase II, aspartate transcarbamoylase, and dihydroorotase—are part of a single trifunctional protein. The protein, known by the acronym CAD, contains three identical polypeptide chains (each of M_r 230,000), each with active sites for all three reactions. This suggests that large, multienzyme complexes may be the rule in this pathway.

Once orotate is formed, the ribose 5-phosphate side chain, provided once again by PRPP, is attached to yield orotidylate (Fig. 22–34). Orotidylate is then decarboxylated to uridylate, which is phosphorylated to UTP. CTP is formed from UTP by the action of **cytidylate synthetase,** by way of an acyl phosphate intermediate (consuming one ATP). The nitrogen donor is normally glutamine, although NH_4^+ can be used directly by the cytidylate synthetases of many species.

figure 22–34

De novo synthesis of pyrimidine nucleotides: synthesis of UTP and CTP via orotidylate. The pyrimidine is constructed from carbamoyl phosphate and aspartate. The ribose 5-phosphate is then added to the completed pyrimidine ring by orotate phosphoribosyltransferase. The first step in this pathway (not shown here; see Fig. 18–10) is the synthesis of carbamoyl phosphate from CO_2 and NH_4^+, catalyzed in eukaryotes by carbamoyl phosphate synthetase II.

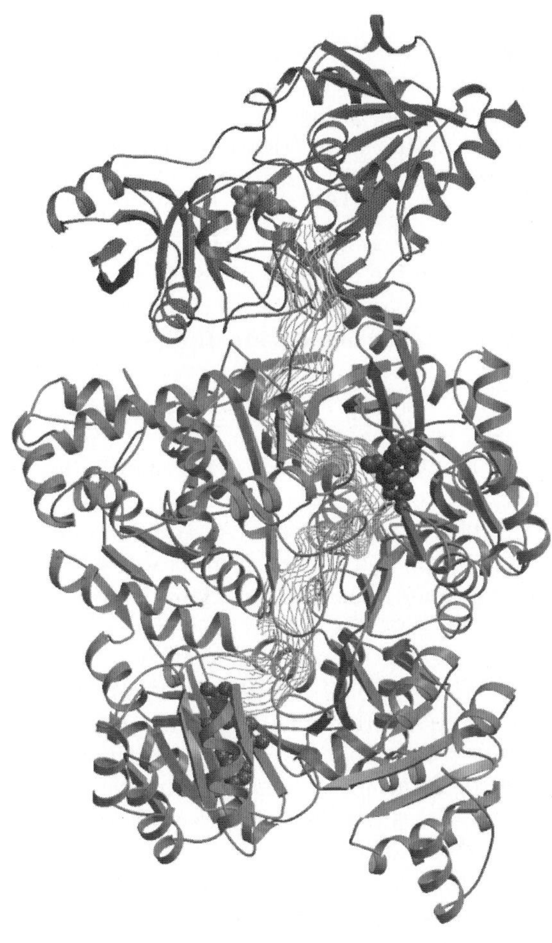

figure 22–35
Channeling of intermediates in bacterial carbamoyl phosphate synthetase. The reaction catalyzed by this enzyme is illustrated in Figure 18–10. The large and small subunits are shown in blue and gray, respectively. A glutamine molecule binds to the small subunit, donating its amido nitrogen as NH_4^+ in a glutamine amidotransferase–type reaction. The NH_4^+ enters a channel (shown as a yellow mesh) that takes it to a second active site, where it combines with bicarbonate in a reaction requiring ATP (bound ADP at this second active site is shown in blue). The carbamate then reenters the channel to reach the third active site, where it is phosphorylated to carbamoyl phosphate (bound ADP shown in red). The entire channel is nearly 100 Å long.

Pyrimidine Nucleotide Biosynthesis Is Regulated by Feedback Inhibition

Regulation of the rate of pyrimidine nucleotide synthesis in bacteria occurs in large part through the enzyme aspartate transcarbamoylase (ATCase), which catalyzes the first reaction in the sequence and is inhibited by CTP, the end product of the sequence (Fig. 22–34). The bacterial ATCase molecule consists of six catalytic subunits and six regulatory subunits (see Fig. 8–24). The catalytic subunits bind the substrate molecules, and the allosteric subunits bind the allosteric inhibitor CTP. The entire ATCase molecule, as well as its subunits, exists in two conformations, active and inactive. When CTP is not bound to the regulatory subunits, the enzyme is maximally active. As CTP accumulates and binds to the regulatory subunits, they undergo a change in conformation. This change is transmitted to the catalytic subunits, which then also shift to an inactive conformation. ATP prevents the changes induced by CTP. Figure 22–36 shows the effects of the allosteric regulators on the activity of ATCase.

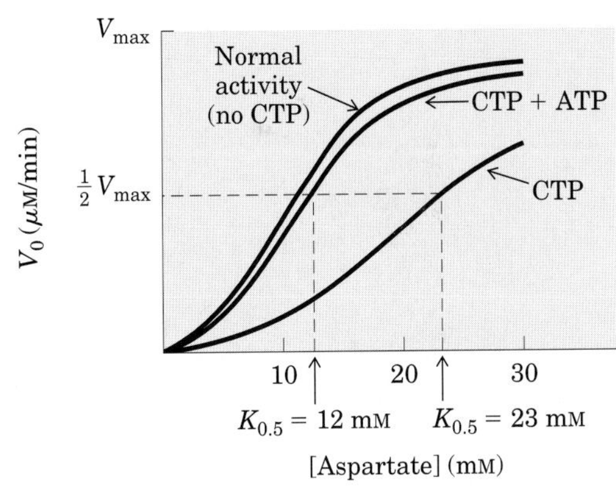

figure 22–36
Allosteric regulation of aspartate transcarbamoylase by CTP and ATP. Addition of 0.8 mM CTP, the allosteric inhibitor of ATCase, increases the $K_{0.5}$ for aspartate (lower curve) and the rate of conversion of aspartate to *N*-carbamoylaspartate. ATP at 0.6 mM fully reverses this effect (middle curve).

Nucleoside Monophosphates Are Converted to Nucleoside Triphosphates

Nucleotides to be used in biosynthesis are generally converted to nucleoside triphosphates. The conversion pathways are common to all cells. Phosphorylation of AMP to ADP is promoted by **adenylate kinase,** in the reaction

$$\text{ATP} + \text{AMP} \rightleftharpoons 2\text{ADP}$$

The ADP so formed is phosphorylated to ATP by the glycolytic enzymes or through oxidative phosphorylation.

ATP also brings about the formation of other nucleoside diphosphates by the action of a class of enzymes called **nucleoside monophosphate kinases.** These enzymes, which are generally specific for a particular base but nonspecific for the sugar (ribose or deoxyribose), catalyze the reaction

$$\text{ATP} + \text{NMP} \rightleftharpoons \text{ADP} + \text{NDP}$$

The efficient cellular systems for rephosphorylating ADP to ATP tend to pull this reaction in the direction of products.

Nucleoside diphosphates are converted to triphosphates by the action of a ubiquitous enzyme, **nucleoside diphosphate kinase,** which catalyzes the reaction

$$\text{NTP}_\text{D} + \text{NDP}_\text{A} \rightleftharpoons \text{NDP}_\text{D} + \text{NTP}_\text{A}$$

This enzyme is notable in that it is not specific for the base (purines or pyrimidines) or the sugar (ribose or deoxyribose). This nonspecificity applies to both phosphate acceptor (A) and donor (D), although the donor (left side of the equation) is almost invariably ATP because it is present in higher concentration than other nucleoside triphosphates under aerobic conditions.

Ribonucleotides Are the Precursors of Deoxyribonucleotides

Deoxyribonucleotides, the building blocks of DNA, are derived from the corresponding ribonucleotides by direct reduction at the 2′-carbon atom of the D-ribose to form the 2′-deoxy derivative. For example, adenosine diphosphate (ADP) is reduced to 2′-deoxyadenosine diphosphate (dADP), and GDP is reduced to dGDP. This reaction is somewhat unusual in that the reduction occurs at a nonactivated carbon; no closely analogous chemical reactions are known. The reaction is catalyzed by **ribonucleotide reductase,** best characterized in *E. coli,* in which its substrates are ribonucleoside diphosphates.

The reduction of the D-ribose portion of a ribonucleoside diphosphate to 2′-deoxy-D-ribose requires a pair of hydrogen atoms, which are ultimately donated by NADPH via an intermediate hydrogen-carrying protein, **thioredoxin.** This ubiquitous protein serves a similar redox function in photosynthesis (see Fig. 20–36) and other processes. Thioredoxin has pairs of —SH groups that carry hydrogen atoms from NADPH to the ribonucleoside diphosphate. Its oxidized or disulfide form is reduced by NADPH in a reaction catalyzed by **thioredoxin reductase** (Fig. 22–37), and reduced thioredoxin is then used by ribonucleotide reductase to reduce the nucleoside diphosphates (NDPs) to deoxyribonucleoside diphosphates (dNDPs). A second source of reducing equivalents for ribonucleotide reductase is glutathione (GSH). Glutathione serves as the reductant for a protein closely related to thioredoxin called **glutaredoxin,** which then transfers the reducing power to ribonucleotide reductase (Fig. 22–37).

Ribonucleotide reductase is notable in that its reaction mechanism provides the best-characterized example of the involvement of free radicals in biochemical transformations, once thought to be rare in biological systems. The enzyme in *E. coli* and most eukaryotes is a dimer, with subunits designated R1 and R2 (Fig. 22–38). The R1 subunit contains two kinds of regulatory sites, as described below. The two active sites of the enzyme are formed at the interface between the R1 and R2 subunits. At each active site, R1 contributes two sulfhydryl groups required for activity and R2 contributes a stable tyrosyl radical. The R2 subunit also has a binuclear iron (Fe^{3+}) cofactor that helps generate and stabilize the tyrosyl radicals (Fig. 22–38). The tyrosyl radical is too far from the active site to interact directly with it, but it generates another radical at the active site that functions in catalysis.

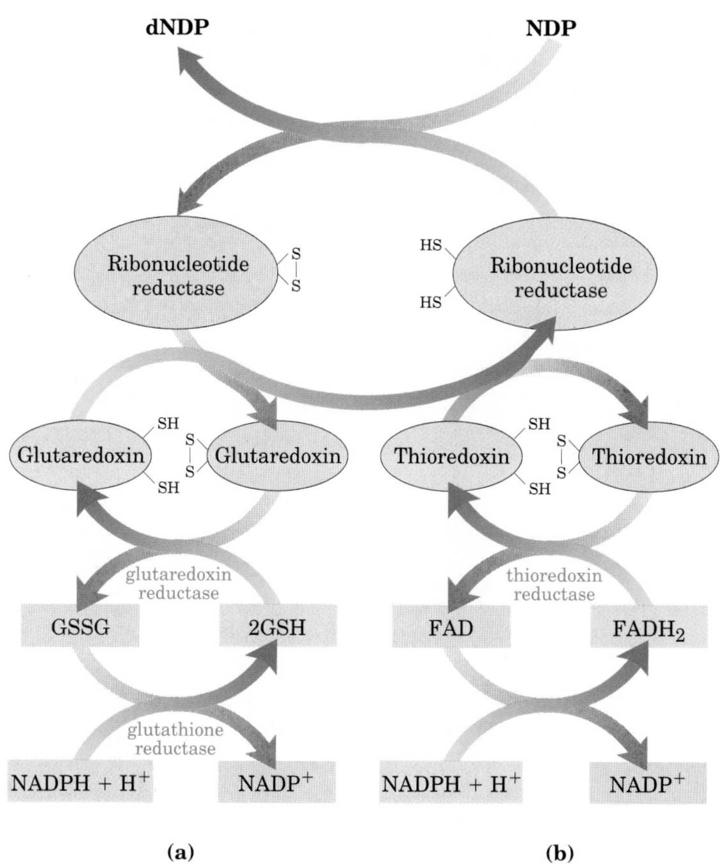

figure 22–37

Reduction of ribonucleotides to deoxyribonucleotides by ribonucleotide reductase. Electrons are transmitted (blue arrows) to the enzyme from NADPH by **(a)** glutaredoxin or **(b)** thioredoxin. The sulfide groups in glutaredoxin reductase are contributed by two molecules of bound glutathione (GSH; GSSG indicates oxidized glutathione). Note that thioredoxin reductase is a flavoenzyme, with FAD as prosthetic group.

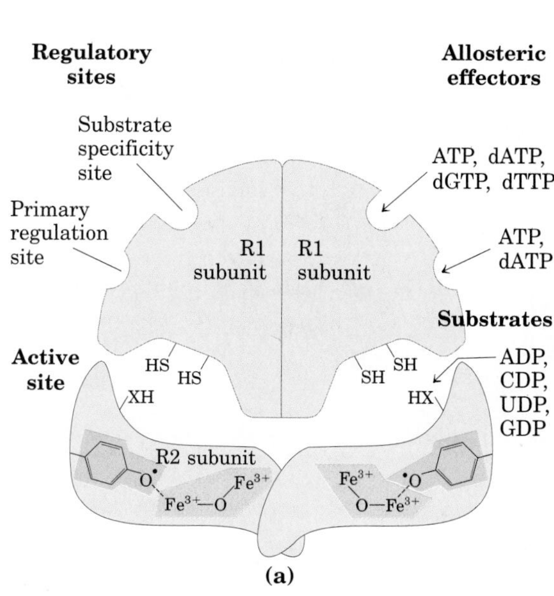

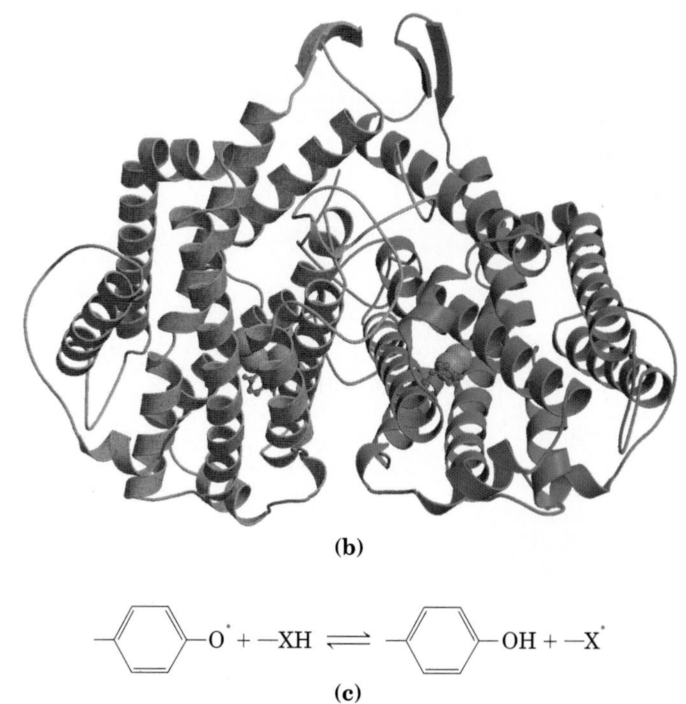

figure 22–38

Ribonucleotide reductase. (a) Subunit structure. The functions of the two regulatory sites are explained in Figure 22–40. The active site contains two thiols and a group (—XH) that can be converted to an active-site radical; this group is probably Cys^{439}, which functions as a thiyl radical. **(b)** The R2 subunits of *E. coli* ribonucleotide reductase. The Tyr residue that acts as the tyrosyl radical is shown in red; the binuclear iron center is orange. **(c)** The tyrosyl radical functions to generate the active-site radical ($—X^{\bullet}$), which is used in the mechanism shown in Figure 22–39.

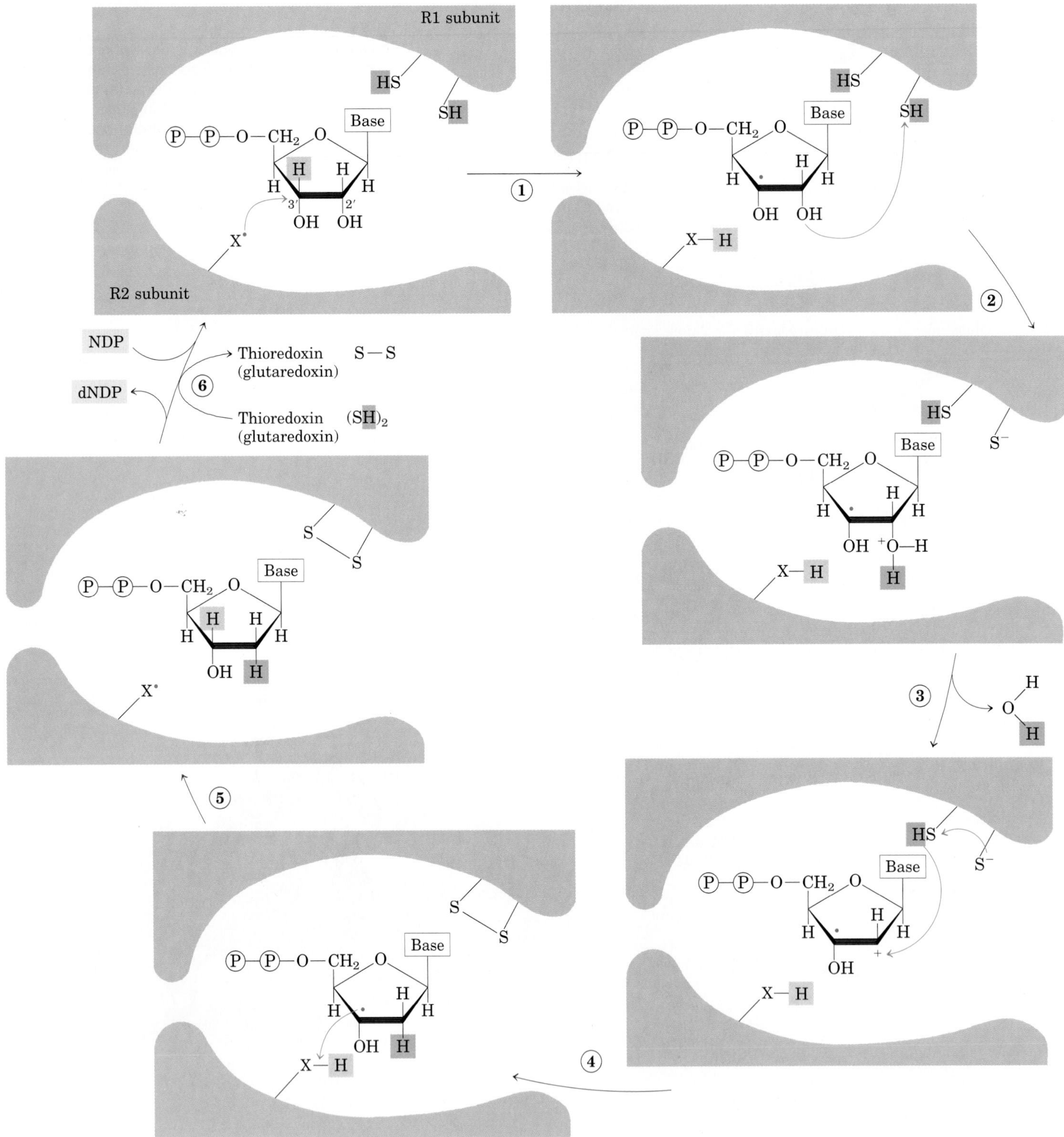

figure 22–39

Proposed mechanism for ribonucleotide reductase. In the enzyme from *E. coli* and most eukaryotes, the active thiol groups are on the R1 subunit; the active-site radical (—X•) is on the R2 subunit and is probably a thiyl radical of Cys[439] in *E. coli* (see Fig. 22–38). Steps ① through ⑥ are described in the text.

A likely mechanism for the ribonucleotide reductase reaction is illustrated in Figure 22–39. The 3′-ribonucleotide radical formed in step ① helps stabilize the cation subsequently formed at the 2′ carbon after the loss of H_2O (steps ② and ③). Two one-electron transfers accompanied by oxidation of the dithiol reduce the radical cation and regenerate the 3′-ribonucleotide radical (step ④). Step ⑤ is the reverse of step ①, regenerating the tyrosyl radical and forming the deoxy product. The oxidized dithiol is reduced to complete the cycle (step ⑥). In *E. coli,* likely sources of the required reducing equivalents for this reaction are thioredoxin and glutaredoxin, as noted above.

Four classes of ribonucleotide reductase have been reported. Their mechanisms (where known) generally conform to the scheme in Figure 22–39, but they differ in the identity of the group supplying the active-site radical and in the cofactors used to generate it. The *E. coli* enzyme (class I) requires oxygen to regenerate the tyrosyl radical if it is quenched and thus functions only in an aerobic environment. Class II enzymes, found in other microorganisms, have 5′-deoxyadenosylcobalamin (see Box 17–2) rather than a binuclear iron center. Class III enzymes have evolved to function in an anaerobic environment. *E. coli* contains a separate class III ribonucleotide reductase when grown anaerobically; this enzyme contains an iron-sulfur cluster (structurally distinct from the binuclear iron center of the class I enzyme) and requires NADPH and *S*-adenosylmethionine for activity. It uses nucleoside triphosphates rather than nucleoside diphosphates as substrates. A class IV ribonucleotide reductase, containing a binuclear manganese center, has been reported in some microorganisms. The evolution of different classes of ribonucleotide reductase to allow production of DNA precursors in different environments reflects the importance of this reaction in nucleotide metabolism.

Regulation of *E. coli* ribonucleotide reductase is unusual in that not only its *activity* but its *substrate specificity* is regulated by the binding of effector molecules. Each R1 subunit has two types of regulatory site (Fig. 22–38). One type affects overall enzyme activity and binds either ATP, which activates the enzyme, or dATP, which inactivates it. The second type alters substrate specificity in response to the effector molecule—either ATP, dATP, dTTP, or dGTP—that is bound there (Fig. 22–40). When ATP or dATP is bound, reduction of UDP and CDP is favored. When dTTP or dGTP is bound, reduction of GDP or ADP, respectively, is stimulated. The scheme is designed to provide a balanced pool of precursors for DNA synthesis. ATP is also a general activator for biosynthesis and ribonucleotide reduction. The presence of dATP in small amounts increases the reduction of pyrimidine nucleotides. An oversupply of the pyrimidine dNTPs is signaled by high levels of dTTP, which shifts the specificity to favor reduction of GDP. High levels of dGTP, in turn, shift the specificity to ADP reduction, and high levels of dATP shut the enzyme down. These effectors are thought to induce several distinct enzyme conformations with altered specificities.

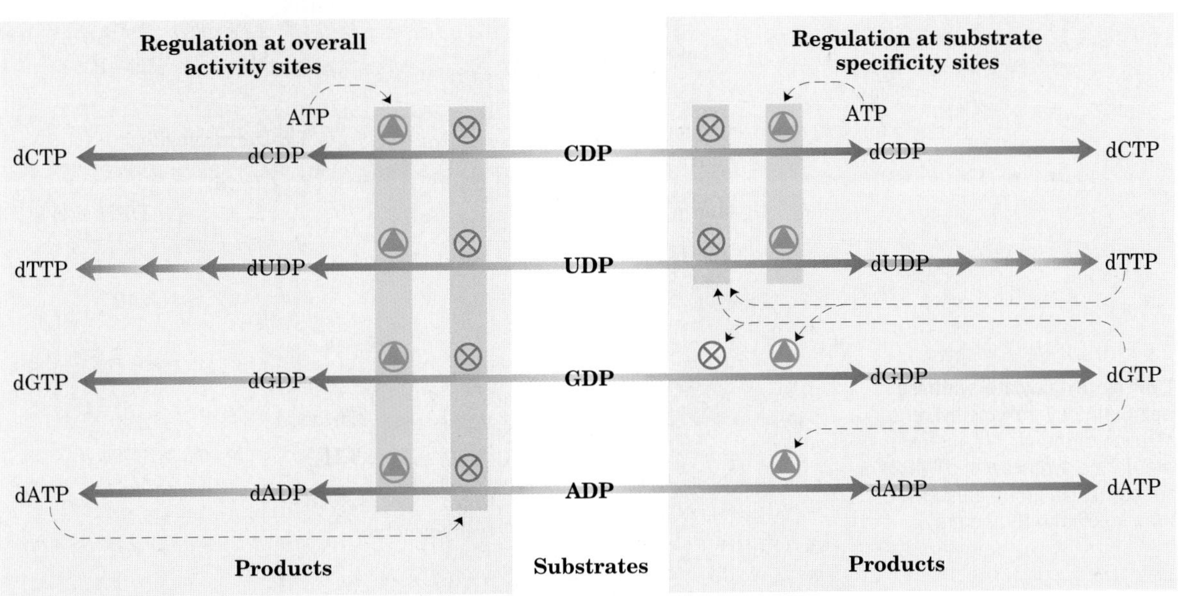

figure 22–40
Regulation of ribonucleotide reductase by deoxynucleoside triphosphates. The overall activity of the enzyme is affected by binding at one type of regulatory site (left). The substrate specificity of the enzyme is affected by the nature of the effector molecule bound at the second type of regulatory site (right). Inhibition or stimulation of the enzyme's activity with the four different substrates is indicated. The pathway from dUDP to dTTP is described later (see Figs 22–41, 22–42).

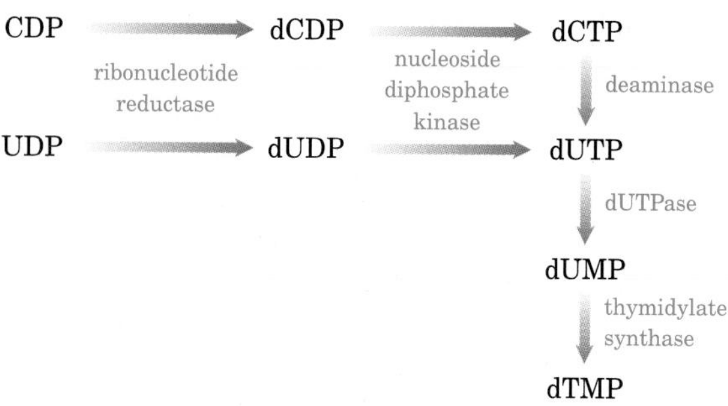

figure 22–41

Synthesis of thymidylate (dTMP). The pathways are shown beginning with the reaction catalyzed by ribonucleotide reductase. Figure 22–42 gives details of the thymidylate synthase reaction.

Thymidylate Is Derived from dCDP and dUMP

DNA contains thymine rather than uracil, and the de novo pathway to thymine involves only deoxyribonucleotides. The immediate precursor of thymidylate (dTMP) is dUMP. In bacteria, the pathway to dUMP begins with formation of dUTP, either by deamination of dCTP or by phosphorylation of dUDP (Fig. 22–41). The dUTP is converted to dUMP by a dUTPase. The latter reaction must be efficient to keep dUTP pools low and prevent incorporation of uridylate into DNA.

Conversion of dUMP to dTMP is catalyzed by **thymidylate synthase.** A one-carbon unit at the hydroxymethyl (—CH_2OH) oxidation level (see Fig. 18–16) is transferred from N^5,N^{10}-methylenetetrahydrofolate to dUMP, then reduced to a methyl group (Fig. 22–42). The reduction occurs at the expense of oxidation of tetrahydrofolate to dihydrofolate, which is unusual in tetrahydrofolate-requiring reactions. (Details of this reaction are shown in Figure 22–48.) The dihydrofolate is reduced to tetrahydrofolate by **dihydrofolate reductase**—a regeneration that is essential for the many processes that depend on tetrahydrofolate. In plants and at least one protist, thymidylate synthase and dihydrofolate reductase are part of a single bifunctional protein.

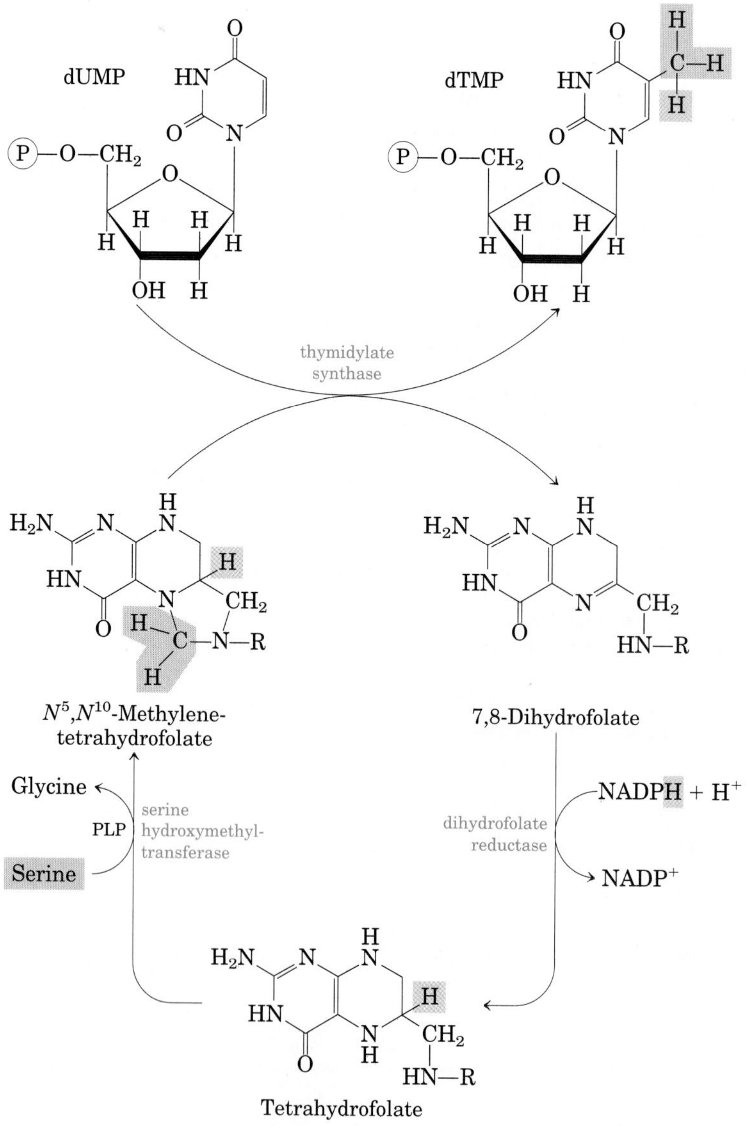

figure 22–42

Conversion of dUMP to dTMP by thymidylate synthase and dihydrofolate reductase. Serine hydroxymethyltransferase is required for regeneration of the N^5,N^{10}-methylene form of tetrahydrofolate. In the synthesis of dTMP, all three hydrogens of the added methyl group are derived from N^5,N^{10}-methylenetetrahydrofolate (red and gray).

Degradation of Purines and Pyrimidines Produces Uric Acid and Urea, Respectively

Purine nucleotides are degraded by a pathway in which the phosphate group is lost by the action of **5′-nucleotidase** (Fig. 22–43). Adenylate yields adenosine, which is deaminated to inosine by **adenosine deaminase,** and inosine is hydrolyzed to hypoxanthine (its purine base) and D-ribose. Hypoxanthine is oxidized successively to xanthine and then uric acid by **xanthine oxidase,** a flavoenzyme with an atom of molybdenum and four iron-sulfur centers in its prosthetic group. Molecular oxygen is the electron acceptor in this complex reaction.

GMP catabolism also yields uric acid as end product. GMP is first hydrolyzed to guanosine, which is then cleaved to free guanine. Guanine undergoes hydrolytic removal of its amino group to yield xanthine, which is converted to uric acid by xanthine oxidase (Fig. 22–43).

Uric acid is the excreted end product of purine catabolism in primates, birds, and some other animals, but in many other vertebrates it is further degraded to **allantoin** by the action of **urate oxidase.** In other organisms

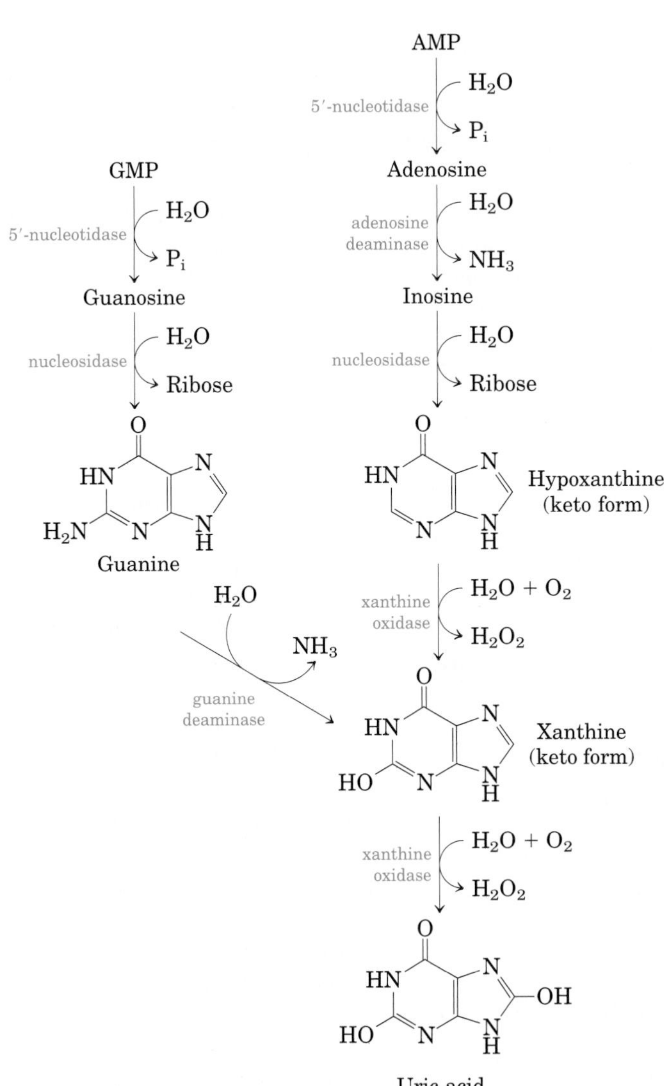

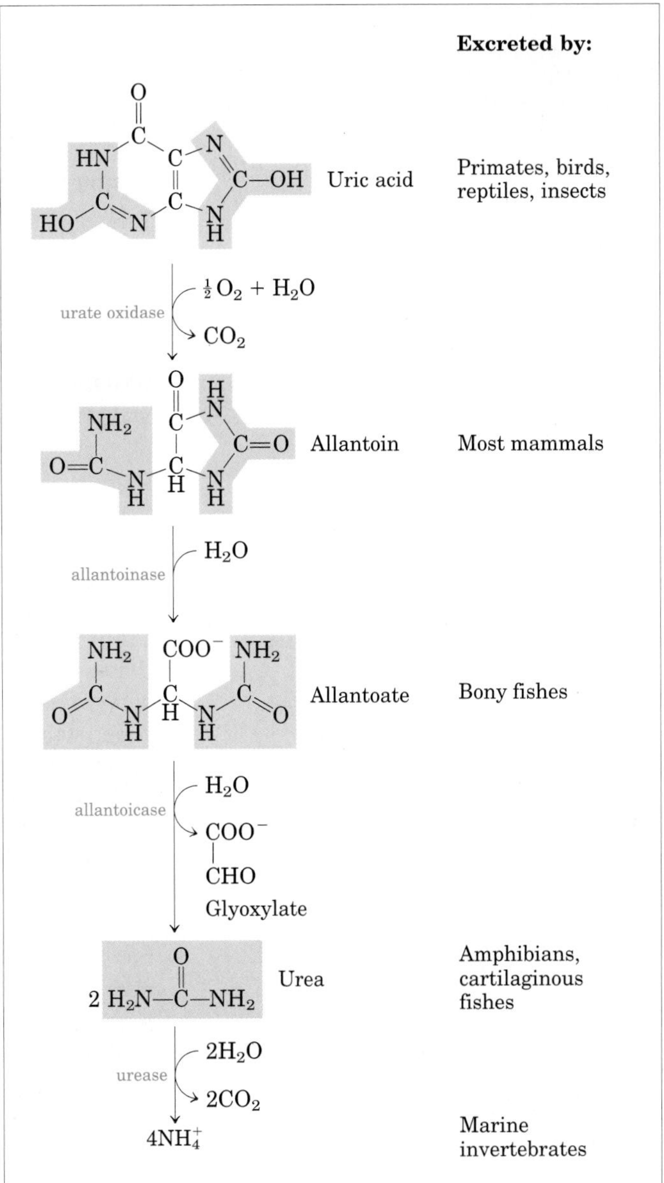

figure 22–43
Catabolism of purine nucleotides. Note that primates excrete much more nitrogen as urea via the urea cycle (Chapter 18) than as uric acid from purine degradation. Similarly, fish excrete much more nitrogen as NH_4^+ than as urea produced by the pathway shown here.

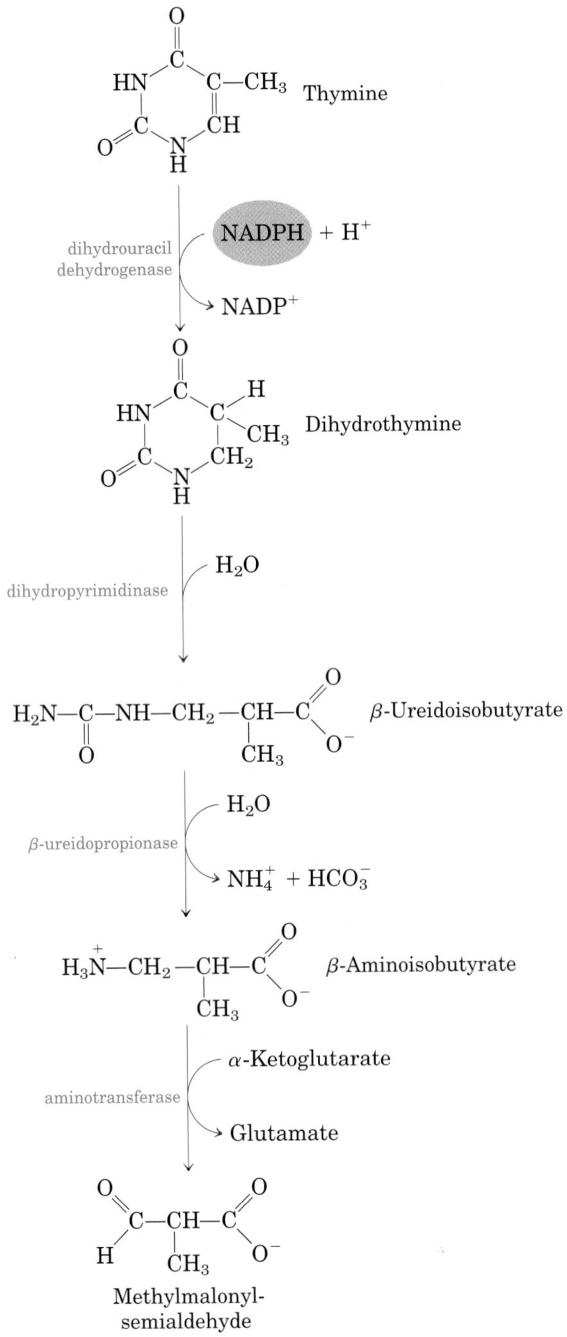

Thymine

dihydrouracil dehydrogenase — NADPH + H^+ → $NADP^+$

Dihydrothymine

dihydropyrimidinase — H_2O

β-Ureidoisobutyrate

β-ureidopropionase — H_2O → NH_4^+ + HCO_3^-

β-Aminoisobutyrate

aminotransferase — α-Ketoglutarate → Glutamate

Methylmalonyl-semialdehyde

figure 22–44
Catabolism of a pyrimidine. Shown here is the pathway for thymine.

the pathway is further extended, as shown in Figure 22–43. A normal adult human excretes uric acid at a rate of about 0.6 g/24 h; the excreted product arises in part from ingested purines and in part from turnover of the purine nucleotides of nucleic acids.

The pathways for degradation of pyrimidines generally lead to NH_4^+ production and thus to urea synthesis. Thymine, for example, is degraded to methylmalonylsemialdehyde (Fig. 22–44), an intermediate of valine catabolism. It is further degraded through propionyl-CoA and methylmalonyl-CoA to succinyl-CoA.

Genetic aberrations in human purine metabolism have been found, some with serious consequences. For example, **adenosine deaminase deficiency** leads to severe immunodeficiency disease in which T lymphocytes and B lymphocytes do not develop properly. Lack of adenosine deaminase leads to a 100-fold increase in the cellular concentration of dATP, a strong inhibitor of ribonucleotide reductase (Fig. 22–40). High levels of dATP produce a general deficiency of other dNTPs in T lymphocytes. The basis for B-lymphocyte toxicity is less clear. Patients with adenosine deaminase deficiency lack an effective immune system and do not survive unless isolated in a sterile "bubble" environment.

Purine and Pyrimidine Bases Are Recycled by Salvage Pathways

Free purine and pyrimidine bases are constantly released in cells during the metabolic degradation of nucleotides. Free purines are in large part salvaged and reused to make nucleotides, in a pathway much simpler than the de novo synthesis of purine nucleotides described earlier. One of the primary salvage pathways consists of a single reaction catalyzed by **adenosine phosphoribosyltransferase,** in which free adenine reacts with PRPP to yield the corresponding adenine nucleotide:

$$\text{Adenine} + \text{PRPP} \longrightarrow \text{AMP} + \text{PP}_i$$

Free guanine and hypoxanthine (the deamination product of adenine; see Fig. 22–43) are salvaged in the same way by **hypoxanthine-guanine phosphoribosyltransferase.** A similar salvage pathway exists for pyrimidine bases in microorganisms, and possibly in mammals.

A genetic lack of hypoxanthine-guanine phosphoribosyltransferase activity, seen almost exclusively in male children, results in a bizarre set of symptoms called **Lesch-Nyhan syndrome.** Children with this genetic disorder, which becomes manifest by the age of 2 years, are sometimes poorly coordinated and mentally retarded. In addition, they are extremely hostile and show compulsive self-destructive tendencies: they mutilate themselves by biting off their fingers, toes, and lips.

The devastating effects of Lesch-Nyhan syndrome illustrate the importance of the salvage pathways. Hypoxanthine and guanine arise constantly from the breakdown of nucleic acids. In the absence of hypoxanthine-guanine phosphoribosyltransferase, PRPP levels rise and purines are therefore overproduced by the de novo pathway, resulting in high levels of uric acid production and goutlike damage to tissue (see below). The brain is especially dependent on the salvage pathways, and this may account for the central nervous system damage in children with Lesch-Nyhan syndrome. This syndrome, and the immunodeficiency disease resulting from adenosine deaminase deficiency, are among the targets of early trials in human gene therapy (see Box 29–2).

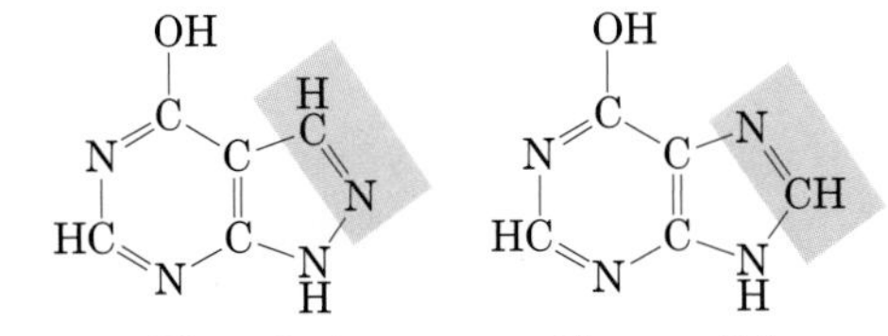

figure 22–45

Allopurinol, an inhibitor of xanthine oxidase. Hypoxanthine is the normal substrate of this enzyme. Only a slight alteration (shaded) in the structure of hypoxanthine yields a medically effective enzyme inhibitor. Allopurinol is an example of a useful drug that was designed as a competitive inhibitor.

Allopurinol

Hypoxanthine (enol form)

Overproduction of Uric Acid Causes Gout

Long thought, erroneously, to be due to "high living," gout is a disease of the joints caused by an elevated concentration of uric acid in the blood and tissues. The joints become inflamed, painful, and arthritic, owing to the abnormal deposition of sodium urate crystals. The kidneys are also affected, because excess uric acid is deposited in the kidney tubules. Gout occurs predominantly in males. Its precise cause is not known but is suspected to be a genetic deficiency of one or another enzyme of purine metabolism.

Gout is effectively treated by a combination of nutritional and drug therapies. Foods especially rich in nucleotides and nucleic acids, such as liver or glandular products, are withheld from the diet. Major alleviation of the symptoms is provided by the drug **allopurinol** (Fig. 22–45), an inhibitor of xanthine oxidase, the enzyme catalyzing conversion of purines to uric acid. When xanthine oxidase is inhibited, the excreted products of purine metabolism are xanthine and hypoxanthine, which are more water-soluble than uric acid and less likely to form crystalline deposits. Allopurinol was developed by Gertrude Elion and George Hitchings, who also developed acyclovir, used in treating people with AIDS, and other purine analogs used in cancer chemotherapy.

Gertrude Elion (1918–1999) and George Hitchings (1905–1998)

Many Chemotherapeutic Agents Target Enzymes in the Nucleotide Biosynthetic Pathways

Cancer cells grow more rapidly than the cells of most normal tissues. They have greater requirements for nucleotides as precursors of DNA and RNA, and consequently are generally more sensitive than normal cells to inhibitors of nucleotide biosynthesis. A growing array of important chemotherapeutic agents act by inhibiting one or more enzymes in these pathways. We describe here several well-studied examples that illustrate productive approaches to treating cancer patients and help us understand how these enzymes work.

The first set of examples includes compounds that inhibit glutamine amidotransferases. Recall that glutamine is a nitrogen donor in at least half a dozen separate reactions in nucleotide biosynthesis. The binding sites for glutamine and the mechanism by which NH_4^+ is extracted are quite similar in many of these enzymes. Most are strongly inhibited by glutamine analogs such as **azaserine** and **acivicin** (Fig. 22–46). Azaserine, characterized by John Buchanan in the 1950s, was one of the first examples of a mechanism-based enzyme inactivator (suicide inactivator; see p. 269 and Box 22–2). Acivicin shows promise as a cancer chemotherapeutic agent.

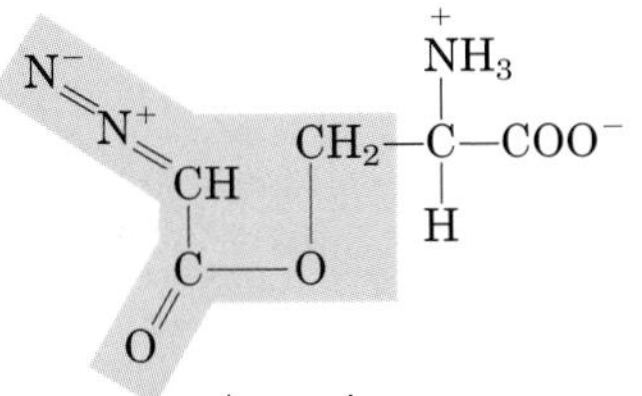

Azaserine

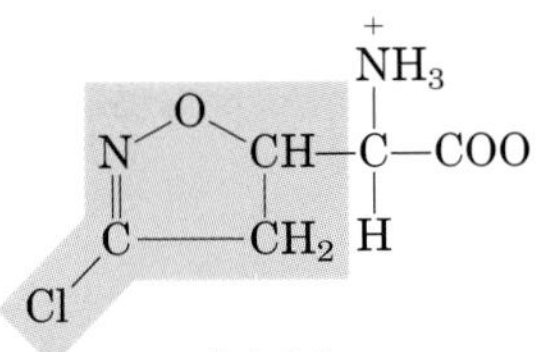

Acivicin

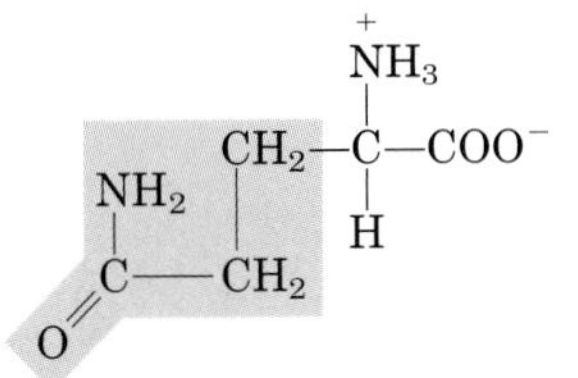

Glutamine

figure 22–46

Azaserine and acivicin, inhibitors of glutamine amidotransferases. These analogs of glutamine interfere in a number of amino acid and nucleotide biosynthetic pathways.

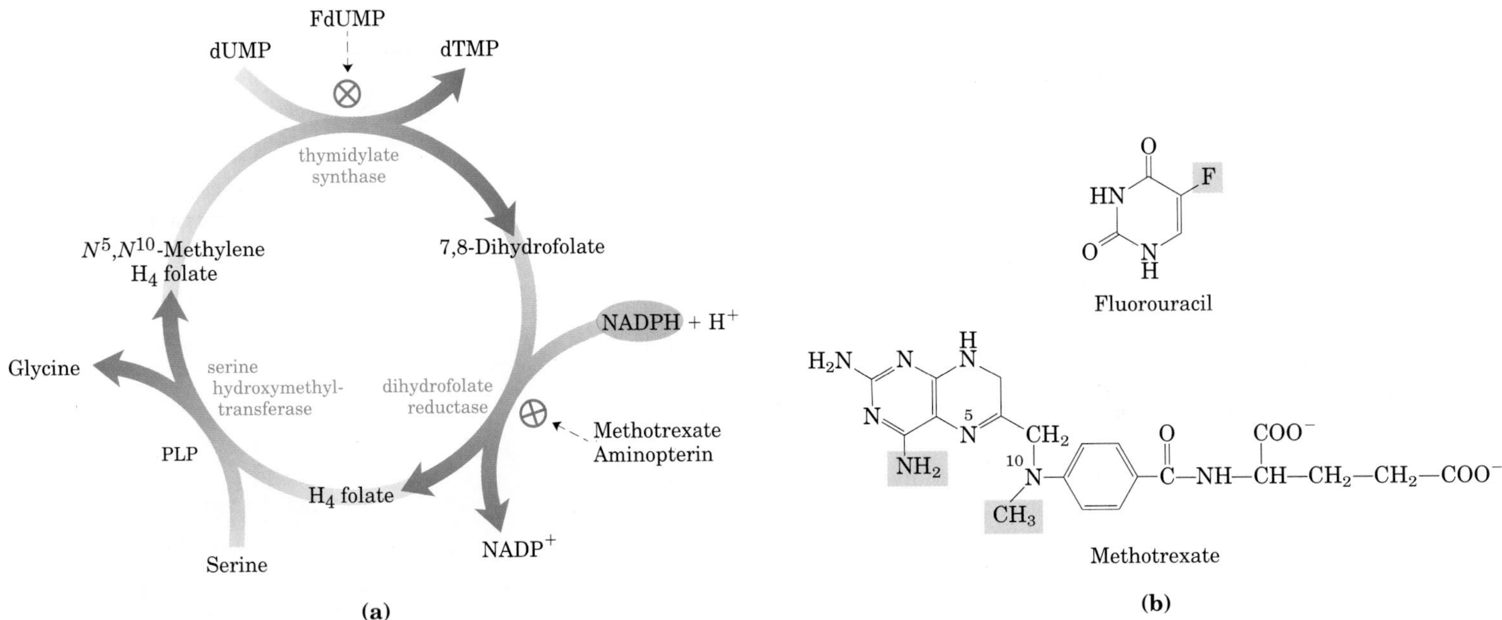

figure 22–47

Thymidylate synthesis and folate metabolism as targets of chemotherapy. (a) During thymidylate synthesis, N^5,N^{10}-methylenetetrahydrofolate is converted to 7,8-dihydrofolate; the N^5,N^{10}-methylenetetrahydrofolate is regenerated in two steps (see Fig. 22–42). The resulting cycle is a major target of several chemotherapeutic agents. **(b)** Fluorouracil and methotrexate, important chemotherapeutic agents. In cells, fluorouracil is converted to FdUMP, which inhibits thymidylate synthase. Methotrexate, a structural analog of tetrahydrofolate, inhibits dihydrofolate reductase; the shaded amino and methyl groups replace a carbonyl oxygen and a proton, respectively, in folate (see Fig. 22–42). Another important folate analog, aminopterin, is identical to methotrexate except that it lacks the shaded methyl group.

Other useful targets for pharmaceutical agents are thymidylate synthase and dihydrofolate reductase, enzymes that provide the only cellular pathway for thymine synthesis (Fig. 22–47). One inhibitor that acts on thymidylate synthase, **fluorouracil,** is an important chemotherapeutic agent. Fluorouracil itself is not the enzyme inhibitor. In the cell, salvage pathways convert it to the deoxynucleoside monophosphate FdUMP, which binds to and inactivates the enzyme. Inhibition by FdUMP (Fig. 22–48) is a classic example of mechanism-based enzyme inactivation. Another prominent chemotherapeutic agent, **methotrexate,** is an inhibitor of dihydrofolate reductase. This folate analog acts as a competitive inhibitor; the enzyme binds methotrexate with about 100 times higher affinity than dihydrofolate. **Aminopterin** also inhibits dihydrofolate reductase.

The medical potential of inhibitors of nucleotide biosynthesis is not limited to cancer treatment. All fast-growing cells (including bacteria and protists) are potential targets. Parasitic protists, such as the trypanosomes that cause African sleeping sickness (African trypanosomiasis) lack pathways for de novo nucleotide biosynthesis and are particularly sensitive to agents that interfere with their scavenging of nucleotides from the surrounding environment using salvage pathways. Allopurinol (Fig. 22–45) and a number of related purine analogs have shown promise for the treatment of African trypanosomiasis and related afflictions. See Box 22–2 for another approach to combating African trypanosomiasis made possible by advances in our understanding of metabolism and enzyme mechanism.

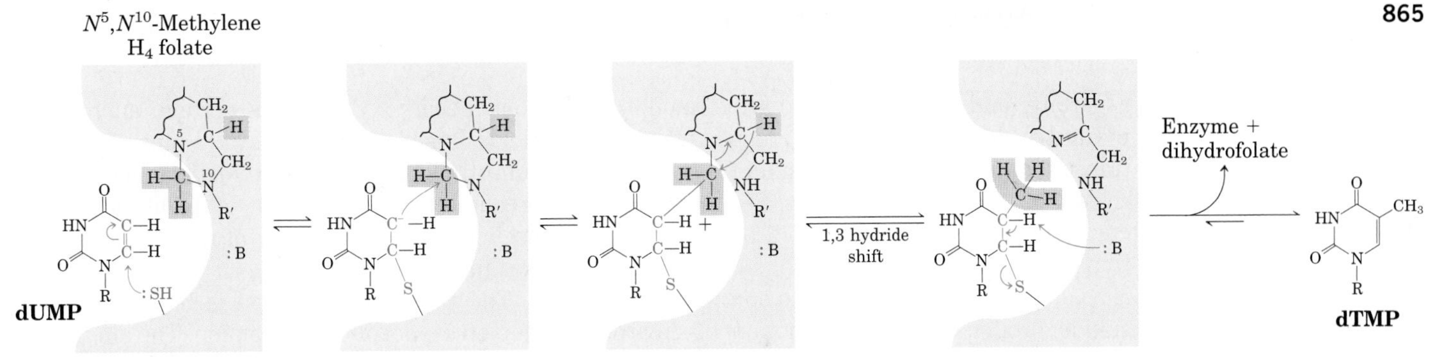

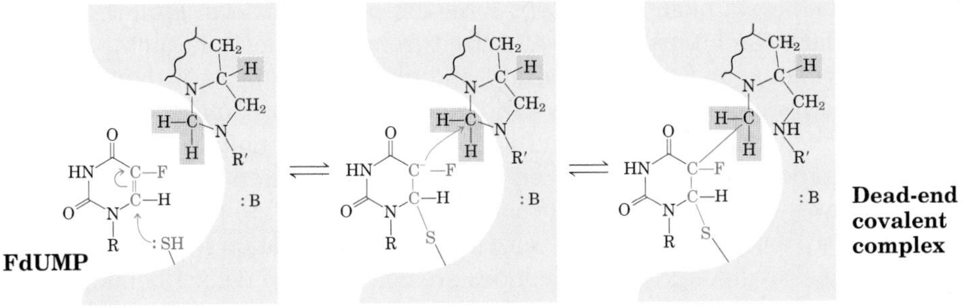

figure 22–48

Conversion of dUMP to dTMP and its inhibition by FdUMP. At the top is the normal reaction mechanism of thymidylate synthase. The nucleophilic sulfhydryl group contributed by the enzyme and the ring atoms of dUMP taking part in the reaction are shown in red; :B denotes an amino acid side chain that acts as a base to abstract a proton in the last step. The hydrogens derived from the methylene group of N^5,N^{10}-methylenetetrahydrofolate are shaded in gray. A novel feature of this reaction mechanism is a 1,3 hydride shift, which moves a hydride ion (shaded red) from C-6 of H_4 folate to the methyl group of thymidine, resulting in the oxidation of tetrahydrofolate to dihydrofolate. It is this hydride shift that apparently does not occur when FdUMP is the substrate (below). The first two steps of the reaction proceed normally, but result in a stable complex, with FdUMP linked covalently to the enzyme and to tetrahydrofolate, that inactivates the enzyme.

summary

The molecular nitrogen that makes up 80% of the earth's atmosphere is unavailable to most living organisms until it is reduced. Fixation of atmospheric N_2 takes place in certain free-living soil bacteria and in symbiotic bacteria in the root nodules of leguminous plants. Formation of ammonia by bacterial fixation of N_2, nitrification of ammonia to nitrate by soil organisms, conversion of nitrate to ammonia by higher plants, synthesis of amino acids from ammonia by all organisms, and conversion of nitrate to N_2 by denitrifying soil bacteria constitute the nitrogen cycle. Fixation of N_2 as NH_3 is carried out by the nitrogenase complex, in a reaction that requires ATP. The nitrogenase complex is very labile in the presence of O_2.

In living systems, reduced nitrogen is incorporated first into amino acids and then into a variety of other biomolecules, including nucleotides. The key entry point is the amino acid glutamate. Glutamate and glutamine are the nitrogen donors in a wide variety of biosynthetic reactions. Glutamine synthetase, which catalyzes the formation of glutamine from glutamate, is a key regulatory enzyme of nitrogen metabolism.

The amino acid and nucleotide biosynthetic pathways make repeated use of the biological cofactors pyridoxal phosphate, tetrahydrofolate, and *S*-adenosylmethionine. Pyridoxal phosphate is required for transamination reactions involving glutamate and for other amino acid transformations. One-carbon transfers require *S*-adenosylmethionine (at the $—CH_3$ oxidation level) and tetrahydrofolate (usually at the —CHO and $—CH_2OH$ oxidation levels). Glutamine amidotransferases catalyze reactions that incorporate nitrogen derived from glutamine.

Plants and bacteria synthesize all 20 standard amino acids. Mammals can synthesize 10 of the 20; the other 10 are required in the diet (essential amino acids). Among the nonessential amino acids, glutamate is formed by reductive amination of α-ketoglutarate and is the precursor of glutamine, proline, and arginine. Alanine and aspartate (and thus asparagine) are formed from pyruvate and oxaloacetate, respectively, by transamination. The carbon chain of serine is derived from 3-phosphoglycerate. Serine is a precursor of glycine; the β-carbon atom of serine is transferred to tetrahydrofolate. In microorganisms, cysteine is produced from serine and from sulfide produced by the reduction of environmental sulfate. Mammals produce cysteine from methionine and serine by a series of reactions requiring *S*-adenosylmethionine and cystathionine. Among the essential amino acids, the aromatic amino acids (phenylalanine, tyrosine, and tryptophan) are formed by a pathway in which chorismate occupies a key branch point. Phosphoribosyl pyrophosphate is a precursor of tryptophan and histidine. The pathway to histidine is interconnected with the purine synthetic pathway. Tyrosine can also be formed by hydroxylation of phenylalanine (and thus is considered nonessential). The pathways for the other essential amino acids are complex. The amino acid biosynthetic pathways are subject to allosteric end-product inhibition; the regulatory enzyme is usually the first in the sequence. Regulation of the various synthetic pathways is coordinated.

Many other important biomolecules are derived from amino acids. Glycine is a precursor of porphyrins. Glycine and arginine give rise to creatine and phosphocreatine. Glutathione, formed from three amino acids, is an important cellular reducing agent. Bacteria synthesize D-amino acids from L-amino acids in racemization reactions requiring pyridoxal phosphate. The PLP-dependent decarboxylation of certain amino acids yields important biological amines, including neurotransmitters. The aromatic amino acids are precursors of a number of plant substances. Arginine is the precursor of nitric oxide, an important signal molecule.

The purine ring system is built up step-by-step on 5-phosphoribosylamine. The amino acids glutamine, glycine, and aspartate furnish all the nitrogen atoms of purines. Two ring-closure steps form the purine nucleus. Pyrimidines are synthesized from carbamoyl phosphate and aspartate, and ribose 5-phosphate is then attached to yield the pyrimidine ribonucleotides. Purine and pyrimidine biosynthetic pathways are regulated by feedback inhibition. Nucleoside monophosphates are converted to their triphosphates by enzymatic phosphorylation reactions. Ribonucleotides are converted to deoxyribonucleotides by ribonucleotide reductase, an enzyme with novel mechanistic and regulatory characteristics. The thymine nucleotides are derived from dCDP and dUMP. Uric acid and urea are the end products of purine and pyrimidine degradation. Free purines can be salvaged and rebuilt into nucleotides by a separate pathway. Genetic deficiencies in certain salvage enzymes cause serious disorders such as Lesch-Nyhan syndrome and immunodeficiency disease. Accumulation of uric acid crystals in the joints, possibly caused by another genetic deficiency, results in gout. Enzymes of the nucleotide biosynthetic pathways are targets for an array of chemotherapeutic agents used to treat cancer and other diseases.

further reading

Nitrogen Fixation

Burris, R.H. (1991) Nitrogenases. *J. Biol. Chem.* **266,** 9339–9342.

A short and well-written summary.

Howard, J.B. & Rees, D.C. (1994) Nitrogenase: a nucleotide-dependent molecular switch. *Annu. Rev. Biochem.* **64,** 235–264.

Leigh, G.J. (1995) The mechanism of dinitrogen reduction by molybdenum nitrogenases. *Eur. J. Biochem.* **229,** 14–20.

Amino Acid Biosynthesis

Abeles, R.H., Frey, P.A., & Jencks, W.P. (1992) *Biochemistry,* Jones and Bartlett Publishers, Boston.

This book includes excellent accounts of reaction mechanisms, including one-carbon metabolism and pyridoxal phosphate enzymes.

Bender, D.A. (1985) *Amino Acid Metabolism,* 2nd edn, Wiley-Interscience, New York.

Crane, B.R., Siegel, L.M., & Getzoff, E.E. (1995) Sulfite reductase structure at 1.6 Å: evolution and catalysis for reduction of inorganic anions. *Science* **270,** 59–67.

Pan P., Woehl, E., & Dunn, M.F. (1997) Protein architecture, dynamics and allostery in tryptophan synthase channeling. *Trends Biochem. Sci.* **22,** 22–27.

Compounds Derived from Amino Acids

Bredt, D.S. & Snyder, S.H. (1994) Nitric oxide: a physiologic messenger molecule. *Annu. Rev. Biochem.* **63,** 175–195.

Meister, A. & Anderson, M.E. (1983) Glutathione. *Annu. Rev. Biochem.* **52,** 711–760.

Rondon, M.R., Trzebiatowski, J.R., & Escalante-Semerena, J.C. (1997) Biochemistry and molecular genetics of cobalamin biosynthesis. *Prog. Nucleic Acid Res. Mol. Biol.* **56,** 347–384.

Stadtman, T.C. (1996) Selenocysteine. *Annu. Rev. Biochem.* **65,** 83–100.

Nucleotide Biosynthesis

Benkovic, S.J. (1980) On the mechanism of action of folate- and biopterin-requiring enzymes. *Annu. Rev. Biochem.* **49,** 227–251.

Blakley, R.L. & Benkovic, S.J. (1985) *Folates and Pterins,* Vol. 2: *Chemistry and Biochemistry of Pterins,* Wiley-Interscience, New York.

Carreras, C.W. & Santi, D.V. (1995) The catalytic mechanism and structure of thymidylate synthase. *Annu. Rev. Biochem.* **64,** 721–762.

Daubner, S.C., Schrimsher, J.L., Schendel, F.J., Young, M., Henikoff, S., Patterson, D., Stubbe, J., & Benkovic, S.J. (1985) A multifunctional protein possessing glycinamide ribonucleotide synthetase, glycinamide ribonucleotide transformylase, and aminoimidazole ribonucleotide synthetase activities in *de novo* purine biosynthesis. *Biochemistry* **24,** 7059–7062.

Eliasson, R., Pontis, E., Sun, X., & Reichard, P. (1994) Allosteric control of the substrate specificity of the anaerobic ribonucleotide reductase from *Escherichia coli. J. Biol. Chem.* **269,** 26,052–26,057.

Hardy, L.W., Finer-Moore, J.S., Montfort, W.R., Jones, M.O., Santi, D.V., & Stroud, R.M. (1987) Atomic structure of thymidylate synthase: target for rational drug design. *Science* **235,** 448–455.

Holmgren, A. (1989) Thioredoxin and glutaredoxin systems. *J. Biol. Chem.* **264,** 13,963–13,966.

Jones, M.E. (1980) Pyrimidine nucleotide biosynthesis in animals: genes, enzymes, and regulation of UMP biosynthesis. *Annu. Rev. Biochem.* **49,** 253–279.

Jordan, A. & Reichard P. (1998) Ribonucleotide reductases. *Annu. Rev. Biochem.* **67,** 71–98.

Kornberg, A. & Baker, T.A. (1991) *DNA Replication,* 2nd edn, W.H. Freeman and Company, New York.

Includes a good summary of nucleotide biosynthesis.

Lee, L., Kelly, R.E., Pastra-Landis, S.C., & Evans, D.R. (1985) Oligomeric structure of the multifunctional protein CAD that initiates pyrimidine biosynthesis in mammalian cells. *Proc. Natl. Acad. Sci. USA* **82,** 6802–6806.

Licht, S., Gerfen, G.J., & Stubbe, J. (1996) Thiyl radicals in ribonucleotide reductases. *Science* **271,** 477–481.

Mueller, E.J., Meyer, E., Rudolph, J., Davisson, V.J., & Stubbe, J. (1994) N^5-carboxyaminoimidazole ribonucleotide: evidence for a new intermediate and two new enzymatic activities in the de novo purine biosynthetic pathway of *Escherichia coli. Biochemistry* **33,** 2269–2278.

Stubbe, J. & Riggs-Gelasco, P. (1998) Harnessing free radicals: formation and function of the tyrosyl radical in ribonucleotide reductase. *Trends Biochem. Sci.* **23,** 438–443.

Villafranca, J.E., Howell, E.E., Voet, D.H., Strobel, M.S., Ogden, R.C., Abelson, J.N., & Kraut, J. (1983) Directed mutagenesis of dihydrofolate reductase. *Science* **222,** 782–788.

Structural studies on this important enzyme.

Genetic Diseases

Scriver, C.R., Beaudet, A.L., Sly, W.S., & Valle, D. (eds) (1995) *The Metabolic and Molecular Bases of Inherited Disease,* 7th edn, McGraw-Hill, Inc., New York.

This book has good chapters on disorders of amino acid, porphyrin, and heme metabolism. See also the chapters on inborn errors of purine and pyrimidine metabolism.

problems

1. ATP Consumption by Root Nodules in Legumes Bacteria residing in the root nodules of the pea plant consume more than 20% of the ATP produced by the plant. Suggest why these bacteria consume so much ATP.

2. Glutamate Dehydrogenase and Protein Synthesis The bacterium *Methylophilus methylotrophus* can synthesize protein from methanol and ammonia. Recombinant DNA techniques have improved the yield of protein by introducing into *M. methylotrophus* the glutamate dehydrogenase gene from *E. coli.* Why does this genetic manipulation increase the protein yield?

3. Transformation of Aspartate to Asparagine There are two routes for transforming aspartate to asparagine at the expense of ATP. Many bacteria have an asparagine synthetase that uses ammonium ion as the nitrogen donor. Mammals have an asparagine synthetase that uses glutamine as the nitrogen donor. Given that the latter requires an extra ATP (for the synthesis of glutamine), why do mammals use this route?

4. Equation for the Synthesis of Aspartate from Glucose Write the net equation for the synthesis of aspartate (a nonessential amino acid) from glucose, carbon dioxide, and ammonia.

5. Phenylalanine Hydroxylase Deficiency and Diet Tyrosine is normally a nonessential amino acid, but individuals with a genetic defect in phenylalanine hydroxylase require tyrosine in their diet for normal growth. Explain.

6. Cofactors for One-Carbon Transfer Reactions Most one-carbon transfers are promoted by one of three cofactors: biotin, tetrahydrofolate, or *S*-adenosylmethionine (Chapter 18). *S*-Adenosylmethionine is generally used as a methyl group donor; the transfer potential of the methyl group in N^5-methyltetrahydrofolate is insufficient for most biosynthetic reactions. However, one example of the use of N^5-methyltetrahydrofolate in methyl group transfer is in methionine formation by the methionine synthase reaction (step ⑨ of Fig. 22–15); methionine is the immediate precursor of *S*-adenosylmethionine (see Fig. 18–17). Explain how the methyl group of *S*-adenosylmethionine can be derived from N^5-methyltetrahydrofolate, even though the transfer potential of the methyl group in N^5-methyltetrahydrofolate is 10^3 times *lower* than that in *S*-adenosylmethionine.

7. Concerted Regulation in Amino Acid Biosynthesis The glutamine synthetase of *E. coli* is independently modulated by various products of glutamine metabolism (see Fig. 22–6). In this concerted inhibition, the extent of enzyme inhibition is greater than the sum of the separate inhibitions caused by each product. For *E. coli* grown in a medium rich in histidine, what would be the advantage of concerted inhibition?

8. Relationship between Folic Acid Deficiency and Anemia Folic acid deficiency, believed to be the most common vitamin deficiency, causes a type of anemia in which hemoglobin synthesis is impaired and erythrocytes do not mature properly. What is the metabolic relationship between hemoglobin synthesis and folic acid deficiency?

9. Nucleotide Biosynthesis in Amino Acid Auxotrophic Bacteria Normal *E. coli* cells can synthesize all the standard amino acids, but some mutants, called amino acid auxotrophs, are unable to synthesize a specific amino acid and require its addition to the culture medium for optimal growth. Besides their role in protein synthesis, some amino acids are also precursors for other nitrogenous cell products. Consider the three amino acid auxotrophs that are unable to synthesize glycine, glutamine, and aspartate, respectively. For each mutant, what nitrogenous products other than proteins would the cell fail to synthesize?

10. Inhibitors of Nucleotide Biosynthesis Suggest mechanisms for the inhibition of (a) alanine racemase by L-fluoroalanine and (b) glutamine amidotransferases by azaserine.

11. Mode of Action of Sulfa Drugs Some bacteria require *p*-aminobenzoate in the culture medium for normal growth, and their growth is severely inhibited by the addition of sulfanilamide, one of the earliest sulfa drugs. Moreover, in the presence of this drug, 5-aminoimidazole-4-carboxamide ribonucleotide (AICAR; see Fig. 22–31) accumulates in the culture medium. These effects are reversed by addition of excess *p*-aminobenzoate.

$H_2N-C_6H_4-C(=O)O^-$

p-Aminobenzoate

$H_2N-C_6H_4-S(=O)_2-NH_2$

Sulfanilamide

(a) What is the role of *p*-aminobenzoate in these bacteria? (Hint: See Fig. 18–15).

(b) Why does AICAR accumulate in the presence of sulfanilamide?

(c) Why are the inhibition and accumulation reversed by addition of excess *p*-aminobenzoate?

12. Pathway of Carbon in Pyrimidine Biosynthesis Predict the locations of ^{14}C in orotate isolated from cells grown on a small amount of uniformly labeled [^{14}C]succinate. Justify your prediction.

13. Nucleotides As Poor Sources of Energy Under starvation conditions, organisms can use proteins and amino acids as sources of energy. Deamination of amino acids produces carbon skeletons that can enter the glycolytic pathway and the citric acid cycle to produce energy in the form of ATP. Nucleotides, on the other hand, are not similarly degraded for use as energy-yielding fuels. What observations about cellular physiology support this statement? What aspect of the structure of nucleotides makes them a relatively poor source of energy?

14. Treatment of Gout Allopurinol (see Fig. 22–45), an inhibitor of xanthine oxidase, is used to treat chronic gout. Explain the biochemical basis for this treatment. Patients treated with allopurinol sometimes develop xanthine stones in the kidneys, although the incidence of kidney damage is much lower than in untreated gout. Explain this observation in the light of the following solubilities in urine: uric acid, 0.15 g/L; xanthine, 0.05 g/L; and hypoxanthine, 1.4 g/L.

15. Inhibition of Nucleotide Synthesis by Azaserine The diazo compound *O*-(2-diazoacetyl)-L-serine, known also as azaserine (see Fig. 22–46), is a powerful inhibitor of glutamine amidotransferases. If growing cells are treated with azaserine, what intermediates of nucleotide biosynthesis would accumulate? Explain.

Integration and Hormonal Regulation of Mammalian Metabolism

chapter 23

In Chapters 14 through 22 we have discussed metabolism at the level of the individual cell, emphasizing those pathways common to almost all cells, prokaryotic and eukaryotic. We have seen how metabolic processes within cells are regulated at the level of individual enzymes by substrate availability, by allosteric mechanisms, and/or by phosphorylation or other covalent modifications of enzymes.

To appreciate fully the significance of individual metabolic pathways and their regulation, we must view these pathways in the context of the whole organism. An essential characteristic of multicellular organisms is cell differentiation and division of labor. In addition to the central pathways of energy-yielding metabolism that occur in all cells, the tissues and organs of complex organisms such as humans have specialized functions. These impose characteristic fuel requirements and patterns of metabolism. Hormonal signals integrate and coordinate the metabolic activities of different tissues and optimize the allocation of fuels and precursors to each organ. In this chapter we focus on mammals, looking at the specialized metabolism of several major organs and tissues and the integration of metabolism in the whole organism.

We begin by examining the distribution of nutrients to various organs—the central role here is played by the liver—and the metabolic cooperation among these organs. To illustrate the integrative role of hormones, we describe the interplay of epinephrine, glucagon, and insulin in coordinating fuel metabolism in muscle, liver, and adipose tissue. The metabolic disturbances in diabetes further show the importance of hormonal regulation of metabolism. Next we broaden our discussion to include the wide variety of hormone types and hormone actions on processes other than fuel metabolism. Finally, we discuss the long-term regulation of body mass by hormones.

These two mice are the same age and have the same genetic makeup, except that the one on the left is genetically defective in the production of a hormone (leptin) that regulates feeding behavior and metabolic activity. As a result, the hormone-deficient mouse accumulates enormous reserves of triacylglycerides in its adipose tissue. In this chapter we discuss the hormonal mechanisms that coordinate and integrate the metabolic activities of all the tissues to produce homeostasis in the animal.

Tissue-Specific Metabolism: The Division of Labor

Each tissue of the human body has a specialized function, reflected in its anatomy and metabolic activity. Skeletal muscle allows directed motion; adipose tissue stores and releases fats, which serve as fuel throughout the body; the brain pumps ions to produce electrical signals. The liver plays a central processing and distributing role in metabolism and furnishes all other organs and tissues with an appropriate mix of nutrients via the bloodstream. The functional centrality of the liver is indicated by the common reference to all other tissues and organs as "extrahepatic" or "peripheral." We therefore begin our discussion of the division of metabolic labor by considering the transformations of carbohydrates, amino acids, and fats in the mammalian liver. This is followed by brief descriptions of the major metabolic functions of adipose tissue, muscle, brain, and the tissue that interconnects all others: the blood.

The Liver Processes and Distributes Nutrients

During digestion in mammals, the three major classes of nutrients (carbohydrates, proteins, and lipids) undergo enzymatic hydrolysis into their monomeric subunits. This breakdown is necessary because the epithelial cells lining the intestinal lumen absorb only relatively small molecules. Many of the fatty acids and monoacylglycerols released by digestion in the intestine are assembled within these epithelial cells into triacylglycerols.

After being absorbed, most of the sugars and amino acids and some triacylglycerols pass via the blood to hepatocytes in the liver; the remaining triacylglycerols enter adipose tissue via the lymphatic system. Hepatocytes transform dietary nutrients into the fuels and precursors required by other tissues, and export them via the blood. The kinds and amounts of nutrients supplied to the liver vary with several factors, including the diet and the time interval between meals. The demand of the extrahepatic tissues for fuels and precursors varies among organs and with the activity of the organism.

To meet these changing circumstances, the liver has remarkable metabolic flexibility. For example, when the diet is rich in protein, hepatocytes contain high levels of enzymes for amino acid catabolism and gluconeogenesis. Within hours after a shift to a high-carbohydrate diet, the levels of these enzymes drop and the synthesis of enzymes essential to carbohydrate metabolism begins. Other tissues also adjust their metabolism to the prevailing conditions, but none is as adaptable as the liver, and none is so central to the organism's overall metabolism. What follows is a survey of the possible fates of sugars, amino acids, and lipids that enter the liver from the bloodstream. To help you recall the metabolic transformations discussed here, Table 23–1 shows the major pathways and processes to which we will refer and indicates the figure in which each pathway is presented in detail.

Sugars The glucose transporter in hepatocytes (GluT2) is so effective that the concentration of glucose within a hepatocyte is essentially that in the blood. Glucose entering hepatocytes is phosphorylated by glucokinase to yield glucose 6-phosphate. Glucokinase has a much higher K_m for glucose (10 mM) than does hexokinase (p. 555); unlike hexokinase, it is not inhibited by its product, glucose 6-phosphate. The presence of glucokinase allows hepatocytes to continue phosphorylating glucose when the glucose concentration rises well above levels that would overwhelm hexokinase. Fructose, galactose, and mannose, all absorbed from the small intestine, are also converted into glucose 6-phosphate by enzymatic pathways examined earlier. Glucose 6-phosphate is at the crossroads of carbohydrate metabolism in the liver. It may take any of five major metabolic routes (Fig. 23–1), depending on the current metabolic needs of the organism. By the action of various allosterically regulated enzymes, and through hormonal regulation of enzyme synthesis and activity, the flow of glucose is directed into one or more of these pathways in the liver.

① Glucose 6-phosphate is dephosphorylated by glucose 6-phosphatase to yield free glucose (p. 729), which is exported to replenish blood

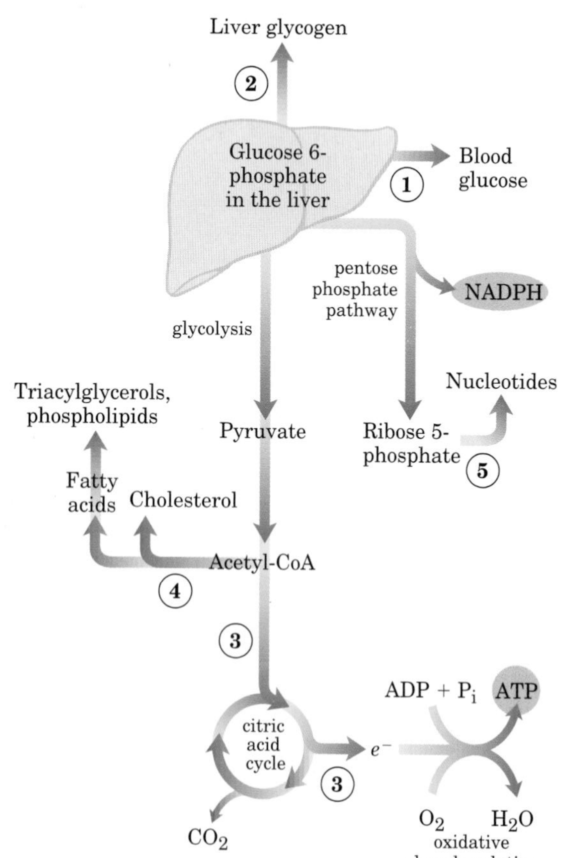

figure 23–1
Metabolic pathways for glucose 6-phosphate in the liver. Here and in the following figures, anabolic pathways are shown leading upward, catabolic pathways leading downward, and distribution to other organs horizontally. The numbered processes correspond to descriptions in the text.

table 23–1

Pathways of Carbohydrate, Amino Acid, and Fat Metabolism Shown in Earlier Figures

Pathway	Figure reference
Citric acid cycle: acetyl-CoA ⟶ $2CO_2$	16–7
Oxidative phosphorylation: ATP synthesis	19–16
Carbohydrate catabolism	
Glycogenolysis: glycogen ⟶ glucose 1-phosphate ⟶ blood glucose	15–11
Hexose entry into glycolysis: fructose, mannose, galactose ⟶ glucose 6-phosphate	15–11
Glycolysis: glucose ⟶ pyruvate	15–2
Pyruvate dehydrogenase reaction: pyruvate ⟶ acetyl-CoA	16–2
Lactic acid fermentation: glycogen ⟶ lactate + 2ATP	15–3
Pentose phosphate pathway: glucose 6-phosphate ⟶ pentose phosphates + NADPH	15–20
Carbohydrate anabolism	
Gluconeogenesis: citric acid cycle intermediates ⟶ glucose	20–2
Glucose-alanine cycle: glucose ⟶ pyruvate ⟶ alanine ⟶ glucose	18–8
Glycogen synthesis: glucose 6-phosphate ⟶ glucose 1-phosphate ⟶ glycogen	20–12
Amino acid and nucleotide metabolism	
Amino acid degradation: amino acids ⟶ acetyl-CoA, citric acid cycle intermediates	18–29
Amino acid synthesis	22–9
Urea cycle: NH_3 ⟶ urea	18–9
Glucose-alanine cycle: alanine ⟶ glucose	18–8
Nucleotide synthesis: amino acids ⟶ purines, pyrimidines	22–31; 22–34
Hormone and neurotransmitter synthesis	22–27
Fat catabolism	
β Oxidation of fatty acids: fatty acid ⟶ acetyl-CoA	17–8
Oxidation of ketone bodies: β-hydroxybutyrate ⟶ acetyl-CoA ⟶ CO_2	17–17
Fat anabolism	
Fatty acid synthesis: acetyl-CoA ⟶ fatty acids	21–5
Triacylglycerol synthesis: acetyl-CoA ⟶ fatty acids ⟶ triacylglycerol	21–18; 21–19
Ketone body formation: acetyl-CoA ⟶ acetoacetate, β-hydroxybutyrate	17–16
Cholesterol and cholesteryl ester synthesis: acetyl-CoA ⟶ cholesterol ⟶ cholesteryl esters	21–32 through 21–36
Phospholipid synthesis: fatty acids ⟶ phospholipids	21–27

glucose. Export is the pathway of choice when glucose 6-phosphate is limited, because the blood glucose concentration must be kept sufficiently high (4 mM) to provide adequate energy for the brain and other tissues. ② Glucose 6-phosphate not immediately needed to form blood glucose is converted into liver glycogen. ③ Glucose 6-phosphate may be oxidized for energy production via glycolysis, decarboxylation of pyruvate (by the pyruvate dehydrogenase reaction), and the citric acid cycle. The ensuing electron transfer and oxidative phosphorylation yield ATP. (Normally, however, fatty acids are the preferred fuel for energy production in hepatocytes.) ④ Glucose 6-phosphate not used to make blood glucose or liver glycogen is degraded via glycolysis and the pyruvate dehydrogenase reaction into acetyl-CoA, which serves as the precursor for the synthesis of lipids: fatty acids, which are incorporated into triacylglycerols and phospholipids, and cholesterol. Much of the lipid synthesized in the liver is transported to other tissues by blood lipoproteins. ⑤ Finally, glucose 6-phosphate is the substrate for the pentose phosphate pathway, yielding both reducing power (NADPH), needed for the biosynthesis of fatty acids and cholesterol, and D-ribose 5-phosphate, a precursor in nucleotide biosynthesis.

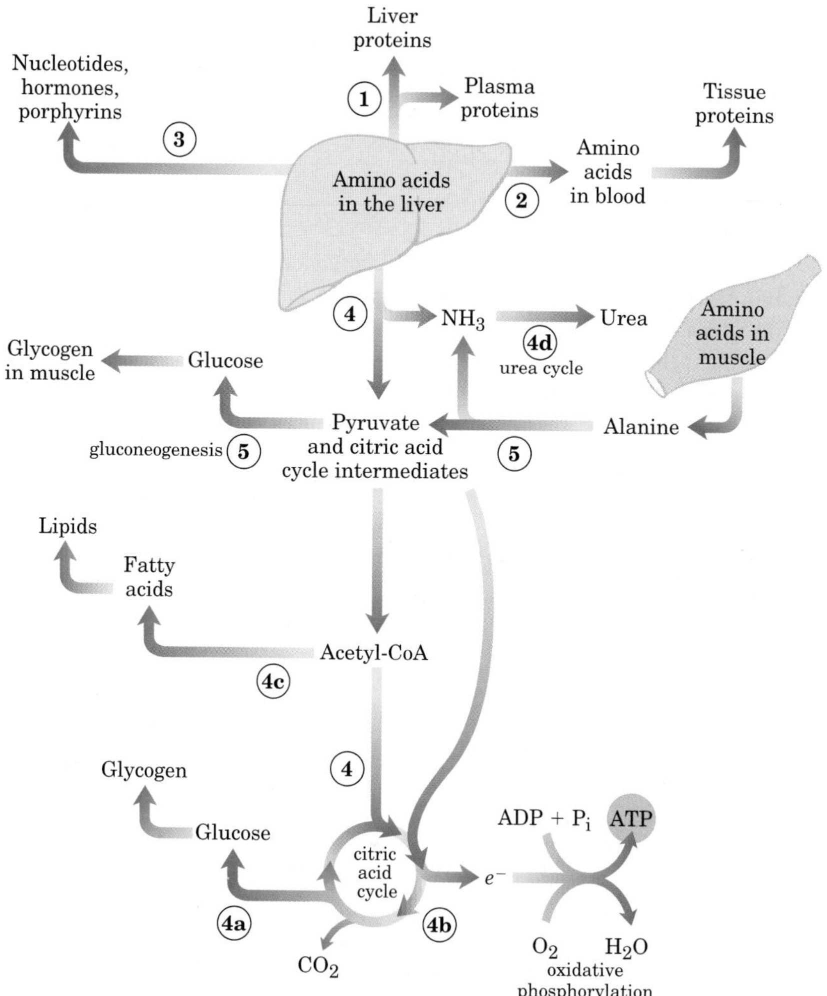

figure 23–2
Metabolism of amino acids in the liver.

Amino Acids Amino acids that enter the liver have several important metabolic routes (Fig. 23–2). ① They act as precursors for protein synthesis in hepatocytes, a process discussed in Chapter 27. The liver constantly renews its own proteins, which have a very high turnover rate, with an average half-life of only a few days. The liver is also the site of biosynthesis of most of the plasma proteins of the blood. ② Alternatively, amino acids may pass from the liver via the blood to other organs, to be used as precursors in the synthesis of tissue proteins. ③ Certain amino acids are precursors in the biosynthesis of nucleotides, hormones, and other nitrogenous compounds in the liver and other tissues.

④ Amino acids not needed as biosynthetic precursors are deaminated and degraded to yield acetyl-CoA and citric acid cycle intermediates. Citric acid cycle intermediates may be converted into glucose and glycogen via the gluconeogenic pathway (4a). Acetyl-CoA may be oxidized via the citric acid cycle for ATP energy (4b), or it may be converted into lipids for storage (4c). The ammonia released on degradation of amino acids is converted into the excretory product urea (4d).

The liver also metabolizes amino acids that arrive intermittently from other tissues. The blood is normally adequately supplied with glucose just after the digestion and absorption of dietary carbohydrate or, between meals, by the conversion of liver glycogen into blood glucose. During a period between meals, especially if prolonged, some muscle protein is degraded to amino acids ⑤. These amino acids donate their amino groups (by transamination) to pyruvate, the product of glycolysis, to yield alanine, which is transported to the liver and deaminated. The resulting pyruvate is converted by hepatocytes into blood glucose (via gluconeogenesis), and the ammonia is converted into urea for excretion. The glucose returns to the skeletal muscles to replenish muscle glycogen stores. One benefit of this cyclic process, the glucose-alanine cycle (see Fig. 18–8), is the smoothing out of fluctuations in blood glucose levels between meals. The amino acid deficit incurred in the muscles is made up after the next meal by incoming dietary amino acids.

Lipids The fatty acid components of the lipids entering hepatocytes also have several different fates (Fig. 23–3). ① Fatty acids are converted into liver lipids. ② Under most circumstances, fatty acids are the major oxidative fuel in the liver. Free fatty acids may be activated and oxidized to yield acetyl-CoA and NADH. The acetyl-CoA is further oxidized via the citric acid cycle to yield ATP by oxidative phosphorylation. ③ Excess acetyl-CoA released on oxidation of fatty acids and not required by the liver is converted into the ketone bodies, acetoacetate and β-hydroxybutyrate; these are circulated in the blood to other tissues, to be used as fuel for the citric acid cycle. Ketone bodies may be regarded as a transport form of acetyl groups. They can supply a significant fraction of the energy in some other tissues, up to one-third in the heart, and as much as 60% to 70% in the brain during prolonged fasting. ④ Some of the acetyl-CoA derived from fatty acids (and from glucose) is used for the biosynthesis of cholesterol, which is required for membrane biosynthesis. Cholesterol is also the precursor of all steroid hormones and of the bile salts, which are essential for the digestion and absorption of lipids.

The final two metabolic fates of lipids involve specialized mechanisms for the transport of insoluble lipids in the blood. ⑤ Fatty acids are converted to the phospholipids and triacylglycerols of the plasma lipoproteins, which carry lipids to adipose (fat) tissue for storage as triacylglycerols. Cholesterol and cholesteryl esters are also transported as lipoproteins. ⑥ Some free fatty acids become bound to serum albumin and are carried in the blood to the heart and skeletal muscles, which absorb and oxidize free fatty acids as a major fuel. Serum albumin is the most abundant plasma protein; one molecule of serum albumin can carry up to 10 molecules of free fatty acid, releasing them at the consuming tissue where they are taken up by passive diffusion.

Thus, the liver serves as the body's distribution center, exporting nutrients in the correct proportions to the other organs, smoothing out fluctuations in metabolism caused by the intermittent nature of food intake, and processing excess amino groups into urea and other products to be disposed of by the kidneys. The liver also detoxifies foreign organic compounds, such as drugs, food additives, preservatives, and other possibly harmful agents with no food value. Detoxification often involves the cytochrome P-450–dependent hydroxylation of relatively insoluble organic compounds to make them sufficiently soluble for further breakdown and excretion (see Box 21–1).

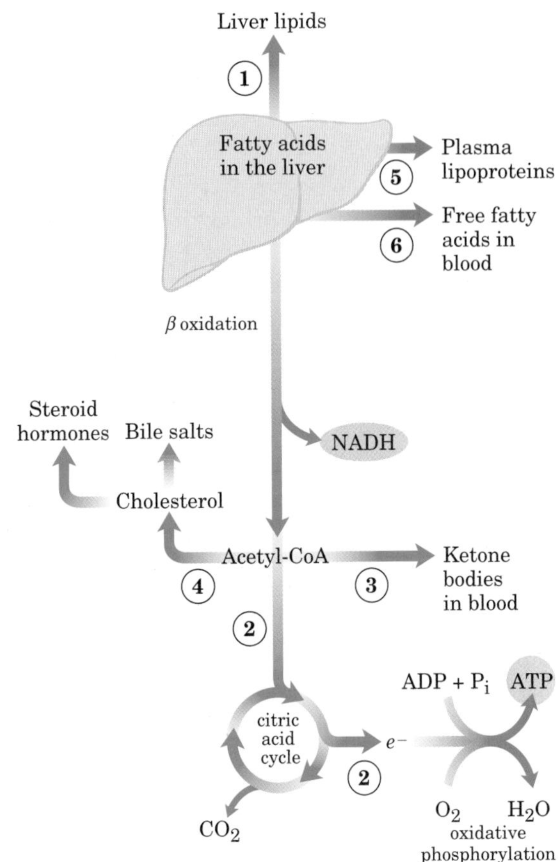

figure 23–3
Metabolism of fatty acids in the liver.

Adipose Tissue Stores and Supplies Fatty Acids

Adipose tissue, which consists of adipocytes (fat cells) (Fig. 23–4), is amorphous and widely distributed in the body: under the skin, around the deep blood vessels, and in the abdominal cavity. It typically makes up about 15% of the mass of a young adult human, with approximately 65% of this mass being in the form of triacylglycerols. Adipocytes are metabolically very active, responding quickly to hormonal stimuli in a metabolic interplay with the liver, skeletal muscles, and the heart.

Like other cell types in the body, adipocytes have an active glycolytic metabolism, they use the citric acid cycle to oxidize pyruvate and fatty

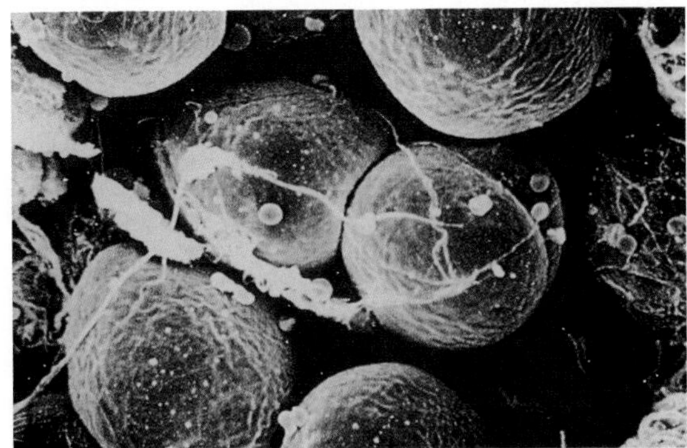

figure 23–4
Scanning electron micrograph of human adipocytes. Capillaries and collagen fibers form a supporting network around adipocytes in fat tissues. Almost the entire volume of these metabolically active cells is taken up by fat droplets. Magnification is ×440.

acids, and they carry out mitochondrial oxidative phosphorylation. During periods of high carbohydrate intake, adipose tissue can convert glucose via pyruvate and acetyl-CoA into fatty acids, from which triacylglycerols are made and stored as large fat globules. In humans, however, most fatty acid synthesis occurs in hepatocytes, not in adipocytes. Adipocytes store triacylglycerols arriving from the liver (carried in the blood as VLDLs; see Fig. 21–38a) and from the intestinal tract, particularly after meals rich in fat.

When fuel is needed, triacylglycerols stored in adipose tissue are hydrolyzed by lipases within the adipocytes to release free fatty acids, which may then be delivered via the bloodstream to skeletal muscles and the heart. The release of fatty acids from adipocytes is greatly accelerated by the hormone epinephrine, which stimulates the activation of triacylglycerol lipase (see Fig. 17–3). Insulin counterbalances this effect of epinephrine, decreasing the activity of triacylglycerol lipase. Humans and many other animals, particularly those that hibernate, have adipose tissue called brown fat, which is specialized to generate heat rather than ATP during oxidation of fatty acids.

Muscle Uses ATP for Mechanical Work

Skeletal muscle accounts for over 50% of the total O_2 consumption in a resting human being and up to 90% during very active muscular work. Metabolism in skeletal muscle is specialized to generate ATP as the immediate source of energy. Moreover, skeletal muscle is adapted to do its mechanical work in an intermittent fashion, on demand. Sometimes skeletal muscles must deliver much work in a short time, as in a 100 m sprint; at other times more extended work is required, as in running a marathon or giving birth.

Skeletal muscle can use free fatty acids, ketone bodies, or glucose as fuel, depending on the degree of muscular activity (Fig. 23–5). In resting muscle, the primary fuels are free fatty acids from adipose tissue and ketone bodies from the liver. These are oxidized and degraded to yield acetyl-CoA, which enters the citric acid cycle for oxidation to CO_2. The ensuing transfer of electrons to O_2 provides the energy for ATP synthesis by oxidative phosphorylation. Moderately active muscle uses blood glucose in addition to fatty acids and ketone bodies. The glucose is phosphorylated, then degraded by glycolysis to pyruvate, which is converted to acetyl-CoA and oxidized via the citric acid cycle.

In maximally active muscles, the demand for ATP is so great that the blood flow cannot provide O_2 and fuels fast enough to produce the necessary ATP by aerobic respiration alone. Under these conditions, stored muscle glycogen is broken down to lactate by fermentation. Each glucose unit

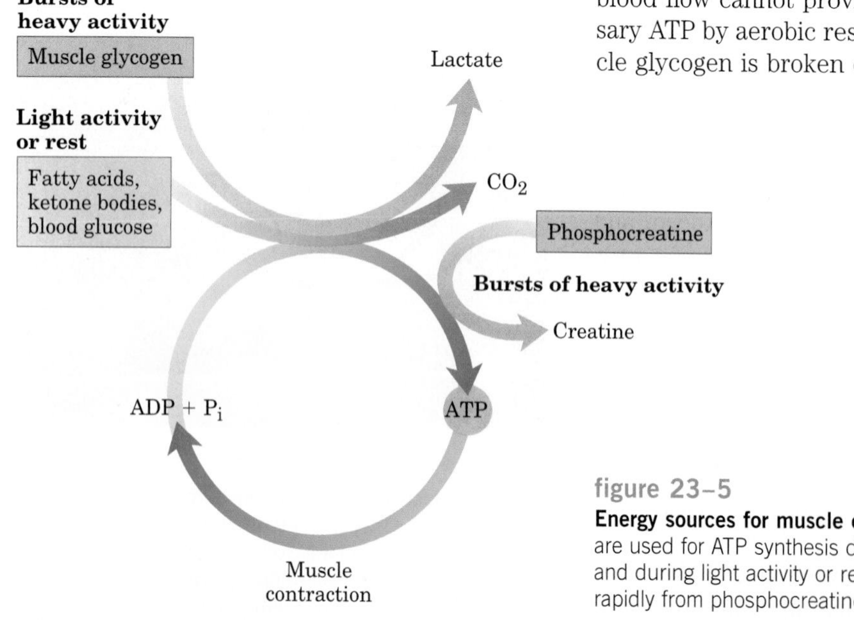

figure 23–5
Energy sources for muscle contraction. Different fuels are used for ATP synthesis during bursts of heavy activity and during light activity or rest. ATP can be obtained rapidly from phosphocreatine.

degraded yields three molecules of ATP because phosphorolysis of glycogen produces glucose 6-phosphate, sparing the ATP normally consumed in the hexokinase reaction. Lactic acid fermentation thus provides extra ATP energy quickly, supplementing basal ATP production resulting from the aerobic oxidation of other fuels via the citric acid cycle. The use of blood glucose and muscle glycogen as fuels for muscular activity is greatly enhanced by the secretion of epinephrine, which stimulates the formation of blood glucose from glycogen in the liver and the breakdown of glycogen in muscle tissue. Skeletal muscle does not contain glucose 6-phosphatase and cannot convert glucose 6-phosphate to free glucose for export to other tissues. Consequently, muscle glycogen is completely dedicated to providing energy in the muscle, via glycolysis.

Because skeletal muscle stores relatively little glycogen (about 1% of its total weight), there is a limit to the amount of glycolytic energy available during all-out exertion. Moreover, the accumulation of lactate and the consequent decrease in pH that occurs in maximally active muscles reduces their efficiency.

After a period of intense muscular activity, heavy breathing continues for some time. Much of the O_2 thus obtained is used for the production of ATP by oxidative phosphorylation in the liver. This ATP is used for gluconeogenesis from lactate, carried in the blood from the muscles to the liver. The glucose thus formed returns to the muscles to replenish their glycogen, completing the Cori cycle (Fig. 23–6; see also Box 15–1).

Skeletal muscle contains 10–30 mM phosphocreatine (see Table 14–5), which can rapidly regenerate ATP from ADP by the creatine kinase reaction. During periods of active contraction and glycolysis, this reaction proceeds predominantly in the direction of ATP synthesis (Fig. 23–5); during recovery from exertion, the same enzyme is used to resynthesize phosphocreatine from creatine at the expense of ATP.

$$\text{ATP} + \underset{\text{Creatine}}{H_2N{-}\overset{+}{C}(=NH_2){-}N(CH_3){-}CH_2{-}COO^-} \underset{\text{creatine kinase}}{\rightleftharpoons} \underset{\text{Phosphocreatine}}{{}^-O{-}P(=O)(O^-){-}NH{-}\overset{+}{C}(=NH_2){-}N(CH_3){-}CH_2{-}COO^-} + \text{ADP}$$

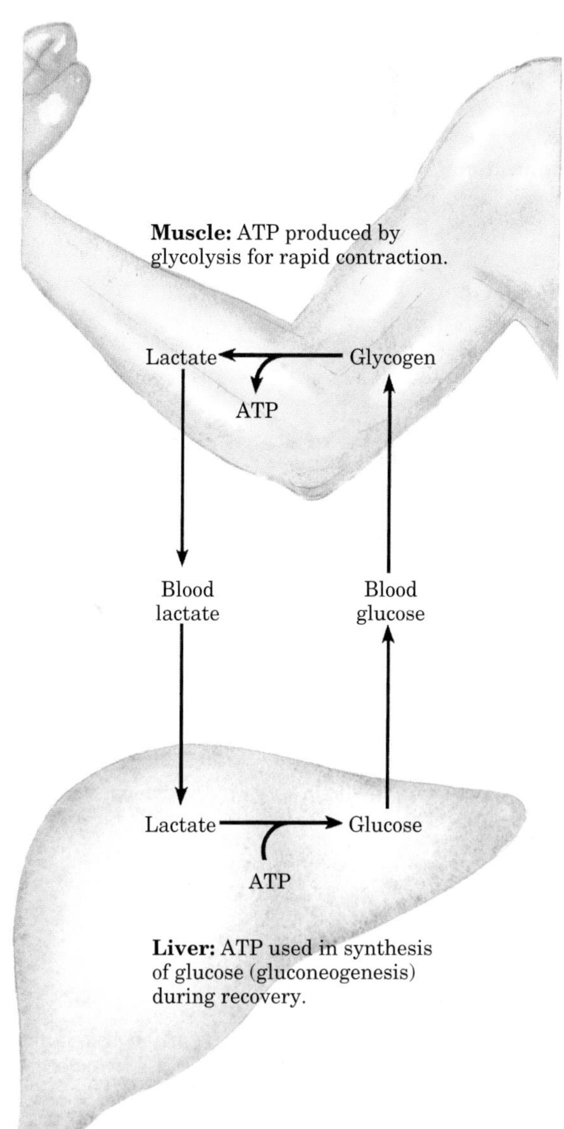

Muscle: ATP produced by glycolysis for rapid contraction.
Lactate
Glycogen
ATP
Blood lactate
Blood glucose
Lactate
Glucose
ATP
Liver: ATP used in synthesis of glucose (gluconeogenesis) during recovery.

figure 23–6
Metabolic cooperation between skeletal muscle and the liver. Extremely active muscules use glycogen as an energy source, generating lactate via glycolysis. During recovery, some of this lactate is transported to the liver and used to form glucose via gluconeogenesis. This glucose is released to the blood and returned to the muscles to replenish their glycogen stores. The overall pathway (glucose ⟶ lactate ⟶ glucose) constitutes the Cori cycle.

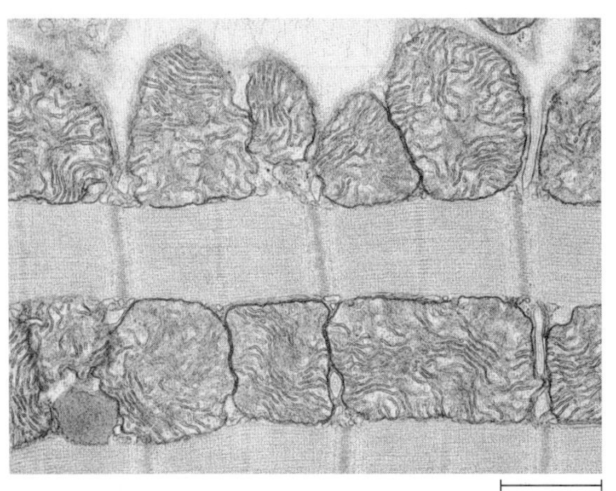

figure 23–7

Electron micrograph of heart muscle. In the profuse mitochondria, pyruvate, fatty acids, and ketone bodies are oxidized to drive ATP synthesis.This steady aerobic metabolism allows the human heart to pump blood at a rate of nearly 6 liters per minute, or about 350 liters per hour, or 200 million liters over 70 years.

Heart muscle differs from skeletal muscle in that it is continuously active in a regular rhythm of contraction and relaxation. In contrast to skeletal muscle, the heart has a completely aerobic metabolism at all times. Mitochondria are much more abundant in heart muscle than in skeletal muscle, making up almost half the volume of the cells (Fig. 23–7). The heart uses as fuel a mixture of blood-borne glucose, free fatty acids, and ketone bodies; these fuels are oxidized via the citric acid cycle to deliver the energy required to generate ATP by oxidative phosphorylation. Like skeletal muscle, heart muscle does not store lipids or glycogen in large amounts. Small amounts of reserve energy are stored in the form of phosphocreatine, enough for a few seconds of contraction. Because the heart is normally aerobic and obtains its energy from oxidative phosphorylation, the failure of O_2 to reach a portion of the heart muscle when the blood vessels are blocked by lipid deposits (atherosclerosis) or blood clots (coronary thrombosis) can cause that region of the heart muscle to die. This process is called myocardial infarction, more commonly known as a heart attack.

The Brain Uses Energy for Transmission of Electrical Impulses

The metabolism of the brain is remarkable in several respects. First, the brain of adult mammals normally uses only glucose as fuel (Fig. 23–8). Second, the brain has a very active respiratory metabolism (Fig. 23–9); it uses O_2 at a fairly constant rate, accounting for almost 20% of the total O_2 consumed at rest. The brain contains very little glycogen, so it is continuously dependent on incoming glucose from the blood. Should blood glucose fall significantly below a critical level for even a short period of time, severe and sometimes irreversible changes in brain function may occur.

Although the brain cannot directly use free fatty acids or lipids from the blood as fuels, it can, when necessary, use β-hydroxybutyrate (a ketone body) formed from fatty acids in hepatocytes. The capacity of the brain to

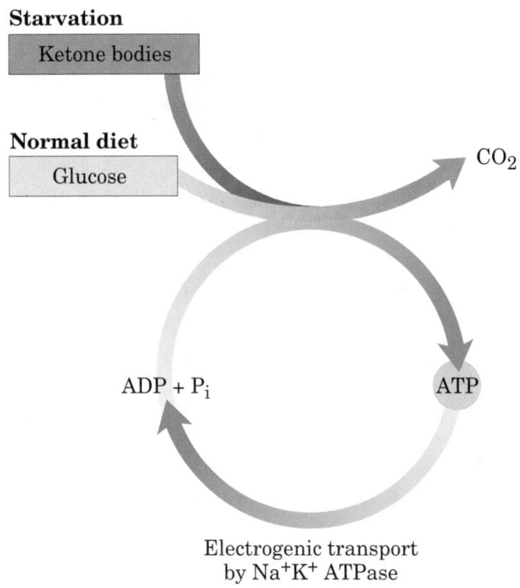

figure 23–8

Energy sources in the brain vary with nutritional state. The ketone body used by the brain is β-hydroxybutyrate.

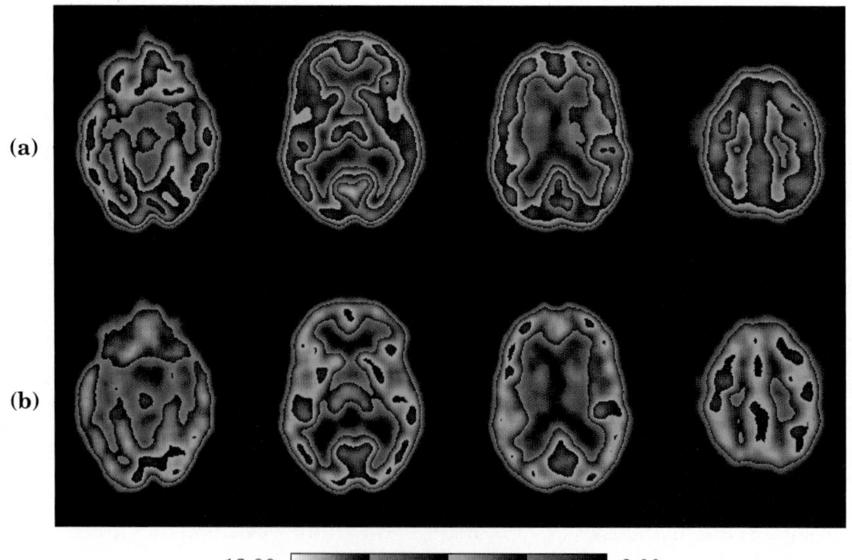

figure 23–9

Glucose metabolism in the brain. The technique of positron emission tomography (PET) scanning shows metabolic activity in specific regions of the brain. PET scans allow visualization of isotopically labeled glucose in precisely localized regions of the brain of a living person, in real time. ^{13}C-enriched glucose (not a radioisotope, and therefore not a health hazard) is injected into the bloodstream; a few seconds later, a PET scan shows how much glucose has been taken up by each region of the brain—a measure of metabolic activity. Shown here are PET scans of front-to-back cross sections of the brain at four levels from top (left) to bottom (right). The scans contrast glucose metabolism when the experimental subject is **(a)** rested and **(b)** deprived of sleep for 48 hours.

oxidize β-hydroxybutyrate via acetyl-CoA becomes important during prolonged fasting or starvation, after liver glycogen has been depleted, because it allows the brain to use body fat as an energy source. This spares muscle proteins, until they become the brain's ultimate source of glucose (via gluconeogenesis in the liver) during severe starvation.

Glucose is oxidized by glycolysis and the citric acid cycle, providing almost all of the ATP used by the brain. Energy is required to create and maintain an electrical potential across the plasma membrane of neurons. The plasma membrane contains an electrogenic ATP-driven antiporter, the Na^+K^+ ATPase, which simultaneously pumps 2 K^+ ions into and 3 Na^+ ions out of the neuron (see Fig. 12–33). The resulting transmembrane potential changes transiently, as an electrical signal (action potential) sweeps from one end of a neuron to the other (see Fig. 13–5). Action potentials are the chief mechanism of information transfer in the nervous system.

Blood Carries Oxygen, Metabolites, and Hormones

Blood mediates the metabolic interactions between all tissues. It transports nutrients from the small intestine to the liver, and from the liver and adipose tissue to other organs; it also transports waste products from the tissues to the kidneys for excretion. Oxygen moves via the blood from the lungs to the tissues, and CO_2 generated by tissue respiration returns via blood to the lungs for exhalation. Blood also carries hormonal signals from one tissue to another. In its role as signal carrier, the circulatory system resembles the nervous system; both regulate and integrate the activities of different organs.

The average adult human has 5 to 6 L of blood. Almost half of this volume is occupied by three types of blood cells (Fig. 23–10): **erythrocytes** (red cells), filled with hemoglobin and specialized for carrying O_2 and CO_2; much smaller numbers of **leukocytes** (white cells) of several types, central to the immune system that defends against infections; and **platelets,** which help to mediate blood clotting. The liquid portion is the **blood plasma,** which is 90% water and 10% solutes. In it are dissolved or suspended a large variety of proteins, lipoproteins, nutrients, metabolites, waste products, inorganic ions, and hormones. Over 70% of the plasma solids are **plasma proteins** (Fig. 23–10), primarily immunoglobulins (circulating antibodies), serum albumin, apolipoproteins involved in the transport of lipids, transferrin (for iron transport), and blood-clotting proteins such as fibrinogen and prothrombin.

The ions and low molecular weight solutes in the blood plasma are not fixed components, but are in constant flux between blood and various tissues. Dietary uptake of inorganic ions is, in general, counterbalanced by their excretion in the urine. For many blood components, something near a dynamic steady state is achieved; the concentration of the component changes little, although a continuous flux occurs between the digestive tract, blood, and urine. The plasma levels of Na^+, K^+, and Ca^{2+} remain close to 140, 5, and 2.5 mM, respectively, with little change in response to dietary intake. Any significant departure from these values can result in serious illness or death. The kidneys play an especially important role in maintaining the ion balance by selectively filtering waste products and excess ions out of the blood while preventing the loss of essential nutrients and ions.

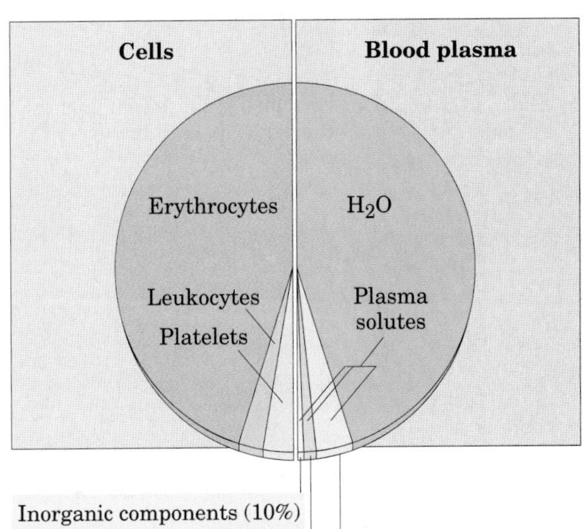

figure 23–10

The composition of blood. Whole blood can be separated into blood plasma and cells by centrifugation. About 10% of blood plasma is solutes, of which about 10% consists of inorganic salts, 20% small organic molecules, and 70% plasma proteins. The major dissolved components are indicated. Blood contains many other substances, often in trace amounts. These include other metabolites, enzymes, hormones, vitamins, trace elements, and bile pigments. Measurements of the concentrations of components in blood plasma are important in the diagnosis and treatment of disease.

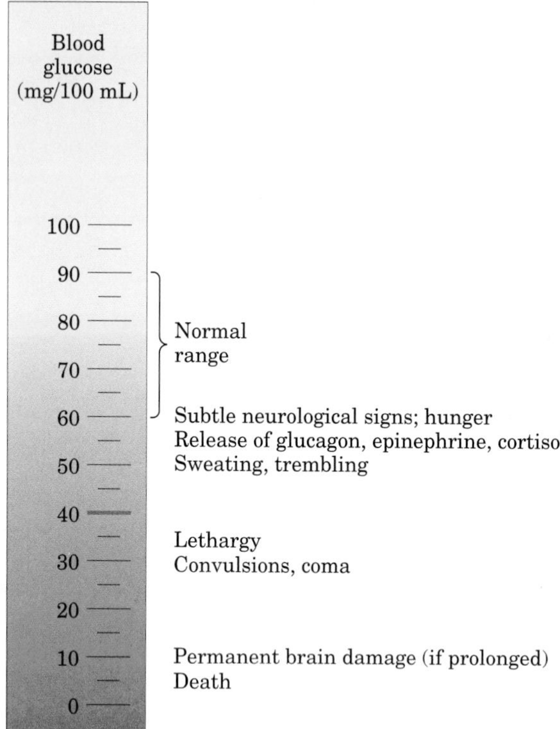

figure 23–11
Physiological effects of low blood glucose in humans. Blood glucose levels of 40 mg/100 mL and below constitute severe hypoglycemia.

The concentration of glucose dissolved in the plasma is also subject to tight regulation. We have noted the requirement of the brain for glucose and the role of the liver in maintaining the glucose concentration in the normal range of 60 to 90 mg/100 mL of blood (about 4.5 mM). When blood glucose in a human drops to 40 mg/100 mL (the hypoglycemic condition), the person experiences discomfort and mental confusion (Fig. 23–11); further reductions lead to coma, convulsions, and in extreme hypoglycemia, death. Maintaining the normal concentration of glucose in the blood is therefore a very high priority of the organism, and a variety of regulatory mechanisms have evolved to achieve that end. Among the most important regulators of blood glucose are the hormones insulin, glucagon, and epinephrine.

Hormonal Regulation of Fuel Metabolism

The minute-by-minute adjustments that keep the blood glucose level near 4.5 mM involve the combined actions of insulin, glucagon, and epinephrine on metabolic processes in many body tissues, but especially in liver, muscle, and adipose tissue. Insulin signals these tissues that the blood glucose concentration is higher than necessary; as a result, excess glucose is taken up from the blood into cells and converted to the storage compounds glycogen and triacylglycerol. Glucagon carries the message that blood glucose is too low; tissues respond by producing glucose through glycogen breakdown and gluconeogenesis, and by oxidizing fats to reduce the use of glucose. Epinephrine is released into the blood to prepare the muscles, lungs, and heart for a burst of activity.

Epinephrine Signals Impending Activity

When an animal is confronted with a stressful situation that requires increased activity—fighting or fleeing, in the extreme case—neuronal signals from the brain trigger the release of **epinephrine** and **norepinephrine** from the adrenal medulla. Both hormones increase the rate and strength of the heartbeat and raise the blood pressure, thereby increasing the flow of O_2 and fuels to the tissues, and the hormones dilate the respiratory passages, facilitating the uptake of O_2 (Table 23–2).

Epinephrine acts primarily on muscle, adipose tissue, and liver. It activates glycogen phosphorylase and inactivates glycogen synthase by cAMP-dependent phosphorylation of the enzymes (see Fig. 20–15), thus stimulating the conversion of liver glycogen into blood glucose, the fuel for anaerobic muscular work. Epinephrine also promotes the anaerobic breakdown of the glycogen of skeletal muscle into lactate by fermentation, thus

table 23–2

Physiological and Metabolic Effects of Epinephrine: Preparation for Action

Effect	Result
Physiological	
↑ Heart rate	Increased delivery of O_2 to tissues (muscle)
↑ Blood pressure	
↑ Dilation of respiratory passages	
Metabolic	
↑ Glycogen breakdown (muscle, liver)	Increased production of glucose for fuel
↓ Glycogen synthesis (muscle, liver)	
↑ Gluconeogenesis (liver)	
↑ Glycolysis (muscle)	Increased ATP production in muscle
↑ Fatty acid mobilization (adipose tissue)	Increased availability of fatty acids as fuel
↑ Glucagon secretion	Reinforce metabolic effects of epinephrine
↓ Insulin secretion	

stimulating glycolytic ATP formation. The stimulation of glycolysis is accomplished by raising the concentration of fructose 2,6-bisphosphate, a potent allosteric activator of the key glycolytic enzyme phosphofructokinase-1 (see Figs. 20–7, 20–8). Epinephrine also stimulates fat mobilization in adipose tissue, activating (by cAMP-dependent phosphorylation) triacylglycerol lipase (see Fig. 17–3). Finally, epinephrine stimulates glucagon secretion and inhibits insulin secretion, reinforcing its effect of mobilizing fuels and inhibiting fuel storage.

Glucagon Signals Low Blood Glucose

Even in the absence of significant physical activity or stress, several hours after the intake of dietary carbohydrate, blood glucose levels fall below 4.5 mM because of the ongoing oxidation of glucose by the brain and other tissues. Lowered blood glucose triggers secretion of glucagon and decreases insulin release (Fig. 23–12). Glucagon causes an increase in blood glucose concentration in several ways (Table 23–3). Like epinephrine, glucagon stimulates the net breakdown of liver glycogen by activating glycogen phosphorylase and inactivating glycogen synthase; both effects are the result of phosphorylation of the regulated enzymes, triggered by cAMP. Unlike epinephrine, glucagon inhibits glucose breakdown by glycolysis in the liver and stimulates glucose synthesis by gluconeogenesis. Both of these effects result from lowering the level of fructose 2,6-bisphosphate, an allosteric inhibitor of the gluconeogenic enzyme fructose 1,6-bisphosphatase (FBPase-1) and an activator of phosphofructokinase-1. Recall that the fructose 2,6-bisphosphate level is ultimately controlled by a cAMP-dependent protein phosphorylation reaction (see Fig. 20–8). Glucagon also inhibits the glycolytic enzyme pyruvate kinase (by promoting its cAMP-dependent phosphorylation), thus blocking the conversion of phosphoenolpyruvate to pyruvate and preventing oxidation of pyruvate via the citric acid cycle; the resulting accumulation of phosphoenolpyruvate favors gluconeogenesis.

By stimulating liver glycogen breakdown, preventing glucose utilization in the liver by glycolysis, and promoting gluconeogenesis, glucagon enables the liver to export glucose to the blood, restoring blood glucose to its normal level (Fig. 23–12).

Although its primary target is the liver, glucagon (like epinephrine) also affects adipose tissue, activating triacylglycerol lipase by causing its cAMP-dependent phosphorylation. This lipase liberates free fatty acids, which are exported to the liver and other tissues as fuel, thus sparing glucose for the brain. The net effect of glucagon is therefore to stimulate glucose synthesis and release by the liver and to cause the mobilization of fatty acids from adipose tissue, to be used instead of glucose as fuel for tissues other than the brain (Table 23–3). All of these effects of glucagon are mediated by cAMP-dependent protein phosphorylation.

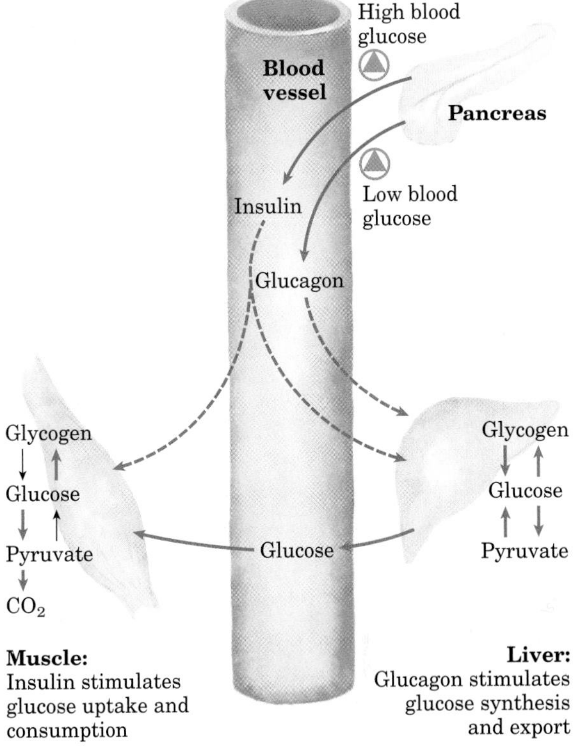

figure 23–12

Regulation of blood glucose by insulin and glucagon. Blue arrows indicate processes stimulated by insulin; red arrows indicate processes stimulated by glucagon. High blood glucose results in insulin secretion by the pancreas, and low blood glucose leads to glucagon release, as described in the text.

table 23–3

Effects of Glucagon on Blood Glucose: Production and Release of Glucose By the Liver

Metabolic effect	Effect on glucose metabolism	Target enzyme
↑ Glycogen breakdown (liver)	Glycogen ⟶ glucose	↑ Glycogen phosphorylase
↓ Glycogen synthesis (liver)	Less glucose stored as glycogen	↓ Glycogen synthase
↓ Glycolysis (liver)	Less glucose used as fuel in liver	↓ Phosphofructokinase-1
↑ Gluconeogenesis (liver)	Amino acids, Glycerol, Oxaloacetate ⟶ glucose	↑ Fructose 1,6-bisphosphatase ↓ Pyruvate kinase
↑ Fatty acid mobilization (adipose tissue)	Less glucose used as fuel by liver, muscle	↑ Triacylglycerol lipase

During Fasting and Starvation, Metabolism Shifts to Provide Fuel for the Brain

The fuel reserves of a normal adult human are of three types: glycogen stored in the liver and in muscle in relatively small quantities; larger quantities of triacylglycerols in adipose tissues; and tissue proteins, which can be degraded when necessary to provide fuel (Table 23–4).

Figure 23–13 shows the changes in fuel metabolism during starvation. After an overnight fast, almost all of the liver glycogen and most of the muscle glycogen have been depleted. Within 24 hours, the blood glucose concentration begins to fall, insulin secretion slows, and glucagon secretion is stimulated. These hormonal signals mobilize triacylglycerols, which become the primary fuels for muscle and liver. To provide glucose for the brain, the liver degrades certain proteins (those most expendable in an organism not ingesting food). Their nonessential amino acids are deaminated (see Chapter 18), and their amino groups are converted into urea in the liver. The urea is exported via the bloodstream to the kidney and is excreted. Also in the liver, the carbon skeletons of glucogenic amino acids (see Table 20–3) are converted into pyruvate or intermediates of the citric acid cycle. These intermediates, as well as the glycerol derived from triacylglycerols in adipose tissue, provide the starting materials for gluconeogenesis in the liver, yielding glucose for the brain. Eventually the use of citric acid cycle intermediates for gluconeogenesis depletes oxaloacetate (step ③ in Fig. 23–13), inhibiting entry of acetyl-CoA into the citric acid cycle. Acetyl-CoA produced by fatty acid oxidation accumulates (steps ⑤ and ⑥), favoring the formation of acetoacetyl-CoA (step ⑦) and ketone bodies in the liver. After a few days of fasting, the levels of ketone bodies in the blood rise as these fuels are exported from the liver to heart and skeletal muscle and the brain, which use them instead of glucose (step ⑧).

Triacylglycerols stored in the adipose tissue of an adult of normal weight could provide enough fuel to maintain a basal rate of metabolism for about three months; a very obese adult has enough stored fuel to endure a fast of more than a year (Table 23–4). However, such a fast would be extremely dangerous; it would almost certainly lead to severe overproduction

table 23–4

Available Metabolic Fuels in a Normal 70 kg Man and in an Obese 140 kg Man at the Beginning of a Fast

Type of fuel	Weight (kg)	Caloric equivalent [thousands of kcal (kJ)]	Estimated survival (months)*
Normal 70 kg man:			
Triacylglycerols (adipose tissue)	15	141 (589)	
Proteins (mainly muscle)	6	24 (100)	
Glycogen (muscle, liver)	0.225	0.90 (3.8)	
Circulating fuels (glucose, fatty acids, triacylglycerols, etc.)	0.023	0.10 (0.42)	
Total		166 (694)	3
Obese 140 kg man:			
Triacylglycerols (adipose tissue)	80	752 (3,140)	
Proteins (mainly muscle)	8	32 (134)	
Glycogen (muscle, liver)	0.23	0.92 (3.8)	
Circulating fuels	0.025	0.11 (0.46)	
Total		785 (3,280)	14

*Survival time is calculated on the assumption of a basal energy expenditure of 1,800 kcal/day.

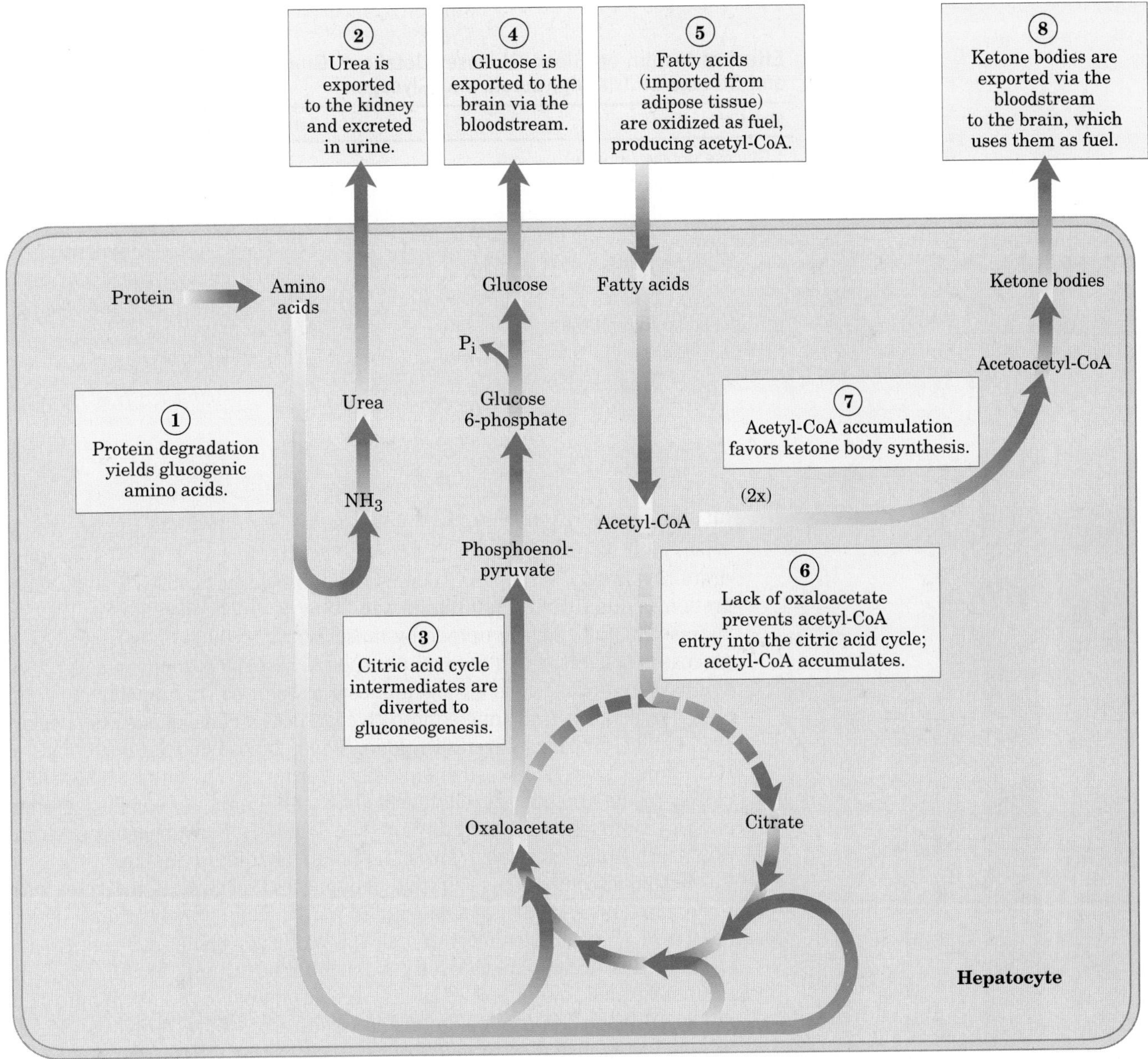

figure 23–13

Fuel metabolism in the liver during prolonged starvation. After depletion of stored carbohydrates, proteins become an important source of glucose, produced from glucogenic amino acids by gluconeogenesis (steps ① through ④). Fatty acids imported from adipose tissue are converted into ketone bodies for export to the brain (steps ⑤ through ⑧). Broken arrows represent reactions through which there is reduced flux during starvation.

of ketone bodies (described below), and perhaps to death. When fat reserves are gone, the degradation of essential proteins begins; this leads to loss of heart and liver function, and eventually death. Stored fat can provide adequate energy (calories) during a fast or rigid diet, but vitamins and minerals must be provided, and sufficient dietary glucogenic amino acids are needed to replace those being used for gluconeogenesis. Diet rations are therefore commonly fortified with vitamins, minerals, and amino acids or proteins.

table 23–5

Effect of Insulin on Blood Glucose: Uptake of Glucose by Cells and Storage as Triacylglycerols and Glycogen

Metabolic effect	Target enzyme
↑ Glucose uptake (muscle)	↑ Glucose transporter
↑ Glucose uptake (liver)	↑ Glucokinase
↑ Glycogen synthesis (liver, muscle)	↑ Glycogen synthase
↓ Glycogen breakdown (liver, muscle)	↓ Glycogen phosphorylase
↑ Glycolysis, acetyl-CoA production (liver, muscle)	↑ Phosphofructokinase-1
	↑ Pyruvate dehydrogenase complex
↑ Fatty acid synthesis (liver)	↑ Acetyl-CoA carboxylase
↑ Triacylglycerol synthesis (adipose tissue)	↑ Lipoprotein lipase

Insulin Signals High Blood Glucose

When glucose enters the bloodstream from the intestine after a carbohydrate-rich meal, the resulting increase in blood glucose causes increased secretion of **insulin** and decreased secretion of glucagon (Fig. 23–12). Insulin stimulates glucose uptake by muscle tissue (Table 23–5), where the glucose is converted to glucose 6-phosphate. Insulin also activates glycogen synthase and inactivates glycogen phosphorylase, so that much of the glucose 6-phosphate is channeled into glycogen. As a consequence of accelerated uptake of glucose from the blood, blood glucose concentration falls to the normal level, slowing insulin release from the pancreas. There is a closely adjusted feedback relationship between the rate of insulin secretion and the blood glucose concentration, which holds blood glucose concentration nearly constant despite large fluctuations in dietary intake.

Insulin also stimulates the storage of excess fuel as fat. It activates both the oxidation of glucose 6-phosphate to pyruvate via glycolysis and the oxidation of pyruvate to acetyl-CoA. If not oxidized further for energy production, this acetyl-CoA is used for fatty acid synthesis in the liver, and these fatty acids are exported as the triacylglycerols of plasma lipoproteins (VLDLs) to the adipose tissue. Insulin stimulates triacylglycerol synthesis in adipocytes, using fatty acids released from the VLDL triacylglycerols. These fatty acids are ultimately derived from the excess glucose taken from the blood by the liver.

In summary, the effect of insulin is to favor the conversion of excess blood glucose into two storage forms: glycogen (in the liver and muscle) and triacylglycerols (in adipose tissue) (Table 23–5).

Cortisol Signals Stress, Including Low Blood Glucose

A variety of stresses (anxiety, fear, pain, hemorrhage, infections, low blood glucose) stimulate release of the corticosteroid hormone **cortisol** from the adrenal cortex. Cortisol acts on muscle, liver, and adipose tissue to supply the organism with fuel for impending intense activity. Cortisol is a relatively slow-acting hormone that alters metabolism by changing the kinds and amounts of certain enzymes that are newly synthesized in its target cell, rather than by regulating existing enzyme molecules.

In adipose tissue, cortisol stimulates the release of fatty acids from stored triacylglycerols. The fatty acids are exported to the blood to serve as fuel for various tissues, and the glycerol resulting from triacylglycerol breakdown is used for gluconeogenesis in the liver. Cortisol stimulates the breakdown of nonessential muscle proteins and the export of amino acids to the liver, where they serve as precursors for gluconeogenesis. In the liver, cortisol promotes gluconeogenesis by stimulating synthesis of the key enzyme PEP carboxykinase (see Fig. 20–3b); glucagon also has this effect, whereas insulin has the opposite effect. Glucose produced in this way is stored in the liver as glycogen, or exported immediately by tissues that need glucose for fuel. The net effect of these metabolic changes is to raise blood glucose back to its normal level and to store glycogen to support the fight-or-flight response commonly associated with stress. The effects of cortisol therefore counterbalance those of insulin.

Diabetes Is a Defect in Insulin Production or Action

Diabetes mellitus, caused by a deficiency in the secretion or action of insulin, is a relatively common disease: nearly 6% of the United States population shows some degree of abnormality in glucose metabolism indicative of diabetes or a tendency toward the condition. Diabetes mellitus is really a group of diseases in which the regulatory activity of insulin is defective. There are two major clinical classes of the disease: type I or insulin-dependent diabetes mellitus (IDDM) and type II or non–insulin-dependent diabetes mellitus (NIDDM). In the former, the disease begins early in life and quickly becomes severe. The latter is slow to develop, milder, and often goes unrecognized. IDDM requires insulin therapy and careful, lifelong control of the balance between glucose intake and insulin dose. Characteristic symptoms of diabetes are excessive thirst and frequent urination (polyuria), leading to the intake of large volumes of water (polydipsia). These changes are due to the excretion of large amounts of glucose in the urine, a condition known as **glucosuria.** The term diabetes mellitus means "excessive excretion of sweet urine."

Another characteristic metabolic change resulting from defective insulin action in diabetes is excessive but incomplete oxidation of fatty acids in the liver, resulting in overproduction of the ketone bodies acetoacetate and β-hydroxybutyrate, which cannot be used by the extrahepatic tissues as fast as they are made in the liver. In addition to β-hydroxybutyrate and acetoacetate, the blood of diabetics also contains acetone, which results from the spontaneous decarboxylation of acetoacetate:

$$\underset{\text{Acetoacetate}}{CH_3{-}\overset{\overset{\displaystyle O}{\|}}{C}{-}CH_2{-}COO^-} + H_2O \longrightarrow \underset{\text{Acetone}}{CH_3{-}\overset{\overset{\displaystyle O}{\|}}{C}{-}CH_3} + HCO_3^-$$

Acetone is volatile and is exhaled, giving the breath of an untreated diabetic a characteristic odor sometimes mistaken for ethanol. A diabetic experiencing mental confusion because of high blood glucose is occasionally misdiagnosed as intoxicated, an error that can be fatal. The overproduction of ketone bodies, called **ketosis,** results in greatly increased concentrations in the blood (ketonemia) and urine (ketonuria; see Table 17–2).

The oxidation of triacylglycerols to form ketone bodies produces carboxylic acids, which ionize, releasing protons. In uncontrolled diabetes this can overwhelm the capacity of the bicarbonate buffering system of blood and produce a lowering of blood pH called **acidosis** or, in combination with ketosis, **ketoacidosis,** a potentially life-threatening condition.

Biochemical measurements on the blood and urine are essential in the diagnosis and treatment of diabetes, which causes profound changes in metabolism. A sensitive diagnostic criterion is provided by the **glucose-tolerance test.** After a night without food, the patient drinks a test dose of 100 g of glucose dissolved in a glass of water. The blood glucose concentration is measured before the test dose and at 30 min intervals for several hours thereafter. A normal individual assimilates the glucose readily, the blood glucose rising to no more than about 9 or 10 mM; little or no glucose appears in the urine. Diabetic individuals assimilate the test dose of glucose poorly; their blood glucose level far exceeds the kidney threshold (about 10 mM), causing glucose to appear in their urine.

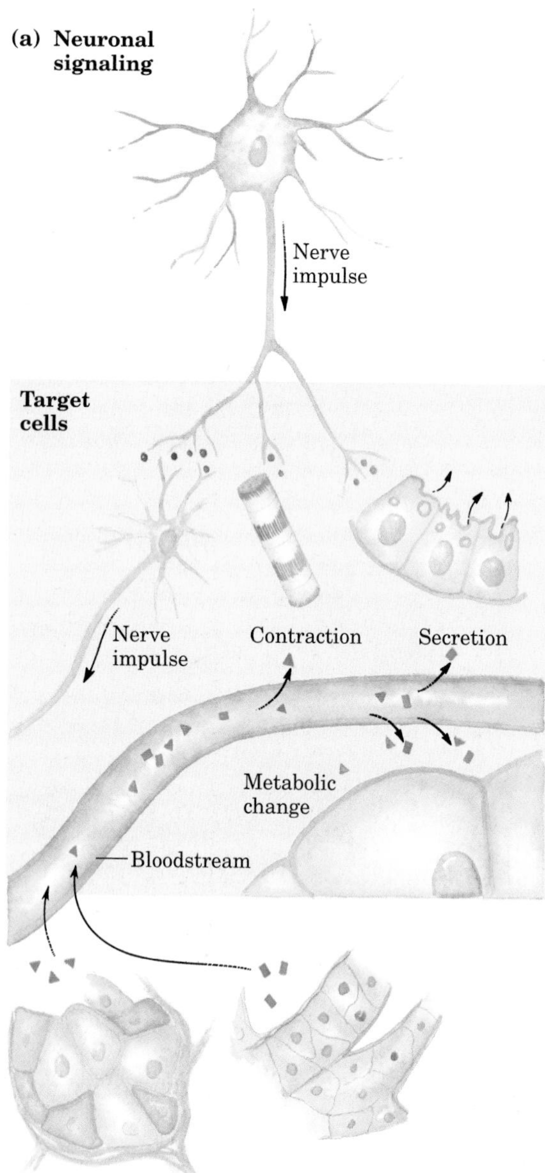

figure 23–14
Signaling by the neuroendocrine system. (a) In neuronal signaling, electrical signals (nerve impulses) originate in the nerve cell body and are carried very rapidly over long distances to the axon tip, where neurotransmitters are released and diffuse to the target cell. The target cell (which may be another neuron, a myocyte, or a secretory cell) is only a fraction of a micrometer or a few micrometers away from the site of neurotransmitter release. **(b)** In the endocrine system, hormones are secreted into the bloodstream, which carries them throughout the body to target tissues that may be a meter or more away from the secreting cell. Both neurotransmitters and hormones interact with specific receptors on or in the target cell, triggering responses.

Hormones: Diverse Structures for Diverse Functions

The regulation of fuel metabolism is only one of many integrative roles of hormones. Virtually every process in complex organisms is regulated by one or more hormones: maintenance of blood pressure, blood volume, and electrolyte balance; embryogenesis; sexual differentiation, development, and reproduction; hunger, eating behavior, and digestion, to name but a few. We now examine methods for detecting and measuring hormones and their interaction with receptors, and we consider a representative selection of hormone types.

The coordination of metabolism in the separate organs of mammals is achieved by the **neuroendocrine system.** Individual cells in one tissue sense a change in the organism's circumstances and respond by secreting an extracellular chemical messenger that passes to another cell, where it binds to a receptor molecule and triggers a change in the second cell.

In neuronal signaling (Fig. 23–14a), the chemical messenger (neurotransmitter; acetylcholine, for example) may travel only a fraction of a micrometer, across the synaptic cleft to the next neuron in a network. In contrast, hormones are carried via the blood to distant organs and tissues; they may travel a meter or more before encountering their target cell (Fig. 23–14b). Except for this anatomical difference, these two chemical signaling mechanisms are remarkably similar. Epinephrine and norepinephrine, for example, serve as neurotransmitters in certain synapses of the brain and smooth muscle and also as hormones that regulate fuel metabolism in the liver and in muscle. The following discussion of cellular signaling emphasizes hormone action, drawing on previous discussions of fuel metabolism, but most of the fundamental mechanisms described here also occur in neurotransmitter action.

Hormone Discovery and Purification Requires a Bioassay

How is a hormone discovered and purified? First a physiological process in one tissue is found to depend on a signal that originates in another tissue. Insulin, for example, was first recognized as a substance produced in the pancreas that affected the volume and composition of urine produced by a dog (Box 23–1). Once a physiological effect of the putative hormone is discovered, a quantitative bioassay for the hormone can be developed. In the case of insulin, the assay consisted of injecting extracts of pancreas (a crude source of insulin) into experimental animals deficient in insulin (Box 23–1), then quantifying the resulting changes in glucose concentrations in blood and urine. Leptin, the hormone that inhibits feeding behavior when the body's supply of stored triacylglycerols is adequate, was discovered as a factor carried in the blood of a normal mouse that reversed the overeating behavior of a mutant mouse (which turned out to lack leptin). To isolate a hormone, extracts containing the putative hormone are fractionated by the same techniques used to purify other biomolecules (solvent fractionation, chromatography, and electrophoresis), and each fraction is then assayed for hormone activity. Once the material has been purified, its chemical composition and structure can be determined.

box 23–1 How Is a Hormone Discovered? The Arduous Pathway to Purified Insulin

Millions of people with type I (insulin-dependent) diabetes mellitus inject themselves daily with pure insulin to compensate for the fact that this critical hormone is not produced by their own pancreatic β cells. Insulin injection is not a cure for diabetes, but it allows people who otherwise would have died young to lead long and productive lives. The discovery of insulin, which began with an accidental observation, illustrates the combination of serendipity and careful experimentation by which many hormones have been discovered.

In 1889, Oskar Minkowski, a young assistant at the Medical College of Strasbourg, and Josef von Mering, at the Hoppe-Seyler Institute in Strasbourg, had a friendly disagreement about whether the pancreas, known to contain lipases, was important in the digestion of fat in the dog. To resolve the issue, they surgically removed the pancreas from a dog. However, before their experiments on fat digestion could be carried out, Minkowski noticed that the dog was now producing far more urine than normal (a common symptom of untreated diabetes). The dog's urine also had glucose levels far above normal (another symptom of diabetes). This implied that the lack of some pancreatic product caused diabetes. Minkowski tried unsuccessfully to prepare an extract of dog pancreas that would reverse the effect of removing the pancreas (that would lower the urinary or blood glucose levels). We know now that insulin is a protein, and that the pancreas is very rich in proteases (trypsin and chymotrypsin), which are normally released directly into the small intestine to aid in digestion. These proteases doubtless degraded the insulin in Minkowski's pancreatic extracts. Despite considerable effort, no significant progress was made in the isolation or characterization of the "antidiabetic factor" until the summer of 1921, when Frederick G. Banting, a young scientist working in the laboratory of J.J.R. MacLeod at the University of Toronto, and a student assistant, Charles Best, took up the problem. Several lines of evidence by then pointed to a group of specialized cells in the pancreas (the islets of Langerhans, see Fig. 23–25) as the source of the antidiabetic factor, which came therefore to be called insulin (from Latin *insula,* "island").

Taking precautions to prevent proteolysis, Banting and Best (later aided by biochemist J.B. Collip) succeeded late in December of 1921 in preparing a purified pancreatic extract that cured the symptoms of experimental diabetes in dogs. On January 25, 1922 (one month later!) their insulin preparation was injected into Leonard Thompson, a 14-year-old boy severely ill from diabetes mellitus. Within days, the levels of ketone bodies and glucose in Thompson's urine dropped dramatically; the extract saved his life. In 1923, Banting and MacLeod won the Nobel Prize for their isolation of insulin. Banting immediately announced that he would share his prize with Best; MacLeod shared his with Collip.

By 1923, pharmaceutical companies were supplying thousands of patients throughout the world with insulin extracted from porcine pancreas. With the development of genetic engineering techniques in the 1980s (see Chapter 29), it became possible to produce unlimited quantities of human insulin by placing the cloned human gene for insulin in a microorganism, which was then cultured on an industrial scale. Some patients are now fitted with implanted insulin pumps, which release adjustable amounts of insulin on demand to meet changing needs at meal times and during exercise. There is a reasonable prospect that, in the future, transplantation of pancreatic tissue will provide diabetic patients with a source of insulin that responds as normal pancreas does, by releasing insulin into the bloodstream only when blood glucose rises.

Frederick G. Banting
1891–1941

J.J.R. MacLeod
1876–1935

Charles Best
1899–1978

J.B. Collip
1892–1965

Pyroglutamate Histidine Prolylamide

pyroGlu–His–Pro–NH_2

figure 23–15

The structure of thyrotropin-releasing hormone (TRH). Purified by heroic efforts from extracts of hypothalamus, TRH proved to be a derivative of the tripeptide Glu–His–Pro. The side-chain carboxyl group of the amino-terminal Glu forms an amide (red) with the α-amino group of Glu, creating pyroglutamate, and the carboxyl group of the carboxyl-terminal Pro is converted to an amide (red). Such modifications are common among the small peptide hormones. In a typical protein of 50 kDa, the charges on the amino- and carboxyl-terminal residues contribute relatively little to the overall charge on the protein but, in a tripeptide hormone, these two charges would dominate the properties of the peptide. Formation of the amide derivatives masks these charges.

This formula for hormone characterization is deceptively simple. Hormones are so potent that they need to be produced only in very small amounts. Obtaining enough of a hormone to allow its chemical characterization often involves biochemical isolations on a heroic scale. When Roger Guillemin and Andrew Schally independently purified and characterized thyrotropin-releasing hormone (TRH) from the hypothalamus, Schally's group processed about 20 tons of hypothalami from nearly two million sheep, while Guillemin's group extracted hypothalami from about one million pigs! TRH proved to be a simple derivative of the tripeptide Glu–His–Pro (Fig. 23–15). Once the structure of the hormone was known, it could be chemically synthesized in large quantities for use in physiological and biochemical studies.

Roger Guillemin

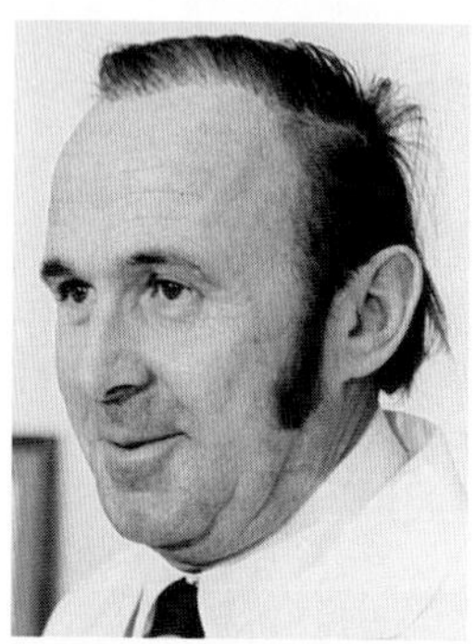

Andrew V. Schally

Rosalyn S. Yalow

For their work on hypothalamic hormones, Schally and Guillemin shared the Nobel Prize in Physiology or Medicine in 1977 with Rosalyn Yalow, who (with Solomon A. Berson) developed the extraordinarily sensitive **radioimmunoassay (RIA)** for peptide hormones and used it to study hormone action. RIA revolutionized hormone research by making possible the rapid, quantitative, and specific measurement of many hormones in minute amounts.

Hormone-specific antibodies are the key to the radioimmunoassay. Purified hormone, injected into rabbits, elicits antibodies that bind to that hormone with very high affinity and specificity. When a constant amount of antibody is incubated with a fixed amount of the radioactively labeled hormone, a certain fraction of the radioactive hormone binds to the antibody (Fig. 23–16). If, in addition to the radiolabeled hormone, unlabeled hormone is also present, the unlabeled hormone will compete with and displace some of the labeled hormone from its binding site on the antibody. This binding competition can be quantified by reference to a standard curve with known amounts of unlabeled hormone. The degree to which labeled hormone is displaced from antibody is a measure of the amount of unlabeled hormone in a sample of blood or tissue extract. By using very highly radioactive hormone, it is possible to make the assay sensitive to picograms of a hormone. A newer variation of this technique, called an enzyme-linked immunosorbent assay (ELISA), is illustrated in Figure 7–28b.

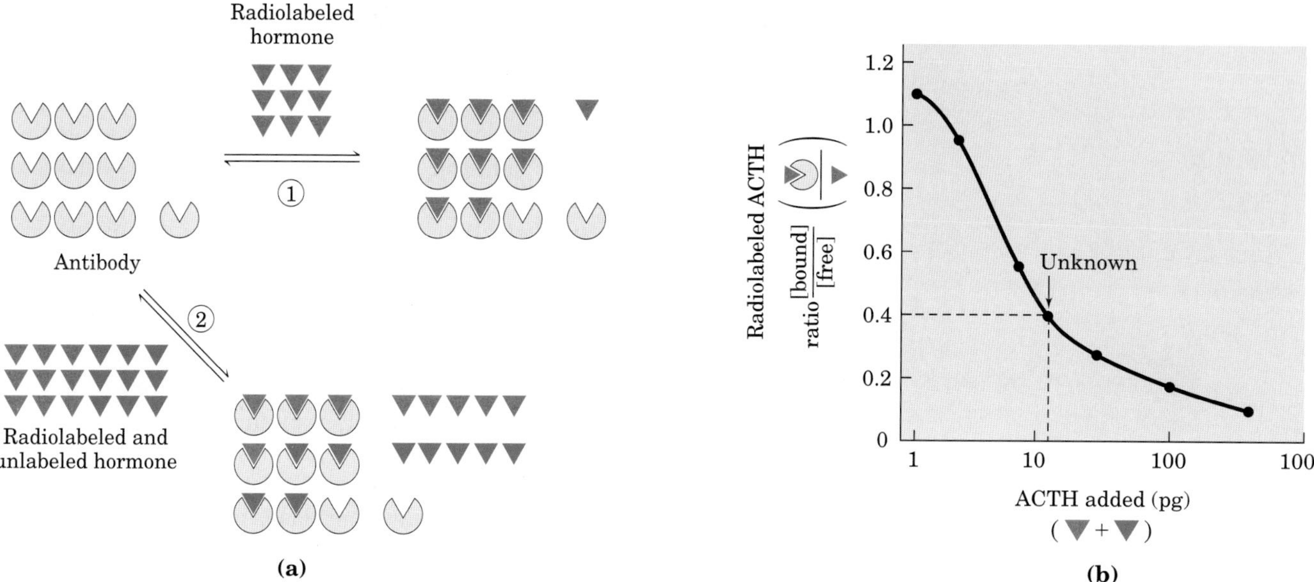

figure 23–16

The principle of the radioimmunoassay (RIA) and a real example. (a) A low concentration of radiolabeled hormone (red) is incubated with a fixed amount of antibody specific for that hormone ①, or in the added presence of various concentrations of unlabeled hormone (blue; ②). Unlabeled hormone molecules compete with labeled ones for binding to the antibody. The amount of labeled hormone bound therefore reflects the concentration of unlabeled hormone present. **(b)** A radioimmunoassay for adrenocorticotropic hormone (ACTH). A standard curve of [bound radiolabeled ACTH] vs. [unlabeled ACTH added] is constructed and used to determine how much (unlabeled) ACTH is in an unknown sample. If the aliquot containing the unknown quantity of unlabeled hormone gives a value of 0.4 for the ratio [bound]/[free] (see arrow), that aliquot must contain about 20 pg of ACTH.

Hormones Act through Specific High-Affinity Cellular Receptors

We saw in Chapter 13 that all hormones act through specific receptors present in hormone-sensitive target cells, to which the hormones bind with high specificity and high affinity (see Fig. 13–2). Each cell type has its own combination of hormone receptors, which define the range of its hormone responsiveness. Moreover, two cell types with the same receptor may have different intracellular targets of hormone action and thus may respond differently to the same hormone. The specificity of hormone action results from structural complementarity between the hormone and its receptor. This interaction is extremely selective, allowing structurally similar hormones to have different effects, and the high affinity of the interaction allows cells to respond to very low concentrations of hormones. In the design of drugs intended to intervene in hormonal regulation, it is essential to know the relative specificity and affinity of the drug and the natural hormone. Recall that the hormone-receptor interaction can be quantified by Scatchard analysis (see Box 13–1) which, under favorable conditions, yields a quantitative measure of affinity (the dissociation constant for the complex) and also gives the number of hormone-binding sites in a preparation of hormone.

The locus of the encounter between hormone and receptor may be extracellular, cytosolic, or nuclear, depending on the hormone type. The intracellular consequences of hormone-receptor interaction are of four general types (see Fig. 13–2): (1) a change in membrane potential results from the opening or closing of a hormone-gated ion channel; (2) a receptor-enzyme is activated by the extracellular hormone; (3) a second messenger, such as cAMP or inositol trisphosphate, is generated inside the cell to act as an allosteric regulator of some key enzyme(s); or (4) there is a change in the level of expression of some gene(s), mediated by a nuclear hormone receptor protein.

Water-soluble peptide and amine hormones (insulin and epinephrine, for example) act from outside their target cells by binding to cell-surface receptors that span the plasma membrane (Fig. 23–17). Upon hormone binding to its extracellular domain, the receptor undergoes a conformational change analogous to that produced in an allosteric enzyme bound by an effector molecule. The conformational change triggers the downstream effects of the hormones.

A single hormone molecule, in forming a hormone-receptor complex, activates a catalyst that produces many molecules of second messenger, so the receptor serves not only as a signal transducer, but also as a signal amplifier. Many hormones act through a signaling cascade, a series of steps in each of which a catalyst activates a catalyst, resulting in very large amplifications of the original signal. An example occurs in the regulation of glycogen synthesis and breakdown by epinephrine (see Fig. 13–15). Epinephrine activates (through its receptor) adenylyl cyclase, which produces many molecules of cAMP for each molecule of receptor-bound hormone. Cyclic AMP in turn activates cAMP-dependent protein kinase, which activates phosphorylase kinase, which activates glycogen phosphorylase. The result is signal amplification: one epinephrine molecule causes the production of many thousands of molecules of glucose 1-phosphate from glycogen.

Water-insoluble hormones (steroid, retinoid, and thyroid hormones) readily pass through the plasma membrane of their target cells to reach their receptor proteins in the nucleus (Fig. 23–17). With this class of hormones, the hormone-receptor complex itself carries the message; it interacts with

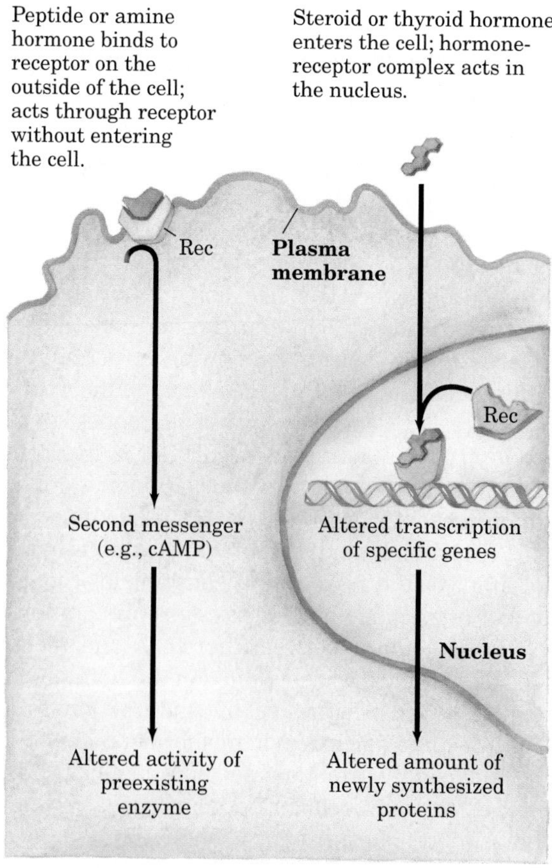

figure 23–17
Two general mechanisms of hormone action. The peptide and amine hormones are faster acting than steroid and thyroid hormones.

DNA to alter the expression of specific genes, changing the enzymatic complement of the cell and thereby its metabolism (see Fig. 13–29).

Those hormones that act through plasma membrane receptors generally trigger very rapid physiological or biochemical responses. Only seconds after epinephrine is secreted into the bloodstream by the adrenal medulla, skeletal muscle responds by accelerating the breakdown of glycogen. By contrast, the thyroid hormones and the sex (steroid) hormones promote maximal responses in their target tissues only after hours or even days. These differences in response time correspond to different modes of action. In general, the fast-acting hormones lead to a change in the activity of one or more preexisting enzyme(s) in the cell, by allosteric mechanisms or by covalent modification of the enzyme(s). The slower-acting hormones generally alter gene expression, resulting in the synthesis of more or less of the regulated protein(s).

Hormones Are Chemically Diverse

There are several distinct classes of hormones in mammals, distinguishable by their chemical structures and their modes of action (Table 23–6). Peptide, amine, and eicosanoid hormones act from outside the target cell via surface receptors. Steroid, vitamin D, retinoid, and thyroid hormones enter the cell and act through nuclear receptors. Nitric oxide also enters the cell, but activates a cytosolic enzyme, guanylyl cyclase (see Fig. 13–9).

Hormones can also be classified by the way they get from the point of their release to their target tissue. **Endocrine** (from the Greek *endon,* meaning "within," and *krinein,* "to release") hormones are released into the blood and carried to target cells throughout the body. **Paracrine** hormones are released into the extracellular space by one cell and diffuse to neighboring target cells. **Autocrine** hormones are released by one cell and affect that same cell by binding to receptors on its own surface.

Mammals are hardly unique in possessing hormonal signaling systems. Insects and nematode worms have highly developed systems for hormonal regulation, whose fundamental mechanisms are similar to those in mammals. Plants, too, use hormonal signals to coordinate the activities of their various tissues. The study of hormone action is not as advanced in plants as in animals, but it is clear that some mechanisms are shared. To illustrate the structural diversity and range of actions of mammalian hormones, let us consider representative examples of each of the major classes in Table 23–6.

table 23–6

Classes of Hormones

Type	Example	Parent /origin	Synthetic path	Mode of action
Peptide	Leu-enkephalin	Tyr–Gly–Gly–Phe–Leu	Proteolytic processing of proenzyme	Plasma membrane receptors; second messengers
Catecholamine	Epinephrine	Tyrosine		
Eicosanoid	PGE_1	20:4 Fatty acid		
Steroid	Testosterone	Cholesterol		
Retinoid	Retinoic acid	Vitamin A		Nuclear receptors; transcriptional regulation
Thyroid	Triiodothyronine (T_3)	Tyr in thyroglobulin		
Vitamin D	1,25-dihydroxycholecalciferol	Cholesterol or vitamin D		
Nitric oxide	Nitric oxide	$NO^{\bullet}$	Arginine + O_2	Cytosolic receptor (guanylate cyclase) and second messenger (cGMP)

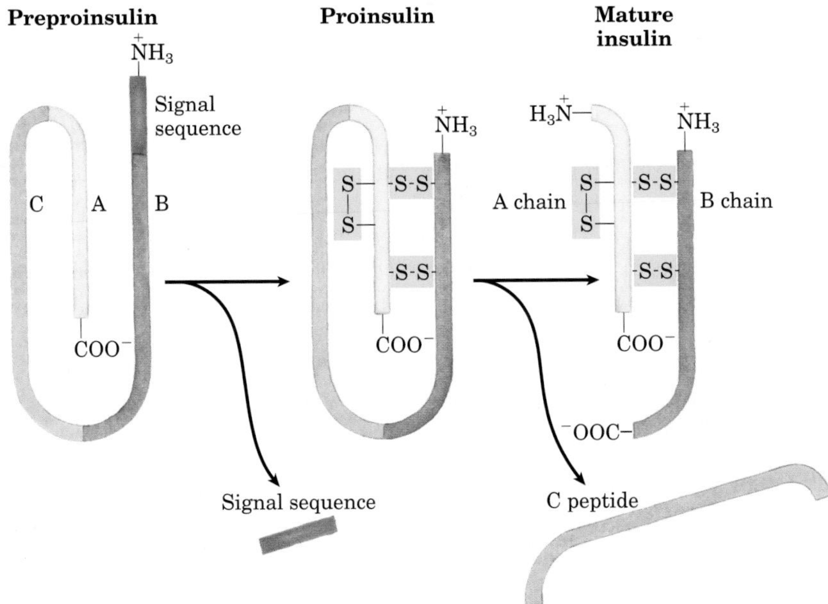

figure 23–18

Insulin. Mature insulin is formed from its larger precursor preproinsulin by proteolytic processing. Removal of 23 amino acids (the signal sequence) at the amino terminus of preproinsulin and formation of three disulfide bonds produces proinsulin. Further proteolytic cuts remove the C peptide from proinsulin, leaving mature insulin, composed of A and B chains. The amino acid sequence of bovine insulin is shown in Figure 5–24.

Peptide Hormones Peptide hormones may have from 3 to over 200 amino acid residues. They include the pancreatic hormones insulin, glucagon, and somatostatin and all of the hormones of the hypothalamus and pituitary (described below). These hormones are synthesized on ribosomes as longer precursor proteins (prohormones), which are packaged into secretory vesicles and proteolytically cleaved to form the active peptides. **Insulin** is a small protein (M_r 5,700) with two polypeptide chains, A and B, joined by two disulfide bonds. It is synthesized in the pancreas as an inactive single-chain precursor, preproinsulin (Fig. 23–18), with an amino-terminal "signal sequence" that directs its passage into secretory vesicles. (Signal sequences are discussed in Chapter 27; see Fig. 27–35). Proteolytic removal of the signal sequence and formation of three disulfide bonds produces proinsulin, which is stored in secretory granules in pancreatic cells. When elevated blood glucose triggers insulin secretion, proinsulin is converted into active insulin by specific proteases, which cleave two peptide bonds to form the mature insulin molecule.

In some cases, prohormone proteins yield a single peptide hormone, but often several active hormones are carved out of the same prohormone. Proopiomelanocortin (POMC) is a spectacular example of multiple hormones encoded in a single gene. The POMC gene encodes a large polypeptide that is progressively carved up into a total of at least nine biologically active peptides (Fig. 23–19). The terminal residues of peptide hormones are often modified, as with TRH (Fig. 23–15).

The concentration of peptide hormones within secretory granules is so high that the contents are virtually crystalline; when the contents of a granule are released by exocytosis, a large amount of hormone is released suddenly. The capillaries that serve peptide-producing endocrine glands are fenestrated (thinner than usual and permeable to peptides), so the secreted hormone molecules readily enter the bloodstream for transport to target cells elsewhere. All peptide hormones act from outside the target cell, through receptors in the plasma membrane. They cause the generation of a second messenger in the cytosol (cAMP, Ca^{2+}, etc.), which changes the activity of some intracellular enzyme, thereby altering the cell's metabolism.

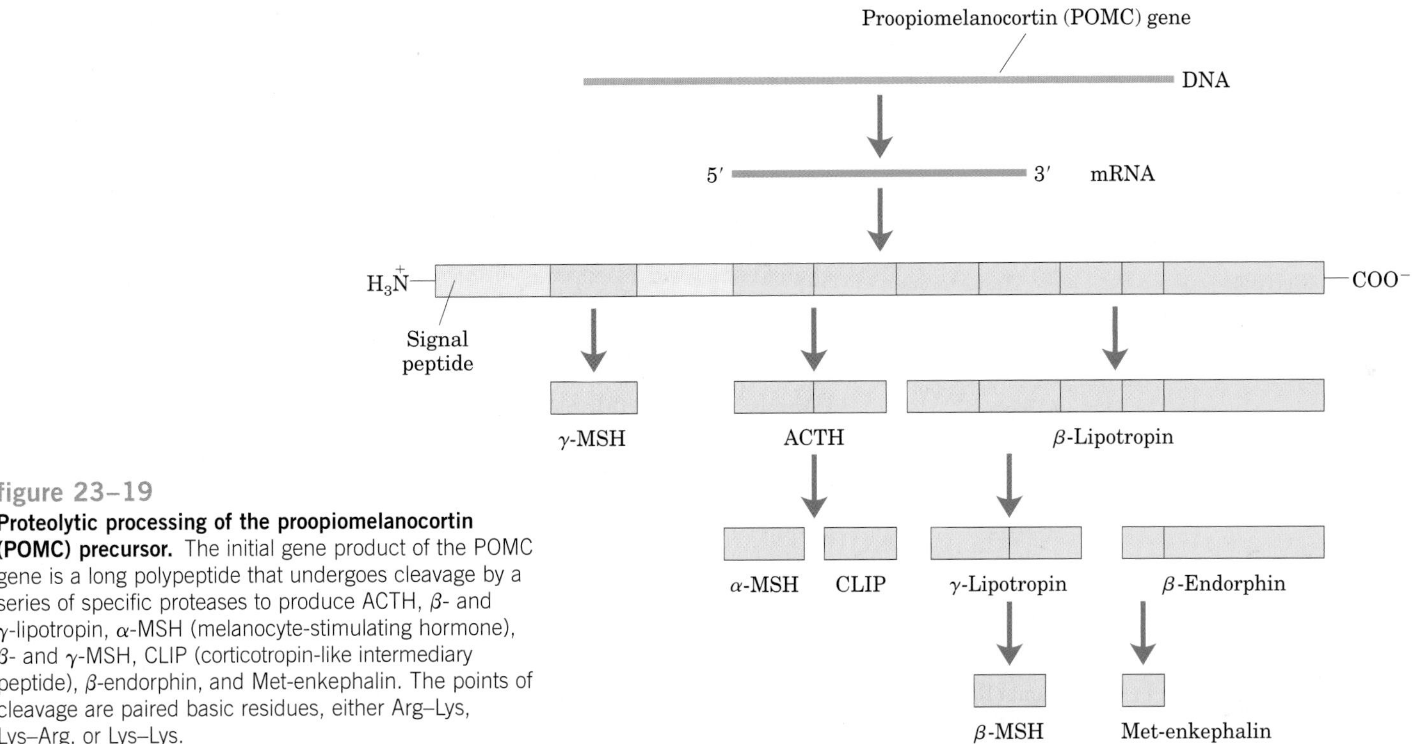

figure 23–19
Proteolytic processing of the proopiomelanocortin (POMC) precursor. The initial gene product of the POMC gene is a long polypeptide that undergoes cleavage by a series of specific proteases to produce ACTH, β- and γ-lipotropin, α-MSH (melanocyte-stimulating hormone), β- and γ-MSH, CLIP (corticotropin-like intermediary peptide), β-endorphin, and Met-enkephalin. The points of cleavage are paired basic residues, either Arg–Lys, Lys–Arg, or Lys–Lys.

Catecholamine Hormones The water-soluble compounds epinephrine and norepinephrine are **catecholamines,** synthesized from tyrosine and named for the structurally related compound catechol. Catecholamines made in the brain and other neural tissue function as neurotransmitters, but **epinephrine** (adrenaline) and **norepinephrine** (noradrenaline) are also made and secreted as hormones by the adrenal glands. Like the peptide hormones, catecholamines are highly concentrated within secretory vesicles and released by exocytosis, and they act through surface receptors to generate intracellular second messengers. They mediate a wide variety of physiological responses to acute stress (Table 23–2).

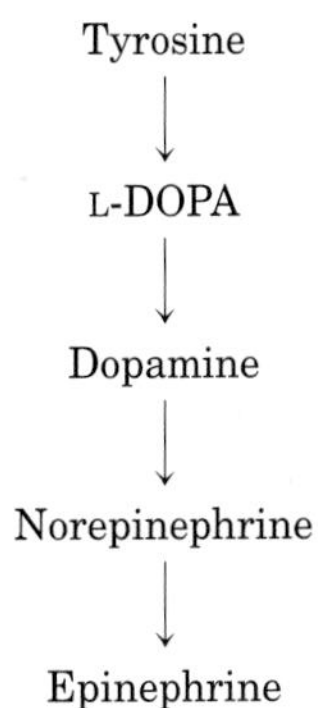

Eicosanoids The eicosanoid hormones (prostaglandins, thromboxanes, and leukotrienes) are derived from the 20-carbon polyunsaturated fatty acid arachidonate. They are not synthesized in advance and stored, but are made when needed from arachidonate that has been enzymatically released from membrane phospholipids by phospholipase A_2 (see Fig. 11–16). The enzymes of the pathway leading to prostaglandins and thromboxanes (see Fig. 21–16) are very widely distributed in mammalian tissues; most cells can produce these signals, and cells of many tissues can respond through specific receptors in the plasma membrane. The eicosanoid hormones are secreted into the interstitial fluid outside a cell (not primarily into the blood) and act in the paracrine fashion on nearby cells. Prostaglandins promote the contraction of smooth muscle, including that of the intestine and uterus (and can therefore be used to induce labor). They also mediate pain and inflammation in all tissues. Antiinflammatory drugs often act by inhibiting steps in the prostaglandin synthetic pathway (see Box 21–2). Thromboxanes regulate platelet function and therefore blood clotting. Leukotrienes LTC_4 and LTD_4 stimulate contraction of smooth muscle in the intestine, pulmonary airways, and trachea, and are mediators of the severe immune response called anaphylaxis.

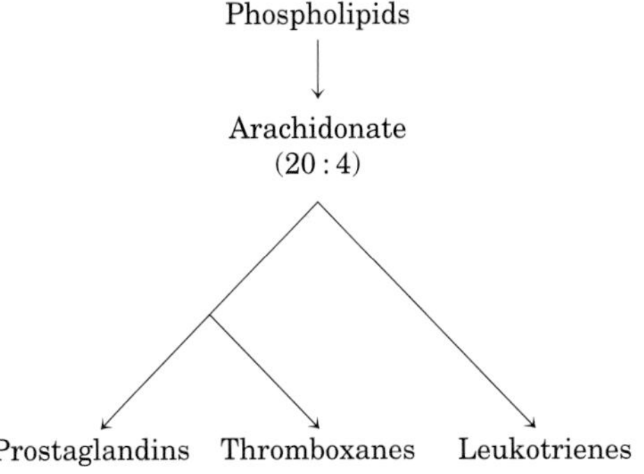

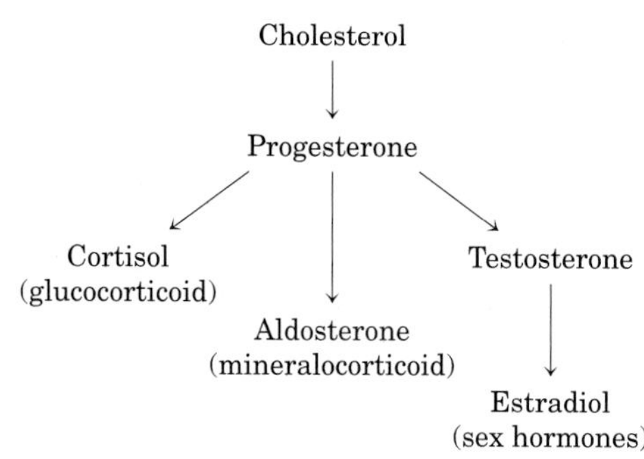

Steroid Hormones The steroid hormones (adrenocortical hormones and sex hormones) are synthesized from cholesterol in several endocrine tissues and carried to their target cells through the bloodstream, bound to carrier proteins. Over 50 corticosteroid hormones of two general types are produced in the adrenal cortex by reactions that remove the side chain from the D ring of cholesterol and introduce oxygen to form keto and hydroxyl groups. Many of these reactions involve cytochrome P-450 enzymes (see Box 21–1, p. 782). All steroid hormones act through nuclear receptors to change the level of expression of specific genes. Glucocorticoids (such as cortisol) primarily affect the metabolism of carbohydrates; mineralocorticoids (such as aldosterone) regulate the concentrations of electrolytes in the blood. Androgens (testosterone) and estrogens (such as estradiol; see Fig. 11–17) are synthesized in the testes and ovaries and, to some extent, in the adrenal cortex. Their synthesis also involves cytochrome P-450 enzymes that cleave the side chain of cholesterol and introduce oxygen atoms. These hormones affect sexual development, sexual behavior, and a variety of other reproductive and nonreproductive functions.

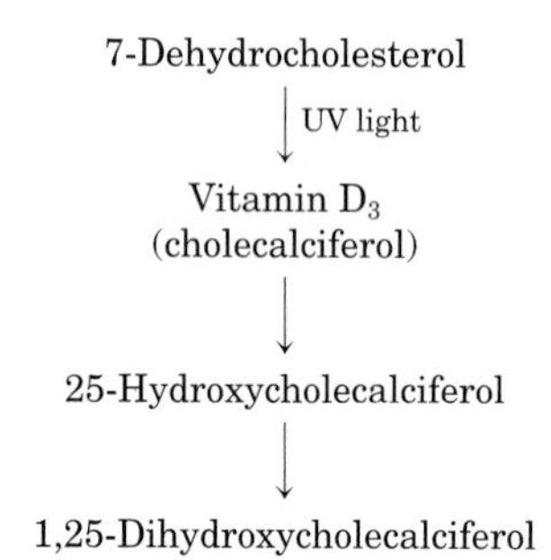

Vitamin D Hormone Calcitriol (1,25-dihydroxycholecalciferol) is produced from vitamin D by hydroxylating enzymes in the liver and kidneys (see Fig. 11–18a). Vitamin D itself is obtained in the diet or by photolysis (by sunlight) of 7-dehydrocholesterol in the skin. Calcitriol works in concert with parathyroid hormone in Ca^{2+} homeostasis, regulating the concentration of Ca^{2+} in the blood and the balance between Ca^{2+} deposition and Ca^{2+} mobilization from bone. Acting through nuclear receptors, calcitriol activates the synthesis of an intestinal Ca^{2+}-binding protein essential for uptake of dietary Ca^{2+}. Inadequate dietary vitamin D, or defects in the biosynthesis of calcitriol, result in serious diseases such as rickets, in which bones are weak and malformed (see Fig. 11–18b).

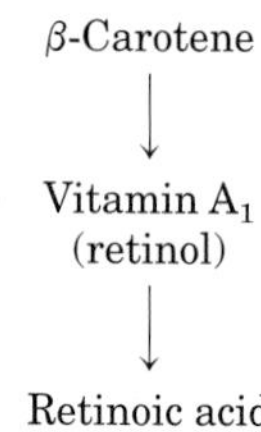

Retinoid Hormones Retinoids are potent hormones that regulate cell growth, survival, and differentiation via nuclear retinoid receptors. The prohormone retinol is synthesized primarily in liver from vitamin A (see Fig. 11–19), and many tissues convert retinol to the hormone retinoic acid (RA). All tissues are retinoid targets, as all cell types have at least one form of nuclear retinoid receptor. In adults, major targets include cornea, skin, epithelia of the lung and trachea, and the immune system. RA regulates the synthesis of proteins essential for growth or differentiation.

Thyroglobulin–Tyr
↓
Thyroglobulin–Tyr–I
(iodinated Tyr residues)
↓ proteolysis
Thyroxine (T_4)
and
triiodothyronine (T_3)

Thyroid Hormones The thyroid hormones T_4 (thyroxine) and T_3 (triiodothyronine) are synthesized in the thyroid gland from the precursor protein thyroglobulin (M_r 650,000). Two Tyr residues in thyroglobulin are enzymatically iodinated and covalently joined; proteolysis then releases free T_4 and T_3. The thyroid hormones stimulate energy-yielding metabolism, especially in liver and muscle, by activating the expression of genes encoding key catabolic enzymes.

Nitric Oxide (NO) Nitric oxide is a relatively stable free radical synthesized from molecular oxygen and the guanidino nitrogen of arginine (see Fig. 22–29) in a reaction catalyzed by **NO synthase.** This enzyme is found in many tissues and cell types: neurons, macrophages, hepatocytes, smooth muscle cells, endothelial cells of the blood vessels, and epithelial cells of the kidney. NO acts near its point of release, entering the target cell and activating the cytosolic enzyme guanylyl cyclase, which catalyzes the formation of the second messenger cGMP (see Fig. 13–9).

What Regulates the Regulators?

The levels of various hormones regulate specific cellular processes, but what regulates the level of each hormone? The brief answer is that the central nervous system receives input from many internal and external sensors—signals about danger, hunger, dietary intake, blood composition and pressure, for example—and orchestrates the production of appropriate hormonal signals by the several endocrine tissues of the body. For a more complete answer, we must look at the major hormone-producing systems of the human body and some of their functional interrelationships.

Figure 23–20 shows the anatomical location of the major endocrine glands in humans, and Figure 23–21 represents the "chain of command" in the hormonal signaling hierarchy. The **hypothalamus** of the brain (Fig. 23–22) is the coordination center of the endocrine system; it receives and integrates messages from the central nervous system. In response to these

Hypothalamus
Pituitary
Thyroid
Adrenals
Pancreas
Kidneys
Ovaries
(female)
Testes (male)

figure 23–20
The major endocrine glands (shaded red).

figure 23–21
The major endocrine systems and their target tissues. Signals originating in the central nervous system (top) are passed via a series of relays to the ultimate target tissues (bottom). In addition to the systems shown, the thymus, pineal glands, and groups of cells in the gastrointestinal tract also secrete hormones.

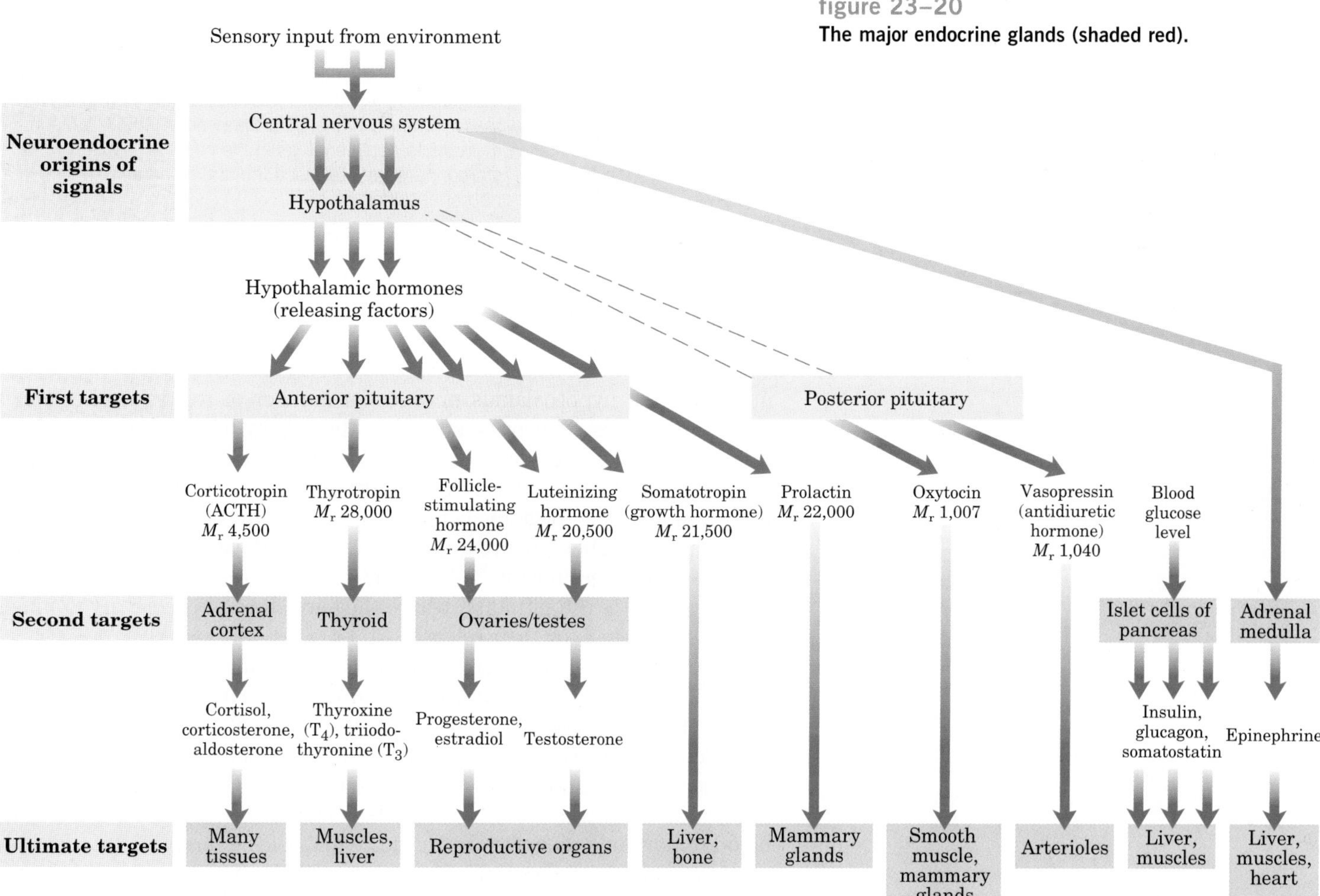

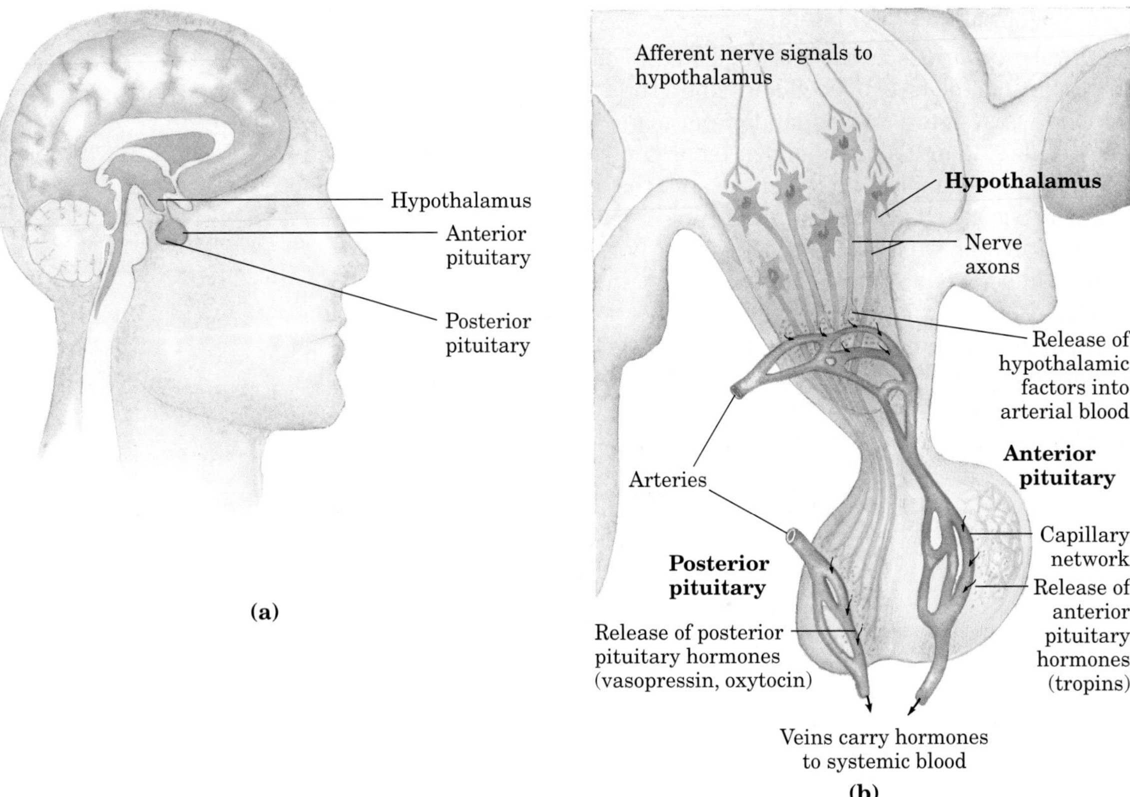

figure 23–22
Neuroendocrine origins of hormone signals. (a) Location of the hypothalamus and pituitary gland. **(b)** Details of the hypothalamus-pituitary system. Signals arriving from connecting neurons stimulate the hypothalamus to secrete releasing factors into a blood vessel that carries the hormones directly to a capillary network in the anterior pituitary. In response to each hypothalamic releasing factor, the anterior pituitary releases its appropriate hormone into the general circulation. Posterior pituitary hormones are made in neurons arising in the hypothalamus, transported in axons to nerve endings in the posterior pituitary, and stored there until released into the blood in response to a neuronal signal.

messages, the hypothalamus produces a number of regulatory hormones (releasing factors) that pass directly to the nearby pituitary gland through special blood vessels and neurons that connect the two glands (Fig. 23–22b). The pituitary gland has two functionally distinct parts. The **posterior pituitary** contains the axonal endings of many neurons that originate in the hypothalamus. These neurons produce the short peptide hormones oxytocin and vasopressin (Fig. 23–23), which then move down the axon to the nerve endings in the pituitary, where they are stored in secretory granules to await the signal for their release.

The **anterior pituitary** responds to hypothalamic hormones carried in the blood by producing **tropic hormones** or **tropins** (from the Greek *tropos,* meaning "turn"). These relatively long polypeptides activate the next rank of endocrine glands (Fig. 23–21), which includes the adrenal cortex, the thyroid gland, the ovaries, and the testes. These glands in turn are stimulated to secrete their specific hormones, which are carried by the blood to hormone receptors on or in the cells of the target tissues. For example, corticotropin-releasing hormone (CRH) from the hypothalamus stimulates the anterior pituitary to release ACTH, which goes to the adrenal

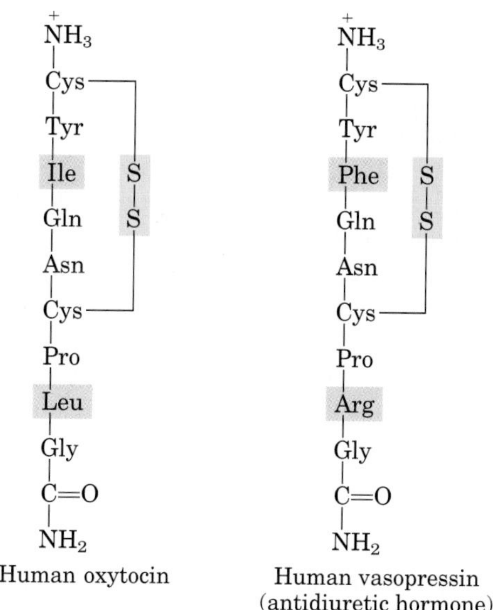

figure 23–23

Two hormones of the posterior pituitary gland. The carboxyl-terminal residues are glycinamide (—NH—CH_2—$CONH_2$); amidation of the carboxyl terminus is common in short peptide hormones. These two hormones, identical in all but two residues (shaded) have very different biological effects. Oxytocin acts on the smooth muscles of the uterus and mammary gland, causing uterine contractions during labor and promoting milk release during lactation. Vasopressin (VP, also called antidiuretic hormone, ADH) increases water reabsorption in the kidney and promotes the constriction of blood vessels, thereby increasing blood pressure.

cortex (the outer layer of the adrenal gland) and triggers the release of cortisol. Cortisol, the ultimate hormone in this cascade, acts through its receptors in many types of target cells to alter their metabolism. In hepatocytes, one of the actions of cortisol is to increase the rate of gluconeogenesis.

Hormonal cascades such as those responsible for the release of cortisol and epinephrine result in large amplifications of the initial signal, and allow for exquisitely fine tuning of the output of the ultimate hormone (Fig. 23–24). At each level in the cascade, a small signal elicits a larger response. The initial electrical signal to the hypothalamus results in the release of a few *nanograms* of corticotropin-releasing hormone, which elicits the release of a few *micrograms* of corticotropin. Corticotropin acts on the adrenal cortex to cause the release of *milligrams* of cortisol, for an overall amplification of at least a millionfold.

At each level of a hormonal cascade, there is the possibility of feedback inhibition of earlier steps in the cascade; an elevated level of the ultimate hormone or of one of the intermediate hormones inhibits release of the earlier hormones in the cascade from the hypothalamus or pituitary. These feedback mechanisms accomplish the same end as the allosteric mechanisms that limit the output of a biosynthetic pathway (compare Fig. 23–24 with Fig. 8–25): a product is made (or released) only until the necessary concentration is reached.

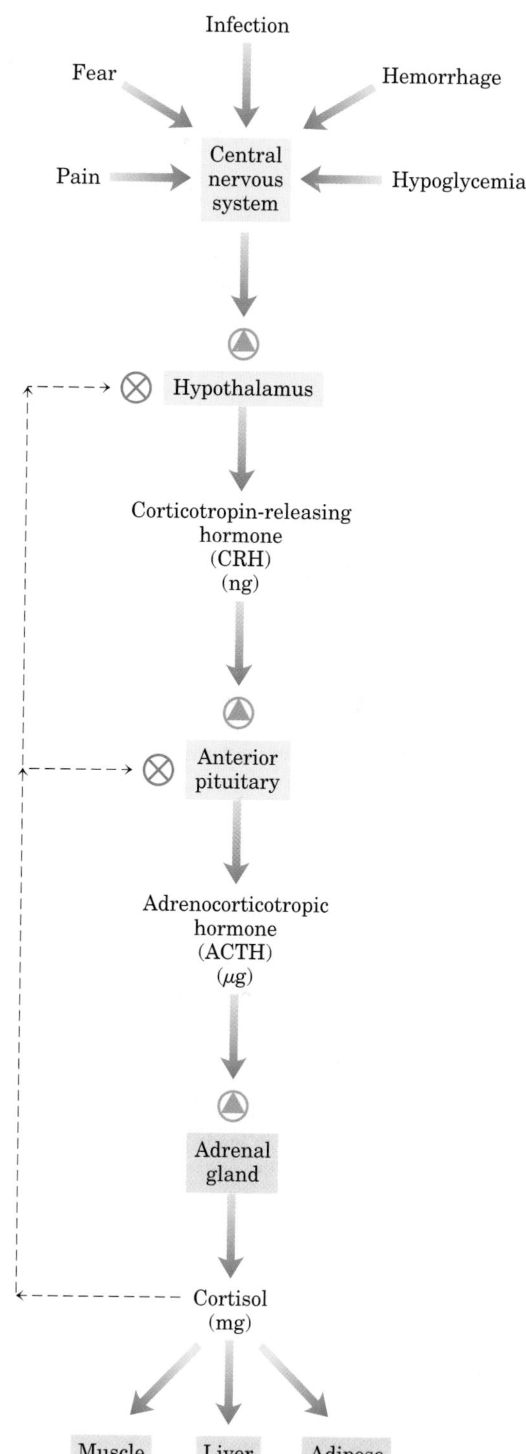

figure 23–24

Cascade of hormone release following central nervous system input to the hypothalamus. In each endocrine tissue along the pathway, a stimulus from the level above is received, amplified, and transduced into the release of the next hormone in the cascade. The cascade is sensitive to regulation at several levels by feedback inhibition by the ultimate hormone. The product therefore controls its own production, as in biosynthetic pathways within a single cell.

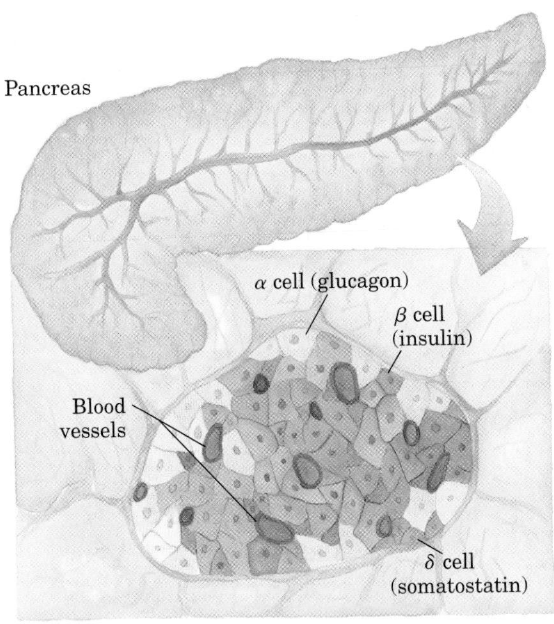

figure 23–25
The endocrine system of the pancreas. In addition to the exocrine or acinar cells (see Fig. 18–3b), which secrete digestive enzymes in the form of zymogens, the pancreas contains endocrine tissue consisting of the islets of Langerhans. The islets contain α, β, and δ cells (also known as A, B, and D cells, respectively), each of which secretes a specific polypeptide hormone.

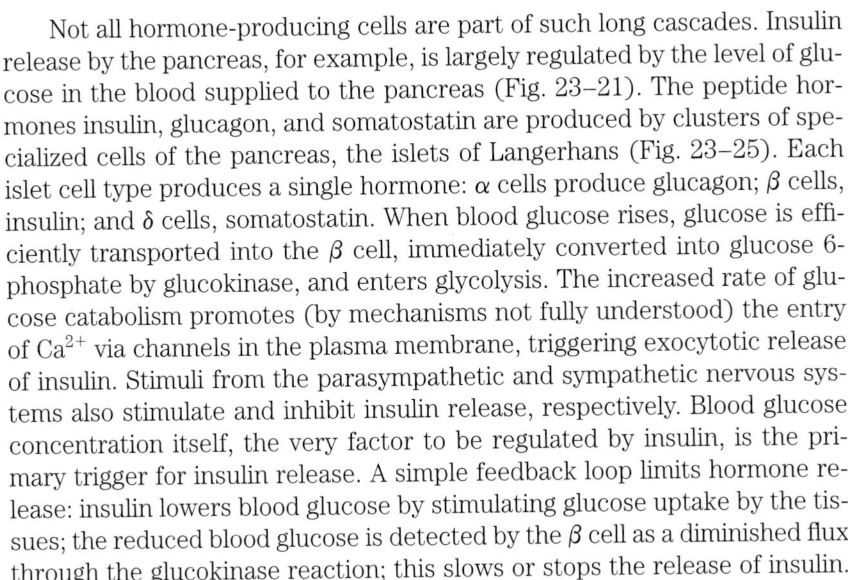

Not all hormone-producing cells are part of such long cascades. Insulin release by the pancreas, for example, is largely regulated by the level of glucose in the blood supplied to the pancreas (Fig. 23–21). The peptide hormones insulin, glucagon, and somatostatin are produced by clusters of specialized cells of the pancreas, the islets of Langerhans (Fig. 23–25). Each islet cell type produces a single hormone: α cells produce glucagon; β cells, insulin; and δ cells, somatostatin. When blood glucose rises, glucose is efficiently transported into the β cell, immediately converted into glucose 6-phosphate by glucokinase, and enters glycolysis. The increased rate of glucose catabolism promotes (by mechanisms not fully understood) the entry of Ca^{2+} via channels in the plasma membrane, triggering exocytotic release of insulin. Stimuli from the parasympathetic and sympathetic nervous systems also stimulate and inhibit insulin release, respectively. Blood glucose concentration itself, the very factor to be regulated by insulin, is the primary trigger for insulin release. A simple feedback loop limits hormone release: insulin lowers blood glucose by stimulating glucose uptake by the tissues; the reduced blood glucose is detected by the β cell as a diminished flux through the glucokinase reaction; this slows or stops the release of insulin.

Long-Term Regulation of Body Mass

The metabolic adjustments triggered by insulin, glucagon, and epinephrine occur on a short time scale—seconds or minutes. Other regulatory mechanisms act on a much longer time scale, controlling feeding and energy expenditure in a way that keeps the mammalian body in homeostasis. Between the ages of 25 and 55, the average American male gains 20 pounds, corresponding to the ingestion over this 30-year period of a few tenths of a percent more calories per day than are expended. This is remarkably precise regulation! Nevertheless, the slight imbalance in favor of weight gain may be health-threatening. Life expectancy drops when adipose tissue becomes too large a fraction of overall body mass. Consequently, there is great interest in understanding how body mass and the storage of fats in adipose tissue are regulated.

Leptin Was Predicted by the Lipostat Theory

The lipostat theory, advanced to account for the relative constancy of body weight, postulates a feedback mechanism that inhibits eating behavior and increases energy consumption whenever body weight exceeds a certain value (the set point); the inhibition is relieved when body weight drops below the set point. This theory predicted that a feedback signal originating in adipose tissue would influence the brain centers that control eating behavior and activity (metabolic and motor). Such a factor was discovered in 1994. **Leptin** (Greek *leptos,* thin), produced in adipocytes, is a small protein (167 amino acid residues) that moves through the blood to the brain, where it acts on the receptors in the hypothalamus to curtail appetite.

figure 23–26
A defect in leptin production leads to obesity. Both of these mice have defects in the obese gene, and are the same age. The mouse on the right was provided with purified leptin by daily injection and weighs 35 grams. The mouse on the left got no leptin, consequently ate more food and was less active, and weighs 67 grams.

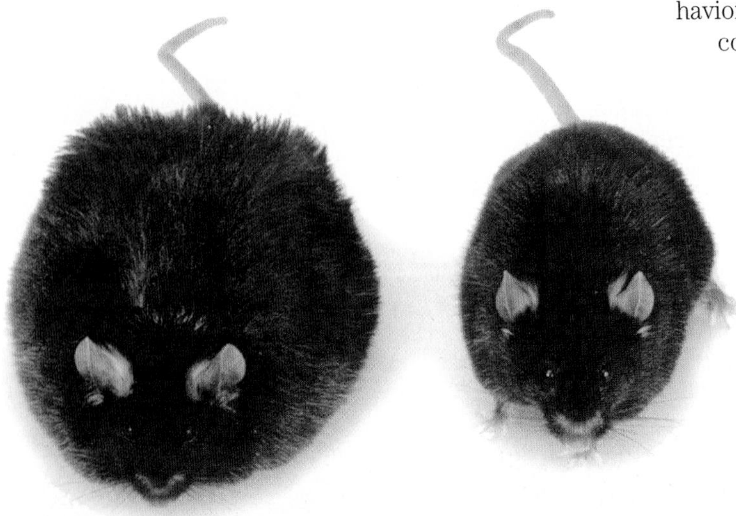

Leptin was identified as the product of a gene designated *OB* (obese) in laboratory mice. Mice with two defective copies (*ob/ob* genotype; lower-case letters signify a mutant form of the gene) show the behavior and physiology of animals in a constant state of starvation: their serum corticosterone levels are elevated; they are unable to stay warm, grow normally, reproduce, or restrain their appetites. As a consequence of the last, they become severely obese, weighing as much as three times more than normal mice (Fig. 23–26). They also have metabolic disturbances very similar to those of diabetic animals, and they are insulin-resistant. When leptin is injected into *ob/ob* mice, they lose weight and increase their locomotor activity and heat production. A second mouse gene, designated *DB* (diabetic), was also found to have a role in appetite regulation. Mice with two defective copies (*db/db*) are obese and diabetic. The *DB* gene was found to encode the **leptin receptor.** When the leptin receptor is defective, the signaling function of leptin is lost.

Leptin is produced only in adipocytes and, to a much lesser extent, in intestinal epithelium and placenta. The leptin receptor is expressed primarily in regions of the brain known to regulate feeding behavior, neurons of the arcuate, ventromedial, and dorsomedial nuclei of the hypothalamus (Fig. 23–27a). The receptor is also expressed in cells of the adrenal cortex and in pancreatic β cells, but at much lower levels.

Leptin carries the message that fat reserves are sufficient and encourages a reduction in fuel intake and an increased expenditure of energy. Leptin's interaction with its receptor in the hypothalamus alters the release of signals that affect appetite. Leptin also stimulates the sympathetic nervous system, increasing blood pressure, heart rate, and thermogenesis (heat production at the expense of metabolic energy) by uncoupling electron transfer from ATP synthesis in the mitochondria of adipose tissue (Fig. 23–27b).

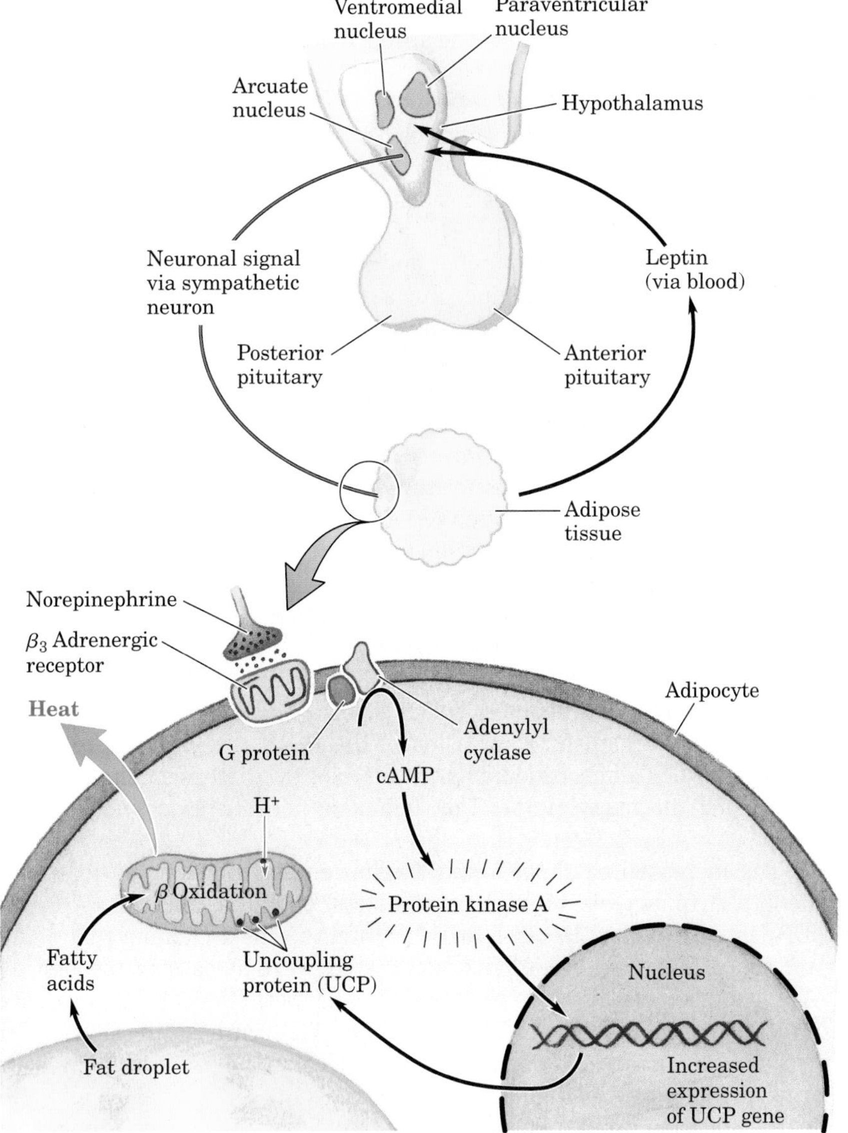

figure 23–27
The hypothalamus in regulation of food intake and energy expenditure. (a) Anatomy of the hypothalamus. **(b)** Interactions between hypothalamus and adipocyte.

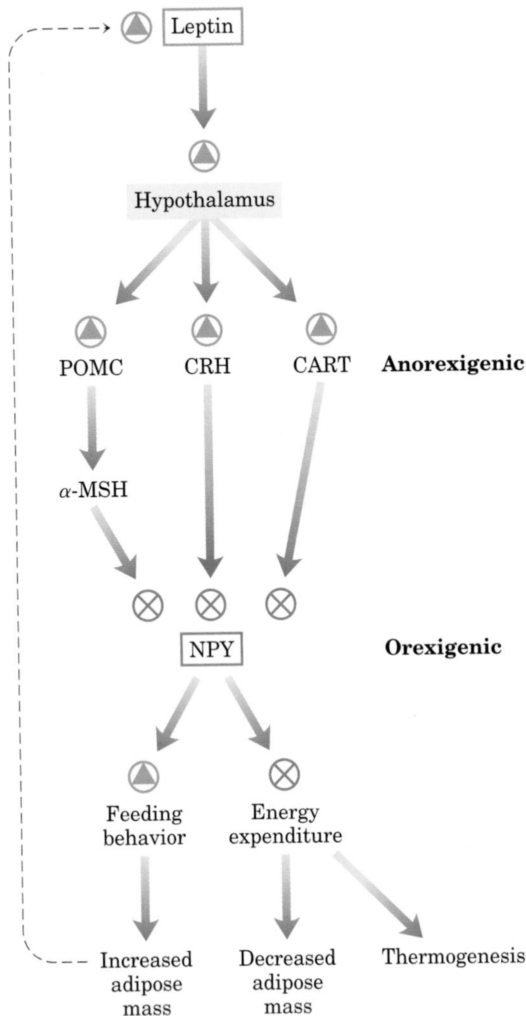

figure 23–28
The leptin cascade. An illustration of the role of the hypothalamus in eating behavior and metabolic activity.

Many Factors Regulate Feeding Behavior and Energy Expenditure

Leptin is not the only hormone that regulates feeding behavior or body weight. Insulin secretion reflects both the size of fat reserves (adiposity) and the current energy balance (blood glucose level). Insulin acts on its receptors in the hypothalamus to inhibit feeding; it also signals muscle, liver, and adipose tissues to increase catabolic reactions, including fat oxidation, which can produce weight loss. Several **anorexigenic** (appetite-suppressing) neuropeptides have been well characterized: α-melanocyte-stimulating hormone (α-MSH), produced in the hypothalamus from the polypeptide precursor proopiomelanocortin (POMC; see Fig. 23–19); corticotropin-releasing hormone (CRH), produced in the paraventricular nucleus; and the hypothalamic peptide CART (for *c*ocaine and *a*mphetamine-*r*egulated *t*ranscript) (Fig. 23–28). Adrenal glucocorticoids are clearly important as well; adrenalectomy (surgical removal of the adrenal gland) reverses or prevents all forms of obesity.

On the opposite side metabolically is neuropeptide Y (NPY), a peptide of 36 amino acid residues, produced in the arcuate nucleus of the hypothalamus. NPY is **orexigenic** (appetite-stimulating) and reduces thermogenesis. Its secretion and action are regulated by leptin and neuropeptides such as melanocortin, CRH, and glucagon-like peptide-1 (GLP-1). The blood level of NPY rises during starvation, and is elevated in both *ob/ob* and *db/db* mice; it presumably underlies the obesity of these mice, rising to levels that are dangerously high in the absence of control by the leptin system.

Leptin Triggers a Regulatory Cascade

The current model for leptin action is a cascade of regulatory events triggered by the interaction of leptin and its receptor and affecting the levels of several downstream hormones that either stimulate or inhibit feeding behavior and energy expenditure. The amount of leptin released by adipose tissue depends on both the number and the size of adipocytes.

The leptin signal is transduced by a mechanism also used by receptors for interferon and growth factors, the so-called JAK/STAT system (Fig. 23–29). The leptin receptor has a single transmembrane segment, and dimerizes when the ligand leptin binds to its extracellular domains. Both monomers of the dimeric receptor are phosphorylated on a Tyr residue of the intracellular domain by a *Ja*nus *k*inase (JAK). The phosphotyrosine residues become docking sites for three proteins that are *s*ignal *t*ransducers and *a*ctivators of *t*ranscription (STATs 3, 5, and 6, sometimes called fat-STATS). The docked STATS are then phosphorylated on Tyr residues by the same Janus kinase (hence the name Janus—like the mythological figure, the kinase has two "faces"). After phosphorylation by JAK, the STATs move to the nucleus, where they bind to specific DNA sequences and stimulate the expression of specific target genes. Expression of the genes for NPY, CRH, and the POMC precursor (which yields α-MSH) is regulated by leptin through this mechanism. Expression of the POMC gene is also regulated (suppressed) by glucocorticoid hormones; clearly, there is a complex interplay among the many factors that influence neuropeptide synthesis and action.

The increased catabolism and thermogenesis triggered by leptin are due in part to increased synthesis of the mitochondrial uncoupling protein UCP-1 in adipocytes. Recall that this protein forms a channel that allows protons to reenter the mitochondrial matrix without passing through the ATP synthase complex (see Fig. 19–28). This permits continual oxidation of fuel (fatty acids in an adipocyte) without ATP synthesis, dissipating energy as heat, potentially consuming dietary calories or stored fats in very large amounts. Leptin stimulates the synthesis of UCP-1 by altering synaptic transmission from neurons in the arcuate nucleus, hyperpolarizing

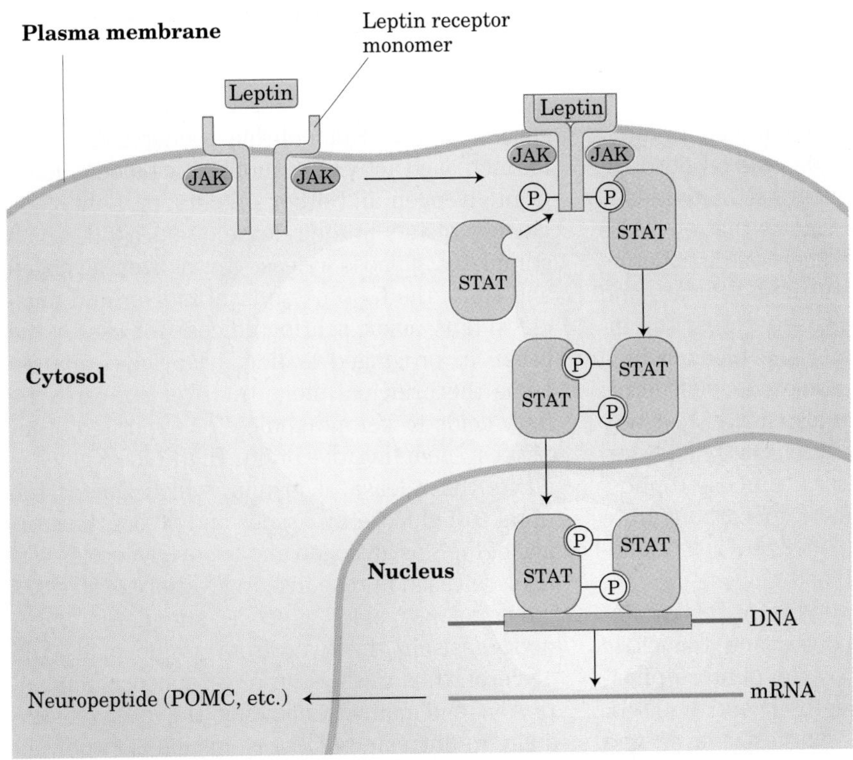

figure 23–29
The JAK-STAT mechanism by which the leptin signal is transduced in the hypothalamus. Leptin binding induces dimerization of the leptin receptor, followed by phosphorylation of Tyr residues of the receptor, catalyzed by Janus kinase (JAK). STATs that bind to the phosphorylated leptin receptor are then phosphorylated on Tyr residues by a separate activity of JAK. The STATs dimerize, binding each other's Ⓟ-Tyr residues, and enter the nucleus, where they bind specific regulatory regions in the DNA and alter expression of specific genes. The products of these genes ultimately influence feeding behavior and energy expenditure.

certain hypothalamic neurons. By stimulating sympathetic neurons originating in the hypothalamus, leptin causes increased release of norepinephrine at the synapse with adipocytes. Norepinephrine acts through β_3-adrenergic receptors to stimulate transcription of the gene for UCP; the resulting uncoupling of electron transfer from oxidative phosphorylation is thermogenic (Fig. 23–27).

Might human obesity be the result of insufficient leptin production, and therefore treatable by the injection of leptin? Blood levels of leptin are in fact usually much *higher* in obese animals (including humans) than in animals of normal body mass (except of course, in *ob/ob* animals, which cannot make leptin). In the very rare cases of extreme obesity in humans where the leptin gene has been found to be defective, leptin injection has resulted in dramatic weight loss. In animals with intact *OB* genes, however, the level of leptin rises with the amount of adipose tissue. In clinical trials, the injection of leptin did not have the dramatic weight-reducing effect that it does on *ob/ob* obese mice in the laboratory. Other possible explanations for morbid obesity include defects in leptin reception and signal transduction, or in the interaction of the leptin system with other systems involved in the maintenance of body mass. All these possibilities are being investigated.

The Leptin System May Have Evolved to Regulate the Starvation Response

Although much of the interest in leptin results from its possible role in preventing obesity, the leptin system probably evolved to adjust an animal's activity and metabolism during periods of fasting and starvation. The *reduction* in leptin level triggered by nutritional deficiency triggers decreased production of thyroid hormone (slowing basal metabolism), decreased production of sex hormones (preventing reproduction), and increased production of glucocorticoids (mobilizing the body's fuel-generating resources). By minimizing energy expenditures and maximizing the use of endogenous reserves of energy, these leptin-mediated responses may allow an animal to survive periods of severe nutritional deprivation.

summary

In mammals there is a division of metabolic labor among specialized tissues and organs. Coordination of the body's diverse metabolic activities is accomplished by hormonal signals that circulate in the blood. The liver is the central distributing and processing organ for nutrients. Sugars and amino acids produced in digestion cross the intestinal epithelium and enter the blood, which carries them to the liver. Some triacylglycerols derived from ingested lipids also make their way to the liver, where the constituent fatty acids are used in a variety of processes. Glucose 6-phosphate is the key intermediate in carbohydrate metabolism. It may be polymerized into glycogen, dephosphorylated to blood glucose, or converted to fatty acids via acetyl-CoA. It may undergo degradation by glycolysis and the citric acid cycle to yield ATP or by the pentose phosphate pathway to yield pentoses and NADPH. Amino acids are used to synthesize liver and plasma proteins, or their carbon skeletons may be converted into glucose and glycogen by gluconeogenesis; the ammonia formed by their deamination is converted into urea. Fatty acids may be converted by the liver into other triacylglycerols, cholesterol, or plasma lipoproteins for transport to and storage in adipose tissue. They may also be oxidized to yield ATP, and to form ketone bodies to be circulated to other tissues.

Skeletal muscle is specialized to produce ATP for mechanical work. During strenuous muscular activity, glycogen is the ultimate fuel and is fermented into lactate, supplying ATP. During recovery the lactate is reconverted (through gluconeogenesis) to glycogen and glucose in the liver. Phosphocreatine is an immediate source of ATP during active contraction. Heart muscle obtains all of its ATP from oxidative phosphorylation. The brain uses only glucose and β-hydroxybutyrate as fuels, the latter being important during fasting or starvation. The brain uses most of its ATP for the active transport of Na^+ and K^+ and the maintenance of the electrical potential of neuronal membranes. The blood links all of the organs, carrying nutrients, waste products, and hormonal signals between them.

The concentration of glucose in the blood is hormonally regulated. Fluctuations in blood glucose (which is normally 60 to 90 mg/100 mL or about 4.5 mM) due to dietary intake or vigorous exercise are counterbalanced by a variety of hormonally triggered changes in the metabolism of several organs. Epinephrine prepares the body for increased activity by mobilizing blood glucose from glycogen and other precursors. Low blood glucose results in the release of glucagon, which stimulates glucose release from liver glycogen and shifts the fuel metabolism in liver and muscle to fatty acids, sparing glucose for use by the brain. In prolonged fasting, triacylglycerols become the principal fuels; the liver converts the fatty acids to ketone bodies for export to other tissues, including the brain. High blood glucose elicits the release of insulin, which speeds the uptake of glucose by tissues and favors the storage of fuels as glycogen and triacylglycerols. Cortisol, released in response to a variety of stresses including low blood glucose, stimulates gluconeogenesis in the liver from amino acids and glycerol, thus raising the blood glucose concentration and counterbalancing the effect of insulin. In untreated diabetes, insulin is either not produced or is not recognized by the tissues, and the utilization of blood glucose is compromised. When blood glucose levels are high, glucose is excreted intact into the urine. Tissues then depend upon fatty acids for fuel (producing ketone bodies) and degrade cellular proteins to make glucose from their glucogenic amino acids. Untreated diabetes is characterized by high glucose levels in the blood and urine and the production and excretion of ketone bodies.

Hormones are chemical messengers secreted by certain tissues into the blood or institial fluid, serving to regulate the activity of other tissues. Radioimmunoassay (RIA) and ELISA are two very sensitive techniques for detecting and quantifying hormones. Hormonal cascades, in which catalysts activate catalysts, amplify the initial stimulus by orders of magnitude, often over very short time scales (seconds). Nerve impulses stimulate the hypothalamus to send specific hormones to the pituitary gland, stimulating (or inhibiting) the release of tropic hormones. The anterior pituitary hormones in turn stimulate other endocrine glands (thyroid, adrenals, pancreas) to secrete their characteristic hormones, which in turn stimulate specific target tissues. Peptide, amine, and eicosanoid hormones act from outside the target cell through specific receptors that alter the level of an intracellular second messenger, such as cAMP. Steroid, vitamin D, retinoid, and thyroid hormones enter target cells and alter gene expression by interacting with specific nuclear receptors.

Adipose tissue produces a hormone, leptin, that regulates feeding behavior and energy expenditure to maintain nearly constant body mass. Leptin production and release rises with the number and size of adipocytes. Leptin acts on receptors in the hypothalamus to cause the synthesis and release of several neuropeptides that modulate food intake and metabolic activity. These include the anorexigenic peptides α-MSH, CRH, and CART. These peptides and others influence the action of the orexigenic (appetite-stimulating) neuropeptide Y (NPY). Leptin also stimulates sympathetic nervous system action on adipocytes, leading to uncoupling of mitochondrial oxidative phosphorylation with consequent thermogenesis. The signal-transduction mechanism for leptin involves phosphorylation of the receptor and of signal-transducing activators of transcription (STATs) by a Janus kinase (JAK). Phosphorylation allows the STATs to bind regulatory regions in nuclear DNA and alter the rates of transcription of genes that encode proteins that set the level of metabolic activity and determine feeding behavior. Leptin release by adipose tissue is reduced during periods of starvation; the resulting changes in neuropeptide release lead to reduced locomotor activity and thermogenesis and curtail reproduction, sparing fuel reserves for functions essential to survival.

further reading

General Background and History

Crapo, L. (1985) *Hormones: The Messengers of Life,* W.H. Freeman and Company, New York.

Short, entertaining account of the history and results of hormone research.

Litwack, G. & Norman, A.W. (1997) *Hormones,* 2nd edn, Academic Press, Orlando.

An excellent, authoritative, well-illustrated, clear introduction to hormone structure, synthesis, and action.

Yalow, R.S. (1978) Radioimmunoassay: a probe for the fine structure of biologic systems. *Science* **200,** 1236–1245.

History of the development of radioimmunoassays; the author's Nobel lecture.

Tissue-Specific Metabolism

Arias, I.M., Boyer, J.L., & Fausto, N. (1994) *The Liver: Biology and Pathobiology,* 3rd edn, Raven Press, New York.

Advanced-level text; includes chapters on the metabolism of carbohydrates, fats, and proteins in the liver.

Nosadini, R., Avogaro, A., Doria, A., Fioretto, P., Trevisan, R., & Morocutti, A. (1989) Ketone body metabolism: a physiological and clinical overview. *Diabetes Metab. Rev.* **5,** 299–319.

Randle, P.J. (1995) Metabolic fuel selection: General integration at the whole-body level. *Proc. Nutr. Soc.* **54,** 317–327.

Hormonal Regulation of Fuel Metabolism

Attie, A.D. & Raines, R.T. (1995) Analysis of receptor-ligand interactions. *J. Chem. Ed.* **72,** 119–123.

Elia, M. (1995) General integration and regulation of metabolism at the organ level. *Proc. Nutr. Soc.* **54,** 213–234.

Harris, R.A. & Crabb, D.W. (1997) Metabolic interrelationships. In *Textbook of Biochemistry with Clinical Correlations,* 4th edn (Devlin, T.M., ed.), pp. 525–562, John Wiley & Sons, Inc., New York.

Description of the metabolic interplay among human tissues during normal metabolism, and the effect on tissue-specific energy metabolism of the stresses of exercise, lactation, diabetes, and renal disease.

Holloszy, J.O. & Kohrt, W.M. (1996) Regulation of carbohydrate and fat metabolism during and after exercise. *Annu. Rev. Nutr.* **16,** 121–138.

Pilkis, S.J. & Claus, T.H. (1991) Hepatic gluconeogenesis/glycolysis: regulation and structure/function relationships of substrate cycle enzymes. *Annu. Rev. Nutr.* **11,** 465–515.

Review at the advanced level.

Snyder, S.H. (1985) The molecular basis of communication between cells. *Sci. Am.* **253** (October), 132–141.

Introductory-level discussion of the human endocrine system.

Zammit, V.A. (1996) Role of insulin in hepatic fatty acid partitioning: emerging concepts. *Biochem. J.* **314,** 1–14.

Hormone Structure and Function

Litwack, G. & Schmidt, R.J. (1997) Biochemistry of hormones I: polypeptide hormones. In *Textbook of Biochemistry with Clinical Correlations,* 4th edn (Devlin, T.M., ed.), pp. 839–891. John Wiley & Sons, Inc., New York.

Litwack, G. & Schmidt, R.J. (1997) Biochemistry of hormones II: steroid hormones. In *Textbook of Biochemistry with Clinical Correlations,* 4th edn (Devlin, T.M., ed.), pp. 893–918. John Wiley & Sons, Inc., New York.

Leptin and the Control of Body Mass

Auwerx, J. & Staels, B. (1998) Leptin. *Lancet* **351,** 737–742.

Brief overview of the leptin system and JAK-STAT signal transductions.

Elmquist, J.K., Maratos-Flier, E., Saper, C.B., & Flier, J.S. (1998) Unraveling the central nervous system pathways underlying responses to leptin. *Nat. Neurosci.* **1,** 445–450.

Short, excellent review.

Flier, J.S. & Maratos-Flier, E. (1998) Obesity and the hypothalamus: novel peptides for new pathways. *Cell* **92,** 437–440.

Freake, H.C. (1998) Uncoupling proteins: beyond brown adipose tissue. *Nutr. Rev.* **56,** 185–189.

Review of the structure, function, and role of uncoupling proteins.

Friedman, J.M. & Halaas, J.L. (1998) Leptin and the regulation of body weight in mammals. *Nature* **395,** 763–770.

Inui, A. (1999) Feeding and body-weight regulation by hypothalamic neuropeptides—mediation of the actions of leptin. *Trends Neurosci.* **22,** 62–67.

Short, intermediate-level review of the leptin system.

Jequier, E. & Tappy, L. (1999) Regulation of body weight in humans. *Physiol. Rev.* **79,** 451–480.

Detailed review of leptin in body weight regulation, the control of food intake, and the roles of white and brown adipose tissues in energy expenditure.

Wood, S.C., Seeley, R.J., Porte, D., Jr., & Schwartz, M.W. (1998) Signals that regulate food intake and energy homeostasis. *Science* **280,** 1378–1383.

Review of the roles of leptin, insulin, and other neuropeptides in the regulation of feeding and catabolism.

problems

1. ATP and Phosphocreatine as Sources of Energy for Muscle In contracting skeletal muscle, the concentration of phosphocreatine drops while the concentration of ATP remains fairly constant. However, in a classic experiment, Robert Davies found that if he first treated muscle with 1-fluoro-2,4-dinitrobenzene (see p. 141), the concentration of ATP declined rapidly, while the concentration of phosphocreatine remained unchanged during a series of contractions. Suggest an explanation.

2. Metabolism of Glutamate in the Brain Glutamate that moves from the blood into the brain is transformed there into glutamine, which is then released into the blood. What is accomplished by this metabolic conversion? How does it take place? Actually, the brain can generate more glutamine than can be made from the glutamate entering from the blood. How does this extra glutamine arise? (Hint: You may want to review amino acid catabolism in Chapter 18; recall that NH_3 is very toxic to the brain.)

3. Absence of Glycerol Kinase in Adipose Tissue Glycerol 3-phosphate is a key intermediate in the biosynthesis of triacylglycerols. Adipocytes, specialized for the synthesis and degradation of triacylglycerols, cannot use glycerol directly because they lack glycerol kinase, which catalyzes the reaction

$$\text{Glycerol} + \text{ATP} \longrightarrow \text{glycerol 3-phosphate} + \text{ADP}$$

How does adipose tissue obtain the glycerol 3-phosphate necessary for triacylglycerol synthesis? Explain.

4. Oxygen Consumption during Exercise A sedentary adult consumes about 0.05 L of O_2 during a 10 s period. A sprinter, running a 100 m race, consumes about 1 L of O_2 during the same time period. After finishing the race, the sprinter will continue to breathe at an elevated but declining rate for some minutes, consuming an extra 4 L of O_2 above the amount consumed by the sedentary individual.

(a) Why do the O_2 needs increase dramatically during the sprint?

(b) Why do the O_2 demands remain high after the sprint has been completed?

5. Thiamine Deficiency and Brain Function Individuals with thiamine deficiency display a number of characteristic neurological signs: loss of reflexes, anxiety, and mental confusion. Why might thiamine deficiency be manifested by changes in brain function?

6. Significance of Hormone Concentration Under normal conditions, the human adrenal medulla secretes epinephrine ($C_9H_{13}NO_3$) at a rate sufficient to maintain a concentration of 10^{-10} M in the circulating blood. To appreciate what that concentration means, calculate the diameter of a round swimming pool, with a water depth of 2 m, that would be needed to dissolve 1 g (about 1 teaspoon) of epinephrine to a concentration equal to that in blood.

7. Regulation of Hormone Levels in the Blood The half-life of most hormones in the blood is relatively short. For example, if radioactively labeled insulin is injected into an animal, half the hormone has disappeared from the blood within 30 min.

(a) What is the importance of the relatively rapid inactivation of circulating hormones?

(b) In view of this rapid inactivation, how can the circulating hormone level be kept constant under normal conditions?

(c) In what ways can the organism make possible rapid changes in the level of circulating hormones?

8. Water-Soluble versus Lipid-Soluble Hormones On the basis of their physical properties, hormones fall into one of two categories: those that are very soluble in water but relatively insoluble in lipids (e.g., epinephrine) and those that are relatively insoluble in water but highly soluble in lipids (e.g., steroid hormones). In their role as regulators of cellular activity, most water-soluble hormones do not penetrate into the interior of their target cells. The lipid-soluble hormones, by contrast, do penetrate into their target cells and ultimately act in the nucleus. What is the correlation between solubility, the location of receptors, and the mode of action of the two classes of hormones?

9. Metabolic Differences in Muscle and Liver in a "Fight or Flight" Situation During a "fight or flight" situation, the release of epinephrine promotes glycogen breakdown in the liver, heart, and skeletal muscle. The end product of glycogen breakdown in the liver is glucose. In contrast, the end product in skeletal muscle is pyruvate.

(a) Why are different products of glycogen breakdown observed in the two tissues?

(b) What is the advantage to the organism during a "fight or flight" condition of having these specific glycogen breakdown routes?

10. Excessive Amounts of Insulin Secretion: Hyperinsulinism Certain malignant tumors of the pancreas cause excessive production of insulin by the β cells. Affected individuals exhibit shaking and trembling, weakness and fatigue, sweating, and hunger.

(a) What is the effect of hyperinsulinism on the metabolism of carbohydrate, amino acids, and lipids by the liver?

(b) What are the causes of the observed symptoms? Suggest why this condition, if prolonged, leads to brain damage.

11. Thermogenesis Caused by Thyroid Hormones Thyroid hormones are intimately involved in regulating the basal metabolic rate. Liver tissue of animals given excess thyroxine shows an increased rate of O_2 consumption and increased heat output (thermogenesis), but the ATP concentration in the tissue is normal. Different explanations have been offered for the thermogenic effect of thyroxine. One is that excess thyroid hormone causes uncoupling of oxidative phosphorylation in mitochondria. How could such an effect account for the observations? Another explanation suggests that the thermogenesis is due to an increased rate of ATP utilization by the thyroxine-stimulated tissue. Is this a reasonable explanation? Why?

12. Function of Prohormones What are the possible advantages in the synthesis of hormones as prohormones?

13. Sources of Glucose in the Human Body The typical human adult uses about 160 g of glucose per day, with 120 g being used by the brain. The available reserve of glucose is adequate for about one day (about 20 g of circulating glucose and 190 g of glycogen). After the reserve becomes depleted, how would a starved body obtain more glucose?

14. Parabiotic *ob/ob* mice By careful surgery, it is possible to connect the circulatory systems of two mice so that the same blood circulates through both animals. In these **parabiotic** mice, products released into the blood by one animal reaches the other animal via the shared circulation. Both animals are free to eat independently. When an *ob/ob* mouse (both copies of *OB* gene are defective) and a normal *OB/OB* mouse (two good copies of the *OB* gene) are made parabiotic, what will happen to the weight of each mouse?

part

IV

Information Pathways

The fourth and final part of this book explores the biochemical mechanisms underlying the apparently contradictory requirements for both genetic continuity and the evolution of living organisms. What is the molecular nature of genetic material? How is genetic information transmitted from one generation to the next with high fidelity? How do the rare changes in genetic material that are the raw material of evolution arise? How is genetic information ultimately expressed in the amino acid sequences of the astonishing variety of protein molecules in a living cell?

The fundamental unit of information in living systems is the **gene.** A gene is defined biochemically as a segment of DNA (or, in a few cases, RNA) that encodes the information required to produce a functional biological product. The final product is usually a protein, so much of the material that follows concerns genes that encode proteins. A functional gene product might also be one of several classes of RNA molecules. The storage, maintenance, and metabolism of these informational units form the focal points of our discussion.

Modern biochemical research on gene structure and function has brought to biology a revolution comparable to that stimulated by the publication of Darwin's theory on the origin of species nearly 150 years ago. An understanding of how information is stored and used in cells has brought penetrating new insights to some of the most fundamental questions about cellular structure and function. A comprehensive conceptual framework for biochemistry is now unfolding.

Today's understanding of information pathways has arisen from the convergence of genetics, physics, and chemistry in modern biochemistry. This was epitomized by the discovery of the double-helical structure of DNA, postulated by James Watson and Francis Crick in 1953 (see Fig. 10–15). Genetic theory contributed the concept of coding by genes. Physics permitted the determination of molecular structure by x-ray diffraction analysis. Chemistry revealed the composition of DNA. The profound impact of the Watson-Crick hypothesis arose from its ability to account for a wide range of observations derived from studies in these diverse disciplines.

facing page

An enlarged image of a DNA microarray. The spots of the microarray each contain DNA from one of the 6,200 genes present in the chromosomes of the yeast *Saccharomyces cerevisia,* with every yeast gene represented in the array. The microarray has been probed with fluorescently labeled nucleic acid derived from the mRNAs present in the cell when it is (1) growing normally in culture and (2) five hours after the cells have begun to form spores. The spots glowing green represent genes that are expressed at relatively higher levels in cells growing normally, while those glowing red represent genes expressed at higher levels during sporulation. The spots that are yellow represent genes that do not change their levels of expression during sporulation. This image is enlarged; the microarray actually measures only 1.8 by 1.8 cm. Microarrays like this are described in Chapter 29. They provide a snapshot of global gene expression in a cell; they also dramatically illustrate the progress made in recent decades in understanding cellular information pathways.

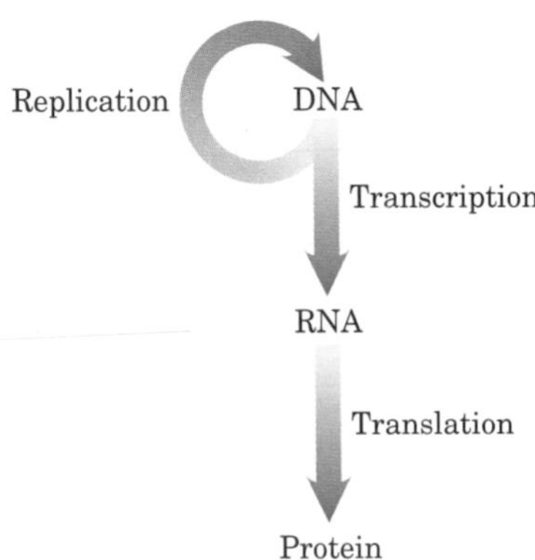

The central dogma of molecular biology, showing the general pathways of information flow via replication, transcription, and translation. The term "dogma" is a misnomer. Introduced by Francis Crick at a time when little evidence supported these ideas, the dogma has become a well-established principle.

This revolution in our understanding of the structure of DNA inevitably stimulated questions about its function. The double-helical structure itself clearly suggested how DNA might be copied so that the information it contains can be transmitted from one generation to the next. Clarification of how the information in DNA is converted into functional proteins came with the discovery of both messenger RNA and transfer RNA and with the deciphering of the genetic code.

These and other major advances gave rise to the central dogma of molecular biology comprising the three major processes in the cellular utilization of genetic information. The first is **replication,** the copying of parental DNA to form daughter DNA molecules with identical nucleotide sequences. The second is **transcription,** the process by which parts of the genetic message encoded in DNA are copied precisely into RNA. The third is **translation,** whereby the genetic message encoded in messenger RNA is translated on the ribosomes into a polypeptide with a particular sequence of amino acids.

Part IV explores these and related processes. In Chapter 24 we examine the structure, topology, and packaging of chromosomes and genes. The processes underlying the central dogma are elaborated in Chapters 25 through 27. Then, as for the biosynthetic pathways, we turn to regulation, examining how the expression of genetic information is controlled (Chapter 28).

A major theme running through these chapters is the added complexity inherent in the biosynthesis of macromolecules that contain information. Assembling nucleic acids and proteins with particular sequences of nucleotides and amino acids, respectively, represents nothing less than preserving the faithful expression of the template upon which life itself is based. The formation of phosphodiester bonds in DNA or peptide bonds in proteins might be expected to be a trivial feat for cells, given the arsenal of enzymatic and chemical tools described in Part III of this book. However, the framework of patterns and rules established in our examination of metabolic pathways thus far must be enlarged considerably when molecular information has to be taken into account. Bonds must be formed between *particular* subunits in informational biopolymers, avoiding either the occurrence or the persistence of sequence errors. This has an enormous impact on the thermodynamics, chemistry, and enzymology of the biosynthetic processes. Formation of a peptide bond requires an input of only about 21 kJ/mol of bonds and can be catalyzed by relatively simple enzymes. In contrast, to synthesize a bond between two specific amino acids at a particular point in a polypeptide, the cell invests about 125 kJ/mol in chemical energy while making use of over 200 enzymes, RNA molecules, and specialized proteins. Information is expensive.

The dynamic interaction between nucleic acids and proteins is another central theme of Part IV. With the important exception of a few catalytic RNA molecules (discussed in Chapter 26), the processes that make up the pathways of cellular information flow are catalyzed and regulated by proteins. An understanding of these enzymes and other proteins can have practical as well as intellectual rewards because they form the basis of recombinant DNA technology. Developments in this field are making possible the prenatal diagnosis of genetic disease; the production of a wide range of potent new pharmaceutical agents; the sequencing of the entire human genome; human gene therapy; the introduction of new traits into bacteria, plants, and animals for industry and agriculture; and many other advances. We finish our tour of the information pathways, and indeed the entire book, in Chapter 29 by examining this technology and its implications for the future.

chapter

24

Genes and Chromosomes

Every cell of a multicellular organism generally contains the same complement of genetic material. Just look at a human being for a hint of the wealth of information contained in each human cell. DNA molecules are the largest macromolecules in the cell and are commonly packaged into structures called **chromosomes.** Most bacteria and viruses have a single chromosome, whereas eukaryotic cells usually contain many. A single chromosome may carry thousands of genes. Together, all of a cell's genes and intergenic DNA (the DNA between genes) form the cellular **genome.**

Measurements carried out in the 1950s suggested that the largest DNAs had a molecular mass of 10^6 daltons or less, equivalent to about 15,000 base pairs. Improved methods for isolation of native DNAs revealed that these initial findings were deceptively low. For example, the 16 chromosomes in the relatively small genome of the yeast *Saccharomyces cerevisiae* have molecular masses ranging from 1.5×10^8 to 1×10^9 daltons, corresponding to DNA molecules with 230,000 to 1,532,000 contiguous base pairs. Today we know that these large native DNA molecules are easily broken by mechanical shear forces, making isolation of the intact molecules somewhat tricky.

The very size of DNA molecules presents an interesting biological puzzle. Chromosomal DNAs are often many orders of magnitude longer than the cells or viruses that contain them (Fig. 24–1; Table 24–1). In this chapter we therefore shift our focus from the secondary structure of DNA (considered in Chapter 10) to the extraordinary degree of organization required for the tertiary packaging of DNA into chromosomes. We first examine the

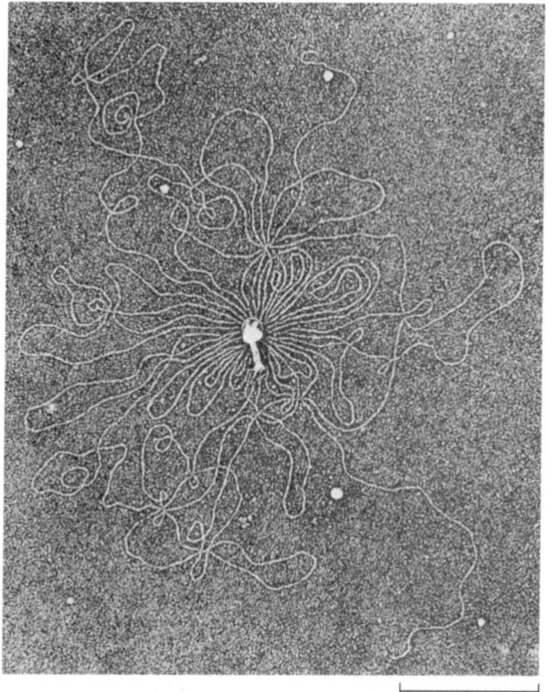

figure 24–1

Bacteriophage T2 protein coat surrounded by its single, linear molecule of DNA. The DNA was released by lysing the bacteriophage particle in distilled water and allowing the DNA to spread on the water surface. An undamaged T2 bacteriophage particle consists of a head structure that tapers to a tail by which the bacteriophage attaches itself to the outer surface a bacterial cell. All the DNA shown in this electron micrograph is normally packaged inside the phage head.

table 24–1

The Sizes of DNA and Viral Particles for Some Bacterial Viruses (Bacteriophages)

Virus	Number of base pairs in viral DNA*	Length of viral DNA (nm)	Long dimension of viral particle (nm)
φX174	5,386†	1,939†	25
T7	39,936	14,377	78
λ (lambda)	48,502	17,460	190
T4	168,889	60,800	210

* The complete base sequences of these bacteriophage genomes have been determined.
†Data are for the replicative form (double-stranded).

elements within viral and cellular chromosomes, then assess their size and organization. We next consider DNA topology, providing a description of the coiling of DNA molecules. Finally, we discuss the protein-DNA interactions that organize chromosomes into compact structures.

Chromosomal Elements

Cellular DNA contains both genes and intergenic regions, both of which may serve functions vital to the cell. More complex genomes, such as those of eukaryotic cells, demand increased levels of chromosomal organization, and this is reflected in the chromosome's structural features. We begin by considering the different types of DNA sequences and structural elements within chromosomes.

Genes Are Segments of DNA That Code for Polypeptide Chains and RNAs

Our understanding of genes has evolved tremendously over the last century. Classically, a gene was defined as a portion of a chromosome that determines or affects a single character or **phenotype** (visible property), such as eye color. A molecular definition was proposed by George Beadle and Edward Tatum in 1940. After exposing spores of the fungus *Neurospora crassa* to x rays and other agents known to damage DNA and cause alterations in DNA sequence **(mutations),** they detected mutant fungal strains that lacked one or another specific enzyme, sometimes resulting in the failure of an entire metabolic pathway. Beadle and Tatum concluded that a gene is a segment of genetic material that determines or codes for one enzyme: the **one gene–one enzyme** hypothesis. Later this concept was broadened to **one gene–one protein,** because many genes code for proteins that are not enzymes.

The modern biochemical definition of a gene is even more precise. Recall that many proteins have multiple polypeptide chains (Chapter 5). In some proteins all the polypeptides are identical, all encoded by the same gene. Other proteins have two or more different kinds of polypeptide chains, each with a distinctive amino acid sequence. Hemoglobin A, the major hemoglobin of adult humans, for example, has two kinds of polypeptide chains, α and β, which differ in amino acid sequence and are encoded by two different genes. Thus the gene-protein relationship is more accurately described by the phrase "one gene–one polypeptide."

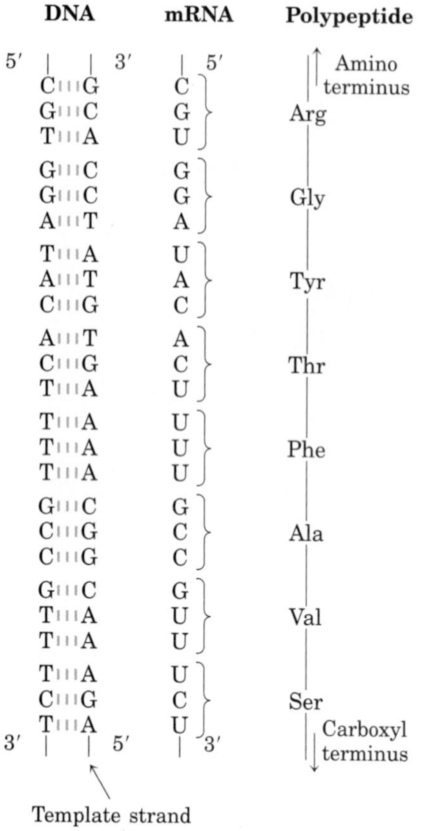

Even this definition is not sufficiently precise because not all genes are ultimately expressed in the form of polypeptide chains. Some, for example, code for tRNAs and rRNAs (Chapters 10 and 26), and a complete definition must include these genes. A gene, therefore, is all of the DNA that encodes the primary sequence of some final gene product, such as a polypeptide or an RNA with a structural or catalytic function. DNA also contains other segments or sequences that have a purely regulatory function. **Regulatory sequences** provide signals that may denote the beginning or the end of genes, or influence the transcription of genes, or function as initiation points for replication or recombination (Chapter 28).

The minimum overall size of genes that encode proteins can be estimated directly. As described in detail in Chapter 27, each amino acid of a polypeptide chain is coded for by a sequence of three consecutive nucleotides in a single strand of DNA (Fig. 24–2), with these coding "triplets" arranged in a sequence that corresponds to the sequence of amino acids in the polypeptide that the gene encodes. A single polypeptide chain may have

DNA	mRNA	Polypeptide
5′ 3′	5′	Amino terminus
C G / G C / T A	C G U	Arg
G C / G C / A T	G G A	Gly
T A / A T / C G	U A C	Tyr
A T / C G / T A	A C U	Thr
T A / T A / T A	U U U	Phe
G C / C G / C G	G C C	Ala
G C / T A / T A	G U U	Val
T A / C G / T A	U C U	Ser
3′ 5′	3′	Carboxyl terminus

Template strand

figure 24–2
Colinearity of the coding nucleotide sequences of DNA and mRNA and the amino acid sequence of a polypeptide chain. The triplets of nucleotide units in DNA determine the amino acids in a protein through the intermediary mRNA. One of the DNA strands serves as a template for synthesis of mRNA, which has nucleotide triplets (codons) complementary to those of the DNA. In some bacterial and many eukaryotic genes, coding sequences are interrupted at intervals by regions of noncoding sequences (called introns).

anywhere from about 50 to several thousand amino acid residues; the gene encoding this polypeptide chain must have at least three times as many base pairs. An average polypeptide chain of 350 amino acid residues corresponds to 1,050 base pairs. Many genes in eukaryotes and a few in prokaryotes are interrupted by noncoding DNA segments and are therefore considerably longer than the simple calculation outlined above would suggest.

How many genes are in a single chromosome? The *Escherichia coli* chromosome, one of the prokaryotic genomes that has been completely sequenced, is a circular DNA molecule (in the sense of an endless loop rather than a perfect circle) with 4,638,858 base pairs. These base pairs encode about 4,300 genes for proteins and another 115 genes for stable RNA molecules. Among eukaryotes, for which our information is not yet so complete, the 3 billion base pairs of the human genome encode 50 to 100 thousand genes on 24 different chromosomes.

George W. Beadle
1903–1989

Edward L. Tatum
1909–1975

Eukaryotic Chromosomes Are Very Complex

Bacteria usually have only one chromosome per cell and, in nearly all cases, each chromosome contains only one copy of each gene. A very few genes, such as those for rRNAs, are repeated several times. Regulatory sequences and genes account for almost all of the DNA in prokaryotes. Moreover, almost every gene is precisely colinear with the amino acid sequence (or RNA sequence) for which it codes (Fig. 24–2).

The organization of genes in eukaryotic DNA is structurally and functionally much more complex, and the study of eukaryotic chromosome structure has yielded many surprises. Tests made of the extent to which segments of mouse DNA occur in multiple copies had an unexpected outcome. About 10% of mouse DNA consists of short lengths of less than 10 base pairs that are repeated millions of times per cell. These are called **highly repetitive** sequences, or **simple-sequence DNA.** Another 20% of mouse DNA was found to occur in lengths of up to a few hundred base pairs that are repeated at least 1,000 times; this DNA is designated **moderately repetitive.** Some of the repetitive DNA may simply be "junk DNA," vestiges of evolutionary sidetracks. However, at least some of it has functional significance, as we shall see. The remaining 70% of mouse DNA consists of unique segments and segments that are repeated only a few times. The unique sequences in eukaryotic chromosomes include most of the genes.

The simple-sequence DNA has also been called **satellite DNA,** so named because its unusual base composition often causes it to migrate as "satellite" bands (separated from the rest of the DNA) when fragmented cellular DNA samples are centrifuged in cesium chloride density gradients. Studies suggest that simple-sequence DNA does not encode proteins or RNAs. Much of it is associated with two important structures in eukaryotic chromosomes: centromeres and telomeres.

A defining feature of a eukaryotic chromosome is its **centromere** (Fig. 24–3), a sequence of DNA that functions during cell division as an attachment point for proteins that link the chromosome to the mitotic spindle. This attachment is essential for the equal and orderly distribution of chromosome sets to daughter cells. The centromeres of yeast chromosomes have been isolated and studied. The sequences essential to centromere function are about 130 base pairs long and are very rich in A=T pairs. The centromeric sequences of higher eukaryotes are much longer and (unlike those of yeast) generally contain simple-sequence DNA. This consists of thousands of tandem copies of one or a few short sequences of 5 to 10 base pairs, in the same orientation. The precise role of simple-sequence DNA in centromere function is not yet understood.

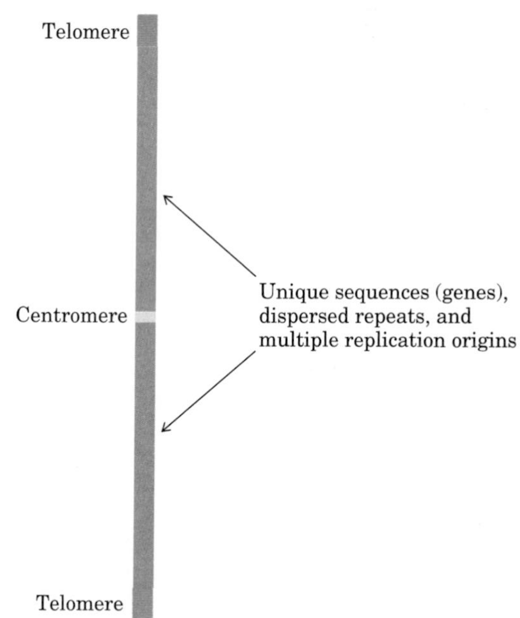

figure 24–3
Important structural elements of a yeast chromosome.

Telomeres are sequences at the ends of eukaryotic chromosomes that help stabilize the chromosome. The best-characterized telomeres are those of simpler eukaryotes. Yeast telomeres end with about 100 base pairs of imprecisely repeated sequences of the form

$$(5')(T_xG_y)_n$$
$$(3')(A_xC_y)_n$$

where x and y are generally between 1 and 4. The number of telomere repeats, n, is in the range of 20 to 100 for the chromosomes of most single-celled eukaryotes and generally over 1,500 in mammals. The ends of a linear DNA molecule cannot be routinely replicated by the cellular replication machinery (which may be one reason why bacterial DNA molecules are circular). Repeated telomeric sequences are added to eukaryotic chromosome ends primarily by the enzyme telomerase, as discussed in Chapter 26. The mechanism that controls the number of repeats in a telomere is not known.

Artificial chromosomes have been constructed as a means of better understanding the functional significance of many structural features of eukaryotic chromosomes. A reasonably stable artificial linear chromosome requires only three components: a centromere, telomeres, and sequences that allow the initiation of DNA replication. Yeast artificial chromosomes (YACs) have been developed as a research tool in biotechnology. Very large segments of foreign (nonyeast) DNA can be included in YACs and then propagated in yeast cells for use in research and biotechnology. Similarly, human artificial chromosomes (HACs) are being developed for the treatment of genetic diseases by somatic gene therapy (Chapter 29).

Most moderately repetitive DNA consists of 150 to 300 base-pair repeats scattered throughout the genome of higher eukaryotes; some of these repeats have been characterized. A number of them are related to (or may be) transposable elements, sequences that move about the genome at very low frequency (Chapter 25). In humans, one class of these repeats (about 300 base pairs long) is the *Alu* family, so named because their sequence generally includes one copy of the recognition sequence for the restriction endonuclease *Alu*I. (Restriction endonucleases are described in Chapter 29.) Hundreds of thousands of *Alu* repeats are dispersed throughout the human genome, comprising 1% to 3% of the total DNA. *Alu* and similar dispersed repeats together make up 5% to 10% of human DNA. No function for this DNA is yet known.

Many Eukaryotic Genes Contain Intervening Nontranscribed Sequences (Introns)

Many, if not most, eukaryotic genes have a distinctive and puzzling structural feature: their nucleotide sequences contain one or more intervening segments of DNA that do not code for the amino acid sequence of the polypeptide product. These nontranslated inserts interrupt the otherwise colinear relationship between the nucleotide sequence of the gene and the amino acid sequence of the polypeptide it encodes. Such nontranslated DNA segments in genes are called **intervening sequences,** or **introns,** and the coding segments are called **exons.** Few prokaryotic genes contain introns.

In the eukaryotic genes examined thus far, investigators have found much variety in the number and position of introns and in the fraction of the total gene length that they occupy. For example, in the gene coding for the single polypeptide chain of the avian egg protein ovalbumin (Fig. 24–4), the introns are much longer than the exons; altogether, seven introns make up 85% of the gene's DNA. In the gene for the β subunit of hemoglobin, a single intron contains over half of the gene's DNA. The gene for the protein

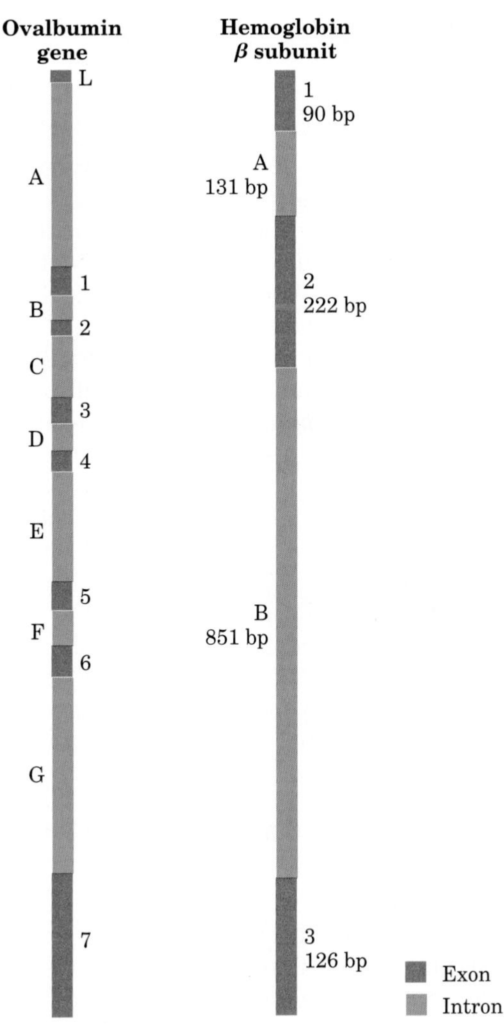

figure 24–4

Intervening sequences, or introns, in two eukaryotic genes. The gene for ovalbumin has seven introns (A to G), splitting the coding sequences into eight exons (L, and 1 to 7). The gene for the β subunit of hemoglobin has two introns and three exons, including one intron that alone contains more than half the base pairs (bp) of the gene.

conalbumin of the chicken egg contains 17 introns, and a collagen gene has been found to have over 50 introns. Genes for histones appear to have no introns. In most cases the function of introns is not clear.

The Size and Sequence Structure of DNA Molecules

Critical to the life of the cell are the packaging and segregation of chromosomes into daughter cells at cell division. Specific DNA sequences and proteins are implicated in these events. In our examination of the organization and structure of chromosomes, we now consider how DNA is organized to fit into prokaryotic and eukaryotic cells.

Viral DNA Molecules Are Relatively Small

Viruses generally contain considerably less genetic information than cells because they use the resources of the host cell for many of the processes required to propagate. An infectious virus particle often consists of no more than its genome (usually a single RNA or DNA molecule) surrounded by a protein coat.

Almost all plant viruses and some bacterial and animal viruses have RNA genomes. These genomes tend to be particularly small. For example, the genomes of mammalian retroviruses such as HIV are about 9,000 nucleotides in length, and that of the bacteriophage Qβ has 4,220 nucleotides. Both types of viruses have single-stranded RNA genomes.

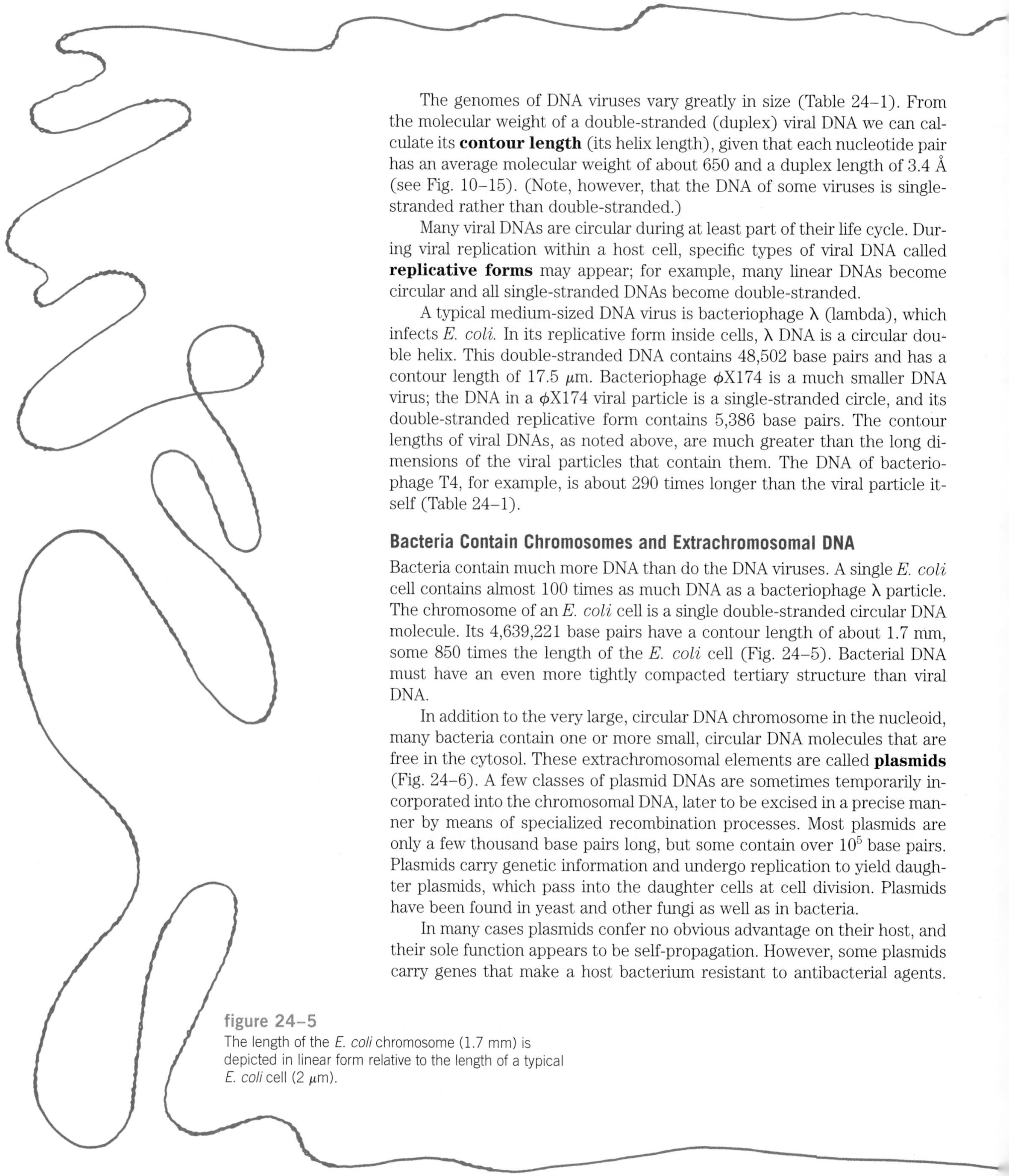

The genomes of DNA viruses vary greatly in size (Table 24–1). From the molecular weight of a double-stranded (duplex) viral DNA we can calculate its **contour length** (its helix length), given that each nucleotide pair has an average molecular weight of about 650 and a duplex length of 3.4 Å (see Fig. 10–15). (Note, however, that the DNA of some viruses is single-stranded rather than double-stranded.)

Many viral DNAs are circular during at least part of their life cycle. During viral replication within a host cell, specific types of viral DNA called **replicative forms** may appear; for example, many linear DNAs become circular and all single-stranded DNAs become double-stranded.

A typical medium-sized DNA virus is bacteriophage λ (lambda), which infects *E. coli.* In its replicative form inside cells, λ DNA is a circular double helix. This double-stranded DNA contains 48,502 base pairs and has a contour length of 17.5 μm. Bacteriophage φX174 is a much smaller DNA virus; the DNA in a φX174 viral particle is a single-stranded circle, and its double-stranded replicative form contains 5,386 base pairs. The contour lengths of viral DNAs, as noted above, are much greater than the long dimensions of the viral particles that contain them. The DNA of bacteriophage T4, for example, is about 290 times longer than the viral particle itself (Table 24–1).

Bacteria Contain Chromosomes and Extrachromosomal DNA

Bacteria contain much more DNA than do the DNA viruses. A single *E. coli* cell contains almost 100 times as much DNA as a bacteriophage λ particle. The chromosome of an *E. coli* cell is a single double-stranded circular DNA molecule. Its 4,639,221 base pairs have a contour length of about 1.7 mm, some 850 times the length of the *E. coli* cell (Fig. 24–5). Bacterial DNA must have an even more tightly compacted tertiary structure than viral DNA.

In addition to the very large, circular DNA chromosome in the nucleoid, many bacteria contain one or more small, circular DNA molecules that are free in the cytosol. These extrachromosomal elements are called **plasmids** (Fig. 24–6). A few classes of plasmid DNAs are sometimes temporarily incorporated into the chromosomal DNA, later to be excised in a precise manner by means of specialized recombination processes. Most plasmids are only a few thousand base pairs long, but some contain over 10^5 base pairs. Plasmids carry genetic information and undergo replication to yield daughter plasmids, which pass into the daughter cells at cell division. Plasmids have been found in yeast and other fungi as well as in bacteria.

In many cases plasmids confer no obvious advantage on their host, and their sole function appears to be self-propagation. However, some plasmids carry genes that make a host bacterium resistant to antibacterial agents.

figure 24–5
The length of the *E. coli* chromosome (1.7 mm) is depicted in linear form relative to the length of a typical *E. coli* cell (2 μm).

E. coli

E. coli DNA

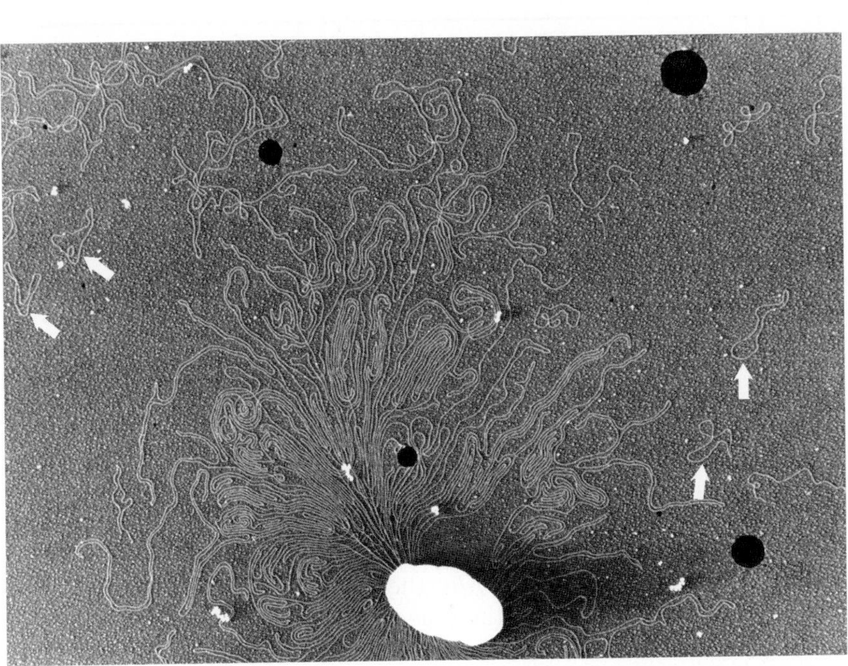

figure 24–6

DNA from a lysed *E. coli* cell. In this electron micrograph several small, circular plasmid DNAs are indicated by white arrows. The black spots and white specks are artifacts of the preparation.

For example, plasmids carrying the gene for the enzyme β-lactamase confer resistance to β-lactam antibiotics such as penicillin and amoxicillin (see Box 20–1). These and similar plasmids may pass from an antibiotic-resistant cell to an antibiotic-sensitive cell of the same or another bacterial species, thus making the recipient cell antibiotic resistant. The extensive use of antibiotics in some human populations has served as a strong selective force, encouraging the spread of transposons (Chapter 25) and plasmids that encode antibiotic resistance in disease-causing bacteria and creating bacterial strains that are resistant to several antibiotics. Physicians are becoming increasingly reluctant to prescribe antibiotics unless a clear clinical need is confirmed. For similar reasons, the widespread use of antibiotics in animal feeds is being curbed.

Plasmids are quite easily isolated intact from bacterial and yeast cells and are useful models for the study of many processes in DNA metabolism. They have also become a central component of the modern technologies used to isolate, clone, and artificially modify genes. Genes from a variety of species can be inserted into isolated plasmids. A modified plasmid with a foreign gene is reintroduced into its normal host cell, is replicated and transcribed, and in the process may cause the host cell to make the protein encoded by the foreign gene, even though it is not part of the cell's normal genome. Chapter 29 describes how such **recombinant DNAs** are made.

table 24–2

Normal Chromosome Number in Some Organisms*

Bacteria	1	Honeybee (female)	32
Fruit fly	8	Fox	34
Red clover	14	Cat	38
Garden pea	14	Mouse	40
Yeast	16[†]	Rat	42
Maize (corn)	20	Rabbit	44
Frog	26	Human	46
Hydra	30	Chicken	78

*The diploid chromosome number is given for all eukaryotes except yeast.

[†]This is the haploid chromosome number for the yeast *Saccharomyces cerevisiae.* Wild yeast strains generally have eight (octoploid) or more sets of these chromosomes.

Eukaryotic Cells Contain More DNA Than Do Prokaryotes

A yeast cell, one of the simplest eukaryotes, has four times more DNA than an *E. coli* cell. Cells of *Drosophila,* the fruit fly used in classical genetic studies, contain more than 25 times as much DNA as *E. coli* cells. Human cells and many other mammalian cells have about 600 times as much DNA as *E. coli*. The cells of many plants and amphibians contain an even greater amount. Although eukaryotic cells contain more DNA than bacterial cells, a eukaryotic genome contains a greater proportion of apparently noncoding DNA. There are roughly 50 genes per millimeter of human DNA, compared with over 2,500 genes per millimeter of *E. coli* DNA. Much of this noncoding DNA may play an important role in the organization of eukaryotic chromosome structure.

The total contour length of all the DNA in a *single* human cell is about 2 m, compared with 1.7 mm for *E. coli* DNA. An adult human body contains approximately 10^{14} cells and thus a total DNA length of 2×10^{11} km. Compare this with the circumference of the earth (4×10^4 km) or the distance between the earth and the sun (1.5×10^8 km)—a dramatic illustration of the extraordinary degree of DNA compaction in our cells.

Microscopic observation of nuclei in dividing eukaryotic cells has shown that the genetic material is subdivided into chromosomes, their diploid number depending upon the species (Table 24–2). A human somatic cell, for example, has 46 chromosomes (Fig. 24–7). Each chromosome of a eukaryotic cell, such as that shown in Figure 24–7a, contains a single, very large, duplex DNA molecule, which may be 4 to 100 times larger than that of an *E. coli* cell. For example, the DNA of one of the smaller human chromosomes has a contour length of about 30 mm, almost 15 times longer than the DNA of *E. coli.* The DNA

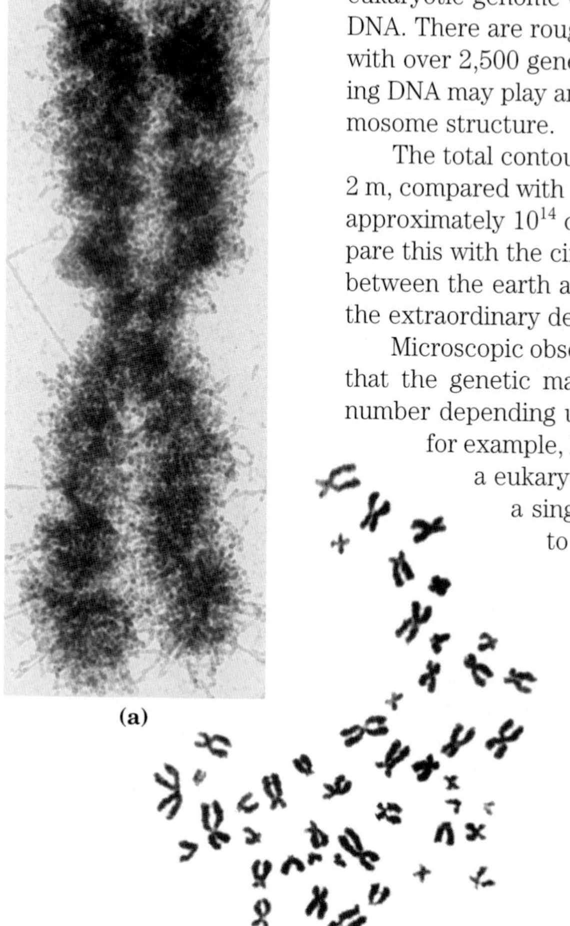

figure 24–7

Eukaryotic chromosomes. (a) A single human chromosome. **(b)** A complete set of chromosomes from a leukocyte from one of the authors. There are 46 chromosomes in every normal human somatic cell.

molecules in the 24 different types of human chromosomes (22 matching pairs plus the X and Y sex chromosomes) vary in length over a 25-fold range. Each type of chromosome in eukaryotes carries a characteristic set of genes.

Organelles of Eukaryotic Cells Also Contain DNA

In eukaryotic cells, DNA differing in base sequence from nuclear DNA is found in mitochondria (Fig. 24–8) and chloroplasts. Less than 0.1% of the cell's DNA is present in mitochondria in typical somatic cells, but in fertilized and dividing egg cells, where mitochondria are much more numerous, the relative amount of mitochondrial DNA is correspondingly larger. Mitochondrial DNA (mtDNA) molecules are much smaller than the nuclear chromosomes. In animal cells, mtDNA contains less than 20,000 base pairs (16,569 base pairs in human mtDNA) and is a circular duplex. Plant cell mtDNA ranges in size from 200,000 to 2,500,000 base pairs. Chloroplast DNA (cpDNA) also exists as circular duplexes and ranges in size from 120,000 to 160,000 base pairs.

The evolutionary origin of mitochondrial and chloroplast DNAs has been the subject of much speculation. A widely accepted view is that they are vestiges of the chromosomes of ancient bacteria that gained access to the cytoplasm of host cells and became the precursors of these organelles (see Fig. 2–15). Mitochondrial DNA codes for the mitochondrial tRNAs and rRNAs and for a few mitochondrial proteins. More than 95% of mitochondrial proteins are encoded by nuclear DNA. Mitochondria and chloroplasts divide when the cell divides. Their DNA is replicated before and during division, and the daughter DNA molecules pass into the daughter organelles.

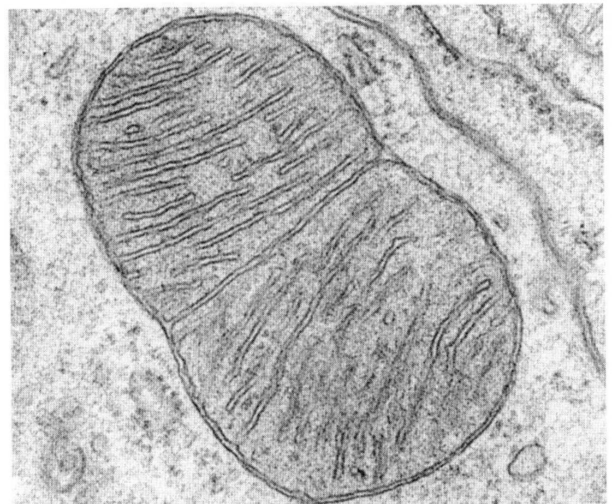

figure 24–8
A dividing mitochondrion. Some mitochondrial proteins and RNAs are encoded by one of the copies of the mitochondrial DNA (none of which are visible here). The DNA (mtDNA) is replicated each time the mitochondrion divides, prior to cell division.

DNA Supercoiling

Cellular DNA, as we have seen, is highly compacted, implying a high degree of structural organization. The folding mechanism must not only pack the DNA but also permit access to the information in the DNA. Before considering how this is accomplished in processes such as replication and transcription, we need to examine an important property of DNA structure known as **supercoiling.**

Supercoiling means the coiling of a coil. A telephone cord, for example, is typically a coiled wire. The path taken by the wire between the base of

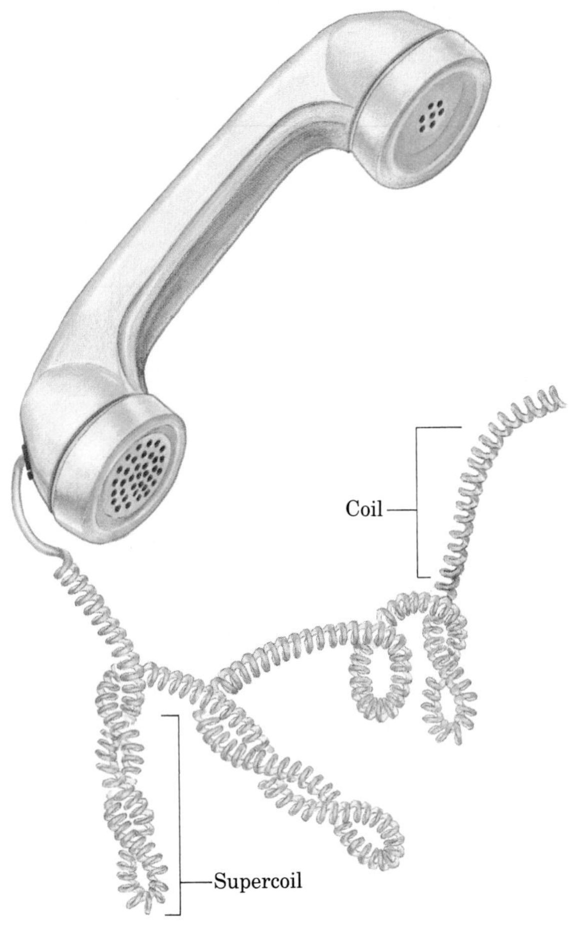

figure 24–9

Supercoils. A typical phone cord is coiled like a DNA helix, and the coiled cord can itself coil in a supercoil. The illustration is especially appropriate because an examination of phone cords helped lead Jerome Vinograd and his colleagues to the insight that many properties of small, circular DNAs can be explained by supercoiling. They first detected DNA supercoiling in small, circular viral DNAs in 1965.

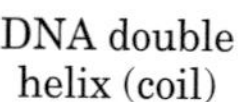

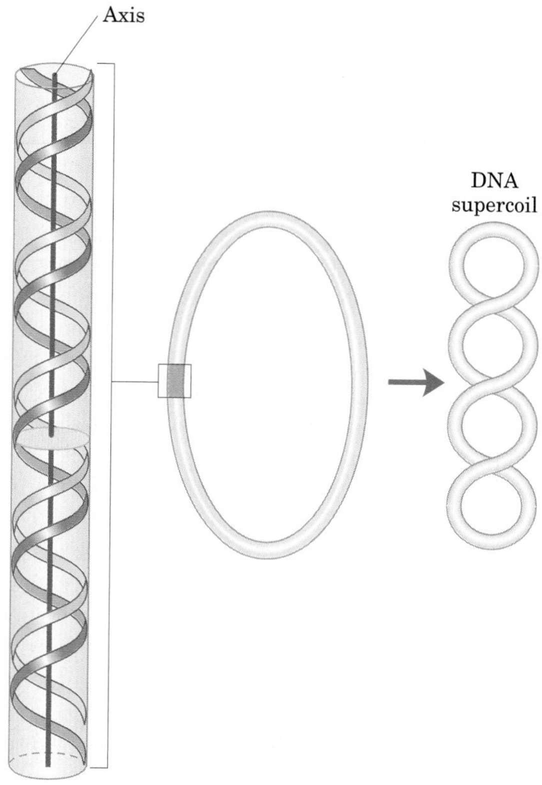

figure 24–10

Supercoiling of DNA. When the axis of the DNA double helix is coiled on itself, it forms a new helix (superhelix). The DNA superhelix is usually called a supercoil.

the phone and the receiver often includes one or more supercoils (Fig. 24–9). DNA is coiled in the form of a double helix, in which both strands of the DNA coil around an axis. The further coiling of that axis upon itself (Fig. 24–10) produces DNA supercoiling. As detailed below, DNA supercoiling is generally a manifestation of structural strain. When there is no net bending of the DNA axis upon itself, the DNA is said to be in a **relaxed** state.

We might predict that DNA compaction involves some form of supercoiling. Perhaps less predictable is the fact that replication and transcription of DNA also affect and are affected by supercoiling. Both processes require a separation of DNA strands—not a simple process in a molecule with two helically interwound strands (Fig. 24–11).

That DNA would bend on itself and become supercoiled in tightly packaged cellular DNA would seem logical and perhaps even trivial were it not for one additional fact: many circular DNA molecules remain highly supercoiled even after they are released from cells and purified, freed from protein and other cellular components. This indicates that supercoiling is an intrinsic property of DNA tertiary structure. It occurs in all cellular DNAs and is highly regulated by each cell.

A number of measurable properties of supercoiling have been established, the study of which has provided many insights into DNA structure and function. This work has drawn heavily on concepts derived from a

figure 24–11

Supercoiling induced by separating the strands of a helical structure. Twist two linear strands of rubber band into a right-handed double helix as shown. Fix one end by having a friend hold onto it, then pull apart the two strands at the other end. The resulting strain will produce supercoiling.

branch of mathematics called **topology,** the study of the properties of an object that do not change under continuous deformations. For DNA, continuous deformations would include conformational changes due to thermal motion or an interaction with proteins or other molecules. Discontinuous deformations involve DNA strand breakage. For circular DNA molecules, a topological property is thus one that is not affected by deformations of the DNA strands as long as no breaks are introduced. Topological properties are changed only by breaking and rejoining of the backbone of one or both DNA strands.

We now examine the fundamental properties and physical basis of supercoiling.

Most Cellular DNA Is Underwound

To understand supercoiling we must first focus on the properties of small, circular DNAs such as plasmids and small viral DNAs. When these DNAs contain no breaks in either strand, they are called **closed-circular DNAs.** If the DNA of a closed-circular molecule conforms closely to the B-form structure (the Watson-Crick structure; see Fig. 10–15), with one turn of the double helix for each 10.5 base pairs, the DNA is relaxed rather than supercoiled (Fig. 24–12). Supercoiling occurs when DNA is subject to some form of structural strain. Purified closed-circular DNA is rarely relaxed, regardless of its biological origin. Furthermore, DNAs derived from a given cellular source have a characteristic degree of supercoiling. DNA structure is therefore strained in a manner that is regulated by the cell to induce the supercoiling.

figure 24–12

Relaxed and supercoiled plasmid DNAs. The molecule in the leftmost electron micrograph is relaxed; the degree of supercoiling increases from left to right.

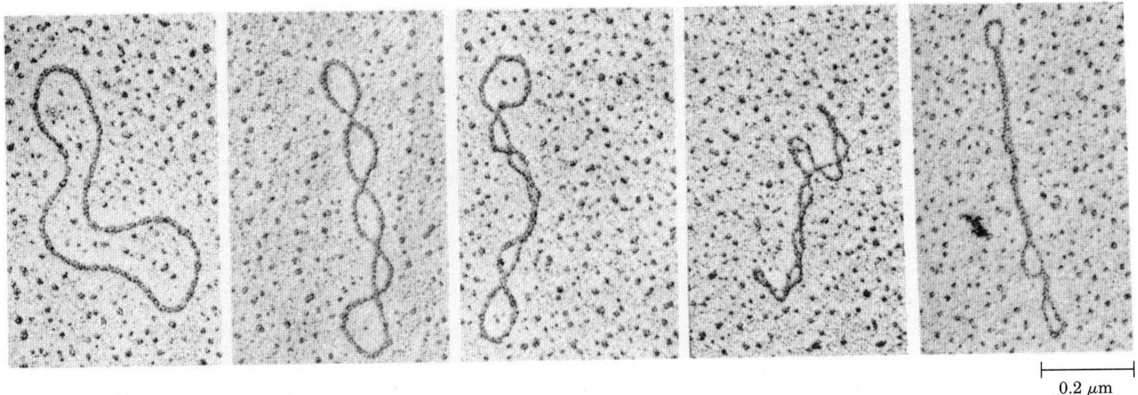

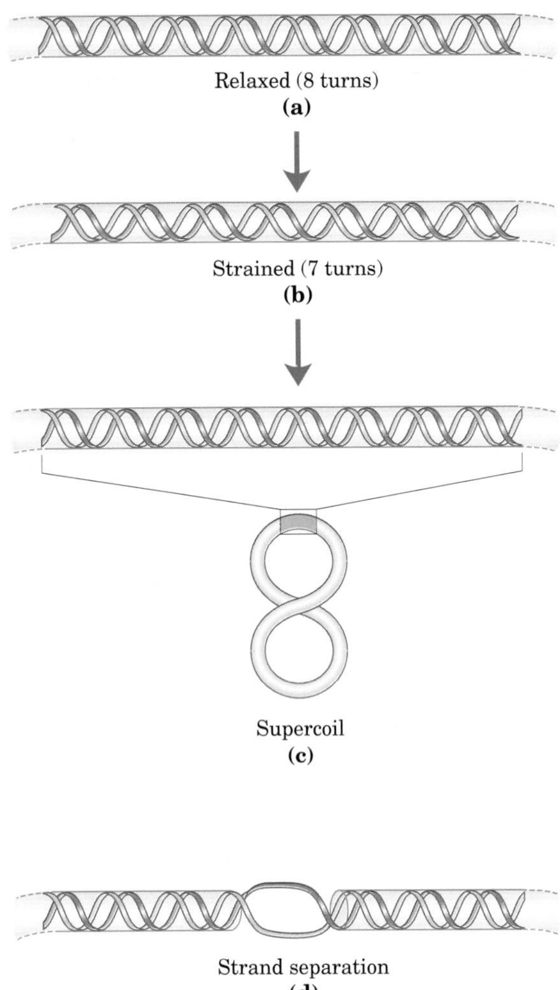

figure 24–13

Effects of DNA underwinding. **(a)** A segment of DNA within a closed-circular molecule, 84 base pairs long, in its relaxed form with eight helical turns. **(b)** Removal of one turn induces structural strain. **(c)** The strain is generally accommodated by formation of a supercoil. **(d)** DNA underwinding also makes it somewhat easier to separate strands. In principle, each turn of underwinding should facilitate strand separation over about 10 base pairs, as shown. However, the hydrogen-bonded base pairs would generally preclude strand separation over such a short distance and the effect becomes important only for longer DNAs and higher levels of DNA underwinding.

In almost every instance, the strain is a result of an **underwinding** of the DNA double helix in the closed circle. In other words, the DNA has *fewer* helical turns than would be expected for the B-form structure. The effects of underwinding are summarized in Figure 24–13. An 84 base pair segment of a circular DNA in the relaxed state would contain eight double-helical turns, or one for every 10.5 base pairs. If one of these turns were removed, there would be 84/7 or 12.0 base pairs per turn rather than the 10.5 found in B-DNA (Fig. 24–13b). This is a deviation from the most stable DNA form, and the molecule is thermodynamically strained as a result. Generally, much of this strain would be accommodated by coiling the axis of the DNA on itself to form a supercoil (Fig. 24–13c; some of the strain would simply become dispersed in the untwisted structure of the larger DNA molecule). In principle, the strain could also be accommodated by separating the two DNA strands over a distance of about 10 base pairs (Fig. 24–13d). In isolated closed-circular DNA, strain introduced by underwinding is generally accommodated by supercoiling rather than strand separation, because coiling the axis of the DNA usually requires less energy than breaking the hydrogen bonds that stabilize paired bases. Note, however, that the underwinding of DNA in vivo makes it easier to separate DNA strands, giving access to the information they contain.

Every cell actively underwinds its DNA with the aid of enzymatic processes (described below), and the resulting strained state represents a form of stored energy. Cells maintain DNA in an underwound state to facilitate its compaction by coiling. In addition, the underwinding of DNA is also important to enzymes of DNA metabolism that must bring about strand separation as part of their function.

The underwound state can be maintained only if the DNA is a closed circle or if it is bound and stabilized by proteins so that the strands are not free to rotate about each other. If there is a break in one strand of an isolated protein-free circular DNA, free rotation at that point will cause the underwound DNA to revert spontaneously to the relaxed state. In a closed-circular DNA without strand breaks, however, the number of helical turns cannot be changed without at least transiently breaking one of the DNA strands. The number of helical turns in a DNA molecule therefore provides a precise description of supercoiling.

DNA Underwinding Is Defined by Topological Linking Number

Topology provides a number of ideas that are useful in this discussion, particularly the concept of **linking number.** Linking number is a topological property because it does not vary when double-stranded DNA is bent or deformed, as long as both DNA strands remain intact. The concept of linking number (Lk) is illustrated in Figure 24–14. We begin by visualizing the separation of the two strands of a double-stranded circular DNA. If the two strands are linked as shown in Figure 24–14a, they are effectively joined by what can be described as a topological bond. Even if all hydrogen bonds and base-stacking interactions were abolished such that the strands were not in physical contact, this topological bond would still link the two strands. Visualize one of the circular strands as the boundary of an imaginary surface (such as a soap film spanning the space framed by a circular wire before you blow a soap bubble). The linking number can be defined as the number of times the second strand pierces this surface. For the molecule in Figure 24–14a, $Lk = 1$; for that in Figure 24–14b, $Lk = 6$. The linking number for a closed-circular DNA is always an integer. By convention, if the links between two DNA strands are arranged so that the strands are interwound in a right-handed helix, the linking number is defined as positive (+); for strands interwound in a left-handed helix, the linking number is negative (−). Nega-

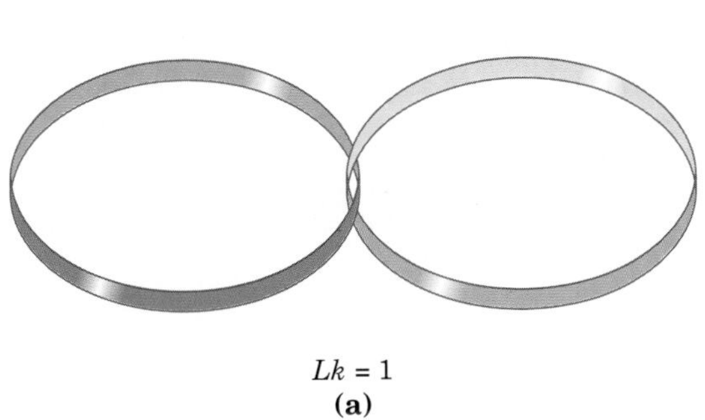

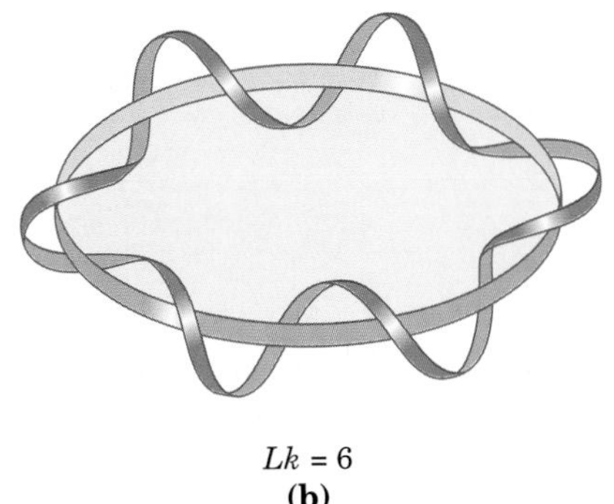

figure 24–14
Linking number, *Lk*. Here, as usual, each blue ribbon represents one strand of a double-stranded DNA molecule. For the molecule in **(a)**, $Lk = 1$. For the molecule in **(b)**, $Lk = 6$. One of the strands in **(b)** is kept untwisted for illustrative purposes to define the border of an imaginary surface (shaded blue). The number of times the twisting strand penetrates this surface provides a rigorous definition of linking number.

tive linking numbers are, for all practical purposes, not encountered in studies of DNA.

We can now extend these ideas to a closed-circular DNA with 2,100 base pairs (Fig. 24–15). When the molecule is relaxed, the linking number is simply the number of base pairs divided by the number of base pairs per turn, which is close to 10.5; in this case, $Lk = 200$. For a circular DNA molecule to have a topological property such as linking number, neither strand may contain a break. If there is a break in either strand, the strands can, in principle, be unraveled and separated completely. In this case, no topological bond exists and Lk is undefined (Fig. 24–15b).

We can now describe DNA underwinding in terms of changes in the linking number. The linking number in relaxed DNA, called Lk_0, is used as a reference. For the molecule shown in Figure 24–15a, $Lk_0 = 200$; if two turns are removed from this molecule, $Lk = 198$. The change can be described by the equation

$$\Delta Lk = Lk - Lk_0 = 198 - 200 = -2$$

It is often convenient to express the change in linking number in terms of a quantity that is independent of the length of the DNA molecule. This quantity, called the **specific linking difference (σ),** or **superhelical density,** is a measure of the number of turns removed relative to the number present in relaxed DNA:

$$\sigma = \frac{\Delta Lk}{Lk_0}$$

In the example in Figure 24–15c, $\sigma = -0.01$, which means that 1% (2 of 200) of the helical turns present in the DNA (in its B form) have been removed. The degree of underwinding in cellular DNAs generally falls in the range of 5% to 7%; that is, $\sigma = -0.05$ to -0.07. The negative sign of σ

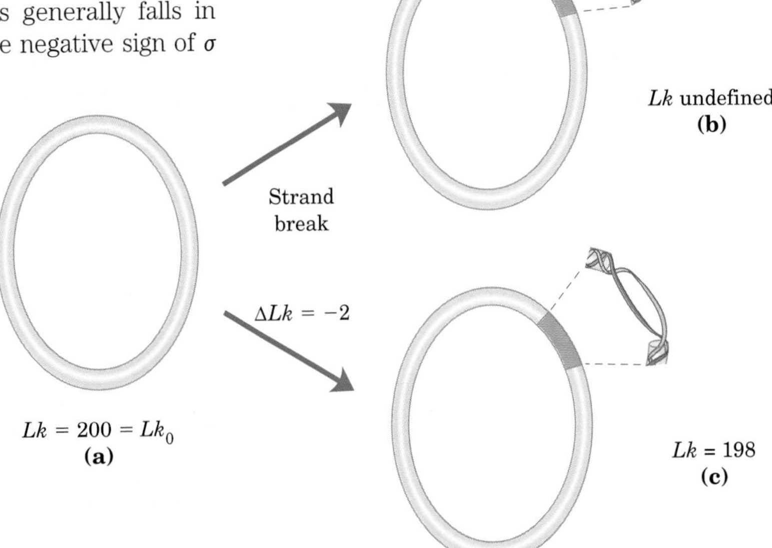

figure 24–15
Linking number applied to closed-circular DNA molecules. A 2,100 base pair circular DNA is shown in three forms: **(a)** relaxed, $Lk = 200$; **(b)** relaxed with a nick (break) in one strand, Lk undefined; **(c)** underwound by two turns, $Lk = 198$. The underwound molecule generally occurs as a supercoiled molecule, but underwinding also facilitates the separation of DNA strands.

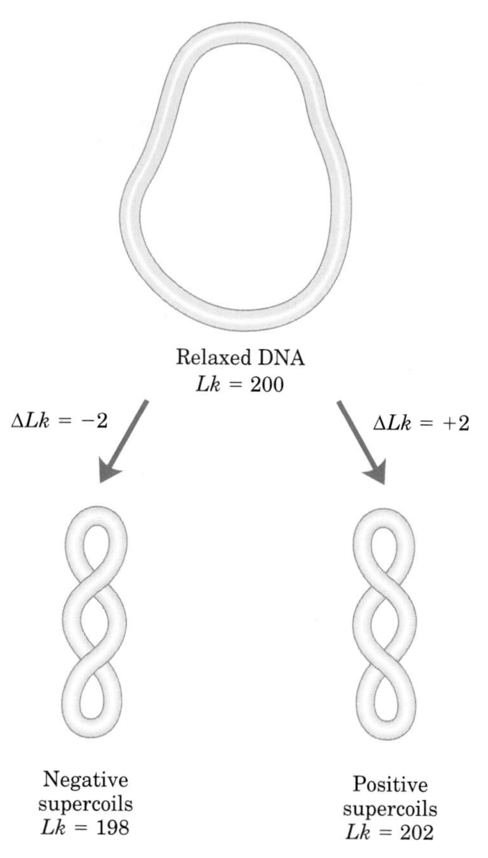

figure 24–16
Negative and positive supercoils. For the relaxed DNA molecule of Figure 24–15a, underwinding or overwinding by two helical turns (Lk = 198 or 202) will produce negative or positive supercoiling, respectively. Note that the DNA axis twists in opposite directions in the two cases.

indicates that the change in linking number is due to underwinding of the DNA. The supercoiling induced by underwinding is therefore defined as negative supercoiling. Conversely, under some conditions DNA can be overwound, resulting in positive supercoiling. Note that the twisting path taken by the axis of the DNA helix when the DNA is underwound (negative supercoiling) is the mirror image of that taken when the DNA is overwound (positive supercoiling) (Fig. 24–16). Supercoiling is not a random process; the path of the supercoiling is largely prescribed by the torsional strain imparted to the DNA by decreasing or increasing the linking number relative to B-DNA.

Linking number can be changed by ±1 by breaking one DNA strand, rotating one of the ends 360° about the unbroken strand, and rejoining the broken ends. This change has no effect on the number of base pairs or the number of atoms in the circular DNA molecule. Two forms of a given circular DNA that differ only in a topological property such as linking number are referred to as **topoisomers.**

Linking number can be broken down into two structural components called writhe (Wr) and twist (Tw) (Fig. 24–17). These are more difficult to describe than linking number, but writhe may be thought of as a measure of the coiling of the helix axis and twist as determining the local twisting or spatial relationship of neighboring base pairs. When the linking number changes, some of the resulting strain is usually compensated for by writhe (supercoiling) and some by changes in twist, giving rise to the equation

$$Lk = Tw + Wr$$

Twist and writhe are geometric rather than topological properties, because they may be changed by deformation of a closed-circular DNA molecule. In addition, Tw and Wr need not be integers.

In addition to causing supercoiling and making strand separation somewhat easier, the underwinding of DNA facilitates a number of structural changes in the molecule. These are of less physiological importance but help illustrate the effects of underwinding. Recall that a cruciform (see Fig. 10–21) generally contains a few unpaired bases; DNA underwinding helps

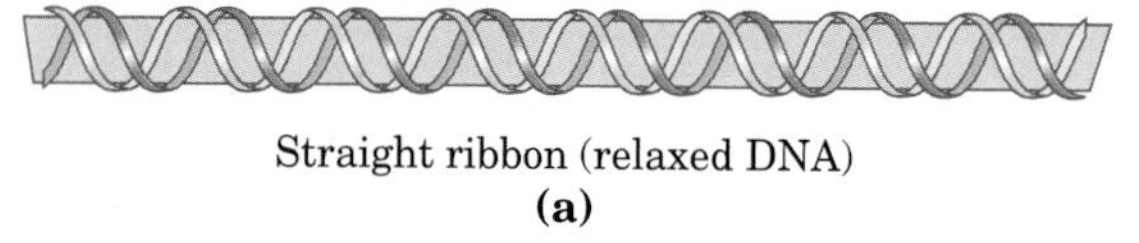

Straight ribbon (relaxed DNA)
(a)

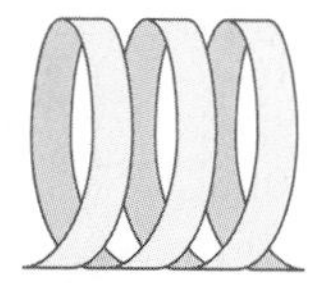

Large writhe, small change in twist
(b)

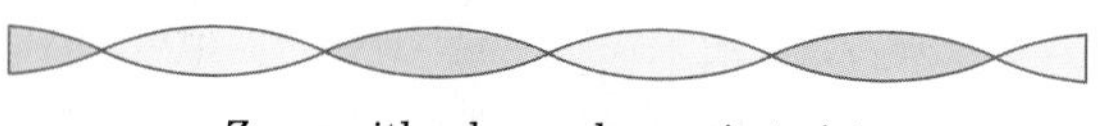

Zero writhe, large change in twist
(c)

figure 24–17
Ribbon model for illustrating twist and writhe. (a) The pink ribbon represents the axis of a relaxed DNA molecule. Strain introduced by twisting the ribbon (underwinding the DNA) can be manifested as **(b)** writhe or **(c)** twist. Changes in linking number are usually accompanied by changes in both writhe and twist.

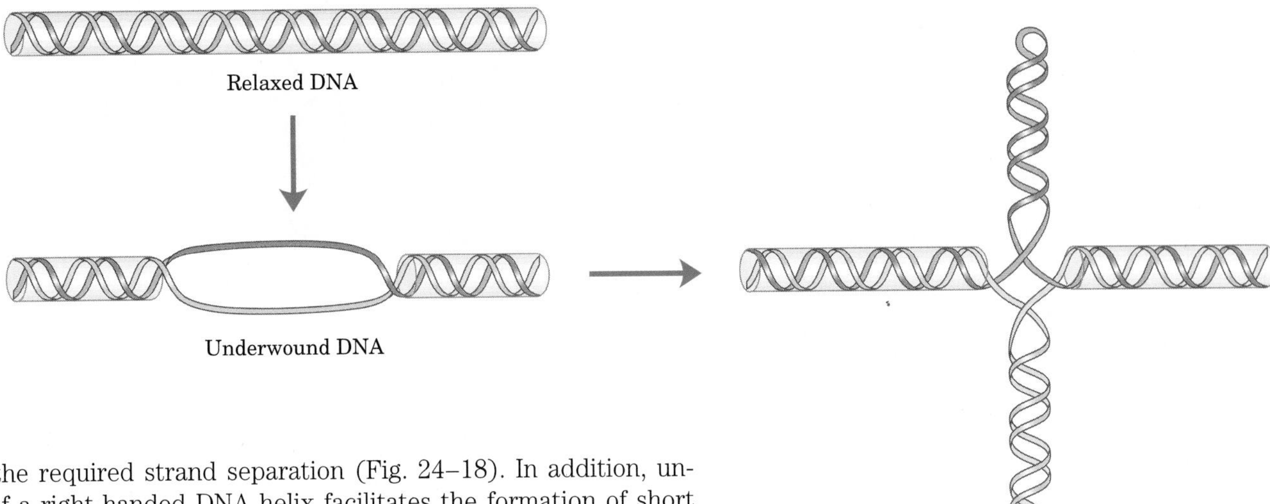

figure 24–18
Promotion of cruciform structures by DNA underwinding. In principle, cruciforms can form at palindromic sequences (see Fig. 10–21), but they seldom occur in relaxed DNA because the linear DNA accommodates more paired bases than does the cruciform structure. Underwinding of the DNA facilitates the partial strand separation needed to promote cruciform formation at appropriate sequences.

to maintain the required strand separation (Fig. 24–18). In addition, underwinding of a right-handed DNA helix facilitates the formation of short stretches of left-handed Z-DNA in regions where the base sequence is consistent with the Z-DNA form (Chapter 10).

Topoisomerases Catalyze Changes in the Linking Number of DNA

DNA supercoiling is a precisely regulated process that influences many aspects of DNA metabolism. Every cell has enzymes whose sole function is to underwind and/or relax DNA. Enzymes that increase or decrease the extent of DNA underwinding are called **topoisomerases;** the property of DNA that they change is the linking number. These enzymes play an especially important role in processes such as replication and DNA packaging. There are two classes of topoisomerases. Type I topoisomerases act by transiently breaking one of the two DNA strands, rotating one of the ends about the unbroken strand, and rejoining the broken ends; they change *Lk* in increments of 1. Type II topoisomerases break both DNA strands and change *Lk* in increments of 2.

The effects of these enzymes can be demonstrated using agarose gel electrophoresis (Fig. 24–19). A population of identical plasmid DNAs with the same linking number migrates as a discrete band during electrophoresis. Topoisomers with *Lk* values differing by as little as 1 can be separated by this method, so changes in linking number induced by topoisomerases can be readily observed.

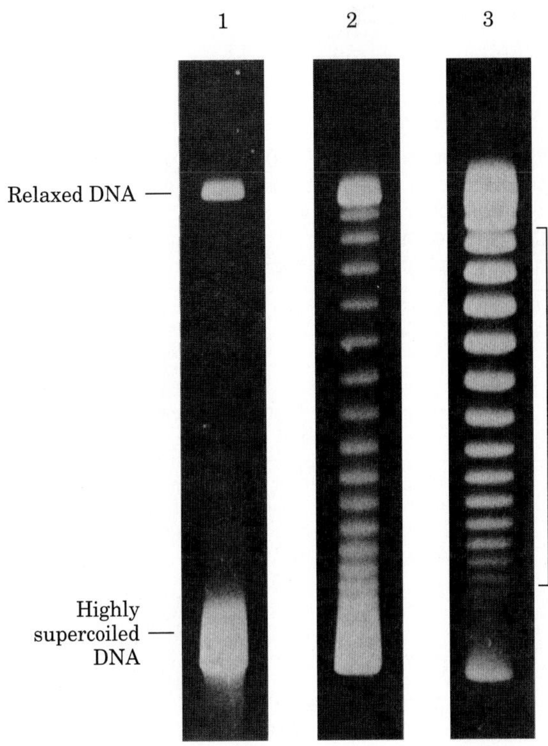

figure 24–19
Changes in linking number catalyzed by topoisomerases. In this experiment, all DNA molecules have the same number of base pairs but exhibit some range in the degree of supercoiling. Because supercoiled DNA molecules are more compact than relaxed molecules, they migrate more rapidly during gel electrophoresis. The gels shown here separate topoisomers (moving from top to bottom) over a limited range of superhelical density. In lane 1, highly supercoiled DNA migrates in a single band even though a number of different topoisomers are probably present. Lanes 2 and 3 illustrate the effect of treating the supercoiled DNA with a type I topoisomerase; the DNA in lane 3 was treated for a longer time than that in lane 2. As the superhelical density of the DNA is reduced to the point where it corresponds to the range in which the gel can resolve individual topoisomers, distinct bands appear. Bands in the region indicated by the bracket next to lane 3 each contain DNA circles with the same linking number.

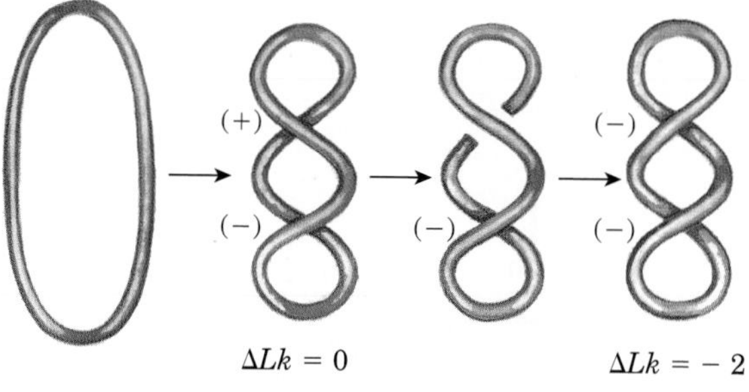

figure 24–20
Mechanism of DNA gyrase. *E. coli* topoisomerase II (DNA gyrase) alters the linking number of circular DNA molecules by an unusual mechanism. The topoisomerase binds to the DNA such that two regions of a DNA molecule are overlaid in a specific configuration in the bound complex, a positive (+) crossover or node. In any circular DNA, a compensating (−) node (a crossover in which the strands are overlaid in the opposite way) forms spontaneously elsewhere in the DNA molecule. Both strands of one DNA segment are broken, the other DNA segment is passed through the break, and the break is then resealed. The product contains two (−) nodes. The change in structure reflects a change in *Lk* of −2.

At least four different individual topoisomerases (I through IV) occur in *E. coli.* Those of type I (topoisomerases I and III) generally relax DNA by removing negative supercoils (increasing *Lk*). A bacterial type II enzyme, called either topoisomerase II or DNA gyrase, can introduce negative supercoils (decrease *Lk*). It uses the energy of ATP and a surprising mechanism to accomplish this (Fig. 24–20). The degree of supercoiling of bacterial DNA is maintained by regulation of the net activity of topoisomerases I and II.

Eukaryotic cells also have type I and type II topoisomerases. Topoisomerases I and III are both type I. The two type II topoisomerases, topoisomerases IIα and IIβ, cannot underwind DNA (introduce negative supercoils), although both can relax both positive and negative supercoils. We consider one probable origin of negative supercoils in eukaryotic cells in our discussion of chromatin later in the chapter.

DNA Compaction Requires a Special Form of Supercoiling

Supercoiled DNA molecules are uniform in a number of respects. The supercoils are right-handed in a negatively supercoiled DNA molecule (Fig. 24–16), and they tend to be extended and narrow rather than compacted, often with multiple branches (Fig. 24–21). At the superhelical densities normally encountered in cells, the length of the supercoil axis, including branches, is about 40% of the length of the DNA involved. This type of supercoiling is referred to as **plectonemic** (from the Greek *plektos,* "twisted," and *nema,* "thread"). This term can be applied to any structure in which strands are intertwined in some simple and regular way, and it well describes the general structure of supercoiled DNA in solution.

figure 24–21
Plectonemic supercoiling. (a) An electron micrograph of plectonemically supercoiled plasmid DNA and **(b)** an interpretation of the observed structure. The purple lines show the axis of the supercoil. Note the branching of the supercoil. **(c)** An idealized representation of this structure.

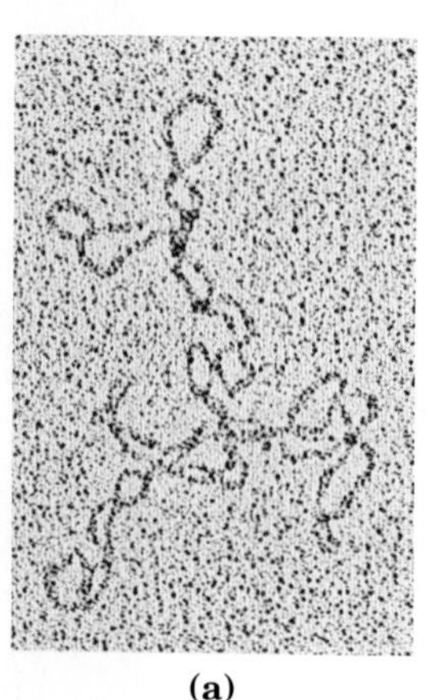
(a)

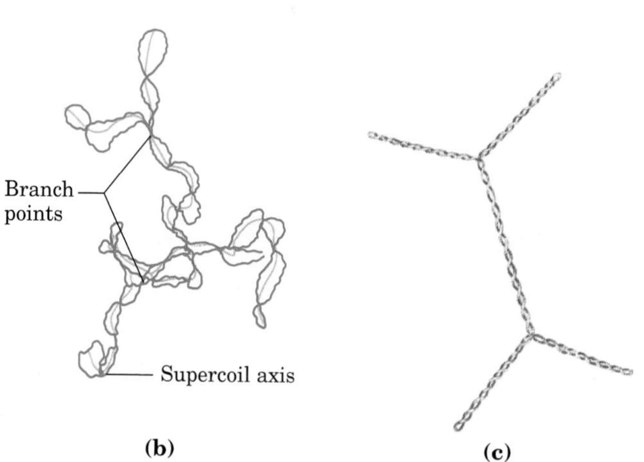

Although plectonemic supercoiling is the form observed in isolated DNAs in the laboratory, it does not give the compaction required to package DNA in the cell. A second form of supercoiling, called **solenoidal** (Fig. 24–22), can be adopted by an underwound DNA. Instead of the extended right-handed supercoils characteristic of the plectonemic form, solenoidal supercoiling involves tight left-handed turns, similar to the shape taken up by a garden hose neatly wrapped on a reel. Although their structures are dramatically different, plectonemic and solenoidal supercoiling are two forms of negative supercoiling that can be taken up by the *same* segment of underwound DNA. The two forms are readily interconvertible. Although the plectonemic form is more stable in solution, the solenoidal form can be stabilized by protein binding and is the form found in chromatin. It provides a much greater degree of compaction (Fig. 24–22b). Solenoidal supercoiling is the mechanism by which underwinding contributes to DNA compaction.

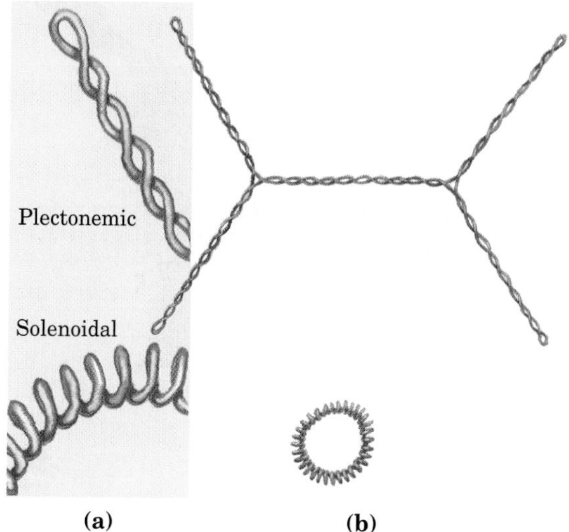

figure 24–22
Plectonemic and solenoidal supercoiling. (a) Plectonemic supercoiling takes the form of extended right-handed coils. Solenoidal negative supercoiling takes the form of tight left-handed turns about an imaginary tube-like structure. The two forms are readily interconverted, although the solenoidal form is generally not observed unless certain proteins are bound to the DNA. **(b)** Plectonemic and solenoidal supercoiling of the same DNA molecule, drawn to scale. Solenoidal supercoiling provides a much greater degree of compaction.

Chromatin and Nucleoid Structure

The term "chromosome" is used to refer to the nucleic acid molecule that is the repository of genetic information in a virus, a bacterium, a eukaryotic cell, or an organelle. It also refers to the densely colored bodies seen in the nuclei of dye-stained eukaryotic cells, as visualized using a light microscope. Eukaryotic chromosomes appear as sharply defined bodies in the nucleus during the period just before and during mitosis, the process of nuclear division in somatic cells. In nondividing eukaryotic cells, the chromosomal material, called **chromatin,** is amorphous and appears to be randomly dispersed throughout the nucleus. As the cells prepare to divide, the chromatin condenses and assembles itself into a species-specific number of well-defined chromosomes (Fig. 24–7).

Chromatin consists of fibers containing protein and DNA in approximately equal masses, along with a small amount of RNA. The DNA in the chromatin is very tightly associated with proteins called **histones,** which package and order the DNA into structural units called **nucleosomes** (Fig. 24–23). Also found in chromatin are many nonhistone proteins, some of which regulate the expression of specific genes (Chapter 28). Beginning with nucleosomes, eukaryotic chromosomal DNA is packaged into a succession of higher-order structures that ultimately yield the compact chromosome seen with the light microscope. We now turn to a description of this structure in eukaryotes and compare it with the packaging of DNA in bacterial cells.

figure 24–23
Nucleosomes. Regularly spaced nucleosomes consist of histone complexes bound to DNA. **(a)** Schematic illustration and **(b)** electron micrograph.

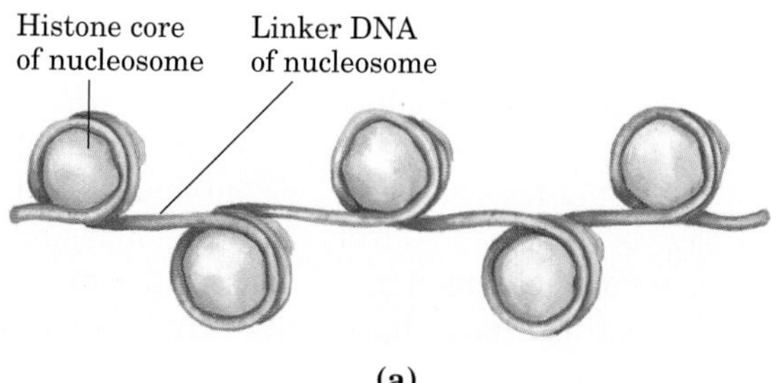

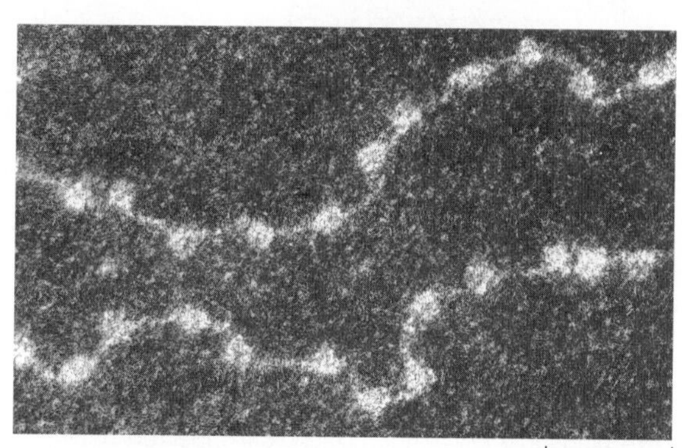

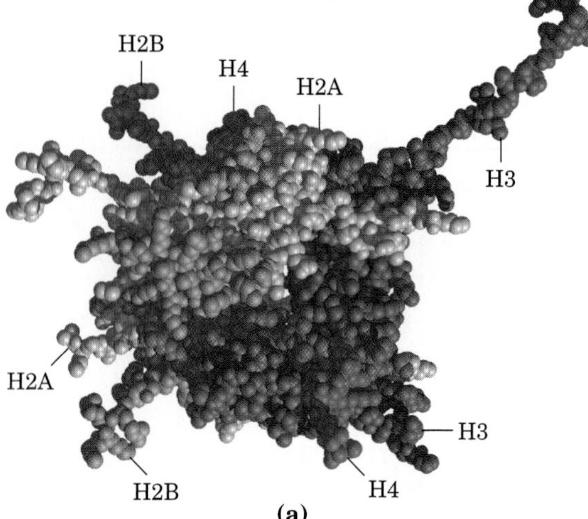

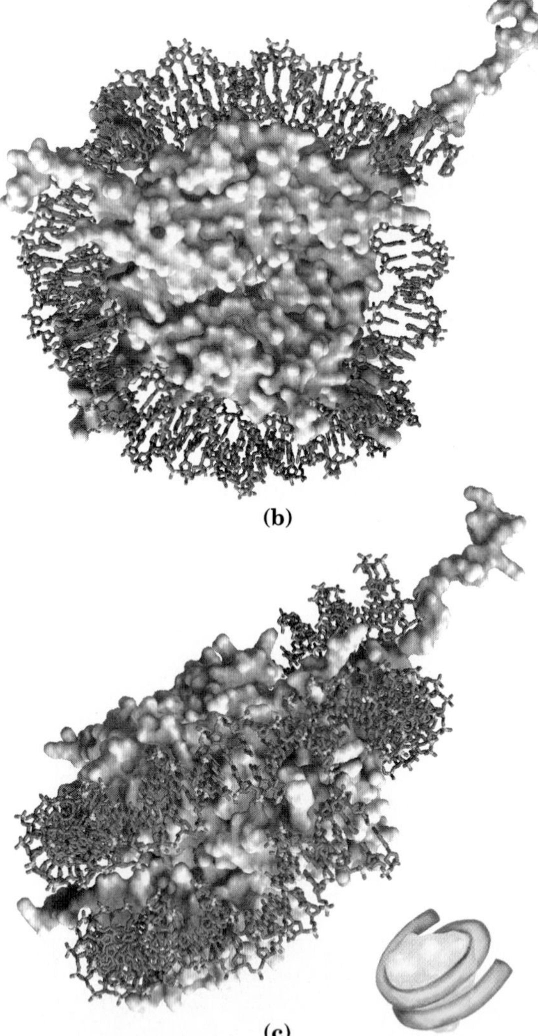

figure 24–24

DNA wrapped around a nucleosome core. (a) Space-filling representation of the nucleosome protein core, with different colors for the different histones. Top **(b)** and side **(c)** views of the crystal structure of a nucleosome with 146 base pairs of bound DNA. The protein is depicted as a gray surface contour and the bound DNA in blue. The DNA binds in a left-handed solenoidal supercoil that circumnavigates the histone complex 1.8 times. A schematic drawing is included in **(c)** for comparison with subsequent figures.

table 24–3

Types and Properties of Histones

Histone	Molecular weight	Number of amino acid residues	Content of basic amino acids (% of total)	
			Lys	Arg
H1*	21,130	223	29.5	1.3
H2A*	13,960	129	10.9	9.3
H2B*	13,774	125	16.0	6.4
H3	15,273	135	9.6	13.3
H4	11,236	102	10.8	13.7

*The sizes of these histones vary somewhat from species to species. The numbers given here are for bovine histones.

Histones Are Small, Basic Proteins

Found in the chromatin of all eukaryotic cells, histones have molecular weights between 11,000 and 21,000 and are very rich in the basic amino acids arginine and lysine (together these make up about one-fourth of the amino acid residues). Five major classes of histones are found in all eukaryotic cells, differing in molecular weight and amino acid composition (Table 24–3). The H3 histones are nearly identical in amino acid sequence in all eukaryotes, as are the H4 histones, suggesting strict conservation of their functions. For example, only two of 102 amino acids differ between the H4 histone molecules of peas and cows, and only eight differ between those of humans and yeast. Histones H1, H2A, and H2B show less sequence similarity among eukaryotic species.

Each of the histones has variant forms because certain amino acid side chains are enzymatically modified by methylation, ADP-ribosylation, phosphorylation, or acetylation. Such modifications affect the net electric charge, shape, and other properties of histones, as well as the structural and functional properties of chromatin, and they play a role in the regulation of transcription, as detailed in Chapter 28.

Nucleosomes Are the Fundamental Organizational Units of Chromatin

The eukaryotic chromosome depicted in Figure 24–7 represents the compaction of a DNA molecule about 10^5 μm long into a cell nucleus that is typically 5 to 10 μm in diameter. This compaction involves several levels of highly organized folding. Subjection of chromosomes to treatments that partially unfold them reveals a structure in which the DNA is bound tightly to beads of protein, often regularly spaced (Fig. 24–23). The beads in this "beads-on-a-string" arrangement are complexes of histones and DNA. The bead plus the connecting DNA that leads to the next bead forms the nucleosome, the fundamental unit of organization upon which the higher-order packing of chromatin is built. The bead of each nucleosome contains eight histone molecules: two copies each of H2A, H2B, H3, and H4. The spacing of the nucleosome beads provides a repeating unit typically of about 200 base pairs, of which 146 base pairs are bound tightly around the eight-part histone core and the remainder serve as linker DNA between nucleosome beads. Histone H1 binds to the linker DNA. Brief treatment of chromatin with enzymes that digest DNA causes preferential degradation of the linker DNA, releasing histone particles containing 146 base pairs of bound DNA that have been protected from digestion. Researchers have crystallized nucleosome cores obtained in this way, and x-ray diffraction analysis has revealed a particle made up of the eight histone molecules with the DNA wrapped around it in the form of a left-handed solenoidal supercoil (Fig. 24–24).

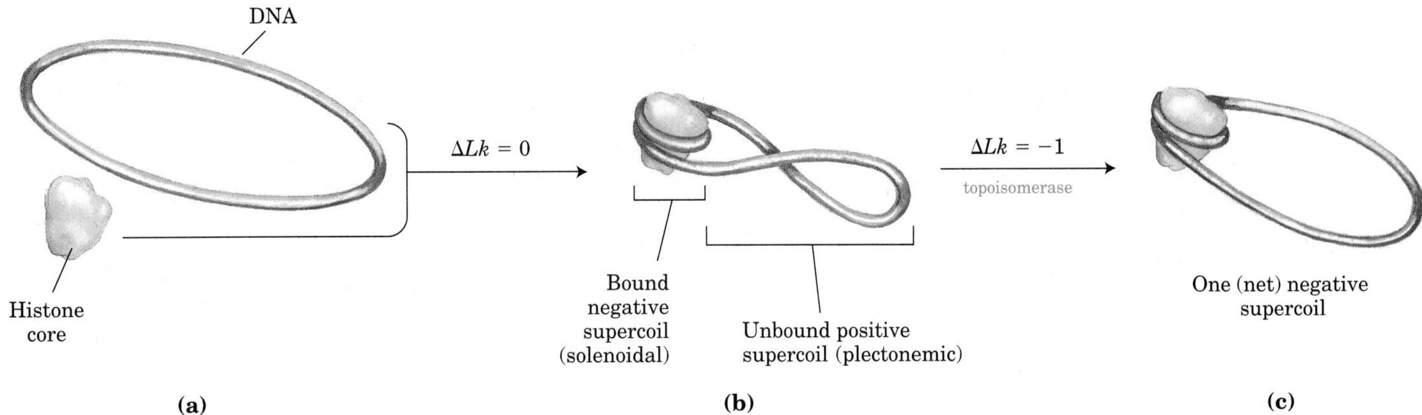

figure 24–25
Chromatin assembly. (a) Relaxed, closed-circular DNA. **(b)** Binding of a histone core to form a nucleosome induces one negative supercoil. In the absence of any strand breaks, a positive supercoil must form elsewhere in the DNA ($\Delta Lk = 0$). **(c)** Relaxation of this positive supercoil by a topoisomerase leaves one net negative supercoil ($\Delta Lk = -1$).

A close inspection of this structure reveals why eukaryotic DNA is underwound even though eukaryotic cells lack enzymes that underwind DNA. Recall that the solenoidal wrapping of DNA in nucleosomes is one form of supercoiling taken up by underwound (negatively supercoiled) DNA. The tight wrapping of DNA around a histone core in nucleosome particles requires the removal of about one helical turn in the DNA. When the protein core of a nucleosome binds in vitro to a relaxed, closed-circular DNA, the binding introduces a negative supercoil. This binding process does not break the DNA or change the linking number, however, so the formation of a negative solenoidal supercoil must be accompanied by a compensatory positive supercoil in the unbound region of the DNA (Fig. 24–25). As mentioned earlier, eukaryotic topoisomerases can relax positive supercoils. Relaxing the unbound positive supercoil leaves the negative supercoil fixed (by virtue of its being bound to the nucleosome histone core) and results in an overall decrease in linking number. Indeed, topoisomerases have proved necessary for assembling chromatin from purified histones and closed-circular DNA in vitro.

Another factor that affects the binding of DNA to histones in nucleosome cores is the sequence of the bound DNA. The histone cores do not bind randomly to DNA; rather, they tend to position themselves at certain locations. This positioning is not fully understood but appears to depend on a local abundance of A=T base pairs in the minor groove of the DNA helix, where it is in contact with the histones (Fig. 24–26). The tight wrapping of the DNA around the nucleosome's histone core requires compression of the minor groove at these points, and a cluster of two or three A=T base pairs makes this compression more likely.

Other proteins are required for the positioning of some nucleosome cores on DNA. In several organisms, proteins have been discovered that bind to a specific DNA sequence and then facilitate the formation of a nucleosome core nearby. Precise positioning of nucleosome cores can play a role in the expression of some eukaryotic genes (Chapter 28).

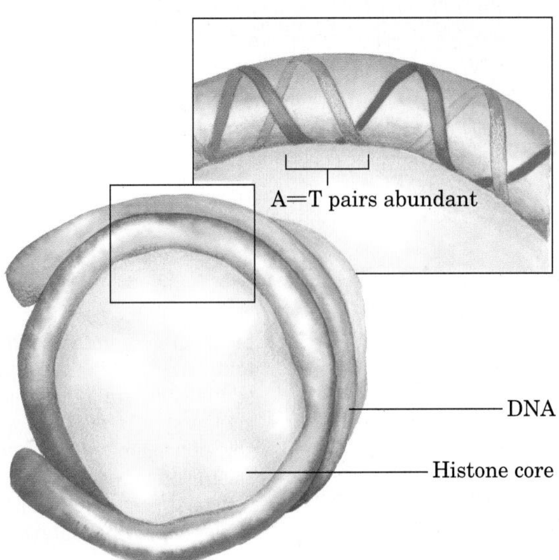

figure 24–26
Positioning of a nucleosome to make optimal use of A=T base pairs where the histone core is in contact with the minor groove of the DNA helix.

(a)

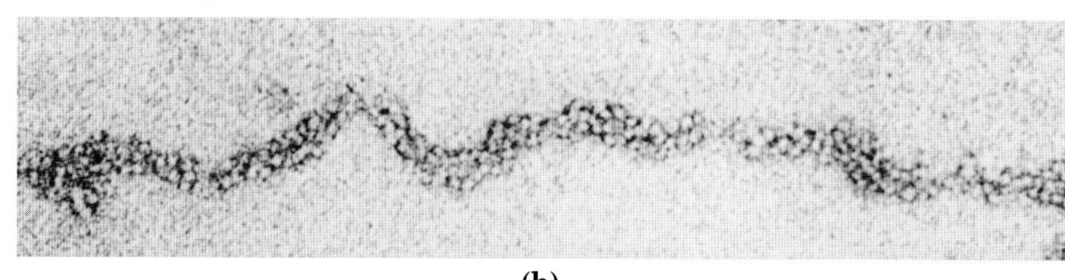

(b)

figure 24–27
The 30 nm fiber, a higher-order organization of nucleosomes. (a) Schematic illustration of the probable structure of the fiber, showing nucleosome packing. **(b)** Electron micrograph.

Nucleosomes Are Packed into Successively Higher-Order Structures

Wrapping of DNA around a nucleosome core compacts the DNA length about sevenfold. The overall compaction in a chromosome is greater than 10,000-fold, however—ample evidence for even higher orders of structural organization. In chromosomes isolated by very gentle methods, nucleosome cores appear to be organized into a structure called the **30 nm fiber** (Fig. 24–27). This packing requires one molecule of histone H1 per nucleosome core. Organization into 30 nm fibers does not extend over the entire chromosome but is punctuated by regions bound by sequence-specific (nonhistone) DNA-binding proteins. The 30 nm structure also appears to depend on the transcriptional activity of the particular region of DNA. Regions in which genes are being transcribed are apparently in a less-ordered state that contains little, if any, histone H1.

The 30 nm fiber, a second level of chromatin organization, provides an approximately 100-fold compaction of the DNA. The higher levels of folding are not yet understood, but it appears that certain regions of DNA associate with a nuclear scaffold (Fig. 24–28). The scaffold-associated regions are separated by loops of DNA with perhaps 20,000 to 100,000 base pairs. The DNA in a loop may contain a set of related genes. For example, in *Drosophila* complete sets of histone-coding genes seem to be clustered together in loops that are bounded by scaffold attachment sites (Fig. 24–29). The scaffold itself appears to contain several proteins, notably large amounts of histone H1 (located in the interior of the fiber) and topoisomerase II. The presence of topoisomerase II further emphasizes the relationship between DNA underwinding and chromatin assembly. Topoisomerase II is so important to the assembly of chromatin that inhibitors of this enzyme can kill rapidly dividing cells. Several drugs used in cancer chemotherapy are topoisomerase II inhibitors that allow the enzyme to promote strand breakage but not the resealing of the breaks.

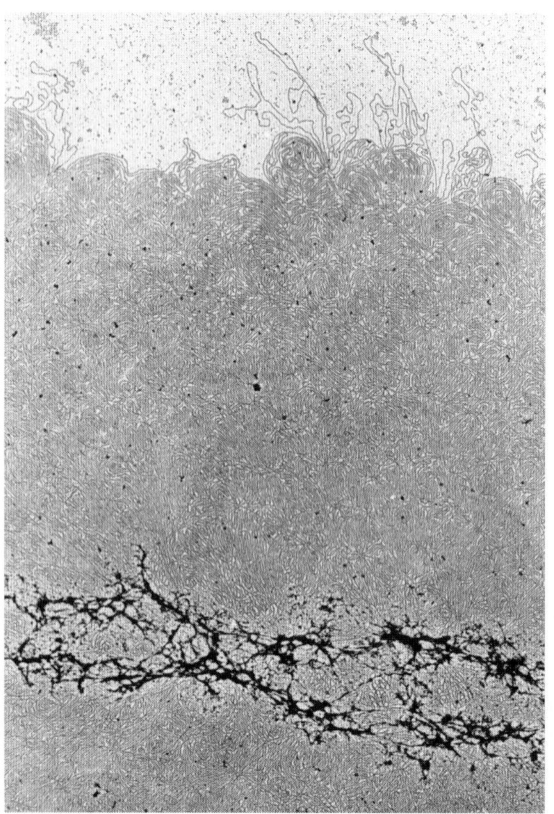

figure 24–28
A partially unraveled human chromosome, revealing numerous loops of DNA attached to a scaffoldlike structure.

figure 24–29
Loops of chromosomal DNA attached to a nuclear scaffold. The DNA in the loops is packaged as 30 nm fibers, so the loops are the next level of organization. Loops often contain groups of genes with related functions. Complete sets of histone-coding genes, as shown in this schematic illustration, appear to be clustered in loops of this kind. Unlike most genes, all histone genes occur in multiple copies in many eukaryotic genomes.

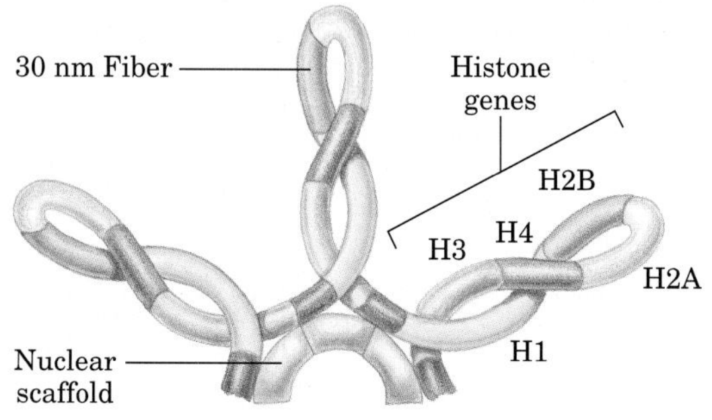

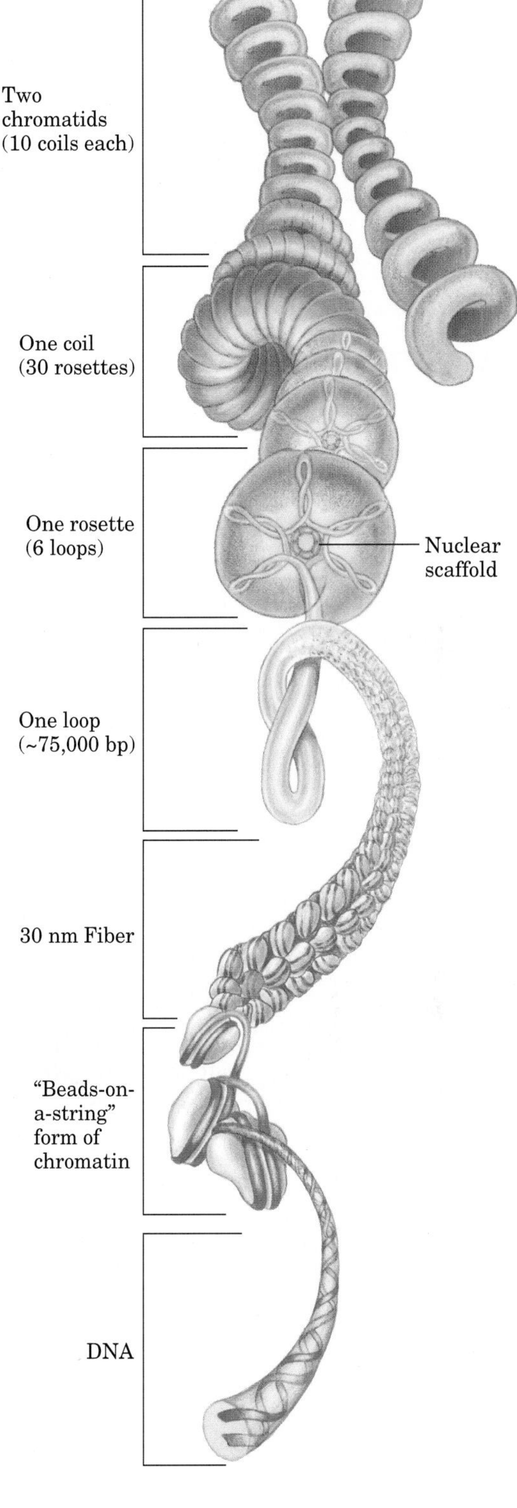

figure 24–30
Model for levels of organization that could provide DNA compaction in a eukaryotic chromosome. The levels take the form of coils upon coils. In cells, the higher-order structures (above the 30 nm fibers) are unlikely to be as uniform as depicted here.

Evidence exists for additional layers of organization in eukaryotic chromosomes, each dramatically enhancing the degree of compaction. One model for achieving this compaction is illustrated in Figure 24–30. Higher-order chromatin structure probably varies from chromosome to chromosome, from one region to the next in a single chromosome, and from moment to moment in the life of a cell. No single model can adequately describe these structures. Nevertheless, the principle is straightforward: DNA compaction in eukaryotic chromosomes is likely to involve coils upon coils upon coils . . .

Bacterial DNA Is Also Highly Organized

We now turn briefly to the structure of bacterial chromosomes. Bacterial DNA is compacted in a structure called the **nucleoid,** which occupies a large fraction of the cell volume (Fig. 24–31). The DNA of bacterial cells appears to be attached at one or more points to the inner surface of the plasma membrane. Much less is known about the structure of the nucleoid than of eukaryotic chromatin. In *E. coli,* a scaffoldlike structure appears to organize the circular chromosome into a series of looped domains, as described above for chromatin. The local organization provided by nucleosomes in eukaryotes does not seem to be duplicated by any comparable structure in bacterial DNA. Histonelike proteins are abundant in *E. coli*—the best-characterized example is a two-subunit protein called HU (M_r 19,000)—but these proteins bind and dissociate within minutes, and no regular, stable structure has been found. The bacterial chromosome is a relatively dynamic structure, possibly reflecting a requirement for more ready access to its genetic information. The bacterial cell division cycle can be as

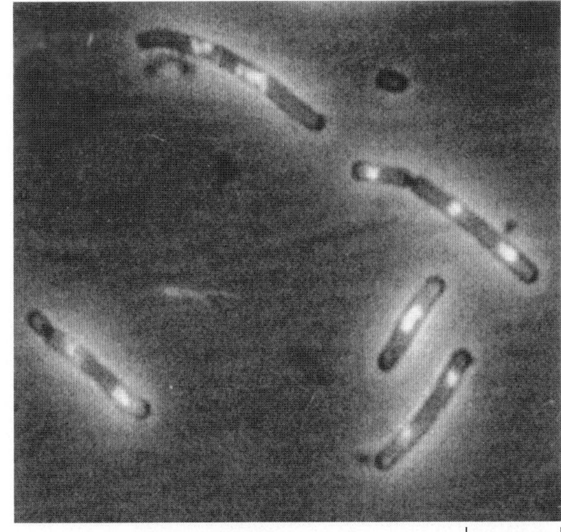

figure 24–31
***E. coli* cells showing nucleoids.** The DNA is stained with a dye that fluoresces when exposed to UV light. The light area defines the nucleoid. Note that some cells have replicated their DNA but have not yet undergone cell division and hence have multiple nucleoids.

short as 15 min, whereas a typical eukaryotic cell may not divide for hours or even months. In addition, a much greater fraction of prokaryotic DNA is used to encode RNA and/or protein products. High rates of cellular metabolism in bacteria mean that a much higher proportion of the DNA is being transcribed or replicated at a given time than in most eukaryotic cells.

With this overview of the complexity of DNA structure, we are now ready to turn, in the next chapter, to a discussion of DNA metabolism.

summary

Genes are segments of a chromosome that contain the information for a functional polypeptide or RNA molecule. In addition to genes, chromosomes contain a variety of regulatory sequences involved in replication, transcription, and other processes. Eukaryotic chromosomes have two important special-function repetitive DNA sequences: centromeres, which are attachment points for the mitotic spindle, and telomeres, which occur at the ends of chromosomes. Many genes in eukaryotic cells, and very occasionally in bacteria, are interrupted by noncoding sequences called introns. The coding segments separated by introns are called exons.

The DNA molecules in chromosomes are the largest macromolecules in cells. Many smaller DNAs also occur in cells, in the form of viral DNAs, plasmids, and (in eukaryotes) mitochondrial and chloroplast DNAs. Many DNAs, especially those in bacteria, mitochondria, and chloroplasts, are circular. Viral and chromosomal DNAs are generally much longer than the viral particles or cells in which they are packaged. The total DNA content of a eukaryotic cell is much greater than that of a bacterial cell.

Most cellular DNAs are supercoiled. Supercoiling is a manifestation of structural strain imparted by the underwinding of the DNA molecule. Underwinding decreases the total number of helical turns in the DNA relative to the relaxed or B form. To maintain an underwound state, DNA must be either a closed circle or bound to protein. Supercoils resulting from underwinding are termed negative supercoils. Underwinding is quantified by a topological parameter called linking number, *Lk*. The linking number of a relaxed, closed-circular DNA is used as a reference (Lk_0) and is equal to the number of base pairs divided by the number of base pairs per turn (approximately 10.5). Underwinding is measured in terms of the specific linking difference, σ, also called the superhelical density, which equals $(Lk - Lk_0)/Lk_0$. For cellular DNAs, σ typically equals -0.05 to -0.07, which means that approximately 5% to 7% of the helical turns in the DNA have been removed. DNA underwinding facilitates strand separation for enzymes involved in DNA metabolism. The plectonemic supercoils in negatively supercoiled DNA in solution are right-handed, and the overall structure is narrow and extended. Solenoidal supercoiling provides a much greater degree of compaction and predominates in the cell.

DNAs that differ only in their linking number are called topoisomers. Enzymes that underwind and/or relax DNA, the topoisomerases, catalyze changes in linking number. The two classes of topoisomerases, type I and type II, change *Lk* in increments of 1 or 2, respectively, per catalytic event. In a bacterial cell, the degree of supercoiling of the DNA is regulated by the activities of topoisomerases that increase and decrease linking number.

The fundamental unit of organization in the chromatin of eukaryotic cells is the nucleosome, which consists of histones and a 200 base pair segment of DNA. A core protein particle containing eight histones (two copies each of histones H2A, H2B, H3, and H4) is encircled by a segment of DNA (about 146 base pairs), which takes the form of a left-handed solenoidal supercoil. Nucleosomes are organized into 30 nm fibers, and the fibers themselves are extensively folded to provide the 10,000-fold compaction required to fit a typical eukaryotic chromosome into a cell nucleus. The higher-order folding involves attachment to a nuclear scaffold that contains large amounts of histone H1 and topoisomerase II. Bacterial chromosomes are also extensively compacted into a structure called a nucleoid, but the chromosome appears to be much more dynamic and irregular in structure than eukaryotic chromatin, reflecting the shorter cell cycle and very active metabolism of a bacterial cell.

further reading

General

Blattner, F.R., Plunkett, G., III, Bloch, C.A., Perna, N.T., Burland, V., Riley, M., Collado-Vides, J., Glasner, J.D., Rode, C.K., Mayhew, G.F., Gregor, J., Davis, N.W., Kirkpatrick, H.A., Goeden, M.A., Rose, D.J., Mau, B., & Shao, Y. (1997) The complete genome sequence of *Escherichia coli* K-12. *Science* **277,** 1453–1474.

New secrets of this common laboratory organism revealed.

Cozzarelli, N.R. & Wang, J.C. (eds) (1990) *DNA Topology and Its Biological Effects*, Cold Spring Harbor Laboratory Press, Cold Spring Harbor, NY.

Kornberg, A. & Baker, T.A. (1991) *DNA Replication,* 2nd edn, W.H. Freeman & Company, New York.

A good place to start for further information on the structure and function of DNA.

Lewin, B. (1997) *Genes VI,* Oxford University Press, New York.

Chapters 21 through 27 are especially relevant.

Lodish, H., Zipursky, S. L., Matsudaira, P., Baltimore, D., & Darnell, J. (1999) *Molecular Cell Biology,* 4th edn, W.H. Freeman & Co., New York.

Another excellent general reference.

Genes and Chromosomes

Goffeau, A., Barrell, B.G., Bussey, H., Davis, R.W., Dujon, B., Feldmann, H., Galibert, F., Hoheisel, J.D., Jacq, C., Johnston, M., Louis, E.J., Mewes, H.W., Murakami, Y., Philippsen, P., Tettelin, H., & Oliver, S.G. (1996) Life with 6000 genes. *Science* **274,** 546, 563–567.

First complete sequence of a eukaryotic genome.

Greider, C.W. & Blackburn, E.H. (1996) Telomeres, telomerase and cancer. *Sci. Am.* **274** (February), 92–97.

Huxley, C. (1997) Mammalian artificial chromosomes and chromosome transgenics. *Trends Genet.* **13,** 345–347.

Jurka, J. (1998) Repeats in genomic DNA: mining and meaning. *Curr. Opin. Struct. Biol.* **8,** 333–337.

Long, M., de Souza, S.J., & Gilbert, W. (1995) Evolution of the intron-exon structure of eukaryotic genes. *Curr. Opin. Genet. Dev.* **5,** 774–778.

Nugent C.I. & Lundblad, V. (1998) The telomerase reverse transcriptase: components and regulation. *Genes Dev.* **12,** 1073–1085.

Schmid, C.W. (1996) Alu: structure, origin, evolution, significance and function of one-tenth of human DNA. *Prog. Nucleic Acid Res. Mol. Biol.* **53,** 283–319.

Sharp, P.A. (1985) On the origin of RNA splicing and introns. *Cell* **42,** 397–400.

Willard, H.F. (1996) Chromosome manipulation: a systematic approach toward understanding human chromosome structure and function. *Proc. Natl Acad. Sci. U. S. A.* **93,** 6847–6850.

Zakian, V.A. (1996) Structure, function, and replication of *Saccharomyces cerevisiae* telomeres. *Annu. Rev. Genet.* **30,** 141–172.

Supercoiling and Topoisomerases

Berger, J.M. (1998) Type II DNA topoisomerases. *Curr. Opin. Struct. Biol.* **8,** 26–32.

Boles, T.C., White, J.H., & Cozzarelli, N.R. (1990) Structure of plectonemically supercoiled DNA. *J. Mol. Biol.* **213,** 931–951.

A study that defines several fundamental features of supercoiled DNA.

Cozzarelli, N.R., Boles, T.C., & White, J.H. (1990) Primer on the topology and geometry of DNA supercoiling. In *DNA Topology and Its Biological Effects* (Cozzarelli, N.R. & Wang, J.C., eds), pp. 139–184, Cold Spring Harbor Laboratory Press, Cold Spring Harbor, NY.

This provides a more advanced and thorough discussion.

Lebowitz, J. (1990) Through the looking glass: the discovery of supercoiled DNA. *Trends Biochem. Sci.* **15,** 202–207.

A short and interesting historical note.

Wang, J.C. (1996) DNA topoisomerases. *Annu. Rev. Biochem.* **65,** 635–692.

Chromatin and Nucleosomes

Felsenfeld, G., Boyes, J., Chung, J., Clark, D., & Studitsky, V. (1996) Chromatin structure and gene expression. *Proc. Natl Acad. Sci. U. S. A.* **93,** 9384–9388.

Filipski, J., Leblanc, J., Youdale, T., Sikorska, M., & Walker, P.R. (1990) Periodicity of DNA folding in higher order chromatin structures. *EMBO J.* **9,** 1319–1327.

Gasser, S.M. (1995) Chromosome structure: coiling up chromosomes. *Curr. Biol.* **5,** 357–360.

Kornberg, R.D. (1974) Chromatin structure: a repeating unit of histones and DNA. *Science* **184,** 868–871.

The classic paper that introduced the subunit model for chromatin.

Travers, A.A. (1994) Chromatin structure and dynamics. *Bioessays* **16,** 657–662.

Wolffe, A.P. (1994) Nucleosome positioning and modification: chromatin structures that potentiate transcription. *Trends Biochem. Sci.* **19,** 240–244.

Zlatanova, J. & van Holde, K. (1996) The linker histones and chromatin structure: new twists. *Prog. Nucleic Acid Res. Mol. Biol.* **52,** 217–259.

problems

1. Packaging of DNA in a Virus Bacteriophage T2 has a DNA of molecular weight 120×10^6 contained in a head about 210 nm long. Calculate the length of the DNA (assume the molecular weight of a nucleotide pair is 650) and compare it with the length of the T2 head.

2. The DNA of Phage M13 The base composition of phage M13 DNA is A, 23%; T, 36%; G, 21%; C, 20%. What does this tell you about the DNA of phage M13?

3. The *Mycoplasma* Genome The complete genome of the simplest bacterium known, *Mycoplasma genitalium,* is a circular DNA molecule with 580,070 base pairs. Calculate the molecular weight and contour length (when relaxed) of this molecule. What is Lk_0 for the *Mycoplasma* chromosome? If $\sigma = -0.06$, what is Lk?

4. Size of Eukaryotic Genes An enzyme isolated from rat liver has 192 amino acid residues and is coded for by a gene with 1,440 base pairs. Explain the relationship between the number of amino acid residues in the enzyme and the number of nucleotide pairs in its gene.

5. Linking Number A closed-circular DNA molecule in its relaxed form has an Lk of 500. Approximately how many base pairs are in this DNA? How is the linking number altered (increases, decreases, doesn't change, becomes undefined) when (a) a protein complex is bound to form a nucleosome, (b) one DNA strand is broken, (c) DNA gyrase and ATP are added to the DNA solution, or (d) the double helix is denatured by heat?

6. Superhelical Density Bacteriophage λ infects *E. coli* by integrating its DNA into the bacterial chromosome. The success of this recombination depends on the topology of the *E. coli* DNA. When the superhelical density (σ) of the *E. coli* DNA is greater than -0.045, the probability of integration is $<20\%$; when σ is less than -0.06, the probability is $>70\%$.

DNA isolated from an *E. coli* culture is found to have a length of 13,800 base pairs and an Lk of 1,222. Calculate σ for this DNA and predict the likelihood that bacteriophage λ will be able to infect this culture.

7. DNA Structure Explain how the underwinding of a B-DNA helix might facilitate or stabilize the formation of Z-DNA.

8. Chromatin Early evidence that helped researchers define nucleosome structure is illustrated by the agarose gel below, in which the thick bands represent DNA. It was generated by briefly treating chromatin with an enzyme that degrades DNA, then removing all protein and subjecting the purified DNA to electrophoresis. Numbers at the side of the gel denote the position to which a linear DNA of the indicated size (in base pairs) would migrate. What does this gel tell you about chromatin structure? Why are the DNA bands thick and spread out rather than sharply defined?

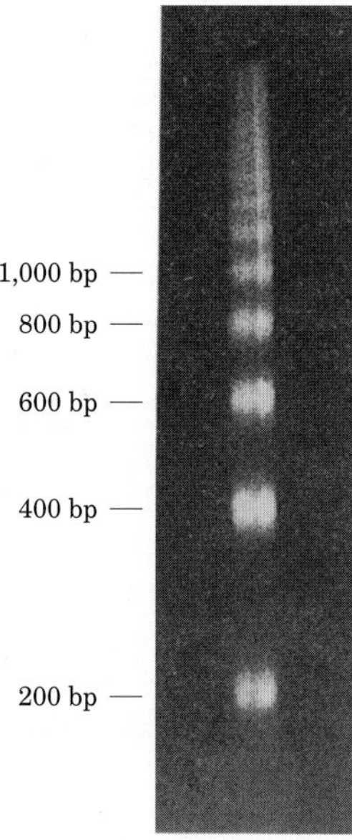

9. Yeast Artificial Chromosomes (YACs) YACs are used to clone large pieces of DNA in yeast cells. What three types of DNA sequences are required to ensure proper replication and propagation of a YAC in a yeast cell?

chapter

25

DNA Metabolism

DNA occupies a unique and central place among biological macromolecules as the repository of genetic information. The nucleotide sequences of DNA encode the primary structures of all cellular RNAs and proteins and, through enzymes, indirectly affect the synthesis of all other cellular constituents. This passage of information from DNA to RNA and protein guides the size, shape, and functioning of every living thing.

DNA is a marvelous device for the stable storage of genetic information. The phrase "stable storage," however, conveys a static and misleading picture. It fails to capture the complexity of processes by which genetic information is preserved in an uncorrupted state and then transmitted from one generation of cells to the next. DNA metabolism comprises both the process by which copies of DNA molecules are faithfully made (replication) and the processes that affect the inherent structure of the information (repair and recombination). Together, these activities are the focus of this chapter.

DNA metabolism is shaped by the requirement for an exquisite degree of accuracy. The chemistry of joining one nucleotide to the next in DNA replication is elegant and simple, almost deceptively so. As we will see, complexity arises in the form of enzymatic devices to ensure that the genetic information is transmitted intact. Uncorrected errors that arise during DNA synthesis can have dire consequences, not only because they can permanently affect or eliminate the function of a gene but also because the change is inheritable.

The enzymes that synthesize DNA may copy DNA molecules that contain millions of bases. They do so with extraordinary fidelity and speed, even though the DNA substrate is highly compacted and bound with other proteins. Formation of phosphodiester bonds to link nucleotides in the backbone of a growing DNA strand is therefore only one part of an elaborate process that requires myriad proteins and enzymes.

Maintaining the integrity of genetic information lies at the heart of our discussion of DNA repair. As detailed in Chapter 10, DNA is susceptible to many types of damaging reactions. Although such reactions occur infrequently, they are significant nevertheless because of the very low biological tolerance for changes in DNA sequence. DNA is the only macromolecule for which repair systems exist; the number, diversity, and complexity of DNA repair mechanisms reflect the wide range of insults that can harm DNA.

Cells often rearrange their genetic information by processes collectively called recombination—seemingly undermining the principle that the stability and integrity of genetic information is paramount. However, most DNA rearrangements in fact play constructive roles in maintaining genomic integrity, contributing in special ways to DNA replication, DNA repair, and chromosome segregation.

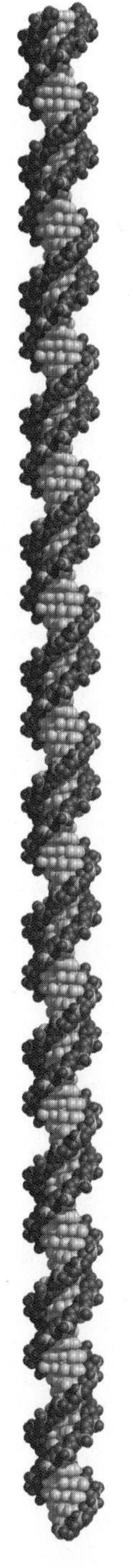

B-form DNA

Special emphasis is given in this chapter to the enzymes of DNA metabolism. They merit careful study, both for their increasing importance in medicine and for their everyday use as reagents in a wide range of modern biochemical technologies. Many of the seminal discoveries in DNA metabolism have been made with *Escherichia coli,* so its well-understood enzymes are generally used to illustrate the ground rules. A quick look at some relevant genes on the *E. coli* genetic map (Fig. 25–1) provides just a hint of what is to come.

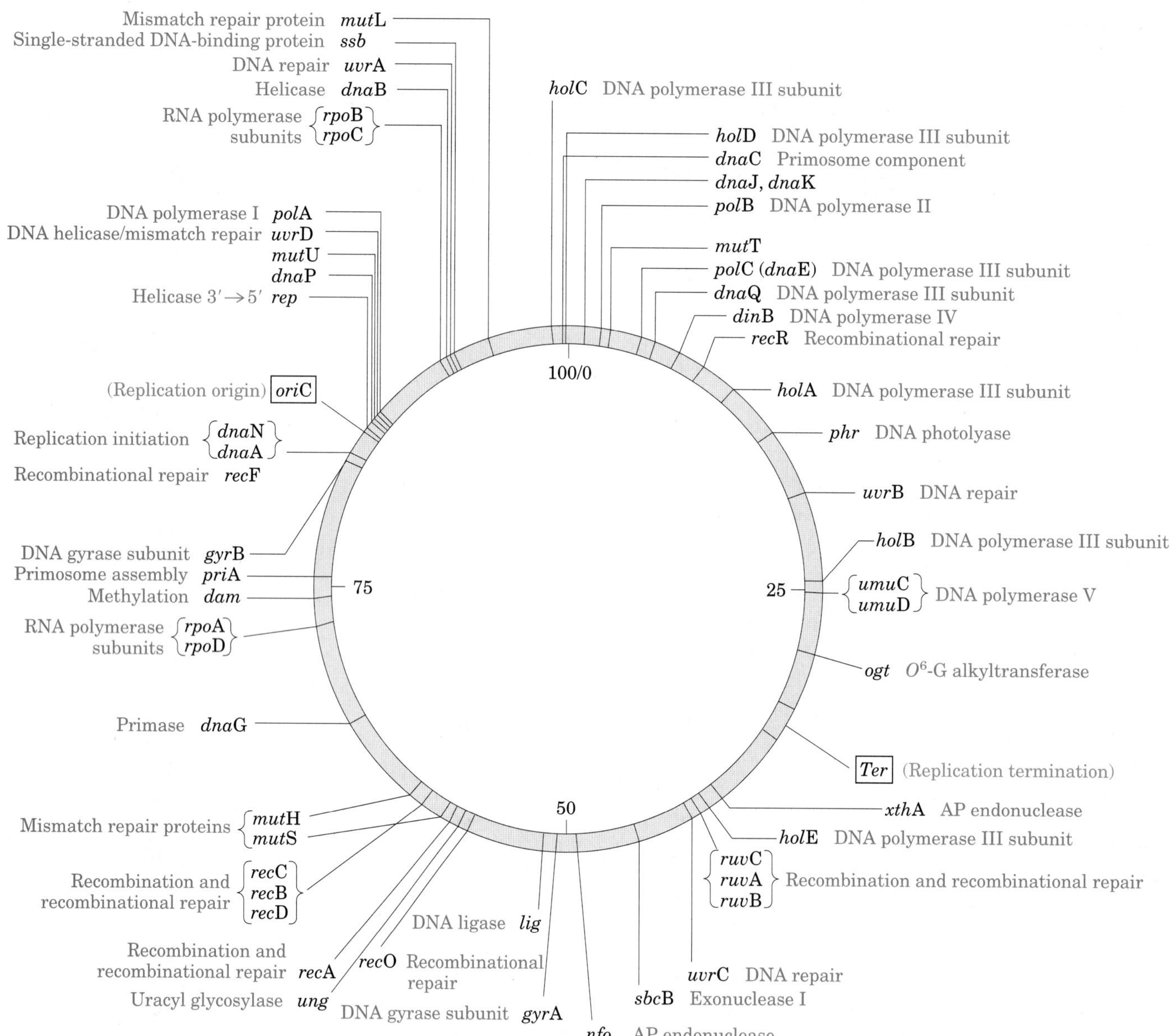

figure 25–1

Map of the *E. coli* chromosome. The map shows the relative positions of genes encoding many of the proteins important in DNA metabolism. The number of genes known to be involved provides a hint of the complexity of these processes. The numbers 0 to 100 inside the circular chromosome denote a genetic measurement called minutes. Each minute corresponds to about 40,000 base pairs along the DNA molecule of *E. coli.* The three-letter names of genes generally reflect some aspect of their function. These include *mut, mut*agenesis; *dna,* DNA replication; *pol,* DNA *pol*ymerase; *rpo,* RNA *po*lymerase; *uvr, UV-r*esistance; *rec, rec*ombination; *Ter, ter*mination of replication; *ori, ori*gin of replication; *dam,* DNA *a*denine *m*ethylation; and *lig,* DNA *lig*ase.

A Word about Terminology

Before beginning to look closely at replication, we must make a short digression into the use of abbreviations in naming genes and proteins. By convention, bacterial genes generally are named using three lowercase, italicized letters that reflect their apparent function. For example, the *dna, uvr,* and *rec* genes affect *DNA* replication, *r*esistance to the damaging effects of *UV* radiation, and *rec*ombination, respectively. Where several genes affect the same process, the letters A, B, C, and so forth, are added—usually reflecting their order of discovery rather than their order in a reaction sequence.

During genetic investigations, the protein product of each gene is usually isolated and characterized. Many bacterial genes have been identified and named before the roles of their protein products are understood in detail. Sometimes the gene product is found to be a previously isolated protein, and some renaming is in order. The *dna*E gene, for example, was found to encode the polymerizing subunit of DNA polymerase III, so this gene was renamed *pol*C to better reflect its function. Often the protein product turns out to be new, with an activity not easily described by a simple enzyme name. In a practice that can be confusing, these bacterial proteins often retain the name of their genes. When referring to the protein, italic type is not used and the first letter is capitalized: the *dna*A and *rec*A gene products are called the DnaA and RecA proteins, respectively. You will encounter many such examples in this chapter.

Similar conventions exist for the naming of eukaryotic genes, although the exact form of the abbreviations may vary and no single convention applies to all eukaryotic systems.

DNA Replication

Long before the structure of DNA was known, scientists wondered at the ability of organisms to create faithful copies of themselves and, later, at the ability of cells to produce many identical copies of large and complex macromolecules. Speculation about these problems centered around the concept of a **template,** a structure that would allow molecules to be lined up in a specific order and joined, to create a macromolecule with a unique sequence and function. The 1940s brought the revelation that DNA was the genetic molecule, but not until James Watson and Francis Crick deduced its structure did it become clear how DNA could act as a template for the replication and transmission of genetic information: *one strand is the complement of the other.* The strict base-pairing rules mean that each strand provides the template for a sister strand with a predictable and complementary sequence (see Figs 10–16, 10–17).

The fundamental properties of the DNA replication process and the mechanisms used by the enzymes that catalyze it have proved to be essentially identical in all organisms. This mechanistic unity is a major theme as we proceed from general properties of the replication process, to *E. coli* replication enzymes, and, finally, to replication in eukaryotes.

DNA Replication Is Governed by a Set of Fundamental Rules

DNA Replication Is Semiconservative Each DNA strand serves as a template for the synthesis of a new strand, producing two new DNA molecules, each with one new strand and one old strand. This is **semiconservative replication.**

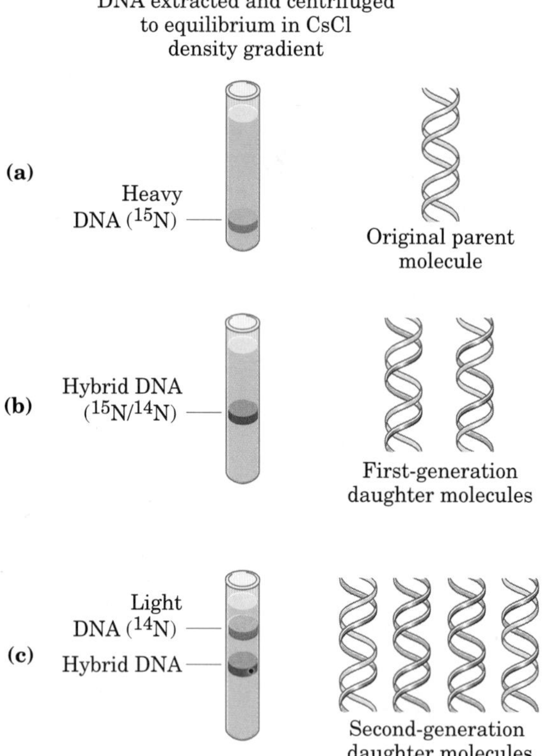

figure 25–2
The Meselson-Stahl experiment. (a) Cells were grown for many generations in a medium containing only heavy nitrogen, ^{15}N, so that all the nitrogen in their DNA was ^{15}N, as shown by a single band (blue) when centrifuged in a CsCl density gradient. **(b)** Once the cells had been transferred to a medium containing only light nitrogen, ^{14}N, cellular DNA isolated after one generation equilibrated at a higher position in the density gradient (purple band). **(c)** Continuation of replication for a second generation yielded two hybrid DNAs and two light DNAs (red), confirming semiconservative replication.

The hypothesis of semiconservative replication was proposed by Watson and Crick soon after publication of their 1953 paper on the structure of DNA, and was proved by ingeniously designed experiments carried out by Matthew Meselson and Franklin Stahl in 1957. Meselson and Stahl grew *E. coli* cells for many generations in a medium in which the sole nitrogen source (NH_4Cl) contained ^{15}N, the "heavy" isotope of nitrogen, instead of the normal, more abundant "light" isotope, ^{14}N. The DNA isolated from these cells had a density about 1% greater than that of normal [^{14}N]DNA (Fig. 25–2a). Although this is only a small difference, a mixture of heavy [^{15}N]DNA and light [^{14}N]DNA can be separated by centrifugation to equilibrium in a cesium chloride density gradient.

The *E. coli* cells grown in the ^{15}N medium were transferred to a fresh medium containing only the ^{14}N isotope, where they were allowed to grow until the cell population had just doubled. The DNA isolated from these first-generation cells formed a *single* band in the CsCl gradient at a position indicating that the double-helical DNAs of the daughter cells were hybrids containing one new ^{14}N strand and one parental ^{15}N strand (Fig. 25–2b).

This result argued against conservative replication, an alternative hypothesis in which one progeny DNA molecule would consist of two newly synthesized DNA strands and the other would contain the two parental strands; this would not yield hybrid DNA molecules in the Meselson-Stahl experiment. The semiconservative replication hypothesis was further supported in the next step of the experiment (Fig. 25–2c). Cells were again allowed to double in number in the ^{14}N medium. The isolated DNA product of this second cycle of replication exhibited *two* bands in the density gradient, one with a density equal to that of light DNA and the other with the density of the hybrid DNA observed after the first cell doubling.

Replication Begins at an Origin and Usually Proceeds Bidirectionally Following the confirmation of a semiconservative mechanism of replication, a host of questions arose. Are the parental DNA strands completely unwound before each is replicated? Does replication begin at random places or at a unique point? After initiation at any point in the DNA, does replication proceed in one direction or both?

An early indication that replication is a highly coordinated process in which the parental strands are simultaneously unwound and replicated was provided by John Cairns, using autoradiography. He made *E. coli* DNA radioactive by growing cells in a medium containing thymidine labeled with tritium (^3H). When the DNA was carefully isolated, spread, and overlaid with a photographic emulsion for several weeks, the radioactive thymidine residues generated "tracks" of silver grains in the emulsion, producing an image of the DNA molecule. These tracks revealed that the intact chromosome of *E. coli* is a single huge circle, 1.7 mm long. Radioactive DNA isolated from cells during replication showed an extra loop (Fig. 25–3a). Cairns concluded that the loop resulted from the formation of two radioactive daughter strands, each complementary to a parental strand. One or both ends of the loop are dynamic points, termed **replication forks,** where parental DNA is being unwound and the separated strands quickly replicated. Cairns's results demonstrated that both DNA strands are replicated simultaneously, and a variation of his experiment (Fig. 25–3b) indicated that replication of bacterial chromosomes is bidirectional: both ends of the loop have active replication forks.

To determine whether the replication loops originate at a unique point in the DNA, landmarks were needed along the DNA molecule. These were provided by a technique called **denaturation mapping,** developed by Ross

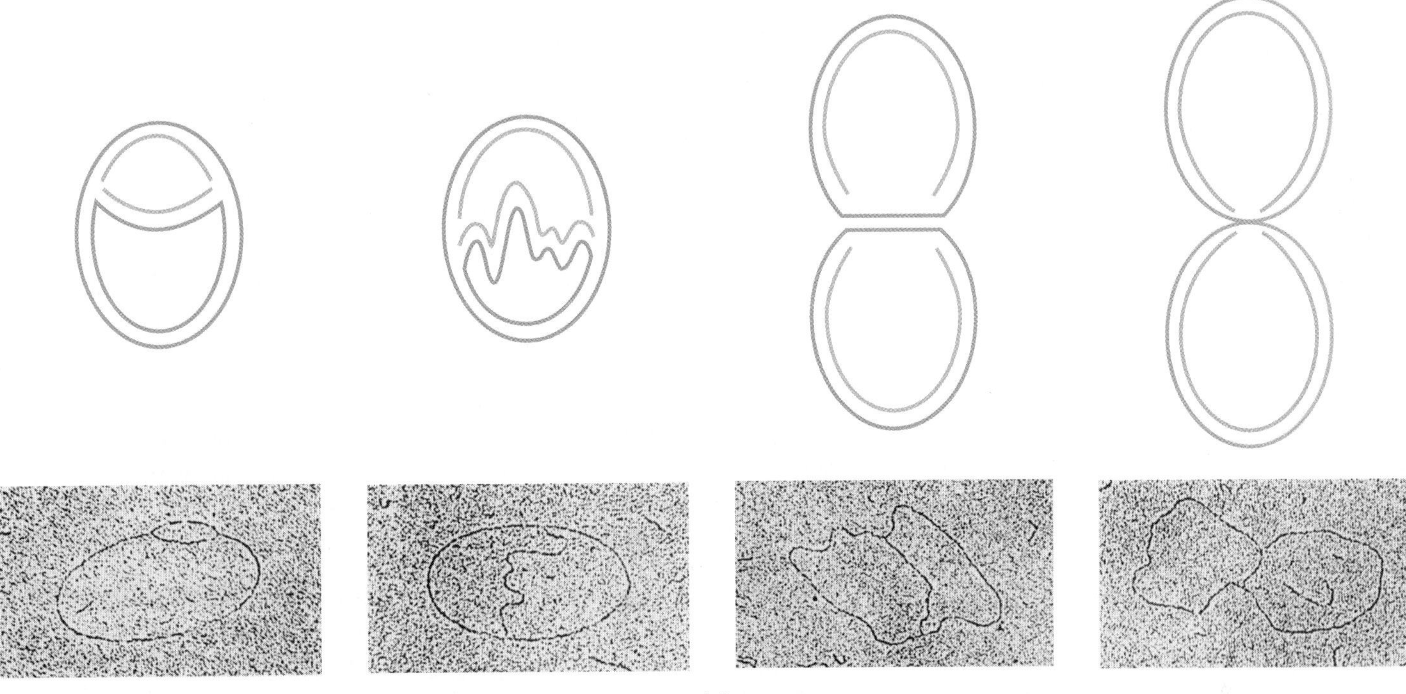

(a)

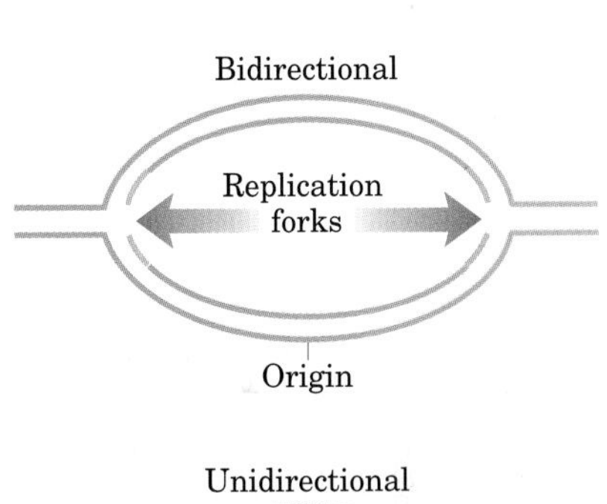

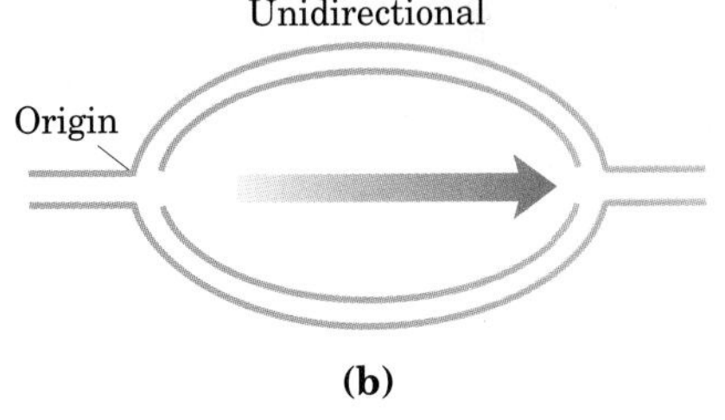

(b)

figure 25–3
Visualization of bidirectional DNA replication. Replication of a circular chromosome produces a structure resembling the Greek letter theta (θ). **(a)** Labeling with tritium (^3H) shows that both strands are replicated at the same time (new strands shown in red). The electron micrographs illustrate the replication of a circular *E. coli* plasmid as visualized by autoradiography. **(b)** Addition of ^3H for a short period just before the reaction is stopped allows a distinction to be made between unidirectional and bidirectional replication, by determining whether label (red) is found at one or both replication forks in autoradiograms. This technique has revealed bidirectional replication in *E. coli, Bacillus subtilis,* and other bacteria.

Inman and colleagues. Using the 48,502 base pair chromosome of bacteriophage λ, Inman showed that DNA could be selectively denatured at sequences unusually rich in A═T base pairs, generating a reproducible pattern of single-stranded bubbles (see Fig. 10–31). Isolated DNAs containing replication loops can be partially denatured in the same way. This allows the position and progress of the replication forks to be measured and mapped, using the denatured regions as points of reference. The technique revealed that the replication loops always initiate at a unique point, called an **origin.** It also confirmed the earlier observation that replication is usually bidirectional. For circular DNA molecules, the two replication forks meet at a point on the side of the circle opposite to the origin.

DNA Synthesis Proceeds in a 5′→3′ Direction and Is Semidiscontinuous

A new strand of DNA is always synthesized in the 5′→3′ direction, with the free 3′ OH being the point at which the DNA is elongated (the 5′ and 3′ ends of a DNA strand are defined in Figure 10–7). Because the two DNA strands are antiparallel, the strand serving as the template is read from its 3′ end toward its 5′ end.

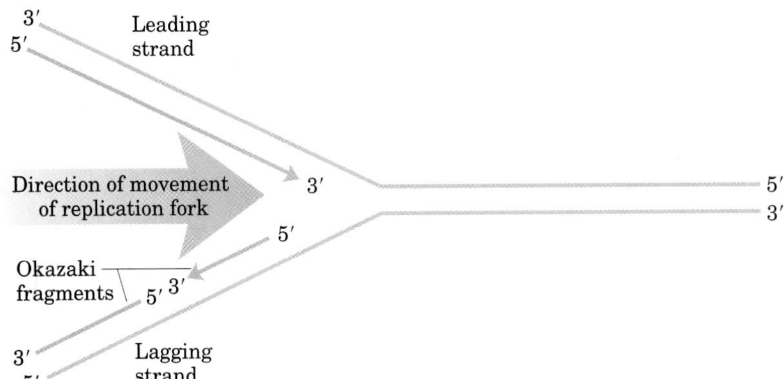

figure 25–4

Defining DNA strands at the replication fork. A new DNA strand (red) is always synthesized in the 5′→3′ direction. The template is read in the opposite direction, 3′→5′. The leading strand is continuously synthesized in the direction taken by the replication fork. The other strand, the lagging strand, is synthesized discontinuously in short pieces (Okazaki fragments) in a direction opposite to that in which the replication fork moves. The Okazaki fragments are then spliced together by DNA ligase. In bacteria, Okazaki fragments are about 1,000 to 2,000 nucleotides long. In eukaryotic cells, they are 150 to 200 nucleotides long.

If synthesis always proceeds in the 5′→3′ direction, how can both strands be synthesized simultaneously? If both strands were synthesized *continuously* while the replication fork moved, one strand would have to undergo 3′→5′ synthesis. This problem was resolved by Reiji Okazaki and colleagues in the 1960s. Okazaki found that one of the new DNA strands is synthesized in short pieces, now called **Okazaki fragments.** This work ultimately led to the conclusion that one strand is synthesized continuously and the other discontinuously (Fig. 25–4). The continuous or **leading strand** is the one in which 5′→3′ synthesis proceeds in the *same* direction as replication fork movement. The discontinuous or **lagging strand** is the one in which 5′→3′ synthesis proceeds in the direction *opposite* to the direction of fork movement. Okazaki fragments range in length from a few hundred to a few thousand nucleotides, depending on the cell type. As we will see later, leading and lagging strand syntheses are tightly coordinated.

DNA Is Degraded by Nucleases

To explain the enzymology of DNA replication, we first introduce the enzymes that degrade DNA rather than synthesize it. These enzymes are called **nucleases,** or **DNases** if they are specific for DNA rather than RNA. Every cell contains several different nucleases, belonging to two broad classes: exonucleases and endonucleases. **Exonucleases** degrade nucleic acids from one end of the molecule. Many operate in only the 5′→3′ or the 3′→5′ direction, removing nucleotides only from the 5′ or the 3′ end, respectively, of one strand of a double-stranded nucleic acid. **Endonucleases** can begin to degrade at any site in a nucleic acid strand or molecule, reducing it to smaller and smaller fragments. A few exonucleases and endonucleases degrade only single-stranded DNA. There are a few important classes of endonucleases that cleave only at specific nucleotide sequences (e.g., the restriction endonucleases that are so important in biotechnology; see Chapter 29). You will encounter many types of nucleases in this and subsequent chapters.

Arthur Kornberg

DNA Is Synthesized by DNA Polymerases

The search for an enzyme that could synthesize DNA began in 1955. Work by Arthur Kornberg and colleagues led to the purification and characterization of DNA polymerase from *E. coli* cells, a single-polypeptide enzyme now called **DNA polymerase I** (M_r 103,000). Much later, investigators found that *E. coli* contains at least four other distinct DNA polymerases, described below.

Detailed studies of DNA polymerase I revealed features of the DNA synthetic process that are now known to be common to all DNA polymerases. The fundamental reaction is a nucleophilic attack by the 3′-hydroxyl group

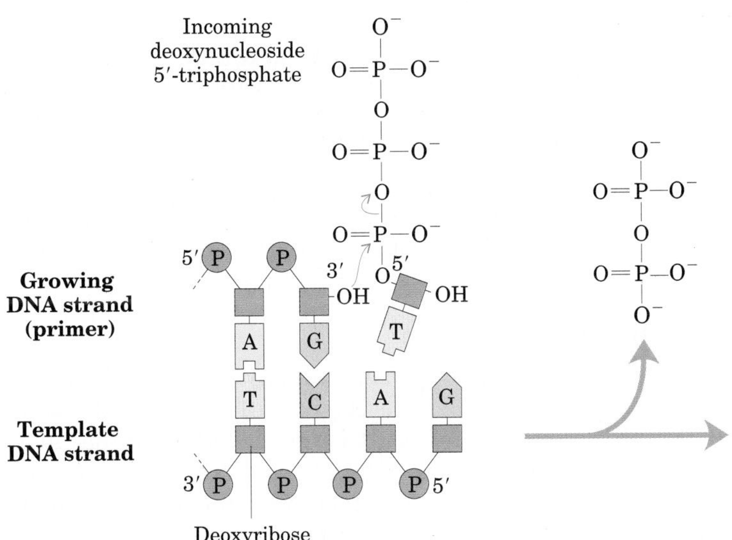

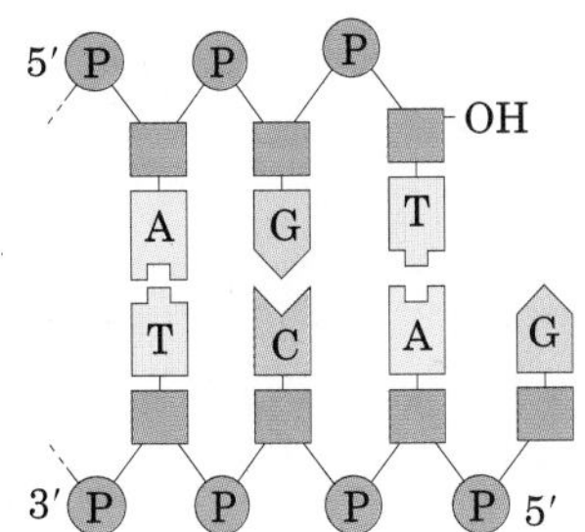

figure 25–5

Elongation of a DNA chain. A single unpaired strand is required to act as template, and a primer strand is needed to provide a free hydroxyl group at the 3′ end to which a new nucleotide unit is added. Each incoming nucleotide is selected in part by base pairing to the appropriate nucleotide in the template strand. The reaction product has a new free 3′ OH, allowing the addition of another nucleotide.

of the nucleotide at the 3′ end of the growing strand on the 5′-α-phosphorus of the incoming deoxynucleoside 5′-triphosphate (Fig. 25–5). Inorganic pyrophosphate is released in the reaction. The general reaction is

$$\underset{\text{DNA}}{(\text{dNMP})_n} + \text{dNTP} \longrightarrow \underset{\text{Lengthened DNA}}{(\text{dNMP})_{n+1}} + \text{PP}_i \qquad (25\text{–}1)$$

where dNMP and dNTP are deoxynucleoside 5′-monophosphate and 5′-triphosphate, respectively. The reaction has an apparent thermodynamic balance in that one phosphodiester bond is formed and one destroyed. However, noncovalent base-stacking and base-pairing interactions help to stabilize the lengthened DNA product relative to the free nucleotide. Also, the formation of products is driven in the cell by the subsequent hydrolysis of the pyrophosphate by the enzyme pyrophosphatase.

Early work on DNA polymerase I led to the definition of two central requirements for DNA polymerization. First, all DNA polymerases require a **template.** The polymerization reaction is guided by a template DNA strand according to the base-pairing rules predicted by Watson and Crick: where a guanine is present in the template, a cytosine is added to the new strand, and so on. This was a particularly important discovery, not only because it provided a chemical basis for accurate semiconservative DNA replication, but because it represented the first example of the use of a template to guide a biosynthetic reaction.

Second, a **primer** is required. A primer is a strand segment (complementary to the template) with a free 3′-hydroxyl group to which a nucleotide can be added. The free 3′ end of the primer is called the **primer terminus.** In other words, part of the new strand must already be in place; all DNA polymerases can only add nucleotides to a preexisting strand. Primers are often oligonucleotides of RNA rather than DNA, and specialized enzymes synthesize primers when and where they are required.

After adding a nucleotide to a growing DNA strand, DNA polymerase either dissociates or moves along the template and adds another nucleotide. Dissociation and reassociation of the polymerase can limit the overall polymerization rate—the process is generally faster if a polymerase adds nucleotides without dissociating from the template. The average number of nucleotides added before a polymerase dissociates defines its **processivity.** DNA polymerases vary greatly in processivity; some add just a few nucleotides before dissociating, others add many thousands.

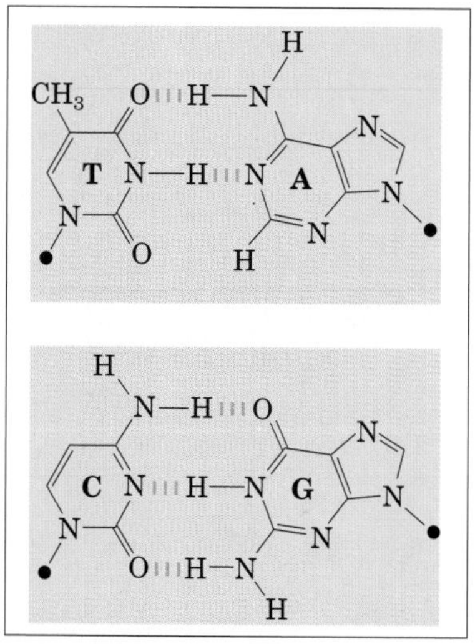

figure 25–6

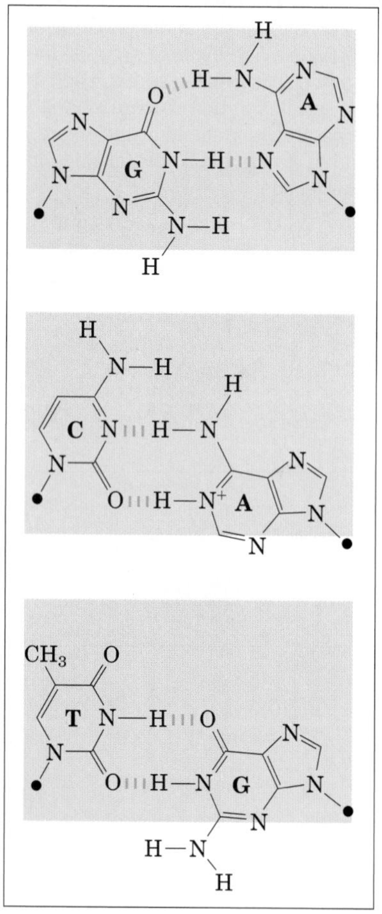

Contribution of base pair geometry to the fidelity of DNA replication. (a) The standard A═T and G≡C base pairs have very similar geometries, and an active site sized to fit one (blue tint) will generally accommodate the other. **(b)** The geometry of incorrectly paired bases can exclude them from the active site. DNA polymerase I has an active site that excludes such incorrect pairing partners.

Replication Is Very Accurate

Replication proceeds with an extraordinary degree of fidelity. In *E. coli,* a mistake is made only once for every 10^9 to 10^{10} nucleotides added. For the *E. coli* chromosome of 4.6×10^6 base pairs, this means that an error occurs only once per 1,000 to 10,000 replications. During polymerization, discrimination between correct and incorrect nucleotides relies not only on the hydrogen bonds that specify the correct pairing between complementary bases, but also on the common geometry of the standard A═T and G≡C base pairs (Fig. 25–6). DNA polymerase I has an active site that accommodates only base pairs with this geometry. An incorrect nucleotide may be able to hydrogen bond with a base in the template, but it generally will not fit into the active site. Incorrect bases can be rejected before the phosphodiester bond is formed.

The accuracy of the polymerization reaction itself, however, is insufficient to account for the high degree of fidelity in replication. Careful measurements in vitro have shown that DNA polymerases insert one incorrect nucleotide for every 10^4 to 10^5 correct ones. These mistakes sometimes occur because a base is briefly in an unusual tautomeric form (see Fig. 10–9), allowing it to hydrogen-bond with an incorrect partner. The error rate is reduced in vivo by additional enzymatic mechanisms.

One mechanism intrinsic to virtually all DNA polymerases is a separate 3′→5′ exonuclease activity that double-checks each nucleotide after it is added. This nuclease activity permits the enzyme to remove a newly added nucleotide and is highly specific for mismatched base pairs (Fig. 25–7). If the polymerase has added the wrong nucleotide, translocation of the enzyme to the position where the next nucleotide is to be added is inhibited. The 3′→5′ exonuclease activity removes the mispaired nucleotide, and the polymerase begins again. This activity, called **proofreading,** is not simply the reverse of the polymerization reaction (Eqn 25–1) because pyrophosphate is not involved. The polymerizing and proofreading activities of a DNA polymerase can be measured separately. Proofreading improves the inherent accuracy of the polymerization reaction 10^2- to 10^3-fold. In the monomeric DNA polymerase I, the polymerizing and proofreading activities have separate active sites within the same polypeptide.

The same base-pairing interactions that guide polymerization are used to discriminate between correct and incorrect bases during proofreading. The use of complementary noncovalent interactions to ensure correct base pairing in two successive steps is common in the synthesis of information-containing molecules. An equivalent strategy ensures the fidelity of protein synthesis (Chapter 27).

When base selection and proofreading are combined, DNA polymerase leaves behind one net error for every 10^6 to 10^8 bases added. Yet the measured accuracy of replication in *E. coli* is higher still. The additional accuracy is accounted for by a separate enzyme system that repairs the mismatched base pairs remaining after replication. This mismatch repair is described with other DNA repair processes later in the chapter.

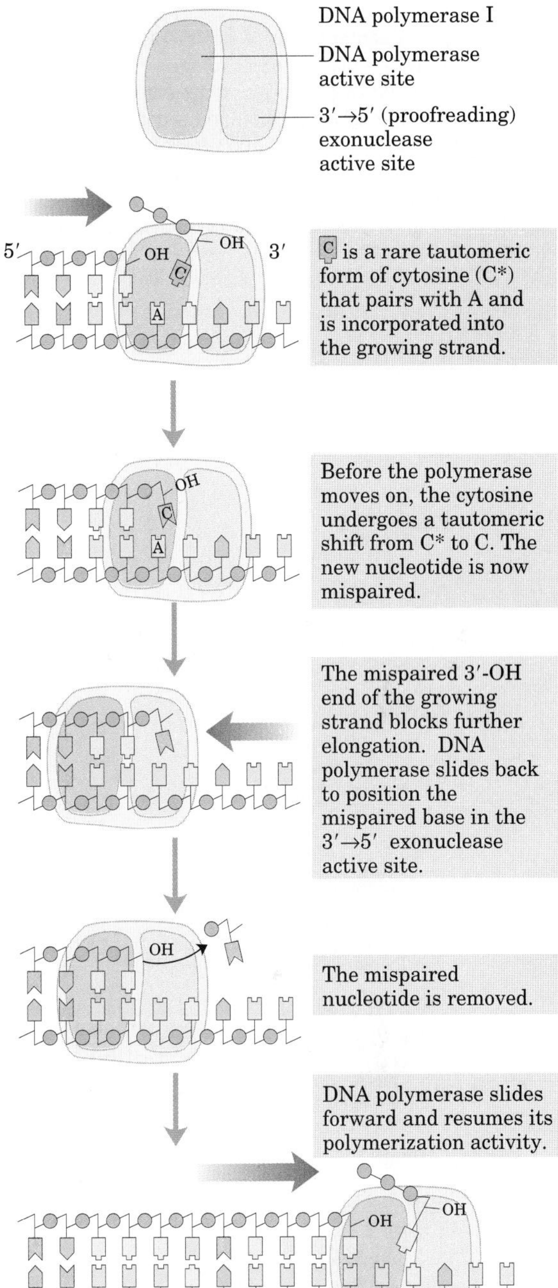

DNA polymerase I
DNA polymerase active site
3′→5′ (proofreading) exonuclease active site
5′
OH
OH
3′
C
A
C is a rare tautomeric form of cytosine (C*) that pairs with A and is incorporated into the growing strand.
Before the polymerase moves on, the cytosine undergoes a tautomeric shift from C* to C. The new nucleotide is now mispaired.
The mispaired 3′-OH end of the growing strand blocks further elongation. DNA polymerase slides back to position the mispaired base in the 3′→5′ exonuclease active site.
The mispaired nucleotide is removed.
DNA polymerase slides forward and resumes its polymerization activity.

figure 25–7

An example of error correction by the 3′→5′ exonuclease activity of DNA polymerase I. Structural analysis has located the exonuclease activity ahead of the polymerase activity as the enzyme is oriented in its movement along the DNA. A mismatched base (here, a C–A mismatch) impedes translocation of DNA polymerase I to the next site. Sliding backward, the enzyme corrects the mistake with its 3′→5′ exonuclease activity, then resumes its polymerase activity in the 5′→3′ direction.

E. coli Has at Least Five DNA Polymerases

More than 90% of the DNA polymerase activity observed in *E. coli* extracts can be accounted for by DNA polymerase I. Soon after the isolation of this enzyme in 1955, however, evidence began to accumulate that it is not suited for replication of the large *E. coli* chromosome. First, the rate at which it adds nucleotides (600 nucleotides/min) is too slow by a factor of 100 or more to account for the rates at which the replication fork is observed to move in the bacterial cell. Second, DNA polymerase I has a relatively low processivity. Third, genetic studies have demonstrated that many genes, and therefore many proteins, are involved in replication: DNA polymerase I clearly does not act alone. Fourth, and most important, in 1969 John Cairns isolated a bacterial strain with an altered gene for DNA polymerase I that produced an inactive enzyme. Although this strain was abnormally sensitive to agents that damaged DNA, it was nevertheless viable!

A search for other DNA polymerases led to the discovery of *E. coli* **DNA polymerase II** and **DNA polymerase III** in the early 1970s. DNA polymerase II is an enzyme involved in DNA repair, discussed later in the chapter. DNA polymerase III is the principal replication enzyme in *E. coli*. Properties of these three DNA polymerases are compared in Table 25–1. DNA polymerases IV and V, identified in 1999, are involved in an unusual form of DNA repair, described later in the chapter.

table 25–1

Comparison of DNA Polymerases of *E. coli*

	DNA polymerase		
	I	II	III
Structural gene*	*pol*A	*pol*B	*pol*C (*dna*E)
Subunits (number of different types)	1	≥4	≥10
M_r	103,000	88,000[†]	830,000
3′→5′ Exonuclease (proofreading)	Yes	Yes	Yes
5′→3′ Exonuclease	Yes	No	No
Polymerization rate (nucleotides/sec)	16–20	40	250–1,000
Processivity (nucleotides added before polymerase dissociates)	3–200	1,500	≥500,000

*For enzymes with more than one subunit, the gene listed here encodes the subunit with polymerization activity. Note that *dna*E is an earlier designation of the gene now referred to as *pol*C.

[†]Polymerization subunit only. DNA polymerase II shares several subunits with DNA polymerase III, including the β, γ, δ, δ', χ, and ψ subunits (see Table 25–2).

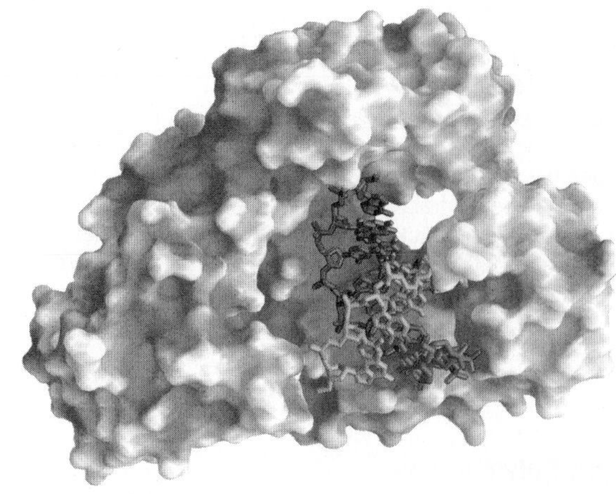

figure 25–8
Large (Klenow) fragment of DNA polymerase I. This polymerase is widely distributed in bacteria. The Klenow fragment, produced by proteolytic treatment of the polymerase, retains the polymerization and proofreading activities of the enzyme. This Klenow fragment is from the thermophilic bacterium *Bacillus stearothermophilus.* The active site for addition of nucleotides is deep in the crevice at the far end of the bound DNA. The dark blue strand is the template.

DNA polymerase I is not the primary enzyme of replication; instead it performs a host of clean-up functions during replication, recombination, and repair. These special functions are enhanced by its 5′→3′ exonuclease activity. This activity, distinct from the 3′→5′ proofreading exonuclease (Fig. 25–7), is located in a structural domain that can be separated from the enzyme by mild protease treatment. When the 5′→3′ exonuclease domain is removed, the remaining fragment (M_r 68,000), called the **large** or **Klenow fragment** (Fig. 25–8), retains the polymerization and proofreading activities. The 5′→3′ exonuclease activity of intact DNA polymerase I can replace a segment of DNA (or RNA) paired to the template strand in a process known as nick translation (Fig. 25–9). Most other DNA polymerases lack a 5′→3′ exonuclease activity.

DNA polymerase III is much more complex than DNA polymerase I, having ten types of subunits (Table 25–2). Its polymerizing and proofreading activities reside in subunits α and ϵ, respectively. The θ subunit associates with α and ϵ to form a core polymerase, which can polymerize DNA but with limited processivity. Two core polymerases can be linked in a complex by a dimer of τ (tau) subunits. This dimeric polymerase complex can then associate with a single clamp-loading complex, which consists of six subunits of five types, $\gamma_2\delta\delta'\chi\psi$. This subassembly of 14 protein subunits (nine different types) is called DNA polymerase III*.

DNA polymerase III* can polymerize DNA, but with a much lower processivity than one would expect for the organized replication of an entire

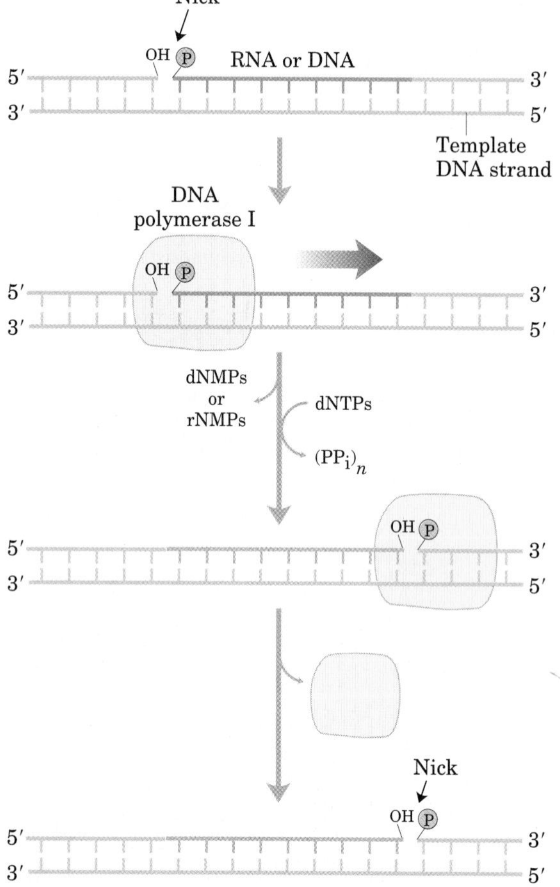

figure 25–9
Nick translation. In this process, an RNA or DNA strand paired to a DNA template is simultaneously degraded by the 5′→3′ exonuclease activity of DNA polymerase I and replaced by the polymerase activity of the same enzyme. These activities have a role in both DNA repair and the removal of RNA primers during replication (both described later). The strand of nucleic acid to be removed (either DNA or RNA) is shown in green, the replacement strand in red. A nick (a broken phosphodiester bond, leaving a free 3′ OH and a free 5′ phosphate) occurs where DNA synthesis is to start. Polymerase I extends the nontemplate DNA strand and moves the nick along the DNA—the process called nick translation. A nick remains where DNA polymerase I dissociates, until it is sealed by another enzyme.

table 25–2

Subunits of DNA Polymerase III of *E. coli*

Subunit	Number of subunits per holoenzyme	M_r of subunit	Gene	Function of subunit	
α	2	132,000	*pol*C (*dna*E)	Polymerization activity	Core polymerase
ϵ	2	27,000	*dna*Q (*mut*D)	3′→5′ Proofreading exonuclease	
θ	2	10,000	*hol*E		
τ	2	71,000	*dna*X	Stable template binding; core enzyme dimerization	
γ	2	52,000	*dna*X*	Clamp-loading complex that loads β subunits on lagging strand at each Okazaki fragment	
δ	1	35,000	*hol*A		
δ'	1	33,000	*hol*B		
χ	1	15,000	*hol*C		
ψ	1	12,000	*hol*D		
β	4	37,000	*dna*N	DNA clamp required for optimal processivity	

*The γ subunit is encoded by a portion of the gene for the τ subunit, such that the amino-terminal 80% of the τ subunit has the same amino acid sequence as the γ subunit. The γ subunit is generated by a translational frameshifting mechanism (see Box 28–1) that leads to premature translational termination.

chromosome. The necessary increase in processivity is provided by the addition of the β subunits, four of which complete the DNA polymerase III holoenzyme. The β subunits associate in pairs to form a donut-shaped structure that encircles the DNA and acts like a clamp (Fig. 25–10). Each dimer associates with a core subassembly of polymerase III* (one dimeric clamp per subassembly) and slides along the DNA as replication proceeds. The β sliding clamp prevents the dissociation of DNA polymerase III from DNA, dramatically increasing processivity to greater than 500,000 (Table 25–1).

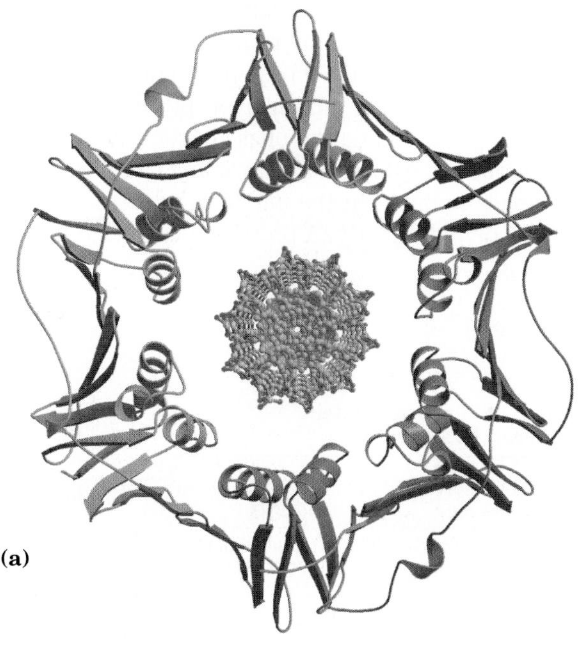

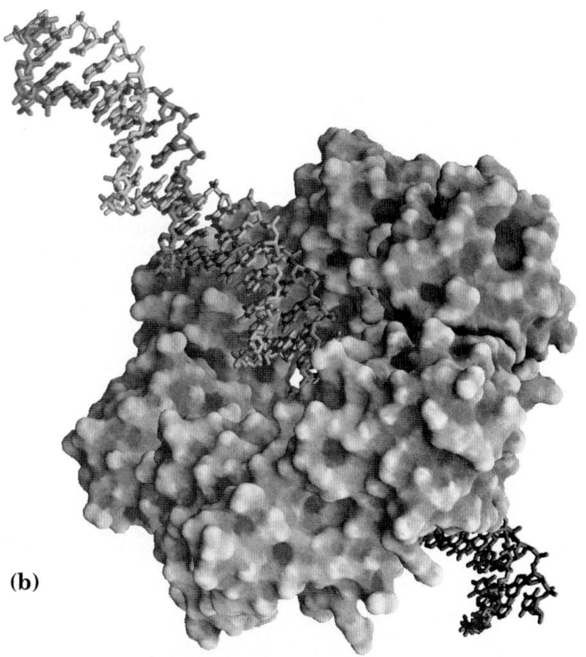

figure 25–10
The two β subunits of *E. coli* polymerase III form a circular clamp that surrounds DNA. The clamp slides along the DNA, enhancing the processivity of the DNA polymerase III holoenzyme to greater than 500,000 by preventing its dissociation. **(a)** End-on view. The two β subunits are shown as gray and light blue ribbon structures surrounding a space-filling model of DNA. **(b)** Side view. Surface contour models of the β subunits (gray) surround a stick representation of a DNA double helix (light and dark blue).

DNA Replication Requires Many Enzymes and Protein Factors

Replication in *E. coli* requires not just a single DNA polymerase but 20 or more different enzymes and proteins, each performing a specific task. The entire complex has been called the **DNA replicase system** or the **replisome.** The enzymatic complexity of replication reflects the constraints imposed by the structure of DNA and by the requirements for accuracy. The major classes of replication enzymes are considered here in terms of the problems that they overcome.

Access to the DNA strands that are to act as templates requires the two parental strands to separate. This is generally accomplished by enzymes called **helicases,** which move along the DNA and separate the strands using chemical energy from ATP. Strand separation creates topological stress in the helical DNA structure (see Fig. 24–11), which is relieved by the action of **topoisomerases.** The separated strands are stabilized by **DNA-binding proteins.** Before DNA polymerases can begin synthesizing DNA, primers must be present on the template—generally short segments of RNA synthesized by enzymes called **primases.** Ultimately, the RNA primers are removed and replaced by DNA—in *E. coli,* this is one of the many functions of DNA polymerase I. After an RNA primer is removed and the gap is filled in with DNA, a nick remains in the DNA backbone in the form of a broken phosphodiester bond. These nicks are sealed by enzymes called **DNA ligases.** All of these processes must be coordinated and regulated, an interplay that has been best characterized in the *E. coli* system.

Replication of the *E. coli* Chromosome Proceeds in Stages

The synthesis of a DNA molecule can be divided into three stages: initiation, elongation, and termination, distinguished both by the reactions taking place and by the enzymes required. As you will find in the next two chapters, synthesis of the other major biological polymers, RNAs and proteins, can be understood in terms of three equivalent stages, with the stages of each pathway having unique characteristics. The events described below reflect information derived from in vitro experiments using purified *E. coli* proteins.

Initiation The *E. coli* bacterial replication origin, called *ori*C, consists of 245 base pairs bearing DNA sequence elements that are highly conserved among bacterial replication origins. The general arrangement of the conserved sequences is illustrated in Figure 25–11. The key sequences for this discussion are two series of short repeats: three repeats of a 13 base pair sequence and four repeats of a 9 base pair sequence.

At least nine different enzymes or proteins (summarized in Table 25–3) participate in the initiation phase of replication. They open the DNA helix at the origin and establish a prepriming complex for subsequent reactions. The key component in the initiation process is the DnaA protein. A single complex of about 20 DnaA protein molecules binds to the four 9 base

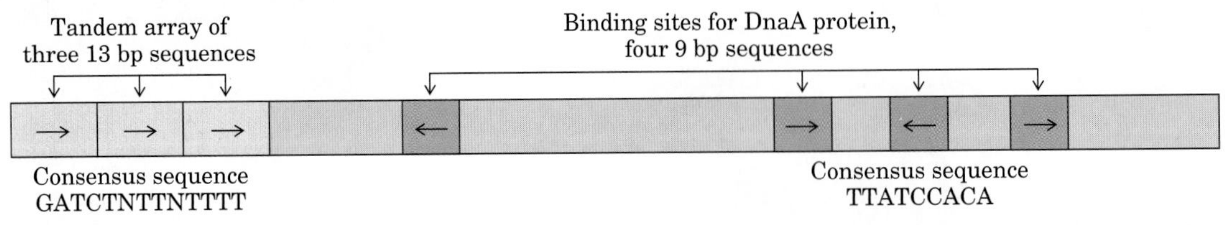

figure 25–11
Arrangement of sequences in the *E. coli* replication origin, *ori*C. Although the repeated sequences (shaded in color) are not identical, certain nucleotides are particularly common in each position, forming a consensus sequence. In positions where there is no consensus, N represents any of the four nucleotides. The arrows indicate the orientations of the nucleotide sequences.

table 25–3

Proteins Required to Initiate Replication at the *E. coli* Origin

Protein	M_r	Number of subunits	Function
DnaA protein	52,000	1	Recognizes origin sequence; opens duplex at specific sites in origin
DnaB protein (helicase)	300,000	6*	Unwinds DNA
DnaC protein	29,000	1	Required for DnaB binding at origin
HU	19,000	2	Histonelike protein; DNA bending protein; stimulates initiation
Primase (DnaG protein)	60,000	1	Synthesizes RNA primers
Single-stranded DNA-binding protein (SSB)	75,600	4*	Binds single-stranded DNA
RNA polymerase	454,000	5	Facilitates DnaA activity
DNA gyrase (DNA topoisomerase II)	400,000	4	Relieves torsional strain generated by DNA unwinding
Dam methylase	32,000	1	Methylates (5′)GATC sequences at *ori*C

*Subunits in these cases are identical.

pair repeats in the origin (Fig. 25–12a), then recognizes and successively denatures the DNA in the region of the three 13 base pair repeats, which are rich in A=T pairs (Fig. 25–12b). This process requires ATP and the bacterial histonelike protein HU. The DnaB protein then binds to the unwound region in a reaction that requires the DnaC protein. Two hexamers of DnaB, one loaded onto each DNA strand, act as helicases, unwinding the DNA bidirectionally and creating two potential replication forks. If the *E. coli* single-stranded DNA-binding protein (SSB) and DNA gyrase (DNA topoisomerase II) are now added in vitro, thousands of base pairs are rapidly unwound by the DnaB helicase, proceeding out from the origin. Many molecules of SSB bind cooperatively to single-stranded DNA, stabilizing the separated strands and preventing renaturation while gyrase relieves the topological stress produced by the DnaB helicase. The DNA unwinding mediated by DnaB is coupled to replication when additional replication proteins are included, as described below.

Initiation is the only phase of DNA replication that is known to be regulated, but it is regulated such that replication occurs only once in each cell cycle. The mechanism of regulation is not yet well understood, but genetic and biochemical studies have provided a few insights.

The timing of replication initiation is affected by DNA methylation and interactions with the bacterial plasma membrane. The *ori*C DNA is methylated by the Dam methylase (Table 25–3), which methylates the N^6 position of adenine within the palindromic sequence (5′)GATC. (Dam is not a biochemical expletive, but stands for *D*NA *a*denine *m*ethylation.) The *ori*C region of *E. coli* is highly enriched in GATC sequences, containing 11 of

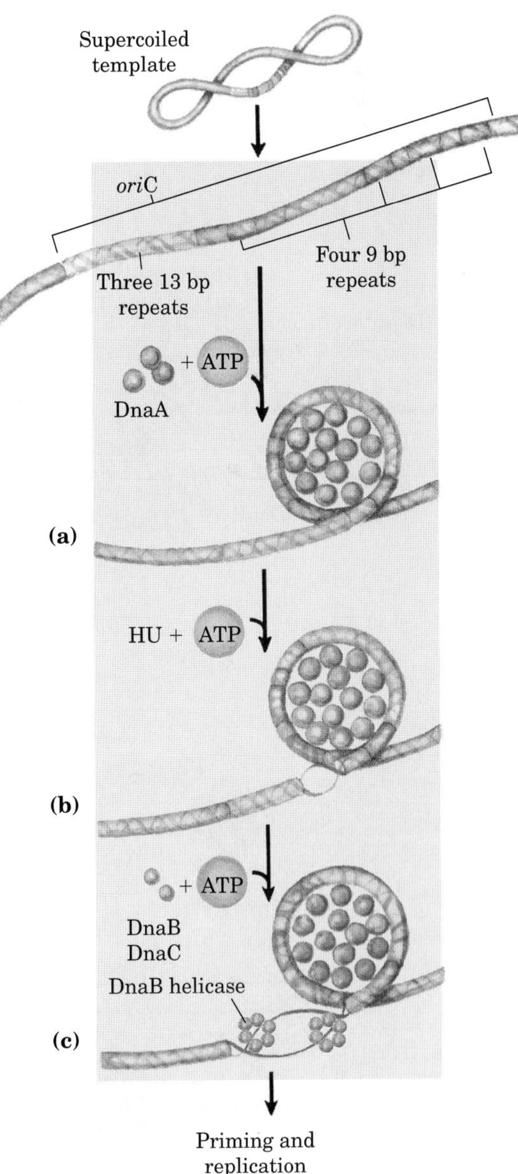

figure 25–12
Model for initiation of replication at the *E. coli* origin, *ori*C. (a) About 20 DnaA protein molecules, each with a bound ATP, bind at the four 9 base pair repeats. The DNA is wrapped around this complex. **(b)** The three A=T-rich 13 base pair repeats are then denatured sequentially. **(c)** Hexamers of the DnaB protein bind to each strand with the aid of DnaC protein. The DnaB helicase activity further unwinds the DNA in preparation for priming and DNA synthesis.

them in its 245 base pairs, whereas the average frequency of GATC in the *E. coli* chromosome is one for every 256 base pairs. Immediately after replication, the DNA is hemimethylated: the parental strands of *ori*C are methylated but the newly synthesized strands are not. The hemimethylated *ori*C sequences are now sequestered for a period by interaction with the plasma membrane (the mechanism is unknown). After a time, *ori*C is released from the plasma membrane, and it must be methylated by Dam methylase before it can again bind DnaA and initiate DNA replication. Regulation of replication initiation may also involve the slow hydrolysis of ATP by DnaA protein, which cycles the protein between active (with bound ATP) and inactive (bound ADP) forms.

Elongation The elongation phase of replication includes two distinct but related operations: leading strand synthesis and lagging strand synthesis. Several enzymes at the replication fork are important to the synthesis of both strands. Parental DNA is first unwound by DNA helicases, and the resulting topological stress is relieved by topoisomerases. Each separated strand is then stabilized by SSB. From this point, synthesis of leading and lagging strands is sharply different.

Leading strand synthesis, the more straightforward of the two, begins with the synthesis by primase (DnaG protein) of a short (10 to 60 nucleotide) RNA primer at the replication origin. Deoxyribonucleotides are added to this primer by DNA polymerase III. Leading strand synthesis then proceeds continuously, keeping pace with the unwinding of DNA at the replication fork.

Lagging strand synthesis, as we have noted, is accomplished in short Okazaki fragments. First, an RNA primer is synthesized by primase and, as in leading strand synthesis, DNA polymerase III binds to the RNA primer and adds deoxyribonucleotides (Fig. 25–13). On this level, the synthesis of each Okazaki fragment seems straightforward, but the reality is quite complex. The complexity lies in the *coordination* of leading and lagging strand synthesis: both strands are produced by a *single* asymmetric DNA polymerase III dimer. This is accomplished by looping the DNA of the lagging strand as shown in Figure 25–14, bringing together the two points of polymerization.

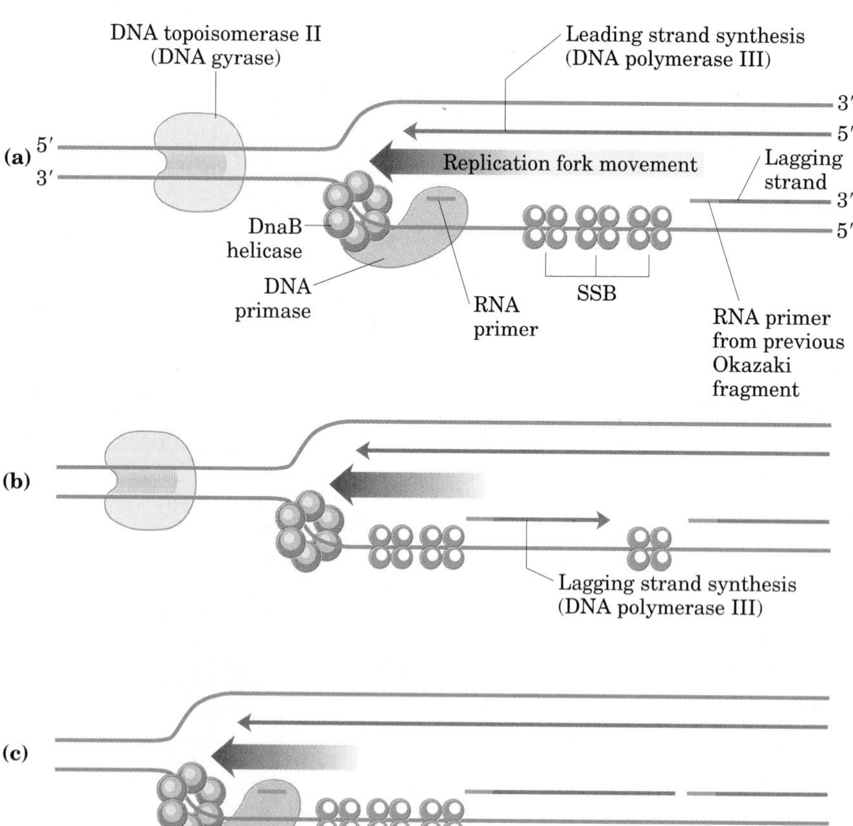

figure 25–13
Synthesis of Okazaki fragments. (a) At intervals, primase synthesizes an RNA primer for a new Okazaki fragment. Note that if we consider the two template strands as lying side by side, lagging strand synthesis formally proceeds in the opposite direction from fork movement. **(b)** Each primer is extended by DNA polymerase III. **(c)** DNA synthesis continues until the fragment extends as far as the primer of the previously added Okazaki fragment. A new primer is synthesized near the replication fork to begin the process again.

The synthesis of Okazaki fragments on the lagging strand entails some elegant enzymatic choreography. The DnaB helicase and DnaG primase constitute a functional unit within the replication complex, called the **primosome.** DNA polymerase III uses one set of its core subunits (the core polymerase) to synthesize the leading strand continuously, while the other set of core subunits cycles from one Okazaki fragment to the next on the looped lagging strand. As the DNA is unwound by the DnaB helicase at the replication fork (Fig. 25–14a), DNA primase occasionally associates with DnaB helicase and synthesizes a short RNA primer (Fig. 25–14b). A new β sliding clamp is then positioned at the primer by the clamp-loading complex of DNA polymerase III (Fig. 25–14c). When synthesis of an Okazaki fragment has been completed, replication halts and the core subunits of DNA polymerase III dissociate from their β sliding clamp (and the completed Okazaki fragment) and associate with the new one (Fig. 25–14d). This initiates synthesis of a new Okazaki fragment. Evidence is accumulating that the complex within which all of this occurs does not itself move. Instead, it is associated with the plasma membrane and the DNA moves through the fixed complex.

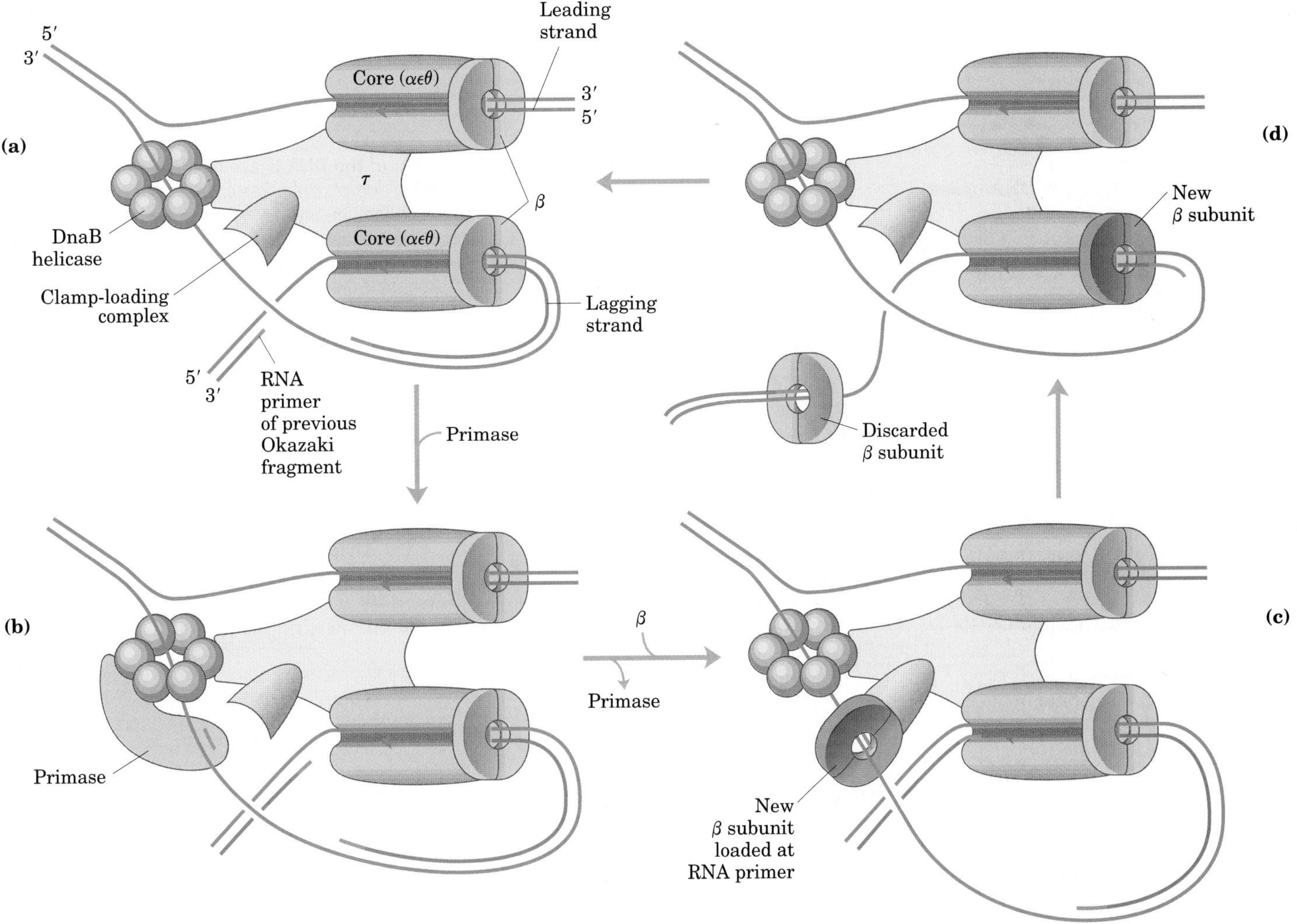

figure 25–14

DNA synthesis on the leading and lagging strands. Events at the replication fork are coordinated by a single DNA polymerase III dimer, in an integrated complex with DnaB helicase. This figure shows the replication process already underway. The lagging strand is looped so that DNA synthesis proceeds steadily on both the leading and lagging strand templates at the same time, catalyzed by the two sets of DNA polymerase III core subunits. Arrows indicate the 3′ end of the two new strands and the direction of DNA synthesis. An Okazaki fragment is being synthesized on the lagging strand.

(a) Continuous synthesis on the leading strand proceeds as DNA is unwound by the DnaB helicase. **(b)** DNA primase binds to DnaB, synthesizes a new primer, then dissociates. **(c)** The DNA polymerase III clamp-loading complex catalyzes the loading of a new β sliding clamp (dark blue) at the new RNA primer. Meanwhile, the Okazaki fragment that was being synthesized is completed. **(d)** The core subunits of DNA polymerase III that synthesized the lagging strand release both the completed Okazaki fragment and the β sliding clamp used in its synthesis. (The clamp-loading complex probably facilitates the recycling of β sliding clamps (not shown).) The same core subunits then bind to the new β sliding clamp and begin synthesis of another Okazaki fragment at its primer.

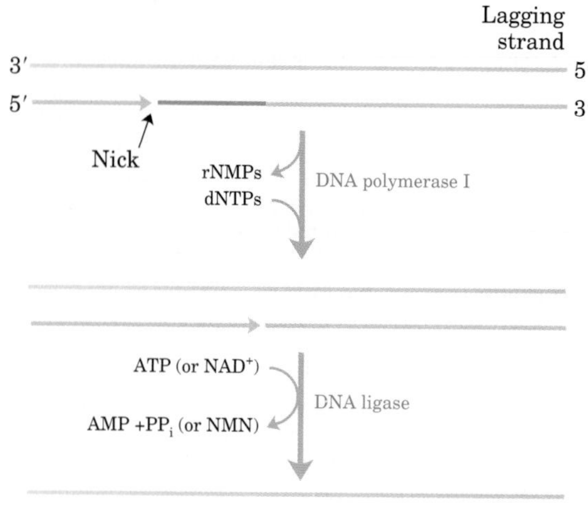

figure 25–15
RNA primers in the lagging strand are removed by the 5′→3′ exonuclease activity of DNA polymerase I and replaced with DNA by the same enzyme. The remaining nick is sealed by DNA ligase. The role of ATP or NAD$^+$ is shown in Figure 25–16.

The process allows rapid synthesis, with about 1,000 nucleotides of new DNA added per second to each strand (leading and lagging). Once an Okazaki fragment has been completed, its RNA primer is removed and replaced with DNA by DNA polymerase I and the remaining nick is sealed by DNA ligase (Fig. 25–15). The proteins acting at the replication fork are summarized in Table 25–4.

DNA ligase catalyzes the formation of a phosphodiester bond between a 3′ hydroxyl at the end of one DNA strand and a 5′ phosphate at the end of another strand. The phosphate must be activated by adenylylation. DNA ligases isolated from viruses and eukaryotes use ATP for this purpose. DNA ligases from bacteria are unusual in that they generally use NAD$^+$—a cofactor that normally functions in hydride transfer reactions (see Fig. 14–15)—as the source of the AMP activating group (Fig. 25–16). DNA ligase is another enzyme of DNA metabolism that has become an important reagent in recombinant DNA experiments (Chapter 29).

Enzyme—NH_3^+
DNA ligase

① Adenylylation of DNA ligase

R—O—P(=O)(O$^-$)—O—Ribose—Adenine
AMP from ATP (R = PP$_i$) or NAD$^+$ (R = NMN)

PP$_i$ (from ATP) or NMN (from NAD$^+$)

Enzyme—$\overset{+}{N}H_2$—P(=O)(O$^-$)—O—Ribose—Adenine
Enzyme-AMP

② Activation of 5′ phosphate in nick

Nick in DNA

Enzyme—NH_3^+

③ Displacement of AMP seals nick

$^-$O—P(=O)(O$^-$)—O—Ribose—Adenine
AMP

DNA ligase

Sealed DNA

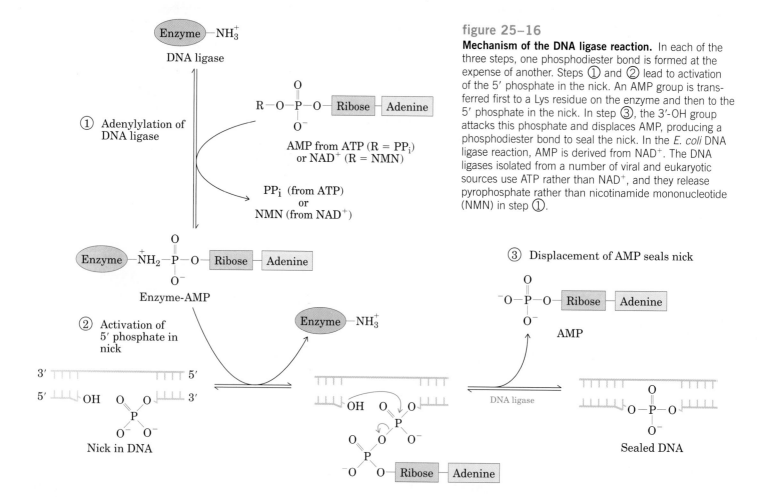

figure 25–16
Mechanism of the DNA ligase reaction. In each of the three steps, one phosphodiester bond is formed at the expense of another. Steps ① and ② lead to activation of the 5′ phosphate in the nick. An AMP group is transferred first to a Lys residue on the enzyme and then to the 5′ phosphate in the nick. In step ③, the 3′-OH group attacks this phosphate and displaces AMP, producing a phosphodiester bond to seal the nick. In the *E. coli* DNA ligase reaction, AMP is derived from NAD$^+$. The DNA ligases isolated from a number of viral and eukaryotic sources use ATP rather than NAD$^+$, and they release pyrophosphate rather than nicotinamide mononucleotide (NMN) in step ①.

table 25–4

Proteins at the *E. coli* Replication Fork

Protein	M_r	Number of subunits	Function
SSB	75,600	4	Binding to single-stranded DNA
DnaB protein (helicase)	300,000	6	DNA unwinding; primosome constituent
Primase (DnaG protein)	60,000	1	RNA primer synthesis; primosome constituent
DNA polymerase III	900,000	18–20	New strand elongation
DNA polymerase I	103,000	1	Filling of gaps, excision of primers
DNA ligase	74,000	1	Ligation
DNA gyrase (DNA topoisomerase II)	400,000	4	Supercoiling

Modified from Kornberg, A. (1982) *Supplement to DNA Replication,* Table S11–2, W.H. Freeman and Company, New York.

Termination Eventually, the two replication forks of the circular *E. coli* chromosome meet at a terminus region containing multiple copies of a 20 base pair sequence called *Ter* (for *ter*minus) (Fig. 25–17a). The *Ter* sequences are arranged on the chromosome to create a sort of trap that a replication fork can enter but cannot exit. *Ter* sequences function as binding sites for a protein called Tus (for *t*erminus *u*tilization *s*ubstance). The Tus-*Ter* complex can arrest a replication fork from only one direction. Only one Tus-*Ter* complex functions per replication cycle—the one first encountered by either replication fork. Given that opposing replication forks generally halt when they collide, *Ter* sequences do not seem essential, but they may prevent overreplication by one replication fork in the event that the other is delayed or halted by an encounter with DNA damage or some other obstacle.

When either replication fork encounters a functional Tus-*Ter* complex, it halts; the other fork halts when it meets the first (arrested) fork. The final few hundred base pairs of DNA between these large protein complexes are then replicated (by an as yet unknown mechanism), completing two topologically interlinked (catenated) circular chromosomes (Fig. 25–17b). DNA circles linked in this way are called **catenanes.** Separation of the catenated circles in *E. coli* requires topoisomerase IV (a type II topoisomerase). The separated chromosomes are then segregated into daughter cells at cell division. The terminal phase of replication of other circular chromosomes, including many eukaryotic DNA viruses, is similar.

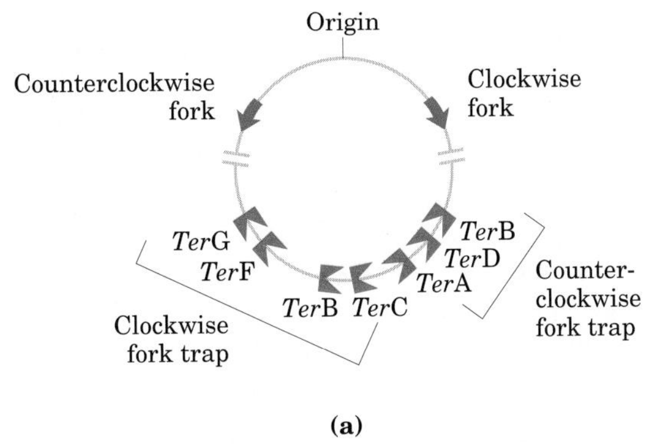

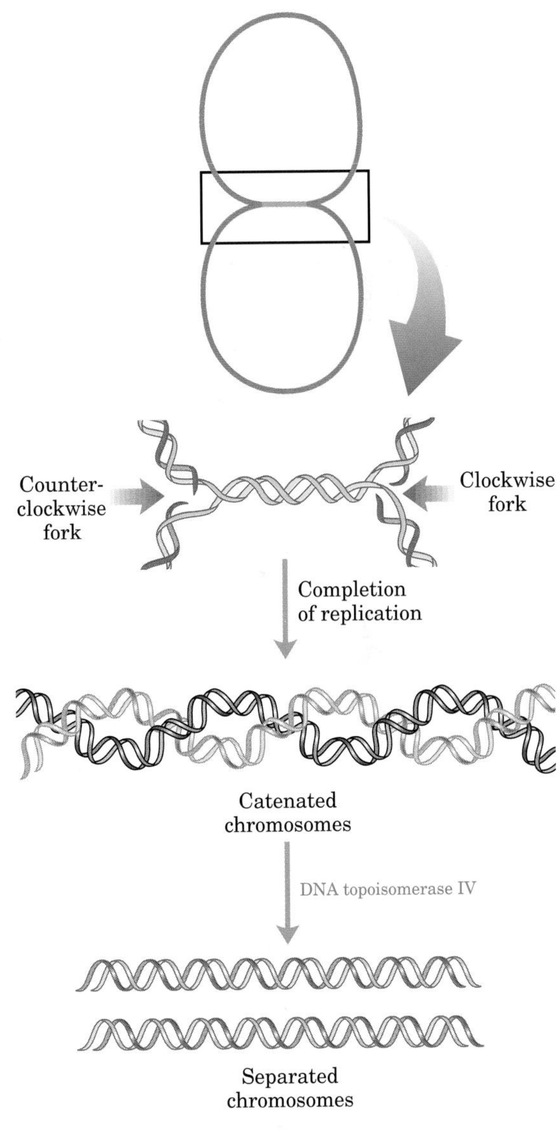

figure 25–17

Termination of chromosome replication in *E. coli*. **(a)** *Ter* sequences are positioned on the chromosome in two clusters with opposite orientations. **(b)** Replication of the DNA separating the opposing replication forks leaves the completed chromosomes joined as catenanes, or topologically interlinked circles. The circles are not covalently linked, but cannot be separated because they are interwound and each is covalently closed. Separating the catenated circles requires the action of topoisomerases. In *E. coli*, a type II topoisomerase called DNA topoisomerase IV plays the major role in the separation of catenated chromosomes, transiently breaking both DNA strands of one chromosome and allowing the other chromosome to pass through the break (see Fig. 24–20).

Replication in Eukaryotic Cells Is More Complex

The DNA molecules in eukaryotic cells are considerably larger than those in bacteria and are organized into complex nucleoprotein structures (chromatin; see Chapter 24). The essential features of DNA replication are the same in eukaryotes and prokaryotes. However, some interesting variations on the general principles discussed above promise new insights into the regulation of replication and its link with the cell cycle.

Origins of replication, called *a*utonomously *r*eplicating *s*equences (ARS) or **replicators,** have been identified and best studied in yeast. Yeast replicators span about 150 base pairs and contain several essential conserved sequences. About 400 replicators are distributed among the 17 chromosomes in a haploid yeast genome. Initiation of replication in all eukaryotes requires a multisubunit protein, the *o*rigin *r*ecognition *c*omplex (ORC), which binds to several sequences within the replicator. ORC interacts with and is regulated by a number of other proteins involved in the control of the eukaryotic cell cycle.

The rate of replication fork movement in eukaryotes (~50 nucleotides/sec) is only one-twentieth that observed in *E. coli.* At this rate, replication of an average human chromosome proceeding from a single origin would take more than 500 hours. Replication of human chromosomes in fact proceeds bidirectionally from multiple origins spaced 30,000 to 300,000 base pairs apart. Eukaryotic chromosomes are almost always much larger than bacterial chromosomes, so multiple origins are probably a universal feature in eukaryotic cells.

Like bacteria, eukaryotes have several types of DNA polymerases. Some have been linked to particular functions, such as the replication of mitochondrial DNA. The replication of nuclear chromosomes involves DNA polymerase α, in association with DNA polymerase δ. **DNA polymerase α** is typically a multisubunit enzyme with similar structure and properties in all eukaryotic cells. One subunit has a primase activity, and the largest subunit (M_r ~180,000) contains the polymerization activity. However, this polymerase has no proofreading 3′→5′ exonuclease activity, making it unsuitable for high-fidelity DNA replication. DNA polymerase α is believed to function only in the synthesis of short primers (containing both RNA and DNA) for Okazaki fragments on the lagging strand. These primers are then extended by the multisubunit **DNA polymerase δ.** This enzyme is associated with and stimulated by a protein called *p*roliferating *c*ell *n*uclear *a*ntigen (PCNA; M_r 29,000), which is found in large amounts in the nuclei of proliferating cells. PCNA has a structure and function analogous to that of the β subunit of *E. coli* DNA polymerase III (Fig. 25–10), forming a circular clamp that greatly enhances the processivity of the polymerase. DNA polymerase δ has a 3′→5′ proofreading exonuclease activity and appears to carry out both leading and lagging strand synthesis in a complex analogous to the dimeric bacterial DNA polymerase III.

Yet another polymerase, **DNA polymerase ϵ,** replaces DNA polymerase δ in some situations, such as in DNA repair. DNA polymerase ϵ may also function at the replication fork, perhaps playing a role analogous to that of the bacterial DNA polymerase I, removing the primers of Okazaki fragments on the lagging strand.

Two other protein complexes also function in eukaryotic DNA replication. RPA is a eukaryotic single-stranded DNA-binding protein equivalent in function to the *E. coli* SSB protein. RFC is a clamp loader (or matchmaker) for PCNA and facilitates the assembly of active replication complexes.

The termination of replication on linear eukaryotic chromosomes involves the synthesis of special structures called **telomeres** at the ends of the chromosome, as discussed in the next chapter.

DNA Repair

A cell generally has only one or two sets of genomic DNA. Damaged proteins and RNA molecules can be quickly replaced using information encoded in the DNA, but DNA molecules themselves are irreplaceable. Maintaining the integrity of the information in DNA is a cellular imperative, supported by an elaborate set of DNA repair systems. DNA can become damaged by a variety of processes, some spontaneous, others catalyzed by environmental agents (Chapter 10). Replication itself can occasionally damage the information content in DNA by leaving mispaired bases.

The chemistry of DNA damage is diverse and complex. The cellular response to this damage includes a wide range of enzymatic systems that catalyze some of the most interesting chemical transformations in DNA metabolism. We first examine the effects of alterations in DNA sequence and then consider specific repair systems.

Mutations Are Linked to Cancer

The best way to illustrate the importance of DNA repair is to consider the effects of *unrepaired* DNA damage (a lesion). The most serious outcome is a change in the base sequence of the DNA, which, if replicated and transmitted to future cell generations, becomes permanent. A permanent change in the nucleotide sequence of DNA is called a **mutation.** Mutations can involve the replacement of one base pair with another (substitution mutation) or the addition or deletion of one or more base pairs (insertion or deletion mutations). If the mutation affects nonessential DNA or if it has a negligible effect on the function of a gene, it is called a **silent mutation.** Rarely, a mutation confers some biological advantage. Most mutations, however, are deleterious.

In mammals there is a strong correlation between the accumulation of mutations and cancer. A simple test developed by Bruce Ames measures the potential of a given chemical compound to promote certain easily detected mutations in a specialized bacterial strain (Fig. 25–18). Few of the chemicals that we encounter day to day score as mutagens in this test. However, of the compounds known to be carcinogenic from extensive animal trials, more than 90% are also found to be mutagenic in the Ames test. Because of this strong correlation between mutagenesis and carcinogenesis, the Ames test for bacterial mutagens is widely used as a rapid and inexpensive screen for potential human carcinogens.

The genome of a typical mammalian cell accumulates many thousands of lesions during a 24-hour period. However, as a result of DNA repair, fewer than one lesion in 1,000 becomes a mutation. DNA is a relatively stable molecule, but in the absence of repair systems, the cumulative effect of many infrequent but damaging reactions would make life impossible.

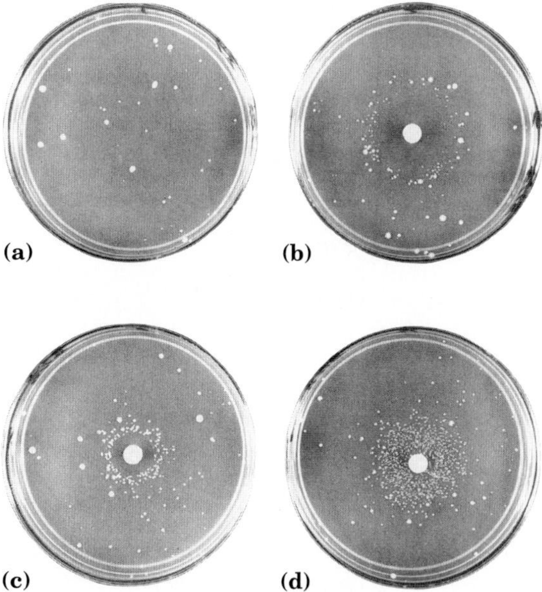

figure 25–18

Ames test for carcinogens, based on their mutagenicity. A strain of *Salmonella typhimurium* having a mutation that inactivates an enzyme of the histidine biosynthetic pathway is plated on a histidine-free medium. Few cells grow. **(a)** The few small colonies of *S. typhimurium* that do grow on a histidine-free medium carry spontaneous back-mutations that permit the histidine biosynthetic pathway to operate. Three identical nutrient plates **(b),** **(c),** and **(d)** have been inoculated with an equal number of cells. Each plate then receives a disk of filter paper containing progressively lower concentrations of a mutagen. The mutagen greatly increases the rate of back-mutation and hence the number of colonies. The clear areas around the filter paper indicate where the concentration of mutagen is so high that it is lethal to the cells. As the mutagen diffuses away from the filter paper, it is diluted to sublethal concentrations that promote back-mutation. Mutagens can be compared on the basis of their effect on mutation rate. Because many compounds undergo a variety of chemical transformations after entering a cell, compounds are sometimes tested for mutagenicity after first incubating them with a liver extract. A number of compounds have been found to be mutagenic only after this treatment.

All Cells Have Multiple DNA Repair Systems

The number and diversity of repair systems reflect both the importance of DNA repair to cell survival and the diverse sources of DNA damage (Table 25–5). Some common types of lesions, such as pyrimidine dimers (see Fig. 10–34), can be repaired by several distinct systems. Many DNA repair processes also appear to be extraordinarily inefficient energetically—an exception to the pattern observed in the metabolic pathways, where every ATP is generally accounted for and used optimally. When the integrity of the genetic information is at stake, the amount of chemical energy invested in a repair process seems to be almost irrelevant.

DNA repair is possible largely because the DNA molecule consists of two complementary strands. DNA damage in one strand can be removed and accurately replaced by using the undamaged complementary strand as a template.

We now consider the principal types of repair systems, beginning with those that repair the rare nucleotide mismatches that are left behind by replication.

Mismatch Repair The correction of mismatches after replication in *E. coli* improves the overall fidelity of replication by a factor of 10^2 to 10^3. The mismatches are nearly always corrected to reflect the information in the old (template) strand, so the repair system must somehow discriminate between the template and the newly synthesized strand. The cell accomplishes this by tagging the template DNA with methyl groups to distinguish it from newly synthesized strands. The mismatch repair system of *E. coli* includes at least 12 protein components (Table 25–5) that function either in strand discrimination or in the repair process itself.

Strand discrimination is based on the action of Dam methylase (Table 25–3), which, you will recall, methylates DNA at the N^6 position of all adenines within (5′)GATC sequences. Immediately after passage of the replication fork, there is a short period (a few seconds or minutes) during which the template strand is methylated but the newly synthesized strand is not (Fig. 25–19). The transient unmethylated state of GATC sequences in the newly synthesized strand permits the new strand to be distinguished from the template strand. Replication mismatches in the vicinity of a hemimethylated GATC sequence are then repaired according to the information in the methylated parental (template) strand. Tests in vitro show that if both strands are methylated at a GATC sequence, few mismatches are repaired; if neither strand is methylated, repair occurs but does not favor either strand. This methyl-directed mismatch repair system efficiently repairs mismatches up to 1,000 base pairs from a hemimethylated GATC sequence.

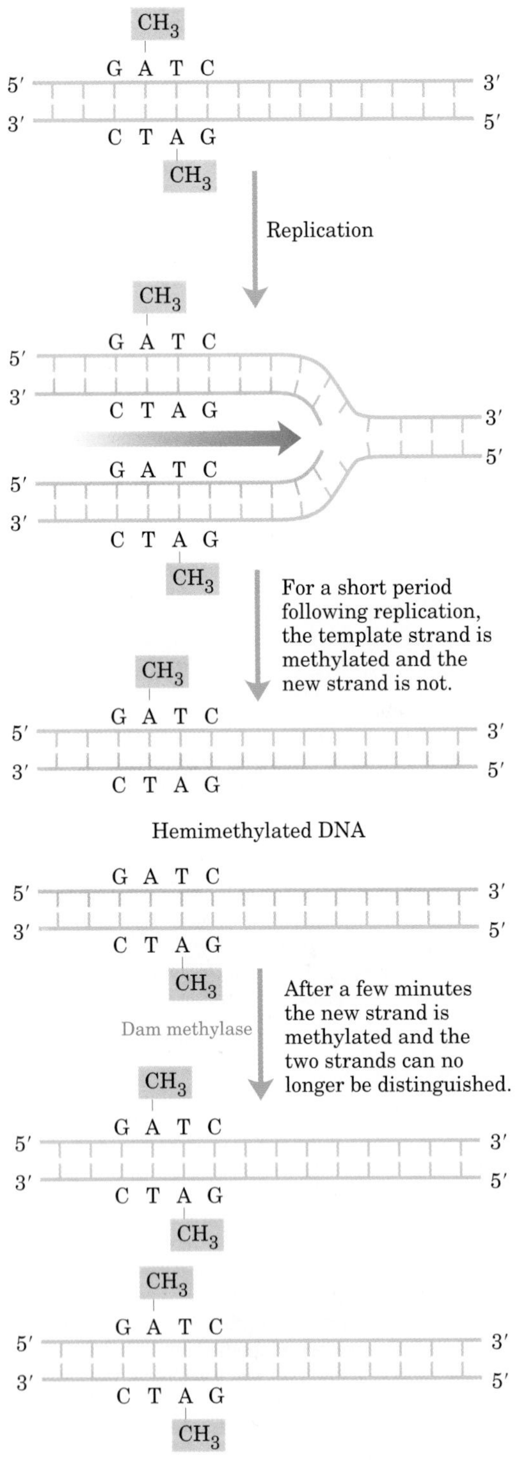

figure 25–19
Methylation and mismatch repair. Methylation of DNA strands can serve to distinguish parental (template) strands from newly synthesized strands in *E. coli* DNA, a function that is critical to mismatch repair (see Fig. 25–20). The methylation occurs at the N^6 of adenines in (5′)GATC sequences. This sequence is a palindrome (see Fig. 10–20) and thus present in opposite orientations on the two strands.

table 25–5

Types of DNA Repair Systems in *E. coli*

Enzymes/proteins	Type of damage
Mismatch repair	
Dam methylase	Mismatches
MutH, MutL, MutS proteins	
DNA helicase II	
SSB	
DNA polymerase III	
Exonuclease I	
Exonuclease VII	
RecJ nuclease	
Exonuclease X	
DNA ligase	
Base-excision repair	
DNA glycosylases	Abnormal bases (uracil, hypoxanthine, xanthine); alkylated bases; pyrimidine dimers in some other organisms
AP endonucleases	
DNA polymerase I	
DNA ligase	
Nucleotide-excision repair	
ABC excinuclease	DNA lesions that cause large structural changes (e.g., pyrimidine dimers)
DNA polymerase I	
DNA ligase	
Direct repair	
DNA photolyases	Pyrimidine dimers
O^6-Methylguanine-DNA methyltransferase	O^6-Methylguanine

The mechanism by which mismatch corrections are directed by relatively distant GATC sequences is illustrated in Figure 25–20. MutL protein forms a complex with MutS protein and binds to all mismatched base pairs (except C–C). MutH protein binds to MutL and to GATC sequences encountered by the MutL-MutS complex. DNA on both sides of the mismatch is threaded through the MutL-MutS complex, creating a DNA loop; movement of both legs of the loop through the complex simultaneously is equivalent to the complex moving in both directions at once along the DNA. MutH has a site-specific endonuclease activity that is inactive until the

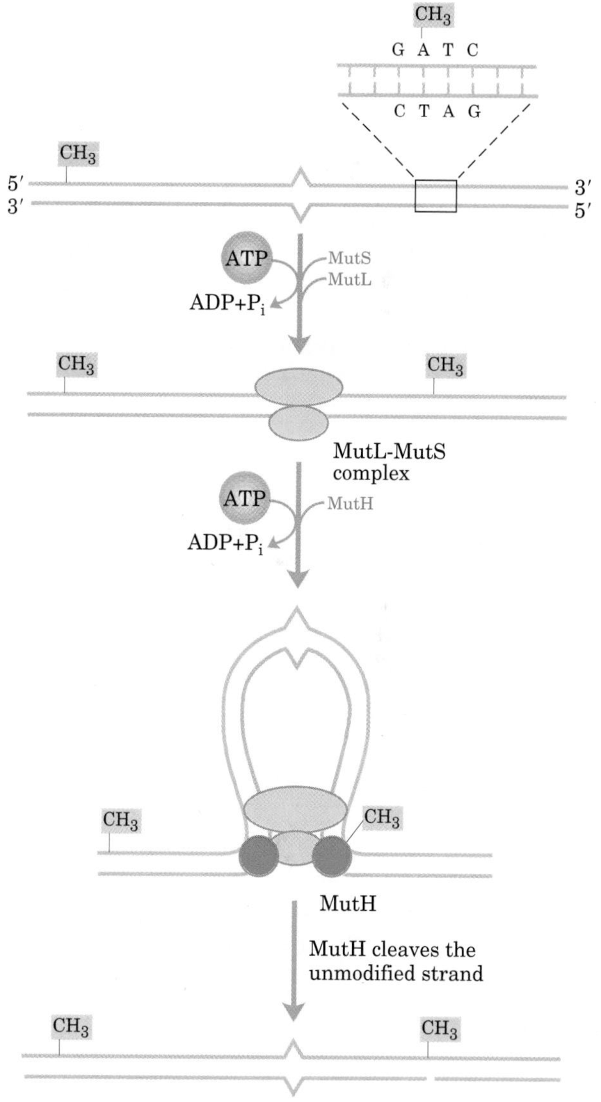

figure 25–20

Model for the early steps of methyl-directed mismatch repair. The proteins involved in this process in *E. coli* have been purified (see Table 25–5). Recognition of the sequence (5′)GATC and of the mismatch are specialized functions of the MutH and MutS proteins, respectively. The MutL protein forms a complex with MutS at the mismatch. DNA is threaded through this complex such that the complex moves simultaneously in both directions along the DNA until it encounters a MutH protein bound at a hemimethylated GATC sequence. MutH cleaves the unmethylated strand on the 5′ side of the G in the GATC sequence. A complex consisting of DNA helicase II and one of several exonucleases then degrades the unmethylated DNA strand from that point toward the mismatch (see Fig. 25–21).

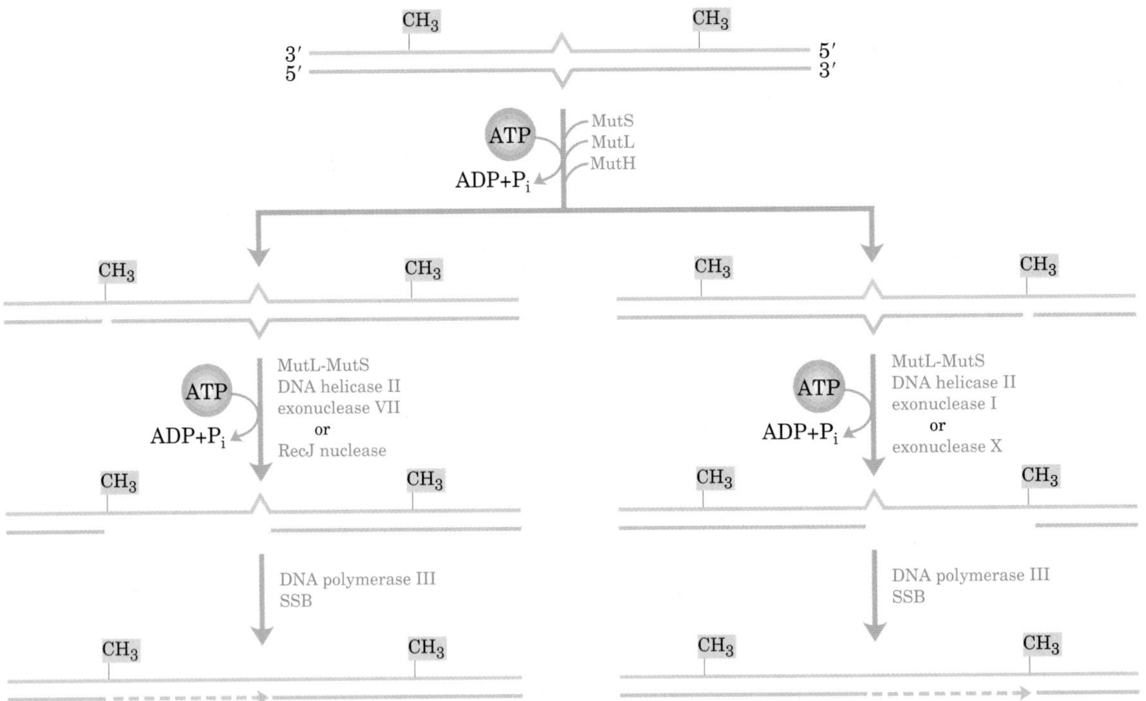

figure 25–21
Completing methyl-directed mismatch repair. The combined action of DNA helicase II, SSB, and one of four different exonucleases removes a segment of the new strand between the MutH cleavage site and a point just beyond the mismatch. The exonuclease used depends on the location of the cleavage site relative to the mismatch. The resulting gap is filled in by DNA polymerase III, and the nick is sealed by DNA ligase (not shown).

complex encounters a hemimethylated GATC sequence. At this site, MutH catalyzes cleavage of the unmethylated strand on the 5′ side of the G in GATC, which marks the strand for repair. Further steps in the pathway depend on where the mismatch is located relative to this cleavage site (Fig. 25–21).

When the mismatch is on the 5′ side of the cleavage site, the unmethylated strand is unwound and degraded in the 3′→5′ direction from the cleavage site through the mismatch, and this segment is replaced with new DNA. This process requires the combined action of DNA helicase II, SSB, and either exonuclease I or exonuclease X (both of which degrade strands of DNA in the 3′→5′ direction), DNA polymerase III, and DNA ligase. The pathway for repair of mismatches on the 3′ side of the cleavage site is similar, except that the exonuclease is replaced by either exonuclease VII (which degrades single-stranded DNA in either the 5′→3′ or 3′→5′ direction) or RecJ nuclease (an exonuclease that degrades single-stranded DNA in the 5′→3′ direction).

Mismatch repair is a particularly expensive process for bacteria in terms of energy expended. The mismatch may be 1,000 base pairs or more from the GATC sequence. The degradation and replacement of a strand segment of this length requires an enormous investment in activated deoxynucleotide precursors to repair a *single* mismatched base. This again underscores the importance to the cell of genomic integrity.

All eukaryotic cells have proteins structurally and functionally analogous to the bacterial MutS and MutL proteins. Alterations in human genes encoding proteins of this type produce some of the most common inherited cancer-susceptibility syndromes (Box 25–1), further demonstrating the value to the organism of DNA repair systems. The detailed mechanism of

box 25–1 DNA Repair and Cancer

Human cancer develops when certain genes that regulate normal cell division (oncogenes and tumor suppressor genes; see Chapter 13) fail or are altered. Cells may grow out of control and form a tumor. The genes controlling cell division can be damaged by spontaneous mutation or overridden by the invasion of a tumor virus (Chapter 26). Not surprisingly, alterations in DNA repair genes that result in an increase in the rate of mutation can greatly increase an individual's susceptibility to cancer. Defects in the genes encoding the proteins involved in nucleotide-excision repair, mismatch repair, and recombinational repair have all been linked to human cancer. Clearly, DNA repair can be a matter of life and death.

Nucleotide-excision repair requires a larger number of proteins in humans than in bacteria, although the overall pathways are very similar. Genetic defects that inactivate nucleotide-excision repair have been associated with several genetic diseases, the best-studied of which is xeroderma pigmentosum, or XP. Because nucleotide-excision repair is the sole repair pathway for pyrimidine dimers in humans, people with XP are extremely light sensitive and readily develop sunlight-induced skin cancers. Most people with XP also have neurological abnormalities, presumably because of their inability to repair certain lesions caused by the high rate of oxidative metabolism in neurons.

One of the most common inherited cancer-susceptibility syndromes is hereditary nonpolyposis colon cancer, or HNPCC. This syndrome has been traced to defects in mismatch repair. Human and other eukaryotic cells have multiple proteins analogous to the bacterial MutL and MutS proteins (see Fig. 25–20a). Defects in at least five different mismatch repair genes can give rise to HNPCC. The most prevalent are defects in the hMLH1 (human MutL homolog 1) and hMSH2 (human MutS homolog 2) genes. In individuals with HNPCC, cancer generally develops at an early age, with colon cancers being most common.

Most human breast cancer occurs in women with no known predisposition. However, about 10% of cases are associated with inherited defects in two genes, *Brca*1 and *Brca*2. The BRCA1 and BRCA2 proteins interact with a protein called Rad 51, the eukaryotic homolog of the RecA protein. This suggests that BRCA1 and 2 are involved in recombinational DNA repair (see Fig. 25–35). Women with defects in either the *Brca*1 or *Brca*2 genes have a greater than 80% chance of developing breast cancer.

DNA mismatch repair in eukaryotes has not been worked out. We do know that the mechanism by which newly synthesized DNA strands are identified does not involve GATC sequences, and it may differ in other ways from the methyl-directed system used by bacteria.

Base-Excision Repair Every cell has a class of enzymes called **DNA glycosylases** that recognize particularly common DNA lesions (such as the products of cytosine and adenine deamination; see Fig. 10–33a) and remove the affected base by cleaving the *N*-glycosyl bond. This cleavage creates an apurinic or apyrimidinic site in the DNA, commonly referred to as an abasic or **AP site.** Each DNA glycosylase is generally specific for one type of lesion.

Uracil glycosylase, for example, found in most cells, specifically removes from DNA the uracil that results from spontaneous deamination of cytosine. Mutant cells that lack this enzyme have a high rate of G≡C to A=T mutations. This glycosylase does not remove uracil residues from RNA or thymine residues from DNA. As explained in Chapter 10, the capacity to distinguish thymine from the product of cytosine deamination

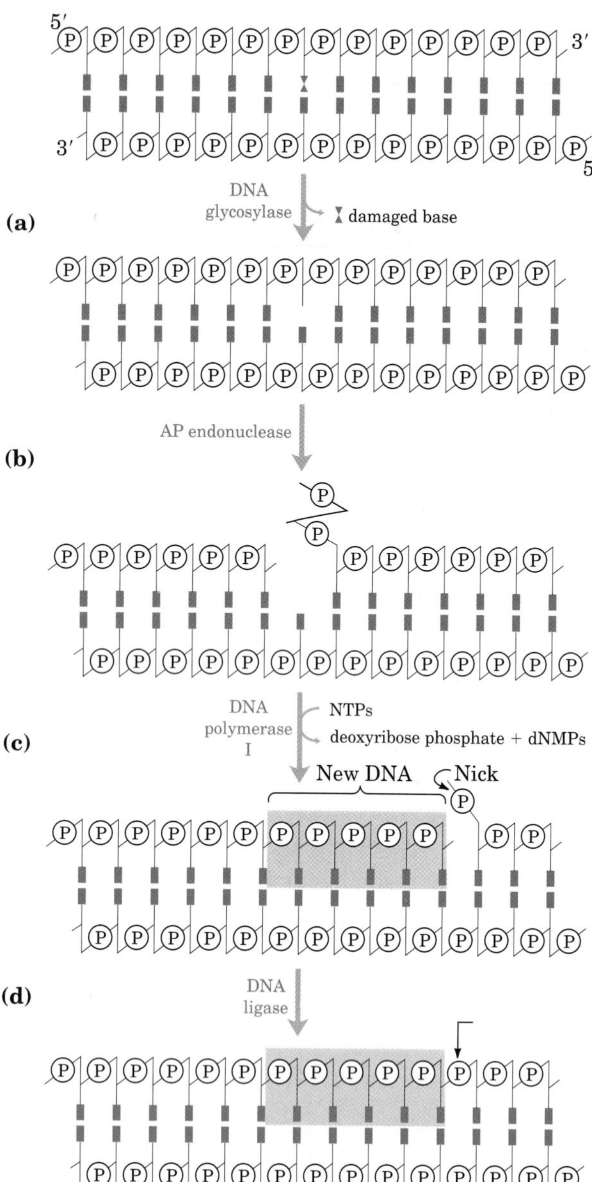

figure 25–22
DNA repair by the base-excision repair pathway. **(a)** A DNA glycosylase recognizes a damaged base and cleaves between the base and deoxyribose in the backbone. **(b)** An AP endonuclease cleaves the phosphodiester backbone near the AP site. **(c)** DNA polymerase I initiates repair synthesis from the free 3′ OH at the nick, removing a portion of the damaged strand (with its 5′→3′ exonuclease activity) and replacing it with undamaged DNA. **(d)** The nick remaining after DNA polymerase I has dissociated is sealed by DNA ligase.

(uracil), which is necessary for the selective repair of the latter, may be one reason why DNA evolved to contain thymine instead of uracil.

Other DNA glycosylases recognize and remove hypoxanthine (arising from adenine deamination) and alkylated bases such as 3-methyladenine and 7-methylguanine. Glycosylases that recognize other lesions, including pyrimidine dimers, have also been identified. Remember that AP sites also arise from the slow, spontaneous hydrolysis of the *N*-glycosyl bonds in DNA (see Fig. 10–33).

Once an AP site has been formed, another group of enzymes must repair it. The repair is *not* made by simply inserting a new base and re-forming the *N*-glycosyl bond. Instead, the deoxyribose 5′-phosphate left behind is removed and replaced with a new nucleotide. This process begins with enzymes called **AP endonucleases,** which cut the DNA strand containing the AP site. The position of the incision relative to the AP site (5′ or 3′) varies with the type of AP endonuclease. A segment of DNA including the AP site is then removed, the DNA is replaced by DNA polymerase I, and the remaining nick is sealed by DNA ligase (Fig. 25–22).

Nucleotide-Excision Repair DNA lesions that cause large distortions in the helical structure of DNA generally are repaired by the nucleotide-excision system, a repair pathway critical to the survival of all free-living organisms. In nucleotide-excision repair (Figure 25–23), a multisubunit enzyme hydrolyzes two phosphodiester bonds, one on either side of the lesion. In *E. coli* and other prokaryotes, the enzyme system hydrolyzes the fifth phosphodiester bond on the 3′ side and the eighth phosphodiester bond on the 5′ side to generate a fragment of 12 to 13 nucleotides (depending on whether the lesion involves one or two bases). In humans and other eukaryotes, the enzyme hydrolyzes the sixth phosphodiester bond on the 3′ side and the twenty-second phosphodiester bond on the 5′ side, producing a fragment of 27 to 29 nucleotides. Following the dual incision, the excised oligonucleotides are released from the duplex and the resulting gap is filled by DNA polymerase I in *E. coli* and DNA polymerase ϵ in humans. The nick is sealed by DNA ligase.

In *E. coli*, the key enzymatic complex is the ABC excinuclease. This activity is carried out by three subunits, UvrA (M_r 104,000), UvrB (M_r 78,000), and UvrC (M_r 68,000). A complex of the UvrA and UvrB proteins (A_2B) scans the DNA and binds to the site of a lesion. The UvrA dimer then dissociates, leaving a tight UvrB-DNA complex. UvrC protein then binds to UvrB, and UvrB makes an incision at the fifth phosphodiester bond on the 3′ side

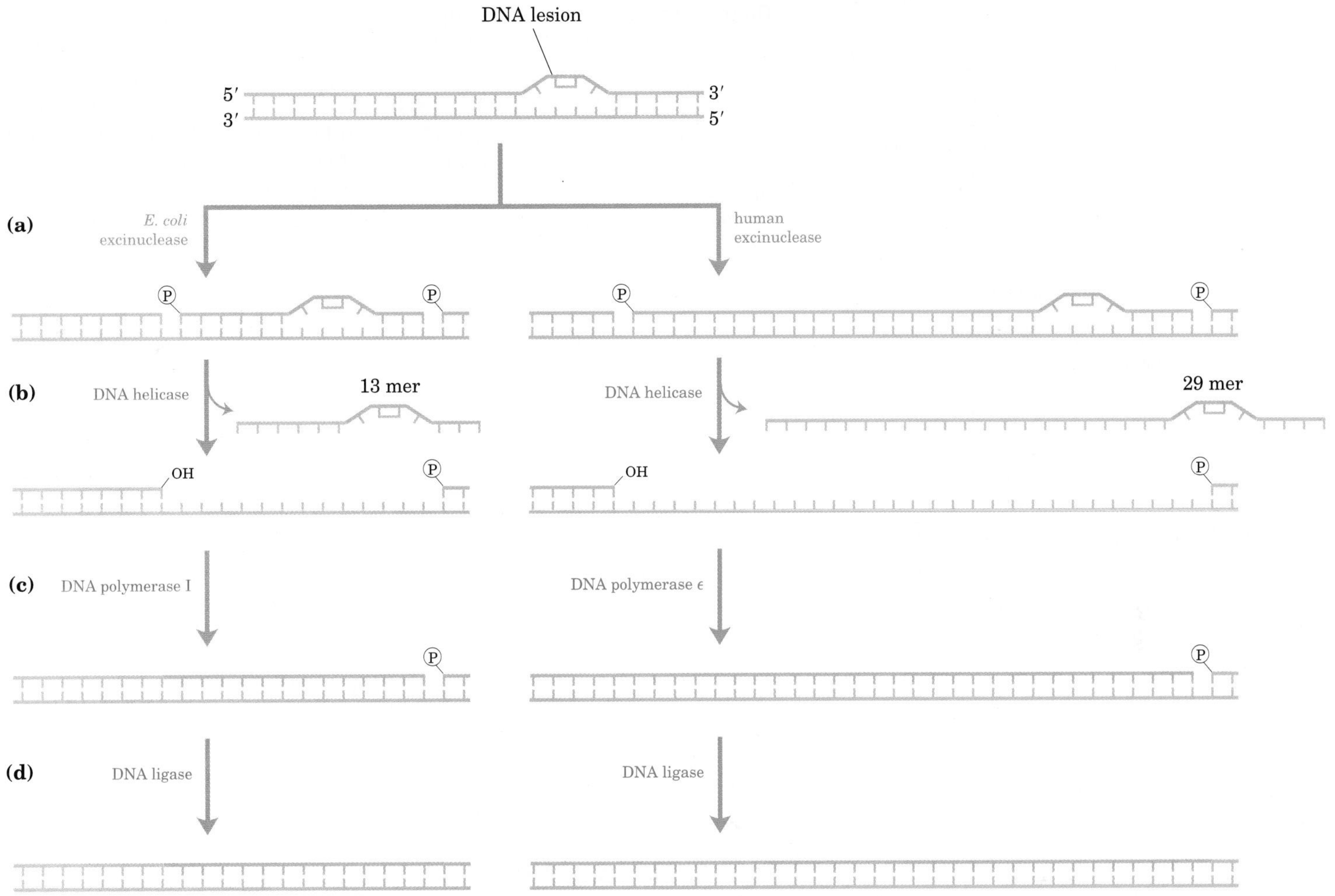

figure 25–23
Nucleotide-excision repair in *E. coli* and humans. The general pathway of nucleotide excision repair is similar in all organisms. **(a)** An excinuclease binds to DNA at the site of a bulky lesion. **(b)** The excinuclease cleaves the damaged DNA strand on either side of the lesion, and the DNA segment is removed with the aid of a helicase. **(c)** The gap is filled in by DNA polymerase, and **(d)** the remaining nick is sealed with DNA ligase.

of the lesion. This is followed by a UvrC-mediated incision at the eighth phosphodiester bond on the 5′ side. The resulting 12 to 13 nucleotide fragment is removed by the UvrD helicase. The short gap thus created is filled in by DNA polymerase I and DNA ligase. This pathway is a primary repair route for many types of lesions, including pyrimidine/cyclobutane dimers, 6-4 photoproducts (see Fig. 10–34), and several other types of base adducts including benzo[*a*]pyrene-guanine, which is formed in DNA by exposure to cigarette smoke. The nucleolytic activity of the ABC excinuclease is novel in the sense that two cuts are made in the DNA (Fig. 25–23). The term "excinuclease" is used to distinguish this activity from that of standard endonucleases.

Eukaryotic excinucleases function, mechanistically, in a manner quite similar to the bacterial enzyme, although 16 polypeptides with no similarity to the *E. coli* excinuclease subunits are required for the dual incision. Genetic deficiencies in nucleotide-excision repair in humans give rise to a variety of serious diseases (Box 25–1).

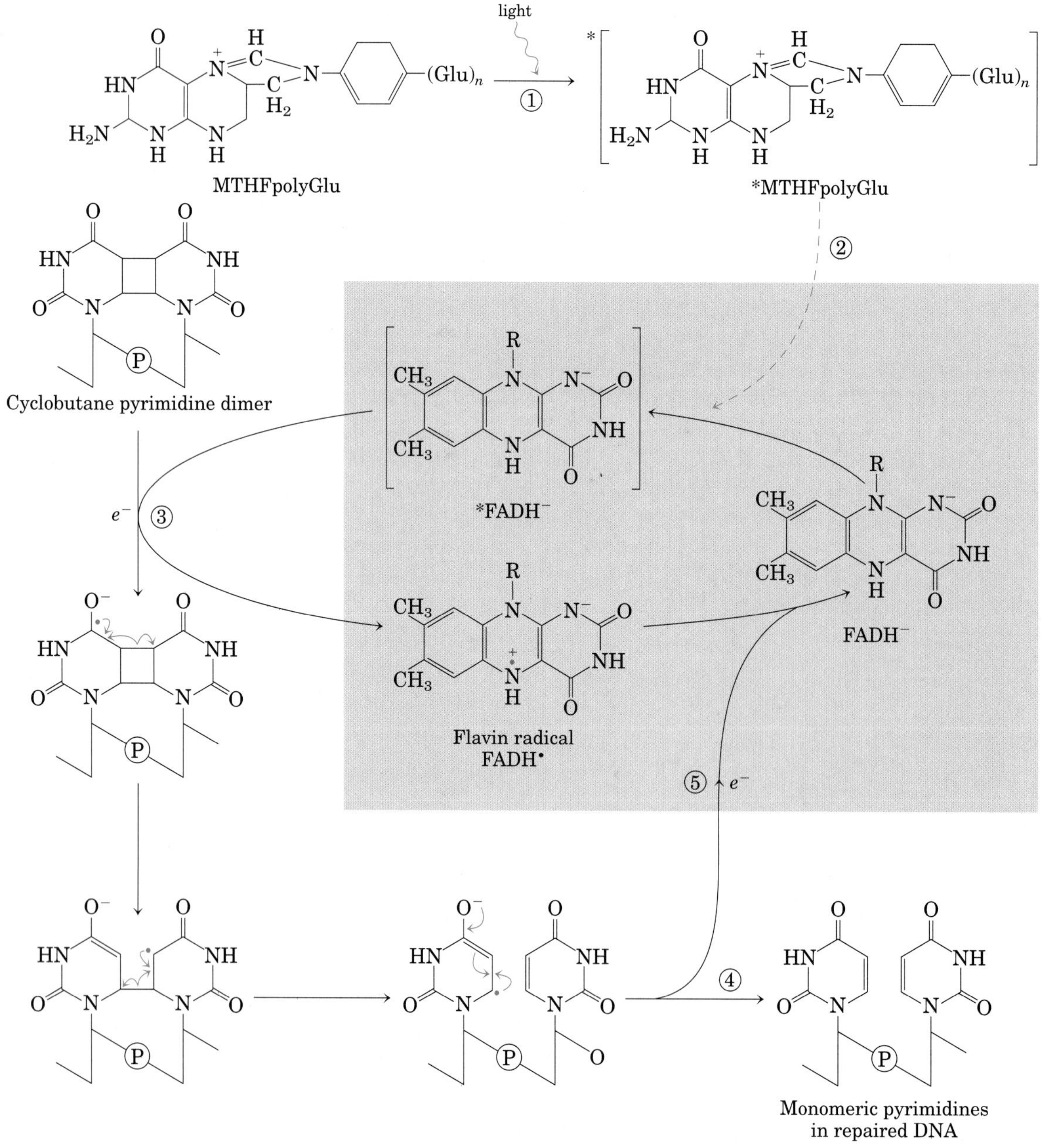

figure 25–24

Repair of pyrimidine dimers with photolyase. Energy derived from absorbed light is used to reverse the photoreaction that caused the lesion. The two chromophores in *E. coli* photolyase (M_r 54,000), N^5,N^{10}-methenyltetrahydrofolylpolyglutamate (MTHFpolyGlu) and $FADH^-$, perform complementary functions. On binding of photolyase to a pyrimidine dimer, repair proceeds as follows. ① A blue-light photon (300–500 nm wavelength) is absorbed by the N^5,N^{10}-methenyltetrahydrofolylpolyglutamate, which functions as a photoantenna. ② The excitation energy is transferred to $FADH^-$ in the active site of the enzyme. ③ The excited flavin ($*FADH^-$) donates an electron to the pyrimidine dimer (shown here in a simplified representation) to generate an unstable dimer radical. ④ Electronic rearrangement restores the monomeric pyrimidines, and ⑤ the electron is transferred back to the flavin radical to regenerate $FADH^-$.

Direct Repair Several types of damage are repaired without removing a base or nucleotide. The best-characterized example is direct photoreactivation of cyclobutane pyrimidine dimers, a reaction promoted by **DNA photolyases.** Pyrimidine dimers result from a light-induced reaction, and photolyases use energy derived from absorbed light to reverse the damage (Fig. 25–24). Photolyases generally contain two cofactors that serve as light-absorbing agents, or chromophores. One of the chromophores is always $FADH^-$. In *E. coli* and yeast, the other chromophore is a folate.

Guanine — Cytosine

methylation and replication

O^6-Methylguanine — Thymine

(a)

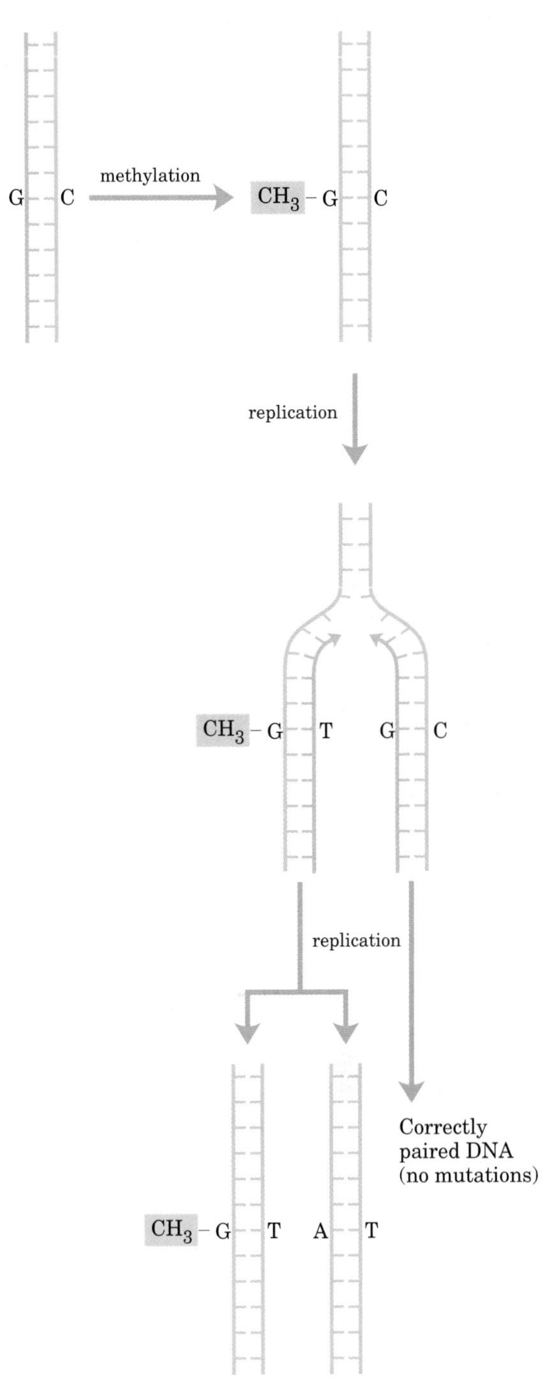

figure 25–25
Example of how DNA damage results in mutations. **(a)** The methylation product O^6-methylguanine pairs with thymine rather than cytosine. **(b)** If not repaired, this leads to a G≡C to A=T mutation after replication.

Another example is the repair of O^6-methylguanine, which forms in the presence of alkylating agents and is a common and highly mutagenic lesion. It tends to pair with thymine rather than cytosine during replication, and therefore causes G≡C to A=T mutations (Fig. 25–25). Direct repair of O^6-methylguanine is carried out by O^6-methylguanine-DNA methyltransferase, a protein that catalyzes transfer of the methyl group of O^6-methylguanine to one of its own Cys residues. This methyltransferase is not strictly an enzyme, because a single methyl transfer event permanently methylates the protein, making it inactive in this pathway. The consumption of an entire protein molecule to correct a single damaged base is another vivid illustration of the priority given to maintaining the integrity of cellular DNA. Although methylated methyltransferase is inactive in repair, it is not discarded; the methylated protein functions as a transcriptional activator (Chapter 28), increasing expression of its own gene and the genes for a few other repair enzymes.

Cys—SH (active) → Cys—S—CH_3 (inactive)

methyltransferase

O^6-Methylguanine nucleotide → Guanine nucleotide

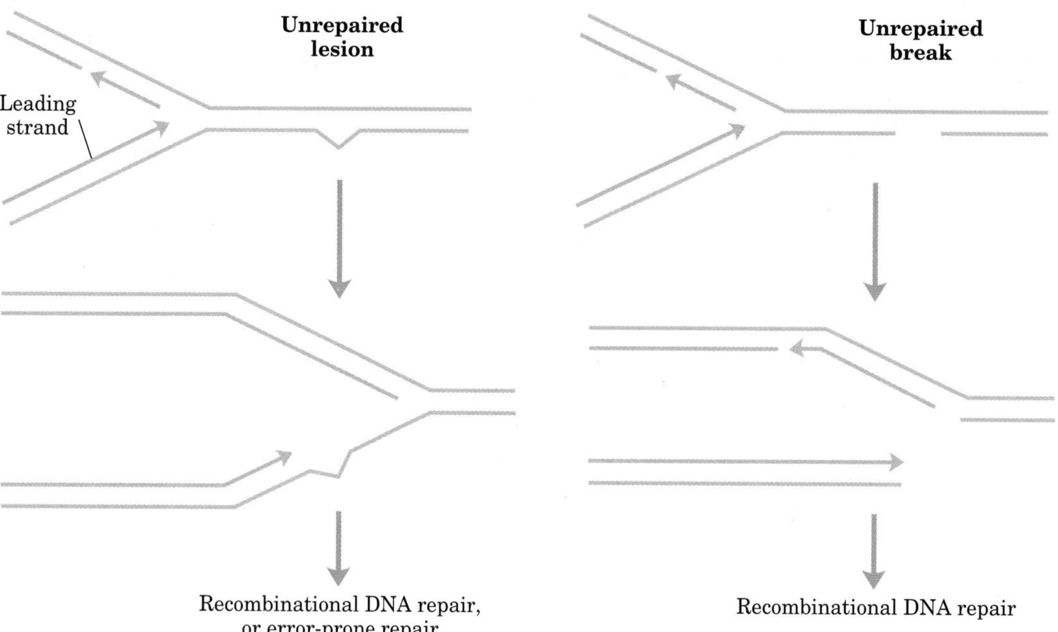

figure 25–26

DNA damage and its effect on DNA replication. If the replication fork encounters an unrepaired lesion or strand break, replication generally halts and the fork may collapse. A lesion is left behind in an unreplicated, single-stranded segment of the DNA; a strand break becomes a double-strand break. In each case, the damage to one strand cannot be repaired by mechanisms described earlier in this chapter because the complementary strand required to direct accurate repair is damaged or absent. There are two possible avenues for repair: recombinational DNA repair (described in Fig. 25–35) or when lesions are unusually numerous, error-prone repair. This latter repair mechanism involves a novel DNA polymerase (DNA polymerase V, encoded by the *umu*C and *umu*D genes) that can replicate, albeit inaccurately, over many types of lesions. It is called error-prone because mutations often result.

The Interaction of Replication Forks with DNA Damage Leads to Recombination or Error-Prone Repair

The repair pathways considered to this point generally work only for lesions in double-stranded DNA, the undamaged strand providing the correct genetic information to restore the damaged strand to its original state. However, in certain types of lesions, such as double-strand breaks, double-strand cross-links, or lesions in single-stranded DNA, the complementary strand is itself damaged or absent. Double-strand breaks and lesions in single-stranded DNA most often arise when a replication fork encounters an unrepaired DNA lesion (Fig. 25–26). Such lesions and DNA cross-links can also result from ionizing radiation and oxidative reactions.

At a stalled bacterial replication fork, there are two avenues for repair. In the absence of a second strand to specify the repair, the information required for accurate repair must come from a separate, homologous chromosome. It therefore involves homologous genetic recombination. This process, called **recombinational DNA repair,** is considered in detail later in this chapter. Whenever DNA damage occurs at unusually high levels (e.g., if the cell is exposed to strong ultraviolet light), a second repair path, called **error-prone repair,** becomes available. When this pathway is active, DNA repair becomes significantly less accurate and a high mutation rate results. Error-prone repair is part of a cellular stress response to extensive DNA damage called, appropriately enough, the **SOS response.**

Some SOS proteins, such as the UvrA and UvrB proteins already described (Table 25–6), are normally present in the cell but are induced to higher levels as part of the SOS response. Additional SOS proteins participate in a novel pathway for error-prone repair. The UmuD protein is cleaved to a shorter form called UmuD′, and this forms a complex with the UmuC protein to create a specialized DNA polymerase (DNA polymerase V) that can replicate past many of the DNA lesions that would normally block replication. Proper base pairing is often impossible at the site of such a lesion,

table 25–6

Genes Induced as Part of the SOS Response in *E. coli*

Gene name	Protein encoded and/or role in DNA repair
Genes of known function	
*pol*B (*din*A)	Encodes polymerization subunit of DNA polymerase II, required for replication restart in recombinational DNA repair
*uvr*A, *uvr*B	Encode ABC excinuclease subunits UvrA and UvrB
*umu*C, *umu*D	Encode DNA polymerase V
*sul*A	Encodes protein that inhibits cell division, possibly to allow time for DNA repair
*rec*A	Encodes RecA protein required for error-prone repair and recombinational repair
*din*B	Encodes DNA polymerase IV
Genes involved in DNA metabolism, but role in DNA repair unknown	
ssb	Encodes single-stranded DNA-binding protein (SSB)
*uvr*D	Encodes DNA helicase II (DNA-unwinding protein)
*him*A	Encodes subunit of integration host factor, involved in site-specific recombination, replication, transposition, regulation of gene expression
*rec*N	Required for recombinational repair
Genes of unknown function	
*din*D	
*din*F	

Note: Some of these genes and their functions are further discussed in Chapter 28.

so this translesion replication is error-prone. Given the emphasis on the importance of genomic integrity throughout this chapter, it may seem incongruous that a system would exist that increases the rate of mutation. However, we can think of this system as a desperation strategy. The resulting mutations kill many cells, but this is the biological price that cells pay to overcome an otherwise insurmountable barrier to replication, as it permits at least a few mutant cells to survive. In addition to DNA polymerase V, translesion replication requires the RecA protein, SSB, and some subunits derived from DNA polymerase III. Yet another DNA polymerase, DNA polymerase IV, is induced during the SOS response. Replication by DNA polymerase IV, a product of the *din*B gene, is also highly error-prone. Some genes induced in the SOS response have as yet unknown functions (Table 25–6).

You will encounter the RecA protein again because it has several distinct functions in the bacterial cell. Its roles in recombination and in the regulation of the SOS response are well characterized. (Regulation of the SOS response is described in Chapter 28.) We now turn to a discussion of genetic recombination.

DNA Recombination

The rearrangement of genetic information within and among DNA molecules encompasses a variety of processes, collectively placed under the heading of genetic recombination. Practical applications of DNA rearrangements in altering the genomes of a variety of organisms are now being explored (Chapter 29).

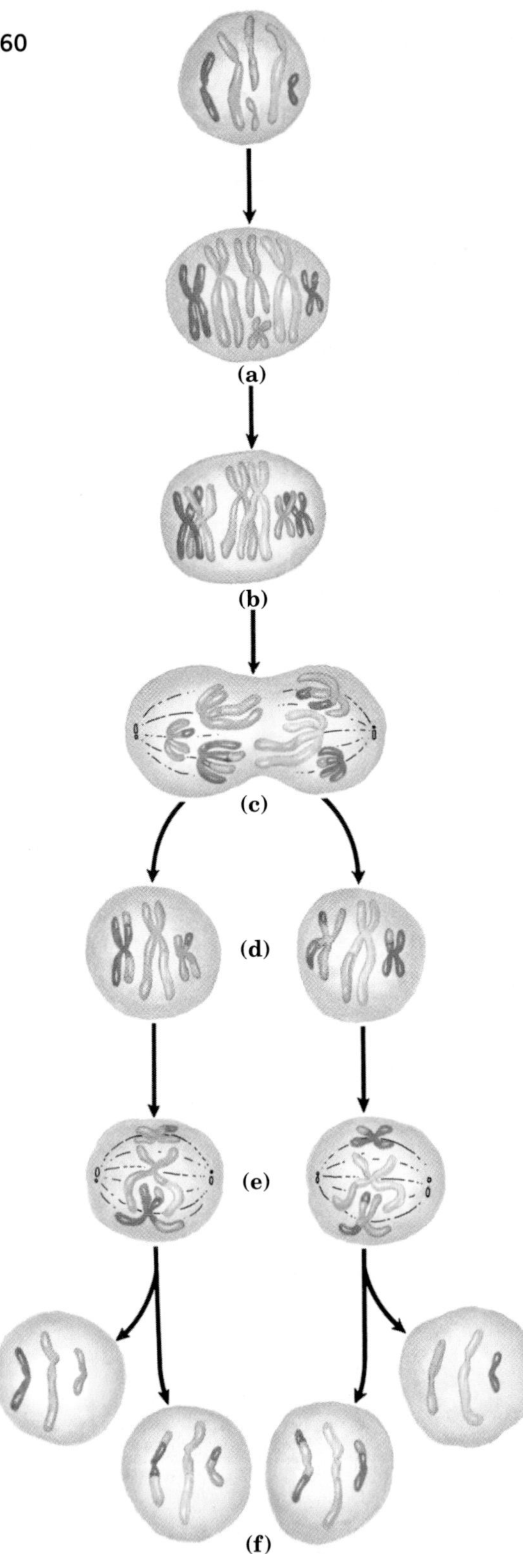

Genetic recombination events fall into at least three general classes. **Homologous genetic recombination** (also called general recombination) involves genetic exchanges between any two DNA molecules (or segments of the same molecule) that share an extended region of nearly identical sequence. The actual sequence of bases is irrelevant, as long as it is similar in the two DNAs. **Site-specific recombination** differs from homologous recombination in that the exchanges occur only at a *particular* DNA sequence. **DNA transposition** is distinct from both other classes in that it usually involves a short segment of DNA with the remarkable capacity to move from one location in a chromosome to another. These "jumping genes" were first observed in maize in the 1940s by Barbara McClintock. There is in addition a wide range of unusual genetic rearrangements for which no mechanism or purpose has yet been proposed, but here we focus on these three general classes.

The functions of genetic recombination systems are as varied as their mechanisms. They include the operation of specialized DNA repair systems and specialized activities in DNA replication, the regulation of expression of certain genes, the facilitation of proper chromosome segregation during cell division in eukaryotes, the maintenance of genetic diversity, and the implementation of programmed genetic rearrangements during embryonic development. In most cases, genetic recombination is closely integrated with other processes in DNA metabolism, and this becomes a theme of our discussion.

Homologous Genetic Recombination Has Multiple Functions

In bacteria, homologous genetic recombination is primarily a DNA repair process, and in this context is referred to as **recombinational DNA repair.** It is usually directed at the reconstruction of replication forks stalled at the site of DNA damage. Homologous genetic recombination can also occur during processes such as conjugation (mating), when chromosomal DNA is transferred from a donor to a recipient bacterial cell. Recombination during conjugation, albeit rare in wild bacterial populations, contributes to genetic diversity.

In eukaryotes, homologous genetic recombination is generally associated with cell division (ensuring the orderly segregation of chromosomes) and with DNA repair. Recombination occurs with the highest frequency during **meiosis,** the process by which diploid germ-line cells with two sets of chromosomes divide to produce haploid gametes—sperm cells or ova in higher eukaryotes—each gamete having only one member of each chromosome pair (Fig. 25–27). In outline, meiosis begins with replication of the DNA in the germ-line cell so that each DNA molecule is present in four copies. The cell then goes through two rounds of cell division without an intervening round of DNA replication. This reduces the DNA content to the haploid level in each gamete.

figure 25–27
Meiosis in eukaryotic germ-line cells. (a) The chromosomes of a hypothetical diploid germ-line cell (six chromosomes; three homologous pairs) are replicated and held together at their centromeres. Each replicated double-stranded DNA molecule is called a chromatid (sister chromatid). **(b)** In prophase I, just before the first meiotic division, the three homologous sets of chromatids align to form tetrads, held together by covalent links at homologous junctions (chiasmata). Crossovers occur within the chiasmata (see Fig. 25–28). These transient associations help ensure that the two tethered chromosomes segregate properly to the opposite poles in the next step. **(c)** The homologous pairs separate and migrate toward opposite poles of the dividing cell. **(d)** This first meiotic division produces two daughter cells, each with three pairs of chromatids. **(e)** The homologous pairs line up across the equator of the cell, in preparation for separation of the chromatids (which are now called chromosomes). **(f)** The second meiotic division produces four haploid daughter cells that can serve as gametes. Each has three chromosomes, half the number of the diploid germ-line cell. The chromosomes have resorted and recombined.

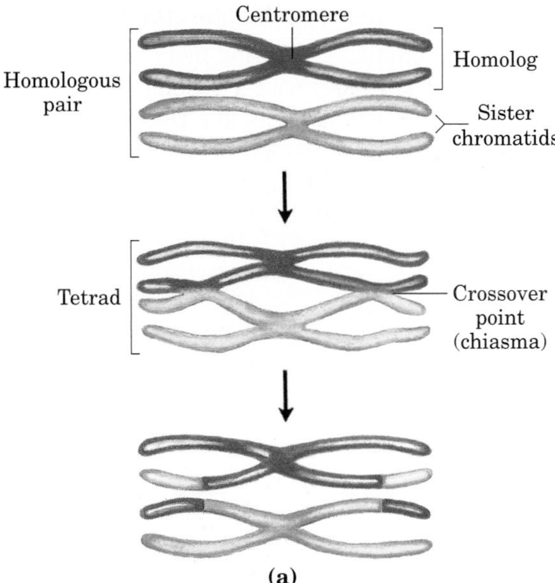

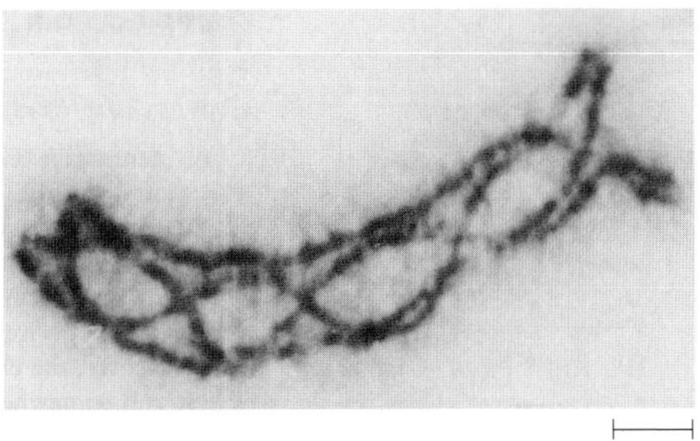

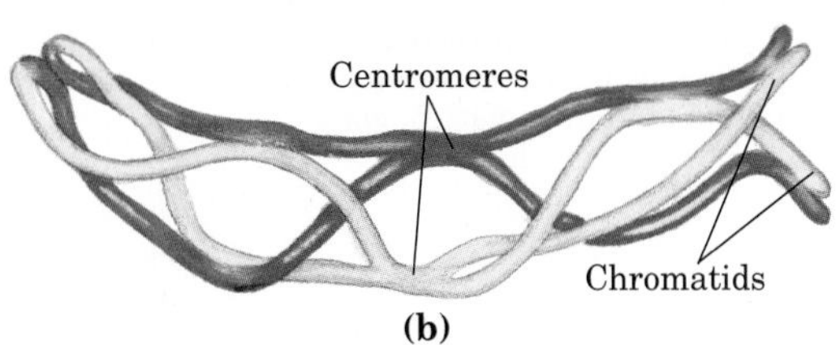

figure 25–28
Crossing over. (a) Crossing over often produces an exchange of genetic material. **(b)** The homologous chromosomes of a grasshopper are shown during prophase I of meiosis. Multiple points of joining (chiasmata) are evident between the two homologous pairs of chromatids. These chiasmata are the physical manifestation of prior homologous recombination (crossing over) events.

After the DNA is replicated during the prophase of the first meiotic division, the resulting DNA copies remain associated at their centromeres and are referred to as sister chromatids. At this stage, each set of four homologous chromosomes exists as two pairs of chromatids. Genetic information is now exchanged between the closely associated homologous chromatids by homologous genetic recombination, a process involving the breakage and rejoining of DNA (Fig. 25–28). This exchange, also called crossing over, can be observed with the light microscope. Crossing over links the two pairs of sister chromatids together at points called chiasmata (singular, chiasma).

Crossing over effectively links together all four homologous chromatids, a linkage that is essential to the proper segregation of chromosomes in the subsequent meiotic cell divisions. Crossing over is not an entirely random process, and "hot spots" have been identified on many eukaryotic chromosomes. However, the assumption that crossing over can occur with equal probability at almost any point along the length of two homologous chromosomes remains a reasonable approximation in many cases, and it is this assumption that permits the genetic mapping of genes. The frequency of homologous recombination in any region separating two points on a chromosome is roughly proportional to the distance between the points, allowing the relative positions of and distances between different genes to be determined.

Homologous recombination thus serves at least three identifiable functions: (1) it contributes to the repair of several types of DNA damage; (2) it provides, in eukaryotic cells, a transient physical link between chromatids that promotes the orderly segregation of chromosomes at the first meiotic cell division; and (3) it enhances the genetic diversity in a population.

Barbara McClintock
1902–1992

Recombination during Meiosis Is Initiated with Double-Strand Breaks

A likely pathway for homologous recombination during meiosis is outlined in Figure 25–29a. The model has four key features. First, homologous chromosomes are aligned. Second, a double-strand break in a DNA molecule is enlarged by an exonuclease, such that a single-strand extension with a free 3′-hydroxyl group is left at the broken end (step ① in Fig. 25–29a). Third, the exposed 3′ ends invade the intact duplex DNA, and this is followed

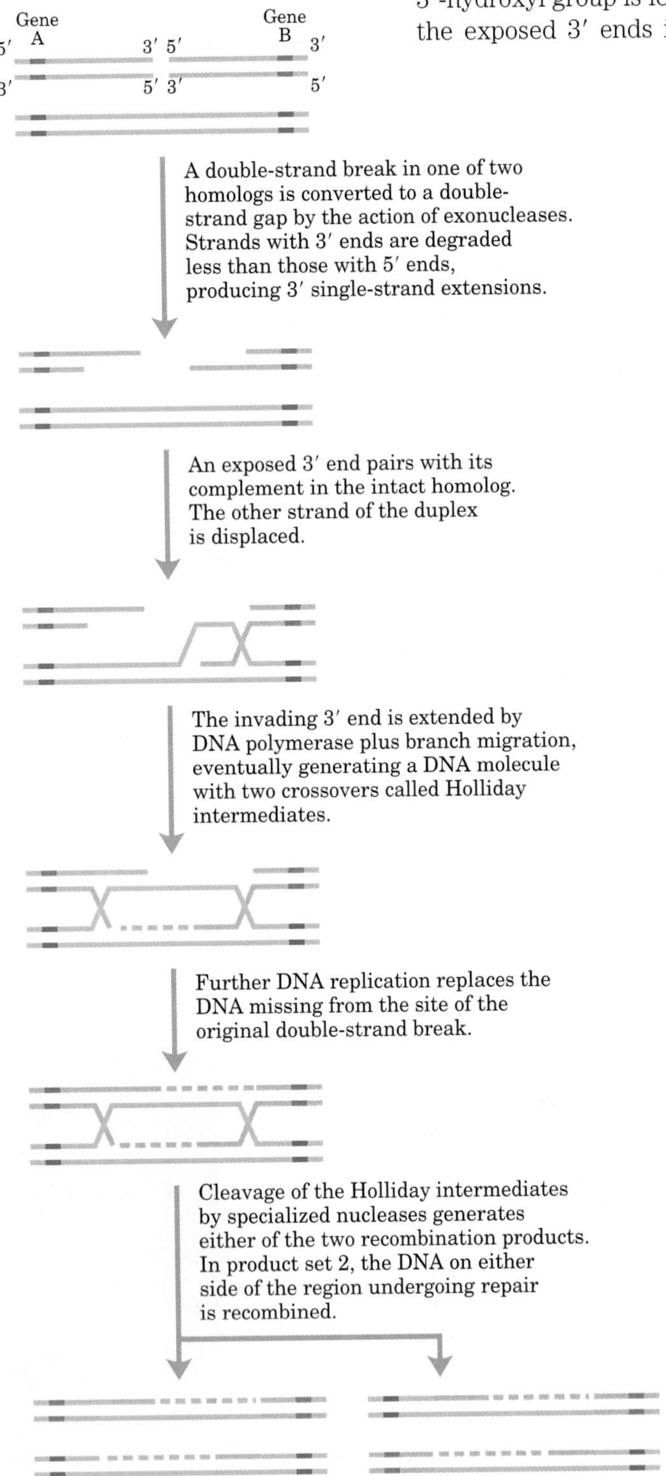

Gene A
Gene B
5′
3′ 5′
3′
3′
5′ 3′
5′
A double-strand break in one of two homologs is converted to a double-strand gap by the action of exonucleases. Strands with 3′ ends are degraded less than those with 5′ ends, producing 3′ single-strand extensions.
An exposed 3′ end pairs with its complement in the intact homolog. The other strand of the duplex is displaced.
The invading 3′ end is extended by DNA polymerase plus branch migration, eventually generating a DNA molecule with two crossovers called Holliday intermediates.
Further DNA replication replaces the DNA missing from the site of the original double-strand break.
Cleavage of the Holliday intermediates by specialized nucleases generates either of the two recombination products. In product set 2, the DNA on either side of the region undergoing repair is recombined.
Product set 1
Product set 2
(a)

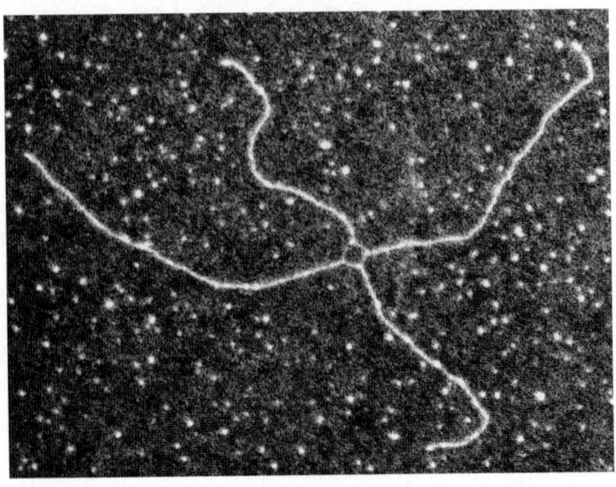

figure 25–29
Recombination during meiosis. (a) Double-strand break repair model for homologous genetic recombination. The two homologous chromosomes involved in this recombination event have similar sequences. Each of the two genes shown has different alleles on the two chromosomes. The DNA strands and alleles are colored differently so that their fate can be followed. The steps are described in the text. **(b)** A Holliday intermediate formed between two bacterial plasmids in vivo, as seen with the electron microscope. The intermediates are named for Robin Holliday, who first proposed their existence in 1964.

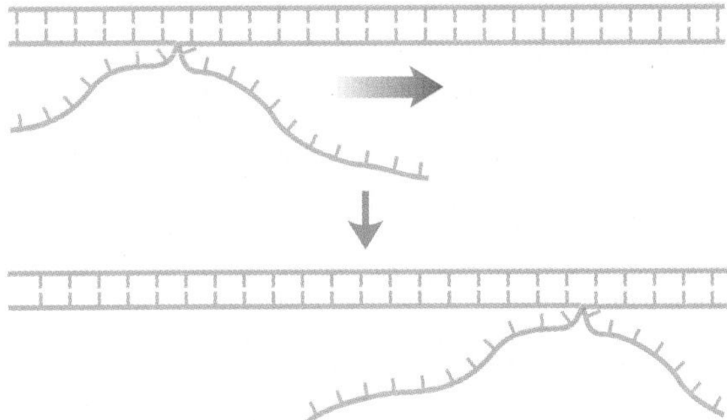

figure 25–30
Branch migration. When a template strand pairs with two different complementary strands, a branch is formed at the point where the three complementary strands meet. The branch "migrates" when a base pair to one of the two complementary strands is broken and replaced with a base pair to the other complementary strand. In the absence of an enzyme to direct it, this process can move the branch spontaneously in either direction. Spontaneous branch migration is blocked wherever one of the otherwise complementary strands has a sequence nonidentical to the other.

by **branch migration** (Fig. 25–30) and/or replication to create a pair of crossover structures, called Holliday junctions (steps ② to ④). Fourth, cleavage of the two crossovers creates two complete recombinant products (step ⑤).

In this **double-strand break repair model** for recombination, the 3′ ends are used to initiate the genetic exchange. Once paired with the complementary strand on the intact homolog, a region of hybrid DNA is created containing complementary strands from two different parental DNAs (the product of step ② in Fig. 25–29a). Each of the 3′ ends can then act as a primer for DNA replication. The structures that are formed, called **Holliday intermediates** (Fig. 25–29b), are a feature of homologous genetic recombination pathways in all organisms.

Homologous recombination can vary in many details from one species to another, but most of the steps outlined above are generally present in some form. There are two ways to cleave or "resolve" the Holliday intermediate so that the two recombinant products carry genes in the same linear order as in the substrates (the original, unrecombined chromosomes). If cleaved one way, the DNA flanking the region containing the hybrid DNA is not recombined; if cleaved the other way, the flanking DNA is recombined (step ⑤ of Fig. 25–29a). Both outcomes are observed in vivo in eukaryotes and in prokaryotes.

The homologous recombination illustrated in Figure 25–29 is a very elaborate process with subtle molecular consequences for the generation of genetic diversity. To understand how this process contributes to diversity, it is important to note that the two homologous chromosomes that undergo recombination are not necessarily *identical.* The linear array of genes may be the same, but the base sequences in some of the genes may differ slightly (in different alleles). In a human, for example, one chromosome may contain the allele for normal hemoglobin (hemoglobin A) while the other contains the allele for hemoglobin S (the sickle-cell mutation). The difference may consist of no more than one base pair among millions. Homologous recombination does not change the linear array of genes, but it can determine which alleles become linked together on a single chromosome.

Recombination Requires Specific Enzymes

Enzymes that promote various steps of homologous recombination have been isolated from both prokaryotes and eukaryotes. In *E. coli,* the *rec*B, C, and D genes encode the RecBCD enzyme, which has both helicase and nuclease activities. The RecA protein promotes all the central steps in the homologous recombination process: the pairing of two DNAs, formation of Holliday intermediates, and branch migration (as described below). The RuvA and B proteins form a complex that binds to Holliday intermediates, displaces RecA protein, and promotes branch migration at higher rates than

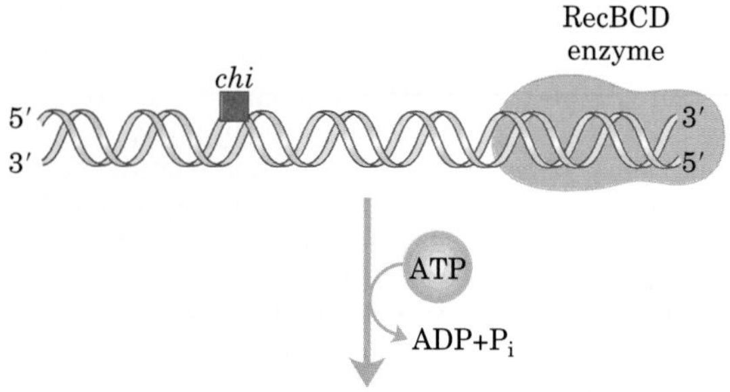

Helicase and nuclease activities of enzyme degrade the DNA.

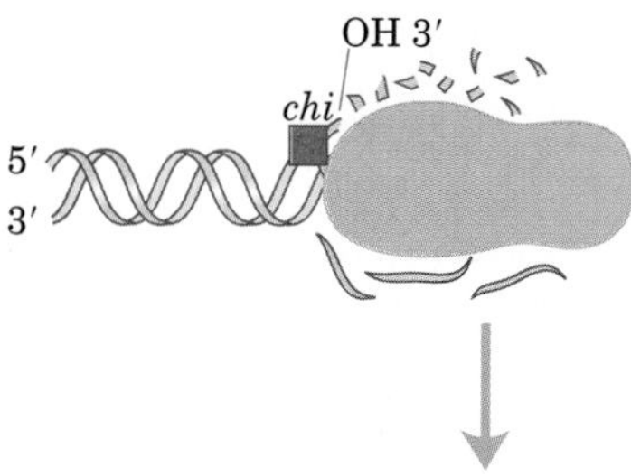

On reaching a *chi* sequence, nuclease activity on the strand with the 3′ end is suppressed. The other strand continues to be degraded, generating a 3′-terminal single-stranded end.

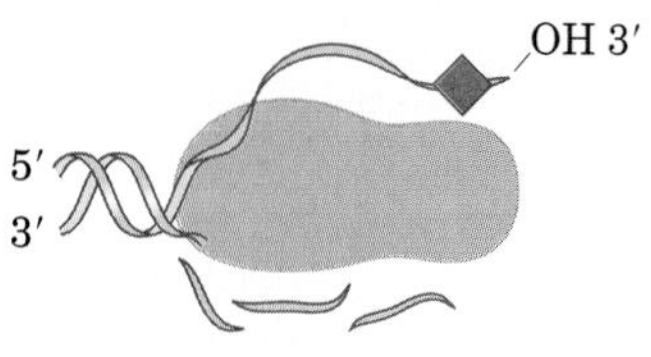

figure 25–31
Helicase and nuclease activities of the RecBCD enzyme. Entering at a double-stranded end, RecBCD unwinds and degrades the DNA until it encounters a *chi* sequence. The interaction with *chi* alters the activity of RecBCD so that it generates a single-stranded DNA with a 3′ end, suitable for subsequent steps in recombination. Movement of the enzyme requires ATP hydrolysis. This enzyme is believed to help initiate homologous genetic recombination in *E. coli.* It is also involved in the repair of double-strand breaks at collapsed replication forks.

does RecA. Nucleases that specifically cleave Holliday intermediates, often called resolvases, have been isolated from bacteria and yeast. The RuvC protein is one of at least two such nucleases in *E. coli*.

The RecBCD enzyme binds to linear DNA at a free (broken) end and moves inward along the double helix, unwinding and degrading the DNA in a reaction coupled to ATP hydrolysis (Fig. 25–31). The activity of the enzyme is altered when it interacts with a sequence called *chi,* (5′)GCTGGTGG. From that point, degradation of the strand with a 3′ terminus is greatly reduced, but degradation of the 5′-terminal strand is increased. A single-stranded DNA with a 3′ end is thus created, a structure used during subsequent steps in recombination (Fig. 25–29). The 1,009 *chi* sequences scattered throughout the *E. coli* genome promote recombination in the regions where they occur. Sequences that enhance recombination frequency have also been identified in several other organisms.

RecA is unusual among the proteins of DNA metabolism in that its active form is an ordered, helical filament of up to several thousand RecA monomers that assemble cooperatively on DNA (Fig. 25–32). This filament normally forms on single-stranded DNA, such as that produced by the RecBCD enzyme. The filament will also form on a duplex DNA with a single-stranded gap. In this case, the first RecA monomers bind to the single-stranded DNA in the gap, after which the assembled filament rapidly envelops the neighboring duplex. The RecF, RecO, and RecR proteins regulate the assembly and disassembly of RecA filaments.

A useful model to illustrate the recombination activities of the RecA filament is the in vitro DNA strand exchange reaction (Fig. 25–33). A homologous duplex DNA is taken up and aligned with a single strand of DNA already bound within the RecA filament. Strands are then exchanged between the two DNAs to create hybrid DNA. The exchange occurs at a rate of six base pairs per second and progresses in the 5′→3′ direction relative to the single-stranded DNA within the RecA filament. This reaction can involve either three or four strands (Fig. 25–33); in the latter case, a Holliday structure forms during the process.

As the duplex DNA is incorporated within the RecA filament and aligned with the bound single-strand DNA over regions of hundreds of base

figure 25–32
RecA. (a) Nucleoprotein filament of RecA protein on single-stranded DNA, as seen with the electron microscope. The striations indicate the right-handed helical structure of the filament. **(b)** Surface contour model of the molecular structure of a 24-subunit RecA filament. The filament has six subunits per turn. One subunit is colored red to provide perspective.

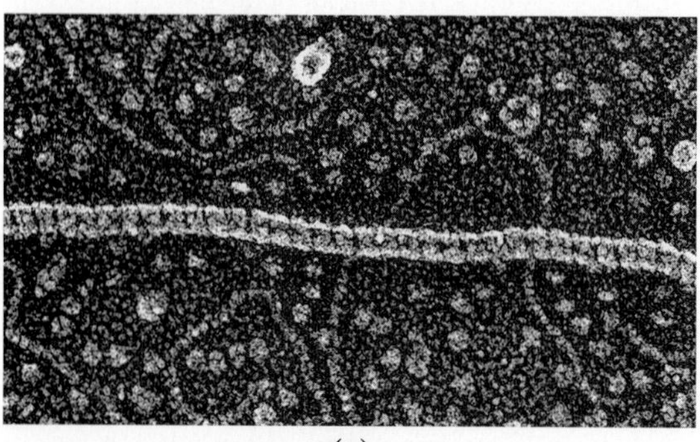

(a)

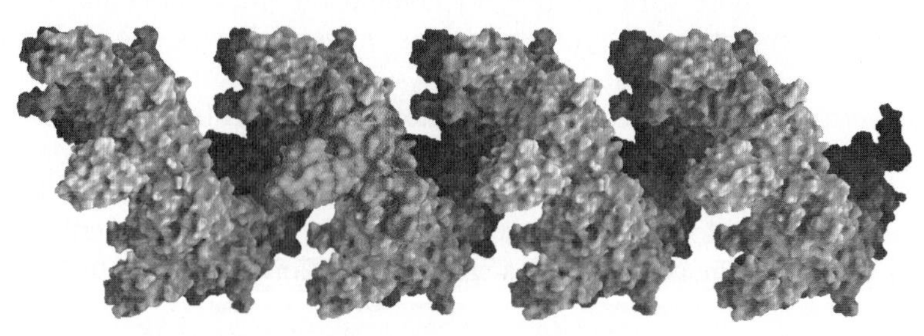

(b)

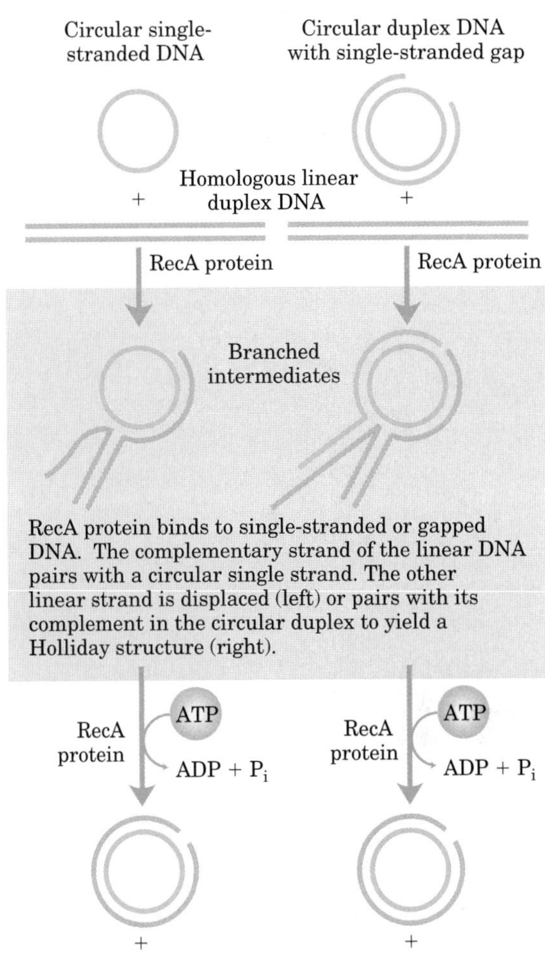

figure 25–33
DNA strand-exchange reactions promoted by RecA protein in vitro. Strand exchange involves the separation of one strand of a duplex DNA from its complement and transfer of the strand to an alternative complementary strand to form a new duplex (heteroduplex) DNA. The transfer forms a branched intermediate. Formation of the final product depends on branch migration, which is facilitated by RecA. The reaction can involve three strands (left) or a reciprocal exchange between two homologous duplexes—four strands in all (right). When four strands are involved, the branched intermediate that results is a Holliday structure. RecA protein promotes the branch migration phases of these reactions using energy derived from ATP hydrolysis.

pairs, one strand of the duplex switches pairing partners (Fig. 25–34b). Because DNA is a helical structure, continued strand exchange requires an ordered rotation of the two aligned DNAs. This brings about a spooling action (Fig. 25–34c,d) that shifts the branch point along the helix. ATP is hydrolyzed by RecA protein during this reaction.

Once a Holliday intermediate has been formed, a host of enzymes—topoisomerases, the RuvAB branch migration protein, a resolvase, other nucleases, DNA polymerase I or III, and DNA ligase—are required to complete recombination. The RuvC protein (M_r 20,000) of *E. coli* cleaves Holliday intermediates to generate full-length, unbranched chromosome products.

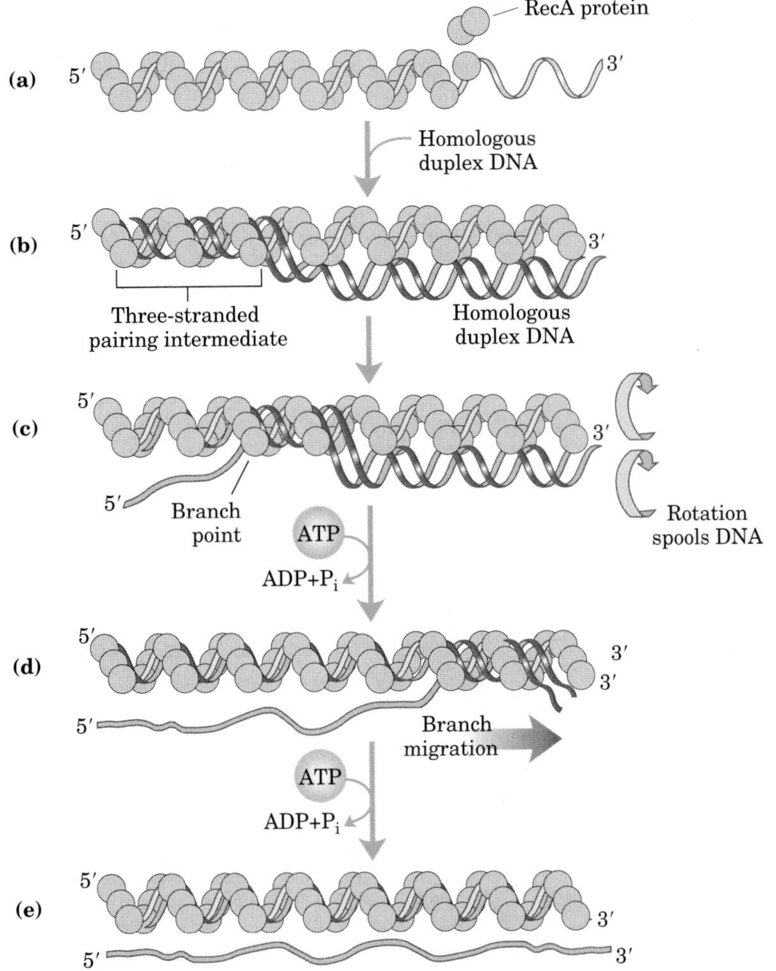

figure 25–34
Model for DNA strand exchange mediated by RecA protein. A three-strand reaction is shown. The balls representing RecA protein are undersized relative to the thickness of DNA to clarify the fate of the DNA strands. **(a)** RecA protein forms a filament on the single-stranded DNA. **(b)** A homologous duplex incorporates into this complex. **(c)** As spooling shifts the three-stranded region from left to right, one of the strands in the duplex is transferred to the single strand originally bound in the filament. The other strand of the duplex is displaced, and a new duplex is formed within the filament. As rotation continues **(d, e)**, the displaced strand separates entirely. ATP is hydrolyzed by RecA protein during this strand exchange.

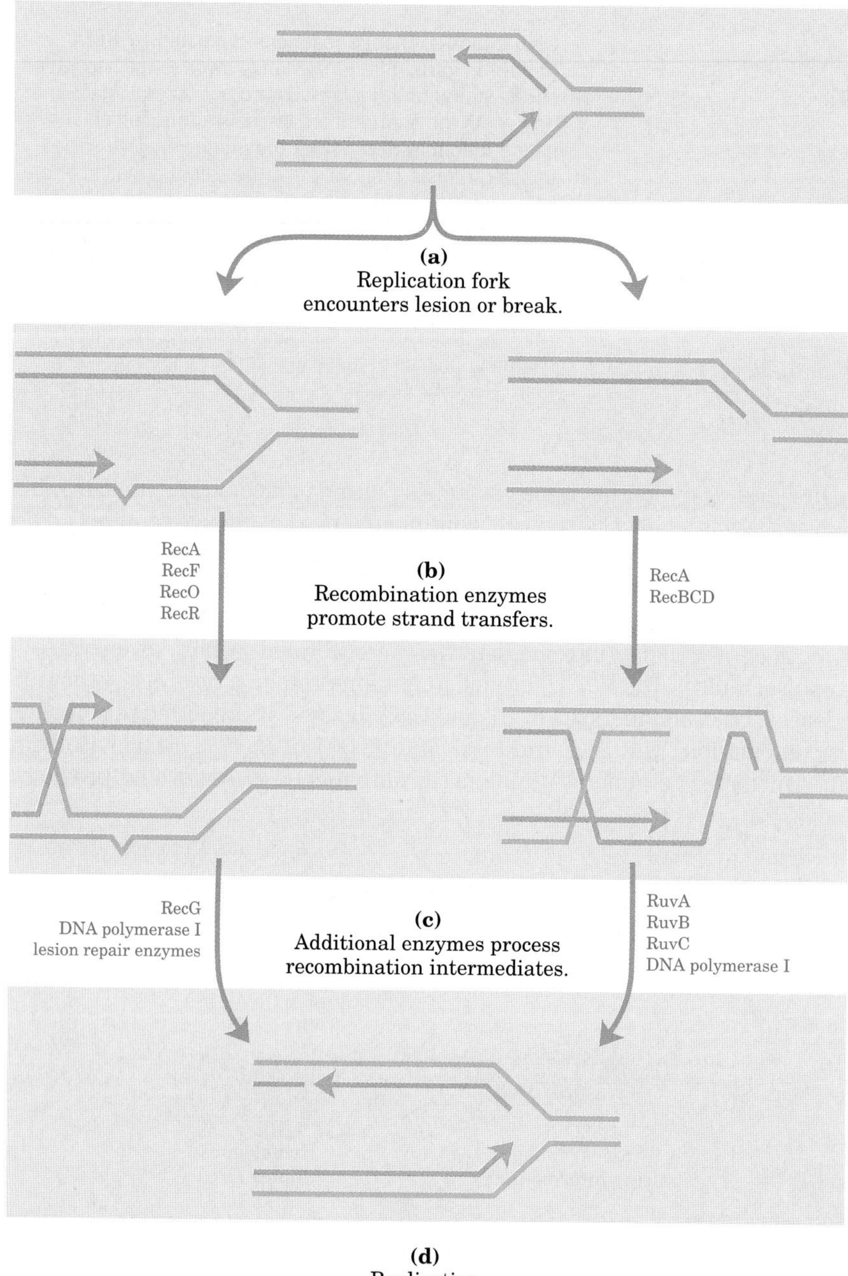

figure 25–35

Models for recombinational DNA repair of stalled replication forks. (a) The replication fork collapses on encountering a DNA lesion (left) or strand break (right). **(b)** Recombination enzymes promote DNA strand transfers needed to repair the branched DNA structure at the replication fork. A lesion in a single-strand gap is repaired in a reaction requiring the RecF, RecO, and RecR proteins. Double-strand breaks are repaired in a pathway requiring the RecBCD enzyme. Both pathways require RecA. **(c)** Recombination intermediates are processed by additional enzymes (e.g., RuvA, RuvB, and RuvC, which process Holliday intermediates). Lesions in double-stranded DNA are repaired by nucleotide-excision repair or other pathways. **(d)** The replication fork is reestablished with the aid of enzymes catalyzing origin-independent replication restart, and chromosomal replication is completed. The overall process requires an elaborate coordination of all aspects of bacterial DNA metabolism.

All Aspects of DNA Metabolism Come Together to Repair Stalled Replication Forks

Like all cells, bacteria sustain high levels of DNA damage even under normal growth conditions. Most DNA lesions are repaired rapidly by base-excision repair, nucleotide-excision repair, and the other pathways already described. Nevertheless, almost every bacterial replication fork encounters an unrepaired DNA lesion or break at some point in its journey from the replication origin to the terminus (Fig. 25–26). DNA polymerase III cannot proceed past many types of DNA lesions, and these encounters tend to leave the lesion in a single-strand gap. An encounter with a DNA strand break creates a double-strand break. Both situations require recombinational DNA repair (Fig. 25–35). Under normal growth conditions, stalled replication forks are reactivated by an elaborate repair pathway encompassing recombinational DNA repair, the restart of replication, and the repair of any lesions left behind. All aspects of DNA metabolism come together in this process.

Once the replication fork has been halted, it can be restored by at least two major paths, both of which require the RecA protein. The repair pathway for lesion-containing DNA gaps also requires the RecF, RecO, and RecR proteins. Repair of double-strand breaks requires the RecBCD enzyme (Fig. 25–35b). Additional recombination steps are followed by a process called origin-independent restart of replication, in which the replication fork is reassembled with the aid of a complex of seven proteins (PriA, B, and C, and DnaB, C, G, and T). This complex, originally discovered as a component required for the replication of ϕX174 DNA in vitro, is now called the replication restart primosome. Replication fork restart also requires DNA polymerase II, in a role not yet defined, giving way to DNA polymerase III for the extensive replication generally required to complete the chromosome.

The repair of stalled replication forks entails a coordinated transition from replication to recombination and back to replication. The recombination steps function to fill the DNA gap or rejoin the broken DNA branch to recreate the branched DNA structure at the replication fork. Lesions left behind in what is now duplex DNA are repaired by pathways such as base-excision or nucleotide-excision repair. Thus a wide range of enzymes encompassing every aspect of DNA metabolism ultimately take part in the repair of a stalled replication fork. This type of repair process is clearly a major function of the homologous recombination system of every cell, and defects in recombinational DNA repair play an important role in human disease (Box 25–1).

Site-Specific Recombination Results in Precise DNA Rearrangements

Homologous genetic recombination can involve any two homologous sequences. In a very different type of process, recombination is limited to specific sequences. These recombination reactions occur in virtually every cell, filling specialized roles that vary greatly from one species to another. Examples include regulating the expression of certain genes and promoting programmed DNA rearrangements during embryonic development or in the replication cycles of some viral and plasmid DNAs. Each site-specific recombination system consists of an enzyme called a recombinase and a short (20 to 200 base pair), unique DNA sequence where the recombinase acts (the recombination site). One or more auxiliary proteins may regulate the timing or outcome of the reaction.

In vitro studies of many site-specific recombination systems have elucidated some general principles, including the fundamental reaction pathway

(Fig. 25–36a). A separate recombinase recognizes and binds to each of two recombination sites on two different DNA molecules or within the same DNA. One DNA strand in each site is cleaved at a specific point within the site, and the recombinase becomes covalently linked to the DNA at the cleavage site through a phosphotyrosine (or phosphoserine) bond (step ①). The transient protein-DNA linkage preserves the phosphodiester bond lost in cleaving the DNA, such that high-energy cofactors such as ATP are unnecessary in subsequent steps. The cleaved DNA strands are rejoined to new partners, forming a Holliday intermediate with new phosphodiester bonds created at the expense of the protein-DNA linkage (step ② of Fig. 25–36a). To complete the reaction, the process must be repeated at a second point within each of the two recombination sites (steps ③ and ④). In some systems, both strands of each recombination site are cut concurrently and rejoined to new partners without the Holliday intermediate. The exchange is always reciprocal and precise, so that the recombination sites are regenerated when the reaction is complete. Recombinase can be viewed as a site-specific endonuclease and ligase in one package.

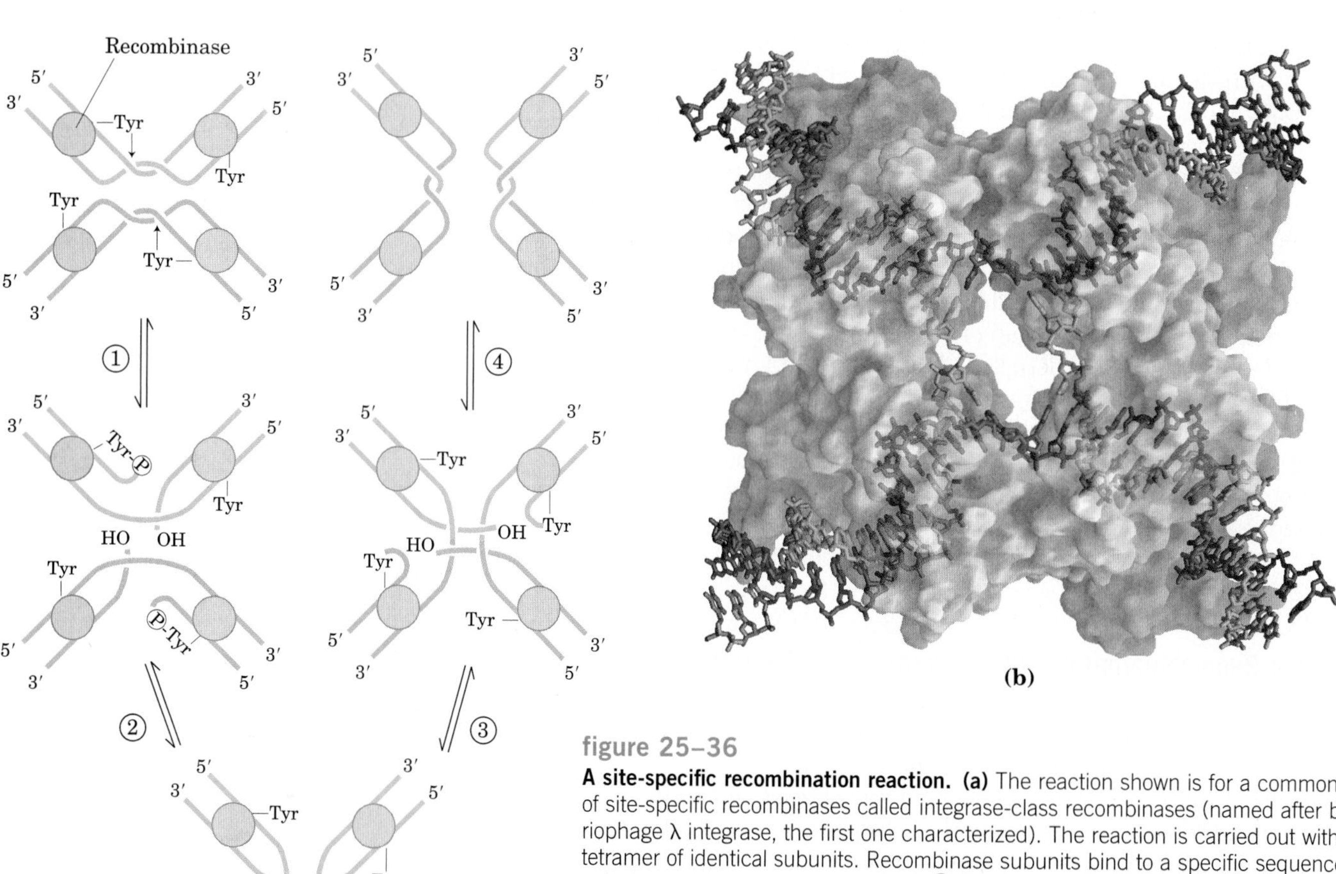

figure 25–36

A site-specific recombination reaction. (a) The reaction shown is for a common class of site-specific recombinases called integrase-class recombinases (named after bacteriophage λ integrase, the first one characterized). The reaction is carried out within a tetramer of identical subunits. Recombinase subunits bind to a specific sequence, often called simply the recombination site. ① One strand in each DNA is cleaved at particular points within the sequence. The nucleophile is the OH group of an active-site Tyr residue, and the product is a covalent phosphotyrosine link between protein and DNA. ② The cleaved strands join to new partners, producing a Holliday intermediate. Steps ③ and ④ complete the reaction by a process similar to the first two steps. The original sequence of the recombination site is regenerated after recombining the DNA flanking the site. These steps occur within a complex of multiple recombinase subunits that sometimes includes other proteins not shown here. **(b)** A surface contour model of a four-subunit integrase-class recombinase called the Cre recombinase, bound to a Holliday structure recombination intermediate shown with light blue and dark blue helix strands. The protein has been rendered transparent so that the bound DNA is visible.

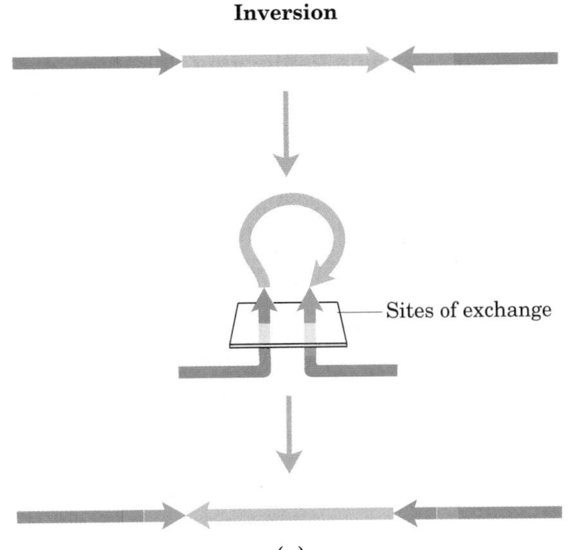

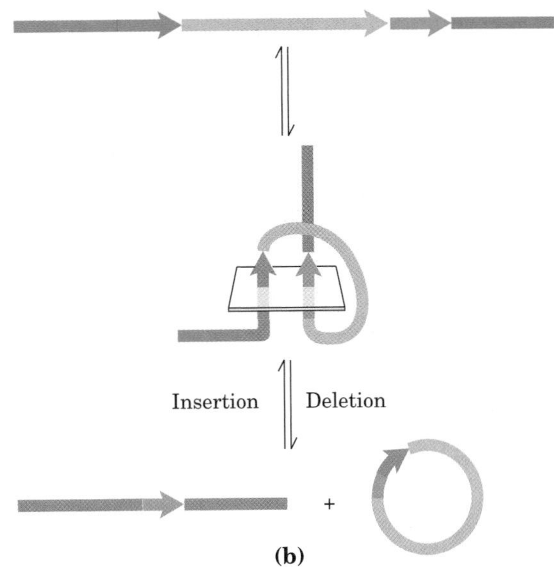

figure 25–37
Effects of site-specific recombination. The outcome of site-specific recombination depends on the location and orientation of the recombination sites (red and blue) in a double-stranded DNA molecule. Orientation here refers to the order of nucleotides in the recombination site, not the 5′→3′ direction. **(a)** Recombination sites with opposite orientation in the same DNA molecule. The result is an inversion. **(b)** Recombination sites with the same orientation, either on one DNA molecule (producing a deletion) or two DNA molecules (producing an insertion).

The sequences of the recombination sites recognized by site-specific recombinases are partially asymmetric (nonpalindromic), and the two recombining sites align in the same orientation during the recombinase reaction. The outcome depends on the location and orientation of the recombination sites (Fig. 25–37). If the two sites are on the same DNA molecule, the reaction inverts or deletes the intervening DNA, depending on whether the recombination sites have the opposite or the same orientation, respectively. If the sites are on different DNAs, the recombination is intermolecular, and an insertion occurs if one or both DNAs are circular. Some recombinase systems are highly specific for one of these reactions and act only on sites with particular orientations.

The first site-specific recombination system studied in vitro was that encoded by the bacteriophage λ. When λ phage DNA enters an *E. coli* cell, a complex series of regulatory events commits the DNA to one of two fates. The λ DNA either replicates and produces more bacteriophages (destroying the host cell) or integrates into the host chromosome, replicating passively along with the chromosome for many cell generations. Integration is accomplished by a phage-encoded recombinase (λ integrase) that acts at recombination sites on the phage and bacterial DNAs, called attachment sites *att*P and *att*B, respectively (Fig. 25–38).

The role of site-specific recombination in regulating gene expression is considered in Chapter 28.

figure 25–38
Integration and excision of bacteriophage λ DNA at the chromosomal target site. The attachment site on the λ phage DNA (*att*P) shares only 15 base pairs of complete homology with the bacterial site (*att*B) in the region of the crossover. The reaction generates two new attachment sites (*att*R and *att*L) flanking the integrated phage DNA. The recombinase is the λ integrase (or INT protein). Integration and excision use different attachment sites and different auxiliary proteins. Excision uses the proteins XIS, encoded by the bacteriophage, and FIS, encoded by the bacterium. Both reactions require the protein IHF (*i*ntegration *h*ost *f*actor), encoded by the bacterium.

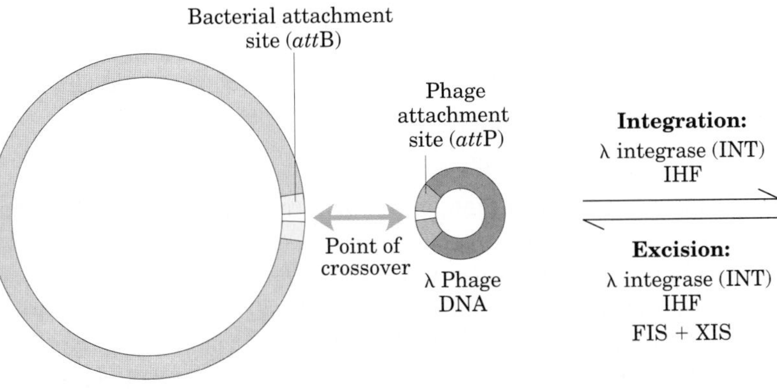

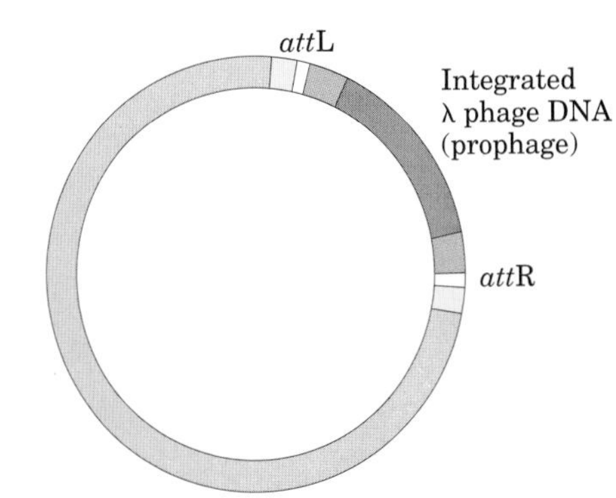

Complete Chromosome Replication Can Require Site-Specific Recombination

Recombinational DNA repair of a circular bacterial chromosome, while essential, sometimes generates deleterious byproducts. The resolution of a Holliday junction at a replication fork by a nuclease such as RuvC, followed by completion of replication, can give rise to one of two products: the usual two monomeric chromosomes or a contiguous dimeric chromosome (Fig. 25–39). In the latter case, the covalently linked chromosomes cannot be segregated to daughter cells at cell division and the dividing cells become "stuck." A specialized site-specific recombination system in *E. coli*, called the XerCD system, converts the dimeric chromosomes to monomeric chromosomes so that cell division can proceed. The reaction is a site-specific deletion reaction (Fig. 25–37b). This is another example of the close coordination between DNA recombination processes and other aspects of DNA metabolism.

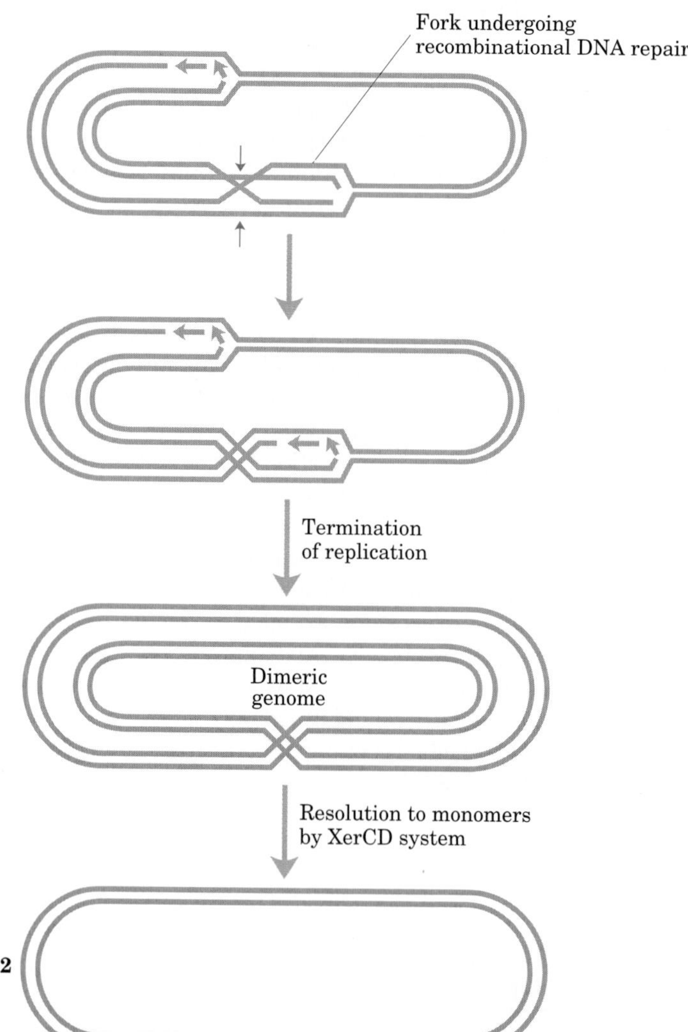

figure 25–39
DNA deletion to undo a deleterious effect of recombinational DNA repair. The resolution of a Holliday intermediate during recombinational DNA repair can (if cut at the points indicated by red arrows) generate a contiguous dimeric chromosome. A specialized site-specific recombinase in *E. coli* called XerCD converts the dimer to monomers, allowing chromosome segregation and cell division to proceed.

Transposable Genetic Elements Move from One Location to Another

We now consider how recombination allows the movement of transposable elements, or **transposons.** These are segments of DNA, found in virtually all cells, that move or "hop" from one place on a chromosome (the donor site) to another on the same or a different chromosome (the target site). DNA sequence homology is not usually required for this movement, called

transposition; the new location is usually chosen more or less randomly. Insertion of a transposon in an essential gene could kill the cell, so transposition is tightly regulated and usually very infrequent. Transposons are perhaps the simplest of molecular parasites, adapted to replicate passively within the chromosomes of host cells. In some cases they carry genes that are useful to the host cell and exist in a kind of symbiosis with the host.

Bacteria have two classes of transposons. **Insertion sequences** (simple transposons) contain only the sequences required for transposition and the genes for proteins (transposases) that promote the process. **Complex transposons** contain one or more genes in addition to those needed for transposition. These extra genes might, for example, confer resistance to antibiotics and thus enhance the survival chances of the host cell. The spread of antibiotic-resistance elements among disease-causing bacterial populations that is rendering some antibiotics ineffectual (pp. 912–913) is mediated in part by transposition.

Bacterial transposons vary in structure, but most have short repeated sequences at each end that serve as binding sites for the transposase. When transposition occurs, a short sequence at the target site (5 to 10 base pairs) is duplicated to form an additional short repeated sequence that flanks each end of the inserted transposon (Fig. 25–40). These duplicated segments result from the cutting mechanism used to insert a transposon into the DNA at a new location.

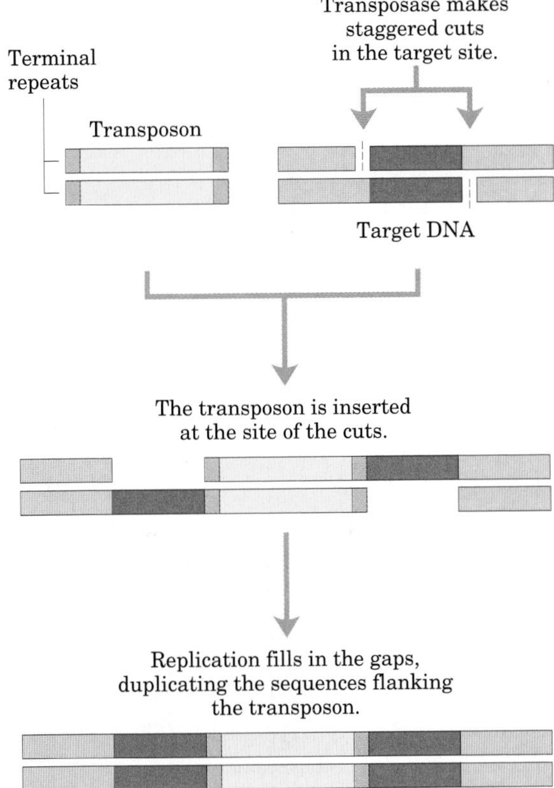

figure 25–40
Duplication of the DNA sequence at a target site when a transposon is inserted. The duplicated sequences are shown in red. These sequences are generally only a few base pairs long, so their size (compared with that of a typical transposon) is greatly exaggerated in this drawing.

There are two general pathways for transposition in bacteria. In direct or simple transposition (Fig. 25–41, left), cuts on each side of the transposon excise it, and the transposon moves to a new location. This leaves a double-stranded break in the donor DNA that must be repaired. At the target site, a staggered cut is made (as in Fig. 25–40), the transposon is inserted into the break, and DNA replication fills in the gaps to duplicate the target site sequence. In replicative transposition (Fig. 25–41, right), the entire transposon is replicated, leaving a copy behind at the donor location. The **cointegrate** is an intermediate in this process, consisting of the donor region covalently linked to DNA at the target site. Two complete copies of

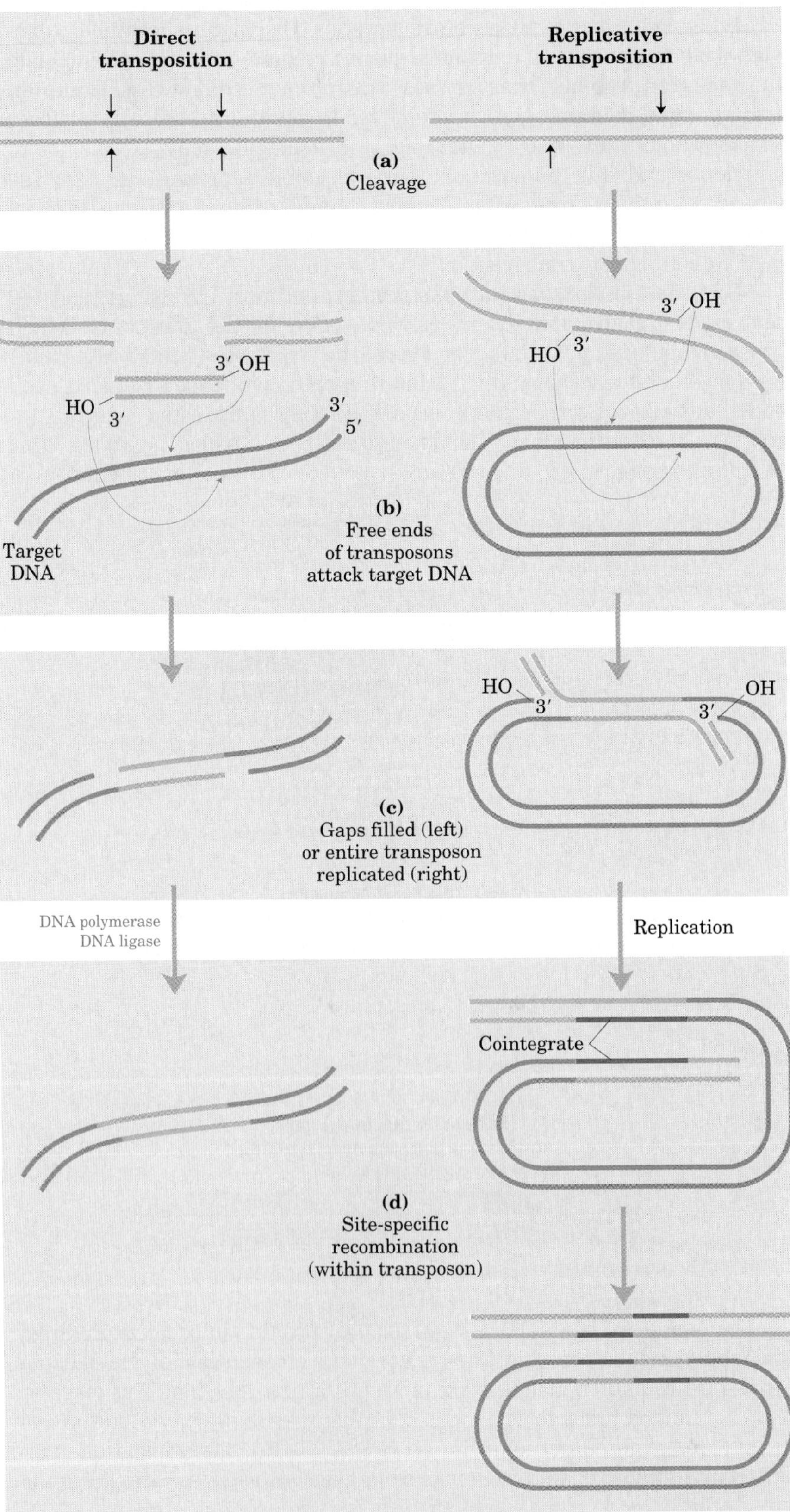

figure 25–41
Two general pathways for transposition: direct (simple) and replicative. (a) The DNA is first cleaved on each side of the transposon, at the sites indicated by arrows. **(b)** The liberated 3′-hydroxyl groups at the ends of the transposon act as nucleophiles in a direct attack on phosphodiester bonds in the target DNA. The target phosphodiester bonds are staggered (not directly across from each other) in the two DNA strands. **(c)** The transposon is now linked to the target DNA. In direct transposition, replication fills in gaps at each end. In replicative transposition the entire transposon is replicated, creating a cointegrate intermediate. **(d)** The cointegrate is often resolved later with the aid of a separate site-specific recombination system. The cleaved host DNA left behind after direct transposition is either repaired by DNA end-joining or degraded (not shown). The latter outcome can be lethal to an organism.

the transposon are present in the cointegrate, both having the same relative orientation in the DNA. In some well-characterized transposons, the cointegrate intermediate is converted to products by site-specific recombination, in which specialized recombinases promote the required deletion reaction.

Eukaryotes also have transposons, structurally similar to bacterial transposons, and some use similar transposition mechanisms. In other cases, however, the mechanism of transposition appears to involve an RNA intermediate. Evolution of these transposons is intertwined with the evolution of certain classes of RNA viruses. Both will be described in the next chapter.

Immunoglobulin Genes Are Assembled by Recombination

Some DNA rearrangements are a programmed part of development in eukaryotic organisms. An important example is the generation of complete immunoglobulin genes from separate gene segments in vertebrate genomes. A mammal, such as a human, is capable of producing *millions* of different immunoglobulins (antibodies) with distinct binding specificities, even though the human genome contains only about 100,000 genes. Recombination allows an organism to produce an extraordinary diversity of antibodies from a limited DNA-coding capacity. Studies of the recombination mechanism reveal a close relationship to DNA transposition and suggest that this system for generating antibody diversity may have evolved from an ancient cellular invasion of transposons.

Human genes encoding proteins of the immunoglobulin G (IgG) class can be used to illustrate how antibody diversity is generated. Immunoglobulins consist of two heavy and two light polypeptide chains (see Fig. 7–23). Each chain has two regions, a variable region, with a sequence that differs greatly from one immunoglobulin to another, and a region that is virtually constant within a class of immunoglobulins. There are also two distinct families of light chains, called kappa and lambda, which differ somewhat in the sequences of their constant regions. For each of the three types of polypeptide chain (heavy chain, and kappa and lambda light chains), diversity in the variable regions is generated by a similar mechanism. The genes for these polypeptides are divided into segments, and the genome contains clusters with multiple versions of each segment. The joining of one version of each of the segments creates a complete gene.

Figure 25–42 depicts the organization of the DNA encoding the kappa light chains of human IgG and shows how a mature kappa light chain is generated. In undifferentiated cells, the coding information for this polypeptide chain is separated into three segments. The V (variable) segment encodes the first 95 amino acid residues of the variable region, the J (joining) segment encodes the remaining 12 residues of the variable region, and the C segment encodes the constant region. The genome contains about 300 different V segments, four different J segments, and one C segment.

As a stem cell in the bone marrow differentiates to form a mature B lymphocyte, one V segment and one J segment are brought together by a

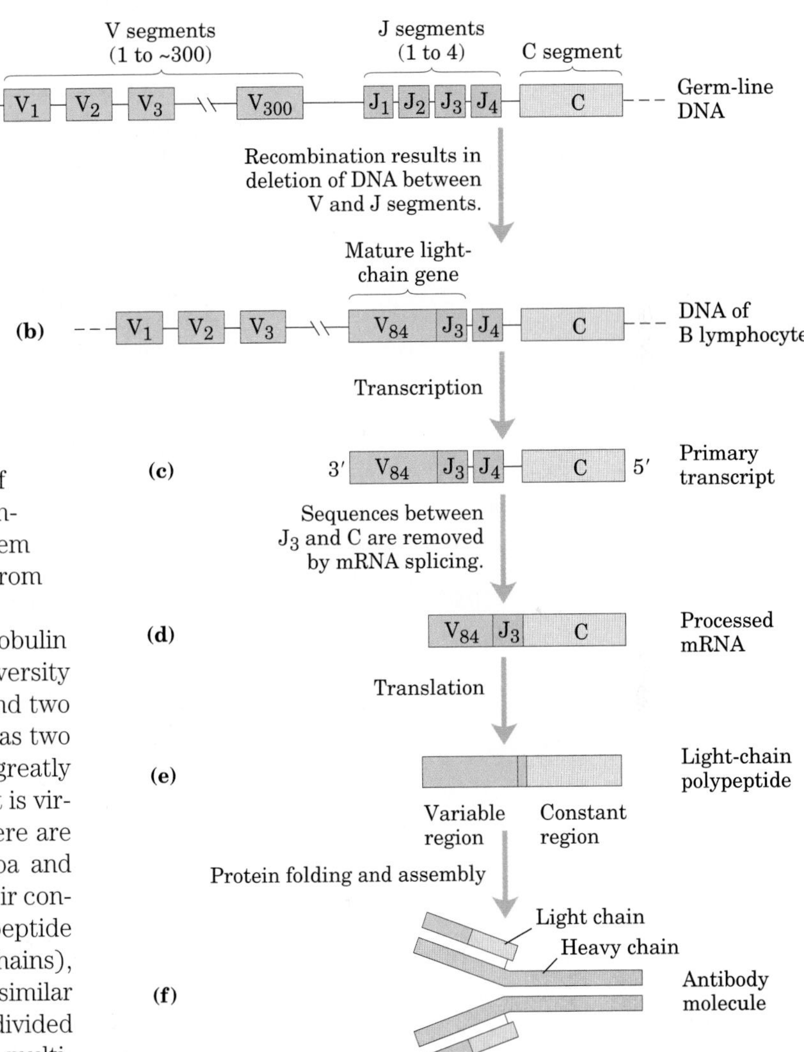

figure 25–42

Recombination of the V and J gene segments of the human IgG kappa light chain. This process is designed to generate antibody diversity. **(a)** The arrangement of IgG-coding sequences in a bone marrow stem cell. **(b)** Recombination deletes the DNA between a particular V segment and a J segment. The transcript **(c)** is processed by RNA splicing **(d)**, as described in Chapter 26, then translated **(e)** to produce the light-chain polypeptide. **(f)** The light chain can combine with any of 5,000 possible heavy chains to produce an antibody molecule.

specialized recombination system (Fig. 25–42). During this programmed DNA deletion, the intervening DNA is discarded. There are $300 \times 4 = 1,200$ possible V-J combinations. The recombination process is not as precise as the site-specific recombination described earlier. Additional variation occurs in the sequence at the V-J junction that increases the overall variation by a factor of at least 2.5: about $2.5 \times 1,200 = 3,000$ different V-J combinations can be generated.

The final joining of the V-J combination to the C region is accomplished by an RNA-splicing reaction after transcription, a process described in Chapter 26. The genes for the heavy chains and the lambda light chains are formed by similar processes. For heavy chains, there are more gene segments and more than 5,000 possible combinations. Because any heavy chain can combine with any light chain to generate an immunoglobulin, each human has at least $3,000 \times 5,000$ or 1.5×10^7 possible IgGs. Additional diversity is generated because the V sequences are subject to high mutation rates (of unknown mechanism) during B-lymphocyte differentiation. Each mature B lymphocyte produces only one type of antibody, but the range of antibodies produced by different cells is clearly enormous.

The recombination mechanism for joining the V and J segments is illustrated in Figure 25–43. Just beyond each V segment and just before each J segment are recombination signal sequences (RSS). These are bound by proteins called RAG1 and RAG2 (*r*ecombination *a*ctivating *g*ene). The RAG proteins catalyze the formation of a double-strand break between the signal sequences and the V (or J) segments to be joined. The V and J segments are then joined with the aid of a second complex of proteins.

Did the immune system evolve in part from ancient transposons? The mechanism for generation of the double-strand breaks by RAG1 and RAG2 does mirror several reaction steps in transposition (Fig. 25–43). In addition, the deleted DNA, with its terminal RSS sequences, has a sequence structure found in most transposons. In the test tube, RAG1 and RAG2 can associate with this deleted DNA and insert it, transposon-like, into other DNA molecules (a reaction that probably happens only rarely in B lymphocytes). Although we cannot know for certain, the properties of the immunoglobulin gene rearrangement system suggest an intriguing origin in which the distinction between host and parasite has been blurred by evolution.

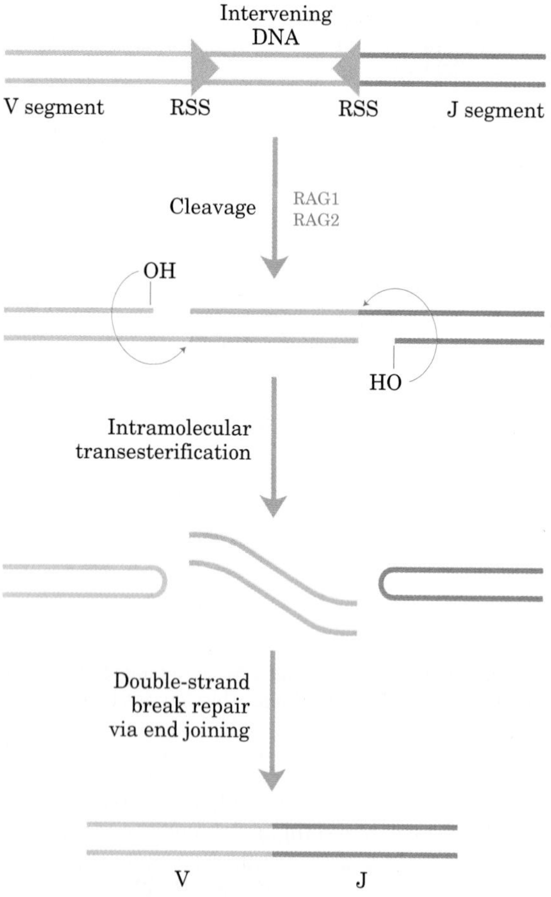

figure 25–43
Mechanism of immunoglobulin gene rearrangement. The RAG1 and RAG2 proteins bind to the recombinational signal sequences (RSS) and cleave one DNA strand between the RSS and the V (or J) segments to be joined. The liberated 3′ hydroxyl then acts as a nucleophile, attacking a phosphodiester bond in the other strand to create a double-strand break. The resulting hairpin ends on the V and J segments are cleaved, and the ends are covalently linked by a complex of proteins specialized for end-joining repair of double-strand breaks. The steps in the generation of the double-strand break catalyzed by RAG1 and RAG2 are chemically related to steps seen in transposition reactions.

summary

The integrity of DNA is of utmost importance to the cell, as reflected in the complexity and redundancy of the enzyme systems that participate in DNA replication, repair, and recombination, and in the energy these systems require.

Replication of DNA occurs with very high fidelity and within a designated time period in the cell cycle. Replication is semiconservative, with each strand acting as a template for a new daughter strand. The reaction starts at a DNA sequence called the origin and usually proceeds bidirectionally. DNA is synthesized in the $5' \rightarrow 3'$ direction by DNA polymerases. At the replication fork, the leading strand is synthesized continuously in the same direction as the progressing replication fork. The lagging strand is synthesized discontinuously as Okazaki fragments, which are subsequently ligated. The fidelity of DNA replication is maintained by (1) base selection by the polymerase, (2) a $3' \rightarrow 5'$ proofreading exonuclease activity that is part of most DNA polymerases, and (3) specific repair systems for mismatches left behind after replication.

Most cells have several DNA polymerases. In *E. coli,* DNA polymerase III is the primary replication enzyme. DNA polymerase I is responsible for special functions during replication, recombination, and repair. Replication of the *E. coli* chromosome involves many enzymes and protein factors organized into complexes. Initiation of replication requires binding of DnaA protein to the origin, strand separation, and the entry of the DnaB and DnaC proteins to set up two replication forks. Initiation is the only phase of replication that is known to be regulated. The process of elongation has different requirements for each strand. DNA strands are separated by the DnaB helicase, with the resulting topological strain relieved by topoisomerases, and single-stranded DNA-binding proteins stabilize the separated strands. In synthesis of the lagging strand, the DnaG primase binds to DnaB at regular intervals and synthesizes short RNA primers. Both the leading and the lagging strands are synthesized by DNA polymerase III, and the syntheses are coupled. RNA primers are removed and replaced with DNA by DNA polymerase I, and nicks are sealed by DNA ligase. DNA synthesis is terminated at specialized sites in a bacterial chromosome and requires several proteins, including topoisomerases. Replication is similar in eukaryotic cells, but eukaryotic chromosomes have multiple replication origins. Several eukaryotic DNA polymerases have been identified.

Cells have multiple systems for DNA repair. Mismatch repair in *E. coli* is directed by transient undermethylation of (5')GATC sequences on the newly synthesized strand. Other systems recognize and repair damage caused by environmental agents such as radiation and alkylating agents, and damage caused by spontaneous reactions of nucleotides. Some repair systems recognize and excise only damaged or incorrect bases, leaving an AP (apurinic or apyrimidinic) site in the DNA. This is repaired by excising and replacing the segment of DNA containing the AP site. Other excision repair systems recognize and remove a variety of bulky lesions and pyrimidine dimers. Some types of DNA damage can also be repaired by direct reversal of the reaction causing the damage: pyrimidine dimers are directly converted to monomeric pyrimidines by photolyase, and the methyl group in O^6-methylguanine is removed by a specific methyltransferase. Error-prone repair is a specialized and mutagenic replication process that occurs when DNA damage is so heavy that the need of cells to replicate outweighs the need to avoid errors.

DNA sequences are rearranged in recombination reactions, usually in processes tightly coordinated with DNA replication or repair. Homologous genetic recombination can take place between any two DNAs that share sequence homology. In meiosis (in eukaryotes), it is one of the processes that creates genetic diversity. Homologous recombination also is needed to repair some types of DNA damage. A Holliday intermediate, a crossover between the strands of two homologous DNAs, is formed during the process. In *E. coli,* the RecA protein promotes both formation of Holliday intermediates and branch migration to extend heteroduplex DNA.

Site-specific recombination occurs only at specific target sequences, and this process also can involve a Holliday intermediate. Recombinases cleave the DNA at specific points and ligate the strands to new partners. This type of recombination is found in virtually all cells, and its many functions include DNA integration and regulation of gene expression. In virtually all cells, small segments of DNA called transposons use recombination to move within or between chromosomes. A programmed recombination reaction related to transposition joins immunoglobulin gene segments to form immunoglobulin genes during B-lymphocyte differentiation in vertebrates.

further reading

General

Kornberg, A. & Baker, T.A. (1991) *DNA Replication,* 2nd edn, W.H. Freeman and Company, New York.

An excellent primary source for all aspects of DNA metabolism.

Friedberg, E.C., Walker, G.C., & Siede, W. (1995) *DNA Repair and Mutagenesis,* American Society for Microbiology, Washington, DC.

A thorough treatment and a good place to start exploring this field.

Replication

Burgers, P.M.J. (1998) Eukaryotic DNA polymerases in DNA replication and DNA repair. *Chromosoma* **107,** 218–227.

Crooke, E. (1995) Regulation of chromosomal replication in *E. coli:* sequestration and beyond. *Cell* **82,** 877–880.

Goodman, M.F. (1997) Hydrogen bonding revisited: geometric selection as a principal determinant of DNA replication fidelity. *Proc. Natl. Acad. Sci. USA* **94,** 10,493–10,495.

Goodman, M.F. & Fygenson, K.D. (1998) DNA polymerase fidelity: from genetics toward a biochemical understanding. *Genetics* **148,** 1475–1482.

Kamada, K., Horiuchi, T., Ohsumi, K., Shimamoto, N., & Morikawa, K. (1996) Structure of a replication-terminator protein complexed with DNA. *Nature* **383,** 598–603.

Kelman, Z. & O'Donnell, M. (1995) DNA polymerase III holoenzyme: structure and function of a chromosomal replicating machine. *Annu. Rev. Biochem.* **64,** 176–200.

Marians, K.J. (1992) Prokaryotic DNA replication. *Annu. Rev. Biochem.* **61,** 673–719.

Stillman, B. (1996) Cell cycle control of DNA replication. *Science* **274,** 1659–1664.

Toyn, J.H., Toone, M.W., Morgan, B.A., & Johnston, L.H. (1995) The activation of DNA replication in yeast. *Trends Biochem. Sci.* **20,** 70–73.

Waga, S. & Stillman, B. (1998) The DNA replication fork in eukaryotic cells. *Annu. Rev. Biochem.* **67,** 721–751.

Woodgate, R. (1999) A plethora of lesion-replicating DNA polymerases. *Genes Develop.* **13,** 2191–2195.

Repair

Kolodner, R.D. (1995) Mismatch repair: mechanisms and relationship to cancer susceptibility. *Trends Biochem. Sci.* **20,** 397–401.

McCullough, A.K., Dodson, M.L., & Lloyd, R.S. (1999) Initiation of base excision repair: glycosylase mechanisms and structures. *Annu. Rev. Biochem.* **68,** 255–286.

Modrich, P. & Lahue, R. (1996) Mismatch repair in replication fidelity, genetic recombination, and cancer biology. *Annu. Rev. Biochem.* **65,** 101–133.

Sancar, A. (1994) Mechanisms of DNA excision repair. *Science* **266,** 1954–1956.

Sancar, A. (1995) DNA repair in humans. *Annu. Rev. Genet.* **29,** 69–105.

Sancar, A. (1996) DNA excision repair. *Annu. Rev. Biochem.* **65,** 43–81.

Seeberg, E., Eide, L., & Bjoras, M. (1995) The base excision repair pathway. *Trends Biochem. Sci.* **20,** 391–397.

Walker, G.C. (1995) SOS-regulated proteins in translesion DNA synthesis and mutagenesis. *Trends Biochem. Sci.* **20,** 416–420.

Wood, R.D. (1996) DNA repair in eukaryotes. *Annu. Rev. Biochem.* **65,** 135–167.

Recombination

Cox, M.M. (1999) Recombinational DNA repair in bacteria and the RecA protein. *Prog. Nucl. Acid Res. Mol. Biol.* **63,** 311–366.

Craig, N.L. (1995) Unity in transposition reactions. *Science* **270,** 253–254.

Eggleston, A.K. & West, S.C. (1996) Exchanging partners: recombination in *E. coli. Trends Genet.* **12,** 20–26.

Grindley, N.D. (1997) Site-specific recombination: synapsis and strand exchange revealed. *Curr. Biol.* **7,** R608–R612.

Hallet, B. & Sherratt, D.J. (1997) Transposition and site-specific recombination: adapting DNA cut-and-paste mechanisms to a variety of genetic rearrangements. *FEMS Microbiol. Rev.* **21,** 157–178.

Kogoma, T. (1996) Recombination by replication. *Cell* **85,** 625–627.

Kowalczykowski, S.C., Dixon, D.A., Eggleston, A.K., Lauder, S.D., & Rehrauer, W.M. (1994) Biochemistry of homologous recombination in *Escherichia coli. Microbiol. Rev.* **58,** 401–465.

Lewis, S.M. & Wu, G.E. (1997) The origins of V(D)J recombination. *Cell* **88,** 159–162.

Lieber, M. (1996) Immunoglobulin diversity: rearranging by cutting and repairing. *Curr. Biol.* **6,** 134–136.

Pâques, F. & Haber, J.E. (1999) Multiple pathways of recombination induced by double-strand breaks in *Saccharomyces cerevisiae. Microbiol. Mol. Biol. Rev.* **63,** 349–404.

Roca, A.I. & Cox, M.M. (1997) RecA protein: structure, function, and role in recombinational DNA repair. *Prog. Nucl. Acid Res. Mol. Biol.* **56,** 129–223.

Roth, D.B. & Craig, N.L. (1998) VDJ recombination: a transposase goes to work. *Cell* **94,** 411–414.

problems

1. Conclusions from the Meselson-Stahl Experiment The Meselson-Stahl experiment (see Fig. 25–2) proved that DNA undergoes semiconservative replication in *E. coli.* In the "dispersive" model of DNA replication, the parental DNA strands are cleaved into pieces of random size, then joined with pieces of newly replicated DNA to yield daughter duplexes. In the Meselson-Stahl experiment, each strand would contain random segments of heavy and light DNA. Explain how the results of Meselson and Stahl's experiment ruled out such a model.

2. Heavy Isotope Analysis of DNA Replication A culture of *E. coli* growing in a medium containing $^{15}NH_4Cl$ is switched to a medium containing $^{14}NH_4Cl$ for three generations (an eightfold increase in population). What is the molar ratio of hybrid DNA (^{15}N–^{14}N) to light DNA (^{14}N–^{14}N) at this point?

3. Replication of the *E. coli* Chromosome The *E. coli* chromosome contains 4,639,221 base pairs.

(a) How many turns of the double helix must be unwound during replication of the *E. coli* chromosome?

(b) From the data in this chapter, how long would it take to replicate the *E. coli* chromosome at 37 °C if two replication forks proceed from the origin? Assume replication occurs at a rate of 1,000 base pairs per second. Under some conditions *E. coli* cells can divide every 20 min. How might this be possible?

(c) In the replication of the *E. coli* chromosome, about how many Okazaki fragments would be formed? What factors guarantee that the numerous Okazaki fragments are assembled in the correct order in the new DNA?

4. Base Composition of DNAs Made from Single-Stranded Templates Predict the base composition of the total DNA synthesized by DNA polymerase on templates provided by an equimolar mixture of the two complementary strands of bacteriophage øX174 DNA (a circular DNA molecule). The base composition of one strand is A, 24.7%; G, 24.1%; C, 18.5%; and T, 32.7%. What assumption is necessary to answer this problem?

5. DNA Replication Kornberg and his colleagues incubated soluble extracts of *E. coli* with a mixture of dATP, dTTP, dGTP, and dCTP, all labeled with ^{32}P in the α-phosphate group. After a time, the incubation mixture was treated with trichloroacetic acid, which precipitates the DNA but not the nucleotide precursors. The precipitate was collected, and the extent of precursor incorporation into DNA was determined from the amount of radioactivity present in the precipitate.

(a) If any one of the four nucleotide precursors were omitted from the incubation mixture, would radioactivity be found in the precipitate? Explain.

(b) Would ^{32}P be incorporated into the DNA if only dTTP were labeled? Explain.

(c) Would radioactivity be found in the precipitate if ^{32}P labeled the β or γ phosphate rather than the α phosphate of the deoxyribonucleotides? Explain.

6. Leading and Lagging Strands Prepare a table that lists the names and compares the functions of the precursors, enzymes, and other proteins needed to make the leading versus lagging strands during DNA replication in *E. coli.*

7. Function of DNA Ligase Some *E. coli* mutants contain defective DNA ligase. When these mutants are exposed to ^{3}H-labeled thymine and the DNA produced is sedimented on an alkaline sucrose density gradient, two radioactive bands appear. One corresponds to a high molecular weight fraction, the other to a low molecular weight fraction. Explain.

8. Fidelity of Replication of DNA What factors promote the fidelity of replication during the synthesis of the leading strand of DNA? Would you expect the lagging strand to be made with the same fidelity? Give reasons for your answers.

9. Importance of DNA Topoisomerases in DNA Replication DNA unwinding, such as that occurring in replication, affects the superhelical density of DNA. In the absence of topoisomerases, the DNA would become overwound ahead of a replication fork as the DNA is unwound behind it. A bacterial replication fork will stall when the superhelical density (σ) of the DNA ahead of the fork reaches $+0.14$, (see Chapter 24).

Bidirectional replication is initiated at the origin of a 6,000 base pair plasmid in vitro, in the absence of topoisomerases. The plasmid initially has a σ of -0.06. How many base pairs will be unwound and replicated by each replication fork before the forks stall? Assume that each fork travels at the same rate and that each includes all components necessary for elongation except topoisomerase.

10. The Ames Test In a nutrient medium that lacks histidine, a thin layer of agar containing about 10^9 *Salmonella typhimurium* histidine auxotrophs (mutant cells that require histidine to survive) produces about 13 colonies over a 2-day incubation period at 37 °C (see Fig. 25–18). How do these colonies arise in the absence of histidine? The experiment is repeated in the presence of 0.4 μg of 2-aminoanthracene. The number of colonies produced over 2 days exceeds 10,000. What does this indicate about 2-aminoanthracene? What can you surmise about its carcinogenicity?

11. DNA Repair Mechanisms Vertebrate and plant cells often methylate cytosine in DNA to form 5-methylcytosine (see Fig. 10–5a). In these same cells, a specialized repair system recognizes G–T mismatches and repairs them to G≡C base pairs. How might this repair system be advantageous to the cell? (Explain in terms of the presence of 5-methylcytosine in the DNA.)

12. DNA Repair in People with Xeroderma Pigmentosum The condition known as xeroderma pigmentosum (XP) arises from mutations in at least seven different human genes. The deficiencies are generally in genes encoding enzymes involved in some part of the pathway for human nucleotide excision repair. The various types of XP are labeled A–G (XP-A, XP-B, etc.), with a few additional variants lumped under the label XP-V.

Cultures of cells from normal individuals and from some patients with XP-G are irradiated with ultraviolet light. The DNA is isolated and denatured, and the resulting single-stranded DNA is characterized by analytical ultracentrifugation.

(a) Samples from the normal fibroblasts show a significant reduction in the average molecular weight of the single-stranded DNA after irradiation, but samples from the XP-G fibroblasts show no such reduction. Why might this be?

(b) If you assume that a nucleotide-excision repair system is operative, which step might be defective in the fibroblasts from the patients with XP-G? Explain.

13. Holliday Intermediates How does the formation of Holliday intermediates in homologous genetic recombination differ from their formation in site-specific recombination?

chapter

26

RNA Metabolism

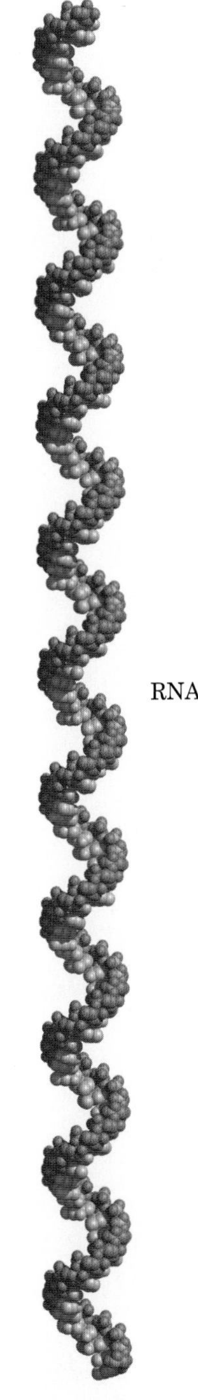

The expression of the information in a gene normally involves production of an RNA molecule transcribed from a DNA template. Strands of RNA and DNA may seem quite similar at first glance, differing only in the hydroxyl group at the 2′ position of the pentose and the substitution of uracil for thymine. However, unlike DNA, most RNAs function as single strands. These strands fold back on themselves with the potential for much greater structural diversity than DNA (Chapter 10). RNA is thus suited to a variety of cellular functions.

RNA is the only macromolecule known to function both in the storage and transmission of information and in catalysis, leading to much speculation about its possible role as an essential chemical intermediate in the development of life on this planet. The discovery of catalytic RNAs, or ribozymes, has changed the very definition of an enzyme, extending it beyond the domain of proteins.

Proteins nevertheless remain important to RNA and its functions. Within the modern cell, all nucleic acids, including RNAs, are complexed with proteins. Some of these complexes are quite elaborate, and RNA can assume both structural and catalytic roles within complicated biochemical machines.

All RNA molecules except the RNA genomes of certain viruses are derived from information permanently stored in DNA. During **transcription,** an enzyme system converts the genetic information in a segment of double-stranded DNA into an RNA strand with a base sequence complementary to one of the DNA strands. Three major kinds of RNA are produced. **Messenger RNA (mRNA)** encodes the amino acid sequence of one or more polypeptides specified by a gene or set of genes. **Transfer RNA (tRNA)** reads the information encoded in the mRNA and transfers the appropriate amino acid to a growing polypeptide chain during protein synthesis. Molecules of **ribosomal RNA (rRNA)** are constituents of ribosomes, the intricate cellular machines where proteins are synthesized. Many additional specialized RNAs have regulatory or catalytic functions or are precursors to the major classes of RNA listed above.

During replication the entire chromosome is normally copied, but transcription is more selective. Only particular genes or groups of genes are transcribed at any one time, and some portions of the DNA genome are never transcribed. The cell restricts the expression of genetic information to the formation of gene products needed at any particular moment. Specific regulatory sequences mark the beginning and end of the DNA segments to be transcribed and designate which DNA strand is to be used as template. The regulation of transcription is described in detail in Chapter 28.

Here we examine the synthesis of RNA on a DNA template and the postsynthetic processing and turnover of RNA molecules. In the process we encounter many of the specialized functions of RNA, including catalytic functions. Interestingly, the substrates for RNA enzymes are often other RNA molecules.

This chapter also introduces systems in which RNA is the template and DNA the product, rather than vice versa. The information pathways thus come full circle, revealing that template-dependent nucleic acid synthesis has standard rules regardless of the nature of template or product (RNA or DNA). This examination of the biological interconversion of DNA and RNA as information carriers leads to a discussion of the evolutionary origin of biological information.

DNA-Dependent Synthesis of RNA

Our discussion of RNA synthesis begins with a comparison between transcription and DNA replication (Chapter 25). Transcription resembles replication in its fundamental chemical mechanism, its polarity (direction of synthesis), and its use of a template. And like replication, transcription has initiation, elongation, and termination phases—though in the literature on transcription, initiation is further divided into discrete phases of DNA binding and initiation of RNA synthesis. Transcription differs from replication in that it does not require a primer and, generally, only limited segments of a DNA molecule are involved. Additionally, within transcribed segments only one DNA strand serves as a template.

RNA Is Synthesized by RNA Polymerases

The discovery of DNA polymerase and its dependence on a DNA template spurred a search for an enzyme that synthesizes RNA complementary to a DNA strand. By 1960, four research groups had independently detected an enzyme in cellular extracts that could form an RNA polymer from ribonucleoside 5′-triphosphates. Subsequent work on the purified *Escherichia coli* RNA polymerase helped to define the fundamental properties of transcription (Fig. 26–1). **DNA-dependent RNA polymerase** requires, in addition to a DNA template, all four ribonucleoside 5′-triphosphates (ATP, GTP, UTP, and CTP) as precursors of the nucleotide units of RNA, as well as Mg^{2+}. The chemistry of RNA synthesis has much in common with DNA synthesis. RNA polymerase elongates an RNA strand by adding ribonucleotide units to the strand's 3′-hydroxyl end and builds RNA in the 5′→3′ direction. The 3′-hydroxyl group acts as a nucleophile, attacking the α-phosphate of the incoming ribonucleoside triphosphate (as illustrated for DNA synthesis in Fig. 25–5) and releasing pyrophosphate. The overall reaction is

$$\underset{\text{RNA}}{(\text{NMP})_n} + \text{NTP} \xrightarrow{\text{RNA polymerase}} \underset{\text{Lengthened RNA}}{(\text{NMP})_{n+1}} + \text{PP}_i$$

RNA polymerase requires DNA for activity and is most active when bound to a double-stranded DNA. As noted above, only one of the two DNA strands serves as a template. The template DNA strand is copied in the 3′→5′ direction (antiparallel to the new RNA strand), just as in DNA replication. Each nucleotide in the newly formed RNA is selected by Watson-Crick base-pairing interactions; uridylate (U) residues are inserted in the RNA to pair with adenylate residues in the DNA template; and so on. Base pair geometry (see Fig. 25–6) may also play a role in base selection.

Unlike DNA polymerase, RNA polymerase does not require a primer to initiate synthesis. Initiation occurs when RNA polymerase binds at specific DNA sequences called promoters (described below). The 5′-triphosphate

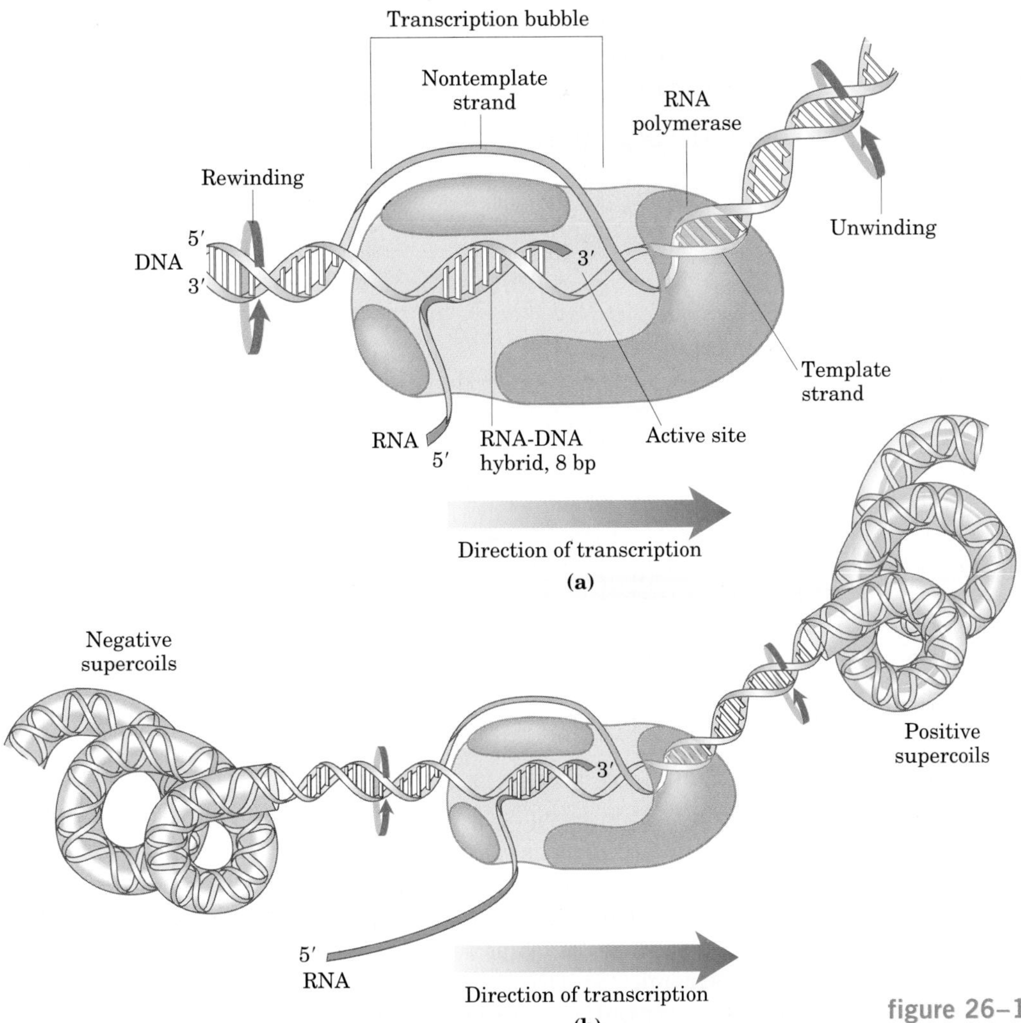

figure 26–1
Transcription by RNA polymerase in *E. coli.* To synthesize an RNA strand complementary to one of two DNA strands in a double helix, the DNA is transiently unwound. **(a)** About 17 base pairs are unwound at any given time. The transcription bubble moves from left to right as shown, keeping pace with RNA synthesis. The DNA is unwound ahead and rewound behind as RNA is transcribed. Red arrows show the direction in which the DNA and the short RNA-DNA hybrid must rotate to permit this process. As the DNA is rewound, the RNA-DNA hybrid is displaced and the RNA strand is extruded. The RNA polymerase is in close contact with the DNA ahead of the transcription bubble, as well as with the separated DNA strands and the RNA within and immediately behind the bubble. The polymerase footprint encompasses about 35 base pairs of DNA during elongation. **(b)** Supercoiling of DNA brought about by transcription. Positive supercoils form ahead of the transcription bubble and negative supercoils form behind.

group of the first residue in a nascent (newly formed) RNA molecule is not cleaved to release PP_i, but instead remains intact throughout transcription. During the elongation phase of transcription, the growth end of the new RNA strand base-pairs temporarily with the DNA template to form a short hybrid RNA-DNA double helix, estimated to be eight base pairs long (Fig. 26–1a). The RNA in this hybrid duplex "peels off" shortly after its formation, and the DNA duplex re-forms.

To enable RNA polymerase to synthesize an RNA strand complementary to one of the DNA strands, the DNA duplex must unwind over a short distance, forming a transcription "bubble." During transcription, the *E. coli* RNA polymerase generally keeps about 17 base pairs unwound. A short RNA-DNA hybrid occurs in the unwound region. Elongation of a transcript by *E. coli* RNA polymerase proceeds at a rate of 50 to 90 nucleotides/sec. Because DNA is a helix, movement of a transcription bubble requires considerable strand rotation of the nucleic acid molecules. DNA strand rotation is restricted in most DNAs by DNA-binding proteins and other structural barriers. As a result, a moving RNA polymerase generates waves of positive supercoils ahead of the transcription bubble and negative supercoils behind (Fig. 26–1b). This has been observed both in vitro and in vivo (in bacteria). In the cell, the topological problems caused by transcription are relieved through the action of topoisomerases.

figure 26–2

Template and nontemplate (coding) DNA strands. The two complementary strands of DNA are defined by their function in transcription. The RNA transcript is synthesized on the complementary template strand and is identical in sequence (with U in place of T) to the nontemplate or coding strand.

(5′) CGCTATAGCGTTT(3′)	DNA nontemplate (coding) strand
(3′) GCGATATCGCAAA(5′)	DNA template strand
(5′) CGCUAUAGCGUUU(3′)	RNA transcript

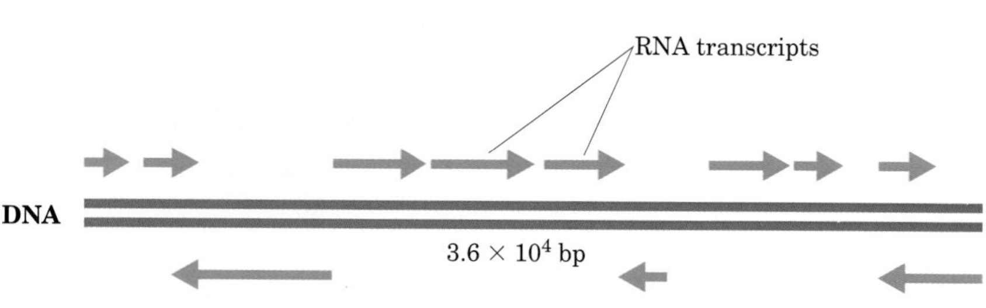

figure 26–3

The genetic information of the adenovirus genome (a conveniently simple example) is encoded by a double-stranded DNA molecule of 36,000 base pairs, both strands of which encode proteins. The information for most proteins is encoded by the top strand (by convention, the strand transcribed left to right), but some is encoded by the bottom strand, which is transcribed in the opposite direction. Synthesis of mRNAs in adenovirus is actually much more complex than shown here. Many of the mRNAs shown for the upper strand are initially synthesized as a long transcript (25,000 nucleotides), which is then extensively processed to produce the separate mRNAs. Adenovirus causes upper respiratory tract infections in some vertebrates.

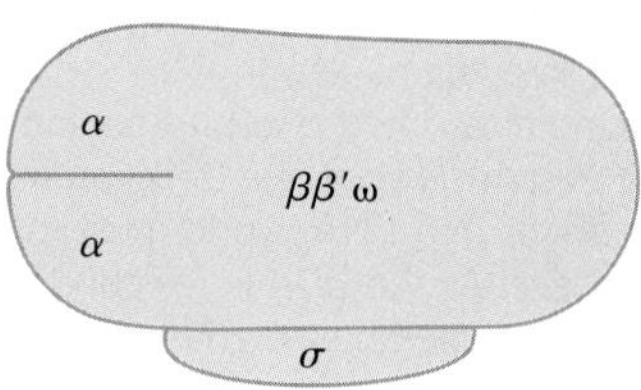

figure 26–4

Structure of *E. coli* RNA polymerase and its function in transcription. The polymerase has six subunits. The α (two copies), β, β', and ω subunits have molecular weights of 36,500, 151,000, 155,000, and 11,000, respectively. The ω subunit is not required for activity in vitro, and its function is not known. The σ subunit functions only in promoter recognition and is not present during the elongation phase of transcription. The several types of σ subunits recognize different regulatory sequences. The major type is σ^{70}, reflecting its M_r of 70,000. The active site for RNA synthesis is believed to be formed by the β and β' subunits.

The two complementary DNA strands have different roles in transcription. The strand that serves as template for RNA synthesis is called the **template strand.** The DNA strand complementary to the template, the **nontemplate strand,** or **coding strand,** is identical in base sequence to the RNA transcribed from the gene, with U in place of T (Fig. 26–2). The coding strand for a particular gene may be located in either strand of a given chromosome (Fig. 26–3). The regulatory sequences that control transcription (described later in this chapter) are by convention designated by the sequences in the nontemplate strand.

The DNA-dependent RNA polymerase of *E. coli* is a large, complex enzyme with five core subunits ($\alpha_2\beta\beta'\omega$; M_r 390,000) and a sixth subunit, one of a group designated σ. The σ subunit, the variants of which are designated by size (molecular weight), binds transiently to the core and directs the enzyme to specific binding sites on the DNA (described below). These six subunits constitute the RNA polymerase holoenzyme (Fig. 26–4). The RNA polymerase holoenzyme of *E. coli* thus exists in several forms, depending on the type of σ subunit. The most common subunit is σ^{70} (M_r 70,000), and the upcoming discussion focuses on the corresponding RNA polymerase holoenzyme.

RNA polymerases lack a proofreading 3′→5′ exonuclease activity (such as that of many DNA polymerases), and about one error is made for every 10^4 to 10^5 ribonucleotides incorporated into RNA during transcription. Because many copies of an RNA are generally produced from a single gene and all RNAs are eventually degraded and replaced, a mistake in an RNA molecule is of less consequence to the cell than a mistake in the permanent information stored in DNA.

	UP element		−35 Region	Spacer	−10 Region	Spacer	RNA start
							+1
Consensus sequence	NNAAA$^{AA}_{TT}$T$^{A}_{T}$TTTTNNAAAANNN	N	TTGACA	N_{17}	TATAAT	N_6	
*rrn*B P1	AGAAAATTATTTTAAATTTCCT	N	GTGTCA	N_{16}	TATAAT	N_8	A
trp			TTGACA	N_{17}	TTAACT	N_7	A
lac			TTTACA	N_{17}	TATGTT	N_6	A
*rec*A			TTGATA	N_{16}	TATAAT	N_7	A
*ara*BAD			CTGACG	N_{18}	TACTGT	N_6	A

figure 26–5

Typical *E. coli* promoters recognized by an RNA polymerase holoenzyme containing σ^{70}. Sequences of the nontemplate strand are shown and are read in the 5′→3′ direction, the convention for representations of this kind. The sequences vary from one promoter to the next, but comparisons of many promoters reveal similarities, particularly in the −10 and −35 regions. The sequence element UP, not present in all *E. coli* promoters, is illustrated in the promoter for the highly expressed rRNA gene *rrn*B P1. UP elements are generally found in the region between −40 and −60 and strongly stimulate transcription at the promoters that contain them. The UP element in the *rrn*B P1 promoter encompasses the region between −38 and −59. The consensus sequence for *E. coli* promoters recognized by σ^{70} is shown second from the top. Spacer regions contain slightly variable numbers of nucleotides (N). Only the first nucleotide coding the RNA transcript (at position +1) is shown.

RNA Synthesis Is Initiated at Promoters

Initiation of RNA synthesis at random points in a DNA molecule would be an extraordinarily wasteful process. Instead, the RNA polymerase binds to specific sequences in the DNA called **promoters,** which direct the transcription of adjacent segments of DNA (genes). The sequences where RNA polymerases bind can be quite variable, and much research has focused on identifying the particular sequences that are critical to promoter function.

In *E. coli,* RNA polymerase binding occurs within a region stretching from about 70 base pairs before the transcription start site to about 30 base pairs beyond it. By convention, the DNA base pairs that correspond to the beginning of an RNA molecule are given positive numbers, and those preceding the RNA start site are given negative numbers. The promoter region thus extends between positions −70 and +30. Analyses and comparisons of the sequences in the most common class of bacterial promoters (those recognized by an RNA polymerase holoenzyme containing σ^{70}) have revealed similarities in two short sequences centered about positions −10 and −35 (Fig. 26–5). These sequences are important interaction sites for the σ^{70} subunit. Although the sequences are not identical for all bacterial promoters in this class, certain nucleotides that are particularly common at each position form a **consensus sequence** (recall the *E. coli ori*C consensus sequence; see Fig. 25–11). The consensus sequence at the −10 region is (5′)TATAAT(3′); the consensus sequence at the −35 region is (5′)TTGACA(3′). A third AT-rich recognition element, called the UP (upstream promoter) element, occurs in the promoters of certain highly expressed genes between positions −40 and −60. The UP element is bound by the α subunit of RNA polymerase. The efficiency with which an RNA polymerase binds to and initiates transcription at a promoter is determined in large measure by these sequences, the spacing between them, and their distance from the transcription start site.

Many independent lines of evidence attest to the functional importance of the sequences in the −35 and −10 regions. Mutations that affect the function of a given promoter often involve a base pair in these regions. Variations in the consensus sequence also affect the efficiency of RNA polymerase binding and transcription initiation. A change in only one base pair can decrease the rate of binding by several orders of magnitude. The promoter sequence thus establishes a basal level of expression for particular *E. coli* genes that can vary greatly from one gene to the next. A method that provides information about the interaction between RNA polymerase and promoters is illustrated in Box 26–1.

The pathway of transcription initiation is becoming increasingly well-defined (Fig. 26–6). It consists of two major parts, each with multiple steps. First, the polymerase binds to the promoter, forming, in succession, a closed complex (in which the bound DNA is intact) and an open complex (in which the bound DNA is intact and partially unwound). Second, transcription is initiated within the complex, leading to a conformational change that converts the complex to the elongation form, followed by movement of the transcription complex away from the promoter (promoter clearance). Any of these steps can be affected by the specific sequence of the promoter sequences.

E. coli has other classes of promoters that are bound by RNA polymerase holoenzymes with different σ factors. An example is the promoters of the heat-shock genes. The products of this set of genes are made at higher levels when the cell has received an insult, such as a sudden increase in temperature. RNA polymerase binds to the promoters of these genes when σ^{70} is replaced with the σ^{32} (M_r 32,000) subunit, which is specific for the heat-shock promoters (see Fig. 28–3). By using different σ factors the cell can coordinate the expression of sets of genes that permit major changes in cell physiology.

Transcription Is Regulated

Requirements for any gene product vary with cellular conditions or developmental stage, and transcription of each gene is carefully regulated to form gene products only in the proportions needed. Regulation can occur at any step in transcription, including elongation and termination. However, much of the regulation is directed at the polymerase binding and transcription initiation steps outlined in Figure 26–6. Differences in promoter sequences are just one of several levels of control.

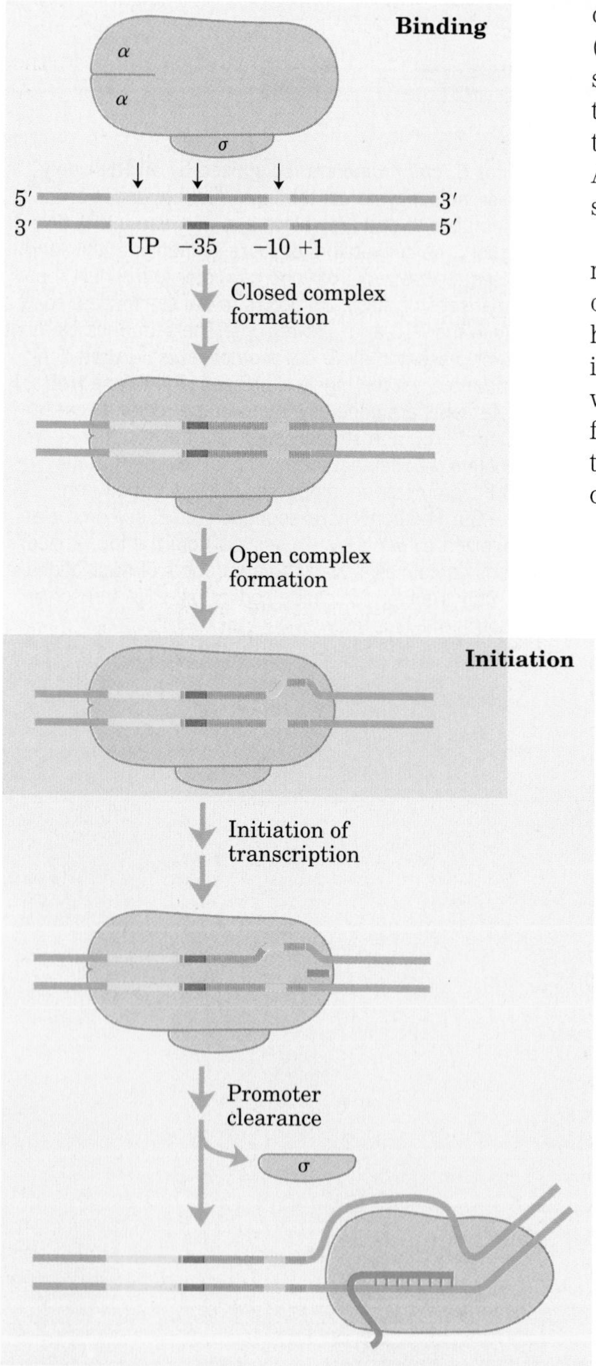

figure 26–6

Steps in the initiation of transcription by *E. coli* RNA polymerase. Initiation requires several steps generally divided into two phases, the binding and initiation phases. In the binding phase, the initial interaction of the RNA polymerase with the promoter leads to the formation of a closed complex, in which the promoter DNA is stably bound but not unwound. A 12 to 15 base pair region of DNA—from within the −10 region to position +2 or +3—is then unwound to form an open complex. Additional intermediates (not shown) have been detected in the pathways leading to the closed and open complexes, along with several changes in protein conformation. The initiation phase encompasses transcription initiation and promoter clearance. Once the first eight or nine nucleotides of a new RNA are synthesized, the σ subunit is released and the polymerase leaves the promoter and becomes committed to productive elongation of the RNA.

box 26–1 RNA Polymerase Leaves Its Footprint on a Promoter

Footprinting, a technique derived from principles used in DNA sequencing, identifies the DNA sequences bound by a particular protein. A DNA fragment thought to contain sequences recognized by a DNA-binding protein is isolated and radiolabeled at one end of one strand (Fig. 1). Chemical or enzymatic reagents are used to introduce random breaks in the DNA fragment (averaging about one per molecule). Separation of the labeled cleavage products (broken fragments of various lengths) by high-resolution electrophoresis will produce a ladder of radioactive bands. In a separate tube, the cleavage procedure is repeated on copies of the same DNA fragment to which the protein is bound, preventing cleavage of the DNA in the bound region. The two sets of cleavage products are then subjected to electrophoresis and compared side by side. A hole ("footprint") occurs in the series of radioactive bands derived from the protein-bound sample because of protection of the DNA by the protein. This identifies the sequences that are bound by the protein.

The precise location of the protein-binding site can be determined by directly sequencing (see Fig. 10–37) copies of the same DNA fragment and including the sequencing lanes (not shown here) on the same gel with the footprint. Footprinting results for the binding of RNA polymerase to a DNA fragment containing a promoter are shown in Figure 2. The polymerase covers 60 to 80 base pairs; protection by the bound enzyme includes the −10 and −35 regions.

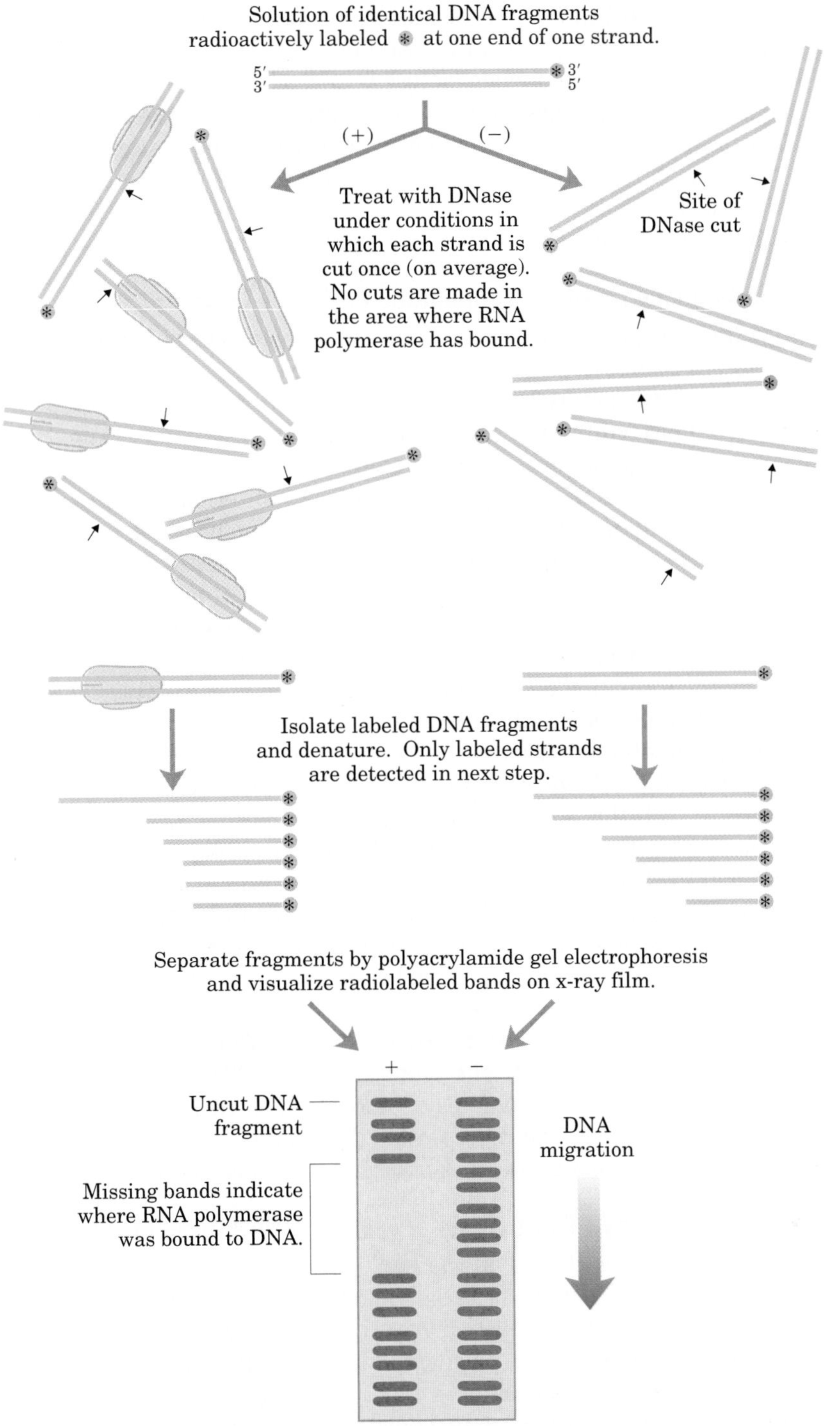

Solution of identical DNA fragments radioactively labeled ✻ at one end of one strand.

5′ 3′ ✻ 3′ 5′

(+) (−)

Treat with DNase under conditions in which each strand is cut once (on average). No cuts are made in the area where RNA polymerase has bound.

Site of DNase cut

Isolate labeled DNA fragments and denature. Only labeled strands are detected in next step.

Separate fragments by polyacrylamide gel electrophoresis and visualize radiolabeled bands on x-ray film.

+ −

Uncut DNA fragment

Missing bands indicate where RNA polymerase was bound to DNA.

DNA migration

figure 1

Footprint analysis of the binding site for RNA polymerase on a DNA fragment. Separate experiments are carried out in the presence (+) and absence (−) of RNA polymerase.

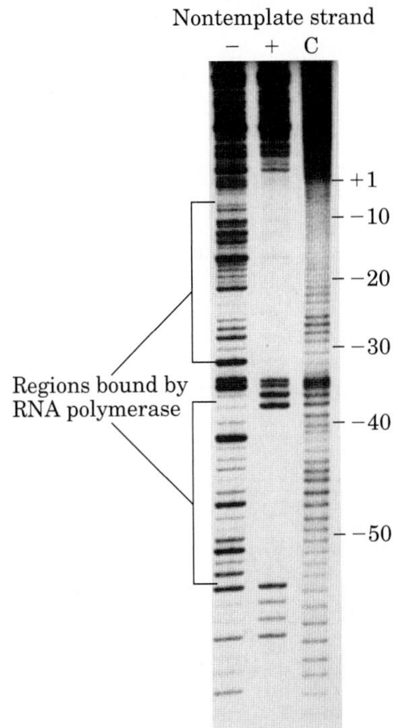

figure 2

Footprinting results of RNA polymerase binding to the *lac* promoter (see Fig. 26–5). In this experiment, the 5′ end of the nontemplate strand was radioactively labeled. Lane C is a control in which the labeled DNA fragments are cleaved with a chemical reagent that produces a more uniform banding pattern.

The binding of proteins to sequences both near to and distant from the promoter can also affect levels of gene expression. Protein binding can either *activate* transcription by facilitating RNA polymerase binding or steps further along in the initiation process, or it can *repress* transcription by blocking the activity of the polymerase. In *E. coli,* one protein that activates transcription is the **cAMP receptor protein (CRP),** which increases the transcription of genes coding for enzymes that metabolize sugars other than glucose when cells are grown in the absence of glucose. **Repressors** are proteins that block the synthesis of RNA at specific genes. In the case of the Lac repressor (Chapter 28), transcription of the genes for enzymes of lactose metabolism is blocked when lactose is unavailable.

Transcription is the first step in the complicated and energy-intensive pathway of protein synthesis, so much of the regulation of protein levels in both bacterial and eukaryotic cells is directed at transcription, particularly its early stages. In Chapter 28 we describe many mechanisms by which this regulation is accomplished.

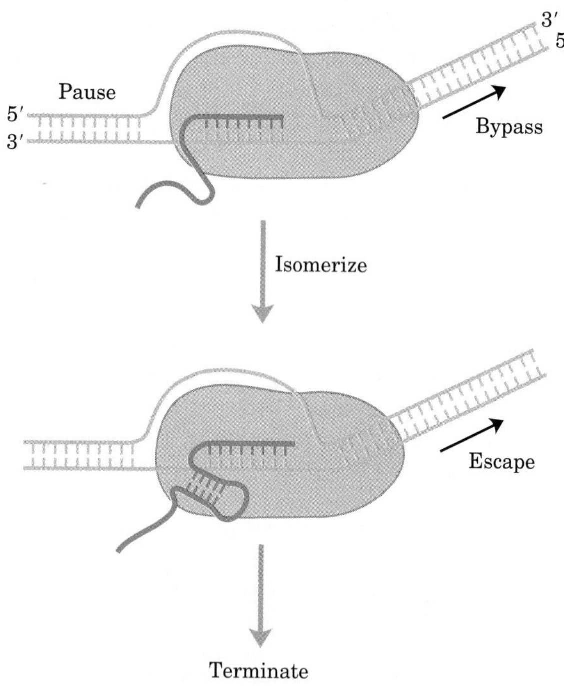

figure 26–7
A model for ρ-independent termination of transcription in *E. coli.* RNA polymerase pauses at a variety of DNA sequences, some of which are terminators. One of two outcomes is then possible: the polymerase bypasses the site and continues on its way or the complex undergoes a conformational change (isomerization). In the latter case, intramolecular pairing of complementary sequences in the newly formed RNA transcript may form a hairpin that disrupts the RNA-DNA hybrid and/or the interactions between the RNA and the polymerase, resulting in isomerization. An A=U hybrid region at the 3′ end of the new transcript is relatively unstable, and the RNA dissociates completely, leading to termination and dissociation of the RNA molecule. This is the usual outcome at terminators. At other pause sites, the complex may escape after the isomerization step to continue RNA synthesis.

Specific Sequences Signal Termination of RNA Synthesis

RNA synthesis is processive (i.e., the RNA polymerase has high processivity; see p. 937)—necessarily so, because if an RNA polymerase releases an RNA transcript prematurely, it cannot resume synthesis of the same RNA but instead must start over. However, an encounter with certain DNA sequences results in a pause in RNA synthesis, and at some of these sequences transcription is terminated. The process of termination is not yet well understood in eukaryotes, so our focus is again on bacteria. *E. coli* has at least two classes of termination signals: one class relies on a protein factor called ρ (rho), and the other is ρ-independent.

Most ρ-independent terminators have two distinguishing features. The first is a region whose RNA transcript has self-complementary sequences, permitting the formation of a hairpin structure (see Fig. 10–21a) centered 15 to 20 nucleotides before the end of the RNA strand. The second feature is a short string of adenylates in the template strand that are transcribed into uridylates at the 3′ end of the RNA. When a polymerase arrives at a termination site with this structure, it pauses (Fig. 26–7). Formation of the hairpin structure in the RNA may disrupt important interactions between the RNA and the RNA polymerase, facilitating dissociation of the transcript.

The ρ-dependent terminators lack the sequence of repeated adenylates in the template strand but sometimes have a short sequence that is transcribed to form a hairpin. The ρ protein loads onto the RNA at specific binding sites and migrates in the 5′→3′ direction until it reaches the transcription complex that is paused at a termination site. Here it contributes to release of the RNA transcript. The ρ protein has an ATP-dependent RNA-DNA helicase activity that promotes translocation of the protein along the RNA, and ATP is hydrolyzed by ρ protein during the termination process. The detailed mechanism by which the protein acts is not known.

Eukaryotic Cells Have Three Kinds of Nuclear RNA Polymerases

The transcriptional machinery in the nucleus of a eukaryotic cell is much more complex than that in bacteria. Eukaryotes have three RNA polymerases, designated I, II, and III, which are distinct complexes but have certain subunits in common. Each polymerase has a specific function and binds to a specific promoter sequence.

RNA polymerase I (Pol I) is responsible for the synthesis of only one type of RNA, a transcript called preribosomal RNA, which contains the precursor for the 18S, 5.8S, and 28S rRNAs (see Fig. 26–24). Pol I promoters vary greatly in sequence from one species to another.

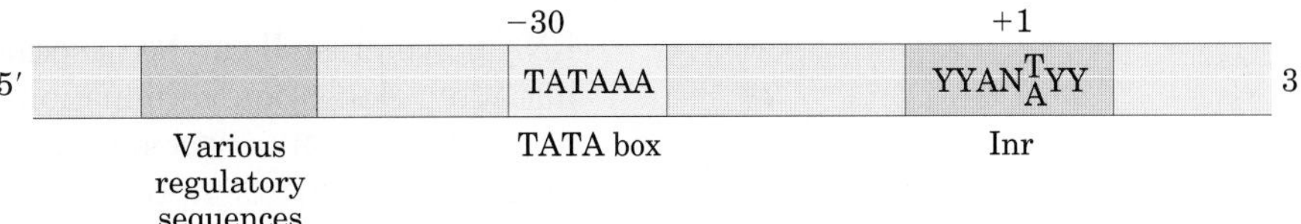

figure 26–8
Common sequences in promoters recognized by eukaryotic RNA polymerase II. The TATA box is the major assembly point for the proteins of the preinitiation complexes of RNA polymerase II. The DNA is unwound at the initiator sequence (Inr), and the transcription start site is usually within or very near this sequence. In the Inr consensus sequence shown, N represents any nucleotide; Y, a pyrimidine nucleotide. Many additional sequences serve as binding sites for a wide variety of proteins that affect the activity of RNA polymerase II. These sequences are important in regulating RNA polymerase II promoters and vary greatly in type and number, but in general they make the eukaryotic promoter much more complex than is shown. Many are located within a few hundred base pairs of the TATA box on the 5′ side; others may be thousands of base pairs away. The sequence elements summarized here are more variable among the RNA polymerase II promoters of eukaryotes than among the *E. coli* promoters (Fig. 26–5). Many RNA polymerase II promoters lack a TATA box or a consensus Inr element or both. Additional sequences around the TATA box and downstream (to the right as drawn) of Inr may be recognized by one or more transcription factors.

The principal function of RNA polymerase II (Pol II) is synthesis of mRNAs and some specialized RNAs. This enzyme must recognize thousands of promoters that can vary greatly in sequence. Many Pol II promoters have a few sequence features in common, including a TATA box (consensus sequence TATAAA) near base pair −30 and an Inr sequence (initiator) near the RNA start site at +1 (Fig. 26–8).

RNA polymerase III (Pol III) makes tRNAs, the 5S rRNA, and some other small specialized RNAs. The promoters recognized by RNA polymerase III are well characterized. Interestingly, some of the sequences required for the regulated initiation of transcription by RNA polymerase III are located within the gene itself, whereas others are in more conventional locations upstream of the RNA start site (Chapter 28).

RNA Polymerase II Requires Many Other Protein Factors for Its Activity

RNA polymerase II is central to eukaryotic gene expression and has been studied extensively. It is strikingly more complex than its bacterial counterpart, although the largest subunit exhibits a high degree of homology to the β' subunit of bacterial RNA polymerase. The polymerase is a huge enzyme with 12 subunits. It requires an array of other proteins called **transcription factors** in order to form the active transcription complex. The **basal transcription factors** required at every RNA polymerase II promoter (factors usually designated TFII with an additional identifier) are highly conserved in all eukaryotes (Table 26–1). The process of transcription by

table 26–1

Proteins Required for Transcription at the RNA Polymerase II Promoters of Eukaryotes

Transcription factor	Number of subunits	Subunit M_r	Functions
Initiation			
RNA polymerase II	12	10,000–220,000	Catalyzes RNA synthesis
TBP (TATA-binding protein)	1	38,000	Specifically recognizes the TATA box
TFIIA	3	12,000, 19,000, 35,000	Stabilizes binding of TFIIB and TBP to the promoter
TFIIB	1	35,000	Binds to TBP; recruits RNA polymerase–TFIIF complex
TFIID	12	15,000–250,000	Interacts with positive and negative regulatory proteins
TFIIE	2	34,000, 57,000	Recruits TFIIH; ATPase and helicase activities
TFIIF	2	30,000, 74,000	Binds tightly to RNA polymerase II; binds to TFIIB and prevents binding of RNA polymerase to nonspecific DNA sequences
TFIIH	12	35,000–89,000	Unwinds DNA at promoter; phosphorylates RNA polymerase; recruits nucleotide-excision repair complex
Elongation*			
ELL†	1	80,000	
P-TEFb	2	43,000, 124,000	
SII (TFIIS)	1	38,000	
Elongin (SIII)	3	15,000, 18,000, 110,000	

*All elongation factors suppress the pausing or arrest of transcription by the RNA polymerase II–TFIIF complex.

†The name is derived from the term *e*leven-nineteen *l*ysine-rich *l*eukemia. The gene for the factor ELL is the site of chromosomal recombination events frequently associated with the cancerous condition known as acute myeloid leukemia.

RNA polymerase II can be described in terms of several phases—assembly, initiation, elongation, termination—each associated with characteristic proteins (Fig. 26–9). The step-by-step assembly pathway described below leads to active transcription in vitro. In the cell, many of the proteins may be present in larger preassembled complexes, simplifying the pathways for assembly on promoters. Figure 26–9 and Table 26–1 can help you keep track of the participants in this process.

Assembly of RNA Polymerase and Transcription Factors at a Promoter The formation of a closed complex begins when the TATA-binding protein (TBP) binds to the TATA box (Fig. 26–9b). TBP is bound in turn by the transcription factor TFIIB, which also binds to DNA on either side of TBP. TFIIA binding, although not always essential, can stabilize the TFIIB-TBP complex on the DNA and can be important at nonconsensus promoters where TBP binding is relatively weak. The TFIIB-TBP complex is next bound by another complex consisting of TFIIF and RNA polymerase II. TFIIF helps target RNA polymerase II to its promoters, both by interacting with TFIIB and by reducing the binding of RNA polymerase to nonspecific sites on the DNA. Finally, TFIIE and TFIIH bind to create the closed complex. TFIIH has DNA helicase activity that promotes the unwinding of DNA near the RNA start site (a process requiring the hydrolysis of ATP), thereby creating an open complex. Counting all the subunits of the various essential factors (excluding TFIIA), this minimal active assembly requires more than 30 polypeptides.

RNA Strand Initiation and Promoter Clearance TFIIH has an additional function during the initiation phase. A kinase activity in one of its subunits phosphorylates RNA polymerase II at several places in the carboxyl-terminal domain of the polymerase's largest subunit (Fig. 26–9). This causes a conformational change in the overall complex, initiating transcription. Phosphorylation of the carboxyl-terminal domain may be important during the subsequent elongation phase, and it affects the interactions between the transcription complex and other enzymes involved in processing the transcript (as described below).

During synthesis of the initial 60 to 70 nucleotides of RNA, first TFIIE and then TFIIH are released, and RNA polymerase II enters the elongation phase of transcription.

Elongation, Termination, and Release TFIIF remains associated with RNA polymerase II throughout elongation. During this stage, the activity of the polymerase is greatly enhanced by proteins called elongation factors (Table 26–1). Once the RNA transcript is completed, transcription is terminated by mechanisms not yet well understood. RNA polymerase II is dephosphorylated and recycled, ready to initiate another transcript (Fig. 26–9).

Regulation of RNA Polymerase II Activity Regulation of transcription at RNA polymerase II promoters is quite elaborate. It involves the interaction of a wide variety of other proteins with the preinitiation complex. Some of these regulatory proteins interact with transcription factors, others with RNA polymerase II itself. Many interact with TFIID, a complex of about 12 proteins, including TBP and certain *T*BP-*a*ssociated *f*actors, or TAFs. The regulation of transcription is described in more detail in Chapter 28.

Diverse Functions of TFIIH In eukaryotes, the repair of damaged DNA (see Table 25–5) within genes that are actively being transcribed is more efficient than the repair of other damaged DNA, and the template strand is

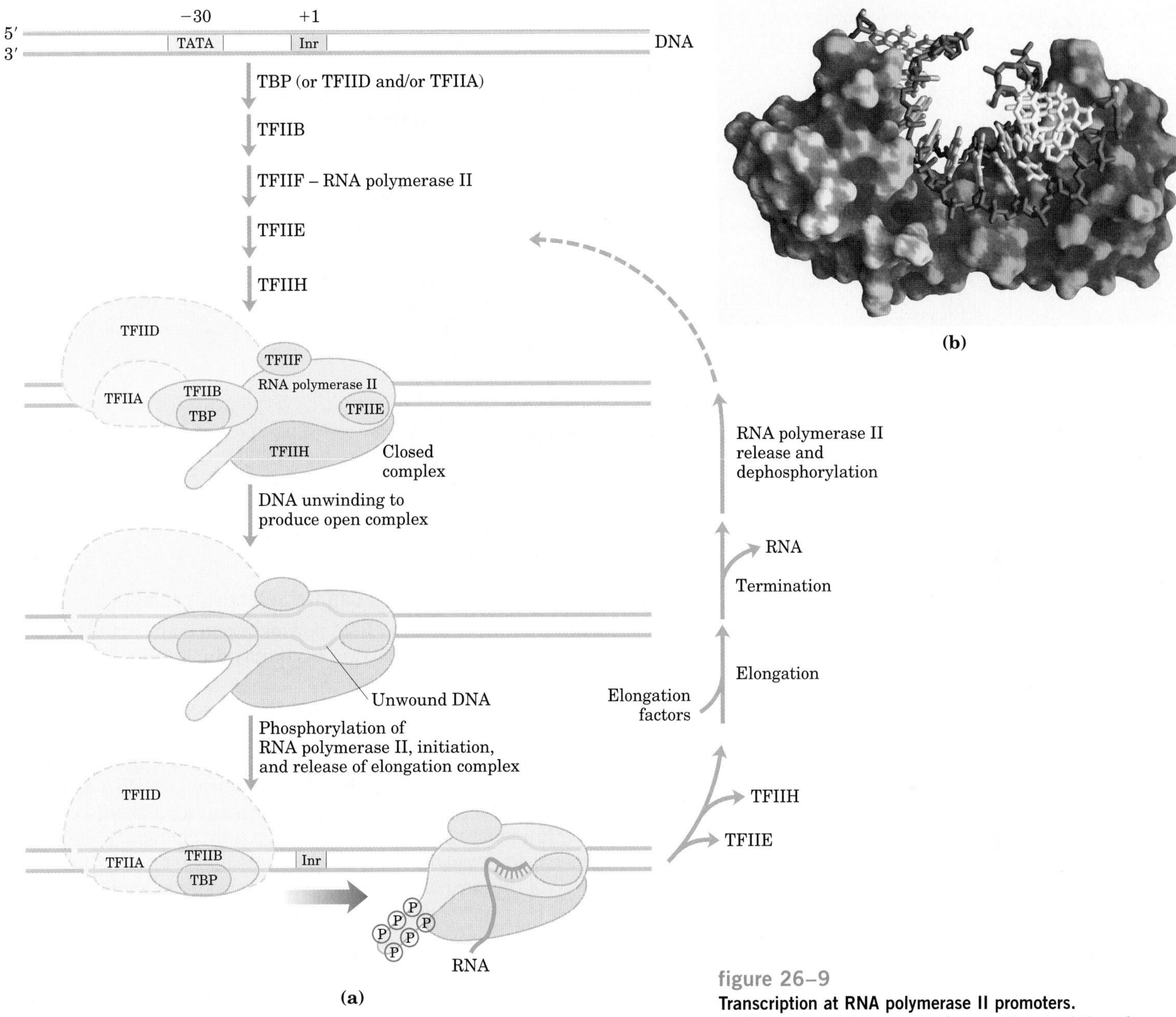

figure 26–9
Transcription at RNA polymerase II promoters. **(a)** The sequential assembly of a complex consisting of TBP (often with TFIIA), TFIIB plus RNA polymerase II, TFIIE, TFIIF, and TFIIH results in a closed complex. TBP often binds as part of a larger complex called TFIID. Some of the TFIID subunits play a role in transcription regulation (see Fig. 28–32). The DNA is unwound at the Inr region by the helicase activities of TFIIH and perhaps TFIIE, creating an open complex. The carboxyl-terminal domain of the largest RNA polymerase II subunit is phosphorylated by TFIIH, and the polymerase then escapes the promoter and begins transcription. Elongation is accompanied by the release of many transcription factors and is also enhanced by elongation factors (see Table 26–1). After termination, the RNA polymerase is released, dephosphorylated, and recycled. **(b)** The structure of TBP (gray) bound to DNA (blue and white).

repaired somewhat more efficiently than the nontemplate strand. These remarkable observations are explained by the alternative roles of the TFIIH subunits. Not only does TFIIH participate in the formation of the closed complex during assembly of a transcription complex (as described above), but some of its subunits are also essential components of the separate nucleotide-excision repair complex (Chapter 25).

When RNA polymerase II transcription halts at the site of a DNA lesion, TFIIH can interact with the lesion and recruit the entire nucleotide-excision repair complex. Genetic loss of certain TFIIH subunits can produce human diseases. Some examples are xeroderma pigmentosum (see Box 25–1) and Cockayne's syndrome, which is characterized by arrested growth, photosensitivity, and neurological disorders.

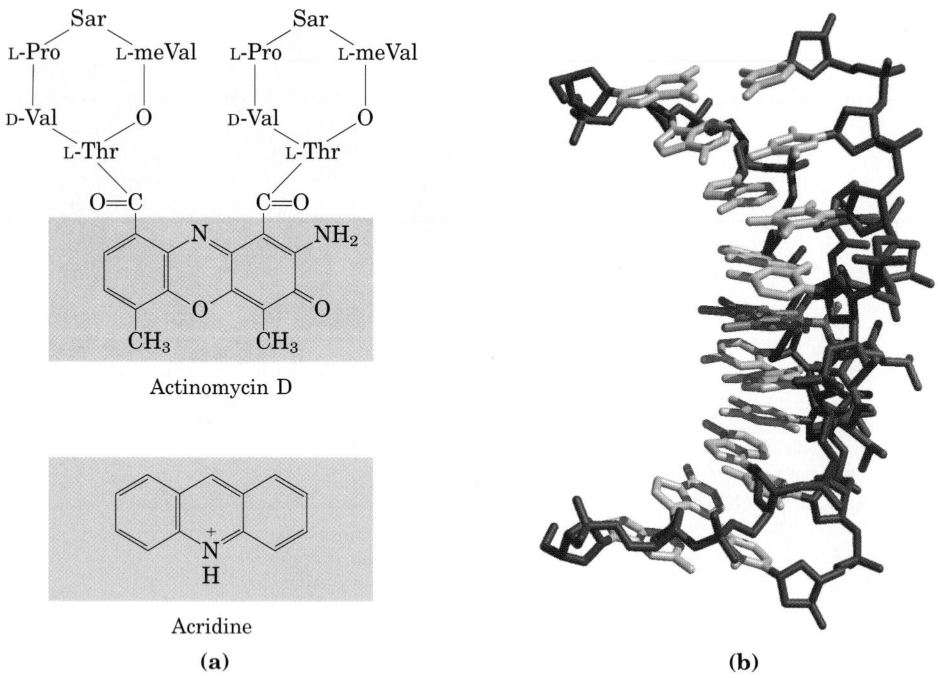

Actinomycin D

Acridine

(a)

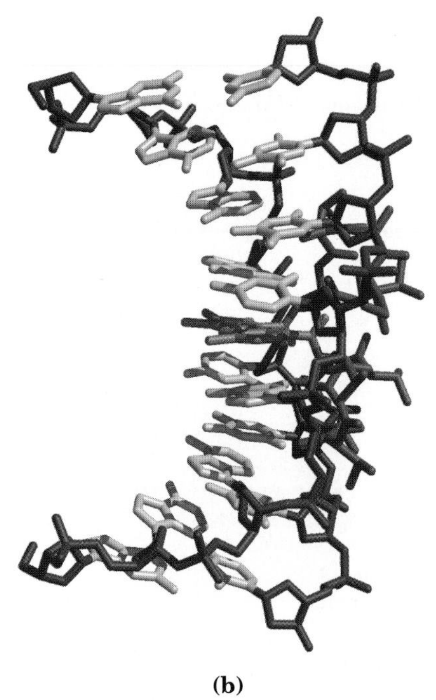

(b)

figure 26–10

Actinomycin D and acridine, inhibitors of DNA transcription. (a) The shaded portion of actinomycin D is planar and intercalates between two successive G≡C base pairs in duplex DNA. The two cyclic peptide structures of actinomycin D bind to the minor groove of the double helix. Sarcosine (Sar) is *N*-methylglycine; meVal is methylvaline. Acridine also acts by intercalation in the DNA. **(b)** A complex of actinomycin D with DNA. The DNA backbone is shown in blue, the bases are white, the intercalated part of actinomycin (shaded in **(a)**) is orange, and the remainder of the actinomycin is red. The DNA is bent as a result of actinomycin binding.

DNA-Dependent RNA Polymerase Can Be Selectively Inhibited

The elongation of RNA strands by RNA polymerase in both bacteria and eukaryotes is inhibited by the antibiotic **actinomycin D** (Fig. 26–10). The planar portion of this molecule inserts itself (intercalates) into the double-helical DNA between successive G≡C base pairs, deforming the DNA. This prevents the movement of the polymerase along the template. Because actinomycin D inhibits RNA elongation in intact cells as well as in cell extracts, it is used to identify cell processes that depend on RNA synthesis. **Acridine** inhibits RNA synthesis in a similar fashion (Fig. 26–10).

Rifampicin inhibits bacterial RNA synthesis by binding to the β subunit of bacterial RNA polymerases (Fig. 26–4), preventing the promoter clearance step of transcription. It is sometimes used as an antibiotic.

The mushroom *Amanita phalloides* has evolved a very effective defense mechanism against predators. It produces **α-amanitin,** which disrupts mRNA formation in animal cells by blocking RNA polymerase II and, at higher concentrations, RNA polymerase III as well. Neither RNA polymerase I nor bacterial RNA polymerase is sensitive to α-amanatin—nor the RNA polymerase II of *A. phalloides* itself!

RNA Processing

Many of the RNA molecules in bacteria and virtually all RNA molecules in eukaryotes are processed to some degree after they are synthesized. Some of the most interesting molecular events in RNA metabolism occur during this postsynthetic processing. Intriguingly, the enzymes that catalyze these reactions sometimes consist of RNA rather than protein. The discovery of catalytic RNAs, or **ribozymes,** has brought a revolution in thinking about RNA function and about the origin of life.

A newly synthesized RNA molecule is called a **primary transcript.** Perhaps the most extensive processing of primary transcripts occurs in eukaryotic mRNAs and in tRNAs of both bacteria and eukaryotes. The primary transcript for a eukaryotic mRNA typically contains sequences encompassing one gene, although the sequences encoding the polypeptide may not be

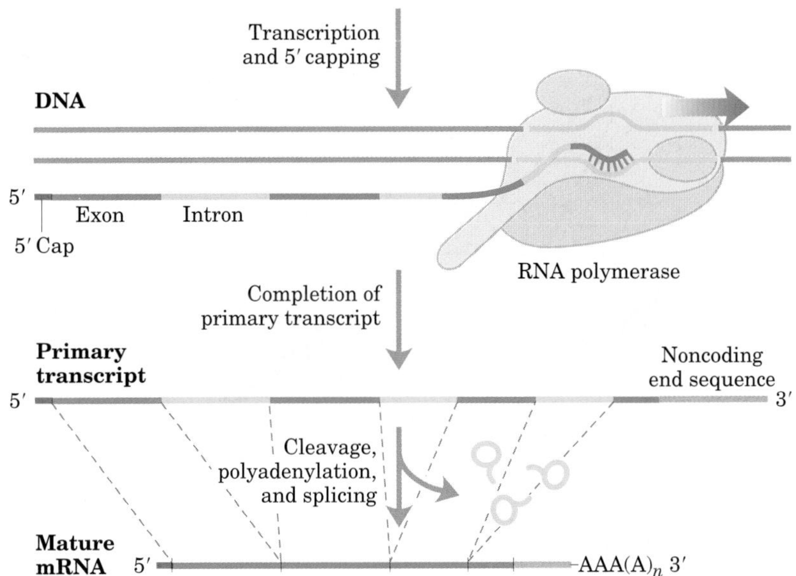

figure 26–11
Formation of the primary transcript and its processing during maturation of mRNA in a eukaryotic cell. The 5′ cap (in red) is added before synthesis of the primary transcript is complete. A noncoding sequence following the last exon is shown in orange. Splicing can occur either before or after the cleavage and polyadenylation steps. All of the processes shown here take place within the nucleus.

contiguous. Noncoding tracts that break up the coding region of the transcript are called introns and the coding segments are called exons (see the discussion of introns and exons in DNA in Chapter 24). In a process called **splicing,** the introns are removed from the primary transcript and the exons are joined to form a continuous sequence that specifies a functional polypeptide. Eukaryotic mRNAs are also modified at each end. A modified residue called a 5′ cap (see p. 997) is added at the 5′ end. The 3′ end is cleaved and 80 to 250 adenylate residues are added to create a poly(A) "tail." These processes are outlined in Figure 26–11 and described in more detail below.

The primary transcripts of prokaryotic and eukaryotic tRNAs are processed by the removal of sequences from each end (called cleavage) and in a few cases by the removal of introns (splicing). Many bases and sugars in tRNAs are also modified; mature tRNAs are replete with unusual bases not found in other nucleic acids (see Fig. 26–26).

The ultimate fate of any RNA is its complete degradation. The rate of turnover of RNAs plays a critical role in determining their steady-state level and the rate at which cells can shut down expression of a gene whose product is no longer needed. During the development of multicellular organisms, for example, it is important that certain proteins be expressed at one stage only, and the mRNA encoding such a protein must therefore be made and destroyed at the appropriate times.

The Introns Transcribed into RNA Are Removed by Splicing

In bacteria, a polypeptide chain is generally encoded by a DNA sequence that is colinear with the amino acid sequence, continuing along the DNA template without interruption until the information needed to specify the polypeptide is complete. However, the notion that all genes are continuous was disproved in 1977 when Phillip Sharp and Richard Roberts independently discovered that the genes for polypeptides in eukaryotes are often interrupted by noncoding sequences (introns). The vast majority of genes in vertebrates contain introns; among the few exceptions are those that encode histones. The occurrence of introns in other eukaryotes varies. Many genes in the yeast *Saccharomyces cerevisiae* lack introns, although in some other yeast species introns are more common. Introns are also found in a few bacterial and archaebacterial genes.

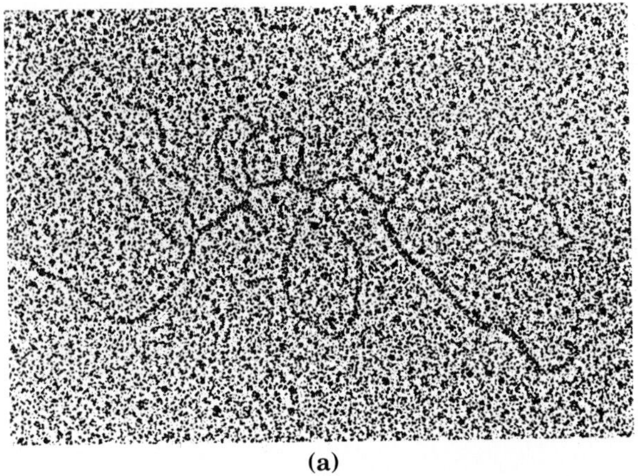
(a)

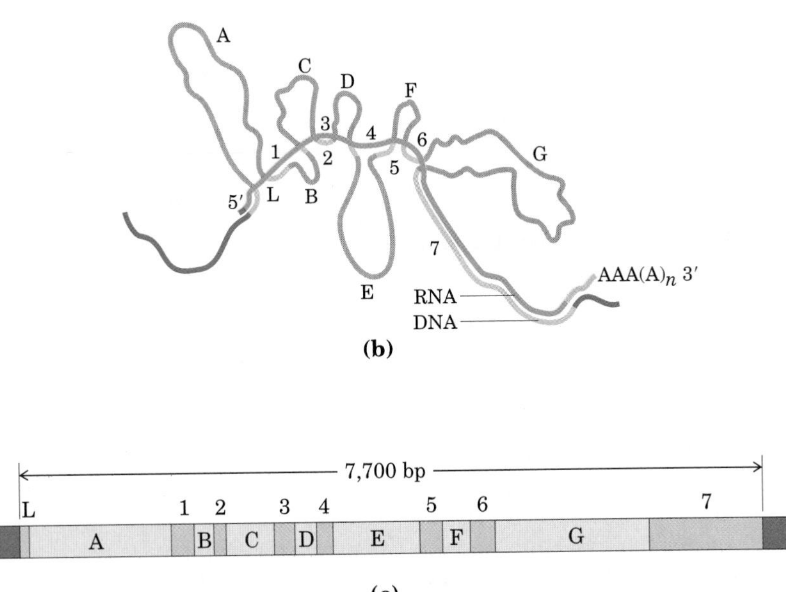

figure 26–12
Demonstration of noncoding sequences in the chicken ovalbumin gene by RNA-DNA hybridization. (a) Mature mRNA was hybridized to its complementary strand in denatured DNA containing the ovalbumin gene, and the resulting molecules were viewed with the electron microscope. Some regions of the DNA have no complement in the mature mRNA (formed by splicing of the primary transcript), producing the single-stranded DNA loops shown in the electron micrograph. The loops define the locations and sizes of introns. **(b)** The introns are labeled A to G and the exons L and 1 to 7; L encodes a signal sequence that targets the protein for export from the cell (see Fig. 27–36). The poly(A) tail defines the 3′ end of the mRNA. **(c)** A linear representation of the ovalbumin gene, showing introns and exons.

Introns in DNA are transcribed along with the rest of the gene by RNA polymerases. The introns in the primary RNA transcript are then spliced, and the exons are joined to form a mature, functional RNA. Introns were discovered when DNA of a known gene was completely denatured and then renatured in the presence of the mature RNA derived from the gene. The RNA-DNA hybrid that formed had looped-out DNA sequences that did not base-pair with any of the RNA (Fig. 26–12). This and other methods have shown that many genes have introns, some genes dozens of them.

In eukaryotic mRNAs, most exons are less than 1,000 nucleotides long, with many in the 100 to 200 nucleotide size range, encoding stretches of 30 to 60 amino acids within a longer polypeptide. Introns vary in size from 50 to 20,000 nucleotides. Genes of higher eukaryotes, including humans, typically have much more DNA devoted to introns than to exons. Many genes have introns; some genes have dozens of them.

RNA Catalyzes Splicing

There are four classes of introns. The first two, called group I and group II introns, share some key characteristics but differ in the details of their splicing mechanisms. Group I introns are found in some nuclear, mitochondrial, and chloroplast genes coding for rRNAs, mRNAs, and tRNAs. Group II introns are generally found in the primary transcripts of mitochondrial or chloroplast mRNAs in fungi, algae, and plants. Group I and group II introns are also found among the rarer examples of introns in bacteria. Neither class requires a high-energy cofactor (such as ATP) for splicing. The splicing mechanisms in both groups involve two transesterification reaction steps (Fig. 26–13). A ribose 2′- or 3′-hydroxyl group makes a nucleophilic attack on a phosphorus and, in each step, a new phosphodiester bond is formed at the expense of the old, maintaining the balance of energy. These reactions are very similar to the DNA breaking and rejoining reactions promoted by topoisomerases (see Fig. 24–20) and site-specific recombinases (see Fig. 25–36).

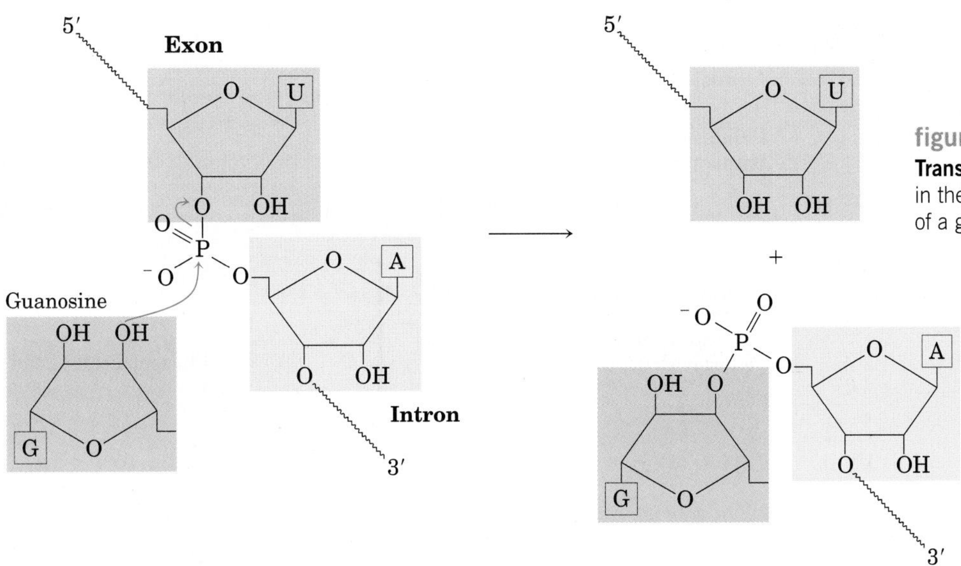

figure 26–13

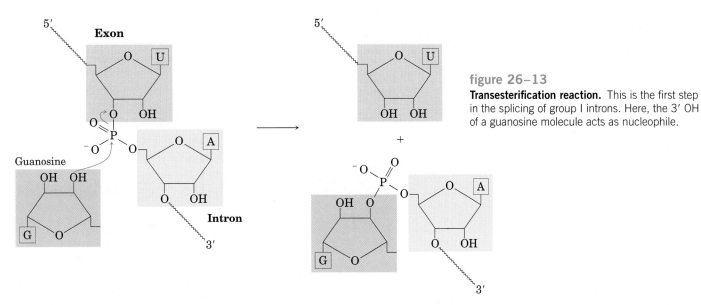

Transesterification reaction. This is the first step in the splicing of group I introns. Here, the 3′ OH of a guanosine molecule acts as nucleophile.

The group I splicing reaction requires a guanine nucleoside or nucleotide cofactor, but the cofactor is not used as a source of energy; instead, the 3′-hydroxyl group of guanosine is used as a nucleophile in the first step of the splicing pathway. The guanosine 3′-hydroxyl group forms a normal 3′,5′-phosphodiester bond with the 5′ end of the intron (Fig. 26–14). The 3′ hydroxyl of the exon that is displaced in this step then acts as a nucleophile in a similar reaction at the 3′ end of the intron. The result is precise excision of the intron and ligation of the exons.

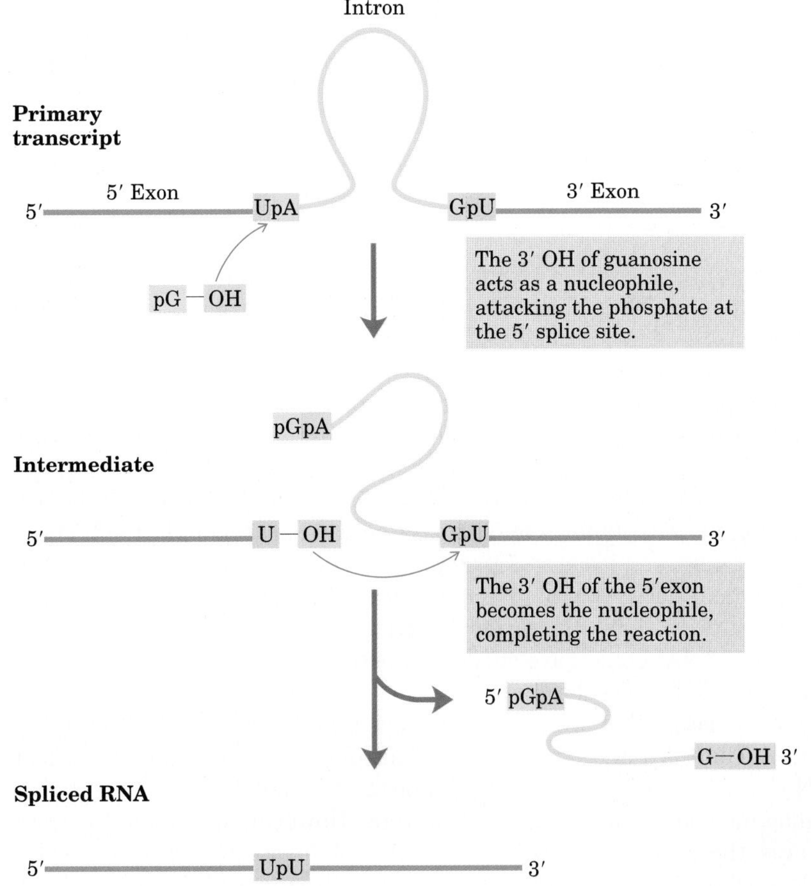

figure 26–14

Splicing mechanism of group I introns. The nucleophile in the first step may be guanosine, GMP, GDP, or GTP. The spliced intron is eventually degraded.

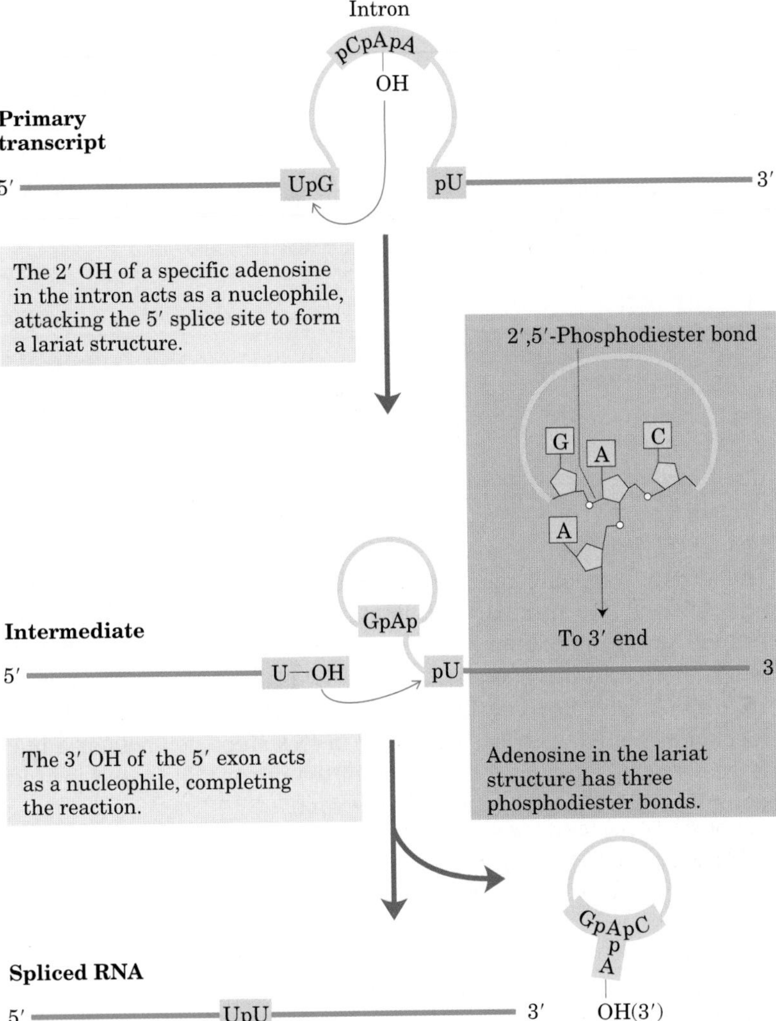

figure 26–15
Splicing mechanism of group II introns. The chemistry is similar to that of group I intron splicing, except for the identity of the nucleophile in the first step and formation of a lariatlike intermediate, in which one branch is a 2′,5′-phosphodiester bond.

In group II introns the pattern is similar except for the nucleophile in the first step, which in this case is the 2′-hydroxyl group of an adenylate residue *within* the intron (Fig. 26–15). A branched lariat structure is formed as an intermediate.

Attempts to identify the enzymes that promote splicing of group I and group II introns produced a major surprise: these introns are *self-splicing*. No protein enzymes are involved. This was first revealed in studies of the splicing mechanism of the group I rRNA intron from the ciliated protozoan *Tetrahymena thermophila* by Thomas Cech and colleagues in 1982. These workers transcribed isolated *Tetrahymena* DNA (including the intron) in vitro using purified bacterial RNA polymerase. The resulting RNA spliced itself accurately without any protein enzymes from *Tetrahymena*. The discovery that RNAs could have catalytic functions was a milestone in our understanding of biological systems.

Thomas Cech

Intron classes that are not self-splicing are not designated with a group number. The third and largest class of introns are those found in nuclear mRNA primary transcripts. They undergo splicing by the same lariat-forming mechanism as the group II introns. However, splicing in this case requires the action of specialized RNA-protein complexes called *s*mall *n*uclear *r*ibo*n*ucleo*p*roteins (snRNPs, often pronounced "snurps"). Each

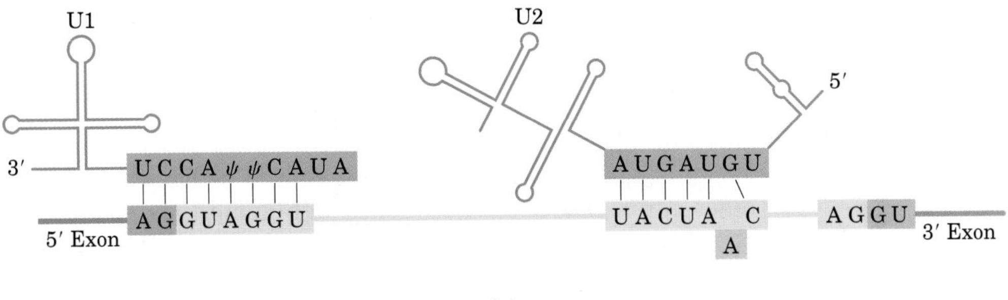

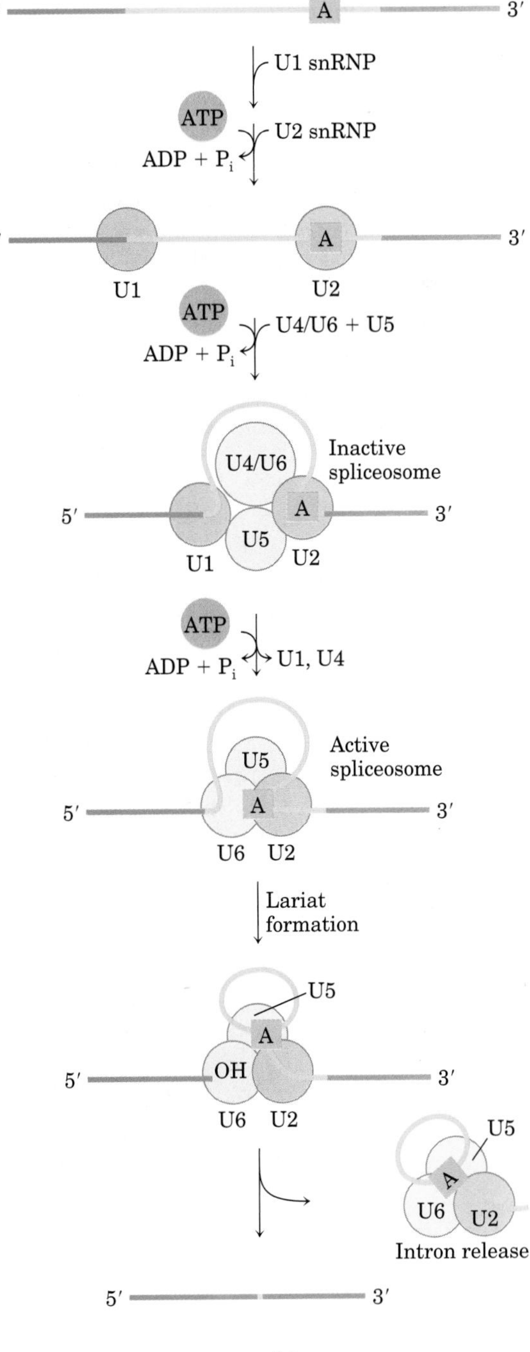

figure 26–16
Splicing mechanism in mRNA primary transcripts. **(a)** RNA pairing interactions in the formation of spliceosome complexes. The U1 snRNA has a sequence near its 5′ end that is complementary to the splice site at the 5′ end of the intron. Base pairing of U1 to this region of the primary transcript helps define the 5′ splice site during spliceosome assembly. (The ψ represents pseudouridine (see Fig. 26–26).) U2 is paired to the intron at a position encompassing the adenosine residue (red) that becomes the nucleophile during the splicing reaction. Base pairing of U2 snRNA causes a bulge that displaces and helps to activate the adenosine, whose 2′ OH will form the lariat structure through a 2′,5′-phosphodiester bond. **(b)** Assembly of spliceosomes occurs in several steps. The U1 and U2 snRNPs bind, then the remaining snRNPs (the U4/U6 complex and U5) bind to form an inactive spliceosome. Internal rearrangements convert this species to an active spliceosome in which U1 and U4 have been expelled and U6 is paired with both the 5′ splice site and U2. This is followed by the catalytic steps, which parallel those of the splicing of group II introns (Fig. 26–15).

snRNP contains one of a class of eukaryotic RNAs 100 to 200 nucleotides long called **small nuclear RNAs (snRNAs).** Five snRNAs (U1, U2, U4, U5, and U6) involved in splicing reactions are generally found in abundance in eukaryotic nuclei. The RNAs and proteins in snRNPs are highly conserved in eukaryotes from yeast to human.

The U1 snRNA contains a sequence complementary to sequences near the 5′ splice site of nuclear mRNA introns (Fig. 26–16a), and the U1 snRNP binds to this region in the primary transcript. Addition of the U2, U4, U5, and U6 snRNPs leads to formation of a complex called the **spliceosome** within which the actual splicing reaction occurs (Fig. 26–16b). The snRNPs together contribute five RNAs and about 50 proteins to the spliceosome, a supramolecular assembly nearly as complex as the ribosome (described in Chapter 27). ATP is required for assembly of the spliceosome, but the RNA cleavage-ligation reactions do not seem to require ATP. Some mRNA introns are spliced by a less common type of spliceosome, in which the U1 and U2 snRNPs are replaced by the U11 and U12 snRNPs. The spliceosomes used in nuclear RNA splicing may have evolved from more ancient group II introns, with the snRNPs replacing the catalytic domains of their self-splicing ancestors.

The fourth class of introns, found in certain tRNAs, is distinguished from the group I and II introns in that the splicing reaction requires ATP and an endonuclease (Fig. 26–17). The splicing endonuclease cleaves the phosphodiester bonds at both ends of the intron, and the two exons are joined by a mechanism similar to the DNA ligase reaction (see Fig. 25–16).

Although spliceosomal introns appear to be limited to eukaryotes, the other intron classes are not. Genes with group I and II introns have now been found in both bacteria and bacterial viruses. Bacteriophage T4, for example, has several protein-encoding genes with group I introns. Introns appear to be more common in archaebacteria (p. 24) than in bacteria.

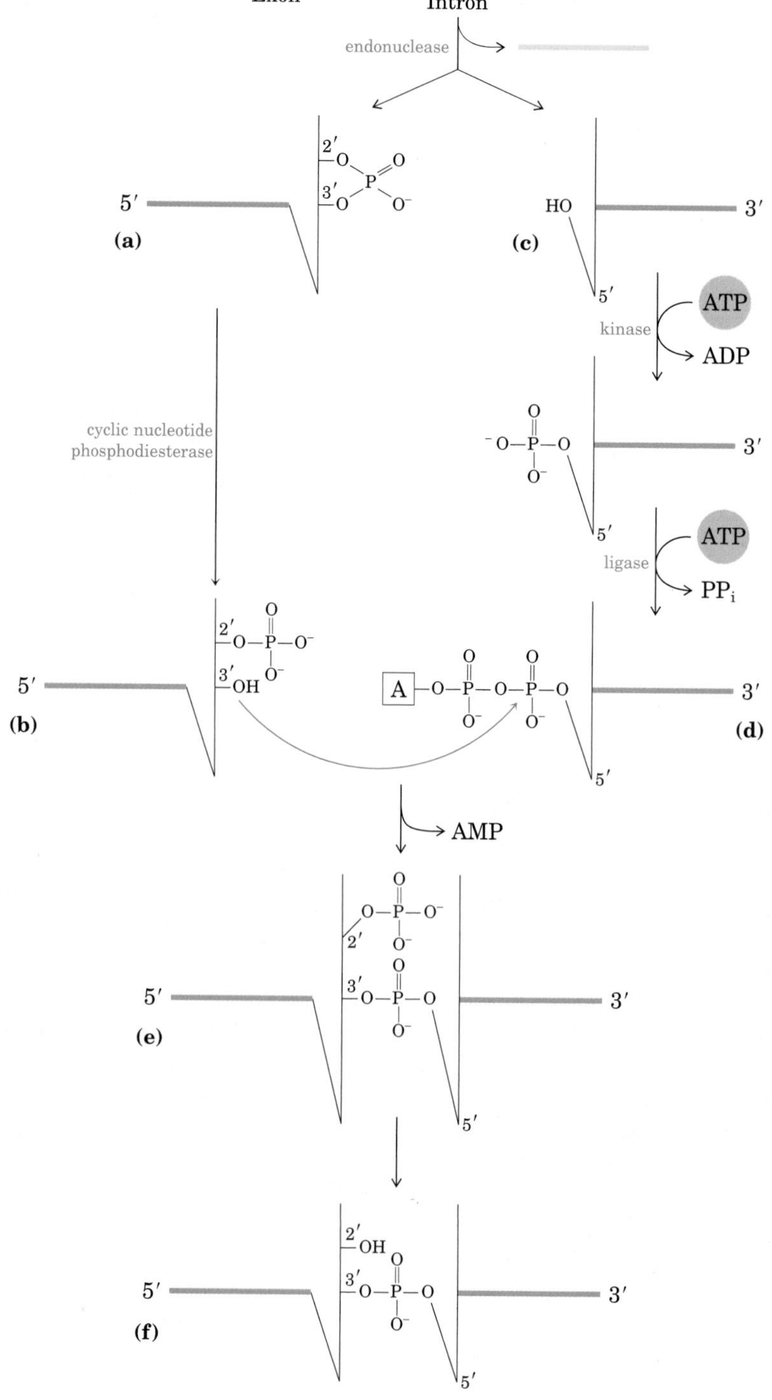

figure 26–17

Splicing of yeast tRNA. The intron is first removed by endonuclease-catalyzed cleavage at both ends. The 2′,3′-cyclic phosphate on the 5′ exon **(a)** is cleaved by a cyclic nucleotide phosphodiesterase, leaving a 2′ phosphate **(b).** The 5′ OH left on the 3′ exon **(c)** is activated in two steps requiring a high-energy cofactor (ATP) to form **(d).** The free 3′ OH of the 5′ exon now acts as a nucleophile to displace AMP from the 3′ exon, joining the two exons with a 3′,5′-phosphodiester bond **(e).** The 2′ phosphate is removed to yield the final product **(f).**

Eukaryotic mRNAs Undergo Additional Processing

Mature eukaryotic mRNAs have distinctive structural features at both ends. Most have a **5′ cap,** a residue of 7-methylguanosine linked to the 5′-terminal residue of the mRNA through an unusual 5′,5′-triphosphate linkage (Fig. 26–18). At the 3′ end, most eukaryotic mRNAs have a string of 80 to 250 adenylate residues called the **poly(A) tail.**

The functions of the 5′ cap and the 3′ poly(A) tail are only partially known. The 5′ cap binds to a specific protein and may participate in the binding of the mRNA to the ribosome to initiate translation (Chapter 27). The poly(A) tail also serves as a binding site for a specific protein. It is likely that the 5′ cap and poly(A) tail and their associated proteins help protect mRNA from enzymatic destruction. Many prokaryotic mRNAs also acquire poly(A) tails, but these tails stimulate mRNA decay rather than protecting the mRNAs from degradation.

Both types of terminal structures are added in several steps. The 5′ cap is formed by condensation of a molecule of GTP with the triphosphate at the 5′ end of the transcript. The guanine is subsequently methylated at N-7, and additional methyl groups often are added at the 2′ hydroxyls of the first and second nucleotides adjacent to the cap (Fig. 26–18). The methyl groups are derived from *S*-adenosylmethionine.

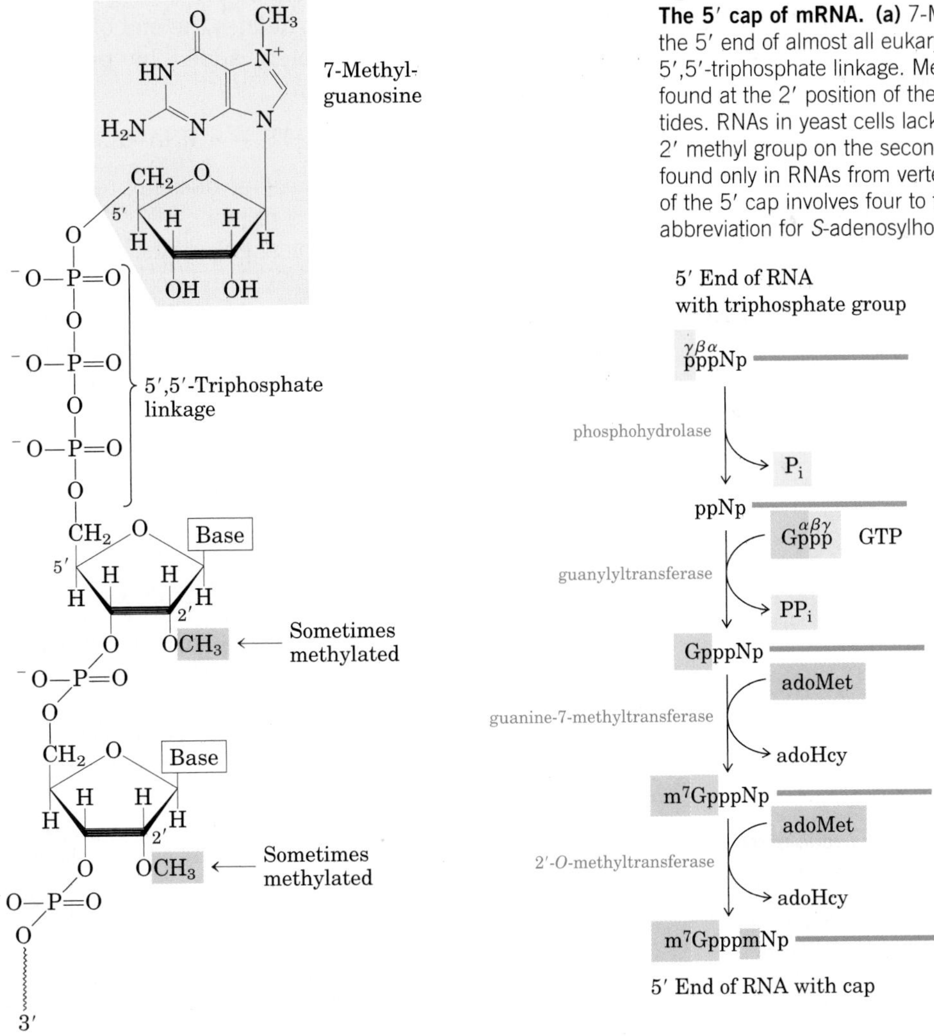

figure 26–18
The 5′ cap of mRNA. (a) 7-Methylguanosine is joined to the 5′ end of almost all eukaryotic mRNAs in an unusual 5′,5′-triphosphate linkage. Methyl groups (red) are often found at the 2′ position of the first and second nucleotides. RNAs in yeast cells lack the 2′-methyl groups. The 2′ methyl group on the second nucleotide is generally found only in RNAs from vertebrate cells. **(b)** Generation of the 5′ cap involves four to five separate steps. (The abbreviation for *S*-adenosylhomocysteine is adoHcy.)

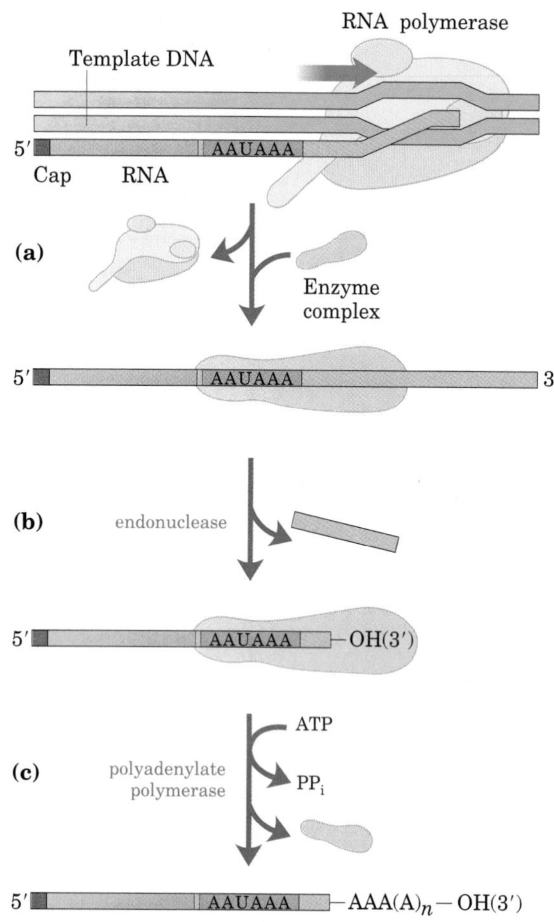

figure 26–19

Addition of the poly(A) tail to the primary RNA transcript of eukaryotes. RNA polymerase synthesizes RNA beyond the segment of the transcript containing the cleavage signal sequences, including the highly conserved upstream sequence (5′)AAUAAA. **(a)** The cleavage signal sequence is bound by an enzyme complex that includes an endonuclease, a polyadenylate polymerase, and several other multisubunit proteins involved in sequence recognition, stimulation of cleavage, and regulation of the length of the polyA tail. **(b)** The RNA is cleaved by the endonuclease at a point 10 to 30 nucleotides 3′ to (downstream of) the sequence AAUAAA. **(c)** The polyadenylate polymerase synthesizes a poly(A) tail 80 to 250 nucleotides in length, beginning at the cleavage site.

The poly(A) tail is added in a multistep process. The transcript is extended beyond the site where the poly(A) tail is to be added, then is cleaved at the poly(A) addition site by an endonuclease component of a large enzyme complex (Fig. 26–19). The mRNA site where cleavage occurs is marked by two sequence elements: the highly conserved sequence (5′)AAUAAA(3′), 10 to 30 nucleotides on the 5′ side (upstream) of the cleavage site, and a less well-defined sequence rich in G and U residues 20 to 40 nucleotides downstream of the cleavage site. Cleavage generates the free 3′-hydroxyl group that defines the end of the mRNA, to which adenylate residues are immediately added by **polyadenylate polymerase,** which catalyzes the reaction

$$\text{RNA} + n\text{ATP} \longrightarrow \text{RNA–(AMP)}_n + n\text{PP}_i$$

where n = 80 to 250. This enzyme does not require a template but does require the cleaved mRNA as a primer.

The processing of a typical eukaryotic mRNA is summarized in Figure 26–20. In some cases the polypeptide-coding region of the mRNA is also modified by RNA "editing" (see Box 27–1). A particularly dramatic example occurs in parasitic protozoa called trypanosomes: large regions of an mRNA are synthesized without any uridylate, and the U residues are inserted later by RNA editing.

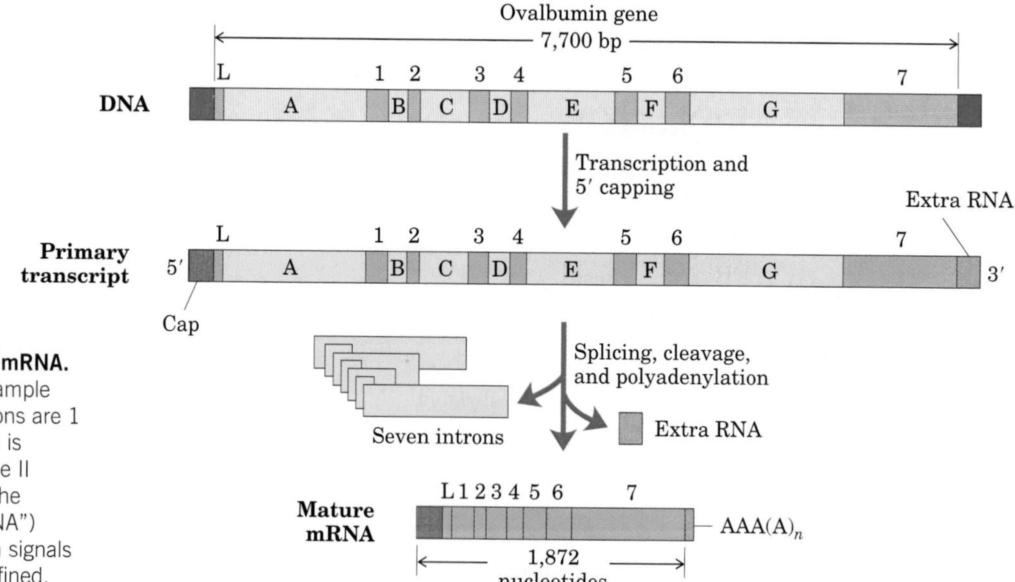

figure 26–20

Overview of the processing of a eukaryotic mRNA. The ovalbumin gene is again used as an example (see Fig. 26–12). Introns are A to G and exons are 1 to 7 and L. About three-quarters of the RNA is removed during processing. RNA polymerase II extends the primary transcript well beyond the cleavage and polyadenylation site ("extra RNA") before terminating transcription. Termination signals for RNA polymerase II have not yet been defined.

Multiple Products Are Derived from One Gene by Differential RNA Processing

The transcription of introns consumes cellular resources and energy without returning any obvious benefit to the organism, but introns may confer an advantage not yet appreciated by scientists. Introns may be vestiges of a molecular parasite not unlike the transposons considered in Chapter 25. Although any benefits of introns are not yet clear in most cases, cells have evolved to take advantage of splicing pathways to alter the expression of certain genes.

Although most eukaryotic mRNA transcripts produce only one mature mRNA and one corresponding polypeptide, some can be processed in more than one way to produce *different* mRNAs and thus different polypeptides. The primary transcript contains molecular signals for all of the alternative processing pathways. The pathway favored in a given cell is determined by processing factors, RNA-binding proteins that promote one particular pathway.

Complex transcripts can have either more than one site for cleavage and polyadenylation, or alternative splicing patterns, or both (Fig. 26–21). If there are two or more sites for cleavage and polyadenylation, use of the one closest to the 5′ end will remove more of the primary transcript sequence (Fig. 26–21a). This mechanism, called poly(A) site choice, is used to generate diversity in the variable domains of immunoglobulin heavy chains. Alternative splicing patterns (Fig. 26–21b) produce, from a common primary transcript, three different forms of the myosin heavy chain at different stages of fruit fly development. *Both* mechanisms come into play

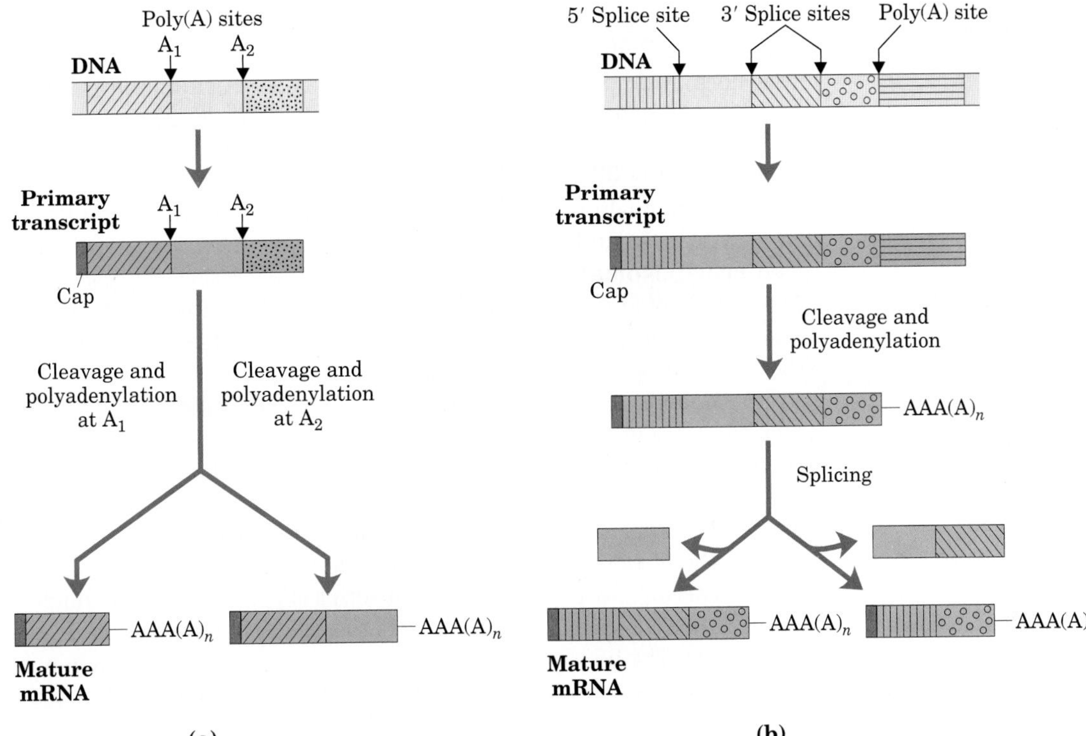

figure 26–21
Two mechanisms for the alternative processing of complex transcripts in eukaryotes. (a) Alternative cleavage and polyadenylation patterns. Two poly(A) sites, A_1 and A_2, are shown. **(b)** Alternative splicing patterns. Two different 3′ splice sites are shown. In both mechanisms, different mature mRNAs are produced from the same primary transcript.

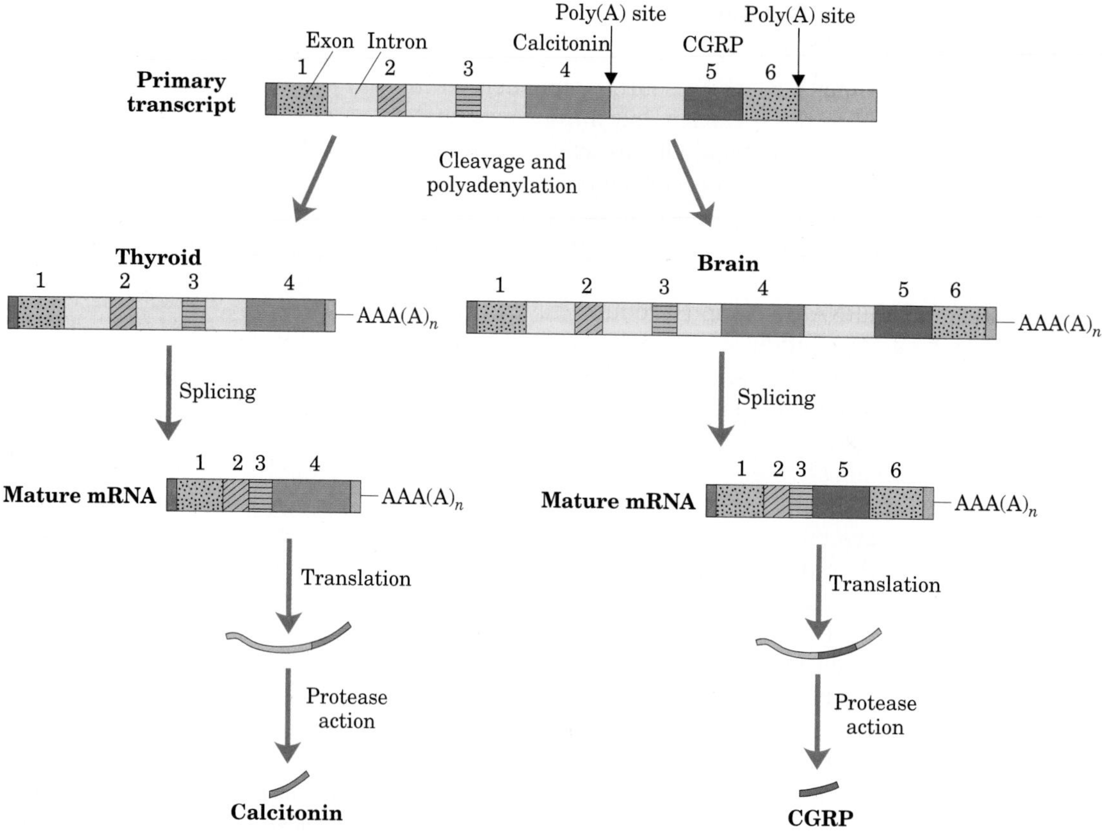

figure 26–22

Alternative processing of the calcitonin gene transcript in rats. The primary transcript has two poly(A) sites; one predominates in the brain, the other in the thyroid. In the brain, splicing eliminates the calcitonin exon (exon 4); in the thyroid, this exon is retained. The resulting peptides are processed further to yield the final hormone products: calcitonin-gene-related peptide (CGRP) in the brain and calcitonin in the thyroid.

when a single RNA transcript is processed differently to produce two different hormones: the calcium-regulating hormone calcitonin in rat thyroid and calcitonin-gene-related peptide in rat brain (Fig. 26–22).

Ribosomal RNAs and tRNAs Also Undergo Processing

Posttranscriptional processing is not limited to mRNA. Ribosomal RNAs of both prokaryotic and eukaryotic cells are made from longer precursors called **preribosomal RNAs,** or pre-rRNAs. In bacteria, 16S, 23S, and 5S rRNAs (and some tRNAs) arise from a single 30S RNA precursor of about 6,500 nucleotides. RNA at both ends of the 30S precursor and between the rRNAs is removed during processing (Fig. 26–23).

The genome of *E. coli* encodes seven pre-rRNA molecules. All these genes have essentially identical rRNA coding regions, but they differ in the regions in between. The region between the genes for the 16S and 23S rRNAs generally encodes one or two tRNAs, with different tRNAs arising from different pre-rRNA transcripts. Coding sequences for tRNAs are also found on the 3′ side of the 5S rRNA in some precursor transcripts.

In eukaryotes, a 45S pre-rRNA transcript is processed in the nucleolus to form the 18S, 28S, and 5.8S rRNAs characteristic of eukaryotic ribosomes (Fig. 26–24). The 5S rRNA of most eukaryotes is made as a completely separate transcript by a different polymerase (RNA polymerase III instead of RNA polymerase I).

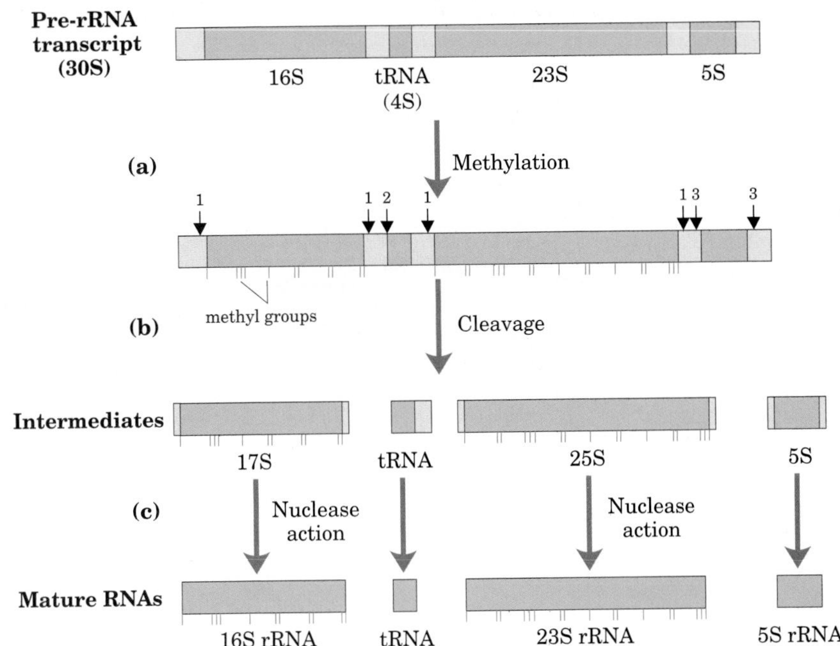

figure 26–23

Processing of pre-rRNA transcripts in bacteria. (a) Before cleavage, the 30S RNA precursor is methylated at specific bases. **(b)** Cleavage liberates precursors of rRNAs and tRNA(s). Cleavage at the points labeled 1, 2, and 3 is carried out by the enzymes RNase III, RNase P, and RNase E, respectively. As discussed later in the text, RNase P is a ribozyme. **(c)** The final 16S, 23S, and 5S rRNA products result from the action of a variety of specific nucleases. The seven copies of the gene for pre-rRNA in the *E. coli* chromosome differ in the number, location, and identity of tRNAs included in the primary transcript. Some copies of the gene have additional tRNA gene segments between the 16S and 23S rRNA segments and at the far 3′ end of the primary transcript.

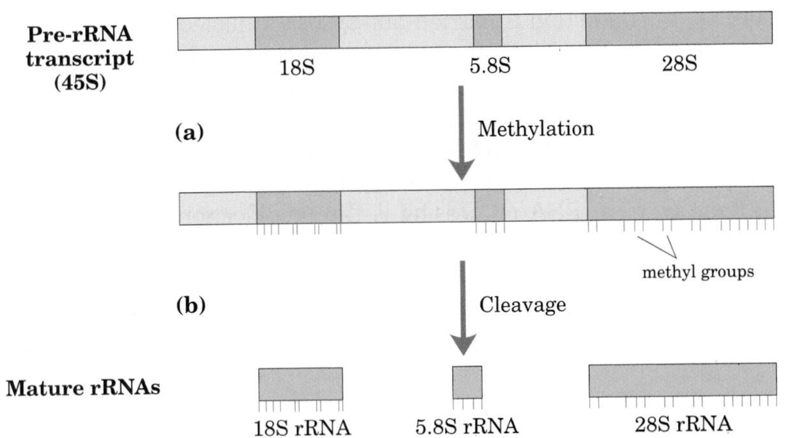

figure 26–24

Processing of pre-rRNA transcripts in vertebrates. **(a)** The 45S precursor is methylated at more than 100 of its 14,000 nucleotides, mostly on the 2′-OH groups of ribose units retained in the final products. **(b)** A series of enzymatic cleavages produces the 18S, 5.8S, and 28S rRNAs. The cleavage reactions require RNAs found in the nucleolus, called small nucleolar RNAs (snoRNAs), within protein complexes reminiscent of spliceosomes. The 5S rRNA is produced separately.

Most cells have 40 to 50 distinct tRNAs, and eukaryotic cells have multiple copies of many of the tRNA genes. Transfer RNAs are derived from longer RNA precursors by enzymatic removal of nucleotides from the 5′ and 3′ ends (Fig. 26–25). In eukaryotes, introns are occasionally present in tRNA transcripts and must be excised. Where two or more different tRNAs are contained in a single primary transcript, they are separated by enzymatic cleavage. RNase P is the endonuclease that removes RNA at the 5′ end of tRNAs. The 3′ end of the tRNA is processed by one or more nucleases, including an exonuclease called RNase D. RNase P, found in all organisms, contains both protein and RNA. The RNA component is essential for activity, and in bacterial cells it can carry out its processing function with precision even in the absence of the protein component. RNase P is therefore another example of a catalytic RNA, as described in more detail below.

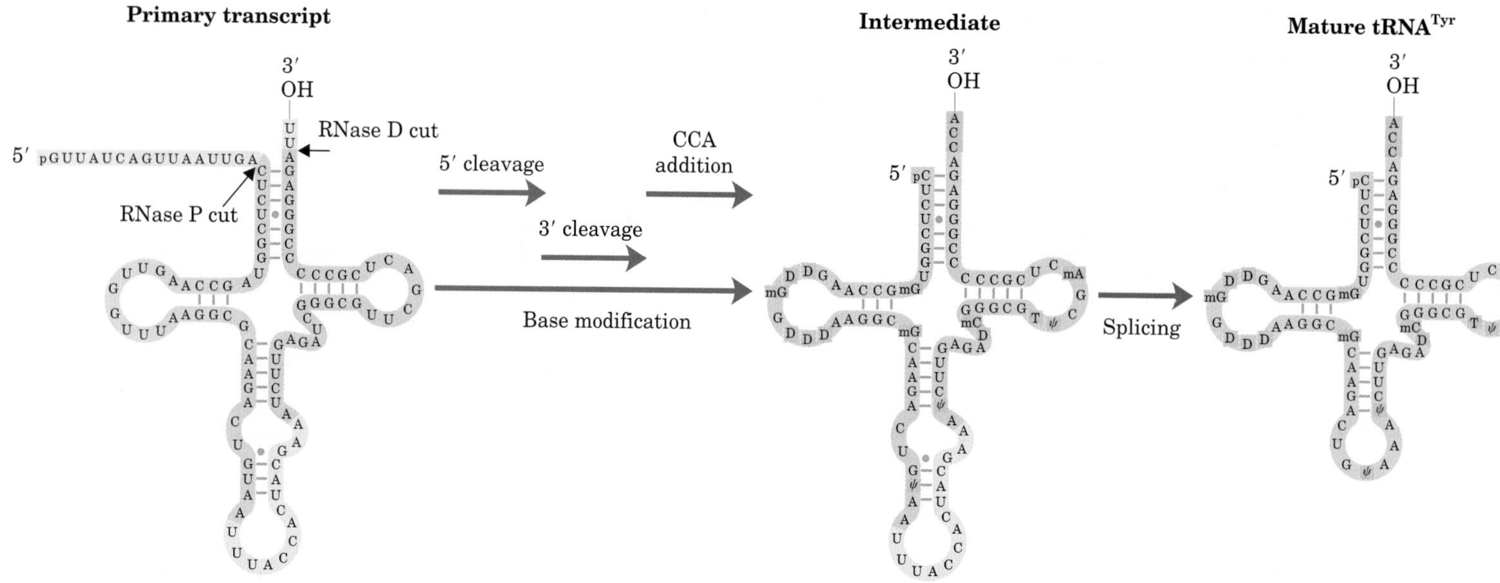

figure 26–25
Processing of tRNAs in bacteria and eukaryotes. The yeast tRNA^{Tyr} (the tRNA specific for tyrosine binding) is used to illustrate the important steps. The nucleotide sequences shown in yellow are removed from the primary transcript. The ends are processed first, the 5′ end before the 3′ end. CCA is then added to the 3′ end, a necessary step in processing eukaryotic tRNAs and those bacterial tRNAs that lack this sequence in the primary transcript. While the ends are being processed, specific bases in the rest of the transcript are modified (see Fig. 26–26). For the eukaryotic tRNA shown here, the final step is splicing of the 14 nucleotide intron. Introns are found in some eukaryotic tRNAs but not in bacterial tRNAs.

Transfer RNA precursors may undergo further posttranscriptional processing. The 3′-terminal trinucleotide CCA(3′) to which an amino acid will be attached during protein synthesis (Chapter 27) is absent from some bacterial and all eukaryotic tRNA precursors and is added during processing (Fig. 26–25). This addition is carried out by tRNA nucleotidyltransferase, an unusual enzyme that binds the three ribonucleoside triphosphate precursors in separate active sites and catalyzes formation of the phosphodiester bonds to produce the CCA(3′) sequence. The creation of this defined sequence of nucleotides is therefore not dependent on a DNA or RNA template—the template is the binding site of the enzyme.

The final type of tRNA processing is the modification of some of the bases by methylation, deamination, or reduction (Fig. 26–26). In the case of pseudouridine (ψ) the base (uracil) is removed and reattached to the sugar through a different N atom. Some of these modified bases occur at characteristic positions in all tRNAs (Fig. 26–25).

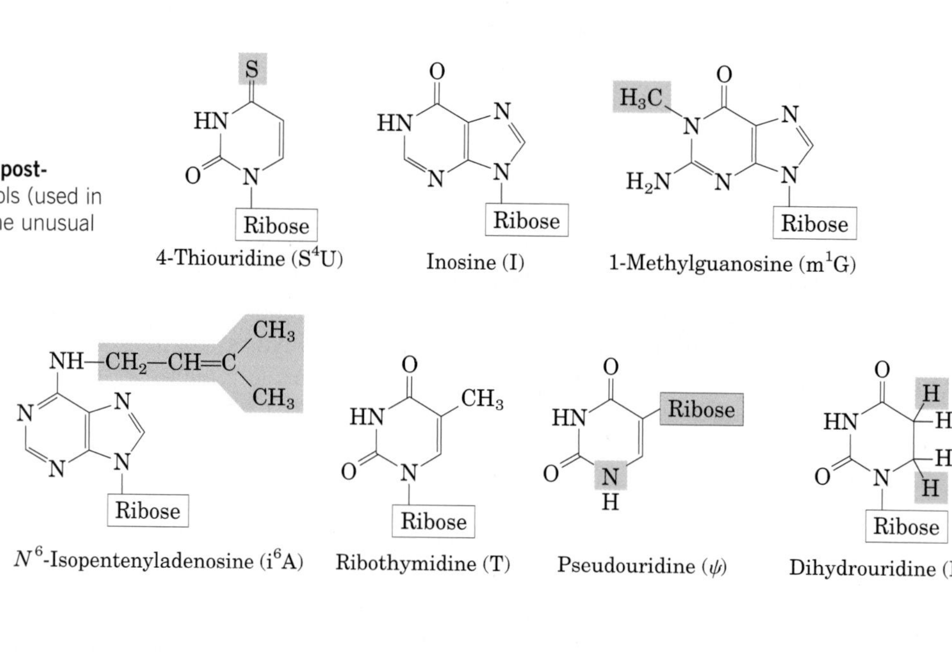

figure 26–26
Some modified bases of tRNAs, produced in posttranscriptional reactions. The standard symbols (used in Fig. 26–25) are shown in parentheses. Note the unusual ribose attachment point in pseudouridine.

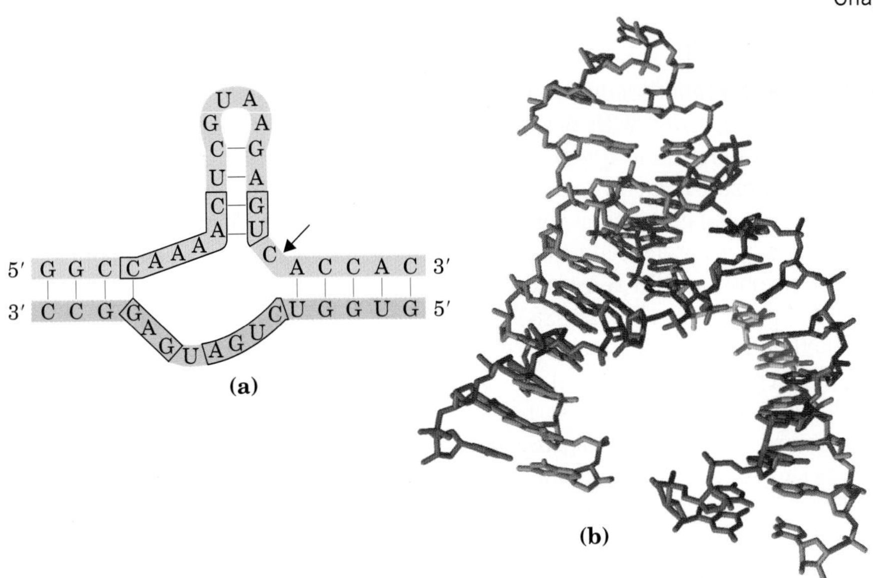

figure 26–27

Hammerhead ribozyme. Certain virus-like elements called virusoids have small RNA genomes and usually require another virus to assist in their replication and/or packaging. Some virusoid RNAs include small segments that promote site-specific RNA cleavage reactions associated with replication. These segments are called hammerhead ribozymes because their secondary structures are shaped like the head of a hammer. Hammerhead ribozymes have been defined and studied separately from the much larger viral RNAs. **(a)** The minimal sequences required for catalysis. The boxed nucleotides are highly conserved and required for catalytic function. The arrow indicates the site of self-cleavage. **(b)** Three-dimensional structure. The strands are colored as in **(a).** The hammerhead ribozyme is a metalloenzyme; Mg^{2+} ions are required for activity. The phosphodiester bond at the site of self-cleavage is between the two nucleotide residues shown in yellow.

Some Events in RNA Metabolism Are Catalyzed by RNA Enzymes

The study of posttranscriptional processing of RNA molecules led to one of the most exciting discoveries in modern biochemistry—the existence of RNA enzymes. The best-characterized ribozymes are the self-splicing group I introns, RNase P, and the hammerhead ribozyme. Most of the activities of these ribozymes are based on two fundamental reactions: transesterification (Fig. 26–13) and phosphodiester bond hydrolysis (cleavage). The substrate for ribozymes is often an RNA molecule, and it may even be part of the ribozyme itself. When its substrate is RNA, an RNA catalyst can make use of base-pairing interactions to align the substrate for the reaction.

Ribozymes vary greatly in size. A self-splicing group I intron may have over 400 nucleotides. The hammerhead ribozyme consists of two RNA strands with a total of only 41 nucleotides (Fig. 26–27). As with protein enzymes, the three-dimensional structure of ribozymes is important for function. Activity is lost if the RNA is heated beyond its melting temperature, if denaturing agents are added, or if complementary oligonucleotides are added, which disrupt normal base-pairing patterns. Ribozymes can also be inactivated if essential nucleotides are changed. The secondary structure of a self-splicing group I intron from the 26S rRNA precursor of *Tetrahymena* is shown in detail in Figure 26–28.

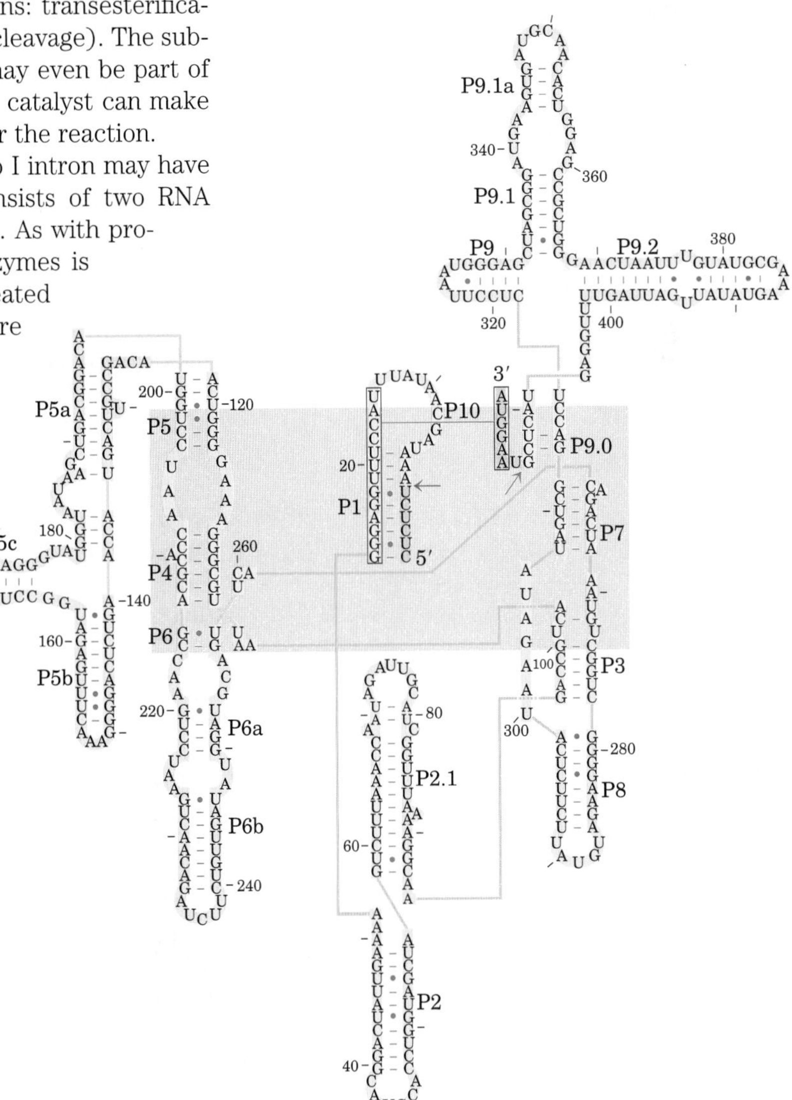

figure 26–28

Secondary structure of the self-splicing rRNA intron from *Tetrahymena*. Intron and exon sequences are shaded yellow and green, respectively. Each thick yellow line represents a bond between neighboring nucleotides in a continuous sequence (necessitated by showing this complex molecule in two dimensions)—all nucleotides are shown. The catalytic core of the self-splicing activity is shaded. Some base-paired regions are labeled (P1, P3, P2.1, P5a, and so forth) according to an established convention for this RNA molecule. The P1 region, which contains the internal guide sequence (boxed), is the location of the 5′ splice site (red arrow). Part of the internal guide sequence pairs with the end of the 3′ exon, bringing the 5′ and 3′ splice sites (red and blue arrows) into close proximity. The three-dimensional structure of a large segment of this intron is illustrated in Figure 10–28d.

Enzymatic Properties of Group I Introns Self-splicing group I introns share several properties with enzymes besides greatly accelerating the reaction rate, including their kinetic behaviors and their great specificity. Binding of the guanosine cofactor to the *Tetrahymena* group I rRNA intron is saturable ($K_m \approx 30\ \mu$M) and can be competitively inhibited by 3′-deoxyguanosine. The intron is very precise in its excision reaction, largely due to a segment called the **internal guide sequence** that can base-pair with exon sequences near the 5′ splice site (Fig. 26–28). This pairing promotes the alignment of specific bonds to be cleaved and rejoined.

Because the intron itself is chemically altered by having its ends cleaved during the splicing reaction, it may appear to lack one key enzymatic property: the ability to catalyze multiple reactions. Closer inspection has shown that after excision, the 414 nucleotide intron from *Tetrahymena* rRNA can, in vitro, act as a true enzyme (in vivo it is quickly degraded). A series of intramolecular cyclization/cleavage reactions in the excised intron leads to the loss of 19 nucleotides from its 5′ end. The remaining 395 nucleotide linear RNA, called L-19 IVS, promotes nucleotidyl transfer reactions in which some oligonucleotides are lengthened at the expense of others (Fig. 26–29). The best substrates are oligonucleotides, for example a synthetic oligomer of five cytidylates, $(C)_5$, which can base-pair with the same guanylate-rich internal guide sequence that held the 5′ exon in place for self-splicing.

The enzymatic activity of L-19 IVS results from a cycle of transesterification reactions mechanistically similar to self-splicing. Each ribozyme can process about 100 substrate molecules per hour and is not altered in the reaction; therefore the intron acts as a catalyst. It follows Michaelis-Menten kinetics, is specific for RNA oligonucleotide substrates, and can be competitively inhibited. The k_{cat}/K_m (the specificity constant) is $10^3\ \text{M}^{-1}\ \text{s}^{-1}$, lower than that of many enzymes, but this ribozyme accelerates hydrolysis by a factor of 10^{10} relative to the uncatalyzed reaction. The ribozyme makes use of substrate orientation, covalent catalysis, and metal-ion catalysis—strategies used by protein enzymes.

Characteristics of Other Ribozymes *E. coli* RNase P has both an RNA component (the M1 RNA, 377 nucleotides) and a protein component (M_r 17,500). In 1983, Sidney Altman and Norman Pace and their coworkers discovered that under some conditions, the M1 RNA alone is capable of catalysis, cleaving tRNA precursors at the correct position. The protein component apparently stabilizes the RNA or facilitates its function in vivo. The RNase P ribozyme recognizes the three-dimensional shape of its pre-tRNA substrate along with the CCA sequence, and thus can cleave the 5′ leaders from diverse tRNAs (Fig. 26–25).

The catalytic repertoire of ribozymes continues to expand. Some virusoids, small RNAs associated with plant RNA viruses, include a structure that promotes a self-cleavage reaction. The hammerhead ribozyme illustrated in Figure 26–27 is one of these, catalyzing the hydrolysis of an internal phosphodiester bond. The splicing reaction that occurs in a spliceosome is believed to rely on a catalytic center formed by the U2, U5, and U6 snRNAs (Fig. 26–16). Also, an RNA component of ribosomes (Chapter 27) may participate in the catalysis of protein synthesis.

Exploring catalytic RNAs has provided new insights into catalytic function in general and has important implications for our understanding of the origin and evolution of life on this planet, a topic discussed at the end of the chapter.

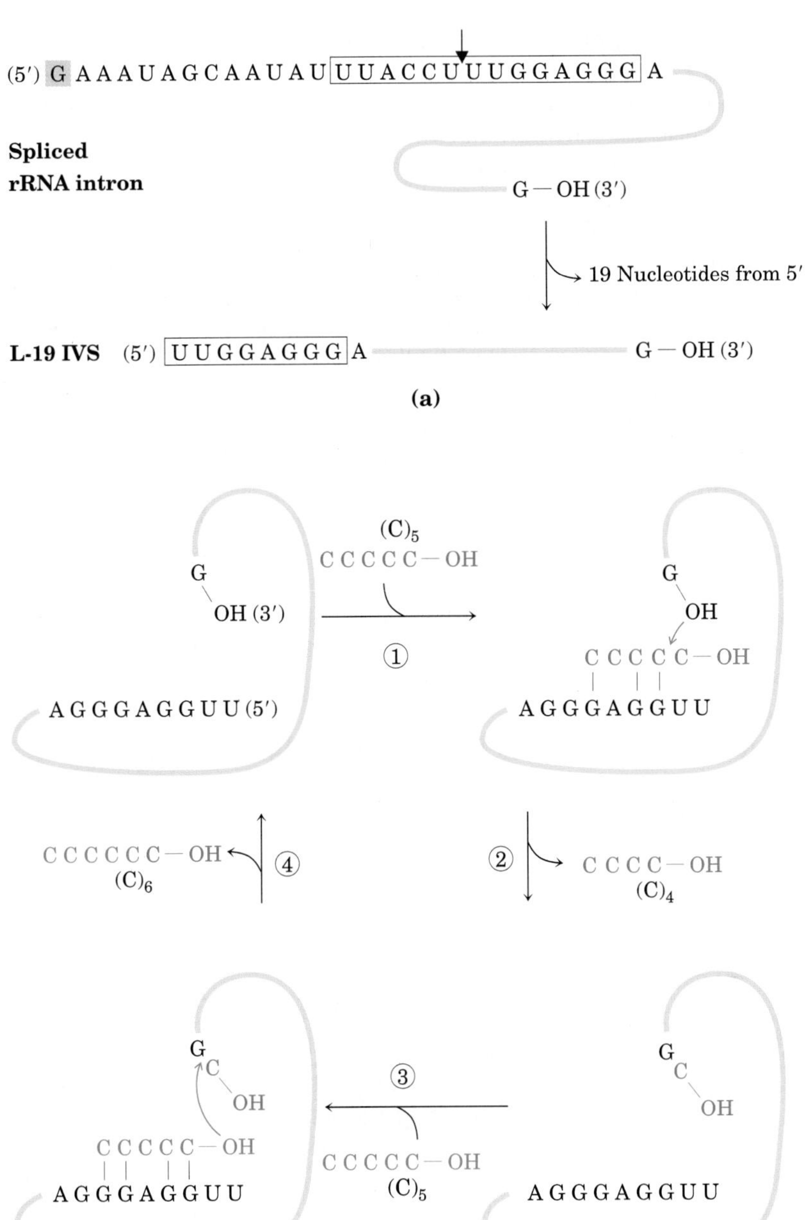

figure 26–29

In vitro catalytic activity of L-19 IVS. (a) L-19 IVS is generated by the autocatalytic removal of 19 nucleotides from the 5′ end of the spliced *Tetrahymena* intron. The cleavage site is indicated by the arrow in the internal guide sequence (boxed). The G residue (shaded in red) added in the first step of the splicing reaction (see Fig. 26–14) is part of the removed sequence. A portion of the internal guide sequence remains at the 5′ end of L-19 IVS. **(b)** L-19 IVS lengthens some RNA oligonucleotides at the expense of others in a cycle of transesterification reactions (steps ① through ④). The 3′ OH of the G residue at the 3′ end of L-19 IVS plays a key role in this cycle of reactions (note that this is *not* the G residue added in the splicing reaction). $(C)_5$ is one of the ribozyme's better substrates because it can base-pair with the guide sequence remaining in the intron. Although this catalytic activity is probably irrelevant to the cell, it has important implications for current hypotheses on evolution, discussed at the end of this chapter.

Cellular mRNAs Are Degraded at Different Rates

The expression of genes is regulated at many levels. A crucial factor governing a gene's expression is the cellular concentration of its associated mRNA. The concentration of any molecule depends on two factors: its rate of synthesis and its rate of degradation. When synthesis and degradation of an mRNA are balanced, the concentration of the mRNA remains in a steady state. A change in either rate will lead to net accumulation or depletion of the mRNA. Degradative pathways ensure that mRNAs do not build up in the cell and direct the synthesis of unnecessary proteins.

The rates of degradation vary greatly for mRNAs from different eukaryotic genes. For a gene product that is needed only briefly, the half-life of its mRNA may be only minutes or even seconds. Gene products needed constantly by the cell may have mRNAs that are stable over many cell generations. The average half-life of a vertebrate cell mRNA is about three hours, with the pool of each type of mRNA turning over about ten times per cell generation. The half-life of bacterial mRNAs is only about 1.5 min, perhaps related to regulatory requirements.

RNA is degraded by ribonucleases present in all cells, usually in a $5' \rightarrow 3'$ direction, although $3' \rightarrow 5'$ exoribonucleases also exist. Stable mRNAs carry a sequence at their $3'$ ends that inhibits exoribonucleases. In bacteria, mRNA degradation often begins almost as soon as transcription is complete, with degradative ribonucleases following close behind the ribosomes that translate the mRNA into protein (Chapter 27). A hairpin structure in bacterial mRNAs with a ρ-independent terminator (Fig. 26–7) confers stability against degradation. Similar hairpin structures can make some parts of a primary transcript more stable, leading to nonuniform degradation of polycistronic transcripts. In eukaryotic cells, the $3'$ poly(A) tail is important to the stability of many mRNAs. A major degradative pathway in eukaryotic cells involves first shortening the poly(A) tail, then decapping the $5'$ end and degrading the RNA in the $5' \rightarrow 3'$ direction.

Polynucleotide Phosphorylase Makes Random RNA-like Polymers

In 1955, Marianne Grunberg-Manago and Severo Ochoa discovered the bacterial enzyme **polynucleotide phosphorylase,** which in vitro catalyzes the reaction

$$(\text{NMP})_n + \text{NDP} \rightleftharpoons \underset{\substack{\text{Lengthened} \\ \text{polynucleotide}}}{(\text{NMP})_{n+1}} + \text{P}_\text{i}$$

Marianne Grunberg-Manago

Severo Ochoa
1905–1993

Polynucleotide phosphorylase was the first nucleic acid–synthesizing enzyme discovered (Arthur Kornberg's discovery of DNA polymerase followed soon thereafter). The reaction catalyzed by polynucleotide phosphorylase differs fundamentally from the polymerase activities discussed so far in that it is not template-dependent. The enzyme uses the $5'$-diphosphates of ribonucleosides as substrates and cannot act on the homologous $5'$-triphosphates or on deoxyribonucleoside $5'$-diphosphates. The RNA polymer formed by polynucleotide phosphorylase contains normal $3',5'$-phosphodiester linkages, which can be hydrolyzed by ribonuclease. The reaction is readily reversible and can be pushed in the direction of breakdown of the polyribonucleotide by increasing the phosphate concentration. The probable function of this enzyme in the cell is the degradation of mRNAs to nucleoside diphosphates.

Because the polynucleotide phosphorylase reaction does not use a template, the polymer it forms does not have a specific base sequence. The reaction proceeds equally well with any or all of the four nucleoside diphosphates, and the base composition of the resulting polymer reflects nothing more than the relative concentrations of the $5'$-diphosphate substrates in the medium.

Polynucleotide phosphorylase can be used in the laboratory to prepare RNA polymers with many different sequences and frequencies of bases. Synthetic RNA polymers of this sort were critical for deducing the genetic code for the amino acids (Chapter 27).

RNA-Dependent Synthesis of RNA and DNA

In our discussion of DNA and RNA synthesis up to this point, the role of the template strand has been reserved for DNA. However, some enzymes use an RNA template for nucleic acid synthesis. With the very important exception of viruses with an RNA genome, these enzymes play only a modest role in information pathways. RNA viruses are the source of most RNA-dependent polymerases characterized so far.

The existence of RNA replication requires an elaboration of the central dogma depicted earlier (Fig. 26–30; contrast with the diagram on p. 906). The enzymes involved in RNA replication are extremely useful in recombinant DNA technology (see Fig. 29–9) and have profound implications for investigations into the nature of self-replicating molecules that may have existed in prebiotic times.

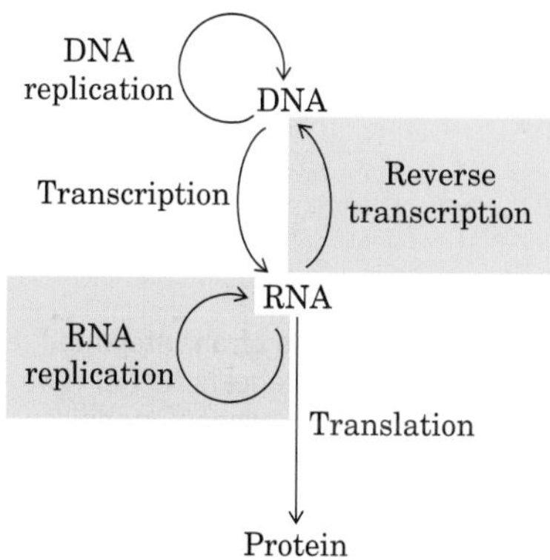

figure 26–30
Extension of the central dogma to include RNA-dependent synthesis of RNA and DNA.

Reverse Transcriptase Produces DNA from Viral RNA

Certain RNA viruses that infect animal cells carry within the viral particle an RNA-dependent DNA polymerase called **reverse transcriptase.** On infection, the single-stranded RNA viral genome (~10,000 nucleotides in length) and the enzyme enter the host cell. The reverse transcriptase first catalyzes the synthesis of a DNA strand complementary to the viral RNA (Fig. 26–31), then it degrades the RNA strand of the viral RNA-DNA hybrid and replaces it with DNA. The resulting duplex DNA often becomes incorporated into the genome of the eukaryotic host cell. These integrated (and dormant) viral genes can be activated and transcribed, and the gene products—viral proteins and the viral RNA genome itself—are packaged as new viruses. The RNA viruses that contain reverse transcriptases are known as **retroviruses** (*retro* is the Latin prefix for "backward").

The existence of reverse transcriptases in RNA viruses was predicted by Howard Temin in 1962, and the enzymes were ultimately detected by Temin and independently by David Baltimore in 1970. Their discovery aroused much attention as dogma-shaking proof that genetic information can flow "backward" from RNA to DNA.

Howard Temin
1934–1994

David Baltimore

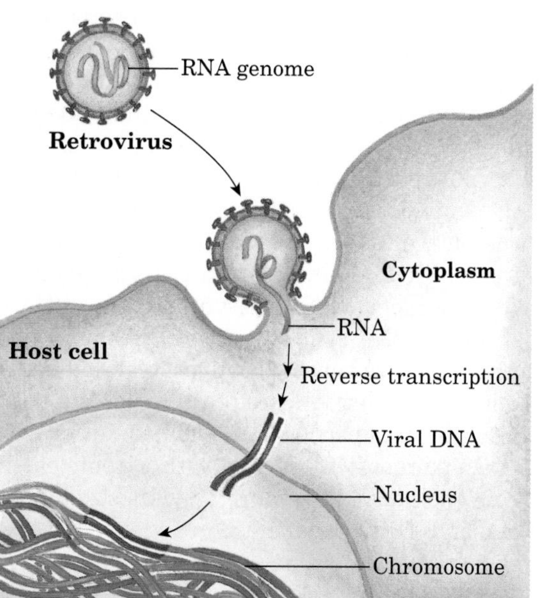

figure 26–31
Retroviral infection of a mammalian cell and integration of the retrovirus into the host chromosome. Integration of viral DNA into host DNA is mechanistically similar to the insertion of transposons in bacterial chromosomes (Fig. 25–41). For example, a few base pairs of host DNA become duplicated at the site of integration, forming short repeats of four to six base pairs at each end of the inserted retroviral DNA (not shown). The viral particles entering the host cell carry viral reverse transcriptase and a cellular tRNA (picked up from a former host cell) already base-paired to the viral RNA. The tRNA facilitates immediate conversion of viral RNA to double-stranded DNA, as described in the text.

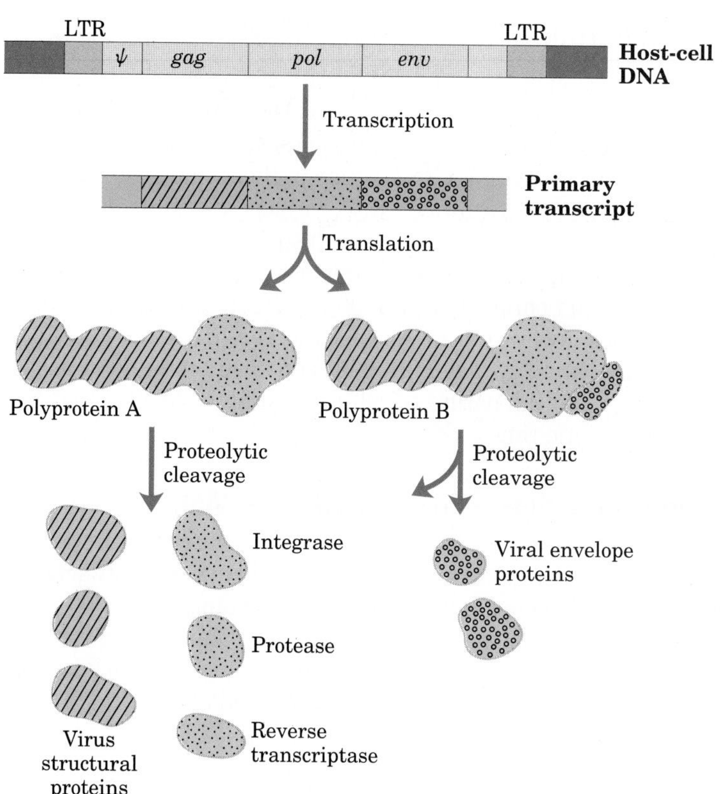

figure 26–32

Structure and gene products of an integrated retrovirus genome. The long terminal repeats (LTRs) have sequences needed for the regulation and initiation of transcription. The sequence denoted ψ is required for packaging of retroviral RNAs into mature viral particles. Transcription of the retroviral DNA produces a primary transcript encompassing the *gag, pol,* and *env* genes. Translation (Chapter 27) produces a polyprotein, a single long polypeptide derived from the *gag* and *pol* genes, which is proteolytically cleaved into six distinct proteins. Splicing of the primary transcript yields an mRNA derived largely from the *env* gene, which is also translated into a polyprotein, then cleaved to generate viral envelope proteins.

Retroviruses typically have three genes: *gag* (derived from the historical designation *g*roup *a*ssociated anti*g*en), *pol,* and *env* (Fig. 26–32). The transcript that contains *gag* and *pol* is translated into a long "polyprotein," a single large polypeptide that is cleaved into six proteins with distinct functions. The proteins derived from the *gag* gene make up the interior core of the viral particle. The *pol* gene encodes the protease that cleaves the long polypeptide, an integrase that inserts the viral DNA into the host chromosomes, and reverse transcriptase. Many reverse transcriptases have two subunits, α and β. The *pol* gene specifies the β subunit (M_r 90,000), and the α subunit (M_r 65,000) is simply a proteolytic fragment of the β subunit. The *env* gene encodes the proteins of the viral envelope. At each end of the linear RNA genome are long terminal repeat (LTR) sequences of a few hundred nucleotides. Transcribed into the duplex DNA, these sequences facilitate the integration of the viral chromosome into the host DNA and contain promoters for viral gene expression.

Viral reverse transcriptases contain Zn^{2+}, as do many DNA and RNA polymerases. Each transcriptase is most active with the RNA of its own virus but can be used experimentally to make DNA complementary to a variety of RNAs. Reverse transcriptases catalyze three different reactions: (1) RNA-dependent DNA synthesis, (2) RNA degradation, and (3) DNA-dependent DNA synthesis. The DNA/RNA synthesis and RNA degradation activities utilize separate active sites on the protein. For DNA synthesis to begin, the reverse transcriptase requires a primer, a cellular tRNA carried within the viral particle but obtained during an earlier infection. This tRNA is base-paired at its 3′ end with a complementary sequence in the viral RNA. The new DNA strand is synthesized in the 5′→3′ direction, as in all RNA and DNA polymerase reactions. Reverse transcriptases, like RNA polymerases, do not have 3′→5′ proofreading exonucleases. They generally have error rates of about one per 20,000 nucleotides added. An error rate this high is extremely unusual in DNA replication and appears to be a feature of most enzymes that replicate the genomes of RNA viruses. A consequence is a faster rate of viral evolution, which is a factor in the frequent appearance of new strains of disease-causing retroviruses.

Reverse transcriptases have become important reagents in the study of DNA-RNA relationships and in cloning DNA. They make possible the synthesis of DNA complementary to an mRNA template, and synthetic DNA prepared in this manner, called **complementary DNA (cDNA),** can be used to clone cellular genes (see Fig. 29-9).

Retroviruses Cause Cancer and AIDS

Retroviruses have featured prominently in recent advances in the molecular understanding of cancer. Most retroviruses do not kill their host cells but remain integrated in the cellular DNA, replicating when the cell divides. Some retroviruses, classified as RNA tumor viruses, contain an oncogene that can cause the cell to grow abnormally (see Fig. 13–36). The first retrovirus of this type to be studied was the Rous sarcoma virus (also called avian sarcoma virus; Fig. 26–33), named for Peyton Rous, who studied chicken tumors now known to be caused by this virus. Since the initial discovery of oncogenes by Harold Varmus and Michael Bishop, many dozens of such genes have been found in retroviruses.

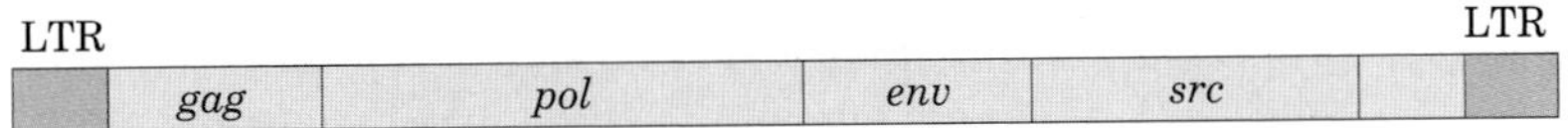

figure 26–33

Rous sarcoma virus genome. The *src* gene encodes a tyrosine-specific protein kinase, one of a class of enzymes known to function in systems that affect cell division, cell-cell interactions, and intercellular communication (see Chapter 13). The same gene is found in the DNA of normal chickens and in the genomes of many other eukaryotes, including humans. When associated with the Rous sarcoma virus, this oncogene is often expressed at abnormally high levels, contributing to unregulated cell division and cancer.

The human immunodeficiency virus (HIV), which causes acquired immune deficiency syndrome (AIDS), is a retrovirus. Identified in 1983, HIV has an RNA genome with standard retroviral genes along with several other unusual genes (Fig. 26–34). Unlike many other retroviruses, HIV kills many of the cells it infects (principally T lymphocytes) rather than causing the formation of tumors. This gradually leads to suppression of the immune system in the host organism. The reverse transcriptase of HIV is even more error prone than other known reverse transcriptases—ten times more so—resulting in high mutation rates in this virus. One or more errors are generally made every time the viral genome is replicated, so that any two viral RNA molecules almost always differ.

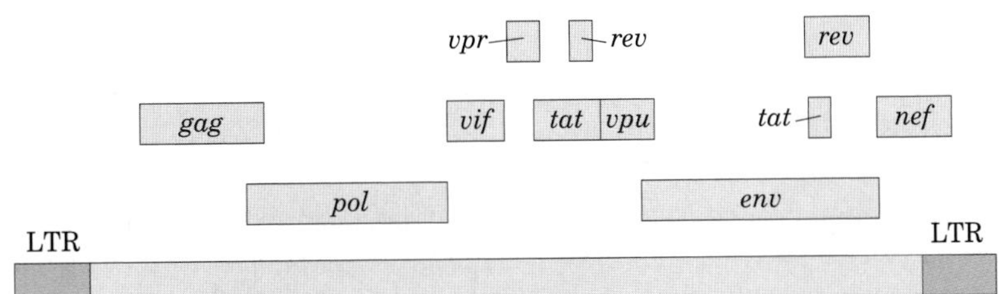

figure 26–34

The genome of HIV, the virus that causes AIDS. In addition to the typical retroviral genes, HIV contains several small genes with a variety of functions (not identified here, and not all known). Some of these genes overlap (Chapter 27). Alternative splicing mechanisms produce many different proteins from this small (9.7×10^6 nucleotides) genome.

Modern vaccines for viral infections often consist of one or more coat proteins of the virus, produced by methods described in Chapter 29. These proteins are not infectious on their own but stimulate the immune system to recognize and resist subsequent viral invasions (Chapter 7). Because of the high error rate of the HIV reverse transcriptase, the *env* gene in this virus (along with the rest of the genome) undergoes very rapid mutation, complicating the development of an effective vaccine. However, repeated cycles of cell invasion and replication are needed to propagate an HIV infection, so inhibition of viral enzymes offers promise as an effective therapy. The HIV protease is targeted by a class of drugs called protease inhibitors (see Box 8–3). Reverse transcriptase is the target of some additional drugs widely used to treat HIV-infected individuals (Box 26–2).

box 26–2 Fighting AIDS with Inhibitors of HIV Reverse Transcriptase

Research into the chemistry of template-dependent nucleic acid biosynthesis, combined with modern techniques of molecular biology, has elucidated the life cycle and structure of the human immunodeficiency virus (HIV), the retrovirus that causes AIDS. A few years after the isolation of HIV, this research resulted in the development of drugs capable of prolonging the lives of those infected by HIV.

The first drug to be approved for clinical use was AZT, a structural analog of deoxythymidine. AZT was first synthesized in 1964 by Jerome P. Horwitz. It failed as an anticancer drug (the purpose for which it was made), but in 1985 it was found to be a useful treatment for AIDS. AZT is taken up by T lymphocytes, immune system cells that are particularly vulnerable to HIV infection, and converted to AZT triphosphate. The triphosphate cannot be given directly because it cannot cross the plasma membrane. HIV's reverse transcriptase has a higher affinity for AZT triphosphate than for dTTP, and binding of AZT triphosphate to this enzyme competitively inhibits dTTP binding. When AZT is added to the 3′ end of the growing RNA strand, its lack of a 3′ hydroxyl means that the RNA strand is terminated prematurely and viral RNA synthesis quickly grinds to a halt.

AZT triphosphate is not as toxic to the T lymphocytes themselves, because *cellular* DNA polymerases have a lower affinity for this compound than for dTTP. At concentrations of 1 to 5 μM, AZT affects HIV reverse transcription but not most cellular DNA replication. Unfortunately, AZT appears to be toxic to the bone marrow cells that are the progenitors of erythrocytes, and patients taking AZT often develop anemia. AZT can increase the survival time of patients with advanced AIDS by about a year, and it delays the onset of AIDS in individuals who are still in the early stages of HIV infection. Some other AIDS drugs, such as dideoxyinosine, have a similar mechanism of action. Newer drugs target and inactivate the HIV protease. Because of the high error rate of HIV reverse transcriptase and the resulting rapid evolution of HIV, effective treatments of HIV infections use a combination of drugs directed at both the protease and the reverse transcriptase.

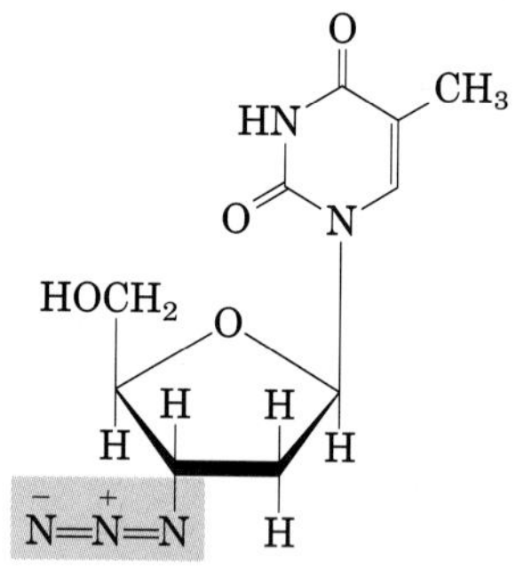

3′-Azido-2′,3′-dideoxythymidine (AZT)

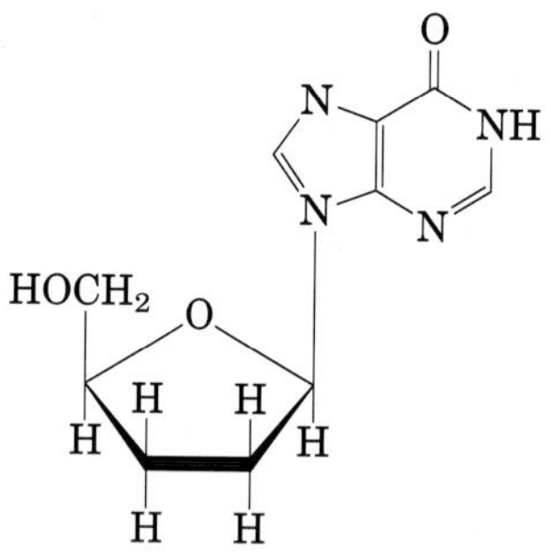

2′,3′-Dideoxyinosine (DDI)

Many Transposons, Retroviruses, and Introns May Have a Common Evolutionary Origin

Some well-characterized eukaryotic DNA transposons from sources as diverse as yeast and fruit flies have a structure very similar to that of retroviruses; these are sometimes called retrotransposons (Fig. 26–35). Retrotransposons encode an enzyme homologous to the retroviral reverse transcriptase, and their coding regions are flanked by LTR sequences. They transpose from one position to another in the cellular genome by means of an RNA intermediate, probably using reverse transcriptase to make a DNA copy of the RNA, followed by integration of the DNA at a new site. Most

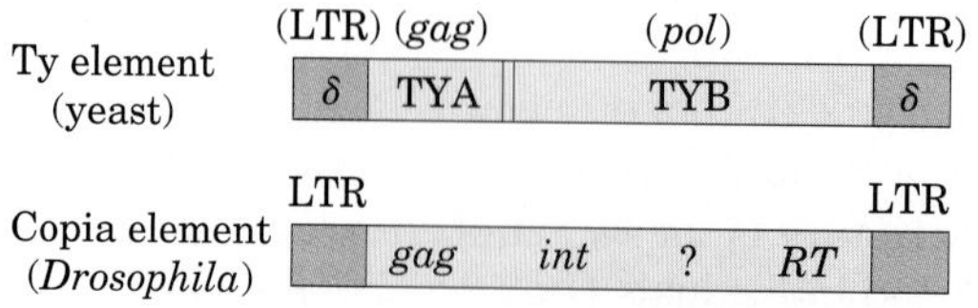

figure 26–35
Eukaryotic transposons. The Ty element of the yeast *Saccharomyces* and the copia element of the fruit fly *Drosophila* serve as examples of eukaryotic transposons, which often have a structure similar to retroviruses but lacking the *env* gene. The δ sequences of the Ty element are functionally equivalent to retroviral LTRs. In the copia element, *int* and *RT* are homologous to the integrase and reverse transcriptase segments, respectively, of the *pol* gene.

transposons in eukaryotes probably use this mechanism for transposition, distinguishing them from bacterial transposons, which move as DNA directly from one chromosomal location to another (see Fig. 25–41).

Retrotransposons lack an *env* gene and so cannot form viral particles. They can be thought of as defective viruses, trapped in cells. Comparisons between retroviruses and eukaryotic transposons suggest that reverse transcriptase is an ancient enzyme that predates the evolution of multicellular organisms.

Interestingly, many group I and group II introns are also mobile genetic elements. In addition to their self-splicing activities, they encode DNA endonucleases that promote their movement. During genetic exchanges between cells of the same species, or when DNA is introduced into a cell by parasites or other means, these endonucleases promote insertion of the intron into an identical site in another DNA copy of a homologous gene that does not contain the intron, in a process termed **homing** (Fig. 26–36). Whereas group I intron homing is DNA-based, group II intron homing occurs via an RNA intermediate. The endonucleases of the group II introns have associated reverse transcriptase activity. The proteins can form complexes with the intron RNAs themselves, after the introns are spliced from the primary transcripts. Because the homing process involves insertion of the RNA intron into DNA and reverse transcription of the intron, the movement of these introns has been called retrohoming. Over time, every copy of a particular gene in a population may acquire the intron. Much more rarely, the intron may insert itself into a new location in an unrelated gene. If this event does not kill the host cell, it can lead to the evolution and distribution of an intron in a new location. The structures and mechanisms used by mobile introns support the idea that at least some introns originated as molecular parasites whose evolutionary past can be traced to retroviruses and transposons.

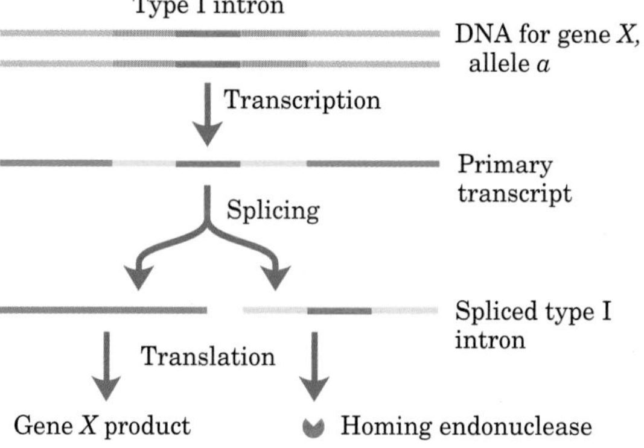

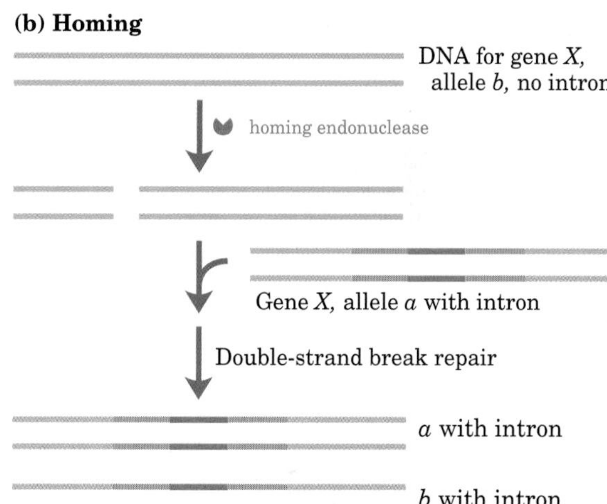

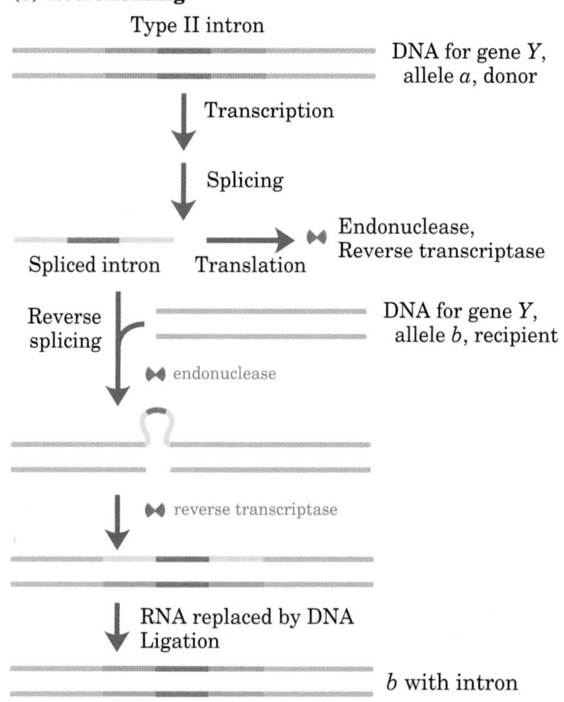

figure 26–36

Introns that move: homing and retrohoming. Certain introns include a gene (shown in red) for enzymes that promote homing (type I introns) or retrohoming (type II introns). **(a)** The gene within the spliced intron is bound by a ribosome and translated. Type I homing introns specify a site-specific endonuclease, called a homing endonuclease. Type II retrohoming introns specify a protein with both endonuclease and reverse transcriptase activities. **(b)** Homing. Allele *a* of a gene *X* containing a type I homing intron is present in a cell containing allele *b* of the same gene, which lacks the intron. The homing endonuclease produced by *a* cleaves *b* at the position corresponding to the intron in *a*, and double-strand break repair (recombination with allele *a*; see Fig. 25–29a) then creates a new copy of the intron in *b*. **(c)** Retrohoming. Allele *a* of gene *Y* contains a retrohoming type II intron; allele *b* lacks the intron. The spliced intron inserts itself into the coding strand of *b* in a reaction that is the reverse of the splicing that excised the intron from the primary transcript (see Fig. 26–15), except that here the insertion is into DNA rather than RNA. The noncoding DNA strand of *b* is then cleaved by the intron-encoded endonuclease/reverse transcriptase. This same enzyme uses the inserted RNA as a template to synthesize a complementary DNA strand. The RNA is then degraded by cellular ribonucleases and replaced with DNA.

Telomerase Is a Specialized Reverse Transcriptase

Telomeres, the structures at the ends of linear eukaryotic chromosomes (see Fig. 24–3), generally consist of many tandem copies of a short oligonucleotide sequence. This sequence usually has the form T_xG_y in one strand and C_yA_x in the complementary strand, where x and y are typically in the range of 1 to 4 (p. 910). Telomeres range in length from a few dozen base pairs in some ciliated protozoans to tens of thousands of base pairs in mammals. The TG strand is longer than its complement, leaving a region of single-stranded DNA of up to a few hundred nucleotides at the 3′ end.

The ends of a linear chromosome are not readily replicated by cellular DNA polymerases. DNA replication requires a template and primer, but beyond the end of a linear DNA molecule no template is available for the pairing of an RNA primer. Without a special mechanism for replicating the ends, chromosomes would be shortened somewhat in each cell generation. The enzyme **telomerase** solves this problem by adding telomeres to chromosome ends.

Although the existence of this enzyme may not be surprising, the mechanism by which it acts is remarkable and unprecedented. Telomerase, like some other enzymes described in this chapter, contains both RNA and protein components. The RNA component is about 150 nucleotides long and contains about 1.5 copies of the appropriate C_yA_x telomere repeat. This region of the RNA acts as a template for synthesis of the T_xG_y strand of the telomere. Telomerase thereby acts as a cellular reverse transcriptase that provides the active site for RNA-dependent DNA synthesis. Unlike retroviral reverse transcriptases, telomerase copies only a small segment of RNA that it carries within itself. Telomere synthesis requires the 3′ end of a chromosome as primer and proceeds in the usual 5′→3′ direction. Having synthesized one copy of the repeat, the enzyme repositions to resume extension of the telomere (Fig. 26–37).

The complementary C_yA_x strand is thought to be synthesized by cellular DNA polymerases starting with an RNA primer. The single-stranded region is protected by specific binding proteins in many lower eukaryotes, especially those species with telomeres of less than a few hundred base pairs. In higher eukaryotes (including mammals) with telomeres many thousands of base pairs long, the single-stranded end is sequestered in a specialized structure called a **T loop.** The single-stranded end is folded back and paired with its complement in the double-stranded portion of the telomere. The formation of a T loop involves invasion of the 3′ end of the telomere's single strand into the duplex DNA, perhaps by a mechanism similar to the initiation of homologous genetic recombination (Chapter 25). In mammals, the looped DNA is bound by two proteins called TRF1 and TRF2, with the latter protein involved in formation of the T loop. T loops protect the 3′ends of chromosomes, making them inaccessible to nucleases and the enzymes that repair double-strand breaks.

In protozoans (such as *Tetrahymena*), loss of telomerase activity results in a gradual shortening of telomeres with each cell division, ultimately leading to the death of the cell line. A similar link between telomere length and cell senescence (cessation of cell division) has been observed in humans. In germ-line cells, which contain telomerase activity, telomere lengths are maintained; in somatic cells, which lack telomerase, they are not. There is a linear, inverse relationship between the length of telomeres in cultured fibroblasts and the age of the individual from whom the fibroblasts were taken: telomeres in human somatic cells gradually shorten as an individual ages. If the telomerase reverse transcriptase is introduced into human somatic cells in vitro, telomerase activity is restored and the cellular life span increases markedly. Is the gradual shortening of telomeres a

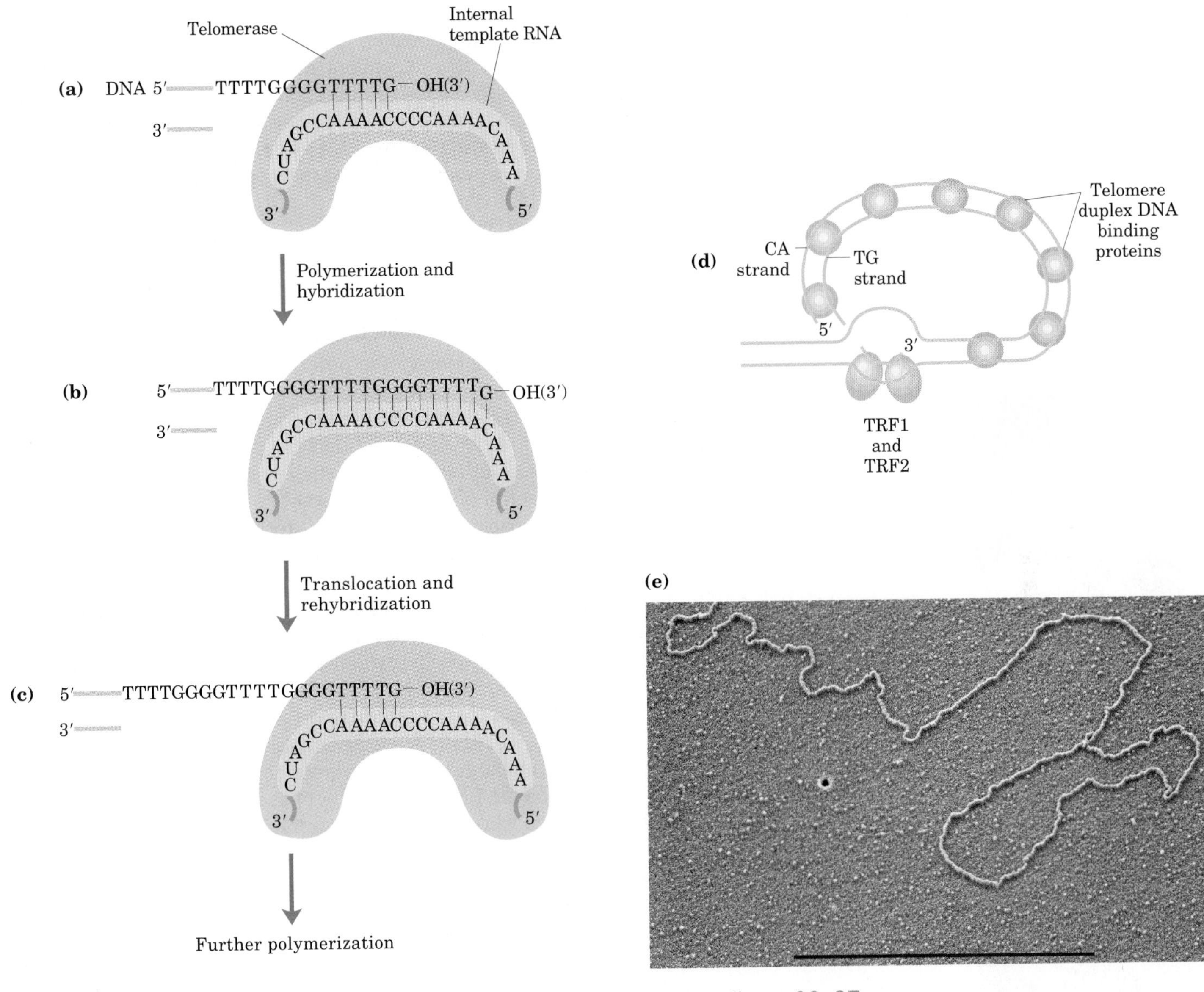

figure 26–37

The TG strand and T loop of telomeres. (a) The internal template RNA of telomerase binds to and base-pairs with the DNA's TG primer (T_xG_y). **(b)** Telomerase adds more T and G residues to the TG primer, then **(c)** repositions the internal template RNA to allow the addition of more T and G residues. **(d)** Proposed structure of T loops in telomeres. The single-stranded tail synthesized by telomerase is folded back and paired with its complement in the duplex portion of the telomere. **(e)** Electron micrograph of a T loop at the end of a chromosome isolated from a mouse hepatocyte. The bar at the bottom of the micrograph represents a length of 5,000 base pairs.

key to the aging process? Is our natural life span determined by the length of the telomeres we are born with? Further research in this area should yield some fascinating insights.

Some Viral RNAs Are Replicated by RNA-Dependent RNA Polymerase

Some *E. coli* bacteriophages, including f2, MS2, R17, and Qβ, have RNA genomes. The single-stranded RNA chromosomes of these viruses, which also function as mRNAs for the synthesis of viral proteins, are replicated in the host cell by an **RNA-dependent RNA polymerase (RNA replicase).** This enzyme (M_r ~210,000) has four subunits. One subunit (M_r 65,000), which has the active site for replication, is the product of the replicase gene encoded by the viral RNA. The other three subunits are host proteins normally involved in host-cell protein synthesis: the *E. coli* elongation factors Tu (M_r 30,000) and Ts (M_r 45,000) (which normally ferry amino acyl–tRNAs to the ribosomes) and the protein S1 (normally an integral part of the 30S ribosomal subunit). These three host proteins may help the RNA replicase locate and bind to the 3′ ends of the viral RNAs.

RNA replicase isolated from Qβ-infected *E. coli* cells catalyzes the formation of an RNA complementary to the viral RNA in a reaction equivalent to that catalyzed by DNA-dependent RNA polymerases. New RNA strand synthesis proceeds in the 5′→3′ direction by a chemical mechanism identical to that used in all other nucleic acid synthetic reactions that require a template. RNA replicase requires RNA as its template and will not function with DNA. It lacks a proofreading endonuclease and has an error rate similar to that of RNA polymerase. Unlike the DNA and RNA polymerases, RNA replicases are specific for the RNA of their own virus; the RNAs of the host cell are generally not replicated. This explains how RNA viruses are preferentially replicated in the host cell, which contains many other types of RNA.

RNA Synthesis Offers Important Clues to Biochemical Evolution

Carl Woese

Francis Crick

Leslie Orgel

The extraordinary complexity and order that distinguish living from inanimate systems are key manifestations of fundamental life processes. Maintaining the living state requires that *selected* chemical transformations occur very rapidly—especially those that use environmental energy sources and that synthesize elaborate or specialized cellular macromolecules. Life depends on powerful and selective catalysts—enzymes—and on informational systems capable of both securely storing the blueprint for these enzymes and accurately reproducing the blueprint for generation after generation. Chromosomes encode the blueprint not for the cell but rather for the enzymes that construct and maintain the cell. The parallel demands for information and catalysis present a classic conundrum: what came first, the information needed to specify structure or the enzymes needed to maintain and transmit the information?

The unveiling of the structural and functional complexity of RNA led Carl Woese, Francis Crick, and Leslie Orgel to propose in the 1960s that this macromolecule might serve as both information carrier and catalyst. The discovery of catalytic RNAs took this proposal from conjecture to hypothesis and has led to widespread speculation that an "RNA world" might have been important in the transition from prebiotic chemistry to life. The parent of all life, in the sense that it could reproduce itself across generations leading to the present, might have been a self-replicating RNA or a polymer with equivalent chemical characteristics.

How might a self-replicating polymer come to be? How might it maintain itself in an environment where the precursors for polymer synthesis are scarce? How could evolution progress from such a polymer to the modern DNA-protein world? These difficult questions can be addressed by careful experimentation, providing clues about how life on Earth began and evolved.

The probable origin of purine and pyrimidine bases can be found in experiments designed to test hypotheses about prebiotic chemistry (pp. 74–75). Beginning with simple molecules thought to be present in the early atmosphere (CH_4, NH_3, H_2O, H_2), electrical discharges such as lightning generate, first, more reactive molecules such as HCN and aldehydes, then an array of amino acids and organic acids (see Table 3–6). When molecules such as HCN become abundant, purine and pyrimidine bases are synthesized in detectable amounts. Remarkably, a concentrated solution of ammonium cyanide, refluxed for a few days, generates adenine in up to 0.5% yield (Fig. 26–38). Adenine may well have been the first and most abundant nucleotide constituent to appear on Earth. Intriguingly, most enzyme cofactors contain adenosine as part of their structure, although it plays no direct role in the cofactor function (see Fig. 10–41). This may suggest an evolutionary relationship, based on the simple synthesis of adenine from cyanide.

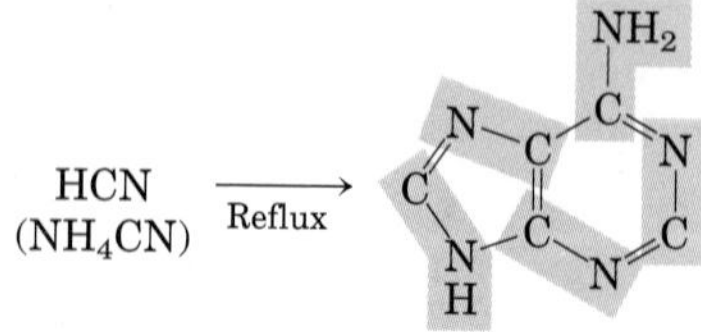

figure 26–38
Possible prebiotic synthesis of adenine from ammonium cyanide. Adenine is derived from five molecules of cyanide, denoted by shading.

The RNA world hypothesis requires a nucleotide polymer to reproduce itself. Can a ribozyme bring about its own synthesis in a template-directed

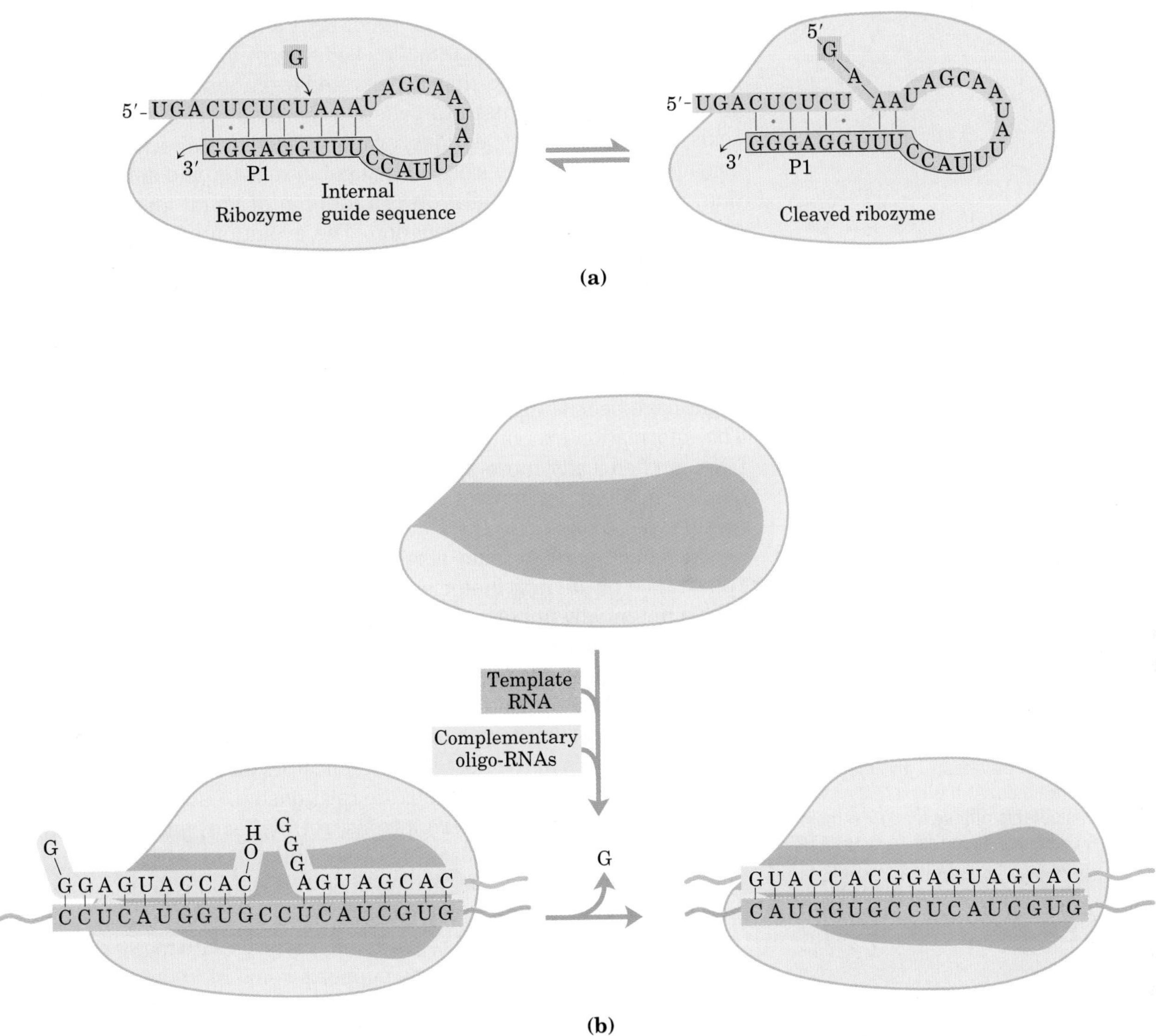

figure 26–39

RNA-dependent synthesis of an RNA polymer from oligonucleotide precursors. (a) The first step in the removal of the self-splicing group I intron of the rRNA precursor of *Tetrahymena* is reversible attack of a guanosine residue on the 5′ splice site. Only P1, the region of the ribozyme that includes the internal guide sequence and the 5′ splice site, is shown in detail; the rest of the ribozyme is represented as a green blob. The complete secondary structure of the ribozyme is shown in Figure 26–28. **(b)** If P1 is removed, the ribozyme retains both its three-dimensional shape and its catalytic capacity. A new RNA molecule added in vitro can bind to the ribozyme in the same manner as does the internal guide sequence of P1 in **(a).** This provides a template for further RNA polymerization reactions when oligonucleotides complementary to the added RNA base-pair with it. The ribozyme can link these oligonucleotides in a process equivalent to the *reversal* of the reaction in **(a).** Although only one such reaction is shown in **(b),** repeated binding and catalysis can result in the RNA-dependent synthesis of long RNA polymers.

manner? The self-splicing rRNA intron of *Tetrahymena* (Fig. 26–28) catalyzes the reversible attack of a guanosine residue on the 5′ splice junction (Fig. 26–39). If the 5′ splice site and the internal guide sequence are removed from the intron, the rest of the intron can bind RNA strands paired with short oligonucleotides. Part of the remaining intact intron effectively acts as a template for the alignment and ligation of the short oligonucleotides. The reaction is in essence a reversal of the attack of guanosine on the 5′ splice junction. However, the result is the synthesis of long RNA polymers from short ones, with the sequence of the product defined by an RNA template.

A self-replicating polymer would quickly use up available supplies of precursors provided by the relatively slow processes of prebiotic chemistry. Thus, from an early stage in evolution, metabolic pathways would be required to generate precursors efficiently. The ribozymes found in nature have a limited repertoire of catalytic functions, but we know that the catalytic potential of RNA is far greater. The search for RNAs with new catalytic functions has been aided by the development of a method that rapidly searches pools of random polymers of RNA and extracts those with particular activities. SELEX (*s*ystematic *e*volution of *l*igands by *ex*ponential enrichment) is nothing less than accelerated evolution in a test tube (Fig. 26–40). SELEX has been used to generate RNA molecules that bind to amino acids, organic dyes, nucleotides, cyanocobalamin, and other molecules. Ribozymes have also been isolated that catalyze ester and amide bond formation, SN2 reactions, metallation of (addition of metal ions to) porphyrins, and carbon–carbon bond formation. The evolution of enzymatic cofactors with nucleotide "handles" that facilitate their binding to ribozymes might have further expanded the repertoire of chemical processes available to primitive metabolic systems.

As we'll see in the next chapter, some natural RNA molecules may catalyze the formation of peptide bonds, offering an idea of how the RNA world might have been transformed by the greater catalytic potential of proteins. The information-carrying role of RNA may have passed to DNA because DNA is chemically more stable. RNA replicase and reverse transcriptase may be modern versions of enzymes that once played important roles in making the transition to the modern DNA-based system.

Molecular parasites may also have originated in an RNA world. With the appearance of the first inefficient self-replicators, transposition could have been a potentially important alternative to replication as a strategy for successful reproduction and survival. Early parasitic RNAs would simply hop into a self-replicating molecule via catalyzed transesterification, and then be passively replicated. Natural selection would have driven transposition to become site-specific, targeting sequences that did not interfere with the catalytic activities of the host RNA. Replicators and RNA transposons could have existed in a primitive symbiotic relationship, each contributing to the evolution of the other. Modern introns, retroviruses, and transposons may all be vestiges of a "piggy-back" strategy pursued by early parasitic RNAs. These elements continue to make major contributions to the evolution of their hosts.

Although the RNA world remains a hypothesis with many gaps yet to be explained, experimental evidence supports some of its key elements. Further experimentation should increase our understanding. Important clues to the puzzle will be found in the workings of fundamental chemistry, within living cells, and perhaps on other planets.

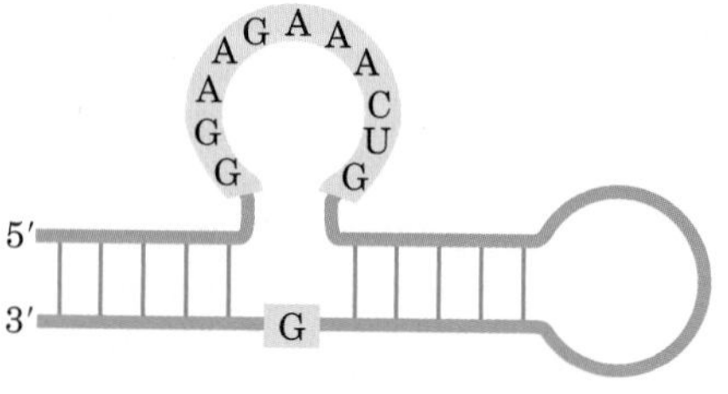

figure 26–40

The SELEX method for generating RNA polymers with new functions. (a) In this example, SELEX is used to select an RNA species that binds tightly to ATP. ① A random mixture of RNA polymers is subjected to "unnatural selection" by passing it through a resin to which ATP is attached. The practical limit for the complexity of an RNA mixture in SELEX is about 10^{15} different sequences, which allows for the complete randomization of 25 nucleotides ($4^{25} = 10^{15}$). RNA polymers that pass through the column are discarded ②; those that bind to ATP are washed from the column with salt solution and collected ③. The collected polymers are amplified ④, often by using reverse transcriptase to make many DNA complements to the selected RNAs, then an RNA polymerase to make many RNA complements of the resulting DNA molecules. ⑤ This new pool of RNA is subjected to the same selection procedure, and the cycle is repeated a dozen or more times. At the end, only a few RNA sequences with considerable affinity for ATP remain. **(b)** Structure of an RNA oligonucleotide that binds to ATP. Nucleotides required for the binding activity are shown. Molecules with this general structure bind to ATP with $K_d < 50$ μM.

summary

Transcription is catalyzed by DNA-dependent RNA polymerase, a complex enzyme that uses ribonucleoside 5′-triphosphates to synthesize RNA complementary to the template strand of duplex DNA. Transcription occurs in several phases, including the binding of RNA polymerase to a DNA site called a promoter, initiation of transcript synthesis, elongation, and termination. Bacterial RNA polymerase requires a special subunit to recognize the promoter. As the first committed step in transcription, binding of RNA polymerase to the promoter and the initiation of transcription are closely regulated. Eukaryotic cells have three different types of RNA polymerases. Binding of eukaryotic RNA polymerase II to its promoters requires the activity of an array of proteins called transcription factors. Transcription stops at sequences called terminators.

Ribosomal RNAs and transfer RNAs are derived from longer precursor RNAs that are trimmed by nucleases. Some bases are also modified enzymatically during the maturation process. Eukaryotic mRNAs are also made from longer precursors. Primary RNA transcripts often contain noncoding regions called introns, which are removed by splicing. Excision of the group I introns found in some rRNAs requires a guanosine cofactor. Some group I and group II introns are capable of self-splicing; no protein enzymes are required. Nuclear mRNA precursors have a third class of introns that are spliced with the aid of RNA-protein complexes called snRNPs, assembled into elaborate structures called spliceosomes. Messenger RNAs are also modified by addition of a 7-methylguanosine residue at the 5′ end, followed by cleavage and polyadenylation at the 3′ end to form a long poly(A) tail. A fourth class of introns, found in some tRNAs, is the only class known to be spliced by protein enzymes.

The self-splicing introns and the RNA component of RNase P (the enzyme that cleaves the 5′ end of tRNA precursors) belong to a class of biological catalysts called ribozymes. These have the properties of true enzymes. They generally promote hydrolytic cleavage and transesterification, using RNA as substrate. Combinations of these reactions can be promoted by the excised group I rRNA intron of *Tetrahymena,* resulting in a type of RNA polymerization reaction.

Polynucleotide phosphorylase can reversibly form RNA-like polymers from ribonucleoside 5′-diphosphates, adding or removing ribonucleotides at the 3′-hydroxyl end of the polymer. It degrades RNA in vivo.

RNA-dependent DNA polymerases, also called reverse transcriptases, are produced in animal cells infected by RNA viruses called retroviruses. These enzymes transcribe the viral RNA into DNA, a process that can be used experimentally to form complementary DNA. Many eukaryotic transposons are related to retroviruses, and their mechanism of transposition includes an RNA intermediate. Telomerase, the enzyme that synthesizes the telomere ends of linear chromosomes, is a specialized reverse transcriptase that contains an internal RNA template.

RNA-dependent RNA polymerases, or replicases, are found in bacterial cells infected with certain RNA viruses. They are template-specific for the viral RNA.

The existence of catalytic RNAs and pathways for the interconversion of RNA and DNA has led to speculation that an important stage in evolution was the appearance of an RNA (or an equivalent polymer) that could catalyze its own replication. The biochemical potential of RNAs can be explored by SELEX, a method for rapidly selecting RNA sequences with particular binding or catalytic properties.

further reading

General

Jacob, F. & Monod, J. (1961) Genetic regulatory mechanisms in the synthesis of proteins. *J. Mol. Biol.* **3,** 318–356.

A classic article that introduced many important ideas.

Lodish, H., Berk, A., Zipursky, S.L., Matsudaira, P., Baltimore, D., & Darnell, J. (1999) *Molecular Cell Biology,* 4th edn, W.H. Freeman and Company, New York.

DNA-Directed RNA Synthesis

Conaway, R.C. & Conaway, J.W. (1997) General transcription factors for RNA polymerase II. *Prog. Nucl. Acid Res. Mol. Biol.* **56,** 327–346.

Particularly good summary of what is known about elongation factors.

Conaway, J.W. & Conaway, R.C. (1999) Transcription elongation and human disease. *Annu. Rev. Biochem.* **68,** 301–320.

DeHaseth, P.L., Zupancic, M.L., & Record, M.T. Jr. (1998) RNA polymerase-promoter interactions: the comings and goings of RNA polymerase. *J. Bacteriol.* **180,** 3019–3025.

Estrem, S.T., Gaal, T., Ross, W., & Gourse, R.L. (1998) Identification of an UP element consensus sequence for bacterial promoters. *Proc. Natl. Acad. Sci. USA* **95,** 9761–9766.

Friedberg, E.C. (1996) Relationships between DNA repair and transcription. *Annu. Rev. Biochem.* **65,** 15–42.

Gross, C.A., Chan, C.L., & Lonetto, M.A. (1996) A structure/function analysis of *Escherichia coli* RNA polymerase. *Philos. Trans. R. Soc. Lond. Biol. Sci.* **351,** 475–482.

Kornberg, R.D. (1996) RNA polymerase II transcription control. *Trends Biochem. Sci.* **21,** 325–327.

Introduces an issue of *Trends in Biochemical Sciences* that is devoted to RNA polymerase II.

Mooney, R.A., Artsimovitch, I., & Landick, R. (1998) Informational processing by RNA polymerase: recognition of regulatory signals during RNA chain elongation. *J. Bacteriol.* **180,** 3265–3275.

RNA Processing

Beelman, C.A. & Parker, R. (1995) Degradation of mRNA in eukaryotes. *Cell* **81,** 179–183.

Colgan, D.F. & Manley, J.L. (1997) Mechanism and regulation of mRNA polyadenylation. *Genes Dev.* **11,** 2755–2766.

Curcio, M.J. & Belfort, M. (1996) Retrohoming: cDNA-mediated mobility of group II introns requires a catalytic RNA. *Cell* **84,** 9–12.

Frank, D.N. & Pace, N.R. (1998) Ribonuclease P: unity and diversity in a tRNA-processing ribozyme. *Annu. Rev. Biochem.* **67,** 153–180.

Michel, F. & Ferat, J.-L. (1995) Structure and activities of group II introns. *Annu. Rev. Biochem.* **64,** 435–461.

Murray, H.L. & Jarrell, K.A. (1999) Flipping the switch to an active spliceosome. *Cell* **96,** 599–602.

Narlikar, G.J. & Herschlag, D. (1997) Mechanistic aspects of enzymatic catalysis: lessons from comparison of RNA and protein enzymes. *Annu. Rev. Biochem.* **66,** 19–59.

Sarkar, N. (1997) Polyadenylation of mRNA in prokaryotes. *Annu. Rev. Biochem.* **66,** 173–197.

Scott, W.G. & Klug, A. (1996) Ribozymes: structure and mechanism in RNA catalysis. *Trends Biochem. Sci.* **21,** 220–224.

Spinelli, S.L., Kierzek, R., Turner, D.H., & Phizicky, E.M. (1999) Transient ADP-ribosylation of a 2′-phosphate implicated in its removal from ligated tRNA during splicing in yeast. *J. Biol. Chem.* **274,** 2637–2644.

Wondering where the phosphoryl group went in the last step in Figure 26–17? This paper will tell you.

Staley, J.P. & Guthrie, C. (1988) Mechanical devices of the spliceosome—motors, clocks, springs, and things. *Cell* **92,** 315–326.

RNA-Directed RNA or DNA Synthesis

Bishop, J.M. (1991) Molecular themes in oncogenesis. *Cell* **64,** 235–248.

A good overview of oncogenes; it introduces a series of more detailed reviews included in the same issue of *Cell.*

Blackburn, E.H. (1992) Telomerases. *Annu. Rev. Biochem.* **61,** 113–129.

Boeke, J.D. & Devine, S.E. (1998) Yeast retrotransposons: finding a nice, quiet neighborhood. *Cell* **93,** 1087–1089.

Collins, K. (1999) Ciliate telomerase biochemistry. *Annu. Rev. Biochem.* **68,** 187–218.

Frankel, A.D. & Young, J.A.T. (1998) HIV-1: fifteen proteins and an RNA. *Annu. Rev. Biochem.* **67,** 1–25.

Greider, C.W. (1996) Telomere length regulation. *Annu. Rev. Biochem.* **65,** 337–365.

Griffith, J.D., Comeau, L., Rosenfield, S., Stansel, R.M., Bianchi, A., Moss, H. & de Lange, T. (1999) Mammalian telomeres end in a large duplex loop. *Cell* **97,** 503–514.

Lingner, J. & Cech, T.R. (1998) Telomerase and chromosome end maintenance. *Curr. Opin. Genet. Dev.* **8,** 226–232.

Temin, H.M. (1976) The DNA provirus hypothesis: the establishment and implications of RNA-directed DNA synthesis. *Science* **192,** 1075–1080.

A discussion of the original proposal for reverse transcription in retroviruses.

Zakian, V.A. (1995) Telomeres: beginning to understand the end. *Science* **270,** 1601–1607.

Ribozymes and Evolution

Gold, L., Brown, D., He, Y.-Y., Shtatland, R., Singer, B.S., & Wu, Y. (1997) From oligonucleotide shapes to genomic SELEX: novel biological regulatory loops. *Proc. Natl. Acad. Sci. USA* **94,** 59–64.

Pace, N.R. (1991) Origin of life—facing up to the physical setting. *Cell* **65,** 531–533.

Wilson, D.S. & Szostak, J.W. (1999) In vitro selection of functional nucleic acids. *Annu. Rev. Biochem.* **68,** 611–648.

problems

1. RNA Polymerase (a) How long would it take for the *E. coli* RNA polymerase to synthesize the primary transcript for the *E. coli* genes encoding the enzymes for lactose metabolism (the 5,300 base pair *lac* operon, considered in Chapter 28)? (b) How far along the DNA would the transcription "bubble" formed by RNA polymerase move in 10 s?

2. Error Correction by RNA Polymerases DNA polymerases are capable of editing and error correction, but RNA polymerases do not appear to have this capacity. Given that a single base error in either replication or transcription can lead to an error in protein synthesis, suggest a possible biological explanation for this striking difference.

3. RNA Posttranscriptional Processing Predict the likely effects of a mutation in the sequence (5′)AAUAAA in a eukaryotic mRNA transcript.

4. Coding vs. Template Strands The RNA genome of phage Qβ is the nontemplate or coding strand, and when introduced into the cell it functions as an mRNA. Suppose the RNA replicase of phage Qβ synthesized primarily template-strand RNA and uniquely incorporated this, rather than nontemplate strands, into the viral particles. What would be the fate of the template strands when they entered a new cell? What enzyme would such a template-strand virus need to include in the viral particles for successful invasion of a host cell?

5. The Chemistry of Nucleic Acid Biosynthesis Describe three properties common to the reactions catalyzed by DNA polymerase, RNA polymerase, reverse transcriptase, and RNA replicase. How is the enzyme polynucleotide phosphorylase similar to and different from these three enzymes?

6. RNA Splicing What is the minimum number of transesterification reactions needed to splice an intron from an mRNA transcript? Why?

7. RNA Genomes The RNA viruses have relatively small genomes. For example, the single-stranded RNAs of retroviruses have about 10,000 nucleotides and the Qβ RNA is only 4,220 nucleotides long. Given the properties of reverse transcriptase and RNA replicase described in this chapter, can you suggest a reason for the small size of these viral genomes?

8. Screening RNAs by SELEX The practical limit for the number of different RNA sequences that can be screened in a SELEX experiment is 10^{15}. (a) Suppose you are working with oligonucleotides 32 nucleotides in length. How many sequences exist in a randomized pool containing every sequence possible? (b) What percentage of these can be screened in a SELEX experiment? (c) Suppose you wish to select an RNA molecule that catalyzes the hydrolysis of a particular ester. From what you know about catalysis (Chapter 8), propose a SELEX strategy that might allow you to select the appropriate catalyst.

9. Slow Death The death cap mushroom, *Amanita phalloides,* contains several dangerous substances, including the lethal α-amanitin. This toxin blocks RNA elongation in consumers of the mushroom by binding to eukaryotic RNA polymerase II with very high affinity; it is deadly in concentrations as low as 10^{-8} M. The initial reaction to ingestion of the mushroom is gastrointestinal distress (caused by some of the other toxins). These symptoms disappear, but about 48 hours later, the mushroom-eater dies, usually from liver dysfunction. Speculate on why it takes this long for α-amanitin to kill.

10. Detection of Rifampicin-Resistant Strains of TB Rifampicin is an important antibiotic used to treat tuberculosis (TB), as well as other mycobacterial diseases. Some strains of *Mycobacterium tuberculosis,* the causative agent of TB, are resistant to rifampicin. These strains become resistant through mutations that alter the *rpo*B gene, which encodes the β subunit of the RNA polymerase. Rifampicin cannot bind to the mutant RNA polymerase and so is unable to block the initiation of transcription. DNA sequences from a large number of rifampicin-resistant *M. tuberculosis* strains have been found to have mutations in a specific 69 base pair region of *rpo*B. One well-characterized strain with rifampicin resistance has a single base-pair alteration in *rpo*B that results in a single amino acid substitution in the β subunit: a His residue is replaced by an Asp residue.

(a) Based on your knowledge of protein chemistry (Chapters 5 and 6), suggest a technique that would allow detection of the rifampicin-resistant strain containing this particular mutant protein.

(b) Based on your knowledge of nucleic acid chemistry (Chapter 10), suggest a technique to identify the mutant form of *rpo*B.

Biochemistry on the Internet

11. The Ribonuclease Gene Human pancreatic ribonuclease has 128 amino acid residues.

(a) What is the minimum number of nucleotide pairs required to code for this protein?

(b) The mRNA expressed in human pancreas cells was copied with reverse transcriptase to create a "library" of human DNA. The sequence of the mRNA coding for human pancreatic ribonuclease was determined by sequencing the "complementary DNA" or cDNA from this library that included an open reading frame for the protein. Use the Entrez database system to find the published sequence of this mRNA (accession number D26129). See the *Principles of Biochemistry* Web site for the current URL and more detailed instructions. What is the length of this mRNA?

(c) How can you account for the discrepancy between the size you calculated in (a) and the actual length of the mRNA?

chapter

27 Protein Metabolism

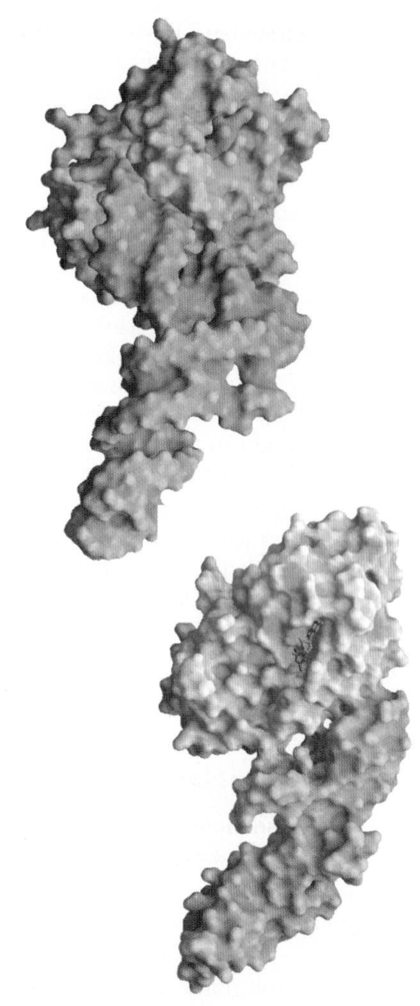

Protein synthesis is an elaborate interplay of protein factors and RNA molecules and, in some cases, proteins that look like RNAs. Two proteins that bind to ribosomes at particular steps in protein synthesis are shown. Elongation factor EF-Tu complexed with tRNA (green) is shown in the top image. Another elongation factor, EF-G, complexed with GDP (red) is shown in the bottom image. Interestingly, the carboxyl-terminal part of EF-G (green) mimics the structure of tRNA in both shape and charge distribution. The structural mimicry of tRNA is just one facet of the function of the molecular machines that synthesize polypeptides in all cells.

Proteins are the end products of most information pathways. A typical cell requires thousands of different proteins at any given moment. These must be synthesized in response to the cell's current needs, transported (targeted) to their appropriate cellular locations, and degraded when no longer needed.

Understanding protein synthesis, the most complex biosynthetic process, has been one of the greatest challenges in biochemistry. Eukaryotic protein synthesis involves over 70 different ribosomal proteins; 20 or more enzymes to activate the amino acid precursors; a dozen or more auxiliary enzymes and other protein factors for the initiation, elongation, and termination of polypeptides; perhaps 100 additional enzymes for the final processing of different proteins; and 40 or more kinds of transfer and ribosomal RNAs. Overall, almost 300 different macromolecules cooperate to synthesize polypeptides. Many of these macromolecules are organized into the complex three-dimensional structure of the ribosome itself.

To appreciate the central importance of protein synthesis, consider the cellular resources devoted to this process. Protein synthesis can account for up to 90% of the chemical energy used by a cell for all biosynthetic reactions. Every prokaryotic and eukaryotic cell contains thousands of copies of each of its proteins and RNAs. The 20,000 ribosomes, 100,000 related protein factors and enzymes, and 200,000 tRNAs in a typical bacterial cell can account for more than 35% of the cell's dry weight.

Despite this great complexity, proteins are made at exceedingly high rates. A polypeptide of 100 residues is synthesized in an *Escherichia coli* cell (at 37 °C) in about 5 s. Synthesis of the thousands of different proteins in a cell is tightly regulated, so that just enough is made to match the metabolic circumstances of the cell. To maintain the appropriate mix and concentration of proteins, the targeting and degradative processes must keep pace with synthesis. Research is gradually uncovering the finely coordinated cellular choreography that guides each protein to its proper cellular location and selectively degrades it when it is no longer required.

The Genetic Code

Three major advances set the stage for our present knowledge of protein biosynthesis. In the early 1950s Paul Zamecnik and his colleagues designed a set of experiments to investigate where in the cell proteins are synthesized. They injected radioactive amino acids into rats and, at different time intervals after the injection, the liver was removed, homogenized, and fractionated by centrifugation. The subcellular fractions were then examined for the presence of radioactive protein. When hours or days were allowed

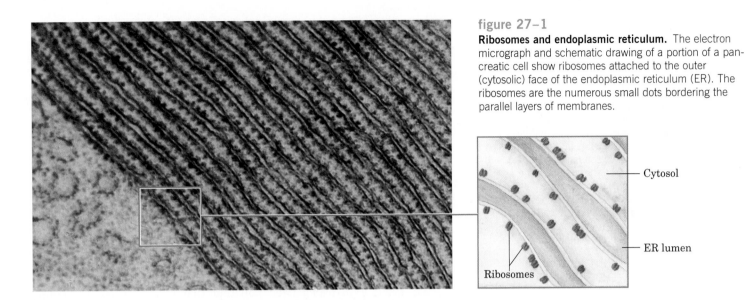

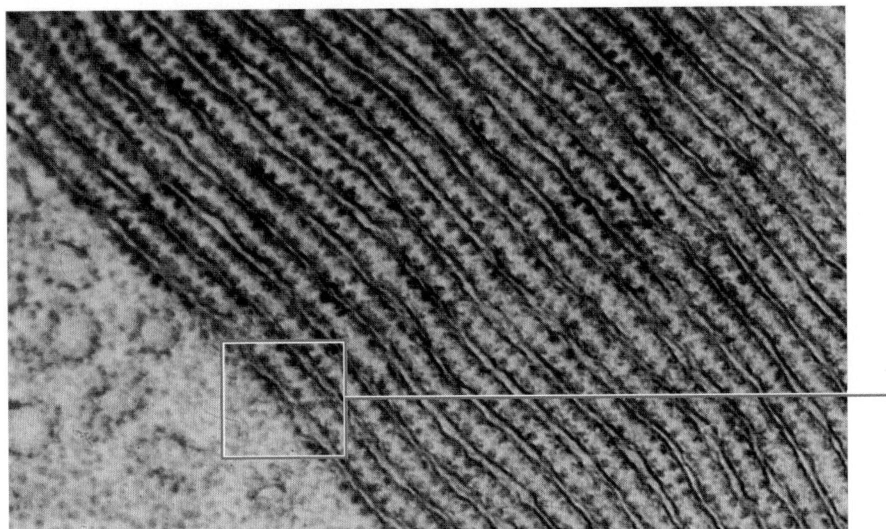

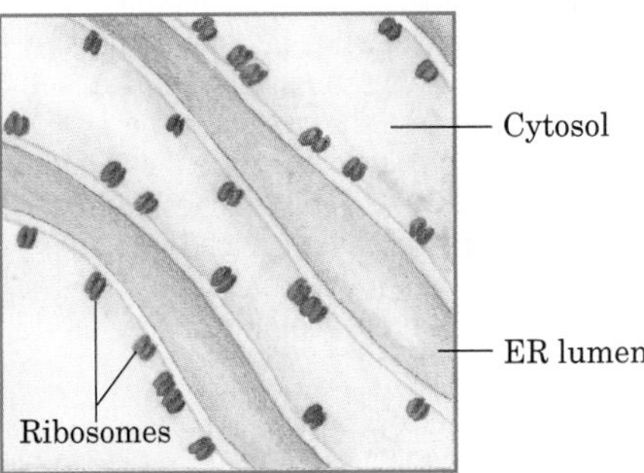

figure 27–1

Ribosomes and endoplasmic reticulum. The electron micrograph and schematic drawing of a portion of a pancreatic cell show ribosomes attached to the outer (cytosolic) face of the endoplasmic reticulum (ER). The ribosomes are the numerous small dots bordering the parallel layers of membranes.

to elapse after injection of the labeled amino acids, *all* the subcellular fractions contained labeled proteins. However, when only minutes had elapsed, labeled protein was found only in a fraction containing small ribonucleoprotein particles. These particles, visible in animal tissues by electron microscopy, were therefore identified as the site of protein synthesis from amino acids, and later were named ribosomes (Fig. 27–1).

The second key advance was made by Mahlon Hoagland and Zamecnik when they found that amino acids were "activated" when incubated with ATP and the cytosolic fraction of liver cells. The amino acids became attached to a heat-stable soluble RNA, later called transfer RNA (tRNA), forming **aminoacyl-tRNAs.** The enzymes that catalyze this process are the **aminoacyl-tRNA synthetases.**

Paul Zamecnik

The third major advance occurred when Francis Crick considered how the genetic information encoded in the four-letter language of nucleic acids could be translated into the 20-letter language of proteins. Crick reasoned that a small nucleic acid (perhaps RNA) could serve the role of an adaptor, one part of the adaptor molecule binding a specific amino acid and another part recognizing the nucleotide sequence encoding that amino acid in an mRNA (Fig. 27–2). This idea was soon verified. The tRNA adaptor "translates" the nucleotide sequence of an mRNA into the amino acid sequence of a polypeptide. The overall process of mRNA-guided protein synthesis is often referred to simply as **translation.**

These three developments soon led to recognition of the major stages of protein synthesis and ultimately to the elucidation of the genetic code that specifies each amino acid.

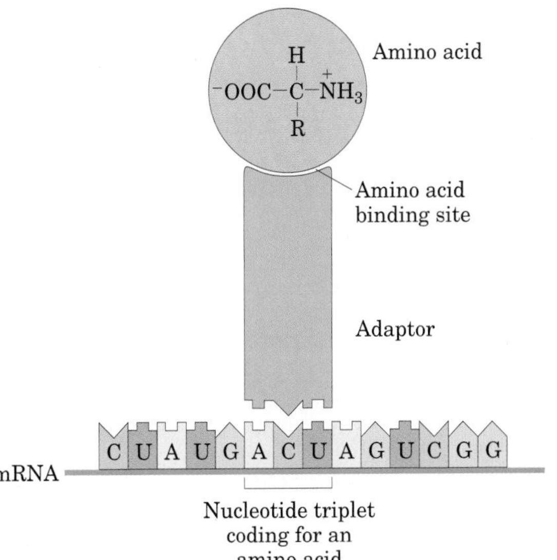

figure 27–2

Crick's adaptor hypothesis. Today we know that the amino acid is covalently bound at the 3′ end of a tRNA molecule and that a specific nucleotide triplet elsewhere in the tRNA interacts with a particular triplet codon in mRNA through hydrogen bonding of complementary bases.

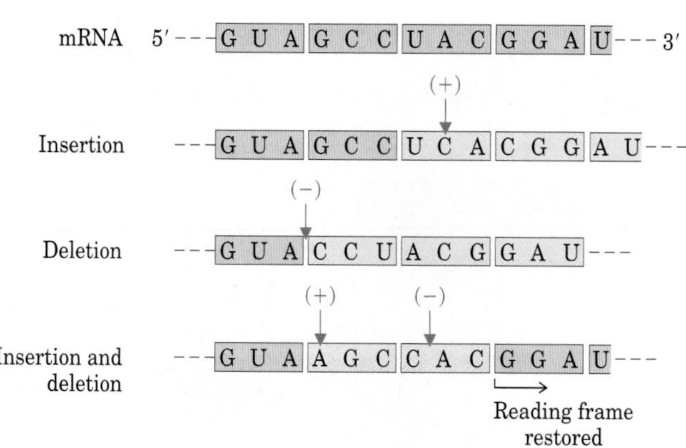

figure 27–3
The triplet, nonoverlapping code. Evidence for the general nature of the genetic code came from many types of experiments, including genetic experiments on the effects of deletion and insertion mutations. Inserting or deleting one base pair (shown here in the mRNA transcript) alters the sequence of triplets in a nonoverlapping code; all amino acids coded by the mRNA following the change are affected. Combining insertion and deletion mutations affects some amino acids but can eventually restore the correct amino acid sequence. Adding or subtracting three nucleotides (not shown) leaves the remaining triplets intact, providing evidence that a codon has three, rather than four or five, nucleotides. The triplet codons shaded in gray are those transcribed from the original gene; codons shaded in blue are new codons resulting from the insertion or deletion mutations.

The Genetic Code Was Cracked Using Artificial mRNA Templates

By the 1960s it had long been apparent that at least three nucleotide residues of DNA are required to encode each amino acid. The four code letters of DNA (A, T, G, and C) in groups of two can yield only $4^2 = 16$ different combinations, insufficient to encode 20 amino acids. Groups of three, however, yield $4^3 = 64$ different combinations.

Several key properties of the genetic code were established in early genetic studies (Figs 27–3, 27–4). A **codon** is a triplet of nucleotides that codes for a specific amino acid. Translation occurs in such a way that these nucleotide triplets are read in a successive, nonoverlapping fashion. A specific first codon in the sequence establishes the **reading frame,** in which a new codon begins every three nucleotide residues. There is no punctuation between codons for successive amino acid residues. The amino acid sequence of a protein is defined by a linear sequence of contiguous triplet codons. In principle, any given single-stranded DNA or mRNA sequence has three possible reading frames. Each reading frame gives a different sequence of codons (Fig. 27–5), but only one is likely to encode a given protein. A key question remained. What were the three-letter code words for each amino acid?

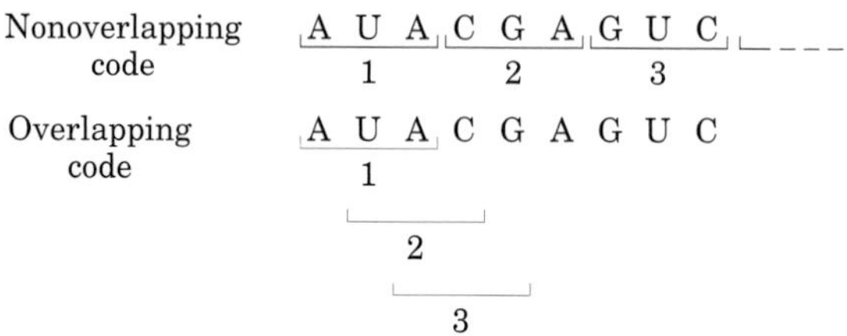

figure 27–4
Overlapping versus nonoverlapping genetic codes. In a nonoverlapping code, codons do not share nucleotides. In the examples shown, the consecutive codons are numbered. In an overlapping code, some nucleotides in the mRNA are shared by different codons. In a triplet code with maximum overlap, many nucleotides, such as the third nucleotide from the left (A), are shared by three different codons. Note that in an overlapping code, the sequence of the first codon limits the possible sequences for the second codon. A nonoverlapping code provides much more flexibility in the sequence of neighboring codons and therefore in the possible amino acid sequences designated by the code. The genetic code used in all living systems is now known to be nonoverlapping.

In 1961 Marshall Nirenberg and Heinrich Matthaei reported the first breakthrough. They incubated synthetic polyuridylate, poly(U), with an *E. coli* extract, GTP, ATP, and a mixture of the 20 amino acids in 20 different tubes, each tube containing a different radioactively labeled amino acid. Because poly(U) mRNA is made up of many successive UUU triplets, it should promote the synthesis of a polypeptide containing only the amino acid encoded by the triplet UUU. A radioactive polypeptide was indeed formed in only one of the 20 tubes, the one containing radioactive phenylalanine. Nirenberg and Matthaei therefore concluded that the triplet codon UUU encodes phenylalanine. The same approach revealed that polycytidylate,

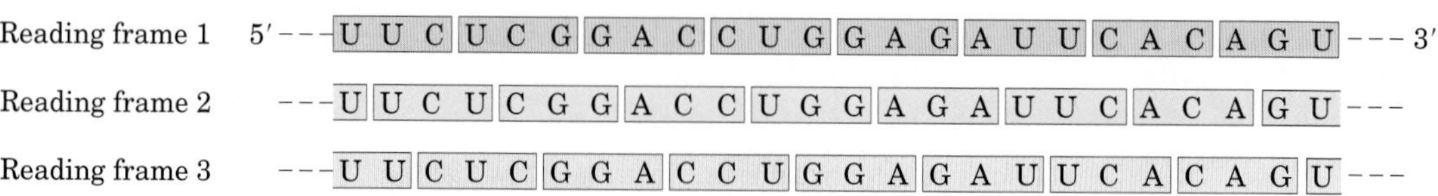

figure 27–5
In a triplet, nonoverlapping code, all mRNAs have three potential reading frames, shaded here in different colors. Note that the triplets, and hence the amino acids specified, are different in each reading frame.

poly(C), encodes a polypeptide containing only proline (polyproline), and polyadenylate, poly(A), encodes polylysine. Polyguanylate did not work in this experiment because it spontaneously forms tetraplexes (see Fig. 10–22) that cannot be bound by ribosomes.

The synthetic polynucleotides used in such experiments were made using polynucleotide phosphorylase (p. 1006), which catalyzes the formation of RNA polymers starting from ADP, UDP, CDP, and GDP. This enzyme requires no template and makes polymers with a base composition that directly reflects the relative concentrations of the nucleoside 5′-diphosphate precursors in the medium. If polynucleotide phosphorylase is presented with UDP only, it makes only poly(U). If it is presented with a mixture of five parts ADP and one part CDP, it makes a polymer in which about five-sixths of the residues are adenylate and one-sixth are cytidylate. This random polymer is likely to have many triplets of the sequence AAA, lesser numbers of AAC, ACA, and CAA triplets, relatively few ACC, CCA, and CAC triplets, and very few CCC triplets (Table 27–1). Using a variety of artificial mRNAs made by polynucleotide phosphorylase from different starting mixtures of ADP, GDP, UDP, and CDP, investigators soon identified the base compositions of the triplets coding for almost all the amino acids. Although these experiments revealed the base composition of the coding triplets, they could not reveal the sequence of the bases.

Marshall Nirenberg

In 1964 Nirenberg and Philip Leder achieved another experimental breakthrough. Isolated *E. coli* ribosomes would bind a specific aminoacyl-tRNA in the presence of the corresponding synthetic polynucleotide messenger. (By convention, the identity of a tRNA is indicated by a superscript and an aminoacylated tRNA is indicated by a hyphenated name: correctly aminoacylated $tRNA^{Ala}$ is alanyl-$tRNA^{Ala}$ or Ala-$tRNA^{Ala}$.) For example, ribosomes incubated with poly(U) and phenylalanyl-$tRNA^{Phe}$ (Phe-$tRNA^{Phe}$)

table 27–1

Incorporation of Amino Acids into Polypeptides in Response to Random Polymers of RNA*

Amino acid	Observed frequency of incorporation (Lys = 100)	Tentative assignment for nucleotide composition† of corresponding codon	Expected frequency of incorporation based on assignment (Lys = 100)
Asparagine	24	A_2C	20
Glutamine	24	A_2C	20
Histidine	6	AC_2	4
Lysine	100	AAA	100
Proline	7	AC_2, CCC	4.8
Threonine	26	A_2C, AC_2	24

*Presented here is a summary of data from one of the early experiments designed to elucidate the genetic code. A synthetic RNA containing only A and C residues in a 5:1 ratio directed polypeptide synthesis. Both the identity and the quantity of amino acids incorporated were determined. Based on the relative abundance of A and C residues in the synthetic RNA, and assigning the codon AAA (the most likely codon) a frequency of 100, there should be three different codons of composition A_2C, each at a relative frequency of 20; three codons of composition AC_2, each at a relative frequency of 4.0; and codon CCC at a relative frequency of 0.8. The CCC assignment was based on information derived from prior studies with poly(C). Where two tentative codon assignments are made, both are proposed to code for the same amino acid.

†Note that these designations of nucleotide composition contain no information on nucleotide sequence (except, of course, AAA and CCC).

bind both RNAs, but if the ribosomes are incubated with poly(U) and some other aminoacyl-tRNA, the aminoacyl-tRNA is not bound because it does not recognize the UUU triplets in poly(U) (Table 27–2). Even trinucleotides could promote specific binding of appropriate tRNAs, allowing the use of chemically synthesized oligonucleotides. Researchers could then determine which aminoacyl-tRNA bound to about 50 of the 64 possible triplet codons. For some codons, either no aminoacyl-tRNAs or more than one would bind. Another method was needed to complete and confirm the entire genetic code.

table 27–2

Trinucleotides Induce Specific Binding of Aminoacyl-tRNAs to Ribosomes

	^{14}C-Labeled aminoacyl-tRNA bound to ribosome*		
Trinucleotide	**Phe-tRNA^{Phe}**	**Lys-tRNA^{Lys}**	**Pro-tRNA^{Pro}**
UUU	4.6	0	0
AAA	0	7.7	0
CCC	0	0	3.1

Source: Modified from Nirenberg, M. & Leder, P. (1964) RNA code words and protein synthesis. *Science* **145,** 1399.

*Each number represents the factor by which the amount of bound ^{14}C increased when the indicated trinucleotide was present, relative to a control in which no trinucleotide was added.

H. Gobind Khorana

At about this time, a complementary approach was provided by H. Gobind Khorana, who developed chemical methods to synthesize polyribonucleotides with defined, repeating sequences of two to four bases. The polypeptides produced by these mRNAs had one or a few amino acids in repeating patterns. These patterns, when combined with information from the random polymers used by Nirenberg and colleagues, permitted unambiguous codon assignments. The copolymer $(AC)_n$, for example, has alternating ACA and CAC codons: ACACACACACACACA. The polypeptide synthesized on this messenger contained equal amounts of threonine and histidine. Given that a histidine codon has one A and two Cs, (Table 27–1), CAC must code for histidine and ACA for threonine.

An RNA with three bases in a repeating pattern should yield three different types of polypeptide, each derived from a different reading frame and containing a single kind of amino acid. An RNA with four bases in a repeating pattern should yield a single type of polypeptide with a repeating pattern of four amino acids (Table 27–3). Consolidation of the results from all such experiments permitted the assignment of 61 of the 64 possible codons. The other three were identified as termination codons, in part because they disrupted amino acid coding patterns when they occurred in a synthetic RNA polymer (Fig. 27–6; Table 27–3).

figure 27–6

Effect of a termination codon in a repeating tetranucleotide. Dipeptides or tripeptides are synthesized, depending on where the ribosome initially binds. The three different reading frames are shown in different colors. Termination codons (in red) are encountered every fourth codon in all three reading frames.

Reading frame 1 5′ --- G U A | A G U | A A G | U A A | G U A | A G U | A A --- 3′

Reading frame 2 --- G | U A A | G U A | A G U | A A G | U A A | G U A | A ---

Reading frame 3 --- G U | A A G | U A A | G U A | A G U | A A G | U A A | ---

Meanings for all the triplet codons (Fig. 27–7) were established by 1966 and have been verified in many different ways. The cracking of the genetic code is regarded as one of the most important scientific discoveries of the twentieth century.

Codons are the key to the translation of genetic information, allowing the synthesis of specific proteins. The reading frame is set when translation of an mRNA molecule begins and is maintained as the synthetic machinery reads sequentially from one triplet to the next. If the initial reading frame is off by one or two bases, or if translation somehow skips a nucleotide in the mRNA, all the subsequent codons will be out of register; the result is a "missense" protein with a garbled amino acid sequence (Box 27–1).

Several codons serve special functions. The **initiation codon,** AUG, signals the beginning of a polypeptide in all cells, in addition to coding for Met residues in internal positions of polypeptides. As noted above, three of the 64 possible nucleotide triplets (UAA, UAG, and UGA; Fig. 27–7) do not code for any known amino acids. These **termination codons** (also called stop codons or nonsense codons) normally signal the end of polypeptide synthesis.

As described later in the chapter, initiation of protein synthesis in the cell is an elaborate process that relies on initiation codons and other signals in the mRNA. In retrospect, the experiments of Nirenberg and Khorana to identify codon function should not have worked in the absence of initiation codons. Serendipitously, experimental conditions caused the normal initiation requirements for protein synthesis to be relaxed. Diligence combined with chance to produce a breakthrough—a common occurrence in the history of biochemistry.

In a random sequence of nucleotides, one in every 20 codons in each reading frame is, on average, a termination codon. In general, a reading frame without a termination codon among 50 or more codons is called an **open reading frame.** Long open reading frames usually correspond to genes that encode proteins. An uninterrupted gene coding for a typical protein with a molecular weight of 60,000 would require an open reading frame with 500 or more codons.

table 27–3

Polypeptides Produced in Response to Synthetic RNA Polymers with Repeating Sequences of Three and Four Bases

Polynucleotide	Polypeptide products
Trinucleotide repeats	
$(UUC)_n$	$(Phe)_n$, $(Ser)_n$, $(Leu)_n$
$(AAG)_n$	$(Lys)_n$, $(Arg)_n$, $(Glu)_n$
$(UUG)_n$	$(Leu)_n$, $(Cys)_n$, $(Val)_n$
$(CCA)_n$	$(Pro)_n$, $(His)_n$, $(Thr)_n$
$(GUA)_n$	$(Val)_n$, $(Ser)_n$, (chain terminator)*
$(UAC)_n$	$(Tyr)_n$, $(Thr)_n$, $(Leu)_n$
$(AUC)_n$	$(Ile)_n$, $(Ser)_n$, $(His)_n$
$(GAU)_n$	$(Asp)_n$, $(Met)_n$, (chain terminator)*
Tetranucleotide repeats	
$(UAUC)_n$	$(Tyr–Leu–Ser–Ile)_n$
$(UUAC)_n$	$(Leu–Leu–Thr–Tyr)_n$
$(GUAA)_n$	Di- and tripeptides*
$(AUAG)_n$	Di- and tripeptides*

*With these polynucleotides, the patterns of amino acid incorporation into polypeptides are affected by the presence of codons that are termination signals for protein biosynthesis. In the repeating trinucleotide sequences, one of the three reading frames includes only termination codons and thus only two homopolypeptides are observed (generated from the remaining two reading frames). In some of the repeating tetranucleotide sequences, every fourth codon is a termination codon in every reading frame, so only short peptides are produced. This is illustrated in Figure 27–6 for $(GUAA)_n$.

Second letter of codon

First letter of codon (5′ end)	U	C	A	G
U	UU**U** Phe UU**C** Phe UU**A** Leu UU**G** Leu	UC**U** Ser UC**C** Ser UC**A** Ser UC**G** Ser	UA**U** Tyr UA**C** Tyr UA**A** Stop UA**G** Stop	UG**U** Cys UG**C** Cys UG**A** Stop UG**G** Trp
C	CU**U** Leu CU**C** Leu CU**A** Leu CU**G** Leu	CC**U** Pro CC**C** Pro CC**A** Pro CC**G** Pro	CA**U** His CA**C** His CA**A** Gln CA**G** Gln	CG**U** Arg CG**C** Arg CG**A** Arg CG**G** Arg
A	AU**U** Ile AU**C** Ile AU**A** Ile AU**G** Met	AC**U** Thr AC**C** Thr AC**A** Thr AC**G** Thr	AA**U** Asn AA**C** Asn AA**A** Lys AA**G** Lys	AG**U** Ser AG**C** Ser AG**A** Arg AG**G** Arg
G	GU**U** Val GU**C** Val GU**A** Val GU**G** Val	GC**U** Ala GC**C** Ala GC**A** Ala GC**G** Ala	GA**U** Asp GA**C** Asp GA**A** Glu GA**G** Glu	GG**U** Gly GG**C** Gly GG**A** Gly GG**G** Gly

figure 27–7

"Dictionary" of amino acid code words as they occur in mRNAs. The codons are written in the 5′→3′ direction. The third base of each codon (in bold type) plays a lesser role in specifying an amino acid than the first two. The three termination codons are shaded in red, the initiation codon AUG in green. All the amino acids except methionine and tryptophan have more than one codon. In most cases, codons that specify the same amino acid differ only at the third base.

box 27–1 Translational Frameshifting and RNA Editing: mRNAs That Change Horses in Midstream

Once the reading frame has been set during protein synthesis, codons are translated without overlap or punctuation until a termination codon is encountered. Usually, the other two possible reading frames contain no useful genetic information. However, a few genes are structured so that ribosomes "hiccup" at a certain point in the translation of their mRNAs, changing the reading frame from that point on. This appears to be a mechanism either to allow two or more related but distinct proteins to be produced from a single transcript or to regulate the synthesis of a protein.

One of the best-documented examples occurs in the translation of the mRNA for the overlapping *gag* and *pol* genes of the Rous sarcoma virus (see Fig. 26–34). The reading frame for *pol* is offset to the left by one base pair (−1 reading frame) relative to the reading frame for *gag* (Fig. 1).

The product of the *pol* gene (reverse transcriptase) is translated as a larger polyprotein using the same mRNA as is used for the *gag* protein alone. (see Fig. 26–33). The polyprotein, or *gag-pol* protein, is then trimmed to the mature reverse transcriptase by proteolytic digestion. Production of the polyprotein requires a translational frameshift in the overlap region to allow the ribosome to bypass the UAG termination codon at the end of the *gag* gene (shown in red in Fig. 1).

Frameshifts occur during about 5% of translations of this mRNA, and the *gag-pol* polyprotein (and ultimately reverse transcriptase) is synthesized at about 1/20 the frequency of the *gag* protein, a level that suffices for efficient reproduction of the virus. In some retroviruses another translational frameshift allows the translation of an even larger polyprotein that includes the product of the *env* gene fused to the *gag* and *pol* gene products (see Fig. 26–33). A similar mechanism produces both the τ and γ subunits of *E. coli* DNA polymerase III from a single *dnaX* gene transcript (see Table 25–2).

This mechanism also occurs in the gene for *E. coli* release factor 2 (RF_2), discussed later in this chapter, which is required for termination of protein synthesis at the termination codons UAA and UGA. The 26th codon in the transcript of the gene for RF_2 is UGA, which would normally halt protein synthesis. The remainder of the gene is in the +1 reading frame (offset one base pair to the right) relative to this UGA codon. Translation pauses at this codon but termination does not occur unless RF_2 is bound to the codon (a binding that is less likely, the lower the level of RF_2). The absence of bound RF_2 prevents the termination of protein synthesis at UGA and allows time for a frameshift to occur. The UGA plus the C that follows it (UGAC) is therefore read as GAC = Asp. Translation then proceeds in the new reading frame to complete synthesis of RF_2. In this way, RF_2 regulates its own synthesis in a feedback loop.

Some mRNAs are edited before translation. The initial transcripts of the genes that encode the cytochrome oxidase subunit II in some protist mitochondria do not correspond precisely to the sequence needed at the carboxyl terminus of the protein product. A posttranscriptional editing process inserts four uridines that shift the translational reading frame of the transcript. Figure 2a

figure 1
The *gag-pol* overlap region in Rous sarcoma virus.

--- Leu — Gly — Leu — Arg — Leu — Thr — Asn — Leu Stop
gag reading frame 5′---C U A G G G C U C C G C U U G A C A A A U U U A U A G G G A G G G C C A---3′
pol reading frame ---C U A G G G C U C C G C U U G A C A A A U U U A U A G G G A G G G C C A---
Ile — Gly — Arg — Ala ---

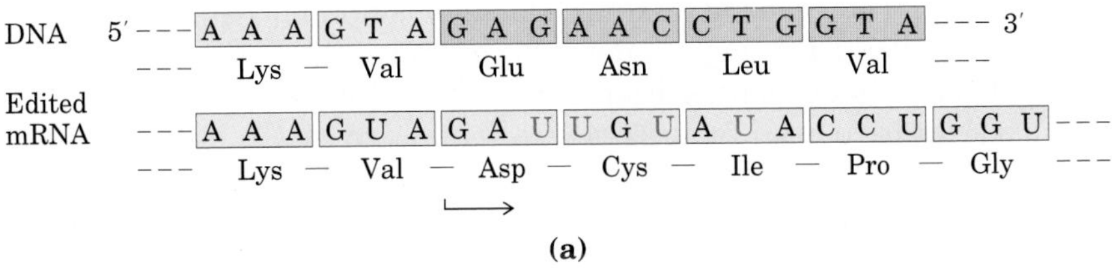

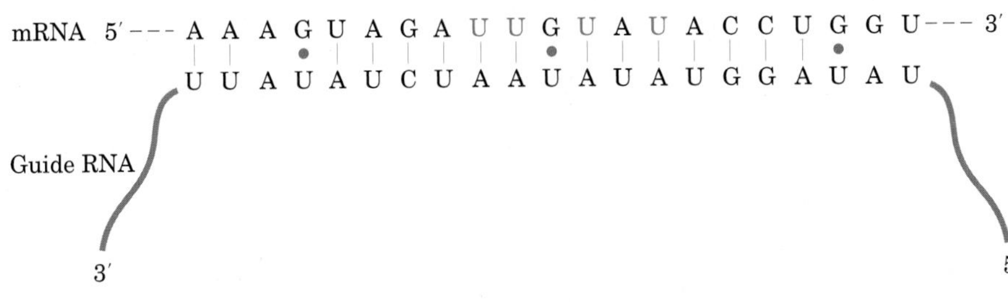

figure 2
RNA editing of the transcript of the cytochrome oxidase subunit II gene from mitochondria of *Trypanosoma brucei.* **(a)** Insertion of four uridine residues produces a revised reading frame. **(b)** A special class of guide RNAs in these mitochondria, complementary to the edited product, appear to act as templates for the editing process.

shows the added uridine residues in the small part of the transcript that is affected by editing. Neither the function nor the mechanism of this editing process is understood. A special class of RNA molecules encoded by these mitochondria has been detected with sequences complementary to the edited mRNAs. These so-called guide RNAs (Fig. 2b) appear to act as templates for the editing process. Note that the base pairing involves a number of G=U base pairs (marked by blue dots), which are common in RNA molecules.

A distinct form of RNA editing occurs in the gene for the apolipoprotein B component of low-density lipoprotein in vertebrates. One form of apolipoprotein B, called apoB-100 (M_r 513,000), is synthesized in the liver; a second form, apoB-48 (M_r 250,000), is synthesized in the intestine. Both are encoded by an mRNA produced from the gene for apoB-100. A cytosine deaminase enzyme found only in the intestine binds to the mRNA at codon 2,153 (CAA = Gln) and converts the C to a U to introduce the termination codon UAA. The apoB-48 produced in the intestine from this modified mRNA is simply an abbreviated form (corresponding to the amino-terminal half) of apoB-100 (Fig. 3). This reaction permits the synthesis of two different proteins from one gene in a tissue-specific manner.

figure 3
RNA editing of the transcript of the gene for the apolipoprotein B-100 component of low-density lipoprotein. Deamination that occurs only in the intestine converts a specific cytosine to uracil, converting a Gln codon to a stop codon and producing a truncated protein.

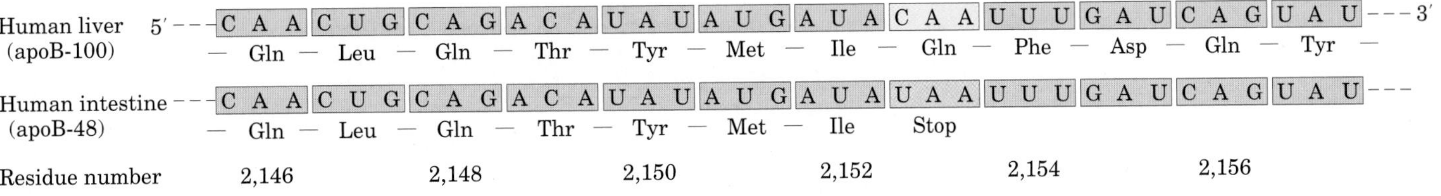

table 27–4

Degeneracy of the Genetic Code

Amino acid	Number of codons
Ala	4
Arg	6
Asn	2
Asp	2
Cys	2
Gln	2
Glu	2
Gly	4
His	2
Ile	3
Leu	6
Lys	2
Met	1
Phe	2
Pro	4
Ser	6
Thr	4
Trp	1
Tyr	2
Val	4

A striking feature of the genetic code is that an amino acid may be specified by more than one codon, so the code is described as **degenerate.** This does *not* suggest that the code is flawed: each codon specifies only one amino acid. The degeneracy of the code is not uniform. Whereas methionine and tryptophan have single codons, for example, three amino acids (Leu, Ser, Arg) have six codons, five amino acids have four, Ile has three, and nine amino acids have two (Table 27–4).

The genetic code is nearly universal. With the intriguing exception of a few minor variations in mitochondria, some bacteria, and some single-celled eukaryotes (Box 27–2), amino acid codons are identical in all species examined so far. Human beings, *E. coli,* tobacco plants, amphibians, and viruses share the same genetic code. Thus it would appear that all life forms have a common evolutionary ancestor whose genetic code has been preserved throughout biological evolution.

When several different codons specify one amino acid, the difference between them usually lies at the third base (at the 3′ end). For example, alanine is coded by the triplets GCU, GCC, GCA, and GCG. The codons for most amino acids can be symbolized by XY^{A}_{G} or XY^{U}_{C}. The first two letters of each codon are the primary determinants of specificity, a feature that has some interesting consequences.

Wobble Allows Some tRNAs to Recognize More than One Codon

Transfer RNAs base-pair with mRNA codons by means of a three-base sequence on the tRNA called the **anticodon.** The first base of the codon in mRNA (read in the 5′→3′ direction) pairs with the third base of the anticodon (Fig. 27–8a).

If the anticodon triplet of a tRNA recognized only one codon triplet through Watson-Crick base pairing, cells would have a different tRNA for each codon of an amino acid. This is not the case. For example, the anticodons in some tRNAs contain the nucleotide inosinate (designated I), which contains the uncommon base hypoxanthine (see Fig. 10–5b). Inosinate can form hydrogen bonds with three different nucleotides (U, C, and A; Fig. 27–8b), although these pairings are much weaker than the hydrogen bonds between the Watson-Crick base pairs G≡C and A═U. In yeast, one $tRNA^{Arg}$ has the anticodon (5′)ICG, which recognizes three arginine codons, (5′)CGA, (5′)CGU, and (5′)CGC. The first two bases of these codons are identical (CG) and form strong Watson-Crick base pairs with the corresponding bases of the anticodon. The third bases of the three arginine codons (A, U, and C) form rather weak hydrogen bonds with the I residue at the first position of the anticodon.

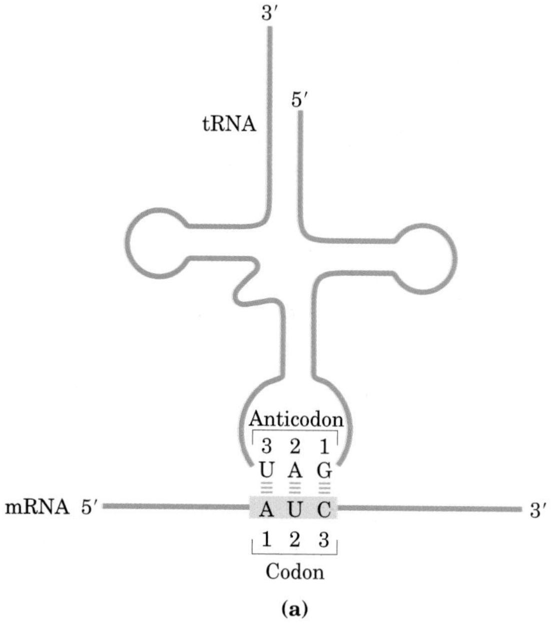

figure 27–8

Pairing relationship of codon and anticodon. (a) Alignment of the two RNAs is antiparallel. The tRNA is shown in the traditional cloverleaf configuration. **(b)** Codon pairing relationships when the tRNA anticodon contains inosinate.

Examination of these and other codon-anticodon pairings led Crick to conclude that the third base of most codons pairs rather loosely with the corresponding base of its anticodons; to use his picturesque word, the third bases of such codons (and the first bases of their corresponding anticodons) "wobble." Crick proposed a set of four relationships called the **wobble hypothesis:**

1. The first two bases of an mRNA codon always form strong Watson-Crick base pairs with the corresponding bases of the tRNA anticodon and confer most of the coding specificity.
2. The first base of the anticodon (reading in the $5' \rightarrow 3'$ direction; this pairs with the third base of the codon) determines the number of codons recognized by the tRNA. When the first base of the anticodon is C or A, base pairing is specific and only one codon is recognized by that tRNA. When the first base is U or G, binding is less specific and two different codons may be read. When inosine (I) is the first (wobble) nucleotide of an anticodon, three different codons can be recognized—the maximum number for any tRNA. These relationships are summarized in Table 27–5.
3. When an amino acid is specified by several different codons, the codons that differ in either of the first two bases require different tRNAs.
4. A minimum of 32 tRNAs are required to translate all 61 codons.

The wobble (or third) base of the codon contributes to specificity but, because it pairs only loosely with its corresponding base in the anticodon, it permits rapid dissociation of the tRNA from its codon during protein synthesis. If all three bases of codons engaged in strong Watson-Crick pairing with the three bases of the anticodons, tRNAs would dissociate too slowly and this would severely limit the rate of protein synthesis. Codon-anticodon interactions balance the requirements for accuracy and speed.

table 27–5

How the Wobble Base of the Anticodon Determines the Number of Codons a tRNA Can Recognize*

1. One codon recognized:			
	Anticodon	(3′) X–Y–**C** (5′)	(3′) X–Y–**A** (5′)
	Codon	(5′) Y–X–**G** (3′)	(5′) Y–X–**U** (3′)
2. Two codons recognized:			
	Anticodon	(3′) X–Y–**U** (5′)	(3′) X–Y–**G** (5′)
	Codon	(5′) Y–X–**A/G** (3′)	(5′) Y–X–**C/U** (3′)
3. Three codons recognized:			
	Anticodon	(3′) X–Y–**I** (5′)	
	Codon	(5′) Y–X–**A/U/C** (3′)	

*X and Y denote complementary bases capable of strong Watson-Crick base pairing with each other. The bases in the wobble positions—the 3′ position of codons and 5′ position of anticodons—are shaded in red.

box 27–2 Natural Variations in the Genetic Code

In biochemistry, as in other disciplines, exceptions to general rules can be problematic for instructors and frustrating for students. At the same time they teach us that life is complex and inspire us to search for more surprises. Understanding the exceptions can even reinforce the original rule in surprising ways.

One would expect little room for variation in the genetic code. Even a single amino acid substitution can have profoundly deleterious effects on the structure of a protein. Nevertheless, variations in the code do occur in some organisms, and they are both interesting and instructive. The rarity and types of variation provide powerful evidence for a common evolutionary origin of all living things.

To alter the code, changes must occur in one or more tRNAs, with the obvious target for alteration being the anticodon. Such a change would lead to the systematic insertion of an amino acid at a codon that does not specify that amino acid, according to the normal code (see Fig. 27–7). The genetic code, in effect, is defined by two elements: (1) the anticodons on tRNAs (which determine where an amino acid is placed in a growing polypeptide) and (2) the specificity of the enzymes — aminoacyl-tRNA synthetases — that charge the tRNAs, which determine the identity of the amino acid attached to a given tRNA.

Most sudden changes in the code would have catastrophic effects on cellular proteins, so code alterations are most likely where relatively few proteins would be affected. This could happen in small genomes encoding only a few proteins. The biological consequences of a code change could also be limited by restricting changes to the three termination codons, because these do not generally occur within genes (see Box 27–1 for exceptions to *this* rule). This pattern is in fact observed.

Most of the very rare known variations in the genetic code occur in mitochondrial DNA (mtDNA), which encodes only 10 to 20 proteins. Mitochondria have their own tRNAs, so their code variations do not affect the much larger cellular genome. The most common changes in mitochondria, and the only code changes observed in cellular genomes, involve termination codons. These changes affect termination in the products of only a subset of genes, and sometimes the effects in those genes are minor because the genes have multiple (redundant) termination codons.

In mitochondria, the changes can be viewed as a kind of genomic streamlining. Vertebrate mtDNAs have genes that encode 13 proteins, two rRNAs, and 22 tRNAs (see Fig. 19–30). An unusual set of wobble rules allows the 22 tRNAs to decode all 64 possible codon triplets—not all the 32 tRNAs required for the normal code are needed. Four codon families (in which the amino acid is determined entirely by the first two nucleotides) are decoded by a single tRNA with a U in the first (or wobble) position in the anticodon. Either the U pairs somehow with any of the four possible bases in the third position of the codon, or a "two out of three" mechanism is used—that is, no base pairing is needed at the third position of the codon. Other tRNAs recognize codons with either A or G in the third position, and yet others recognize U or C, so that virtually all the tRNAs recognize either two or four codons.

In the normal code, only two amino acids are specified by single codons, methionine and tryptophan (see Table 27–4). If all mitochondrial tRNAs recognize two codons, then additional Met and Trp codons might be expected in mitochondria. And so we find that the single most common code variation observed is that UGA specifies not termination but tryptophan. $tRNA^{Trp}$ recognizes and inserts a Trp residue at either the termination codon UGA or the normal Trp codon UGG. Converting AUA from an Ile codon to a Met codon has a similar effect; the normal Met codon is AUG, and a single tRNA can recognize both codons. This is the second most common mitochondrial code variation. The known coding variations in mitochondria are summarized in Table 1.

Turning to the much rarer changes in the codes for cellular (as distinct from mitochondrial) genomes, we find that the only known variation in a prokaryote is again the use of UGA to encode Trp residues, occurring in the simplest free-living cell, *Mycoplasma capricolum*. Among eukaryotes, the only known extramitochondrial coding changes occur in a few species of ciliated protists, in which both termination codons UAA and UAG can specify glutamine.

Changes in the code need not be absolute—a codon need not always encode the same amino acid. In *E. coli* we find two examples of amino acids being inserted at positions not specified in the general code. The first is the occasional use of the codon GUG (Val) as an initiating codon. This occurs only for those genes in which the

table 1

Known Variant Codon Assignments in Mitochondria

	Codons*				
	UGA	AUA	AGA AGG	CUN	CGG
Normal code assignment	**Stop**	**Ile**	**Arg**	**Leu**	**Arg**
Animals					
Vertebrates	Trp	Met	Stop	+	+
Drosophila	Trp	Met	Ser	+	+
Yeasts					
Saccharomyces cerevisiae	Trp	Met	+	Thr	+
Torulopsis glabrata	Trp	Met	+	Thr	?
Schizosaccharomyces pombe	Trp	+	+	+	+
Filamentous fungi	Trp	+	+	+	+
Trypanosomes	Trp	+	+	+	+
Higher plants	+	+	+	+	Trp
Chlamydomonas reinhardtii	?	+	+	+	?

*A question mark Indicates that the codon has not been observed in the indicated mitochondrial genome; N, any nucleotide; +, the codon has the same meaning as in the normal code.

GUG is properly located relative to particular mRNA sequences that affect the initiation of translation (as discussed later in this chapter).

Contextual signals that alter coding patterns also occur in the second *E. coli* example. A few proteins in all cells (e.g., formate dehydrogenase in bacteria and glutathione peroxidase in mammals) require the element selenium for their activity. It is generally present in the form of the modified amino acid selenocysteine. Although modified amino acids are generally produced in posttranslational reactions (described in this chapter), in *E. coli*, selenocysteine is introduced into formate dehydrogenase during translation in response to an in-frame UGA codon. A specialized type of serine tRNA, present at lower levels than other serine tRNAs, recognizes UGA and no other codons. This tRNA is charged with serine, and the serine is then enzymatically converted to selenocysteine prior to its use at the ribosome. This charged tRNA will not recognize just any UGA codon; instead, some contextual signal in the mRNA, yet to be identified, permits the tRNA to recognize only those few UGA codons within certain genes that specify selenocysteine. In effect, there are 21 standard amino acids in *E. coli*, and UGA doubles as a codon for both termination and (sometimes) for selenocysteine.

These variations tell us that the code is not quite as universal as once believed, but that its flexibility is severely constrained. The variations are obviously derivatives of the general code, and no example of a completely different code has ever been found. The limited scope of code variants strengthens the principle that all life on this planet evolved on the basis of a single (slightly flexible) genetic code.

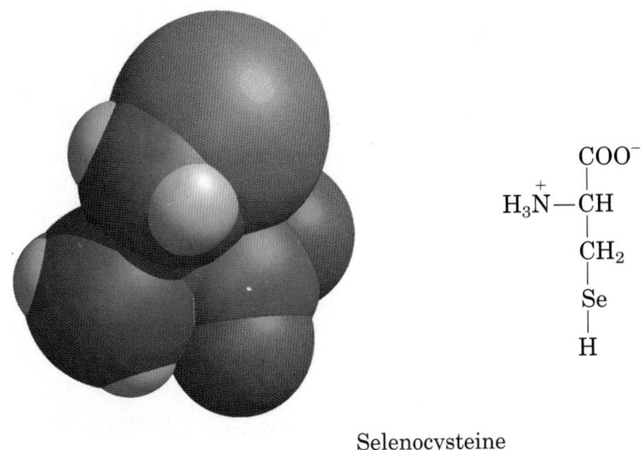

Selenocysteine

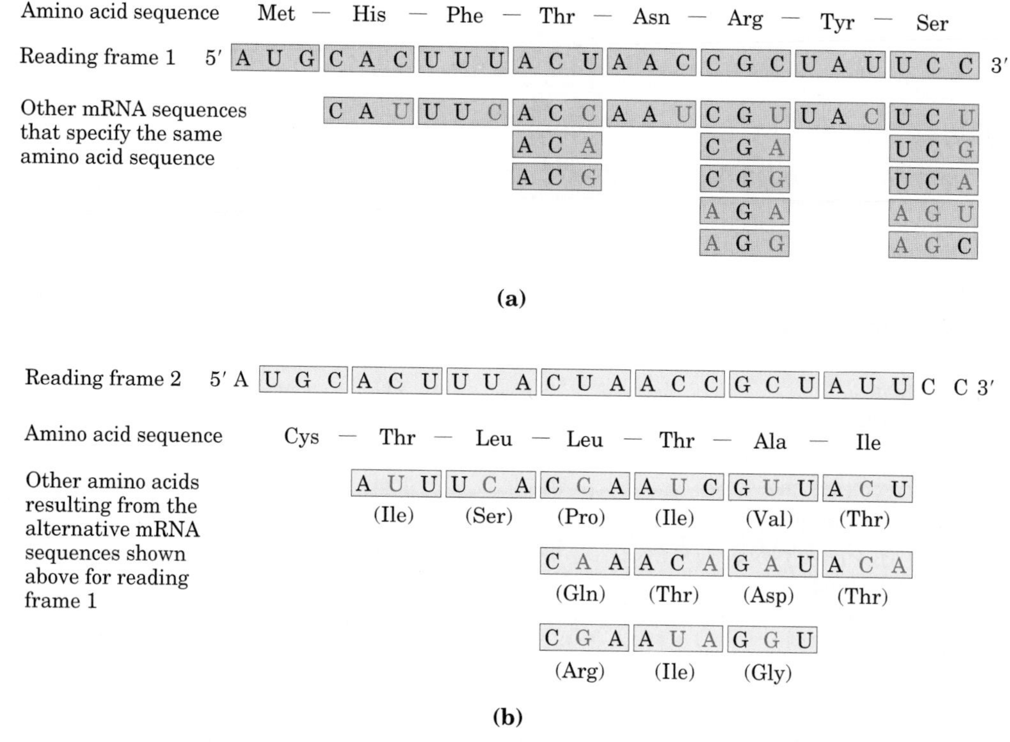

figure 27–9

Reading frame and amino acid sequence. An amino acid sequence specified by one reading frame severely limits the potential amino acids encoded by the other two reading frames. **(a)** The codons of reading frame 1 that can produce the indicated amino acid sequence. Most of the permitted nucleotide changes (red) are in the third (wobble) position of each codon. **(b)** At the top are shown the codons that can exist in reading frame 2 without changing the amino acid sequence encoded by reading frame 1. Below are shown the alternative codons that correspond to the alternative mRNA sequences listed in **(a)**. The possible amino acids that can be encoded by reading frame 2 without changing the sequence encoded by reading frame 1 are in parentheses.

Overlapping Genes in Different Reading Frames Are Found in Some Viral DNAs

Although a given nucleotide sequence can, in principle, be read in any of its three reading frames, most DNA sequences encode a protein product in only one reading frame. To generate proteins from different reading frames, termination codons must not crop up in the new sequence and each codon must correspond to a suitable amino acid. The genetic code imposes strict limits on the numbers of amino acids that can be encoded by reading frame 2 without changing the amino acids specified by reading frame 1 (Fig. 27–9). Sometimes one amino acid (and its corresponding codon) may be substituted for another in reading frame 1 and still retain the function of the encoded protein, making it more likely that reading frames 2 or 3 might also encode a useful protein; but even taking these factors into account, the flexibility in other reading frames is very limited.

Generally, only one reading frame is used to encode a protein and genes do not overlap, but there are a few interesting exceptions. In several viruses, the same DNA sequence codes for two different proteins by employing two different reading frames. The DNA of bacteriophage ϕX174, which contains 5,386 nucleotide residues, is not long enough to code for the

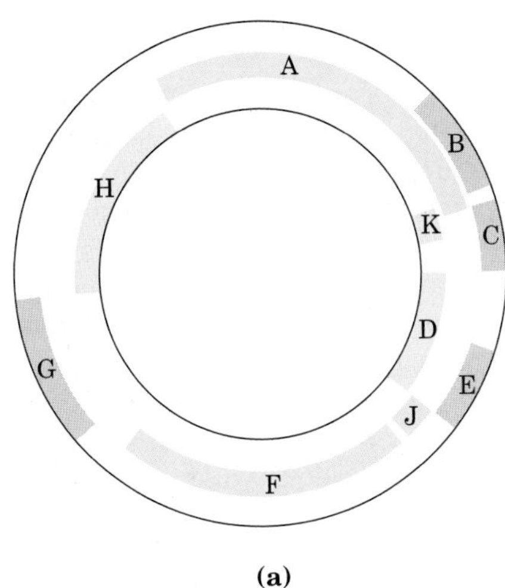

(a)

Gene C 133
AUG AGA AAA UUC GAC CUA U--- 3′
Met – Arg – Lys – Phe – Asp – Leu ---

Gene A 5′--- GGC GGA AAA UGA
--- Gly – Gly – Lys – Stop

Gene K 5′--- G GCG GAA AAU GAG AAA AUU CGA CCU AU--- 3′
--- Ala – Glu – Asn – Glu – Lys – Ile – Arg – Pro – ---

(b)

figure 27–10
Genes within genes. (a) The circular DNA of ϕX174 contains ten genes (A to K), shown here in concentric arcs corresponding to their reading frames. Gene B lies within the sequence of gene A but uses a different reading frame. Similarly, gene E lies primarily within gene D but also overlaps with gene J, each gene using a different reading frame. Gene K overlaps with genes A, B, and C. The unshaded segments are untranslated spacer regions. **(b)** Portion of the nucleotide sequence of the mRNA transcript of genes C, A, and K of ϕX174 DNA. Four bases are used in all three genes, in different reading frames. One region of K on the 5′ side (not shown) also overlaps a third gene, B.

ten different proteins produced by the ϕX174 DNA genome. A comparison of the nucleotide sequence of the ϕX174 chromosome with the amino acid sequences of the proteins it encodes revealed several overlapping gene sequences. Figure 27–10a shows gene B nested within gene A and most of gene E within gene D; Figure 27–10b diagrams the overlap of genes K, A, and C. In several cases, the initiation codon of one gene overlaps the termination codon of the other gene. These nested and overlapping sequences account for the surprisingly small size of the ϕX174 genome compared with the number of amino acid residues in the ten proteins it encodes.

Similar observations have been made in viral DNAs of phage λ and the cancer-causing simian virus 40 (SV40) and in RNA phages such as Qβ, Q17, and phage G4, a close relative of ϕX174. Overlapping genes or genes within genes may give viruses a selective advantage because the small size of the viral capsid encourages economical use of a limited amount of DNA to encode the variety of proteins needed to infect and replicate within a host cell. As viruses reproduce (and therefore evolve) faster than their host cells, they may represent a limit in biological streamlining.

The genetic code tells us how protein sequence information is stored in nucleic acids and provides some clues about how that information is translated into protein. We now turn to the molecular mechanisms of the translation process.

Protein Synthesis

As we have seen for DNA and RNA, the synthesis of polymeric biomolecules can be considered in terms of initiation, elongation, and termination stages. These fundamental processes are typically bracketed by two additional stages: the activation of precursors prior to synthesis and the postsynthetic processing of the completed polymer. Protein synthesis follows this pattern. The activation of amino acids before their incorporation into polypeptides and posttranslational processing of the completed polypeptide play particularly important roles in assuring both the fidelity of synthesis and the proper function of the protein product. The cellular components involved in the five stages of protein synthesis in *E. coli* and other bacteria are listed in Table 27–6; the requirements in eukaryotic cells are quite similar, although the components are in some cases more numerous.

An overview of the five stages of protein synthesis provides a useful outline for the discussion that follows.

Stage 1: Activation of Amino Acids Synthesis of a polypeptide with a defined sequence demands that two fundamental chemical requirements be met: (1) the carboxyl group of each amino acid must be activated to facilitate formation of a peptide bond, and (2) a link must be established between each new amino acid and the information that encodes it in the mRNA. Both of these requirements are met by attaching the amino acid to a tRNA in the first stage of protein synthesis. Attaching the right amino acid to the right tRNA is critical. This reaction takes place in the cytosol, not on the ribosome. Each of the 20 amino acids is covalently attached to a specific tRNA at the expense of ATP energy, using Mg^{2+}-dependent activating

table 27–6

Components Required for the Five Major Stages of Protein Synthesis in *E. coli*

Stage	Essential components
1. Activation of amino acids	20 amino acids 20 aminoacyl-tRNA synthetases 20 or more tRNAs ATP Mg^{2+}
2. Initiation	mRNA *N*-Formylmethionyl-tRNA Initiation codon in mRNA (AUG) 30S ribosomal subunit 50S ribosomal subunit Initiation factors (IF-1, IF-2, IF-3) GTP Mg^{2+}
3. Elongation	Functional 70S ribosome (initiation complex) Aminoacyl-tRNAs specified by codons Elongation factors (EF-Tu, EF-Ts, EF-G) GTP Mg^{2+}
4. Termination and release	Termination codon in mRNA Polypeptide release factors (RF_1, RF_2, RF_3) ATP
5. Folding and posttranslational processing	Specific enzymes, cofactors, and other components for removal of initiating residues and signal sequences, additional proteolytic processing, modification of terminal residues, and attachment of phosphate, methyl, carboxyl, carbohydrate, or prosthetic groups

enzymes called aminoacyl-tRNA synthetases. Aminoacylated tRNAs are said to be "charged."

Stage 2: Initiation Next, the mRNA bearing the code for the polypeptide to be made binds to the smaller of two ribosomal subunits and to the initiating aminoacyl-tRNA. The large ribosomal subunit then binds to form an initiation complex. The initiating aminoacyl-tRNA base-pairs with the mRNA codon AUG that signals the beginning of the polypeptide. This process, which requires GTP, is promoted by cytosolic proteins called initiation factors.

Stage 3: Elongation The nascent polypeptide is lengthened by covalent attachment of successive amino acid units, each carried to the ribosome and correctly positioned by its tRNA, which base-pairs to its corresponding codon in the mRNA. Elongation requires cytosolic proteins called elongation factors. The binding of each incoming aminoacyl-tRNA and movement of the ribosome along the mRNA are facilitated by the hydrolysis of GTP as each residue is added to the growing polypeptide.

Stage 4: Termination and Release Completion of the polypeptide chain is signaled by a termination codon in the mRNA. The new polypeptide is then released from the ribosome, aided by proteins called release factors.

Stage 5: Folding and Posttranslational Processing In order to achieve its biologically active form, the new polypeptide must fold into its proper three-dimensional conformation. Before or after folding, the new polypeptide may undergo enzymatic processing, including removal of one or more amino acids from the amino terminus; addition of acetyl, phosphoryl, methyl, carboxyl, or other groups to certain amino acid residues; proteolytic cleavage; or attachment of oligosaccharides or prosthetic groups.

Before examining these five stages in detail, we must examine two key components in protein biosynthesis: the ribosome and tRNAs.

The Ribosome Is a Complex Supramolecular Machine

Each *E. coli* cell contains 15,000 or more ribosomes, making up almost a quarter of the dry weight of the cell. Bacterial ribosomes contain about 65% rRNA and about 35% protein; they have a diameter of about 18 nm and are composed of two unequal subunits with sedimentation coefficients of 30S and 50S and a combined sedimentation coefficient of 70S. Both subunits contain ribosomal proteins and at least one rRNA (Table 27–7).

table 27–7

RNA and Protein Components of the *E. coli* Ribosome

Subunit	Number of different proteins	Total number of proteins	Protein designations	Number and type of rRNAs
30S	21	21	S1–S21	1 (16S rRNA)
50S	33	36	L1–L36*	2 (5S and 23S rRNAs)

*The L1 to L36 protein designations do not correspond to 36 different proteins. The protein originally designated L7 is in fact a modified form of L12, and L8 is a complex of three other proteins. Also, L26 proved to be the same protein as S20 (and not part of the 50S subunit). This gives 33 different proteins in the large subunit. There are four copies of the L7/L12 protein, with the three extra copies bringing the total protein count to 36.

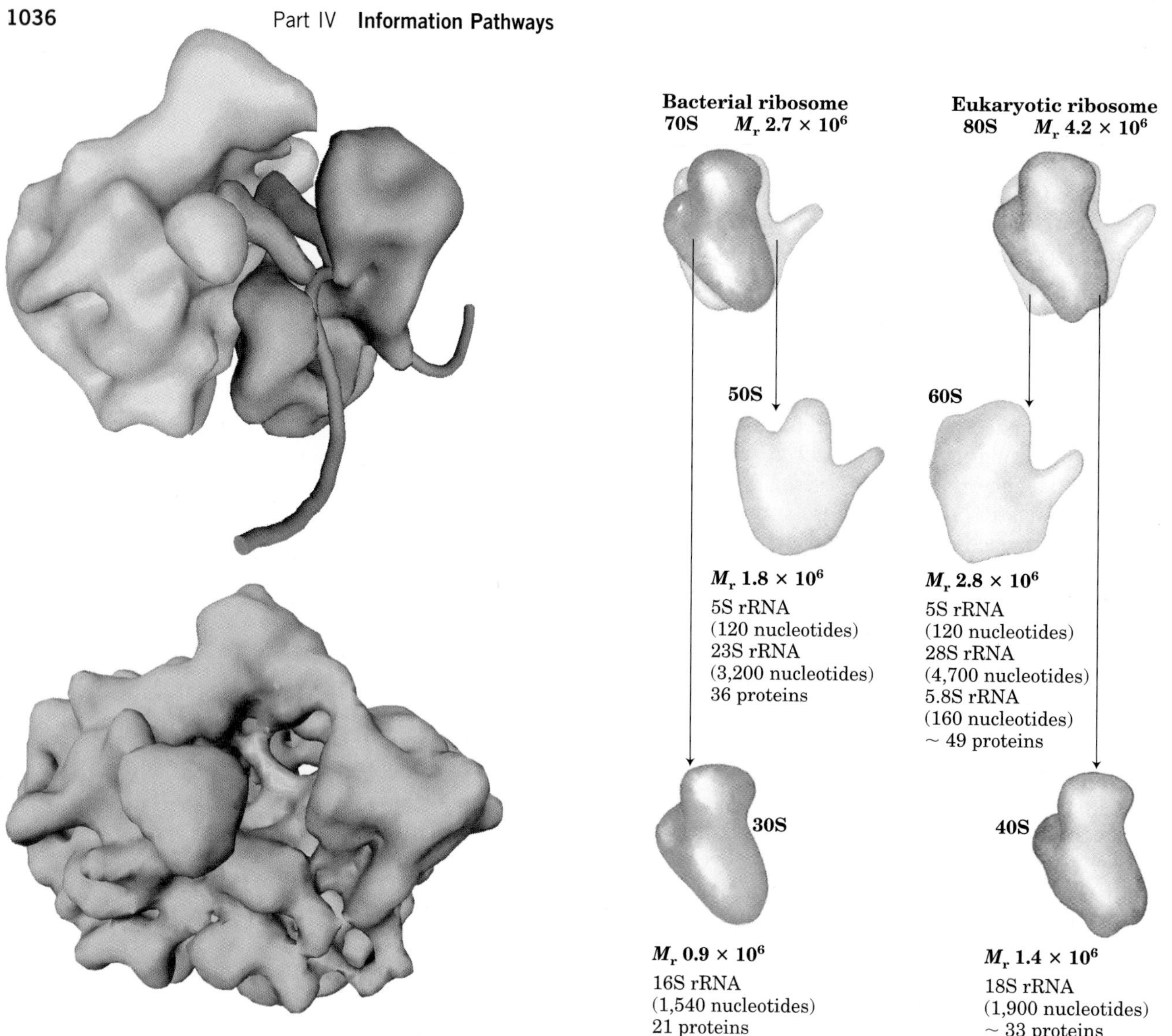

figure 27–11

Ribosomes. (a) Structure of the bacterial ribosome at near-molecular resolution. The model at the top shows the locations of the mRNA (green rope) and the bound tRNAs (shades of green). The large subunit is on the left, the small subunit on the right. The image at the bottom is the complete ribosome at about 15 Å resolution. The subunits are not distinguished, but the orientation is similar to that of the image at the top. Some density corresponding to a tRNA is shown in green. **(b)** Ribosomal subunits are identified by their S (Svedberg unit) values, sedimentation coefficients that refer to their rate of sedimentation in a centrifuge. The S values are not necessarily additive when subunits are combined, because rates of sedimentation are affected by shape as well as mass.

The three-dimensional structures of the ribosomal subunits of *E. coli* have been deduced (although not quite to molecular resolution) from a combination of x-ray diffraction, electron microscopy, and other structural methods (Fig. 27–11). The two irregularly shaped subunits fit together to form a cleft through which the mRNA passes as the ribosome moves along the mRNA during translation.

The variety of proteins in ribosomes is enormous, with molecular weights ranging from about 6,000 to 75,000. The structures of many of these proteins are known, and the structure of the entire ribosome is coming into focus (Fig. 27–11a). The sequences of the rRNAs of many organisms have been determined. Each of the three single-stranded rRNAs of *E. coli* has a specific three-dimensional conformation featuring extensive intrachain base pairing (Fig. 27–12).

The rRNAs appear to serve as a framework to which ribosomal proteins are bound. In the late 1960s, Masayasu Nomura and colleagues demonstrated that both ribosomal subunits can be broken down into their RNA and protein components, then reconstituted in vitro. Under appropriate experimental conditions, they spontaneously reassemble to form 30S or 50S subunits nearly identical in structure and activity to native subunits. The 55

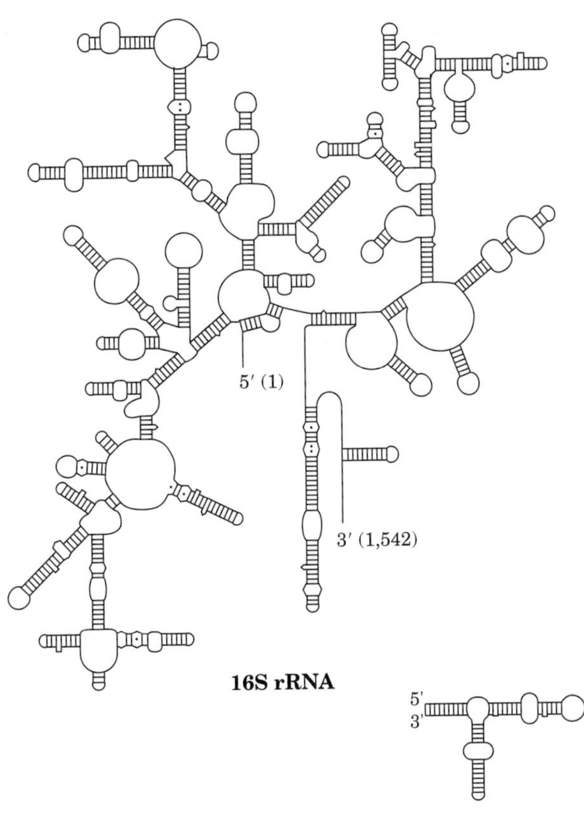

figure 27–12
Bacterial rRNAs. Shown here are models for the secondary structure of *E. coli* 16S and 5S rRNAs. The first (5′ end) and final (3′ end) ribonucleotide residues of the 16S rRNA are numbered.

proteins in the bacterial ribosome serve as either enzymes or structural components in protein synthesis, although the detailed functions of most of these proteins are yet to be elucidated.

The ribosomes of eukaryotic cells (other than mitochondrial and chloroplast ribosomes) are larger and more complex than bacterial ribosomes (Fig. 27–11b), with a diameter of about 23 nm and a sedimentation coefficient of about 80S. They also have two subunits, which vary in size among species but on average are 60S and 40S. Altogether, eukaryotic ribosomes contain over 80 different proteins. The ribosomes of mitochondria and chloroplasts are somewhat smaller and simpler than bacterial ribosomes. Nevertheless, ribosomal structure and function are strikingly similar in all organisms and organelles.

Masayasu Nomura

Transfer RNAs Have Characteristic Structural Features

To understand how tRNAs can serve as adaptors in translating the language of nucleic acids into the language of proteins, we must first examine their structure in more detail. Transfer RNAs are relatively small and consist of a single strand of RNA folded into a precise three-dimensional structure (see Fig. 10–28a). The tRNAs in bacteria and in the cytosol of eukaryotes have between 73 and 93 nucleotide residues, corresponding to molecular weights of 24,000 to 31,000. Mitochondria and chloroplasts contain distinctive, somewhat smaller tRNAs. Cells have at least one kind of tRNA for each amino acid; at least 32 tRNAs are required to recognize all the amino acid codons (some recognize more than one codon), but some cells use more than 32.

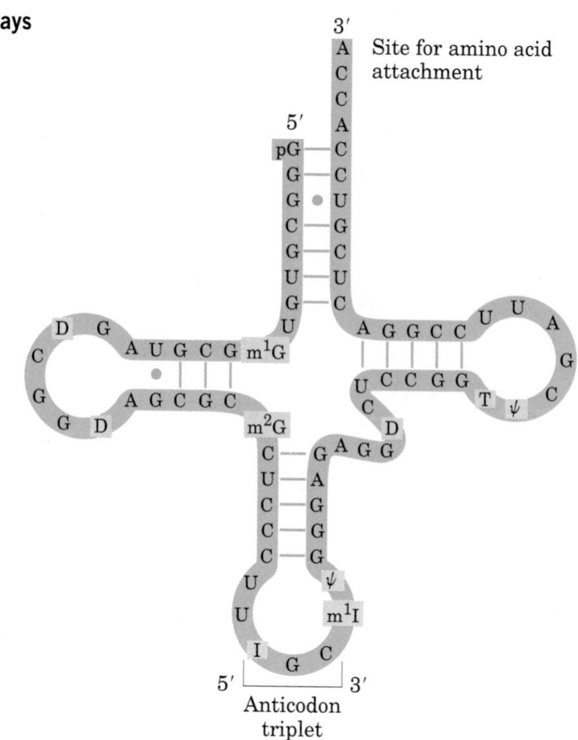

figure 27–13

Nucleotide sequence of yeast tRNA^{Ala}. The structure, deduced in 1965 by Robert W. Holley and his colleagues, is shown in the cloverleaf conformation in which intrastrand base pairing is maximal. The following symbols are used for the modified nucleotides: ψ, pseudouridine; I, inosine; T, ribothymidine; D, 5,6-dihydrouridine; m^{1}I, 1-methylinosine; m^{1}G, 1-methylguanosine; m^{2}G, N^2-dimethylguanosine. The modified bases are shaded in red, and most are illustrated in Figure 26–26. Blue lines between parallel sections indicate base pairs. The anticodon can recognize three codons for alanine (GCA, GCU, and GCC). Other features of tRNA structure are shown in Figures 27–14 and 27–15. Note the presence of G=U base pairs in both the amino acid arm (top) and the D arm (left), signified by a blue dot to indicate a non–Watson-Crick pairing. In RNAs, guanosine is often found base-paired with uridine, although the G=U pair is not as stable as the Watson-Crick G≡C pair (Chapter 10).

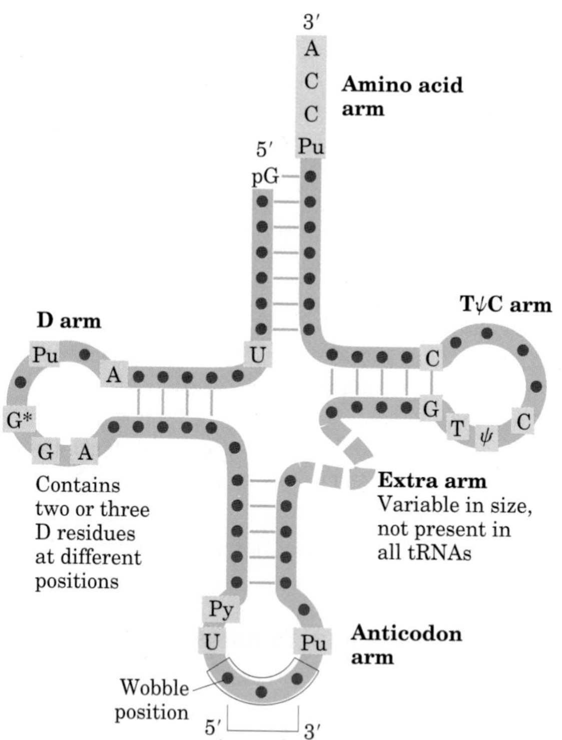

figure 27–14

General cloverleaf secondary structure of all tRNAs. The large dots on the backbone represent nucleotide residues, and the blue lines represent base pairs. Characteristic and/or invariant residues common to all tRNAs are shaded in red. Transfer RNAs vary in length from 73 to 93 nucleotides. Extra nucleotides occur in the extra arm or in the D arm. At the end of the anticodon arm is the anticodon loop, which always contains seven unpaired nucleotides. The D arm contains two or three D residues, depending on the tRNA. In some tRNAs, the D arm has only three hydrogen-bonded base pairs. In addition to the symbols explained in Figure 27–13: Pu, purine nucleotide; Py, pyrimidine nucleotide; G*, guanylate *or* 2′-*O*-methylguanylate.

Robert W. Holley
1922–1993

Yeast alanine tRNA (tRNA^{Ala}), the first nucleic acid to be completely sequenced (Fig. 27–13), contains 76 nucleotide residues, ten of which have modified bases. Comparisons of many tRNAs from various species revealed many common denominators of structure (Fig. 27–14). Eight or more of the nucleotide residues have modified bases and sugars, many of which are methylated derivatives of the principal bases. Most tRNAs have a guanylate (pG) residue at the 5′ end, and all have the trinucleotide sequence CCA(3′)

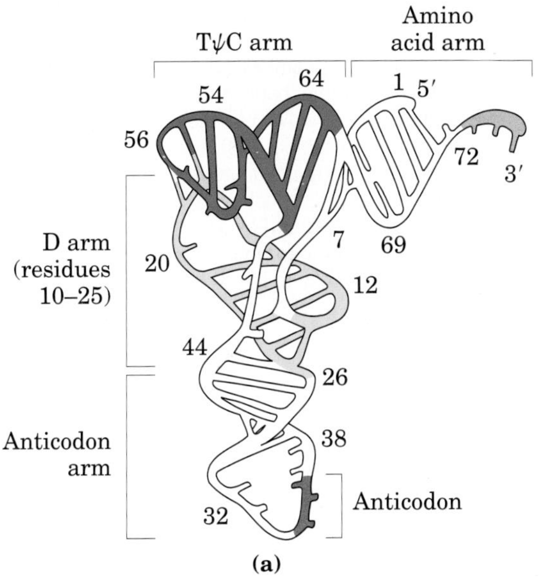

(a)

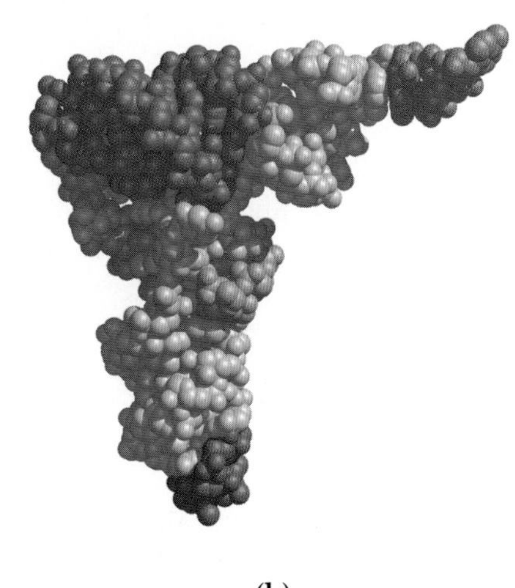
(b)

figure 27–15
Three-dimensional structure of yeast tRNA^{Phe} deduced from x-ray diffraction analysis. The shape resembles a twisted L. **(a)** A schematic diagram with the various arms identified in Figure 27–14 shaded in different colors. **(b)** A space-filling model. Color coding is the same in both representations. The three bases of the anticodon are shown in red and the CCA sequence at the 3′ end (the attachment point for the amino acid) in orange. The TψC and D arms are blue and yellow, respectively.

at the 3′ end. When drawn in two dimensions, the hydrogen-bonding pattern of all tRNAs forms a cloverleaf structure with four arms; the longer tRNAs have a short fifth or extra arm (Fig. 27–14). In three-dimensions, a tRNA has the form of a twisted L (Fig. 27–15).

Two of the arms of a tRNA are critical for the adaptor function. The **amino acid arm** can carry a specific amino acid esterified by its carboxyl group to the 2′- or 3′-hydroxyl group of the adenosine residue at the 3′ end of the tRNA. The **anticodon arm** contains the anticodon. The other major arms are the **D arm,** which contains the unusual nucleotide dihydrouridine, and the **TψC arm,** which contains ribothymidine (T), not usually present in RNAs, and pseudouridine (ψ), which has an unusual carbon–carbon bond between the base and ribose (see Fig. 26–26). The D and TψC arms contribute important interactions for the overall folding of tRNA molecules, but specific functions for these regions have not yet been determined.

Having looked at the structures of ribosomes and tRNAs, we now consider the details of the five stages of protein synthesis.

Stage 1: Aminoacyl-tRNA Synthetases Attach the Correct Amino Acids to Their tRNAs

During the first stage of protein synthesis, the 20 different amino acids are esterified in the cytosol to their corresponding tRNAs by aminoacyl-tRNA synthetases. Each enzyme is specific for one amino acid and one or more corresponding tRNAs. Most organisms have one aminoacyl-tRNA synthetase for each amino acid. For amino acids with two or more corresponding tRNAs, the same enzyme usually aminoacylates all of them.

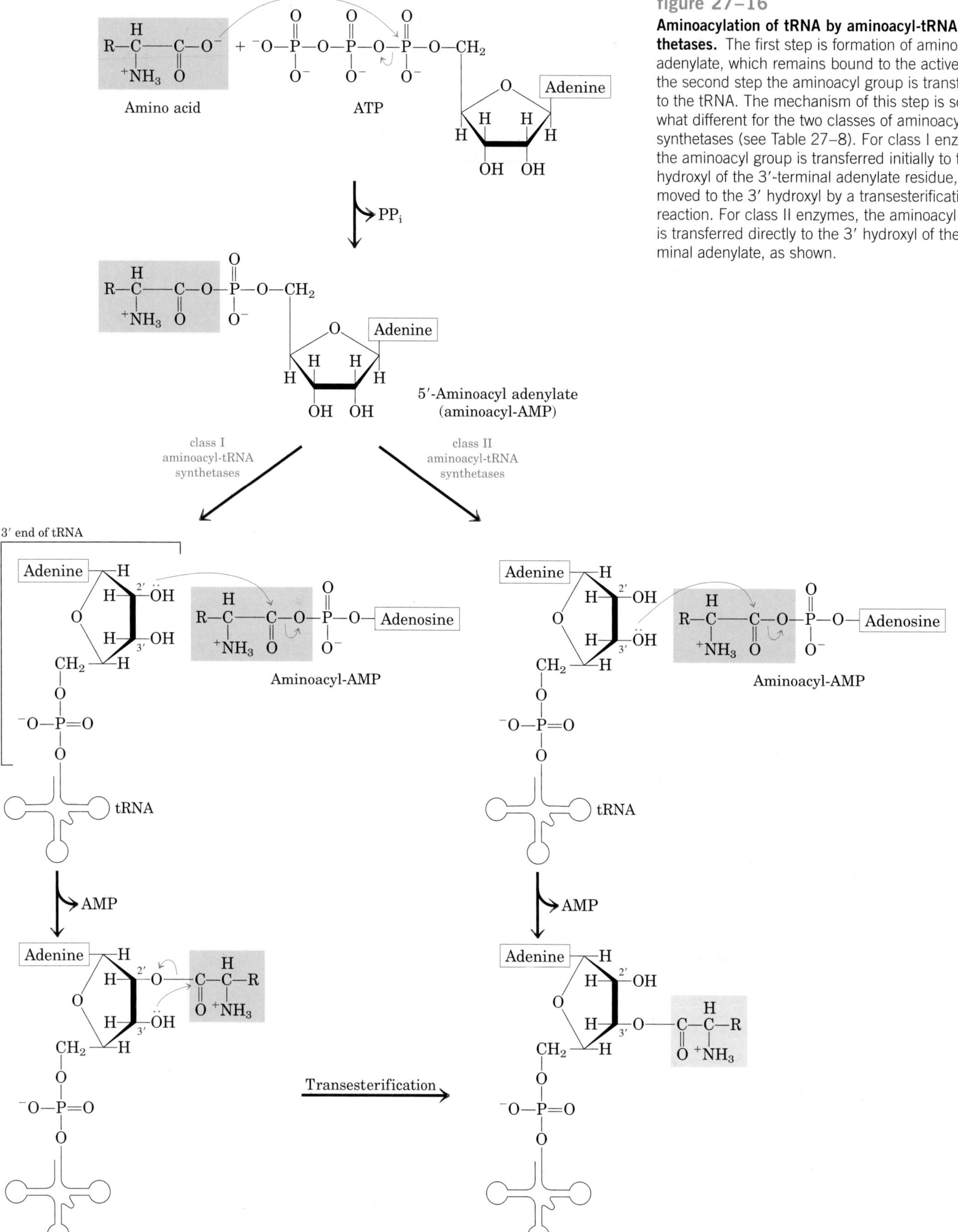

figure 27–16

Aminoacylation of tRNA by aminoacyl-tRNA synthetases. The first step is formation of aminoacyl adenylate, which remains bound to the active site. In the second step the aminoacyl group is transferred to the tRNA. The mechanism of this step is somewhat different for the two classes of aminoacyl-tRNA synthetases (see Table 27–8). For class I enzymes, the aminoacyl group is transferred initially to the 2′ hydroxyl of the 3′-terminal adenylate residue, then moved to the 3′ hydroxyl by a transesterification reaction. For class II enzymes, the aminoacyl group is transferred directly to the 3′ hydroxyl of the terminal adenylate, as shown.

Nearly all the aminoacyl-tRNA synthetases of *E. coli* have been isolated and some have been crystallized. They have been divided into two classes (Table 27–8) based on substantial differences in primary and tertiary structure and in reaction mechanism (Fig. 27–16). The two classes appear in all organisms and there is no evidence for a common ancestor. The biological, chemical, or evolutionary reasons for two enzyme classes involved in essentially identical processes remain obscure. The reaction catalyzed by an aminoacyl-tRNA synthetase is

$$\text{Amino acid} + \text{tRNA} + \text{ATP} \xrightleftharpoons[\text{aminoacyl-tRNA synthetase}]{Mg^{2+}} \text{aminoacyl-tRNA} + \text{AMP} + PP_i$$

The activation reaction occurs in two steps in the enzyme's active site. In the first step an enzyme-bound intermediate, aminoacyl adenylate (aminoacyl-AMP), forms when the carboxyl group of the amino acid reacts with the α-phosphoryl group of ATP to form an anhydride linkage, with displacement of pyrophosphate. In the second step the aminoacyl group is transferred from enzyme-bound aminoacyl-AMP to its corresponding specific tRNA. The course of this second step depends on the class to which the enzyme belongs. The resulting ester linkage between the amino acid and the tRNA (Fig. 27–17) has a highly negative standard free energy of hydrolysis ($\Delta G'^{\circ} = -29$ kJ/mol). The pyrophosphate formed in the activation reaction undergoes hydrolysis to phosphate by inorganic pyrophosphatase. Thus *two* high-energy phosphate bonds are ultimately expended for each amino acid molecule activated, rendering the overall reaction for amino acid activation essentially irreversible:

$$\text{Amino acid} + \text{tRNA} + \text{ATP} \xrightarrow[\substack{\text{aminoacyl-tRNA synthetase}\\ \text{inorganic pyrophosphatase}}]{Mg^{2+}} \text{aminoacyl-tRNA} + \text{AMP} + 2P_i \qquad \Delta G'^{\circ} \approx -29 \text{ kJ/mol}$$

table 27–8

Two Classes of Aminoacyl-tRNA Synthetases*

Class I	Class II
Arg	Ala
Cys	Asn
Gln	Asp
Glu	Gly
Ile	His
Leu	Lys
Met	Phe
Trp	Pro
Tyr	Ser
Val	Thr

*Here, Arg represents arginyl-tRNA synthetase, and so forth. The classification applies to all organisms for which tRNA synthetases have been analyzed and is based on protein structural distinctions and on the mechanistic distinction outlined in Figure 27–16.

Proofreading by Aminoacyl-tRNA Synthetases The aminoacylation of tRNA accomplishes (1) activation of an amino acid for peptide bond formation and (2) attachment of the amino acid to an adaptor tRNA that ensures appropriate placement of the amino acid in a growing polypeptide. The identity of the amino acid attached to a tRNA is not checked on the ribosome, so attachment of the correct amino acid to the tRNA is essential to the fidelity of protein synthesis.

Enzyme specificity is limited by the binding energy available from enzyme-substrate interactions (Chapter 8). Discrimination between two similar amino acid substrates has been studied in detail in the case of Ile-$tRNA^{Ile}$ synthetase, which distinguishes between valine and isoleucine, differing by only a single methylene ($—CH_2$) group. Ile-$tRNA^{Ile}$ synthetase favors activation of isoleucine (to form Ile-AMP) over valine by a factor of 200—as expected given the amount that a methylene group could enhance the binding of isoleucine over valine. Yet valine is erroneously incorporated into proteins in positions normally occupied by isoleucine at a frequency of only about 1 in 3,000. This greater than tenfold increase in accuracy is brought about by a proofreading function of Ile-tRNA synthetase, also present in some other aminoacyl-tRNA synthetases. Recall a general principle from the discussion of proofreading by DNA polymerases (p. 938): if available binding interactions do not provide sufficient discrimination between two substrates, the necessary specificity can be achieved by substrate-specific binding in *two successive* steps. The effect of forcing the system

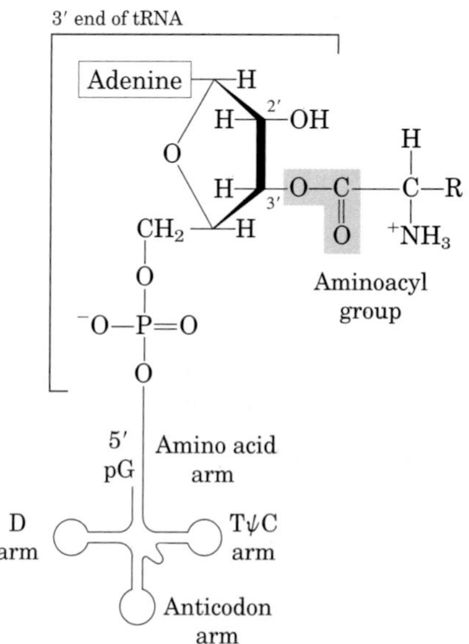

figure 27–17
General structure of aminoacyl-tRNAs. The aminoacyl group is esterified to the 3′ position of the terminal adenylate residue. The ester linkage that both activates the amino acid and joins it to the tRNA is shaded red.

$$\begin{array}{c} COO^- \\ | \\ H_3\overset{+}{N}-C-H \\ | \\ H-C-CH_3 \\ | \\ CH_3 \end{array} \qquad \begin{array}{c} COO^- \\ | \\ H_3\overset{+}{N}-C-H \\ | \\ H-C-CH_3 \\ | \\ CH_2 \\ | \\ CH_3 \end{array}$$

Valine Isoleucine

through two successive filters is multiplicative. In the case of Ile-tRNA synthetase, the first filter is the initial binding of the amino acid to the enzyme and its activation to aminoacyl-AMP. The second filter is the binding of *incorrect* aminoacyl-AMP products to a separate active site on the enzyme. Substrate that binds in this second active site is hydrolyzed. The R group of valine is slightly smaller than that of isoleucine; Val-AMP fits the hydrolytic (proofreading) site of the Ile-tRNA synthetase, but Ile-AMP does not. Thus Val-AMP is hydrolyzed to valine and AMP in the proofreading active site, and tRNA bound to the synthetase does not become aminoacylated to the wrong amino acid.

In addition to proofreading after formation of the aminoacyl-AMP intermediate, most aminoacyl-tRNA synthetases can also hydrolyze the ester linkage between amino acids and tRNAs in aminoacyl-tRNAs. This hydrolysis is greatly accelerated for incorrectly charged tRNAs, providing yet a third filter to enhance the fidelity of the overall process. The few aminoacyl-tRNA synthetases that activate amino acids with no close structural relatives demonstrate little or no proofreading activity; in these cases, the active site for aminoacylation can sufficiently discriminate between the proper substrate and incorrect amino acids.

The overall error rate of protein synthesis ($\sim$1 mistake per 10^4 amino acids incorporated) is not nearly as low as for DNA replication. Flaws in a protein are eliminated when the protein is degraded and are not passed on to future generations, so they have less biological significance. The degree of fidelity in protein synthesis is sufficient to ensure that most proteins contain no mistakes and that the large amount of energy required to synthesize a protein is rarely wasted.

Interaction between an Aminoacyl-tRNA Synthetase and a tRNA: A "Second Genetic Code" An individual aminoacyl-tRNA synthetase must be specific not only for a single amino acid but for certain tRNAs as well. Discriminating among dozens of tRNAs is just as important for the overall fidelity of protein biosynthesis as distinguishing among amino acids. The interaction between aminoacyl-tRNA synthetases and tRNAs has been referred to as the "second genetic code," reflecting its critical role in maintaining the accuracy of protein synthesis. The "coding" rules appear to be more complex than those in the "first" code.

Figure 27–18 summarizes what we know about the nucleotides involved in recognition by some aminoacyl-tRNA synthetases. Some nucleotides are conserved in all tRNAs and therefore cannot be used for discrimination. Nucleotide positions that are involved in discrimination by the aminoacyl-tRNA synthetases have been identified by observing that changes at these nucleotides alter the enzyme's substrate specificity. These interactions seem to be concentrated in the amino acid arm and the anticodon arm, including the nucleotides of the anticodon itself, but are also located in other parts of the molecule. Determination of the crystal structures of aminoacyl-tRNA synthetases complexed with their cognate tRNAs and ATP has added a great deal to our understanding of these interactions (Fig. 27–19).

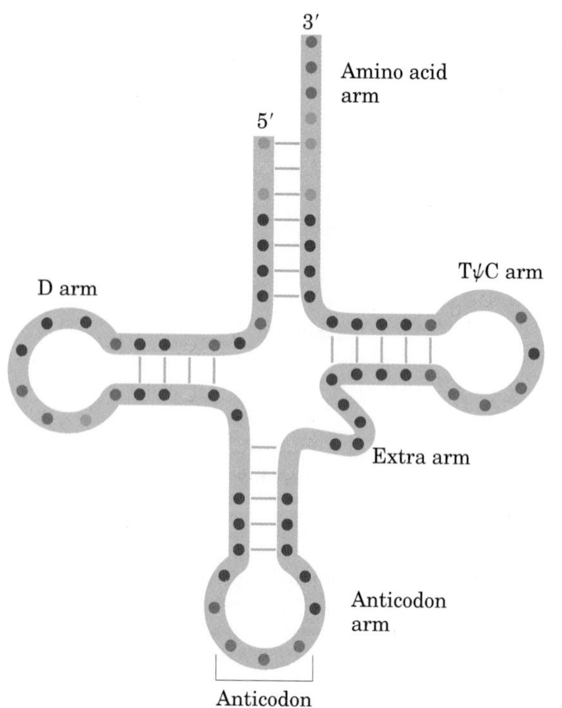

figure 27–18

Known positions in tRNAs recognized by aminoacyl-tRNA synthetases. Positions in blue are the same in all tRNAs and therefore cannot be used to discriminate one from another. Other positions are known recognition points for one (orange) or more (green) aminoacyl-tRNA synthetases. Structural features other than sequence are important for recognition by some of the synthetases.

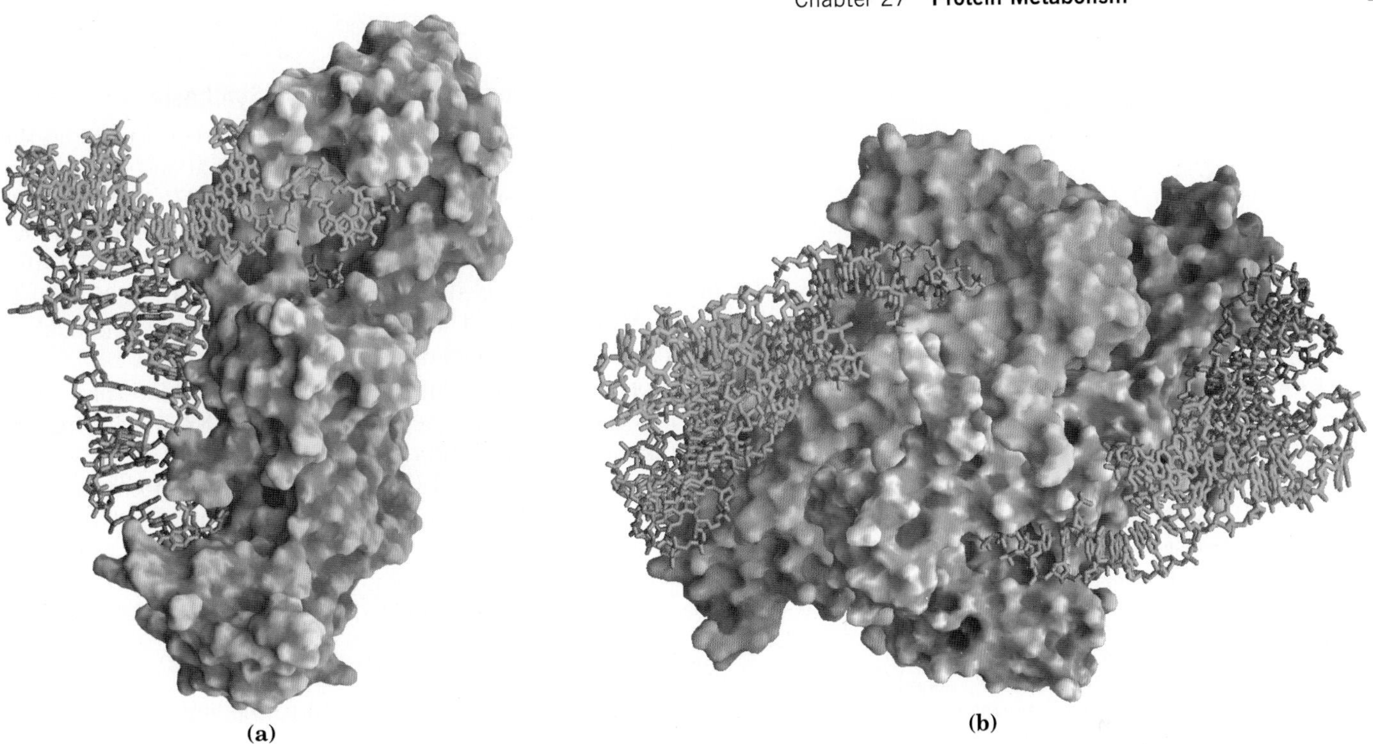

figure 27–19
Aminoacyl-tRNA synthetases. Both synthetases are complexed with their cognate tRNAs (green stick structures). Bound ATP (red) pinpoints the active site near the end of the aminoacyl arm. **(a)** This Gln-tRNA synthetase from *E. coli* is a typical monomeric type I synthetase. **(b)** Asp-tRNA synthetase from yeast is a typical dimeric type II synthetase.

Ten or more specific nucleotides may be involved in recognition of a tRNA by its specific aminoacyl-tRNA synthetase. In contrast, across a range of organisms from bacteria to humans, the primary determinant of tRNA recognition by the Ala-tRNA synthetases is a single G=U base pair in the amino acid arm of $tRNA^{Ala}$ (Fig. 27–20a). A short RNA with as few as seven base pairs arranged in a simple hairpin minihelix is efficiently aminoacylated by the Ala-tRNA synthetase as long as the RNA contains the critical G=U (Fig. 27–20b). This relatively simple alanine system may be an evolutionary relic of a period when RNA oligonucleotides, ancestors to tRNA, were aminoacylated in a primitive system for protein synthesis.

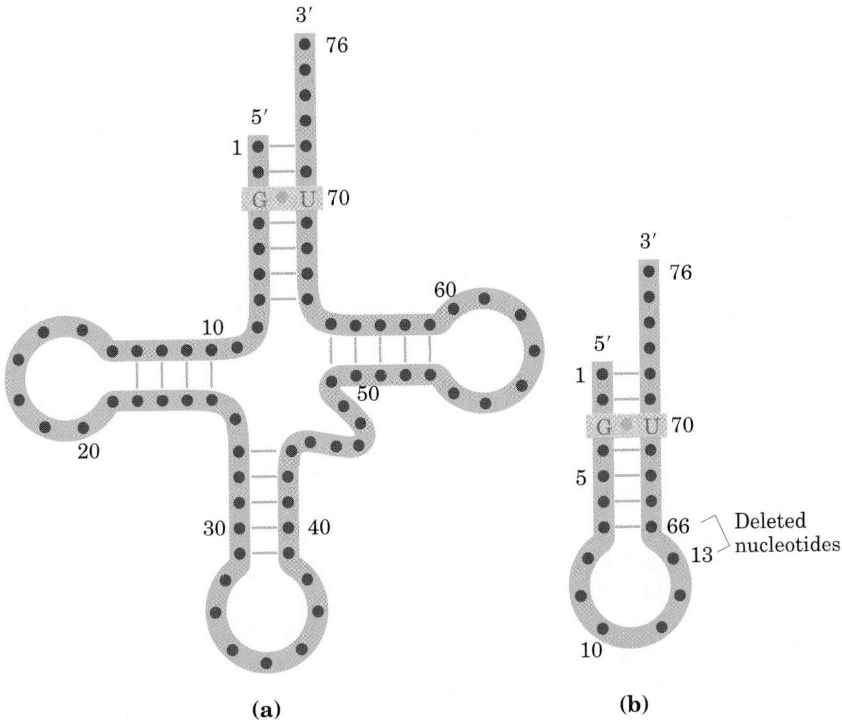

figure 27–20
(a) The $tRNA^{Ala}$ structural elements recognized by the Ala-tRNA synthetase are unusually simple. A single G=U base pair (red) is the only element needed for specific binding and aminoacylation. **(b)** A short synthetic RNA minihelix, which has the critical G=U base pair but lacks most of the remaining tRNA structure, is specifically aminoacylated with alanine almost as efficiently as the complete $tRNA^{Ala}$.

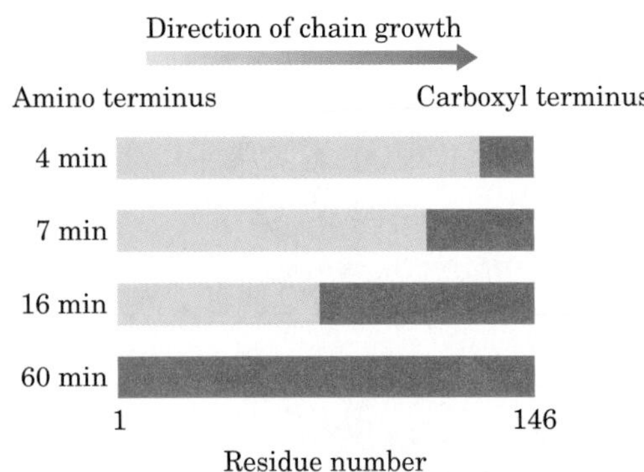

figure 27–21

Proof that polypeptides grow by addition of new amino acid residues to the carboxyl end—the Dintzis experiment. Reticulocytes (immature erythrocytes) actively synthesizing hemoglobin were incubated with radioactive leucine (selected because it occurs frequently in both the α- and β-globin chains). Samples of completed α chains were isolated from the reticulocytes at various times afterward, and the distribution of radioactivity was determined. The dark red zones show the portions of completed α-globin chains containing radioactive Leu residues. At 4 min, only a few residues at the carboxyl end of α-globin were labeled, because the only *complete* globin chains that incorporated label after 4 min were those that had nearly completed synthesis at the time the label was added. With longer incubation times, successively longer segments of the polypeptide contained labeled residues, always in a block at the carboxyl end of the chain. The unlabeled end of the polypeptide (the amino terminus) was thus defined as the initiating end, which means that polypeptides grow by successive addition of amino acids to the carboxyl end.

Stage 2: A Specific Amino Acid Initiates Protein Synthesis

Protein synthesis begins at the amino-terminal end and proceeds by the stepwise addition of amino acids to the carboxyl-terminal end of the polypeptide, as determined by Howard Dintzis in 1961 (Fig. 27–21). The AUG initiation codon thus specifies an *amino-terminal* methionine residue. Although methionine has only one codon, (5′)AUG, all organisms have two tRNAs for methionine. One is used exclusively when (5′)AUG is the initiation codon for protein synthesis. The second is used to code for methionine in an internal position in a polypeptide.

The distinction between an initiating (5′)AUG and an internal one is straightforward. In bacteria, the two separate types of tRNA specific for methionine are designated $tRNA^{Met}$ and $tRNA^{fMet}$. The amino acid residue incorporated in response to the (5′)AUG initiation codon is *N*-formylmethionine (fMet). It arrives at the ribosome as *N*-formylmethionyl-$tRNA^{fMet}$ (fMet-$tRNA^{fMet}$), which is formed in two successive reactions. First, methionine is attached to $tRNA^{fMet}$ by the Met-tRNA synthetase (which in *E. coli* aminoacylates both $tRNA^{fMet}$ and $tRNA^{Met}$):

$$\text{Methionine} + \text{tRNA}^{\text{fMet}} + \text{ATP} \longrightarrow \text{Met-tRNA}^{\text{fMet}} + \text{AMP} + \text{PP}_i$$

Next, a transformylase transfers a formyl group to the amino group of the Met residue from N^{10}-formyltetrahydrofolate:

$$N^{10}\text{-Formyltetrahydrofolate} + \text{Met-tRNA}^{\text{fMet}} \longrightarrow \text{tetrahydrofolate} + \text{fMet-tRNA}^{\text{fMet}}$$

The transformylase is more selective than the Met-tRNA synthetase; it is specific for Met residues attached to $tRNA^{fMet}$, presumably recognizing some unique structural feature of that tRNA. By contrast, Met-$tRNA^{Met}$ inserts methionine in interior positions in polypeptides.

Addition of the *N*-formyl group to the amino group of methionine by the transformylase prevents fMet from entering interior positions in a polypeptide while also allowing fMet-$tRNA^{fMet}$ to be bound at a specific initiation site on the ribosome that accepts neither Met-$tRNA^{Met}$ nor any other aminoacyl-tRNA.

H COO⁻ | H—C(=O)—N—C—H | CH₂ | CH₂ | S | CH₃

N-Formylmethionine

In eukaryotic cells, all polypeptides synthesized by cytosolic ribosomes begin with a Met residue (rather than fMet), but, again, a specialized initiating tRNA is used that is distinct from the $tRNA^{Met}$ used at (5′)AUG codons at interior positions in the mRNA. Polypeptides synthesized by the ribosomes in the mitochondria and chloroplasts, however, begin with *N*-formylmethionine. This strongly supports the view that mitochondria and chloroplasts originated from bacterial ancestors that were symbiotically incorporated into precursor eukaryotic cells at an early stage of evolution (see Fig. 2–15).

How can the single (5′)AUG codon identify both the starting *N*-formylmethionine (or methionine, in eukaryotes) and Met residues that occur in interior positions in polypeptides? The answer becomes clear in the following discussion.

The Three Steps of Initiation The **initiation** of polypeptide synthesis in bacteria requires (1) the 30S ribosomal subunit, (2) the mRNA coding for the polypeptide to be made, (3) the initiating fMet-$tRNA^{fMet}$, (4) a set of three proteins called initiation factors (IF-1, IF-2, and IF-3), (5) GTP, (6) the 50S ribosomal subunit, and (7) Mg^{2+}. Formation of the initiation complex takes place in three steps (Fig. 27–22).

In step ① the 30S ribosomal subunit binds two initiation factors, IF-1 and IF-3. IF-3 prevents the 30S and 50S subunits from combining prematurely. The mRNA then binds to the 30S subunit. The initiating (5′)AUG is guided to its correct position by the **Shine-Dalgarno sequence** in the

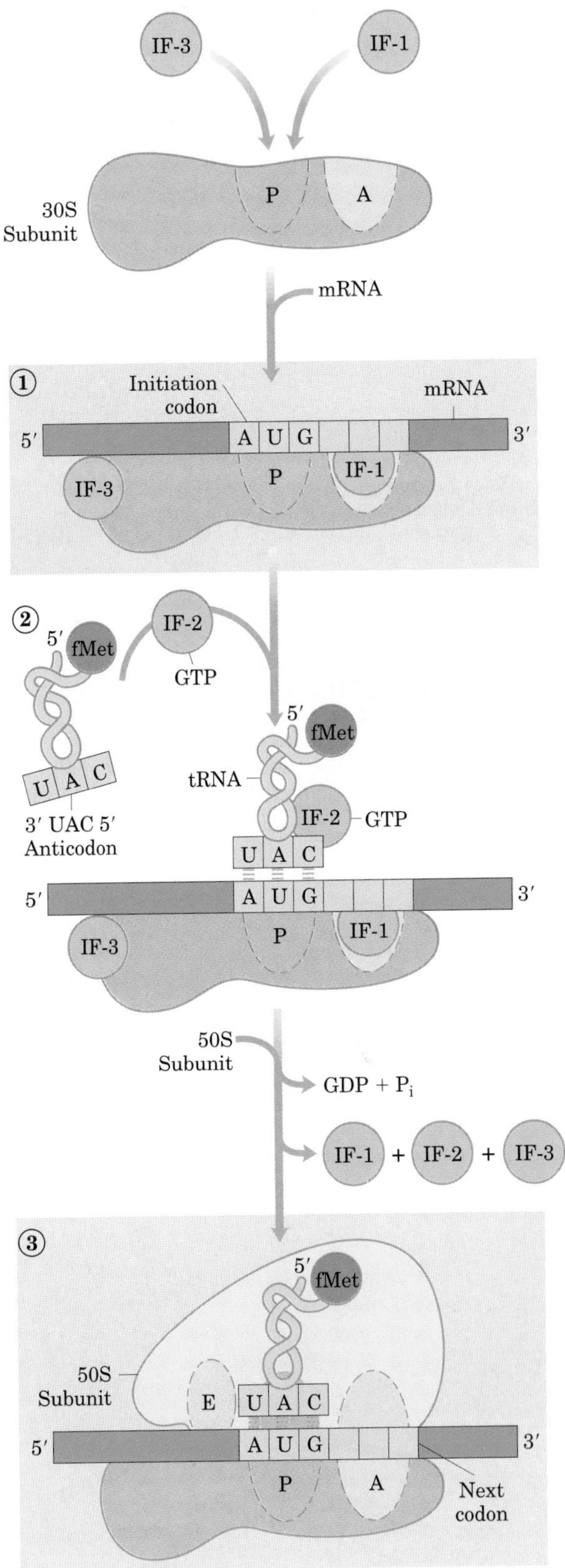

figure 27–22

Formation of the initiation complex. The complex forms in three steps (described in the text) at the expense of the hydrolysis of GTP to GDP and P_i. IF-1, IF-2, and IF-3 are initiation factors. P designates the peptidyl site, A the aminoacyl site, and E the exit site. Here the anticodon of the tRNA is oriented 3′ to 5′, left to right, as in Figure 27–8 but opposite to the orientation in Figures 27–18 and 27–20.

mRNA. This consensus sequence is an initiation signal of four to nine purine residues, eight to 13 base pairs to the 5′ side of the initiation codon (Fig. 27–23a). It base-pairs with a complementary pyrimidine-rich sequence near the 3′ end of the 16S rRNA of the 30S ribosomal subunit (Fig. 27–23b). This mRNA-rRNA interaction positions the initiating (5′)AUG sequence of the mRNA in the precise position on the 30S subunit where it is required for initiation of translation. The particular (5′)AUG where fMet-$tRNA^{fMet}$ is to be bound is distinguished from other methionine codons by its proximity to the Shine-Dalgarno sequence in the mRNA.

Bacterial ribosomes have three sites that bind aminoacyl-tRNAs, the **aminoacyl** or **A site,** the **peptidyl** or **P site,** and the **exit** or **E site.** Both the 30S and the 50S subunits contribute to the characteristics of the A and P sites, whereas the E site is largely confined to the 50S subunit. The initiating (5′)AUG is positioned at the P site, the only site to which fMet-$tRNA^{fMet}$ can bind (Fig. 27–22). The fMet-$tRNA^{fMet}$ is the only aminoacyl-tRNA that binds first to the P site; during the subsequent elongation stage, all other incoming aminoacyl-tRNAs (including the Met-$tRNA^{Met}$ that binds to interior AUGs) bind first to the A site and only subsequently to the P and E sites. The E site is the site from which the "uncharged" tRNAs leave during elongation. The initiation factor IF-1 binds at the A site and prevents tRNA binding at this site during initiation.

In step ② of the initiation process (Fig. 27–22), the complex consisting of the 30S ribosomal subunit, IF-3, and mRNA is joined by both GTP-bound IF-2 and the initiating fMet-$tRNA^{fMet}$. The anticodon of this tRNA now pairs correctly with the mRNA's initiation codon.

In step ③ this large complex combines with the 50S ribosomal subunit; simultaneously, the GTP bound to IF-2 is hydrolyzed to GDP and P_i, which are released from the complex. All three initiation factors depart from the ribosome at this point.

Completion of the steps in Figure 27–22 produces a functional 70S ribosome called the **initiation complex,** containing the mRNA and the initiating fMet-$tRNA^{fMet}$. The correct binding of the fMet-$tRNA^{fMet}$ to the P site in the complete 70S initiation complex is assured by at least three points of recognition and attachment: the codon-anticodon interaction involving the initiating AUG fixed in the P site; interaction between the Shine-Dalgarno sequence in the mRNA and the 16S rRNA; and binding interactions between the ribosomal P site and the fMet-$tRNA^{fMet}$. The initiation complex is now ready for elongation.

Initiation in Eukaryotic Cells Eukaryotic translation is generally similar to translation in bacterial cells, with most of the significant differences appearing in the mechanism of initiation. Eukaryotic mRNAs are bound to the ribosome as a complex with a number of specific binding proteins. Several of these are thought to tie together the 5′ and 3′ ends of the message. At the 3′ end, the mRNA is bound by a protein called poly(A) binding protein (PAB). Eukaryotic cells have at least nine initiation factors. A complex

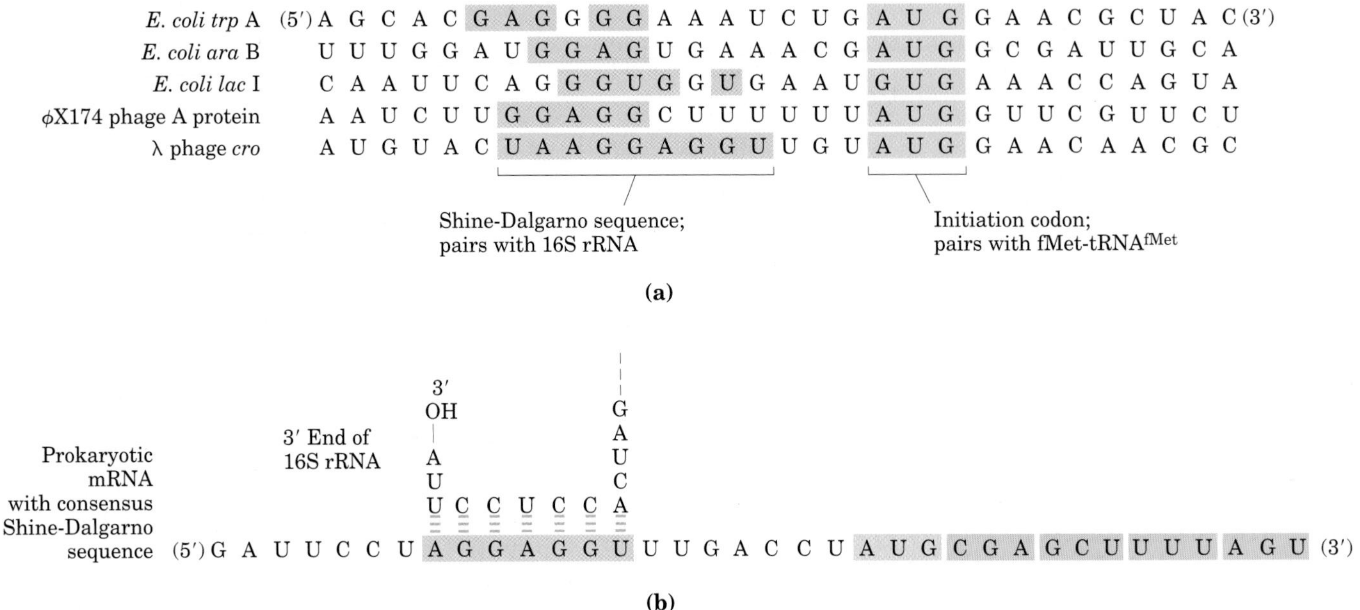

figure 27–23

Sequences on the mRNA that serve as signals for initiation of protein synthesis in bacteria. (a) Alignment of the initiating AUG (shaded in green) at its correct location on the 30S ribosomal subunit depends in part on upstream Shine-Dalgarno sequences (shaded in red). Portions of the mRNA transcripts of five prokaryotic genes are shown. **(b)** The Shine-Dalgarno sequences pair with a sequence near the 3′ end of the 16S rRNA.

called eIF4F, which includes the proteins eIF4E, eIF4G, and eIF4A, binds to the 5′ cap through eIF4E. The protein eIF4G binds to both eIF4E and PAB, effectively tying them together (Fig. 27–24). The protein eIF4A has an RNA helicase activity. It is the eIF4F complex that associates with another factor, eIF3, and with the 40S ribosomal subunit. The efficiency of translation is affected by many properties of the mRNA and proteins in this complex, including the length of the 3′ poly(A) tract (in most cases, longer is better). The end-to-end arrangement facilitates translational regulation of gene expression by mechanisms that we consider in the next chapter.

The initiating (5′)AUG is located within the mRNA not by its proximity to a Shine-Dalgarno–like sequence, but by a scan of the mRNA from the 5′ end until the first AUG is encountered, signaling the beginning of the reading frame. The eIF4F complex is probably involved in the scanning process, perhaps using the RNA helicase activity of eIF4A to eliminate secondary structure in the 5′ untranslated portion of the mRNA. Scanning is facilitated by another protein, eIF4B.

The roles of the various bacterial and eukaryotic initiation factors in the overall process are summarized in Table 27–9. The mechanism by which these proteins act remains an important area of investigation.

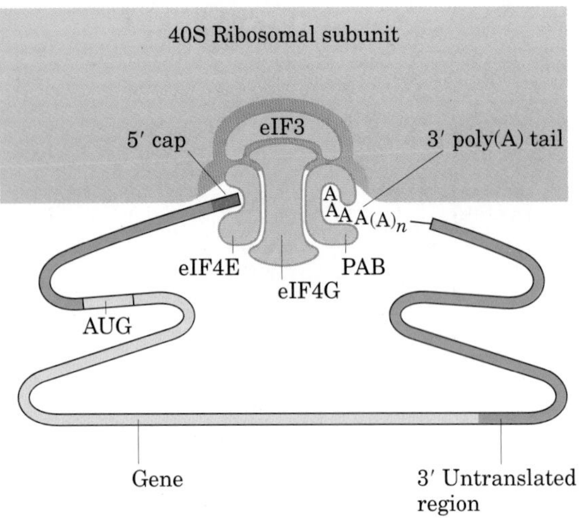

figure 27–24

Protein complexes in the formation of a eukaryotic initiation complex. The 3′ and 5′ ends of eukaryotic mRNAs are linked by a complex of proteins that includes several initiation factors and the poly(A) binding protein (PAB). The factors eIF4E and eIF4G are part of a larger complex called eIF4F. This complex binds to the 40S ribosomal subunit.

table 27–9

Protein Factors Required for Initiation of Translation in Bacterial and Eukaryotic Cells

Bacterial

Factor	Function
IF-1	Prevents premature binding of tRNAs to A site
IF-2	Facilitates binding of fMet-$tRNA^{fMet}$ to 30S ribosomal subunit
IF-3	Binds to 30S subunit; prevents premature association of 50S subunit; enhances specificity of P site for fMet-$tRNA^{fMet}$

Eukaryotic

Factor*	Function
eIF2	Facilitates binding of initiating Met-$tRNA^{Met}$ to 40S ribosomal subunit
eIF2B, eIF3	First factors to bind 40S subunit; facilitate subsequent steps
eIF4A	RNA helicase activity removes secondary structure in the mRNA to permit binding to 40S subunit; part of the eIF4F complex
eIF4B	Binds to mRNA; facilitates scanning of mRNA to locate the first AUG
eIF4E	Binds to the 5′ cap of mRNA; part of the eIF4F complex
eIF4G	Binds to eIF4E and to poly(A) binding protein (PAB); part of the eIF4F complex
eIF5	Promotes dissociation of several other initiation factors from 40S subunit as a prelude to association of 60S subunit to form 80S initiation complex
eIF6	Facilitates dissociation of inactive 80S ribosome into 40S and 60S subunits

*The prefix "e" identifies these as eukaryotic factors.

Stage 3: Peptide Bonds Are Formed in the Elongation Stage

The third stage of protein synthesis is **elongation.** Again, our initial focus is on bacterial cells. Elongation requires (1) the initiation complex described above, (2) aminoacyl-tRNAs, (3) a set of three soluble cytosolic proteins called **elongation factors** (EF-Tu, EF-Ts, and EF-G in bacteria), and (4) GTP. Three steps are needed to add each amino acid residue, and they are repeated as many times as there are residues to be added.

Elongation Step 1: Binding of an Incoming Aminoacyl-tRNA In the first step of the elongation cycle (Fig. 27–25), the appropriate incoming aminoacyl-tRNA is first bound to a complex of GTP-bound EF-Tu. The resulting aminoacyl-tRNA–EF-Tu·GTP complex binds to the A site of the 70S initiation complex. The GTP is hydrolyzed and an EF-Tu·GDP complex is released from the 70S ribosome. The EF-Tu·GTP complex is regenerated in a process involving EF-Ts and GTP.

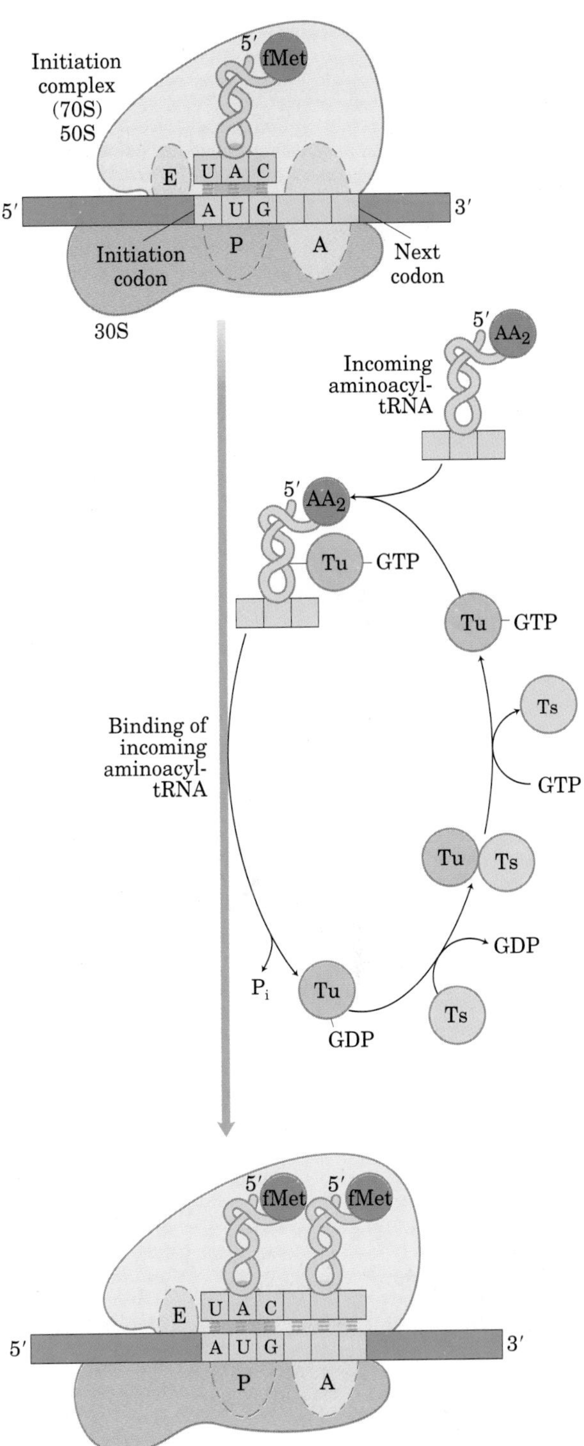

figure 27–25
First step in elongation (bacteria): binding of the second aminoacyl-tRNA. The second aminoacyl-tRNA enters the A site of the ribosome bound to EF-Tu (shown here as Tu), which also contains GTP. Binding of the second aminoacyl-tRNA to the A site is accompanied by hydrolysis of the GTP to GDP and P_i and release of the EF-Tu·GDP complex from the ribosome. The bound GDP is released when the EF-Tu·GDP complex binds to EF-Ts, and EF-Ts is subsequently released when another molecule of GTP binds to EF-Tu. This recycles EF-Tu and makes it available to repeat the cycle.

Elongation Step 2: Peptide Bond Formation A peptide bond is now formed between the two amino acids that are still bound by their tRNAs to the A and P sites on the ribosome. This occurs by the transfer of the initiating *N*-formylmethionyl group from its tRNA to the amino group of the second amino acid, now in the A site (Fig. 27–26). The α-amino group of the amino acid in the A site acts as a nucleophile, displacing the tRNA in the P site to form the peptide bond. This reaction produces a dipeptidyl-tRNA in the A site, and the now "uncharged" (deacylated) $tRNA^{fMet}$ remains bound to the P site. The tRNAs then shift to a hybrid binding state, with elements of each spanning two different sites on the ribosome, as shown in Figure 27–26.

The enzymatic activity that catalyzes peptide bond formation has historically been referred to as **peptidyl transferase** and was widely assumed to be intrinsic to one or more of the proteins in the large ribosomal subunit. This reaction may in fact be catalyzed by the 23S rRNA, possibly adding to the known catalytic repertoire of ribozymes.

Elongation Step 3: Translocation In the final step of the elongation cycle, **translocation,** the ribosome moves one codon toward the 3′ end of the mRNA (Fig. 27–27). This movement shifts the anticodon of the dipeptidyl-tRNA, which is still attached to the second codon of the mRNA, from the A site to the P site, and shifts the deacylated tRNA from the P site to the E site. The tRNA is then released from the E site into the cytosol. The third codon of the mRNA now lies in the A site and the second codon in the P site. Movement of the ribosome along the mRNA requires EF-G (also called translocase) and the energy provided by hydrolysis of another molecule of GTP. A change in the three-dimensional conformation of the entire ribosome enables its movement along the mRNA. The structure of EF-G mimics the structure of the EF-Tu/tRNA complex (see p. 1020), suggesting that EF-G may bind at the A site and displace the peptidyl-tRNA.

The ribosome, with its attached dipeptidyl-tRNA and mRNA, is now ready for the next elongation cycle and attachment of a third amino acid residue. This process occurs in the same way as addition of the second residue (as shown in Figs 27–25, 27–26, and 27–27). For each amino acid residue correctly added to the growing polypeptide, two GTPs are hydrolyzed to GDP and P_i as the ribosome moves from codon to codon along the mRNA toward the 3′ end.

The polypeptide remains attached to the tRNA of the most recent amino acid to be inserted. This association maintains the functional connection between the information in the mRNA and its decoded polypeptide output. At the same time, the ester linkage between this tRNA and the carboxyl terminus of the growing polypeptide activates the terminal carboxyl group for nucleophilic attack by the incoming amino acid to form a new peptide bond (Fig. 27–26). As the existing ester linkage between the polypeptide and tRNA is broken during peptide bond formation, the linkage between the polypeptide and the information in the mRNA persists because each newly added amino acid is itself attached to its tRNA.

The elongation cycle in eukaryotes is quite similar to that in prokaryotes. Three eukaryotic elongation factors (eEF1α, eEF1$\beta\gamma$, and eEF2) have functions analogous to those of the bacterial elongation factors EF-Tu, EF-Ts, and EF-G, respectively. Eukaryotic ribosomes do not have an E site. Uncharged tRNAs are expelled directly from the P site.

Proofreading on the Ribosome The GTPase activity of EF-Tu during the first step of elongation in bacterial cells (Fig. 27–25) makes an important contribution to the rate and fidelity of the overall biosynthetic process. Both

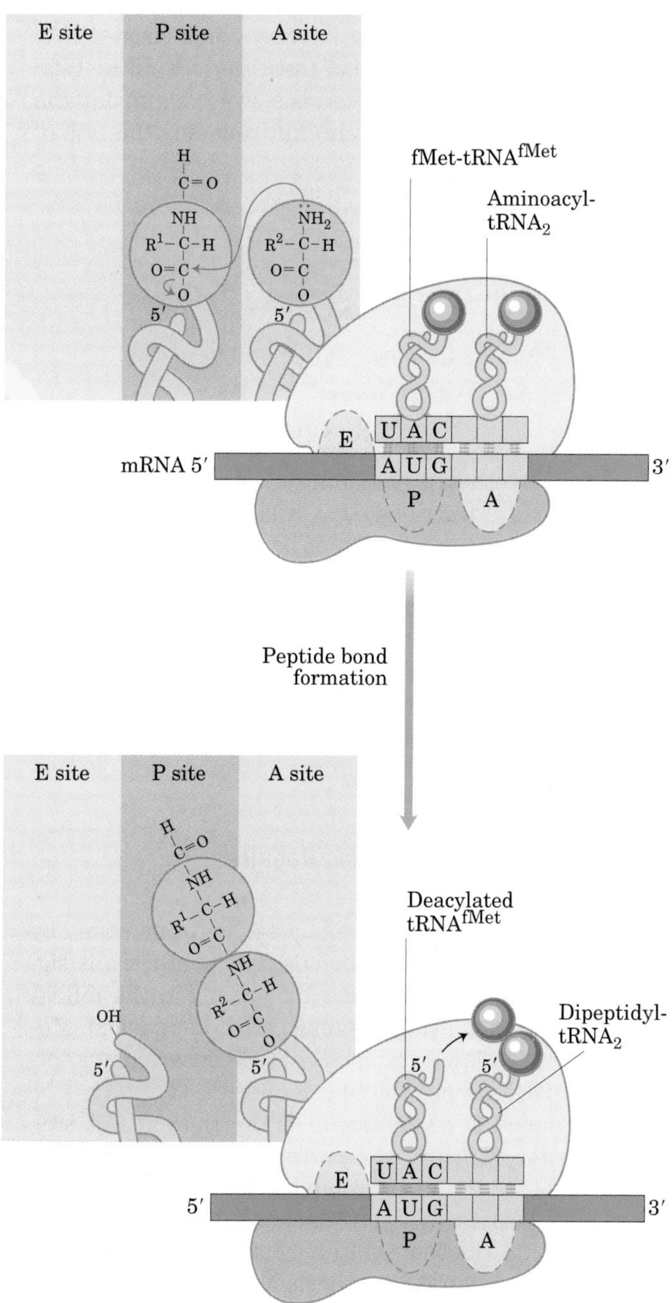

figure 27–26

Second step in elongation (bacteria): formation of the first peptide bond. The peptidyl transferase catalyzing this reaction is probably the 23S rRNA ribozyme. The *N*-formylmethionyl group is transferred to the amino group of the second aminoacyl-tRNA in the A site, forming a dipeptidyl-tRNA. At this stage, both tRNAs bound to the ribosome shift position in the 50S subunit to take up a hybrid binding state. The uncharged tRNA shifts so that its 3′ and 5′ ends are in the E site. Similarly, the 3′ and 5′ ends of the peptidyl tRNA shift to the P site. The anticodons remain in the A and P sites.

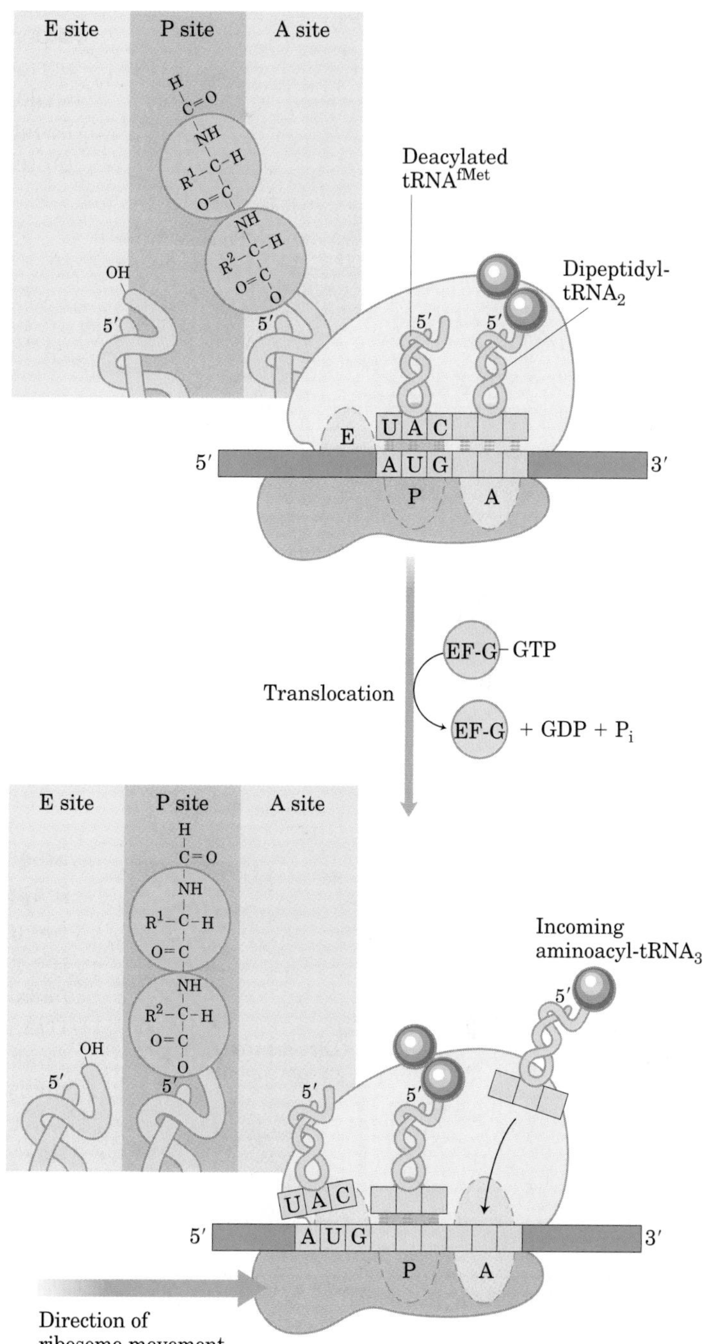

figure 27–27

Third step in elongation (bacteria): translocation. The ribosome moves one codon toward the 3′ end of mRNA, using energy provided by hydrolysis of GTP bound to EF-G (translocase). The dipeptidyl-tRNA is now entirely in the P site, leaving the A site open for the incoming (third) aminoacyl-tRNA. The uncharged tRNA dissociates from the E site, and the elongation cycle begins again.

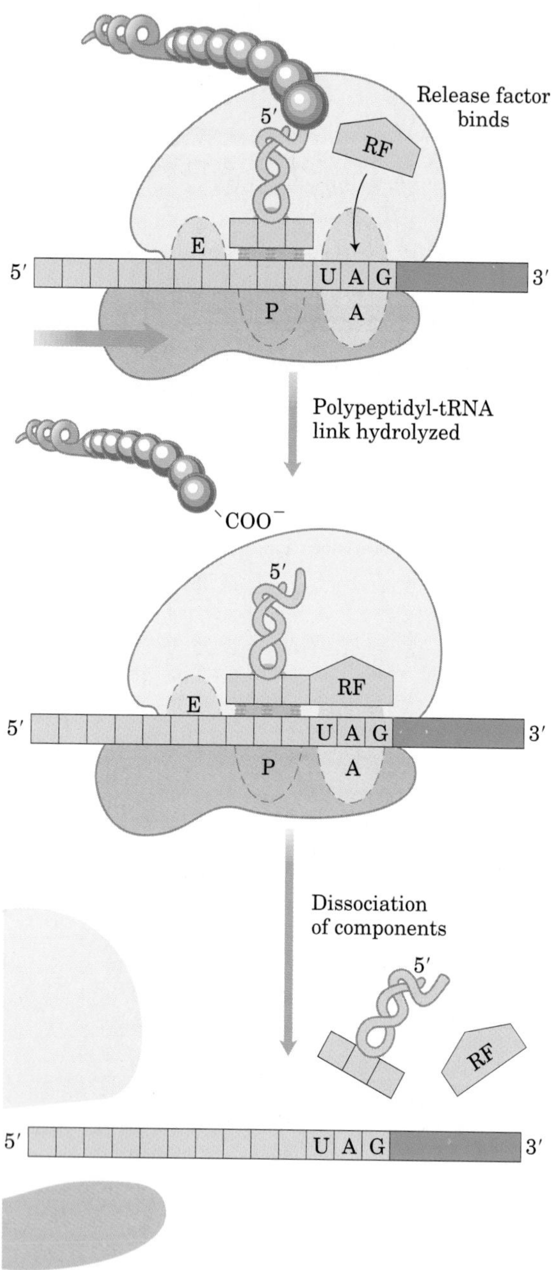

figure 27–28
Termination of protein synthesis in bacteria. Termination occurs in response to a termination codon in the A site. First, a release factor (RF_1 or RF_2 depending on which termination codon is present) binds to the A site. This leads to hydrolysis of the ester linkage between the nascent polypeptide and the tRNA in the P site and release of the completed polypeptide. Finally, the mRNA, deacylated tRNA, and release factor leave the ribosome, and the ribosome dissociates into its 30S and 50S subunits.

the EF-Tu·GTP and EF-Tu·GDP complexes exist for a few milliseconds before they dissociate. These two intervals provide opportunities for the codon-anticodon interactions to be proofread. Incorrect aminoacyl-tRNAs normally dissociate from the A site during one of these periods. If the GTP analog GTPγS is used in place of GTP, hydrolysis is slowed, improving the fidelity (by increasing the proofreading intervals) but reducing the rate of protein synthesis.

Guanosine 5′-*O*-(3-thiotriphosphate) (GTPγS)

The process of protein synthesis (including the characteristics of codon-anticodon pairing already described) has clearly been optimized through evolution to balance the requirements of both speed and fidelity. Improved fidelity might diminish speed, while increases in speed would probably compromise fidelity. Note that the proofreading mechanism on the ribosome establishes only that the proper codon-anticodon pairing has taken place. The identity of the amino acids attached to tRNAs is not checked on the ribosome. If a tRNA is aminoacylated with the wrong amino acid (as can be done experimentally), this incorrect amino acid is efficiently incorporated into a protein in response to whatever codon is normally recognized by the tRNA.

Stage 4: Termination of Polypeptide Synthesis Requires a Special Signal

Elongation continues until the ribosome adds the last amino acid coded by the mRNA. Termination, the fourth stage of polypeptide synthesis, is signaled by the presence of one of three termination codons in the mRNA (UAA, UAG, UGA), immediately following the final coded amino acid. Mutations in a tRNA anticodon that allow an amino acid to be inserted at a termination codon are generaly deleterious to the cell (Box 27–3).

In bacteria, once a termination codon occupies the ribosomal A site, three **termination** or **release factors,** the proteins RF_1, RF_2, and RF_3, contribute to (1) hydrolysis of the terminal peptidyl-tRNA bond, (2) release of the free polypeptide and the last tRNA, now uncharged, from the P site, and (3) dissociation of the 70S ribosome into its 30S and 50S subunits, ready to start a new cycle of polypeptide synthesis (Fig. 27–28). RF_1 recognizes the termination codons UAG and UAA, and RF_2 recognizes UGA and UAA. Either RF_1 or RF_2 (depending on which codon is present) binds at a termination codon and induces peptidyl transferase to transfer the growing polypeptide to a water molecule rather than to another amino acid. These release factors have domains thought to mimic the structure of tRNA, as shown for the elongation factor EF-G on p. 1020. The specific function of RF_3 has not been firmly established, although it is thought to release the ribosomal subunit. In eukaryotes, a single release factor called eRF recognizes all three termination codons.

Energy Cost of Fidelity in Protein Synthesis Synthesizing a protein true to the information specified in its mRNA requires energy. Formation of each aminoacyl-tRNA uses two high-energy phosphate groups. ATP is consumed each time an incorrectly activated amino acid is hydrolyzed by the

box 27–3 Induced Variation in the Genetic Code: Nonsense Suppression

When a mutation introduces a termination codon in the interior of a gene, translation is prematurely halted and the incomplete polypeptide is usually inactive. These are called nonsense mutations. The gene can be restored to normal function if a second mutation either (1) converts the misplaced termination codon to a codon specifying an amino acid, or (2) suppresses the effects of the termination codon. Such restorative mutations are called **nonsense suppressors;** they generally involve mutations in tRNA genes that produce altered (suppressor) tRNAs that can recognize the termination codon and insert an amino acid at that position. Most known suppressor tRNAs have single base substitutions in their anticodons.

Suppressor tRNAs constitute an experimentally induced variation in the genetic code to allow the reading of what are usually termination codons, similar to the naturally occurring code variations described in Box 27–2. Nonsense suppression does not completely disrupt normal information transfer in a cell because the cell usually has several copies of each tRNA gene; some of these duplicate genes are weakly expressed and account for only a minor part of the cellular pool of a particular tRNA. Suppressor mutations usually involve a "minor" tRNA, leaving the major tRNA to read its codon normally.

For example, *E. coli* has three identical genes for $tRNA^{Tyr}$, each producing a tRNA with the anticodon (5′)GUA. One of these genes is expressed at relatively high levels and thus its product represents the major $tRNA^{Tyr}$ species; the other two genes are transcribed in only small amounts. A change in the anticodon of the tRNA product of one of these duplicate $tRNA^{Tyr}$ genes, from (5′)GUA to (5′)CUA, produces a minor $tRNA^{Tyr}$ species that will insert tyrosine at UAG stop codons. This insertion of tyrosine at UAG is carried out inefficiently, but can permit production of enough full-length protein from a gene with a nonsense mutation to allow the cell to survive. The major $tRNA^{Tyr}$ continues to translate the genetic code normally for the majority of proteins.

The mutation that leads to the creation of a suppressor tRNA does not always occur in the anticodon. The suppression of UGA nonsense codons generally involves the $tRNA^{Trp}$ that normally recognizes UGG. The alteration that allows it to read UGA (and insert Trp residues at these positions) does not occur in the anticodon. Instead, a G to A change at position 24 (in an arm of the tRNA somewhat removed from the anticodon) alters the tRNA so that it can recognize *both* UGG and UGA. A similar change is found in tRNAs involved in the most common naturally occurring variation in the genetic code (UGA = Trp; see Box 27–2).

Suppression should lead to many abnormally long proteins, but this does not always occur. We do not yet understand many details of the molecular events that occur during translation termination and nonsense suppression.

deacylation activity of an aminoacyl-tRNA synthetase (p. 1041). GTP is cleaved to GDP and P_i during the first elongation step and during the translocation step. More than four high-energy bonds are required for the formation of each peptide bond of a polypeptide.

This represents an exceedingly large thermodynamic "push" in the direction of synthesis: at least 4×30.5 kJ/mol = 122 kJ/mol of phosphodiester bond energy is required to generate a peptide bond, which has a standard free energy of hydrolysis of only about -21 kJ/mol. The net free-energy change in peptide bond synthesis is thus -101 kJ/mol. Proteins are information-containing polymers. The biochemical goal is not simply the formation of a peptide bond, but the formation of a peptide bond between two *specified* amino acids. Each of the high-energy bonds expended in this process plays a critical role in maintaining proper alignment between each new codon in the mRNA and its associated amino acid at the growing end of the polypeptide. This energy permits very high fidelity in the biological translation of the genetic message of mRNA into the amino acid sequence of proteins.

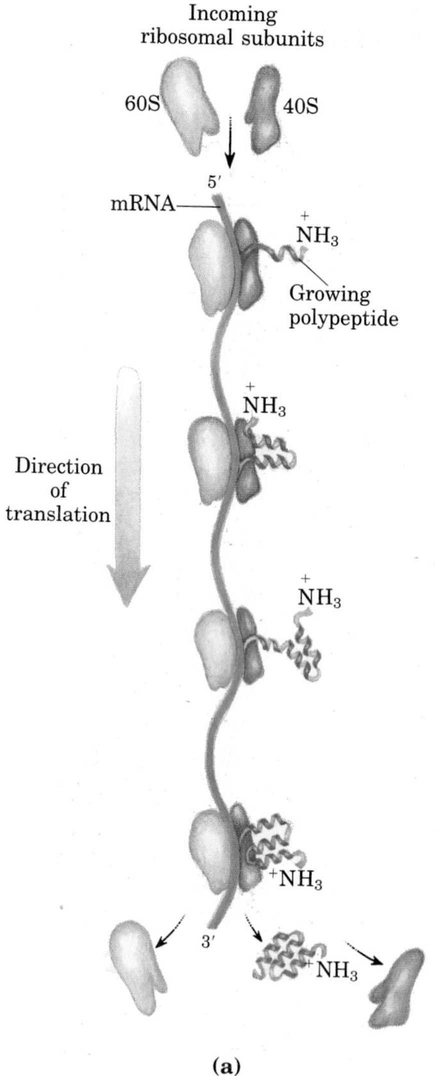

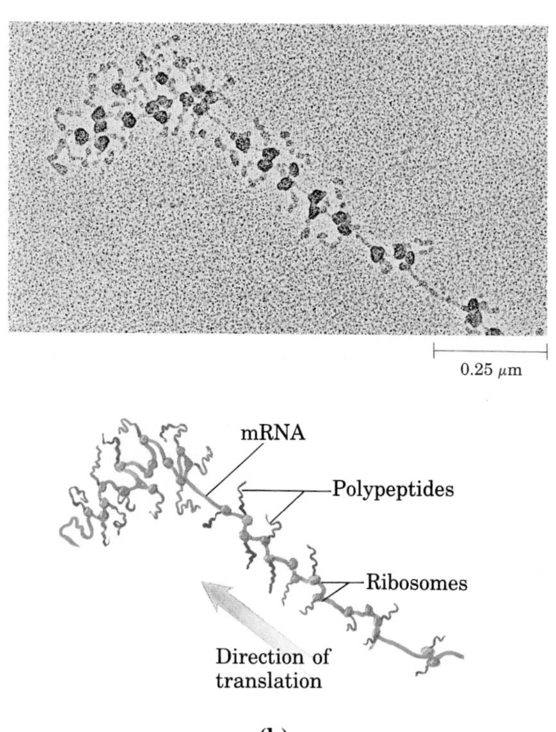

figure 27–29
Polysome. (a) Four ribosomes are shown translating a eukaryotic mRNA molecule simultaneously, moving from the 5′ end to the 3′ end and synthesizing a polypeptide from the amino terminus to the carboxyl terminus. **(b)** Electron micrograph and explanatory diagram of a polysome from the silk gland of a silkworm larva. The mRNA is being translated by many ribosomes simultaneously. The nascent polypeptides become longer as the ribosomes move toward the 3′ end of the mRNA. The final product of this process is silk fibroin.

Rapid Translation of a Single Message by Polysomes Large clusters of 10 to 100 ribosomes that are very active in protein synthesis can be isolated from both eukaryotic and bacterial cells. A fiber between adjacent ribosomes in the cluster (called a **polysome**) is visible in electron micrographs (Fig. 27–29). The connecting strand is a single molecule of mRNA that is being translated simultaneously by many closely spaced ribosomes, allowing the highly efficient use of the mRNA.

In bacteria, transcription and translation are tightly coupled. Messenger RNAs are synthesized and translated in the same 5′→3′ direction. Ribosomes begin translating the 5′ end of the mRNA before transcription is complete (Fig. 27–30). The situation is quite different in eukaryotic cells, where newly transcribed mRNAs must be transferred out of the nucleus before they can be translated.

Bacterial mRNAs generally exist for only a few minutes (p. 1006) before they are degraded by nucleases. In order to maintain high rates of protein synthesis, the mRNA for a given protein or set of proteins must be made

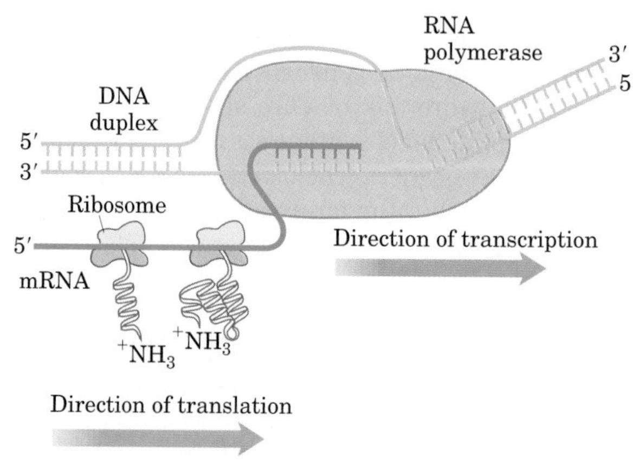

figure 27–30
Coupling of transcription and translation in bacteria. The mRNA is translated by ribosomes while it is still being transcribed from DNA by RNA polymerase. This is possible because the mRNA in bacteria does not have to be transported from a nucleus to the cytoplasm before encountering ribosomes. In this schematic diagram the ribosomes are depicted as smaller than the RNA polymerase. In reality the ribosomes (M_r 2.7 × 10^6) are an order of magnitude larger than the RNA polymerase (M_r 3.9 × 10^5).

continuously and translated with maximum efficiency. The short lifetime of mRNAs in bacteria allows synthesis of a protein to cease rapidly when the protein is no longer needed by the cell.

Stage 5: Newly Synthesized Polypeptide Chains Undergo Folding and Processing

In the fifth and final step of protein synthesis, the nascent polypeptide chain is folded and processed into its biologically active form. During or after its synthesis, the polypeptide progressively assumes its native conformation, with the formation of appropriate hydrogen bonds and van der Waals, ionic, and hydrophobic interactions. In this way the linear or one-dimensional genetic message in the mRNA is converted into the three-dimensional structure of the protein. Some newly made proteins, both prokaryotic and eukaryotic, do not attain their final biologically active conformation until they have been altered by one or more processing reactions called **posttranslational modifications.**

Amino-Terminal and Carboxyl-Terminal Modifications Initially, all polypeptides begin with a residue of *N*-formylmethionine (in bacteria) or methionine (in eukaryotes). However, the formyl group, the amino-terminal Met residue, and often additional amino-terminal (and, in some cases, carboxyl-terminal) residues may be removed enzymatically and thus do not appear in the final functional protein. In as many as 50% of eukaryotic proteins, the amino group of the amino-terminal residue is *N*-acetylated after translation. Carboxyl-terminal residues are also sometimes modified.

Loss of Signal Sequences As we shall see later in this chapter, the 15 to 30 residues at the amino-terminal end of some proteins play a role in directing the protein to its ultimate destination in the cell. Such **signal sequences** are ultimately removed by specific peptidases.

Modification of Individual Amino Acids The hydroxyl groups of certain Ser, Thr, and Tyr residues of some proteins are enzymatically phosphorylated by ATP (Fig. 27–31a); the phosphate groups add negative charges to these polypeptides. The functional significance of this modification varies from one protein to the next. For example, the milk protein casein has many phosphoserine groups that bind Ca^{2+}. Calcium, phosphate, and amino acids are all valuable to suckling young, so casein efficiently provides three essential nutrients. And as we have seen in numerous instances, phosphorylation-dephosphorylation cycles regulate the activity of many enzymes and regulatory proteins.

Extra carboxyl groups may be added to Glu residues of some proteins. For example, the blood-clotting protein prothrombin contains a number of γ-carboxyglutamate residues (Fig. 27–31b) in its amino-terminal region, introduced by an enzyme that requires vitamin K. These carboxyl groups bind Ca^{2+}, required to initiate the clotting mechanism.

Monomethyl- and dimethyllysine residues (Fig. 27–31c) occur in some muscle proteins and in cytochrome *c*. The calmodulin of most organisms contains one trimethyllysine residue at a specific position. In other proteins, the carboxyl groups of some Glu residues undergo methylation, removing their negative charge.

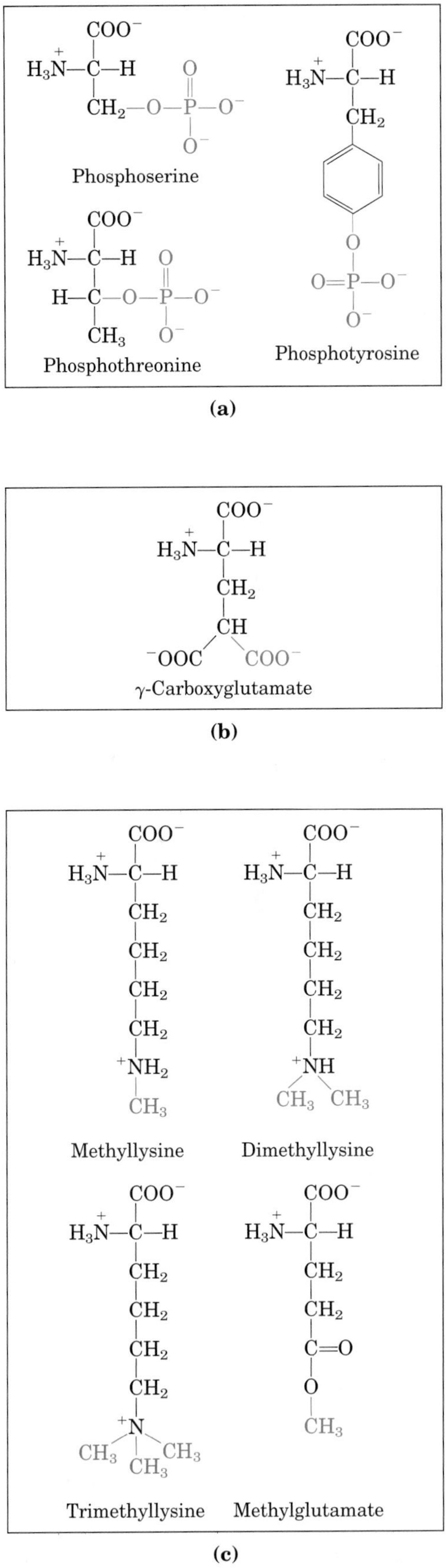

Phosphoserine

Phosphothreonine

Phosphotyrosine

(a)

γ-Carboxyglutamate

(b)

Methyllysine

Dimethyllysine

Trimethyllysine

Methylglutamate

(c)

figure 27–31
Some modified amino acid residues. (a) Phosphorylated amino acids. **(b)** A carboxylated amino acid. **(c)** Some methylated amino acids.

Attachment of Carbohydrate Side Chains The carbohydrate side chains of glycoproteins are attached covalently during or after the synthesis of the polypeptide. In some glycoproteins, the carbohydrate side chain is attached enzymatically to Asn residues (*N*-linked oligosaccharides), in others to Ser or Thr residues (*O*-linked oligosaccharides) (see Fig. 9–25). Many proteins that function extracellularly, as well as the lubricating proteoglycans that coat mucous membranes, contain oligosaccharide side chains (see Fig. 9–23).

Addition of Isoprenyl Groups A number of eukaryotic proteins are modified by the addition of groups derived from isoprene (isoprenyl groups). A thioether bond is formed between the isoprenyl group and a Cys residue of the protein (see Fig. 12–13). The isoprenyl groups are derived from pyrophosphorylated intermediates of the cholesterol biosynthetic pathway (see Fig. 21–34), such as farnesyl pyrophosphate (Fig. 27–32). Proteins modified in this way include the Ras proteins, products of the *ras* oncogenes and proto-oncogenes, and G proteins (both discussed in Chapter 13), and proteins called lamins, found in the nuclear matrix. In some cases the isoprenyl group helps to anchor the protein in a membrane. The transforming (carcinogenic) activity of the *ras* oncogene is lost when isoprenylation of the Ras protein is blocked, stimulating interest in identifying inhibitors of this posttranslational modification pathway for use in cancer chemotherapy.

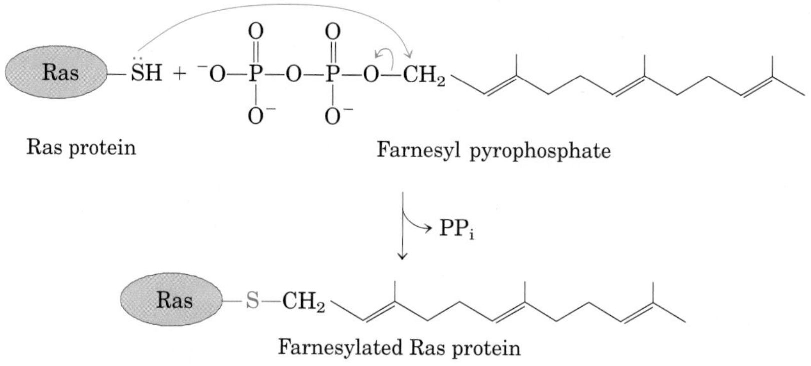

Ras —SH + ⁻O—P(=O)(O⁻)—O—P(=O)(O⁻)—O—CH_2—(farnesyl)

Ras protein

Farnesyl pyrophosphate

→ PP_i

Ras —S—CH_2—(farnesyl)

Farnesylated Ras protein

figure 27–32

Farnesylation of a Cys residue on a protein. The thioether linkage is shown in red. The Ras protein is the product of the *ras* oncogene.

Addition of Prosthetic Groups Many prokaryotic and eukaryotic proteins require for their activity covalently bound prosthetic groups. Two examples are the biotin molecule of acetyl-CoA carboxylase and the heme group of cytochrome *c*.

Proteolytic Processing Many proteins are initially synthesized as large, inactive precursor proteins that are proteolytically trimmed to produce their smaller, active forms. Examples include insulin, some viral proteins, and proteases such as trypsin and chymotrypsin (see Fig. 8–31).

Formation of Disulfide Cross-Links After folding into their native conformations, some proteins form intrachain or interchain disulfide bridges between Cys residues. Disulfide bonds are common in proteins to be exported from eukaryotic cells. The cross-links formed in this way help to protect the native conformation of the protein molecule from denaturation in an extracellular environment that can differ greatly from intracellular conditions and is generally oxidizing.

Protein Synthesis Is Inhibited by Many Antibiotics and Toxins

Protein synthesis is a central function in cellular physiology and is the primary target of many naturally occurring antibiotics and toxins. Except as noted, these antibiotics inhibit protein synthesis in bacteria. The differ-

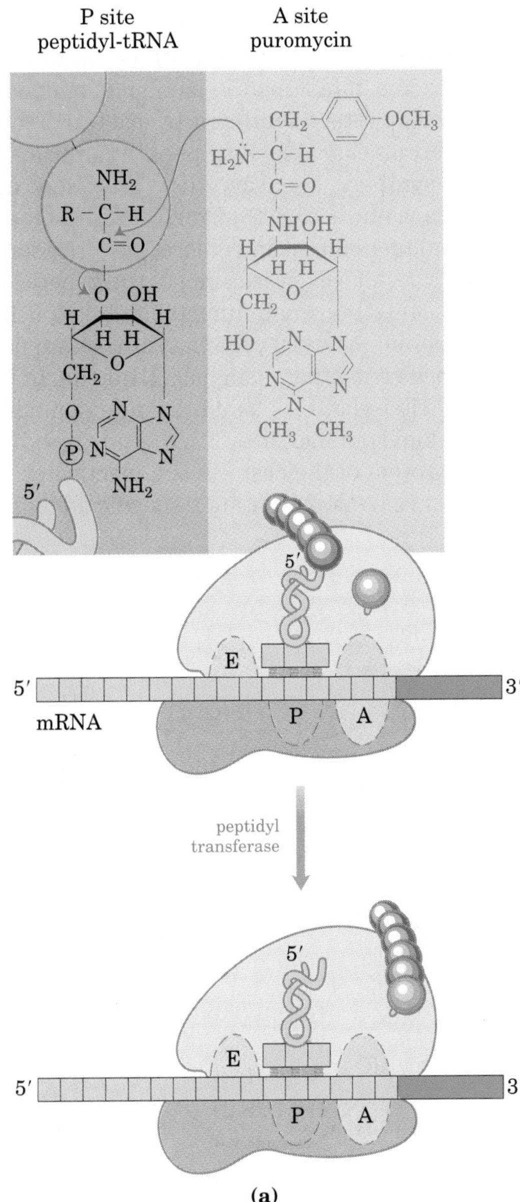

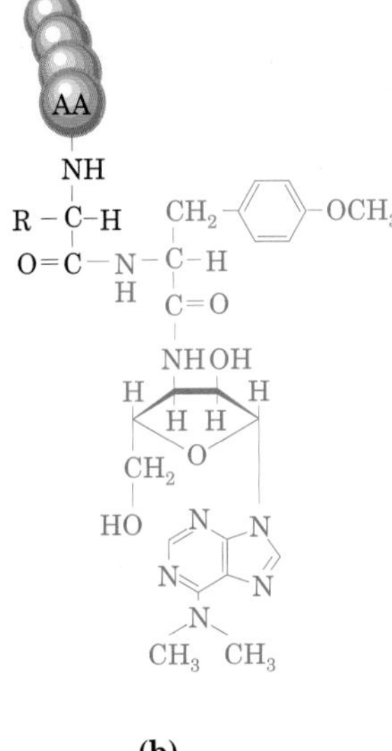

figure 27–33

Disruption of peptide bond formation by puromycin. (a) The antibiotic puromycin resembles the aminoacyl end of a charged tRNA, and it can bind to the ribosomal A site and participate in peptide bond formation. The product of this reaction, instead of being translocated to the P site, dissociates from the ribosome, causing premature chain termination. **(b)** Peptidyl puromycin.

ences between bacterial and eukaryotic protein synthesis, though in some cases subtle, are sufficient that most of the compounds discussed below are relatively harmless to eukaryotic cells. Natural selection has favored the evolution of compounds that exploit those minor differences in order to affect bacterial systems selectively. These important biochemical weapons are synthesized by some microorganisms and are extremely toxic to others. Because nearly every step in protein synthesis can be specifically inhibited by one antibiotic or another, antibiotics have become valuable tools in the study of protein synthesis.

Puromycin, made by the mold *Streptomyces alboniger*, is one of the best-understood inhibitory antibiotics. Its structure is very similar to the 3′ end of an aminoacyl-tRNA, enabling it to bind to the ribosomal A site and participate in peptide bond formation, producing peptidyl-puromycin (Fig. 27–33). However, because puromycin resembles only the 3′ end of the tRNA, it does not engage in translocation and dissociates from the ribosome shortly after it is linked to the carboxyl terminus of the peptide. This prematurely terminates polypeptide synthesis.

Tetracyclines inhibit protein synthesis in bacteria by blocking the A site on the ribosome, inhibiting the binding of aminoacyl-tRNAs. **Chloramphenicol** inhibits protein synthesis by bacterial (and mitochondrial and chloroplast) ribosomes by blocking peptidyl transfer; it does not affect cytosolic protein synthesis in eukaryotes. Conversely, **cycloheximide** blocks the peptidyl transferase of 80S eukaryotic ribosomes but not that of 70S bacterial (and mitochondrial and chloroplast) ribosomes. **Streptomycin,** a basic trisaccharide, causes misreading of the genetic code in bacteria at relatively low concentrations and inhibits initiation at higher concentrations.

Several other inhibitors of protein synthesis are notable because of their toxicity to humans and other mammals. **Diphtheria toxin** (M_r 65,000) catalyzes the ADP-ribosylation of a diphthamide (a modified histidine) residue of eukaryotic elongation factor eEF2, thereby inactivating it. **Ricin,** an extremely toxic protein of the castor bean, inactivates the 60S subunit of eukaryotic ribosomes by depurinating a specific adenosine in 23S rRNA.

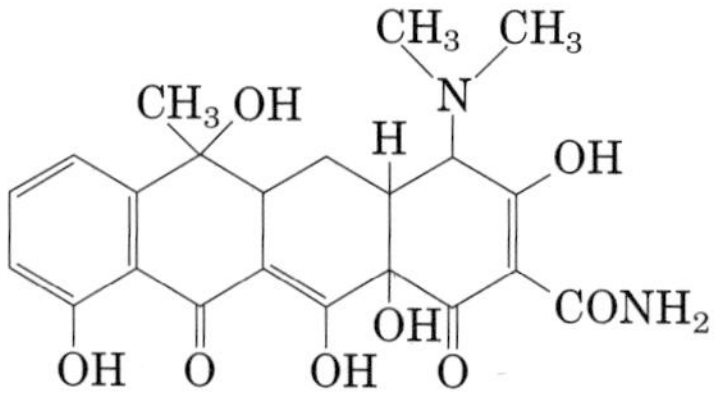

Tetracycline

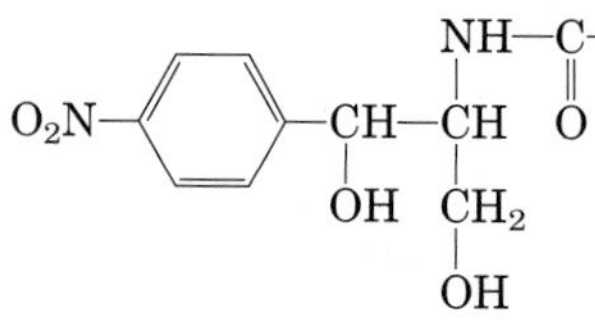

Chloramphenicol

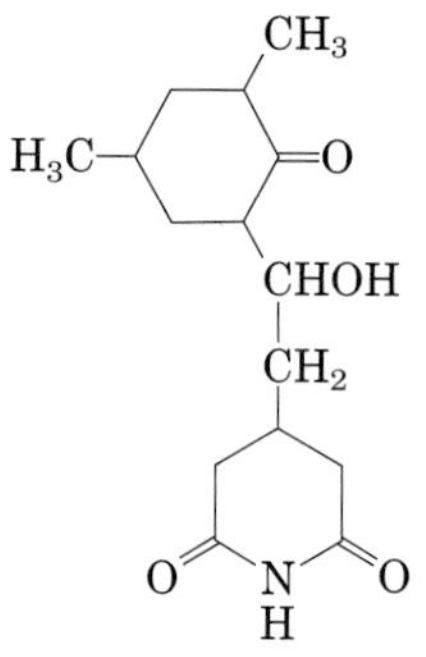

Cycloheximide

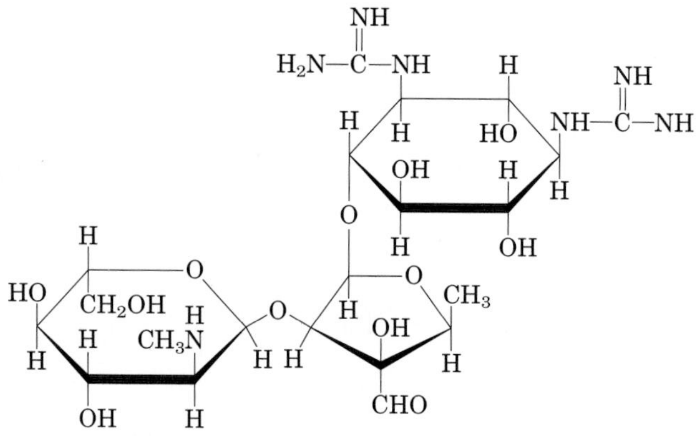

Streptomycin

Protein Targeting and Degradation

The eukaryotic cell is made up of many structures, compartments, and organelles, each with specific functions that require distinct sets of proteins and enzymes. Almost all these proteins are synthesized on ribosomes in the cytosol, so how are they directed to their final cellular destinations?

We are now beginning to understand this complex and fascinating process. Proteins destined for secretion, integration in the plasma membrane, or inclusion in lysosomes generally share the first few steps of a pathway that begins in the endoplasmic reticulum. Proteins destined for mitochondria, chloroplasts, or the nucleus use three separate mechanisms, and proteins destined for the cytosol simply remain where they are synthesized.

The most important element in many of these targeting pathways is a short sequence of amino acids called a **signal sequence,** whose function was first postulated by David Sabatini and Günter Blobel in 1970. The signal sequence directs a protein to its appropriate location in the cell and often is removed during transport or after the protein has reached its final destination. For proteins slated for transport into mitochondria, chloroplasts, or the ER, the signal sequence is at the amino terminus of a newly synthesized polypeptide. In many cases, the targeting capacity of particular signal sequences has been confirmed by fusing the signal sequence from one protein to a second protein and showing that the signal directs the second protein to the location where the first protein is normally found. The selective degradation of proteins no longer needed by the cell also relies largely on a set of molecular signals embedded in each protein's structure.

Günter Blobel

In this concluding section we examine protein targeting and degradation, emphasizing the underlying signals and molecular regulation so crucial to cellular metabolism. Except where noted, the focus is on eukaryotic cells.

Posttranslational Modification of Many Eukaryotic Proteins Begins in the Endoplasmic Reticulum

Perhaps the best-characterized targeting system begins in the ER. Most lysosomal, membrane, or secreted proteins have an amino-terminal signal sequence (Fig. 27–34) that marks them for translocation into the lumen of the ER; hundreds of such signal sequences have been determined. The carboxyl terminus of the signal sequence is defined by a cleavage site, where protease action removes the signal sequence after the protein is imported into the ER. Signal sequences vary from 13 to 36 amino acid residues in length, but all have the following features: (1) about 10 to 15 hydrophobic amino acid residues, (2) one or more positively charged residues, usually near the amino terminus preceding the hydrophobic sequence, and (3) a short sequence at the carboxyl terminus (near the cleavage site) that is relatively polar, typically having amino acid residues with short side chains (especially Ala) at the positions closest to the cleavage site.

	Sequence (↓ = cleavage site)
Human influenza virus A	Met Lys Ala Lys Leu Leu Val Leu Leu Tyr Ala Phe Val Ala Gly ↓ Asp Gln --
Human preproinsulin	Met Ala Leu Trp Met Arg Leu Leu Pro Leu Leu Ala Leu Leu Ala Leu Trp Gly Pro Asp Pro Ala Ala Ala ↓ Phe Val --
Bovine growth hormone	Met Met Ala Ala Gly Pro Arg Thr Ser Leu Leu Leu Ala Phe Ala Leu Leu Cys Leu Pro Trp Thr Gln Val Val Gly ↓ Ala Phe --
Bee promellitin	Met Lys Phe Leu Val Asn Val Ala Leu Val Phe Met Val Val Tyr Ile Ser Tyr Ile Tyr Ala ↓ Ala Pro --
Drosophila glue protein	Met Lys Leu Leu Val Val Ala Val Ile Ala Cys Met Leu Ile Gly Phe Ala Asp Pro Ala Ser Gly ↓ Cys Lys --

figure 27–34
Translocation into the ER directed by amino-terminal signal sequences of some eukaryotic proteins. The hydrophobic core (yellow) is preceded by one or more basic residues (blue). Note the polar and short-side-chain residues immediately preceding the cleavage sites (indicated by red arrows).

As originally demonstrated by George Palade, proteins with these signal sequences are synthesized on ribosomes attached to the ER. The signal sequence itself helps to direct the ribosome to the ER (Fig. 27–35). The targeting pathway begins with the initiation of protein synthesis on free ribosomes. The signal sequence appears early in the synthetic process, because it is at the amino terminus, which is synthesized first. As it emerges from the ribosome, this signal sequence and the ribosome itself are bound by the large **signal recognition particle** (SRP). SRP then binds GTP and halts the elongation of the polypeptide when it is about 70 amino acids long and the signal sequence has completely emerged from the ribosome.

George Palade

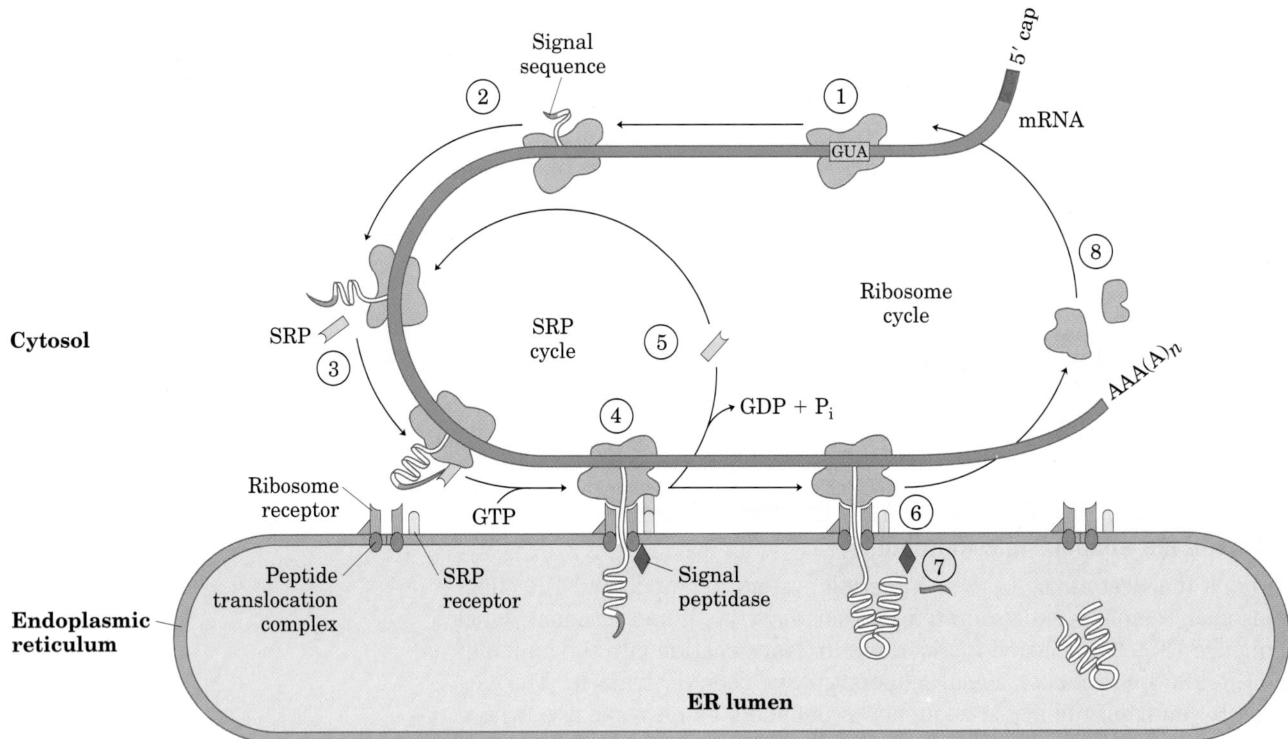

figure 27–35

Directing eukaryotic proteins with the appropriate signals to the endoplasmic reticulum. This process involves the SRP cycle and translocation and cleavage of the nascent polypeptide. ① The ribosomal subunits assemble in an initiation complex at the initiation codon and begin protein synthesis. ② If an appropriate signal sequence appears at the amino terminus of the nascent polypeptide, ③ SRP binds to the ribosome, binds GTP, and halts elongation. ④ The ribosome-SRP complex is bound by receptors on the ER that also bind GTP. ⑤ SRP dissociates and is recycled, accompanied by the hydrolysis of GTP in both SRP and the receptor. ⑥ Protein synthesis resumes, coupled to translocation of the polypeptide chain into the ER lumen through the peptide translocation complex. ⑦ The signal sequence is cleaved from the nascent polypeptide by a signal peptidase within the lumen. ⑧ The ribosomal subunits dissociate from the mRNA and are recycled.

SRP is a rod-shaped complex containing a 300 nucleotide RNA (7SL-RNA) and six different proteins, with a combined molecular weight of 325,000. One protein subunit of SRP binds directly to the signal sequence, inhibiting elongation by sterically blocking the entry of aminoacyl-tRNAs and inhibiting peptidyl transferase. Another protein subunit binds and hydrolyzes GTP. The SRP receptor is a heterodimer of α (M_r 69,000) and β (M_r 30,000) subunits, both of which bind and hydrolyze GTP in this targeting pathway.

The GTP-bound SRP directs the ribosome (still bound to the mRNA) and the incomplete polypeptide to GTP-bound SRP receptors in the cytosolic face of the ER. The nascent polypeptide is delivered to a **peptide translocation complex** in the ER, which may interact directly with the ribosome. SRP dissociates from the ribosome, accompanied by hydrolysis of GTP in both SRP and the SRP receptor.

Elongation of the polypeptide now resumes, with the ATP-driven translocation complex feeding the growing polypeptide into the ER lumen until the complete protein has been synthesized. The signal sequence is then removed by a signal peptidase within the ER lumen. The ribosome dissociates and is recycled.

Glycosylation Plays a Key Role in Protein Targeting

In the ER lumen, newly synthesized proteins are further modified in several ways. Following the removal of signal sequences, polypeptides are folded,

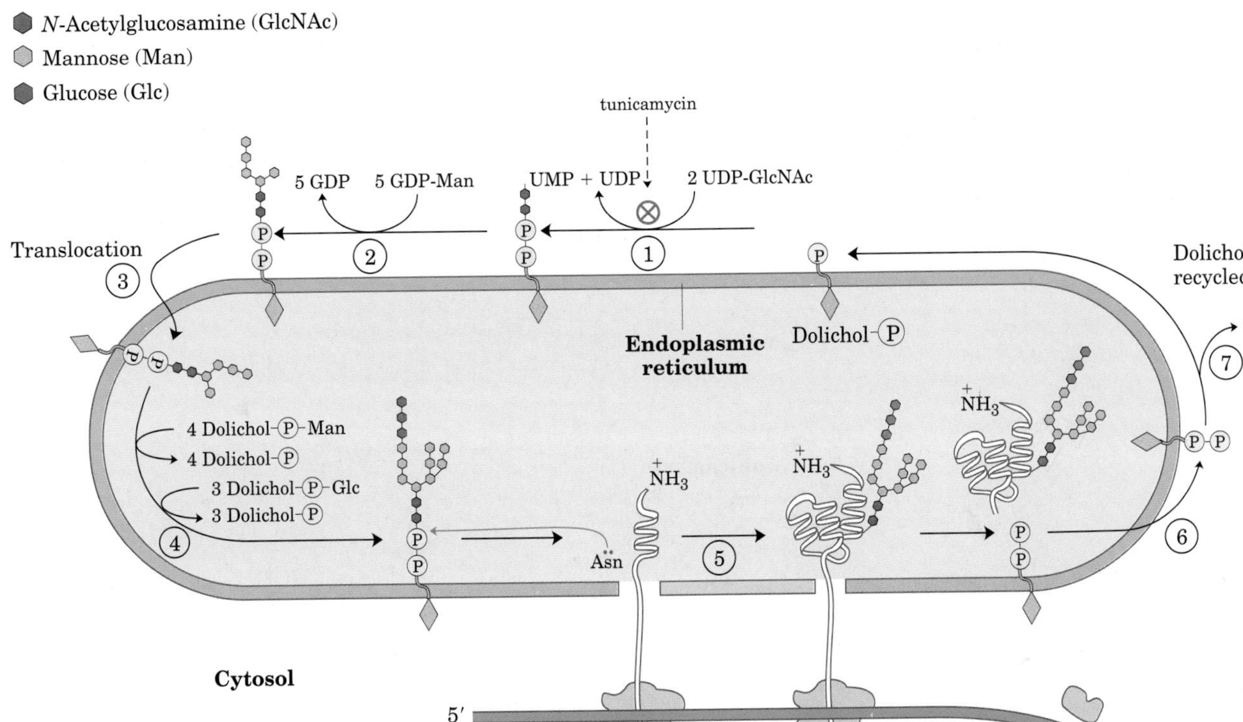

figure 27–36
Synthesis of the core oligosaccharide of glycoproteins. The core oligosaccharide is built up by the successive addition of monosaccharide units. (1), (2) The first steps occur on the cytosolic face of the ER. (3) Translocation moves the incomplete oligosaccharide across the membrane (mechanism not shown), and (4) completion of the core oligosaccharide occurs within the lumen of the ER. The synthetic precursors that contribute additional mannose and glucose residues to the growing oligosaccharide in the lumen are dolichol phosphate derivatives. (5) In the first step in the construction of the *N*-linked oligosaccharide moiety of a glycoprotein, the core oligosaccharide is transferred from dolichol phosphate to an Asn residue of the protein within the ER lumen. The core oligosaccharide is then further modified in the ER and the Golgi complex in pathways that differ for different proteins. The five sugar residues surrounded by a beige screen (lower right) are retained in the final structure of all *N*-linked oligosaccharides. (6) The released dolichol pyrophosphate is again translocated so that the pyrophosphate is on the cytosolic face of the ER, then (7) a phosphate is hydrolytically removed to regenerate dolichol phosphate.

disulfide bonds are formed, and many proteins are glycosylated. Glycosylated proteins, or glycoproteins, are often linked to their oligosaccharides through Asn residues. These *N*-linked oligosaccharides are diverse (Chapter 9), but the pathways by which they form have a common first step. A 14 residue core oligosaccharide is built up in a stepwise fashion, then transferred from a dolichol phosphate donor molecule to certain Asn residues in the protein (Fig. 27–36). The transferase is on the lumenal face of the ER

$$^{-}O-\overset{\displaystyle O}{\overset{\|}{\underset{\underset{\displaystyle O^-}{|}}{P}}}-O-CH_2-CH_2-\overset{\displaystyle CH_3}{\overset{|}{\underset{\underset{\displaystyle H}{|}}{C}}}-CH_2-\left(CH_2-CH=\overset{\displaystyle CH_3}{\overset{|}{C}}-CH_2\right)_n-CH_2-CH=\overset{\displaystyle CH_3}{\overset{|}{C}}-CH_3$$

Dolichol phosphate
(n = 9–22)

and thus cannot catalyze glycosylation of cytosolic proteins. After the transfer, the core oligosaccharide is trimmed and elaborated in different ways on different proteins, but all *N*-linked oligosaccharides retain a pentasaccharide core derived from the original 14 residue oligosaccharide. Several antibiotics act by interfering with one or more steps in this process and have aided in elucidating the steps of protein glycosylation. The best-characterized is **tunicamycin,** which mimics the structure of UDP–*N*-acetylglucosamine and blocks the first step of the process.

Tunicamycin

N-Acetylglucosamine

Uracil

Fatty acyl side chain

(n = 8–11)

Tunicamine

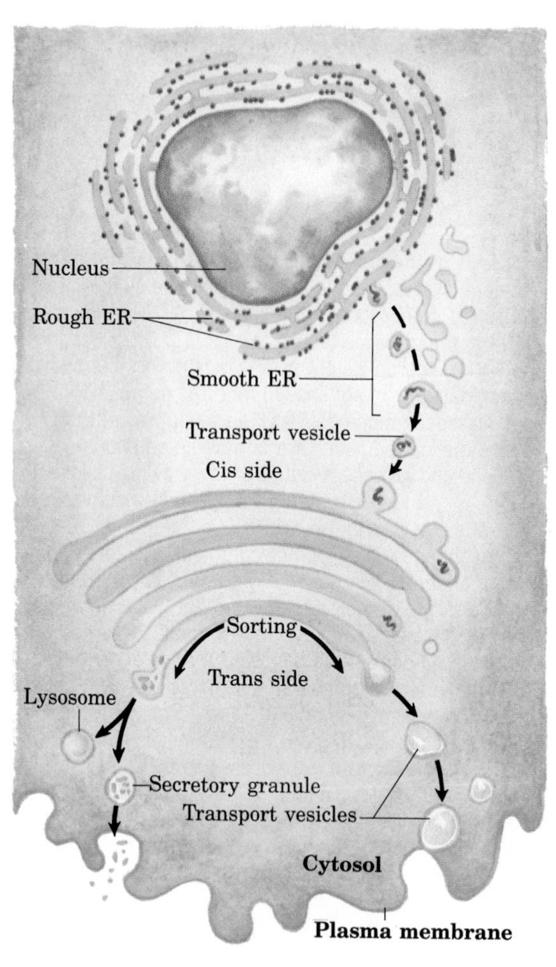

figure 27–37

Pathway taken by proteins destined for lysosomes, the plasma membrane, or secretion. Proteins are moved from the ER to the cis side of the Golgi complex in transport vesicles. Sorting occurs primarily in the trans side of the Golgi complex.

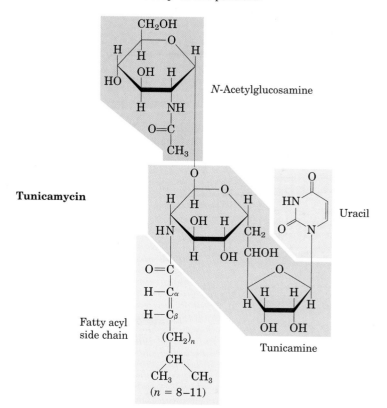

Next, suitably modified proteins can be moved to a variety of intracellular destinations. Proteins are moved from the ER to the Golgi complex in transport vesicles (Fig. 27–37). In the Golgi complex, *O*-linked oligosaccharides are added and *N*-linked oligosaccharides are further modified. By mechanisms not yet fully understood, proteins are also sorted here and are sent to their final destinations. Within the Golgi complex, the processes that segregate proteins targeted for secretion from those targeted for the plasma membrane or lysosomes must distinguish between these proteins on the basis of structural features other than the signal sequence, which was removed in the ER lumen.

This sorting process is best understood in the case of hydrolases destined for transport to lysosomes. On arrival of a hydrolase in the Golgi complex from the ER, an as yet undetermined feature (sometimes called a signal patch) of the three-dimensional structure of the hydrolase is recognized by a phosphotransferase that phosphorylates certain mannose residues in the oligosaccharide (Fig. 27–38). The presence of one or more mannose 6-phosphate residues in its *N*-linked oligosaccharide is the structural signal that targets the protein to lysosomes. A receptor protein in the membrane of the Golgi complex recognizes the mannose 6-phosphate signal and binds the hydrolase so marked. Vesicles containing these receptor-hydrolase complexes bud from the trans side of the Golgi complex and make their way to sorting vesicles. Here, the receptor-hydrolase complex dissociates in a process facilitated by the lower pH in the sorting vesicle and by phosphatase-catalyzed removal of phosphate groups from the mannose 6-phosphate residues. The receptor is then recycled to the Golgi complex, and vesicles containing the hydrolases bud from the sorting vesicles and move

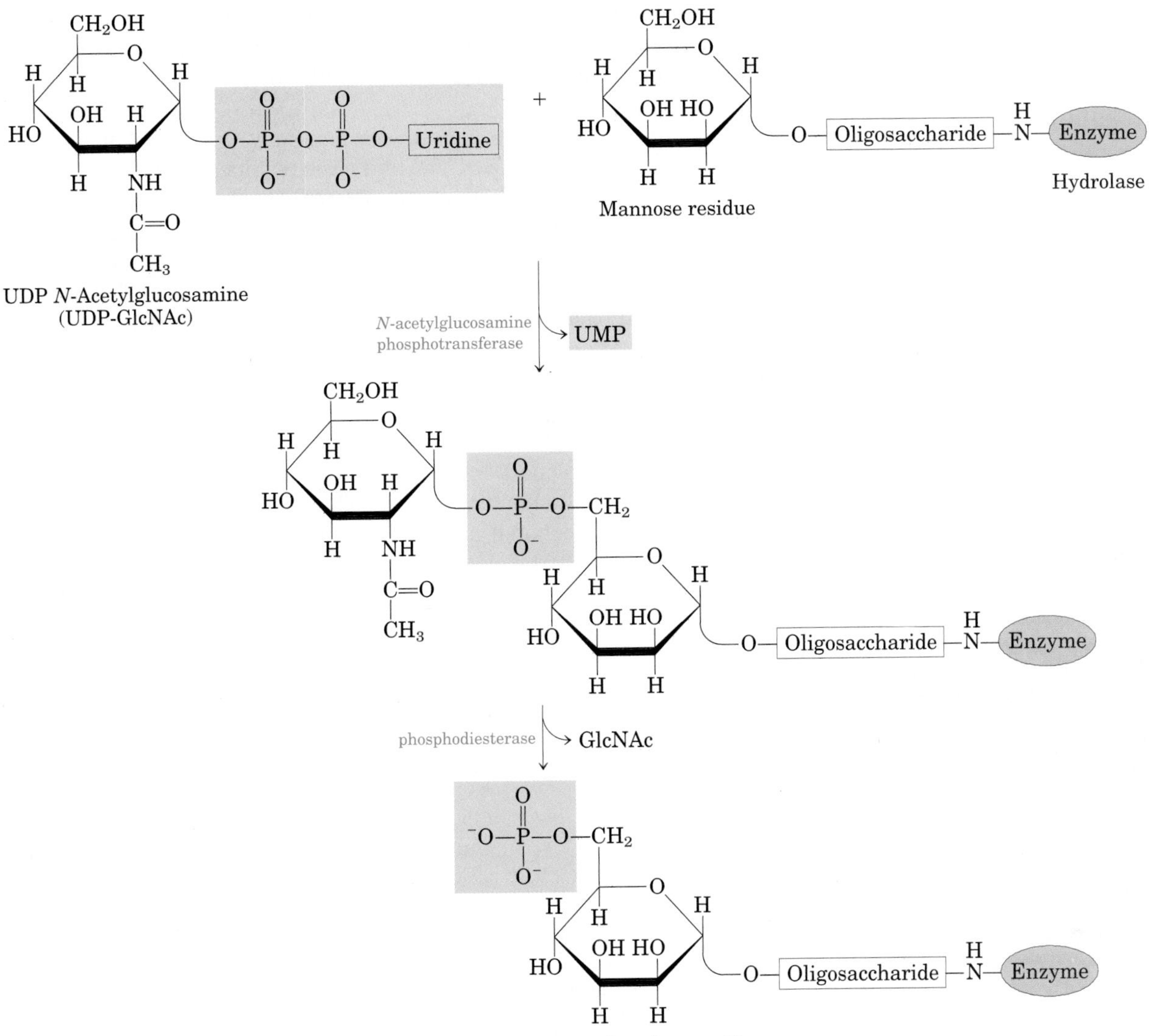

figure 27–38
Phosphorylation of mannose residues on lysosome-targeted enzymes. *N*-Acetylglucosamine phosphotransferase recognizes some as yet unidentified structural feature of hydrolases destined for lysosomes.

to the lysosomes. In cells treated with tunicamycin, hydrolases that should be targeted for lysosomes are instead secreted, confirming that the *N*-linked oligosaccharide plays a key role in targeting these enzymes to lysosomes.

Proteins Are Targeted to Mitochondria and Chloroplasts by Similar Pathways

Although mitochondria and chloroplasts contain DNA, most of their proteins are encoded by nuclear DNA and must be targeted to the appropriate organelle. Unlike the targeting of proteins to the ER, the mitochondrial and chloroplast targeting pathways can begin *after* a precursor protein has been completely synthesized and released from the ribosome. Nevertheless, there are similarities among the three targeting pathways.

Precursor proteins destined for mitochondria or chloroplasts have amino-terminal signal sequences that are bound by cytosolic chaperone proteins (p. 197). The precursors are delivered to receptors on the exterior surface of the target organelle, then to a protein channel that usually spans the inner and outer membranes of the organelle. Translocation through the channel is facilitated by hydrolysis of ATP or GTP and, in some cases, by a transmembrane electrochemical potential. Inside the organelle, the signal sequence of the precursor is removed, and the mature protein is refolded (Fig. 27–39).

Mitochondrial signal sequences are generally 20 to 35 amino acid residues long and are rich in Ser, Thr, and basic residues. Mitochondrial protein precursors become bound by either Hsp70 (*h*eat-shock *p*rotein 70) or MSF (*m*itochondrial import *s*timulation *f*actor). These molecular chaperones stabilize unfolded conformations of the protein precursors. MSF is specific to mitochondria; Hsp70 belongs to a common class of chaperone proteins found in bacteria and eukaryotic cells (see Fig. 6–30).

The bound mitochondrial protein precursor is delivered to a protein complex called Tom (for *t*ransport across the *o*uter *m*embrane), the mitochondrial receptor, and then to another Tom complex that provides a channel through the outer membrane (Fig. 27–39a). Another protein complex, known as Tim (for *t*ransport across the *i*nner *m*embrane), translocates the precursor into the mitochondrial matrix. The translocation is fueled by interactions with a mitochondrial Hsp70 protein (mHsp70) coupled to ATP hydrolysis, as well as by the electrochemical gradient across the inner membrane. Within the matrix, folding of the mitochondrial protein is facilitated by a number of different chaperone proteins (including mHsp70), and the signal sequence is removed by a processing protease in the matrix.

Some complex mitochondrial signal sequences include "stop transfer" signals. When this signal is recognized during transport of the mitchondrial protein precursor through the membrane channel, translocation is halted at

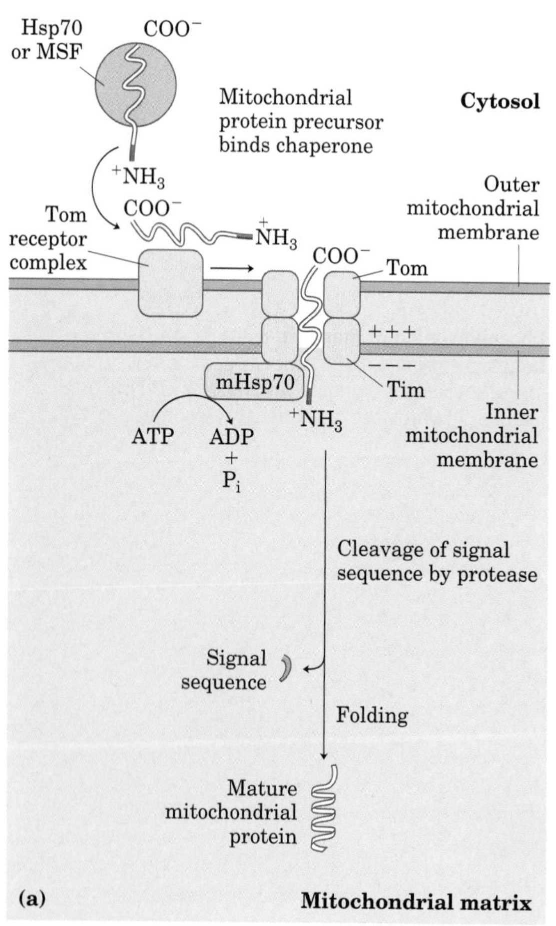

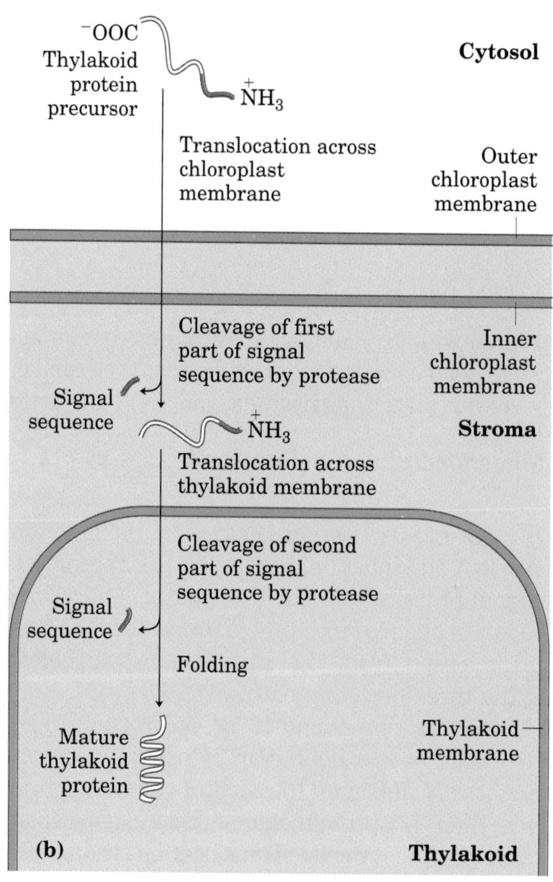

figure 27–39

Protein targeting to mitochondria and chloroplasts. **(a)** Targeting to the mitochondrial matrix. A mitochondrial protein precursor is bound by a cytosolic chaperone (Hsp70 or MSF) and delivered to a Tom receptor complex, then to a membrane channel made up of specialized Tom and Tim channel complexes. Translocation of the precursor protein is driven both by the ATPase of mitochondrial Hsp70 protein (mHsp70) on the matrix side of the inner mitochondrial membrane and by the electrochemical potential across the inner membrane. The signal sequence is cleaved in the mitochondrial matrix, and the protein folds into its functional conformation. **(b)** Proteins targeted to the chloroplast thylakoids have bipartite signal sequences. The first part of the sequence (red) mediates targeting to the chloroplast stroma, where it is removed; the second part (blue) mediates translocation into the thylakoid, where it, in turn, is removed.

a point when a hydrophobic stretch of amino acid residues in the precursor spans either the inner or the outer mitochondrial membrane. The protein is then released to generate an integral membrane protein. In some cases, a sequence spanning the inner membrane is cleaved to create a soluble protein of the intermembrane space.

Targeting of precursor proteins to chloroplasts follows a similar pattern. An interesting variation occurs in proteins targeted to the thylakoids (Fig. 27–39b). These proteins have a bipartite signal sequence. The amino-terminal portion of the signal sequence targets the protein to the stroma before being cleaved; the rest of the sequence then targets the protein to the thylakoids.

Signal Sequences for Nuclear Transport Are Not Cleaved

Molecular communication between the nucleus and the cytosol requires the movement of macromolecules through nuclear pores (see Fig. 2–11). RNA molecules synthesized in the nucleus are exported to the cytoplasm. Ribosomal proteins synthesized on cytosolic ribosomes are imported into the nucleus and assembled into 60S and 40S ribosomal subunits in the nucleolus; completed subunits are then exported back to the cytosol. A variety of nuclear proteins (RNA and DNA polymerases, histones, topoisomerases, proteins that regulate gene expression, and so forth) are synthesized in the cytosol and imported into the nucleus. This traffic is modulated by a complex system of molecular signals and transport proteins that is gradually being elucidated.

In most multicellular eukaryotes, the nuclear envelope breaks down at each cell division. Once cell division is completed and the nuclear envelope is reestablished, the dispersed nuclear proteins must be reimported. To allow repeated nuclear importation, the signal sequence that targets a protein to the nucleus (the nuclear localization sequence, NLS) is not removed. Rather than occurring at the amino terminus, an NLS may be located almost anywhere along the primary sequence of a nuclear protein.

Nuclear importation is mediated by a number of proteins that cycle between the cytosol and the nucleus (Fig. 27–40), including importin α and β and a small GTPase called Ran. A heterodimer of importin α and β functions as a soluble receptor for proteins targeted to the nucleus, with the α subunit binding NLS-bearing proteins in the cytosol. The complex of the NLS-bearing protein and the importin docks at a nuclear pore and is translocated through the pore by an energy-dependent mechanism that requires the Ran GTPase. The two importin subunits separate during the translocation, and the NLS-bearing protein dissociates from importin α inside the nucleus. Importin α and β are then exported from the nucleus to repeat the process. How importin α remains dissociated from the many NLS-bearing proteins inside the nucleus is not yet known.

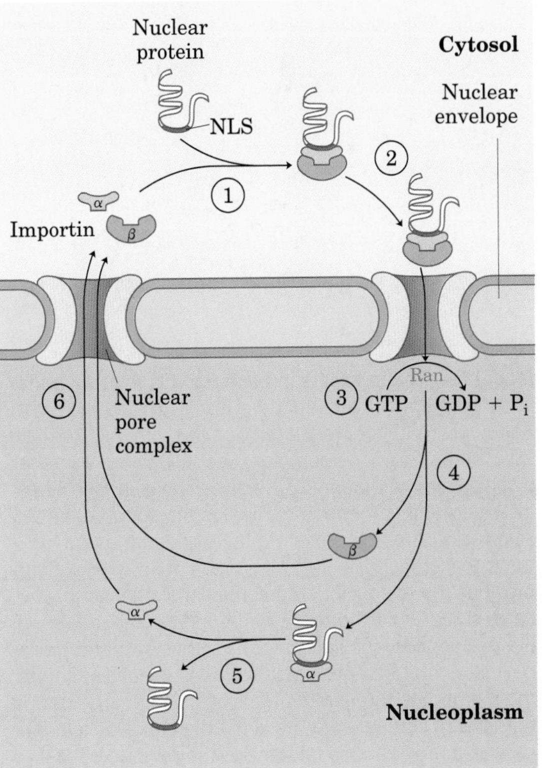

figure 27–40
Targeting of nuclear proteins. ① A protein with an appropriate nuclear localization signal (NLS) is bound by a complex of importin α and β. ② The resulting complex binds to a nuclear pore, and ③ translocation is mediated by the Ran GTPase. ④ Inside the nucleus, importin β dissociates from importin α, and ⑤ importin α then dissociates from the nuclear protein. ⑥ Importin α and β are transported out of the nucleus and recycled.

Protein	Signal sequence (cleavage site ↓)
Inner membrane proteins	
Phage fd, major coat protein	Met Lys Lys Ser Leu Val Leu Lys Ala Ser Val Ala Val Ala Thr Leu Val Pro Met Leu Ser Phe Ala↓Ala Glu --
Phage fd, minor coat protein	Met Lys Lys Leu Leu Phe Ala Ile Pro Leu Val Val Pro Phe Tyr Ser His Ser↓Ala Glu --
Periplasmic proteins	
Alkaline phosphatase	Met Lys Gln Ser Thr Ile Ala Leu Ala Leu Leu Pro Leu Leu Phe Thr Pro Val Thr Lys Ala↓Arg Thr --
Leucine-specific binding protein	Met Lys Ala Asn Ala Lys Thr Ile Ile Ala Gly Met Ile Ala Leu Ala Ile Ser His Thr Ala Met Ala↓Asp Asp --
β-Lactamase of pBR322	Met Ser Ile Gln His Phe Arg Val Ala Leu Ile Pro Phe Phe Ala Ala Phe Cys Leu Pro Val Phe Ala↓His Pro --
Outer membrane proteins	
Lipoprotein	Met Lys Ala Thr Lys Leu Val Leu Gly Ala Val Ile Leu Gly Ser Thr Leu Leu Ala Gly↓Cys Ser --
LamB	Leu Arg Lys Leu Pro Leu Ala Val Ala Val Ala Ala Gly Val Met Ser Ala Gln Ala Met Ala↓Val Asp --
OmpA	Met Met Ile Thr Met Lys Lys Thr Ala Ile Ala Ile Ala Val Ala Leu Ala Gly Phe Ala Thr Val Ala Gln Ala↓Ala Pro --

figure 27–41

Signal sequences targeting different locations in bacteria. Basic amino acids (blue) near the amino terminus and hydrophobic core amino acids (yellow) are highlighted. The cleavage sites marking the ends of the signal sequences are indicated by red arrows. Note that the inner bacterial cell membrane (see Fig. 2–5) is where phage fd coat proteins and DNA are assembled into phage particles. OmpA is outer membrane protein A; LamB is a cell surface receptor protein for bacteriophage lambda.

Bacteria Also Use Signal Sequences for Protein Targeting

Bacteria can target proteins to their inner or outer membranes, to the periplasmic space between these membranes, or to the extracellular medium. They use signal sequences at the amino terminus of the proteins (Fig. 27–41) much like those on eukaryotic proteins targeted to the ER, mitochondria, and chloroplasts.

Most proteins exported from *E. coli* make use of the pathway shown in Figure 27–42. Following translation, a protein to be exported may fold only slowly, the amino-terminal signal sequence impeding the folding. The soluble chaperone protein SecB binds to the protein's signal sequence or other features of its incompletely folded structure. The bound protein is then delivered to SecA, a protein associated with the inner surface of the cytoplasmic membrane. SecA acts as both a receptor and a translocating ATPase.

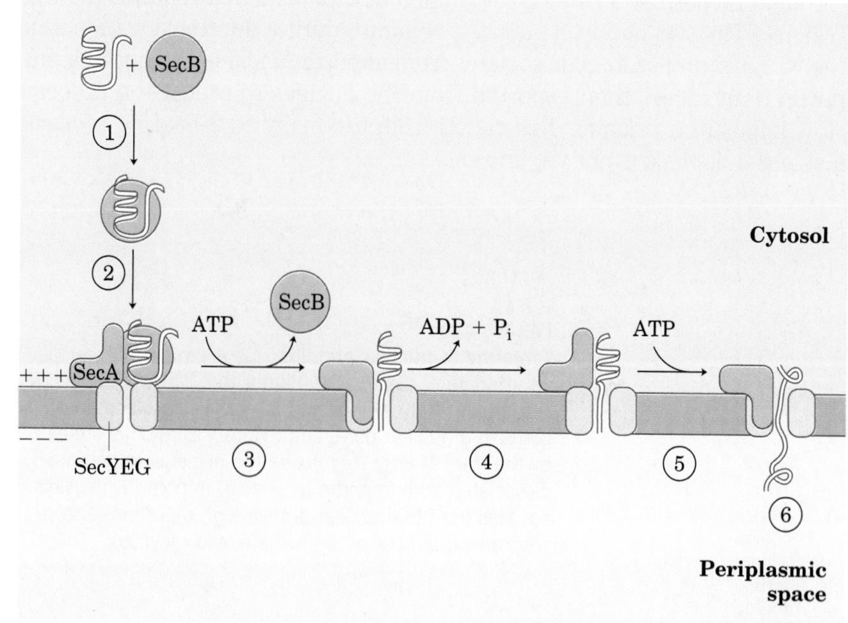

figure 27–42

Model for protein export in bacteria. ① A newly translated polypeptide binds to the cytosolic chaperone protein SecB, which ② delivers it to SecA, a protein associated with the translocation complex (SecYEG) in the bacterial cytoplasmic membrane. ③ SecB is released, and SecA inserts itself into the membrane, forcing about 20 amino acid residues of the protein to be exported through the translocation complex. ④ Hydrolysis of an ATP by SecA provides the energy for a conformational change that causes SecA to withdraw from the membrane, releasing the polypeptide. ⑤ SecA binds another ATP, and the next stretch of 20 amino acid residues is pushed across the membrane through the translocation complex. Steps ④ and ⑤ are repeated until ⑥ the entire protein has passed through and is released to the periplasm. The electrochemical potential across the membrane also provides some of the driving force required for protein translocation.

Released from SecB and bound to SecA, the protein is delivered to a translocation complex in the membrane, made up of SecY, E, and G, and is translocated stepwise through the membrane at the SecYEG complex in lengths of about 20 amino acid residues. Each step is facilitated by the hydrolysis of ATP, catalyzed by SecA.

An exported protein is thus pushed through the membrane by a SecA protein located on the cytoplasmic surface, rather than being pulled through the membrane by a protein (akin to mitochondrial Hsp70) located on the periplasmic surface. This difference may simply reflect the need for the translocating ATPase to be where the ATP is. The transmembrane electrochemical potential can also provide energy for the translocation of a protein being exported, by a mechanism not yet understood.

Although most exported bacterial proteins use this pathway, some follow an alternative pathway that uses signal recognition and receptor proteins homologous to components of the eukaryotic SRP and SRP receptor (Fig. 27–35).

Cells Import Proteins by Receptor-Mediated Endocytosis

Some proteins are imported into cells from the surrounding medium; examples include low-density lipoprotein (LDL), the iron-carrying protein transferrin, peptide hormones, and circulating proteins destined for degradation. The proteins bind to receptors in invaginations of the membrane called **coated pits,** which concentrate receptors destined for endocytosis in preference to other cell-surface proteins. The pits are coated on their cytosolic side with a lattice of the protein **clathrin,** which forms closed polyhedral structures (Fig. 27–43). As more of the receptors are occupied by target proteins, the clathrin lattice grows until a complete membrane-bounded endocytic vesicle buds off the plasma membrane and enters the cytoplasm. The clathrin is quickly removed by uncoating enzymes, and the vesicle fuses with an endosome. The activity of ATPases in the endosomal membranes reduces the pH, facilitating dissociation of receptors from their target proteins.

The imported proteins and receptors then go their separate ways, their fates varying with the cell and protein type. Transferrin and its receptor are eventually recycled. Some hormones, growth factors, and immune complexes, after eliciting the appropriate cellular response, are degraded along with their receptors. LDL is degraded after the associated cholesterol has been delivered to its destination, but the LDL receptor is recycled (see Fig. 21–40).

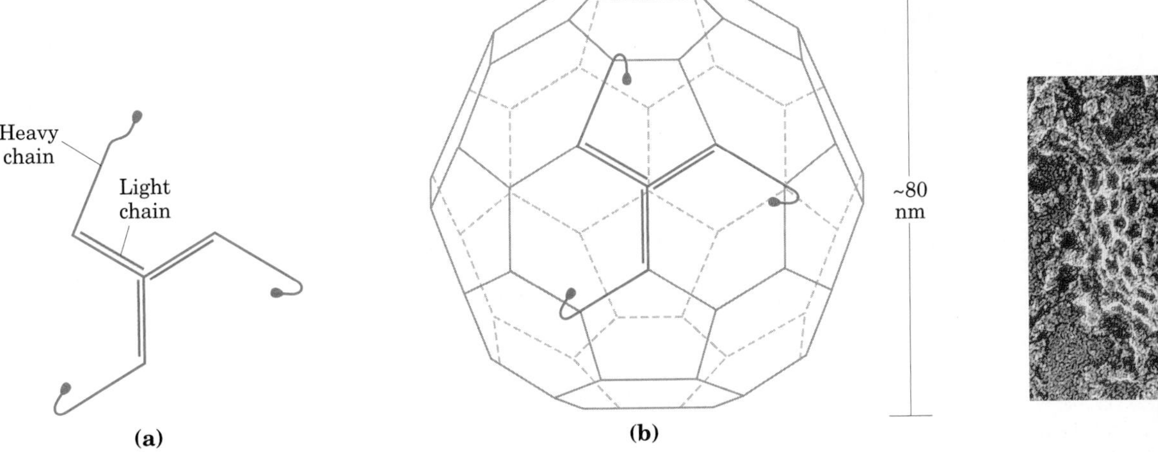

figure 27–43
Clathrin. (a) Three light (L) chains (M_r 35,000) and three heavy (H) chains (M_r 180,000) of the $(HL)_3$ clathrin unit, organized as a three-legged structure called a triskelion. **(b)** Triskelions tend to assemble into polyhedral lattices. **(c)** Electron micrograph of a coated pit on the cytosolic face of the plasma membrane of a fibroblast.

Receptor-mediated endocytosis is exploited by some toxins and viruses to gain entry to cells. Diphtheria toxin, cholera toxin, and influenza virus all enter cells in this way.

Protein Degradation Is Mediated by Specialized Systems in All Cells

Protein degradation prevents the buildup of abnormal or unwanted proteins and permits the recycling of amino acids. The half-lives of eukaryotic proteins vary from 30 s to many days. Most proteins turn over rapidly relative to the lifetime of a cell, although a few (such as hemoglobin) can last for the life of the cell (about 110 days for an erythrocyte). Rapidly degraded proteins include those that are defective because of incorrectly inserted amino acids or because of damage accumulated during normal functioning. And enzymes that act at key regulatory points in metabolic pathways often turn over rapidly.

Defective proteins and those with characteristically short half-lives are generally degraded in both bacterial and eukaryotic cells by selective ATP-dependent cytosolic systems. A second system in vertebrates, operating in lysosomes, recycles the amino acids of membrane proteins, extracellular proteins, and proteins with characteristically long half-lives.

In *E. coli,* many proteins are degraded by an ATP-dependent protease called Lon (the name refers to the "long form" of proteins, observed only when this protease is absent). The protease is activated in the presence of defective proteins or those slated for rapid turnover; two ATP molecules are hydrolyzed for every peptide bond cleaved. The precise role of this ATP hydrolysis is not yet clear. Once a protein has been reduced to small inactive peptides, other ATP-independent proteases complete the degradation process.

The ATP-dependent pathway in eukaryotic cells is quite different, involving the protein **ubiquitin,** which occurs throughout the eukaryotic kingdoms. One of the most highly conserved proteins known, ubiquitin (76 amino acid residues) is essentially identical in organisms as different as yeasts and humans. Ubiquitin is covalently linked to proteins slated for destruction via an ATP-dependent pathway involving three separate enzymes (E1, E2, and E3 in Fig. 27–44). How ubiquitination targets proteins for proteolysis is not yet known. The eukaryotic ATP-dependent proteolytic system is a large complex (M_r 1×10^6) called the **proteasome.** The mode of action of the protease component of the system and the role of ATP are now being elucidated.

Although all the signals that trigger ubiquitination are not understood, a simple one has been found. For many proteins, the identity of the first residue that remains after removal of the amino-terminal Met residue, and any other posttranslational proteolytic processing of the amino-terminal

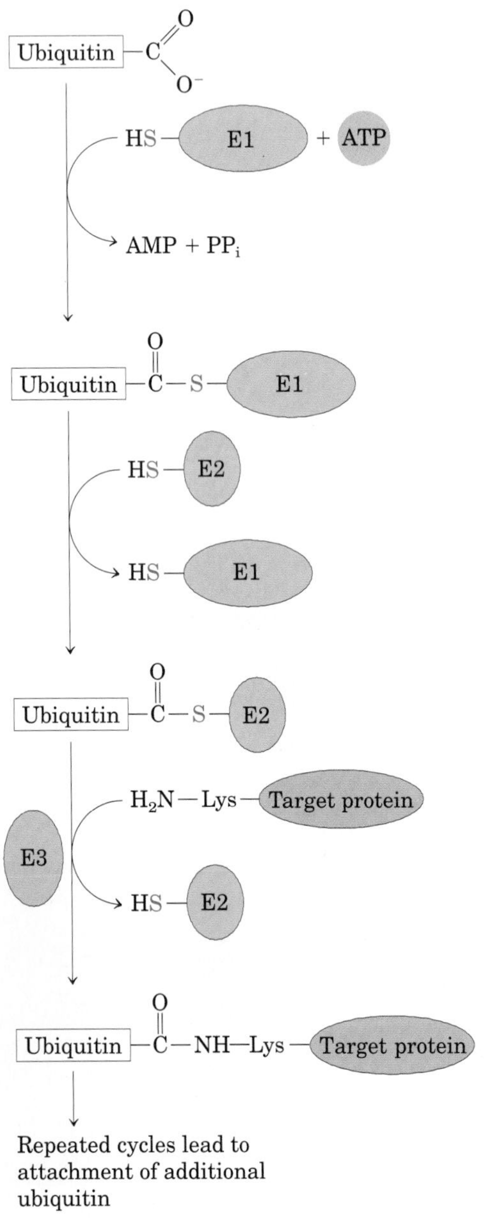

Ubiquitin—COO^-
HS—E1 + ATP
AMP + PP_i
Ubiquitin—C(=O)—S—E1
HS—E2
HS—E1
Ubiquitin—C(=O)—S—E2
H_2N—Lys—Target protein
E3
HS—E2
Ubiquitin—C(=O)—NH—Lys—Target protein
Repeated cycles lead to attachment of additional ubiquitin

figure 27–44
Three-step cascade pathway by which ubiquitin is attached to a protein. Two different enzyme-ubiquitin intermediates are involved. The free carboxyl group of ubiquitin's carboxyl-terminal Gly residue is ultimately linked through an amide (isopeptide) bond to an ϵ-amino group of a Lys residue of the target protein. Additional cycles produce polyubiquitin, a covalent polymer of ubiquitin subunits that targets the attached protein for destruction in eukaryotes.

end, has a profound influence on half-life (Table 27–10). These amino-terminal signals have been conserved over billions of years of evolution, being the same in both bacterial protein degradation systems and in the human ubiquitination pathway. More complex signals, such as the destruction box discussed in Chapter 13 (p. 472), are also being identified.

Ubiquitin-dependent proteolysis is as important for regulation of cellular processes as for the elimination of defective proteins. Many proteins required at only one stage of the eukaryotic cell cycle are rapidly degraded by the ubiquitin-dependent pathway after their function has been completed. The same pathway is required for the processing and presentation of class I MHC antigens (see Fig. 7–20). Ubiquitin-dependent destruction of cyclin is critical to cell-cycle regulation (see Fig. 13–33b). The E2 and E3 components of the ubiquitination cascade pathway (Fig. 27–44) are in fact two large families of proteins. Different E2 and E3 enzymes exhibit different specificities for target proteins and thus regulate different cellular processes. Some E2 and E3 enzymes are highly localized in certain cellular compartments, reflecting a specialized function. In a changing environment, protein degradation is as important to a cell's survival as protein synthesis, and much remains to be learned about these interesting pathways.

table 27–10

Relationship between Protein Half-Life and Amino-Terminal Amino Acid Residue

Amino-terminal residue	Half-life*
Stabilizing	
Met, Gly, Ala, Ser, Thr, Val	>20 h
Destabilizing	
Ile, Gln	~30 min
Tyr, Glu	~10 min
Pro	~7 min
Leu, Phe, Asp, Lys	~3 min
Arg	~2 min

Source: Modified from Bachmair, A., Finley, D., & Varshavsky, A. (1986) In vivo half-life of a protein is a function of its amino-terminal residue. *Science* **234,** 179–186.

*Half-lives were measured in yeast for a single protein modified so that in each experiment it had a different amino-terminal residue. (See Chapter 29 for a discussion of techniques used to engineer proteins with altered amino acid sequences.) Half-lives may vary for different proteins and in different organisms, but this general pattern appears to hold for all organisms: amino acids listed here as stabilizing when present at the amino terminus have a stabilizing effect on proteins in all cells.

summary

Proteins are synthesized with a particular amino acid sequence through the translation of information encoded in mRNA carried out by an RNA-protein complex called a ribosome. Amino acids are specified by mRNA codons consisting of nucleotide triplets. Translation requires adaptor molecules, the tRNAs, which recognize codons and insert amino acids into their appropriate sequential positions in the polypeptide.

The base sequences of the codons were deduced from experiments using synthetic mRNAs of known composition and sequence. The genetic code is degenerate: it has multiple code words for nearly all the amino acids. The third position in each codon is much less specific than the first and second and is said to wobble. The standard genetic code words are universal in all species, with some minor deviations in mitochondria and a few single-celled organisms.

The initiating amino acid, *N*-formylmethionine in bacteria, methionine in eukaryotes, is coded by AUG. Recognition of a particular AUG as the initiation codon requires a purine-rich initiating signal (the Shine-Dalgarno sequence) on the 5′ side of the AUG. The triplets UAA, UAG, and UGA are signals for termination of translation. Protein synthesis occurs on the ribosomes. Bacteria have 70S ribosomes, with a large (50S) and a small (30S) subunit. Eukaryotic ribosomes are significantly larger (80S) and contain more proteins.

Protein synthesis occurs in five stages. In stage 1, amino acids are activated by specific aminoacyl-tRNA synthetases in the cytosol. These enzymes catalyze the formation of aminoacyl-tRNAs, with simultaneous cleavage of ATP to AMP and PP_i. The fidelity of protein synthesis depends on the accuracy of this reaction, and some of these enzymes carry out proofreading steps at separate active sites.

Transfer RNAs have 73 to 93 nucleotide residues, some of which have modified bases. They have an amino acid arm with the terminal sequence CCA(3′) to which an amino acid is esterified, an anticodon arm, a TψC arm, and a D arm; some tRNAs have a fifth arm. The anticodon of a tRNA is responsible for the specificity of interaction between the aminoacyl-tRNA and the complementary mRNA codon. The growth of polypeptides on ribosomes begins with the amino-terminal amino acid and proceeds by successive additions of new residues to the carboxyl-terminal end.

In bacteria, the initiating aminoacyl-tRNA in all proteins is *N*-formylmethionyl-tRNA^{fMet}. Initiation of protein synthesis (stage 2) involves formation of a complex between the 30S ribosomal subunit, mRNA, GTP, fMet-tRNA^{fMet}, three initiation factors, and the 50S subunit; GTP is hydrolyzed to GDP and P_i. In the elongation steps (stage 3), GTP and elongation factors are required for binding the incoming aminoacyl-tRNA to the A site on the ribosome. In the first peptidyl transfer reaction, the fMet residue is transferred to the amino group of the incoming aminoacyl-tRNA. Movement of the ribosome along the mRNA then translocates the dipeptidyl-tRNA from the A site to the P site, a process requiring hydrolysis of GTP. Deacylated tRNAs dissociate from the ribosomal E site. After many such elongation cycles, synthesis of the polypeptide is terminated (stage 4) with the aid of release factors. At least four high-energy phosphate bonds are required to generate each peptide bond, an energy investment required to guarantee fidelity of translation.

In stage 5 of protein synthesis, polypeptides fold into their active, three-dimensional forms. Many proteins are further processed by posttranslational modification reactions.

After synthesis, many proteins are directed to particular locations in the cell. One targeting mechanism involves a peptide signal sequence, generally found at the amino terminus of a newly synthesized protein. In eukaryotic cells, one class of signal sequences is recognized and bound by a large protein-RNA complex called the signal recognition particle (SRP). SRP binds the signal sequence as soon as it appears on the ribosome and transfers the entire ribosome and incomplete polypeptide to the endoplasmic reticulum. Polypeptides with these signal sequences are moved into the ER lumen as they are synthesized; there they may be modified and moved to the Golgi complex, then sorted and sent to lysosomes, the plasma membrane, or transport vesicles. Proteins targeted to mitochondria and chloroplasts in eukaryotic cells, and for export in bacteria, also make use of an amino-terminal signal sequence. Other known targeting signals include carbohydrates and signal patches. Some proteins are imported into the cell by receptor-mediated endocytosis. The same receptors are also used by some toxins and viruses to gain entry into cells.

Proteins are eventually degraded by specialized proteolytic systems present in all cells. Defective proteins and those slated for rapid turnover are generally degraded by an ATP-dependent proteolytic system. In eukaryotic cells, the proteins are first tagged by linking them to a highly conserved protein called ubiquitin. Ubiquitin-dependent proteolysis is critical to the regulation of many cellular processes.

further reading

The Genetic Code

Cedergren, R. & Miramontes, P. (1996) The puzzling origin of the genetic code. *Trends Biochem. Sci.* **21,** 199–200.

Crick, F.H.C. (1966) The genetic code: III. *Sci. Am.* **215** (October), 55–62.

An insightful overview of the genetic code at a time when the code words had just been worked out.

Fox, T.D. (1987) Natural variation in the genetic code. *Annu. Rev. Genet.* **21,** 67–91.

Hatfield, D. & Oroszlan, S. (1990) The *where, what* and *how* of ribosomal frameshifting in retroviral protein synthesis. *Trends Biochem. Sci.* **15,** 186–190.

Nirenberg, M.W. (1963) The genetic code: II. *Sci. Am.* **208** (March), 80–94.

A description of the original experiments.

Stadtman, T.C. (1996) Selenocysteine. *Annu. Rev. Biochem.* **65,** 83–100.

Stuart, K. (1991) RNA editing in mitochondrial mRNA of trypanosomatids. *Trends Biochem. Sci.* **16,** 68–72.

Protein Synthesis

Arnez, J.G. & Moras, D. (1997) Structural and functional considerations of the aminoacylation reaction. *Trends Biochem. Sci.* **22,** 211–216.

Björk, G.R., Ericson, J.U., Gustafsson, C.E.D., Hagervall, T.G., Jönsson, Y.H., & Wikström, P.M. (1987) Transfer RNA modification. *Annu. Rev. Biochem.* **56,** 263–288.

Burbaum, J.J. & Schimmel, P. (1991) Structural relationships and the classification of aminoacyl-tRNA synthetases. *J. Biol. Chem.* **266,** 16,965–16,968.

Chapeville, F., Lipmann, F., von Ehrenstein, G., Weisblum, B., Ray, W.J., Jr., & Benzer, S. (1962) On the role of soluble ribonucleic acid in coding for amino acids. *Proc. Natl. Acad. Sci. USA* **48,** 1086–1092.

Classic experiments providing proof for Crick's adaptor hypothesis and showing that amino acids are not checked after they are linked to tRNAs.

Dintzis, H.M. (1961) Assembly of the peptide chains of hemoglobin. *Proc. Natl. Acad. Sci. USA* **47,** 247–261.

A classic experiment establishing that proteins are assembled beginning at the amino terminus.

Giege, R., Sissler, M., & Florentz, C. (1998) Universal rules and idiosyncratic features in tRNA identity. *Nucleic Acids Res.* **26,** 5017–5035.

Gingras, A.-C., Raught, B., & Sonenberg, N. (1999) eIF4 initiation factors: effectors of mRNA recruitment to ribosomes and regulators of translation. *Annu. Rev. Biochem.* **68,** 913–964.

Gray, N.K., & Wickens, M. (1998) Control of translation initiation in animals. *Ann. Rev. Cell Dev. Biol.* **14,** 399–458.

Green, R. & Noller, J.F. (1997) Ribosomes and translation. *Annu. Rev. Biochem.* **66,** 679–716.

Maden, B.E.H. (1990) The numerous modified nucleotides in eukaryotic ribosomal RNA. *Prog. Nucleic Acid Res. Mol. Biol.* **39,** 241–303.

Moore, P.B. (1998) The three-dimensional structure of the ribosome and its components. *Annu. Rev. Biophys. Biomol. Struct.* **27,** 35–58.

Sachs, A.B., Sarnow, P., & Hentze, M.W. (1997) Starting at the beginning, middle, and end: translation initiation in eukaryotes. *Cell* **89,** 831–838.

Schimmel, P. (1993) GTP hydrolysis in protein synthesis: two for Tu? *Science* **259,** 1264–1265.

Sprinzl, M. (1994) Elongation factor Tu: a regulatory GTPase with an integrated effector. *Trends Biochem. Sci.* **19,** 245–250.

Stark, H., Orlova, E.V., Rinke-Appel, J., Jünke, N., Mueller, F., Brimacombe, R., & van Heel, M. (1997) Arrangement of tRNAs in pre- and post-translocational ribosomes revealed by electron cryomicroscopy. *Cell* **88,** 19–28.

Protein Targeting and Secretion

Görlich, D. & Mattaj, I.W. (1996) Nucleocytoplasmic transport. *Science* **271,** 1513–1518.

Neupert, W. (1997) Protein import into mitochondria. *Annu. Rev. Biochem.* **66,** 863–917.

Pryer, N.K., Wuestehube, L.J., & Schekman, R. (1992) Vesicle-mediated protein sorting. *Annu. Rev. Biochem.* **61,** 471–516.

Rapoport, T.A., Jungnickel, B., & Kutay, U. (1996) Protein transport across the eukaryotic endoplasmic reticulum and bacterial inner membranes. *Annu. Rev. Biochem.* **65,** 271–303.

Schatz, G. & Dobberstein, B. (1996) Common principles of protein translocation across membranes. *Science* **271,** 1519–1525.

Schekman, R. & Orci, L. (1996) Coat proteins and vesicle budding. *Science* **271,** 1526–1532.

Schmid, S.L. (1997) Clathrin-coated vesicle formation and protein sorting: an integrated process. *Annu. Rev. Biochem.* **66,** 511–548.

Varshavsky, A. (1997) The ubiquitin system. *Trends Biochem. Sci.* **22,** 383–387.

Voges, D., Zwickl, P., & Baumeister, W. (1999) The 26S proteasome: a molecular machine designed for controlled proteolysis. *Annu. Rev. Biochem.* **68,** 1015–1057.

Ward, W.H.J. (1987) Diphtheria toxin: a novel cytocidal enzyme. *Trends Biochem. Sci.* **12,** 28–31.

problems

1. Messenger RNA Translation Predict the amino acid sequences of peptides formed by ribosomes in response to the following mRNA sequences, assuming that the reading frame begins with the first three bases in each sequence.

(a) GGUCAGUCGCUCCUGAUU
(b) UUGGAUGCGCCAUAAUUUGCU
(c) CAUGAUGCCUGUUGCUAC
(d) AUGGACGAA

2. How Many Different mRNA Sequences Can Specify One Amino Acid Sequence? Write all the possible mRNA sequences that can code for the simple tripeptide segment Leu–Met–Tyr. Your answer will give you some idea about the number of possible mRNAs that can code for one polypeptide.

3. Can the Base Sequence of an mRNA Be Predicted from the Amino Acid Sequence of Its Polypeptide Product? A given sequence of bases in an mRNA will code for one and only one sequence of amino acids in a polypeptide, if the reading frame is specified. From a given sequence of amino acid residues in a protein such as cytochrome *c*, can we predict the base sequence of the unique mRNA that coded it? Give reasons for your answer.

4. Coding of a Polypeptide by Duplex DNA The template strand of a segment of double-helical DNA contains the sequence

(5′)CTTAACACCCCTGACTTCGCGCCGTCG

(a) What is the base sequence of the mRNA that can be transcribed from this strand?

(b) What amino acid sequence could be coded by the mRNA in (a), starting from the 5′ end?

(c) If the complementary strand of this DNA were transcribed and translated, would the resulting amino acid sequence be the same as in (b)? Explain the biological significance of your answer.

5. Methionine Has Only One Codon Methionine is one of two amino acids with only one codon. How does the single codon for methionine specify both the initiating residue and interior Met residues of polypeptides synthesized by *E. coli*?

6. Synthetic mRNAs The genetic code was elucidated with polyribonucleotides synthesized either enzymatically or chemically in the laboratory. Given what we now know about the genetic code, how would you make a polyribonucleotide that could serve as an mRNA coding predominantly for many Phe residues and a small number of Leu and Ser residues? What other amino acid(s) would be coded for by this polyribonucleotide, but in smaller amounts?

7. Energy Cost of Protein Biosynthesis Determine the minimum energy cost, in terms of high-energy phosphate groups expended, required for the biosynthesis of the β-globin chain of hemoglobin (146 residues), starting from a pool including all necessary amino acids, ATP, and GTP. Compare your answer with the direct energy cost of the biosynthesis of a linear glycogen chain of 146 glucose residues in ($\alpha 1 \rightarrow 4$) linkage, starting from a pool including glucose, UTP, and ATP (Chapter 20). From your data, what is the *extra* energy cost of making a protein, in which all the residues are ordered in a specific sequence, compared to the cost of making a polysaccharide containing the same number of residues, but lacking the informational content of the protein?

In addition to the direct energy cost for the synthesis of a protein, there are indirect energy costs—those required for the cell to make the necessary enzymes for protein synthesis. Compare the magnitude of the indirect costs to a eukaryotic cell of the biosynthesis of linear ($\alpha 1 \rightarrow 4$) glycogen chains and the biosynthesis of polypeptides, in terms of the enzymatic machinery involved.

8. Predicting Anticodons from Codons Most amino acids have more than one codon and attach to more than one tRNA, each with a different anticodon. Write all possible anticodons for the four codons of glycine: (5′)GGU, GGC, GGA, and GGG.

(a) From your answer, which of the positions in the anticodons are primary determinants of their codon specificity in the case of glycine?

(b) Which of these anticodon-codon pairings has/have a wobbly base pair?

(c) In which of the anticodon-codon pairings do all three positions exhibit strong Watson-Crick hydrogen bonding?

9. Effect of Single-Base Changes on Amino Acid Sequence Much important confirmatory evidence on the genetic code has come from the nature of single-residue changes in the amino acid sequence of mutant proteins. Which of the following single-residue replacements would be consistent with the genetic code? Which cannot be the result of a single-base mutation? Why?

(a) Phe → Leu
(b) Lys → Ala
(c) Ala → Thr
(d) Phe → Lys
(e) Ile → Leu
(f) His → Glu
(g) Pro → Ser

10. Basis of the Sickle-Cell Mutation Sickle-cell hemoglobin has a Val residue at position 6 of the β-globin chain, instead of the Glu residue found in normal hemoglobin A. Can you predict what change took place in the DNA codon for glutamate to account for its replacement by valine?

11. Importance of the "Second Genetic Code" Some aminoacyl-tRNA synthetases do not recognize and bind the anticodon of their cognate tRNAs but instead use other structural features of the tRNAs to impart binding specificity. The tRNAs for alanine apparently fall into this category.

(a) What features of $tRNA^{Ala}$ are recognized by Ala-tRNA synthetase?

(b) Describe the consequences of a C → G mutation in the third position of the anticodon of $tRNA^{Ala}$.

(c) What other kinds of mutations might have similar effects?

(d) Mutations of these types are never found in natural populations of organisms. Why? (Hint: Consider what might happen both to individual proteins and to the organism as a whole.)

12. Maintaining the Fidelity of Protein Synthesis The chemical mechanisms used to avoid errors in protein synthesis are different from those used during DNA replication. DNA polymerases use a 3′→5′ exonuclease proofreading activity to remove mispaired nucleotides incorrectly inserted into a growing DNA strand. There is no analogous proofreading function on ribosomes; and, in fact, the identity of an amino acid attached to an incoming tRNA and added to the growing polypeptide is never checked. A proofreading step that hydrolyzed the previously formed peptide bond after an incorrect amino acid had been inserted into a growing polypeptide (analogous to the proofreading step of DNA polymerases) would be impractical. Why? (Hint: Consider how the link between the growing polypeptide and the mRNA is maintained during elongation; see Figs 27–26 and 27–27.)

13. Predicting the Cellular Location of a Protein The gene for a eukaryotic polypeptide 300 amino acid residues long is altered so that a signal sequence recognized by SRP occurs at the polypeptide's amino terminus and a nuclear localization signal (NLS) occurs internally, beginning at residue 150. Where is the protein likely to be found in the cell?

14. Requirements for Protein Translocation across a Membrane The secreted bacterial protein OmpA has a precursor, ProOmpA, which has the amino-terminal signal sequence required for secretion. If purified ProOmpA is denatured with 8 M urea and the urea is then removed, (e.g., by running the protein solution rapidly through a gel filtration column) the protein can be translocated across isolated bacterial inner membranes in vitro. However, translocation becomes impossible if ProOmpA is allowed to incubate for a few hours in the absence of urea. Furthermore, the capacity for translocation is maintained for an extended period if ProOmpA is incubated in the presence of another bacterial protein called trigger factor. Describe the probable function of this factor.

15. Protein-Coding Capacity of a Viral DNA The 5,386 base pair genome of bacteriophage ϕX174 (see Fig. 27–10) includes genes for 10 proteins, designated A to K, with sizes given in the table below. How much DNA would be required to encode these 10 proteins? How can you reconcile the size of the ϕX174 genome with its protein-coding capacity?

Protein	Number of amino acid residues
A	455
B	120
C	86
D	152
E	91
F	427
G	175
H	328
J	38
K	56

chapter

28 Regulation of Gene Expression

(a)

(b)

Defects in the regulation of gene expression can alter the developmental program of an organism. An example is this *bithorax* mutant of *Drosophila melanogaster* with an extra set of wings. **(a)** Normal body structure. **(b)** Homeotic mutant *(bithorax)* in which a segment has developed incorrectly to produce an extra set of wings.

Of the 4,000 or so genes in the typical bacterial genome or the estimated 100,000 genes in the human genome, only a fraction are expressed at any given time. Some gene products are present in very large amounts in cells: the elongation factors required for protein synthesis, for example, are among the most abundant proteins in bacteria, and ribulose 1,5-bisphosphate carboxylase/oxygenase (rubisco) of plants and photosynthetic bacteria is the most abundant known enzyme in the biosphere. Other gene products occur in much smaller amounts; for instance, a cell may contain only a few molecules of the enzymes that repair rare DNA lesions. Requirements for some gene products change over time. The need for enzymes in certain metabolic pathways may wax or wane as food sources change or are depleted. During development of a multicellular organism, some proteins that influence cellular differentiation are present for only a brief time in a few cells. Specialization of cellular function can dramatically affect the need for various gene products; an example is the uniquely high concentration of hemoglobin in erythrocytes. Given the high cost of protein synthesis, regulation of gene expression is essential to making optimal use of available energy.

The cellular concentration of a protein is determined by a delicate balance of at least seven processes, each having several potential points of regulation:

1. Synthesis of the primary RNA transcript.
2. Posttranscriptional processing of mRNA.
3. mRNA degradation.
4. Protein synthesis (translation).
5. Posttranslational modification of proteins.
6. Protein degradation.
7. Protein targeting and transport.

These processes are summarized in Figure 28–1. We have examined several of these mechanisms in previous chapters. Posttranscriptional modification of mRNA by processes such as alternative splicing patterns (see Fig. 26–21b) or RNA editing (see Box 27–1) can affect which proteins are produced from mRNA transcripts and in what amounts. A variety of nucleotide sequences in an mRNA can affect the rate at which it is degraded (p. 1006). Many factors affect the rate at which an mRNA is translated into a protein, as well as influencing the posttranslational modification, targeting, and eventual degradation of that protein (Chapter 27).

This chapter focuses primarily on the regulation of transcription initiation, although some aspects of the regulation of translation are also described. Of the regulatory processes illustrated in Figure 28–1, those operating at the level of transcription initiation are the best documented and

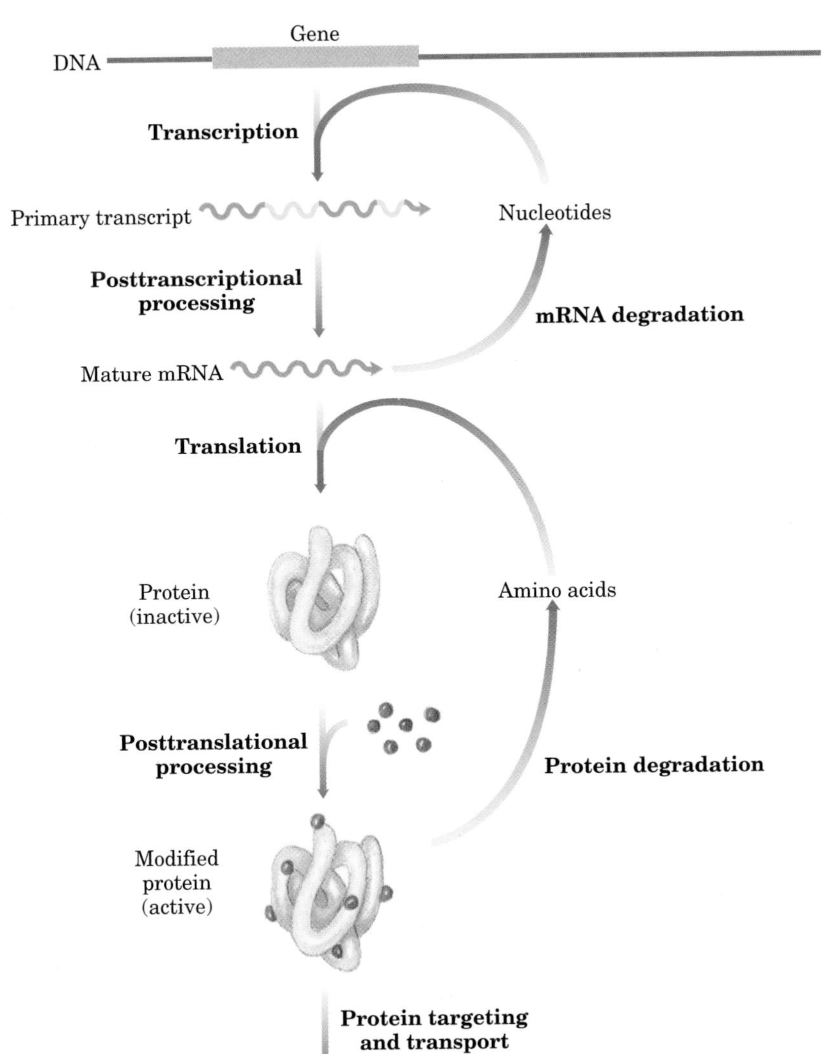

figure 28–1
Seven processes that affect the steady-state concentration of a protein. Each process has several potential points of regulation.

the most common. As in all biochemical processes, the most efficient place for regulation is at the beginning of the pathway. However, synthesizing informational macromolecules is so extraordinarily expensive that cells have evolved elaborate mechanisms to regulate the process. The regulatory processes themselves can involve a considerable investment of chemical energy.

Control of transcription initiation permits the synchronized regulation of multiple genes encoding products with interdependent activities. For example, when their DNA is heavily damaged, bacterial cells require a coordinated increase in the levels of the many enzymes involved in DNA repair. Perhaps the most sophisticated form of coordination occurs in the complex regulatory circuits that guide the development of multicellular eukaryotes.

We begin by examining the interactions between proteins and DNA that are the key to transcriptional regulation. We next discuss the specific proteins that influence the expression of specific genes, first in prokaryotic and then in eukaryotic cells. Included in the discussion are several different mechanisms that allow cells to regulate gene expression and coordinate the expression of multiple genes.

Principles of Gene Regulation

Genes for products that are required at all times are expressed at a more or less constant level in virtually every cell of a species or organism. Genes for the enzymes of central metabolic pathways, such as the citric acid cycle, fall into this category and are often referred to as **housekeeping genes.** Unvarying expression of a gene is called **constitutive gene expression.**

The cellular levels of some gene products rise and fall in response to molecular signals; this is **regulated gene expression.** Gene products that increase in concentration under particular molecular circumstances are referred to as **inducible;** the process of increasing their expression is called **induction.** The expression of many genes encoding DNA repair enzymes, for example, is induced by high levels of DNA damage. Conversely, gene products that decrease in concentration in response to a molecular signal are referred to as **repressible,** and the process is called **repression.** For example, ample supplies of tryptophan lead to repression of the genes for the enzymes that catalyze tryptophan biosynthesis in bacteria.

Transcription is mediated and regulated by protein-DNA interactions, especially those involving the protein components of RNA polymerase (Chapter 26). We first consider how the activity of RNA polymerase is regulated, and proceed to a general description of the proteins participating in this process. We then examine the molecular basis for the recognition of specific DNA sequences by DNA-binding proteins.

RNA Polymerase Binds to DNA at Promoters

RNA polymerases bind to DNA and initiate transcription at sites called promoters (see Fig. 26–5), generally found near points at which RNA synthesis begins on the DNA template. The regulation of transcription initiation often entails changes in how RNA polymerase interacts with a promoter.

The nucleotide sequences of promoters vary considerably, influencing the binding affinity of RNA polymerases and thus the frequency of transcription initiation. Some *Escherichia coli* genes are transcribed once per second, others less than once per cell generation. Much of this variation is due to differences in promoter sequence. In the absence of regulatory proteins, differences in the sequences of two promoters may affect the frequency of transcription initiation by a factor of 1,000 or more. Most *E. coli* promoters have a sequence close to a consensus (Fig. 28–2). Mutations away from the consensus sequence usually decrease promoter function; conversely, mutations toward consensus usually enhance promoter function.

Although housekeeping genes are expressed constitutively, the cellular concentrations of the proteins they encode nevertheless vary widely. For

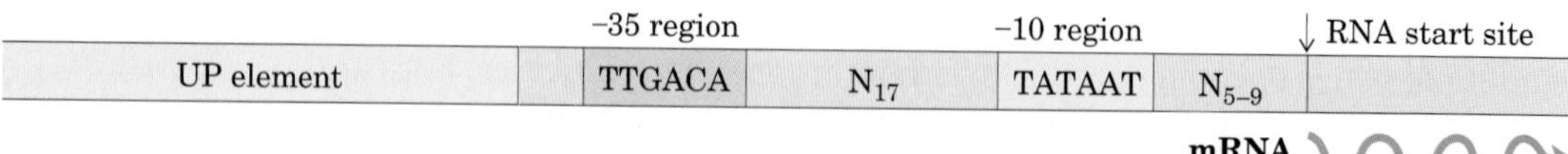

figure 28–2

Consensus sequence for many *E. coli* promoters. N indicates any nucleotide. Most base substitutions in the −10 and −35 regions have a negative effect on promoter function. Some promoters also include the UP element (see Fig. 26–5). By convention, DNA sequences are shown as they occur in the nontemplate strand, with the 5′ terminus on the left. Nucleotides are numbered from the transcription start site, with positive numbers to the right (in the direction of transcription) and negative numbers to the left.

these genes, the RNA polymerase–promoter interaction strongly influences the rate of transcription initiation; differences in promoter sequence allow the cell to synthesize the appropriate level of each housekeeping gene product.

The basal rate of transcription initiation at the promoters of nonhousekeeping genes is also determined by the promoter sequence, but expression of these genes is further modulated by regulatory proteins. These proteins often work by enhancing or interfering with the interaction between RNA polymerase and the promoter.

The sequences of eukaryotic promoters are more variable than their prokaryotic counterparts (see Fig. 26–8). The three eukaryotic RNA polymerases usually require an array of general transcription factors in order to bind to a promoter. Yet, as with prokaryotic gene expression, the basal level of transcription is determined by the effect of promoter sequence on the function of RNA polymerase and its associated transcription factors.

Transcription Initiation Is Regulated by Proteins That Bind to or near Promoters

At least three types of proteins regulate transcription initiation by RNA polymerase: **specificity factors** alter the specificity of RNA polymerase for a given promoter or set of promoters; **repressors** impede access of RNA polymerase to the promoter; and **activators** enhance the RNA-promoter interaction.

We introduced prokaryotic specificity factors in Chapter 26 (see Fig. 26–4), although we did not refer to them by that name. The σ subunit of the *E. coli* RNA polymerase holoenzyme is a specificity factor that mediates promoter recognition and binding. Most *E. coli* promoters are recognized by a single σ subunit (M_r 70,000) called σ^{70}. Under some conditions, notably when the bacteria are subjected to heat stress, some of the σ^{70} subunits are replaced by another specificity factor (M_r 32,000) called σ^{32} (p. 982). When bound to σ^{32}, RNA polymerase is directed to a specialized set of promoters with a different consensus sequence (Fig. 28–3). These promoters control the expression of a set of genes that encode the heat-shock response. Thus, through changes in the binding affinity of the polymerase that direct it to different promoters, a set of related genes is coordinately regulated. In eukaryotic cells, some of the general transcription factors, in particular the TATA binding protein, may be considered specificity factors.

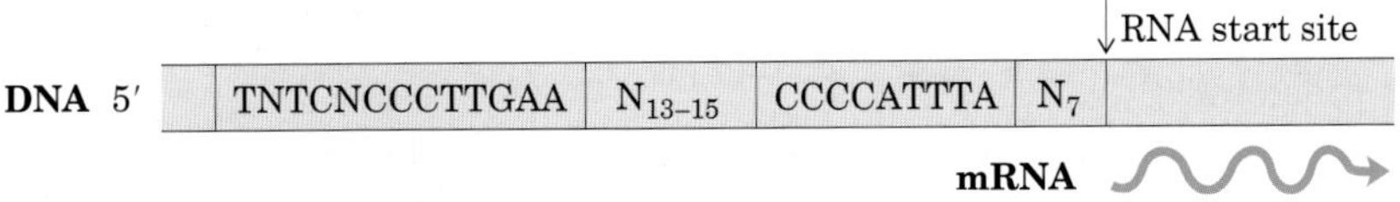

figure 28–3
Consensus sequence for promoters that regulate expression of the heat-shock genes of *E. coli*. This system responds to temperature increases as well as some other environmental stresses, and it involves the induction of a set of proteins. Binding of RNA polymerase to heat-shock promoters is mediated by a specialized σ subunit of the polymerase, σ^{32}, which replaces σ^{70} in the RNA polymerase initiation complex.

Repressors bind to specific sites on the DNA. In prokaryotic cells, such binding sites, called **operators,** are generally near a promoter. RNA polymerase binding, or its movement along the DNA after binding, is blocked when the repressor is present. Regulation by means of a repressor protein that blocks transcription is referred to as **negative regulation.** Repressor binding to DNA is regulated by a molecular signal (sometimes called an **effector**), usually a small molecule or a protein, that binds to the repressor and causes a conformational change. The interaction between repressor and signal molecule either increases or decreases transcription. In some

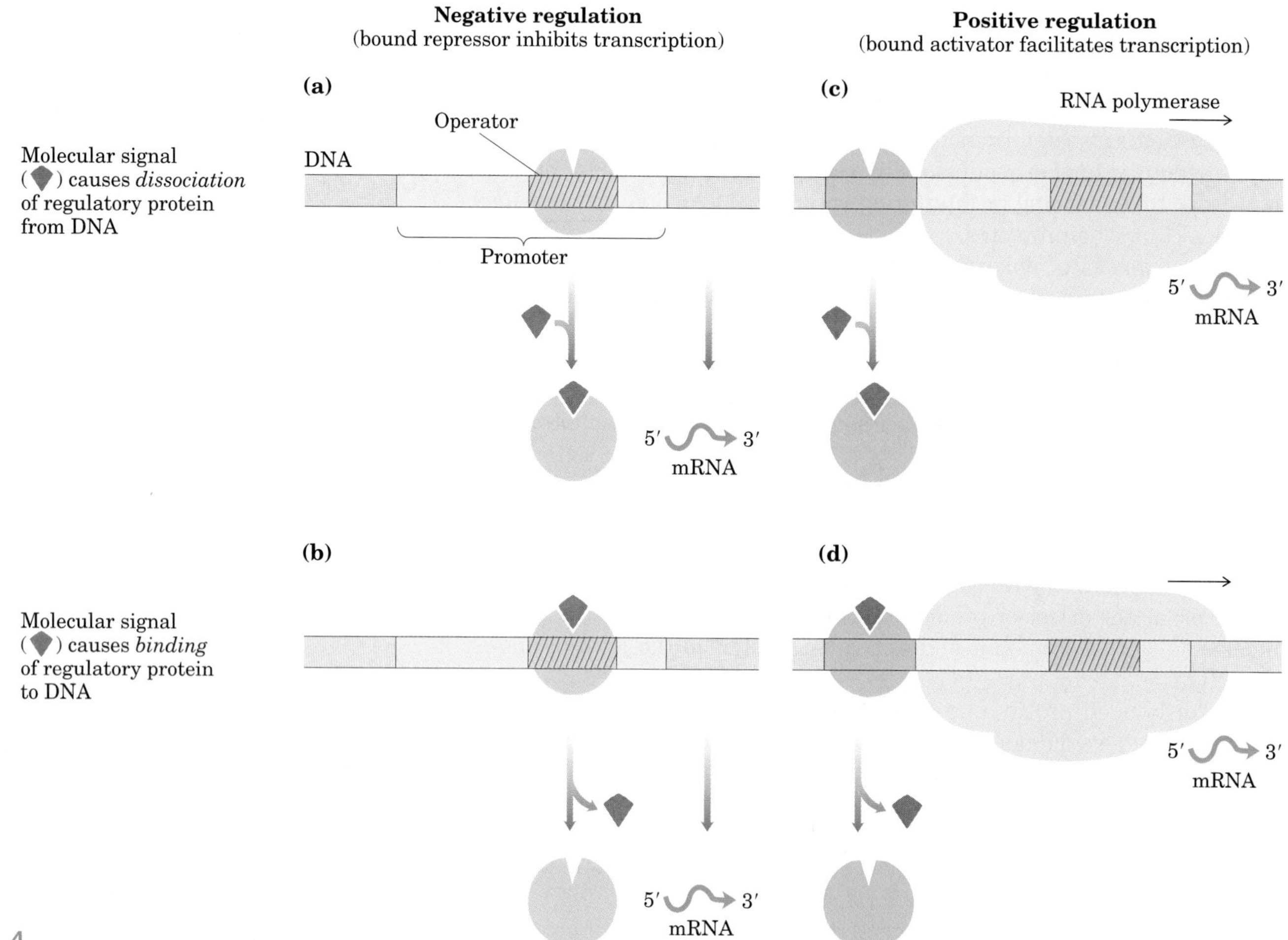

figure 28–4

Common patterns of regulation of transcription initiation. Two types of negative regulation are illustrated. **(a)** Repressor (red) is bound to the operator in the absence of the molecular signal; the external signal causes dissociation of the repressor to permit transcription. **(b)** Repressor is bound in the presence of the signal; the repressor dissociates and transcription ensues when the signal is removed. Positive regulation is mediated by gene activators. Again, two types are shown. **(c)** Activator (green) binds in the absence of the molecular signal and transcription proceeds; when the signal is added, the activator dissociates and transcription is inhibited. **(d)** Activator binds in the presence of the signal; it dissociates only when the signal is removed. Note that "positive" and "negative" regulation refer to the type of regulatory protein involved: the bound protein either facilitates or inhibits transcription. In either case, addition of the molecular signal may increase or decrease transcription, depending on its effect on the regulatory protein.

cases, the conformational change results in dissociation of a DNA-bound repressor from the operator (Fig. 28–4a). Transcription initiation can then proceed unhindered. In other cases, the interaction between an inactive repressor and the signal molecule causes the repressor to bind to the operator (Fig. 28–4b). In eukaryotic cells, the binding site for a repressor may be some distance from the promoter; binding has the same effect as in bacterial cells: inhibiting the assembly or activity of a transcription complex at the promoter.

Activators provide a molecular counterpoint to repressors; they bind to DNA and *enhance* the activity of RNA polymerase at a promoter—**positive regulation.** Activator binding sites are often adjacent to promoters that are bound weakly or not at all by RNA polymerase alone, such that little transcription occurs in the absence of the activator. Some eukaryotic activators bind to DNA sites called **enhancers** that are quite distant from the promoter, influencing the rate of transcription at a promoter that may be located thousands of base pairs away. Some activators are normally bound to DNA, enhancing transcription until dissociation of the activator is triggered by the binding of a signal molecule (Fig. 28–4c). In other cases the activator binds DNA only after interaction with a signal molecule (Fig. 28–4d). Signal molecules can therefore increase or decrease transcription depending on how they affect the activator. Positive regulation is particularly common in eukaryotes, and is more complex than in prokaryotes, as we shall see.

Most Prokaryotic Genes Are Regulated in Units Called Operons

Bacteria have a simple general mechanism for coordinating the regulation of genes that encode products involved in a set of related processes: these genes are clustered on the chromosome and are transcribed together. Most prokaryotic mRNAs are polycistronic—multiple genes on a single transcript—and the single promoter that initiates transcription of the cluster is the site of regulation for expression of all the genes in the cluster. The gene cluster and promoter, plus additional sequences that function together in regulation, are called an **operon** (Fig. 28–5). Operons that include two to six genes transcribed as a unit are common; some operons contain 20 or more genes.

Many of the principles of prokaryotic gene expression were first defined by studies of lactose metabolism in *E. coli,* which can use lactose as its sole carbon source. In 1960, François Jacob and Jacques Monod published a short paper in the *Proceedings* of the French Academy of Sciences that described how two adjacent genes involved in lactose metabolism were coordinately regulated by a genetic element located at one end of the gene cluster. The genes were those for β-galactosidase, which cleaves lactose to galactose and glucose, and galactoside permease, which transports lactose into the cell (Fig. 28–6). The terms "operon" and "operator" were first introduced in this paper. With the operon model, gene regulation could, for the first time, be considered in molecular terms.

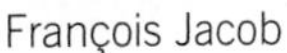

François Jacob

Jacques Monod

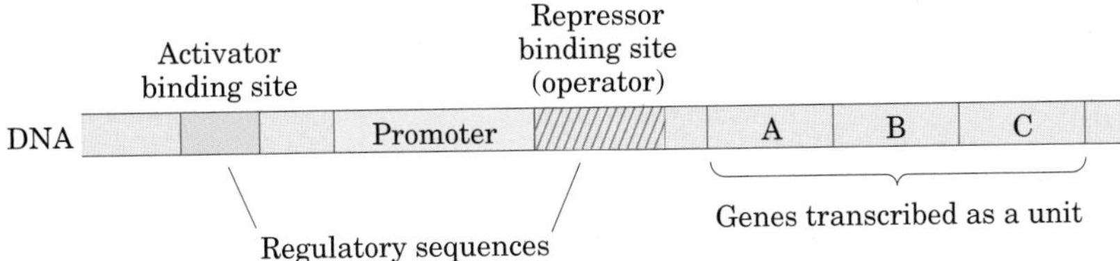

figure 28–5

Representative prokaryotic operon. Genes A, B, and C are transcribed on one polycistronic mRNA. Typical regulatory sequences include binding sites for proteins that either activate or repress transcription from the promoter.

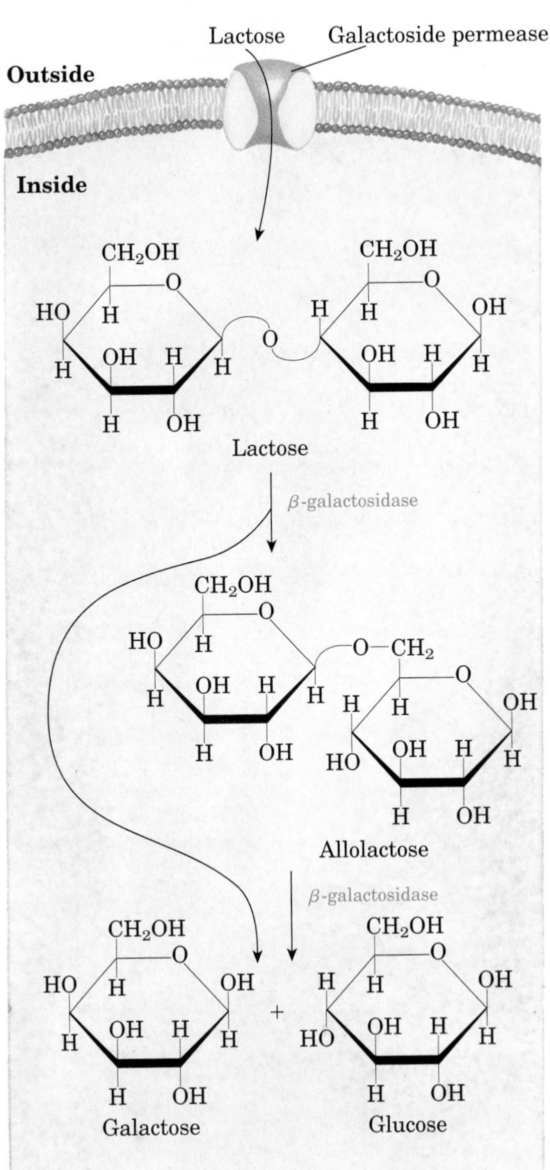

figure 28–6

Lactose metabolism in *E. coli*. Uptake and metabolism of lactose require the activities of galactoside permease and β-galactosidase. Conversion of lactose to allolactose by transglycosylation is a minor reaction also catalyzed by β-galactosidase.

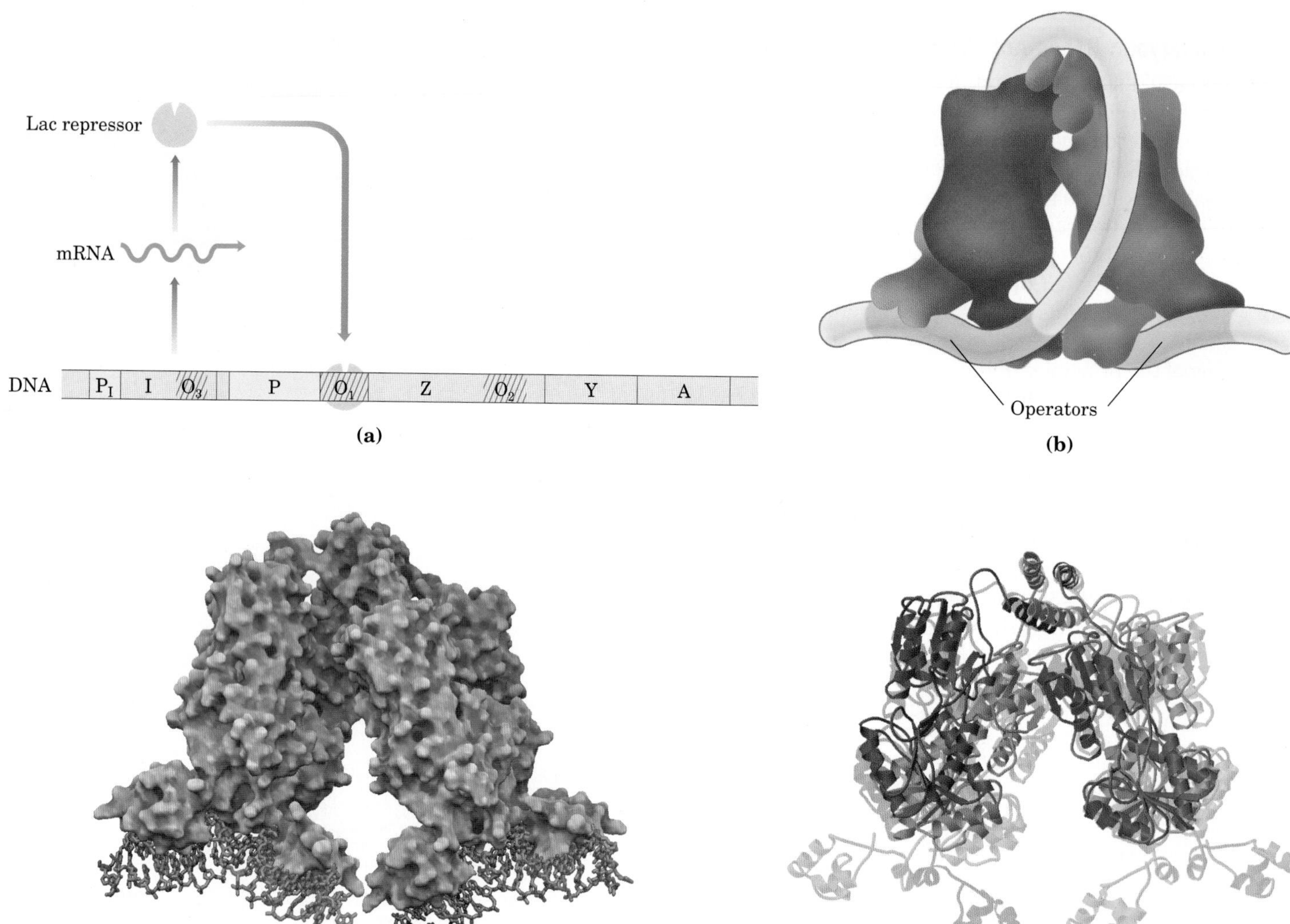

figure 28–7

The *lac* operon. (a) The *lac* operon in the repressed state. The I gene encodes the Lac repressor. The *lac* Z, Y, and A genes encode β-galactosidase, galactoside permease, and transacetylase, respectively. P is the promoter for the *lac* genes, and P_I is the promoter for the I gene. O_1 is the main operator for the *lac* operon; O_2 and O_3 are secondary operator sites called pseudooperators. **(b)** The Lac repressor binds the main operator and one of the pseudooperators, apparently forming a loop in the DNA that might wrap around the repressor as shown. **(c)** Lac repressor bound to DNA. This shows the protein (gray) bound to only short, discontinuous segments of DNA (blue). **(d)** Conformational change in the Lac repressor caused by binding of the artificial inducer isopropylthiogalactoside, IPTG. The tetrameric repressor is shown without IPTG bound (transparent image) and with IPTG bound (bound IPTG not shown; overlaid solid image). DNA bound to the transparent structure is not shown. When IPTG is bound and DNA is not bound, the DNA-binding domains are no longer defined in the crystal structure.

The *lac* Operon Is Subject to Negative Regulation

The lactose (*lac*) operon (Fig. 28–7a) includes the genes for β-galactosidase (Z), galactoside permease (Y), and thiogalactoside transacetylase (A), whose physiological function is not yet certain. Each of the three genes is preceded by ribosome binding sites (not shown in Fig. 28–7) that independently direct the translation of that gene (Chapter 27). Regulation of the *lac* operon follows the pattern outlined in Figure 28–4a.

The study of *lac* operon mutants has revealed some details of the workings of the operon's regulatory system. In the absence of lactose, the *lac* operon genes are repressed. Mutations in the operator or in another gene, the I gene, result in constitutive synthesis of the gene products. When the I gene is defective, repression can be restored by introducing a functional I gene into the cell on another DNA molecule, demonstrating that the I gene encodes a diffusible molecule that causes gene repression. This molecule proved to be a protein, now called the Lac repressor, a tetramer of identical monomers. The operator to which it binds most tightly (called O_1) abuts the transcription start site (Fig. 28–7a). The I gene is transcribed from its own promoter (P_I) independently of the *lac* operon genes. The *lac* operon has two secondary binding sites for the Lac repressor, sometimes referred to as pseudooperators because they are not absolutely required for operator function. One (O_2) is centered near position +410, within the gene encoding β-galactosidase (Z); the other (O_3) is near position −90, within the I gene. To repress the operon, the Lac repressor appears to bind to both the main operator and one of the two secondary sites, with the intervening DNA looped out (Fig. 28–7b, c). Either binding arrangment blocks transcription initiation.

Despite this elaborate binding complex, repression is not absolute. Binding of the Lac repressor reduces the rate of transcription initiation by a factor of 1,000. If the O_2 and O_3 sites are eliminated by deletion or mutation, the binding of repressor to O_1 alone reduces transcription about 100-fold. Even in the repressed state, each cell has a few molecules of β-galactosidase and galactoside permease, presumably synthesized on the rare occasions when the repressor transiently dissociates from the operators. This basal level of transcription is essential to operon regulation.

When cells are provided with lactose, the *lac* operon is induced. An inducer (signal) molecule binds to a specific site on the Lac repressor, causing a conformational change (Fig. 28–7d) that results in dissociation of the repressor from the operator. The inducer in the *lac* operon system is not lactose itself but an isomer of lactose called allolactose (Fig. 28–6). After entry into the *E. coli* cell (via the existing molecules of permease), lactose is converted to allolactose by one of the few existing β-galactosidase molecules. Release of the Lac repressor caused by allolactose allows the *lac* operon genes to be expressed and leads to a 1,000-fold increase in the concentration of β-galactosidase.

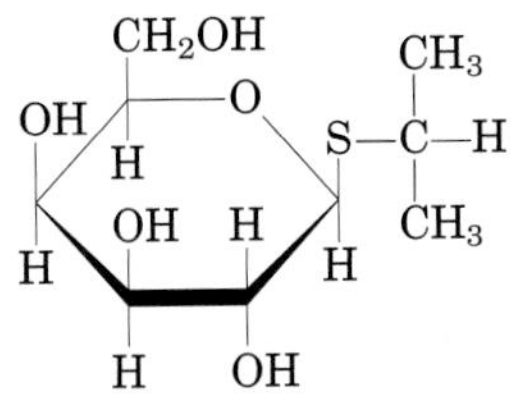

Several β-galactosides structurally related to allolactose are inducers of the *lac* operon but are not substrates for β-galactosidase; others are substrates but not inducers. One particularly effective and nonmetabolizable inducer of the *lac* operon that is often used experimentally is isopropylthiogalactoside (IPTG). An inducer that cannot be metabolized allows researchers to explore the physiological function of lactose as a carbon source for growth separate from its function in the regulation of gene expression.

Isopropylthiogalactoside (IPTG)

In addition to the multitude of operons now known in bacteria, a few polycistronic operons have been found in the cells of lower eukaryotes. In the cells of higher eukaryotes, however, almost all genes are transcribed separately.

The mechanisms by which operons are regulated can vary significantly from the simple model presented in Figure 28–7. Even the *lac* operon is more complex than has been indicated here, with an activator also contributing to the overall scheme, as described later in this chapter. Before any further discussion of the layers of regulation of gene expression, we examine the critical molecular interactions between DNA-binding proteins (such as repressors and activators) and the DNA sequences to which they bind.

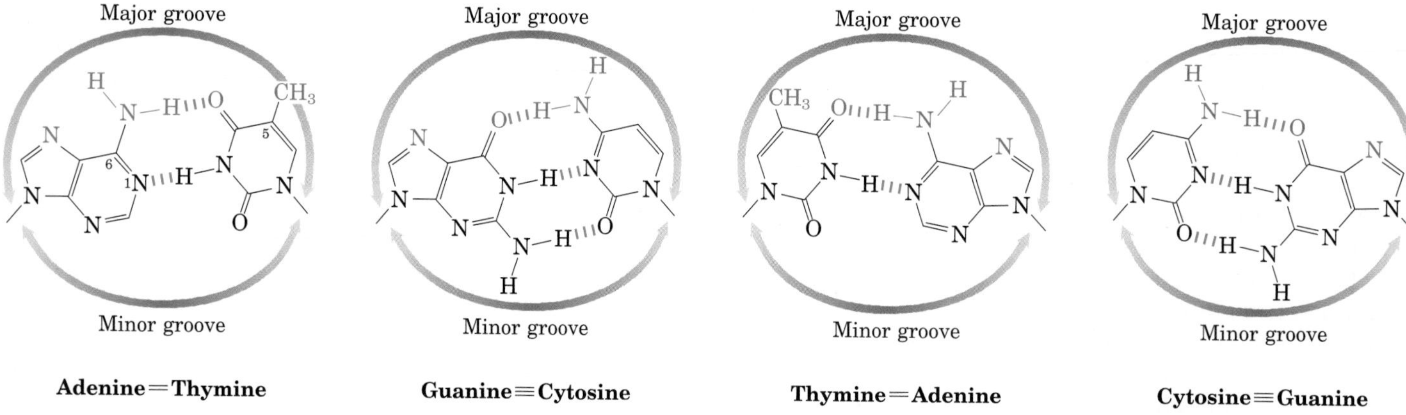

figure 28–8
Groups in DNA available for protein binding. Shown here are functional groups on all four base pairs that are displayed in the major and minor grooves of DNA. The groups that can be used for base-pair recognition by proteins are shown in red.

Regulatory Proteins Have Discrete DNA-Binding Domains

Regulatory proteins generally bind to specific DNA sequences. Their affinity for these target sequences is roughly 10^4 to 10^6 times higher than their affinity for other DNA sequences. Most regulatory proteins have discrete DNA-binding domains containing substructures that interact closely and specifically with the DNA. These binding domains usually include one or more of a relatively small group of recognizable and characteristic structural motifs.

To bind specifically to DNA sequences, regulatory proteins must recognize surface features on the DNA. Most of the chemical groups that differ among the four bases and permit discrimination between base pairs are hydrogen-bond donor and acceptor groups exposed in the major groove of DNA (Fig. 28–8), and most of the protein-DNA contacts that impart specificity are hydrogen bonds. A notable exception is the nonpolar surface near C-5 of pyrimidines, where thymine can be readily distinguished from cytosine by its protruding methyl group. Protein-DNA contacts are also possible in the minor groove of the DNA, but the hydrogen-bonding patterns here generally do not allow ready discrimination between base pairs.

Within regulatory proteins, the amino acid side chains most often hydrogen-bonded to bases in the DNA are those of Asn, Gln, Glu, Lys, and Arg residues. Is there a simple recognition code in which a particular amino acid always pairs with a particular base? The two hydrogen bonds that can form between Gln or Asn and the N^6 and N-7 positions of adenine cannot form with any other base. An Arg residue can similarly form two hydrogen bonds to N-7 and O^6 of guanine (Fig. 28–9).

Examination of the structures of many DNA-binding proteins has shown, however, that a protein can recognize each base pair in multiple ways, leading to the conclusion that there is no simple amino acid–to–base code. For some proteins, the Gln-adenine interaction can specify A=T base pairs, but in others a van der Waals pocket for the methyl group of thymine can recognize A=T base pairs. Researchers cannot yet examine the structure of a DNA-binding protein and infer the DNA sequence to which it binds.

To interact with bases in the major groove of DNA, a protein requires a relatively small structure that can stably protrude from the protein surface. The DNA-binding domains of regulatory proteins tend to be small (60 to 90 amino acid residues), and the structural motifs within these domains that are actually in contact with the DNA are smaller still. Small proteins are often unstable because of their limited capacity to form layers of structure to

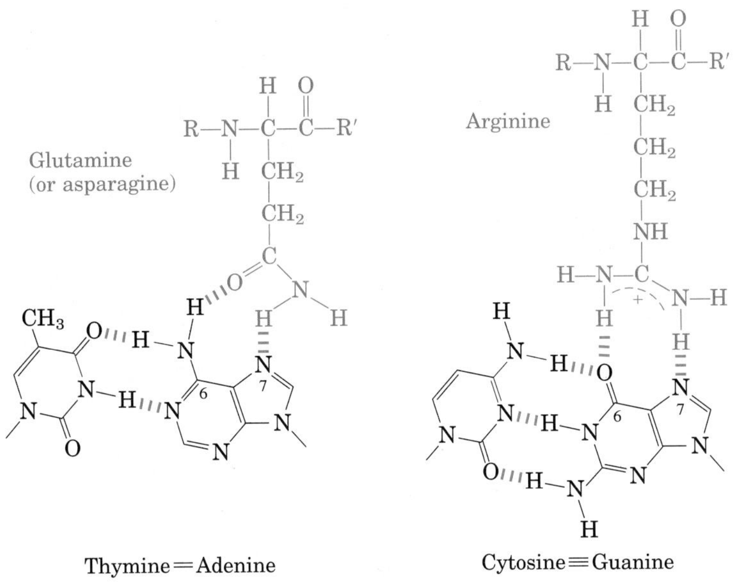

Glutamine (or asparagine)

Arginine

Thymine=Adenine

Cytosine≡Guanine

figure 28–9
Two examples of specific amino acid–base pair interactions that have been observed in DNA-protein binding.

bury hydrophobic groups (Chapter 6; see p. 161). The DNA-binding motifs provide either a very compact stable structure or a means to allow a segment of protein to protrude from the protein surface.

The DNA binding sites for regulatory proteins are often inverted repeats of a short DNA sequence (a palindrome) at which multiple subunits (usually two) of a regulatory protein bind cooperatively. The Lac repressor is unusual in that it functions as a tetramer, with two dimers tethered together at the end away from the DNA-binding sites (Fig. 28–7b). An *E. coli* cell normally contains about 20 tetramers of the Lac repressor. Each of the tethered dimers separately binds to a palindromic operator sequence, in contact with 17 base pairs of a 22 base pair region in the *lac* operon (Fig. 28–10). And each of the tethered dimers can independently bind to an operator sequence, with one generally binding to O_1 and the other to O_2 or O_3 (as in Fig. 28–7b). The symmetry of the O_1 operator sequence corresponds to the twofold axis of symmetry of two paired Lac repressor subunits. The tetrameric Lac repressor binds to its operator sequences with an estimated dissociation constant in vivo of about 10^{-10} M. The repressor discriminates between the operators and other sequences by a factor of about 10^6, so that binding to these few base pairs among the 4.7 million or so of the *E. coli* chromosome is highly specific.

Several DNA-binding motifs have been described, but here we focus on two that play major roles in the binding of DNA by regulatory proteins: the **helix-turn-helix** and the **zinc finger.** We also consider a type of DNA-binding domain—the homeodomain—found in some eukaryotic proteins.

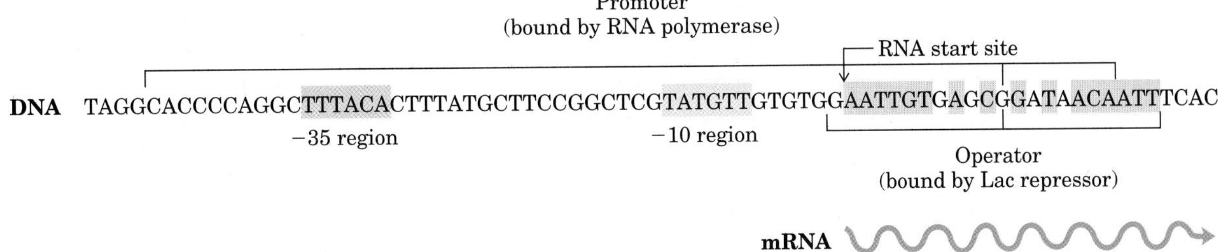

figure 28–10
Relationship between the *lac* operator sequence and the *lac* promoter. The bases shaded beige exhibit twofold (palindromic) symmetry about the axis indicated by the vertical line.

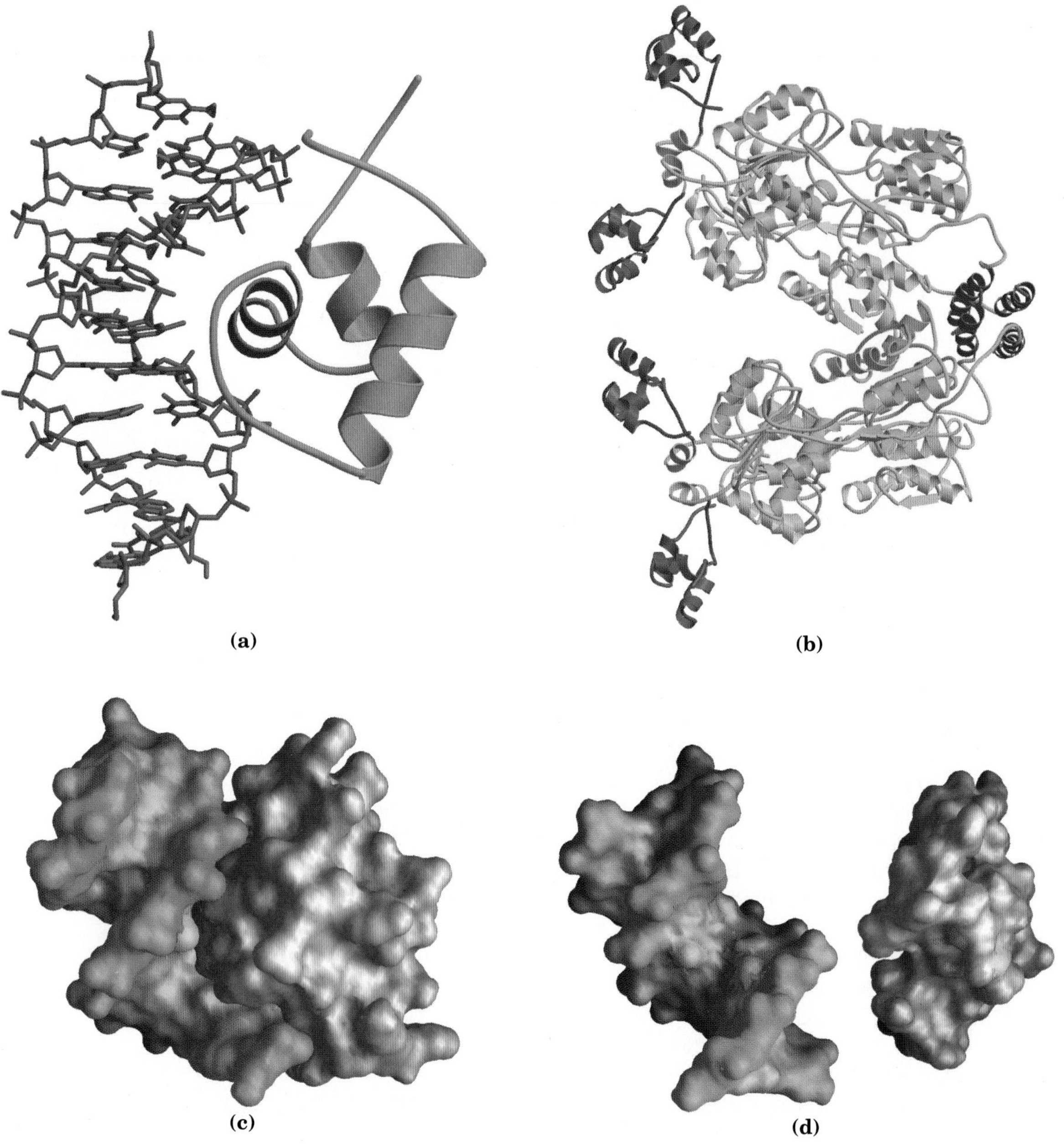

figure 28–11
Helix-turn-helix. (a) DNA-binding domain of the Lac repressor, with the helix-turn-helix motif shown in red and orange. The DNA recognition helix is red. **(b)** Entire Lac repressor, with the DNA-binding domains in blue, and the α helices involved in tetramerization in red. The remainder of the protein (shades of green) has the binding sites for allolactose. The allolactose-binding domains are linked to the DNA-binding domains through linker helices (yellow). **(c)** Surface rendering of the DNA-binding domain of the Lac repressor (gray) bound to DNA (blue). **(d)** The same DNA-binding domain as in **(c)**, but separated from the DNA, with the binding interaction surfaces shown. Some groups on the protein and DNA that interact through hydrogen-bonding are shown in red; some groups that interact through hydrophobic interactions are in orange. Only a few of the groups involved in sequence recognition are shown. The complementary nature of the two surfaces is evident.

Helix-Turn-Helix This DNA-binding motif is crucial to the interaction of many prokaryotic regulatory proteins with DNA, and similar motifs occur in some eukaryotic regulatory proteins. The helix-turn-helix motif comprises about 20 amino acids in two short α-helical segments, each seven to nine amino acid residues long and separated by a β turn (Fig. 28–11). This structure generally is not stable by itself; it is simply the reactive portion of a somewhat larger DNA-binding domain. One of the two α-helical segments is called the recognition helix because it usually contains many of the amino acids that interact with the DNA in a sequence-specific way. This α helix is stacked on other segments of the protein structure so that it protrudes from the protein surface. When bound to DNA, the recognition helix is positioned in or nearly in the major groove. The Lac repressor has this DNA-binding motif (Fig. 28–11).

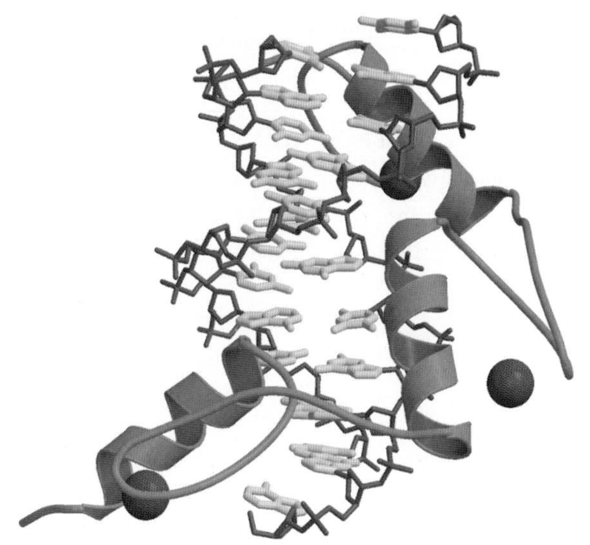

figure 28–12
Zinc fingers. Three zinc fingers (gray) from the regulatory protein Zif 268, complexed with DNA (blue and white). Each Zn^{2+} (maroon) coordinates with two His and two Cys residues.

Zinc Finger Zinc fingers consist of about 30 amino acid residues, four of which (four Cys, or two Cys and two His) coordinate a single Zn^{2+} ion. The zinc does not itself interact with DNA; rather, the coordination with zinc stabilizes this small structural motif. Several hydrophobic side chains in the core of the structure also lend stability. Figure 28–12 shows the interaction between DNA and three zinc fingers of a single polypeptide from the mouse regulatory protein Zif 268.

Zinc fingers occur in many eukaryotic DNA-binding proteins. The interaction of a single zinc finger with DNA is typically weak, and many DNA-binding proteins, like Zif 268, have multiple zinc fingers that enhance binding substantially when they interact simultaneously with DNA. One DNA-binding protein of the frog *Xenopus* has 37 zinc fingers. There are few known examples of the zinc-finger motif in prokaryotic proteins.

The precise manner in which proteins with zinc fingers bind to DNA varies from one protein to the next. Sometimes zinc fingers contain the amino acid residues that are important in sequence discrimination, but others appear to bind DNA nonspecifically (the amino acids required for specificity occurring elsewhere in the protein). Zinc fingers also function as RNA-binding motifs—for example, in certain proteins that bind eukaryotic mRNAs and act as translational repressors. This role is discussed later in the chapter.

Homeodomain A DNA-binding domain has been identified in a number of proteins that function as transcriptional regulators, especially during eukaryotic development. This domain of 60 amino acids—called the **homeodomain** because it was discovered in homeotic genes, genes that regulate the development of body patterns—is highly conserved and has been identified in proteins from a wide variety of organisms, including humans (Fig. 28–13). The DNA-binding segment of the domain is related to the helix-turn-helix motif. The DNA sequence encoding this domain is called the **homeobox**.

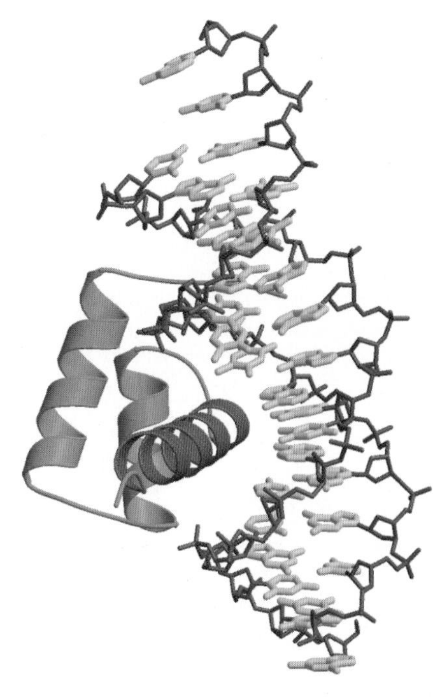

figure 28–13
Homeodomain. Shown here is a homeodomain bound to DNA. One of the α helices, stacked on two others, can be seen protruding into the major groove. This is only a small part of the much larger protein Ultrabithorax (Ubx), involved in the regulation of development in fruit flies.

Source	Regulatory protein	Amino acid sequence: DNA-binding region	6-Amino acid connector	Leucine zipper
Mammal	C/EBP	DKNSNEYRVRRERNNIAVRKSRDKA	KQRNVE	TQQKVLELTSDNDRLRKRVEQLSRELDTLRG–
	Jun	SQERIKAERKRMRNRIAASKCRKRK	LERIAR	LEEKVKTLKAQNSELASTANMLTEQVAQLKQ–
	Fos	EERRRIRRIRRERNKMAAAKCRNRR	RELTDT	LQAETDQLEDKKSALQTEIANLLKEKEKLEF–
Yeast	GCN4	PESSDPAALKRARNTEAARRSRARK	LQRMKQ	LEDKVEELLSKNYHLENEVARLKKLVGER
	Consensus molecule	RR R R RR ---------- – – – – N – – – – – – – – R – – – KK K K KK	------	L------L------L------L------L---

Invariant Asn

(a)

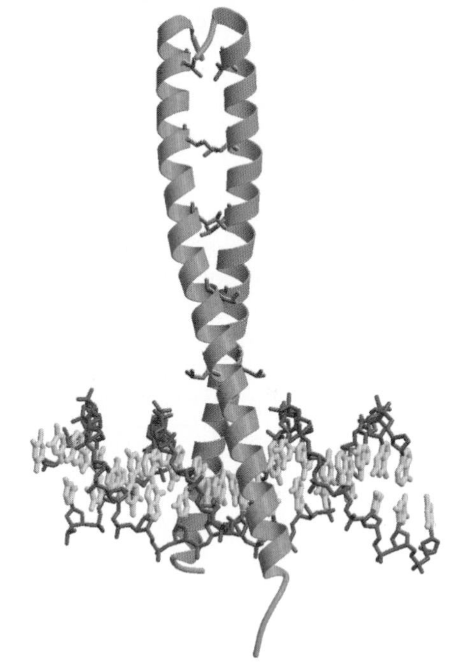

(b)

figure 28–14

Leucine zippers. (a) Comparison of amino acid sequences of several leucine zipper proteins. Note the Leu (L) residues occurring at every seventh position in the zipper region, and the number of Lys (K) and Arg (R) residues in the DNA-binding region. **(b)** Leucine zipper from the yeast activator protein GCN4. Only the "zippered" α helices (white and light blue), derived from different subunits of the dimeric protein, are shown. The two helices wrap around each other in a gently coiled coil. The interacting Leu residues are shown in red.

Regulatory Proteins Also Have Protein-Protein Interaction Domains

Regulatory proteins contain domains not only for DNA binding but also for protein-protein interactions—with RNA polymerase, other regulatory proteins, or other subunits of the same regulatory protein. Examples include many eukaryotic transcription factors that function as gene activators, which often bind to DNA as dimers, using DNA-binding domains that contain zinc fingers. Some structural domains are devoted to the interactions required for dimer formation, which is generally a prerequisite for DNA binding. Like DNA-binding motifs, the structural motifs that mediate protein-protein interactions tend to fall within one of a few common categories. Two important examples are the **leucine zipper** and the **basic helix-loop-helix.** Structural motifs of this type are the basis for classifying some regulatory proteins into structural families.

Leucine Zipper This motif is an amphipathic α helix with a series of hydrophobic amino acid residues concentrated on one side (Fig. 28–14), with the hydrophobic surface forming the area of contact between the two polypeptides of the dimer. A striking feature of these α helices is the occurrence of Leu residues at every seventh position, forming a straight line along the hydrophobic surface. Although researchers initially thought the Leu residues interdigitated (hence the name "zipper"), we now know that they line up side by side as the interacting α helices coil around each other (forming a coiled coil; Fig. 28–14b). Regulatory proteins with leucine zippers often have a separate DNA-binding domain with a high concentration of basic (Lys or Arg) residues that can interact with the negatively charged phosphates of the DNA backbone. Leucine zippers have been found in many eukaryotic and a few prokaryotic proteins.

Basic Helix-Loop-Helix Another common structural motif occurs in some eukaryotic regulatory proteins implicated in the control of gene expression during the development of multicellular organisms. These proteins share a conserved region of about 50 amino acid residues important in both DNA binding and protein dimerization. This region can form two short amphipathic α helices linked by a loop of variable length, the helix-loop-helix (distinct from the helix-turn-helix motif associated with DNA binding). The helix-loop-helix motifs of two polypeptides interact to form dimers (Fig. 28–15). In these proteins, DNA binding is mediated by an adjacent short amino acid sequence rich in basic residues, similar to the separate DNA-binding region in proteins containing leucine zippers.

Subunit Mixing in Eukaryotic Regulatory Proteins Several families of eukaryotic transcription factors have been defined based on close structural similarities. Within each family, dimers can sometimes form between two identical proteins (a homodimer) or between two different members of the family (a heterodimer). A hypothetical family of four different leucine-zipper proteins may thus form up to ten different dimeric species. In many cases the different combinations appear to have distinct regulatory and functional properties.

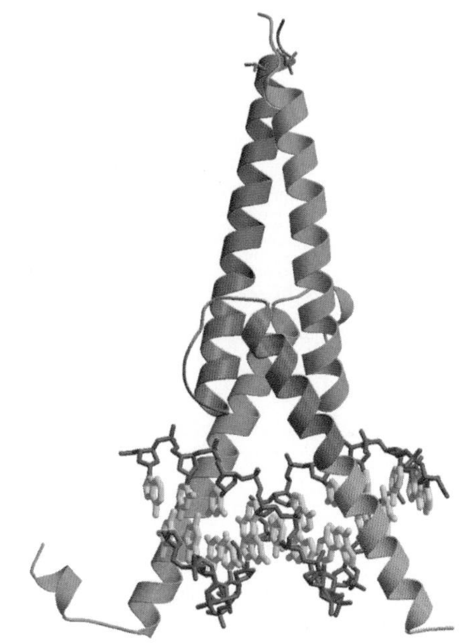

figure 28–15
Helix-loop-helix. Shown here is the human transcription factor Max, bound to its DNA target site. The protein is dimeric; one subunit is colored. The DNA-binding segment (pink) merges with the first helix of the helix-loop-helix (red). The second helix merges with the carboxyl-terminal end of the subunit (purple). Interaction of the carboxyl-terminal helices of the two subunits describes a coiled coil very similar to that of a leucine zipper (see Fig. 28–14b), but with only one pair of interacting Leu residues (red side chains near the top) in this particular example. The overall structure is sometimes called a helix-loop-helix/leucine zipper motif.

In addition to structural domains devoted to DNA binding and dimerization (or oligomerization), many regulatory proteins must interact with RNA polymerase, with other, unrelated regulatory proteins, or with both. At least three different types of additional domains for protein-protein interaction have been characterized (primarily in eukaryotes): glutamine-rich, proline-rich, and acidic domains, the names reflecting the amino acid residues that are especially abundant.

Protein-DNA binding interactions are the basis of the intricate regulatory circuits fundamental to gene function. We now turn to a closer examination of the gene regulatory schemes, first in prokaryotic then in eukaryotic systems.

Regulation of Gene Expression in Prokaryotes

As in many other areas of biochemical investigation, the study of the regulation of gene expression advanced earlier and faster in bacteria than in other experimental organisms. The examples of bacterial gene regulation presented here are chosen from among scores of well-studied systems, partly for their historical significance, but primarily because they provide a good overview of the range of regulatory mechanisms employed in prokaryotes. Many of the principles of prokaryotic gene regulation are also relevant to understanding gene expression in eukaryotic cells.

We begin by examining the lactose, arabinose, and tryptophan operons. Each system has regulatory proteins, but the overall mechanisms of regulation are very different. A short discussion of the SOS response in *E. coli* follows, illustrating how genes scattered throughout the genome can be coordinately regulated. We then describe two prokaryotic systems of quite different types, illustrating the diversity of gene regulatory mechanisms: regulation of ribosomal protein synthesis at the level of translation, with many of the regulatory proteins binding to RNA (rather than DNA), and regulation of a process called phase variation in *Salmonella* by genetic recombination.

First let's return to the *lac* operon to examine its features in greater detail.

The *lac* Operon Is Subject to Positive Regulation

The operator-repressor-inducer interactions described earlier for the *lac* operon (Fig. 28–7) provide an intuitively satisfying model for an on/off switch in the regulation of gene expression. In truth, operon regulation is rarely so simple. A bacterium's environment is too complex for its genes to be controlled by a single signal. Other factors besides lactose affect the expression of the *lac* genes, such as the availability of glucose. Glucose, metabolized directly by glycolysis, is *E. coli's* preferred energy source. Other sugars can serve as the main or sole nutrient, but extra steps are required to prepare them for entry into glycolysis, necessitating the synthesis of additional enzymes. Expressing the genes for proteins that metabolize sugars such as lactose or arabinose is wasteful if glucose is abundant.

What happens to the expression of the *lac* operon when both glucose and lactose are present? Another regulatory mechanism, called **catabolite repression,** prevents expression of the genes for catabolism of lactose, arabinose, and other sugars in the presence of glucose, even when these secondary sugars are also present. The effect of glucose is mediated by cAMP and a protein called **cAMP receptor protein,** or **CRP** (the protein is sometimes called CAP, for *c*atabolite gene *a*ctivator *p*rotein). This homodimer (subunit M_r 22,000) has binding sites for DNA and cAMP. Binding is mediated by a helix-turn-helix motif within the DNA-binding domain of the protein (Fig. 28–16). When glucose is absent, CRP binds to a site near the *lac* promoter (Fig. 28–17a) and stimulates RNA transcription 50-fold. CRP is therefore a positive regulatory element responsive to glucose levels, whereas the Lac repressor is a negative regulatory element responsive to lactose. The two act in concert; CRP has little effect on the *lac* operon when the Lac repressor is blocking transcription, and dissociation of the repressor from the *lac* operator has little effect on transcription of the *lac* operon unless CRP is present to facilitate transcription—when CRP is not bound, the wild-type *lac* promoter is a relatively weak promoter (Fig. 28–17b). The open complex of RNA polymerase and the promoter (see Fig. 26–6) does not form readily unless CRP is present. CRP interacts directly with RNA polymerase through the latter's α subunit (Fig. 28–16).

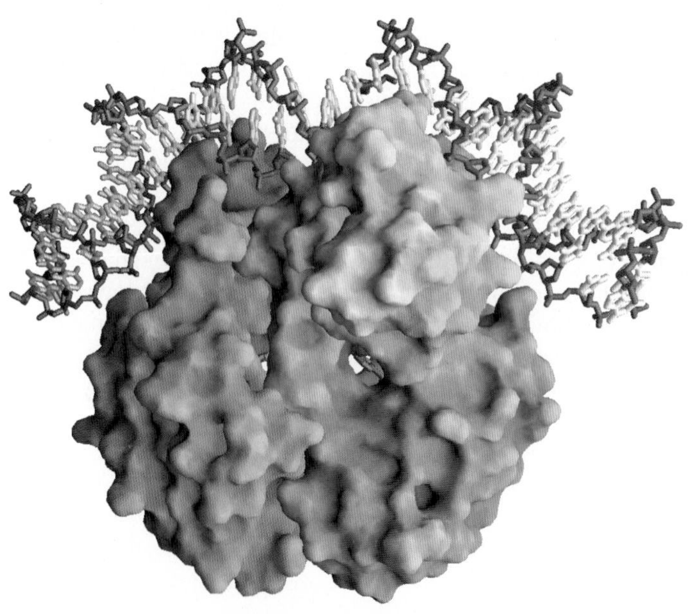

figure 28–16
CRP homodimer. Bound molecules of cAMP are shown in red. Note the bending of the DNA around the protein. The region that interacts with RNA polymerase is shaded yellow.

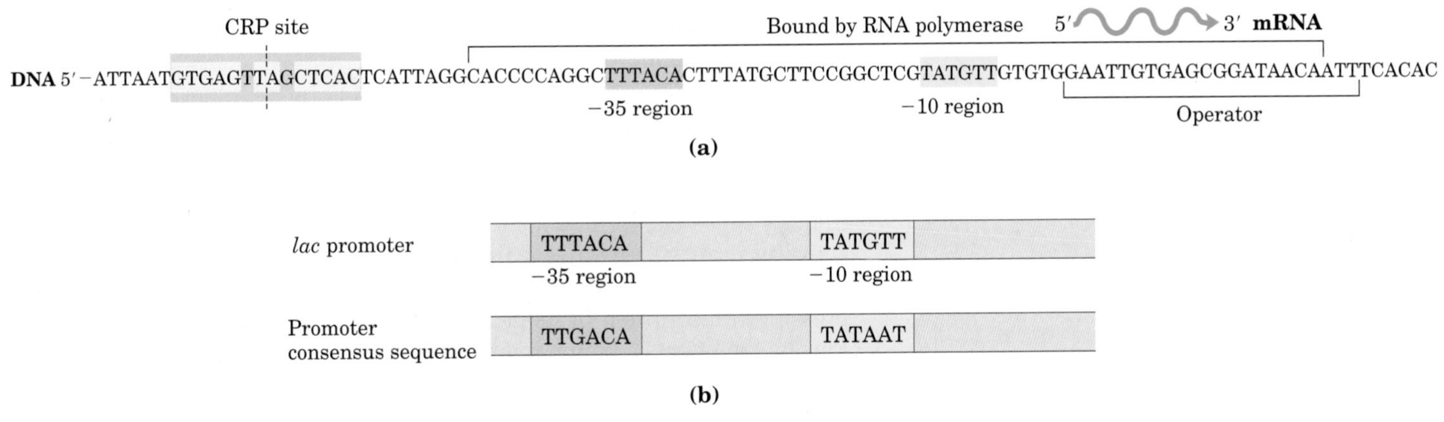

figure 28–17
Activation of transcription of the *lac* operon by CRP. **(a)** The binding site for CRP is near the promoter. As in the case of the *lac* operator, the CRP site has twofold symmetry (bases shaded beige) about the axis indicated by the dashed line. **(b)** Sequence of the *lac* promoter compared with the promoter consensus sequence. The differences mean that RNA polymerase binds relatively weakly to the *lac* promoter until the polymerase is activated by CRP.

The effect of glucose on CRP is mediated by cAMP (Fig. 28–18). CRP has a cAMP-binding site, and it binds to DNA most avidly when cAMP concentrations are high. In the presence of glucose, the synthesis of cAMP is inhibited and efflux of cAMP from the cell is stimulated. As [cAMP] declines, CRP binding to DNA declines, thereby decreasing the expression of the *lac* operon. Strong induction of the *lac* operon therefore requires both lactose (to inactivate the *lac* repressor) and a lowered concentration of glucose (to trigger an increase in [cAMP] and increased binding of cAMP to CRP).

CRP and cAMP are involved in the coordinated regulation of many operons, primarily those that encode enzymes for the metabolism of other secondary sugars such as lactose and arabinose. A network of operons with a common regulator is called a **regulon.** This arrangement allows for coordinated shifts in cellular functions that can involve the action of hundreds of genes. This is a major theme in the regulated expression of dispersed networks of genes in eukaryotes. Other bacterial regulons include the heat-shock gene system that responds to changes in temperature (p. 1075) and the genes induced in *E. coli* as part of the SOS response to DNA-damage, described later.

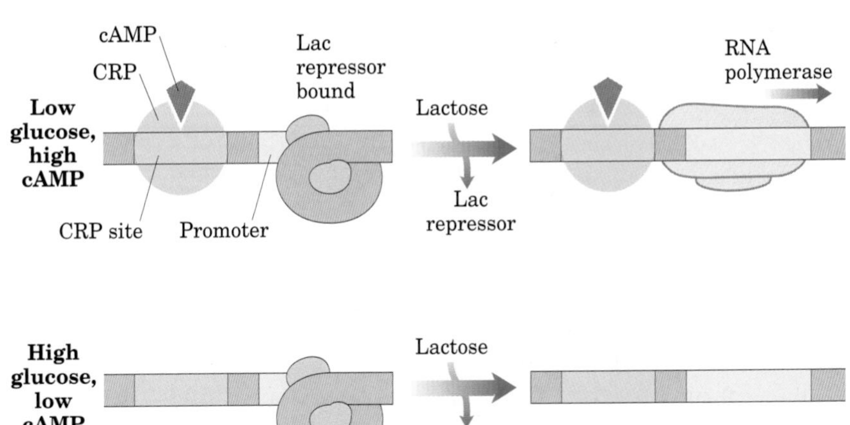

figure 28–18
Combined effects of glucose and lactose on expression of the *lac* operon. Transcription occurs only when glucose concentrations are low (so that cAMP levels are high and CRP is bound) and lactose concentrations are high (so that the Lac repressor is not bound).

The *ara* Operon Undergoes Both Positive and Negative Regulation by a Single Regulatory Protein

A more complex regulatory scheme is found in the arabinose *(ara)* operon of *E. coli,* a system that illustrates several additional regulatory mechanisms. First, the regulatory protein AraC exerts both positive and negative control. The binding of a signal molecule alters the conformation of AraC from its repressor form, which binds one DNA regulatory sequence, to its activator form, which binds a different DNA sequence, with radically different effects. Second, the AraC protein regulates its own synthesis by repressing transcription of its gene, a phenomenon called **autoregulation.** Finally, the effects of some regulatory DNA sequences can be exerted from a distance; that is, these sequences are not always contiguous with promoters. Distant DNA sequences are brought into proximity by **DNA looping,** through specific protein-protein and protein-DNA interactions. In this respect, the *ara* system is a useful model for eukaryotic gene expression, in which regulation by relatively distant sites in the DNA is common.

E. coli can use arabinose as a carbon source by first converting it to xylulose 5-phosphate, an intermediate in the pentose phosphate pathway (see Fig. 15–21). This conversion requires the enzymes ribulose kinase, arabinose isomerase, and ribulose 5-phosphate epimerase, encoded by the genes *ara*B, *ara*A, and *ara*D, respectively (Fig. 28–19). The *ara* operon includes these three genes (collectively called *ara*BAD), a regulatory site with two operators (*ara*O_1 and *ara*O_2), another binding site for AraC called *ara*I (I for inducer), and a promoter adjacent to *ara*I (P_{BAD}). The operator *ara*O_2 has a single binding site for AraC protein; *ara*I and *ara*O_1 each have two binding sites (called half-sites) in the same orientation. The nearby *ara*C gene is transcribed from its own promoter (P_C) in the opposite direction from the *ara*B, A, and D genes. A CRP-binding site is adjacent to the *ara* operon promoter; transcription of the *ara* operon is modulated by CRP-cAMP, in the same way as the *lac* system. At this point the similarities to the *lac* regulatory system largely end.

figure 28–19

The *ara* operon. Genes and regulatory sites are described in the text. The metabolism of arabinose, as catalyzed by the enzymes encoded by the *ara* genes, is diagrammed at the bottom. The end product of this pathway, D-xylulose 5-phosphate, is an intermediate in the pentose phosphate pathway (Chapters 15 and 20).

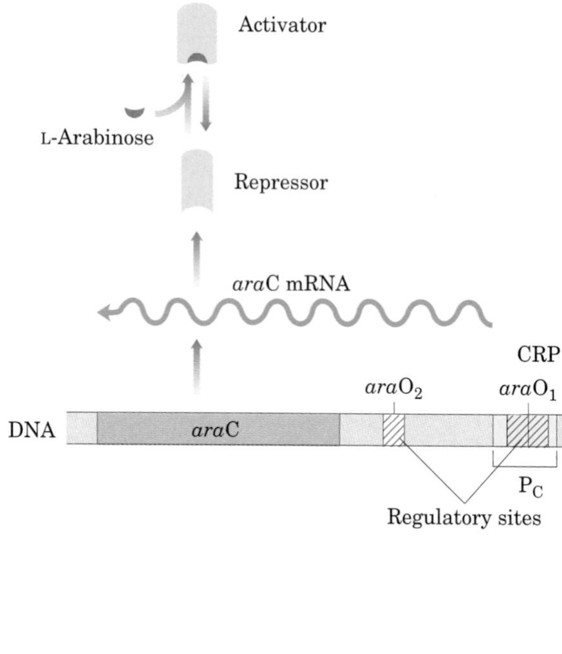

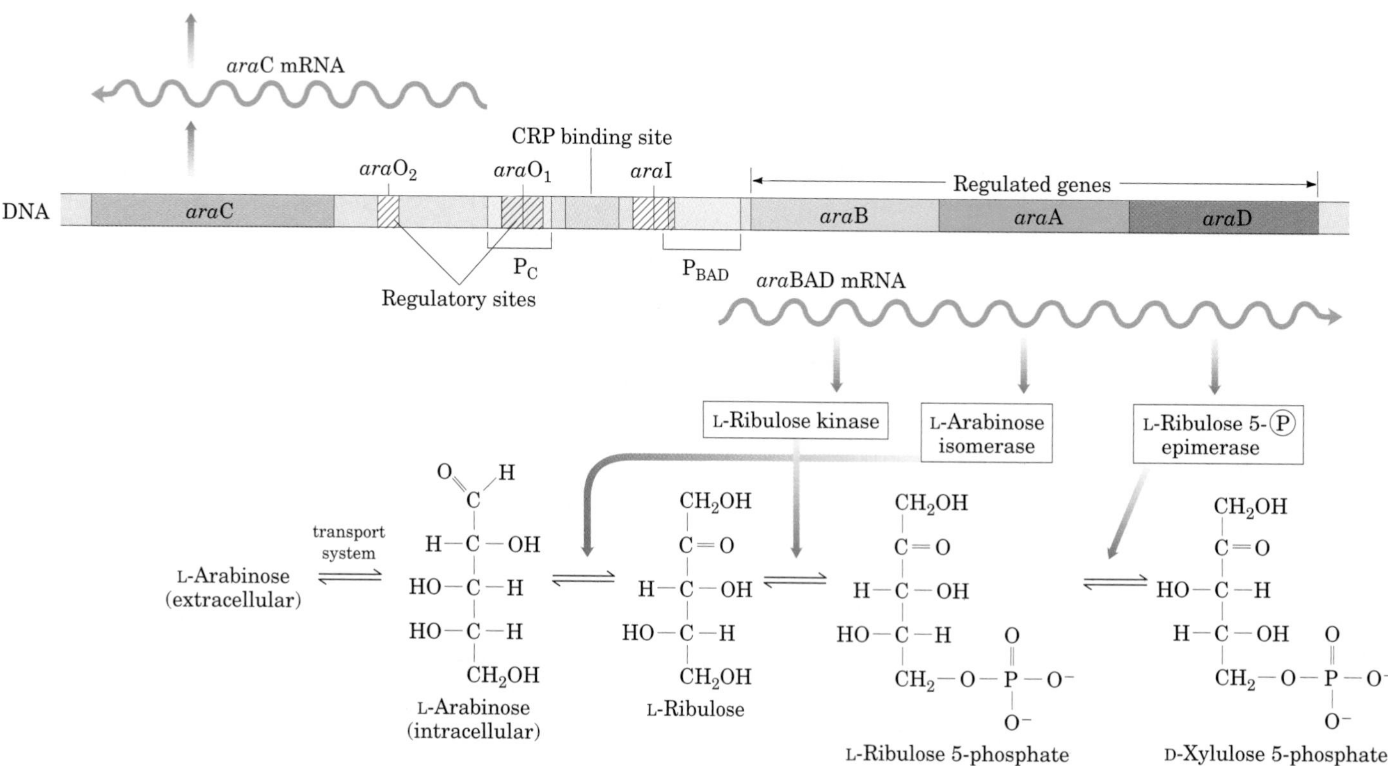

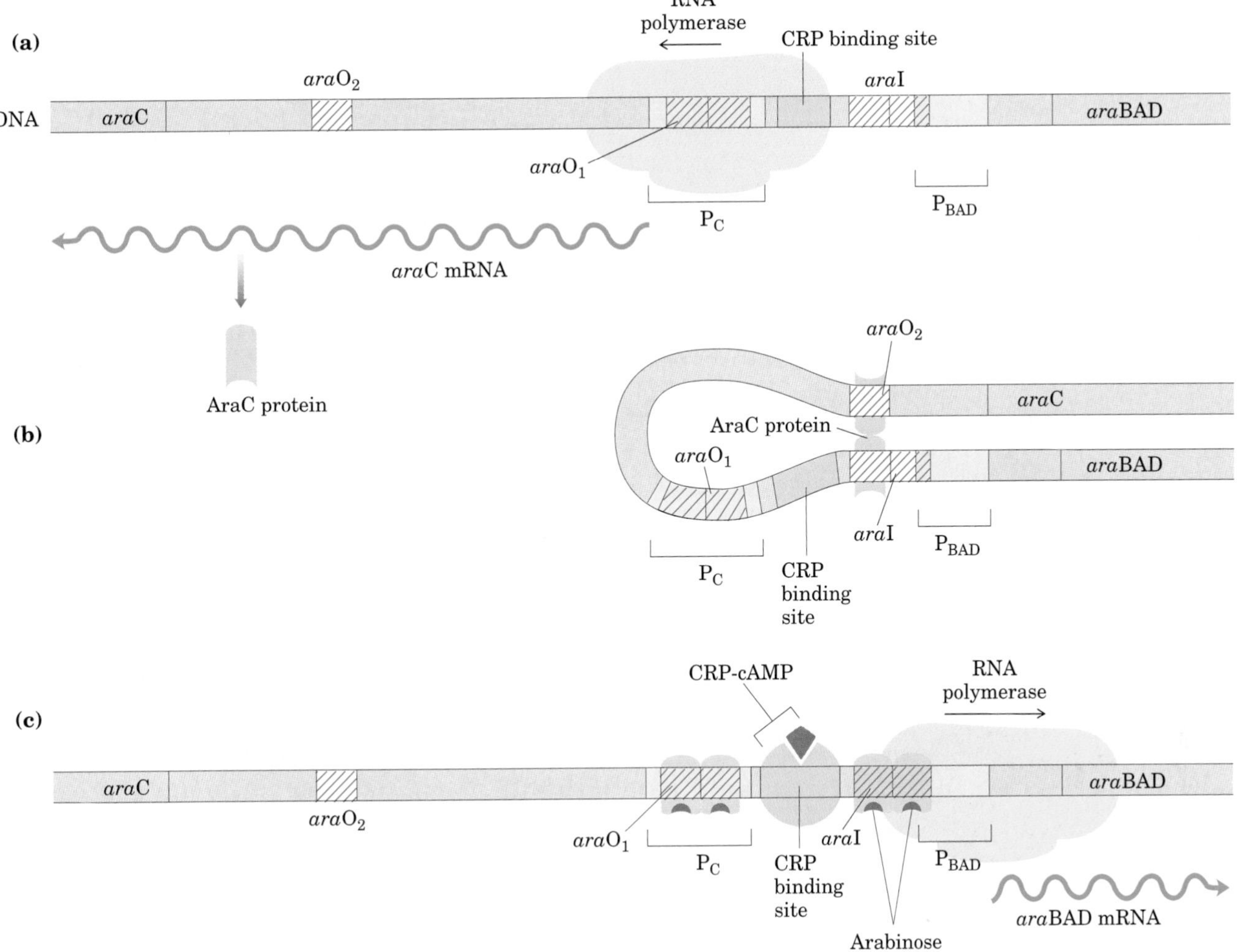

figure 28–20

Regulation of the *ara* operon. (a) When the AraC protein is depleted, the *ara*C gene is transcribed from its own promoter. **(b)** When glucose levels are high and arabinose levels are low, AraC binds to both *ara*O_2 and one half-site of *ara*I and brings these sites together to form a DNA loop, repressing *ara*BAD. **(c)** When arabinose is present and the glucose concentration is low, AraC binds arabinose and changes conformation to become an activator. The DNA loop is opened, and AraC binds to each half-site of *ara*I and *ara*O_1. The tandem proteins interact with each other and act in concert with CRP-cAMP to facilitate transcription of the *ara*BAD genes.

The role of the AraC protein in the regulation of the *ara* operon is complex (Fig. 28–20). First, when its concentration exceeds about 40 copies per cell, it regulates its own synthesis by binding at *ara*O_1 and repressing transcription of the *ara*C gene. Second, it is both a positive and a negative regulator of the *ara*BAD genes, binding in this capacity to both *ara*O_2 and *ara*I.

Regulation by AraC can be summarized by considering two possible metabolic scenarios:

1. *Glucose is abundant and arabinose is not.* Under these conditions, the dimeric AraC protein binds to both *ara*O_2 and half of *ara*I, forming a DNA loop of about 210 base pairs (Fig. 28–20b), repressing transcription of the *ara*BAD genes.
2. *Glucose is present only at low levels, but arabinose is available.* Under these conditions, CRP-cAMP becomes abundant and occupies the CRP-binding site adjacent to *ara*I. Arabinose also binds to AraC and alters its conformation. The DNA loops formed by the AraC homodimer opens. AraC protein now binds to both half-sites at *ara*I and becomes an activator, acting in concert with CRP-cAMP to induce transcription of the *ara*BAD genes (Fig. 28–20c).

When both arabinose and glucose are abundant, or both are absent, the *ara* operon remains repressed, although the detailed status of the various regulatory proteins and their DNA-binding sites under these circumstances is not clear. Regulation of the *ara* operon is rapid and reversible, an example of a swift response at the level of gene regulation to changes in environmental conditions.

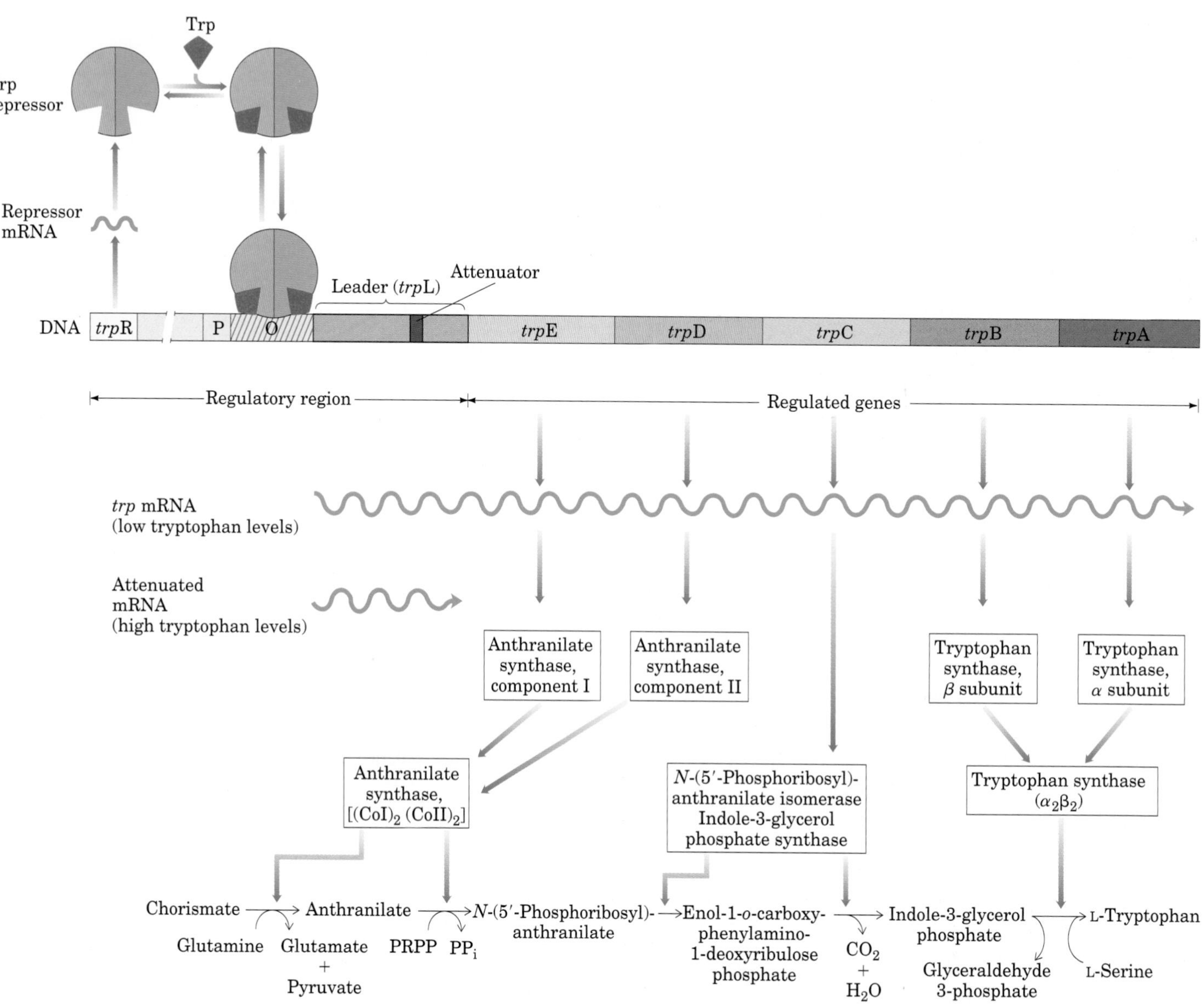

figure 28–21

The *trp* operon. This operon is regulated by two mechanisms: when tryptophan levels are high, (1) the repressor (upper left) binds its operator and (2) transcription of *trp* mRNA is attenuated (see Fig. 28–23). The biosynthesis of tryptophan by the enzymes encoded in the *trp* operon is diagrammed at the bottom.

Many Genes for Amino Acid Biosynthesis Are Regulated by Transcription Attenuation

The 20 standard amino acids are required in large amounts for protein synthesis, and *E. coli* can synthesize all of them. The genes for the enzymes needed to synthesize a given amino acid are generally clustered in an operon and expressed whenever existing supplies of that amino acid are inadequate for cellular requirements. When the amino acid is abundant, the biosynthetic enzymes are not needed and the operon is repressed.

The *E. coli* tryptophan *(trp)* operon (Fig. 28–21) includes five genes for the enzymes required to convert chorismate to tryptophan. Note that two of the enzymes catalyze more than one step in the pathway. The mRNA from the *trp* operon has a half-life of only about 3 min, allowing the cell to respond rapidly to changing needs for this amino acid. The Trp repressor is a homodimer, each subunit containing 107 amino acid residues (Fig. 28–22). When tryptophan is abundant, it binds to the Trp repressor, causing a conformational change that permits the repressor to bind to the *trp* operator and inhibit expression of the *trp* operon. The *trp* operator site overlaps the promoter, so binding of the repressor may block binding of RNA polymerase.

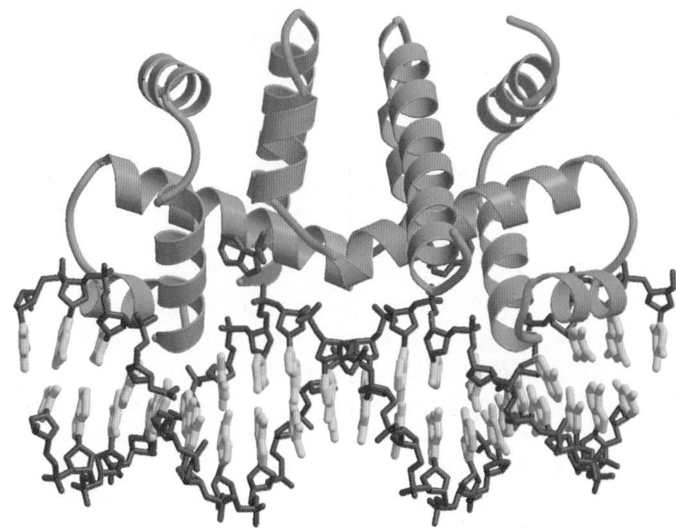

figure 28–22
Trp repressor. The repressor is a dimer, with both subunits (gray and light blue) binding the DNA at helix-turn-helix motifs. Bound molecules of tryptophan are in red.

Once again, this simple on/off circuit mediated by a repressor is not the entire regulatory story. Different cellular concentrations of tryptophan can vary the rate of synthesis of the biosynthetic enzymes over a 700-fold range. Once repression is lifted and transcription begins, the rate of transcription is fine-tuned by a second regulatory process called **transcription attenuation,** in which transcription is initiated normally but is abruptly halted *before* the operon genes are transcribed. The frequency with which transcription is attenuated is regulated by the available tryptophan and relies on the very close coupling of transcription and translation in bacteria.

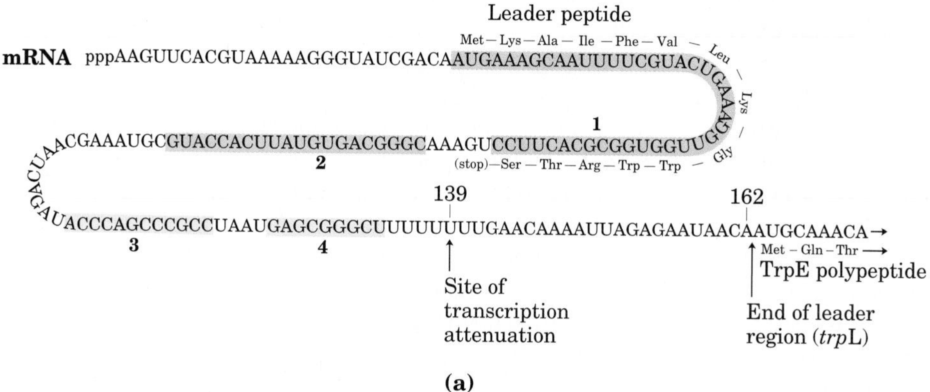

figure 28–23

Transcriptional attenuation in the *trp* operon. Transcription is initiated at the beginning of the 162 nucleotide mRNA leader encoded by a DNA region called *trp*L (see Fig. 28–21). A regulatory mechanism determines whether transcription is attenuated at the end of the leader or continues into the structural genes. **(a)** The *trp* mRNA leader (*trp*L). The attenuation mechanism in the *trp* operon involves sequences 1 to 4 (highlighted). **(b)** Sequence 1 encodes a small peptide, the leader peptide, containing two Trp residues (W); it is translated immediately after transcription begins. Sequences 2 and 3 are complementary, as are sequences 3 and 4. The attenuator structure forms by the pairing of sequences 3 and 4 (top). Its structure and function are similar to those of a transcription terminator (see Fig. 26–7). Pairing of sequences 2 and 3 (bottom) prevents the attenuator structure from forming. Note that the leader peptide has no other cellular function. Translation of its open reading frame has a purely regulatory function that determines which complementary sequences (2 and 3 or 3 and 4) are paired. **(c)** Base-pairing schemes for the complementary regions of the *trp* mRNA leader.

The *trp* operon attenuation mechanism uses signals encoded in four sequences within a 162 nucleotide **leader** region at the 5′ end of the mRNA that precedes the initiation codon of the first gene (Fig. 28–23a). Within the leader lies a region called the **attenuator,** made up of sequences 3 and 4. These sequences base-pair to form a G≡C-rich stem-and-loop structure closely followed by a series of uridylate residues. This attenuator structure acts as a transcription terminator (Fig. 28–23b). Sequence 2 is an alternative complement for sequence 3 (Fig. 28–23c). If sequences 2 and 3 base-pair, the attenuator structure cannot form and transcription continues into the *trp* biosynthetic genes; the loop formed by the pairing of sequences 2 and 3 does not obstruct transcription.

Regulatory sequence 1 is crucial to a tryptophan-sensitive mechanism that determines whether sequence 3 pairs with sequence 2 (allowing transcription to continue) or with sequence 4 (attenuating transcription). Formation of the attenuator stem-and-loop structure depends on events that occur during *translation* of regulatory sequence 1, which encodes a leader peptide (so called because it is encoded by the leader region of the mRNA) of 14 amino acids, two of which are Trp residues. The leader peptide has no other known cellular function; its synthesis is simply an operon regulatory device. This peptide is translated immediately after it is transcribed, by a ribosome that follows closely behind RNA polymerase as transcription proceeds.

When tryptophan concentrations are high, concentrations of charged tryptophan tRNA (Trp-tRNA^{Trp}) are also high. This allows translation to proceed rapidly past the two Trp codons of sequence 1 into sequence 2, before sequence 3 is synthesized by RNA polymerase. In this case sequence 2

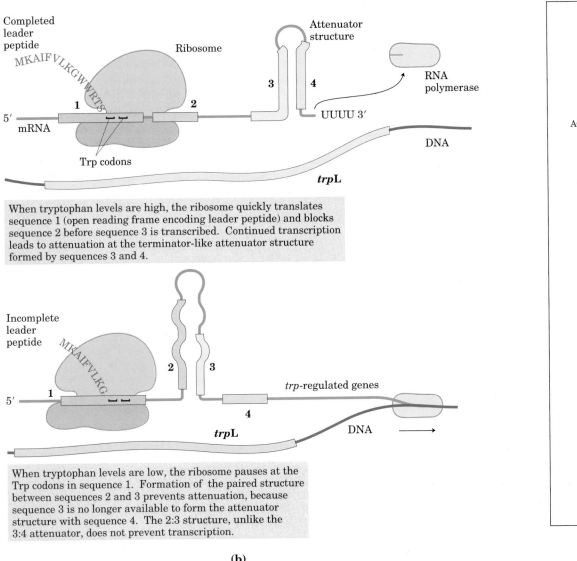

Completed leader peptide

Ribosome

Attenuator structure

RNA polymerase

mRNA

Trp codons

DNA

*trp*L

When tryptophan levels are high, the ribosome quickly translates sequence 1 (open reading frame encoding leader peptide) and blocks sequence 2 before sequence 3 is transcribed. Continued transcription leads to attenuation at the terminator-like attenuator structure formed by sequences 3 and 4.

Incomplete leader peptide

trp-regulated genes

*trp*L

DNA

When tryptophan levels are low, the ribosome pauses at the Trp codons in sequence 1. Formation of the paired structure between sequences 2 and 3 prevents attenuation, because sequence 3 is no longer available to form the attenuator structure with sequence 4. The 2:3 structure, unlike the 3:4 attenuator, does not prevent transcription.

(b)

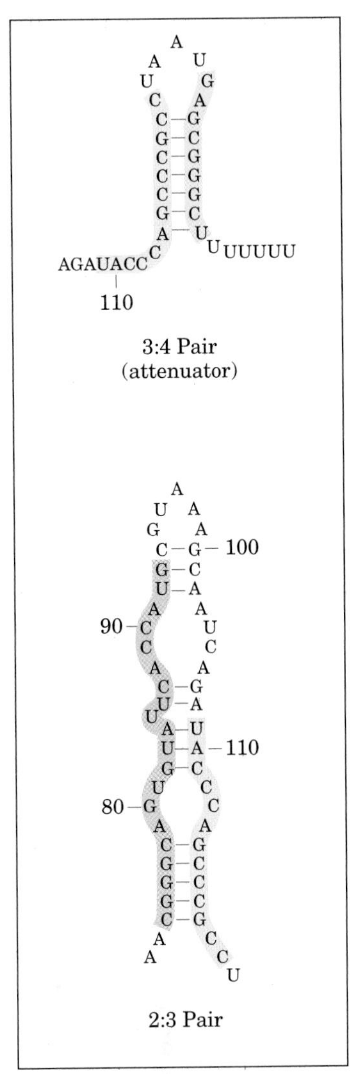

(c)

is covered by the ribosome and unavailable for pairing to sequence 3 when sequence 3 is synthesized; the attenuator structure (sequences 3 and 4) forms and transcription is halted (Fig. 28–23b, top). When tryptophan concentrations are low, however, the ribosome stalls at the two Trp codons in sequence 1 because charged $tRNA^{Trp}$ is less available. Sequence 2 remains free while sequence 3 is synthesized, allowing these two sequences to base-pair and permitting transcription to proceed (Fig. 28–23b, bottom). In this way, the proportion of transcripts that are attenuated increases as tryptophan concentration increases.

Many other amino acid biosynthetic operons use a similar attenuation strategy to fine-tune biosynthetic enzymes to meet the prevailing cellular requirements. The 15 amino acid leader peptide produced by the *phe* operon contains seven Phe residues. The *leu* operon leader peptide has four contiguous Leu residues. The leader peptide for the *his* operon contains seven contiguous His residues. In fact, in the *his* operon and a number of others (*trp* excepted), attenuation is sufficiently sensitive to be the only regulatory mechanism.

Induction of the SOS Response Requires Destruction of Repressor Proteins

Extensive DNA damage in the bacterial chromosome triggers the induction of many distantly located genes. This response, called the SOS response (p. 958), provides another good example of coordinated gene regulation. Many of the induced genes are involved in DNA repair and mutagenesis (see Table 25–6). The key regulatory elements are the RecA protein and the LexA repressor.

The LexA repressor (M_r 22,700) inhibits transcription of all the SOS genes (Fig. 28–24), and induction of the SOS response requires removal of LexA. This is not a simple dissociation from DNA in response to binding of

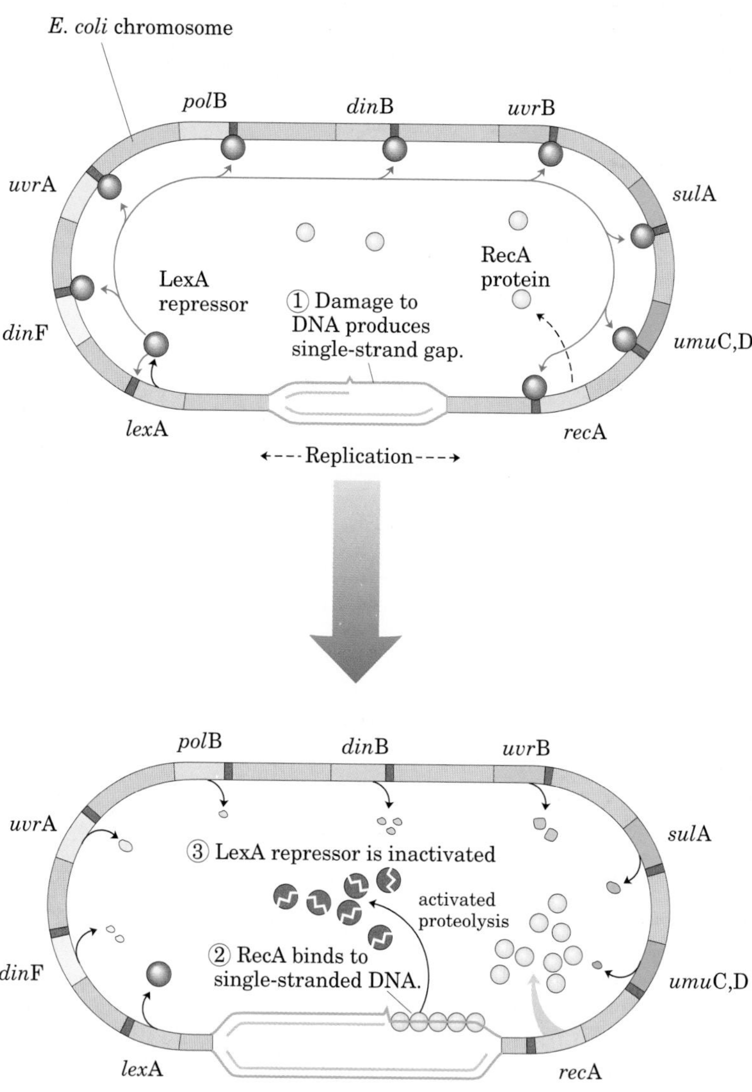

figure 28–24

SOS response in *E. coli*. See Table 25–6 for the functions of these genes. The LexA protein is the repressor in this system, with an operator site (red) near each gene. The normal cell, because the *recA* gene is not entirely repressed by the LexA repressor, contains about 1,000 RecA monomers. ① When DNA is extensively damaged (e.g., by UV light), DNA replication is halted and the number of single-strand gaps in the DNA increases. ② RecA protein binds to this damaged, single-stranded DNA, activating the protein's coprotease activity. ③ While bound to DNA, the RecA protein facilitates cleavage and inactivation of the LexA repressor. When the repressor is inactivated, the SOS genes, including *recA*, are induced; RecA levels increase 50- to 100-fold.

a small molecule, as in the regulation of the *lac, ara,* and *trp* operons described above. Instead, the LexA repressor is inactivated when it catalyzes its own cleavage at a specific Ala–Gly peptide bond, producing two roughly equal protein fragments. At physiological pH, this autocleavage reaction requires the RecA protein. RecA is not a protease in the classical sense, but its interaction with LexA facilitates the self-cleavage reaction of the repressor. This function of RecA is sometimes called a coprotease activity.

The RecA protein provides the functional link between the biological signal (DNA damage) and induction of the SOS genes. Heavy DNA damage leads to numerous single-strand gaps in the DNA, and RecA facilitates cleavage of the LexA repressor only when RecA is bound to single-stranded DNA (Fig. 28–24, bottom). Binding at the gaps eventually activates RecA, leading to cleavage of the LexA repressor and SOS induction.

During induction of the SOS response in a severely damaged cell, RecA also cleaves and thus inactivates the repressors that otherwise allow certain viruses to be propagated in a dormant lysogenic state within the bacterial host. This provides a remarkable illustration of evolutionary adaptation. These repressors, like LexA, also undergo self-cleavage at a specific Ala–Gly peptide bond, so induction of the SOS response permits replication of the virus and lysis of the cell, releasing new viral particles. Thus the bacteriophage can make a hasty exit from a compromised bacterial host cell.

Synthesis of Ribosomal Proteins Is Coordinated with rRNA Synthesis

In bacteria, an increased cellular demand for protein synthesis is met by increasing the number of ribosomes rather than altering the activity of individual ribosomes. In general, the number of ribosomes increases as the cellular growth rate increases. At high growth rates, ribosomes make up approximately 45% of the cell's dry weight. The proportion of cellular resources devoted to making ribosomes is so large, and the function of ribosomes so important, that cells must coordinate the synthesis of ribosomal proteins (r-proteins) and rRNAs. This regulation is distinct from the mechanisms described so far because it occurs largely at the level of *translation*.

The 52 genes that encode the r-proteins occur in at least 20 operons, each with 1 to 11 genes. Some of these operons also contain the genes for the subunits of DNA primase (see Fig. 25–14), RNA polymerase (see Fig. 26–4), and protein synthesis elongation factors (see Fig. 27–26)—explaining the close coupling of replication, transcription, and protein synthesis during cell growth.

The r-protein operons are regulated primarily through a translational feedback mechanism. One r-protein encoded by each operon also functions as a **translational repressor,** which binds to the mRNA transcribed from that operon and blocks translation of all the genes that the messenger encodes (Fig. 28–25). In general, the r-protein that plays the role of repressor also binds directly to an rRNA. Each translational repressor r-protein binds with higher affinity to the appropriate rRNA than to its mRNA, so the mRNA is bound and translation repressed only if the level of the r-protein exceeds that of the rRNA. This ensures that translation of the mRNAs encoding r-proteins is repressed only when synthesis of these r-proteins exceeds that needed to make functional ribosomes. In this way the rate of r-protein synthesis is kept in balance with rRNA availability.

The mRNA binding site for the translational repressor is near the translational start site of one of the genes in the operon, usually the first gene (Fig. 28–25). In other operons this would affect only that one gene, because in bacterial polycistronic mRNAs most genes have independent translation

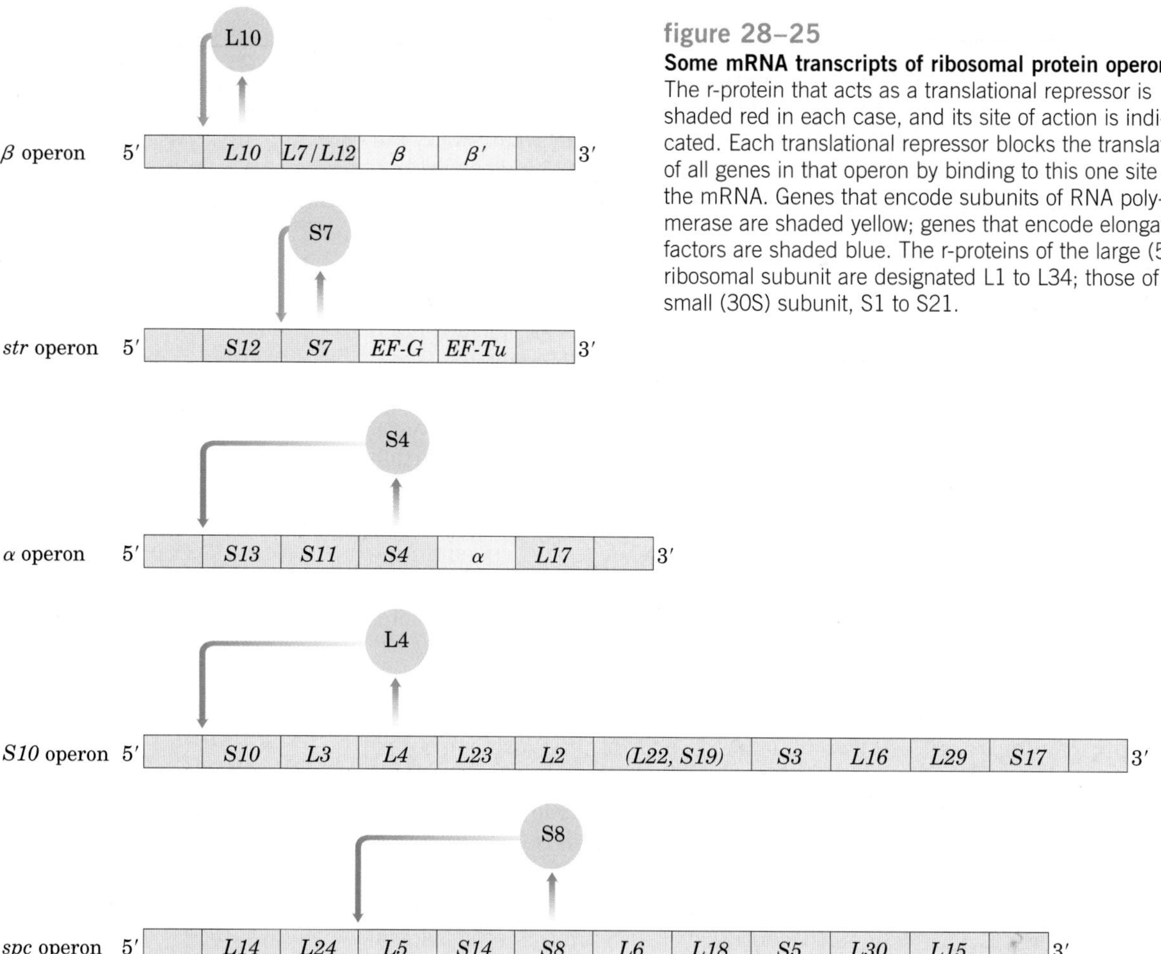

figure 28–25
Some mRNA transcripts of ribosomal protein operons. The r-protein that acts as a translational repressor is shaded red in each case, and its site of action is indicated. Each translational repressor blocks the translation of all genes in that operon by binding to this one site on the mRNA. Genes that encode subunits of RNA polymerase are shaded yellow; genes that encode elongation factors are shaded blue. The r-proteins of the large (50S) ribosomal subunit are designated L1 to L34; those of the small (30S) subunit, S1 to S21.

signals. In the r-protein operons, however, the translation of one gene depends on the translation of all the others. The mechanism of this translational coupling is not yet understood in detail. However, in some cases the translation of multiple genes appears to be blocked by folding of the mRNA into an elaborate three-dimensional structure that is stabilized both by internal base-pairing (as in Fig. 10–25) and by binding of the translational repressor protein. When the translational repressor is absent, ribosome binding and translation of one or more of the genes disrupts the folded structure of the mRNA and allows all the genes to be translated.

The r-protein operons also seem to be regulated at the level of transcription initiation, because transcription increases with increasing cellular growth rates. We do not as yet understand the mechanism of transcriptional regulation or the detailed relationship between transcriptional and translational regulation in this system.

Because the synthesis of r-proteins is coordinated with the available rRNAs, the regulation of ribosome production reflects the regulation of rRNA synthesis. In *E. coli,* rRNA synthesis from the seven rRNA operons responds to cellular growth rate and to changes in the availability of crucial nutrients, particularly amino acids.

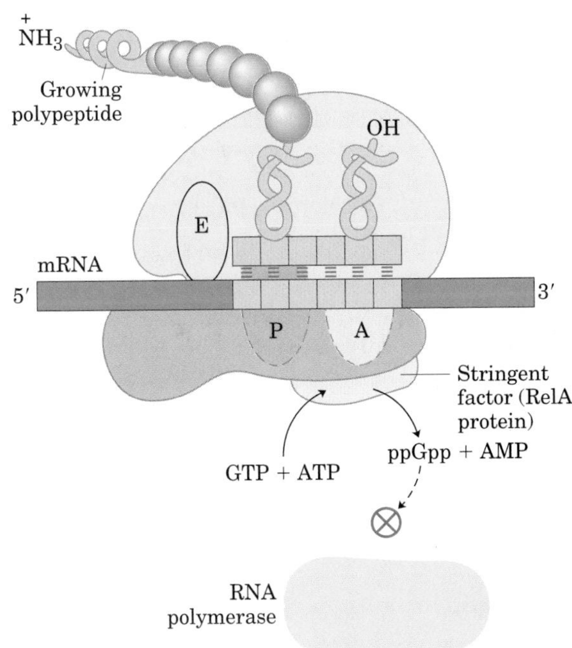

figure 28–26
Stringent response in *E. coli*. This response to amino acid starvation is triggered by binding of an uncharged tRNA in the ribosomal A site. A protein called stringent factor binds to the ribosome and catalyzes the synthesis of ppGpp. The signal ppGpp reduces transcription of some genes and increases that of others, in part by binding to the β subunit of RNA polymerase and altering the enzyme's promoter specificity. Synthesis of rRNA is reduced when ppGpp levels increase.

The regulation coordinated with amino acid concentrations is called the **stringent response** (Fig. 28–26). When amino acid concentrations are low, rRNA synthesis is halted. Amino acid starvation leads to the binding of uncharged tRNAs to the A site on ribosomes. This triggers a sequence of events that begins with the binding of an enzyme called **stringent factor** (RelA protein) to the ribosome. When thus bound to the ribosome, stringent factor catalyzes formation of the unusual nucleotide guanosine tetraphosphate (ppGpp; see Fig. 10–42), adding pyrophosphate to the 3′ position of GTP in the reaction

$$\text{GTP} + \text{ATP} \longrightarrow \text{ppGpp} + \text{AMP}$$

The abrupt rise in ppGpp in response to amino acid starvation leads to a great reduction in rRNA synthesis, mediated at least in part by the binding of ppGpp to RNA polymerase.

The nucleotide ppGpp, along with cAMP, belongs to a class of modified nucleotides that act as cellular second messengers (p. 358). In *E. coli*, these two nucleotides serve as starvation signals; they cause large changes in cellular metabolism by increasing or decreasing the transcription of hundreds of genes. In eukaryotic cells, similar nucleotide second messengers also have multiple regulatory functions. The coordination of cellular metabolism with cell growth is highly complex, and further regulatory mechanisms undoubtedly remain to be discovered.

Some Genes Are Regulated by Genetic Recombination

Salmonella bacteria live in the intestines of mammals and move by rotating flagella on their cell surfaces (Fig. 28–27). The many copies of the protein flagellin (M_r 53,000) that make up the flagella are prominent targets of mammalian immune systems. In a mechanism that evades the immune response, *Salmonella* cells switch between two distinct flagellin proteins (FljB and FliC) roughly once every 1,000 generations, using a process called **phase variation.**

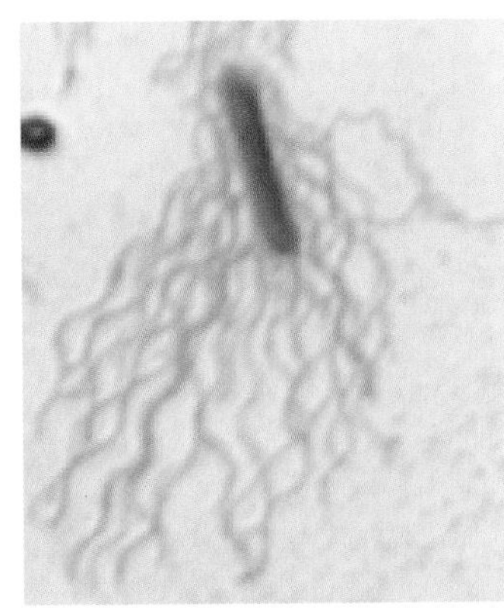

figure 28–27
Salmonella typhimurium, with flagella evident.

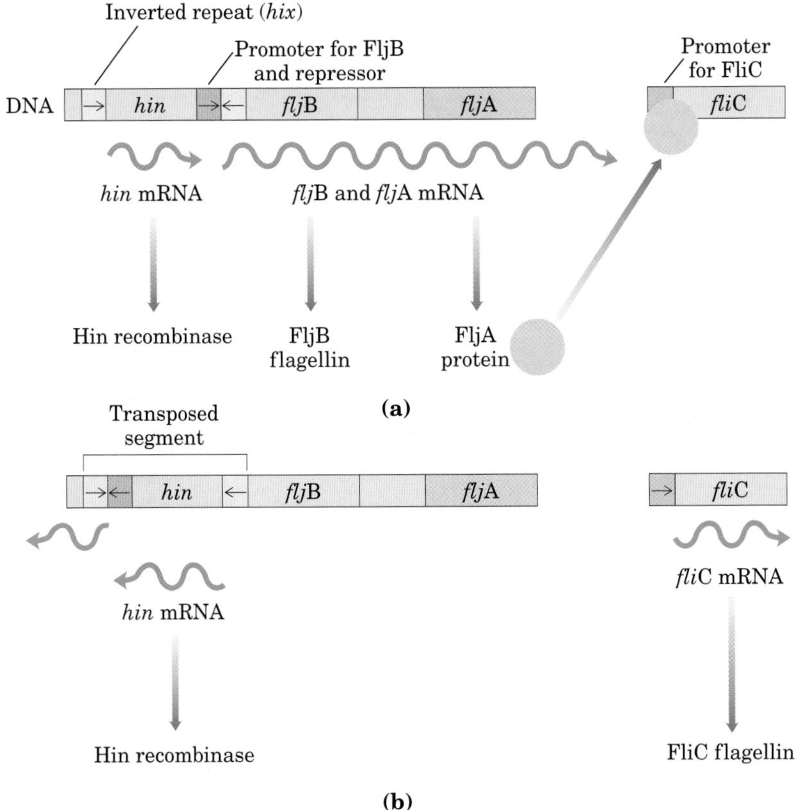

figure 28–28

Regulation of flagellin genes in *Salmonella:* phase variation. The products of genes *fli*C and *flj*B are different flagellins. The *hin* gene encodes the recombinase that catalyzes inversion of the DNA segment containing the *flj*B promoter and the *hin* gene. The recombination sites (inverted repeats) are called *hix* (yellow). **(a)** In one orientation, *flj*B is expressed along with a repressor protein (product of the *flj*A gene) that represses transcription of the *fli*C gene. **(b)** In the opposite orientation only the *fli*C gene is expressed; the *flj*A and *flj*B genes cannot be transcribed. The interconversion between these two states, called phase variation, also requires two other nonspecific DNA binding proteins, HU (*h*istonelike protein from *U*13, a strain of *E. coli*) and FIS (*f*actor for *i*nversion *s*timulation).

The switch is accomplished by periodic inversion of a segment of DNA containing the promoter for a flagellin gene. The inversion is a site-specific recombination reaction (p. 967) mediated by the recombinase called Hin at specific sequences of 14 base pairs (*hix* sequences) at either end of the DNA segment. When the DNA segment is in one orientation, the gene for FljB flagellin and the gene encoding a repressor (FljA) are expressed (Fig. 28–28a). The repressor shuts down expression of the gene for FliC flagellin. When the DNA segment is inverted (Fig. 28–28b), the *flj*A and *flj*B genes are no longer transcribed, and the *fli*C gene is induced as the repressor becomes depleted. The Hin recombinase, encoded by the *hin* gene in the DNA segment that undergoes inversion, is expressed when the DNA segment is in either orientation, so it is always possible to switch from one state to the other.

This type of regulatory mechanism has the advantage of being absolute: gene expression is impossible when the gene is physically separated from its promoter (see the position of the *flj*B promoter in Fig. 28–28b). An absolute on/off switch may be important in this system (even though it affects only one of the two flagellin genes), because a flagellum with just one copy of the wrong flagellin might be vulnerable to host antibodies directed against that protein. The *Salmonella* system is by no means unique.

Similar regulatory systems occur in a number of bacteria and bacteriophages, and recombination systems with similar functions have been found in eukaryotes (Table 28–1). Gene regulation by DNA rearrangements that move genes and/or promoters is particularly common in pathogens that benefit by changing their host range or changing their surface proteins, thereby fooling host immune systems.

table 28–1

Examples of Gene Regulation by Recombination

System	Recombinase/ recombination site	Type of recombination	Function
Phase variation (*Salmonella*)	Hin/*hix*	Site-specific	Alternative expression of two flagellin genes allows evasion of host immune response.
Host range (bacteriophage μ)	Gin/*gix*	Site-specific	Alternative expression of two sets of tail fiber genes affects host range.
Mating type switch (yeast)	HO endonuclease, RAD52 protein, other proteins/*MAT*	Nonreciprocal gene conversion*	Alternative expression of two mating types of yeast, a and α, creates cells of different mating types that can mate and undergo meiosis.
Antigenic variation (trypanosomes)†	Varies	Nonreciprocal gene conversion*	Successive expression of different genes encoding the variable surface glycoproteins (VSGs) allows evasion of host immune response.

*Nonreciprocal gene conversion is a class of recombination events not discussed in Chapter 25. Genetic information is moved from one part of the genome (where it is silent) to another (where it is expressed) in a reaction similar to replicative transposition (see Fig. 25–41).

†Trypanosomes cause African sleeping sickness and other diseases (see Box 22–2). The outer surface of a trypanosome is made up of multiple copies of a single VSG, the major surface antigen. A cell can change surface antigens to more than 100 different forms, precluding an effective defense by the host immune system. Trypanosome infections are chronic and, if untreated, result in death.

Regulation of Gene Expression in Eukaryotes

The initiation of transcription is a major regulation point for both prokaryotic and eukaryotic gene expression. Although some of the same regulatory mechanisms are used in both systems, there is a fundamental difference in the mechanism of transcription in eukaryotes and bacteria. The inherent activity of promoters and transcriptional machinery in vivo in the absence of regulatory sequences can be defined as the transcriptional ground state. In bacteria, RNA polymerase generally has access to every promoter and can bind and initiate transcription at some level of efficiency in the absence of activators or repressors. The transcriptional ground state is therefore nonrestrictive. In contrast, strong eukaryotic promoters are generally inactive in vivo in the absence of regulatory sequences; that is, the transcriptional ground state is restrictive. This fundamental difference gives rise to at least four important features that distinguish the regulation of gene expression in eukaryotes from that in bacteria.

First, access to eukaryotic promoters is restricted by the structure of chromatin, and activation of transcription is associated with multiple changes in chromatin structure in the transcribed region. Second, although both positive and negative regulatory elements are found in eukaryotic cells, positive mechanisms predominate in all systems characterized to date. Given that the transcriptional ground state is restrictive, virtually every eukaryotic gene requires activation to be transcribed. Third, eukaryotic cells have larger, more complex multimeric regulatory proteins than do bacteria. Finally, transcription in the eukaryotic nucleus is separated in both space and time from translation in the cytoplasm.

As the following discussions demonstrate, the complexity of regulatory circuits in eukaryotic cells is extraordinary. We conclude with an illustrated description of the elaborate regulatory cascade that controls development in fruit flies.

Transcriptionally Active Chromatin Is Structurally Distinct from Inactive Chromatin

The effects of chromosome structure on gene regulation in eukaryotes have no clear parallel in prokaryotes. Transcription of a eukaryotic gene is strongly repressed when its DNA is condensed within chromatin. Transcriptionally active chromosomal regions have an increased sensitivity to nuclease-mediated degradation. Nucleases such as DNase I tend to cleave the DNA of carefully isolated chromatin into fragments with multiples of about 200 base pairs, reflecting the regular repeating structure of the nucleosome (see Fig. 24–24). In actively transcribed regions, the fragments produced by nuclease activity are smaller and more heterogeneous in size. These regions contain sequences that are especially sensitive to DNase I, called **hypersensitive sites,** consisting of about 100 to 200 base pairs within the 1,000 base pairs flanking the 5′ ends of transcribed genes. In some genes, hypersensitive sites are found farther from the 5′ end, near the 3′ end, or even within the gene itself.

Modifications Increase the Accessibility of DNA

Many hypersensitive sites correspond to binding sites for known regulatory proteins, and the relative absence of nucleosomes in these regions may facilitate binding of the proteins. Nucleosomes are entirely absent in some regions that are very active in transcription, such as in the rRNA genes. Transcriptionally active chromatin tends to be deficient in histone H1, and the other core histones are more likely to be modified by acetylation (see below) or by the attachment of ubiquitin (see Fig. 27–44).

5′-Methylation of cytosine residues of CpG sequences is common in eukaryotic DNA (p. 351), but DNA in transcriptionally active chromatin tends to be undermethylated. Furthermore, CpG sites in particular genes are more often undermethylated in cells from tissues where those genes are expressed than in those where they are not expressed. The overall pattern suggests that active chromatin is prepared for transcription by the removal of potential structural barriers.

Chromatin Is Remodeled by Acetylation and Nucleosomal Displacements

The detailed mechanisms for transcription-associated structural changes in chromatin, called **chromatin remodeling,** are now coming to light, including identification of a variety of enzymes directly implicated in the process. These include enzymes that acetylate and deacetylate the core histones of the nucleosome and others that use the chemical energy of ATP to remodel nucleosomes on the DNA (Table 28–2).

Each of the core histones (H2A, H2B, H3, H4; see Fig. 24–24) has two distinct structural domains. A central domain is involved in histone-histone interaction and the wrapping of DNA around the nucleosome. A second, lysine-rich amino-terminal domain is generally positioned near the exterior of the assembled nucleosome particle; the Lys residues are acetylated by histone acetyltransferases (HATs). Cytosolic (type B) HATs acetylate newly

table 28–2

Some Enzyme Complexes Catalyzing Chromatin Structural Changes during Transcription

Enzyme complex*	Oligomeric structure	Source	Activities
GCN5-ADA2-ADA3	3 polypeptides	Yeast	GCN5 has type A HAT activity
SAGA/PCAF	>20 polypeptides	Eukaryotes	Includes GCN5-ADA2-ADA3
SWI/SNF	>11 polypeptides; M_r 2 × 10^6	Eukaryotes	ATP-dependent nucleosome remodeling
NURF	4 polypeptides; M_r 500,000	*Drosophila*	ATP-dependent nucleosome remodeling
CAFI	>2 polypeptides	Humans; *Drosophila*	Responsible for binding histones H3 and H4 to DNA
NAP1	1 polypeptide; M_r 125,000	Widely distributed in eukaryotes	Responsible for binding histones H2A and H2B to DNA

*The abbreviations used to identify eukaryotic genes and proteins are often more confusing or obscure than those used for bacteria. The complex of GCN5 (*g*eneral *c*ontrol *n*onderepressible) and ADA (*a*lteration/*d*eficiency *a*ctivation) proteins was discovered during investigation of the regulation of the genes of nitrogen metabolism in yeast. These proteins can be part of the larger SAGA complex. SAGA (*S*PF, *A*DA2,3, *G*CN5, *a*cetyltransferase) is a yeast complex; its equivalent in humans is PCAF (*p*300/*C*BP-*a*ssociated *f*actor). SWI was discovered as a protein required for expression of certain genes involved in mating type *swi*tching in yeast, and SNF (sucrose *n*on*f*ermenting) as a factor for expression of the yeast gene for sucrase. Subsequent studies revealed multiple SWI and SNF proteins that acted in a complex. The SWI/SNF complex has a role in the expression of a wide range of genes and has been found in many eukaryotes, including humans. NURF is *nu*clear *r*emodeling *f*actor; CAF1, *c*hromatin *a*ssembly *f*actor; and NAP1, *n*ucleosome *a*ssembly *p*rotein.

synthesized histones before the histones are imported into the nucleus. The subsequent assembly of the histones into chromatin is facilitated by additional proteins: CAF1 for H3 and H4, and NAP1 for H2A and H2B.

Where chromatin is being activated for transcription, the nucleosomal histones are further acetylated by nuclear (type A) HATs. The acetylation of multiple Lys residues in the amino-terminal domains of histones H3 and H4 can reduce the affinity of the entire nucleosome for DNA. Acetylation may also prevent or promote interactions with other proteins involved in transcription or its regulation. When transcription of a gene is no longer required, the acetylation of nucleosomes in that vicinity is reduced by histone deacetylases, as part of a general gene-silencing process that restores the chromatin to a transcriptionally inactive state.

Chromatin remodeling also requires protein complexes that actively move or displace nucleosomes, hydrolyzing ATP in the process. The enzyme complex SWI/SNF (Table 28–2), found in all eukaryotic cells, contains at least 11 polypeptides (total M_r 2 × 10^6), which together create hypersensitive sites in the chromatin and stimulate the binding of transcription factors. SWI/SNF is not required for the transcription of every gene. NURF (Table 28–2) is another ATP-dependent enzyme complex that remodels chromatin in ways that complement and overlap the activity of SWI/SNF. These enzyme complexes play an important role in preparing a region of chromatin for active transcription.

Many Eukaryotic Promoters Are Positively Regulated

As already noted, eukaryotic RNA polymerases have little or no intrinsic affinity for their promoters; initiation of transcription is almost always dependent on the action of multiple activator proteins. One major reason for the apparent predominance of positive regulation seems obvious: the storage of DNA within chromatin effectively renders most promoters inaccessible, so genes are normally silent in the absence of other regulation. The structure of chromatin affects access to some promoters more than others, but repressors that bind the DNA so as to preclude access of RNA polymerase (negative regulation) would often be simply redundant. Other factors are at play in the use of positive regulation, and speculation generally centers around two: the large size of eukaryotic genomes and the greater efficiency of positive regulation.

First, nonspecific DNA binding of regulatory proteins becomes a more important problem in the much larger genomes of higher eukaryotes. And the chance that a single specific binding sequence will occur randomly at an inappropriate site also increases with genome size. Specificity for transcriptional activation can be improved if each of several positive regulatory proteins must bind specific DNA sequences and then form a complex. The average number of regulatory sites for genes in a multicellular organism is probably at least five. The requirement for multiple positive regulatory proteins binding to specific DNA sequences vastly reduces the probability of the random occurrence of functional juxtaposition of all the necessary binding sites. In principle, a similar strategy could be used by multiple negative regulatory elements, but this brings us directly to the second reason for the use of positive regulation: it is simply more efficient. If the 100,000 genes in the human genome were negatively regulated, each cell would have to synthesize, at all times, almost 100,000 different repressors (or many times that number if multiple regulatory elements were used at each promoter) in sufficient concentrations to permit specific binding to each of the "unwanted" genes. In positive regulation, most of the genes are normally inactive (i.e., RNA polymerases do not bind the promoters) and the cell has to synthesize only the activator proteins needed to promote transcription of the subset of genes required in that cell.

These arguments notwithstanding, there are examples of negative regulation in eukaryotes from yeast to humans.

DNA-Binding Transactivators and Coactivators Facilitate Assembly of the General Transcription Factors

To begin our exploration of the regulation of gene expression in eukaryotes we start with a further discussion of RNA polymerase II, the enzyme responsible for the synthesis of eukaryotic mRNAs. Although most RNA polymerase II promoters include the TATA box and Inr (initiator) sequences with their standard spacing (see Fig. 26–8), they vary greatly in both the number and the location of additional sequences required for the regulation of transcription. These additional regulatory sequences are usually called **enhancers** in higher eukaryotes and **upstream activator sequences** (UASs) in yeast. A typical enhancer may be found hundreds or even thousands of base pairs upstream from the transcription start site, or may even be downstream of the transcription start site within the gene itself. When bound by the appropriate regulatory proteins, an enhancer increases transcription at nearby promoters regardless of its orientation in the DNA. The UASs of yeast function in a similar way, although generally they must be positioned upstream of and within a few hundred base pairs of the transcription start site. An average RNA polymerase II promoter may be affected by a half-dozen of these regulatory sequences, and even more complex promoters are quite common.

Three Classes of Proteins Are Involved in Transcriptional Activation

Successful binding of active RNA polymerase II holoenzyme at one of its promoters usually requires the action of other proteins (Fig. 28–29), of three types. The first class encompasses the **basal transcription factors** (see Fig. 26–9, Table 26–1) required at every RNA polymerase II promoter. The second class includes the **DNA-binding transactivators,** which bind to enhancers or UASs and facilitate transcription. The third class is a group of proteins called **coactivators,** which act indirectly, not by binding to the DNA, and are required for essential communication between the DNA-binding transactivators and the complex composed of RNA polymerase II and the general transcription factors. Furthermore, a variety of repressor proteins can interfere with communication between the RNA polymerase and the DNA-binding transactivators, resulting in repression of transcription (Fig. 28–29b). More details on the protein complexes shown in Figure 28–29, and on the ways in which they interact to activate transcription, now follow.

TATA-Binding Protein The TATA-binding protein (TBP) is the first component to bind in the assembly of a preinitiation complex at the TATA box of a typical RNA polymerase II promoter. This complex includes the general transcription factors TFIIB, TFIIE, TFIIF, TFIIH; RNA polymerase II; and perhaps TFIIA. Although required for transcription, this minimal preinitiation complex is relatively insensitive to regulation and may not form at all if the promoter is obscured within chromatin. Regulation is imposed by the transactivators and coactivators.

DNA-Binding Transactivators The requirements for transactivators vary greatly from one promoter to another. A few transactivators are known to facilitate transcription at hundreds of promoters, while others are specific for a few promoters. Many transactivators are sensitive to the binding of signal molecules, providing the capacity to activate or deactivate transcription in response to a changing cellular environment. Some enhancers with which DNA-binding transactivators associate are quite distant from the promoter's TATA box. How do DNA-binding transactivators function at a distance? The answer in most cases seems to be that, as indicated earlier, the intervening DNA is looped so that the various protein complexes can interact directly. The looping is facilitated by certain nonhistone proteins that are abundant in chromatin and bind nonspecifically to DNA. These high mobility group (HMG) proteins (the name referring to their electrophoretic mobility in polyacrylamide gels) play an important structural role in chromatin remodeling and transcriptional activation.

Coactivator Protein Complexes Most transcription requires the presence of additional protein complexes. Some major regulatory protein complexes that interact with RNA polymerase II have been defined both genetically and biochemically. These coactivator complexes act as intermediaries between the DNA-binding transactivators and the RNA polymerase.

The best-characterized coactivator is the transcription factor TFIID. In most eukaryotes, TFIID is a large complex that includes TBP and ten or more TBP-associated factors (TAFs). Some TAFs resemble histones and may play a role in displacing nucleosomes during the activation of transcription. Many DNA-binding transactivators may facilitate transcription initiation by interacting with one or more TAFs.

Another major coactivator found in yeast consists of about 20 polypeptides in a protein complex called **mediator.** This binds tightly to the carboxyl-terminal domain (CTD) of yeast RNA polymerase II. Homologs of some mediator proteins have been identified in eukaryotes from yeast to humans. We do not know whether the functions of TFIID and mediator are distinct, complementary, or overlapping. Like TFIID, some DNA-binding transactivators are believed to interact with one or more components of the mediator complex. Coactivator complexes function at or near the promoter's TATA box.

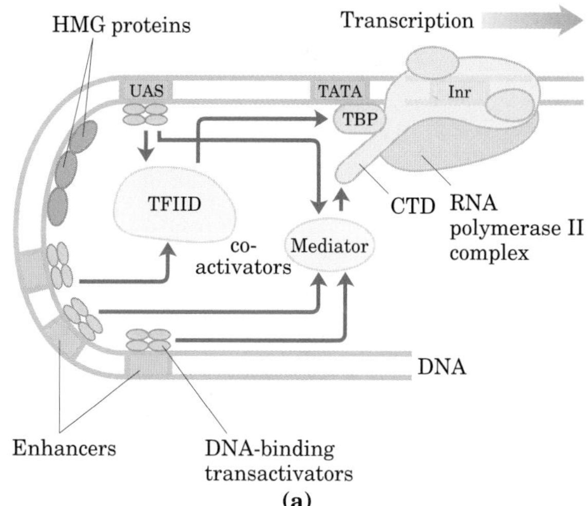

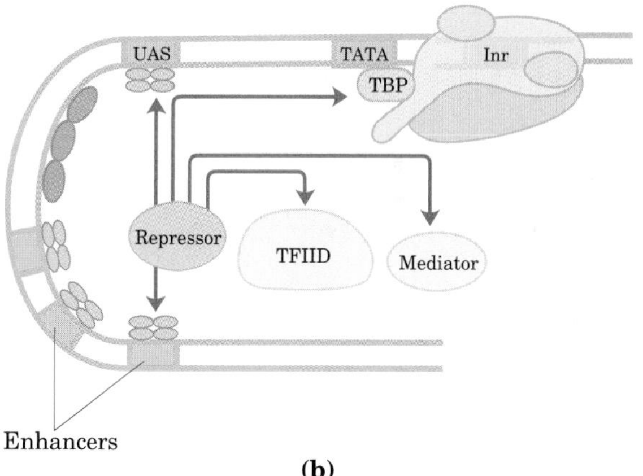

figure 28–29

Eukaryotic promoters and regulatory proteins. RNA polymerase II and its associated general transcription factors form a preinitiation complex at the TATA box and Inr site of the cognate promoters, a process facilitated by DNA-binding transactivators, acting through TFIID and/or mediator. **(a)** Shown here is a composite promoter with typical sequence elements and protein complexes found in both yeast and higher eukaryotes. The carboxyl-terminal domain (CTD) of RNA polymerase II (see Fig. 26–9) is an important point of interaction with mediator and other protein complexes. Not shown are the protein complexes involved in histone acetylation and chromatin remodeling. For the DNA-binding transactivators, DNA-binding domains are shown in green, activation domains in red. The interactions symbolized by blue arrows are discussed in the text. **(b)** A wide variety of eukaryotic transcriptional repressors function by a range of mechanisms. Some bind directly to DNA, displacing a protein complex required for activation; others interact with various parts of the transcription or activation complexes to prevent activation. Possible points of interaction are indicated with red arrows.

Choreography of Transcriptional Activation We can now begin to piece together the sequence of transcriptional activation events at a typical RNA polymerase II promoter. First, crucial remodeling of the chromatin takes place in stages. Some DNA-binding transactivators have significant affinity for their binding sites even when the sites are within condensed chromatin. Binding of one transactivator may facilitate the binding of others, gradually displacing some nucleosomes.

The bound transactivators can then interact directly with HATs and/or enzyme complexes such as SWI/SNF, accelerating the remodeling of the surrounding chromatin. In this way a bound transactivator can draw in other components necessary for further chromatin remodeling to facilitate the transcription of specific genes. The bound transactivators, generally acting through complexes such as TFIID or mediator, stabilize the binding of RNA polymerase II and its associated transcription factors and greatly facilitate formation of the preinitiation transcription complex. Complexity in these regulatory circuits is the rule rather than the exception, with multiple DNA-bound transactivators promoting transcription.

Reversible Transcriptional Activation Although rarer, some eukaryotic regulatory proteins that bind to RNA polymerase II promoters can act as repressors, inhibiting the formation of active preinitiation complexes (Fig. 28–29b). Some transactivators can adopt multiple conformations, enabling them to serve as transcriptional activators or repressors. For example, some steroid hormone receptors (described later in the chapter) function in the nucleus as DNA-binding transactivators, stimulating the transcription of certain genes when a particular steroid hormone signal is present. When the hormone is absent, the receptor proteins revert to a conformation that acts as a repressor, *preventing* the formation of preinitiation complexes. In some cases, this repression involves interaction with histone deacetylases and other proteins that help restore the surrounding chromatin to its transcriptionally inactive state.

The Genes Required for Galactose Metabolism in Yeast Are Subject to Both Positive and Negative Regulation

Some of the general principles described above can be illustrated by one well-studied eukaryotic regulatory circuit (Fig. 28–30). The enzymes required for the importation of galactose and its metabolism in yeast are encoded by genes scattered over several chromosomes (Table 28–3). Each of the *GAL* genes is transcribed separately, and yeast cells have no operons like those in bacteria. However, all the *GAL* genes have similar promoters and are regulated coordinately by a common set of proteins. The promoters for the *GAL* genes consist of the TATA box and Inr sequences, as well as an

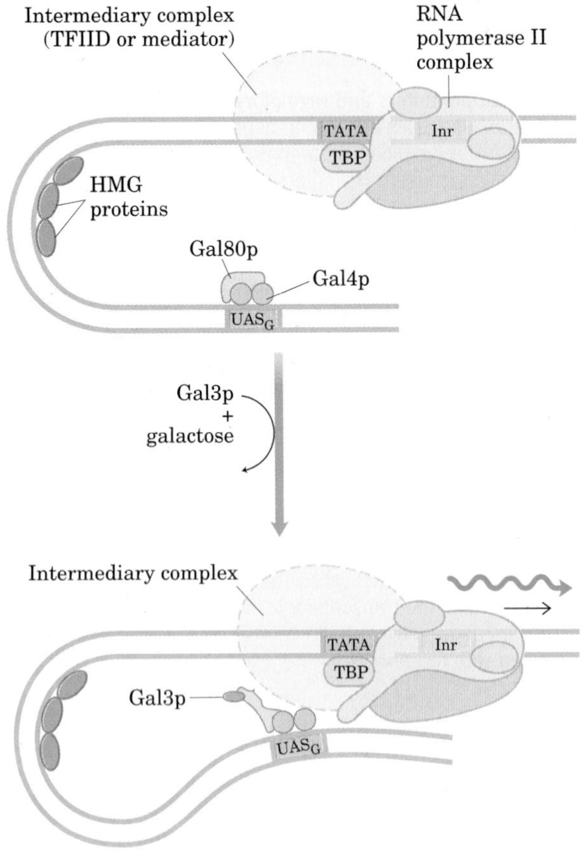

figure 28–30

Regulation of transcription at genes of galactose metabolism in yeast. Galactose is imported into the cell and converted to galactose 6-phosphate by a pathway involving six enzymes whose genes are scattered over three chromosomes (Table 28–3). Transcription of these genes is regulated by the combined actions of the proteins Gal4p, Gal80p, and Gal3p, with Gal4p playing the central role of DNA-binding transactivator. The Gal4p/Gal80p complex is inactive in gene activation. Binding of galactose to Gal3p and its interaction with Gal80p produce a conformational change in Gal80p that allows Gal4p to function in transcription activation.

table 28–3

Genes of Galactose Metabolism in Yeast

	Protein function	Chromosomal location	Protein size (number of residues)	Relative protein expression in different carbon sources		
				Glucose	Glycerol	Galactose
Regulated genes						
GAL1	Galactokinase	II	528	–	–	+++
GAL2	Galactose permease	XII	574	–	–	+++
PGM2	Phosphoglucomutase	XIII	569	+	+	++
GAL7	Galactose 1-phosphate uridylyltransferase	II	365	–	–	+++
GAL10	UDP-glucose 4-epimerase	II	699	–	–	+++
MEL1	α-Galactosidase	II	453	–	+	++
Regulatory genes						
GAL3	Inducer	IV	520	–	+	++
GAL4	Transcriptional activator	XVI	881	+/–	+	+
GAL80	Transcriptional inhibitor	XIII	435	+	+	++

Adapted from Reece, R. & Platt, A. (1997) Signaling activation and repression of RNA polymerase II transcription in yeast. *Bioessays* **19**, 1001–1010.

upstream activator sequence (UAS_G) recognized by a DNA-binding transcriptional activator called Gal4 protein (Gal4p). Regulation of gene expression by galactose involves an interplay between Gal4p and two other proteins, Gal80p and Gal3p (Fig. 28–30).

Gal80p forms a complex with Gal4p, preventing Gal4p from functioning as an activator of the *GAL* promoters. When galactose is present, it binds Gal3p, which then interacts with Gal80p, releasing Gal4p to function as an activator at the various *GAL* promoters.

Other protein complexes also have a role in the activation of transcription of the *GAL* genes. These may include the SAGA complex for histone acetylation, the SWI/SNF complex for histone remodeling, and the mediator complex. Figure 28–31 provides an idea of the complexity of protein interactions in the overall process of transcriptional activation in eukaryotic cells.

Glucose is the preferred carbon source for yeast, as it is for bacteria. When glucose is present, most of the *GAL* genes are repressed whether galactose is present or not. The *GAL* regulatory system described above is effectively overriden by a complex catabolite repression system that includes several proteins, not depicted in Figure 28–31.

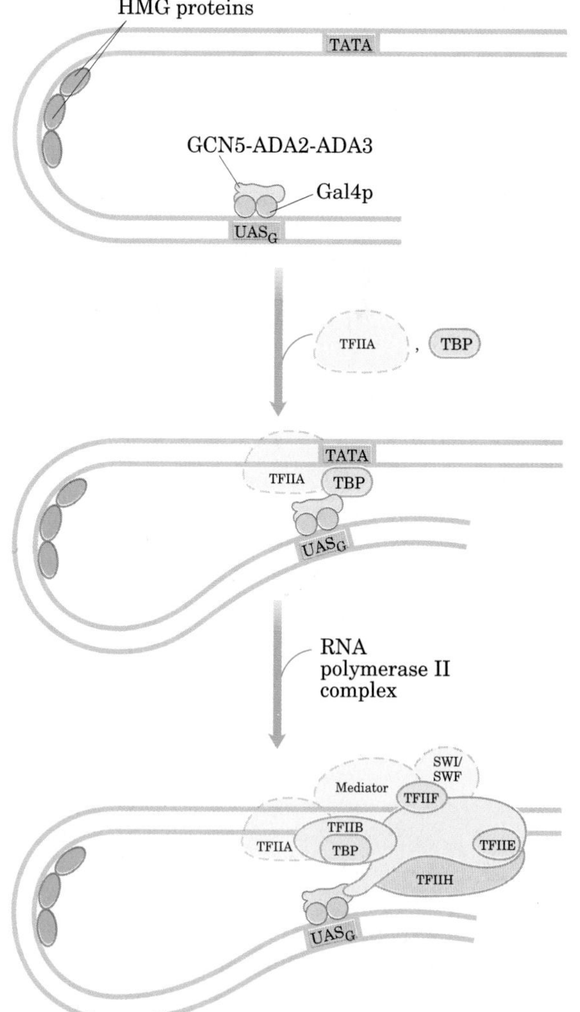

figure 28–31

Protein complexes involved in transcription activation of a group of related eukaryotic genes. The *GAL* system is shown to illustrate the complexity of this process, but not all of these protein complexes are yet known to affect *GAL* gene transcription. Note that many of the complexes (such as SWI/SNF, GCN5-ADA2-ADA3, mediator) affect the transcription of many genes. The complexes assemble stepwise. DNA-binding transactivators bind first, then the additional protein complexes needed to remodel the chromatin and allow transcription to begin.

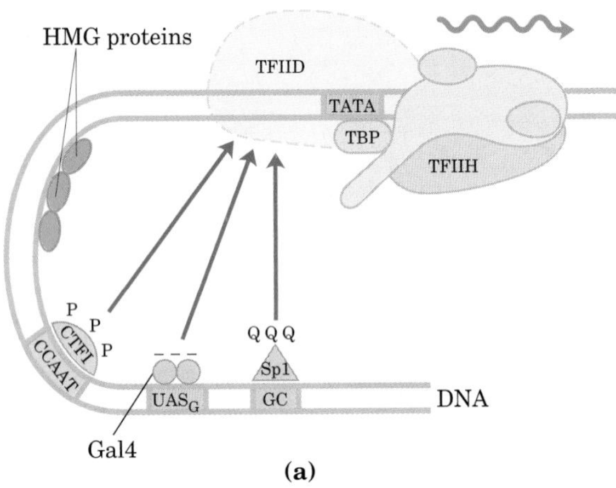

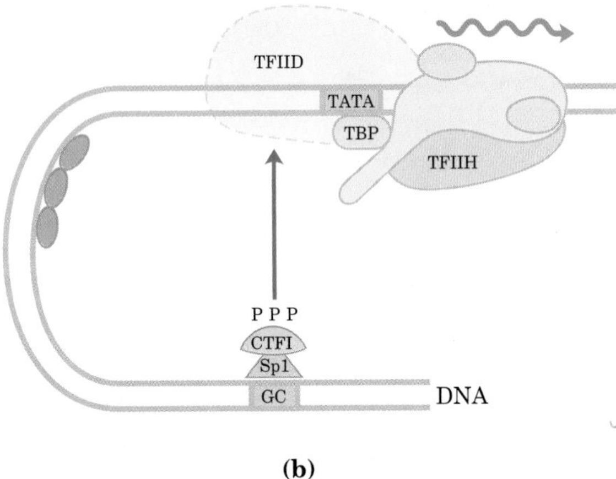

figure 28–32

DNA-binding transactivators. (a) Typical DNA-binding transactivators such as CTF1, Gal4p, and Sp1 have a DNA-binding domain and an activation domain. The nature of the activation domain is indicated by symbols: – – –, acidic; QQQ, glutamine-rich; PPP, proline-rich. Some or all of these proteins may activate transcription by interacting with intermediary complexes such as TFIID or mediator. Note that the binding sites illustrated here are not generally found together near a single gene. **(b)** A chimeric protein containing the DNA-binding domain of Sp1 and the activation domain of CTF1 activates transcription if a GC box is present.

DNA-Binding Transactivators Have a Modular Structure

DNA-binding transactivators typically have a distinct structural domain for specific DNA binding and one or more additional domains for transcriptional activation or for interaction with other regulatory proteins. Interaction of two regulatory proteins is often mediated by domains containing leucine zippers (Fig. 28–14) or helix-loop-helix motifs (Fig. 28–15). We consider here three distinct types of structural domains used in activation by DNA-binding transactivators (Fig. 28–32a): Gal4p, Sp1, and CTF1.

Gal4p contains a zinc finger–like structure in its DNA-binding domain near the amino terminus; this domain has six Cys residues that coordinate two Zn^{2+}. The protein functions as a homodimer (with dimerization mediated by interactions between two coiled coils) and binds to UAS_G, a palindromic DNA sequence about 17 base pairs in length. Gal4p has a separate activation domain with many acidic amino acid residues. Experiments substituting a variety of different peptide sequences for the **acidic activation domain** of Gal4p suggest that the acidic nature of this domain is critical to its function, although its precise amino acid sequence can vary considerably.

Sp1 (M_r 80,000) is a DNA-binding transactivator for a large number of genes in higher eukaryotes. Its DNA binding site, the GC box (consensus sequence GGGCGG), is usually quite near the TATA box. The DNA-binding domain of the Sp1 protein is near its carboxyl terminus and contains three zinc fingers. Two other domains in Sp1 function in activation. These are notable in that 25% of their amino acid residues are Gln. A wide variety of other activator proteins also have these **glutamine-rich domains.**

CCAAT-binding transcription factor 1 (CTF1) belongs to a family of DNA-binding transactivators that bind a sequence called the CCAAT site (its consensus sequence is $TGGN_6GCCAA$, where N is any nucleotide). The DNA-binding domain of CTF1 contains many basic amino acid residues, and the binding region is probably arranged as an α helix. This protein has neither a helix-turn-helix nor a zinc-finger motif; its DNA-binding mechanism remains to be clarified. CTF1 has a **proline-rich activation domain,** with Pro accounting for more than 20% of the amino acid residues.

The discrete activation and DNA-binding domains of regulatory proteins often act completely independently, as has been demonstrated in "domain-swapping" experiments. The proline-rich activation domain of CTF1 can be joined through genetic engineering (Chapter 29) to the DNA-binding domain of Sp1 to create a protein that, like normal Sp1, binds to GC boxes on the DNA and activates transcription at a nearby promoter (Fig. 28–32b). The DNA-binding domain of Gal4p has similarly been replaced experimentally with the DNA-binding domain of the prokaryotic LexA repressor (of the SOS response; Fig. 28–24). This chimeric protein neither binds at UAS_G nor activates the yeast *GAL* genes (as would normal Gal4p) unless the UAS_G sequence in the DNA is replaced by the LexA recognition site.

Eukaryotic Gene Expression Can Be Regulated by Intercellular and Intracellular Signals

The effects of steroid hormones (and of thyroid and retinoid hormones, which share their mode of action) provide additional well-studied examples of the modulation of eukaryotic regulatory proteins by direct interaction with molecular signals. Unlike other types of hormones, hormones of the steroid type do not bind to plasma membrane receptors. Instead, they interact with intracellular receptors that are themselves transcriptional transactivators. Steroid hormones too hydrophobic to dissolve readily in the blood (estrogen, progesterone, and cortisol, for example) travel on specific carrier proteins from the point of their release to their target tissues. In the target tissue, the hormone passes through the plasma membrane by simple

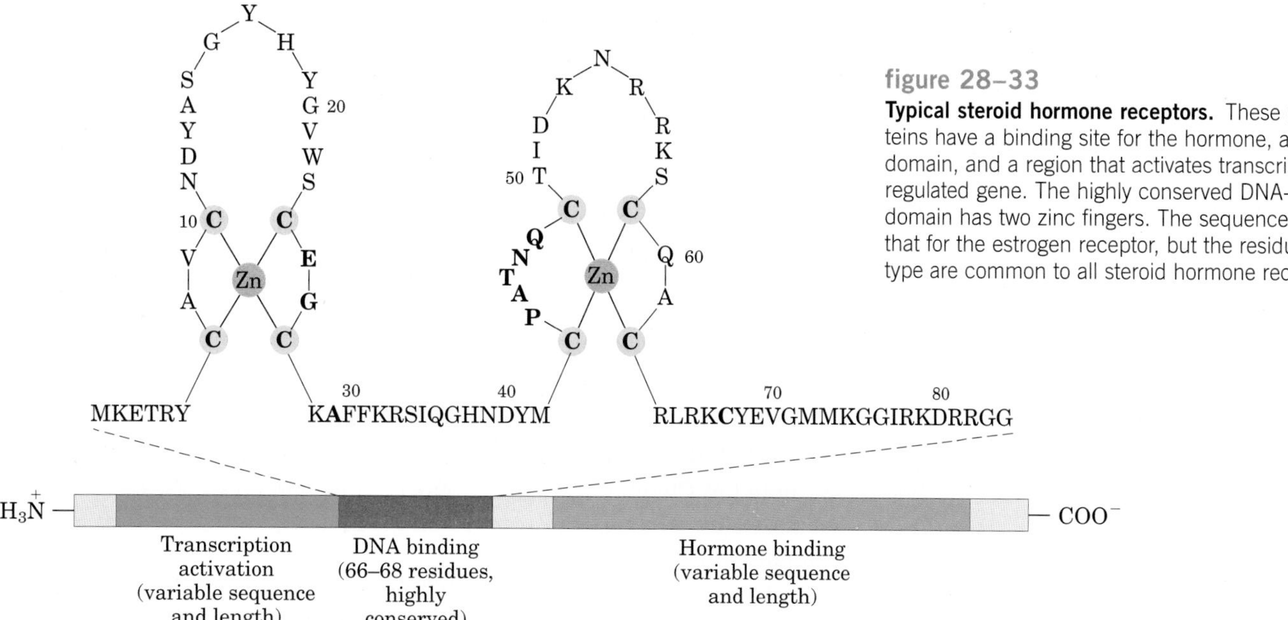

figure 28–33

Typical steroid hormone receptors. These receptor proteins have a binding site for the hormone, a DNA-binding domain, and a region that activates transcription of the regulated gene. The highly conserved DNA-binding domain has two zinc fingers. The sequence shown here is that for the estrogen receptor, but the residues in bold type are common to all steroid hormone receptors.

diffusion and binds to its specific receptor protein in the nucleus (Fig. 28–33). The hormone-receptor complex acts by binding to highly specific DNA sequences called **hormone response elements** (HREs) and altering gene expression. Hormone binding triggers changes in the conformation of the receptor proteins so that they become capable of interacting with additional transcription factors. The bound hormone-receptor complex can either enhance or suppress the expression of adjacent genes.

The DNA sequences (HREs) to which hormone-receptor complexes bind are similar in length and arrangement, but different in sequence, for the various steroid hormones. Each receptor has a consensus HRE sequence (Table 28–4) to which the hormone-receptor complex binds well, each consensus sequence consisting of two six-nucleotide sequences, either contiguous or separated by three nucleotides, in tandem or in a palindromic arrangement. The hormone receptors have a highly conserved DNA-binding domain with two zinc fingers (Fig. 28–33). The hormone-receptor complex binds to the DNA as a dimer, with the zinc finger domains of each monomer recognizing one of the six-nucleotide sequences. The ability of a given hormone to act through the hormone-receptor complex to alter the expression of a specific gene depends on the exact sequence of the HRE, its position relative to the gene, and the number of HREs associated with the gene.

table 28–4

Hormone Response Elements Bound by Steroid-Type Hormone Receptors

Receptor	Consensus sequence bound*
Androgen	$GG^{A}/_{T}ACAN_2TGTTCT$
Glucocorticoid	$GGTACAN_3TGTTCT$
Retinoic acid (some)	$AGGTCAN_5AGGTCA$
Vitamin D	$AGGTCAN_3AGGTCA$
Thyroid hormone	$AGGTCAN_3AGGTCA$
RX[†]	AGGTCANAGGTCANAGGTCANAGGTCA

*N represents any nucleotide.

[†]Forms dimer with retinoic acid receptor or vitamin D receptor.

Unlike the DNA-binding domain, the ligand-binding region of the receptor protein, always at the carboxyl terminus, is quite specific to the particular receptor. In the ligand-binding region, the glucocorticoid receptor is only 30% similar to the estrogen receptor and 17% similar to the thyroid hormone receptor. The size of the ligand-binding region varies dramatically; in the vitamin D receptor it has only 25 amino acid residues, whereas in the mineralocorticoid receptor it has 603 residues. Mutations that change one amino acid in these regions can result in loss of responsiveness to a specific hormone. Some humans unable to respond to cortisol, testosterone, vitamin D, or thyroxine have mutations of this type.

Regulation Can Occur through Phosphorylation of Nuclear Transcription Factors

We noted in Chapter 13 that the effects of insulin on gene expression are mediated by a series of steps leading ultimately to the activation of a protein kinase in the nucleus that phosphorylates specific DNA-binding proteins and thereby alters their ability to act as transcription factors. This general mechanism mediates the effects of many nonsteroid hormones. For example, the β-adrenergic pathway that leads to elevated levels of cytosolic cAMP, which acts as a second messenger in eukaryotes as it does in prokaryotes (see Figs 13–11, 28–18), also affects the transcription of a set of genes, each of which is located near a specific DNA sequence called a cAMP response element (CRE). The catalytic subunit of protein kinase A, released when cAMP levels rise (see Fig. 13–14), enters the nucleus and phosphorylates a nuclear protein, the CRE-binding protein (CREB). When phosphorylated, CREB binds to CREs near certain genes and acts as a transcription factor, turning on the expression of these genes.

Many Eukaryotic mRNAs Are Subject to Translational Repression

Regulation at the level of translation assumes a much more prominent role in eukaryotes than in bacteria and is observed in a range of cellular situations. In contrast to the tight coupling of transcription and translation that occurs in bacteria, the transcripts generated in a eukaryotic nucleus must be processed and transported to the cytoplasm before translation. This can impose a significant delay on the appearance of a protein. When a rapid increase in protein production is needed, a translationally repressed mRNA already in the cytoplasm can be activated for translation without delay. Translational regulation may play an especially important role in regulating certain very long eukaryotic genes (a few are measured in the millions of base pairs), for which transcription and mRNA processing can require many hours. Some genes are regulated at both the transcriptional and translational stages, with the latter playing a role in the fine-tuning of cellular protein levels. In some anucleate cells, such as reticulocytes (immature erythrocytes), transcriptional control is entirely unavailable and translational control of stored mRNAs becomes essential. As described later in this chapter, translational controls can also have spatial significance during development, where the regulated translation of prepositioned mRNAs creates a local gradient of the resulting protein product.

Eukaryotes have at least three major mechanisms of translational regulation.

1. Initiation factors are subject to phosphorylation by a number of protein kinases. The phosphorylated forms are often less active and cause a general depression of translation in the cell.

2. Some proteins bind directly to mRNA and act as translational repressors, many of them binding at specific sites in the 3′ untranslated region (3′UTR). So positioned, these proteins interact with other translation initiation factors bound to the mRNA or with the 40S ribosomal subunit to prevent translation initiation (Fig. 28–34; compare this with Fig. 27–24).
3. Binding proteins, present in eukaryotes from yeast to mammals, disrupt the interaction between eIF4E and eIF4G (see Fig. 27–24). The mammalian versions are called 4E-BPs (short for eIF4E binding proteins). When cell growth is slow, these proteins limit translation by binding to the site on eIF4E that normally interacts with eIF4G. When cell growth resumes or increases in response to growth factors or other stimuli, the binding proteins are inactivated by protein kinase–dependent phosphorylation.

The variety of translational regulation mechanisms provides flexibility, allowing focused repression of a few mRNAs or global regulation of all cellular translation.

Translational regulation has been particularly well-studied in reticulocytes. A major mechanism in these cells involves eIF2, the initiation factor that binds to the initiator tRNA and conveys it to the ribosome; when Met-tRNA has been bound in the P site, the factor eIF2B binds to eIF2, recycling it with the aid of GTP binding and hydrolysis. The maturation of reticulocytes includes destruction of the cell nucleus, leaving behind a plasma membrane packed with hemoglobin. Messenger RNAs deposited in the cytoplasm before the loss of the nucleus allow for the replacement of hemoglobin as required. When reticulocytes become deficient in iron or heme, the translation of globin mRNAs is repressed. A protein kinase called HCR (hemin-controlled repressor) is activated, catalyzing the phosphorylation of eIF2. In its phosphorylated form, eIF2 forms a stable complex with eIF2B that sequesters the eIF2, making it unavailable for participation in translation. In this way, the reticulocyte coordinates the synthesis of globin with the availability of heme.

Many additional examples of translational regulation have been found in studies of the development of multicellular organisms, a topic to which we now turn.

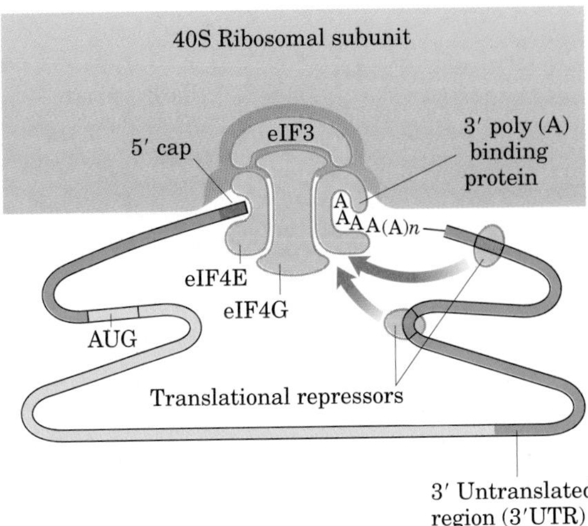

figure 28–34

Translational regulation of eukaryotic mRNA. A major mechanism for translational regulation in eukaryotes involves the binding of translational repressors (RNA-binding proteins) to specific sites in the 3′ untranslated region (3′UTR) of the mRNA. These proteins interact with eukaryotic initiation factors or with the ribosome (see Fig. 27–24) to prevent or slow translation.

Development Is Controlled by Cascades of Regulatory Proteins

For sheer complexity and intricate coordination, the patterns of gene regulation that bring about the development of a zygote into a multicellular animal or plant have no peers. Development requires transitions in morphology and protein composition that depend on tightly coordinated changes in the expression of the organism's genome. More genes are expressed during early development than in any other part of the life cycle. For example, the sea urchin oocyte has about 18,500 *different* mRNAs, compared with about 6,000 different mRNAs in the cells of typical differentiated tissues in the same organism. The mRNAs in the oocyte give rise to a cascade of events that regulate the expression of many genes in both space and time.

Several animals have emerged as important model systems for the study of development, because they are easy to maintain in a laboratory and have relatively short generation times. These include nematodes, fruit flies, zebra fish, mice, and the plant *Arabidopsis.* This discussion focuses on the development of fruit flies. Our understanding of the molecular events during development of *Drosophila melanogaster* is particularly well-advanced and can be used to illustrate patterns and principles of general significance.

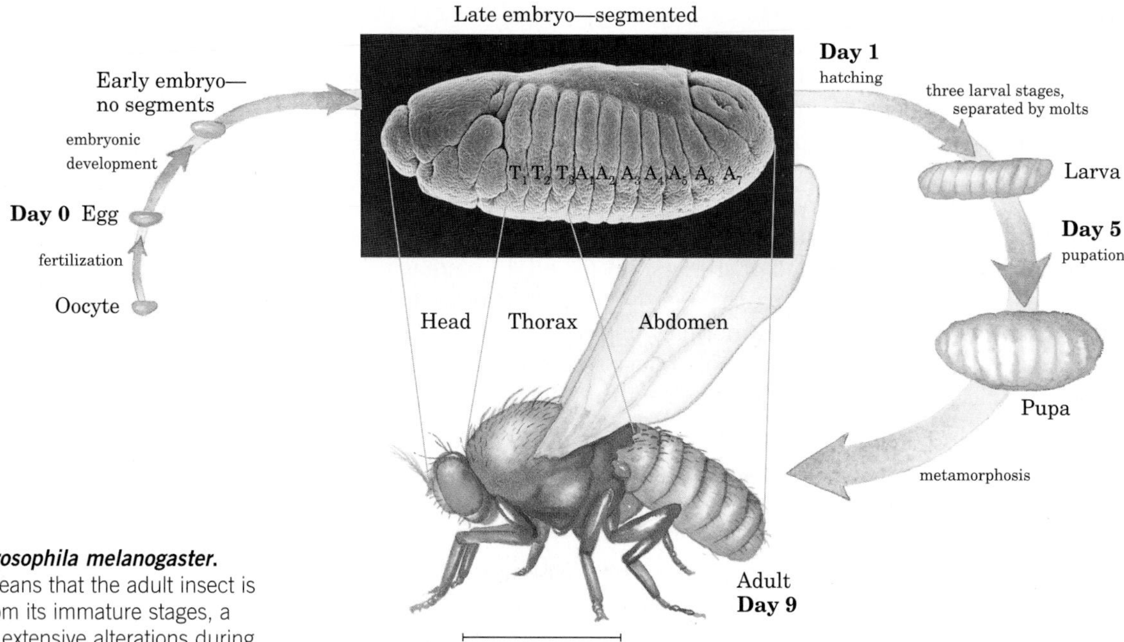

figure 28–35
Life cycle of the fruit fly *Drosophila melanogaster*. Complete metamorphosis means that the adult insect is radically different in form from its immature stages, a transformation that requires extensive alterations during development. By the late embryonic stage, segments have formed, each containing specialized structures from which the various appendages and other features in the adult fly will develop.

The life cycle of the fruit fly includes complete metamorphosis during its progression from an embryo to an adult (Fig. 28–35). Among the most important characteristics of the embryo are its **polarity** (the anterior and posterior parts of the animal are readily distinguished, as are its dorsal and ventral parts) and its **metamerism** (the embryo body is made up of serially repeating segments, each with characteristic features). During development, these segments become organized into a head, thorax, and abdomen. Each segment of the adult thorax has a different set of appendages. Development of this complex pattern is under genetic control, and a variety of pattern-regulating genes have been discovered that dramatically affect the organization of the body.

The *Drosophila* egg, along with 15 nurse cells, is surrounded by a layer of follicle cells (Fig. 28–36a). As the egg cell is formed (before fertilization), mRNAs and proteins originating in the nurse and follicle cells are deposited in the cell; some of these play a critical role in development.

Once a fertilized egg is laid, its nucleus divides and the nuclear descendants continue to divide in synchrony every 6 to 10 min (Fig. 28–36b). Plasma membranes are not formed around these nuclei, and they are distributed within the egg cytoplasm (or syncytium). Between the 8th and 11th rounds of nuclear divisions, the nuclei migrate to the outer layer of the egg forming a monolayer of nuclei surrounding the common yolk-rich cytoplasm; this is the syncytial blastoderm. After a few additional divisions, membrane invaginations surround the nuclei to create a layer of cells that form the cellular blastoderm. At this stage, the mitotic cycles in the various cells lose their synchrony. The developmental fate of the cells is determined by the mRNAs and proteins originally deposited in the egg by the nurse and follicle cells.

Proteins whose local concentration or activity causes the surrounding tissue to take up a particular shape or structure are sometimes called **morphogens;** they are the products of pattern-regulating genes. As defined by Christiâne Nüsslei̇n-Volhard, Edward B. Lewis, and Eric F. Wieschaus, three major classes of pattern-regulating genes function in successive stages of development to specify the basic features of the *Drosophila* embryo's body: maternal, segmentation, and homeotic genes. **Maternal genes** are expressed

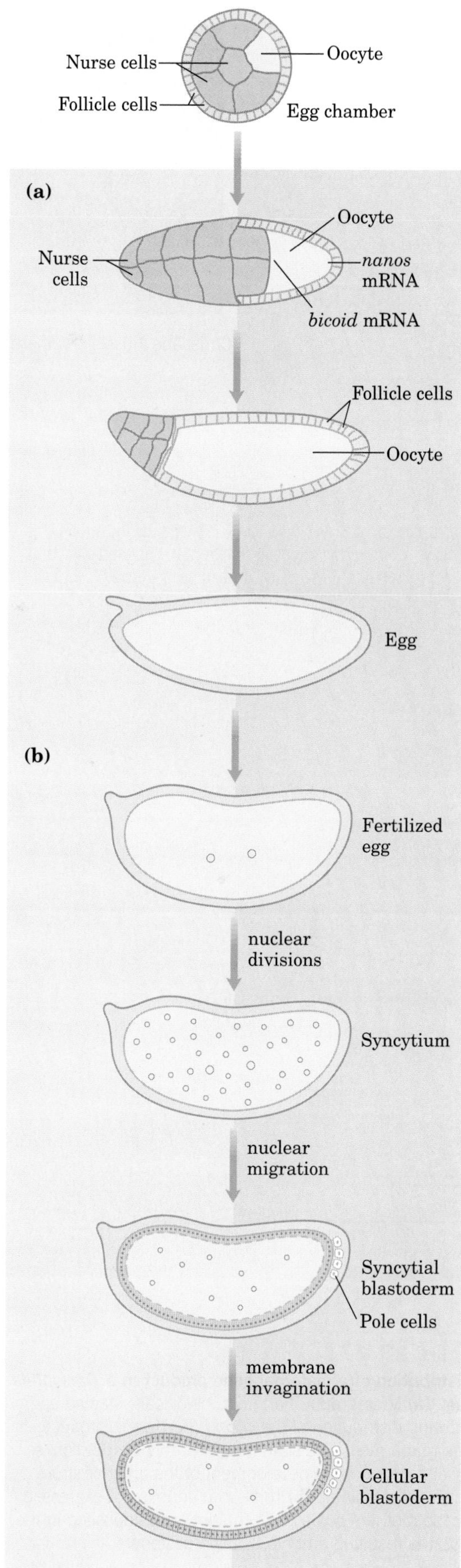

figure 28–36

Early development in *Drosophila*. (a) Development of the egg. Maternal mRNAs (including the *bicoid* and *nanos* gene transcripts, discussed in the text) and proteins are deposited in the developing oocyte (unfertilized egg cell) by nurse cells and follicle cells. (**b**) Early embryonic development. The two nuclei of the fertilized egg divide in synchrony in the common cytoplasm (syncytium), then migrate to the periphery. Membrane invaginations surround the nuclei to create a monolayer of cells at the periphery; this is the cellular blastoderm stage. During the early nuclear divisions, several nuclei at the far posterior become pole cells, which later become the germ line cells.

in the unfertilized egg, and the resulting **maternal mRNAs** remain dormant until fertilization. These provide most of the proteins needed in very early development, until the cellular blastoderm is formed. Some of the proteins encoded by maternal mRNAs direct the spatial organization of the developing embryo at early stages, establishing its polarity. **Segmentation genes** are transcribed after fertilization and direct the formation of the proper number of body segments. At least three subclasses of segmentation genes act at successive stages: **gap genes** divide the developing embryo into several broad regions, and **pair-rule genes** together with **segment polarity genes** define 14 stripes that become the 14 segments of a normal embryo. **Homeotic genes** are expressed still later. They specify which organs and appendages will develop in particular body segments.

The many regulatory genes in these three classes direct the development of an adult fly with a head, thorax, and abdomen, with the proper number of segments, and with the correct appendages on each segment. Although embryogenesis takes about a day to complete, all of these genes are activated during the first four hours. Some mRNAs and proteins are present for only a few minutes at specific points during this period. Some of the genes code for transcription factors that affect the expression of other genes in a kind of developmental cascade. Regulation at the level of translation also occurs, and many of the regulatory genes encode translational repressors, most of which bind to the 3′UTR of the mRNA (Fig. 28–34). Because many mRNAs are deposited in the egg well before their translation is required, translational repression provides an especially important avenue for regulation in developmental pathways.

Maternal Genes Some maternal genes are expressed within the nurse and follicle cells, and some in the egg itself. Within the unfertilized *Drosophila* egg, the maternal gene products establish two axes: anterior-posterior and dorsal-ventral. These genes define which regions of the radially symmetric egg will develop into the head and abdomen and top and bottom of the adult fly. A key event in very early development is establishment of mRNA and protein gradients along these body axes. Some maternal mRNAs have protein products that diffuse through the cytoplasm to create an asymmetric distribution in the egg. Different cells in the cellular blastoderm therefore inherit different amounts of these proteins, setting the cells on different developmental paths. The products of the maternal mRNAs include transcriptional activators or repressors as well as translational repressors, all regulating the expression of other pattern-regulating genes. The resulting specific patterns and sequences of gene expression therefore differ between cell lineages, ultimately orchestrating the development of each adult structure.

Christiâne Nüsslein-Volhard

The anterior-posterior axis is defined at least in part by the products of the *bicoid* and *nanos* genes of *Drosophila.* The *bicoid* gene product is a major anterior morphogen, and the *nanos* gene product is a major posterior morphogen. The mRNA from the *bicoid* gene is synthesized by nurse cells and deposited in the unfertilized egg near its anterior pole. Christiâne Nüsslein-Volhard found that this mRNA is translated soon after fertilization, and the Bicoid protein diffuses through the cell to create a concentration gradient radiating out from the anterior pole by the seventh nuclear division (Fig. 28–37a). The Bicoid protein is a transcription factor that activates the expression of a number of segmentation genes; the protein contains a homeodomain (p. 1082). Bicoid is also a translational repressor that inactivates certain mRNAs. The amounts of Bicoid protein in various parts of the embryo affect the subsequent expression of a number of other genes in a threshold-dependent manner. Genes are transcriptionally activated or translationally repressed only where the Bicoid protein concentration exceeds the threshold. Changes in the shape of the Bicoid concentration gradient have dramatic effects on the body pattern. Lack of Bicoid protein results in development of an embryo with two abdomens but neither head nor thorax (Fig. 28–37b); however, these embryos develop normally if an adequate amount of *bicoid* mRNA is injected into the egg at the appropriate end. The *nanos* gene has an analogous role, but its mRNA is deposited at the posterior end of the egg and the anterior-posterior protein gradient peaks at the posterior pole. The Nanos protein is a translational repressor.

(a)

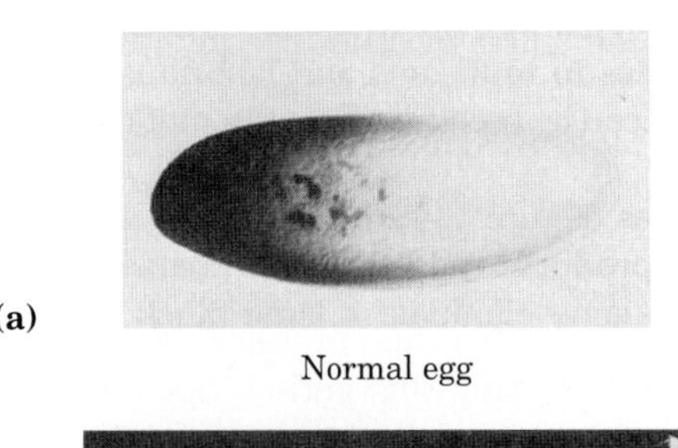

Normal egg

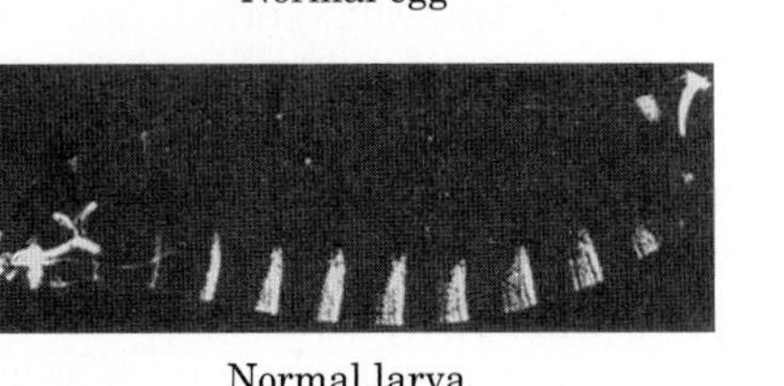

Normal larva

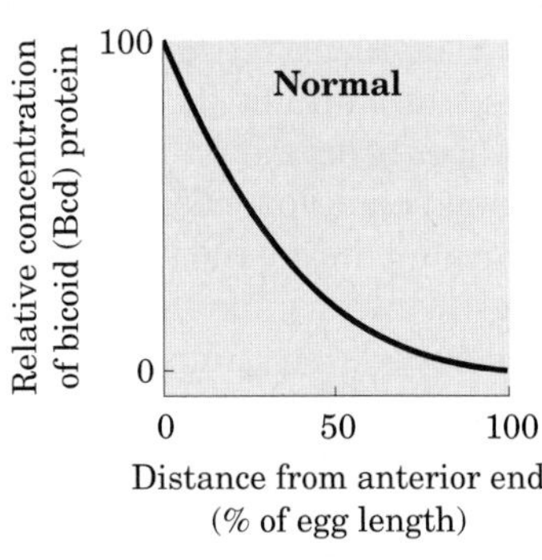

(b)

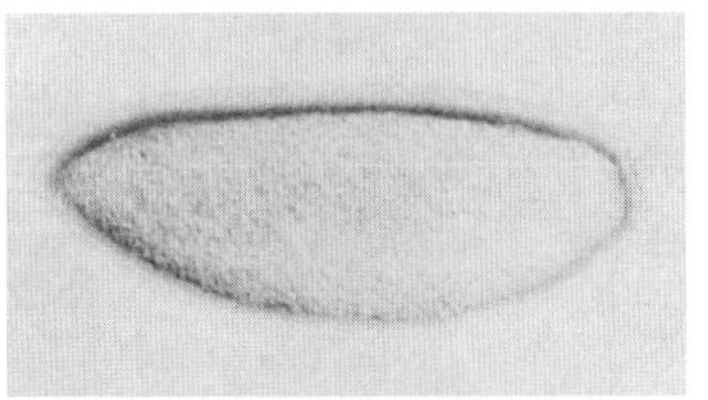

bcd⁻/bcd⁻ egg

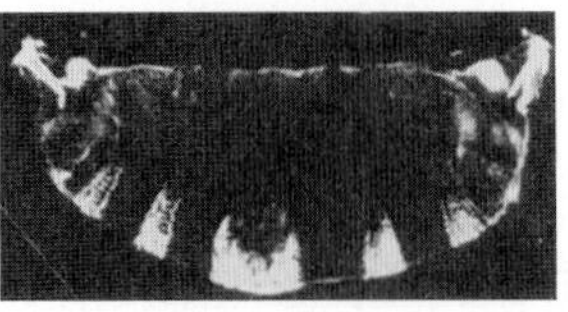

Double-posterior larva

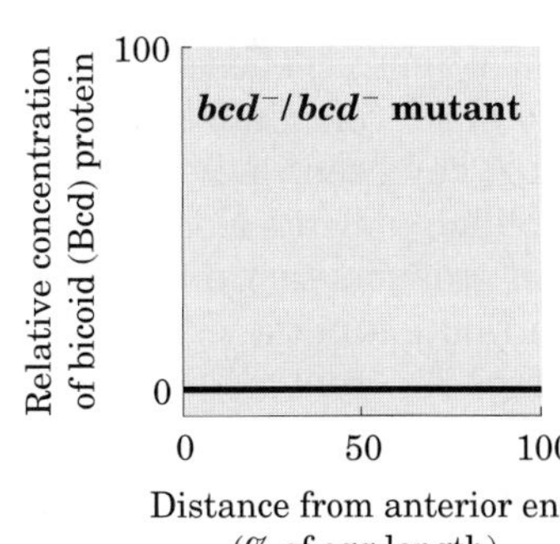

figure 28–37

Distribution of a maternal gene product in a *Drosophila* egg. (a) Micrograph of an immunologically stained egg, showing distribution of the *bicoid (bcd)* gene product. The graph measures stain intensity. This distribution is essential for normal development of the anterior structures of the animal. **(b)** If the *bcd* gene is not expressed by the mother (bcd^-/bcd^- mutant) and deposited in the egg, the resulting embryo has two posteriors.

A broader look at the effects of maternal genes reveals the outline of a developmental circuit. In addition to the *bicoid* and *nanos* mRNAs, which are deposited in the egg asymmetrically, a number of other maternal mRNAs are deposited uniformly throughout the egg cytoplasm. Three of these mRNAs encode the Pumilio, Hunchback, and Caudal proteins, all affected by *nanos* and *bicoid* (Fig. 28–38). Caudal and Pumilio are involved in development of the posterior end of the fly. Caudal is a transcriptional activator with a homeodomain; Pumilio is a translational repressor. Hunchback protein plays an important role in the development of the fly anterior and is also a transcriptional regulator of a variety of genes, in some cases a positive regulator, in other cases negative. Bicoid suppresses translation of *caudal* in the anterior and also acts as a transcriptional activator of *hunchback* in the cellular blastoderm. Since *hunchback* is expressed both from maternal mRNAs and from genes in the developing egg, it is considered both a maternal and a segmentation gene. The result of the activities of Bicoid is an increased concentration of Hunchback at the anterior end of the egg. The Nanos and Pumilio proteins act as translational repressors of *hunchback*, suppressing synthesis of its protein near the posterior end of the egg. Pumilio does not function in the absence of the Nanos protein, and the gradient of Nanos expression confines the activity of both proteins to the posterior region. Translational repression of the *hunchback* gene leads to degradation of *hunchback* mRNA near the posterior end. However, lack of Bicoid protein in the posterior leads to expression of *caudal*. In this way, the Hunchback and Caudal proteins become asymmetrically distributed in the egg.

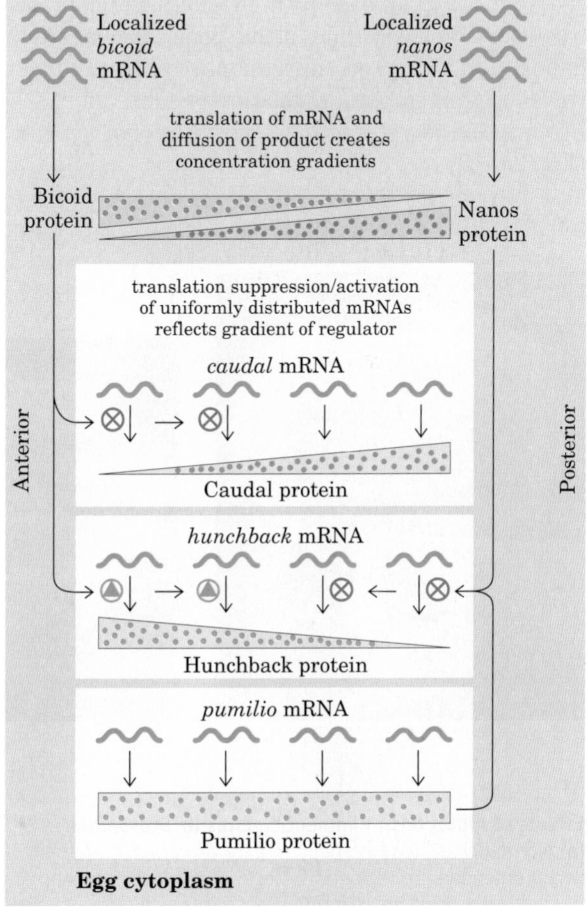

figure 28–38

Regulatory circuits of the anterior-posterior axis in a *Drosophila* egg. The *bicoid* and *nanos* mRNAs are localized near the anterior and posterior poles, respectively. The *caudal, hunchback,* and *pumilio* mRNAs are distributed throughout the egg cytoplasm. The gradients of Bicoid and Nanos proteins lead to accumulation of Hunchback protein in the anterior and Caudal protein in the posterior of the egg. Because Pumilio protein requires Nanos protein for its activity as a translational repressor of *hunchback*, it functions only at the posterior end.

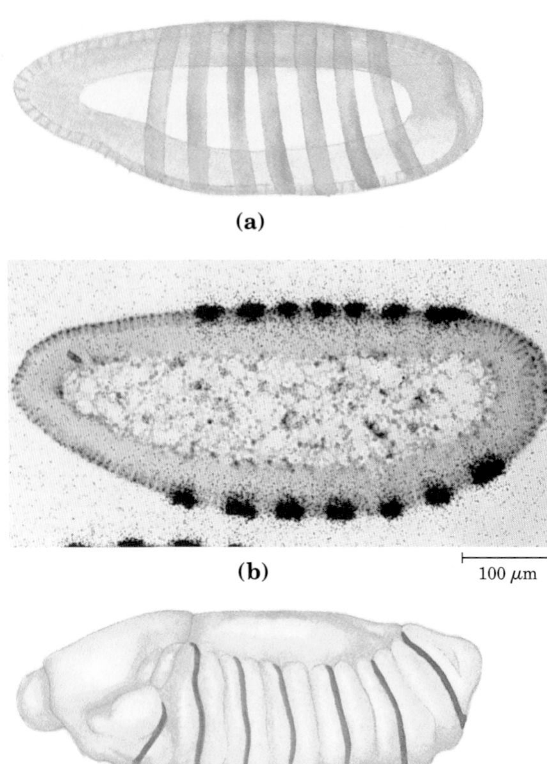

figure 28–39

Distribution of the *fushi tarazu (ftz)* gene product in early *Drosophila* embryos. (a) In the normal embryo, the gene product can be detected in seven bands around the circumference of the embryo (shown schematically). **(b)** These bands appear as dark spots (generated by a radioactive label) in a cross-sectional autoradiograph, and **(c)** demarcate the anterior margins of the segments in the late embryo (marked in red).

Segmentation Genes Gap genes, pair-rule genes, and segment polarity genes, three subclasses of segmentation genes in *Drosophila,* are activated at successive stages of embryonic development. Expression of the gap genes is generally regulated by the products of one or more maternal genes. At least some of the gap genes encode transcription factors that affect the expression of other segmentation or (later) homeotic genes.

One well-characterized segmentation gene is *fushi tarazu (ftz),* of the pair-rule subclass. When *ftz* is deleted, the embryo develops seven segments instead of the normal 14, each segment twice the normal width. The Fushi-tarazu protein (Ftz) is a transcriptional activator with a homeodomain. The mRNAs and proteins derived from the normal *ftz* gene accumulate in a striking pattern of seven stripes that encircle the posterior two-thirds of the embryo (Fig. 28–39). The stripes demarcate the positions of segments that develop later; these segments are eliminated if *ftz* function is lost. The Ftz protein and a few similar regulatory proteins directly or indirectly regulate the expression of vast numbers of genes in the continuing developmental cascade.

Homeotic Genes Loss of homeotic genes by mutation or deletion causes the appearance of a normal appendage or body structure at an inappropriate body position. An important example is the *ultrabithorax (ubx)* gene. When Ubx function is lost, the first abdominal segment develops incorrectly, having the structure of the third thoracic segment. Other known homeotic mutations cause the formation of an extra set of wings (p. 1072), or two legs at the position in the head where the antennae are normally found (Fig. 28–40).

The homeotic genes often span long regions of DNA. The *ubx* gene, for example, is 77,000 base pairs in length and contains introns of up to 50,000 base pairs. Transcription of this gene takes nearly one hour. The delay this imposes on *Ubx* gene expression is believed to be a timing mechanism involved in the temporal regulation of subsequent steps in development. The Ubx protein is yet another transcriptional activator with a homeodomain (Fig. 28–13).

(a)

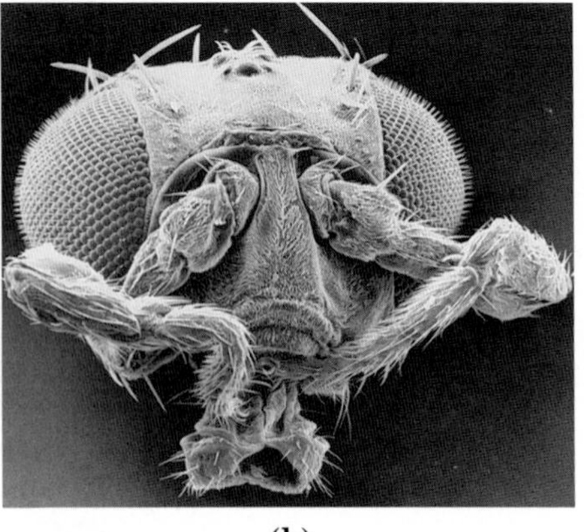

(b)

figure 28–40

Effects of mutations in homeotic genes in *Drosophila*. (a) Normal head. **(b)** Homeotic mutant *(antennapaedia)* in which antennae are replaced by legs. In another homeotic mutant, *bithorax,* a segment develops incorrectly to produce an extra set of wings (see p. 1072).

Many of the principles of development outlined above apply to eukaryotes from nematodes to humans. Some of the regulatory proteins themselves are conserved. For example, the products of the homeobox-containing genes *HOX* 1.1 in mouse and *antennapedia* in fruit fly differ at only one amino acid residue. Of course, although the molecular regulatory mechanisms may be similar, many of the ultimate developmental events are not conserved (humans do not have wings or antennae). The discovery of structural determinants with identifiable molecular functions is the first step in understanding the molecular events underlying development. As more genes and their protein products are discovered, the biochemical side of this vast puzzle will be elucidated in increasingly rich detail.

summary

The expression of genes is regulated by a number of processes that affect the rates at which gene products are synthesized and degraded. Much of this regulation occurs at the level of initiation of transcription, mediated by regulatory proteins that either repress or activate transcription at specific promoters. The effect of repressors is called negative regulation; activation is positive regulation.

Regulatory proteins are DNA-binding proteins that recognize specific DNA sequences; most have distinct DNA-binding domains. Within these domains, common structural motifs that bind DNA are the helix-turn-helix, zinc finger, and homeodomain. Regulatory proteins also contain domains for protein-protein interactions, including the leucine zipper and helix-loop-helix involved in dimerization and other domains involved in activation of transcription.

In bacteria, genes that encode products with interdependent functions are often clustered as a single transcriptional unit called an operon, and transcription of the genes is generally blocked by binding of a specific repressor protein at a DNA site called an operator. Dissociation of the repressor from the operator is mediated by a specific small molecule, an inducer. These principles were first elucidated in studies of the lactose *(lac)* operon. The Lac repressor dissociates from the *lac* operator when the repressor binds to its inducer, allolactose.

The lactose operon of *E. coli* also exhibits positive regulation by the cAMP receptor protein (CRP). When [cAMP] is high (i.e., when [glucose] is low), CRP binds a specific site on the DNA, stimulating transcription of the *lac* operon and production of lactose-metabolizing enzymes. The presence of glucose depresses [cAMP], decreasing expression of *lac* genes and other genes and preventing the metabolism of secondary sugars. A group of coordinately regulated operons is referred to as a regulon.

In the arabinose *(ara)* operon, the AraC protein acts as both activator and repressor. Some repressors, as in the *ara* operon, regulate their own synthesis (autoregulation). Some regulatory proteins in the *ara* system bind DNA sites many base pairs apart and interact by DNA looping.

Operons that produce the enzymes of amino acid synthesis have a regulatory circuit called attenuation that uses a transcription termination site (the attenuator). Formation of the attenuator structure in the mRNA is modulated by a mechanism that couples transcription and translation while responding to small changes in amino acid concentration.

In the SOS system, multiple unlinked genes that are repressed by a single type of repressor protein are induced simultaneously when DNA damage triggers RecA protein–facilitated autocatalytic proteolysis of the repressor. Some prokaryotic genes are regulated by genetic recombination processes that move promoters relative to the genes being regulated. Prokaryotic regulation can also occur at the level of translation. In the synthesis of ribosomal proteins, one protein in each r-protein operon acts as a translational repressor. The mRNA is bound by the repressor and translation thus blocked only when the r-protein is present in excess of available rRNA. These diverse mechanisms permit very sensitive cellular responses to changes in environmental conditions.

Eukaryotic cells use many of the same regulatory schemes, although positive regulation is more common and transcription is accompanied by large changes in chromatin structure. Promoters for RNA polymerase II typically have a TATA box and Inr sequence, as well as multiple binding sites for DNA-binding transactivators.

The latter sites, sometimes located hundreds or thousands of base pairs away from the TATA box, are called upstream activator sequences in yeast and enhancers in higher eukaryotes. Large complexes of proteins are generally required to regulate transcriptional activity. The effects of DNA-binding transactivators on RNA polymerase II are mediated by coactivator protein complexes such as TFIID or mediator. The modular structures of the transactivators have distinct activation and DNA-binding domains. Other protein complexes, including histone acetyltransferases such as GCN5-ADA2-ADA3 and ATP-dependent complexes such as SWI/SNF and NURF, reversibly remodel chromatin structure.

Hormones affect the regulation of gene expression in one of two ways. Steroid or thyroid hormones interact directly with intracellular receptors that are regulatory proteins; binding of the hormone has either positive or negative effects on the transcription of genes targeted by the hormone. Nonsteroid hormones bind to cell-surface receptors, triggering a signaling pathway that can lead to phosphorylation of a regulatory protein, affecting its activity.

Development of a multicellular organism presents the most complex regulatory challenge. The fate of cells in the early embryo is determined by establishment of anterior-posterior and dorsal-ventral gradients of proteins that act as transcriptional transactivators or repressors, regulating the genes required for the development of structures appropriate to a particular part of the organism. Sets of regulatory genes operate in temporal and spatial succession, transforming given areas of an egg cell into predictable structures in the adult organism.

further reading

General

Hershey, J.W.B., Mathews, M.B., & Sonenberg, N. (1996) *Translational Control,* Cold Spring Harbor Laboratory Press, Cold Spring Harbor, NY.

Many detailed reviews cover all aspects of this topic.

Lewin, B. (2000) *Genes VII*, Oxford University Press, New York.

The latest edition of an authoritative text.

Müller-Hill, B. (1996) *The* lac *Operon: A Short History of a Genetic Paradigm*, Walter de Gruyter, New York.

An excellent detailed account of the investigation of this important system.

Neidhardt, F.C. (ed.) (1996) Escherichia coli *and* Salmonella typhimurium, 2nd edn, Vol. 1: *Cellular and Molecular Biology* (Curtis, R., Ingraham, J.L., Lin, E.C.C., Magasanik, B., Low, K.B., Reznikoff, W.S., Riley, M., Schaechter, M., & Umbarger, H.E., eds), American Society for Microbiology, Washington, DC.

An excellent source for reviews of many bacterial operons.

Pabo, C.O. & Sauer, R.T. (1992) Transcription factors: structural factors and principles of DNA recognition. *Annu. Rev. Biochem.* **61,** 1053–1095.

Schleif, R. (1993) *Genetics and Molecular Biology,* 2nd edn, The Johns Hopkins University Press, Baltimore, MD.

Provides an excellent account of the experimental basis of major concepts of gene regulation in prokaryotes.

Regulation of Gene Expression in Prokaryotes

Blumenthal, R.M., Borst, D.W., & Matthews, R.G. (1996) Experimental analysis of global gene regulation in *Escherichia coli. Prog. Nucl. Acid Res. Mol. Biol.* **55,** 1–86.

Condon, C., Squires, C., & Squires, C.L. (1995) Control of rRNA transcription in *Escherichia coli. Microbiol. Rev.* **59,** 623–645.

Gourse, R.L., Gaal, T., Bartlett, M.S., Appleman, J.A., & Ross, W. (1996). rRNA transcription and growth rate–dependent regulation of ribosome synthesis in *Escherichia coli. Annu. Rev. Microbiol.* **50,** 645–677.

Jacob, F. & Monod, J. (1961) Genetic regulatory mechanisms in the synthesis of proteins. *J. Mol. Biol.* **3,** 318–356.

The operon model and the concept of messenger RNA were proposed in this historic paper.

Johnson, R.C. (1991) Mechanism of site-specific DNA inversion in bacteria. *Curr. Opin. Genet. Dev.* **1,** 404–411.

Kolb, A., Busby, S., Buc, H., Garges, S., & Adhya, S. (1993) Transcriptional regulation by cAMP and its receptor protein. *Annu. Rev. Biochem.* **62,** 749–795.

Landick, R. & Roberts, J.W. (1996) The shrewd grasp of RNA polymerase. *Science* **273,** 202–203.

Yanofsky, C., Konan, K.V., & Sarsero, J.P. (1996) Some novel transcription attenuation mechanisms used by bacteria. *Biochimie* **78,** 1017–1024.

Regulation of Gene Expression in Eukaryotes

Bashirullah, A., Cooperstock, R.L., & Lipshitz, H.D. (1998) RNA localization in development. *Annu. Rev. Biochem.* **67,** 335–394.

Beardsley, T. (1991) Smart genes. *Sci. Am.* **265** (August), 86–95.

A good overview of gene regulation during development.

DeRobertis, E.M., Oliver, G., & Wright, C.V.E. (1990) Homeobox genes and the vertebrate body plan. *Sci. Am.* **263** (July), 46–52.

Edmondson, D.G. & Roth, S.Y. (1996) Chromatin and transcription. *FASEB J.* **10,** 1173–1182.

Gingras, A.-C., Raught, B., & Sonenberg, N. (1999) eIF4 initiation factors: effectors of mRNA recruitment to ribosomes and regulators of translation. *Annu. Rev. Biochem.* **68,** 913–963.

Gray, N.K. & Wickens, M. (1998) Control of translation initiation in animals. *Annu. Rev. Cell Dev. Biol.* **14,** 399–458.

Johnson, A.D. (1995) The price of repression. *Cell* **81,** 655–658.

Kornberg, R.D. (1996) RNA polymerase II transcription control. *Trends Biochem. Sci.* **21,** 325–326.

The lead article in an issue devoted to RNA polymerase II and its regulation.

Mannervik, M., Nibu, Y., Zhang, H., & Levine, M. (1999) Transcriptional coregulators in development. *Science* **284,** 606–609.

McKnight, S.L. (1991) Molecular zippers in gene regulation. *Sci. Am.* **264** (April), 54–64.

A good description of leucine zippers.

Melton, D.A. (1991) Pattern formation during animal development. *Science* **252,** 234–241.

Muller, W.A. (1997) *Developmental Biology,* Springer, New York.

A good elementary text.

Pugh, B.F. (1996) Mechanisms of transcription complex assembly. *Curr. Opin. Cell Biol.* **8,** 303–311.

Reece, R. & Platt, A. (1997) Signaling activation and repression of RNA polymerase II transcription in yeast. *Bioessays* **19,** 1001–1010.

Rivera-Pomar, R. & Jackle, H. (1996) From gradients to stripes in *Drosophila* embryogenesis: filling in the gaps. *Trends Genet.* **12,** 478–483.

Steger, D.J. & Workman, J.L. (1996) Remodeling chromatin structures for transcription: what happens to the histones? *Bioessays* **18,** 875–884.

Struhl, K. (1995) Yeast transcriptional regulatory mechanisms. *Annu. Rev. Genet.* **29,** 651–674.

Struhl, K. (1999) Fundamentally different logic of gene regulation in eukaryotes and prokaryotes. *Cell* **98,** 1–4.

Thummel, C.S. (1992) Mechanisms of transcriptional timing in *Drosophila. Science* **255,** 39–40.

Travers, A. (1999) An engine for nucleosome remodeling. *Cell* **96,** 311–314.

Wade, P.A., Pruss, D., & Wolffe, A. (1997) Histone acetylation: chromatin in action. *Trends Biochem. Sci.* **22,** 128–132.

Wickens, M., Kimble, J., & Strickland, S. (1996) Translational control of developmental decisions. In *Translational Control* (Hershey, J.W.B., Mathews, M.B., & Sonenberg, N., eds), pp. 411–450, Cold Spring Harbor Laboratory Press, Cold Spring Harbor, NY.

Zlatanova, J. (1990) Histone H1 and the regulation of transcription of eukaryotic genes. *Trends Biochem. Sci.* **15,** 273–276.

problems

1. Effect of mRNA and Protein Stability on Regulation *E. coli* cells are growing in a medium with glucose as the sole carbon source. Tryptophan is suddenly added. The cells continue to grow, and divide every 30 min. Describe (qualitatively) how the amount of tryptophan synthase activity in the cells changes under the following conditions:

(a) The *trp* mRNA is stable (degraded slowly over many hours).

(b) The *trp* mRNA is degraded rapidly, but tryptophan synthase is stable.

(c) The *trp* mRNA and tryptophan synthase are both degraded more rapidly than normal.

2. Negative Regulation Describe the probable effects on gene expression in the *lac* operon of mutations in:

(a) the *lac* operator that delete most of O_1

(b) the *lac*I gene that inactivate the repressor

(c) the promoter that eliminate the region around position −10.

3. Specific DNA Binding by Regulatory Proteins A typical prokaryotic repressor protein discriminates between its specific DNA-binding site (operator) and nonspecific DNA by a factor of 10^4 to 10^6. About ten molecules of repressor per cell are sufficient to ensure a high level of repression. Assume that a very similar repressor existed in a human cell, with a similar specificity for its binding site. How many copies of the repressor would be required to elicit a level of repression similar to that in the prokaryotic cell? (Hint: The *E. coli* genome contains about 4.7 million base pairs; the human haploid genome has about 2.4 billion base pairs.)

4. Repressor Concentration in *E. coli* The dissociation constant for a particular repressor-operator complex is very low, about 10^{-13} M. An *E. coli* cell (volume 2×10^{-12} mL) contains 10 copies of the repressor. Calculate the cellular concentration of the repressor protein. How does this value compare with the dissociation constant of the repressor-operator complex? What is the significance of this result?

5. Catabolite Repression *E. coli* cells are growing in a medium containing lactose but no glucose. Indicate whether each of the following changes or conditions would increase, decrease, or not change the expression of the *lac* operon. It may be helpful to draw a model depicting what is happening in each situation.

(a) Addition of a high concentration of glucose

(b) A mutation that prevents dissociation of the Lac repressor from the operator

(c) A mutation that completely inactivates β-galactosidase

(d) A mutation that completely inactivates galactoside permease

(e) A mutation that prevents binding of CRP to its binding site near the *lac* promoter

6. Transcription Attenuation How would transcription of the *E. coli trp* operon be affected by the following manipulations of the leader region of the *trp* mRNA?

(a) Increasing the distance (number of bases) between the leader peptide gene and sequence 2

(b) Increasing the distance between sequences 2 and 3

(c) Removing sequence 4

(d) Changing the two Trp codons in the leader peptide gene to His codons

(e) Eliminating the ribosome-binding site for the gene that encodes the leader peptide

(f) Changing several nucleotides in sequence 3 so that it can base-pair with sequence 4 but not with sequence 2

7. Repressors and Repression How would the SOS response in *E. coli* be affected by a mutation in the *lex*A gene that prevented autocatalytic cleavage of the LexA protein?

8. Regulation by Recombination In the phase variation system of *Salmonella,* what would happen to the cell if the Hin recombinase became more active and promoted recombination (DNA inversion) several times in each cell generation?

9. Initiation of Transcription in Eukaryotes A new RNA polymerase activity is discovered in crude extracts of cells derived from an exotic fungus. The RNA polymerase initiates transcription only from a single, highly specialized promoter. As the polymerase is purified its activity declines, and the purified enzyme is completely inactive unless crude extract is added to the reaction mixture. Suggest an explanation for these observations.

10. Functional Domains in Regulatory Proteins A biochemist replaces the DNA-binding domain of the yeast Gal4 protein with the DNA-binding domain from the Lac repressor and finds that the engineered protein no longer regulates transcription of the *GAL* genes in yeast. Draw a diagram of the different functional domains you would expect to find in the Gal4 protein and in the engineered protein. Why does the engineered protein no longer regulate transcription of the *GAL* genes? What might be done to the DNA-binding site recognized by this chimeric protein to make it functional in activating transcription of *GAL* genes?

11. Inheritance Mechanisms in Development A *Drosophila* egg that is bcd^-/bcd^- may develop normally but as an adult will not be able to produce viable offspring. Explain.

Biochemistry on the Internet

12. TATA Binding Protein and the TATA Box To examine the interactions between transcription factors and DNA, go to the Protein Data Bank and download the PDB file 1TGH. This file models the interactions between a human TATA binding protein and a segment of double-stranded DNA. Use the Noncovalent Bond Finder found at the Chime Resources Web site to examine the roles of hydrogen bonds and hydrophobic interactions involved in the binding of this transcription factor to the TATA box in DNA. (For the current URLs of these Web sites and for further instructions on how to use the Noncovalent Bond Finder, go to http://www.worthpublishers.com/lehninger.)

Within the Noncovalent Bond Finder program, load the PDB file and display the protein in Spacefill mode and the DNA in Wireframe mode. Go to the Lehninger Web site for more detailed instructions in order to answer the following questions.

(a) Which of the base pairs in the DNA form hydrogen bonds with the protein? Which of these contribute to the specific recognition of the TATA box by this protein? (Hydrogen bond lengths range from 2.5 to 3.3 Å between the hydrogen donor and hydrogen acceptor.)

(b) Which amino acid residues in the protein interact with these base pairs? On what basis did you make this determination? Do these observations agree with the information presented in the text?

(c) What is the sequence of the DNA in this model and which portions of the sequence are recognized by the TATA binding protein?

(d) Can you identify any hydrophobic interactions in this complex? (Hydrophobic interactions usually occur with interatomic distances of 3.3 to 4.0 Å.)

chapter 29

Recombinant DNA Technology

This final chapter describes a technology that is now fundamental to the advance of modern biological sciences, defining present and future biochemical frontiers and illustrating many important principles of biochemistry. Elucidation of the laws governing enzymatic catalysis, macromolecular structure, cellular metabolism, and information pathways allows research to be directed at ever more complex biochemical processes. Cell division, immunity, embryogenesis, vision, taste, oncogenesis, cognition—all are orchestrated in an elaborate symphony of molecular and macromolecular interactions that are being understood with increasing clarity. The real implications of the biochemical journey begun in the nineteenth century are found in the ever-increasing power to understand and alter living systems.

Paul Berg

To understand a complex biological process, a biochemist will isolate and study the individual components in vitro, then piece together the parts to get a coherent picture of the overall process. A major source of molecular insights lies in the cell's own information archive, its DNA. The sheer size of chromosomes, however, presents an enormous challenge: how does one find and study a particular gene among the 100,000 genes nested in the billions of base pairs of a mammalian genome? Solutions began to emerge during the 1970s.

Herbert Boyer

Stanley N. Cohen

Decades of advances by thousands of scientists working in genetics, biochemistry, cell biology, and physical chemistry came together in the laboratories of Paul Berg, Herbert Boyer, and Stanley Cohen to yield techniques for locating, isolating, preparing, and studying small segments of DNA derived from much larger chromosomes. Techniques for **DNA cloning** have opened hitherto unimaginable opportunities to identify and study the genes involved in almost every known biological process. These new methods are transforming basic research, agriculture, medicine, ecology, forensics, and many other fields, while presenting society with bewildering choices and significant ethical dilemmas.

The first two parts of this chapter outline the fundamental biochemical principles that support this revolutionary technology, drawing on an understanding of material discussed in the previous five chapters. Genetic selection and screening are then considered, and the final section of this chapter illustrates the range of applications and the potential of this technology.

DNA Cloning: The Basics

To clone means to make identical copies. This term originally applied to the procedure of isolating one cell and allowing it to reproduce, creating a population of identical cells for study. DNA cloning involves separating a specific gene or DNA segment from a larger chromosome, attaching it to a small molecule of carrier DNA, and then replicating this modified DNA thousands or millions of times through both the increase in cell number and the creation of multiple copies of the cloned DNA in each cell. The result is selective amplification of a particular gene or DNA segment. Cloning of DNA from any organism entails five general procedures:

1. Cutting DNA at precise locations. Sequence-specific endonucleases (restriction endonucleases) provide the necessary molecular scissors.
2. Joining two DNA fragments covalently. DNA ligase does this.
3. Selecting a small molecule of DNA capable of self-replication. Segments of DNA to be cloned can be joined to plasmids or viral DNAs (**cloning vectors;** a vector is a delivery agent). Composite DNA molecules comprising covalently linked segments from two or more sources are called **recombinant DNAs.**
4. Moving recombinant DNA from the test tube to a host cell that will provide the enzymatic machinery for DNA replication.
5. Selecting or identifying host cells that contain recombinant DNA.

The methods used to accomplish these and related tasks are collectively referred to as **recombinant DNA technology,** or more informally as **genetic engineering.**

Much of this initial discussion will focus on DNA cloning in the bacterium *Escherichia coli,* the first organism used for recombinant DNA work and still the most common host cell. *E. coli* has many advantages: its DNA metabolism (and many other biochemical processes) are well understood; many naturally occurring cloning vectors associated with *E. coli,* such as bacteriophages and plasmids, are well characterized; and effective techniques are available for moving DNA from one bacterial cell to another. DNA cloning in other organisms is addressed later in the chapter.

Restriction Endonucleases and DNA Ligase Yield Recombinant DNA

A set of enzymes now available thanks to decades of research on nucleic acid metabolism are particularly important to recombinant DNA technology (Table 29–1). Two classes of enzymes lie at the heart of the general ap-

table 29–1

Some Enzymes Used in Recombinant DNA Technology

Enzyme(s)	Function
Type II restriction endonucleases	Cleave DNAs at specific base sequences
DNA ligase	Joins two DNA molecules or fragments
DNA polymerase I (*E. coli*)	Fills gaps in duplexes by stepwise addition of nucleotides to 3′ ends
Reverse transcriptase	Makes a DNA copy of an RNA molecule
Polynucleotide kinase	Adds a phosphate to the 5′-OH end of a polynucleotide to label it or permit ligation
Terminal transferase	Adds homopolymer tails to the 3′-OH ends of a linear duplex
Exonuclease III	Removes nucleotide residues from the 3′ ends of a DNA strand
Bacteriophage λ exonuclease	Removes nucleotides from the 5′ ends of a duplex to expose single-stranded 3′ ends
Alkaline phosphatase	Removes terminal phosphates from either the 5′ or 3′ end (or both)

proach to generating and propagating a recombinant DNA molecule (Fig. 29–1). First, **type II restriction endonucleases** cleave DNA at specific DNA sequences to generate a set of smaller fragments. Second, the DNA fragment to be cloned can be joined to a suitable cloning vector by using **DNA ligases** to seal the DNA molecules together. The recombinant vector is then introduced into a host cell, which amplifies the fragment while the cell undergoes many generations of cell divisions.

Restriction endonucleases are found in a wide range of bacterial species. Werner Arber discovered that their biological function is to recognize and cleave foreign DNA (e.g., the DNA of an infecting virus); such DNA is said to be *restricted.* In the host cell's DNA, the sequence that would be recognized by the restriction endonuclease is protected from digestion by methylation of the DNA, catalyzed by a specific DNA methylase. The restriction endonuclease and the corresponding methylase in a bacterium are sometimes referred to as a **restriction-modification system.**

There are three types of restriction endonucleases, designated I, II, and III. Types I and III are generally large, multisubunit complexes containing both the endonuclease and methylase activities. Type I restriction endonucleases cleave DNA at random sites that can be more than 1,000 base pairs from the recognition sequence. Type III restriction endonucleases cleave the DNA about 25 base pairs from the recognition sequence. Both types move along the DNA in a reaction that requires the energy of ATP. Type II restriction endonucleases, first isolated by Hamilton Smith, are simpler, require no ATP, and cleave the DNA within the recognition sequence itself. The extraordinary utility of these enzymes was demonstrated by Daniel Nathans, who first used them to develop novel methods for mapping and analyzing genes and genomes.

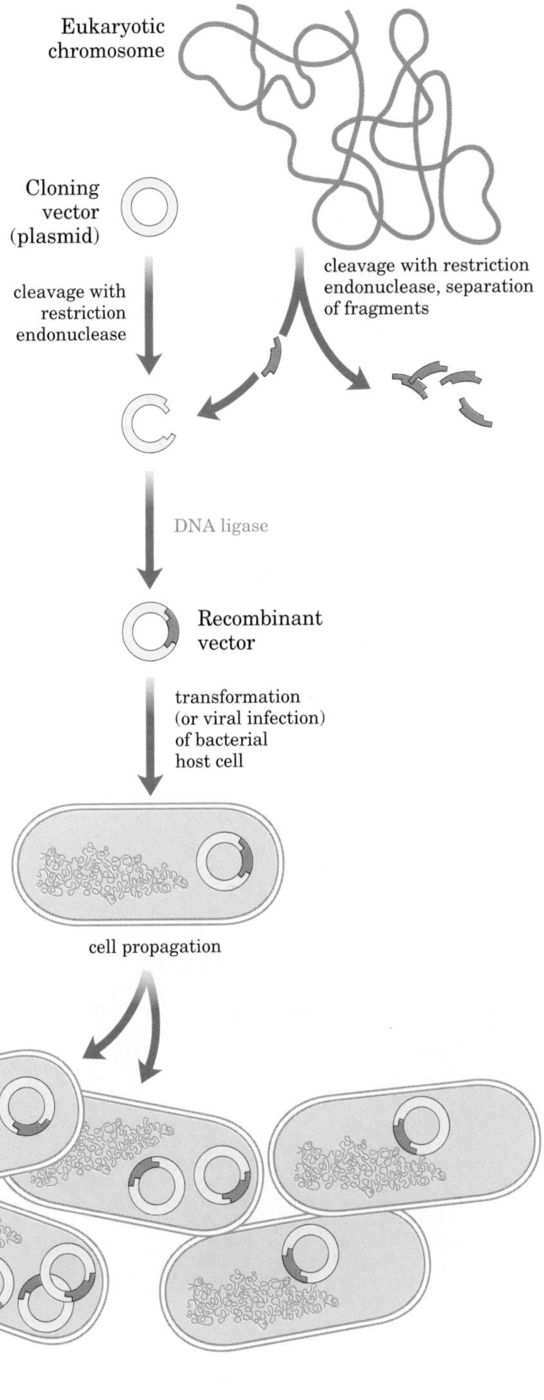

figure 29–1
Schematic illustration of DNA cloning. A fragment of DNA of interest to the researcher is obtained by cleaving eukaryotic chromosomes with a restriction endonuclease. After isolating the fragment and ligating it to a cloning vector that has also been cleaved with a restriction endonuclease, the resulting recombinant DNA is introduced into a host cell where it can be propagated (cloned). Note that the size of the *E. coli* chromosome relative to that of a typical cloning vector (such as a plasmid) is much greater than depicted here.

table 29–2

Recognition Sequences for Some Type II Restriction Endonucleases

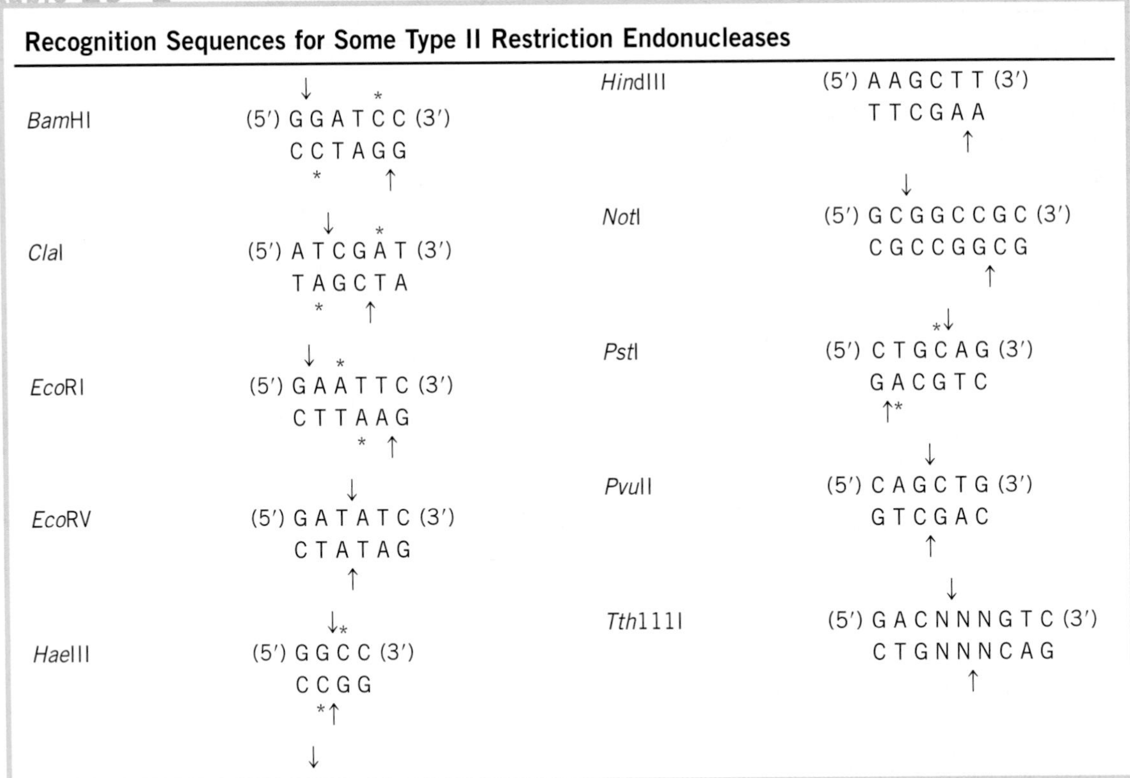

Enzyme	Recognition sequence	Enzyme	Recognition sequence
*Bam*HI	↓ * (5′) G G A T C C (3′) C C T A G G * ↑	*Hin*dIII	(5′) A A G C T T (3′) T T C G A A ↑
*Cla*I	↓ * (5′) A T C G A T (3′) T A G C T A * ↑	*Not*I	↓ (5′) G C G G C C G C (3′) C G C C G G C G ↑
*Eco*RI	↓ * (5′) G A A T T C (3′) C T T A A G * ↑	*Pst*I	*↓ (5′) C T G C A G (3′) G A C G T C ↑*
*Eco*RV	↓ (5′) G A T A T C (3′) C T A T A G ↑	*Pvu*II	↓ (5′) C A G C T G (3′) G T C G A C ↑
*Hae*III	↓* (5′) G G C C (3′) C C G G *↑ ↓	*Tth*111I	↓ (5′) G A C N N N G T C (3′) C T G N N N C A G ↑

Arrows indicate the phosphodiester bonds cleaved by each restriction endonuclease. Asterisks indicate bases that are methylated by the corresponding methylase (where known). N denotes any base. Note that the name of each enzyme consists of a three-letter abbreviation of the bacterial species from which it is derived (e.g., *Bam* for B*acillus* *am*y*loliquefaciens*, *Eco* for *E*s*cherichia* *co*l*i*). The Roman numerals included in the enzyme names (e.g., *Bam*HI) distinguish different restriction endonucleases isolated from the same bacterial species rather than the type of restriction enzyme.

Thousands of restriction endonucleases have been discovered in different bacterial species, and over 100 different DNA sequences are recognized by one or more of these enzymes. The recognized sequences are usually four to six base pairs in length and palindromic (see Fig. 10–20). A few sequences recognized by some type II restriction endonucleases are presented in Table 29–2. In a few cases, the interaction between a restriction endonuclease and its target sequence has been elucidated in exquisite molecular detail. The complex comprising the type II restriction endonuclease *Eco*RV and its target sequence is illustrated in Figure 29–2.

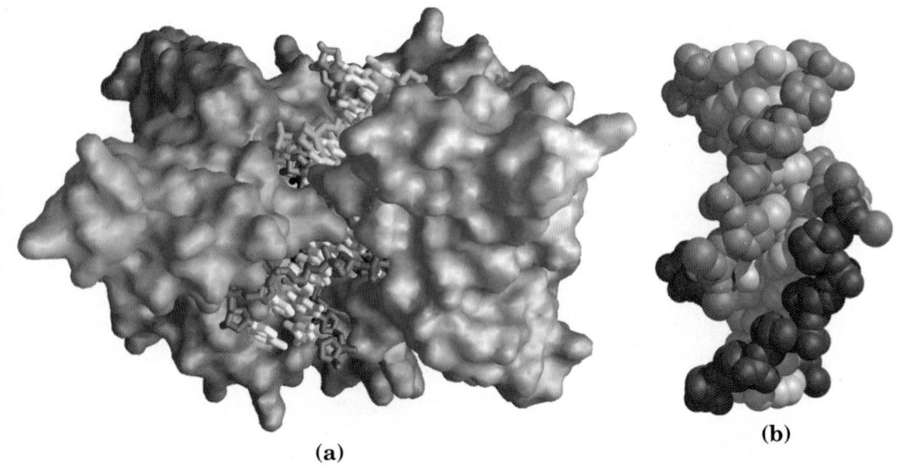

figure 29–2
Interaction of *Eco*RV restriction endonuclease with its target sequence. **(a)** The dimeric enzyme (with its two subunits in light blue and gray) is shown bound to the products of DNA cleavage at the sequence recognized by the EcoRV endonuclease. The DNA backbone is shown in two shades of blue to distinguish the segments separated by cleavage. **(b)** In this view, the protein has been removed and the DNA has been turned 180°. The cleavage points are staggered on the two DNA strands so the enzyme generates sticky ends. Bound magnesium ions, shown in orange, play a role in catalysis of the cleavage reaction.

Some restriction endonucleases make staggered cuts on the two DNA strands, leaving two to four nucleotides of one strand unpaired at each resulting end. These are referred to as **sticky ends** (Fig. 29–3a) because they can base-pair with each other or with complementary sticky ends of other DNA fragments. Other restriction endonucleases cleave both strands of DNA at the opposing phosphodiester bonds, leaving no unpaired bases on either end; these are often called **blunt ends** (Fig. 29–3b).

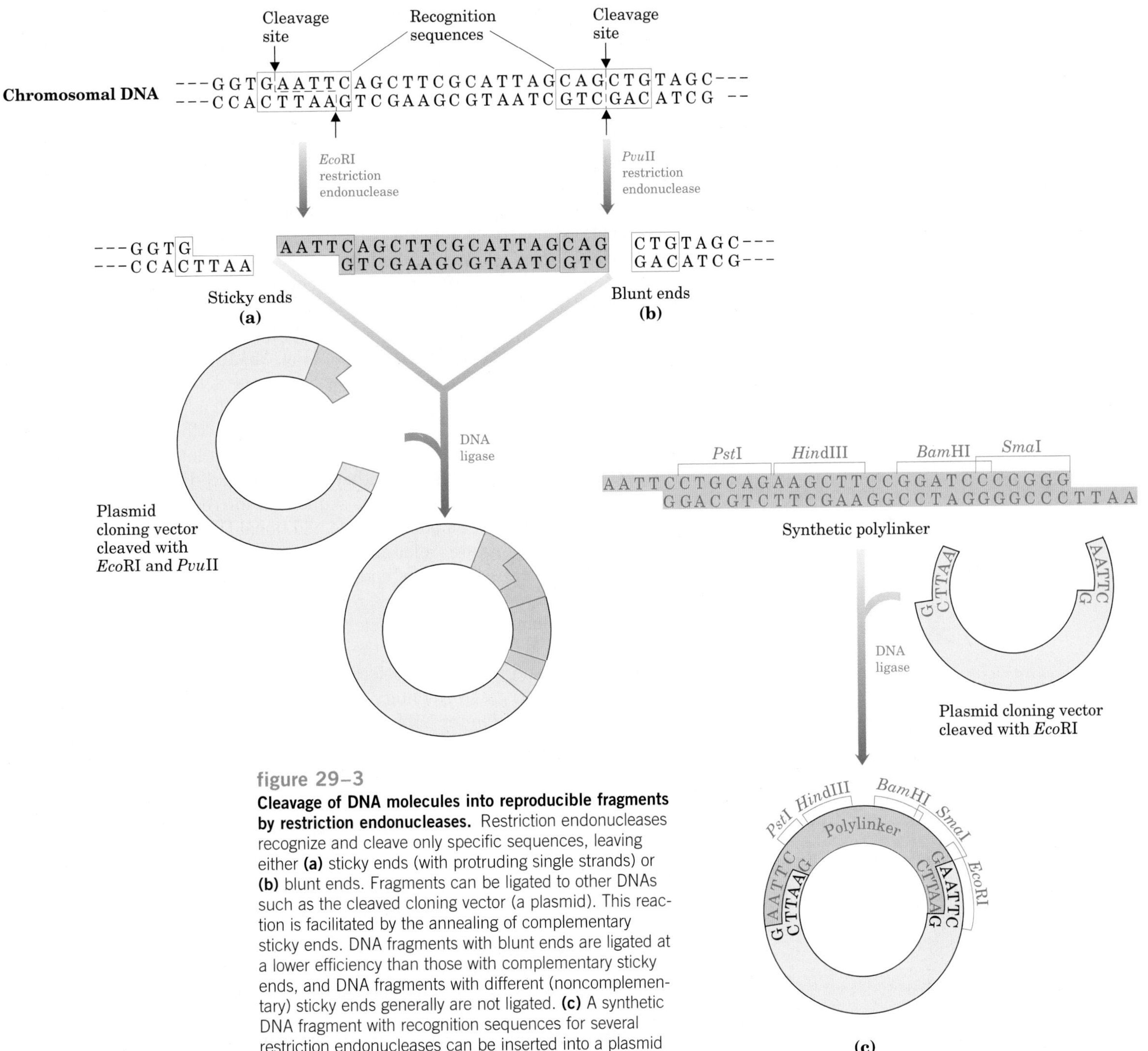

figure 29–3

Cleavage of DNA molecules into reproducible fragments by restriction endonucleases. Restriction endonucleases recognize and cleave only specific sequences, leaving either **(a)** sticky ends (with protruding single strands) or **(b)** blunt ends. Fragments can be ligated to other DNAs such as the cleaved cloning vector (a plasmid). This reaction is facilitated by the annealing of complementary sticky ends. DNA fragments with blunt ends are ligated at a lower efficiency than those with complementary sticky ends, and DNA fragments with different (noncomplementary) sticky ends generally are not ligated. **(c)** A synthetic DNA fragment with recognition sequences for several restriction endonucleases can be inserted into a plasmid that has been cleaved by a restriction endonuclease. The insert is called a linker; an insert with multiple restriction sites is called a polylinker.

The average size of the DNA fragments produced by cleaving genomic DNA with a restriction endonuclease depends on the frequency with which a particular restriction site occurs in the DNA molecule; this in turn depends largely on the size of the recognition sequence. In a DNA molecule with a random sequence in which all four nucleotides are equally abundant, a six base pair sequence recognized by a restriction endonuclease such as *Bam*HI will occur on average once every 4^6 (4,096) base pairs. Enzymes that recognize a four base pair sequence would produce smaller DNA fragments in a DNA molecule with a random sequence; a recognition sequence of this size would be expected to occur about once every 4^4(256) base pairs. Particular recognition sequences tend to occur less frequently than this because nucleotide sequences in DNA are not random and the four nucleotides are not equally abundant. The average size of the fragments produced by restriction endonuclease cleavage of a large DNA can be increased by simply terminating the reaction prior to completion; the result is called a partial digest.

Once a DNA molecule has been cleaved into fragments, a particular fragment of known size can be enriched by agarose or acrylamide gel electrophoresis (p. 134) or HPLC (p. 131). Cleavage of a typical mammalian genome by a restriction endonuclease usually yields too many different fragments to make isolation of a particular DNA fragment by electrophoresis or HPLC practical. An intermediate step in the cloning of a specific gene or DNA segment of interest is often the construction of a DNA library (as described later in this chapter).

When the target DNA fragment is isolated, it is joined to a similarly digested cloning vector using DNA ligase (Fig. 25–16). The base-pairing of complementary sticky ends greatly facilitates the ligation reaction (Fig. 29–3a). A fragment generated by *Eco*RI generally will not link to a fragment generated by *Bam*HI. Blunt ends can also be ligated, albeit less efficiently. New DNA sequences can be created by inserting synthetic DNA fragments (called **linkers**) between the ends that are being ligated (Fig. 29–3c). DNA fragments with multiple recognition sequences for restriction endonucleases (often useful later as points where additional DNA can be inserted by cleavage and ligation) are called **polylinkers.**

The effectiveness of sticky ends in selectively joining two DNA fragments was apparent in the earliest recombinant DNA experiments. Before restriction endonucleases were widely available, some workers found that sticky ends could be generated by the combined action of the bacteriophage λ exonuclease and terminal transferase (Table 29–1). The fragments to be joined were given complementary homopolymeric tails (Fig. 29–4). This method was used by Peter Lobban and Dale Kaiser in 1971 in the first experiments to join naturally occurring DNA fragments. Similar methods were used soon after in the laboratory of Paul Berg to join DNA segments from simian virus 40 (SV40) to DNA derived from bacteriophage λ, thereby creating the first recombinant DNA molecule involving DNA segments from different species.

Cloning Vectors Allow Amplification of Inserted DNA Segments

The principles that govern the delivery of recombinant DNA in clonable form to a host cell, and its subsequent amplification, can be seen by considering three popular cloning vectors commonly used in experiments with *E. coli*—plasmids, bacteriophages, and bacterial artificial chromosomes.

Plasmids Plasmids (see Fig. 24–6) are circular DNA molecules that replicate separately from the host chromosome. Naturally occurring bacterial

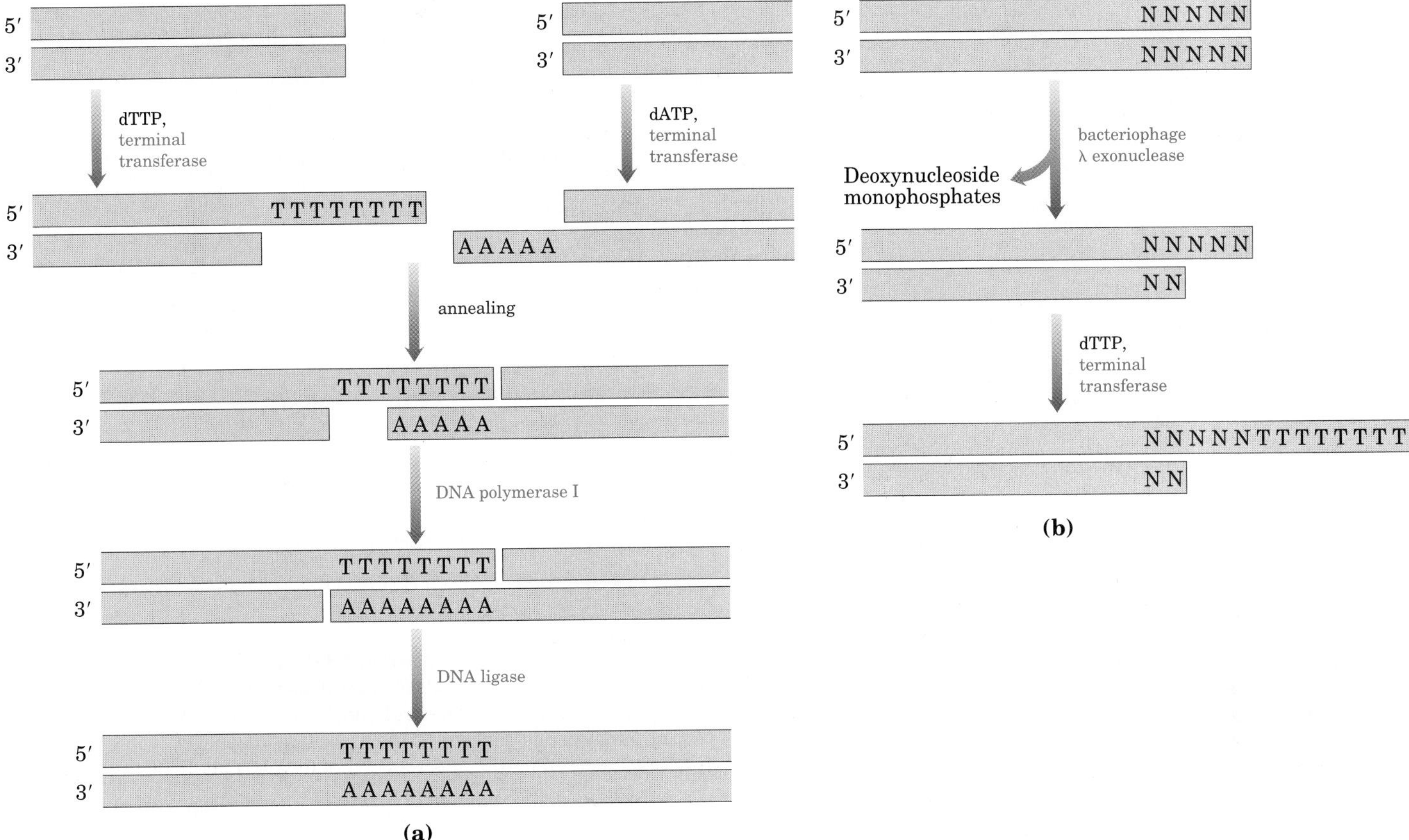

figure 29–4
Terminal transferase can be used to generate sticky ends for joining two DNA fragments. (a) Complementary homopolymeric tails are added to the ends of the two fragments to be joined, forming sticky ends. After annealing, the gaps are filled and the nicks sealed by the sequential action of DNA polymerase I and DNA ligase. **(b)** The optimal substrate for terminal transferase is the 3′ OH at the end of a single strand of DNA at least three nucleotides long. If the ends of the duplex DNA have a 5′ protruding single strand or are blunt ends, the bacteriophage λ exonuclease (which degrades DNA strands in the 5′→3′ direction) can create a good substrate for terminal transferase. N denotes any base.

plasmids range in size from 5,000 to 400,000 base pairs. They can be introduced into bacterial cells by a process called **transformation.** To get the cells to take up the DNA, the cells and plasmid DNA are incubated together at 0 °C in a calcium chloride solution, then subjected to a shock by rapidly shifting the temperature to 37–43 °C. For reasons not well understood, some of the cells treated in this way take up the DNA. Alternatively, cells incubated with the plasmid DNA can be subjected to a high-voltage pulse. This approach, called **electroporation,** transiently renders the cell membrane permeable to large molecules.

Regardless of the approach used, few cells actually take up the plasmid DNA, so a method is needed to select those that do. The usual strategy is to ensure that the plasmid includes a gene that the host cell requires for growth under specific conditions, such as a gene that confers resistance to an antibiotic. Only cells that have been transformed by the recombinant plasmid can grow in the presence of that antibiotic, making any cell that contains the plasmid "selectable" under those conditions. Such a gene is sometimes called a selectable marker.

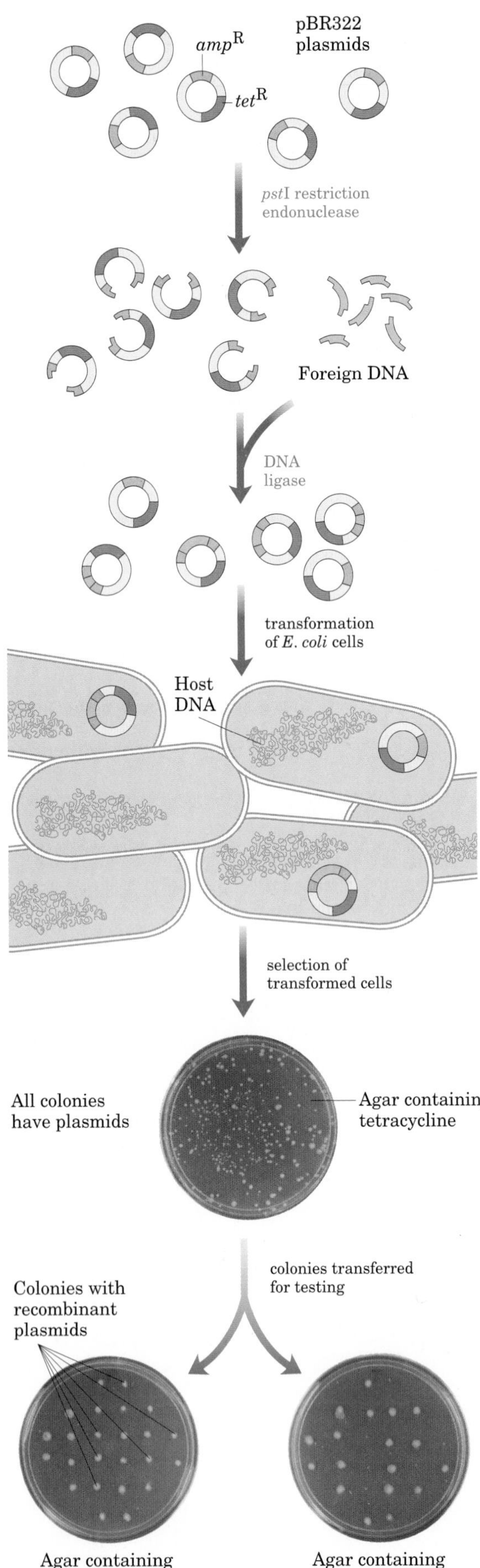

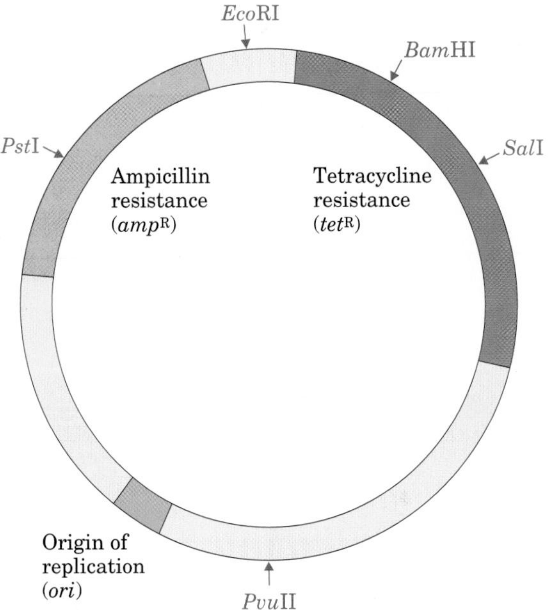

figure 29–5

The constructed *E. coli* plasmid pBR322. Note the location of some important restriction sites for *Pst*I, *Eco*RI, *Bam*HI, *Sal*I, and *Pvu*II; ampicillin- and tetracycline-resistance genes; and the replication origin *(ori).* Constructed in 1977, this was one of the early plasmids designed expressly for cloning in *E. coli.*

Many different plasmid vectors suitable for cloning have been developed by modifying naturally occurring plasmids. The *E. coli* plasmid pBR322 offers a good example of the features useful in a cloning vector (Fig. 29–5):

1. An origin of replication is required to propagate the plasmid and maintain it at a level of 10 to 20 copies per cell.
2. Two genes that confer resistance to different antibiotics allow the identification of cells that contain the intact plasmid, or a recombinant version of the plasmid (Fig. 29–6).

figure 29–6

Cloning foreign DNA in *E. coli* with pBR322. If foreign DNA is inserted at the *Pst*I restriction site, the ampicillin-resistance element is disrupted and inactivated. After ligation of the DNA and transformation of *E. coli* cells, the cells are grown on agar plates containing tetracycline, to select for those that have taken up a plasmid. Individual colonies from these agar plates are transferred using sterile toothpicks to matching positions within a grid on two additional plates; one plate contains tetracycline (a control) and the other contains both tetracycline and ampicillin. Those cells that grow in the presence of tetracycline, but do not form colonies on the plate containing tetracycline plus ampicillin, contain recombinant plasmids (recall that the ampicillin-resistance element becomes nonfunctional with the insertion of foreign DNA). Cells that contain pBR322 that was ligated without the insertion of a foreign DNA fragment retain ampicillin resistance and grow on both plates. Identification of recombinant clones requires selection followed by screening.

3. Several unique recognition sequences for different restriction endonucleases provide sites where the plasmid can later be cut to insert foreign DNA.
4. An overall small size facilitates the plasmid's entry into cells and the biochemical manipulation of the DNA.

Transformation of typical bacterial cells with purified DNA (never a very efficient process) decreases as plasmid size increases, so it is difficult to clone DNA segments longer than about 15,000 base pairs when plasmids are used as the vector.

Bacteriophages Bacteriophage λ has a very efficient mechanism for delivering its 48,502 base pairs of DNA into a bacterium and can be used as a vector to clone somewhat larger DNA segments (Fig. 29–7). Two key features contribute to its utility:

1. About one-third of the λ genome is nonessential and can be replaced with foreign DNA.
2. DNA will be packaged into infectious phage particles only if it is between 40,000 and 53,000 base pairs long, a constraint that can be used to direct packaging of recombinant DNA only.

Bacteriophage λ vectors have been developed that can be readily cleaved into three pieces, two of which contain essential genes but which together are only about 30,000 base pairs long. The third piece of "filler" DNA in the vector is discarded when the vector is to be used for cloning. Additional DNA must therefore be inserted between the two essential segments to generate ligated DNA molecules long enough to produce viable phage particles. In effect, the packaging mechanism selects for recombinant viral DNAs. Bacteriophage λ vectors permit the cloning of DNA fragments of up to 23,000 base pairs. Once the bacteriophage λ fragments are ligated to foreign DNA fragments of suitable size, the resulting recombinant DNAs can be packaged into phage particles by adding them to crude bacterial cell extracts that contain all the proteins needed to assemble a complete phage. This is called **in vitro packaging** (Fig. 29–7). All viable phage particles will contain a foreign DNA fragment. The subsequent transmission of the recombinant DNA into *E. coli* cells is highly efficient.

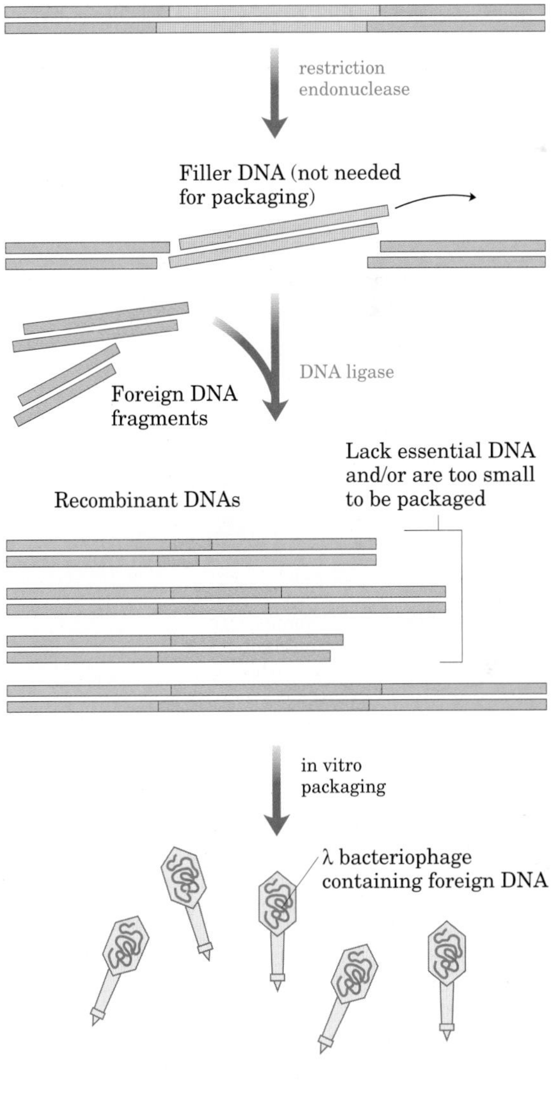

figure 29–7
Bacteriophage λ cloning vectors. Recombinant DNA methods are used to remove from the bacteriophage λ genome those genes not needed for phage production, replacing them with "filler" DNA to make the phage vector DNA large enough for packaging into phage particles. This filler is replaced with foreign DNA in cloning experiments. Recombinants are packaged into viable phage particles in vitro only if they include an appropriately sized foreign DNA fragment as well as both of the essential λ DNA end fragments.

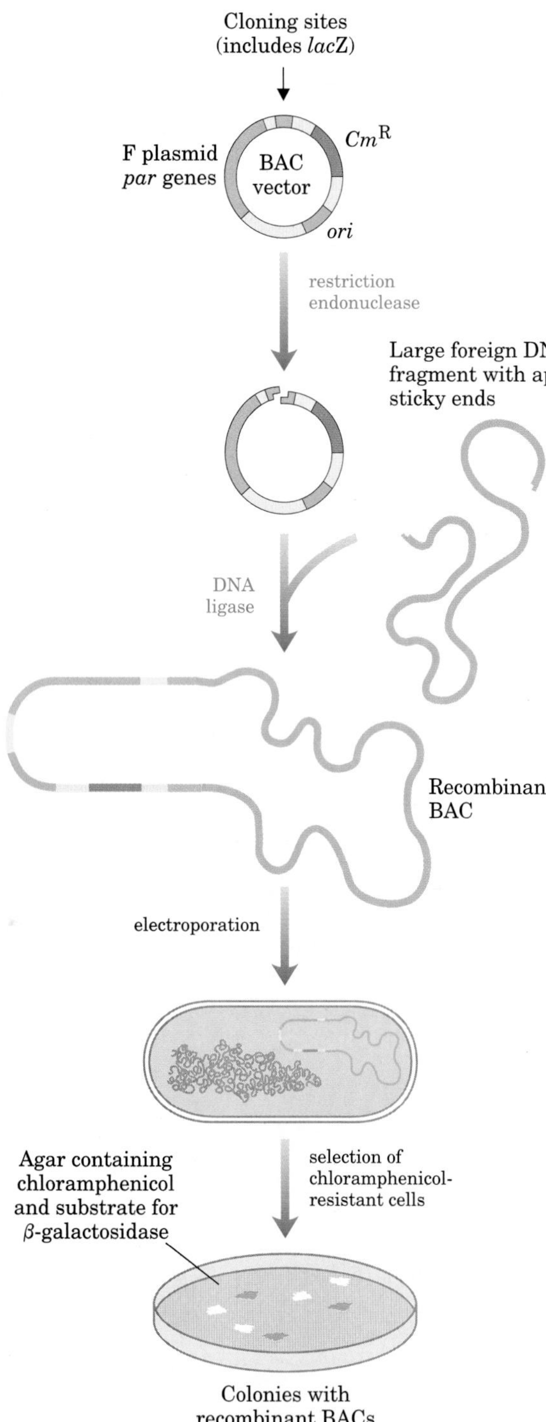

figure 29–8

Cloning with bacterial artificial chromosomes (BACs). The vector is a relatively simple plasmid, with a replication origin *ori* that directs replication. The *par* genes, derived from a type of plasmid called an F plasmid, facilitate the even distribution of plasmids to daughters cells at cell division. This increases the likelihood of each daughter cell carrying one copy of the plasmid, even when few copies are present. The low copy number is useful in cloning large segments of DNA because it limits the opportunities for unwanted recombination reactions that can unpredictably alter large cloned DNAs over time. Selectable markers are included in the BAC. A *lacZ* gene is situated in the cloning region such that it is inactivated by cloned DNA inserts. Introduction of recombinant BACs into cells by electroporation is facilitated by the use of cells with an altered (more porous) cell wall. Recombinant DNAs are screened for resistance to the antibioitic chloramphenicol (Cm). Plates also contain an artificial substrate for β-galactosidase that yields a colored product. Colonies with active β-galactosidase (and hence no DNA insert in the BAC vector) turn blue; colonies with the desired DNA inserts are white.

Bacterial Artificial Chromosomes Bacterial artificial chromosomes **(BACs),** are simply plasmids designed for the cloning of very long segments of DNA (Fig. 29–8). They generally include selectable markers such as chloramphenicol resistance (Cm^R), as well as a very stable replication origin (*ori*) that maintains the plasmid at one or two copies per cell. DNA fragments several hundred thousand base pairs in length are cloned into the BAC vector. The large circular DNAs are then introduced into host bacteria by electroporation. The bacteria used as hosts for recombinant BACs have mutations that compromise the structure of the bacterial cell wall, facilitating the uptake of these large DNA molecules.

Isolating a Gene from a Cellular Chromosome

Because a single gene is only a very small part of any chromosome, isolating a DNA fragment containing a particular gene often requires two procedures. First, a DNA library is constructed that contains many thousands of DNA fragments derived from a genome. Second, the DNA fragment containing the gene of interest is identified by taking advantage of the key property that distinguishes it from other DNA fragments—its sequence.

Cloning a Gene Often Requires a DNA Library

DNA libraries can take a variety of forms, depending on the source of the DNA. Among the most common is a **genomic library**—produced when the complete genome of a particular organism is cleaved into thousands of fragments, and *all* of them are cloned by insertion into a cloning vector. In the first step, DNA to be cloned is partially digested using a restriction endonuclease such that any given sequence will appear in fragments of a variety of sizes. A fragment size range is chosen that is both compatible with the cloning vector, and ensures that virtually all sequences will be represented among the clones within the library. Fragments that are too large or too small for cloning are removed by centrifugation or electrophoresis. The

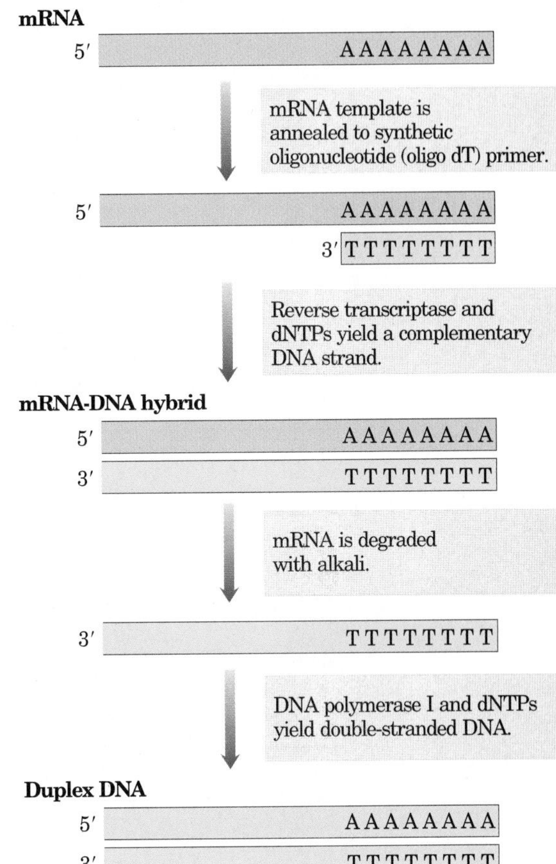

figure 29–9
Constructing a cDNA library from mRNA. In practice, the mRNA from a cell will include transcripts from thousands of genes, and the cDNAs generated will be correspondingly heterogeneous. The duplex DNA produced by this method is inserted into an appropriate cloning vector.

cloning vector is also cleaved with the same restriction endonuclease and ligated to the genomic DNA fragments. The ligated DNA mixture is used to transform bacterial cells or is packaged into bacteriophage particles (as described in Figs 29–6 to 29–8) to generate bacteria or bacteriophages each harboring a different recombinant DNA molecule. Ideally, all of the DNA in the genome will be represented in the library.

Each transformed bacterium grows into a colony or "clone" of identical cells, each bearing the same recombinant plasmid. When a bacteriophage vector has been used, each type of recombinant phage creates a clear region of lysed cells (a plaque) within a lawn of bacteria distributed evenly on an agar plate; all of the recombinant bacteriophages within a plaque are identical. The challenge then is to identify the clone that contains the particular gene of interest from among the thousands of clones in the library. To get an idea of the scale of the problem, consider the case of a mammal with a genome of 3×10^9 base pairs of DNA. If BACs are used as cloning vectors, and if the objective is a 99% probability that any desired gene of unique sequence will be represented in the library, then the library must contain about 50,000 recombinant BACs, each with a different insert of 300,000 base pairs.

The fragments in a genomic library derived from a higher eukaryote include not only genes but the noncoding DNA that makes up a large portion of many eukaryotic genomes. A more specialized and exclusive DNA library can be constructed that includes only those genes that are *expressed* in a given organism or even in certain cells or tissues. Expressed genes are those that are transcribed into RNA. The mRNA from an organism or certain cells derived from the organism is first extracted, and **complementary DNAs (cDNAs)** are then produced from the RNA in a multistep reaction catalyzed by reverse transcriptase (Fig. 29–9). The resulting double-stranded DNA fragments are then inserted into a suitable vector and cloned, creating a population of clones called a **cDNA library.** The search for a particular gene can be facilitated by focusing on a cDNA library generated from the mRNAs of a cell that expresses that gene. For example, the cloning of globin genes can be facilitated by first generating a cDNA library from erythrocyte precursor cells, where about half the mRNAs code for globins.

Specific DNA Sequences Can Be Amplified

The human genome project, along with the many associated efforts to sequence the genomes of organisms of every type, is providing unprecedented access to gene sequence information. Whereas the creation of one or more DNA libraries has often been an intermediate step in the sequencing of a genome, a gene can readily be cloned without the aid of a library once the genomic sequence is completed. If one knows the sequence of at least part of a DNA segment to be cloned, the number of copies of that DNA segment can be hugely amplified using the **polymerase chain reaction (PCR),** conceived by Kary Mullis in 1983. The amplified DNA can be cloned directly, or used in a variety of analytical procedures.

PCR has an elegant simplicity. Two synthetic oligonucleotides are synthesized, each complementary to sequences on opposite strands of the

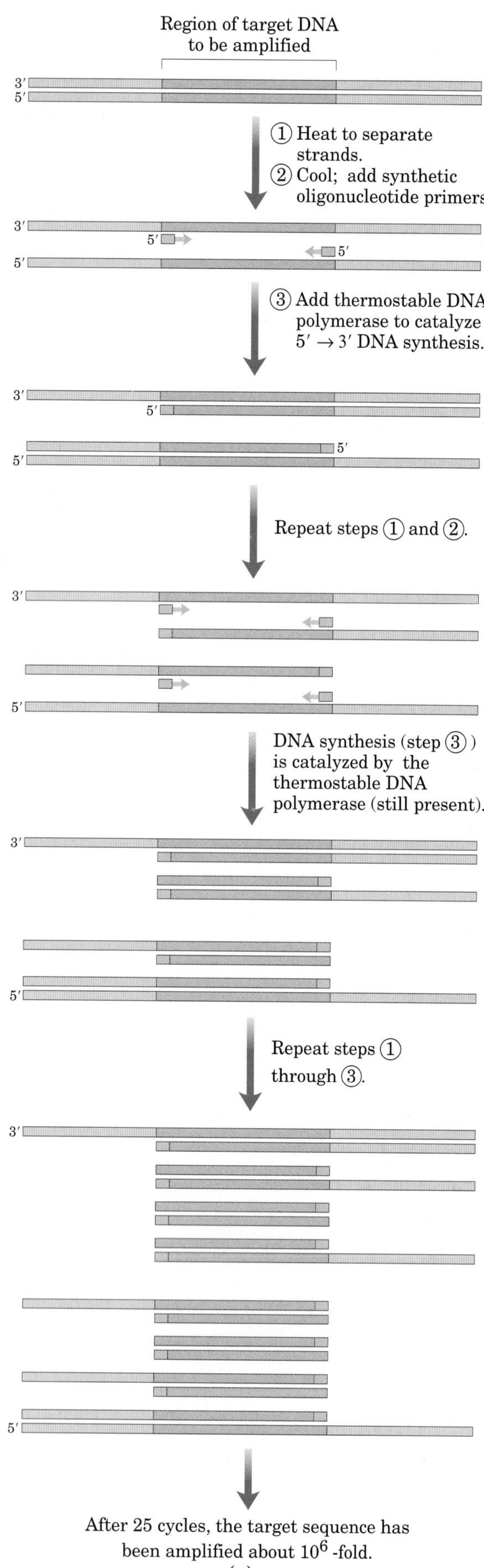

Region of target DNA to be amplified
3′
5′
① Heat to separate strands.
② Cool; add synthetic oligonucleotide primers.
③ Add thermostable DNA polymerase to catalyze 5′ → 3′ DNA synthesis.
Repeat steps ① and ②.
DNA synthesis (step ③) is catalyzed by the thermostable DNA polymerase (still present).
Repeat steps ① through ③.
After 25 cycles, the target sequence has been amplified about 10^6-fold.
(a)

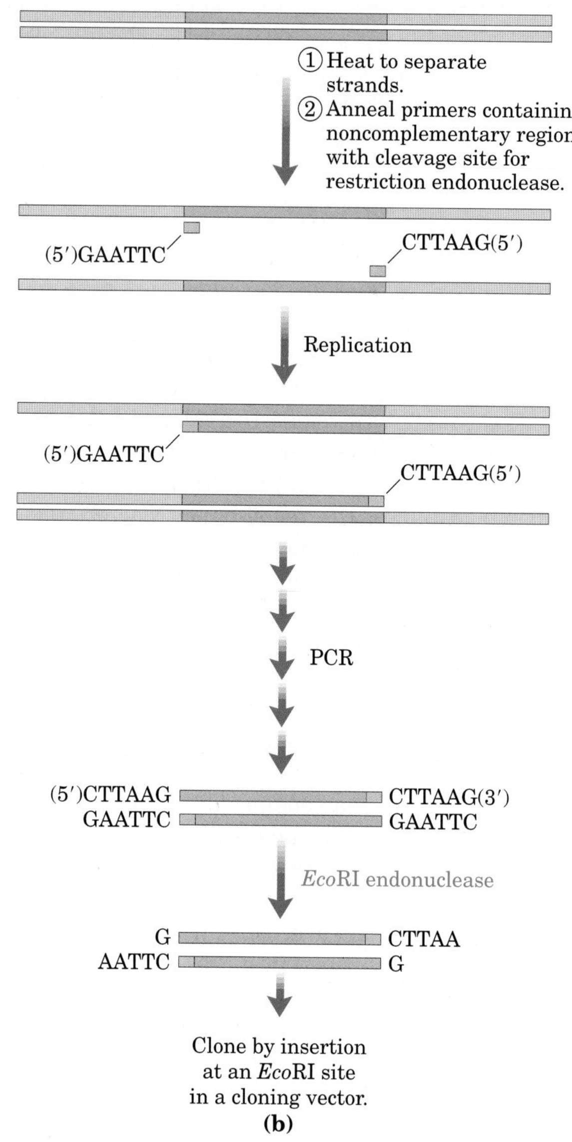

① Heat to separate strands.
② Anneal primers containing noncomplementary regions with cleavage site for restriction endonuclease.
(5′)GAATTC
CTTAAG(5′)
Replication
PCR
(5′)CTTAAG
GAATTC
CTTAAG(3′)
GAATTC
EcoRI endonuclease
G
AATTC
CTTAA
G
Clone by insertion at an EcoRI site in a cloning vector.
(b)

figure 29–10

Amplifying a specific DNA segment using the polymerase chain reaction. (a) The procedure has three steps. ① DNA strands are separated by heating, then ② annealed to an excess of short synthetic DNA primers (blue) that flank the region to be amplified. After polymerization (3), the process is repeated for 25 or 30 cycles. The thermostable DNA polymerase *TaqI* (from *Thermus aquaticus,* bacteria that grow in hot springs) is not denatured by the heating steps. **(b)** DNA amplified by PCR can be cloned. The primers can include noncomplementary DNA at the ends that have a site for cleavage by a restriction endonuclease. Although these parts of the primers do not anneal to the target DNA, the PCR process incorporates them into the DNA that is amplified. Cleavage of the amplified fragments at these sites creates sticky ends that facilitate the ligation of the amplified DNA into a cloning vector.

target DNA at positions just beyond the ends of the segment to be amplified. The oligonucleotides serve as replication primers, with the 3′ ends of the hybridized probes oriented toward each other and positioned to prime DNA synthesis across the desired DNA segment (Fig. 29–10).

Isolated DNA containing the segment to be amplified is heated briefly to denature it and then cooled in the presence of a large excess of the synthetic oligonucleotide primers. The four deoxynucleoside triphosphates are then added, and the primed DNA segment is replicated selectively. The cycle of heating, cooling, and replication is repeated 25 or 30 times over a few hours in an automated process, amplifying the DNA segment flanked by the primers until it can be readily analyzed and/or cloned. Heat-stable DNA polymerases, such as the *Taq*I polymerase (derived from a bacterium that lives at 90 °C) are used in PCR. The enzyme remains active after every heating step and does not have to be replenished. Careful design of the primers used for PCR, such as including restriction endonuclease cleavage sites, can facilitate the subsequent cloning of the amplified DNA (Fig. 29–10b).

The PCR method is sensitive enough to detect and amplify as little as one DNA molecule in almost any type of sample. Although DNA degrades slowly over time (p. 348), DNA has been cloned successfully with the aid of PCR from samples over 40,000 years old. The technique has been used to clone DNA fragments from the mummified remains of humans and extinct animals such as the woolly mammoth, creating the new fields of molecular archaeology and molecular paleontology. DNA from burial sites has been amplified by PCR and used to trace ancient human migrations. Epidemiologists can use PCR-enhanced DNA samples from human remains to trace the evolution of human pathogenic viruses. In addition to its usefulness for cloning DNA, PCR is a potent new tool in forensic medicine (Box 29–1). It is also being used for detection of viral infections before they cause symptoms and in prenatal diagnosis of a wide array of genetic diseases.

Hybridization Allows the Detection of Specific Sequences

DNA hybridization, introduced in Chapter 10 (Fig. 10–32), is the most common sequence-based process for detecting a particular gene or segment of nucleic acid. There are many variations of the basic method, most making use of a labeled (e.g., radioactive) DNA or RNA fragment, a **probe,** complementary to the DNA being sought. In one classic approach used to detect a particular DNA sequence within a DNA library, nitrocellulose paper is pressed onto an agar plate which has many individual bacterial colonies from the library, each containing a different recombinant DNA. Some cells from each colony adhere to the paper, forming a replica of the plate. The paper is treated with alkali to disrupt the cells and denature the DNA within, which remains bound to the region of the paper around the colony from which it came. The radioactive DNA probe is then added to the paper, where it anneals only to complementary DNA. After washing away any unannealed probe DNA, the hybridized DNA can be detected by autoradiography (Fig. 29–11).

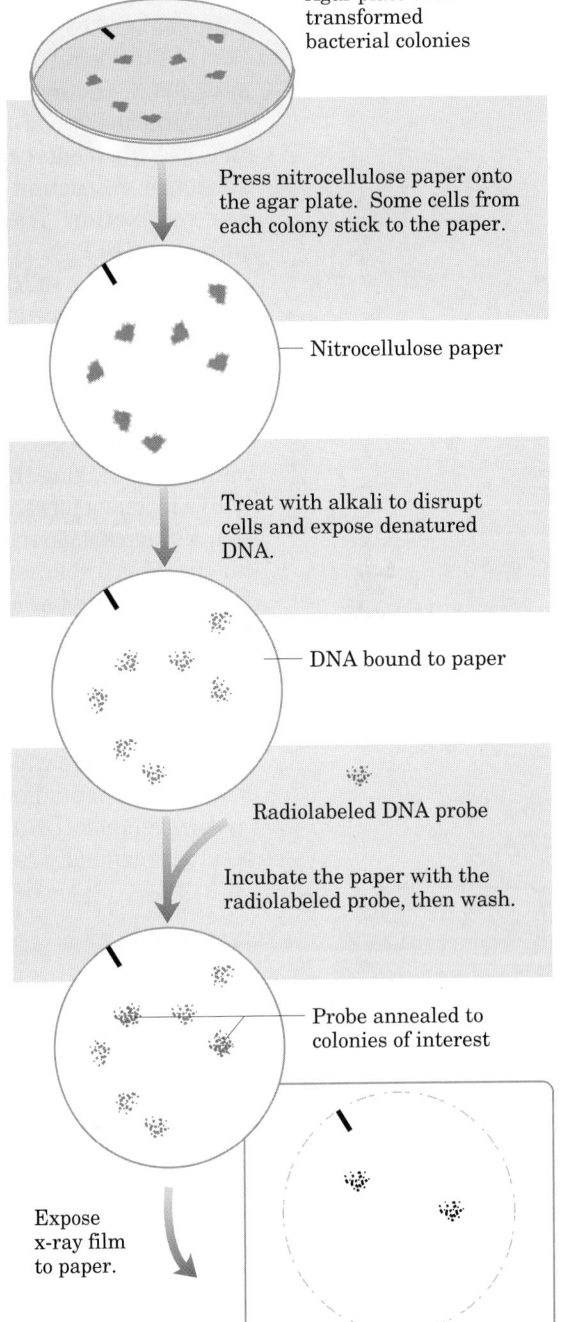
Agar plate with transformed bacterial colonies
Press nitrocellulose paper onto the agar plate. Some cells from each colony stick to the paper.
Nitrocellulose paper
Treat with alkali to disrupt cells and expose denatured DNA.
DNA bound to paper
Radiolabeled DNA probe
Incubate the paper with the radiolabeled probe, then wash.
Probe annealed to colonies of interest
Expose x-ray film to paper.

figure 29–11
Identifying a clone with a particular DNA segment using hybridization. The radioactive DNA probe hybridizes to complementary DNA, and is revealed by autoradiography. Once the labeled colonies have been identified, the corresponding colonies on the original agar plate can be used as a source of cloned DNA for further study.

box 29–1 A Potent Weapon in Forensic Medicine

Traditionally, one of the most accurate methods for placing an individual at the scene of a crime has been a fingerprint. A technique that is based on methods developed for recombinant DNA technology, **DNA fingerprinting** (also called DNA typing or DNA profiling), can be more powerful than any other identification method.

DNA fingerprinting is based on **sequence polymorphisms.** These are slight sequence differences (usually single base-pair changes) that occur from individual to individual once every 500 to 1,000 base pairs, on average. Each difference from the consensus human genome sequence occurs in some fraction of the human population; every individual has some of them. Some of the sequence changes affect recognition sites for restriction enzymes, resulting in variation from individual to individual in the size of the DNA fragments produced by digestion with a particular restriction enzyme. These variations are **restriction fragment length polymorphisms (RFLPs).**

The detection of RFLPs relies on a specialized hybridization procedure called **Southern blotting** (Fig. 1). DNA fragments from digestion of genomic DNA by restriction endonucleases are first separated according to size by electrophoresis in an agarose gel. The DNA fragments are denatured by soaking the gel in alkali, then blotted onto a nylon membrane to reproduce the distribution of fragments in the gel. The membrane is then immersed in a solution containing a radioactively labeled DNA probe. A probe for a sequence that is repeated several times in the human genome generally identifies a few of the thousands of DNA fragments generated when the human genome is digested with a restriction endonuclease. Fragments to which the probe hybridizes are revealed by autoradiography, as in Figure 29–11.

The genomic DNA sequences used in these tests are generally regions containing repetitive DNA (short sequences repeated thousands of times in tandem; see p. 909), which are common in the genomes of higher eukaryotes. The number of repeated units in such DNA varies between individuals (unless they are identical twins). If a suitable probe is chosen, the pattern of bands produced by such an experiment is distinctive for each individual. Combining the use of several probes makes the test so selective that it can positively identify a single individual in the entire human population. However, the Southern blot procedure requires relatively fresh DNA samples and larger amounts of DNA than are generally present at a crime scene. RFLP analysis sensitivity is augmented by using PCR (see Fig. 29–10a) to permit vanishingly small amounts of DNA to be amplified. This allows DNA fingerprints to be obtained from a single hair, a drop of blood, a small semen sample from a rape victim, or from samples that might be months or even many years old.

These methods are now proving decisive in court cases worldwide. In the example in Figure 1, the DNA from a semen sample obtained from a rape and murder victim was compared to DNA samples from the victim and two suspects. Each of the DNA samples was cleaved into fragments and separated by gel electrophoresis. Radioactive DNA probes were used to identify a small subset of these fragments that contained sequences complementary to the probe. The sizes of the fragments identified varied from one individual to the next, as seen here in the different patterns for the three individuals (victim and two suspects) tested. One suspect's DNA exhibits a banding pattern identical to that of a semen sample taken from the victim. A single probe was used here, but three or four different probes would be used (in separate experiments) to achieve an unambiguous positive identification.

Frequently, the limiting step in detecting and/or cloning a gene is the generation of a complementary strand of nucleic acid to use as a probe. The origin of a probe depends on what is known about the gene under investigation. Sometimes a homologous gene cloned from another species can be used as a probe. Alternatively, if the protein product of a gene has been purified, probes can be designed and synthesized on the basis of its amino acid sequence and a knowledge of the genetic code (Fig. 29–12). Increasingly, the necessary DNA sequence information can be obtained from sequence databases that detail the structure of millions of genes from a wide range of organisms.

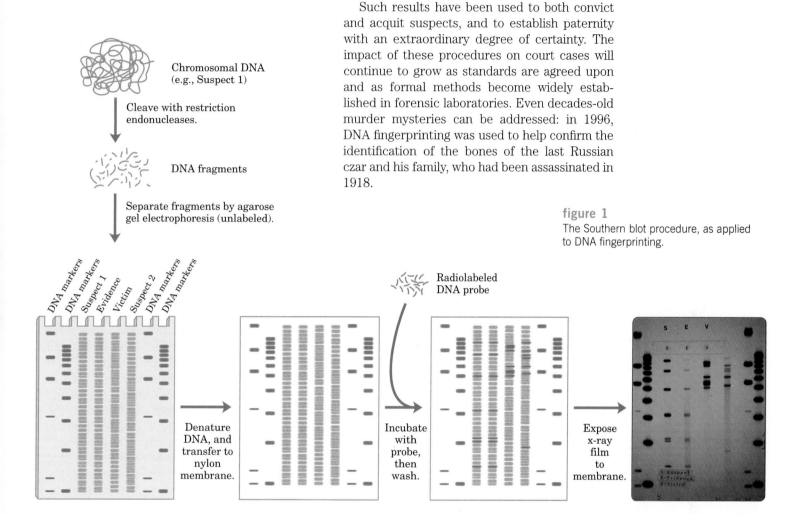

Such results have been used to both convict and acquit suspects, and to establish paternity with an extraordinary degree of certainty. The impact of these procedures on court cases will continue to grow as standards are agreed upon and as formal methods become widely established in forensic laboratories. Even decades-old murder mysteries can be addressed: in 1996, DNA fingerprinting was used to help confirm the identification of the bones of the last Russian czar and his family, who had been assassinated in 1918.

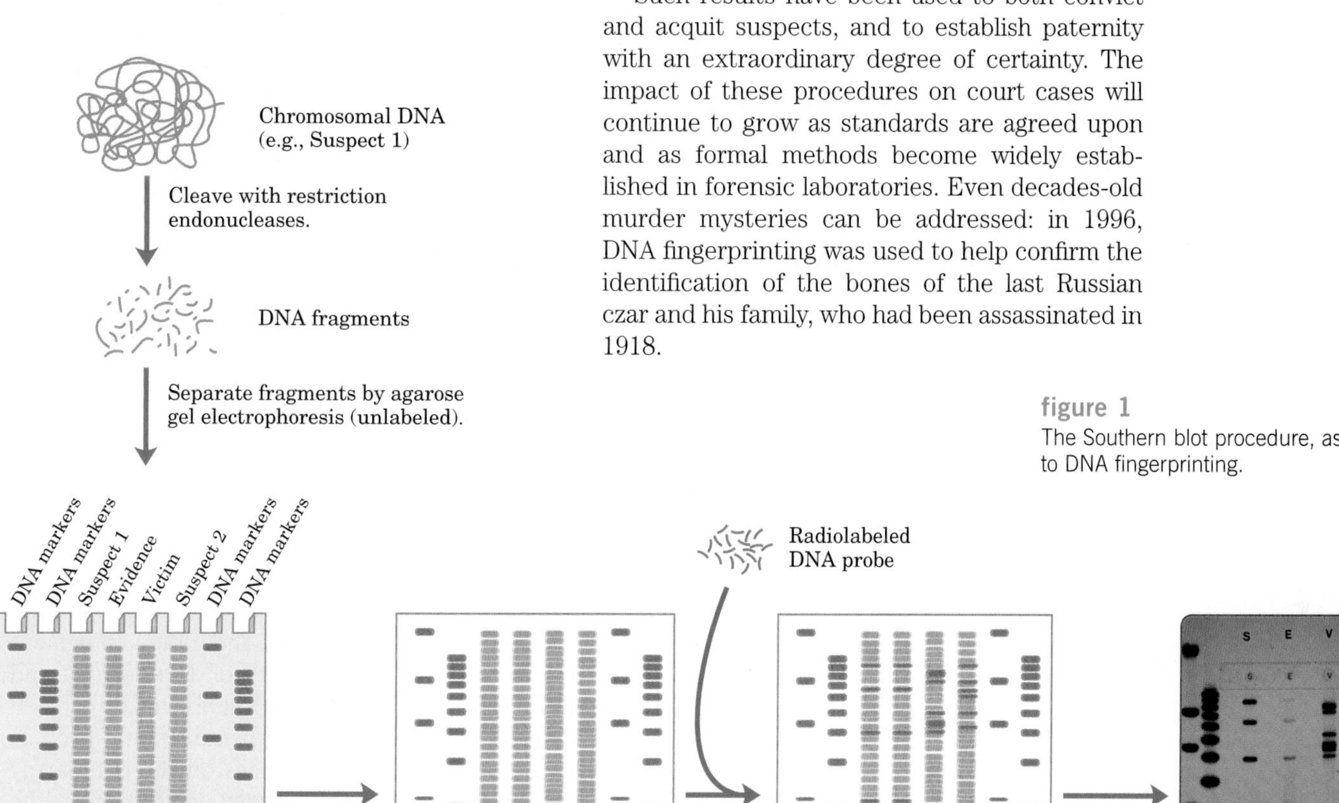

figure 1
The Southern blot procedure, as applied to DNA fingerprinting.

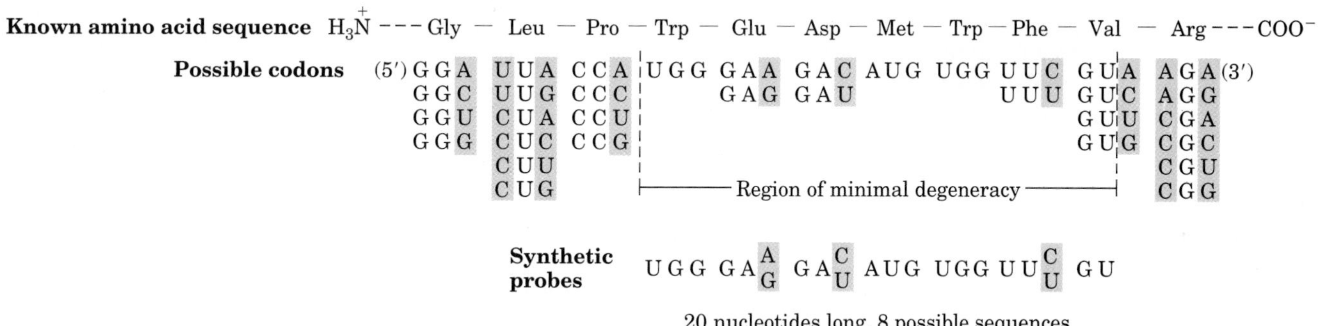

figure 29–12
Designing a probe to detect the gene for a protein of known amino acid sequence. The degeneracy of the genetic code means that more than one DNA sequence can code for any given amino acid sequence. As the correct DNA sequence cannot be known in advance, the probe is designed to be complementary to a region of the gene with minimal degeneracy (fewest possible codons). Oligonucleotides are synthesized with selectively randomized sequences, so that they contain either of the two possible nucleotides at each position of potential degeneracy (shaded in red). In the example shown, the synthesized oligonucleotide is actually a mixture of eight different sequences: one of the eight will complement the gene perfectly, and all eight will match at least 17 of 20 positions.

DNA Microarrays Provide Compact Libraries for Studying Genes and Their Expression

The explosion of DNA sequence information from genome sequencing projects has revealed a sobering truth. Despite many years of biochemical advances, there are still thousands of genes in every eukaryotic cell (and quite a few even in bacteria) that we know nothing about. New methods for rapidly screening genomic sequence data and obtaining clues about gene function are needed. Major refinements of the technology underlying DNA libraries, PCR, and hybridization have come together in the development of **DNA microarrays** (sometimes called DNA chips) that allow the rapid and simultaneous screening of many thousands of genes. DNA segments from known genes, a few dozen to hundreds of nucleotides in length, are placed on a solid surface, using robotic devices that accurately deposit nanoliter quantities of DNA solution. Many thousands of such spots are deposited in a predesigned array on a surface measuring only a few square centimeters. A dramatic example of this technique appears at the beginning of Part IV of this book (p. 904). Segments from each of the more than 6000 genes in the completely sequenced yeast genome were separately amplified by PCR, and each was deposited in a defined pattern to create that microarray. An alternative strategy is to synthesize DNA directly on the solid surface.

A microarray can answer such questions as which genes are expressed at a given stage in the development of an organism. The total complement of mRNA is isolated from cells at two different stages of development and converted to cDNA using reverse transcriptase and fluorescently labeled deoxynucleotides. The fluorescent cDNAs can then be mixed and used as probes, each hybridizing to complementary sequences on the microarray (Fig. 29–13). In this example, the labeled nucleotides used to make the cDNA for each sample fluoresce in two different colors. The cDNA from the two samples is mixed and used to probe the microarray. Spots that fluoresce green represent mRNAs more abundant at the single cell stage; those that fluoresce red represent sequences more abundant later in development. Using a mixture of two samples to measure relative rather than absolute abundance of sequences corrects for variations in the amounts of DNA originally deposited in each spot on the grid and other inconsistencies that might exist from spot to spot in a microarray. The spots that fluoresce provide a snapshot of all the genes being expressed in the cells at the moment they were harvested—gene expression examined on a genome-wide scale. For a gene of unknown function, the time and circumstances of its expression can provide important clues about its role in the cell.

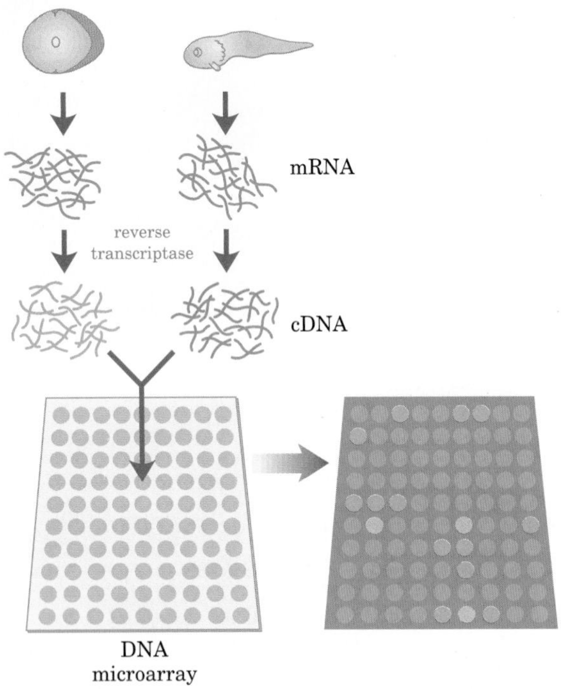

figure 29–13
Constructing a DNA microarray. Any known DNA sequence, from any source, can be used in a microarray. The DNA can be generated by chemical synthesis or PCR. The DNA is positioned on a solid surface (usually specially treated glass slides) with the aid of a robotic device capable of depositing very small drops (nanoliters) in precise patterns. UV light is then used to cross-link the DNA to the glass slides. Once the DNA is attached to the surface, the microarray can be probed with other fluorescently labeled nucleic acids. For example, the mRNA isolated from a cell (representing all the genes being expressed in that cell) can be converted to cDNA probes by reverse transcriptase, using fluorescently labeled dNTPs. The fluorescent cDNAs anneal to complementary sequences on the microarray. After removal of unhybridized probe, each spot that flouresces represents a gene that was being expressed in the sample.

Here, mRNA samples are collected from cells at two different stages in the development of a frog. The cDNA probes for each sample are made with nucleotides that fluoresce in different colors; a mixture of the cDNAs is used to probe the microarray. Spots that fluoresce green represent mRNAs more abundant at the single-cell stage, whereas spots that fluoresce red represent sequences more abundant later in development.

Applications of Recombinant DNA Technology

Cloning a gene is normally only the first step in a much grander design, such as the production of large amounts of its protein product. The amino acid sequence of the protein in question can be altered by introducing base-pair changes in the gene, a strategy that can be very powerful in exploring protein folding, structure, and function. Increasingly sophisticated methods for moving DNA into and out of cells of all types provide new avenues for studying gene function and regulation, and for the introduction of new traits into plants and animals.

Our focus now turns to applications of DNA cloning, beginning with the proteins produced by cloned genes. We then describe cloning procedures used for a variety of eukaryotic cells, before finishing with an overview of the potential and implications of this technology.

Cloned Genes Can Be Expressed

Frequently it is the product of the cloned gene, rather than the gene itself, that is of primary interest—particularly when the protein has commercial, therapeutic, or research value. Understanding the fundamentals of DNA, RNA, and protein metabolism and their regulation in *E. coli* has made it possible to express cloned genes in order to study their protein products.

Most eukaryotic genes lack the DNA sequence elements (promoters, etc.) required for their expression in *E. coli* cells, so bacterial regulatory sequences for transcription and translation must be inserted at appropriate positions relative to the eukaryotic gene in the vector DNA. Cloned genes can be so efficiently expressed in some cases that their protein product may represent 10% or more of the cellular protein. Some foreign proteins can kill an *E. coli* cell when the concentration is that high, requiring gene expression to be limited to the few hours before the planned harvest of the cells.

Cloning vectors with the transcription and translation signals needed for such regulated expression of a cloned gene are often called **expression vectors.** The rate of expression of the cloned gene is controlled by replacing the gene's own promoter and regulatory sequences with more efficient and convenient versions supplied by the vector. Generally, a well-characterized promoter and its regulatory elements are positioned near several unique restriction sites for cloning, so that genes inserted at the restriction sites will be expressed from the regulated promoter (Fig. 29–14). Some of these vectors incorporate other features, such as a bacterial ribosome binding site to enhance translation of the mRNA derived from the gene, and/or a transcription termination sequence. Overexpression of cloned genes in bacteria and other cells can provide large quantities of specific proteins, which can simplify the work of researchers or provide high yields for industry.

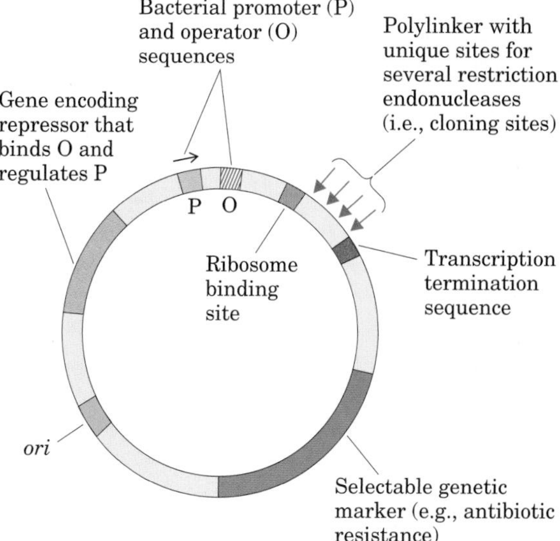

figure 29–14

Types of DNA sequences found in a typical *E. coli* expression vector. The gene to be expressed is inserted into one of the restriction sites in the polylinker, near the promoter, with the end encoding the amino terminus proximal to the promoter. The promoter allows efficient transcription of the inserted gene, and the transcription termination sequence sometimes improves the amount and stability of the mRNA produced. The operator permits regulation by means of a repressor that binds to it (p. 1075). The ribosome-binding site provides sequence signals needed for efficient translation of the mRNA derived from the gene. The selectable marker allows the selection of cells containing the recombinant DNA.

Cloned Genes Can Be Altered

Cloning techniques can be used not only to overproduce proteins but to produce protein products subtly altered from their native forms. Specific amino acids may be replaced individually by **site-directed mutagenesis.** This powerful approach to studying protein structure and function changes the amino acid sequence of a protein by altering the DNA sequence of the cloned gene.

If appropriate restriction sites flank the sequence to be altered, a change can be made by simply removing a DNA segment and replacing it with a synthetic one that is identical to the original except for the desired change (Fig. 29–15a). When suitably located restriction sites are not present, an approach called **oligonucleotide-directed mutagenesis** (Fig. 29–15b) can be used to create a specific DNA sequence change. A short synthetic DNA strand with a specific base change is annealed to a single-stranded copy of the cloned gene within a suitable vector. The mismatch of a single base pair out of 15 to 20 does not prevent annealing if it is done at an appropriate temperature. This annealed strand serves as a primer for the synthesis of a strand complementary to the plasmid vector. This slightly mismatched duplex recombinant plasmid is used to transform bacteria, where the mismatch is repaired by cellular DNA repair enzymes. About half of the repair events will remove and replace the altered base and restore the gene to its original sequence; the other half will remove and replace the *normal* base, retaining the desired mutation. Transformants are screened (often simply by sequencing their plasmid DNA) until a bacterial colony containing a plasmid with the altered sequence is found.

Changes can also be introduced that involve more than one base pair. Large parts of a gene can be deleted by cutting out a segment with restriction endonucleases and ligating the remaining portions to form a smaller gene. Parts of two different genes can be ligated to create new combinations. The product of such a fused gene is called a **fusion protein** (unrelated to the fusion proteins introduced in Chapter 12 that participate in membrane fusion.)

Ingenious methods exist to bring about virtually any genetic alteration in vitro. Reintroduction into the cell allows the consequences of the alteration to be investigated. Site-directed mutagenesis has greatly facilitated research on proteins by allowing investigators to make specific changes in the primary structure of a protein and to examine the effects of these changes on the folding, three-dimensional structure, and activity of the protein. These methods are being used commercially to create proteins with enhanced activity, or with the ability to function in extremes of temperature and pH or in harsh environments such as organic solvents.

figure 29–15

Two approaches to site-directed mutagenesis. (a) A DNA segment is synthesized and is used to replace a DNA fragment that has been removed by cleavage with a restriction endonuclease. **(b)** An oligonucleotide is synthesized with a desired sequence change at one position. This is hybridized to a single-stranded copy of the gene to be altered, and acts as primer for synthesis of a duplex DNA (with one mismatch) that is then used to transform cells. Cellular mismatch repair will generally convert about 50% of the mismatches to reflect the desired sequence change.

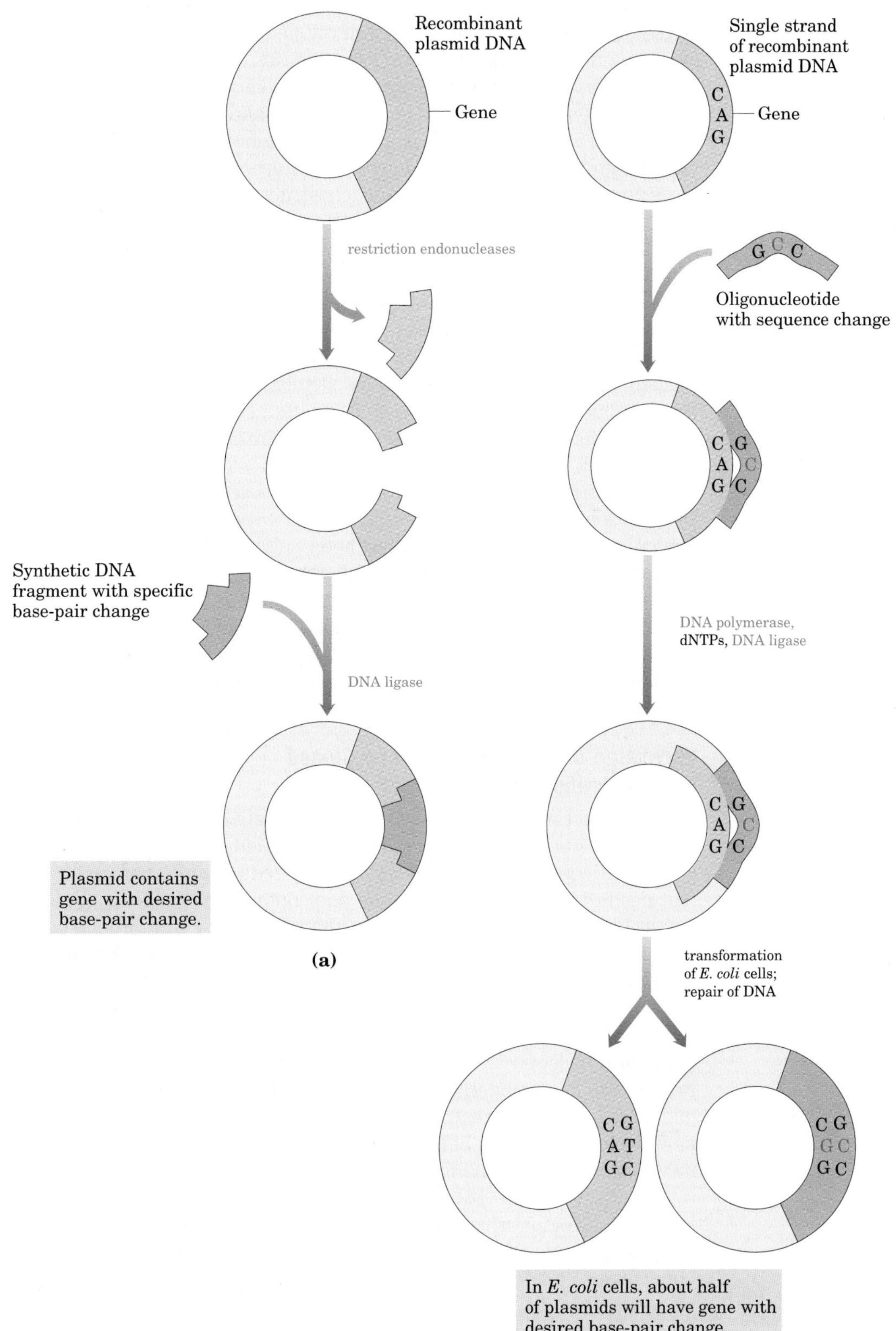
Recombinant
plasmid DNA
Gene
restriction endonucleases
Synthetic DNA
fragment with specific
base-pair change
DNA ligase
Plasmid contains
gene with desired
base-pair change.
(a)
Single strand
of recombinant
plasmid DNA
C
A
G
Gene
G C C
Oligonucleotide
with sequence change
C G
A C
G C
DNA polymerase,
dNTPs, DNA ligase
C G
A C
G C
transformation
of E. coli cells;
repair of DNA
C G
A T
G C
C G
G C
G C
In E. coli cells, about half
of plasmids will have gene with
desired base-pair change.
(b)

Yeast Is an Important Eukaryotic Host for Recombinant DNA

E. coli cells are by no means the only host cell for genetic engineering. Yeasts are particularly convenient eukaryotic organisms for this work: as with *E. coli,* yeast genetics is a well-developed discipline; the genome of the most commonly used yeast, *Saccharomyces cerevisiae,* contains only 14×10^6 base pairs (a simple genome by eukaryotic standards, less than fourfold larger than the *E. coli* chromosome) and its sequence is known in its entirety; and finally, yeast is a microorganism that is very easy to maintain and grow on a large scale in the laboratory.

Expression vectors have been constructed for yeast employing the same principles that govern use of *E. coli* vectors described above. Expression of eukaryotic genes in yeast can be preferable to expression in *E. coli.* Many eukaryotic proteins are normally modified after translation by enzymes not present in bacteria, so eukaryotic proteins produced in bacteria may lack modifications essential for their activity. Convenient methods now available for moving DNA into and out of yeast cells facilitate the study of many aspects of eukaryotic cell biochemistry. In transformed yeast cells, the introduced DNA may be integrated into a cellular chromosome by homologous recombination (Chapter 25). This **integrative transformation** occurs at low frequency.

Transformation efficiencies can be increased by introducing cloned DNA on a plasmid capable of replication in yeast. The naturally occurring 2 micron (2 μm) yeast plasmid has been engineered to create a variety of cloning vectors that include a replication origin and other sequences needed for plasmid maintenance in yeast. Some recombinant plasmids incorporate multiple replication origins and other elements that allow them to be maintained in more than one species (for example, yeast or *E. coli*). Plasmids that can be propagated in cells of two or more different species are called **shuttle vectors.**

Very Large DNA Segments Can Be Cloned in Yeast Artificial Chromosomes

A cloning vector with the capacity to incorporate larger DNA segments can reduce the number of individual clones needed to archive an organism's genome in a genomic DNA library, thereby reducing the number of clones that need to be screened to identify one containing a particular gene of interest. Work with large genomes and the associated need for high capacity cloning vectors led to the development of **yeast artificial chromosomes** (**YACs,** Fig. 29–16).

YAC vectors contain all of the elements needed to maintain a eukaryotic chromosome in the yeast nucleus (Fig. 24–3). These include a yeast origin of replication, a centromere, telomeres, and two selectable markers. Prior to use in cloning, the vector is propagated as a circular bacterial plasmid. Cleavage with a restriction endonuclease (*Bam*H1 in Fig. 29–16) removes a length of DNA between two telomere sequences, leaving the telomeres at the ends of the linearized DNA. Cleavage at another internal site (at *Eco*RI in Fig. 29–16) divides the vector into two DNA segments, referred to as vector arms, each with a different selectable marker.

The genomic DNA is prepared by partial digestion with restriction endonucleases (*Eco*RI in Fig. 29–16). Genomic fragments are then separated by **pulsed field gel electrophoresis,** a variation of gel electrophoresis (Fig. 5–19) that allows the separation of very large DNA segments. The DNA fragments of appropriate size (up to about 2 million base pairs) are mixed with the prepared vector arms and ligated. The ligation mixture is then used to transform treated yeast cells with very large DNA molecules. Culture on a medium that requires the presence of both selectable marker genes ensures that a yeast cell grows only if it contains an artificial chro-

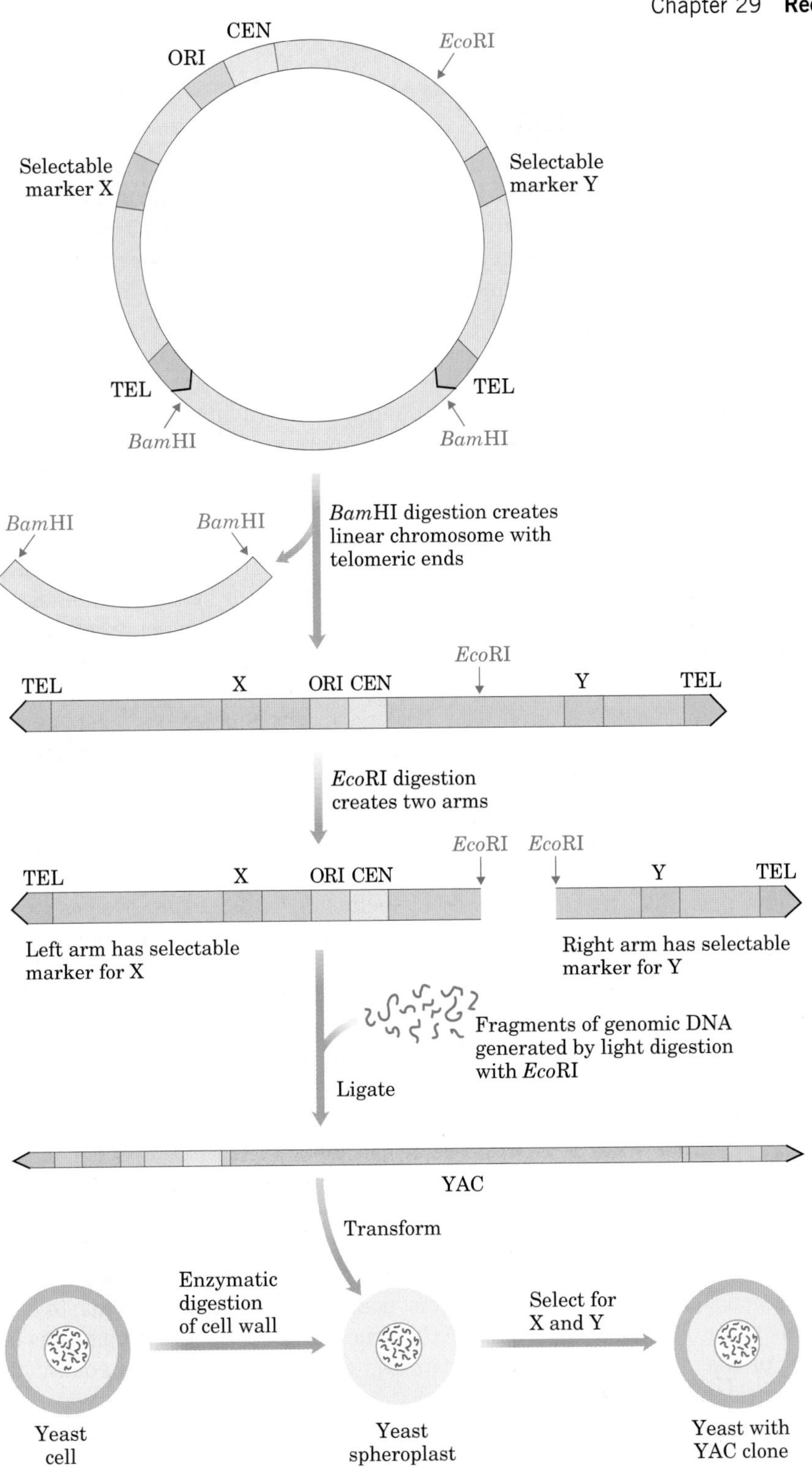

figure 29–16

Constructing a yeast artificial chromosome (YAC). A YAC vector includes an origin of replication (ORI), a centromere (CEN), two telomeres (TEL), and selectable markers (X and Y). Digestion with both *Bam*H1 and *Eco*RI generates two separate DNA arms, each with a telomeric end and one selectable marker. A large segment of DNA (for example, up to 2×10^6 base pairs from the human genome) can be ligated to the arms to create a yeast artificial chromosome. The YAC can be used to transform prepared yeast cells, which are selected for X and Y. The surviving yeast cells propagate the DNA insert.

mosome with a large insert sandwiched between the two vector arms (Fig. 29–16). The stability of YAC clones increases with size up to a point. Those with inserts of more than 150,000 base pairs are nearly as stable as normal cellular chromosomes, whereas those with inserts of less than 100,000 base pairs are gradually lost during mitosis (so yeast cell clones carrying only the two vector ends ligated together or with only short inserts are generally not found). YACs that lack a telomere at either end are rapidly degraded.

YAC DNA libraries and clones are widely used to study large genomes. They have become particularly important in the effort to sequence the entire human genome and the associated effort to isolate genes that are at the heart of many human diseases (Box 29–2).

box 29–2 The Human Genome and Human Gene Therapy

Sometime in the not too distant future, a human patient will be diagnosed with adenosine deaminase deficiency, a deleterious genetic condition that affects the immune system. The treatment will be innoculation with a genetically engineered variant of HIV (the virus that causes AIDS) that has been stripped of the genes it normally requires to reproduce itself. This engineered virus will be completely benign, revealing its HIV origin only by its protein coat.

Once in the bloodstream, however, that protein coat will facilitate the targeted uptake of the engineered virus by immune system cells, where the coat will be shed to expose a recombinant DNA molecule with a functional adenosine deaminase gene. An efficient genetic recombination system packaged in the engineered virus particle will integrate this DNA into a specific location on a chromosome, without disrupting any other cellular function. A powerful promoter will then command the expression of the integrated gene, restoring the immune system to normal function. A genetic disease will have been cured.

This scenario, or something quite like it, is destined to become part of everyday medicine.

The first human gene therapy trial was carried out at the National Institutes of Health in Bethesda, Maryland in 1990. The patient was a four-year-old girl whose immune system had been crippled by the adenosine deaminase deficiency used in our fictional example. This rare human condition was chosen as a target of early trials because it met some important criteria. The genetic defect is well characterized and, in principle, could be addressed by correcting a single gene. The functional gene can operate even in the presence of the dysfunctional gene, so it is not necessary to eliminate the latter. Finally, and most important, the effects of the disease are sufficiently serious that the considerable risks inherent in a new technology are outweighed by the potential for a cure.

In the 1990 trial, as in many subsequent ones, bone marrow cells were removed from the patient and transformed in a laboratory with an engineered retrovirus containing a functional adenosine deaminase gene, then reintroduced into the patient's marrow. Four years later, the child was leading an apparently normal life, going to school and even testifying about her experiences before Congress. In spite of this apparent success, unambiguous positive results have been elusive and even the adenosine deaminase trials have not been decisive. The children who have received gene therapy for this condition are also given regular injections of synthetic adenosine deaminase, making it unclear which treatment is primarily responsible for the positive clinical outcome.

The trials continue, fueled by ever more-imaginative protocols for treating new conditions, and encouraging the development of a new biotechnology industry dedicated to human gene therapy. The many obstacles remaining to routine and effective gene therapy are being addressed by technological advances, even as more diseases are identified that could in principle be treated with these emerging techniques. The genes involved in many human diseases have already been tracked down, including those responsible for Huntington's disease, cystic fibrosis, familial hypercholesterolemia, and various genetic conditions conferring an inherited predisposition to cancer (see Box 25–1). A major engine for progress is the ongoing human genome project. Scheduled for completion shortly after the turn of the millenium, this worldwide effort to sequence the entire human genome is complemented by programs to sequence the genomes of many other organisms of interest to the research and medical communities.

A formidable obstacle to gene therapy is the efficient delivery of DNA to affected cells. Retro-

Cloning in Plants Is Aided by a Bacterial Plant Parasite

The introduction of recombinant DNA into plants has enormous implications for agriculture, making it possible to alter the nutritional profile or yield of crops or their resistance to environmental stresses, such as insect pests, disease, cold, salinity, and drought. Fertile plants of some species may be generated from a single transformed cell, enabling an introduced gene to be transmitted to progeny through seed.

viral vectors can introduce DNA to cells efficiently, but this DNA reaches and integrates into host chromosomes only when the nuclear envelope breaks down transiently during cell division. Cells that divide infrequently or not at all, such as mature neurons and skeletal muscle cells, cannot be transformed easily by these vectors. Furthermore, retroviral vectors inevitably introduce DNA into many cell types that undergo regular division but are not appropriate targets for the therapy, and since the DNA is integrated at random locations on the host chromosomes, there is a possibility that some critical gene will be disrupted. These factors limit the scope for administering the engineered virus directly to the patient, as opposed to transforming cells in vitro and then returning them to the treated individual. However, some vectors can be targeted to particular tissues by viral coat proteins that are bound by the receptors of target cells only.

Alternatives to retrovirus vectors are being developed. Engineered adenoviruses can carry large segments of recombinant DNA, but they lack a recombination system for integrating the DNA into the genome. Hence, the introduced genes tend to be expressed only transiently. This could be useful when an expressed protein, such as a coat protein from a pathogenic bacterium, is needed only transiently to induce a strong immune response to the pathogen. Unfortunately, adenoviruses themselves trigger a strong immune response, which can blunt the effectiveness of repeated treatments. Methods are also needed to target adenoviruses to particular tissues. Synthetic liposomes have no inherent capacity to cause disease, but they are less efficient than viruses in transferring recombinant DNA to cells. The range of options for gene therapy vectors will grow as new viruses with useful properties are investigated and harnessed, and as liposome technology is improved. Advances in understanding genetic recombination and gene expression will also enhance the design of treatment vectors.

Human gene therapy will not be limited to genetic diseases. Treatment could be targeted at cancer cells, delivering genes for proteins that might destroy the cell or restore the normal control of cell division. Immune system cells associated with tumors, called tumor-infiltrating lymphocytes, can be genetically modified to produce tumor necrosis factor (TNF; see p. 477). When these lymphocytes are taken from a cancer patient, modified, and then reintroduced, the engineered cells target the tumor, and the TNF they produce facilitates tumor shrinkage. AIDS could also potentially be treated with gene therapy; DNA that encodes an RNA molecule complementary to a vital HIV mRNA could be introduced into immune system cells (the target of HIV). This RNA would be transcribed from the introduced DNA and pair with the HIV mRNA, preventing its translation and interfering with the virus life cycle. Alternatively, a gene could be introduced that encodes an inactive form of a multisubunit HIV enzyme; the entire enzyme might be inactivated if one subunit is nonfunctional.

Our growing understanding of the human genome and the genetic basis for some diseases brings the promise of early diagnosis and constructive intervention. It also brings new ethical quandaries. Tests for a wide range of genetic conditions exist that can predict our future health, yet often there are no effective treatment protocols with which to respond meaningfully to this information. Adequate safeguards are needed to protect against genetic discrimination in access to insurance or employment. The advance of this technology places increasing responsibility on both individuals and society as a whole to find ways to protect the privacy of citizens and to avoid the establishment of a genetic underclass of individuals known to carry potentially undesirable genetic attributes.

Naturally occurring plant cell plasmids have not yet been found to facilitate cloning in plants, so the biggest technical challenge is getting DNA into plant cells. An important and adaptable ally in this effort is the soil bacterium *Agrobacterium tumefaciens.* This bacterium can invade plants at the site of a wound, transforming nearby cells and inducing them to form a tumor called a crown gall. *Agrobacterium* contains the large (~200,000 base pair) **Ti plasmid** (Fig. 29–17a). When the bacterium contacts a damaged plant cell, a 23,000 base pair segment of the Ti plasmid called the T DNA is transferred from the plasmid and integrated at a random position in one of the plant cell's chromosomes. The transfer of T DNA from *Agrobacterium* to the plant cell chromosome relies on two 25 base pair repeats that flank the T DNA and on the products of the virulence *(vir)* genes, also on the Ti plasmid (Fig. 29–17a).

The T DNA encodes enzymes that convert plant metabolites to two classes of compounds that benefit the bacterium (Fig. 29–18). The first class produces plant growth hormones (auxins and cytokinins) that stimulate growth of the transformed plant cells to form the crown gall tumor. The second class generates a series of unusual amino acids called opines, which serve as a food source for the bacterium. The opines are produced in high concentrations in the tumor cells and secreted to the surroundings, where they can be metabolized only by *Agrobacterium* using enzymes encoded elsewhere on the Ti plasmid. The bacterium thereby diverts plant resources by converting them to a form that benefits only itself.

This rare example of DNA transfer from a prokaryote to a eukaryotic cell is a natural genetic engineering process—one that can be harnessed to transfer recombinant DNA (instead of T DNA) to the plant genome. A common cloning strategy employs an *Agrobacterium* with two different recombinant plasmids. The first is a Ti plasmid from which the T DNA segment has been removed in the laboratory (Fig. 29–19a). The second is an *Agrobacterium–E. coli* shuttle vector in which the 25 base pair repeats of the T DNA flank the foreign gene that a researcher wants to introduce into the plant cell, along with a selectable marker such as kanamycin resistance (Fig. 29–19b). The engineered *Agrobacterium* is used to infect a leaf, but crown galls are not formed because the T DNA genes for auxin, cytokinin, and opine biosynthetic enzymes are absent from both plasmids. Instead, the *vir* gene products from the altered Ti plasmid direct the transformation of the plant cells by the gene flanked by the T DNA 25 base pair repeats in the second plasmid. The transformed plant cells can be selected by growth on agar plates that contain kanamycin and are induced with growth hormones to form new plants that contain the foreign gene in every cell.

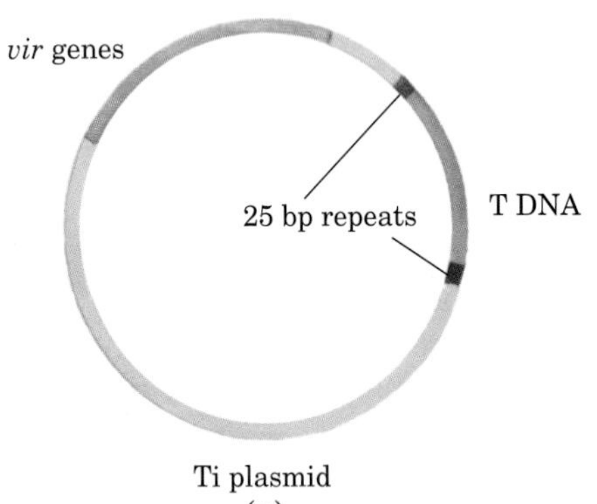
vir genes
25 bp repeats
T DNA
Ti plasmid
(a)

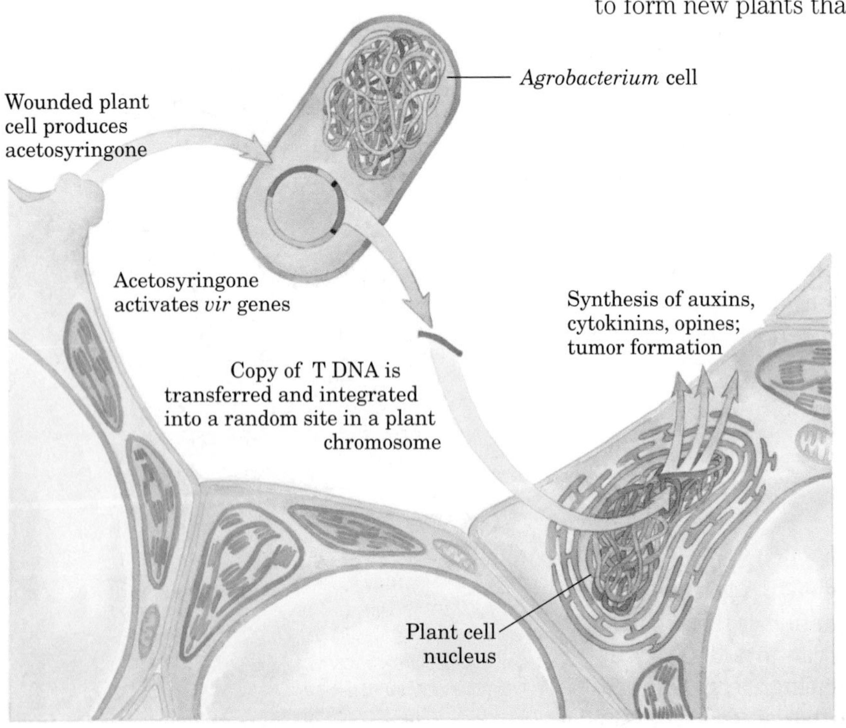
Agrobacterium cell
Wounded plant cell produces acetosyringone
Acetosyringone activates *vir* genes
Synthesis of auxins, cytokinins, opines; tumor formation
Copy of T DNA is transferred and integrated into a random site in a plant chromosome
Plant cell nucleus
(b)

figure 29–17
A bacterial plant parasite transfers DNA to plant cells. **(a)** The Ti (tumor-inducing) plasmid of *Agrobacterium tumefaciens.* **(b)** Wounded plant cells produce and release the phenolic compound acetosyringone. When *Agrobacterium* detects this compound, the virulence *(vir)* genes on the Ti plasmid are expressed. The *vir* genes encode enzymes needed to introduce the T DNA segment of the Ti plasmid into the genome of nearby plant cells. A single-stranded copy of the T DNA is synthesized and transferred to the plant cell, where it is converted to duplex DNA and integrated into a plant cell chromosome. The T DNA encodes enzymes that synthesize not only growth hormones but also opines (see Fig. 29–18), compounds that can be metabolized only by *Agrobacterium,* which uses them as a nutrient source. Expression of the T DNA genes by transformed plant cells thus leads to aberrant plant cell growth (tumor formation) and the diversion of plant cell nutrients to the invading bacteria.

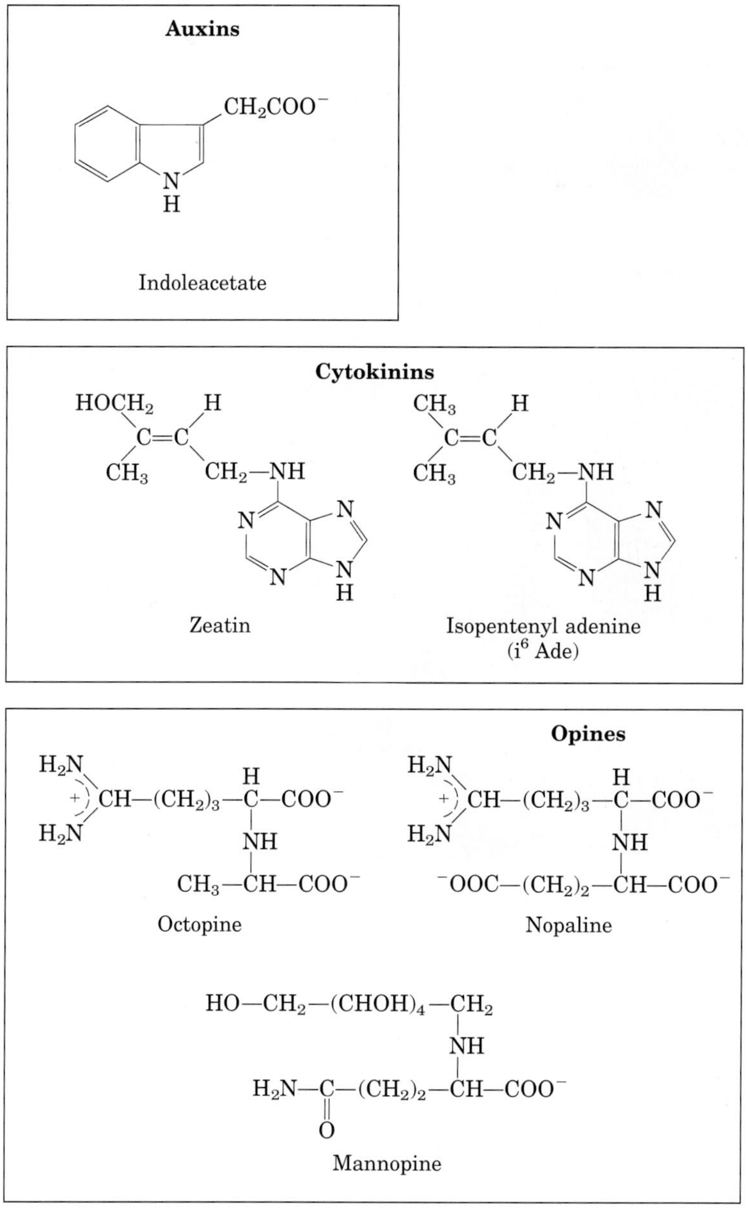

Auxins

CH_2COO^- (indole ring, N–H)

Indoleacetate

Cytokinins

$HOCH_2(CH_3)C{=}CH{-}CH_2{-}NH{-}$(adenine)

Zeatin

$(CH_3)_2C{=}CH{-}CH_2{-}NH{-}$(adenine)

Isopentenyl adenine (i^6 Ade)

Opines

$(H_2N)_2C^+{-}NH{-}(CH_2)_3{-}CH(COO^-){-}NH{-}CH(CH_3){-}COO^-$

Octopine

$(H_2N)_2C^+{-}NH{-}(CH_2)_3{-}CH(COO^-){-}NH{-}CH(COO^-){-}(CH_2)_2{-}COO^-$

Nopaline

$HO{-}CH_2{-}(CHOH)_4{-}CH_2{-}NH{-}CH(COO^-){-}(CH_2)_2{-}C(=O){-}NH_2$

Mannopine

figure 29–18

Metabolites produced in *Agrobacterium*-infected plant cells. Auxins and cytokinins are plant growth hormones. The most common auxin, indoleacetate, is derived from tryptophan. Cytokinins are adenine derivatives. Opines generally are derived from amino acid precursors; at least 14 different opines are produced by enzymes encoded by the Ti plasmids of different species of *Agrobacterium*.

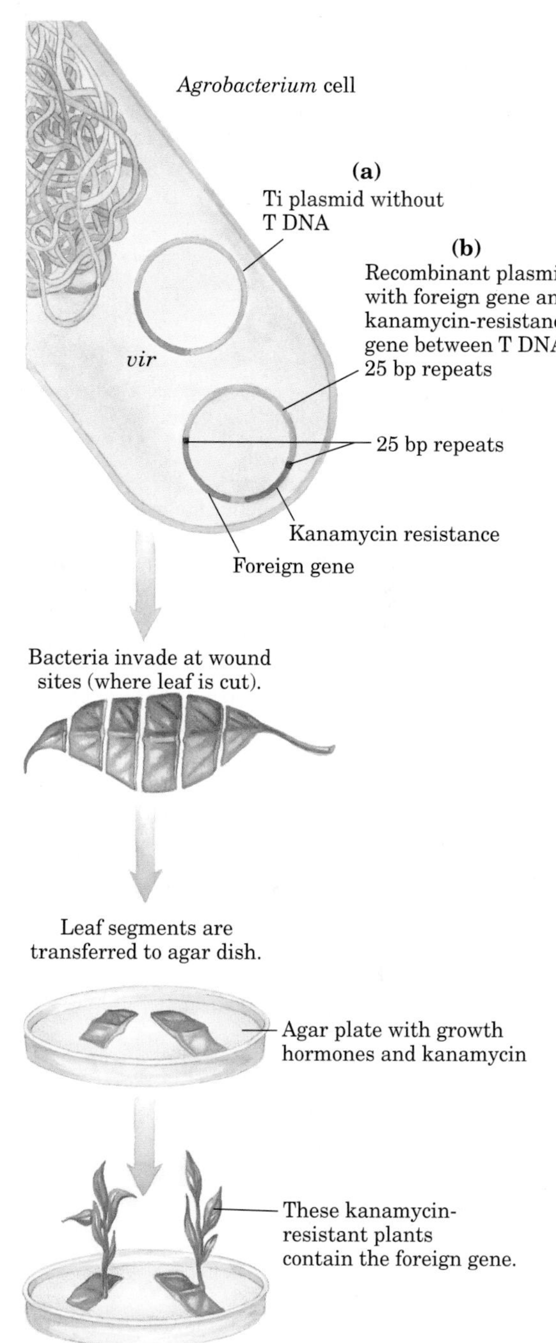

figure 29–19

A two-plasmid strategy to create a recombinant plant. **(a)** One plasmid is a modified Ti plasmid that lacks T DNA. **(b)** The other plasmid lacks the usual T DNA genes but contains a segment of DNA that bears both a foreign gene of interest (e.g., the gene for the insecticidal protein described in Figure 29–21) and an antibiotic-resistance element (here, kanamycin resistance), flanked by the two T-DNA 25 base pair repeats. The second plasmid also contains the replication origin needed for propagation in *Agrobacterium.* When bacteria invade at the site of a wound (the edge of the cut leaf), the *vir* genes on the first plasmid mediate transfer into the plant genome of the segment of the second plasmid that is flanked by the 25 base pair repeats. New plants are generated when leaf segments with the transformed cells are placed on an agar dish that contains both kanamycin and appropriate levels of plant growth hormones. Nontransformed plant cells are killed by the kanamycin. The gene of interest and the antibiotic-resistance element are normally transferred together, so plant cells that grow in this medium generally contain the foreign gene of interest.

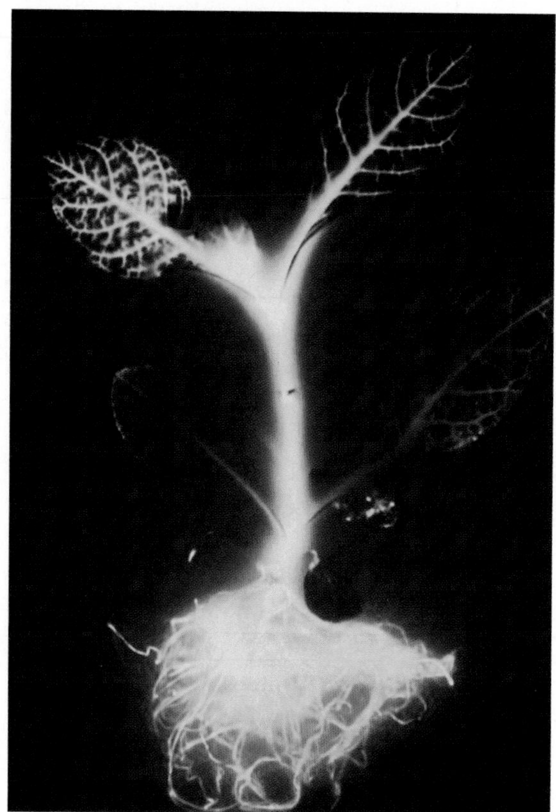

figure 29–20
A tobacco plant in which the gene for firefly luciferase is expressed. Light was produced after the plant was watered with a solution containing luciferin, the substrate for this light-producing enzyme (see Box 14–3). Don't expect glow-in-the-dark ornamental plants at your local nursery anytime soon; the light is actually quite weak; this photograph required a 24-hour exposure. The real point—that this technology allows the introduction of new traits into plants—is nevertheless elegantly made.

The successful transfer of recombinant DNA into plants is vividly illustrated by an experiment in which the luciferase gene from fireflies was introduced into the cells of a tobacco plant (Fig. 29–20), which is often used experimentally because its cells are particularly easy to transform with *Agrobacterium*. The potential of this technology is not limited to the production of glow-in-the-dark plants. The same approach has been used to produce crop plants that are resistant to herbicides, plant viruses, and insect pests (Fig. 29–21). Potential benefits include increased yields and a reduction in the need for environmentally harmful agricultural chemicals.

Biotechnology can introduce new traits into a plant much faster than traditional methods of plant breeding. A prominent example is the development of soybeans that are resistant to the general herbicide glyphosate (the active ingredient in RoundUp). Glyphosate breaks down rapidly in the environment (sensitive plants can be planted in a treated area after as little as 48 hours), and its use does not generally lead to contamination of

figure 29–21
Tomato plants engineered to be resistant to some insect larvae. Two tomato plants were exposed to equal numbers of moth larvae. The plant on the left has not been genetically altered. The plant on the right expresses a gene for a protein toxin derived from the bacterium *Bacillus thuringiensis*. This protein, introduced by a protocol similar to that depicted in Figure 29–19, is toxic to the larvae of some moth species while being harmless to humans and other organisms. Insect resistance has also been genetically engineered in cotton and other plants.

(a)

(b)

groundwater or carryover from one year to the next. A field of these soybeans can be treated once during a summer growing season to eliminate essentially all weeds in the field, while leaving the soybeans unaffected (Fig. 29–22). Potential pitfalls of the technology, such as the evolution of glyphosate-resistant weeds or the escape of difficult-to-control recombinant plants, remain a concern of researchers and the public.

$$\mathrm{O}^- - \overset{\overset{\displaystyle \mathrm{O}^-}{\|}}{\underset{\underset{\displaystyle \mathrm{O}^-}{|}}{\mathrm{P}}} - \mathrm{CH_2} - \mathrm{NH} - \mathrm{CH_2} - \mathrm{COO}^-$$

Glyphosate

(c)

figure 29–22

Glyphosate-resistant soybean plants growing in a field in Wisconsin. (a) Without glyphosate treatment, this part of the soybean field is overrun with weeds. **(b)** Glyphosate-resistant soybean plants thrive in the glyphosate-treated section of the field. **(c)** Glyphosate breaks down rapidly in the environment. Agricultural use of engineered plants proceeds with considerable deliberation, balancing the extraordinary promise of the technology with the need to select new traits with care. Both science and society as a whole have a stake in ensuring that the use of the resultant plants has no adverse impact on the environment or human health.

Cloning in Animal Cells Points the Way to Human Gene Therapy

The transformation of animal cells with foreign genetic material offers an important mechanism for expanding our knowledge of the structure and function of animal genomes, as well as for the generation of animals with new traits. This potential has stimulated intensive research into increasingly sophisticated means of cloning animals.

Most work of this kind requires a source of cells into which DNA can be introduced. Although intact tissues are often difficult to maintain and manipulate in vitro, many types of animal cells can be isolated and grown in the laboratory if their growth requirements are carefully met. Cells derived from a particular animal tissue and grown under appropriate **tissue culture** conditions can maintain their differentiated properties (for example, a liver cell remains a liver cell) for weeks or even months.

No suitable plasmidlike vector is available for introducing DNA into an animal cell, so transformation usually requires the integration of the DNA into a host-cell chromosome. The efficient delivery of DNA to a cell nucleus and its integration into a chromosome without disrupting any critical genes remains the major technical problem in the genetic engineering of animal cells.

Available methods for bringing DNA into an animal cell vary in efficiency and convenience. Some success has been achieved using spontaneous uptake of DNA or electroporation, techniques that are roughly comparable to the common methods used to transform bacteria. They are inefficient, however, transforming only one in 10^2 to 10^4 cells. **Microinjection** entails the injection of DNA directly into a cell's nucleus using a very fine needle. For skilled practitioners, this method has a high success rate, but the total number of cells that can be treated is small because each must be injected individually.

The most efficient and widely used methods for transforming animal cells rely on liposomes or viral vectors. Liposomes are small vesicles consisting of a lipid bilayer that encloses an aqueous compartment (see Fig. 12–4). Liposomes that enclose a recombinant DNA molecule can be fused with the

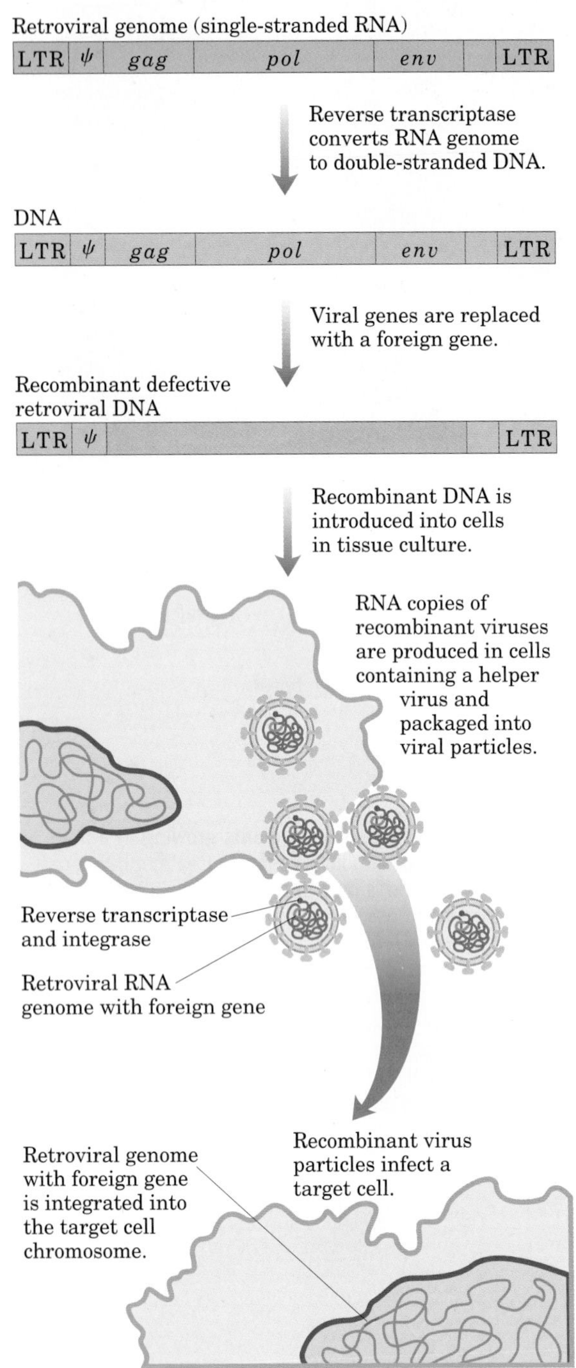
Retroviral genome (single-stranded RNA)
LTR ψ gag pol env LTR
Reverse transcriptase converts RNA genome to double-stranded DNA.
DNA
LTR ψ gag pol env LTR
Viral genes are replaced with a foreign gene.
Recombinant defective retroviral DNA
LTR ψ LTR
Recombinant DNA is introduced into cells in tissue culture.
RNA copies of recombinant viruses are produced in cells containing a helper virus and packaged into viral particles.
Reverse transcriptase and integrase
Retroviral RNA genome with foreign gene
Retroviral genome with foreign gene is integrated into the target cell chromosome.
Recombinant virus particles infect a target cell.

figure 29–23
Cloning in mammalian cells with retroviral vectors. A typical retroviral genome (somewhat simplified here) is engineered to carry a desired foreign gene and added to a host cell tissue culture. The helper virus (not shown) lacks the packaging sequence, ψ, so its RNA transcripts cannot be packaged into virus particles. However, it provides the *gag, pol,* and *env* gene products needed to package the engineered retrovirus into functional virus particles, enabling the foreign gene on the recombinant retroviral genome to be introduced efficiently into the target cells.

membranes of target cells to deliver DNA into the cell. The DNA sometimes reaches the nucleus, where it can integrate into a chromosome (mostly at random locations). **Viral vectors** are even more efficient at delivering DNA. Animal viruses have evolved effective mechanisms for introducing their nucleic acids into cells, and several types also have mechanisms to integrate their DNA into a host-cell chromosome. Some of these, such as retroviruses (p. 1007) and adenoviruses, have been modified to act as viral vectors to introduce foreign DNA into mammalian cells.

Work on retroviral vectors illustrates some of the strategies being used (Fig. 29–23). When an engineered retrovirus enters a cell, its RNA genome is converted to DNA by reverse transcriptase and then integrated into the host genome by viral integrase. Long terminal repeat (LTR) sequences are required for integration of retroviral DNA in the host chromosome (see Fig. 26–32), and the ψ sequence is required to package the viral RNA in viral particles.

The *gag, pol,* and *env* genes of the retroviral genome, required for retroviral replication and assembly of viral particles, can be replaced with foreign DNA. To assemble viruses that contain the recombinant genetic information, the DNA must be introduced into cultured cells that are simultaneously infected with a helper virus that has the genes to produce virus particles but lacks the ψ sequence required for packaging. This allows the recombinant DNA to be transcribed, and its RNA to be packaged into virus particles. These particles can act as vectors to introduce the recombinant RNA into target cells. Viral reverse transcriptase and integrase enzymes (produced by the helper virus) are also packaged in the viral particle and introduced into the target cells. Once the engineered viral genome is inside a cell, these enzymes create a DNA copy of the recombinant viral RNA genome and integrate it into a host chromosome. The integrated recombinant DNA then becomes a permanent part of the target cell chromosome and is replicated with the chromosome at every cell division. The cell itself is not endangered by the integrated viral DNA because the recombinant virus lacks the genes needed to produce RNA copies of its genome and package them into new virus particles. The use of recombinant retroviruses is often the best method for introducing DNA into large numbers of mammalian cells.

Each type of virus has different attributes, so several classes of animal viruses are being engineered as vectors to transform mammalian cells. Adenoviruses, for example, lack a mechanism for integrating DNA into a chromosome. Recombinant DNA introduced via an adenoviral vector is therefore expressed for only a short time and then destroyed. This can be useful if the objective is transient expression of a gene.

Transformation of animal cells by any of the above techniques can be problematic. Introduced DNA is generally integrated into chromosomes at random locations. Even when the foreign DNA contains a sequence similar to a sequence on a host chromosome, allowing targeting to that position, nonhomologous integrants still outnumber the targeted ones 100 to 100,000 fold. If these integration events disrupt essential genes, they can kill the cell. Integration sites can also determine the level of expression of an integrated gene, because integrated genes are not transcribed equally well everywhere in the genome.

Despite these challenges, the transformation of animal cells has been used extensively to study chromosome structure, as well as the function, regulation, and expression of genes in animal cells. The successful introduction of recombinant DNA into an animal can be illustrated by an experiment that permanently altered an easily observable heritable physical trait.

Microinjection of DNA into the nuclei of fertilized mouse eggs can produce efficient transformation (chromosomal integration). When the injected eggs are introduced into a female mouse and allowed to develop, the new gene is often expressed in some of the newborn mice. Those in which the germ line has been altered can be identified by testing their offspring. By careful breeding, a **transgenic** mouse line can be established in which all the mice are homozygous for the new gene or genes. This technology was used to introduce into mice the gene for human growth hormone, under the control of an inducible promoter. When the mice were fed a diet that included the inducer, some of the mice that developed from injected embryos grew to an unusually large size (Fig. 29–24). Transgenic mice have now been produced with a wide range of genetic variations, including many relevant to human disease and its control, pointing the way to human gene therapy (Box 29–2).

figure 29–24
Cloning in mice. The gene for human growth hormone was introduced into the genome of the mouse on the right. Expression of the gene resulted in the unusually large size of this mouse.

Recombinant DNA Technology Yields New Products and Choices

The products of recombinant DNA technology range from proteins to engineered organisms. Large amounts of commercially useful proteins can be produced by these techniques. Microorganisms can be designed for special tasks; plants or animals can be engineered with traits that are useful in agriculture or medicine. Some products of this technology have been approved for consumer or professional use and many more are in development. Genetic engineering has been transformed over a few years from a promising technology to a multibillion dollar industry, with much of the growth occurring in the pharmaceuticals industry. Some major classes of new products are listed in Table 29–3.

table 29–3

Some Recombinant DNA Products in Medicine

Product category	Examples/uses
Anticoagulants	Tissue plasminogen activator (TPA) activates plasmin, an enzyme involved in dissolving clots; effective in treating heart attack victims.
Blood factors	Factor VIII promotes clotting and is deficient in hemophiliacs; use of factor VIII produced by recombinant DNA technology eliminates infection risks associated with blood transfusions.
Colony stimulating factors	Immune system growth factors that stimulate leukocyte production; used to treat immune deficiencies and to fight infections.
Erythropoietin	Stimulates erythrocyte production; used to treat anemia in patients with kidney disease.
Growth factors	Stimulate differentiation and growth of various cell types; used to promote wound healing.
Human growth hormone	Used to treat dwarfism.
Human insulin	Used to treat diabetes.
Interferons	Interfere with viral reproduction; used to treat some cancers.
Interleukins	Activate and stimulate different classes of leukocytes; possible uses in treating wounds, HIV infection, cancer, and immune deficiencies.
Monoclonal antibodies	Extraordinary binding specificity is used in: diagnostic tests; targeted transport (of drugs, toxins, or radioactive compounds to tumors as a cancer therapy); many other applications.
Superoxide dismutase	Prevents tissue damage from reactive oxygen species when tissues briefly deprived of O_2 during surgery suddenly have blood flow restored.
Vaccines	Proteins derived from viral coats are as effective in "priming" an immune system as the killed virus more traditionally used for vaccines, but are safer; first developed was the vaccine for hepatitis B.

Erythropoietin is typical of the newer products. Erythropoietin (M_r 51,000) is a protein hormone that stimulates erythrocyte production. People with kidney disease often have a deficiency of this protein, resulting in anemia. Erythropoietin produced by recombinant DNA technology can be used to treat these individuals, reducing the need for repeated blood transfusion.

Other applications of this technology continue to appear. Enzymes produced by recombinant DNA technology are already used to produce detergents, sugars, and cheese. Engineered proteins are being used as food additives to supplement nutrition, flavor, and fragrance. Microorganisms are being engineered with altered or entirely novel metabolic pathways to extract oil and minerals from ground deposits, to digest oil spills, and to detoxify hazardous waste dumps and sewage. Engineered plants with improved resistance to drought, frost, pests, and disease are increasing crop yields and reducing the need for agricultural chemicals. Complete animals can be cloned by moving an entire nucleus and all of its genetic material to a prepared egg from which the nucleus has been removed. The full range of long-term consequences of this technology on our species and on the global environment are impossible to foresee, but will certainly demand ever-improving understanding of both cellular metabolism and ecology.

summary

The study of gene structure and function has been greatly facilitated by recombinant DNA technology. Isolation of a gene from a large chromosome requires methods for cutting and joining DNA fragments, the availability of small DNA vectors that can replicate autonomously and into which the gene can be inserted, methods to introduce the vector with its foreign DNA into a cell in which it can be propagated to form clones, and methods to identify the cells containing the DNA of interest. Advances in this technology are revolutionizing many aspects of medicine, agriculture, and other industries.

The first organism used for DNA cloning was *E. coli.* Bacterial restriction endonucleases and DNA ligases provide the most important instruments for cutting DNA at specific sequences and joining DNA fragments. Bacterial cloning vectors include plasmids, bacteriophages, and bacterial artificial chromosomes (BACs). These permit the cloning of DNA fragments of different sizes. In each case, the vectors provide a replication origin for propagation in the bacterial host, and a selectable genetic trait (such as antibiotic resistance) to facilitate the identification of cells that harbor the recombinant vector. DNA is introduced into target cells using viral vectors or via methods that make the cell wall artificially permeable.

The first step in cloning a gene is often the construction of a DNA library that includes fragments representing virtually the entire genome of a given species. A cDNA library is limited to expressed genes by cloning only complementary DNA copies of isolated mRNAs. A specific segment of DNA can be amplified and cloned using the polymerase chain reaction. Clones containing a specific gene in a large library can be detected by hybridization, using a radioactive probe that contains the complementary nucleotide sequence. DNA microarrays provide defined libraries for the investigation of genome-wide gene expression patterns.

Expression vectors provide the DNA sequences required for transcription, translation, and regulation of cloned genes. They allow the production of large amounts of cloned proteins for research and commercial purposes. Cloned genes also can be finely altered by site-directed mutagenesis, which is useful in studies of protein structure and function.

Yeast is sometimes used for cloning eukaryotic DNA, offering many of the same advantages as *E. coli.* YAC vectors permit the cloning of

2×10^6 base pairs of DNA. Cloning of plants and animal cells is generating a variety of organisms with altered traits. Plants that are resistant to disease, insects, herbicides, and drought are being produced with the aid of the natural gene-transfer process promoted by the Ti plasmid of the parasitic bacterium *Agrobacterium tumefaciens.* Engineered DNA can be introduced into animal cells by electroporation, microinjection, liposomes, or viral vectors, resulting in animals with new inheritable genetic traits. The technology extends to humans, and human gene therapy is being directed at treating human genetic diseases. The economic, social, and environmental implications of recombinant DNA technology are widespread and profound.

further reading

General

Jackson, D.A., Symons, R.H., & Berg, P. (1972) Biochemical method for inserting new genetic information into DNA of simian virus 40: circular SV40 DNA molecules containing lambda phage genes and the galactose operon of *Escherichia coli. Proc. Natl. Acad. Sci. USA* **69,** 2904–2909.

The first recombinant DNA experiment linking DNA from two different organisms.

Lobban, P.E. & Kaiser, A.D. (1973) Enzymatic end-to-end joining of DNA molecules. *J. Mol. Biol.* **78,** 453–471.

Report of the first recombinant DNA experiment.

Sambrook, J., Fritsch, E.F., & Maniatis, T. (1989) *Molecular Cloning: A Laboratory Manual,* 2nd edn, Cold Spring Harbor Laboratory Press, Cold Spring Harbor, NY.

Although supplanted by more recent manuals, this three-volume set includes much useful background information on the biological, chemical, and physical principles underlying both classic and still-current techniques.

Libraries and Gene Isolation

Arnheim, N. & Erlich, H. (1992) Polymerase chain reaction strategy. *Annu. Rev. Biochem.* **61,** 131–156.

Audic, S. & Béraud-Colomb, E. (1997) Ancient DNA is thirteen years old. *Nat. Biotech.* **15,** 855–858.

Brown, T. & Brown, K. (1994) Ancient DNA: using molecular biology to explore the past. *Bioessays* **16,** 719–726.

Ivanov, P.L., Wadhams, M.J., Roby, R.K., Holland, M.M., Weedn, V.W., & Parsons, T.J. (1996) Mitochondrial DNA sequence heteroplasmy in the Grand Duke of Russia Georgij Romanov establishes the authenticity of the remains of Tsar Nicholas II. *Nat. Genet.* **12,** 417–420.

Lindahl, T. (1997) Facts and artifacts of ancient DNA. *Cell* **90,** 1–3.

Products of Recombinant DNA Technology

Bailey, J.E. (1991) Toward a science of metabolic engineering. *Science* **252,** 1668–1675.

An overview of efforts to reengineer entire metabolic pathways in microorganisms for commercial purposes.

Blaese, R.M. (1997) Gene therapy for cancer. *Sci. Am.* **276** (June), 111–115.

Brown, P.O. & Botstein, D. (1999) Exploring the new world of the genome with DNA microarrays. *Nat. Genet.* **21,** 33–37.

Collins, F.S., Patrinos, A., Jordan, E., Chakravarti, A., Gesteland, R., Walters, L., and the members of the Department of Energy and National Institutes of Health planning groups. (1998) New goals for the US human genome project: 1998–2003. *Science* **282**, 682–689.

Culver, K.W., Osborne, W.R.A., Miller, A.D., Fleisher, T.A., Berger, M., Anderson, W.F., & Blaese, R.M. (1991) Correction of ADA deficiency in human T lymphocytes using retroviral-mediated gene transfer. *Transplant Proc.* **23,** 170–171.

Davies, S.W., Turmaine, M., Cozens, B.A., Difiglia, M., Sharp, A.H., Ross, C.A., Scherzinger, E., Wanker, E.E., Mangianni, L., & Bates, G.P. (1997) Formation of neuronal intranuclear inclusions underlies the neurological dysfunction in mice transgenic for the HD mutation. *Cell* **90,** 537–548.

Gene therapy of adenosine deaminase deficiency.

Foster, E.A., Jobling, M.A., Taylor, P.G., Donnelly, P., de Knijff, P., Mieremet, R., Zerjal, T., & Tyler-Smith, C. (1999) The Thomas Jefferson paternity case. *Nature* **397**, 32.

Last article of a series in an interesting case study of the uses of biotechnology to address historical questions.

Friedmann, T. (1997) Overcoming the obstacles to gene therapy. *Sci. Am.* **276** (June), 96–101.

Hansen, G. & Wright M.S. (1999) Recent advances in the transformation of plants. *Trends Plant Sci.* **4,** 226–231.

Hooykaas, P.J.J. & Schilperoort, R.A. (1985) The Ti-plasmid of *Agrobacterium tumefaciens:* a natural genetic engineer. *Trends Biochem. Sci.* **10,** 307–309.

The Huntington's Disease Collaborative Research Group (1993) A novel gene containing a trinucleotide repeat that is expanded and unstable on Huntington's disease chromosomes. *Cell* **72,** 971–983.

Koopman, P., Gubbay, J., Vivian, N., Goodfellow, P., & Lovell-Badge, R. (1991) Male development of chromosomally female mice transgenic for *Sry. Nature* **351,** 117–121.

Recombinant DNA technology is used to demonstrate that a single gene directs the development of chromosomally female mice into males.

Lapham, E.V., Kozma, C., & Weiss, J. (1996) Genetic discrimination: perspectives of consumers. *Science* **274,** 621–624.

Mahowald, M.B., Verp, M.S., & Anderson, R.R. (1998) Genetic counseling: clinical and ethical challenges. *Annu. Rev. Genet.* **32,** 547–559.

Marshall, E. (1995) Gene therapy's growing pains. *Science* **269,** 1050–1055.

Murray, T.H. (1991) Ethical issues in human genome research. *FASEB J.* **5,** 55–60.

This issue also contains a number of other useful papers on the human genome project.

Palmiter, R.D., Brinster, R.L., Hammer, R.E., Trumbauer, M.E., Rosenfeld, M.G., Birnberg, N.C., & Evans, R.M. (1982) Dramatic growth of mice that develop from eggs microinjected with metallothionein–growth hormone fusion genes. *Nature* **300,** 611–615.

A description of how to make giant mice.

Peterson, K.R., Clegg, C.H., Li, Q.L., & Stamatoyannopoulos, G. (1997) Production of transgenic mice with yeast artificial chromosomes. *Trends Genet.* **13,** 61–66.

Ramsay, M. (1994) Yeast artificial chromosome cloning. *Molec. Biotech.* **1,** 181–201.

Scherzinger, E., Lurz, R., Turmaine, M., Mangiarini, L., Hollenbach, B., Hasenbank, R., Bates, G.P., Davies, S.W., Lehrach, H., & Wanker, E.E. (1997) Huntington-encoded polyglutamine expansions form amyloid-like protein aggregates in vitro and in vivo. *Cell* **90,** 549–558.

Schuler, G.D., Boguski, M.S., et al. (1996) A gene map of the human genome. *Science* **274,** 540–546.

Thompson, J. & Donkersloot, J.A. (1992) *N*-(Carboxyalkyl)amino acids: occurrence, synthesis, and functions. *Annu. Rev. Biochem.* **61,** 517–557.

A good summary of the structure and biological functions of opines.

Verma, I. & Somia, N. (1997) Gene therapy—promises, problems and prospects. *Nature* **389,** 239–242.

problems

1. Cloning When joining two or more DNA fragments, a researcher can adjust the sequence at the junction in a variety of subtle ways, as seen in the following exercises.

(a) Draw the structure of each end of a linear DNA fragment that was produced by an *Eco*RI restriction digest (include those sequences remaining from the *Eco*RI recognition sequence).

(b) Draw the structure resulting from the reaction of this end sequence with DNA polymerase I and the four deoxynucleoside triphosphates.

(c) Draw the sequence produced at the junction that arises if two ends with the structure derived in (b) are ligated.

(d) Draw the structure produced if the structure derived in (a) is treated with a nuclease that degrades only single-stranded DNA.

(e) Draw the sequence of the junction produced if an end with structure (b) is ligated to an end with structure (d).

(f) Draw the structure of the end of a linear DNA fragment that was produced by a *Pvu*II restriction digest (as in (a)).

(g) Draw the sequence of the junction produced if an end with structure (b) is ligated to an end with structure (f).

(h) Suppose you can synthesize a short duplex DNA fragment with any sequence you desire. With such a synthetic fragment and the procedures described in (a) through (g), design a protocol that will remove an *Eco*RI restriction site from a DNA molecule and incorporate a new *Bam*HI restriction site at approximately the same location. (Hint: see Fig. 29–3.)

(i) Design four different short synthetic double-stranded DNA fragments that would permit ligation of structure (a) with a DNA fragment produced by a *Pst*I restriction digest. In one of these synthetic fragments, design the sequence so that the final junction contains the recognition sequences for both *Eco*RI and *Pst*I. In the second and third synthetic fragments, design the sequence so that the junction contains only the *Eco*RI or the *Pst*I recognition sequence, respectively. Design the sequence of the fourth fragment so that neither the *Eco*RI nor the *Pst*I sequence appears in the junction.

2. Selecting for Recombinant Plasmids When cloning a foreign DNA fragment into a plasmid, it is often useful to insert the fragment at a site that interrupts a selectable marker (such as the tetracycline-resistance gene of pBR322). The loss of function of the interrupted gene can be used to identify clones containing recombinant plasmids with foreign DNA. With a bacteriophage λ vector it is not necessary to do this, yet one can easily distinguish vectors that incorporate large foreign DNA fragments from those that do not. How are these recombinant vectors identified?

3. DNA Cloning The plasmid cloning vector pBR322 (see Fig. 29–5) is cleaved with the restriction endonuclease *Pst*I. An isolated DNA fragment from a eukaryotic genome (also produced by *Pst*I cleavage) is added to the prepared vector and ligated. The mixture of ligated DNAs is then used to transform bacteria, and plasmid-containing bacteria are selected by growth in the presence of tetracycline.

(a) In addition to the desired recombinant plasmid, what other types of plasmids might be found among the transformed bacteria that are tetracycline resistant? How can they be distinguished?

(b) The cloned DNA fragment is 1,000 bp in length and has an *Eco*RI site 250 bp from one end. Three different recombinant plasmids are cleaved with *Eco*RI and analyzed by gel electrophoresis, giving the patterns shown. What does each pattern say about the cloned RNA? Note that in pBR322, the *Pst*I and *Eco*RI restriction sites are about 750 bp apart. The entire plasmid with no cloned insert is 4,361 bp. Size markers in lane 4 have the number of nucleotides noted.

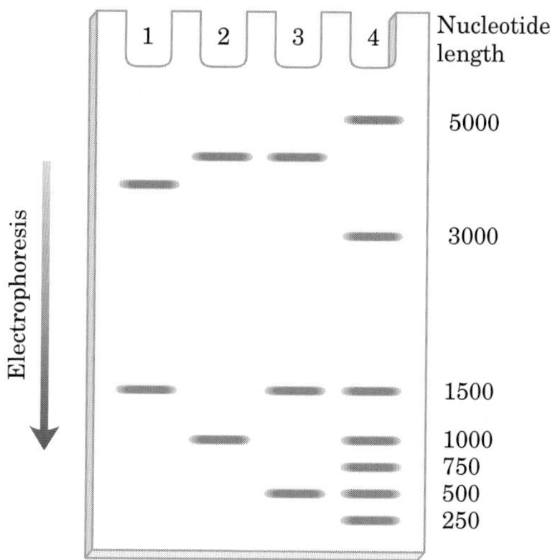

4. Expressing a Cloned Gene You have isolated a plant gene that encodes a protein in which you are interested. On the drawing below, indicate sequences or sites that you will need to get this gene transcribed, translated, and regulated in *E. coli.*

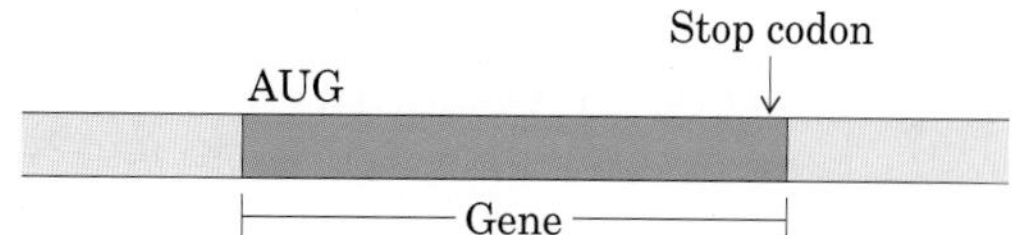

5. Identifying the Gene for a Protein with a Known Amino Acid Sequence Design a DNA probe that would allow you to identify the gene for a protein with the following amino-terminal amino acid sequence. The probe should be 18 to 20 nucleotides long, a size that provides adequate specificity if there is sufficient homology between the probe and the gene.

$H_3\overset{+}{N}$–Ala–Pro–Met–Thr–Trp–Tyr–Cys–
Met–Asp–Trp–Ile–Ala–Gly–Gly–Pro–
Trp–Phe–Arg–Lys–Asn–Thr–Lys–

6. Cloning in Plants The strategy outlined in Figure 29–19 employs *Agrobacteria* that contain two separate plasmids. Suggest a reason why the sequences on the two plasmids are not combined on one plasmid.

7. Cloning in Mammals The retroviral vectors described in Figure 29–23 make it possible to integrate foreign DNA efficiently into a mammalian genome. Explain how these vectors, which lack genes for replication and viral packaging *(gag, pol, env),* are assembled into infectious viral particles. Suggest why it is important that these vectors lack the replication and packaging genes.

8. Designing a Diagnostic Test for a Genetic Disease Huntington's disease (HD) is an inherited neurodegenerative disorder, characterized by the gradual, irreversible impairment of psychological, motor, and cognitive functions. Symptoms typically appear in middle age, but onset can occur at almost any age. The course of the disease can last 15 to 20 years. The molecular basis of the disease is increasingly well understood. The genetic mutation underlying HD has been traced to a gene encoding a protein (M_r 350,000) of unknown function. In individuals who will not develop HD, a region of the gene that encodes the amino terminus of the protein has a sequence of CAG codons (for glutamine) that is repeated 6 to 39 times in succession. In individuals with adult-onset HD, this codon is typically repeated 40 to 55 times. In individuals with childhood-onset HD, this codon is repeated more than 70 times. The length of this simple trinucleotide repeat indicates whether an individual will develop HD, and at approximately what age the first symptoms will occur.

A small portion of the amino-terminal coding sequence of the 3,143-codon HD gene is given below. The nucleotide sequence of the DNA is shown in black, the amino acid sequence of the open reading frame in blue, and the CAG repeat is shaded. Outline a PCR-based test for HD that could be carried out using a blood sample. Assume the PCR primer must be 25 nucleotides long. You may wish to refer to Chapter 26 (p. 982) for conventions for displaying gene sequences.

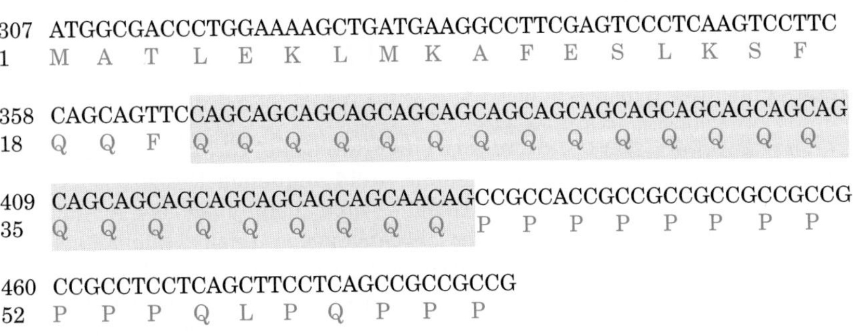

Source: The Huntington's Disease Collaborative Research Group (1993) A novel gene containing a trinucleotide repeat that is expanded and unstable on Huntington's disease chromosomes. *Cell* **72,** 971–983.

9. Using PCR to Detect Circular DNA Molecules A segment of genomic DNA in a ciliated protozoan is sometimes deleted. The deletion is a genetically programmed reaction associated with cellular mating. A researcher proposes that the DNA is deleted in a site-specific recombination event (Fig. 25–36), with the deleted DNA ending up as a circular DNA reaction product as shown below. Suggest how the researcher might use the polymerase chain reaction (PCR) to detect the presence of the circular form of the deleted DNA in an extract of the protozoan.

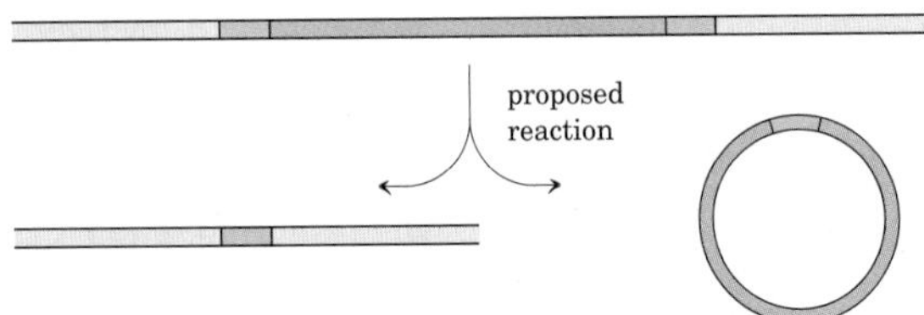

10. DNA Fingerprinting and RFLP Analysis DNA is extracted from the blood cells of two different humans. In separate experiments, the DNA from each individual is cleaved by restriction endonucleases A, B, and C, and the fragments separated by electrophoresis. A hypothetical map of a 10,000 base pair segment of a human chromosome is shown. Individual #2 has point mutations that eliminate restriction recognition sites B* and C*. You probe the gel with a radioactive oligonucleotide complementary to the indicated sequence and expose a piece of x-ray film to the gel. Indicate where you would expect to see bands on the film. The lanes of the gel are marked in the accompanying figure.

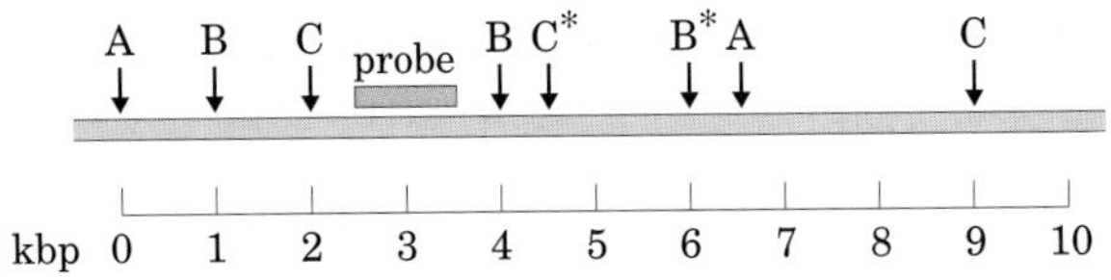

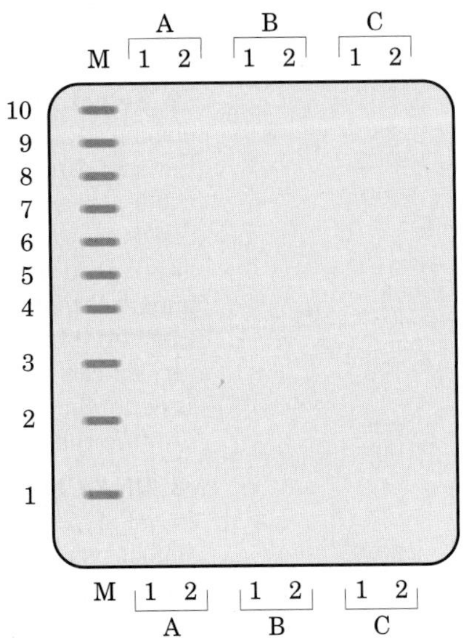

11. RFLP Analysis for Paternity Testing DNA fingerprinting and RFLP analysis are often used to test paternity. A child inherits chromosomes from both mother and father, so DNA from each child displays restriction fragments derived from each parent. In the gel shown here, which child, if any, can be excluded as being the biological offspring of the father? Explain your reasoning. Lane M is the sample from the mother, F from the father, and C1, C2, and C3 from the children.

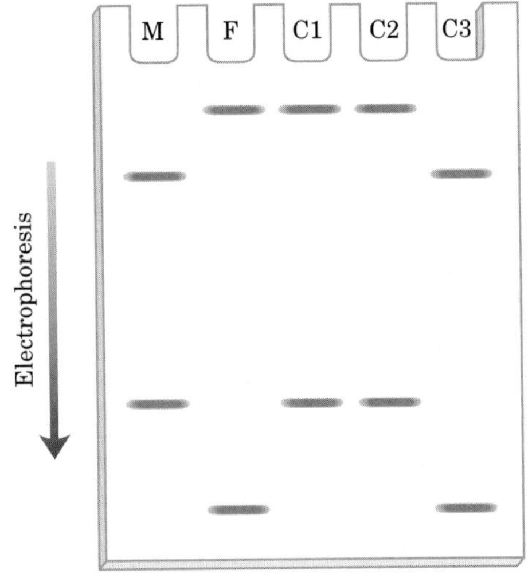

Appendix A Common Abbreviations in the Biochemical Research Literature

Å	angstrom
A	adenine, adenosine, or adenylate
A	absorbance
Ab	antibody
ACh	acetylcholine
ACP	acyl carrier protein
ACTH	adrenocorticotropic hormone
Acyl-CoA	acyl derivatives of coenzyme A (also, acyl-S-CoA)
ADH	alcohol dehydrogenase
adoHcy	*S*-adenosylhomocysteine
adoMet	*S*-adenosylmethionine (also, SAM)
Ag	antigen
AIDS	acquired immunodeficiency syndrome
AKAP	PKA-anchoring protein
Ala	alanine
$[\alpha]_D^{25°C}$	specific rotation
AMP, ADP, ATP	adenosine 5′-mono-, di-, triphosphate
AQP	aquaporin
Arg	arginine
ARS	autonomously replicating sequence
Asn	asparagine
Asp	aspartate
ATCase	aspartate transcarbamoylase
atm	atmosphere
ATPase	adenosine triphosphatase
B_{12}	coenzyme B_{12}, cobalamin
BAC	bacterial artificial chromosome
BMR	basal metabolic rate
bp	base pair
1,3-BPG	1,3-bisphosphoglycerate
C	cytosine, cytidine, or cytidylate
CaM	calmodulin
cAMP, cGMP	3′,5′-cyclic AMP, 3′,5′-cyclic GMP
CDK	cyclin-dependent protein kinase
cDNA	complementary DNA
CFTR	cystic fibrosis transmembrane conductance regulator
Chl	chlorophyll
CL	cardiolipin
CMP, CDP, CTP	cytidine 5′-mono-, di-, triphosphate
CoA	coenzyme A (also, CoASH)
CoQ	coenzyme Q (ubiquinone; also, UQ)
CRP	cAMP receptor protein
Cys	cysteine
D	dihydrouridine
D	diffusion coefficient
d	density
dADP, dGDP, etc.	deoxyadenosine 5′-diphosphate, deoxyguanosine 5′-diphosphate, etc.
dAMP, dGMP, etc.	deoxyadenosine 5′-monophosphate, deoxyguanosine 5′-monophosphate, etc.

dATP, dGTP, etc.	deoxyadenosine 5′-triphosphate, deoxyguanosine 5′-triphosphate, etc.
DEAE	diethylaminoethyl
DFP (DIFP)	diisopropylfluorophosphate
DHAP	dihydroxyacetone phosphate
DHF	dihydrofolate (also, H_2 folate)
DMS	dimethyl sulfate
DNA	deoxyribonucleic acid
DNase	deoxyribonuclease
DNP	2,4-dinitrophenol
Dol	dolichol
DOPA	dihydroxyphenylalanine
E	electrical potential
E.C.	Enzyme Commission (followed by numbers indicating the formal classification of an enzyme)
EDTA	ethylenediaminetetraacetate
EF	elongation factor
EGF	epidermal growth factor
ELISA	enzyme-linked immunosorbent assay
EM	electron microscopy
ϵ	molar absorption coefficient
ER	endoplasmic reticulum
η	viscosity
$\mathcal{F}$	faraday
f	frictional coefficient
FA	fatty acid
FAD, $FADH_2$	flavin adenine dinucleotide, and its reduced form
FBPase-1	fructose-1,6-bisphosphatase
FBPase-2	fructose-2,6-bisphosphatase
Fd	ferredoxin
FDNB (DNFB)	1-fluoro-2,4-dinitrobenzene
FFA	free fatty acid
FH	familial hypercholesterolemia
fMet	*N*-formylmethionine
FMN, $FMNH_2$	flavin mononucleotide, and its reduced form
F_oF_1	mitochondrial ATP synthase
FP	flavoprotein
F1P	fructose 1-phosphate
F6P	fructose 6-phosphate
Fru	D-fructose
ΔG	free-energy change
$\Delta G'^{\circ}$	standard free-energy change
$\Delta G^{\ddagger}$	activation energy
ΔG_B	binding energy
ΔG_p	free-energy change of ATP hydrolysis under cellular conditions
G	guanine, guanosine, or guanylate
g	acceleration due to gravity
GABA	δ-aminobutyric acid
Gal	D-galactose
GalN	D-galactosamine

GalNAc N-acetyl-D-galactosamine
GAP glyceraldehyde-3-phosphate; G protein–activating factor
GDH glutamate dehydrogenase
GH growth hormone
GLC gas-liquid chromatography
Glc D-glucose
GlcA D-gluconic acid
GlcN D-glucosamine
GlcNAc N-acetyl-D-glucosamine (also, NAG)
GlcUA D-glucuronic acid
Gln glutamine
Glu glutamate
Gly glycine
GMP, GDP, GTP guanosine 5′-mono-, di-, triphosphate
G1P glucose 1-phosphate
GPCR G protein–coupled receptor
GPI glycosylated derivatives of phosphatidylinositol
GRK G protein–coupled receptor kinase
G6P glucose 6-phosphate
GSH, GSSG glutathione, and its oxidized form
ΔH enthalpy change
Hb, HbO_2, HbCO hemoglobin, oxyhemoglobin, carbon monoxide hemoglobin
HDL high-density lipoprotein
H_2 folate dihydrofolate (also, DHF)
H_4 folate tetrahydrofolate (also, THF)
His histidine
HIV human immunodeficiency virus
HMG-CoA β-hydroxy-β-methylglutaryl-CoA
hnRNA heterogeneous nuclear RNA
HPLC high-performance liquid chromatography
HRE hormone response element
HTH helix-turn-helix
Hyp hydroxyproline
I inosine
IF initiation factor
Ig immunoglobulin
IgG immunoglobulin G
Ile isoleucine
IMP, IDP, ITP inosine 5′-mono-, di-, triphosphate
IP_3 inositol 1,4,5-trisphosphate
IR infrared
IS insertion sequence
j joule
JAK Janus kinase
K dissociation constant
K_a acid dissociation constant
K_{eq} equilibrium constant
K'_{eq} equilibrium constant under standard conditions
K_I inhibition constant
K_m Michaelis constant
K_S dissociation constant
k rate constant
k_{cat} turnover number
kb kilobase
kbp kilobase pair
α-KG α-ketoglutarate
lac lactose
λ wavelength
LDH lactate dehydrogenase
LDL low-density lipoprotein
Leu leucine
LH luteinizing hormone
Lk linking number
ln logarithm to the base *e*
log logarithm to the base 10
LT leukotriene
LTR long terminal repeat
Lys lysine
M_r relative molecular mass
Man D-mannose
MAPK mitosis-associated protein kinase
Mb, MbO_2 myoglobin, oxymyoglobin
Met methionine
mRNA messenger RNA
MSH melanocyte-stimulating hormone
mDNA mitochondrial DNA
μ electrophoretic mobility
Mur muramic acid
Mur2Ac N-acetylmuramic acid (also, NAM)
NAD^+, NADH nicotinamide adenine dinucleotide, and its reduced form
$NADP^+$, NADPH nicotinamide adenine dinucleotide phosphate, and its reduced form
NAG N-acetylglucosamine (also, GlcNAc)
NAM N-acetylmuramic acid (also, MurNAc)
Neu5Ac N-acetylneuraminic acid (sialic acid)
NMN^+, NMNH nicotinamide mononucleotide, and its reduced form
NMP, NDP, NTP nucleoside mono-, di-, and triphosphate
NMR nuclear magnetic resonance
NO nitric oxide
OAA oxaloacetate
P pressure
P_i inorganic orthophosphate
pO_2 partial pressure of oxygen
PAB or PABA *p*-aminobenzoate
PAGE polyacrylamide gel electrophoresis
PC plastocyanin; phosphatidylcholine
PCR polymerase chain reaction
PDE cyclic nucleotide phosphodiesterase
PE phosphatidylethanolamine
PEP phosphoenolpyruvate
PFK phosphofructokinase
PG prostaglandin
2PGA 2-phosphoglycerate
3PGA 3-phosphoglycerate
pH log $1/[H^+]$
Phe phenylalanine
pI isoelectric point
PI phosphatidylinositol
PIP_2 phosphatidyinositol 4,5-bisphosphate

PK	protein kinase; pyruvate kinase
p*K*	log 1/*K*
PKA	cAMP-dependent protein kinase
PKC	Ca^{2+}/calmodulin-dependent protein kinase
PKG	cGMP-dependent protein kinase
PLC	phospholipase C
PLP	pyridoxal-5-phosphate
Pol	polymerase (DNA or RNA)
PP_i	inorganic pyrophosphate
PQ	plastoquinone
pRb	retinoblastoma protein
Pro	proline
PRPP	5-phosphoribosyl-1-pyrophosphate
PS	phosphatidylserine
$\Delta\psi$	transmembrane electrical potential
PUFA	polyunsaturated fatty acid
Rb	retinoblastoma gene, a tumor supressor gene
RC	photoreaction center
RER	rough endoplasmic reticulum
RF	release factor; replicative form
RFLP	restriction-fragment length polymorphism
RGS	regulator of G protein signaling, a protein
RIA	radioimmunoassay
Rib	D-ribose
RNA	ribonucleic acid
RNase	ribonuclease
rRNA	ribosomal RNA
RSV	Rous sarcoma virus
RTK	receptor tyrosine kinase
ΔS	entropy change
SAM	*S*-adenosylmethionine (also, adoMet)
SDS	sodium dodecyl sulfate
SER	smooth endoplasmic reticulum
Ser	serine
SNARE	synaptosome-associated protein receptor
snRNA	small nuclear RNA
SRP	signal recognition particle
STAT	signal transducer and activator of transcription
STP	standard temperature and pressure
T	thymine, thymidine, or thymidylate
T	absolute temperature
THF	tetrahydrofolate (also, H_4 folate)
Thr	threonine
TLC	thin layer chromatography
TMP, TDP, TTP	thymidine 5′-mono-, di-, triphosphate
TMV	tobacco mosaic virus
TPI	triose phosphate isomerase
TPP	thiamine pyrophosphate
tRNA	transfer RNA
Trp	tryptophan
TX	thromboxane
Tyr	tyrosine
U	uracil, uridine, or uridylate
UCP	uncoupling protein of the mitochondrial inner membrane
UDP-Gal	uridine diphosphate galactose (also, UDP-galactose)
UDP-Glc	uridine diphosphate glucose (also, UDP-glucose)
UMP, UDP, UTP	uridine 5′-mono-, di-, triphosphate
UQ	coenzyme Q (ubiquinone; also, CoQ)
UV	ultraviolet
V_m	transmembrane electrical potential
V_{max}	maximum velocity
V_0	initial velocity
Val	valine
VLDL	very low-density lipoprotein
YAC	yeast artificial chromosome
Z	net charge

Appendix B Abbreviated Solutions to Problems

Chapter 2

1. **(a)** Diameter of magnified cell = 500 mm
   (b) 2.7×10^{12} actin molecules
   (c) 3.7×10^{4} mitochondria
   (d) 3.91×10^{10} glucose molecules
   (e) 50 glucose molecules per hexokinase molecule
2. **(a)** 1.1×10^{-12} g = 1.1 pg **(b)** 5% **(c)** 4.8%
3. **(a)** 1.6 mm; 800 times longer than the cell; DNA must be tightly coiled. **(b)** 3,900 proteins
4. **(a)** Metabolic rate is limited by diffusion, which is limited by surface area. **(b)** $1.2 \times 10^{7}\ m^{-1}$ for the bacterium; $4 \times 10^{4}\ m^{-1}$ for the amoeba; the surface-to-volume ratio is 300 times higher in the bacterium.
5. **(a)** 7,900 **(b)** $3.1 \times 10^{-10}\ m^2$ **(c)** $2.7 \times 10^{-9}\ m^2$ **(d)** 765% **(e)** Mitochondria, chloropolasts, gill lamellae of fish, root hairs of plants
6. **(a)** 2×10^{6} s (about 23 days!) **(b)** Fast, directed transport allows unidirectional movement of molecules, larger cell size, localization of cellular functions to discrete regions, and regulation of molecular movement. **(c)** Proteolytic or oxidative enzymes released from osmotically lysed organelles (lysosomes, peroxisomes) would cause breakdown of cellular molecules that one is attempting to isolate.

Chapter 3

1. The vitamin molecules from the two sources are identical; the body cannot distinguish the source; only associated impurities might vary with the source.
2. **(a)** $H_2N-CH_2-CH_2-OH$ (Amino: H_2N; Hydroxyl: OH)
   (b) $HOCH_2-CH(OH)-CH_2OH$ (Hydroxyls)

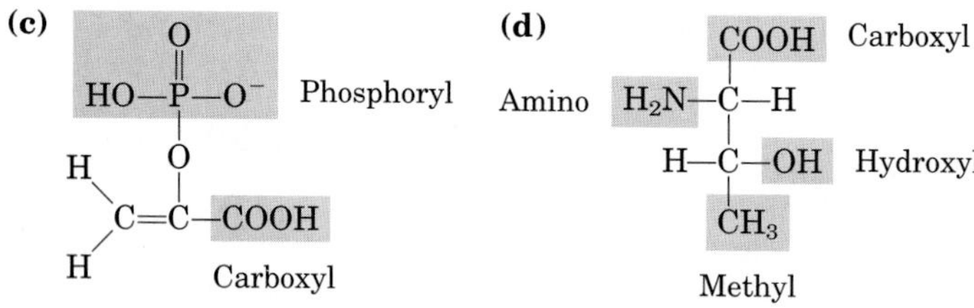

   (e) $HOOC-CH_2-CH_2-NH-C(=O)-CH(OH)-C(CH_3)_2-CH_2OH$ (Carboxyl: COOH; Amido: NH—C=O; Hydroxyl: OH; Methyl: CH_3, CH_3; Hydroxyl: CH_2OH)
   (f) $OHC-CH(NH_2)-CH(OH)-CH(OH)-CH(OH)-CH_2OH$ (Aldehyde: CHO; Amino: NH_2; Hydroxyl: HO—C—H; Hydroxyls: OH, OH, CH_2OH)
3. $HO(HO)C_6H_3-C^{*}H(OH)-CH_2-NH-CH(CH_3)-CH_3$

   The two enantiomers have different interactions with a chiral biological "receptor" (a protein).
4. Only one enantiomer of the drug was physiologically active. Dexedrine consisted of the single enantiomer; benzedrine consisted of a racemic mixture.
5. **(a)** 3 Phosphoric acid molecules; α-D-ribose; guanine **(b)** Choline; phosphoric acid; glycerol; oleic acid; palmitic acid **(c)** Tyrosine; 2 glycines; phenylalanine; methionine
6. **(a)** CH_2O; $C_3H_6O_3$
   (b)
   1. $HOCH_2-CH=C(OH)_2$
   2. $CH_3-C(OH)=C(OH)_2$
   3. $HOCH_2-C(OH)=CH(OH)$
   4. $(HO)_2CH-C(OH)=CH_2$
   5. $(HO)_2CH-CH=CH(OH)$
   6. $CH_3-CH(OH)-COOH$
   7. $HOCH_2-CH_2-COOH$
   8. $HOCH_2-CH(OH)-CHO$
   9. $CH_3-C(OH)_2-CHO$
   10. $(HO)_2CH-CH_2-CHO$
   11. $HOCH_2-CO-CH_2OH$
   12. $CH_3-CO-CH(OH)_2$

   (c) X contains a chiral center; eliminates all but **6** and **8.**
   (d) X contains an acid functional group; eliminates **8;** structure **6** is consistent with all data.
   (e) Structure **6;** we cannot distinguish between the two possible enantiomers.
7. **(a)** Only the amino acids have amino groups; separation could be based on the charge or binding affinity of these groups. Fatty acids are less soluble in water than amino acids, and the two types of molecules also differ in size and shape—either of these property differences could be the basis for separation. **(b)** Glucose is a smaller molecule than a nucleotide; separation could be based on size. The nitrogenous base and/or the phosphate group also endow nucleotides with characteristics (solubility, charge) that could be used for separation from glucose.

8. (a) It is improbable that silicon could serve as the central organizing element for life, especially in an oxygen-containing atmosphere such as that of Earth. Long chains of silicon atoms are not readily synthesized; the polymeric macromolecules necessary for more complex functions would not readily form. Oxygen disrupts bonds between silicon atoms, and silicon–oxygen bonds are extremely stable and difficult to break, preventing the breaking and making of bonds, which is essential to life processes.

Chapter 4

1. 3.32 mL

2. pH 1.1

3. 1.7×10^{-9} mol

4. (a) 460 mM Na^+, 10 mM K^+, 590 mM Cl^-, 10 mM Ca^{2+}, 50 mM Mg^{2+}; the seawater contains 1,120 mosmol/L. **(b)** Cells contain 235 mosmol/L. **(c)** Ions (notably Na^+ and Cl^-) diffuse down their concentration gradients into the cells; water moves down its concentration gradient out of the cells. The cells become severely dehydrated. **(d)** 2.59 g **(e)** 5.31 g

5. (a) pH 8.6 to 10.6 **(b)** 4/5 **(c)** 10 mL **(d)** pH = $pK_a - 2$

6. (a) 0.1 M HCl **(b)** 0.1 M NaOH **(c)** 0.1 M NaOH

7. (d) Bicarbonate, a weak base, titrates —OH to —O^-, making the compound more polar and more water-soluble.

8. Stomach; the neutral form of aspirin present at the lower pH is less polar and passes through the membrane more easily.

9. $NaH_2PO_4{\cdot}H_2O$, 5.80 g; Na_2HPO_4, 8.23 g

10. (a) Blood pH is controlled by the carbon dioxide–bicarbonate buffer system,

$$CO_2 + H_2O \rightleftharpoons H^+ + HCO_3^-$$

During *hypoventilation*, [CO_2] increases in the lungs and arterial blood, driving the equilibrium to the right, raising [H^+] and lowering pH. **(b)** During *hyperventilation*, [CO_2] decrease in the lungs and arterial blood, reducing [H^+] and increasing pH above the normal 7.4 value. **(c)** Lactate is a moderately strong acid, completely dissociating under physiological conditions and thus lowering the pH of blood and muscle tissue. Hyperventilation removes hydrogen ions, raising the pH of the blood and tissues in anticipation of the acid buildup.

Chapter 5

1. L; determine the absolute configuration at the α carbon and compare it with D- and L-glyceraldehyde.

2. (a) I **(b)** II **(c)** IV **(d)** II **(e)** IV **(f)** II and IV **(g)** III **(h)** III **(i)** V **(j)** III **(k)** V **(l)** II **(m)** III **(n)** V **(o)** I, III, and V

3. (a) pI > pK_a of the α-carboxyl group and pI < pK_a of the α-amino group, so both groups are charged (ionized).
(b) 1 in 2.19×10^7. The pI of alanine is 6.01. From Table 5–1 and the Henderson-Hasselbalch equation, 1/4,680 carboxyl groups and 1/4,680 amino groups are uncharged. The fraction of alanine molecules with both groups uncharged is the product of these fractions.

4. (a)–(c)

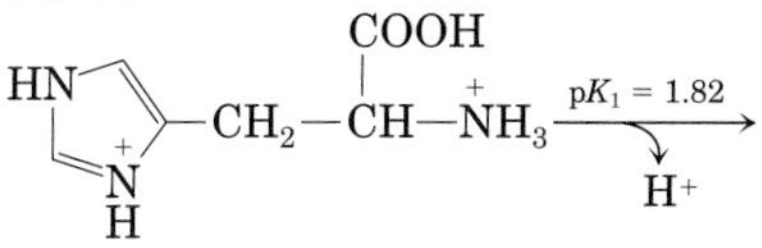

1: imidazolium(HN, $\overset{+}{N}$H)–CH_2–CH(COOH)–$\overset{+}{N}H_3$ —pK_1 = 1.82→ H^+

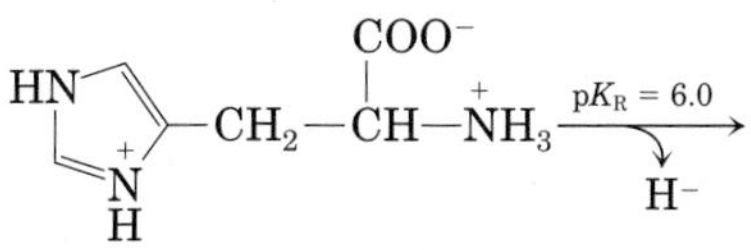

2: imidazolium(HN, $\overset{+}{N}$H)–CH_2–CH(COO^-)–$\overset{+}{N}H_3$ —pK_R = 6.0→ H^+

3: imidazole(HN, N)–CH_2–CH(COO^-)–$\overset{+}{N}H_3$ —pK_2 = 9.17→ H^+

4: imidazole(HN, N)–CH_2–CH(COO^-)–NH_2

pH	Structure	Net charge	Migrates toward:
1	**1**	+2	Cathode (+)
4	**2**	+1	Cathode (+)
8	**3**	0	Does not migrate
12	**4**	−1	Anode (−)

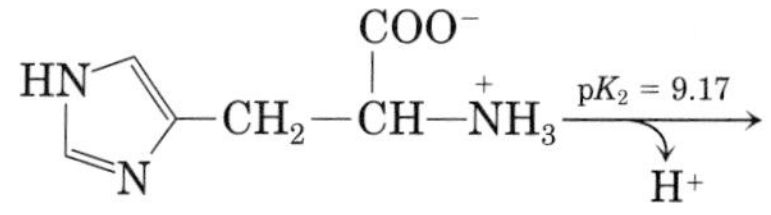

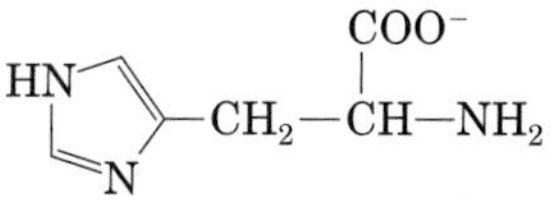

5. (a) Asp **(b)** Met **(c)** Glu **(d)** Gly **(e)** Ser

6. (a) 2 **(b)** 4
(c)

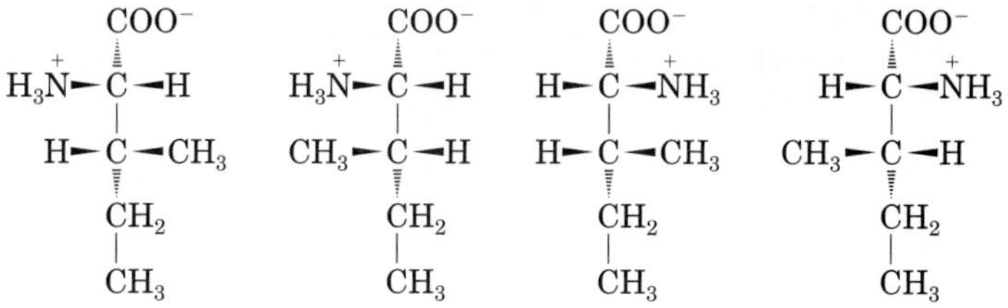

Four stereoisomers (COO^- / C / C / CH_2 / CH_3):
1. $H_3\overset{+}{N}$–C–H; H–C–CH_3
2. $H_3\overset{+}{N}$–C–H; CH_3–C–H
3. H–C–$\overset{+}{N}H_3$; H–C–CH_3
4. H–C–$\overset{+}{N}H_3$; CH_3–C–H

7. (a) Structure at pH 7:

$H_3\overset{+}{N}$–CH(CH_3)–C(=O)–N(H)–CH(CH_3)–C(=O)–N(H)–CH(NH_3)–C(=O)–O^-

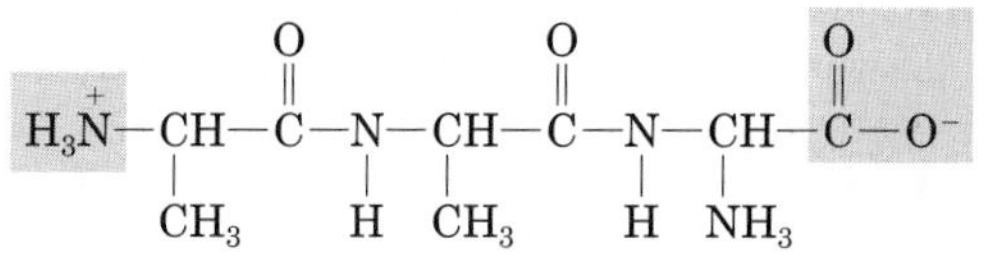

$pK_2 = 9.69$ $pK_1 = 2.34$

(b) Electrostatic interaction between the carboxylate anion and the protonated amino group of the alanine zwitterion favorably affects the ionization of the carboxyl group. This favorable electrostatic interaction decreases as the length of the poly-Ala increases, resulting in an increase in pK_1.
(c) Ionization of the protonated amino group destroys the favorable electrostatic interaction. With increasing distance between the charged groups, removal of the proton from the amino group in poly-Ala is easier and thus pK_2 is lower.

8. 75,020

9. (a) 32,100 **(b)** 2

10. (a) at pH 3, +2; at pH 8, 0; at pH 11, −1 **(b)** pI = 7.8

11. pI ≈ 1; carboxylate groups; Asp and Glu

12. Lys, His, Arg; negatively charged phosphate groups in DNA interact with positively charged side groups in histones.

13. (a) $(Glu)_{20}$ **(b)** $(Lys–Ala)_3$ **(c)** $(Asn–Ser–His)_5$ **(d)** $(Asn–Ser–His)_5$

14. (a) Specific activity after step 1 is 200 units/mg; step 2,600 units/mg; step 3,250 units/mg; step 4, 4,000 units/mg; step 5, 15,000 units/mg; step 6, 15,000 units/mg **(b)** Step 4 **(c)** Step 3 **(d)** Yes. Specific activity did not increase in step 6; polyacrylamide gel electrophoresis in the presence of SDS

15. Tyr–Gly–Gly–Phe–Leu

16.

```
         →Leu↘
    Orn          Phe
   ↗                ↘
 Val                 Pro
  ↑                   ↓
 Pro                 Val
   ↖                ↙
    Phe          Orn
        ↖Leu↙
```

The arrows correspond to the orientation of the peptide bonds, —CO → NH—.

17. 88%, 97%

18. **(a)** The protein to be isolated (citrate synthase, CS) is a relatively small fraction of the total cellular protein. Cold temperatures reduce protein degradation; sucrose provides an isotonic environment that preserves the integrity of organelles during homogenization. **(b)** This step separates organelles on the basis of relative size. **(c)** The first addition of ammonium sulfate removes some unwanted proteins from the homogenate. Additional ammonium sulfate precipitates CS. **(d)** To resuspend (solubilize) CS, ammonium sulfate must be removed under conditions of pH and ionic strength that support the native conformation. **(e)** CS molecules are larger than the pore size of the chromatographic gel. Protein is detectable at 280 nm because of absorption at this wavelength by Try and Trp residues. **(f)** CS has a positive charge and thus binds to the negatively charged cation-exchange column. After the neutral and negatively charged proteins pass through, CS is displaced from the column using the washing solution of higher pH, which alters the charge on CS. **(g)** Different proteins can have the same pI. The SDS gel confirmed that only a single protein was purified. SDS is difficult to remove completely from a protein, and its presence distorts the acid-base properties of the protein, including pI.

Chapter 6

1. **(a)** Shorter bonds are stronger and have a higher bond order (are multiple rather than single). The C—N bond is stronger than a single bond and is midway between a single and a double bond in character. **(b)** Rotation about the peptide bond is difficult at physiological temperatures because of its partial double-bond character.

2. **(a)** The principal structural units in the wool fiber polypeptides are successive turns of the α helix, which are spaced at 5.4 Å intervals. Steaming and stretching the fiber yields an extended polypeptide chain with the β conformation, in which the distance between adjacent R groups is about 7.0 Å. As the polypeptide reassumes an α-helical structure, the fiber shrinks. **(b)** Processed wool shrinks when polypeptide chains are converted from an extended β conformation to the native α-helical conformation in the presence of moist heat. The β-pleated sheets of silk, with their small, closely packed amino acid side chains, are more stable than those of wool.

3. About 43 peptide bonds per second

4. At pHs above 6, the carboxyl groups of poly(Glu) become deprotonated; repulsion among the negatively charged carboxylate groups leads to unfolding. Similarly, at pHs below 9, the amino groups of poly(Lys) become protonated; repulsion among these positively charged groups also leads to unfolding.

5. **(a)** Disulfide bonds are covalent bonds, which are much stronger than the noncovalent interactions that stabilize most proteins. They cross-link protein chains, increasing their stiffness, mechanical strength, and hardness. **(b)** Cystine residues (disulfide bonds) prevent the complete unfolding of the protein.

6. **(a)** Bends are most likely at residues 7 and 19; Pro residues in the cis configuration accommodate turns well. **(b)** The Cys residues at positions 13 and 24 can form disulfide bonds. **(c)** External surface: polar and charged residues (Asp, Gln, Lys); interior: nonpolar and aliphatic residues (Ala, Ile); Thr, though polar, has a hydropathy index near zero and thus can be found either on the external surface or interior of the protein.

7. 30 amino acid residues; 89%

8. The bacterial enzyme (a collagenase) can destroy the connective-tissue barrier of the host, allowing the bacterium to invade the host tissues. Bacteria do not contain collagen.

9. **(a)** Calculating the number of moles of DNP-valine formed per mole of protein gives the number of amino termini and thus the number of polypeptide chains. **(b)** 4 **(c)** Different chains would likely run as discrete bans on an SDS-polyacrylamide gel.

10. **(a)** NF-κ-B transcription factor **(b)** You will obtain similar results, but with additional related proteins listed. **(c)** The protein has two subunits, a p65 form (A chain) and a p50 form (B chain), which interact to form a heterodimer (quaternary structure). **(d)** The NF-κ-B transcription factor is a heterodimer that binds specific DNA sequences, enhancing transcription of nearby genes. One such gene is the immunoglobulin κ light chain, from which it gets its name.

Chapter 7

1. Protein B has a higher affinity for ligand X; it will be half-saturated at a much lower concentration of X than will protein A. Protein A has a K_a of 10^6 M^{-1}. Protein B has a K_a of 10^9 M^{-1}.

2. All three have $n_H < 1.0$. Apparent negative cooperativity in ligand binding can be caused by the presence of two or more types of ligand-binding sites with different affinities for the ligand on the same or different proteins in the same solution. Apparent negative cooperativity is also commonly observed in heterogeneous protein preparations. There are few well-documented cases of true negative cooperativity.

3. Effect on myoglobin's affinity for O_2: **(a)** none; **(b)** none; **(c)** none. Effect on hemoglobin's affinity for O_2: **(a)** decreases; **(b)** increases; **(c)** decreases.

4. The cooperative behavior of hemoglobin arises from subunit interactions.

5. **(a)** The observation that hemoglobin A (HbA; maternal) is only 33% saturated when the pO_2 is 4 kPa, while hemoglobin F (HbF; fetal) is 58% saturated under the same physiological conditions, indicates that the O_2 affinity of HbF is higher than that of HbA.
(b) The higher O_2 affinity of HbF assures that oxygen will flow from maternal blood to fetal blood in the placenta. Fetal blood approaches full saturation where the O_2 affinity of HbA is low.
(c) The observation that the O_2-saturation curve of HbA undergoes a larger shift on BPG binding than does that of HbF suggests that HbA binds BPG more tightly than does HbF. Differential binding of BPG to the two hemoglobins may determine the difference in their O_2 affinities.

6. **(a)** Hb Memphis **(b)** HbS, Hb Milawukee, Hb Providence, maybe Hb Cowtown **(c)** Hb Providence

7. **(a)** 1.25×10^{-8} M **(b)** 5×10^{-8} M **(c)** 7.5×10^{-8} M **(d)** 2×10^{-7} M. Note that a rearrangement of Eqn 7–8 gives $[L] = \theta K_d/(1 - \theta)$.

8. The epitope is likely to be a structure that is buried when G-actin polymerizes to form F-actin.

9. Many pathogens, including HIV, have evolved mechanisms by which they can repeatedly alter the surface proteins to which immune system components initially bind. Thus the host organism regularly faces new antigens and requires time to mount an immune response to each one. As the immune system responds to one variant, new variants are created.

10. Binding of ATP to myosin triggers dissociation of myosin from the actin thin filament. In the absence of ATP, actin and myosin bind tightly to each other.

11.

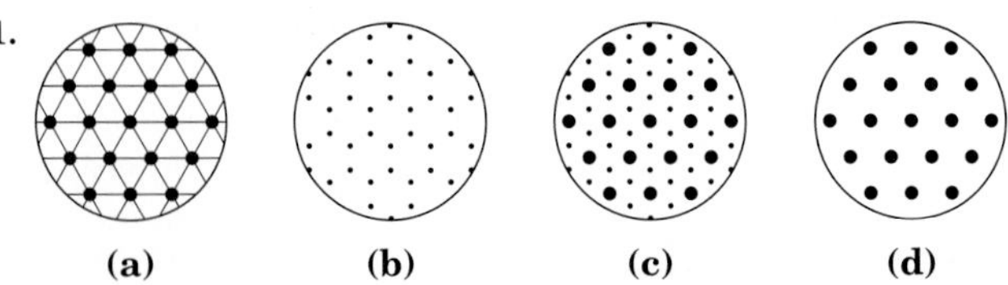

12. **(a)** Chain L is the light chain and chain H is the heavy chain of the Fab fragment of this antibody molecule. Chain Y is lysozyme.
(b) β Structure is predominant in the variable and constant regions of the fragment.
(c) Fab heavy-chain fragment, 218 amino acid residues; light-chain fragment, 214 residues; lysozyme, 129 residues. Less than 15% of the lysozyme molecule is in contact with the Fab fragment.
(d) In the H chain, the residues that appear to be in contact with lysozyme include Gly^{31}, Tyr^{32}, Arg^{99}, Asp^{100}, and Tyr^{101}. In the L chain the residues that appear to be in contact with lysozyme include Tyr^{32}, Tyr^{49}, Tyr^{50}, and Trp^{92}. In lysozyme, residues Asn^{19}, Gly^{22}, Tyr^{23}, Ser^{24}, Lys^{116}, Gly^{117}, Thr^{118}, Asp^{119}, Gln^{121}, and Arg^{125} appear to be situated at the antigen-antibody interface. Not all these residues are adjacent in the primary structure. Folding of the polypeptide chain into higher levels of structure brings the nonconsecutive residues together to form the antigen-binding site.

Chapter 8

1. The activity of the enzyme that converts sugar to starch is destroyed by heat denaturation.

2. 2.4×10^{-6} M

3. 9.5×10^{8} yr

4. The enzyme-substrate complex is more stable than the enzyme alone.

5. **(a)** 188 Å **(b)** Three-dimensional folding of the enzyme brings these amino acid residues into close proximity.

6. The reaction rate can be measured by following the decrease in the absorption of NADH (at 340 nm) as the reaction proceeds. Determine the K_m value; using substrate concentrations well above the K_m, measure the initial rate (rate of NADH disappearance with time, measured spectrophotometrically) at several known enzyme concentrations, and make a plot of the initial rates at increasing concentrations of enzyme. The plot should be linear, with a slope that provides a measure of LDH concentration.

7. **(a)** 1.7×10^{-3} M **(b)** 0.33; 0.67; 0.91

8. $V_{max} \approx 140$ μM/min; $K_m \approx 1 \times 10^{-5}$ M

9. **(a)** $V_{max} = 51.55$ m M /min, $K_m = 0.598$ mM **(b)** Competitive inhibition

10. $K_m = 2.2$ mM; $V_{max} = 0.51$ μmol/min

11. Curve A

12. $k_{cat} = 2.0 \times 10^{7}$ min^{-1}

13. The basic assumptions of the Michaelis-Menten equation still hold. The reaction is at steady state, and the rate is determined by:

$$V_0 = k_2\,[\mathrm{ES}]$$

The equations needed to solve for [ES] are:

$$[\mathrm{E_t}] = [\mathrm{E}] + [\mathrm{ES}] + [\mathrm{EI}]$$

and

$$[\mathrm{EI}] = \frac{[\mathrm{E}][\mathrm{I}]}{K_\mathrm{I}}$$

The term [E] can be obtained by rearranging Eqn 8–19. The rest follows the pattern established by the Michaelis-Menten equation derivation in the text.

Chapter 9

1. Yes; the empirical formula is CH_2O, typical of a carbohydrate.

2. With reduction of the carbonyl oxygen to a hydroxyl group, the chemistry at C-1 and C-3 is the same; the glycerol molecule is not chiral.

3. Osazone formation destroys the configuration around C-2 of aldoses, so aldoses differing only at the C-2 configuration give the same derivative, with the same melting point.

4. **(a)**

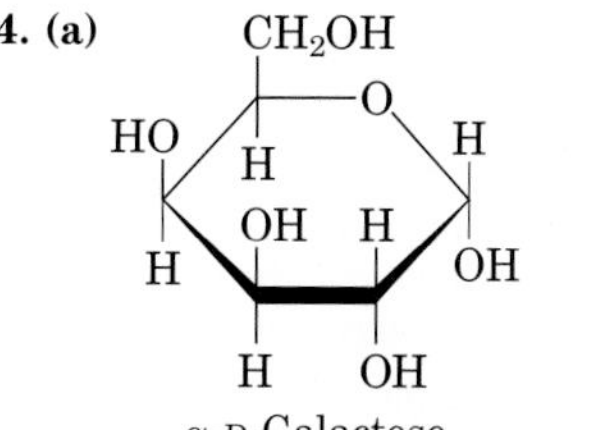

α-D-Galactose

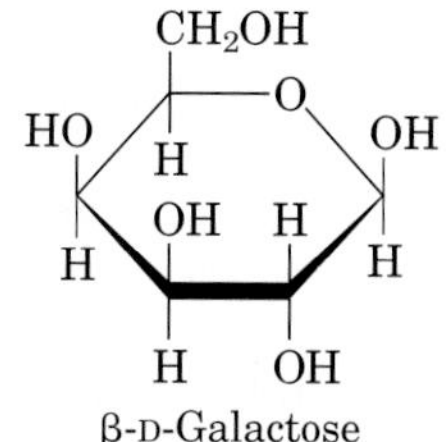

β-D-Galactose

(b) A fresh solution of α-D-galactose, or of β-D-galactose, undergoes mutarotation to an equilibrium mixture of the α and β forms.
(c) 28% α form; 72% β form

5. Fructose cyclizes to either the pyranose or the furanose structure. Increasing the temperature shifts the equilibrium in the direction of the furanose, the less sweet form.

6. The rate of mutarotation (interconversion of α and β anomers) is sufficiently high that as the enzyme consumes β-D-glucose, more α-D-glucose is converted to the β form, and eventually, all the glucose is oxidized. Glucose oxidase is specific for glucose and does not detect other reducing sugars (such as galactose) that react with Fehling's reagent.

7. **(a)** Measure the change in optical rotation with time.
(b) The optical rotation of the mixture is negative (inverted) relative to that of the sucrose solution.
(c) 0.63

8. Prepare a slurry of sucrose and water for the core; add a small amount of invertase; immediately coat with chocolate.

9. Sucrose has no free anomeric carbon to undergo mutarotation.

10. Native cellulose consists of glucose units linked by (β1→4) glycosidic bonds, which force the polymer chain into an extended conformation (see Fig. 9–17). Parallel series of these extended chains form intermolecular hydrogen bonds, aggregating into long, tough, insoluble fibers. Glycogen consists of glucose units linked by (α1→4) glycosidic bonds, which cause bends in the chain and prevent formation of long fibers. In addition, glycogen is highly branched and, because many of its hydroxyl groups are exposed to water, is highly hydrated and disperses in water.

Cellulose is a structural material in plants, consistent with its side-by-side aggregation into insoluble fibers. Glycogen is a storage fuel in animals. Highly hydrated glycogen granules with their many nonreducing ends are rapidly hydrolyzed by glycogen phosphorylase to release glucose 1-phosphate.

11. 7,700 residues/s

12. 11 s

13. The negative charges on chondroitin sulfate repel each other and force the molecule into an extended conformation. The polar molecule attracts many water molecules, increasing the molecular volume.

14. Positively charged amino acid residues would bind the highly negatively charged groups on heparin. In fact, Lys residues of antithrombin III interact with heparin.

15. Oligosaccharides; their subunits can be combined in more ways than the amino acid subunits of oligopeptides. Each hydroxyl group can participate in glycosidic bonds, and the configuration of each glycosidic bond can be either α or β. The polymer can be linear or branched.

16. **(a)** Branch-point residues yield 2,3-di-*O*-methylglucose; the unbranched residues yield 2,3,6-tri-*O*-methylglucose. **(b)** 3.75%

17. Chains of (1→6)-linked D-glucose residues with occasional (1→3)-linked branches, with about one branch for every 20 residues

Chapter 10

1. 0.35 mg/ml

2. N-3 and N-7

3. **(a)** (5′)GCGCAATATTTTGAGAAATATTGCGC(3′); it contains a palindrome. The individual strands can form hairpin structures; the two strands can form a cruciform.

4. 0.00094 g

5. **(a)** 40° **(b)** 0°

6. The RNA helix is in the A conformation; the DNA helix is generally in the B conformation.

7. In eukaryotic DNA, about 5% of C residues are methylated. 5-Methylcytosine can spontaneously deaminate to form thymine; the resulting G–T pair is one of the most common mismatches in eukaryotic cells.

8. Base stacking in nucleic acids tends to reduce the absorption of UV light. Denaturation involves loss of base stacking, and UV absorption increases.

9. One DNA contains 32% A, 32% T, 18% G, and 18% C; the other 17% A, 17% T, 33% G, and 33% C. This assumes that both are double-stranded. The DNA with 33% G and 33% C most likely came from the thermophilic bacterium; its higher G≡C content makes it much more stable to heat.

10.

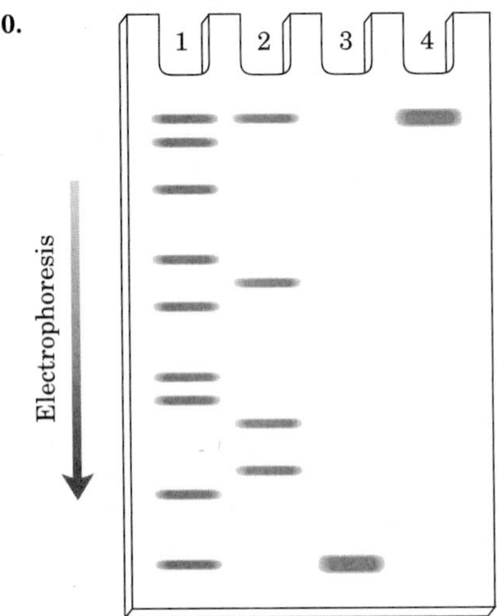

11.

Deoxyribose   Guanine   Phosphate

Solubilities: phosphate > deoxyribose > guanine. The highly polar phosphate groups and sugar moieties are on the outside of the double helix, exposed to water; the hydrophobic bases are in the interior of the helix.

12. (5′)P—GCGCCAUUG(3′)—OH
(5′)P—GCGCCAUU(3′)—OH
(5′)P—GCGCCAU(3′)—OH
(5′)P—GCGCCA(3′)—OH
(5′)P—GCGCC(3′)—OH
(5′)P—GCGC(3′)—OH
(5′)P—GCG(3′)—OH
(5′)P—GC(3′)—OH
and the nucleoside 5′-phosphates

13. **(a)** Water is a participant in most biological reactions, including those that cause mutations. The low water content in endospores reduces the activity of mutation-causing enzymes and slows the rate of nonenzymatic depurination reactions, which are hydrolysis reactions.
(b) UV light induces formation of cyclobutane pyrimidine dimers. Because *B. subtilis* is a soil organism, spores can be lofted to the top of the soil or into the air, where they may be subject to prolonged UV exposure.

14. **(a)** The DNA backbone can be identified using **Backbone** display in Chime. Individual bases are named at the bottom of the frame by clicking on bases in the ball-and-stick mode. The base at the 5′ end is adenine. This is a right-handed helix.
(b) This is a left-handed helix.
(c) See the Study Guide or the *Principles of Biochemistry* Web site for additional tips on viewing structures in stereo.

Chapter 11

1. The term "lipid" does not specify a particular chemical structure. Compounds are categorized as lipids based on their greater solubility in organic solvents than in water.

2. **(a)** The number of cis double bonds. Each cis double bond causes a bend in the hydrocarbon chain, lowering the melting temperature.
(b) Six different triacylglycerols can be constructed. Melting points OOO < OOP = OPO < PPO = POP < PPP, where O = oleic and P = palmitic acid. The greater the content of saturated fatty acid, the higher is the melting point.
(c) Branched-chain fatty acids increase the fluidity of membranes because they decrease the extent of membrane lipid packing.

3. Lecithin, an amphipathic compound, is an emulsifying agent, facilitating the solubilization of butter.

4. *Hydrophobic units:* **(a)** 2 fatty acids; **(b)**, **(c)**, and **(d)** 1 fatty acid and the hydrocarbon chain of sphingosine; **(e)** the hydrocarbon backbone. *Hydrophilic units:* **(a)** phosphoethanolamine; **(b)** phosphocholine; **(c)** D-galactose; **(d)** several sugar molecules; **(e)** alcohol group (OH)

5. The triacylglycerols of animal fats (grease) are hydrolyzed by NaOH (saponified) to form soaps, which are much more soluble in water than are triacylglycerols.

6. **(a)** The free OH group on C-2 and the phosphorylcholine head group on C-3 are hydrophilic; the fatty acid on C-1 of lysolecithin is hydrophobic.
(b) Certain steroids such as prednisone inhibit the action of phospholipase A_2, inhibiting the release of arachidonic acid from C-2. Arachidonic acid is converted to a variety of eicosanoids, some of which cause inflammation and pain.
(c) Phospholipase A_2 releases arachidonic acid, a precursor of other eicosanoids with vital protective functions in the body; it also breaks down dietary glycerophospholipids.

7. Diacylglycerol is hydrophobic and remains in the membrane lipid. Inositol 1,4,5-trisphosphate is highly polar, very soluble in water, and more readily diffusible in the cytosol. Both are second messengers.

8. Water-soluble vitamins are more rapidly excreted in the urine and are not stored effectively. Fat-soluble vitamins have very low solubility in water and are stored in body lipids.

9. **(a)** Glycerol and the sodium salts of palmitic and stearic acids.
(b) D-Glycerol 3-phosphorylcholine and the sodium salts of palmitic and oleic acids.

10. Solubilities in water: monoacylglycerol > diacylglycerol > triacylglycerol.

11. First eluted to last eluted: cholesteryl palmitate and triacylglycerol; cholesterol and *n*-tetradecanol; phosphatidylcholine and phosphatidylethanolamine; sphingomyelin; and phosphatidylserine and palmitate.

12. **(a)** Subject acid hydrolysates of each compound to chromatography (GLC or silica gel TLC) and compare the result with known standards. Sphingomyelin hydrolysate: sphingosine, fatty acids, phosphocholine, choline, and phosphate; cerebroside hydrolysate: sphingosine, fatty acids, sugars, but no phosphate.
(b) Strong alkaline hydrolysis of sphingomyelin yields sphingosine; phosphatidylcholine yields glycerol. Detect hydrolysate components on thin-layer chromatograms by comparing with standards or by their differential reaction with 2,4-DNP (only sphingosine reacts to form a colored product). Treatment with phospholipase A_1 or A_2 releases free fatty acids from phosphatidylcholine, but not from sphingomyelin.

13. Phosphatidylethanolamine and phosphatidylserine

Chapter 12

1. From the known amount of lipid used, its molecular weight, and the area occupied by a monolayer (determined as shown), the area per molecule can be calculated.

2. The data support a bilayer of lipid in the dog erythrocytes. The surface area of 8 × 10^9 dog cells is 62 m^2, or 190 μm^2, and the total area of a lipid monolayer from one cell is 98 μm^2. In the case of sheep and human erythrocytes, the data suggest a monolayer, not a bilayer. In fact, there were significant experimental errors in these early experiments; recent, more accurate measurements support a bilayer in all cases.

3. 63 SDS molecules per micelle

4. **(a)** Lipids that form bilayers are amphipathic molecules: they contain a hydrophilic and a hydrophobic unit. To minimize the hydrophobic area that is exposed to the water surface, these lipids form two-dimensional sheets with the hydrophilic units exposed to water and the hydrophobic units buried in the interior of the sheet. Furthermore, to avoid exposing the hydrophobic edges of the sheet to water, lipid bilayers close upon themselves. Similarly, if the sheet is perforated, the hole will seal because the membrane is semifluid. **(b)** These sheets form the closed membrane surfaces that envelop cells and compartments within cells (organelles).

5. 2.1 nm. In the straight-chain form, all bond angles are about 109°. The distance between the first and third carbons is about 0.26 nm, so each —CH_2— group contributes about 0.13 nm to the chain length. For palmitate (16:0), the length is 16 × 0.013 nm = 2.1 nm. Two palmitates placed end to end span about 4.2 nm, approximately the width of a typical bilayer.

6. A decrease. Because cell-cell fusion requires movement of individual lipids in the bilayers, the process will occur much faster at 37 °C, when the lipids are in the "fluid" phase, than at 10 °C, when they are in the "solid" phase. The same holds for membrane protein mixing.

7. 35 kJ/mol, neglecting the effects of transmembrane electrical potential; 0.60 mol

8. 13 kJ/mol

9. Most of the oxygen consumed by a tissue is for oxidative phosphorylation, the source of most of the ATP. Therefore, about two-thirds of the ATP synthesized by the kidney is used for pumping K^+ and Na^+.

10. No; the symporter may carry more than one equivalent of Na^+ for each mole of glucose transported.

11. Salt extraction indicates a peripheral location, and inaccessibility to protease in intact cells indicates an internal location. X resembles a peripheral membrane protein.

12. The interactions among membrane lipids are noncovalent and reversible, and they form spontaneously.

13. The temperature of body tissues at the extremities, such as near the hooves, is generally lower than that of tissues closer to the center of the body. If lipid is to remain fluid at this lower temperature, as required by the fluid mosaic model, it must contain a higher proportion of unsaturated fatty acids; unsaturated fatty acids lower the melting point of lipid mixtures.

14. The energetic cost of moving the highly polar, sometimes charged, head group through the hydrophobic interior of the bilayer is prohibitive.

15. At pH 7, tryptophan bears a positive and a negative charge, but indole is uncharged. The movement of the less polar indole through the hydrophobic core of the bilayer is energetically more favorable.

16. 2.2 × 10^{-3} seconds

17. Treat a suspension of cells with unlabeled NEM in the presence of excess lactose, remove the lactose, then add radiolabeled NEM. Use SDS-PAGE to determine the M_r of the radiolabeled band (the transporter).

18. Construct a hydropathy plot; hydrophobic regions of 20 or more residues suggest transmembrane segments. Determine whether the protein in intact erythrocytes reacts with a membrane-impermeant reagent specific for primary amines; if so, the transporter is of type I.

19. The leucine transporter is specific for the L isomer, but the binding site can accommodate either L-leucine or L-valine. Reduction of V_{max} in the absence of Na^+ indicates that leucine (or valine) is transported by symport with Na^+.

20. V_{max} reduced; K_t unaffected

21. ~1%; estimated by calculating the surface area of the cell and of 10,000 transport molecules (using dimensions of hemoglobin (p. 210) as a model globular protein).

22. **(a)** The rise-per-residue for an α helix (Chapter 6) is about 1.5 Å = 0.15 nm. To span a 4 nm bilayer, an α helix must contain about 27 residues; thus for seven spans, about 190 residues are required. A protein of M_r 64,000 has about 580 residues.
(b) To locate transmembrane regions, a hydropathy plot is used.
(c) Because about half of this portion of the epinephrine receptor consists of charged residues, it is likely that it represents an intracellular loop that connects two adjacent membrane-spanning regions of the protein.
(d) Because this helix is composed mostly of hydrophobic residues, it is likely that this portion of the receptor is one of the membrane-spanning regions of the protein.

Chapter 13

1. Albuterol raises [cAMP], leading to relaxation and enlargement of the bronchi. Because β-adrenergic receptors control many other processes, this drug would have undesirable side effects. To minimize them, find an agonist specific for the subtype of β-adrenergic receptors found in the smooth muscle of bronchi.

2. The amplification results from catalysts activating catalysts, as when an insulin receptor phosphorylates many molecules of IRS-1, when one molecule of Raf phosphorylates many molecules of MEK, each of which phosphorylates and activates many molecules of MAPK.

3. All of the following contribute to response termination: hormone degradation, hydrolysis of GTP bound to a G protein, degradation, metabolism, or sequestration of second messenger, receptor desensitization, and removal of receptor from the cell surface.

4. Two cells expressing the same surface receptor may have different complements of target proteins for protein phosphorylation.

5. **(a)** If the V_m were set by permeability primarily to K^+, the Nernst equation would predict a V_m of −86 mV, not the observed −95 mV, so some other conductance must contribute to V_m. **(b)** Chloride ion is probably the determinant of V_m; the predicted E_{Cl^-} is −95 mV.

6. **(a)** The V_m of the oocyte membrane goes from −65 mV to −10 mV—that is, it is depolarized. **(b)** The effect of KCl depends on an influx of Ca^{2+} from the extracellular medium.

7. Hyperpolarization of rod cells results in the closing of voltage-dependent Ca^{2+} channels in the presynaptic region of the rod cell. The resulting decrease in $[Ca^{2+}]_{in}$ diminishes the release of an inhibitory neurotransmitter, which suppresses the activity in the next neuron of the visual circuit. When this inhibition is removed in response to a light stimulus, the circuit becomes active and visual centers in the brain are excited.

8. **(a)** Because adenylyl cyclase is a membrane-bound protein, centrifugation sediments it into the particulate fraction. **(b)** Epinephrine stimulates the production of cAMP, a soluble substance that stimulates glycogen phosphorylase. **(c)** The heat-stable substance is cAMP; it can be prepared by treating ATP with barium hydroxide.

9. **(a)** It increases the level of cAMP. **(b)** The observations suggest that cAMP regulates Na^+ permeability. **(c)** Replace lost body fluids and electrolytes.

10. Unlike cAMP, dibutyryl-cAMP passes readily through the cell membrane.

11. G_s remains in its activated form once it has bound the nonhydrolyzable analog. Analog injection would therefore prolong the effect of epinephrine on the injected cell.

12. Shared properties of Ras and G_s: both can bind either GDP or GTP; both are activated by GTP; both can, when active, activate a downstream enzyme; both have intrinsic GTPase activity that shuts them off after a short period of activation. Differences between Ras and G_s: Ras is a small, monomeric protein; G_s is heterotrimeric. Functional differences: G_s activates adenylyl cyclase, G_i inhibits it.

13. Vasopressin acts by elevating cytosolic [Ca^{2+}] to 10^{-6} M, activating protein kinase C. EGTA injection will block vasopressin action, but should not affect the response to glucagon, which uses cAMP, *not* Ca^{2+}, as its intracellular messenger.

14. Individuals with Oguchi disease might have a defect in rhodopsin kinase or in arrestin.

15. (a) The mutation makes R unable to bind and inhibit the C subunit, so C is constantly active. **(b)** The mutation prevents cAMP binding to R, leaving C inhibited by bound R.

16. PKA (cAMP); PKG (cGMP); PKC (Ca^{2+}, DAG); Ca^{2+}/CaM kinase (Ca^{2+}, CaM); cyclin-dependent kinase (cyclin); protein tyrosine kinase (ligand for the receptor, such as insulin); MAP kinase (Raf); Raf (Ras); glycogen phosphorylase kinase (PKA).

17. An oncogene encodes a regulator protein, which, when defective, leads to unregulated cell growth and division. The normal product of this gene triggers cell division, but only when the appropriate signal is present. The mutant version of the oncogene product constantly sends the signal to divide, whether or not growth factors are present. A tumor suppressor gene in its normal cellular form encodes a protein that restrains cell division. Mutant forms of the protein fail to suppress cell division, but if either of the two alleles of the gene present encodes a normal protein, the normal function will continue.

18. Children who develop multiple tumors in both eyes were born with every cell in the retina already having one defective copy of the *Rb* gene. Early in their lives, as retinal cells divided, several cells each independently suffered a second mutation that damaged the one remaining good copy of the *Rb* gene, producing a tumor. The single-tumor form of the disease occurs in children who at birth had two good copies of the *Rb* gene in every cell. In a single cell, two mutations make both copies of the *Rb* gene defective. It is very unlikely that this will happen twice in the same person, and these children therefore develop only one tumor, and in only one eye.

19. A mutation in Ras that inactivates its GTPase activity creates a protein that, once activated by the binding of GTP, continues to give the signal via Raf to divide.

Chapter 14

1. Consider the developing chick as the system; the nutrients, egg shell, and outside world are the surroundings. Transformation of the single cell into a chick drastically reduces the entropy of the system. Initially, the parts of the egg outside the embryo (the surroundings) contain complex fuel molecules (a low-entropy condition). During incubation, some of these complex molecules are converted into large numbers of CO_2 and H_2O molecules (high entropy). This increase in the entropy of the surroundings is larger than the decrease in entropy of the chick (the system).

2. (a) −4.75 kJ/mol **(b)** 7.6 kJ/mol **(c)** −13.7 kJ/mol

3. (a) 261 m **(b)** 609 M **(c)** 0.29

4. $K'_{eq} = 21$; $\Delta G'^\circ = -7.6$ kJ/mol

5. −30.7 kJ/mol

6. (a) −1.7 kJ/mol **(b)** −4.4 kJ/mol
(c) At a given temperature, the value of $\Delta G'^\circ$ for any reaction is fixed and is defined for standard conditions (both fructose 6-phosphate and glucose 6-phosphate at 1 M). In contrast, ΔG is a variable that can be calculated for any set of reactant and product concentrations.

7. Less. The overall equation for ATP hydrolysis can be approximated as:

$$ATP^{4-} + H_2O \longrightarrow ADP^{3-} + HPO_4^{2-} + H^+$$

(This is only an approximation because the ionized species shown here are the major, but not the only, forms present.) Under standard conditions (i.e., [ATP] = [ADP] = [P_i] = 1 M), the concentration of water is 55 M and does not change during the reaction. Because H^+ ions are produced in the reaction, at a higher [H^+] (pH 5.0) the equilibrium would be shifted to the left and less free energy would be released.

8. 9.6

9. (a) 3.8×10^{-3} M; [glucose 6-phosphate] = 8.7×10^{-8} M; no
(b) 14 M; because the maximum solubility of glucose is less than 1 M, this is not a reasonable step.
(c) 843 ($\Delta G'^\circ = -17$ kJ/mol); [glucose] = 1.2×10^{-7} M; yes
(d) No; this would require such high [P_i] that the phosphate salts of divalent cations would precipitate.
(e) By directly transferring the phosphoryl group from ATP to glucose, the phosphoryl group transfer potential ("tendency" or "pressure") of ATP is utilized without generating high concentrations of intermediates. The essential part of this transfer is, of course, the enzymatic catalysis.

10. (a) −12.5 kJ/mol **(b)** −14.6 kJ/mol

11. (a) 3.1×10^{-4} **(b)** 69.4 **(c)** 7.5×10^4

12. −10.0 kJ/mol

13. 46.0 kJ/mol

14. (a) 46.0 kJ/mol **(b)** 46 kg; 68% **(c)** ATP is synthesized as it is needed then broken down to ADP and P_i; its concentration is maintained at a steady state.

15. The ATP system is in a dynamic steady state; [ATP] remains constant because the rate of ATP consumption equals its rate of synthesis. ATP consumption involves release of the terminal (γ) phosphoryl group; synthesis of ATP from ADP involves replacement of this phosphoryl group. Hence, the terminal phosphate undergoes rapid turnover. In contrast, the central (β) phosphate undergoes only relatively slow turnover.

16. (a) 1.7 kJ/mol
(b) Inorganic pyrophosphatase catalyzes the hydrolysis of pyrophosphate and drives the net reaction toward the synthesis of acetyl-CoA.

17. 34.2 kJ/mol

18. (a) NAD^+/NADH **(b)** Pyruvate/lactate **(c)** Lactate formation
(d) −26 kJ/mol **(e)** 3.59×10^4

19. (a) 1.14 V **(b)** −220 kJ/mol **(c)** About 4

20. (a) −0.35 V **(b)** −0.32 V **(c)** −0.29 V

21. In order of increasing tendency: **(a)**; **(d)**; **(b)**; **(c)**

22. (c) and **(d)**

Chapter 15

1. Net equation: Glucose + 2ATP →
2 glyceraldehyde 3-phosphate + 2ADP + $2H^+$
$\Delta G'^\circ = 2.1$ kJ/mol

2. Net equation:
2 Glyceraldehyde 3-phosphate + 4ADP + $2P_i$ →
2 lactate + $2NAD^+$
$\Delta G'^\circ = -113.6$ kJ/mol

3. (a) $^{14}CH_3CH_2OH$ **(b)** [3-^{14}C] glucose or [4-^{14}C] glucose

4. Soybeans and wheat contain starch, a polymer of glucose, which is broken down to glucose by the microorganisms. The glucose is then broken down to pyruvate via glycolysis. Because the process is carried out in the absence of oxygen (i.e., it is a fermentation), pyruvate is reduced to lactic acid and ethanol by the microorganisms. If oxygen were present, pyruvate would be oxidized to acetyl-CoA and then to CO_2 and H_2O. Some of the acetyl-CoA, however, would also be hydrolyzed to acetic acid (vinegar) in the presence of oxygen.

5. At C-1. This experiment demonstrates the reversibility of the aldolase reaction. The C-1 of glyceraldehyde 3-phosphate is equivalent to C-4 of fructose 1,6-bisphosphate (see Fig. 15–4). The starting glyceraldehyde 3-phosphate must have been labeled at C-1. The C-3 of dihydroxyacetone phosphate becomes labeled through the triose phosphate isomerase reaction, thus giving rise to the labeling of C-3 in fructose 1,6-bisphosphate.

6. There would be no anaerobic production of ATP; aerobic ATP production would be diminished only slightly.

7. No; lactate dehydrogenase is required to recycle NAD^+ from the NADH formed during the oxidation of glyceraldehyde 3-phosphate.

8. The transformation of glucose to lactate occurs when muscle cells are low in oxygen and provides a means of generating ATP under oxygen-deficient conditions. Because lactate can be transformed back to pyruvate, glucose is not wasted; pyruvate can be oxidized by aerobic reactions when oxygen becomes plentiful. This metabolic flexibility gives the organism a greater capacity to adapt to its environment.

9. It rapidly removes the 1,3-bisphosphoglycerate in a favorable subsequent step, catalyzed by phosphoglycerate kinase.

10. **(a)** 1.4×10^{-9} M **(b)** The physiological concentration (0.023 mM) is 16,000 times higher than the equilibrium concentration; this reaction does not reach equilibrium in the cell. Many reactions in the cell are not at equilibrium.

11. **(a)** 3-Phosphoglycerate would be the product. **(b)** In the presence of arsenate there is no net ATP synthesis under anaerobic conditions.

12. **(a)** The stoichiometry of alcohol fermentation requires 2 mol of P_i per mole of glucose.
(b) Ethanol is the reduced product formed during reoxidation of NADH to NAD^+, and CO_2 is the byproduct of the conversion of pyruvate into ethanol. Yes; pyruvate must be converted to ethanol, to produce a continuous supply of NAD^+ for the oxidation of glyceraldehyde 3-phosphate. Fructose 1,6-bisphosphate accumulates; it is formed as an intermediate in glycolysis.
(c) Arsenate replaces P_i in the glyceraldehyde 3-phosphate dehydrogenase reaction to yield an acyl arsenate, which spontaneously hydrolyzes. This prevents the formation of ATP, but 3-phosphoglycerate continues through the pathway.

13. Dietary niacin is used to synthesize NAD^+. Oxidations carried out by NAD^+ are part of cyclic processes, with NAD^+ as an electron carrier (reducing agent). Because of this cycling, one molecule can oxidize many thousands of molecules of glucose, and thus the dietary requirement for the precursor vitamin (niacin) is relatively small.

14. The phosphate group on glucose 6-phosphate is completely ionized at pH 7, giving the molecule an overall negative charge. Because membranes are generally impermeable to electrically charged molecules, glucose 6-phosphate could not pass from the bloodstream into cells and hence could not enter the glycolytic pathway and generate ATP. (This is why glucose, once phosphorylated, cannot escape from the cell.)

15. Net equation: Glycerol + $2NAD^+$ + ADP + $P_i \rightarrow$
pyruvate + 2NADH + ATP + $2H^+$

16. **(a)** 0.029 **(b)** 307 **(c)** No. The value of the mass-action ratio is much lower than K'_{eq}, indicating that the PFK-1 reaction is far from equilibrium in cells; this reaction is slower than the subsequent reactions in glycolysis. Flux through the glycolytic pathway is largely determined by the activity of PFK-1.

17. In the absence of O_2, the ATP needs are met by anaerobic glucose metabolism (fermentation to lactate). Because aerobic oxidation of glucose produces far more ATP than does fermentation, less glucose is needed to produce the same amount of ATP.

18. **(a)** There are two binding sites for ATP: a catalytic site and a regulatory site. Binding of ATP to an allosteric site inhibits PFK-1, either by reducing V_{max} or increasing K_m for ATP at the catalytic site.
(b) Glycolytic flux is reduced when ATP is plentiful.
(c) The graph indicates that the addition of ADP suppresses the inhibition of ATP. Because the adenine nucleotide pool is fairly constant, consumption of ATP leads to an increase in ADP levels. The data indicate that the activity of PFK-1 may be regulated by the [ATP]/[ADP] ratio.

19. **(a)** *In muscle,* glycogen is broken down to supply energy (ATP) via glycolysis. Glycogen phosphorylase catalyzes the conversion of stored glycogen to glucose 1-phosphate, which is converted to glucose 6-phosphate, an intermediate in glycolysis. During strenuous activity, skeletal muscle requires large quantities of glucose 6-phosphate. In *the liver,* the breakdown of glycogen is used to maintain a steady level of blood glucose between meals (glucose 6-phosphate is converted to free glucose).
(b) In actively working muscle, ATP flux requirements are very high and glucose 1-phosphate needs to be produced rapidly, requiring a high V_{max}.

20. **(a)** $[P_i]$/[glucose 1-phosphate] $\approx$ 3.5/1
(b) The value of this ratio in the cell (>100:1) indicates that [glucose 1-phosphate] is far below the equilibrium value. The rate at which glucose 1-phosphate is removed (through entry into glycolysis) is greater than its rate of production (by the glycogen phosphorylase reaction). This indicates that metabolite flow is from glycogen to glucose 1-phosphate, and that the glycogen phosphorylase reaction is probably the regulatory step in glycogen breakdown.

21. **(a)** Increases **(b)** Decreases **(c)** Increases

22. *Case A:* (f), (3); *Case B:* (c), (3); *Case C:* (d), (1)

23. In galactokinase deficiency, galactose accumulates; in galactose 1-phosphate uridylyltransferase deficiency, galactose 1-phosphate accumulates. The latter is more toxic.

Chapter 16

1. **(a)**
① *Citrate synthase:*
Acetyl-CoA + oxaloacetate + $H_2O \rightarrow$ citrate + CoA + H^+
② *Aconitase:* Citrate $\rightarrow$ isocitrate
③ *Isocitrate dehydrogenase:*
Isocitrate + $NAD^+ \rightarrow \alpha$-ketoglutarate + CO_2 + NADH
④ *α-Ketoglutarate dehydrogenase:*
α-Ketoglutarate + NAD^+ + CoA $\rightarrow$ succinyl-CoA + CO_2 + NADH
⑤ *Succinyl-CoA synthetase:*
Succinyl-CoA + P_i + GDP $\rightarrow$ succinate + GTP + CoA
⑥ *Succinate dehydrogenase:*
Succinate + FAD $\rightarrow$ fumarate + $FADH_2$
⑦ *Fumarase:* Fumarate + $H_2O \rightarrow$ malate
⑧ *Malate dehydrogenase:*
Malate + $NAD^+ \rightarrow$ oxaloacetate + NADH + H^+
(b), (c) Step ① CoA, condensation; ② none, isomerization; ③ NAD^+, oxidative decarboxylation; ④ NAD^+, CoA, and thiamine pyrophosphate, oxidative decarboxylation; ⑤ CoA, phosphorylation; ⑥ FAD, oxidation; ⑦ none, hydration; ⑧ NAD^+, oxidation
(d) Acetyl-CoA + $3NAD^+$ + FAD + GDP + P_i + $2H_2O \rightarrow$
$2CO_2$ + 3NADH + $FADH_2$ + GTP + $2H^+$ + CoA

2. **(a)** Oxidation; methanol $\rightarrow$ formaldehyde + [H—H]
(b) Oxidation; formaldehyde $\rightarrow$ formate + [H—H]
(c) Reduction; CO_2 + [H—H] $\rightarrow$ formate + H^+
(d) Reduction; glycerate + [H—H] $\rightarrow$ glyceraldehyde + OH^-
(e) Oxidation; glycerol $\rightarrow$ dihydroxyacetone + [H—H]
(f) Oxidation; $2H_2O$ + toluene $\rightarrow$ benzoate + 3 [H—H] + H^+
(g) Oxidation; succinate $\rightarrow$ fumarate + [H—H]
(h) Oxidation; pyruvate + $H_2O \rightarrow$ acetate + [H—H] + CO_2

3. From the structural formulas, we see that the carbon-bound H/C ratio of hexanoic acid (11/6) is higher than that of glucose (7/6). Hexanoic acid is more reduced and yields more energy upon complete combustion to CO_2 and H_2O.

4. **(a)** Oxidized; ethanol + $NAD^+ \rightarrow$ acetaldehyde + NADH + H^+
(b) Reduced; 1,3-bisphosphoglycerate + NADH + $H^+ \rightarrow$
glyceraldehyde 3-phosphate + NAD^+ + HPO_4^{2-}
(c) Unchanged; pyruvate + $H^+ \rightarrow$ acetaldehyde + CO_2

(d) Oxidized; pyruvate + $NAD^+ \rightarrow$ acetate + NADH + H^+ + CO_2
(e) Reduced; oxaloacetate + NADH + $H^+ \rightarrow$ malate + NAD^+
(f) Unchanged; acetoacetate + $H^+ \rightarrow$ acetone + CO_2

5. **(a)** Oxygen consumption is a measure of the activity of the first two stages of cellular respiration: glycolysis and the citric acid cycle. The addition of oxaloacetate or malate stimulates the citric acid cycle and thus stimulates respiration. **(b)** The added oxaloacetate or malate serves a catalytic role, because it is regenerated in the latter part of the citric acid cycle.

6. **(a)** 5.6×10^{-6} **(b)** 1.1×10^{-8} M **(c)** 28 molecules

7. The terminal phosphoryl group in GTP can be transferred to ADP in a reaction that is catalyzed by nucleoside diphosphate kinase and has an equilibrium constant of unity:

$$\text{GTP} + \text{ADP} \longrightarrow \text{GDP} + \text{ATP}$$

8. **(a)** $^-OOC—CH_2—CH_2—COO^-$ (succinate) **(b)** Malonate is a competitive inhibitor of succinate dehydrogenase. **(c)** A block in the citric acid cycle stops NADH formation, which stops electron transfer, which stops respiration. **(d)** A large excess of succinate (substrate) overcomes the competitive inhibition.

9. **(a)** Add uniformly labeled $[^{14}C]$ glucose and check for the release of $^{14}CO_2$. **(b)** Equally distributed in C-2 and C-3 of oxaloacetate; an infinite number

10. **(a)** C-1 **(b)** C-3 **(c)** C-3 **(d)** Methyl group **(e)** Equally distributed in the $—CH_2—$ groups **(f)** C-4 **(g)** Equally distributed in C-2 and C-3

11. Thiamine is required for the synthesis of thiamine pyrophosphate (TPP), a prosthetic group in the pyruvate dehydrogenase and α-ketoglutarate dehydrogenase enzyme complexes. A thiamine deficiency reduces the activity of these enzyme complexes and causes the observed accumulation of precursors.

12. No. For every two carbons that enter as acetate, two leave the cycle as CO_2; thus there is no net synthesis of oxaloacetate. Net synthesis of oxaloacetate occurs by the carboxylation of pyruvate, an anaplerotic reaction.

13. **(a)** Inhibition of aconitase **(b)** Fluorocitrate; competes with citrate; by a large excess of citrate **(c)** Citrate and fluorocitrate are inhibitors of PFK-1. **(d)** All catabolic processes necessary for ATP production are shut down.

14. In glycolysis,
Glucose + $2P_i$ + 2ADP + $2NAD^+ \rightarrow$
2 pyruvate + 2ATP + 2NADH + $2H^+$ + $2H_2O$
Then pyruvate carboxylase catalyzes the reaction
2 Pyruvate + $2CO_2$ + 2ATP + $2H_2O \rightarrow$
2 oxaloacetate + 2ADP + $2P_i$ + $4H^+$
In the citric acid cycle, malate dehydrogenase then catalyzes the reaction
2 Oxaloacetate + 2NADH + $2H^+ \rightarrow$ 2 L-malate + $2NAD^+$
which recycles nicotinamide coenzymes under anaerobic conditions.
The overall reaction is
Glucose + $2CO_2 \rightarrow$ 2 L-malate + $4H^+$
The overall reaction produces four H^+, which increases the acidity and thus the tartness of the wine.

15. Net reaction: 2 Pyruvate + ATP + $2NAD^+$ + $H_2O \rightarrow$
α-ketoglutarate + CO_2 + ADP + P_i + 2NADH + $3H^+$

16. **(a)** Decreases **(b)** Increases **(c)** Decreases

17. **(a)** Citrate is produced through the action of citrate synthase on oxaloacetate and acetyl-CoA. Citrate synthase can be used for net synthesis of citrate when (1) there is a continuous influx of new oxaloacetate and acetyl-CoA and (2) isocitrate synthesis is restricted, as it is in a medium low in Fe^{3+}. Aconitase requires Fe^{3+}, so an Fe^{3+}-deficient medium restricts the synthesis of aconitase.
(b) Sucrose + $H_2O \rightarrow$ glucose + fructose
Glucose + $2P_i$ + 2ADP + $2NAD^+$ + $\rightarrow$
2 pyruvate + 2ATP + 2NADH + $2H^+$ + $2H_2O$
Fructose + $2P_i$ + 2ADP + $2NAD^+$ + $\rightarrow$
2 pyruvate + 2ATP + 2NADH + $2H^+$ + $2H_2O$
2 Pyruvate + $2NAD^+$ + 2CoASH $\rightarrow$ 2 acetyl-SCoA + 2NADH + $2CO_2$
2 Pyruvate + $2CO_2$ + 2ATP + $2H_2O \rightarrow$
2 oxaloacetate + 2ADP + $2P_i$ + $4H^+$
2 Acetyl-SCoA + 2 oxaloacetate + $2H_2O \rightarrow$ 2 citrate + 2CoASH + $2H^+$
The overall reaction is
Sucrose + H_2O + $2P_i$ + 2ADP + $6NAD^+ \rightarrow$
2 citrate + 2ATP + 6NADH + $10H^+$
(c) Note that the overall reaction consumes NAD^+. Because the cellular pool of this oxidized coenzyme is limited, it must be recycled by the electron-transfer chain with consumption of oxygen. Consequently, the overall conversion of sucrose to citric acid is an aerobic process and requires molecular oxygen.

18. Succinyl-CoA is an intermediate of the citric acid cycle; its accumulation signals reduced flux through the cycle, calling for reduced entry of acetyl-CoA into the cycle. Citrate synthase, by regulating the primary oxidative pathway of the cell, regulates the supply of NADH and, thereby, the flow of electrons from NADH to oxygen.

19. Fatty acid catabolism increases [acetyl-CoA], which stimulates pyruvate carboxylase. The resulting increase in [oxaloacetate] stimulates acetyl-CoA consumption by the citric acid cycle, and [citrate] rises, inhibiting glycolysis at the level of PFK-1. In addition, increased [acetyl-CoA] inhibits the pyruvate dehydrogenase complex, slowing the utilization of pyruvate from glycolysis.

20. Oxygen is needed to recycle NAD^+ from the NADH produced by the oxidative reactions of the citric acid cycle. Reoxidation of NADH occurs during mitochondrial oxidative phosphorylation.

21. *Resting:* [ATP] high; [AMP] low; [acetyl-CoA] and [citrate] intermediate. *Running:* [ATP] intermediate; [AMP] high; [acetyl-CoA] and [citrate] low. Glucose flux through glycolysis increases during the anaerobic sprint because: (1) the ATP inhibition of glycogen phosphorylase and PFK-1 is partially relieved, (2) AMP stimulates both enzymes, and (3) lower [citrate] and [acetyl-CoA] relieves their inhibitory effects on PFK-1 and pyruvate kinase, respectively.

22. The migrating bird relies on the highly efficient aerobic oxidation of fats, rather than the anaerobic metabolism of glucose used by a sprinting rabbit. The bird reserves its muscle glycogen for short bursts of energy during emergencies.

23. Toward citrate; ΔG for the citrate synthase reaction under these conditions is about -8.5 kJ/mol.

24. Steps ④ and ⑤ are essential in the reoxidation of the reduced lipoamide cofactor.

Chapter 17

1. The fatty acid portion; the carbons in fatty acids are more reduced than those in glycerol.

2. **(a)** 4.0×10^5 kJ (9.5×10^4 kcal) **(b)** 48 days **(c)** 0.5 lb/day

3. The first step in fatty acid oxidation is analogous to the conversion of succinate to fumarate; the second step, to the conversion of fumarate to malate; the third step, to the conversion of malate to oxaloacetate.

4. **(a)** $R—COO^-$ + ATP $\rightarrow$ acyl-AMP + PP_i
Acyl-AMP + CoA $\rightarrow$ acyl-CoA + AMP
(b) Irreversible hydrolysis of PP_i to $2P_i$ by cellular inorganic pyrophosphatase

5. Yes; some of the tritium is removed from palmitate during the dehydrogenation reactions of β oxidation. The removed tritium appears as tritiated water.

6. Fatty acyl groups condensed with CoA in the cytosol are first transferred to carnitine, releasing CoA, then transported into the mitochondrion, where they are again condensed with CoA. The cytosolic and mitochondrial pools of CoA are thus kept separate, and no radioactive CoA from the cytosolic pool enters the mitochondrion.

7. **(a)** In the pigeon, β oxidation predominates; in the pheasant, anaerobic glycolysis of glycogen predominates. **(b)** Pigeon muscle would consume more oxygen. **(c)** Fat contains more energy per gram than glycogen does. In addition, the anaerobic breakdown of glycogen is limited by the tissue's tolerance to lactate buildup. Thus, the pigeon, operating on the oxidative catabolism of fats, is the long-distance flyer. **(d)** These enzymes are the regulatory enzymes of their respective pathways and thus limit ATP production rates.

8. **(a)** The carnitine-mediated entry of fatty acids into mitochondria is the rate-limiting step in fatty acid oxidation. Carnitine deficiency slows fatty acid oxidation; added carnitine increases the rate.
(b) All of these increase the metabolic need for fatty acid oxidation.
(c) Carnitine deficiency might result from a deficiency of Lys, its precursor, or from a defect in one of the enzymes in the biosynthetic path to carnitine.

9. Oxidation of fats releases metabolic water; 1.4 L of water per kg of tripalmitoylglycerol (ignores the small contribution of glycerol to the mass)

10. The bacteria can be used to completely oxidize hydrocarbons to CO_2 and H_2O. However, contact between hydrocarbons and bacterial enzymes may be difficult to achieve. Bacterial nutrients such as nitrogen and phosphorus may be limiting and inhibit growth.

11. **(a)** M_r 136; phenylacetic acid **(b)** Even

12. Because the mitochondrial pool of CoA is small, CoA must be recycled from acetyl-CoA via the formation of ketone bodies. This allows the operation of the β-oxidation pathway, necessary for energy production.

13. **(a)** Glucose yields pyruvate via glycolysis, and pyruvate is the main source of oxaloacetate. Without glucose in the diet, [oxaloacetate] drops and the citric acid cycle slows.

(b) Odd-numbered; propionate conversion to succinyl-CoA provides intermediates for the citric acid cycle and four-carbon precursors for gluconeogenesis.

14. β Oxidation of ω-fluorooleate forms fluoroacetyl-CoA, which enters the citric acid cycle and produces fluorocitrate, a powerful inhibitor of aconitase. Inhibition of aconitase shuts down the citric acid cycle. Without reducing equivalents from the citric acid cycle, oxidative phosphorylation (ATP synthesis) is fatally slowed.

15. Enz-FAD, having a more positive standard reduction potential, is a better electron acceptor than NAD^+, and the reaction is driven in the direction of fatty acyl–CoA oxidation. This more favorable equilibrium is obtained at the cost of 1 ATP; only 1.5 ATP are produced per $FADH_2$ oxidized in the respiratory chain (versus 2.5 per NADH).

16. 9 turns; arachidic acid, a 20-carbon saturated fatty acid, yields 10 molecules of acetyl-CoA, the last two both formed in the ninth turn.

17. See Fig. 17–11. [3-^{14}C]Succinyl-CoA is formed, which gives rise to oxaloacetate labeled at C-2 and C-3.

18. ATP hydrolysis in the energy-requiring reactions of a cell takes up water in the reaction ATP + $H_2O \rightarrow$ ADP + P_i; thus, there is no net production of H_2O.

19. Methylmalonyl-CoA mutase requires the cobalt-containing cofactor formed from vitamin B_{12} (see Box 17–2).

20. Mass lost per day is about 0.66 kg, or about 138 kg in 7 months. Ketosis could be avoided by degradation of nonessential body proteins to supply amino acid skeletons for gluconeogenesis.

Chapter 18

1.

(a) $^-OOC{-}CH_2{-}C({=}O){-}COO^-$ Oxaloacetate

(b) $^-OOC{-}CH_2{-}CH_2{-}C({=}O){-}COO^-$ α-Ketoglutarate

(c) $CH_3{-}C({=}O){-}COO^-$ Pyruvate

(d) $C_6H_5{-}CH_2{-}C({=}O){-}COO^-$ Phenylpyruvate

2. This is a coupled-reaction assay. The product of the slow transamination (pyruvate) is rapidly consumed in the subsequent "indicator reaction" catalyzed by lactate dehydrogenase, which consumes NADH. Thus the rate of disappearance of NADH is a measure of the rate of the aminotransferase reaction. The indicator reaction is monitored by observing the decrease in absorption of NADH at 340 nm with a spectrophotometer.

3. No; the nitrogen in Ala can be transferred to oxaloacetate via transamination, to form Asp.

4. **(a)** Phenylalanine hydroxylase; a low-phenylalanine diet **(b)** The normal route of Phe metabolism via hydroxylation to Tyr is blocked, and Phe accumulates. **(c)** Phe is transformed to phenylpyruvate by transamination, and then to phenyllactate by reduction. The transamination reaction has an equilibrium constant of 1.0, and phenylpyruvate is formed in significant amounts when phenylalanine accumulates. **(d)** Because of the deficiency in production of Tyr, which is a precursor of melanin, the pigment normally present in hair

5. Catabolism of the carbon skeletons of Val, Met, and Ile is impaired because of the absence of functional methylmalonyl-CoA mutase (a coenzyme B_{12} enzyme). The physiological effects of loss of this enzyme are described in Table 18–2 and Box 18–2.

6. 15 moles of ATP per mole of lactate; 13 ATP per Ala, when nitrogen removal is included

7. **(a)** $^{15}NH_2{-}CO{-}^{15}NH_2$

(b) $^-OO^{14}C{-}CH_2{-}CH_2{-}^{14}COO^-$

(c) $R{-}NH{-}C({=}^{15}NH){-}^{15}NH_2$

(d) $R{-}NH{-}C({=}O){-}^{15}NH_2$

(e) No label

(f) $^-OO^{14}C{-}CH(^{15}NH_2){-}CH_2{-}^{14}COO^-$

8. **(a)** Ile $\xrightarrow{①}$ II $\xrightarrow{②}$ IV $\xrightarrow{③}$ I $\xrightarrow{④}$ V $\xrightarrow{⑤}$ III $\xrightarrow{⑥}$ acetyl-CoA + propionyl-CoA **(b)** Step ① transamination, no analogous reaction, PLP; ② oxidative decarboxylation, analogous to the pyruvate dehydrogenase reaction, NAD^+ ; ③ oxidation, analogous to the succinate dehydrogenase reaction, FAD; ④ hydration, analogous to the fumarase reaction, no cofactor required; ⑤ oxidation, analogous to the malate dehydrogenase reaction, NAD^+; ⑥ thiolysis (reverse aldol condensation), analogous to the thiolase reaction, CoA.

9. **(a)** Fasting resulted in low blood glucose; subsequent administration of the experimental diet led to rapid catabolism of glucogenic amino acids. **(b)** Oxidative deamination caused the rise in ammonia levels; the absence of Arg (an intermediate in the urea cycle) prevented the conversion of ammonia to urea; Arg is not synthesized in sufficient quantities in the cat to meet the needs imposed by the stress of the experiment. This suggests that Arg is an essential amino acid in the cat's diet.
(c) Ornithine is converted to Arg by the urea cycle.

10. H_2O + glutamate + NAD^+ ⟶
α-ketoglutarate + NH_4^+ + NADH + H^+
NH_4^+ + 2ATP + H_2O + CO_2 ⟶
carbamoyl phosphate + 2ADP + P_i + $3H^+$
Carbamoyl phosphate + ornithine ⟶ citrulline + P_i + H^+
Citrulline + aspartate + ATP ⟶
argininosuccinate + AMP + PP_i + H^+
Argininosuccinate ⟶ arginine + fumarate
Fumarate + H_2O ⟶ malate
Malate + NAD^+ ⟶ oxaloacetate + NADH + H^+
Oxaloacetate + glutamate ⟶ aspartate + α-ketoglutarate
Arginine + H_2O ⟶ urea + ornithine

2 Glutamate + CO_2 + $4H_2O$ + $2NAD^+$ + 3ATP ⟶
2 α-ketoglutarate + 2NADH + $7H^+$ + urea +
2ADP + AMP + PP_i + $2P_i$ (1)

Additional reactions that need to be considered are:
AMP + ATP ⟶ 2ADP (2)
O_2 + $8H^+$ + 2NADH + 6ADP + $6P_i$ ⟶
$2NAD^+$ + 6ATP + $8H_2O$ (3)
H_2O + PP_i ⟶ $2P_i$ + H^+ (4)
Summing equations (1) through (4),
2 Glutamate + CO_2 + O_2 + 2ADP + $2P_i$ ⟶
2 α-ketoglutarate + urea + $3H_2O$ + 2ATP

11. A likely mechanism is:

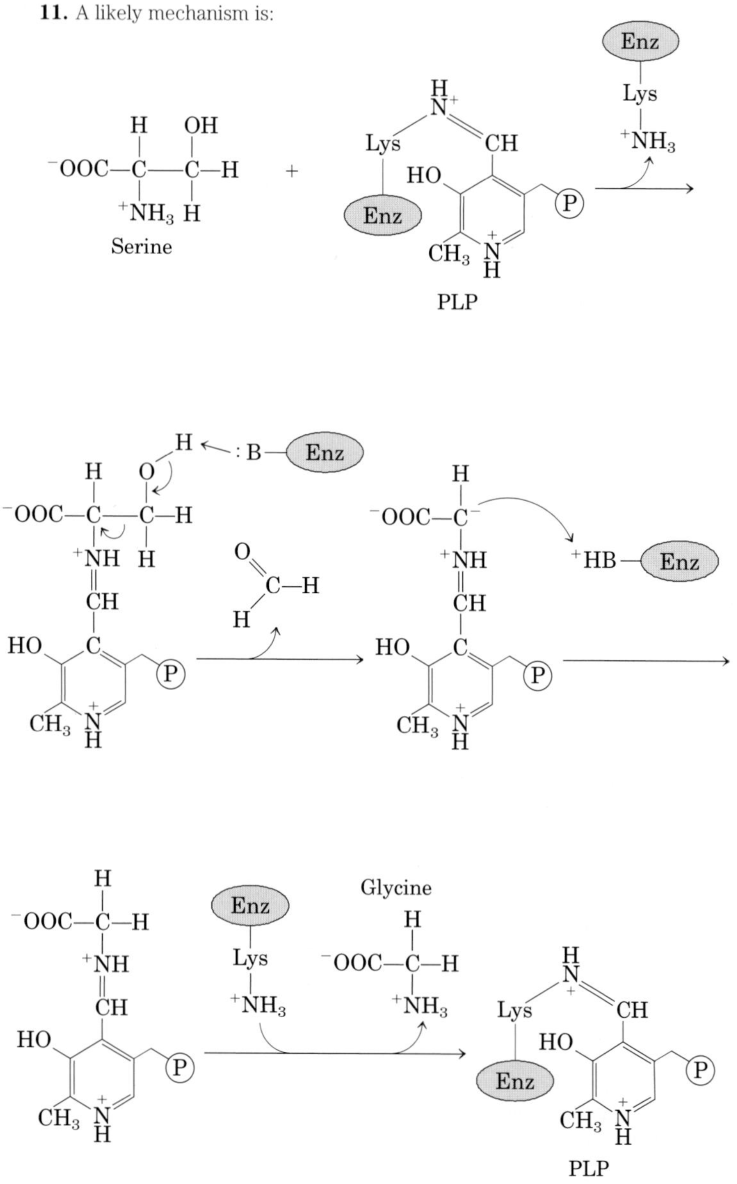

Serine

PLP

Glycine

PLP

The formaldehyde (HCHO) produced in the second step reacts rapidly with tetrahydrofolate at the enzyme active site to produce N^5,N^{10}-methylenetetrahydrofolate (see Fig. 18–16).

12. **(a)** Transamination; no analogies in either pathway; cofactor: PLP
(b) Oxidative decarboxylation; analogous to oxidative decarboxylation of pyruvate to acetyl-CoA prior to entry into citric acid cycle and of α-ketoglutarate to succinyl-CoA in citric acid cycle; cofactors: NAD^+, FAD, lipoate, and thiamine pyrophosphate
(c) Dehydrogenation (oxidation); analogous to dehydrogenation of succinate to fumarate in citric acid cycle and of fatty acyl–CoA to enoyl-CoA in β oxidation; cofactor: FAD
(d) Carboxylation; analogous to carboxylation of pyruvate to oxaloacetate in citric acid cycle; cofactors: ATP and biotin
(e) Hydration; analogous to hydration of fumarate to malate in citric acid cycle and of enoyl-CoA to 3-hydroxyacyl-CoA in β oxidation; cofactors: none
(f) Reverse aldol reaction; analogous to reverse of citrate synthase reaction in citric acid cycle and identical to cleavage of β-hydroxy-β-methylglutaryl-CoA in formation of ketone bodies; cofactors: none

13. The second amino group introduced into urea is transferred from Asp, which is generated during the transamination of Glu to oxaloacetate, a reaction catalyzed by aspartate aminotransferase. Approximately one half of all the amino groups excreted as urea must pass through the aspartate aminotransferase reaction, making this the most highly active aminotransferase.

14. **(a)** A person on a diet consisting only of protein must use amino acids as the principal source of metabolic fuel. Because the catabolism of amino acids requires the removal of nitrogen as urea, the process consumes abnormally large quantities of water to dilute and excrete the urea in the urine. Furthermore, electrolytes in the "liquid protein" must be diluted with water and excreted. If the daily water loss through the kidney is not balanced by a sufficient water intake, a net loss of body water results. **(b)** When considering the nutritional benefits of protein, one must keep in mind the total amount of amino acids needed for protein synthesis and the distribution of amino acids in the dietary protein. Gelatin contains a nutritionally unbalanced distribution of amino acids. As large amounts of gelatin are ingested and the excess amino acids are catabolized, the capacity of the urea cycle may be exceeded, leading to ammonia toxicity. This is further complicated by the dehydration that may result from excretion of large quantities of urea. A combination of these two factors could produce coma and death.

15. Ala and Gln play special roles in the transport of amino groups from muscle and from other nonhepatic tissues, respectively, to the liver.

Chapter 19

1. *Reaction (1):* **(a), (d)** NADH; **(b), (e)** E–FMN; **(c)** NAD^+/NADH and E–FMN/$FMNH_2$
Reaction (2): **(a), (d)** E–$FMNH_2$; **(b), (e)** Fe^{3+}; **(c)** E–FMN/$FMNH_2$ and Fe^{3+}/Fe^{2+}
Reaction (3): **(a), (d)** Fe^{2+}; **(b), (e)** Q; **(c)** Fe^{3+}/Fe^{2+} and Q/QH_2

2. The side chain makes ubiquinone soluble in lipids and allows it to diffuse in the semifluid membrane.

3. From the difference in standard reduction potential ($\Delta E'^\circ$) for each pair of half reactions, one can calculate $\Delta G'^\circ$. The oxidation of succinate by FAD is favored by the negative standard free-energy change ($\Delta G'^\circ = -3.86$ kJ/mol). Oxidation by NAD^+ would require a large, positive, standard free-energy change ($\Delta G'^\circ = 67.6$ kJ/mol).

4. **(a)** All carriers reduced; CN^- blocks the reduction of O_2 catalyzed by cytochrome oxidase. **(b)** All carriers reduced; in the absence of O_2, the reduced carriers are not reoxidized. **(c)** All carriers oxidized **(d)** Early carriers more reduced; later carriers more oxidized

5. **(a)** The inhibition of NADH dehydrogenase by rotenone decreases the rate of electron flow through the respiratory chain, which in turn decreases the rate of ATP production. If this reduced rate is unable to meet the organism's ATP requirements, the organism dies.

(b) Antimycin A strongly inhibits the oxidation of Q in the respiratory chain, reducing the rate of electron transfer and leading to the consequences described in (a).
(c) Because antimycin A blocks *all* electron flow to oxygen, it is a more potent poison than rotenone, which blocks electron flow from NADH but not from $FADH_2$.

6. (a) The rate of electron transfer necessary to meet the ATP demand increases, and thus the P/O ratio decreases.
(b) High concentrations of uncoupler produce P/O ratios near zero. The P/O ratio decreases, and more fuel must therefore be oxidized to generate the same amount of ATP. The extra heat released by this oxidation raises the body temperature.
(c) Increased activity of the respiratory chain in the presence of an uncoupler requires the degradation of additional fuel. By oxidizing more fuel (including fat reserves) to produce the same amount of ATP, the body loses weight. When the P/O ratio approaches zero, the lack of ATP results in death.

7. Valinomycin acts as an uncoupler. It combines with K^+ to form a complex that passes through the inner mitochondrial membrane, dissipating the membrane potential. ATP synthesis decreases, which causes the rate of electron transfer to increase. This results in an increase in the H^+ gradient, in O_2 consumption, and in the amount of heat released.

8. (a) The formation of ATP is inhibited. **(b)** The formation of ATP is tightly coupled to electron transfer: 2,4-dinitrophenol is an uncoupler of oxidative phosphorylation. **(c)** Oligomycin

9. Cytosolic malate dehydrogenase plays a key role in the transport of reducing equivalents across the inner mitochondrial membrane via the malate-aspartate shuttle.

10. (a) Glycolysis will become anaerobic. **(b)** Oxygen consumption will cease. **(c)** Lactate formation will increase. **(d)** ATP synthesis will decrease to 2ATP/glucose.

11. The steady-state concentration of P_i in the cell is much higher than that of ADP. P_i released by ATP hydrolysis changes total $[P_i]$ very little.

12. (a) NADH is reoxidized via electron transfer instead of lactic acid fermentation. **(b)** Oxidative phosphorylation is more efficient. **(c)** The high mass-action ratio of the ATP system inhibits phosphofructokinase-1.

13. The fermentation to ethanol could be accomplished in the presence of O_2, an advantage because strict anaerobic conditions are difficult to maintain. The Pasteur effect is not observed because the citric acid cycle and electron-transfer chain are inactive.

14. (a) External medium: 4.0×10^{-8} M; matrix: 2.0×10^{-8} M **(b)** 2:1; $\approx$ 1.7 kJ/mol **(c)** 21 **(d)** No **(e)** From the transmembrane potential

15. (a) 0.9 μmol/s·g **(b)** About 5.5 s; to provide a constant level of ATP, regulation of ATP production must be tight and rapid.

16. About 53 μmol/s·g. With a steady state [ATP] of 7 μmol/g, this is equivalent to 10 turnovers of the ATP pool per second; the reservoir would last about 0.13 s.

17. The inner mitochondrial membrane is impermeable to NADH, but the reducing equivalents of NADH are transferred (shuttled) through the membrane indirectly: they are transferred to oxaloacetate in the cytosol, the resulting malate is transported into the matrix, and mitochondrial NAD^+ is reduced to NADH.

18. Pyruvate dehydrogenase is located in mitochondria; glyceraldehyde phosphate dehydrogenase is located in cytosol. The NAD pools are separated by the inner mitochondrial membrane.

19. For the maximum photosynthetic rate, photosystem I (which absorbs light of 700 nm) and photosystem II (which absorbs light of 680 nm) must be operating simultaneously.

20. The extra weight comes from the water consumed in the overall reaction.

21. Purple sulfur bacteria use H_2S as the hydrogen donor in photosynthesis. No oxygen is evolved because the single photosystem that is present lacks the manganese-containing water-splitting complex.

22. 0.45

23. During illumination, a proton gradient is established. When ADP and P_i are added, ATP synthesis is driven by the gradient, which becomes exhausted in the absence of light.

24. DCMU blocks electron transfer between photosystem II and the first site of ATP production.

25. (a) 56 kJ/mol **(b)** 0.29 V

26. From the difference in reduction potentials, one can calculate that $\Delta G'^{\circ} = 17.4$ kJ for the redox reaction. From Figure 19–36 it is apparent that the energy of photons in any region of the visible spectrum is more than sufficient to drive this endergonic reaction.

27. 1.2×10^{-77}; the reaction is highly unfavorable! In chloroplasts, the input of light energy overcomes this barrier.

28. –920 kJ/mol

29. No; the electrons from H_2O flow to the artificial electron acceptor Fe^{3+}, not to $NADP^+$.

30. About once every 0.1 s; 1 in 10^8 is excited.

31. Light of 700 nm excites photosystem I but not photosystem II; electrons flow from P700 to $NADP^+$, but no electrons flow from P680 to replace them. When light of 680 nm excites photosystem II, electrons tend to flow to photosystem I, but the electron carriers between the two photosystems quickly become completely reduced.

32. No; the excited electron from P700 returns to refill the electron "hole" created by illumination. Photosystem II is not needed to supply electrons, and no O_2 is evolved from H_2O. NADPH is not formed because the excited electron returns to P700.

Chapter 20

1. No; glucose synthesis requires the input of energy (4ATP + 2GTP) and reducing power (2NADH), obtainable only through the citric acid cycle/oxidative phosphorylation pathway.

2. (a) In the pyruvate carboxylase reaction, $^{14}CO_2$ is added to pyruvate, but phosphoenolpyruvate carboxykinase removes the *same* CO_2 in the next step. Thus, ^{14}C is not (initially) incorporated into glucose.
(b)

CH_3–C(=O)–$^{14}COO^-$ (1-^{14}C-pyruvate) ⟶ COO^-–CH_2–C(=O)–$^{14}COO^-$ (Oxaloacetate) ⟶ CH_2=C(OPO_3^{2-})–$^{14}COO^-$ (Phosphoenolpyruvate)

⟶ CH_2OH–CH(OPO_3^{2-})–$^{14}COO^-$ (2-Phosphoglycerate) ⟶ $CH_2OPO_3^{2-}$–CH(OH)–$^{14}COO^-$ (3-Phosphoglycerate) ⟶ $CH_2OPO_3^{2-}$–CH(OH)–^{14}C(=O)–OPO_3^{2-} (1,3-Bisphosphoglycerate)

⟶ $CH_2OPO_3^{2-}$–CH(OH)–^{14}CH=O (Glyceraldehyde 3-phosphate) ⇌ $CH_2OPO_3^{2-}$–C(=O)–$^{14}CH_2OH$ (Dihydroxyacetone phosphate)

↓

$$\begin{array}{cc} {}^{2-}O_3POH_2C & CH_2OPO_3^{2-} \\ | & | \\ HOHC & C{=}O \\ | & | \\ HO{-}{}^{14}C & {}^{14}C{-}OH \\ | & | \\ H & H \end{array} \longrightarrow \longrightarrow \text{3,4-}^{14}\text{C-Glucose}$$

Fructose 1,6-bisphosphate

3. Pyruvate carboxylase is a mitochondrial enzyme. The [^{14}C]oxaloacetate formed mixes with the oxaloacetate pool of the citric acid cycle and is equilibrated with the citric acid cycle intermediates to form a mixture of [1-^{14}C] and [4-^{14}C]oxaloacetate. Oxaloacetate labeled at C-1 leads to the formation of [3,4-^{14}C]glucose (see Problem 2).

4. 4 ATP equivalents per glucose molecule

5. The flow of glucose to pyruvate and back to glucose would constitute a futile cycle. The pathways are reciprocally regulated—as the flow through one increases, the flow through the other decreases. Glycolysis is stimulated and gluconeogenesis inhibited by high [AMP]; glycolysis is inhibited and gluconeogenesis stimulated when [AMP] is low.

6. PFK-1 is activated by AMP and inhibited by ATP; it regulates glycolysis. FBPase-1 is activated by ATP and inhibited by AMP; it regulates gluconeogenesis.

7. (a), (b), and (d) are glucogenic; (c) and (e) are not.

8. **(a)** The rapid increase in glycolysis; the rise in pyruvate and NADH results in a rise in lactate. **(b)** Lactate is transformed to glucose via pyruvate; this is a slower process because formation of pyruvate is limited by NAD^+ availability, the LDH equilibrium is in favor of lactate, and conversion of pyruvate to glucose is energy-requiring. **(c)** The equilibrium for the LDH reaction is in favor of lactate formation.

9. Lactate is transformed to glucose in the liver by gluconeogenesis (see Fig. 20–2). A defect in FBPase-1 would prevent the entry of lactate into the gluconeogenic pathway in hepatocytes, causing lactate to accumulate in the blood.

10. Consumption of alcohol forces a competition for NAD^+ between ethanol metabolism and gluconeogenesis. The problem is compounded by strenuous exercise and lack of food because at these times the level of blood glucose is already low.

11. Succinate is transformed to oxaloacetate, which passes into the cytosol and is transformed to phosphoenolpyruvate by PEP carboxykinase. Two moles of PEP are then required to produce a mole of glucose by the route outlined in Figure 20–2.

12. If the catabolic and anabolic pathways of glucose metabolism are operating simultaneously, futile cycling of ATP occurs, with extra O_2 consumption.

13. Glycogen synthase has the lowest measured activity of the enzymes in glycogen synthesis, and is thus a likely regulatory point. This is confirmed by the observation (b) that the stimulation of glycogen synthesis by the activation of the regulatory enzyme leads to a decrease in the concentration of the substrate of the glycogen synthase reaction (UDP-glucose) and an increase in the concentration of the product of the reaction (UDP).

14. Storage consumes 1 mol of ATP per mole of glucose 6-phosphate; this represents 0.026 (2.6%) of the total ATP available from glucose 6-phosphate metabolism (i.e., the efficiency of storage is 97.4%).

15. **(c)** UDP-glucose pyrophosphorylase

16. The diversion of glucose and its precursor oxaloacetate to milk production under conditions of extensive fatty acid catabolism results in ketosis. Ruminants can readily transform propionate to succinyl-CoA (via the intermediates propionyl-CoA, D-methylmalonyl-CoA, and L-methylmalonyl-CoA) and thus into oxaloacetate to avert the ketosis.

17. This observation suggests that ATP and NADPH are generated in the light and are essential for CO_2 fixation; conversion stops as the supply of ATP and NADPH becomes exhausted.

18. X is 3-phosphoglycerate; Y is ribulose 1,5-bisphosphate.

19. In maize, CO_2 is fixed by the C_4 pathway worked out by Hatch and Slack, in which phosphoenolypyruvate is carboxylated rapidly to oxaloacetate (some of which undergoes transamination to aspartate) and reduced to malate. Only after subsequent decarboxylation does the CO_2 enter the Calvin cycle.

20. The isocitrate dehydrogenase reaction

21. **(a)** The equation for the lengthening of dextran by one glucose residue is:

$$\text{Sucrose} + (\text{glucose})_n \longrightarrow (\text{glucose})_{n+1} + \text{fructose}$$

(b) The fructose generated in the synthesis of dextran is readily imported and metabolized by the bacteria.

22. **(a)** Low levels of P_i in the cytosol and high levels of triose phosphate in the chloroplast **(b)** High levels of triose phosphate in the cytosol **(c)** The P_i–triose phosphate antiport system

Chapter 21

1. **(a)** The 16 carbons of palmitate are derived from 8 acetyl groups of 8 acetyl-CoA molecules. The ^{14}C-labeled acetyl-CoA gives rise to malonyl-CoA labeled at C-1 and C-2.
(b) The metabolic pool of malonyl-CoA, the source of all palmitate carbons except the first two (C-16 and C-15), does not become labeled with small amounts of ^{14}C-labeled acetyl-CoA. Hence, only [15,16-^{14}C]palmitate is formed.

2. Both glucose and fructose are degraded to pyruvate in glycolysis. The pyruvate is converted to acetyl-CoA by the pyruvate dehydrogenase complex. Some of this acetyl-CoA enters the citric acid cycle, which produces reducing equivalents (NADH and NADPH). Mitochondrial electron transfer to O_2 yields ATP.

3. $$8\ \text{Acetyl-CoA} + 15\ \text{ATP} + 14\text{NADPH} + 9H_2O \longrightarrow \text{palmitate} + 8\ \text{CoA} + 15\text{ADP} + 15P_i + 14\text{NADP}^+ + 2H^+$$

4. **(a)** 3 Deuteriums per palmitate; all located on C-16; all other two-carbon units are derived from unlabeled malonyl-CoA. **(b)** 7 Deuteriums per palmitate; all *even*-numbered carbons except C-16

5. By using the three-carbon unit malonyl-CoA, the activated form of acetyl-CoA (recall that malonyl-CoA synthesis requires ATP), metabolism is driven in the direction of fatty acid synthesis by the exergonic release of CO_2.

6. The rate-limiting step in the biosynthesis of fatty acids is the carboxylation of acetyl-CoA catalyzed by acetyl-CoA carboxylase. High [citrate] and [isocitrate] indicate that conditions are favorable for fatty acid synthesis: an active citric acid cycle is providing a plentiful supply of ATP, reduced pyridine nucleotides, and acetyl-CoA. Citrate stimulates (increases the V_{max} of) acetyl-CoA carboxylase. Furthermore, because citrate binds more tightly to the filamentous form of the enzyme (the active form), high [citrate] drives the protomer $\rightleftharpoons$ filament equilibrium in the direction of the active form. In contrast, palmitoyl-CoA (the end product of fatty acid synthesis) drives the equilibrium in the direction of the inactive (protomer) form. Hence, when the end product of fatty acid synthesis accumulates, the biosynthetic path is slowed down.

7. **(a)** $$\text{Acetyl-CoA}_{(mit)} + \text{ATP} + \text{CoA}_{(cyt)} \longrightarrow \text{acetyl-CoA}_{(cyt)} + \text{ADP} + P_i + \text{CoA}_{(mit)}$$
(b) 1 ATP per acetyl group **(c)** Yes

8. The double bond in palmitoleate is introduced by an oxidation catalyzed by fatty acyl–CoA desaturase, a mixed-function oxidase that requires O_2 as a cosubstrate.

9. $$3\ \text{Palmitate} + \text{glycerol} + 7\text{ATP} + 4H_2O \longrightarrow \text{tripalmitin} + 7\text{ADP} + 7P_i + 7H^+$$

10. In adult rats, stored triacylglycerols are maintained at a steady-state level through a balance of the rates of degradation and biosynthesis. Hence, the triacylglycerols of adipose (fat) tissue are constantly turned over, explaining the incorporation of ^{14}C label from dietary glucose.

11. Net reaction:
Dihydroxyacetone phosphate + NADH + palmitate + oleate + 3ATP + CTP + choline + $4H_2O \longrightarrow$ phosphatidylcholine + NAD^+ + 2AMP + ADP + H^+ + CMP + $5P_i$;
7 ATP per molecule of PC

12. Methionine deficiency reduces the level of *S*-adenosylmethionine, which is required for the de novo synthesis of PC. The salvage pathway does not employ *S*-adenosylmethionine, but uses available choline. Thus PC can be synthesized even when the diet is deficient in Met, as long as choline is available.

13. ^{14}C label appears in three places in the activated isoprene:

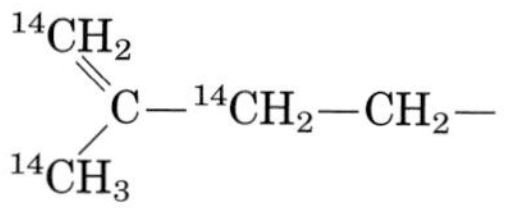

$$\begin{matrix} {}^{14}CH_2 \\ \| \\ C-{}^{14}CH_2-CH_2- \\ | \\ {}^{14}CH_3 \end{matrix}$$

14. **(a)** ATP **(b)** UDP-glucose **(c)** CDP-ethanolamine **(d)** UDP-galactose **(e)** Fatty acyl–CoA **(f)** *S*-Adenosylmethionine **(g)** Malonyl-CoA **(h)** Isopentenyl pyrophosphate

15. Linoleate is required in the synthesis of prostaglandins. Mammals are unable to transform oleate to linoleate, so linoleate is an essential fatty acid in animals. However, plants can transform oleate to linoleate and thus they provide animals with the required linoleate (Fig. 21–13).

16. The rate-determining step in the biosynthesis of cholesterol is the synthesis of mevalonate catalyzed by hydroxymethylglutaryl-CoA reductase. This enzyme is allosterically regulated by mevalonate and cholesterol derivatives. High levels of intracellular cholesterol also reduce transcription of the gene encoding HMG-CoA reductase.

17. Without ApoE, LDLs cannot be recognized and taken up by LDL receptors in the liver and other tissues. The result is high LDL levels in blood and, sometimes, atherosclerosis.

Chapter 22

1. In their symbiotic relationship with the plant, bacteria supply ammonium ion by reducing atmospheric nitrogen, which requires large quantities of ATP.

2. The transfer of carbon skeletons can be catalyzed by (1) glutamine synthetase and (2) glutamate dehydrogenase. The latter enzyme produces glutamate, the amino group donor in all transamination reactions.

3. Toxic ammonium ions are transformed to glutamine in the mammalian route, reducing toxic effects on the brain.

4. Glucose + $2CO_2$ + $2NH_3 \longrightarrow$ 2 aspartate + $2H^+$ + $2H_2O$

5. If phenylalanine hydroxylase is defective, the biosynthetic route to Tyr is blocked and Tyr must be obtained from the diet.

6. In *S*-adenosylmethionine synthesis, triphosphate is released from ATP. Hydrolysis of the triphosphate renders the reaction thermodynamically more favorable.

7. If the inhibition of glutamine synthase were not concerted, saturating concentrations of histidine would shut down the enzyme and cut off production of glutamine, which the bacterium needs to synthesize other products.

8. Folic acid is a precursor of tetrahydrofolate (Fig. 18–15), required in the biosynthesis of glycine (Fig. 22–12), a precursor of porphyrins. A folic acid deficiency therefore impairs hemoglobin synthesis.

9. For Asp and Glu auxotrophs: adenine, guanine, uridine, and cytosine; for Gly auxotrophs: adenine and guanine.

10. **(a)** See Figure 18–6 for the reaction mechanism of amino acid racemization. The F atom of fluoroalanine is an excellent leaving group. Fluoroalanine causes irreversible (covalent) inhibition of alanine racemase. One plausible mechanism is:

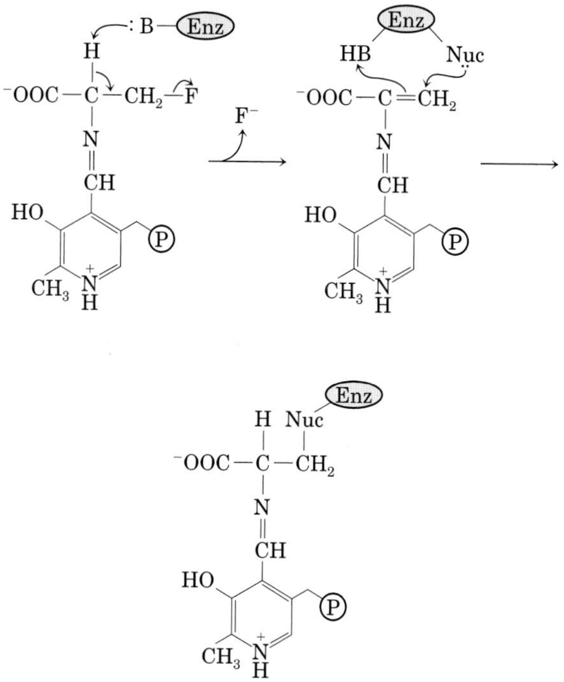

:B—Enz; H; $^-OOC-C-CH_2-F$; N; CH; HO; P; CH_3; $\overset{+}{N}$; H $\xrightarrow{F^-}$ Enz; HB; Nuc; $^-OOC-C{=}CH_2$; N; CH; HO; P; CH_3; $\overset{+}{N}$; H $\longrightarrow$ Enz; H Nuc; $^-OOC-C-CH_2$; N; CH; HO; P; CH_3; $\overset{+}{N}$; H

Nuc denotes any nucleophilic amino acid side chain in the enzyme active site.

(b) Azaserine (see Fig. 22–46) is an analog of Gln. The diazoacetyl group is highly reactive and forms covalent bonds with nucleophiles at the active site of a glutamine amidotransferase.

11. **(a)** As shown in Figure 18–15, *p*-aminobenzoate is a component of N^5,N^{10}-methylenetetrahydrofolate, the cofactor involved in the transfer of one-carbon units.
(b) In the presence of sulfanilamide, a structural analog of *p*-aminobenzoate, bacteria are unable to synthesize tetrahydrofolate, a cofactor necessary for converting AICAR to FAICAR, causing AICAR to accumulate.
(c) The competitive inhibition by sulfanilamide of the enzyme involved in tetrahydrofolate biosynthesis is overcome by the addition of excess substrate (*p*-aminobenzoate).

12. The ^{14}C-labeled orotate arises from the following pathway:

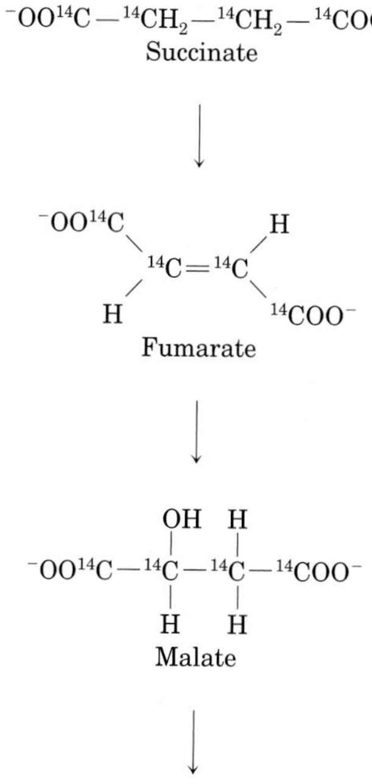

$$^-OO^{14}C-{}^{14}CH_2-{}^{14}CH_2-{}^{14}COO^-$$
Succinate

↓

$^-OO^{14}C$, H; ${}^{14}C{=}{}^{14}C$; H, ${}^{14}COO^-$
Fumarate

↓

$$^-OO^{14}C-\underset{H}{\overset{OH}{{}^{14}C}}-\underset{H}{\overset{H}{{}^{14}C}}-{}^{14}COO^-$$
Malate

↓

$$^{-}OO^{14}C-{}^{14}C(=O)-{}^{14}CH_2-{}^{14}COO^{-}$$

Oxaloacetate

↓ transamination

$$^{-}OO^{14}C-{}^{14}CH(\overset{+}{N}H_3)-{}^{14}CH_2-{}^{14}COO^{-}$$

Aspartate

↓

O ‖ ^{14}C ^{-}O $^{14}CH_2$ H_2N C ^{14}C—$^{14}COO^{-}$ O N H H

↓

H O ‖ ^{14}C N $^{14}CH_2$ C ^{14}C—$^{14}COO^{-}$ O N H H

↓

H O ‖ ^{14}C H N ^{14}C C ^{14}C—$^{14}COO^{-}$ O N H

Orotate

The first three steps are part of the citric acid cycle.

13. Organisms do not store nucleotides to be used as fuel and do not completely degrade them, but rather hydrolyze them to release the bases, which can be recovered in salvage pathways. The low C:N ratio of nucleotides makes them poor sources of energy.

14. Treatment with allopurinol has two biochemical consequences. (1) Conversion of hypoxanthine to uric acid is inhibited, causing accumulation of hypoxanthine, which is more soluble and more readily excreted. This alleviates the clinical problems associated with AMP degradation. (2) Conversion of guanine to uric acid is also inhibited, causing accumulation of xanthine, which is, unfortunately, even less soluble than uric acid. This is the source of xanthine stones. Because the amount of GMP degradation is low relative to AMP degradation, the kidney damage caused by xanthine stones is less than that caused by untreated gout.

15. 5-Phosphoribosyl-1-pyrophosphate; this is the first ammonia acceptor in the purine biosynthetic pathway.

Chapter 23

1. Steady-state levels of ATP are maintained by phosphoryl transfer to ADP from phosphocreatine. 1-Fluoro-2,4-dinitrobenzene inhibits creatine kinase.

2. Ammonia is very toxic to nervous tissue, especially the brain. Excess NH_3 is removed by the transformation of glutamate to glutamine, which travels to the liver and is subsequently transformed to urea. The additional glutamate arises from the transformation of glucose to α-ketoglutarate. Additional NH_3 is removed by transamination of α-ketoglutarate to glutamate, and conversion of glutamate to glutamine.

3. From glucose, by the following route:

$$\text{Glucose} \xrightarrow{\text{glycolysis}} \text{dihydroxyacetone phosphate}$$

$$\text{Dihydroxyacetone phosphate} + \text{NADH} + \text{H}^+ \longrightarrow \text{glycerol 3-phosphate} + \text{NAD}^+$$

4. (a) Increased muscular activity increases the demand for ATP, which is met by increased O_2 consumption. **(b)** After the sprint, lactate produced by anaerobic glycolysis is converted back to glucose and glycogen, which requires ATP and therefore O_2.

5. Glucose is the primary fuel of the brain. TPP-dependent oxidative decarboxylation of pyruvate to acetyl-CoA is essential to complete glucose metabolism.

6. 1.86×10^2 m

7. (a) Inactivation provides a rapid means to change hormone concentrations. **(b)** A constant insulin level is maintained by equal rates of synthesis and degradation. **(c)** Other means of varying hormone concentration include changes in the rates of release from storage, of transport, and of conversion from prohormone to active hormone.

8. Water-soluble hormones bind to receptors on the outer surface of the cell, triggering the formation of a second messenger (cAMP) *inside* the cell. Lipid-soluble hormones can pass through the cell membrane, then act on their target molecules or receptors directly.

9. (a) Heart and skeletal muscle lack the enzyme glucose 6-phosphate phosphatase. Any glucose 6-phosphate that is produced enters the glycolytic pathway, and under O_2-deficient conditions is converted into lactate via pyruvate.
(b) Phosphorylated intermediates cannot escape from the cell, because the membrane is not permeable to charged species. In a "fight or flight" situation, the concentration of glycolytic precursors needs to be high in preparation for muscular activity. The liver, on the other hand, must release the glucose necessary to maintain the blood glucose level. Glucose is formed from glucose 6-phosphate and passes from the liver cells to the bloodstream.

10. (a) Excessive utilization of blood glucose by the liver, leading to hypoglycemia; shutdown of amino acid and fatty acid catabolism **(b)** Little circulating fuel is available for ATP requirements. Brain damage results because glucose is the main source of fuel for the brain.

11. Thyroxine acts as an uncoupler of oxidative phosphorylation. Uncouplers lower the P/O ratio, and the tissue must increase respiration to meet the normal ATP demands. Thermogenesis could also be due to the increased rate of ATP utilization by the thyroid-stimulated tissue, because the increased ATP demands are met by increased oxidative phosphorylation and thus respiration.

12. Because prohormones and preprohormones are inactive, they can be stored in quantity in secretory granules. Rapid activation is achieved by enzymatic cleavage in response to an appropriate signal.

13. In animals, many precursors lead to the synthesis of glucose (Fig. 20–1). In humans, the principal precursors are glycerol from triacylglycerols and glucogenic amino acids from protein.

14. The *ob/ob* mouse, which is initially obese, will lose weight. The *OB/OB* mouse will retain its normal body weight.

Chapter 24

1. 62,769 nm. Dividing the approximate molecular weight of T2 DNA by 650 gives a length of 184,615 base pairs; multiplying this by 3.4 Å per base pair (see Chapter 10) gives 627,691 Å = 62,769 nm. Thus the DNA is about 300 times longer than the T2 phage head.
2. Because neither the numbers of A and T residues nor the numbers of G and C residues are equal, the DNA cannot be a base-paired double helix; the M13 DNA is single-stranded.
3. $M_r = 3.77 \times 10^8$; length = 197 μm; Lk_0 = 55,245; Lk = 51,930
4. The exons of this gene contain 3 × 192 = 576 base pairs. The remaining 864 base pairs are present in introns and possibly a leader or signal sequence.
5. 5,250 base pairs. In relaxed DNA, *Lk* is equivalent to the number of turns in the DNA helix. Multiplying 500 by 10.5 base pairs per turn gives a length of 5,250 base pairs. **(a)** Doesn't change; *Lk* cannot change without breaking and re-forming the covalent backbone of the DNA. **(b)** Becomes undefined; a circular DNA with a break in one strand has, by definition, no *Lk*. **(c)** Decreases; in the presence of ATP, gyrase underwinds DNA. **(d)** Doesn't change; this assumes that neither of the DNA strands was broken during the heating process.
6. $\sigma = -0.07$; there is a >70% chance that the phage DNA will be incorporated into the *E.coli* DNA.
7. A right-handed helix has a positive *Lk*; a left-handed helix (such as Z-DNA) has a negative *Lk*. Decreasing the *Lk* of a closed circular B-DNA molecule by underwinding it facilitates the formation of regions of Z-DNA within certain sequences. (See Chapter 10 for a description of sequences that permit the formation of Z-DNA.)
8. The results demonstrate a fundamental structural unit in chromatin that repeats about every 200 base pairs; the DNA is accessible to the nuclease only at 200 base pair intervals. The brief treatment was insufficient to cleave the DNA at every accessible point, so that a ladder of DNA bands was created in which the sizes of the DNA fragments were in multiples of 200 base pairs (further treatment would have caused most of the DNA to be in the lowest band). The thickness of the DNA bands suggests that the distance between cleavage sites varies somewhat. For instance, not all of the fragments in the lowest band are exactly 200 base pairs long.
9. A centromere, telomeres, and an autonomous replicating sequence or replication origin

Chapter 25

1. If random, dispersive replication had taken place in the second generation, all the DNAs would have had the same density and would have appeared as a single band, not the two bands that were observed in the Meselson-Stahl experiment.
2. In this extension of the Meselson-Stahl experiment, after three generations the molar ratio of ^{15}N–^{14}N DNA to ^{14}N–^{14}N DNA is 2/6 = 0.33.
3. **(a)** 4.42×10^5 turns; the length of the chromosome (in base pairs) is divided by 10.5 **(b)** 39 min; When the cells are dividing every 20 min, a replicative cycle is inititiated every 20 min, each cycle beginning before the prior one is complete. **(c)** About 2,320 to 4,640 Okazaki fragments are formed by DNA polymerase III from an RNA primer and a DNA template. These fragments are 1,000 to 2,000 bases long, and are firmly bound to the template strand by base pairing. Each fragment is quickly joined to the lagging strand, thus preserving the correct order of the fragments.
4. Composition of total DNA synthesized would be A, 28.7%; G, 21.3%; C 21.3%; T 28.7%. The DNA strand made from the given template strand would have A, 32.7%; G, 18.5%; C, 24.1%; T, 24.7%. The DNA strand made from the complementary template strand would have A, 24.7%; G, 24.1%; C, 18.5%; T, 32.7%. It is assumed that the two template strands are replicated completely.
5. **(a)** The incorporation of ^{32}P label into DNA results from the synthesis of new DNA. The synthesis of DNA requires the presence of all four nucleotide precursors.
   (b) Yes. Although all four nucleotide precursors must be present for DNA synthesis to occur, only one of them has to be radioactive in order for radioactivity to be observed in the new DNA.
   (c) No radioactivity would be incorporated if the ^{32}P label were not in the α phosphate because DNA polymerase, which catalyzes this reaction, cleaves off pyrophosphate—i.e., the β and γ phosphate groups.
6. *Leading strand:* precursors—dATP, dGTP, dCTP, dTTP; enzymes—DNA gyrase, helicase, single-strand DNA-binding protein, DNA polymerase III, topoisomerases, and pyrophosphatase.
   Lagging strand: precursors—ATP, GTP, CTP, UTP, dATP, dGTP, dCTP, dTTP; enzymes—DNA gyrase, helicase, single-strand DNA-binding protein, primase, DNA polymerase III, DNA polymerase I, DNA ligase, topoisomerases, and pyrophosphatase. NAD^+ is also required as a cofactor for DNA ligase.
7. Mutants with defective DNA ligase produce DNA duplex in which one of the strands remains in pieces (as Okazaki fragments). When this duplex is denatured, sedimentation results in one fraction containing the intact single strand (the high molecular weight band) and one fraction containing the unspliced fragments (the low molecular weight band).
8. Watson-Crick base pairing between the template and leading strand; proofreading and removal of wrongly inserted nucleotides by the 3′-exonuclease activity of DNA polymerase III. Yes—maybe; because the factors ensuring fidelity of replication are operative in both the leading and the lagging strands, the lagging strand would probably be made with the same fidelity. However, the greater number of distinct chemical operations involved in making the lagging strand compared with the leading strand might provide a greater opportunity for errors to arise.
9. About 1,200 base pairs will be unwound (about 600 in each direction)
10. A small fraction (13 of 10^9 cells) of the histidine-requiring mutants spontaneously undergo a back-mutation and regain their capacity to synthesize histidine. The addition of 2-aminoanthracene increases the rate of back-mutations about 1,000-fold, and is therefore mutagenic. Since most carcinogens are mutagenic, 2-aminoanthracene is probably carcinogenic.
11. Spontaneous deamination of 5-methylcytosine (see p. 348) produces thymine, and thus a G–T mismatched pair. Such G–T mismatches are among the most common mismatches in the DNA of eukaryotes. The specialized repair system restores the G≡C pair.
12. **(a)** Ultraviolet irradiation produces pyrimidine dimers, which are excised in normal fibroblasts in a process involving cleavage of the damaged strand by a special excinuclease. Thus the denatured single-stranded DNA contains the many fragments caused by the cleavage and the average molecular weight is lowered.
    (b) The absence of fragments in the single-stranded DNA from the diseased cells (no change in average molecular weight) after irradiation suggests that the special excinuclease is defective or missing in these cells.
13. During homologous genetic recombination, a Holliday intermediate may be formed almost anywhere within the two paired, homologous chromosomes. Once formed, the branch point of the intermediate may move extensively by branch migration. In site-specific recombination, the Holliday intermediate is formed between two specific sites, and branch migration is generally restricted by heterologous sequences on either side of the recombination sites.

Chapter 26

1. **(a)** 59–106 s; the RNA has 5,300 bases and is transcribed at a rate of 50–90 nucleotides/s (see p. 981). **(b)** 500–900 nucleotides/s
2. A single base error in DNA replication, if not corrected, would cause one of the two daughter cells, and all its progeny, to have a mutated chromosome. A single base error in RNA transcription would not affect the chromosome; it would lead to formation of some defective copies of one protein, but because mRNAs turn over rapidly, most copies of the protein would not be defective. The progeny of this cell would be normal.
3. Normal posttranscriptional processing at the 3′ end (cleavage and polyadenylation) would be inhibited or blocked.
4. Because the template-strand RNA does not encode the enzymes needed to initiate viral infection, it would probably be inert or simply degraded by cellular ribonucleases. Replication of the template-strand RNA and propagation of the virus could occur only if intact RNA replicase (RNA-dependent RNA polymerase) were introduced into the cell along with the template strand.
5. (1) Use of a template strand of nucleic acid, which is copied in the 3′ → 5′ direction; (2) use of nucleoside triphosphate substrates, with displacement of PP_i
6. Generally two: one to cleave the phosphodiester bond at one intron-exon junction, and the other to link the resulting free exon end to the exon at the other end of the intron. If the nucleophile in the first step were water, this step would be a hydrolytic event and only one transesterification step would be required to complete the splicing process.
7. These enzymes lack a 3′ → 5′ proofreading exonuclease and have a high error rate; the likelihood of a replication error that would inactivate the virus is much less in a small genome than in a large one.
8. **(a)** $4^{32} = 1.8 \times 10^{19}$ **(b)** 0.005% **(c)** For the "unnatural selection" step, use a chromatographic resin to which is bound a molecule that is a transition-state analog of the ester hydrolysis reaction (e.g., an appropriate phosphonate compound; see Box 8–3).
9. Though RNA synthesis is quickly halted by α-amanitin toxin, it takes several days for the critical mRNAs and the proteins in the liver to degrade, causing liver dysfunction and death.
10. **(a)** After lysis of the cells and partial purification, the protein extract could be subjected to isoelectric focusing. The β subunit could be detected by an antibody-based assay. The difference in amino acid residues between the normal β subunit and the mutated form (i.e., the different charges on the amino acids) would alter the electrophoretic mobility of the mutant protein in an isoelectric focusing gel relative to the protein from a nonresistant strain. **(b)** Direct DNA sequencing (by the Sanger method)
11. **(a)** 384 nucleotide pairs **(b)** 1,620 nucleotide pairs **(c)** Most of the nucleotides are untranslated regions at the 3′ and 5′ ends of the mRNA. Also, most mRNAs code for a signal sequence (see Chapter 27) in their protein products, which is eventually cleaved off to produce the mature and functional protein.

Chapter 27

1. **(a)** Gly–Gln–Ser–Leu–Leu–Ile **(b)** Leu–Asp–Ala–Pro **(c)** His–Asp–Ala–Cys–Cys–Tyr **(d)** Met–Asp–Glu in eukaryotes; fMet–Asp–Glu in prokaryotes
2. UUAAUGUAU, UUGAUGUAU, CUUAUGUAU, CUCAUGUAU, CUAAUGUAU, CUGAUGUAU, UUAAUGUAC, UUGAUGUAC, CUUAUGUAC, CUCAUGUAC, CUAAUGUAC, CUGAUGUAC
3. No; because nearly all the amino acids have more than one codon (e.g., Leu has six), any given polypeptide can be coded for by a number of different base sequences (see Problem 2). However, because some amino acids are encoded by only one codon and those with multiple codons often share the same nucleotide at two of the three positions, *certain parts* of the mRNA sequence encoding a protein of known amino acid sequence can be predicted with high certainty.
4. **(a)** (5′)CGACGGCGCGAAGUCAGGGGUGUUAAG(3′)
(b) Arg–Arg–Arg–Glu–Val–Arg–Gly–Val–Lys
(c) No; the complementary antiparallel strands in double-helical DNA do not have the same base sequence in the 5′ → 3′ direction. RNA is transcribed from only one specific strand of duplex DNA. The RNA polymerase must therefore recognize and bind to the correct strand.
5. There are two tRNAs for methionine: $tRNA^{fMet}$, the initiating tRNA, and $tRNA^{Met}$, which can insert Met in interior positions in a polypeptide. Only $fMet\text{-}tRNA^{fMet}$ is recognized by the initiation factor IF-2 and is aligned with the initiating AUG positioned at the ribosomal P site in the initiation complex. AUG codons in the interior of the mRNA can bind and incorporate only $Met\text{-}tRNA^{Met}$.
6. Allow polynucleotide phosphorylase to act on a mixture of UDP and CDP in which UDP has, say, five times the concentration of CDP. The result would be a synthetic RNA polymer with many UUU triplets (coding for Phe), a smaller number of UUC (also Phe), UCU (Ser), and CUU (Leu), and a much smaller number of UCC (also Ser), CUC (also Leu), and CCU (Pro).
7. A minimum of 583 high-energy phosphate groups (based on 4 per amino acid residue added, except that there are only 145 translocation steps). Correction of any errors requires 2 high-energy phosphate groups each. For glycogen synthesis, 292 high-energy phosphate groups would be used. The extra energy cost for β-globin synthesis reflects the cost of the information content of the protein.

   At least 20 activating enzymes, 70 ribosomal proteins, 4 rRNAs, 32 or more tRNAs, an mRNA, and 10 or more auxiliary enzymes must be made by the eukaryotic cell in order to synthesize a protein from amino acids. The synthesis of an (α1 → 4) chain of glycogen from glucose requires only 4 or 5 enzymes (see Chapter 20).
8.

Glycine codons	Anticodons
(5′)GGU	(5′)ACC, GCC, ICC
GGC	GCC, ICC
GGA	UCC, ICC
GGG	CCC, UCC

(a) The 3′ end and the middle position **(b)** The pairings with anticodons (5′)GCC, ICC, and UCC **(c)** The pairings with anticodons (5′)ACC and CCC

9. (a), (c), (e), and (g); mutations (b), (d), and (f) cannot be the result of single-base mutations; (b) and (f) would require substitutions of two bases, and (d) would require substitutions of all three bases.
10. The two DNA codons for Glu are GAA and GAG, and the four DNA codons for Val are GTT, GTC, GTA, and GTG. A single-base change in GAA to form GTA or in GAG to form GTG could account for the Glu → Val replacement in sickle-cell hemoglobin. Much less likely are two-base changes from GAA to GTG, GTT, or GTC; and from GAG to GTA, GTT, or GTC.
11. **(a)** The Ala-tRNA synthetase recognizes the G3–U70 base pair in the amino acid arm of $tRNA^{Ala}$. **(b)** The mutant $tRNA^{Ala}$ would insert Ala residues at codons encoding Pro. **(c)** Another type of mutation that might have similar effects is an alteration in $tRNA^{Pro}$ that allowed it to be recognized and aminoacylated by Ala-tRNA synthetase. **(d)** Most of the proteins in the cell would be inactivated, making these lethal mutations and hence never observed. This represents a powerful selective pressure for maintaining the genetic code.
12. The amino acid most recently added to a growing polypeptide chain is the only one covalently attached to a tRNA and hence is the only link between the polypeptide and the mRNA that is encoding it. A proofreading activity would sever this link, halting synthesis of the polypeptide and releasing it from the mRNA.

13. The protein will be directed into the endoplasmic reticulum, and from there the targeting will depend on additional signals. The SRP will bind the amino-terminal signal early in protein synthesis and direct the nascent polypeptide and ribosome to receptors in the ER. Because the protein is translocated into the lumen of the ER as it is synthesized, the NLS will never be accessible to the proteins involved in nuclear targeting.

14. Trigger factor is a molecular chaperone that stabilizes an unfolded and translocation-competent conformation of ProOmpA.

15. DNA with a minimum of 5,784 bp; some of the coding sequences must be nested or overlapping.

Chapter 28

1. **(a)** Tryptophan synthase levels remain high in spite of the presence of Trp. **(b)** Levels again remain high. **(c)** Levels rapidly decrease, preventing wasteful synthesis of Trp.

2. **(a)** Constitutive expression of the operon; most mutations in the operator would make the repressor less likely to bind. **(b)** Either constitutive expression, as in (a), or constant repression, if the mutation destroyed the capability to bind to lactose and related compounds and hence the response to inducers **(c)** Either increased or decreased expression of the operon (under conditions in which it is induced), depending on whether the mutation made the promoter more or less similar to the consensus *E. coli* promoter (see Fig. 28–2)

3. About 5,000 copies

4. 8×10^{-9} M, about 10^5 times greater than the dissociation constant. With 10 copies of active repressor in the cell, we can conclude that the operator site is always bound by the repressor molecule.

5. Each condition would decrease expression of *lac* operon genes.

6. **(a)** The ribosome completing the translation of sequence 1 would no longer overlap and block sequence 2. Attenuation would become much less effective because sequence 2 would always be available to pair with sequence 3, preventing formation of the attenuator structure. **(b)** Sequence 2 would pair less efficiently with sequence 3. Attenuation would increase because the attenuator structure would be formed more often, even when sequence 2 was not blocked by a ribosome. **(c)** Attenuation would not occur, and the only regulation would be that afforded by the Trp repressor. **(d)** The attenuation would lose its sensitivity to tryptophan tRNA and might become sensitive to His tRNA. **(e)** Attenuation would rarely, if ever, occur because sequences 2 and 3 would always block formation of the attenuator. **(f)** The attenuator would always form regardless of the availability of tryptophan.

7. Induction of the SOS response could not occur, making the cells more sensitive to high levels of DNA damage.

8. Each cell would have flagella made up of both types of flagellar protein, and the cell would be vulnerable to antibodies generated in response to either protein.

9. A dissociable factor necessary for activity (for example, a specificity factor similar to the σ subunit of the *E. coli* enzyme) may have been purified away from the polymerase.

10. **Gal4 protein**

Gal4 DNA-binding domain	Gal4 activator domain

Engineered protein

Lac repressor DNA-binding domain	Gal4 activator domain

The engineered protein can't bind to the Gal4 binding site in the *GAL* gene (UAS_G) because it lacks the Gal4 DNA-binding domain. Modify the Gal4p DNA binding site to give it the nucleotide sequence to which the Lac repressor normally binds (using methods described in Chapter 29).

11. The *bcd* mRNA needed for development is contributed to the egg by the mother. The egg will then develop normally, even if its genotype is bcd^-/bcd^-, as long as the mother has one normal *bcd* gene and the bcd^- allele is recessive. However, the adult bcd^-/bcd^- female will be sterile because she has no normal *bcd* mRNA to contribute to her own eggs.

12. **(a)** Hydrogen bonds form between the protein and DNA backbone at A106, A110, A118, T119, and A122, and between the protein and DNA bases at A106, T107, A118, and T119. The latter four nucleotides contribute directly to DNA sequence recognition. **(b)** Using the Noncovalent Bond Finder, it is possible to identify which base pairs hydrogen bond with which amino acid residue. DNA backbone: A106–Arg^{290}, A110–Ser^{212}, A118–Arg^{199}, T119–Arg^{204}, A122–Ser^{303}. DNA bases: A106–Asn^{253}, T107–Asn^{253}, A118–Asn^{163}, T119–Asn^{163}. Asn, Gln, Glu, Lys, and Arg are commonly found hydrogen-bonded to bases in DNA. The majority of residues in the TATA binding protein that are involved in hydrogen bonds are Arg and Asn.

(c) TATATATA (residues 103 to 110)
ATATATAT (residues 122 to 115)

The TATA binding protein recognizes A106,T107/T119, A118.

(d) Use the Noncovalent Bond Finder to find interactions between carbons only that occur between 3.3–4.0 Å.

Chapter 29

1. **(a)** (5′)---G(3′) and (5′)AATTC---(3′)
---CTTAA G---

(b) (5′)---GAATT(3′) and (5′)AATTC---(3′)
---CTTAA TTAAG---

(c) (5′)---GAATTAATTC---(3′)
---CTTAATTAAG---

(d) (5′)---G(3′) and (5′)C---(3′)
---C G---

(e) (5′)---GAATTC---(3′)
---CTTAAG---

(f) (5′)---CAG(3′) and (5′)CTG---(3′)
---GTC GAC---

(g) (5′)---CAGAATTC---(3′)
---GTCTTAAG---

(h) First, cut the DNA with *Eco*RI as in (a). At this point, one could treat the DNA as in (b) or (d), then ligate a synthetic DNA fragment with the *Bam*HI recognition sequence between the two resulting blunt ends. Another (more efficient) approach would be to synthesize a DNA fragment with the structure:

(5′)AATTGGATCC
CCTAGGTTAA

This would ligate efficiently to the sticky ends generated by *Eco*RI cleavage, would introduce a *Bam*HI site, but would not regenerate the *Eco*RI site.

(i) The four fragments (with N = any nucleotide), in order of discussion in the problem, are:

(5′)AATTCNNNNCTGCA
GNNNNG

(5′)AATTCNNNNGTGCA
GNNNNC

(5′)AATTGNNNNCTGCA
CNNNNG

(5′)AATTGNNNNGTGCA
CNNNNC

2. λ phage DNA can be packaged into infectious phage particles only if it is between 40,000 and 53,000 base pairs in length. Since bacteriophage vectors generally include about 30,000 base pairs (in two pieces), they will not be packaged into phage particles unless they contain a sufficient length of inserted DNA (10,000 to 23,000 base pairs).
(b) The clones in lanes 1 and 2 each have one DNA fragment inserted in different orientations. The clone in lane 3 has two DNA fragments, ligated such that the *Eco*RI proximal ends are joined.

3. **(a)** Plasmids in which the original pBR322 was regenerated without insertion of a foreign DNA fragment; these would retain resistance to ampicillin. Also, two or more molecules of pBR322 might be ligated together with or without insertion of foreign DNA.
(b) The clones in lanes 1 and 2 each have one DNA fragment inserted in different orientations. The clone in lane 3 has two DNA fragments, ligated such that the *Eco*RI proximal ends are joined.

4. You will need a suitable bacterial promoter, a ribosome-binding site for translation, and regulatory site(s) such as operators placed on the 5′ side of the gene (on the coding strand). The ribosome-binding site should be immediately left (5′) of the gene, and the promoter further to the left. The regulatory elements must be located where they could affect the promoter. Many possibilities for expression and regulation are described in Chapters 26–28.

5. Focus on the amino acids with the fewest codons: Met and Trp. The best possibility is the span of DNA from the codon for the first Trp residue to the first two nucleotides of the codon for Ile. The sequence of the probe would be:

(5′)UGGUA(U/C)UG(U/C)AUGGA(U/C)UGGAU

The synthesis would be designed to incorporate either U or C where indicated, producing a mixture of eight 20-nucleotide probes that differ only at one or more of these positions.

6. Simply for convenience; the 200,000 base pair Ti plasmid, even when the T DNA is removed, is too large to isolate in quantity and manipulate in vitro. It is also too large to reintroduce into a cell by standard transformation techniques. The *vir* genes will facilitate the transfer of any DNA between the T DNA repeats, even if they are on a separate plasmid. The second plasmid in the two-plasmid system, because it requires only the T DNA repeats and a few sequences necessary for plasmid selection and propagation, is relatively small, easily isolated, and easily manipulated (foreign DNA easily added and/or altered).

7. The vectors must be introduced into a cell infected with a helper virus that can provide the necessary replication and packaging functions but cannot itself be packaged. Once recombinant DNA is integrated into the chromosome of the target cell with these vectors, the lack of recombination and packaging functions makes the integration very stable by preventing the deletion or replication of the integrated DNA.

8. Your test would require DNA primers, a heat-stable DNA polymerase, deoxynucleoside triphosphates, and a PCR machine (thermal cycler). The primers would be designed to simplify a DNA segment encompassing the CAG repeat. The DNA strand shown is the coding strand, oriented 5′ → 3′ left to right. The primer targeted to DNA to the left of the repeat would be identical to any 25-nucleotide sequence shown in the region to the left of the CAG repeat. The primer on the right side must be *complementary* and *antiparallel* to a 25-nucleotide sequence to the right of the CAG repeat. Using the primers, DNA including the CAG repeat would be amplified by PCR, and its size would be determined by comparison to size markers after electrophoresis. The length of the DNA would reflect the length of the CAG repeat, providing a simple test for the disease.

9. Design PCR primers complementary to DNA in the deleted segment, but which would direct DNA synthesis away from each other. No PCR product will be generated unless the ends of the deleted segment are joined to create a circle.

10.

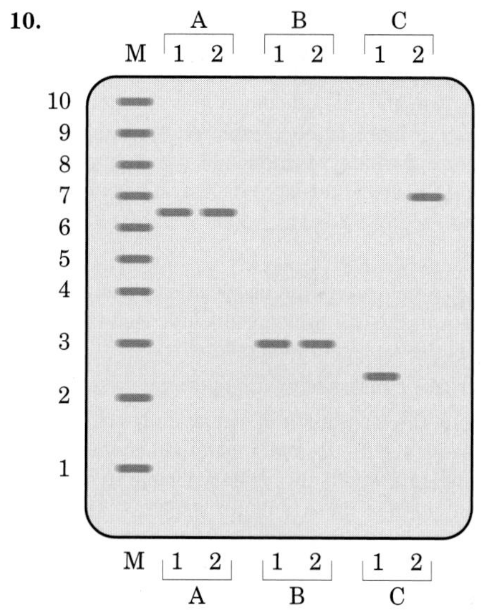

11. None of the children can be excluded. All have one band that could be derived from the father.

Illustration Credits

Photos and Line Art

PART I Opener Lee C. Coombs/Phototake NYC

CHAPTER 1 **Figure 1–1 (a)** Ron Boardman/Frank Lane Agency/Corbis; **(b)** W. Perry Conway/Corbis; **(c)** Karen Tweedy-Holmes/Corbis; **p. 4** Corbis/UPI/Bettmann; **Figure 1–2** The Bridgeman Art Library; **Figure 1–16 (a)** Erich Lessing/Art Resource, NY; **(b)** Dr. Gopal Murti-CNRI/Phototake NYC; **p. 15** (Linnaeus, Darwin) Corbis/Bettmann.

CHAPTER 2 **Figure 2–2 (a)** John Hansen/Photo Researchers; **(b)** Christine Gall, University of California, Irvine; **p. 23 (top)** CNRI/Science Photo Library/Photo Researchers; **(bottom)** David Scharf/Peter Arnold; **p. 24** Sinclair Stammers/Science Photo Library/Photo Researchers; **Figure 2–5 (top left)** T.J. Beveridge/Biological Photo Service; **(top right and bottom left)** Biological Photo Service; **(bottom right)** Norma J. Lange/Biological Photo Service; **Figure 2–9 (top and center)** Don W. Fawcett/Photo Researcher; **(bottom)** Science Source/Photo Researchers; **Figure 2–10 (bottom)** Biological Photo Service; **Figure 2–11 (a)** Don W. Fawcett/Photo Researchers; **(b)** Ursula Goodenough, Washington University, Department of Biology; **Figure 2–12** Adapted from Becker, W.M & Deamer, D.W. (1991) *The World of the Cell,* 2nd edn, Fig. 13–20, Benjamin/Cummings Publishing Company, Menlo Park, CA; **(photo)** G.F. Bahr/Biological Photo Service; **Figure 2–13** Don W. Fawcett & Keith Porter/Photo Researchers; **Figure 2–14** Biological Photo Service; **Figure 2–16 (a) (top)** Robert Goldman, Northwestern University Medical School, Department of Cell Biology & Anatomy; **(bottom)** M. Schliwa/Visuals Unlimited; **(b) (top)** Dr. Gopal Murti/Science Photo Library/Photo Researchers; **(bottom)** K.G. Murti/Visuals Unlimited; **(c) (top)** Dr. Peter Dawson/Science Photo Library/Photo Researchers; **(bottom)** Ueli Aebi, Maurice E. Müller-Institut für Hochauflösende Elektronenmikroskopie am Biozentrumder Universität Basel; **Figure 2–20** Adapted from Alberts, B., Bray, D., Lewis, J., Raff, M., Roberts, K., & Watson, J.D. (1989) *Molecular Biology of the Cell,* 2nd edn, pp. 165–166, Garland Publishing, Inc., New York; **Figure 2–21 (a)** Don W. Fawcett/Photo Researchers; **(b)** Don. W. Fawcett/Visuals Unlimited; **(c)** Biophoto Associates/Photo Researchers; **(d)** Pr. S. Cinti/Ancona U/CNRI/Phototake NYC; **(e)** Dr. Dennis Kunkel/Phototake NYC; **(f)** Account Phototake/Phototake NYC; **Figure 2–22 (a)** D.M. Phillips/Visuals Unlimited; **(b)** Don W. Fawcett/Photo Researchers; **(c)** D. Allbertini-D. Fawcett/Visuals Unlimited; **(d)** Courtesy Ray F. Evert from Raven, Evert, Eichhorn, *Biology of Plants*, 6th edition, Freeman/Worth, 1999; **Figure 2–23 (a)** John Finch, MRC Laboratory of Molecular Biology; **(b)** Dauquet, Institut Pasteur/Photo Researchers.

CHAPTER 3 **p. 53** Bill Curtsinger/Photo Researchers; **Figure 3–10** Adapted from Carroll, F. (1998) *Perspectives on Structure and Mechanism in Organic Chemistry,* p. 63, Brooks/Cole Publishing Co., Pacific Grove, CA; Bill Curtsinger/Photo Researchers, Inc.; **Box 3–1** (Pasteur) The Granger Collection; **Figure 3–13** PDB ID: 1BRZ; Caldwell, J.E., Abildgaard, F., Dzakula, Z., Ming, D., Hellekant, G., & Markley, J.L. (1998) Solution structure of the thermostable sweet-tasting protein brazzein. *Nat. Struct. Biol.* **5,** 427–431; **Figure 3–26** Adapted from Becker, W.M. & Deamer, D.W. (1991) *The World of the Cell,* 2nd edn, Fig. 2–15, Benjamin/Cummings Publishing Company, Menlo Park, CA; **p. 72** Historical Pictures Service/Stock Montage; **Figure 3–29 (a)** François Gohier/Photo Researchers; **(b and c)** J. William Schopf.

CHAPTER 4 **p. 82** NASA; **Box 4–1 Figure 1 (a and b)** Runk/Schoenberger/Grant Heilman Photography; **Figure 2 (a and b)** Adrian p. Davies/Bruce Coleman; **p. 107** Jon Bertsch/Visuals Unlimited.

CHAPTER 5 **Figure 5–1 (a)** Runk/Schoenburger/Grant Heilman Photography; **(b)** Bill Longcore/Photo Researchers; **(c)** Animals Animals; **Figure 5–16** Adapted from Dickerson, R.E. & Geis I. (1969) *The Structure and Action of Proteins,* Benjamin/Cummings Publishing Company, Menlo Park, CA. Copyright © 1969 by Irving Geis; **Figure 5–19 (b)** Richard R. Burgess & Jerome J. Jendrias, University of Wisconsin–Madison, Biotechnology Center; **Figure 5–22 (b)** Patrick H. O'Farrell, University of California Medical Center, San Francisco, Department of Biochemistry and Biophysics; **p. 142** UPI/Corbis-Bettmann; **Box 5–3, Figures 1 and 2** Adapted from Mann, M. & Wilm, M. (1995) Electrospray mass spectrometry for protein characterization. *Trends Biochem Sci.* **20,** 219–224; **p. 150** Corbis/Bettmann.

CHAPTER 6 **Figures 6–3, 6–9 (a)** Adapted from Creighton, T.E. (1984) *Proteins,* p. 166. Copyright © 1984 by W.H. Freeman and Company. Reprinted by permission; **Figure 6–9 (b)** Personal communication from Hazel Holden, Department of Biochemistry and Enzyme Institute, University of Wisconsin–Madison; **Figure 6–10** From Zubay, G. (1988) *Biochemistry,* p. 34. Copyright © 1988 by Macmillan Publishing Company. Reprinted by permission; **p. 161** (Corey) AP/Wide World Photos; (Pauling) Corbis/Bettmann; **Figure 6–13** Science Source/Photo Researchers; **Figure 6–14 (b)** Dr. Dennis Kunkel/Phototake NYC; **Figure 6–16** Data provided by Lai-Su L. Yeh, Protein Identification Resource, National Biomedical Research Foundation, Georgetown University Medical Center; **Box 6–3 Figure 1 (a, b, c)** George N. Phillips, Jr., Rice University, Department of Biochemistry; **Figures 2 and 3 (a)** Volkman, B.F., Alam, S.L., Satterlee, J.D., & Markley, J.L. (1998) Solution structure and backbone dynamics of component IV-glycera dibranchiata monomeric hemoglobin-CO. *Biochemistry* **37,** 10,906–10,919; **p. 189 (top)** Corbis/Hulton Deutsch Collection; **Figure 6–25 (b)** Science Source/Photo Researchers; **Figure 6–26** Personal communication from Yong Duan and Peter A. Kollman, University of California at San Francisco; **(a)** data from Sendak et al. (1996) *Biochemistry* **35,** 12,978–12,992, Nishii et al. (1995) *J. Mol. Biol.* **250,** 223–238; **(b)** data from Houry et al. (1996) *Biochemistry* **35,** 10,125–10,133; **Figure 6–29** Adapted from Wolynes, P.G., et al. (1995) Navigating the folding routes, *Science* **267,** 1619–1620; **Box 6–4, Figure 1** Stephen J. DeArmond.

CHAPTER 7 **Figure 7–18 (a)** Andrew Syred/Science Photo Library/Custom Medical Stock Photo; **(b)** Custom Medical Stock Photo; **p. 231** (Kohler and Milstein) Corbis/UPI/Bettmann; **Figure 7–28 (b)** State of Wisconsin Lab of Hygiene, Madison, WI; **(c)** Son, M., Gundersen, R.E., & Nelson, D.L. (1993) *J. Biol. Chem.* **268,** 5940–5948; **Figure 7–29 (a)** David Shotton/University of Oxford, Department of Zoology; **Figure 7–30 (a)** Eisaku Katayama, Institute of Medical Science, The University of Tokyo, Department of Fine Morphology; **(b)** Roger Craig, University of Massachusetts Medical School; **Figure 7–31 (b and c)** James E. Dennis/Phototake NYC.

CHAPTER 8 **p. 244** (Sumner and Haldane) AP/Wide World Photos; (Buchner) Science Photo Library/Photo Researchers; **p. 258** (Michaelis) Rockefeller University Archive Center; (Menten) Courtesy Dorothy C. Craig; **Box 8–3, Figure 1** Adapted from Fersht, A. (1985) *Enzyme Structure and Mechanism,* 2nd edn, p. 318. Copyright © 1977, 1985 by W.H. Freeman and Company. Reprinted by permission.

CHAPTER 9 **p. 293** Richard Hamilton Smith/Corbis; **Figure 9–14 (a)** Biophoto Associates/Photo Researchers; **(b)** Barry King/Biological Photo Service; **Figures 9–16b, 9–17b** Adapted from Lehninger, A. (1982) *Principles of Biochemistry,* p. 290, Worth Publishers, New York. Copyright © by Irving Geis; **Figure 9–28** R. M. Genta and D. Y. Graham, Veterans Affairs Medical Center, Houston, TX; **Figure 9–29** Adapted from Sharon, N. & Lis, H. (1993) Carbohydrates in Cell Recognition, *Sci. Am.* **268** (January), 82–89.

CHAPTER 10 **p. 332** (Watson and Crick) Corbis/UPI/Bettmann; **Figure 10–14** Science Source/Photo Researchers; **p. 335** (Franklin) Science Photo Library/Photo Researchers; (Wilkins) Corbis/UPI/Bettmann; **Figure 10–23 (a)** Htun, H. & Dahlberg, J.E. (1988) Single strands, triple strands, and kinks in H-DNA. *Science* **241,** 1793. Copyright © 1988 by the American Association for the Advancement of Science; **(b)** Htun, H. & Dahlberg, J.E. (1989) Topology and formation of triple-stranded H-DNA. *Science* **243,** 1571. Copyright © 1989 by the American Association for the Advancement of Science; **Figure 10–27** James, B., Olsen, G.J., Liu, J., & Pace, N.R. (1988) The secondary structure of ribonuclease P RNA, the catalytic element of a ribonucleoprotein enzyme. *Cell* **52,** 19–26; **Figure 10–30 (b)** After Marmur, J. & Doty, p. (1962) Determination of the base composition of deoxyribonucleic acid from its thermal denaturation temperature. *J. Mol. Biol.* **5,** 109–118; **Figure 10–31** Ross B. Inman, University of Wisconsin–Madison, Department of Molecular Biology; **Figure 10–36 (d)** Lloyd Smith, University of Wisconsin–Madison, Department of Chemistry; **Figure 10–37** Data provided by Lloyd Smith, Department of Chemistry, University of Wisconsin–Madison.

CHAPTER 11 **Figure 11–3 (a)** Biological Photo Service; **(b)** Courtesy Howard Goodman, Department of Genetics, Harvard Medical School; **Figure 11–5 (b)** William J. Weber/Visuals Unlimited; **p. 373** From D. L. Drabkin, *Thudichum: Chemist of the Brain,* University of Pennsylvania Press, 1958, credited to J. L. W. Thudichum, Tubingen, F. Pietzcker (1898); **Box 11–2, Figure 2** Herbert A. Fischler, Isaac Albert Research Institute of the Kingsbrook Jewish Medical Center; **p. 379** Ira Wyman/Sygma; **Figure 11–18 (b) (left and right)** Courtesy Dr. Russell Chesney, Chairman of Pediatrics, University of Tennessee, Memphis in collaboration with Professor H. F. DeLuca, Department of Biochemistry, University of Wisconsin; **p. 382** (Dam and Doisy) AP/Wide World Photos.

CHAPTER 12 **Figure 12–1 (a–f)** Joan Peterson, University of Wisconsin–Madison; **Figure 12–5** Data from Zachowski, A. (1993) Phospholipids in animal eukaryotic membranes: transverse asymmetry and movement. *Biochem. J.* **294,** 1–14; **Box 12–1, Figure 2** Margaritis, L.H., Elgsaeter, A., and Branton , D. (1977) *J. Cell Biol.* **72,** 47–56; **Figure 12–10** Adapted from Marchesi, V.T., Furthmayr, H., & Tomita, M. (1976) The red cell membrane. *Annu. Rev. Biochem.* **45,** 667–698; **Figure 12–25** Adapted from Mueckler, M. (1994) Facilitative glucose transporters, *Eur. J. Biochem.* **219,** 713–725; **Box 12–2, Figure 1** Adapted from Lienhard, F.E., Slot, J.W., James, D.E., & Mueckler, M.M. (1992) How cells absorb glucose, *Sci. Am.* **266** (January),

86–91; **Box 12–3, Figure 2** Tom Moninger, University of Iowa, Ames; **p. 420** (Skou) Courtesy Information Office, University of Aarhus, Denmark; **Figure 12–39** Adapted from Changeux, J.p. (1993) Chemical signaling in the brain, *Sci. Am.* **269** (November), 58–62; **Figure 12–40** Adapted from Taylor, R. (1994) Evolutions: the voltage-gated sodium channel, *J NIH Res.* **6,** 112.

CHAPTER 13 **p. 450** (Gilman) Office of News and Publications, The University of Texas Southwestern Medical Center at Dallas; (Rodbell) Courtesy Andrew Rodbell; **Figure 13–18 (a)** Courtesy Michael D. Cahalan, Department of Physiology and Biophysics, University of California, Irvine; **(b)** Rooney, T.A., Sass, E.J., & Thomas, A.p. (1989) Characterization of cytosolic calcium oscillations induced by phenylephrine and vasopressin in single fura-2-loaded hepatocytes. *J. Biol. Chem.* **264,** 17,131–17,141; **Figure 13–20** Adapted from Curtis, H. & Barnes, *Biology,* 5th edn., Fig. 42–13, p. 869, Worth Publishers, Inc., New York; **Figure 13–24** Adapted from Nathans, J. (1989) The genes for color vision. *Sci. Am.* **260** (February), 42–49; **Box 13–2, Figure 1** Courtesy Professor J.D. Mollon, Department of Experimental Psychology, Cambridge; **p. 467** (Fischer and Krebs) Mary Levin/University Photography, University of Washington; **Figure 13–29** O'Malley, B.W. & Tsai, M.-J. (1992) Molecular pathways of steroid receptor action. *Biol. Reprod.* **46,** 163–167; **Figure 13–32** Data from Pines, J. (1999) Four-dimensional control of the cell cycle. *Nat. Cell Biol.* **1,** E73; **Figure 13–38** Adapted from Kinzler, K.W. & Vogelstein, B. (1996) Lessons from hereditary colorectal cancer. *Cell* **87,** 159–170.

CHAPTER 14 **p. 490** The Granger Collection; **p. 491** © Sidney Harris; **Box 14–3** E.R. Degginger/Photo Researchers.

CHAPTER 15 **p. 527** (Lipman) Corbis/UPI/Bettmann; (Kalckar) Boston University Photo Services; **p. 532** (Harden) Hulton Getty/Liaison; (Young) Courtesy Medical History Museum, The University of Melbourne; **Box 15–1** Animals Animals; **Figure 15–10** New Brunswick Scientific Co., Inc.; **Figure 15–15 (a and b)** Sterchi, E., et al. (1990) Biogenesis of intestinal lactase-phlorizin hydrolase in adults with lactose intolerance. *J. Clin. Invest.* **86,** 1329–1334.

CHAPTER 16 **p. 568** Courtesy Public Relations Office, University of Sheffield; **Figure 16–5** Lester J. Reed, University of Texas at Austin, Clayton Foundation Biochemical Institute; **Figure 16–17** Richard N. Trelease, Arizona State University, Department of Botany.

CHAPTER 17 **p. 598** AP/Wide World Photos/Gretchen Ertl; **Box 17–1** Stouffer Productions/Animals Animals; **p. 611** The Nobel Foundation.

CHAPTER 18 **p. 623** Bernard G. Silberstein/Photo Researchers.

CHAPTER 19 **p. 660** (Lehninger) Alan Mason Chesney Medical Archives of Johns Hopkins Medical Institutions; **p. 664** (Beinert) Courtesy Helmut Beinert, University of Wisconsin; **Box 19–1, Figure 1** D. Cavagnaro/Visuals Unlimited; **Figure 19–13** Adapted from Williams, R.J.P. (1995) *Nature* **378,** 235; a correction to *Nature* **376,** 643 (1995); **p. 675** (Mitchell) AP/Wide World Photos; **p. 678** (Racker) Courtesy E. Racker; **p. 680** (Walker) The Nobel Foundation; **p. 682** (Boyer) The Nobel Foundation; **Figure 19–22 (f)** Adapted from Fillingame, R.H (1999) Protein structure: Molecular rotary motors. *Science* **286,** 1687; **Figure 19–24** Adapted from Sambongi, Y., et al. (1999) Mechanical rotation of the c subunit oligomer in ATP synthase (F^oF^1): Direct observation. *Science* **286,** 1722–1724; **(photo, right)** Courtesy Ryohei Yasuda and Kazuhiko Kinosita, from Yasuda, et al. (1998) *Cell* **93,** 1117–1124; **Figure 19–30 (a)** Morris, M.A. (1990) Mitomutations in neuro-opthalmological diseases: a review. *J. Clin. Neuro-opthalmology* **10,** 159–166; **(b)** From Wallace, D., et al. (1988) Familial mitochondrial encephalomyopathy (MERRF): genetic, pathophysiological, and biochemical characterization of a mitochondrial DNA disease. *Cell* **55,** 601–610; **Figure 19–31** Adapted from Harris, D.A. (1995) *Bioenergetics at a Glance*. Blackwell Science, London, p. 36; **Figure 19–35 (b)** Biological Photo Service; **Figure 19–39** Kühlbrandt, W., Wang, D.N., & Fujiyoshi, Y. (1994) Atomic model of plant light-harvesting complex by electron crystallography. *Nature* **367,** 614–620; **Figures 19–40, 19–44 (a and b), 19–47, 19–48, 19–50** Adapted from Heldt, H.-W. (1997) *Plant Biochemistry and Molecular Biology,* Oxford University Press, Oxford, pp. 57, 62, 63, 100, 101, 133; **Figure 19–54 (a)** Adapted from Luecke, H., et al. (1999) Structural changes in bacteriorhodopsin during ion transport at 2 angstrom resolution. *Science* **286,** 255–260; **(b)** Adapted from Gennis, R.B. & Ebrey, T.G. (1999) Proton pump caught in the act. *Science* **286,** 252–253; **p. 708** University of California, Berkeley; **p. 709** (Jagendorf) Cornell University.

CHAPTER 20 **p. 722** Jonathan Blair/Corbis; **p. 735** (Leloir) The Nobel Foundation; **p. 747** (Calvin) Lawrence Berkeley National Laboratory; **Figure 20–14** Smythe, C. & Cohen, p. (1991) The discovery of glycogenin and the priming mechanism for glycogen biogenesis. *Eur. J. Biochem.* **200,** 625–631; **Figure 20–35** Halliwell, B. (1984) *Chloroplast Metabolism: The Structure and Function of Chloroplasts in Green Leaf Cells,* p. 97, Clarendon Press, Oxford; **Figure 20–40** Ray Evert, University of Wisconsin–Madison, Department of Botany.

CHAPTER 21 **Figure 21–12 (b)** Daniel Lane, Johns Hopkins University, School of Medicine; **p. 792** (Kennedy) Harvard Medical School; **p. 804** (Bloch, Lynen, Cornforth) AP/Wide World Photos; (Popjak) Corbis/UPI/Bettmann; **Figure 21–37 (b)** Courtesy Robert L. Hamilton and the Arteriosclerosis Specialized Center of Research, University of California, San Francisco; **Figure 21–38 (b)** Alan Attie, Department of Biochemistry, University of Wisconsin–Madison; **p. 809** Corbis/UPI/Bettmann.

Chapter 22 **p. 818** Ric Ergenbright/Corbis; **Figure 22–4 (a)** C.p. Vance/Visuals Unlimited; **(b)** Jeremy Burgess/Photo Researchers; **Box 22–2, Figure 1** John Mansfield, University of Wisconsin–Madison, Department of Veterinary Science; **p. 849** (Buchanan) Massachusetts Institute of Technology Museum Collection; **Figure 22–38 (a)** Thelander, L. & Reichard, p. (1979) Reduction of ribonucleotides. *Annu. Rev. Biochem.* **48,** 133–158; **p. 863** Courtesy Kathy Bendo Hitchings.

CHAPTER 23 **p. 869** Courtesy Jackson Laboratory; **Figure 23–4** Fred Hossler/Visuals Unlimited; **Figure 23–7** D.W. Fawcett/Photo Researchers; **Figure 23–9** Courtesy M.L. Thomas, H.C. Sing, G. Belenky, Division of Neuropsychiatry, Walter Reed Army Institute of Research, U.S. Army Medical Research Materiel Command; **Box 23–1** (McCleod, Collip, Banting, Best) Corbis/UPI/Bettmann; **Figure 23–16** Adapted from Orth, D.N. (1975) General considerations for radioimmunoassay of peptide hormones. *Meth. Enzymol.* **37,** 22–38. **p. 886** (Guillemin) Corbis/UPI/Bettmann; (Schally) AP/Wide World Photos; (Yalow) Courtesy Rosalyn S. Yalow, Veterans Affairs Medical Center, NYC; **Figure 23–26** John Sholtis, The Rockefeller University, New York; **Figure 23–27** Adapted from Ezzell, C. (1995) Fat times for obesity research: Tons of new information, but how does it all fit together? *J. NIH Res.* **7,** 39–43; **Figure 23–29** Adapted from Auwerx, J & Staels, B. (1998) Leptin. *Lancet* **351,** 737–742.

PART IV Opener (p. 904) Patrick O. Brown, Stanford University School of Medicine, Department of Biochemistry.

CHAPTER 24 **Figure 24–1** From Kleinschmidt, A.K., Land, D., Jackerts, D., & Zahn, R.K. (1962) *Biochem. Biophys. Acta* **61,** pp. 857–864; **p. 909** (Beadle) Archive Photos; (Tatum) Corbis/UPI/Bettmann; **Figure 24–6** Huntington Potter and David Dressler, Harvard Medical School, Department of Neurobiology; **Figure 24–7 (a and b)** G.F. Bahr/Biological Photo Service; **Figure 24–8** D.W. Fawcett/Photo Researchers; **Figures 24–10, 24–13, 24–14** Cozzarelli, N.R., Boles, T.C., & White, J.H. (1990) Primer on the topology and geometry of DNA supercoiling. In *DNA Topology and Its Biological Effects* (Cozzarelli, N.R. & Wang, J.C., eds.), pp. 139–184, Cold Spring Harbor Laboratory Press, Cold Spring Harbor, New York; **Figure 24–11** Saenger, W. (1984) *Principles of Nucleic Acid Structure,* p. 452, Springer-Verlag, New York; **Figure 24–12** Laurien Polder from Kornberg, A. (1980) *DNA Replication,* p. 29, W.H. Freeman, New York; **Figure 24–19** Keller, W. (1975) Characterization of purified DNA-relaxing enzyme from human tissue culture cells. *Proc. Nat. Acad. Sci. USA* **72,** p. 2553; **Figure 24–21 (a)** James H. White, T. Christian Boles, & N.R. Cozzarelli, University of California, Berkeley, Department of Molecular and Cell Biology; **Figure 24–23 (b)** Ada L. Olins and Donald E. Olins, Oak Ridge National Laboratory; **Figure 24–27 (b)** Barbara Hamkalo, University of California, Irvine, Department of Molecular Biology and Biochemistry; **Figure 24–28** D.W. Fawcett/Visuals Unlimited; **p. 930, Problem 24–10** Roger Kornberg, MRC Laboratory of Molecular Biology.

CHAPTER 25 **Figure 25–3 (a)** Bernard Hirt, Institut Suisse de Recherches Experimentales sur le Cancer; **(c)** Cairns, J. (1963) Cold Spring Harbor *Symp. Quant. Biol.* **28,** 44; **p. 936** (Kornberg) AP/Wide World Photos; **Figure 25–12** Bramhill, D. & Kornberg, A. (1988) A model for initiation at origins of DNA replication. *Cell* **54,** 915–918; **Figure 25–17** Wake, R.G. & King, G.F. (1997) A tale of two terminators… *Structure* **5,** 1–5; **Figure 25–18** Bruce N. Ames, University of California, Berkeley, Department of Biochemistry & Molecular Biology; **Figure 25–20** Adapted from a figure provided by Paul Modrich; **Figure 25–21** Adapted from Grilley, M., Griffith, J., & Modrich, p. (1993) Bidirectional excision in methyl-directed mismatch repair. *J. Biol. Chem.* **268,** 11,830–11,837; **Figure 25–22** Watson, J.D., et al. (1987) *Molecular Biology of the Gene,* 4th edn, p. 350, Benjamin/Cummings Publishing Company, Menlo Park, CA; **Figure 25–23** Adapted from a figure provided by Aziz Sancar; **Figure 25–28 (b)** Bernard John, The Australian National University; **p. 961** (McClintock) AP/Wide World Photos; **Figure 25–29 (b)** Huntington Potter and David Dressler, Harvard Medical School, Department of Neurobiology; **Figure 25–32 (a)** John Heuser, Washington University Medical School, Department of Biochemistry; **Figure 25–34** Roca, A.I. & Cox, M.M. (1990) The RecA protein: structure and function. *Crit. Rev. Biochem. Mol. Biol.* **25,** 415–456; **Figure 25–36 (a)** Adapted from Guo, F., Gopaul, D.N., & Van Duyne, G.D. (1997) Structure of Cre recombinase complexed with DNA in a site-specific recombination synapse. *Nature* **389,** 40–46;

CHAPTER 26 **Figure 26–7** R. Landick (1997) RNA polymerase slides home: pause and termination site recognition, *Cell* **88,** 741–744; **Box 26–1, Figure 2** Carol Gross, University of Wisconsin–Madison, Department of Stomatology; **Figure 26–12 (a)** Pierre Chambon, Laboratorie de Génétique Moléculaire des Eucaryotes, Faculté de Médecine (CNRS); **(b and c)** Chambon, p. (1981) Split genes. *Sci. Am.* **244** (May), 60–71; **Figure 26–15** Cech, T.R. (1986) RNA as an enzyme. *Sci. Am.* **255** (November), 64–75; **p. 994** (Cech) Corbis/UPI/Bettmann; **Figure 26–16a** Kramer, A. (1996) The structure and function of proteins involved in mammalian pre-mRNA splicing. *Annu. Rev. Biochem.* **65,** 367–440; **Figure 26–28** Cech, T.R., Damberger, S.H., & Gutell, R.R. (1994) Representation of the secondary and tertiary structure of group I introns, *Struct. Biol.* **1,** 273–280; **Figure 26–29** Cech, T.R. (1986) RNA as an enzyme. *Sci. Am.* **255** (November), 64–75; **p. 1006** (Grunberg-Manago) Courtesy Marianne Grunberg-Manago; (Ochoa) AP/Wide World Photos; **p. 1007** (Temin) Corbis/UPI/Bettmann; (Baltimore) AP/Wide World Photos; **Figure 26–34** Haseltine,

W.A. & Wong-Staal, F. (1988) The molecular biology of the AIDS virus. *Sci. Am.* **259** (October), 52–62; **Figure 26–35** Kingsman, A.J. & Kingsman, S.M. (1988) Ty: a retroelement moving forward. *Cell* **53,** 333–335; **Figure 26–37 (a–c)** Boeke, J.D. (1990) Reverse transcriptase, the end of the chromosome, and the end of life. *Cell* **61,** 193–195; **(e)** Jack Griffith, University of North Carolina at Chapel Hill, Comprehensive Cancer Center; **p. 1014** (Woese, Crick, Orgel) AP/Wide World Photos.

CHAPTER 27 **Figure 27–1** D.W. Fawcett/Visuals Unlimited; **p. 1021** (Zamecnik) News Office, Massachusetts General Hospital; **p. 1023** AP/Wide World Photos; **Figure 27–12** Gutell, R.R., et al. (1985) Comparative anatomy of 16-S-like ribosomal DNA. *Prog. Nucleic Acid Res. Mol. Biol.* **32,** 155–216; **p. 1024** (Khorana) Courtesy Archives, University of Wisconsin, Madison; **p. 1037** (Nomura) Courtesy Masayasu Nomura; **p. 1038** (Holley) Corbis/UPI/Bettmann; **Figure 27–15** Kim, S.-H., et al. (1974) Three-dimensional tertiary structure of yeast phenylalanine transfer RNA. *Science* **185,** 435–440; **Figure 27–29 (b)** Steven L. McKnight and Oscar L. Miller, University of Virginia, Department of Biology; **p. 1057** (Blobel) Courtesy Günter Blobel, Rockefeller University; (Palade) AP/Wide World Photos; **Figure 27–43 (c)** John Heuser, Washington University Medical School, Department of Biochemistry.

CHAPTER 28 **p. 1072 (top and bottom)** E.B. Lewis, California Institute of Technology, Division of Biology; **p. 1077** (Jacob) Corbis/Bettmann; (Monod) Corbis/Bettmann; **Figure 28–7 (b)** Personal communication from M. Thomas Record, Department of Biochemistry, University of Wisconsin–Madison; **Figure 28–14 (a)** McKnight, S.L. (1991) Molecular zippers in gene regulation. *Sci. Am.* **264** (April), 54–64; **Figure 28–23 (a)** Watson, J.D., et al. (1987) *Molecular Biology of the Gene,* 4th edn, p. 487, Benjamin/Cummings Publishing Company, Menlo Park, CA; **Figure 28–25** Nomura, M., Gourse, R., & Baughman, G. (1984) Regulation of the synthesis of ribosomes and ribosomal components. *Annu. Rev. Biochem.* **53,** 75–117; **Figure 28–27** John D. Cunningham/Visuals Unlimited; **Figure 28–33** Schwabe, J.W.R. & Rhodes, D. (1991) Beyond zinc fingers: steroid hormone receptors have a novel structural motif for DNA recognition. *Trends Biochem. Sci.* **16,** 291–296; **Figure 28–35** F.R. Turner, University of Indiana, Bloomington, Department of Biology; **Figure 28–37 (a)** Wolfgang Driever and Christiane Nüsslein-Volhard, Max-Planck-Institut; **p. 1112** (Nüsslein-Volhard) AP/Wide World Photos/Mark Lennihan; **Figure 28–38** Adapted from a figure provided by Marv Wickens, University of Wisconsin–Madison, Department of Biochemistry; **Figure 28–39 (b)** Courtesy Phillip Ingham, Imperial Cancer Research Fund, Oxford University; **Figure 28–40 (a and b)** F.R. Turner, University of Indiana, Bloomington, Department of Biology.

CHAPTER 29 **p. 1119** (Berg) Courtesy Stanford Visual Art Services; (Boyer) Courtesy Genentech, Inc.; (Cohen) Courtesy Stanford Visual Art Services; **Figure 29–6** Elizabeth A. Wood, University of Wisconsin–Madison, Department of Biochemistry; **Box 29–1, Figure 1** Phil Borden/PhotoEdit; **Figure 29–20** Keith Wood, University of California, San Diego, Department of Biology; **Figure 29–21** From Fischhoff, D.A. & Bowdish, K.S. (1987) Insect tolerant transgenic tomato plants. *BioTechnology* **5,** 807–812. Photo courtesy David Fischhoff, Monsanto Company; **Figure 29–22 (a and b)** Chris Boerboom, University of Wisconsin–Madison, Department of Agronomy; **Figure 29–24** R.L. Brinster and R.E. Hammer, University of Pennsylvania, School of Veterinary Medicine.

Molecular Modeling

MOLECULAR GRAPHICS Unless indicated, all molecular graphics were produced by Jean-Yves Sgro, Ph.D., at the Department of Biochemistry, University of Wisconsin–Madison, using a Silicon Graphics Unix workstation and the following software.

GRASP software from Anthony Nicholls. See Nicholls, A., Sharp, K.A., & Honig, B. (1991) Protein folding and association: insights from interfacial and thermodynamic properties of hydrocarbons. *Proteins Struct. Funct. Genet.* **11,** 281–296. http://trantor.bioc.columbia.edu/grasp/

MidasPlus software from the Computer Graphics Laboratory, University of California, San Francisco. See Ferrin, T.E., et al. (1988) The MIDAS display system. *J. Mol. Graphics* **6,** 13. http://www.cgl.ucsf.edu/Outreach/midasplus/

MOLMOL See Koradi, R., Billeter, M., & Wüthrich, K. (1996) MOLMOL: a program for display and analysis of macromolecular structures. *J. Mol. Graphics* **14,** 51–55. http://www.mol.biol.ethz.ch/wuthrich/software/molmol/

Molscript (versions 1.4 and 2.1) from Per Kraulis. See Kraulis, P.J. (1991) MOLSCRIPT: a program to produce both detailed and schematic plots of protein structures. *J. Appl. Crystallogr.* **24,** 946–950. http://www.avatar.se/molscript/

RasMol 2.6b for Unix (©Roger Sayle, 1992–1994) was used to create template scripts for Molscript/Raster3D renderings subsequently altered to add specific colors or renderings. See http://www.umass.edu/microbio/rasmol/

Raster3D (v. 2.0) See Merritt, E.A. & Murphy, M.E.P. (1994) Raster3D version 2.0, a program for photorealistic molecular graphics. *Acta Cryst.* **D50,** 869–873.

WebLabViewerLite3.2 software from Molecular Simulation, Inc. http://www.msi.com

In some cases images were superimposed with PhotoShop 4.0 or 5.0 (Adobe, http://www.adobe.com/) on a Macintosh computer or with Image Magick (http://www.wizards.dupont.com/cristy/ImageMagick.html) on a Silicon Graphics Unix workstation.

ATOMIC COORDINATES Unless indicated, all atomic coordinates were obtained from the Protein Data Bank at the Research Collaboratory for Structural Bioinformatics at http://www.rcsb.org/pdb/. See Berman, H.M., Westbrook, J., Feng, Z., Gilliland, G., Bhat, T.N., Weissig, H., Shindyalov, I.N., Bourne, P.E. (2000) The Protein Data Bank *Nucl. Acids Res.* **28,** 235–242; Bernstein, F.C., Koetzle, T.F., Williams, G.J.B., Meyer, E.F., Jr., Brice, M.D., Rodgers, J.R., Kennard, O., Shimanouchi, T., & Tasumi, M. (1977) Protein data bank. A computer-based archival file for macromolecular structures. *J. Mol. Biol.* **112,** 535; and Abola, E.E., Bernstein, F.C., Bryant, S.H., Koetzle, T.F., & Weng, J. (1987) Protein data bank. In *Crystallographic Databases—Information Content, Software Systems, Scientific Applications* (Allen, F.H., Bergerhoff, G., & Sievers, R., eds), p. 107, Data Commission of the International Union of Crystallography, Bonn.

Some structures (noted below) were generated using Sybyl 6.2, Tripos Inc. http://www.tripos.com/

Cover In descending order: **PDB ID: 1AON;** Xu, Z., Horwich, A.L., & Sigler, P.B. (1997) The crystal structure of the asymmetric GroEL-GroES-(ADP)$_7$ chaperonin complex. *Nature* **388,** 741; **PDB ID: 1VTM;** Pattanayek, R. & Stubbs, G. (1992) Structure of the U2 strain of tobacco mosaic virus refined at 3.5 angstroms resolution using x-ray fiber diffraction. *J. Mol. Biol.* **228,** 516; **PDB ID: 1LXA;** Raetz, C.R.H., & Roderick, S.L. (1995) A left-handed parallel β-helix in the structure of UDP *N*-acetylglucosamine acyltransferase. *Science* **270,** 997–1000; **PDB ID: 1BMF;** Abrahams, J.P., Leslie, A.G., Lutter, R., & Walker, J.E. (1994) Structure at 2.8 Å resolution of F_1-ATPase from bovine heart mitochondria. *Nature* **370,** 621; **PDB ID: 1MBO;** Phillips, S.E.V. (1980) Structure and refinement of oxymyoglobin at 1.6 angstroms resolution. *J. Mol. Biol.* **142,** 531.

PART I

Figure 2–23c PDBID: 1IFD; Marvin, D.A. (1990) Model-building studies of inovirus: genetic variations on a geometric theme. *Int. J. Biol. Macromol.* **12,** 125; **Figure 2–23d** PDBID: 2DPV; Tsao, J., Chapman, M.S., Agbandje, M., Keller, W., Smith, K., Wu, H., Luo, M., Smith, T.J., Rossmann, M.G., Compans, R.W., & Parrish, C.R. (1991) The three-dimensional structure of canine parvovirus and its functional implications. *Science* **251,** 1456; **Figure 2–23e** PDBID: 1EAH; Lentz, K.N., Smith, A.D., Geisler, S.C., Cox, S., Buontempo, P., Skelton, A., Demartino, J., Rozhon, E., Schwartz, J., Girijavallabhan, V., O'Connell, J., & Arnold, E. (1997) Structure of poliovirus type 2 Lansing complexed with antiviral agent Sch48973: comparison of the structural and biological properties of three poliovirus serotypes. *Structure* **5,** 961; **Figure 2–23f** PDBID: 2BPA; McKenna, R., Xia, D., Willingmann, P., Ilag, L.L., Krishnaswamy, S., Rossmann, G., Olson, N.H., Baker, T.S., & Incardona, N.L. (1992) Atomic structure of single-stranded DNA bacteriophage phiX174 and its functional implications. *Nature* **355,** 137; **Figure 3–7b,c** coordinates generated by Sybyl; **Figure 3–12** PDBID: 1AKX; Brodsky, A.S. & Williamson, J.R. (1997) Solution structure of the HIV-2 Tar-argininamide complex. *J. Mol. Biol.* **267,** 624.

PART II

Opener (p. 112) PDB ID: 7GCH; Brady, K., Wei, A., Ringe, D., & Abeles, R.H. (1990) Structure of chymotrypsin-trifluoromethyl ketone inhibitor complexes. Comparison of slowly and rapidly equilibrating inhibitors. *Biochemistry* **29,** 7600; **Figure 6–1** PDB ID: 6GCH; Brady, K., Wei, A., Ringe, D., & Abeles, R.H. (1990) Structure of chymotrypsin-trifluoromethyl ketone inhibitor complexes. Comparison of slowly and rapidly equilibrating inhibitors. *Biochemistry* **29,** 7600; Glycine coordinates from Sybyl compounds database; **Figure 6–4c,d, 6–5** PDB ID: 4TNC; Satyshur, K.A., Rao, S.T., Pyzalska, D., Drendel, W., Greaser, M., & Sundaralingam, M. (1988) Refined structure of chicken skeletal muscle troponin C in the two-calcium state at 2-angstroms resolution. *J. Biol. Chem.* **263,** 1628; **Figure 6–12** PDB ID: 1CDG (modified); Lawson, C.L., Van Montfort, R., Strokopytov, B., Rozeboom, H.J., Kalk, K.H., De Vries, G.E., Penninga, D., Dijkhuizen, L., & Dijkstra, B.W. (1994) Nucleotide sequence and x-ray structure of cyclodextrin glycosyltransferase from *Bacillus circulans* strain 251 in a maltose-dependent crystal form. *J. Mol. Biol.* **236,** 590; **Figure 6–16** PDB ID: 1MBO; Phillips, S.E.V. (1980) Structure and refinement of oxymyoglobin at 1.6 angstroms resolution. *J. Mol. Biol.* **142,** 531; **Box 6–3 Figure 1d** PDB ID: 2MBW; Brucker, E.A., Olson, J.S., Phillips, G.N., Jr., Dou, Y., & Ikeda-Saito, M. (1996) High resolution crystal structures of the deoxy-, oxy-, and aquomet-forms of cobalt myoglobin. *J. Biol. Chem.* **271,** 25,419–25,422; **Box 6–3 Figures 3b,c** Created by Brian Volkman, National Magnetic Resonance Facility at Madison, using MOLMOL; PDB ID: 1VRF **(part b)** and 1VRE **(part d)**; Volkman, B.F., Alam, S.L., Satterlee, J.D., & Markley, J.L. (1998) Solution structure and backbone dynamics of component IV-glycera dibranchiata monomeric hemoglobin-CO. *Biochemistry* **37,** 10,906–10,919; **Figure 6–18 (left)** PDB ID: 1CCR; Ochi, H., Hata, Y., Tanaka, N., Kakudo, M., Sakurai, T., Aihara, S., & Morita, Y. (1983) Structure of rice ferricytochrome *c* at 2.0 angstroms resolution. *J. Mol. Biol.* **166,** 407; **(center)** PDB ID: 3LYM; Kundrot, C.E. & Richards, F.M. (1987) Crystal structure of hen egg-white lysozyme at a hydrostatic pressure of 1000 atmospheres. *J. Mol. Biol.* **193,** 157; **(right)** PDB ID: Howlin, B., Moss, D.S., & Harris, G.W. (1989) Segmented anisotropic refinement of bovine ribonuclease A by the application of the rigid-body TLS model. *Acta Crystallogr., Sect. A* **45,** 851; **Figure 6–19** PDB ID: 4TNC; See citation at Fig. 6–4c; **Figure 6–20d (left)** PDB ID: 7AHL; Song, L., Hobaugh, M.R., Shustak, C., Cheley, S., Bayley, H., & Gouaux, J.E. (1996) Structure of staphylococcal αhemolysin, a heptameric transmembrane pore. *Science* **274,** 1859; **Figure 6–20d (right)** PDB ID: 1DNP; Park, H.W., Kim, S.T., Sancar, A., & Deisenhofer, J. (1995) Crystal structure of DNA photolyase from *Escherichia coli. Science* **268,** 1866;

Figure 6–21 PDB ID: 1PKN; Larsen, T.M., Laughlin, L.T., Holden, H.M., Rayment, I., & Reed, G.H. (1994) Structure of rabbit muscle pyruvate kinase complexed with Mn^{2+}, K^+, and pyruvate. *Biochemistry* **33,** 6301; **Figure 6–22 (all α)** PDB ID: 1A06; Sugio, S., Kashima, A., Mochizuki, S., Noda, M., & Kobayashi, K. (1999) Crystal structure of human serum albumin at 2.5 angstrom resolution. *Protein Eng.* **12,** 439; PDB ID: 1BCF; Frolow, F., Kalb (Gilboa), A.J., & Yariv, J. (1994) The structure of a unique, two-fold symmetric, haem-binding site. *Nat. Struct. Biol.* **1,** 453; PDB ID: 1GAI; Aleshin, A.E., Stoffer, B., Firsov, L.M., Svensson, B., & Honzatko, R. B. (1996) Crystallographic complexes of glucoamylase with maltooligosaccharide analogs: relationship of stereo-chemical distortions at the nonreducing end to the catalytic mechanism. *Biochemistry* **35,** 8319–8328; PDB ID: 1ENH; Clarke, N.D., Kissinger, C.R., Desjarlais, J., Gilliland, G.L., & Pabo, C.O. (1994) Structural studies of the engrailed homeodomain. *Protein Sci.* **3,** 1779; **(all β)** PDB ID: 1HOE; Pflugrath, J.W., Wiegand, G., Huber, R., & Vertesy, L. (1986) Crystal structure determination, refinement and the molecular model of the a-amylase inhibitor Hoe-467A. *J. Mol. Biol.* **189,** 383; PDB ID: 1LXA; Raetz, C.R.H. & Roderick, S.L. (1995) A left-handed parallel β-helix in the structure of UDP *N*-acetyl-glucosamine acyltransferase. *Science* **270,** 997–1000; PDB ID: 1PEX; Gomis-Ruth, F.X., Gohlke, U., Betz, M., Knauper, V., Murphy, G., Lopez-Otin, C., & Bode, W. (1996) The helping hand of collagenase-3 (Mmp-13): 2.7 Å crystal structure of its C-terminal haemopexin-like domain. *J. Mol. Biol.* **264,** 556–566; PDB ID: 1JPC; Wright, C.S. & Hester, G. (1996) 2Å Structure of a cross-linked complex between snowdrop lectin and a branched penta-mannoside. Evidence for two unique binding modes. *Structure* **4,** 1339–1352; PDB ID: 1CD8; Leahy, D.J., Axel, R., & Hendrickson, W.A. (1992) Crystal structure of a soluble form of the human T cell co-receptor Cd8 at 2.6 angstroms resolution. *Cell* **68,** 1145; **(α/β)** PDB ID: 1DEH; Davis, G.J., Stone, C.J., Bosron, W.F., & Hurley, T.D. (1996) X-ray structure of human $\beta_3\beta_3$ alcohol dehydrogenase: the contribution of ionic interactions to coenzyme binding. *J. Biol. Chem.* **271,** 17,057–17,061; PDB ID: 1DUB; Engel, C.K., Mathieu, M., Zeelen, J.P., Hiltunen, J.K., & Wierenga, R. K. (1996) Crystal structure of enoyl-coenzyme A (CoA) hydratase at 2.5 angstroms resolution: a spiral fold defines the CoA-binding pocket. *EMBO J.* **15,** 5135; PDB ID: 1PFK; Shirakihara, Y., & Evans, P.R. (1988) Crystal structure of the complex of phosphofructokinase from *Escherichia coli* with its reaction products. *J. Mol. Biol.* **204,** 973; **(α+β)** PDB ID: 2PIL; Forest, K.T., Dunham, S.A., Koomey, M., & Tainer, J.A. (1999) Crystallographic structure of phosphorylated pilin from *Neisseria:* phosphoserine sites modify type IV pilus surface chemistry and morphology. *Mol. Bicrobiol.* **31,** 743; PDB ID: 1SYN; Stout, T.J. & Stroud, R.M. (1996) The complex of the anti-cancer therapeutic, BW1843U89, with thymidylate synthase at 2.0 Å resolution: implications for a new mode of inhibition. *Structure* **4,** 67–77; PDB ID: 1EMA; Ormo, M., Cubitt, A.B., Kallio, K., Gross, L.A., Tsien, R.Y., & Remington, S.J. (1996) Crystal structure of the *Aequorea victoria* green fluorescent protein. *Science* **273,** 1392; PDB ID: 1U9A; Tong, H., Hateboer, G., Perrakis, A., Bernards, R., & Sixma, T.K. (1997) Crystal structure of murine/human Ubc9 provides insight into the variability of the ubiquitin-conjugating system. *J. Biol. Chem.* **272,** 21,381–21,387; **Figure 6–23** PDB ID: 2HHB; Fermi, G., Perutz, M.F., Shaanan, B., & Fourme, R. (1984) The crystal structure of human deoxyhaemoglobin at 1.74 angstroms resolution. *J. Mol. Biol.* **175,** 159; **Figure 6–25a** PDB ID: 2PLV; Filman, D.J., Syed, R., Chow, M., Macadam, A.J., Minor, P.D., & Hogle, J.M. (1989) Structural factors that control conformational transitions and serotype specificity in type 3 poliovirus. *EMBO J.* **8,** 1567; **Figure 6–25b** PDB ID: 1VTM; Pattanayek, R. & Stubbs, G. (1992) Structure of the U2 strain of tobacco mosaic virus refined at 3.5 angstroms resolution using x-ray fiber diffraction. *J. Mol. Biol.* **228,** 516; **Figure 6–28** Personal communication from Yong Duan and Peter A. Kollman, University of California, San Francisco, Department of Pharamaceutical Science; **Figure 6–31** PDB ID: 1AON; Xu, Z., Horwich, A.L., & Sigler, P.B. (1997) The crystal structure of the asymmetric GroEL-GroES-$(ADP)_7$ chaperonin complex. *Nature* **388,** 741; **Figure 7–1c** Heme extracted from PDB ID: 1CCR; See citation for Fig. 6–18 (left); **Figures 7–3, 7–5, 7–6 (left)** PDB ID: 1MBO; Phillips, S.E.V. (1980) Structure and refinement of oxymyoglobin at 1.6 angstroms resolution. *J. Mol. Biol.* **142,** 531; **Figures 7–6 (right), 7–8, 7–9a, 7–10, 7–11 (T state), 7–17** PDB ID: 1HGA; Liddington, R., Derewenda, Z., Dodson, E., Hubbard, R., & Dodson, G. (1992) High resolution crystal structures and comparisons of T state deoxyhaemoglobin and two liganded T-state haemoglobins: T(α-oxy)haemoglobin and T(met) haemoglobin. *J. Mol. Biol.* **228,** 551. In Fig. 7–17, coordinates for 2,3-bisphosphoglycerate were obtained from Paul Reisberg, Wellesley College, Department of Chemistry; **Figures 7–10, 7–11 (R state)** PDB ID: 1BBB; Silva, M.M., Rogers, P.H., & Arnone, A. (1992) A third quaternary structure of human hemoglobin a at 1.7-angstroms resolution. *J. Biol. Chem.* **267,** 17,248. In Fig. 7–11 R-state was modified to represent O_2 instead of CO; **Figure 7–21** PDB ID: 1DDH; Li, H., Natarajan, K., Malchiodi, E.L., Margulies, D.H., & Mariuzza, R.A. (1998) Three-dimensional structure of H-2Dd complexed with an immunodominant peptide from human immunodeficiency virus envelope glycoprotein 120. *J. Mol. Biol.* **283,** 179; **Figure 7–23b** PDB ID: 1IGT; Harris, L.J., Larson, S.B., Hasel, K.W., & McPherson, A. (1997) Refined structure of an intact IgG2A monoclonal antibody. *Biochemistry* **36,** 1581; **Figure 7–27a** PDB ID: 1GGC; Stanfield, R.L., Takimoto-Kamimura, M., Rini, J.M., Profy, A.T., & Wilson, I.A. (1993) Major antigen-induced domain rearrangements in an antibody. *Structure* **1,** 83; **Figure 7–27b,c** PDB ID: 1GGI; Rini, J.M., Stanfield, R.L., Stura, E.A., Salinas, P.A., Profy, A.T., & Wilson, I.A. (1993) Crystal structure of an human immunodeficiency virus type 1 neutralizing antibody, 50.1, in complex with its V3 loop peptide antigen. *Proc. Natl. Acad. Sci. USA* **90,** 6325; **Figure 7–29** Personal communication from Ivan Rayment, Enzyme Institute and Department of Biochemistry, University of Wisconsin–Madison (See also PDB ID: 2MYS); **Figure 7–30** Personal communication from Ivan Rayment, Enzyme Institute and Department of Biochemistry, University of Wisconsin–Madison; **Figure 8–1** PDB ID: 7GCH; Brady, K., Wei, A., Ringe, D., & Abeles, R.H. (1990) Structure of chymotrypsin-trifluoromethyl ketone inhibitor complexes. Comparison of slowly and rapidly equilibrating inhibitors. *Biochemistry* **29,** 7600; **Figure 8–4** PDB ID: 1RA2; Sawaya, M.R. & Kraut, J. (To be published) Loop and subdomain movements in the mechanism of *Escherichia coli* dihydrofolate reductase crystallographic evidence; **Figure 8–18b,c,d** PDB ID: 7GCH; see citation for Fig. 8–1; **Figure 8–21a** PDB ID: 2YHX; Anderson, C.M., Stenkamp, R.E., & Steitz, T.A. (1978) Sequencing a protein by x-ray crystallography. II. Refinement of yeast hexokinase B coordinates and sequence at 2.1 angstroms. *J. Mol. Biol.* **123,** 15; **Figure 8–21b** PDB ID: 1HKG; Steitz, T.A., Shoham, M., & Bennett, W.S., Jr. (1981) Structural dynamics of yeast hexokinase during catalysis. *Philos. Trans. R. Soc. London, Ser. B* **293,** 43. Glucose (GLC) coordinates were transformed from PDB ID: 1GLK; St. Charles, R., Harrison, R.W., Bell, G.I., Pilkis, S.J., & Weber, I.T (1994) Molecular model of human beta-cell glucokinase built by analogy to the crystal structure of yeast hexokinase B. *Diabetes* **43,** 784; **Figure 8–22b** PDB ID: 1ONE; Larsen, T.M., Wedekind, J.E., Rayment, I., & Reed, G.H. (1996) A carboxylate oxygen of the substrate bridges the magnesium ions at the active site of enolase: structure of the yeast enzyme complexed with the equilibrium mixture of 2-phosphoglycerate and phosphoenolpyruvate at 1.8 Å resolution. *Biochemistry* **35,** 4349; **Figure 8–24** PDB ID: 2AT2; Stevens, R.C., Reinisch, K.M., & Lipscomb, W.N. Molecular structure of *Bacillus subtilis* aspartate transcarbamoylase at 3.0 angstroms resolution. *Proc. Natl. Acad. Sci. USA* **88,** 6087 (1991). Thank you to Raymond C. Stevens (University of California, Berkeley) for providing symmetry-related coordinates; **Figure 8–29** PDB ID: 3AMV; Oikonomakos, N.G., Tsitsanou, K.E., Zographos, S.E., Skamnaki, V.T., Goldmann, S., & Bischoff, H. (1999) Allosteric inhibition of glycogen phosphorylase a by the potential antidiabetic drug 3-isopropyl 4-(2-chlorophenyl)-1,4-dihydro-1-ethyl-2-methyl-pyridine-3,5,6-tricarboxylate. *Protein Sci.* **8,** 1930. IMP (inosine monophosphate) extracted from PDB ID 2SKE; Oikonomakos, N.G., Zographos, S.E., Tsitsanou, K.E., Johnson, L.N., & Acharya, K.R. (1996) Activator anion binding site in pyridoxal phosphorylase B. The binding of phosphite, phosphate and fluorophosphate in the crystal. *Protein Sci.* **5,** 2416; **Figure 9–26** Lipopolysaccharide (LPS) from PDB ID: 2FCP; Ferguson, A.D., Hofmann, E., Coulton, J.W., Diederichs, K., & Welte, W. (1998) Siderophore-mediated iron transport: crystal structure of FhuA with bound lipopolysaccharide. *Science* **282,** 2215–2220; **Figure 10–15** Coordinates generated by Sybyl; **Figure 10–19** Coordinates generated by Sybyl; **Figure 10–22b** PDB ID: 1BCE; Asensio, J.L., Brown, T., & Lane, A.N. (1998) Comparison of the solution structures of intramolecular DNA triple helices containing adjacent and non-adjacent CG.C+ triplets. *Nucl. Acids Res.* **26,** 3677–3686; **Figure 10–2d** PDB ID: 1QDG; Marathias, V.M., Wang, K.Y., Kumar, S., Swaminathan, S., & Bolton, P.H. (1996) Determination of the number and location of the manganese binding sites of DNA quadruplexes in solution by EPR and NMR in the presence and absence of thrombin. *J. Mol. Biol.* **260,** 378–394; **Figure 10–25** Coordinates generated by Sybyl; **Figure 10–28a** PDB ID: 1TRA; Westhof, E. & Sundaralingam, M. (1986) Restrained refinement of the monoclinic form of yeast phenylalanine transfer RNA. Temperature factors and dynamics, coordinated waters, and base-pair propeller twist angles. *Biochemistry* **25,** 4868; **Figure 10–28c** PDB ID: 1MME; Scott, W.G., Finch, J.T., & Klug, A. (1995) The crystal structure of an all-RNA hammerhead ribozyme: a proposed mechanism for RNA catalytic cleavage. *Cell* **81,** 991; **Figure 10–28d** PDB ID: 1GRZ; Golden, B.L., Gooding, A.R., Podell, E.R., & Cech, T.R. (1998) A preorganized active site in the crystal structure of the *Tetrahymena* ribozyme. *Science* **282,** 259; **Figures 11–1a,b, 11–2, 11–11** coordinates from Sybyl compounds database; **Figure 11–14** PDB coordinates from Dave Woodcock, Okanagan University College, Kelowna, British Columbia, Department of Chemistry; **Figure 12–12** Protein coordinates from PDB ID: 1A8A; Swairjo, M.A., Concha, N.O., Kaetzel, M.A., Dedman, J.R., & Seaton, B.A. (1995) Ca^{2+}-bridging mechanism and phospholipid head group recognition in the membrane-binding protein annexin V. *Nat. Struct. Biol.* **2,** 968. Manually docked onto lipid bilayer coordinates obtained from Heller, H., Schaefer, M. & Schulten, K. (1993) Molecular dynamics simulation of a bilayer of 200 lipids in the gel and in the liquid-crystal phases. *J. Phys Chem.* **97,** 8343–8360; **Figure 12–15** PDB ID: 2AT9; Mitsuoka, K., Hirai, T., Murata, K., Miyazawa, A., Kidera, A., Kimura, Y., & Fujiyoshi, Y. (1999) The structure of bacteriorhodopsin at 3.0 Å resolution based on electron crystallography: implication of the charge distribution. *J. Mol. Biol.* **286,** 861; **Figure 12–16** PDB ID: 1PRC; Deisenhofer, J., Epp, O., Sinning, I., & Michel, H. (To be published) Crystallographic refinement at 2.3 angstroms resolution and refined model of the photosynthetic reaction center from *Rhodopseudomonas viridis*; **Figure 12–18** PDB ID: 2FCP; Ferguson, A.D., Hofmann, E., Coulton, J.W., Diederichs, K., & Welte, W. (1998) Siderophore-mediated iron transport: crystal structure of FhuA with bound lipopolysaccharide. *Science* **282,** 2215–2220; **Figure 12–37** Coordinates prepared for The Virtual Museum of Minerals and Molecules, http://www.soils.wisc.edu/virtual_museum by Phillip Barak, University of Wisconsin–Madison, Department of Soil Science, using data from Neupert-Laves, K. & Dobler, M. (1975) The crystal structure of a K^+ complex of valinomycin. *Helv. Chim. Acta* **58,** 432; **Figure 12–38a,b,c** PDB ID: 1BL8; Doyle, D.A., Cabral, J.M., Pfuetzner, R.A., Kuo, A., Gulbis, J.M., Cohen, S.L., Chait, B.T., & Mackinnon, R. (1998) The structure of the potassium channel: molecular basis of K^+ conduction and selectivity. *Science* **280,** 69; **Figure 12–43a,b** PDB ID: 2FCP; see citation for Fig. 12–18;

Figure 13–12 PDB ID: 1AZS; Tesmer, J.J., Sunahara, R.K., Gilman, A.G., & Sprang, S.R. (1997) Crystal structure of the catalytic domains of adenylyl cyclase in a complex with $G_{s\alpha}$•GTPγS. *Science* **278,** 1907; **Figure 13–14b** PDB ID: 1YDS; Engh, R.A., Girod, A., Kinzel, V., Huber, R., & Bossemeyer, D. (1996) Crystal structures of catalytic subunit of cAMP-dependent protein kinase in complex with isoquinolinesulfonyl protein kinase inhibitors H7, H8, and H89. Structural implications for selectivity. *J. Biol. Chem.* **271,** 26,157; **Figure 13–19a** PDB ID: 1CLL; Chattopadhyaya, R., Meador, W.E., Means, A.R., Quiocho, F. (1992) A calmodulin structure refined at 1.7 angstroms resolution. *J. Mol. Biol.* **228,** 1177; **Figure 13–19b** PDB ID: 1CDL; Meador, W.E., Means, A.R., & Quiocho, F.A. (1992) Target enzyme recognition by calmodulin: 2.4 angstroms structure of a calmodulin-peptide complex. *Science* **257,** 1251; **Figure 13–22** PDB ID: 1BAC; Chou, K.-C., Carlacci, L., Maggiora, G.M., Parodi, L.A., & Schulz, M.W. (1992) An energy-based approach to packing the 7-helix bundle of bacteriorhodopsin. *Protein Sci.* **1,** 810; **Figure 13–31a** PDB ID: 1HCK; Schulze-Gahmen, U., De Bondt, H.L., & Kim, S.-H. (1996) High-resolution crystal structures of human cyclin-dependent kinase 2 with and without ATP: bound waters and natural ligand as guides for inhibitor design. *J. Med. Chem.* **39,** 4540–4546; **Figure 13–31b** PDB ID: 1FIN; Jeffrey, P.D., Russo, A.A., Polyak, K., Gibbs, E., Hurwitz, J., Massague, J., & Pavletich, N.P. (1995) Mechanism of Cdk activation revealed by the structure of a cyclin a-Cdk2 complex. *Nature* **376,** 313; **Figure 13–31c** PDB ID: 1JST; Russo, A.A., Jeffrey, P.D., & Pavletich, N.P. (1996) Structural basis of cyclin-dependent kinase activation by phosphorylation. *Nat. Struct. Biol.* **3,** 696.

PART III

Opener (p. 484) PDB ID: 1QO1; Stock, D., Leslie, A. G.W., & Walker, J.E. (1999) Molecular architecture of the rotary motor in ATP synthase. *Science* **286,** 1700–1705. Protein backbones were calculated from Ca PDB entry with using MIDAS-PLUS and Sybyl; **Figure 15–18** PDB ID: 1PFK; Shirakihara, Y. & Evans, P.R. (1988) Crystal structure of the complex of phosphofructokinase from *Escherichia coli* with its reaction products. *J. Mol. Biol.* **204,** 973; **Figure 16-8a** PDB ID: 5CSC; Liao, D.-I., Karpusas, M., & Remington, S.J. (1991) Crystal structure of an open conformation of citrate synthase from chicken heart at 2.8-Å resolution. *Biochemistry* **30,** 6031–6036; **Figure 16–8b** PDB ID: 5CTS; Karpusas, M., Branchaud, B., & Remington, S.J. (1990) Proposed mechanism for the condensation reaction of citrate synthase. 1.9-angstroms structure of the ternary complex with oxaloacetate and carboxymethyl coenzyme A. *Biochemistry* **29,** 2213; **Figure 16–10** PDB ID: 1SCU; Wolodko, W.T., Fraser, M.E., James, M.N.G., & Bridger, W.A. (1994) The crystal structure of succinyl-CoA synthetase from *Escherichia coli* at 2.5 angstroms resolution. *J. Biol. Chem.* **289,** 10,883; **Figure 18-5c,d,e** PDB ID: 1AJS; Rhee, S., Silva, M.M., Hyde, C.C., Rogers, P.H., Metzler, C.M., Metzler, D.E., & Arnone, A. (1997) Refinement and comparisons of the crystal structures of pig cytosolic aspartate aminotransferase and its complex with 2-methylaspartate. *J. Biol. Chem.* **272,** 17,293; **Figure 19-5d** PDB ID: 1FRD; Jacobson, B.L., Chae, Y.K., Markley, J.L., Rayment, I., & Holden, H.M. (1993) Molecular structure of the oxidized, recombinant, heterocyst (2Fe-2S) ferredoxin from Anabaena 7120 determined to 1.7 angstroms resolution. *Biochemistry* **32,** 6788; **Figure 19–10** PDB ID: 1BJY; Iwata, S., Lee, J.W., Okada, K., Lee, J.K., Iwata, M., Rasmussen, B., Link, T.A., Ramaswamy, S., & Jap, B.K. (1998) Complete structure of the 11-subunit bovine mitochondrial cytochrome bc_1 complex. *Science* **281,** 64; **Figure 19–12** PDB ID: 1OCC; Tsukihara, T., Aoyama, H., Yamashita, E., Tomizaki, T., Yamaguchi, H., Shinzawa-Itoh, K., Nakashima, R., Yaono, R., & Yoshikawa, S. (1996) The whole structure of the 13-subunit oxidized cytochrome *c* oxidase at 2.8 Å. *Science* **272,** 1136; **Figure 19–22b,c** PDB ID: 1BMF; Abrahams, J.P., Leslie, A.G., Lutter, R., & Walker, J.E. (1994) Structure at 2.8 Å resolution of F_1-ATPase from bovine heart mitochondria. *Nature* **370,** 621; **Figure 19–22e,f** PDB ID: 1QO1; Stock, D., Leslie, A. G.W., & Walker, J.E. (1999) Molecular architecture of the rotary motor in ATP synthase. *Science* **286,** 1700–1705. Protein backbones were calculated from the C_α PDB entry using MIDAS-PLUS and Sybyl; **Figure 19–45** PDB ID: 1PRC; Deisenhofer, J., Epp, O., Sinning, I., & Michel, H. (1995) Crystallographic refinement at 2.3 angstroms resolution and refined model of the photosynthetic reaction center from *Rhodopseudomonas viridis*. *J. Mol. Biol.* **246,** 429–457; **Figure 19–54** PDB ID: 1C8R; Luecke, H., Schobert, B., Richter, H.-T., Cartailler, J.-P., & Lanyi, J.K. (1999) Structural changes in bacteriorhodopsin during ion transport at 2 Å resolution. *Science* **286,** 255; **Figure 20–24a,b** PDB ID: 8RUC; Andersson, I. (1996) Large structures at high resolution: the 1.6 angstroms crystal structure of spinach ribulose-1,5-bisphosphate carboxylase/oxygenase complexed with 2-carboxyarabinitol bisphosphate. *J. Mol. Biol.* **259,** 160–174. **Figure 20–24c** PDB ID: 9RUB; Lundqvist, T. & Schneider, G. (1991) Crystal structure of activated ribulose-1,5-bisphosphate carboxylase complexed with its substrate, ribulose-1,5-bisphosphate. *J. Biol. Chem.* **266,** 12,604; **Figure 20–33b** PDB ID: 8RUC; see citation for Fig. 20–24a,b; **Box 21–2, Figure 2** PDB ID: 3PGH; Kurumbail, R.G., Stevens, A.M., Gierse, J.K., McDonald, J.J., Stegeman, R.A., Pak, J.Y., Gildehaus, D., Miyashiro, J.M., Penning, T.D., Seibert, K., Isakson, P.C., & Stallings, W.C. (1996) Structural basis for selective inhibition of cyclooxygenase-2 by anti-inflammatory agents. *Nature* **384,** 644; **Figure 22–3** PDB ID: 1N2C; Schindelin, H., Kisker, C., Schlessman, J.L., Howard, J.B., & Rees, D.C. (1997) Structure of ADP X AlF_4^--stabilized nitrogenase complex and its implications for signal transduction. *Nature* **387,** 370; **Figure 22–5** PDB ID: 2GLS; Yamashita, M.M., Almassy, R.J., Janson, C.A., Cascio, D., & Eisenberg, D. (1989) Refined atomic model of glutamine synthetase at 3.5 angstroms resolution. *J. Biol. Chem.* **264,** 17,681; **Figure 22–35** Coordinates and assistance provided by Jim Thoden and Hazel Holden, University of Wisconsin–Madison, Department of Biochemistry and Enzyme Institute. See also Thoden, J.B., Huang, X., Raushel, F.M., & Holden, H.M. (1999) The small subunit of carbamoyl phosphate synthetase: snapshots along the reaction pathway. *Biochemistry* **38,** 16,158–16,166; **Figure 22–38b** PDB ID: 1PFR; Logan, D.T., Su, X.D., Aberg, A., Regnstrom, K., Hajdu, J., Eklund, H., & Nordlund, P. (1996) Crystal structure of reduced protein r2 of ribonucleotide reductase: the structural basis for oxygen activation at a dinuclear iron site. *Structure* **4,** 1053.

PART IV

Figure 24–24 PDB ID: 1AOI; Luger, K., Maeder, A.W., Richmond, R.K., Sargent, D.F., & Richmond, T.J. (1997) Crystal structure of the nucleosome core particle at 2.8 Å resolution. *Nature* **389,** 251; **p. 931** Coordinates generated by Sybyl; **Figure 25–8** PDB ID: 3BDP; Kiefer, J.R., Mao, C., Braman, J.C., & Beese, L.S. (1998) Visualizing DNA replication in a catalytically active *Bacillus* DNA polymerase crystal. *Nature* **391,** 304; **Figure 25–10** PDB ID: 2POL; Kong, X.-P., Onrust, R., O'Donnell, M., & Kuriyan, J. (1992) Three-dimensional structure of the β subunit of *Escherichia coli* DNA polymerase III holoenzyme: a sliding DNA clamp. *Cell* **69,** 425; **Figure 25–32b** PDB ID: 2REB; Story, R.M., Weber, I.T., & Steitz, T.A. (1992) The structure of the *E. coli* RecA protein monomer and polymer. *Nature* **355,** 318; **Figure 25–36b** PDB ID: 3CRX; Gopaul, D.N., Guo, F., & Van Duyne, G.D. (1998) Structure of the Holliday junction intermediate in Cre-Loxp site-specific recombination. *EMBO J.* **17,** 4175; **p. 979** Coordinates generated by Sybyl; **Figure 26–9b** PDB ID: 1TGH; Juo, Z.S., Chiu, T.K., Leiberman, P.M., Baikalov, I., Berk, A.J., & Dickerson, R.E. (1996) How proteins recognize the TATA-box. *J. Mol. Biol.* **261,** 239; **Figure 26–10b** PDB ID: 1DSC; Lian, C., Robinson, H., & Wang, A.H.-J. (1996) Structure of actinomycin D bound with $(GAAGCTTC)_2$ and $(GATGCTTC)_2$ and its binding to the (CAG)N:(CTG)N triplet sequence by NMR analysis. *J. Am. Chem. Soc.* **118,** 8791; **Figure 26–27b** PDB ID: 1MME; see citation for Fig. 10–28c; **p. 1020** PDB ID: 1B23; Nissen, P., Thirup, S., Kjeldgaard, M., Nyborg, J. (1999) The crystal structure of Cys-tRNA-EF-TU-GDPNP reveals general and specific features in the ternary complex and in tRNA. *Structure* **7,** 143; **(bottom)** PDB ID: 1DAR; Al-Karadaghi, S., Aevarsson, A., Garber, M., Zheltonosova, J., & Liljas, A. (1996) The structure of elongation factor G in complex with GDP: conformational flexibility and nucleotide exchange. *Structure* **4,** 555; **Box 27–2** Coordinates generated by Sybyl and modified with MIDAS-PLUS; **Figure 27–11a** modified from a density map provided by Joachim Frank, New York State Department of Health, Wadsworth Center; **Figure 27–15b** PDB ID 4TRA; Westhof, E., Dumas, P., & Moras, D. (1988) Restrained refinement of two crystalline forms of yeast aspartic acid and phenylalanine transfer RNA crystals. *Acta Crystallogr. Sect. A* **44,** 112; **Figure 27–19a** PDB ID: 1QRT; Arnez, J.G. & Steitz, T.A. (1996) Crystal structures of three misacylating mutants of *E. coli* glutaminyl-tRNA synthetase complexed with $tRNA^{Gln}$ and ATP. *Biochemistry* **35,** 14,725; **Figure 27–19b** PDB ID: 1ASZ; Cavarelli, J., Eriani, G., Rees, B., Ruff, M., Boeglin, M., Mitschler, A., Martin, F., Gangloff, J., Thierry, J.C., & Moras, D. (1994) The active site of yeast aspartyl-tRNA synthetase: structural and functional aspects of the aminoacylation reaction. *EMBO J.* **13,** 327; **Figure 28–7c,d** PDB ID: 1LBG and **(for part d)** 1LBH; Lewis, M., Chang, G., Horton, N.C., Kercher, M.A., Pace, H.C., Schumacher, M.A., Brennan, R.G., & Lu, P. (1996) Crystal structure of the lactose operon repressor and its complexes with DNA and inducer. *Science* **271,** 1247. Side chains were added with software programs MIDAS-PLUS and Sybyl to the C_α-only PDB entry; **Figure 28–11a,c,d** PDB ID: 1LCC; Chuprina, V.P., Rullmann, J.A.C., Lamerichs, R.M.J.N., Van Boom, J.H., Boelens, R., & Kaptein, R. (1993) Structure of the complex of Lac repressor headpiece and an 11 base-pair half-operator determined by nuclear magnetic resonance spectroscopy and restrained molecular dynamics. *J. Mol. Biol.* **234,** 446; **Figure 28–11b** PDB ID: 1LBG; see citation for Fig. 28–7c; **Figure 28–12** PDB ID: 1A1L; Elrod-Erickson, M., Benson, T.E., & Pabo, C.O. (1998) High-resolution structures of variant Zif268-DNA complexes: implications for understanding zinc finger-DNA recognition. *Structure* **6,** 451; **Figure 28–13** PDB ID: 1B8I; Passner, J.M., Ryoo, H.-D., Shen, L., Mann, R.S., & Aggarwal, A.K. (1999) Structure of a DNA-bound ultrabithorax-extradenticle homeodomain complex. *Nature* **397,** 714; **Figure 28–14b** PDB ID: 1YSA; Ellenberger, T.E., Brandl, C.J., Struhl, K., & Harrison, S.C. (1992) The GCN4 basic region leucine zipper binds DNA as a dimer of uninterrupted α helices: crystal structure of the protein-DNA complex. *Cell* **71,** 1223; **Figure 28–15** PDB ID: 1HLO; Brownlie, P., Ceska, T.A., Lamers, M., Romier, C., Theo, H., & Suck, D. (1997) The crystal structure of an intact human max-DNA complex: new insights into mechanisms of transcriptional control. *Structure* **5,** 509; **Figure 28–16a** PDB ID: 1RUN; Parkinson, G., Gunasekera, A., Vojtechovsky, J., Zhang, X., Kunkel, T.A., Berman, H., & Ebright, R.H. (1996) Aromatic hydrogen bond in sequence-specific protein-DNA recognition. *Nat. Struct. Biol.* **3,** 837; **Figure 28–22** PDB ID: 1TRO; Otwinowski, Z., Schevitz, R.W., Zhang, R.-G., Lawson, C.L., Joachimiak, A.J., Marmorstein, R., Luisi, B.F., & Sigler, P.B. (1988) Crystal structure of Trp repressor operator complex at atomic resolution. *Nature* **335,** 321; **Figure 29–2** PDB ID: 1RVC; Kostrewa, D. & Winkler, F.K. (1995) Mg^{2+} binding to the active site of EcoRV endonuclease: a crystallographic study of complexes with substrate and product DNA at 2 angstroms resolution. *Biochemistry* **34,** 683.

Glossary

a

absolute configuration: The configuration of four different substituent groups around an asymmetric carbon atom, in relation to D- and L-glyceraldehyde.

absorption: Transport of the products of digestion from the intestinal tract into the blood.

acceptor control: The regulation of the rate of respiration by the availability of ADP as phosphate group acceptor.

accessory pigments: Visible light-absorbing pigments (carotenoids, xanthophyll, and phycobilins) in plants and photosynthetic bacteria that complement chlorophylls in trapping energy from sunlight.

acidosis: A metabolic condition in which the capacity of the body to buffer H^+ is diminished; usually accompanied by decreased blood pH.

actin: A protein making up the thin filaments of muscle; also an important component of the cytoskeleton of many eukaryotic cells.

activation energy ($\Delta G^\ddagger$): The amount of energy (in joules) required to convert all the molecules in 1 mole of a reacting substance from the ground state to the transition state.

activator: (1) A DNA-binding protein that positively regulates the expression of one or more genes; that is, transcription rates increase when an activator is bound to the DNA. (2) A positive modulator of an allosteric enzyme.

active site: The region of an enzyme surface that binds the substrate molecule and catalytically transforms it; also known as the catalytic site.

active transport: Energy-requiring transport of a solute across a membrane in the direction of increasing concentration.

activity: The true thermodynamic activity or potential of a substance, as distinct from its molar concentration.

activity coefficient: The factor by which the numerical value of the concentration of a solute must be multiplied to give its true thermodynamic activity.

acyl phosphate: Any molecule with the general chemical form $R{-}\underset{\underset{O}{\|}}{C}{-}OPO_3^{2-}$.

adenosine 3′,5′-cyclic monophosphate: *See* cyclic AMP.

adenosine diphosphate: *See* ADP.

adenosine triphosphate: *See* ATP.

adipocyte: An animal cell specialized for the storage of fats (triacylglycerols).

adipose tissue: Connective tissue specialized for the storage of large amounts of triacylglycerols.

ADP (adenosine diphosphate): A ribonucleoside 5′-diphosphate serving as phosphate group acceptor in the cell energy cycle.

aerobe: An organism that lives in air and uses oxygen as the terminal electron acceptor in respiration.

aerobic: Requiring or occurring in the presence of oxygen.

alcohol fermentation: The anaerobic conversion of glucose to ethanol via glycolysis. *See also* fermentation.

aldose: A simple sugar in which the carbonyl carbon atom is an aldehyde; that is, the carbonyl carbon is at one end of the carbon chain.

alkalosis: A metabolic condition in which the capacity of the body to buffer OH^- is diminished; usually accompanied by an increase in blood pH.

allosteric enzyme: A regulatory enzyme, with catalytic activity modulated by the noncovalent binding of a specific metabolite at a site other than the active site.

allosteric protein: A protein (generally with multiple subunits) with multiple ligand-binding sites, such that ligand binding at one site affects ligand binding at another.

allosteric site: The specific site on the surface of an allosteric enzyme molecule to which the modulator or effector molecule is bound.

α helix: A helical conformation of a polypeptide chain, usually right-handed, with maximal intrachain hydrogen bonding; one of the most common secondary structures in proteins.

Ames test: A simple bacterial test for carcinogens, based on the assumption that carcinogens are mutagens.

amino acid activation: ATP-dependent enzymatic esterification of the carboxyl group of an amino acid to the 3′-hydroxyl group of its corresponding tRNA.

amino acids: α-Amino-substituted carboxylic acids, the building blocks of proteins.

amino-terminal residue: The only amino acid residue in a polypeptide chain with a free α-amino group; defines the amino terminus of the polypeptide.

aminoacyl-tRNA: An aminoacyl ester of a tRNA.

aminoacyl-tRNA synthetases: Enzymes that catalyze synthesis of an aminoacyl-tRNA at the expense of ATP energy.

aminotransferases: Enzymes that catalyze the transfer of amino groups from α-amino to α-keto acids; also called transaminases.

ammonotelic: Excreting excess nitrogen in the form of ammonia.

amphibolic pathway: A metabolic pathway used in both catabolism and anabolism.

amphipathic: Containing both polar and nonpolar domains.

ampholyte: A substance that can act as either a base or an acid.

amphoteric: Capable of donating and accepting protons, thus able to serve as an acid or a base.

anabolism: The phase of intermediary metabolism concerned with the energy-requiring biosynthesis of cell components from smaller precursors.

anaerobe: An organism that lives without oxygen. Obligate anaerobes die when exposed to oxygen.

anaerobic: Occurring in the absence of air or oxygen.

anaplerotic reaction: An enzyme-catalyzed reaction that can replenish the supply of intermediates in the citric acid cycle.

angstrom (Å): A unit of length (10^{-8} cm) used to indicate molecular dimensions.

anhydride: The product of the condensation of two carboxyl or phosphate groups in which the elements of water are eliminated to form a compound with the general structure $R{-}\underset{\underset{O}{\|}}{X}{-}O{-}\underset{\underset{O}{\|}}{X}{-}R$, where X is either carbon or phosphorus.

anion-exchange resin: A polymeric resin with fixed cationic groups; used in the chromatographic separation of anions.

anomers: Two stereoisomers of a given sugar that differ only in the configuration about the carbonyl (anomeric) carbon atom.

antibiotic: One of many different organic compounds that are formed and secreted by various species of microorganisms and plants, are toxic to other species, and presumably have a defensive function.

antibody: A defense protein synthesized by the immune system of vertebrates. *See also* immunoglobulin.

anticodon: A specific sequence of three nucleotides in a tRNA, complementary to a codon for an amino acid in an mRNA.

antigen: A molecule capable of eliciting the synthesis of a specific antibody in vertebrates.

antiparallel: Describing two linear polymers that are opposite in polarity or orientation.

antiport: Cotransport of two solutes across a membrane in opposite directions.

apoenzyme: The protein portion of an enzyme, exclusive of any organic or inorganic cofactors or prosthetic groups that might be required for catalytic activity.

apolipoprotein: The protein component of a lipoprotein.

apoprotein: The protein portion of a protein, exclusive of any organic or inorganic cofactors or prosthetic groups that might be required for activity.

apoptosis: (app′-a-toe′-sis) Programmed cell death, in which a cell brings about its own death and lysis, signaled from outside or programmed in its genes, by systematically degrading its own macromolecules.

arrestins: A family of proteins that bind to the phosphorylated carboxyl-terminal region of serpentine receptors, preventing their interactions with G proteins and thereby terminating the signal through those receptors.

asymmetric carbon atom: A carbon atom that is covalently bonded to four different groups and thus may exist in two different tetrahedral configurations.

ATP (adenosine triphosphate): A ribonucleoside 5′-triphosphate functioning as a phosphate group donor in the cell energy cycle; carries chemical energy between metabolic pathways by serving as a shared intermediate coupling endergonic and exergonic reactions.

ATP synthase: An enzyme complex that forms ATP from ADP and phosphate during oxidative phosphorylation in the inner mitochondrial membrane or the bacterial plasma membrane, and during photophosphorylation in chloroplasts.

ATPase: An enzyme that hydrolyzes ATP to yield ADP and phosphate; usually coupled to some process requiring energy.

attenuator: An RNA sequence involved in regulating the expression of certain genes; functions as a transcription terminator.

autotroph: An organism that can synthesize its own complex molecules from very simple carbon and nitrogen sources, such as carbon dioxide and ammonia.

auxin: A plant growth hormone.

auxotrophic mutant (auxotroph): A mutant organism defective in the synthesis of a given biomolecule, which must therefore be supplied for the organism's growth.

Avogadro's number (*N*): The number of molecules in a gram molecular weight (a mole) of any compound (6.02×10^{23}).

b

back-mutation: A mutation that causes a mutant gene to regain its wild-type base sequence.

bacteriophage (phage): A virus capable of replicating in a bacterial cell.

basal metabolic rate: The rate of oxygen consumption by an animal's body at complete rest, long after a meal.

base pair: Two nucleotides in nucleic acid chains that are paired by hydrogen bonding of their bases; for example, A with T or U, and G with C.

β conformation: An extended, zigzag arrangement of a polypeptide chain; a common secondary structure in proteins.

β oxidation: Oxidative degradation of fatty acids into acetyl-CoA by successive oxidations at the β-carbon atom.

β turn: A type of secondary structure in polypeptides consisting of four amino acid residues arranged in a tight turn so that the polypeptide turns back on itself.

bilayer: A double layer of oriented amphipathic lipid molecules, forming the basic structure of biological membranes. The hydrocarbon tails face inward to form a continuous nonpolar phase.

bile salts: Amphipathic steroid derivatives with detergent properties, participating in digestion and absorption of lipids.

binding energy: The energy derived from noncovalent interactions between enzyme and substrate or receptor and ligand.

binding site: The crevice or pocket on a protein in which a ligand binds.

biocytin: The conjugate amino acid residue arising from covalent attachment of biotin, through an amide linkage, to a Lys residue.

biomolecule: An organic compound normally present as an essential component of living organisms.

biopterin: An enzymatic cofactor derived from pterin and involved in certain oxidation-reduction reactions.

biosphere: All the living matter on or in the earth, the seas, and the atmosphere.

biotin: A vitamin; an enzymatic cofactor involved in carboxylation reactions.

bond energy: The energy required to break a bond.

branch migration: Movement of the branch point in branched DNA formed from two DNA molecules with identical sequences. *See also* Holliday intermediate.

buffer: A system capable of resisting changes in pH, consisting of a conjugate acid–base pair in which the ratio of proton acceptor to proton donor is near unity.

c

calorie: The amount of heat required to raise the temperature of 1.0 g of water from 14.5 to 15.5 °C. One calorie (cal) equals 4.18 joules (J).

Calvin cycle: The cyclic pathway used by plants to fix carbon dioxide and produce triose phosphates.

cAMP: *See* cyclic AMP.

cAMP receptor protein (CRP): A specific regulatory protein that controls initiation of transcription of the genes producing the enzymes required for a bacterial cell to use some other nutrient when glucose is lacking. Also called catabolite gene activator protein (CAP).

CAP: *See* catabolite gene activator protein.

capsid: The protein coat of a virion or virus particle.

carbanion: A negatively charged carbon atom.

carbocation: A positively charged carbon atom; also called a carbonium ion.

carbon-assimilation reactions: Reaction sequences in which atmospheric CO_2 is converted into organic compounds.

carbon-fixation reaction: The reaction catalyzed by rubisco during photosynthesis, or by other carboxylases, in which atmospheric CO_2 is initially incorporated into an organic compound.

carboxyl-terminal residue: The only amino acid residue in a polypeptide chain with a free α-carboxyl group; defines the carboxyl terminus of the polypeptide.

carotenoids: Lipid-soluble photosynthetic pigments made up of isoprene units.

catabolism: The phase of intermediary metabolism concerned with the energy-yielding degradation of nutrient molecules.

catabolite gene activator protein (CAP): *See* cAMP receptor protein.

catalytic site: *See* active site.

catecholamines: Hormones, such as epinephrine, that are amino derivatives of catechol.

catenane: Circular polymeric molecules with a noncovalent topological link resembling the links of a chain.

cation-exchange resin: An insoluble polymer with fixed negative charges; used in the chromatographic separation of cationic substances.

cDNA: *See* complementary DNA.

central dogma: The organizing principle of molecular biology: genetic information flows from DNA to RNA to protein.

centromere: A specialized site within a chromosome, serving as the attachment point for the mitotic or meiotic spindle.

cerebroside: Sphingolipid containing one sugar residue as a head group.

channeling: The direct transfer of a reaction product (common intermediate) from the active site of one enzyme to the active site of a different enzyme catalyzing the next step in a sequential pathway.

chemiosmotic coupling: Coupling of ATP synthesis to electron transfer via an electrochemical H^+ gradient across a membrane.

chemotaxis: A cell's sensing of and movement toward, or away from, a specific chemical agent.

chemotroph: An organism that obtains energy by metabolizing organic compounds derived from other organisms.

chiral center: An atom with substituents arranged so that the molecule is not superimposable on its mirror image.

chiral compound: A compound that contains an asymmetric center (chiral atom or chiral center) and thus can occur in two nonsuperimposable mirror-image forms (enantiomers).

chlorophylls: A family of green pigments functioning as receptors of light energy in photosynthesis; magnesium-porphyrin complexes.

chloroplasts: Chlorophyll-containing photosynthetic organelles in some eukaryotic cells.

chromatin: A filamentous complex of DNA, histones, and other proteins, constituting the eukaryotic chromosome.

chromatography: A process in which complex mixtures of molecules are separated by many repeated partitionings between a flowing (mobile) phase and a stationary phase.

chromosome: A single large DNA molecule and its associated proteins, containing many genes; stores and transmits genetic information.

chylomicron: A plasma lipoprotein consisting of a large droplet of triacylglycerols stabilized by a coat of protein and phospholipid; carries lipids from the intestine to the tissues.

cis and trans isomers: *See* geometric isomers.

cistron: A unit of DNA or RNA corresponding to one gene.

citric acid cycle: A cyclic system of enzymatic reactions for the oxidation of acetyl residues to carbon dioxide, in which formation of citrate is the first step; also known as the Krebs cycle or tricarboxylic acid cycle.

clones: The descendants of a single cell.

cloning: The production of large numbers of identical DNA molecules, cells, or organisms, from a single ancestral DNA molecule, cell, or organism.

closed system: A system that exchanges neither matter nor energy with the surroundings. *See also* system.

cobalamin: *See* coenzyme B_{12}.

codon: A sequence of three adjacent nucleotides in a nucleic acid that codes for a specific amino acid.

coenzyme: An organic cofactor required for the action of certain enzymes; often contains a vitamin as a component.

coenzyme A: A pantothenic acid–containing coenzyme serving as an acyl group carrier in certain enzymatic reactions.

coenzyme B_{12}: An enzymatic cofactor derived from the vitamin cobalamin, involved in certain types of carbon skeletal rearrangements.

cofactor: An inorganic ion or a coenzyme required for enzyme activity.

cognate: Describing two biomolecules that normally interact; for example, an enzyme and its normal substrate, or a receptor and its normal ligand.

cohesive ends: *See* sticky ends.

cointegrate: An intermediate in the migration of certain DNA transposons in which the donor DNA and target DNA are covalently attached.

colligative properties: Properties of solutions that depend on the number of solute particles per unit volume; for example, freezing-point depression.

common intermediate: A chemical compound common to two chemical reactions, as a product of one and a reactant in the other.

competitive inhibition: A type of enzyme inhibition reversed by increasing the substrate concentration; a competitive inhibitor generally competes with the normal substrate or ligand for a protein's binding site.

complementary: Having a molecular surface with chemical groups arranged to interact specifically with chemical groups on another molecule.

complementary DNA (cDNA): A DNA used in DNA cloning, usually made by reverse transcriptase; complementary to a given mRNA.

configuration: The spatial arrangement of an organic molecule that is conferred by the presence of either (1) double bonds, about which there is no freedom of rotation, or (2) chiral centers, around which substituent groups are arranged in a specific sequence. Configurational isomers cannot be interconverted without breaking one or more covalent bonds.

conformation: The spatial arrangement of substituent groups that are free to assume different positions in space, without breaking any bonds, because of the freedom of bond rotation.

conformation, β: *See* β conformation.

conjugate acid-base pair: A proton donor and its corresponding deprotonated species; for example, acetic acid (donor) and acetate (acceptor).

conjugate redox pair: An electron donor and its corresponding electron acceptor form; for example, Cu^+ (donor) and Cu^{2+} (acceptor), or NADH (donor) and NAD^+ (acceptor).

conjugated protein: A protein containing one or more prosthetic groups.

consensus sequence: A DNA or amino acid sequence consisting of the residues that occur most commonly at each position within a set of similar sequences.

conservative substitution: Replacement of an amino acid residue in a polypeptide by another residue with similar properties; for example, substitution of Glu by Asp.

constitutive enzymes: Enzymes required at all times by a cell and present at some constant level; for example, many enzymes of the central metabolic pathways. Sometimes called housekeeping enzymes.

contour length: The length of a helical polymeric molecule as measured along the molecule's helical axis.

corticosteroids: Steroid hormones formed by the adrenal cortex.

cotransport: The simultaneous transport, by a single transporter, of two solutes across a membrane. *See* antiport, symport.

coupled reactions: Two chemical reactions that have a common intermediate and thus a means of energy transfer from one to the other.

covalent bond: A chemical bond that involves sharing of electron pairs.

cristae: Infoldings of the inner mitochon-drial membrane.

CRP: *See* cAMP receptor protein.

cyclic AMP (cAMP): A second messenger within cells; its formation by adenylyl cyclase is stimulated by certain hormones or other molecular signals.

cyclic electron flow: In chloroplasts, the light-induced flow of electrons originating from and returning to photosystem I.

cyclic photophosphorylation: ATP synthesis driven by cyclic electron flow through photosystem I.

cyclin: One of a family of proteins that activate cyclin-dependent protein kinases and thereby regulate the cell cycle.

cytochromes: Heme proteins serving as electron carriers in respiration, photosynthesis, and other oxidation-reduction reactions.

cytokine: One of a family of small secreted proteins (such as interleukins or interferons) that activate cell division or differentiation by binding to plasma membrane receptors in sensitive cells.

cytokinesis: The final separation of daughter cells following mitosis.

cytoplasm: The portion of a cell's contents outside the nucleus but within the plasma membrane; includes organelles such as mitochondria.

cytoskeleton: The filamentous network providing structure and organization to the cytoplasm; includes actin filaments, microtubules, and intermediate filaments.

cytosol: The continuous aqueous phase of the cytoplasm, with its dissolved solutes; excludes the organelles such as mitochondria.

d

dalton: The weight of a single hydrogen atom (1.66×10^{-24} g).

dark reactions: *See* carbon-assimilation reactions.

de novo pathway: Pathway for synthesis of a biomolecule, such as a nucleotide, from simple precursors; as distinct from a salvage pathway.

deamination: The enzymatic removal of amino groups from biomolecules such as amino acids or nucleotides.

degenerate code: A code in which a single element in one language is specified by more than one element in a second language.

dehydrogenases: Enzymes catalyzing the removal of pairs of hydrogen atoms from their substrates.

deletion mutation: A mutation resulting from the deletion of one or more nucleotides from a gene or chromosome.

denaturation: Partial or complete unfolding of the specific native conformation of a polypeptide chain, protein, or nucleic acid.

denatured protein: A protein that has lost its native conformation by exposure to a destabilizing agent such as heat or detergent.

deoxyribonucleic acid: *See* DNA.

deoxyribonucleotides: Nucleotides containing 2-deoxy-D-ribose as the pentose component.

desaturases: Enzymes that catalyze the introduction of double bonds into the hydrocarbon portion of fatty acids.

desolvation: In aqueous solution, the release of bound water surrounding a solute.

dextrorotatory isomer: A stereoisomer that rotates the plane of plane-polarized light clockwise.

diabetes mellitus: A metabolic disease resulting from insulin deficiency; characterized by a failure in glucose transport from the blood into cells at normal glucose concentrations.

dialysis: Removal of small molecules from a solution of a macromolecule, by allowing them to diffuse through a semipermeable membrane into water.

differential centrifugation: Separation of cell organelles or other particles of different size by their different rates of sedimentation in a centrifugal field.

differentiation: Specialization of cell structure and function during embryonic growth and development.

diffusion: The net movement of molecules in the direction of lower concentration.

digestion: Enzymatic hydrolysis of major nutrients in the gastrointestinal system to yield their simpler components.

diploid: Having two sets of genetic information; describing a cell with two chromosomes of each type.

dipole: A molecule having both positive and negative charges.

diprotic acid: An acid having two dissociable protons.

disaccharide: A carbohydrate consisting of two covalently joined monosaccharide units.

dissociation constant: (1) An equilibrium constant (K_d) for the dissociation of a complex of two or more biomolecules into its components; for example, dissociation of a substrate from an enzyme. (2) The dissociation constant (K_a) of an acid, describing its dissociation into its conjugate base and a proton.

disulfide bridge: A covalent cross link between two polypeptide chains formed by a cystine residue (two Cys residues).

DNA (deoxyribonucleic acid): A polynucleotide having a specific sequence of deoxyribonucleotide units covalently joined through 3′, 5′-phosphodiester bonds; serves as the carrier of genetic information.

DNA chimera: A DNA containing genetic information derived from two different species.

DNA cloning: *See* cloning.

DNA library: A collection of cloned DNA fragments.

DNA ligase: An enzyme that creates a phosphodiester bond between the 3′ end of one DNA segment and the 5′ end of another.

DNA looping: The interaction of proteins bound at distant sites on a DNA molecule so that the intervening DNA forms a loop.

DNA microarray: A collection of DNA sequences immobilized on a solid surface, with individual sequences laid out in patterned arrays that can be probed by hybridization.

DNA polymerase: An enzyme that catalyzes template-dependent synthesis of DNA from its deoxyribonucleoside 5′-triphosphate precursors.

DNA replicase system: The entire complex of enzymes and specialized proteins required in biological DNA replication.

DNA supercoiling: The coiling of DNA upon itself, generally as a result of bending, underwinding, or overwinding of the DNA helix.

DNA transposition: *See* transposition.

domain: A distinct structural unit of a polypeptide; domains may have separate functions and may fold as independent, compact units.

double helix: The natural coiled conformation of two complementary, antiparallel DNA chains.

double-reciprocal plot: A plot of $1/V_0$ versus 1/[S], which allows a more accurate determination of V_{max} and K_m than a plot of V_0 versus [S]; also called the Lineweaver-Burk plot.

e

***E*′°:** *See* standard reduction potential.

***E. coli (Escherichia coli)*:** A common bacterium found in the small intestine of vertebrates; the most well-studied organism.

electrochemical gradient: The sum of the gradients of concentration and of electric charge of an ion across a membrane; the driving force for oxidative phosphorylation and photophosphorylation.

electrochemical potential: The energy required to maintain a separation of charge and of concentration across a membrane.

electrogenic: Contributing to an electrical potential across a membrane.

electron acceptor: A substance that receives electrons in an oxidation-reduction reaction.

electron carrier: A protein, such as a flavoprotein or a cytochrome, that can reversibly gain and lose electrons; functions in the transfer of electrons from organic nutrients to oxygen or some other terminal acceptor.

electron donor: A substance that donates electrons in an oxidation-reduction reaction.

electron transfer: Movement of electrons from substrates to oxygen via the carriers of the respiratory (electron transfer) chain.

electrophile: An electron-deficient group with a strong tendency to accept electrons from an electron-rich group (nucleophile).

electrophoresis: Movement of charged solutes in response to an electrical field; often used to separate mixtures of ions, proteins, or nucleic acids.

electroporation: Introduction of macromolecules into cells after rendering the cells transiently permeable by the application of a high-voltage pulse.

elongation factors: Specific proteins required in the elongation of polypeptide chains by ribosomes.

eluate: The effluent from a chromato-graphic column.

enantiomers: Stereoisomers that are nonsuperimposable mirror images of each other.

end-product inhibition: *See* feedback inhibition.

endergonic reaction: A chemical reaction that consumes energy (that is, for which ΔG is positive).

endocrine glands: Groups of cells specialized to synthesize hormones and secrete them into the blood to regulate other types of cells.

endocytosis: The uptake of extracellular material by its inclusion within a vesicle (endosome) formed by an invagination of the plasma membrane.

endonuclease: An enzyme that hydrolyzes the interior phosphodiester bonds of a nucleic acid; that is, it acts at points other than the terminal bonds.

endoplasmic reticulum: An extensive system of double membranes in the cytoplasm of eukaryotic cells; it encloses secretory channels and is often studded with ribosomes (rough endoplasmic reticulum).

endothermic reaction: A chemical reaction that takes up heat (that is, for which ΔH is positive).

energy charge: The fractional degree to which the ATP/ADP/AMP system is filled with high-energy phosphate groups.

energy coupling: The transfer of energy from one process to another.

enhancers: DNA sequences that facilitate the expression of a given gene; may be located a few hundred, or even thousand, base pairs away from the gene.

enthalpy (*H*): The heat content of a system.

enthalpy change (ΔH): For a reaction, is approximately equal to the difference between the energy used to break bonds and the energy gained by the formation of new ones.

entropy (*S*): The extent of randomness or disorder in a system.

enzyme: A biomolecule, either protein or RNA, that catalyzes a specific chemical reaction. It does not affect the equilibrium of the catalyzed reaction; it enhances the rate of a reaction by providing a reaction path with a lower activation energy.

enzyme cascade: A series of reactions, often involved in regulatory events, in which one enzyme activates another (often by phosphorylation), which activates a third, and so on. The effect of a catalyst activating a catalyst is a large amplification of the signal that initiated the cascade.

epimerases: Enzymes that catalyze the reversible interconversion of two epimers.

epimers: Two stereoisomers differing in configuration at one asymmetric center, in a compound having two or more asymmetric centers.

epithelial cell: Any cell that forms part of the outer covering of an organism or organ.

epitope: An antigenic determinant; the particular chemical group or groups within a macromolecule (antigen) to which a given antibody binds.

equilibrium: The state of a system in which no further net change is occurring; the free energy is at a minimum.

equilibrium constant (K_{eq}): A constant, characteristic for each chemical reaction; relates the specific concentrations of all reactants and products at equilibrium at a given temperature and pressure.

erythrocyte: A cell containing large amounts of hemoglobin and specialized for oxygen transport; a red blood cell.

***Escherichia coli*:** *See E. coli.*

essential amino acids: Amino acids that cannot be synthesized by humans (and other vertebrates) and must be obtained from the diet.

essential fatty acids: The group of polyunsaturated fatty acids produced by plants, but not by humans; required in the human diet.

ethanol fermentation: *See* alcohol fermentation.

eukaryote: A unicellular or multicellular organism with cells having a membrane-bounded nucleus, multiple chromosomes, and internal organelles.

excited state: An energy-rich state of an atom or molecule; produced by the absorption of light energy.

exergonic reaction: A chemical reaction that proceeds with the release of free energy (that is, for which ΔG is negative).

exocytosis: The fusion of an intracellular vesicle with the plasma membrane, releasing the vesicle contents to the extracellular space.

exon: The segment of a eukaryotic gene that encodes a portion of the final product of the gene; a portion that remains after posttranscriptional processing and is transcribed into a protein or incorporated into the structure of an RNA. *See* intron.

exonuclease: An enzyme that hydrolyzes only those phosphodiester bonds that are in the terminal positions of a nucleic acid.

exothermic reaction: A chemical reaction that releases heat (that is, for which ΔH is negative).

expression vector: *See* vector.

f

facilitated diffusion: Diffusion of a polar substance across a biological membrane through a protein transporter; also called passive diffusion or passive transport.

facultative cells: Cells that can live in the presence or absence of oxygen.

FAD (flavin adenine dinucleotide): The coenzyme of some oxidation-reduction enzymes; it contains riboflavin.

fatty acid: A long-chain aliphatic carboxylic acid found in natural fats and oils; also a component of membrane phospholipids and glycolipids.

feedback inhibition: Inhibition of an allosteric enzyme at the beginning of a metabolic sequence by the end product of the sequence; also known as end-product inhibition.

fermentation: Energy-yielding anaerobic breakdown of a nutrient molecule, such as glucose, without net oxidation; yields lactate, ethanol, or some other simple product.

fibroblast: A cell of the connective tissue that secretes connective tissue proteins such as collagen.

fibrous proteins: Insoluble proteins that serve in a protective or structural role; contain polypeptide chains that generally share a common secondary structure.

fingerprinting: *See* peptide mapping.

first law of thermodynamics: The law stating that in all processes, the total energy of the universe remains constant.

Fischer projection formulas: *See* projection formulas.

5′ end: The end of a nucleic acid that lacks a nucleotide bound at the 5′ position of the terminal residue.

flagellum: A cell appendage used in propulsion. Bacterial flagella have a much simpler structure than eukaryotic flagella, which are similar to cilia.

flavin-linked dehydrogenases: Dehydrogenases requiring one of the riboflavin coenzymes, FMN or FAD.

flavin nucleotides: Nucleotide coenzymes (FMN and FAD) containing riboflavin.

flavoprotein: An enzyme containing a flavin nucleotide as a tightly bound prosthetic group.

fluid mosaic model: A model describing biological membranes as a fluid lipid bilayer with embedded proteins; the bilayer exhibits both structural and functional asymmetry.

fluorescence: Emission of light by excited molecules as they revert to the ground state.

FMN (flavin mononucleotide): Riboflavin phosphate, a coenzyme of certain oxidation-reduction enzymes.

footprinting: A technique for identifying the nucleic acid sequence bound by a DNA-or RNA-binding protein.

fractionation: The process of separating the proteins or other components of a complex molecular mixture into fractions based on differences in their physical properties, such as size, net charge, and solubility.

frame shift: A mutation caused by insertion or deletion of one or more paired nucleotides, changing the reading frame of codons during protein synthesis; the polypeptide product has a garbled amino acid sequence beginning at the mutated codon.

free energy (G): The component of the total energy of a system that can do work at constant temperature and pressure.

free energy of activation ($\Delta G^{\ddagger}$): *See* activation energy.

free-energy change (ΔG): The amount of free energy released (negative ΔG) or absorbed (positive ΔG) in a reaction at constant temperature and pressure.

free radical: *See* radical.

functional group: The specific atom or group of atoms that confers a particular chemical property on a biomolecule.

furanose: A simple sugar containing the five-membered furan ring.

fusion protein: (1) A family of proteins that facilitate membrane fusion. (2) The protein product of a gene created by the fusion of two distinct genes or portions of genes.

futile cycle: A set of enzyme-catalyzed cyclic reactions that results in release of thermal energy by the hydrolysis of ATP.

g

G proteins: A family of heterotrimeric GTP-binding proteins that act in intracellular signaling pathways. Commonly, ligand binding to a serpentine receptor induces the exchange of GTP for bound GDP, enabling the G protein to activate a downstream enzyme in a signaling pathway. G proteins have intrinsic GTPase activity, and therefore self-inactivate.

$\Delta G'^{\circ}$: *See* standard free-energy change.

gametes: Reproductive cells with a haploid gene content; sperm or egg cells.

gangliosides: Sphingolipids, containing complex oligosaccharides as head groups; especially common in nervous tissue.

gel filtration: *See* size-exclusion chromatography.

gene: A chromosomal segment that codes for a single functional polypeptide chain or RNA molecule.

gene expression: Transcription, and in the case of proteins, translation, to yield the product of a gene; a gene is expressed when its biological product is present and active.

gene splicing: The enzymatic attachment of one gene, or part of a gene, to another.

general acid-base catalysis: Catalysis involving proton transfer(s) to or from a molecule other than water.

genetic code: The set of triplet code words in DNA (or mRNA) coding for the amino acids of proteins.

genetic information: The hereditary information contained in a sequence of nucleotide bases in chromosomal DNA or RNA.

genetic map: A diagram showing the relative sequence and position of specific genes along a chromosome.

genome: All the genetic information encoded in a cell or virus.

genomic library: A DNA library containing DNA segments representing all (or most) of the sequences in an organism's genome.

genotype: The genetic constitution of an organism, as distinct from its physical characteristics, or phenotype.

geometric isomers: Isomers related by rotation about a double bond; also called cis and trans isomers.

germ-line cell: A type of animal cell that is formed early in embryogenesis and may multiply by mitosis or may produce, by meiosis, cells that develop into gametes (egg or sperm cells).

globular proteins: Soluble proteins with a globular (somewhat rounded) shape.

glucogenic amino acids: Amino acids with carbon chains that can be metabolically converted into glucose or glycogen via gluconeogenesis.

gluconeogenesis: The biosynthesis of a carbohydrate from simpler, noncarbohydrate precursors such as oxaloacetate or pyruvate.

glycan: Another term for polysaccharide; a polymer of monosaccharide units joined by glycosidic bonds.

glycerophospholipid: An amphipathic lipid with a glycerol backbone; fatty acids are ester-linked to C-1 and C-2 of glycerol, and a polar alcohol is attached through a phosphodiester linkage to C-3.

glycoconjugate: A compound containing a carbohydrate component bound covalently to a protein or lipid, forming a glycoprotein or glycolipids.

glycolipid: A lipid containing a carbohydrate group.

glycolysis: The catabolic pathway by which a molecule of glucose is broken down into two molecules of pyruvate.

glycoprotein: A protein containing a carbohydrate group.

glycosaminoglycan: A heteropolysaccharide of two alternating units: one is either *N*-acetylglucosamine or *N*-acetylgalactosamine; the other is a uronic acid (usually glucuronic acid). Formerly called mucopolysaccharide.

glycosidic bonds: Bonds between a sugar and another molecule (typically an alcohol, purine, pyrimidine, or sugar) through an intervening oxygen.

glyoxylate cycle: A variant of the citric acid cycle, for the net conversion of acetate into succinate and, eventually, new carbohydrate; present in bacteria and some plant cells.

glyoxysome: A specialized peroxisome containing the enzymes of the glyoxylate cycle; found in cells of germinating seeds.

Golgi complex: A complex membranous organelle of eukaryotic cells; functions in the posttranslational modification of proteins and their secretion from the cell or incorporation into the plasma membrane or organellar membranes.

gram molecular weight: The weight in grams of a compound that is numerically equal to its molecular weight; the weight of 1 mole.

grana: Stacks of thylakoids, flattened membranous sacs or disks, in chloroplasts.

ground state: The normal, stable form of an atom or molecule; as distinct from the excited state.

group transfer potential: A measure of the ability of a compound to donate an activated group (such as a phosphate or acyl group); generally expressed as the standard free energy of hydrolysis.

h

half-life: The time required for the disappearance or decay of one-half of a given component in a system.

haploid: Having a single set of genetic information; describing a cell with one chromosome of each type.

hapten: A small molecule which, when linked to a larger molecule, elicits an immune response.

Haworth perspective formulas: A method for representing cyclic chemical structures so as to define the configuration of each substituent group; the method commonly used for representing sugars.

helicase: An enzyme that catalyzes the separation of strands in a DNA molecule before replication.

helix, α: *See* α helix.

heme: The iron-porphyrin prosthetic group of heme proteins.

heme protein: A protein containing a heme as a prosthetic group.

hemoglobin: A heme protein in erythrocytes; functions in oxygen transport.

Henderson-Hasselbalch equation: An equation relating the pH, the pK_a, and the ratio of the concentrations of the proton-acceptor (A^-) and proton-donor (HA) species in a solution.

hepatocyte: The major cell type of liver tissue.

heteroduplex DNA: Duplex DNA containing complementary strands derived from two different DNA molecules with similar sequences, often as a product of genetic recombination.

heteropolysaccharide: A polysaccharide containing more than one type of sugar.

heterotroph: An organism that requires complex nutrient molecules, such as glucose, as a source of energy and carbon.

heterotropic: Describes an allosteric modulator that is distinct from the normal ligand.

heterotropic enzyme: An allosteric enzyme requiring a modulator other than its substrate.

hexose: A simple sugar with a backbone containing six carbon atoms.

high-energy compound: A compound that on hydrolysis undergoes a large decrease in free energy under standard conditions.

high-performance liquid chromatography (HPLC): Chromatographic procedure, often conducted at relatively high pressures, using automated equipment that permits refined and highly reproducible profiles.

Hill reaction: The evolution of oxygen and the photoreduction of an artificial electron acceptor by a chloroplast preparation in the absence of carbon dioxide.

histones: The family of five basic proteins that associate tightly with DNA in the chromosomes of all eukaryotic cells.

Holliday intermediate: An intermediate in genetic recombination in which two double-stranded DNA molecules are joined by virtue of a reciprocal crossover involving one strand of each molecule.

holoenzyme: A catalytically active enzyme including all necessary subunits, prosthetic groups, and cofactors.

homeobox: A conserved DNA sequence of 180 base pairs encoding a protein domain found in many proteins that play a regulatory role in development.

homeodomain: The protein domain encoded by the homeobox.

homeostasis: The maintenance of a dynamic steady state by regulatory mechanisms that compensate for changes in external circumstances.

homeotic genes: Genes that regulate the development of the pattern of segments in the *Drosophila* body plan; similar genes are found in most vertebrates.

homologous genetic recombination: Recombination between two DNA molecules of similar sequence, occurring in all cells; occurs during meiosis and mitosis in eukaryotes.

homologous proteins: Proteins having sequences and functions similar in different species; for example, the hemoglobins.

homopolysaccharide: A polysaccharide made up of only one type of monosaccharide unit.

homotropic: Describes an allosteric modulator that is identical to the normal ligand.

homotropic enzyme: An allosteric enzyme that uses its substrate as a modulator.

hormone: A chemical substance synthesized in small amounts by an endocrine tissue and carried in the blood to another tissue, where it acts as a messenger to regulate the function of the target tissue or organ.

hormone receptor: A protein in, or on the surface of, target cells that binds a specific hormone and initiates the cellular response.

hormone response element (HRE): A short (12 to 20 bp) DNA sequence to which receptors for steroid, retinoid, thyroid, and vitamin D hormones bind, altering the expression of the contiguous genes. For each hormone, there is a consensus sequence preferred by the cognate receptor.

hyaluronic acid: A high molecular weight, acidic polysaccharide typically composed of the alternating disaccharide GlcUA(β1$\rightarrow$3)GlcNAc. Hyaluronic acid is a major component of the extracellular matrix, and forms larger complexes (proteoglycans) with proteins and other acidic polysaccharides.

hydrogen bond: A weak electrostatic attraction between one electronegative atom (such as oxygen or nitrogen) and a hydrogen atom covalently linked to a second electronegative atom.

hydrolases: Enzymes (proteases, lipases, phosphatases, nucleases, for example) that catalyze hydrolysis reactions.

hydrolysis: Cleavage of a bond, such as an anhydride or peptide bond, by the addition of the elements of water, yielding two or more products.

hydronium ion: The hydrated hydrogen ion (H_3O^+).

hydropathy index: A scale that expresses the relative hydrophobic and hydrophilic tendencies of a chemical group.

hydrophilic: Polar or charged; describing molecules or groups that associate with (dissolve easily in) water.

hydrophobic: Nonpolar; describing molecules or groups that are insoluble in water.

hydrophobic interactions: The association of nonpolar groups, or compounds, with each other in aqueous systems, driven by the tendency of the surrounding water molecules to seek their most stable (disordered) state.

hyperchromic effect: The large increase in light absorption at 260 nm occurring as a double-helical DNA is melted (unwound).

hypoxia: The metabolic condition in which the supply of oxygen is severely limited.

i

immune response: The capacity of a vertebrate to generate antibodies to an antigen, a macromolecule foreign to the organism.

immunoglobulin: An antibody protein generated against, and capable of binding specifically to, an antigen.

in vitro: "In glass"; that is, in the test tube.

in vivo: "In life"; that is, in the living cell or organism.

induced fit: A change in the conformation of an enzyme in response to substrate binding that renders the enzyme cataly-tically active; also used to denote changes in the conformation of any macromolecule in response to ligand binding such that the binding site of the macromolecule better conforms to the shape of the ligand.

inducer: A signal molecule that, when bound to a regulatory protein, produces an increase in the expression of a given gene.

induction: An increase in the expression of a gene in response to a change in the activity of a regulatory protein.

informational macromolecules: Biomolecules containing information in the form of specific sequences of different monomers; for example, many proteins, lipids, polysaccharides, and nucleic acids.

initiation codon: AUG (sometimes GUG in prokaryotes); codes for the first amino acid in a polypeptide sequence: *N*-formylmethionine in prokaryotes, and methionine in eukaryotes.

initiation complex: A complex of a ribosome with an mRNA and the initiating Met-$tRNA^{Met}$ or fMet-$tRNA^{fMet}$, ready for the elongation steps.

inorganic pyrophosphatase: An enzyme that hydrolyzes a molecule of inorganic pyrophosphate to yield two molecules of (ortho) phosphate; also known as pyrophosphatase.

insertion mutation: A mutation caused by insertion of one or more extra bases, or a mutagen, between successive bases in DNA.

insertion sequence: Specific base sequences at either end of a transposable segment of DNA.

integral proteins: Proteins firmly bound to a membrane by hydrophobic interactions; as distinct from peripheral proteins.

integrin: One of a large family of heterodimeric transmembrane proteins that mediate adhesion of cells to other cells or to the extracellular matrix.

intercalating mutagen: A mutagen that inserts itself between successive bases in a nucleic acid, causing a frame-shift mutation.

intercalation: Insertion between stacked aromatic or planar rings; for example, the insertion of a planar molecule between two successive bases in a nucleic acid.

interferons: A class of glycoproteins with antiviral activities.

intermediary metabolism: In cells, the enzyme-catalyzed reactions that extract chemical energy from nutrient molecules and utilize it to synthesize and assemble cell components.

intron (intervening sequence): A sequence of nucleotides in a gene that is transcribed but excised before the gene is translated.

ion channel: An integral protein that provides for the regulated transport of a specific ion, or ions, across a membrane.

ion-exchange resin: A polymeric resin that contains fixed charged groups; used in chromatographic columns to separate ionic compounds.

ion product of water (K_W): The product of the concentrations of H^+ and OH^- in pure water: $K_W = [H^+][OH^-] = 1 \times 10^{-14}$ at 25 °C.

ionizing radiation: A type of radiation, such as x rays, that causes loss of electrons from some organic molecules, thus making them more reactive.

ionophore: A compound that binds one or more metal ions and is capable of diffusing across a membrane, carrying the bound ion.

iron-sulfur center: A prosthetic group of certain redox proteins involved in electron transfers; Fe^{2+} or Fe^{3+} is bound to inorganic sulfur and to Cys groups in the protein.

isoelectric focusing: An electrophoretic method for separating macromolecules on the basis of their isoelectric pH.

isoelectric pH (isoelectric point): The pH at which a solute has no net electric charge and thus does not move in an electric field.

isoenzymes: *See* isozymes.

isomerases: Enzymes that catalyze the transformation of compounds into their positional isomers.

isomers: Any two molecules with the same molecular formula but a different arrangement of molecular groups.

isoprene: The hydrocarbon 2-methyl-1, 3-butadiene, a recurring structural unit of the terpenoid biomolecules.

isoprenoid: Any of a large number of natural products synthesized by enzymatic polymerization of two or more isoprene units; also called terpenoids.

isothermal: Occurring at constant temperature.

isotopes: Stable or radioactive forms of an element that differ in atomic weight but are otherwise chemically identical to the naturally abundant form of the element; used as tracers.

isozymes: Multiple forms of an enzyme that catalyze the same reaction but differ from each other in their amino acid sequence, substrate affinity, V_{max}, and / or regulatory properties; also called isoenzymes.

k

keratins: Insoluble protective or structural proteins consisting of parallel polypeptide chains in α-helical or β conformations.

ketogenic amino acids: Amino acids with carbon skeletons that can serve as precursors of the ketone bodies.

ketone bodies: Acetoacetate, D-β-hydroxybutyrate, and acetone; water-soluble fuels normally exported by the liver but overproduced during fasting or in untreated diabetes mellitus.

ketose: A simple monosaccharide in which the carbonyl group is a ketone.

ketosis: A condition in which the concentration of ketone bodies in the blood, tissues, and urine is abnormally high.

kinases: Enzymes that catalyze the phosphorylation of certain molecules by ATP.

kinetics: The study of reaction rates.

Krebs cycle: *See* citric acid cycle.

l

lagging strand: The DNA strand that, during replication, must be synthesized in the direction opposite to that in which the replication fork moves.

law of mass action: The law stating that the rate of any given chemical reaction is proportional to the product of the activities (or concentrations) of the reactants.

leader: A short sequence near the amino terminus of a protein or the 5′ end of an RNA that has a specialized targeting or regulatory function.

leading strand: The DNA strand that, during replication, is synthesized in the same direction in which the replication fork moves.

leaky mutant: A mutant gene that gives rise to a product with a detectable level of biological activity.

leaving group: The departing or displaced molecular group in a unimolecular elimination or a bimolecular substitution reaction.

lectin: A protein that binds a carbohydrate, commonly an oligosaccharide, with very high affinity and specificity, mediating cell-cell interactions.

lethal mutation: A mutation that inactivates a biological function essential to the life of the cell or organism.

leucine zipper: A protein structural motif involved in protein-protein interactions in many eukaryotic regulatory proteins; consists of two interacting α helices in which Leu residues in every seventh position are a prominent feature of the interacting surfaces.

leukotrienes: A family of molecules derived from arachidonate; muscle contractants that constrict air passages in the lungs and are involved in asthma.

levorotatory isomer: A stereoisomer that rotates the plane of plane-polarized light counterclockwise.

ligand: A small molecule that binds specifically to a larger one; for example, a hormone is the ligand for its specific protein receptor.

ligases: Enzymes that catalyze condensation reactions in which two atoms are joined using the energy of ATP or another energy-rich compound.

light reactions: The reactions of photosynthesis that require light and cannot occur in the dark; also known as the light-dependent reactions.

Lineweaver-Burk equation: An algebraic transform of the Michaelis-Menten equation, allowing determination of V_{max} and K_m by extrapolation of [S] to infinity.

linking number: The number of times one closed circular DNA strand is wound about another; the number of topological links holding the circles together.

lipases: Enzymes that catalyze the hydrolysis of triacylglycerols.

lipid: A small water-insoluble biomolecule generally containing fatty acids, sterols, or isoprenoid compounds.

lipoate (lipoic acid): A vitamin for some microorganisms; an intermediate carrier of hydrogen atoms and acyl groups in α-keto acid dehydrogenases.

lipoprotein: A lipid-protein aggregate that serves to carry water-insoluble lipids in the blood. The protein component alone is an apolipoprotein.

liposome: A small, spherical vesicle composed of a phospholipid bilayer, which forms spontaneously when phospholipids are suspended in an aqueous buffer.

low-energy phosphate compound: A phosphorylated compound with a relatively small standard free energy of hydrolysis.

lyases: Enzymes that catalyze the removal of a group from a molecule to form a double bond, or the addition of a group to a double bond.

lymphocytes: A subclass of leukocytes involved in the immune response. B lymphocytes synthesize and secrete antibodies; T lymphocytes either play a regulatory role in immunity or kill foreign and virus-infected cells.

lysis: Destruction of a cell's plasma membrane or of a bacterial cell wall, releasing the cellular contents and killing the cell.

lysosome: A membrane-bounded organelle in the cytoplasm of eukaryotic cells; it contains many hydrolytic enzymes and serves as a degrading and recycling center for unneeded components.

m

macromolecule: A molecule having a molecular weight in the range of a few thousand to many millions.

mass-action ratio: For the reaction $aA + bB \rightleftharpoons cC + dD$, the ratio: $\frac{[C]^c\,[D]^d}{[A]^a\,[B]^b}$.

matrix: The aqueous contents of a cell or organelle (the mitochondrion, for example) with dissolved solutes.

meiosis: A type of cell division in which diploid cells give rise to haploid cells destined to become gametes.

membrane potential (V_m): The difference in electrical potential across a biological membrane, commonly measured by the insertion of a microelectrode. Typical membrane potentials vary from −25 mV (by convention, the negative sign indicates that the inside is negative relative to the outside) to greater than −100 mV across some plant vacuole membranes.

membrane transport: Movement of a polar solute across a membrane via a specific membrane protein (a transporter).

messenger RNA (mRNA): A class of RNA molecules, each of which is complementary to one strand of DNA; carries the genetic message from the chromosome to the ribosomes.

metabolism: The entire set of enzyme-catalyzed transformations of organic molecules in living cells; the sum of anabolism and catabolism.

metabolite: A chemical intermediate in the enzyme-catalyzed reactions of metabolism.

metalloprotein: A protein having a metal ion as its prosthetic group.

metamerism: Division of the body into segments; in insects, for example.

micelle: An aggregate of amphipathic molecules in water, with the nonpolar portions in the interior and the polar portions at the exterior surface, exposed to water.

Michaelis constant (K_m): The substrate concentration at which an enzyme-catalyzed reaction proceeds at one-half its maximum velocity.

Michaelis-Menten equation: The equation describing the hyperbolic dependence of the initial reaction velocity, V_0, on substrate concentration, [S], in many enzyme-catalyzed

reactions: $V_0 = \frac{V_{max}[S]}{K_m + [S]}$.

Michaelis-Menten kinetics: A kinetic pattern in which the initial rate of an enzyme-catalyzed reaction exhibits a hyperbolic dependence on substrate concentration.

microbodies: Cytoplasmic, membrane-bounded vesicles containing peroxide-forming and peroxide-destroying enzymes; include lysosomes, peroxisomes, and glyoxysomes.

microfilaments: Thin filaments composed of actin, found in the cytoplasm of eukaryotic cells; serve in structure and movement.

microsomes: Membranous vesicles formed by fragmentation of the endoplasmic reticulum of eukaryotic cells; recovered by differential centrifugation.

microtubules: Thin tubules assembled from two types of globular tubulin subunits; present in cilia, flagella, centrosomes, and other contractile or motile structures.

mismatch: a base pair in a nucleic acid that cannot form normal Watson-Crick pairs.

mismatch repair: an enzymatic system for repairing base mismatches in DNA.

mitochondrion: Membrane-bounded organelle in the cytoplasm of eukaryotes; contains the enzyme systems required for the citric acid cycle, fatty acid oxidation, electron transfer, and oxidative phosphorylation.

mitosis: The multistep process in eukaryotic cells that results in the replication of chromosomes and cell division.

mixed-function oxidases: Enzymes (a monooxygenase, for example) that catalyze reactions in which two reductants, one of which is generally NADPH, the other the substrate, are oxidized. One oxygen atom is incorporated into the product, the other is reduced to H_2O. These enzymes often employ cytochrome P-450 to carry electrons from NADPH to O_2.

mixed inhibition: The reversible inhibition pattern resulting when an inhibitor molecule can bind to either the free enzyme or to the enzyme-substrate complex (not necessarily with the same affinity).

modulator: A metabolite that, when bound to the allosteric site of an enzyme, alters its kinetic characteristics.

molar solution: One mole of solute dissolved in water to give a total volume of 1,000 mL.

mole: One gram molecular weight of a compound. *See* Avogadro's number.

monoclonal antibodies: Antibodies produced by a cloned hybridoma cell, which therefore are identical and directed against the same epitope of the antigen.

monolayer: A single layer of oriented lipid molecules.

monoprotic acid: An acid having only one dissociable proton.

monosaccharide: A carbohydrate consisting of a single sugar unit.

mRNA: *See* messenger RNA.

mucopolysaccharide: An older name for a glycosaminoglycan.

multienzyme system: A group of related enzymes participating in a given metabolic pathway.

mutarotation: The change in specific rotation of a pyranose or furanose sugar or glycoside accompanying the equilibration of its α- and β-anomeric forms.

mutases: Enzymes that catalyze the transposition of functional groups.

mutation: An inheritable change in the nucleotide sequence of a chromosome.

myofibril: A unit of thick and thin filaments of muscle fibers.

myosin: A contractile protein; the major component of the thick filaments of muscle and other actin-myosin systems.

n

NAD, NADP (nicotinamide adenine dinucleotide, nicotinamide adenine dinucleotide phosphate): Nicotinamide-containing coenzymes functioning as carriers of hydrogen atoms and electrons in some oxidation-reduction reactions.

native conformation: The biologically active conformation of a macromolecule.

negative cooperativity: A phenomenon of some multisubunit enzymes or proteins in which binding of a ligand or substrate to one subunit impairs binding to another subunit.

negative feedback: Regulation of a biochemical pathway achieved when a reaction product inhibits an earlier step in the pathway.

neuron: A cell of nervous tissue specialized for transmission of a nerve impulse.

neurotransmitter: A low molecular weight compound (usually containing nitrogen) secreted from the terminal of a neuron and bound by a specific receptor in the next neuron; serves to transmit a nerve impulse.

nicotinamide adenine dinucleotide, nicotinamide adenine dinucleotide phosphate: *See* NAD, NADP.

ninhydrin reaction: A color reaction given by amino acids and peptides on heating with ninhydrin; widely used for their detection and estimation.

nitrogen cycle: The cycling of various forms of biologically available nitrogen through the plant, animal, and microbial worlds, and through the atmosphere and geosphere.

nitrogen fixation: Conversion of atmospheric nitrogen (N_2) into a reduced, biologically available form by nitrogen-fixing organisms.

nitrogenase complex: A system of enzymes capable of reducing atmospheric nitrogen to ammonia in the presence of ATP.

noncyclic electron flow: The light-induced flow of electrons from water to $NADP^+$ in oxygen-evolving photosynthesis; it involves both photosystems I and II.

nonessential amino acids: Amino acids that can be made by humans and other vertebrates from simpler precursors, and are thus not required in the diet.

nonheme iron proteins: Proteins, usually acting in oxidation-reduction reactions, containing iron but no porphyrin groups.

nonpolar: Hydrophobic; describing molecules or groups that are poorly soluble in water.

nonsense codon: A codon that does not specify an amino acid, but signals the termination of a polypeptide chain.

nonsense mutation: A mutation that results in the premature termination of a polypeptide chain.

nonsense suppressor: A mutation, usually in the gene for a tRNA, that causes an amino acid to be inserted into a polypeptide in response to a termination codon.

nuclear magnetic resonance (NMR) spectroscopy: A technique that utilizes certain quantum mechanical properties of atomic nuclei to study the structure and dynamics of the molecules of which they are a part.

nucleases: Enzymes that hydrolyze the internucleotide (phosphodiester) linkages of nucleic acids.

nucleic acids: Biologically occurring polynucleotides in which the nucleotide residues are linked in a specific sequence by phosphodiester bonds; DNA and RNA.

nucleoid: In bacteria, the nuclear zone that contains the chromosome but has no surrounding membrane.

nucleolus: A densely staining structure in the nucleus of eukaryotic cells; involved in rRNA synthesis and ribosome formation.

nucleophile: An electron-rich group with a strong tendency to donate electrons to an electron-deficient nucleus (electrophile); the entering reactant in a bimolecular substitution reaction.

nucleoplasm: The portion of a cell's contents enclosed by the nuclear membrane; also called the nuclear matrix.

nucleoside: A compound consisting of a purine or pyrimidine base covalently linked to a pentose.

nucleoside diphosphate kinase: An enzyme that catalyzes the transfer of the terminal phosphate of a nucleoside 5′-triphosphate to a nucleoside 5′-diphosphate.

nucleoside diphosphate sugar: A coenzymelike carrier of a sugar molecule, functioning in the enzymatic synthesis of polysaccharides and sugar derivatives.

nucleoside monophosphate kinase: An enzyme that catalyzes the transfer of the terminal phosphate of ATP to a nucleoside 5′-monophosphate.

nucleosome: Structural unit for packaging chromatin; consists of a DNA strand wound around a histone core.

nucleotide: A nucleoside phosphorylated at one of its pentose hydroxyl groups.

nucleus: In eukaryotes, a membrane-bounded organelle that contains chromosomes.

o

oligomer: A short polymer, usually of amino acids, sugars, or nucleotides; the definition of "short" is somewhat arbitrary, but usually less than 50 subunits.

oligomeric protein: A multisubunit protein having two or more identical polypeptide chains.

oligonucleotide: A short polymer of nucleotides (usually less than 50).

oligopeptide: A few amino acids joined by peptide bonds.

oligosaccharide: Several monosaccharide groups joined by glycosidic bonds.

oncogene: A cancer-causing gene; any of several mutant genes that cause cells to exhibit rapid, uncontrolled proliferation. *See also* proto-oncogene.

open reading frame: A group of contiguous nonoverlapping nucleotide codons in a DNA or RNA molecule that do not include a termination codon.

open system: A system that exchanges matter and energy with its surroundings. *See also* system.

operator: A region of DNA that interacts with a repressor protein to control the expression of a gene or group of genes.

operon: A unit of genetic expression consisting of one or more related genes and the operator and promoter sequences that regulate their transcription.

optical activity: The capacity of a substance to rotate the plane of plane-polarized light.

optimum pH: The characteristic pH at which an enzyme has maximal catalytic activity.

organelles: Membrane-bounded structures found in eukaryotic cells; contain enzymes and other components required for specialized cell functions.

origin: The nucleotide sequence or site in DNA where DNA replication is initiated.

osmosis: Bulk flow of water through a semipermeable membrane into another aqueous compartment containing solute at a higher concentration.

osmotic pressure: Pressure generated by the osmotic flow of water through a semipermeable membrane into an aqueous compartment containing solute at a higher concentration.

oxidases: Enzymes that catalyze oxidation reactions in which molecular oxygen serves as the electron acceptor, but neither of the oxygen atoms is incorporated into the product. Compare **oxygenases.**

oxidation: The loss of electrons from a compound.

oxidation, β: *See* β oxidation.

oxidation-reduction reaction: A reaction in which electrons are transferred from a donor to an acceptor molecule; also called a redox reaction.

oxidative phosphorylation: The enzymatic phosphorylation of ADP to ATP coupled to electron transfer from a substrate to molecular oxygen.

oxidizing agent (oxidant): The acceptor of electrons in an oxidation-reduction reaction.

oxygen debt: The extra oxygen (above the normal resting level) consumed in the recovery period after strenuous physical exertion.

oxygenases: Enzymes that catalyze reactions in which oxygen atoms are directly incorporated into the product, forming a hydroxyl or carboxyl group. In reactions catalyzed by a monooxygenase, only one of the two O atoms is incorporated; the other is reduced to H_2O; in reactions catalyzed by a dioxygenase, both O atoms are incorporated into the product. Compare **oxidases.**

p

palindrome: A segment of duplex DNA in which the base sequences of the two strands exhibit twofold rotational symmetry about an axis.

paradigm: In biochemistry, an experimental model or example.

partition coefficient: A constant that expresses the ratio in which a given solute will be partitioned or distributed between two given immiscible liquids at equilibrium.

pathogenic: Disease-causing.

pentose: A simple sugar with a backbone containing five carbon atoms.

pentose phosphate pathway: A pathway that serves to interconvert hexoses and pentoses and is a source of reducing equivalents and pentoses for biosynthetic processes; present in most organisms. Also called the phosphogluconate pathway.

peptidases: Enzymes that hydrolyze peptide bonds.

peptide: Two or more amino acids covalently joined by peptide bonds.

peptide bond: A substituted amide linkage between the α-amino group of one amino acid and the α-carboxyl group of another, with the elimination of the elements of water.

peptide mapping: The characteristic two-dimensional pattern (on paper or gel) formed by the separation of a mixture of peptides resulting from partial hydrolysis of a protein; also known as peptide fingerprinting.

peptidoglycan: A major component of bacterial cell walls; generally consists of parallel heteropolysaccharides cross-linked by short peptides.

peripheral proteins: Proteins that are loosely or reversibly bound to a membrane by hydrogen bonds or electrostatic forces; generally water-soluble once released from the membrane.

permeases: *See* transporters.

peroxisome: Membrane-bounded organelle in the cytoplasm of eukaryotic cells; contains peroxide-forming and peroxide-destroying enzymes.

pH: The negative logarithm of the hydrogen ion concentration of an aqueous solution.

phage: *See* bacteriophage.

phenotype: The observable characteristics of an organism.

phosphatases: Enzymes that hydrolyze a phosphate ester or anhydride, releasing inorganic phosphate, P_i.

phosphodiester linkage: A chemical grouping that contains two alcohols esterified to one molecule of phosphoric acid, which thus serves as a bridge between them.

phosphogluconate pathway: An oxidative pathway beginning with glucose 6-phosphate and leading, via 6-phosphogluconate, to pentose phosphates and yielding NADPH. Also called the pentose phosphate pathway.

phospholipid: A lipid containing one or more phosphate groups.

phosphorolysis: Cleavage of a compound with phosphate as the attacking group; analogous to hydrolysis.

phosphorylases: Enzymes that catalyze phosphorolysis (defined above).

phosphorylation: Formation of a phosphate derivative of a biomolecule, usually by enzymatic transfer of a phosphoryl group from ATP.

phosphorylation potential (ΔG_p): The actual free-energy change of ATP hydrolysis under the nonstandard conditions prevailing within a cell.

photochemical reaction center: The part of a photosynthetic complex where the energy of an absorbed photon causes charge separation, initiating electron transfer.

photon: The ultimate unit (a quantum) of light energy.

photophosphorylation: The enzymatic formation of ATP from ADP coupled to the light-dependent transfer of electrons in photosynthetic cells.

photoreduction: The light-induced reduction of an electron acceptor in photosynthetic cells.

photorespiration: Oxygen consumption occurring in illuminated temperate-zone plants, largely due to oxidation of phosphoglycolate.

photosynthesis: The use of light energy to produce carbohydrates from carbon dioxide and a reducing agent such as water.

photosynthetic phosphorylation: *See* photophosphorylation.

photosystem: In photosynthetic cells, a functional set of light-absorbing pigments and its reaction center.

phototroph: An organism that can use the energy of light to synthesize its own fuels from simple molecules such as carbon dioxide, oxygen, and water; as distinct from a chemotroph.

pK_a: The negative logarithm of an equilibrium constant.

plasma membrane: The exterior membrane surrounding the cytoplasm of a cell.

plasma proteins: The proteins present in blood plasma.

plasmalogen: A phospholipid with an alkenyl ether substituent on the C-1 of glycerol.

plasmid: An extrachromosomal, independently replicating, small circular DNA molecule; commonly employed in genetic engineering.

plastid: In plants, a self-replicating organelle; may differentiate into a chloroplast.

platelets: Small, enucleated cells that initiate blood clotting; they arise from cells called megakaryocytes in the bone marrow. Also known as thrombocytes.

pleated sheet: The side-by-side, hydrogen-bonded arrangement of polypeptide chains in the extended β conformation.

plectonemic: A structure in a molecular polymer in which there is a net twisting of strands about each other in some simple and regular way.

polar: Hydrophilic, or "water-loving"; describing molecules or groups that are soluble in water.

polarity: (1) In chemistry, the nonuniform distribution of electrons in a molecule; polar molecules are usually soluble in water. (2) In molecular biology, the distinction between the 5′ and 3′ ends of nucleic acids.

poly(A) tail: A length of adenosine residues added to the 3′ ends of many mRNAs in eukaryotes (and sometimes in bacteria).

polycistronic mRNA: A contiguous mRNA with more than two genes that can be translated into proteins.

polyclonal antibodies: A heterogeneous pool of antibodies produced in an animal by a number of different B lymphocytes in response to an antigen. Different antibodies in the pool recognize different parts of the antigen.

polylinker: A short, often synthetic, fragment of DNA containing recognition sequences for several restriction endonucleases.

polymerase chain reaction (PCR): A repetitive procedure that results in a geometric amplification of a specific DNA sequence.

polymorphic: Describing a protein for which amino acid sequence variants exist in a population of organisms, but the variations do not destroy the protein's function.

polynucleotide: A covalently linked sequence of nucleotides in which the 3′ hydroxyl of the pentose of one nucleotide residue is joined by a phosphodiester bond to the 5′ hydroxyl of the pentose of the next residue.

polypeptide: A long chain of amino acids linked by peptide bonds; the molecular weight is generally less than 10,000.

polyribosome: *See* polysome.

polysaccharide: A linear or branched polymer of monosaccharide units linked by glycosidic bonds.

polysome (polyribosome): A complex of an mRNA molecule and two or more ribosomes.

P/O ratio: The number of moles of ATP formed in oxidative phosphorylation per $\frac{1}{2}O_2$ reduced (thus, per pair of electrons passed to O_2). Experimental values used in this text are 2.5 for passage of electrons from NADH to O_2, and 1.5 for passage of electrons from FADH to O_2. Some textbooks use the integral values of 3.0 and 2.0.

porphyria: Genetic condition resulting from the lack of one or more enzymes required to synthesize porphyrins.

porphyrin: Complex nitrogenous compound containing four substituted pyrroles covalently joined into a ring; often complexed with a central metal atom.

positive cooperativity: A phenomenon of some multisubunit enzymes or proteins in which binding of a ligand or substrate to one subunit facilitates binding to another subunit.

posttranscriptional processing: The enzymatic processing of the primary RNA transcript, producing functional mRNA, tRNA, and/or rRNA molecules.

posttranslational modification: Enzymatic processing of a polypeptide chain after translation from its mRNA.

primary structure: A description of the covalent backbone of a polymer (macromolecule), including the sequence of monomeric subunits and any interchain and intrachain covalent bonds.

primary transcript: The immediate RNA product of transcription before any posttranscriptional processing reactions.

primase: An enzyme that catalyzes the formation of RNA oligonucleotides used as primers by DNA polymerases.

primer: A short oligomer (of sugars or nucleotides, for example) to which an enzyme adds additional monomeric subunits.

primer terminus: The end of the primer to which monomeric subunits are added.

primosome: An enzyme complex that synthesizes the primers required for lagging strand DNA synthesis.

probe: A labeled fragment of nucleic acid containing a nucleotide sequence complementary to a gene or genomic sequence that one wishes to detect in a hybridization experiment.

processivity: For any enzyme that catalyzes the synthesis of a biological polymer, the property of adding multiple subunits to the polymer without dissociating from the substrate.

prochiral molecule: A symmetric molecule that can react asymmetrically with an enzyme having an asymmetric active site, generating a chiral product.

projection formulas: A method for representing molecules to show the configuration of groups around chiral centers; also known as Fischer projection formulas.

prokaryote: A bacterium; a unicellular organism with a single chromosome, no nuclear envelope, and no membrane-bounded organelles.

promoter: A DNA sequence at which RNA polymerase may bind, leading to initiation of transcription.

proofreading: The correction of errors in the synthesis of an information-containing biopolymer by removing incorrect monomeric subunits after they have been covalently added to the growing polymer.

prostaglandins: A class of lipid-soluble, hormonelike regulatory molecules derived from arachidonate and other polyunsaturated fatty acids.

prosthetic group: A metal ion or an organic compound (other than an amino acid) that is covalently bound to a protein and is essential to its activity.

proteasome: Supramolecular assembly of enzymatic complexes that function in the degradation of damaged or unneeded cellular proteins.

protein: A macromolecule composed of one or more polypeptide chains, each with a characteristic sequence of amino acids linked by peptide bonds.

protein kinases: Enzymes that transfer the terminal phosphoryl group of ATP or another nucleoside triphosphate to a Ser, Thr, Tyr, Asp, or His side chain in a target protein, thereby regulating the activity or other properties of that protein.

protein targeting: The process by which newly synthesized proteins are sorted and transported to their proper locations in the cell.

proteoglycan: A hybrid macromolecule consisting of a heteropolysaccharide joined to a polypeptide; the polysaccharide is the major component.

proto-oncogene: A cellular gene, usually encoding a regulatory protein, that can be converted into an oncogene by mutation.

proton acceptor: An anionic compound capable of accepting a proton from a proton donor; that is, a base.

proton donor: The donor of a proton in an acid-base reaction; that is, an acid.

proton-motive force: The electrochemical potential inherent in a transmembrane gradient of H^+ concentration; used in oxidative phosphorylation and photophosphorylation to drive ATP synthesis.

protoplasm: A general term referring to the entire contents of a living cell.

purine: A nitrogenous heterocyclic base found in nucleotides and nucleic acids; containing fused pyrimidine and imidazole rings.

puromycin: An antibiotic that inhibits polypeptide synthesis by being incorporated into a growing polypeptide chain, causing its premature termination.

pyranose: A simple sugar containing the six-membered pyran ring.

pyridine nucleotide: A nucleotide coenzyme containing the pyridine derivative nicotinamide; NAD or NADP.

pyridoxal phosphate: A coenzyme containing the vitamin pyridoxine (vitamin B_6); functions in reactions involving amino group transfer.

pyrimidine: A nitrogenous heterocyclic base found in nucleotides and nucleic acids.

pyrimidine dimer: A covalently joined dimer of two adjacent pyrimidine residues in DNA, induced by absorption of UV light; most commonly derived from two adjacent thymines (a thymine dimer).

pyrophosphatase: *See* inorganic pyrophosphatase.

q

quantum: The ultimate unit of energy.

quaternary structure: The three-dimensional structure of a multisubunit protein; particularly the manner in which the subunits fit together.

r

R group: (1) Formally, an abbreviation denoting any alkyl group. (2) Occasionally, used in a more general sense to denote virtually any organic substituent (the R groups of amino acids, for example).

racemic mixture (racemate): An equimolar mixture of the D and L stereoisomers of an optically active compound.

radical: An atom or group of atoms possessing an unpaired electron; also called a free radical.

radioactive isotope: An isotopic form of an element with an unstable nucleus that stabilizes itself by emitting ionizing radiation.

radioimmunoassay: A sensitive and quantitative method for detecting trace amounts of a biomolecule, based on its capacity to displace a radioactive form of the molecule from combination with its specific antibody.

rate constant: The proportionality constant that relates the velocity of a chemical reaction to the concentration(s) of the reactant(s).

rate-limiting step: (1) Generally, the step in an enzymatic reaction with the greatest activation energy or the transition state of highest free energy. (2) The slowest step in a metabolic pathway.

reaction intermediate: Any chemical species in a reaction pathway that has a finite chemical lifetime.

reading frame: A contiguous and nonoverlapping set of three-nucleotide codons in DNA or RNA.

recombinant DNA: DNA formed by the joining of genes into new combinations.

recombination: Any enzymatic process by which the linear arrangement of nucleic acid sequences in a chromosome is altered by cleavage and rejoining.

recombinational DNA repair: recombinational processes that are directed at the repair of DNA strand breaks or cross-links, especially at inactivated replication forks.

redox pair: An electron donor and its corresponding oxidized form; for example, NADH and NAD^+.

redox reaction: *See* oxidation-reduction reaction.

reducing agent (reductant): The electron donor in an oxidation-reduction reaction.

reducing end: The end of a polysaccharide having a terminal sugar with a free anomeric carbon; the terminal residue can act as a reducing sugar.

reducing equivalent: A general or neutral term for an electron or an electron equivalent in the form of a hydrogen atom or a hydride ion.

reducing sugar: A sugar in which the carbonyl (anomeric) carbon is not involved in a glycosidic bond and can therefore undergo oxidation.

reduction: The gain of electrons by a compound or ion.

regulatory enzyme: An enzyme having a regulatory function through its capacity to undergo a change in catalytic activity by allosteric mechanisms or by covalent modification.

regulatory gene: A gene that gives rise to a product involved in the regulation of the expression of another gene; for example, a gene coding for a repressor protein.

regulatory sequence: A DNA sequence involved in regulating the expression of a gene; for example, a promoter or operator.

regulon: A group of genes or operons that are coordinately regulated even though some, or all, may be spatially distant within the chromosome or genome.

relaxed DNA: Any DNA that exists in its most stable and unstrained structure, typically B form under most cellular conditions.

release factors: *See* termination factors.

releasing factors: Hypothalamic hormones that stimulate release of other hormones by the pituitary gland.

renaturation: Refolding of an unfolded (denatured) globular protein so as to restore native structure and protein function.

replication: Synthesis of daughter nucleic acid molecules identical to the parental nucleic acids.

replication fork: The Y-shaped structure generally found at the point where DNA is being synthesized.

replicative form: Any of the full-length structural forms of a viral chromosome that serve as distinct replication intermediates.

replisome: The multiprotein complex that promotes DNA synthesis at the replication fork.

repressible enzyme: In bacteria, an enzyme whose synthesis is inhibited when its reaction product is readily available to the cell.

repression: A decrease in the expression of a gene in response to a change in the activity of a regulatory protein.

repressor: The protein that binds to the regulatory sequence or operator for a gene, blocking its transcription.

residue: A single unit within a polymer; for example, an amino acid within a polypeptide chain. The term reflects the fact that sugars, nucleotides, and amino acids lose a few atoms (generally the elements of water) when incorporated in their respective polymers.

respiration: Any metabolic process that leads to the uptake of oxygen and the release of CO_2.

respiration-linked phosphorylation: ATP formation from ADP and P_i, driven by electron flow through a series of membrane-bound carriers, with a proton gradient as the direct source of energy driving rotational catalysis by ATP synthase.

respiratory chain: The electron transfer chain; a sequence of electron-carrying proteins that transfer electrons from substrates to molecular oxygen in aerobic cells.

restriction endonucleases: Site-specific endodeoxyribonucleases causing cleavage of both strands of DNA at points within or near the specific site recognized by the enzyme; important tools in genetic engineering.

restriction fragment: A segment of double-stranded DNA produced by the action of a restriction endonuclease on a larger DNA.

restriction fragment length polymorphisms (RFLPs): Variations, among individuals in a population, in the length of certain restriction fragments within which certain genomic sequences occur. These variations result from rare sequence changes that create or destroy restriction sites in the genome.

retrovirus: An RNA virus containing a reverse transcriptase.

reverse transcriptase: An RNA-directed DNA polymerase in retroviruses; capable of making DNA complementary to an RNA.

ribonuclease: A nuclease that catalyzes the hydrolysis of certain internucleotide linkages of RNA.

ribonucleic acid: *See* RNA.

ribonucleotide: A nucleotide containing D-ribose as its pentose component.

ribosomal RNA (rRNA): A class of RNA molecules serving as components of ribosomes.

ribosome: A supramolecular complex of rRNAs and proteins, approximately 18 to 22 nm in diameter; the site of protein synthesis.

ribozymes: Ribonucleic acid molecules with catalytic activities; RNA enzymes.

Rieske iron-sulfur protein: A type of iron-sulfur protein in which two of the ligands to the central iron ion are His side chains. These proteins act in many electron-transfer sequences, including oxidative phosphorylation and photophosphorylation.

RNA (ribonucleic acid): A polyribonucleotide of a specific sequence linked by successive 3′, 5′-phosphodiester bonds.

RNA polymerase: An enzyme that catalyzes the formation of RNA from ribonucleoside 5′-triphosphates, using a strand of DNA or RNA as a template.

RNA splicing: Removal of introns and joining of exons in a primary transcript.

rRNA: *See* ribosomal RNA.

S

***S*-adenosylmethionine (adoMet):** An enzymatic cofactor involved in methyl group transfers.

salvage pathway: Synthesis of a biomolecule, such as a nucleotide, from intermediates in the degradative pathway for the biomolecule; a recycling pathway, as distinct from a de novo pathway.

saponification: Alkaline hydrolysis of triacylglycerols to yield fatty acids as soaps.

sarcomere: A functional and structural unit of the muscle contractile system.

satellite DNA: Highly repeated, nontranslated segments of DNA in eukaryotic chromosomes; most often associated with the centromeric region. Its function is not clear.

saturated fatty acid: A fatty acid containing a fully saturated alkyl chain.

second law of thermodynamics: The law stating that in any chemical or physical process, the entropy of the universe tends to increase.

second messenger: An effector molecule synthesized within a cell in response to an external signal (first messenger) such as a hormone.

secondary metabolism: Pathways that lead to specialized products not found in every living cell.

secondary structure: The residue-by-residue conformation of the backbone of a polymer.

sedimentation coefficient: A physical constant specifying the rate of sedimentation of a particle in a centrifugal field under specified conditions.

selectins: A large family of membrane proteins, lectins that bind oligosaccharides on other cells tightly and specifically, and serve to carry signals across the plasma membrane.

SELEX: A method for rapid experimental identification of nucleic acid sequences (usually RNA) that have particular catalytic or ligand-binding properties.

serpentine receptors: A large family of membrane receptor proteins with seven transmembrane helical segments. These receptors often associate with G proteins to transduce an extracellular signal into a change in cellular metabolism.

Shine-Dalgarno sequence: A sequence in an mRNA required for binding prokaryotic ribosomes.

SH2 domain: A protein domain that binds tightly to a phosphotyrosine residue in certain proteins such as the receptor tyrosine kinases, initiating the formation of a multiprotein complex that acts in a signaling pathway.

shuttle vector: A recombinant DNA vector that can be replicated in two or more different host species. *See also* vector.

sickle-cell anemia: A human disease characterized by defective hemoglobin molecules; caused by a homozygous allele coding for the β chain of hemoglobin.

sickle-cell trait: A human condition recognized by the sickling of erythrocytes when exposed to low oxygen tension; occurs in individuals heterozygous for the allele responsible for sickle-cell anemia.

signal sequence: An amino acid sequence, often at the amino terminus, that signals the cellular fate or destination of a newly synthesized protein.

signal transduction: The process by which an extracellular signal (chemical, mechanical, or electrical) is amplified and converted to a cellular response.

silent mutation: A mutation in a gene that causes no detectable change in the biological characteristics of the gene product.

simple diffusion: The movement of solute molecules across a membrane to a region of lower concentration, unassisted by a protein transporter.

simple protein: A protein yielding only amino acids on hydrolysis.

site-directed mutagenesis: A set of methods used to create specific alterations in the sequence of a gene.

site-specific recombination: A type of genetic recombination that occurs only at specific sequences.

size-exclusion chromatography: A procedure for the separation of a mixture of molecules on the basis of size, based on the capacity of porous polymers to exclude solutes above a certain size. Also called gel filtration.

small nuclear RNA (snRNA): Any of several small RNA molecules in the nucleus; most have a role in the splicing reactions that remove introns from mRNA, tRNA, and rRNA molecules.

somatic cells: All body cells except the germline cells.

SOS response: In bacteria, a coordinated induction of a variety of genes as a response to high levels of DNA damage.

Southern blot: A DNA hybridization procedure in which one or more specific DNA fragments are detected in a larger population by means of hybridization to a complementary, labeled nucleic acid probe.

specific acid-base catalysis: Acid or base catalysis involving the constituents of water (hydroxide or hydronium ions).

specific activity: The number of micromoles (μmol) of a substrate transformed by an enzyme preparation per minute per milligram of protein at 25 °C; a measure of enzyme purity.

specific heat: The amount of energy (in joules or calories) needed to raise the temperature of 1 g of a pure substance by 1 °C.

specific rotation: The rotation, in degrees, of the plane of plane-polarized light (D-line of sodium) by an optically active compound at 25 °C, with a specified concentration and light path.

specificity: The ability of an enzyme or receptor to discriminate among competing substrates or ligands.

sphingolipid: An amphipathic lipid with a sphingosine backbone to which are attached a long-chain fatty acid and a polar alcohol.

spliceosome: A complex of RNAs and proteins involved in the splicing of mRNAs in eukaryotic cells.

splicing: *See* gene splicing; RNA splicing.

standard free-energy change ($\Delta G°$): The free-energy change for a reaction occurring under a set of standard conditions: temperature, 298 K; pressure, 1 atm or 101.3 kPa; and all solutes at 1 M concentration. $\Delta G'°$ denotes the standard free-energy change at pH 7.0.

standard reduction potential ($E'°$): The electromotive force exhibited at an electrode by 1 M concentrations of a reducing agent and its oxidized form at 25 °C and pH 7.0; a measure of the relative tendency of the reducing agent to lose electrons.

steady state: A nonequilibrium state of a system through which matter is flowing and in which all components remain at a constant concentration.

stem cells: The common, self-regenerating cells in bone marrow that give rise to differentiated blood cells such as erythrocytes and lymphocytes.

stereoisomers: Compounds that have the same composition and the same order of atomic connections, but different molecular arrangements.

sterols: A class of lipids containing the steroid nucleus.

sticky ends: Two DNA ends in the same DNA molecule, or in different molecules, with short overhanging single-stranded segments that are complementary to one another, facilitating ligation of the ends; also known as cohesive ends.

stop codons: *See* termination codons.

stroma: The space and aqueous solution enclosed within the inner membrane of a chloroplast, not including the contents within the thylakoid membranes.

structural gene: A gene coding for a protein or RNA molecule; as distinct from a regulatory gene.

substitution mutation: A mutation caused by the replacement of one base by another.

substrate: The specific compound acted upon by an enzyme.

substrate-level phosphorylation: Phosphorylation of ADP or some other nucleoside 5′-diphosphate coupled to the dehydrogenation of an organic substrate; independent of the electron-transfer chain.

suicide inhibitor: A relatively inert molecule that is transformed by an enzyme, at its active site, into a reactive substance that irreversibly inactivates the enzyme.

supercoil: The twisting of a helical (coiled) molecule on itself; a coiled coil.

supercoiled DNA: DNA that twists upon itself because it is under- or overwound (and thereby strained) relative to B-form DNA.

superhelical density: In a helical molecule such as DNA, the number of supercoils (superhelical turns) relative to the number of coils (turns) in the relaxed molecule.

suppressor mutation: A mutation that totally or partially restores a function lost by a primary mutation; located at a site different from the site of the primary mutation.

Svedberg (S): A unit of measure of the rate at which a particle sediments in a centrifugal field.

symbionts: Two or more organisms that are mutually interdependent; usually living in physical association.

symport: Cotransport of solutes across a membrane in the same direction.

synthases: Enzymes that catalyze condensation reactions in which no nucleoside triphosphate is required as an energy source.

synthetases: Enzymes that catalyze condensation reactions using ATP or another nucleoside triphosphate as an energy source.

system: An isolated collection of matter; all other matter in the universe apart from the system is called the surroundings.

t

telomere: Specialized nucleic acid structure found at the ends of linear eukaryotic chromosomes.

template: A macromolecular mold or pattern for the synthesis of an informational macromolecule.

template strand: A strand of nucleic acid used by a polymerase as a template to synthesize a complementary strand.

terminal transferase: An enzyme that catalyzes the addition of nucleotide residues of a single kind to the 3′ end of DNA chains.

termination codons: UAA, UAG, and UGA; in protein synthesis, signal the termination of a polypeptide chain. Also known as stop codons.

termination factors: Protein factors of the cytosol required in releasing a com-pleted polypeptide chain from a ribosome; also known as release factors.

termination sequence: A DNA sequence that appears at the end of a transcriptional unit and signals the end of transcription.

terpenes: Organic hydrocarbons or hydrocarbon derivatives constructed from recurring isoprene units. They produce some of the scents and tastes of plant products.

tertiary structure: The three-dimensional conformation of a polymer in its native folded state.

tetrahydrobiopterin: The reduced coenzyme form of biopterin.

tetrahydrofolate: The reduced, active coenzyme form of the vitamin folate.

thiamine pyrophosphate: The active coenzyme form of vitamin B_1; involved in aldehyde transfer reactions.

thioester: An ester of a carboxylic acid with a thiol or mercaptan.

3′ end: The end of a nucleic acid that lacks a nucleotide bound at the 3′ position of the terminal residue.

thrombocytes: *See* platelets.

thromboxanes: A class of molecules derived from arachidonate and involved in platelet aggregation during blood clotting.

thylakoid: Closed cisterna, or disk, formed by the pigment-bearing internal membranes of chloroplasts.

thymine dimer: *See* pyrimidine dimer.

tissue culture: Method by which cells derived from multicellular organisms are grown in liquid media.

titration curve: A plot of the pH versus the equivalents of base added during titration of an acid.

tocopherols: Forms of vitamin E.

topoisomerases: Enzymes that introduce positive or negative supercoils in closed, circular duplex DNA.

topoisomers: Different forms of a covalently closed, circular DNA molecule that differ only in their linking number.

topology: The study of the properties of an object that do not change under continuous deformations such as twisting or bending.

toxins: Proteins produced by some organisms and toxic to certain other species.

trace element: A chemical element required by an organism in only trace amounts.

transaminases: *See* aminotransferases.

transamination: Enzymatic transfer of an amino group from an α-amino acid to an α-keto acid.

transcription: The enzymatic process whereby the genetic information contained in one strand of DNA is used to specify a complementary sequence of bases in an mRNA chain.

transcriptional control: The regulation of a protein's synthesis by regulation of the formation of its mRNA.

transduction: (1) Generally, the conversion of energy or information from one form to another. (2) The transfer of genetic information from one cell to another by means of a viral vector.

transfer RNA (tRNA): A class of RNA molecules (M_r 25,000 to 30,000), each of which combines covalently with a specific amino acid as the first step in protein synthesis.

transformation: Introduction of an exogenous DNA into a cell, causing the cell to acquire a new phenotype.

transgenic: Describing an organism that has genes from another organism incorporated within its genome as a result of recombinant DNA procedures.

transition state: An activated form of a molecule in which the molecule has undergone a partial chemical reaction; the highest point on the reaction coordinate.

translation: The process in which the genetic information present in an mRNA molecule specifies the sequence of amino acids during protein synthesis.

translational control: The regulation of a protein's synthesis by regulation of the rate of its translation on the ribosome.

translational repressor: A repressor that binds to an mRNA, blocking translation.

translocase: (1) An enzyme that catalyzes membrane transport. (2) An enzyme that causes a movement, such as the movement of a ribosome along an mRNA.

transpiration: Passage of water from the roots of a plant to the atmosphere via the vascular system and the stomata of the leaves.

transporters: Proteins that span a membrane and transport specific nutrients, metabolites, ions, or proteins across the membrane; sometimes called permeases.

transposition: The movement of a gene or set of genes from one site in the genome to another.

transposon (transposable element): A segment of DNA that can move from one position in the genome to another.

triacylglycerol: An ester of glycerol with three molecules of fatty acid; also called a triglyceride or neutral fat.

tricarboxylic acid cycle: *See* citric acid cycle.

triose: A simple sugar with a backbone containing three carbon atoms.

tRNA: *See* transfer RNA.

tropic hormone (tropin): A peptide hormone that stimulates a specific target gland to secrete its hormone; for example, thyrotropin produced by the pituitary stimulates secretion of thyroxine by the thyroid.

turnover number: The number of times an enzyme molecule transforms a substrate molecule per unit time, under conditions giving maximal activity at substrate concentrations that are saturating.

U

ubiquitin: A small, highly conserved protein that targets an intracellular protein for degradation by proteasomes. Several ubiquitin molecules are covalently attached in tandem to a Lys residue in the target protein by a specific ubiquitinating enzyme.

ultraviolet (UV) radiation: Electromagnetic radiation in the region of 200 to 400 nm.

uncompetitive inhibition: The reversible inhibition pattern resulting when an inhibitor molecule can bind to the enzyme-substrate complex but not to the free enzyme.

uncoupling agent: A substance that uncouples phosphorylation of ADP from electron transfer; for example, 2,4-dinitrophenol.

uniport: A transport system that carries only one solute, as distinct from cotransport.

unsaturated fatty acid: A fatty acid containing one or more double bonds.

urea cycle: A metabolic pathway in vertebrates, for the synthesis of urea from amino groups and carbon dioxide; occurs in the liver.

ureotelic: Excreting excess nitrogen in the form of urea.

uricotelic: Excreting excess nitrogen in the form of urate (uric acid).

V

V_{max}: The maximum velocity of an enzymatic reaction when the binding site is saturated with substrate.

vector: A DNA molecule known to replicate autonomously in a host cell, to which a segment of DNA may be spliced to allow its replication; for example, a plasmid or an artificial chromosome.

vectorial metabolism: Metabolic transformations in which the location (not the chemical composition) of a substrate changes relative to a cellular membrane dividing two compartments. Transporters catalyze vectorial reactions, as do the proton pumps of oxidative and photophosphorylation.

viral vector: A viral DNA altered so that it can act as a vector for recombinant DNA.

virion: A virus particle.

virus: A self-replicating, infectious, nucleic acid–protein complex that requires an intact host cell for its replication; its genome is either DNA or RNA.

vitamin: An organic substance required in small quantities in the diet of some species; generally functions as a component of a coenzyme.

W

wild type: The normal (unmutated) phenotype.

wobble: The relatively loose base pairing between the base at the 3′ end of a codon and the complementary base at the 5′ end of the anticodon.

X

x-ray crystallography: The analysis of x-ray diffraction patterns of a crystalline compound, used to determine the molecule's three-dimensional structure.

Z

zinc finger: A specialized protein motif involved in DNA recognition by some DNA-binding proteins; characterized by a single atom of zinc coordinated to four Lys residues or to two His and two Lys residues.

zwitterion: A dipolar ion, with spatially separated positive and negative charges.

zymogen: An inactive precursor of an enzyme; for example, pepsinogen, the precursor of pepsin.

Index

Bold page number indicates term is bold in text. Configuration/position designations disregarded in alphabetizing. Greek letter adjectives alphabetized as if spelled out.

t = table; f = figure; s = structural formula; p = photo or portrait; e = electron micrograph

a

b

d

e

f

g

h

i

j

k

m

n

O

p

q

r

S

X

y

z

Abbreviations for Amino Acids

A	Ala	Alanine	N	Asn	Asparagine
B	Asx	Asparagine or aspartate	P	Pro	Proline
			Q	Gln	Glutamine
C	Cys	Cysteine	R	Arg	Arginine
D	Asp	Aspartate	S	Ser	Serine
E	Glu	Glutamate	T	Thr	Threonine
F	Phe	Phenylalanine	V	Val	Valine
G	Gly	Glycine	W	Trp	Tryptophan
H	His	Histidine	X	—	Unknown or nonstandard amino acid
I	Ile	Isoleucine			
K	Lys	Lysine	Y	Tyr	Tyrosine
L	Leu	Leucine	Z	Glx	Glutamine or glutamate
M	Met	Methionine			

Asx and Glx are used in describing the results of amino acid analytical procedures in which Asp and Glu are not readily distinguished from their amide counterparts, Asn and Gln.

The Standard Genetic Code

UUU	Phe	UCU	Ser	UAU	Tyr	UGU	Cys
UUC	Phe	UCC	Ser	UAC	Tyr	UGC	Cys
UUA	Leu	UCA	Ser	UAA	Stop	UGA	Stop
UUG	Leu	UCG	Ser	UAG	Stop	UGG	Trp
CUU	Leu	CCU	Pro	CAU	His	CGU	Arg
CUC	Leu	CCC	Pro	CAC	His	CGC	Arg
CUA	Leu	CCA	Pro	CAA	Gln	CGA	Arg
CUG	Leu	CCG	Pro	CAG	Gln	CGG	Arg
AUU	Ile	ACU	Thr	AAU	Asn	AGU	Ser
AUC	Ile	ACC	Thr	AAC	Asn	AGC	Ser
AUA	Ile	ACA	Thr	AAA	Lys	AGA	Arg
AUG	Met*	ACG	Thr	AAG	Lys	AGG	Arg
GUU	Val	GCU	Ala	GAU	Asp	GGU	Gly
GUC	Val	GCC	Ala	GAC	Asp	GGC	Gly
GUA	Val	GCA	Ala	GAA	Glu	GGA	Gly
GUG	Val	GCG	Ala	GAG	Glu	GGG	Gly

* AUG also serves as the initiation codon in protein synthesis.

1 **H** 1.008																		2 **He** 4.003
3 **Li** 6.94	4 **Be** 9.01												5 **B** 10.81	6 **C** 12.011	7 **N** 14.01	8 **O** 16.00	9 **F** 19.00	10 **Ne** 20.18
11 **Na** 22.99	12 **Mg** 24.31												13 **Al** 26.98	14 **Si** 28.09	15 **P** 30.97	16 **S** 32.06	17 **Cl** 35.45	18 **Ar** 39.95
19 **K** 39.10	20 **Ca** 40.08		21 **Sc** 44.96	22 **Ti** 47.90	23 **V** 50.94	24 **Cr** 52.00	25 **Mn** 54.94	26 **Fe** 55.85	27 **Co** 58.93	28 **Ni** 58.71	29 **Cu** 63.55	30 **Zn** 65.37	31 **Ga** 69.72	32 **Ge** 72.59	33 **As** 74.92	34 **Se** 78.96	35 **Br** 79.90	36 **Kr** 83.30
37 **Rb** 85.47	38 **Sr** 87.62		39 **Y** 88.91	40 **Zr** 91.22	41 **Nb** 92.91	42 **Mo** 95.94	43 **Te** 98.91	44 **Ru** 101.07	45 **Rh** 102.91	46 **Pd** 106.4	47 **Ag** 107.87	48 **Cd** 112.40	49 **In** 114.82	50 **Sn** 118.69	51 **Sb** 121.75	52 **Te** 126.70	53 **I** 126.90	54 **Xe** 131.30
55 **Cs** 132.91	56 **Ba** 137.34	57–70 *	71 **Lu** 174.97	72 **Hf** 178.49	73 **Ta** 180.95	74 **W** 183.85	75 **Re** 186.2	76 **Os** 190.2	77 **Ir** 192.2	78 **Pt** 195.09	79 **Au** 196.97	80 **Hg** 200.59	81 **Tl** 204.37	82 **Pb** 207.19	83 **Bi** 208.98	84 **Po** (209)	85 **At** (210)	86 **Rn** (222)
87 **Fr** (223)	88 **Ra** 226.03	89–102 **	103 **Lr** 262.11	104 **Rf** 261.11	105 **Db** 262.11	106 **Sg** 263.12	107 **Bh** 264.12	108 **Hs** 265.13	109 **Mt** 268	110 **Uun** 269	111 **Uuu** 272	112 **Uub** 277		114 **Uuq** 289		116 **Uuh** 289		118 **Uuo** 293

*Lanthanides	57 **La** 138.91	58 **Ce** 140.12	59 **Pr** 140.91	60 **Nd** 144.24	61 **Pm** 144.91	62 **Sm** 150.36	63 **Eu** 151.96	64 **Gd** 157.25	65 **Tb** 158.93	66 **Dy** 162.50	67 **Ho** 164.93	68 **Er** 167.26	69 **Tm** 168.93	70 **Yb** 173.04
**Actinides	89 **Ac** 227.03	90 **Th** 232.04	91 **Pa** 231.04	92 **U** 238.03	93 **Np** 237.05	94 **Pu** 244.06	95 **Am** 243.06	96 **Cm** 247.07	97 **Bk** 247.07	98 **Cf** 251.08	99 **Es** 252.08	100 **Fm** 257.10	101 **Md** 258.10	102 **No** 259.10

Some Conversion Factors	
Length	1 cm = 10 mm = 10^4 μm = 10^7 nm = 10^8 Å = 0.394 in 1 in = 2.54 cm
Mass	1 g = 10^{-3} kg = 10^3 mg = 10^6 μg = 3.53×10^{-2} oz 1 oz = 28.3 g
Temperature	°C = 5/9(°F − 32) K = °C + 273
Energy	1 J = 10^7 erg = 0.239 cal 1 cal = 4.184 J
Pressure	1 torr = 1 mm Hg = 1.32×10^{-3} atm = 1.333×10^2 Pa 1 atm = 758 torr = 1.01×10^5 Pa
Radioactivity	1 Ci = 3.7×10^{10} dps = 37 GBq 1,000 dpm = 16.7 Bq

Some Physical Constants, with Symbols and Values		
Atomic mass unit (dalton)	amu	1.661×10^{-24} g
Avogadro's number	N	6.022×10^{23}/mol
Becquerel	Bq	1 dps
Boltzmann constant	**k**	1.381×10^{-23}J/K; 3.298×10^{-24} cal/K
Curie	Ci	3.70×10^{10} dps
Electron volt	eV	1.602×10^{-19} J; 3.828×10^{-20} cal
Faraday constant	$\mathcal{F}$	96,480 J/V • mol
Gas constant	R	1.987 cal/mol • K; 8.315 J/mol • K
Planck's constant	h	1.584×10^{-34} cal • s; 6.626×10^{-34} J • s
Speed of light (in vacuum)	c	2.998×10^{10} cm/s